S0-BBU-879

ENCYCLOPEDIA OF

FOOD SCIENCE AND TECHNOLOGY

VOLUME 1

WILEY ENCYCLOPEDIA OF FOOD SCIENCE AND TECHNOLOGY, Second Edition

ENCYCLOPEDIA OF

FOOD SCIENCE AND TECHNOLOGY Second Edition

VOLUME 1

Frederick J. Francis
University of Massachusetts
Amherst, Massachusetts

A Wiley-Interscience Publication
John Wiley & Sons, Inc.
New York / Chichester / Weinheim / Brisbane / Singapore / Toronto

This book is printed on acid-free paper. ♾

Published simultaneously in Canada.

For ordering and customer service, call 1-800-CALL-WILEY.

Library of Congress Cataloging-in-Publication Data:

Wiley encyclopedia of food science and technology.—2nd ed. / [edited by] Frederick J. Francis.
p. cm.
Rev. ed. of: Encyclopedia of food science and technology / Y.H. Hui, editor-in-chief. c1992.
Includes bibliographical references.
ISBN 0-471-19285-6 (set : cloth : alk. paper).—ISBN 0-471-19255-4 (v. 1 : cloth : alk. paper).—ISBN 0-471-19256-2 (v. 2 : cloth : alk. paper).—ISBN 0-471-19257-0 (v. 3 : cloth : alk. paper).—ISBN 0-471-19258-9 (v. 4 : cloth : alk. paper)
1. Food industry and trade Encyclopedias. I. Francis, F. J. (Frederick John), 1921- . II. Encyclopedia of food science and technology.
TP368.2.E62 2000
664′.003—dc21 99-29003
CIP

Printed in the United States of America.

10 9 8 7 6 5 4 3 2 1

PREFACE

The preface to the First Edition of this encyclopedia provided a detailed account of its breadth and aims. This breadth is maintained in the Second Edition and enlarged to include more of the associated areas of food science. A number of these areas are of current prime importance, including food safety, functional foods, and neutraceuticals. Seventy years ago, when the discipline of food science and technology was first introduced, the accepted definition was the science concerned with food "from the farm gate to the consumer." It later became obvious that many of the conditions associated with conventional agricultural production practices had a definite impact on the quality and quantity of the resultant food, so the lines between agricultural production and food science and technology became blurred.

The above definition of food science and technology was appropriate for food production techniques in the early part of the twentieth century. We were primarily dependent on fresh fruits and vegetables in season and meats, augmented by commodities which could be easily stored such as grains, root crops, dried fruits and fermented foods, and beverages. The "New England boiled dinner," composed of corned beef, potatoes, carrots, cabbage, turnips, onions, and parsnips, was a good example. A major change from a seasonal concept of food supply to a processed system occurred in the 1920s with the development of food preservation techniques such as canning, freezing, dehydration, and chemical preservation. When this was combined with developments in packaging, storage, transportation, and consumer marketing, the rate of change to a processed-food society was speeded up. Today we have the advantages of a stable processed food supply combined with fresh fruits and vegetables produced locally and also shipped in from other countries. This has produced the best of both systems, but it has some problems. The concentration of production, packaging, storage, and transportation into large units has increased the possibility of contamination with chemicals and microorganisms, a major concern of regulatory officials; it may also be a source of concern for readers of this encyclopedia. Hopefully, the sections on food safety, estimation of risk, interpretation of risk, and management of risk will allay the fears for the safety of the global food supply.

Another change in the concept of food science and technology occurred in the 1950s with increased concern over adequate nutrition in food and the nutrition delivery system. The science of nutrition at that point had developed to the point where meaningful predictions could be made on the effect of components in the food supply on the health and well-being of the populace. This led to increased emphasis on food processors to maintain the nutritive value of their products in order to supply the "Acceptable Daily Intake" recommended by governmental and other agencies. However, maintenance of nutrients already present in food was not enough, and the science of fortification was developed primarily to prevent the development of conditions which could lead to a deterioration in the overall health of the individual. This concept was enlarged to encompass components of the diet which were not necessarily related to deficiencies and disease but were related to overall well-being. This led to the concept of "Neutraceuticals," components which combine nutritive and pharmaceutical properties. The term "functional foods" was also introduced to describe components which have a technological function in foods as well as a nutritive or health benefit. Both areas are rapidly expanding and may be of interest to readers of this encyclopedia.

ACKNOWLEDGMENTS

There are three groups of people who make possible the completion of a work of this size: the Editorial Board, the contributors, and the production personnel. The members of the Editorial Board were chosen for their understanding of particular areas of food science and technology that I thought should be given more consideration in the Second Edition. I believed that the First Edition had an excellent set of papers, primarily in technology, and I wanted to build on that base. All of the editors enjoy a superb reputation in their respective fields and I am grateful to them for their willingness to share opinions and expertise. The contributors supplied the real foundation for the encyclopedia and I am most grateful for their willingness to contribute their time and energy to write the papers. This expression of professionalism is reassuring for the future of the discipline. Finally, I would like to thank the group of professionals at Wiley for their administrative support in what otherwise would be an overwhelming assignment.

Frederick J. Francis
Amherst, Massachusetts

PREFACE TO THE FIRST EDITION

Although this encyclopedia is designed for food scientists and technologists, it also contains information useful to food engineers, chemists, biologists, ingredient suppliers, and other professionals involved in the food chain.

The topics selected are based on the following frame of reference:

A. Basic and Applied Sciences
   1. Biology: botany, bacteriology, microbiology, mycology, eg, photosynthesis, food microbiology, mycotoxins.
   2. Chemistry: biochemistry, physical chemistry, analytical chemistry, organic chemistry, radiochemistry, forensic chemistry, eg, food chemistry, food analysis, interesterification, emulsification.
   3. Physics: rheology, thermodynamics, cryogenics, radiophysics, ultrasonics, eg, food rheology, thermodynamics, microwave, irradiation.
   4. Nutrition: basic, applied, and clinical nutrition, eg, nutrition, protein, energy, dietetics, nutritional quality and food processing.
   5. Psychology: sensory behaviors, eg, food and color, sensory sciences, taste and odor.
   6. Medicine: metabolisms, toxicology, heart diseases, eg, food utilization, cholesterol, foodborne diseases.
   7. Economics: marketing, development, eg, food marketing, food industry economic development.
   8. Integrated sciences: history of foods, food structure.

B. Processing Technology, Engineering, and the Twenty-three Unit Operations
   1. Raw material preparation, eg, apricots and apricot processing, milk and milk products.
   2. Size reduction, eg, potato and potato processing, crabs and crab processing.
   3. Mixing and forming, eg, confectionery, biscuits and cracker technology.
   4. Mechanical separations, eg, meat processing technology, cheeses, poultry product and technology.
   5. Membrane concentration, eg, milk and membrane technology
   6. Biotechnology, eg, genetic engineering
   7. Irradiation, eg, irradiation
   8. Blanching, eg, blanching, vegetables processing
   9. Pasteurization, eg, yogurt
   10. Heat sterilization, eg, aseptic processing
   11. Evaporation, eg, evaporation, distillation, heat transfer, heat exchangers
   12. Extrusion, eg, extrusion and extrusion cooking, snack foods
   13. Dehydration, eg, fruit dehydration, vegetable dehydration, drier technology and engineering
   14. Baking and roasting, eg, baking science and technology, breakfast cereals
   15. Frying, eg, frying technology, fats and oils
   16. Microwave and infrared radiation, eg, microwave science and technology
   17. Chilling and controlled- (modified-) atmosphere storage, eg, chilled foods, controlled atmospheres for fresh fruits and vegetables, packaging and modified atmospheres.
   18. Freezing, eg, food freezing, freezing systems, refrigerated foods
   19. Freeze drying and freeze concentration, eg, freeze drying, driers technology and engineering.
   20. Coating or enrobing, eg, fish and shellfish products, poultry processing technology,
   21. Packaging, eg, packaging materials, packaging and labeling
   22. Filling and sealing of containers, eg, aseptic processing and packaging
   23. Materials handling and process control, eg, computer and food processing, artificial intelligence, expert systems

C. Food Laws and Regulations
   1. Basic principles, eg, food laws and regulations
   2. Food identity, eg, standards and grades
   3. Food chemicals, eg, food additives
   4. Health and safety, eg, sanitation
   5. Food processing plants, eg, food plant design and construction, food processing: standard industrial classifications.
   6. Food plant inspection, eg, food laws and regulations.

Apart from members of the editorial advisory board, hundreds of professionals in the academic, industry, and government have personally counseled me on many aspects of the work. Their combined effort is reflected in the articles in this encyclopedia: appropriateness of the selection of topics; diverse expertise and background of each contributor; the excellent treatment of each article. The users are the best judge of these claims.

A small portion of the text is devoted to cover nontechnical data:

1. Descriptions of selected scientific institutions, eg, Institute of Food Technologists.
2. Food science and technology in selected countries in the world, eg, International Development: South Africa.

3. Research and development in major U.S. government agencies, eg, U.S. Department of Agriculture.

ACKNOWLEDGMENTS

Basically, there are three groups of people who make the completion of this large work possible: the advisors, the contributors, and the production personnel.

The advisors include the members of the editorial board, my friends, colleagues, and other experts. They helped me to formulate the subject matters, select contributors and reviewers, and solve scientific, technical, and engineering matters. Although it is not possible to name them all here, they are the foundation of the project. And I am grateful to them.

The contributors are the components and parts and give substance to the encyclopedia. Their contribution is most appreciated.

The production team puts the parts together and you are the best judge of their professionalism. Personally, I am specially grateful to Michalina Bickford, the managing editor for the encyclopedia.

The support of my family will always be appreciated.

Y. H. Hui
Cutten, California

CONTRIBUTORS

An entry that appears without a contributor's signature is based on the corresponding entry taken from the First Edition of this Encyclopedia and updated by the Editors of this Second Edition.

APV Crepaco, Inc., *Lake Mills, Wisconsin,* Corrosion and Food Processing; Distillation: Technology and Engineering; Dryers: Technology and Engineering; Evaporators: Technology and Engineering; Heat Exchangers: Fouling; Heat Exchangers: Paraflow; Heat Exchangers: Plate versus Tubular; Heat Exchangers: Scraped Surface

Judith A. Abbott, *USDA/ARS, Beltsville, Maryland,* Food Crops: Nondestructive Quality Evaluation

G. Abraham, *USDA, Southern Regional Research Center,* Oilseeds and Vegetable Oils

Raija Ahvenainen, *VTT Biotechnology and Food Research, VTT, Finland,* International Development: Finland

Robert E. Altomare, *General Foods USA, Tarrytown, New York,* Snack Food Technology

S. M. Alzamora, *Washington State University, Pullman, Washington,* Water Activity: Microbiology

Burton A. Amernick, *Pollock, Vande Sande, and Amernick, RLLP, Washington, D.C.,* Patents

Elizabeth Andress, *University of Georgia, Athens, Georgia,* Canning: Regulatory and Safety Considerations

Cathy Y. W. Ang, *North Carolina State University, Raleigh, North Carolina,* Poultry Flavors

Frank H. Arthur, *USDA/ARS, Manhattan, Kansas,* Grains and Protectants

Bonnie Atwood, *U.S. Army Natick, Natick, Massachusetts,* Military Food

Alex A. Avery, *Hudson Institute, Churchville, Virginia,* Food Regulations: International, Codex Alimentarius

Dennis T. Avery, *Hudson Institute, Churchville, Virginia,* Food Regulations: International, Codex Alimentarius

Jerry K. Babbitt, *National Marine Fisheries Service, Kodiak, Alaska,* Crabs and Crab Processing

T. Sudhakar Babu, *University of Waterloo, Waterloo, Ontario, Canada,* Photosynthesis

Robert J. Baer, *South Dakota State University, Brookings, South Dakota,* Milk and Milk Products

M. O. Balaban, *State University of Campinas—UNICAMP, São Paulo, Brazil,* Supercritical Fluid Technology

G. V. Barbosa-Cánovas, *Washington State University, Pullman, Washington,* Water Activity: Microbiology

Shai Barbut, *University of Guelph, Guelph, Ontario, Canada,* Poultry Meat Processing and Product Technology

Steven A. Barker, *Louisiana State University, Baton Rouge, Louisiana,* Antibiotics in Foods of Animal Origin

Harold J. Barnett, *National Oceanic and Atmospheric Administration, Seattle, Washington,* Seafood: Flavors and Quality

Roger G. Bates, *University of Florida, Gainesville, Florida,* Hydrogen-Ion Activity (pH)

J. N. BeMiller, *Purdue University, West Lafayette, Indiana,* Carbohydrates: Classification, Chemistry, and Labeling; Gums; Lactose; Sweeteners: Nutritive; Starch; Sugar: Sucrose

Richard C. Berberet, *Oklahoma State University, Stillwater, Oklahoma,* Postharvest Integrated Pest Management

Carolyn Berdanier, *University of Georgia, Athens, Georgia,* Minerals: Micronutrients

R. G. Berger, *Institut für Lebensmittelchemie der Universität Hannover, Hannover, Germany,* Genetic Engineering: Food Flavors

Daniel Berkowitz, *U.S. Army Natick, Natick, Massachusetts,* Military Food

Mrinal Bhattacharva, *University of Minnesota, St. Paul, Minnesota,* Extrusion Processing: Texture and Rheology

Darlena K. Blucher, *Humboldt State University, Arcata, California,* History of Foods

Michael M. Blumenthal, *Libra Technologies, Inc., Metuchen, New Jersey,* Frying Technology

Joseph Borsa, *MDS Nordion, Inc., Kanata, Ontario, Canada,* Irradiation of Foods

J. R. Botta, *Moncton Food Quality Investigations, Inc., Moncton, New Brunswick, Canada,* Lobster: Biology and Technology; Seafood: Sensory Evaluation and Freshness

Malcolm C. Bourne, *Cornell University, Geneva, New York,* Water Activity: Food Texture

Elizabeth A. E. Boyle, *Kansas State University, Manhattan, Kansas,* Meat Products

Francine Bradley, *University of California, Davis, California,* Poultry: Meat From Avian Species

J. W. Brady, *Cornell University, Ithaca, New York,* Molecular Modeling: Computer Modeling of Functional Properties

D. E. Briggs, *University of Birmingham, Birmingham, England,* Malts and Malting

Aaron L. Brody, *Rubbright•Brody, Inc., Duluth, Georgia,* Packaging: Part II—Labeling; Packaging: Part IV—Controlled/Modified Atmosphere/Vacuum Food Packaging

B. Brückner, *Institute of Vegetable and Ornamental Crops, Grossbeeren/Erfurt, Germany,* International Development: Germany

Christine M. Bruhn, *University of California, Davis, California,* Food Safety Education: Consumer

John Bumbalough, *Land O'Lakes, Inc., Arden Hills, Minnesota,* Margarine

David Burdick, *Hoffman-La Roche, Inc.,* Vitamins: Pyridoxine (B_6); Vitamins: Thiamine (B_1)

Charles J. Cante, *General Foods USA, White Plains, New York,* Breakfast Cereals; Snack Food Technology

Armand Cardello, *U.S. Army Natick, Natick, Massachusetts,* Military Food

Marijana Caric, *Novi Sad University, Novi Sad, Yugoslavia,* Processed Cheese

Robert Casani, *Hoffman-La Roche, Inc.,* Vitamins: Vitamin E

Robert G. Cassens, *University of Wisconsin, Madison, Wisconsin,* Muscle as Food

Domenico R. Cassone, *Nabisco, Inc., East Hanover, New Jersey,* Biscuit and Cracker Technology

S. Cenkowski, *University of Manitoba, Winnipeg, Manitoba, Canada,* Cleaning-in-Place (CIP); Energy Usage in Food Processing Plants

Tuu-jyi Chai, *University of Maryland, Cambridge, Maryland,* Fish and Shellfish Microbiology

James V. Chambers, *Purdue University, West Lafayette, Indiana,* Waste Management: Animal Processing

Roberto S. Chamul, *Mississippi State University, Mississippi State, Mississippi,* Machine Vision Systems: Process and Quality Control

D. Chandarana, *National Food Processors Association, Dublin, California,* Aseptic Processing and Packaging Systems

Ping Yang Chang, *Food Industry Research and Development Institute, Hsinchu, Taiwan,* Sulfites and Food

Muhammad Munir Chaudry, *Cornell University, Ithaca, New York,* Muslim Dietary Laws: Food Processing and Marketing

Chee-Jen Chen, *Food Industry Research and Development Institute, Hsinchu, Taiwan,* Mushrooms: Cultivation

T. C. Chen, *University of Arkansas, Fayetteville, Arkansas,* Poultry Meat Microbiology

Wen-Lian Chen, *Food Industry Research and Development Institute, Hsinchu, Taiwan,* Soy Foods, Fermented

Munir Cheryan, *University of Illinois, Urbana, Illinois,* Membrane Processing

Young-Meng Chiang, *National Taiwan University, Taipei, Taiwan,* Seaweed Aquaculture: Taiwan

H. Chiba, *Kobe Women's University, Kobe, Japan,* Peptides

Yoon Hin Chong, *Palm Oil Research Institute of Malaysia, Selangor, Malaysia,* Palm Oil

B. B. Chrystall, *Massey University–Albany, Auckland, New Zealand,* International Development: New Zealand; Meat and Electrical Stimulation; Meat Science

J. Peter Clark, *Oak Park, Illinois,* Food Plant Design and Construction

Warren S. Clark, Jr., *American Dairy Products Institute, Chicago, Illinois,* Dry Milk; Evaporated Milk

J. T. Clayton, *University of Massachusetts, Amherst, Massachusetts,* Food Engineering

John Clemmings, *Häagen-Dazs Research and Development, Edina, Minnesota,* Ice Cream and Frozen Dessert

Dean Cliver, *University of California, Davis, California,* Bovine Spongiform Encephalopathy (BSE)

G. Coccodrilli, *General Foods USA, White Plains, New York,* Breakfast Cereals

Bernard Cole, *Leiner Davis Gelatin, West Krugersdorp, South Africa,* Gelatin

David Collins-Thompson, *Nestlé Research and Development Center, Inc., New Milford, Connecticut,* Chilled Foods

Chris Combs, *Dow Corning Corporation, Midland, Michigan,* Foams and Silicones in Food Processing

Richard Cotter, *Whitehall-Robins, Madison, New Jersey,* Osteoporosis

Roger A. Coulombe, Jr., *Utah State University, Logan, Utah,* Toxicants, Natural

Thomas Crosby, *Bunge Foods Corporation, Bradley, Illinois,* Fats and Oils: Properties, Processing Technology, and Commercial Shortenings

Gerrit W. Cuperus, *Oklahoma State University, Stillwater, Oklahoma,* Postharvest Integrated Pest Management

Ashim K. Datta, *Cornell University, Ithaca, New York,* Thermal Sterilization of Canned Liquid Foods

P. Michael Davidson, *University of Tennessee, Knoxville, Tennessee,* Antimicrobial Compounds

Robert V. Decareau, *Microwave Consulting Services, Amherst, New Hampshire,* Microwave Science and Technology

Eric A. Decker, *University of Massachusetts, Amherst, Massachusetts,* Antioxidants; Phytochemicals: Antioxidants

Marnie L. DeGregorio, *Kraft Foods USA, Tarrytown, New York,* Fruit Preserves and Jellies

David Deis, *Abbott Laboratories, Columbus, Ohio,* Enteral Formulas and Feeding Systems

J. M. de Man, *University of Guelph, Guelph, Ontario, Canada,* Fats and Oils: Chemistry, Physics, and Applications

C. E. Devine, *Technology Development Group, HortResearch, Hamilton, New Zealand,* Meat and Electrical Stimulation; Meat Science

D. E. Dixon-Holland, *Neogen Corporation, Lansing, Michigan,* Immunological Methodology

Cindy Dominguez, *Whitehall-Robins, Madison, New Jersey,* Osteoporosis

Mark L. Dreher, *Mead Johnson Nutritionals, Bristol-Myers Squibb, Evansville, Indiana,* Carbohydrates: Functionality and Physiological Significance

Grant E. DuBois, *The Coca-Cola Company, Atlanta, Georgia,* Sweeteners: Nonnutritive

Muri Dueppen, *AOAC INTERNATIONAL, Gaithersburg, Maryland,* AOAC INTERNATIONAL

C. Patrick Dunne, *U.S. Army Natick, Natick, Massachusetts,* Military Food

Richard A. Durst, *University of Florida, Gainesville, Florida,* Hydrogen-Ion Activity (pH)

Phyllis Entis, *QA Life Sciences, San Diego, California,* Membrane Filtration Systems

Marilyn Erickson, *University of Georgia, Griffin, Georgia,* Food Spoilage

Ralph Ernst, *University of California, Davis, California,* Poultry: Meat From Avian Species

Felix E. Escher, *Swiss Federal Institute of Technology (ETH), Zurich, Switzerland,* International Development: Switzerland

N. A. M. Eskin, *University of Manitoba, Winnipeg, Manitoba, Canada,* Canola Oil

Terry D. Etherton, *Penn State University, University Park, Pennsylvania,* Lipids: Nutrition

David A. Evans, *University of Massachusetts, Amherst, Massachusetts,* Disinfectants

James Faller, *University of Illinois, Urbana, Illinois,* Food Processing

Daniel F. Farkas, *Oregon State University, Corvallis, Oregon,* High-Pressure Processing

Paul Fazio, *Häagen-Dazs Research and Development, Edina, Minnesota,* Ice Cream and Frozen Dessert

P. J. Fellers, *State of Florida Department of Citrus, Lake Alfred, Florida,* Limonin

P. Fellows, *Intermediate Technology Development Group, Rugby, United Kingdom,* Dehydration

Robin M. Fenwick, *New Zealand Dairy Board, Wellington, New Zealand,* Dairy Ingredients for Foods

Mario P. de Figueiredo, *Consultant, Chesterfield, Missouri,* Quality Assurance

Marshall Fishman, *USDA / ARS, Wyndmoor, Pennsylvania,* Pectic Substances

Dick Fizzell, *General Foods USA, White Plains, New York,* Breakfast Cereals

Henry Fleming, *USDA / ARS, Raleigh, North Carolina,* Vegetables, Pickling

John D. Floros, *Purdue University, West Lafayette, Indiana,* Mass Transfer and Diffusion in Foods; Optimization Methods in Food Processing and Engineering

F. J. Francis, *Editor-in-Chief, University of Massachusetts, Amherst, Massachusetts,* Aromatherapy; Biologically Stable Intermediates; Color and Food; Colorants Section; Fruits, Semi-Tropical; Fruits, Temperate; Fruits, Tropical

Daniel Y. C. Fung, *Kansas State University, Manhattan, Kansas,* Food Fermentation; Foodborne Diseases; Microbiology of Foods; Rapid Methods of Microbiological Analysis

Carol A. Gallagher, *Saint Joseph's University, Philadelphia, Pennsylvania,* Food Marketing

Elisabeth L. Garcia, *University of California, Davis, California,* Food Safety and Risk Communication; Food Safety and Risk Management; Pesticide Residues in Food; Toxicology and Risk Assessment

David R. Gard, *Solutia, Inc., St. Louis, Missouri,* Phosphates and Food Processing

Marianne H. Gellette, *McCormick & Co., Inc., Hunt Valley, Maryland,* Vanilla Extract

Aristippos Gennadios, *Banner Pharmacaps, Inc., High Point, North Carolina,* Edible Films and Coatings

Richard J. George, *Saint Joseph's University, Philadelphia, Pennsylvania,* Food Marketing

Shibani A. Ghosh, *University of Massachusetts, Amherst, Massachusetts,* Affluence, Food Excess, and Nutritional Disorders; Hunger, Food Deprivation, and Nutritional Disorders

Linda Gilbert, *HealthFocus, Inc., Des Moines, Iowa,* Vegetarian and Vegetarian-Aware Eating Trends

H. Douglas Goff, *University of Guelph, Guelph, Ontario, Canada,* Food Preservation

M. M. Góngora-Nieto, *Washington State University, Pullman, Washington,* Water Activity: Microbiology

J. Richard Gorham, *Uniformed Services University of the Health Sciences, Bethesda, Maryland,* Filth and Extraneous Matter in Food

Harold N. Graham, *New York City, New York,* Caffeine; Tannins; Tea

Bob Graves, *Balchem Corporation, Slate Hill, New York,* Encapsulation Techniques

Bruce Greenberg, *University of Waterloo, Waterloo, Ontario, Canada,* Photosynthesis

Scott M. Griffith, *Admix, Inc., Manchester, New Hampshire,* Beverages: Carbonated; Beverages: Noncarbonated

James A. Guzinski, *Kalsec, Inc., Kalamazoo, Michigan,* Spices and Herbs: Natural Extractives

Norman F. Haard, *University of California, Davis, California,* Marine Enzymes

Leigh Hadden White, *Häagen-Dazs Research and Development, Edina, Minnesota,* Ice Cream and Frozen Dessert

R. L. Hall, *Baltimore, Maryland,* International Union of Food Science and Technology

Yong D. Hang, *Cornell University, Geneva, New York,* Waste Management: Fruits and Vegetables

Milford A. Hanna, *University of Nebraska, Lincoln, Nebraska,* Edible Films and Coatings

Robert Harkins, *North Brunswick, New Jersey,* Sugar: Substitutes, Bulk, Reduced Calorie

Ronald D. Harris, *Kraft General Foods, Inc., Glenview, Illinois,* Cheese

Robert C. Hastert, *Hastech Corporation, Omaha, Nebraska,* Hydrogenation

James G. Hawkes, *Continental Colloids, Inc., Chicago, Illinois,* Food Processing: Effect on Nutritional Quality

Jim Heber, *Helsingborg, Sweden,* Freezing Systems for the Food Industry

Susan L. Hefle, *University of Nebraska, Lincoln, Nebraska,* Food Allergy

Barbara B. Heidolph, *Solutia, Inc., St. Louis, Missouri,* Phosphates and Food Processing

C. P. Herman, *University of Toronto, Toronto, Ontario, Canada,* Appetite

David Hettinga, *Land O' Lakes, Arden Hills, Minnesota,* Butter and Butter Products

S. Hill, *International Food Information Service, Reading, United Kingdom,* International Food Information Service (IFIS)

John T. Hines, *Kraft General Foods, Inc., Glenview, Illinois,* Cheese

Alan R. Hirsch, *Smell and Taste Treatment and Research Foundation, Chicago, Illinois,* Aphrodisiacs and Stimulants

Arnold L. Hirsch, *Alpharma, Inc.,* Vitamins: Vitamin D

Edward Hirsch, *U.S. Army Natick, Natick, Massachusetts,* Military Food

Y. C. Ho, *Rutgers University, New Brunswick, New Jersey,* Packaging: Part I—General Considerations; Packaging: Part III—Materials

Patrick G. Hoffman, *McCormick & Co., Inc., Hunt Valley, Maryland,* Vanilla Extract

Stefan Hogekamp, *Universität Karlsruhe, Karlsruhe, Germany,* Agglomeration and Agglomerator Systems

David N. Holcomb, *Technical University of Nova Scotia, Halifax, Nova Scotia, Canada,* Rheology

Stanley Holgate, *U.S. Army Natick Research, Development and Engineering Center, Natick, Massachusetts,* United States: Armed Forces Food Research and Development

D. E. Hood, *Baltimore, Maryland,* International Union of Food Science and Technology

R. Carl Hoseney, *R & R Research Services, Manhattan, Kansas,* Cereals Science and Technology

R. J. Hron, *USDA, Southern Regional Research Center,* Oilseeds and Vegetable Oils

Fu-hung Hsieh, *University of Missouri, Columbia, Missouri,* Extrusion; Extrusion Cooking; Ultrafiltration and Reverse Osmosis

Victor Huang, *Häagen-Dazs Research and Development, Edina, Minnesota,* Ice Cream and Frozen Dessert

Y. H. Hui, *American Food and Nutrition Center, Cutten, California,* Alkaloids; Cultural Nutrition; Food Processing: Standard Industrial Classification; Food Utilization; Shellfish; Soybeans and Soybean Processing

W. Jeffrey Hurst, *Hershey Foods Technical Center, Hershey, Pennsylvania,* Laboratory Robotics and Automation

Robert L. Jackman, *University of Guelph, Guelph, Ontario, Canada,* Proteins: Denaturation and Food Processing

Lauren Jackson, *FDA-NCFST, Argo, Illinois,* Mycotoxin Analysis

Claude Janelle, *Agriculture and AgriFood Canada, Ottawa, Ontario, Canada,* International Development: Canada

D. S. Jayas, *University of Manitoba, Winnipeg, Manitoba, Canada,* Cleaning-in-Place (CIP); Energy Usage in Food Processing Plants

P. Jelen, *University of Alberta, Edmonton, Alberta, Canada,* Whey: Composition, Properties, Processing, and Uses

B. Jessen, *Danish Meat Research Institute, Roskilde, Denmark,* Meat Starter Cultures and Meat Product Manufacturing

Mark Kantor, *University of Maryland, College Park, Maryland,* Fresh-Cut Fruits and Vegetables: Modified Atmosphere Packaging

George Katsushi Iwama, *University of British Columbia, Vancouver, British Columbia, Canada,* Fishes: Anatomy and Physiology; Fishes: Species of Economic Importance

Frances Katz, *Institute of Food Technologists, Chicago, Illinois,* Institute of Food Technologists (IFT)

Tetsuya Kawakita, *Ajinomoto Company, Inc., Kawasaki, Japan,* Nucleotides; Proteins: Amino Acids

Michael Keane, *University College Cork, Cork, Ireland,* International Development: Ireland

Patricia Kearney, *PMK Associates, Arlington, Virginia,* Nuts

J. F. Kefford, *North Ryde, New South Wales, Australia,* International Development: Australia

D. Mark Kettunen, *General Foods USA, Tarrytown, New York,* Snack Food Technology

George G. Khachatourians, *University of Saskatchewan, Saskatoon, Saskatchewan, Canada,* Genetic Engineering: Principles and Applications

J. W. Kiceniuk, *Fernleigh House, Halifax, Nova Scotia, Canada,* Lobster: Biology and Technology

Seok Joong Kim, *The University of Texas Health Science Center, San Antonio, Texas,* Phytochemicals: Melatonin

K. J. Kirkpatrick, *New Zealand Dairy Board, Wellington, New Zealand,* Dairy Ingredients for Foods

Donald Klingborg, *University of California, Davis, California,* Bovine Spongiform Encephalopathy (BSE)

William H. Knightly, *Emulsion Technology, Inc., Wilmington, Delaware,* Emulsifier Technology in Foods

Hannu Korkeala, *VTT Biotechnology and Food Research, VTT, Finland,* International Development: Finland

Penny M. Kris-Etherton, *Penn State University, University Park, Pennsylvania,* Lipids: Nutrition

David Kritchevsky, *The Wistar Institute, Philadelphia, Pennsylvania,* Cancer Risk and Diet

Donald H. Kropf, *Kansas State University, Manhattan, Kansas,* Meat, Modified Atmosphere Packaging

Volker Kuellmer, *Hoffman-La Roche, Inc.,* Vitamins: Ascorbic Acid

Rauno Lampi, *Westboro, Massachusetts,* Retort Pouch

Duane K. Larick, *North Carolina State University, Raleigh, North Carolina,* Poultry Flavors

A. K. Lau, *University of British Columbia, Vancouver, British Columbia, Canada,* Evaporation

Chang Y. Lee, *Cornell University, Geneva, New York,* Browning Reaction, Enzymatic; Phenolic Compounds

Chong M. Lee, *University of Rhode Island, Kingston, Rhode Island,* Surimi: Science and Technology

Lothar Leistner, *Federal Centre for Meat Research, Kulmbach, Germany,* Hurdle Technology

D. Paul Leitch, *U.S. Army Natick Research, Development and Engineering Center, Natick, Massachusetts,* United States: Armed Forces Food Research and Development

Eunice C. Y. Li-Chan, *University of British Columbia, Vancouver, British Columbia, Canada,* Hydrophobicity in Food Protein Systems

Hanhua Liang, *Purdue University, West Lafayette, Indiana,* Mass Transfer and Diffusion in Foods

I Chiu Liao, *Taiwan Fisheries Research Institute, Keelung, Taiwan,* Eel

Rong C. Lin, *U.S. Food and Drug Administration, Washington, D.C.,* Water Activity: Good Manufacturing Practice

Rudy R. Lin, *Swift-Eckrich, Inc., Downers Grove, Illinois,* Elastins and Meat Ligaments

Miia Lindström, *VTT Biotechnology and Food Research, VTT, Finland,* International Development: Finland

Ming-Sai Liu, *Food Industry Research and Development Institute, Hsinchu, Taiwan,* Bamboo Shoots

John K. Lodge, *University of California, Berkeley, California,* Phytochemicals: Lipoic Acid

Goran Löndahl, *Helsingborg, Sweden,* Freezing Systems for the Food Industry

Austin R. Long, *Louisiana State University, Baton Rouge, Louisiana,* Antibiotics in Foods of Animal Origin

G. R. Longdill, *Meat Industry Research Institute of New Zealand, Hamilton, New Zealand,* Meat Slaughtering and Processing Equipment

A. López-Malo, *Washington State University, Pullman, Washington,* Water Activity: Microbiology

B. S. Luh, *University of California, Davis, California,* Vegetable Processing

Ted M. Lupina, *Kalsec, Inc., Kalamazoo, Michigan,* Spices and Herbs: Natural Extractives

Bernadene Magnuson, *University of Idaho, Moscow, Idaho,* Food Toxicology

Raymond R. Mahoney, *University of Massachusetts, Amherst, Massachusetts,* Enzymology; Immobilized Enzymes

Charles H. Manley, *Takasago International Corporation, USA, Inc., Teterboro, New Jersey,* Dairy Flavors

Robert A. Martin, *Hershey Foods Technical Center, Hershey, Pennsylvania,* Laboratory Robotics and Automation

Paul R. Mathewson, *Food Technology Resource Group, Park City, Utah,* Enzymes in Food Production

Donald N. Maynard, *University of Florida, Bradenton, Florida,* Vegetable Production

G. Mazza, *Pacific Agri-Food Research Centre, Summerland, British Columbia, Canada,* Functional Foods

Terry L. McAninch, *Birko Corporation, Westminster, Colorado,* Detergents

Julian McClements, *University of Massachusetts, Amherst, Massachusetts,* Food Analysis

B. E. McDonald, *University of Manitoba, Winnipeg, Manitoba, Canada,* Canola Oil

Richard R. McFeaters, *Nabisco, Inc., East Hanover, New Jersey,* Biscuit and Cracker Technology

Martina McGloughlin, *University of California, Davis, California,* Genetic Engineering: Animals

John U. McGregor, *Louisiana State University Agricultural Center, Baton Rouge, Louisiana,* Cultured Milk Products

Kristen McNutt, *Consumer Choices, Inc., Winfield, Illinois,* Food and Nutrition Science Alliance Organizations

M. Medina, *National Institute for Agricultural and Food Research and Technology (INIA), Madrid, Spain,* International Development: Spain

M. A. A. Meireles, *State University of Campinas—UNICAMP, São Paulo, Brazil,* Supercritical Fluid Technology

H. H. Meissner, *ARC—Animal Nutrition and Animal Products Institute, Irene, South Africa,* International Development: South Africa

Alice Meyer, *U.S. Army Natick, Natick, Massachusetts,* Military Food

Suzette Middleton, *Procter & Gamble Co., Cincinnati, Ohio,* Fats and Oils: Substitutes

A. J. Miller, *United States Department of Agriculture, Philadelphia, Pennsylvania,* Sausages

J. David Miller, *Carleton University, Ottawa, Ontario, Canada,* Mycotoxins

David B. Min, *The Ohio State University, Columbus, Ohio,* Fats and Oils: Flavors

Enrico Miniati, *Perugia University, Perugia, Italy,* Phytochemicals: *Vaccinium*

Vikram V. Mistry, *South Dakota State University, Brookings, South Dakota,* Milk and Milk Products

Peter L. Montagne, *Sharples, Inc., Warminster, Pennsylvania,* Centrifuges: Principles and Applications

Alanna J. Moshfegh, *USDA/ARS, Beltsville, Maryland,* Food Consumption Surveys in the U.S. Department of Agriculture

Howard R. Moskowitz, *Moskowitz Jacobs Inc., White Plains, New York,* Sensory Science: Principles and Applications

K. Darwin Murrell, *USDA/ARS/NRRC, Peoria, Illinois,* Parasitic Organisms

Eric C. Mussen, *University of California Entomology Extension, Davis, California,* Beekeeping

S. Nakai, *University of British Columbia, Vancouver, British Columbia, Canada,* Proteins: Structure and Functionality

Ananth Narayan, *U.S. Army Natick, Natick, Massachusetts,* Military Food

Jennifer Near, *University of California, Davis, California,* Poultry: Meat From Avian Species

A. C. Noble, *University of California, Davis, California,* Wine

John Norback, *University of Wisconsin, Madison, Wisconsin,* Food Processing: Technology, Engineering, and Management

M. Nuñez, *National Institute for Agricultural and Food Research and Technology (INIA), Madrid, Spain,* International Development: Spain

K. Rajinder Nuath, *Kraft, General Foods, Inc., Glenview, Illinois,* Cheese

John O'Brien, *University of Surrey, Guildford, United Kingdom,* International Development: United Kingdom

Robert Olsen, *Schreiber Foods, Inc., Tempe, Arizona,* Artificial Intelligence; Computer Applications in the Food Industry

B. Dave Oomah, *Pacific Agri-Food Research Centre, Summerland, British Columbia, Canada,* Functional Foods

Jose M. Ordovas, *Tufts University, Boston, Massachusetts,* Olives and Olive Oil

Thomas M. Ott, *Purepulse Technologies, Westwood, New Jersey,* Pulsed Light Processing

Robert A. Outten, *Hoffmann-La Roche, Inc.,* Vitamins: Biotin

Joseph L. Owades, *Consultant, Sonoma, California,* Alcoholic Beverages and Human Responses; Beer; Distilled Beverage Spirits

Lester Packer, *University of California, Berkeley, California,* Phytochemicals: Lipoic Acid

Mahesh Padmanabhan, *University of Minnesota, St. Paul, Minnesota,* Extrusion Processing: Texture and Rheology

Patti Pagliuco, *Institute of Food Technologists, Chicago, Illinois,* Institute of Food Technologists: Awards

William D. Pandolfe, *APV Americas-Homogenizers, Wilmington, Massachusetts,* Colloid Mills; Homogenizers

Robert Parker, *Cornell University, Ithaca, New York,* Phytochemicals: Carotenoids

John G. Parsons, *South Dakota State University, Brookings, South Dakota,* Milk and Milk Products

Kenneth G. Payie, *University of Guelph, Guelph, Ontario, Canada,* Proteins: Denaturation and Food Processing

Peter L. Pellett, *University of Massachusetts, Amherst, Massachusetts,* Affluence, Food Excess, and Nutritional Disorders; Hunger, Food Deprivation, and Nutritional Disorders

Per Oskar Persson, *Helsingborg, Sweden,* Freezing Systems for the Food Industry

J. F. Pfeiffer, *University of California, Davis, California,* Wine

Thomas W. Phillips, *Oklahoma State University, Stillwater, Oklahoma,* Postharvest Integrated Pest Management

Raul H. Piedrahita, *University of California, Davis, California,* Aquaculture: Engineering and Construction

George M. Pigott, *Sea Resources Engineering, Inc., Kirkland, Washington,* Extraction; Fish and Shellfish Products; Heat; Heat Exchangers; Heat Transfer; Thermodynamics

J. Terry Pitts, *USDA/ARS, Manhattan, Kansas,* Grains and Protectants

Y. Pomeranz, *Washington State University, Pullman, Washington,* Wheat Science and Technology

Dorothy Pond-Smith, *Washington State University, Pullman, Washington,* Foodservice Systems

William Porter, *U.S. Army Natick, Natick, Massachusetts,* Military Food

John J. Powers, *University of Georgia, Athens, Georgia,* Food Science and Technology: Definition and Development; Food Science and Technology: The Profession; Institute of Food Technologists: History and Perspectives; Sensory Science: Standardization and Instrumentation

J. H. Prentice, *Axminster, Devon, United Kingdom,* Cheese Rheology

R. L. Preston, *Texas Tech University, Lubbock, Texas,* Livestock Feeds

George Purvis, *Purvis Consulting, Inc., Fremont, Michigan,* Infant Foods

Jeff Rattray, *Purdue University, West Lafayette, Indiana,* Mass Transfer and Diffusion in Foods; Optimization Methods in Food Processing and Engineering

Thimma R. Rawalpally, *Hoffmann-La Roche, Inc.,* Vitamins: Folic Acid; Vitamins: Pantothenic Acid

Carrie E. Regenstein, *University of Rochester, Rochester, New York,* Kosher Foods and Food Processing

Joe M. Regenstein, *Cornell Information Technologies, Ithaca, New York,* Kosher Foods and Food Processing; Muslim Dietary Laws: Food Processing and Marketing

D. S. Reid, *University of California, Davis, California,* Food Freezing

Russel J. Reiter, *The University of Texas Health Science Center, San Antonio, Texas,* Phytochemicals: Melatonin

Mikelle Roeder, *University of Idaho, Moscow, Idaho,* Animal Science and Livestock Production

Richard A. Roeder, *University of Idaho, Moscow, Idaho,* Animal Science and Livestock Production

Debra Rooney, *Abbott Laboratories, Columbus, Ohio,* Enteral Formulas and Feeding Systems

Diane Rosenwald, *Häagen-Dazs Research and Development, Edina, Minnesota,* Ice Cream and Frozen Dessert

Leif Rynnel, *Helsingborg, Sweden,* Freezing Systems for the Food Industry

Raymond G. Saba, *Rutgers University, New Brunswick, New Jersey,* Packaging: Part I—General Considerations; Packaging: Part III—Materials

Fred W. Schenck, *Sun City, Arizona,* Syrups (Starch Sweeteners and Other Syrups)

Donald Schlimme, *University of Maryland, College Park, Maryland,* Fresh-Cut Fruits and Vegetables: Modified Atmosphere Packaging

Glenn R. Schmidt, *Colorado State University, Fort Collins, Colorado,* Meat Products

Helmar Schubert, *Universität Karlsruhe, Karlsruhe, Germany,* Agglomeration and Agglomerator Systems

Gerald L. Schulz, *U.S. Army Natick Research, Development and Engineering Center, Natick, Massachusetts,* Military Food; United States: Armed Forces Food Research and Development

Henry G. Schwartzberg, *University of Massachusetts, Amherst, Massachusetts,* Coffee; Freeze Concentration; Freeze Drying

John W. Scott, *Hoffmann-La Roche, Inc.,* Vitamins: Survey; Vitamins: Vitamin B_{12}

V. N. Scott, *National Food Processors Association, Washington, D.C.,* Low-Acid and Acidified Foods

Jeremy D. Selman, *Campden Food & Drink Research Association, Campden, Gloucestershire, United Kingdom,* Blanching

Amaral Sequeira-Munoz, *University of California, Davis, California,* Marine Enzymes

P. Shatadal, *University of Manitoba, Winnipeg, Manitoba, Canada,* Cleaning-in-Place (CIP)

Kalidas Shetty, *University of Massachusetts, Amherst, Massachusetts,* Phytochemicals: Biotechnology of Phenolic Phytochemicals for Food Preservatives and Functional Food Applications; Solid-State Fermentation and Value–Added Utilization of Fruit and Vegetable Processing By-Products

Robert L. Shewfelt, *University of Georgia, Athens, Georgia,* Food Crops: Postharvest Deterioration; Food Crops: Varietal Differences, Maturation, Ripening, and Senescence

John M. Siddle, *Nestlé Research and Development Center, Inc., New Milford, Connecticut,* Chilled Foods

Joseph B. Sieczka, *Cornell University, Riverhead, New York,* Potatoes and Potato Processing

Juan L. Silva, *Mississippi State University, Mississippi State, Mississippi,* Machine Vision Systems: Process and Quality Control

Gerald Silverman, *U.S. Army Natick, Natick, Massachusetts,* Military Food

Benjamin K. Simpson, *University of California, Davis, California,* Marine Enzymes

Kenneth L. Simpson, *University of Rhode Island, Kingston, Rhode Island,* Carotenoid Pigments

Durand Smith, *Cyclopss Corporation, Salt Lake City, Utah,* Ozone and Food Processing

Jeff L. Smith, *University of Guelph, Guelph, Ontario, Canada,* Proteins: Denaturation and Food Processing

J. F. Smullen, *Hershey Foods Corporation, Hershey Park, Pennsylvania,* Chocolate and Cocoa; Licorice Confectionery

Laszlo P. Somogyi, *Consulting Food Scientist, Kensington, California,* Food Additives; Fruit Dehydration; Vegetable Dehydration

Peter Sporns, *University of Alberta, Edmonton, Alberta, Canada,* Food Chemistry and Biochemistry

William J. Stadelman, *Purdue University, West Lafayette, Indiana,* Eggs and Egg Products

John L. Stanton, *Saint Joseph's University, Philadelphia, Pennsylvania,* Food Marketing

Aliza Stark, *Hebrew University of Jerusalem, Rehovot, Israel,* Fiber, Dietary

Clyde E. Stauffer, *Technical Food Consultants, Cincinnati, Ohio,* Emulsifiers, Stabilizers, and Thickeners; Enzyme Assays for Food Scientists; Bakery Leavening Agents; Bakery Specialty Products

Donald F. Steenson, *The Ohio State University, Columbus, Ohio,* Fats and Oils: Flavors

Keith H. Steinkraus, *Food Industry Research and Development Institute, Hsinchu, Taiwan,* Soy Foods, Fermented

K. Stevenson, *National Food Processors Association, Dublin, California,* Aseptic Processing and Packaging Systems

Robert R. Stickney, *Texas Sea Grant College Program, Bryan, Texas,* Aquaculture

William Stoddard, *Cyclopss Corporation, Salt Lake City, Utah,* Ozone and Food Processing

G. S. Stoewsand, *Cornell University, Geneva, New York,* Phytochemicals: Wine

Jean Storlie, *Nutritional Labeling Solutions, Maple Grove, Minnesota,* Packaging: Part II—Labeling

Jong-Ching Su, *National Taiwan University, Taipei, Taiwan,* International Development: Taiwan

Ellen J. Sullivan, *Institute of Food Technologists, Chicago, Illinois,* Food and Nutrition Science Alliance: FANSA

Taneko Suzuki, *Nihon University, Kanagawa, Japan,* Krill Protein Processing

J. E. Swan, *University of Waikato, Hamilton, New Zealand,* Animal By-Product Processing

Hugh Symons, *Kingston Upon Thames, England,* Refrigerated Foods: Food Freezing and Processing; Refrigerated Foods: Food Freezing and World Food Supply; Refrigerated Foods: Handling and Inventory

Bernard F. Szuhal, *Central Soya Company, Inc., Fort Wayne, Indiana,* Lecithins

Donna R. Tainter, *Tone Brothers, Inc., Ankeny, Iowa,* Spices and Seasonings

Takuji Tanaka, *University of Guelph, Guelph, Ontario, Canada,* Proteins: Denaturation and Food Processing

M. S. Tapia, *Washington State University, Pullman, Washington,* Water Activity: Microbiology

Irwin Taub, *U.S. Army Natick, Natick, Massachusetts,* Military Food

Steve L. Taylor, *University of Nebraska, Lincoln, Nebraska,* Food Allergy

A. A. Teixeira, *University of Florida, Gainesville, Florida,* Thermal Processing of Food

A. R. Thicker, *Institute for Toxicology and Chemotherapy, Heidelberg, Germany,* Nitrosamines

Gordon Timbers, *Agriculture and AgriFood Canada, Ottawa, Ontario, Canada,* International Development: Canada

Katherine S. Tippett, *USDA/ARS, Beltsville, Maryland,* Food Consumption Surveys in the U.S. Department of Agriculture

Ross Tomaino, *University of Massachusetts, Amherst, Massachusetts,* Phytochemicals: Antioxidants

Cindy B. S. Tong, *USDA/ARS, Beltsville, Maryland,* Food Crops: Storage

Susan Trimbo, *Whitehall-Robins, Madison, New Jersey,* Osteoporosis

M. Tsoubeli, *Campbell Soup Co., Camden, New Jersey,* Food Microstructure

Marvin A. Tung, *Technical University of Nova Scotia, Halifax, Nova Scotia, Canada,* Rheology

F. J. Vaccarino, *University of Toronto, Toronto, Ontario, Canada,* Appetite

Susan D. Van Arnum, *Westfield, New Jersey,* Vitamins: Niacin, Nicotinamide, and Nicotinic Acid; Vitamins: Vitamin A; Vitamins: Vitamin K

Martien van den Hoven, *Creamy Creations, Rijkevoort, The Netherlands,* Dairy Ingredients: Applications in Meat, Poultry, and Seafoods

Y. Vodovotz, *NASA Johnson Space Center, Houston, Texas,* Food Microstructure

A. L. Waldroup, *University of Arkansas, Fayetteville, Arkansas,* Poultry Meat Microbiology

Michael Sanderson Walker, *P&O Nedlloyd, London, United Kingdom,* Refrigerated Foods: Transportation

Chien Yi Wang, *USDA/ARS, Beltsville, Maryland,* Controlled Atmospheres for Fresh Fruits and Vegetables

Pie Yi Wang, *Armour Swift-Eckrich, Downers Grove, Illinois,* Meat Processing: Technology and Engineering

Shaw Wang, *Rutgers University, Piscatawav, New Jersey,* Kinetics; Surface (Interfacial) Tension

Alley E. Watada, *USDA/ARS, Beltsville, Maryland,* Controlled Atmospheres for Fresh Fruits and Vegetables; Food Crops: Storage

L. M. Weddig, *National Food Processors Association, Washington, D.C.,* Low-Acid and Acidified Foods

Herb Weiss, *Balchem Corporation, Slate Hill, New York,* Encapsulation Techniques

John C. Wekell, *National Oceanic and Atmospheric Administration, Seattle, Washington,* Seafood: Flavors and Quality

Curtis L. Weller, *University of Nebraska, Lincoln, Nebraska,* Edible Films and Coatings

Phillip Wells, *Bestfoods North America, Somerset, New Jersey,* Nutritional Labeling

J. Welti-Chanes, *Washington State University, Pullman, Washington,* Water Activity: Microbiology

R. C. Whiting, *United States Department of Agriculture, Philadelphia, Pennsylvania,* Sausages

W. W. Widmer, *State of Florida Department of Citrus, Lake Alfred, Florida,* Limonin

Cecilia Wilkinson Enns, *USDA/ARS, Beltsville, Maryland,* Food Consumption Surveys in the U.S. Department of Agriculture

Carl K. Winter, *University of California, Davis, California,* Food Safety and Risk Communication; Food Safety and Risk Management; Pesticide Residues in Food; Toxicology and Risk Assessment

Dominic W. S. Wong, *United States Department of Agriculture, Albany, California,* Food Chemistry: Mechanism and Theory; Oxidation

Adrienne E. Woytowich, *University of Saskatchewan, Saskatoon, Saskatchewan, Canada,* Genetic Engineering: Principles and Applications

Bih Keng Wu, *Food Industry Research and Development Institute, Hsinchu, Taiwan,* Mushrooms: Processing

Y. Victor Wu, *USDA/ARS, Peoria, Illinois,* Cereals, Nutrients, and Agricultural Practices

Rickey Y. Yada, *University of Guelph, Guelph, Ontario, Canada,* Proteins: Denaturation and Food Processing

Kit L. Yam, *Rutgers University, New Brunswick, New Jersey,* Packaging: Part I—General Considerations; Packaging: Part III—Materials

T. C. S. Yang, *U.S. Army Soldier and Biological Chemical Command, Natick, Massachusetts,* Aseptic Processing: Ohmic Heating

V. A. Yaylayan, *McGill University, St. Anne de Bellevue, Quebec, Canada,* Flavor Chemistry

Fumio Yoneda, *Fujimoto Pharmaceutical Corporation,* Vitamins: Riboflavin (B_2)

M. Yoshikawa, *Kobe Women's University, Kobe, Japan,* Peptides

Gideon Zeidler, *University of California, Davis, California,* Poultry: Meat From Avian Species

Zuoxing Zheng, *University of Massachusetts, Amherst, Massachusetts,* Solid-State Fermentation and Value–Added Utilization of Fruit and Vegetable Processing By-Products

B. L. Zoumas, *Hershey Foods Corporation, Hershey, Pennsylvania,* Chocolate and Cocoa

CONVERSION FACTORS, ABBREVIATIONS, AND UNIT SYMBOLS

SI Units (Adopted 1960)

The International System of Units (abbreviated SI), is being implemented throughout the world. This measurement system is a modernized version of the MKSA (meter, kilogram, second, ampere) system, and its details are published and controlled by an international treaty organization (The International Bureau of Weights and Measures) (1).

SI units are divided into three classes:

BASE UNITS	
length	meter† (m)
mass	kilogram (kg)
time	second (s)
electric current	ampere (A)
thermodynamic temperature‡	kelvin (K)
amount of substance	mole (mol)
luminous intensity	candela (cd)

SUPPLEMENTARY UNITS	
plane angle	radian (rad)
solid angle	steradian (sr)

DERIVED UNITS AND OTHER ACCEPTABLE UNITS

These units are formed by combining base units, supplementary units, and other derived units (2–4). Those derived units having special names and symbols are marked with an asterisk in the list below.

Quantity	Unit	Symbol	Acceptable equivalent
*absorbed dose	gray	Gy	J/kg
acceleration	meter per second squared	m/s^2	
*activity (of a radionuclide)	becquerel	Bq	1/s
area	square kilometer	km^2	
	square hectometer	hm^2	ha (hectare)
	square meter	m^2	
concentration (of amount of substance)	mole per cubic meter	mol/m^3	
current density	ampere per square meter	A/m^2	
density, mass density	kilogram per cubic meter	kg/m^3	g/L; mg/cm^3
dipole moment (quantity)	coulomb meter	C · m	
*dose equivalent	sievert	Sv	J/kg
*electric capacitance	farad	F	C/V
*electric charge, quantity of electricity	coulomb	C	A · s
electric charge density	coulomb per cubic meter	C/m^3	
*electric conductance	siemens	S	A/V
electric field strength	volt per meter	V/m	
electric flux density	coulomb per square meter	C/m^2	
*electric potential, potential difference, electromotive force	volt	V	W/A
*electric resistance	ohm	Ω	V/A

†The spellings "metre" and "litre" are preferred by ASTM; however, "-er" is used in the encyclopedia.

‡Wide use is made of Celsius temperature (t) defined by

$$t = T - T_0$$

where T is the thermodynamic temperature, expressed in kelvin, and T_0 = 273.15 K by definition. A temperature interval may be expressed in degrees Celsius as well as in kelvin.

Quantity	Unit	Symbol	Acceptable equivalent
*energy, work, quantity of heat	megajoule	MJ	
	kilojoule	kJ	
	joule	J	N · m
	electronvolt†	eV†	
	kilowatt-hour†	kW · h†	
energy density	joule per cubic meter	J/m^3	
*force	kilonewton	kN	
	newton	N	$kg \cdot m/s^2$
*frequency	megahertz	MHz	
	hertz	Hz	1/s
heat capacity, entropy	joule per kelvin	J/K	
heat capacity (specific), specific entropy	joule per kilogram kelvin	J/(kg · K)	
heat-transfer coefficient	watt per square meter kelvin	$W/(m^2 \cdot K)$	
*illuminance	lux	lx	lm/m^2
*inductance	henry	H	Wb/A
linear density	kilogram per meter	kg/m	
luminance	candela per square meter	cd/m^2	
*luminous flux	lumen	lm	cd · sr
magnetic field strength	ampere per meter	A/m	
*magnetic flux	weber	Wb	V · s
*magnetic flux density	tesla	T	Wb/m^2
molar energy	joule per mole	J/mol	
molar entropy, molar heat capacity	joule per mole kelvin	J/(mol · K)	
moment of force, torque	newton meter	N · m	
momentum	kilogram meter per second	kg · m/s	
permeability	henry per meter	H/m	
permittivity	farad per meter	F/m	
*power, heat flow rate, radiant flux	kilowatt	kW	
	watt	W	J/s
power density, heat flux density, irradiance	watt per square meter	W/m^2	
*pressure, stress	megapascal	MPa	
	kilopascal	kPa	
	pascal	Pa	N/m^2
sound level	decibel	dB	
specific energy	joule per kilogram	J/kg	
specific volume	cubic meter per kilogram	m^3/kg	
surface tension	newton per meter	N/m	
thermal conductivity	watt per meter kelvin	W/(m · K)	
velocity	meter per second	m/s	
	kilometer per hour	km/h	
viscosity, dynamic	pascal second	Pa · s	
	millipascal second	mPa · s	
viscosity, kinematic	square meter per second	m^2/s	
	square millimeter per second	mm^2/s	
volume	cubic meter	m^3	
	cubic diameter	dm^3	L (liter) (5)
	cubic centimeter	cm^3	mL
wave number	1 per meter	m^{-1}	
	1 per centimeter	cm^{-1}	

†This non-SI unit is recognized by the CIPM as having to be retained because of practical importance or use in specialized fields (1).

In addition, there are 16 prefixes used to indicate order of magnitude, as follows:

Multiplication factor	Prefix	Symbol
10^{18}	exa	E
10^{15}	peta	P
10^{12}	tera	T
10^{9}	giga	G
10^{6}	mega	M
10^{3}	kilo	k
10^{2}	hecto	h[a]
10	deka	da[a]
10^{-1}	deci	d[a]
10^{-2}	centi	c[a]
10^{-3}	milli	m
10^{-6}	micro	μ
10^{-9}	nano	n
10^{-12}	pico	p
10^{-15}	femto	f
10^{-18}	atto	a

[a]Although hecto, deka, deci, and centi are SI prefixes, their use should be avoided except for SI unit-multiples for area and volume and nontechnical use of centimeter, as for body and clothing measurement.

For a complete description of SI and its use the reader is referred to ASTM E380 (4).

A representative list of conversion factors from non-SI to SI units is presented herewith. Factors are given to four significant figures. Exact relationships are followed by a dagger. A more complete list is given in the latest editions of ASTM E380 (4) and ANSI Z210.1 (6).

CONVERSION FACTORS TO SI UNITS

To convert from	To	Multiply by
acre	square meter (m^2)	4.047×10^{3}
angstrom	meter (m)	1.0×10^{-10}†
are	square meter (m^2)	1.0×10^{2}†
astronomical unit	meter (m)	1.496×10^{11}
atmosphere, standard	pascal (Pa)	1.013×10^{5}
bar	pascal (Pa)	1.0×10^{5}†
barn	square meter (m^2)	1.0×10^{-28}†
barrel (42 U.S. liquid gallons)	cubic meter (m^3)	0.1590
Bohr magneton (μ_B)	J/T	9.274×10^{-24}
Btu (International Table)	joule (J)	1.055×10^{3}
Btu (mean)	joule (J)	1.056×10^{3}
Btu (thermochemical)	joule (J)	1.054×10^{3}
bushel	cubic meter (m^3)	3.524×10^{-2}
calorie (International Table)	joule (J)	4.187
calorie (mean)	joule (J)	4.190
calorie (thermochemical)	joule (J)	4.184†
centipoise	pascal second ($Pa \cdot s$)	1.0×10^{-3}†
centistokes	square millimeter per second (mm^2/s)	1.0†
cfm (cubic foot per minute)	cubic meter per second (m^3/s)	4.72×10^{-4}
cubic inch	cubic meter (m^3)	1.639×10^{-5}
cubic foot	cubic meter (m^3)	2.832×10^{-2}
cubic yard	cubic meter (m^3)	0.7646
curie	becquerel (Bq)	3.70×10^{10}†
debye	coulomb meter ($C \cdot m$)	3.336×10^{-30}
degree (angle)	radian (rad)	1.745×10^{-2}

†Exact.

CONVERSION FACTORS TO SI UNITS

To convert from	To	Multiply by
denier (international)	kilogram per meter (kg/m)	1.111×10^{-7}
	tex‡	0.1111
dram (apothecaries')	kilogram (kg)	3.888×10^{-3}
dram (avoirdupois)	kilogram (kg)	1.772×10^{-3}
dram (U.S. fluid)	cubic meter (m^3)	3.697×10^{-6}
dyne	newton (N)	1.0×10^{-5}†
dyne/cm	newton per meter (N/m)	1.0×10^{-3}†
electronvolt	joule (J)	1.602×10^{-19}
erg	joule (J)	1.0×10^{-7}†
fathom	meter (m)	1.829
fluid ounce (U.S.)	cubic meter (m^3)	2.957×10^{-5}
foot	meter (m)	0.3048†
footcandle	lux (lx)	10.76
furlong	meter (m)	2.012×10^{-2}
gal	meter per second squared (m/s^2)	1.0×10^{-2}†
gallon (U.S. dry)	cubic meter (m^3)	4.405×10^{-3}
gallon (U.S. liquid)	cubic meter (m^3)	3.785×10^{-3}
gallon per minute (gpm)	cubic meter per second (m^3/s)	6.309×10^{-5}
	cubic meter per hour (m^3/h)	0.2271
gauss	tesla (T)	1.0×10^{-4}
gilbert	ampere (A)	0.7958
gill (U.S.)	cubic meter (m^3)	1.183×10^{-4}
grade	radian	1.571×10^{-2}
grain	kilogram (kg)	6.480×10^{-5}
gram force per denier	newton per tex (N/tex)	8.826×10^{-2}
hectare	square meter (m^2)	1.0×10^{4}†
horsepower (550 ft · lbf/s)	watt (W)	7.457×10^{2}
horsepower (boiler)	watt (W)	9.810×10^{3}
horsepower (electric)	watt (W)	7.46×10^{2}†
hundredweight (long)	kilogram (kg)	50.80
hundredweight (short)	kilogram (kg)	45.36
inch	meter (m)	2.54×10^{-2}†
inch of mercury (32°F)	pascal (Pa)	3.386×10^{3}
inch of water (39.2°F)	pascal (Pa)	2.491×10^{2}
kilogram-force	newton (N)	9.807
kilowatt hour	megajoule (MJ)	3.6†
kip	newton (N)	4.448×10^{3}
knot (international)	meter per second (m/S)	0.5144
lambert	candela per square meter (cd/m^3)	3.183×10^{3}
league (British nautical)	meter (m)	5.559×10^{3}
league (statute)	meter (m)	4.828×10^{3}
light year	meter (m)	9.461×10^{15}
liter (for fluids only)	cubic meter (m^3)	1.0×10^{-3}†
maxwell	weber (Wb)	1.0×10^{-8}†
micron	meter (m)	1.0×10^{-6}†
mil	meter (m)	2.54×10^{-5}†
mile (statute)	meter (m)	1.609×10^{3}
mile (U.S. nautical)	meter (m)	1.852×10^{3}†
mile per hour	meter per second (m/s)	0.4470
millibar	pascal (Pa)	1.0×10^{2}
millimeter of mercury (0°C)	pascal (Pa)	1.333×10^{2}†
minute (angular)	radian	2.909×10^{-4}
myriagram	kilogram (kg)	10

†Exact.
‡See footnote on p. xviii.

CONVERSION FACTORS TO SI UNITS

To convert from	To	Multiply by
myriameter	kilometer (km)	10
oersted	ampere per meter (A/m)	79.58
ounce (avoirdupois)	kilogram (kg)	2.835×10^{-2}
ounce (troy)	kilogram (kg)	3.110×10^{-2}
ounce (U.S. fluid)	cubic meter (m^3)	2.957×10^{-5}
ounce-force	newton (N)	0.2780
peck (U.S.)	cubic meter (m^3)	8.810×10^{-3}
pennyweight	kilogram (kg)	1.555×10^{-3}
pint (U.S. dry)	cubic meter (m^3)	5.506×10^{-4}
pint (U.S. liquid)	cubic meter (m^3)	4.732×10^{-4}
poise (absolute viscosity)	pascal second (Pa · s)	0.10†
pound (avoirdupois)	kilogram (kg)	0.4536
pound (troy)	kilogram (kg)	0.3732
poundal	newton (N)	0.1383
pound-force	newton (N)	4.448
pound force per square inch (psi)	pascal (Pa)	6.895×10^{3}
quart (U.S. dry)	cubic meter (m^3)	1.101×10^{-3}
quart (U.S. liquid)	cubic meter (m^3)	9.464×10^{-4}
quintal	kilogram (kg)	1.0×10^{2}†
rad	gray (Gy)	1.0×10^{-2}†
rod	meter (m)	5.029
roentgen	coulomb per kilogram (C/kg)	2.58×10^{-4}
second (angle)	radian (rad)	4.848×10^{-6}†
section	square meter (m^2)	2.590×10^{6}
slug	kilogram (kg)	14.59
spherical candle power	lumen (lm)	12.57
square inch	square meter (m^2)	6.452×10^{-4}
square foot	square meter (m^2)	9.290×10^{-2}
square mile	square meter (m^2)	2.590×10^{6}
square yard	square meter (m^2)	0.8361
stere	cubic meter (m^3)	1.0†
stokes (kinematic viscosity)	square meter per second (m^2/s)	1.0×10^{-4}†
tex	kilogram per meter (kg/m)	1.0×10^{-6}†
ton (long, 2240 pounds)	kilogram (kg)	1.016×10^{3}
ton (metric) (tonnee)	kilogram (kg)	1.0×10^{3}†
ton (short, 2000 pounds)	kilogram (kg)	9.072×10^{2}
torr	pascal (Pa)	1.333×10^{2}
unit pole	weber (Wb)	1.257×10^{-7}
yard	meter (m)	0.9144†

†Exact.

BIBLIOGRAPHY

1. The International Bureau of Weights and Measures, BIPM (Parc de Saint-Cloud, France) is described in Appendix X2 of Ref. 4. This bureau operates under the exclusive supervision of the International Committee for Weights and Measures (CIPM).
2. *Metric Editorial Guide (ANMC-78-1),* latest ed., American National Metric Council, 5410 Grosvenor Lane, Bethesda, Md. 20814, 1981.
3. *SI Units and Recommendations for the Use of Their Multiples and of Certain Other Units (ISO 1000-1981),* American National Standards Institute, 1430 Broadway, New York, 10018, 1981.
4. Based on *ASTM E380-89a (Standard Practice for Use of the International System of Units (SI)),* American Society for Testing and Materials, 1916 Race Street, Philadelphia, Pa. 19103, 1989.
5. *Fed. Reg.,* Dec. 10, 1976 (41 FR 36414).
6. For ANSI address, see Ref. 3.

R. P. Lukens
ASTM Committee E-43 on SI Practice

ACIDULANTS

Food acidulants are categorized either as general purpose acids or as specialty acids (1). General purpose acids are those that have a broad range of functions and can be used in most foods where acidity is desired or necessary. Specialty acids are those that are limited in their functionality and/or range of application.

Citric acid and malic acid are the predominant general purpose acidulants with tartaric and fumaric acids. Fumaric acid is included in this category even though its low solubility limits its potential range of application. However, it is used in popular and widely consumed food products.

All other acidulants fall into the specialty acid category. The most commonly used specialty acids are acetic acid (vinegar), cream of tartar (potassium acid tartrate), phosphoric acid, glucono-delta-lactone, acid phosphate salts, lactic acid, and adipic acid.

CITRIC ACID

Citric acid is the premier acid for the food and beverage industry because it offers a unique combination of desirable properties, ready availability in commercial quantities, and competitive pricing. It is estimated that citric acid worldwide accounts for more than 80% of general purpose acidulants used. It is found naturally in almost all living things, both plant and animal. It is the predominant acid, in substantial quantity, in citrus fruits (oranges, lemons, limes, etc), in berries (strawberries, raspberries, currants) and in pineapples. Citric acid is also the predominant acid in many vegetables, such as potatoes, tomatoes, asparagus, turnips, and peas, but in lower concentrations. The citrate ion occurs in all animal tissues and fluids. The total circulating citric acid in the serum of man is approximately 1 mg/kg of body weight (2).

Citric acid is manufactured by fermentation, a natural process using living organisms. The acid is recovered in pure crystalline form either as the anhydrous or monohydrate crystal depending on the temperature of crystallization. The transition temperature is 36.6°C. Crystallization above this temperature produces an anhydrous product while the monohydrate forms at lower temperatures. With an occasional exception for technological reasons, the anhydrous form is preferred for its physical stability and is the more widely available commercial form.

The salts of citric acid that are used in the food industry are sodium citrate dihydrate and anhydrous, monosodium citrate, potassium citrate monohydrate, calcium citrate tetrahydrate, and ferric ammonium citrate, in both brown and green powder. These salts are used either for functional purposes such as buffering or emulsification, as a source of cation for technological purposes, such as calcium to aid the gelation of low methoxyl pectin, or as a mineral source for food supplementation.

Citric acid is a hydroxy tribasic acid that is a white granule or powder. It is odorless with no characteristic taste other than tartness. Its physical and chemical properties are listed in Table 1. Citric acid and its salts are used across the broad spectrum of food and beverage products:

- Beverages
- Gelatin desserts
- Baked goods
- Jellies, jams, and preserves
- Candies
- Fruits and vegetables
- Dairy products
- Meats
- Seafood
- Fats and oils

MALIC ACID

This is the second most popular general purpose food acid although less than one-tenth the quantity of citric acid used. Malic acid is a white, odorless, crystalline powder or granule with a clean tart taste with no characterizing flavor of its own. The properties of malic acid are listed in Table 1.

Malic acid is made through chemical synthesis by the hydration of maleic acid. An inspection of its structure in Table 1 reveals that malic acid has an asymmetric carbon which provides for the existence of isomeric forms. The synthetic procedure for malic acid produces a mixture of the D and L isomers. The item of commerce, racemic D, L-malic acid does not occur in nature (3) although the acid is affirmed as GRAS for use in foods.

L-Malic acid is the isomeric form that is found in nature. It is the predominant acid in substantial quantity in apples and cherries and in lesser quantities in prunes, watermelon, squash, quince, plums, and mushrooms. Like citric acid, L-malic acid plays an essential part in carbohydrate metabolism in man and other animals.

There are no salts of malic acid that are items of commerce for the food and beverage industries. The application for malic and citric acid cover the same broad range of food categories.

TARTARIC ACID

While tartaric acid has the characteristics of a general purpose acid, fluctuating availability and price have caused users to reformulate where possible. Tartaric acid is a dibasic dihydroxy acid. The product is a white odorless granule or powder that has a tart taste and a slight characteristic flavor of its own. The properties of tartaric acid are listed in Table 1.

Table 1. Properties of General Purpose Food Acidulants

	Citric	Malic	Tartaric	Fumaric
Structure	COOH–CH$_2$–C(OH)(COOH)–CH$_2$–COOH (HO—C—COOH central carbon)	COOH–HOCH–CH$_2$–COOH	COOH–HCOH–HOCH–COOH	COOH–CH=HC–COOH
Formula	$C_6H_5O_7$	$C_4H_6O_5$	$C_4H_6O_6$	$C_4H_4O_4$
Molecular weight	192.12	134.09	150.09	116.07
Melting point, °C	153	131	169	286
Solubility at 25°C (g/100 mL of water)	181	144	147	0.63
Caloric value, kcal/g	2.47	2.39	1.84	2.76
Hygroscopicity (Resistance to moisture pick up)				
At 66% RH	Fair	Fair	Fair	Good
At 86% RH	Poor	Poor	Poor	Good
Ionization constants				
K_1	8.2×10^{-4}	4×10^{-4}	1.04×10^{-3}	1×10^{-3}
K_2	1.77×10^{-5}	9×10^{-6}	5.55×10^{-5}	3×10^{-5}
K_3	3.9×10^{-6}			

Tartaric acid has two asymmetric carbons which permit the formation of a dextro (+) rotatory form, a levo (−) rotatory form, and a meso form which is inactive due to internal compensation. A racemic mixture of dextro- and levo-rotatory forms is also a possibility.

Apart from very limited synthetic production in South Africa, tartaric acid is extracted from the residues of the wine industry. Therefore the natural form, L-(+)tartaric acid, is the article of commerce. The structure in Table 1 is L-(+)tartaric acid.

Potassium acid tartrate, also known as cream of tartar, is the major salt used in the food industry. It is occasionally used as the acid portion of a chemical leavening system for baked goods. It is also used as a doctor in candy making to prevent sugar crystallization by inverting a portion of the sucrose.

Tartaric acid can be used in most food categories and functions that require acidification but in practice it is limited to grape flavored products, particularly beverages and candies. It is also used where high tartness is desired in a highly soluble acid.

FUMARIC ACID

Fumaric acid is also a naturally occurring, organic, general purpose food acid. Although not found ubiquitously or in the concentrations of citric acid, fumaric nevertheless is found in all mammals as well as rice, sugar cane, wine, plant leaves, bean spouts, and edible mushrooms.

Fumaric acid is made synthetically by the isomerization of maleic acid. It is also produced by fermentation of glucose or molasses with *Rhizopus* spp. (4). It is a white crystalline powder that has the clean tartness necessary for a food acidulant. Table 1 shows that fumaric acid is the strongest of the food acids and is also the least soluble. The low solubility of fumaric acid limits its usefulness in foods. In processes where 50% stock solutions are mandatory such as carbonated beverages and jams and jellies, fumaric acid cannot be used.

Fumaric acid is extensively used in noncarbonated fruit juice drinks. Its greater acid strength allows lower use levels than citric acid and its solubility and rate of solution are sufficient for this process. Fumaric is also used in consumer packed gelatin desserts because of its low hygroscopicity. The only salt of fumaric acid that has had any application in the food industry is ferrous fumarate for iron fortification.

FUNCTIONS OF FOOD ACIDS

Food acidulants and their salts perform a variety of functions. These functions are as antioxidants, curing and pickling agents, flavor enhancers, flavoring agents and adjuvants, leavening agents, pH control agents, sequestrants, and synergists. The definitions for these functions are contained in the *U.S. Code of Federal Regulations* (5). Some of the functions overlap and in any given application an acidulant will often perform two or more functions.

Flavor Enhancer and Flavor Adjuvant

These functions are performed by a majority of the acidulants consumed by the food and beverage industries. Acids provide a tang or tartness that compliments and enhances many flavors but do not impart a characteristic flavor of their own. The acid itself should have a clean taste and be free of off notes that are foreign to foods. Some acids, such as succinic acid, have a distinctive taste which is incompatible with most food products; and hence, these acids have achieved very little use.

The need for tartness is obvious. Citrus and berry flavors would be flat and lifeless without at least a touch of acidity. However, not all fruit flavors require the same degree of tartness. Lemon candies and beverages are traditionally very sour, while orange and cherry are a little less tart. Flavors like strawberry, watermelon, and tropical fruits require only a trace of acidity for flavor enhancement.

In noncola carbonated beverages, beverage mixes, candies and confections, syrups and toppings, and any application where high solubility is required, citric and malic acids are used extensively. Fumaric acid is used in all ready constituted still beverages for economic reasons. Several acids are suitable for some flavors but not for others. For example, phosphoric acid is used in cola beverages but not in fruit flavored ones. Tartaric acid presents still another category. It has traditionally been used in grape flavored products even though it is suitable for other flavors.

The general purpose acids impart different degrees of tartness that are in part a result of their different acid strengths. Table 2 summarizes the tartness equivalence of the general purpose acids. The relationship shown is based only on tartness intensity and not character of flavor. This relationship can change depending on the formulation ingredients and the particular flavor system being studied. Malic acid, for example, has been claimed to be 10–15% more tart than citric acid in juice based, fruit flavored still beverages. In fruit and berry carbonated beverages, both acids have been perceived as being of equal tartness. Tartness is a difficult property to measure precisely and it must be determined by a trained and experienced taste panel.

Acids have also been used for their effects on masking undesired flavors in foods and food ingredients. Both citric and malic acids and citrate salts are known for their ability to mitigate the unpleasant aftertaste of saccharin. Gluconate salts and glucono-delta-lactone (GDL) have been patented for this function (6,7). Claims of enhanced benefits for malic acid over citric acid when used with the new intense sweeteners have been made but definitive advantages have not yet been demonstrated.

pH Control

Control of acidity in many food products is important for a variety of reasons. Precise pH control is important in the manufacture of jams, jellies, gelatin desserts, and pectin jellied candies in order to achieve optimum development of gel character and strength. Precise pH control is also important in the direct acidification of dairy products to achieve a smooth texture and proper curd formation. Increasing acidity enhances the activity of antimicrobial food preservatives, decreases the heat energy required for sterilization, inactivates enzymes, aids the development of cure color in processed meats, and aids the peelability of frankfurters.

Gelatin desserts are generally adjusted to an average pH of 3.5 for proper flavor and good gel strength. However, the pH can range from 3.0–4.0. Adipic and fumaric acids are used in gelatin desserts that are packaged for retail sales. Their low hygroscopicity allows use of packaging materials that are less moisture resistance and less expensive.

In jams and jellies, the firmness of pectin gel is dependent on rigid pH control. Slow set pectin attains maximum firmness at pH 3.05–3.15 while rapid set pectin reaches maximum firmness at pH 3.35–3.45 (8). The addition of buffer salts such as sodium citrate and sodium phosphate assist in maintaining the pH within the critical pH range for the pectin type. These salts also delay the onset of gelation by lowering the gelation temperature. The acid should be added as late as possible in the process. Premature acid addition will result in some pectin hydrolysis and weakening of the gel in the finished product. The acid is added as a 50% stock solution and thus soluble acids are required. Citric is generally used in this application but malic and tartaric are also satisfactory.

The United States Federal Standards of Identity (9) provide for the direct acidification of cottage cheese by the addition of phosphoric, lactic, citric, or hydrochloric acid as an alternate procedure to production with lactic acid producing bacteria. Milk is acidified to a pH of 4.5–4.7 without coagulation, and then after mixing, is heated to a maximum of 120°F without agitation to form a curd. Glucono-delta-lactone is also permitted for this application. It is added in such amounts as to reach a final pH value of 4.5–4.8 and is held until it becomes coagulated. GDL is preferred for this application because it must undergo hydrolysis to gluconic acid before it can lower pH. Thus the rate at which the pH is lowered is slowed, avoiding local denaturation.

The activity of antimicrobial agents (benzoic acid, sorbic acid, propionic acid) is due primarily to the undissociated acid molecule (10). On the basis of undissociated acid concentration, Giannuzzi et al. (11) have shown that citric acid is more effective than ascorbic or lactic acids in inhibiting *Listeria monocytogenes* in a trypticase soya broth containing yeast extract. Under refrigerated temperatures, higher inhibition indices were obtained in the presence of lower concentrations of citric acid. Activity is therefore pH dependent and theoretical activity at any pH can be calculated. Table 3 shows the effect of pH on dissociation. It can be seen why acidification improves preservative perfor-

Table 2. Tartness Equivalence

General-purpose acids	Weight equivalence
Citric	1.0
Fumaric	0.6–0.7
Tartaric	0.8–0.9
Malic	0.85–1.0

Table 3. Effect of pH on Dissociation[a]

pH	Sorbic	Benzoic	Propionic
3	98	94	99
4	86	60	88
5	37	13	42
6	6	1.5	6.7
7	0.6	0.15	0.7
(pK_a)	4.67	4.19	4.87

[a]Percentage of undissociated acid.

mance and why benzoates are not generally recommended above pH 4.5.

The use of acid to make heat preservation more effective, especially against spore-forming food spoilage organisms, is an established part of food technology. Under U.S. Federal Standards of Identity (12), the addition of a suitable organic acid or vinegar is required in the canning of artichokes (to reduce the pH to 4.5 or below) and is optional in the canning of the vegetables listed in Table 4. Vinegar is not permitted in mushrooms. Citric acid is specifically permitted as an optional ingredient in canned corn and canned field corn.

The advantage of acidification is especially well illustrated in the canning of whole tomatoes. When the pH of these is greater than 4.5, there is increased incidence of spoilage in the cans. When tomatoes of pH 3.9 are processed at 212°F, only 34 min are required to kill a normal or high spore load without decreases in color and flavor and deterioration of structure. In contrast, at pH 4.8 the cooking must be 110 min (13).

In the processing of fruits and vegetables, whether for canning, freezing, or dehydration, the prevention of discoloration in the fresh cut tissue is a major concern. Reactions in which polyphenolic compounds are changed by oxidation into colored materials play an important part in this discoloration which may be accompanied by undesirable flavors. The ascorbic acid naturally present in many fruits and vegetables offers some protection, but this is of relatively short duration because of destruction of ascorbic acid by natural enzymes and air. Heating, as applied in blanching, destroys the oxidative enzymes which cause discoloration but may alter flavor and texture if continued sufficiently to completely inactivate oxidative enzymes.

Lowering pH by addition of acid substantially decreases the activity of natural color producing enzymes in fruits. Citric acid also sequesters traces of metals which may accelerate oxidation. Even greater protection is obtained by using citric acid in conjunction with a reducing agent such as ascorbic or erythorbic acid. Some processors have found that a combination of sodium erythorbate and citric acid best serves their needs.

Leavening Agent

The basis for the formulation of effervescent beverage powders, effervescent compressed tablet products, and chemically leavened baked goods is the reaction of an acidulant with a carbonate or bicarbonate resulting in the generation of carbon dioxide. The physical state of some food acids as dry solids is a property appropriate for beverage mixes and chemical leavening systems. In the absence of water, there is essentially no interaction between such acids and sodium bicarbonate. Thus these dry mixes can be stored for long periods.

A desired property for an acidulant in a chemical leavening system is that it react smoothly with the sodium bicarbonate to assure desirable volume, texture, and taste. Leavening acids and acid salts vary quantitatively in their neutralizing capacity. This relationship is shown in Table 5. A new heat-activated leavening agent, dimagnesium phosphate, was recently reported for use in finished baked products (14).

The various acids differ in their rate of reaction in response to elevation of temperature. This must be taken into consideration in selecting an acidulant for a particular condition. Under some conditions, a mixture of acidulants may be most suitable to achieve desired reaction times. Table 6 compares the reaction times of GDL and cream of tartar.

Glucono-delta-lactone is an inner ester of gluconic acid that is produced commercially by fermentation involving *Aspergillus niger* or *A. suboxydans*. When it hydrolyzes, gluconic acid forms and this reacts with sodium bicarbonate. Although GDL is relatively expensive, there are certain specialized types of products such as pizza dough and cake doughnuts for which it is eminently suited as an acid component of the leavening system. Cream of tartar (potassium acid tartrate) has limited solubility at lower temperatures. There is a limited evolution of gas during the initial stages of mixing in reduced temperature bat-

Table 4. Canned Vegetables in which Use of Acids is Optional

Asparagus	Celery	Potatoes
Bean sprouts	Greens, collard	Rutabagas
Beans, butter	Greens, dandelion	Salsify
Beans, lima	Greens, mustard	Spinach
Beans, shelled	Kale	Sweet peppers, green
Beet greens	Mushrooms	Sweet peppers, red
Beets	Okra	Sweet potatoes
Broccoli	Onions	Swiss chard
Brussels sprouts	Parsnips	Tomatoes
Cabbage	Peas, black-eyed	Truffles
Carrots	Peas, field	Turnip greens
Cauliflower	Pimientos	Turnips

Table 5. Neutralizing Value of Various Acidulants Used in Chemical Leavening

Acid	Parts acid to neutralize one part sodium bicarbonate
Fumaric	0.69
Glucono-delta-lactone	2.12
Cream of tartar	2.24
Sodium acid pyrophosphate	1.39
Anhydrous monocalcium phosphate	1.20
Monocalcium phosphate monohydrate	1.25
Sodium aluminum sulfate	1.00
Sodium aluminum phosphate	1.00

Table 6. Comparison of Carbon Dioxide Evaluation in GDL and Cream of Tartar at Room Temperature

Time (min)	GDL (%)	Cream of tartar (%)
0.5	12.7	53.7
2	19.5	80.0
5	32.3	92.0
20	69.0	99.4
60	92.4	100.0

ters. At room temperature and above, the rate of reaction increases. Because of these characteristics, and its pleasant taste, cream of tartar is used in some baking powders and in the leavening systems of a number of baked goods and dry mixes.

Antioxidants, Sequestrants, and Synergists

Oxidation is promoted by the catalytic action of certain metallic ions present in many foods in trace quantities. If not naturally present in a food, minute quantities of these metals, particularly iron and copper, can be picked up from processing equipment. Oxidation is the cause of rancidity, an off flavor development in fat. It is also responsible for off color development that renders a food unappetizing in appearance. Hydroxy-polycarboxylic acids such as citric acid sequester these trace metals and render them unavailable for reaction. In this regard the acids function as antioxidants.

Hydroxy-polycarboxylic acids are often used in combination with antioxidants such as ascorbates or erythorbates to inhibit color and flavor deterioration caused by trace metal catalyzed oxidation. The ascorbates and erythorbates as well as BHA, BHT, and other approved antioxidants and reducing agents are oxygen scavengers and are effective when used alone. The effect of the combination of a sequestrant, such as a hydroxy-polycarboxylic acid, and an antioxidant is synergistically greater than the additive effect of either component used alone.

Citric acid is the most prominent antioxidant synergist although malic and tartaric acid have been used. In meat products, U.S. Department of Agriculture regulations permit citric acid in dry sausage (0.003%), fresh pork sausage (0.01%), and dried meats (0.01%). A short dip in a bath containing 0.25% citric and 0.25% erythorbic acid improves quality retention in frozen fish. This treatment is also applicable to shellfish to sequester iron and copper that catalyze complex blueing and darkening reactions.

Untreated fats and oils, both animal and vegetable, are likely to become rancid in storage. Oxidation is promoted by the catalytic action of certain metallic ions such as iron, nickel, manganese, cobalt, chromium, copper, and tin. Minute quantities of these metals are picked up from processing equipment. Adding citric acid to the oil sequesters these trace ions, thereby assisting antioxidants to prevent development of off flavors. Although the oil solubility of citric acid is limited, this can be overcome by first dissolving it in propylene glycol. The antioxidant can be dissolved in the same solvent so that the two can be added in combination.

Curing Accelerator

The acids approved by the U.S. Department of Agriculture for this function in meat products (1) must be used only in combination with curing agents. In addition to ascorbates and erythorbates, the approved acids are:

1. Fumaric acid to be used at a maximum of 0.065% (1 oz–100 lb) of the weight of the meat before processing.
2. GDL to be used at 8 oz to each 100 lb of meat in cured, comminuted meat products and at 16 oz per 100 lb of meat in Genoa salami.
3. Sodium acid pyrophosphate not to exceed, alone or in combination with other curing accelerators, 8 oz per 100 lb of meat nor 0.5% in the finished product.
4. Citric acid or sodium citrate to replace up to 50% of the ascorbate or erythorbates used or in 10% solution to spray surfaces of cured cuts.

In conjunction with sodium erythorbate or related reducing compounds, GDL accelerates the rate of development of cure color in frankfurters during smoking. This permits shortening smokehouse time by one half or more and products have less shrinkage and better shelf life.

The special property of GDL upon which these advantages depend is its lactone structure at room temperature. In this form there is no free acid group and the GDL can thus be safely added during the emulsifying stage of sausage making without fear of shorting out the emulsion. Under the influence of heat in the smoking process, the ester hydrolyzes rapidly and is converted in part to gluconic acid. This lowers the pH of the emulsion during smoking, providing conditions under which sodium erythorbate or other reducing compounds (erythorbic acid, ascorbic acid, and sodium ascorbate) react with greater speed to convert the nitrite of the cure mixture into nitric oxide. The nitric oxide, in turn, acts upon the meat pigment to form the desired red nitrosomyoglobin.

CONCLUSION

The many functions and broad range of applications of food acidulants makes the selection of the most suitable acid for a food product a matter of serious concern. The physical and chemical properties of food approved acidulants must be an essential part of the knowledge of those food technologists who make the decision.

BIBLIOGRAPHY

1. J. D. Dziezak, "Acidulants: Ingredients That Do More Than Meet the Acid Test," *Food Technol.* **44**, 75–83 (1990).
2. E. F. Bouchard and E. G. Merritt in M. Grayson, ed., Kirk-Othmer, *Encyclopedia of Chemical Technology*, Vol. 6, 3rd ed., John Wiley & Sons, Inc., New York, 1979, pp. 150–179.
3. *Code of Federal Regulations*, Title 21 §184.1069.
4. D. E. Blenford, "Food Acids," *Food Flavor Ingredients Processing and Packaging* **8**, 11 (1986).
5. *Code of Federal Regulations*, Title 21 §170.3(o).
6. U.S. Pat. 3,285,751 (Nov. 15, 1966) Artificial Sweetner Composition, P. Kracauer (to Cumberland Packing Company).
7. U.S. Pat. 3,684,529 (Aug. 15, 1972) Sweetening Compositions, J. J. Liggett (to William E. Hoerres).
8. M. Glicksman, *Gum Technology in the Food Industry*, Academic Press, Inc., Orlando, Fla., 1969, p. 179.
9. *Code of Federal Regulations*, Title 21 §133.129.
10. Davidson and Juneja, 1990.

11. L. Giannuzzi and N. E. Zaritsky, "Effect of Ascorbic Acid in Comparison to Citric Acid and Lactic Acid on *Listeria monocytogenes* Inhibition at Refrigerated Temperatures," *Lebensmittel Wissenschaft und Technologie* **29**, 278–285 (1996).
12. *Code of Federal Regulations*, Title 21 §155.200.
13. C. T. Townsend, cited in S. Leonard, R. M. Pangborn, and B. S. Luh, *Food Technology* **13**, 418 (1959).
14. D. R. Gard et al., "Evaluation of a New Leavening Acid, Dimagnesium Phosphate, With a Traditional Chemical Leavening System," *Proc. IFT Ann. Meet.*, 1996, p. 114.
15. *Code of Federal Regulations*, Title 9 §318.7(c)(4).

AFFLUENCE, FOOD EXCESS, AND NUTRITIONAL DISORDERS

Nutrition research in the first half of the present century focused mainly on identifying and assessing nutrients in foods and recognizing deficiency diseases. Since essential nutrients were unevenly distributed in individual foods, consumption of a variety of foods was recommended to provide a more nutritionally balanced diet (1). With undernutrition still a problem even in the richer countries, greater consumption of meat and dairy products was recommended. Increasing wealth in society allowed these recommendations to match producer and consumer wishes and acknowledged, at the same time, the perceived high nutritional values of these foods of animal origin. By the 1950s, however, evidence was accumulating that the high rates of premature deaths from some of the major chronic diseases could be related to diet. Originally this was thought to be a problem only in the industrialized countries but, as medicine conquered infectious diseases, the phenomenon became recognized as existing worldwide.

The major nutrition-related degenerative diseases include obesity, diabetes mellitus, cardiovascular diseases, and certain cancers. The still continuing Framingham Heart Study, which began in 1948, has been responsible for demonstrating many associations between these diseases and diet, notably excess intake of food energy and fat, especially saturated fat and cholesterol, as well as lifestyle factors such as smoking, emotional stress, and lack of physical exercise. Consequently, current diet/health recommendations differ dramatically from those made earlier.

OBESITY

Obesity affects many millions of Americans and is a major public health problem. It is characterized by excess body fat caused by an imbalance between energy intake and energy expenditure (2). The specific reasons behind such an imbalance, however, remain the subject of much debate.

Genetic, environmental and behavioral variables all influence the risk of becoming overweight (2,4), but the relative importance of each remains unclear. The prevalence of obesity for different age groups in the United States is shown in Table 1. Direct assessment of body fat can be used to evaluate obesity but the procedure is mainly limited to research. For public health studies and clinical practice, simple measures such as height and weight tables, body mass index (BMI), or skinfold measurements are used. The BMI (weight in kilograms divided by the square of height in meters) is an indicator that shows the best correlation with independent measures of body fat (2,5). The use of BMI was first proposed in 1871 and was long known as the Quetelet index. It is now widely used for assessing the degree of obesity or overweight (Table 2).

Obesity is more than a problem in its own right since it is closely linked to other diseases such as hypertension, diabetes, and cardiovascular disease. Many long-term studies have shown a greater risk of these diseases with increasing levels of obesity, even when other risk factors are present (5). Later studies have refined these conclusions and have demonstrated that the distribution of fat, especially in the abdominal area, is an additional factor (5). A waist circumference of over 40 in. (102 cm) in men and over 35 in. (89 cm) in women signifies increased risk in those who have a BMI of 25.0 to 34.9.

With increasing BMI, average blood pressure and total cholesterol levels rise while average high-density lipoprotein (HDL—a protective indicator) levels fall (6). Men in the highest obesity category have more than twice the risk of hypertension and elevated cholesterol when compared with men of normal weight. Women in the highest obesity category have four times the risk of either or both of these risk factors (6). Obese individuals are also at increased risk for several other problems, including lipid disorders, type II diabetes, coronary heart disease, stroke, gallbladder disease, osteoarthritis, sleep apnea, respiratory problems, and certain cancers.

A number of efforts have been initiated to educate the public on altering behavioral risk factors such as improper dietary habits and lack of exercise. Examples include the Dietary Guidelines for Americans (Table 3), revised editions of which are periodically issued by the U.S. Department of Agriculture (USDA) (7). Recently the first federal guidelines on the identification, evaluation, and treatment of overweight and obesity in adults were released (6). These clinical practice guidelines are designed to help physicians in their care of overweight and obesity and include the scientific background for assessment as well as the principles of safe and effective weight loss. Three key indicators are recommended BMI, waist circumference, and the patient's risk factors for diseases and conditions associated with obesity. Overweight is defined as a BMI of 25.0 to 29.9 while obesity is a BMI greater than 30.0. These values are consistent with the definitions used in many other countries and support the Dietary Guidelines for Americans. A BMI of 30.0, for example, a weight of 221 lb for a 6-ft person (100 kg: 1.83 m) or 186 lb for a person of 5 ft, 6 in. (84 kg: 1.68 m) indicates about 30 lb (13.6 kg) overweight for both men and women. Highly muscular people may have a high BMI without increased health risks. For the majority, however, BMI is an excellent indicator of overall risk.

The most successful strategies for weight loss include reduction of food energy intake, increased physical activity,

Table 1. Prevalence of Overweight, Severe Overweight, and Morbid Obesity (NHANES II)

	Males		Females		Total
	Millions	Prevalence %	Millions	Prevalence %	Millions
Overweight	15.4	24.2	18.6	27.1	34.0
Severe overweight	5.1	8.0	7.4	10.6	12.5
Morbid obesity	0.327	0.6	1.7	2.5	2.0

Source: Ref. 3.

Table 2. Classification of Overweight and Obesity Based on Body Mass Index

Degree of Obesity	BMI[a]
Grade III (morbid obesity)	>40
Grade II (obese)	30–39.9
Grade I (overweight)	25–29.9
Grade 0 (normal)	20–24.9

[a]BMI = weight in kilograms divided by the square of the height in meters.
Source: Ref. 2.

Table 3. Dietary Guidelines for Americans

- Eat a variety of foods.
- Balance the food you eat with physical activity; maintain or improve your weight.
- Choose a diet with plenty of grain products, vegetables, and fruits
- Choose a diet low in fat, saturated fat, and cholesterol.
- Choose a diet moderate in sugars.
- Choose a diet moderate in salt and sodium.
- If you drink alcoholic beverages, do so in moderation.

Source: Ref. 7.

and behavior therapy designed to improve eating and physical activity habits. Other recommendations are listed in Table 4.

One in five children in the United States is now overweight. Care and treatment differ from adults, and therefore pediatric guidelines have been proposed by an expert committee (8). Clinicians who care for these children and their families are urged not to express blame but to show "sensitivity, compassion and a conviction that obesity is an important chronic medical condition that can be treated." Balanced nutritional intakes with avoidance of excess together with adequate physical activity are important goals for all family members.

HYPERTENSION

In the adult, hypertension (high blood pressure) is defined as a pressure greater than, or equal to, 140 mmHg systolic, or greater than or equal to, 90 mmHg diastolic pressure (9). In 90 to 95% of the cases of high blood pressure, the specific cause may be unknown (10). Hypertension is a risk factor for both coronary heart disease and stroke. Although it can occur in children and adolescents, it is more prevalent in the middle-aged and elderly, especially African Americans and the obese. Heavy drinkers and women who are taking oral contraceptives (11) are also at increased risk. Individuals with diabetes mellitus, gout, or kidney disease also have a higher frequency of hypertension. Salt consumption can increase blood pressure for some. High blood pressure is related to obesity and to increases in body weight over time (9,12). Factors increasing the risk of developing high blood pressure are listed in Table 5.

Weight loss, an active lifestyle, reduction in sodium intake, and moderation of alcohol consumption are recommended for prevention and management (9). For many, however, medical intervention with antihypertensive drugs is required to maintain acceptable blood pressure.

Table 4. Selected Recommendations from the First Federal Obesity Guidelines Panel

- Engage in moderate physical activity (30 min or more) on most/all days of the week.
- Reduce dietary fat and calories. Cutting back on dietary fat can help reduce calories and is heart-healthy.
- The initial goal of treatment should be to reduce body weight by about 10% from baseline, an amount that reduces obesity-related risk factors. With success, and if warranted, further weight loss can be attempted.
- A reasonable time line for a 10% reduction in body weight is six months of treatment (weight loss of 1 to 2 lb per week).
- Weight maintenance should be a priority after the first 6 months of weight-loss therapy.
- Overweight and obese patients who do not wish to lose weight, or are otherwise not candidates for weight-loss treatment, should be counseled on strategies to avoid further weight gain.
- Age alone should not preclude weight-loss treatment in older adults. A careful evaluation of potential risks and benefits in the individual patient should guide management.

Source: Ref. 6.

Table 5. Risk Factors for Hypertension (High Blood Pressure)

- Heredity
- Male gender
- Sodium or salt sensitivity
- Heavy alcohol consumption
- Sedentary and inactive lifestyle
- Race (African Americans are at greater risk)
- Age
- Obesity and overweight
- Use of oral contraceptives and some other medications

Source: Ref. 11.

DIABETES MELLITUS

What can now be recognized as diabetes was described in the ancient civilizations of Egypt, Greece, and India. The sweet taste of urine in those with the condition was noted in the 1600s and the term "mellitus" meaning honeylike was introduced (13). Diabetes mellitus is a major public health problem worldwide. It ranks sixth as a primary cause of death in the United States, but when its complications are included, it ranks third. These complications can be very serious and involve, in the United States, 50% of the amputations of all lower extremities in adults and 25% of all kidney failure and are also a leading cause of blindness.

Non-insulin-dependent diabetes (NIDDM, or type II) is the form of diabetes characterized as a chronic nutritionally related condition (Table 6) and is a disorder showing abnormalities in glucose, fat, and protein metabolism. The onset of type II diabetes can be triggered by dietary and lifestyle factors similar to those associated with cardiovascular diseases. Diabetes and heart disease in later life appear to be linked to weight at birth (13).

CARDIOVASCULAR DISEASE

Cardiovascular disease (CVD) has been for many years the leading single cause of death in the United States (14): it includes both coronary heart disease (CHD) and stroke. CHD is the most common form of cardiovascular disease.

In 1987 nearly 1 million deaths in the United States, half of the total number, occurred due to some form of CVD (15). In 1998, CVD remains the leading cause of death in the United States (14), although there have been some reductions in the rate. Cardiovascular disease is often thought to affect mainly men and the elderly, but it is also a major killer of women and people in the prime of life (14). An estimated 58 million Americans live with some form of the disease, and almost 10 million Americans aged 65 years and older report disabilities caused by heart disease. Stroke is also a leading cause of disability in the United States, affecting more than 1 million people nationwide (14). The health burden of this condition is rivaled by the economic burden, which has a profound impact on the health care system.

Extensive clinical and epidemiological studies have identified several major and contributing risk factors of heart disease and stroke (16). The major risk factors are listed in Table 7. Some such as increasing age, male gender, and genetic background cannot be changed, treated, or modified (16). Others, for example, smoking, high serum cholesterol, high blood pressure, physical inactivity, obesity, and overweight, are under some control by the individual. Smokers have twice the risk of heart disease compared with nonsmokers. Nearly one-fifth of all deaths from cardiovascular diseases (180,000 deaths per year) are attributable to smoking (14). Surveillance data indicate that an estimated 1 million young people become regular smokers each year (14).

The risk of heart disease increases with a rise in cholesterol levels especially when other risk factors are present (17–19). Plasma total cholesterol was accepted as a causal factor (among multiple factors) by the World Health Organization (WHO) expert committee in 1982 and by the U.S. National Institute of Health Consensus Development Conference in 1985 (17). Diet and its effects on plasma cholesterol levels are discussed in the next section. Plasma triglyceride levels have also been correlated with increased risk of heart disease (17) and are associated with increased low-density lipoprotein (LDL) cholesterol levels. High blood pressure increases the risk of a stroke, heart attack, kidney failure, and congestive heart failure. When obesity, smoking, high blood cholesterol levels, or diabetes are also present, high blood pressure increases the risk of a heart attack or stroke severalfold.

Regular moderate-to-vigorous exercise plays a significant role in preventing heart and blood vessel disease (16). Exercise helps control blood cholesterol, diabetes, and obesity as well as maintaining blood pressure. However, surveys have shown that more than half of American adults do not practice the recommended level of physical activity. Obese people are at a greater risk of heart disease, high blood pressure, high cholesterol, and other chronic diseases and diabetes (14,16). Diabetes is also a serious risk factor for heart disease with more than 80% of diabetics succumbing to some form of heart or blood vessel disease (16).

ROLE OF DIET IN CARDIOVASCULAR DISEASE

Improper eating habits accompanied by the lack of exercise increase the risk of gaining excess weight, a major risk factor for heart disease, high blood pressure, and diabetes (14). Diet also affects plasma cholesterol levels. Cholesterol is carried in the blood associated with two major types of lipoproteins; LDL and HDL. LDL cholesterol has been

Table 6. Classification of Diabetes Mellitus

Spontaneous Diabetes Mellitus (DM)
Insulin-dependent (IDDM, or type I)
Non-insulin-dependent (NIDDM or type II)
  Nonobese NIDDM
  Obese NIDDM
  Maturity onset diabetes of young people
Secondary diabetes
Gestational diabetes

Source: Ref. 13.

Table 7. Major Risk Factors for Cardiovascular Disease

- Increasing age
- Heredity
- High blood cholesterol levels
- Physical inactivity
- Diabetes mellitus
- Male gender
- Smoking
- High blood pressure
- Obesity and overweight

Source: Ref. 16.

correlated with increased risk of cardiovascular disease. For many years it has been recognized that dietary cholesterol has only a limited effect on plasma cholesterol levels (17). Absorption of ingested cholesterol is poor, and part of the cholesterol in plasma is synthesized in the liver.

Total lipid intake, and the type of fat consumed, have more effect in raising plasma cholesterol than does dietary cholesterol (17,20). Saturated fatty acids were found to raise cholesterol, polyunsaturated fatty acids lowered plasma cholesterol, and monounsaturated fatty acids had an intermediate effect. In the classic seven-country prospective study where lipid intake was correlated with CHD (21,22), the disease incidence was also related to the intake of saturated fat. Most dietary prevention trials have reduced total fat, saturated fat, and cholesterol intakes along with moderately increased polyunsaturated fat levels.

Similar recommendations are also inclusive in many of the guidelines for health in the United States. Recent research indicates that monounsaturated fatty acids such as oleic acid, which were thought to be neutral, have, in fact, a substantial cholesterol lowering effect (23). *Trans* fatty acids formed during hydrogenation of certain edible oils may, perhaps, increase LDL cholesterol and decrease HDL cholesterol (24). A high fish intake has been associated with beneficial effects on the prevention of CHD. Fish oils contain DHA (docosahexaenoic acid), which has a plasma triglyceride-lowering effect (25).

High levels of dietary carbohydrate, especially complex carbohydrate, are associated with a decreased risk of cardiovascular disease (17). A recent study found rice bran as well as oat bran to have a hypocholesterolemic effect (26). Increasing intakes of a number of vitamins have also been shown to be protective toward cardiovascular disease. These include vitamins B_6, C, E, and folate. Vitamin C and E are antioxidants and have been hypothesized as preventing damage to coronary arteries. Elevated serum vitamin C has also been correlated with increased HDL cholesterol levels in women (27). Increased consumption of folate and vitamin B_6 have also been shown to reduce risk of CHD in women (28). Current dietary recommendations (Table 3) incorporate many of these proven relationships between diet and health.

CANCER

Cancer is a disease condition characterized by excessive growth of cells due to abnormal multiplication and replication. The biological process shows several experimentally distinct phases following exposure to a carcinogen. These include: initiation, tumor promotion, and tumor progression (29). After CVD, cancer is the next highest cause of death in the United States. Many adverse effects that occur in patients are due, not only to the cancer itself, but also to the treatment. Loss of appetite (anorexia) is the most common side effect. When advanced and persistent, this along with other metabolic and physiologic changes can eventually lead to severe undernutrition termed cancer cachexia.

Anorexia is not unique to cancer; however, it is persistent and severe in certain cancers such as the carcinoma of the stomach, breast, and large bowel. Onset of anorexia is insidious and may not be accompanied by any obvious manifestations other than progressive weight loss. Energy expenditure is high in cancer patients (29) while glucose intolerance is found frequently and may be due to the increased insulin resistance or inadequate insulin release. Abnormalities also occur in fat metabolism and include excess body fat depletion, protein loss, increased lipolysis, changes in free fatty acid and glycerol turnover, as well as decreased lipogenesis. Abnormalities in protein metabolism include increased whole body protein turnover, increased hepatic protein synthesis, persistent muscle protein breakdown, and decreased levels of plasma branched-chain amino acids. Malabsorption as well as protein loss through the gastrointestinal tract may also occur.

DIET AND CANCER RISK

Diet may have either positive and negative influences on cancer risk (30). About 35% of cancer occurrence is related to dietary factors (31). The role of diet in cancer etiology is summarized in Table 8. Dietary fat intake was associated with breast cancer in animals as early as 1942 (32,33). More recently, higher intake levels of dietary fat have been related to increased risk of colon cancer (34). Positive correlations between per capita fat intakes and breast cancer rates have been described (35,36). It has been argued, however, that it may be the high food energy intake that is causative rather than the percentage of the food energy coming from dietary fat (37). Nevertheless, diets high in fat, particularly saturated fat, have also been associated with a higher incidence of cancer of the colon, prostate, and breast (17).

Increased consumption of fruits and vegetables has been recommended to reduce cancer risk as these foods contain protective factors. Many are also high in fiber, an increased consumption of which has been associated with decreased risk of cancer (especially colon cancer). Diets high in plant foods, starches, fiber, and various carotenes are commonly associated with a lower incidence of alimentary tract cancers. A number of dietary and nondietary factors have been found to decrease the incidence of various cancers. Lycopene found in tomatoes appears to be protective against colorectal cancer (38) as it can scavenge peroxyl radicals and quench singlet oxygen. Increased intakes of plants of the cabbage family appear to be protective against certain cancers (39). More recently, broccoli sprouts have been shown to contain high levels of the anticarcinogenic chemical sulforaphane. This compound has been known to help mobilize the body's natural cancer-fighting resources and reduces the risk of developing cancer (40).

ANOREXIA NERVOSA AND BULIMIA NERVOSA

Although not included within the category of nutritionally related chronic diseases, the eating disorders anorexia nervosa and bulimia nervosa are important. These diseases are primarily disorders of perception of body image and are characterized by an excessive concern over being fat. They

Table 8. Dietary Factors and Cancer Etiology

Carcinogenic dietary factors	Anticarcinogenic factors
Energy excess associated with increased cancer mortality in men and women	Energy deficit inhibits tumor growth
Amount and type of fat in the diet also related to increased cancer risk; high saturated fat, cholesterol, and low polyunsaturated fat are risk factors	High levels of monounsaturated fat in the diet show decreased incidence of certain cancers
High protein intake associated with increased risk of enhanced tumorigenesis	High levels of fiber from fruits and vegetables are associated with low levels of colon and rectal cancer.
Zinc deficiency associated with increased risk of tumors	Vitamin A and its analogues and precursor (carotenids) are possible inhibitors of carcinogenesis; β-carotene may be protective in a mechanism independent of its role as a vitamin A precursor
Excess alcohol intake	Vitamin C has antioxidant properties that may influence tumorigenesis
High intake of coffee is a possible risk factor	Vitamin E as an intracellular antioxidant may protect against carcinogens
Artificial sweeteners such as saccharin increase risk of bladder cancer	Calcium intake has a inverse association with colon cancer risk
Nitrates, nitrites, and nitrosamines may be causative factors of gastric cancer	Selenium intake has been associated with decreased tumor growth in animal models
Methods of food preparation, such as charcoal broiling, smoking food, and frying, may increase risk	

Source: Ref. 31.

are often regarded as modern disorders despite the fact that similar conditions have been recognized in medicine for more than a century.

Anorexia nervosa is a condition of self-engendered weight loss whose occurrence was originally thought to be restricted to young women. It also occurs in young men who are concerned with their body image such as dancers and models. The diseases appear to be largely confined to affluent societies that espouse Western cultural ideals.

Diagnostic criteria include: refusal to maintain minimally normal body weight for age and height; intense fear of gaining weight or becoming fat, even though already underweight; undue influence of body weight or shape on self-evaluation; and denial of the seriousness of the current low body weight with amennorhea often occurring in postmenarchal females (41). Associated symptoms include: depressed mood, irritability, social withdrawal, loss of sexual libido, preoccupation with food and rituals, as well as reduced alertness and concentration (42).

One form of the disease invokes restrictive feeding behavior commonly associated with normal dieting, such as undereating, refusal to take high-energy foods, and strenuous exercise. This behavior is abnormal only in the degree to which it is pursued. Restlessness is very common once emaciation sets in and continues until physical deterioration leads to weakness and lassitude. The "purging" form involves more dangerous behaviors, such as self-induced vomiting, and laxative or diuretic use.

Bulimia nervosa is a variant of anorexia nervosa and shares many of its clinical and demographic features. It is closely related to the purging form of anorexia nervosa. One of the major differences is that bulimic patients maintain normal weight. The condition generally involves persistent dietary restriction that is eventually interrupted by episodes of binge eating with compensatory behaviors such as vomiting and laxative abuse. Behavioral disturbances often become the focus of intense guilt feelings. In the early stages of the disease, all patients attempt to control their weight by dieting and abstaining from high-energy foods. They are constantly preoccupied by thoughts of food, but their pattern of eating alternates between fasting and gorging. Patients are extremely secretive about their bulimic episodes. It is this secrecy that makes the condition difficult to diagnose.

Both conditions occur predominantly in industrialized, developed countries and are rare elsewhere (43). Immigrants are more likely to develop eating disorders than their peers in their country of origin, probably indicating the importance of sociocultural factors in the etiology and distribution of these disorders (43).

CONCLUSION

The chronic, nutritionally related diseases just described are major causes of death and disability in rich industrialized countries. American and North European diets have tended to be high in animal foods (meat, dairy, fish, eggs) and low in foods of plant origin (grains, fruits, and vegetables). It is claimed by Garrow (44) that most of the chronic diseases in the Western society are the manifestation of the high availability and variety of foods leading to overconsumption. Only a small proportion of income is now required to be spent on food in the industrialized countries (45). Excessive intake of animal foods leads to a dietary pattern that is high in saturated fat and cholesterol and low in fiber.

In contrast, the southern European or the Mediterranean diet comprises fruits, vegetables, and grains with smaller amounts of meat, fish, eggs, and dairy products (46,47). Olive oil is often the major lipid, so that the diet is low in cholesterol and saturated fat and high in monounsaturated fatty acids. A comparative study between Italians and Americans was performed in the early 1950s. It was found that Italian diets were remarkably low in fat (20% of energy) or just half of the proportion observed in the diets of comparable American groups. The typical American diet, rich in meat and dairy fats was thus, together with higher concentrations of blood cholesterol, identified with increased risk of coronary heart disease (48). A seven-country study performed over 20 years confirmed these relationships (22). Recommendations for the "Mediterranean Diet" have become popular within the United States. This diet plan is indicated in Table 9 (49).

Ironically, while such diets are now being consumed by the affluent, recent dietary surveys carried out on the island of Crete have reported an increase in intake of meat, fish, and cheese and a decrease in intakes of bread, fruit, potatoes, and olive oil (50). Similar changes have been observed in Italy (51). An increased availability of animal foods throughout the Mediterranean area has also been documented (47). These dietary changes have been accompanied by increases in chronic disease risk factors such as higher concentrations of serum cholesterol, hypertension, and obesity as well as reduced levels of physical activity (50,52).

Chronic disease risk is increasing, not only in the Western society, but also in the more affluent classes of the developing countries (53). The rich in poor countries often have a similar pattern of food consumption to that observed in the affluent countries. They are also subject to many of the same lifestyle factors, including smoking and reduced physical activity.

Dietary Guidelines (Table 3) can help in reducing both heart disease and cancer risk. The guidelines now emphasize moderation in intake, especially of saturated fat, along with increased physical activity. Increasing intakes of fruits, vegetables, and complex carbohydrates are also recommended.

Paradoxically, these present recommendations for the affluent define diets and lifestyles closer to those common in the past for the less affluent. As a further paradox, these latter societies, as their wealth increases, are often attempting to emulate the diets and lifestyles of the West. Consequently, they are now increasingly subject to the same pattern of disease.

Table 9. The Mediterranean Diet Plan

Frequency of consumption	Foods
	In significant amounts
Daily	Whole grains and grain products (breads, pasta, rice, couscous, polenta, bulgur) and potatoes
	Fruits and vegetables
	Beans, other legumes, and nuts
	In small or minimal amounts
Daily	Cheese and yogurt
A few times a week	Fish, poultry, eggs, and sweets
A few times a month	Red meat (or in small amounts more often)

In addition, regular physical activity is important. Moderate wine consumption is optional.

Source: Ref. 49.

BIBLIOGRAPHY

1. P. James, "The Nature of Food: Essential Requirements," in B. Harriss and Sir R. Hoffenberg, eds., *Food: Multidisciplinary Perspectives*, Blackwell Publishers, Oxford, UK, 1994, pp. 27–40.
2. F. X. Pi-Sunyer, "Obesity," in M. E. Shils, J. A. Olson, and M. Shike, eds., *Modern Nutrition in Health and Disease*, 8th ed, no. 2, Lea & Febiger, Philadelphia, Pa., 1994, pp. 984–1006.
3. D. F. Williamson, "Prevalence and Demographics of Obesity," in K. D. Brownell and C. G. Fairburn, eds., *Eating disorders and Obesity: A Comprehensive Handbook*, The Guilford Press, New York, 1995, pp. 391–395.
4. National Institutes of Health, "NIH Consensus Development Conference Statement: Health Implications of Obesity" (Feb. 11–13, 1985). URL: *http://text.nlm.nih.gov/nih/cdc/www/49txt.html* (last accessed Aug. 18, 1998).
5. National Institutes of Health and National Heart, Lung and Blood Institute, "First Federal Obesity Guidelines Released," URL: *http://www.nih.gov/news/pr/jun98/nhlbi-17.html* (last accessed June 17, 1998: 1–4).
6. Heart Information Network, "Most Americans Are Overweight," URL: *http://www.heartinfo.org/mosamfat197.html* (last accessed Aug. 18, 1998: 1–2)
7. U.S. Department of Agriculture, U.S. Department of Health and Human Services, *Nutrition and Your Health: Dietary Guidelines for Americans*, 4th ed., Home and Garden Bulletin No. 23, Hyattsville, Md.: U.S. Government Printing Office, Washington, D.C., 1995.
8. S. E. Barlow and W. H. Dietz, "Obesity Evaluation and Treatment: Expert Committee Recommendations," *Pediatrics* **102**(3) (1998). URL: *http://www.pediatrics.org/cgi/content/full/102/3/e29* (last accessed Sept. 23, 1998: 1–11).
9. American Heart Association, "High Blood Pressure: AHA Recommendation," URL: *http://www/amhrt.org/Heart_and_Stroke_A_Z_Guide/hbp.html* (last accessed Aug. 18, 1998: 1–2).
10. American Heart Association, "High Blood Pressure Causes," URL: *http://www/amhrt.org/Heart_and_Stroke_A_Z_Guide/hbpc.html* (last accessed Aug. 18, 1998: 1).
11. American Heart Association, "High Blood Pressure, Factors That Contribute To," URL: *http://www/amhrt.org/Heart_and_Stroke_A_Z_Guide/hbpf.html* (last accessed Aug. 18, 1998: 1–2).
12. T. A. Kotchen and J. M. Kotchen, "Nutrition, Diet and Hypertension," in M. E. Shils, J. A. Olson, and M. Shike, eds., *Modern Nutrition in Health and Disease*, 8th ed., no 2, Lea & Febiger, Philadelphia, Pa., 1994, pp. 1287–1297.
13. J. W. Anderson and P. B. Geil, "Nutritional Management of Diabetes Mellitus," in M. E. Shils, J. A. Olson, and M. Shike, eds., *Modern Nutrition in Health and Disease*, 8th ed., no. 2, Lea & Febiger, Philadelphia, Pa., 1994, pp. 1259–1286.
14. U.S. Department of Health and Human Services: Public Health Service, Centers for Disease Control and Prevention, "Preventing Cardiovascular Disease," URL: *http://www.cdc.gov/nccdphp/cvd/cvdaag.html* (last accessed Aug. 18, 1998: 1–8).
15. D. J. McNamara, "Cardiovascular Disease," in M. E. Shils, J. A. Olson, and M. Shike, eds., *Modern Nutrition in Health and Disease*, 8th ed., no. 2, Lea & Febiger, Philadelphia, Pa., 1994, pp. 1533–1544.
16. American Heart Association, "Risk Factors and Coronary Heart Disease," URL: *http://www/amhrt.org/Heart_and_Stroke_A_Z_Guide/riskfact.html* (last accessed Aug. 18, 1998: 1–3).

17. A. S. Truswell, "The Evolution of Diets for Western Diseases," in B. Harriss and Sir R. Hoffenberg, eds., *Food: Multidisciplinary Perspectives*, Blackwell Publishers, Oxford, UK, 1994, pp. 41–62.

18. A. Keys et al., "Lessons from Serum Cholesterol Studies in Japan, Hawaii and Los Angeles," *Ann. Intern. Med.* **48**, 83–94 (1958).

19. W. B. Kannel et al., "Serum Cholesterol, Lipoproteins and the Risk of Coronary Heart Disease. The Framingham Study," *Ann. Intern. Med.* **74**, 1–12 (1971).

20. S. M. Mellinkoff, T. A. Marchella, and J. G. Reinhold, "The Effect of a Fat Free Diet in Causing Low Serum Cholesterol," *Amer. J. Med. Sci.* **220**, 203–207 (1950).

21. A. Keys, "Coronary Heart Disease in Seven Countries," *Circulation* **41** (Suppl. 1), 1–211 (1970).

22. A. Keys, *Seven Countries. A Multivariate Analysis of Death and Coronary Heart Disease*, Harvard Univ. Press, Cambridge, Mass., 1980.

23. R. P. Mensink and M. B. Katan, "Effect of a Diet Enriched with Monounsaturated or Polyunsaturated Fatty Acids on the Levels of Low-Density and High-Density Lipoprotein Cholesterol in Healthy Men and Women," *New Engl. J. Med.* **321**, 436–441 (1989).

24. R. P. Mensink and M. B. Katan, "Effect of Dietary Trans Fatty Acids on the High and Low Density Lipoprotein Cholesterol Levels in Healthy Subjects," *New Engl. J. Med.* **323**, 436–445 (1990).

25. H. Gerster, "Can Adults Adequately Convert Alpha-Linolenic Acid to Eicosapentaenoic Acid and Docosahexaenoic Acid," *Int J. Vitamin Nutr. Res.* **68**(3), 159–173 (1998).

26. A. L. Gerhardt and N. B. Gallo, "Full Fat Rice Bran and Oat Bran Similarly Reduce Hypercholsterolemia in Humans," *J. Nutr.* 5 **128**(5), 865–869 (1998).

27. J. A. Simon and E. S. Hudes, "Relation of Serum Ascorbic Acid to Serum Lipids and Lipoproteins in US Adults," *J. Amer. College Nutr.* **17**(3), 250–255 (1998).

28. E. B. Rimm et al., "Folate and Vitamin B6 from Diet and Supplements in Relation to Risk of Coronary Heart Disease among Women," *J. Amer. Med. Assoc.* **279**(5), 359–364 (1998).

29. M. E. Shils, "Nutrition and Diet in Cancer Management," in M. E. Shils, J. A. Olson, and M. Shike, eds., *Modern Nutrition in Health and Disease*, 8th ed., no 2, Lea & Febiger, Philadelphia, Pa., 1994, pp. 1317–1348.

30. M. W. Pariza, "Diet, Cancer and Food Safety," in M. E. Shils, J. A. Olson, and M. Shike, eds., *Modern Nutrition in Health and Disease*, 8th ed., no. 2, Lea & Febiger, Philadelphia, Pa., 1994, pp. 1545–1558.

31. C. L. Chenney and S. N. Aker, "Nutritional Care in Neoplastic Disease," in L. K. Mahan and M. Arlin, eds., *Krause's Food Nutrition and Diet Therapy*, 8th ed., W.B. Saunders Company, Harcourt Brace Jovanovich Inc., Philadelphia, Pa., 1992, pp. 625–642.

32. A. Tannenbaum, "The Genesis and Growth of Tumors. III Effects of a High Fat Diet," *Cancer Res.* **2**, 460 (1942).

33. T. L. Dao and R. Hilf, "Dietary Fat and Breast Cancer: A Search for Mechanisms," in Maryce M. Jacobs, ed., *Exercise, Calories, Fat and Cancer, Advances in Experimental Medicine and Biology 322*, Plenum Press, New York, 1992, pp. 223–238.

34. K. K. Carroll and H. T. Khor, "Dietary Fat in Relation to Tumorigenesis," *Prog. Biochem. Pharmacol.* **10**, 308 (1975).

35. S. Graham et al., "Diet in the Epidemiology of Breast Cancer," *Amer. J. Epidemiol.* **116**, 68–75 (1982).

36. G. R. Howe et al., "Dietary Factors and the Risk of Breast Cancer: Combined Analysis of 12 Case-Control Studies," *J. Nat. Cancer Inst.* **82**, 561–569 (1990).

37. R. K. Boutwell, "Caloric Intake, Dietary Fat Level and Experimental Carcinogenesis," in Maryce M. Jacobs, ed., *Exercise, Calories, Fat and Cancer, Advances in Experimental Medicine and Biology 322*, Plenum Press, New York, 1992, pp. 95–101.

38. C. La. Vecchia, "Mediterranean Epidemiological Evidence on Tomatoes and the Prevention of Digestive Tract Cancers," *Proc. Soc. Exp. Biol. Med.* **218**(2), 125–128 (1998).

39. U.S. Department of Health and Human Services, *The Surgeon General's Report on Nutrition and Health*, U.S. Government Printing Office, Washington, D.C., Publication No. 88-50210, 1988, 712 pp.

40. Health News, "Study Finds Broccoli Sprouts Have High Levels of Anticancer Chemical," URL: *http://www.canoe.ca/HealthNutrition/sep15-broccoli.html* (last accessed Sept. 15, 1997: 1–2).

41. P. E. Garfinkel, "Classification and Diagnosis of Eating Disorders," in K. D. Brownell and C. G. Fairburn, eds., *Eating and Obesity: A Comprehensive Handbook*, The Guilford Press, New York, 1995, pp. 125–134.

42. P. J. V. Beumont, "The Clinical Presentation of Anorexia and Bulimia Nervosa," in K. D. Brownell and C. G. Fairburn, eds., *Eating Disorders and Obesity: A Comprehensive Handbook*, The Guilford Press, New York, 1995, pp. 151–158.

43. H. W. Hoek, "The Distribution of Eating Disorders," in K. D. Brownell and C. G. Fairburn, eds., *Eating Disorders and Obesity: A Comprehensive Handbook*, The Guilford Press, New York, 1995, pp. 207–211.

44. J. Garrow, "Diseases of Diet in Affluent Societies," in B. Harris and Sir R. Hoffenberg, eds., *Food: Multidisciplinary Perspectives*, Blackwell Publishers, Oxford, UK, 1994.

45. J. Waterlow, "Diet in the Classical Period of Greece and Rome," *Eur. J. Clin. Nutr.* **43** (Suppl. 2), 3–12 (1989).

46. A. Keys, "Mediterranean Diet and Public Health: Personal Reflections," *Amer. J. Clin. Nutr.* **61** (Suppl.), 1321S–1323S (1995).

47. M. Nestle, "Mediterranean Diets: Historical and Research Overview," *Amer. J. Clin. Nutr.* **61** (Suppl.), 1313S–1320S (1995).

48. A. Keys et al., "Studies on Serum Cholesterol and Other Characteristics on Clinically Healthy Men in Naples," *Arch. Intern. Med.* **93**, 328–335 (1954).

49. D. Schardt, B. Liebman, and S. Schmidt, "Going Mediterranean," *Nutr. Action Health Lett.* **21**, 1, 5–7 (1994).

50. A. Kafatos et al., "Coronary Heart Disease Risk Factor Status of the Cretan Urban Population in the 1980s," *Amer. J. Clin. Nutr.* **54**, 591–598 (1991).

51. A. Ferro-Luzzi and F. Branca, "The Mediterranean Diet, Italian Style: Prototype of a Healthy Diet," *Amer. J. Clin. Nutr.* **61** (Suppl.), 1338S–1345S (1995).

52. G. A. Spiller, ed., *The Mediterranean Diets in Health and Disease*, AVI, Van Nostrand Reinhold, New York, 1991.

53. World Health Organization, "Diet, Nutrition, and the Prevention of Chronic Diseases," *Technical Report Series* No. 797, World Health Organization, Geneva, Switzerland, 1990.

S. Ghosh
P. L. Pellett
University of Massachusetts
Amherst, Massachusetts

AFLATOXINS. See MYCOTOXINS.

AGGLOMERATION AND AGGLOMERATOR SYSTEMS

GENERAL ASPECTS

Numerous food powders experience significant changes in their properties during storage, transportation, or processing, which are related to the particle size distribution. Attrition causes reduction in average particle size while aggregation increases it. Fines generated by attrition may either form clusters or coat larger particles (plating). Interparticle adhesion is decisively influenced by particle size, the ratio between adhesion and weight usually being inversely proportional to the square of the particle size (1). As a result, this ratio is two orders of magnitude higher for particles of 10 μm than for particles of 100 μm. Dry food powders with average sizes of 80 to 100 μm are usually free flowing, whereas powders having sizes below 20 to 30 μm become cohesive, form secondary particles (clusters) of larger size, and form lumps when rewetted.

Adhesion without formation of bridges between the adjacent particles occurs as a result of either van der Waals or electrostatic forces and causes the formation of comparatively weak agglomerates. Adhesion associated with the formation of bridges produces much stronger agglomerates. Free flowing and cohesive food powders may undergo segregation during their storage, transportation, and handling. Primarily because of the differences in particle size and also in density, shape, and resilience, fine particles migrate to the bottom while large particles find themselves at the top of the vessel. As a result, some minor components of beverage blends (colors, flavors, vitamins) may become unevenly distributed between packages.

The purpose of particle size enlargement by agglomeration is to improve powder properties like bulk density, flowability, meterability, dusting, powder mix homogeneity, storage stability, and optical appearance. Powdered foods, which are in most cases intended to be dispersed in liquid, should also have good wettability, sinkability, dispersibility, and (for soluble materials) solubility, that is good "instant properties." A powder layer spread on a liquid surface should imbibe the liquid, submerge, disperse, and dissolve within a few seconds with little mechanical aid and without forming lumps. A powder treated by a technical process to have such properties is called "instantized." Agglomeration is the predominant method for instantizing powdered foods, and an example of the dependency of the wetting time of a powder layer on the average agglomerate size is shown in Figure 1.

Another major quality factor for instant foods is the preservation of flavor components. Instant beverages containing, for example, coffee extract are particularly susceptible to flavor loss caused by high temperature or excessive contact with air, as, for example, in a fluidized bed.

Obviously, simultaneous improvement of all powder properties is impossible. Increasing agglomerate stability, for example, results in most cases in a decline of the instant properties. The way the agglomerates are formed in the production process determines their properties, and comprehension of the basic physical principles of particle adhesion and the mechanisms likely to predominate in a given agglomeration process is helpful.

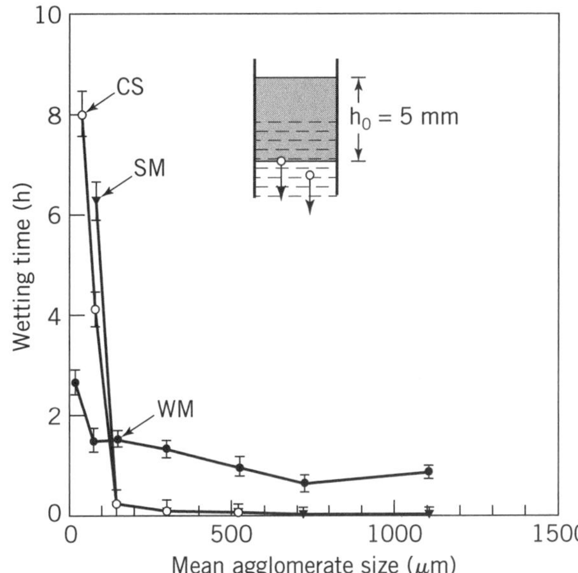

Figure 1. Wetting time (including standard deviations) of a powder layer of height h_0 = 5 mm for three commercial instantized food powders. CS, cocoa-sugar mix (21°C); SM, skim milk powder (21°C); WM, whole milk powder (50°C).

UNDESIRED AGGLOMERATION—CAKING

Agglomeration via caking may occur unintentionally since blends of particles are always exposed for some time to the ambient environmental conditions (temperature and/or humidity). For example, food powders that include lipids (soups, sauces, baking mixes) may undergo caking if the temperature exceeds the melting point of the lipids. As a result, sticky liquid bridges are formed. Once cooled, the lipids recrystallize, liquid bridges between particles become solid, and caking is reinforced. Although starchy and proteinaceous components are relatively insensitive to the environmental conditions, the soluble components of food powders (sugars, salts) absorb moisture and eventually change their state from solid to liquid.

The ability of sugars to soften depends on the conditions under which they were produced and stored. These conditions are responsible for the formation of areas of crystalline or amorphous structure. Amorphous sugars absorb much more moisture at a given water activity (relative humidity) and have lower glass-transition temperatures than crystalline sugars (2). Whereas a stable, crystalline structure is formed at equilibrium conditions, the amorphous one is created at nonequilibrium conditions. Relatively slow moisture withdrawal during carefully controlled crystallization (nuclei formation and crystal growth) leads to the development of a crystalline structure. Fast moisture withdrawal from a solution of carbohydrate via spray drying, roller drying, or freeze drying helps to produce mainly the amorphous form; even the mechanical impact of milling of sugar crystals produces an amorphous surface capable of recrystallization after absorbing water (3). Upon recrystallization, amorphous sucrose releases water, which facilitates formation of bridges between particles and initiates caking. Adding high molecular weight components (eg, maltodextrin) to a blend containing sugars may reduce caking (4).

Caking may be effectively suppressed by adding anticaking agents like tricalcium phosphate, magnesium oxide, calcium silicate, and so on, which absorb a portion of moisture from the blend and thus reduce the amount of available moisture. Although total moisture content of the blend with or without anticaking agent stays virtually unchanged, it is relative humidity generated by the blend in a sealed chamber that reflects the amount of available moisture: blend with added anticaking agent generates lower RH than blend without anticaking agent. The effectiveness of the anticaking agents depends largely on their water-holding capacity, so that with an unlimited source of humidity (open storage), their impact is lessened.

Even packaged food powders may undergo caking influenced by the environment inside their packages. Being relatively isolated, the headspace inside the package is affected not only by the surface moisture of the particles and temperature in the warehouse, but by the permeability and heat conductivity of the package film. Variations in the temperature and humidity outside of the packaged material often accelerate an exchange in surface moisture between the ingredients and initiate caking.

AGGLOMERATION METHODS

The first step in any agglomeration process (except drying methods starting with a slurry) is to make the primary particles contact each other, which is frequently achieved by external force. Powders for instant products usually consist of primary particles smaller than 200 μm to facilitate solubility. In a second step, permanent adhesion forces stronger than any possibly existing disruptive forces must be established between these particles. For food powders, this is usually achieved by wetting (which causes partial dissolution and the development of liquid bridges) and subsequent drying (which leaves solid bonds in place of the liquid bridges).

The duration and intensity of the forces acting among the particles during agglomerate formation and stabilization have an important influence on agglomerate porosity and stability. For example, an agglomeration process in which high forces act on the particles and agglomerates will turn out dense, smooth, and stable agglomerates that are easy to handle and dispense. However, instant properties would be poor owing to low agglomerate porosity and strong bonds between the primary particles. Such a process, like compaction, would be inappropriate for instantizing.

The final product should have the following properties:

- sufficient agglomerate porosity for fast liquid suction by capillary action, although a critical porosity must not be exceeded (5);
- particle size in the range of 0.2 to 2 mm; and
- sufficient agglomerate strength to withstand handling and transportation.

Agglomeration processes suitable for producing instantized food powders can be divided into three groups:

- moist agglomeration,
- agglomeration by drying, and
- combined methods.

Moist Agglomeration

Moist agglomeration, using capillary and liquid bridge forces to achieve sufficient interparticle adhesion during agglomeration, is the most important process for the production of instantized powders. This method starts from dry powder, which is moistened either by condensing vapor, atomized liquid, or a mixture of both. The material is then dried and solid bridges between the primary particles provide the necessary strength. A large variety of equipment is available for moist agglomeration. Schematics of some typical examples are shown in Figure 2.

Except for the static process Figure 2f, all methods are "dynamic"; that is, agglomerates are formed due to the collision and subsequent adhesion of the particles. Common features of all processes shown in Figure 2 are the moistening of the dry powder, the size enlargement of the wet particles, and subsequent drying and cooling, if required. The drying rate and temperature considerably influence the strength of the dry agglomerates because of different crystal structures and distributions of the interparticle solid bridges formed by crystallization of dissolved substances during drying (6).

Fluidized-bed agglomerators for batchwise or continuous production, Figure 2a and 2b, are provided by numerous manufacturers. Units for continuous operation (eg, APV Anhydro) utilize either a moisture product feed or a rewet system so that sufficient moisture is present to agglomerate the product. A typical unit has three fluidized zones. These include the entry or wetting zone, a drying zone, and a cooling zone.

Also, several mechanical agglomerators utilize mechanical mixers to provide the liquid addition, product interaction, and mixing to facilitate the agglomeration process.

The "SCHUGI" mixer (Bepex Corp.), shown in Figure 2d, has extremely short residence times (approximately 1 s) and has considerable flexibility in the types and amounts of feedstock. The system employs a flexible housing so that product buildup on the interior walls of the agglomerator is minimized, if not eliminated (7).

Jet agglomeration (Fig. 2e) has been used in the food industry for several years to produce agglomerates with favorable instant properties from fine powders. In a jet agglomeration plant, freely moving, wetted particles are made to collide with each other to form agglomerates. The solid material fed to the agglomerator consists of individ-

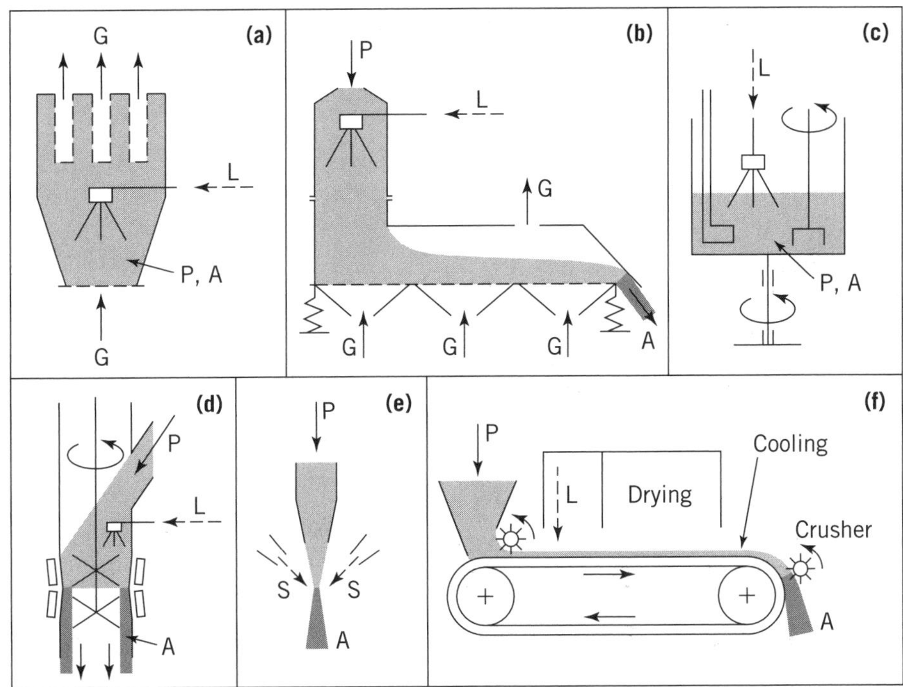

Figure 2. Principles of moist agglomeration processes for the production of instantized powdered foods. (**a**) Batch fluidized-bed agglomerator, (**b**) continuous mixing chamber with integrated fluidized-bed agglomeration, (**c**) batch mixer agglomerator, (**d**) continuous mixer agglomerator with a perpendicular balling region ("SCHUGI" mixer), (**e**) continuous steam fusion process ("jet agglomeration"), (**f**) continuous belt aggomerator. A, agglomerates; G, gas; L, liquid; P, powder; S, steam.

ual particles and dry preagglomerates bound mainly by van der Waals forces. The effectiveness of this method depends on a variety of process parameters influencing interparticle collision frequency, relative velocities of the particles, interparticle contact forces between wetted particles, and strain on the agglomerates (8).

Drying Processes

The second main group of processes for agglomerating food powders comprises special drying processes. Two examples are shown schematically in Figure 3, a spray drier with fluidized bed and a freeze drier.

Spray drying is one of the most widely used processes in food powder technology. The concentrate/slurry, either in the form of a suspension or a solution, is finely distributed using a nozzle or an atomizer disc, dried, and cooled in a connected fluidized bed from which the product is withdrawn in agglomerated form. The fluidized bed also serves the purpose of removing the fines, which are collected in a cyclone and recycled into the spray drier. Agglomeration occurs in the vicinity of the nozzle or atomizer, where the fine dry particles collide with the slurry droplets. In many cases the product is also after-dried in the fluidized bed, a process referred to as two-stage drying.

Freeze drying is relatively expensive but especially useful for products sensitive to high temperatures. Another advantage is the possibility to vary the porosity of the agglomerates over a wide range by foaming the concentrate before freezing (this can be achieved in a spray drier, too, by gassing the slurry immediately before atomization). As a novel technology, microwave drying in a vacuum chamber can be used instead of freeze drying.

Combined Methods

Two examples for combined agglomeration methods are presented in Figure 4. These are special spray driers, the "Filtermat" by GEA Niro AIs (Soeborg, Denmark), with integrated perforated belt drier (Fig. 4a), and spray drying with integrated fluidized bed and fines recycling (Fig. 4b).

In the "Filtermat" process, the product is partially dried in the spray-dry section of the machine to a moisture content of approximately 16 to 25%, whereupon it is deposited by gravity onto the perforated belt and further dried to approximately 5 to 10% moisture. Sufficient moisture is present in the intermediate dried product so that agglomerates are formed on the belt. These are then classified and/or size-reduced to the desired size (9).

These methods allow production of very loose, but still sufficiently strong agglomerates (5) with good instant properties and have been widely used in the food industry.

Further Methods

Pressure agglomeration is rarely applied to the production of redispersible products, because the resulting agglomerates are so compact that they show insufficient dispersibility. Substances dispersing readily—due to a bursting effect of special additives—can be agglomerated using die pressing or low pressure extrusion processes. If no great demands are set upon the redispersibility, roll pressing is a useful and inexpensive process. Mostly low-pressure ring-roller presses are used in which the moistened powder is pressed through holes and thereby shaped into agglomerates.

A special kind of agglomeration is coating, to improve the wetting behavior of the particles. Usually the particles are coated by pure liquids, solutions, or suspensions that harden on the surface of the solid material. A novel technology is coating with submicron particles by "mechanofusion" (10) to improve the wettability and for other applications (11).

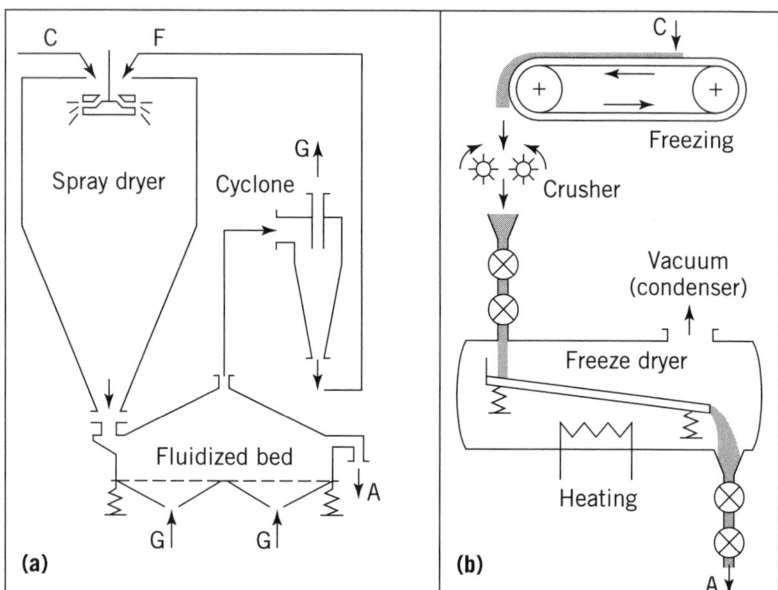

Figure 3. Principles of production of instantized powdered foods by spray drying (**a**) and freeze drying (**b**). A, agglomerates; C, concentrate; F, fines; G, gas.

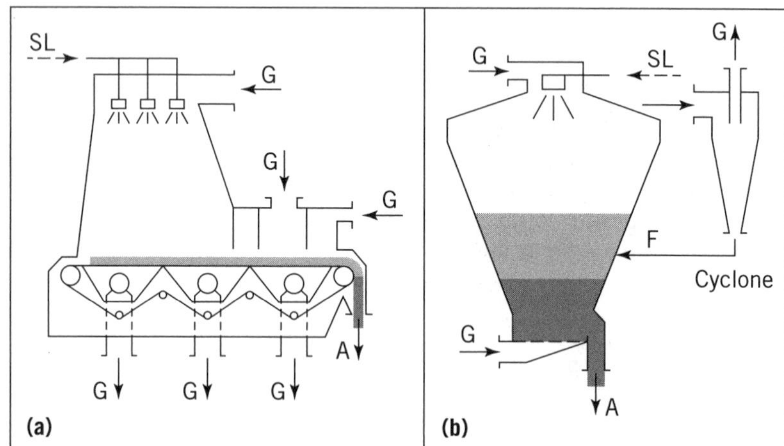

Figure 4. Principles of combined agglomeration processes for the production of instantized powdered foods: "Filtermat" spray drier with perforated belt drier (**a**) and spray drier with integrated fluidized bed and fines recycling (**b**). A, agglomerates; F, fines; G, gas; SL, slurry.

CHARACTERIZATION OF AGGLOMERATED POWDERS

Particle size distribution of agglomerated products should be measured using an adequately representative sample and with minimum disruption. Care should be taken to prevent swelling, dissolution, or disintegration. For sufficiently stable particles, sieving is still widely in use (dry for particles larger than 30 to 40 μm and wet for smaller particles), but over the past few years, laser diffraction has become a standard method due to the availability of a large number of different devices (Coulter, Malvern, Sympatec, etc). The last method is more versatile because it allows the measurement of larger particles with a dry feeder in a relatively nonviolent way and is also applicable for fines in a dispersing liquid. Automatic image analysis systems are a recent development employing charge-coupled device (CCD) camera imaging of particles falling out of a dry feeder.

Agglomerated particles may partially disintegrate when subjected to a more or less violent impact (vibration, shaking) during storage, transportation, and handling. In the case of agglomerated coffee, attrition is the primary cause in separation of fines from the outer surface of agglomerates (12). Formation of so-called secondary particles (shattering) from the disintegrated pieces is also taking place. The strength or resistance of agglomerated products to attrition may be measured by comparing its size distribution before and after rotation for a limited time inside a cylinder (friabilator). Other properties of agglomerates (bulk or free flow density, flowability, cohesion, angles of spatula and repose) are measured, for example, with the Hosokawa (MicroPul) tester.

For powdered foods, information about wetting and dispersing behavior is of special interest to the manufacturer. Wetting of a powder is easily tested by preparing a sample of defined height (5–8 mm) in a cylindrical testing vessel with a slide covering the liquid reservoir. When the slide is pulled out sideways, the powder sits on the liquid surface and is wetted. The wetting time measured in this way is useful for quality control purposes and for product property comparison.

Powder dispersion measurement is more difficult and depends on the ability of measuring the amount of material actually dispersed after the mixture/dispersion has been prepared in a defined way. For many products, this can be achieved by, for example, photometry (milk powder, cocoa beverages), conductometry (powdered extracts of coffee, tea) or refraction index measurement (sugars). The dispersed powder mass divided by the total powder mass in the sample is called degree of dispersibility (which varies between 0 and 1). Milk powder can reach a value of up to 0.8 (depending on fat content and age), instant cocoa beverages up to 0.98 and instant coffee up to 1.

PRODUCT EXAMPLES

Agglomerated food products are inseparable from some specific processes that were used to obtain these products. Steam fusion/steam jet agglomeration is used for agglomerating water-soluble instant beverage powders with high sugar content like cocoa drinks or products containing coffee extract (Fig. 5). After grinding, the particles with an average particle size of 25 to 75 μm are fed into an agglomeration chamber in the shape of dry clusters (average size up to the mm range). A uniformly distributed curtain of powder moves downward where it interacts with jets of steam. Steam wets the particles and fuses them into agglomerates. The wetted agglomerates pass from the top section of the agglomerating tower into a drying zone in the bottom portion of the tower, which is supplied with hot air. The agglomerates are dried and then cooled and screened. Critical parts of this agglomeration process are the feed port section, where initial cluster formation occurs, and the steam jet zone, where further agglomeration by collision of sticky clusters takes place (8).

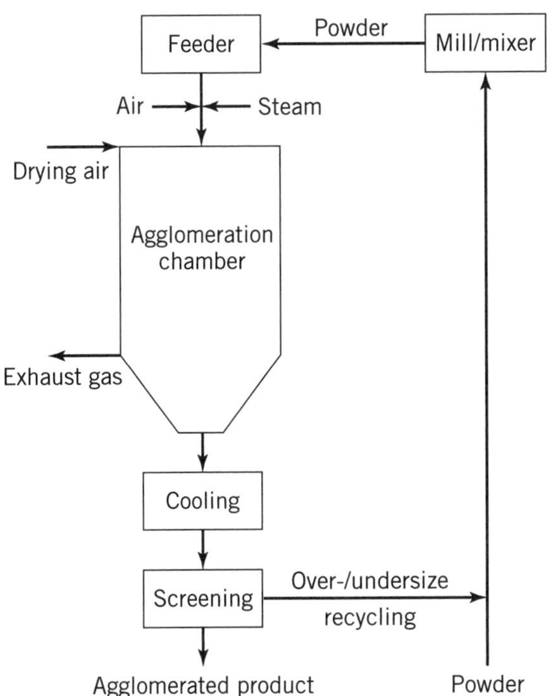

Figure 5. Principle flow diagram of a steam fusion/steam jet agglomeration process.

A method for making agglomerated bits containing aspartame includes preblending aspartame and a bulking agent (maltodextrin) to form a premix (13). The latter is then mixed with other dry ingredients (flavors, starch binders, dispersing agents, and vitamins) to form a dry mix. Liquid ingredients (vegetable oil and water) are blended into a dry mix with a ribbon blender or a paddle mixer to form moistened clumps. These granules must be dried in a forced-air convection oven and then screened to obtain a desirable particle size distribution. While starches (tapioca, corn, potato, modified wheat) and gums are used as the binders in forming agglomerates, baking soda and maltodextrin assist in the dispersion of the agglomerated product. These bits are suitable for use in home-cooked grain cereals and other foods.

Agglomerated potato granules can be prepared from a mixture of potato granules, egg white solids, and water (14). After the wet premix is formed, a gentle sieving, drying, and crushing are used to obtain agglomerates with the desired size and density.

Agglomerated bread crumbs may be produced from a starch containing raw material (flour, meal) and water in a continuous pellet mixer (15); subsequent baking in a humidified atmosphere (to control the desired gelatinization) and sizing (cutting) of the agglomerates are utilized. A controlled retrogradation (recrystallization of starch) occurs due to the controlled cooling process.

Agglomerated beverage blends having aspartame as a sweetener may be manufactured to prevent clumping of aspartame and to improve its water dispersibility (16). Agglomeration has been conducted in a jacketed blender so that a heating or cooling fluid may be passed through the jacket while the blender is rotated to provide adequate mixing. The blending time and temperature (but not, however, humidity) were controlled to obtain a desired agglomerate size distribution.

Aqueous, sugary syrups (honey, high fructose corn syrup, invert sugar, corn syrup, etc) were dehydrated using thin film drying in the presence of binders (soy protein and ungelatinized starch that was partially gelatinized *in situ*) (17). A spray of water was added during tumbling. The resultant agglomerates were dried and then slightly coated with a high melting point fat, apparently to prevent caking.

Agglomerates of garlic, onion, and their mixtures were produced in an upright chamber (18). Wetting of particles was provided by atomized water in the upper part of this chamber, followed by drying of the agglomerated product with air in the lower part of the chamber.

An agglomerated milk product was prepared by spraying a concentrate of milk into a stream of drying gas directed against the surface of a fluidized layer of already spray-dried particles (19). Adjustment of temperatures, flow rates of drying air, and residence times allowed better efficiency for skim milk, whole milk, and whey particles.

Agglomerates of meat analogues were prepared by extrusion cooking of soy concentrate, comminution of the extrudate, mixing it with a water slurry of binder, frying the mixture in edible fat or oil to produce an agglomerated mat, and sizing (20). Particles with the desired size distribution were used for meat-type sauces.

A porous and pelletized food product may be formed by premixing two or more ingredients, one of which is capable of forming sticky bonds after being moistened by an aqueous medium (21). This interaction occurs when the mixture of particles is tumbled and rolled on a pelletizing disc; the adhered particles form pellets or wet aggregates. Examples of a dry mix may include sugars, starches, dried milk products, proteinaceous materials, dehydrated juices, and powdered coffee concentrates. Some of these products (sugars, starches) may become self-adherent in contact with water and are used to form agglomerates. To provide pellets with controlled porosity, the particulate mixture also includes a chemical leavening system (sodium bicarbonate and leavening acid). Once moist agglomerates are in contact with hot air, two processes take place: drying and formation of gaseous carbon dioxide (due to a reaction between the bicarbonate and leavening acid). The resulting pellets have a porous, cellular structure with a crisp, crunchy, and friable texture.

The vast majority of gelatin dessert mixes require the use of hot water to dissolve the gelatin and an extended time (3–4 h) to prepare the meal. However, if gelatin-containing mix with a limited moisture content of 1 to 3% is agitated and slowly heated to 190 to 195°F, it forms agglomerates (22). Subsequent cooling helps to form so-called cold-water-soluble gelatin, which dissolves and disperses in water at 40 to 50°F. Agglomeration of gelatin mixes (ie, sucrose, gelatin, citric acid) was conducted in a jacketed rotating blender.

Other agglomerated (instantized) products include maltodextrin and dextrose, which may be used as carriers for flavors, colors, and nonnutritive sweeteners in instant beverages and desserts (23,24), soy protein isolates for high-protein beverage blends and for better dispersibility in meat emulsions, prejelled starches and gums as soup thickeners, whey protein concentrates and calcium caseinates for dairy blends, all developed by IFT, Inc. (25), and coated animal feed with improved nutritional value (26). Powders such as egg proteins, cocoa, or various fibers will have improved dispersibility after agglomeration in the presence of maltodextrin or surfactants (27).

BIBLIOGRAPHY

1. H. Schubert, "Food Particle Technology. Part 1: Properties of Particles and Particulate Food Systems," *J. Food Technol.* **6**, 1–32 (1987).
2. B. Makower and W. B. Dye, "Equilibrium Moisture Content and Crystallization of Amorphous Sucrose and Glucose," *Agric. Food Chem.* **4**, 72–77 (1984).
3. J. M. Flink, "Transition in Dried Carbohydrates," in M. Peleg and E. B. Bagley, eds., *Physical Properties of Foods*, AVI Publishing, Westport, Conn., 1983.
4. E. C. To and J. M. Flink, "Collapse," *J. Food Technol.* **13**, 551–594 (1978).
5. H. Schubert, "Instantization of powdered food products" *Int. Chem. Eng.* **33**, 28–45 (1993).
6. W. Pietsch, *Size Enlargement by Agglomeration*, Wiley, New York, 1991.
7. P. Koenig et al., *Bepex Corp. Product Literature*, Bepex Corp., Minneapolis, Minn.
8. S. Hogekamp, "Über eine modifizierte Strahlagglomerationsanlage zur Herstellung schnell dispergierbarer Pulver," Ph.D. dissertation, Universität Karlsruhe, 1997.
9. S. Mignat, "Spray Drying and Instantization—A Survey," Dragoco Report, Factoring Information Service 4, Holzminden, Germany, 1988, pp. 105–116.
10. T. Yokoyama, "Mechanofusion," Micromeritics Laboratory, Hosokawa Micron Corp., Osaka, Japan, 1988.
11. D. J. Harron, "Particle Fabrication: New Method of Agglomerating Heterogeneous Powders," *Proc. Fifth International Symposium on Agglomeration*, Brighton, UK, Inst. Chem. Engs., Rugby, UK, 1989, pp. 175–185.
12. J. Malave-Lopez and M. Peleg, "Patterns of Size Distribution Changes during the Attrition of Instant Coffees," *J. Food Sci.* **51**(3), 691–694 (1986).
13. U.S. Pat. 4,741,910 (May 3, 1988), J. Karwowski and A. M. Magliacano (to Nabisco Brands, Inc.).
14. U.S. Pat. 4,797,292 (Jan. 10, 1989), J. DeWit (to Nestec S.A.).
15. U.S. Pat. 4,344,975 (Aug. 17, 1982), W. Seiler (to Gebruder Buhler AG).
16. U.S. Pat. 4,554,167 (Nov. 19, 1985), R. M. Sorge et al. (to General Foods Corp.).
17. U.S. Pat. 3,941,893 (Mar. 2, 1976), E. F. Glabe et al. (to Food Technology, Inc.).
18. U.S. Pat. 4,394,394 (July 19, 1983), L. J. Nava and N. L. Ewing (to Foremost-McKesson, Inc.).
19. U.S. Pat. 4,490,403 (Dec. 25, 1984), J. Pisecky et al. (to A/S Niro Atomizer).
20. U.S. Pat. 4,441,461 (May 8, 1984), P. J. Loos et al. (to the Procter and Gamble Co.).
21. U.S. Pat. 4,310,560 (Jan. 12, 1982), R. C. Doster and S. L. Nelson (to Carnation Co.).
22. U.S. Pat. 4,571,346 (Feb. 18, 1986), D. M. Lehmann et al. (to General Foods Corp.).
23. "New Ingredients," *Food Technol.* **43**(1), 169 (1989).
24. "A Sweetener for Specialized Applications," *Food Eng.* **59**(9), 56 (1986).
25. "Instantizing Ingredients," *Food Eng.* **57**(3), 58–59 (1984).
26. U.S. Pat. 5,786,008 (July 28, 1997), R. Humphry et al. (to Moorman Manufacturing Co.).
27. D. D. Duxbury, "Innovative Ingredient Properties Enhanced, Customized by Agglomeration, Coating Process," *Food Processing* **49**(5), 96–100 (May 1988).

STEFAN HOGEKAMP
HELMAR SCHUBERT
Universität Karlsruhe
Karlsruhe, Germany

ALCOHOL, POLYHYDRIC. See SUGAR: SUBSTITUTES, BULK, REDUCED CALORIE; SWEETENERS: NUTRITIVE.

ALCOHOLIC BEVERAGES AND HUMAN RESPONSES

Alcoholic beverages are, in essence, flavored solutions of ethanol. The flavors may come from grains, as in beer; or from grapes and other fruit, as in wine; or from any source of carbohydrates, grains, sugar, or grapes, as in whiskey, rum, and brandy. In addition, consumers may add their own flavors, as lime with some beers or fruits with some wine or carbonated sodas with distilled spirits. The spectrum of flavors is wide indeed. But the purpose of drinking any of these is to supply ethanol in measured doses to the user.

Ethanol is as unique as humanity itself. It is a food but requires no digestion. It acts on many organs in the body but has no cellular receptors as do all other drugs. It is stable in the atmosphere to any chemical change, whereas all other foods will undergo some kind of decomposition. It is the only food produced solely by microbial action. It enters any cell in the body, freely, without any transport mechanism. All other foods (and all other substances except water) require a transport mechanism to enter any cell. It provides energy more rapidly than any other food.

This article inquires more closely into these and other aspects of ethanol. Although alcohol is a generic term for a large group of related substances, so common is ethanol that the term alcohol has been usurped for it and will be used here from now on to mean ethanol. So common is the drinking of alcoholic beverages that the word drink or drinker implies the drinking of alcoholic beverages and not any others.

People have always needed a release from reality. From earliest recorded history this release has come quite effectively from alcohol. It must have been discovered by accident, and probably in more than one place. It is readily produced from any *saccharous* source, is pleasant tasting, and not prone to any pathogenic divergence.

Whether alcohol appeared first from grapes as wine or from grain as beer or from honey as mead is not known. The catalyst that converts any of these into alcohol is ubiquitous. A recipe for beer has been found on a clay tablet from Mesopotamia some 4000 yr old. It was probably known during the new Stone Age, some 6000 yr ago. All but three or four of the many cultures that have survived to modern times knew alcohol. It is absent from polar people and Australian aborigines.

Probably the nature of the alcohol in any culture depended on the prevalence of the source. In cool northern Europe it was likely to be beer or mead. In the Near East it may have been beer or wine. In the Far East it was probably beer. In early cultures the making of alcohol was so cherished that it fell under the domain of the priest and clergy. Vestiges of this still remain in many monasteries in Europe.

GENERAL METABOLISM OF ALCOHOL

The first step in the metabolism of alcohol is a dehydrogenation to acetaldehyde.

$$\begin{array}{ccc} \text{H H} & & \text{H H} \\ \text{HC-C-OH} & \rightleftarrows & \text{HC-C=O} \\ \text{H H} & & \text{H} \end{array}$$

This is mediated by the enzyme alcohol dehydrogenase, with nicotine adenine dinucleotide (NAD^+) as hydrogen acceptor. The reaction is reversible, and the reverse reaction is the last step in the process by which alcohol is produced by yeast.

This reaction is followed by the oxidation of the aldehyde to acetate, brought about by another enzyme, aldehyde dehydrogenase, again with NAD^+. This reaction has never been reversed. The acetate in turn joins with coenzyme A to form the ever-present acetyl CoA. This can take part in the citric acid cycle and be oxidized to CO_2 and H_2O. This scheme is shown in Figure 1.

Alcohol dehydrogenase (ADH) exists in about 20 forms, each with differing activity toward ethanol and to other alcohols. These isozymes vary in concentration among diverse ethnic groups, no doubt accounting for different sensitivities to alcohol by different peoples. All forms have zinc as the core metallic element. ADH is found in all tissues, including red and white blood cells and the brain. Before 1970, it was thought that ADH existed only in the liver, but that is certainly not the case. That it is present in many isosteric forms is probably rooted in the many functions it performs and the many needs it satisfies in metabolism.

Aldehyde dehydrogenase (ALDH) also exists widely in humans. Cytoplasmic ALDH is the same in all people, whereas the mitochondrial ALDH does differ among people, with that found in Asians being less active than the form found in whites. But it is probable that mitochondrial ALDH is not nearly as important in oxidizing acetaldehyde as is cytoplasmic ALDH. Because very little acetaldehyde is found circulating in the blood even after high alcohol intake, it is assumed that the rate-limiting step in alcohol metabolism is the first step—its dehydrogenation to acetaldehyde.

HOW ALCOHOL IS CONSUMED

Wine

The fermentation of the juice of grapes produces a wine containing about 12% alcohol by volume (10% by weight). The stoichiometry of fermentation,

$$\begin{array}{ccccc} C_6H_{12}O_6 & \rightarrow & 2C_2H_5OH & + & 2CO_2 \\ 180 & & 92 & & 88 \end{array}$$

dictates that a 22°Brix grape juice will give an alcohol solution of somewhat more than 10% by weight. Perhaps by evolutionary coincidence, a 10% alcohol solution is close to the limit that most yeasts can produce.

Because most countries forbid the addition of water to grape juice before fermentation, wines worldwide are very similar in alcohol content. Champagnes, which are fermented twice, may be a little higher, say 14% by volume. Fortified wines, such as port and sherry, are wines to which brandy has been added at some stage. These may contain as much as 20% alcohol by volume.

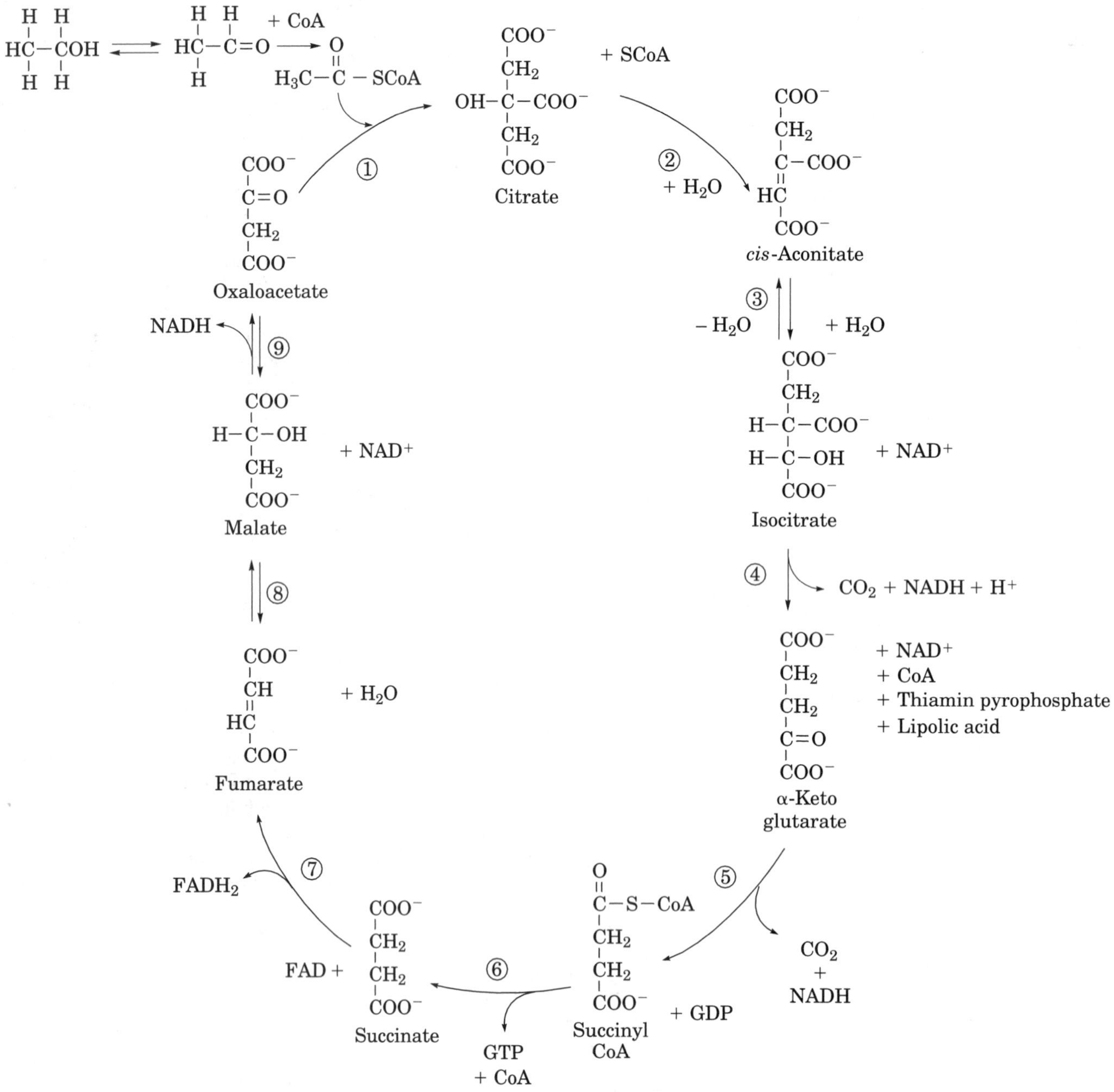

Figure 1. The citric acid cycle.

Other fruits may also be fermented to wine, but only apple juice, as cider, has found even limited acceptance.

Beer

The solution from which beer is made, called wort, is made at the brewery and not in nature. As such, its concentration and fermentability (see BEER) are designed for the beer being made and may vary widely. But the concentration of the wort only varies between 12 and 18°Brix, and the fermentability between about 60% and 90%. So beer is never as high in alcohol as wine. Typically, beer contains 5% alcohol by volume (4% by weight). So-called malt liquors contain about 6% alcohol by volume.

Mead

As honey is about 82% solids, it must be diluted to about 20°Brix and then fermented. The high price of honey and the rather weak flavor of its fermented solution has not permitted mead to be more than a historical curiosity.

Distilled Spirits

These spirits are distilled at a rather high proof, be they from grains (whiskey or vodka), molasses (rum), or wine (brandy), and are diluted to a marketable strength. The normal concentration is about 80 proof or 40% by volume or 32% by weight. Whiskeys are usually consumed after mixing with water, often carbonated, and also flavored. As purchased, wine is 2 stronger than beer, and whiskey is 3 or more times stronger than wine.

ABSORPTION OF ALCOHOL

Alcohol is absorbed into the bloodstream mainly from the upper small intestine. Although it enters the stomach, there is only a limited exit into the blood from that organ.

Emptying of the stomach's contents is controlled by the pyloric sphincter muscle at the base of the stomach. It opens when the pH of the contents falls below 3. The presence of proteins, which act as buffers, or fats, which delay access of the stomach's enzymes and acid to their substrates, delay the fall in pH and thus also the exit of food or drink from the stomach.

All the blood that nourishes the entire digestive tract, and that in turn carries all the digested foodstuffs from the tract, first enters the liver via the portal vein. Thus alcohol absorbed from the stomach or the small intestine goes first to the liver. This probably accounts for the old and incorrect belief that alcohol is oxidized to acetaldehyde only in the liver.

Alcohol consumed at high concentrations is absorbed more rapidly than when consumed in diluted forms. The presence of carbon dioxide also accelerates the absorption. Fatigue and exercise delay the absorption.

It has been recently found that a considerable amount of alcohol is metabolized in the stomach to acetaldehyde, and that the quantity oxidized is higher in men than in women and lower in alcoholic men and women (1). The blood alcohol curves for alcohol administered by mouth or intravenously for nonalcoholic and alcoholic men and women are shown in Figure 2. Of course, the alcohol metabolized in the stomach can have no effect elsewhere. This so-called first-pass metabolism accounts for the more ready effect that alcohol has on women than on men.

A comparison among whiskey, wine, and beer showing the maximum blood alcohol concentrations for the three beverages is shown in Table 1 (2). Wine and whiskey reach about the same maximum concentration, but whiskey is faster. Beer achieves a lower maximum, and it takes longer to get there.

At a lower alcohol level, the same relative relations are maintained, but the maximum concentrations attained are lower (Fig. 3) (3).

Table 2 shows the relationship between body weight, amount of beer and whiskey to reach certain blood alcohol levels, and the time at which the maximum concentration is reached. For example, an average person weighing 175 lb (79.5 kg) who consumes four 12-oz bottles of beer will have a maximum blood alcohol level of 0.08% in about 95 min (4).

Another set of experiments compared the sequential blood alcohol levels in a group of 50 men weighing between 145 and 175 lb who consumed 44 g of alcohol (four bottles of beer or four shots of whiskey) at once. The results in

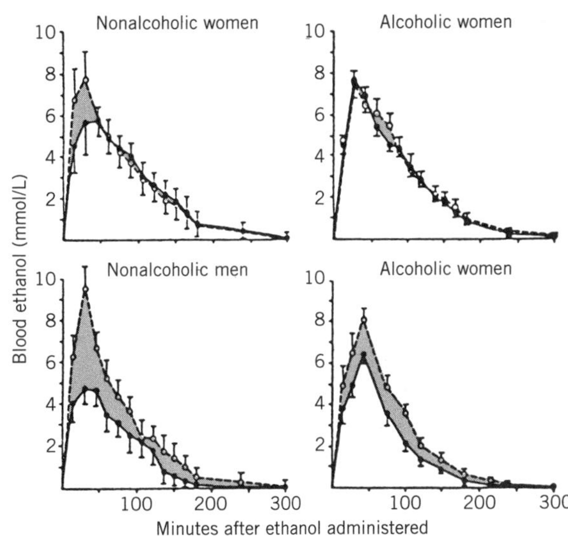

Figure 2. Effects of gender and chronic alcohol abuse on blood ethanol concentrations. Ethanol was administered orally (solid lines) or intravenously (dashed lines) in a dose of 0.3 g/body weight. The shaded area represents the difference between the curves for the two routes of administration (the first-pass metabolism).

Table 1. Effects of Alcoholic Beverages on the Maximum Blood Alcohol Levels and the Time at Which These Levels Are Reached

	c_{max}, %	t_{max}, min
Whiskey neat	0.108	57
Whiskey and water	0.107	71
Whiskey and soda water	0.104	51
Table wine	0.111	75
Beer	0.086	103

Note: Alcohol administered at a rate of 0.75 g/kg body weight. Equal to an all-at-once consumption of either four 12-oz. bottles of beer (for a 160-lb person) or two-thirds of a 750 mL bottle of 12% alcohol wine, or 3-1/2 shots (1-1/2 oz each) of 80-proof whiskey.

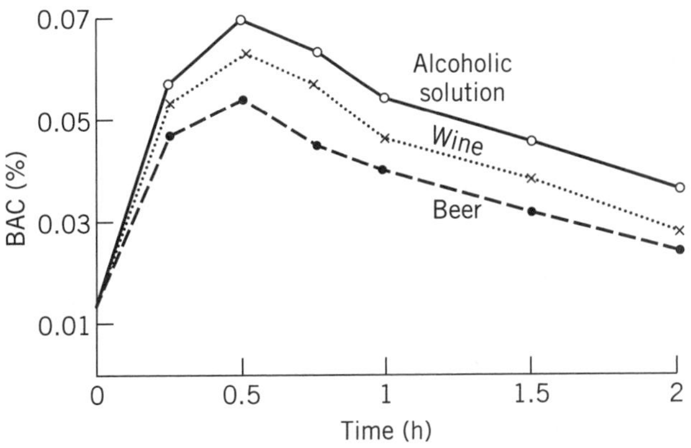

Figure 3. Blood alcohol curve after ingestion of 0.5 g alcohol per kg body weight in form of a diluted alcoholic solution (12.5%), of wine (11%), and of beer (5.5%) (mean value of 13 subjects).

Table 2.

Body weight, lb (kg)	g Alcohol per kg wt	To provide given amount of alcohol: No. of 12-oz bottles of beer (5%)	To provide given amount of alcohol: No. of 1-1/2-oz shots of whiskey (80 Proof)	Time to reach max. alc, min: Beer	Time to reach max. alc, min: Whiskey	Max. blood alcohol level, %: Beer	Max. blood alcohol level, %: Whiskey
120 (54.5)	0.25	<1	1	30	15	0.03	0.04
	0.50	2	2	50	35	0.06	0.075
	0.75	2-7/8	3	95	60	0.08	0.11
	1.0	3-3/8	4	150	95	0.10	0.14
	1.5	5-3/4	6	(240)	165	0.11	0.17
150 (68.2)	0.25	1-1/4	1-1/4	30	15	0.03	0.04
	0.50	2-1/2	1-1/2	50	35	0.06	0.075
	0.75	3-1/2	3-1/2	95	60	0.08	0.11
	1.0	4-3/4	5	150	95	0.10	0.14
	1.5	7-1/4	7-1/2	(240)	165	0.12	0.17
175 (79.5)	0.25	1-1/2	1-1/2	30	15	0.03	0.04
	0.50	2-4/5	3	50	35	0.06	0.075
	0.75	4	4-1/2	95	60	0.08	0.11
	1.0	5-1/2	6	150	95	0.10	0.14
	1.5	8-1/4	9	(240)	165	0.12	0.17
200 (90.9)	0.25	1-5/8	1-3/4	30	15	0.03	0.04
	0.50	3-1/4	3-1/2	50	35	0.06	0.75
	0.75	4-3/4	5	95	60	0.08	0.11
	1.0	6-1/2	6-3/4	150	95	0.10	0.14
	1.5	9-1/2	10-1/4	(240)	165	0.12	0.17

Figure 4 show that while whiskey produces a maximum blood alcohol level of 0.085, beer only gives a maximum of 0.045 (5).

Fig. 5 shows the maximum blood alcohol concentration, in percent of various alcoholic beverages. The effect of dilution on whiskey action is very marked (6).

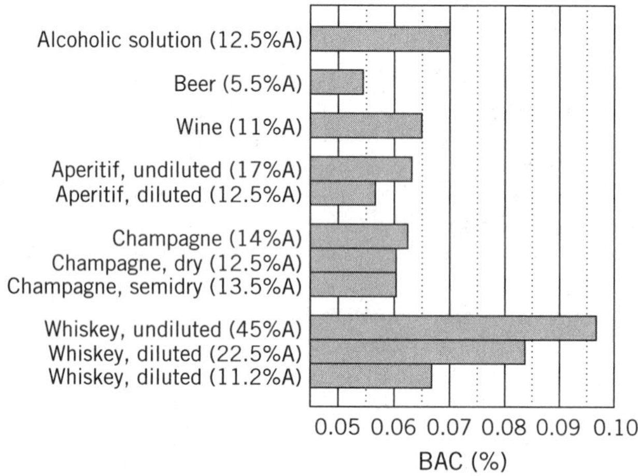

Figure 5. Blood alcohol maximum after ingestion of 0.5 g alcohol per kg body weight in form of several alcoholic beverages with different dilutions.

PHYSIOLOGICAL ACTIONS OF ALCOHOL

Alcohol is one of only two substances to which human cells have never found a need to bar entrance. The only other freely moving substance is water. Because there is no barrier to entrance, there are no known cellular receptors to alcohol. This free movement indicates that the body does not consider alcohol toxic or in any way foreign. It is this facile movement that permits alcohol to be excreted in the urine and in breath and also permits it to be found and measured in the breath.

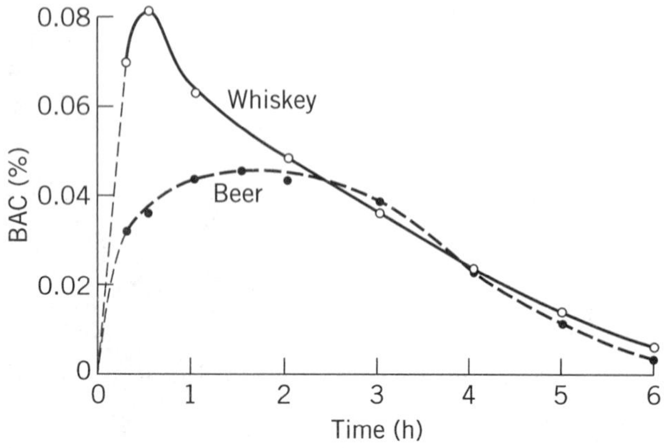

Figure 4. Blood alcohol curve after ingestion of 44 g alcohol in form of whiskey and of beer by 50 subjects, weighing between 65 and 79 kg.

There is many-faceted evidence that alcohol is a product of normal metabolism in humans; it has even been found in teetotalers. The amount is low, about 0.001%.

Effect of Alcohol on Fluid Balance

The initial effect of alcohol is to produce a slight retention of water, followed by a more marked diuresis of water and sodium. This latter effect is due to an inhibition by alcohol of the pituitary antidiuretic hormone (vasopressin). This diuresis occurs only when the concentration of blood alcohol content is rising. The concerted drinking of wine or whiskey, but not beer, requires the imbibition of other fluids to maintain homeostasis (7).

Effect of Alcohol on Nutrition

Alcohol is readily metabolized and yields 6.9 kcal/g. The rate of metabolism varies from person to person and ranges from 50 to 100 mg/kg of body weight per hour. Alcohol does not affect the use of dietary protein and is as effective as carbohydrates in protecting the body from loss of protein.

Alcohol does not raise heat production or CO_2 production, so it has no thermogenic effect. There is no increase in alcohol metabolism with exercise, nor a decrease with rest.

Alcohol has only a minor effect on appetite. Distilled spirits consumed in concentrated form decrease gastric motility, whereas dilute alcohol increases gastric motion. These will appear to affect appetite.

In addition, during a meal, the palate-cleansing action of wine and beer will promote eating and lessen the feeling of satiety. It is probably the tannin in wine and the hops in beer that are responsible for this action. Yet heavy drinkers are far less likely to be obese than nondrinkers. Among a large group of low-income women in their 30s, 41% of nondrinkers were obese, whereas only 22% of heavy drinkers were so rated. Among men, the ratio is 31% obese for nondrinkers and 16% for drinkers.

Effects of Alcohol on the Central Nervous System

The primary effect of alcohol in the body is a sequential depressant action on the central nervous system. From the spinal cord rises the medulla oblongata, the pons, and the midbrain, the control centers for the autonomic nervous system, which regulates reflexes and those functions beyond control of the will. Above the midbrain lies the thalamus and hypothalamus, whose dominion includes functions not willfully regulated, such as body temperature, metabolism of fats, and blood sugar level.

Above all this, and more recent in the evolution of the species, lie the cerebellum and the cerebral cortex. The former has sway over voluntary muscular movements and manual skills, as well as balance and speech. The cerebral cortex, the corolla of the mind, is master of all that distinguishes humans and makes them distinguished.

Alcohol, which has ready access to all cells and crosses the almost impenetrable blood-brain barrier as if it were a sieve, exerts its earliest action on the cerebral cortex. Its action is not as a stimulant but as a tranquilizer or depressant. It lessens inhibitions and increases confidence. It replaces discontent with discovery, bashfulness with bravado, and cowardice with courage. Alcohol negates a no and affirms a yes.

These effects, which people had sought long before the tensions of modern life were imposed on them, occur at blood levels between 0.02 and 0.05%. Whatever pressures and problems were mankind's lot in Biblical days, they were surely different in scope and pervasiveness than those we confront today. But alcohol was used then and was considered of priestly stature. In many cultures, its production was limited to the clergy.

Apparently, the evolution of the cerebral cortex, with its awesome powers of the mind, brings with it a need to lessen its potent sway over our actions. This alcohol does with speed and thoroughness.

At blood alcohol levels between 0.05 and 0.10%, the sphere of its influence expands to the cerebellum. Balance and speech become less than normal. The gait becomes unsteady and speech uncertain. It is at the higher of these levels that ability to operate machinery is impaired.

At 0.2%, balance is more seriously affected and speech nearly incoherent. Neurotransmission in the cerebellum is disturbed. At increasing blood alcohol concentrations, the effects become deeper and parts of the brain developed earlier in evolution are also effected.

At 0.3–0.4%, most of the behavior that we call human is sharply curtailed if not lost. Anesthesia sets in. In fact, alcohol was the earliest anesthetic, before ether was discovered.

At 0.5–0.6% a remarkable shut-off valve operates and wakefulness is not maintained. This normally prevents any further action, because even the autonomic nervous system, controlling basic functions of respiration and heart beat, are denervated at a level above 0.6% and death ensues. But sleep normally prevents further inhibition, so fatal levels are not attained except with rapid drinking of strong whiskey. Fatal levels cannot be achieved with wine or beer.

Interaction of Alcohol and Other Nutrients

Alcohol and its primary metabolite, acetate, serve as excellent sources of energy. Alcohol and acetate penetrate the blood–brain barrier almost as efficiently as glucose. Everywhere, alcohol is used in preference to all other energy sources. Acetyl CoA, the entry point of all energy sources into the citric acid cycle, is most easily produced from alcohol.

Also, while all other energy sources—fatty acids, glucose, and amino acids—can be stored in polymolecular forms, alcohol and acetate cannot be stored. Neither do they act as substrates for conversion to glucose, amino acids, or lipids. Alcohol must thus be metabolized as it becomes available (8).

Effect on the Digestive Tract

Alcohol slightly increases gastric and intestinal mobility. It also has a slight positive but very transient effect on appetite. It has no effect on the absorption of fats or cholesterol from the intestinal tract. Some of the very slight appetite stimulation may not even be due to alcohol, but to other factors present in some alcoholic beverages—juniper in gin, bitters in some aperitifs, hops in beer.

Effect on Cardiovascular System

Alcohol slightly dilates peripheral capillaries, leading to blushing and a feeling of warmth. The effects are transitory and meager.

Moderate alcohol consumption reduces the risk of coronary heart disease. It also reduces the risk of cardiac mortality. Its action is probably mainly due to its effect on raising the HDL levels in the blood, as well as lowering the triglyceride levels. Physical activity has the same effect, and the effects are additive.

Moderate drinking is taken to be between 25 and 80 g of alcohol a day (two to six 12-oz. bottles of beer, 5% alcohol by volume; 250–750 ml of table wine [12% alcohol]; or 70–250 ml of whiskey [80 proof]).

The LDL in the blood (the "bad" cholesterol) is now believed to have its malign effect on atherosclerosis after it is oxidized. And so antioxidants have been shown to have a protective effect. This is the so-called French Paradox: the French consume diets high in cholesterol and saturated fats and yet have very low coronary artery disease. The explanation is their relatively high consumption of alcohol, mainly in the form of red wine. Many of the polyphenals, gallic acid, rutin, epicatechin, and quercitin present in red wine are potent antioxidants (9). Such phenolics are also present in oak-aged whiskey and in full-bodied and darker beers.

Alcoholic beverages therefore have a two-pronged effect on reducing coronary heart disease, raising the "good" HDL and keeping the LDL from becoming "bad."

Effect of Alcohol on Men versus Women

Alcohol per se has the same effects on either sex. However, a given amount of alcohol has a greater effect on women than on men for several reasons.

First, women are usually smaller than men, so a given volume of an alcoholic beverage will have a greater effect on the smaller person.

Second, women have a lower water content in their bodies than do men. Thus the alcohol, which is only soluble in water and not in fat, will reach a higher concentration in women. This difference in water content, on an equal weight basis, is about 12%.

And last, and probably most important, women have a lower gastric activity of alcohol dehydrogenase than do men. They therefore do not destroy as much alcohol in their stomach as do men, allowing more of it to be absorbed into the blood. This phenomenon, only recently discovered (1), is known as first-pass metabolism. It may differ by 20–30% between men and women. Thus the bioavailability of alcohol is considerably greater in women than in men. However, men over 50 years of age begin to lose the high alcohol dehydrogenase activity in their stomachs and become equal to women in their first-pass metabolism (10).

Alcohol as Medicine

Before the advent of the modern pharmacopoeia, alcohol was prescribed for many human ills. It dulls pain, summons sleep, soothes the spirit, and tames the teeming mind. Each of these functions can now be accomplished by a coterie of drugs, so alcohol is rarely recommended for these ills. But no one substance does all of what alcohol does, and people in all cultures have found alcohol to be the preferred, and possibly less noxious, means to tranquility. Furthermore, as a controlled way to peace and pleasure, alcohol has no peer.

Regulations in the United States prohibit any declaration of therapeutic value for any alcoholic beverage, so each generation must rediscover the value of alcohol. Many countries permit some therapeutic claim for alcohol, such as a bedtime relaxer or tension reducer.

As medicine, alcohol differs from drugs. All drugs are detoxified in the liver, ie, the liver so chemically modifies the drug as to render it inactive and able to be excreted, usually by the kidneys. This detoxification usually involves oxidation, hydrolysis, or sulfation. But alcohol is not detoxified, it is metabolized, just as is any foodstuff. No drug supplies energy; alcohol does. It may be more valid to consider alcohol as a food that has effects on the body rather than as a drug. As was pointed out earlier, alcohol has free entrance to every cell in the body; no drug has. All drugs operate by attaching themselves to a specific cellular receptor; alcohol needs none and has none.

Alcohol has at least one clinically proven and prescribed medicinal function. It is the only antidote to poisoning with ethylene glycol (antifreeze). Quick administration of alcohol, usually intravenously, makes all alcohol dehydrogenase sites act on the alcohol, harmlessly, instead of acting on the glycol, which produces very toxic oxalic acid. The glycol, not oxidized, is slowly excreted by the kidneys. The preferential reaction of alcohol dehydrogenase with alcohol permits this therapeutic action.

There are foods other than alcohol that affect the brain and human behavior. The amino acid tryptophane is a soporific and has been used as such. Meals high in carbohydrates produce high tryptophane blood levels. This they do by producing an influx of insulin into the blood, which in turn mobilizes some amino acids from plasma to muscle, leaving a higher concentration of tryptophane in the plasma. Histidine, threonine, tyrosine, and choline also are neuroactive, but their action is compromised by complex factors.

There are no confirmed allergies to alcohol per se. Its free entrance to all cells and ubiquitous presence mitigate against any allergic reaction.

Alcohol the Day After

The day after a serious session of drinking is often one of mental and physical fatigue and headache. In the United States, these symptoms are lumped together and called a hangover. In other languages the term is more descriptive, a wooden mouth in French, for example. The symptoms relate to the known constituents of alcoholic beverages and the known physiology of alcohol.

Alcohol masks fatigue, so a night of drinking will lead to excessive activity, which will be revealed by fatigue and malaise the next day. Rest always allows return to the normal state. The dry mouth or excessive thirst is the result of the diuretic effect of alcohol (and possibly the very salty nibbles that often accompany drinking). Drinking water or fruit juice, and a little time, alleviate this condition.

The headache may result from the fatigue and the dehydration but may also result from the congeners of distilled spirits and of wine. The congeners are simple compounds produced by yeast concomitant with alcohol. They range from ethyl acetate to amyl alcohol. Also, particularly in some wines, and in brandy, there may be measurable amounts of methyl alcohol. These levels of methyl alcohol are not toxic per se, but when metabolized to formaldehyde may produce headaches. The drinking of a small amount of some alcoholic beverage the next day, possibly beer or vodka, will induce the liver to return to oxidizing alcohol and allow the methanol to be eliminated by the kidneys.

Alcohol and Alcoholism

Alcohol, although it effects many organs and systems, is singularly responsible for only one disease, alcoholism. Those who have this disease are called alcoholics.

An alcoholic has been defined as a person who has a compulsion to drink and cannot stop drinking once begun, to the point where that person's social and economic life are adversely affected. There is no certain cause of this disease, although a genetic factor may well be involved. An alcoholic is a heavy drinker, but not all heavy drinkers are alcoholics. Drinking because they want to characterizes drinkers, heavy or light, but drinking because they have to is the hallmark of alcoholics. In an alcoholic drinking sets up a chain reaction, which terminates only with sleep. There is no clear evidence that an alcoholic metabolizes alcohol any differently than others, but there appears to be, in alcoholics, a strong predilection toward alcohol metabolism, rather than carbohydrate or fat metabolism, as the major energy source.

Alcoholism may lead to several other pathological conditions. It is important to understand that alcoholics may derive up to or even above 50% of their daily caloric requirements in the form of alcohol. The human system can only tolerate such univalent load with carbohydrates. Either fat or protein in such excess will lead to problems of a different nature, but none the less severe.

The major problem facing alcoholics is a series of metabolic and structural changes in the liver. The initial change is the accumulation of fats in the liver, resulting in a fatty liver. This is easily reversible. It occurs because, in the presence of alcohol, the liver uses it as a preferential source of energy and the fatty acids, which it would otherwise use, accumulate. This change is followed, after some years, by structural changes in the liver, including scarring or necrosis of the liver tissue, the mark of cirrhosis. About 20% of long-term alcoholics develop cirrhosis.

Alcohol consumption equivalent to about 50% of the daily caloric intake, for from 5 to 20 yr, is necessary to induce cirrhosis in susceptible persons. Cessation of drinking often leads to reversal of the cirrhotic condition.

The male to female ratio of alcoholics in the United States is now 4:1. Although alcohol is a necessary component in alcoholism, it is no more correct to say that alcohol is the cause of alcoholism than it would be to say that marriage is the cause of divorce. The roots of alcoholism are social, metabolic, and genetic. No medical treatment exists. The use of drugs that produce unpleasant symptoms if alcohol is consumed is not widespread because of toxic effects. Societal treatment, in groups of peers such as Alcoholics Anonymous, has proved most beneficial.

Prenatal alcohol abuse can lead to birth defects known as fetal alcohol syndrome. The most benign symptom is low birth weight, and the most serious is impaired brain development and mental retardation. The most severe effects are known to occur in chronic alcoholics, but because it is neither known what the safe lower limit of alcohol intake is nor when during the pregnancy alcohol is most likely to harm the fetus, it is generally recommended to avoid alcohol during pregnancy (11,12).

The Salubrity of Alcohol

The vast majority of consumers of alcohol—as beer, wine, or distilled spirits—do so because it makes them feel better. It relieves stress and smooths the way from work to winding down. Not for naught has the time in a tavern after work and before the trip home come to be called the happy hour.

In addition to general relief from stress and care, moderate drinkers have been found to show genuine improvement in health. Recall that the ill effects of alcohol are confined to those who consume, regularly and for many years, alcohol in excess of 80 g a day. This amounts to almost two cases (twenty-four 12 oz packages) of beer a week, or more than a 750 mL bottle of wine a day, or six 1-1/2 oz shots of whiskey a day.

The consumption of moderate amounts of alcohol has been found to decrease stepwise the risk of coronary heart disease. The lowest level of consumption produces the least salutary effect, and the highest levels, more than 31 g but less than 80 g a day, reduce the risk to only 20–75% of that in nondrinkers. These results have been found in many studies, in many lands, among men and women, and among various socioeconomic strata. The Framingham Study, the longest and largest research of its kind, has always found this correlation between alcohol consumption and coronary heart disease—moderate alcohol use lowers the risk of coronary heart disease.

The recent (1989) Physicians Health Study, in which 22,000 physicians were studied prospectively, to determine the effect of aspirin on cardiovascular mortality showed incidentally that daily alcohol consumption significantly reduced the overall mortality rate over that produced by aspirin alone.

Physicians using alcohol daily and taking aspirin had a myocardial infarction rate of 1% whereas those taking aspirin alone had a 1.4% rate. This 40% improvement was among the best of any other risk factor studied. The synergistic effect of moderate alcohol consumption was not expected.

Alcohol and Drugs

Alcohol has a potentiating effect on some drugs, particularly sedatives. This may be drug specific, but also in many cases is due to the fact that the liver, where all drugs are detoxified, may prefer to metabolize alcohol and the drugs may persist in the bloodstream longer than they would normally. Repeating the drug dosage, in, say, 4 or 6 h, may lead to an overdosage if the drug has not been removed by the liver.

Those taking drugs for any chronic ailments should find out if alcohol interferes with the action or metabolism of those drugs.

BIBLIOGRAPHY

1. M. Frezza et al., *New Eng. J. Med.* **322**, 95–99 (1990).
2. C. Leake and M. Silverman, *Alcoholic Beverages in Clinical Medicine*, World Publishing Co., Cleveland, 1966, p. 54.
3. A. Piendl, *How the Blood Alcohol Curve Develops After Drinking Beer*, Brauwelt International, 1993, pp. 154–157.
4. K. Crow and R. Batt, *Human Metabolism of Alcohol*, Vol. I, CRC Press, Inc., Boca Raton, Fla., 1989, p. 53.
5. C. F. Gastineau et al., *Fermented Food Beverages in Nutrition*, Academic Press, New York, 1979.
6. B. Rouche, *The Netural Spirit*, Little, Brown, Boston, 1960.
7. L. Goodman and A. Gilman, *The Pharmacological Basis of Therapeutics*, Macmillan & Co., New York, 1970, p. 139.
8. L. Bisson et al., *Amer. J. Enol. Vit.* **46**, 449–462 (1995).
9. E. Frankel et al., *J. Agric. Food Chem.* **43**, 890–894 (1995).
10. H. Seitz et al., *New Eng. J. Med.* **323**, 58 (1990).
11. *The Merck Manual*, 16th ed., 1992, p. 2009.
12. T. M. Roebuck et al., *Alcohol Clin. Exp. Res.* **22**, 339–344 (1998).

GENERAL REFERENCES

C. Leake and M. Silverman, *Alcoholic Beverages in Clinical Medicine*, World Publishing Co., Cleveland, 1966.

National Research Council (U.S.), *Diet and Health*, National Academy of Sciences, National Academy Press, Washington, D.C., 1989.

D. A. Roe, *Alcohol and the Diet*, AVI Publishing Co., Westport, Conn., 1979.

E. Rubin, "Alcohol and the Cell," *Annals of the New York Academy of Sciences* **492**, p. 1–7, 181–190 (1987).

F. A. Seixas, K. Williams and S. Eggleston, "Medical Consequences of Alcoholism," *Annals of the New York Academy of Sciences* **252**, p. 11–19, 63–78, 85–100 (1975).

F. A. Seixas, and S. Eggleston, "Alcohol and the Central Nervous System," *Annals of the New York Academy of Sciences* **215**, p. 10–36; 325–330 (1973).

JOSEPH L. OWADES
Consultant
Sonoma, California

ALKALOIDS

CLASSIFICATION AND PROPERTIES

Alkaloids are naturally occurring substances with a particularly wide range of structures and pharmacologic activities. They may be conveniently divided into three main categories: the true alkaloids, the protoalkaloids, and the pseudoalkaloids.

The true alkaloids have the following characteristics: they show a wide range of physiological activity, are usually basic, normally contain nitrogen in a heterocyclic ring, are biosynthesized from amino acids, are of limited taxonomic distribution, and occur in the plant as the salt of an organic acid. Exceptions are colchicine, aristolichic acid, and the quaternary alkaloids. The protoalkaloids, also known as the biological amines, include mescaline and *N,N*-dimethyltryptamine. They are simple amines synthesized from amino acids in which the nitrogen is not in a heterocyclic ring. The pseudoalkaloids, those not derived from amino acids, include two major series of compounds: the steroidal and terpenoid alkaloids (eg, cones-sine) and the purines (eg, caffeine). Most alkaloids occur in the Angiosperms, the flowering plants, but they are also found in animals, insects, marine organisms, microorganisms, and the lower plants. The only common characteristic of alkaloid names is that they end in "ine" (except camptothecin).

Alkaloids are preferably grouped by their biosynthetic origin from amino acids (eg, ornithine, lysine, phenylalanine, tryptophan, histidine, and anthranilic acid) rather than their heterocyclic nucleus. The pseudoalkaloids, which are not derived from amino acids and clearly cannot be classified this way, are best organized in terms of their parent terpenoid class, eg, diterpenoid and steroidal alkaloids, or as purines or ansamacrolides.

Physical Properties

Most alkaloids are colorless crystalline solids with a defined melting point or decomposition range (eg, vindoline and morphine). Some alkaloids are amorphous gums and some are liquids (eg, nicotine and conine) and some are colored (eg, berberine is yellow and betanidine is red).

The free base of the alkaloid is normally soluble in an organic solvent; however, the quaternary bases are only water soluble, and some of the pseudo- and protoalkaloids show substantial solubility in water. The salts of most alkaloids are soluble in water.

The solubility of alkaloids and their salts is of considerable significance in the pharmaceutical industry, both in the extraction of the alkaloid from plant or fungus and in the formulation of the final pharmaceutical preparation. Solubility is also of considerable significance in the clinical distribution of an alkaloidal drug.

Chemical Properties

Most alkaloids are basic, which makes them extremely susceptible to decomposition, particularly by heat and light.

Ornithine-Derived Alkaloids

Ornithine-derived alkaloids include the tropanes (atropine, *l*-hyoscyamine, *l*-scopolamine, and cocaine), the Senecio alkaloids, and nicotine.

Tropane. Tropane alkaloids are derived from plants in the Solanaceae, Erythroxylaceae, and Convolvulaceae families. These alkaloids comprise two parts: an organic acid and an alcohol (normally a tropan-3α-ol). The pharmacologically active members of this group include atropine, the optically inactive form of *l*-hyoscyamine, which is isolated from deadly nightshade (*Atropa belladonna*); *l*-hyoscyamine and *l*-scopolamine, which are found in the leaves of *Duboisia metel* L., *D. meteloides* L., and *D. fastuosa* var. *alba*.

The tropane alkaloids are parasympathetic inhibitors. For example, atropine acts through antagonism of muscarinic receptors, the receptors responsible for the slowing of the heart, constriction of the eye pupil, vasodilation, and stimulation of secretions. Atropine prevents secretions (eg, sweat, saliva, tears, and pancreas) and dilates the pupil. Atropine is used to reduce pain of renal and intestinal cholic and other gastrointestinal tract disorders, to prolong mydriasis when necessary, and as an antidote to poisoning by cholinesterase inhibitors. Small doses produce respiratory and myocardial stimulation and decrease nasal secretion, and the drug has little local anesthetic action. Hyoscyamine and scopolamine have mydriatic effects. They are also used in combination as sedatives, in anti-motion-sickness drugs, and in antiperspirant preparations.

Cocaine. Cocaine is a potent central nervous system (CNS) stimulant and adrenergic blocking agent. It is extracted from South American cocoa leaves or prepared by converting ester alkaloids to exgonine, followed by methylation and benzoylation. It is too toxic to be used as an anesthetic by injection, but the hydrochloride is used as a topical anesthetic. It has served as a model for a tremendous synthetic effort to produce an anesthetic of increased stability and reduced toxicity.

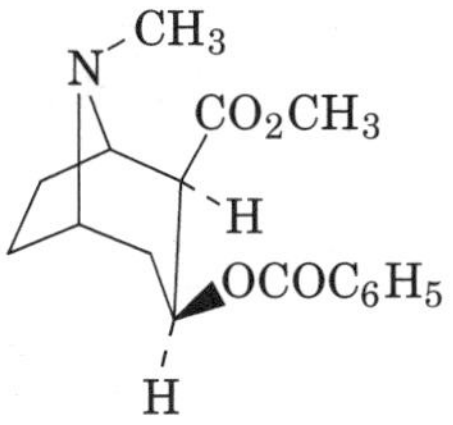

CH_3 N CO_2CH_3 H $OCOC_6H_5$ H

Cocaine

Senecio Alkaloids. Senecio alkaloids possess a pyrrolizidine nucleus and occur in the genera Senecio (Compositae), Heliotropium and Trichodesma, and Crotalaria. They are biosynthetically derived from two units of ornithine in a manner similar to some of the lupin alkaloids. Certain of the alkaloids having an unsaturated nucleus are potent hepatotoxins.

Nicotine. Nicotine is toxic, soluble in water, and a constituent of tobacco. The lethal human dose is ca 40 mg/kg. Pharmacologically, there is an initial stimulation followed by depression and paralysis of the autonomic ganglia. The biosynthesis of nicotine is well established.

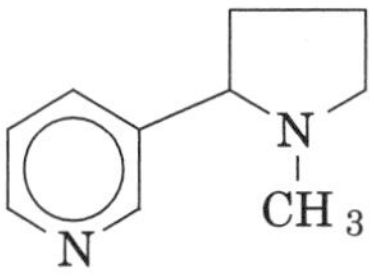

N N CH_3

Nicotine

Lysine-Derived Alkaloids

Lysine-derived alkaloids contain the pyridine nucleus or its reduced form, piperidine. They include alkaloids derived from the Areca or Betel nut, Lobelia alkaloids, and those derived from pomegranate or club mosses.

Arecoline, a colorless liquid alkaloid that has a pronounced stimulant action (in large doses, paralysis may occur) is found in the Areca or Betal nut. As the hydrobromide it is used as a diaphoretic and anthelmintic.

Lobelia inflata, known as Indian tobacco, contains lobeline, which is similar to, but less potent than, nicotine in pharmacologic action and is used as an emetic; the sulfate salt is used in antismoking tablets.

The root of *Punica granatum* contains alkaloids such as pelletierine and pseudopelletierine, which are formed from lysine and acetate. Pelletierine, is toxic to tapeworms and is used as an anthelmintic.

The club mosses, *Lycopodium* spp., produce polycyclic alkaloids such as lycopodine, whereas *Hydrangea* spp. yield febrifugine, an active antimalarial agent. Anabasine is found in *Haloxylon persicum* Bunge; this alkaloid has antismoking and respiratory muscle stimulation action similar to lobeline and is also used as a metal anticorrosive.

A host of complex alkaloids such as lupinine, sparteine, cytisine, and matrine are found in the lupins, a large plant family of the Leguminosae. Sparteine paralyzes motor nerve endings and sympathetic ganglia. The sulfate is used as an oxytoxic and the adenylate derivative is used to treat cardiac insufficiencies. Cytisine, found in the seeds of the highly toxic plant *Cytisus laburnum* L., is a strongly basic alkaloid that produces convulsions and death by respiratory failure.

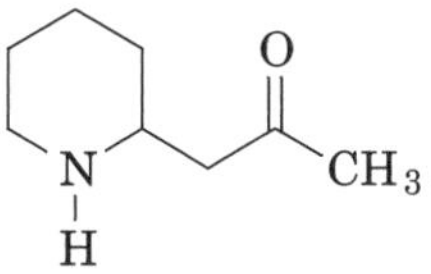

O N H CH_3 CH_2OH N

Pelletierine 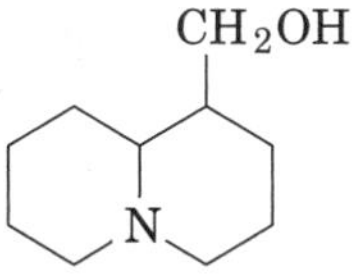 Lupinine

Anthranilic Acid-Derived Alkaloids

Anthranilic acid-derived alkaloids exhibit great structural diversity. This group includes dictamnine, platydesmine, vasicine, cusparine, and rutecarpine. Vasicine has oxytocic activity.

Phenylalanine- and Tyrosine-Derived Alkaloids

Phenylalanine- and tyrosine-derived alkaloids are by far the most numerous group of alkaloids, ranging from simple phenethylamines to the very complex dimeric benzylisoquinolines and the highly rearranged Cephalotaxus alkaloids.

Ephedrine. Ephedrine, from the Chinese drug Ma Huang, is soluble in water, alcohol, chloroform, and diethyl ether. It melts over the range of 33–42°C, depending on the water content. Little of the ephedrine of United States commerce is obtained from natural sources. Ephedrine is produced commercially through a biosynthetic process. Ephedrine has mydriatric effects and demonstrates adrenaline-like activity. It causes a rise in arterial blood pressure, increased secretions, and dilated pupils; it is a monoamine oxidase inhibitor and is used in nasal decongestants.

Peyote. Peyote, the small cactus of the Indians of north central Mexico, contain over 60 constituents. It is used as a hallucinogen in religious ceremonies and for medicinal purposes. A principal constituent is mescaline, a simple trimethoxyphenethylamine.

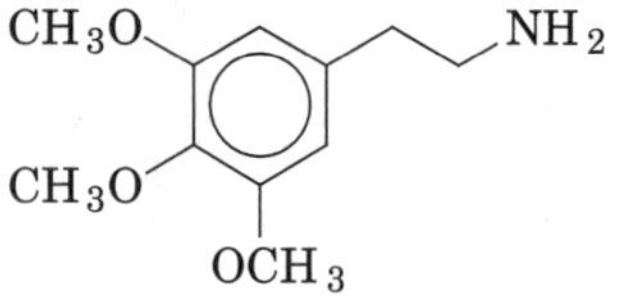

CH_3O NH_2
CH_3O
OCH_3

Mescaline

Ipecac. Ipecac, derived from *Cephaelis ipecacuanha* (native to Brazil), contains emetine and cephaeline, both of commercial significance. Emetine exhibits profound pharmacologic effects including clinical antiviral activity and is used in the treatment of amebic dysentery. Its side effects are cardiotoxicity, muscle weakness, and gastrointestinal problems including diarrhea, nausea, and vomiting. Two synthetic routes to emetine are of commercial importance: the Roche synthesis, which produces dehydroemetine, and the Burroughs-Wellcome synthesis. In handling emetine and its products, exposure should be limited as it can cause severe conjunctivitis, epidermal inflammation, and asthma attacks in susceptible individuals.

Isoquinoline and Related Alkaloids

Isoquinoline and related alkaloids are the largest group of alkaloids derived from phenylalanine (or its hydroxylated derivatives) and corresponding β-phenylacetaldehydes. The alkaloids are prevalent in the plant families Fumariaceae, Papaveraceae, Ranunculaceae, Rutaceae, and Berberidaceae. There are many structural types, including the simple tetrahydroisoquinolines; the benzylisoquinolines; the bisbenzylisoquinolines, such as the *dl*- and *d*-isomers of tetrandrine, which exhibit anticancer activity; the proaporphines, such as glaziovine, which is an antidepressant; the aporphines, such as glaucine; the aporphine–benzylisoquinoline dimers, such as thalicarpine, which shows cytotoxic and antitumor activity; the oxoaporphines; the protoberberines, a group of over 70 alkaloids such as xylopinine, berberine, canadine, and corydaline, known for such diverse pharmaceutical uses as tranquilizers, CNS depressants, antibacterial and antiprotozoal agents, anticancer agents, and alpha adrenergic blockers; the benzophenanthridines, a group of 30 alkaloids such as fagaronine and nitidine, which exhibit antitumor properties; the protopines; the phthalideisoquinolines, a group that includes narcotine, which possesses antitussive activity, depresses smooth muscles, and is not narcotic, and hydrastine, which is used as an astringent in mucous membrane inflammation; and the homoaporphines.

The Opium Alkaloids

The opium alkaloids number over 25, some of which are of commercial importance and major significance. Opium is the air-dried milky exudate from incised, unripe capsules of *Papaver somniferum* L. or *P. albumen* Mill (Papaveraceae). Notable opium-derived alkaloids include morphine, codeine, thebaine, noscapine, and papaverine.

Morphine is the most important alkaloid. It is isolated from opium. Along with its salts, it is classified as a narcotic analgesic and is strongly hypnotic. Side effects include constipation, nausea, and vomiting in addition to habituation, reduced power of concentration, and reduction in fear and anxiety. Respiration is also deepened.

Codeine is the methyl ether of morphine, and thebaine is one of the methyl enol ethers of codeinone. Codeine pharmacologically resembles morphine but is weaker, less toxic, and exhibits less depressant action (it does not depress respiration in normal therapy). It is used in the treatment of minor pain and as an antitussive. Codeine is available as a free base or the sulfate or phosphate salt. Thebaine is a convulsant poison rather than a narcotic.

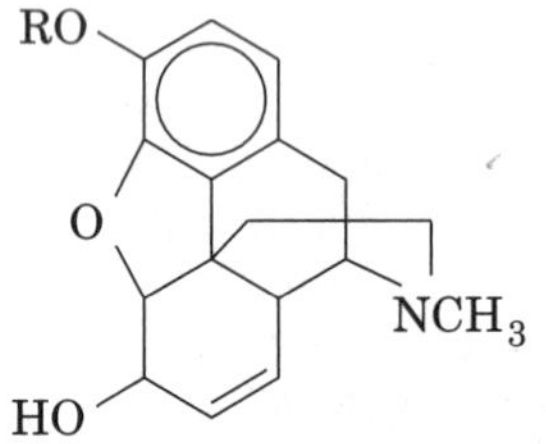

RO
O
NCH_3
HO

Morphine R = H
Codeine R = CH_3

Papaverine, a smooth-muscle relaxant that is neither narcotic nor additive, occurs in *P. somniferum* to the extent of 0.8–1.0%, commonly accompanied by noiscopine (narcotine). Most papaverine used is synthetically produced. The glyoxylate salt is used in treating arterial and venous disorders.

Amaryllidaceae

Amaryllidaceae alkaloids include galanthamine, margetine, and narciprimine. Galanthamine is a water-insoluble crystalline alkaloid that exhibits powerful cholinergic activity and analgesic activity comparable to morphine. It has been used to treat diseases of the nervous system. Its derivatives show anticholinesterase, antibacterial, and CNS depressant activity. Narciclasine, margetine, and narciprimine exhibit anticancer activity.

Colchicine

Colchicine, also known as hermodactyl, surinjan, and ephemeron, has some of the most unusual solubility characteristics of any alkaloid: it is soluble in water, alcohol, and chloroform, but only slightly soluble in ether or petroleum ether. Colchicine-type alkaloids are present in ten other genera of the Liliaceae and 19 species of Colchicum. Reviews of the chemistry of colchicine and related compounds and their history and pharmacology are available. Colchicine has the ability to artificially induce polyploidy or multiple chromosome groups. It is also used to suppress gout.

Cephalotaxus

Cephalotaxus alkaloids are found in the Japanese plum yews, *Cephalotoxus* spp. The esters, such as harringtonine and homoharringtonine, are potent antileukemic agents. The absolute configuration of the ester moiety has been determined. The α-hydroxy ester is essential for *in vivo* antileukemic activity.

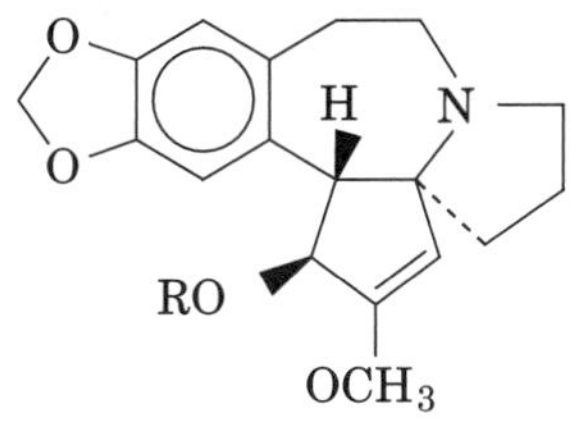

Cephalotaxine R = H

Harringtonine R = $-\mathrm{COC(OH)(CH_2CO_2CH_3)(CH_2)_2C(OH)(CH_3)_3}$

Deoxyharringtonine R = $-\mathrm{COC(OH)(CH_2CO_2CH_3)(CH_2)_2CH(CH_3)}$

Securinine

Securinine, isolated from *Securinega suffruticosa* Rehd, is similar to strychnine in action, but exhibits lower toxicity, stimulates respiration, raises blood pressure, and increases cardiac output. The chemistry, pharmacology, and biosynthesis of securinine and related compounds have been reviewed.

Tryptophan-Derived Alkaloids

Tryptophan-derived alkaloids, which occur in the families Apocynaceae, Rubiaceae, and Loganiaceae, have recently become of great interest, particularly those derived from tryptamine and a monoterpene unit. There are many diverse structural types in this group. Attempts to determine the important details of the biosynthetic interconversions of these compounds have been reported.

The simplest indole alkaloids are derived from tryptamine itself. These include indole-3-acetic acid, a potent plant-growth stimulator; serotonin (5-hydroxytryptamine), a vital mammalian product that inhibits or stimulates smooth muscles and nerves; *N*-acetyl-5-methoxytryptamine (melotonin), a constituent of the pineal gland with melanophase-stimulating properties; 5-methoxy-*N,N*-dimethyltryptamine, a constituent of the hallucinogenic Virola snuffs; psilocybine, a hallucinogenic found in the mushroom *Psilocybe mexicana* Heim.

The harmala alkaloids, such as harmine and harmaline, are powerful monoamine oxidase inhibitors, previously used in the treatment of Parkinsonism. Harmine is the active ingredient of the narcotic drug yage.

Ellipticine and derivatives show anticancer activity.

Several alkaloids of *Calycanthus* spp. are the products of dimerization or trimerization of simple tryptamine residues, such as folicanthine.

The main indole alkaloid skeletons are derived from tryptamine and a C_{10} unit. Examples include corynanthine; yohimbine, which has hypotensive and cardiostimulant properties and is used to treat rheumatic disease; ajmaline, which has coronary dilating and antiarrhythmic properties; decarbomethoxydihydrovobasine, which shows vasodilating and hypotensive activities whereas related compounds exhibit antiviral activity; akuammicine; tabersonine; catharanthine; rhynchophylline; vindoline; dihydrovobasine; 10-methoxyibogamine, whose acyl derivatives exhibit analgesic and anti-inflammatory activity; and strychnine. This structural diversity has been the source of intense biosynthetic interest.

Physostigmine

Physostigmine, found in the perennial West African woody climber, *Physostigma venenosum* Balfour, is pharmacologically similar to pilocarpine and is a reversible cholinesterase inhibitor used to treat glaucoma.

The Ergot Alkaloids

The ergot alkaloids are obtained from ergot, the dried sclerotium of the fungus *Claviceps purpurea* (Fries) Tulsane (Hypocreaceae). Ergot alkaloids are produced by isolation from the crude drug grown in the field, by extraction from saprophytic cultures, and by partial and total synthesis.

The ergot alkaloids act pharmacologically to produce peripheral, neurohormonal, and adrenergic blockage, and to produce smooth-muscle contraction as well. The two medicinally important ergot alkaloids are ergotamine and ergonovine.

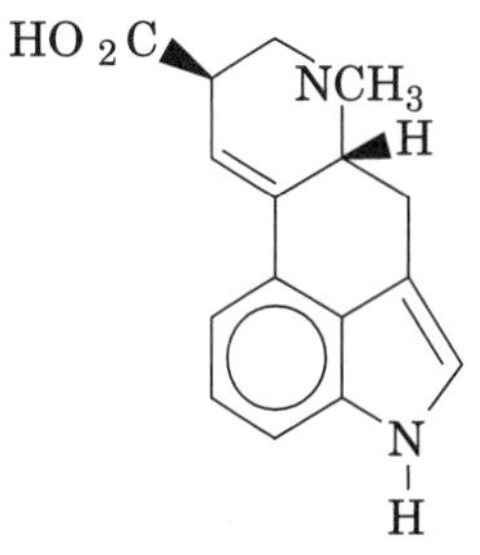

Lysergic acid

Three main groups of ergot alkaloids exist:

1. The clavine type, a group of over 20 alkaloids, which is water insoluble and does not give lysergic acid on hydrolysis. This group includes elymoclavine, agroclavine (a potent uterine stimulant), and chanoclavine-I.

2. The aqueous lysergic acid derivatives such as ergonovine, which in its maleate salt (or as methyl ergonovine maleate) is the drug of choice to treat postpartum hemorrhage.

3. The peptide ergot alkaloids, a group of water insoluble lysergic acid derivatives. This group includes ergotamine, ergocornine, and ergocryptine. Ergotoxine, a mixture of three peptide ergot alkaloids, possesses strong sympatholytic action and is used as a peripheral vasodilator and antihypertensive. Dihydroergotoxine is used for vascular disorders in the aged.

Some ergot alkaloids (eg, 2-bromo-α-ergocryptine) stimulate prolactin release and are being evaluated for treatment of breast cancer.

Catharanthus and Vinca Alkaloids

Catharanthus and Vinca alkaloids, usually discussed together, are quite distinct. The most important alkaloids of the Catharanthus genus are vincaleukoblastine, leurocristine, and leurosine, all antileukemic agents. Vincaleukoblastine and leurocristine are used clinically. The most important alkaloid of Vinca is vincamine, used to treat hypertension, angina, and migraine headaches. Alkaloids of this type produce marked hypotensive effects and curare-like action. The ethers of vincaminol are potent muscle relaxants.

Rauwolfia

Rauwolfia alkaloids include reserpine, the first tranquilizer, rescinnamine, and deserpidine. Reserpine is a sedative and tranquilizer useful in treating hypertension. It is also used as a rodenticide.

Strychnine

Strychnine, from the seeds of many *Strychnos* species, is a widely known poison (although, in fact, it is only moderately toxic). Pharmacologically, strychnine excites all portions of the CNS; it is a powerful convulsant and death results from asphyxia. It has no therapeutic uses in Western medicine, although its nitrate is used in treating chronic aplastic anemia.

Cinchona Alkaloids

Cinchona alkaloids, derived from the dried stem or root bark of various Cinchona species, include quinine and quinidine. These alkaloids are bitter tasting white crystalline solids, sparingly soluble in water. Quinine is toxic to many bacteria and other unicellular organisms and was the only specific antimalarial remedy until the Second World War. It is a local anesthetic of considerable duration. Quinine is commonly used as the sulfate and dihydrochloride. Quinidine, produced by the isomerization of quinine or found in Cuprea bark, is more effective on cardiac muscle than quinine and is used to prevent or abolish certain cardiac arrhythmias.

Camptothecine

Camptothecine, isolated from the Chinese tree *Camptotheca acuminata* Decsne, is used to treat cancer in the People's Republic of China.

Histidine-Derived Alkaloids

Histidine-derived alkaloids include pilocarpine and saxitoxin. Pilocarpine stimulates parasympathetic nerve endings and is used to treat glaucoma. The main commercial source of pilocarpine is *Pilocarpus microphyllus* Stapf., known as Marnham jaborandi. Saxitoxin is an extremely toxic neuromuscular blocking agent found in the so-called coastal red tides of North America.

Monoterpenoid Alkaloids

Monoterpenoid alkaloids include chaksine, a guanidine alkaloid from *Cassia lispikula* Vahl, which induces respiratory paralysis in mice; β-skytanthine, which is tremorigenic; cantleyine, derived from a monoterpene before loganin; and those derived from secologanin, such as gentianine, which exhibits hypotensive, anti-inflammatory, and muscle-relaxant actions, gentioflavine, gentiatibetine, pedicularine, and actinidine, a potent feline attractant.

Diterpene Alkaloids

Diterpene alkaloids are not of commercial or therapeutic significance, but some have potent pharmacological activity, eg, aconitine and Erythrophleum alkaloids.

Steroidal and Triterpene Alkaloids

Steroidal and triterpene alkaloids are found in the plant families Solanaceae, Liliaceae, Apocynaceae, and Buxaceae. There are four main groups based on the botanical source: the Veratrum, Solanum, Holarrhena and Funtumia, and Buxus alkaloids.

The Veratrum alkaloids include jervine, protoveratrine A, and protoveratrine B; the latter two produce pronounced bradycardia and a fall of blood pressure by stimulation of vagal afferents. The Solanum alkaloids are of interest as potential sources of steroids. Examples of these alkaloids are tomatidine and solanidine. Some Solanum alkaloids exhibit fungistatic activity. Biosynthetically, the alkaloids are derived from acetate and mevalonate.

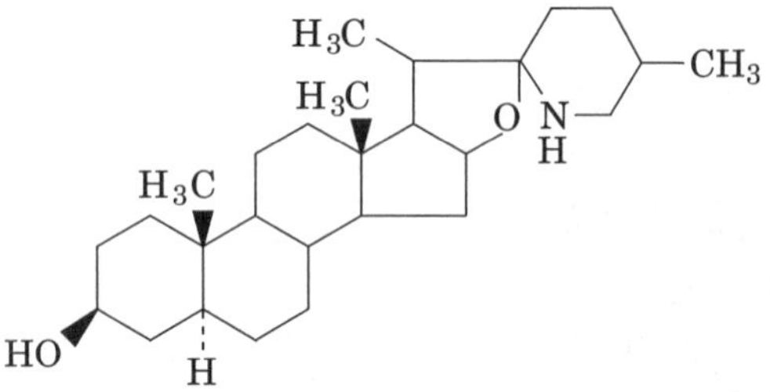

Tomatidine

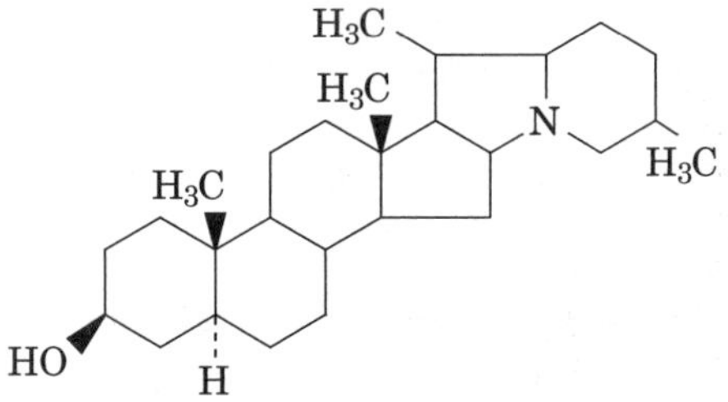

Solanidine

Table 1. Effects of Natural Food Alkaloids on Humans and Animals

Alkaloid(s); identified or potential food sources
Clinical effects
Toxic symptoms

Amatoxins (α-, β-, γ-amanitins, amanin, amanullin); toxic mushrooms [Death Cup (*Amanita phalloides, A. verna, A. virosa*)]
Liver and kidney damage (man)
Delayed vomiting, abdominal pain and diarrhea, coma, death; LD^a (man, oral) = 0.11 mg/kg

Arecoline; Areca nut or Betal nut (*Areca catechu*)
Parasympathomimetic agent, cathartic (horses); cholinergic (man)
Flushing, perspiration, bronchial spasms, contraction of pupils, diarrhea, dyspnea, and collapse; LD^a (mouse, subcutaneous) = 102 mg/kg

Caffeine; coffee, tea, cola nuts, guarana
Cardiac, respiratory, and psychic stimulation, diuresis; vascular cephalalgesia (man)
Nausea, restlessness, vomiting, insomnia, tremors, tachycardia, cardiac arrhythmia (man); LD_{50}^b (rat, oral) = 19.6 mg/kg

Capsaicine; red peppers
Skin irritant
Sweating, salivation (mammals)

Carapine; papaya (*Carica papaya*)
CNS depression (mammals)
Bradycardia (mammals)

Caulerpicin; marine algae (*Caulerpa sertulariodes C. racemosa* var. *clavifera*)
Anesthesia (mammals)
Numbness of lips and tongue (man)

Chavicine; black and white pepper (*Piper nigrum*)
Sharp peppery taste (man)
No information

Choline; brain, egg yolk, bivalve mollusks (*Callista brevisiphonata*), areca nut
Lipotrophic agent, antihypercholesterolemic agent
Nausea, vomiting, diarrhea (man); LD^a (rabbit, subcutaneous) = 450 mg/kg

Convicine; fava bean (*Vicia sativa* or *V. fava*)
Inhibition of glucose-6-phosphate dehydrogenase, decreased or reduced glutathione levels in red blood cells
Hemolysis, growth retardation (rat), hemoglobinemia (dog)

Curry alkaloids (murrayanine, mukoeic acid, girinimbine murrayacine, koenimbine, koenine, koeniginine, koenigine, mahanimbine, Cyclomahanimbine, curryanine); curry plant (*Murraya koenigii* Spreng)
For skin eruptions
No information

Cystisine; milk [contaminated from *Cystisus laburnum, Sophora secundiflora* (mountain laurel)
No information
Excitement, sweating, incoordination, convulsion, death (man, pig, cattle, horse); LD^a (as nitrate, dog, subcutaneous) = 4.3 mg/kg

3,4-Dihydroxy-L-phenylalanine (DOPA); fava bean (*Vicia fava*)
Decrease in reduced glutathione in red blood cells (man)
No information

Dopamine; banana, avocado, cephalopods
Hypertensive agent, increased cardiac output
No information

Dioscorine; wild yam (*Dioscorea hirsutus, D. hispida*)
CNS stimulation: analeptic, diuretic, expectorant
Sialorrhea, nausea, vomiting, diarrhea, confusion, cold sweat, pallor, clonic convulsion, paralysis, asphyxia (man), emetic hemolytic agent; LD^a (mouse, intraperitoneal) = 130 mg/kg

Eptatretin; Pacific hagfish (*Polistotrena stouti*)
Cardiac stimulant (frog, dog)
None, due to rapid detoxication

Ergot alkaloids (Eorgonovine, ergotamine, ergosine, ergocristine, ergocyptine, ergocornine, ergosinine, ergocristinine, etc); Barley, rye (contaminated from ergot produced by the fungi *Claviceps paspali* and *C. pupurea*)
Uterine stimulation, analgesic for migraine headache
Tachycardia, hypertension, vomiting, diarrhea, mental confusion, hallucinations, convulsions, gangrene, gastrointestinal (GI) disturbance; LD^a ergotamine, rat, intravenous) = 60 mg/kg

Gelsemine; Honey [contaminated from yellow jasmine] (*Gelsemium sempervirens*) nectar]
Uterine stimulation, CNS stimulant
Dizziness, dimness of vision, mydriasis, nausea, muscular debility, unusual prostration, weak pulse, dyspnea (man); minimum LD (rabbit, subcutaneous) = 0.15 mg/kg Glucobrassicin; broccoli, brussels sprouts, cabbage, cauliflower, kohlrabi, radish, rutabaga
Goitrogenesis (rabbit, rat)
Hyperplasia of the thyroid; kidney and liver enlargement (rabbit, rat)

Table 1. Effects of Natural Food Alkaloids on Humans and Animals *(continued)*

Alkaloid(s); identified or potential food sources	Clinical effects	Toxic symptoms
Goitrin; cabbage, rape, rutabaga	Goitrogenesis (rat)	Hyperplasia of thyroid, liver and kidney enlargement (rat)
Gramine; barley (*Hordeum vulgare*)	Vasoactivity, stimulation of intestines and uterus (rabbit)	Hypertension, clonic convulsion, and excitation of respiratory center
Histamine; derived from beer, chocolate, fish, sauerkraut, wine, and yeasts	Vasoactivity (mammals)	Headache (man), hypotension (man, mammals); LD_{50}[b] (mouse, intraperitoneal) = 12.7 g/kg, LD[a] (monkey, intravenous) = 52 mg/kg
Hordenine; germinated barley (*Hordeum vulgare*), sorghum (*Sorghum vulgare*), millet (*Panicum miliaceum*)	Sympathomimetic intestinal stimulation, respiratory stimulation (cat, dog); uterine stimulation (guinea pig)	Hypertension (man, dog), psychostimulation, convulsion, respiratory inhibition (man), LD[a] (as sulfate, dog, oral) = 1.9 g/kg
Islanditoxin; rice (contaminated from *Penicillium islandicum*)	Carcinogenic hepatotoxin (mouse)	Liver damage, liver cancer (mouse, rat); LD_{50}[b] (mouse, oral) = 5.0 mg/10 g
Laminine; marine algae (*Laminaria* spp) and others	Hypotension (rabbit); depression of contraction of smooth muscles (mouse, guinea pig)	Hypotension (rabbit)
Methyl pyrazine (various derivatives); green peas (*Pisum sativum*)	No information	No information
Methyl pyrroline; black pepper (*Piper nigrum*)	No information	No information
Mimosine; *Leucaena glauca*	No information	Alopecia (mammals)
Murexine (urocanyl choline); shellfish (*Muricidae* spp.)	Excitation of respiratory center, neuromuscular blocking, muscle relaxation (vertebrates and invertebrates)	Muscular and respiratory paralysis (vertebrates and invertebrates); LD_{100} (as oxalate, mouse, subcutaneous) = 310 mg/kg
Muscaridine; mushroom fly agaric (*Amanita muscaria*)	Uterine stimulation (rabbit, guinea pig)	Hashish or alcohol-like intoxication (man)
Muscarine; mushroom fly agaric (*Amanita muscaria* and other mushrooms)	Uterine stimulation (rabbit, guinea pig)	Sialorrhea, lacrimation, diaphoresis, nausea, vomiting, bradycardia, convulsions, coma, death; LD_{100} (mouse) = 16 μg/day
Nicotine; tomato	Respiratory stimulation (man); uterine stimulation (cat, rabbit, pig); cerebral and visceral ganglial stimulation (man)	Hypotension (dog); respiratory depression and paralysis (cat); hyperglycemia, convulsion, dizziness, nausea (man), etc; LD (man, oral) = 40 mg/70 kg; LD_{50}[b] (rat, oral) = 60 mg/kg
Norepinephrine; banana (particularly the pulp), orange (trace), potato	Vasoactivity sympathomimetic, adrenergic	Hypertension
Pahutoxin; fishes [Hawaiian boxfish (Pahu), Ostraciontidae family]	Neuromuscular effects (mammals)	Hemolysis (mammals); ataxia, respiratory distress, coma (mouse); minimum LD[a] (mouse, intraperitoneal) = 0.25 mg/g
Phallotoxins (phalloidin, phalloin, phallisin, phallicidin); toxic mushrooms [Death Cup, (*Amanita phalloides, A. verna, A virosa*)]	Liver and kidney damage (man)	Vomiting, abdominal pain, diarrhea, coma, and death (man); LD (man, oral) = 3 mg/kg
β-Phenylethylamine; Mushrooms, bitter almonds	Respiratory stimulation, intestinal relaxation, symphatomimetic (man)	Hyperventilation, hypertension, hypotension (cat); LD_{50} (mouse, intraperitoneal) = 350 mg/kg
Piperine; black pepper (*Piper nigrum*)	No information	No information
Pyrrolizidine alkaloids; wheat and other cereals, legume (contaminated from *Crotalaria laburnifolia, C. striata*)	No information	Liver damage, carcinogenesis, venoocclusive disease (mammals); glaucoma (rat, mouse); LD_{50}[b] (mouse) = 20 mg/kg
Sanguinarine; mustard oil and cereal grains (contaminated from *Argemone mexicana*)	Expectorant; for chronic eczema and skin cancers	LD[a] (as sulfate, intraperitoneal, rat) = 450 mg/kg

Table 1. Effects of Natural Food Alkaloids on Humans and Animals *(continued)*

Alkaloid(s); identified or potential food sources	Clinical effects	Toxic symptoms
Saxitoxin; shellfish, bivalve mollusks, and other pelecypods (contaminated from the dinoflagellates *Gonyaulax catenella* and *G. tamarensis*)	Neural stimulation (mammals); hypotension (cat and rabbit); myocardial and respiratory depression (mammals)	Peripheral paralysis, tingling and numbness of lips, respiratory failure (man and other mammals); $LD_{50}{}^{b}$ (man, oral) = 10–20 μg/kg
Serotonine; avocado, banana, pineapple, octopus, papaya, plantain, passion fruit, red plum tomato	Respiratory stimulation (dog); respiratory inhibition (cat)	Hypertension (man); CNS depression (most animals); inhibition of ovulation (rabbit); teratogenesis (mouse)
Solanine (solanidine); Irish potato (*Solanum tuberosum*), tomato (*Lycopersicon esculentum*)	Acetylcholine esterase inhibition (mammals)	Drowsiness, hyperesthesia, dyspnea, vomiting, diarrhea (man); LD^{a} (rabbit, intravenous) = 25 mg/kg
Tetrodotoxin; Pufferfish (Tetradontidae and Diodontidae families)	Neuromuscular blocking (mammals); hypotension (cat); respiratory inhibition, hypothermia (dog)	Weakness, dizziness, pallor and paresthesia, nausea, vomiting, sweating, salivation, muscular paralysis, cyanosis, respiratory paralysis, death (man); minimum LD (cat, subcutaneous) = 11 mg/kg; $LD_{50}{}^{b}$ (mouse, oral) = 325 mg/kg
Tetramine; shellfish (Buccinidae and Cymatidae families)	Curare-like effects, hypotension, bradycardia (mammals, frog)	Salivation, lacrimation, miosis, peristalsis (mouse); motor paralysis (mammals)
Threobromine; cacao bean; cola nuts; tea	Diuresis, myocardial stimulation, vasodilation, respiratory stimulation (cat)	GI distress; $LD_{50}{}^{b}$ (cat, oral) = 220 mg/kg
Theophylline; tea	Diuresis, hypertensive cephaloanalgesia (man)	Nausea and vomiting, vertigo, insomnia, flushing, convulsions, death; $LD_{50}{}^{b}$ (mouse, oral) = 550 mg/kg
Tomatine; tomato juice	No information	LD (rat, oral) 1 g/kg
Toxoflavin; Indonesian bongkrek (contaminated from *Pseudomonas cocovenenans*)	No information	$LD_{50}{}^{b}$ (mouse, oral) = 800 mg/kg $LD_{50}{}^{b}$ (mouse, intravenous) = 2.0 mg/kg
Trigonelline; green peas (*Pisum sativum*), coffee, soybeans, potatoes	No information	LD^{b} (rat, subcutaneous) = 4.9 g/kg
Tryptamine; plum (red and blue), orange, tomato	Vasoactivity, musculotropic (rabbit); intestinal and uterine contraction (rabbit, guinea pig)	Hypertension, headache (man)
Tyramine; avocado, banana, octopus, orange, spinach, potato, tomato	Respiratory stimulation (rat, cat); vasoconstriction (mammals)	Paroxysmal hypertension, intracerebral hemorrhage (man); mydriasis, bradycardia (dog); hypothermia (cat); LD (rabbit, intravenous) = 250 mg/kg
Vicine; fava bean (*Vicia sativa* or *V. fava*)	Inhibition of glucose-6-phosphate of dehydrogenase; decrease in reduced glutathione	Hemolysis, growth retardation (rat); hemoglobinemia (dog)

Source: Ref. 1; see also reference under S. W. Pettetier.
[a]LD = lethal dose.
[b]LD_{50} = median lethal dose.

Purine Alkaloids

Purine alkaloids are derivatives of the xanthine nucleus and include caffeine, theophylline, and theobromine, the principal constituents of plants used throughout the world as stimulating beverages.

Caffeine has the structure 1,3,7-trimethylxanthine. It is derived from cola, coffee (qv), tea (qv), guarana, and maté. Theophylline, 1,3-dimethylxanthine, is found in tea. Theobromine, 3,7-dimethylxanthine, is found in cocoa and tea. The xanthine derivatives have pharmacological properties in common: central nervous system (CNS) and respiratory stimulation; skeletal-muscle stimulation; diuresis; cardiac stimulation; and smooth-muscle relaxation. Caffeine is used to increase CNS activity; it acts on the cortex to produce clear thought and to reduce drowsiness and fatigue. Theophylline is used in smooth-muscle relaxants. Theophylline, as the ethylenediamine salt, is used in preference to caffeine in cardiac edema and in angina pectoris.

Miscellaneous Alkaloids

Coniine is an extremely toxic alkaloid that induces paralysis of the motor nerve endings and is the primary toxic constituent of poison hemlock. It was the first alkaloid to be synthesized. Carpaine, a crystalline macrocyclic alkaloid that induces bradycardia, depresses the CNS and is a potent amoebicide. Alkaloids are found in the poisonous Amanita species of mushrooms, such as α- and β-amanita toxins, ibotenic acid, muscimol, and muscazone. Maytansine and related ansamacrolides are potent antileukemic agents. Surugatoxin, found in the carnivorous gastropod *Babylonia japonica*, produces a pronounced mydriatic effect, sometimes resulting in death.

Ingestion and Human Health

The effects of natural alkaloids on humans and animals vary. This article is concerned mainly with food alkaloids and their clinical effects when ingested intentionally or accidentally. Table 1 summarizes the information.

BIBLIOGRAPHY

The section on classification and properties has been adapted from reference 1. Check original article for specific reference citations.

1. G. A. Cordell, "Alkaloids," in M. Grayson, ed., *Kirk-Othmer Concise Encyclopedia of Chemical Technology*, Wiley-Interscience, New York, 1985.

GENERAL REFERENCES

M. S. Abdel Kader et al., "DNA-Damaging Steroidal Alkaloids From *Eclipta alba* From the Suriname Rainforest," *Journal of Natural Products* **61**, 1202–1208 (1998).

Atta ur Rahman and M. I. Choudhary, "Diterpenoid and Steroidal Alkaloids," *Natural Products Report* **14**, 191–203 (1997).

J. M. Betz et al., "Chiral Gas Chromatographic Determination of Ephedrine Alkaloids in Dietary Supplements Containing Ma Huang," *J. AOAC Int.* **80**, 303–315 (1997).

B. Beumann et al., "The Effect of Ephedrine Plus Caffeine on Plasma Lipids and Lipoproteins During a 4.2 MJ/day Diet," *International Journal of Obsesity Related Metabolism Disorders* **18**, 329–332 (1994).

E. F. Domino, E. Hornbach and T. Demana, "The Nicotine Content of Common Vegetables," *N. Engl. J. Med.* **329**, 437 (1993).

K. Frach and G. Blaschke, "Separation of Ergot Alkaloids and Their Epimers and Determination in Sclerotia by Capillary Chromatography," *J. Chromatogr.* **808**, 247–252 (1998).

J. R. Gallon, K. N. Chit and E. G. Brown, "Biosynthesis of the Tropane-Related Cyanobacterial Toxin—A Role of Ornithine Decarboxylase," *Phytochemistry* **29**, 1107–1111 (1990).

C. Giroud et al., "Thermospray Liquid Chromatography/Mass Spectrometry (TSP LC/MS) Analysis of the Alkaloids From Cinchona *in vitro* Cultures," *Planta Med.* 57 (2), 142–148 (1991).

B. Gutsche and M. Herderich, "HPLC-MS/MS Profiling of Tryptophan-Derived Alkaloids in Food. Identification of Tetrahydro-beta-carbolinedicarboxylic Acids," *J. Agric. Food Chem.* **45**, 2458–2462 (1997).

J. F. Hanock, "The Colchicine Story," *HortScience* **32**, 1011–1012 (1997).

X. J. Hao et al., "Dieterpene Alkaloids From Roots of *Spitaea japonica*," *Phytochemistry* **38**, 545–547 (1995).

J. N. Hathcock, ed., *Nutritional Toxicology*, Vols. 1 and 2, Academic Press, Orlando, Fla., 1987.

J. Hong et al., "Enhancement of Catharanthine Production by the Addition of Paper Pulp Wastes Liquors to *Catharanthus roseus* in Chemostat Cultivation," *Biotechnol. Lett.* **19**, 967–969 (1997).

P. J. Houghton et al., "Two Securinega-type Alkaloids From *Phyllanthus amarus*," *Phytochemistry* **43**, 715–717 (1996).

A. Itoh, T. Tanahashi and N. Nakagura, "Six Tetrahydroisoquinoline-Monoterpene Glucosides From *Cephalis ipecacuanha*," *Phytochemistry* **30**, 3117–3123 (1991).

B. A. Kimball and C. A. Furcolow, "Ion-pair High Performance Liquid Chromatographic Determination of Strychnine Alkaloid in an Animal Tissue," *J. Agric. Food Chem.* **43**, 700–703 (1995).

Y. M. Liu and S. J. Sheu, "Determination of Ephedrine and Pseudoephedrine by Capillary Electrophoresis," *J. Chromatogr.* **637**, 219–223 (1993).

P. R. Martens and K. Vandevelde, "A Near Lethal Case of Combined Strychnine and Aconitine Poisoning," *J. Toxicol. Clin. Toxicol.* **31**, 133–138 (1993).

A. Montagnac et al., "Isoquinoline Alkaloids from *Ancistrocladus tectorius*," *Phytochemistry* **39**, 701–704 (1995).

P. M. Newberne, "Naturally Occurring Food-Borne Toxicants," in M. E. Shils and V. R. Young, eds., *Modern Nutrition in Health and Disease*, 7th ed., Lea & Febiger, Philadelphia, Pa., 1988.

A. J. Parr, "The Production of Secondary Metabolites by Plant Cell Cultures," *J. Biotechnol.* **10**, 1–26 (1989).

S. W. Pelletier, ed., *Alkaloids: Chemical and Biological Perspectives*, 6 Vols., Wiley-Interscience, New York, 1983–1988.

O. Poobrasert et al., "Cytotoxic Degradation Product of Physostigmine," *Journal of Natural Products* **59**, 1087–1089 (1996).

A. Rosso and S. Zuccaro, "Determination of Alkaloids From the Colchicine Family by Reversed-Phase High-Performance Liquid Chromatography," *J. Chromatogr. A* **825**, 96–101 (1998).

J. E. Saxton, "Recent Progress in the Chemistry of the Monoterpenoid Indole Alkaloids," *Natural Products Report* **14**, 559–590 (1997).

L. A. Skorupa and M. C. Assis, "Collecting and Preserving Ipecac (*Psychotria ipecacuanha, Rubiacceae*) Germplasm in Brazil," *Economic Botany* **52**, 209–210 (1998).

A. W. Smallwood, C. S. Tschee and R. D. Satzger, "Basic Drug Screen and Quantitation of Five Toxic Alkaloids in Milk, Chocolate Milk, Orange Juice, and Blended Vegetable Juice," *J. Agric. Food Chem.* **45**, 3976–3979 (1997).

L. Takano et al., "Alkaloids from *Cephalotaxus harringtonia*," *Phyochemistry* **43**, 299–303 (1996).

L. Takano et al., "Cephalotaxidine, a Novel Dimeric Alkaloid From *Cephalotaxus harringtonia* var. *Drupacea*," *Tetrahedron Lett.* **37**, 7053–7054 (1996).

S. L. Taylor and R. A. Scanlon, *Food Toxicity*, Marcel Dekker, New York, 1990.

A. Ulubelen et al., "Diterpene Alkaloids From *Delphinium peregrinum*," *Phytochemistry* **31**, 1019–1022 (1992).

S. J. P. van Belle et al., "Determination of Vinca Alkaloids in Mouse Tissues by High-Performance Liquid-Chromatography," *J. Chromatogr.* **578**, 223–229 (1992).

Y. H. Hui
American Food and Nutrition Center
Cutten, California

AMERICAN DIETETIC ASSOCIATION. See FOOD AND NUTRITION SCIENCE ALLIANCE (FANSA); FOOD AND NUTRITION SCIENCE ALLIANCE ORGANIZATIONS.

AMINO ACIDS. See PROTEINS: AMINO ACIDS.

ANIMAL BY-PRODUCT PROCESSING

Domesticated animals are grown and slaughtered for meat for human consumption. Normally, only 30 to 40% of the animal is utilized for human food (meat cuts, edible offals, processed meat products). In monetary terms this represents by far the most important product of meat processing plants. However, economics of operating the meat slaughter plant, together with the cost of pollution abatement and/or disposal of inedible material from the slaughter operation, demands that the rest of the material produced from the slaughter operation be used profitably. In most countries everything produced by or from the animal other than the dressed carcass is considered a by-product. Another name for by-products is coproducts. The term "coproduct" is coming into common acceptance, to indicate that these materials are contributing to process profitability. In some countries the terms "offal" and "by-product" are interchangeable. Hides and pelts are obvious by-products and are discussed elsewhere. A partial, but not all-encompassing, list of animal by-products is given in Refs. 1–4.

Many products can be made from the nonmeat parts of the animal (Fig. 1), and various by-products often contribute significantly to the meat plant's profitability. In some instances the commercial value of these by-products is often higher than the sum of the running expenses and the margin required for the meat plant to operate profitably. Also, because there is a worldwide shortage of animal protein, it is essential to maximize the use of these raw materials. This applies both to products used directly for human consumption and to protein-containing material that can be processed and fed back to animals.

By-products are usually classified as "edible" or "inedible." Edible by-products, such as heart, liver, tongue, oxtail, kidney, brain, sweetbreads, and tripe, that are segregated, chilled, and processed under sanitary conditions are called "variety meats" (or in some countries, offals). Chitterlings and natural casings (intestines) and fries (lamb or calf testicles) can also be eaten. In some countries, the blood and/or blood fractions from healthy animals, processed hygienically under conditions specified by the appropriate regulatory authority, can be an edible by-product. Some by-products must be processed or refined before they can be eaten. Examples include stomachs for tripe, bones and skin pieces for gelatin manufacture, and fatty tissue for edible fat. Most noncarcass material, if cleaned, handled, and processed in an appropriate manner, could be edible.

There is a sizable international trade in edible by-products because they are a very economical source of high-quality protein and fat. Custom, religion, palatability, and reputation of these products usually limit their use for human consumption. How a meat processor classifies a specific product depends both on the possible utilization of that product and the availability of a potential market. For example, many potentially edible by-products are downgraded to an inedible use because a profitable market does not exist. Many animal by-products are often underused. Their effective utilization depends on having a practical commercial process to convert the animal by-product into a usable commodity, actual or potential markets for the commodity produced, a large enough volume of economically priced raw material in one location for processing, a method for storing perishable material before and after

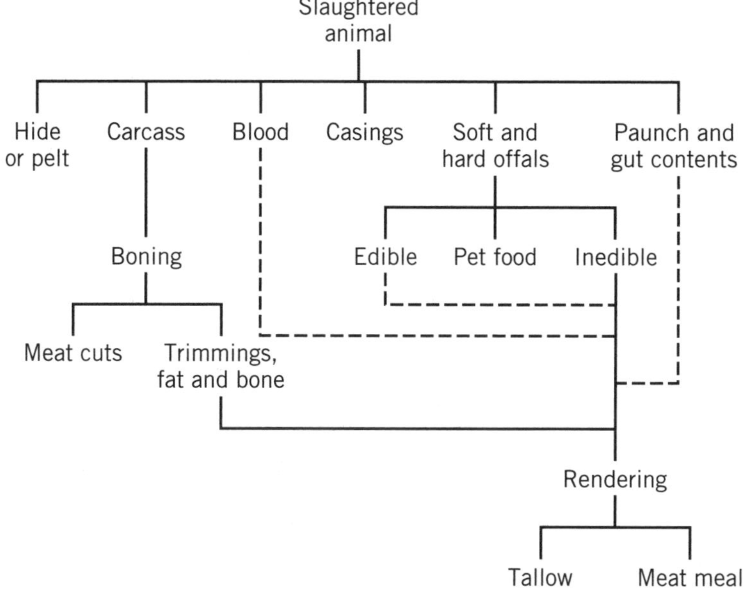

Figure 1. Flow diagram of material from a slaughtered animal.

processing, and, often, the availability of highly trained technical operators.

By-product yields (as a percentage of the animal's slaughter weight) depend on the species and category of animals slaughtered, the degree of animal processing, and on the end form of the by-product. Edible offals, material for pet food, and blood for edible use can represent up to 20% of carcass weight. When by-products are further processed (eg, rendered into meal and processed fat), one- to two-thirds of the original weight of raw material is lost as water.

BLOOD

Whole blood is a liquid with a dry matter content of 18 to 20%. Animal type and age may slightly influence blood composition. Blood undergoes complex biochemical reactions (clotting, coagulation, syneresis) once it has been released. Uncoagulated blood can be fractionated into plasma (a pale yellow fluid) and red blood cells (a fraction containing hemoglobin). Serum can be obtained when clotted blood synereses.

About 3 to 4% of the animal's live weight is blood, so large quantities are produced daily at slaughter facilities, even small ones. As 90 to 95% of the dry matter in blood is protein, blood is an obvious source of high-quality protein. Whole blood and the processed fractions (especially plasma) can be used as food ingredients. The dark red color of whole blood and red blood cells usually limits the use of these products in foods (5–7), although several processes exist for manufacturing decolorized blood fractions. However, the high costs and the logistics of collecting blood hygienically usually preclude extensive use of blood for edible products, so it is usually processed into an inedible blood meal. Although this processing into blood meal produces some income, it is done mainly to reduce pollution treatment costs: whole blood has a high biological oxygen demand (250,000 mg/L) and therefore can be a major source of pollution. Blood is usually not collected hygienically, in which case it may contain urine and ingesta. As well, it usually has been diluted with wash water. As a result, the dry matter content of the whole blood can be reduced to as low as 10%, especially if there is poor water management.

Whole blood can be stabilized chemically by adding urea, ammonia, metabisulfite or sulfuric acid, and the preserved blood can then be held without refrigeration and fed to animals (8). However, in most commercial processes, the blood is heated to coagulate the proteins, and these coagulated solids are then separated and dried to produce blood meal with a moisture content of 8 to 10%. This meal is a cheap source of animal protein high in lysine. The three most common methods of processing inedible blood are as follows:

1. Apply indirect heat to the whole blood to boil off most of the water. This process is very energy inefficient and produces a denatured product with very low solubility. This method is often used at smaller older plants that have not updated their equipment.
2. Inject live steam to bring the temperature to about 90°C to coagulate the blood, then remove most of the water in a decanter and dry the solids (eg, in a ring dryer with hot circulating gases, a rotating drum cooker, or a batch dryer). This method uses less energy than the previously mentioned technique because about half of the water is removed mechanically. A denatured protein powder is produced, the solubility of which depends on the type of dryer used: powders produced in ring dryers usually are more soluble. The high temperatures and/or long times used during processing lower the nutritional value of the blood meal.
3. Concentrate whole blood by ultrafiltration, then dry the concentrate in a spouted-bed drier (9). A very soluble powder is produced.

The yields of dried blood depend on the processing parameters (10). Aging the blood (leaving for 12 or more hours before processing) can increase yields. Any water added to the raw blood reduces yields. If the steam coagulator is not working efficiently, the losses of protein in the "blood water" from the centrifuge can increase.

Blood and its components (serum, albumin, red blood cells, hemoglobin) can be used in food, animal feeds, laboratory reagents, medical preparations, and industrial uses and as fertilizer. Blood albumin can be a substitute for egg albumen. Dried blood meal is used as a protein supplement in livestock feed. It is deficient in the amino acids tryptophan and isoleucine but is high in lysine, although lysine availability is affected by the drying method. Laboratory uses for blood include as a nutrient for tissue culture media and as a necessary ingredient in some agars for bacteriological use. Many blood components are isolated from whole blood and used in chemical and medical analyses or as nutritive supplements. Industrial uses of blood include as an adhesive and for its film-forming properties in the paper, lithographic, plywood, veneering, fiber, plastics, and glue industries. As a fertilizer, blood meal contributes nitrogen, aids humus formation, and improves soil structure.

Blood char, or blood charcoal, is the carbon component of whole blood or blood meal. Blood char is produced by treating 20% of the weight of whole blood or 50% of the weight of blood meal with activating agents and heating in airtight containers to 650 to 750°C for 6 to 8 h. Blood char contains 80% carbon and is used for absorption of gases, as an industrial decolorant, and as an antidote for chemical poisoning.

RENDERING

Introduction

Animal tissues are composed essentially of water, fat, and protein, with some minerals. The term "rendering" refers to a variety of processes that are used to separate the water, fat, and protein components, as far as is practicable, into commercial products. Until the 1850s, the highly perishable by-products from meat slaughtering were considered waste and were buried. The rendering industry developed to convert these materials into farm fertilizer and realize a profit from the operation. Proteinaceous by-products yielded nitrogen fertilizer, whereas bone produced phosphate fertilizer. Today, the rendering industry produces many useful products that can be broadly classified as edible and inedible fats, fine chemicals, meat meals, and bone meals.

Animal by-products used by renderers consist of fatty tissue, trimmings, bones, hooves, and soft offals (viscera). Processing carcasses into cuts and boneless meat increases the amount of material for rendering through the bones and trimmings that result. The material for rendering can represent 30 to 60 percent of the weight of a slaughtered animal. In developed countries, with centralized slaughter of large numbers of animals, a large volume of material is available for renderers. This has led to the development of sophisticated equipment and processes.

Basic Principles

The basic purpose of rendering is to produce stable products of commercial value, free of disease-bearing organisms, from raw material that is often unsuitable or unfit for human consumption. Most of the fat comes from fatty tissue, which can be located anywhere within connective tissue, and is made up of cells containing fat. This fat is deposited when animals have a surplus of dietary energy. The fat cells are surrounded by reticular fibers, and the fat cannot be released from animal tissue until the supporting structure has been broken. The two basic processes in rendering are separation of the fat and drying of the residue. The most common method used to rupture fat cells is heat, although enzymic and solvent-extraction rendering processes also exist.

A large proportion of the raw material for rendering is viscera (soft offal) and its associated contents. Because paunch and gut contents contain chemicals that can adversely affect fat quality, and solids that can downgrade meal quality, paunches are usually opened and emptied. The viscera are then cut and may be washed. Size reduction increases mass and heat transfer, although in some rendering processes it may be difficult to get even heat transfer in the processing vessel if the particles are too small. Both hard and soft offals may undergo size reduction. This size reduction is usually done in devices with rotating knives or anvils (termed hogors or prebreakers when producing large particles, grinders or mincers when producing fine-particle material) or with rotating hammer devices (hammer mills). Hooves and bones (hard offals) need to be reduced in size but usually do not require washing. The raw material is then processed to separate the fat from the nonfat phase.

Processes

A rendering process can generally be classified as wet or dry, depending on whether the fat is removed from the raw material before or after the drying operation. Processes can operate in a batch, semicontinuous, or continuous mode, and some of the newer rendering systems are classified as low-temperature-rendering (LTR) systems because of their milder heat treatment. Table 1 lists the best-known and most-used rendering systems, divided according to process type. Connected systems produce products of similar quality when processing similar raw material. There is no one "best" process for all applications; the most appropriate method will often depend on the application.

Profitability of any rendering system can be maximized by ensuring maximum yields and by obtaining the best possible product quality. Regular and planned maintenance of equipment will minimize repairs and maintenance costs. Processing additional water (from excessive washing and hosing) is wasteful, because more energy and a larger rendering capacity is required for a given throughput of solid material.

Wet Rendering. Wet rendering is an old processing method. In older systems, the preground raw material is cooked in a closed vertical tank (termed a digestor, autoclave, or cooker) under pressure by direct steam injection, usually to 380 to 500 kPa for 3 to 6 h (Fig. 2). Operators use past experience to gauge the end point. The pressure is then slowly released, and the liquid and solid phases are allowed to settle. The fat that has floated to the top is drawn off and can be "polished" in disc centrifuges to remove residual water and particles (fines). The water phase (liquor) in the cooker is drained off, then the solid material (greaves) is removed and can be pressed or centrifuged to remove additional liquid (stickwater) before being dried. The liquor and stickwater may be further processed to remove fines, which can be recycled to the greaves, and residual fat. Wet rendering produces good-quality fat (but only if the viscera are cut and washed). However, it requires very long cook times, is very labor intensive, and has significant losses (up to 25% of the solids may be lost in the stickwater). It is energy intensive, although heat can be recovered from the vent steam.

A more modern wet-rendering system is semicontinuous and involves cooking the raw material in a conventional dry-rendering cooker under pressure (to ensure sterilization) for a short time, then processing the cooked material in decanters to separate the liquor from the wet solids (Fig. 3). The meal is dried in continuous driers, and the fat is separated from the liquor in disc centrifuges. Process water is evaporated in multiple-effect evaporators and the concentrate added to the wet solids. The system produces high-quality fat and low-fat meal and uses less energy than conventional wet- or dry-rendering systems. However, capital costs, repair costs, and maintenance costs are high.

Dry Rendering. Both batch and continuous processes exist (Fig. 4). The material is heated in a horizontal, steam-jacketed vessel until most of the water has evaporated. The vessel has an agitator, which also may be steam heated. The evaporated water is usually condensed to recover heat and reduce atmospheric pollution. In batch systems, the raw material can be subjected to 200 to 500 kPa for some specified time to sterilize the material and/or hydrolyze wool and hair. It may take up to 3 h to produce a "dry" material, and the end-point temperature is often 120 to 140°C. The stage at which pressure is applied can influence the ease of further processing. Cooking times in continuous

Table 1. Best-Known and Most-Used Rendering Processes

			Cooker temperature, °C
Dry	Batch	Cookers	105–130
	Continuous	Keith	105–140
		Stord Bartz with centrifuges	110–140
		Duke	120–140
	Continuous with evaporator	Carver-Greenfield	90–105
		TM-1	120–140
		Retrofit	120–140
		Stord Barts with presses	120–140
Wet	Batch	Digesters	90–110
	Semicontinuous	Centrimeal	125–140
		Instant meal	125
	Continuous low temperature	Atlas Balfour Centribone MLTR Pfaudler Rendertech Stord Bartz wet pressing	60–95

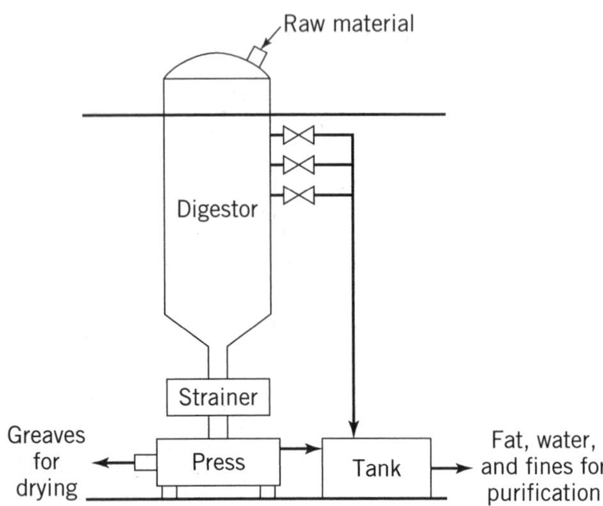

Figure 2. Wet-rendering system.

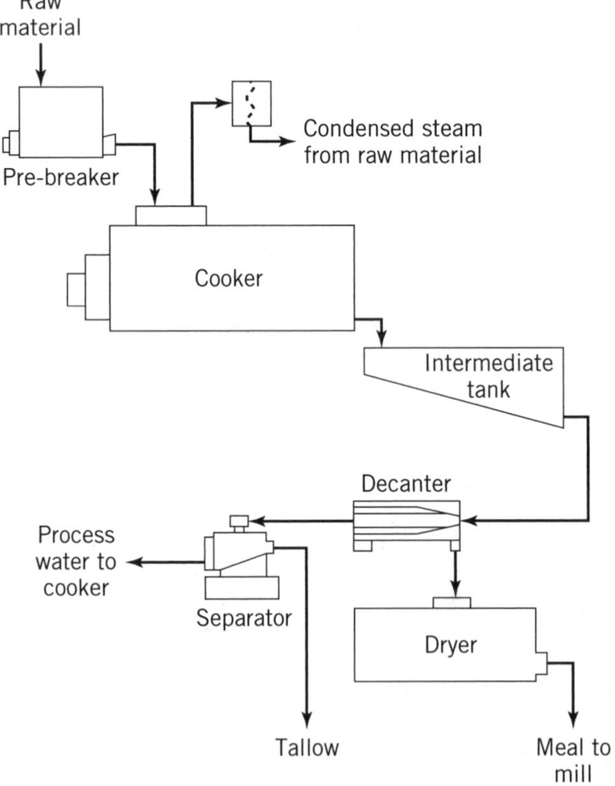

Figure 3. Semicontinuous wet-rendering system.

systems depend on cooker volume, cooker heat-transfer capability, and the characteristics and feed-rate of the raw material. Cooker contents are discharged into a percolator to remove free-draining fat; then the solid material is pressed (a continuous operation) or centrifuged (a batch operation) to remove additional fat. The material is then ground into meat meal. The fat is polished in disc centrifuges to remove fines and moisture.

Batch dry rendering has the following advantages:

- Little material is lost.
- Cooking, pressurizing, and sterilizing can be carried out in the same vessel.
- Different cookers can be assigned for processing different raw material, and hence for producing different grades of fat.
- Heat can be recovered from the vent steam.

The disadvantages are as follow:

- The fat is usually of poorer quality than that from wet or LTR systems.
- The high temperatures used produce fines, which can pass into the fat, degrading its quality, and which can be lost in the effluent from the polishing centrifuges.

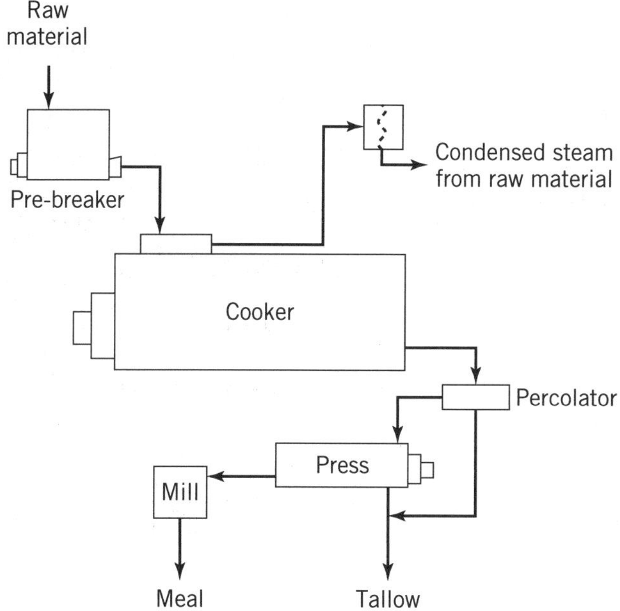

Figure 4. Dry-rendering system.

- The meal has a higher fat content than meal from wet and LTR systems.
- Raw material must be washed (producing effluent) to produce good-quality fat.
- Indiscriminate hosing and inadequate draining of washed material add extra water, increasing evaporation loadings and, hence, energy requirements.
- High-protein, low-fat material, such as slaughter wastes from calves and young lambs, is difficult to process.
- The end point of the cooks is difficult to control, so fat quality can be variable.
- It is difficult to keep the processing area clean and tidy; the process is not completely enclosed, so cooked products can be recontaminated.
- Energy usage is high, especially if vent steam is not recovered as hot water; dry-rendering cookers are not efficient dryers.
- The process is labor intensive.

Continuous dry rendering has most of the advantages of batch dry rendering, uses less labor, and requires less space. It has the disadvantages that the system cannot be pressurized, so sterilization and hydrolysis cannot be done, and tallow quality tends to be lower than with batch operations because end-point temperatures are often higher.

Low-Temperature Rendering (LTR). In the wet- and dry-rendering systems discussed so far, the raw material is subjected to high temperatures for long times. In dry rendering, near the end of the cooking process, when most of the free water has evaporated, the solids (cracklings) are essentially "frying" in the fat. With such high temperatures, the raw material must be washed to ensure that paunch contents and other "dirt" does not downgrade the color of the fat. Operators often overdry the cracklings to make pressing easier, so dry-rendered meal can be high in fat and low in moisture.

LTR systems were developed in the late 1970s to overcome some of the disadvantages of dry-rendering systems. Heat treatment is minimized and phase separation is carried out at low temperatures (70 to 100°C). Systems are usually continuous, so material flow through the size-reduction equipment, heating units, phase separating units, evaporators, and dryers must be balanced using surge bins and/or variable speed drives. Raw material, which can be unwashed, is minced, then heated to 70 to 95°C in a unit called a coagulator, preheater, melting section, or rendering vessel. The phases are then mechanically separated in decanters or presses.

In a MIRINZ low-temperature-rendering (MLTR) system (Fig. 5) processing 5 t of raw material per hour, the rendering vessel volume is only 1 m^3. Despite this, the residence time for the raw material, along with some recycled fat, to reach 80 to 90°C using indirect heat (in coils in the vessel) is just 6 to 8 min (11). The resulting liquor and solids are separated in a decanter, and the wet solids are dried, often in a direct-fired rotary drier, to give a low-fat meal. The liquor is separated into high-quality fat and an aqueous phase (stickwater). Product loss in the stickwater should be low.

In other low-temperature systems (eg, the Atlas or Stord Bartz wet pressing systems), raw material is heated indirectly to about 90 to 95°C over a 30- to 60-min period in a preheater, then pressed in a twin-screw press (Fig. 6). Drainings from the preheater and press are further heated to coagulate any soluble protein, then centrifuged to remove fines. The fines and pressed solids are dried, usually in contact driers.

The stickwater from LTR systems can be concentrated by ultrafiltration or in evaporators, then dried. The steam side of the evaporator is often supplied with waste vapors from the cooker/drier to save energy. Heat can be recovered from the stickwater and from the vapors from the drier and used to preheat incoming raw material, or to provide hot water.

The advantages of LTR systems are as follow:

- Energy requirements are usually about half those of dry-rendering systems.
- The raw material may not need to be washed.
- A low heat treatment is used, so high-quality fat is produced.
- The meal has a low fat content and has a high nutritional value because heat treatment is minimal.
- Labor requirements are low.
- The system can be easily automated.

However, the systems have high capital costs, may have high repair and maintenance costs, and usually require highly trained technical operators.

Other methods of treating waste by-products from the meat industry (and also the fish and other protein indus-

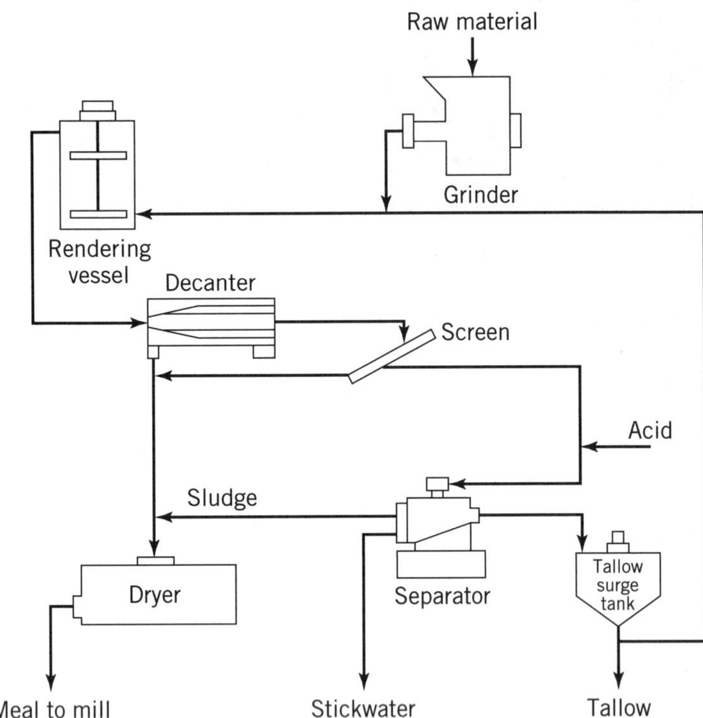

Figure 5. MIRINZ low-temperature rendering (MLTR) system.

tries) include hydrolysis using proteolytic enzymes or ensiling (12).

Uses of Fat

Animal fats are composed of triglycerides—three fatty acids esterified to glycerol. Fat quality is measured by titer; free fatty acid (FFA); FAC color (standard set by the Fat Analysis Committee of the American Oil Chemists Society); bleach color; and moisture, insoluble impurities, and unsaponifiable matter (MIU). Other tests that can be specified include saponification number, iodine value, peroxide value, and smoke point.

The fatty acid chain length and degree of saturation of the carbon bonds affect the fat's hardness or melting point (titer); the longer the chain length and the more saturated the fatty acid, the higher the titer. Fats of different species of animals and from different sites in the body have different titers. Type of feed can affect the titer, but rendering method does not.

When fat molecules break down, free fatty acids are released, so FFA content, which is usually expressed as percentage of free oleic acid, indicates the degree of spoilage that has occurred. To minimize FFA values, poor-quality raw material should be segregated from good-quality material, and material should be processed promptly. If material cannot be processed promptly, the raw material should be kept whole (unbroken) as long as practicable, to minimize microbial and enzymic activity, and preserved by cooling, or by adding acid. Processing equipment and storage tanks should be kept clean.

The factors affecting the color of fats include: animal breed, feed, age, and condition; source of the raw material; and presence of contaminants (feces, gut contents, etc). Fats can be almost white, yellow, or green (from contact with chlorophyll in digested plant material), or red or brown (from overheating or from contact with blood). Processing parameters during rendering (temperature, time) influence fat color. Some color components can be removed by bleaching with activated clay, and the color of the bleached sample read in a Lovibond tintometer. Fat bleachability indicates the temperature and handling conditions that the tallow has been subjected to; the cleaner the raw material and the lower the temperatures used, the lighter the bleach color.

The MIU value indicates fat purity. Moisture content should be as low as possible, because microbial and enzymic activity can hydrolyze fat at fat–water interfaces. Insoluble impurities such as protein fines, ground bone, manure, and so on that were not completely removed during processing may form colloidal fines that are not removed by settling or centrifuging. Trace amounts of copper, tin, and zinc can cause fat oxidation, and any polyethylene that has melted during rendering can adversely affect industrial processes that utilize fats. Unsaponifiable matter is fatty components such as cholesterol and gums that cannot be converted into soap by the use of alkali. Such matter reduces soap yields and affects catalyst efficiency. Unsaponifiable matter can also impart objectionable odors.

The saponification number indicates the average length of fatty acid chains, and the iodine number indicates the degree of unsaturation. These values can be used to identify types of fats and oils. The peroxide value indicates the degree of rancidity. Tallow that is not rancid and with good oxidative stability will have a low peroxide value. The smoke point is directly related to FFA and is the temperature to which the fat can be heated before it begins to smoke.

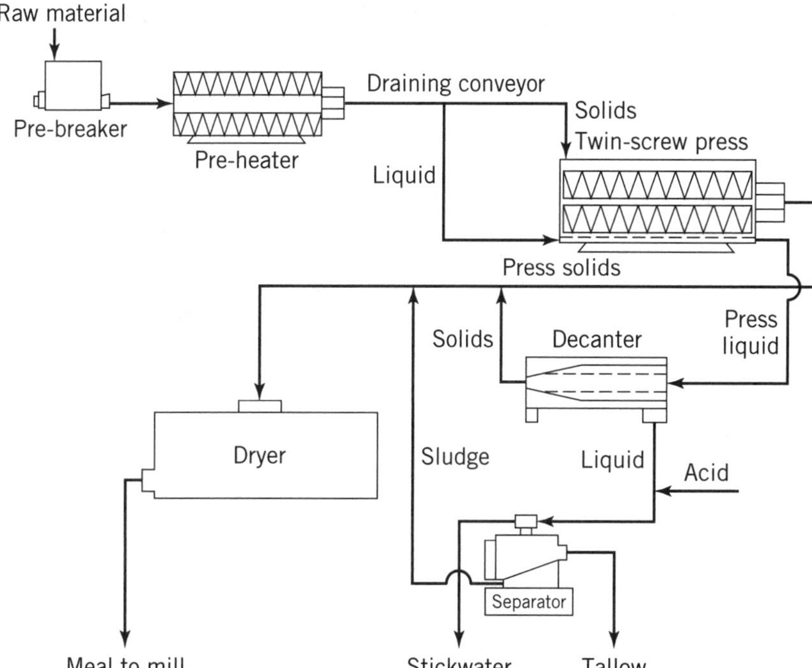

Figure 6. Wet pressing low-temperature-rendering system.

The two important fats of (land) animals used commercially are lard (fat from pigs) and tallow (fat from sheep and beef animals). Lard is made from specified, clean, sound tissues of healthy pigs, whereas rendered pork fat can be made from any fatty tissue. Edible tallow (dripping) is manufactured from specified edible fatty tissue. Oleo stock (premier jus) is a high-grade tallow prepared by low-temperature wet rendering of the fresh internal fat from beef. Inedible tallows (usually defined as fat with a melting point >40°C) and greases (melting point <40°C) are produced in many grades from inedible fatty tissue and dead stock.

Many different grades of tallow can be produced, and most countries have their own trading standards for tallow and grease. Tallow price reflects the prices obtained for other oils and fats, especially soy and palm oil. Vegetable oils are usually used for edible purposes and tend to be higher priced than tallow, which is used mainly for inedible processes such as soap manufacture, oleochemical production, and as an energy source in animal feed. Beef tallow is the largest component of bath bar soaps because of its economics and availability and its chemical composition. Fats are a useful component of animal feeds, as they have about twice the energy content of protein and carbohydrates. They also reduce dust, improve color and texture, enhance palatability, increase pelleting efficiency, and reduce machinery wear during feed production. Tallow can be used as an oleochemical feedstock. The resulting derivatives can be used to make abrasives, candles, cement additives, cleaners, cosmetics, paints, polishes, perfumes, detergents, plastics, synthetic rubber and water-repellent compounds. Tallow can be fractionated into its various components by crystallization, supercritical extraction, solvent extraction, or detergent fractionation. Tallow and its methyl esters can also be used as a fuel oil. Considerable amounts of tallow are still used for soap making. Glycerine, a coproduct of soap or fatty acid production, can be purified and used as a chemical in its own right.

Edible tallow and lard can be used in margarine, shortenings, and cooking fats. Tallow tends to give a better flavor to fried foods and is more stable during the cooking process than vegetable oils. However, because hard fats such as tallow can be associated with heart disease, the overall consumption of edible tallow has declined.

Uses of Meals

The material remaining after water and fat have been removed can have one of many names, depending on the rendering method and/or raw material used. These names include tankage, meat-meal tankage, digester tankage, wet-rendered tankage, or feeding tankage (finely ground, dried residue from wet-rendering material low in hair, hoof, horn, manure, and paunch contents), digester tankage with bone, meat and bone meal digester tankage, meat and bone meal tankage, or feeding tankage with bone (of higher phosphorus content than feeding tankage); meat meal (usually from dry rendering processes); and meat and bone meal or meat and bone scrap (of higher phosphorus content than meat meal). Feather meal is made from finely ground, wet-rendered feathers. This material is very digestible, but not well balanced nutritionally. Steamed bone meal (from wet rendering), or bone meal (from dry rendering), is defatted, dried, and ground bones suitable for animal feeding. Poultry meal has a similar composition, appearance, and nutritional value to meat meal but is made from poultry by-products. Fish meal is another high-quality meal similar to meat meal.

The type of raw material rendered and the rendering process used influence the composition of the meal produced. Meals made from proteinaceous material are high in nitrogen (the crude protein content is usually >50%) and also contain calcium, phosphorus, and fat. Bone meal and low-quality meat meals are made from material with a high bone content and have a relatively small market. One main use is as a fertilizer, but these meals are now being replaced by mineral fertilizers. Meals with a higher protein content are sold as meat and bone meal for use in animal feed formulations. They have a typical composition of 50% minimum crude protein, 4 to 10% moisture, 8 to 16% fat, and 20 to 30% ash. The price received for a given meal usually depends on its crude protein content, although some buyers may specify a minimum digestibility and availability of amino acids.

Bovine spongiform encephalopathy (BSE), commonly known as mad cow disease, was identified in the 1980s. Because of the belief that BSE transmission is via feeding-rendered ruminant protein to other ruminants, some countries introduced legislation banning the use of ruminant-derived animal proteins in animal feeds. This can have a great economic impact on the rendering industry. Researchers are developing economic sterilization processes and/or developing methods to monitor the sterilization effect of rendering processes (13,14).

FINE CHEMICALS AND PHARMACEUTICALS

Animals have ductless (endocrine) glands that secrete hormones, and ducted glands and organs that release enzymes and other biologically active molecules. Many biological chemicals can be recovered. A description of the animal glands and the medicinal and pharmaceutical uses of by-products is given in References 15 to 17.

The glands and tissues used for producing fine chemicals represent a small portion of the animal's live weight. Age, sex, and species of the animal determine the content of the active material. Most glands are very perishable and must be processed quickly to limit autolysis and microbial activity. The logistics of collecting sufficient raw material to operate a fine chemical plant or process usually limit operations either to larger meat plants or to centralized processing of material from many small plants.

SAUSAGE CASINGS

Intestines can be processed into natural casings used in sausage manufacture (18). Casing quality is affected by handling and cleaning procedures. Factors such as the age and species of the animal, breed, fodder consumed, and the conditions in which the animals are raised also affect casing quality and value. Animal casings naturally come in a wide variety of different shapes and sizes. The preference for a particular type of casing varies from country to country.

Reconstituted casings are manufactured from hide pieces (19), and artificial casings are often made from cellulose. Animal intestines are also used for the manufacture of surgical sutures, and strings for musical instruments and tennis rackets. Intestinal tract not used for these purposes is converted into pet food or is rendered to yield meat meal and tallow.

PET FOOD

Inedible animal tissue and rendered material can be used in both wet and dry pet foods. Pet foods require high-quality ingredients, which means that renderers may have to segregate raw material. Generally, pet-food manufacturers are using less fatty tissue and tallow and more protein-rich tissue and/or meat meal in their formulations.

BIBLIOGRAPHY

1. American Meat Institute Committee on Textbooks, *By-Products of the Meat Packing Industry*, Institute of Meat Packing, University of Chicago, Chicago, 1958.
2. A. Levie, *The Meat Handbook*, 2nd ed., AVI Publishing Co., Westport, Conn., 1967.
3. F. Gerrard, *Meat Technology*, Leonard Hill, London, 1971, pp. 65–77.
4. P. Filstrup, *Handbook for the Meat By-Products Industry*, Alfa-Laval Slaughterhouse By-products Department, Denmark, 1976.
5. C. W. Dill and W. A. Landmann, "Food Grade Proteins from Edible Blood," in A. M. Pearson and T. R. Dutson, eds., *Edible Meat By-Products*, Vol. 5, *Advances in Meat Research*, Elsevier Applied Science, New York, 1988, pp. 127–145.
6. C. L. Knipe, "Production and Use of Animal Blood and Blood Proteins for Human Food," A. M. Pearson and T. R. Dutson, eds., *Edible Meat By-Products*, Vol. 5, *Advances in Meat Research*, Elsevier Applied Science, New York, 1988, pp. 147–165.
7. V. M. Gorbatov, "Collection and Utilization of Blood and Blood Proteins for Edible Purposes in the USSR," A. M. Pearson and T. R. Dutson, eds., *Edible Meat By-Products*, Vol. 5, *Advances in Meat Research*, Elsevier Applied Science, New York, 1988, pp. 231–274.
8. J. Vrchlabsky and J. Budig, "Animal Blood as Feed–Chemical Preservation," *Meat Focus Int.* **5**, 82–83 (1996).
9. Q. T. Pham, "Behaviour of a Conical Spouted-bed Dryer for Animal Blood," *Canadian J. Chem. Eng.* **61**, 426–434 (1983).
10. J. E. Swan, "*Maximising Yields from Conventional Blood Processing*," MIRINZ Food Technology and Research Ltd. Bulletin No. 10, Hamilton, New Zealand, 1985.
11. T. Fernando, "The MIRINZ Low-Temperature Rendering System," *Proc. of the 22nd Meat Industry Research Conference*, MIRINZ Food Technology and Research Ltd. Publ. No 816, Hamilton, New Zealand, 1982, pp. 79–84.
12. B. Urlings, *Fermentation of Animal By-Products*, ADDIX, Wijk Bij Duurstede, The Netherlands, 1992.
13. H. Heubel, "The Continuous Sterilization Process for Slaughterhouse By-Products," in B. A. P. Urlings, ed., *Upgrading of Slaughter By-Products for Animal Nutrition*, ECCEAMST, Utrecht, The Netherlands, 1995, pp. 33–37.
14. D. M. Taylor, S. L. Woodgate, and M. J. Atkinson, "Inactivation of the Bovine Spongiform Encephalopathy Agent by Rendering Procedures," *Veterinary Record* **137**, 605–610 (1995).
15. H. W. Ockerman and C. L. Hansen, *Animal By-Product Processing*, Ellis Horwood, Chichester, England, 1988.

16. R. J. Banis, "Pharmaceutical and Diagnostic By-Products," in J. F. Price and B. S. Schweigert, eds., *The Science of Meat and Meat Products*, 3rd ed., Food and Nutrition Press, Inc., Westport, Conn., 1987.
17. A. M. Pearson and T. R. Dutson, eds., *Inedible Meat By-Products*, Vol. 8, *Advances in Meat Research*, Elsevier Applied Science, New York, 1992.
18. R. E. Rust, "Production of Edible Casings," A. M. Pearson and T. R. Dutson, eds., *Edible Meat By-Products*, Vol. 5, *Advances in Meat Research*, Elsevier Applied Science, New York, 1988, pp. 147–165.
19. L. L. Hood, "Collagen in Sausage Casings," A. M. Pearson, T. R. Dutson, and A. J. Bailey, eds., *Collagen as a Food*, Vol. 4, *Advances in Meat Research*, AVI Publishing Co., New York, 1987, pp. 109–129.

GENERAL REFERENCES

References 11 and 17 give a good overview.
T. Fernando, "Blood Meal, Meat and Bone Meal and Tallow," in A. M. Pearson and T. R. Dutson, eds., *Inedible Meat By-Products*, Vol. 8, *Advances in Meat Research*, Elsevier Applied Science, New York, 1992, pp. 81–112.
J. F. Price and B. S. Schweigert, eds., "Meat Animal By-Products and Their Utilization," *The Science of Meat and Meat Products*, 3rd ed., Food and Nutrition Press, Inc., Westport, Conn., 1987.
D. Swern, ed., *Bailey's Industrial Oil and Fat Products*, Vols. 1–3, 4th ed., Wiley, New York, 1982.
I. A. Wolff, ed., "Red Meat and Meat Products," *CRC Handbook of Processing and Utilization in Agriculture*, Vol. 1, *Animal Products*, CRC Press, Boca Raton, Fla., 1982.

J. E. Swan
University of Waikato
Hamilton, New Zealand

ANIMAL SCIENCE AND LIVESTOCK PRODUCTION

The discipline of animal science is concerned with the application of the biological and physical sciences to the efficient production of livestock. Food and fiber are the primary products of livestock production in the United States. Examples of products derived from animal agriculture include meat, milk, eggs, wool, mohair, leathers, pharmaceuticals, and products used in research laboratories throughout the world.

BUSINESS ASPECTS OF LIVESTOCK PRODUCTION

Livestock are raised on a wide variety of terrains, climates, and management systems. However, concentrations of livestock raising often border on regions of grain raising (cattle and pigs are heavily concentrated in the Midwestern states of Iowa, Kansas, Nebraska, and Texas) or located near population centers (dairy cattle are concentrated in the Northeastern states, Wisconsin, Idaho, and California) (1). Fifty percent of the land area of the United States is classified as grazing or range area that cannot be used to cultivate crops (2). Ruminant livestock (cattle and sheep) are efficient utilizers of the millions of acres of land in the United States too rough or dry to grow crops.

Agriculture is the largest industry in the United States, accounting for 15% of the gross national product. About 2 million people are directly involved in production agriculture with ca 1.37 million agricultural operations involving beef or dairy cattle (2).

Beef Cattle

The beef cattle industry is the largest segment of American agriculture, accounting for almost 25% of all farm marketings (3). Indeed, nearly 26 billion lb of beef were produced in 1997 (4). Cattle are raised in every state, including Hawaii and Alaska. Commercial cattle operations tend to be large and diverse enterprises that involve many types and breeds of cattle. The beef cattle industry in the United States is distinctly segmented into three production schemes: cow-calf (both commercial and purebred), stocker (intensive grazing), and feedlot. The term feeders is sometimes used to describe cattle, pigs, and lambs going into intensive feeding programs. Feeder cattle production is the goal of commercial (nonpurebred) cow-calf and stocker operations. Feeder cattle may be either steers or heifers and are the animals of highest value in terms of beef production. The singular goal of purebred cattle production is the selective breeding and selection of superior offspring to attain distinguishable characteristics of the breed, such as carcass merit, production efficiency, or reproductive traits.

Stocker cattle are weaned calves grazed on grass for inexpensive growth. This period is characterized by growth of lean tissues and frame size. In the Midwest, stocker cattle may be grazed on small grain pastures, grain stubble, or legume pastures before being placed in feedlots for fattening. Stocker cattle are usually owned for less than 1 year.

Fattening steers and heifers in feedlots involves feeding nearly market-sized cattle moderately high to high energy rations (grain) until they have reached a sufficient finish (fattening) to produce a carcass that will grade USDA choice. Large feedlots predominate in the Western states and account for a large percentage of fed-beef marketings. Midwestern feedlots tend to be more plentiful, but are generally smaller and produce a smaller percentage of fed-beef marketings. Most fat steers and heifers are slaughtered at 15–24 months of age.

Bulls are sometimes fed for slaughter and are usually referred to as bullocks. Bulls generally gain weight more rapidly and are more feed efficient than steers or heifers. Bulls also tend to be leaner. Older cows culled from beef or dairy cow herds are also sometimes fed grain for short periods of time before slaughter to increase the palatability (juiciness and tenderness) characteristics of the meat.

Dairy Cattle

Dairy farms are highly specialized agricultural enterprises, with considerable financial and labor input and returns. Dairying as a secondary enterprise on farms and ranches is uncommon. The number of dairy farms and dairy cattle in the United States has steadily declined during the past 20 years. The number of farms with dairy cattle decreased about 92% from 2 million in 1950 to 164,000 in 1992 (1,5,6). However, the decrease has been offset by a significant increase in milk production (Fig. 1). Modern dairy cattle produce an average of 16,400 lb of milk per lactation (7).

The price dairy farmers receive for whole (fluid) milk has been, historically, on hundred weight basis according to the percentage of butterfat. However, as the value of butterfat has declined as the most valuable component of whole milk, the price differential has increased less than the price per pound. The decrease in demand for butterfat is directly attributable to the introduction of margarine (containing plant oils), increased consumption of cheeses (made from whole milk), and processed food containing nonfat dry milk. Also, there has been a strong trend for health-conscious people to eat and drink only low fat and non-fat dairy products such as low fat cottage cheese, low fat yogurt and skim milk.

Locations of dairy farms have traditionally been close to major markets. This is because of the high cost of transporting whole milk, which contains a high proportion of water, and because milk is relatively perishable. However, in recent years location of dairies near consumers has been a less important consideration than the high cost of real estate near major markets. Also, conflicts between large dairies and home owners over water quality, dust, flies, and odors have forced dairy farms to relocate to less populated areas. Improved processing techniques and more rapid transportation networks also have contributed to decreased costs in moving milk to markets.

Swine

Swine production units are located in every state of the United States, but are generally concentrated in or near feed-grain-growing regions of the central United States. The corn belt states of Iowa, Illinois, Minnesota, Indiana, and Nebraska are leaders in market hog production (1,9). Swine farrowing was traditionally a spring and fall operation, but with the development of improved housing and nutrition, production of swine is a year-round business. This has tended to equalize the monthly marketings of pigs and reduced the seasonal fluctuations in market price.

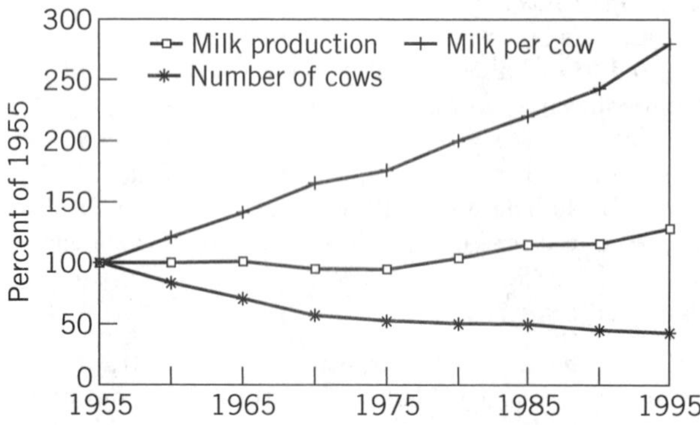

Figure 1. Index of milk production, milk per cow, and cow numbers in the United States. *Source:* Refs. 5, 6, and 8.

The trend in pork production, as in other agricultural operations, has been toward larger and more specialized operations. There is also a trend toward more confinement units in the production of pigs. Swine production systems usually can be described by one of three basic systems: farrowing to finishing, feeder pig production, and finishing feeder pigs.

Finished pigs are usually sold directly to the processor by the producer. Selecting the market and timing the presentation of pigs for slaughter are two of the most important decisions made by the swine producer. Most pigs are sold on a live-weight basis, but some are sold based on carcass yield and grade. Computerized hog-marketing systems are also being used in some parts in the United States.

Sheep

The products of sheep production businesses are primarily lambs and mutton for meat and wool. Most of the sheep in the United States are raised in the Western states. Texas, California, Colorado, South Dakota, Wyoming, and Utah account for more than 50% of the sheep produced (10). Sheep production systems are generally classified as farm flocks or range flocks. Predators, such as coyotes and wild dogs, and disease are some of the common problems associated with raising sheep. Lamb mortality rates as high as 20% are not uncommon. Most feeder lambs are sold to feedlots weighing between 65–80 lb. Ewes culled from flocks are also marketed for slaughter. The seasonality of breeding for sheep unfortunately causes a shortage of fresh lamb during certain times of the year.

THE DISCIPLINE OF ANIMAL SCIENCE

The multiple and interdisciplinary needs of farm animal production that focus on growth, reproduction, and lactation are at the center of animal science programs in research, teaching, and extension. Animal science is primarily an applied science that incorporates advances in science and technology from all disciplines. Nearly all land-grant universities, as well as several state and private colleges with agricultural programs, offer a bachelor of science degree in animal science. Most animal science departments in land-grant universities also offer graduate programs leading to master of science (M.S.) and doctor of philosophy degrees (Ph.D.). The Ph.D. degree is a research-based science degree.

Curriculum at the undergraduate level is designed to prepare students for professional careers in animal agriculture. The student majoring in animal science usually takes specialized courses, which include nutrition, breeding and genetics, biochemistry, animal reproduction, agricultural economics, animal health, meat science, dairy science, and animal production and products courses.

The Journal of Animal Science. The Journal of Dairy Science, and *Meat Science* serve as the repository for advances in the science of livestock production as well as their products.

PRINCIPLES OF ANIMAL REPRODUCTION AND BREEDING

Productivity of livestock depends on reproductive efficiency and is usually measured by the number of offspring produced by an animal or herd per unit of time. Therefore, management of reproductive cycles is critical for obtaining maximum efficiency of reproducing animals.

Animal Reproduction

Dairy and beef cattle are derived from common ancestors and thus have common reproductive characteristics. The period of time from one ovulation to the next ovulation is called the estrous cycle. The estrous cycle in cattle is on average 21 days in length and is characterized as being one in which the cow does not permit mating to occur except at the time of ovulation. This period of time in which mating is permitted is relatively short, being in the order of 16–18 h. The estrous cycle in cattle is continuous (polyestrous) throughout the year and is not seasonal as is the case in some other ruminant-type animals such as sheep, deer, antelope, and elk. Cattle ovulate 10–14 h after the end of estrus, and normally one follicle ovulates per estrous cycle. The gestation period is 283 days, and twins occur only 1% of the time.

Sheep are generally seasonally polyestrous, with recurring estrous cycles during the fall of the year followed by a prolonged quiescent period. Some breeds of sheep that originated in equatorial regions and are subject to less variations in temperature and photoperiod have longer breeding seasons. The period of sexual receptivity in sheep lasts for 24–36 h, but may vary widely. The length of the estrous cycle is 14–19 days. Seasonal breeding in animals is associated with an increased frequency and size of episodic releases of luteinizing hormone (11). Ovulation normally occurs near the end of estrus. Ovulation rate is influenced by breed and nutrition, but twinning is extremely common in sheep. Gestation is 150 days in length.

The pig, like the cow, is polyestrous. The period of sexual receptivity in pigs lasts about 40–60 h and may be influenced by season, breed, and endocrine dysfunction (12). The length of the estrous cycle is 21 days. Ovulation rates are strongly influenced by weight of the gilt at breeding and the amount of inbreeding. Gestation in the pig is 114 days and average litter size is in the order of 9–12 pigs per litter.

The estrous cycle in all livestock is characterized by profound changes in behavioral patterns and blood hormone profiles. Cyclic changes during the estrous period reflect the secretory functions and interdependence of the ovary, uterus, hypothalamus, and pituitary gland. The estrous cycle is controlled by ovarian hormone secretions (estrogens and progesterone) and may be subdivided into a follicular phase, an ovulatory phase, and a luteal phase (Fig. 2). The ovary functions to provide fertilizable ova and a balanced ratio of steroid hormones to facilitate development of the reproductive tract for migration of the early embryo and successful implantation in the uterus.

The male reproductive system in livestock is comprised of two testes (producing both sperm and the male sex hormone testosterone), excretory ducts (epididymis, vas deferens, and ejaculatory duct), and accessory structures (prostate, seminal vesicles, bulbourethral glands, and penis). The scrotum, containing two testes, provides for efficient regulation of testicular temperature. Descent of both testes from the abdomen to their proper scrotal location is necessary for maximal fertility. Failure of one or both testes to descend is a common reproductive organ defect in livestock, especially in swine, where the condition is a hereditary defect transmitted by the male and is referred to as cryptorchidism (13).

Reproductive management of the male primarily involves maintenance of health and nutrition to optimize sperm production and libido. Methods for assessing individual male fertility, such as breeding soundness evaluations, are subjective and poorly correlated with pregnancy rates.

Breeds of Livestock

Decisions about breeding are some of the most important decisions a livestock producer must make. The livestock breeder must consider the heritability of a characteristic, such as prolificacy, feed conversion, milk yield, or carcass merit, for example. The livestock breeder also must consider the merits of inbreeding, outbreeding, crossbreeding, and the relative merits of different breeds of animals.

The development of modern breeds of livestock began in the 1700s. The origin of cattle and sheep breeds can be traced to Europe and the British Isles. For classification purposes there are four basic categories of modern beef and dairy cattle. The classification system for cattle is not consistent because two of the categories reflect geographical origin of the cattle and the other two reflect the purpose of the cattle.

1. British and continental breeds (beef): Includes Angus, Hereford, Shorthorn, Charolais, Limousin, Chianina, Gelbvieh
2. North American breeds (beef): Brahman, Brangus, Beefmaster, and Santa Gertrudis
3. Dual-purpose breeds (beef and dairy): Ayrshire, Milking Shorthorn, Red Poll, and Brown Swiss
4. Dairy breeds: Guernsey, Holstein, and Jersey

The foundation of modern-day pigs can be traced to European and Asiatic strains of pigs. Modern swine are raised to produce lean high-quality pork. However, not long ago, breeds of swine were classified as lard-type and bacon-type. Although the individual breeds of swine continue today, all swine producers strive for the same high-quality meat-type pig.

As in the case of cattle, the classification system for breeds of sheep is inconsistent. The Rambouillet and Merino breeds were developed for the production of the fine wool characteristic, but much lamb and mutton comes from these breeds. Likewise, income derived from wool represents a significant portion of the value of the meat-type breeds.

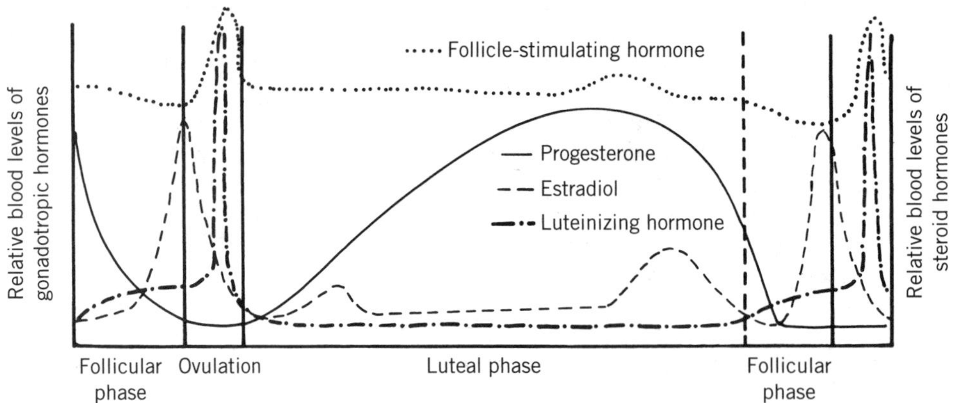

Figure 2. Cyclic changes in ovarian steroid and gonadotropic hormones in the ewe.

Animal Breeding

Crossbreeding is one of the quickest and most economical methods for increasing total beef or pork production. Heterosis, also sometimes commonly referred to as hybrid vigor, is the result of crossbreeding. Increased performance and vigor is not limited to animal breeding, but also results in hybrid plant production. Hybrid increases are seen in the first generation (F1) crossbred animals. F1 animals are the result of crossbreeding two purebred lines of livestock. Increased performance benefits may be as high as 25% for F1 calves. Hybrid chickens and crossbred pigs have contributed to the increased economics of poultry and pork production.

Three general crossbreeding systems are used to produce market animals: (1) Mating females of one breed with males of another to produce a F1 market animal; (2) three-breed terminal crossbreeding, where F1 females produced according to (1) are mated to bulls of a third breed to produce a three-breed market animal with a maximal level of heterosis (alternatively, the F1 female could be mated to one of the original parent breed to produce a backcross market animal); (3) rotational crossbreeding, where the breed of the sire is rotated among the three original breeds. Rotational breedings usually only use three breeds because the heterosis gained by the addition of a fourth breed is small.

PRINCIPLES OF ANIMAL NUTRITION AND FEEDING

The proper feeding of livestock is a matter of supplying them with the correct amount of nutrients essential for reproduction, growth, or lactation. Nutrients are chemical elements and compounds required by the cells of the animal's body in support of the three basic functions: (1) structural matter for building and maintaining the body; (2) a source of energy for work, thermogenesis (heat production), and fat deposition; and (3) regulating body processes or the synthesis of body regulators. An auxiliary function of nutrients would be their use in milk production.

The Nutrients

Nutrients required for livestock, just like for humans, can be categorized into six functional or chemical classes: carbohydrates, proteins, fats, minerals, vitamins, and water. All carbohydrates or saccharides are related structurally and chemically and contain the same amount of gross energy. Carbohydrates are comprised primarily of hexose ($C_6H_{12}O_6$) and pentose ($C_5H_{10}O_5$) molecules. Tetrose and triose molecules are present in small quantities but are generally not important in animal nutrition. Carbohydrates are usually categorized as monosaccharides, disaccharides, and polysaccharides based on how many hexose and pentose molecules are linked together. Common monosaccharides, also called simple sugars, consist of glucose, fructose, and galactose. Disaccharides consist of two monosaccharides linked together with hydroxyl groups of each sugar unit. The common monosaccharides include sucrose (table sugar), maltose, and lactose. Polysaccharides have the empirical formula $(C_6H_{10}O_5)_n$ and contain large polymers of monosaccharides. Polysaccharides important in livestock nutrition include starch, glycogen, and fiber (hemicellulose, cellulose, and lignin).

Carbohydrates are important energy sources in livestock feeds, comprising 65–80% of plant dry weight. Carbohydrate as a class of nutrients is usually divided into two groups: nitrogen free extract (NFE) and fiber. Fiber is what remains after a feed has been boiled in dilute acid or alkali and roughly approximates the amount of carbohydrate poorly digested in the animal's intestinal tract. NFE represents the soluble, readily digestible carbohydrate portion of a feedstuff. Corn grain, eg, contains only 2.2% fiber and 70% NFE. Good-quality alfalfa hay contains 26% fiber and 46% NFE. Cellulose is not an efficient energy source for nonruminant livestock (swine and poultry) but can be readily digested by the bacteria of the rumen. Because of this characteristic, ruminant livestock occupy an important niche in utilizing a potentially wasted feed resource.

Lipids are water-insoluble organic molecules that can be extracted from plant and animal tissues with nonpolar solvents such as benzene and ether. Lipids contain 2.25 times more energy per unit weight than either carbohydrate or protein. The term lipid is used in a general sense and used interchangeably with fat and oils. Livestock rations are generally low in fat, with most grains and roughages containing less than 5% lipid. Fats may sometimes be added to a ration, especially for swine and poultry, and high producing dairy cattle when higher energy rations are desired.

Lipids are composed of carbon, hydrogen, and oxygen, as are carbohydrates, but contain less oxygen. Lipids have several important biological functions, including storage and transport forms of energy and components of cell surface membranes. The fatty acid composition of body fat may be altered by the composition of dietary fats. This is especially true for swine and poultry.

Protein constitutes the most expensive portion of livestock feed. Protein primarily consists of 20 α amino acids linked together by peptide bonds. Both plant and animal tissues contain a diversity of proteins with variable amounts of amino acids. Ten amino acids cannot be synthesized by animal tissues or can be synthesized only to a limited extent. These 10 amino acids are referred to as dietary essential amino acids (Table 1). The primary function of dietary protein in livestock rations is to supply amino acids, the building blocks of proteinaceous body tissues. Ruminant livestock do not have dietary amino acid requirements as such, but depend on ruminal bacterial protein synthesis to supply these nutrients. Because of this, ruminant animals are able to utilize nonprotein nitrogen, such as urea, to replace part of the natural protein in the ration. The use of rumen bypass protein supplements, such as corn gluten meal and blood meal, has been shown to increase growth in beef cattle and milk production in dairy cattle (14,15). Swine rations are usually balanced for the first two or three most-limiting amino acids as well as for balancing the ration for crude protein.

Sixteen or more minerals are required by livestock. Of these, seven are classified as macrominerals (Ca, P, K, Na, S, Mg, Fe), with the remaining constituting microminerals (Co, Cu, F, I, Fe, Mn, Mo, Se, Zn). A description of the minerals' functions and deficiency characteristics in livestock nutrition has been reviewed in Ref. 16.

Vitamins are dietary-essential organic molecules that are required in minute quantities by livestock. Vitamins are usually classified into two groups: water soluble and fat soluble (Table 2). Approximately 15 vitamins are known to function in animal metabolism, but only several of these are needed in the ration of livestock because synthesis of certain vitamins occurs in the animal. Animal nutritionists formulating livestock rations are especially concerned about vitamins A and D for ruminants and certain B vitamins for nonruminants. Ruminants do not generally require B vitamin supplementation because the bacteria of the rumen synthesize sufficient quantities. Vitamin E is normally present in ample quantities in natural feeds, but may be supplemented to ruminants and nonruminant livestock receiving processed grain diets. Vitamin E supplementation of cattle diets also has been shown to stabilize the red color of fresh beef and help prevent oxidative rancidity (17). Livestock normally do not need ascorbic acid supplementation where adequate quantities are synthesized in the tissues. Ref. 16 outlines a review of vitamins in livestock nutrition.

Table 1. Dietary Essential Amino Acids Required by Swine and Poultry

Arginine	Lysine	Tryptophan
Histidine	Methionine	Valine
Isoleucine	Phenylalanine	Glycine (poultry)
Leucine	Threonine	Proline (poultry)

Table 2. Fat-Soluble and Water-Soluble Vitamins Required by Livestock

Fat-soluble	Water-soluble
Vitamin A	Thiamine
Vitamin D	Riboflavin
Vitamin E	Niacin
Vitamin K	Pyridoxine
	Biotin
	Pantothenic acid
	Vitamin B_{12}
	Choline
	Folic acid
	Inositol
	Para-aminobenzoic acid

Digestion

Digestion involves the physical and chemical preparation of feed in the gastrointestinal tract to absorption-ready nutrients. The digestive systems of livestock are anatomically and functionally similar, with the very important distinction of the ruminant having a large, four-compartment stomach (Fig. 3). The abomasum or the true stomach is functionally similar to the stomach of monogastrics. The rumen, the largest of the four compartments, with a capacity approaching 50 gal, functions as a fermentation vat and contains billions of bacteria and protozoa. The presence of the bacteria and the enzymes that they secrete enables the ruminant to efficiently use cellulose from fibrous plants and feedstuffs. The microorganisms also synthesize amino acids to make bacterial protein and supply the ruminant animal with essential amino acids.

RED MEAT PRODUCTION

The slaughtering procedure for beef and pork usually involves immobilization, bleeding, removing hair (pigs) or skin (cattle and sheep), and removing viscera—trachea, esophagus, lungs, heart, stomach, intestines, and reproductive organs. Immobilization of meat animals, except those killed for kosher meat, is accomplished according to humane, accepted methods (captive bolt or electric shock, eg) regulated by state and federal laws. Most commercial slaughtering plants use a movable rail where immobilized animals are individually hung by their shanks and moved along a deassembly line to dress the slaughtered animal. Electrical stimulation of fresh beef, pork, veal, and lamb carcasses is sometimes employed to improve tenderness

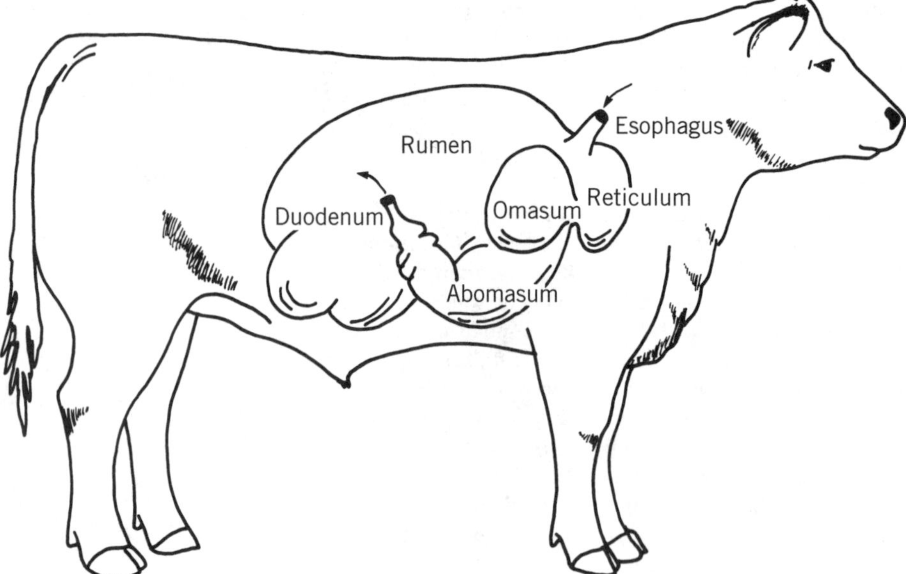

Figure 3. The digestive system of the polygastric ruminant.

(18). To facilitate rapid chilling of the carcass, beef carcasses are entirely split along the backbone, while pig carcasses are usually split to the neck.

The term carcass is what remains after dressing and usually refers to both sides of the same slaughtered animal. A side is one of the two halves of a split carcass. Beef quarters are the portions remaining after splitting a side of beef between the 12th and 13th rib. Wholesale cuts are divisions of the quarters including the round, full loin or short loin and sirloin, flank, rib, short plate, chuck, brisket, and fore shank. Primal cuts are any of the wholesale cuts except for short plates, briskets, flanks, and fore shanks.

The principal cuts of beef, pork, and lamb are shown in Figures 4, 5, and 6. Most beef is sold as carcass sides or quarter or wholesale cuts to retailers, wholesalers, chain-store breaking plants, and restaurants. However, an increasing portion of beef is sold as 60–80 lb saw- or knife-ready subprimal cuts shipped in boxes. Processors nearly always reduce pork carcasses to wholesale and retail cuts at the plant because hams and bellies, and sometimes other cuts, are smoked and cured. Most lamb is sold as intact carcasses, but some carcasses are reduced to wholesale cuts.

Significant quantities of beef are ground, tested for lean content, and blended to meet federal and state quality requirements for fat content. Low-quality carcasses, such as cutter and canner grades of beef, pale and soft pork carcasses, and cull grades of lamb and mutton, are not sold as wholesale or retail cuts but are instead boned-out at the plant. The edible meat from these carcasses may be sold as hamburger, or used in the preparation of a variety of meat products, such as canned meat products or sausage.

The USDA through the Food Safety and Inspection Service (FSIS) is responsible for ensuring the wholesomeness of the meat supply. FSIS assurance includes antemortem and postmortem inspection and compliance with sanitation and temperature, ingredients, and label requirements. Meat bearing the U.S. "inspected and passed" stamp informs the buyer the meat was processed in a federally approved plant and is wholesome and safe for human consumption. Additionally, the carcass may be graded for meat yield and quality. The yield grade is an estimate of the portion of the carcass that is edible meat. Quality grades are intended to identify meat on the basis of palatability and cooking attributes. Factors used to establish quality grades include kind or species of animal, class or physical attributes, maturity, marbling, firmness of flank, color and structure of lean, and conformation, fleshing, and finish of the carcass. USDA quality grades for various kinds of meat are given in Table 3.

The Food and Drug Administration (FDA) is responsible for ensuring the safety of the drugs used in livestock production and food additives. Both the FDA and the Federal Trade Commission have been involved with monitoring and regulating claims made during advertising of food products.

MILK PRODUCTION

Dairy cattle are normally milked two or three times per 24-h day. Following freshening (parturition), maximum milk production is normally reached after 45 days. A slow, but steady decline in milk produced by the cow occurs after peak yield until she is dried. Normal lactation periods for commercial dairy farms are 305 days lactation and 60 days dry (Fig. 7). The dry period allows for the cow to replenish her body reserves and rest her mammary tissue. Because a calf must be produced every 12.5 to 13.5 months in order to meet this idealized production schedule, cows are bred at approximately 80–90 days of lactation.

Feeding dairy cattle represents approximately 50% of the cost of milk production. The percentage of protein and fat may be influenced by feeding and management practices on dairy farms. Dairies may select feeding programs to produce fat and produce percentages that best suit milk

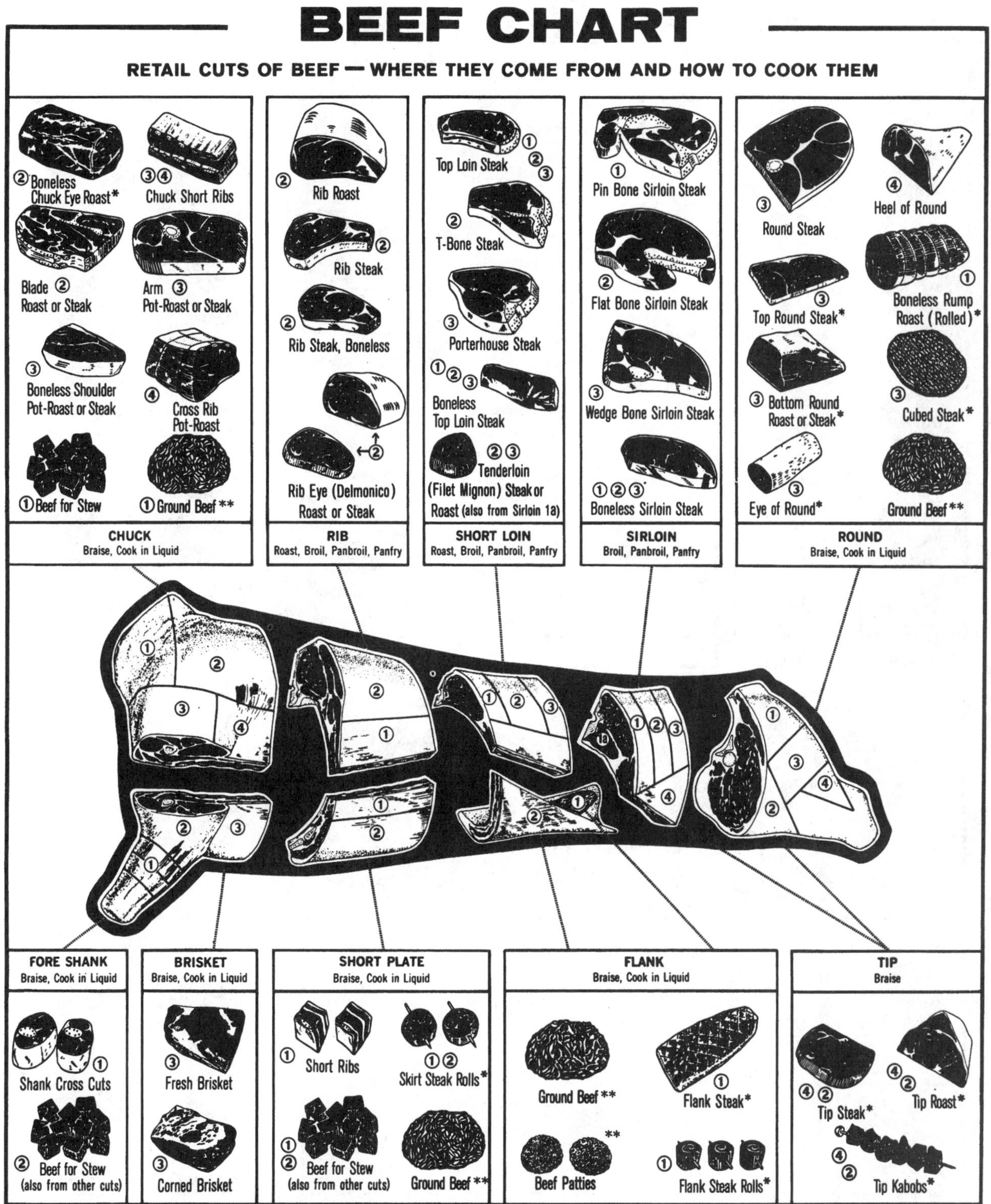

Figure 4. Wholesale and retail cuts of beef.

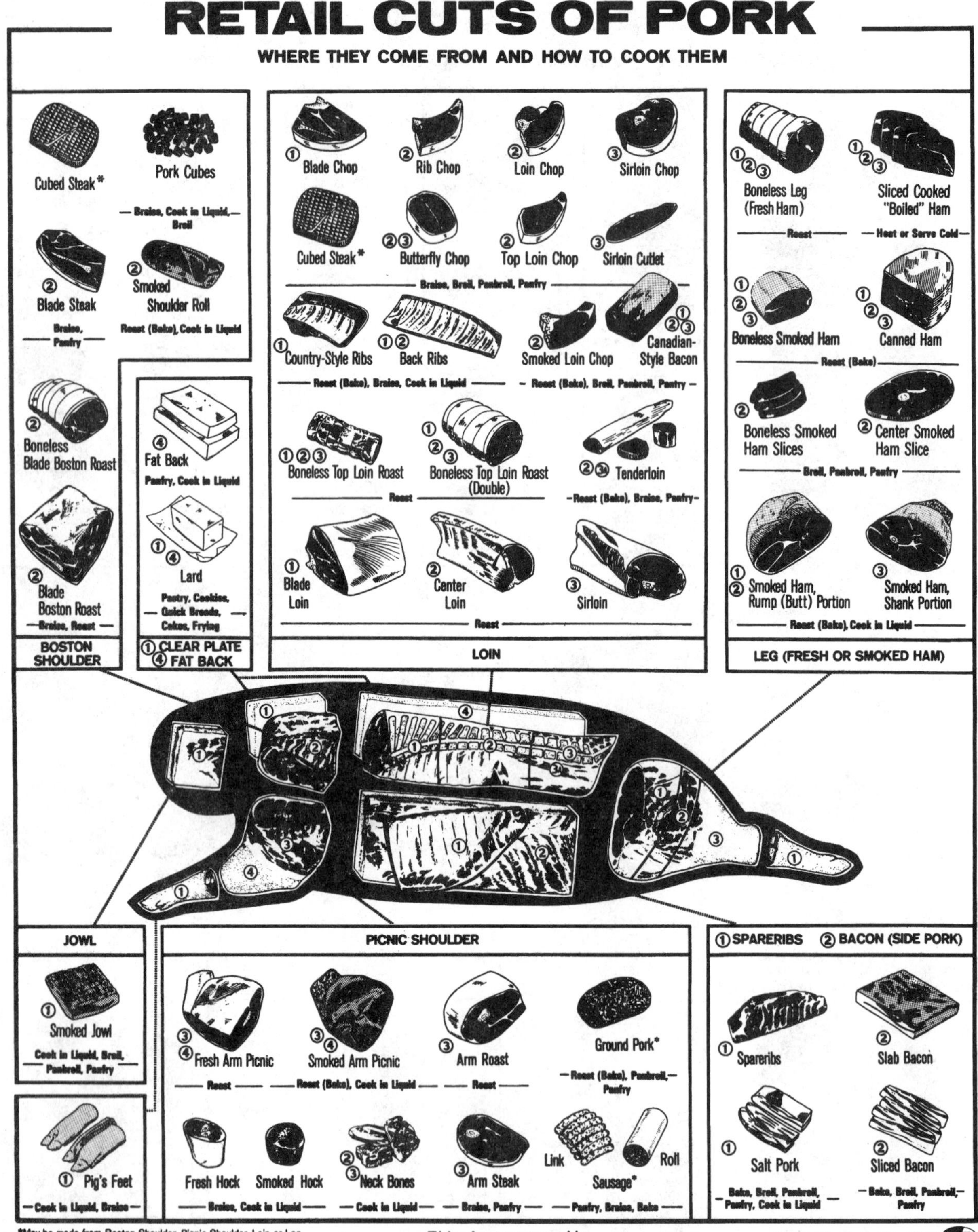

Figure 5. Wholesale and retail cuts of pork.

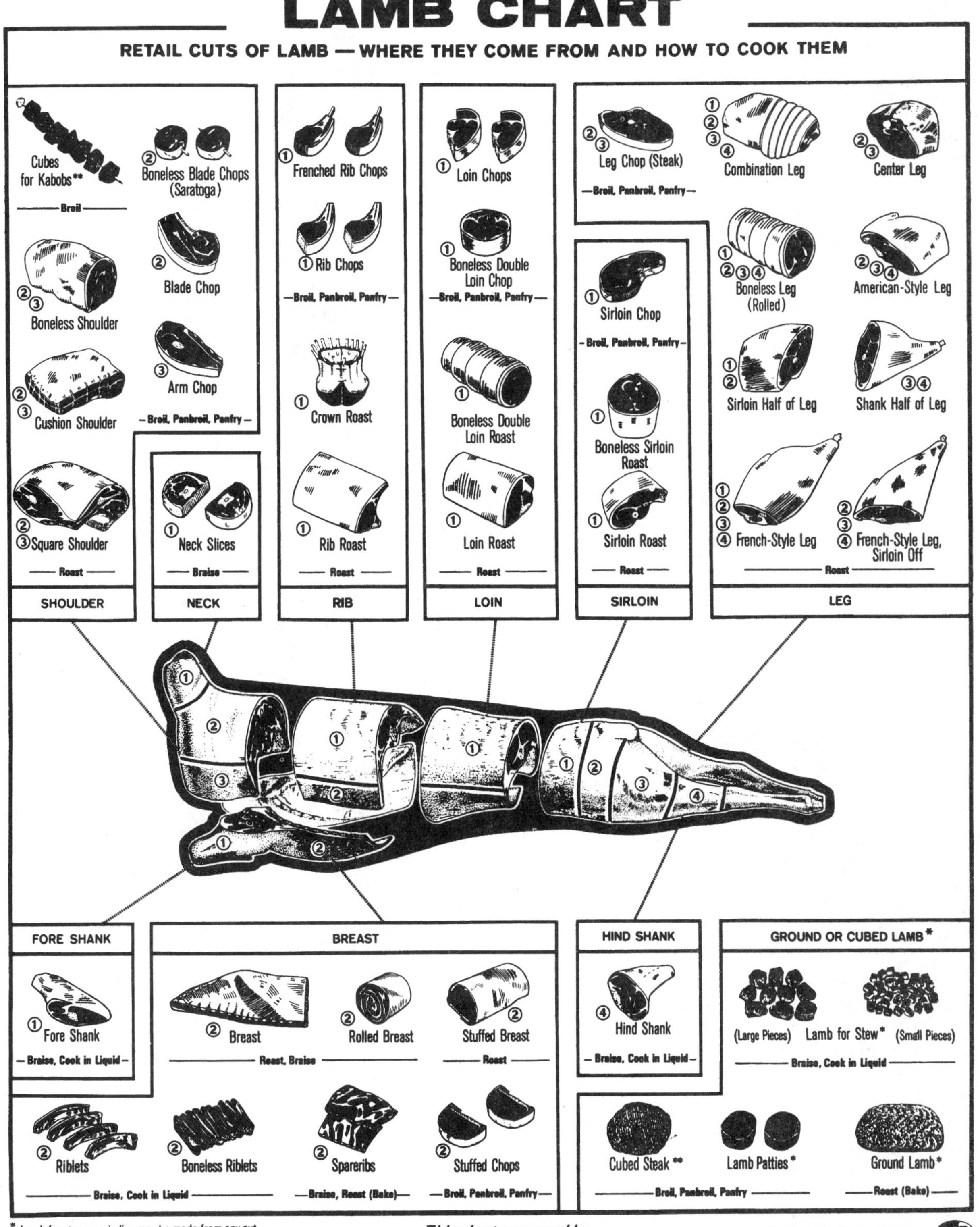

Figure 6. Wholesale and retail cuts of lamb.

Table 3. USDA Grades for Meat

Kind	Class	Grade names
Beef quality grades	Steer, heifer, cow	Prime, choice, good, standard, commercial, utiltiy, cutter, canner
	Bullock	Prime, choice, good, standard, utility
Beef yield grades	All classes	1, 2, 3, 4, 5
Calf quality grades		Prime, choice, good, standard, utility, cull
Veal quality grades		Same as calf
Lamb and mutton quality grades	Lamb, yearling lamb	Prime, choice, good, utility
Lamb yield grades	All classes	1, 2, 3, 4, 5
Pork carcasses	Barrow, gilt	U.S. No. 1, U.S. No. 2, U.S. No. 3, U.S. No. 4, U.S. utility
	Sows	U.S. No. 1, U.S. No. 2, U.S. No. 3, U.S. No. 4, U.S. utility, cull

Source: Ref. 19.

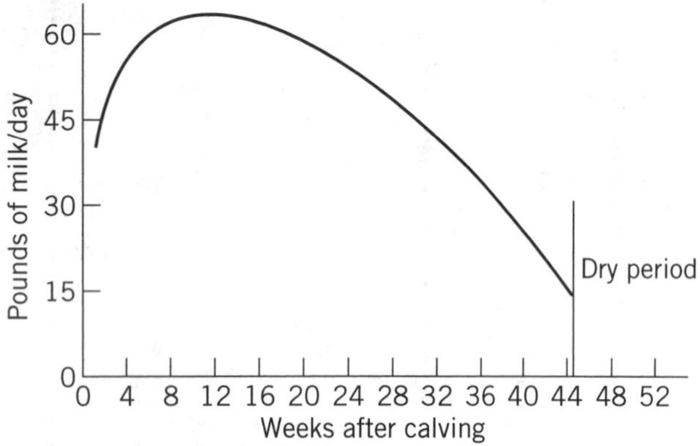

Figure 7. Typical lactation curve.

pricing schemes. However, because milk flavor is contained in the lipid component of milk, feed flavors are the most common causes of off-flavor in milk. Color of milk may also vary among breeds. Guernsey cattle milk, eg, has a more yellow color than milk from Holstein cattle.

The term for the facilities used for the milking of cows is the milking parlor. Modern milking parlors must meet rigid sanitation requirements, keep cows comfortable, and be labor efficient. In comparison to the Federal Meat Inspection Service, state and local laws regulate cow health, sanitation, and milk-handling equipment. Dairy cows are no longer milked by hand but instead by modern machines that rapidly move milk from the milking parlor to refrigerated storage tanks. Milking machines function to remove milk from the cow's udder by applying partial vacuum to the teat end. This vacuum is alternately applied by pulsation consisting of milk and rest cycle 45–60 times per minute.

Milk moves from the cow through a series of stainless-steel pipes to a bulk cooling-storage tank. Milk must be cooled rapidly to approximately 40°F to prevent bacterial growth. Some states allow for in-line rapid cooling coils. However, thorough cleaning of the coils must be ensured.

Most dairies sell their milk directly to processors or through cooperative milk marketing associations. Fluid milk is usually marketed and priced under federal milk marketing orders, in accordance with the Agricultural Marketing Act of 1937. The objective of the federal order is to promote and maintain orderly milk marketing conditions for dairy farmers and assure consumers of an adequate supply of milk.

Milk production is somewhat seasonal, with more milk being produced in the spring. The price of milk paid to dairy farmers is usually on the hundred-weight basis adjusted for percentage of components. Milk protein and fat are usually considered. The pricing scheme will consider local market factors and final use of the milk. For example, cheese plants usually will consider milk protein in establishing prices paid to producers. Milk purchased for the manufacture of cheese may be priced on a cheese yield formula that estimates the quantity of final product.

Almost 50% of the milk produced by farms is processed for consumption as whole, low-fat, or skim milk. All milk is pasteurized to destroy disease-producing microorganisms that may be present in unprocessed milk. The pasteurization process involves heating the milk to 161°F for 15 s in a continuous flow process.

Bovine somatotropin (bST) administration to dairy cattle to enhance milk production has been shown to be efficacious. Depending on the dosage of somatotropin administered, milk production will be enhanced by 10–40% without significant changes in the composition of the milk (20). Concern regarding the impact of this technology on the small dairy farmer has been shown to be unfounded. Since its 1994 approval and sale for use in dairy cattle, bST has been shown to have no deleterious effects on reproduction (21), hematological profiles (22), general health (23), or incidence of mastitis (24). This information confirms earlier work done in Europe (25). The success of bST in increasing milk production and the development of a slow-release product good for 14 days has made it one of the most popular products for use in dairy cattle in recent history.

FUTURE DEVELOPMENTS

The productivity of modern livestock production was made possible through basic and applied research. The application of the principles of genetics, discovery of vitamins and other nutrients necessary for health and productivity of livestock, and the control of reproduction have made large contributions to the economy of producing food. However, for livestock production and its products to remain economically competitive the emergence of new technologies must continue.

Several new technologies and discoveries with applications in livestock production hold promise for enhancing the competitiveness of United States livestock production. Recombinant DNA technology is largely responsible for the production of commercially abundant supplies of somatotropin, also commonly known as growth hormone, to enhance milk and meat production. Porcine somatotropin (PST) administration in pigs increases rate of gain, feed efficiency, and carcass muscling, while significantly reducing carcass fat deposition (26). Similar, though less dramatic, results are seen in beef cattle given bST (27). This technology, along with other methods to regulate lean tissue growth, such as β agonists, will help the pork and beef industry produce meat that is leaner and thus help to reduce fat consumption in the American diet.

Artificial insemination and embryo transplantation of superior lines of swine, sheep, and dairy and beef cattle will continue to increase in popularity among livestock breeders. The ability of scientists to develop improved embryo freezing techniques will greatly enhance the efficacy of this technology. The successful introduction of transgenic livestock with improved rates of growth, carcass characteristics, lactation, and disease resistance will influence the methods and efficiency of livestock production (28).

Transgenic animals provide opportunities to greatly decrease genetic lag in breeding schemes (29) and enhance transfer of superior genes (30). Transgenics may make it possible to alter milk composition in ways that enhance human health, such as decreasing lactose for lactose-intolerant individuals and increasing the resemblance to human milk to improve neonatal nutrition (31). There is even research into the use of the transgenic mammary gland as a mammalian bioreactor that could produce mammalian therapeutic proteins, thus avoiding the disadvantages of pharmaceutical proteins produced by microbial fermentors (lack of bioactivity, presence of allergenic prokaryotic proteins) (32). This process has been dubbed "pharming." Currently, transgenic technology is still inefficient, expensive, and socially controversial; however, it holds promise for enhancing the productivity of livestock and the quality of life for humans.

BIBLIOGRAPHY

1. *Change in U.S. Livestock Production, 1969–1992*, U.S. Department of Agriculture Economic Report, Economic Research Service, ERS-AER-754.
2. *Factsheet, "The Beef Industry,"* National Cattlemen's Association, Englewood, Colo., 1990.
3. C. Lambert, *Beef Industry Facts*, National Cattlemen's Association, Englewood, Colo., 1990.
4. *Livestock Slaughter 1997 Summary*. USDA National Agricultural Statistics Service, U.S. Government Printing Office, Washington, D.C. March, 1998.
5. *Dairy Producer Highlights*, National Milk Producers Federation, Arlington, Va., 1982.
6. U.S. Department of Agriculture, *Agricultural Statistics*, U.S. Government Printing Office, Washington, D.C., 1990.
7. U.S. Department of Agriculture Economics and Statistics System, *Dairy Outlook*, U.S. Government Printing Office, Washington, D.C., December, 1996.
8. *The Structure of Dairy Markets: Past, Present, Future*, U.S. Department of Agriculture Economic Report, Economic Research Service, ERS-AER-757.
9. *Fact Book of Agriculture*, U.S. Department of Agriculture, Miscellaneous Publication No. 1063, 1989.
10. U.S. Department of Agriculture, *Agricultural Statistics*, U.S. Government Printing Office, Washington, D.C., 1989.
11. P. Yuthasastrakosal et al., "Release of LH in Anoestrous and Cyclic Ewes," *Journal of Reproduction and Fertility* **50**, 319–321 (1977).
12. L. Anderson, "Pigs," in E. S. E. Hafez, ed., *Reproduction in Farm Animals*, 4th ed., Lea and Febiger, Philadelphia, Penn., 1980.
13. R. Ashdown and J. Hancock, "Functional Anatomy of Male Reproduction," in E. S. E. Hafez, ed., *Reproduction in Farm Animals*, 4th ed., Lea and Febiger, Philadelphia, Penn., 1980.
14. National Research Council, *Ruminant Nitrogen Usage*, National Academy Press, Washington, D.C., 1985.
15. R. Preston and S. Bartle, "Quantification of Escape Protein and Amino Acid Needs for New Feedlot Cattle," *J. Animal Sci.* **68**, 531 (1990).
16. M. Jurgens, *Animal Feeding and Nutrition*, Kendall/Hunt, Dubuque, Iowa, 1982.
17. Q. Liu et al., "Titration of Fresh Meat Color Stability and Malondialdehyde Development with Holstein Steers Fed Vitamin E-Supplemented Diets," *J. Animal Sci.* **74**, 117–126 (1996).
18. Riley et al., "Palatability of Beef from Steer and Bull Carcasses as Influenced by Electrical Stimulation, Subcutaneous Fat, Thickness and Marbling," *J. Animal Sci.* **56**, 592–597 (1983).
19. M. Judge et al., *Principles of Meat Science*, 2nd ed., Kendall/Hunt, Dubuque, Iowa, 1989.
20. P. Eppard et al., "Effect of Dose of Bovine Growth Hormone on Lactation of Dairy Cows," *J. Dairy Sci.* **68**, 1109–1115 (1985).
21. C. Kirby et al., "Follicular Function in Lactating Dairy Cows Treated with Sustained-Release Bovine Somatotropin," *J. Dairy Sci.* **80**, 273–285 (1997).
22. P. Eppard et al., "Effect of Exogenous Somatotropin on Hematological Variables of Lactating Cows and Their Offspring," *J. Dairy Sci.* **80**, 1582–1591 (1997).
23. G. Hartnell et al., "Effect of Bovine Somatotropin Administered to Periparturient Dairy Cows on the Incidence of Metabolic Disease," *J. Dairy Sci.* **79**, 2170–2181 (1996).
24. L. Judge et al., "Recombinant Bovine Somatotropin and Clinical Mastitis: Incidence, Discarded Milk Following Therapy, and Culling," *J. Dairy Sci.* **80**, 3212–3218 (1997).
25. J. Skarda and H. Mader, "Impact of Bovine Somatotropin on Dairying in Eastern Europe," *J. Dairy Sci.* **74** (Suppl 2), 72–82 (1991).
26. E. Kanis et al., "Effect of Recombinant Porcine Somatotropin on Growth and Carcass Quality of Growing Pigs: Interaction With Genotype, Gender and Slaughter Weight," *J. Animal Sci.* **68**, 1193–1200 (1990).
27. B. Dalke et al., "Dose-Response Effects of Recombinant Bovine Somatotropin Implants on Feedlot Performance in Steers," *J. Animal Sci.* **70**, 2123–2137 (1992).

28. G. Seidel, "Characteristics of Future Agricultural Animals," in J. W. Evans and A. Hollaender, eds., *Genetic Engineering of Animals: An Agricultural Perspective*, Plenum Press, New York, 1986.
29. S. Bishop, "A Comparison of Bonus and Quota Mating Systems for Utilising the Sex-Determining Region Y Gene in Terminal Sire Beef Cattle Breeding," *Theoret. Appl. Genet.* **90**, 487–491 (1995).
30. L. Cundiff et al., "Challenges and Opportunities for Integrating Genetically Modified Animals Into Traditional Animal Breeding Plans," *J. Animal Sci.* **71** (Suppl 3), 20–25 (1993).
31. C. Karatzas and J. Turner, "Toward Altering Milk Composition by Genetic Manipulation: Current Status and Challenges," *J. Dairy Sci.* **80**, 2225–2232 (1997).
32. R. Wall et al., "Transgenic Dairy Cattle: Genetic Engineering on a Large Scale," *J. Dairy Sci.* **80**, 2213–2224 (1997).

GENERAL REFERENCES

D. Acker, *Animal Science and Industry*, 3rd ed., Prentice-Hall, Englewood Cliffs, N.J., 1983.
J. Blakely and D. Bade, *The Science of Animal Husbandry*, 4th ed., Reston Publishing, Reston, Va, 1985.
M. Jurgens, *Animal Feeding and Nutrition*, 6th ed., Kendall Hunt, Dubuque, Iowa, 1988.
National Research Council, *Designing Foods: Animal Product Options in the Market Place*, National Academy Press, Washington, D.C., 1988.

Richard A. Roeder
Mikelle Roeder
University of Idaho
Moscow, Idaho

ANTIBIOTICS IN FOODS OF ANIMAL ORIGIN

In 1877 Pasteur and Joubert discovered that growth of *Bacillus anthracis* could be inhibited by the presence of other microorganisms. This discovery led to the isolation of pyacyanase, the first antibiotic. In 1928 penicillin was discovered by Fleming when he observed that the outgrowth of an agar culture of *Staphylococcus aureus* was inhibited by mold. In 1932 Domagk discovered that the dye prontosil red, the first sulfonamide, demonstrated antimicrobial properties. Since these early discoveries the number of antimicrobials either isolated from natural sources or synthesized in the laboratory has grown tremendously. Antibiotics, in the narrowest sense, are products produced by living organisms that are not toxic to the producing organism but are capable of inhibiting the growth of or terminating other organisms. However, there are many synthetically manufactured compounds that inhibit or kill organisms as effectively as natural antibiotics. The chemistry and mode of action of these synthetic antimicrobial compounds are well documented. While there are virtually thousands of known antimicrobials, only a few are marketed for use in present-day food animal production.

Treatment of food animals with therapeutic and subtherapeutic dosage forms of antibiotic-antimicrobial drugs has increased in the last decade. This has been due to the advent of modern mass production operations that involve the maintenance of thousands of animals simultaneously and requires the utilization of carefully formulated and medicated diets that serve to maximize growth, minimize production costs, and provide an acceptable consumer product that is wholesome and affordable. Animal feeds that contain antibiotics and other antimicrobials as a prophylactic are now used routinely in beef, swine, chicken, and turkey production. In this way the antibiotics—antimicrobials help to maintain the optimal health of the animals so treated. Antibiotics may be coadministered with antimicrobial drugs resulting in a net increase in drug effectiveness compared to individual drug administrations. Strictly speaking, sulfonamides are not antibiotics. However, for purposes of this article sulfonamide antimicrobials will be collectively termed antibiotics.

The aminoglycoside, beta-lactam, ionophore, macrolide, sulfonamide, tetracycline, and other antibiotics are an integral part of food animal production. Regulations governing the use, dosage, and withdrawal times for many of the members of these antibiotic classes in animal production have been established in the United States by federal law, as outlined in the *Code of Federal Regulations*, Title 21, for each compound (1). The U.S. Food and Drug Administration (FDA) regulates the use of antibiotics in animal production while the U.S. Department of Agriculture (USDA) monitors residue levels by testing animal-derived products for antibiotics. Specific information regarding manufacturers, new animal drug application (NADA) codes, approved dosage forms and directions for proper use can be found in the *Code of Federal Regulations*, Title 21 and other published sources (2).

Antibiotics may manifest themselves as residues in animal-derived human foods if improperly used or if withdrawal times have not been observed for treated animals. Animal-derived human foods that contain violative antibiotic residues may pose a potential human health hazard. These potential health hazards can be broken down into three broad categories (3): toxicological, microbiological, and immunopathological. Toxicological concerns relate to the direct toxic effect of the compound on the consumer, resulting in physiological abnormalities, an example being sulfamethazine, which has recently been shown to produce cancer in laboratory animals (4). Microbiological concerns relate to the transmittance of antibiotic resistance. Antibiotics have the potential to act as a selective force that favors the emergence of resistant pathogenic bacteria in the natural flora of meat consumers. An example is a reported outbreak of salmonella poisoning in humans believed to be associated with a resistant strain in hamburger obtained from culled cattle that had been treated routinely with chloramphenicol (5). Finally, there are immunopathological mechanisms, whereby the drug serves as an antigen, demonstrates allergenic properties, and may result in hypersensitivity reactions to the drug that has sensitized some individuals. Penicillin is a prime example. Antibiotics are the principal compounds of concern in the federal residue control strategy (6). Such routine monitoring for antibiotic residues in animal-derived foods is intended to minimize the occurrence of violative antibiotic levels in the food supply. Such a capability is dependent on the analytical methods utilized.

Methods for the determination of antibiotics can include but are not limited to bioassay (ba), thin layer chromatography (tlc), gas chromatography (gc), liquid chromatography (lc), and various immunoassay (ia) techniques. Each of these methods has found utility in antibiotic determinations and many have been shown to be precise, specific, or sensitive, depending on which analytical technique, detector, visualization method, and compound are used. In this regard, a method utilized to detect an antibiotic residue at low nanogram antibiotic per gram of sample (ng/g) or even picogram antibiotic per gram of sample (pg/g) levels may be confounded by the presence of interferences found in the sample matrix or in the sample extract. Thus analytical capability is governed ultimately by the sample preparation and extraction steps. The isolation of such residues from a complex biological matrix poses unique problems to the analyst and because of the number of antibiotics being utilized in animal production the need for multiresidue determinations exists. An emerging trend is the development of rapid multiresidue—multidrug class isolation techniques that result in clean, interference-free extracts and that minimize cost, time, and expendable material requirements, enabling the analyst to test for multiple drugs isolated from one sample. In conjunction with improved isolation methods the need to screen samples rapidly and accurately for antibiotic residues exists. Such antibiotic screening protocols and analytical capability can be enhanced by rapid, reproducible, efficient, and cost-effective residue isolation techniques.

SAMPLE PREPARATION AND RESIDUE ISOLATION

Sample preparation requirements are dependent on the particular analysis that is to be performed. For example, ba or ia procedures for antibiotics may require little or no sample preparation, whereas sample preparation for tlc, lc, and gc procedures can be a major limiting factor of the analyses.

Liquid samples such as milk, blood, urine, saliva, or other body fluids can be assayed directly for antibiotics by ba or ia techniques, or the samples may require only minor cleanup steps (centrifugation, pH adjustment, or a protein precipitation) prior to the assay. In some cases the antibiotic residues that may be present in the sample at low concentrations may require an analyte-concentration step prior to the analysis.

Residue enrichment methods that concentrate the antibiotic residue prior to an analysis may involve solvent extractions, column chromatography, or solid-phase extraction (spe) techniques. The utility of solvent-solvent extractions is limited because of the polar characteristics of many antibiotics such as beta-lactams, aminoglycosides, macrolides, polyether ionophores, and tetracyclines. Ion-exchange column chromatography techniques utilized to isolate polar ionizable antibiotics are only marginally effective. Solid-phase extraction appears to hold more promise as a routine residue enrichment approach. The application of this technique proved very effective in cleaning and concentrating penicillin residues in animal tissue prior to HPLC analysis (7). Methods that attempt to circumvent cleanup steps by assaying samples for antibiotics directly are only rarely efficient, except for ba or ia techniques of some liquid samples such as milk.

Supercritical fluid extraction (SFE), a relatively new extraction technique, is also superior to conventional sample preparation methods as no cleaning is required prior to analysis. SFE minimized sample manipulation and eliminated the use of organic solvents when used to extract sulfamethazine in chicken eggs (8). This advantage was also evident when supercritical CO_2 modified with acetonitrile was used to recover sulfonamide from chicken liver, beef liver, and egg yolk (9).

Sample-preparation steps may not be necessary in some cases for antibiotic determinations but their use can often facilitate more accurate analyses. Therefore, it is incumbent on the analyst to obtain the cleanest sample extract possible. Clean extracts from some liquids and most non-liquid matrices such as muscle or organ tissues are more difficult to obtain and generally require an extraction step and multiple manipulations to isolate the antibiotic residue from the sample with high percentage recoveries. Unfortunately, sample extracts obtained in this manner may contain naturally occurring inhibitors or interferences that could affect the analysis. Thus further cleanup of extracts is usually required prior to performing the analysis. Because of the polar nature of many antibiotics, buffered aqueous solutions are routinely employed for antibiotic extractions and it is difficult, if not impossible, to isolate antibiotic residues from these aqueous extracts by partitioning with organic solvents. Thus drying of the aqueous extract, a time-consuming process, or the use of other preparative steps such as column chromatography or spe techniques may be required before an analytical determination can be made. A direct assay of the aqueous extract may be possible if the concentration of the antibiotic in the extract is sufficiently high or if the analytical method is particularly sensitive. As the complexity of the sample and the number of sample cleanup steps increases, the utility of techniques such as ba, ia, and tlc for quick antibiotic screening purposes decreases. In addition, extracts suitable for ba or ia procedures may not be sufficiently clean for more sophisticated tlc, lc, or gc determinations, thereby requiring additional and different residue isolation procedures for confirmatory techniques.

Sophisticated techniques based on lc, gc, lc-mass spectrometry (ms), and gc-ms usually require more rigorous sample preparation to isolate the antibiotic free from interferences found in the sample extract. These classic isolation techniques can be laborious and time-consuming. Classic isolation techniques for aminoglycoside (10), β-lactam (11–13), chloramphenicol (14), ionophore (15), macrolide (13,16), sulfonamide (17–20), and tetracycline (21) antibiotics generally involve their extraction from biological matrices with large volumes of extracting solvents, chemical manipulations such as pH adjustments and protein precipitations, centrifugations, back washing, and the evaporation of large volumes of organic solvents. This approach limits the usefulness of many classic isolation techniques for multiresidue determinations by lc and gc.

Analytical capability is, therefore, limited to a large extent by interferences present in the sample extract. While state-of-the-art analytical techniques can detect picogram levels of pure antibiotic standard compounds, this same level of detection may not be achievable for antibiotics obtained from the extraction of biological samples. Specifically, coextracted interferences may hinder antibiotic determinations by tlc, lc, or gc because they may have similar detector and chromatographic characteristics or they may exist in such a large quantity that they overwhelm the detection method used. Interferences in ba or ia techniques can contribute to cross-reactions that may lead to false-positive or false-negative determinations. Thus a major limiting factor associated with antibiotic determinations is not the available analytical capability but the sample-preparation steps required to extract the antibiotics. The sample-preparation steps should ideally result in clean biological extracts that contain the antibiotic residue with high percentage recoveries and that have minimal interferences that might limit the choice of analysis.

The need to test for more antibiotic residues in more foods requires rapid, rugged, and multiresidue isolation techniques allowing the analyst to test for multiple drugs in the same sample. Classic residue isolation techniques have not been able to meet this challenge. Matrix solid-phase dispersion (mspd) techniques (22–28), recently developed for the isolation of drug residues from animal matrices, have the potential to greatly enhance many antibiotic residue isolation protocols. In mspd the sample (0.5 g) is dispersed onto octadecasilyl polymeric—derived silica beads [C-18 reversed-phase packing material (2 g), 1,000 m^2 surface area, theoretical]. The dispersion mechanism, utilizing a mortar and pestle, involves the disruption, unfolding, and rearrangement of matrix constituents by mechanical and hydrophobic forces onto the C-18 beads. Lipid and lipophilic materials associate with the lipophilic C-18, allowing the more hydrophilic components and protein regions to extend outward away from the nonpolar, inner C-18—lipid region. Water and more polar constituents preferentially associate with the hydrophilic ends. A column fashioned from the C-18—sample matrix blend can be eluted sequentially with solvents (8 mL) of different polarities to effectively remove interferences in one solvent and elute the target residue in a different solvent. The process can be envisioned as an exhaustive extraction process whereby a large volume of solvent is passed over a thin layer of sample. Mspd has been utilized for the isolation of beta-lactams from beef tissue (23); sulfonamides from milk (24), infant formula (25), and pork muscle tissue (26); as well as for chloramphenicol (27) and tetracycline (28) isolations from milk. The theoretical aspects describing the disruption, unfolding, and rearrangement of matrix constituents onto the C-18 have been published (22–28).

Advances in multiclass—multiresidue isolation procedures, which are rapid, rugged, generic in nature, and free from interferences and which facilitate the isolation of multiple antibiotic residues from one sample, will greatly enhance analytical determinations of antibiotics.

ANALYTICAL METHODOLOGY

Bioassays

Bioassays are used routinely to test for violative levels of antibiotics in milk, animal tissues, and feeds. Bioassays involve the inhibition of growth of specific bacterial spores or viable bacteria in the presence of a sample or sample extract that contains antibiotic residues. Bioassays may also utilize the measurement of labeled antibiotic analyte bound to receptors, a ligand assay, on vegetative bacterial cells. Bioassays have been utilized to detect aminoglycosides (29–32), β-lactams (30,32–34), chloramphenicol (32,35–37), ionophore (15), sulfonamide (29,30), and tetracycline (29,30,32,33,38) antibiotics in food.

For example, the swab test on premises (stop) procedure (30) is used by federal meat inspectors to test for chloramphenicol, tetracycline, aminoglycoside, penicillin G, and sulfonamide antibiotics. The stop method involves taking a sterile cotton swab and macerating the target organ or tissue with the noncotton end. The size of the macerated zone should be slightly larger in diameter and deeper than the cotton end of the swab. The cotton end is then inserted into the disrupted area and allowed to absorb fluids (30 min). The fluid-soaked cotton swab is then placed onto a nutrient agar plate that has been inoculated with a lawn of specific bacterial spores and the plates are then incubated (16–22 h). Standards of known concentration are run in parallel. A zone of microbial growth inhibition on the sample plate is an indication of the presence of an antibiotic and the size of the zone of inhibition can be semicorrelated with the size of the zone of inhibition for a given concentration of pure antibiotic standard. The sample is positive if the size of the zone of inhibition is similar to that for a known pure standard. Unfortunately, the stop procedure detects only the presence of inhibitors, not their specific identity. Other antibiotics or naturally occurring inhibitors may contribute to the size of the zone of growth inhibition observed, indicating a positive sample although the amount of the individual antibiotic compounds present may be less than the violative level.

The classic disk assay procedure (34) is similar to the stop procedure for the same antibiotics except that a filter paper disk is placed on the inoculated nutrient agar, the liquid sample or suitable extract is added to the disk and then the plate is incubated for a minimum of 2.5 h or until a zone of inhibition can be observed. The zone of inhibition may be enhanced by dyeing techniques (39,40), which aid in its detection. Semiquantitative determinations by the disk assay can be accomplished, provided incubation times are increased (16–22 h). The disk assay procedure is the official method described in the Pasteurized Milk Ordinance (PMO) (41) for antibiotic testing in milk. The disk assay procedure suffers limitations similar to those of the stop procedure. However, the PMO allows for the use of any method that gives results equivalent to the disk assay method; therefore, many states utilize alternative techniques such as the color reaction test (crt) to test for antibiotics in milk. The Delvo test (39) is an example of a crt technique used to determine antibiotics, specifically beta-lactams, in milk. The test involves placing the milk sample onto agar containing a viable strain of *Bacillus*, nutrients, and pH indicators. If the color of the agar changes from purple (basic) to yellow (acid) after incubation (1.5 h) then no penicillin is present to prevent the outgrowth of the acid-producing bacteria. Color reaction tests can be rapid and simple to perform.

The microbial receptor assay (mra) (Charm test) technique (38) can be used to detect beta-lactam, macrolide, and aminoglycoside antibiotic classes. The mra method involves the use of C-14 or tritium isotopically labeled analyte to displace nonlabeled analyte from the bacterial receptors located on vegetative cells under a standard set of conditions. An equilibrium condition between unlabeled and isotopically labeled analyte results, allowing quantitative measurement of the amount of antibiotic present by an appropriate radiometric method. However, the identity of the compound is not known and must be determined by other methodology.

The thin layer chromatography—bioautography of tlb bioassay technique (15,32,40) uses traditional thin layer chromatography to separate sample constituents on silica or microcrystalline cellulose chromatography (mcc) plates. Different antibiotics will migrate on the plates according to their chemical characteristics and the developing solvent utilized. The developed plates are covered with a spore-inoculated nutrient agar and then incubated (16–24 h). Zones of growth inhibition observed for different locations on the plate after incubation indicate the presence of antibiotics. Aminoglycoside (32), β-lactam (32), chloramphenicol (32), ionophore (15,32), macrolide (32,40), and tetracycline (32) antibiotics have been assayed by tlb. The tlb method is more specific and sensitive (36) than either the stop or classical disk assay procedures because it takes advantage of the ability of tlc to separate 14 different antibiotics from each other as well as from other impurities that may be present in the sample. Therefore, zones of inhibition can be more closely correlated to different antibiotics. The tlc-developing solutions are designed to optimize separations between similar antibiotics. However, the separation of up to 14 different antibiotic residues can only be accomplished by utilizing three separate tlc plates. Furthermore, extensive sample-preparation and extraction steps are employed in the tlb procedure that are similar to those employed for more sophisticated analytical determinations. Because of the incubation time required the tlb procedure is not as rapid as the crt or the disk assay methods for screening purposes. However, the tlb procedure has an advantage over the stop or disk assay procedures for screening purposes because it can provide for more precise determinations between different antibiotics. Absolute confirmation of the antibiotics is not possible by tlb and time requirements are in excess of many sophisticated, more definitive, liquid and gas chromatographic techniques.

The stop, disk assay, crt, and tlb procedures are valuable screening tools. The first three techniques are relatively cheap, easy to perform, and require a minimal amount of equipment and technician training. However, they lack specificity and sensitivity and/or they may be subject to interpretive errors in the presence of naturally occurring inhibitors. Furthermore, each of these four methods requires confirmational testing of positives. The mra method using isotopically labeled antibiotics can be expensive and requires special handling and equipment; but it can provide for quantitative determinations. These bioassays, if used judiciously, can minimize the number of samples screened by more costly analytical methods, but they cannot replace such methods for absolute confirmations.

Immunoassay

Immunoassay techniques are based on classic antibody-antigen reactions whereby the antibody will bind with its corresponding antigen (antibiotic) and result in visible turbidity if reacted in solution or form visible immunoprecipitation in a gel at the location where the antigen and antibody meet. Visible end point immunoprecipitation that occurs at low antibiotic concentrations may be difficult to see and may require detection by more sensitive nonvisual means.

Immunoassay techniques such as enzyme immunoassay (eia), radio immunoassay (ria), enzyme-linked immunosorbent assay (elisa), fluorescence polarization immunoassay (fpia), particle-concentration immunoassay (pcia), particle-concentration fluorescence immunoassay (pcfia), quenching fluoroimmunoassay (fia), and latex-agglutination inhibition immunoassay (laia) require the measurement of by-products produced by linked enzyme systems or the measurement of radioactive, fluorescence, metal, or latex labels that have been attached to one of the reactants (42). The displacement between the unlabeled and labeled antigen or antibody allows for true measurement of the immunoprecipitate concentration, which corresponds to the concentration and type of analyte present in the sample extract. Ria determinations of chloramphenicol (43,44) as a residue in eggs, milk, and meat were comparable to values obtained by gc. The eia determination of monensin (45), a polyether antibiotic in urine, serum, and fecal extracts, may be applicable to food extracts as well. A review of ia techniques such as ria, fia, fpia, and laia for aminoglycoside determinations in body fluids has been published (46). The recent use of a monoclonal-antibody—based agglutination test (spot test) for beta-lactams in milk (47) and elisa for chloramphenicol determinations in swine tissue (48), milk (49), and sulfamethazine in swine tissue (50) has been adapted to other antibiotics including spectinomycin (51,52), enrofloxacin, (53) and gentamicin (54).

Immunoassays can be sensitive, class specific, accurate, and provide a means for rapidly screening samples for antibiotics; the future development of monoclonal antibodies will allow for more specificity in immunoassay determinations. The growth in the use of immunoassay-based detection has lead to the development of commercial kits allowing for on-site detection of chemical contaminants (55–57). Radioimmunoassay techniques that require radioisotopes may be subject to specific regulations, require expensive counting equipment, and pose disposal problems. Reagent kits can be relatively expensive and have a limited shelf life. In this regard, nonisotopic immunoassays such as elisa, fpia, pcia, or pcfia, and monoclonal-based ias will in all likelihood play an increasingly important role in antibiotic screening immunoassay determinations.

At present, ia techniques for antibiotics approved for use in animal production are limited and have minimal utility for aminoglycoside, β-lactam, chloramphenicol, and sulfonamide determinations, but they have the potential for use as screens for all the major antibiotic classes. Immunoassay techniques have been limited to liquid samples such as milk because of the difficulties associated with the isolation of antibiotic residues from matrices such as muscle or organ tissues. Increased use of ia techniques to screen for antibiotics in tissues is directly dependent on the development of rapid residue isolation techniques that isolate the antibiotic from the tissue and provide for an extract that is free from cross-reacting components and nonspecific binding (NSB) factors. The development of such isolation techniques will further the use of ia techniques for the screening of antibiotics in tissues.

Immunoassay techniques are more specific than microbiological based bioassays and would allow for the rapid screening of large numbers of samples. With the need to screen more samples for more drugs, the ia techniques will become more important in residue control protocols. Immunoassay techniques can greatly reduce the number of samples that are presently screened by more sophisticated analytical techniques and have the potential to replace many bioassay and tlc methods, provided rapid and efficient antibiotic extraction procedures can be developed. As a regulatory tool, ia techniques for antibiotic screening coupled to highly specific and accurate techniques, such as lc, gc, lc-ms, or gc-ms for confirmations of positives, mark the future with respect to a rapid and reliable residue control strategy.

Thin Layer Chromatography

Thin layer chromatography techniques have been used in chemical separations for decades. Thin layer chromatography is rapid and inexpensive, can be highly sensitive depending on the compound examined and the visualization technique employed, and is easy to use and versatile. It can be adapted to separations of all classes of antibiotics by utilizing different sorbents and solvent-developing systems. Aminoglycoside (31,32), β-lactam (32,48,58), chloramphenicol (32), ionophore (15,32), macrolide (16,32,40), sulfonamide (59,60), and tetracycline (38,61,62) antibiotics have been successfully assayed by tlc. However, the utility of tlc for the detection of picogram levels of antibiotics is limited and thus tlc techniques have come to play only a minor role in antibiotic analyses.

The sulfa on site (sos) tlc procedure for sulfonamide determinations in swine urine was developed by the USDA–FSIS and is presently in use in 100 of the largest swine-slaughtering facilities in the United States (63). This method was adopted in 1988 as the official method for in-plant testing of swine for sulfamethazine (64). Thin layer chromatography techniques can complement other antibiotic assay techniques such as tic–bioautography bioassays (15,32,40). TLC/bioautography qualitative and semiquantitative methods are routinely used in Austria for the analysis of antibiotic residues in poultry, meat, and fish (65). This method is limited to a large degree by the cleanliness of the sample that is to be analyzed. Impurities in the sample can interfere with the chromatography of the antibiotic by altering the migration of the antibiotic on the tlc plates compared to pure standards, and these impurities can negate visualization techniques. Thin layer chromatography determinations also require confirmations of suspect residues isolated from the tlc plates to determine if the residue is indeed an antibiotic. In this regard bioassay and immunoassay techniques provide for a more precise determination of the presence of antibiotics in a sample extract. Thin layer chromatography will perhaps play a decreasing role in antibiotic residue control strategies where low-level detection is required for some antibiotics, but should not be eliminated from the analyst's tools utilized for the isolation and purification of compounds from animal-derived matrices.

Liquid Chromatography

The convenience and versatility of liquid chromatography has led to its adoption as the analytical method of choice for the determination of many drugs, especially antibiotics. Liquid chromatography, using selective detectors, can give reproducible results that are specific, sensitive, and precise. The polar characteristics of antibiotics make them well suited to lc procedures, in which mobile phase solvent systems and columns can be varied to facilitate specific antibiotic determinations.

Recent publications describing lc methods for the analysis of aminoglycoside (10,31,66), β-lactam (11–13,16,23,67–69), chloramphenicol (14,27,70,71), ionophore (15,16), macrolide (13,16,72), sulfonamide (17,19,24–26,73–75), and tetracycline (21,36,61,68,76–80) antibiotics underscore the utility of lc determinations for the analysis of these compounds as residues in foods and other biological matrices.

Food extracts containing residues of sulfonamide (15,18,24–27), β-lactam (11,16,23,67), chloramphenicol (14,27), tylosine (16) and spiramycin (13) macrolides, and tetracycline (21,36,61,68,78) antibiotics may be analyzed directly by ultraviolet (uv) detection at picogram—nanogram levels because they have characteristic uv absorbances and large extinction coefficients. Photodiode array uv detection of antibiotics can provide the analyst with uv spectra of suspect peaks (81) and thus serve as a preconfirmational screening tool. However, some beta-lactams and macrolides, which have a maximum absorbance in the low uv (210–240 nm) range or relatively small extinction coefficients, can be more difficult to analyze by lc—uv because of coextracted interferences that absorb readily in this range. Optimizing lc chromatographic conditions to separate a particular antibiotic residue from interferences that may be present in the extract severely limits lc techniques in terms of multiresidue antibiotic determinations.

Aminoglycoside and most macrolide antibiotics that have low or nonexistent uv-absorbing properties may require the formation of derivatives to aid in their detection or the use of alternative detection methods. Benzene sulfonyl chloride (82) and 1-fluoro-2,4-dinitrobenzene (83) have been used to prepare uv derivatives of aminoglycosides for analyses by lc. Detection methods for antibiotics that do not require making derivatives are ideal because reaction conditions for many derivatizations can be difficult to optimize. However, until improved or new detection methods are developed for antibiotics, derivatives will continue to be used to facilitate sensitive and selective detections, especially for aminoglycosides.

The analyses of aminoglycoside (10), β-lactam (12), macrolide (16), and tetracycline (38) antibiotics can also be facilitated by the formation of fluorescent derivatives. The ionophore lasalocid has native fluorescence due to its salicyclic acid-type aromatic moiety and, therefore, can be analyzed directly in food extracts at nanogram levels by fluorescence detection (15). Fluorescent derivatives of tetracyclines (38) can be made by simply complexing the tetracycline with different metal ions. The tetracycline—metal ion complexes have unique excitation and emission wavelengths that can be measured to quantitatively determine tetracycline concentrations in the food extract. In the case of aminoglycoside antibiotics, which do not have native fluorescence, fluorescent derivatives must be made to facilitate their detection. The preparation of fluorescent derivatives can require exacting reaction conditions, additional equipment in the form of reactors and delivery pumps, and in many cases the removal of reaction reagents before an analysis can be made. However, this cannot be avoided in the case of aminoglycosides because no other suitable chromatographic method presently exists for their detection at nanogram levels. Fluorescence detection of *o*-phthalaldehyde (OPA) or fluorescamine derivatives has been shown to give accurate and reliable values for aminoglycoside residues in animal tissue (10,31). The derivatization of the aminoglycoside amino group with a fluorogenic agent can be accomplished precolumn or postcolumn. The added cost of the postcolumn reactor and solvent pump needed to deliver the derivatization solution may be disadvantageous. Alternatively, precolumn derivatization reaction mixtures may require the removal of derivatizing reagents prior to the analyses. Specific applications may require either approach and are dependent on the type of food that is to be extracted and the inherent interferences present.

Fluorescent derivatives (12) of penicilloaldehyde products obtained by enzymatic hydrolysis of beta-lactam rings and reaction with dansyl hydrazine have been reported for eight neutral beta-lactams. This is a novel approach for beta-lactam determinations and might be applicable to other antibiotics that are inherently unstable and difficult to analyze. Techniques that serve to characterize unique enzymatic or degradative products of antibiotics may be a useful tool for many antibiotic determinations.

Liquid chromatography of antibiotics will continue to be the method of choice for most antibiotic determinations. Coupling of photodiode array uv, fluorescence, electrochemical and other yet to be developed detectors in tandem can provide information in terms of retention times, structures, and characteristic uv and fluorescent spectra. Information obtained by this approach may be sufficient to confirm the presence of specific antibiotics if coupled with positive results obtained by immunoassay techniques for the specific antibiotic in question. Liquid chromatography—mass spectrometry is not presently suitable for low-level detection of the major antibiotic classes, but advances in this area, the use of tandem detection methods, and specific monoclonal-based immunoassay screening techniques would contribute significantly to overall antibiotic residue control strategies needed to insure a safe and wholesome food supply.

Gas Chromatography

Gas chromatography coupled to specific detectors can provide valuable information about the retention time, structure, chemical characteristics, and identity of compounds. Gas chromatography has been utilized for antibiotic determinations but is limited to some extent by the molecular weight, high polarity and the relative lack of thermal stability of many antibiotics. Chemical derivatives can serve to impart greater stability and volatility to antibiotics. However, controlling reaction conditions to insure the formation of consistent derivative products with high yields can be difficult to accomplish. These difficulties have limited the usefulness of gc for routine antibiotic determinations.

Presently there are no suitable gc methods for the routine determination of aminoglycoside, beta-lactam, and most ionophore, macrolide, and tetracycline antibiotics. Gas chromatography of trimethylsilyl (TMS) derivatives of aminoglycoside (84), lasalocid ionophore (14), and tetracycline (85) antibiotics have been reported. However, the usefulness of this technique for low-level determinations of these antibiotics in food extracts is limited because of the need to control silylating reaction conditions carefully, the formation of multiple TMS antibiotic derivative products, the abundance of TMS interferences contributed by the sample, and the lack of stability of TMS derivatives.

Gas chromatography methods for chloramphenicol have been reviewed (14). Gas chromatography utilizing electron capture (ECD), flame ionization (FID) and thermionic (TID) detection of TMS and heptafluorobutyl derivatives of chloramphenicol isolated from muscle, liver, kidney, and milk resulted in picogram—nanogram detection giving results comparable to radioimmunoassay determinations (43). Gas chromatography of chloramphenicol residues in food extracts can provide for confirmations of suspect residues and complement lc determinations.

Gas chromatography methods for sulfonamides have also been reviewed (17). Sulfonamide determinations in animal tissues (20,86) by gc have been accomplished by analyzing volatile methylated or acylated derivatives. Methylation of sulfonamides at the *N*-1 position followed by acylation at the *N*-4 position can provide for stable volatile derivatives suitable for gc analysis. However, it can be difficult to control reaction conditions necessary to optimize this two-step derivative reaction. Low yields and the formation of multiple derivative products may result in nonrepresentative and misleading data. Although methods for the quantitative determination of sulfonamides (18,20,75) have been reported, it has been noted (18) that some sulfonamides determined in this manner resulted in low and variable recoveries. Because of this variability, it is necessary to be careful when quantitatively evaluating data obtained by this technique.

Advances in supercritical fluid chromatography (sfc) may eventually lead to applications involving antibiotic determinations. At present, sfc is limited as a routine analytical tool to nonpolar substances (87). Analysis of polar antibiotics by sfc will require advances in sfc hardware and columns. In addition, the solubility characteristics of different antibiotics in different supercritical media will require extensive laboratory investigation to define optimal conditions in terms of pressures and polar mobile phase modifiers that may be required:

Thus gas chromatography methods have minimal utility for antibiotic determinations. Gas chromatography will remain a supplemental analytical tool for antibiotic residue determinations until new derivatization schemes, reagents, and refined reaction conditions are developed. The development of such innovations can, however, advance present marginal gc antibiotic techniques to full-fledged quantitative antibiotic confirmatory methods, which will contribute significantly to an integrated antibiotic residue control strategy.

Lc–ms and Gc–ms

Mass detectors coupled to chromatographic techniques such as lc and gc can provide for the unequivocal identification of compounds. Sensitivity limitations associated with lc–mc precluded its use in antibiotic residue analyses when low-level determinations were required. However, in recent years lc–ms has been developed to permit such determinations. Keever, Voysker, and Tyczkowska (88) reported a quantitative procedure using lc–electrospray ms to detect ceftiofur in milk at levels down to 10 ppb. An interlaboratory study to monitor pirlimycin residues in bovine milk using lc–thermospray ms was able to detect as low as 0.4 ppm in milk and 0.5 ppm in liver tissue (89). Gc–ms techniques for antibiotics have been predominantly applied to colatile sulfonamide derivatives (17,20,75,86) but may provide for confirmation of suspect residues, although it may be inadequate as a regulatory tool. However, gc–ms procedures were reported suitable for identifying and quantifying chloramphenicol residues with detection limits of 0.1–1.0 ng/kg (90). This method compared well with results obtained using radioimmunoassay and enzyme immunoassay and was considered to be a good screening test. Nagota and Oka (91) reported a capillary gc–ms procedure capable of measuring antibiotic residues in yellowtail fish including chloramphenicol, florfenicol, and thamphenicol. Recovery of each antibiotic was greater than 65% with a detection limit of 5 ppb. Further refinement of lc–ms and gc–ms techniques, advances in ms sensitivity, more efficient lc–ms interfaces, and future developments may allow ms to become a routine confirmatory tool that will greatly enhance antibiotic residue determinations.

FUTURE TRENDS

Microbiological and bioassays have historically served to screen foods for the presence of antibiotics. While these assays have proven to be quite sensitive and inexpensive they are far too nonspecific and cause too lengthy a delay in obtaining data. This delay may allow antibiotic contaminated animal-derived foods to be marketed and consumed with drug detection being accomplished after the fact. This policy is inherently inadequate but has arisen from the limitations of the available analytical technologies. In this sense, bioassays and other methodologies that require lengthy waiting periods and prolonged and complex analytical procedures cannot meet the increasing need to perform rapid drug screening and confirmation on more samples for more drugs.

The advent of more specific, sensitive, and simple to perform immunoassays should overcome many of the problems of performing rapid screening and early detection of antibiotic contamination in animal-derived foods. In this regard, simple elisa, card, or test-strip assays that are suitable for use at production sites, slaughterhouses, and packing plants could dramatically reduce the occurrence of antibiotic violations and preclude the problems that can arise from antibiotic contamination of the food supply.

These and other more complex immunoassay methods could be performed on milk, urine, blood, or other body fluids. However, new methods for the rapid isolation of drugs from tissues may also make such immunoassays more directly applicable to the screening of this complex sample material. Matrix solid-phase dispersion (mspd) techniques have shown promise in providing a simple, rapid, and generic method for performing drug isolations from tissues and milk. Thus the combination of ia and mspd may prove to be a useful approach to the more rapid screening and analysis of tissues and animal derived products for antibiotics as well as other drugs. A further advantage of mspd is the capability to isolate a class of drugs or several classes of drugs from a single sample. In this regard the advances in immunoassay screening technology must be matched by advances in extraction procedures. Extraction methodology that is too narrow in extraction capability leads to a need to develop a specific method for each antibiotic drug. Extraction methods that are multidrug and multidrug class specific will have the greatest utility for performing screening analyses in the future, whether such screens are conducted by immunoassay, tlc, hplc, or gc.

Of these analytical techniques ia and tlc will be best utilized for rapid screening, whereas hplc and gc may be the best methods for subsequent confirmation and quantitation of antibiotic residues. However, hplc methods that are capable of separating and detecting several drug residues obtained, perhaps, from a single multidrug class extraction may also prove to be useful screening techniques. The usefulness of such screens is directly related to the efficiency of recovery of a multidrug extraction method, the elimination of background interferences, and the ability of various detectors to indicate the presence or absence of a drug with adequate sensitivity.

For the antibiotics, gc will remain of limited value but should always be considered and applied where feasible. It is expected that sfc will eventually play a major role in the extraction and analysis of compounds such as the antibiotics. However, current results are less than promising and the awaited applications may be long in development.

BIBLIOGRAPHY

1. *Code of Federal Regulations*, 21 CFR "Food and Drugs," U.S. Government Printing Office, Washington, D.C., 1989.
2. S. F. Sundlof, J. E. Riviere, and A. L. Craigmill. *Food Animal Residue Avoidance Databank. Trade Name File. A Compre-*

hensive Compendium of Food Animal Drugs, 3rd ed, Institute of Food and Agricultural Sciences, University of Florida, Gainesville, Fla., 1989.

3. Food and Agriculture Organization of the United Nations. *Residues of Veterinary Drugs in Foods*, FAO Food and Nutrition Paper **32**. Food and Agriculture Organization of the United Nations, Rome, Italy, 1985.
4. National Center for Toxicological Research, *Chronic Toxicity and Carcinogenicity of Sulfamethazine in B6C3F1 Mice*, technical report, National Center for Toxicological Research, Jefferson, Ark., 1988.
5. J. S. Spika et al., "Chloramphenicol-Resistant *Salmonella newport* Traced through Hamburgers to Dairy Farms," *New Eng. J. Med.* **316**, 565–570 (1987).
6. J. Brown, ed., U.S. Department of Agriculture/Food Safety and Inspection Service. *Compound Evaluation and Analytical National Residue Program Plan*, USDA, FSIS Science and Technology Program, Washington, D.C., 1988.
7. J. Boison and L. Keng, "Application of Solid Phase Technology for the Multiresidue Analysis of Penicillin Residues in Animal Tissues by High Performance Liquid Chromatography with Ultraviolet-Visible Detection," *Seminars in Food Analysis* **1**, 33–38 (1996).
8. J. W. Pensabene, W. Fiddler, and D. J. Donoghue, "Supercritical Fluid Extraction Compared with Solvent Method for Incurred Sulfamethazine in Chicken Eggs," *J. Food Sci.* **63**, 25–26 (1998).
9. M. T. Combs et al., "Quantitative Recovery of Sulfonamides from Chicken Liver, Beef Liver, and Egg Yolk Via Modified Supercritical Carbon Dioxide," *J. Agric. Food Chem.* **45**, 1779–1783 (1997).
10. V. K. Agarwal, "High Performance Liquid Chromatographic Determinations of Gentamicin in Animal Tissue," *J. Liquid Chromatography* **12**, 613–628 (1989).
11. J. O. Miners, "The Analysis of Penicillins in Biological Fluids and Pharmaceutical Preparations by High Performance Liquid Chromatography: A Review," *J. Liquid Chromatography* **8**, 2827–2843 (1985).
12. R. K. Munns et al., "Multiresidue Method for Determination of Eight Neutral β-Lactam Penicillins in Milk by Fluorescence-Liquid Chromatography," *Journal of the Association of Official Analytical Chemists* **68**, 968–971 (1985).
13. T. Nagata and M. Saeki, "Determination of Ampicillin Residues in Fish Tissues by Liquid Chromatography," *Journal of the Association of Official Analytical Chemists* **69**, 448–450 (1986).
14. E. H. Allen, "Review of Chromatographic Methods for Chloramphenicol Residues in Milk, Eggs and Tissues from Food Producing Animals," *Journal of the Association of Official Analytical Chemists* **68**, 990–999 (1985).
15. G. Weiss and A. MacDonald, "Methods for Determination of Ionophore Type Antibiotic Residues in Animal Tissues," *Journal of the Association of Official Analytical Chemists* **68**, 971–980 (1985).
16. W. A. Moats, "Chromatographic Methods for Determination of Macrolide Antibiotic Residues in Tissues and Milk of Food-Producing Animals, *Journal of the Association of Official Analytical Chemists* **68**, 980–984 (1985).
17. W. Horwitz, "Analytical Methods for Sulfonamides in Foods and Feeds. II. Performance Characteristics of Sulfonamide Methods," *Journal of the Association of Official Analytical Chemists* **64**, 814–824 (1981).
18. A. J. Manuel and W. A. Steller, "Gas-Liquid Chromatographic Determination of Sulfamethazine in Swine and Cattle Tissues," *Journal of the Association of Official Analytical Chemists* **64**, 794–799 (1981).
19. B. L. Cox and A. Krzeminski, "High Pressure Liquid Chromatographic Determination of Sulfamethazine in Pork Tissue," *Journal of the Association of Official Analytical Chemists* **65**, 1311–1315 (1982).
20. R. M. Simpson, F. B. Suhre, and J. W. Shafer, "Quantitative Gas Chromatographic Mass Spectrometric Assay of Five Sulfonamide Residues in Animal Tissue," *Journal of the Association of Official Analytical Chemists* **68**, 23–26 (1985).
21. R. B. Ashworth, "Liquid Chromatographic Assay of Tetracyclines in Tissues of Food Producing Animals," *Journal of the Association of Official Analytical Chemists* **68**, 1013–1018 (1985).
22. S. A. Barker, A. R. Long, and C. R. Short, "A New Approach to the Isolation of Drug Residues From Animal Tissues," in W. Huber, ed., *Proceedings of the Sixth Symposium on Veterinary Pharmacology and Therapeutics*, Blacksburg, Va., 1988, pp. 55–56.
23. S. A. Barker, A. R. Long, and C. R. Short, "Isolation of Drug Residues from Tissues by Solid Phase Dispersion," *J. Chromatography* **475**, 353–361 (1989).
24. A. R. Long, C. R. Short, and S. A. Barker, "Multiresidue Method for the Isolation and Liquid Chromatographic Determination of Eight Sulfonamides in Milk," *J. Chromatography* **502**, 87–94 (1990).
25. A. R. Long et al., "A Multiresidue Method for the Isolation and Liquid Chromatographic Determination of Seven Sulfonamides in Infant Formula," *J. Liquid Chromatography* **12**, 1601–1612 (1989).
26. A. R. Long et al., "Multiresidue Method for the Determination of Sulfonamides in Pork Tissue," *J. Agric. Food Chem.* **38**, 423–426 (1990).
27. A. R. Long et al., "Method for the Isolation and Liquid Chromatographic Determination of Chloramphenicol in Milk," *J. Agric. Food Chem.* **38**, 427–429 (1990).
28. A. R. Long et al., "Matrix Solid Phase Dispersion (MSPD) Isolation and Liquid Chromatographic Determination of Oxytetracycline, Tetracycline and Chlortetracycline in Milk," *Journal of the Association of Official Analytical Chemists* **73**, 379–384 (1990).
29. R. B. Read, Jr. et al., "Detection of Sulfa Drugs and Antibiotics in Milk," *Applied Microbiology* **21**, 806–808 (1971).
30. R. W. Johnston et al., "A New Screening Method for the Detection of Antibiotic Residues in Meat and Poultry Tissues," *J. Food Protection* **44**, 828–831 (1981).
31. B. Shaikh and E. H. Allen, "Overview of Physical Chemical Methods for Determining Aminoglycoside Antibiotics in Tissues and Fluids of Food Producing Animals," *Journal of the Association of Official Analytical Chemists* **68**, 1007–1013 (1985).
32. E. Neidert, P. W. Saschenbrecker, and F. Tittiger, "Thin Layer Chromatographic/Bioautographic Method for Identification of Antibiotic Residues in Animal Tissues," *Journal of the Association of Official Analytical Chemists* **70**, 197–200 (1987).
33. J. R. Bishop et al., "Quantitative Assay for Antibiotics Used Commonly in Treatment of Bovine Infections," *J. Dairy Sci.* **68**, 3031–3036 (1985).
34. American Public Health Association, *Standard Methods for the Examination of Dairy Products*, Port City Press Inc., Baltimore, Md., 1985.

35. C. J. Singer and S. E. Katz, "Microbiological Assay for Chloramphenicol Residues," *Journal of the Association of Official Analytical Chemists* **68**, 1037–1041 (1985).

36. G. O. Korsrud and J. D. MacNeil, "A Comparison of Three Bioassay Techniques and High Performance Liquid Chromatography for the Detection of Chlortetracycline Residues in Swine Tissues," *Food Additives and Contaminants* **5**, 149–153 (1987).

37. C. D. C. Salisbury et al., "Survey of Chloramphenicol Residues in Diseased Swine," *Can. J. Veterinary Res.* **52**, 15–17 (1988).

38. M. Riaz, "The Quantitative Analysis of Tetracyclines," *Journal of the Chemical Society of Pakistan* **8**, 571–583 (1986).

39. S. Williams, ed., *Official Methods of Analysis*, 14th ed., Association of Official Analytical Chemists, Inc., Arlington, Va., 1984.

40. M. Petz et al., "Thin-Layer Chromatographic Determination of Erythromycin and Other Macrolide Antibiotics in Livestock Products," *Journal of the Association of Official Analytical Chemists* **70**, 691–697 (1987).

41. Food and Drug Administration, *PMO, Grade A Pasteurized Milk Ordinance*, Public Health Service Publication No. **229**, Washington, D.C., 1985.

42. B. A. Morris, M. N. Clifford, and R. Jackman, eds., *Immunoassays for Veterinary and Food Analysis—Part 1*, Elsevier Science Publishing Co., Inc., New York, 1988.

43. D. Arnold et al., "Radioimmunological Determination of Chloramphenicol Residues in Musculature, Milk and Eggs," *Archiv für Lebensmittelhygiene* **35**, 131–139 (1984).

44. D. Arnold and A. Somogyi, "Trace Analysis of Chloramphenicol Residues in Eggs, Milk and Meat: Comparison of Gas Chromatography and Radioimmunoassay," *Journal of the Association of Official Analytical Chemists* **68**, 984–990 (1985).

45. M. E. Mount, D. L. Failla, and S. Wie, "Enzyme Immunoassay for the Feed Additive Monensin," *Journal of the Chemical Society of Pakistan* **8** (1986).

46. M. C. Rouan, "Antibiotic Monitoring in Body Fluids," *J. Chromatography* **340**, 361–400 (1985).

47. J. J. Ryan et al., "Detection of Penicillin, Cephapirin and Cloxacillin in Commingled Raw Milk by the Spot Test," *J. Dairy Sci.* **69**, 1510–1517 (1986).

48. C. Van de Water et al., "An Enzyme Linked Immunosorbent Assay for the Determination of Chloramphenicol Using Monoclonal Antibody," *Zeitschrift für Lebensmittel-Untersuchang und-Forschung* **185**, 202–207 (1987).

49. J. F. M. Nouws, J. Laurensen, and M. M. L. Aerts, "Monitoring Milk for Chloramphenicol Residues by an Immunoassay (Quick-Card)," *Veterinary Quarterly* **10**, 270–272 (1988).

50. D. E. Dixon-Holland and S. E. Katz, "Competitive Direct Enzyme-Linked Immunosorbent Assay for Detection of Sulfamethazine Residues in Swine Urine and Muscle Tissues," *Journal of the Association of Official Analytical Chemists* **71**, 1137–1140 (1988).

51. T. Tanaka et al., "Easy Enzyme-linked Immunosorbent Assay for Spectinomycin in Chicken Plasma," *J. AOAC Int.* **79**, 426–430 (1996).

52. J. D. Thacker, E. S. Casale, and C. M. Tucker, "Immunoassays (ELISA) for Rapid, Quantitative Analysis in the Food Processing Industry," *J. Agric. Food Chem.* **44**, 2680–2685 (1996).

53. P. Hammer and W. Heeschen, "Antibody-Capture-Immunoassay for the Detection of Enrofloxacin in Raw Milk," *Milchwissenschaft* **50**, 513–514 (1995).

54. J. Ara et al., "Dot-ELISA for the Rapid Detection of Gentamicin in Milk," *J. Clin. Laboratory Analysis* **9**, 320–324 (1995).

55. B. M. Medina et al., "Evaluation of Commercial Immunochemical Assays for Determination of Sulfamethazine in Milk," *J. Food Protection* **55**, 330–332 (1992).

56. U. Samarajeewa et al., "Application of Immunoassays in the Food Industry," *CRC Reviews in Food Science and Nutrition* **29**, 403–434 (1991).

57. Y. Takahashi et al., "Enzyme Immunoassay of Sulfamethoxazole in Chicken Tissues: Interlaboratory Study," *J. Veterinary Medical Sci.* **56**, 1207–1208 (1994).

58. T. Saesmaa, "Identification and Purity Determination of Benzathine and Embonate Salts of Some Beta-Lactam Antibiotics by Thin Layer Chromatography, *J. Chromatography* **463**, 469–473 (1989).

59. J. Sherma et al., "Spectrometric and Thin-Layer Chromatographic Quantification of Sulfathiazole Residues in Honey," *J. Chromatography* **463**, 229–233 (1989).

60. B. Wyhowski de Bukanski, J. M. Degroodt, and H. Beernaert, "A Two-Dimensional High Performance Thin-Layer Chromatographic Screening Method for Sulphonamides in Animal Tissues," *Zeitschrift für Lebensmittel-Untersuchung und-Forschung* **187**, 242–245 (1988).

61. H. Oka et al., "Improvement of Chemical Analysis of Antibiotics X. Determination of Eight Tetracyclines Using Thin Layer and High Performance Liquid Chromatography," *J. Chromatography* **393**, 285–296 (1987).

62. Y. Ikai et al., "Improvement of Chemical Analysis of Antibiotics. XIII. Systematic Simultaneous Analysis of Residual Tetracyclines in Animal Tissues Using Thin-Layer and High Performance Liquid Chromatography," *J. Chromatography* **411**, 313–323 (1987).

63. R. L. Ellis, "Less Expensive and Faster Than Laboratory Tests," *Association of Official Analytical Chemists Newsletter, The Referee* **13**, 9 (1989).

64. S. F. Sunderlof, "Drug and Chemical Residues in Livestock," *Veterinarian Clinical North American Food Animal Plant* **4**, 411–419 (1989).

65. I. Schwaiger, R. Schuch, and H. Hoertner, "TLC/Bioautographic Analysis of Antibiotic Residues in Poultry Meat and Milk," *Current Status and Future Trends, Proceedings of Euro Food Chem VIII*, Vol. 3, Vienna, 1995, pp. 727–731.

66. C. Stubbs and I. Kanfer. "High Performance Liquid Chromatography of Erythromycin Propionyl Ester and Erythromycin Base in Biological Fluids," *J. Chromatography* **427**, 93–101 (1988).

67. M. Margosis, "Quantitative Liquid Chromatography of Ampicillin: Collaborative Study," *Journal of the Association of Official Analytical Chemists* **70**, 206–212 (1987).

68. K. Tyczkowska and A. L. Aronson, "Ion-Pair Liquid Chromatographic Determination of Some Penicillins in Canine and Equine Sera," *Journal of the Association of Official Analytical Chemists* **71**, 773–775 (1988).

69. W. A. Moats and R. Harik-Khan, "Liquid Chromatographic Determination of β-Lactam Antibiotics in Milk: A Multiresidue Approach," *J. AOAC Int.* **78**, 49–54 (1995).

70. G. Knupp, G. Bugl-Kreickmann, and C. Commichaw, "A Method for the Verification of Chloramphenicol Residues in Animal Tissues and Eggs by High Performance Liquid Chromatography with Radioimmunological Detection (HPLC-RIA)," *Zeitschrift für Lebensmittel-Untersuchung und-Forschung* **184**, 390–391 (1987).

71. M. F. Pochard, G. Burger, M. Chevalier, and E. Gleizes, "Determination of Chloramphenicol Residues with Reverse-Phase High Pressure Liquid Chromatography. Use of a Pharmacokinetic Study in Rainbow Trout with Confirmation by Mass Spectrometry," *J. Chromatography* **490**, 315–323 (1987).
72. B. Delepine, D. Hurtaud-Pessel, and P. Saunders, "Multiresidue Method for Confirmation of Macrolide Antibiotics in Bovine Muscle by Liquid Chromatography/Mass Spectrometry," *J. AOAC Int.* **79**, 397–404 (1996).
73. M. M. Aerts, W. M. Beek, and U. A. Brinkman, "Monitoring of Veterinary Drug Residues by a Combination of Continuous Flow Techniques and Column-Switching High Performance Liquid Chromatography. I. Sulfonamides in Egg, Meat and Milk Using Post-Column Derivatization with Dimethylaminobenzaldehyde," *J. Chromatography* **435**, 97–112 (1988).
74. R. Malisch, "Multi-Method for the Determination of Residues of Chemotherapeutics, Antiparasitics and Growth Promoters in Foodstuffs of Animal Origin. Part I. General Procedure and Determination of Sulfonamides," *Zeitschrift für Lebensmittel-Untersuchung und-Forschung* **182**, 385–399 (1986).
75. G. D. Paulson, A. D. Mitchell, and R. G. Zaylskie, "Identification and Quantitation of Sulfamethazine Metabolites by Liquid Chromatography and Gas Chromatography-Mass Spectrometry," *Journal of the Association of Official Analytical Chemists* **68**, 1000–1006 (1985).
76. W. A. Moats, "Determination of Tetracycline Antibiotics in Tissues and Blood Serum of Cattle and Swine by High Performance Liquid Chromatography," *J. Chromatography* **358**, 253–259 (1986).
77. F. Kondo, S. Morikawa, and S. Tateyama, "Simultaneous Determination of Six Tetracyclines in Bovine Tissue, Plasma and Urine by Reverse-Phase High Performance Liquid Chromatography," *J. Food Protection* **52**, 41–44 (1988).
78. I. Norlander, H. Johnsson, and B. Österdahl, "Oxytetracycline Residues in Rainbow Trout Analyses by a Rapid HPLC Method," *Food Additives and Contaminants* **4**, 291–295 (1987).
79. W. An-Bang, Y.-M. Chen, and C.-Y. Chen, "Simultaneous Determination of Sulfa Drugs and Tetracyclines in Feeds and Animal Tissues by High Performance Liquid Chromatography," *J. Food and Drug Analysis* **2**, 297–310 (1994).
80. S. Kawata et al., "Liquid Chromatographic Determination of Oxytetracycline in Swine Livers," *J. AOAC Int.* **79**, 1463–1465 (1997).
81. D. W. Hill and K. J. Langner, "HPLC Photodiode Array UV Detection for Toxicological Drug Analysis," *J. Liquid Chromatography* **10**, 377–409 (1987).
82. N. E. Larsen, K. Marinelli, and A. M. Heilesen, "Determination of Gentamicin in Serum Using Liquid Column Chromatography," *J. Chromatography* **221**, 182–187 (1980).
83. K. Tsuji, J. F. Goetz, W. Van Meter, and K. A. Gusciora, "Normal-Phase High Performance Liquid Chromatographic Determination of Neomycin Sulfate Derivatized with 1-Fluoro-2,4-dinitrobenzene," *J. Chromatography* **175**, 141–152 (1979).
84. K. Tsuji and J. H. Robertson, "Gas-Liquid Chromatographic Determination of Neomycins B and C," *Analytical Chem.* **41**, 1332–1335 (1969).
85. K. Tsuji and J. H. Robertson, "Formation of Trimethylsilyl Derivatives of Tetracyclines for Separation and Quantitation by Gas-Liquid Chromatography," *Analytical Chem.* **45**, 2136–2140 (1973).
86. S. J. Stout, W. A. Stellar, A. J. Manuel, M. O. Poeppel, and A. R. DaCumha, "Confirmatory Method for Sulfamethazine Residues in Cattle and Swine Tissues Using Gas Chromatography-Chemical Ionization Mass Spectrometry," *Journal of the Association of Official Analytical Chemists* **67**, 142–144 (1984).
87. M. V. Novotny, "Recent Developments in Analytical Chromatography," *Science* **246**, 51–57 (1989).
88. J. Keever, R. D. Voykser, and K. L. Tyczkowska, "Quantitative Determination of Ceftiofur in Milk by Liquid Chromatography–Mass Spectrometry," *J. Chromatography* **794**, 57–62 (1998).
89. D. N. Helller, "Determination of Pirlimycin Residue in Milk by Liquid Chromatographic Analysis of the 9-Fluorenylmethyl Chloroformate Derivative," *J. AOAC Int.* **80**, 975–981 (1997).
90. L. A. van Ginkel et al., "Development and Validation of a Gas Chromatographic–Mass Spectrometric Procedure for the Identification and Quantification of Residues of Chloramphenicol," *Analytica Chimica Acta* **237**, 61–69 (1990).
91. T. Nagata and H. Oka, "Detection of Residual Chloramphenicol, Florfenicol, and Thiamphenicol in Yellowtail Fish Muscles by Capillary Gas Chromatography–Mass Spectrometry," *J. Agric. and Food Chem.* **44**, 1280–1284 (1996).

AUSTIN R. LONG
STEVEN A. BARKER
Louisiana State University
Baton Rouge, Louisiana

ANTIMICROBIAL COMPOUNDS

Food antimicrobials are chemical compounds added to or present in foods for the purpose of retarding microbial growth or killing microorganisms. The major targets for antimicrobials are bacteria, molds, and yeasts that are either pathogenic or cause spoilage of foods. The effectiveness of food antimicrobials against viruses and parasites carried by foods is less well characterized. Food antimicrobials are sometimes referred to as food preservatives; however, the latter include food additives that are antimicrobials, antibrowning agents, and antioxidants. Under normal use conditions, food antimicrobials are bacteriostatic or fungistatic rather than bactericidal or fungicidal. The former indicates inhibition of growth of cells while the latter indicates killing of a population. Bacteriostasis is often reversible. Because food antimicrobials are generally static in nature, they will not preserve a food indefinitely. Depending on storage conditions, the food product eventually spoils or becomes hazardous. In addition, food antimicrobials are normally not capable of concealing spoilage of a food product. Rather, the food remains wholesome during its extended shelf life. Food antimicrobials are often used in combination with other food preservation procedures such as heat or refrigeration.

The effectiveness of food antimicrobials depends on many factors, including those related to the target microorganisms, characteristics of the food product, the storage environment, and processing of the food (1). Microbial factors that affect antimicrobial activity include inherent resistance of a microorganism, initial number, growth stage, cellular composition (eg, gram reaction), previous exposure to stress, and injury. Important factors affecting activity associated with the food product include pH, oxidation reduction potential, and water activity. pH is the most important factor influencing the effectiveness of many food antimicrobials. Antimicrobials that are weak acids are most effective in their undissociated or protonated form because they are able to penetrate the cytoplasmic membrane of a microorganism more effectively in this form. Therefore, the pK_a value of these compounds is important in selecting a particular compound for an application. The lower the pH of a food product, the greater the proportion of acid in the undissociated form and the greater the antimicrobial activity. Storage factors affecting antimicrobial activity include time, temperature, and atmosphere. Processing of foods may lead to shifts in microflora and reduction in microbial numbers. Most of the preceding factors influence microbial inhibition or inactivation in an interactive manner.

Cellular targets of food antimicrobials may include the cell wall, cytoplasmic membrane, metabolic enzymes, protein synthesis, or genetic systems. The exact mechanisms for most food antimicrobials are not known or are not well defined. These compounds likely have multiple targets with concentration-dependent thresholds for inactivation or inhibition. A given target is important in an inhibitor's overall mechanism only when its sensitivity is within the range of the antimicrobial concentration that inhibits growth (2).

Following are discussions of compounds that are used traditionally in foods as antimicrobials including salt (sodium chloride), organic acids, nitrites, parahydroxybenzoic acid esters, sulfites, and dimethyl dicarbonate. In addition, selected compounds that are approved as food additives for other uses but have antimicrobial activity as well as compounds that occur naturally in animals, plants, and from microorganisms and have a potential for use as food antimicrobials are addressed.

SODIUM CHLORIDE

Sodium chloride (NaCl), or common salt, is probably the oldest known food preservative. Few present-day foods are preserved directly by high concentrations of NaCl. Rather salt is used primarily as an adjunct to other processing methods such as canning or curing (3). In general, foodborne pathogenic bacteria are inhibited by a water activity of 0.92 or less (equivalent to a NaCl concentration of 13% w/v). The exception is *Staphylococcus aureus*, which has a minimum water activity for growth of 0.83 to 0.86. Another relatively salt-tolerant foodborne pathogen is *Listeria monocytogenes*, which can survive in saturated salt solutions at low temperatures. Fungi are more tolerant to low water activity than bacteria. The minimum for growth of xerotolerant fungi is 0.61 to 0.62, but most are inhibited by 0.85 or lower (4).

The antimicrobial activity of sodium chloride is related to its ability to reduce water activity (a_w). As the water activity of the external medium is reduced, cells are subjected to osmotic shock and rapidly lose water through plasmolysis. During plasmolysis, a cell ceases to grow and either dies or remains dormant. To resume growth, the cell must reduce its intracellular water activity (5). In addition to osmotic influence on growth, other possible mechanisms of sodium chloride inhibition include limiting oxygen solubility, alteration of pH, toxicity of sodium and chloride ions, and loss of magnesium ions (6).

ORGANIC ACIDS AND ESTERS

Many organic acids are used as food additives, but not all have antimicrobial activity. The most effective antimicrobials are acetic, lactic, propionic, sorbic, and benzoic acids. The activity of organic acids is related to pH and the undissociated form of the acid is primarily responsible for antimicrobial activity (7). The use of organic acids is generally limited to foods with a pH less than 5.5, since most organic acids have pK_a's of pH 3.0 to 5.0 (7). Another factor affecting potential activity is polarity. This relates both to the ionization of the molecule and contribution of any alkyl side groups or hydrophobic parent molecules. Antimicrobials must be lipophilic to attach and pass through the cell membrane but also be soluble in the aqueous phase (8).

The mechanism of action of organic acids and their esters has some common elements. As already stated, in the undissociated form, organic acids can penetrate the cell membrane lipid bilayer more easily. Once inside the cell, the acid dissociates because the cell interior has a higher pH than the exterior (9). Bacteria maintain internal pH near neutrality to prevent conformational changes to the cell structural proteins, enzymes, nucleic acids, and phospholipids. Protons generated from intracellular dissociation of the organic acid acidify the cytoplasm and must be extruded to the exterior. This requires energy in the form of ATP, which will eventually deplete cellular energy.

Acetic Acid/Acetates

Acetic acid (pK_a = 4.75), the primary component of vinegar, and its sodium, potassium, and calcium salts are some of the oldest food antimicrobials. Derivatives of acetic acid including diacetate salts and dehydroacetic acid have also been used as food antimicrobials. In contrast to most organic acids, acetic acid is generally more effective against yeasts and bacteria than against molds (10). Bacteria inhibited include *Bacillus*, *Clostridium*, *Listeria monocytogenes*, *Staphylococcus aureus*, and *Salmonella*. Only *Acetobacter* species (microorganisms involved in vinegar production), lactic acid bacteria, and butyric acid bacteria are tolerant to acetic acid (7,11). Some yeasts and molds are sensitive to acetic acid and include species of *Aspergillus*, *Penicillium*, *Rhizopus*, and some strains of *Saccharomyces* (7).

Acetic acid and its salts have shown variable success as antimicrobials in food applications. Acetic acid can increase poultry shelf life and decreases the heat resistance of *Salmonella newport*, *Salmonella typhimurium*, and *Campylobacter jejuni* in poultry scald tank water (12,13). The compound has been used as a spray sanitizer at 1.5 to 2.5% on meat carcasses and as a dip for beef or lamb (14).

Two-percent acetic acid affected large-scale reductions in the number of viable *Escherichia coli* O157:H7 on beef after seven days at 5°C (15). Sodium acetate has been shown to increase the shelflife of meat and fish products and inhibits the molds, *Aspergillus flavus*, *A. fumigatus*, *A. niger*, *A. glaucus*, *Penicillium expansum*, and *Mucor pusillus* (7). Sodium diacetate inhibits mold growth in cheese spread and is an effective inhibitor of rope-forming bacteria (*Bacillus subtilis*) in baked goods (7). It is also inhibitory to *L. monocytogenes*, *E. coli*, *Pseudomonas fluorescens*, *S. enteritidis*, and *Shewanella putrefaciens*, but not *S. aureus*, *Yersinia enterocolitica*, *P. fragi*, *Enterococcus faecalis*, or *Lactobacillus* (7,16). Dehydroacetic acid has a high pK_a of 5.27 and is therefore active at higher pH values. It is inhibitory to bacteria at 0.1 to 0.4% and fungi at 0.005 to 0.1% (7).

Acetic acid is used commercially in baked goods, cheeses, condiments and relishes, dairy product analogues, fats and oils, gravies and sauces, and meats. The sodium and calcium salts are used in breakfast cereals, candy, cheeses, fats and oils, gelatin, jams and jellies, meats, snack foods, sauces, and soup mixes (7). Sodium diacetate is used in baked goods, candy, cheese spreads, gravies, meats, sauces and soup mixes (7).

Benzoic Acid/Benzoates

Benzoic acid and sodium benzoate were the first antimicrobial compounds permitted in foods by the U.S. Food and Drug Administration (FDA). Benzoic acid occurs naturally in cranberries, plums, prunes, apples, strawberries, cinnamon, cloves, and most berries. The undissociated form of benzoic acid ($pK_a = 4.19$) is the most effective antimicrobial agent; therefore, the most effective pH range is 2.5 to 4.5.

Benzoic acid and sodium benzoate are used primarily as antifungal agents. The inhibitory concentration of benzoic acid at pH less than 5.0 against most yeasts ranges from 20 to 700 μg/ml, whereas for molds it is 20 to 2000 μg/mL (11,17). Fungi including *Byssochlamys nivea*, *Pichia membranaefaciens*, *Talaromyces flavus*, and *Zygosaccharomyces bailii* are resistant to benzoic acid (18). While bacteria associated with food poisoning including *Bacillus cereus*, *Listeria monocytogenes*, *Staphylococcus aureus*, and *Vibrio parahaemolyticus* are inhibited by 1000 to 2000 μg/ml undissociated acid, the control of many spoilage bacteria requires much higher concentrations (11). Benzoic acid at 0.1% is effective in reducing viable *E. coli* O157:H7 in apple cider (pH 3.6–4.0) by 3 to 5 logs in seven days at 8°C (19).

Sodium benzoate is used as an antimicrobial at up to 0.1% in carbonated and still beverages, syrups, cider, margarine, olives, pickles, relishes, soy sauce, jams, jellies, preserves, pie and pastry fillings, fruit salads, and salad dressings and in the storage of vegetables (17).

Lactic Acid/Lactates

Lactic acid ($pK_a = 3.79$) is produced naturally during fermentation of foods by lactic acid bacteria including *Lactococcus*, *Lactobacillus*, *Streptococcus*, *Pediococcus*, *Leuconostoc*, and *Carnobacterium*. While the acid and salts act as preservatives in food products, their primary uses are as pH control agents and flavorings. Lactic acid inhibits *Clostridium botulinum*, *C. perfringens*, *C. sporogenes*, *Listeria monocytogenes*, *Salmonella*, *Staphylococcus aureus*, and *Yersinia enterocolitica* (7,20,21). Lactic acid at 1–2% reduces *Enterobacteriaceae* and aerobic mesophilic microorganisms on beef, veal, pork, and poultry and delays growth of spoilage microflora during long-term storage of products (7). Sodium lactate (2.5–5.0%) inhibits *C. botulinum*, *C. sporogenes*, *L. monocytogenes*, and spoilage bacteria in various meat products (7,22,23).

Propionic Acid/Propionates

Up to 1% propionic acid ($pK_a = 4.87$) is produced naturally in Swiss cheese by *Propionibacterium freudenreichii* ssp. *shermanii*. Propionic acid and sodium, potassium, and calcium propionates are used primarily against molds; however, some yeasts and bacteria are also inhibited. Propionates (0.1–5.0%) retard the growth of *E. coli*, *Staphylococcus aureus*, *Sarcina lutea*, *Salmonella*, *Proteus vulgaris*, *Lactobacillus plantarum*, and *Listeria monocytogenes*, and the yeasts *Candida* and *Saccharomyces cerevisiae* (7). Rope-forming bacteria in bread dough (*B. subtilis*) are inhibited by propionic acid at pH 5.6 to 6.0 (7).

Propionic acid and propionates are used as antimicrobials in baked goods and cheeses. Propionates may be added directly to bread dough because they have no effect on the activity of baker's yeast (7). There is no limit to the concentration of propionates allowed in foods, but amounts used are generally less than 0.4% due to sensory changes that might occur (7).

Sorbic Acid/Sorbates

Sorbic acid occurs naturally in the berries of the mountain ash tree (rowanberry) (24). As with other organic acids, the antimicrobial activity of sorbic acid is greatest when the compound is in the undissociated state. With a pK_a of 4.75, activity is greatest at a pH less than 6.0 to 6.5.

Sorbates are probably the most well characterized of all food antimicrobials as to their spectrum of action and inhibit bacteria, yeasts, and molds at concentrations of 0.05 to 0.3% (24). Food-related yeasts and molds inhibited by sorbates include *Byssochlamys*, *Candida*, *Saccharomyces*, and *Zygosaccharomyces* and *Aspergillus*, *Fusarium*, *Geotrichum*, and *Penicillium*, respectively (24). Sorbates inhibit the growth of yeasts and molds in microbiological media, cheeses, fruits, vegetables and vegetable fermentations, sauces, and meats. Sorbates inhibit growth and mycotoxin production by *Aspergillus flavus*, *A. parasiticus*, *Byssochlamys nivea*, *Penicillium expansum*, and *P. patulum* (24). Some of the genera of bacteria inhibited by sorbate include *Acinetobacter*, *Aeromonas*, *Bacillus*, *Campylobacter*, *Clostridium*, *Escherichia*, *Lactobacillus*, *Listeria*, *Pseudomonas*, *Salmonella*, *Staphylococcus*, *Vibrio*, and *Yersinia* (24).

A number of *Penicillium*, *Saccharomyces*, and *Zygosaccharomyces* species can grow in the presence of and degrade potassium sorbate (24). Sorbates may be degraded through a decarboxylation reaction resulting in the formation of 1,3-pentadiene, a compound having a kerosene-like or hydrocarbon-like odor.

Sorbate inhibits the growth of many pathogenic bacteria in foods, including *Salmonella* and *S. aureus* in sausage; *S. aureus* in bacon; *Vibrio parahaemolyticus* in seafood; *Salmonella*, *S. aureus*, and *E. coli* in poultry; *Yersinia enterocolitica* in pork; and *Salmonella typhimurium* in milk and cheese (24,25). In addition, the compound inhibits growth of the spoilage bacteria, *Pseudomonas putrefaciens* and *P. fluorescens*, histamine production by *Proteus morgani* and *K. pneumoniae* and listeriolysin O production by *L. monocytogenes* (24,26). Sorbates are effective anticlostridial agents in cured meats and other meat and seafood products. The compound prevents spores of *C. botulinum* from germinating and forming toxin in beef, pork, poultry, and soy protein frankfurters and emulsions and bacon (24).

Sorbate is applied to foods by direct addition, dipping, spraying, dusting, or incorporation into packaging. Baked goods, icing, fruit, and cream fillings can be protected from yeast and molds through the use of 0.05 to 0.10% potassium sorbate applied either as a spray after baking or by direct addition (24). Sorbates may be used in or on beverages, jams, jellies, preserves, margarine, chocolate syrup, salads, dried fruits, dry sausages, salted and smoked fish, cheeses, and in various lactic acid fermentations (24).

Fumaric Acid/Fumarates

Fumaric acid has been used to prevent the malolactic fermentation in wines and as an antimicrobial agent in wines (7). The compound is lethal to ascospores of *Talaromyces flavus* and *Neosartorya fischeri* (7). Esters of fumaric acid (monomethyl, dimethyl, and ethyl) at 0.15 to 0.2% have been tested as a substitute or adjunct for nitrate in bacon. Fumaric acid esters inhibit *Clostridium botulinum* toxin formation in bacon, and the methyl, dimethyl, ethyl, and diethyl fumarates inhibit fungal growth in tomato juice and on bread (7).

Citric Acid/Citrates

Although citric acid is not used directly as an antimicrobial, it has activity against some molds and bacteria. Citric acid retards growth and toxin production by *Aspergillus parasiticus* and *A. versicolor*, but not *Penicillium expansum* (7). It is inhibitory to *Salmonella* in media and on poultry carcasses, to growth and toxin production by *C. botulinum* in shrimp and tomato products, to *S. aureus* in microbiological medium and to flat sour bacteria isolated from tomato juice (7). The mechanism of citrate inhibition has been theorized to be related to its ability to chelate metal ions (27). However, Buchanan and Golden (28) found that, while undissociated citric acid is inhibitory against *Listeria monocytogenes*, the dissociated molecule protects the microorganism. They theorized that this protection is due to chelation by the anion.

Fatty Acid Esters

Certain fatty acid esters have antimicrobial activity in foods. One of the most effective of the fatty acid esters is glyceryl monolaurate (monolaurin) (29,30). Monolaurin is effective as an antimicrobial against Gram-positive bacteria including *Bacillus*, *Listeria monocytogenes*, *Micrococcus*, and *Staphylococcus aureus* but generally ineffective against Gram-negative bacteria (31–33). Presence of ethylenediaminetetraacetic acid (EDTA) expands the activity spectrum of monolaurin to include Gram-negative bacteria such as *Salmonella, E. coli* O157:H7, and *Vibrio* and decreases the minimum inhibitory concentrations against Gram-positive strains (29,34). Monolaurin is inhibitory to molds and yeasts including *Aspergillus*, *Alternaria*, *Candida*, *Cladosporium*, *Penicillium*, and *Saccharomyces*. Monolaurin is effective against *L. monocytogenes* in cottage cheese, Camembert cheese, yogurt, and meat products (32,33).

NITRITES

Sodium ($NaNO_2$) and potassium (KNO_2) nitrite have a specialized use in cured meat products. Meat curing utilizes salt, sugar, spices, and ascorbate or erythorbate along with nitrite. In addition to serving as an antimicrobial, nitrite has many functions in cured meats, including formation of characteristic cured meat color, contribution to the flavor and texture, and as an antioxidant. The primary use for sodium nitrite as an antimicrobial is to inhibit *Clostridium botulinum* growth and toxin production in cured meats. In association with other components in the curing mix, such as salt and reduced pH, nitrite inhibits outgrowth of spores of *C. botulinum* and other clostridia. Nitrite inhibits bacterial spore formers by inhibiting outgrowth of the germinated spore (35). The effectiveness of nitrite depends on reduced pH, salt concentration, presence of ascorbate and isoascorbate, storage and processing temperatures, and initial inoculum size.

Meat products that may contain nitrites include bacon, bologna, corned beef, frankfurters, luncheon meats, ham, fermented sausages, shelf-stable canned cured meats, and perishable canned cured meat (eg, ham). Nitrite is also used in a variety of fish and poultry products. The concentration used in these products is specified by governmental regulations but is generally limited to 156 ppm (mg/kg) for most products and 100 to 120 ppm (mg/kg) in bacon. Sodium erythorbate or isoascorbate is required in products containing nitrites as a cure accelerator and as an inhibitor to the formation of nitrosamines, carcinogenic compounds formed by reactions of nitrite with secondary or tertiary amines. Sodium nitrate is used in certain European cheeses to prevent spoilage by *C. tyrobutyricum* or *C. butyricum* (35).

PARAHYDROXYBENZOIC ACID ESTERS (PARABENS)

Alkyl esters of *p*-hydroxybenzoic acid (parabens) have been known to possess antimicrobial activity since the 1920s. Esterification of the carboxyl group of benzoic acid allows the molecule to remain undissociated up to pH 8.5, giving the parabens an effective range of pH 3.0 to 8.0 (36). In most countries, the methyl, propyl, and heptyl parabens are allowed for direct addition to foods as antimicrobials while the ethyl and butyl esters are approved in a limited number of countries.

The antimicrobial activity of *p*-hydroxybenzoic acid esters is, in general, directly proportional to the chain length of the alkyl component (8,36). As the alkyl chain length of the parabens increases, inhibitory activity generally increases. Increasing activity with decreasing polarity is more evident against Gram-positive than against Gram-negative bacteria (8). Parabens are generally more active against molds and yeast than against bacteria (8).

To take advantage of their respective solubility and increased activity, methyl and propyl parabens are normally used in a combination of 2–3:1 (methyl:propyl). The compounds may be incorporated into foods by dissolving in water, ethanol, propylene glycol, or the food product itself. The *n*-heptyl ester is used in fermented malt beverages (beers) and noncarbonated soft drinks and fruit-based beverages. Parabens are used in a variety of foods including baked goods, beverages, fruit products, jams and jellies, fermented foods, syrups, salad dressings, and wine (8).

SULFITES

Sulfur dioxide (SO_2) and its salts (potassium sulfite [K_2SO_3], sodium sulfite [Na_2SO_3], potassium bisulfite [$KHSO_3$], sodium bisulfite [$NaHSO_3$], potassium metabisulfite [$K_2S_2O_5$], and sodium metabisulfite [NaS_2O_5]) have been used as disinfectants since the time of the ancient Greeks and Romans (37). Although sulfites now have multiple uses as food additives, their original purpose was as an antimicrobial. Sulfites are used primarily in fruit and vegetable products to control spoilage and fermentative yeasts, and molds on fruits and fruit products (eg, wine), acetic acid bacteria, and malolactic bacteria (37). In addition to use as antimicrobials, sulfites act as antioxidants and inhibit enzymatic and nonenzymatic browning in a variety of foods. Their primary application is in fruits and vegetable products, but they are also used to a limited extent in meats.

The most important factor impacting the antimicrobial activity of sulfites is pH. The inhibitory effect of sulfites is most pronounced when the acid or $SO_2 \cdot H_2O$ is in the undissociated form. As the pH decreases, the proportion of $SO_2 \cdot H_2O$ increases and the bisulfite (HSO_3^-) ion concentration decreases. Therefore, their most effective pH range is less than 4.0. Sulfites show increased effectiveness at low pH because the unionized sulfur dioxide can pass across the cell membrane in this form.

Sulfites, especially as the bisulfite ion, are very reactive. These reactions not only determine the mechanism of action of the compounds, they also influence antimicrobial activity. For example, sulfites form addition compounds (α-hydroxysulfonates) with aldehydes and ketones.

Sulfur dioxide is fungicidal even in low concentrations against yeasts and molds. The inhibitory concentration range of sulfur dioxide is 0.1 to 20.2 μg/ml for *Saccharomyces*, *Zygosaccharomyces*, *Pichia*, *Hansenula*, and *Candida* species (37). Sulfur dioxide at 25 to 100 μg/ml inhibits *Byssochlamys nivea* growth and patulin production in grape and apple juices (38). Against bacteria, sulfur dioxide is more inhibitory to Gram-negative rods than to Gram-positive rods. Gram-negative bacteria susceptible to sulfites include *Salmonella*, *E. coli*, *Citrobacter*, *Yersinia enterocolitica*, *Enterobacter*, *Serratia marcescens*, and *Hafnia* (39).

Sulfur dioxide is used to control the growth of undesirable microorganisms in fruits, fruit juices, wines, sausages, fresh shrimp, and acid pickles and during extraction of starches (37). It is added at 50 to 100 mg/L to expressed grape juices used for making wines to inhibit molds, bacteria, and undesirable yeasts. At appropriate concentrations, sulfur dioxide does not interfere with wine yeasts or with the flavor of wine. The optimum level of sulfur dioxide (50–75 mg/L) is maintained to prevent postfermentation changes by microorganisms. In a few countries, sulfites may be used to inhibit the growth of microorganisms on fresh meat and meat products. Sulfite or metabisulfite added in sausages is effective in delaying the growth of molds, yeast, and salmonellae during storage at refrigerated or room temperature (39). Sulfur dioxide restores a bright color but may give a false impression of freshness.

DIMETHYL DICARBONATE

Dimethyl dicarbonate (DMDC) is a colorless liquid that is slightly soluble in water. The compound is very reactive with many substances, including water, ethanol, alkyl and aromatic amines, and sulfhydryl groups (40). The primary target microorganisms for DMDC are yeasts, including *Saccharomyces*, *Zygosaccharomyces*, *Rhodotorula*, *Candida*, *Pichia*, *Torulopsis*, *Torula*, *Endomyces*, *Kloeckera*, and *Hansenula*. The compound is also bactericidal at 30 to 400 mg/L to a number of species including *Acetobacter pasteurianus*, *E. coli*, *Pseudomonas aeruginosa*, *Staphylococcus aureus*, several *Lactobacillus* species, and *Pediococcus cerevisiae* (40). The compound has been shown to be bactericidal against *E. coli* O157:H7 in apple cider (41). Molds are generally more resistant to DMDC than yeasts or bacteria.

PHENOLIC ANTIOXIDANTS

Phenolic antioxidants including, butylated hydroxyanisole (BHA), butylated hydroxytoluene (BHT), propyl gallate (PG) and tertiary butylhydroquinone (TBHQ), are used in foods primarily to delay autoxidation of unsaturated lipids. The first report on the antibacterial effectiveness of BHA was that of Chang and Branen (42) in which *E. coli*, *Salmonella typhimurium*, and *Staphylococcus aureus* were inhibited in nutrient broth. Subsequent studies generally demonstrate that Gram-positive bacteria are more susceptible to BHA than Gram-negative bacteria. BHT is generally less effective than other phenolic antioxidants (43). TBHQ is an extremely effective inhibitor of Gram-positive bacteria including *S. aureus* and *Listeria monocytogenes* at concentrations generally less than 64 μg/mL (43).

BHA inhibits *Aspergillus flavus*, *A. parasiticus*, *Penicillium*, *Geotrichum*, *Byssochlamys*, and *Saccharomyces cerevisiae* in microbiological media (43). In addition to growth inhibition, BHA inhibits production of the mycotoxins (43).

A number of studies have been carried out to determine the antimicrobial effectiveness of phenolic antioxidants in foods (43). In nearly all studies, the concentration of phenolic antioxidants required for inhibition in a food, especially a meat product, is significantly higher than that needed for *in vitro* inhibition. This is probably because the presence of lipid or protein dramatically decreases the activity of phenolic antioxidants due to binding (44). Application studies in lower fat and protein products have shown more promise.

PHOSPHATES

Some phosphate compounds, including sodium acid pyrophosphate (SAPP), tetrasodium pyrophosphate (TSPP), sodium tripolyphosphate (STPP), sodium tetrapolyphosphate, sodium hexametaphosphate (SHMP), and trisodium phosphate (TSP), have variable levels of antimicrobial activity in foods (45). Gram-positive bacteria are generally more susceptible to phosphates than Gram-negative bacteria. TSPP, SAPP, STPP, and SHMP have been shown to inhibit *Bacillus subtilis*, *Enterococcus faecalis*, *Clostridium sporogenes*, *C. bifermentans*, and *Staphylococcus aureus* (45,46). Sodium polyphosphates at 1% inhibited lag and generation times of *Listeria monocytogenes* in BHI broth, especially in the presence of NaCl (47). Wagner and Busta (48) found that SAPP has no effect on the growth of *C. botulinum* but delayed or prevented toxicity to mice. It was theorized that this was due to binding of the toxin molecule or inactivation of the protease responsible for protoxin activation.

Phosphate derivatives also have antimicrobial activity in food products (49). SAPP, SHMP, or polyphosphates enhance the effect of nitrite, pH, and salt against *C. botulinum* (45). Phosphates, sodium chloride, reduced water activity, water content, reduced pH, and lactic acid interact to prevent the outgrowth of *C. botulinum* in pasteurized process cheese (50). A 10% sodium tetrapolyphosphate dip preserved cherries against the fungal growth by *Penicillium*, *Rhizopus*, and *Botrytis* (45). Various phosphate salts have antimicrobial activity against rope-forming *Bacillus* in bread and *Salmonella* in pasteurized egg whites (51). Trisodium phosphate (TSP) at levels of 8 to 12% reduces pathogens, especially *Salmonella*, on poultry most likely due to a physical removal process or high pH (11–12) (52,53).

Several mechanisms have been suggested for bacterial inhibition by polyphosphates. The ability of polyphosphates to chelate metal ions, such as magnesium, appears to play an important role in their antimicrobial activity (45). Knabel et al. (54) stated that the chelating ability of polyphosphates is responsible for growth inhibition of *B. cereus*, *L. monocytogenes*, *S. aureus*, *Lactobacillus*, and *Aspergillus flavus*. In addition, inhibition is reduced at lower pH due to protonation of the chelating sites on the polyphosphates. It was concluded that polyphosphates inhibited Gram-positive bacteria and fungi by removal of essential cations from binding sites on the cell walls of these microorganisms (54).

NATURALLY OCCURRING COMPOUNDS AND SYSTEMS

Nisin

Nisin, a polypeptide produced by *Lactococcus lactis* spp. *lactis*, was isolated, characterized, and named by Mattick and Hirsh (55). The peptide has 34 amino acids and a molecular weight of 3,500 Da, however it usually occurs as a dimer. The solubility of the compound depends on the pH of the solution. At pH 2.2, the solubility of nisin is 56 mg/mL, at pH 5.0, 3 mg/mL, and it is less soluble at neutral and alkaline pH. Nisin solution in dilute HCl at pH 2.5 is stable to autoclaving (121°C) with no marked loss of antimicrobial activity. At pH 7.0, inactivation occurs even at room temperature. Nisin remains stable for years in the dry form, but activity is gradually lost in foods. The effectiveness of nisin increases as pH decreases. Nisinase from *Streptococcus thermophilus*, *Lactobacillus plantarum*, other lactic acid bacteria, and certain *Bacillus* species inactivate nisin (55). In addition, resistance to nisin may develop in cells exposed to the compound through alterations of the cell surface or cell membrane (18).

Nisin has a narrow spectrum affecting only Gram-positive bacteria, including lactic acid bacteria, streptococci, bacilli, and clostridia. By itself, it does not generally inhibit Gram-negative bacteria, yeasts, or molds. Nisin is inhibitory to the spore formers *Bacillus* and *Clostridium*, including *Clostridium botulinum*. Nisin concentrations necessary to inhibit *Clostridium botulinum* in brain heart infusion broth were 200, 80, and 20 μg/mL for types A, B, and E, respectively (56). In contrast, the concentration required to inhibit *C. botulinum* in cooked meat medium (CMM) was beyond the highest tested for types A (>200 μg/mL) and B (>80 μg/mL). It was theorized that the higher levels required in CMM were due to binding of the nisin by meat particles. Nisin reduces the heat resistance of spore formers. The sensitivity of vegetative bacteria to nisin varies. *Staphylococcus*, *Enterococcus*, *Pediococcus*, *Leuconostoc*, *Lactobacillus*, and *Listeria monocytogenes* have all been shown to be sensitive to nisin (57–60). Gram-negative bacteria are resistant to nisin activity, but they can be sensitized by disruption of the outer membrane either chemically with chelators, such as EDTA, or mechanically. This expands the spectrum of nisin to Gram-negative pathogens such as *Escherichia coli*, *Salmonella*, *Yersinia* (61).

The application of nisin as a food preservative has been studied extensively. Nisin-producing starter cultures were first used to prevent gas or "blowing" of Swiss-type cheese caused by *Clostridium tyrobutyricum* and *C. butyricum* (62). Nisin has been recommended for use in canned vegetable products to prevent the outgrowth of *Clostridium botulinum* when less severe sterilization conditions are desired or required (61). The compound has been shown to have potential benefit in some meat products, although, as already stated, binding to meat may be a problem. Nisin has been suggested as an adjunct to nitrite in cured meats to prevent the growth of clostridia (61). The compound has been tested as a preservative in seafood, dairy products, vegetables, soups, sauces, beer and ale (61). Nisin was less active against *L. monocytogenes* in milk and ice cream with increasing fat concentrations (63,64). This was probably due to binding of nisin to fat globules.

Nisin is permitted for use in foods in many countries including the U.S. It was approved by the U.S. FDA for use in pasteurized cheese spreads and pasteurized process cheese spread to inhibit the growth of *C. botulinum* at a maximum of 250 ppm in 1988.

Natamycin

Natamycin, a polyene macrolide antibiotic, was first isolated in 1955 from a culture of *Streptomyces natalensis*, a microorganism found in soil from Natal, South Africa. The name natamycin is synonymous with pimaricin, a name used in earlier literature. In the United States, natamycin was approved for use in cheese making as a mold spoilage inhibitor in 1982. Natamycin may be applied to the surface of cuts and slices of cheese by dipping or spraying an aqueous solution containing 200 to 300 ppm. Natamycin does not readily migrate into the cheese and does not adversely affect flavor or appearance.

Natamycin is active against nearly all molds and yeasts but has little or no effect on bacteria or viruses (65). Most molds are inhibited at concentrations of natamycin from 0.5 to 25 μg/mL. Most yeasts are inhibited at natamycin concentrations from 1.0 to 5.0 μg/mL. In addition to fungal growth inhibition, natamycin has been shown to inhibit aflatoxin B_1 production of *Aspergillus flavus* and penicillic acid production by *Penicillium cyclopium* as well as eliminate ochratoxin production by *A. ochraceus* and patulin production by *P. patulum* (66).

Several factors affect the stability and resulting antimycotic activity of natamycin. While pH has no apparent effect on antifungal activity, it does influence stability of the compound. In the pH range of most food products (pH 5–7), natamycin is very stable. Under normal storage conditions, temperature has little effect on natamycin activity when in neutral aqueous suspension. Sunlight, contact with certain oxidants (eg, organic peroxides and sulfhydryl groups), and heavy metals all adversely effect stability of natamycin solutions or suspensions (67).

Natamycin added in the wash water was effective in increasing the shelf life of cottage cheese up to 13.6 days at 4.4°C (68). Adding natamycin to the cottage cheese dressing was even more effective in extending shelf life. At 500 or 1000 μg/mL, natamycin delayed mold growth on cheese for up to six months but did not prevent it completely (69). Natamycin was shown to be effective in preserving Italian cheeses with no detrimental effect on ripening (70).

In addition to dairy products, early work on natamycin suggested it might be useful to inhibit fungal growth on fruit products and poultry. Natamycin was shown to be an effective antifungal agent on strawberries, raspberries, and cranberries and in orange juice (71,72). Natamycin, alone and in combination with chlortetracycline, inhibited yeast growth on chicken stored 12 to 15 days at 4.4°C (73).

Bacteriocins

It has long been known that lactic acid bacteria are antagonistic to other microorganisms through production of antimicrobial compounds such as hydrogen peroxide, acetic acid, lactic acid, and diacetyl. However, these and other food fermentation microorganisms also produce peptide antimicrobial compounds known as bacteriocins. Bacteriocins are bactericidal to susceptible genera or species and are often produced under plasmid control. These compounds are not considered to be antibiotics as they are not used for human disease control, are larger than medical antibiotics, and usually have a narrow spectrum. Potentially useful bacteriocins have been demonstrated from *Lactococcus*, *Lactobacillus*, *Pediococcus*, *Leuconostoc*, *Propionibacterium*, *Bifidobacterium*, and *Carnobacterium*. Although none of the compounds are approved for use in foods, there is great potential for their use in the future. Most of the bacteria that produce bacteriocins being considered as food antimicrobials are used or associated with fermented dairy, meat, and vegetables products. For that reason, the antimicrobials they produce would be assumed to be safe from a toxicological standpoint. Many bacteriocins have been isolated and their effectiveness characterized in microbiological media, but much work remains to be done on determining their effectiveness in foods.

Excluding nisin, the most well characterized bacteriocins are those produced by *Pediococcus* species. *Pediococcus acidilactici* produces pediocin AcH and PA-1, and *P. pentosaceus* produces pediocin A. Pediocin AcH and PA-1 inhibit *Bacillus*, *Brochothrix*, *Clostridium*, *Enterococcus*, *Lactobacillus*, *Leuconostoc*, *Listeria monocytogenes*, other *Pediococcus*, *Propionibacterium*, and *Staphylococcus aureus*. Pediocin A is inhibitory to other lactic acid bacteria and the pathogens *Bacillus cereus*, *Clostridium botulinum*, *C. perfringens*, *L. monocytogenes*, and *S. aureus*. Gram-negative bacteria can be made to be susceptible to pediocins following exposure to sublethal stress treatments such as freezing or heat (61). In foods, pediocin AcH inhibits spoilage bacteria in meats, dairy products, salads, and salad dressings, and it has a variable effect on growth of pathogens inoculated into foods (61). Limitations on the use of pediocins in foods include: (1) it may cause resistance development in some microorganisms; (2) it may select for resistant strains in foods; and (3) food components, especially lipids, could interfere with antimicrobial activity (61).

Several species of *Lactobacillus* have been shown to elucidate bacteriocins. These include *L. acidophilus*, *L. brevis*, *L. casei*, *L. delbrueckii* ssp. *bulgaricus*, *L. fermentum*, *L. helveticus*, *L. plantarum*, and *L. sake* (18,61). Bacteriocins produced by *Leuconostoc* include those produced by *L. carnosum*, *L. dextranicum*, *L. gelidum*, *L. mesenteroides*, and *L. paramesenteroides*. Various of these bacteriocins have shown activity against *C. botulinum*, *L. monocytogenes*, *S. aureus*, and *Yersinia enterocolitica* (61). *Carnobacterium piscicola* and *Bifidobacterium* have shown to produce bacteriocins with activity primarily against gram positive bacteria including *L. monocytogenes*. Milk fermented by *Propionibacterium freudenreichii* ssp. *shermanii* is effective against Gram-negative psychrotrophic bacteria in cottage cheese and spoilage microorganisms in a variety of other foods. Among the inhibitors in this product are propionic acid, lactic acid, and a low-molecular-weight proteinaceous compound (74).

Lysozyme

Lysozyme (1,4-β-*N*-acetylmuramidase) is a 14,600 Da enzyme present in avian eggs, mammalian milk, tears and other secretions, insects, and fish. It is most stable under acidic conditions. The enzyme catalyzes hydrolysis of the β-1,4 glycosidic bonds between *N*-acetylmuramic acid and *N*-acetylglucosamine of the peptidoglycan of bacterial cell walls. This causes cell wall degradation and eventual lysis.

Lysozyme is most active against Gram-positive bacteria, most likely because of the exposed peptidoglycan in the cell wall. The enzyme inhibits *Bacillus stearothermophilus*, *B. cereus*, *Clostridium botulinum*, *C. thermosaccharolyticum*, *C. tyrobutyricum*, *Listeria monocytogenes*, and *Micrococcus lysodeikticus* (18). Variation in susceptibility of Gram-positive bacteria is likely due to the presence of teichoic acids or other materials that bind the enzyme and that certain species have greater proportions of 1,6 or 1,3 glycosidic linkages in the peptidoglycan that are more resistant than the 1,4 linkage (75). For example, certain strains of *L. monocytogenes* are not inhibited by lysozyme unless EDTA is present (76). Lysozyme is less effective against Gram-negative bacteria due to a reduced peptidoglycan content and presence of outer membrane of lipopolysaccharide and lipoprotein. Gram-negative cell susceptibility can be increased by pretreatment with chelators (eg, EDTA) or if cells are subjected to shock (pH, heat, osmotic), drying or freeze–thaw cycling (75).

The most common use of lysozyme in foods is in hard cheeses such as Edam and Gouda to prevent a defect known as late blowing. This defect is characterized by formation of holes in the cheese and the production of unacceptable flavors caused by *C. tyrobutyricum*. In foods other than cheese, lysozyme has been evaluated as an antimicrobial in seafoods, sake, potato salad, sushi, Chinese noodles, creamed custard, and sausage. Lysozyme and lysozyme combined with EDTA may have potential for reducing *Salmonella* on poultry and spoilage microflora from shrimp. Lysozyme was bactericidal to *L. monocytogenes* in shredded cabbage and lettuce, fresh green beans, corn, and carrots, and this activity was enhanced by EDTA (76). The compound was less effective in refrigerated meats and soft cheese. In some studies, no inhibition was demonstrated with EDTA and lysozyme against *Salmonella typhimurium* or *Pseudomonas fluorescens* (77).

Lysozyme is one of the few naturally occurring antimicrobials approved by regulatory agencies for use in foods. In the United States and Europe, lysozyme is allowed to prevent gas formation in cheeses such as Edam and Gouda by *C. tyrobutyricum*. Lysozyme is used to a great extent in Japan to preserve seafood, vegetables, pasta, and salads.

Lactoperoxidase System

Lactoperoxidase (MW = 78,000 Da) is an enzyme that occurs in raw milk, colostrum, saliva and other biological secretions. Bovine milk naturally contains 10 to 60 mg/L of lactoperoxidase (78). This enzyme reacts with thiocyanate (SCN^-) in the presence of hydrogen peroxide and forms antimicrobial compound(s). The components are called the lactoperoxidase system (LPS). Fresh milk contains 1 to 10 mg/L of thiocyanate, which is not always sufficient to activate the LPS (79). Hydrogen peroxide, the third component of the LPS, is not present in fresh milk due to the action of natural catalase, peroxidase, or superoxide dismutase. Approximately 8 to 10 mg/L hydrogen peroxide are required to activate the LPS, which can be added directly, through the action of lactic acid bacteria or through the enzymatic action of xanthine oxidase, glucose oxidase, or sulfhydryl oxidase. In LPS reaction, thiocyanate is oxidized to the antimicrobial hypothiocyanate ($OSCN^-$), which also exists in equilibrium with hypothiocyanous acid (78).

The LPS is more effective against active against Gram-negative bacteria, including *Pseudomonas*, than Gram-positive bacteria. However, it does inhibit both Gram-positive and Gram-negative foodborne pathogens, including *Salmonella*, *Staphylococcus aureus*, *Listeria monocytogenes*, and *Campylobacter jejuni* (78). There is variable activity against catalase-negative microorganisms, including the lactic acid bacteria such as *Lactococcus*, *Lactobacillus*, and *Streptococcus*.

The LPS system can increase the shelf life of raw milk. This could be useful in countries that have poorly developed refrigerated storage systems (78). LPS has also been used as a preservation process in cream, cheese, liquid whole eggs, ice cream, and infant formula.

Lactoferrin and Other Iron-Binding Proteins

Lactoferrin is an iron-binding protein that occurs in milk and colostrum and has potential for use as a food antimicrobial (78). Lactoferrin is a glycoprotein with a molecular weight of around 76,500 Da. Lactoferrin has two iron-binding sites per molecule and, for each Fe^{3+} bound, requires one bicarbonate (HCO_3^-). Lactoferrin is potentially active in milk due to the low iron concentration and presence of bicarbonate (79). The exact biological role of lactoferrin is unknown; however, it may act as a barrier to infection of the nonlactating mammary gland and protect the gastrointestinal tract of the newborn against infection (78).

Lactoferrin is inhibitory to a number of microorganisms including *Bacillus subtilis*, *B. stearothermophilus*, *Listeria monocytogenes*, *Micrococcus*, *E. coli*, and *Klebsiella* (18). There is evidence that the most effective form of lactoferrin is the iron-unsaturated, or apolactoferrin, form (80). Some Gram-negative bacteria may be resistant because they adapt to low-iron environments by producing siderophores such as phenolates and hydroxamates. Microorganisms with a low iron requirement, such as lactic acid bacteria, would not be inhibited by lactoferrin (78).

Lactoferricin B is a small peptide (25 amino acids) produced by acid-pepsin hydrolysis of bovine lactoferrin (81). The compound is inhibitory to *Shigella*, *Salmonella*, *Yersinia enterocolitica*, *E. coli* O157:H7, *Staphylococcus aureus*, *L. monocytogenes*, and *Candida* species at concentrations ranging from 1.9 to 125 mg/mL (82). Another iron-binding molecule, ovotransferrin or conalbumin, occurs in egg albumen. Each ovotransferrin molecule has two iron-binding sites and, like lactoferrin, it binds an anions, such as bicarbonate or carbonate with each ferric iron bound. Ovotransferrin is inhibitory against both Gram-positive and Gram-negative bacteria, but the former are generally more sensitive, with *Bacillus* and *Micrococcus* species being particularly sensitive (75). Some yeasts are also sensitive.

Avidin

Avidin is a glycoprotein present in egg albumen. The concentration varies with hen's age, but the mean is 0.05% of the total egg albumen protein (75). The protein is 66,000 to 69,000 Da and has four identical subunits of 128 amino acids each. Avidin strongly binds the cofactor biotin at a ratio of four molecules of biotin per molecule of avidin. Biotin is a cofactor for enzymes in the tricarboxylic acid cycle and fatty acid biosynthesis. Avidin inhibits growth of some bacteria and yeasts that have a requirement for biotin with the primary mechanism being nutrient deprivation.

Spices and Their Essential Oils

Spices are roots, bark, seeds, buds, leaves, or fruit of aromatic plants added to foods as flavoring agents. It has been known since ancient times that spices and their essential oils have varying degrees of antimicrobial activity. Cloves, cinnamon, oregano, thyme, and, to a lesser extent, sage and rosemary have the strongest antimicrobial activity among spices.

The major antimicrobial components of clove and cinnamon are eugenol (2-methoxy-4-(2-propenyl)-phenol) and cinnamic aldehyde (3-phenyl-2-propenal), respectively. Cinnamon and clove extracts or their essential oils inhibit *Aeromonas hydrophila*, *Bacillus*, *Enterobacter aerogenes*, lactic acid bacteria, and *Staphylococcus aureus* in microbiological systems and in foods (8). Bullerman (83) observed that cinnamon in raisin bread inhibited growth and aflatoxin production by *Aspergillus parasiticus*. Cinnamon and clove were the most effective of 16 ground herbs and spices tested at 2% w/v against nine mycotoxin-producing *Aspergillus* and *Penicillium* species (84).

The antimicrobial activity of oregano and thyme has been attributed to their essential oils which contain the terpenes carvacrol (2-methyl-5-(1-methylethyl)phenol) and thymol (5-methyl-2-(1-methylethyl)phenol), respectively. These compounds have inhibitory activity against a number of bacterial species, molds, and yeasts including *Bacillus subtilis*, *E. coli*, *Lactobacillus plantarum*, *Pediococcus cerevisiae*, *Pseudomonas aeruginosa*, *Proteus*, *Salmonella enteritidis*, *S. aureus*, *Vibrio parahaemolyticus*, and *Aspergillus parasiticus* (8).

The active fraction of sage and rosemary has been suggested to be the terpene fractions of the essential oils. Rosemary contains primarily borneol (endo-1,7,7-trimethylbicyclo[2.2.1] heptan-2-ol) along with pinene, camphene, camphor while sage contains thujone (4-methyl-1-(1-methylethyl)bicyclo[3.1.0]-hexan-3-one). Sage and rosemary are more active against gram-positive than gram-negative bacterial strains (85). Sensitivity of *Bacillus cereus*, *S. aureus*, and *Pseudomonas* to sage is greatest in microbiological medium and significantly reduced in foods (86). It is theorized that loss of activity was due to solubilization of the antimicrobial fraction in the lipid of the foods.

Vanillin (4-hydroxy-3-methoxybenzaldehyde) is a major constituent of vanilla beans, the fruit of an orchid (*Vanilla planifola*, *Vanilla pompona*, or *Vanilla tahitensis*). Vanillin is most active as against molds and nonlactic Gram-positive bacteria (87). Vanillin at 1500 μg/mL significantly inhibited strains of *Aspergillus* in fruit-based agars containing mango, papaya, pineapple, apple, and banana (88). The compound was least effective in banana and mango agars. This was attributed to binding of the vanillin by protein or lipid in these fruits, a phenomenon demonstrated for other antimicrobial phenolic compounds (44).

Many other spices have been tested and shown to have limited or no activity. They include allspice, anise, bay leaf, black pepper, cardamom, celery seed, chili powder, coriander, cumin, curry powder, dill, fenugreek, ginger, juniper oil, mace, marjoram, mustard, nutmeg, orris root, paprika, parsley, red pepper, sesame, spearmint, tarragon, and white pepper (89).

ONIONS AND GARLIC

Probably the most well characterized antimicrobial system in plants is that found in the juice and vapors of onions (*Allium cepa*) and garlic (*Allium sativum*). Growth and toxin production of many microorganisms have been shown to be inhibited by onion and garlic, including the bacteria *Bacillus cereus*, *Clostridium botulinum* type A, *E. coli*, *Lactobacillus plantarum*, *Salmonella*, *Shigella*, and *Staphylococcus aureus*, and the fungi *Aspergillus flavus*, *A. parasiticus*, *Candida albicans*, *Cryptococcus*, *Rhodotorula*, *Saccharomyces*, *Torulopsis*, and *Trichosporon* (18). The major antimicrobial component from garlic is allicin (diallyl thiosulfinate; thio-2-propene-1-sulfinic acid-5-allyl ester) (18), which is formed by the action of the enzyme allinase on the substrate alliin (*S*-(2-propenyl)-L-cysteine sulfoxide). The reaction only occurs when cells of the garlic are disrupted, releasing the enzyme to act on the substrate. A similar reaction occurs in onion except the substrate is *S*-(1-propenyl)-L-cysteine sulfoxide and one of the major products is thiopropanal-*S*-oxide. The products apparently responsible for antimicrobial activity are also responsible for the flavor of onions and garlic. In addition to antimicrobial sulfur compounds, onions contain the phenolic compounds protocatechuic acid and catechol, which could contribute to their antimicrobial activity (90). The mechanism of action of allicin is most likely inhibition of sulfhydryl-containing enzymes (18).

Other Plant Extracts

Isothiocyanates (R-N=C=S) are derivatives from glucosinolates in cells of plants of the Cruciferae or mustard family (cabbage, kohlrabi, Brussel sprouts, cauliflower, broccoli, kale, horseradish, mustard, turnips, rutabaga). They are potent antifungal and antimicrobial agents (91). These compounds are formed from the action of the enzyme myrosinase (thioglucoside glucohydrolase) on the glucosinolates when the plant tissue is injured or mechanically disrupted. Common isothiocyanate side groups include allyl (AIT), ethyl, methyl, benzyl, and phenyl. The compounds are inhibitory to fungi, yeasts, and bacteria in the range of 0.016 to 0.062 μg/mL in the vapor phase (92). Inhibition depends on the compound type against bacteria but generally Gram-positive bacteria are less sensitive to AIT than are Gram-negative bacteria. The mechanism by which isothiocyanates inhibit cells may involve enzymes by direct reaction with disulfide bonds or through thiocyanate (SCN^-) anion reaction to inactivate sulfhydryl enzymes (91). These compounds have very low sensory thresholds, but they may be useful as food antimicrobials due to their low inhibitory concentrations.

Phenolic Compounds

Phenolic compounds occurring in foods are classified as simple phenols and phenolic acids, hydroxycinnamic acid derivatives, and the flavonoids (93). The various classes of phenolics and individual compounds within each class have various antimicrobial activities.

Smoking of foods, such as meats, cheeses, fish, and poultry, has a preservative effect through drying and the deposition of chemicals. The major antimicrobial components of wood smoke are phenol and cresol. Smoke distillates and liquid smoke are inhibitory to *Staphylococcus aureus*, *E. coli*, *Listeria monocytogenes*, and *Saccharomyces cerevisiae* (43). Consistent with results using other phenolic compounds as antimicrobials, smoke is more effective against Gram-positive than Gram-negative bacteria. Isoeugenol is one of the most effective antimicrobial compounds in liquid smoke, followed by *m*-cresol and *p*-cresol (94).

The phenolic acids, including derivatives of *p*-hydroxybenzoic acid (protocatechuic, vanillic, gallic, syringic, ellagic) and *o*-hydroxybenzoic acid (salicylic) may be found in plants and foods. Esters of gallic acid including butyl, isobutyl, isoamyl, *p*-octyl, and *p*-dodecyl have been shown to be effective against *Clostridium botulinum* in microbiological media at 37°C (95). Another ester of gallic acid, propyl gallate, inhibited growth of *Alcaligenes faecalis*, *Enterobacter aerogenes*, *E. coli*, *Klebsiella pneumoniae*, *L. monocytogenes*, *Proteus vulgaris*, *Salmonella enteritidis*, *Salmonella paratyphi*, and *Shigella flexneri* (96). Tannic acid has antimicrobial activity against *Aeromonas hydrophila*, *E. coli*, *L. monocytogenes*, *S. enteritidis*, *S. aureus*, and *Streptococcus faecalis* (43). Stead (97) demonstrated stimulation of growth of the lactic acid spoilage bacteria, *Lactobacillus collinoides* and *Lactobacillus brevis*, by gallic, quinic, and chlorogenic acids. It was suggested that the compounds were being metabolized by the microorganisms and that presence of these compounds in beverages may potentially result in antagonistic reactions with other food-grade antimicrobials.

Resin from the flowers of the hop vine (*Humulus lupulus* L.) used in the brewing industry for imparting a desirable bitter flavor to beer contain α-bitter acids (humulone, cohumulone, adhumulone) and β-bitter acids (lupulone, colupulone, xanthohumol, adlupulone). Both types of acids possess antimicrobial activity primarily against Gram-positive bacteria and fungi at low water activity (98). A compound isolated from green olives and identified as the phenolic glycoside oleuropein, or its aglycone, inhibits *Lactobacillus plantarum*, *Leuconostoc mesenteroides*, *Pseudomonas fluorescens*, *Bacillus subtilis*, *Rhizopus* sp., and *Geotrichum candidum* (43).

Hydroxycinnamic acids, including caffeic, *p*-coumaric, ferulic, and sinapic acids, are found in plants and occur as esters and less often as glucosides (93). Hydroxycinnamic acids have been shown to inhibit bacteria, including *B. cereus*, *Erwinia carotovora*, *Lactobacillus collinoides*, *Lactobacillus brevis*, *S. aureus*, and fungi, including *Aspergillus flavus*, *A. parasiticus*, and *Saccharomyces cerevisiae* (43). Cinnamic acid was found to be inhibitory to *L. monocytogenes* Scott A growth and listeriolysin O activity at pH 5.5 (99). The degree of inhibition is generally inversely related to the polarity of the compounds.

The furocoumarins are related to the hydroxycinnamates. These compounds, including psoralen and its derivatives, are phytoalexins in citrus fruits, parsley, carrots, celery, and parsnips at levels of 2–6 μg/mL. They inhibit many Gram-positive bacteria including, *L. monocytogenes*, *E. coli* O157:H7 and *Micrococcus luteus*, following irradiation with long wave ultraviolet light (100). These compounds inhibit growth by interfering with DNA replication.

The flavonoids consist of catechins, proanthocyanidins, anthocyanidins and flavons, flavonols, and their glycosides (93). Proanthocyanidins or condensed tannins are polymers of favan-3-ol and are found in apples, grapes, strawberries, plums, sorghum, and barley (93). The tannins have been tested for their antimicrobial effectiveness against molds, yeasts, bacteria, and viruses. For example, proanthocyanidins and flavonols accounted for the majority of inhibition of *Saccharomyces bayanus* by cranberries (101). Tannins have been reported to be inhibitory to *Aeromonas*, *Bacillus*, *C. botulinum*, *C. perfringens*, *Enterobacter*, *Klebsiella*, *Proteus*, *Pseudomonas*, *Shigella*, *S. aureus*, *Streptococcus*, and *Vibrio* (102). Extracts of blueberries, crabapples, strawberries, red wines, grape juice, apple juice, and tea inactivated poliovirus, coxsackievirus, echovirus, reovirus, and herpes simplex virus (103,104). It was concluded that the primary viral inhibitors in these products are condensed or hydrolyzable tannins.

Glucose Oxidase

Glucose oxidase (β-D-glucose:oxidoreductase) catalyzes the oxidation of glucose to gluconic acid with the production of hydrogen peroxide (89,105). Glucose oxidase acting on glucose in honey produces hydrogen peroxide as an inhibitory agent. Bacteria inhibited by glucose oxidase include *Acinetobacter*, *Bacillus cereus*, *Campylobacter jejuni*, *Clostridium perfringens*, *Listeria monocytogenes*, *Pseudomonas fluorescens*, *Salmonella*, *Staphylococcus aureus*, and *Yersinia* (106,107). In most cases, it is hydrogen peroxide produced by the enzyme that is responsible for inhibition of these microorganisms. Glucose oxidase in the presence of added glucose was bactericidal against *Salmonella enteritidis*, *Micrococcus luteus*, and *B. cereus* and bacteriostatic against *P. fluorescens* in liquid whole egg stored at 7°C (105). In contrast, glucose oxidase had no effect on *Salmonella* or *Pseudomonas* on chicken breast meat (108).

Utilization of Naturally Occurring Antimicrobials in Foods

With the exception of nisin, natamycin, and lysozyme, no isolated naturally occurring antimicrobials are approved for use in foods by regulatory agencies. For acceptance by regulatory agencies, more information will be needed concerning the antimicrobial activity of the compounds in foods, their stability, and their toxicological effects on humans. In addition, prior to acceptance by the food industry, they must be shown to have no adverse effects on sensory qualities of foods and be cost-effective. These are huge barriers to the use of isolated compounds. It is more likely that the future of naturally occurring antimicrobials is in their use as crude or partially purified extracts of their component source.

BIBLIOGRAPHY

1. G. W. Gould, ed., *Mechanisms of Action of Food Preservation Procedures*, Elsevier Applied Science, London, 1989.
2. T. Eklund, "Organic Acids and Esters," in G. W. Gould, ed., *Mechanisms of Action of Food Preservation Procedures*, Elsevier Applied Science, London, 1989, pp. 161–200.
3. J. N. Sofos, "Antimicrobial Effects of Sodium and Other Ions in Foods," *J. Food Safety* **6**, 45–78 (1983).
4. J. E. L. Corry, "Relationships of Water Activity to Fungal Growth," in L. R. Beuchat, ed., *Food and Beverage Mycology*, 2nd ed., AVI/Van Nostrand Reinhold, New York, 1987, pp. 51–99.
5. W. H. Sperber, "Influence of Water Activity on Foodborne Bacteria—A Review," *J. Food Prot.* **46**, 142–150 (1983).
6. G. J. Banwart, *Basic Food Microbiology*, AVI Publishing, Westport, Conn., 1979.
7. S. Doores, "Organic Acids," in P. M. Davidson and A. L. Branen, eds., *Antimicrobials in Foods*, 2nd ed., Marcel Dekker, New York, 1993, pp 95–136.
8. P. M. Davidson, "Chemical Preservatives and Natural Antimicrobial Compounds," in M. P. Doyle, L. R. Beuchat, and T. J. Montville, eds., *Food Microbiology: Fundamentals and Frontiers*, American Society for Microbiology, Washington, D.C., 1997, pp. 520–556.
9. D. Hunter and I. H. Segel, "Effect of Weak Acids on Amino Acid Transport by *Penicillium chrysogenum*. Evidence for a Proton or Charge Gradient as the Driving Force," *J. Bacteriol.* **113**, 1184–1192 (1973).
10. M. Ingram, F. J. H. Ottoway, and J. B. M. Coppock, "The Preservative Action of Acid Substances in Food," *Chem. Ind. (London)* **42**, 1154–1163 (1956).
11. A. C. Baird-Parker, "Organic Acids," International Commission on Microbiological Specifications for Foods, ed., *Microbial Ecology of Foods, Vol. I. Factors Affecting Life and Death of Microorganisms*, Academic Press, New York, 1980, pp. 135–141.
12. G. J. Mountney and J. O'Malley, "Acids as Poultry Meat Preservatives," *Poultry Sci.* **44**, 582 (1985).
13. A. J. Okrend, R. W. Johnston, and A. B. Moran, "Effect of Acetic Acid on the Death Rates at 52°C of *Salmonella newport*, *Salmonella typhimurium*, and *Campylobacter jejuni* in Poultry Scald Water," *J. Food Prot.* **49**, 500–503 (1986).
14. M. E. Anderson et al., "Counts of Six Types of Bacteria on Lamb Carcasses Dipped or Sprayed with Acetic Acid at 25°C or 55°C and Stored Vacuum Packaged at 0°C," *J. Food Prot.* **51**, 874–877 (1988).
15. G. R. Siragusa and J. S. Dickson, "Inhibition of *Listeria monocytogenes*, *Salmonella typhimurium* and *Escherichia coli* O157:H7 on Beef Muscle Tissue by Lactic or Acetic Acid Contained in Calcium Alginate Gels," *J. Food Safety* **13**, 147–158 (1993).
16. L. A. Shelef, and L. Addala, "Inhibition of *Listeria monocytogenes* and Other Bacteria by Sodium Diacetate," *J. Food Safety* **14**, 103–115 (1994).
17. J. R. Chipley, "Sodium Benzoate and Benzoic Acid," in P. M. Davidson and A. L. Branen, eds., *Antimicrobials in Foods*, 2nd ed., Marcel Dekker, New York, 1993, pp. 11–48.
18. J. N. Sofos et al., *Naturally Occurring Antimicrobials in Food*, Task Force Report No. 132, Council for Agricultural Science and Technology, Ames, Iowa, 1998.
19. T. Zhao, M. P. Doyle, and R. E. Besser, "Fate of Enterohemorrhagic *Escherichia coli* O157:H7 in Apple Cider With and Without Preservatives," *Appl. Env. Microbiol.* **59**, 2526–2530 (1993).
20. T. E. Minor and E. H. Marth, "Growth of *Staphylococcus aureus* in Acidified Pasteurized Milk," *J. Milk Food Technol.* **33**, 516–520 (1970).
21. R. E. Brackett, "Effect of Various Acids on Growth and Survival of *Yersinia enterocolitica*," *J. Food Prot.* **50**, 598–601 (1987).
22. N. Chen and L. A. Shelef, "Relationship Between Water Activity, Salts of Lactic Acid, and Growth of *Listeria monocytogenes* in a Meat Model System," *J. Food Prot.* **55**, 574–578 (1992).
23. G. A. Pelroy et al., "Inhibition of *Listeria monocytogenes* in Cold-Process (Smoked) Salmon by Sodium Lactate," *J. Food Prot.* **57**, 108–113 (1994).
24. J. N. Sofos and F. F. Busta, "Sorbic Acid and Sorbates," in P. M. Davidson and A. L. Branen, eds., *Antimicrobials in Foods*, 2nd ed., Marcel Dekker, New York, 1993, pp. 49–94.
25. M. C. Robach, "Use of preservatives to control microorganisms in food," *Food Technol.* **34**, 81–84 (1980).
26. R. C. McKellar, "Effect of Preservatives and Growth Factors on Secretion of Listeriolysin O by *Listeria monocytogenes*," *J. Food Prot.* **56**, 380–384 (1993).
27. A. L. Branen and T. W. Keenan, "Growth Stimulation of *Lactobacillus casei* by Sodium Citrate," *J. Dairy Sci.* **53**, 593 (1970).
28. R. L. Buchanan and M. H. Golden, "Interaction of Citric Acid Concentration and pH on the Kinetics of *Listeria monocytogenes* Inactivation," *J. Food Prot.* **57**, 567–570 (1994).
29. J. J. Kabara, "Medium-Chain Fatty Acids and Esters," in P. M. Davidson and A. L. Branen, eds., *Antimicrobials in Foods*, 2nd ed., Marcel Dekker, New York, 1993, pp. 307–342.
30. J. J. Kabara and T. Eklund, "Organic Acids and Esters," in N. J. Russell and G. W. Gould, eds., *Food Preservatives*, Blackie, Glasgow, United Kingdom, 1991, pp. 44–71.
31. S. M. Razavi-Rohani and M. W. Griffiths, "The Effect of Mono and Polyglycerol Laurate on Spoilage and Pathogenic Bacteria Associated With Foods," *J. Food Safety* **14**, 131–151 (1994).
32. D. H. Oh and D. L. Marshall, "Enhanced Inhibition of *Listeria monocytogenes* by Glycerol Monolaurate With Organic Acids," *J. Food Sci.* **59**, 1258–1261 (1994).
33. L.-L. Wang and E. A. Johnson, "Control of *Listeria monocytogenes* by Monoglycerides in Foods," *J. Food Prot.* **60**, 131–138 (1997).
34. L. R. Beuchat, "Comparison of Anti-Vibrio Activities of Potassium Sorbate, Sodium Benzoate and Glycerol and Sucrose Esters of Fatty Acids," *Appl. Environ. Microbiol.* **39**, 1178–1182 (1980).
35. R. B. Tompkin, "Nitrite," in P. M. Davidson and A. L. Branen, eds., *Antimicrobials in Foods*, 2nd ed., Marcel Dekker, New York, 1993, pp. 191–262.
36. T. R. Aalto, M. C. Firman, and N. E. Rigier, "*P*-hydroxybenzoic Acid Esters as Preservatives. I. Uses, Antibacterial and

Antifungal Studies, Properties and Determination," *J. Am. Pharm. Assoc. (Sci. Ed.)* **42**, 449–458 (1953).

37. C. S. Ough, "Sulfur Dioxide and Sulfites," in P. M. Davidson and A. L. Branen, eds., *Antimicrobials in Foods*, 2nd ed., Marcel Dekker, New York, 1993, pp. 137–190.
38. J. O. Roland et al., "Effects of Sorbate, Benzoate, Sulfur Dioxide and Temperature on Growth and Patulin Production by *Byssochlamys nivea* in Grape Juice," *J. Food Prot.* **47**, 237–241 (1984).
39. J. G. Banks and R. G. Board, "Sulfite-inhibition of *Enterobacteriaceae* Including *Salmonella* in British Fresh Sausage and in Culture Systems," *J. Food Prot.* **45**, 1292–1297 (1982).
40. C. S. Ough, "Dimethyl Dicarbonate and Diethyl Dicarbonate," in P. M. Davidson and A. L. Branen, eds., *Antimicrobials in Foods*, 2nd ed., Marcel Dekker, New York, 1993, pp. 343–368.
41. T. L. Fisher and D. A. Golden, "Survival of *Escherichia coli* O157:H7 in Apple Cider as Affected by Dimethyl Dicarbonate, Sodium Bisulfite, and Sodium Benzoate," *J. Food Sci.* **63**, 904–906 (1998).
42. H. C. Chang and A. L. Branen, "Antimicrobial Effects of Butylated Hydroxyanisole (BHA)," *J. Food Sci.* **40**, 349–351 (1975).
43. P. M. Davidson, "Parabens and Phenolic Compounds," in P. M. Davidson and A. L. Branen, eds., *Antimicrobials in Foods*, 2nd ed., Marcel Dekker, New York, 1993, pp. 263–306.
44. E. Rico-Muñoz and P. M. Davidson, "The Effect of Corn Oil and Casein on the Antimicrobial Activity of Phenolic Antioxidants," *J. Food Sci.* **48**, 1284–1288 (1983).
45. L. A. Shelef and J. A. Seiter, "Indirect Antimicrobials," in P. M. Davidson and A. L. Branen, eds., *Antimicrobials in Foods*, 2nd ed., Marcel Dekker, New York, 1993, pp. 325–328.
46. F. Kelch, "Effect of Commercial Phosphates on the Growth of Microorganisms," *Fleischwirtschaft* **10**, 325–328 (1958).
47. L. L. Zaika and A. H. Kim, "Effect of Sodium Polyphosphates on Growth of *Listeria monocytogenes*," *J. Food Prot.* **56**, 577–580 (1993).
48. M. K. Wagner and F. F. Busta, "Inhibition of *Clostridium botulinum* 52a Toxicity and Protease Activity by Sodium Acid Pyrophosphate in Media Systems," *Appl. Environ. Microbiol.* **50**, 16–20 (1985).
49. M. K. Wagner, "Phosphates as Antibotulinal Agents in Cured Meats. A Review," *J. Food Prot.* **49**, 482–487 (1986).
50. N. Tanaka, "Challenge of Pasteurized Process Cheese Spreads with *Clostridium botulinum* Using In-process and Post-process Inoculation," *J. Food Prot.* **45**, 1044–1050 (1982).
51. R. B. Tompkin, "Indirect Antimicrobial Effects in Foods: Phosphates," *J. Food Safety* **6**, 13–27 (1983).
52. J. Giese, "Experimental Process Reduces *Salmonella* on Poultry," *Food Technol.* **46**, 112 (1992).
53. H. S. Lillard, "Effect of Trisodium Phosphate on Salmonellae Attached to Chicken Skin," *J. Food Prot.* **57**, 465–469 (1994).
54. S. J. Knabel, H. W. Walker, and P. A. Hartman, "Inhibition of *Aspergillus flavus* and Selected Gram-positive Bacteria by Chelation of Essential Metal Cations by Polyphosphates," *J. Food Prot.* **54**, 360–365 (1991).
55. A. Hurst and D. G. Hoover, "Nisin," in P. M. Davidson and A. L. Branen, eds., *Antimicrobials in Foods*, 2nd ed., Marcel Dekker, New York, 1993, pp. 369–394.
56. V. N. Scott and S. L. Taylor, "Effect of Nisin on the Outgrowth of *Clostridium botulinum* Spores," *J. Food Sci.* **46**, 117–120 (1981).
57. N. Benkerroum and W. E. Sandine, "Inhibitory Action of Nisin Against *Listeria monocytogenes*," *J. Dairy Sci.* **71**, 3237–3245 (1988).
58. A. Laukova, "Inhibition of Ruminal *Staphylococci* and *Enterococci* by Nisin *in Vitro*," *Lett. Appl. Microbiol.* **20**, 34–36 (1995).
59. F. Radler, "Possible Use of Nisin in Winemaking. I. Action of Nisin Against Lactic Acid Bacteria and Wine Yeasts in Solid and Liquid Media," *Am. J. Enol. Vitic.* **41**, 1–6 (1990).
60. D. O. Ukuku and L. A. Shelef, "Sensitivity of Six Strains of *Listeria monocytogenes* to Nisin," *J. Food Prot.* **60**, 867–869 (1997).
61. B. Ray and M. Daeschel, *Food Biopreservatives of Microbial Origin*, CRC Press, Boca Raton, Fla., 1992.
62. A. Hirsch et al., "A Note on the Inhibition of an Anaerobic Sporeformer in Swiss-type Cheese by a Nisin-producing *Streptococcus*," *J. Dairy Res.* **18**, 205–206 (1951).
63. D. Jung, F. W. Bodyfelt, and M. A. Daeschel, "Influence of Fat and Emulsifiers on the Efficacy of Nisin in Inhibiting *Listeria monocytogenes* in Fluid Milk," *J. Dairy Sci.* **75**, 387–393 (1992).
64. J. P. Dean and E. A. Zottola, "Use of Nisin in Ice Cream and Effect on the Survival of *Listeria monocytogenes*," *J. Food Prot.* **59**, 476–480 (1996).
65. P. M. Davidson and C. H. Doan, "Natamycin," P. M. Davidson and A. L. Branen, eds., *Antimicrobials in Foods*, 2nd ed., Marcel Dekker, New York, 1993, pp. 395–408.
66. L. L. Ray and L. B. Bullerman. "Preventing Growth of Potentially Toxic Molds Using Antifungal Agents," *J. Food Prot.* **45**, 953–963 (1982).
67. H. Brik, "Natamycin," in K. Flory, ed., *Analytical Profiles of Drug Substances*, Academic Press, New York, 1981, pp. 513.
68. K. M. Nilson et al., "Pimaricin and Mycostatin for Retarding Cottage Cheese Spoilage," *J. Dairy Sci.* **58**, 668–671 (1975).
69. H. Lück and C. E. Cheeseman, "Mould Growth on Cheese as Influenced by Pimaricin or Sorbate Treatments," *S. Afr. J. Dairy Technol.* **10**, 143–146 (1978).
70. R. Lodi, R. Todesco, and V. Bozzetti, "New Applications of Natamycin with Different Types of Italian Cheese," *Microbiol. Aliments Nutr.* **7**, 81 (1989).
71. J. C. Ayres and E. L. Denisen. "Maintaining Freshness of Berries Using Selected Packaging Materials and Antifungal Agents," *Food Technol.* **12**, 562–567 (1958).
72. R. J. Shirk and W. L. Clark, "The Effect of Pimaricin in Retarding the Spoilage of Fresh Orange Juice," *Food Technol.* **17**, 1062–1066 (1963).
73. J. C. Ayres et al., "Use of Antibiotics in Prolonging Storage Life of Dressed Chicken," *Food Technol.* **10**, 563–568 (1956).
74. D. G. Hoover, "Bacteriocins with Potential for Use in Foods," in P. M. Davidson and A. L. Branen, eds., *Antimicrobials in Foods*, 2nd ed., Marcel Dekker, New York, 1993, pp. 409–440.
75. H. S. Tranter, "Lysozyme, Ovotransferrin and Avidin," in V. M. Dillon and R. G. Board, eds., *Natural Antimicrobial Systems and Food Preservation*, CAB International Wallingford, United Kingdom, 1994, pp. 65–98.
76. V. L. Hughey, R. A. Wilger, and E. A. Johnson, "Antibacterial Activity of Hen Egg White Lysozyme Against *Listeria monocytogenes* Scott A in Foods," *Appl. Env. Microbiol.* **55**, 631–638 (1989).
77. K. D. Payne, P. M. Davidson, and S. P. Oliver, "Comparison of EDTA and Apo-lactoferrin With Lysozyme on the Growth of Foodborne Pathogenic and Spoilage Bacteria," *J. Food Prot.* **57**, 62–65 (1994).

78. B. Ekstrand, "Lactoperoxidase and Lactoferrin," in V. M. Dillon and R. G. Board eds., *Natural Antimicrobial Systems and Food Preservation*, CAB International, Wallingford, United Kingdom, 1994, pp. 15–64.
79. K. M. Wilkins and R. G. Board, "Natural Antimicrobial Systems," in G. W. Gould, ed., *Mechanisms of Action of Food Preservation Procedures*, Elsevier Applied Science, London, 1989, pp. 285–362.
80. K. D. Payne et al., "Influence of Bovine Lactoferrin on the Growth of *Listeria monocytogenes*," *J. Food Prot.* **53**, 468–472 (1990).
81. W. Bellamy et al., "Antibacterial Spectrum of Lactoferricin B, a Potent Bactericidal Peptide Derived from the *N*-terminal Region of Bovine Lactoferrin," *J. Appl. Bacteriol.* **73**, 472–479 (1992).
82. E. M. Jones et al., "Lactoferricin, a New Antimicrobial Peptide," *J. Appl. Bacteriol.* **77**, 208–214 (1994).
83. L. B. Bullerman, "Inhibition of Aflatoxin Production by Cinnamon," *J. Food Sci.* **39**, 1163–1165 (1974).
84. M. A. Azzouz and L. B. Bullerman, "Comparative Antimycotic Effects of Selected Herbs, Spices, Plant Components and Commercial Fungal Agents," *J. Food Sci.* **45**, 1298–1301 (1982).
85. L. A. Shelef, O. A. Naglik, and D. W. Bogen, "Sensitivity of Some Common Foodborne Bacteria to the Spices Sage, Rosemary, and Allspice," *J. Food Sci.* **45**, 1042–1044 (1980).
86. L. A. Shelef, E. K. Jyothi, and M. A. Bulgarelli, "Growth of Enteropathogenic and Spoilage Bacteria in Sage-containing Broth and Foods," *J. Food Sci.* **49**, 737–740 (1984).
87. J. M. Jay and G. M. Rivers, "Antimicrobial Activity of Some Food Flavoring Compounds," *J. Food Safety* **6**, 129–139 (1984).
88. A. López-Malo, S. M. Alzamora, and A. Argaiz, "Effect of Natural Vanillin on Germination Time and Radial Growth of Moulds in Fruit-based Agar Systems," *Food Microbiol.* **12**, 213–219 (1995).
89. E. H. Marth, "Antibiotics in Foods—Naturally Occurring, Developed and Added," *Residue Rev.* **12**, 65–161 (1966).
90. J. C. Walker and M. A. Stahmann, "Chemical Nature of Disease Resistance in Plants," *Ann. Rev. Plant Physiol.* **6**, 351–366 (1955).
91. P. J. Delaquis and G. Mazza, "Antimicrobial Properties of Isothiocyanates in Food Preservation," *Food Technol.* **49**, 73–84 (1995).
92. K. Isshiki et al., "Preliminary Examination of Allyl Isothiocyanate Vapor for Food Preservation," *Biosci. Biotechnol. Biochem.* **56**, 1476–1477 (1992).
93. C. T. Ho, "Phenolic Compounds in Food. An Overview," in C. T. Ho, C. Y. Lee, and M. T. Huang, eds., *Phenolic Compounds in Food and Their Effects on Health I. Analysis, Occurrence, and Chemistry*, ACS Symposium Series 506., American Chemical Society, Washington, D.C., 1992, pp. 2–7.
94. N. G. Faith, A. E. Yousef, and J. B. Luchansky, "Inhibition of *Listeria monocytogenes* by Liquid Smoke and Isoeugenol, a Phenolic Component Found in Smoke," *J. Food Safety* **12**, 303–314 (1992).
95. N. R. Reddy and M. D. Pierson, "Influence of pH and Phosphate Buffer on Inhibition of *Clostridium botulinum* by Antioxidants and Related Phenolic Compounds," *J. Food Prot.* **45**, 925–927 (1982).
96. K.-T. Chung et al., "Antimicrobial Properties of Tannic Acid, Propyl Gallate and Related Compounds," ASM General Meeting, New Orleans, La., 1992.
97. D. Stead, "The Effect of Hydroxycinnamic Acids on the Growth of Wine-spoilage Lactic Acid Bacteria," *J. Appl. Bacteriol.* **75**, 135–141 (1993).
98. L. R. Beuchat, "Antimicrobial Properties of Spices and Their Essential Oils," in V. M. Dillon and R. G. Board eds., *Natural Antimicrobial Systems and Food Preservation*, CAB International, Wallingford, United Kingdom, 1994, pp. 167–180.
99. Y. Kouassi and L. A. Shelef, "Listeriolysin O Secretion by *Listeria monocytogenes* in the Presence of Cysteine and Sorbate," *Lett. Appl. Microbiol.* **20**, 295–299 (1995).
100. J. Ulate-Rodriguez et al., "Inhibition of *Listeria monocytogenes, Escherichia coli* O157:H7 and *Micrococcus luteus* by Linear Furocoumarins in Culture Media," *J. Food Prot.* **60**, 1046–1049 (1997).
101. A. G. Marwan and C. W. Nagel, "Microbial Inhibitors of Cranberries," *J. Food Sci.* **51**, 1009–1013 (1986).
102. K.-T. Chung, C.-I. Wei, and M. G. Johnson, "Are Tannins a Double-edged Sword in Biology and Health," *Trends Food Sci. Technol.* **9**, 168–175 (1998).
103. J. Konowalchuk and J. I. Speirs, "Antiviral Effect of Commercial Juices and Beverages," *Appl. Env. Microbiol.* **35**, 1219–1220 (1978).
104. J. Konowalchuk and J. I. Speirs, "Antiviral Effect of Apple Beverages," *Appl. Env. Microbiol.* **36**, 798–801 (1978).
105. D. Dobbenie, M. Uyttendaele, and J. Debevere, "Antibacterial Activity of the Glucose Oxidase/Glucose System in Liquid Whole Egg," *J. Food Prot.* **58**, 273–279 (1995).
106. C. A. Kantt and J. A. Torres, "Growth Inhibition by Glucose Oxidase of Selected Organisms Associated with the Microbial Spoilage of Shrimp (*Pandalus jordani*): In Vitro Model Studies," *J. Food Prot.* **56**, 147–152 (1993).
107. M. Tiina and M. Sandholm, "Antibacterial Effect of the Glucose Oxidase-Glucose System on Food-poisoning Organisms," *Int. J. Food Microbiol.* **8**, 165–174 (1989).
108. D. K. Jeong et al., "Trials on the Antibacterial Effect of Glucose Oxidase on Chicken Breast Skin and Muscle," *J. Food Safety* **13**, 43–50 (1992).

P. Michael Davidson
University of Tennessee
Knoxville, Tennessee

ANTIOXIDANTS

Oxidative reactions cause damage to lipids and proteins, thus influencing food quality. For example, oxidation of lipids results in the formation of volatile compounds that cause rancidity, oxidation of pigments (eg, carotenoids and myoglobin) leading to color changes, and oxidation of vitamins (eg, A, C, and E) leading to alterations in nutritional composition. The biological tissue from which we derive foods contains several distinctively different mechanisms to control oxidation catalysts, reactive oxygen species, and free radicals. In addition, many antioxidant additives are available to increase the oxidative stability of foods. Utilization of antioxidant additives and protection of endogenous antioxidants can be effective methods to increase the quality and shelf life of foods.

CHAIN-BREAKING ANTIOXIDANTS

Control of free radicals is an effective method to prevent both autoxidation and oxidation accelerated by pro-oxidants, since free radicals are a universal reaction intermediate for all lipid oxidation reactions. Chain-breaking antioxidants inhibit free-radical reactions by scavenging radicals resulting in inactivation of the original free radical and formation of a more stable antioxidant radical (1). The effectiveness of a chain-breaking antioxidant is a function of its chemical and physical properties. For a chain-breaking antioxidant to be effective, it must possess a hydrogen, which is weakly bound to the antioxidant so that it can be freely donated to the free radical. As the bond energy of the hydrogen on the chain-breaking antioxidant decreases, the transfer of the hydrogen to the free radical is more rapid (1,2). The effectiveness of a chain-breaking antioxidant is also related to its ability to decrease radical energy so that the antioxidant radical cannot promote further oxidation of unsaturated fatty acids. Effective antioxidants decrease free radical energy by resonance delocalization (2,3). Effective chain-breaking antioxidants also do not react rapidly with oxygen to form peroxides. Degradation of antioxidant peroxides can deplete the system of the antioxidant and can result in the formation of additional free radicals (1).

The most common functional groups that inactivate free radicals are hydroxyl (eg, phenolics), sulfhydryl (eg, cysteine and glutathione) and amino (eg, uric acid, spermine, and proteins) groups (4). Phenolic compounds are the most common forms of synthetic and natural chain-breaking antioxidants in foods. Examples of naturally occurring phenolics that inhibit lipid oxidation include α-tocopherol, epicatechin, ferulic acid, and carnosic acid (Fig. 1). Many of the natural antioxidants are used in foods as plant extracts such as rosemary and mixed tocopherol isomers. Synthetic phenolics approved as food additives include butylated hydroxytoluene, butylated hydroxyanisole, tertiary butylhydroquinone, and propyl gallate (Fig. 2). These phenolics are effective because of their ability to inactivate free radicals and form low-energy phenolic radicals. Natural phenolic extracts are typically added to foods at concentrations $\leq$500 ppm whereas the synthetic antioxidants are normally limited to concentrations $\leq$200 ppm of the fat content (5,6).

Effective utilization of phenolic antioxidants in foods often depends on the physical characteristics of the food and the solubility characteristics of the phenolics. Porter (7) originally described the "antioxidant paradox" as a phenomenon where the ability of a chain-breaking antioxidant to inhibit oxidation in bulk oils increased as the polarity of the antioxidant increased. Conversely in oil-in-water lipid emulsions, the effectiveness of lipid-soluble antioxidants increased as their polarity decreased. This phenomenon is due to the ability of polar, lipid-soluble chain-breaking antioxidants to concentrate at the oil–air interface of bulk oils (where oxidation was most prevalent) and the ability of nonpolar lipid-soluble chain-breaking antioxidants to be retained in the droplet of emulsified lipids. The structural diversity of naturally occurring phenolic antioxidants means that they vary greatly in polarity. Trolox (a water-soluble analog of tocopherol) and gallic acid inhibit lipid oxidation more effectively in bulk oils than their nonpolar counterparts, α-tocopherol and propyl gallate. In an oil-in-water emulsion, α-tocopherol is more effective at inhibiting lipid oxidation than Trolox, and methyl carnosate and carnosol are more effective than their more polar counterpart, carnosoic acid. In lipid emulsions, the phenolic antioxidants can partition into the continuous phase, the emulsifier and droplet interior. As the polarity of phenolic antioxidants decreases, the amount of the phenolic partitioning in the lipid droplet increases as does the ability of the phenolic to inhibit lipid oxidation (8–12).

Ascorbic acid and glutathione (Fig. 1) can also inactivate free radicals and act as chain-breaking antioxidants. Glutathione and ascorbate are endogenous food antioxidants, and ascorbate is often used as an antioxidant food additive. Since ascorbate and glutathione are highly water soluble, they would primarily interact with water-soluble free radicals and reactive oxygen species (4). However, ascorbate and glutathione are strong reducing compounds that can convert transition metals to their reduced state that, in turn, can promote the formation of free radicals from hydrogen and lipid peroxides (13). Therefore, ascorbate and glutathione can exhibit pro-oxidative activity if the activity of transition metals is not controlled.

Chain-breaking antioxidants will go through a series of reactions with free radicals leading to their eventual inactivation (1). However, two or more chain-breaking antioxidants can interact synergistically, resulting in antioxidant regeneration. Synergistic antioxidant interactions can occur when one chain-breaking antioxidant preferentially reacts with free radicals due to its lower bond dissociation energies or physical location (eg, it is the same environment as the free radical) (3). Such a chain breaking is very effective at inhibiting oxidation; however, it is consumed rapidly during oxidation. However, it is possible that the highly reactive chain-breaking antioxidant can interact with another chain-breaking antioxidant resulting in antioxidant regeneration. An example of such a system is α-tocopherol and ascorbic acid (14). α-Tocopherol reacts more readily with lipid radicals due to its presence in the lipid phase. Ascorbic acid can then reduce the oxidized α-tocopherol back to its active form. In turn, the resulting dehydroascorbate is regenerated by enzymes that utilize reduced nicotinamide adenine dinucleotide (NADH) or reduced nicotinamide adenine dinucleotide phosphate (NADPH) as reducing equivalents (1).

INHIBITING LIPID OXIDATION CATALYSTS

The presence and activity of lipid oxidation catalysts often determines the shelf life of foods. Therefore, control of lipid oxidation catalysts can be very important in preventing the oxidative deterioration of foods.

Transition Metals

Transition metals promote the formation of free radicals by accelerating the decomposition of peroxides (15). The activity of pro-oxidant metals can be controlled by chelators that prevent metal redox cycling, occupy the reactive

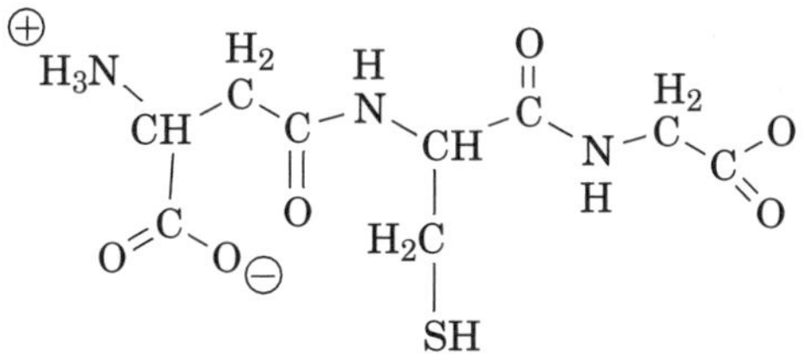

Glutathione

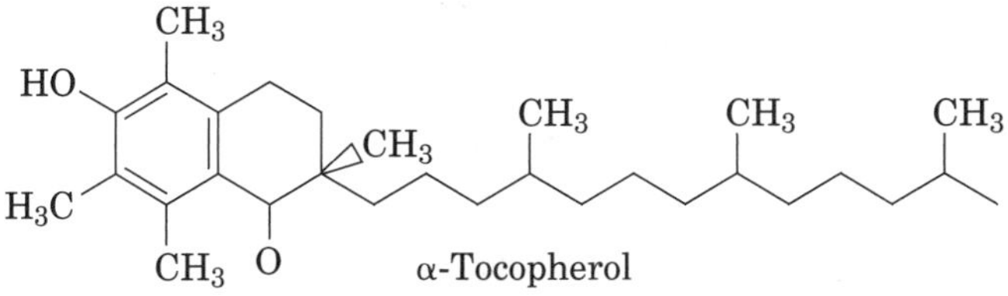

α-Tocopherol

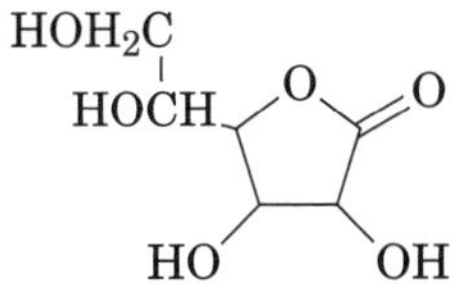

Ascorbic acid

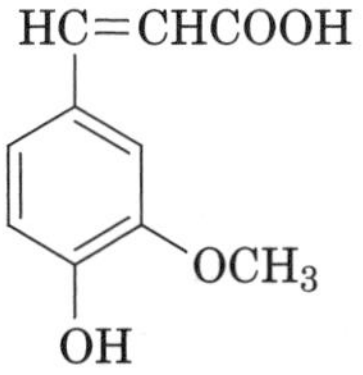

Ferulic acid

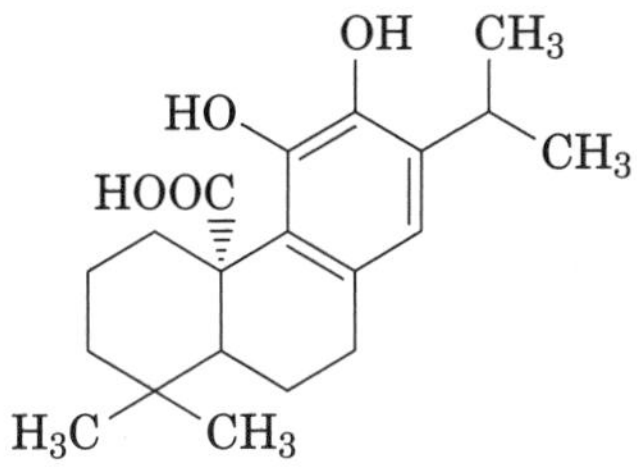

Carnosoic acid

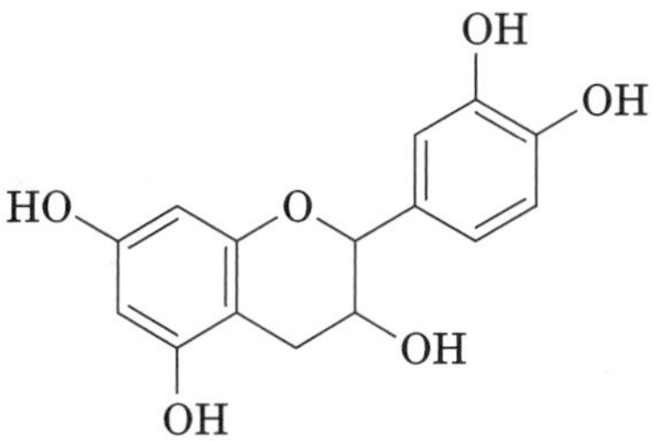

Epicatechin

Figure 1. Examples of natural chain-breaking antioxidants in foods.

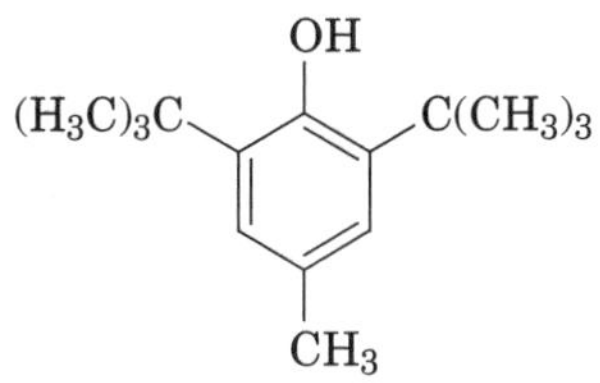

Butylated hydroxytoluene

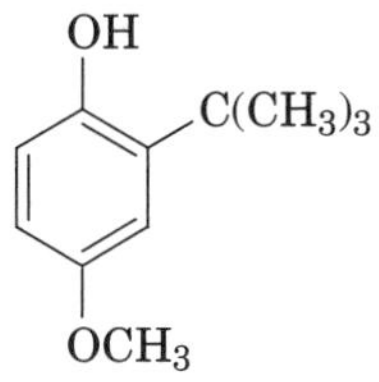

2-Butylated hydroxyanisole

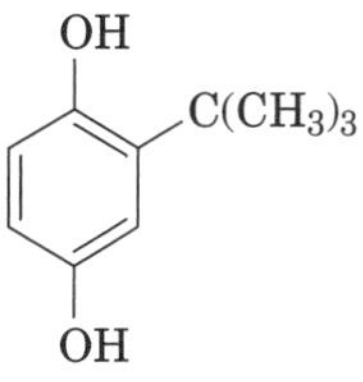

Tertiary butylhydroquinone

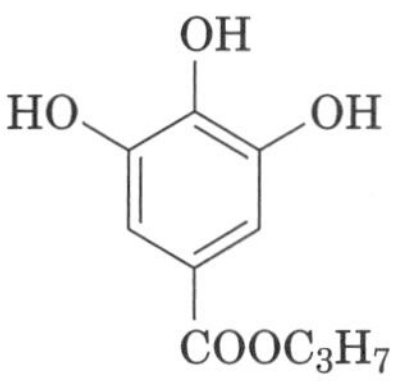

Propyl gallate

Figure 2. Synthetic chain-breaking antioxidants approved for food use.

metal coordination sites, form insoluble metal complexes, and stearically hinder interactions between metals and lipids (16). Common chelators added to foods include ethylenediaminetetraacetic acid (EDTA), polyphosphates, and citric acid (4). In addition, biological tissues from which we derive foods contain proteins and peptides that bind and inactivate metals such as serum albumin (Cu), transferrin (Fe), lactoferrin (Fe), ferritin (Fe), carnosine (Cu), and histidine (Cu) (4).

Combinations of chelators and chain-breaking antioxidants often result in increased inhibition of lipid oxidation (3). The ability of chelators to decrease metal-catalyzed free-radical generation decreases the rate at which chain-breaking antioxidants are consumed, thus making the concentration of the chain-breaking antioxidant greater at any given time.

Singlet Oxygen

Singlet oxygen is a high-energy form of oxygen that can directly interact with unsaturated fatty acids to form lipid peroxides (17). Singlet oxygen is most commonly formed from the interaction of oxygen with light-activated photosensitizers such as chlorophyll and riboflavin. Singlet oxygen can be inactivated by both chemical and physical mechanisms. Chemical inactivation occurs when singlet oxygen attacks the double bonds of compounds whose oxidation does not result in the development of rancidity (eg, carotenoids, tocopherols, amino acids, peptides, proteins, phenolics, and ascorbate) (17,18). Carotenoids can also inactivate singlet oxygen by physically quenching singlet oxygen through a transfer of energy from singlet oxygen to the carotenoid. The resulting excited state of the carotenoid then returns to its ground state by the slow dissipation of energy to the surrounding media (17). Similarly, carotenoids can absorb energy from photoactivated sensitizers, thereby decreasing their ability to promote the conversion of oxygen to singlet oxygen. For a carotenoid to be capable of the physical quenching of singlet oxygen it must possess a minimum of nine double bonds (19).

Lipoxygenases

Lipoxygenases catalyze the formation of lipid peroxides. Little is known about compounds that can directly inhibit lipoxygenase activity. Therefore lipoxygenases are most commonly inhibited by heat inactivation.

CONTROL OF REACTIVE OXYGEN SPECIES

The biological tissues from which we derive foods contain enzymes that control reactive oxygen species. The concentration and activity of the antioxidant enzymes can vary greatly in foods as a function of genetic and tissue variability and due to enzyme inactivation during food processing operations (4).

Superoxide Anion

Addition of an electron to molecular oxygen results in the formation of superoxide anion. Superoxide accelerates oxidative reactions by reducing transition metals to their more reactive states and by its conversion to the highly reactive perhydroxyl radical at low pH (15). Superoxide dismutase is an antioxidant enzyme found in foods that catalyzes the conversion of superoxide anion to hydrogen peroxide (4).

Peroxides

Both lipid peroxides and hydrogen peroxide will decompose into free radicals in the presence of transition metals, irradiation, and elevated temperatures (3). Catalase, glutathione peroxidase, and ascorbate peroxidase are antioxidant enzymes in food that convert peroxides into stable compounds. Catalase and ascorbate peroxidase are active against hydrogen peroxide, whereas glutathione peroxidase is active against both hydrogen and lipid peroxides (4,15). Hydrogen peroxide is converted to water, and lipid peroxides are converted to lipid alcohols (4).

EFFECTIVE UTILIZATION OF ANTIOXIDANTS IN FOODS

Utilization of antioxidant additives and preservation of endogenous antioxidants can be effective methods for improving food quality. Maximizing protection by endogenous antioxidants can be accomplished by minimizing antioxidant removal during the processing of raw materials, minimizing inactivation of antioxidants during processing (eg, inactivation of antioxidant enzymes during thermal processing operations) (20), selecting food sources that are naturally high in antioxidants, and including antioxidants in the diet of livestock (21). When antioxidant additives are used, several factors should be considered, including the antioxidant's mechanism of action in relation to the pro-oxidants in the food, the solubility characteristics of the antioxidant in relation to the physical location where oxidative reactions are most prevalent, how the antioxidant will interact with other food components (eg, prooxidative nature of ascorbate in the presence of transition metals), and how food processing operations will influence antioxidant stability.

BIBLIOGRAPHY

1. D. C. Liebler, "The Role of Metabolism in the Antioxidant Function of Vitamin E," *Crit. Rev. Toxicol.* **23**, 147–169 (1993).
2. F. Shahidi and J. P. K. Wanasundara, "Phenolic Antioxidants," *Crit. Rev. Food Sci. Nutr.* **32**, 67–103 (1992).
3. W. W. Nawar, "Lipids" in O. Fennema, ed., *Food Chemistry*, 3rd ed., Marcel Dekker, New York, 1996, pp. 225–320.
4. E. A. Decker, "Antioxidant Mechanisms," in C. C. Akoh and D. B. Min, eds., *Lipid Chemistry*, Marcel Dekker, New York, 1997.
5. J. T. R. Nickerson and L. J. Ronsivalli, *Elementary Food Science*, 2nd ed., AVI Publishing Company, Westport, Conn., 1980, pp. 62–63.
6. G. R. Schmidt, "Processing and Fabrication," in P. J. Bechtel, ed., *Muscle as Food*, Academic Press, Orlando, Fla., 1986.
7. W. L. Porter, "Paradoxical Behavior of Antioxidants in Food and Biological Systems," in G. M. William, ed., *Antioxidants: Chemical Physiological, Nutritional and Toxicological Aspects*, Princeton Scientific, Princeton, N.J. 1993, pp. 93–122.

8. E. N. Frankel et al., "Interfacial Phenomena in the Evaluation of Antioxidants: Bulk Oils vs. Emulsions," *J. Agric. Food Chem.* **42**, 1054–1059 (1994).
9. S.-W. Huang et al., "Antioxidant Activity of α-Tocopherol and Trolox in Different Lipid Substrates: Bulk Oils vs. Oil-in-Water Emulsions," *J. Agric. Food Chem.* **44**, 444–452 (1996).
10. S.-W. Huang et al., "Antioxidant Activity of Carnosic Acid and Methyl Carnosate in Bulk Oils and Oil-in-Water Emulsions," *J. Agric. Food Chem.* **44**, 2951–2956 (1996).
11. E. N. Frankel et al., "Antioxidant Activity of a Rosemary Extract and Its Constituents, Carnosic Acid, Carnosol, and Rosmarinic Acid in Bulk Oil and Oil-in-Water Emulsions," *J. Agric. Food Chem.* **44**, 131–135 (1996).
12. S.-W. Huang et al., "Partitioning of Selected Antioxidants in Corn Oil-in-Water Model Systems," *J. Agric. Food Chem.* **45**, 1991–1994 (1997).
13. E. A. Decker and H. O. Hultin, (1992) "Lipid Oxidation in Muscle Foods via Redox Iron," in A. J. St. Angelo, ed., *Lipid Oxidation in Foods*, American Chemical Society Symposium Series Vol. 500, American Chemical Society Books, Washington, D.C., 1992, pp. 33–54.
14. P. B. McCay, "Vitamin E: Interactions with Free Radicals and Ascorbate," *Ann. Rev. Nutr.* **5**, 323–340 (1985).
15. J. Kanner, J. B. German, and J. E. Kinsella, "Initiation of Lipid Peroxidation in Biological Systems," *Crit. Rev. Food Sci. Nutr.* **25**, 317–364 (1987).
16. E. Graf and J. W. Eaton, "Antioxidant Functions of Phytic Acid," in *Free Rad. Biol. Med.* **8**, 61–69 (1990).
17. D. G. Bradley and D. B. Min, "Singlet Oxygen Oxidation of Foods," *Crit. Rev. Food Sci. Nutr.* **31**, 211–236 (1992).
18. T. A. Dahl, W. R. Midden, and P. E. Hartman, "Some Prevalent Biomolecules as Defenses against Singlet Oxygen Damage," *Photochem. Photobiol.* **47**, 357–362 (1988).
19. P. DiMascio, P. S. Kaiser, and H. Sies, "Lycopene as the Most Efficient Biological Carotenoid Singlet Oxygen Quencher," *Arch. Biochem. Biophys.* **274**, 532–538 (1989).
20. L. Mei, A. D. Crum, and E. A. Decker, "Development of Lipid Oxidation and Inactivation of Antioxidant Enzymes in Cooked Beef and Pork," *J. Food Lipids* **1**, 273–283 (1994).
21. C. Faustman, "Food from Supplement-Fed Animals," in J. Smith, ed., *Technology for Reduced Additive Food*, Blackie Academic & Professional, New York, pp. 160–194, 1993.

ERIC A. DECKER
University of Massachusetts
Amherst, Massachusetts

AOAC INTERNATIONAL

AOAC INTERNATIONAL is a scientific association whose main purpose is to promote methods validation and quality measurements in the analytical sciences. AOAC also develops and validates chemical and microbiological analytical methods used in laboratories worldwide to analyze foods, feeds, agricultural and industrial chemicals, pharmaceuticals and cosmetics, water, soil, disinfectants, forensic materials, and many other substances.

HISTORY

AOAC INTERNATIONAL, originally known as the Association of Official Agricultural Chemists, was formed in 1884 to adopt uniform methods of analysis for fertilizers. Over the years, the number of areas of analytical sciences represented in AOAC increased steadily. A name change in 1965 to the Association of Official Analytical Chemists partially reflected the expansion of AOAC's analytical interests. By 1991 the Association had long ceased to be confined to regulatory (official) analytical chemists from North America—having a majority of members working in the private sector, a significant number involved in microbiology and other forms of analysis, and more than one-fourth living outside the United States. Consequently, in that year, the name was changed to AOAC INTERNATIONAL, thereby retaining the initials by which the Association had been known for more than 100 years, eliminating reference to a specific scientific discipline or profession, and reflecting the international membership and focus of AOAC.

MEMBERSHIP

Individual Membership

Individual members work in private industry, government, and academia in more than 85 countries worldwide. They include analytical chemists, microbiologists and other biologists, biochemists, toxicologists, forensic scientists, and other scientists in laboratory, administrative, and management positions. To qualify, they must have an interest in the purpose and goals of the Association, have a degree in science, and be engaged (directly or indirectly) in analysis or analytical science in fields relevant to the purpose of AOAC INTERNATIONAL.

Sustaining Membership

Sustaining member organizations are private firms, institutes, associations, and local, state, provincial, and national governmental agencies with an interest in AOAC INTERNATIONAL's mission. They are engaged, directly or indirectly, in analysis or research in the areas of interest to AOAC.

Sections of AOAC INTERNATIONAL

The AOAC INTERNATIONAL sections program was initiated in 1981. Geographically organized, it provides a forum in which members of AOAC and other analytical scientists can more easily and regularly share information. Most sections have an annual meeting and several produce newsletters and/or have their own websites.

Awards Programs

AOAC provides annual awards for scientific achievement, scientific potential, and service to the association. AOAC's most prestigious award, The Harvey W. Wiley Award, is presented each year to a scientist (or group of scientists) who has made an outstanding contribution to analytical methodology in an area of interest to the Association. AOAC's annual scholarship award is given to a worthy upper-level undergraduate or graduate student majoring in a field of study that supports the mission of the Association. The Fellow of AOAC INTERNATIONAL Award recognizes the dedication of the volunteers who serve the Association.

TECHNICAL SERVICES

The AOAC® Official Methods℠ Program

The AOAC® *Official Methods*℠ program is designed to provide high-quality validated methods that can be used with confidence by regulatory agencies, regulated industry, product testing laboratories, academic institutions, and others. Proposed AOAC® *Official Methods*℠ are subjected to a multilaboratory collaborative study according to internationally recognized standards and receive rigorous scientific review of performance results. Adopted methods are published in the *Journal of AOAC INTERNATIONAL* and in *Official Methods of Analysis of AOAC INTERNATIONAL*.

The AOAC® Peer-Verified Methods℠ Program

This AOAC INTERNATIONAL method validation program, initiated in 1993, is the program of choice when speed of validation is essential and a lesser degree of confidence is acceptable. The performance characteristics of these methods are checked in one or more independent laboratories and reviewed against specific performance standards. Accepted *Peer-Verified Methods*℠ are listed in AOAC's magazine and journal, and copies are available for purchase.

The AOAC® Performance Tested Methods℠ Program

The AOAC® *Performance Tested Methods*℠ Program provides an independent third-party review for proprietary methods to confirm manufacturers' performance claims. This program is administered by the AOAC Research Institute, a subsidiary of AOAC INTERNATIONAL. Successfully validated test kits are licensed to use the AOAC Research Institute "Performance Tested" seal.

The AOAC® Laboratory Proficiency Testing Program

Developed in 1998, the AOAC® Laboratory Proficiency Testing Program provides a means for laboratories to test the quality and accuracy of their analytical results and benchmark them against those of other participating laboratories. The program supports laboratory accreditation and international standards through compliance with ISO guides.

TECHNICAL DIVISIONS

Technical Division on Reference Materials

The Technical Division on Reference Materials (TDRM) promotes the use of reference materials to verify the accuracy of methods. Liaisons from TDRM work closely with the AOAC methods committees to define needs for reference materials. TDRM plans and conducts educational and informational activities on the use of reference materials.

Technical Division on Laboratory Management

The Technical Division for Laboratory Management (TDLM) provides a forum for networking, information sharing, and problem solving on issues of mutual interest for managing an efficient, cost-effective, quality laboratory. It offers opportunities for leadership and mentoring as well as opportunities for learning and professional enhancement. TDLM maintains a listserv to discuss topics such as quality assurance, training, personnel management, safety, downsizing, budgeting, laboratory information management systems (LIMS), proficiency testing, and accreditation.

MEETINGS AND EDUCATION

Meetings

The annual meeting and exposition is the focal point of AOAC INTERNATIONAL's yearly work. The scientific/technical program includes symposia on a variety of topics of current interest as well as presentations of contributed papers, collaborative studies, and poster sessions organized to correspond with the topic areas of the methods committees. The program also includes workshops on methodology, forums on specialized subjects, and a panel presentation and round table discussion on regulatory topics. At its exposition, the annual meeting offers an opportunity for attendees to be introduced to and acquire information about new products and services offered by vendors of analytical equipment, services, and publications.

Training Courses

AOAC INTERNATIONAL has assumed a leadership role in providing continuing education opportunities for all members of the analytical community, including the scientists who use or develop methods, quality assurance personnel, and laboratory managers. In addition to courses scheduled by AOAC, the Association can also make arrangements with organizations to present AOAC training courses at their location.

PUBLICATIONS

Official Methods of Analysis of AOAC INTERNATIONAL

Considered to be the analytical methods "bible" by thousands of analysts who use and depend on it, *Official Methods of Analysis of AOAC INTERNATIONAL* (*OMA*) is AOAC's premiere publication. Methods validated and adopted by AOAC are published in *OMA*, and the book is updated annually via revisions available on a subscription basis to purchasers of the book.

Journal of AOAC INTERNATIONAL

The *Journal of AOAC INTERNATIONAL* publishes original research articles and reports on the development, validation, and interpretation of analytical methods for agricultural commodities, foods, drugs, cosmetics, and the environment. A limited number of invited reviews and feature articles on selected subjects are also published, as well as the proceedings from symposia. The *Journal of AOAC INTERNATIONAL* publishes fully refereed reports on developing, improving, and testing methods, including reports on the collaborative studies that precede and support adoption of AOAC® *Official Methods*SM.

Inside Laboratory Management

Inside Laboratory Management (ILM) features articles on laboratory management, regulatory news, safety concerns, new instrumentation, and emerging technologies; publishes news of AOAC activities; disseminates information to and among methods volunteers; provides notice on stages of methods in progress in the AOAC® methods validation programs; offers a vehicle for communication among the AOAC constituencies; and generally promotes the purposes, activities, and accomplishments of the Association and its members.

MURI DUEPPEN
AOAC INTERNATIONAL
Gaithersburg, Maryland

APHRODISIACS AND STIMULANTS

When one thinks of aphrodisiacs, the image of oysters comes to mind. Yet there is no scientific evidence (short of possible pheromonal components of truffles [5α-androst-16-en-3α-ol] and celery [5α-androst-16-en-3α-one]) (1,2) that most of the traditional foods considered as aphrodisiacs have any aphrodisiac qualities at all. Alternatively, there is considerable evidence that foods can act to inhibit sexual arousal. In particular, alcohol has a marked inhibiting effect on sexual functions in both men and women (3,4). Shakespeare described, "It provokes the desire, but it takes away the performance" (5). Scientific research has just begun to investigate the aphrodisiacal effects of food on sexual arousal.

COMPONENTS OF FOOD

When one thinks of giving chocolates on Valentine's Day to a loved one, we hope this acts as an aphrodisiac . . . but does it? To understand this better, the sensory components of food need to be delineated. These include texture, visual appearance, viscosity, tactile properties, and gustatory and olfactory components.

A Twinkie, a cucumber, and a doughnut are particularly Freudian based on their shape and visual appearance. The texture and viscosity of oysters, likewise, may induce reminiscence of sexual organs. Alternatively, the sexual experience surrounding the ingestion of a food may also cause the food itself to become endowed with a sexual valence, which may be aphrodisiacal in nature. Another possible mechanism is not through achieving a direct sexual meaning, but rather, a food can become sexually charged by becoming indirectly associated with sexual function, for example, by being considered a "guilty pleasure." Since in our puritan society, sex and other hedonic pleasures are frowned upon and repressed, one pleasure can be unconsciously substituted for another, while retaining the same underlying meaning. Eating chocolate or strawberries and whipped cream may be viewed in a sinful way as a reward of "badness for good behavior" (as a dessert after having eaten the meal of brussels sprouts), or as a violation of an unwritten moral code that elevates these foods into a nonessential, potentially dangerous, risky (ie, exciting) food, and hence strongly desired. Thus, eating these foods is perceived on some level as a violation of social morals and revolt against hedonic repression, which may be manifested by sexual arousal. Another nonfood example of this repressed hedonic affective overflow is the Las Vegas effect of gambling, being perceived as a "naughty" thing to do, and hence becoming imbued with components of sexual arousal. Likewise, sinful foods may come to be endowed with other perceived sinful qualities, and this would include sexual arousal.

Alternatively, the taste of the food, that is, sweet, sour, bitter, or salty, if endowed with a high degree of sugar, may induce release of endorphins and norepinephrine, which may act to stimulate sexual arousal. Alternatively, sweet-induced reactive hypoglycemia may act to inhibit the associated sexual arousal. Sexual arousal in the male is manifested by erection and is predominantly parasympathetically mediated (6). Thus foods that induce parasympathetic activation (and inhibit sympathetic discharge) may have the greatest effect of inducing sexual arousal.

Last, the smell of the food can influence sexual arousal. Since approximately 90% of what is perceived as taste is really smell, the smell of food may have a far greater effect than is traditionally given credit in assessing food's overall effect on behavior (7). This can be demonstrated by holding one's nose while eating chocolates: it produces a taste of chalk. This is a manifestation of an olfactory synesthesia. The concept of synesthesia is when one sensation is misperceived as another sensation. For instance, pressing on the eyeballs induces an illusion of the presence of light, yet it is the pressure that is misperceived as light. In the olfactory realm, odors that are sniffed from outside the nose going up to the olfactory epithelium at the top of the nose are interpreted as smell. Alternatively, odors that do not go through the orthonasal route rather through the retronasal route, from the mouth up through the nasopharynx in the back of the throat to the top of the nose, are interpreted as taste.

SMELL AND SEXUAL AROUSAL

Throughout history aromas have been linked with sexual response. Perfume jars were preserved in the chambers designed for sexual relations in volcanic remnants of ancient Pompeii. Ancient Sumarians used perfumes to entice women. Traditional Chinese rituals highlight the relationship between smell and affection. Hieroglyphics also relate the two. Virtually all cultures use perfume in their marriage rites. In the modern world as well, the connection is pervasive. Use of perfumes and colognes as romantic enticements are a multibillion dollar business (8). The prominence of the historical connection between odors and sex among diverse cultures implies a high level of evolutionary importance. Could it be inbred within the brain of humans?

Freud, almost a hundred years ago, in *Civilization and Its Discontents* recognized the relationship between olfaction and sexual arousal. He suggested that to remain a civilized society we had to repress our olfactory instincts, otherwise humans would walk around sexually excited all the time (9). Freud, followed by Bieber and later analysts, delineated the connection between smell and sexual development (10). This linkage indicates a specific point in time in the development of Oedipal conflicts. When children recognize the smell of the same-sex parent and the scent of the opposite-sex parent, they learn to dislike the smell of the same sex and transfer positive feelings toward the smell of the opposite-sex parent (11). This establishes a psychodynamic interrelationship between smell and sexual identity. The transfer of emotions to a parent's odor imbues sexual meaning to otherwise inert odors (12). In those who are unable to work through this phase of Oedipal development, fetishes develop as in Freud's rat man (13).

Freud's analysis followed a millennia of pervasive cultural myth and practice that linked sexuality and olfaction. In Western culture the court jester wore a mask with a phalliclike nose (14). Native cultures practiced the Eskimo kiss (15). The Pacific Islanders engaged in a similar type of nasal kiss (16).

This cultural awareness may derive from the unique characteristics of human neuroanatomy. Humans have the largest limbic system compared with body size of any primate (17). The olfactory lobe directly connects with the limbic system (18). The relative size of the limbic system and the extensive olfactory/limbic interconnection implies olfaction has a significant effect on human behavior. The effect of external stimuli on the olfactory system and subsequent effect on behavior is well documented (19–23). Erectile tissue exists in both human genitals and within structures in the nose (24). With sexual arousal and excitation there is engorgement of tissue not only in the genitals but also in the nose (25). Other anatomical evidence is that common neurotransmitters facilitate both olfactory and sexual functioning (26,27).

Besides purely anatomical similarities, the linkage between olfactory function and sexual function is recognized in a clinical setting. More than 17% of individuals with chemosensory dysfunction develop impaired sexual desire or other sexual dysfunction (28). Genetic disorders can affect both systems; for example, those with Kalliman's syndrome have both olfactory deficit and impaired sexual drive and functioning (29). Other diseases impair olfactory ability and sexual functioning concomitantly, including cerebral vascular disorders (18,30), Parkinson's disease (31,32), senile dementia of the Alzheimer's type (33), hypothyroidism (34,35), and vitamin deficiency states including B_{12} deficiency (36).

Research into the olfactory–sexual link has reached the laboratory as well as the physician's office. Olfactory-related sexual behavior has been well documented in laboratory animals. When the olfactory bulb is lesioned in hamsters, it causes an impaired sex drive (37). The lesion does not have to be in the brain itself; for example, lesion of the olfactory mucosa external to the brain, which prevents odorant molecule binding, similarly causes impaired sexual function and reduced sex drive (38). Homologous lesions in mice cause impaired copulatory ability (39,40). Alternatively, in the female rat, lesion of the olfactory bulb leads to an increase in mating behavior suggesting indiscriminate sexual activity (41). This olfactory–sexual connection is further highlighted in lab animals since castration of the animal leads to not only impaired sexual drive but also impaired olfactory functioning (42). Alternatively, ovariectomy leads to not only impaired sexual functioning but also reduced olfactory ability (43,44). Lesions of the olfactory bulb, or even of the nasal cartilage alone, cause both olfactory deficit (45,46) and malformation of the developing animal's sex organs. These studies suggest a strong link among olfaction, olfactory organs, and sexual functioning.

Following the strong evidence from laboratory animals, significant research has been accomplished in humans, especially regarding the sexual olfactory linkage in women. This could be because women's ability to smell is greater than that of men (47). There exists a fluctuation in olfactory ability concurrent with menstrual cycle (48), such that at the time of ovulation a woman's ability to smell a musk-like substance is about 100,000 times better than that of men (49). This increased olfactory ability at the time of ovulation is even prominent in women who are taking progesterone (50). A woman's superior ability to smell, as compared with man, does not occur until puberty, around the time when apocrine glands obtain their distinct "scents" smell (51). Unlike other sensory modalities, which do not increase at time of puberty (51), olfactory ability does, demonstrating the importance of interrelationship between sexual development and olfaction.

Many explanations exist to describe this olfactory–sexual connection; the first to be discussed here are pheromones. While pheromones have been well characterized extensively in animals, they have not been fully characterized in humans, although much evidence suggests that pheromones also exist in humans. For example, pheromones exist throughout the animal kingdom, including insects; mechanisms of such evolutionary importance that exist throughout so many species usually exist in humans as well (52).

Anatomic evidence suggesting pheromones may exist in humans is the presence of apocrine glands in the axilla and surrounding the genitals. Unlike eccrine, or sweat glands, apocrine glands secrete a high-density steroid substance (53). In nonhuman primates, pheromones are released from these glands (54). In these sites we find hair tufts, creating greater surface area, which would allow greater dispersion of any pheromone that may be present (55). Apocrine glands are larger in men than in women (56), which may act in harmony with women's greater olfactory ability; for example, it would do no good to secrete pheromones that cannot be detected. Further anatomical evidence is the existence of the vomeronasal organ (VNO). The VNO in some mammals is where pheromones act (57). In humans this organ is of unknown significance and until recently was thought to be only of vestigial origin (58,59).

It is now postulated to be the site where human pheromones act.

Physiologically one sees specific changes suggesting the presence of pheromones. For example, change in olfactory ability is associated with the ovulatory cycle as already described (50). Another change is the nasal engorgement at time of sexual excitation (60). This engorgement allows for greater olfactory ability, since the odorants form eddy currents in the nose, which do not go directly down to the lung, but rather reach the olfactory epithelium at the top of the nose (61). This mechanism would enhance an individual's ability to detect any potential pheromones that may be present.

Further evidence for the existence of pheromones is based on clinical observations. McClintock described the development of menstrual synchrony in an all-women dormitory (62). While entering as freshmen the women's menstrual cycles were asynchronous but became synchronized by midterm, suggesting that there was a pheromone present acting to synchronize the women. Likewise, increases in copulatory behavior occur in married couples associated with time of ovulation, suggesting that a pheromone may be present that stimulates the male to respond or self-stimulates arousal in the women and the men respond to this associated behavior (63). Michael et al. demonstrated that when a potential pheromone (copulant) was placed on the chest of women, as compared with a placebo odorant, there was an increase in copulation associated with the use of the copulant (64).

Pheromones that have received the most attention have been androstenol and androstenediol, both pig pheromones (1). Androstenol is reported to be less active than androstendiol, and there is an equilibrium whereby androstendiol is easily converted to androstenol. Kirk-Smith et al. placed androstendiol on surgical masks and had individuals rating slides, pictures of men and women (65). Both men and women tended to rate the figures of women as prettier, and so on, in the presence of the androstendiol. When androstendiol was placed on surgical masks in another experiment, and participants were asked to rate other individuals, they tended to rate the individuals in a more sexually oriented matter (66).

That women have a greater ability to detect potential pheromones is of significance since the propagation of the species is based on their sexual activity, and, hence, birth rates are maximized by women's appropriate response to pheromones. Existence of pheromones is also suggested based on territorial markings. Gustavson et al. placed androstenol and androstenediol underneath bathroom stalls in a men's room and watched where men sat (67). They tended to avoid those stalls with the androstenol, suggesting that it may act not only as a pheromone but also as a repellent or territorial marker for other males.

While pheromones are one mechanism used to explain the olfactory-sexual link, there are several others. For example, odors could act on sexual arousal through a learned conditioned response (68). If the odor is one that is associated in the past with being sexually aroused, it might induce a sexually arousing mechanism. This could be a primary conditioned response or through secondary effects, for instance by inducing a more positive mood state or relaxed state. Inducing a more positive or relaxed state might cause a reduction in inhibitions. Positive moods and relaxed states can be achieved either directly through a conditioned response, through a learned response, or through the phenomenon of olfactory-evoked nostalgia where an odor induces a positive mood state in an individual as a result of recalling the past (69).

Alternatively, odors may act to enhance sexual arousal by actually acting directly on areas of the brain that induce sexual arousal (eg, the septal nucleus), by acting directly, almost as a drug, influencing them through dopaminergic, cholinergic, or serotinergic mechanisms (70). Odors induce a change in dopamine in the olfactory bulb and its connections, and medications that have been said to induce sexual arousal include those that are known to affect dopamine (this side effect is seen in L-dopa treated Parkinson's patients) (71).

Alternatively odors could affect sexual arousal by inducing a state of risk taking or of generalized pleasure seeking, as in seeking food or other pleasure-oriented responses (72). Perhaps odors could inhibit associated cortical functioning, allowing a release of the id or the underlying limbic system functioning, thus manifesting more primitive responses (73). This release is seen in decorticate animals as in the Kluver-Bucy syndrome or in children with developmental cortical deficits, as in Down syndrome patients who become obese in response to their relatively uninhibited appetite. Similar responses are seen in individuals who became more tired, thus cortically suppressed, and hence more easily induced to sexual arousal or eating. Similarly, cortical suppression with alcohol leads to initially a lack of discrimination for sexual partners. Alternatively, this may be because of alcohol-induced inhibition of olfactory reception (74). This is seen in olfactory-deprived rats, which engage in indiscriminate mating.

Environmental stimuli that affect one sex in a species also usually affect the opposite sex. In a study by Hirsch and Gruss, odors were found to affect male sexual arousal (75). Each of the 30 odors produced an increase in penile blood flow. The combined odor of lavender and pumpkin pie had the greatest effect, increasing median penile blood flow by 40%.

Table 1 shows the median percentage increases in penile blood flow produced in the 31 participants by each of the odors. Second in effectiveness was the combination of black licorice and doughnut, which increased the median penile blood flow 31.5%, followed by the combination of pumpkin pie and doughnut, which produced a 20% increase. Least effective was cranberry, which increased penile blood flow by 2%. None of the odors tested were found to reduce penile blood flow.

Among individuals with normal olfactory ability, there were several significant correlations: higher brachial penile indices correlated with greater age and with greater responses to the odor of vanilla ($p = 0.05$); self-assessed level of sexual satisfaction correlated with greater responses to the odor of strawberry ($p = 0.05$); and frequency of sexual intercourse correlated with greater responses to the odors of lavender ($p = 0.03$), oriental spice ($p = 0.02$), and cola ($p = 0.03$).

Table 1. Increases in Penile Blood Flow Produced by Various Odors on 31 Individuals

Odor	Median % increase
Lavender and pumpkin pie	40.0
Doughnut and black licorice	31.5
Pumpkin pie and doughnut	20.0
Orange	19.5
Lavender and doughnut	18.0
Black licorice and cola	13.0
Black licorice	13.0
Doughnut and cola	12.5
Lily of the valley	11.0
Buttered popcorn	9.0
Vanilla	9.0
Pumpkin pie	8.5
Lavender	8.0
Musk	7.5
Cola	7.0
Doughnut	7.0
Peppermint	6.0
Cheese pizza	5.0
Roasting meat	5.0
Parsley	4.5
Cinnamon buns	4.0
Green apple	3.8
Rose	3.5
Strawberry	3.5
Oriental spice	3.5
Baby powder	3.3
Floral	3.0
Chocolate	2.8
Pink grapefruit	2.5
Cranberry	2.0

Table 2. Summary of Results in Study of the Effect of Odors on Female Sexual Arousal

Average change in vaginal blood flow (%)	Change in vaginal blood flow (%)
Good and Plenty and cucumber	+13
Baby powder	+13
Pumpkin pie and lavender	+11
Baby powder and chocolate	+4
Perfume	+0–1
Cologne	−1
Good and Plenty	−12
Charcoal barbeque meat	−14
Cherry	−18
Women who find masturbation extremely arousing	
Good and Plenty and banana nut bread	+28
Good and Plenty and cucumber	+22
Perfume	+18
Women who find masturbation not arousing or repulsive	
Baby powder	+16
Pumpkin pie and lavender	+10
Good and Plenty	−20
Women who are positively aroused by genital finger manipulation by a partner	
Good and Plenty and cucumber	+18
Pumpkin pie and lavender	+12
Women who are not aroused or are inhibited by finger manipulation by a partner	
Perfume	−14
Cologne	−14
Good and Plenty and cucumber	−13
Women who are not usually multiorgasmic in sexual encounters with partner	
Baby powder	+15
Women who are usually multiorgasmic in sexual encounters with partner	
Baby powder	−8

Odors found that consistently increased penile blood flow were those of food. If odors can increase penile blood flow, the ramifications extend beyond mere aphrodisiacs. Might odors be used to treat vasculogenic impotence, as in diabetics? Or if an odor could be found that decreases penile blood flow, might it be used to treat sexual deviants, such as pedophiles, as part of their aversion training?

In a homologous manner, odors' effect on women's sexual arousal were tested with results summarized in Table 2. Several questions remain as to why food odors enhance female sexual arousal. Possible mechanisms by which odors can induce female sexual arousal include Pavlovian conditioned response olfactory-evoked nostalgia, or physiological. Odors may act to induce the Pavlovian conditioned response (69) and may act to release sexual instincts or basic appetites for hedonic experiences including sexual arousal. Alternatively, there may be a conditioning response to the smell of food if this was something that frequently preceded enjoyable sexual experiences in the past, that induced arousal. Hence, in a conditioned response paradigm, reintroduction to food may induce the phenomena of sexual arousal. It's likely it is smell rather than the taste that elicits this response since approximately 90% of taste is actually smell (76). By eating food prior to sexual arousal, the women were primarily experiencing smell rather than taste, thus odors acted as the conditioning stimulus. Presentation of these odors may have acted to focus these women's attentions on the stimulus rather than on the external environment within the laboratory.

Another explanation is that the odors may have acted through a distractor effect. The women might think about the odor rather than think about all the negative aspects of being inside a laboratory. This would then remove associated negative emotional tone and hence remove inhibitions, allowing sexual arousal to occur.

Another potential mechanism is the phenomenon of olfactory-evoked nostalgia. Approximately 86% of women described a strong olfactory-evoked nostalgic response where odor induced a vivid memory of one's childhood (77). The odors might have induced an olfactory-evoked response of these women's childhood. Childhood memory is often associated with positive emotions. By inducing the women to be in a more positive mood state, odors may have reduced inhibitions, therefore allowing sexual arousal.

Memories of one's childhood might induce not only a positive mood state but also a sense of security and safety. These feelings would then be reintroduced to the women on exposure to the odor, allowing them to be less inhibited. By feeling safe and secure, urges would come forth as manifested by enhanced vaginal blood flow. Alternatively, the odors may have induced recall of sexual experiences in one's adolescence, which then would have a direct induction of sexual arousal.

Another possible mechanism by which odors could impact female sexual arousal is by directly acting in a physiological manner. The odors could have affected perception of another sensation, for example, touch or pain. Touch perception threshold may have been reduced by the odors. Hence, the women were able to feel the vaginal probe to a greater degree than they were without the odor present. This increased touch perception may have induced the vaginal probe to become a sexually arousing stimulus. Studies suggest that odors can enhance tactile perception (78), as well as affect perceptional thresholds in other forms, that is, perception of external space (79). Or the odors might not affect the touch threshold so much as the awareness of the touch threshold. Another possible physiological mechanism by which odors may enhance sexual arousal is by reducing pain. By reducing pain, odors may have acted to remove inhibitions. A low level of pain may have existed in these women as a result of the vaginal probe and odors may have reduced this pain. Evidence that odors can influence pain has been demonstrated in migraine headaches (21).

Another possible mechanism arises from the olfactory link to the limbic system. Odors, by inducing a strong emotional response associated with limbic system firing, may induce a neighborhood effect. Overflow limbic system discharges may project to all limbic system structures, including the septal nuclei, the sexual arousal center. Hence, the generalized emotional response to odors may have acted physiologically to induce arousal.

The effects of specific odors must also be considered. For example, an evolutionary hypothesis may be entertained in trying to understand why primarily food odors affect sexual arousal. Primitive humans congregated around food kills (80). Perhaps there they had the most opportunities to procreate. An increase in sexual arousal in response to food odors, therefore, may have conferred an evolutionary advantage for survival.

Humans can detect approximately 10,000 odors (81). Only 46 were examined. Studies have suggested that many odors affect behavior; for example, floral smells affect learning (82) and buying behavior (83), green apple affects claustrophobia, barbecue smoke induces agoraphobia (79), and food odors affect weight loss (84). Quite possibly other food odors would affect vaginal and penile blood flow more than those we examined.

BIBLIOGRAPHY

1. R. Claus, H. O. Hoppen, and H. Karg, "The Secret of Truffles: A Steroidal Pheromone?" *Experientia* **37**, 1178–1179 (1981).
2. R. Claus and H. O. Hoppen, "The Boar-Pheromone Steroid Identified in Vegetables," *Experientia* **35**, 1674–1675 (1979).
3. G. M. Farkas and R. C. Rosen, "Effect of Alcohol on Elicited Male Sexual Response," *J. Stud. Alcohol* **37**, 265–272 (1976).
4. G. T. Wilson and D. M. Lawson, "Effects of Alcohol on Sexual Arousal in Women," *J. Abnorm. Psychol.* **85**, 489–497 (1976).
5. W. Shakespeare, *Macbeth*, Act 2, Scene 3.
6. H. D. Weiss, "The Physiology of Human Penile Erection," *Ann. Intern. Med.* **76**, 792–799 (1972).
7. A. R. Hirsch, "Scentsation, Olfactory Demographics and Abnormalities," *Int. J. Aromatherapy* **4**(1), 16–17 (1992).
8. P. Dicther, "Second Hand Scents," *Drug and Cosmetic Industry* **156**, 72–75 (1995).
9. S. Freud, "Bemerkunger uber einen Fall von Zwangs Neurosa," *Ges. Schr.* **VIII**, 350 (1908).
10. I. Bieber, "Olfaction in Sexual Development and Adult Sexual Organization," *Am. J. Psychother.* **13**, 851–859 (1959).
11. I. Bieber, T. B. Breber, and R. C. Friedman, "Olfaction and Human Sexuality: A Psychoanalytic Approach," in M. J. Serby and K. L. Chobor, eds., *Science of Olfaction*, Springer-Verlag, New York, 1992, pp. 396–409.
12. M. Kalogerakis, "The Role of Olfaction in Sexual Development," *Psychosom. Med.* **25**, 420–432 (1963).
13. S. Freud, "Notes Upon a Case of Obsessional Neurosis," (1909) in J. Strachey, ed. and trans., *The Standard Edition of the Complete Psychological Works of Sigmund Freud*, Vol. 10, Hogarth Press, London, United Kingdom, 1962, p. 247.
14. S. Freud, "Report on My Studies in Paris and Berlin 1886," (1892–1899), in J. Strachey, ed. and trans., *The Standard Edition of the Complete Psychological Works of Sigmund Freud*, Vol. 1, Hogarth Press, London, United Kingdom, 1962, pp. 173–280.
15. A. A. Brill, "The Sense of Smell in the Neuroses and Psychoses," *Psychoanal. Q.* **1**, 7–42 (1932).
16. I. Eibl-Eibesfeldt, *Human Ethology*, Aldine de Gruyter, New York, 1989.
17. E. Armstrong and M. Onge, "Number of Neurones in the Anterior Thalamic Complex of the Primate Limbic System," *Am. J. Phys. Anthropol.* **54**, 197 (Abstract) (1981).
18. R. B. Goodspeed, J. F. Gent, and F. A. Catalanotto, "Chemosensory Dysfunction: Clinical Evaluation Results from a Taste and Smell Clinic," *Postgrad. Med.* **81**, 251–260 (1987).
19. A. R. Hirsch and R. Gomez, "Weight Reduction Through Inhalation of Odorants," *J. Neurol. Orthop. Med. Surg.* **16**, 28–31 (1995).
20. A. R. Hirsch, "Effects of Ambient Odors on Slot Machine Usage in a Las Vegas Casino," *Psychology and Marketing* **12**, 585–594 (1995).
21. A. R. Hirsch and C. Kang, "The Effects of Green Apple Fragrance on Migraine Headache," *Headache* **37**, 312 (1997).
22. A. R. Hirsch and L. H. Johnson, "Odors and Learning," *J. Neurology Ortho Med Surg* **17**, 119–126 (1996).
23. A. R. Hirsch, "Negative Health Effects of Environmental Odor Pollution: A Review," in J. Lee and R. Phalen, eds., *The Proceedings of the Second Colloquium on Particulate Air Pollution and Human Health*, 1996, pp. 4-207 to 4-215.
24. H. E. Hamlin, "Working Mechanisms for the Liquid and Gaseous Intake and Output of the Jacobson's Organ," *Am. J. Physiol.* **91**, 201–205 (1929).

25. P. Cole, *"Respiratory Role of the Upper Airways,"* Mosby-Yearbook, St. Louis, Mo., 1993, pp. 8–93.
26. N. Halasz and G. M. Shepherd, "Neurochemistry of the Vertebrae Olfactory Bulb," *Neurosci.* **10**, 579–619 (1983).
27. J. G. Herndon, Jr., "Effects of Midbrain Lesions on Female Sexual Behavior in the Rat," *Physiol. Behav.* **17**, 143–148 (1976).
28. A. R. Hirsch and T. J. Trannel, "Chemosensory Disorders and Psychiatric Diagnoses," *J Neurol Orthop Med Surg* **17**, 25–30 (1996).
29. F. J. Kallann, W. A. Schoefeld, and S. E. Barrera, "The Genetic Aspects of Primary Eunuchoidism," *Am. J. Ment. Defic.* **48**, 203–236 (1943).
30. E. D. Goddess, N. N. Wagner, and D. R. Silverman, "Post-stroke Sexual Activity of CVA Patients," *Med. Aspects Hum. Sex.* **13**, 16–30 (1979).
31. R. L. Doty, D. Deems, and S. Stellar, "Olfactory Dysfunction in Parkinson's Disease: A General Deficit Unrelated to Neurologic Signs, Disease Stage or Disease Duration," *Neurology* **38**, 1234–1237 (1988).
32. F. Boller and E. Frank, *Sexual Dysfunction in Neurological Disorders, Diagnosis, Management, and Rehabilitation*, Raven Press, New York, 1982, p. 50.
33. R. L. Doty, P. F. Reyes, and T. Gregor, "Presence of Both Odor Identification and Detection Deficits in Alzheimer's Disease," *Brain Res. Bull.* **18**, 597–600 (1987).
34. R. J. McConnell et al., "Defects of Taste and Smell in Patients with Hypothyroidism," *Am. J. Med.* **59**, 354–364 (1975).
35. R. C. Kolodny, W. H. Masters, and V. E. Johnson, *Textbook of Sexual Medicine*, Little, Brown, Boston, 1979, pp. 142–143.
36. R. W. Rundles, "Prognosis in the Neurologic Manifestations of Pernicious Anemia," *Blood* **1**, 209–219 (1946).
37. M. Devor and M. Murphy, "The Effects of Peripheral Olfactory Blockade on the Social Behavior of the Male Golden Hamster," *Behav. Biol.* **9**, 31–42 (1973).
38. P. D. MacLean, *Triune Concept of the Brain and Behavior*, University of Toronto Press, Toronto, 1973.
39. K. Larsson, "Impaired Mating Performances in Male Rats After Anosmia Induced Peripherally or Centrally," *Brain Behav. Evol.* **4**, 463–471 (1971).
40. D. A. Edwards, M. L. Thompson, and K. G. Burge, "Olfactory Bulb Removal vs. Peripherally Induced Anosmia: Differential Effects on the Aggressive Behavior of Male Mice," *Behav. Biol.* **7**, 823–828 (1972).
41. R. L. Moss, "Modification of Copulatory Behavior in the Female Rat Following Olfactory Bulb Removal," *J. Comp. Physiol. Psychol.* **74**, 374–382 (1971).
42. H. Mortimer, R. P. Wright, and J. B. Collip, "The Effect of Oestrogenic Hormones on the Nasal Mucosa; Their Role in the Nasosexual Relationship; and Their Significance in Clinical Rhinology," *Can. Med. Assoc. J.* **35**, 615–621 (1936).
43. R. P. Michael, J. Herbert, and J. Welegalla, "Ovarian Hormones and the Sexual Behaviour of the Male Rhesus Monkeys after Administration of Progesterone to Their Female Partners: Preliminary Communication," *Lancet* **1**, 1015–1016 (1966).
44. J. LeMagnen, "Physiologie des sensations: nouvelles donnees sur la phenomene de l'exaltolide," *Compt. Rend.* **230**, 1103 (1950) as referenced in M. G. Kalogerakis, "The Role of Olfaction in Sexual Development," *Psychosom. Med.* **25**, 420–432 (1963).
45. C. H. Best and N. B. Taylor, *The Physiological Basis of Medical Practice*, 6th ed., Williams & Wilkins, Baltimore, Md., 1955, p. 895, as referenced in M. G. Kalogerakis, "The Role of Olfaction in Sexual Development," *Psychosom. Med.* **25**, 420–432 (1963).
46. W. K. Whitten, "The Effect of the Removal of the Olfactory Bulbs on the Gonads of Mice," *J. Endocrinol.* **14**, 160–163 (1956).
47. R. L. Doty et al., "Sex Differences in Odor Identification Ability: A Cross Cultural Analysis," *Neurophysiologia* **23**, 667–672 (1985).
48. R. A. Schneider and S. Wolf, "Olfactory Perception Thresholds for Citral Using a New Type of Olfactorium," *J. Appl. Physiol.* **8**, 337–342 (1955).
49. J. LeMagnen, "Physiologie des sensations-nouvelles données sur le phenomene de l'exaltolide," *C.R. Acad. Sci.* **230**, 1103–1105 (1950).
50. R. L. Doty et al., "Cyclical Changes in Olfactory and Auditory Sensitivity During the Menstrual Cycle: No Attenuation by Oral Contraceptives," in W. Briepohl, ed., *Olfaction and Endocrine Regulation*, IRL, London, 1982, pp. 35–42.
51. W. Velle, "Sex Differences in Sensory Functions," *Perspectives in Biology and Medicine* **30**, 491–522 (1987).
52. M. J. Rogel, "A Critical Evaluation of the Possibility of Higher Primate Reproductive and Sexual Pheromones," *Psychol. Bull.* **85**, 810–830 (1978).
53. B. W. L. Brooksbank, R. Brown, and R. Gustafson, "The Detection of 5α-androst-16-en-3α-ol in Human Male Axillary Sweat," *Experientia* **30**, 864–865 (1974).
54. B. H. Kingston, "The Chemistry and Olfactory Properties of Musk, Civet and Castoreum," Proceedings of the 2nd International Congress on Endocrinology, *Int. Congr. Ser.* **83**, 209–214 (1965).
55. A. M. Kligman and N. Shehadek, *"Pubic Apocrine Glands and Odour," Arch. Dermatol.* **89**, 461 (1964).
56. S. Bird and D. B. Gower, "The Validation and Use of a Radioimmunoassay for 5α-androst-16-en-3α-one in Human Axillary Collections," *J. Steroid Biochem.* **14**, 213–219 (1981).
57. M. Meredity et al., "Vomeronasal Pump; Significance for Male Hamster Sexual Behaviours," *Science* **207**, 1224–1226 (1980).
58. E. C. Crosby and T. Humphrey, "Studies of the Vertebrae Telencephalon," *J. Comp. Neurol.* **71**, 121 (1938).
59. J. Garcia-Velasco and M. Mondragon, "The Incidence of the Vomeronasal Organ in 1000 Human Subjects and Its Possible Clinical Significance," *J. Steroid Biochem. Mol. Biol.* **39**, 561–563 (1991).
60. N. D. Fabricant, "Sexual Functions and the Nose," *Am. J. Med. Sci.* **239**, 156–160 (1960).
61. M. Mozell et al., "Nasal Airflow," in T. V. Getchell et al., eds., *Smell and Taste in Health and Disease*, Raven Press, New York, 1991, pp. 481–492.
62. M. K. McClintock, "Menstrual Synchrony and Suppression," *Nature* **229**, 244–245 (1971).
63. D. B. Adams, A. R. Gold, and A. D. Burt, "Rise in Female Initiated Sexual Activity at Ovulation and Its Suppression by Oral Contraceptives," *N. Engl. J. Med.* **299**, 1145–1150 (1978).
64. R. P. Michael, R. W. Bonsall, and P. Warner, "Human Vaginal Secretions: Volatile Fatty Acid Contents," *Science* **186**, 1217–1219 (1974).
65. M. Kirk-Smith et al., "Human Social Attitudes Affected by Androstenol," *Research Communications in Psychology, Psychiatry and Behavior* **3**, 379–381 (1978).

66. E. E. Filsinger, J. J. Braun, and W. C. Monte, "An Examination of the Effects of Putative Pheromones on Human Judgments," *Ethology and Sociobiology* **6**, 227–236 (1985).

67. R. Gustavson, E. Dawson, and G. Bonnett, "Androstenol, A Putative Human Pheromone, Affects Human (Homo Sapiens) Male Choice Performance," *J. Comp. Psychol.* **101**, 210–212 (1987).

68. I. P. Pavlov, *"Conditioned Reflexes,"* (G. V. Anrep, trans.) Oxford University Press, London, 1927.

69. A. R. Hirsch, "Nostalgia, The Odors of Childhood and Society," *Psychiatric Times* **9**(8), 29 (1992).

70. M. B. Bowers, M. Van Woert, and L. Davis, "Sexual Behavior During L-dopa Treatment for Parkinsonism," *Am. J. Psychiatry* **127**, 6191–6193 (1971).

71. M. Hyyppa, U. K. Rinne, and V. Sonninen, "The Activating Effect of L-dopa Treatment on Sexual Functions and Its Experimental Background," *Acta Neurol. Scand.* **46**(Suppl. 43), 223–224 (1970).

72. H. R. Arkes, L. T. Herren, and A. M. Isen, "The Role of Potential Loss in the Influence of Affect on Risk-Taking Behavior," *Organizational Behavior and Human Decision Processes* **42**, 181–193 (1988).

73. H. Kluver and P. C. Bucy, "Psychic Blindness and Other Symptoms Following Bilateral Temporal Lobectomy in Rhesus Monkeys," *Am. J. Physiol.* **119**, 352–353 (1937).

74. R. L. Moss, "Modification of Copulatory Behaviour in the Female Rat Following Olfactory Bulb Removal," *J. Comp. Physiol. Psychol.* **74**, 374–382 (1971).

75. A. R. Hirsch and J. J. Gruss, "Olfactory Stimuli and Sexual Response in the Human Male," in J. Rose, ed., *The Proceedings of the World of Aromatherapy*, San Francisco, Calif., 1996, pp. 26–76.

76. T. Engen, *The Perception of Odors*, Academic Press, New York, 1982.

77. A. R. Hirsch, "Nostalgia: A Neuropsychiatric Understanding," *Advances in Consumer Research* **19**, 390–395 (1992).

78. J. Fox, "Improving Tactile Stimulation of the Blind," *Am. J. Ther.* **19**, 5 (1965).

79. A. R. Hirsch and J. J. Gruss, "Odors and Perception of Room Size," presented at the 148th Annual Meeting of the American Psychiatric Association, Miami, Fla. 1995.

80. J. Diamond, *Third Chimpanzee: Evolution and Future of the Human Animal*, Harper Collins, New York, 1992, p. 68.

81. D. Ackerman, *Natural History of the Senses*, Random House, New York, 1990, p. 26.

82. A. R. Hirsch and L. H. Johnston, "Odors and Learning," *J. Neurol. Orthop. Med. Surg.* **17**, 119–126 (1996).

83. A. R. Hirsch and S. Gay, "Effect of Ambient Olfactory Stimuli on the Evaluation of a Common Consumer Product," *Chem. Senses* **16**, 535 (1991).

84. A. R. Hirsch and R. Gomez, "Weight Reduction Through Inhalation of Odorants," *J. Neurol. Orthop. Med. Surg.* **16**, 28–31 (1995).

ALAN R. HIRSCH
Smell and Taste Treatment and Research Foundation
Chicago, Illinois

APPETITE

The term appetite does not have a single widely accepted meaning. Colloquially, it refers to a motivational state or desire, ordinarily for food but more generally for almost any reinforcer. Often the desire is not for food per se but for a specific food or nutrient (eg, salt). Hunger and appetite are often regarded as synonyms, although some investigators have distinguished between hunger as an objective depletion of energy or a specific nutrient and appetite as a subjective desire reflecting psychological factors more than physiological depletion. When people report a loss of appetite in conjunction with, say, an episode of clinical depression, the decline in the subjective desire for food is assumed to correspond to a suppression of the physiological correlates implicated in objective hunger processes; but it remains possible that subjective appetite and objective hunger might diverge under certain circumstances. Certainly the expression of an intense desire for an attractive dessert that appears unexpectedly at the end of a large meal cannot easily be reconciled with a simple depletion model of hunger.

People may report on appetite as a subjective state, but in accepting such reports at face value, one must proceed at one's own risk. First, people do not readily distinguish between their objective need for calories or nutrients (hunger) and their subjective desires (appetite). Second, few if any people can make sophisticated quantitative distinctions between different gradations of appetite or hunger. Indeed, it has been argued that people often infer the intensity of these internal states only retrospectively, by observing the amount of food they have consumed (1). Direct intro-(or viscero-)spection of hunger or appetite, except at the extremes of deprivation or satiation, may not be possible for the untrained subject. Thus, people may assume that they ate as much as they did because they were correspondingly hungry, rather than accept the less obvious or plausible explanation that they ate because of social influence or simply because the food was there (2). The possibility that intake might not be directly reflective of hunger or appetite also eludes many physiological researchers (behavioral neuroscientists), who accept as an axiom that all eating is driven by hunger and that the cessation is eating is necessarily a matter of the onset of physiological satiety.

To the extent that we rely on the amount consumed as an index of hunger or appetite, we may be misled by the fact that people often desire food but do not consume it even when it is available (as happens with dieters) and, obversely, people often consume food for which they have little or no reported desire (as occasionally happens under conditions of social pressure). Finally, much of the research on appetite is conducted on nonhuman animals, necessitating the substitution of physiological or consummatory assessments for verbal assessments. We readily assume the amount that an animal eats corresponds to its "subjective desire" for food, but of course we cannot be sure. Like humans, nonhuman animals often eat more in the presence of conspecifics (3), and we cannot simply infer that the presence of others increases hunger/appetite; it may well be that the presence of others increases intake directly, without affecting hunger/appetite.

Although we cannot complacently assume a direct correspondence between appetite and hunger, or between appetite and intake, the paucity of research on appetite (or hunger) as a phenomenon separate from intake demands that we suspend our concerns about the correspondence between desire and behavior and examine the factors that influence eating. Whether all of these factors operate through the mediator of subjective appetite is doubtful, but at this stage in the development of the research, we have little choice but to make that assumption. None of the research on nonhuman animals has connected appetite to intake, and even in humans, the majority of the research has focused on what is consumed, with a relative neglect of the corresponding subjective state. We are thus forced to shift our attention to the determinants of intake, on which a significant body of research has accumulated. These determinants may be classified as arising from the internal (physiological) and external environments. The role of appetite qua subjective state in this research is largely unexplored, but we may indulge ourselves with the probably only partly incorrect belief that eventually researchers will connect the subjective state of appetite to the more objective determinants of intake, both in humans and (somehow) even in nonverbal animals.

The factors that affect eating have not been systematically cataloged, let alone integrated into a complete systems approach to eating. We have elected to divide the influences on eating into internal and external categories, corresponding roughly to physiological versus environmental factors. Some factors, such as the palatability of the available food, depend jointly on the properties of food, the history of the organism, and the acute physiological state of the organism; such factors, obviously, do not fit neatly into the internal-external categorization. The internal and external distinction is also threatened by findings that demonstrate that through conditioning, external signals can stimulate hunger, or at least feeding (4). Likewise, the perception of attractive food triggers (cephalic) anticipatory digestive reflexes that in turn may affect feeding (5). By the same token, deprivation or other physiological manipulations can alter receptivity to external influences (6). Any comprehensive analysis of eating must contend with these mutual, interactive influences, but current research only hints at some of these complexities. In the absence of such a fully integrated view of eating, this article will be confined to a less satisfactory and somewhat arbitrary listing of influences.

PHYSIOLOGICAL INFLUENCES

As indicated, most laypeople and many researchers subscribe to an essentially physiological view of feeding. The feeding system is activated by various physiological events, most notably the progressive depletion over time of energy or more specific nutrients. This system is usually conceptualized as involving a central regulation of signals from the periphery, with eating as a tool in the service of homeostasis.

METABOLIC AND HORMONAL INFLUENCES

Metabolic factors governing the control of eating are generally separated into those associated with turning on eating (hunger) and those associated with turning off eating (satiety). Hunger (usually accompanied by eating) develops during the fasting phase of metabolism, after the ingestion and absorption of nutrients has been completed. Satiety (usually accompanied by cessation of eating) develops during the absorptive phase of metabolism, when a meal is being ingested and while nutrients are being absorbed from the intestine.

The principal sources of energy for all tissues are glucose and fatty acids, which may be drawn directly from the bloodstream. Energy reserves are stored in three different forms: protein, fat, and glycogen, which must be converted to glucose and fatty acids to be used. During the absorptive phase, metabolism is directed at the accumulation of these reserves; glucose and amino acids are converted to glycogen and fat, and amino acids are also stored as protein.

In the fasting phase, metabolic processes are aimed at utilizing energy. After a meal, blood glucose is used for energy needs, but as the supply of available glucose diminishes, energy reserves must be converted to usable forms. Glycogen is converted back to glucose and fat to fatty acid. These processes are associated with hunger. In the event of long-term deprivation or starvation, there is an increasing demand for energy derived from the conversion of amino acids to glucose, and fatty acids to ketones. These conversion processes provide the substrate for internal influences on eating.

Initiating Eating

Low levels of circulating glucose (and a corresponding higher rate of glycogen to glucose conversion) are important in the initiation of eating. Experimental manipulations that decrease glucose availability are a powerful means of stimulating eating (7). It is important to note, however, that a low level of circulating glucose is not an entirely accurate characterization of the metabolic trigger for eating. Consider untreated diabetics, who have a high level of circulating glucose but who are chronically hungry, because the glucose is not metabolically available owing to insufficient insulin, the hormone necessary for glucose uptake into cells. Thus, it is the unavailability of glucose that is the critical signal. Accordingly, because of its importance in clearing glucose from the bloodstream, insulin is considered an important hormone for the initiation of eating.

Recent research attempting to characterize more precisely the connection between blood glucose level and feeding indicates that meal onsets are preceded by small transient declines in blood glucose (8). If this decline in blood glucose is prevented, feeding will not occur. Feeding will not begin until a transient decline in blood glucose is experienced, supporting the notion that a drop in glucose availability is causally related to eating. Of special interest is the finding that the glucose-drop signal that initiates eating is of surprisingly short duration and small magnitude.

Whereas glucose appears to be important for regulating eating in the short term (ie, meal to meal), lipids appear to play a more important role in regulating food intake in the long term (9). During fasting, the body's fat reserves are converted to fatty acids for energy; this process probably provides an additional signal for turning on eating.

Terminating Eating

Physiological factors responsible for the cessation of eating can be divided into stomach and hormonal factors. Each of these contributes to what is typically referred to as satiety.

Stomach Factors. The stomach produces two types of satiety signal: one associated with the nutrient content of the meal and the other with the volume of the meal (10). It seems likely that these two meal-related signals are detected by separate mechanisms in the gut, because rats can regulate caloric intake independently from meal volume and meal volume can produce satiety cues that do not depend on caloric content.

Nutrient receptors in the stomach respond to the caloric density of a meal. A large but low-calorie meal would activate a large number of nutrient receptors but at a low rate, whereas a small but calorically dense meal would activate a small number of nutrient receptors at a high rate. In addition to this indirect effect on nutrient receptor activation rate, meal volume also induces satiety directly by activating stretch receptors in the stomach.

Current research indicates that stomach distension cues (which activate stretch receptors) and caloric cues (which activate nutrient receptors) are transmitted through separate channels. Stomach distension cues are transmitted to the brain by way of the vagus nerve, whereas stomach nutrient cues do not require the integrity of the vagus nerve and may therefore involve hormonal messengers (10).

Hormonal Factors. Although a number of different hormones have been implicated in satiety, the present discussion will briefly describe the influence of glucagon and cholecystokinin (CCK), the two hormones that have received the most attention as likely satiety agents. Glucagon is produced by islet cells of the pancreas as well as by cells in the gut. Animals treated with glucagon eat smaller than normal meals and animals treated with antibodies to glucagon eat larger than normal meals, suggesting that endogenous glucagon may contribute to meal termination. Because glucagon stimulates the conversion of glycogen (the stored form of glucose that cannot be directly utilized for energy) to glucose, the resultant increased levels of glucose released into the hepatic vein from the liver may stimulate glucose receptors and thereby decrease hunger signals.

Cholecystokinin has also received considerable attention as a signal associated with postprandial satiety (11). CCK is released from the duodenum (the upper portion of the small intestine) when food reaches the stomach. Animal and human studies indicate that systemic administration of CCK inhibits food intake through the normal behavioral sequence that characterizes satiety, without affecting water intake, implying that the behavioral effects of CCK are specific to feeding. Physiological studies indicate that CCK exerts its inhibitory effects on feeding through activation of vagus nerve input to the brain.

Brain Systems Controlling Eating

The Hypothalamus. Current notions regarding the organization of the brain's feeding system derive from early animal studies aimed at identifying brain regions responsible for turning on or turning off feeding. One of the original findings in this regard was that destruction of the medial hypothalamus, in particular the ventromedial hypothalamic nucleus (VMH), produced a behavioral syndrome characterized by severe overeating and ultimately obesity (12). Humans with VMH tumors display a similar behavioral profile, confirming across species the importance of the VMH for eating.

Initially, the VMH was viewed as the brain's satiety center, but it is now recognized that both the psychological and neural characteristics of satiety are far too complex to be accounted for solely by the VMH. However, it is generally agreed that the VMH and its connections are important for relaying satiety signals from the periphery to neural systems involved in turning off feeding. Recent evidence indicates that VMH lesions also destroy fibers projecting to the pituitary stalk and brain stem, both of which are important in the control of various hormones. Of relevance here are findings showing that VMH lesions increase insulin secretion and decrease glucagon secretion, a combination associated with overeating. Thus the overeating associated with VMH lesions may be secondary to the peripheral hormonal and metabolic effects of the lesions rather than a direct result of the missing VMH.

In contrast to the inhibitory role ascribed to the VMH, the lateral hypothalamus (LH) has become associated with stimulation of eating. Animal experiments reveal that lesions of the LH produce aphagia and adipsia, whereas stimulation of the LH produces eating responses (12). Although LH lesions produce aphagia and adipsia, artificially fed animals can eventually recover and maintain close to normal eating habits; this recovery indicates that the central nervous system is sufficiently plastic to enable other brain areas to take over, to some extent, the feeding functions of the LH. It should be noted, however, that although eating deficits can eventually recover, LH-lesioned rats seldom return to their prelesion weight. This finding suggests that the LH contributes to the establishment and maintenance of a set point for body weight, which may be altered permanently by the lesion.

Other Hypothalamic Structures Involved in Eating. Another important structure implicated in eating behavior is the paraventricular nucleus (PVN) of the hypothalamus. In fact, many of the effects associated with VMH manipulations have more recently been interpreted as PVN effects, owing to the discovery that VMH lesions destroy fibers originating from the PVN, some of which are associated with control of hormonal functions already described and others of which project to brain stem mechanisms involved in the control of eating. Moreover, it has been demonstrated that PVN lesions have effects on eating similar to those of VMH lesions. For these reasons, as well as for reasons to be discussed next, researchers have begun to view the PVN as a critical structure for the control of feeding.

Neurotransmitters and Eating. Any description of the neural basis of eating would be incomplete without a discussion of the neurotransmitter systems implicated in eating. Neurotransmitters are responsible for interneuronal communication. Thus, the activity of cells within particular brain regions is determined by the neurotransmitters that stimulate or inhibit them.

Norepinephrine, endorphins, galanin, and neuropeptide Y are transmitters that act in the PVN to facilitate food intake (13). Of special interest here is the finding that the stimulatory effects of norepinephrine in the PVN appear to be specific to carbohydrate intake (13). Elsewhere, growth hormone-releasing factor in the medial preoptic area of the hypothalamus stimulates eating (14). Neurotransmitter signals involved in inhibiting eating include dopamine and norepinephrine in the perifornical region of the hypothalamus as well as serotonin in the PVN (13). Interestingly, serotonin's inhibitory effect, like norepinephrine's stimulatory effect, is carbohydrate selective (13).

Although exactly how these transmitter signals are integrated during normal eating is still unclear, it is assumed that these chemical signals reflect food-relevant information derived from the periphery. For example, consider the regulation of carbohydrate intake. A high-carbohydrate meal increases tryptophan uptake in the brain and ultimately the synthesis of serotonin, for which tryptophan is the precursor. Increased serotonin activity, as discussed earlier, suppresses food intake, particularly carbohydrates. Thus increased carbohydrate intake eventually feeds back negatively on further carbohydrate intake (probably through serotonin's action in the PVN), providing a mechanism for the short-term regulation of carbohydrate intake.

Extrahypothalamic Influences on Feeding. Nonhypothalamic brain systems have also been implicated in the control of feeding. In particular, activation of the nigrostriatal and mesolimbic dopamine pathways, which use dopamine as a neurotransmitter, is associated with stimulatory effects on feeding. These pathways originate in the midbrain and project rostrally to striatal and limbic structures in the forebrain, with some of the axons involved coursing through the lateral hypothalamus en route. This anatomical fact has been taken to suggest that some of the feeding decrements associated with LH lesions may in fact be due to destruction of ascending dopamine neurons. Although the role of the nigrostriatal and mesolimbic dopamine pathways in feeding is not entirely understood, it is of interest to note that these pathways are crucial for reward and reinforcement functions in general (15), suggesting that activation of these systems may reinforce the sensory-motor associations involved in eating and also increase the rewarding properties of food or eating (16,17). Thus contrary to the conventional homeostatic view of eating regulation, these particular signals may be associated with nonregulatory influences on eating such as taste hedonics or learned feeding patterns.

CONDITIONED HUNGER

Ordinarily, the physiological substrate of appetite is thought to respond to energy considerations as previously outlined. However, these reactions are also subject to conditioning, such that particular environmental stimuli may, through repeated pairing, come to elicit the physiological states associated with the onset or offset of eating. For instance, it has been demonstrated that previously neutral stimuli repeatedly paired with the presentation of food can induce sated animals to continue eating (4). Visual, olfactory, or social cues associated with hunger may induce feeding in people who are not otherwise hungry; comparable cues that have been paired with satiety may likewise terminate eating in individuals who are not otherwise sated. It has been suggested that such external, conditioned cues do not control behavior directly, but rather operate on the physiological substrate of hunger and satiety; for example, the mere sight of attractive food cues (of the sort that have reliably predicted eating in the past) may induce physiological reactions (eg, insulin release) associated with hunger (18). Even temporal cues may stimulate appetite; people become hungry as mealtime approaches, but if for some reason the meal is skipped, hunger may subside (C. P. Herman, J. Polivy, and I. Biernacka, unpublished data, 1998).

SPECIFIC APPETITES AND NUTRIENT DEPLETIONS

The organism must concern itself not only with overall energy balance, but also with its inventory of the nutrients crucial to its well-being. Many well-known diseases (eg, scurvy and rickets) may appear in otherwise well-fed individuals who lack a particular vitamin or mineral. Specific appetites refer to hardwired cravings for certain substances the absence of which is detected and, wherever the food supply permits, rectified. It should be noted that whereas some deficits (eg, salt) produce clear cravings, not all deficiencies are detectable by the suffering organism.

Recently, speculation has arisen that the organism may regulate nutrients (especially macronutrients) not so much for their medicinal value as for their psychoactive effects. Most notably, a carbohydrate-craving hypothesis has been championed (19). It is argued that a deficiency of carbohydrates affects neurotransmitter levels to create a negative affective state; accordingly, individuals may learn to (over)eat carbohydrates in an attempt to bolster their mood, especially when they are feeling upset or depressed. This hypothesis has been offered to explain the overeating characteristic of bulimia nervosa and some forms of obesity.

SENSORY AND PERCEPTUAL FACTORS

Flavor

The term "flavor" is intended to encompass both gustatory (taste) and olfactory (smell) properties of food. Palatability is a matter of both aspects of flavor; when people say that a food tastes good, they tend to underestimate the contribution of smell. The palatability of food, which corresponds to how positively its flavor is rated and also includes certain secondary properties of the food such as its texture, is a major determinant of intake. Organisms are basically finicky, avoiding indiscriminate consumption in favor of selective consumption (where possible) of preferred foods. The finding that nonhuman animals consume more highly palatable food is compromised by the fact that palatability must often be inferred from intake itself; but humans also show a strong tendency to consume more of foods that they independently rate as more pleasant.

Palatability has a direct effect on intake, but whether palatability overrides other potentially competing considerations in controlling intake remains an active research question. It has been observed that the duration of the meal is largely a matter of the palatability of the available food but the likelihood of a rat initiating feeding is more closely tied to caloric deprivation (20). In short, palatability may affect some meal parameters more than others. In humans, extremes of food deprivation or satiety mute the effects of flavor; starving people tend to ignore the flavor of food and eat a great deal of unpalatable food if that is all that is available, whereas fully sated people eat very little of even quite palatable food. Interestingly, research indicates that at least under certain qualifying conditions hungry animals (including humans) do not display reduced finickiness; indeed, finickiness may be exaggerated both with respect to acceptance of good-tasting food and rejection of bad-tasting food (21,22). An apparent paradox has been identified in which "the animal eats . . . for taste when he needs calories [ie, when deprived]" (23). This paradox may be more apparent than real, in that the flavor of food is ordinarily confounded with its caloric density (ie, good-tasting foods are calorically denser, leading one humorist to describe the calorie as a term that scientists use to measure how good something tastes). Accordingly, the hungry animal may become more finicky as a way of maximizing energy intake in an energy deficiency crisis.

The potency of palatability as a determinant of eating may vary in different types of individuals. Obese people, for instance, have long been regarded as prone to overeat highly palatable foods (rather than food per se); this view was crystallized in the externality theory of obesity, which argues that obese people's eating is controlled virtually entirely by environmental cues, including taste, whereas normal-weight people's eating is more responsive to physiological cues of hunger and satiety (24). The observation that obese people are especially responsive to taste (24) may be meshed with the observation that hungry organisms are especially responsive to taste (23), if it is postulated that most obese people are chronically hungry as a result of their partially successful attempts to reduce their weight (25).

Other eating problems have likewise been associated with extreme partiality toward highly palatable foods. Although the evidence is conflictual (26), binge eating in normal-weight individuals is often interpreted as targeted specifically on highly palatable carbohydrates (27), which elevate serotonin levels and, ultimately, mood (19). Conceivably, this view might be extended in the direction of proposing that some flavors are deemed palatable precisely because of their central hedonic consequences.

Sensory Factors

If eating offset were associated solely with the accumulation of particular nutrient stores, then it would be difficult to account for the "dessert effect." That is, despite a satiating meal (fully repleting nutrient stores), people can still find room for dessert. One way of accounting for this effect is in terms of sensory specific satiety, which refers to an animal becoming satiated to a particular flavor without showing satiation to other normally preferred flavors (28). Thus the individual may become satiated to the sensory characteristics of the main course while not being satiated to the sensory characteristics of the dessert. (It should be noted that there may be other contributors to the "dessert effect," including social facilitation, as described later.)

Collateral support for the role of nongut sensory factors in satiety is found in studies of sham-feeding (esophagotomized) animals. In these animals, food is chewed and swallowed but does not reach the gut, and the nutrients are therefore not absorbed. Of special interest here is the fact that although these animals eat larger than normal meals (providing evidence for the involvement of post-ingestional factors in regulating eating offset), they do eventually stop eating. This has been interpreted to indicate that sensory aspects of the food (taste, olfaction, and texture) are sufficient to induce satiety-like effects on eating. A full explication of satiety will eventually have to incorporate and reconcile sensory and postingestional factors.

Perceptual Salience

Does the presence of food act merely to allow individuals to satisfy their preexisting hunger, or might the sight of food actually stimulate the appetitive drive, which would not have existed in the absence of food? One experiment found that normal-weight individuals ate the same amount (just under two sandwiches) regardless of whether they were initially presented with one sandwich (and had to obtain additional sandwiches from a refrigerator) or three sandwiches (29). This study suggests that the visual salience of food has a minimal impact on consumption. Obese subjects in this study, however, ate considerably more when presented with three sandwiches than when presented with only one, suggesting that visual cues have a strong effect on them. This finding was interpreted as support for the externality theory of obesity, with the visual salience of food representing an external cue par excellence. (One important implication of this study is that in the absence of excitatory external cues, obese people should eat less than do normal-weight people.) Another experiment found that normal-weight subjects ate the same amount of cashews regardless of whether the cashews were brightly illuminated or difficult to see (and regardless of whether they were instructed to think about the cashews or about something else), whereas obese subjects ate above-normal amounts of the visually salient cashews and below-normal amounts of the dimly illuminated cashews (30). Obese subjects also ate more when instructed to think about the cashews, reinforcing the point that it is awareness of food that matters and not simply optical effects. It should be noted that some researchers interpret these findings to mean that obese subjects eat in direct response to external cues and not because such cues stimulate hunger, which in turn drives eating. However, it has been suggested that exposure to food cues may trigger anticipatory (cephalic phase) reflexes at the physiological level, such that the sight or smell of palatable food may literally make the fat person hungry (18). Recently, we have shown that directing people's attention toward or away from food-related stimuli can increase or decrease their reported hunger (C. P. Herman, J. Ostovich, and J. Polivy, unpublished data, 1998); this effect was obtained in normal-weight nondieters, suggesting that the effects of food stimuli are not confined to the obese. Our study did not measure eating, however, so we cannot conclude that attentional or perceptual manipulations will affect intake.

Appetizer Effects

Does intake itself sometimes stimulate more intake? The term "appetizer" would suggest so, and common experience would seem to confirm that appetite can be stimulated by small delicacies. No compelling research has been conducted on this topic in humans, but if strong appetizer effects were confirmed, then the presumptive physiological mediation of such effects (ie, the appetizer triggers an anticipatory digestive reflex, which in turn demands satisfaction in the form of a complete meal) should be distinguished from alternative explanations (eg, the appetizer initiates a behavioral response sequence that is not hunger driven but that perseverates until satiety occurs). The role of behavioral momentum in feeding and other repetitive action patterns has not been adequately explored, although it has been argued that obese subjects overeat because of an inability to terminate repetitive behavior sequences (especially eating) once they have begun (31). This would suggest that appetizer effects might be stronger in the obese.

Preload studies, in which subjects are initially given a fixed amount of food and then permitted to continue ad lib, show a complex pattern. When subjects are classified as dieters (obese or normal weight) or nondieters, opposite reactions are observed, with nondieters eating ad lib in inverse proportion to preload size and dieters eating more following large preloads than following small preloads. The nondieter pattern would seem to contradict expectations based on an appetizer effect, and the data from dieters likewise do not fit in well, because a small preload (appetizer) would be expected to trigger the greatest amount of subsequent eating. Interpretations of these effects in dieters emphasize deliberate cognitive control of intake and its disruption, rather than the sort of analysis ordinarily invoked to explain the hypothetical appetizer effect (32).

EMOTIONAL FACTORS

It is readily agreed that emotional disturbance affects appetite and eating, but the precise nature of this effect is variable. In nonhuman animals, arousal tends to promote feeding. Nonspecific activation may render them more responsive to external, food-related stimuli, which direct behavior toward eating (33). (Such external responsiveness is presumably not specifically directed at food-related stimuli; rather, it focuses on whatever powerful behavior-relevant stimuli may dominate the animal's immediate perceptual environment, and might yield sexual, aggressive, or other stimulus-bound behavior depending on the environmental configuration that obtains when activation occurs.) It has been suggested that any sort of activation or drive induction might accentuate stimulus-bound eating in humans (C. P. Herman et al., unpublished data). Such stimulus-bound eating would not necessarily increase intake but rather make it more responsive to environmental conditions; if the food were bad-tasting, it might suppress eating. Such a mechanism might explain why hungry organisms sometimes eat less of a bad-tasting food than nondeprived organisms do (22,23).

Studies of arousal in humans tend to focus on particular valenced emotions (eg, anxiety, depression, and anger). In general, such negative emotions have developed a reputation for suppressing eating, even when palatable food is available. This effect is usually attributed to the sympathomimetic effects of emotional distress; these effects presumably act to suppress peripheral hunger signals. Clinical depression is indexed by a loss of appetite, among other criteria. Still, some depressed individuals tend to eat more and gain weight (34). Indeed, emotional agitation tends to increase eating among many people, to the point where emotional disturbance is often cited as a prime cause of weight gain and failure to maintain dietary restraint. How can these contradictory effects be reconciled?

It does appear to be the case that strong negative emotions produce physiological effects that oppose appetite. (It seems conceivable that minor agitation or disturbances might stimulate eating in humans much as seems to be the case in other animals, but such weak manipulations have not been explored.) Powerful distress (eg, intense clinical depression) probably suppresses appetite for virtually everyone, but short of overwhelming distress, emotional agitation appears to have opposite effects depending on the type of individual involved. Normal eaters, whose behavior is at least partially governed by the sort of hunger and satiety signals discussed previously, tend to respond to distress by eating less. Individuals who ordinarily attempt to suppress their eating (ie, dieters, including perhaps most of the obese), however, tend to eat more when they are upset (and possibly when elated, although studies of the effects of positive moods are so few as to defy conclusiveness). Dieters, because they must overcome normal physiological controls on eating if they are to succeed in reducing their intake, eventually dissociate their eating from such controls in favor of cognitive calculations of appropriate intake (see "Cognitive Factors"). Their intake is deliberately inhibited, but such inhibitions are vulnerable to disruption by (among other factors) emotional agitation. The emotionally distressed dieter, then, becomes disinhibited and overeats, unconstrained by the sort of normal satiety considerations that depend on physiological signals and conditioning, both of which are functionally atrophied in the dieter. The effects of distress may be exacerbated or attenuated depending on the type of distress involved; for example, physical threats tend to have a stronger suppressive effect, whereas ego threats tend to have a more facilitative effect (35; C. P. Herman et al., unpublished data, 1998). The exploration of overeating under conditions of distress remains a prominent concern of researchers studying obesity and especially bulimia nervosa.

SOCIAL FACTORS

Other people influence an individual's eating in a variety of ways. Social facilitation refers to increased intake in the presence of others; this usually occurs when these other people are eating substantial amounts as well (36). More typically, social inhibition of eating is encountered; for example, being observed by noneating others (D. A. Roth, et al., unpublished data, 1998) or eating along with others who eat very little (37) tends to reduce consumption. More generally, people seem to be highly vulnerable to social influences on eating, from indirect pressure (such as conformity to a model or experimental confederate) to more direct influences (such as requests or demands that a certain amount of food be eaten). These social effects are relatively powerful compared with other experimental manipulations of factors affecting eating and lend support to a nonhomeostatic view of eating, at least in the short term. Eating seems to be acutely responsive to social influences that bear little relation to an individual's physiological state or needs. Homeostatic considerations may eventually correct for such short-term deviations; for example, if a person undereats in a public spotlight, he or she may compensate by eating more later in a private setting. Another consideration is the likelihood that such social effects as copying others or not gorging when being watched serve a broader biological survival purpose. It is possible to benefit from others' experiences, and people may look to social guides when the appropriate amount or type of food to be consumed is ambiguous, as it often is. Furthermore, social norms may demand some sort of equity in the distribution of the food supply. Still it is clear that such social factors may have a profound effect on eating, and this effect may act in direct opposition to purely physiological needs and signals. Even 24-hour food-deprived experimental subjects, who ought to eat in response to powerful internal signals, are strongly influenced by the eating patterns of a model (38).

Another sort of social influence on eating occurs when people eat in a particular way to convey a certain impression. This impression-management view of eating focuses on eating as a self-presentational strategy (39). For example, women eat less when they are paired with a man than when paired with another woman, presumably because they wish to convey a more feminine impression to the man, and eating lightly is part of the feminine stereotype. Again, this sort of influence on eating seems to have little to do with the exigencies of internal state.

COGNITIVE FACTORS

The impact of both internal (eg, hunger) and external (eg, food salience) factors on eating is frequently mediated by cognition; people may decide to eat (or not eat) upon consideration of (or despite) the allure of such pressures. Because such internal and external factors have seemingly direct effects on eating (ie, because these factors can be interpreted as affecting eating directly, without the mediation of decisions that add little in the way of explanation) most analyses of eating do not bother to include deliberation or conscious choice as important elements. When we find that people eat in a manner opposed to both internal and external pressures, however, some sort of deliberative element seems necessary to account for the behavior. If someone were to continue eating an unpalatable food despite being sated, it would probably be necessary to have access to that individual's phenomenology to help provide an explanation. More commonly, people fail to eat despite their evident hunger and the availability of palatable food. Such dietary restraint demands an analysis that adds a set of mental factors to the control of eating in addition to physiological and environmental stimuli to which the person responds more or less reflexively (40). Such cognitive dietary calculations do not arise spontaneously, of course; they themselves are learned and may be reinforced or extinguished. But the prevalence (in humans, at least) of eating patterns that depart radically from the straightforward application of our knowledge of physiology and environmental stimulus control requires that attention be paid to willful opposition to such signals. One of the most serious problems arising from dieting is that the chronic overriding of normal controls on eating renders the dieter confused about whether or when she or he is feeling hunger or appetite. This problem is exacerbated in the recognized eating disorders (anorexia nervosa, bulimia nervosa, binge eating disorder).

INTERACTIONS AMONG RELEVANT FACTORS

A variety of examples have been presented of the nonadditive influence of factors affecting eating. For instance, hunger does not simply add to palatability as a determinant of eating, hungry people may eat more good-tasting food, as expected, but they may eat less bad-tasting food, depending on circumstances. Likewise, social influences may combine with, mask, or oppose hunger and satiety considerations in eating. An attempt has been made to reconcile the various influences on eating in a boundary model (41). The basic premise is that physiological influences will predominate when the person is particularly hungry or sated and that environmental and cognitive considerations will be most influential when the person is indifferent (neither hungry nor sated). This model has not been successful in predicting behavior, but its emphasis on the need to assess the influence of one factor (eg, social influence) in the context of other factors (eg, hunger-satiety) has proven fruitful experimentally. Research on the influence of one particular factor in isolation seems likely to misrepresent the importance of that influence; under certain conditions, its influence will be exaggerated, whereas under others it may be suppressed (42).

CONCLUSIONS

The scientific study of appetite has yielded a wide range of interesting phenomena. Still, the field is a painfully long way from achieving any sort of satisfying integration, in which the mutual and combined influences of various factors acting simultaneously might be predicted or explained. This conclusion is especially warranted in the case of human eating behavior, where social and cognitive factors are particularly evident. Such factors may apply as well to other animals. It can only be hoped that human and animal research, often seemingly at odds with each other, will eventually be synthesized into a comprehensive analysis of the determinants of feeding and appetite.

BIBLIOGRAPHY

1. D. J. Bem, "Self-perception Theory," in L. Berkowitz, ed., *Advances in Experimental Social Psychology*, Vol. 6, Academic Press, New York, 1972.
2. R. E. Nisbett and T. D. Wilson, "Telling More Than We Can Know: Verbal Reports on Mental Processes," *Psychol. Rev.* **84**, 231–259 (1977).
3. K. T. Hoyenga and S. Aeschleman, "Social Facilitation of Feeding in the Rat," *Psychonomic Sci.* **14**, 239–241 (1969).
4. H. P. Weingarten, "Conditioned Cues Elicit Feeding in Sated Rats: A Role for Learning in Meal Initiation," *Science* **220**, 431–433 (1983).
5. T. L. Powley, "The Ventromedial Hypothalamic Syndrome, Satiety, and a Cephalic Phase Hypothesis," *Psychol. Rev.* **84**, 89–126 (1977).
6. M. Cabanac, "Physiological Role of Pleasure," *Science* **173**, 1103–1107 (1971).
7. G. A. Bray and L. A. Campfield, "Metabolic Factors in the Regulation of Feeding and Body Energy Storage," *Metabolism* **24**, 99–117 (1975).
8. L. A. Campfield and F. J. Smith, "Functional Coupling between Transient Declines in Blood Glucose and Feeding Behavior: Temporal Relationships," *Brain Res. Bull.* **17**, 427–433 (1986).
9. N. Mrosovsky, "Body Fat: What Is Regulated?" *Physiol. Behav.* **38**, 407–414 (1986).
10. M. F. Gonzalez and J. A. Deutsch, "Vagotomy Abolishes Cues of Satiety Produced by Gastric Distention," *Science* **212**, 1283–1284 (1981).
11. G. P. Smith, "Gut Hormone Hypothesis of Postprandial Satiety," in A. J. Stunkard and E. Stellar, eds., *Eating and Its Disorders*, Raven, New York, 1984, pp. 67–76.
12. S. P. Grossman, "Contemporary Problems Concerning Our Understanding of Brain Mechanisms That Regulate Food Intake and Body Weight," in A. J. Stunkard and E. Stellar, eds., *Eating and Its Disorders*, Raven, New York, 1984, pp. 5–14.
13. S. Liebowitz, "Brain Monoamines and Peptides: Role in the Control of Eating Behavior," *Federation Proc.* **45**, 1396–1403 (1986).
14. F. J. Vaccarino and M. Hayward, "Microinjections of Growth Hormone-Releasing Factor into the Medial Preoptic Area/Suprachiasmatic Nucleus Region of the Hypothalamus Stimulate Food Intake in Rats," *Regulatory Peptides* **21**, 21–28 (1988).
15. F. J. Vaccarino, B. B. Schiff, and S. E. Glickman, "Biological View of Reinforcement," in S. B. Klein and R. R. Mowrer, eds., *Contemporary Learning Theories*, Erlbaum, Hillsdale, N.J., 1989, pp. 111–142.
16. K. R. Evans and F. J. Vaccarino, "Amphetamine and Morphine-Induced Feeding: Evidence for Involvement of Reward Mechanisms," *Neurosci. Biobehav. Rev.* **14**, 1–14 (1990).
17. N. M. White, "Control of Sensorimotor Function by Dopaminergic Nigroatriatal Neurons: Influence on Eating and Drinking," *Neurosci. Biobehav. Rev.* **10**, 15–36 (1989).
18. J. Rodin, "Has the Distinction between Internal Versus External Control of Feeding Outlived Its Usefulness?" in G. Bray, ed., *Recent Advances in Obesity Research*, Vol. II, Newman Publishing, London, 1978, pp. 75–85.
19. R. J. Wurtman, "Neurotransmitters, Control of Appetite, and Obesity," in M. Winick, ed., *Control of Appetite*, Wiley, New York, 1988, pp. 27–34.
20. J. Le Magnen, "Advances in Studies of the Physiological Control and Regulation of Food Intake," in E. Stellar and J. M. Sprague, eds., *Progress in Physiological Psychology*, Vol. 4, Academic Press, Orlando, Fla., 1971.
21. P. Pliner, C. P. Herman, and J. Polivy, "Palatability as a Determinant of Eating, Finickiness as a Function of Taste, Hunger and the Prospect of Good Food," in E. D. Capaldi and T. L. Powley, eds., *Taste, Experience, and Feeding*, American Psychological Association, Washington, D.C., 1990, pp. 210–226.
22. N. Kauffman, C. P. Herman, and J. Polivy, "Hunger-Induced Finickiness in Humans," *Appetite* **24**, 203–218 (1995).
23. H. L. Jacobs and K. N. Sharma, "Taste Versus Calories: Sensory and Metabolic Signals in the Control of Food Intake," *Ann. New York Acad. Sci.* **157**, 1084–1125 (1969).
24. S. Schachter, "Some Extraordinary Facts about Obese Humans and Rats," *Amer. Psychol.* **26**, 129–145 (1971).
25. R. E. Nisbett, "Hunger, Obesity, and the Ventromedial Hypothalamus," *Psychol. Rev.* **79**, 433–453 (1972).
26. C. P. Herman and J. Polivy, "What Does Abnormal Eating Tell Us about Normal Eating?" in H. L. Meiselman and H. J. H. MacFie, eds., *Food Choice, Acceptance and Consumption*, Blackie Academic and Professional, London, 1996, pp. 207–238.
27. J. C. Rosen et al., "Binge Eating Episodes in Bulimia Nervosa: The Amount and Type of Food Consumed," *Int. J. Eating Disorders* **5**, 255–267 (1986).
28. B. J. Rolls, E. T. Rolls, and E. A. Rowe, "The Influence of Variety on Human Food Selection and Intake," in L. M. Barker, ed., *The Psychobiology of Human Food Selection*, AVT Publishing, Westport, Conn., 1982, pp. 101–122.
29. R. E. Nisbett, "Determinants of Food Intake in Obesity," *Science* **159**, 1254–1255 (1968).
30. L. Ross, "Effects of Manipulating Salience of Food upon Consumption by Obese and Normal Eaters," in S. Schachter and J. Rodin, eds., *Obese Humans and Rats*, Lawrence Erlbaum Assoc., Potomac, Md., 1974, pp. 43–52.
31. D. Singh, "Role of Response Habits and Cognitive Factors in Determination of Behavior of Obese Humans," *J. Personality Soc. Psychol.* **27**, 220–238 (1973).
32. C. P. Herman and J. Polivy, "Studies of Eating in Normal Dieters," in B. T. Walsh, ed., *Eating Behavior in Eating Disorders*, American Psychiatric Association Press, Washington, D.C., 1988, pp. 95–111.
33. T. W. Robbins and P. J. Fray, "Stress-Induced Eating: Fact, Fiction or Misunderstanding?" *Appetite* **1**, 103–133 (1980).

34. J. Polivy and C. P. Herman, "Clinical Depression and Weight Change: A Complex Relation," *J. Abnormal Psychol.* **85**, 338–340 (1976).
35. T. F. Heatherton, C. P. Herman, and J. Polivy, "Effects of Physical Threat and Ego Threat on Eating Behavior," *J. Personality Soc. Psychol.* **60**, 138–143 (1991).
36. J. M. de Castro and E. S. de Castro, "Spontaneous Meal Patterns of Humans: Influence of the Presence of Other People," *Amer. J. Clin. Nutr.* **50**, 237–247 (1989).
37. J. Polivy et al., "Effects of a Model on Eating Behavior: The Induction of a Restrained Eating Style," *J. Personality* **47**, 100–112 (1979).
38. S. J. Goldman, C. P. Herman, and J. Polivy, "Is the Effect of a Social Model on Eating Attenuated by Hunger?" *Appetite* **17**, 129–140 (1991).
39. D. Mori and P. L. Pliner, "'Eating Lightly' and the Self-Presentation of Femininity," *J. Personality Soc. Psychol.* **53**, 693–702 (1987).
40. C. P. Herman and J. Polivy, "Restrained Eating," in A. J. Stunkard, ed., *Obesity*, Saunders, Philadelphia, 1980, 208–225.
41. C. P. Herman and J. Polivy, "A Boundary Model for the Regulation of Eating," in A. J. Stunkard and E. Stellar, eds., *Eating and Its Disorders*, Raven, New York, 1984, pp. 141–156.
42. C. P. Herman, "Human Eating: Diagnosis and Prognosis," *Neurosci. Biobehav. Rev.* **20**, 107–111 (1996).

C. P. HERMAN
F. J. VACCARINO
University of Toronto
Toronto, Ontario
Canada

APPLES. See FRUITS, TEMPERATE.

APRICOTS. See FRUITS, TEMPERATE.

AQUACULTURE

The Food and Agriculture Organization (FAO) of the United Nations has defined aquaculture as, "the farming of aquatic organisms, including fish, molluscs, crustaceans, and aquatic plants" (1). Other definitions include "the rearing of aquatic organisms under controlled or semi-controlled conditions" and "underwater agriculture" (2). The list of organism groups mentioned in the FAO definition is not complete. Aquaculturists have also been involved in the production of echinoderms (sea urchins), cephalopods (octopus, squid, cuttlefish), reptiles (alligators, sea turtles, freshwater turtles), and amphibians (frogs). Aquaculture is an inclusive term that encompasses fresh, brackish, marine, and even hypersaline waters. The term mariculture is more restrictive in that it is usually defined as referring to aquaculture in saline environments.

Public sector aquaculture has historically involved production of aquatic animals to augment or establish recreational and commercial fisheries. More recently, it has also come to include attempts to recover threatened or endangered species by maintaining them in captivity and producing new generations that will ultimately be used to stock the waters where the species has become depleted. Public sector aquaculture is widely practiced in North America and to a lesser extent in other parts of the world. The FAO definition of aquaculture indicates that farming implies ownership of the organisms being cultured, which would seem to exclude public sector aquaculture.

Going hand in hand with attempts to recover endangered species are enhancement stocking programs aimed at releasing juvenile animals to build back stocks of aquatic animals that have been reduced due to overfishing. Examples of enhancement programs currently in existence include the stocking of cod in Norway, flounders in Japan, and red drum in the United States.

The bulk of global production from aquaculture is utilized directly as human food, with public aquaculture playing a minor role or being absent in many nations. Private aquaculture is not only about human food production, however. In some regions, well-developed private sector aquaculture is involved in the production of bait and ornamental fishes and invertebrates. The birth of aquaculture may have, in fact, been associated with attempts to produce ornamental fish to please royalty. Such might be the case with the development of colorful koi carp in China (koi are a colorful form of common carp, *Cyprinus carpio*).

Aquatic plants are cultured in many regions of the world. In fact, aquatic plants, primarily seaweeds, account for nearly 25% of the world's aquaculture production (3). Seaweed is widely used as food in the orient and elsewhere and can be found in many items that people use in their daily lives. Extracts from seaweed are important components of many pharmaceuticals, automobile tires, toothpaste, ice cream, and a wide variety of other products.

As implied, the origins of aquaculture may be rooted in China and could, in fact, go back some 4000 years. The ancient Egyptians may also have been early, if not the first aquaculturists. Pictographs in the tombs of the pharaohs can be interpreted to indicate that tilapia (fish native to North Africa and the Middle East) were being raised, possibly as food or display fish for the royal court. Regardless, Asia dominates the world in aquaculture production today (4). In North America, the culture of fish began in the nineteenth century and grew rapidly in the public sector after the establishment of the U.S. Fish and Fisheries Commission in 1871 (5). Private aquaculture existed as a minor industry for many decades, coming into prominence in the 1960s. Since then the United States has become one of the leaders in aquaculture research and development, although production, while significant at nearly 400,000 metric tons by 1996 (6), amounted to only about 2% of the world's total of 34 million metric tons.

The U.S. commercial aquaculture industry is dominated by channel catfish, trout, salmon, minnows, oysters, mussels, clams, and crawfish. A number of other fish and invertebrates are also being reared. Included are tilapia, striped bass and hybrid striped bass, red drum, goldfish, tropical fish, and shrimp. In the public sector, hatcheries produce large numbers of such species as salmon, trout, largemouth and smallmouth bass, sunfish, crappie, northern pike, muskellunge, walleye, and catfish for stocking or growout.

Global aquaculture production continues to grow annually, but increasing competition for suitable land and water, problems associated with wastewater from aquaculture facilities, disease outbreaks, and potential shortages of animal protein for aquatic animal feeds are becoming increasingly important. New technology, including the application of molecular genetic approaches to improving performance and disease resistance in aquaculture species, along with the development of water reuse (recirculating) systems, the establishment of offshore facilities, and the expansion of enhancement programs may provide the impetus for a resurgence in growth of the industry.

GLOBAL AQUACULTURE PRODUCERS AND PRODUCTION

Aquaculture production around the world grew at an average annual rate of 9.6% for the period 1984 through 1995 (3). In 1995, total production approached 28 million metric tons (3). Production jumped to 34.11 million metric tons in 1996 (6). A breakdown by type of organisms grown is presented in Table 1 for each year. The total value of the 1995 aquaculture crop exceeded $42 billion (3) and rose to $46.5 billion in 1996 (6).

The People's Republic of China is far and away the world's largest producer of aquaculture species. In 1995, production in China exceeded 17.5 million metric tons. India, in second place, produced slightly over 1.6 million metric tons, or less than 10% of China's production (3). In 1996, China's production exceeded 23 million metric tons, with India climbing modestly to 1.8 million metric tons (6). The top 20 aquaculture-producing nations during 1995 and 1996, in descending order, are shown in Table 2. The same countries are on both lists, but some changes in ranking are apparent. Most notable, perhaps, is the movement of Chile from ranking 16 to 12, reflecting the emphasis that country has been placing on aquaculture, particularly in the rearing of salmon.

U.S. aquaculture production actually dropped from 1995 to 1996 by 4.9%, and the United States dropped by one position in the top 20 from eighth to ninth (Table 2).

Table 1. Percentages of Total World Aquaculture Crop (by Weight) Associated with Different Types of Organisms during 1995 and 1996

Type of organism	1995 (%)	1996 (%)
Aquatic plants	24.5	22.7
Mollusks	18.3	24.9
Crustaceans	4.1	3.4
Finfish	52.8	48.8
Others	0.3	0.2
Total	*100.0*	*100.0*

Source: Refs. 3 and 6.

Table 2. Top 20 Aquaculture-Producing Nations of the World in 1995 and 1996

Rank	1995	1996
1	People's Republic of China	People's Republic of China
2	India	India
3	Japan	Japan
4	South Korea	Philippines
5	Philippines	South Korea
6	Indonesia	Indonesia
7	Thailand	Thailand
8	United States of America	North Korea
9	Bangladesh	United States of America
10	Taiwan	Bangladesh
11	Norway	Norway
12	France	Chile
13	Italy	France
14	Viet Nam	Taiwan
15	North Korea	Spain
16	Chile	Italy
17	Spain	Viet Nam
18	Malaysia	United Kingdom
19	United Kingdom	Ecuador
20	Ecuador	Malaysia

Source: Refs. 3 and 6.

Of some importance is the fact that unlike most nations, where the bulk of the animals produced in aquaculture are used directly as human food, a considerable portion of U.S. aquaculture is associated with the production of bait minnows and ornamental fish. In 1997, aquaculture sales of tropical fish in the state of Florida were $57 million. Aquatic plant sales for species used in wetland restoration and aquaria netted an additional $13.2 million (7). The total value of U.S. aquaculture produce in 1996 was slightly more than $735 million (6).

SPECIES UNDER CULTIVATION

The emphasis here is on aquatic animal production, but as previously indicated, aquatic plants comprise nearly 25% of global aquaculture production by weight. Many hundreds of thousands of people are involved, worldwide, in aquatic plant production. The quantity of brown seaweeds, red seaweeds, green seaweeds, other types of algae, and miscellaneous aquatic plants such as watercress, water chestnuts, and ornamental plants produced in 1996 was estimated at more than 7.7 million metric tons (6). Several species of macroalgae are listed in Table 3. Microscopic algae and cyanobacteria are sometimes marketed as food or as a nutritional supplement (eg, *Spirulina* sp.). In addition, an undocumented quantity of algae (mostly of the single-celled variety) is produced for use as food for filter-feeding aquatic animals (primarily mollusks and zooplankton). Planktonic organisms such as rotifers are reared on algae and then used to feed the young of crustaceans and fish that do not accept prepared feeds.

Animal aquaculture is concentrated on finfish, mollusks, and crustaceans. Sponges, echinoderms, tunicates, turtles, frogs, and alligators are all being cultured, but production is insignificant in comparison with the three prin-

Table 3. Examples of Common and Scientific Names of Selected Aquaculture Species

Common name	Scientific name
Finfish	
African catfish	*Clarias gariepinus*
Atlantic halibut	*Hippoglossus hippoglossus*
Atlantic salmon	*Salmo salar*
Bighead carp	*Aristichthys nobilis*
Bigmouth buffalo	*Ictiobus bubalus*
Black crappie	*Pomoxis nigromaculatus*
Blue catfish	*Ictalurus furcatus*
Blue tilapia	*Oreochromis aureus*
Bluegill	*Lepomis macrochirus*
Brook trout	*Salvelinus fontinalis*
Brown trout	*Salmo trutta*
Catla	*Catla catla*
Channel catfish	*Ictalurus punctatus*
Chinook salmon	*Oncorhynchus tshawytscha*
Chum salmon	*Oncorhynchus keta*
Coho salmon	*Oncorhynchus kisutch*
Common carp	*Cyprinus carpio*
Fathead minnow	*Pimephales promelus*
Gilthead sea beam	*Sparus aurata*
Goldfish	*Carassius auratus*
Grass carp	*Ctenopharyngodon idella*
Largemouth bass	*Micropterus salmoides*
Milkfish	*Chanos chanos*
Mozambique tilapia	*Oreochromis mossambicus*
Mrigal	*Cirrhinus mrigala*
Mud carp	*Cirrhina molitorella*
Muskellunge	*Esox masquinongy*
Nile tilapia	*Oreochromis niloticus*
Northern pike	*Esox lucius*
Pacu	*Colossoma metrei*
Pink salmon	*Oncorhynchus gorbuscha*
Plaice	*Pleuronectes platessa*
Rabbitfish	*Siganus* spp.
Rainbow trout	*Oncorhynchus mykiss*
Red drum	*Sciaenops ocellatus*
Rohu	*Labeo rohita*
Sea bass	*Dicentrarchus labrax*
Shiners	*Notropis* spp.
Silver carp	*Hypophthalmichthys molitrix*
Smallmouth bass	*Micropterus dolomieui*
Sole	*Solea solea*
Steelhead	*Oncorhynchus mykiss*
Striped bass	*Morone saxatilis*
Walking catfish	*Clarias batrachus*
Walleye	*Stizostedion vitreum*
White crappie	*Pomoxis annularis*
Yellow perch	*Perca flavescens*
Yellowtail	*Seriola quinqueradiata*
Mollusks	
American oyster	*Crassostrea virginica*
Bay scallop	*Aequipecten irradians*
Blue mussel	*Mytilus edulis*
Northern quahog	*Mercenaria mercenaria*
Pacific oyster	*Crassostrea gigas*
Southern quahog	*Mercenaria campechiensis*
Crustaceans	
Blue shrimp	*Litopenaeus stylirostris*
Giant river prawn	*Macrobrachium rosenbergii*
Kuruma shrimp	*Marsupenaeus japonicus*
Marron crayfish	*Cherax tenuimanus*
Pacific white shrimp	*Litopenaeus vannamei*
Red claw crayfish	*Cherax quadricarinatus*
Red swamp crawfish	*Procambarus clarkii*
Tiger shrimp	*Penaeus monodon*
White river crawfish	*Procambarus acutus acutus*
White shrimp	*Penaeus setiferus*
Yabby crayfish	*Cherax destructor*
Algae (seaweeds)	
California giant kelp	*Macrocystis pyrifera*
Eucheuma	*Eucheuma cottoni*
False Irish moss	*Gigartina stellata*
Gracilaria	*Gracilaria* sp.
Irish moss	*Chondrus crispus*
Laminaria	*Laminaria* spp.
Nori or laver	*Porphyra* spp.
Wakame	*Undaria* spp.

cipal groups. Common and scientific names of many of the species of the finfish, mollusks, and crustaceans currently under culture are presented in Table 3. Included are examples of bait, recreational, and food animals.

Various species of carp and other members of the family Cyprinidae lead the world in terms of quantity of animals produced. In 1996, the total was more than 9.6 million metric tons (6). China is the leading carp-producing nation and is the world's leading aquaculture nation overall. Significant amounts of carp are also produced in India and parts of Europe. Species of carp where production exceeded 1 million metric tons in 1996 were silver carp, grass carp, common carp, and bighead carp.

Fish in the family Salmonidae (trout and salmon) are in high demand, with the interest in salmon being greatest in developed nations. Salmon, mostly Atlantic salmon, are produced in Canada, Chile, Norway, New Zealand, Scotland, and the United States. Fish in the family Cichlidae, which includes several cultured species of tilapia, are reared primarily in the tropics, but have been widely introduced throughout both the developed and developing world.

Catfish are not a major contributor to aquaculture production globally, but the channel catfish industry dominates U.S. aquaculture. Catfish production in the United States, primarily channel catfish, was 209,090 metric tons in 1992 (4) and 206,078 metric tons in 1995 (3). Total world catfish production in 1995, which includes walking catfish, was 312,000 metric tons.

Among the invertebrates, most of the world's production is associated with mussels, oysters, shrimp, scallops, and clams. Crawfish culture is of considerable importance in the United States but amounted to only 24,211 metric tons in 1992 (4), insignificant compared with some other invertebrate species. Globally, crawfish production was 29,350 tons, of which more than 28,000 tons was attributable to the red swamp crawfish (3).

Small amounts of crabs, lobsters, and abalone are being cultured in various nations. All three bring good prices in the marketplace but have drawbacks associated with their culture. Lobsters are highly cannibalistic. Rearing them separately to keep them from consuming one another during molting has precluded economic culture in most instances. Spiny lobster (*Panulirus* spp.) culturists produced a total of 69 metric tons in 1995 (3). Crab culture has also failed to develop in some parts of the world. In the United States, blue crab (*Callinectes sapidus*) has been attempted but has not been economically feasible, largely due to problems with cannibalism. Global production of crabs in the family Portunidae (of which *C. sapidus* is a member) during 1995 was only 20 metric tons. Chinese river crabs and various other species of marine crabs are being successfully cultured, with 1995 world production at more than 89,000 metric tons (3).

The dominant mollusks being cultured are oysters, clams, mussels, and to a lesser extent, scallops. Mollusk culture ranges from highly extensive to intensive. An example of one of those extremes is the placement of shell material (cultch) on the substrate (in public waters or a leased site) to provide additional substrate on which naturally produced oyster larvae (spat) can settle. That approach is highly extensive. At the other (intensive) extreme is producing and settling spat in a hatchery and rearing them on strings suspended from rafts or in trays.

Among the mollusks of aquaculture importance, abalones eat seaweeds and can only be reared in conjunction with a concurrent seaweed culture facility or in regions where large supplies of suitable seaweeds are available from nature. In some instances the value of the seaweed for direct human consumption may make the highest and best use of the plants.

Well over 100 species of aquatic animals are being cultured primarily for human food. That number expands to several hundred if ornamental, bait, and recreational species produced by aquaculturists are included. Many researchers have turned their attention to species for which there is demand by consumers, but for which the technology required for commercial production is marginal or not completely developed. Examples are dolphin, also known as mahimahi (*Coryphaena hippurus*), cobia (*Rachycentron canadum*), red snapper (*Lutjanus campechanus*), Pacific halibut (*Hippoglossus stenolepis*), southern flounder (*Paralichthys lethostigma*), winter flounder (*Pseudopleuronectes americanus*), American lobster (*Homarus americanus*), and blue crab (*Callinectes sapidus*). Each of the species mentioned is marine and has small eggs and larvae. Providing the first feeding stages with acceptable food has been a common problem, as has the fragility of the early life stages of many species, and the problem of cannibalism.

Fish with large eggs, such as trout, salmon, catfish, and tilapia, were among the first to be economically successful in modern times. Small eggs do not necessarily mean that sophisticated research is required to develop the technology required for successful culture. Carp, which have been cultured in China for millennia, have extremely small eggs. At the time the methodology for carp culture was developed, there were no research scientists, although there must have been dedicated farmers who used their common sense and trial-and-error methods to establish aquaculture.

In contrast, relatively large eggs do not necessarily mean that culture will be easy. Fish with very large eggs (eg, such marine catfish as the gaftopsail catfish, *Bagre marinus*) have eggs well over a centimeter in diameter but such low fecundity that the number of broodstock required to maintain a culture facility would doom the venture. Atlantic and Pacific halibut have eggs of about 3-mm diameter, much smaller than those of trout and salmon, but relatively large relative to many marine species, including other species of flatfish. Halibut eggs are relatively easy to hatch, but the larvae are extremely fragile so providing the nearly static conditions for development without causing water quality degradation is difficult. Nearly one month is required for egg incubation, followed by another approximately 35 days until first feeding, and then several additional weeks before metamorphosis. Following metamorphosis, halibut postlarvae become easy to maintain, but mortality rate from egg to metamorphosis is generally very high.

ECONOMICS

In nations where fish are produced for recreational fishing, funding for hatcheries is often raised from user fees such as fishing licenses. Such hatcheries are usually government facilities. For most private aquaculture companies to get started, outside funding is required. Funding may come through banks and other commercial lending sources or from venture capitalists. The high risks associated with aquaculture have made it difficult for many firms to obtain bank loans, although that situation is changing as bankers become more knowledgeable and comfortable with underwriting aquaculture ventures. Large corporations have also entered the aquaculture field. Some have abandoned aquaculture after a few years, but others continue to diversify into the arena. Highly profitable corporations may have an easier time obtaining the requisite funding for establishing aquaculture ventures than small entrepreneurs.

A key factor in obtaining funding support for aquaculture is development of a sound business plan. The plan needs to demonstrate that the prospective culturist has identified all costs associated with establishment of the facility and its day-to-day operation. One or more suitable sites should be identified and the species to be cultured selected before the business plan is submitted. Cost estimates should be verifiable. Having actual bids for specific tasks at specific locations, for example, pond construction, well drilling, building construction, and vehicle costs, in a particular geographic region will strengthen the business plan.

Land costs vary enormously both between and within countries. Compare, for example, the cost of coastal land in south Florida, where it might be possible to consider rearing shrimp, with that of Mississippi farmland suitable for catfish farming. The former might be thousands of dollars for every meter of oceanfront, while the latter may be obtained for one or two thousand dollars per hectare.

The amount of land required will also vary, not only as a function of the amount of anticipated production, but also on the type of culture system that is used. It may take several hectares of static culture ponds to produce the same biomass of animals as one modest size raceway through which large volumes of water are constantly flowing, for example. Construction costs vary from one location to another. Local labor and fuel costs must be factored into the equation. The experience of contractors in building aquaculture facilities is another factor to be considered.

Much of the world's aquaculture is conducted in tropical or subtropical regions that feature long growing seasons (often year-round). Many such areas lie within developing nations where labor and land are typically inexpensive and where governments sometimes offer tax incentives to prospective aquaculturists. Problems with obtaining good quality feed, parts, dependable energy supplies, and other infrastructure are common and represent a trade-off that needs to be considered. Skilled technical personnel, such as hatchery managers, may have to be brought into facilities sited in developing countries because individuals with the required credentials may not be available. Expatriate labor can be quite expensive but may be necessary.

The need for redundancy in the culture system needs to be assessed in conjunction with each facility. Failure of a well pump that brings up water to supply a static pond system may not be a serious problem in countries where new pumps can be purchased in a nearby town. However, it can be disastrous in developing countries where new pumps and pump parts are often not available but must be ordered from another country. Several weeks or months may pass before the situation can be remedied unless the culturist maintains a selection of spares. If the culture system requires constantly flowing water (eg, an open raceway system or a recirculating system), loss of flow due to pump failure for even a few minutes may result in disaster. A pump may break down, or there may be a power failure that results in pump failure. In either case, having a backup that will automatically come on to keep the system operational will help ensure against tragic losses. In the case of power failures, a backup diesel generator with an automatic switching mechanism is a popular choice.

Redundancy is not restricted to pumps but should be considered for all types of equipment where failure can lead to rapid loss of environmental control. The up-front costs of backup systems can be quite high, but loss of a crop may cause a business failure.

The business plan should provide projections of annual production. Based on those estimates and assumed food conversion ratios, an estimate of feed costs can be made. Food conversion ratio (FCR) is calculated by determining the amount of feed consumed for each unit of weight gain. For example, if a fish grows from 1 to 2 kg on 2 kg of feed, the FCR is 2 kg of feed consumed divided by 1 kg gain or 2.0. For many aquaculture ventures, between 40 and 50% of the variable costs involved in aquaculture can be attributed to feed.

Aquaculturists may elect to purchase animals for stocking or maintain their own broodstock and operate a hatchery. The decision may rest on such factors as the availability and cost of fry fish, postlarval fish, oyster spat, or other early life history stages in the location selected for the aquaculture venture. In regions where the species selected is being produced in large quantity (eg, channel catfish in Mississippi or penaeid shrimp in Thailand), sufficient local hatchery production may be available to supply a new venture. A catfish farmer in Nebraska or shrimp farmer in Alabama might find it more economical to establish a hatchery than to ship in young animals from long distances.

For aquaculturists working with banks and other types of lending institutions, money for land purchase, site preparation, and facility construction can be obtained through loans of 15 to 30 years. Equipment such as pumps and trucks are usually depreciated over a few years and are funded with shorter-term loans (often seven years). Operating expenses for such items as feed, chemicals, fuel, utilities, salaries, taxes, and insurance may require periodic short-term loans to keep the business solvent. The projected income should be based on a realistic estimate of farmgate value of the product and an accurate assessment of anticipated production. Each business plan should project income and expenses over the term of all loans to demonstrate to the lending agency or venture capitalist that there is a high probability the investment will be repaid. Realistic business plans often show losses for up to several years while the facility is being constructed and put into full production. Some aquaculturists even plan for an occasional crop loss when developing their long-range projections of profitability.

REGULATION

The extent to which governments regulate aquaculture varies greatly from one nation to another. In some parts of the world, particularly in developing nations, there has historically been little or no regulation. Inexpensive land and labor, low taxes, excellent climates, and a lack of government interference have drawn many aquaculturists to underdeveloped countries, most of which are in the tropics. Unregulated expansion of aquaculture in some countries has led to pollution problems and destruction of valuable habitats such as mangrove swamps and has enhanced the spread of disease from one farm to another. The need for imposing more restrictive regulations is now becoming evident around the world. Response to that need varies considerably from one nation to another.

Environmental problems associated with aquaculture have become a global phenomenon. In Japan, previously unrestricted development of net-pen culture of various kinds of marine fish in bays led to what was termed *self-pollution*. The waste feed and excretory products from the fish degraded water quality to the point that the aquacultured fish were unable to grow adequately or, in some cases, were unable to survive. Strict controls on the number of net-pens that could be established within each bay were established and the problem was adequately addressed. Similarly, strict regulations on aquaculture have been imposed throughout Europe in an attempt to prevent environmental deterioration. Inland aquaculture ventures on private lands have come under much less scrutiny and governmental control than those established in coastal regions in or adjacent to public waters.

The United States is an example of a mixture of local, state, and federal regulations. Permits from a county, state, or federal agency may be required for drilling wells, pumping water, releasing water, employing exotic species, constructing facilities, and so on. In the United States, most permits can be obtained at the local or state level. In some instances the federal government has devolved permitting authority to the states when state regulations are as rigorous or more so than national regulations. Federal agencies become involved when aquaculture projects are conducted in navigable waters (U.S. Army Corps of Engineers) or might impact threatened or endangered species (National Biological Service). For aquaculture ventures established in the marine environment outside of state waters, only federal permits are required. Once again, a U.S. Army Corps of Engineers permit will be required. With its mandate to regulate the nation's fisheries, the National Marine Fisheries Service will also become involved in the aquaculture permitting process in federal waters. Interest in establishing aquaculture operations in conjunction with operating or out-of-production oil and gas platforms has developed in recent years. For such a platform to be used for aquaculture, a permit from the Minerals Management Service will be required. During the period when the various permit applications are being reviewed, public hearings may be held and various other federal agencies may be provided with the opportunity to comment. The process can be long, arduous, and expensive whether in state or federal waters. It is not uncommon for the prospective aquaculturist to have to deal with more than a dozen agencies when seeking state permits.

In general, it is easier to establish an aquaculture facility on private land than in public waters such as a lake or coastal embayment. Prospective aquaculturists who want to establish facilities in public waters may be confronted at public hearings by outraged citizens who do not want to see an aquaculture facility in what they consider to be their water. The issue is highly contentious in some nations (eg, the United States). In other countries (eg, Japan), when properly conducted, aquaculture in public waters not only is seen as a good use of natural resources, but also can be considered an amenity.

Obtaining permits is often not simple. Few states have one office that can accommodate the prospective aquaculturist. In most cases it is necessary to contact a number of state agencies to apply for permits. As mentioned, public hearings may be required before permits are approved. The process can take months or even years to complete. The costs involved in going through the process may be prodigious. After the expenditure of considerable amounts of time and money, there is no guarantee that the permits will ultimately be granted.

Most states in the United States now have an aquaculture coordinator, usually housed in the state department of agriculture, who can assist prospective aquaculturists in finding a path through the permitting process. In other countries the process can vary from highly complicated to virtually unregulated. Anyone considering development of an aquaculture facility should become educated on the permitting process of the state or nation in which the facility will be developed. In cases where the process is involved, it should be initiated well in advance of the anticipated time of actual facility construction and, if possible, before land acquisition is completed.

CULTURE SYSTEMS

At one extreme, aquaculture can be conducted with a small amount of intervention from humans and the employment of little technology. At the other extreme, aquaculture involves total environmental control and the use of computers, molecular genetics, and complex modern technology. Many aquaculturists operate between the extremes. The range of culture approaches can be described as running from extensive to intensive, or even hyperintensive, with extensive systems being relatively simple and intensive systems being complex to very complex. In general, as the level of culture intensity increases, stocking density and, as a consequence, production per unit area of culture system or volume of water increase.

The most extensive types of aquaculture involve minimal human intervention to promote increases in natural productivity. One good example is the scattering of cultch material to provide substrates for oyster spat settlement as previously described. A second is the stocking of water bodies used in conjunction with recreational fishing.

The stocking of ponds, lakes, and reservoirs to increase the production of desirable fish that depend on natural productivity for their food supply and are ultimately captured by recreational anglers is practiced primarily in developed nations. In developing nations, community ponds are sometimes stocked to provide food for people living at the subsistence level. Some would consider the practice of stocking fish for recreation or subsistence as lying outside of the realm of aquaculture, but since it involves human intervention and often employs fish produced in hatcheries, such programs are certainly associated with, if not an actual part of, aquaculture. Since the stocked animals are expected to live off available natural food supplies in most instances (agricultural waste products are sometimes thrown into ponds by subsistence farmers to enhance productivity), production levels are often less than 100 kg/ha/yr.

Most of the aquaculture practiced around the world is conducted in ponds (Fig. 1). Ponds range in size, but production units are generally 0.1 to 10 ha in area. The intensity of aquaculture in ponds can range from a few kilograms per hectare to thousands of kilograms per hectare of annual production. Production levels in the U.S. channel catfish industry were only a few hundred kilograms per hectare in the early 1960s and rose to about 3000 kilograms per hectare by the end of that decade. A decade

Figure 1. Catfish culture pond in Texas with paddlewheel aerator used to increase dissolved oxygen level, as necessary.

Figure 2. Concrete raceways at a trout farm in Idaho.

later, some farmers were able to produce 8000 to 10,000 kg/ha/yr in open ponds. Technological advances in feed quality, control of diseases, and water quality management were among the factors that allowed for the increased production rates.

The first step up from the recreational stocking or subsistence level of culture is to fertilize the water to increase natural productivity. Fertilization encourages algae blooms, which in turn stimulate the production of plankton and benthic organisms, all of which can provide a source of nutrients to the target culture species. Moving up in intensity from fertilization is the provision of supplemental feeds, which are those that provide some additional nutrition but cannot be depended upon to supply all the required nutrients. Provision of complete feeds, those that do provide all of the nutrients required by the culture species, translates to another increase in intensity. Associated with one or more of the stages described might be the application of techniques that lead to the maintenance of good water quality. Examples are continuous water exchange, mechanical aeration, and the use of various chemicals to adjust such factors as pH, alkalinity, and hardness.

With the application of increased technology and control over the culture system, intensity continues to increase. Utilization of specific pathogen-free animals, provision of nutritionally complete feeds, careful monitoring and control of water quality, and the use of animals bred for good performance can lead to impressive production levels.

Where water is plentiful and inexpensive, raceway culture is an attractive option and one that allows for production levels well in excess of what is possible in ponds. Trout are frequently reared in linear raceways from hatching to market size. Linear raceways are longer than they are wide and are usually no deeper than 1 to 2 in. (Fig. 2). High-density raceways used in production facilities are commonly constructed of poured concrete. Small raceways of the type used in hatcheries and research facilities may be constructed of fiberglass or other resilient materials. Water is introduced at one end and flows by gravity through the raceway to exit the other end. Circular raceways, called tanks (Fig. 3), are also used by aquaculturists. Tanks are usually no more than 2 in. deep and may be from less than 1 m to as much as 10 m in diameter. Concrete tanks can be found, but most are constructed of fiberglass, metal, or wood that is sealed and covered with epoxy or some other waterproof material. Plastic liners are commonly used in metal or wood tanks to prevent leakage, and in the case of metal, to avoid exposing the aquaculture animals to trace element toxicity.

Figure 3. Circular tanks for the commercial production of tilapia in geothermal water in Idaho.

Linear raceways are commonly used by trout and salmon culturists both for commercial production and for hatchery programs conducted by government agencies to produce fish for stocking. Large numbers of state and federal salmon hatcheries in Washington and Oregon, along with governmental and private hatcheries in Alaska, collect and fertilize eggs, hatch them, and rear the young fish to the smolt stage at which time they become physiologically adapted to enter seawater. It is following smoltification that the fish are transported to release sites. Migrating adults will have imprinted on the hatchery water as juveniles and can be counted upon to return to their hatcheries of origin with few exceptions.

Commercial salmon culturists can rear their fish to market size in freshwater raceways, although most salmon are grown from smolt to market size or adulthood in the marine environment, either as free roaming fish or in confinement. Releasing smolts into the open marine environment is called ocean ranching and is a technique that takes advantage of the homing instinct of salmon. When the fish that had been released as smolts return to spawn, sufficient numbers of adults are collected for use as broodfish to continue the cycle. The remainder may be harvested by the aquaculturist or by commercial fishers after which the fish are processed and marketed.

Salmon, steelhead trout, and a variety of marine fish are currently being reared in net-pens (Fig. 4). The typical salmon net-pen is several meters (sometimes as much as 20 m) on each side and may be 10 m or more deep (1). Smaller units, called cages, are sometimes used by freshwater culturists. Cages tend to have volumes of no more than a few cubic meters (Fig. 5).

Net-pen technology was developed in the 1960s but has only been widely employed commercially for salmon production since the 1980s, when the Norwegian salmon farming industry was developed. The Japanese began producing large numbers of sea bream and yellowtail in net-pens during the 1960s. Other nations have used the technology as well. Most of the net-pens currently in production are located in protected waters since they are easily damaged or destroyed by storms.

Competition by various user groups for space in protected coastal waters in much of the world has led to strict controls and, in some cases, prohibitions against the establishment of inshore net-pen facilities. As a result, there is growing interest in developing the technology to move offshore. Various designs for offshore net-pens have been developed and a few have been tested (Fig. 6). A number of different designs, including systems that are semisubmersible or totally submersible, have been able to withstand storm waves of at least 6 m, but the costs of those systems are very high compared with inshore net-pens, so commercial viability may be difficult to achieve unless the species being reared is of very high value. Tuna, for example, are being cultured to a limited extent and can bring very high prices, particularly if they are of high quality, for the Japanese sushi trade. Some success in spawning and larval rearing of tuna has been achieved, but in most instances juvenile fish are captured and placed in net-pens for growout to market size. Depending on the size at which the fish are stocked, the net-pen growout period may require only a few months.

The highest level of intensity that can be found in aquaculture is associated with reuse systems, often called recirculating systems. In these systems, the bulk of the wa-

Figure 4. A marine net-pen facility in Norwegian fjord.

Figure 5. Marine cages in Malaysia.

Figure 6. A salmon net-pen designed for use in areas where wave heights during storms may reach as high as 6 m.

ter passing through the chambers in which the finfish or shellfish are held is continuously treated and reused. Once filled initially, reuse systems can theoretically be operated for long periods of time without much water replacement. It is necessary to add some water to such systems to make up for that lost to evaporation and splashout and in conjunction with solids removal.

Recirculating systems can be used for all phases of culture, from broodstock maintenance through spawning, hatching, larval rearing, to growout. High energy requirements because of the requirement to run pumps and aerators 24 hours a day, and sometimes greatly increased if water must be heated or chilled, have rendered many recirculating systems uneconomical. Exceptions include systems used for the production of high-priced products (such as ornamental species), those where water and/or heat is basically free (such as systems that employ power plant cooling water), or when only part of the rearing cycle employs recirculating technology (eg, fingerling production in a reuse system followed by pond growout).

Many of the recirculating systems in use today are operated in a mode between entirely closed and completely open (open systems typically have sufficient flow rates to exchange the water in the culture chambers every hour or less). In most systems there is a significant percentage of replacement water added either continuously or intermittently on a daily basis. Such partial recirculating systems may exchange from a few percent to several hundred percent of system volume each day.

The heart of a recirculating water system is the biofilter, a device that contains a medium on which bacteria that help purify the water become established (Fig. 7). Fish and aquatic invertebrates produce ammonia as a primary metabolite. If not removed or converted to a less toxic chemical, ammonia can quickly reach lethal levels. Two genera of bacteria are responsible for ammonia removal in biofilters. The first, *Nitrosomonas*, converts ammonia (NH_3) to nitrite (NO_2^-). The second, *Nitrobacter*, converts nitrite to nitrate (NO_3^-). Nitrite is highly toxic to aquatic animals, although nitrate can be allowed to accumulate to relatively high levels. If both genera of bacteria are active, the conversion from ammonia through nitrite to nitrate is so rapid that nitrite levels remain within the safe range.

In addition to the biofilter and culture chambers, recirculating systems typically also employ one or more settling chambers or mechanical filters to remove solids such as unconsumed feed, feces, and mats of bacteria that slough from surfaces within the system. Each recirculating system requires a mechanical means of moving water from component to component. That usually means mechanical pumping, though air-lifts can also be used. Most systems incorporate one pump to lift water and rely on gravity to provide flow through the various system components and back to the pump (Fig. 8). That approach reduces the need to balance flow rates through two or more pumps and reduces operating costs.

Control of circulating bacteria and oxidation of organic matter can be obtained through ozonation of the water. Ozone (O_3) is highly toxic to aquatic organisms. Ozone must be allowed to dissipate prior to exposing the water to the aquaculture animals. With time, and with the assis-

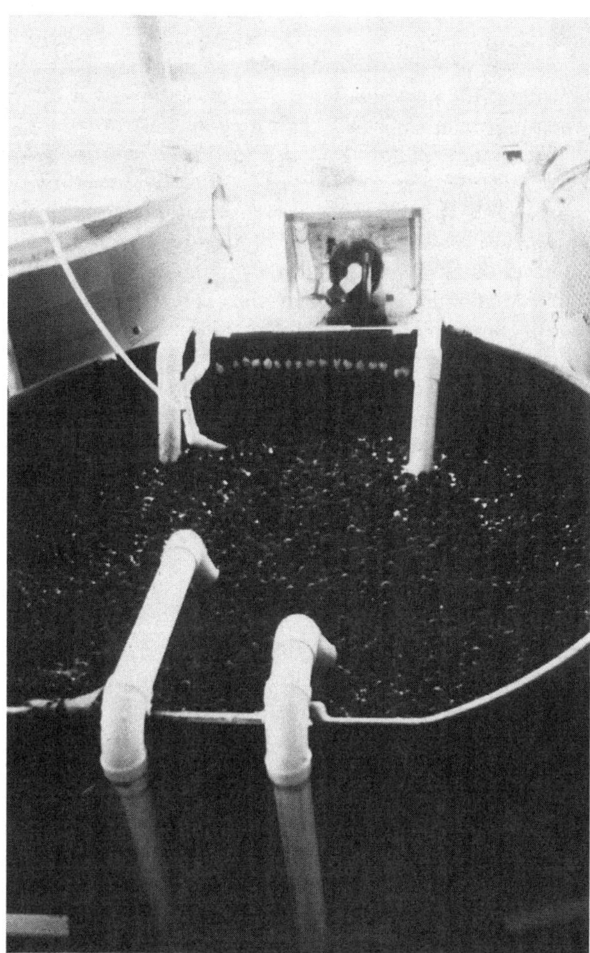

Figure 7. A circular tank filled with plastic balls to which bacteria attach is one type of biofilter that can be used by aquaculturists; in this case, the biofilter helps maintain water quality at a baitstand in Texas.

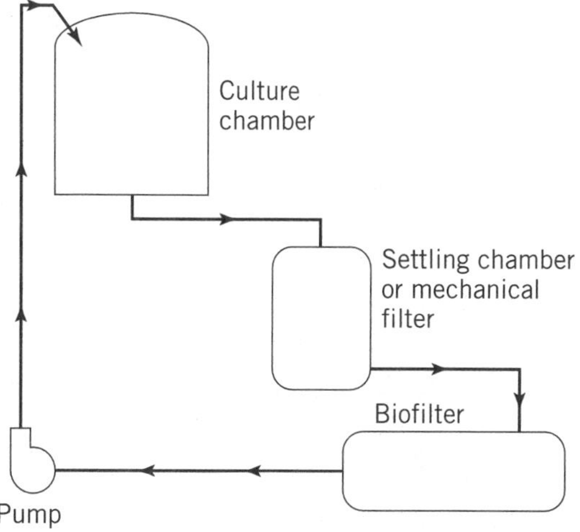

Figure 8. Schematic representation of a typical closed water system design.

tance of aeration, ozone can be driven off or converted to molecular oxygen. Various commercial firms market ozone generators and can assist aquaculturists in selecting the proper equipment to meet system needs.

Ultraviolet (uv) light has also been used to sterilize the water in aquaculture systems. The effectiveness of uv decreases with the thickness of the water column being treated, so the water is usually flowed past uv lights as a thin film (alternatively, the water may flow through a tube a few centimeters in diameter that is surrounded by uv lights). Ultraviolet light systems require more routine maintenance than ozone systems. The bulbs emit lower and lower quantities of uv light with time and need to be changed periodically. In addition, organic material exposed to uv light will foul the surface of the transparent quartz or plastic tubes separating the bulbs from the water, thereby creating a film between the water and the uv source that reduces the effectiveness of the uv light.

Recirculating systems often feature other types of apparatus, such as foam strippers and supplemental aeration. Devices for denitrifying nitrate to nitrogen gas have been developed to the point that they may find a place in commercial culture systems in the near future.

The technology associated with recirculating systems is expensive. Redundancy in the system, that is, providing backups for all critical components, and automation are important considerations. When a pump fails, for example, the failure must be instantly communicated to the culturist, who must have the ability to keep the system operating while the problem is being addressed. Loss of a critical component for even a few minutes can result in the loss of all animals within the system. Computerized water-quality monitoring systems that will sound alarms and dial emergency telephone numbers to report system failures are finding increased use among culturists who employ recirculating systems. The computer can also be used to put backup systems on line in the event of a primary system failure.

Recirculating systems can make aquaculture feasible in locations where conditions might otherwise not be conducive to successful operations. Tropical and subtropical animals such as tilapia and penaeid shrimp have, for example, been reared in recirculating systems in the northern tier of states in the United States. Recirculation systems can also be used to reduce transportation costs by making it possible to grow animals near markets. In areas where there are concerns about pollution from aquaculture system effluents or with the use of exotic species, recirculating systems provide an alternative approach to more extensive types of operations.

WATER SOURCES AND QUALITY

Sources of water for aquaculture include municipal supplies, wells, springs, streams, lakes, reservoirs, estuaries, and the ocean. The water may be used directly from the source or it may be treated in some fashion prior to use.

Most municipal water sources are chlorinated and contain sufficiently high levels of chlorine so as to be toxic to aquatic life. Chlorine can be removed by passing the water through activated charcoal filters or through the use of sodium thiosulfate metered into the incoming water at the proper rate. Municipal water is usually not used in aquaculture operations that require large quantities of water, either continuously or periodically, because of the initial high cost for the water and the cost of pretreatment to remove chlorine.

Most aquaculturists, if asked, would probably indicate a preference for well water over other sources. Both freshwater and saline wells are common sources of water for aquaculture. The most commonly used pretreatments for well water include temperature alteration (either heating or cooling); aeration to add oxygen or to drive off or oxidize such substances as carbon dioxide, hydrogen sulfide, and iron; and modification of salinity (in mariculture systems). Pretreatment may also include adjusting hardness and alkalinity through the application of appropriate chemicals. For example, limestone ($CaCO_3$) is commonly used to increase both hardness and alkalinity.

To heat or cool water requires large amounts of energy. A major consideration in locating an aquaculture facility is to have not only a sufficient supply of water but also water at or near the optimum temperature for growing the species that has been selected. The vast supply of spring water of almost perfect temperature in the Hagerman Valley of Idaho supports the majority of the rainbow trout production in the United States. Where geothermal water is available, tropical species can be grown in locations where ambient winter temperatures would otherwise not allow them to survive. Tilapia farmers have been successful in parts of Idaho and other states where geothermal water of suitable quality is available. Not all geothermal wells produce water of suitable quality, however. High levels of sulfur and other toxic chemicals are commonly found in geothermal water, so thorough testing should precede any attempt to use such water sources for aquaculture.

Another large cost associated with incoming water is associated with its movement. Many aquaculture facilities that utilize surface waters and those that obtain their water from wells other than artesian wells are required to pump the water into their facilities. Pumping costs can be a major expense, particularly when the facility requires continuous inflow or high volumes of makeup water to replace evaporative losses from ponds.

Surface water can sometimes be obtained through gravity flow by locating aquaculture facilities at elevations below those of adjacent springs, streams, lakes, or reservoirs. Coastal facilities may be able to obtain water through tidal flow.

The most common treatment of incoming surface water is removal of particulate matter. This can be effected through the use of settling basins or mechanical filters. Particle removal may involve the reduction or elimination of suspended inorganic material such as clay, silt, and sand. It may also involve removal of organic material, including living organisms. Organisms that will enter aquaculture facilities if not filtered from the incoming water include phytoplankton and zooplankton, plants and plant parts, macroinvertebrates, and fish. Some of the organisms, if not removed, can be expected to survive and grow to become predators on, or competitors with, the target aquaculture species. Very small organisms, such as bacteria, can be removed mechanically though filtering particles of a few microns in size; such removal becomes inordinately expensive if there are large volumes involved. In most cases, filtration of incoming water involves passing the water through screens of various sizes, the smallest of which is often no smaller than a window screen. If a situation calls for elimination of microorganisms, other forms of water treatment, such as ozonation and the use of uv radiation, are more efficient and effective than filtration.

For many freshwater species that can be characterized as warmwater (such as channel catfish and tilapia with an optimum of about 30°C) or coldwater (such as trout with an optimum of about 20°C), the conditions outlined in Table 4 should provide an acceptable environment. So-called midrange species are those with an optimum temperature for growth of about 25°C (examples are walleye, northern pike, muskellunge, and yellow perch). Typically, midrange species do well under the conditions, other than temperature, specified in Table 4 for coldwater species. Some species have higher or lower tolerances than others. For example, tilapia can tolerate temperatures in excess of 34°C but have poor tolerance for low temperature. Most tilapia species die when the temperature falls below about 12°C. Tilapia have a remarkably high tolerance for ammonia compared with such species as trout and salmon, which have a high tolerance for cold water but cannot tolerate water temperatures much above 20°C. Marine fish may be able to tolerate a wide range of salinity (such euryhaline species include many species of flounder, red drum, salmon, and some species of shrimp), or they may have a narrow tolerance range (they are called stenohaline species, examples of which are dolphin, halibut, and lobsters). Recommended water quality conditions for marine fish production systems are presented in Table 5.

The water quality criteria for each species should be determined from the literature or through experimentation when literature information is unavailable. Synergistic effects that occur among water quality variables can have an influence on the tolerance a species has under any given set of circumstances. Ammonia is a good example. Ionized ammonia (NH^+) is not particularly lethal to aquatic animals, but unionized ammonia (NH^3) can be toxic even when present at a fraction of a part per million (depending on species). The percentage of unionized ammonia in the water at any given total ammonia concentration changes in relation to such factors as temperature and pH. As either temperature or pH increases so does the percentage of unionized ammonia relative to the level of total ammonia. Thus, in warmwater and in seawater (which generally has a higher pH than freshwater), ammonia toxicity will occur at a lower total ammonia concentration than in cool or lower pH water.

Table 4. General Water Quality Requirements for Trout and Warmwater Aquatic Animals in Fresh Water

Variable	Acceptable level or range	
	Coldwater	Warmwater
Temperature, °C	<20	26–30
Alkalinity, mg/L	10–400	50–400
Dissolved oxygen, mg/L	>5	~5
Hardness, mg/L	10–400	50–400
pH	6.5–8.5	6.5–8.5
Total ammonia, mg/L	<0.1	<1.0
Ferrous iron, mg/L	0	0
Ferric iron, mg/L	0.5	0–0.5
Carbon dioxide, mg/L	0–10	0–15
Hydrogen sulfide, mg/L	0	0
Cadmium, μg/L	<10	<10
Chromium, μg/L	<100	<100
Copper, μg/L	<25	<25
Lead, μg/L	<100	<100
Mercury, μg/L	<0.1	<0.1
Zinc, μg/L	<100	<100

Source: Refs. 2, 8, and 9.

Table 5. Suggested Water Quality Conditions for Marine Fish Production Facilities

Variable	Acceptable level or range
Temperature, °C	1–40 (depends on species)
Salinity, g/kg	1–40 (depends on species)
Dissolved oxygen, mg/L	>6
pH	<7.9–8.2
Total ammonia, μg/L as NH_3	<10
Iron, μg/L	100
Carbon dioxide, mg/L	<10
Hydrogen sulfide, μg/L	<1
Cadmium, μg/L	<3
Chromium, μg/L	<25
Copper, μg/L	<3
Mercury, μg/L	<0.1
Nickel, μg/L	<5
Lead, μg/L	<4
Zinc, μg/L	<25

Source: Ref. 10.

Another example is dissolved oxygen (DO). The amount of DO that water can hold at saturation is affected by both temperature and salinity. The warmer and/or saline the water, the lower the saturation DO level. Atmospheric pressure also affects oxygen saturation. The saturation oxygen level decreases as elevation increases.

Biocides should not be present in water used for aquaculture. Sources of herbicides and pesticides include runoff from agricultural land, contamination of the water table, and spray drift from crop-dusting activity. Excessive levels of phosphorus and nitrogen may occur where runoff from fertilized land enters an aquaculture facility either from surface runoff or groundwater contamination. Trace metal levels should be low as indicated in Tables 4 and 5.

Most aquaculture facilities release water constantly or periodically into the environment without passing it through a municipal sewage treatment plant. The effects of aquaculture effluents on natural systems have become a subject of intense scrutiny in recent years and have, in some instances, resulted in opposition to further development of aquaculture facilities in some locales, particularly in public waters. There have even been demands that some existing operations should be shut down.

Regulation of aquaculture varies greatly both between and within nations. Some governmental agencies with jurisdiction over aquaculture have placed severe restrictions on the levels of such nutrients as phosphorus and nitrogen that can be released into receiving waters. Regulations on suspended solids levels in effluent water are also common. The installation of settling ponds or created wetlands, exposure of the water to filter feeding animals that will remove solids, and mechanical filtration have been used to treat effluents. Reduction or removal of dissolved nutrients through tertiary treatment is possible but is generally not economically feasible with current technology. Research is currently under way to develop feeds containing reduced levels of nutrients or to provide nutrients in forms that the culture animals can better utilize. The goal in both approaches is to reduce losses of nutrients to the environment through excretion.

NUTRITION AND FEEDING

Problems associated with excessive levels of nutrients and unwanted nuisance species have already been mentioned. In some cases, aquaculturists use intentional fertilization to produce desirable types of natural food for the species under culture. Examples of this approach include inorganic fertilizer applications in ponds to promote phytoplankton and zooplankton blooms that provide food for young fish such as channel catfish in the United States, the development of algal mats (called lab-lab) through fertilization of milkfish ponds in Asia, and the use of organic fertilizers (from livestock and human excrement) in Chinese carp ponds to encourage the growth of phytoplankton, macrophytes, and benthic invertebrates. In the latter instance, various species of carp with different food habits are stocked to ensure that all the types of natural foods produced as a result of fertilization are consumed.

Provision of live foods is currently necessary for survival of the early stages of many aquaculture species because acceptable prepared feeds have yet to be developed. Algae is routinely cultured for the early stages of mollusks produced in hatcheries. Once the mollusks are placed in grow-out areas, natural productivity is depended on to provide the algae upon which the shellfish feed.

In cases where zooplankton are reared as food for predatory larvae or fry, it may be necessary to maintain three cultures (algae, zooplankton, and the desired aquaculture species). Though wild zooplankton have been used successfully in some instances (eg, in Norway, wild zooplankton have been collected and fed to larval Pacific halibut), it is much more common practice to provide rotifers or brine shrimp nauplii as zooplanktonic food for first-feeding stages of animals being reared for marketing. After being fed rotifers and/or brine shrimp nauplii for periods ranging from several days to several weeks, depending on the species being reared, the aquaculture animals can usually be trained to accept pelleted feeds, though a weaning process during which both natural and prepared feeds are offered is sometimes required. Problems associated with utilizing prepared feeds from first feeding include difficulty in providing very small particles that contain all the required nutrients, loss of soluble nutrients into the water from small particles before the animals consume the feed, and in some cases, the fact that prepared feeds do not behave the same as live foods when placed in the water. For species that are sight feeders, behavior of the food is often an important factor. Feed color may also be a factor, as is texture, in whether a food item is accepted. Some fish, for example, reject hard pellets but will accept pellets that have a more spongy texture.

Some of the most popular aquaculture species accept prepared feeds from first feeding. Included are catfish, tilapia, salmon, and trout. All of the fish listed have relatively large eggs (several millimeters in diameter) that develop into fry that have concomitantly large yolk sacs. The nutrients in the yolk sac lead to production of first-feeding fry with well-developed digestive tracts that produce the enzymes required to efficiently digest diets that contain the same types of ingredients used for larger animals. First-feeding fry of catfish and so on have mouth gapes of sufficient size that they can engulf particles that do contain all essential nutrients.

Over the last few decades, fish nutritionists have successfully determined the nutritional requirements of many aquaculture species and have developed practical feed formulations based on those requirements. For species such as Atlantic salmon, various species of Pacific salmon and trout, common carp, channel catfish, and tilapia, sufficient information exists to design diets precisely suited to each species. Aquaculturists are always interested in the development of new species. In each instance, the nutritional requirements of the new species must be investigated. Although there are many similarities among aquatic animals with respect to nutritional requirements, diets that produce the best growth at the least cost will vary significantly from species to species and can only be formulated when precise nutritional requirements are known. Determination of those requirements may require several years of research, although diets based on existing formulations may be at least adequate to produce reasonable growth and survival and can be employed while the research is being conducted.

Requirements for energy, protein, carbohydrates, lipids, vitamins, and minerals have been determined for the species commonly cultured (11). As a rule of thumb, trout and salmon diets will, if accepted, support growth and survival in virtually any aquaculture species. Such diets often serve as the control against which experimental diets are compared.

Since feeds contain substances other than those required by the animals of interest, studies have also been conducted on antinutritional factors in feedstuffs and on the use of additives. Certain feed ingredients contain chemicals that retard growth or may actually be toxic. Examples are gossypol in cottonseed meal and trypsin inhibitor in soybean meal. Restricting the amounts of the feedstuffs used in a particular diet is one way to avoid problems. In some cases, as is true of trypsin inhibitor, proper processing can destroy the antinutritional factor. In this case, heating of soybean meal is effective.

Animals that do not readily accept pelleted feeds may be enticed to do so if the feed carries an odor that induces ingestion. Color development is an important consideration in aquarium species and some animals produced for human food. External coloration is desired in aquarium species. Pink flesh in cultured salmon is desired by much of the consuming public. Coloration, whether external or of the flesh, can be achieved by incorporating ingredients that contain pigments or by adding extracts or synthetic compounds. One class of additives that impart color is the carotenoids.

Prepared feeds are marketed in various forms from very small crumbles to flakes and pellets of various sizes. Pelleted rations may be hard, semimoist, or moist. Hard pellets typically contain less than 10% water and can be stored under cool, dry conditions for at least 90 days without deterioration of quality. Dry pellets should, however, always be stored in a cool, dry place. Semimoist pellets are chemically stabilized to protect them from degradation and mold if they are properly stored (either refrigerated or in a cool, dry place), while moist pellets that contain high percentages of water must be frozen if they are not used immediately after manufacture.

Hard pellets are the type preferred if the species under culture will accept them. Semimoist feeds are most commonly used in conjunction with feeding young fish and species that find hard pellets unpalatable. Semimoist diets tend to be very expensive and can often only be used economically in conjunction with early life history stages where only small amounts of feed are required. Moist feeds, which contain high percentages of fresh fish, are usually available only in the vicinity of fish-processing plants.

Moist and semimoist feed pellets are produced in machines similar to sausage grinders. The most widely used types of prepared feeds, hard pellets, are produced by pressure pelleting or extrusion (crumbles are produced by grinding larger pellets). Pressure pelleting involves pushing the ground and mixed feed ingredients through holes in a die that is a few centimeters thick to produce spaghetti-like strands of the desired diameter. The strands are cut to length as they exit the die. Steam is often injected into the pellet mill in a location that exposes the feed mixture to moist heat just before the mix enters the die. Exposure to steam improves binding and extends pellet water stability. In general, a pelleted diet should remain stable in water for at least 10 minutes so the fish have sufficient opportunity to consume the feed before it dissolves.

Extruded pellets are produced by exposing the ground and mixed ingredients to much higher heat and pressure and for a longer time than is the case with pressure pellets. In the extrusion process the ingredients undergo some cooking that can be beneficial in reducing the levels of certain antinutritional factors, such as trypsin inhibitor. There may be concomitant losses of heat labile nutrients such as vitamin C, so overfortification to obtain the desired level in the final product may be required. In some cases, a heat labile nutrient is dissolved in oil and sprayed on the pellets after they have exited the extruder.

Specialty feeds such as flakes can be made by running newly manufactured pellets through a press or through use of a double drum dryer. The latter type of flakes begin as a slurry of feed ingredients and water. When the slurry is pressed between the hot rollers of the double drum dryer, wafer thin sheets of dry feed are produced that are then broken into smaller pieces. The different colors observed in some tropical fish foods represent a mixture of flakes, with each of the different colored flakes containing one or more different additives that impart color.

Pressure pellets sink when placed in water, whereas under the proper conditions, floating pellets can be produced through the extrusion process. That is accomplished when the feed mixture contains high levels of starch. When heated, starch expands and the pellets produced will trap air as they leave the barrel of the extruder. The trapped air gives the pellets a density of less than 1.0. Floating pellets are desirable for species that come to the surface to feed, because the aquaculturist can visually determine that the fish are actively feeding and can control daily feeding rates based on observed consumption.

Sinking extruded pellets are used for shrimp and other species that will not surface to obtain food and require pellets with high water stability. Shrimp consume very small particles, so they will nibble pieces from a pellet over an extended period of time. Unless heavily fortified with binders, pressure pellets will dissolve before shrimp have adequate time to fully consume them. Extruded feeds, whether sinking or floating, may remain intact for up to 24 hours after being placed in the water. Pressure pellets begin to disintegrate after a few minutes, unless, as mentioned, supplemental binders are incorporated into the feed mixture.

Nearly all aquaculture feeds contain at least some animal protein since the amino acid levels in plant proteins cannot meet the requirements of most aquatic animals. Fish meal is the most commonly used source of animal protein in aquaculture feeds, though blood meal, poultry byproduct meal, and meat and bone meal have also been successfully used. In 1995 more than 15,500,000 tons of prepared feeds were manufactured for use by the world's aquaculture producers. To make that amount of feed, 1,728,000 tons of fish meal were used. That amounted to 25.6% of the world's fish meal production (12).

Commonly used plant proteins include cornmeal, cottonseed meal, peanut meal, rice, soybean meal, and wheat. A number of other ingredients have also been used, many of which are only locally available. Most formulations contain a small percentage of added fat from such sources as fish oil, beef tallow, or more commonly, oilseed oils such as corn oil and soybean oil.

Complete rations contain added vitamins and minerals. Purified amino acids, binders, carotenoids, and antioxidants are other components found in many feeds. Growth hormone and antibiotics are sometimes used. Regulations on the incorporation of hormones along with other chemicals and drugs into aquatic animal feeds are in place in the United States and some other countries (Table 6). Few such regulations have been promulgated in developing nations.

Feeding practices vary from species to species. It is important not to overfeed since waste feed not only means wasted money, it can also lead to degradation of water quality. Most species require only 3 to 4% of body weight

Table 6. Therapeutants and Disinfecting Agents Approved for Use in U.S. Aquaculture

Name of compound	Use of compound
Therapeutants	
Copper	Antibacterial for shrimp
Formalin	Parasiticide for various species
Furanace (Nifurpyrinol)	Antibiotic for aquarium fishes
Oxytetracycline (Terramycin)	Antibiotic for fish and lobsters
Sodium chloride	Osmoregulatory enhancer for fish
Sulfadimethoxine (Romet)	Antibacterial for salmonids and catfish
Trichlorofon (Masoten)	Parasiticide for baitfish and goldfish
Disinfectants	
Calcium hypochlorite (HTH)	Used in raceways and on equipment
Didecyl dimethyl ammonium chloride (Sanaqua)	Used in aquaria and on equipment
Povidone-iodine compounds (Argentyne, Betadine, Wescodyne)	Disinfection of fish eggs

Source: Ref. 13.

in dry feed daily for optimum growth. Very young and adult animals are exceptions. Young animals are fed at higher rates because they are growing rapidly and consume a greater percentage of body weight daily than older animals. In addition, it is important to have food readily available to them. Food should be spread evenly over the culture chamber area so the young animals do not have to expend a great deal of energy searching for a meal. That is easily possible in tanks and small raceways, but large culture units do not lend themselves to having feed spread across the entire water surface. In those instances the feed can be spread along the pond or raceway's sides. Feeding rates as high as 50% of body weight daily are not uncommon for young animals. Since total biomass is small, even in intensively stocked units such as raceways, the economic cost is not high. Water quality in raceways can be maintained by siphoning out waste feed periodically. In ponds, any unconsumed feed acts as fertilizer, and the quantities used are not high enough to affect water quality adversely. Over time as the animals become increasingly mobile, the feeding rate can be reduced to satiation, which will often be about 10% of body weight daily for early juveniles and will fall to the previously mentioned 3 to 4% as the animals grow. Broodfish and other adult animals, such as ornamental species, being held in captivity are often fed at a maintenance level, which is usually about 1% of body weight daily.

Young animals may be fed several times daily. Examples include the standard practices of feeding fry channel catfish every three hours and young northern pike as frequently as every few minutes. Keeping carnivorous species such as northern pike satiated helps reduce the incidence of cannibalism. Animals stocked in growout culture chambers may be fed several times a day, but it is common practice to feed only once or twice to satiation. In warm weather, feed should be provided early in the day (but after making certain that the dissolved oxygen level is within the optimum range) and late in the afternoon. As the water cools in the fall in temperate climates, feeding rate is often reduced. Channel catfish farmers often feed during winter on warm days or every three days (unless there is ice cover) at a rate of no more than 1% of body weight at each feeding. The fish will overwinter without being fed but will lose weight. Weight loss can be averted by providing a modest amount of feed during winter.

REPRODUCTION AND GENETICS

Species such as carp, salmon, trout, channel catfish, and tilapia have been bred for many generations in captivity, though they usually differ little in appearance or genetically from their wild counterparts. A few exceptions exist, such as the leather carp, a common carp strain selectively bred to produce only one row of scales, and the Donaldson trout, a strain of rainbow trout developed over numerous generations to grow more rapidly to larger size and with a stouter body than its wild cousins.

Selective breeding has been long practiced as a means of improving aquaculture stocks. In some instances it has not been possible or is at least quite difficult and expensive to produce broodstock and spawn them in captivity, so culturists continue to rear animals obtained from nature. Most of the species that are being reared in significant quantities around the world are produced in hatcheries using either captured or cultured broodstock. Milkfish is a notable exception. That species has been spawned in captivity, but most of the fish reared in confinement are collected as juveniles in seines and sold to fish culturists. Wild shrimp postlarvae continue to be used to stock ponds in some parts of the world, though hatcheries may also be available in the event sufficient numbers of wild postlarvae are unavailable in a given year. In the United States, where shrimp culture involves the use of exotic species, all the animals reared come from hatcheries, both in the United States and in Latin America.

Spawning techniques vary widely from one species to another. Tilapia and catfish are typically allowed to spawn in ponds. Artificial nests are provided for catfish while tilapia dig their own nests in the pond bottom. Fertilized eggs can be collected from the mouths of female tilapia, but it is common practice to collect schools of fry after they are released from the mother's mouth to forage on their own. Catfish lay eggs in adhesive masses. Spawning chambers such as milk cans and grease cans are placed in ponds and may be examined every few days for the presence of egg masses. Some catfish farmers allow the eggs to hatch in the pond, though most farmers collect eggs and incubate them in a hatchery.

Adult Pacific salmon die after spawning. Females are usually sacrificed by cutting open the abdomen to release the eggs. Milt is obtained by squeezing the belly of males. Trout and Atlantic salmon can be reconditioned to spawn annually. Eggs are usually obtained from those species in the same fashion as from male Pacific salmon. Banks of egg-hatching trays, called *Heath trays*, through which water is flowed are typically used to hatch trout and salmon eggs (Fig. 9).

Unlike catfish, tilapia, trout, and salmon that produce from several hundred to several thousand relatively large eggs per female, many marine species produce large numbers of very small eggs. Hundreds of thousands to millions of eggs are produced by such species as halibut, flounder, red drum, striped bass, and shrimp. Catfish, salmon, and trout spawn once a year, while tilapia and some marine species spawn repeatedly at intervals of a few days to a few weeks for as long as several months if the proper environmental conditions are maintained (2).

Fish breeders have worked with varying degrees of success to improve growth and disease resistance in several species. As genetic engineering techniques are adapted to aquatic animals, dramatic and rapid changes in the genetic makeup of aquaculture species may be possible. However, since it is virtually impossible to prevent the escape of animals into the natural environment from aquaculture facilities, potential negative impacts of such organisms on wild populations cannot be ignored. Maintaining genetic diversity like that of the wild population in cultured stocks makes good sense, particularly in conjunction with enhancement and when escape from confinement systems would produce a high likelihood of escapees intermingling with wild counterparts.

Figure 9. A bank of Heath trays in a salmon hatchery in Washington State.

For some species, one sex may grow more rapidly than the other. A prime example is tilapia, which mature at an early age (often within six months of hatching). At maturity, submarketable females divert large amounts of food energy to egg production. Also, since they are mouth brooders (holding the eggs and fry within their mouths for about two weeks) and repeat spawners (spawning about once a month if the water temperature is suitable), the females grow very slowly once they mature. Males, on the other hand, continue to grow rapidly and can become marketable within a few months after reaching maturity. All-male, or predominantly male, populations of tilapia can be produced by feeding androgens to fry, which are undifferentiated sexually. Various forms of testosterone have been used effectively in sex reversing tilapia and other fishes.

In species such as flatfish, females may grow more rapidly than males and ultimately reach much larger sizes. For them, producing all-female populations for growout might be beneficial.

DISEASES AND THEIR CONTROL

Aquatic animals are susceptible to a variety of diseases, including those caused by viruses, bacteria, fungi, and parasites. A range of chemicals and vaccines has been developed for treating the known diseases, although some conditions have resisted all control attempts to date, and severe restrictions on the use of therapeutants in some nations has impaired the ability of aquaculturists to control disease outbreaks. The United States is a good example of a nation in which the variety of treatment chemicals is limited (Table 6). In many instances when a drug is cleared for use on aquatic animals in the United States, the species on which the drug can be used is limited. Clearing a drug for catfish, for example, does not necessarily mean it can be used on trout. The cost of obtaining clearance for drug and chemical use can be in the millions of dollars, and it is often uneconomical to attempt gaining clearance for species that are not major contributors to overall aquaculture production.

Maintenance of conditions in the culture environment that minimize stress is one of the best methods of avoiding diseases. Vaccines have been developed against several diseases and more are under development. Selective breeding of animals with disease resistance has met with only limited success. Good sanitation and disinfection of contaminated facilities are important avoidance and control measures. Some disinfectants are listed in Table 6. Pond soils can be sterilized with burnt lime (CaO), hydrated lime ($Ca[OH]_2$), or chlorine compounds (14).

When treatment chemicals have to be used, they may be incorporated in the food; dissolved in water for use in dips, flushes, and baths that allow relatively short exposure periods; or allowed to remain in the water until the chemical breaks down. Higher dosages can be used in dips, flushes, and baths than with chronic exposure. Since one of the first responses of aquatic animals to disease is reduction or cessation of feeding, treatments with medicated feeds must be initiated as soon as development of an outbreak is suspected and before the animals quit feeding. Antibiotics, such as oxytetracycline, can be dissolved in the water but may be less effective than when given orally.

Vaccines can be administered through injections, orally, or by immersion. Injection is the most effective means of vaccinating aquatic animals, but it is stressful, time-consuming, and expensive. The time and expense may be acceptable for use in conjunction with broodfish and other valuable animals. Oral administration of vaccines may be ineffective, because many vaccines are deactivated in the digestive tract of the animals the vaccines are intended to protect. Dip treatment by which the vaccines enter the animals through diffusion from the water are not generally as effective as injection but can be used to vaccinate large numbers of animals in short periods of time. Vaccines have been developed for a number of fish diseases, but scientists have not achieved much success with invertebrates, which have a primitive immune system.

HARVESTING, PROCESSING, AND MARKETING

Harvesting techniques vary depending on the type of culture system involved. Seines are often used to capture fish from ponds, or the majority of the animals can be collected by draining the pond through netting. Fish pumps are available that can physically remove aquatic animals directly onto hauling trucks from ponds, raceways, cages, and net-pens without causing damage to the animals.

Aquaculturists may harvest and even process their own crops, although custom harvesting and hauling companies are often available in areas where the aquaculture industry is sufficiently developed to support them. Some processing plants also provide harvesting and live-hauling services.

Some species, with channel catfish being a good example, can develop off-flavors. A characteristic off-flavor in catfish is often described as an earthy-musty, or muddy, flavor. The problem is associated with the chemical geosmin and related compounds that are produced by certain types of algae (2). Processors often require that a sample fish from each pond scheduled for harvest be brought to the plant about two weeks prior to harvest for a testing. Subsequent samples are taken to the processor three days before and during the day of processing. A portion of the sample fish is cooked and tasted. If off-flavor is detected during any of the tests, the plant will reject the fish from that pond until the problem is resolved. Once the source of geosmin is no longer present, the fish will metabolize the compound. Purging fish of geosmin may involve moving them into clean well water or merely waiting until the algae bloom dissipates, after which the geosmin will be rapidly metabolized. Within a few days after the source of the problem has been eliminated, the fish can be retested and, if no off-flavor is detected, harvested and processed.

Centralized processing plants specifically designed to handle regional aquaculture crops are established in areas where production is sufficiently high. In coastal regions, aquacultured animals are often processed in plants that also serve capture fisheries.

Marketing can be done by aquaculturists who operate their own processing facilities. Most aquaculture operations depend on a regional processing plant to market the final product. In all cases aquaculturists should remember that their job is not complete until the product reaches the consumer in prime condition.

BIBLIOGRAPHY

1. FAO, *Agriculture Production Statistics, 1974–1993*, Food and Agriculture Organization of the United Nations, 1995.
2. R. R. Stickney, *Principles of Aquaculture*, John Wiley & Sons, New York, 1994.
3. FAO, *Aquaculture Production Statistics 1986–1995*, Food and Agriculture Organization of the United Nations, Circular 815, Rev. 9, Rome (1997).
4. *Aquaculture Buyer's Guide '95 and Industry Directory, Vol. 8*, 1995, Aquaculture Magazine, Asheville, N. Car.
5. R. R. Stickney, *Aquaculture in the United States: A Historical Review*, John Wiley & Sons, New York, 1996.
6. A. Tacon, *International AquaFeed* **2**, 13–16 (1998).
7. Anonymous, *Aquaculture News* **6**, 1, 12 (1998).
8. R. G. Piper, I. B. McElwain, L. E. Orme, J. P. McCraren, L. G. Fowler, and J. R. Leonard, *Fish Hatchery Management*, U.S. Fish and Wildlife Service, Washington, D.C. 1982.
9. C. E. Boyd, *Water Quality in Ponds for Aquaculture*, Alabama Agricultural Experiment State, Auburn University, 1990.
10. J. Huegenin and J. Colt, *Design and Operating Guide for Aquaculture Seawater Systems*, Elsevier, New York, 1989.
11. National Research Council, *Nutrient Requirements of Fish*, National Academy Press, Washington, D.C., 1993.
12. A. Tacon, *International Aquafeed Directory and Buyers' Guide 1997/98* (1998).
13. F. P. Meyer and R. A. Schnick, "A Review of Chemicals Used for the Control of Fish Diseases," *Rev. Aquatic Sci.* **1**, 693 (1989).
14. C. E. Boyd, *Bottom Soils, Sediment, and Pond Aquaculture*, Chapman and Hall, New York, 1995.

ROBERT R. STICKNEY
Texas Sea Grant College Program
Bryan, Texas

AQUACULTURE: ENGINEERING AND CONSTRUCTION

BACKGROUND

Aquaculture's origins are ancient, but the explicit application of engineering to aquaculture is very recent. Only during the last 20 to 30 years has there been visible research activity in this field. This activity has accelerated recently fueled in part by the newly formed Aquacultural Engineering Society (AES, c/o The Freshwater Institute, P.O. Box 1746, Shepherdstown, WV 25443). The AES is an international organization bringing together individuals interested in the engineering applications to aquaculture. Although the number of aquacultural engineering practitioners is still small, interest in the field is very high, and engineering sessions are typically some of the best attended at meetings of aquaculture societies.

The development of aquacultural engineering has followed a path analogous to that followed by agricultural engineering 50 to 75 years earlier. Typically, engineers trained in related disciplines have acquired enough understanding of the biological components of aquaculture production systems to be able to design, evaluate, and manage these systems. These early activities have led to the training of students, graduate and undergraduate, in the combination of engineering and biological principles that would allow them to function effectively as aquacultural engineers. However, the number of educational programs that offer aquacultural engineering courses and degree programs continues to be small, and there continue to be all too frequent cases of engineers and other designers working in aquaculture without adequate training in the discipline, pointing to the continuing need to expand educational offerings in aquacultural engineering.

Aquaculture encompasses a wide range of activities, culture techniques, and species. The application of engineering principles to the different forms of aquaculture varies widely from the sophisticated, explicit engineering needed for an intensive system that incorporates water reuse, to the simple, nonformal engineering used for extensive, subsistence-type aquacultural operations. Some of the processes described here, then, will be applicable only to certain types of production systems.

DEFINITION

Aquacultural engineering is the application of engineering principles to the design, construction, and management of systems for the production of aquatic animals or plants. Specific areas covered include the overall system design and management; the physical system used to hold the animals or plants; the supporting structures; processes, monitoring equipment, and techniques needed to ensure an adequate environmental quality in the production system; and the facilities and equipment needed to handle fish* and other materials such as feed. As a research discipline, aquacultural engineering also deals with the evaluation of the effects of water quality and other environmental conditions on fish, as well as the effect that fish have on the water. Effects on the fish that are of concern include growth rates, susceptibility to disease, and reproductive success. An area of increasing concern is the potential environmental impact of aquaculture operations, and aquacultural engineers are being called upon to help address this topic. In addition, traditional engineering disciplines ranging from environmental to mechanical to electrical engineering are used in aquaculture.

SYSTEM DESIGN

An aquaculture system must be designed to provide a healthy environment for the target fish or plant, and this must be achieved within constraints inherent in the choice of location where the facility is to be constructed. Other considerations that need to be taken into account include: marketing, economics, and regulatory restrictions. The procedure for the development of an aquaculture system is summarized in Figure 1 and follows stages of data collection, evaluation, design, and construction (1,2). Background information needed includes data on the biological (or bioengineering as they are sometimes referred to) characteristics of the target species, as well as data on the site, possible markets for the product, and regulations affecting aquaculture at the chosen site. The biological data needed to quantify the relationship between the target organism and its environment include information on tolerances and optimum levels for various water quality parameters, the effect of target animal activity on water quality, space and water velocity requirements, reproductive characteristics, feeding behavior, and others. Site data are primarily related to water availability and to the quality of the water, but other factors are important, such as soil characteristics (soil permeability, physical properties affecting possible construction, and fertility), topography, climate, other land uses, transportation, and other infrastructure. Site characteristics may be considered at the level of a specific aquaculture site, or at the regional planning level, where attempts have been made at incorporating satellite imaging and geographic information system (GIS) techniques into the site characterization process (3). Desirable marketing information includes not only the type of species that is salable but also the particular requirements imposed by the market on product size, level of processing, seasonal fluctuations in demand, and so on. Last, regulatory aspects of aquaculture must not be overlooked. Restrictions on land and water use are augmented by considerations of species approvals, and in some cases by the lack of familiarity of regulators with aquaculture, and their unwillingness to consider it as a form of agriculture, rather than as an industrial enterprise. The Food and Agriculture Organization (FAO) of the United Nations has recently released a "code of practice" for aquaculturists (4). The intent of the code is to promote the safe and environmentally sound development of aquaculture by encouraging aquaculture practitioners to consider the full scope of possible impacts of aquaculture installations.

A last topic on which background information should be gathered is an overall review of the methods and procedures used for the culture of the species of interest. Wherever possible this should focus on conditions that are similar to those of the planned facility.

From this point on, the design process follows fairly well established overall guidelines for engineering design. A preliminary evaluation is carried out where the background information is reviewed and an initial decision is made about the possible viability of the operation, or of the need to make changes and collect additional background information.

Once a decision has been made to proceed with the planning of a facility, detailed review of the practices used by others in the culture of the target species is undertaken. This review should be based on published information as well as on personal contacts with other producers, with

*In the context of this article, fish may be a finfish such as trout or catfish, or a crustacean such as shrimp or lobsters, or a mollusk such as oysters or clams.

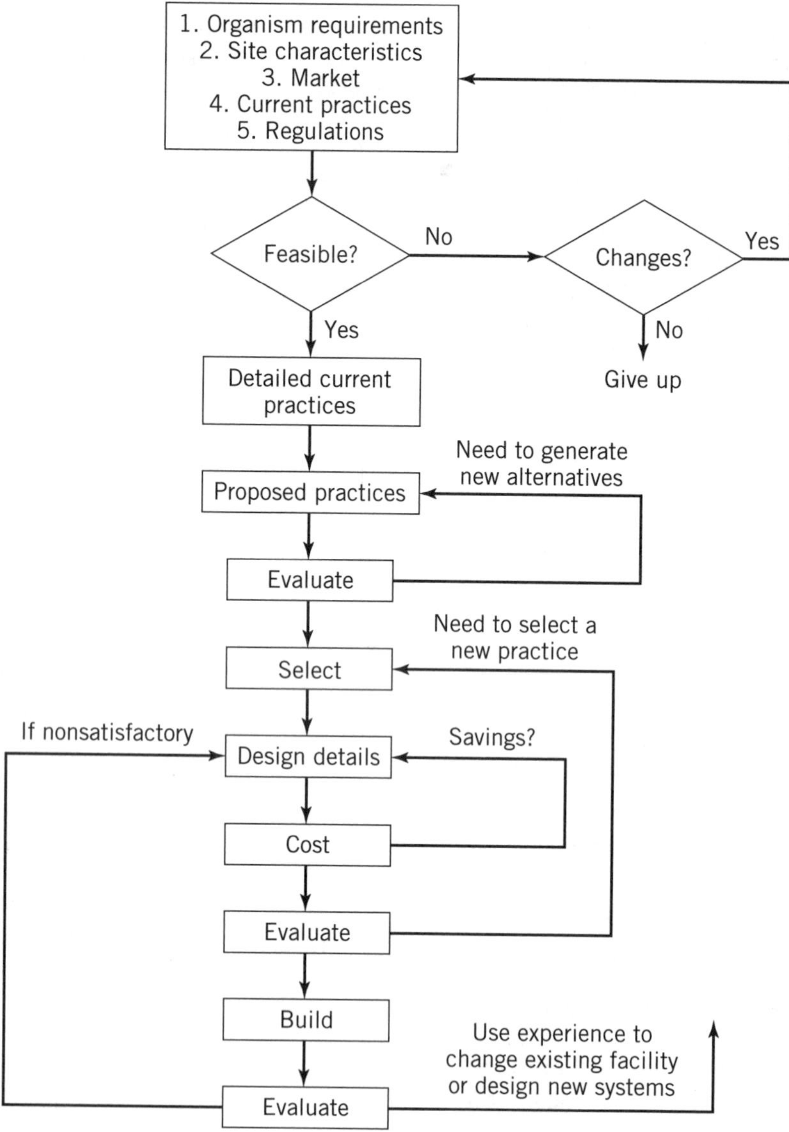

Figure 1. Flow diagram of the design process followed for aquaculture operations.

extension agents, and with researchers. As a first step, several alternative designs may be generated and evaluated to select the most promising one for further development. After evaluation and selection, the details of the design are completed, including estimated costs and preparation of construction documents. Evaluation of the plans and facilities should be carried out at the various stages of the process to minimize the probability of errors, as well as to ensure maximum functionality and operational efficiency for the investment. Evaluation of the completed installation will serve in the preparation of future designs or in the possible expansion of the facility.

HOLDING SYSTEMS

The types of impoundments or holding systems used in aquaculture are determined by the requirements of the particular species and by the production and management practices being employed. Typical systems include ponds, raceways, tanks, and cages. Although construction materials for the various types of holding systems differ depending on specific requirements, they should have general characteristics, including being corrosion resistant, nontoxic, with smooth and nonabrasive surfaces, strong and lightweight, resistant to ultraviolet radiation, and low in cost (5,6). Typically, no material satisfies all criteria optimally, and selection is based on the particular constraints of a given project.

Ponds

Pond size, shape, and construction technique varies widely (7,8). Ponds tend to be shallow, less than 1.5 m deep, and vary in area between less than 100 m^2 to more than 10 ha (Fig. 2). The pond bottom is sometimes covered with concrete or plastic liners to prevent water losses to infiltration. The ponds may be built by excavation, by construction of dikes, or more commonly by a combination of the two techniques. In all cases, pond design should incorporate struc-

Figure 2. Pond being used for the culture of shrimp with supplemental mechanical aeration.

tures for water inflow for filling and maintenance flows, and for effluents including flows during pond drainage and emptying.

Raceways

Raceways are relatively narrow, long channels with water constantly entering through one end of the channel and leaving through the opposite end (Fig. 3). Raceways are often built of concrete, and the normal dimensions are in a ratio of approximately 30:3:1 (length:width:depth). Raceways are the most common holding system for the production of trout and are usually approximately 30 m long. Flow in a raceway approaches plug flow characteristics, creating a gradient of water quality along the raceway; water quality is best near the head of the raceway where the water enters and degrades toward the drain side of the raceway. Raceways are often constructed in series of up to six raceways with reaeration between raceways to replenish some of the oxygen used by the fish.

Tanks

Aquaculture tanks are normally designed to optimize the movement of water around the center of the tank. As such, common shapes are round, square with rounded corners, hexagonal, and octagonal (5,9,10) (see Fig. 4). The choice is often made on the basis of space availability and construction materials, as there aren't large differences in the hydraulic properties of the various shapes. Tank size ranges from an equivalent diameter of less than 1 m to more than 10 m. Water depth is normally maintained between 0.5 m for the smaller tanks and 1 to 2 m for the larger tanks. The two most common materials used for tanks are fiberglass and concrete.

The size and shape of tanks used in aquaculture vary greatly, but the common characteristic is that flow approaches the ideal continuously stirred tank reactor (CSTR) model in which water quality within the tank is approximately uniform. To achieve this uniformity, water is introduced along the periphery of the tank through one or more inlets that normally inject the water tangentially, and the water effluent is collected from the center of the tank (10,11). Contrary to raceways, it is possible to achieve a certain degree of independence between water flow rate and water velocity in a tank by manipulating the inlet and outlet characteristics and location. An important feature is that by adjusting the circulation pattern and water velocity to maintain particulate matter in suspension, tanks may be made to be "self-cleaning" (10,11).

A recent development in tank culture is the introduction of "double drains" in which two water streams are created in the drain, one constituting a large percentage of the flow containing a very small fraction of the solids released to the tank water (fish feces and uneaten feed), and the second stream conveying no more than 5 to 10% of the total flow from the tank, but containing up to 90 or 95% of the tank solids (10,11). These new double drains offer opportunities for improved water utilization efficiency and water treatment systems, with the high flow–low solids stream being relatively easy to treat for reuse, and the low flow–high solids stream being suitable for treatment prior to disposal. Flow rates in tanks are normally selected to result in residence times ranging from 10 to 15 min to several hours, and the value used depends on the level of water treatment (normally aeration) taking place within the tank and on the fish biomass being held.

Figure 3. Farm that uses raceways for the culture of trout. The raceways are laid out in series and in parallel to optimize land and water use.

Figure 4. Round concrete tanks used for the culture of striped bass. Water in the tanks is reoxygenated using pure oxygen injection systems and aerated for carbon dioxide removal.

Cages

Although cages have been used in aquaculture for many years, they have received a great deal of attention recently, especially with the development of the salmon industry in Norway and other countries (Fig. 5). Conventional cage technology is based on a rigid frame that is either floating or fixed, and to which a cage made of some type of netting material is attached. Traditionally, cages had been rectangular and relatively small, with total volumes of less than 10 m^3. The recent growth of the industry, the availability of new materials and technologies, and the presence of competing users for coastal resources have resulted in the development of large cage systems suitable for open sea deployment (6,12). Some of these cages may be more than 50 m in diameter, 30 m deep, with production volumes of more than 50,000 m^3, and are designed to hold up to 1,000,000 kg of fish (6,12). Engineering contributions into the design, construction, and anchoring of cages becomes very specialized as the cage size increases and as cages are placed in more exposed locations. Cages are subjected to static and dynamic loads. Static loads are relatively easy to quantify and are caused by the weight of the fish and the structure itself, but transient loads caused by wind, waves, and currents may be orders of magnitude greater than static loads and are very difficult to determine accurately (6). Frequent cage maintenance is required to prevent fouling of the net from restricting the flow of water. Exchange of water through the net must be sufficient to maintain adequate water quality for the fish.

QUANTIFYING PROCESSES

As was mentioned in the definition of aquacultural engineering, many engineering disciplines can be applied to the design of aquaculture systems. In this section, some of the critical processes that are unique in an aquaculture system and that define design parameters will be described. The key processes in the design of aquatic systems have to do with the primary water quality factors that affect the cultured organism and with how the organism affects water quality.

Figure 5. Cages used for the culture of Atlantic salmon in Norway.

As a starting point in the description of important processes in an aquaculture operation, one can consider how a fish relates to its environment. Fish consume oxygen from the water and release water products from their metabolism back into the water. The main waste products of concern are ammonia, carbon dioxide, and particulate and dissolved organics. Depletion of dissolved oxygen below a certain level, which is species and size dependent, results in increased stress to the fish and may ultimately result in suffocation. Similarly, the accumulation of waste products may reach toxic levels, as in the case of ammonia and carbon dioxide; it may alter the pH of the water, again as a result of carbon dioxide accumulation; it may affect growth and reproductive behavior as a result of the accumulation of dissolved organics such as hormones; or it may affect the overall oxygen balance of the system by the oxygen demand created by the decomposition of particulate and dissolved organics.

A major difference between aquaculture and land-based food production systems is that in an aquaculture system the animal or plant takes its nutrients and oxygen and releases its waste products into the same environment: the water. Maintaining water quality in an aquaculture system is based on providing sufficient oxygen to satisfy the needs of the fish, and in removing metabolites to prevent their buildup to toxic levels. Design of systems to achieve those goals requires the quantification of processes that take place within the system, especially those related directly to the target species. The relative importance of the various consumption and production processes depends on the type of aquatic production system being considered. In general, the shorter the residence time of the water in the production system, the more important are the fish-related processes with respect to other biological processes taking place. If, on the other hand, the residence time is long (more than one day and as long as one year), as is the case in most ponds, biological and chemical processes that are not directly associated with the target product's metabolism become very important in determining water quality. As an example, dissolved oxygen in a tank with a short residence time will be determined by the concentration in the influent water, the fish biomass being held in the tank, the amount of oxygen being added to the system through aeration mechanisms, and the oxygen consumption by the fish. In a pond, on the other hand, many processes can have significant effects on dissolved oxygen concentration. These processes include photosynthetic oxygen production, respiration by all the organisms in the ponds (phytoplankton, zooplankton, fish, bacteria, and sediment decomposition processes), and gas exchange between the water and the atmosphere.

Quantification, design, and management of aquaculture production systems are most predictable in those systems with short residence times, where the majority of water quality changes are associated with fish activity. Pond systems, often with long residence times, are difficult to quantify and their management remains very much an art rather than a science.

Design of systems is ultimately based on mass balances on the critical water quality factors already mentioned. The general form of the mass balance equation for a particular substance (eg, dissolved oxygen or ammonia) in a control volume can be written as

$$\text{ACCUMULATION} = \text{INFLOW} + \text{PRODUCTION} - \text{OUTFLOW} - \text{CONSUMPTION} \quad (1)$$

For simplicity in calculations and because of incomplete information on the time dependence of production and consumption rates, the mass balance is often simplified to consider steady-state conditions in which ACCUMULATION becomes zero, and equation 1 can be written as:

$$\text{INFLOW} + \text{PRODUCTION} = \text{OUTFLOW} + \text{CONSUMPTION} \quad (2)$$

And the INFLOW and OUTFLOW terms may be expressed as the product of the concentration of the substance being observed and the flow rate:

$$\text{INFLOW} = Q \times C_i \quad (3)$$

$$\text{OUTFLOW} = Q \times C_o \quad (4)$$

where Q is the flow rate ($m^3\ d^{-1}$), C_i is the influent concentration ($g\ m^{-3}$), and C_o is the effluent concentration ($g\ m^{-3}$).

In a fully mixed rearing impoundment, the effluent concentration is approximately equal to the concentration at any point within the impoundment and is the concentration to which the target species is exposed. The production and consumption functions depend on the species, size, and condition of the fish and on the environmental conditions to which they are exposed. In addition, these rates fluctuate through diel cycles in response to activity levels associated primarily with feeding and digestion. In general, there is little quantitative information on how the various factors affect the rates of oxygen consumption and metabolite production, and for purposes of design, it is common to relate these rates to feeding levels (13,14):

$$\text{DO}_{\text{cons}} = \text{FEED} \times \text{DO}_{\text{rc}} \quad (5)$$

$$\text{CO}_{2\text{prod}} = \text{FEED} \times \text{CO}_{2\text{rc}} \quad (6)$$

$$\text{TAN}_{\text{prod}} = \text{FEED} \times \text{TAN}_{\text{rc}} \quad (7)$$

$$\text{SOLIDS}_{\text{prod}} = \text{FEED} \times \text{SOLIDS}_{\text{rc}} \quad (8)$$

where DO_{cons} is the rate of oxygen consumption by fish within the impoundment ($g\ d^{-1}$), FEED is the amount of feed applied to the impoundment ($g_{\text{feed}}\ d^{-1}$), DO_{rc} is the oxygen consumption per unit of feed applied ($g\ g_{\text{feed}}^{-1}$), and similarly for carbon dioxide (CO_2), total ammonia nitrogen (TAN), and solids productions. Feeding rate is often expressed as a proportion of the fish biomass applied per day, and tables are available from feed manufacturers for various fish species and ages. Approximate values used for the consumption of oxygen and production of carbon dioxide, ammonia, and solids per unit mass of feed ($g\ \text{gfeed}^{-1}$) are 0.20, 0.28, 0.03, and 0.30, respectively (13,14). These values can be useful as a first approximation, but detailed calculations must be based on values specifically developed for the species of interest and for the feed and culture system being used.

WATER CONDITIONING

Water conditioning operations may be used to treat the water and increase the fish biomass that can be held for a given water flow rate. The rates of oxygen consumption and metabolite production and the tolerances of fish to concentrations of these substances are such that oxygen is the first limiting factor encountered. If aeration is used to replenish oxygen, then the second limiting factor reached may be either carbon dioxide or ammonia accumulation, depending on characteristics of the water supply (in particular alkalinity and pH of the water) and on fish tolerances.

Oxygenation

Oxygen sources differ between pond and other types of holding systems. Possible sources of oxygen in a pond are the production by aquatic plants, including phytoplankton, mechanical reaeration, and reaeration across the water surface with minimum contributions by the water exchange, with the relative importance of the sources depending on the type of production system being used. In raceways and tanks, oxygen sources are limited to the water supply and some form of mechanical reaeration.

Oxygen Production by Phytoplankton. Phytoplankton produce oxygen through photosynthesis. The rate of oxygen production by phytoplankton is very difficult to quantify but conceptually may be expressed as

$$\text{DOPHYTO} = \text{DO}_{\text{equi}} \times \text{PHYTO} \times \text{MAX} \times \text{NUTRI} \times \text{LIGHT} \times \text{TEMP} \quad (9)$$

where DOPHYTO is the oxygen production rate by phytoplankton ($g\ m^{-3}\ h^{-1}$), DO_{equi} is the oxygen production per unit of phytoplankton biomass growth ($g\ g^{-1}$), PHYTO is the phytoplankton concentration ($g\ m^{-3}$), MAX is the maximum phytoplankton growth rate under optimum conditions (h^{-1}), NUTRI is the nutrient limitation on growth (0–1), LIGHT is the light limitation on growth (0–1), and TEMP is the temperature limitation on growth (0–1).

A multitude of expressions have been proposed for the various limitation terms (eg, Ref. 15).

The overall effect of phytoplankton on oxygen concentration in a pond is further complicated by their respiration at night. This combination of oxygen production by phytoplankton during the day and consumption at night by phytoplankton, fish, and other organisms results in wide variations in dissolved oxygen over diurnal cycles (8). Maintaining a balance between the phytoplankton production and consumption terms in ponds becomes the primary goal of water quality management in most pond production systems.

A new production system has been developed recently that makes use of the water-treating and oxygen-producing capabilities of ponds. The system has been called the Partitioned Aquaculture System (PAS) (16) and consists of a shallow pond used for water treatment and oxygen production, and a small raceway in which fish are held. Water is continuously recirculated between the raceway and the pond. A secondary fish crop is raised in a section of the raceway downstream from the primary fish crop. The species to be raised as the secondary crop is selected such that it has the ability to filter the water and remove particulates such as algae and uneaten feed from the primary crop. These new systems are still in the experimental stage but offer great promise given their high production rates per unit area and the fact that the water is recycled and no effluent is produced.

Oxygen Production by Mechanical Aeration. Oxygen transfer into the water takes place as a result of a driving force equal to the difference between the saturation concentration and the actual concentration found in solution, a mass-transfer coefficient, and the area of contact between the liquid and gas phases. The mass-transfer coefficient and the area of contact are normally combined into an overall gas-transfer coefficient, K_{La}, that may be determined experimentally. The equation describing the rate of oxygen transfer can be written as

$$DO_{Trans} = (DO_{Sat} - DO)K_{La}V \qquad (10)$$

where DO_{Trans} is the dissolved oxygen transfer rate ($g\,h^{-1}$), DO_{Sat} is the saturation concentration of dissolved oxygen ($g\,m^{-3}$), DO is the dissolved oxygen concentration in solution ($g\,m^{-3}$), K_{La} is the overall gas-transfer coefficient (h^{-1}), and V is the volume over which the aeration is taking place (m^3).

Saturation concentration is determined by the composition of the gas phase, by temperature, by atmospheric pressure, and by the presence of dissolved substances in the water (eg, salinity) (17). Although more complete estimation procedures are available that consider the factors just mentioned (17), saturation concentration for fresh water as a function of temperature may be expressed as a regression equation (18):

$$\begin{aligned} DO_{Sat} = {} & 14.652 - 0.41022 \times TEMP \\ & + 0.007991 \times TEMP^2 \\ & - 0.000077774 \times TEMP^3 \end{aligned} \qquad (11)$$

where TEMP is the water temperature (°C).

Equation 12 and tables of dissolved oxygen saturation are normally prepared for an atmosphere of 20.9% oxygen. Oxygen enrichment of the atmosphere will result in a corresponding increase in saturation dissolved oxygen concentration. Enrichment may be achieved by aerating with "pure" oxygen.

An important difference between aeration systems used for aquaculture and those used in water and wastewater treatment is that the "driving force" ($DO_{Sat} - DO$) in aquaculture tends to be lower, reducing the rate of oxygen transfer for a given aerator. The lower driving force is caused by the requirement of most fish for dissolved oxygen concentrations above 3 to 4 $g\,m^{-3}$, compared with the usual limit of just above zero in wastewater treatment.

Common types of aerators used in aquaculture may be classified as surface, gravity, and diffuser types (5,9,19). Surface aerators spray water into the air or beat the water, increasing the turbulence and area of contact between the water and air. Gravity aerators rely on the fall of water for aeration. Different types of structures such as plates, baffles, and so on may be used in gravity aerators to increase transfer rates. A particularly useful and common gravity aerator is the packed column aerator (PCA), where water is introduced at the top, and air or another gas is introduced at the bottom of a column filled with a medium designed to maximize turbulence and transfer area. Design equations for the use of PCAs in aquaculture have been proposed by Hackney and Colt (20). In diffuse aeration, gas bubbles are introduced into the water, and transfer takes place between the bubble and the water. The simplest type of diffuse aerator is an airstone where bubbles are released at the bottom of a tank or pond. This type of aerator tends to be inefficient in conventional aquaculture systems due to the shallow water depths used and the resulting short contact times between the bubble and the water. Variations on the basic concept of diffuse aeration have been developed, including some systems designed to be used as part of a pipeline and that pressurize and trap bubbles allowing for longer contact times and, in some cases, complete dissolution of the bubbles into the liquid (5,9).

The use of pure oxygen in aquaculture operations has become commonplace over the last few years, especially for intensive systems that include some form of water reuse. Pure oxygen systems rely on either liquid bottled oxygen or on the on-site production of oxygen. The use of pure oxygen makes possible the addition of large amounts of oxygen to the water, increasing fish biomass held in a given water supply.

Oxygen Production by Surface Reaeration. The transfer of oxygen by reaeration is important in ponds where it can result in substantial net gains or losses of oxygen depending on whether the dissolved oxygen concentration is below or above saturation. Estimates of reaeration rates may be obtained from the following (21):

$$DO_{Reaer} = (DO_{Sat} - DO)K_{Reaer}/D \qquad (12)$$

where DO_{Reaer} is the reaeration rate ($g\,m^{-3}\,h^{-1}$),

$$\begin{aligned} K_{Reaer} = {} & 0.03 \times v_w^{0.5} - 0.0132 \times v_w + 0.0015 \\ & \times v_w^2 (m\,h^{-1}) \end{aligned} \qquad (13)$$

v_w is the wind velocity ($m\,s^{-1}$), and D is the pond depth (m).

An implicit assumption in equation 13 is that conditions in the pond are uniform and there is a minimum of stratification.

Ammonia Removal

Ammonia is toxic to aquatic animals and is also the waste product of their protein metabolism. The sensitivity of aquatic animals to ammonia concentration depends on species, live stage, level of stress, and on other environmental conditions. Ammonia exists in water as the equilibrium product of ammonium ion and unionized ammonia:

$$NH_4^+ \leftrightarrow NH_3 + H^+ \quad pK_a = 9.25 \text{ at } 25°C \tag{14}$$

Unionized ammonia (NH_3) is the toxic form of ammonia (5,14). Approximate estimates of the concentration of unionized ammonia in fresh water can be obtained form measures of total ammonia (NH_3 plus NH_4^+ concentrations) and pH as (5):

$$[NH_3] = [TA] \times (1/10^{(pK_a - pH)}) \tag{15}$$

where $[NH_3]$ is the unionized ammonia concentration (mol/L) and [TA] is the total ammonia concentration (mol/L).

Removal of ammonia from aquaculture systems may be achieved by ion exchange or by biological filtration. Ion exchange is carried out with a natural zeolite (clinoptilolite) that has a high affinity for ammonium ions. Removal capacity of the resin varies widely, but a common design value is around 1 mg NH_4^+ (g clino)$^{-1}$ (22). Clinoptilolite resin may be regenerated with brackish or salt water, making the use of the resin suitable for fresh water applications only. Ion exchange columns need to be designed to incorporate the downtime involved with the recharging and reconditioning. In addition, unless properly maintained, these columns have a tendency to be colonized by bacteria and become biological filters.

The biological removal of ammonia by the process of nitrification is carried out by two groups of bacteria: nitrosomonas, which take ammonia to nitrite; and nitrobacter, which complete the reaction to nitrate. Nitrate is toxic to fish only at very high concentration and therefore is a suitable end product for ammonia and nitrite, forms of nitrogen toxic to fish. The nitrification process is relatively slow and sensitive to temperature, pH, and ammonia concentration. To achieve practical nitrification rates, high bacterial biomass must be maintained, which is normally achieved with some form of attached growth filter (23,24). Many types of filters are used for ammonia removal: downflow, upflow, rotating disks, submerged, trickling, fluidized beds, and soon (23,24). Removal rates are highly variable, and this is an area of active research at the present time.

Solids Removal

Removal of solids produced by aquaculture may be attempted for making the water suitable for reuse, or to meet discharge guidelines. Solids account for a high percentage of the biochemical oxygen demand (BOD) in aquaculture effluents; they tend to be highly variable in size and break up easily if subjected to mechanical forces. Solids are formed from uneaten feed and fecal material. Their density tends to be very close to that of water.

Removal is achieved by filtration with particulate (eg, sand) or microscreen (eg, stainless steel) filters, or by sedimentation (25). The high organic content of the particulates make them highly biodegradable, and filters need to be backwashed frequently to prevent clogging by a mat of biological slime. Screen filters are common in aquaculture and a number of filters have been developed that incorporate some form of automated cleaning (eg, drum or disk filters). Simple settling tanks or ponds with overflow rates between 40 and 80 m d^{-1} can result in total suspended solids removal rates of 65 to 85% (25). Aquaculture settling ponds traditionally have been operated with infrequent sludge removal, relying on biological decomposition to prevent excessive accumulation of sludge.

Recent concerns with the possible release of nutrients, especially phosphorus, from aquaculture operations to natural waters have caused a reevaluation of techniques available for solids removal. Given the technical difficulties and expense associated with treating aquaculture effluents, a combination of nutritional and engineering approaches are being studied to reduce the solids (and nutrient) content in effluents.

FUTURE DEVELOPMENTS

The aquaculture farm of the future will take many forms. The difference with today's farms will be in the level of intensity of water use, in the degree of control over the culture environment exerted by the operator, and in the predictability of growth and overall production rates. The application of engineering will become more and more important as the aquaculture industry continues to grow and diversify. Developments will be in new culture systems, materials, equipment, and water quality monitoring, treatment, and control. The efficient use of natural resources, especially water, will be a primary constraint on new aquaculture operations. To make more efficient use of water, higher fish biomass will be maintained per unit of water volume and per unit of water use by the implementation of water treatment and recirculation systems. The safe rearing of these large biomasses will require continuous monitoring of water quality, and the use of alarms and backup systems. In most cases these will be all computer controlled.

Improvements in water treatment are also needed to make possible reductions in water use and the commercial viability of culture systems based on water reuse. New, more efficient and reliable filtration systems for the removal of particulate and dissolved organics and of ammonia will be developed.

Systems for inventorying and handling fish are needed. Obtaining accurate estimates of number and size of fish in a culture system are practically impossible today as they are based on sampling or on the use of growth functions and initial stocking values. Equipment and techniques that can be used to count animals and estimate their size with a minimum of disruption and stress to the fish will have to be developed.

BIBLIOGRAPHY

1. R. D. Mayo, "A Format for Planning a Commercial Model Aquaculture Facility," paper presented at Northeast Fish and Wildlife Conference, February 25–28, 1974, Grate Gorge, N.J. Technical Reprint No. 30. Kramer, Chin and Mayo, Inc., Consulting Engineers, Seattle, Wash., 1974.

2. C. M. Brown and C. E. Nash, "Planning an Aquaculture Facility—Guidelines for Bioprogramming and Design," Aquaculture Development and Coordination Programme Report ADCP/REP/87/24, United Nations Development Programme, Food and Agriculture Organization of the United Nations, Rome, 1988.
3. S. S. Nath et al., "Applications of Heat Balance and Fish Growth Models for Continental-Scale Assessment of Aquaculture Potential in Latin America," in D. Burke et al., eds., *Pond Dynamics / Aquaculture Collaborative Research Support Program, Fourteenth Annual Technical Report*, PD/A CRSP Office of International Research and Development, Oregon State University, Corvallis, Oreg., 1996.
4. Food and Agriculture Organization of the United Nations (FAO) Fisheries Department, Aquaculture Development, FAO Technical Guidelines for Responsible Fisheries, No. 5, Rome, 1997.
5. T. B. Lawson, *Fundamentals of Aquacultural Engineering*, Chapman & Hall, New York, 1995.
6. M. C. M. Beveridge, *Cage Aquaculture*, 2nd ed. Fishing News Books Ltd., Surrey, England, 1996.
7. C. R. dela Cruz, "Fishpond Engineering: A Technical Manual for Small- and Medium-Scale Coastal Fish Farms in Southeast Asia," *South China Sea Fisheries Development and Coordinating Programme Manual No. 5*, United Nations Development Programme, Food and Agriculture Organization of the United Nations. Manila, Philippines, 1983.
8. H. S. Egna, and C. E. Boyd, eds. *Dynamics of Pond Aquaculture*, CRC Press, Boca Raton, Fla., 1997.
9. F. W. Wheaton, *Aquacultural Engineering*, Wiley, New York, 1977.
10. S. Skybakmoen, "Fish Rearing Tanks," *Aquaculture Series Brochure*, AGA AB, Lidingö, Sweden, 1993.
11. M. B. Timmons and S. T. Summerfelt, "Advances in Circular Culture Tank Engineering to Enhance Hydraulics, Solids Removal, and Fish Management," in M. B. Timmons and T. M. Losordo, eds., *Advances in Aquacultural Engineering*, Aquacultural Engineering Society Proceedings III. Northeast Regional Agricultural Engineering Service, NRAES-105, Ithaca, New York, 1997, pp. 66–84.
12. J. Gunnarsson, "Bridgestone HI-SEAS Fish Cage: Design and Documentation," in H. Reinertsen et al., eds., *Fish Farming Technology, Proc. of the First International Conference on Fish Farming Technology*, Trondheim, Norway, August 9–12, 1993, A. A. Balkema, Rotterdam, 1993, pp. 199–201.
13. J. Colt and K. Orwicz, "Modeling Production Capacity of Aquatic Culture Systems under Freshwater Conditions," *Aquacultural Eng.* **10**, 1–29 (1991).
14. T. M. Losordo and H. Westers, "System Carrying Capacity," in M. B. Timmons and T. M. Losordo, eds., *Aquaculture Water Reuse Systems: Engineering Design and Management*, Elsevier, Amsterdam, 1994, pp. 9–60.
15. S. E. Jørgensen, *Fundamentals of Ecological Modelling*, Elsevier, Amsterdam, 1986.
16. D. E. Brune and J. K. Wang, "Recirculation in Photosynthetic Aquaculture Systems," *Aquaculture Magazine* **24**(3), 63–71 (1998).
17. J. Colt, "Computation of Dissolved Gas Concentrations in Water as Functions of Temperature, Salinity, and Pressure," American Fisheries Society Special Publication No. 14, Bethesda, Md., 1984.
18. H. L. Elmore and T. W. Hayes, "Solubility of Atmospheric Oxygen in Water," *J. Sanitary Eng. Div. ASCE* **86**(SA4), 41 (1960).
19. B. J. Watten, "Aeration and Oxygenation," in M. B. Timmons and T. M. Losordo, eds., *Aquaculture Water Reuse Systems: Engineering Design and Management*, Elsevier, Amsterdam, 1994, pp. 173–208.
20. G. Hackney and J. Colt, "The Performance and Design of Packed Column Aeration Systems for Aquaculture," *Aquacultural Eng.* **1**, 275–295 (1982).
21. R. B. Banks and F. F. Herrera, "Effects of Wind and Rain on Surface Reaeration," *J. Environ. Eng. Div. ASCE* **103**(EE3), 489–504 (1977).
22. W. Bruin, J. W. Nightingale, and L. Mumaw, "Preliminary Results and Design Criteria from and On-line Zeolite Ammonia Removal Filter in a Semi-closed Recirculating System," in L. J. Allen and E. C. Kinney, eds., *Proc. of the Bio-Engineering Symposium for Fish Culture*, American Fisheries Society, Bethesda, Md., 1981, pp. 92–96.
23. F. W. Wheaton, et al., "Nitrification Filter Principles," in M. B. Timmons and T. M. Losordo, eds., *Aquaculture Water Reuse Systems: Engineering Design and Management*, Elsevier, Amsterdam, 1994, pp. 101–126.
24. F. W. Wheaton, et al., "Nitrification Filter Design Methods," in M. B. Timmons and T. M. Losordo, eds., *Aquaculture Water Reuse Systems: Engineering Design and Management*, Elsevier, Amsterdam, 1994, pp. 127–171.
25. S. Chen, D. Stechey, and R. F. Malone, "Suspended Solids Control in Recirculating Aquaculture Systems," in M. B. Timmons and T. M. Losordo, eds., *Aquaculture Water Reuse Systems: Engineering Design and Management*, Elsevier, Amsterdam, 1994, pp. 61–100.

RAUL H. PIEDRAHITA
University of California
Davis, California

AROMATHERAPY

Aromatherapy is a branch of the alternate medicine approach that uses herbal remedies to improve an individual's health and appearance and to alter one's mood. The alleged benefits from aromatherapy range from stress relief to enhancement of immunity and the unlocking of emotions from past experience. The concept has ancient roots but is primarily used today by the cosmetics, fragrance, and alternative-medicine industries.

The proponents of aromatherapy claim that the tools of the trade, such as wood-resin distillates and flower, leaf, stalk, root, grass, and fruit extracts, contain antibiotics, antiseptics, hormones, and vitamins. Some proponents go even further and have characterized the essential oils, which are volatile, aromatic, and flammable, as the soul or spirit of plants. One of aromatherapy's promises is that the essential oils have a "spiritual dimension" and can restore "balance" and "harmony" to one's body and to one's life. One of its principles is the "doctrine of signatures," which claims that a plant's visible and olfactory characteristics reveal its secret qualities. For example, the violet suggests shyness so the proponents conclude that the scent of violets engenders shyness and modesty. Aromatherapy encompasses topical application of essential oils, bathing in water containing essential oils, sniffing them, or actually ingesting them. Products supporting this concept include shaving gels, aftershaves, facial cleansers, bath salts, shower gels, shampoos, hair conditioners, body masks, moisturizers, sunscreen preparations, lipsticks, deodorants, candles, lamps, diffusers, pottery, massage oils, massage devices, and jewelry such as lockets and pendants for carrying essential oils.

The most common aromatherapy applications are aesthetic, where a sense of well-being is derived from enjoying perfumes, scented candles and baths, and other fragrances. At the opposite end of the spectrum is medical aromatherapy, also known as aromatic medicine, which includes massage therapists, naturopaths, some nurses, and some medical doctors. The alleged beneficial effects are numerous: essential oil from bergamot normalizes emotions; essential oil from roses or sandalwood increases confidence; essential oil from eucalyptus alleviates sorrow, and oil from patchouli creates a desire for peace. Eucalyptus oil and peppermint oil have been used to treat respiratory diseases. But it is conceded that some oils can be harmful. Concentrated oils from cloves, cinnamon, nutmeg, and ginger can burn the skin and ingestion of oil from pennyroyal can cause miscarriages. One of the obvious applications of aromatherapy involves the sense of smell, and one researcher believes that the lack of ability to smell is correlated with a gain in weight. Others believe that the ability to smell or even the ability of the body to produce certain odors indicates certain diseases or impairments. If this proves to be true, it would be an interesting diagnostic tool.

The concept behind aromatherapy is accepted in parts of Europe. For example, in France, medical students are taught how to prescribe essential oils, and in Britain, hospital nurses use aromatherapy to treat patients suffering anxiety and depression and to make terminal-care patients more comfortable. Dorene Peterson, principal of the Australasian College of Herbal Studies, commented, "There is a philosophical difference between hard-core science and the approach that believes there's vibrational energy that's part of the healing process. Alternative medicine is offered now in quite a number of medical schools. I think a lot of hard-core scientists and doctors who have been trained in the data-oriented scientific approach are realizing there's more to heaven and earth than we really know about." However, she admits that empirical evidence is necessary for widespread acceptance. In the United States, there are no legal standards concerning the education in aromatherapy, or the certification or occupational practice of aromatherapy. Several nonaccredited organizations offer short courses or correspondence courses, but no accredited educational institutions offer majors in aromatherapy. It is safe to say that aromatherapy remains outside the mainstream of medical therapy, but it is well ensconced in the cosmetic area.

GENERAL REFERENCE

C. A. Sweet, "Scents and Nonsense," *Priorities* **9**, 30–33 (1997).

F. J. Francis
Editor-in-Chief
University of Massachusetts
Amherst, Massachusetts

ARTIFICIAL INTELLIGENCE

BACKGROUND

Artificial intelligence is defined as the application for computer technology to resemble human thought and action. It is also the branch of computer science concerned with the development of machines having this ability. Artificial intelligence was widely acclaimed in the 1980s as revolutionary technology. However, when expectations went unrealized, its popularity faded. More recently it is reemerging with revamped development tools and languages and advanced flexibility. The restrictions of expensive, large machines and obscure languages have been overcome. Many of the lofty early claims have been dismissed and replaced with more pragmatic and profitable applications. The use of artificial intelligence is subtly in place for a variety of industrial, business, and consumer applications.

Areas of artificial intelligence include robotics, machine vision, voice recognition, natural language processing, expert systems, neural networks, and fuzzy logic. The term artificial intelligence was first used in 1956 at a conference held at Dartmouth College in Hanover, New Hampshire. The basis of the conference was the conjecture that every feature of intelligence can be so precisely described that a machine can be made to simulate it (1). Since that time the field has grown, evolved, fragmented, developed, and in many ways become incorporated into mainstream software.

Robotics

Robotics, along with computer vision, have been applied in food-processing operations for sorting and evaluation. Robots are used in manufacturing, performing such functions as welding and painting. Labor-intensive, repetitive tasks such as those found in packaging have been prime applications for robots. Even routine lab testing has incorporated robotics when the volume of samples justifies the expense. Through artificial intelligence techniques robots can function in unstructured situations, performing tasks such as identifying objects and selecting and assembling them into a unit.

Machine Vision

Machine vision is the ability of a machine to look at or see objects and to differentiate between those objects intelligently. A vision system can observe defects even at microscopic levels. Because of its success, the implementation of machine vision is extensive. In the food industry, machine vision is used for sorting purposes in quality control applications. Machine vision, coupled with neural network-based products, is being used to recognize patterns, optical characters, and bar codes.

Voice Recognition

Voice recognition systems respond to the human voice instead of a keyboard or other input. In certain industrial settings such as package routing, voice recognition has been successful. Initially, because of the wide variety in human voice patterns, voice recognition was constrained to a narrow range of applications or subjects. The technology has since advanced to common usage in telecommunications for automated information exchange.

Natural Language Processing

Natural language processing allows computer software to understand normal conversational commands. Natural language is gaining strength in foreign language translation. Although syntax often suffers, at least approximate translations can be commercially provided and are readily available by means of the Internet. In database applications, natural language programs translate queries from natural language to some database query language. Natural language processing is steadily developing in order to facilitate the user/computer interface.

Expert Systems

Expert systems are computer programs designed to solve problems requiring experts. In the field of artificial intelligence, knowledge-based expert systems received most of the early attention. An expert system includes a knowledge base and an inference engine. The knowledge base is the accumulation and representation of knowledge specific to a particular task. It is often constructed in the form of rules, but methods such as frames and semantic nets are also used. Rule-based expert systems rule statements normally include a premise and a conclusion, using If and Then statements. The inference engine is the system's machinery for selecting and applying knowledge from the knowledge base to the specific problem. An inference engine uses the knowledge base along with responses from the user to solve problems. The inference engine takes information and proceeds forward through the knowledge base looking for a valid path. The engine may either take information and proceed forward through the knowledge base looking for a valid path, or start with goals and work backward until a clear path is found. In conventional programs the knowledge base is part of the program, but with an expert system it is separate. This distinction makes it possible to substitute a new knowledge base for a new task in place of the existing knowledge base. Expert systems can also be updated readily as new knowledge is discovered. Expert systems take advantage of an expert's heuristics or knowledge from experience. The combination of heuristics and learned principles helps account for the value of an expert. Usually an expert system is developed through an iterative process where an initial program is prepared and then developed and improved with additional items of knowledge. Expert systems cannot exactly model human problem-solving processes, but rather they attempt to interact with expert thought concerning specific areas of knowledge. Most expert systems allow confidence factors in order to express uncertainty. Uncertainty may arise from either the stored knowledge base or user responses.

Although an expert system is useful for arriving at a solution from numerous options, its range of focus needs to be narrow to avoid an excessively complex system. The economic advantages of going to the time and expense of developing an expert system need to be established. If the problem is very simple or if ample resources are already available to solve the problem within the company, then the investment may not be worthwhile. Some problems are too broad or general to be solved by an expert system. To be solved by an expert system, the problem would have to be solvable to start with by a human expert in the field. Expert systems work well with programs where symbolic logic and rules are required.

Expert systems can benefit problem-solving situations in which humans suffer from cognitive overload or fail to monitor all available information. It can be difficult to simultaneously manipulate all relevant information to obtain the optimal solution. An expert system may be more consistent in certain situations in which there is little time to think. Expert systems can often deliver solutions more quickly than their human counterparts. They are also readily available at any time. Human experts are then able to better use their skills on more difficult problems beyond the ability of the expert system. The process of building a knowledge base can help to clarify and organize an expert's logic and thinking process. A disadvantage of an expert system is that it can be costly to develop. It also needs to be maintained and updated to remain current.

Neural Networks

Neural networks attempt to figuratively model the neurological processes of the brain, in a stimulus/response structure. Neural networks essentially recognize patterns in data leading from initial events to results. These programs are well suited to systems in which traditional statistics are not applicable. The technology's power lies in its ability to analyze large combinations of variables very quickly. The neural network is trained by observing a multitude of patterns. For many processes indications of product quality parameters, such as moisture, color, texture, concentration, or taste, cannot be measured at the point of control. Quality labs check samples from the end of the line, which is too late for process control, resulting in considerable scrap or costly rework. Neural networks can be trained to predict quality parameters from on-line process data at the point of control. This information can then be used either manually or automatically for controlling the process. Whenever product quality needs to be controlled based on measurements taken after the process, neural networks are useful.

Fuzzy Systems

Fuzzy systems are distinguished primarily by their support for a gradual and continuous transition rather than a sharp and abrupt change between binary values. Information obtained in food process control may be noisy, incomplete, scrambled, erroneous, or generally uncertain (2). Fuzzy systems are most suitable for uncertain or approximate reasoning, particularly systems with a mathematical model that is difficult to derive. Designers are able to represent descriptive or qualitative expressions such as "slow" or "moderately fast" and incorporate them with symbolic statements. Integrated circuit manufacturers have fabricated chips with built-in fuzzy logic capabilities.

EXAMPLES

The addition of an expert system can advise a process controller when to make adjustments. These devices use pattern recognition algorithms, calculate control loop performance on-line, and readjust operator intervention. Expert systems can help monitor a process and notify operators of an out-of-specification situation. Self-tuning controllers can also use model-based references where the difference in response between the process and the model is used to tune the controller and also update the model. A classic expert system for process control was designed for advisory control of thermal processing in a major soup company (3). The expertise of a retiring staff member for troubleshooting, start-up, and shutdown of hydrostatic and rotary sterilizers was captured in an expert system program, saving the company up to US$3 million per year. An expert system can respond to user questions, query the user for additional information, and provide a report of the interaction along with a process and instrumentation diagram. Fault analyzers respond to alarms and other process inputs to diagnose abnormal conditions. They can find a previously determined pattern in a set of alarm messages. Rather than receiving a number of alarms, the operator will receive a simplified message stating the recommendation. Corrective action can then be taken.

A widely used rule-based expert system from Gensym Corp (Cambridge, Mass.), among other applications, has been useful in controlling wastewater treatment. The arrangement of equipment and controls is a complex, time-consuming assignment and many factors need to be considered. With control-system configuration, operating conditions and parameters, objectives, and constraints such as equipment limitations can be entered into a process model. The configuration best satisfying all requirements is then presented to the user.

Neural networks can predict patterns in process data, which in turn can be used to control the process as desired. With neural network pattern recognition software, responses are observed and necessary control adjustments are made. Applications have been reported across a wide variety of industries. Pattern recognition methods have been used for the identification of flours (4). BrainMaker (Nevada City, Calif.) has reported a wide variety of applications for its neural network product including investment advising, heart attack diagnosis, beverage quality testing, chemical structure recognition, and protein sequence pattern recognition. The application of neural networks in the dairy industry was recently discussed in an Australian seminar (5). Reports included neural networks in ultrahigh temperature (UHT) operations, natural language computer control of crucial steps in cheese making, and fermentation processes. The effects of transmembrane pressure and cross-flow velocity on cross-flow filtration were modeled using a neural network approach. Modeling was obtained after only five experimental trials for either raw cane sugar remelt or natural gum solution (6). The use of neural networks for modeling instrumental-sensory relationships has been investigated. The advantages and disadvantages of using neural networks were compared with multivariate linear methods of principal components regression (PCR) and partial least-squares (PLS) regression. It was concluded that neural networks cannot replace PCR and PLS for linear relationships but do offer potential for modeling nonlinear relationships (7).

Questions regarding expansion of neural networks when necessary have prompted research on network configuration. Learning rates for neural networks, which were expanded by coupling modules, rather than simply directly expanding the existing network were improved. Coupled neural networks were found to improve food starch identification (8).

The relationships between process parameters and flow rates through ultrafiltration membranes have been examined using fuzzy mathematics (9). Main parameters in the permeate flow were identified. Fuzzy logic has been used in many familiar applications including, home appliances, video cameras, automobiles, robots, and aircraft. One company has developed a hybrid product combining neural networks, fuzzy logic, genetic algorithms, and chaos theory (Neuristics, Baltimore, Md.). The company maintains this design overcomes limitations of neural networks, explaining the logic behind its forecasts.

Regarding business applications, neural networks have been successfully used for bank loan applications. A program called Countrywide Loan Underwriting Expert System, or CLUES (Brightware, Inc, Novato, CA), evaluates a loan file's strengths and weaknesses in human terms. Loans can be approved in less than 2 min compared with 50 min for human underwriters. Benefits include personnel cost savings and better quality and consistency in underwriting. Neural networks can be used to set up intelligent agents for information services. These agents are able to automatically locate and retrieve competitive intelligence information. Consumer electronics industries have developed software programs with the ability to operate autonomously, performing operations ranging from electronic shopping to getting data over the Internet. The growth of the Internet may create new artificial intelligence applications.

ADVANTAGES/FUTURE DEVELOPMENT

The many facets of artificial intelligence are able to reduce labor costs, improve productivity and quality, and remove human bias. Expert systems can act as training tools, both during the process of development, as well as during implementation. Advanced computer-aided systems can improve food safety, reduce processing times, and achieve higher quality and yields at minimal costs if applied correctly to the food processing system (10). Human talents and abilities can be better utilized and even assisted as these programs are made available for process control and analysis.

The large number of computing techniques within the very broad terms of artificial intelligence will continue to drive many of the advanced applications in science, industry, and consumer products. Products will continue to improve in data assimilation and communication, replacing an increasing number of tasks previously requiring human intervention.

Artificial intelligence will continue in importance as it becomes embedded in mainstream systems. The different components of artificial intelligence are each finding their own particular embedded applications.

BIBLIOGRAPHY

1. E. Charniak and D. McDermott, *Introduction to Artificial Intelligence*, Addison-Wesley, Reading, Mass., 1985.
2. T. Eerikainen et al., "Fuzzy Logic and Neural Network Applications in Food Science and Technology," *Trends Food Sci. Technol.* **4**, 237–242 (1993).
3. R. K. Tyson and R. A. Herrod, "Capturing 32 Years on Computer," *Chilton's Food Eng.* **57**, 69–71 (Dec. 1985).
4. P. Zagrodzki et al., "Characterisation of Flour by Means of Pattern Recognition Methods," *Food Chem.* **53**, 295 (1995).
5. J. A. Hourigan, "Artificial Neural Networks in the Dairy Industry," *Aust. J. Dairy Technol.* **49**, 110–121 (1994).
6. M. Dornier et al., "Interest of Neural Networks for the Optimization of the Crossflow Filtration Process," *Lebensmittel-Wissenschaft Technol.* **28**, 300–309 (1995).
7. C. Wilkenson and D. Yuksel, "Using Artificial Neural Networks to Develop Prediction Models for Sensory-Instrumental Relationships: An Overview," *Food Quality Preference* **8**, 439–445 (1997).
8. Y. W. Huang et al., "Modular Neural Networks for Identification of Starches in Manufacturing Food Products," *Biotechnol. Progr.* **9**, 401–410 (1993).
9. P. Vaija et al., "Prediction of the Flux through an Ultrafiltration Membrane Using Fuzzy Mathematics," *J. Membrane Sci.* **83**, 173–179 (1993).
10. S. Linko, "Expert Systems: What Can They Do for the Food Industry?" *Trends Food Sci. Technol.* **9**, 3–12 (1998).

ROBERT OLSEN
Schreiber Foods, Inc.
Tempe, Arizona

ASEPTIC PROCESSING AND PACKAGING SYSTEMS

In aseptic processing, packages and food product are sterilized in separate systems. The sterile package is then filled with sterile product, closed, and sealed in a sterile chamber. Because aseptic processing is a continuous operation, the behavior of one part of the system can affect the overall performance of the entire system. As a result, there are numerous critical factors associated with aseptic processing and packaging often requiring automated control systems. Process establishment for aseptic systems must consider not only the sterilization of the product, the processing equipment, and downstream piping, but also the sterilization of the packaging material and the packaging equipment and the maintenance of sterile conditions throughout the aseptic system.

DEFINITIONS

To assist in the discussion of aseptic processing and packaging systems, some definitions are presented here.

Aseptic describes a condition in which there is an absence of microorganisms, including viable spores. In the food industry, the terms aseptic, sterile, and commercially sterile are often used interchangeably.

Aseptic system refers to the entire system necessary to produce a commercially sterile product contained in a hermetically sealed container. This term includes the product processing system and the packaging system.

Aseptic processing system refers only to the system that processes the product and delivers it to a packaging system.

Aseptic packaging system refers to any piece of equipment that fills a sterile package or container with sterile product and hermetically seals it under aseptic conditions. These units or systems may also form and sterilize the package.

BASIC ASEPTIC SYSTEM

Figure 1 is a diagram of a simplified aseptic system. Raw or unprocessed product is heated, sterilized by holding at high temperature for a predetermined amount of time, then cooled and delivered to a packaging unit for packaging. Commercial sterility is maintained throughout the system, from product heating to the discharge of hermetically sealed containers.

DESCRIPTION OF THE ASEPTIC PROCESSING SYSTEM

Although the equipment for aseptic processing systems varies, all systems have certain common features:

1. A pumpable product
2. A means to control and document the flow rate of product through the system
3. A method of heating the product to sterilizing temperatures
4. A method of holding product at an elevated temperature for a time sufficient for sterilization
5. A method of cooling product to filling temperature
6. A means to sterilize the system before production and to maintain sterility during production
7. Adequate safeguards to protect sterility and prevent nonsterile product from reaching the packaging equipment

Preproduction Sterilization

Producing a commercially sterile product cannot be assured unless the processing system and filler have been adequately sterilized before starting production. It is important that the system be thoroughly cleaned before sterilization; otherwise the process may not be effective. Some systems, or portions thereof, use saturated steam for sterilization. However, for most systems, equipment sterilization is accomplished by circulation of hot water through the system for a sufficient length of time to render it commercially sterile. When water is used, it is heated in the

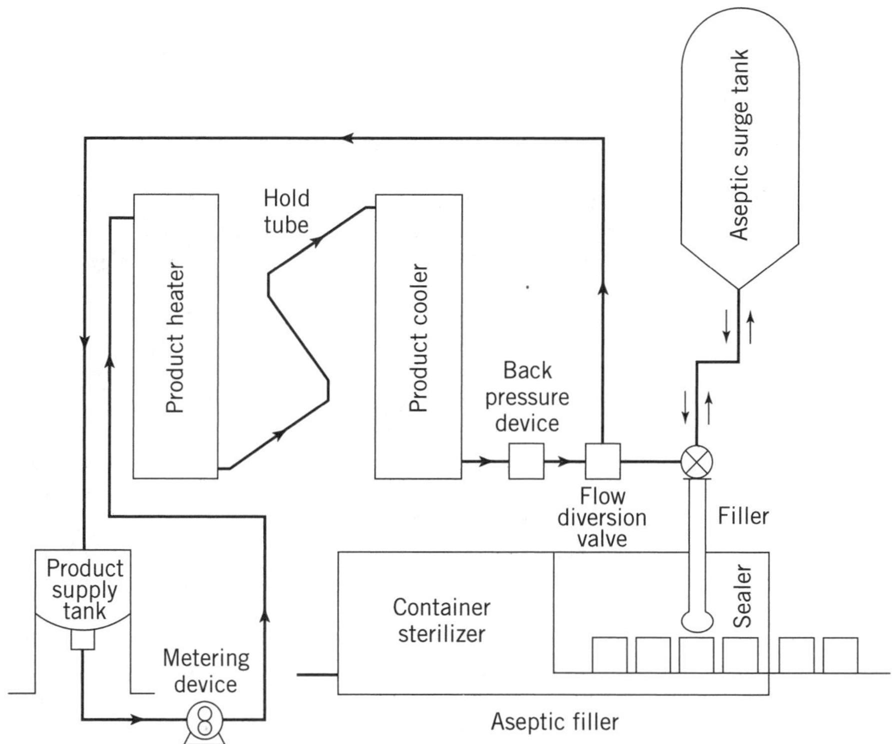

Figure 1. Simplified diagram of an aseptic processing system.

product heater and then pumped through all downstream piping and equipment up to (and generally past) the filler valve on the packaging unit. All product contact surfaces downstream from the product heater must be maintained at or above a specified temperature by continuously circulating the hot water for a required period of time.

Surge tanks are generally sterilized with saturated steam rather than hot water owing to their large capacity. Although sterilization of surge tanks may occur separately, it is usually conducted simultaneously with hot water sterilization of the other equipment.

To control aseptic system sterilization properly, it is necessary that a thermometer or thermocouple be located at the coldest point(s) in the system to ensure that the proper temperature is maintained throughout. Thus, the temperature-measuring device is generally located at the most distant point from the heat exchangers. Timing of the sterilization cycle begins when the proper temperature is obtained at this remote location. If this temperature should fall below the minimum, the cycle should be restarted after the sterilization temperature is reestablished.

Flow Control

Sterilization time or residence time, as indicated in the scheduled process, is directly related to the rate of flow of the fastest-moving particle through the system. The fastest-moving particle is a function of the flow characteristics of the food. Consequently, a process must be designed to ensure that product flows through the system at a uniform and constant rate so that the fastest-moving particle of food receives at least the minimum amount of heat for the minimum time specified by the scheduled process. This constant flow rate is generally achieved with a pump, called a timing or metering pump.

Timing pumps may be variable speed or fixed rate. The pumping rate of the fixed-rate pump cannot be changed without dismantling the pump. Variable-speed pumps are designed to provide flexibility and allow for easy rate changes. When a variable-speed pump is used, it must be protected against unauthorized changes in the pump speed that could affect the rate of product flow through the system.

Product Heating

A product heater brings the product to sterilizing temperature. There are two major categories of product heaters in aseptic food processing: direct and indirect.

Direct heating, as the name implies, involves direct contact between the heating medium (steam) and the product. Direct heating systems can be one of two types: steam injection and steam infusion.

Steam injection introduces steam into the product in an injection chamber as product is pumped through the chamber (Figure 2). Steam infusion introduces product through a steam-filled infusion chamber (Fig. 3). These systems are currently limited to homogeneous, low-viscosity products.

Direct heating has the advantage of very rapid heating, which minimizes organoleptic changes in the product. The problems of fouling or burn-on of product in the system may also be reduced in direct heating systems compared with indirect systems.

There are also some disadvantages. The addition of water (from the condensation of steam in the product) in-

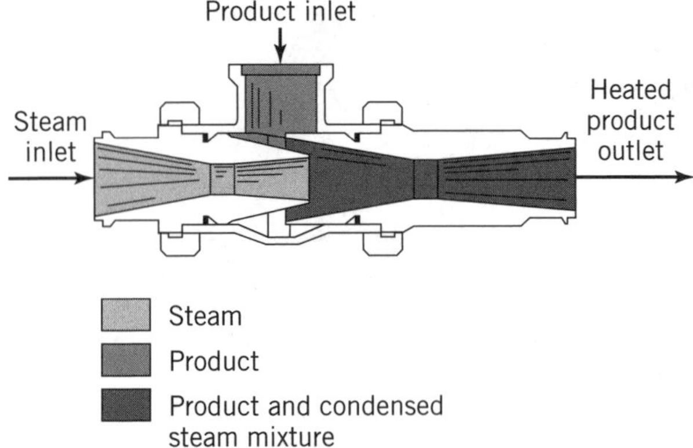

Figure 2. Steam injection.

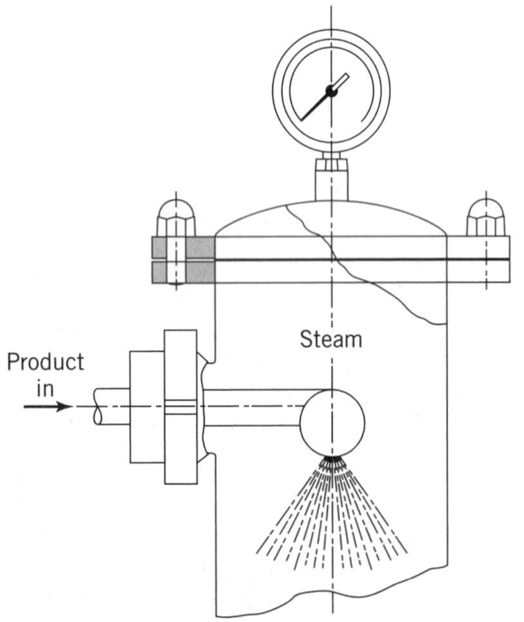

Figure 3. Steam infusion.

creases product volume. Because this change in volume increases product flow rate through the hold tube, it must be accounted for when establishing the scheduled process. Depending on the product being produced, water that was added as steam may need to be removed. Water removal is discussed under product cooling. Steam used for direct heating must be of culinary quality and must be free of noncondensable gases. Thus, strict controls on boiler feed water additives must be followed.

The other major category of product heaters is indirect heating units. *Indirect heating* units have a physical separation between the product and the heating medium. There are three major types of indirect heating units: plate, tubular, and swept-surface heat exchangers.

Plate heat exchangers are used for homogeneous liquids of relatively low viscosity. The plates serve as both a barrier and a heat-transfer surface with product on one side and the heating medium on the other. Each plate is gasketed, and a series of plates are held together in a press. The number of plates can be adjusted to meet specific needs.

Tubular heat exchangers employ either two or three concentric tubes instead of plates as heat-transfer surfaces. Product flows through the inner tube of the double-tube style and through the middle tube of the triple-tube style, with the heating medium in the other tube(s) flowing in the opposite direction to the product. In a shell and tube heat exchanger (considered a type of tubular exchanger), the tube is coiled inside a shell. Product flows through the tube while the heating medium flows in the opposite direction through the shell. As with plate heat exchangers, tubular heat exchangers are used for homogeneous products of low viscosity.

Scraped-surface heat exchangers are normally used for processing more viscous products (Fig. 4). The scraped-surface heat exchanger consists of a mutator shaft with scraper blades concentrically located within a jacketed, insulated heat exchange tube. The rotating blades continuously scrape the product off the wall. This scraping reduces buildup of product and burn-on. The heating medium flowing on the opposite side of the wall is circulating water or steam.

Some systems incorporate the use of product-to-product heat exchangers. These devices are either plate or tubular heat exchangers with product flowing on both sides of the plates or through both sets of tubes. This process allows the heat from the hot, sterile product to be transferred to the cool incoming, nonsterile product. Energy and cost savings can be significant by recycling the heat from sterile product.

When a product-to-product regenerator is used, the regenerator must be designed, operated, and controlled so that the pressure of the sterilized product in the regenerator is at least 1 psi greater than the pressure of any nonsterilized product in the regenerator. This pressure differential helps ensure that any leakage in the regenerator will be from sterilized product into nonsterilized product.

Hold Tube

Once the product has been brought to sterilizing temperature in the heater, it flows to a hold tube. The time required for the fastest product particle to flow through the hold tube is referred to as the residence time. The residence time must be equivalent to or greater than the time necessary at a specific temperature to sterilize the product and is specified in the scheduled process. Hold-tube volume, which is determined by hold-tube diameter and length, combined with the flow rate and flow characteristics of the product, determines the actual residence time of the product in the hold tube. Because the hold tube is essential for ensuring that the product is held at sterilization temperatures for the proper time, certain precautions must be followed:

1. The hold tube must have an upward slope in the direction of product flow at least 0.25 in./ft to assist in eliminating air pockets and prevent self-draining.

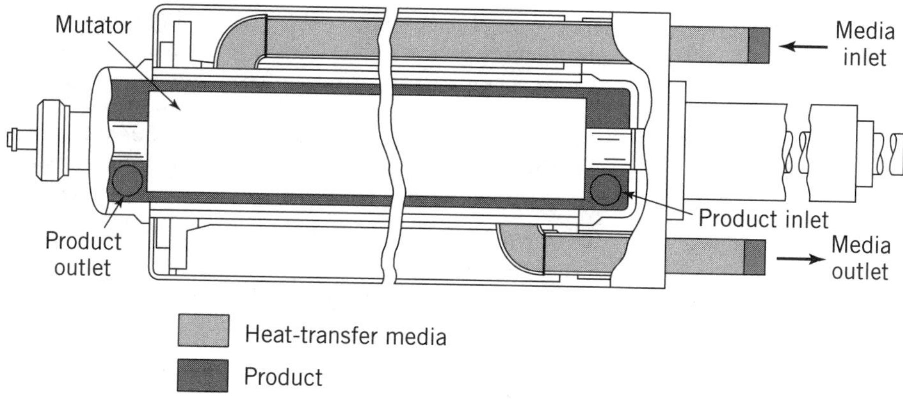

Figure 4. Scraped-surface heat exchanger.

2. If the hold tube can be taken apart, care should be taken that all parts are replaced and that no parts are removed or interchanged to make the tube shorter or different in diameter. Such accidental alterations could shorten the time the product remains in the tube.
3. If the hold tube can be taken apart, care should be exercised when reassembling to ensure that the gaskets do not protrude into the inner surface. The tube interior should be smooth and easily cleanable.
4. There must be no condensate drip on the tube, and the tube should not be subjected to drafts or cold air, which could affect the product temperature in the hold tube.
5. Heat must not be applied at any point along the hold tube.
6. The product in the hold tube must be maintained under a pressure sufficiently above the vapor pressure of the product at the process temperature to prevent flashing or boiling because flashing can decrease the product residence time in the hold tube. The prevention of flashing is usually accomplished by use of a back-pressure device.

The temperature of the food in the hold tube must be monitored at the inlet and outlet of the tube. The temperature at the inlet of the tube is monitored with a temperature recorder-controller sensor that must be located at the final heater outlet and must be capable of maintaining process temperature in the hold tube. A mercury-in-glass thermometer or other acceptable temperature-measuring device (such as an accurate thermocouple recorder) must be installed in the product sterilizer hold tube outlet between the hold tube and the cooler inlet. An automatic recording thermometer sensor must also be located in the product at the hold tube outlet between the hold tube and cooler to indicate the product temperature. The temperature-sensing device chart graduations must not exceed 2°F (1°C) within a range of 10°F (6°C) of the required product sterilization temperature.

Product Cooling

Product flows from the hold tube into a product cooler that reduces product temperature before filling. In systems that use indirect heating, the cooler will be a heat exchanger that may be heating raw product while cooling sterile product. Those systems that use direct heating will typically include a flash chamber or vacuum chamber. The hot product is exposed to a reduced pressure atmosphere within the chamber, resulting in product boiling or flashing. The product temperature is lowered, and a portion or all of the water that was added to the product during heating is removed by evaporation. On discharge from the flash chamber, product may be further cooled in some type of heat exchanger.

Maintaining Sterility

After the product leaves the hold tube, it is sterile and subject to contamination if microorganisms are permitted to enter the system. One of the simplest and best ways to prevent contamination is to keep the sterile product flowing and pressurized. A back-pressure device is used to prevent product from boiling or flashing and maintains the entire product system under elevated pressure.

Effective barriers against microorganisms must be provided at all potential contamination points, such as rotating or reciprocating shafts and the stems of aseptic valves. Steam seals at these locations can provide an effective barrier, but they must be monitored visually to ensure proper functioning. If other types of barriers are used, there must be a means provided to permit the operator to monitor the proper functioning of the barrier.

Aseptic Surge Tanks

Aseptic surge tanks have been used in aseptic systems to hold sterile product before packaging. These vessels, which range in capacity from about 100 gal to several thousand gallons, provide flexibility, especially for systems in which the flow rate of a product sterilization system is not compatible with the filling rate of a given packaging unit. If the valving that connects a surge tank to the rest of the system is designed to allow maximum flexibility, the packaging and processing functions can be carried out independently, with the surge tank acting as a buffer between the two systems. A disadvantage of the surge tank is that all sterile product is held together, and if there is a contamination problem, all product is lost. A sterile air or other sterile gas supply system is needed to maintain a protective positive pressure within the tank and to displace the contents. This positive pressure must be monitored and controlled to protect the tank from contamination.

Automatic Flow Diversion

An automatic flow-diversion device may be used in an aseptic processing system to prevent the possibility of potentially unsterile product from reaching the sterile packaging equipment. The flow-diversion device must be designed so that it can be sterilized and operated reliably. Past experience has shown that flow-diversion valves of the gravity drain type should not be used in aseptic systems owing to the possibility of recontamination of sterile product. Because the design and operation of a flow-diversion system is critical, it should be done in accordance with the recommendations of an aseptic processing authority.

NONTHERMAL PROCESSING

The major disadvantage of an aseptic processing (and any thermal processing system) is that heat treatment results in destruction of nutrients and organoleptic properties of foods. An alternative to aseptic processing is nonthermal preservation of food. Methods of nonthermal preservation include Pulsed Electric Field (PEF), High-Pressure Processing (HPP), and Ultraviolet Processing (UV). PEF treatment and UV are inherently adaptable to a continuous processing system, whereas HPP was developed as a batch system, although there is one continuous HPP system.

PEF is a process that uses very short duration (~2μs) of high electric field pulses (~30–50 kV) to destroy microorganisms in liquid and other pumpable foods. Research has shown that PEF is very effective against vegetative microorganisms; however, its effect against spore-forming microorganisms is still under study. PEF causes an electroporation of the semipermeable microbial cell envelope causing cell injury and eventual death.

HPP involves the application of high pressures (up to 140,000 psi) to food. The high pressure disrupts microbial cells and deactivates certain enzymes. Up until recently HPP has been a batch operation. However, one company, Flow International, has developed a semicontinuous high-pressure system that is currently undergoing testing. HPP has also been shown to be quite effective on vegetative microorganisms; however, for HPP to be effective against spore formers, moderate heat (60–70°C) is necessary in combination with the high pressure.

UV processing, as the name implies, uses ultraviolet electromagnetic rays to affect microbial destruction. Currently, UV has been shown to be effective only against vegetative microorganisms. UV has also been used effectively for water treatment.

ASEPTIC PACKAGING SYSTEMS

Basic Requirements

Aseptic packaging units are designed to combine sterile product with a sterile package, resulting in a hermetically sealed shelf-stable product. As with aseptic processing systems, certain features are common to all aseptic packaging systems. The packaging units must do the following:

1. Create and maintain a sterile environment in which the package and product can be brought together
2. Sterilize the product contact surface of the package
3. Aseptically fill the sterile product into the sterilized package
4. Produce hermetically sealed containers
5. Monitor and control critical factors

Sterilization Agents

Sterilization agents are used in aseptic packaging units to sterilize the packaging material and the internal equipment surfaces to create a sterile packaging environment. In general, these agents involve heat, chemicals, high-energy radiation, or a combination of these. For aseptic packaging equipment the sterilization agents used must effectively provide the same degree of protection in terms of microbiological safety that traditional sterilization systems provide for canned foods. This requirement applies to both the food contact surface of the packaging material and the internal machine surfaces that constitute the aseptic or sterile zone within the machine. The safety and effectiveness of these agents must be proven and accepted or approved by regulatory agencies for packaging commercially sterile low-acid or acidified foods in hermetically sealed containers. Food processors considering use of an aseptic packaging unit should request written assurances that the equipment has passed such testing and that the equipment and sterilizing agents are acceptable to the regulatory agencies for their intended use.

Heat is the most widely used method of sterilization. Steam or hot water is commonly used and referred to as moist heat. Superheated steam or hot air may also be used in certain situations and is referred to as dry heat. Dry heat is a much less effective sterilization agent than moist heat at the same temperature. Systems that use moist heat operate at elevated pressures compared with dry heat systems, which operate at atmospheric pressures. Other methods may be used to generate heat, such as microwave radiation or infrared light. As new methods are developed, they will have to be evaluated by aseptic processing authorities.

Chemical agents, primarily hydrogen peroxide, are often used in combination with heat as sterilization agents. The Food and Drug Administration (FDA) regulations specify that a maximum concentration of 35% hydrogen peroxide may be used for food contact surfaces. If hydrogen peroxide is used as a sterilant, the packaging equipment must be capable of producing finished packages that also meet FDA requirements for residuals. Not more than 0.5 ppm hydrogen peroxide may be present in tests done with distilled water packaged under production conditions. These regulations apply also to products regulated by the U.S. Department of Agriculture (USDA).

Other sterilants, such as high-energy radiation (UV-light, γ radiation, or electron-beam radiation), could be used alone or in combination with existing methods. Completely new alternative sterilants may be developed in the future. Whatever methods are developed, they will have to be proven effective in order to protect the public health and will be compared with existing methods.

Aseptic Zones

The aseptic zone is the area within the aseptic packaging machine that is sterilized and maintained sterile during production. This is the area in which the sterile product is filled and sealed in the sterile container. The aseptic zone begins at the point where the package material is sterilized or where presterilized package material is introduced into the machine. The area ends after the seal is placed on the package and the finished package leaves the sterile area. All areas between these two points are considered as part of the aseptic zone.

Before production, the aseptic zone must be brought to a condition of commercial sterility analogous to that achieved on the packaging material or other sterile product contact surfaces. This area may contain a variety of surfaces, including moving parts composed of different materials. The sterilant(s) must be uniformly effective and their application controllable throughout the entire aseptic zone.

Once the aseptic zone has been sterilized, sterility must be maintained during production. The area should be constructed in a manner that provides sterilizable physical barriers between sterile and nonsterile areas. Mechanisms must be provided to allow sterile packaging materials and hermetically sealed finished packages to enter and leave the aseptic zone without compromising the sterility of the zone.

The sterility of the aseptic zone can be protected from contamination by maintaining the aseptic zone under positive pressure of sterile air or other gas. As finished containers leave the sterile area, sterile air flows outward, preventing contaminants from entering the aseptic area. The sterile air pressure within the aseptic zone must be kept at a level proven to maintain sterility of the zone. Air or gases can be sterilized using various sterilization agents, but the most common methods are incineration (dry heat) and/or ultrafiltration.

Production of Aseptic Packages

A wide variety of aseptic packaging systems are in use today. These are easily categorized by package type:

1. Preformed rigid and semirigid containers, including
   a. Metal cans
   b. Composite cans
   c. Plastic cups
   d. Glass containers
   e. Drums
2. Web-fed paperboard laminates and plastic containers
3. Partially formed laminated paper containers
4. Thermoform-fill-seal containers
5. Preformed bags or pouches
6. Blow-molded containers

A number of different packaging systems are represented in these categories. Not all of these systems, however, are being used in the United States for aseptic applications.

Containers in these categories may be sterilized by a variety of means. For example, one system utilizing metal cans uses superheated steam to sterilize the containers. In other systems, preformed plastic cups may be sterilized by hydrogen peroxide and heat or by saturated steam. Systems using containers formed from paperboard laminates also utilize hydrogen peroxide and heat or hydrogen peroxide and ultraviolet irradiation to sterilize packages. Thermoform-fill-seal containers may be sterilized by the heat of extrusion (dry heat) or by hydrogen peroxide and heat. Plastic pouches or bags may be sterilized by gamma irradiation, by the heat of extrusion, or by chemical means such as hydrogen peroxide.

Research is now being conducted to explore alternative sterilization methods for most categories of packaging. Thus, it can be said with some certainty that currently familiar equipment may not be state of the art tomorrow. Nevertheless, whatever equipment or sterilant or packaging material is used, the monitoring and control of critical factors will be vital to successful operation.

SUMMARY

1. Aseptic is a term that describes the absence of microorganisms and may be used interchangeably with commercial sterility.
2. An aseptic processing system consists of a timing pump, a means to heat the product, a hold tube, and a means to cool the product.
3. The processing system from the hold tube past the fillers is brought to a condition of commercial sterility before the introduction of product into the system.
4. The thermal process occurs in the hold tube where flow rate, or residence time, and temperature are critical factors.
5. In the event of a process deviation, flow-diversion devices are used to prevent potentially inadequately processed product from reaching the filler.
6. Aseptic packaging machines create and maintain an aseptic zone (sterile environment) in which sterilized containers are filled and sealed.
7. Sterilization agents such as heat, chemicals, high-energy radiation, or a combination of these, can be used to sterilize packages or machine surfaces.

GENERAL REFERENCES

Adapted from *Canned Foods—Principles of Thermal Process Control, Acidification and Container Closure Evaluation*, 6th ed., The Food Processors Institute, Washington, D.C., 1995.

K. Stevenson
D. Chandarana
National Food Processors Association
Dublin, California

ASEPTIC PROCESSING: OHMIC HEATING

The recent development of the ohmic heater represents a major advance in the continuous processing of particulate food products. The ohmic heater uses resistance heating within the flow of electrically conducting liquid and particulates to provide heat and is capable of handling food products containing particulates of up to 25 mm.

When combined with an aseptic container or bag-in-box filling system, this new process offers the food processor with the possibility of producing high-quality, ready-prepared meals containing large quantities of undamaged chunks of meat and vegetables, which can be stored at ambient temperatures for long periods of time. The process has also been successfully applied to the sterilization of diced and sliced fruit at solids concentrations of up to 80% drained weight.

PRINCIPLE OF OHMIC HEATING

The ohmic heating effect occurs when an electric current is passed through a conducting product (Fig. 1). In practice, low-frequency alternating current (50 or 60 Hz) from the public main supply is used to eliminate the possibility of adverse electrochemical reactions and minimize power supply complexity and cost.

In common with microwave heating, electrical energy is transformed into thermal energy. However, unlike microwave heating, the depth of penetration is virtually unlimited and the extent of heating is governed by the spatial uniformity of electrical conductivity throughout the product and its residence time in the heater. For most practical purposes the product does not experience a large temperature gradient within itself as it heats, and liquid and particulate are heated virtually simultaneously. The requirement to overprocess the liquid to ensure sterility at the center of a large particulate as with scraped surface and tubular heat exchangers is therefore reduced. This results in less heat damage to the liquid phase and prevents overcooking of the outside of the particulates.

Another major advantage is that there are no heat transfer surfaces, which reduces the possibility of deposit formation and the transfer of burnt particles to the finished product. Nor is there a need for mechanical agitation of the product, which can cause loss of true particulate identity.

The applicability of ohmic heating depends on product electrical conductivity. Most food preparations contain a moderate percentage of free water with dissolved ionic salts and hence conduct sufficiently well for the ohmic effect to be applied. The system will not directly heat fats, oils, alcohol, bone, or crystalline structures such as ice.

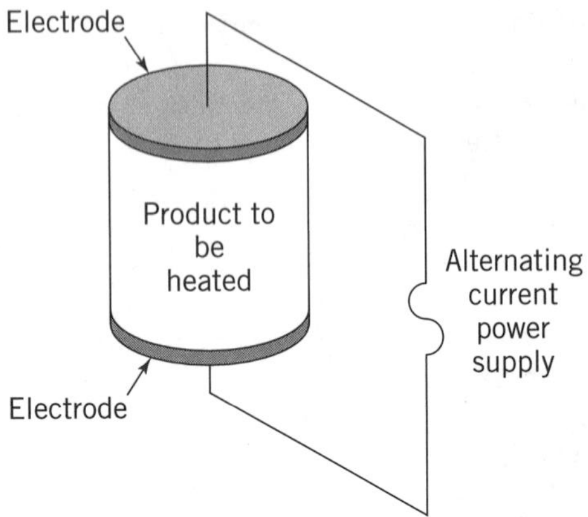

Figure 1. Principle of ohmic heating.

DESIGN OF THE OHMIC HEATER

The ohmic heater column typically consists of four or more electrode housings machined from a solid block of poly(tetrafluoroethylene) (PTFE) and encased in stainless steel, each containing a single cantilever electrode (Fig. 2). The electrode housing are connected using stainless steel interconnecting tubes lined with an electrically insulating plastics liner. Suitable lining materials include poly(vinylidene difluoride) (PVDF) and glass. These flanged tube sections are bolted together and sealed with flat food-grade rubber gaskets.

The column is mounted in a vertical or near vertical position with the flow of product in an upward direction. A vent valve positioned at the top of the heater ensures the column is always full. The column is configured such that each heating section has the same electrical impedance; hence, the interconnecting tubes generally increase in length toward the outlet. This is because the electrical

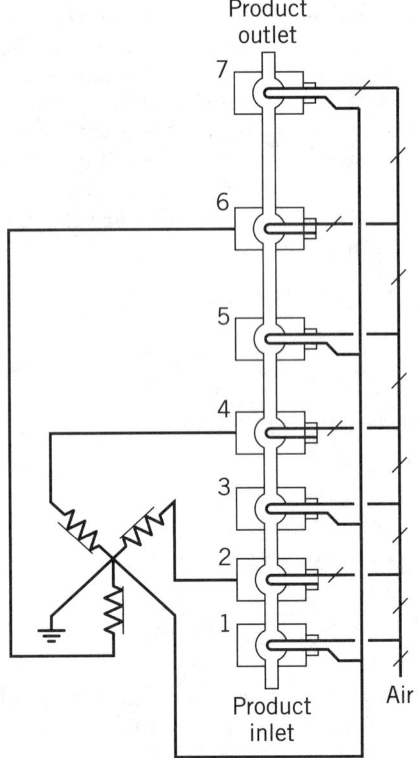

Figure 2. Design of an ohmic heater column.

conductivity of food products usually increases with increase in temperature; indeed, for aqueous solutions of ionized salts there is a linear relationship between temperature and electrical conductivity. This phenomenon is attributed to increased ionic mobility with increase in temperature and also applies to most food products. Exceptions could be products in which viscosity increases markedly at higher temperatures such as those containing ungelled starches.

MEASUREMENT OF ELECTRICAL CONDUCTIVITY OF PARTICULATE FOOD PRODUCTS

Conductivity measurements of products with large particles are often impossible using commercially available cells because of their small internal clearances.

As consistent results can only be obtained using a cell having a cross-sectional area considerably greater than that of the largest particle, APV has developed a special cell to overcome this problem, and can now accurately measure the conductivity of products containing large particulates.

TEMPERATURE CONTROL OF THE OHMIC HEATER

Ohmic heating plants are supplied with a fully automatic feed-forward temperature control system. Pure feedback is not satisfactory due to the time constant in the heater column.

Inlet changes, which will affect the final product outlet temperature, are changes in inlet temperature, mass flow rate and product specific heat capacity. In the feed-forward control system, a microprocessor scans these variables including the latter, which is an operator-entered or computer-generated variable. It continuously computes the electrical power required to heat the product and compares this value with the signal from a power transducer on the output side of the transformer. Feedback monitoring is used to prevent any long-term drift in outlet temperature.

ASEPTIC PROCESSING USING THE OHMIC HEATER

The ohmic heater assembly can be visualized in the context of a complete product sterilization or cooking process. A typical line diagram of such a plant is shown in Figure 3. In common with other types of continuous systems, careful design of the ancillary process equipment is necessary, and once heated, the product must be cooled by more conventional means such as scraped surface or tubular heat exchangers. The latter is generally preferred for particulate processing to maximize the greatest advantage of ohmic heating, that of a minimal structural damage to the particulates.

Presterilization of the ohmic heater assembly, holding tube and coolers, is carried out by recirculation of a solution of sodium sulfate at a concentration that approximates the electrical conductivity of the food material that will subsequently be processed. Sterilization temperatures are achieved through the passage of electrical current and backpressure is controlled using a backpressure valve. The aseptic storage reservoir, interface catch tank, and connecting pipework to the filler are sterilized by traditional steam methods.

The use of sterilizing solution of similar product electrical conductivity minimizes adjustment of electrical power during subsequent change to product, thus ensuring a smooth and efficient changeover period with little temperature fluctuation.

Once the plant has been sterilized, the recycled solution is cooled using a heat exchanger in the recirculation line. When steady-state conditions are reached, sterilizing solution is run to drain and product introduced into the hopper of a positive displacement feed pump. Typically, this might be an auger-fed mono or rotary or a Marlen reciprocating piston pump. Product is usually prepared in premix vessels, which can incorporate preheating, or blanching operations.

Backpressure during the changeover period is controlled by regulation of top pressure in a catch tank using sterile compressed air or nitrogen. This tank serves to collect the sodium sulfate/product interface. Once the interface has been collected, product is diverted to the main aseptic storage vessel where top pressure is similarly used to control backpressure in the system.

Backpressure is maintained at a constant 1 bar when sterilizing high-acid food products at temperatures of 90 to 95°C. A 4-bar backpressure is used for low-acid food products where sterilization temperatures of 120 to 140°C are necessary. Safety features are incorporated to ensure power is automatically switched off should there be any loss of pressure.

The use of pressurized storage vessels has proven to be an extremely effective method of controlling backpressure in pilot-plant heaters where throughputs are typically less than 750 kg/h. In larger systems operating in excess of 2 t/h, aseptic positive displacement pump downstream of the cooler can be used as an alternative depending on the composition of the food product. This alternative overcomes the requirement for two aseptic storage vessels in order for the system to continuously feed product to the filling machine.

Product is progressively heated to the required sterilization temperature as it rises through the ohmic heater assembly. It then enters an air-insulated holding tube before being cooled in a series of tubular heat exchangers. More rapid cooling can be achieved with some products in which the final particulate level will be less than 40%. This system involves a combination of ohmic and traditional heat treatment and takes advantage of the capability of the ohmic heater in being able to process products containing up to 80% particulates. During preparation, the product is formulated into a high concentration particulate stream and a separate liquid stream. The liquid stream is conventionally sterilized and cooled in a plate or tubular heat exchanger system before being injected into the particulate stream leaving the holding tube of the ohmic heater. Such operation has the advantage that it reduces the capital and operating cost for a given throughput. It also allows the electrical conductivity of the carrier fluid in the ohmic heater to be more closely matched to that of

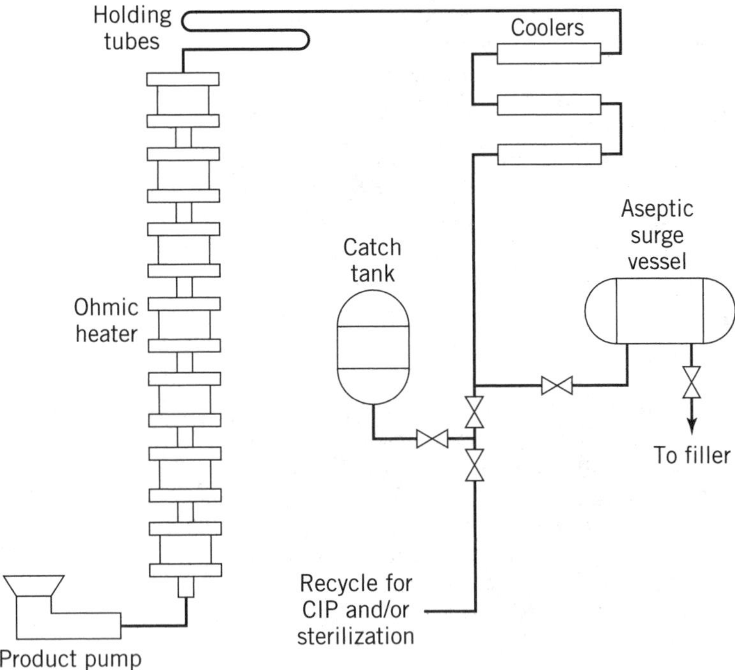

Figure 3. Ohmic plant line diagram.

the particulate, should they be markedly different, by manipulation of the dissolved ionic contents of the two processing streams. This assists in ensuring that both particulates and carrier fluid are heated at the same rate.

After cooling, product enters the main storage reservoir prior to aseptic filling. It is not usually necessary to carry out intermediate cleaning and/or resterilization of the plant when several different products are processed. This is due to the almost complete absence of fouling with most food materials. After one product is processed, the plant is flushed with a food-compatible liquid or base sauce before introducing the next product. The catch tank is used to collect product/sauce interfaces.

When food processing is complete, the power is turned off and the plant rinsed with water. This is followed by cleaning with a 2% (w/v) solution of caustic soda recirculated at 60 to 70°C for 30 min. Heating of the cleaning solution is by standard methods and not using the ohmic principle. This is because the electrical conductivity of the cleaning solution is too high for the ohmic principle to be applied.

PRODUCT QUALITY

Included in the list of perceived advantages of continuous aseptic processing (Fig. 4) is that of improved product quality, with the basic standard being set by products sterilized by traditional canning methods.

These quality standards include microbiological safety (ie, process lethality), cooking effects, and nutrient/vitamin retention. Unlike other processes requiring heat penetration, either from the outside to the center of a can (in-can sterilization) or from the outside of the particulate to the center (tubular or swept surface), ohmic has the ability to provide exceptionally fast heating of all of the product virtually simultaneously; this provides a high degree of microbiological security, as there is a very narrow range of lethality from the surface to the center of the particulates.

- Fresher tasting, more nutritional products can be produced containing large particulates.
- There is ability to heat food product in continuous flow without the need for any hot heat transfer surfaces.
- Process is ideal for shear-sensitive products.
- Risk of fouling is considerably reduced.
- Heat can be generated in the product solids without reliance on thermal conductivity through liquid.
- System is quiet in operation.
- Maintenance costs are lower.
- Process is easy to control and can be started or shut down instantly.
- There are potential savings in processing and packaging costs.

Figure 4. Advantages of electrical resistance heating.

In a 1996 publication by Kim et al. (2), a team involving the U.S. Army Soldier and Biological Chemical Command and APV UK Ltd., had demonstrated that not only the center temperature of a particulate rises faster than the surface, but also particulates could heat faster than fluid in ohmic heating. A higher temperature of the particulates provided additional heat to the surrounding fluid as both phases exited the ohmic column and entered the holding tube. Therefore a higher temperature reading was registered at the outlet than that at the inlet of holding tube (Fig. 5). They also suggested that lethality estimation of microorganisms based on fluid temperature at the ohmic column outlet would be very conservative. This is a very

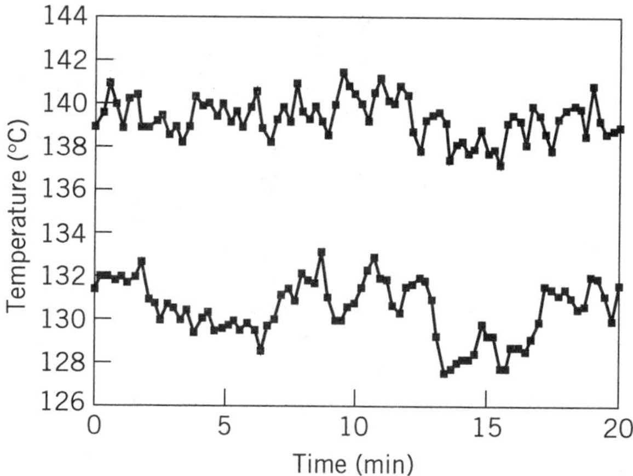

Figure 5. Fluid temperature increases in the holding tube in ohmic heating. Bottom: ohmic column outlet; top: holding tube exit.

important information for the application of regulatory approval of such novel technology.

Challenge tests carried out (on now commercially available products) by the Campden Food and Drink Research Association have involved particulates of meat and vegetable up to 19 mm cube. An average of 10 million spores of *Bacillus stearothermophilus* per particle embedded in a food/alginate mixture were included in batches of mince and mushrooms in bolognese sauce, and in kidney, peas, and gravy, in order to determine the lethality at the center and throughout particles of various sizes from 3 mm to 19 mm.

The results obtained (Table 1) demonstrate the high level of microbiological security of the process and the very narrow range of lethality from the surface to the center of the particulates.

The cook value (C_0) of a process is also a function of time and temperature (dependent upon the viscosity of the product). For small cans, F_0 values for viscous products can typically range from 5 to 300 resulting in high cook values. Larger cans will show even greater variations in F_0 values and higher degrees of overcook to achieve satisfactory F_0 values at the center.

Continuous systems using heat penetration will also have wide variations in F_0 values from the center of the particulate to the surface, the variation depending greatly on the size of the particulate, the ratio of solids to liquid (as the carrier fluid must provide the excess heat required to penetrate the particulate), and the sterilization temperature selected.

Fig. 6 and Fig. 7 show the F_0 value given to the carrier fluid to achieve a minimal F_0 value of 5 in the center of the particulate, and the processing times required at various temperatures. It can be seen that for particulates larger than 10 mm, a considerable wide range of F_0 values result

Table 1. Lethality Tests of APV Ohmic System on *Bacillus stearothermophilus*

Particle type	Number of particles	F_s range, min	Mean	Standard deviation
		Run 1		
Mince (5 mm)	41	23.6–34.3	27.9	2.2
Mushroom (19 mm)	16	34.9–40.3	37.9	1.4
Mushroom (3 mm)	6	31.9–41.8	36.0	4.1
		Run 2		
Pea (5 mm)	28	33.7–46.4	38.5	2.8
Kidney (19 mm)	30	47.–62.8	54.8	3.8
Kidney (3 mm)	20	45.6–56.2	50.2	3.2

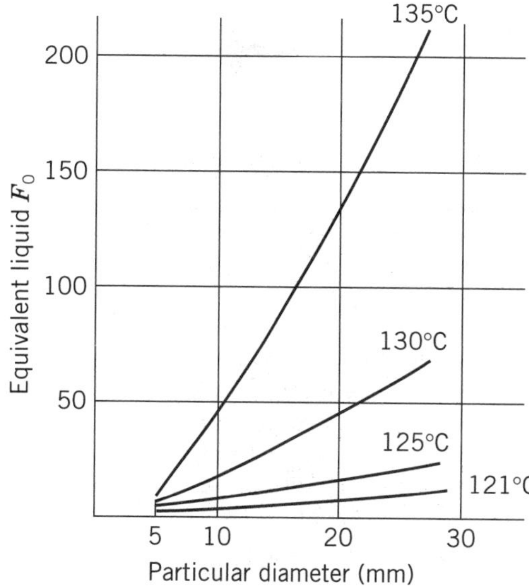

Figure 6. Liquid F_0 necessary to obtain $F_0 = 5$ in center of particulate.

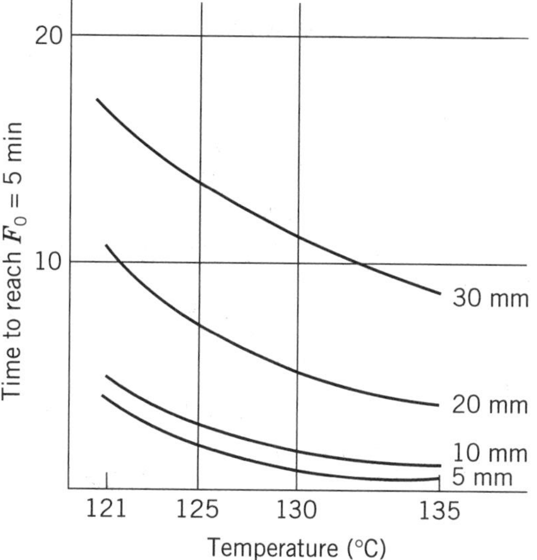

Figure 7. Time taken to reach $F_0 = 5$ in center of particulate.

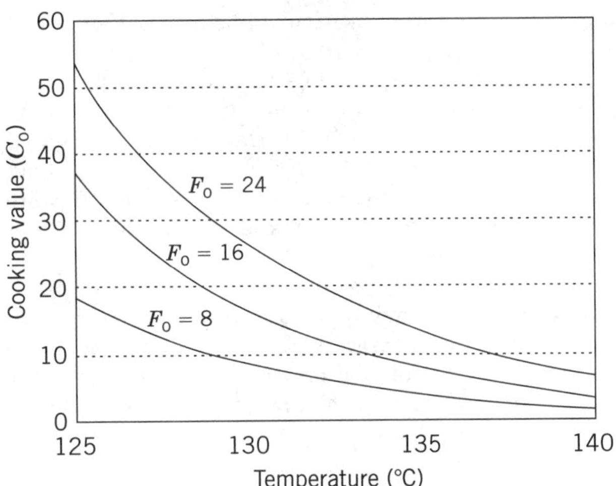

Figure 8. Cooking values versus temperature

irrespective of the sterilization temperature. This results in undesirable overcooking of the carrier fluid and the outside surface of the particulates. For example, processing a 19-mm particle to a minimum F_0 value of 5 at a sterilization temperature of 135°C will require the carrier fluid to have had an equivalent F_0 treatment of 108, as the heat penetration time will have been the order of 4 min. In contrast, ohmic heating can achieve a high F_0 level with a low value of cook and nutrient/vitamin destruction. Figure 8 shows the effect of processing temperature on F_0 and C_0. The C_0 value is based on a Z value of 33 and is calculated as

$$10^y[T]$$

where y is the [(°C − 100)/Z] and T is the time, min. Sterilization at 140°C to a virtually uniform F_0 of 24 equates to a likewise uniform C_0 value of less than 5.

This relationship allows such products as diced potato to be subjected to an F_0 value greater than 24 and still not be fully cooked; thus, the particle retains its sharp edges throughout the process. Indeed such products processed through the ohmic system therefore require further cooking (as part of the reheat process) by the end user.

The ability to heat particulates uniformly, without mechanical damage, with lower nutrient and vitamin loss, and with no fouling of heat transfer surfaces leaves no doubt that ohmic heating will play a major role in the growing requirement for aseptic food products containing large particulates, as it provides the food processor with an opportunity to produce new high-quality, long-life products previously unseen with alternative sterilization techniques.

Typical of such high-quality added-value products are those already fully tested that contain meat chunks, shrimps, baked beans, mushrooms, diced or sliced vegetables, pasta, whole strawberries, black currants, blackberries and sliced kiwi fruit.

BIBLIOGRAPHY

1. D. F. Dinnage, "Continuous Aseptic Processing Using the Ohmic Heating Process," *dfi News* **11**(3), 6–13 (1989).
2. H.-J. Kim et al., "Validation of Ohmic Heating for Quality Enhancement of Food Products," *Food Technol.* **50**(5), 253–261 (1996).

T. C. S. Yang
U.S. Army Soldier and Biological Chemical Command
Natick, Massachusetts

B

BAKERY LEAVENING AGENTS

All baked foods are leavened. Even unleavened bread such as matzoh is raised by the steam generated during baking. Normally, leavening is thought of as being due to the products of fermentation (carbon dioxide, alcohol) or to CO_2 generated by chemical reactions. However, the expansion of air incorporated into a dough during mixing and the formation of steam, from water in the dough, also cause the baking piece to expand. If this does not happen, the product is dense and unappetizing. This article discusses leavening obtained by chemical reactions as well as from yeast fermentation.

CHEMICAL LEAVENING

The common form of chemical leavening involves a food acid and soda (sodium bicarbonate). The soda dissolves readily in the aqueous phase of the dough or batter. When the acid dissolves, the hydrogen ion reacts with bicarbonate ion, releasing carbon dioxide (CO_2), which expands the baking piece. Water-vapor pressure increases as the internal temperature of the bakery item increases. Air incorporated during dough mixing also expands. Finally, certain materials, notably ammonium bicarbonate, decompose on heating and generate gases that leaven the product.

Air

During dough mixing air is incorporated as numerous small bubbles (1,2). The actual amount varies, but with an ordinary horizontal mixer a bread dough contains about 15 to 20% air by volume. If yeast fermentation is occurring, this volume shrinks by 20% owing to oxygen depletion (3), and during baking, the nitrogen remaining expands by 17% owing to temperature rise, so the net leavening contribution due to air is nil. In the absence of yeast, thermal expansion is still only 17%, so the overall effect is small.

In cookie doughs and cake batters the amount of air retained is variable. In a standard sugar cookie dough, for example, the dough contains about 10% air by volume (4). Cake batter specific gravity may vary from 0.60 to 1.10 g/mL, depending on the formulation and emulsifier system used. The density of air-free batter is about 1.30 g/mL (5), so the amount of air in the batter may range from 15 to 55% by volume. Even at the highest level, however, the thermal expansion of this air contributes only slightly to the final volume of the cake.

Nevertheless, it is important to get good air incorporation in the dough or batter during mixing, because the air bubbles serve as nuclei for other leavening gases. The internal pressure produced by the surface tension of a gas bubble is inversely related to bubble diameter. If a small bubble and a large bubble are near each other, the gas diffuses from the small one to the large one, and the small bubble disappears. If no gas bubbles are already present in, say, a cake batter when baking is begun, then the CO_2 that is formed by the chemical reactions in the batter has no place to go, because a new bubble cannot form (with a zero radius the surface tension is infinite). If the cake batter contains relatively few air bubbles, the CO_2 diffuses to these, and the resulting cake has a coarse, open grain, often accompanied by tunnels. With many small air nuclei, the grain is fine and close.

Surfactants aid incorporation and subdivision of air into dough (2). Emulsifiers help incorporate air into shortening during the preparation of cookie doughs and cake batters. In all cases the presence of many air bubbles for nucleation is an important part of any leavening action.

Steam

The formation of steam by evaporation of water would seem to be important only when the temperature in the baking piece exceeds 100°C in items such as cookies, crackers, and snacks such as tortilla chips. Also, it has been suggested that the water present in roll-in margarine, used in making puff pastry, evaporates and helps to give the characteristic open structure of this product. Extruded goods, of course, depend almost entirely on steam for their expansion.

The vapor pressure of water increases with increasing temperature, and this increased pressure can expand the gas bubbles in cake batter and in bread. In calculations of the expansion of sponge-cake batter (which contained no soda for chemical leavening), it was found that the theoretical curves (which took into account the thermal expansion of the air) fitted the experimental measurements within a few percent (5). When theoretical expansion for bread dough in the oven (see "Oven Spring") was calculated, it was concluded that water evaporation accounted for 60% of the total volume increase (6).

Ammonium Bicarbonate

Ammonium bicarbonate is often used as a supplementary leavening agent for cookies and snack crackers. At room temperature, dissolved in the dough water, it is stable, but when the temperature reaches 40°C (104°F) in the early stages of the oven, it decomposes:

$$NH_4HCO_3 \rightarrow NH_3 + CO_2 + H_2O$$

One mole of ammonium bicarbonate (79 g, or 2.8 oz) gives 2 mol of gas (44.8 L, or ca 1.6 ft^3).

This potential for extensive gas formation means that ammonium bicarbonate must be uniformly distributed throughout the dough; the presence of small undissolved pieces would give rather large blowouts in the product. Ammonium bicarbonate is usually dissolved in a few quarts of warm water and then added to the mixer along with the other water.

Because ammonia is water soluble, this leavener is applicable only in low-moisture products. If the finished moisture is above 5% (say, in a soft cookie), the water will retain some of the NH_3 and the product has a characteristic ammonia taste. In contrast to this general rule, ammonium bicarbonate is sometimes added to eclair and popover doughs (*pâté de choux*). In this case the combination of high baking temperatures, thin walls, and a large internal cavity allows enough space for ammonia to diffuse out of the baked good.

Sodium Bicarbonate

Baking soda (sodium bicarbonate) has been the workhorse chemical leavener of baked goods for well over a century. In most cases it is combined with a leavening acid to form CO_2, but soda itself undergoes thermal decomposition:

$$2NaHCO_3 + \text{heat} \rightarrow Na_2CO_3 + CO_2 + H_2O$$

This reaction usually requires rather high temperatures (>120°C) to happen at an appreciable rate, so using soda as a sole leavening agent is restricted mainly to cookies and snack crackers in which the internal temperatures can approach this range.

The more usual reaction is with hydrogen ions from leavening acids:

$$NaHCO_3 + H^+ \rightarrow Na^+ + CO_2 + H_2O$$

Soda is readily soluble in water (saturation solubility of 6.5% at 0°C, 14.7% at 60°C) and dissolves in the dough or batter water during mixing. The rate of reaction is governed by the rate of dissolution of the leavening acid (see "Leavening Acids"); no H^+ is present until the acid dissolves and ionizes. Of course, the soda can react with acidic ingredients in the formula such as buttermilk and even flour itself. Chlorinated cake flour, which is more acidic than ordinary pastry flour, will neutralize about 0.27 g of soda per 100g of flour. Soda is a mild alkali; the pH of an aqueous solution is about 8.2. Increasing the amount of soda in a formula raises the pH of the dough, for example, in raising the dough pH of saltine crackers to 7.4. Excess soda (more than that neutralized by available leavening acid) is added to devil's food cake batter to raise the pH of the finished cake crumb to about 7.8 and generate the dark chocolate color desired (cocoa is lighter colored in mild acid, darker colored in mild alkali).

Sodium bicarbonate is available in various granulations (Table 1); it is important to choose the correct one for the application. If the ingredient is added at the mixer, and the dough or batter will be processed fairly quickly, then rapid dissolution is wanted and No. 3 (fine powdered) or No. 1 (powdered) would be appropriate. At the other extreme is the inclusion of soda in a dry mix (eg, cake, cake doughnut) that is packaged and stored for a period of time before use. In this case thermal decomposition of the dry powder becomes a factor. Grade No. 1 (powdered) decomposes 50% faster than grade No. 5 (coarse granular) in this situation. A loss of 2 to 4% per week at 50°C (122°F) and 0.5 to 1% at 30°C has been reported. A common practice is to add 5 to 10% more than needed when the dry mix is blended, in the hope that when the mix is used (with uncontrolled variables of age and storage temperatures), the amount of soda will be adequate to perform as expected. It would be preferable to use grade No. 4 (granular) or No. 5 (coarse granular) for this application and reduce the amount of overage in the dry mix.

Potassium Bicarbonate

Recently this material has become commercially available as a substitute for sodium bicarbonate; it is functionally equivalent but lowers the sodium content of the finished product. As shown in Table 1 the granulation grade currently offered is intermediate between No. 4 (granular) and No. 5 (coarse granular) sodium bicarbonate. The effect on pH and the reactions to CO_2 are the same. Because the molecular weight is greater (100.11 for $KHCO_3$ vs 84.01 for $NaHCO_3$), 19% more is required to get the same effect; 10 lb of sodium bicarbonate is replaced with 11.9 lb (11 lb 14-1/4 oz) of potassium bicarbonate.

A review of bicarbonates for leavening has been published (7).

Leavening Acids

The traditional acidulants for baking were vinegar (acetic acid), lemon juice (citric acid), cream of tartar (potassium acid tartrate), or sour milk (lactic and acetic acids). In all cases the acid reacted with the soda as soon as they were mixed and the baker relied on batter viscosity to retain the CO_2 until the cake or cookie was baked in the oven. In 1864 monocalcium phosphate hydrate was patented for use in making a commercial baking powder. Later (ca 1885) sodium aluminum sulfate began to be used; this compound has a low solubility in water at room temperature and essentially none in the aqueous phase of a batter, so it does not release CO_2 until late in the baking cycle. The combi-

Table 1. Sodium Bicarbonate Granulation

Grade number	Cumulative % retained, minimum-maximum							
USS sieve:	60	70	80	100	140	170	200	325
Microns:	250	210	177	149	105	88	74	44
1. Powdered				0–2			20–45	60–100
2. Fine granular			0–Tr	0–2			70–100	90–100
3. Fine powdered				0–Tr	0–5		0–20	20–50
4. Granular			0–Tr	0–2			80–100	93–100
5. Coarse granular	0–8	0–35		65–100		95–100		
6. Potassium bicarbonate	0–5			40–60			80–100	

nation of the two leavening acids gave rise to double-acting baking powder. During the first third of the twentieth century a number of phosphate compounds were studied, and the range of leavening acids available today to formulators was developed. These acids have the advantage of controlled and predictable rates of CO_2 release and account for essentially all commercial leavening acids sold today. Several good reviews on the properties and uses of leavening acids have been published (8–12) that give more details about various specific applications.

Table 2 lists the leavening acids and the chemical reactions that occur when they dissolve in water to liberate H^+. Of use to the formulator are the neutralizing value (NV) and the equivalence value (EV). Suppliers characterize their products by NV, which is defined as the pounds of sodium bicarbonate that are neutralized by 100 lb of the acid. When developing product formulas, it is more convenient to use EV; for example, if 2% soda is included in the formula, 2.78% sodium acid pyrophosphate (SAPP) is added to achieve exact equivalence.

Neutralizing value is determined by experimental titration (9,10). The yield of H^+ shown in Table 2 corresponds to NV, although there are some discrepancies. For instance, dicalcium phosphate dihydrate (DCP.Di) should yield 2.15 H^+ to have an NV of 35. One that is even more puzzling is the reaction of SAPP. The pyrophosphate ion has four ionization constants, the first two of which are neutralized during manufacture of disodium pyrophosphate. The additional ionization constants for the dihydrogen dianion formed on solution in water are 5.77 for pK_1 and 8.22 for pK_2. In a cake batter at pH 7 the dianion should yield only 1.1 H^+, equivalent to an NV of 42, but in actual fact NV is 72 as shown in the table. Obviously other, unrecognized, reactions occur that increase the acidifying potential of SAPP. Neutralizing value (and EV) are for practical guidance; the reactions are for interest and edification.

Reaction Rates. Two kinds of reaction rates are of interest to bakers: (*1*) immediate, at room temperature; and (*2*) during baking. Both are functions of the solubility of the leavening acid. The first rate, called the dough reaction rate (DRR), is measured by mixing all the dry ingredients (cake flour, nonfat dry milk, shortening, salt, soda, and leavening acid), adding water, stirring briefly, and then measuring the evolution of gas at 27°C for 10 min (13). In the absence of leavening acid there is a blank reaction of about 20% of the soda due to acidity of the flour and dry milk. Monocalcium phosphate monohydrate ($MCP.H_2O$) reacts about as quickly as cream of tartar; that is, it is a very fast acting acid. Anhydrous MCP is about 80% as reactive. Dicalcium phosphate dihydrate (DCP.Di) is essentially unreactive under these conditions, as are sodium aluminum phosphate (SAlP) and sodium acid pyrophosphate for refrigerated dough (SAPP-RD). The various grades of SAPP (see the following discussion) show increasing extents of reaction, with the reaction rate for sodium aluminum sulfate (SAS) being somewhere in the middle of the SAPPs (9) and dimagnesium phosphate (DMP) slightly slower than SAlP.

Baking (temperature-dependent) reaction rates measure the rate of CO_2 release from a dough at various temperatures. The rate is actually governed by the solubility of the acid. A slow acid such as SAlP is nearly insoluble at room temperature, so it cannot ionize to give H^+ for reaction with soda; as the temperature rises, the material dissolves and starts to perform its function. Figure 1 shows the rates for several leavening acids (8,9,14). The ranking of acids is the same as that given by the DRR test, but Figure 1 is useful in choosing acids for giving lift at various points in the baking cycle. For instance, MCP hydrate generates CO_2 in a cake batter during mixing, whereas SAlP starts to leaven the cake only midway through the baking cycle. DCP.Di is effectively unreactive (insoluble) until the temperature reaches about 80°C, near the end of the bake cycle. However, it has value in triggering a late release of CO_2, which helps prevent dipped centers or fallen layer cakes.

Sodium acid pyrophosphate is provided in various grades ranging from fast to slow baking reaction rates. Al-

Table 2. Reactions of Leavening Agents

Leavening acid	NV[a]	EV[b]	Reaction
Monobasic calcium phosphate (MCP)	80[c]	1.25[d]	$3Ca(H_2PO_4)_2 \rightarrow Ca_3(PO_4)_2 + 3HPO_4^{2-} + H_2PO_4^- + 7H^+$
Sodium acid pyrophosphate (SAPP)	72	1.39	$Na_2H_2P_2O_7 \rightarrow 2Na^+ + P_2O_7^{2-} + 2H^+$
Sodium aluminum phosphate (SAlP)	100	1.00	$NaH_{14}Al_3(PO_4)_8 \cdot 4H_2O + 5H_2O \rightarrow 3Al(OH)_3 + Na^+ + 4H_2PO_4^- + 4HPO_4^{2-} + 11H^+$
Sodium aluminum sulfate (SAS)	104	0.96	$NaAl(SO_4)_2 + 3H_2O \rightarrow Al(OH)_3 + Na^+ + 2SO_4^{2-} + 3H^+$
Dimagnesium phosphate (DMP)	40	2.50	$MgHPO_4 \cdot 3H_2O \rightarrow Mg(OH)_2 + H_2PO_4^- + H^+ + H_2O$
Dicalcium phosphate · $2H_2O$ (DCP.Di)	35	2.86	$3CaHPO_4 \cdot 2H_2O \rightarrow Ca_3(PO_4)_2 + HPO_4^{2-} + 6H_2O + 2H^+$

[a]Neutralizing value, grams of $NAHCO_3$ neutralized by 100 g of the leavening acid.
[b]Equivalence value = 100/NV; grams of leavening acid to neutralize 1 g of $NaHCO_3$.
[c]NV is 83 for anhydrous MCP, 80 for the monohydrate form.
[d]EV is 1.20 for anhydrous MCP, 1.25 for the monohydrate form.

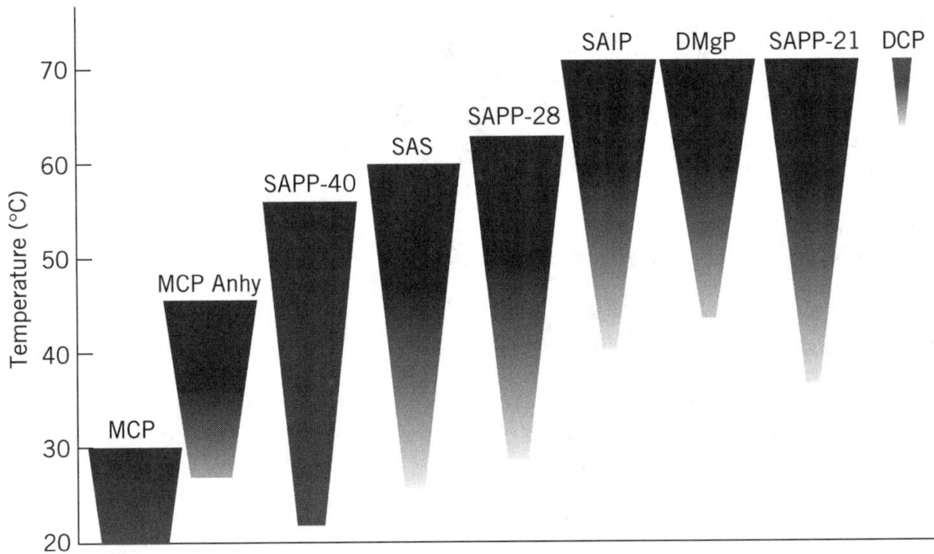

Figure 1. The rate of reaction of various leavening acids as a function of temperature. The width of the triangles indicates the rate of reaction. MCP is the fastest-acting acid, while DCP.Di is the slowest leavening acid. *Source:* Ref. 9 (data for MGP from B. Heidolph, personal communication).

though the basic molecule is the same in all cases, Ca^{++} slows the dissolution rate of SAPP granules. Thus the inclusion of milk in a SAPP-leavened formulation (eg, a pancake batter) delays the time of leavening slightly. The various reaction grades are made by intentionally adding a certain amount of Ca^{++} to the disodium salt during manufacture (adjustments in granule size also control dissolution rate). The grade designations used by Monsanto Chemical Co. range from SAPP-43 (the fastest) through 40, 37, and 28 to 21 (formerly called SAPP RD-1, the slowest). The corresponding grades marketed by Stauffer Chemical Co. (now a division of Rhone-Poulenc) are named Perfection, Donut Pyro, Victor Cream, BP Pyro, and SAPP #4. The slowest grade is used in making refrigerated biscuits and doughs, whereas for other uses (cakes, cake doughnuts, biscuits, and pancakes) a combination of SAPP acids is used to get continuous leavening throughout the baking cycle.

The high reaction rate of MCP may be moderated by coating it with a somewhat insoluble material. This coating dissolves slowly with time, allowing the MCP to then dissolve and form H^+. Instead of 70% of the reaction occurring during the first 10 min after mixing, coated MCP typically gives 20% reaction immediately and then the other 50% over the next 30 min.

The slow-reacting SAlP and DMP leavening acids are useful in pancake and waffle batter for restaurant use, where little or no evolution of CO_2 is desired during the holding period between uses.

Formula Balance. Some of the ions found in leavening acids may influence other properties of certain baked goods. The pyrophosphate ion gives a certain taste that is detectable by some people. Aluminum ion plays a role in developing optimum layer cake structure (15), and SAlP and/or SAS is useful as part of the leavening system. Sodium aluminum sulfate is reported to contribute to rancidity in dry cake mixes stored for a long time. Calcium ion also helps set protein structure and contributes to fine grain in cakes and cake doughnuts. One leavening acid may be selected over another with an equivalent reaction rate just to take advantage of these ion effects.

Developing a formulation with a properly balanced chemical leavening system is partly science and partly art—preplanning followed by trial and adjustment. The selection of timing of the leavening acids, the ratio between the various ones that might be used, the total amount of CO_2 required, and the amount of excess of sodium bicarbonate desired are all factors that interplay with other ingredients and structural functionalities in the baked goods. Achieving the correct leavening balance in a particular formula requires a certain amount of trial and error.

Baking Powder

In the latter part of the nineteenth century, baking powders were developed that combined both parts of the chemical leavening system—baking soda and a leavening acid. The first baking powder contained only one leavening acid, anhydrous MCP, and reacted fairly quickly in cake batters, biscuits, and so on. Soon afterward, sodium aluminum sulfate was incorporated; this gives leavening action during the middle of the baking process, in addition to the early action due to the MCP. This double-acting baking powder was a great success. In Table 3 some typical compositions of baking powders are shown. In addition to the soda and acid, various fillers and calcium salts are also included. The level of potential leavening gas is nearly constant: it is the amount of CO_2 equivalent to 28% sodium bicarbonate in the single-acting powder, and carbon dioxide equivalent to 30% soda in all the double-acting types. Thus these powders are interchangeable from the standpoint of the amount of leavening achieved during baking.

Table 3. Baking Powder Compositions

	Single-acting	Double-acting types					
		Household			Commercial		
Sodium bicarbonate, granular	28.0	30.0	30.0	30.0	30.0	30.0	30.0
Monocalcium phosphate, hydrate		8.7	12.0	5.0	5.0		5.0
Monocalcium phosphate, anhydrous	34.0						
Sodium aluminum sulfate		21.0	21.0	26.0			
Sodium acid pyrophosphate					38.0	44.0	38.0
Cornstarch, redried	38.0	26.6	37.0	19.0	24.5	26.0	27.0
Calcium sulfate		13.7					
Calcium carbonate				20.0			
Calcium lactate					2.5		

Note: Composition given in percentage of each ingredient in the baking powder.

Household double-acting baking powder is based on MCP.H_2O and SAS, whereas the commercial types contain MCP.H_2O and SAPP as the leavening acids. The use of a combination of different grades of SAPP in place of SAS allows better control of the timing of CO_2 release in the oven, which is sometimes an advantage in commercial bakery production.

BIOLOGICAL LEAVENING

The use of yeasts in food processing is prehistoric in origin. Baking, brewing, and wine-making all depend on the ability of yeasts to carry out anaerobic fermentation of sugars, yielding CO_2 and ethanol. In brewing and wine-making, alcohol is the prime product of interest; in baking, the leavening effect of CO_2 and ethanol is more important. In this article only the leavening action of yeast in baked goods is discussed.

Oven Spring

The use of yeast in bread production gives a product containing many small gas cells—leavened bread. By the end of the proofing period the aqueous phase of the bread is saturated with CO_2 and the volume has roughly doubled owing to the pressure of CO_2 that has diffused to air cells or bubble nuclei. During the first part of the bake cycle the loaf expands; then at some temperature the matrix sets, expansion stops, and the rest of the baking time is for the purpose of starch gelatinization, crust coloring, flavor development, and so on. The increase in loaf volume during baking is called oven spring (16). The magnitude of oven spring depends on two factors: (*1*) generation and expansion of gases; and (*2*) the amount of time available for loaf expansion before the structure sets. The first factor is primarily a function of yeast fermentation; the second one is affected by dough components such as shortening, surfactants, gluten protein, and flour lipids.

The influence of several dough components on finished volume has been explored (17,18). A key observation is shown in Figure 2 (18,19). Dough with or without shortening expands at the same rate initially, but at 55°C the structure of the fat-free dough begins to set, slowing the expansion rate. In the presence of 3% shortening, dough continues to expand up to 80°C; the longer time at the initial expansion rate results in a larger final volume. Similar results were found in the presence of 0.6% monoglyceride or 0.5% sodium stearoyl lactylate (17).

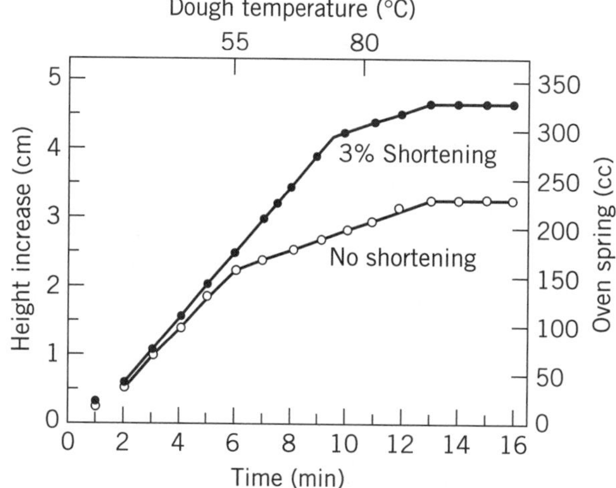

Figure 2. The increase in volume of bread dough in a resistance-heating oven. Doughs contained no shortening or 3% (flour basis) shortening. Initial rate of expansion is the same for both doughs. In the absence of shortening some (as yet unknown) event takes place at 55°C that results in a decreased rate of expansion. This slowdown occurs at about 75°C in the presence of shortening. Both loaves cease all expansion at about 90°C. *Source:* Ref. 19.

Much of the CO_2 formed during fermentation is lost to the atmosphere (20). If the internal gases in a bread were at equilibrium with the atmosphere, there would be no excess CO_2 pressure inside the gas cells. The fact that there is such a pressure (as shown by increased dough volume) is due to inhibition of diffusion of CO_2 from the gas cells to the surrounding air. The causes of this inhibition are not known, but increased viscosity due to water-soluble pentosans and/or the developed gluten phase has been suggested (21). At the end of proof, about 35% of the CO_2 formed during fermentation is present in gas cells, about 5% is dissolved in the aqueous phase, and 60% has been lost to the atmosphere (22). When the solubility of CO_2 in bread dough was measured experimentally, a value of 0.81

mL (equivalent) dissolved per gram of dough was found (21). Others calculated 0.32 mL (equivalent) dissolved per gram of dough, based on the known solubility of CO_2 in water (20).

A mass balance calculation for a pup loaf based on 100 g of flour (173.5 g of dough) found that CO_2 and air occluded during mixing accounted for 370 mL of loaf volume at the end of proofing, and in the oven this gas phase expanded by 62 mL (20). Dissolved CO_2, driven out of the aqueous phase owing to rising temperature, contributed 43 mL to oven expansion, and CO_2 generated by increased yeast activity during early stages of baking amounted to 4 mL. The total volume expansion in the oven from CO_2 and air was 109 mL, but the actual increase in loaf volume was 360 mL. (If the solubility data of Ref. 22 are used, the volume of CO_2 expelled from the aqueous phase would be 109 mL, and the total volume expansion due to CO_2 and air would be 175 mL.)

The difference, 251 mL (or 185 mL, if Ref. 23 data are used), must be supplied by the evaporation of ethanol and water. The amount of dough expansion that could be expected owing to evaporation of water, based on the vapor pressure of water as internal dough temperature rises was calculated (6). This is a slight overestimate, as the aqueous phase of the dough is roughly 0.5 M in salt and 0.3 M in sugar, as well as containing water-soluble proteins, flour pentosans, and fermentation acids. These solutes lower the vapor pressure of the water by roughly 3%.

The amount of ethanol formed during fermentation and proofing usually amounts to 1% of total dough weight (24), or about 2.5% by weight of the liquid phase. About half this ethanol is vaporized during baking. The boiling point of the ethanol azeotrope (95% ethanol, 5% water) is 78°C, so all of it would be expected to vaporize by the end of the bake (although half is apparently trapped and does not escape to the surrounding atmosphere). At 1% of the weight of the pup loaf, 1.7 g of ethanol vaporizes to 830 mL. Vaporization of only half the ethanol is more than sufficient to supply the volume of oven spring over and above that from CO_2 and air.

To summarize, CO_2 contributes about 25% of the gas pressure responsible for oven spring in the standard white loaf considered; expansion of air accounts for about 15%; and ethanol vaporization makes up the rest. The actual magnitude of oven spring depends on other dough components that govern the temperature at which rapid expansion ceases (the foam structure of leavened dough converts to a porous sponge, allowing leavening gases to escape).

Proofing Temperature

Another factor that influences the amount of oven spring is the temperature of proofing. In a study using a sponge and dough variety bread (containing wheat bran) it was found that lower proofing temperatures produced larger final loaf volumes (25). The 1-lb loaves were proofed to a height of 0.5 in. above the pan rim in 50 to 55 min. Additional yeast was added on the dough side to maintain this proof time at each experimental temperature (75–125°F, in 10-degree steps). The bread proofed at 75° had a loaf volume of 3058 cc, while that proofed at 125° had a final volume of 2365 cc. Further, the crumb was progressively more open and overall bread quality was poorer as proofing temperatures increased.

The probable mechanism for this enhanced oven spring was analyzed in some detail (26). Expansion of CO_2 and air accounted for only a small part of the difference. The main contributor to better bread quality is probably twofold: (*1*) improved gluten strength (less weakening effect from yeast cells dying at the higher temperature), and (*2*) a longer loaf expansion time until the foam (dough) converts to a sponge (bread) structure.

BIBLIOGRAPHY

1. J. C. Baker and M. D. Mize, "The Origin of the Gas Cell in Bread Dough," *Cereal Chemistry* **18**, 19–34 (1941).
2. R. C. Junge, R. C. Hoseney, and E. Variano-Marston, "Effect of Surfactants on Air Incorporation in Dough and the Crumb Grain of Bread," *Cereal Chemistry* **58**, 338–342 (1981).
3. N. Chamberlain, "The Use of Ascorbic Acid in Breadmaking," in J. N. Councell and D. H. Homig, eds., *Vitamin C*, Applied Science, London, 1981, pp. 87–104.
4. J. L. Vetter et al., "Effect of Shortening Type and Level on Cookie Spread," *Technical Bulletin of the American Institute of Baking* **6**(10), 1–5 (1984).
5. M. Mizukoshi, H. Maeda, and H. Amano, "Model Studies of Cake Baking. II. Expansion and Heat Set of Cake Batter During Baking," *Cereal Chemistry* **57**, 352–355 (1980).
6. A. H. Bloksma, "Rheology of the Breadmaking Process," *Cereal Foods World* **35**, 228–236 (1990).
7. M. S. Lajoie and M. C. Thomas, "Versatility of Bicarbonate Leavening Bases," *Cereal Foods World* **36**, 420–424 (1991).
8. T. P. Kichline and J. F. Conn, "Some Fundamental Aspects of Leavening Agents," *Bakers Digest* **44**(4), 36–40 (1970).
9. J. F. Conn, "Chemical Leavening Systems in Flour Products," *Cereal Foods World* **26**, 119–123 (1981).
10. H. M. Reiman, "Chemical Leavening Systems," *Bakers Digest* **51**(4), 33–34, 36, 42 (1977).
11. D. K. Dubois, "Chemical Leavening," *Technical Bulletin of the American Institute of Baking* **3**(9), 1–6 (1981).
12. B. B. Heidolph, "Designing Chemical Leavening Systems," *Cereal Foods World* **41**, 118–126 (1996).
13. J. R. Parkes et al., "Method for Measuring Reactivity of Chemical Leavening Systems," *Cereal Chemistry* **37**, 503–518 (1960).
14. C. E. Stauffer, "Precise Leavening," *Baking & Snack* **19**(1), 46, 48, 50, 52, 54 (1997).
15. N. B. Howard, "The Role of Some Essential Ingredients in the Formation of Layer Cake Structures," *Bakers Digest* **46**(5), 28–30, 32, 34, 36–37, 64 (1972).
16. J. C. Baker, "Function of Yeast in Baked Products. Effects of Yeast on Bread Flavor," in C. S. McWilliams and M. S. Peterson, eds., *Yeast—Its Characteristics, Growth, and Function in Baked Products*, National Academy of Science, National Research Council, Washington, D.C., 1957.
17. W. R. Moore and R. C. Hoseney, "Influence of Shortening and Surfactants on Retention of Carbon Dioxide in Bread Dough," *Cereal Chemistry* **63**, 67–70 (1986).
18. W. R. Moore and R. C. Hoseney, "The Effects of Flour Lipids on the Expansion Rate and Volume of Bread Baked in a Resistance Oven," *Cereal Chemistry* **63**, 172–174 (1986).

19. R. C. Junge and R. C. Hoseney, "A Mechanism by Which Shortening and Certain Surfactants Improve Loaf Volume in Bread," *Cereal Chemistry* **58**, 408–412 (1981).
20. W. R. Moore and R. C. Hoseney, "The Leavening of Bread Dough," *Cereal Foods World* **30**, 791–792 (1985).
21. C. E. Stauffer, "Principles of Dough Formation," in S. Cauvain and L. Young, eds., *Technology of Breadmaking*, Chapman & Hall, London, 1998.
22. R. C. Hoseney, "Gas Retention in Bread Doughs," *Cereal Foods World* **29**, 306–308 (1984).
23. G. E. Hibberd and N. S. Parker, "Gas Pressure–Volume–Time Relationships in Fermenting Doughs. I. Rate of Production and Solubility of Carbon Dioxide in Dough," *Cereal Chemistry* **53**, 338–346 (1976).
24. J. W. Stitley et al., "Bakery Oven Ethanol Emissions. Experimental and Plant Survey Results," *Technical Bulletin of the American Institute of Baking* **9**(12), 1–11 (1987).
25. K. Siffring and B. L. Bruinsma, "Effects of Proof Temperature on the Quality of Pan Bread," *Cereal Chemistry* **70**, 351–353 (1993).
26. C. E. Stauffer, "Low-Temperature Proofing," *Baking & Snack* **19**(9), 50, 52–55 (1997).

Clyde E. Stauffer
Technical Food Consultants
Cincinnati, Ohio

BAKERY SPECIALTY PRODUCTS

The main items produced by commercial bakeries, in terms of total tonnage, are bread and rolls, cakes, cookies, and crackers, with bread (either pan bread or hearth-baked types) representing the largest part of this production. Over many centuries many other types of food products, using wheat flour as their main ingredient, have been developed. This article discusses several of these products that are available to consumers today.

NATIVE FLAT BREADS

The earliest form of wheat bread is made by grinding cereal grain(s), adding water, mixing to a coherent mass, and allowing this mass to rest and ferment for a period of time. Then small portions are taken, flattened into a sheet, and baked quickly on a hot surface, usually either the floor or the wall of a heated oven. The process is usually a sourdough type; that is, a portion of the fermented dough is retained at the end of the day's baking and mixed in with the next day's dough, thus inoculating it with the yeast and other fermentative microorganisms that produce ethanol, carbon dioxide, and organic acids as they metabolize and grow.

These breads are produced widely throughout the Middle East and the Indian subcontinent. Depending on details of formulation and production, they are called chappati, nan, roti, tamouri, tamees, korsan, shamuf, and several other names. The simplest formula consists of whole wheat meal (frequently stone-ground by hand in the kitchen), salt, water, and sourdough starter. At the other extreme, the dough also contains ground pulses, sesame seeds, shortening (either ghee or sesame seed paste), honey or other sweetener, and spices (anise, cardamom, curry powder, dill, etc). Although wheat is the major cereal grain used, ground rice, barley, maize, or sorghum are also sometimes included at up to 20% of the weight of wheat meal or flour.

Production may be either in the home or in small commercial establishments. Home production is done strictly by hand. Commercial bakeries have a mechanical slow-speed mixer and perhaps a set of motorized sheeting rollers. The oven is usually heated by building a wood fire inside it, raking out the coals when it is hot, and baking on the floor (hearth) of the oven. In the home, ovens have been seen in the shape of an urn, about 3 ft high with an 18-in. opening in the top and a small opening in the bottom for feeding the fire. When the sides of the oven are hot, the sheeted dough is inserted through the top and plastered (by hand) against the oven wall. Dexterity and practice is required for this, because the oven temperature approaches 500°F. The baked bread is removed with wooden tongs. In the commercial hearth oven, insertion and removal is performed using a wooden peel.

As is common practice in most parts of the world, the bread is produced, purchased, and consumed on a daily basis, so staling and development of mold is not a problem. The meat and vegetable dishes, in the countries where flat breads are common, tend to be stews and purées, and freshly baked flat bread is an ideal and flavorful accompaniment to such a meal.

A similar sort of bread called tortilla, based on maize, is common in Latin American countries and recently in the United States. The maize (corn, in U.S. terminology) is soaked in dilute lime (calcium hydroxide) for up to 24 h to remove the hull. The endosperm is ground, while wet, to make masa (dough), then shaped into balls, flattened, and baked as described above. These differ from the flat breads described previously in the way in which the grain is processed and in having no fermentation step. The leavening of the tortilla, such as it is, is by steam. Tortillas can also be made using wheat flour. Commercial flour tortillas are usually leavened by the addition of a small amount of sodium bicarbonate. Commercial production of tortillas, to accompany Mexican food specialties, incorporates a number of refinements; space does not allow a full discussion of these factors in this article.

PITA BREAD

Another type of Middle Eastern bread, related to the flat breads, is pita (pocket, balady, burr) bread. A wheat flour is used, with an extraction rate of 75 to 82%, depending on the specific type of bread being made. The strength of this flour is greater than that of the flour (or whole wheat meal) used in flat breads. Usually the bread is made by a sourdough process, but with a rather large (20%) portion of sourdough, so that fermentation is more rapid than with the usual flat bread. (Pita doughs can be made adding yeast to each dough, as for regular bread production.) The absorption is somewhat greater than for regular bread dough. The dough is allowed to ferment for about 1 h after mixing, then it is divided into pieces weighing about 3 to 4 oz each. These receive a short intermediate proof, are flattened into circles about 1/2 in. thick and 6 to 9 in. in diameter, and allowed to proof for 30 to 45 min.

The proofed pieces are baked for a short time (1.5–2 min) in an extremely hot oven (700–900°F). The high oven heat causes formation of a top and bottom crust almost instantaneously after the piece is inserted. As the heat penetrates the dough the ethanol and water rapidly evaporate and the vapors, trapped by the crust, cause the bread to balloon. As the bread cools, after removal from the oven, the internal pressure dissipates and it collapses to the familiar round, flat shape, but with the internal "pocket" still present.

PIZZA DOUGH

Pizza is basically an open-faced sandwich made on a leavened flat bread, in which the topping and the bread are baked together. It has become a popular baked food worldwide, and although the familiar tomato sauce/sausage/cheese topping is the most popular, fish, ground lamb, purėed legumes, and a wide variety of chopped vegetables are also used. The possible variety of pizzas is limited only by the availability of ingredients and the ingenuity of the chef.

Pizza crusts are of two types—thin cracker-type crusts and thick bread-type crusts—and are mixed shortly before use or are mixed and retarded in a commissary for later use. These variations lead to differences in dough formulas and handling. The flour used for pizza dough should be a good grade of bread flour with 12.0 to 12.5% protein. The thin-crust dough for immediate use requires even more protein, so rather than carry a 14% protein flour for this dough and a 12% flour for thick-crust doughs, it is more convenient to add 2 to 3% of vital wheat gluten to the 12% flour. Gluten generally is not used in commissary doughs unless flour of an adequate strength is not available. The gluten protein forms a seal at the crust/topping interface, minimizing formation of a soggy layer during baking.

Dough for immediate use receives little fermentation time beyond a 10 to 15-min floor time. It should be somewhat undermixed (just beyond the "cleanup" stage) and quite warm (33–38°C, 90–100°F). These doughs often resist sheeting and tend to shrink back in the oven. To counteract this effect, a reductant, either L-cyteine or sodium bisulfite, is used. Cornmeal also reduces dough elasticity. Adding it does away with the need for reductants and improves crust color, flavor, and eating properties. Sodium stearoyl lactylate improves the crust volume of thick-crust pizzas, giving closer grain and a more tender crust.

The variations in water absorption between the two types of crusts are consonant with the descriptions of cracker-type crust and bread-type crust. If absorption is too high, the dough is sticky, whereas if it is too low, the dough tears when it is sheeted.

Commissary pizza doughs are mixed, divided, rounded, and immediately placed in the retarder at 5 to 8°C (40–45°F). The dough ball should double its volume in 24 h at this temperature. If the volume increase is greater than this, the dough temperature (usually around 25°C, 78°F) should be reduced; if the volume increase is less than double, mix the doughs to a warmer temperature. Slight adjustments in yeast levels may be made to get the amount of fermentation to the desired point, but too much yeast in the dough may result in large blisters on the crust when it is sheeted and baked in the pizza shop.

Frozen pizza is a common supermarket food item. The consumer does not thaw the pizza and allow it to proof; rather, the piece is simply placed directly in the oven and baked. For this application, chemical leavening rather than yeast is used to give the desired texture to the crust. The leavening is of the double-acting baking powder type.

BAGELS

Bagels supposedly originated in Vienna, during the period (late seventeenth century) when the city was fighting off the Turks. Jan Sobieski, the king of Poland and a famous horseman, was a major factor in the defeat of the Turks, and a hard roll in the shape of a stirrup (bügel in German), immersed briefly in a boiling-water bath before being baked to preserve the shape, was developed in his honor. In later times the shape was simplified and the name corrupted to bagel, but the same production procedures are still followed.

Bagels traditionally are made from a lean, low-absorption straight dough, using a high-protein (ca 13.5–14%), clear, hard-red spring-wheat flour. In recent years numerous adjustments have been made to the basic formula, in the direction of making the bagel richer (eg, adding whole eggs), softer (by the addition of shortening or enzymes for longer shelf life), or sweeter (by adding honey and/or raisins). The main feature that separates bagels from other bakery items (besides the short boiling-water treatment) is the type of fermentation schedule used.

A combination of proofing (35–40°C, 95–105°F) and retarding (5–10°C, 40–50°F) is used to obtain the desired product volume as well as the internal and external characteristics. If the bagels are overproofed, they collapse in the oven; if they are underretarded, they tend to ball, that is, expand and close the center hole. Also, tiny blisters on the surface of the finished bagel are considered desirable. These blisters are due to the presence of lactic acid formed during retarding, and if they are absent (but the bagels are otherwise acceptable), 0.1 to 0.5% lactic acid may be added to the dough.

The sequence and timing of the fermentation steps is open to experimentation. Proofing may be done before retarding, after retarding, or both. Retarding times may be as short as 8 h or as long as 24 h. Specific process parameters must be worked out by the baker, depending on the formula and equipment being used, to make the product that is most acceptable to customers.

Just before baking, the proofed bagels are placed in a boiling-water bath; they float on the top of the bath and are boiled for about 1 min, being flipped halfway through. This treatment gelatinizes the starch on the surface of the bagel and causes the formation of a hard, shiny crust when the bagel is baked. The usual baking procedure is directly in a hearth oven, although they also may be baked on sheet pans. Various toppings (poppy seed, sesame seed) are applied to the boiled (and still moist) bagel; other flavorings such as dried onion are mixed in with the dough.

PRETZELS

Considered to be the world's oldest snack foods, pretzels reputedly originated around the year 600 in a monastery in the Italian Alps. Supposedly, the monk in charge of the bakery formed strips of bread dough into a shape intended to resemble arms folded across the breast in prayer. When baked, these pretiola (little reward in Italian) were given to children who learned their prayers properly. Later, these baked bread snacks became popular north of the Alps, where the name was germanicized to bretzels, a spelling still used by some pretzel bakers.

Soft pretzels, presumably identical with those made by the Italian monk, are still popular today. They have a moisture content similar to that of bread and a short shelf life. Hard pretzels were supposedly discovered when a baker's apprentice forgot to remove the last trays of pretzels from the oven at the end of the day, and the overnight residence in the cooling oven dried them, giving them a hard, crisp texture. The rest of this discussion concerns itself with the low-moisture (2–3%) hard pretzel.

Pretzel dough is extremely stiff, made with a low-protein flour (that from soft white winter wheat is preferred), and only 38 to 42% water. The only other ingredients are yeast (ca 0.25%) and roughly 1% each of salt, shortening, and dry malt. This dough is mixed in a horizontal sigma-arm mixer, allowed to ferment for up to 4 h, then processed.

The dough is divided into small pieces, elongated into a thin strip, and twisted into the typical pretzel shape by a special machine. The formed pretzels receive a short intermediate proof, then are carried through a caustic bath, which contains 1.25 ± 0.25% sodium hydroxide, held at about 190°F. After exiting this bath, salt is sprinkled on the tops and the pretzels enter the oven. Here they are baked at about 450°F for 4 to 5 min. The caustic bath gelatinizes the starch on the surface of the pretzel, giving the crisp, shiny crust. In the oven all residual caustic is converted to sodium bicarbonate. The baked pretzels are then piled onto slower-moving conveyors that carry them through a drying stage, where they remain for 30 to 90 min (depending on the size of the piece) at a temperature of approximately 250°F. This treatment slowly reduces the moisture to about 2% and also allows equilibration so that the finished pretzel does not check (form small cracks on its surface).

Variations on this production scheme are many. A machine is available that extrudes the dough through a die already shaped in the twisted form, so tying is not necessary. By changing the die the machine will also make straight rods, which are a popular form for this snack. Also, the exact dough formulation and extent of fermentation varies among manufacturers, the details being closely held as proprietary information, each company sure that its particular formula makes the best possible pretzel. As with any low-moisture snack, the shelf life is governed by the quality of the packaging; the loss in consumer acceptance is directly related to moisture uptake, with about 3.5% moisture marking the upper limit of acceptability.

ENGLISH MUFFINS

The nomenclature of this item is anomalous, as it was developed in the United States; the term *English* applied to them is a highly successful marketing ploy. They are a disc-shaped baked product, usually around 3 to 4 in. diameter and about three-quarters in. high. A good general description is: "Good eating muffins are relatively tough, chewy, and honeycombed with medium to large size holes (1/8–1/4 in. diameter). Flavor is bland and somewhat sour. Side walls are straight and light colored. The edge between the sides and the flat, dark brown top crust is gently rounded, not sharp." They are baked in a covered griddle cup on a highly automated line. The main features of importance are that the dough must be soft so it flows to fill the cup, and the formulation must lead to the formation of the large internal holes.

A typical English muffin dough contains 83 to 87% water (flour basis equals 100%), 2% sugar, 1.5% salt, 5 to 8% yeast, and 0.5 to 0.7% calcium propionate. Vital wheat gluten at 1 to 2% is sometimes added to increase the chewiness of the finished muffin. At this absorption level the dough is very soft, and it must be cold (20°C, 68°F) when taken from the mixer so that it does not stick when it is divided, rounded, and deposited from the intermediate proofer into the griddle cups. Dusting material, usually a blend of corn flour and cornmeal, is used rather liberally (3–4% of the dough weight) to facilitate the various transfers.

Fermentation of the dough piece after it is divided and rounded is rather short, usually about 30 min. Three factors—short time, cold dough, and high calcium propionate levels—all tend to retard fermentation, so a rather high yeast level is required. Leavening in the griddle cups is due to CO_2, ethanol, and steam; in addition, some bakers add 0.5% baking powder to get an early kick in the cups so that the mold is quickly filled by the dough. The finished English muffin has a rather high moisture content (about 45%) and is prone to mold formation. Calcium propionate is used at the elevated level mentioned to help overcome this problem, and it is also added to the dusting flour. Another method is to spray the muffin with a potassium sorbate solution as it leaves the oven. If this is done, the propionate may be omitted from the dough, which in turn allows a decrease of about 1% in the amount of yeast.

In some instances the internal grain porosity is still less than desired, and the addition of protease enhances the openness of the crumb. This additive also increases pan flow, so absorption might be reduced by 1 to 2%, which might or might not be desirable. Sometimes a tangy, acidic taste is wanted (sourdough muffins), and a dry sour base or vinegar (acetic acid) is added to the dough, The increased acidity (lower pH) of the finished product enhances the antimold action of calcium propionate. Another popular variety contains raisins, added in amounts equal to 25 to 50% of the flour weight.

DOUGHNUTS

Doughnuts are fried in deep fat rather than baked in an oven as are most other bakery products. They are related to another product, fritters, which are made with a soft, rich dough containing pieces of fruit (apples) or vegetables (corn), dropped by spoonfuls into hot fat. An apocryphal story relates that a nineteenth-century Yankee ship captain loved apple fritters but had trouble holding onto them when at the ship's wheel during high seas. His cook, anxious to please, got the idea of making the fritters with a hole in the center so the captain could stick them on the handspokes of the wheel during busy moments. Although there is a certain implausibility about this story, the fact remains that doughnuts, in one form or another, are extremely popular sweet snacks or desserts.

There are three basic types of doughnuts: cake doughnuts, yeast-raised doughnuts, and French crullers. Cake doughnuts are chemically leavened, yeast-raised doughnuts use a sweet yeasted dough, and French crullers are steam leavened. These will be discussed separately.

Cake Doughnuts

Batters for this product are a leaner, lower-sugar version of layer cake batters. A typical regular cake doughnut mix contains (on 100% flour basis) about 40% sugar and 13% nonemulsified shortening. Other ingredients frequently found in doughnut mixes are defatted soy flour, nonfat dry milk, potato starch, and dried egg yolk. These ingredients give a tender eating quality to the finished product, as well as limiting fat absorption during frying. The protein ingredients decrease fat absorption, while potato starch helps retain moisture in the finished doughnut and gives a longer shelf life. Lecithin is sometimes used to enhance the wetting of the dry mix during batter preparation, and occasionally monoglyceride is added to make a more tender final product. Other emulsifiers are not used, as they increase fat absorption without conferring any quality enhancement.

Most cake doughnuts are produced by extruding (or depositing) the batter from a reservoir directly into the hot fat fryer. Bench-cut doughnuts are made using the same recipe, but with a reduced amount of water. This gives a dough (rather than a batter) that can be rolled out on the workbench and cut into rings using a doughnut cutter. This process takes much more hand labor than depositing and is usually used only in home baking or in small, traditional doughnut shops.

In commercial bakeries and most doughnut shops, doughnut batter is deposited through a cutter into the frying fat. Proper batter viscosity is extremely important to attain uniformity of flow through the cutter and uniform spread in the fryer. Batter viscosity is governed by the amount of water used in mixing (usually about 70% based on flour, or 38–44% based on total dry mix weight) and the way that water is partitioned between flour, sugar, soy flour, and the other ingredients present. Sometimes high-viscosity gums (guar, locust bean, sodium carboxymethylcellulose) are added to the dry mix at levels of 0.1 to 0.25%. The gum helps give uniform viscosity from batch to batch and also acts as a water binder that reduces fat absorption and lengthens shelf life.

Leavening in cake doughnuts is by 1.5% soda (flour basis) and 2.1% sodium acid pyrophosphate (SAPP). The fast-reacting grades of SAPP are required, from SAPP-37 to SAPP-43 (the fastest grade). When the doughnut is first deposited in the hot (190°C, 375°F) frying fat, the ring of dough sinks and is supported on a bottom plate. Leavening begins, and after 3 to 7 s the doughnut rises to the surface and frying of the first side commences. As heat penetrates the dough piece it slowly expands; after the first side is fried, the doughnut is flipped and the second side is fried. During this time expansion continues, but the diameter of the doughnut has been fixed because the first side has set (the starch has gelatinized and proteins have denatured). Thus, second-stage expansion tends to be into the center hole and the typical star is formed (or the doughnut balls and fills the center hole if conditions are incorrect).

When the finished doughnut is cut crosswise, three regions are readily seen. The central core is surrounded by a more porous leavened ring, but this porosity tends to be greater on the second side than the first. If the core is inadequately leavened, it will be dense and gummy, and the doughnut will not have a good eating quality. Part of the art of the doughnut formulator is in choosing leavening acids that strike the best compromise to achieve moderate porosity of the two sides along with at least some opening up of the core.

Yeast-Raised Doughnuts

These doughnuts are made using a rather lean, sweet dough. After full development of the dough, it is sheeted, cut into rings, and the doughnuts are proofed for 30 to 45 min. The raised doughnuts are then introduced into the fryer, where they are fried for about 1 min on each side. (Because they are already leavened, they float on the fat throughout the frying cycle.) They are then removed, the surface fat is drained, and the fried doughnuts are (most often) glazed with a simple sugar/water glaze. If they are to be iced or otherwise coated, they are usually cooled before these operations are carried out.

Yeast-raised doughnuts can be made in many different shapes. In addition to rings, the dough is also frequently cut into strips and twisted into various fanciful shapes before proofing and frying. In the form of rectangles about 2 by 4 in., they are iced and topped with chopped nuts or fruit to make Long Johns. In the form of discs, the fried piece can be filled with jelly or custard. This product, filled with raspberry-flavored jelly, is known as a Berliner in Austria and Germany. (This explains why the Germans roared with laughter when, in 1963, President Kennedy proclaimed, "Ich bin ein Berliner." He was telling them that he was a jelly-filled doughnut.)

French Crullers

These doughnuts are made using eclair dough (see the next section). The dough is deposited in a ring shape onto a sheet of parchment paper, and after a few moments to allow it to set, the parchment is dipped into the frying fat and the rings of dough are loosened from the paper. The dough can also be deposited using a specially shaped cutter that produces a fluted surface on the ring of dough. The dough piece expands to a much greater extent than does cake doughnut batter, and the finished piece is much less dense than other doughnuts. It is frequently glazed with a butter/rum flavored glaze.

ECLAIRS

The dough for eclairs and French crullers (*paté de choux*) is a paste made in a rather unusual procedure. Water (two parts) and shortening (one part) are placed in a kettle and brought to a rolling boil. After the fat is melted, a pinch of salt is added, flour (1–1.5 parts) is added, and the mixture is stirred briskly. The gelatinized flour paste is cooled to about 150°F, and two parts whole egg are added in several increments, with continuous stirring. A small amount of ammonium bicarbonate, dissolved in one-half part milk, is stirred in. The final paste should be soft enough to deposit from a pastry bag, yet firm enough to hold its shape once deposited. If it is too stiff, it may be softened by the addition of a small amount of milk or water.

The finished dough is deposited on sheet pans lined with parchment paper, then baked in a hot (425°F) oven until the eclairs are crispy. Leavening action is due to the formation of steam in the interior of the piece. If inadequate leavening occurs, a small amount of ammonium bicarbonate (no more than 1% of the weight of the water used) may be dissolved in a little water and mixed in at the end of the dough-making procedure. The finished piece should be crisp on the outside, hollow on the inside, and with no soggy layer on the interior wall.

Eclairs are usually filled with a chilled custard, either plain or a French custard (butter-flavored). They are then iced, with chocolate being the most popular flavor, or dusted with confectioners' (very finely powdered) sugar. To prevent bacterial growth, custard-filled eclairs should be refrigerated until served.

LAYERED DOUGH PRODUCTS

Numerous bakery products have a unique texture, in that the baked dough is present in many thin layers rather than one continuous matrix. This effect is obtained by placing a layer of fat between two layers of dough, rolling out this sandwich to a thin sheet, folding the sheet over onto itself into three or four layers, and repeating this rolling and folding process two or three more times. If done properly, the final sheet consists of 50 to 100 layers of dough interspersed with an equal number of layers of fat. When baked, the dough layers set up more or less independently, giving the flaky texture of the finished product.

Puff Pastry

Puff pastry dough is used for a large number of culinary items, from basic pastries such as fruit turnovers and napoleons, to patty shells for holding a variety of fillings for appetizers, to dough for enclosing fish or meat, such as beef Wellington. It needs no fermentation but does require some care in the roll-in process to obtain a flaky final product.

The dough for this product is a flour/water dough (no yeast), containing about 6% shortening and 6% whole eggs. After it is fully developed it is sheeted out, fat is spread over two-thirds of the dough, which is then folded to give a sandwich having three layers of dough separated by two layers of fat. This is sheeted out and folded as already described. Usually the dough is refrigerated after the second sheeting and folding, to keep the fat in the plastic state. If the dough becomes too warm, the fat melts and gets incorporated into the dough itself, and the layering effect is lost. The fat usually used is a special puff-paste margarine, having a broad plastic range, and containing about 15% water. In the oven the water (which is finely emulsified in the fat) forms steam, which helps give the puff to puff pastry. Thus, puff pastry is a steam-leavened product.

Croissants

Croissants are another legacy of the siege of Vienna. The story is that the Turks were busy digging a tunnel under the city walls but were overheard by a baker in the early hours of the morning, as they were digging under his shop. He alerted the authorities, and a countertunnel foiled the plot. After the siege was lifted, the baker made rolls in the shape of a crescent (the emblem on the Turkish flag), which were sold to celebrate the victory, and which became a Viennese tradition. Marie Antoinette introduced these Kipfeln to the French court, and they became popular under the French name croissant.

A typical croissant dough is basically a straight dough, containing 10% sugar, 1% salt, and 4% margarine (flour basis), with a fairly high (9%) yeast level and slightly below normal absorption. Croissant dough is mixed cold (18–20°C, 64–68°F). This practice, together with the slight underabsorption, retards fermentation. Proofing temperatures are also relatively low, not exceeding the melting point of the fat (ca 37°C, 98°F). For these reasons a rather high level of yeast is used in the dough. Numerous variations on the basic mixing, sheeting, and roll-in process are used, most of which seem to be based more on tradition than on actually yielding superior product. After mixing, the dough is allowed to ferment for up to 1 hr, then the fat is rolled in, essentially as described for puff pastry. The traditional fat for croissants is unsalted butter, although many commercial companies today use a bakers' margarine that does not get as hard as butter at retarder temperatures and so spreads more uniformly during the sheeting and folding process. The usual fat-to-dough ratio is about 25%, that is, 1 lb of butter to 4 lb of dough.

The rolled-in dough is sheeted out, then cut into triangles having a height about twice the width of the base. This triangle is rolled up, starting from the base, the shaped dough piece is placed on a sheet pan, and the ends are pulled down to give the typical crescent shape. After proofing for 1 to 3 h, it is baked and cooled. Care must be taken that the piece is not underproofed, otherwise the finished product loses its flakiness and is tough, chewy, and unpalatable. Croissants are frequently sliced horizontally and used to make sandwiches, with a wide variety of fillings being offered by various specialty shops.

Danish Pastry

Danish dough is a rich, sweet, yeasted dough made with a higher than usual absorption. The dough is quite soft, mixed cold (62–64°F), and not fully developed in the mixer. Gluten development takes place during the roll-in process. The dough is divided into pieces of a suitable size (eg, 16 lb) and then retarded for several hours. Some fermentation takes place during this time. The cold dough is then sheeted out and fat is rolled in, as already described. Specialty shops use butter or bakers' margarine for the roll-in fat, but all-purpose shortening works equally well. The dough sheet is usually given the first two sheetings and folds, then returned to the retarder for 6 to 12 h. It is then given one more sheeting and three-fold, and returned to the retarder for 24 to 48 h for further fermentation.

Danish pastry is almost invariably made by sheeting out the rolled-in dough, cutting into various shapes, then adding some sort of sweet filling (fruit, nuts, cinnamon, etc) and folding or rolling up the dough to retain this filling during proofing and baking. The variety of shapes and fillings is limited only by the ingenuity of the baker. The proofed pieces are sometimes washed lightly with diluted egg just before baking, to give a shine to the finished piece. Alternatively, the piece may be sprayed with a light sugar syrup as it exits the oven, to give the same effect. If a Danish coffee cake, for example, is to receive an icing, this is usually a rather thin icing that is drizzled over the top of the coffee cake in a random pattern, rather than applied in a solid layer (as on sweet rolls). A good Danish pastry is one of the foundation stones of any small retail bakery today.

PIES

The term *pie* (like biscuit) means something rather different in the United States than it does on the other side of the Atlantic Ocean. In Europe a pie usually contains meat, potatoes, and other vegetables and may have a crust of puff pastry, of mashed potatoes, or of a flour dumpling. It is normally the main dish of a meal. In the United States, on the other hand, a pie is for dessert, or is a sweet snack. It need not be a baked food; for example, ice cream pies are now popular. In at least one instance (Boston cream pie), pie is more properly a filled, layered sponge cake. Having said that, this section discusses pies primarily in terms of the traditional one-crust or two-crust baked American dessert pie, and, briefly, the fried pie, a related, traditional American dessert.

A pie crust is one of the simplest of baked doughs, comprising only flour, shortening, salt, and water. The ability to combine these few ingredients properly to produce a flaky, tender pie crust, however, is rare, as attested by the number of leathery, soggy, nearly inedible pie crusts that are served to diners in U.S. restaurants every day. The first step in making a pie crust is to cut the shortening into the flour; that is, mix the two using a fork or similar mixing tool, reducing the particle size of the shortening to small lumps, perhaps one-quarter the size of a pea. Then a measured amount of cold water (even ice water, in hot weather), with the salt dissolved in it, is added, and the dough is gently mixed until the flour is wetted and the dough forms a lumpy, slightly sticky ball. The dough should then be allowed to rest for a time, preferably in a refrigerator, to allow the water to equilibrate with the flour protein.

An amount of this dough just sufficient to form the crust (either top or bottom) is then rolled out into a circle. An absolute minimum of dusting flour is used, and it is best to use cloth sleeve on the rolling pin and a flat cloth on the bench top, both of which have been rubbed with flour and the excess flour removed. Once the desired diameter of crust is achieved, the circle of dough is loosely rolled up on the rolling pin and transferred to the pie pan, where it is gently shaped and the edges sealed (if it is the top crust, and the filling is in place).

The main mistake in making a flaky, tender pie crust lies in excessive working of the dough and consequent development of the gluten in the flour. This can happen if too much water (or water that is warm) is added to the dough, and it is mixed until the water disappears. Also, excessive dusting flour, and rolling the dough more than once, will contribute to gluten formation. In large-scale pie production (either in a commercial bakery, or in an institutional setting) the dough just described may be too tender for the kind of handling it receives, and a certain amount of gluten development will be done on purpose, in the interests of production efficiency. Although this speeds up the operating line, it produces the leathery crusts that are abandoned on dessert plates across the country.

Fillings for pies are usually either some kind of fruit in a starch-stabilized juice matrix or a custard. The formulation of a good pie filling deserves an entire article to itself and will not be explored here. However, the stabilizing system should be such that during baking the filling will not boil or foam excessively, to the point where it overflows the side of the pie. If the pie crust at the filling/crust interface does not bake rapidly enough, it will soak up moisture from the filling, and the final crust will be soggy. This may be due to using a cold filling in the pie or having too low an oven temperature.

One type of pie avoids this pitfall, by adding the filling after the bottom crust is baked. An example is strawberry pie; a single crust is baked in the pie pan, then the fruit is placed in the cooled crust and a strawberry-flavored starch glaze is poured over the top. The pie is refrigerated until served.

Fried Pies

These desserts are traditional in the southeastern part of the United States, although they are now produced commercially and sold throughout the country. A pie crust with a slightly developed gluten is used. A circle is rolled out, fruit filling is placed on one half the circle, the other half of the crust is folded over, and the edges are sealed. This is then fried in a half inch of hot fat, in a heavy skillet. In the commercial version, the shaped and filled piece is fried in a deep fat (doughnut) fryer. A top conveyor belt keeps the pie submerged throughout its journey through the hot fat. The crust absorbs some of the frying fat, tenderizing it so that it is not leathery as it would be if baked. The commercial fried pie is given a sugar glaze before being cooled, wrapped, and dispatched to the consumer. The traditional fried pie, on the other hand, is usually eaten while still warm, fresh from the kitchen, and any glaze or other topping would be superfluous.

GENERAL REFERENCES

Technical Bulletins of the American Institute of Baking: *Bagels*, **8**(11) 1986; **10**(4), 1988; *Croissants*, **8**(10), 1986; *Donuts*, **1**(6,7) 1979; **2**(7), 1980; **4**(9), 1982; *English Muffins*, **1**(1), 1979; *Pizza Dough*, **1**(11), 1979; **8**(12), 1986; *Chemically Leavened Pizza Dough*, **19**(11), 1997.

A. Matz, *Bakery Technology and Engineering*, AVI Publishing, Westport, Conn., 1972.

E. J. Pyler, *Baking Science & Technology*, 3rd ed., Sosland Publishing, Merriam, Kans. 1988.

W. J. Sultan, *Practical Baking*, 5th ed., Van Nostrand Reinhold, New York, 1990.

CLYDE E. STAUFFER
Technical Food Consultants
Cincinnati, Ohio

BAMBOO SHOOTS

FAMILY AND RELATED SPECIES

Bamboo belongs to the family Gramineae. The emerging shoots of several species, of the genera *Dendrocalamus, Bambusa, Phyllostachys*, and others are used as a vegetable (1).

PRACTICAL DESCRIPTION

The members of the family Gramineae are tall woody shoots of perennial grasses. The reproductive organs resemble the grasses, particularly the oat (*Avena sativa*). The anthers are borne on long filaments resembling those of corn (*Zea mais*) (2).

ORIGIN, DISTRIBUTION, AND PRODUCTION

The majority of the edible bamboos originated in China, Japan, and Southeast Asia. Some are wild and others are cultivated. The use of bamboo shoots outside of this area is restricted (1). The important edible species include the tropical clump bamboos, *Bambusa oldhami* Nakai and *Dendrocalamus latiflorus* Munro (Fig. 1), and the spreading bamboos. *Phyllostachys edulis* (Fig. 2), *P. pubescens*, and *P. makinoi*, which are not confined to the tropics.

The planted area of edible bamboos in Taiwan is ca 30,000 ha with an average yield of 12 t/ha. The annual production of bamboo shoots in 1989 was 401,152 t (3).

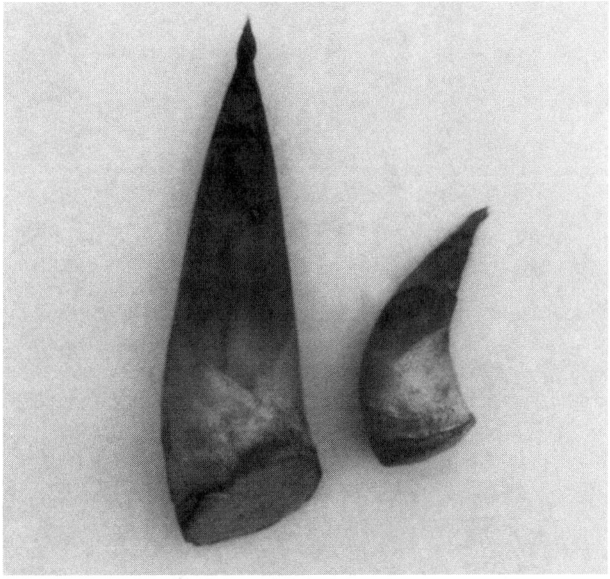

Figure 1. Shoots of *Bambusa oldhami* (right) and *Dendrocalamus latiflorus* (left).

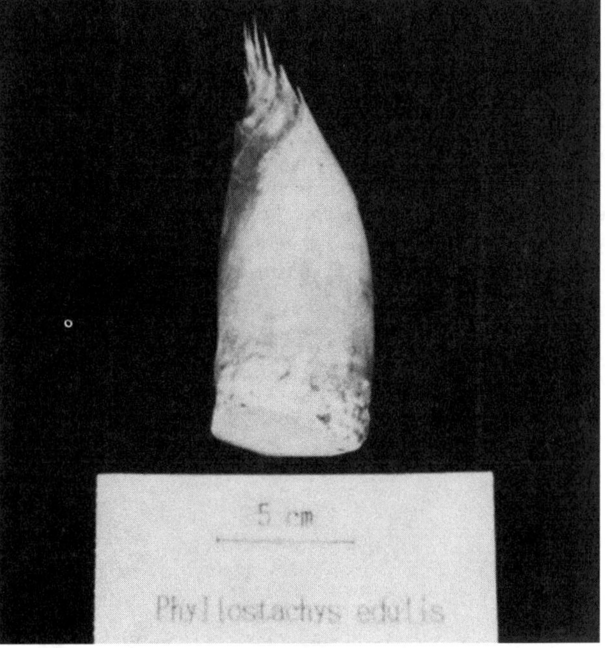

Figure 2. Shoot of *Phyllostachys edulis.*

CULTURE

The cultivation of bamboo has been practiced in the Orient for thousands of years. Bamboo is propagated by the transfer of clumps or rhizomes as shown in Figure 3 (4). Some bamboos can be propagated from culm cuttings. In the life cycle of a clone, the bamboo flowers once, and the top dies. Parts of the same clone, even when transferred to different areas and conditions, will flower at the same time (2). Very few bamboo flowers produce viable seeds, but the collection of seeds is important for the breeding of an entirely new seedling, which can start a new life cycle of 50–60 yr (5).

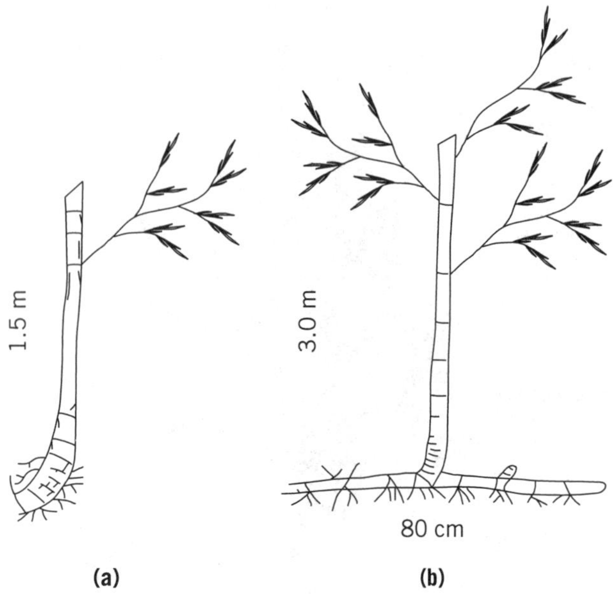

Figure 3. Rhizomes transferred for propagation of bamboos (**a**) *Bambusa* sp, or *Dendrocalamus* sp; (**b**) *Phyllostachys* sp.

HARVEST

The harvest of bamboo for its edible shoots is an ancient and highly specialized act (6). The grower must know what proportion of the young shoots may be cut and when to cut without endangering the vigor of the parent plant. The usual practice is to cover and mound up the clumps with soil, rice hull, rice straw, bamboo leaves, etc so that the emerging shoots are in the dark. Polyethylene (PE) or poly(vinyl chloride) (PVC) sheets have recently been used on top of the mounds (7). Exposure of shoots to light causes greening and bitterness and these shoots are mainly used for processing. The shoots are ready for cutting when the tips are just emerging from the surface of the soil in the mound, if left much later they will become woody. First the mud is removed by hand and then the shoot is cut off with a sharp knife or spade designed for the purpose (Fig. 4) (5).

In Japan, bamboo shoots are forced by growing under mulch heated with electric cables placed 6–8 cm below the soil. About 2–3 cm of rice straw are placed on the soil surface, an additional 4–5 cm layer of soil is put on top, and finally plastic sheeting is laid on top of all. The soil temperature is kept between 13 and 15°C. By the mulching procedure, bamboo shoots can be harvested almost a month earlier than those not mulched (2).

STORAGE

Harvested bamboo shoots are highly perishable with a respiration rate over 100 mg CO_2/kg·h at 20°C (8). The fiber content increases quickly from the cut end toward the tip. The quality and yield of canned bamboo shoots depend greatly on the holding condition of raw materials before processing (9,10). Rapid precooling and storage at low temperature and high humidity conditions can prevent quality deterioration and extend the shelf life of the bamboo shoot from 1–2 days to 7–10 days (11). In Taiwan, harvested bamboo shoots are washed and packed in plastic or bamboo baskets for nearby markets or in PE-lined cartons, which are shipped to the wholesale fresh markets in Taipei.

PRODUCTION SEASON

In Taiwan markets shoots of *B. oldhami* Nakai are available from May to early October. They are called green bamboo shoots and are sweet and tender. Green bamboo shoots

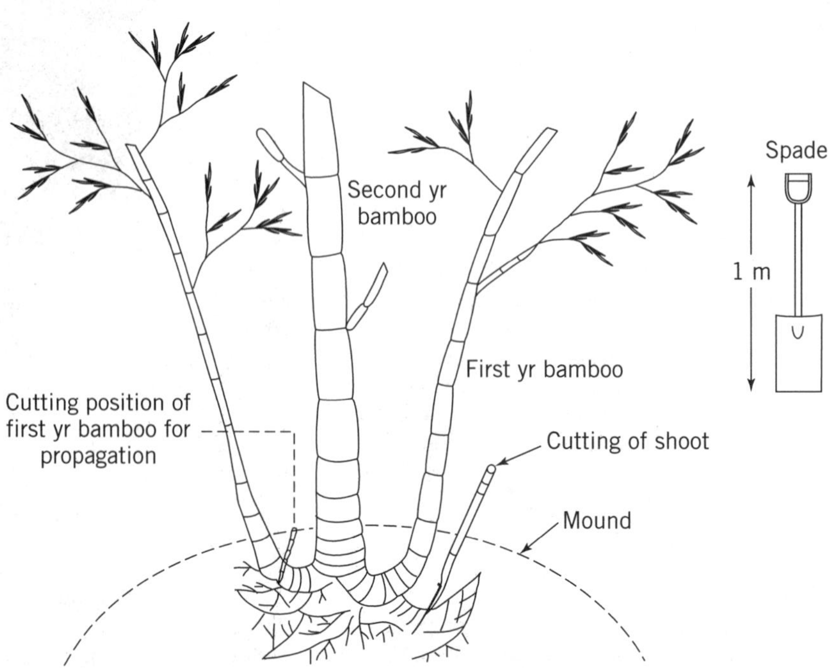

Figure 4. Shooting and harvesting position of *Bambusa oldhami*.

Figure 5. Harvest of *Dendrocalamus latiflorus* shoot.

Figure 6. Sun drying of fermented shoots of *D. latiflorus*.

are popular in summer when other vegetables are in shortage. *Dendrocalamus latiflorus* Munro has a longer production season from late March to early November. The average weight of each shoot is 2 kg. Green bamboo shoots weigh 0.65 kg. The shoots of *D. latiflorus* are cut when 50–100 cm above the ground (Fig. 5) and are mainly used for processing because they are relatively bitter and tough.

Phyllostachys edulis yields winter shoots, available from November to February, which are smaller, sweeter, and of better quality than spring shoot, available from March to May. Winter shoots of *P. edulis* are considered the best flavored shoots and are precious among winter vegetables due to small production.

FOOD USES

In the preparation of shoots for use, the sheaths and tough basal portions are removed, and the tender shoots are then sliced or cut lengthwise and boiled in water. The still crispy slices can be used in many dishes. Upon tasting, if the shoot is bitter, it is boiled again to remove the bitterness. Raw bamboo shoot is acrid, and boiling removes the acridity. The bitterness is from cyanogenic glucosides, which, when ingested, is poisonous because cyanide is released. Parboiling in water leaches the compound to render the shoots nontoxic (2). Green bamboo shoots and the winter shoots of *P. edulis* are not bitter because they are harvested before they emerge from the soil surface. Both shoots can be directly cooked with other food without precooking.

In Taiwan, shoots of *D. latiflorus* Munro, which are much larger in size, are used for canning or fermentation, fewer go to fresh market. Canned bamboo shoots are mainly for export and the amount is decreased year on year due to high labor cost. Dried fermented shoots are prepared from fresh bamboo shoots of ca 1 m long, the green portion is removed before processing. After blanching and fermenting for about one month the shoots are sun dried and sliced (Fig. 6). The final product has a crispy, tender texture and good flavor from the lactic acid fermentation (12).

Table 1. Approximate Composition of *Bambusa oldhami* and *Phyllostachys edulis* in a 100-g Edible Portion

	Composition	
Constituent	*B. oldhami*	*P. edulis*
Energy (cal/100 g)	60	22
Macroconstituent, 100%		
Water	83	92
Protein	3.9	2.4
Fat	1.1	0.3
Ash	0.7	1.0
Fiber	1.3	0.9
N-free extract	10.3	3.2.
Minerals, (mg %)		
Ca	44	43
Mg	21	45
P	39	38
Fe	3.4	2.2
Na	98	26
K	330	435
Vitamins, (mg %)		
A (IU)	30	20
B_1	0.02	0.08
B_2	0.03	0.07
Niacin	2.23	1.58
C	3	3

NUTRITIONAL VALUE

Bambusa oldhami and *Phyllostachys edulis* are two varieties sold in Taiwan fresh markets. The approximate nutritional composition in a 100-g edible portion is given in Table 1 (13).

BIBLIOGRAPHY

1. R. M. Ruberté, "Leaf and Miscellaneous Vegetables," in F. W. Martin, ed., *Handbook of Tropical Food Crops*, CRC Press, Inc., Boca Raton, Fl., 1984.
2. M. Yamaguchi, "Bamboo Shoot," *World Vegetables*, AVI Publishing Co., Westport, Conn., 1983.
3. Department of Agriculture and Forestry, "Crop Production," in *Taiwan Agricultural Yearbook*, Taiwan Provincial Government, Taichun, Taiwan, 1990.

4. B. N. Lee, "Bamboo Shoot," *Farmers Guide. Harvest Farm Magazine* Ltd., Taipei, Taiwan, 1980 pp. 885–888.
5. G. M. Leu, "Disappearing of Bamboos" *Harvest* **35**, 16–17 (June 28, 1985).
6. G. A. C. Herklots, *Vegetables in South-East Asia*, George Allen & Unwin Ltd., London, 1972.
7. M. T. Chang, "Cultivation of Bamboo Shoot," *Harvest* **36**, 42–45 (Aug. 1, 1986).
8. M. L. Liao et al., "Respiration Measurement of Some Fruits and Vegetables of Taiwan," *FIRDI Technical Report* **E-66**, Hsinchu, Taiwan, 1982.
9. T. R. Yen et al., "Quality Improvement of Canned Bamboo Shoot," *FIRDI Technical Report* **237**, Hsinchu, Taiwan, 1982.
10. R. Y. Chen et al., "Postharvest Handling and Storage of Bamboo Shoots," *Acta Horticulture* 258:309–316, 1989.
11. R. Y. Chen et al., "Control of Postharvest Quality of Bamboo Shoot," *FIRDI Technical Report* **423**, Hsinchu, Taiwan, 1986.
12. G. T. Huang et al., "Improvement on the Quality of Dried Fermented Bamboo Shoots," *FIRDI Technical Report* E-121, Hsinchu, Taiwan, 1985.
13. FIRDI, "Table of Taiwan Food Composition," *FIRDI Technical Report* **24**, Hsinchu, Taiwan, 1971.

MING-SAI LIU
Food Industry Research and Development Institute
Hsinchu, Taiwan

BANANAS. See FRUITS, TROPICAL.

BARLEY. See CEREALS SCIENCE AND TECHNOLOGY.

BEEKEEPING

Honey bees represent less than 12 of the thousands of bee species that exist around the world. Honey bees are unique in that they live in perennial social societies, and they produce a number of "hive products" that are used by humans in various ways. However, their greatest value to humans is their role as pollinators of cultivated crops. Approximately one-third of the diet consumed by individuals living in advanced societies is the product of honey bee pollination.

HISTORY

Honey bees (genus *Apis*) are believed to have originated in Southeast Asia (1). New species of *Apis* are still being described from that region today. Four species that naturally inhabit large areas of the world include the Western honey bee (*Apis mellifera*) from Europe and Africa, and the Eastern honey bee (*Apis cerana*), the giant honey bee (*Apis dorsata*), and the dwarf honey bee (*Apis florea*) from Southeast Asia. Only *A. mellifera* and *A. cerana* are cavity nesting species and thus have been collected by humans and successfully transferred and managed in artificial domiciles or hives.

It is likely that the first beekeepers were honey hunters or honey gathers, as depicted in a Paleolithic rock painting found in eastern Spain dating from around 7000 B.C. (2). There, gatherers collected honey by robbing wild bee nests. Honey hunting is still widely practiced in Asia to obtain honey from *A. dorsata* and to a lesser extent from *A. florea* nests. It can be postulated that an early stage in the development of beekeeping involved the cutting of a log section enclosing a bee colony and transporting the log to the dwelling of the gatherer. There the log and bees would provide a convenient source of honey and brood for periodic robbing by the new owner. The use of log sections as beehives persists in some African regions. The first hives made by humans probably were composed of clay, straw, or tree bark. In the Middle East, honey bees are still kept in clay pottery, mud, and straw hives. Some of these hive materials were used some 6000 years ago.

The advent of what is considered the "modern era" of beekeeping came about with the discovery of movable-frame beehives. The person generally credited with publicizing the significance of the "bee space" was Lorenzo L. Langstroth who, in 1851, discovered that bees would refrain from building wax connections between neighboring combs, or combs and the side of the beehive, if the space separating them was approximately 3/8 in. (3). This seemingly modest discovery led directly to the evolution of hives that had fully removable frames that enclose and support the beeswax combs. These frames could be manipulated easily by the beekeeper and removed for the purpose of examining the brood or extracting honey. The latter part of the nineteenth century saw the development of an amazing array of inventions based on this technology, including machines for removing wax (uncapping) and extracting honey from the new removable frames. Eventually, the Langstroth hive became established as the standard-size beehive that remains largely unchanged to the present day.

BIOLOGY OF HONEY BEES

All members of the genus *Apis* are eusocial, which means truly social. To entomologists this term indicates that there is cooperative care of the young by individuals of the same species, an overlap in generations, and production of a sterile or nonreproducing worker caste. This type of social development is uncommon considering that there are over 1 million named insect species. Living as a colony, eusocial species like honey bees provide more food and protection for themselves and their young than can any solitary individual. It is because honey bees store food for future needs that humans have been able to exploit them.

Scientists subdivide the wide-ranging species *A. mellifera* and *A. cerana* into a number of races (4). These races differ in behavioral and physical characteristics that are the result of adaptations of the species over time, following isolation in geographical regions. In fact, such races are sometimes called geographic races. In *A. mellifera* some races can be physically distinguished only through the use of computer-assisted morphometric (body measurement) analysis. To describe honey bee races, specialists use a trinomial system of nomenclature as opposed to the traditional binomial. For example, the Italian honey bee race's trinomial is *A. mellifera ligustica*. Other examples are *A. mellifera mellifera*, a race from northern Europe and the first honey bee race known to have been introduced into the New World, and *A. mellifera carnica*, the Carniolan honey bee from southern Austria and Yugoslavia.

A colony of honey bees may contain more than 45,000 adult individuals during the summer when the population peaks. Four life stages present in the colony are the eggs, larvae, pupae, and adults. These stages can be further subdivided by the two sexes, females and males (drones). The females may be divided further into two castes, workers and queens.

WORKERS

Workers, as their name implies, do the work of the colony. The worker caste builds the combs used to house developing eggs, larvae, and pupae (collectively called brood) and to store honey and pollen. They forage outside the colony for nectar (carbohydrate) and pollen (proteins, vitamins, minerals, and fats) to provide food for the colony. Workers also feed the brood, queen, and drones and defend the colony against intruders. When a worker bee stings, the barbed stinger usually pulls out of her abdomen and remains attached to her victim, where its odor attracts more bees to sting. Bees that have stung die within a day.

Sexually, workers are females with 32 chromosomes and usually lack functional ovaries. However, in certain situations, such as the absence of a queen, it is possible for workers to lay eggs. Laying workers are incapable of laying fertile eggs since they do not mate with drones to obtain sperm. Consequently, they lay eggs that only produce haploid drones. This leads to eventual death of the colony. In rare situations, laying workers can produce females impaternate (without fathers). This parthenogenic means of reproduction is found in the highest frequency in the race *A. mellifera capensis*.

During the busy foraging season, the life span of a worker is about six weeks. Workers literally work themselves to death. However, workers reared late in the fall often live through the winter into the following spring.

QUEENS

Queens are also females. They have the capacity to mate and lay fertilized eggs. Usually each colony has a single queen. However, under certain conditions colonies may have more than one queen. A few commercial beekeepers intentionally manipulate beehives so they can maintain a two-queen colony that is sometimes more productive than a single-queen colony. Queens live for as long as five years but most commercial beekeepers replace them every two years. The queen is the egg-laying female of the colony. The maximum number of eggs a queen lays varies throughout the season. It is estimated that a good queen can lay 1500 to 2000 eggs a day. A colony will produce new queens if its queen becomes a drone layer (depletes her stored sperm) or is injured or if the colony is about to swarm. New queens have to be raised in specially constructed cells. The workers feed the developing larva in such a way that a queen, rather than a worker, is produced.

DRONES

The drones are males and their primary function is to mate with virgin queens. They are raised in specially constructed cells that are larger than the worker cells. Genetically, drones are haploid, having a chromosome number of 16, in contrast to the diploid status of workers and queens, which have a chromosome number of 32. In temperate areas, drones are produced in large numbers, usually beginning in March, and are maintained until the first killing frost. A typical colony in summer will have 1000 to 2000 drones. Populous colonies sometimes will keep their drones into late November. Some scientists suspect that drones have functions other than mating, but no evidence has been forthcoming to support this viewpoint.

MATING

In nature, mating of drones and queens occurs in flight. The queen mates only at one early period of her life, during the so-called nuptial flight(s). About seven days after she emerges from her cell as an adult, she is considered sexually mature and will leave the hive to mate. A queen mates with about 17 drones (5). Queens are believed to fly to a location known as a drone congregation area (DCA) to mate. To this day, it is not possible to predict with certainty what geographical condition or other attributes are needed to establish a DCA. The results of recent work done with radar suggest that DCAs may be busy intersections in a complex network of drone pathways. Drones injure themselves mortally during mating, after which they fall to the ground and die.

SWARMING

The process of swarming represents the natural method of colony reproduction. Under certain conditions, often associated with the seasonal growth cycle of the colony, workers begin to raise additional queens in the colony. At some point before the virgin queens emerge from their cells, the old queen will depart from the nest with approximately half of the workers and drones and attempt to establish a colony at a new home site. The propensity for swarming and the conditions leading up to this behavior vary within and among races of honey bees. Sometimes one or more of the newly emerged virgin queens will also depart with a group of workers and drones in an "after swarm," also attempting to establish a new colony.

DISEASES AND PESTS

Larval Diseases

Like all living animals, the honey bee is subject to the ravages of diseases and pests. The diseases range from viral to fungal in origin. One troublesome brood disease is called

American foulbrood disease. This disease brought about the establishment of many state bee inspection programs (6). American foulbrood disease is caused by *Paenibacillus larvae larvae* (formerly *Bacillus larvae*). Only the spore stage of this bacterium is infective and only in honey bee larvae less than three days of age. Another important brood disease is European foulbrood disease, caused by *Melissococcus pluton*. In both cases, beekeepers sometimes use an antibiotic (oxytetracycline) for disease prevention and treatment. Another bacterial disease of minor importance is powdery scale disease, which is caused by *Paenibacillus larvae pulvifaciens*.

All the known fungal diseases that affect the brood are of minor significance except for chalkbrood disease, which is caused by *Ascosphaera apis*. Until 1968, no chalkbrood disease had been found in the United States. Currently, chalkbrood is reported present in most of the states, and many beekeepers report reductions in their honey crop due to this disease. There is no effective chemical control for chalkbrood.

A number of viruses affect honey bees. Most virus diseases of bees are considered of minor importance and no treatment is necessary. Two brood diseases of viral origin are sacbrood and filamentous virus disease (formerly believed to be of rickettsial origin).

Adult Diseases

Nosema disease is caused by the microsporidian *Nosema apis*. This protozoan shortens the life span of adult bees and reduces their production of royal jelly. Consequently, brood and honey production are reduced by 30 to 40%. The subtle effects of this disease are difficult to recognize and most beekeepers do not realize that their honey bees are infected. Fumagillin is used by some beekeepers to treat this disease.

Chronic bee paralysis is a virus-induced disease that is occasionally seen in honey bee colonies. This disease mimics the effects of pesticide kills, and many beekeepers may misdiagnose the condition. Fortunately, the disease is not serious. It is generally believed that requeening can abate the disease.

Other minor diseases affect the health of honey bees but are of little economic significance. Noninfectious conditions caused by pesticides and poisonous plants, such as California buckeye (*Aesculus californica*), cornlily (*Veratrum califonicum*), death camas (*Zygadenus venenosus*), and some locoweeds (*Astragalus* spp) can be serious problems in some years. A more detailed discussion of this subject has been published (7).

Parasitic Mites

The first of two problematic, then "exotic," parasitic mites of honey bees was detected in the United States in 1984 (8,9). *Acarapis woodi*, the tracheal mite, lives and reproduces in the thoracic tracheae (breathing tubes) of adult honey bees. During warmer times of the year, mites remain in young worker hosts only during the three weeks that they perform chores in the hive. When the workers begin to forage, the mites leave their original hosts at night, when the bees are relatively motionless, and seek newly emerged bees to infest. All castes of bees are susceptible, and the mites tend to remain in drones and queens. Tracheal mites feed by pushing their mouthparts through membranous areas of the tracheal tubes and sucking hemolymph (bee blood). Mite feeding causes hemolymph chemistry changes, destruction of flight muscles, darkening (healing?) of tracheal tubes, transmission of diseases, and reduced life expectancy. Colony losses are common in areas of colder climates, where new bees are not reared for months during the winter.

Menthol fumigation inside the beehive currently is the only registered and legal means of reducing the mite population. Treatments are temperature sensitive, with high heat driving bees from the boxes resulting in some queen losses. Once mite populations are reduced, they often can be prohibited from increasing by placing thin patties, made of two parts sugar to one part vegetable shortening, in contact with the bees in the hive.

The second important mite to be detected in the United States was the varroa mite, *Varroa jacobsoni*, found in 1987 (10,11). *Varroa* is an ectoparasite that feeds through membranous portions of the exoskeleton of adult and larval bees. Reproduction takes place only in capped cells containing worker or drone pupae. All developmental stages of the mite feed on the pupa. Physiologically immature male offspring mate with their sisters, delivering sperm that complete maturation in their sisters. When the bee emerges from its cell, the mites are liberated.

As with tracheal mites, *Varroa* feeding reduces protein content in the host's hemolymph and reduces its life span. As mite populations increase, two or more *Varroa* infest the same cell. The host pupa is apt to shrink and have malformed wings and legs. Such individuals are carried out of their cells by worker bees and discarded outside the hive. Without replacement workers, the colony perishes. *Varroa* also are disease vectors. In 1995 and 1996 there were so many *Varroa* in the country that we lost nearly 90% of our feral (unmanaged) colonies. Gardeners and commercial growers realized these losses when they noticed a total lack of honey bees on crops requiring pollination.

Introduction of plastic strips containing a time-release acaricide is the only currently registered, legal chemical control for *Varroa* mites. The selective acaricide kills the mites on contact without killing the bees. In a number of locations around the world, scientists are searching for new and different chemistries to control the mite. Other researchers are screening U.S. and foreign bee stocks for populations demonstrating tolerance or resistance to *Varroa* and the tracheal mite. Results look much better for the tracheal mite than for *Varroa*, which is an interspecific transplant from its original bee host, *A. cerana* (12).

AFRICANIZED HONEY BEES

An especially defensive, man-made, hybrid race of honey bees reached southern Texas in 1990. The result of intentional crossing of European and African stocks of honey bees in Brazil, the Africanized honey bees (AHBs) expanded their population and the territory in which they live at a phenomenal rate. The leading edge of the population expansion progressed about 250 to 300 miles per year as the bees moved from Brazil to Mexico. Fortunately they slowed down considerably as they entered more temperate climates. Population expansion to the east seems to have stopped west of Houston, Texas, but is continuing to the west in New Mexico, Arizona, Nevada, and California. Various opinions exist, but it appears that AHBs may not move, on their own, too much further north than their current locations (13).

Away from their hives, AHBs forage similarly to European honey bees (EHBs), can be used for pollinating crops, and produce good honey yields if managed carefully. However, their propensity to defend their nests extremely quickly, with very large number of stinging bees, makes them inappropriate for the type of beekeeping that is practiced in the United States. Fortunately, there is a segment of the commercial beekeeping industry that produces new queens, mated outside of areas "colonized" by AHBs. These new queens may be substituted for old queens or for queens of unsuitable stocks, but only after the old queen has been found and removed from the colony. The worker bee population will be converted fully to offspring of the new queen in about six weeks.

POLLINATION

The contribution of honey bees to agriculture far exceeds the value of honey production. Whereas honey bees produce an average of 200 million lb of honey a year valued at $150 million, the annual value of crops pollinated by honey bees exceeds $10 billion (14,15).

PRODUCTS OF THE HIVE

Most people think only of honey when honey bees are mentioned. Other useful products of the hive, on a worldwide basis, include bee venom, propolis, royal jelly, pollen, and beeswax (16). The larvae and pupae also can be eaten. However, in most cases, bee brood is a novelty food or is crushed in the process of extracting honey and consumed with the honey.

Bee Venom

Bee venom is used in the United States primarily for desensitizing individuals allergic to honey bee stings. In some countries bee venom is also used for treating arthritis. Interest is increasing in the United States in using bee venom therapy for other autoimmune conditions, such as multiple sclerosis (17). Specific devices are used by beekeepers for the collection of bee venom. One device generates a mild electric current to irritate the honey bees enough to sting a plastic surface. The surface material must allow withdrawal of the barbed stinger after venom is deposited. After air drying, the bee venom is scraped off the underside of the plastic material, diluted to known concentrations, sterilized, and packaged for medical use.

Propolis

Propolis or "bee glue" is a resinous material the bees collect from plants. This material is used by honey bees to seal small openings or cracks and for mummification of foreign objects that are too large for the bees to remove from the hive (such as a dead mouse). Because the bees also use propolis to cement together wooden parts of man-made beehives, most beekeepers consider propolis a nuisance and maintain bees that use less of the substance. Propolis has been shown to possess antimicrobial properties and is used in some European and Asian countries as the active ingredient in tinctures and throat lozenges. Propolis can be scraped from wooden hive parts and can also be harvested from a perforated plastic mat that is placed between the supers (boxes that make up a beehive). After the bees deposit propolis on it, the mat is removed and frozen previous to breaking off the substance. Periodically, bee journals carry advertising for propolis, but the demand for this material is quite limited.

Royal Jelly

Royal jelly (bee milk) is fed to young honey bee larvae, and it is this food that differentiates queens from workers. Only female larvae selected to be queens are fed royal jelly voluminously throughout their developmental period. Royal jelly is used in some cosmetic products and food supplements. In some countries it is believed that royal jelly possesses medicinal properties. Collection of royal jelly is labor intensive. In commercial production, larvae are grafted (moved into a queen cell cup) as if to produce queens. Nurse bees are induced to feed the larvae liberal amounts of royal jelly. Three days after grafting, the larvae are removed from the cells and the royal jelly harvested.

Pollen

Pollen is consumed by humans primarily as a dietary supplement. Its protein, vitamin, mineral, and fat content is variable, depending on the season and plant source. In a survey of a mixed sample of bee-collected pollen from seven states, the average protein content of pollen on a dry weight basis was 24.1%. However, when individual plant sources were analyzed, the range was 7.02 to 29.87% protein.

No precise data are available on the amount of pollen needed by one colony over an entire year. This would depend on the population and brood rearing activities. Nevertheless, a good estimate is 44 lb. Because most colonies collect pollen in excess of their immediate needs, beekeepers can supplement their income by collecting pollen.

Pollen is collected with a pollen trap, a device with a screen that literally scrapes the pollen off the hind legs of returning foragers. Pollen traps must be emptied frequently, because pollen is an attractive food source for many insects and mites, and it will absorb moisture and become moldy. Some of the pollen collected by beekeepers is used to feed honey bee colonies during dearth periods. Processing of pollen is generally a simple matter after its removal from the hive. The first step is to dry the pollen, usually by air drying. Then foreign debris, such as insect parts and the remains of diseased bees, is removed. In some cases, pollen pellets are also separated by color.

The market for pollen is not especially lucrative. Some foreign pollen is imported into the United States for sale in health food stores. There are no reliable figures on the size of this trade.

Beeswax

One of the most useful products of the hive is beeswax. Honey bees produce beeswax from honey or sugar syrup through internal body chemical processes. It is estimated that it takes 8 lb of honey to produce 2 lb of beeswax. In terms of costs of both products, it is economically more profitable for a beekeeper to produce honey. Because beekeepers must remove the beeswax cappings from honeycombs prior to extracting honey, beeswax is usually considered to be a by-product of honey production. Cappings and old combs can be liquefied by heat to extract the beeswax, which floats on the surface and hardens when the mixture is allowed to cool. Beeswax ranges in color from almost white through light green or tan to golden yellow. Like honey, the color of beeswax from cappings varies according to the floral source, while beeswax from old combs is usually darkest.

The majority of beeswax produced in the United States is used in candle making and cosmetics. Other uses include dental wax, furniture polish, lubricants, jewelry making ("lost-wax casting"), batik, egg coloring, sewing, and honeycomb foundation (used by beekeepers).

Honey

Honey can be defined simply as a plant exudate of floral or extrafloral origin that is modified by honey bees and is stored in honeycombs. Nectars collected by honey bees are processed into honey. Principally, this ripening procedure involves enzymatic inversion of sucrose into fructose and glucose and reduction in the water content of nectar. Other changes occur in the process, but not much is known about them. The color, odor, and taste of honey are strongly influenced by floral source (18). Depending on the floral source and location, honey is traditionally harvested whenever the combs are filled and capped with wax. Both the stage of the ripening process, achieved before the honey is harvested, and the ambient relative humidity affect the moisture content of the final product. Honey contains spores of yeast that will germinate and ferment the honey if the moisture content is much above 18%.

Honey yields are extremely variable, and two colonies placed side by side seldom produce identical honey crops. Honey depends on a number of factors, in addition to differences in nectar production among plant species. Honey production also is influenced by ground moisture, air temperature (day and night), wind conditions, and rainfall, because they affect both nectar production and bee flight. Honey bees exhibit preferences among plant species and may collect nectar from many sources. Consequently, honey from only one plant source is extremely rare. For instance, honey that is advertised as tulip poplar frequently contains honey from a number of floral sources, though the predominant floral flavor is that of tulip poplar. As a result, many beekeepers label their honey as mountain honey, wildflower honey, and so forth.

Honey is available in many forms, although the most popular and well known is extracted honey in the familiar queen-line jars. Extracted honey is harvested from the combs by first removing the beeswax cappings and then extracting the cell contents by centrifugal force. Some honey packers heat liquid honey (72°C for 4–5 min) to destroy yeasts and to delay granulation. When honey granulates, the remaining liquid portion of the honey increases in moisture content and the likelihood of fermentation also increases. Other beekeepers only heat their honey enough (34–38°C) to aid in the process of filtration, and still others do not heat their honey at all. The filtration process is somewhat variable, ranging from merely straining out dead bees, comb, and other large debris, to filtration that removes bits of beeswax and even pollen.

The U.S. Department of Agriculture (USDA) has established a system of grading honey based on percentage of sugars, clarity, flavor, aroma, and absence of defects (beeswax, propolis, etc). In addition to grades, honey is also classified by color: water white, extra white, white, extra light amber, light amber, amber, and dark amber. Honey color varies according to floral source. There have been attempts to develop worldwide honey standards. Meanwhile, in the United States, the USDA Standards for Grades of Extracted Honey and Comb Honey are currently being used.

In addition to liquid, extracted honey, many products using honey are available on grocery store shelves. These include honey-roasted nuts, cereals, baked goods, ice cream, and candy. Honey is widely used in baked goods because of its hygroscopic properties. The fructose in honey attracts moisture and prolongs freshness in baked products. To the purist, however, comb honey is the ultimate form for eating. Within this category of packaging there are several subdivisions, such as whole combs, cut combs, chunk honey, and comb honey sections (square and round). Honey sold in combs is unheated, unfiltered, and unlikely to be adulterated.

The typical beekeeper removes only the honey that is surplus to the colony's needs. In the United States, the commercial beekeeper's colonies produce an average of 80 lb of honey annually. In good years, beekeepers report honey yields of 200 lb or more in some areas of the United States and Canada.

FUTURE DEVELOPMENT

In recent years the honey bee industry in the United States has faced many difficult problems. Foreign honey imports and lower honey prices, coupled with increased cost of production, have created a considerable financial challenge. Because of the recent establishment of the two parasitic mites and Africanized honey bees in the United States, Canada and Mexico have banned imports of U.S. package bees and queens, an action severely damaging to the queen-producing segment of the industry. The reliance of U.S. agriculture on honey bees, however, makes it imperative that these problems be solved. Many crops cannot be effectively pollinated without the placement of honey bee colonies in the orchards or fields. Fortunately, the demand for one of the direct products of the insects, honey, shows signs of increasing. The formation of the National Honey Board, with its stated goal of expanding domestic and foreign markets for honey and to develop and improve markets, bodes well for the future of commercial and hobby beekeeping in the United States.

BIBLIOGRAPHY

1. E. Crane, "The World's Beekeeping—Past and Present," in J. M. Graham, ed., *The Hive and the Honey Bee*, revised ed., Dadant & Sons, Hamilton, Ill., 1992.
2. E. Crane, *The Archaeology of Beekeeping*, Duckworth, London, 1983.
3. L. L. Langstroth, *Langstroth on the Hive and the Honey-bee, A Bee Keeper's Manual*, Hopkins, Bridgman & Co., Northampton, Mass., 1853.
4. A. Dietz, "Honey Bees of the World," in J. M. Graham, ed., *The Hive and the Honey Bee*, revised ed., Dadant & Sons, Hamilton, Ill., 1992.
5. R. Page, "Sperm Utilization in Social Insects," *Ann. Rev. Entomol.* **31**, 297–320 (1986).
6. A. S. Michael, D. A. Knox, and H. Shimanuki, "U.S. Federal and State Laws Relating to Honey Bees," in R. A. Morse and R. Nowogrodzki, eds., *Honey Bee Pests, Predators, and Diseases*, 2nd ed., Comstock Publishing, Cornell University Press, Ithaca, New York, 1990.
7. E. L. Atkins, "Injury to Honey Bees by Poisoning," in J. M. Graham, ed., *The Hive and the Honey Bee*, revised ed., Dadant & Sons, Hamilton, Ill., 1992.
8. M. Bruner, "Acarine Mites Found in Texas," *Gleanings in Bee Culture* **112**, 402 (1984).
9. M. Delfinado-Baker, "*Acarapis woodi* in the United States," *Amer. Bee J.* **124**, 805–806 (1984).
10. J. M. Graham, "Varroa Mites Found in the United States," *Amer. Bee J.* **127**, 745–746 (1987).
11. K. Flottom, "Of Mites and Men (Again!)," *Gleanings in Bee Culture* **115**, 613 (1987).
12. M. Delfinado-Baker and C. Y. S. Peng, "*Varroa jacobsoni* and *Tropilaelaps clareae*—A Perspective of Life History and Why Asian Bee-mites Preferred European Honey Bees," *Amer. Bee J.* **135**, 415–420 (1995).
13. O. R. Taylor, "African Bees: Potential Impact in the United States," *Bull. Entoml. Soc. of Amer.* **31**, 14–24 (1985).
14. M. Levin, "Value of Bee Pollination to United States Agriculture," *Amer. Bee J.* **124**, 184–186 (1984).
15. W. S. Robinson, R. Nowogrodzki, and R. A. Morse, "The Value of Honey Bees as Pollinators of U.S. Crops, (Two part series)," *Amer. Bee J.* **129**, 411–423 and 478–487 (1989).
16. A. Mizrahi and Y. Lensky, *Bee Products—Properties, Applications, and Apitherapy*, Plenum Press, New York, 1997.
17. M. Simics, *Bee Venom: Exploring the Healing Power*, Apitronic Publishing, Calgary, Alberta, Canada, 1994.
18. E. Crane, *Honey: A Comprehensive Survey*, Heinemann, London, 1976.

ERIC C. MUSSEN
University of California Entomology Extension
Davis, California

BEER

Beer is traditionally, and in many countries, legally, defined as an alcoholic beverage derived from barley malt, with or without other cereal grains, and flavored with hops.

Barley malt is barley that has been purposely wetted, allowed to partly germinate, and then dried or kilned. This three-step process is called malting. Because barley is virtually the only cereal that is malted, it is often just referred to as malt. The drying stabilizes the malt and allows beer to be produced year round and in places far removed from where barley is grown. It thus differs from wine, which is only produced seasonally and only near grape-growing areas.

Other cereal grains that are used to make beer are rice, corn, sorghum, and wheat. The choice, depends on the individual brewer and often on the availability of the different cereals.

HISTORY

Beer was likely a casual discovery when a bowl of grain that got wet, and then very wet, accidentally fermented with airborne yeast. It has been made intentionally for more than 6,000 years and is indigenous to most temperate and tropical cultures.

The oldest known written recipe was found on a 4,000-year-old Mesopotamian clay tablet, and it was for beer (Fig. 1). The Babylonians made 16 kinds of beer, using barley, wheat, and honey. The Egyptians described a "beer of truth" and a "beer of eternity".

Beer making was an important trade in ancient Mesopotamia, Egypt, Greece, and Rome. In less-settled cultures in Europe, Africa, China, India, and South America some kind of fermented grain beverages were made on a small scale for home consumption. It may well have been discovered that brewing beer makes any water safe for human consumption.

Large-scale brewing began in medieval Europe as a result of increased travel and trade and the attendant increase in inns and taverns offering food and lodging. Monasteries began brewing beer, as well as wine and liqueurs, and accepted voyagers as guests. This helped spread knowledge about beer.

In North America, beer was introduced with the early English and Dutch settlers. The Pilgrims landed in Plymouth, instead of going farther south, because they had run out of beer. Many of the founders of the United States were active brewers or maltsters.

All the beer of the early settlers was actually ale, fermented by strains of yeast, *Saccharomyces cerevisiae*, that are airborne and are also used in making bread and wine. Ales are produced at ambient temperatures.

In the 1840s German brewers came to the United States and began brewing lager beer, fermented by strains of a yeast *S. carlsbergensis* that exists only in breweries and is used only for making beer. A characteristic of this yeast is that it can only function at temperatures below 70°F. For this reason, the early lager breweries were built near riv-

Figure 1. Mesopotamian clay tablet. *Source:* Courtesy of the Beer Institute.

ers or lakes that froze in the winter. Ice was cut from these and stored in cellars for refrigeration during the summer. Thus New York, St. Louis, and Philadelphia became centers of lager brewing. By the 1870s mechanical refrigeration was invented and used first in a brewery in Brooklyn.

Lager beer has almost completely replaced ale as the beverage of choice, and today ale is only produced in the United Kingdom and parts of Canada, with only token amounts elsewhere. In the United Kingdom ale is referred to as beer and what the rest of the world calls beer is called lager. In what follows, beer will be used to refer to both ale and beer.

The Making of Beer

Beer production involves three distinct but interrelated stages. First is the preparation of an extract of the malt and the grains selected. This extract is called wort (pronounced to rhyme with "hurt"). This step takes place in the brewhouse and is often referred to as brewing. It takes about 4–10 h. A photograph of a typical brewhouse is shown in Figure 2.

The next stage is fermentation, the conversion of this liquid by yeast into beer. This is a temperature-controlled process that takes 3–10 d.

Fermentation is usually conducted in large stainless steel vessels that hold the volume of an entire brew, or several brews. Their size may be between 7,000 and 375,000 gal, and they may be horizontal rectangular in shape, or vertical cylindrical, with or without a conical bottom. The conical bottom allows simple gravity expulsion of yeast. A set of cylindroconical tanks are shown in Figure 3. The rectangular tanks require manual raking of yeast for reuse in subsequent brews.

The final stage is finishing or the refining of this liquid into salable beer by the brewery. It may take 2–25 d. It takes place in tanks equal in size to the fermentation tanks

Figure 2. Brewhouse, showing lauter tubs above and kettles below. *Source:* Courtesy of Anheuser-Busch, Inc.

Figure 3. Cylindroconical tanks. *Source:* Courtesy of Paul Mueller Company.

or multiples of the fermenter size. Aging tanks, preassembled on site, are shown in Figure 4.

Each stage offers the brewery many choices. It is for these reasons that there are not many beers, perhaps none, that are exactly alike. The finished beer may be packed in large containers, holding 2–31 gal, or into small ones, holding 7–40 oz.

RAW MATERIALS OF BREWING

Water

Water is, only quantitatively, the major component of beer. Its requirements for brewing are fairly simple. First, it must meet local or international standards for potable water. The World Health Organization's standards for drinking water are shown in Table 1.

The second requirement is that it not be too alkaline. A maximum alkalinity of 50 ppm (as calcium carbonate) is acceptable.

A third requirement is that it be hard and contain calcium. The level of calcium considered most desirable is 100 ppm (as calcium), but lesser amounts may also be used. The water in Pilsner, Czechoslovakia, had these attributes, neutrality and hardness, and soon became the standard by which all beer is made.

All water supplies are either surface water or groundwater. Surface water is derived from rainfall or snow and may be quite pure. If it is collected in reservoirs near the source of rain or snow, it is usually soft and neutral. If it is collected in rivers, its purity will depend on the distance between the collection and distribution points and the uses it suffers in between. Such water may need purification with sand and charcoal.

Figure 4. Aging tanks assembled on site. *Source:* Courtesy of Paul Mueller Company.

Table 1. World Health Organization Standards for Drinking Water

	Maximum permissible concentrations (mg/L)
Chlorides	60
Sulfates	400
Calcium	200
Magnesium	150
Total dissolved solids	1,500

Groundwater comes from wells or springs and is usually tasteless but may be quite alkaline or hard, or both. Such water may need acidification to remove alkalinity and, less often, softening if it is too hard. Such water rarely needs treatment with activated charcoal.

Barley Malt

Malted barley is the principal ingredient of beer. It furnishes almost all the required elements for making beer. Malt provides starch and sugars, which will, partly or completely, become alcohol during fermentation; and protein and amino acids, which will supply nutrition for the yeast, as well as color, foam, and flavor in the finished beer.

Barley used to make malt is almost always selected from special strains that have been developed and grown specifically for this purpose. There are two major types of barley, distinguished by the number of rows on the stalk. Six-rowed barley is grown mainly in the midwest of the United States and Canada, but now also in Europe and Australia. Two-rowed barely is grown in the far west, and also in Europe and other barley growing regions. The two-rowed barleys are considered by some to give a smoother beer. A drawing of a barley kernel is shown in Figure 5.

The husk, which protects the seed and is composed of the palea and lemma, is, in barley, alone among the cereals, fused to the testa and the pericarp. It is this fusion that permits barley to be malted efficiently and to be the universal basis for making beer. Grains that lose their husks at germination tend to become moldy when malted in bulk.

The malting of barley begins with steeping, in which the barley, in tanks, is soaked in aerated water. The steeped water is changed several times. The steeping is complete when the barley reaches a moisture content of 45%. It is then transferred to long, horizontal compartments, through which moist air is blown. In 4–6 d, the barley germinates and starts to grow as if it were to become a barley plant. During this time the barley produces starch-splitting enzymes, notably α-amylase and β-amylase, and also proteinases and cellulases. The amylases break down some of the starch in the barley; the proteinases degrade some of the proteins to amino acids, and the cellulases soften the cell walls. All these enzymes are secreted by the aleurone layer and migrate into the endosperm. These changes, which collectively are called modification, prepare the barley for its use in brewing.

At the time when the modification is considered complete, the malt is transferred to a kiln and dried. The temperature at which the malt is dried determines the color and, partly, the flavor of the beer from which it will be made. Kilning stabilizes the malt and allows it to be shipped widely and to be stored for a year or more.

The analysis of a typical malt is shown in Table 2.

UNMALTED CEREALS (ADJUNCTS)

Cereals such as rice, corn, and sorghum are often used to complement or attenuate the malt. These grains do not

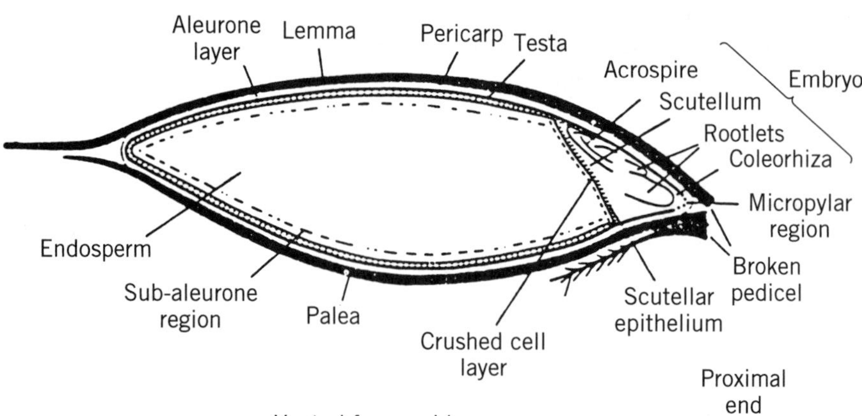

Figure 5. Schematic longitudinal section of a barley grain, taken to one side of the furrow, showing the disposition of the parts.

Table 2. Analysis of Barley Malt (%)

Moisture	4.0
Starch and dextrins	52.5
Simple sugars	9.5
Total protein	13.0
Soluble protein	5.4
Cellulose	6.0
Other fiber	10.0
Fats	2.5
Minerals	2.5

contribute to flavor, color, or foam but serve only to supply carbohydrates to the wort. Their use probably arose when malt was in short supply. They are widely used throughout the world. Only in parts of Germany, in Switzerland, and in Greece is their use prohibited by law. This prohibition, which dates to 1516 in Bavaria, as the *Reinheitsgebot*, originally was intended as a means to police the taxation of beer because malt could only be made in a malthouse. The prohibition on the use of other sources of carbohydrates precluded the surreptitious increase in beer production. It remains to this day as a vestigial reminder of days past that has served to prevent the importation into those three countries of beers made with unmalted cereals. However, the European Common Market has overruled this law and such beers may now be imported into the all-malt countries.

The quantity of unmalted cereals may vary from 10 to 50% of the total mash, depending on the beer to be produced. Increased use gives a lighter, less bodied beer. If rice or sorghum is used, it must be ground before mashing. If corn is chosen, then a physically separated fraction of the ground corn kernel that contains mainly starch and protein and very little corn oil is used.

Starch is the major component of both malt and the other cereals. It is a complex polysaccharide made up of about 35% of a straight-chain polymer of glucose known as amylose and about 65% of a branched polymer, amylopectin. The links in all of amylose are 1,4′ bonds between neighboring glucose units. The bonds in amylopectin are both 1,4′ and 1,6′,—the latter at the branch points. Schematic diagrams of amylose and amylopectin are shown in Figure 6 and 7, respectively.

SPECIAL MALTS

Some malts are made with special properties, mainly differing in color and flavor. A caramel, or crystal, malt is made by heating a wet malt with steam under pressure. The resulting malt is used with regular malt to produce dark beers. A so-called chocolate malt is a variant of crystal malt. These malts provide color and special flavor to the beer. Malt may also be darkened by kilning at higher temperatures (200°F or higher) or for longer times. Such malt only imparts more color to the beer.

Wheat may be malted, with difficulty, but it produces an ordinary beer unless a yeast is used that by itself produces a spicy flavor. This has found some favor in parts of Germany and Belgium.

HOPS

Hops, *Humulus lupulus*, is a truly remarkable plant. Although it originally grew wild, and still does, it is intensively cultivated in a few countries for use only in beer. The major hop-growing countries are Germany, the United States, the Soviet Union, Czechoslovakia, Yugoslavia, and the United Kingdom. Figure 8 shows hops growing in Germany. Hop picking by hand in Czechoslovakia is shown in Figure 9.

The use of hops in beer only goes back some 500 years. Many plants had been used to flavor beer, until it was discovered that hops not only makes beer pleasantly flavored but also controls the growth of spoilage bacteria. The use of hops is now universal in beer.

Hops contain a group of compounds, humulones, which are very insoluble in water, but which undergo a chemical rearrangement during the brewing process to form an isomeric group of compounds called isohumulones. The isohumulones are soluble in water and impart to beer a

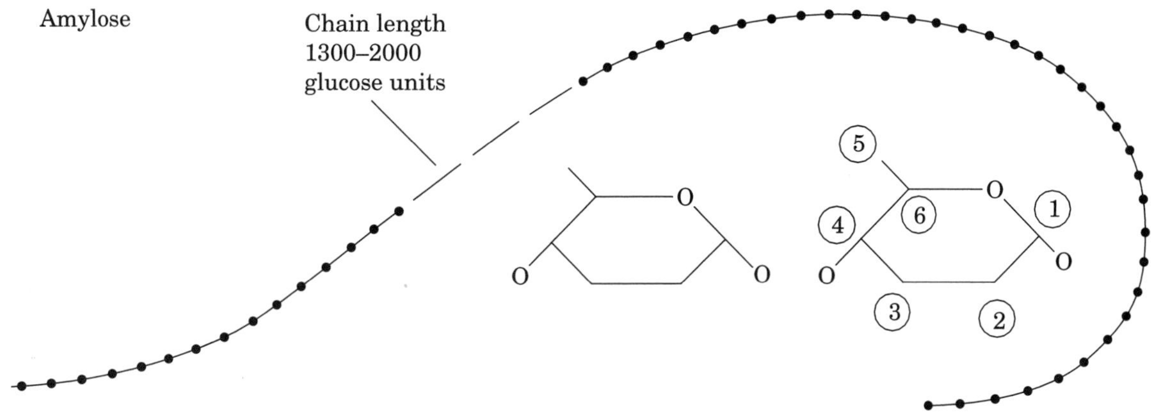

Figure 6. Structure of the linear starch fraction. Each black circle represents a glucose unit, and the aldehydic terminus of the chain is indicated by an asterisk. Interglucose bonding is α-1,4, as shown in the inset, with carbon atoms in the glucose unit numbered in conventional fashion.

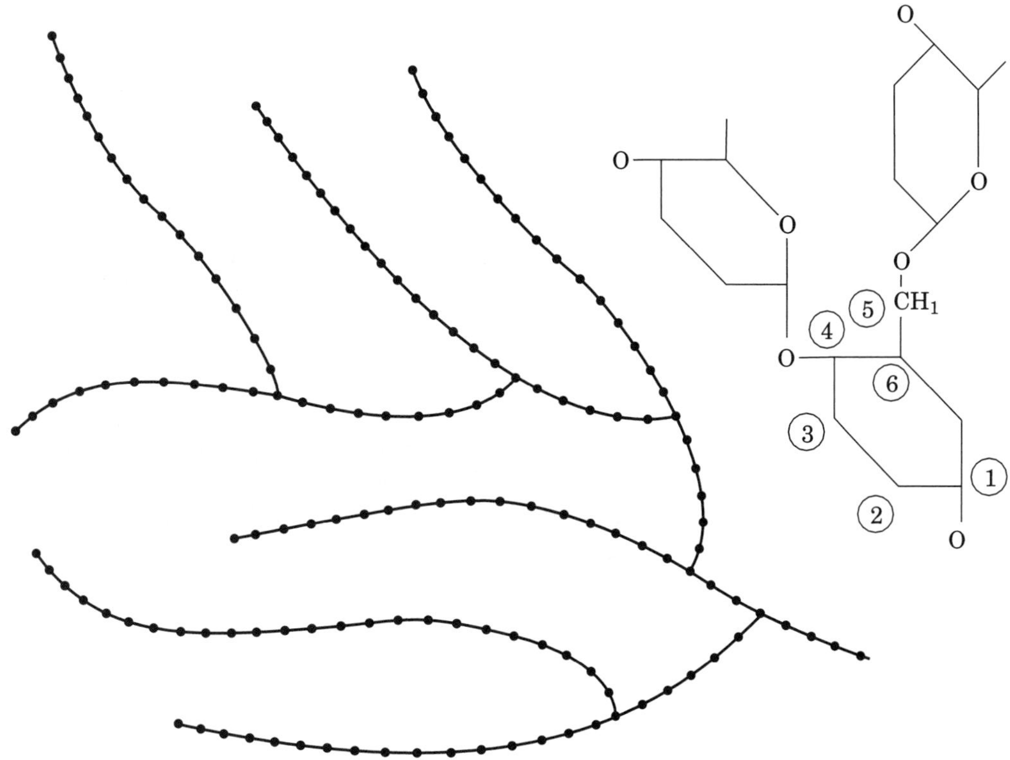

Figure 7. Structure of a small section of the branched starch fraction. Branching is effected through an α-1,6 linkage, as shown in the inset.

palate-cleansing bitterness that provides beer with its unusual property of drinkability. The isohumulones in beer can be readily analyzed, permitting a quantitative measure of their presence and thus of the beer's bitterness. Not many flavor compounds in other foods are so easily measured. A diagram of humulone and its conversion product isohumulone is shown in Figure 10. The analogues of humulone (and isohumulone)—cohumulone, adhumulone, and prehumulone—differ only in the number of carbon atoms in the side chains.

In addition, hops contain a volatile oil containing many odoriferous compounds, some of which survive into the finished beer. Some varieties of hops, such as Clusters and Galena, are used primarily for bitterness, whereas others, such as Hallertau and Cascade, are used for their aroma.

The essential oil of hops is, as in all plant materials, of terpenic origin. The major terpenes are myrcene (1), a diterpene; humulene (2), and caryophyllene (3) sesquiterpenes (Fig. 11). Oxidation products of these hydrocarbons, such as humaladienone and caryophyllene epoxide have

Figure 8. Hops growing in Germany. *Source:* Courtesy of S. S. Steiner.

Figure 9. Hand picking of hops. *Source:* Courtesy of S. S. Steiner.

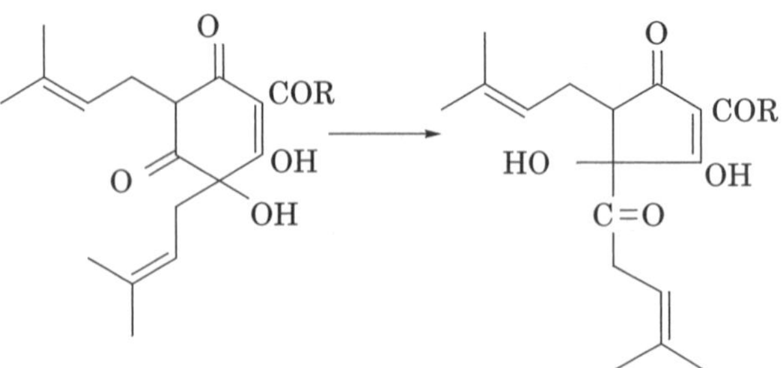

Figure 10. Humulone and isohumulone.

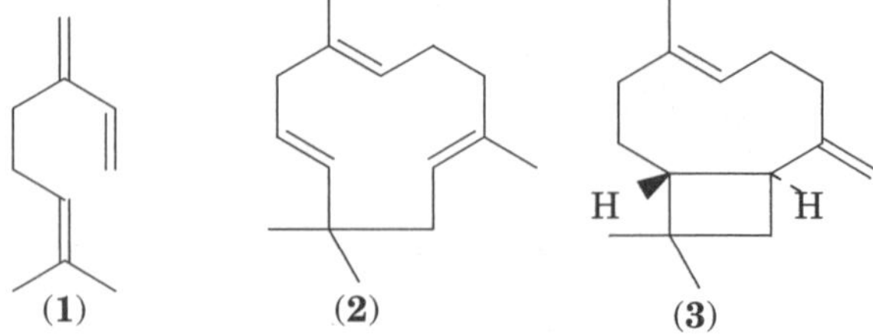

Figure 11. The terpenes (1) myrcene, (2) humulene, and (3) caryophyllene.

been found and are probably more important to the flavor of beer.

Also present in hop oil are alcohols, such as linalool and geraniol; ketones, such as undecanone-2; and esters, such as geranyl butyrate. These survive into beer and are important components of the hoppy aroma of some beers.

Hops is a perennial vine that grows to more than 20 ft in one season. They are trained to high wires held by long poles. The flowers are mechanically picked and dried in a kiln. Hops are used in several forms. The dried cones, compressed into bales weighing 200 lb, are used by many brewers. In this state, hops must be kept refrigerated. Alternatively, the hop cones can be milled to a powder and then recompressed into pellets. These are easier to use at a brewery, but also need refrigeration. Lastly, the hops may be extracted, with an organic solvent such as hexane or alcohol, or with liquid carbon dioxide, and the extract used. These do not need refrigeration and are popular in tropical countries. A cluster of Fuggle hops is shown in Figure 12. Some varieties of hops grown commercially are listed in Table 3, with the country producing the major quantities indicated.

A typical analysis of hops is shown in Table 4.

YEAST

Yeast is a unicellular microscopic organism that is fairly distinctive in being able to metabolize sugars either to carbon dioxide and water, in the presence of air; or to alcohol and carbon dioxide, in the absence of air. It is this latter trait that is used in all alcoholic fermentations.

One genus of yeast, Saccharomyces is used in brewing. But there are two species. One is *S. cerevisiae*, used in producing ales, and also for making bread, wine, and whiskey. It is a fairly hardy yeast and survives in the atmosphere.

The other is *S. carlsbergensis* (uvarum) used only for lager beers and found no where else but in breweries. It is temperature-sensitive and does not survive in the atmosphere.

Figure 12. A cluster of Fuggle hops. *Source:* Courtesy of S. S. Steiner.

Table 3. Hop varieties (In Order of Acreage Devoted to Each Variety)

United States	Europe	Australia
Washington	Perle (G)	Pride of Ringwood
Galena	Hallertau Magnum (G)	
Nugget	Hersbrucker (G)	
Columbus	Hallertau Tradition (G)	
Willamette	Spalter Select (G)	
Cluster	Hallertauer (G)	
Chinook	Nugget (G)	
Cascade	Tettnanger (G)	
Mt. Hood	Northern Brewer (UK & G)	
Perle	Fuggle (UK)	
Tettnanger	Golding (UK & SU)	
Horizon	Brewers Gold (UK & G)	
Olympic	Styrian (Y & SU)	
Golding	Saaz (Cz)	
	Hallertau Tarus (G)	
Idaho	Spalter (G)	
	Target (G)	
Galena	Huller (G)	
Cluster	Orion (G)	
Chinook	Record (G)	
Willamette	Bullion (G)	
Nugget		
Mt. Hood		
Oregon		
Willamette		
Nugget		
Perle		
Golding		
Mt. Hood		
Fuggle		
Tettnanger		

Note: G, Germany; Y, Yugoslavia; SU, Soviet Union; Cz, Czechoslovakia; and UK, United Kingdom.

Table 4. Chemical Composition of HopS

Moisture	10%
Total resins	17–20%
Volatile oils	0.3–1.2%
Polyphenols	2–5%
Waxes and lipids	3%
Ash	7%
Cellulose	55%

Yeast reproduces asexually and multiplies severalfold during a normal fermentation. Because yeast enters a commercially sterile liquid, wort, it may be reused many times in successive brews without danger of contamination. Beer is the only fermented product that starts with a sterile medium. A photograph through an electron microscope of a dividing yeast cell is shown in Figure 13.

The two species of yeast differ in many biochemical characteristics. Table 5 gives some of these differences.

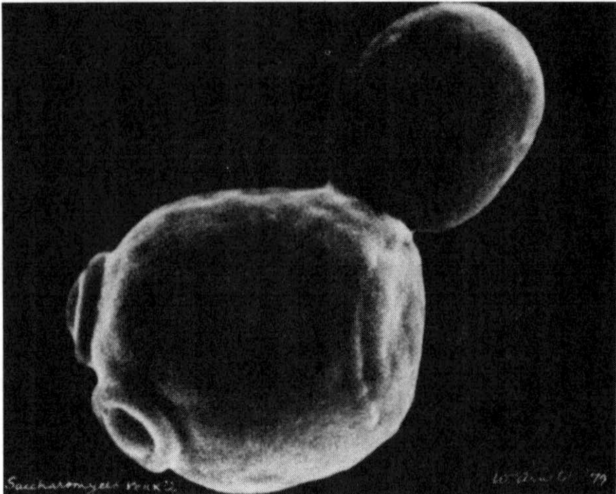

Figure 13. Dividing yeast cell.

Table 5. Biochemical Characteristaics of Yeast Used in Beer Making

	S. carlsbergensis	*S. cerevisiae*
Ferment melibiose	+	−
Ferment raffinose	+ + +	+
Utilize ethanol aerobically	−	+
Produce hydrogen sulfide	+ + +	+
Attitude after fermentation	Settles to bottom	Rises to top

Most breweries have their own strains of yeast and continue to use them indefinitely. Very often a pure yeast culture is maintained and propagated in each plant, or, more often, in the parent plant if the brewery has more than one plant.

THE BREWING PROCESS

The Brewhouse

Milling. Grains are normally received in breweries in bulk and transferred to silos. Malt needs to be crushed before it is used. This is done just before use. The aim in milling is to crush the endosperm of the barley, but leave the husk as intact as possible. These husks act as the filtration medium during later processing. A six-roller malt mill is shown in Figure 14.

Corn as received in a brewery does not need any treatment before use. It has already been crushed by a corn miller and separated pneumatically. The lighter germ of the corn is conveyed in a different airstream from the heavier, starchy endosperm. It is this latter fraction that is used by breweries.

Rice needs merely to be crushed before use.

Mashing. This step determines the ultimate structure of the finished beer. A typical mashing cycle requires the use of four vessels, a cooker, mash tub (tun in Europe), lauter tub, and kettle. Examples of some brewhouse control panels are shown in Figure 15.

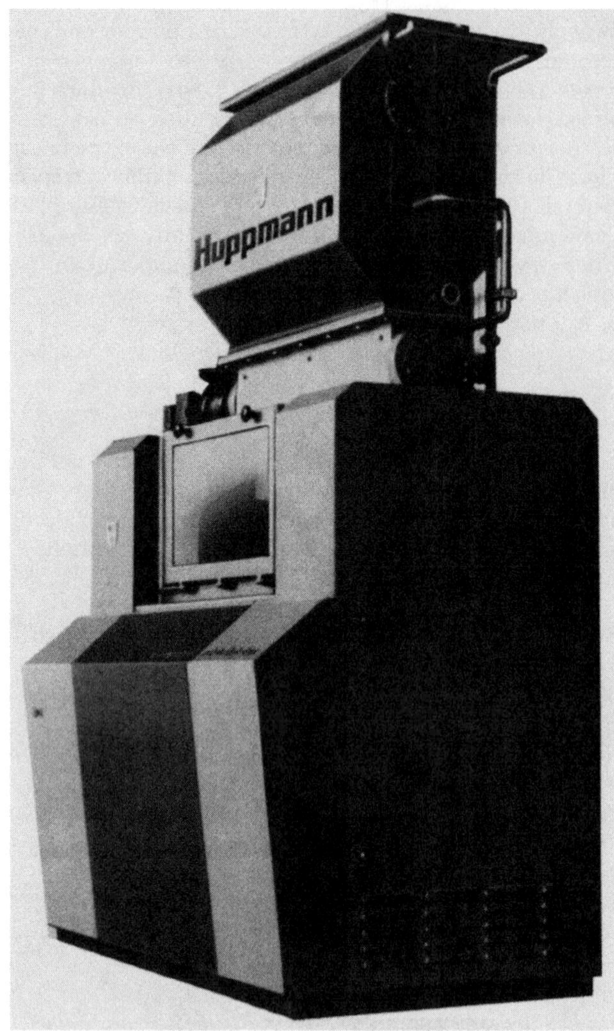

Figure 14. Six-roller malt mill. *Source:* Courtesy of Wittemann Hasselberg.

The cooker is a simple vessel, with an agitator and some means of being heated. The lauter tub is a complex vessel with a specially perforated false bottom through which the mash is strained or filtered. It has variable speed rakes, with adjustable flights that can be raised and lowered. The husks of the malt serve as the filtration medium in the lauter tub. A cut-out diagram of a lauter tub is shown in Figure 16. The kettle is also complex. It is heated internally with steam passing through coils and a percolator, or directly with a flame; or externally in a steam boiler. Steam-heated kettles are shown in Figure 17

There are several systems of mashing, each with certain advantages and each setting the requirements for the brewhouse equipment. The most common system in the United States and most other countries is a double-mash system. The corn or rice is boiled in the cooker, with a small amount of malt to lower the viscosity. The main portion of malt is admitted to the mash tub, and, after a stand at about 50°C for proteolysis, the temperature is raised by

Figure 15. Brewhouse control panels. *Source:* Courtesy of Wittemann Hasselberg.

adding the contents of the boiling cooker. The combined mash is allowed to rest at the temperature that results from this mixing for a variable time, and then the temperature is raised selectively until the final mashing off temperature, usually 75°C, is reached.

In the United Kingdom and its brewing satellites the most common mashing system is a simple infusion method. In this system, the mashing process takes place at a single temperature. This system developed with mash tubs that could not be heated. It is not nearly as versatile as the double-mash system. Mashing, with or without corn or rice, boiled separately in a small vessel, takes place at about 65°C.

In Germany, and its brewing followers, still a third system, called decoction brewing, is common. Here too the mash tub cannot be heated, but a smaller decoction vessel can. The malt is transferred into the mash tub at a low temperature. A small portion of this mash is pumped to the decoction vessel and boiled. The boiled mash is pumped back into the main mash vessel. This serves to raise its temperature to a level that depends on the relative volumes in both vessels. This transfer is repeated once or twice (double decoction or triple decoction), leading to a series of rising temperatures in the mash tub. This process is rather lengthy and has not found wide use.

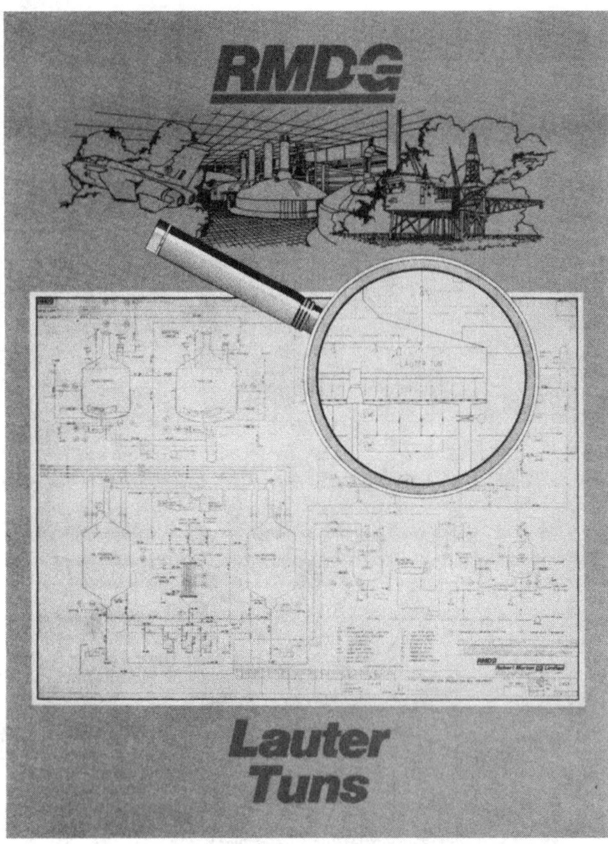

Figure 16. Cut-out diagram of lauter tub. *Source:* Courtesy of Robert Morton.

The reactions that occur in the mash tub are very complex. At a low temperature, about 50°C, proteolytic and cytolytic enzymes in the malt are operative. The corn or rice are not present at this stage and don't need to be. Proteolysis is necessary to supply amino acids for the growth of yeast and also to break down large, insoluble proteins to simple soluble peptides for foam enhancement. Cytolysis is required to reduce the viscosity of the barley gums (soluble fiber) to accelerate the later filtration. After some 10 to 60 min, the temperature is raised by the addition of the boiled cooker (or decoction vessel). The resulting temperature may be held for 10–60 min, and the temperature raised again, finally ending at 75°C. During these elevated temperatures, two amylolytic enzymes, α-amylase and β-amylase, cooperate to hydrolyze most of the starch to fermentable sugars. Beta-amylase, which operates best at about 63°C, attacks starch from its nonreducing end and splits off the disaccharide, maltose. It cannot proceed past a 1,6′ linkage, and so depends on the concurrent action of α-amylase. Alpha-amylase, which operates best at 70°C, splits any 1,4′ linkage at random, and so provides a supply of nonreducing ends for the β-amylase to operate on. Cooperatively, the action of these two enzymes can result in the splitting of about 65–75% of the starch into mono-, di-, and trisaccharides.

The mash temperature is raised to 75–77°C when the desired action of the amylolytic enzymes has been achieved. This arrests any further enzymatic action and prepares the mash for the next step, lautering.

Lautering (Straining or Filtration). The converted mash is transferred to a vessel that will permit separation of the liquid (wort) from the insoluble solids (husk, fiber) of the malt and grains. Three different types of vessels accomplish this separation. The one most widely used is a lauter tub, a large cylindrical vessel with a slotted false bottom and a set of movable and adjustable flights. The mash is pumped into this tub, allowed to settle for a few minutes, and then liquid is allowed to flow to the next stage, the kettle. The first runnings are always cloudy and are run back onto the mash. When it runs clear, it is run to the kettle. After the clear liquid, called first wort, stops run-

Figure 17. A row of brew kettles. *Source:* Courtesy of Miller Brewing Co.

ning, fresh water (sparge water) is sprayed onto the grain bed to elute the wort adhering to the grains.

Sparging continues until the next vessel is full, the concentration of the wort has reached the desired value, or the solids in the effluent are too dilute to be worthwhile. In some breweries, these last runnings are recovered and used for a succeeding brew. At the end of sparging, the rakes are lowered, their angle changed, and the spent grains are pushed into now-open large holes at the bottom of the lauter tub. These grains have a ready market as a valuable animal feed.

Another device used to separate wort from grain is a mash filter, a plate-and-frame press with plastic cloth as the barrier. These are faster than lauter tubs, but lack flexibility to grain depth. They have found some use in the United States, but wider use outside that country.

A third device, patented by Anheuser-Busch, is a Strainmaster, a modified lauter tub with perforated tubes projecting into the grain bed that serve to collect the wort. It has fallen out of favor and is being replaced by the lauter tub. This also is more rapid than a lauter tub. They are used almost exclusively in the United States.

The Kettle. The clear wort running from any of the grain-separation vessels is collected in a large vessel, the kettle, and boiled. Heat is normally supplied by steam in a set of coils and a center-mounted percolator. Alternatively, the vessel may be heated by direct fire or by an external collandria heated by steam.

During the heating period hops are added, in one or several portions, and at varying times. The varieties, the quantity, and the duration of their boiling time, all affect the flavor of the finished beer.

The boiling of the wort serves many vital functions:

1. It extracts the resin from the hops.
2. It isomerizes the humulones to soluble isohumulones.
3. It volatilizes most, but not all of the volatile oils in the hops.
4. It stops all enzymatic action.
5. It sterilizes the wort.
6. It concentrates the wort.
7. It removes grainy odors from the malt and other cereals.
8. It darkens the wort and produces flavor-inducing melanoidin reactions between the amino acids and simple sugars.

At the end of the boil, the wort is passed through a strainer to remove hop leaves, if whole hops were used, or directly to a tank in which the wort is allowed to settle. Recently, tangential entrance to this tank was found to produce a rapid settling of the precipitate that always forms in the kettle. This precipitate, called trub, contains hop bitter resins, proteins, and tannins and needs to be eliminated from the beer.

After settling some 20–30 min the wort is passed through a heat exchanger and cooled to the temperature desired for fermentation. Immediately after cooling, the wort is aerated to provide oxygen for the yeast.

Yeast, from a preceding brew, is added (pitched) in measured amounts to the aerated wort. The usual quantity is 1 lb of liquid yeast per barrel of wort. This will give a count of about 12 million yeast cells per milliliter. The second stage of the brewing process now begins.

Fermentation

Yeast enters a cool solution containing oxygen, fermentable sugars, and various nutrients. The yeast quickly absorbs the oxygen as well as minor nutrients that it requires, such as phosphate, potassium, magnesium, and zinc. It then begins to metabolize sugars and amino acids. The glycolytic pathway that yeast uses is common to many higher organisms, and only the ability to split pyruvic acid into acetaldehyde and carbon dioxide makes yeast distinctive. The scheme for the fermentation of glucose is shown in Figure 18.

The temperature during fermentation is very rigidly controlled. Because fermentation itself is exothermic, the fermentation vessels in all beer fermentations are cooled by a refrigerated fluid circulating through coils in the tanks or through jackets surrounding the tanks. The process usually takes 5–9 d.

During this time the yeast also metabolizes amino acids and makes three to four times as much yeast as was pitched. The excess yeast, over that used for succeeding brews, is collected and sold for use in dietary supplements and flavors for soups and snack foods.

The metabolism of amino acids is orderly and proceeds in a specific sequence. Certain amino acids are metabolized first, followed by the others, as shown in Table 6.

S. carlsbergensis has a marked ability to flocculate when it completes fermentation. This tendency allows the yeast to settle quickly when fermentable sugars are exhausted and makes collection and reuse of the yeast fairly simple. Some brewers hasten this settling by using centrifuges to collect their yeast.

S. cerevisiae, on the other hand, first rises to the top of its fermenting tank, where it may be collected by skimming, but then settles to the bottom and may also be collected there. Certain strains of *S. cerevisiae* settle more quickly and have been chosen by many brewers of ales and stouts.

Finishing

The fermented beer may now be finished in one of several ways. The simplest and most widely used is merely to transfer the beer to another tank, chilling it en route. This stage, called ruh (rest), allows much of the still suspended yeast to settle and also removes some harsh, sulfury notes. Furthermore, the yeast removes some undesirable flavor compounds, notably diacetyl, which were produced during the earlier fermentation. Ruh normally takes 7–14 d.

An alternative way of finishing beer is to move it to another tank sometime before it is completely fermented. Again, it is chilled en route, and the so-called secondary fermentation is allowed to proceed at a much lower tem-

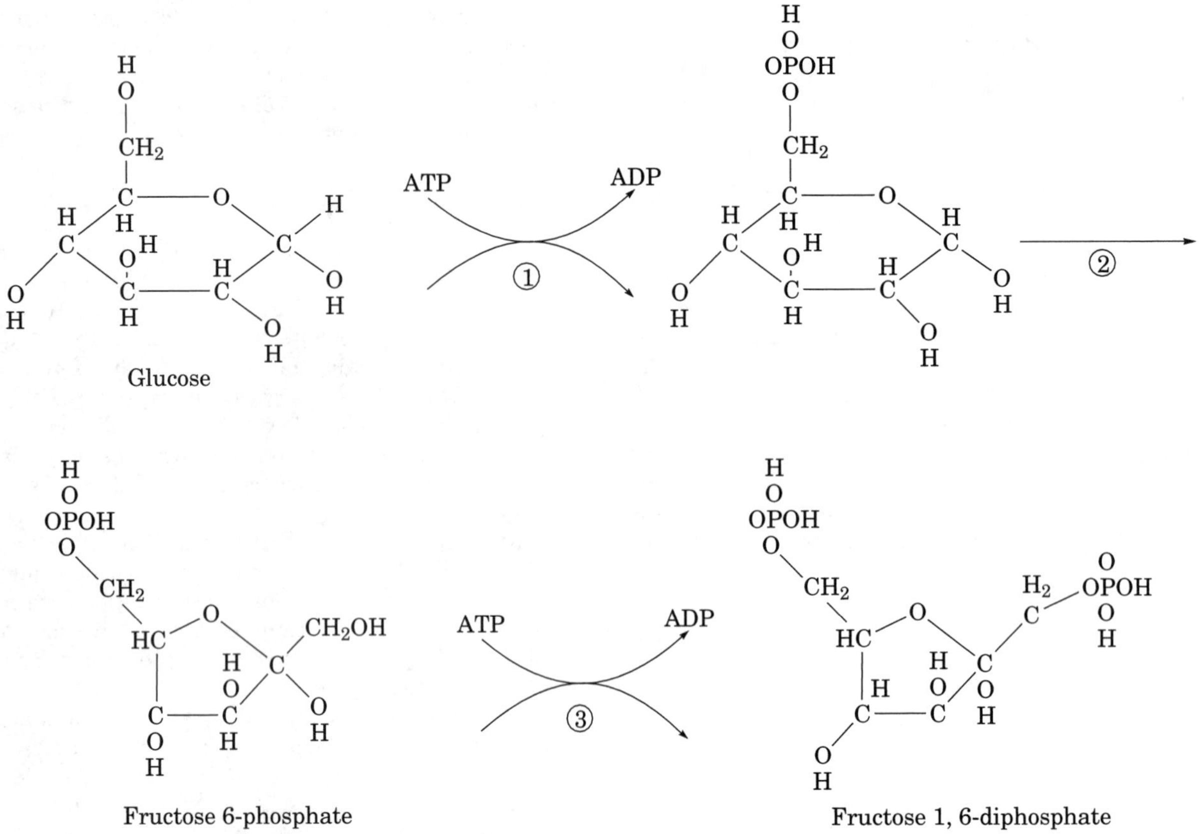

(1) Needs hexokinase, ATP, Mg^{2+} (or Mn^{2+})
(2) Needs glucosephosphate isomerase
(3) Needs phosphofructokinase—rate limiting

(**a**) Phosphorylation

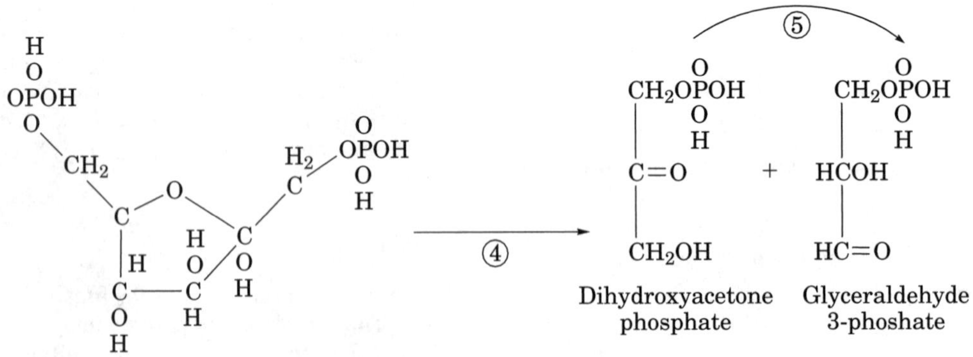

(4) Needs aldolase, Zn^{2+}, is a reverse aldol condensation
(5) Needs triose isomerase

(**b**) Splitting of hexose

Figure 18. Chemistry of alcoholic fermentation.

Glyceraldehyde 3-phosphate $\xrightarrow[\text{(6)}]{NAD^+ \rightarrow NADH + H^+}$ 1,3-Diphosphoglyceric acid

$\xrightarrow[\text{(7)}]{ADP \rightarrow ATP}$ 3-Phosphoglyceric acid $\xrightarrow{\text{(8)}}$ 2-Phosphoglyceric acid

(6) Needs glyceraldehyde phosphate dehydrogenase and NAD^+
(7) Needs phosphoglyceric kinase
(8) Needs phosphoglyceromutase

(**c**) Oxidation without oxygen

2-Phosphoglyceric acid $\xrightarrow{\text{(9)}}$ Phospho-enolpyruvic acid $\xrightarrow[\text{(10)}]{ATP \rightarrow ADP}$ Pyruvic acid

(9) Needs enolase and Mg^{2+}
(10) Needs pyruvate kinase

(**d**) Formation of pyruvic acid

Figure 18. Chemistry of alcoholic fermentation (*continued*).

perature, 0–3°C. Much the same cleansing action occurs, but the early transfer ensures the presence of sufficient yeast to cleanse the beer more efficiently. It takes 2–4 w. It has not found wide favor, probably because of the need to transfer the beer at times that could be inconvenient.

A third, and most elegant way, is to transfer the fermented beer to another tank, chilling it less, and treating it with a small quantity of beer that just started to ferment a day earlier. This subjects the fermented beer not just to a specific quantity of yeast, but to very active yeast that very effectively cleanses the fermented beer. This process, called krauesening, is used by very fastidious brewers. It normally takes 3–5 w.

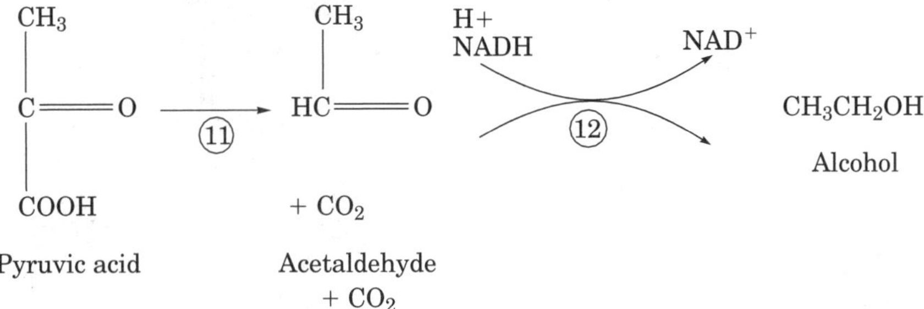

(11) Needs carboxylase, Mg^{2+} and thiamin
(12) Needs alcohol dehydrogenase and Zn^{2+}

(**e**) Production of alcohol and carbon dioxide

Figure 18. Chemistry of alcoholic fermentation (*continued*).

Table 6. Metabolism of Amino Acids

Extent of absorption	Acid
Group A	
Almost completely absorbed within 24 hours	Glutamic acid Glutamine Aspartic acid Asparagine Serine Threonine Lysine Arginine
Group B	
Absorbed gradually during fermentation	Valine Methionine Leucine Isoleucine Histidine
Group C	
Absorbed only after Group A amino acids are gone	Glycine Phenylalanine Tyrosine Tyrptophane Alanine Ammonia
Group D	
Absorbed only slightly	Proline

After any of these finishing processes, the beer is filtered, always in the cold, to remove insoluble particles and yeast. This filtration uses diatomaceous earth as the filter medium. The beer may be filtered again, and this filtration may be a sterile filtration to remove all yeast and lactobacilli or another diatomaceous earth filtration. A sterile filtration may involve cotton fibers, a porous plastic sheet, or a ceramic filter as the retaining barrier.

Beer Packaging

Beer is packaged in either cans or bottles or in large containers or kegs for use in restaurants and taverns. The packaging machinery for beer has become very sophisticated and fast. Because oxygen changes the flavor of beer (see later), modern fillers evacuate the bottle before it is filled and replace the air with CO_2. Then the beer enters against a counter pressure of CO_2. Bottles are being filled at more than 20/s, and cans at more than 30/s. A high-speed can filler is shown Figure 19.

If such packaged beer has not been sterile-filtered, it must be pasteurized, as beer is a fertile medium for many microbes. The pasteurization may be done just before filling, a so-called bulk pasteurization, or after filling in long tunnels with hot-water sprays. Bulk pasteurization takes about a minute, tunnel pasteurization about an hour.

Kegs may be filled with diatomaceous earth-filtered beer, in which case they are kept refrigerated; or they may be filled with sterile filtered or bulk pasteurized beer, in which case they do not need refrigeration.

Problems of Beer

The complexity of beer makes it prone to several problems. First and foremost is that the delicate flavor of beer is adversely affected by oxygen, so great pains are taken in the brewery and in the packaging operation to minimize the pick up of oxygen. When yeast is present, it will scavenge oxygen and keep the beer oxygen-free. But after filtration beer is very prone to pick up oxygen. At every transfer, steps are taken to avoid oxygen pickup, which is monitored closely. The filling operation has been mentioned earlier.

Correctly packaged beer has a flavor shelf life of about 4 m: if kept cold, it will last longer. The change in flavor with oxidation is difficult to describe because it is unique to beer. It has been described as papery, bready, or cardboard. But mainly it loses the fine appeal of fresh beer.

Another problem of beer is that it develops a peculiar, skunky aroma if packaged in clear or green bottles and is exposed to light. The problem has been eliminated with the

Figure 19. High-speed can filler. *Source:* Courtesy of H&R Inc.

discovery that brown bottles filter the offending wavelengths of light. The aroma is caused by the action of light on isohumulone, which splits off an unsaturated hydrocarbon, which in turn reacts with a trace of hydrogen sulfide in the beer to give a mercaptan.

A third problem is that beer is a good medium for some nonpathogenic microbes and so must be sterile-filtered or pasteurized. Either of these treatments effectively eliminate the problem.

Another shortcoming that has largely been eliminated is the tendency for beer to become cloudy when it is chilled. This haze is caused by a reversibly insoluble protein-polyphenol complex. It is removed in one of several ways. The protein may be solubilized by the enzyme papain, or the polyphenols may be removed by absorption onto silica gel. These treatments are effective and widely used.

Light Beer

The enzymes present in barley malt cannot completely degrade starch to fermentable sugars. So all regular beers contain dextrins, fragments of starch held by 1,6′ linkages and too large to be fermented by yeast. There is an enzyme, however, that can break up all the starch—amyloglucosidase.

A true light beer is one in which all the starch has been made fermentable. This is accomplished by the use of amyloglucosidase, either in the brewhouse to work with the α-amylase and β-amylase or during fermentation to work after the α-amylase and β-amylase.

Some light beers are made by merely extending the action of the α-amylase and β-amylase. This produces a beer reduced in dextrins but not free of them. The legal definition of a light beer in this country permits this.

Analysis of Beer

Typical analysis of some American beers are shown in Table 7.

Table 7. Analysis of American Beers

	Standard beer	All malt beer	Light beer
Water	92	91	6
Alcohol, wt%	3.8	3.9	3.3
Carbohydrates, %	3.4	3.8	<1.0
Protein, %	0.3	0.6	0.2
Acids, as lactic, %	0.1	0.1	0.1
Ash, %	0.1	0.2	0.1
Isohumulones, ppm	15	25	15

Table 8. Vitamins in Beer

Vitamin	Vitamin content of beer (mg/L)	% Minimum daily requirements supplied by 1 L
Thiamin	0.02	<2
Riboflavin	0.3	20%
Pantotenic acid	1.0	25%
Pyridoxine	0.5	20%
Biotin	0.007	7%
Cyanocobalamin	0.1	3%

Types of Beer

Beer may be classified in many ways:

- By type yeast
  1. Lager yeast, S. carlsbergensis—By color: Light to amber color—lager beer; Dark color—porter; Dark and strong—bock beer. Or by process: Ruh beer; Krauesened beer.
  2. Ale yeast, S. cerevisiae—By color: Light to amber color—ale; Dark colored—stout.
- By properties of the water
  1. Alkaline, hard water—Munich type

2. Neutral, hard water—Pilsen type (All beers made now are Pilsen type)

- By alcohol content
  1. No-alcohol beer
  2. Low-alcohol beer (shank beer)
  3. Beer, ale
  4. Malt liquor
  5. Barley wine
- By market category
  1. Super premium
  2. Premium
  3. Standard, regional
  4. Generic

Dietary Aspects

Beer, being made from grains, has considerable nutritional value. Table 8 shows the vitamin content of a typical beer.

Beer also contains the trace minerals zinc, copper, and manganese. It is also very low in sodium, 12–18 mg sodium per liter, and high in potassium, 110–125 mg potassium per liter. This ratio of sodium to potassium has made for the inclusion of beer in the diets of patients with fluid retention problems. The well-known diuretic effect of beer is also a factor in many diets.

PLANT OPERATIONS

Breweries are operated most efficiently on round-the-clock schedules. The limiting factor in production is almost always capacity of the brewhouse, mainly the lauter tub and kettle. Some breweries will have one malt mill, one cereal cooker, one mash tub, two lauter tubs, and two kettles. With a mash filter or Strainmaster, only one may be required. Large breweries will have more than one brewing line. The maximum volume of beer that can be produced from the largest practical kettle (31,000 gal) is about 3–4 million barrels (90–120,000,000 gal). Any plant making more than this has more than one brewing line.

Round-the-clock brewhouse operation also demands 24-h manning in the fermentation cellars. Although fermenting and aging tanks are always above street level, the rooms they are in are referred to as cellars, from the days when the absence of mechanical refrigeration made underground storage necessary to keep the fermenting beer cool.

The speed of packaging equipment allows the so-called bottleshop section of a brewery to run on less than round-the-clock operation, even if the rest of the brewery is on a 24-h schedule.

The many choices involved in the brewing process, and the actual day-to-day running of the brewery (but not the bottleshop) are under the domain of the brewmaster. All else in a brewery, including the bottleshop, are the domain of the plant manager.

QUALITY CONTROL

Quality are those attributes of a product that makes it desirable to a consumer. In the production of beer there are many opportunities for undesirable properties—in aroma, in taste, and in appearance—to develop. Brewers have learned to be very careful about their products.

Quality control begins with the raw materials used in brewing. The water used is monitored for taste, alkalinity, hardness, and trace metals.

Barley malt is controlled very closely. The variety of barley is always specified, and sometimes the region in which it is grown. The malting process is controlled indirectly, by the analysis of the finished malt. The kilning cycle is specified, and the temperatures on the kiln are controlled.

Analyses exist for malt that reveal the size and vitality of the barley used, the extent to which the barley is germinated, how well it was modified or malted, and how well it will perform in the brewery. A typical analysis of a two-row and six-row malt is shown in Table 9.

The assortment indicates the size and grade of the barley used to make this malt. Uniformity in size is as important as absolute values. The growth shows how many dead kernels there were in the barley and how extensively and how uniformly the barley germinated.

The extract values, which give the amount of soluble matter in the malt, is an important economic factor. The difference in extract between a coarsely ground and a finely ground malt indicates how well the barley was modified; the smaller the difference the better the modification.

The diastatic power and α-amylase value measure the β-amylase and α-amylase, respectively. The ratio of soluble protein over total protein indicates how well the barley grew during malting and can predict yeast behavior in the brewery.

Hops are specified by variety and country or state in which they are grown. Samples from lots that could be purchased are examined visually by the brewer. Acceptance

Table 9. Analyses of Two-Row and Six-Row Malts

	Two-row malt	Six-row malt
Assortment		
On 7/64 in., %	61.4	47.5
6/64	30.7	43.3
5/64	7.0	9.2
Through 5/64	0.9	1.0
Growth		
0–1/4	0	0
1/4–1/2	1	2
1/2–3/4	5	6
3/4–1	84	83
Over 1	10	9
Moisture, %	4	4
Extract, fine grind, as is, %	77.3	75.2
Extract, coarse grind, as is, %	76.0	73.8
Diastatic power	133	142
α-Amylase	51	55
Total protein, %	13.7	13.8
Soluble protein, %	5.8	5.6

depends on aroma and appearance. Analysis for humulones accompanies the sample and may be a criterion also.

Corn and rice are judged by extract values and low oil content. In the brewery, control begins with the wort produced in the kettle. This is judged for solids content, color, bitterness, pH, degree of fermentability, and sterility.

The fermentation process is monitored for rate of fermentation (decrease in specific gravity) and temperature and biologically for the presence of lactobacilli. The aging process is checked for gravity and microbiologic state.

After filtration, the oxygen content is monitored, as is the carbon dioxide level. Just before packaging, and possibly again in the package, the beer is analyzed completely to maintain standards and uniformity.

Tasting is an integral part of quality control. In many breweries beer is tasted just before it is transferred from one tank to another. It is again tasted just before packaging, and again, critically, in the package. Taste panels are selected from brewery workers and are trained to detect certain flavors in beer and to describe them. These panels operate daily, or twice daily, with rotating membership. A view of a taste panel is shown in Figure 20.

WORLD BEER PRODUCTION

Beer is made in almost every country in the world. It probably ranks next to bread in the ubiquity of its production. Table 10 lists the production, in 1,000 hectoliters, in these countries.

GLOSSARY

Apparent extract. The density, usually measured with a hydrometer or a densitometer of a fermenting or fermented liquid. In beer it measures the extent to which the wort has fermented. The value reflects the residual solids in the beer, modified by the alcohol present (which lowers the reading).

Figure 20. A view of a taste panel. *Source:* Courtesy of Miller Brewing Co.

Table 10. World Beer Production in 1997 (in 1,000 hL)

The Americas	
United States	236,430
Brazil	88,200
Mexico	51,949
Canada	22,355
Colombia	20,000
Venezuela	17,232
Argentina	12,063
Peru	8,500
Chile	3,900
Dominican Republic	2,400
Ecuador	2,270
Bolivia	1,800
Paraguay	1,600
Cuba	1,500
Panama	1,350
Costa Rica	1,200
Honduras	1,184
Guatemala	1,000
Uruguay	800
El Salvador	700
Jamaica	574
Nicaragua	400
Trinidad	320
Puerto Rico	314
Guyana	180
Haiti	140
Bahamas	139
Netherlands Antilles	135
Barbados	112
Surinam	89
Belize	70
Martinique	70
Santa Lucia	49
St. Vincent	35
Grenada	34
Antigua	22
St. Kitts	18
Dominica	12
Total	479,146
Far East	
China	170,000
Japan	67,695
South Korea	16,740
Philippines	13,475
Thailand	8,360
Vietnam	5,638
India	4,250
Taiwan	3,900
Indonesian	1,722
Malaysia	1,477
Hong Kong	890
Singapore	804
Nepal	350
Laos	286
Sri Lanka	266
Cambodia	175
Lebanon	126
Mongolia	100
Burma	60
Pakistan	20
Total	296,334

Table 10. World Beer Production in 1997 (in 1,000 hL) *(continued)*

Near East	
Israel	800
Syria	102
Jordan	56
Iraq	50
Total	1,008
South Pacific	
Australia	17,349
New Zealand	3,214
Papua-New Guinea	390
Figi	161
Tahiti	146
New Caledonia	125
Samoa	45
Solomon Islands	19
Total	22,457
Europe	
Germany	114,800
UK	58,139
USSR	25,249
Spain	24,879
Netherlands	24,701
France	19,483
Poland	18,800
Czech Republic	18,649
Belgium	14,168
Italy	11,455
Austria	9,366
Denmark	9,180
Ireland	8,152
Romania	7,506
Turkey	7,448
Hungary	7,168
Portugal	6,623
Yugoslavia	6,106
Ukraine	6,090
Sweden	4,810
Finland	4,793
Slovak Republic	4,394
Greece	3,945
Croatia	3,607
Switzerland	3,563
Bulgaria	3,004
Norway	2,298
Slovenia	2,123
Lithuania	1,356
Kazakhstan	730
Latvia	687
Bosnia-Herzegovina	620
Macedonia	600
Estonia	585
Uzbekistan	500
Luxembourg	481
Georgia	450
Cyprus	320
Albania	152
Malta	132
Armenia	52
Iceland	51
Azerbaijan	15
Total	437,230

Table 10. World Beer Production in 1997 (in 1,000 hL) *(continued)*

Africa	
South Africa	25,000
Nigeria	4,300
Kenya	3,000
Cameroon	3,253
Zimbabwe	1,649
Tanzania	1,615
Zaire	1,525
Ivory Coast	1,240
Burundi	1,161
Angola	980
Ethiopia	856
Namibia	888
Uganda	830
Rwanda	808
Gabon	801
Marocco	800
Tunisia	780
Guinea	773
Malawi	760
Mozambique	698
Zanbua	558
Upper Volta	458
Botswana	419
Egypt	400
Lesotho	398
Benin	358
Algeria	350
Madagascar	350
Congo	347
Mauritius	347
Swaziland	294
Togo	292
Reunion	239
Eritrea	218
Central African Republic	207
Senegal	162
Chad	157
Guinea	131
Sierra Leone	100
Niger	72
Seychelles	70
Mali	65
Liberia	60
Cape Verde	54
Gambia	20
Total	57,868

Barley malt. A legally required ingredient in beer; it is barley that has been wetted and induced to germinate and then heated to stop the germination. The germination partly digests the starch and protein in the barley and produces some enzymes. The temperature at which the terminal drying is done determines the color of the resulting beer. Barley may be either two-rowed or six-rowed and may be either blue or white. Malt provides carbohydrates, protein, and yeast nutrients to the brewing process.

Bitterness unit or isohumulones. A quantitative measure of the amount of isohumulones—the major bitter compound from hops—in beer, in milligrams per liter.

Corn grits, corn flakes, corn starch, corn syrup, rice, milo, and sorghum. Ingredients used as adjuncts to dilute the barley malt; they are only a source of carbohydrates.

Enzyme. A protein, made by a living cell or organism, that catalyzes or promotes a chemical reaction. Most of the processes of life, and most of those in brewing, are mediated by enzymes.

Hops. The flower of a perennial vine Humulus lupulus grown in the United States only in Washington, Oregon, Idaho, and California. Many varieties exist, with differing aromas and bittering qualities. The United States grows the varieties called Clusters, Northern Brewer, Fuggles, Bullion, and Cascades.

Krauesen. A stage after primary fermentation in which about 15% by volume of freshly fermenting beer is added to beer that has completed its fermentation. The term, as a noun, is also used to describe the foamy head that appears on top of a fermenting tank at the height of fermentation.

Original gravity. The concentration of solids in the wort from which beer is made. In the United States and most other countries, the concentration is expressed in percent solids, or °Plato or °Balling (after the workers who laboriously determined these values). Thus a wort with 12% solids would be said to have a gravity of 12°P (or 12°B), and the beer made from it would be said to have an original gravity of 12°P. An analysis of beer permits you to calculate the original gravity of the wort from which it was made, without reference to the wort. In the United Kingdom, the concentration of solids in the wort is expressed as the actual specific gravity. A wort with 12% solids would have a specific gravity of 1.048 and would be said to have a gravity of ten forty-eight.

Real extract. The actual percentage of solids in a beer, determined after removing the alcohol present.

Ruh. The period of rest between fermentation and filtration during which the beer clarifies itself and loses some harsh notes.

Wort. The liquid from which beer is made; a warm-water extract of malt and other grains.

Yeast. The microorganism that converts simple sugars to alcohol and carbon dioxide. Two types of yeast are used in brewing—Saccharomyces carlsbergensis for lager beers and S. cerevisiae for ales. Most brewers have their own strain of these yeasts. During a normal brewers fermentation, yeast multiplies three- to fourfold, yielding more than enough yeast for successive brews. Yeast also produces many compounds that have a marked influence on the flavor of the beer.

GENERAL REFERENCES

Briggs, D. E., J. S. Hough, R. Stevens, and T. W. Young, *Malting and Brewing Science*, Vol. 1, *Malt and Sweet Wort*; Vol. 2, *Hopped Wort and Beer*, Chapman and Hall, New York, 1981.

Broderick, H. M., *The Practical Beer*, Master Brewers Association of the Americas, Madison, Wis., 1977.

Broderick, H. M., *Beer Packaging*, Master Brewers Association of the Americas, Madison, Wis., 1982.

Burgess, A. H., *Hops*, Interscience Publishers, New York, 1964.

European Brewery Convention, Fermentation and Storage Symposium, *E.B.C. Monograph V*, European Brewery Convention, Amsterdam, The Netherlands, 1978.

European Brewery Convention, Symposium on the Relationship Between Malt and Beer, Helsinki, *E.B.C. Monograph VI*, European Brewery Convention, Amsterdam, The Netherlands, 1980.

European Brewery Convention, Flavour Symposium, Copenhagen November 1981, *E.B.C. Monograph VII*, European Brewery Convention, Amsterdam, The Netherlands, 1981.

Findlay, W. P. K., *Modern Brewing Technology*, Macmillan, London, 1971.

Pollock, J. R. A., *Brewing Science*, Vol. 1, 1979; Vol. 2, 1981; Vol. 3, 1987, Academic Press, Orlando, Fla.

Preece, I. A., *The Biochemistry of Brewing*, Oliver & Boyd, London. 1954.

Rose, A. H., and J. S. Harrison, *The Yeasts*, Vol. 2, Academic Press, Orlando Fla., 1987.

Strausz, D. A., *The Hop Industry of Eastern Europe and the Soviet Union*, Washington State University Press, Pullman, Wash., 1969.

Joseph L. Owades
Consultant
Sonoma, California

BEVERAGES: CARBONATED

Carbonated beverages, or soft drinks, are the most popular beverages in North America and a growing category worldwide. Per capita consumption in the United States reached 54 gallons per year in 1997. Worldwide consumption is estimated at almost 7.5 gallons per year for every man, woman, and child alive. This has created an industry with estimated annual U.S. sales of over $54 billion, employing more than 136,000 workers, producing hundreds of different flavors.

Few products are as closely associated with American ingenuity, enterprise, and culture as soft drinks. The trademark of the market leader, the Coca-Cola Company, is the most recognized trademark in the world. That one company served an estimated 1 billion customers in 1997. Although a few large corporations dominate sales, there has always been a market for hundreds of small businesses offering unique alternatives to the mass-marketed brand names.

The emergence and growth of this category resulted from the ideas of individuals in the United States and Europe, who found better flavors, better marketing, or better ways of producing their beverages. Although descriptions of naturally carbonated mineral waters date back 2,000 years, the production and distribution of flavored soft drinks began little more than 200 years ago. The development of soft drinks follows (see "History of Carbonated Beverages"). What is a carbonated beverage and how is it different from other beverages?

DESCRIPTION

The favored term for carbonated beverages is now carbonated soft drink, or simply soft drink, to differentiate them from beverages containing alcohol. They are characterized by containing carbon dioxide (CO_2) gas, which effervesces (bubbles) when pressure is released, as when opening a bottle. Soft drinks have some added flavor and usually a sweetener to distinguish them from seltzer water. The primary ingredient of course is water, comprising from 85 to 99% of the formula.

INGREDIENTS

The water used in soft drinks must be free of objectionable tastes, odors, or colors. It must be safe to drink and low in organic and mineral content. Most soft drink bottlers employ extensive water treatment to assure this primary ingredient is pure and suitable for their products. Although the water must meet federal drinking water standards, most bottlers go well beyond those specifications to assure stability in a complex beverage system that must be shelf stable for six months or more.

Carbon dioxide is delivered to the bottling plant as a liquid compressed under high pressure. As the gas is needed, the liquid CO_2 is metered through a regulator where pressure is decreased and the liquid is vaporized to a gas. The gas is then dissolved in water in direct proportion to the pressure and temperature, following the universal gas laws. The amount of CO_2 in the product is measured in gas volumes, where *one gas volume equals the volume of gas that will dissolve in water under standard conditions*. The treated water is chilled to absorb more CO_2 and carbonated by exposing it to a countercurrent flow of CO_2 gas under pressure. Different flavors specify different levels of carbonation, up to a practical maximum of 5.0 gas volumes.

Most soft drinks are sweetened with sugars, predominantly in recent years by a liquid corn sweetener known as high fructose corn syrup (HFCS). Corn syrup is refined from cornstarch, which is a polymer of the simple sugar dextrose (*d*-glucose). Starch is hydrolyzed to dextrose and then a portion of the dextrose is enzymatically converted to fructose to enhance sweetness. HFCS is transported across North America by rail tank car or by tank truck and delivered in bulk. Prior to the 1950s, granulated sugar (sucrose) was most common, delivered in bags or bulk, and this sweetener still predominates in most areas of the world.

Sugar, HFCS, and honey are nutritive sweeteners, meaning they supply some food energy or calories. A significant portion of the U.S. market, about 15%, prefers diet or sugar-free soft drinks made with nonnutritive sweeteners. These are typically intense sweeteners, hundreds of times sweeter than sugar, and are therefore used in small amounts measured in parts per million (ppm or μg/L).

Where sugars may constitute 10 to 12% by weight of the beverage, an intense sweetener makes up a fraction of 1%. Sugars also contribute to the mouth-feel of the beverage and help carry and smooth flavors across the tongue. This presents a challenge to the flavor chemist formulating sugar-free products. The balance of flavors found in a sugar-sweetened beverage is difficult to duplicate in the diet version. Examples of nonnutritive sweeteners include aspartame, saccharin, sucralose acesulfame potassium, and alitame. Table 1 lists several of the sweeteners used in the beverage industry with relative sweetness compared with sucrose, solubility at 20°C, and approximate usage levels in beverages.

Another important ingredient is an acidulent or acid, typically citric acid or phosphoric acid, which serves several purposes. The primary function is to modify the flavor by contributing a bite or tang sensation while balancing the sweetness in a sweet–sour relationship. The acid also lowers the beverage pH (a measure of acidity or hydrogen ion concentration) increasing the solubility of some ingredients and helping to prevent spoilage. Carbon dioxide gas forms carbonic acid in solution and therefore is itself a minor acidulent. Other common acid ingredients include malic acid, found in apples and cherries; tartaric acid, primarily found in grapes; and acetic acid, found in vinegar.

The defining ingredient of soft drinks is the flavor system, which differentiates one product from another. Flavors can be simple or very complex combinations of ingredients blended together to achieve a desired balance. Some soft drink flavors have hundreds of ingredients, some of which may be added just to prevent analysis and unauthorized duplication. The secrecy of such formulas is legendary. Fruit juices or juice extractives are common in fruit flavors. Extracts of spices, nuts, and herbs characterize colas, ginger ales, and root beers.

Flavors are categorized by law as "natural" or "artificial" depending on their source and means of manufacture. The U.S. Code of Federal Regulations (21 CFR 101.22) states:

> Natural flavors include: essential oils, oleoresins, essence or extractive. . . . or product of roasting, heating or enzymolysis, which contains the flavoring constituents derived from a spice, fruit or fruit juice, vegetable or vegetable juice, edible yeast, herb, bark, bud, root, leaf. . . . meat, seafood or dairy product . . . whose significant function is flavoring rather than nutritional.

Artificial flavors are not derived from this list of foods, but typically are synthesized chemicals that may or may not be identical to natural flavor constituents. A great amount of research has been done and is continuing to find new flavoring materials in nature and in the laboratory. All in-

Table 1. Beverage Sweeteners

Sweetener	Relative sweetness	20°C (%) solubility	Approx. (mg/L) usage level
Acesulfame potassium	200	27	500
Alitame	2,000	14	50
Aspartame	180	1	600
Crystalline fructose	1.5	80	90,000
Cyclamate	30	13	(Blend)
HFCS 42	0.9		111,000 (solids)
HFCS 55	1		100,000 (solids)
Saccharin	300		350
Sorbitol	0.6	230	(Blend)
Sucralose	600	28	180
Sucrose	1	60	100,000

gredients used in foods must be approved by the U.S. Food and Drug Administration (FDA). Most common soft drink flavoring materials are on the generally recognized as safe (GRAS) list.

It is well established that our perception and memory for flavor is primarily based on our sense of smell. Taste receptors on the tongue play a role but are not as sensitive as the olfactory receptors found in our nasal passages. This is why the volatile essences or "top notes" are so important to successful flavors.

Colors are also important ingredients in beverages. Consumer acceptance and even flavor perception is dependent on the appearance of the product. Until recently, the choices of stable color ingredients have been limited to about a dozen useful in soft drinks. As our scientific understanding of chemistry advances, more choices are becoming available, allowing a rainbow of colors to be offered. The regulation of colors can present some contradictions: the natural color component of red cranberries can be used in a cranberry drink and labeled "natural"; however, the same natural color ingredient used in cherry flavor must be labeled "artificial color."

Preservatives are found in most soft drinks to prevent spoilage, protect flavors, and extend shelf life. Soft drinks are generally packaged cold to maintain carbonation, and they cannot be heat processed to eliminate spoilage yeast and mold spores. Although conditions in modern bottling plants are sanitary, the total elimination of microorganisms is nearly impossible. Agents that prevent microbial growth in soft drinks include CO_2, acids, and some flavoring materials. Beyond these, many formulas include a preservative to prevent spoilage and assure product safety. The most common are sodium or potassium benzoate, salts of the common benzoic acid found in fruits and berries. Another type of preservative is an antioxidant (such as vitamin C), which protects sensitive flavor ingredients from oxidation and the resulting development of off-flavors. Sequestrants (such as citric acid) prevent chemical reactions by binding with the metals that act as catalysts.

Other functional ingredients include foaming agents, emulsifiers, clouding agents, and gums. Foaming agents stabilize the foam "head" on a glass of root beer or sarsaparilla. Emulsifiers stabilize oil-in-water emulsions critical to oil-based flavors, such as citrus flavors. Clouding agents make beverages cloudy or opaque to appear more natural, like orange juice. Gums can have a variety of functions including improving the mouth-feel of diet drinks.

Table 2 compares the ingredients in popular soft drink flavors.

Table 2. Typical Soft Drink Ingredients

Flavor	Brix (% sugar)	Acid	pH	Co_2 gas volumes	Calories/8 oz.
Cola	10	Phosphoric	2.3	3.0	93
Cream	11–13	Citric/none	5.5	2.9	115
Grape	14	Tartaric	3.0	2.8	125
Lemon-lime	11	Citric	2.7	3.5	108
Orange	12–14	Citric	3.5	2.5	120
Ginger ale	8	Citric	2.6	4.0	80
Root beer	11–12	Varies	5.0	3.2	113

NUTRITIONAL VALUE

Soft drinks are usually consumed for refreshment and enjoyment, not for nutritional value, but they do have value. They are a significant source of essential water in the diet. Many people who drink soft drinks on the job or while travelling may have limited choices of safe or appealing liquid alternatives. Soft drinks supply quick energy in the form of readily absorbable carbohydrates (sugars), which can be valuable to athletes, workers, and those with calorie deficient diets. Soft drinks have often provided much needed relief to victims of floods and natural disasters, who have no clean source of water for drinking or preparing meals. Recent lifting of government restrictions has allowed fortification of soft drinks with vitamins and nutritive additives. Now we have "value added" products enriched with vitamin C or calcium to increase nutritional benefits.

AVAILABILITY

One important reason for the increasing consumption of soft drinks is availability. Historically, soft drinks were made at home, or close to home, and by the early 1800s, many cities had choices of several purveyors. Even before the advent of bottled product, soda fountains were common in drugstores, candy shops, and saloons. As the industry grew in the later nineteenth and early twentieth centuries, mobility increased, bottling "pop" became common, and bottlers did their best to make it available in grocery stores, gas stations, and corner stands. Now you can buy soft drinks from convenience stores, fast food counters, and the ubiquitous vending machines. Of course they are available on the Internet! Increasing availability is a corporate marketing priority, and the competition is fierce.

TERMINOLOGY

Carbonated beverage. Preferred term adopted by manufacturers in 1919 to favorably disassociate the packaged effervescent beverages from other types of soft drink.

Soft drink. Commonly adopted early in the twentieth century to disassociate flavored refreshment waters from alcoholic beverages. Includes all varieties of carbonated and still beverages in bottles and cans and also those prepared at the fountain.

Soda. Originally a specific type of carbonated water, which included sodium bicarbonate for its medicinal properties. Term became generic in America for the broad category of flavored carbonated soft drinks. Because soda is not an ingredient in modern beverages, the term "soda water" is disfavored by the industry, although "soda" persists as a popular designation for carbonated beverages, particularly in the northeastern United States.

Pop. Popular term in the late nineteenth century for bottled flavored soda waters and still widely used by consumers in the central parts of the United States. Derived from popping noise made when the cork or other closure was removed from the bottle.

Tonic. A New England generic name for packaged carbonated beverages derived from the health values attributed to early carbonated waters. Also an English term now used in the United States to designate carbonated quinine water.

HISTORY OF CARBONATED BEVERAGES

Ancient Greeks and Romans knew the virtues of bathing in naturally occurring carbonated mineral waters. By the sixteenth century, extensive efforts were made by physicians and scientists to characterize and reproduce these physical benefits. The Swiss alchemist Phillipus Aureolus Paracelsus, recorded the earliest written observations. A century later, the Belgian physician Jean Baptiste Van Helmont (1577–1644) applied the name "gas" to mineral spring vapors. In 1757, Joseph Black, a Scottish physician, established that "fixed air" (CO_2 gas) could be extracted from limestone with oil of vitriol (sulfuric acid). In the 1780s Thomas and William Henry discovered that increasing pressure increased solubility of gases (Henry's law). By 1823, Sir Humphry Davy and Michael Faraday were liquifying CO_2 gas in England.

Simultaneously in America, scientists were studying mineral springs near Saratoga, New York, and developing ways to duplicate their properties. In 1798 the term "soda water" was first introduced. In 1807, Dr. Benjamin Silliman began selling carbonated mineral waters in New Haven, Connecticut. The first U.S. patent for the manufacture of imitation mineral waters was issued to Joseph Hawkins of Philadelphia. Developments followed rapidly on both sides of the Atlantic Ocean. Bottled soda water was first sold in the United States in 1835. Flavors were added soon after that in Germany and the United States, the earliest being fruit juices. Artificial flavors appeared in the 1850s along with ginger ale, root beer, and vanilla flavors sold through pharmacies. These early "tonics" were compounded and dispensed by chemists in apothecaries for consumption on the premises. The health benefits of mineral waters were expanded with the addition of herbs and sodium bicarbonate to soothe the stomach. By the late 1800s, the availability of liquid CO_2 in steel cylinders allowed the transition from soda fountains to bottling shops, and the industry took off. Some of the earliest products that grew into businesses include:

- Cantrell & Cochrane (C&C) ginger ale was exported from Belfast, Ireland, in the 1850s.
- Vernors ginger ale was developed in Detroit, Michigan, by James Vernor when he returned from the Civil War in the late 1860s.
- Hires root beer was introduced at the Philadelphia Centennial in 1876 by Charles Hires.
- In 1881, Henry Millis started selling Clicquot Club ginger ale in Boston.
- In 1884, Dr. Augustin Thompson developed Moxie in Lowell, Massachusetts.
- Dr Pepper formula developed by young Charles Alderton, was sold to Robert S. Lazenby, proprieter of the Circle "A" Ginger Ale Company in Waco, Texas, in 1885.
- In 1886, Dr. John Pemberton, an Atlanta druggist, produced the first 25 gallons of what became Coca-Cola, in a three-legged pot in his backyard. By 1904, 123 franchised bottlers sold one million gallons of Coke.
- Caleb Bradham of New Bern, North Carolina, named his new beverage Pepsi-Cola in 1898. By 1910 he had sold franchises to 280 bottlers.
- In 1890, John McLaughlin opened a small plant near Toronto to sell carbonated water to local drug stores. He introduced Canada Dry pale dry ginger ale in 1904.
- The Royal Crown name was trademarked by Hatcher Grocery Company in Columbus, Georgia, in 1905.

By the turn of the twentieth century the soft drink industry was well on its way to success. In 1900, 2,800 bottling plants produced over 1 billion 6 oz. bottles. Coca-Cola's first newspaper advertisement appeared in 1902. By 1921 there were over 6,000 bottlers helped in part by Prohibition, which closed many beer breweries. The number of bottlers peaked in 1929 at 7,920 and then tapered down during the Great Depression. After World II, the number of plants increased to 6,900 in 1949, but has been decreasing ever since, to less than 500 at the end of the twentieth century. But throughout the century, except for a few periods of war rationing restriction, the volume of soft drinks has increased as the plants became more efficient, more sanitary, and extended their reach through better packaging, transportation, and marketing.

SOFT DRINK INDUSTRY

An important factor in the sales growth of soft drinks has been innovation. The earliest successful purveyors had better ideas, better products, and better locations. To expand their businesses, they needed to develop better packaging, better production methods, and improved distribution. The important milestones on the road to success are acknowledged as (*1*) addition of flavors to mineral waters; (*2*) development of the fluted crown closure (bottle cap), capable of withstanding the pressures exerted by carbonated beverages; and (*3*) mass-produced glass bottles of uniform size, allowing development of faster bottling machinery. With a uniform package, the next innovation was the sale of franchise agreements, which allowed a franchisee to bottle and sell product exclusively within a prescribed geographic territory. Franchising the product quickly expanded its sales area with little capital cost to the franchisor. The Coca-Cola Company (Coke) and the Pepsi-Cola Company (Pepsi) quickly had large networks of hundreds of bottling plants each, few of which were owned by the parent company. Bottlers were independent businessmen operating within their territories as they wished with some minor constraints by the franchise contract. The most significant restriction was they must buy the flavor syrup from the parent company and follow the formula specifications.

Bottlers, being independent businessmen, became very competitive with bottlers of rival brands. Early "turf" battles for space in neighborhood stores and gas stations gradually evolved into the corporate marketing competition known as the cola wars. The advertising and distribution rivalry between Coke and Pepsi is legendary for its competition for priority in shelf space, sales outlets, stadium rights, and the bragging rights of being the "Official Soft Drink" of any and every event.

Bottlers also came to recognize the value of cooperation with their rivals on certain issues that adversely affected their business. The issues that finally brought them together were taxes and bottle "misappropriation," the stealing of empty bottles by a competitor to be refilled with the rival brand. The second issue led to the development of distinctive bottle designs for each brand and the controversial imposition of bottle return deposits. Bottlers formed regional and statewide organizations to discuss and resolve common issues.

SOFT DRINK BOTTLER ASSOCIATIONS

The first attempt at national organization was the U.S. Bottlers Protective Association in 1882. This group dissolved within five years and was replaced by the American Bottlers Protective Association (ABPA) in 1889. The ABPA held meetings to oppose taxes and resolve differences between members, but was largely ineffective. Finally, in 1919, the American Bottlers of Carbonated Beverages (ABCB) was formed under the strong leadership of James Vernor of Detroit. The group raised money to fight tax proposals that unfairly burdened the industry. They held regular meetings and formed committees to address industry issues. Soon, the annual meetings included equipment displays and the presentation of technical papers for the education of the membership.

The 1920s were boom years for soft drinks with amazing growth in sales and production technology. The ABCB hastened the spread of modern technology and equipment. One of the first accomplishments was the establishment of a scientific research program at Iowa State University to study industry needs and develop better testing methods. In 1923 they adopted the slogan "Quality—Purity—Service" to focus the image of the industry as a safe and reliable source of refreshment for consumers. The group introduced the Sanitary Code for Bottling Plants in 1929, establishing benchmarks for sanitary operations. In 1933 ABCB opened a beverage testing lab in Washington, D.C., to work with bottlers on product and ingredient specifications. This was supplemented by a mobile laboratory, which traveled the nation visiting bottling plants, upgrading local understanding and technical standards. This mobile lab continued for 27 years until supplanted by the growth of corporate field programs.

The ABCB also published important guidelines on the operations and financial controls of bottling plants. The industry faced many challenges from state and federal regulations and taxes that held the organization together, despite the competitive interests of the membership. The Great Depression of the 1930s drove many businesses out of existence, and bottling plants closed, but the five cent price of a bottle of cold soda made it an affordable treat. The advent of the rationing of sugar, trucks, and equipment in World War II also affected many bottlers, but the industry grew rapidly after the war. Some companies benefited from the war as American GIs spread their preference for soft drinks across Europe and Africa. Major brands were becoming international businesses.

ABCB was instrumental in the formation of the scientific association Society of Soft Drink Technologists in 1953. This group, now known as the International Society of Beverage Technologists, has greatly advanced the technical understanding of beverage chemistry and production. Annual technical meetings are attended by bottler, ingredient, and equipment members from around the world with the presentation and publishing of peer-reviewed technical papers.

In 1967, the ABCB changed its name to the National Soft Drink Association (NSDA), headquartered in Washington, D.C. A permanent staff of lawyers and technical managers coordinates activities of various volunteer committees to promote the industry and protect it from excessive legislation and taxes. The original reasons bottlers came together over 100 years ago, taxes and bottle return deposits, are still hot issues today.

SOFT DRINK PRODUCTION

The reason more beverage units are produced each year while the number of producing plants decreases is the advancement of beverage manufacturing technology. The industry has modernized in all aspects as it has developed larger and more efficient factories. Many related advances have combined to improve water purification, material handling, refrigeration, packaging speeds, warehousing, and transportation. The bottler of 1950 mixed a few bags of sugar to make 100 cases of 7 oz. bottles with a 32-station bottle filler filling about 150 bottles per minute. In 1998, the bottler mixes a tank truck of liquid sweetener to make 30,000 cases of 12 oz. cans with a 120-station can filler running at 2,000 cans per minute. He has automatic equipment to unload truckloads of empty cans to feed the conveyors and automatic equipment to load the finished product onto pallets, keeping several forklifts busy loading 15 semitrailer trucks to carry the product to market. Thanks to the interstate highway system, these trucks can now travel a couple hundred miles in a few hours to a distribution warehouse for local delivery. Table 3 summarizes the steps in the production of the modern soft drink can.

The modern beverage plant uses the latest in management tools and techniques to increase efficiencies, productivity, and product quality. Inventories are kept low by scheduling incoming materials to arrive hours before needed with "just-in-time" deliveries from suppliers. Production runs are scheduled to fill weekly orders from customers, assuring rapid warehouse turnover to keep fresh

Table 3. Steps in Soft Drink Can Production

Step	Description
Water treatment	Coagulation, sand filtration, carbon filtration, and polishing
Ingredients	Flavors, acidulent, colors, and preservative measured
Liquid sweetener	Filtered and metered into syrup tank
Syrup blending	Syrup compounded from ingredients, sweetener, and water
Water	Chilled and carbonated
Beverage proportioning	Beverage = 1 part syrup + 5 parts carbonated water
Can receiving	Empty cans depalletized, inspected, and cleaned
Can filling	Cans gravity filled through open end on revolving filler
Can seaming	Filled cans have lid applied and sealed with double seam.
Date coding	Each can has a date code printed on the side or bottom.
Fill check	Net contents measured, low fill cans rejected to scrap
Can warmer	Cans warmed above dew point to prevent condensation in cases
Multipacker	Cans grouped and wrapped in 6, 12, 18, or 24 packs
Case packer	Six or 12 packs combined into cases of 24
Palletizer	One hundred cases oriented into alternating stack patterns on wood pallets
Stretch wrap	Pallets wrapped in clear plastic film for protection and stability
Warehousing	Product stored in warehouse until needed to fill orders
Bulk delivery	Semitrailer loads transported to customer warehouses
Local delivery	Smaller trucks deliver to stores, restaurants, vending machines, etc.

product on the store shelves. Heating and cooling energy is recycled to minimize waste. Packaging materials are recycled, solid waste is minimized, and liquid wastes are pretreated before release to municipal treatment facilities.

Many plants run 16- to 20-hour daily production schedules with a sanitation shift during downtime. Quality improvement and worker safety programs are constantly reinforced by management. Strict quality programs are monitored by parent company representatives, who provide materials, training, and technical support to each plant in their system. Every facility has a modern testing laboratory with calibrated instrumentation to assure accurate quality test results. Many now have on-line instrumentation providing constant data logging of critical measurements.

MARKETING

The soft drink story begins with marketing and ends with consumer satisfaction. Carbonated beverage marketing is big business and possibly the most visible product promotion campaign existing. Direct advertising on signs, motor vehicles, store displays, print ads, television, radio, clothing, and anywhere else a logo or jingle can be found constantly reminds us we are thirsty. From subtle placements in movies to giant skywriting over sports events, soft drink advertising is ubiquitous. From the early patent medicine purveyors to the latest extreme-sport computer-graphic commercials, the point is the same, sell more soft drinks. These are fun refreshing products, so the ads tend to be more attractive and enjoyable than most. Not many products have more appealing packaging graphics or vending machine decoration to be sure.

Although good quality advertising and packaging increase desire to purchase, the product must also consistently deliver the promised refreshment at a reasonable price. Competitive promotion pressures have kept the price of soft drinks quite reasonable. There are times during holiday promotions when products are sold at or below cost to increase market share and meet volume targets. This discounting has contributed to the demise of the small, inefficient bottler and the growth of the larger franchisors. Another factor in that evolution is the consolidation of supermarket chains, giving stores more control over which products get shelf space and end-of-aisle displays.

Soft drinks are marketed primarily to the young, from teenage years through the twenties, because early preferences often continue as people age. The brands want to be associated with the "in crowd" and good times. The brand confers status to those that are seen with a can. As the population demographics are aging, more ads will be seen with older people enjoying soft drinks, perhaps with nostalgia.

Not only do the advertising campaigns adjust to changing times, but new product offerings will be available for changing tastes and lifestyles. The recent proliferation of sports drinks, flavored waters, and good-for-you beverages are keeping up with the trends. The future is unpredictable, but you can be sure soft drinks will evolve to fill the niche for liquid refreshment.

GENERAL REFERENCES

American Bottlers of Carbonated Beverages, *Plant Operation Manual*, Washington, D.C., 1940.

American Bottlers of Carbonated Beverages, *Bottlers Tentative Sugar Standard*, Washington, D.C., 1953.

American Bottlers of Carbonated Beverages, *Bottlers Tentative Liquid Sugar Standard*, Washington, D.C., 1956.

American Can Company, *Carbonated Beverages in the United States—Historical Review*, Greenwich, Conn., 1972.

Beverage Digest Online home page, URL: *www.beverage-digest.com* (last accessed April 19, 1999).

BEVnet, The Beverage Network home page, URL: *www.bevnet.com* (last accessed April 19, 1999).

Beverage World home page, URL: *www.beverageworld.com* (last accessed June 25, 1999).

Cadbury Schweppes plc. home page, URL: *www.cadbury schweppes.com* (last accessed April 19, 1999).

Coca-Cola Company, *Portrait of a Business*, Atlanta, Ga., 1971.

Dr. Pepper Museum and Free Enterprise Institute home page, URL: *www.drpeppermuseum.com* (last accessed April 19, 1999).

M. B. Jacobs, *Manufacture and Analysis of Carbonated Beverages*, Chemical Publishing, New York, 1959.

C. H. Lofland, *The National Soft Drink Association—A Tradition of Service*, National Soft Drink Association Washington, D.C., 1986.

National Soft Drink Association home page, URL: *www.nsda.org* (last accessed April 19, 1999).

T. Oliver, *The Real Coke, The Real Story*, Random House, New York, 1986.

J. J. Riley and the American Bottlers of Carbonated Beverages, *Organization in the Soft Drink Industry*, Washington, D.C., 1946.

J. G. Woodroof and G. F. Phillips, *Beverages: Carbonated and Noncarbonated*, revised edition, AVI Publishing, Westport, Conn., 1981.

SCOTT M. GRIFFITH
Admix, Inc.
Manchester, New Hampshire

BEVERAGES: NONCARBONATED

How many types of beverages can you name? There are thousands of different flavors, categories, brands, and packages of beverages available in the United States today. They all have one thing in common: They are all predominantly water. Our bodies are composed primarily of water and water-soluble compounds. We must drink liquids to maintain a water balance and replenish the water our bodies expel constantly. The amount of liquid required will vary with our physical condition, exertion level, and environment. We can only survive for a week or two without some water intake. Fortunately, most of us consume plenty of beverages to meet our biological need for water.

Our water balance controls the natural desire for a drink, called *thirst*. When our bodies sense an excess of salts, called *electrolytes*, in our blood, we feel the sensation of thirst. We can satisfy our thirst by restoring the water balance through beverage consumption, but we have many other reasons to consume beverages beyond our physical need.

Natural water was undoubtedly the first beverage consumed by prehistoric people, and it remains the most popular beverage in areas where it is safe and available to drink. Unfortunately, as the world becomes more developed and crowded, those areas are shrinking fast. Throughout history, we have developed beverage alternatives with desirable flavors, physical sensations, and nutritional value. We have learned to identify and purify unsafe water and change it into the rainbow of beverages available today.

Several other beverages are found in nature, the most obvious and important to our development is milk. As mammalian babies, mother's milk is our first beverage and source of nutrition following birth. Until the relatively recent development of nutritional baby formula, milk was essential to a baby's survival. As we grow, we can experience other beverages and eventually solid food. Frequently the next beverage after milk and water, is fruit juice, another naturally occurring drink derived by squeezing ripe fruit. Of course, the fruit juice we enjoy today is much improved over what was available from fruit growing in the wild. Other naturally occurring beverages would also be liquids derived from plants such as nectars, saps, and vegetable juices.

BEVERAGE TYPES

Webster's defines a beverage simply as a drinkable liquid. We drink beverages to satisfy our thirst, for enjoyment, for psychological reward, for ceremony or celebration, and for nutrition. The proliferation of beverage choices was most likely driven by curiosity at first, but now seems to be more profit driven. We have beverages for all occasions: to stimulate us in the morning, to calm us down at night, to sooth our physical maladies, to celebrate holidays and special occasions, and to perform religious ceremonies. There are ethnic drinks identified with specific cultures, sports drinks for strenuous exercise, health drinks for improved nutrition, and even astronaut drinks developed for space travel. See Table 1 for a list of the popular categories available in the U.S. market. There are two common measures of beverage sales and consumption: liquid volume and retail sales in dollars. These measures can rank beverages quite differently because some beverages are very expensive per unit volume. For instance, alcoholic beverages including beer, wine, and distilled spirits, account for almost half of the beverage dollars spent by consumers, but only comprise about 15% of the volume sold.

HISTORY

Early beverages of commerce include milk, mineral waters, tea, beer, and wine. Chapters could be written on the histories of each of these, with roots in early civilizations. Explorers constantly expanded the list with discoveries of herbs, spices, and fruits from distant lands. Beverage purveyors developed preservation methods, processing techniques, and improved packaging to expand their markets. New machines such as grinders, presses, distilleries, and dehydrators improved yields and quality. Pasteurization and natural preservatives, along with better packaging to reduce spoilage, greatly extended shelf life and distribution range. As availability of supplies increased, marketing

Table 1. Beverage Categories

Water—tap, purified, spring, sparkling, flavored, caffeinated.
Carbonated soft drinks
Alcoholic—beer, cider, wine, distilled spirits, low-calorie beer
Milk and dairy-based beverages
Coffee and tea—hot and cold, cappucino, latte
Fruit juices and juice drinks (less than 100% juice)
Fruit flavored (no juice)
Powdered drink mixes
Vegetable juices
Sports drinks—isotonics
Frozen drinks—shakes, slushes
Health drinks—nutrition fortified, herbal, nutraceuticals
Diet drinks—low-calorie, low-fat, weight control aids
Chocolate based—hot cocoa, cold sweetened
New Age—Innovative new blends that cross categories

and promotion of beverages accelerated the trial and proliferation of new drinks. The beverage market grew to become huge. Beverages make up the largest portion of our daily food intake. Proliferation of products and packages has made drinks conveniently available almost everywhere we go in the Western world. Packages range from little tea bags and powder packets up to 5-gal bulk dispensers and kegs.

U.S. BEVERAGE MARKET

What are the most popular beverages available in the United States in the late 1990s? By far the volume leader is carbonated soft drinks, with a per capita consumption of more than 55 gal per year and growing. They are also the largest single category in retail dollars, estimated at more than $50 billion. Tap water may be second in volume, but declining and contributing negligible dollar sales. Beer is in third place, with steady annual volume of about 22 gal per person, followed by coffee, milk, bottled water, juice, and tea, in that order. The fastest growing segment is bottled water.

RECENT DEVELOPMENTS

The decades of the 1980s and 1990s saw a tremendous proliferation of new products in ready-to-drink teas, juice blends, New Age, and nutraceuticals. Ready-to-drink teas (bottles or cans) were pioneered by the big dry-tea producers such as Lipton, Nestlé, Red Rose, and Tetley. These were cold-filled 12-oz cans with preservatives, offered as an alternative to carbonated soft drinks. Drinking cold or iced tea was an American innovation very popular in the South. The packaged product did not really catch on until a small New York company introduced Snapple, all natural flavored teas. These were hot-filled products in 16-oz bottles with fruit flavors. With some clever marketing, Snapple grew at an amazing rate for several years. It spawned many imitators and created a new market segment. Snapple sales plummeted when a large food conglomerate bought the company and tried to change its successful distribution system.

At about the same time, imported bottled waters became popular in small, fancy bottles appealing to upscale consumers. Perrier and Clearly Canadian expanded their offerings with flavored waters and created a new category of innovative products, dubbed New Age, that could not be defined by the existing beverage categories. New Age products were attractively packaged, very expensive, and very profitable when consumers made them a trend. But as more products tried to ride the coattails of their success, the market was diluted, and many new products failed to last long. Many industry watchers felt too many new products were being introduced; sales outlets and distributors had no room to take on so many products. An industry shakeout was inevitable as supermarkets raised the cost of new product introductions by charging for shelf space and quickly dropping products that did not sell well. Many smaller companies failed or were sold to larger companies, but innovation continued at the more successful firms.

Sports drinks are a recent innovation that has grown steadily but is still dominated by the pioneer, Gatorade. Sports drinks are designed to replace the water, energy, and electrolytes lost during strenuous exercise. Gatorade was developed in the 1960s by researchers led by Dr. Robert Cade to help the University of Florida football team (Gators) maintain stamina while playing in the hot and humid Florida climate. It was successful with the athletes and soon was used by other teams. Clever marketing got it placed on professional football team benches with large logos on coolers and cups plainly visible on national television. This rapid brand identification made it popular with athletes and sports fans across the country. Soon, Gatorade was sponsoring youth sports events with associated promotion of branded accessories like water bottles, hats, and headbands. Its marketing clout kept other sports drinks from making any inroads until Coke and Pepsi developed products and used their distribution power to challenge the market leader. Their market shares are slowly gaining on Gatorade.

Juice companies have been livening up their product lines with exotic juice blends, featuring tropical fruits like guava, papaya, mango, and kiwi. Juices are expensive, so juice drinks with 5 or 10% juice were introduced to compete at a lower price. Lately the trend has swung back toward 100% juice blends, which allow some cost savings with lower-priced juices predominating in the blend. Juice sales are not growing overall, but certain segments such as orange juice and the new blends are doing well.

Coffee sales have been reinvigorated by specialty stores like Starbucks and Dunkin Donuts and upscale specialty products like cappuccino and latte. New cold and frozen products from these stores may help even out the seasonal sales cycles of hot products.

The latest trend in beverages is the healthy or "good for you" segment. This seemed to start with herbal teas making claims to have physiological affects. There has been a sharp increase in fortification of products with calcium, beginning with orange juice. This was based on the health benefits of calcium in preventing Osteoporosis, a degenerative bone condition common in older women.

Now a new class of nutraceutical beverages has emerged claiming health benefits derived from herbs, trace minerals, and amino acids. A list of some of these esoteric ingredients follows in Table 2. The term nutraceutical comes from combining nutrition and pharmaceutical, as

Table 2. Some Interesting New Beverage Ingredients

Ginseng root	Gotu kola
Guarana	Echinacea
Ginko biloba	Kava kava
Dong quai	Bee pollen
Fo-ti	Carnitine
Damiana	Proline
Agavé nectar	Taurine
Yerba maté	Yohimbe

these products are targeted toward better health through diets rich in these ingredients with curative properties.

OTHER TRENDS AFFECTING BEVERAGES

Consumer trends that affect beverage developments include the healthier lifestyle movement, desire for more "natural" products with fewer additives, the increase in meals eaten away from home, and the desire for convenience. Natural products and products with fewer additives have become more popular. Improvements in processing have made these products possible while consumer demand has allowed producers to raise prices to cover increased costs. Changes in packaging technology have also had an impact. Aseptic packaging allows less heat processing and resulting flavor loss for shelf-stable juice products. Better oxygen and light barrier materials have improved shelf life and reduced vitamin losses in dairy products. Easy-opening spouts on water and sports drink bottles have become popular. Convenience trends have hurt the sales of powder mix beverages and frozen juice concentrates prepared at home. At the same time, convenience store and vending machine packaged product sales have grown. Bottled water is a good example of convenience packaging increasing sales and margins for a commodity product. The convenience trend is making beverages of all kinds more readily available where we work, shop, and recreate.

GENERAL REFERENCES

Beverage Digest LLC home page, URL: *www.beverage-digest.com* (last accessed June 25, 1999).

Bill Communications home page, URL: *www.beverageworld.com* (last accessed June 25, 1999).

Hoovers, Inc. home page, URL: *www.hoovers.com* (last accessed June 25, 1999).

Kraft Foods, Inc. home page, URL: *www.kraftfoods.com* (last accessed June 25, 1999).

Quaker Oats Company home page, URL: *www.quakeroats.com* (last accessed June 25, 1999).

J. G. Woodroof and G. F. Philips, *Beverages: Carbonated and Noncarbonated*, rev. ed., AVI Publishing, Westport, Conn., 1981.

Scott M. Griffith
Admix, Inc.
Manchester, New Hampshire

BIOLOGICALLY STABLE INTERMEDIATES

The ability to preserve food from one harvest to another has been vital to the development of civilization and has played a very important role in discovery and survival. Methods of food preservation, in one form or another, have been employed for thousands of years and are usually defined as a procedure that delays spoilage and makes food items available for consumption at a later date. Often the food item can be eaten without further preparation as, for example, when the inhabitants of the hot dry desert areas used to bury dates, figs, and grapes in the hot dry sands to dehydrate them. With partial removal of water, the sugar content of the fruit was high enough to prevent the growth of microorganisms, and the fruit would keep for long periods of time. The fruit could be consumed at a later date without the need for further processing. A number of other technologies were also developed to accomplish this. These include simple dehydration, chemical preservation, pickling, fermentation, canning, freezing, and, more recently, a number of more-sophisticated approaches such as irradiation, hypobaric applications, pulsed light, and so on. All of these were concerned with processing food in a manner that would make it edible at a later date. But with the increasing importance of food fabrication, another concept emerged. Food commodities were essential as ingredients in many formulated foods, and a need arose to have food ingredients available in a "biologically stable form" for future formulation. With the limited harvesting schedules available for many fruits and vegetables, it was economically prohibitive to install sufficient formulation equipment to make all the desired product at time of harvest. The manufactures wanted stable ingredients such that they could run their formulation lines on a year-round basis. The following passages describe some food commodities available in biologically active form that are not usually consumed in that form.

Nature has provided many foods in a biologically stable form. The most obvious example is cereal grains. It is no surprise that cereal grains became the main source of food for humans from the dawn of civilization. Cereal grains, when properly matured, dried, and stored under good conditions, will last almost forever. They are subject to loss from insects, birds, and animals but seldom from microorganisms. Cereal grains are the ideal biologically stable food, but they do require processing such as grinding, soaking, and cooking prior to consumption. Many vegetables are normally biologically stable. Root crops such as potatoes, carrots, parsnips, turnips, rutabagas, and beets can be stored throughout the winter season. The early settlers in New England depended on storage of root crops in "root cellars," which were simply excavations in the side of a hill, to store root vegetables and cabbage over the winter. The natural humidity of the earthen surroundings in the cave provided sufficient moisture to prevent dehydration and also sufficient insulation to maintain a cool storage temperature without freezing. The availability of these vegetables in the winter months led to the development of the well-known "New England boiled dinner" (1), which was composed of potatoes, carrots, onions, cabbage, turnips, parsnips, and chunks of meat, usually corned beef, but sometimes salt pork or codfish. Very few fruits are biologically stable, with the possible exception of the banana. Bananas are the world's largest fresh fruit crop. Its popularity, aside from its delicious flavor, may be partially due to the fact that it requires a three-week ripening period after harvesting, which was long enough to allow bananas to be shipped by boat from the tropical producing areas to nearly all parts of the world. The central Asian nomads had their own biologically stable food. They simply opened a vein in their horse's neck and drank the blood (2). In the following examples, biologically stable foods will be confined to those that are processed to form an intermediate commodity that is not normally consumed as such and requires further processing into a consumer item.

DRIED CODFISH

Probably the oldest, and certainly the most successful, biologically stable product is dried codfish (3). The earliest record is attributed to the Vikings who traveled from Norway to Iceland to Greenland to Canada, which coincidentally is the range of the Atlantic cod (*Gadus morhua*). In the tenth century, Thorwald and his son Erik the Red were expelled from Norway for murder and traveled to Iceland where they were again expelled for murder and in the year 965 traveled to Greenland. The Vikings were able to travel such long distances because they dried cod by splitting the fish and exposing the carcasses to the dry freezing air. The low temperatures prevented the frozen fish from decaying and the dry air was effective in removing moisture. The dried product resembled a wooden plank, but it did not decay. This was the same process developed by the natives in the high mountains of South America to produce "chofa" from potatoes. The Vikings had a thriving trade in dried cod with the Romans many centuries earlier, but it really flourished in the trade with Europe in the Middle Ages. Erik the Red colonized Greenland and his son Leif Eriksson sailed on to what he called Stoneland but was probably the coast of Labrador. The Norsemen made five expeditions to Canada between 985 and 1011 and all were possible because they had rations of dried cod.

In the Middle Ages in Europe, dried cod was an important food commodity, and the supply was coming from Scandinavia and Iceland. The dried cod being sold in Europe was a superior product to that being used in the Norse exploratory journeys because the Europeans had salt, and dried salt cod lasted longer and simply tasted better. The Basques in northern Spain entered the European market with large quantities of dried salt cod and for centuries no one knew where it was coming from. Actually, the Basques had discovered the limitless cod stocks of the eastern shore of America. Europe, in the Middle Ages, was a big market for whale meat, and the Basques were providing a large portion of it. In their whaling expeditions, they discovered the cod fishing grounds. When John Cabot landed in Newfoundland in 1497, he claimed the land for England. Jacques Cartier arrived in 1534 and reported seeing 1000 Basque ships fishing for cod. But the Basques, in the interests of secrecy, never claimed the land, and it remained for Jacques Cartier to claim the land for Canada. The Pilgrims landed in 1621 and the wealth of the New England cod fishery began to unfold.

The eighteenth century saw a race to exploit the New England cod fishery, and the European countries of Spain, Portugal, France, England, and Scandinavia rushed to supply the huge European demand for dried cod. The Americans soon learned that this was a lucrative world trade and developed the fishing industry. The French colonies on Haiti, Martinique, and Guadalupe, and the Dutch Republic of Suriname (then Dutch Guiana) became good customers because salt cod provided a nutritious and cheap source of food for slaves. Ships sailed for Africa loaded with salt cod and used half the cargo to purchase slaves and then set sail for the Carribbean where they exchanged the slaves and the rest of the cargo for molasses. The molasses was used to produce rum, which had become a naval necessity. A tot of rum was issued each day to each sailor, and we are led to believe that it was appreciated. Rum became a generic term for alcoholic beverages. The trade in dried cod became intertwined with the slave and molasses trade, and the Boston "Cod Aristocracy" was born. Today a model of a codfish hangs from the ceiling of the state house in Boston.

Codfish was an ideal candidate for biologically stable products for several reasons. First, it was very low in fat, which reduces the tendency to become rancid when stored, as opposed to salmon, which the early New Englanders also had in abundance. Salmon has a high oil content and cannot be preserved by drying with the same efficiency as codfish. Second, codfish was available in limitless quantities and could be easily caught. Third, abundant rocky shorelines were available to allow for drying in the open air. The Vikings reported that the dried cod without salt could be eaten directly by chipping off pieces, but this must have been very difficult. They probably soaked the dry slabs in water to soften the product. Dried salt cod cannot be eaten directly but must be soaked to lower the salt content. A number of changes of water were necessary to lower the salt content, and some ingenious methods were developed. The French reported that the journey by boat from Bordeaux to Aveyron took about two days, which was approximately the soaking time required for salt cod, so they towed it in a mesh bag behind the boat. Pollution of the Lot River discouraged this practice. In the meantime, flush toilets were developed and the old "water closet" had a tank of water elevated above the toilet. Storing cod in the tank provided an efficient way to remove the salt.

It is ironic that the "limitless" harvest of cod that was responsible for so much international trade turned out to be anything but limitless. Ships became bigger and more powerful for dragging trawls and more efficient in finding fish with electronic gear, and the capacity to process fish at sea was developed. The development of large factory ships made it possible for any country to exploit the cod fishing grounds, and soon ships from Europe, Russia, Korea, China, and many others were appearing in the Atlantic off the New England coast. The fisheries could not sustain the fishing effort, and cod became commercially extinct. In 1994 Canada closed the fisheries off Newfoundland, and in 1999 the United States closed a large portion of the cod fishing grounds. Fishery biologists say that the fisheries will need a number of old spawners to replenish the cod population. This means fish up to 15 years of age, so the fisheries will be closed for a long time.

SULFITED FRUITS

The use of sulfur dioxide and its salts to preserve fruits and some vegetables is a very old process. Most fruits were harvested over a very short period and had to be preserved in bulk until they could be processed into jams and jellies. Sulfur dioxide served this role very well. Fruit could be cleaned, placed into barrels, and covered with a solution of sulfur dioxide in water, or a solution of sodium bisulfite or one of its salts. After mixing in the barrels, the fruit could be kept for years. The fruit could be removed from the barrels, placed in a kettle, and boiled to remove the sulfur dioxide. After addition of pectin and sugars and boiling to produce the required moisture level, the product could be filled into consumer-sized packages. Strawberries were an important sulfited product because of the large market for strawberry jam but the sulfite treatment bleached out the red color. The chemical reaction was a simple addition reaction between the red anthocyanin pigment and the sulfite ion that was easily broken such that the boiling process reduced the sulfite content and restored the desirable red color.

The sulfite process is important today in the manufacture of maraschino cherries. Nearly all the cherries used for maraschino cherries are mechanically harvested and suffer some mechanical damage in the process of being shaken off the tree. Since cherries discolor very quickly, they are conveyed in a matter of seconds into a brine tank on the mechanical harvester. The simplest of the sulfite brines is a solution of sodium bisulfite in water, but a number of other formulations are used. The purpose of the sulfite brine is to preserve the fruit by inhibiting the enzymes that produce discoloration and to bleach out the red color. Sometimes a secondary bleach in sodium chlorite is used. A source of calcium ions is usually added to the brine to firm the cherries. The brined cherries are leached to remove the sulfite ions and any other soluble material and conveyed to a tank containing a red colorant dissolved in a sugar syrup. The cherries readily absorb the red colorant and are then drained and packaged. Maraschino, candied, and glaceéd cherries are made the same way except that the sugar content is different.

Around the beginning of the twentieth century, sulfiting was the major way to preserve fruits in bulk for further processing in Europe, particularly in England because of the English people's liking for strawberry jam. The introduction of freezing provided a method of preservation that produced a superior final product from both an appearance and a flavor point of view and sulfiting decreased in importance. Sulfiting is still used in some parts of the world but the technology has been essentially displaced except for maraschino cherries.

TOMATO PRODUCTS

Tomato juice and tomato concentrate are more recent examples of biologically stable products and are ideal candidates for preserving. The tomato harvesting season is relatively short, the crop is huge in size, the demand for tomato products in formulated foods was growing rapidly, and nearly all the tomatoes for processing are mechanically harvested. Mechanical harvesting is an economic necessity, but it does result in more mechanical damage to the crop than the old-fashioned hand harvesting. This means that the crop must be processed as soon after harvesting as possible. Minimal processing to produce a biologically stable intermediate is a virtual necessity to minimize the costs of production and ensure a high-quality product.

Biogically stable intermediate products became feasible with the concept of aseptic packaging developed by W. M. Martin in 1935 (4). He developed the concept of sterilizing a product, sterilizing the container, and bringing the two together, with a sterile cap, in a sterile atmosphere. Conventional canning at that time consisted of filling and capping a container and heat treating the closed container to ensure sterility. This means that the center point of the container needed to receive sufficient heat treatment to ensure sterility; however, the outer portions of the container would be overheated and the quality of the food would be lowered. This effectively limited the size of the container to a No. 10 can, approximately 1 gal. If the product could be sterilized in a thin film outside the container, and packaged aseptically, the size of the container was limitless. I remember visiting a large food processor in the 1940s. They required a large quantity of tomato paste to formulate into products like pork and beans, soup, pizza sauce, sardines in tomato sauce, barbecue sauce, and many others. During the tomato season they would pack their year's quota of tomato juice, tomato ketchup, and tomato soup and make the remainder into tomato concentrate or paste. The concentrate would be filled into No. 10 cans, capped, and stored in a huge warehouse. The concentration process would make the puree essentially sterile, and the hot fill would sterilize the cans and lids. This process worked very well, but it was labor intensive and it was not perfect. Some of the cans leaked, and one leaky can makes quite a mess.

Bulk storage of tomato juice in the United States started essentially with the work of Nelson at Purdue University (5–7). He experimented with aseptic storage of tomato juice in tanks up to 100,000 gal, and even larger, and now numerous large storage systems are in place worldwide. The concept is simple, but the implementation is more complicated. Tomatoes can be mechanically harvested into large containers and brought immediately to the factory for processing. After washing, the tomatoes are chopped and screened to remove skins and seeds and pumped to a heat exchanger that sterilizes the juice. A number of systems are available to sterilize the juice, but nearly all employ highly efficient heat exchangers that use the high temperature/short time (HTST) or the ultrahigh temperature (UHT) approach to deliver a quantity of heat lethal to microorganisms. The short heating time ensures maximum retention of nutrients and flavor. A deaeration system is necessary to remove dissolved air in order to minimize loss of quality during subsequent storage. Large tanks are sterilized with hot water and chlorine, and the juice is pumped into them. The whole system, including the pumps, pressure reduction valves, heat exchangers, and the valve system, is closed and operated aseptically, which ensures that sterile air enters the tank to replace the product that is removed on demand for later formulation. The tanks can be refrigerated with interior coils or simply stored in a refrigerated building, if desired, depending on the quality demands of the final product. The tanks can also be mounted on a rail car or truck for shipment to any area. The same system can be operated with tomato concentrate, depending on the requirements of the formulator.

GRAPE JUICE

The development of bulk storage of grape juice paralleled the experience with tomato juice, but the products are different primarily due to the higher sugar content and flavor lability of grape juice. The aims are similar. The "Concord" grape juice industry primarily in the eastern United States is used for juice, jams, jellies, and fruit juice drinks. The western U.S. grape industry is about 10 times as big and is used for wines, brandies, raisins, juices, and fruit juices. Both have a need for bulk storage of grape juice. Friedman (8) attributed the development of the Concord grape juice industry in the last 70 years to three major technological developments: (*1*) bulk storage of juice, (*2*) continuous processing, and (*3*) mechanical harvesting. Continuous processing was introduced about 1955 to replace the batch process being used by the apple and grape industry prior to that time. The batch process consisted of crushing the heated grapes and building a pad of grapes on a filter cloth on a wooden rack. The racks and layers of grapes were alternated to build a large pile and the whole stack was placed in a filter press to remove the juice. A number of types of continuous presses were developed that removed the juice in one pass. The batch process was labor intensive and slow, whereas the new presses enabled some plants to process 2000 tons of grapes per day. The introduction of mechanical harvesters, at approximately the same time as continous pressing, enabled the operators to harvest a large quantity of fruit in a short time. Most harvesters consist of a machine that straddles the row of grapes, and beater blades separate the grapes from the vine. The grapes are collected and transferred to a tank in the vineyards for transport to the processing factory. Today, nearly all grapes for processing are mechanically harvested.

A typical bulk storage operation would be as follows. Grapes are crushed and destemmed, heated to about 60°C, and treated for 30 to 40 min with a depectinizing enzyme to disintegrate the pulp and release the juice. A filter aid, usually a cellulose derivative, is added, and the skins and seeds are separated in a continuous press. The juice is filtered, flash-pasteurized, cooled, and pumped directly to the sterile storage tanks with the same sterility precautions as described for tomato juice. The storage tanks may be made of stainless steel or, more usually, of mild steel lined with a phenolic (epoxy) resin. Tanks holding 320,000 gal were standard in the industry for years, but modern installations usually hold 714,000 gal. Apparently grape juice can be held at 30°F for 12 months or more with little loss in quality. The large tanks are usually freestanding with refrigerated coils inside the tank, but smaller tanks could be stored in a refrigerated room. Bulk storage of juice has several important advantages: (*1*) The "argols," crystals of potassium bitartrate, precipitate out, which prevents the accumulation of tartrates as unsightly sediment in consumer packages; (*2*) a substantial amount of pooling takes place with a resulting beneficial standardization and uniformity; and (*3*) the juice can be withdrawn, as desired, to maintain year-round production lines. The same procedure may be used to produce and store juice concentrates. Juice can be pumped to the concentrators at 15 to 20° Brix and stored at 60° Brix. The concentrate is more stable, microbiologically, because of the high sugar content. Alternatively, single strength juice can be stored for later concentration, but this is only done with white juices because two heat treatments would cause too much color degradation for colored grape juice.

Many of the fruit drinks on the market use grape juice as the main ingredient for several reasons. It is inexpensive, available in large quantities, provides sweetness, and has a desirable fruity flavor that is compatible with other added flavors. Some grape concentrates have another advantage in that they can provide a very attractive color in addition to the sugar and flavor.

FROZEN FOODS

Frozen foods do not qualify under the definition of biologically stable foods that are not normally consumed in that form since they obviously can be consumed, but the packages are much bigger. Frozen foods are one of the best biologically stable foods from a quality standpoint and play a very important role in the reformulation of convenience foods. Products such as peas, green beans, lima beans, corn, sliced carrots, and almost any vegetable can be frozen individually and packed in containers up to 2000 lb for future reformulation into entrees, pies, and so on. Many fruits can be prepared, frozen individually, and stored in large containers for reformulation into fruit cocktail, specialty desserts, and so on.

BIBLIOGRAPHY

1. *American Cooking*. Time-Life Series on Foods of the World, New York, 1968, p. 90.
2. R. Tannahill, Food in History Stein and Day, New York, 1973.
3. M. Kurlanski, *Cod*. Penguin Books, New York, 1997.
4. American Can Co., *The Story of Canco Research*, American Can Co., Chicago, 1955.
5. P. E. Nelson, "Technical Developments in Bulk Storage Processing," *Hortscience* **6**(3), 222–223 (1971).
6. U.S. Pat. 3,714,956 (1973), P. E. Nelson (to Nelson).
7. P. E. Nelson, "Aseptic Bulk Processing of Fruits and Vegetables," *Food Technol.* **44**(2), 96–97 (1990).
8. I. E. Friedman, "Industrial Application of Bulk Storage in the "Concord" Grape Industry," *Hortscience* **6**(3), 228–229 (1971).

F. J. Francis
Editor-in-Chief
University of Massachusetts
Amherst, Massachusetts

BISCUIT AND CRACKER TECHNOLOGY

If a definition of a biscuit were to encompass the wide range of products as understood by consumers and the baking industry, that definition would be different from the one contained in most standard dictionaries. In the United States, the common biscuit is a chemically leavened bread/roll that is generally circular in outline and flat in profile; it is rather similar in composition to some British scones. This product is sometimes called a baking powder biscuit.

In the United Kingdom, and most of the rest of the English-speaking world, biscuits are the two types of products called crackers and cookies. Crackers are nonsweet products used like bread. Cookie items include a vast array of dessert foods, characterized mainly by being baked in small pieces and having a texture or consistency that is drier, chewier (or crisper), and denser than most cakes, and are usually sweet. An important characteristic of crackers and cookies is that they usually have a much longer shelf life than baked products such as bread and cake. However, intermediate moisture cookies (8–12% moisture) are being sold that confound this distinction (eg, soft-filled cookies, brownies, and fruit-filled bars).

Except for their lower moisture content, crackers and cookies are similar in composition to breads and cakes. The range of ingredient percentages in cracker and cookie formulas will be found to overlap with formulas for breads and cakes. Manufacturing processes, procedures, and equipment are also similar, yet different, for many parts of the process (ingredient handling, mixing, and baking), although the forming, depositing, and packaging equipment may be highly specialized.

INGREDIENTS

Flour

The principal structure-forming ingredient used in biscuit doughs is wheat flour. Very few cookies or crackers are made without any wheat flour, and those are usually quite atypical in organoleptic characteristics. To provide for flour strength in weaker flours, vital wheat gluten can replace wheat flour to a limited extent.

There are many different kinds of wheat flour, and the specifications of this ingredient must be carefully chosen, if it is to impart satisfactory machining properties to the dough and have a desirable appearance with good eating qualities. Cracker doughs typically use soft wheat flour fortified with relatively high protein content flour made from hard red winter wheat or spring wheat. Some cracker doughs and most cookies will be based on soft wheat flour, sometimes with a small proportion of flour from hard wheat.

Soft wheat flours suitable for biscuits may vary in protein content (mostly gluten) from 7.0 to 9.5% (for cookies) to 10% or more (for crackers). The flour may be unbleached or heavily bleached to a pH range of 4.4 to 4.8 for certain specialty items, such as soft cookies that contain higher levels of sugar and shortening. Cookies using bleached flour will rise higher with little spread; in contrast, unbleached flour will typically spread more than it rises. Cracker sponge flours may contain added malt (wheat or barley). Malt flours are made by wetting whole grains and allowing the enzymes to become active. The malted grains are then dried at low heating temperatures to remove moisture but not deactivate the enzymes. Such malted flour is then added to the cracker sponge for its enzymatic properties (1).

Ingredient specifications for cookie flour will usually include protein content, moisture content, ash content, particle size, starch damage, pH, odor, and flavor. Microbiological tests will be required to reveal contamination by insects, fungi, pesticides, and other unwanted materials. It is also common to specify certain rheological tests (eg, Alveograph), which are expected to correlate with dough response to processing conditions. Finally, the flour must yield cookies of specified characteristics (spread/height), when it is used in dough that is prepared and baked under standardized conditions. One relevant test that gives an indication of flour function is the bake-spread test. Written procedures for this test can be found in AACC Method 10-53 (2). The diameter and height of the finished test cookie are measured. Analyzing the results indicates the amount of rise or spread performance of a flour. Sometimes this measurement is expressed as a ratio of the height to diameter, but that approach does not fully characterize flour performance. It is better to analyze and compare both results for trends.

Flours other than wheat can be used in crackers and cookies for color, texture, and flavor. Prime examples of these ingredients would be rye and corn flour. Sorghum flour, cottonseed meal, soybean meal, triticale flour, barley flour, and other cereal and noncereal powders have been suggested as additives for enhancing nutritional or other properties. These nonwheat ingredients have structure-forming properties that do not function as well as typical wheat varieties and must be supplemented with a strong wheat flour or vital wheat gluten for most purposes. Rice flour has been used to make cookies suitable for persons who are allergic to wheat gluten; it is incapable of forming typical dough. Rice flour has been used successfully for the production of rice crackers, but the process and procedures are very different from standard baking practices. A modified form of wheat flour starch, called resistant starch, passes easily through the digestive system, thus reducing its caloric contribution to the body.

Shortenings

Shortenings are essential components of most crackers and cookies. The type and amount of shortenings and emulsifiers in a formula affect both the machining response of a dough or batter and the quality of a finished product. Coatings (such as chocolate) and fillings (such as sandwich cremes) depend on specific fats and oils to furnish the structural part of such components.

Natural fats and oils suitable for shortenings include butter, lard, beef fats, and vegetable oils. The latter material includes refined and modified soybean, cottonseed, coconut, palm, and corn oils. The processing of shortenings may include some or all of the steps of refining (to remove contaminants), deodorizing, winterizing (to remove high-melting-point fats), bleaching, hydrogenation, fractionation, and blending. The purpose of these operations is to yield a shortening that is bland in flavor, essentially colorless, and free of aromas; has a melting point and solid fat content within the desired ranges for the application for which it is destined; and has other characteristics desired by the purchaser. Butter, and to a lesser extent lard and beef fats, have natural flavors and physical qualities that are considered desirable in some applications.

Recently, new fat systems that are non- or poorly digestible have been developed. One of these fat systems is a combination of sugar esters that is nondigestible in the body. Another fat system, which contributes a reduced calorie load, modifies the triglyceride composition through esterification with nondigestible fatty acids.

The function of shortenings in baked goods is primarily to modify the physical properties or texture of the finished product, making it more tender or flakier and, in some cases, giving it a glossier, more appealing appearance. Shortenings may also affect the rheological characteristics of doughs. Most crackers have a topping or spray oil applied after baking, to add flavors or solids (cheese, spices and/or herbs) that improve the eating qualities.

Emulsifiers are specialized fats or fat systems that act as surfactants between various systems (eg, oil and water). Emulsifiers are specialized fats that can be composed of mono-, di-, or triglycerides or their combinations. Chemically, an emulsifier is a molecule composed of a water-soluble or hydrophilic portion and a water-insoluble or hydrophobic portion. Different solubility tendencies thus exist within the same molecule. This leads to the phenomenon of incomplete solubility in both water and oils. Emulsifiers partition, or orient, themselves at the interface between oil and water phases (3). Naturally occurring emulsifiers include egg yolk or fluid lecithin. Fluid lecithin is derived from either soybeans or corn during the milling process.

Sweeteners

All cookies and many crackers contain some form of sweetener. The quantity of sweetener is usually such that it has significant effects on the texture and appearance of the product, as well as on its flavor. Machining properties and response of the dough piece to oven conditions are also related to the type and quantity of sweetener employed. In fermented goods, sugars serve as substrates from which yeast and other microorganisms form carbon dioxide and the flavoring substances characteristic of these products. Commercial sucrose (cane or beet sugar) functions not only as a nutritive sweetener but also as a texturizer, coloring agent, and as a means of controlling spread during baking. However, differences in the crystallization properties between cane and beet sugars can be quite significant. In marshmallow, jellies, and fruit jams with relatively high moisture content, sucrose has the valuable property of delaying microbiological spoilage, when it is present in a high-enough concentration.

Sucrose can be obtained in various particle sizes or in the form of syrups. When added to a dough as granulated or powdered sugar, its particle size influences the dimensions of the finished cookie, by controlling the extent to which the dough piece spreads as it is baking.

Syrups are easier to handle, being adaptable to fluid-transfer systems of pumps, pipes, valves, and meters. Syrups can crystallize in handling systems, causing problems. Sucrose will not form a stable aqueous solution of greater than 67% concentration at room temperature, and this is not high enough to ensure resistance to all forms of microbiological growth. Treating a solution of sucrose with an acid or an enzyme known as invertase hydrolyzes the sucrose into fructose and dextrose (3). Invert sugar solutions are sweeter than pure sugar solutions of the same concentration. Invert syrups are less likely to crystallize and are more resistant to spoilage.

Corn syrups are sweeteners prepared by hydrolyzing cornstarch with acids or enzymes. The standard types contain glucose and varying amounts of maltose, larger oligomers, polymers, and other carbohydrates, plus minor amounts of impurities. In preparing the high-fructose corn syrup (HFCS) varieties, manufacturers isomerize part of the glucose to fructose. Because fructose is sweeter than glucose, a substantial increase in sweetening power results. HFCS can be substituted for invert sugar syrups in many applications, but they are not identical.

Regular corn syrup is categorized by type, according to dextrose equivalent (DE), which is a measure of reducing sugar content. A second parameter is total solids content of a syrup. Cookies made with syrups with higher-reducing sugar contents have a greater sweetness that those made with a similar content of low-DE syrup. In addition, the browning of the crust and, for that matter, the interior of the cookie is affected by high contents of reducing sugar from any source. Low-DE corn syrups can increase the apparent viscosity of doughs and batters and make a finished product chewier.

Molasses is less-refined, concentrated sugar syrup containing some of the impurities, flavoring and coloring materials from the sugar-refining process. Molasses is available in various degrees of darkness; the darker the color, the more bitter and less sweet the syrup is. True blackstrap, the final residue of the refining process, is not suitable for human consumption and is used primarily as an animal feed additive.

Commercial brown sugar is granulated sugar that has been coated with a small percentage of molasses. No commercial brown sugar is produced by removing partially refined granulated sugar from the refining process. There are, however, less-refined sugars available from refiners (eg, Demarara) that occasionally provide some economic advantages over pure cane or beet sugar.

Honey is primarily invert syrup with various impurities that give typical flavors and color. Various concentrated fruit juices are used, as alternatives to sucrose and corn syrups, as natural sweeteners. These fruit juices are principally from grape, pear, or apple sources that have been treated to have low characterizing flavor. These fruit syrups are usually more expensive than sucrose or corn syrup.

Caramel color is a widely used brown pigment made from acid- or alkali-treated corn syrup (or sugar syrup). Caramel color is used in many types of cookies, in liquid or powder form, for its colorant properties and its low cost. It contributes no flavor to a finished product.

Synthetic sweeteners, such as saccharin and aspartame, can be used in certain dietetic cookies. However, sugar serves as a primary bulking agent in cookies. When significant reductions in sugar content are made, texture and other physical characteristics are affected. Most synthetic sweeteners are not heat stable, but recent developments are changing that limitation. New synthetic sugars are under development, which have up to 2000 times the sweetening power of sugar. Of course, most of these synthetic sweeteners are regulated by law and must pass stringent food and drug tests. Nonnutritive and reduced-calorie sweeteners are playing a more important role, as consumer demand requires such specialized products.

Leavening Agents

A leavening agent is a substance or system that expands or lightens a dough or batter at some stage of its processing. Leavening can be achieved in three ways: mechanically, biologically, or chemically. The leavening effect is absolutely essential to the formation of a finished product having the appearance and eating qualities that are required by consumers. Leaveners familiar to every baker are baking powder and yeast. Air and water vapor (steam) are also leaveners. They cause expansion of the product during baking, providing the dough or batter has a structure capable of retaining the gas. Ammonium bicarbonate is a chemical additive that has leavening and other effects on cookie and cracker doughs.

Yeast alters the physical properties of dough and its handling characteristics during mixing and fermentation. Yeast acts on certain sugars to form alcohol and carbon dioxide. Gluten from the flour absorbs water and forms extensible membranes that trap carbon dioxide and expand to decrease the density of the dough mass. The diffusion and accumulation of carbon dioxide throughout the dough mass generates powerful stretching action. These changes in gluten are described as the "mellowing," maturing, or conditioning of the dough (3).

An egg white meringue, which is the basis of angel food cake, relies on egg albumen to initially entrap the air that has been beaten into the mixture. The addition of flour strengthens the dough mass in the latter stages of mixing and baking.

Few cookies depend on yeast leavening. Crackers made by the sponge-and-dough method (eg, soda crackers, saltines, and certain snack crackers) require a fermentation period to develop their textures and characterizing flavors. Typical sponges contain flour, water, and yeast mixtures; the doughs usually contain the remaining ingredients, including sodium bicarbonate. The acids produced during sponge fermentation will react with sodium bicarbonate to neutralize the acid and yield additional leavening.

The great majority of cookies contain some sort of chemical leavening system such as baking powder, which is a mixture of sodium bicarbonate and a reactive acidic compound such as cream of tartar (potassium acid tartrate). In addition, fillers such as starch and other additives will be included in commercial baking powders to standardize their strength and delay storage deterioration.

When formulating cookie doughs, most food technologists add sodium bicarbonate in combination with a reactive acidic material, in proportions appropriate to the pH of the dough, rather than rely on premixed baking powder. Other types of chemical leaveners (eg, monocalcium phosphate, sodium acid pyrophosphate) have specific characteristics that can be tailored to the requirements of a given dough formula. The goal is usually to obtain a finished product with a pH close to neutrality, although this generality may not apply to certain products (eg, chocolate doughs containing Dutched cocoa, with a resulting higher pH).

Ammonium bicarbonate is a self-reacting leavener that decomposes as the dough is heated during the early stages of baking, giving off ammonia, carbon dioxide, and water. It is acceptable only for products baked to lower moistures (3–9%). In high-moisture products, the ammonia dissolves into the water, producing an unpleasant aroma and/or taste that most people find offensive. However, ammonium bicarbonate can be very effective in altering finished product characteristics (height, spread, and texture) for both cookies and crackers.

Many cookies undergo an increase in volume (decrease in density due to leavening) that is quite small compared with that of bread. For example, traditional shortbread cookies may contain no added soda and do not seem to increase in size during baking. An examination of their interior shows, however, that some expansion has taken place, as a result of air and water (steam) that have expanded during the baking process.

Other Ingredients

Ingredient water can have a significant effect on dough properties, but most production plants are able to develop formula modifications that offset any undesirable effects that might result from use of the water from their particular supply system. The water from any potable source, such as the municipal pipelines or a well, can be regarded as legally suitable for incorporation in a biscuit formula. However, it should not be necessary to say that any water that has an undesirable odor or flavor should be subjected to a purification process before it is mixed into a dough. The pH of the ingredient water supply, especially from municipal sources, can in some cases vary widely throughout the week or day, and this can cause the physical properties of the dough to change, adversely affecting the machining of this material.

Fresh milk and eggs are not common ingredients in most commercial cookies partly because they are relatively costly. When present, they are likely to constitute only a small percentage of the product. Eggs can have a substantial effect on the physical properties of a dough or batter, acting as a structure former, leavener (due to entrapment of air and water vapor), emulsifier, and lubricants. Egg whites are better as a structure former and leavener; yolks are more effective as an emulsifier and lubricant. Eggs do affect flavor, not always positively, and yolks add color, usually a desirable result. Nonfat milk has ambivalent results on structure while whole milk generally has a weakening effect due to the fat. Both milk and eggs add good-quality protein to a biscuit and are sometimes used as a wash on the surface to color or glaze the finished piece, but for some consumers, cholesterol concerns may be a negative attribute.

Salt is an important ingredient. Not only is it an essential flavor-enhancing ingredient in any baked product, but it has effects on the physical characteristics of doughs. Some of these effects may be favorable, such as the slight toughening effect it has on some doughs, and others are considered unfavorable, such as its retarding effect on yeast-leavened systems, when incorporated at high levels.

Crackers and especially cookies are susceptible to rancidity, because of their high fat content and their fairly long shelf life. Antioxidants can counteract this tendency to a considerable degree. Common antioxidants are butylated hydroxyanisol (BHA), butylated hydroxytoluene (BHT), and *tert*-butylhydroquinone (TBHQ). Amounts that can be added are restricted by federal regulations, and antioxidants must be labeled as cancer-causing agents in some states and may be banned outright in certain countries. Citric acid and phosphoric acid are sometimes useful for chelating metal ions that would otherwise accelerate rancidity. Research has shown that some natural tocopherols (vitamin E) and other antioxidants (rosemary extract) can prevent oxidation.

FORMULAS AND PROCEDURES

In general, formulas can be divided into different groups. Saltines and other fermented products use a sponge-and-dough system that is normally formulated with yeast. In contrast, straight cracker doughs and the like will have small amounts of sugar, fat, and other characterizing additives and are chemically leavened. Cookie formulas will contain moderate to high amounts of sugar and shortening and are primarily leavened with sodium bicarbonate and ammonium bicarbonate in combination with a leavening acid to produce a baked good with high volume. Cookies may also contain other ingredients such as icings, fillings, fruits, nut pastes or pieces, flavors, and chocolate to give them distinctive value.

Cookie formulas are generally classified according to the kind of equipment used to form the individual pieces. Stamping machines, rotary cutters, rotary molders, wire-cut machines and depositors are used for more than 90% of the cookies commercially produced in the United States. The type of equipment being used for a product sets limits on the dough rheological properties or the final composition of the dough.

Deposit cookies are the machine-made counterparts of hand-dropped cookies, and many published (1) formulas for the latter can be easily adapted to factory requirements. General rules are that deposit cookies should contain (on the basis of flour as 100%), about 35 to 40% sugar, 65 to 75% shortening, and 15 to 25% whole eggs. The flour should be milled from soft wheat, and it should be unbleached.

Although wire-cut cookie equipment is rather tolerant as far as dough consistency is concerned, there are still some rheological requirements that must be observed. The dough should be sufficiently cohesive enough to hold together as it is extruded through orifices. Yet it must be relatively nonsticky and short enough, so that it separates cleanly when cut by the wire. Such formulas may contain several times as much sugar as flour, shortening up to 100% of the flour, and are generally the type of doughs to which particulates (nuts, chocolate chips, or candy pieces) are added. Doughs may be almost as soft as some cake batters or too stiff to be easily molded by hand. The softest wire-cut doughs overlap deposit doughs in consistency, while the other extreme is close to the consistency of some rotary-molded doughs.

Whatever the mixing procedure, it must be sufficient to produce a uniform distribution of ingredients. Many cookie doughs are rather tolerant in the amount and type of mixing that will yield satisfactory performance and products. Saltines and the like must be developed (or brought to a stage of near-maximum dough strength) by the mixing process. For high-sugar-fat formulas, a preliminary creaming stage, in which the sugar and shortening are mixed together for a time before other ingredients are added, is said to ensure uniformity and air incorporation. This creaming step also produces a finer texture in the finished cookie. Most particulates (eg, chocolate chips or nuts) are best added at a late stage of mixing, either with or after the addition of flour.

Brownies are one of the few kinds of cookies that can be baked in a continuous sheet on an oven band, then cut into pieces during or after cooling. Brownies are made with high proportions of sugar, invert syrup and other hygroscopic ingredients, so that they are soft and chewy when fresh, and retain this texture fairly well over a period of weeks or months when properly packaged. The batter, which is often quite soft, is extruded directly onto the oven band as a sheet of uniform thickness, and no other forming operations are performed.

The manufacturing process for sponge-and-dough (eg, saltines) consists of the following steps: weighing the sponge ingredients, mixing the sponge, fermenting the sponge in troughs, mixing the sponge with the remainder of the dough ingredients, fermenting the dough, laminating the dough, sheeting the dough to a required thickness, cutting and embossing the dough sheet, removing the scrap, sprinkling salt on the dough sheet, baking, breaking the baked dough sheet into pieces, and packaging. The manufacturing steps for yeast-leavened or chemically leavened snack crackers resemble the foregoing sequence to a considerable extent, with variations depending on the need for special shapes or content of additional ingredients.

Reciprocating cutters (stamping type) and rotary-cutting machines must be fed, by conveyor belts, with a continuous sheet of dough. An important requirement of these machines is that the scrap dough must be removed in one piece and reincorporated, if the operation is to be efficient. The scrap is returned to a prelamination stage in an operation that will uniformly reincorporate the scrap back into the fresh dough. Care must be taken, in that excessive scrap or rework may affect machining and finished product attributes. Also, the sheet thickness must be maintained within a narrow range, so that the weight of the dough pieces will not vary significantly. The dough must have some elasticity and be cohesive enough to bear its weight, if it is to form a sheet that will retain its continuity and not tear. Excessive elasticity creates problems with shrinkage of the pieces and difficulties in maintaining uniform piece weight. To obtain the desired characteristics, the dough must contain a substantial amount of wheat flour, to provide the gluten that will give the dough strength and elasticity. In processing, the dough must be developed by a mixing operation that orients the gluten molecules, or it must be repeatedly sheeted and layered. The content of ingredients that weaken or "shorten" the dough, such as sweeteners and shortenings, must be kept relatively low. Moisture content, for all practical purposes, is the amount of ingredient water added and should be sufficient enough to allow full hydration of the gluten, without weakening the dough excessively. Nearly all cookies and crackers are baked after forming. A few recipes for fried cookies and crackers can be found in the literature, but these do not appear to have been commercialized to an appreciable extent.

EQUIPMENT

The type of equipment that can be used to process a given biscuit dough is highly dependent on the consistency of the dough. The types of dough that can be processed satisfactorily with one or more of the kinds of available equipment include everything from soft, fluffy batters and wet, flowable doughs to very stiff, claylike masses and tough but elastic doughs resembling bread doughs in many respects.

Mixers

Cookie and cracker doughs are mixed in two main types of mixers: upright and spindle mixers. The upright mixer is used mainly for cookie and some cracker doughs. Various arrangements of the mixing arms can be used, including sigma, z-type, and double-armed. Depending on the product to be made, the characteristics of the formula will determine what mixing arm arrangement is used. Spindle mixers are much more aggressive and are used for saltine and like doughs or other high-strength cracker doughs.

Other specialized mixers, called aerators, are used to produce batters for most sugar wafer production and some drop cookies (eg, vanilla wafers). These machines take a liquid to semipaste batter and incorporate pressurized air into the mixture. Through formulation and mechanical agitation, the air is incorporated into the batter, thus reducing the specific gravity of the resultant batter.

Laminators

Many cracker doughs, processed through cutting machines, first undergo multiple sheeting and folding steps. The basic principle is to form a thin, endless dough sheet by passing the dough between two rollers (or multiple pairs of rollers may be used in sequence), folding this thin dough sheet into a stack of multiple layers, then incorporating together the multiple layers. This laminating step gives the dough internal structure, which enhances its performance in the cutting machine, and gives the final product desired characteristics. If a saltine is examined closely, traces of this layering process can be found.

Forming Equipment

Each type of forming equipment has been designed to process a dough or batter with a particular range of physical characteristics. The action or method required for shaping the dough necessarily sets certain limits on the appearance of the product. It is convenient to separate the following discussion into cookie equipment and cracker equipment, because the machines used in forming these two types of product are quite different.

Cookie Equipment

The following three broad types of forming or shaping devices, as described afterward, are used to manufacture the great majority of commercial cookies:

1. Extruders push the dough or batter through a constricting orifice; they are exemplified by deposit machines, bar presses, and wire-cut equipment.
2. Rotary molders shape the dough in die cavities cut into the surface of a metal cylinder.
3. Stamping machines or rotating cutters cut shaped pieces from a continuous sheet of dough.

Extruders. Extruders vary widely in complexity, from simple equipment consisting of a hopper fitted with feed rollers that press the dough through adjustable slits, to complicated devices that extrude deposit-cookie batters through orifices moving in predetermined patterns. The most common type of machine consists of a hopper and one or more feed rollers that force the dough through an array of tubes called die cups. The dies may have orifices of different shapes: square, round, oval, scalloped, and so on. On wire-cut machines, disks are sliced from a continuously extruded cylinder of dough and allowed to drop onto the oven band or transfer belt. Other, more complicated extruder-type machines can extrude multiple doughs or doughs with fillers. Such machines are sometimes called encrusters. A dough casing and filler are coextruded and either sliced by a wire device or pinched closed via a mechanical device. Sometimes, the dough and filler combination is applied directly to the oven band, forming a continuous strip (eg, fruit bar).

Rotary Molders. A simple rotary-molding machine consists of a hopper, a feed roll, a cylindrical die, a knife or scraper, a cloth or woven-plastic belt (also called a web or apron), and a rubber-covered compression roller. There will also be a frame, motors, controls, and so on. Such machines may be permanently affixed to the oven frame or they may be constructed as movable attachments for cutting machine lines.

The curved surface of the metal cylinder is covered with engraved cavities having the shape desired for the dough piece. Alternatively, plastic die cavities may be fastened to the surface. Pressure is exerted on the dough in the hopper. This causes the dough to be forced into the die cavities, as the die roller rotates beneath a slit at the bottom of the hopper. The cylinder rotates past a knife or scraper (which forms the flat bottom on the dough piece), until it contacts the cloth belt passing beneath it. The belt is forced against the die by a rubber-covered roller, which can be adjusted to vary the pressure. As the cylinder lifts off the belt in its continuing rotation, the dough piece adheres to the belt more strongly than it does to the cavity and is, therefore, drawn out of the die. The belt carries the dough piece to the oven band, where transfer is completed by various kinds of simple mechanisms. There are limits to the adjustments that can be made on the machine to achieve these operational essentials, so the physical characteristics of the dough can vary only slightly.

Stamping Machines. In deposit machines, the batter is extruded intermittently through shaped nozzles. Separation into cookie-size portions is achieved by lowering the oven band during the time the extrusion is stopped. This causes the batter on the band to be pulled away and separated from the batter still in the extrusion orifice. Then the band is moved up again toward the nozzle, and extrusion begins again. Such machines are relatively simple, but they are also quite demanding with regard to dough characteristics. In more advanced machines, the nozzles can be moved to form various patterns such as curves, wavy figures, swirls, and circles. A second depositor can be synchronized with the first to put jelly or some other filling on top of the cookie.

A bar-press machine (sometimes called rout presses) extrudes continuous strings or strips of dough directly onto the oven band. Separation of these strips into individual cookie pieces can be made, either before or after baking, by a cutting device. The die plate may be inclined in the direction of the extrusion, so that the dough ribbon is supported for a longer period of time, an arrangement that reduces breaking or thinning of the dough strand caused by gravitational pull. The die orifices are usually slots with a straight lower edge to give a flat bottom to the cookie, and a grooved top edge, to give a ribbed upper surface. Height of the strip can be varied by moving one of the slot edges.

Wire-cut machines represent an advance in complexity over bar-press and deposit machines in that they include a device that cuts off pieces of the extruded dough as it emerges from the die orifice. The cutoff device consists of a wire or blade that is quickly drawn through the dough by a reciprocating harp. The harp is simply a frame or support attached to a mechanism that moves it back and forth beneath the die cups. The dough emerges from the cups in approximately cylindrical form, if round cookies are being made. However, changing the shape of the orifice can modify the cross section of the dough strand. Several of the die cups are held on a bar that fits snugly into a channel at the bottom of the hopper. Wire-cut machines can probably handle more types of dough than any other cookie-forming apparatus, but they do not operate well on elastic, extensible doughs.

Cracker Equipment

Vertically reciprocating cutters, or stamping machines, are used mostly for crackers, although such devices are also used to make embossed cookies. Rotary cutters employ metal cylinders with cutting dies attached to the curved surface. Two stamping machines may be placed in tandem to perform two different operations on the same piece of dough (eg, cutting the outline and docking the piece).

Ovens

It is frequently said that the biscuit factory must be designed around the oven, because the limitations of this device affect nearly every other operation in the bakery. The bakery equipment found in nearly all large biscuit factories include some form of band oven, a type of oven that is ideally suited for rapid and uniform processing of small pieces of dough. Small operations may use reel ovens, deck ovens, rotary ovens, or just about any other kind of baking equipment. The following discussion will concentrate on the band oven.

The band oven consists of an endless steel belt passing through a baking chamber that can be heated directly or indirectly. The band may be a solid strip of steel or a belt of woven steel wire. Perforated steel or wire-mesh bands allow steam to escape from the bottom of the dough pieces. This helps to prevent gas pockets and retard spread and other distortions in the finished baked good.

The baking chamber consists of a series of modular units (eg, 10 ft long, 7 ft wide, and 5 ft high). Heating means are supplied above and below the band. Thick insulation covers all surfaces that are not occupied by air input and output ducts, ports, and other fixtures. In direct-heated ovens, heat is often provided by ribbon burners above and below the band. Open-flame or surface combustion (ceramic) elements have been used, but the former type is more common. Gas (methane or propane), oil, or even diesel fuel is burned, depending on the limitations of the heating system and the market price of fuel. Electrically heated ovens function satisfactorily but are economically impractical in most locations.

Indirect-heated ovens burn the fuel in a chamber separate from the baking tunnel, and hot air is transferred to the band by fans and convection. The length of the oven is divided into zones, which permits different temperatures to be applied to the product during different stages of baking.

Electronic ovens, of both the dielectric and microwave varieties, have been used to bake cookies and crackers. Microwave heating has demonstrated considerable success in finishing or drying, after baking in a conventional oven. This form of energy application enables the manufacturer to reduce moisture in a cracker, without undue browning of the cracker's surface. New designs in oven technology have led to the development of hybrid ovens, which combine conventional hot air and microwave technology.

Other Equipment

Sugar wafers are made from wafer sheets that are baked in a specialized wafer oven. The oven consists of an enclosed chamber through which pairs of plates, which can be brought together and separated, as required, are transported. At the beginning of the process, a fluid batter is deposited on the bottom plate, and then the other plate in the pair is immediately brought near the first to form a closed chamber within which the batter rapidly expands under the influence of heat. The plates are patterned on their contiguous surfaces so the baked wafer emerges with a sort of waffle pattern or any other desired pattern. In fact, there are many points where the process resembles home baking of waffles. The baked wafers are removed from the griddle, cooled, spread with a filling, topped with another wafer, cut into desired shapes, and packed.

Sandwiching machines assemble two cookie wafers (usually of the molded type, but sometimes wire-cut) and a layer of filling to form a sandwich cookie. Many variations in shapes and flavors are possible; in fact, crackers can be sandwiched with cheese or peanut butter filling to make a nonsweet snack. The cookies, called basecakes, arrive at the sandwiching machines from cooling conveyors leading from the oven or from chutes. Vibrating conveyors align the basecakes into a stacked-on-edge orientation. Basecakes are fed by the conveyor into magazines or chutes in the proper orientation for sandwiching; they are removed one at a time by means of double pins on double chains. As the bottom cakes travel through the machine with their embossed side downward, they receive a deposit of soft creme extruded through a rotating sleeve having shaped orifices. As the bottom basecake with its creme deposit reaches the second set of magazines, the top cakes are dropped onto the creme. Then the sandwich is gently pressed together to ensure adherence of the components, and to establish uniform thickness of the finished cookies. A much more complete discussion of biscuit equipment has been published elsewhere (4).

PACKAGING, STORAGE, AND DISTRIBUTION

Most biscuits will have moisture contents of less than 5%, with some of the crisper varieties, such as saltines, averaging around 2%. Exceptions are the chewy cookies such as brownies and cakelike cookies, which, when fresh, will have moisture contents of more than 10%. Within the range of normal ambient relative humidities experienced in the United States, drier cookies and crackers will tend to absorb water vapor. Chewier cookies will lose moisture, so it is highly desirable to package all biscuits in containers having low rates of moisture vapor transfer, if extended shelf lives are to be achieved. A more detailed discussion of packaging technology has been published elsewhere (5).

Deterioration of cookies and crackers generally takes the form of loss of flavor or acquisition of stale flavors, due to chemical changes (such as oxidation and loss of volatiles by evaporation), and undesirable texture changes. These changes are caused by loss or uptake of moisture, depending on whether a given biscuit is expected to be chewy or crisp, respectively. Sometimes, the appearance also changes because the surface reflects and refracts light differently, due to physical changes such as crystallization of sugars, solidification or liquefaction of fat fractions, or formation of empty vacuoles as moisture evaporates.

Microbiological changes are usually not particularly important causes of storage deterioration of cookies and crackers, because the low "water activity" (actually, relative humidity) of the substrate strongly inhibits the growth of yeasts, bacteria, and molds. The water activity (denoted as A_w) should be less than 0.70, which is the point below which most yeast, mold, and bacteria cannot survive. Sometimes, mold inhibitors such as potassium sorbate or calcium propionate are used to control spoilage, but these additives are regulated and are pH dependent. Occasionally, fermentation by osmophilic yeasts will take place in marshmallow and jellies, but formula adjustments to decrease the water activity of these materials, plus improvements in sanitation, can often remedy such problems. An extensive discussion of packaging and preservation methods for biscuits and other baked goods has been published elsewhere (5).

CURRENT AND FUTURE TRENDS

New product introductions in the United States in recent years have emphasized health and nutritional claims (not always the same things) and high-priced gourmet variations. Health claims are faddish in nature and vary according to the enthusiasm of the moment. High-fiber cookies and crackers can be produced through heavy additions of wheat, oat, or corn bran or more exotic ingredients such as psyllium seed husks, purified cellulose (from wood, sugar beet pulp, etc), and pectin-based materials. No-fat and reduced-fat cookies and crackers have also been popular. Consumers realize there are health benefits in such products, but it is difficult for manufacturers to predict such product trends.

It can be expected that future trends will follow new dietary fads, as they become entrenched in the popular imagination. The exact direction cannot be predicted, but products of reduced caloric content or those meeting health requirements of an aging population will probably be popular.

Gourmet cookies are achieved by formula variations that are heavy on chocolate, fruit preserves, nuts, butter, and other highly indulgent ingredients. Pricing structure usually restricts such products to select market niches. Other gourmet cookies can be made by the addition of perishable components such as whipped cream, butter-cream fillings, and other adjuncts that are practical only for freshly prepared items or for cookies that are distributed in frozen form.

BIBLIOGRAPHY

1. P. Ellis, *Biscuit and Cracker Technology and Manufacturing*, Crompton Press, Morris Plains, New Jersey, 1988.
2. American Association of Cereal Chemists (AACC), 1995, *Approved Methods of the AmericanAmerican Association of Cereal Chemists*, Method 10-53 (approved October 15, 1997).
3. P. Ellis, *Cookie and Cracker Manufacturing*, Vols. 1 and 2, The Biscuit and Cracker Manufacturers Association, Washington, D.C., 1990.
4. S. A. Matz, *Ingredients for Bakers*, Pan-Tech International, Inc., McAllen, Tex., 1987.
5. M. Bakker, *Encyclopedia of Packaging Technology*, John Wiley and Sons, New York, 1986.

GENERAL REFERENCES

H. Faridi, *The Science of Cookie and Cracker Production*, Chapman & Hall, New York, 1994.

D. J. R. Manley, *Technology of Biscuits, Crackers and Cookies*, Ellies Heorrwood Ltd., Chichester, UK, 1983.

S. A. Matz, *Formulas and Process for Bakers*, Pan-Tech International, Inc., McAllen, Tex., 1988.

S. A. Matz, *Equipment for Bakers*, Pan-Tech International, Inc., McAllen, Tex., 1988.

S. A. Matz, *Bakery Technology*, Pan-Tech International, Inc., McAllen, Tex., 1989.

S. A. Matz and T. D. Matz, *Cookie and Cracker Technology*, 2nd ed., AVI Publishing Co., Inc. Westport, Conn., 1978.

W. H. Smith, *Biscuits, Crackers, and Cookies*, Magazines for Industry, Inc., New York, 1972.

P. R. Whiteley, *Biscuit Manufacture*, Elsevier Publishing Co., London, UK, 1971.

RICHARD R. MCFEATERS
DOMENICO R. CASSONE
Nabisco, Inc.
East Hanover, New Jersey

BLANCHING

The blanching process typically utilizes temperatures around 75–95°C for times of about 1–10 min, depending on the product requirements. It is a necessary pretreatment for many vegetables in order to achieve satisfactory quality in dehydrated, canned, and frozen products. A blanching process may be needed if there is likely to be a delay in reaching enzyme inactivation temperatures or, as in freezing preservation, if such temperatures are never achieved. The process should ensure the required reduction of enzyme activity that otherwise might cause undesirable changes in odor, flavor, color, texture, and nutritive value during frozen storage. Another major effect is the removal of intercellular gases. This reduces the potential for oxidative changes in the food and allows the achievement of suitable headspace vacua within cans. As a heat process, blanching may result in some reduction in microbial load, and texture may be improved. Vegetable matter tends to shrink because of loss of turgor, which can aid the achievement of the required fill weight. Undesirable losses of heat sensitive nutrients may be caused, and, in water blanching, soluble constituents may be leached, resulting in large volumes of effluent (1).

BLANCHING METHODS

Water Blanching

Water blanching is the most widely used method. There are four basic designs of water blancher.

1. The tubular blancher in which particulate vegetable matter is transported by pumped water through a system of tubes for the required residence time. Direct steam injection is used to heat the water to blanch temperature, while tube length and product flow rate govern the blanch time.

2. The rotary screw blancher in which a central screw rotates, at a speed that moves the product forward to give the required blanch time, within a static drum containing the blanch water.

3. The rotary blancher, which comprises a drum with a scroll attached to the inner wall, rotating on trunnions.

4. The thermascrew blancher in which the rotating central screw is hollow and contains the heating medium.

Other designs have been developed for various products, such as the countercurrent blanching tower in which product and hot water are moved in opposite directions to promote rapid heat transfer. Product characteristics vary considerably and types (b) and (c) have been used widely. For rotary water blanchers, by a relatively simple scale-up of the basic design, ie, increasing blancher length, blancher capacity is increased for a range of blanch times. Estimates of product outputs can be predicted from data based on trials, as shown by the example in Figure 1 and Table 1.

Capital and running costs are relatively low for water blanchers, and thermal efficiency has been reported to be as high as 60% (2). Improvements in design have been directed at reducing energy and water consumption to minimize leaching and effluent. Mean heat utilization for a blancher with an average spinach capacity of 58,000 kg/h has been estimated at 39%, assuming a value of 758 kJ/kg for raw spinach leaves free from adhering water (3). Energy consumptions of a tubular water blancher and a water blancher with a screw conveyor have been measured as a basis for suggesting energy conserving modifications. The former required 0.54 MJ/kg and the latter 0.91 MJ/kg, indicating the importance of complete steam condensation.

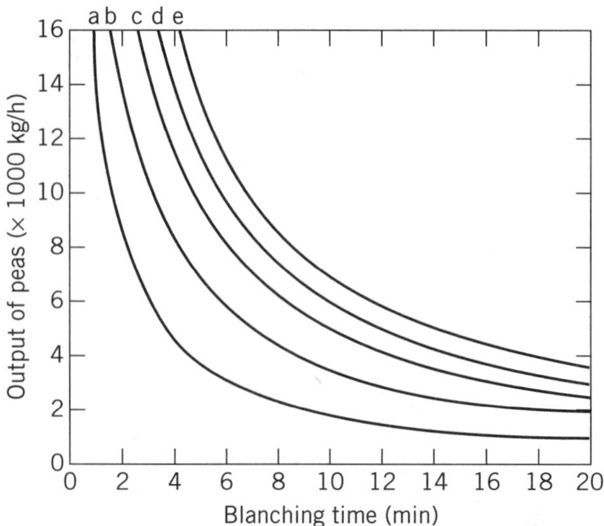

Figure 1. Typical example of relation between product output and blanch time for various products blanched in various sizes of rotary water blancher. For a blancher diameter = 2 m, blancher length is approximately a = 3.1 m, b = 4.5 m, c = 5.9 m, d = 7.4 m, and e = 8.8 m.

Table 1. Multiply Graph Outputs by Appropriate Factor for These Products

Product	Factor
Potatoes	
whole	1.15
french fried	1.00
Carrots, whole	1.10
Beans, soaked	1.08
Peas	
soaked	1.08
fresh	1.00
Green beans	
cross cut	0.70
whole	0.60
Cauliflower, florets	0.50
Spinach, leaf	0.30
Brussels sprouts	0.30
Cabbage, leaf	0.30

Note: See Figure 1.

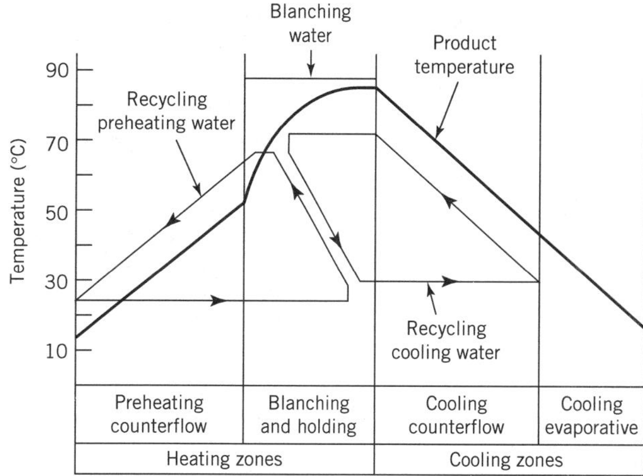

Figure 2. Representation of flow of water and temperatures in a Cabinplant integrated blancher–cooler with heat exchanger.

A study of water heating by heat exchange (68% efficient) and by steam injection (17% efficient) confirmed the energy savings potential of heating by using a heat exchanger to minimize the escape of steam rather than by steam injection (4). Energy consumption can be reduced by minimizing the flow rate of water to the blancher (5), and a comparison of energy consumptions and effluent volumes of various water blanchers is shown in Table 2.

Loss of solubles may be reduced by maximizing product to water ratio, by recycling the blanch water, and by minimizing fresh water addition. This allows the solute concentration in the blanch water to rise toward that of the vegetable cell sap, the isotonic condition thus minimizing net solute loss (6). Tissue damage may promote more rapid loss of solutes and oxidative changes, for example, in peas (7), and hence mechanically induced product movement may be significant both to nutrient quality and product break-up of the more delicate vegetables such as broccoli. The Cabinplant integrated blancher–cooker (Denmark) utilizes minimal water volumes, heat exchangers, and a design to optimize many of the above points, and is the most sophisticated water blancher available (Fig. 2) (8).

One advantage of water blanching is that processing aids such as sodium chloride and sodium bicarbonate may be added to the blanch water to obtain the preferred quality of, for example, cabbage (9). However, blanch waters can also provide a good microbial growth medium. Total microbial loads may be reduced on heating by 10^4–10^5 per gram of product, but numbers of thermoduric organisms may not be reduced. It is therefore important that blanchers be easily accessible for cleaning, be constructed of stainless steel, and that dead spaces where food splashes may collect be designed out wherever possible.

Steam Blanching

As a means of reducing leaching losses, much attention has been given to the development of steam blanchers. The first major development of steam blanching was that of the Individual Quick Blanch (IQB) system (10). This consisted of a 25-s exposure to steam of a single layer of diced carrot, followed by 50 s in a deep bed to allow equilibriation of temperature throughout the carrot pieces. A venturi thermocycle blancher was described, which was claimed to reduce steam consumption by 50%, maintain a uniform temperature throughout the blanching area, and, because of

Table 2. Comparison of Water Blanchers

Type of blancher	Product	Energy efficiency (%)	Energy (MJ/kg product)	Effluent volume or mass/ton product
Ordinary water blancher, calculated values	Peas	60		240–384 L
Ordinary water blancher	Peas		4.8	1,080–3,600 kg
Ordinary water blancher			4.5–6.8	400–1,800 kg
Tank water blancher	Spinach	30.7	1.5	
Ordinary screw blancher	Cauliflower	31.3	0.9	
Tubular blancher	Beans	34.7	0.54	
Pilot screw blancher with steam injection	Cauliflower	16.8	2.09	
Pilot screw blancher with heat exchanger	Cauliflower	67.6	0.51	
Screw blancher	Peas		2.09	4.000 L
Cabinplant blancher and cooling section	Peas		0.27	240 L

Source: Ref. 13.

the water steam traps, provide a vapor-free plant environment (11). Further development of the IQB concept involved an integrated blanching–cooling system comprising vertical helical vibratory conveyors and condensate spray cooling, although problems were encountered when handling nonparticulate material such as leafy green vegetables (12).

In spite of the improved performance of such blancher designs regarding energy and leaching, capital costs and other variable costs have been so high that the conventional water blancher has remained the most competitive system. For many large-scale operations this may no longer be true, and Table 3 compares the performance of various steam blanchers (13). The most recently developed energy efficient steam blancher is that manufactured by the Atlantic Bridge Company (ABCO), Canada (14). The principle of operation of the ABCO K2 was developed from the earlier IQB concept.

Hot-Gas Blanching

Hot-gas blanching has been studied because of the potential for reducing leaching and, in particular, for reducing effluent. One report indicated that for spinach, green beans, corn on the cob, and leeks the retention of water-soluble vitamins was much the same as for other commercial blanching operations. Excessive surface drying in some cases and the presence of atmospheric oxygen, which could promote oxidative changes, were disadvantages. Also, it was shown that operational costs could be much higher than water blanching for some vegetables (15).

Microwave and Electroconductive Blanching

A number of studies of microwave blanching have been carried out to explore the potential for reduced leaching and rapid heating, particularly for vegetables of large cross section such as potatoes, corn on the cob, and brussels sprouts (16). On an industrial scale, processing advantages have been identified, but in the past it was concluded that these were outweighed by the high electricity costs (17). Until recently microwave generators were limited to about 30 kW, with the microwave blancher incorporating steam preheating and humidity control for optimum efficiency. However, with the recent development of 60-kW generators, which also permit more uniform heating, the cost effectiveness for blanching operations needs reviewing (18). There has been limited interest to date in electroconductive blanching (19).

OBJECTIVES FOR BLANCHER CONTROL

Close control of blanch time and temperature influences uniformity of product quality as well as energy consumption. For example, where a vegetable does require blanching prior to freezing, it is now being established that the blanch treatment required depends largely on the heat stability of those enzymes directly responsible for the main deteriorative changes in a given product during frozen storage. Hence energy may be wasted if the blanching conditions are loosely controlled to inactivate peroxidase enzymes, whereas the less heat stable lipoxygenases are the relevant enzymes to inactivate (20).

Accepting that there will always be variations in the vegetable raw material in terms of the physical and thermal properties, consistently uniform blanching conditions will result in consistent blanching effects. This may be less easy to achieve where one blancher is to be used to meet the requirements for a range of vegetable products. Consistently controlled blanching also permits the prediction of the degree of enzyme inactivation (21) and of leaching losses of various solutes using suitable models for water blanching (22).

EFFECTS OF BLANCHING

Water blanchers have been shown to be the main operation contributing to the solids content of cannery wastes, and some recent studies have considered the treatment and concentration of waste blanch water. The use of water as the blanch medium also allows the addition of certain chemicals as processing aids, such as citric acid, and this is reviewed elsewhere (1).

Table 3. Comparison of Steam Blanchers

Type of blancher	Product	Energy efficiency (%)	Energy (MJ/kg product)	Effluent volume or mass/ton product
Ordinary steam blancher, calculated values	Peas	5		191–313 L
Ordinary steam blancher	Peas		6.7–9.0	200 kg
Ordinary steam blancher	Spinach	13	2.12	
Ordinary steam blancher with water cooling	Peas			5.151 L
Steam blancher with end seals	Spinach	27	0.95	
Steam blancher with water curtains	Spinach	19	1.56	
Vibratory spiral blancher	Beans	85		28 L
Individ. quick blancher	Peas			225 L
Steam blancher with end seals and steam recycling	Spinach	31	0.91	
Steam blancher with thermocompression			0.32–0.49	
IQB-prototype K1 & evap. cooling	Peas		0.27–0.32	130–150 kg
IQB-prototype K2 & evap. cooling			0.32–0.56	75–81 kg

Source: Ref. 13.

Weight

In some instances weight losses can be excessive; for example, when blanching mushrooms, weight losses in excess of 19% and volume losses between 11 and 15% have been recorded (23). It has been shown that weight loss from vegetable tissue during water blanching occurs by two main mechanisms. In the typical temperature range of 50 to 55°C the cytoplasmic membranes that enclose the cell contents become disorganized, and on loss of turgor the cells contract and express some cell solution. Simultaneously, the damaged cell membranes allow free diffusion of solutes out of the cells. Because of continued diffusion of solutes out of the tissue during the blanch time, the net tissue weight loss also increases. The example of whole peas blanched in the laboratory at 85°C is shown in Figure 3. The kinetics of mushroom shrinkage have been described by three apparently first-order reactions (24). Artificially altering the water content of the vegetable prior to blanching will also influence subsequent weight loss, although the effect on the diffusional loss of solutes is small (25).

The blanching of dried vegetables has been studied since it was found that the standard rehydration step of around 18 h in cold water could be avoided by an extended blanch of 10–70 min at 71–85°C, depending on the vegetable. During this time sufficient water was taken up to compensate for lack of soaking, with resulting improved overall quality (26).

Nutrients

Some loss of nutrients occurs in all cases of blanching, but water blanching may promote excessive losses, typically up to 40% for minerals and certain vitamins, 35% for sugars, and 20% for protein (1). The most commonly measured nutrient is ascorbic acid because of its high water solubility, susceptibility to oxidation, and ease of analysis. Loss of vitamin C from spinach has been reported as being greater than 70% after 3–5-min blanching (27). Spinach blanched for a range of times at different temperatures was found to lose vitamin C and thiamin following first-order reaction kinetics (28). The decimal reduction times decreased as temperature increased, and Z values (temperature range over which the decimal reduction time changes tenfold) were about 65°C for both these vitamins.

Losses of vitamin C from peas during water blanching have been studied extensively. Temperatures of 35–97°C for times of 15 s to 25 min were used, and the changes in ascorbic acid and dehydroascorbic acid levels were measured (7,29). It was also found that vitamin C concentration was higher in the seed coat (30). A typical 1-min blanch at 97°C resulted in 28% loss of ascorbic acid from Dark Skinned Perfection peas. A broader study showed that at 94–99°C the K1 blanch–cool system (see also Table 3) had a profound effect on ascorbic acid content of some vegetables but not others. For example, large losses were observed for peas and broccoli but not green beans. An analysis of the beans showed that ascorbic acid was concentrated in the seeds (38 mg/100 g fresh wt), which are protected by the pod (8.4 mg/100 g fresh wt) (31). This highlights the need to study the location and distribution of nutrients within various tissues and to study how the concentrations vary with maturity and variety.

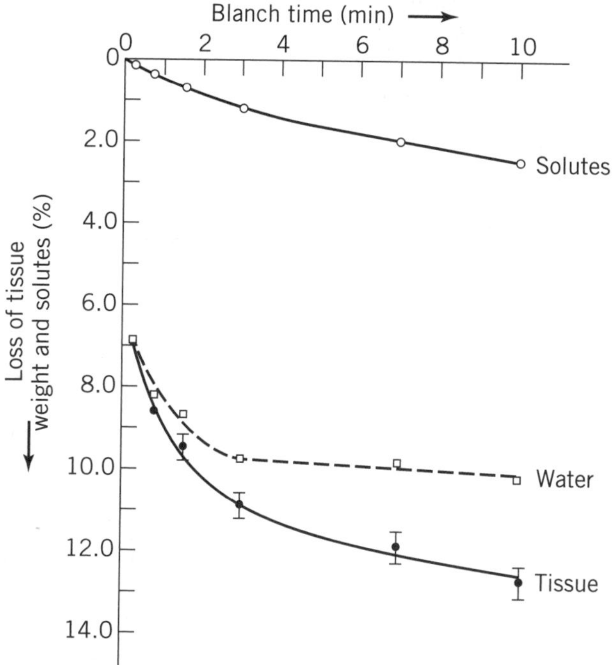

Figure 3. Percentage loss of tissue weight, solutes, and water (by difference) from whole pea samples after the given blanch time at 85°C (means of two duplicated replicates). *Source:* Ref. 25.

Toxic Constituents

In addition to the leaching of nutrients, toxic constituents naturally present in the vegetable may also be leached. The level of nitrates in foods has caused concern because of potential toxicity, and high levels of nitrites may cause methaemoglobinaemia in infants. Nitrate levels in spinach petioles have been shown to be as much as 5 to 10 times greater than in leaves, but water blanching subsequently reduced levels by more than 50% (32). Any nitrite formed by reduction was eliminated, and it was concluded that a combination of leaf selection and blanching could reduce nitrate levels to 25% of the original content. Nitrate levels in carrots were reported for raw carrot, 339 mg nitrate/kg; water washed at 17°C, 226 mg/kg; blanched at 65°C, 187 mg/kg; blanched at 80°C, 181 mg/kg; and blanched at 95°C, 165 mg/kg (33).

The calcium uptake from blanch water influences the ratio of soluble and insoluble oxalates (32). Although excess calcium chloride in the blanch water reduced soluble oxalate in spinach, it also adversely affected color. The influence of washing and blanching on the cadmium content of several vegetables has been studied. Contents initially were 0.121–0.379 mg % for spinach, 0.029 mg % for green beans, and 0.038 mg % for peas. Although 0.1 ppm cadmium was recommended as the maximum in the fresh weight, some spinach still contained higher levels after blanching (34).

Contaminants

Blanching can significantly reduce the level of contaminating microorganisms, which is especially important prior to freezing or dehydration. A 3-min blanch of soy sprouts in boiling water reduced the total bacteria to ≤14,000/g, coliforms to <10/g, and salmonella to zero (35). However, the numbers of thermoduric organisms may not be reduced, and *Bacillus stearothermophilus* was isolated from blanch water used at 90°C for 5-min treatments prior to canning peas in an investigation of the cause of spoiled packs (36). Because blanching may render the vegetable more easily infected by microorganisms, hygienic handling is important.

The content of pesticide residues in vegetables has been studied recently. Commercial blanching was shown to remove 50% of DDT (dichlorodiphenyltrichloroethane) and 68–73% of carbonyl residue from green beans (37). The fate of di-syston has been studied during the processing of potatoes (38), and a 15-min blanch at 100°C reduced the levels of Aldrin, heptachlor epoxide, and Endrin in Irish and sweet potatoes (39). Blanching for 3 min at 100°C reduced levels of DDT isomers in turnip greens (40). The effects of water cooking on the pesticide residues in spinach have been studied (41).

Laboratory canning operations were highly effective in reducing strontium and cesium concentrations in beans and kale. However, blanching of sweet potatoes appeared to result in a transfer of radioactivity from the peel to the core, suggesting that skins of contaminated potatoes should be removed prior to thermal treatment (42).

Enzymes

Peroxidase has often been used as a blanching efficiency indicator enzyme because it is the most heat resistant enzyme and is easy to measure (43). Recent work has shown that a significant but not well-defined proportion of active peroxidase can be left in many vegetables after blanching and a long storage life can still be achieved (44). Less than 5% residual peroxidase activity did not affect quality during storage of carrot, cauliflower, french bean, onion, leek, and swede stored at −20 or −30°C for 15 months (20). Good quality was retained in carrots after storing at −20°C when palmitoyl-CoA hydrolase was inactivated by blanching, even though catalase and peroxidase activity was present. Although this hydrolase was a better indicator enzyme, there was no cheap, easy method of measurement. The complete inactivation of peroxidase does correlate well with the achievement of best quality in peas, but the best sensory quality of green beans was achieved prior to complete peroxidase inactivation.

A study on brussels sprouts found that the activities of polyphenoloxidase and peroxidase were related to the size of the sprouts, and increased from the center to the outer layers. Reducing activity to about 1% of the initial level required 7 min at 98°C for large sprouts, 5 min for medium, and 4 min for small. Residual activity increased toward the center of the sprout (45). The inactivation of lipid-degrading enzymes has also been shown to be very important to final sensory quality (46).

Color, Flavor, and Odor

The blanching operation promotes the thermal degradation of the blue-green chlorophyll pigments to the yellow-green pheophytins (47,48). This has been demonstrated in peas, and it has been shown that the addition of sodium carbonate or calcium oxide or alkalinization with sodium hydroxide during a 3-min blanch at 90°C increased chlorophyll stability. Addition of calcium chloride had the opposite effect. Iron and tin ions promoted, and copper and tin inhibited, chlorophyll decomposition; chlorophyll *b* was more resistant to heat than chlorophyll *a* (49).

Steam blanching of ground red peppers gave a major improvement in color in the first 10 min. This effect was greatest in material that had not been stored, in which the ripening process was not terminated. Because of changes in the conjugated system of carotenes, the red proportion increased and the yellow decreased (50). The effect of different types of blanch on the color of sliced carrots using hot water, in-can steam, direct steam, and microwaves resulted in similar color changes. These arose from changes in ultrastructure, especially in chromoplasts, from which liberated carotenes dissolved into lipids, and some were lost into the water (51).

The use of sodium acid pyrophosphate has been reported in blanch solutions, especially for potato and cauliflower to prevent discoloration during storage of the processed products. A cause of the discoloration may be the reaction between *ortho*-dihydroxyphenols with ferrous ions, forming a pigment in the ferric form on oxidation. Pyrophosphate is unique among condensed phosphates because of its strong ferric ion binding ability.

When peaches were microwaved a uniform browning of the skin resulted because the high levels of polyphenoloxidase in the outer fruit parts received the least heating effect. To prevent this a combination of microwave heating to blanch the inner parts and lye peeling to scald the outer parts was proposed (52).

Soluble or volatile flavor components may be lost during blanching. However, the resulting enzyme inactivation and oxygen removal may aid greater flavor retention during subsequent frozen storage. Unblanched carrot, cauliflower, and french bean developed off-odor after 9, 3, and 6 months, respectively, at −30°C (53). Unblanched onion, leek, and swede did not develop detectable off-flavor or off-odor, and no changes in total lipid content were found. In onion, no lipoxygenase or peroxidase has been found.

Structure and Texture

Blanching may induce both physical and biochemical changes in the structure and textural properties, depending on blanch time and temperature, and type and state of the vegetable. Blanching green beans at 90°C for 30 to 240 s did not seem to influence the degree of cell damage (54), and no tissue disruption in carrots was observed after blanching for 3 min at 100°C or cooking for 10 min at 100°C. This suggested that physiological and chemical rather than physical changes may have occurred to cause softness of the tissues. Subsequent freezing did cause cell disruption (55).

It has been observed that the effects of moisture and heating during blanching lead to a swelling of the cell walls, for example, of green bean pods (150–200%), and to a beginning of separation of the individual cells. This is attributed to the extraction of protopectin from the middle lamella. It was found that not only short blanching but also boiling to doneness preserved the histological structure well. No evidence of any cell wall rupture was observed during cooking, and it was concluded that earlier suggestions that cell wall rupture did occur probably arose through poor methodology in preparing sections (56).

The effects of boiling, steaming, and pressure cooking on the loss of pectic substances, and also the effect of water hardness on potato structure, have been reported (57). Steaming and pressure cooking gave similar results, but boiling resulted in significantly greater losses followed by a characteristic leveling off with time. Steaming probably caused less cell wall damage than boiling because of the lack of water to wash away degraded pectic substances. Boiled potatoes retained significantly more pectic substances when cooked with hard water. Surface to volume ratio was increased by slicing, thus increasing the rate of calcium diffusion into the tissue and of degraded pectic substance out of the tissue.

French fries have been preblanched in disodium dihydrogen pyrophosphate to sequester ions, such as calcium, and prevent discoloration, and this treatment did not reduce protopectin stability or cause deterioration of French fry texture. Low-specific-gravity potatoes were also studied because these gave a soft texture after frying. After 15 min at 70°C in 0.5% calcium chloride the best firming results were obtained, although still not equal to the firmness of high-specific-gravity potatoes (58). It was shown that blanching increased shear value and springiness but decreased hardness of asparagus. The optimum blanch times at 100°C were 135 s for stalk diameters of 16–25 mm, 120 s for 12–15 mm, 105 s for 9–11 mm, and 90 s for 6–8 mm (59).

The texture and color of beans were improved by a stepwise blanch treatment involving high and low temperatures (60). Others found that low-temperature blanching at 74°C for 20–30 min resulted in firmer carrots. Their studies confirmed the conclusion that the increase in firmness was caused by the effects of pectin methyl esterase, which was activated at low temperature and inactivated at a higher temperature (61). The rate of thermal softening of these vegetables follows first-order kinetics (62). Mung bean shoota were blanched at 75°C for 30 s to activate pectinesterase and then held at 55°C for 30 min to maximize deesterification. Such firming has been applied to tomato, potato, and cauliflower. Subsequent addition of calcium salts to the canning liquor caused insoluble calcium pectate gels in the cell walls, thus firming the cell walls (63). The noncooking effects of salts added after cooking on the texture of canned snap beans has been studied (64). Blanching may also affect the dietary fiber content of vegetables (65).

Intercellular Gases

Blanching induces the expansion and removal of intercellular gases from within vegetable tissues. This reduces the potential for subsequent oxidative changes and permits the attainment of adequate headspace vacua in cans. Air removal from the surfaces of a vegetable may change the apparent color shade. When peas of various sizes were blanched between 66 and 95°C for a few seconds up to 10 min it was found that removal of internal gases had largely occurred within the first minute. Gas volume varied from 0.47 mL/100 g grade 2 peas to 1.21 mL/100 g grade 7 peas (66).

The gas and volume changes in sliced green peppers and beans were measured during blanching in steam at 1.5 atm and in water at 100, 90, and 80°C. Initially the peppers contained 14% gas, which was reduced to 2.8–3.8% by steam or water at 100°C, but only to 3.2–5.8% at the lower temperatures. Beans initially contained 19.3%, which was reduced to 1.1–4.1% by steam or water at 100°C, and to 4.2–6.4% at the lower temperatures. Maximum gas reduction was achieved after 2 min with beans and after 1 min with peppers (67). Further data are available for red cabbage and carrots (68).

Several early blanching studies reported that steam blanching was slower than water blanching. Since the heat transfer coefficient of condensing steam is at least an order of magnitude greater than that of hot water to the same surface, it was postulated that the noncondensable gases mixed with steam accumulated at the surfaces being heated, thus interfering with the heat transfer from the steam. Blancher designs where the noncondensable gases are not swept away continually and removed from the system might be prone to a stagnant layer of gas accumulating at the surfaces, thus reducing heat transfer rate. Studies using pure steam revealed that the most important source of noncondensable gases was from the interior of the vegetables being blanched. Foaming was observed, demonstrating that heat transfer had to occur through an insulating layer of foam from the steam to the vegetable. A surface heat transfer coefficient for carrot in steam at atmospheric pressure was given as 1136 $W/m^2 \cdot K$ (69). The work showed that the rate of heat transfer was increased by vacuum pretreatment and decreased by pressure pretreatment. Results indicated that some means of degassing vegetable particles prior to steam blanching would reduce the heating time much more effectively than increases in steam velocity.

Heat and Mass Transfer

The calculation of heat transfer rates during blanching requires a knowledge of the geometry of the vegetable piece, the blanching conditions, including the initial food temperature and blanch medium temperature, and the thermal properties of the vegetable. For unsteady-state heat conduction, numerical solutions to Fourier's general law are available for the central temperature history of the three elementary shapes; an infinite slab, an infinite cylinder, and a sphere (70). Thermokinetic models for enzyme inactivation have also been studied using unsteady-state heat conduction procedures (71).

The two contributions to the total resistance to mass transfer in the blanching of vegetables are the surface resistance due to convection and the internal resistance due to mass diffusion. These may be represented by Fick's first and second laws, together with a mass balance at the interface. When there is sufficient agitation of the blanching liquid, the surface resistance becomes small, and it can be assumed that the total resistance is due to only the internal resistance. Then only the solution to Fick's second law is required, and this is available for the infinite slab, infinite cylinder, and sphere, with the mean concentration obtained after integration with respect to position as a function of time, again in nondimensionalized form (72).

These solutions were originally developed for drying applications over a wide range of concentrations; however, in the case of blanching vegetables, the concentrations are much smaller and near to zero. The solutions have been recalculated for smaller increments in this range of concentrations using the first ten terms for the three series (6). Where the apparent diffusion coefficient for a solute is known at various temperatures it is possible to predict the changes in concentration that will occur under defined conditions of blanching (73,74).

BIBLIOGRAPHY

1. J. D. Selman, "The Blanching Process," in S. Thorne, ed., *Developments in Food Preservation*, Vol. 4, Elsevier Applied Science, London, 1986.
2. J. L. Bomben, "Effluent Generation, Energy Use and Cost of Blanching," *Journal of Food Process Engineering* **1**, 329–341 (1977).
3. A. Schaller, "Heat Utilisation in Water Blanching of Spinach Leaves," *Confructa* **17**, 264–273 (1973).
4. T. R. Rumsey, E. P. Scott, and P. A. Carroad, "Energy Consumption in Water Blanching," *Journal of Food Science* **47**, 295–298 (1982).
5. M. A. Rao, H. J. Cooley, and A. A. Vitali, "Thermal Energy Consumption for Blanching and Sterilisation of Snap Beans," *Journal of Food Science* **51**, 378–380 (1986).
6. J. D. Selman, P. Rice, and R. K. Abdul-Rezzak, "A Study of the Apparent Diffusion Coefficients for Solute Losses from Carrot Tissue during Blanching in Water," *Journal of Food Technology* **18**, 427–440 (1983).
7. J. D. Selman and E. J. Rolfe, "Effects of Water Blanching on Pea Seeds, II: Changes in Vitamin C Content," *Journal of Food Technology* **17**, 219–234 (1982).
8. M. Togeby, N. Hansen, E. Mosekilde, and K. P. Poulsen, "Modelling Energy Consumption, Loss of Firmness and Enzyme Inactivation in an Industrial Blanching Process," *Journal of Food Engineering* **5**, 251–267 (1986).
9. C. Srisangnam, D. K. Salunkhe, N. R. Reddy, and G. G. Dull, "Quality of Cabbage, I: Effects of Blanching, Freezing and Freeze-Dehydration on the Acceptability and Nutrient Retention of Cabbage," *Journal of Food Quality* **3**, 217–231 (1980).
10. M. E. Lazar, D. B. Lund, and W. C. Dietrich, "A New Concept in Blanching," *Food Technology* **25**, 684 (1971).
11. C. R. Havighorst, "Venturi Tubes Recycle Heat in Blancher," *Food Engineering* **45**, 89–90 (1973).
12. J. L. Bomben, "Vibratory Blancher-Cooler Saves Water, Heat and Waste," *Food Engineering International* **1**, 37–38 (1976).
13. K. P. Poulsen, "Optimisation of Vegetable Blanching," *Food Technology* **40**, 122–129 (1986).
14. D. Atherton and J. B. Adams, "New Blanching Systems," *Technical Memorandum No. 319*, Campden Food and Drink Research Association, Chipping Campden, England, 1983.
15. J. W. Ralls, H. J. Maagdenberg, N. L. Yacoub, M. E. Zinnecker, J. M. Reiman, H. D. Karnath, D. M. Homnick, and W. A. Mercer, "In-Plant Hot-Gas Blanching of Vegetables," *Proceedings of the 4th National Symposium of Food Processing Wastes*, Environmental Protection Agency, Syracuse, N.Y., pp. 178–222, 1973.
16. W. C. Dietrich, C. C. Huxsoll, and D. C. Guadagni, "Comparison of Microwave, Conventional and Combination Blanching of Brussels Sprouts for Frozen Storage," *Food Technology* **24**, 613–617 (1970).
17. A. C. Cross and D. Y. C. Fung, "The Effect of Microwaves on Nutrient Value of Foods," *Critical Reviews in Food Science and Nutrition* **16**, 355–381 (1982).
18. W. Stolp, D. J. van Zuilichem, and F. Westbrook, "Microwave Blanching of Fresh Mushrooms," *Vaedingsmiddelentechnologie* **22**, 45–47 (1989).
19. R. L. Garrote, E. R. Silva, and R. A. Bertone, "Blanching of Small Whole Potatoes in Boiling Water and by Electroconductive Heating," *Lebensmittel-Wissenschaft und Tecnologie* **21**, 41–45 (1988).
20. P. Baardseth, "Quality Changes in Frozen Vegetables," *Food Chemistry* **3**, 271–282 (1978).
21. R. P. Singh and G. Chen, "Lethality-Fourier Method to Predict Blanching," in P. Linko, Y. Malkki, J. Olkku, and J. Larinkari, eds., *Food Process Engineering*, Vol. 1, Applied Science, London, 1980, pp. 70–74.
22. S. M. Alzamora, G. Hough, and J. Chirife, "Mathematical Prediction of Leaching Losses of Water Soluble Vitamins during Blanching of Peas," *Journal of Food Technology* **20**, 251–262 (1985).
23. T. R. Gormley and C. MacCanna, "Canning Tests on Mushroom Strains," *Irish Journal of Food Science and Technology* **4**, 57–64 (1980).
24. M. Konanayakam and S. K. Sastry, "Kinetics of Shrinkage of Mushrooms during Blanching," *Journal of Food Science* **53**, 1406–1411 (1988).
25. J. D. Selman and E. J. Rolfe, "Effects of Water Blanching on Pea Seeds, I: Fresh Weight Changes and Solute Loss," *Journal of Food Technology* **14**, 493–507 (1979).
26. UK Patent Application GB2088695 A (1982), I. J. Farrow, Cleary & Co. Ltd, "Blanching Dried Vegetables."
27. H. Koaze, T. Okamura, K. Ishibashi, K. Hironaka, and S. Kato, "Studies on Frozen Vegetables, IV: Effect of Blanching on Retention of Reduced Ascorbic Acid, Chlorophyll and Peroxidase Activity of Frozen Spinach," *Refrigeration (Reito)* **55**, 771–776 (1980).
28. K. Paulus, R. Duden, A. Fricker, K. Heintze, and H. Zohm, "Influence of Heat Treatment of Spinach at Temperatures up to 100°C, II: Changes of Drained Weight and Contents of Dry Matter, Vitamin C, Vitamin B1 and Oxalic Acid," *Lebensmittel Wissenschaft Technologie* **8**, 11–16 (1975).
29. J. D. Selman, "Review—Vitamin C Losses from Peas during Blanching in Water," *Food Chemistry* **3**, 189–197 (1978).
30. J. D. Selman and E. J. Rolfe, "Studies on the Vitamin C Content of Developing Pea Seeds," *Journal of Food Technology* **14**, 157–171 (1979).
31. D. B. Cumming, R. Stark, and K. A. Sanford, "The Effect of an Individual Quick Blanching Method on Ascorbic Acid Retention in Selected Vegetables," *Journal of Food Processing and Preservation* **5**, 31–37 (1981).
32. B. L. Bengtsson, "Effect of Blanching on Mineral and Oxalate Content of Spinach," *Journal of Food Technology* **4**, 141–145 (1969).
33. T. Mari and I. Binder, "Reduction of the Nitrate Content of Carrots by Various Blanching Methods," *Hütoipar* **25**, 7–9 (1978).

34. H. J. Beilig and H. Treptow, "Influence of Processing on Cadmium Content of Spinach, Green Beans and Peas," *Berichte über Landwirtschaft* **55**, 809–816 (1977).

35. G. Marcy and W. Adam, "Soy Sprout Salad as a Health Hazard," *Archiv für Lebensmittelhygiene* **28**, 197–198 (1977).

36. A. C. Georgescu and M. Bugulescu, "Thermophilic Organisms in Insufficiently Sterilized Canned Peas," *Industria Alimentara* **20**, 251–253 (1969).

37. E. R. Elkins, F. C. Lamb, R. P. Farrow, R. W. Cook, M. Kawai, and J. R. Kimball, "Removal of DDT, Malathion and Carbaryl from Green Beans by Commercial and Home Preparation Procedures," *Journal of Agricultural and Food Chemistry* **16**, 962–966 (1968).

38. M. G. Kleinschmidt, "Fate of Di-syston-in Potatoes during Processing," *Journal of Agricultural and Food Chemistry* **19**, 1196–1197 (1971).

39. J. M. Solar, J. A. Liuzzo, and A. F. Novak, "Removal of Aldrin, Heptachlor Epoxide and Endrin from Potatoes during Processing," *Journal of Agricultural and Food Chemistry* **19**, 1008–1010 (1971).

40. J. G. Fair, J. L. Collins, M. R. Johnston, and D. L. Coffey, "Levels of DDT Isomers in Turnip Greens after Blanching and Thermal Processing," *Journal of Food Science* **38**, 189–191 (1973).

41. G. Melkebeke, M. van Assche, W. Dejonckheere, W. Steurbaut, and R. H. Kips, "Effect of Some Culinary Treatments on Pesticide Residue Contents of Spinach," *Revue de l'Agriculture* **36**, 369–378 (1983).

42. C. M. Weaver and N. D. Harris, "Removal of Radioactive Strontium and Caesium from Vegetables during Laboratory Scale Processing," *Journal of Food Science* **44**, 1491–1493 (1979).

43. D. C. Williams, M. H. Lim, A. O. Chen, R. M. Pangborn, and J. R. Whitaker, "Blanching of Vegetables for Freezing—Which Indicator Enzyme to Choose," *Food Technology* **40**, 130–140 (1986).

44. H. Böttcher, "The Enzyme Content and the Quality of Frozen Vegetables, II: Effects on the Quality of Frozen Vegetables," *Nahrung* **19**, 245–253 (1975).

45. E. Pogorzelski, J. Rotsztejn, and J. Berdowski, "Effect of Blanching Conditions on Polyphenoloxidase and Peroxidase Activities of Brussels Sprouts," *Przemysl Fermentauyjny i Owocowo-Warszawny* **25**, 38–39 (1981).

46. P. Baardseth and E. Naesset, "The Effect of Lipid-Degrading Enzyme Activities on Quality of Blanched and Unblanched Frozen Stored Cauliflower, Estimated by Sensory and Instrumental Analysis," *Food Chemistry* **32**, 39–46 (1989).

47. J. Abbas, M. A. Rouet-Mayer, and J. Philippon, "Comparison of the Kinetics of Two Pathways of Chlorophyll Degradation in Blanched or Unblanched Frozen Green Beans," *Lebensmittel Wissenschaft und Technologie* **22**, 68–72 (1989).

48. N. Muftugil, "Effect of Different Types of Blanching on the Color and the Ascorbic Acid and Chlorophyll Contents of Green Beans," *Journal of Food Processing and Preservation* **10**, 69–76 (1986).

49. B. Segal, "Chlorophyll Decomposition in Green Peas during Processing," *Industria Alimentara* **21**, 493–497 (1970).

50. K. Klyamov, "Effect of Blanching on Colour of Ground Red Peppers," *Bilgarski Plodove Zelenchutsi i, Konservi* **8**, 25–27 (1975).

51. S. Mirza and I. D. Morton, "Blanching of Carrots and Nutrient Loss," *Journal of the Science of Food and Agriculture* **25**, 1043 (1974).

52. C. Avisse and P. Varoquaux, "Microwave Blanching of Peaches," *Journal of Microwave Power* **12**, 73–77 (1977).

53. P. Baardseth, "Meat Products and Vegetables (Thermal Processing), in *Physical, Chemical and Biological Changes in Food Caused by Thermal Processing*, Applied Science Publishers, London, 1977, pp. 280–289.

54. C. Buonocore and G. Crivelli, "Structural Changes in Vegetable Products as a Function of Freezing Time, V: Observations on Green Beans," *Industrie Agrarie* **8**, 225–230 (1970).

55. A. R. Rahman, W. L. Henning, and D. E. Westcott, "Histological and Physical Changes in Carrots as Affected by Blanching, Cooking, Freezing, Freeze-drying and Compression," *Journal of Food Science* **36**, 500–502 (1971).

56. M. Grote and H. G. Fromm, "Fine Structural Analysis of the Morphological Changes Involved in the Blanching, Cooking, Dehydration and Rehydration of Green Bean Pod Tissue," *Zeitschrift für Lebensmittel Untersuchung und Forschung* **166**, 203–207 (1978).

57. D. E. Johnston, D. Kelly, and P. P. Dornan, "Losses of Pectic Substances during Cooking and the Effect of Water Hardness," *Journal of the Science of Food and Agriculture* **34**, 733–736 (1983).

58. A. S. Jaswal, "Effects of Various Chemical Blanchings on the Texture of French Fries," *American Potato Journal* **47**, 13–18 (1970).

59. T. Motohiro and N. Inoue, "Manufacture of Canned Asparagus, IV: Changes in Shear Value, Springiness and Hardness of Asparagus on Heat Treatment," *Journal of Food Science and Technology (Japan)* **20**, 1–4 (1973).

60. E. Steinbuch, "Technical Note: Improvement of Texture of Frozen Vegetables by Stepwise Blanching Treatments II," *Journal of Food Technology* **12**, 435–436 (1977).

61. C. Y. Lee, M. C. Bourne, and J. P. van Buren, "Effect of Blanching Treatments on the Firmness of Carrots," *Journal of Food Science* **44**, 615–616 (1979).

62. M. C. Bourne, "Effect of Blanch Temperature on Kinetics of Thermal Softening of Carrots and Green Beans," *Journal of Food Science* **52**, 667–668, 690 (1987).

63. A. J. Taylor, J. M. Brown, and L. M. Downie, "The Effect of Processing on the Texture of Canned Mung Bean Shoots," *Journal of the Science of Food and Agriculture* **32**, 134–138 (1981).

64. J. P. van Buren, "Effects of Salts Added after Cooking on the Texture of Canned Snap Beans," *Journal of Food Science* **49**, 910–912 (1984).

65. M. Nyman, K. E. Palsson, and N. G. Asp, "Effects of Processing on Dietary Fibre in Vegetables," *Lebensmittel Wissenschaft und Technologie* **20**, 29–36 (1987).

66. R. S. Mitchell, D. J. Casimir, and L. J. Lynch, "Blanching of Green Peas, II: Puncturing to Improve Efficiency of Gas Removal," *Food Technology (Champaign)* **23**, 819–822 (1969).

67. S. I. Jankov, "Physical Changes in Sliced Green Peppers, Beans and Peas during Blanching, *Confructa* **15**, 88–92 (1970).

68. C. Kluge, K. Melhardt, and L. Linke, "Studies on Gas Escape during the Thermal Treatment of Vegetables," *Lebensmittelindustrie* **33**, 234–237 (1986).

69. C. C. A. Ling, J. L. Bomben, D. F. Farkas, and C. J. King, "Heat Transfer from Condensing Steam to Vegetable Pieces," *Journal of Food Science* **39**, 692–695 (1974).

70. P. J. Schneider, *Conduction Heat Transfer*, 6th ed., Addison-Wesley, Reading, Mass., 1974.

71. J. A. Luna, R. L. Garrote, and J. A. Bressan, "Thermo-kinetic Modeling of Peroxidase Inactivation during Blanching–Cooling of Corn on the Cob," *Journal of Food Science* **51**, 141–145 (1986).
72. A. B. Newman, "The Drying of Porous Solids: Diffusion Calculations," *Transactions of the AIChE* **27**, 310 (1931).
73. G. Hough, S. M. Alzamora, and J. Chirife, "Effect of Piece Shape and Size on Leaching of Vitamin C during Water Blanching of Potato," *Journal of Food Engineering* **8**, 303–310 (1988).
74. P. Rice, J. D. Selman, and R. K. Abdul-Rezzak, "Nutrient Loss in the Hot Water Blanching of Potatoes," *International Journal of Food Science and Technology* **25**, 61–65 (1990).

GENERAL REFERENCES

"Steaming into the Future," *Potato Business World* **6**, 19–25 (1998).

S. Boyes et al., "Microwave and Water Blanching of Corn Kernels: Control of Uniformity of Heating During Microwave Blanching," *Journal of Food Processing and Preservation* **21**, 461–484 (1997).

J. I. Mate et al., "Effect of Blanching on Structural Quality of Dried Potato Slices," *J. Agric. Food Chem.* **46**, 676–681 (1998).

S. Meenakshi, B. Raghavan, and K. O. Abraham, "Processing of Marjoram (*Marjorana hortensis Moensch.*) and Rosemary (*Rosmarinus officinalis L.*). Effect of Blanching Methods on Quality," *Nahrung* **40**, 264–266 (1996).

G. Prestemo, C. Fuster, and M. C. Risueno, "Effects of Blanching and Freezing on the Structure of Carrot Cells and Their Implications for Food Processing," *J. Sci. Food Agric.* **77**, 223–229 (1998).

A. R. Quinero et al., "Optimization of Low Temperature Blanching of Frozen Jalapeno Pepper (*Capsicum annuum*) Using Response Surface Methodology," *J. Food Sci.* **63**, 519–522 (1998).

H. S. Raswamy and M. O. Fakhouri, "Microwave Blanching: Effect on Peroxidase Activity, Texture and Quality of Frozen Vegetables," *Journal of Food Science and Technology India* **35**, 216–222 (1998).

U.S. Patents 752,431,730,717 (1998), D. R. Zittel.

JEREMY D. SELMAN
Campden Food & Drink Research Association
Campden, Gloucestershire
United Kingdom

BLOOD. See ANIMAL BY-PRODUCT PROCESSING; KOSHER FOODS AND FOOD PROCESSING.

BLUEBERRIES. See FRUITS, TEMPERATE.

BOTULISM. See FOODBORNE DISEASES.

BOVINE SPONGIFORM ENCEPHALOPATHY (BSE)

Bovine spongiform encephalopathy (BSE, mad cow disease), a progressive degenerative disorder of the central nervous system of cattle, was first recognized in the United Kingdom (UK) in November 1986 (1). Believed to be caused by a newly classified proteinaceous infectious particle, a prion, BSE is one of several recognized fatal transmissible spongiform encephalopathies (TSEs) found in various species. The epizootic of BSE in the UK affected more than 170,000 cattle, 70 domestic cats, and some zoo animals and is believed to have resulted in 40 human cases of variant Creutzfeldt-Jakob disease (v-CJD, also known as new variant or nv-CJD). The human cases resulted from the ingestion of beef contaminated with the BSE agent.

OVERVIEW

TSEs disrupt the function of nerve cells. In TSEs there are characteristic vacuolar degeneration of neurons in the brain and spinal cord, giving the tissue a spongelike appearance (Fig. 1). TSEs evoke no immune response and are inevitably fatal (2–5).

The incubation period (ie, the time between exposure and the onset of clinical signs) in cattle is very long, rang-

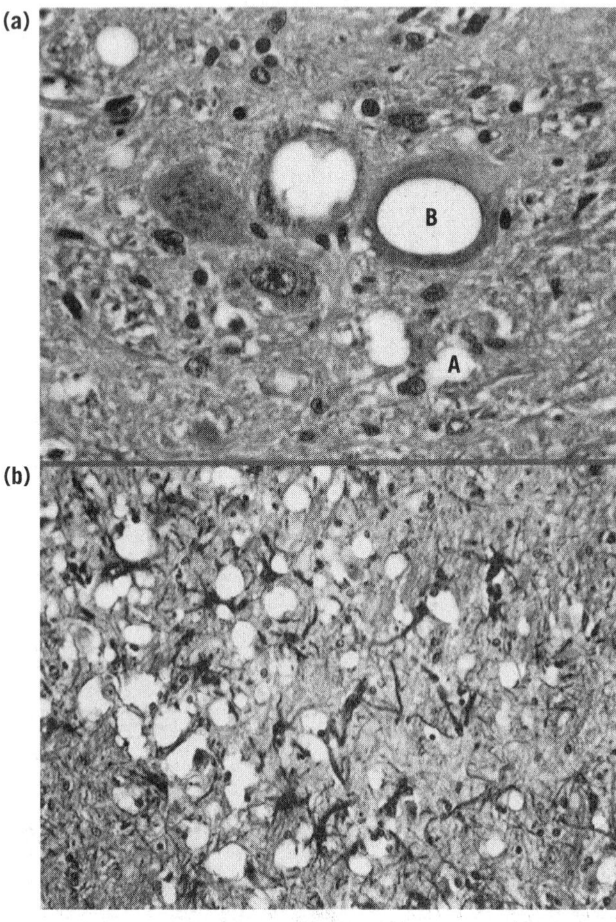

Figure 1. Lesions in the gray matter of the brain of a sheep with scrapie: (A) typical spongiform change in neurons; (B) spongiform change and astrocytic hypertrophy and hyperplasia. (**a**) Hematoxylin and eosin stain; (**b**), glial fibrillar acid protein (GFAP) stain. Magnification ×500. *Source:* Courtesy of Dr. Robert Higgins, School of Veterinary Medicine, University of California, Davis.

ing from two to eight years. Most animals show clinical signs between three and five years of age and demonstrate a slowly progressive disease with nonspecific neurological signs, including incoordination, weakness, tremors, and anxiety. Once signs develop, animals usually die within two weeks to six months.

Epidemiologically, BSE has been considered an "expanded point source epizootic"; that is, it resulted from the consumption of contaminated protein feed supplements (6). Meat and bone meal (MBM) produced from rendered carcasses and offal from dead and butchered livestock has been a common protein supplement used for decades in livestock production. Changes in rendering practices in the late 1970s and early 1980s were considered a risk factor in amplifying the disease through the resulting manufacture of infectious MBM feed supplements (6–10).

About 2000 cattle have been diagnosed with the disease in other countries, some as the result of exportation from the UK of live cattle incubating the disease, and the majority from exportation of contaminated MBM subsequently fed to native cattle (7–9).

More than 170,000 cattle from more than 33,500 British herds (67% of all dairy herds, 16% of all beef herds) have been confirmed with the disease. The low average within-herd incidence, with 35% of the farms experiencing only one case, and 69% with four or fewer cases, is attributed to uneven exposure to the agent in widely spaced batches of feed (6).

In 1996 a possible link was made between BSE in cattle and 10 cases of Creutzfeldt-Jakob disease (CJD) in humans (11). To date 29 humans have been confirmed with v-CJD (12), shown to share identical lesion patterns and chemical characteristics with BSE (13–15). Evidence strongly supports the hypothesis that v-CJD in people and BSE in cattle are the same disease, with transmission to people occurring through the ingestion of infected beef products (16,17).

Perishable foods are especially vulnerable to acute crises in consumer confidence. Following the March 1996 announcement of the possible link between BSE and human disease, British beef consumption fell for more than a month to 25% of its preannouncement level, and the European Union (EU) banned British beef imports. Global markets, including those in countries without BSE, have been impacted by this epizootic.

Beginning in July 1988 the UK responded to the cattle epizootic with a series of control measures. Mandatory programs were enacted requiring euthanasia of affected cattle, prohibiting the use of affected cattle carcasses or specified offal from sheep or cattle in feeds, and eventually banning MBM from all agricultural uses. A selective cull program was instituted targeting herds that had BSE cases and removing nonaffected cattle that might have been shared exposure to the agent. Current projections are for BSE eradication between 2001 and 2005.

Measures to protect human health began in 1989 with the Specified Offal Ban (SBO), prohibiting the use of brain, spinal cord, tonsil, thymus, spleen, and intestinal tissues of cattle origin in foods intended for human consumption. Additional measures, requiring meat from animals over 6 months of age to be deboned prior to sale, and banning most cattle over 30 months of age from use as human food, were enacted.

By-products from slaughtering cattle are used in horticultural applications, pet foods, cosmetics, and pharmaceuticals. Those tissues not meeting specified standards are prohibited from use in manufacturing these products and are now incinerated.

TRANSMISSIBLE SPONGIFORM ENCEPHALOPATHIES

TSEs are caused by similar, but not identical, agents and produce specific and characteristic lesions in the brain of affected hosts. The apparent resistance/susceptibility factors in animals and humans are not yet fully understood, and multiple "strains" have been identified in some TSEs, including scrapie (2,3,13,14,17). All TSEs have long incubation periods, evoke no immune response, and are inevitably fatal.

TSEs in animals include cattle BSE, scrapie of sheep and goats, transmissible mink encephalopathy, and chronic wasting disease of mule deer and elk (6,9,18). Confirmed cases in kudus, a gemsbok, an eland, a nyala, an oryx, an Ankole cow, cheetahs, pumas, a tiger, ocelots, and more than 70 domestic cats are also believed to be BSE caused by the feeding of contaminated protein supplements (19).

Human TSEs include kuru, Creutzfeldt-Jakob disease, Gerstmann–Straussler–Scheinker syndrome, and fatal familial insomnia. Human TSEs are heritable in about 15% of the cases and are considered sporadic in about 85% of the cases. Worldwide a few hundred cases have been associated with the use of tissues from infected donors (corneal transplants, cadaveric dura mater, and human pituitary extracts used to prepare growth and gonadotropin hormones), or contaminated intracranial electrodes and neurosurgical instruments. Kuru, a TSE associated with the ritual handling and cannibalistic consumption of the brain of deceased tribal members in a region of New Guinea, with a known incubation period up to 30 years, has all but disappeared due to changes in local customs (2,3).

BSE LINK TO HUMAN DISEASE

In 1990 the UK established the Spongiform Encephalopathy Advisory Committee to advise on the potential implications TSEs have on advised the government that it had become concerned about 10 cases of Creutzfeldt-Jakob disease with onset of symptoms in 1994 and 1995. These cases were unusual because they included patients younger than the expected (average age 27 vs 63 for sporadic CJD cases). In addition, the neuropathology and clinical course in these patients was distinguishable from that in other CJD patients. A review of medical histories, genetic analysis, and consideration of other possible causes of CJD "failed to explain these cases adequately" (4,5,11).

As many as 900,000 cattle might have been infected between 1984 and 1995, with 450,000 entering the human food chain prior to 1989 and 283,000 between 1989 and 1995 (6). Cattle are slaughtered at young ages and BSE's long incubation period would preclude most infected animals from showing any clinical signs. Cattle in the earlier stages of incubation are expected to have had much lower accumulations of infectious agent in their brain and spinal cord.

There have been no confirmed cases of v-CJD in people with occupations expected to have the highest risk of exposure, including farmers, abattoir workers, butchers, and veterinarians. The number of new cases of all types of CJD in the UK remains stable, at about one new case per million people per year. This rate is about the same in all countries, including Australia and New Zealand, where neither scrapie nor BSE have been identified. Currently 29 people, one in France and the remainder in the UK, are believed to have v-CJD (13). The proposed link between BSE and v-CJD was eventually substantiated (13,14). There are no predictions of the number of human victims that might eventually result.

THE CAUSATIVE AGENT

Prions and the "Protein Only" Hypothesis

Prions, the name coined by Stanley Prusiner for proteinaceous infectious particles, have changed our understanding of infectious agents. All previously recognized pathogenic agents use nucleic acid, either DNA or RNA, to reproduce. Prions are abnormal structural variants of normal cellular proteins found in the membranes of neurons and other tissues (2,3).

Although scientific opinion generally supports Prusiner's 1982 hypothesis that TSEs are caused by abnormal prion proteins, other research groups continue looking for an unrecognized organism with nucleic acid (3), and one group continues to promote *Spiroplasma* sp. as a potential causative agent (20).

Prions are single molecules of about 250 amino acids and contain no nucleic acid. Abnormal prion protein is formed within the host by genetic mutation (eg, familial CJD), or by unknown means (eg, sporadic CJD), or enters the body via oral or other routes (eg, BSE, kuru, scrapie, etc), and proceeds to convert by contact normal protein into structurally aberrant forms. The conversion of normal cellular prion protein to the distorted prion conformation is characterized by a structural change from predominantly alpha helices to beta sheets (alpha helix:beta sheet 42:3 in normal prion protein, 30:43 in abnormal) (3,21). The distorted form has increased resistance to natural proteases and, as a result, accumulates in tissues, contributing to neuronal death and spongiform changes in the brain. Amyloid plaques are formed from remnants of prion protein in some, but not all, forms of TSE disease.

TSEs such as scrapie have been shown to exist as different "strains," evidenced by different incubation periods, lesion patterns, and chemical characteristics. Historically, strain differences in pathogenic organisms have always been associated with genetic differences represented by varying nucleic acid sequences (3). Strain differences in prion diseases appear to be from conformational differences of the same protein molecule. Functional behavior of the molecule is linked to the conformation and may influence susceptibility/resistance, infectious dose, virulence, disease pattern, and incubation period (13,14,17).

Transmission

Variant CJD is believed to have been transmitted by consuming infected beef products such as sausages, franks, and hamburgers that contained portions of brain, spinal cord, dorsal root ganglia, retina, intestines, or bone marrow as components. A list of Specified Risk Materials (SRM) has been adopted by the European Union (EU) Scientific Steering Committee as a result of calf assay experiments using various tissues from known infected animals (Table 1; 22).

Prions are extremely resistant to heat, ultraviolet light, ionizing radiation, and common disinfectants that normally inactivate viruses and bacteria, and destroy nucleic acid (2,3,23–25). Contaminated surgical instruments have been shown to retain infectivity following standard hospital sterilization techniques. Transplanted tissues from humans with CJD, including dura mater grafts, extracts from pituitary glands, and corneas, have been demonstrated vehicles of transmission. Strategies to protect the blood supply, including provisions to deplete blood products of leukocytes because these white blood cells may play a role in the transport of ingested abnormal prions to the nervous system, are currently under consideration in the UK (26).

Kuru was transmitted by ingestion and through open cuts or abrasions. Scrapie in sheep has been transmitted through open wounds and new infections have been associated with lambing of normal animals with infected animals. Although the majority of cattle infections resulted from ingestion of contaminated protein feed supplements, studies on potential maternal transmission report a small risk of infection for calves born shortly before or after the onset of clinical signs of the disease in infected dams. Experts indicate this rate is insufficient to perpetuate the epizootic in cattle but will delay the eradication effort (6,27).

Table 1. Levels of TSE Infectivity Associated with Various Ruminant Tissues

Category	Organs
High infectivity	Bovine brain, eyes, spinal cord and dorsal root ganglia; *dura mater,* pituitary, skull and bovine vertebral column, lungs ovine/caprine brain, eyes, and spinal cord, dorsal root ganglia and vertebral columns; ovine and caprine spleens, lungs
Medium infectivity	Total intestine from duodenum to rectum, tonsils Bovine and caprine spleen, placenta, uterus, fetal tissue, adrenal, cerebrospinal fluid, lymph nodes
Low infectivity	Liver, pancreas, thymus, bone marrow, other bones, nasal mucosa, peripheral nerves
No detected infectivity	Skeletal muscle, heart, kidney, colostrum, milk, discrete adipose tissues, salivary gland, saliva, thyroid, mammary gland, ovary, testis, seminal testes, cartilaginous tissue, skin, hair, blood clot, serum, urine, bile, feces

Source: Ref. 22.

The persistence of chronic wasting disease of mule deer and elk raises questions about possible methods of lateral transmission between unrelated animals. There is a small risk of transmission to uninfected lambs if they are born in the immediate physical vicinity of infected ewes also giving birth.

Susceptibility

The amino acid sequence in the prion protein of cattle differs from that of sheep at 7 positions, and that of humans at 30 positions. It is thought that greater differences provide a more formidable "species barrier" to transmission. Scrapie strains have not been shown to transmit to cattle; however, BSE has been experimentally transmitted to sheep and several other species. Prions from cattle and humans, but not sheep, share specific amino acid sequences in two areas of the protein, possibly accounting for the transmission in spite of the amino acid sequence differences (2,3,21,28).

All cases of v-CJD have been recorded in people homozygous for methionine at prion protein codon 129. This genotype is found in 38% of the UK population. Genotypes that are heterozygous for methionine/valine (UK frequency 51%), or homozygous valine (UK frequency 11%), may have different incubation periods or increased resistance to infection (15,29).

BSE has a characteristic pattern of lesions in all species examined when changes are mapped by region of the brain. Also, protein digestion patterns of BSE prion are identical between species and differ from other TSEs. Other TSEs, however, demonstrate "strain" differences using these tests and in clinical progression of the disease. Strain differences are believed to represent differing molecular conformations of identical proteins (13,14).

Prevention and Treatment

Both normal and distorted prion proteins fail to stimulate any detectable host immune response, and no inflammatory reaction is associated with accumulations of abnormal prion protein in neurons. The agent presents special challenges to those attempting to develop effective drugs for treatment and vaccines for prevention.

A reliable antemortem test is an area of special need, and a number of promising methods are under evaluation. The long incubation period contributes to the difficulty of validating the various tests being developed. Ultimately the determination of the risk from medical products incorporating organs, cells, sera, and bovine by-products, and that from various food products, depends on sensitive and specific antemortem testing methods.

THE BSE EPIZOOTIC

First recognized in November of 1986, with retrospective histological evidence of the first case dating from November 1985, the BSE epizootic peaked in 1992 with more than 1000 cattle cases confirmed per week (see Fig. 2). Reviews of archived tissue samples from cattle with undiagnosed disease have failed to identify any unrecognized cases prior to 1985 (1,6).

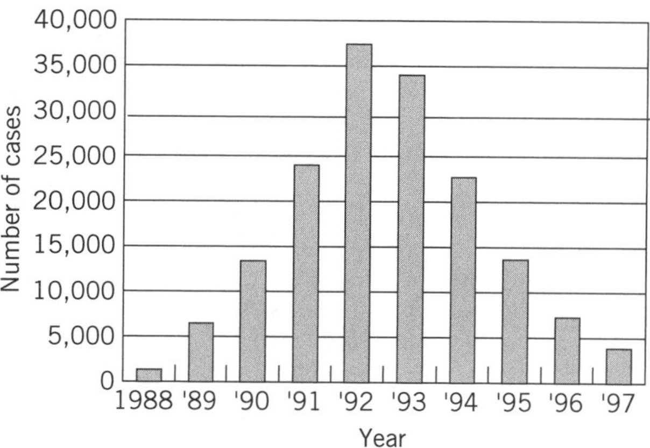

Figure 2. Confirmed cases of BSE in UK cattle by year.

Epidemiological studies suggest that UK cattle became infected through exposure to contaminated MBM feed supplements in 1981 to 1982 (6). Although the original source of the contamination is unknown, the most accepted theory proposes that sheep scrapie prions completed a species jump and induced normal cattle prion protein to become distorted (6–9). This distorted cattle prion protein distorts others, eventually resulting in clinical disease. Affected animals were slaughtered or died, rendered, and the resulting MBM containing abnormal cattle prion protein was fed to other cattle. The UK sheep population has a high enzootic rate of scrapie, and a high ratio of sheep to cattle (45 million to 12 million). This resulted in sheep representing 14% of rendered products, compared with 0.6% in the United States. The higher level of MBM supplements fed to young calves in the UK (2–4% MBM in rations) may explain the high incidence of BSE as compared with other EU countries (30).

A second theory suggests that cattle have always had a low incidence of BSE, perhaps at a rate similar to the 1:1,000,000/year incidence of CJD found in humans, which was unrecognized (31). Changes in rendering practices in the late 1970s and early 1980s no longer destroyed infectivity sufficiently to prevent accumulation of the necessary infectious dose.

A third hypothesis, offered in an effort to explain the rapid appearance and spread of the epizootic, theorizes importation of the agent to the UK in MBM containing abnormal proteins from an unknown ruminant other than cattle (32). The original source would have to be a country or countries lacking the infrastructure necessary to identify clinical cases and with rendering systems inadequate to destroy the agent. The imported material would need to be in sufficient quantity, or commingled with domestic MBM, to result in the volumes of contaminated product with infectious dose levels necessary to account for the epizootic.

Whatever the original source, changes in British rendering methods in the late 1970s and early 1980s are proposed to have played a role in the widespread amplification of the disease (6–8). The solvent extraction step with steam recovery was eliminated, and a liquid slurry process was introduced (33). Both changes may have resulted in decreased exposure of materials to heat during rendering. Critics note that TSEs such as scrapie have been demonstrated to be resistant to solvents, and the temperatures used (about 100°C) are too low to inactivate the agent (23,24,32). The impacts from these changes in the rendering system on the epizootic are unknown. The use of MBM in cattle rations as a source of supplemental protein had been a common practice for decades prior to the emergence of BSE, and scrapie is known to have been endemic in the UK for more than 250 years.

BSE has been confirmed in native cattle in Ireland, France, Portugal, the Netherlands, and Switzerland. Exported British cattle have been found with the disease in Oman, the Falkland Islands, Germany, Denmark, Canada, and Italy (1). The export of rendered animal feed products from the UK is considered the source of the disease outbreaks in native cattle in other countries. There are concerns that the number of cattle, and the amount of feed, exported should have resulted in significantly more disease than reported (30). There is confusion about whether the expected cases failed to occur for unknown reasons, went unrecognized within the surveillance systems in place, or were unreported.

ACTIVITIES TO PROTECT THE FOOD SUPPLY

In the UK

In July 1988, with about 2000 cattle cases confirmed, and in response to a just completed epidemiological investigation implicating MBM (6), the UK made BSE a notifiable disease. Euthanasia of affected cattle became mandatory, and their carcasses were banned from use in feed supplements. A few months later the carcass ban was expanded to include offal from sheep and cattle, although strict enforcement was not achieved. In 1989 the Specified Bovine Offal (SBO) ban prohibited the use of brain, spinal cord, tonsil, thymus, spleen, and intestinal tissues of cattle origin in foods intended for human consumption and fertilizer manufacturing.

Beginning in 1991 controls on MBM went through a number of iterations, starting with prohibitions from use in ruminant feeds, and eventually resulting in a ban from all agricultural use. In August 1996 mammalian MBM was made illegal to possess or use in any part of the food chain including on-farm, in transit, or at feed mills. A surveillance system was imposed including statistical sampling of protein supplements for the presence of ruminant and porcine proteins, and sheep and goat heads, spinal cords, and spleens were banned from rendered use.

Special programs for calf processing were initiated to prevent male calves less than 20 days old from being used for human food. Data showing that cattle did not harbor significant amounts of abnormal prion protein until later in life led to the "Over Thirty Month scheme" (OTM, OTMS), which prohibited the slaughter of cattle over 30 months of age for human food. Beef cattle from BSE-free herds, and reared on grass, are considered at very low risk of the disease and can be exempted and slaughtered for human consumption up to 42 months of age.

In 1997, the selective cull program was initiated, requiring all cattle born on farms six months before and after a confirmed case of BSE to be removed from the herd, euthanized, and handled as if they are infected. No replacements are allowed onto the farm until all such animals have been removed. Additional restrictions required all meat from cattle more than 6 months of age to be deboned as a precaution to prevent any nervous tissue or bone marrow from being consumed.

By-products from slaughtering cattle approved for use as human food are considered "clean" and approved for use in horticultural applications, pet food, cosmetics, and pharmaceuticals, or they can be buried in landfills. All other by-products are considered "dirty" and are stained to prevent accidental use, incinerated at 1000°C for sufficient time to result in a protein-free ash, and then disposed of in landfills. No rendered ruminant products can be used for agricultural purposes.

The number of new cases of BSE has fallen from its peak in 1992 of 900 to 1000 confirmed diagnoses per week, to less than 200 per week in 1998. The UK expects to eradicate the disease between the years 2001 and 2005 (1,9,27).

In the European Union

On March 27, 1996, the EU imposed a ban on exports from the UK of beef, live bovine animals, semen, embryos, and bovine products (including MBM) that might enter the human or animal food chains or be used in pharmaceuticals or cosmetics. In December of 1996, the UK published the Export Certified Herds Scheme (ECHS) to enable documentation that an animal was from a BSE-free herd. In March of 1998, EU farm ministers approved the ECHS, permitting legal exports of beef from Northern Ireland from June 1, 1998. The European Commission then proposed to lift the export ban for the rest of the UK, as applied to animals born after August 1, 1996 (the date the MBM ban became fully effective).

In the United States

The United States has not imported processed beef, live cattle, semen, embryos, MBM, or other beef products from the UK since 1989. In 1990 the U.S. Department of Agriculture (USDA) began tracing the 496 UK cattle imported to the United States between 1981 and 1989 and found no traces of BSE among those they were able to locate. A surveillance program was begun that has examined thousands of brain tissue samples from slaughtered animals across 43 states without finding any evidence of BSE.

In 1991 the USDA began enforcing stringent restrictions on importations from other countries either known to have had cases of BSE or considered to have inadequate surveillance systems to detect the disease. The rendering of sheep for use in animal feed was discouraged, and risk analysis activities were pursued. In 1993 the slaughter surveillance program was expanded to include "downer" cattle in response to reports suggesting disabled cattle, those that are unable to stand without assistance, represent unrecognized cases of BSE. Downer cattle cannot stand for many different reasons including metabolic diseases, injured musculoskeletal systems, or systemic illness resulting in general weakness. BSE-affected cattle may be downers, but downers are not necessarily related to BSE.

On March 29, 1996, livestock and professional animal health organizations in the United States requested an American ban on the use of ruminant protein in ruminant feeds. In addition, these groups requested increased national surveillance, enhanced BSE educational activities, and expanded cooperative research efforts to provide extra measures to prevent the introduction of BSE into the United States. Replacing animal origin proteins for feeds in the United States will require about 3 million additional acres of soybean production (25).

In 1997 USDA banned the importation of live ruminants and most ruminant products from European countries, and the Food and Drug Administration announced a permanent prohibition of specified substances for use in ruminant feed.

GLOSSARY

Amino acids. Organic compounds serving as the building blocks of proteins.

BSE. Bovine spongiform encephalopathy.

Chronic wasting disease. A TSE of mule deer and elk.

CJD. Creutzfeld-Jakob disease of humans.

Codon. The sequence of three chemical units in nucleic acid that determines the specific amino acid formed in protein synthesis.

Dura matter. The membranous covering of the brain.

Enzootic. Diseases continuously prevalent in animals.

Epizootic. Diseases spreading widely among animals.

Fatal familial insomnia. A TSE of humans.

Genotype. The genetic makeup of an organism, as distinguished from its appearance.

GSS. Gerstmann–Strausser–Sheinkers syndrome, a TSE of humans.

Heterozygous. Differing genes for one characteristic.

Homozygous. Identical genes for one characteristic.

Kuru. A TSE of humans.

Leukocytes. White blood cells.

Maternal transmission. Spread of a disease from dam to offspring before birth.

MBM. Meat and bone meal.

Nucleic acid. DNA or RNA.

Offal. Waste products of animals slaughtered for food.

OTM. Over thirty months.

OTMS. Over thirty months scheme.

Prions. Proteinaceous infectious particles.

Rendering. To melt fat or other animal matter.

Scrapie. A TSE of sheep and goats.

SBO. Specified bovine offal.

Species barrier. A hypothetical barrier delaying a TSE of one species from inducing disease in another species.

Species jump. A TSE of one species inducing disease in another species.

SRM. Specified risk material.

SEAC. Spongiform Encephalopathy Advisory Committee.

Transmissable mink encephalopathy. A TSE of mink.

TSE. Transmissible spongiform encephalopathy.

BIBLIOGRAPHY

1. British Ministry of Agriculture, Fisheries and Food, *A Progress Report on Bovine Spongiform Encephalopathy in Great Britain*, 1996, 25 pp.
2. S. B. Prusiner, "Prions," *Sci. Am.* **251**, 50–59 (1984).
3. S. B. Prusiner, "The Prion Diseases," *Sci. Am.* **272**, 47–57 (1995).
4. J. G. Collee and R. Bradley, "BSE—A Decade on—Part 1," *Lancet* **349**, 636–641 (1997).
5. J. G. Collee and R. Bradley, "BSE—A Decade on—Part 2," *Lancet* **349**, 715–721 (1997).
6. R. M. Anderson et al., "Transmission Dynamics and Epidemiology of BSE in British Cattle," *Nature* **382**, 779–788 (1996).
7. J. W. Wilesmith et al., "Bovine Spongiform Encephalopathy: Epidemiological Studies," *Vet. Rec.* **123**, 638–644 (1988).
8. J. W. Wilesmith et al., "BSE: Epidemiological Studies on the Origin," *Vet. Rec.* **128**, 199–203 (1991).
9. J. W. Wilesmith et al., "Bovine Spongiform Encephalopathy: Epidemiological Features 1985 to 1990," *Vet. Rec.* **130**, 90–94 (1992).
10. I. H. Pattison, "Origins of BSE," *Vet. Rec.* **128**, 262–263 (1991).
11. R. G. Will et al., "A New Variant of Creutzfeldt-Jacob Disease in the UK," *Lancet* **347**, 921–925 (1996).
12. National CJD Surveillance Unit, Creutzfeldt-Jakob Disease Surveillance in the UK, Seventh Annual Report," URL: *www.CJD.ed.ac.uk / rep98.html* (last accessed April 19, 1999).
13. J. Collinge et al., "Molecular Analysis of Prion Strain Variation and the Aetiology of 'New Variant' CJD," *Nature* **383**, 685–690 (1996).
14. M. Bruce et al., "Transmissions to Mice Indicate That 'New Variant' CJD Is Caused by the BSE Agent," *Nature* **389**, 498–501 (1997).
15. J. Almond and J. Pattison, "Human BSE," *Nature* **389**, 437–438 (1997).
16. World Health Organization, *Report of a WHO Consultation on Public Health Issues Related to Human and Animal TSE With the Participation of FAO and OIE*, World Health Organization, Geneva, Switzerland, 1996.
17. G. J. Raymond et al., "Molecular Assessment of the Potential Transmissibility of BSE and Scrapie to Humans," *Nature* **388**, 285–288 (1997).
18. T. R. Spraker et al., "Spongiform Encephalopathy in Free-Ranging Mule Deer (*Odocoileus hemionus*), White-Tailed Deer (*Odocoileus virginianus*), and Rocky Mountain Elk (*Cervus elaphus nelsoni*) in North Central Colorado," *Journal of Wildlife Diseases* **33**, 1–6 (1997).
19. J. K. Kirkwood and A. A. Cunningham, "Epidemiological Observations on Spongiform Encephalopathies in Captive Wild Animals in the British Isles," *Vet. Rec.* **135**, 296–303 (1994).

20. F. O. Bastian, "Bovine Spongiform Encephalopathy: Relationship to Human Disease and Nature of the Agent," *ASM News* **59**, 235–240 (1993).
21. S. B. Prusiner and M. R. Scott, "Genetics of Prions," *Annu. Rev. Genet.* **31**, 139–175 (1997).
22. EU Scientific Steering Committee, "Listing of Specified Risk Materials: A Scheme for Assessing Relative Risks to Man. Opinion of the Scientific Steering Committee," adopted Dec. 9, 1997, URL: *http://www.easynet.co.uk/ifst/hottop5c.htm.*
23. D. M. Taylor, "Survival of Mouse-Passaged Bovine Spongiform Encephalopathy Agent After Exposure to Paraformaldehyde-Lysine-Periodate and Formic Acid," *Vet. Microbiol.* **44**, 111–112 (1995).
24. D. M. Taylor, "Deactivation of BSE and Scrapie Agents: Rendering and Other UK Studies," in Editor(s) name(s), *Transmissible Spongiform Encephalopathies*, Scientific Veterinary Committee of European Communities, Brussels, Belgium, 1995, pp. 205–223.
25. D. Harlan, "Risk Reduction—An Ounce of Prevention," *International Symposium on Spongiform Encephalopathies: Generating Rational Policy in the Face of Public Fears*, The Ceres Forum, Georgetown University, December 12–13, 1996.
26. World Health Organization, *Report of an International Consultation of Experts Convened in Geneva from 24 to 26 March 1997 on the Issue of Human and Animal Spongiform Encephalopathies and Medical Products*, World Health Organization, Geneva, Switzerland, 1997.
27. D. J. Stekel et al., "Prediction of Future BSE Spread," *Nature* **381**, 119 (1996).
28. D. C. Krakauer et al., "Phylogenesis of Prion Protein," *Nature* **380**, 675 (1996).
29. R. C. Moore et al., "Mice With Gene Targeted Prion Protein Alterations Show That *Prnp, Sinc* and *Prni* Are Congruent," *Nature Genet.* **18**, 118–125 (1998).
30. D. Butler, "Creutzfeldt-Jakob Researchers Seek Greater Access to Data on Cases," *Nature* **381**, 453 (1996).
31. C. J. Gibbs, "Is BSE Endemic?" *International Symposium on Spongiform Encephalopathies: Generating Rational Policy in the Face of Public Fears*, The Ceres Forum, Georgetown University, December 12–13, 1996.
32. O. Hotz de Baar, "BSE: A Hypothetical Genesis," *PROMED* communication, 12 December 1996.
33. "Origins of BSE [comment]," *Vet. Rec.* **128**, 310 (1991).

DONALD KLINGBORG
DEAN CLIVER
University of California
Davis, California

BREAKFAST CEREALS

Breakfast cereals may be classified as hot cereals, requiring some preparation such as the addition of hot water, or ready-to-eat (RTE) cereals, usually consumed with cold milk. RTE cereals are so-called because they are precooked using one of a number of processes. Corn, wheat, rice, oats, or other grains may be puffed, flaked, shredded, or extruded; coated with sugars; enriched with micronutrients; or enhanced with fruits to provide a convenient, nutritious, and readily digestible breakfast food.

The history of breakfast cereals parallels the growth of the Kellogg Company and the Postum Cereal Company, the latter becoming the nucleus of General Foods (1,2). Health and nutrition were the driving forces in the development of breakfast cereals. At the turn of the century, people were looking for lighter, more digestible foods as they moved into less physically demanding work environments. The heavy fried and starchy foods of the pioneer days became a major cause of stomach disorders among the sedentary segment of the population (2).

About this time Dr. J. H. Kellogg was superintendent of the Battle Creek sanitarium. Kellogg, with his brother W. K. Kellogg, developed the first flaked cereal as a health food to feed the sanitarium patients. C. W. Post was a patient at the sanitarium and was most intrigued at the commercial possibilities of the health foods being served there. When he regained his health, Post went on his own to develop the cereal products that are still known today as Grape Nuts and Post Toasties.

From humble beginnings, breakfast cereals became a large industry in the United States. In 1987 it was a $5.3 billion business, with 90% coming from the RTE segment and 10% from the hot cereal category. The per capita consumption of breakfast cereals in 1987 was 11.7 lb for RTE and 3.6 lb for hot cereals (3).

READY-TO-EAT BREAKFAST CEREALS

Ready-to-eat breakfast cereals can be described by grouping them into a few basic cereal types: puffed, flaked, shredded, baked, and agglomerated.

Puffed Cereals

Puffed cereals are whole grains, grain parts, or a grain dough that has been formed into a specific shape and then expanded by subjecting it to heat and pressure to produce a very light and airy grain product. The two types of process that are normally used for puffed cereals are gun puffing and extrusion.

Gun Puffing. Gun puffing consists of a chamber (gun) that is charged with a quantity of grain material. The chamber is subjected to high-pressure steam (100–250 psig) for 5–45 s. The pressure is then released very rapidly giving a sound reminiscent of a gun. The shot of grain is propelled into an expansion vessel. The change from the high-pressure steam to atmospheric conditions causes the grain piece to expand to about 3–10 times its original volume. The puff is then dried to a final moisture content of 1–3% (4,5).

Extrusion. Extrusion employs either a single- or twin-screw extruder. A single flour or a mixture of grain flours is fed into the extruder. These flours may have been preconditioned with water or steam before extrusion. Other ingredients may be added to the extruder such as water or sugar syrups. The design of the screw, barrel elements, and die of the extruder impart desired characteristics to the finished product. Heating or cooling is accomplished through mechanical energy from the screw (6) and through jacketing the extruder barrel. Pressures in the extruder range from 500 to 2500 psig and temperatures range from

250 to 400°F. As the material exists from the end of the extruder it expands to release steam from the product. The produce is then dried to 1–4% moisture.

Puffed cereals are often coated with a sugar solution that may be flavored and dried to give a presweetened cereal. This coating is normally done in a rotating drum or screw auger where the puffed cereal is fed at a controlled rate. Sugar syrup is sprayed onto the cereal in the desired amount. The coated cereal is dried to remove excess moisture.

Flaked Cereals

Flaked cereals are cooked whole grains, parts of grains, or flours that have been pelletized or agglomerated, flattened into a flake, and dried or toasted. A normal process involves blending the grain with a flavoring syrup that may contain sugar, water, malt, or salt as well as other ingredients or combinations thereof. The blend is cooked in a steam atmosphere in a batch or continuous cooker. Cooking conditions involve moistures between 25 and 40%, time between 30 min and 3 h, and steam pressures of 15–40 psig. The cooked material is dried (pelletized and sized if required) to the desired flaking moisture, normally between 12 and 22% moisture. The material is tempered for 1–24 h to equilibrate moisture throughout the grain piece. After tempering, the pieces are fed through flaking rolls to get the desired flake thickness, normally between 0.010 and 0.035 in. After flaking, the wet flakes go to toasting ovens where the product moistured is quickly brought down to 2–3%. At times, grain flakes (especially rice) are oven puffed. This requires that the wet flakes be subjected to a very high temperature (up to 600°F) for a short period of time, often 1 min or less (7). This is done effectively in a fluidized bed dryer.

Shredded Products

Shredded products are whole grains or mixtures of grains that have been shaped into thin strands or a crosshatch shape, then layered, cut into biscuits or geometric shapes, and finally dried or toasted. For a whole grain such as wheat, the grain is cooked either in atmospheric conditions with an excess of water or in a pressurized steam vessel. Moisture content at this point is approximately 40–43% (8). After cooking, the grain is allowed to temper for several hours to equilibrate; it is then fed into shredding rolls consisting of one smooth roll and one grooved roll. The smooth roll forces the cooked grain into the grooved roll. A comb scrapes the grain material out of the grooves and deposits the strands onto a conveyer belt. Some 18–20 layers of shreds are deposited one on top of another from a bank of shredding rolls, and then cut into bite-size pieces. The pieces go through an oven to bake and dry (4,6).

Crosshatched types of products (chex) are normally made with rice or corn plus a flavoring syrup. Cooking is done in a batch or extrusion cooker and the shredding rolls are basically the same except for the addition of a horizontal groove on the grooved roll which is slightly less deep than the vertical groove. The layered web is then cut into bite-size pieces and puffed in a fluidized bed dryer to develop the characteristic light texture.

Baked Cereals

Baked cereals are the nugget type or rotary mold type. They are normally made in a typical bakery process. The nugget type involves blending grain flours, yeast, salt, and water together in a dough mixer, allowing the dough to proof, forming the dough into loaves or sheets, baking in an oven, then breaking the baked grain material into chunks, and drying to a final moisture of approximately 2%. The chunks are ground into finished food size (8). The rotary mold type utilizes a cookie baking line where the ingredients—grain, water, sugar, and shortening—are made into a dough. Pieces are formed by a cookie mold with a shape and size appropriate for RTE cereal. These are baked and dried in a typical baking oven.

Agglomerated Cereals

Agglomerated cereals are blends of grain products such as rolled oats, crisp rice, and toasted rolled wheat that have been glued together using a sugar or maltodextrin syrup as a binding agent. Vegetable oil is added to prevent sugar absorption into the agglomerated piece and to improve crispness retention in milk. The process involves blending the ingredients together in a ribbon blender or continuously in a mixer auger. The product is spread onto the belt of a traveling screen dryer (4) and moisture is reduced to approximately 2–3%. The sheets of agglomerated grain are sized off the dryer through the use of a grinder and then readied for packaging. Ingredient addition and sequencing are important to insure good agglomeration of the product without causing sugar lumps or bricking of the product in the package.

HOT BREAKFAST CEREALS

Oats

Hot breakfast cereals are usually made of oat flakes. The two principal factors that make oats different from other cereals are its retention of its whole grain identity and its nutlike flavor. Products vary from rolled oats (whole oat flakes), which require five or more minutes to prepare, to instant products, which can be prepared in a bowl with hot water.

In making oat flakes, the oat grain is first heated with dry steam, which reduces its moisture content to approximately 6%. Enzymes that produce rancid flavors during storage are deactivated during steaming. A high-speed impact operation separates the dry friable hull from the groat (oat berries). Even a small percentage of whole oats remaining is unacceptable because the hull fraction in oat flakes is unpalatable. Whole rolled oats are produced by flaking the whole groat.

Steel cutting reduces the groats for further processing. Another steaming step softens the cut groats for the rolling step. The size of the flake is determined by the type of rollers used to flatten the groats. Instant cook-in-the-bowl oats are produced from highly polished rollers with a narrow gap.

Oat flakes vary in sizes from 0.02 to 0.03 in. for rolled oats to 0.011 to 0.013 in. for instant oat flakes. Thicker flakes require longer cooking and retain more flake integrity (6,9).

Wheat

Hot wheat breakfast cereal is made with farina, which is a fraction of middlings from hard wheat. Soft wheat is not suitable because it produces a product that loses particle integrity too rapidly during cooking. Particle size is critical for quality. United States standards require that 100% of the product pass through a 20-mesh sieve and not more than 10% pass through a 45-mesh sieve. Farina must be cooked in boiling water for several minutes to insure complete starch gelatinization.

Instant farina cooks in about 1 min with boiling water. Proteolytic enzymes are used to instantize farina by opening avenues for easier penetration of water into particles to facilitate hydration during cooking (6).

Corn

In the southern portion of the United States a breakfast cereal made from corn called grits or hominy grits is popular. Grits are produced by dry milling white corn and are essentially small pieces of endosperm (6).

NUTRITIONAL CONSIDERATIONS

Breakfast Cereals and Dietary Recommendations

Dietary recommendations from the government and major health organizations suggest that the intake of dietary fat and cholesterol should decrease and the caloric intake from complex carbohydrates should increase (10). One way this can be accomplished is to increase the amount of cereals and grains in the diet. A RTE breakfast cereal is right in line with these dietary recommendations.

A recent General Foods study showed that adults who consumed breakfast, especially one that incorporated RTE cereal, had a better overall diet quality than those who skipped breakfast. In addition, RTE cereal users usually have lower daily intake of fat and cholesterol (11).

Another survey showed that 95% of the survey population consumes some kind of food item in the morning. Eighty-nine percent of these people eat breakfast. RTE cereals were the second most frequently consumed meal, behind a food group that included rolls, muffins, and toast (12). A third survey showed that 88% of the survey population consumed breakfast, and approximately 32% consumed cereal, primarily RTE cereal (13).

Breakfast cereals are, therefore, an excellent vehicle for delivery of calories from complex carbohydrates and, at the same time, reduce fat calories. Figure 1 shows the distribution of calories in a complete cereal breakfast (1 oz cereal, 12 fl oz lowfat milk, two slices toast with spread, and 6 fl oz orange juice) compared with a bacon and egg breakfast (one egg, two slices bacon, one slice toast with spread, 6 fl oz orange juice, and 8 fl oz lowfat milk). The cereal breakfast is lower in fat and higher in carbohydrates than the bacon and egg breakfast.

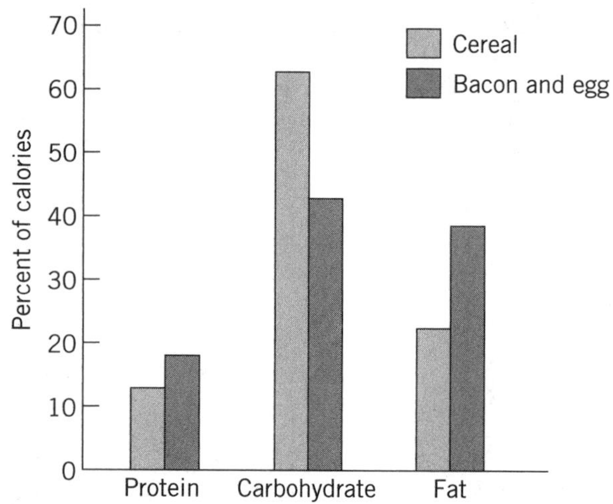

Figure 1. Percent calories in a cereal breakfast versus a bacon and egg breakfast.

Fortification

Cereal fortification has been practiced in the industry for decades and most RTE cereals today are fortified with vitamins and minerals. In 1974 the Food and Drug Administration (FDA) proposed guidelines on cereal fortification that were later incorporated into a general fortification regulation (14). This regulation, along with other guidelines (15), serves as the cornerstone for cereal fortification. Cereal manufacturers today use the following guidelines for specific nutrient fortification.

1. The intake of the nutrient is below the desirable level in the diets of a significant number of people.
2. The food used to supply the nutrient is likely to be consumed in quantities that will make a significant contribution to the diet of the population in need.
3. The addition of the nutrient is not likely to create an imbalance of essential nutrients.
4. The nutrient added is stable under conditions of storage.
5. The nutrient is physiologically available from the food.
6. There is reasonable assurance against excessive intake to a level of toxicity.

Other factors considered when fortifying include the estimated frequency of consumption of the food, the target consumer group, current health issues, and the dietary role of the product.

Breakfast cereals therefore make a significant contribution to the overall diet quality of an individual through their ingredient and fortification profiles. They provide a logical way to translate dietary recommendations to increase carbohydrate intake and decrease fat consumption; at the same time they provide a means for enhancement of vitamins and minerals in the diet.

STABILITY OF BREAKFAST CEREALS

Breakfast cereals should remain stable for up to one year when stored under reasonably cool and dry conditions. RTE cereals should remain crisp without developing off-flavors in storage. They should stay crisp in milk for at least 3–5 min.

Texture Stability

Loss of crispness (staling) in RTE cereals is associated with moisture pickup. Fresh RTE cereals have a moisture content of 2–3% and are very crisp. A good indicator of sensory crispness is water activity (A_w), which is the ratio of the vapor pressure of water over the food material to the vapor pressure of pure water, both measured at the same temperature. Fresh cereals generally have an A_w of about 0.20. As the moisture content goes up, the A_w increases until it reaches a critical value; for most cereals, this is around 0.45 (16). Beyond the critical A_w, the cereal becomes stele and unacceptable.

In the case of fruited cereals such as raisin bran, the raisin, which contains up to 18% moisture, will transfer its moisture during storage to the cereal, which has about 2–3% moisture. As long as the critical A_w of the cereal is not exceeded, the cereal will remain crisp. But the raisin could become unacceptably hard.

Some remedies to keep fruit soft while retaining cereal crispness include controlled addition of moisture to the cereal to modulate moisture migration from the fruit. In some cases this could make the cereal lose crispness. Infusion of fruits with edible humectants such as glycerol will keep the fruits soft while preventing the cereal from becoming stale (16,17). Coating RTE cereals with hydrophobic ingredients such as fats and oils, or incorporation of ingredients like magnesium stearate in the formulation, can significantly improve the bowl life of the cereal (18,19).

Flavor Stability

Breakfast cereals can develop off-flavors during storage. Unsaturated fatty acids and other compounds containing unsaturated linkages, such as vitamins, may undergo autooxidation, resulting in oxidative rancidity. Other reactions that can cause off-flavors are fat hydrolysis and reversion (8,20). Susceptibility to oxidation depends on the type of cereal grain used. Breakfast cereals made from oats are more susceptible to oxidative rancidity because oats contain higher levels of oil, about 7%. Other breakfast cereals, such as those made from rice, wheat, and corn, contain much less oil and thus are more stable in storage.

Antioxidants are used to retard autooxidation in breakfast cereals. Butylated hydroxyanisole (BHA) and butylated hydroxytoluene (BHT) are the most common phenolic antioxidants used. Federal regulations allow BHT and BHA to be added directly to breakfast cereals up to 50 ppm. The more common practice is to add antioxidants in the wax liner of the cereal box. These phenolic antioxidants, volatile at room temperature, diffuse from the liner to the product, thus providing protection from oxidation (21).

Nutrient Stability

Breakfast cereals are usually fortified with vitamins and minerals up to 25% of the U.S. recommended daily allowance or higher. These micronutrients, especially vitamins A, D, and C, are not stable during cereal processing. Hence, micronutrient fortification is done after processing, just before packaging (8,22). Properly maintained temperature and humidity will protect micronutrients during storage.

BREAKFAST CEREAL PACKAGING

The forerunner of today's RTE cereal packaging was introduced about 1900 by Nabisco in packaging the Uneeda Biscuit. For the consumer, a folding carton with an internal liner became a convenient alternative to cracker-barrel–type bulk packaging. Although today's packaging resembles that of the early 1900s, technological changes have occurred in packaging materials and equipment as a result of more demanding product specifications, consumer needs, distribution, merchandising, and regulatory and production requirements.

The major functions of the RTE cereal package are containment, preservation, identification, and integrity. Most RTE cereal packaging systems comprise three components: primary (liner), secondary (folding carton), and tertiary (corrugated shipping container).

Primary Package

The cereal liner's principal function is preservation. The liner evolved over time from an uncoated, bleached sulfate paper to a more protective waxed glassine paper. Plastic films were introduced in the mid-1970s. These coextruded films provide superior product protection via improved moisture and grease barriers and seal integrity. In 1985 films were made with bidirectional barrier properties to retain product aroma while preventing external contamination (23). Cereal liners can be modified to preserve other product attributes. Phenolic antioxidants can be incorporated into both waxed paper liners and plastic film coextrusions to minimize oxidative rancidity, which may impart off-flavors to the product (21). Titanium dioxide is generally added to plastic films to provide varying degrees of opacity where transparency is not desired.

Secondary Package

The principal functions of the folding carton are liner containment and communication (graphics). Most cereal cartons are produced from a machine clay-coated newboard comprising multiple layers of recycled paper fiber and are printed by either offset lithography or rotogravure. A second method prints clay-coated sulfite paper before lamination to an uncoated, special-bending newsboard using adhesives or waxes. Other less-economical, higher quality constructions using virgin fiberboard include coated, solid, bleached sulfate and coated, solid, unbleached sulfate.

Tertiary Package

Maintenance of the integrity of the retail package through the distribution cycle is the main function of the corrugated shipping case. Shipping cases are generally regular slotted containers able to withstand a burst test ranging from 150–275 lb. Top-to-bottom compression strength of the case is important to prevent crushing during handling and storage. Minimum standards for corrugated containers have been established for rail and truck shipments and are covered under specific uniform classification codes (24–26).

Cereal Packaging Line

The bag-in-the-box (BNB) packaging system is replacing the double package liner packing lines. BNB packaging uses vertical form–fill–seal baggers coupled with horizontal cartoners. Product is deposited in the bag (liner) during bag formation and is sealed prior to carton insertion. In contrast, traditional packing lines form a liner and carton around a mandrel prior to product fill. BNB equipment is more efficient and flexible, operates with wider liner material selection, and produces a higher quality cereal liner.

BIBLIOGRAPHY

1. G. Carson, *The Cornflake Crusade*, Rinehart & Co., New York, 1957, pp. 147–175.
2. J. Berry, *Inside Battle Creek*, rev. ed., Battle Creek Public Schools, Brochure No. 6, Battle Creek, Michigan, 1989.
3. The Staff, "State of the Food Industry, Breakfast Cereals," *Food Engineering* **60**, 86 (1988).
4. R. Fast, "Breakfast Cereals: Processed Grains for Human Consumption," *Cereal Foode World* **32**, 241–244 (1987).
5. R. Daniels, *Breakfast Cereal Technology*, Noyes Data Corp., Park Ridge, N.J., 1974.
6. R. Hosney, *Principles of Cereal Science and Technology*, American Association of Cereal Chemists, St. Paul, Minn., 1986, pp. 293–304.
7. D. Tressler and W. Sultan, *Food Products Formulary*, Vol. 2, AVI Publishing Co., Westport, Conn., 1975.
8. N. Kent, *Technology of Cereals*, Pergamon Press, Elmsford, NY, 1983, pp. 144–153.
9. F. H. Webster, "Oat Utilization: Past, Present, and Future," in F. H. Webster, ed., *Oats: Chemistry and Technology*, American Association of Cereal Chemists, Inc., St. Paul, Minn., 1986, pp. 413–426.
10. Committee on Diet and Health, *Diet & Health: Implications for Reducing Chronic Disease Risk*, National Academy Press, Washington, D.C., 1989.
11. K. Morgan, M. Zabik, and G. Stampley, "The Role of Breakfast in Diet Adequacy of the U.S. Adult Population," *Journal of the American College of Nutrition* **5**, 551–563 (1986).
12. J. Markle, F. Vellucci, and G. Coccodrilli, *Food Selection Patterns of Individuals Who Consume a Food Item in the Morning*, Internal Research Report, General Foods USA, White Plains, N.Y., 1985.
13. J. Heyback and F. Vellucci, *Clustering Patterns of Food Intake at Breakfast*, Internal Research Report, General Foods USA, White Plains, N.Y., 1985.
14. 21 *CFR* Part 104, *Federal Register* **45**, 6314–6324 (1980).
15. AMA Council on Foods and Nutrition and the Food and Nutrition Board of the National Academy of Science—National Research Council "Improvement of Nutritive Quality of Foods," *Journal of the American Medical Association* **205**, 160 (1968).
16. U.S. Pat. 4,256,772 (Mar. 17, 1981), A. Shanbhag and A. Szczesniak (to General Foods Corp.).
17. U.S. Pat. 4,103,035 (July 25, 1978), C. Fulger and T. Morfee (to Kellogg Co.).
18. U.S. Pat. 3,484,250 (Dec. 16, 1969), W. Vollink and M. Steeg (to General Foods Corp.).
19. U.S. Pat. 4,588,596 (May 13, 1986), D. Bone and co-workers (to the Quaker Oats Co.).
20. R. B. Coulter, "Extending Shelf Life by Using Traditional Phenolic Antioxidants," *Cereal Foods World* **33**, 207–210 (1988).
21. R. Sims and J. Fioriti, "Antioxidants as Stabilizers for Fats, Oils, and Lipid Containing Foods," in T. E. Furia, ed., *CRC Handbook of Food Additives*, Vol. 11, 2nd ed., CRC Press, Inc., Boca Raton, Fl., 1980, pp. 13–56.
22. R. H. Anderson and co-workers, "Effects of Processing and Storage on Micronutrients in Breakfast Cereals," *Food Technology* **30**, 110–114 (1976).
23. W. Drennan, "Barrier Resins Key New Package Development," *Plastics Packaging*, 17–21 (July–Aug. 1988).
24. *Uniform Freight Classification*, Uniform Freight Classification Committee, Chicago, Ill., annual.
25. *National Motor Freight Classification*, American Trucking Associations, Washington, D.C., annual.
26. *Coordinated Freight Classification*, The New England Motor Rate Bureau, Inc., Burlington, Mass., annual.

GENERAL REFERENCES

A. Brody and K. Marsh, *The Wiley Encyclopedia of Packaging Technology*, 2nd ed., John Wiley & Sons, Inc., New York, 1997.

S. A. Matz, *The Chemistry and Technology of Cereals as Food & Feed*, AVI Publishing Co., Westport, Conn., 1959.

S. A. Matz, *Cereal Technology*, AVI Publishing Co., Westport, Conn., 1970.

DICK FIZZELL
G. COCCODRILLI
CHARLES J. CANTE
General Foods USA
White Plains, New York

BROWNING REACTION, ENZYMATIC

The change in color following mechanical or physiological injury of fruits and vegetables is due to oxidative reactions of phenolic compounds by polyphenol oxidase (PPO) and the reaction products, *o*-quinones, to various polymerized products. In food processing, such color change is commonplace during preparation for canning, dehydration, freezing, or storage. Formation of the dark brown color causes the product to become unattractive and is accompanied by undesirable changes in flavor and a reduction in nutritive value. However, in some other cases such as the manufacture of tea, coffee, or cocoa, this enzymatic browning reaction is an essential part of the processing. Enzyme-catalyzed oxidative browning has been recognized since 1895, when the change in color of fresh apple cider was recognized first as a result of oxidation of tannin by the oxidase present in apple tissue (1).

POLYPHENOL OXIDASE

PPO (phenolase or catechol oxidase) is a ubiquitous group of copper metalloproteins and it is found in microorganisms (bacteria and fungi), plants, and animals, where it is involved in the biosynthesis of melanins and other polyphenolic compounds. PPO is of particular importance in the processing and marketing of horticultural products because of the brown discoloration of bruised fruits and vegetables resulting from the action of this enzyme. Not only does an undesirable color form, but browning is accompanied by a loss of nutrient quality and the development of undesirable flavors. In some tropical fruits, the action of PPO leads to major economic losses (up to 50%). Black spots in shrimp are caused by PPO-catalyzed browning; the browned shrimp are not acceptable to the consumer and they are downgraded in quality. Millions of dollars are spent each year on attempts to control PPO oxidation; to date none of the control methods are entirely successful. On the other hand, browning by PPO is a desired reaction in certain foods such as tea, coffee, cocoa, prunes, and dates.

PPO was first discovered in mid 1800s in mushrooms (2). PPO is a generic term for the group of enzymes that catalyze the oxidation of phenolic compounds at the expense of oxygen to quinones. Quinones are highly reactive, electrophilic molecules, which covalently modify and cross-link a variety of cellular constituents including nucleophiles of proteins such as sulfhydryl, amine, amide, indol, and imidazole substituents. These quinone adducts produce brown or black color on cut surfaces of fruits and vegetables, and they are the major detrimental effect of the enzyme in postharvest physiology and food processing.

PPO is a copper-containing enzyme that belongs to a group of oxidoreductases. Although the function of PPO in plant metabolism is not well understood yet, the recent cloning of PPO presents opportunities to explore its function in plants as well as to explore the extent to which PPO expression can be manipulated. Various aspects of PPO have been reviewed recently (3–8).

Depending on the substrate specificity, International Enzyme Nomenclature has designated monophenol monooxygenase or tyrosinase as EC 1.14.18.1, which hydroxylates monophenols to *o*-diphenols; diphenol oxidase or catechol oxidase or diphenol oxygen oxidoreductase as 1.10.3.2, which dehydrogenates *o*-dihydroxyphenols to *o*-quinones; and laccase or *p*-diphenol oxygen oxidoreductase as 1.10.3.1. The first two enzymes, tyrosinase and catechol oxidase, which many authors refer to as PPOs, occur in practically all plants.

PPO was first described by G. Bertrand in 1895 who demonstrated that the darkening of mushroom was due to the enzymatic oxidation of tyrosine (9). In addition to its general occurrence in plants, PPO is also found in some bacteria, fungi, algae, bryophytes, and gymnosperms (10–13). Laccase was first observed by H. Yoshida in the latex of the Japanese lacquer tree (14). PPO catalyzes two distinct reactions: (*1*) hydroxylation of monophenols to give *o*-diphenols. This monophenol oxidase activity often requires priming with reducing agents or trace amounts of *o*-diphenol indicated as BH_2,

$$\text{phenol} + O_2 + BH_2 \xrightarrow{\text{Monophenol oxidase}} \text{catechol} + B + H_2O$$

and (*2*) the removal of hydrogens from *o*-diphenol to give *o*-quinone.

$$2\ \text{catechol} + O_2 \xrightarrow{\text{Catechol oxidase}} 2\ o\text{-quinone} + 2\ H_2O$$

It was reported in 1944 that PPO preparations from mushrooms had the ability to catalyze oxidation of monophenols in addition to catalyzing oxidation of polyphenols (15). However, the ratio of rate of monophenol oxidation to that of polyphenol oxidation varied considerably from different mushroom sources. Since then many reports have shown a wide range of PPO activities in many different fruits and vegetables in different ratios. Today, it is known that enzyme preparations from many other sources possess these two activities in different ratios and that the ratios may change during isolation and purification. Most PPO preparations from potato, apple, mushroom, and bean possess both activities, whereas those from tea leaf, tobacco, mango, banana, pear, clingstone peach, and sweet cherry have been reported not to act on monohydroxy phenols (16).

Laccase is less frequently encountered as a cause of browning in fruits and vegetables than is catechol oxidase. Laccase oxidizes *o*- and *p*-diphenols,

$$p\text{-hydroquinone or catechol} \xrightarrow[\frac{1}{2}O_2]{\text{Laccase}} p\text{-quinone or } o\text{-quinone} + H_2O$$

In addition, it oxidizes various other *p*-diphenols, *m*-diphenols and *p*-phenylenediamine, but laccase does not oxidize tyrosine.

PPO appears in almost all tissues of plants; however, significant differences in both the level of the activity and in the concentration of its substrates have been observed among different cultivars of fruits and vegetables. The enzyme has been found in a variety of cell fractions, both in organelles and in the soluble fraction of the cell (17,18). The level of PPO often changes markedly during the development of the plant (19–21) and may be significantly affected by growing conditions.

Recent emerging new scientific development in the field of molecular biology offers not only the possibility of understanding the essential function of PPO in plants, which has remained a puzzle for years, but it also promises to control PPO and its substrates to prevent browning in foods.

PPO ANALYSIS

Qualitative and quantitative colorimetric tests and manometric determinations have been used to detect and measure enzyme activity. Because PPOs use molecular oxygen to oxidize phenolic substrates, their activity can be determined by measuring the rate of substrate disappearance or the rate of product formation. When based on the rate of substrate disappearance, oxygen absorption is usually measured either manometrically in a Warburg respirometer or polarographically with an oxygen electrode. The most popular method of assay is to follow the initial rate of formation of the quinone spectrophotometrically by measuring the optical density. Care must be taken to restrict measurement to the initial phase of the oxidation, because the reaction soon slows down. Some researchers object to the use of spectrophotometric analysis because it measures the secondary reaction products of PPO, and the secondary reactions are influenced by many factors difficult to control. Thus the presence of ascorbic acid creates a lag phase and lowers the values obtained, whereas autooxidation products of polyphenols may increase the levels of enzyme activity. It is important to note that the reaction changes with time and temperature and depends on the substrate type and concentration, on the pH of the reaction mixture, and on the buffer used (22,23). A wide variety of substrates can be used with spectrophotometric methods: catechol (24), pyrogallol (25), or chlorogenic acid (26). It has to be taken into account that the reaction products formed from the oxidation of various phenols have absorption maxima at different wavelengths, that the substrate may undergo autooxidation, and that an excess of some substrates causes strong inhibition of the enzyme (26). Other methods for PPO activity are based on disappearance of ascorbic acid, a reaction that is directly proportional to enzyme activity (27) and spectrophotometric methods using Besthorn's hydrazone (28), or 2-nitro-5-thiobenzoic acid (29).

It is important to differentiate between catechol oxidase and laccase on the one hand and peroxidase on the other. Because peroxidative oxidation of phenols is often mistaken for PPO or laccase, peroxides should be removed from the reaction mixture by addition of catalase and alcohol to prevent any oxidation of phenols (30). For the determination of laccase, syringaldehyde (31) and 2,6-dimethoxyphenol (32) have been used as substrates of laccase. The PPO and laccase activities can be differentiated by the use of cinnamic acid derivatives to inhibit PPO and cationic detergents to inhibit laccase (33).

PPO LOCATION IN PLANT

The vascular elements of many fruits and vegetables darken rapidly when they are cut and exposed to air. The relative PPO activity of different parts of apricots, apples, or potato plants vary. PPO is considered to be an intracellular enzyme in plants except in a few cases where it is found in the cell wall fraction. The enzyme is located in a variety of cell fractions, both in organelles where it may be tightly bound to membranes and in the soluble fraction (17,18). Histochemical studies using the electron microscope and the density gradient centrifugation support the idea that PPO is bound within chloroplast lamellae and grana and in mitochondria (34). Recent study suggested more specifically that PPO is located in the grana lamellae loosely associated with photosynthetic photo system II (35). Many studies indicated that PPO is membrane-bound in plastides of nonsenescing tissues (3,5). It was suggested that PPO is located exclusively in plastids and is only released to the cytosol upon wounding, senescence, or deterioration of the organelle (36). So far, no PPO genes encoding a nonplastidic enzyme have been isolated (8). Conversion of particulate forms of the enzyme to soluble forms occurs in fruits when they are exposed to stress conditions or during ripening and storage. Browning reactions that take place after the disruption of tissues rich in PPO may cause binding of soluble PPO to a particulate fraction. The microbial extracellular laccase has been studied extensively but the location of laccase in higher plants is not well known. Although it is believed that the enzyme is found in a cytoplasmic soluble form, more information is needed on its location and distribution in plant tissues.

EXTRACTION AND PURIFICATION

Due to extremely low concentrations of PPO and laccase in plants and their instability, preparation and purification of PPO and laccase are not easy. PPO has been isolated from a variety of sources, but pigment contamination and the occurrence of multiple forms have frequently hampered its characterization. Oxidation reactions taking place during the isolation of the enzyme, due to natural substrates within the plant, result in changes in enzyme properties as well as in apparent multiplicity. Such reactions can be partially prevented by isolation under nitrogen gas, by using reducing agents such as ascorbic acid or cysteine, or by using phenol-adsorbing agents, such as poly(vinylpyrrolidone) or poly(ethylene glycol) (37,38). The binding of PPO to membranes in many tissues further complicates its isolation. Solubilization of acetone powder preparations or extracting with detergents also result in modification in structure and properties of the enzyme. To minimize these changes, all extraction steps should be carried out at temperature below 0°C, preferably at −20 to −30°C. Communition and homogenization are often carried out in liquid nitrogen or under a nitrogen atmosphere (39). Acetone precipitation followed by buffer extraction is one of the methods most often applied (16).

For the purification of the enzyme, several methods have been used that vary according to the enzyme source and the degree of purity to be attained. Most often, precipitation with ammonium sulfate of different saturation, gel chromatography on Sephadex columns, ion exchange chromatography, DEAE-cellulose or DEAE-Sephadex are applied. Recently, hydrophobic gel chromatography using Phenyl-Sepharose CL-4B has been used effectively (40,41). The separation of enzymes by electrophoresis is now a standard procedure. Sodium dodecylsulfate (SDS) techniques and electroblotting of catechol oxidase also have been successfully used (42,43).

PHYSIOCHEMICAL PROPERTIES

Molecular Weight

Most reports on the molecular weight of PPO are based on estimates on partially purified preparations using gel filtration or acrylamide gel electrophoresis. Therefore, a wide range of values are reported. Often crude or partially purified preparations show a multiplicity of forms, which may have resulted from association–dissociation reactions. Depending on the source of the enzyme, the molecular weights vary between 29,000 and 200,000 with subunit molecular weight from 29,000 to 67,000. Until recently, the enzyme from the bacterium *Streptomyces glaucescens* was found to have the smallest functional unit with one copper pair per polypeptide chain of 29,000 (44). The molecular weight of mushroom PPO has been reported to be between 116,000 and 128,000 daltons (37). The enzyme consisted of four identical subunits of M_W ca 30,000, each containing one copper atom (45). Two types of polypeptide chains, heavy (M_W 43,000) and light (M_W 13,400) were reported in mushroom (46). Avocado, banana, tea leaf, and other sources also showed multiple molecular weights in a wide range. The predominant form of PPO in grape had a molecular weight of 55,000 to 59,000 but upon storage it dissociated into 31,000–33,000 and 20,000–21,000 subunits (47). Based on recent knowledge regarding multiplicity of PPO, it is believed that many previous observations on multiplicity were due to secondary reactions, such as tanning, during enzyme preparation.

With the advent of recombinant DNA technology in recent years, numerous amino acid sequences have become available. These include PPO from the bacterium *Streptomyces glaucescens* (48) and *Streptomyces antibioticus* (49), tomato (50,51), broad bean (52), and potato (53). All PPO molecules sequenced are single-chain proteins with calculated M_W of 30,900 (273 amino acid residues) for *S. glaucescens* and *S. antibioticus*, 46,000 (407 amino acid residues) for *N. crassa*, 68,000 (606 amino acid residues) for the broad bean *Vicia faba*, and 66,300 (587 amino acid residues) for tomato.

The molecular weight of laccase was also reported to vary according to their sources. *Rhus vernicifera* laccase has been reported to vary between 101,000 and 140,000 depending on the species and variety. The M_W of the fungal laccase varies between 56,000 and 390,000; peach laccase was reported to have a M_W of 73,500. These data suggest that laccases are a heterogeneous group of glycoproteins, having a basic structural unit of between 50,000 and 70,000 M_W, which can undergo aggregation to give larger units.

pH and Temperature Optima

The optimum pH for the activity of an enzyme preparation from any one source usually varies with different substrates and is characteristic of the substrate as well as the enzyme preparation. The optimum pH of PPO activity has a relative wide range, pH 5.0 and 7.0. Some preparations were reported to be inactive below 4.0. The type of buffer and the purity of the enzyme affect the pH optimum; also, isoenzymes have different pH optima. Enzyme preparations obtained from the same fruit or vegetable at various stages of maturity have been reported to differ in optimum pH of activity as well. The pH optimum of laccase activity depends on the enzyme source and on the substrate. The stability of the enzyme is reported to be dependent upon buffer and other factors in the medium. Optima between pH 4.5 and 7.5 have been reported for enzymes from various fungi and peaches (54).

The optimum temperature of PPO depends essentially on the same factors as the pH optimum. Generally, the activity of PPO increases as the temperature rises from cold to temperatures of maximum activity (25–35°C); activity declines rapidly at temperatures of 35 to 45°C, depending on the enzyme source and substrate (16). Like most enzymes, the application of heat (70–90°C) to plant tissues or enzyme solutions for a short period causes rapid inactivation. Exposures to temperatures below zero may also affect activity. Concentrated solutions of the enzyme in dilute phosphate buffer at a pH near neutrality can be held without detectable loss of activity for several months at 1°C or when frozen at −25°C. However, the enzyme slowly losses activity even in the frozen state. Thermotolerance of PPO also depends on the source of the enzyme. It is difficult to compare heat stability data of purified enzyme preparations with that of the enzyme when in tissues or juice. Different molecular forms of PPO from the same source may have different thermostabilities. Laccase is usually more sensitive to inactivation by heat than is catechol oxidase.

Activators and Inhibitors

The activation mechanism of PPO from different sources is not well understood, but it has been suggested that conformational changes of protein and perhaps protein association or dissociation are involved in the process. Anionic detergents, such as sodium dodecyl sulfate, were found to reactivate PPO in crude and partially purified preparations from different avocado cultivars (55). Short exposures to acid pH or urea also resulted in a severalfold reversible activation of grape and bean PPO.

Many compounds inhibit PPO activity but only a limited number of PPO inhibitors are considered acceptable on grounds of safety and expense for use to control enzymatic browning during food processing. Practical aspects of browning inhibition are discussed later. There are two principal types of PPO inhibitor: compounds that interact with the copper in the enzyme and compounds that affect the site for the phenolic substrate. The metal-chelating agents such as cyanide, diethyldithiocarbamate, carbon monoxide, dimercaptopropanol, sodium azid, phenylthiourea, and potassium methyl xanthate are all inhibitory. In general, there are five major types of PPO inhibitors based on their chemical characters (56,57): (*1*) *reducing agents*, such as ascorbic acid and sulfur dioxide; (*2*) *copper-chelating agents*, such as carbon monoxide, tropolone, and diethyl dithiocarbamate; (*3*) *quinone couplers*, such as cysteine and glutathione; (*4*) *substrate analogues*, such as cinnamic acid, *p*-coumaric acid, and benzoic acid and some substituted cinnamic acids; and (*5*) *miscellaneous compounds*, such as 4-hexylresorcinol and poly(vinylpyrrolidone). Some other natural inhibitors of PPO from various sources have been reported (57). The activity of laccase is not inhibited by carbon monoxide or phenylhydrazine, both of which inhibit PPO. This response permits an easy differentiation between laccase and PPO.

Substrate Specificity

Whereas PPOs from animal tissues are relatively specific for tyrosine and dopa, those from fungi and higher plants act on a wide range of monophenol and *o*-diphenols. There are individual differences among PPOs from different sources. The rate of oxidation of *o*-diphenols by PPO increases with increasing electron withdrawing power of substituents in the *para* position. *o*-Diphenol substitution (-CH_3) at one of the positions adjacent to the -OH groups prevents oxidation. These positions should remain free for oxidation to take place. As mentioned earlier, some PPO preparations from some plants lack cresolase activity, which may be due to changes in the structure of the protein during preparation. Some arguments regarding the physical relationships between the cresolase and catechol oxidase functions suggested that both functions are catalyzed by a single site, whereas others imply the participation of two sites, either on the same enzyme molecule or different ones. Although PPO could oxidize a wide range of phenolics, the individual enzymes tend to prefer a particular substrate or certain type of phenolic substrates. The affinity of plant PPOs of the phenolic substrates is relatively low. The K_m is high, usually around 1 mM, which is higher than most of fungi and bacteria, 0.1 mM. The affinity to oxygen is also relatively low, ranging from 0.1 to 0.5 mM. Monophenols and *o*-diphenols have been considered as the exclusive substrates of PPO for a long time, but a recent study showed that aromatic amines and *o*-aminophenols also undergo the same catalytic reactions (ortho hydroxylation and oxidation), as documented by product analysis and kinetic studies (44).

Cloning of PPO Genes

Scientific interest in molecular biology of PPO is that the molecular weight and the form of PPO have long been a matter of speculation because of the various forms of the PPO as described earlier. Identification of PPO genes offers a great potential to produce transgenic plants having regulated levels of PPO. Because PPO is responsible for significant decreases in postharvest quality of many crops due to enzymatic browning associated with injury or senescence, downregulation of PPO using antisense could significantly increase quality attributes of most fruits and vegetables. However, overexpression of PPO may potentially minimize pest damage to crop plants. This plant defense system is one of many theories of PPO function in plant. Enzymatic browning generated by PPO in a multitude of plant–pest interactions and its wound inducibility have led some scientists to the view that the primary role of PPO is in plant defense (58). Therefore, the challenge here is to selectively modify PPO expression in specific tissue to minimize its effects on browning of fruits and vegetables, while maximizing PPO expression in other parts of the plant to maintain pathogen or herbivore resistance. Much progress has been made recent years. PPO genes have been characterized from broad bean, tomato, potato, and grape, and all PPOs cloned have been found to encode mature peptides of 57,000 to 62,000 Da with 8,000 to 11,000 Da putative transit peptides, which posttranslationally directs the protein to the chloroplast envelope (7,35,51–53). The catalytic centers of the copper binding regions for *V. faba* and tomato PPOs have been elucidated. The further characterization of PPO in various plants is being made, and there is a great potential in near future to manipulate the levels of PPO in plant foods to overcome the adverse effects of PPO.

Mechanism of Oxidation

In spite of extensive studies on PPOs, its reaction mechanism is still unclear. One of the reasons is that the mechanism of PPOs on phenolic compounds is complicated. As stated earlier, the cresolase reaction apparently involves a hydroxylation reaction, which differs from the oxidation of the *o*-diphenols by the catecholase reaction. For a long time there was a difference of opinions as to whether hydroxylation of monophenols and oxidation of *o*-dihydroxy phenols were catalyzed by the same enzyme. The oxidation mechanism of PPO from *N. crassa* has been extensively investigated (44), and a plausible theory of its catalytic activation is shown in Figure 1.

The proposed mechanisms for hydroxylation and dehydrogenation reactions with phenols probably occur by separate pathways but are linked by a common *deoxy* PPO intermediate (middle in Fig. 1). In dehydrogenation of catechol oxidase (cycle A in Fig. 1), oxygen is bound first to the Cu(I) groups of *deoxy* PPO to give *oxy* PPO in which a Cu-Cu distance of the two Cu(II) groups is about 3.4 to 3.6 Å. This would allow two Cu(II) groups of *oxy* PPO to bind to the oxygen atom of the two hydroxyl groups of catechol (an efficient 2e-transfer from catechol to the binuclear site) and from the O_2-PPO-catechol complex (at about the 1 o'clock position on the A cycle of Fig. 1). The catechol is then oxidized to *o*-benzoquinone (at about 11 o'clock position), and the enzyme is reduced to *met* PPO. Another molecule of catechol binds to *met* PPO, is oxidized to *o*-benzoquinone (and water) and the enzyme reduced to *deoxy* PPO, completing the A cycle.

The mechanism of *o*-hydroxylation of a monophenol by PPO is shown in B cycle in Figure 1. In vitro, the reaction begins with *met* PPO of A cycle. *Met* PPO must be reduced by a reducing compound BH_2 (shown previously) if a lag period is to be avoided, to give *deoxy* PPO. *Deoxy* PPO binds O_2 to give *oxy* PPO, the monophenol is bound to one of the Cu(II) groups via the oxygen atom of the hydroxyl group to give the O_2-PPO-monophenol complex (at about 5 o'clock

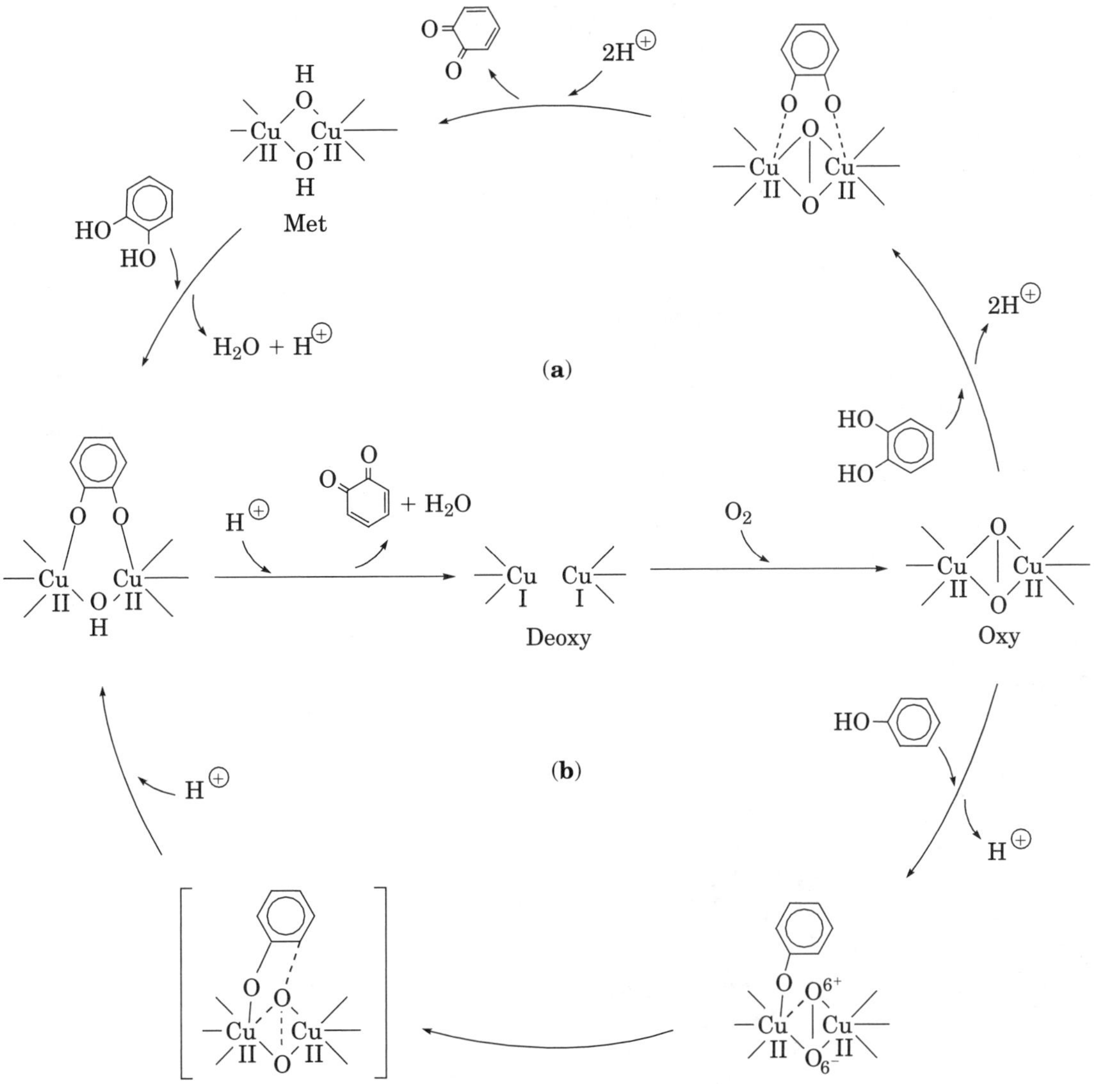

Figure 1. Proposed mechanisms of oxidation of *o*-diphenol (cycle A, for catechol) and monophenol (cycle B) for *Neurospora crassa* PPO. *Source:* Adapted from Refs. 2 and 44.

on the B cycle). Subsequently, the *o*-position of the monophenol is hydroxylated by an oxygen atom of the O_2 of the O_2-PPO-monophenol complex to give catechol, which then dissociates to give *deoxy* PPO, to complete the cycle. Only the first cycle of hydroxylation on a monophenol requires starting at the *met* PPO; all subsequent cycles begin with *deoxy* PPO (2).

Because copper is the prosthetic group, the catalytic activity of PPOs is based on the cupric-cuprous valency change. On isolation of the enzyme in its natural state, the copper is in the cuprous form, but in the presence of *o*-dihydroxy phenols, the copper would be oxidized to cupric form. The substrate is oxidized by losing two electrons and two protons. The two electrons are taken up by the copper of the enzyme, which then passes into the cuprous state. The cuprous enzyme rapidly transfers the two electrons to oxygen, which immediately forms water with the two protons that were liberated, and the enzyme returns to the cupric state. This two-state reaction can be represented by

$$4\,Cu^{2+}(\text{enzyme}) + 2\ \text{catechol} \rightarrow 4\,Cu^{+}(\text{enzyme}) + 2\ o\text{-quinone} + 4\,H$$

$$4\,Cu^{+}(\text{enzyme}) + 4\,H^{+} + O_2 \rightarrow 4\,Cu^{2+}(\text{enzyme}) + 2\,H_2O$$

Recent understanding of the complete amino acid sequence of fungal PPO and the site of interaction with the phenolic substrate made it possible to create models that include a binuclear center. Histidine residues appear to lignand at least one of the two active sites of the copper atoms. Monophenols bind to one of the Cu^{2+} atoms, whereas diphenols bind to both of them, as shown in Figure 1 (59,60). Catechol oxidase can undergo conformational changes induced by the substrate O_2 and by pH. The changes in conformation are accompanied by changes in the K_m of the enzyme for both its substrates, oxygen and diphenol.

A two-step reaction mechanism by laccase has been reported (61). The first step in the oxidation of quinol by laccase was the formation of the semiquinone, with transfer of an electron from substrate to the copper in the enzyme. The second step was a nonenzymatic disproportionation reaction between two semiquinone molecules to give one molecule of quinone and one of quinol. The function of copper and electron transfer in the reaction mechanism has been studied by using inorganic ions, electron paramagnetic resonance, and spectrophotometric methods (3,62).

The Brown Pigment

Although the mechanism of pigment formation from phenolic substrates is not understood completely, the general course of pigmentation is known to involve enzymatic oxidation, nonenzymatic oxidation, nonoxidative transformations, and polymerizations. The *o*-quinones formed from phenolic compounds by PPOs are the precursors of the brown color. The *o*-quinones themselves possess little color, but they are among the most reactive intermediates occurring in plants. They take part in a secondary reaction, bringing about the formation of more intensely colored secondary products. The most important secondary reactions are coupled oxidation of substrates oxidized, complexing with amino compounds and proteins, and condensation and polymerization. The principal reaction of *o*-quinones in browning reaction is the one leading to the formation of the unstable hydroxyquinones. The hydroxyquinones polymerize readily and are easily further oxidized nonenzymatically to a dark brown pigment.

As a typical oxidation reaction, tyrosine is oxidized to dihydroxyphenylalanine (dopa) quinone by PPO and then proceeds to melanin (Scheme 1). Tyrosine is first converted into dopa, which is then oxidized to the corresponding dopa quinone. The dopa quinone, on intramolecular rearrangement, is converted into 5,6-dihydroxy indole-2-carboxylic acid, which on further oxidation is converted into red 5,6-quinone indole carboxylic acid; then finally the black melanins are formed. Quinones also react readily with simple amines, such as

o-Benzoquinone + glycine → 4-*N*-glycyl-*o*-benzoquinone

This reaction product is the intermediate responsible for the deamination of glycine with the concomitant formation of deeply colored pigments. *o*-Quinones also react with proteins, sulfhydryl compounds, such as cysteine, and produce dark-colored, insoluble products.

Dimerization or polymerization of *o*-quinones that lead to the colored products is common. The formation of dimers or oligomers of *o*-quinone by condensation of a hydroxy quinone with a quinone was recently demonstrated. The enzymatically generated caffeoyl tartaric acid *o*-quinones were shown to oxidize other phenols, such as 2-*S*-glutathionyl caffeoyl tartaric acid and flavans, by coupled oxidation mechanisms, with reduction of the cafeoyl tartaric acid quinones back to caffeoyl tartaric acid (63). The *o*-quinones formed by enzymatic or coupled oxidation can also react with a hydroquinones to yield a condensation product (64–66). It is possible to regenerate the original phenolic from an intermediate by reduction provided that oxidation and subsequent transformation had not gone too far. In later stages the browning is no longer reversible.

CONTROL OF BROWNING IN FOODS

Extensive studies have been carried out to control enzymatic browning ever since Lindet (1) recognized in 1895 that the change in color occurring in freshly pressed cider was enzymatic. The practical control of enzymatic browning in foods has been carried out by several methods. The method of choice is dependent upon the food product and the intended use. The enzymatic browning in fruits and vegetables can be controlled at the beginning by selecting cultivars that are least susceptible to discoloration either because of the absence of the certain phenolic substrate or because the substrate or enzyme is present at low concentration. It is also possible to select fruits and vegetables at stages of maturity where discoloration is at minimum. Other methods include the removal of oxygen from fruit and vegetable tissues as well as from the atmosphere surrounding the food; the addition of acids to reduce the pH and thus reduce PPO activity; the addition of antioxidants or reducing substances; the addition or treatment with permissible inhibitors; and heat inactivation of the enzyme. The use of antioxidants during processing and heat inactivation of the enzyme are widely used practices that have been moderately successful.

All fruits susceptible to browning should be processed as quickly as possible. Heating destroys the enzymes responsible for the reaction; thus when fruit is canned or made into jams or jellies, browning stops as soon as the fruit is heated sufficiently to denature the enzyme. The exact temperature necessary varies with enzyme, rate of heating, pH, and other factors such as size of the fruits. Unfortunately some undesirable effects on quality often result from an adequate heat process. In general, cherry, peach, and apricot take 2 to 3 minutes in boiling water and apple, pear, and plum take 4 to 5 minutes to inactivate enzymes responsible for browning on surface. Vegetables such as beans and peas also take 3 to 5 minutes. The optimum blanching conditions for strawberry, black currant, sour cherry, and prune range from 1.5 to 3 minutes at 85°C.

In the preparation of fruit for freezing, sugars and sugar solution have been used to sweeten and to exclude direct contact of the fruit tissue with molecular oxygen. The sugar solutions inhibit discoloration by reducing the concentration of dissolved oxygen and by retarding the rate of diffusion of the oxygen from the air into the fruit tissues. Concentrated sugar solutions also exert an inhibiting effect on fruit PPO. The pH also affects the rate of the browning reaction: acid dips are sometimes used to lower the pH and by this method delay or retard browning. The effect of many salts and compounds such as sulfur dioxide, hydrogen sulfide, hydrocyanic acid, and thiourea on PPO has been studied extensively.

In the home preparation of fruits, pineapple juice and lemon juice have long been used to prevent browning. Pineapple juice has a relatively high concentration of sulfhydryl compounds, which are active antioxidants, whereas lemon juice contains relatively high amounts of both citric acid and ascorbic acid.

Heat treatments and the application of sulfur dioxide or sulfites are common commercial methods for inactivating PPO. PPO activity can be inhibited by the addition of

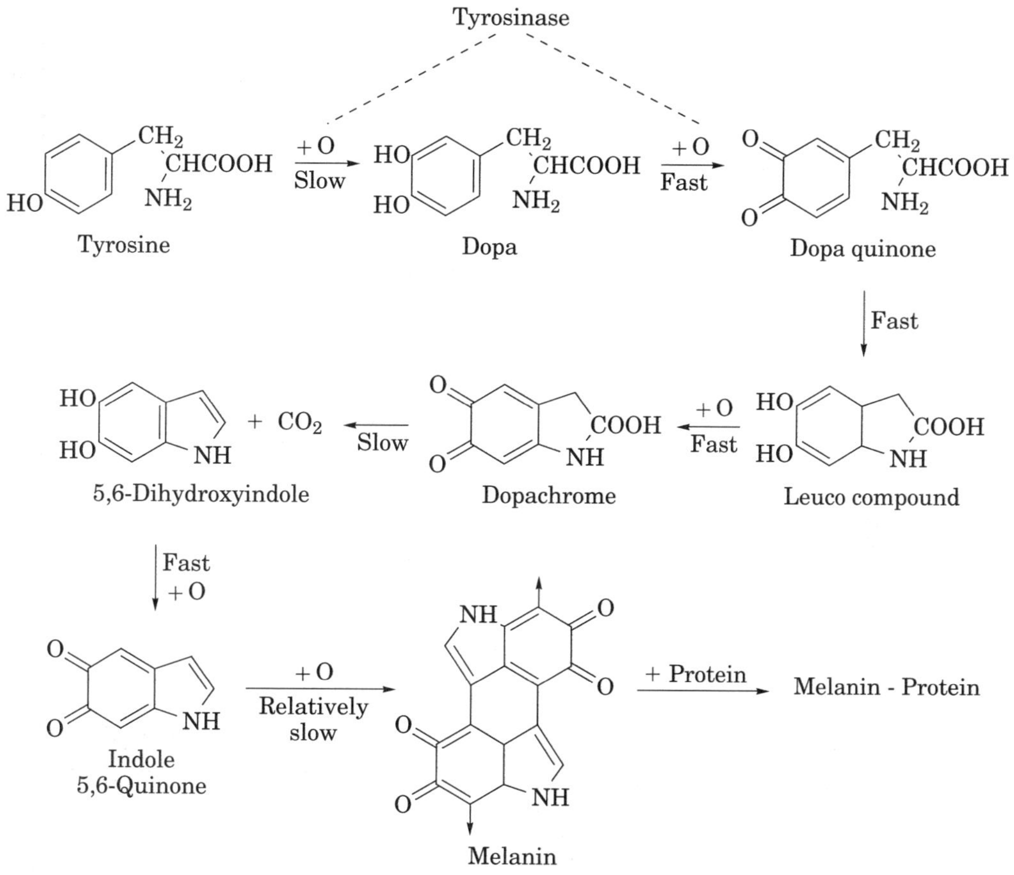

Tyrosinase

Tyrosine $\xrightarrow[\text{Slow}]{+O}$ Dopa $\xrightarrow[\text{Fast}]{+O}$ Dopa quinone $\xrightarrow{\text{Fast}}$ Leuco compound $\xrightarrow[\text{Fast}]{+O}$ Dopachrome $\xrightarrow{\text{Slow}}$ 5,6-Dihydroxyindole + CO_2 $\xrightarrow[\text{Fast}]{+O}$ Indole 5,6-Quinone $\xrightarrow[\text{Relatively slow}]{+O}$ Melanin $\xrightarrow{+\text{Protein}}$ Melanin - Protein

Scheme 1.

sufficient amounts of acidulants, such as citric, malic, or phosphoric acids to yield a pH of 3 or lower. Oxygen can be eliminated by vacuum or by immersing the plant tissues in a brine or syrup. Phenolic substrates can be protected from oxidation by reaction with borate salts, but these are not approved for food use.

For products such as vegetables, which ultimately end up cooked, heat inactivation by steam or hot water blanching is the most practical method of inactivating PPOs. Overblanching should be avoided due to loss of firmness and nutrients. Most commercial blanching of vegetables has been based on the residual activity of the peroxidase because peroxidase is one of the most stable enzymes in plants. It has been generally accepted that if peroxidase is destroyed it is quite unlikely that other enzymes including PPOs will have survived. There are, however, problems associated with the use of peroxidase as an indicator of an adequate blanch in that several enzymes other than peroxidase have been shown to be largely responsible for quality deterioration during frozen storage of different vegetables (67,68). Therefore, optimization of blanching of individual products has to be established in order to tailor the process to each type of raw material and product desired. Heat inactivation of PPO in fruit products has been applied to fruit juices and purees and to fruit intended for such products. Because the enzyme is very labile to heat at 85°C and above, the higher the temperature, the shorter the time required for inactivation. To minimize the undesirable changes due to excess heating, optimum temperature–time requirements for PPO must be established. Rapid cooling after enzyme inactivation is also necessity for best quality retention.

Sulfur dioxide is a very effective inhibitor of PPO and has been used for many years. It readily reacts with compounds such as aldehyde and other carbonyl-containing molecules. These reaction products are ineffective against PPO, and therefore, a sufficient amount of free SO_2 must be maintained. In order to be effective in preventing PPO activity, SO_2 must penetrate throughout the tissues. Sulfurous acid penetrates better than the bisulfite form. The use of excess SO_2 produces both undesirable flavors and an excessively soft product. In recent years the safety of sulfites in foods has been questioned because of their hazard to certain asthmatics. Since August 9, 1986, the U.S. FDA has banned the use of sulfur dioxide in fresh fruits and vegetables and has required a label declaration on those food products (such as dehydrated fruits, frozen potato products, wine) containing more than 10 ppm of sulfiting agent. Because of these restrictions, food processors have turned to a number of sulfite alternatives, mostly formulations effective against enzymatic browning, with varying success. The demand for more effective browning inhibitors has stimulated considerable research activity in this area during the last decade.

Ascorbic acid and its isomer, erythorbic acid, and derivatives such as ascorbic acid-2-phosphate, ascorbic acid-triphosphate, ascorbic acid-6-fatty acid esters are reported to control browning. Ascorbic acid is a very effective reducing agent, as it reduces the *o*-quinones formed by PPO to the original *o*-dihydroxy phenolic compounds. Ascorbic acid alone or in combination with citric acid have been used widely by the food industry. Its prevention of browning lasts as long as any residual ascorbic acid remains, and, therefore, stabilized forms of ascorbic acid should be effective sulfite substitutes. One should keep in mind that the excess amounts of oxidized ascorbic acid can produce brown pigments by nonenzymatic browning reactions. Browning inhibitor penetration can be enhanced by vacuum infiltration, which also removes air from the products' void spaces.

Cinnamic acid and benzoic acid were found to be effective for apple products, and carbon monoxide has been applied to control browning in mushrooms. 4-Hexylresorcinol (Everfresh™) is known to be effective on shrimp. Kojic acid and salicylhydroxamic acid were also known as PPO inhibitors. Compounds that bind or complex PPO substrates also may be potential inhibitors. Poly(vinylpolypyrollidone) and cyclodextrins can bind polyphenols and prevent their participation in enzymatic browning. Glutathione and *N*-acetylcystein were also reported to be effective in controlling browning in apple, potato, and fresh juices. Protease enzymes as well as low molecular weight peptides were claimed to be effective browning inhibitors (69,70). Selecting raw materials for processing that have a low tendency to brown is another way to control browning. Empire and Granny Smith apples and Atlantic potato are examples of cultivars that brown slowly (21,71).

Another approach involves enzymes that transform the substrates of PPOs by methylation or oxidative cleavage of the benzene nucleus. *o*-Methyl transferase methylates the 3-position of 3,4-dihydroxy aromatic compounds in the presence of a methyl donor, converting PPO substrates into inhibitors of PPOs such as caffeic acid to ferulic acid. The exclusion of oxygen as a means of controlling enzymatic browning is generally used in combination with other methods. For example, retail packs of frozen peach slices maintain their high quality when packaged in ascorbic acid-containing syrup combined with hermetically sealed containers wherein the oxygen has been removed from the head space. Oxygen concentrations in the atmosphere of packaged produce can be controlled by modified atmosphere packaging. Although this method can delay browning, low oxygen may induce an anaerobic condition that entails a risk of undesirable quality change and anaerobic microbial growth. Edible coating materials such as xanthan were reported to prevent enzymatic browning and extend the shelf life of fruits and vegetables used in salad bars and prepared salads. Ultrafiltration removes PPO and prevents browning. Osmotic dehydration using sugar or syrup also inhibits enzymatic browning and protect flavors.

Most of chemicals mentioned here are not fail-safe, not acceptable to some consumers, and cannot be used to prevent browning in intact fruits and vegetables. Through better understanding of the mechanism of action of PPO and its essential metabolic roles in plants, it is expected that genetic engineering techniques will be the most important tools in preventing unwanted enzymatic browning. A molecular biology approach will provide a more precise method of decreasing PPO expression, while retaining the desirable genetic traits of plants.

BIBLIOGRAPHY

1. M. Lindet, "Sur l'oxydation du tannin de la pomme a cidre," *Comptes Rendus des Seances de l'Academie des Sciences* **120**, 370–372 (1895).
2. J. R. Whitaker and C. Y. Lee, "Recent Advances in Chemistry of Enzymatic Browning," in C. Y. Lee and J. R. Whitaker, eds., *Enzymatic Browning and Its Prevention*, ACS Symposium Series 600, Washington, D.C., 1995, pp. 2–7.
3. A. M. Mayer, "Polyphenol Oxidases in Plants—Recent Progress," *Phytochemistry* **26**, 11–20 (1987).
4. J. Macheix, J. Sapis, and A. Fleuriet, "Phenolic Compounds and Polyphenol Oxidase in Relation to Browning in Grapes and Wines," *Crit. Rev. Food Sci. Nutr.* **30**, 441–486 (1991).
5. A. M. Mayer and E. Harel, "Phenoloxidases and Their Significance in Fruit and Vegetables," in P. F. Fox, ed., *Food Enzymology*, Elsevier, New York, 1991, pp. 373–398.
6. J. J. Nicolas et al., "Enzymatic Browning Reactions in Apple and Apple Products," *Crit. Rev. Food Sci. Nutr.* **34**, 109–157 (1994).
7. J. C. Steffen, E. Harel, and M. D. Hunt, "Polyphenol Oxidase" B. E. Ellis, ed., in *Genetic Engineering of Secondary Metabolism*, Plenum Press, New York, 1994, pp. 275–312.
8. J. C. Steffen et al., "Polyphenol Oxidase," in *Proceedings of 18th International Conference on Polyphenols*, INRA, Paris, 1998, pp. 223–250.
9. G. Bertrand, "Sur la recherche et la presence de la laccase dans les vegetaux," *Comptes Rendus Hebdomadaires des Sciences de l'Academie des Sciences* **121**, 166–168 (1895).
10. K. Prabhakaran, "Properties of Phenoloxidase in *Mycobacterium leprae*," *Nature* **218**, 973–974 (1968).
11. R. D. Tochner and B. J. D. Meeuse, "Enzymes of Marine Algae. 1. Studies on Phenolase in the Green Algae, *Monostroma fuscum*," *Can. J. Botany* **44**, 551–561 (1966).
12. M. von Poucke, "Copper Oxidase in Thalli *Marchantia pollumorpha*," *Physiological Plantarum* **20**, 932–945 (1967).
13. R. C. Cambie and S. M. Bocks, "A *p*-Diphenol Oxidase from Gymnosperms," *Phytochemistry* **5**, 391–396 (1966).
14. H. Yoshida, "Chemistry of Lacquer (Urushi). Part I," *J. Chem. Soc.* **43**, 472–486 (1883).
15. J. M. Nelson and C. R. Dawson, "Tyrosinase," *Advances in Enzymology* **4**, 99–152 (1944).
16. L. Vamos-Vigyazo, "Polyphenol Oxidase and Peroxidase in Fruits and Vegetables," *Crit. Rev. Food Sci. Nutr.* **15**, 49–127 (1981).
17. E. Harel, A. M. Mayer, and Y. Shain, "Catechol Oxidases from Apples; Their Properties, Subcellular Location and Inhibition," *Physiological Plantarum* **17**, 921–930 (1964).
18. A. M. Mayer, "Nature and Location of Phenolase in Germinating Lettuce," *Physiological Plantarum* **14**, 322–331 (1961).
19. E. Harel, A. M. Mayer, and Y. Shain, "Catechol Oxidases, Endogenous Substrates and Browning in Developing Apples," *J. Sci. Food Agric.* **17**, 389–392 (1966).

20. K. W. Wissemann and C. Y. Lee, "Polyphenoloxidase Activity During Grape Maturation and Wine Production," *Am. J. Enology Viticulture* **31**, 206–211 (1980).

21. Y. CoSeteng and C. Y. Lee, "Changes in Apple Polyphenoloxidase and Polyphenol Concentration in Relation to Degree of Browning," *J. Food Science* **52**, 985–989 (1987).

22. H. S. Mason and C. I. Wright, "The Chemistry of Melanin: V. Oxidation of Dihydroxyphenylalanine by Tyrosinase," *J. Biol. Chem.* **180**, 235–247 (1949).

23. W. S. Pierpoint, "The Enzymatic Oxidation of Chlorogenic Acid and Some Reactions of the Quinone Produced," *Biochem. J.* **98**, 567–580 (1966).

24. V. Khan, "Some Biochemical Properties of Polyphenol Oxidase from Two Avocado Varieties Differing in Their Browning Rates," *J. Food Sci.* **42**, 38–43 (1977).

25. L. Vamos-Vigyazo, K. Vas, and N. Kiss-Kutz, "Studies into the *o*-Diphenol Oxidase Activity of Potatoes: I. Development of an Enzyme Activity Assay Method," *Acta Alimentaria Academiae Scientarum Hungaricae* **2**, 413–429 (1973).

26. K. Mihalyi and L. Vamos-Vigyazo, "A Method for Determining Polyphenol Oxidase Activity in Fruits and Vegetables Applying a Natural Substrate," *Acta Alimentaria Academiae Scientarum Hungaricae* **5**, 69–85 (1976).

27. M. El-Bayoumi and E. Frieden, "A Spectrophotometric Method for the Determination of the Catecholase Activity of Tyrosinase and Some of Its Applications," *J. Am. Chem. Soc.* **79**, 4854–4859 (1957).

28. F. Mazzocco and P. G. Pifferi, "An Improvement of the Spectrophotometric Method for the Determination of Tyrosinase Catecholase Activity by Besthorn's Hydrozone," *Anal. Biochem.* **72**, 643–647 (1976).

29. H. Esterbauer, E. Schwarzl, and M. Hayn, "A Rapid Assay for Catechol Oxidase and Laccase Using 2-Nitro-5-thiobenzoic Acid," *Anal. Biochem.* **77**, 486–494 (1977).

30. A. M. Mayer, E. Harel, and Y. Shain, "2,3-Naphthalenediol, a Specific Competitive Inhibitor of Phenolase," *Phytochemistry* **3**, 447–451 (1964).

31. R. J. Petroski, W. Peczynska-Czoch, and J. P. Rosazza, "Analysis, Production, and Isolation of an Extracellular Laccase from *Polyporus anceps*," *Appl. Environ. Microbiol.* **40**, 1003–1006 (1980).

32. V. Kovac, "Effet des moisissures du genre penicillium sur l'activite laccase du *Botrytis cinerea*," *Annales de Technologie Agricole* **28**, 345–355 (1979).

33. J. R. L. Walker and R. F. McCallion, "The Selective Inhibition of ortho- and para-Diphenol Oxidases," *Phytochemistry* **19**, 373–377 (1980).

34. A. M. Mayer and E. Harel, "Polyphenol Oxidases in Plants," *Phytochemistry* **18**, 193–215 (1979).

35. A. R. Lax and J. W. Cary, "Biology and Molecular Biology of Polyphenol Oxidase," in C. Y. Lee and J. R. Whitaker, eds., *Enzymatic Browning and Its Prevention*, ACS Symposium Series 600, Washington, D.C., 1995, pp. 120–128.

36. K. C. Vaughn, A. R. Lax, and S. O. Duke, "Polyphenol Oxidase: The Chloroplast Oxidase With No Established Function," *Physiology Plantarum* **72**, 659–665 (1988).

37. D. Kertesz and R. Zito, "Mushroom Polyphenol Oxidase. I. Purification and General Properties," *Biochim. Biophys. Acta* **96**, 447–462 (1965).

38. W. D. Loomis and J. Battaile, "Plant Phenolic Compounds and the Isolation of Plant Enzymes," *Phytochemistry* **5**, 423–438 (1966).

39. N. O. Benjamin and M. W. Montgomery, "Polyphenol Oxidase of Royal Ann Cherries: Purification and Characterization," *J. Food Sci.* **38**, 799–806 (1973).

40. K. W. Wisseman and M. W. Montgomery, "Purification of d'Anjou Pear (*Pyrus communis L.*) Polyphenol Oxidase," *Plant Physiology* **78**, 256–262 (1985).

41. K. W. Wisseman and C. Y. Lee, "Purification of Grape Polyphenoloxidase With Hydrophobic Chromatography," *J. Chromatgr.* **192**, 232–235 (1980).

42. E. L. Angleton and W. H. Flurkey, "Activation and Alteration of Plant and Fungal Polyphenoloxidase Isoenzymes in Sodium Dodecylsulfate Electrophoresis," *Phytochemistry* **23**, 2723–2725 (1984).

43. M. Hruskocy and W. H. Flurkey, "Detection of Polyphenoloxidase Isoenzymes by Electroblotting and Photography," *Phytochemistry* **25**, 329–332 (1986).

44. K. Lerch, "Tyrosinase: Molecular and Active-Site Structure," in C. Y. Lee and J. R. Whitaker, eds., *Enzymatic Browning and Its Prevention*, ACS Symposium Series 600, Washington, D.C., 1995, pp. 64–80.

45. S. Bouchilloux, P. McMahill, and H. S. Mason, "The Multiple Forms of Mushroom Tyrosinase. Purification and Molecular Properties of the Enzymes," *J. Biol. Chem.* **238**, 1699–1707 (1963).

46. K. G. Strothkamp, R. L. Jolley, and H. S. Mason, "Quaternary Structure of Mushroom Tyrosinase," *Biochem. Biophys. Res. Commun.* **70**, 519–524 (1976).

47. E. Harel, A. M. Mayer, and E. Lehman, "Multiple Forms of *Vitis vinifera* Catechol Oxidase," *Phytochemistry* **12**, 2649–2654 (1973).

48. M. Huber, G. Hintermann, and K. Lerch, "Primary Structure of Tyrosinase from *Streptomyces glaucescens*," *Biochemistry* **24**, 6038–6044 (1985).

49. V. Bernan et al., "The Nucleotide Sequence of the Tyrosinase Gene from *Streptomyces antibioticus* and Characterization of the Gene Product," *Gene* **37**, 101–110 (1985).

50. T. Shahar et al., "The Tomato 66.3-kD Polyphenoloxidase Gene: Molecular Identification and Development Expression," *Plant Cell* **4**, 135–147 (1992).

51. S. M. Newman et al., "Organization of the Tomato Polyphenol Oxidase Gene Family," *Plant Mol. Biol.* **21**, 1035–1051 (1993).

52. J. W. Cary, A. R. Lax, and W. H. Flurkey, "Cloning and Characterization of cDNA Coding for *Vicia faba* Polyphenol Oxidase," *Plant Mol. Biol.* **20**, 245–253 (1992).

53. M. D. Hunt et al., "cDNA Cloning and Expression of Potato Polyphenol Oxidase," *Plant Mol. Biol.* **21**, 59–68 (1993).

54. A. M. Mayer and E. Harel, "Laccase-like Enzyme in Peaches," *Phytochemistry* **7**, 1253–1256 (1968).

55. V. Kahn, "Latency Properties of Polyphenol Oxidase in Two Avocado Cultivars Differing in Their Rate of Browning," *J. Sci. Food Agric.* **28**, 233–239 (1977).

56. J. R. L. Walker, "Enzymatic Browning in Fruits," in C. Y. Lee and J. R. Whitaker, eds., *Enzymatic Browning and Its Prevention*, Symposium Series 600, Washington, D.C., 1995, pp. 8–22.

57. C. Y. Lee and J. R. Whitaker, *Enzymatic Browning and Its Prevention*, ACS Symposium Series 600, Washington, D.C., 1995.

58. S. Duffy and G. Felton, "Enzymatic Antinutritive Defenses of the Tomato Plant Against Insects," in P. Hedin, ed., *Naturally Occurring Pest Bioregulators*, American Chemical Society, Washington, D.C., 1991, pp. 166–197.

59. M. E. Winkler, K. Lerch, and E. I. Solomon, "Competitive Inhibitor Binding to the Binuclear Copper Active Site in Tyrosinase," *J. Am. Chem. Soc.* **103**, 7001–7003 (1981).
60. J. R. Whitaker, *Principles of Enzymology for the Food Sciences*, 2nd ed., Marcel Dekker, New York, 1994.
61. T. Nakamura, "On the Process of Enzymatic Oxidation of Hydroquinone," *Biochem. Biophys. Res. Commun.* **2**, 111–113 (1960).
62. R. A. Holwerda, S. Wherland, and H. B. Gray, "Electron Transfer Reactions of Copper Protein," *Ann. Rev. Biophys. Bioeng.* **5**, 363–396 (1976).
63. V. Cheynier, C. Osse, and J. Rigaud, "Oxidation of Grape Juice Compounds in Model Solutions," *J. Food Sci.* **53**, 1729–1732 (1988).
64. V. L. Singleton, "Oxygen with Phenols and Related Reactions in Musts, Wines, and Model Systems: Observations and Practical Implications," *Am. J. Enology Viticulture* **38**, 69–77 (1987).
65. V. Chenyier, N. Basire, and J. Rigaud, "Mechanism of trans-Caffeoyltartaric acid and Catechin Oxidation in Model Solutions Containing Grape Polyphenoloxidase," *J. Agric. Food Chem.* **37**, 1069–1071 (1989).
66. J. Oszmianski and C. Y. Lee, "Enzymatic Oxidation Reaction of Catechin and Chlorogenic Acid in a Model System," *J. Agric. Food Chem.* **38**, 1202–1204 (1990).
67. M. H. Lim et al., "Enzyme Indicators of Adequate Blanching," in *Proceedings International Conference on Technical Innovations in Freezing and Refrigeration of Fruits and Vegetables*, University of California–Davis, Calif., July 9–12, 1989, pp. 67–72.
68. C. Y. Lee, N. L. Smith, and D. E. Hawbecker, "An Improved Blanching Technique for Frozen Sweet Corn-on-the-cob," in *Proceedings International Conference on Technical Innovations in Freezing and Refrigeration of Fruits and Vegetables*, University of California–Davis, Calif., July 9–12, 1989, pp. 85–90.
69. T. P. Labuza, J. H. Lillemo, and P. S. Taoukis, "Inhibition of Polyphenoloxidases by Proteolytic Enzymes," *Flussiges Obst.* **59**, 15–20 (1992).
70. J. Oszmianski and C. Y. Lee, "Inhibition of Polyphenol Oxidase Activity and Browning by Honey," *J. Agric. Food Chem.* **38**, 1892–1895 (1990).
71. G. M. Sapers and F. W. Douglas, "Measurement of Enzymatic Browning in Cut Surfaces and in Juice of Raw Apple and Pear Fruits," *J. Food Sci.* **52**, 1258–1262 (1987).

CHANG Y. LEE
Cornell University
Geneva, New York

BULKING AGENTS. See SUGAR: SUBSTITUTES, BULK, REDUCED CALORIE.

BUTTER AND BUTTER PRODUCTS

The art of butter making dates back to times immemorial. Reference to the use of butter for sacrificial worship, for medicinal and cosmetic purposes, and as a human food may be found long before the Christian era. Documents indicate that, at least in the Old World, the taming and domestication of animals constituted the earliest beginnings of human civilization and culture. There is good reason to believe, therefore, that the milking of animals and the origin of butter making aforedate the beginning of organized and permanent recording of human activities.

The evolution of the art of butter making has been intimately associated with the development and use of equipment. With the close of the eighteenth century the construction and use of creaming and butter-making equipment, other than that made of wood, began to receive consideration, and the barrel churn made its appearance.

By the middle of the nineteenth century attention was given to improvement in methods of creaming. These efforts gave birth to the deep-setting system. Up to that time, creaming was done by a method called shallow pan. The deep-setting system shortened the time for creaming and produced a better-quality cream. An inventive Bavarian brewer in 1864 conceived the idea of adapting the principle of the laboratory centrifuge. In 1877 a German engineer succeeded in designing a machine that, although primitive, was usable as a batch-type apparatus. In 1879 engineers in Sweden, Denmark, and Germany succeeded in the construction of cream separators for fully continuous operation (1).

In 1870, the last year before introduction of factory butter making, butter production in the United States was 514 million lb, practically all farm made. Authentic records concerning the beginning of factory butter making are meager. It appears that the first butter factory was built in Iowa in 1871. This beginning also introduced the pooling system of creamery operation (1).

Other inventions that assisted the development of the butter industry included the Babcock test (1890), which accurately determines the percentage of fat in milk and cream, the use of pasteurization to maintain milk and cream quality; and the use of pure cultures of lactic acid bacteria and refrigeration to help preserve cream quality.

BUTTER CONSUMPTION

Butter production and consumption has recently increased from a long decline Table 1–4. A dramatic shift occurred, starting in 1985, to the table spreads category of products (less than 80% fat) from full-fat butter, margarine, and blends (2, Land O'Lakes, unpublished data, 1985). The spreads category encompasses all non-Standard-of-Identity table spreads (ie, from 0 to 79.9% fat). Major market forces have changed this production consumption turnaround. As noted in a 1984 survey, the most important

Table 1. Market Shares for Butter, Margarine, Blends, and Spreads for 1980 to 1995

Year	Butter	Light butter	Margarine	Blends	Spreads
1980	25	—	74	1	—
1985	22	—	73	3	2
1990	17	—	48	2	33
1995	21	1	23	11	44

Source: Ref. 2 and Land O'Lakes unpublished data, 1996.

Table 2. Composition of Lipids in Whole Bovine Milk

Lipid	Weight percent
Hydrocarbons	Trace
Sterol esters	Trace
Triglycerides	97–98
Diglycerides	0.28–0.59
Monoglycerides	0.016–0.038
Free fatty acids	0.10–0.44
Free sterols	0.22–0.41
Phospholipids	0.2–1.0

Source: Refs. 7 and 8.

Table 3. Composition of Lipids from Milk Fat Globule Membrane

Lipid component	Percentage of membrane lipids
Carotenoids (pigment)	0.45
Squalene	0.61
Cholesterol esters	0.79
Triglycerides	53.4
Free fatty acids	6.3[a]
Cholesterol	5.2
Diglycerides	8.1
Monoglycerides	4.7
Phospholipids	20.4

Source: Refs. 9 and 10.
[a]Contained some triglycerides.

Table 4. Per Capita Consumption of Butter for the Years 1950 to 1995

Year	Pounds per capita
1950	10.7
1960	7.7
1970	5.4
1980	4.5
1990	4.0
1995	4.5

Source: Refs. 3, 4, and Land O'Lakes unpublished data, 1996.

barriers to increased butter sales were listed in the following order (2):

1. Price (opinion of an overwhelming majority when butter is compared to margarine)
2. Health (negative consumer attitudes toward cholesterol and saturated fats are increasing)
3. Poor spreadability
4. Inadequate promotional spending
5. Product innovation in margarine and spreads
6. Legislation and regulatory restrictions

The forces that have had a positive impact on butter are as follows:

1. The U.S. Department of Agriculture (USDA) has eliminated all price supports for butter. Prices follow demand/consumption economics with the world market price (which has traditionally been significantly below the U.S. support price) operating as a base floor price. When price dips to world market price, butter becomes very price competitive to margarine/blends/spreads.
2. Health issues associated with butter have been greatly negated, for the public has become apathetic to negative cholesterol messages. In addition, the issue of trans fatty acids in fairly high concentration in margarines, and the negative health impacts of these trans fats has given butter a positive health message.

Butter manufacture continues to serve as the safety valve for the dairy industry. It absorbs surplus milk supply above market requirements for other dairy products. Milk not required by the demand for these products overflows into the creamery, is skimmed, and the cream is converted to butter. When the milk supply for other products runs short of their demand, milk normally intended for butter making is diverted into the channels where needed. Even though consumption patterns have dramatically changed over the years, the butter industry thus provides a never-failing balance wheel that takes up the slack in the relationship of supply and demand for all other dairy products.

DEFINITIONS AND GRADING OF BUTTER

Standards for butter in the United States were established by an Act of Congress and are supported by the USDA standards for grades of butter. In the revised standards the following definitions apply. Butter refers to the food product usually known as butter, which is made exclusively from milk, cream, or both, with or without common salt; and with or without additional coloring matter. The milk fat content of butter is not less than 80% by weight, allowing for all tolerances. Cream refers to the cream separated from milk produced by healthy cows. Cream is pasteurized at a temperature of not less than 73.9°C for not less than 30 min; or it can be pasteurized at a temperature of not less than 89°C for not less than 15 s. Other approved methods of pasteurization give equivalent results (5).

The cream may be cultured by the addition of harmless lactic acid bacteria to enhance flavor, natural flavors obtained by distilling a fermented milk, or cream may be added to the finished butter. In addition, color, derived from a Food and Drug Administration (FDA) approved source, may be used.

There are three U.S. grades of butter: AA, A, and B. Butter is graded by first classifying its flavor organoleptically. In addition to the overall quality of the butter flavor itself, the standards list 17 flavor defects and the degree to which they may be present for each grade. If more than one off-flavor is noted, assignment is made on the basis of the flavor resulting in the lowest grade. This grade is then lowered by defects in the workmanship and the degree to which they are apparent. Disratings are characterized by negative body, flavor, or salt attributes, which are fully described in the standards. Butter that does not meet the requirements for U.S. Grade B is not graded. To bear the USDA seal, the finished product must fall within the following microbiological specifications: proteolytic count not more than 100 per gram, yeast and mold not more than 20 per gram, and coliform not more than 10 per gram. Butter should be stored at 4.4°C or lower or at less than −17.8°C, if it is to be held for more than 30 days (6).

Legal requirements for butter vary considerably in different countries. For example, in Europe butter must contain 82% fat and in France it may contain a maximum of 16% moisture (6). In some tropical parts of the world, milk fat is used in nearly anhydrous form because it is less susceptible to bacterial spoilage. This product is known as ghee. In the Middle East and India ghee is prepared from heated cow or buffalo milk.

COMPOSITION

The composition of milk fat is somewhat complex. Although dominated by triglycerides, which constitute some 98% of milk fat (with small amounts of diglycerides, monoglycerides, and free fatty acids), various other lipid classes are also present in measurable amounts. It is estimated that about 500 separate fatty acids have been detected in milk lipids; it is probable that additional fatty acids remain to be identified. Of these, about 20 are major components; the remainder are minor and occur in small or trace quantities (7,8). The other components include phospholipids, cerebrosides, and sterols (cholesterol and cholesterol esters). Small amounts of fat-soluble vitamins (mainly A, D, and E), antioxidants (tocopherol), pigments (carotene), and flavor components (lactones, aldehydes, and ketones) are also present.

The composition of the lipids of whole bovine milk is given in Table 2 (7,8). The structure and composition of the typical milk fat globule is exceeding complex. The globule is probably 2 to 3 μ in diameter with a 90-Å-thick membrane surrounding a 98 to 99% triglyceride core. The composition of the milk fat membrane is quite different from milk fat itself in that approximately 60% triglycerides are present, much less than in the parent milk fat (Table 3) (9,10).

It has been generally recognized that butterfat consists of about 15 major fatty acids with perhaps 12 or so minor (trace quantity) acids. Triglycerides are normally defined with respect to their carbon number (CN), that is, the number of fatty acid carbon atoms present in the molecule; the three carbon atoms of the glycerol moiety are ignored. Because the fatty acid spectrum of milk fat is dominated by acids containing an even number of carbon atoms, so is the triglyceride spectrum. However, the proportion of triglycerides with an odd carbon number is about three times greater than the proportion of odd-numbered fatty acids.

Although obvious correlations exist between fatty acid composition and triglyceride distribution, detailed information is lacking that would enable the triglyceride distribution to be predicted from the fatty acid composition. Much more needs to be understood of the strategy used in the bovine mammary gland in assembling a complex array of fatty acids into triglycerides. This is not an arcane study; it is necessary if processes such as fractionation are to yield products with consistent qualities throughout the year. In effect, the detailed structure of milk fat is not yet understood. Perhaps this is not surprising if we consider only the 15 major fatty acids; there are 15^3 (3375) possible triglyceride structures using a purely random model.

The concentration of milk fat results in a dairy product called butter. The concentration is a two-stage process with the first stage being effected in a mechanical separator, which causes a 10-fold concentration of the fat into an intermediate product, cream. The second stage is the churning process, which results in a further twofold concentration of the fat.

In butter manufacture, the preparation of the cream is effected quite independently of the butter-manufacturing stage. Today most butter plants receive their cream from other operations rather than directly from the farm, as was the case in the past. Because of general improvement in sanitation practices, the receipt of fresher milk and cream, and advances in the knowledge and understanding of deleterious handling conditions, the quality of cream is constantly improving.

The composition of milk fat is the most important factor affecting the firmness of butter and, therefore, its spreadability. The composition of milk fat changes primarily according to the feed; therefore, the entire problem is connected to the animal's diet. Today, the fatty acid composition of milk fat produced in various countries is rather accurately known, along with its seasonal variations. In Europe the amount of saturated fatty acids is generally highest in winter and lowest in summer or fall (Table 5) (11). Green fodder decreases the amount of saturated fatty acids and correspondingly increases the amount of unsaturated fatty acids (12). The differences between the maximum and minimum values can be fairly large. For palmitic and oleic acids, the quantitatively most important fatty acids, a difference of more than 10% between the maximum and minimum values was found in some cases. This makes it understandable that there are also significant differences in the physical characteristics of the butter. The structure of the triglycerides in the milk fat, along with the fatty acid composition, is important in determining the physical characteristics of the fat, because the softening point of fat has been found to rise as the result of interesterification (13).

Protected oils are hydrolyzed in the abomasum, and the fatty acids are absorbed in the small intestine, thereby avoiding hydrogenation. The 18:2 content in the milk fat was increased about fivefold, and the 14:0, 16:0, and 18:0 were decreased accordingly. Plasma and depot fats were also increased in 18:2 content by this program (14).

Results at the USDA are similar: cow's milk can be increased in 18:2 acid from 3 to 35% by feeding protected safflower oil (15,16). However, at high 18:2 levels, milk develops an oxidized off-flavor, usually after about 24 h, and creams require a longer aging time for satisfactory churning. As expected, butter that contains more than 16% linoleic acid is soft and sticky (8).

Table 5. Compositional Characteristics of Summer and Winter Milk Fat, 1970

Samples investigated (N = 140)	Fatty acids				
	Volatile	Saturated	Monounsaturated	Polyunsaturated	Iodine number
Total	10.98	56.50	29.81	2.50	32.2
Summer	9.49	58.82	33.53	3.14	36.8
Winter	12.45	59.15	26.15	1.86	27.7

Source: Ref. 11.

There are also differences in the butter fat of different cows on identical rations, and the age of the animal and duration of lactation have some influence on butter fat composition. Much of the dairy literature provides information relating dairy animal species and the composition of the butter fat from them.

Textural characteristics of butter are significantly dependent on milk fat composition and method of manufacture. It is possible, therefore, knowing the chemical composition of milk fat, to select the appropriate technological parameters of butter making to improve the texture. To obtain butter with constant rheological characteristics and to control the parameters of the butter-making process it is necessary to consider the difference in the chemical compositions and the properties of the milk fat in various seasons.

From the standpoint of nutritional value, the vitamin A content of butter is important. Because the source of vitamin A in butter is β-carotene or other carotenoid pigments in the feed of the cows, the content of this vitamin varies considerably, being highest in the summer when the dairy herds are in pasture and lowest in winter when there are no green feedstuffs in the rations. A portion of the carotene in the feed that is transferred by the cow into the butter fat varies with the feeding regimen parallel to variations in the production of vitamin A, so that the intensity of the yellow color of butter to some extent serves to indicate its vitamin A content.

The vitamin A potency of butter is in part due to vitamin A as such and in part to carotene, which is partially converted to the vitamin in the human body. The vitamin A content of butter is usually within the range of 6 to 12 mg/g, and the carotene content is in the range of 2 to 10 mg/g (17); 1 IU of vitamin A is defined as the amount possessing the biological activity of 0.6 μg of pure β-carotene.

The vitamin D content of butter is much less significant than that of vitamin A, but it is nevertheless appreciable. It varies from about 0.1 to 1.0 IU/g, being highest in the summer and lowest in the winter (17).

MODIFICATION OF MILK FAT

Melting and Crystallization of Milk Fat Triglycerides

The complex fatty acid composition of milk fat is reflected in its melting behavior. Melting begins at −30°C and is complete only at 37°C. At any intermediate temperature, milk fat is a mixture of solid and liquid. To a large extent, the solid: liquid ration determines the rheological properties of the fat. For example, at refrigeration temperature, butter has a higher solids content than does a tub margarine. Hence the latter product is more easily spread (18).

As crystallization proceeds, the growing crystals impinge to form aggregates. A network results, in which both the solid and liquid phases may be regarded as continuous. Formation of the network greatly increases the firmness of the fat.

As a liquid fat is cooled, crystallization begins. There are two parts to the crystallization: (*1*) nucleation and (*2*) growth. In a bulk fat, nucleation occurs at the surfaces of impurities, a phenomenon described as heterogeneous nucleation. A considerable degree of supercooling is necessary to initiate nucleation. Subsequent growth of the nuclei tends to be slow in natural fats because of competitive inhibition. In materials of low molecular weight, impurities are rejected at the face of growing crystal. In fats, however, the various triglyceride species are so closely related that the term *impurity* tends to lose its meaning (19).

Hydrogenation

Hydrogenation of various fats and oils is used extensively in industry but is not generally applied to butter fat (the high cost of the raw material argues against its use as a feedstock). The process reduces the degree of unsaturation of the fat and increases its melting point.

Given the criticism directed at milk fat because of its saturated nature, there appears to be little future in increasing the degree of saturation by means of hydrogenation. The reverse procedure, desaturation or dehydrogenation, offers more attractive prospects.

Cream Quality

Raw cream should be processed without delay to minimize deterioration of quality. If the cream is to be held for more than two hours between separation and processing, it should be cooled below 5°C. For holding periods exceeding one day it may be advantageous to heat, treat, or pasteurize the cream.

To make high-quality butter, the cream must be of high quality. The flavor of butter is the most important organoleptic property, and it depends in no small part on the flavor quality of the cream from which it is made. There is no real alternative to tasting in assessing flavor and this simple but important procedure should be conducted routinely as part of the overall quality-control program.

The application of hazard analysis of critical control points (HACCP) to butter production is one of the newer techniques to ensure improved butter safety and quality (20).

Fat Content

The percentage of fat in the cream must be known and controlled. It influences fat losses in churning. Knowledge of the fat content assists in yield estimations for operational conditions in continuous manufacture. A number of satisfactory procedures are available, with the Babcock test being the most common.

The chemical composition of the triglycerides, which comprise milk fat, varies throughout the year depending on the stage of lactation and the cow's diet. This causes a cyclic change in the melting properties of the fat. In the control of the butter-making process and the physical properties of the finished butter this factor must be monitored. The term melting property is used rather than softness or hardness because these more correctly refer to an altogether different attribute of solids. A number of procedures have been used to follow the seasonal change in the melting properties. The iodine number, refractive index and differential scanning calorimetry, or pulsed nmr spectroscopy can be used to prepare a melting curve. However, the expense and complexity of these melting curve techniques precludes the approach in most quality-control situations (21). The traditional chemical determinations for fat, saponification value and Polenske value, are of limited value. They are scarcely relevant for quality control, and the information they represent can be more usefully quantified by the determination of the fatty acid profile using gas chromatography.

Today the use of stainless steel has essentially eliminated the exposure of the fat to copper and iron. The presence of copper and, to a lesser extent, iron can catalyze oxidative deterioration of butter during storage, particularly in the presence of salt and a low pH.

Many procedures have been used or proposed to assess the microbiological quality of the milk or cream. Generally, microbiological tests are performed to determine the hygiene of production and storage conditions or for safety reasons. Tests include total counts and counts for specific classes of microorganisms such as yeasts and molds, coliforms, psychotrophs, and pathogens such as *Salmonella*. Rapid-screening tests based on dye reduction or direct observation using a microscope or automatic total counters are also in use.

Lipase Activity

Lipolysis in butter after manufacturing caused by thermoresistant lipase enzymes created in milk or cream by psychotrophic bacteria is an increasing problem. Based on a termination of the lipase activity in cream, the keeping quality of manufactured butter in regard to lipolysis can be predicted with reasonable accuracy. A similar prediction for sweet cream butter can be based on activity determination in the serum phase (22).

Oxidation

The flavor of dairy products is largely determined by the fat component. Consequently, it is particularly important to restrict the development of oxidized off-flavors in the fat source before use. Oxidation is the chief mode of deterioration of fats and a major factor in determining the shelf life of fat-containing foods (23). Unsaturated fatty acid esters react with oxygen to form peroxides. Although flavorless themselves, peroxides are unstable and readily decompose to yield flavorful carbonyl compounds. The latter are the source of the characteristic oxidized flavors that are detectable at low concentrations. The rate of oxidation depends on the concentration of dissolved oxygen, the temperature, the presence of prooxidants such as copper and iron, the degree of unsaturation of the fat, and the presence of antioxidants that may retard the onset of oxidation. Compared with many fats, milk fat has a good oxidative stability, because it is high in total saturates, low in polyunsaturates, and contains natural antioxidants, principally α-tocopherol.

The development of oxidative rancidity in milk fat is the major determinant of the stability of the fat on storage. Dissolved air in the milk fat can give dissolved oxygen levels of up to 40 ppm at 30°C. In practice, the dissolved oxygen level in the freshly processed milk fat would be about 5 ppm at 45°C, a level sufficient to permit the development of oxidative rancidity, but if the milk fat were allowed to equilibrate with the air, then this level could increase to 33 ppm with a consequent increase in the rate of development of oxidative rancidity.

The oxygen level in the milk fat may be limited by either active or passive actions. For passive control, processing procedures and plant design are established to minimize oxygen (air) exposure. Deaeration devices (24), the use of antioxidants, effective destruction of lipases, and nitrogen spanning of container headspace are examples of active control of product quality.

Consistency

Consistency has been defined as "that property of the material by which it resists permanent change of shape and is defined by the complete force flow relation" (8). This implies that the concept of consistency includes many aspects and cannot be expressed by one parameter. Today more importance is attached to spreadability than anything else in the evaluation of the consistency of butter. The general consensus is that butter should be spreadable at refrigerator temperatures. There is no suitable method available to measure such a subjective criteria as the spreadability of butter. For this reason, the firmness of butter, which should correlate well with spreadability, was selected as the parameter to be measured. It was recommended that the use of the cone penetrometer, along with other methods, could give good results (25).

Flavor

One of the most important consumer attributes of butter is the pleasing flavor. Butter flavor is made up of many volatile and nonvolatile compounds. Researchers have identified more than 40 neutral volatiles, of which the most prominent are lactones, ethyl esters, ketones, aldehydes, and free fatty acids (26). The nonvolatiles, of which salt (sodium chloride) is the most prominent, contribute to a balanced flavor profile. Diacetyl and dimethyl sulfide also contribute, especially in cultured butter flavor (27).

Ripened or cultured creams use lactic acid bacterial cultures selected for their ability to produce volatile acid and diactyl flavors. The NIZO (Dutch Research Institute for Dairy) process was developed to incorporate concentrated lactic starters or to directly inject starter distillates to produce the flavor and characteristics of cultured butter (25). In the United Kingdom, Canada, Ireland, and the United States, sweet cream, salted butter is primarily used, but in France, Federal Republic of Germany, and Scandinavia, the taste of cultured butter is preferred (6).

Body and Texture

The physical properties of butter that are noted by the senses are described by butter graders by means of appropriate qualifications of the terms body and texture. The exact meaning of these terms has not been clearly defined. Frequently they are used as if they had the same meaning. Certain properties such as hardness and softness refer to the body of butter, whereas properties such as openness refer to texture. But some of the properties, such as leakiness or crumbliness, lead to confusion. Usually, most body and texture terms are used to describe a defect such as gritty, gummy, sticky, and the like.

Good butter should be of fine and close texture; have a firm, waxy body; and be sufficiently plastic to be spreadable at cold temperatures.

Color

The color of butter may vary from a light creamy white to a dark creamy yellow or orange yellow. Differences in butter color are due to variations in the color of the butter fat, which is affected by the cows' feed and season of the year; variations in the size of the fat globule; presence or absence of salt; conditions of working the butter; and the type and amount of natural coloring added.

Butter colorings are oil soluble and most often are natural annatto, an extraction of the seeds of the tropical tree *Bixa orellana*, or natural carotenes, extractions from various carotene-rich plants. Because they are oil soluble, colorings are added to the cream to obtain the most uniform dispersion.

PROCESSING

Milk and Cream Separation

The most basic and oldest processing method is cream separation. Ancient people are known to have used milk freely. It is probable that they used the cream that rose to the top of milk that had been held for some time in containers, although there is little in ancient literature to suggest that such use was common. It is well established that in early times butter was produced by churning milk.

The principal questions concerning separation in a butter-manufacturing facility are the choice of cream fat content and the choice of the separation technique, milk separation before or after pasteurization, temperature of separation, and regulation of the fat content. Separation of cream from milk is possible because of a difference in specific gravity between the fat and the liquid portion, or serum. Whether separation is accomplished by gravity or centrifugal methods, the result depends on this difference (28).

Crystallization

The crystal structure of fat and the resulting physical properties of butter made by both conventional and alternative processes have received considerable study. When conventionally churned or Fritz butter is made, most of the milk fat is contained within the fat globule in cream during the cooling and crystallization process. The fat globule provides a natural limit to the growth of fat crystals. Cooling and holding of cream is normally carried out overnight, and thus sufficient time exists for the crystallization process to approach equilibrium (29).

The principles of crystallization of plastic fats in the type of equipment used for margarine manufacture have been described (30). It is important to develop small fat crystals that remain substantially discrete and do not form a strong interlocking structure. Small crystals (eg, 5-μm diameter) have a greater total surface area than large crystals and will bind water and free liquid fat by absorption more effectively (29). Large crystals impart a gritty texture to the product. When fats are cooled rapidly in a scraped surface heat exchanger, fat crystallization commences, but the fat is substantially supercooled on exiting. If crystallization is then permitted to continue under quiescent conditions, crystals will grow together and form a lattice structure. The product will thus be hard and brittle and may tend to leak moisture. If, however, crystallization is permitted to occur under agitated conditions (eg, for 1–3 min in a pin worker), the formation of small, independent crystals will be favored and the product will have a fine, smooth texture. If crystallization under agitated conditions is permitted to continue for too long, the product will be too soft for most patting or bulk filling operations, and it is likely to be too soft and greasy at warm room temperatures.

Neutralization

When lactic acid has developed in the raw, unpasteurized cream by microbial activity to a degree considered excessive, neutralizer may be added to return the cream acidity to a desirable level. Sodium carbonates have been found suitable in practice for batch neutralization. For continuous neutralization by pH control, sodium hydroxide is more suitable. These chemicals should be of food grade.

Fractionation

Fractionation by thermal crystallization, steam stripping, short-path distillation, supercritical fluid extraction, or crystallization using solvents can achieve fat alterations of significance to the dairy industry. Milk fat is a mixture of triglycerides with a range of molecular weights, degrees of unsaturation, melting points, and other physical properties. Milk fat is an important component of most dairy products, but it has been consumed traditionally for the most part as butter.

Fractionation of butter fat for the selective removal of cholesterol has been recently reported using supercritical fluid extraction (31), steam stripping or deodorization (32), short-path molecular distillation (33), and absorption processes (28). Various enzymatic investigations are under way to isolate cholesterol reductase (34) and to insert the gene for cholesterol reductase into lactic and acid bacteria biogenetically (35), resulting in the reduction of the dairy fat cholesterol to its major by-product, coprostanol.

Heat Treatment

The heat treatment of cream plays a decisive role in the butter-manufacturing process and the eventual quality of the butter. It is important that milk and cream be handled in the gentlest possible way to avoid mechanical damage to the fat, a serious problem in continuous manufacture (Fritz process) of butter (36). Cream is pasteurized or heat treated for the following reasons: to destroy pathogenic microorganisms and reduce the number of bacteria; to deactivate enzymes, to liquefy the fat for subsequent control of crystallization, and to provide partial elimination of undesirable volatile flavors.

Batch Butter Manufacture

Today batch processing is not used to any extent for the production of large quantities of butter. Batch systems are still encountered in small butter plants, primarily in less industrially developed countries. Continuous systems are more efficient and cost effective for large outputs; batch systems have low capital intensity.

The processing of cream by a batch churn requires filling to approximately 30 to 50% capacity at a cream temperature of 4.4 to 12.8°C. Cream temperature will vary depending on seasonably of the cream, the butter characteristics desired, and the desired rate of fat inversion. Churning is accomplished by rotation of the churn at approximately 35 rpm until small butter granules appear. The process usually requires about 45 min for coalescence of the fat globules and clean separation of the buttermilk so that it can be drained (37). The granules may be washed with cold water to remove surface buttermilk. Salt is then added with water if the fat content standardization is required. The butter is worked to ensure uniformity and desirable body and texture characteristics.

Continuous Butter Manufacture

Between 1930 and 1960 a number of continuous processes were developed. In the Alfa, Alfa-Laval, New Way, and Meleshin processes phase inversion takes place by cooling and mechanical treatment of the concentrated cream. In the Cherry-Burrel Gold'n Flow and Creamery Package processes, phase inversion takes place during or immediately after concentration, producing a liquid identical to melted butter, prior to cooling and working. The Alfa, Alfa-Laval, and New Way processes were unsuccessful commercially. The Meleshin process, however, was adopted successfully in the Republic of Russia. The Cherry-Burrel Gold'n Flow process appears to have been the more successful of the two American processes (29).

The Fritz continuous butter-making process, which is based on the same principles as traditional batch churning, is now the predominant process for butter manufacture in most butter-producing countries. In the churning process crystallization of milk fat is carried out in the cream, with phase inversion and milk fat concentration taking place during the churning and draining steps. However, because of the discovery that cream could be concentrated to a fat content equal to or greater than that of butter, methods have been sought for converting the concentrated or plastic cream directly into butter. Such methods would carry out the principal butter-making steps essentially in reverse order, with concentration of cream in a centrifugal separator, followed by a phase inversion, cooling, and crystallizing of the milk fat (37).

Increased demands on the keeping qualities of butter require that, in addition to careful construction, operation, and cleaning of the milk and cream processing equipment, research to develop machines that will ensure butter production and packing under conditions eliminating contamination and air admixture must be carried out. It has been demonstrated that butter produced under closed conditions has a better keeping quality than butter produced in open systems (38).

There are two classes of continuous processes in use: one using 40% cream, such as the Fritz process, and the other using 80% cream, such as the Cherry-Burrell Gold'n Flow. As much as 85% of the butter in France is made by the Fritz process. In this process 40% fat cream is churned as it passes through a cylindrical beater in a matter of seconds. The butter granules are fed through an auger where the buttermilk is drained and the product is squeeze dried to a low moisture content. It then passes through a second working stage where brine and water are injected to standardize the moisture and salt contents. As a result of the efficient draining of the buttermilk, this process is suitable for the addition of lactic acid bacteria cultures at this point. The process then becomes known as the NIZO method when the lactic starter is injected (Land O'Lakes, unpublished data, 1985). Advantages of the NIZO method over traditional culturing are improved flavor development; acid values as a result of lower pH; more flexible temperature treatment of the cream, because culturing and tempering often are accomplished concurrently; and, most important, sweet cream buttermilk is produced.

The latest in developments in butter making have been outlined in the *European Dairy Magazine* (39), with the article centering on dosing systems, and for the addition of salt and culture. In addition, a number of patents have been issued for improved butter-making processes (40), making butter directly from milk (41), and for flexible integration of added ingredients, and molding and demolding (42).

The Cherry-Burrell Gold'n Flow process is similar to margarine manufacture (43). The process starts with 18.3°C cream that is pumped through a high-speed destabilizing unit and then to a cream separator from which a 90% fat plastic cream is discharged. It is then vacuum pasteurized and held in agitated tanks to which color, flavor, salt, and milk are added. Then this 80% fat–water emulsion, which is maintained at 48.9°C, is cooled by use of scraped surface heat exchangers to 4.4°C. It then passes through a crystallizing tube, followed by a perforated plate that works the butter. Prior to chilling, 5% nitrogen gas is injected into the emulsion.

Although the Meleshin process continues to be in widespread use in the former USSR, the use of alternative continuous butter-making processes based on high-fat cream has declined in Western countries during the past 20 years (29). The principal reasons for this decline appear to be the economics and butter quality, particularly when compared with the Fritz process. A Fritz manufacturing process can be installed in existing batch churn factories with almost no modification to cream-handling or butter-packing equipment. The churns could be retained in case the Fritz breaks down. However, very little batch plant equipment could be reused in the alternative systems (ie, Gold'n Flow). When a completely new plant is being bought, the alternative systems still tend to be more expensive, and operational advantages over the Fritz are not significant. Butter from the Fritz process is nearly identical in its physical and flavor characteristics to batch-churned butter, whereas butter produced by the alternative processes tends to be different. These differences may be perceived as defects by the consumer, and manufacturers have been reluctant to alter a traditional product.

The alternative systems have a number of advantages compared with a modern Fritz line: (*1*) The most attractive advantage is the flexibility to produce a wide range of products with fat contents ranging from 30 to 95% butter–vegetable oil blends and the incorporation of fractionated fats; (*2*) these processes also present the possibility of a number of operational advantages by use of an efficient centrifuge during the cream concentration stage; fat losses in the buttermilk can be substantially reduced; and (*3*) the composition of the butter can be more accurately controlled, either by including a batch standardization step or by the use of accurate continuous metering systems.

Reworking Butter

Since significant amounts of butter are made during high milk production periods of the year (spring) and highest consumption tends to be during the late fall, butter by necessity needs to be stored under conditions to ensure continuing quality. These conditions are usually in bulk frozen form. Thawing and the subsequent reworking have resulted in development of new and superior processes and equipment. The equipment is in a closed system with clean-in-place, and the process includes butter blocks at temperatures $> -5°C$ that are transported to a shredder, from which the butter chips pass via large-diameter screw conveyors to the silo; from here they are pumped through the reworker where optimal moisture distribution is achieved by two counterrotating kneading rollers, and reworked butter emerges at 6 to 14°C. Equipment for reworking of butter blocks that have a temperature of $> -25°C$ incorporates a butter tempering system after the shredder, and temperature of the emerging butter can be regulated (44).

BUTTER PRODUCTS

Cultured Butter Manufacture

There are several ways of making cultured butter from sweet cream. Pasilac-Danish Turnkey Dairies, Ltd. developed the IBC method (45). The main principles of the IBC method are as follows. After sweet cream churning and buttermilk drainage, a starter culture mixture is worked into the butter, which produces both the required lowering of butter pH and, because of the diacetyl content of the starter culture mixture, the required aroma. The starter mixture consists of two types of starter culture. (*1*) *Lactococcus lactis* and (*2*) *L. cremoris* and *L. lactis* ssp. *diacetylactis*. With respect to production costs, the experience with this method shows that for the manufacture of mildly cultured butter, the direct costs are only about one-third of the costs of other methods (45).

Butter-Vegetable Oil Blends

In the early 1980s blends of butter and vegetable oil products appeared in the U.S. market. These blends generally were 40% butter and 60% vegetable oil for a total fat content of 80%, within the margarine Standard of Identity and designation. With the increasing popularity of reduced-fat (less than 80%), spreads, starting in the mid-1980s, other blends with butter fat contents of 2 to 25% were introduced. As noted in Table 1, the full 80% fat blends have declined in preference to the lower-fat spreads; the blend spreads have taken a portion of this market share (Land O'Lakes, unpublished data, 1996).

A number of processes were developed using continuous churns (46) and alternative systems similar to the Cherry-Burrell Gold'n Flow process (47). The major disadvantage to churning, either batch or continuous, was that the resultant buttermilk would be adulterated with some vegetable fat and would be less valuable than standard buttermilk. An advantage of alternative processing systems is their ability to easily accommodate the manufacture of reduced-fat spread blends.

With the popularity of blends in the United States, several products were introduced in Europe and Australia and have subsequently expanded the market for butter fat usage (Land O'Lakes, unpublished data, 1996). An even earlier entry into the blend category was Swedish Bregott in Europe.

Whipped Butter

Whipped butter is a product produced to improve spreadability; air or nitrogen is incorporated. With the incorporation of gas, the butter's volume increases by approximately 33%. Generally, whipped butter is sold in tub rather than stick form.

Spreadable Butter

In 1937 a process known as the Alnarp method was developed to increase the spreadability of butter made from winter milk fat. The apparent reason for its effectiveness is that it increases the liquid fat content of the milk fat (Fig. 1) (48).

The consistency of butter is determined by the percentage of solid fat present, which is directly influenced by the fat composition, the thermal treatment given to cream prior to churning, the mechanical treatment given to butter after manufacture, and the temperature at which the

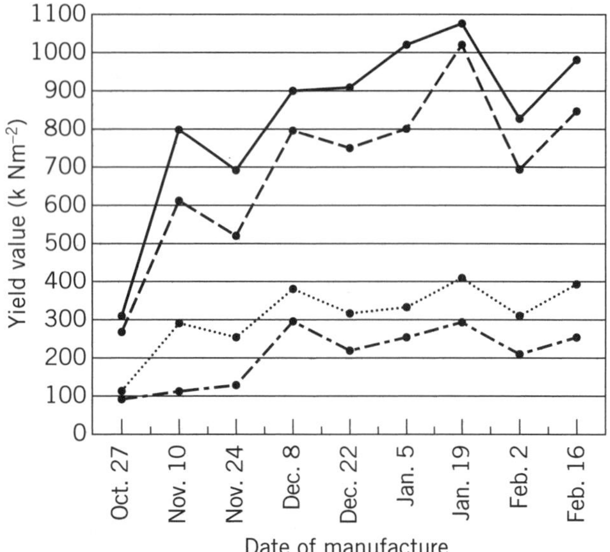

Figure 1. Effect of Alnarp-type treatment on firmness of winter butter. —●— Control 5°C; – –●– – Alnarp; --●-- control 15°C; – · ● · – Alnarp; ----- upper limit of spreadability. *Source:* Ref. 49.

butter is held (50). The European butter market demands that butter be softer and more spreadable in winter and harder in summer. With information on the changes in fat composition from gas–liquid chromatography analysis and the use of nuclear magnetic resonance to estimate solid fat, suitable tempering procedures can be selected to modify the fat composition and produce the most acceptable product for the consumer. A spreadable consistency of butter can be achieved by either varying the fatty acid composition or varying heat-step cream-ripening times and temperatures (51).

In efforts to improve the spreading properties of butter in relation to hard butter fat, one alternative put forward is the use of soft fat fractions obtained in the fractionation of anhydrous milk fat. Although several practical methods of fractionation have been presented, the use of soft fat fractions in butter making has not become general practice. This is evidently because fractionation in all cases significantly raises the cost of the butter produced. In addition, a common problem has been to find suitable uses for the hard fat fractions. Furthermore, in fractionation methods that use solvents or additives, fractionation should be linked to fat refining; and in this process butter also loses its natural food classification. In studies that have used soft butter fat fractions, a substantial softening of the butter has been obtained; however, this butter, like normal butter, hardens as the temperature increases and again decreases. One of the new ways to improve butter spreadability is to add vegetable oil during manufacturing. The best known of these preparations, which cannot be called butter, is the Swedish Bregott, in which 20% of the fat used is vegetable oil. A similar product is made in many other countries, Finland among them, where it has been well received, possibly just because of its better spreadability compared with normal butter.

Two industrial processes in practice for the fractionation of milk fat are the Tirtiaux system, and the DeSmet, which is a semicontinuous bulk crystallization process. They are dry fractionation processes enabling one- or two-step fractionation of butter oil at any temperature from 50 to 2°C. The milk fat fractions thus obtained can be used as such or they can be blended in different proportions for use as ingredients in various food fat formulations or in preparing spreadable butter (52).

Figure 2 depicts the solid fat content profiles of the control and anhydrous milk fat: the low melting, high melting, and 20% very high melting milk fat; the milk fat fractions used in the low melting, high melting, and 20% very high melting butter: the 30S fraction; the 13L fraction; and the 15S fraction. The fraction number includes the fractionation temperature (°C) and its physical form (solid or liquid) (53).

The major shortcoming inherent in this system is the long residence time (8–12 h) for nucleation and crystal growth.

Butter samples made from low melting liquid fractions and from a combination of primarily low melting liquid fractions and a small amount of high melting solid fractions exhibited good spreadability at refrigerator temperature (4°C) but were almost melted at room temperature (21°C). Butters made with a high proportion of low melting liquid fraction, a small proportion of high melting solid, and a small proportion of very high melting solid fractions were spreadable at refrigerator temperature and maintained their physical form at room temperature (53).

Some newer modifications and improvements to processes for the enhancement of spreadability have recently been patented (54,55). They include the use of a hydropho-

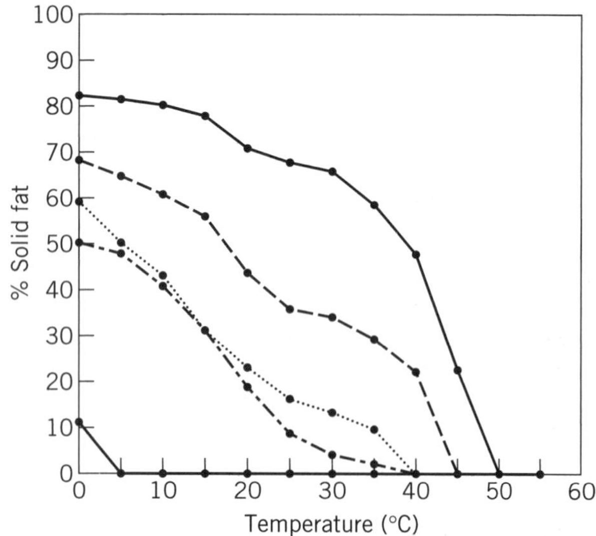

Figure 2. The fraction number includes the fractionation temperature (°C) and its physical form (solid or liquid). —— 15S Fraction (top line); ---- and the milk fat fractions used in the low melting, high melting, and 20% very high melting butter: 30S fraction; ···· the low melting, high melting, and 20% very high melting milk fat; ----- solid fat content profiles of the control and anhydrous milk fat; —— 13L fraction (bottom line).

bic membrane to separate high and low melting point fractions and improved blending techniques.

Reduced-Fat Butter

The first reduced-fat butter (50% fat), called Light Butter, was introduced in the United States by the Lipton Co. in the mid 1980s. The product was withdrawn due to FDA objections of not meeting Standards of Identity for nomenclature. Also, the product contained stabilizers not allowed in the Standard of Identity for butter. In the late 1980s, Ault, Inc. introduced a reduced-fat butter (39% fat) called Pure and Simple, which contained no unusual additives (56). Unfortunately, this all-natural product had severe negatives: it had a short shelf life, experienced moisture seepage, and lacked the highly desirable butter notes. In 1990, Land O'Lakes, Inc. launched its Light Butter (52% fat), which contained emulsifiers, added vitamin A, and preservatives. The FDA was in the process of establishing standards for reduced-fat products at this time and no objection was registered. The new standards were established in 1993 (57), which automatically required Land O'Lakes to reformulate to a 40% butterfat content; it did so and relaunched. The product was a success and has established dominance in the U.S. market.

Manufacturers have experienced many problems with the production of low-fat butter (58). Low-fat butter cannot be manufactured in conventional continuous butter makers. The technology of producing low-fat butter and margarine products is similar to that of ordinary margarine production, and it has nothing in common with modern butter production. These low-calorie water-in-fat emulsions have such a dense package of water droplets that unwanted phase inversion during processing and/or structural weak points in the product can occur, which may, for example, severely limit the microbiological shelf life. The scraped-surface heat exchanger type of machine is preferred for production of low-fat products.

Butterlike products with reduced-fat content are manufactured in several countries. Stabilizers, milk and soy proteins, sodium albumin or caseinate, fatty acids, and other additives are used. A product is now available on a commercial scale in Russia that has the following composition: 45% milk fat, 10% nonfat solids, and 45% moisture. It has a shelf life of 10 days at 5°C (59).

Decreasing the energy content of a diet has clearly been the motive behind those milk fat products in which the fat content is approximately half that of normal butter. Although these products can no longer be called butter according to international standards, they are nonetheless often called low-calorie butter, half-butter, or similar names. A large number of patents have been obtained for these products, because raising the water content to nearly 50% in the manufacture of butterlike spreads requires considerably higher emulsifying properties than the manufacture of normal butter. In addition, the emulsion must often be stabilized with additives. In some countries, a low-fat butter (40% total fat) containing vegetable oil has been designated as Minarine, but Minarine can also be prepared using only butter fat (6).

Anhydrous Milk Fat Manufacture

The quality assurance program for manufacture of butter oil, or anhydrous milk fat (AMF), also focuses on the quality of the raw materials. Naturally, many of the same considerations apply to handling raw cream for AMF manufacture that apply to butter, except that vacreation is not used. Because it is stored under ambient conditions, care against oxidation is essential. Oxidation is perhaps the most important mechanism by which milk fat deteriorates in quality. Because the oxidation reaction is autocatalytic (ie, the products of the reaction act as catalysts to promote further reaction), the normal quality control tests, peroxide value and free fat acidity, could give misleading results when applied to stored butter. Methods of deaeration have been developed that could reduce potential oxidation (24).

Milk fat is present in milk or cream as part of a stable oil-in-water emulsion. The emulsion is stabilized as a result of the protein and phospholipid-rich milk fat globule membrane (MFGM) surrounding the milk fat. During AMF manufacture the aim is to break the emulsion and to separate out all of the nonfat solids and water. To achieve this, the MFGM must first be disrupted mechanically or chemically. Homogenization, an example of mechanical treatment, disrupts the membrane, destroying the membrane layer. For chemical destruction of the membrane layer, an acid such as citric acid can be added to lower the pH of milk or cream to about 4.5 (60). The protein will precipitate, removing a component to maintain an intact fat globule. Direct-from-cream AMF plants usually has three separators. The first concentrates cream from 40% to about 75% fat before phase inversion in a homogenizing device. The oil separator then separates the liberated butter oil to about 99% purity. The oil is washed with water before the third (polishing) separator and the final traces of moisture are removed in a dehydrator at 95°C and under a vacuum of 35 to 50 torr. The dehydrator is usually a simple vessel, and the butter oil is introduced either as a thin film on to the walls or as a spray, to maximize the surface area exposed to the vacuum. Such a device will not remove significant off-flavors, because the vapor flows, temperatures, and pressures are inappropriate for flavor stripping. When producing AMF from butter, fresh or block butter is softened just to a pumpable stage (approximately 50°C) and transferred to a plate heat exchanger to increase the temperature (70–80°C). The oil phase is concentrated through separators and dried under vacuum. Some washing is possible before the final separator removes the last traces of nonfat components (60).

In terms of the preparation of products and their appearance and texture, AMF has several advantages over traditional butter. The latter is in fact subject to seasonal variations, which affect its physical properties. The advantages of AMF are linked to the possibility of standardizing its physical properties (by the selection and mixture of the raw materials used in production) and the possibility of adapting its properties using the fractionation technique.

Ghee

By definition, ghee is a product obtained exclusively from milk and/or fat-enriched milk products of various animal species by means of processes that result in the near total removal of water and nonfat solids (similar to anhydrous milk fat) and in the development of a characteristic flavor and texture. Even so, most ghee contains some nonfat solids to enhance the flavor.

Typically, ghee is manufactured by heating butter to temperatures well above those used during AMF manufacture. The high temperature treatment of the nonfat milk solids and milk fat leads to the development of a strong buttery flavor. However, traditional ghee, as produced in the Middle East and Asia, has a more rancid taste due to less-sophisticated methods of preparation and storage. Manufacturers in the European community are also producing ghee by adding ethyl butyrate to AMF (60,61). Alternate synthetic flavors have been developed to add ghee flavor notes to butter oil.

Technologies continue to advance and processes have been developed to create or enhance AMF flavor. For instance, the continuous production of lipolyzed butter oil or AMF has been developed utilizing pregastric esterases immobilized in a hollow fiber reactor (62).

Butter Flavors

Technologies for the hydrolysis of butterfat to produce and concentrate the free fatty acids to enhance the butter flavor of products have been available for decades. More recently, biotechnologists have developed methods for producing a variety of fairly pure enzymes, economically and in large quantities. The increased availability of lipases (glycerol ester hydrolases) from microbial sources has made it possible for researchers to employ the catalytic properties of these enzymes in innovative ways. One application in which the use of lipases has become well established is the production of lipolyzed flavor from feedstocks of natural origin. Immobilization of lipases on hydrophobic supports has the potential to (*1*) preserve, and in some cases enhance, the activity of lipases over their free counterparts; (*2*) increase their thermal stability; (*3*) avoid contamination of the lipase-modified product with residual activity; (*4*) increase system productivity per unit of lipase employed; and (*5*) permit the development of continuous butter flavor.

Decholesterolized Butter

Fractionation by crystallization, supercritical fluid extraction, and other technology are methods being applied commercially to milk fat to create desirable new products, such as decholesterolized butter. The essential purpose of milk fat application development is to adapt products to fit user demands.

In the 1980s, there was significant research and market activity in developing decholesterolized milk fat. All this activity was for naught, for the hypothesis of creating a "healthier" fat (for butter or milk or other dairy product) was not sound. The nutrition community had long recognized that the link between dietary cholesterol and serum cholesterol was weak and that the ratio of total fat–saturated fat had a greater impact on health. In addition, the FDA issued new standards in 1993 (57) that effectively negated the value of decholestering milk fat. The new law required that to be called low cholesterol, the fat must contain no more than 2 g of saturation fat per serving. Butterfat is approximately 65% saturated. Since the technology to desaturate milk fat is not cost-effective, decholesterization has no economic value.

Desaturated Milk Fat

In addition to chemical and enzymatic means of desaturation, there have been extensive studies on feeding cows specific diets to change butterfat saturation as well as increasing the ratio of potentially desirable fatty acids (19,63,64). In general, good progress in the understanding of rumen physiology, digestion, and function has occurred, but economic potential remains unacceptable. The most promising technologies are the use of protected fats in a feeding regimen. These fats are protected in a way that they pass through the rumen (point of fat hyrogenation) into the remaining digestive system for absorption and subsequently into the mammary glands. Unsaturated fats that are fed to cows have a great opportunity to remain unsaturated as they are synthesized into milk fat. Biotechnology may offer alternatives in the modification of edible fats and milk fat. Research has led to new methods of lipolysis and esterification, but the developments are still at the laboratory level. Nevertheless, commercial application may emerge from these interesting areas of research.

Part of the reason for consumer popularity of the butter–vegetable oil mixtures may be seen in the emphasis on saturated animal fats in current nutritional debates as well as the alleged cause-and-effect relation between butter and heart disease. For these reasons, people interested in good nutrition willingly change over to products that contain a certain amount of vegetable oil but also have the natural butter aroma.

The rise in concern for fat and cholesterol in the U.S. market overshadowed the concern for chemicals and preservatives in the 1980s. By 1990, the concern for cholesterol started to significantly decline, whereas concern for fat, particularly saturated fat, remained high (Fig. 3) (65). The U.S. Nutritional Labeling and Education Act of 1990 has provided the consumer with product label information that has influenced dietary decisions and has educated the consumer, particularly for fat and saturated fat contents.

Because the physical properties of milk fat influence the rheological properties of dairy products, especially butter,

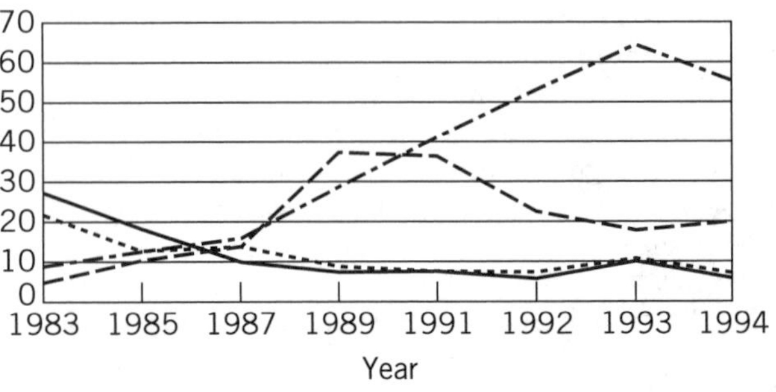

Figure 3. Rise in concern over fats, cholesterols. – – Fat content; —— cholesterol; —— chemicals/additives; ---- preservatives.

there has been considerable interest in the modification of milk fat by physical and chemical means. Economical fractionation of milk fat into oil and plastic fat fractions will facilitate an increased utilization of milk fat in many food applications, such as chocolate, confectionery and bakery products, and in developing new convenient (spreadable) and dietetic (reduced cholesterol, variable fatty acid composition) butter or butter fat–containing products.

BIBLIOGRAPHY

1. O. F. Hunziker, *The Butter Industry*, 3rd ed., Printing Products Corp., Chicago, Ill., 1940.
2. "The World Market for Butter," *International Dairy Federation Bulletin*, Document **170**, 1984.
3. *Statistical Abstract of the United States*, 117th ed., U.S. Department of Commerce, Bureau of the Census, Washington, D.C., 1997.
4. R. Rizek, N. Raper, and K. Tippett, "Trends in U.S. Fat and Oil Consumption," *J. Am. Oil Chem. Soc.* **65**(5), 723 (1988).
5. *U.S. Code of Federal Regulations*, Title 7, Part. 58, Subpart P, U.S. Government Printing Office, Washington, D.C., 1983.
6. M. M. Chrysam, "Table Spreads and Shortenings," in T. H. Applewhite, ed., *Bailey's Industrial Oil and Fat Products*, Vol. 3, John Wiley & Sons, New York, 1985, pp. 65–106.
7. W. R. Morrison, "Milk Lipids," in F. D. Gunstone, ed., *Topics in Liquid Chemistry*, Logos Press, London, United Kingdom, 1970, pp. 51–106.
8. R. G. Jensen, "Symposium: Milk Lipids," *J. Am. Oil Chem. Soc.* **50**, 186–192 (1973).
9. J. R. Brunner, "Milk Lipoproteins," in E. Tria and A. M. Scann, eds., *Structural and Functional Aspects of Lipoproteins in Living Systems*, Academic Press, London, United Kingdom, 1969, pp. 545–578.
10. U. Bracco, J. Hidalgo, and H. Bohren, "Lipid Consumption of the Fat Globule Membrane of Human and Bovine Milk," *J. Dairy Sci.* **55**, 165 (1972).
11. V. Antila, "The Consistency of Butter and the Possibilities to Manipulate It," *Proceedings of the XXI International Dairy Congress, Moscow* **2**, 159 (1982).
12. J. Kiska et al., "Butter Texture and Consistency as a Function of Milk Fat Composition and Method Manufacture," *Brief Communication* **1**(1), 352 (1982).
13. P. W. Parodi, "Relationship Between Triglyceride Structure and Softening Point of Milk Fat," *J. Dairy Res.* **48**, 131 (1981).
14. T. W. Scott, L. J. Cook, and S. C. Mills, "Protection of Dietary Polyunsaturated Fatty Acids Against Microbial Hydrogenation in Ruminants," *J. Am. Oil Chem. Soc.* **48**, 358 (1972).
15. L. F. Edmondson et al., "Feeding Encapsulated Oils to Increase the Polyunsaturation in Milk and Meat Fat," *J. Am. Oil Chem. Soc.* **51**, 72 (1974).
16. R. D. Plowman et al., "Milk Fat with Increase Polyunsaturated Fatty Acids," *J. Dairy Sci.* **55**, 204 (1972).
17. C. R. Brewington, E. A. Caress, and D. P. Schwartz, "Isolation and Identification of New Constituents in Milk Fat," *J. Lipid Res.* **11**, 355 (1970).
18. A. Boudreau and J. Aurl, "Utilizations of Milkfat: Physical and Chemical Modification of Milkfat," *International Dairy Federation Bulletin* **260**, 7–10, 13–16 (1991).
19. W. Banks, "Utilizations of Milkfat: Milk Lipids," *International Dairy Federation Bulletin* **260**, 3–6, 13, 22 (1991).
20. A. Albertini et al., "Applications of the Hazard Analysis Critical Control Point System to Butter Production," *Mondo del Lattle* **50**(2), 95 (1996).
21. B. D. Dixon, "Continuous Butter Manufacture: Raw Materials," *International Dairy Federation Bulletin* **204**, 3–5 (1986).
22. B. Schaffer and S. Szakaly, "Heat-Step Cream Ripening for Producing Spreadable Butter in Hungary," *Brief Communication* **1**(1), 337 (1982).
23. R. Norris, "Utilizations of Milkfat: Uses of Milkfat," *International Dairy Federation Bulletin* **260**(5), 23–25 (1991).
24. German Patent DF 36 23 313 Al (1988), M. Schroder (to Schroeder).
25. G. Van den Berg, "Developments in Buttermaking," *Proceedings of the XXI International Dairy Congress, Moscow* **2**, 153 (1982).
26. T. Siek and J. Lindsay, "Semiquantitative Analysis of Fresh Sweet-Cream Butter Volatiles," *J. Dairy Sci.* **53**, 700 (1970).
27. J. E. Kinsella, "Butter Flavor," *Food Technol.* **29**(5), 82–98 (1975).
28. D. H. Hettinga, *Altering Fat Composition of Dairy Products*, CAST Publication, Ames, Iowa, 1991.
29. M. Kimenai, "Continuous Butter Manufacture," *International Dairy Federation Bulletin* **204**, 16 (1986).
30. L. H. Wiedermann, "Margarine and Margarine Oil, Formulation and Control," *J. Am. Oil Chem. Soc.* **55**, 823 (1978).
31. D. H. Hettinga, "Processing Technologies for Improving the Nutritional Value of Dairy Products," in Board of Agriculture, ed., *Designing Foods*, National Academy Press, Washington, D.C., 1988, p. 292.
32. R&D Applications, "Cholesterol Reduced Fat: They're Here," Prepared Foods, Chicago, Ill., 1989.
33. "Utilizations of Milk Fat," IDF Report B-Doc **164**, in *International Dairy Federation Annual Sessions*, Copenhagen, Denmark, 1989.
34. B. J. Spalding, "Combating Cholesterol via Enzyme Research," *Chemish Weekblad*, June 1, 1988, p. 42.
35. S. Harlander, "Biotechnology: Applications in the Dairy Industry," *J. Am. Oil Chem. Soc.* **65**, 1727 (1988).
36. M. Schweizer, "Continuous Butter Manufacture: Heat Treatment," *International Dairy Federation Bulletin* **204**, 10 (1986).
37. F. H. McDowall, *The Butter Maker's Manual*, New Zealand University Press, Wellington, 1953.
38. P. Friis, "Advantages of Butter Production in Closed Systems," *Brief Communication* **1**(1), 326 (1982).
39. "New Generation of Butter Making Equipment," *European Dairy Magazine* **4**, 39 (1996).
40. PCT International Patent Application WO95/27392 A1 (1995), W. Vennewald (to Westfalia Separator).
41. PCT International Patent Application WO94/26123 A1 (1994), N. Shtukarin (to U.L. Karbysheva).
42. French Patent Application FR 2 743 311 A1 (1997), J. Peneveyre and P. Y. A. R. Guyard (to Peneveyre and Guyard).
43. *Gold'n Flow Continuous Buttermaking Method*, Bulletin **G-493**, Cherry-Burrell Corp., Chicago, Ill., 1954.
44. R. Egli, "Butter Reworking," *Food Tech Europe* **2**(3), 110 (1995).
45. P. Bjerre, in *Milk Fat*, Society of Dairy Technology, Huntingdon, U.K., 1991, pp. 63–73.
46. U.S. Pat. 4,425,370 (Jan. 10, 1984), F. Graves (to Land O'Lakes, Inc.).
47. U.S. Pat 4,447,463 (May 8, 1984), D. Antenore, D. Schmadeke, and R. Stewart (to Land O'Lakes, Inc.).

48. B. Schaffer and S. Szakaly, "The Effect of Temperature Treatment of Cream on the Liquid Fat Ratio," *Brief Communication* **1**(1), 363 (1982).

49. A. M. Fearon and D. E. Johnston, "Improving Butter Spreadability—the Scientific Approach," *Dairy Industry International* **53**(10), 25–29 (1988).

50. N. Cullinane and D. Eason, "Variation in Percentage Solid Fat of Irish Butter," *Brief Communication* **1**(1), 349 (1982).

51. B. K. Mortensen and K. Jansen, "Lipase Activity in Cream and Butter," *Brief Communication* **1**(1), 334 (1982).

52. U.S. Pat. 4,839,190 (June 13, 1989), J. Bumbalough (to Wisconsin Milk Marketing Board, Inc.).

53. K. E. Kaylegian and R. C. Lindsey, "Performance of Selected Milk Fat Fractions in Cold-Spreadable Butter," *J. Dairy Sci.* **75**, 3307–3317 (1992).

54. PCT International Patent Application WO 97/10719 A1 (1997), K. Glugovsky (to Glugovsky and Lassnig Vertriebsges).

55. French Patent Application FR 2 713 656 A1 (1995), M. Parmentier, S. Bomaz, and B. Journet (to Union Beurriere SA).

56. I. Macnab, "Ault Achieves Technological Breakthrough with World's First Pure Light Butter," *Modern Dairy* **68**(3), 23 (1989).

57. R. Wood et al., "Effect of Butter Mono and Polyunsaturated-Enriched Butter Trans Fatty Acid Margarine and Zero Trans Fatty Acid Margarine on Serum Lipids and Lipoproteins in Healthy Men," *J. Lipid Res.* **34**, 1–11 (1993).

58. O. Gerstenberg, "Pitfalls in Low Butter Production," *Dairy Industry International* **53**(11), 28–29 (1988).

59. S. Gulyayev-Zaitsev et al., "Low Calorie Butter," *Brief Communication* **1**, 328 (1982).

60. K. K. Rajah and K. J. Burgess, in *Milk Fat*, Society of Dairy Technology, Huntingdon, U.K., 1991, pp. 37–43.

61. B. K. Wadhwa and M. K. Jain, "Chemistry of Ghee Flavour—a Review," *Indian J. Dairy Sci.* **44**, 372–374 (1992).

62. H. S. Garcia and C. G. Hill, Jr., "Improving the Continuous Production of Lipolyzed Butter Oil With Pregastric Esterases Immobilized in a Hollow Fiber Reactor," *Biotechnol. Tech.* **9**, 467 (1995).

63. W. Banks and W. W. Christie, "Feeding Cows for the Production of Butter With Good Spreadability at Refrigeration Temperatures," *Outlook Agriculture* **19**, 43–47 (1990).

64. R. P. Middaugh et al., "Characteristics of Milk and Butter from Cows Fed Sunflower Seeds," *J. Dairy Sci.* **71**, 3179–3187 (1988).

65. *Trends in the U.S. Consumer Attitudes and the Supermarket*, Food Marketing Institute, Washington, D.C., 1997.

David Hettinga
Land O' Lakes
Arden Hills, Minnesota

C

CACAO BEAN.

See CHOCOLATE AND COCOA.

CAFFEINE

Caffeine (1,3,7-trimethylxanthine) was isolated from coffee in 1820 (1). Thein was isolated from tea in 1827 and later shown to be identical with the coffee isolate. It was subsequently identified in cocoa, maté, kola nuts, and other plants; structure was determined in 1875. Caffeine is one of several methylxanthines that occur naturally, primarily in plant matter that is used to prepare beverages. Other methylxanthines that usually accompany it include theobromine and theophylline. The methylxanthines, as dioxypurines, are related to two nucleic acids: adenine and guanine. This is relevant to their biosynthesis and physiological effects. Their relationship to uric acids accounts for their metabolic fate. The structures of caffeine and related compounds are shown in Figure 1.

Total synthesis was first achieved in 1895 starting with dimethylurea. Dimethyluric acid, chlorotheophylline, and chlorocaffeine were intermediates (2).

PROPERTIES

Physical Properties

Caffeine is a white, odorless powder with a slightly bitter taste. The anhydrous form obtained by crystallization from nonaqueous solvents is a crystalline solid that melts at 235 to 237°C. At atmospheric pressure it begins to sublime without decomposition at 120°C and at 80°C under high vacuum. Its crystals are hexagonal and form parallel plates. When crystallized from water, the monohydrate forms long, silky, white needles. It becomes anhydrous at 80°C. Caffeine is soluble to the extent of 0.6% in water at 0°C, 2.13% at 25°C and 66.7% at 100°C. The pH of dilute solutions is 6.9. In aqueous solution, caffeine forms dimers and higher polymers by base stacking (3). It is fairly stable in dilute bases and acids, forming salts with the latter. Caffeine is considerably more soluble in chlorinated solvents: 8.67% in methylene chloride and 12.20% in chloroform at 25°C. Its ultraviolet absorption spectrum shows a maximum at 274 nm with no variation over the pH range 2 to 14.

Chemical Properties

Complexation occurs with many acid radicals, notably chlorogenate, salicylate, citrate, benzoate, and cinnamate. Chlorogenic acid forms a tightly bound 1:1 complex. It is soluble in chloroform and can be recrystallized from aqueous alcohol, attesting to its stability. In plant matter much of the caffeine exists in this form (4). Complexes with phenolic molecules are only stable below pH 6. Caffeine is a weak base; its acid salts hydrolyze readily. It is converted

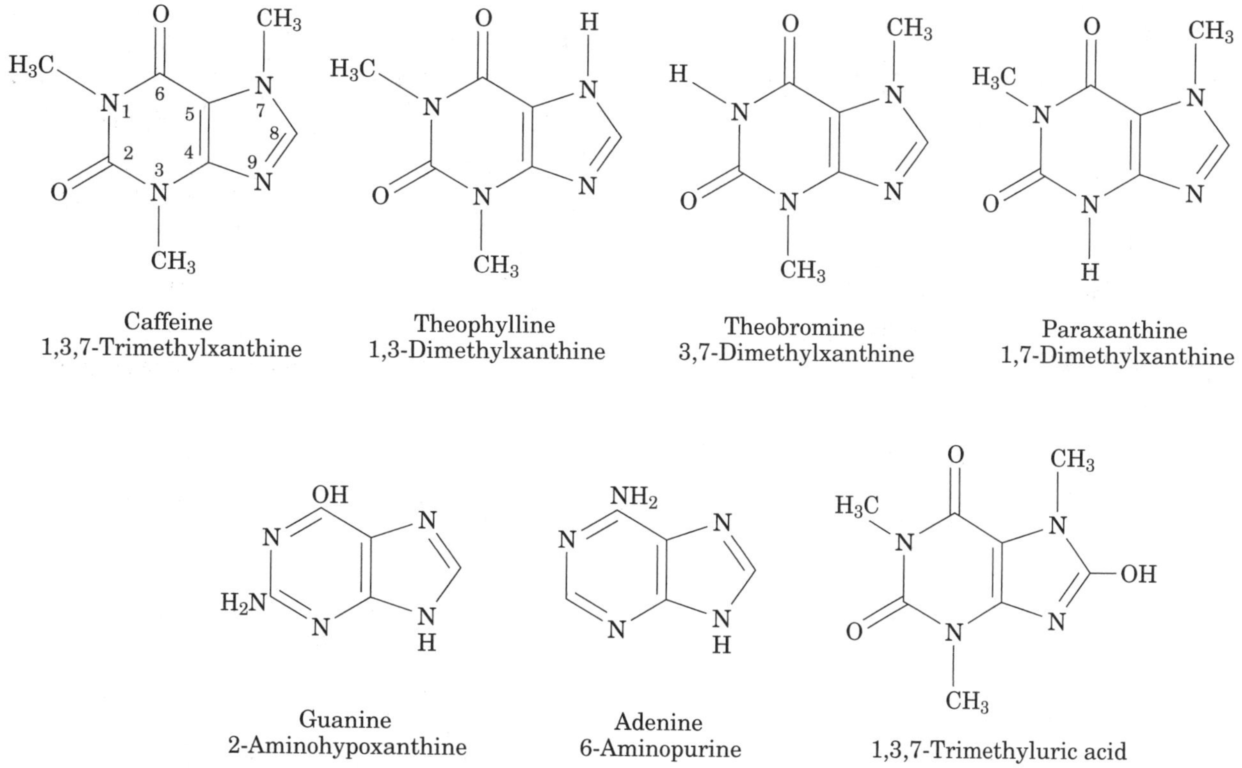

Caffeine
1,3,7-Trimethylxanthine

Theophylline
1,3-Dimethylxanthine

Theobromine
3,7-Dimethylxanthine

Paraxanthine
1,7-Dimethylxanthine

Guanine
2-Aminohypoxanthine

Adenine
6-Aminopurine

1,3,7-Trimethyluric acid

Figure 1. Methylxanthines and related compounds.

to caffeidine by treatment with concentrated NaOH (5). Alloxans are formed by oxidation with chlorine and other oxidants (6). Reaction products are shown in Figure 2.

ANALYSIS

Concern with the physiological effects of caffeine has stimulated research on its analysis in food and biological systems. High-performance liquid chromatographic (hplc) analysis of caffeine in food products requires only the filtration of aqueous extracts and injection onto a suitable column. Detection is frequently based on the measurement of μv absorption at 280 nm (7). Determination of caffeine at the low concentrations present in body fluids is accomplished by direct injection onto reversed-phase hplc columns (8). Gas chromatographic determination of caffeine in beverages using a flame ionization detector has also been described (9). Capillary electrophoresis has also been used for caffeine determination (10).

OCCURRENCE

Caffeine is present in various tissues of about 60 plant species, several of which are used to prepare beverages (11). The most important are the following:

- Coffee beans (*Coffea arabica* and *Coffea canephora* var. *robusta*)
- Tea leaves (*Camellia sinensis* vars. *assamica* and *sinensis*)
- Cocoa beans (*Theobroma cacao*)
- Maté stems and leaves (*Ilex paraguariensis*)
- Guarana seeds (*Paullinia cupana*)
- Kola nuts (*Cola acuminata* and *Cola verticillata*)

BIOGENESIS

The biogenesis of caffeine is probably similar in all plants, but with differences in the rates of various steps. It has its origin in the cell purine pool. Precursors include 7-methylxanthosine, 7-methylxanthine, and theobromine. Synthesis of caffeine from adenine occurs readily in young tea leaf. 7-Methylguanilic acid and 7-methylguanasine are also on the biosynthetic path. A 7-methyl-*N*-nucleoside hydrolase mediates the removal of ribose from 7-methylxanthosine. *S*-adenosylmethionine is the methyl donor for the methylation of the xanthines in reactions catalyzed by *N*-methyltransferase (12,13). A possible reaction path is shown in Figure 3. Caffeine biosyntheses in the coffee plant (12–14) and in the tea plant (14–16) have been studied in detail.

CAFFEINE IN THE HUMAN DIET

Natural Products

A high proportion of the human population has consumed caffeine-containing beverages for many centuries. Authenticated usage of tea in China dates from A.D. 350. The beverage reached Europe in 1600. Coffee came into use as a beverage in Arabia around A.D. 1000 and was brought to Europe in the seventeenth century. Cocoa was consumed by the Aztecs in Mexico. In many parts of South America maté has been a source of substantial caffeine intake. Caffeine content of several contemporary food products is shown in Table 1.

Variations in caffeine content in specific plant species result from varietal diversity, climatic changes in the growing areas, and horticultural techniques. In tea, youngest leaf has the highest concentration (17). Processing conditions also affect caffeine content. High coffee roasting temperatures result in caffeine loss by sublimation (18). There is a higher level of caffeine in tea than in coffee beans, but 200 cups of tea beverage are obtained per pound of tea leaves, whereas only about 40 to 60 cups of coffee are usually prepared per pound of coffee beans.

By far, the greatest sources of dietary caffeine are coffee and tea, but cola beverages provide an increasing portion of the intake. The total per capita intake of caffeine by adults in the United States is about 3 mg/kg of body weight with about 2 mg/kg coming from coffee and most of the rest from tea. For caffeine users, average intake is approximately 4 mg/kg (19). The consumption by heavy users is of interest because of concern with possible health effects. Mean daily intake (mg/kg body weight) in the United States by consumers in the 90th to 100th percentiles of consumption varies with age level as follows:

1–5 yr	4.7	12–17 yr	2.9
6–11 yr	3.2	18 yr+	7.0

Tea and soft drinks are the major caffeine sources for users under the age of 18. Coffee becomes the major source for those over that age (20).

Caffeine as a Food Additive

In addition to its natural occurrence in plants, caffeine is added to the very widely consumed cola drinks and other

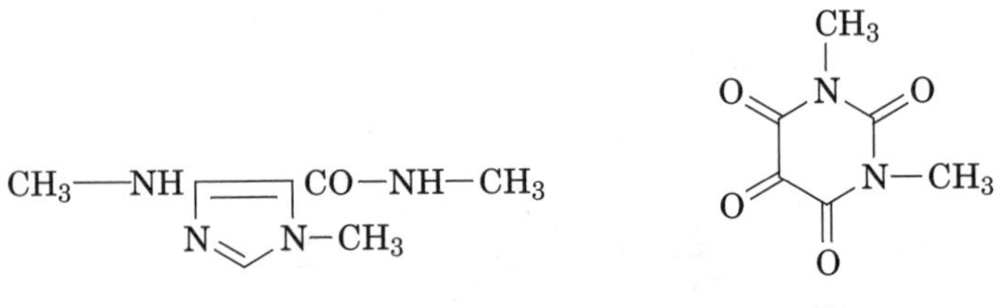

(a) (b)

Figure 2. Reaction products of caffeine: (**a**) caffeidine and (**b**) dimethylalloxan.

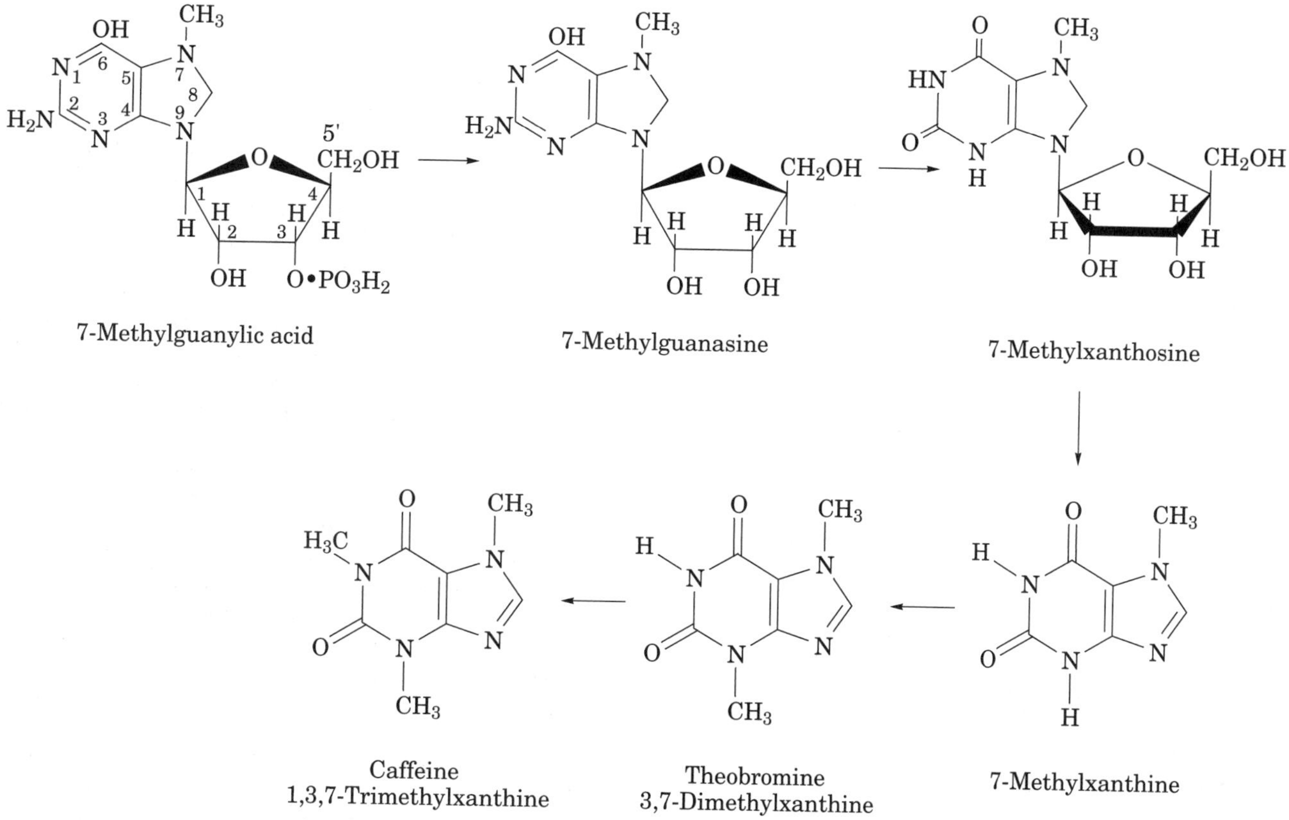

7-Methylguanylic acid

7-Methylguanasine

7-Methylxanthosine

Caffeine
1,3,7-Trimethylxanthine

Theobromine
3,7-Dimethylxanthine

7-Methylxanthine

Figure 3. Biosynthesis of caffeine.

Table 1. Caffeine in Food Products

Food product	Caffeine content (%) Range	Mean	Caffeine per serving[a] (mg) Range	Mean
Coffee, arabica	0.58–1.7	1.2	35–80	60
Coffee, robusta	1.2–3.3	2.2	60–140	120
Coffee, instant	2.2–5.0	3.3	45–105	65
Coffee, decaf.	0.03–0.09	0.06	2–6	3
Tea, black	1.2–4.6	3.0	20–48	35
Tea, green	1.0–2.4	1.9	17–40	24
Tea, instant	4.0–5.0	4.8	20–30	22
Cocoa, powder	0.08–0.35	0.2	1–8	4
Chocolate milk	0.01–0.025	0.02	2–7	5
Milk chocolate	0.005–0.04	0.02	2–14	6
Dark chocolate	0.02–0.10	0.08	6–28	23
Maté	0.9–2.2	2.0	36–42	40
Guarana paste	3.9–5.8	4.7	—	—
Kola nut	0.8–2.2	1.5	—	—
Cola drinks	0.08–0.16	0.09	20–30	28

[a]Serving size is assumed to be 6 oz for hot beverages, 8 oz for cold beverages, and 1 oz for chocolate products.

soda products. This usage is based on its stimulatory properties and the slight degree of bitterness that it imparts. In 1978 the U.S. Select Committee on GRAS Substances reaffirmed the safety of caffeine as a beverage additive at levels current at that time 10 to 12 mg/100 mL). A very few soft drinks have a higher level. No-caffeine colas have become widely consumed.

Prescriptions and Over-the-Counter Medications

In the United States prescriptions and over-the-counter (OTC) medications use a total of about 50 metric tons of caffeine annually. Caffeine-containing medications are used for certain types of pain relief, treatment of infant apnea, alertness increase, and weight loss.

CAFFEINE PRODUCTION

Approximately 450 tons of caffeine are required annually in the United States for addition to beverages and for pharmaceutical use. This need is partially met by the decaffeination of coffee (21). The methylation of theobromine obtained by extraction of cocoa hulls provides part of the remainder of the requirements. Dimethyl sulfate, diazomethane, and methyl iodide can be used to prepare caffeine from xanthine or any of the related monomethyl- or dimethylxanthines (22). Patented synthetic processes are also based on classical organochemical ring closure techniques and do not require the availability of xanthines as starting materials.

In 1995 the United States imported about 1000 tons of caffeine. About 500 tons came from China. Domestic synthetic caffeine was list-priced at $7.25/lb. The product from China was approximately $4.50/lb (23).

METABOLISM OF CAFFEINE

The caffeine molecule is sufficiently hydrophobic to pass freely through biological membranes (24). It is completely absorbed from the gastrointestinal tract and rapidly attains peak plasma levels. Caffeine is equally well absorbed from each of its beverage sources. There are no effective barriers to penetration of any tissue, including placenta and fetus. This characteristic prevents efficient excretion of caffeine by the human kidney since it is readily reabsorbed from the renal tubules.

The half-life of caffeine in plasma varies not only with specie differences but also with age and condition of the individual. The half-life in rodent plasma is 1 to 2 h but it is 6 h in that of the healthy adult human. There is variation depending on smoking habits and the use of some medications. During pregnancy the half-life is increased to 18 h. The immature liver of the newborn human is limited in its ability to metabolize caffeine so that its half-life is three to four days, similar to that of adults with severe liver damage. These differences are significant in extrapolating safety data from animal studies and also in considering diet during pregnancy (24).

The metabolic pathways of caffeine in those mammalian species that have been studied exhibit many similarities (24). The same major reactions take place: demethylation, oxidation at the 8-position on the xanthine ring to form uric acids, and ring cleavage between the 8- and 9-positions to form diaminouricils. *N*-acetylation also occurs. The three dimethyl- and the three monomethylxanthines that can be formed are all found in human urine after caffeine ingestion along with all of the corresponding methylated uric acids. Uracils derived from caffeine, theobromine, and paraxanthine as well as large amounts of acetylated products are also present. No xanthine is formed. The human metabolic pathway is shown in Figure 4.

Specie variations in the metabolic pathways of caffeine were first observed in 1900 (1). In humans, about 70% of ingested caffeine is initially converted to paraxanthine, 25% to a mixture of theobromine and theophylline, and about 5% is oxidized without demethylation to form the corresponding uric acid and uracil compounds. Some primates produce theophylline as the predominant initial metabolite. In the human the final product mix is the result of competing reactions, the rates of which vary with gender, dosage, medications in the diet, and individual differences. Minor amounts of uracils derived from caffeine, theobromine, and paraxanthine also occur. Averaged values for caffeine metabolites found in human urine are shown in Table 2.

The acetylated product (A) in Table 2 is believed to be a true metabolite. Product (B) is probably an artifact of analytical procedures.

Reaction mechanisms are not well known. The biochemical changes take place in the liver. Xanthine oxidase mediates the formation of the uric acid derivatives. It is interesting to note that in the human infant, theophylline, often used for treatment of apnea, is methylated to caffeine as the first step in its metabolism (1).

PHARMACOLOGY

Effects on Cardiovascular System

Caffeine produces minor, transitory increases in blood pressure. Habitual users are less prone to exhibit this effect. Its significance is not known (26). Caffeine has also been reported to cause cardiac arrhythmias, but there is conflicting evidence (26). There appears to be a caffeine-intolerant population that is susceptible to this effect. Caffeine may exacerbate an existing tendency toward arrhythmias. Low dosages may decrease heart rate slightly; high dosages may cause tachycardia in sensitive subjects (27). Cerebral blood flow is decreased, and this effect is the basis for its inclusion in drug preparations for the treatment of migraine headaches (28).

Some studies indicate a positive correlation between caffeine intake and the development of hypercholesterolemia, but many very large efforts to confirm this effect, such as the Framingham study, show no correlation between atherosclerotic cardiovascular disease and coffee intake. Other large studies have shown no correlation between coffee drinking and any form of coronary heart disease (27). The disparate results concerning the effects of caffeine on the cardiovascular system probably relate to the size of the studies, lack of control of unrecognized factors and, very importantly, to the acceptance of "cups of coffee" as a quantitative measure of caffeine content. Coffee and tea contain many physiologically active compounds other than caffeine. Coffee oils, especially kahweol, have been shown to be cholesterogenic (29).

Effects on the Central Nervous System

Although caffeine is the stimulatory drug most widely used throughout the world, it is difficult to quantify its behavioral effects. This is primarily due to the differences in the form of its ingestion, that is, coffee, tea, or the pure substance. Another confounding factor is the varying degrees of tolerance that develop among its users.

Studies of the effects of caffeine on information processing (30), vigilance performance (31), ease of distraction (32), reaction time performance (30), decision making (33), and cognitive functioning (34) have been reported. Results are difficult to assess, especially with regard to cognitive functioning, but in general, low or intermediate dosages of caffeine are beneficial while high dosages impair performance. Gender, age, and task difficulty have notable effects on results.

The effects of caffeine on mood have also been studied. Again, improvements are experienced until dosage levels become high (30). At a dosage of 2 mg/kg, which generates a peak plasma concentration of 5 to 10 μM, individuals generally feel more alert and better able to carry out routine tasks after having become bored or fatigued (35). Caffeine may even counteract mood deficits caused by profound sleep deprivation (36). Conversely, omission of a habitual morning dosage of caffeine often results in nervousness, irritability, and poor work performance. The most widely noticed effect of caffeine is the prolongation of the sleep latency period. Its effect on the quality of sleep shows wide individual variation and is dose, time, and age

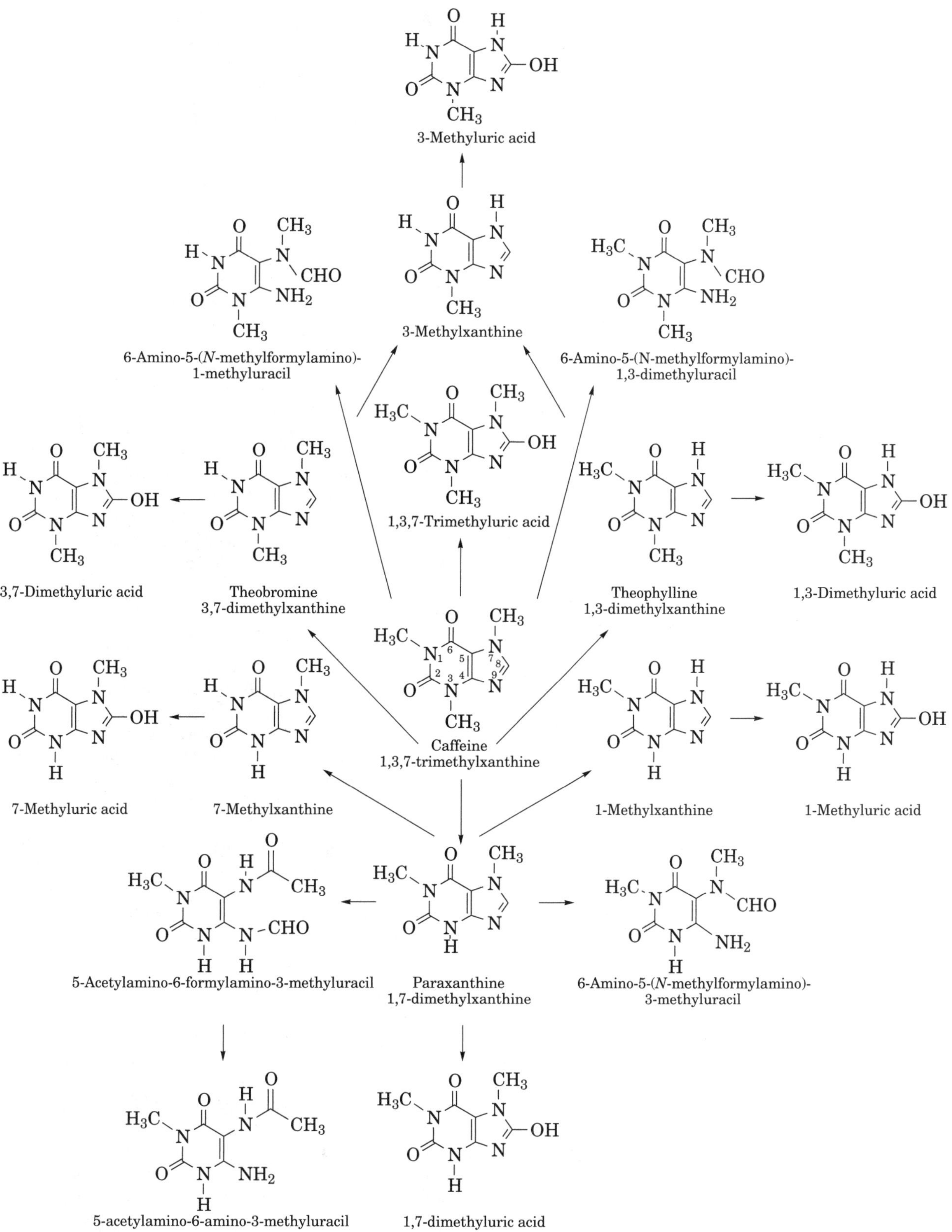

Figure 4. Human metabolism of caffeine.

Table 2. Caffeine Metabolites in Human Urine as % of a 5 mg/kg Dosage

Caffeine	1.2	1,3,7-trimethyluric acid	1.2
Paraxanthine	5.9	1,7-dimethyluric acid	5.4
Theobromine	1.9	3,7-dimethyluric acid[1]	
Theophylline	0.9	1,3-dimethyluric acid	1.8
1-methylxanthine	18.1	1-methyluric acid	14.8
7-methylxanthine	7.7	7-methyluric acid	1.0
3-methylxanthine	4.1	3-methyluric acid	
5-acetylamino-6-amino-3-methyluracil (A)	3.2		
5-acetylamino-6-formylamino-3-methyluracil (B)	14.7		

Source: Ref. 26.

dependent. It is markedly subject to the development of tolerance.

The arousal model as an explanation for the foregoing effects has been suggested (37). Positive response to the arousal effect of caffeine plus those of any other biological and background impacts, increases with an increase in stimuli but is reversed as the total arousal level exceeds the tolerance level.

Other Effects

Caffeine has been reported to increase work output and endurance in long-term exercise regimens. Ergogenic properties of caffeine have been observed in many studies of athletic performance but are subject to confounding factors: nature of activity, degree of training of athletes, level of caffeine tolerance, and habitual caffeine usage. There are also many "no effect" reports. An interesting interpretation of these varying results is that the mood-elevating properties of caffeine may be responsible for some of the positive effects reported (38).

Caffeine produces a mild diuresis in humans and increases the excretion of sodium, potassium, and chloride ions (28). The thermogenic properties of caffeine cause an increase in the resting metabolic rate. Consideration has been given to its use in managing obesity (39). Respiratory rate is increased by caffeine at a plasma level of 5 mg/L. The mechanisms for this stimulation are complex and involve many separate effects. They include increased pulmonary blood flow through vasodilation, increased sensitivity of the respiratory centers to carbon dioxide, and improved skeletal muscle contraction (28). This multiple response is utilized in the treatment of infantile apnea.

Mechanism of Physiological Activities

The effects of caffeine on the cardiovascular, neurological, and renal systems and on lipolysis are primarily triggered by the blockage of adenosine receptors through competitive antagonism (40). The structural relationship between caffeine and adenine (and therefore adenosine) was shown previously. Other suggested mechanisms such as the inhibition of phosphodiesterases and modification of calcium metabolism may also be operative but are inadequate to explain these effects at reasonable caffeine concentrations. Adenosine and caffeine show contrasting effects on blood pressure, activity of the central nervous system, urine output, and lipolysis. The well-known induction of tolerance to caffeine also becomes explicable by this mechanism since caffeine is known to induce the formation of additional adenosine receptors in the rat (40).

Physical Dependence

Addiction to caffeine may develop even from short-term (seven days) administration of high dosages. Headache and fatigue are the most common withdrawal symptoms but depression, irritability, anxiety, and vomiting may also occur. Severity of symptoms that may last for up to one week range from mild to incapacitating. Many caffeine consumers exhibit no withdrawal symptoms. The physiology of caffeine withdrawal is not well understood (41).

Toxicity

The physiological effects of caffeine become greatly intensified at high dosages. Caffeine affects the cortex, the medulla, and eventually the spinal cord as dosages are increased (28). The LD50 of caffeine is similar for many mammals—about 200 mg/kg (42). This would suggest an LD50 of about 3.5 g for humans based on normal metabolic weight corrections. It is more generally accepted that the lethal oral dose is about 10 g. Lethal plasma concentration is in the range of 0.5 to 1.0 mM. To achieve this concentration would require the consumption of about 75 cups of coffee over a very short period of time (40). Lethality resulting from beverage consumption is most unlikely. The few recorded cases of death from caffeine consumption have involved the pure substance or medications containing it.

Caffeine as a Drug

The primary uses of caffeine as a drug are based on its effects on the respiratory, cardiovascular, and central nervous systems. Premature infants, for example, are subject to apnea, a transient but potentially dangerous cessation of breathing. Caffeine has been used to control this syndrome. It decreases apneic episodes and regularizes breathing patterns (28). It is also used for the treatment of bronchiospastic disease in asthmatic patients. Caffeine is used widely in drug mixtures designed for relief from migraine-type headaches because of its vasoconstrictor effects on the cerebral circulation. Many OTC preparations of this type are available. They contain 30 to 200 mg of caffeine per tablet. A common application of caffeine's stimulatory effects on the central nervous system is the use of 100- to 200-mg tablets to prevent drowsiness when driving or whenever this condition is undesirable. There are about 2000 nonprescription and about 1000 prescription drugs containing caffeine.

MUTAGENICITY

Caffeine attains the same concentration in gonadal tissue as in the circulating plasma and may thus reach levels of up to 0.05 mM for very heavy consumers (43). *In vitro* mutagenicity studies in which positive results are obtained generally involve concentrations at least 100 times greater and are therefore at or above the lethal concentration for humans. At these high concentrations, DNA repair is inhibited, an effect attributable to its purine analog structure. Caffeine may thus be considered a weak mutagen with no significance for humans.

TERATOGENICITY

Caffeine is teratogenic to rodents at high dosage levels (44). Effects such as cleft palates and ectrodactyly result from administration of 100mg/kg/d or more, but only when given in a single dose by gavage or injection. Increased fetal absorption and decreased fetal weight also occur at these levels, which approach lethality. True teratogenic responses have usually been associated with dosages that result in maternal toxicity. Delayed bone ossification is observed at much lower levels, but this effect is rapidly compensated postnatally.

The role of caffeine in the outcome of human pregnancy is generally determined by retrospective questioning. Studies have been based on beverage consumption recall and are therefore inexact with regard to the amount of caffeine ingested or the effects of other beverage components. There is no clear evidence of teratogenic effects due to caffeine (45). Conflicting data exist concerning the relationship between maternal caffeine consumption and infant birth weight (46). Since caffeine crosses the placental barrier, it is possible for high concentrations to occur in the neonate (47). Most medical advice cautions against high caffeine intake during pregnancy.

CARCINOGENICITY

The carcinogenetic potential of caffeine has been studied intensively in rodents (48). The results of about a dozen well-controlled studies indicate no carcinogenic effects attributable to caffeine. Epidemiological studies regarding the risk of cancers of the bladder, colon, rectum, pancreas, breast, ovaries, and liver to human coffee drinkers have been carried out in different parts of the world (49). Conflicting results have been obtained. In almost all instances, dose-response relationships have not been observed, casting doubt on any causal relationships between the effect and the consumption of coffee. The fact that tea consumption often produces different effects, sometimes converse, makes it unlikely that caffeine itself is a causative agent of increased cancer risk. An extensive survey of human studies, based primarily on coffee and tea consumption, show little correlation between consumption of the beverages and cancer incidence (50).

Theoretical considerations suggest a lack of carcinogenic potential of caffeine based on the facts that it does not alkylate or combine covalently with DNA, it does not undergo metabolic activation, and it lacks some of the structural features of known purine carcinogens (47).

Determination of the effect of caffeine on fibrocystic disease of the breast has also been pursued. Results of the studies are mixed. There may be some correlation between caffeine consumption (based on dietary coffee) and fibrocystic breast disease (51).

DECAFFEINATION

Concern with the physiological effects of coffee prompted early consideration of ways to decaffeinate the product. The first commercial plant was established by the firm Kaffee HAG in Germany in 1906 and in the United States a few years later. The U.S. plant was expropriated by the government during World War I, terminating production of decaffeinated coffee in the United States until 1932. Only one brand was available until the mid-1950s and solely as the instant product. Consumption later increased rapidly and in 1997 accounted for 35% of the instant product and 12 to 14% of ground roasted coffee. Decaffeinated tea appeared on the market in 1978. In 1997 it accounted for approximately 15% of leaf tea consumed in the United States and 20% of pure instant tea.

Coffee

Chlorinated compounds were the first caffeine extractants used with coffee. To expedite removal of caffeine, coffee beans are steam treated to bring about swelling for greater solvent permeability and to break down caffeine complexes. In all extraction processes it is necessary to wet the coffee for efficient caffeine removal. Green beans are decaffeinated to prevent extraction and loss of the aromatic components that are generated only during roasting. After solvent removal by steam distillation, the beans are roasted.

Trichlorethylene was commonly used as the solvent until it was eliminated in 1977 because of its suspected carcinogenicity. Methylene chloride then became the decaffeinating solvent of choice (36). A residue level of 10 ppm is approved by the U.S. Food and Drug Administration. An unofficial standard providing for the removal of 97% of the original caffeine has come into use for decaffeinated coffee. The beverage therefore contains 2 to 4 mg of caffeine per cup. To eliminate residues of chlorinated compounds, other decaffeinating solvents are now used. Ethyl acetate is approved in the United States. The process has some disadvantages: 5 to 6% solids loss and more difficult desolventization.

Decaffeination using water as the only solvent to contact the coffee is also carried out. A concentrated coffee extract freed of caffeine contacts the green beans. Noncaffeine solids are not removed under these conditions. The extract is then regenerated for use by adsorbing the caffeine on activated carbon (52). Coffee oil expressed from beans is used as a decaffeinating solvent along with other vegetable oils in another operating process. Caffeine is removed from the oils by a liquid–liquid extraction system with water as the second solvent (53).

Supercritical CO_2 extraction is now widely used for decaffeination (54). The process is carried out at 300 bars and at 60 to 90°C. Caffeine solubility increases as temperature and pressure are raised. The extracted caffeine may be removed from the supercritical fluid by reducing the pressure, thereby decreasing its solubility or by passing the mixture over activated carbon. Techniques that avoid the necessity for repressurizing the gas are preferable. Supercritical processes are capital intensive but result in products of superior quality with no solvent residues. In 1997 supercritical extraction was used to prepare about 50% of the decaffeinated coffee in the United States.

Tea

The commercial procedures for the decaffeination of tea are, in principle, similar to those used for coffee. For practical reasons it is necessary to utilize manufactured black tea rather than fresh green leaf. Conservation of aroma during processing is desirable. Decaffeinated tea appeared on the market in 1978. Similar process evolution took place as with coffee. Decaffeination with supercritical CO_2 became a commercial process in 1984 (55). Aroma may be previously stripped with dry supercritical CO_2 or with the moist gas at atmospheric pressure for later addition to the product (56).

BIBLIOGRAPHY

1. M. J. Arnaud, "Products of Metabolism of Caffeine," in P. B. Dews, ed., *Caffeine*, Springer-Verlag, Berlin, 1984, pp. 3–38.
2. E. Fischer and L. Ach, "Synthese des Caffeins," *Chem. Ber.* **28**, 3135–3142 (1895).
3. A. L. Thakkar et al., "Self-association of Caffeine in Aqueous Solution: A Nuclear Magnetic Resonance Study," *J. Chem. Soc. Chem. Commun.* **9**, 524–525 (1970).
4. I. Horman and R. Viani, "The Caffeine-Chlorogenate Complex of Coffee, an NMR Study," ASIC, 5th Colloque, Lisbon, 1971, pp. 102–111.
5. H. Biltz and H. Rakett, "Kaffeiden und Kaffeiden-carbonsaure," *Chem. Ber.* **61**, 1409–1422 (1928).
6. A. C. Cope et al., "1,3-Dimethyl-5-alkyl Barbituric Acids," *J. Am. Chem. Soc.* **63**, 356–358 (1941).
7. W. J. Hurst, R. A. Martin, Jr., and S. M. Tarka, Jr., "Analytical Methods for the Quantitation of Methylxanthines," in G. A. Spiller, ed., *Caffeine*, CRC Press, New York, 1998, pp. 13–31.
8. R. Dorizzi and F. Tagliaro, "Direct HPLC Determination of Theophylline, Theophylline Analogs and Caffeine in Plasma with an ISRP (Pinkerton) Column," *Clin. Chem.* **36**, 566 (1989).
9. E. D. Conte and E. F. Barry, "Gas Chromatographic Determination of Caffeine in Beverages by Alkali Aerosol Flame Ionization Detection," *Microchem. J.* **48**, 372–376 (1993).
10. W. J. Hurst and R. A. Martin, Jr., "The Quantitative Determination of Caffeine in Beverages Using Capillary Electrophoresis Analysis," *Analysis* **21**, 389–391 (1993).
11. B. A. Kihlman, in *Caffeine and Chromosomes*, Elsevier, Amsterdam, 1977, pp. 11–13.
12. E. Looser, T. W. Bauman, and H. Warner, "The Biosynthesis of Caffeine in the Coffee Plant," *Phytochemistry* **13**, 2515–2517 (1974).
13. P. Mazzafera et al., "*S*-Adenosyl-L-Methionine:Theobromine I-*N*-Methyltransferase, an Enzyme Catalyzing the Synthesis of Caffeine in Coffee," *Phytochemistry* **37**, 1577–1584 (1994).
14. T. Suzuki, H. Ashihara, and G. R. Wallace, "Purine and Purine Alkaloid Metabolism in Camellia and Coffea Plants," *Phytochemistry* **31**, 2575–2582 (1992).
15. H. Ashihara and H. Kubota, "Biosynthesis of Purine Alkaloids in Camellia Plants," *Plant Cell Physiol.* **28**, 535–539 (1987).
16. O. Negishi, T. Ozzawa, and H. Imagawa, "Biosynthesis of Caffeine from Purine Nucleotides in Tea Plant," *Biosci. Biotechnol. Biochem.* **56**, 499–503 (1992).
17. M. A. Bokuchava and N. I. Skobeleva, "The Biochemistry and Technology of Tea Manufacture," *Crit. Rev. Food Sci. Nutr.* **12**, 348 (1980).
18. M. Dong et al., "The Occurrence of Caffeine in the Air of New York City," *Atmos. Environ.* **11**, 651–653 (1977).
19. J. J. Barone and H. R. Roberts, "Caffeine Consumption," *Food Chem. Toxicol.* **34**, 119–129 (1996).
20. H. R. Roberts and J. J. Barone, "Biological Effects of Caffeine; History and Use," *Food Technol.* **37**, 32–39 (1983).
21. U.S. Pat. 4,818,552 (Apr. 4, 1989), L. Kaper (to Douwe Egberts).
22. U.S. Pat. 2,534,331 (Dec. 19, 1950), D. W. Woodward (to E.I. du Pont de Nemours & Co.)
23. Anonymous communication, *Chemical Marketing Reporter* **247**, no. 8 Feb. 20 (1995).
24. R. W. Von Borstel, "Biological Effects of Caffeine; Metabolism," *Food Technol.* **37**, 40–45 (1983).
25. D. W. Yesair, A. R. Branfman, and M. M. Callahan, "Human Disposition and Some Biochemical Aspects of Methylxanthines," *Phytochemistry* **13**, 215–233 (1974).
26. M. G. Meyers, "Effects of Caffeine on Blood Pressure," *Arch. Int. Med.* **148**, 1189–1193 (1988).
27. T. K. Leonard, R. R. Watson, and M. E. Mohs, "The Effects of Caffeine on Various Body Systems: A Review," *J. Am. Diet. Assoc.* **87**, 1048–1053 (1987).
28. M. J. Arnaud, "The Pharmacology of Caffeine," *Progress in Drug Research* **31**, 273–313 (1987).
29. R. Urgert et al., "Separate Effects of the Coffee Diterpenes Cafestol and Kahweol on Serum Lipids and Liver Aminotransferases," *Am. J. Clin. Nutr.* **65**, 519–524 (1997).
30. M. Hasenfratz and K. Battig, "Acute Dose–Effect Relationships of Caffeine and Mental Performance, EEG, Cardiological and Subjective Parameters," *Psychopharmacology (Berlin)* **114**, 281–287 (1994).
31. D. Fagan, C. G. Swift, and B. Tiplady, "Effects Of Caffeine on Vigilance and Other Performance Tests in Normal Subjects," *J. Pharmopsychol.* **2**, 19–25 (1988).
32. L. M. Giambra et al., "The Influence of Caffeine Arousal on the Frequency of Task-Unrelated Image and Thought Intrusions," *Imagination, Cognition and Personality* **13**, 215–223 (1994).
33. R. Lubit and B. Russett, "The Effects of Drugs on Decision Making," *Conflict Resolution* **28**, 85–102 (1984).
34. B. D. Smith and K. Tola, "Caffeine: Effects on Psychological Functioning and Performance," in G. A. Spiller ed., *Caffeine*, CRC Press, New York, 1998, pp. 263–266.

35. A. Leviton, "Biological Effects of Caffeine; Behavioral Effects," *Food Technol.* **37**, 44–46 (1983).
36. D. Penetar et al., "Caffeine Reversal of Sleep Deprivation Effects on Alertness and Mood," *Psychopharmacology (Berlin)* **112**, 359–365 (1993).
37. B. Smith, "Effects of Acute and Habitual Caffeine Ingestion in Physiology and Behavior; Tests of a Biobehavioral Arousal Theory," *Pharmopsychoecologia* **7**, 151–167 (1994).
38. R. J. Lamarine, "Caffeine as an Ergogenic Aid," in G. A. Spiller, ed., *Caffeine*, CRC Press, New York, 1998, pp. 244–245.
39. A. G. Dulloo et al., "Normal Caffeine Consumption: Influence on Thermogenesis and Daily Energy Expenditure in Lean and Postobese Human Volunteers," *Am. J. Clin. Nutr.* **49**, 44–50 (1989).
40. R. Von Borstel and R. J. Wurtman, "Caffeine and the Cardiovascular Effects of Physiological Levels of Adenosine," in P. B. Dews, ed., *Caffeine*, Springer-Verlag, Berlin, 1984, pp. 142–150.
41. R. R. Griffiths and P. P. Woodson, "Caffeine Physical Dependence: A Review of Human and Laboratory Animal Studies," *Psychopharmacology (Berlin)* **94**, 437–451 (1988).
42. B. Stavric, "Toxicity to Humans. 2. Caffeine," *Food Chem. Toxicol.* **26**, 645–662 (1988).
43. R. H. Haynes and J. D. B. Collins, "The Mutagenic Potential of Caffeine" in P. B. Dews, ed., *Caffeine*, Springer-Verlag, Berlin, 1984, pp. 231–238.
44. J. G. Wilson and W. J. Scott, Jr., "The Teratogenic Potential of Caffeine in Laboratory Animals," in P. B. Dews, ed., *Caffeine*, Springer-Verlag, Berlin, 1984, pp. 165–187.
45. A. Berger, "Effects of Caffeine Consumption on Pregnancy Outcome," *J. Reproduct. Med.* **33**, 945–956 (1988).
46. J. Nash and T. V. N. Persaud, "Reproductive and Teratological Risks of Caffeine," *Anatomischer Anzeiger* **167**, 265–270 (1988).
47. W. D. Parsons and A. H. Niems. "Prolonged Half-Life of Caffeine in Healthy Term Newborn Infants," *J. Pediatrics* **98**, 640–641 (1981).
48. H. C. Grice, "The Carcinogenic Potential of Caffeine," in P. B. Dews, ed., *Caffeine*, Springer-Verlag, Berlin, 1984, pp. 153–164.
49. Committee on Diet and Health, National Research Council, *Diet and Health*, National Academy Press, Washington, D.C., 1989.
50. G. A. Spiller and B. Bruce, "Coffee, Tea, Methyl Xanthines, Human Cancer, and Fibrocystic Breast Disease," in G.A. Spiller, ed., *Cancer*, CRC Press, New York, 1998, pp. 326–336.
51. U.S. Pat. 3,671,263 (June 20, 1972), J. M. Patel and A. B. Wolfson (to Proctor and Gamble Co.).
52. Anonymous, "Coffex Avails Swift Water Process to N. America," *Tea and Coffee Trade Journal* **160**, 45–47 (1988).
53. U.S. Pat. 4,837,038 (June 6, 1989), J. C. Proudley (to Nestea SA).
54. U.S. Pat. 4,820,537 (Apr. 11, 1989), S. N. Katz (to General Foods Corp.).
55. W. Ger. Pat. 3,515,740 (May 2, 1985), H. Klima, E. Schütz, and H. Vollbrecht (to SKW Trostberg AG).
56. Eur. Pat. 269,970 (June 8, 1988), E. Schütz and H. Vollbrecht (to SKW Trostberg AG).

HAROLD GRAHAM
Consultant
New York, New York

CANCER RISK AND DIET

Cancer can be regarded as a mass of cells that are no longer under normal regulatory growth control. Cancer develops in several stages, namely, initiation, promotion, and progression. The initiation stage is rapid and irreversible and begins with attack of a cancer-inducing agent on a target site in the DNA of the cell. In this process a normal cell is converted into a cancer cell. The transformation is irreversible. During the promotion stage the transformed cell may grow into a tumor; not all cells do. Progression is the stage wherein a benign tumor may become malignant.

Nutritional studies of the influence of diet on cancer risk can be carried out in animals by direct experimentation and in humans by using epidemiological methods. Both methods provide useful data that can then be examined further to determine influences of diet on causation and, possibly, inhibition of tumor growth.

Animal studies are relatively easy to carry out. The animal of choice (almost always a mouse or rat) is subject to treatment with a carcinogenic agent and observed for a specified length of time before being subjected to necropsy. The advantages of this system are an experimental consistency and total control over diet and life span. But even here different strains of mice or rats do not always give the same result. Additionally, the experimental animal is caged, which may produce a certain amount of stress; a lifetime effect of treatment is never seen; and there are virtually no studies in which the diet is varied. Still, animal studies provide useful data concerning mechanisms of carcinogenesis and of effects of dietary and drug treatment. The results of animal studies are not directly applicable to humans.

Epidemiological studies of cancer in humans use varied approaches. Ecological or correlation studies are studies in which differences between populations are examined. These may be examinations of specific nutrients or of patterns of food intake. Generally investigation of food disappearance data is not as useful as studies of individual dietary components. Studies in humans may include migrant studies (cancer incidence in the new location vs cancer incidence in the old location); case-control studies, in which cases are compared with age-matched controls; and follow-up studies, in which appearance of cancer is compared between exposed or unexposed individuals. Studies involving assessment of past dietary intake are subject to errors in recall. Many foods are important sources of more than one nutrient—meat, for instance, is a source of both protein and fat—which adds to the complexity of diet assessment.

A good example of the complexity of epidemiological studies may be found in two studies of diet effects on colon cancer carried out in Australian cities that are about 400 miles apart. In Adelaide, Potter and McMichael (1) found colon cancer risk to be associated with protein and energy intake, but fiber intake was, if anything, a risk for colon cancer. Kune et al. (2) in Melbourne found dietary fiber to be protective.

Armstrong and Doll (3) reviewed the effects of diet on cancer worldwide and suggested that the correlations found in their survey be regarded as guides to further research rather than as evidence of causation. In 1987 a review by Doll and Peto (4) raised interest in diet and cancer to a new level. They suggested that 30 to 70% of cancer in the United States might be related to diet and stated that this value was "highly speculative and chiefly refers to dietary factors which are not yet reliably identified."

In 1982 a committee of the National Academy of Sciences (U.S.) (5) reviewed influences of diet on cancer. Their conclusions were (in part):

1. Epidemiological and experimental data suggest an association between a high-fat diet and cancers of the breast, large bowel, and prostate. The data are not unanimous, however.
2. Dietary restriction reduces tumor incidence in experimental animals.
3. Foods rich in vitamin A and beta-carotene reduce risk of lung and bladder cancer.
4. Selenium may reduce cancer risk, but the results are inconclusive.
5. Diets high in vitamin C may reduce the risk of esophageal or gastric cancer.
6. Studies relating cancer incidence to carbohydrate, fiber, vitamin E, and the B vitamins are inconclusive.
7. Alcohol acts synergistically with smoking to increase risk of cancers of the upper GI and respiratory tracts.

Few new revelations have occurred in the 17 years since that report, except that the role of fat in breast cancer has been downgraded and importance of fruits and vegetables somewhat increased.

One can review dietary data related to cancer by nutrient or by type of cancer. The former model will be used in this review.

MACRONUTRIENTS

Protein

In general there is virtually no relationship between protein intake and either colorectal or breast cancer (6). The data relating to meat intake and the risk of colorectal cancer overall suggest little association if any. Recently, an American multicenter case-control study has concluded, "This study provides little support for an association between meat consumption and colon cancer risk" (7). The finding that cooking meat resulted in formation of heterocyclic amines that are carcinogenic for mice (8) was used as a point of departure for many inquiries into the effects of meat, but a recent report by Augustsson et al. (9) finds that heterocyclic amine intake within the usual dietary range is not likely to increase risk of cancer of the colon, rectum, kidney, or bladder in humans.

CARBOHYDRATE

There is little evidence to support a role for carbohydrates per se in carcinogenesis. Drasar and Irving (10) found a correlation between incidence of breast cancer and intake of simple sugars as did Hems (11), but an inverse correlation was seen with starch intake (12). Longtime feeding of sucrose to rats (13) or mice (14) did not result in increased tumorigenesis. The available evidence is scanty, and it is not possible to form any conclusion except for the safe one; namely, more research is needed. Most dietary fiber is carbohydrate in nature, but that represents a special case that is discussed later.

FAT AND ENERGY

Overweight has been associated with increased risk of a number of cancers (15). Since overweight is often linked to fat intake, the search for dietary correlations with cancer incidence has led to investigation of fat intake and cancer. There is some positive correlation between fat intake and colon cancer risk, but it is not very strong (16). Studies in Great Britain (17) and Sweden (18) found no relation between fat intake and either colon or breast cancer. One of the earliest epidemiologic studies of diet and colon cancer was carried out in 1933 by Stocks and Karn (19) who found negative correlations between colon cancer and intake of bread, vegetables, and dairy food. Similarly, a study comparing diets of Danes (high colon cancer) and Finns (low colon cancer) found positive correlations only with alcohol intake and fecal bile and concentration; among the negative correlations was intake of saturated fat (20).

Data from the National Health and Nutrition Examination Study (NHANES) show no correlation between fat intake and breast cancer (21). The same finding has emerged from broad reviews of the literature (22,23). A study by Willett et al. (24) of more than 89,000 American nurses has revealed no link between fat intake and risk of breast cancer. As with virtually all diet-cancer data the findings are not unanimous.

A new nutritional entity in the diet-cancer arena is conjugated linoleic acid (CLA). Linoleic acid contains double bonds that are separated by a methylene group. A microorganism that occurs in the rumen of ruminants such as the cow or sheep moves the double bonds into conjugation, meaning they are contiguous and not methylene separated. What is now termed CLA is a mixture of positional and geometric isomers of octadecadienoic acid. CLA has been shown to inhibit formation of chemically induced skin and forestomach tumors in mice and mammary tumors in rats (25). CLA is present in dairy foods and meat of ruminant animals. Efforts are being made to increase its content in the common sources to determine if increased dietary intake can deter human carcinogenesis. In rats CLA is effective at levels of 0.25% of the diet. This extrapolates to more than the current intake in humans.

Berg (26) suggested in 1975 that cancers prevalent in developed countries could be due to overnutrition. Almost a century ago Moreschi (27) found that transplanted tumors grew poorly in underfed mice. The effects of caloric (energy) restriction on growth of spontaneous, chemically induced or transplanted tumors has confirmed Moreschi's observation almost unanimously. Studies of chemically induced mammary tumors in rats have shown that at 25% caloric restriction, rats fed 25% fat have fewer tumors than freely fed rats ingesting only 5% fat (28). This area of investigation is very active at the present time, and emerging data show that energy restriction affects enzyme activity and gene expression.

A number of studies have shown that breast cancer risk is related positively to age at menarche, body weight or body mass index, and height (29). These factors can be influenced by nutrition. A role of caloric intake on human cancer has been suggested by several groups of investigators (1,30,31). Energy utilization may also play an important part in carcinogenesis. Women who engaged in organized athletics while in college (32) or men who have spent most of their working life at energy-intensive jobs (33) have less breast or colon cancer, respectively. In 1945 Potter (34) recommended exercise and reduction of caloric intake as means of preventing cancer. These activities are cheap and effective but unpopular.

Alcohol consumption has been suggested to be a risk for breast cancer. The studies that have been reported indicate a bias in that direction, but it is not overly strong (23). Alcohol, as a source of calories, lies between carbohydrate (4 cal/g) and fat (9 cal/g). Alcohol provides about 7 cal/g and can be an important contributor to daily caloric intake.

FIBER

Dietary fiber is a generic term that describes dietary components that are not broken down during the digestive process. Except for lignin, fiber is carbohydrate in nature and is often classified as soluble (pectin, guar gum) or insoluble (cellulose, wheat bran). The many substances that are included under the designation "dietary fiber" are unique chemical compounds and exhibit individual physiological effects.

Interest in fiber was stimulated by the work of Burkitt (35) who attributed the freedom from colon cancer of native African populations to their high fiber intake. Earlier, Higginson and Oettlé (36) made a similar observation, but they did not pursue it with Burkitt's missionary fervor. Armstrong and Doll (3) had observed a negative correlation between cereal intake and cancer risk. In a comprehensive review of the literature up to 1986 (37), data from 18 ecological and 22 case-control studies were considered. Protection from colon cancer by a high-fiber diet was adduced from 66.7% of the ecological studies and 36.4% of the case-control studies; no effect was seen in 27.8% of the ecological studies and 40.9% of the case-control studies; and enhancement was found in the rest (5.6% of the ecological studies and 22.7% of the case control studies). Later reviews also find a preponderance of studies that show a protective effect of fiber but without unanimity.

Fiber fulfills a number of functions that may contribute to its protective effects. It increases fecal bulk, which has been suggested as a protective factor (38); it dilutes colonic contents, which may reduce contact between carcinogens and colonic mucosa; and it reduces fecal pH (39), levels of fecal mutagens (40), and concentration of fecal bile acids (20). Degradation of fiber by colonic microflora also produces water, methane, carbon dioxide, and short-chain fatty acids (SCFA). The major SCFA are acetate, propionate, and butyrate and, of these, butyrate increases differentiation of colonic cells. Butyrate exerts many actions at the cellular and molecular level, all of which work toward protection against carcinogenesis (41).

A summary of epidemiological trials that support a preventive role for dietary fiber has been published recently (41). Reviews by Hill (42) and Jacobs et al. (43) have shown that cereals and cereal fiber do indeed protect against colon cancer. The comparison of colon cancer in Danes and Finns (20) found negative correlations with carbohydrate, cereals, protein, saturated fat, starch, and total dietary fiber. A recent study by Fuchs et al. (44) found that dietary fiber did not protect against colorectal cancer or adenomas. The median intake of cereal fiber in the 88,757 women studied was rather low, ranging from 1.0 to 4.5 g/day going from the first to fifth quintile of intake. An intake of 4.5 g of cereal fiber is on the low side. In their review Rogers and Longnecker (6) state, "most epidemiologic studies of fiber or fiber-containing food intake in relation to the risk of colorectal cancer are consistent with a very small inverse association or no association." However, specific fiber sources such as cereals may offer protection.

A problem in interpretation of data for review purposes is that all fiber is viewed to be the same. In fact dietary fiber from different sources may vary in fiber content and structure and in the accompanying nutrients, which may also have an effect. The epidemiologic data are derived from populations ingesting a fiber-rich diet, not a single source of fiber. A high-fiber diet will also contain carotenoids and other substances unique to specific plant sources, plant sterols, n-3 and n-6 polyunsaturated fatty acids, and possibly trace minerals. The reductionist approach to the data tends to single out a particular substance and assume that to be the protective substance. The substance chosen may be one currently in vogue or may be a specific interest of the investigator. One case in point—many studies have suggested that diets rich in beta-carotene are protective against lung cancer. But adding beta-carotene to the diets of heavy smokers actually increased the incidence of lung cancer (45). In a similar vein total dietary fiber did not lower the risk of colorectal adenomas in women, but fiber from fruits and vegetables did (46).

In assessing the risk of colon cancer in 47,949 (205 cases) as a function of fiber source, fruit fiber ingestion was most strongly correlated with negative risk (47). In an earlier study of a cohort of 7284 men (170 cases) the same authors found that all sources of fiber reduced significantly the risk of colorectal adenoma (48). The data beg the question of why most sources of fiber no longer protect when the adenoma progress to carcinoma. Is it due to the presence of other factors in fruit? Fiber from fruits and vegetables implies other components of fruits and vegetables were present and may have played a role. Several large prospective studies of the effect of wheat bran on colon cancer will be reported in 1999 and 2000.

FRUITS AND VEGETABLES

Substances found in fruits and vegetables may exert anticarcinogenic action by inhibiting formation of carcinogens, by inducing detoxifying enzymes, or by blocking the carcinogen from reaching target tissue. Tumor promotion may be inhibited by retinol and beta-carotene (green/yellow vegetables and fruits), tocopherol (nuts, wheat germ), vitamin C (vegetables, fruits), organosulfur compounds (garlic, onions), curcumin (tumeric, curry), and capsaicin (chili peppers). Covalent DNA binding is inhibited by phenylisothiocyanate (broccoli, cabbage), ellagic acid (fruits, nuts, berries), and flavonoids (fruits, vegetables). Biotransformation of potential carcinogens is inhibited by indole-3-carbinol (cruciferous vegetables). Thus, citrus fruit contains carotenoids, but it also contains flavonoids, glucarates, terpenes, and phenolic acids. Does protection require one or more of these plus carotenoids? Studies along this line of reasoning are virtually unknown. Perhaps studies of combinations of protectants will yield an answer where studies of single substances haven't. Block et al. (49) compiled an exhaustive review of the roles of fruit and vegetables in cancer prevention. A statistically significant protective was found in 128 of 156 dietary studies (82%). Significant protection against lung cancer was found in 24 of 25 studies and against cancers of the oral cavity, larynx, or esophagus in 28 of 29 studies. They also found significant protection against pancreatic or stomach cancer (26 of 30 studies), colorectal and bladder cancer (23 of 38 studies), and cancers of the ovary, cervix, and endometrium in 11 of 13 studies. Protection against breast cancer was adduced from metanalysis. Which particular components of fruit and vegetables or which combination of components is active remains to be elucidated.

MINERALS

Tumorigenesis due to deficiencies of various minerals has been reported, among them selenium (50), iron (51), zinc (52), and molybdenum (53). Data regarding others are sparse but available (5). Antitumor effects of selenium in experimental carcinogenesis have also been reported (54,55). A recently reported clinical trial designed to test the efficacy of dietary selenium treatment for prevention of skin cancer found no effect with regard to the stated objective (56). However, selenium treatment proved efficacious against other forms of cancer. A total of 1312 patients with basal or squamous cell carcinomas were treated with a selenium-containing preparation. Cancer incidence in the selenium-treated group was 77 cases compared with 119 cases in the controls, a highly significant difference. All-cause mortality was lower in the selenium-treated individuals (108 vs 129 in controls). Total cancer deaths numbered 29 in the treated group compared with 57 in the placebo group. Over the total observation period of treatment (4.5 ± 2.8 years) and follow-up (6.4 ± 2.0 years), mortality from colorectal or prostate cancer was significantly lower in the treatment group. Total cancer incidence and mortality were reduced significantly. Considering the original aims of the study, the results were serendipitous and most welcome but illustrate some of the vagaries of nutrition-cancer research. A difficulty with studies of trace minerals is that there is a fine line between necessity and toxicity. Efforts to synthesize mineral-containing compounds with lower toxicity are under way.

SUMMARY

Most dietary guidelines, whether general or aimed at a particular disease state, recommend eating a variety of foods and maintaining desirable weight. The foregoing discussion shows that these suggestions might apply to cancer as well. The risk of overweight has been alluded to. The presence of anticancer substances in fruits and vegetables is well documented. A trend toward reductionism in assessing data has led us to choose a particular component of a food when that same food contains many substances that may act synergistically. The interaction of foods or components of foods is addressed rarely. A study of Japanese men in Hawaii showed a significant association between fiber intake and risk in men who ingested less than 61 g/day of fat, but no association if daily fat intake exceeded 61 (57). Lee et al. (58) found the dietary ratio of meat to vegetables influenced risk of colon cancer in Singapore. More studies of food or food component interactions are needed to clarify the relationship of diet to cancer risk. In the meantime, the dietary suggestion of balance, variety, and moderation may be the most useful.

BIBLIOGRAPHY

1. J. D. Potter and A. J. McMichael, *J. Natl. Cancer Inst.* **76**, 557–569 (1986).
2. S. Kune, G. A. Kune, and L. F. Watson, *Nutr. Cancer* **9**, 21–42 (1987).
3. B. Armstrong and R. Doll, *Int. J. Cancer* **15**, 617–631 (1975).
4. R. Doll and R. Peto, *J. Natl. Cancer Inst.* **66**, 1191–1308 (1981).
5. Committee on Diet, Nutrition, and Cancer, *Diet, Nutrition and Cancer*, National Academy Press, Washington, D.C., 1982.
6. A. E. Rogers and M. P. Longnecker, *Lab. Invest.* **59**, 729–759 (1988).
7. E. Kampman et al., *Cancer Epidemiol. Biomarkers, Prev.* **8**, 15–24 (1999).
8. T. Sugimura, *Mutat. Res.* **376**, 211–219 (1997).
9. K. Augustsson et al., *Lancet* **353**, 703–707 (1999).
10. B. S. Drasar and D. Irving, *Br. J. Cancer* **27**, 167–172 (1973).
11. G. Hems, *Br. J. Cancer* **37**, 974–982 (1978).
12. G. Hems and A. Stuart, *Br. J. Cancer* **31**, 118–123 (1975).
13. L. Friedman et al., *J. Natl. Cancer Inst.* **49**, 751–764 (1972).
14. F. J. C. Roe, L. S. Levy, and R. L. Carter, *Food Cosmet. Toxicol.* **8**, 135–145 (1970).
15. E. A. Lew and L. Garfinkel, *J. Chronic Dis.* **32**, 563–576 (1979).
16. C. LaVecchia et al., *Int. J. Cancer* **41**, 492–498 (1988).
17. L. J. Kinlen, *Lancet* **1**, 946–949 (1982).
18. M. Rosen, L. Nystrom, and S. Wall, *Am. J. Epidemiol.* **127**, 42–49 (1988).
19. P. Stocks and M. K. Karn, *Ann. Eugen. (London)* **5**, 237–280 (1933).
20. O. M. Jensen, R. MacLennon, and J. Wahrendorf, *Nutr. Cancer* **4**, 5–19 (1982).
21. D. Y. Jones et al., *J. Natl. Cancer Inst.* **79**, 465–471 (1987).
22. P. J. Goodwin and N. F. Boyd, *J. Natl. Cancer Inst.* **79**, 473–485 (1987).

23. T. Byers, *Cancer* **62**, 1713–1724 (1988).
24. W. C. Willett et al., *N. Engl. J. Med.* **316**, 22–28 (1987).
25. M. A. Belury, *Nutr. Rev.* **53**, 83–89 (1995).
26. J. W. Berg, *Cancer Res.* **35**, 3345–3350 (1975).
27. C. Moreschi, *Z. Immunitätsforsch* **2**, 651–675 (1909).
28. D. M. Klurfeld et al., *Int. J. Cancer* **43**, 922–925 (1989).
29. D. Kritchevsky, *Cancer* **66**, 1321–1325 (1990).
30. J. B. Bristol et al., *Br. Med. J.* **291**, 1467–1470 (1985).
31. J. L. Lyon et al., *J. Natl. Cancer Inst.* **78**, 853–861 (1987).
32. R. E. Frisch et al., *Br. J. Cancer* **52**, 885–891 (1985).
33. D. H. Garabrant et al., *Am. J. Epidemiol.* **119**, 1005–1014 (1984).
34. V. R. Potter, *Science* **101**, 105–109 (1945).
35. D. P. Burkitt, *Cancer* **28**, 3–13 (1971).
36. J. Higginson and A. G. Oettlé, *J. Natl. Cancer Inst.* **24**, 589–671 (1960).
37. S. M. Pilch, ed., *Physiological Effects and Health Consequences of Dietary Fiber*, FASEB, Bethesda, Md., 1987.
38. J. H. Cummings et al., *Gastroenterology* **103**, 1783–1789 (1992).
39. A. R. P. Walker, B. F. Walker, and A. J. Walker, *Br. J. Cancer* **53**, 489–495 (1986).
40. B. S. Reddy et al., *Cancer Res.* **47**, 644–648 (1987).
41. D. Kritchevsky, *Eur. J. Cancer Prev.* **7**(Suppl. 2), S33–S39 (1998).
42. M. J. Hill, *Eur. J. Cancer Prev.* **6**, 219–225 (1997).
43. D. R. Jacobs, Jr., et al., *Nutr. Cancer* **30**, 85–96 (1998).
44. C. S. Fuchs et al., *N. Engl. J. Med.* **340**, 169–176 (1999).
45. The Alpha Tocopherol, Beta Carotene, Cancer Prevention Study Group, *N. Engl. J. Med.* **330**, 1029–1035 (1994).
46. R. S. Sandler et al., *J. Natl. Cancer Inst.* **85**, 884–891 (1993).
47. E. Giovannucci et al., *Cancer Res.* **54**, 2390–2397 (1994).
48. E. Giovannucci et al., *J. Natl. Cancer Inst.* **84**, 91–98 (1992).
49. G. Block, B. Patterson, and A. Subar, *Nutr. Cancer* **18**, 1–29 (1992).
50. R. J. Shamberger, *J. Natl. Cancer Inst.* **44**, 931–936 (1970).
51. L. G. Larsson, A. Sandström, and P. Westling, *Cancer Res.* **35**, 3308–3316 (1975).
52. H. J. Lin et al., *Nutr. Rep. Int.* **15**, 635–643 (1977).
53. C. S. Yang, *Cancer Res.* **40**, 2633–2644 (1980).
54. J. R. Harr et al., *Clin. Toxicol.* **5**, 187–194 (1972).
55. C. Ip and D. K. Sinha, *Cancer Res.* **41**, 31–34 (1981).
56. L. C. Clark et al., *JAMA, J. Am. Med. Assoc.* **276**, 1957–1963 (1996).
57. L. K. Heilbrun et al., *Int. J. Cancer* **44**, 1–6 (1989).
58. H. P. Lee et al., *Int. J. Cancer* **43**, 1007–1016 (1989).

GENERAL REFERENCES

R. B. Alfin-Slater and D. Kritchevsky, eds., *Human Nutrition. A Comprehensive Treatment*, Vol. 7, Plenum Press, New York 1991.

C. Ip et al., eds., *Dietary Fat and Cancer*, Alan R. Liss, New York, 1986.

M. J. Hill, A. Giacosa, and C. P. J. Caygill, eds., *Epidemiology of Diet and Cancer*, Ellis Horwood Ltd., Chuchester, W. Sussex, England, 1994.

F. L. Meyskins, Jr., and K. N. Prasad, eds., *Vitamins and Cancer*, Humana Press, Clifton, N. J., 1986.

I. R. Rowland, ed., *Nutrition, Toxicity, and Cancer*, CRC Press, Boca Raton, Fla., 1991.

DAVID KRITCHEVSKY
The Wistar Institute
Philadelphia, Pennsylvania

CANNING: REGULATORY AND SAFETY CONSIDERATIONS

Food safety and spoilage have been the major issues of the canning industry since its beginnings in the early 1900s. Preventing food from spoilage during storage was indeed a major reason that the canning industry came to be. However, at the turn of the twentieth century there was little or no scientific knowledge, as we know it today, about canning processes. In the late 1700s, Nicolas Appert in France found a "new" way to preserve food for the armed forces that involved corking bottles filled with food and heating them in boiling water. Not until Pasteur's work in 1864 was the relationship between food spoilage and microorganisms discovered. Several other developments in following decades then brought necessary components together to make use of this new knowledge. The science of bacteriology in the late 1800s to early 1900s provided information about specific bacteria and their growth characteristics. Inventions related to thermocouple technology by 1920 and subsequent determination of mathematical methods for calculating sterilization processes laid the foundation for today's safety standards and regulations in the canning industry.

REGULATORY CONTROLS

Regulations are in place to ensure the control of critical parameters in the canning of foods. The U.S. Federal Food, Drug and Cosmetic Act prohibits the distribution and sale of foods that carry disease-causing contaminants as well as a list of undesirable elements categorized as "filth." Two episodes with botulism in the canning industry in 1971 led to a petition from the National Canners Association, now the National Food Processors Association (NFPA), to the U.S. FDA proposing a statement of policy that was handled as a proposal for regulation. After receiving comments, the FDA published minimum good manufacturing practice (GMP) regulations. After various revisions and comment periods since that time, the FDA now has published several GMP regulations that pertain to canned foods in the *Code of Federal Regulations* (CFR).

"Current Good Manufacturing Practice in Manufacturing, Packing, or Holding Human Food" (21 CFR 110) sets forth general requirements for maintaining sanitary conditions in food establishments, such as design and maintenance of facilities and equipment, and pest control. More specific requirements are set forth in an additional series of regulations in Title 21 of the CFR. "Thermally Processed Low-Acid Foods Packaged in Hermetically Sealed Containers" (Part 113) defines low-acid foods, requires design of a scheduled process by a process authority, defines commercial sterility for low-acid canned foods, and specifies design, control, and instrumentation for retorting systems. This regulation also specifies record keeping requirements, procedures for process deviations, and container closure inspection procedures. "Acidified Foods" (Part 114) defines acidified foods and procedures for acidification; this part describes requirements for process determination, corrections for process deviations, and record keeping. "Emergency Permit Control" (Part 108) describes procedures for exemptions from and compliance with registering an establishment when requirements for operation are not being met.

The U.S. Department of Agriculture (USDA) has published regulations for canning of meat (21 CFR 318.300) and poultry (21 CFR 381.300) products, "Canning and Canned Products." The U.S. Meat Inspection Act and the Poultry Products Inspection Act authorize regulation of canned products by the USDA Food Safety and Inspection Service. The canning regulations for the two products are essentially the same and contain requirements similar to those for low-acid canned foods regulated by the FDA.

Personnel and Training

Canning regulations recognize the importance of people in maintaining the integrity of the canning process, in addition to food, process design, and equipment concerns. All retort (canner), container closure, and inspection operations must be performed under the supervision of trained supervisors. Supervisors of low-acid and acidified foods operations are required by regulations in Parts 108, 113, and 114 to satisfactorily complete instruction in a school or training approved by the FDA. These courses have become known as Better Process Control Schools, developed in cooperation with the NFPA's educational arm, the Food Processors Institute (FPI). Instructors for these schools are selected from the FDA, selected universities, the FPI, the NFPA, and industry. Regulations for canned meat and poultry products (Parts 318 and 381) specify that the supervisor of canning operations must attend a school generally recognized as adequate for properly training supervisors of canning operations.

Imported Canned Foods

Food regulations also allow for protection from unsafe imported thermally processed foods. Foreign food processors of low-acid canned foods and acidified foods must register their establishments with the FDA before exporting products to the United States. The FDA assigns each establishment a number that helps track the registration and processing records. Imported low-acid canned foods and acidified foods are subject to all the requirements under the U.S. Federal Food, Drug, and Cosmetic Act and the Fair Packaging and Labeling Act. The processing of low-acid and acidified foods must also comply with GMP regulations in 21 CFR Parts 113 and 114, respectively. Canning processes must be determined by a thermal process authority and filed with the FDA before being exported.

SAFETY CONCERNS

Microbiology of the Canning Process

The microbiology of the can environment during storage is an extremely important consideration in the safety of canned foods. The objective in food canning is to produce a food and container interior that are commercially sterile. This is accomplished through the application of heat alone or in combination with adjusted parameters of water activity, chemicals, or acidity (usually measured as pH). Commercial sterility results from destruction of all viable microorganisms of public health significance as well as those capable of reproduction at normal nonrefrigerated temperatures during storage and distribution.

Generally, the residual oxygen content of canned foods is minimized. Microorganisms that require an environment with oxygen (obligate aerobes) would not be expected to present spoilage or health hazards. (Special consideration would have to be given to products where oxygen is not completely removed and mild heat treatments are used in conjunction with other preservation means, such as curing salts in canned, cured meat products.) Of greatest concern, then, in the anaerobic environment of canned foods is the pH of the food, as acidity does affect the ability of microorganisms to reproduce and survive destruction by heat.

The foremost health concern with canned foods is survival of the pathogenic *Clostridium botulinum* bacteria. *C. botulinum* is a spore-forming anaerobe. Destruction of the spores in canned low-acid foods is essential to prevent germination and toxin formation during nonrefrigerated storage. It is an accepted generalization that spores of *C. botulinum* can germinate into vegetative cells that multiply, with the ultimate production of toxin if the pH of the canned food product is higher than 4.6. Therefore, the pH value of 4.6 is one delineating factor between low-acid and acid or acidified categories of canned foods. A low-acid food means any food, other than alcoholic beverages, with a finished equilibrium pH greater than 4.6 and a water activity greater than 0.85.

There are some extremely heat-resistant spore-forming bacteria that may survive a process for commercial sterility. These bacteria do not present a health hazard, however, and do not present a spoilage problem under normal, recommended conditions of storage. These are thermophilic, or heat-loving, bacteria that require high temperatures to germinate. If this happens, they may produce acid with little or no gas, or gases that swell cans. It is generally recommended that canned foods be quickly cooled below 105°F in thermal processing and then stored below 95°F to prevent problems from these surviving thermophiles.

Process Design

Critical factors influencing the process needed to obtain commercial sterility include:

1. The composition and nature of the food (eg, pH, sauce viscosity, starch content, particle tendency to mat or clump)

2. Initial number of microorganisms on the food, also called bacterial load
3. The microorganism(s) of concern with the particular food and the heat resistance of those organisms
4. Pack style, or preparation of the food as it affects particle size, viscosity, solids-to-liquid ratio, acid-to-low-acid ratio (for acidified foods) and others depending on the product
5. Jar size and conformation
6. Fill weight
7. Initial (or fill) temperature of the food
8. Retort temperature and style (eg, rotating or still)

Designing a thermal process (canning schedule) that takes into account all the critical factors that can influence safety of the finished product requires the use of specific scientific methods. It requires knowledge and data about the microorganisms of concern in that food and the rate at which they are destroyed by heat in the product and container. Process design also requires and involves collecting data showing heat penetration into the food product in its container, under the conditions used during production. These pieces of information are used to calculate a theoretical process that is then tested by inoculated pack studies, to test that the calculated process does provide commercial sterility.

Regulations for all canned foods state that the food processor must obtain scheduled processes established by a competent processing authority. The regulations do not specifically define acceptable process authorities; there are generally recognized organizations, such as the NFPA, members of the Institute for Thermal Processing Specialists, faculty at certain universities, and a number of consultants recognized as having this expertise. The regulations also do not specify actual procedures for thermal process design. Again, the NFPA provides guidance in how to conduct heat penetration tests as well as performing the tests itself. The Institute for Thermal Processing Specialists has also developed protocols for thermal process design.

Process Filing

Critical details of the process used to can a low-acid or acidified food product must be filed with the FDA. The process time, temperature, and other factors critical to adequacy of the process must be filed for each product and product style in each container type and size. Critical factors that influence the integrity of the process in addition to processing time and temperature include: type of processing equipment used; minimum initial temperature of the food; sterilizing value of the process (F_0) or other evidence of process adequacy; pH; water activity; and factors affecting heat penetration such as formulation, fill weight, drained weight, and headspace. The filing of a scheduled process with all necessary parameters delineated shall be done not later than 60 days after registration of the establishment and before any new product is packed. The FDA then edits all processing information filed for questionable, incomplete, or incorrectly filed information, and returns forms needing clarification or completion. The FDA does not certify or approve processes; it is the responsibility of the processor and thermal processing authority to produce a safe process.

Equipment and Procedures

Scheduled processes determined through testing must be applied under equivalent conditions in commercial practice. The regulations contain specifications for equipment design and management to ensure that critical controls important to a thermal process are in place. The regulations regarding equipment and procedures are detailed, and for actual requirements, the reader should consult the CFR. However, the following is a summary of the controls required to ensure the safety of canned foods.

Equipment design features are prescribed for various types of retorts (equipment used for the canning process). They include type and placement of temperature- and pressure-indicating devices and of temperature-recording devices. Because the steam supply affects maintenance of the temperature in the retort, requirements for steam controllers, the steam inlet to the retort, and steam spreaders are also described. Size and positioning of steam bleeders are specified for certain retorts. Air and water supply and drainage are also critical components discussed. Much attention is given to the placement of retort vents and venting processes; removal of air from the retort is critical to the temperature obtained for steam pressure processing.

Raw Materials

All raw materials and ingredients that are susceptible to microbiological contamination must be suitable for use in processing. Specifications for suppliers should be in place to ensure their wholesomeness and quality. Receiving procedures should ensure adequate examination, especially so as to determine their microbiological condition. If stored before use, they should be examined again immediately prior to processing.

Raw agricultural products should be cleaned to remove excess soil from the surfaces. All raw materials should be stored or held in a manner that prevents contamination, increases in microbiological loads, and loss of quality. Physical characteristics of the product may also affect the thermal process. Product size, texture, and condition can affect heat penetration. Raw product controls important to maintain are sorting and grading for the purposes of size and quality.

The water supply must be of sufficient quality for the intended use and from an acceptable source. Any water that contacts food or food-contact surfaces shall be safe and sanitary. Water must be able to be delivered at appropriate temperatures suitable for its use at different locations in the establishment. The residual chlorine content of water needs to be checked and adjusted at various steps in the process and plant operations.

Product Preparation

Blanching is often done. It helps inhibit enzymatic action, which may later alter the characteristics of the goods.

Blanching also expels respiratory gases and oxygen, a step that assists in control of headspace in the container and allows a greater vacuum to be achieved. Softening of food to enable easier packing, setting of flavors and colors, and cleaning are some additional benefits from blanching. Finally, blanching may allow peeling, dicing, cutting, and other steps to be done more easily.

Blanchers must be operated in such a way as to minimize thermophilic bacterial growth and contamination. Rapid heating is preferred. Blanchers should be operated at temperatures in the 190 to 200°F range to control thermophilic growth. Lower temperatures may be used, but the procedures must be evaluated for the growth of organisms. If the product is not to be immediately packed and processed after blanching, it should be rapidly cooled after heating. Potable water must be used on blanched foods washed before filling.

Product formulation must be controlled; starches, gums, sugars, and other thickening agents are particularly critical factors in heat penetration. Inspectors should determine if changes have been made, and if so, if the scheduled process was tested and changed accordingly if needed. Rehydration of dry ingredients, such as beans, rice, peas, and potatoes also affects thickening and heat penetration. Rehydration should be controlled. Viscosity of packing mediums must also be controlled, and inspectors should check for deviations from the original formulations.

Filling Operations

Containers must be appropriate to the product being processed. They must be appropriately cleaned and even sanitized, if required. The filling operation should be inspected to ensure that no contamination occurs during the filling, such as might occur from a ventilator used to prevent trapped air in the product.

Filling of the container prior to the canning process may also be a critical control. Filling requirements specified in the process must be followed. Fill weights and ratio of solids and liquids must be controlled. Headspace may be critical, especially for agitated products that need a gas bubble to achieve movement of the product. Headspace is normally controlled by the fill weight or volume. Too little headspace may result in underprocessing of the product as well as a low vacuum. After filling, exhausting of containers to remove air from the headspace shall be controlled to meet conditions specified in the scheduled process. This is usually accomplished by hot filling, mechanical methods, steam injection, or gas flushing.

Container closing operations are another area requiring safety considerations. Intact, properly formed seams and seals are essential to successful canning and food safety. They are so essential, in fact, that careful examination of can seaming operations and frequent inspection of seals at the end of retorting and after cooling are required by regulations (see "Container Evaluations"). To avoid container leaks, can seaming equipment or other sealing equipment must be properly maintained and operated. Unacceptable practices at this stage are overfilling and using cans with defective flanges or jars with damaged lugs or threads. Food must not overhang or rest in seal areas.

Holding filled containers in the growth range of microorganisms for long periods of time before processing can lead to spoilage before canning takes place. This is known as incipient spoilage. It may increase the microbiological load before processing and reduce the vacuum obtained. Steps to eliminate incipient spoilage include controlling microbial growth on the raw material, during production, and in the container prior to processing. Holding time before processing should be minimized or carefully controlled to prevent growth. Incipient spoilage does not always result in a health hazard, but may occur, particularly with contamination from staphylococci which produces an enterotoxin.

Processing Operations

Operations in the thermal processing area, often called the "cook room," are vital to safety of the canned food and are heavily regulated. Retort operating procedures and venting procedures to be used for each food and container size must be readily available to the retort operator and the inspector. As discussed earlier, the design of the retorts and their critical operating components are described in detail in regulations for low-acid foods. The operator should be sure all aspects are operating properly before processing food. A system must be established and used that ensures that no product moves through the processing room without being retorted. Labels or other visual indicators should be used to provide indication that the containers have been exposed to the retort process.

All sealed containers must be marked with a code that is permanently visible. The code must identify the establishment; the product; and the year, day, and period when packed. Coding is critical to having a means for isolating and/or recalling questionable product.

The initial temperature of the container contents is a critical safety factor. These temperatures should be measured and recorded with sufficient frequency to be sure that the minimum initial temperature specified in the scheduled process is obtained. This is another reason why lag time between filling and processing should be minimized. Initial temperature is normally determined using a handheld thermometer. (Glass stem thermometers are not used due to potential breakage.) For batch retort loads, a container may be removed from the load going into a retort and set aside where it is not subject to extreme temperature change. When the retort lid is closed and steam turned on, the initial temperature is determined. When retort systems use water to cushion or hold containers, the temperature of that water is a critical concern so the initial food temperature is not lowered. The initial temperature is determined by shaking or stirring the container contents prior to measuring; it is the average temperature of the contents.

Other critical factors during processing are measurements, meters, and devices. Equipment must be accurate, and pressure gauges and thermometers must be located for ease in reading. Timing devices that record process information are to be in place and accurate. Venting of the steam retort is extremely important to achieving the desired processing temperature. Placement of containers in the retort must be consistent with conditions of the scheduled process and must not block venting of the retort.

Deviations from the intended process may occur. A line breakdown may occur at several locations and for many reasons. Procedures should be in place to provide direction for continued use or disposition of the product. Full records must be kept of any reprocessing. Product may be set aside for evaluation by a competent processing authority as to whether it may be used or should be destroyed. Records of evaluation procedures, reasons for the deviation, and actions taken must all be recorded and filed.

During inspection, all these factors of processing operations must be determined, and at least one complete cycle of retort operation should be observed.

Cooling and Storage Operations

Cooling of containers after the retort processing must be controlled to prevent recontamination and damage to seams or seals. It also should be done quickly to prevent growth of thermophilic organisms, as well as overcooking, although the latter is usually detrimental only to quality and not to safety.

Containers may be air or water cooled. Water used for cooling must be of good sanitary quality. Cooling water must be chlorinated or otherwise sanitized. The level of free chlorine in the water must be checked and care given to where the samples are taken for this testing. After processing, the sealing compound used in containers is softened, and as it cools, small amounts of cooling water may be drawn into the container. The sealing compound may then set and seal the leak without loss of vacuum. If contaminated cooling water is used, spoilage and possible danger to the consumer may result.

The time required and the temperature to which retorted foods are cooled must be checked. Cans are generally cooled to 95 to 105°F to leave enough heat in the cans to dry them (when water cooled) but not enough to permit the growth of thermophilic organisms. At this temperature the cook is also stopped, preventing overcooking of the product.

Some cooling procedures may result in damaged seams. Precautions should be taken to prevent buckling of metal containers or damage to seals in glass and other packages. If pressure is reduced too quickly during cooling, buckling may occur from the inability of seams to withstand the change. Buckling is a problem due to the potential for the can seams to be pulled apart enough to leak, and contamination of the contents may occur.

Steps to maintain the integrity of container seals must be continued into postcooling container handling. Bacterial contamination can occur in seam and seal areas if handling lines are not sanitary. Moisture present on containers or the can handling equipment can suspend the organisms and allow them to grow. Rough handling can also damage seals. Crates and baskets used to transfer containers should not have sharp edges or protrusions that damage packages. Leaks from dents or defective seams can result in postprocessing contamination of the commercially sterile product. Cleaning and sanitizing procedures should be in place for all crates, racks, conveyors, belts, and machinery.

Before product is stored prior to distribution, it must be sufficiently cooled to prevent growth of thermophilic organisms (105°F). The storage area must be clean with no product stored directly on the floor. Storage temperatures should not exceed 95°F.

Rust should be prevented or it may lead to damaged seams and seals. If the temperature of the warehouse is too high, the cans may sweat and the moisture will lead to rusting. Rust on the can exteriors may be due to other reasons, including improper operation of retort vents, too long come-up time in retort, or steam containing moisture. Failure to remove the surface water or chemical composition of the cooling water may cause rust.

Container Evaluations

Regular observations and examinations are made to determine that container closures are not defective. The establishment's personnel responsible for these procedures and the inspector need equipment that is used to conduct "tear-down" examinations of the can and its seams. Teardown of cans to determine seam integrity must be done by a trained individual. Samples shall be taken from each seaming station not more than every 4 hours. During teardown, measurements are made on cover and body "hooks" of the double seam. Thickness, tightness, and width are measured or observed. Can seam terminology and procedures are the subject of specialized training in the canning industry, due to the importance of producing and maintaining hermetic seals for food safety.

Sampling of containers for closure evaluation is also done during production runs. At this time, gross closure defects can be observed and corrected. Any defects and corrective actions must be recorded. Measurements and recordings should be made no more than every 30 minutes. Additional inspections of closures are made immediately following a jam in a closing machine, after any adjustment to closing machines, or following a prolonged shutdown before a machine is started up.

SAFEGUARDS

Inspections

One of the protections provided for by canning regulations is inspection of a canning establishment. The regulations recognize that there are certain *critical* elements or steps common to every canning process, and low-acid canning process in particular, that must be controlled to prevent a health hazard. It is the responsibility of the inspection authority (eg, FDA for low acid canned foods) to monitor a firm's compliance with these preventive principles.

The FDA and the USDA publish inspection guidance documents that explain procedures to be followed. These documents may be consulted for detailed information about standards. Changes are being made as needed to keep these procedures updated; for example, they must be consistent with Hazard Analysis Critical Control Points (HACCP) regulatory systems being implemented in various segments of the food processing industry in the 1990s. Although the HACCP system has its origin in the industry's quality control functions of the 1960s, its emphasis is food safety and prevention of health hazards. The use of HACCP systems for regulatory purposes is growing in the 1990s; however, the current GMPs covering low-acid canned foods and acidified foods already have their origins in the HACCP principles presented at the 1971 Conference on Food Protection.

Before an inspection, the investigators should become familiar with the firm's scheduled processes and all critical control factors. A *critical factor* is defined in 21 CFR 113.3(f) as any property, characteristic, condition, aspect, or other parameter, variation of which may affect the scheduled process and the attainment of commercial sterility. Regulations state that critical factors specified in the scheduled process *must* be measured and recorded in processing records. The inspector must review documentation at the firm to determine all critical factors are being controlled.

Recalls

Another safeguard in regulation of the canning industry is the provision for recalls. The FDA and USDA have the authority and responsibility to issue recalls for potentially harmful foods. A company may discover defective product on its own and issue a recall. In other instances, the FDA or USDA finds a problem, informs the company, and suggests or requests a recall. Recalls may then be done voluntarily by the processor in cooperation with the governing agency. However, it the company does not comply with a requested recall, the agency can seek a court order authorizing seizure of the product.

The FDA and USDA have guidelines on how to conduct a recall for companies to follow. There are three classes of recalls according to the level of hazard involved. The strictest procedures are in place to ensure appropriate follow-through for recalls presenting the greatest hazard.

GENERAL REFERENCES

D. L. Downing, ed., *A Complete Course in Canning and Related Processes*, 13th ed., Vols. 1–3, CTI Publications, Baltimore, Md., 1996.

A. Gavin and L. M. Weddig, *Canned Foods. Principles of Thermal Process Control, Acidification and Container Closure Evaluation*, 6th ed., The Food Processors Institute, Washington, D.C., 1995.

Y. H. Hui, "Canning: Regulatory and Safety Considerations," in Y. H. Hui *Encyclopedia of Food Science and Technology*, 1st ed., John Wiley & Sons, New York, 1992, pp. 260–264.

Protocol for Carrying Out Heat Penetration Studies, Institute for Thermal Processing Specialists, Fairfax, Va., 1995.

Thermal Processes for Low-Acid Foods in Metal Containers, Bulletin 26-L, 13th ed., National Food Processors Association, Washington, D.C., 1996.

U.S. Code of Federal Regulations, Title 9, Part 318, Subpart G (.300–.311), Canning and Canned Products (Meat), U.S. Government Printing Office, Washington, D.C., 1996.

U.S. Code of Federal Regulations, Title 9, Part 318, Subpart X (.300–.311), Canning and Canned Products (Poultry), U.S. Government Printing Office, Washington, D.C., 1996.

U.S. Code of Federal Regulations, Title 21, Part 114, Acidified Foods, U.S. Government Printing Office, Washington, D.C., 1996.

U.S. Code of Federal Regulations, Title 21, Part 110, Current Good Manufacturing Practice in Manufacturing, Packing or Holding Human Food, U.S. Government Printing Office, Washington, D.C., 1996.

U.S. Code of Federal Regulations, Title 21, Part 108, Emergency Permit Control, U.S. Government Printing Office, Washington, D.C., 1996.

U.S. Code of Federal Regulations, Title 21, Part 113, Thermally Processed Low-Acid Foods Packaged in Hermetically Sealed Containers, U.S. Government Printing Office, Washington, D.C., 1996.

U.S. Food and Drug Administration, *Guide to Inspections of Acidified Food Manufacturers*, 1998.

U.S. Food and Drug Administration, *Guide to Inspections of Low Acid Canned Food Manufacturers*, Part 1 (PB97-196141), 1996; Part 2 (PB97-196158), 1997; Part 3, 1998.

ELIZABETH ANDRESS
University of Georgia
Athens, Georgia

CANOLA OIL

BACKGROUND

The term *canola* refers to cultivars of an oilseed crop, known in many countries as rapeseed, that is a major source of food and feed throughout the world. The crop has become the world's third most important edible oil source after soybeans and palm (1). The development by plant breeders of low erucic acid and low glucosinolate (double-low) varieties of rapeseed has proved pivotal to the rapid expansion in production and use of rapeseed worldwide. The rapeseed industry in Canada adopted the name canola in 1979 to distinguish those cultivars of *Brassica napus* and *Brassica rapa* (formerly, *B. campestris*) that are genetically low in both erucic acid and glucosinolates. Since the oil and meal from the double-low cultivars are nutritionally superior to those of earlier-grown varieties, the generic term *canola* also was used to identify the products from these varieties. The definition of canola was last updated in 1997 to take account of the development of mustard seed, *Brassica juncea*, varieties with low levels of erucic acid and glucosinolates. The definition adopted by the Canola Council of Canada in 1997 reads: "The oilseeds shall be the seed of the genus *Brassica* which shall contain less than 18 micromoles of total glucosinolates per gram of whole seed at a moisture content of 8.5%; and the oil component of which the seed shall contain less than one percent of all fatty acids as erucic acid" (2). Canola is not an exclusive Canadian trademark; the oilseed industry in other countries, such as Australia, Japan, the United States, and the United Kingdom, also have accepted canola to describe double-low rapeseed.

Canola has become Canada's second most valuable crop after wheat. Its rapid rise in economic importance is due to some positive agronomic advantages and the marketing alternatives it offers producers. Canada is the leading exporter of seed in the world, exporting an average volume of 2.8 million tons annually over the period 1992–1997 (3); Japan is the major importer of Canadian rapeseed/canola. In 1996–1997, the domestic crush of seed in Canada exceeded seed export for the first time (2.71 vs 2.52 million tons). Domestic utilization of canola oil in 1996–1997 was 6.82×10^5 t, up from 4.33×10^5 t in 1992–1993, while oil exports increased from 3.67×10^5 to 5.07×10^5 over the same period (4). Among the processed oils and fats, canola has made substantial gains into the shortening and salad and cooking oil segments of the Canadian market during the past decade. Canola accounted for 70.4% of all deodorized fats and oils, 72.8% of the vegetable oils, and more than 80% of the salad oils produced in Canada in the calendar year 1997 (4). Although current canola production in the United States is relatively small, there is a growing demand for canola products; for 1996–1997 alone, Canada exported 4.24×10^5 t of oil and 8.49×10^5 t of meal to the United States, which accounted for more than 80% of the total oil and meal exports (5).

Canola oil is characterized by a low level of saturated fatty acids and a relatively high content of oleic acid; canola oil is second to olive oil in oleic acid content among common vegetable oils. The blood cholesterol-lowering effect observed in humans, when canola oil made up a major portion of the total dietary fat (6), has been an important factor in the increased use of canola products in the world's markets. Rapeseed/canola now ranks third, behind soybean oil and palm oil, in the global disappearance of fats and oils.

Botanical Origin

Although rapeseed production in Canada commenced after World War II, the cultivation of *Brassica* oilseeds has a long history in Europe and Asia (7). Unlike other oilseed crops, rapeseed is not a product of a single species, but comes from two species of the genus *Brassica: B. napus* and *B. rapa*. The cytogenic relationships between rapeseed and its close relatives have been discussed extensively in several reports and are important in understanding the origin, evolution, and plant-breeding strategies of the *Brassica* species (8–11). The two rapeseed species along with *Brassica juncea* (mustard), commonly referred to as oilseed *Brassicas*, collectively provide more than 13.2% of the world's edible oil supply (11). The small, round *Brassica* seeds contain 40 to 44% oil (dry weight basis) and produce a high-protein content (38–41%) oil-free meal. Within the rapeseed species, both spring and winter cultivars exist; the latter are higher yielding than the spring varieties but are less winter hardy than cereals. Winter cultivars of *B. napus* predominate in Europe, whereas spring cultivars of *B. napus* and *B. rapa* are grown in Western Canada. *B. napus* varieties have a generally higher seed-yielding potential (15–20%) than those of *B. rapa*, but they require an additional 8 to 15 frost-free days to mature; in Western Canada, cultivars of *B. napus* require 95 to 105 days for maturity (9,12).

Development of Double-Low Cultivars

Rapeseed production in Canada began with the introduction of seed of *B. napus* from Argentina and *B. rapa* from Poland; these materials, highly heterogenous, constituted the seed stock for the establishment of breeding programs. Although improvements in seed yield and oil content were the first objectives, particular emphasis was also given to oil and meal quality. The elimination of nutritionally undesirable components—erucic acid, found to cause cardiac lipidosis in several animal species (13), from the oil and sulfur-containing glucosinolates from the meal—became a target at the early stages of the breeding programs. Changes in fatty acid composition were made by introducing the inherited traits of low erucic acid into adapted lines of *B. napus* and *B. rapa*. Canadian breeders also accomplished the transfer of genes responsible for the low glucosinolate characteristic into rapeseed varieties. These sulfur-containing constituents are hydrolyzed by an endogenous enzyme, called *myrosinase* (thioglucosidase, EC 3.2.3.1), into thiocyanates, isothiocyanates, and under some conditions nitriles (Fig. 1) when seed cells are ruptured (eg, during seed crushing). The characteristic odor and flavor of *Brassica* vegetables and condiments (mustard) are largely due to the presence of these compounds. Glucosinolate breakdown products also reduce the palatability and, in nonruminant animals, adversely affect the iodine uptake by the thyroid gland (14,15). Breeding for lower glucosinolate content of the seed has resulted in nutritional upgrading of rapeseed meal for all classes of livestock and poultry. Further reduction in glucosinolates content is required, particularly of the indoyl glucosinolates (16), if proteins from canola meal are to be used in human food formulations. This has been partly accomplished by extraction processes employed in preparation of protein concentrates and isolates (17). The ultimate solution would be to breed varieties free of both alkenyl and indolyl glucosinolates. Total elimination of glucosinolates in canola is a future possibility since a strain of *B. rapa* essentially free of alkenyl glucosinolates has been produced (16).

HORTICULTURE-POSTHARVEST HANDLING

Cultivation-Agronomic Considerations

Canola is a cool season crop that requires more available moisture than wheat, as well as cool night temperatures to recover from hot and dry weather. Hence, canola is usually grown at the wetter and cooler margins of the cereal-growing regions of Canada and other countries.

Because *B. juncea* (mustard) is more heat and drought tolerant than the rapeseed cultivars *B. napus* and *B. rapa*, a double-low germplasm of *B. juncea* has been developed in Canada and is being tested for its potential as an oilseed crop in the dry prairie region of western Canada. This approach is seen as the most promising means of increasing the production area for "canola-type" oilseeds in Canada.

Although rapeseed grows well on a variety of soils, it does best on loamy soils that do not crust and impede seedling emergence (18). Good yields are also obtained on light and heavy soils if rainfall, fertilization, and drainage are adequate (12). Canola is moderately tolerant to saline soils and has greater moisture requirements than cereals. Canola also requires more intensive management practices; for a good crop, the need for nitrogen is 20% higher than for a comparable cereal crop. Certified seed is usually used for sowing to ensure absence of mustard or weeds and the quality characteristics of the cultivar.

Canola is grown in Canada on summer fallow or cereal stubble land. From an agronomic viewpoint, the crop fits

$R-C(=N-O-SO_3^-)-S-C_6H_{11}O_5$ (Glucosinolates) $\xrightarrow[H_2O]{\text{Myrosinase}}$ $[R-C(SH)=NOH]$ + glucose + sulfate

$R-S-C\equiv N$ Thiocyanates | $R-N=C=S$ Isothiocyanates | $R-C\equiv N$ Nitriles + sulfur

Figure 1. General structure of glucosinolates and their enzymic hydrolysis products.

well as a crop rotation with cereals; wheat or barley following canola is the preferred sequence. Most fields are not cropped with canola more than once every four years. This practice permits a break in disease and insect pest cycles, while preserving the soil structure. Canola is susceptible to several fungal diseases such as blackleg (*Leptosphaeria maculans*), stem rot (*Sclerotinia sclerotiorum*), damping-off (*Fusarium* spp.), root rot (*Rhizoctonia* spp.), and black spot or leaf spot (*Alternaria* spp.); disease severity varies with year and location (19). A number of different insects can also cause serious damage to canola (eg, flea beetle) (20). Crop rotation is effective in controlling fungi and insects. Several herbicides are also available for a variety of weed species (12,21).

Harvest and Storage Practices

The date and method of canola harvesting are important determinants of yield and seed quality (12,18,22). Immature seeds have higher chlorophyll and free fatty acids contents. On the other hand, if the harvest is delayed, canola is susceptible to shattering of the pods, particularly in windy areas. To avoid this, swathing (leaving cut plants to dry in the field) is applied when the majority of the seeds are at a firm doughlike stage (moisture content 35–40%). Swathing facilitates uniform maturation and yields seeds of good color and high in oil and protein. Premature swathing (seed moisture >45%) results in reduced yields, whereas late swathing (moisture <20%) increases shattering losses and the possibility of frost damage. Canola is ready to combine when the moisture content is below 15%. Freshly harvested seeds normally require up to six weeks to become dormant; storage in dry and cool environments minimizes respiration losses. Canola is usually transported after harvest to dryer-stocker companies and therefrom to oil-processing plants. For storage, silos of various sizes are used, while on the farms the seed is stored in bins. Immature and field-germinated seeds should not be stored for long periods because they deteriorate rapidly. A high percentage of damaged or fragmented seeds also gives rise to a rapid increase in free fatty acids (22,23).

The moisture content of the seed, because of its importance to enzyme activity and the growth of microflora, is the main criterion for safe storage. In the case of oilseeds, including canola, moisture is held by the nonfatty components, thus bringing the critical level of overall seed moisture content much below that required for safe storage of cereal grains (15%). The maximum moisture content for storage of clean canola seeds in a cool environment (<15°C) is 10% (24); others sources (22,25) even suggest lower optimum levels of moisture (8–9%) for storage over periods of several months. Above this level, the seeds must be air dried; air temperatures must be kept below 38°C to prevent heat damage. Aeration during storage, although at much lower airflow rates than drying, is also required to prevent moisture gradients and to cool the stored grain. Figure 2 shows how storage is affected by both temperature and moisture content. Canola is particularly prone to microbial spoilage, with severe losses in seed and oil quality. The rate of spoilage rapidly increases if nonuniform distributions of moisture and temperature exist in the stored grain. An important consideration regarding moisture content of canola is that before solvent extraction of the oil, the seed must be cooked rapidly at temperatures between 80 and 100°C and at water contents between 6 and 10% to inactivate myrosinase; above 10% moisture, hydrolysis of glucosinolates would occur; below 6%, heat inactivation of the enzyme is difficult.

Chlorophyll is another important quality parameter of canola. Besides imparting an undesirable color to the oil, chlorophyll promotes oxidative rancidity in salad oils. Several environmental factors and agronomic practices, including length of growing season, temperature during seed

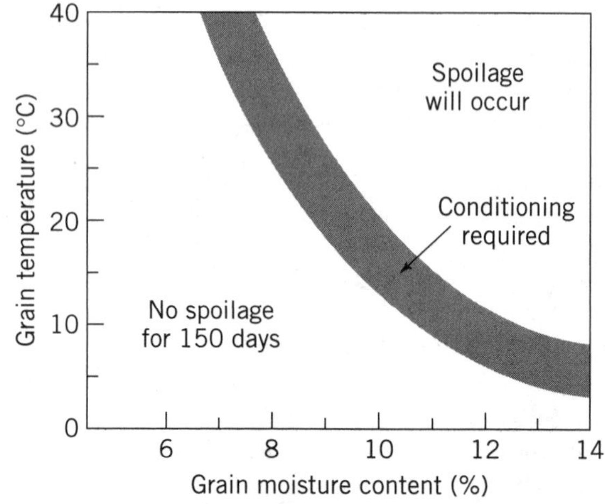

Figure 2. Effects of moisture and temperature on storage of canola. *Source:* Ref. 24.

ripening, cultivar and seeding rate, have been shown to affect the chlorophyll level (26–28). Swathing at 35 to 37% seed moisture followed by drying is effective in reducing substantially the chlorophyll content of canola seeds (29).

BREEDING CANOLA FOR IMPROVED YIELD AND QUALITY—ROLE OF BIOTECHNOLOGY

Important objectives in canola breeding are both agronomic (seed yield, winter and frost hardiness, disease resistance, early maturity, herbicide tolerance, and resistance to shattering) and quality (oil and protein content and composition) traits (9–11,30,31). Although seed yield is one of the most desirable attributes, improvement through selection is a difficult task because of the strong influence from environmental factors.

Oil, being the most valuable component of the seed, is deposited in the form of small lipid droplets in the cytoplasm of embryonal cells. Protein bodies (storage proteins) are also stored in the cytoplasm of these cells (32). In *Brassica* spp., yellow-seeded cultivars have thinner seed coats and thus higher oil and protein content than seeds of darker color (31). It is, therefore, of great interest to producers and crushers to develop yellow-seeded forms or cultivars with increased seed size.

Oil quality is determined by the fatty acid composition. Fatty acid biosynthesis in oilseeds is carried out by assemblies of enzymes located in specific cell compartments (viz., plastids and endoplasmic reticulum of the cytoplasm) (33). Fatty acid synthesis via the fatty acid synthetase (FAS) complex in the plastids produce saturated aliphatic fatty acids (lauric, C12:0; palmitic, C16:0; stearic, C18:0) (Fig. 3). Production of the various saturated fatty acids depends on the presence and activity of specific thioesterases. In most oilseed plants, C18:0 is the primary product of the FAS pathway and it, in turn, is largely desaturated to oleic acid (C18:1n-9) by a specific (stearoyl-ACP) desaturase. Further desaturation of C18:1n-9 to linoleic acid (C18:2n-6) and linolenic acid (18:3n-3) and chain elongation to eicosaenoic acid (C20:1n-9) and erucic acid (C22:1n-9) occurs in the cytoplasm (8,33). Hence, changes in the fatty acid profile in *Brassica* can be achieved by modification of the biosynthetic pathways (Fig. 3). Changes in single genes can have profound effects on the fatty acid composition of the oil. For example, the amount of erucic acid depends on a series of alleles (at two gene loci in *B. napus* and a single locus in *B. rapa*) that influence the elongation of C18:1n-9; by varying the alleles present, it is possible to have erucic acid levels between 0 and over 50%. Interest in erucic acid for use as a feedstock in the oleochemical industry has stimulated interest in producing rapeseed cultivars with very high levels of erucic acid. In general, *Brassica* spp. do not incorporate erucic acid into the *sn*-2 position of the triacylglycerol molecule. However, identification of a genotype of *B. oleracea*, which can accumulate erucic acid in the *sn*-2 position, provides a possible route to *B. napus* lines with enhanced erucic acid levels (34). Another approach being investigated is the cloning of genes for *sn*-2 acyltransferases that recognize erucic acid, for example, from *Limnanthes* spp., and then inserting these genes into rapeseed (34).

Another oil quality breeding objective of canola has been to reduce the level of linolenic acid (C18:3n-3) from 8 to 10% to less than 2% for the purpose of improving oxidative stability of the oil. The first low-linolenic acid cultivar (Stellar), which was introduced in 1987, contained less than 3% linolenic acid (9,35). The oil from this cultivar exhibited improved stability at 60°C and showed negligible changes in sensory and chemical indices of rancidity, compared with regular canola oil (36); improved odor scores from frying tests were also reported for the low-linolenic canola oil (37). Cultivars with even lower levels of linolenic acid (eg, Apollo, avg. 1.7% C18:3n-3) have recently been released (9). Other modifications also have been made to the fatty acid composition of *Brassica* spp., for example, the development of cultivars with 86% oleic acid (C18:1n-9). These modified canola oils were produced in response to the demand for a frying oil with a low level of saturated and *trans* fatty acids and a high stability to oxidative changes without having to be hydrogenated. Another example is the development of cultivars with 10 to 12% palmitic (C16:0) plus palmitoleic (C16:1n-7) acids (15) in order to make margarines exclusively from low erucic acid rapeseed oils. For such products, the tendency to develop large crystals on storage would be diminished.

The application of biotechnology techniques (genetic engineering) to *Brassica* spp. has resulted in some novel oils. Scientists at Calgene Inc. have produced both a high stearic acid and a high lauric acid cultivar of rapeseed/canola using transgenic techniques. The high-lauric canola (>40% C12:0), which is intended for use in the manufacture of detergents and surfactants, was produced by adding the lauroyl-ACP thioesterase gene from the California Bay laurel plant into canola (38,39). The high-stearic canola was produced by inserting the antisense gene construct for stearoyl-ACP desaturase into canola (40). This high-stearic canola could be used in the production of margarine or the manufacture of cocoa butter substitutes (39). Hence, it is reasonable to expect the development of a broad range of rapeseed/canola oils having fatty acid composition appropriate for specific applications (Table 1).

In contrast to oil, relatively little attention has been paid to the protein quality of the meal. Indeed, the reduction of glucosinolate levels in the meal has been the primary objective in the breeding programs. Recessive alleles, responsible for the low-glucosinolate characteristics, have been introduced from *B. napus* var. Bronowski into cultivars of *B. napus* and *B. rapa* by backcrossing (43); several gene loci are involved in the inheritance of glucosinolate content (30,31). Additional improvement in the nutritional value and palatability of canola meal would also result if the 1.0 to 1.5% sinapine and phytic acids present in *Brassica* oilseeds are reduced or eliminated by breeding.

There is great interest among breeders and producers in developing hybrid cultivars of canola for increased yield. Cytoplasmic male sterility (CMS), self-incompatability (SI), and transgenic techniques are among the systems being investigated as means of producing hybrid seeds (10,11,31,39,44). For the CMS system, three inbred lines are used: (*1*) the female parent (male-sterile line owing to cytoplasmic components—incompatibility of "foreign" mitochondria and the nuclear material), (*2*) a maintainer line

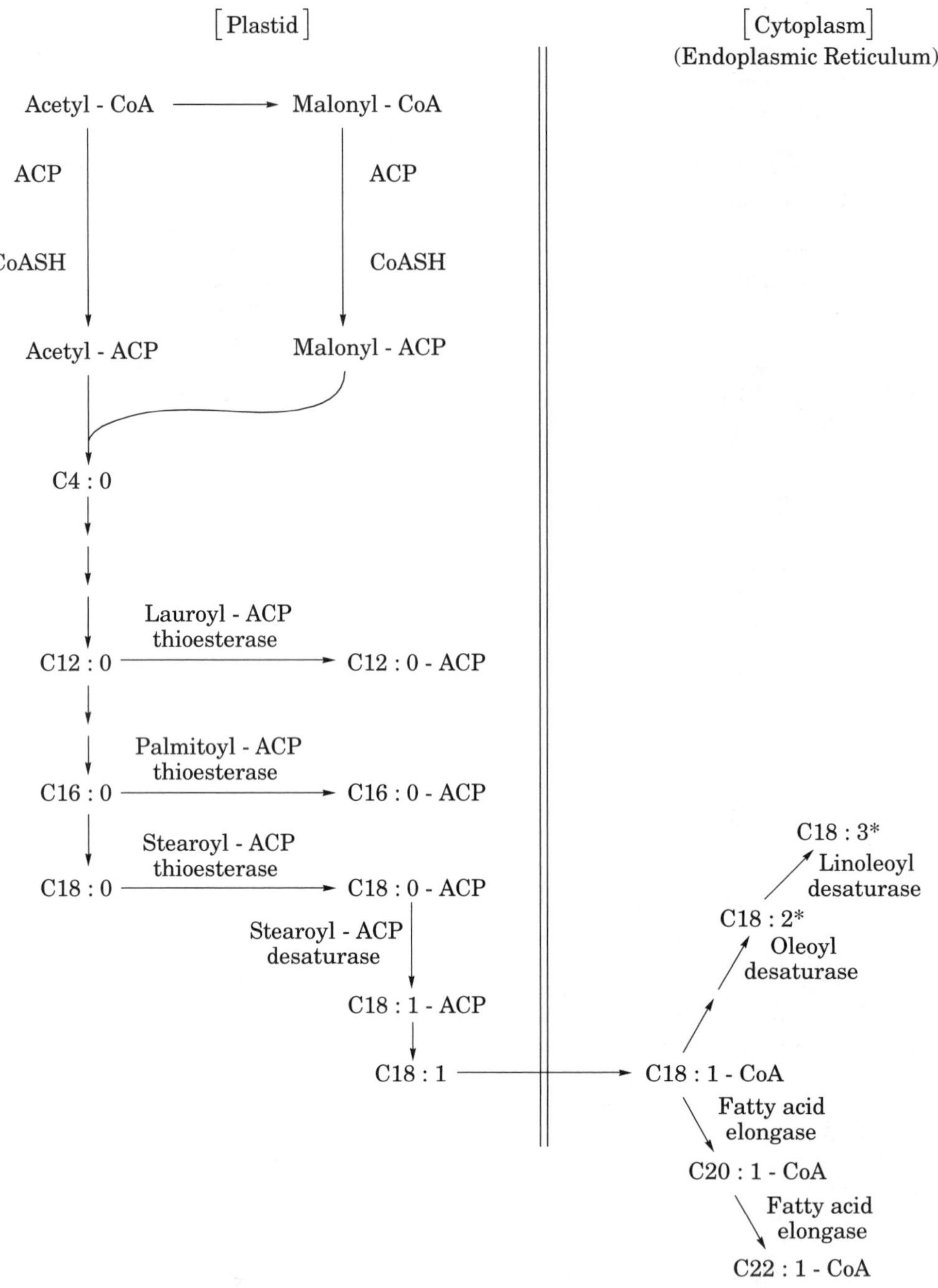

Figure 3. A simplified scheme for the synthesis of fatty acids in plants seeds. Abbr.: ACP—acyl carrier protein; CoASH—coenzyme A. * The oleoyl and linoleoyl desaturases act on C18:1 and C18:2 moieties of phosphatidylcholine to produce C18:2 and C18:3, respectively.

(possess cytoplasm that allows normal production of pollen), and (*3*) a restorer or male parent line that carries nuclear genes that will restore normal fertility to the hybrid cultivars. The expected level of heterosis (up to 40% seed yield increase) is sufficient to offset the cost associated with the development of an effective cytoplasmic male-sterile, genetic restorer system (10,31). Although SI hybrids have been released in Canada, a limitation of the self-incompatibility system for field crops, even though it is used for vegetable crops, is the relatively high cost.

The *Brassica* oilseeds have responded to genetic manipulation better than any other crop. The rapidly growing interest for crop improvement and modification by nonconventional breeding approaches as a result of advances in molecular biology and tissue culture techniques is currently focused on several areas of research (39,45,46).

1. Production of haploids by culture of reproductive organs to reduce the time required for new cultivar development.

Table 1. Fatty Acid Composition of Normal and Modified Rapeseed Oil Resulting from Selection

Oil Type	Fatty acid							
	16:0	16:1	18:0	18:1	18:2	18:3	20:1	22:1
				High erucic				
Normal	3	<1	1	11	13	8	8	50
Selected	2	<1	1	11	12	9	9	57
				Canola				
Low erucic	4	<1	1.5	61	27	9.5	21	<1
High-oleic	3.5	<1	2.5	77	8	2.5	1.5	<1
Low-linolenic	4	<1	1	61	27	2	1	<1
High-lauric[a]	3	<1	1.5	33	11	6.5	1	<1

[a]C12:0, 39%; C14:0, 4%.
Source: Refs. 41 and 42.

2. Protoplast fusion to transfer desirable genetic traits, particularly those carried by cytoplasmic organelles (chloroplast, mitochondria). Transfer of cytoplasmic traits such as male sterility and herbicide resistance was made possible via protoplast fusion and microinjection techniques (39).
3. Selection of cell variants with desirable agronomic characteristics (eg, herbicide resistance, drought tolerance, winter hardiness) after treatment with mutagenic agents and selection pressure of treated cells.
4. Gene transfer using suitable vectors (*Agrobacterium tumefaciens*) or by microinjection techniques.

Consideration is primarily given to genes responsible for herbicide and fungal disease resistance, osmotolerance, manipulation of triglyceride composition, and production of novel high value chemicals (eg, jojoba-type waxes having as precursors eicosenoic and erucic acids).

Most prevalent among the novel-trait canola cultivars are herbicide-tolerant cultivars. Several *B. napus* varieties that are tolerant to the herbicides glyphosate (Round-up Ready), glufosinolate-ammonium (Liberty Link), and imidozolonone (Smart) are presently being grown in Canada. In fact, herbicide-tolerant varieties were estimated to account for 55 to 60% of the total acreage devoted to canola production in 1998 (47). By contrast, it was estimated that only 25 to 30% of the total acreage was devoted to transgenic herbicide-tolerant varieties in 1997. In addition to the development and testing of *B. napus* lines, several lines of herbicide-tolerant *B. rapa* also have been developed and are being tested for agronomic properties. The rapid development of transgenic canola varieties in Canada has been the result of effective collaboration among industry, government, and university researchers and the fact that canola breeding in Canada is conducted primarily by private industry, companies such as Pioneer Hi-Bred, Zeneca Seeds, Plant Genetic Systems (Canada) Lyd., Saskatchewan Wheat Pool, AgrEvo, Limagrain Canada, and Svalof AB (48). The impact of genetic engineering and conventional breeding is expected to result in even more valuable and versatile *Brassica* oilseeds in the future. Specialty oils and by-products for both food and industrial uses, which are within reach of today's breeders, are expected to influence the production and marketing of rapeseed/canola in the years ahead.

COMPARISON OF CANOLA WITH OTHER OILSEEDS

The reduction of erucic acid in rapeseed oil resulted in a marked increase in oleic acid along with smaller increases in linoleic and linolenic acids. Canola oil has a fatty acid composition similar to peanut and olive oil with the exception of the lower palmitic and higher linolenic acid contents. Canola oil contains the lowest level of saturated fatty acids (mainly palmitic and stearic acids) and the highest content of unsaturated fat (mainly monounsaturated) of the common edible vegetable oil sources (Fig. 4). The total quantity of polyunsaturated fatty acids (C18:2n-6 and C18:3n-3) in canola oil is intermediate among the vegetable oils, although it contains an excellent balance between linoleic acid and linolenic acid compared with other widely used vegetable oils. The relatively high linolenic acid content of regular canola oil (10%) makes it susceptible to oxidative rancidity. This problem has been alleviated in the newly developed low linolenic acid cultivars (35,36), although the presence of lipoxygenase in canola seeds has been reported (48). In most respects, there is no obvious physical property distinguishing canola oil from other common vegetable oils. Rapeseed oils (both high and low erucic acid), however, contain a C_{28} brassicasterol, which is not found in other edible vegetable oils (49); brassicasterol is thus a suitable marker to identify *Brassica* oils in vegetable oil blends.

The protein content of canola meal varies with the variety from which it is produced; it ranges between 36 and 42%. Canola meal is higher in fiber than soybean meal (Table 2). The amino acid composition compares well with that of soybean meal (50,51). Although the latter has a greater lysine content, canola meal contains more of the sulfur-containing amino acids (methionine and cystine). Consequently, the two meals tend to complement each other when used as feedstuffs in rations for livestock and poultry. Canola meal is a richer source of minerals than soybean meal (50,51) (Table 3). However, the presence of higher amounts of phytic acid and fiber in canola meal reduces the availability of calcium, magnesium, manganese, phosphorus, zinc, and copper for nonruminant animals (viz., poultry and swine) (51,52). Nevertheless, canola meal is a better source of available calcium, iron, manganese, phosphorus, selenium, and most vitamins than is soybean meal. Canola meal has been shown to have less metabolizable energy than soybean meal (50); on a nutrient content basis alone, canola meal is equivalent to 70 to 75% of 44% protein-content soybean meal for feeding poultry and to about 75 to 80% of the same for feeding swine and ruminants. A reduction in fiber content of the meal would improve the metabolizable energy value. Seeds with yellow color have a lower hull content and, in turn, fiber content than dark seeds, and the hull is more digestible (50). Nonetheless, the available energy in canola meal is a critical factor in its use in animal feeds.

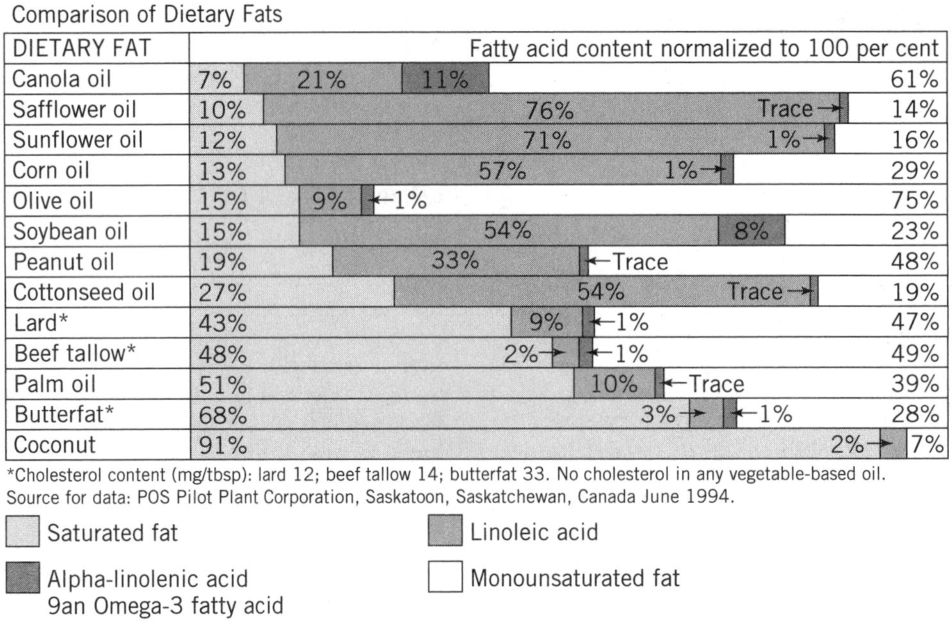

Comparison of Dietary Fats

DIETARY FAT	Fatty acid content normalized to 100 per cent			
Canola oil	7%	21%	11%	61%
Safflower oil	10%	76%	Trace	14%
Sunflower oil	12%	71%	1%	16%
Corn oil	13%	57%	1%	29%
Olive oil	15%	9%	1%	75%
Soybean oil	15%	54%	8%	23%
Peanut oil	19%	33%	Trace	48%
Cottonseed oil	27%	54%	Trace	19%
Lard*	43%	9%	1%	47%
Beef tallow*	48%	2%	1%	49%
Palm oil	51%	10%	Trace	39%
Butterfat*	68%	3%	1%	28%
Coconut	91%	2%		7%

*Cholesterol content (mg/tbsp): lard 12; beef tallow 14; butterfat 33. No cholesterol in any vegetable-based oil.
Source for data: POS Pilot Plant Corporation, Saskatoon, Saskatchewan, Canada June 1994.

Saturated fat | Linoleic acid | Alpha-linolenic acid 9an Omega-3 fatty acid | Monounsaturated fat

Figure 4. Fatty acid composition of common dietary fats. *Source:* Canola Council of Canada.

Table 2. Proximate Analysis and Amino Acid Composition of Canola and Soybean Meals

	Canola meal		Soybean meal	
	As fed, %	In protein, %	As fed, %	In protein, %
Proximate composition				
Moisture	8.0		11.0	
Crude fiber	12.6		7.3	
Ether extract	3.4		0.8	
Protein (N × 6.25)	39.0		45.2	
Amino acid composition				
Alanine	1.73	4.65	1.95	4.31
Arginine	2.32	6.15	3.03	6.71
Aspartic acid	3.03	8.16	5.27	11.66
Cystine	1.05	2.80	0.71	1.61
Glutamic acid	6.46	17.38	8.43	18.65
Glycine	1.92	5.18	2.06	4.55
Histidine	1.26	3.34	1.12	2.48
Isoleucine	1.66	4.38	1.82	4.03
Leucine	2.68	7.07	3.36	7.44
Lysine	2.21	5.95	2.82	6.24
Methionine	0.76	2.03	0.70	1.59
Phenylalanine	1.52	4.04	2.13	4.72
Proline	2.48	6.67	2.27	5.03
Serine	1.69	4.55	2.38	5.27
Threonine	1.68	4.52	1.74	3.85
Tryptophan	0.44	1.19	0.52	1.15
Tyrosine	1.14	3.02	1.33	2.95
Valine	2.14	5.68	1.89	4.18

Source: Refs. 50 and 51.

PROCESSING

Processing of edible oils, including canola oil, is designed to produce the purest extract of triacylglycerols possible. This necessitates removing minor components, naturally present in the oil, that are detrimental to the quality of the finished product (52).

Cleaning Seeds

Before processing, canola seeds undergo cleaning to reduce the presence of any foreign materials. This material, referred to as dockage, was shown to consist mainly of damaged canola seed together with weed seeds (53). A number of operations are involved, including fanning and sieving mills to remove pods and weed seeds; indent machines to remove any large noncanola seeds; destoners to remove dirt and small stones; and finally gravity tables to eliminate anything not removed in the previous processes. Removal of foreign material is designed to ensure less than 2.5% remains in the seed. The presence of damaged canola seed has been shown to be detrimental to the quality of the extracted oil and should be reduced as much as possible before oil extraction (54).

Oil Extraction

Extracting oil from canola seeds is carried out with as little chemical alteration to the oil or meal as possible (55). Once cleaned, canola seeds are rolled or flaked to fracture the seed coat and rupture the oil cells. The production of thin flakes, 0.2- to 0.3-mm thick, is extremely important as a high surface-to-volume ratio is critical during oilseed processing. To facilitate good oil release, the flakes are cooked at 77 to 100°C for 15 to 20 min, to rupture any intact oil cells remaining (56). Flaked and cooked canola seeds gen-

Table 3. Mineral and Vitamin Content of Canola and Soybean Meals

	Canola meals as fed	Soybean meal as fed
	Minerals	
Calcium, %	0.66	0.29
Magnesium, %	0.58	0.27
Phosphorus, %	1.10	0.65
Potassium, %	1.29	2.00
Copper, mg/kg	5.8	23.0
Iron, mg/kg	159	120
Manganese, mg/kg	52	30
Selenium, mg/kg	1.0	0.1
Zinc, mg/kg	70	52
	Vitamins	
Choline, %	0.67	0.28
Biotin, mg/kg	1.0	0.32
Folic acid, mg/kg	2.3	0.6
Niacin, mg/kg	160	29
Pantothenic acid, mg/kg	5.9	16.0
Riboflavin, mg/kg	5.8	2.9
Thiamin, mg/kg	5.2	6.0

Source: Refs. 50 and 51.

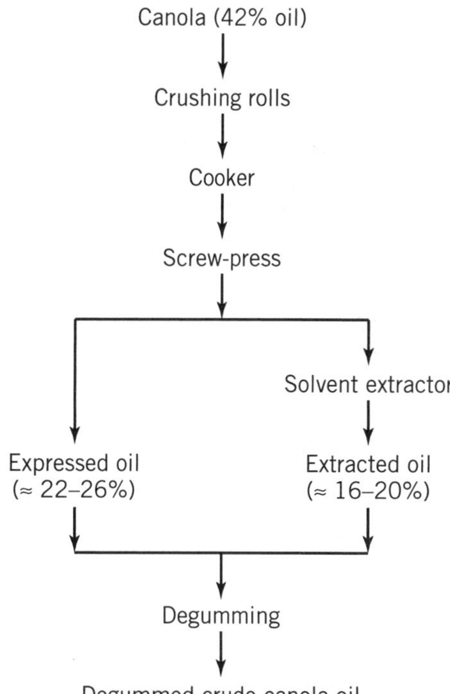

Figure 5. Outline of the primary processing of canola.

erally undergo mild pressing or prepress to reduce oil content from 42 to 16–20%, while compressing the thin flakes into large cake fragments. Canola cake fragments are solvent extracted with normal hexane to remove the remaining oil. This is achieved by countercurrent movement of the cells of pressed canola cake and hexane, thus interfacing the oil in the flake or cake with a rich solvent-oil solution (57). The solvent is recovered from the oil-hexane solution by the conventional distillation system that ensures the oil and meal are solvent free.

The solvent-extracted oil (43–47% of the total oil) is combined with the prepressed oil (53–56% of the total oil) to form the crude oil fraction, as outlined in Figure 5. The crude oil contains a variety of minor constituents (Table 4) detrimental to oil quality that are removed by a series of unit processing steps including degumming, alkali refining, bleaching, hydrogenation, and deodorization.

Degumming

Conventional degumming is carried out in most plants by treating the crude oil with either hot water (85–90°C) or steam (2–5%) while mixing intensively from 1 to 30 min (59). This precipitates the water-hydrated phospholipids, which are then removed by centrifugation. The major drawback to this type of degumming process is that it only removes hydratable phospholipids and still leaves 100 to 200 ppm of phosphorus (0.25–0.5% phosphatides) in the oil. To achieve better results, most Canadian processing plants now carry out acid degumming using citric, phosphoric, or malic acid. This is followed by steam, which results in an oil with phosphorus levels of 50 ppm or less. This product is referred to as acid degummed or super degummed oil.

Table 4. Minor Constituents in Crude, Degummed Canola Oil

Constituents	Amount
Free fatty acids, %	0.5–0.8
Phospholipids (gums), %	0.5–0.8
Unsaponifiables (sterols and sterol esters), %	0.1–0.8
Color bodies, ppm	20–30
Sulfur compounds, ppm	2–10
Iron, ppm	2
Copper, ppm	5

Source: Ref. 58.

Refining

The crude degummed oil is then subjected to refining, which removes free fatty acids, phospholipids, color bodies, iron and copper, as well as some sulfur compounds. The major type of refining in Canada is alkali refining, although there is a shift towards physical refining due to the fewer environmental problems associated with the latter process.

Five major steps are involved in alkali refining.

1. Initial pretreatment of the crude degummed oil with phosphoric acid (0.2–0.5% of 85% phosphoric acid) at 40°C for a minimum of 30 min. This conditions or acidifies the nonhydratable phospholipids (gums) and permits their separation during the refining process. In addition to precipitating nonhydratable phosphatides, chlorophyll as well as pro-oxidant

metals such as iron and copper also are removed (60,61).

2. The second and major step in alkali refining, neutralization with sodium hydroxide (8–12%), depends on the free fatty acid content of the oil and the amount of phosphoric acid used during pretreatment. A slight excess is normally added to ensure complete saponification of these fatty acids.
3. The addition of sodium hydroxide in step 2 produces a soap-stock phase that contains precipitated nonhydratable phospholipids as well as free fatty acids and other insoluble impurities. This soap stock is separated from the oil by centrifugation.
4. Any residual soap stock in the oil is further reduced to 50 ppm or less by washing with water at 86 to 95°C. Citric or phosphoric acid may be added to the washed oil to remove remaining traces of soap stock.
5. During this stage the oil is heated at 105°C with agitation until dry.

Most oil processors use a continuous alkali refining process, although batch refining is still used by some processors. The overall result of alkali refining is a marked reduction in free fatty acids, phosphorus, chlorophyll, and sulfur, as summarized in Table 5.

An alternative method to alkali refining is physical refining, which removes free fatty acids from canola oil by steam distillation. This avoids the production of a soap stock and attendant disposal problems. The oil is first acid degummed with citric or phosphoric acid and then bleached to remove phosphatides and trace metals. This method reduces the phosphorus content to less than 5 ppm, as higher levels cause a darkening of the oil during steam distillation. The latter process (distillative deacidification) is carried out in a specially designed deodorizer. An alternative method to alkali refining is physical or steam refining. This method is used widely in Europe and has been only recently introduced into North America.

Bleaching

Before hydrogenation or deodorization, canola oil is bleached with acid-activated bleaching clay (at 0.125–2.0%) under vacuum at 100 to 125°C for 15 to 30 min. This is an adsorptive process attracting polar substances to active sites of clay surface particles. These include pigments such as chlorophylls and carotenes, residual soaps, phospholipids, and trace metals, as well as primary and secondary oxidation products in the oil. Of particular importance is the presence of very high levels of chlorophylls, which are extremely detrimental to oil quality and must be removed before deodorization. Unlike carotenes, which are heat bleachable in the deodorizer, chlorophylls are not and must be removed by bleaching (63). Most modern oil-processing plants use a continuous bleaching system.

Table 5. Effect of Alkali Refining on the Quality of Canola Oil

Constituents	Crude degummed	Alkali refined
Free fatty acids, %	0.4–1.0	0.05
Phosphorus, ppm	150–250	0–5
Chlorophyll, ppm	5–25	0–25
Sulfur, ppm	3–10	2–7

Source: Ref. 62.

Hydrogenation

Canola oil, like other edible oils, is hydrogenated to improve oxidative stability and to modify the physical properties of the fat or oil. The basic principle of this method is to add hydrogen at the double bonds of the polyunsaturated fatty acids (those containing two or more double bonds), thus eliminating these sites from reaction with oxygen. Only a small portion of these fatty acids may be affected in partially hydrogenated oil stocks, depending on the type of hydrogenation system used. This process requires a catalyst (usually nickel), which may be poisoned by the presence of low levels of sulfur compounds (3–5 ppm) originating from breakdown products of glucosinolates (64). Once hydrogenated, the oil becomes physically harder and more resistant to oxidation. In Canada, canola oil is selectively hydrogenated for the production of margarines and shortenings.

Selective hydrogenation of canola oil is carried under conditions of high temperature, low pressure (200°C, 6 psig), and agitation and affects the more unsaturated fatty acids first. This is illustrated by the fatty acid profiles during the course of hydrogenation of soybean oil shown in Figure 6. All the unsaturated fatty acids are hydrogenated at the same time but at different rates, as indicated by the reaction rate constants. Linolenic acid is hydrogenated 2.3 times faster than linoleic acid, where the latter is hydrogenated 12 times faster than oleic acid (65). This method results in the formation of higher levels of *trans* fatty acids, steeper solid fat index curves, and higher melting points at lower iodine values. The formation of *trans* fatty acids changes the physical properties of the oil as a *trans* double bond is equivalent in physical properties to a saturated

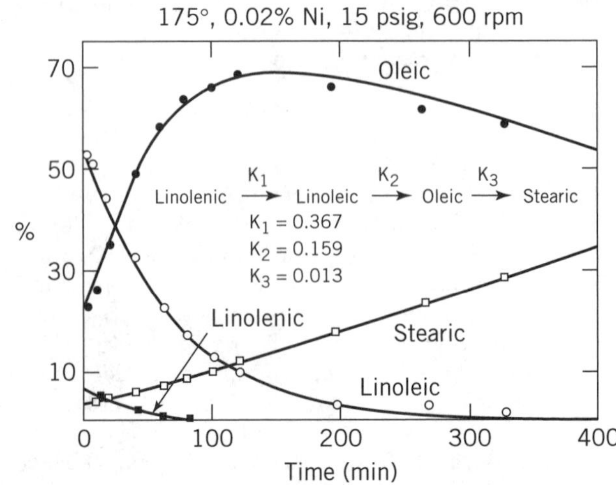

Figure 6. Time-dependent changes in fatty acid composition during hydrogenation of soybean oil. *Source:* Ref. 65.

single bond fatty acid; for example, *cis, trans*-linoleyic acid is equivalent in physical properties to oleic acid. Hardness of margarine oils is due to the higher levels of *trans* fatty acids present rather than to a marked decrease in unsaturation of the fatty acids (66). Because different degrees of hydrogenation are often required by the industry, hydrogenation is a batch process.

Deodorization

Free fatty acids or odiferous or flavor degradation products remaining in the oil are removed by deodorization. This involves steam distillation carried out at very high temperatures (240–270°C) under vacuum (3–8 mmHg). Most plants use a semicontinuous or continuous deodorizing system that is comprised of a large cylindrical tank or shell through which oil is pumped in and passed through a series of trays where it is deaerated and successively deodorized with sparging steam. In the case of canola oil, it must be essentially free of chlorophylls, phosphatidic material, and heavy metals prior to deodorization. Following deodorization, the oil is cooled, passed through a polishing filter, and sparged with nitrogen. The final product is a bland oil that is treated with citric acid (0.005–0.01%) to sequester any trace metals still remaining and other antioxidants to prevent oxidation.

EDIBLE BY-PRODUCTS

Two major edible by-products obtained by processing canola seeds are the oil and the meal. Although canola oil remains the primary edible product, the meal provides an important source of animal feed.

Canola Oil

The major salad oil in Canada is canola, which accounts for more than 80% of the total market. It remains clear and liquid under refrigerated conditions and has acceptable shelf life if protected from direct light. In recent years canola oil has been recognized in the United States as a premium quality oil because of the health implications afforded by its unique fatty acid composition (6,9). This is attributed to the low saturated fatty acid content, the hypocholesterolemic effect of oleic acid (C18:1n-9), a monounsaturated fatty acid present at high levels (ca 60%) in canola oil (Fig. 4), and the favorable balance between linoleic acid and linolenic acid. Oleic acid has been shown equivalent to polyunsaturated fatty acids in lowering low-density lipoprotein (LDL) cholesterol levels (67,68). Studies in Canada (69,70), the United States (71), and Finland (72) have shown canola equally as effective as sunflower oil, soybean oil, and safflower oil in lowering plasma total and LDL cholesterol levels. Because of its low level of saturated fatty acids (<6%) and its relatively high oleic acid content, canola oil has become a premium salad oil in North America. Canola oil, because of its relatively high linolenic acid content and the favorable linolenic acid to linoleic acid balance (viz., C18:3/C18:2 approx. 1:2.5), also may have a favorable effect on thrombosis (clot formation); like atherosclerosis, another major event in coronary heart disease (6,9). Several groups have found canola oil resulted in higher levels of very long chain n-3 polyunsaturated fatty acids (viz., C20:5n-3 and C22:5n-3) in plasma and platelet phospholipids than sunflower oil, soybean oil, or customary mixed fat diets (6,9). The very long chain n-3 polyunsaturated fatty acids in fish oils have been shown to inhibit platelet aggregation (73).

The stability of canola oil is comparable to other edible oils, although its high linolenic acid content (8–10%), like that of soybean oil, makes it susceptible to oxidative rancidity (9). Among eight different vegetable oils stored at 60°C for 8 to 16 days, canola oil had lower peroxide values and volatiles than corn, cottonseed, olive, peanut, safflower, soybean, and sunflower oils (74). A recent study showed a low linolenic acid canola oil product (<2.0%) to be remarkably stable to both heat-accelerated and photochemical oxidation (36,75).

In addition to use as a salad oil, canola oil is used in the production of margarines and shortenings. In North America, margarine is a strictly defined alternative to butter with 80% fat. Although some margarines are made from 100% canola oil, it is generally blended with soybean oil or palm oil. This is carried out to ensure the development of small crystals (β') essential for smooth creaming properties. The unacceptable sandy mouth-feel (due to large fat crystals) for margarines originates from $\beta' \rightarrow \beta$ polymorphic transitions; one way to alleviate this problem is by increasing the heterogeneity of acyl chain lengths of the constituent fatty acids in the oil. Canola oil is selectively hydrogenated and the fat and aqueous phases are mixed together in a crystallizer system or votator to form an emulsion. A variety of different soft and stick margarines are available on the North American market. Shortenings are made in a similar manner to margarines, although in this case, the fat is aerated with air or nitrogen (12%) to improve whiteness and opacity.

High Erucic Acid Oil

As discussed earlier, rapeseed oil has been systematically modified by plant breeders from high erucic acid rapeseed (HEAR) (60%) to a low erucic acid oil (<2%). Nevertheless, HEAR is still permitted in the United States as a stabilizer and thickener component in peanut butter at a maximum level of 2% (76). Superglycerinated, 100% hydrogenated HEAR has been allowed in cake-mix formulations since 1957 in the United States as an emulsifier in the shortening (76).

Canola Meal

After extraction of the oil and removal of hexane, the meal contains approximately 1.5% residual oil and 8 to 10% moisture. It may be granulated to uniform consistency and pelletized or sent directly to storage for marketing (77). Gums removed during degumming are usually added back to the meal to a level of 1.5%; some of the acidulated soap stock is added back as well. The final meal produced is used primarily for the animal feed industry. Canola meal has found much wider acceptance in North America over the past 20 years due primarily to the substantial reduction in glucosinolates accomplished by plant breeding.

Canola meal has, as yet, not found wide acceptance in human nutrition despite the high quality of its protein. This is due to a combination of high levels of phytate and polyphenols as well as the presence of some glucosinolates. A dual solvent process has successfully removed all glucosinolates as well as phytate and hulls (17); the resulting meal was reported to be bland, free-flowing, free from glucosinolates, low in polyphenols, and light in color, with a protein content of 50%. Protein isolates containing 90% protein produced from this meal could have considerable potential for use in human foods.

NONEDIBLE BY-PRODUCTS

The majority of rapeseed currently grown in Canada is canola, although there still remains some acreage devoted to HEAR. This is due to the industrial applications of erucic acid and its cleavage products.

Canola Oil

Canola oil–based printing ink has been found to be superior to petroleum-based ink by reducing the rub problem. Switching to canola-based ink also reduces environmental problems, as 75% of the ink is biodegradable (78). The superior rub properties of canola oil permit both black and color inks to be made with this oil. A number of major Canadian newspapers are now printed with canola oil–based ink.

High Erucic Acid Oil

Erucic acid, a long chain fatty acid ($C_{22:1}$) containing a single double bond, exhibits high fire and smoke points (271°C). These properties allow erucic acid to withstand high temperatures and still remain a liquid at room temperature (79). Consequently, oils containing high levels of erucic acid are used as lubricants or in lubricant formulations (80). Because of their lubricant properties, HEAR oils have found applications as spinning lubricants in the textile, steel, and shipping industries, as drilling oils, and as marine lubricants (81). High erucic acid oil has also been used for the clinical treatment of adrenoleukodistrophy, a rare children's disease. Hydrogenated HEAR oils have industrial applications by producing hard, glossy waxes (82).

Erucic Acid Derivatives

Erucic acid and its hydrogenated derivative, behenic acid, and erucyl and behenyl alcohols as well as other derivatives including esters, amides, amines, and their salts are used extensively in industry as slip, softening, antifoaming, and release agents; emulsifiers; processing aids; conditioners; antistatic agents; stabilizers; and corrosion inhibitors (80).

$$CH_3(CH_2)_7CH{=}CH(CH_2)_{11}COOH$$

erucic acid

$$CH_3(CH_2)_7CH_2CH_2(CH_2)_{11}COOH$$

behenic acid

Erucamide, the amide derivative of erucic acid, has been used for many years as a processing aid and antiblock agent in plastic films. It facilitates the production of plastic parts by acting as a lubricant and forming a thin layer on the surface of the plastic, thus preventing the sheets from sticking to each other.

$$CH_3(CH_2)CH{=}CH(CH_2)_{11}CONH_2$$

erucamide

Behenic acid and its esters are used to enhance the performance of a wide range of pharmaceutical, cosmetic, fabric softener, and hair conditioner products. As antifriction coatings they soften and improve the texture and sewability of cotton and synthetic fabrics (80). In addition to the compounds discussed, there are a large number of HEAR oil-derived alcohols and esters that are used in the cosmetic and pharmaceutical industries.

Erucic acid cleavage products have considerable potential in the production of plastics, resins, and nylons. For example, the oxidative cleavage of erucic acid yields brassylic acid, a 13-carbon dicarboxylic acid, and pelargonic acid, a 9-carbon monocarboxylic acid. A number of long chain nylons prepared from brassylic acid are important in automotive parts and products (79, 83,84).

$$CH_3(CH_2)_6CH{=}CH(CH_2)_{11}COOH \xrightarrow{\text{oxidative cleavage}} \underset{\text{pelargonic acid}}{CH_3(CH_2)_7COOH} + \underset{\text{brassylic acid}}{HOOC(CH_2)_{11}COOH}$$

Alkyl esters of brassylic acid are excellent plasticizers. The corresponding allyl and vinyl esters form polymers and copolymers that may be used in molding compounds, reinforced plastics, laminates, sealants, and coatings (85–88).

STANDARDS AND SAFETY

The standards for crude and refined, bleached, deodorized canola oil, as outlined in Table 6, were established in 1987 by the Canadian General Standard Board (CGSB). Such standards are used in purchasing, consumer protection, health and safety, international trade, and regulatory reference (90). In addition, there is also the Canada Agricultural Products Standard Act, which defines the quality of products, including canola oil (Table 4), for regulating the grading, packing, and marketing of processed canola oil in Canada. These come under the jurisdiction of Agriculture Canada (90).

Canola oil has been approved by Health and Welfare Canada under the Food and Drugs Act and Regulations as safe for human consumption (91). The Canadian legislation favors consumption of canola oil over that of rapeseed oil by limiting the amount of C_{22} monoenoic fatty acids such as erucic acid ($C_{22:1}$) permitted in dietary fats. The three sections of this act, outlined in Table 7, refer to this particular restriction. For example, Section B.09.022 refers to a restriction of less than 5% C_{22} monoenoic fatty acids in fats and oils, in shortenings, salad and cooking oils, as well as table spreads. A similar restriction is cited in Section B.07.043 for salad dressings, while Section

Table 6. Canadian General Standards Board Requirements for Canola Oil

Characteristics	*A. Crude canola oil* Super degummed	Degummed
Free fatty acids (as oleic acid), % by mass, maximum	1.0	1.0
Moisture and impurities, combined % by mass, maximum	0.3	0.3
Phosphorus, ppm, maximum	50	200
Chlorophyll, ppm, maximum	30	30
Sulfur, ppm, maximum	10	10
Refined, bleached color, Lovibond (133.4 mm cell), maximum	1.5 red	1.5 red
Erucic acid, % by mass, maximum	2.0	2.0

Characteristics	*B. Refined, bleached, and deodorized canola oil* Minimum	Maximum
Free fatty acids (as oleic acid), % by mass	—	0.05
Moisture and impurities, combined % by mass	—	0.05
Lovibond color (133.4 mm cell)	—	1.5 red, 15 yellow
Peroxide value, mEq/kg	—	2.0
Cold test, h	12	—
Smoke, point, °C	232	—
Unsaponifiable matter, g/kg	—	15
Saponification value, milligrams potassium hydroxide per gram oil	182	193
Refractive index (n_v 40°C)	1.465	1.467
Iodine value (Wija)	110	126
Crismer value, °C	67	70
Relative density (20°C/water 20°C)	0.914	0.920
Erucic acid, % by mass	—	2.0

Source: Ref. 89.

Table 7. Regulations in Canada's Food and Drugs Act Relating to Canola and Rapeseed Oils

B.09.001 (S)	Vegetable fats and oils shall be fats and oils obtained entirely from the botanical source after which they are named, shall be dry and sweet in flavour and odour and, with the exception of olive oil, may contain Class IV preservatives, an antifoaming agent, and β-carotene in a quantity sufficient to replace that lost during processing, if such addition is shown on the label.
B09.022	No person shall sell cooking oil, margarine, salad oil, simulated dairy products, shortening or food that resembles margarine or shortening, if the product contains more than 5 percent. C_{22} Monoenoic Fatty Acids calculated as a proportion of the total fatty acids contained in the product.
B07.043	No person shall sell a dressing that contains more than 5 percent C_{22} Monoenoic Fatty Acids calculated as a proportion of the total fatty acids contained in the dressing.
B25.054 (1)	Except as otherwise provided in this Division, no person shall sell or advertise for sale of human milk substitute unless it contains, when prepared according to directions for use. (a) per 100 available kilocalories (i) not less than 3.3 and not more than 6.0 grams of fat. (ii) not less than 500 milligrams of linoleic acid in the form of a glyceride. (iii) not more than 1 kilocalories from C_{22} Monoenoic Fatty Acids.

Source: Ref. 92.

B.25.054(a)(iii) limits C_{22} monoenoic acids in infant formula to a maximum of 1% of the calories. In Canada, only 2% erucic acid is now permitted in canola oil, which is much lower than the 5% indicated in the Codex standard for low erucic acid oils by Codex Alimentarious Commission FAO/WHO 1982 (89). Erucic acid oil with a maximum of 2% was subsequently granted GRAS (generally regarded as safe) status in the United States (1985) by the FDA (92). This oil, now identified as canola oil, has since been recognized by the American Health Foundation as a healthful product in the consumer market. A comprehensive review on the safety and health aspects of canola oil (93) concluded that, because of its fatty acid composition, canola oil can be a good substitute for saturated fat in diets intended to reduce the risk of hypercholesterolemic and coronary heart disease. Current nutritional recommendations by many health organizations suggest a reduction of the calories derived from fat in the diet to about 30% and no more than 10% of the caloric intake as saturated fat.

FUTURE OF CANOLA

The discovery of cytoplasmic male sterility (CMS) in *Brassica* spp. a decade ago led to considerable research in developing new hybrids. Because *B. napus* spp. of canola are mainly self-pollinated, the development of hybrids requires cross-pollination. The CMS system has been perfected and is now used for the production of commercial hybrids. Future research should result in improved hybrids for commercial production. The application of biotechnology (and genetic engineering) will assist in the development of improved canola lines that mature early, are frost resistant, are resistant to pesticides, are blackleg tolerant, as well as higher yielding. *B. napus* varieties require long days to flower, so future research will develop lines that mature earlier without any loss in yield (8,31,35,94,95). New high protein lines will also be developed for *B. napus* varieties, whereas higher yields and oil content and modified fatty acid composition are the long-term goals for *B. rapa* breeding research programs. Biotechnology should also result in new lines with cold temperature hardiness and frost resistance that can germinate at much lower soil temperatures and permit early seeding. As breeders and genetic engineers manage to tailor rapeseed/canola, segregating the crop on the basis of specific characteristics will evolve. Processors may develop contracts with farmers to plant varieties that provide products with the desired end-use properties. In addition to the breeding programs, modifications in canola processing technologies will be sought to improve the finished products while being environmentally safe.

BIBLIOGRAPHY

1. "Fats and Oils Industry Changes," *J. Am. Oil Chem. Soc.* **65**, 702–713 (1988).
2. J. Daun and D. Adolphe, "A Revision of the Canola Definition," *Bulletin of Groupe Consultatif International de Recherche sur la Colza* **14**, 134–141 (1997).
3. Canola Council of Canada, URL: *http://www.canola-council.org/Stats/seedsupplydemand.htm.*
4. Canola Council of Canada, URL: *http://www.canola-council.org/Stats/oilmealsupplydemand.htm.*
5. Canola Council of Canada, URL: *http://www.canola-council.org/Stats/oilmealexports.htm.*
6. B. E. McDonald, "Oil Properties of Importance in Human Nutrition," in D. S. Kimber and D. I. McGregor, eds., *Brassica Oilseeds: Production and Utilization*, CAB International, Wallingford, UK, 1995, pp. 291–299.
7. L. A. Appelqvist, "Historical Background," in L. A. Appelqvist and R. Ohlson, eds., *Rapeseed: Cultivation, Composition, Processing and Utilization*, Elsevier, Amsterdam, 1972, pp. 1–8.
8. D. S. Kimber and D. I. McGregor, "The Species and Their Origin, Cultivation and World Production," in D. S. Kimber and D. I. McGregor, eds., *Brassica Oilseeds: Production and Utilization*, CAB International, Wallingford, UK, 1995, pp. 1–7.
9. N. A. M. Eskin et al. "Canola Oil," in Y. H. Hui ed., *Bailey's Industrial Oil & Fat Products Edible Oil and Fat Products*, 5th ed., Vol. 2, *Edible Oil and Fat Products: Oils and Oil Seeds*, John Wiley & Sons, New York, 1995, pp. 1–95.
10. R. K. Downey and G. F. K. Rakow, "Rapeseed and Mustard," in W. R. Fehr, ed., *Principles of Cultivar Development*, Vol. 2, Macmillan, London, 1987, pp. 437–486.
11. R. K. Downey and G. Robbelen, "Brassica Species," in G. Robbelen, R. K. Downey, and A. Ashri, eds., *Oil Crops of the World: Their Breeding and Utilization*, McGraw-Hill, New York, 1989, pp. 339–362.
12. *Rapeseed-Canola*, Canola Council of Canada, Winnipeg, Canada, 1981, pp. 12–14.
13. F. D. Sauer and J. K. G. Kramer, "The Problem Associated with the Feeding of High Erucic Acid Rapeseed Oils and Some Fish Oils to Experimental Animals," in J. K. G. Kramer, F. D. Sauer, and W. J. Pigden, eds., *High and Low Erucic Acid Rapeseed Oils: Production, Usage, Chemistry and Toxicological Evaluation*, Academic Press, New York, 1983, pp. 253–292.
14. G. R. Fenwick, R. K. Heaney, and W. J. Mullin, "Glucosinolates and Their Breakdown Products in Food and Food Plants," *Crit. Rev. Food Sci. Nutr.* **18**, 123–201 (1983).
15. B. Uppstrom, "Seed Chemistry," in D. S. Kimber and D. I. McGregor, eds., *Brassica Oilseeds: Production and Utilization*, CAB International, Wallingford, UK, 1995, pp. 217–242.
16. D. S. Hutcheson, "Development of Improved *B. campestris* Cultivars Essentially Free from Both Alkenyl and Indolyl Glucosinolates," *Canola Research Summary 1985–1989*, Canola Council of Canada, Winnipeg, Manitoba, 1989, p. 54.
17. L. L. Diosady, L. J. Rubin, C. R., Phillips, and M. Naczk, "Effect of Alkanol-Ammonia-Water Treatment on the Glucosinolate Content of Rapeseed Meal," *Can. Inst. Food Sci. Technol. J.* **18**, 311–315 (1985).
18. B. Loof, "Cultivation of Rapeseed," L. A. Appelqvist and R. Ohlson, eds., *Rapeseed: Cultivation, Composition, Processing and Utilization*, Elsevier, Amsterdam, 1972, pp. 49–59.
19. S. R. Rimmer and L. Buchwaldt, "Diseases," in D. S. Kimber and D. I. McGregor eds., *Brassica Oilseeds: Production and Utilization*, CAB International, Wallingford, UK, 1995, pp. 111–140.
20. B. Ekbom, "Insect Pests," in D. S. Kimber and D. I. McGregor, eds., *Brassica Oilseeds: Production and Utilization*, CAB International, Wallingford, UK, 1995, pp. 141–152.
21. J. Orson, "Weeds and Their Control," in D. S. Kimber and D. I. McGregor, eds., *Brassica Oilseeds: Production and Utilization*, CAB International, Wallingford, UK, 1995, pp. 93–109.
22. A. Pouzet, "Agronomy," in D. S. Kimber and D. I. McGregor, eds., *Brassica Oilseeds: Production and Utilization*, CAB International, Wallingford, UK, 1995, pp. 65–92.
23. L. A. Appelqvist and B. Loof, "Postharvest Handling and Storage of Rapeseed," in L. A. Appelqvist and R. Ohlson, eds., *Rapeseed: Cultivation, Composition, Processing and Utilization*, Elsevier, Amsterdam, 1972, pp. 60–100.
24. *Guide to Farm Practice in Saskatchewan*, University of Saskatchewan, Saskatoon, 1987, pp. 60–61.
25. H. B. W. Patterson, *Handling and Storage of Oilseeds, Oils, Fats and Meal*, Elsevier Applied Science, London, 1989, pp. 133–140.
26. K. M. Clear and J. K. Daun, "Chlorophyll in Canola Seed and Oil," in *8th Progress Report on Canola Seed, Oil and Meal Fractions*, Canola Council of Canada, Winnipeg, Manitoba, 1987, pp. 277–279.
27. K. Ward, et al., "Genotypic and Environmental Effects on Seed Chlorophyll Levels in Canola (*Brassica napus*)," in D. I. McGregor, ed., *Proceedings of the 8th International Rapeseed Congress*, Saskatoon, SK, 1991, pp. 1241–1245.
28. K. Ward et al., "Effects of Genotype and Environment on Seed Chlorophyll Degradation During Ripening in Four Cultivars of Oilseed Rape (*Brassica napus*)," *Can. J. Plant Sci.* **72**, 643–649 (1992).
29. S. Cenkowski, S. Sokhansanj, and F. W. Sosulski, "The Effect of Drying Temperature on Green Color and Chlorophyll Content of Canola Seeds," *Can. Inst. Food Sci. Technol. J.* **22**, 383–386 (1989).
30. B. R. Stefansson, "The Development of Improved Rapeseed Cultivars," *Bulletin of Groupe Consultatif International de Recherche sur la Colza* **14**, 143–159 (1997).
31. G. C. Buzza, "Plant Breeding," in D. S. Kimber and D. I. McGregor, eds., *Brassica Oilseeds: Production and Utilization*, CAB International, Wallingford, UK, 1995, pp. 153–175.
32. L. Bengtsson, A. Von Hofsten, and B. Loof, "Botany of Rapeseed," in L. A. Appelqvist and R. Ohlson, eds., *Rapeseed: Cul-*

tivation, Composition, Processing and Utilization, Elsevier, Amsterdam, 1972, pp. 36–48.

33. C. H. Foyer and W. P. Quick, *A Molecular Approach to Primary Metabolism in Higher Plants*, Taylor and Francis, London, 1997, pp. 311–326.
34. D. C. Taylor et al., "Stereospecific Analysis of Seed Triacylglycerols from High-Erucic Acid Brassicaceae: Detection of Erucic Acid at the *sn*-2 Position in *Brassica oleracea* L. Genotype," *J. Am. Oil Chem. Soc.* **71**, 163–167 (1994).
35. R. Scarth, et al., "Stellar: Low Linolenic-High Linoleic Acid Summer Rape," *Can. J. Plant Sci.* **68**, 509–511 (1988).
36. N. A. M. Eskin et al., "Stability of Low Linolenic Acid Canola Oil to Frying Temperature," *J. Am. Oil Chem. Soc.* **66**, 1081–1084 (1989).
37. A. Prevot et al., "A New Variety of Low-Linolenic Rapeseed Oil: Characteristics and Room-Odor Tests," *J. Am. Oil Chem. Soc.* **67**, 161–164 (1990).
38. Anonymous, "Calgene Seeks Approval for Modified Canola," *INFORM* **5**, 716 (1994).
39. D. J. Murphy and R. Mithen, "Biotechnology," in D. S. Kimber and D. I. McGregor, eds., *Brassica Oilseed: Production and Utilization*, CAB International, Wallingford, UK, 1995, pp. 177–193.
40. D. S. Knutzon et al., "Modification of Brassica Seed Oil by Antisense Expression of a Stearoyl-Acyl Carrier Protein Desaturase Gene," *Proc. Natl. Acad. Sci.* **89**, 2624–2628 (1992).
41. B. E. McDonald and K. Fitzpatrick, "Designer Vegetable Oils," in G. Mazza, ed. *Functional Foods: Biochemical and Processing Aspects*. Technomic Publishing, Lancaster, Pa., U.S.A., 1998.
42. R. K. Downey, "It's All in the Breeding," *Proceedings of the 19th Annual Convention of the Canola Council of Canada Meeting*, San Francisco, Calif., March 24–26, 1986, pp. 25–36.
43. R. K. Downey, "The Origin and Description of the Brassica Oilseed Crops," in J. K. G. Kramer, F. D. Sauer, and W. J. Pigden, eds., *High and Low Erucic Acid Rapeseed Oils: Production, Usage, Chemistry and Toxicological Evaluation*, Academic Press, New York, 1983, pp. 1–20.
44. K. F. Thompson and W. G. Hughes, "Breeding and Varieties," in D. H. Scarisbrick and R. W. Daniels, eds., *Oilseed Rape*, Collins, London, 1986, pp. 32–82.
45. W. A. Keller, "The Application of Biotechnology to Canola Improvement," *Crit. Rev. Food Sci. Nutr.* **18**, 56 (1983).
46. Anonymous, "Breeding Canola for Success," *J. Am. Oil Chem. Soc.* **65**, 1567–1568 (1988).
47. Canola Council of Canada, URL: *http://www.canola-council org/board/biotech/31.html* (last accessed Feb. 17, 1998).
48. A. Khalyfa, S. Kermasha, and I. Alli, "Partial Purification and Characterization of Lipoxygenase of Canola Seed (*Brassica napus* var. Westar)," *J. Agric. Food Chem.* **38**, 2003–2008 (1990).
49. R. G. Ackman, "Chemical Composition of Rapeseed Oil," Bulletin of *Groupe Consultatif International de Recherche sur la Colza* **14**, 85–129 (1997).
50. D. R. Clandinin et al., "Composition of Canola Meal," in R. Salmon and D. R. Clandinin, eds., *Canola Meal for Livestock and Poultry*, Canola Council of Canada, Winnipeg, Manitoba, 1989, pp. 5–7.
51. J. M. Bell, "Meal and By-product Utilization in Animal Nutrition," in D. S. Kimber and D. I. McGregor, eds., *Brassica Oilseed: Production and Utilization*, CAB International, Wallingford, UK, 1995, pp. 301–337.
52. N. A. M. Eskin and R. Bacchus, "Processing of Canola Oil," in M. Vaisey-Genser and N. A. M. Eskin, eds., *Canola Oil: Properties and Performance*, Canola Council of Canada, Winnipeg, Canada, 1987.
53. F. W. Hougen, J. K. Daun, and D. C. Durnin. "The Composition and Quality of Canola Dockage," in D. S. Kimber and D. I. McGregor, eds., *Brassica Oilseed: Production and Utilization*, CAB International, Wallingford, UK, 1995, pp. 287–297.
54. F. Ismail, N. A. M. Eskin, and M. Vaisey-Genser, "The Effect of Dockage on the Stability of Canola Oil," in *6th Progress Report on Canola Seed, Oil, Meal and Meal Fractions*, Canola Council of Canada, Winnipeg, Manitoba, 1980, pp. 234–239.
55. R. Simmons, "The Effect of Rapeseed Quality on Processing Procedures," in *Proceedings of the International Conference on Science, Technology and Marketing of Rapeseed and Rapeseed Products*, Rapeseed Association of Canada, St. Adele, Quebec, 1970, p. 121–126.
56. D. L. Beach, "Primary Processing of Vegetable Oils," in J. T. Harapiak, ed., *Oilseed and Pulse Crops in Western Canada*, Western Cooperative Fertilizers Ltd., Calgary, Alberta, 1975, pp. 541–550.
57. D. L. Beach, "Rapeseed Crushing and Extraction," *Bulletin of Groupe Consultatif International de Recherche sur la Colza* **14**, 181–195 (1997).
58. B. F. Teasedale, "Processing of Vegetable Oils," in D. H. Scarisbrick and R. W. Daniels, eds., *Oilseed Rape*, Collins, London, 1986, pp. 551–585.
59. T. K. Mag, "Canola Processing in Canada," *J. Am. Oil Chem. Soc.* **60**, 332A–336A (1983).
60. R. Ohlson and C. Scensson, "Comparison of Oxalix Acid and Phosphoric Acid as Degumming Agents for Vegetable Oils," *J. Am. Oil Chem. Soc.* **53**, 8–11 (1976).
61. G. R. List et al., "Steam-refined Soybean Oil 1. Effect of Refining and Degumming Methods on Oil Quality," *J. Am. Oil Chem. Soc.* **55**, 277–279 (1978).
62. B. F. Teasedale and T. K. Mag, "The Commercial Processing of Low and High Erucic Acid Rapeseed Oils," *Bulletin of Groupe Consultatif International de Recherche sur la Colza* **14**, 199–228 (1997).
63. H. Niewiadomski, "Progress in the Technology of Rapeseed Oil for Edible Purposes," *Chem. Ind.*, 883–888 (1970).
64. J. K. Daun and F. W. Hougen, "Sulfur Content of Rapeseed Oils," *J. Am. Oil Chem. Soc.* **53**, 169–171 (1976).
65. R. R. Allen, "Hydrogenation," *J. Am. Oil Chem. Soc.* **58**, 166–169 (1981).
66. T. J. Weiss, *Foods, Oils and Their Uses*, AVI Publishing, Westport, Conn., 1970.
67. R. S. Mattson and S. M. Grundy, "Comparison of Effects of Dietary Saturated, Monounsaturated and Polyunsaturated Fatty Acids on Plasma Lipids and Lipoproteins in Man," *J. Lipid Res.* **26**, 194–202 (1985).
68. R. P. Mensink and M. B. Katan, "Effect of a Diet Enriched with Monounsaturated or Polyunsaturated Fatty Acids on Level of Low-Density Lipoprotein Cholesterol in Healthy Women and Men," *N. Engl. J. Med.* **321**, 436–441 (1989).
69. B. E. McDonald et al., "Comparison of the Effect of Canola Oil and Sunflower Oil on Plasma Lipids and Lipoproteins and on in vivo Thromboxane A_2 and Prostacyclin Production in Healthy Young Men," *Am. J. Clin. Nutr.* **50**, 1382–1388 (1989).
70. J. K. Chan, V. M. Bruce, and B. E. McDonald, "Dietary α-Linolenic Acid Is as Effective as Oleic Acid and Linoleic Acid in Lowering Blood Cholesterol in Normolipidemic Men," *Am. J. Clin. Nutr.* **53**, 1230–1240 (1991).
71. G. M. Wardlaw et al., "Serum Lipid and Apolipoprotein Concentration in Healthy Men on Diets Enriched in Either Canola Oil or Safflower Oil," *Am. J. Clin. Nutr.* **54**, 104–110 (1991).

72. L. M. Valsta et al., "Effects of a Monounsaturated Rapeseed Oil and a Polyunsaturated Sunflower Oil on Lipoprotein Levels in Humans," *Arter. Thromb.* **12**, 50–57 (1992).
73. P. M. Herold and J. E. Kinsella, "Fish Oil Consumption and Decreased Risk of Cardiovascular Disease: a Comparison of Findings from Animal and Human Feeding Trials," *Am. J. Clin. Nutr.* **43**, 556–598 (1986).
74. J. M. Snyder, E. N. Frankel, and E. Selke, "Capillary Gas Chromatographic Analyses of Headspace Volatiles from Vegetable Oils," *J. Am. Oil Chem. Soc.* **62**, 1675–1679 (1985).
75. R. Przybylski et al., Stability of low linolenic acid canola oil to accelerated storage at 60°C, *Lebensm.-Wiss. Technol.* **26**, 205–209 (1993).
76. Federal Register, **42**, 44335–48336 (1977).
77. M. D. Pickard et al., "Processing of Canola Seed for Quality Meal," in R. Salman and D. R. Clandinin, eds., *Canola Meal for Livestock and Poultry*, Canola Council of Canada, Winnipeg, Canada, 1989, pp. 3–4.
78. Anonymous, "What's Black and White and Read All Over?" *Canola Digest* **22**, 1 (1988).
79. H. J. Nieschlag et al., "Synthetic Wax Esters and Diesters from Crambe and Limnanthes Seed Oils," *Ind. Eng. Chem. Prod. Res. Dev.* **16**, 202–207 (1977).
80. D. L. van Dyne, M. G. Blase, and K. D. Carlson, *Industrial Feedstocks and Products from High Erucic Acid Oil: Crambe and Industrial Rapeseed*, University of Missouri, Columbia, 1990.
81. H. J. Nieschlag and I. A. Wolff, "Industrial Uses of High Erucic Oils," *J. Am. Oil Chem. Soc.* **41**, 1–5 (1971).
82. T. K. Miwa and I. A. Wolff, "Fatty Acids, Fatty Alcohols, Wax Esters and Methyl Esters from *Crambe abyssinica* and *Lunaria annua* Seed Oils," *J. Am. Oil Chem. Soc.* **40**, 742–744 (1963).
83. J. L. Greene, Jr. R. Perkins, Jr., and I. A. Wolff, "Aminotridecanoic Acid from Erucic Acid," *Ind. Eng. Chem. Prod. Res. Dev.* **8**, 171–176 (1969).
84. R. B. Perkins, Jr. et al., "Nylons from Vegetable Oils: −13, −13/13 and −6/13," *Mod. Plastics* **46**, 136–139 (1969).
85. S. P. Chang, T. K. Miwa, and I. A. Wolff, "Alkyl Vinyl Esters of Brassylic Acid," *J. Polym. Sci.*, Part A-1, **5**, 2457–2556 (1967).
86. S. P. Chang, T. K. Miwa, and W. H. Tallent, "Allylic Prepolymers from Bassylic and Azelaic Acids," *J. Appl. Polym. Sci.* **18**, 319–334 (1974).
87. H. J. Nieschlag et al., "Brassylic Acid Esters as Plasticizers for Polyvinylchloride," *Ind. Eng. Chem. Prod. Res. Dev.* **3**, 146–149 (1964).
88. H. J. Nieschlag et al., "Diester Plasticizers from Mixed Crambe Dibasic Acids," *Ind. Eng. Chem. Prod. Res. Dev.* **6**, 201–204 (1967).
89. FAO/WHO, *Codex Standards for Edible Low Erucic Acid Rapeseed Oil*. Codex Standard 123-1981. Codex Alimentarious Commission, Joint Food and Agriculture Organization/World Health Organization Food Standards Programme, April 1982.
90. M. Vaisey-Genser, "Regulations Related to Canola Oil," in M. Vaisey-Genser and N. A. M. Eskin, eds., *Canola Oil: Properties and Performance*, Canola Council of Canada, Winnipeg, Canada, 1987, pp. 10–15.
91. Health and Welfare Canada, *The Food and Drugs Act and Regulations*, Supply and Services Canada, Ottawa, 1953, amended to September 1986.
92. Food and Drug Administration, *Direct Food Substances Affirmed as Generally Regarded as Safe; Low Erucic Acid Rapeseed Oil*. Federal Food, Drug and Cosmetic Acit 21 CRF Part 184, 1555 (c), amended 9 Jan. Department of Health and Human Services, Food and Drug Administration, Washington, D.C., 1985.
93. J. Dupont et al., "Food Safety and Health Effects of Canola Oil," *J. Am. College Nutr.* **8**, 360–375 (1989).
94. Anonymous, "Contiseed Banks on Hybrids," *Canola Digest* **22**, 8 (1988).
95. Anonymous, "Improving Oil/Protein Important at U. of A.," *Canola Digest* **22**, 4 (1988).

GENERAL REFERENCES

References 9, 25, 78, and 89 are also good general references.

L. A. Appelqvist and R. Ohlson, *Rapeseed Cultivation, Composition, Processing and Utilization*, Elsevier, Amsterdam, 1972.

D. S. Kimber and D. I. McGregor, *Brassica Oilseeds: Production and Utilization*, CAB, International, Wellingford, UK, 1995.

J. K. G. Kramer, F. D. Sauer, and W. J. Pigden, eds., *High and Low Erucic Rapeseed Oils—Production, Usage, Chemistry and Toxicological Evaluation*, Academic Press, New York, 1983.

G. Robbelen, R. K. Downey, and A. Ashri, *Oil Crops of the World*, McGraw-Hill, New York, 1989.

R. E. Salmon and D. R. Clandinin, *Canola Meal for Livestock and Poultry*, Canola Council of Canada, Winnipeg, Manitoba.

D. H. Scarisbrick and R. W. Daniels, *Oilseed Rape*, Collins, London, 1986.

D. Swern, *Bailey's Industrial Oil and Fat Products*, 4th ed., John Wiley & Sons, New York, 1985.

M. Vaisey-Genser and N. A. M. Eskin, *Canola Oil: Properties and Performance*, Canola Council of Canada, Winnipeg, Manitoba 1987.

N. A. M. Eskin
B. E. McDonald
University of Manitoba
Winnipeg, Manitoba, Canada

CARAMEL. See Colorants: caramel.

CARBOHYDRATES: CLASSIFICATION, CHEMISTRY, AND LABELING

Carbohydrates can be, and are, classified in several different ways. They are often classified according to functionality in the minds of food scientists, for example, as sweeteners, nonsweet (or only slightly sweet) bulking agents, texturizing materials, suspending agents, and so on. While tables generally have not been constructed in this way, that is the approach taken later in this article. First the more traditional method will be presented. Both types of classification are included in Table 1.

MONOSACCHARIDES

Monosaccharides, often called simple sugars, are individual carbohydrate molecules that cannot be hydrolyzed to smaller, simpler carbohydrate units. The only ones found in food, beverage, and confectionery products to any appreciable extent are D-glucose and D-fructose. Most monosaccharides present in prepared food, beverage, and con-

Table 1. Major Carbohydrate Ingredients

Natural	Derived
Monosaccharides	
Glucose/dextrose[a]	Sorbitol[a]
Fructose[a]	Xylitol[a]
	Mannitol
	Glucono-δ-lactone (glucono-delta-lactone)
Oligosaccharides	
Sucrose[b]	Hydrogenated glucose syrup (maltitol syrup)
Lactose[c]	
Maltodextrins[a]	
Carbohydrate sweeteners	
Sugar[b]	
Glucose (corn) syrups[a]	
High-fructose (corn) syrups[a]	
Honey[a]	
Molasses[a]	
Sorbitol[a]	
Sugar products[b]	
Granulated sugar	
Coarse	
Extrafine	
Fine	
Baker's special, etc	
Brown sugar/soft sugar (light yellow → dark brown)	
Powdered sugar	
Fondant sugar	
Transformed sugar/agglomerated sugar	
Liquid sugar	
Molasses	
Polysaccharides	
Starches[d] (cookup, pregelatinized, and cold-water-swelling)	
Food gums[e]	
Dietary fiber	
Modified food starches[d]	
Crosslinked	cookup, pregelatinized, and cold water swelling
Stabilized	cookup, pregelatinized, and cold water swelling
Crosslinked and stabilized	cookup, pregelatinized, and cold water swelling
Dextrins	
Bulking agents	
Starches[e] and modified food starches[d]	
Food fiber[f]	
Polydextrose	
Fat replacers	
Microcrystalline cellulose	
Dextrins[d]	
Maltodextrins[a]	
Pectin microgel particles	
Inulin oligosaccharides	
β-glucan products	
Sources of fiber	
Cellulose (powdered, microcrystalline, microfibrillated, microreticulated)	
Brans (wheat, rice, barley, corn, oat)	
β-glucan preparations	
Ordinary food gums[e]	
Psyllium gum	
Resistant starch	
Polydextrose	
Sugar beet fiber	
Fruit fiber preparations	

[a]See SWEETENERS: NUTRITIVE.
[b]See SUGAR: SUCROSE.
[c]See LACTOSE.
[d]See STARCH.
[e]See GUMS.
[f]See FIBER, DIETARY.

fectionery products have been added as ingredients, primarily as high-fructose syrups or invert sugar.

D-Glucose is a six-carbon sugar (a hexose) whose carbonyl function is an aldehydo group. Aldehyde-based sugars are known as aldoses. Therefore, D-glucose is an aldohexose (Fig. 1). D-Glucose may appear on labels as glucose or dextrose.

The chemistry of D-glucose is that of aldehydo groups and hydroxyl groups (1,2). The aldehydo group reacts with the hydroxyl group on carbon atom 5 to form a cyclic, hemiacetal structure (a pyranose ring), which is by far the predominate form of D-glucose in solution. Ring formation involves formation of a new chiral carbon atom (carbon atom 1 [C-1], the one on the far right in the structure of β-D-glucopyranose (Fig. 1), so two configurations are formed at that position, the one with the hydroxyl group on C-1 down being the α form and the one with the hydroxyl group up being the β form.

Oxidation of the aldehydo or hemiacetal carbon atom (C-1) produces, respectively, the corresponding acid, D-gluconic acid (or its salt, D-gluconate), or the cyclic ester (lactone) (Fig. 1), the latter being produced directly in commercial production, which employs the enzyme glucose oxidase. This food acidulant is identified on ingredient labels as gluconolactone or glucono-delta-lactone. The product of reduction, D-glucitol (Fig. 2), is commonly known as sorbitol, which is how it most often appears on labels. Glucitol (sorbitol) belongs to the class of carbohydrates known as polyols, polyhydroxy alcohols or, scientifically, alditols. Other monosaccharidic polyols used as food ingredients are mannitol, xylitol, and erythritol (Fig. 3), which are labeled as such. A classification of these and other nutritive sweeteners can be found in the article SWEETENERS: NUTRITIVE.

D-Fructose is a six-carbon sugar (hexose) (Fig. 4) whose carbonyl function is a keto group, making it a ketose, more specifically a ketohexose. D-Fructose may appear on labels as fructose, levulose, or fruit sugar.

The keto group of D-fructose is also involved in ring formation. In solution, such as in honey and high-fructose

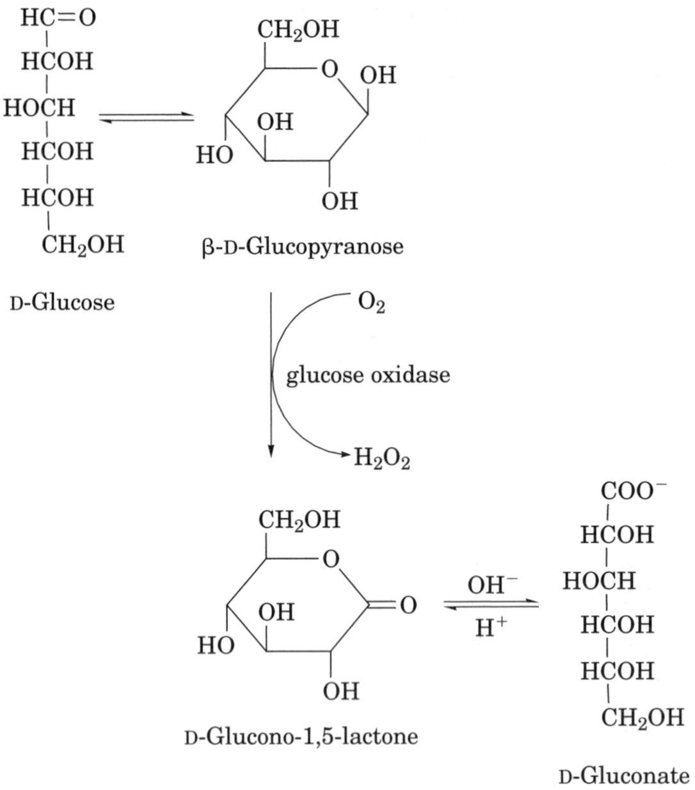

Figure 1. The open-chain (acyclic) and one of the two six-membered (pyranose) ring forms of D-glucose, its enzyme-catalyzed oxidation, and the open-chain (salt) and lactone ring forms of D-gluconic acid.

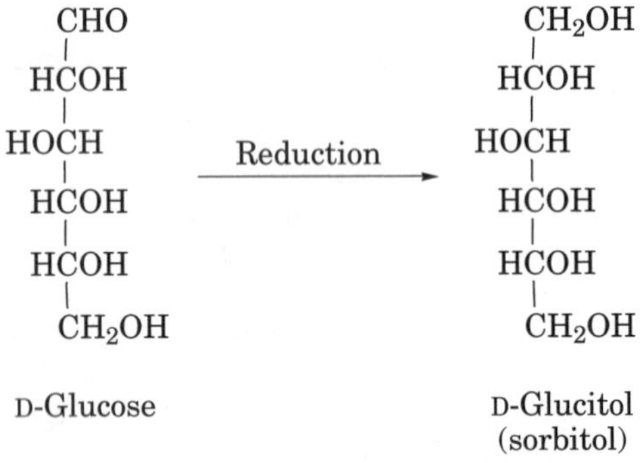

Figure 2. Reduction of D-glucose to sorbitol.

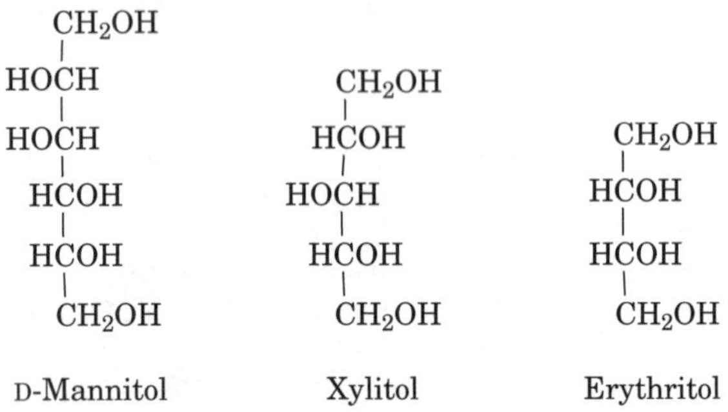

Figure 3. Structures of other food alditols.

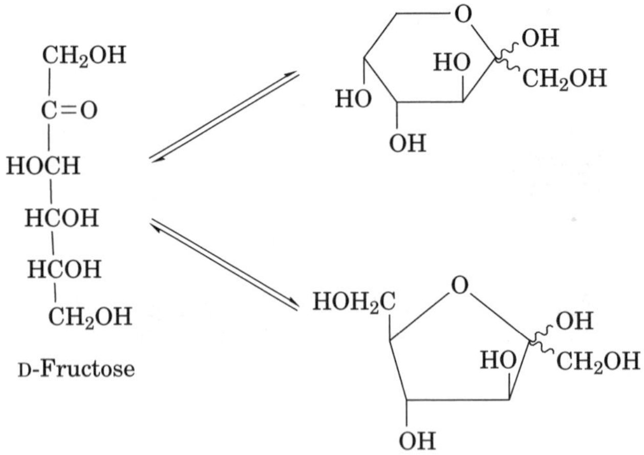

Figure 4. Cyclization of D-fructose to pyranose and furanose ring forms.

syrups, D-fructose occurs as the more stable six-membered (hemiacetal/hemiketal, pyranose) ring (Fig. 4); in the disaccharide sucrose (described in the next section), D-fructose occurs in its five-membered (furanose) ring form (lower ring structure in Fig. 4). (The squiggly lines on carbon 2 in this figure and at the reducing end of Figure 5 indicate a mixture of two configurations—the α and β configurations.) Reduction of D-fructose produces a mixture of D-glucitol (sorbitol) and D-mannitol. Fructose cannot be oxidized to an acid or lactone.

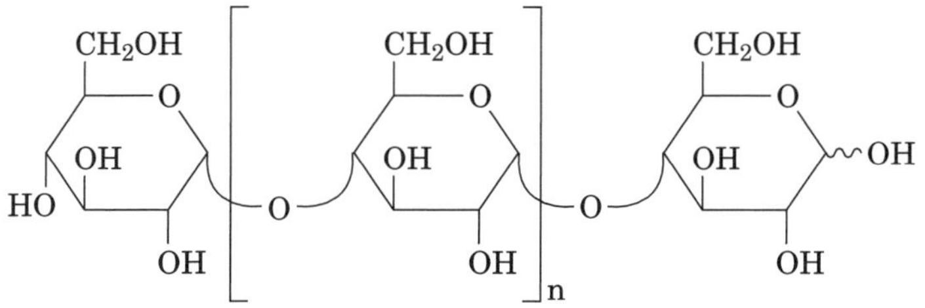

Figure 5. Structures of molecules related to the two starch polymers, amylose and amylopectin: $n = 0$, maltose: $n = 1$, maltotriose; $n = 2$, maltotetraose; $n = 3$, maltopentaose; and so on. In amylose, $n = {\sim}1000$–15,000.

Glucose is isomerized to fructose by the enzyme glucose oxidase, which is the basis for the production of high-fructose syrups (high-fructose corn syrups, HFCS). First starch is completely enzyme-converted to D-glucose, which is then isomerized to D-fructose. Generally, two fructose syrups are made, a 42% fructose and a 55% fructose syrup. Crystalline fructose is also available from this process.

In addition to starch being completely converted to glucose (crystalline and solution/syrup form), various glucose syrups (corn syrups) are also produced from it. Such syrups are described by their dextrose (glucose) equivalence (DE), which is defined as the percent reducing sugar calculated as D-glucose on a dry-weight basis. Thus, the DE is a measure of the degree of hydrolysis. D-Glucose has a DE value of 100 and starch a DE value of 0. Glucose syrups normally have DE values ranging from ~25% to >95%. A DE 25 syrup contains ~5% glucose, ~6% maltose, ~11% maltotriose, and ~78% higher oligosaccharides. A DE 80 syrup contains about 80% glucose.

Reactions of hydroxyl groups are of little significance in monosaccharides of foods, except as they are involved in hemiacetal and lactone ring formation. However, an important process that can occur with any carbohydrate containing a free, or potentially free, aldehydo or keto group also involves the remainder of molecule. Nonenzymic browning begins with a reaction known as the Maillard reaction, which is a chemical reaction between an amino group of an amino acid or protein molecule and reducing sugars, which are any carbohydrate molecules with a carbonyl group (free or in the form of a hemiacetal). The end products of a series of steps in several related pathways are flavors, aromas, and dark-colored soluble and insoluble polymers.

Heating sugars in the absence of compounds containing amino groups to a sufficient temperature effects a complex series of pyrolytic reactions resulting in the production of caramel colors and flavors.

OLIGOSACCHARIDES

Oligosaccharides are short chains of a few sugar units joined in acetal linkages; that is, the hemiacetal groups of a ring form (most often the pyranose ring form) of monosaccharides have been joined during biosynthesis to a hydroxyl group of another monosaccharide unit to create an acetal. In carbohydrate chemistry, the half of the acetal bond joining two sugar units is called a glycosidic linkage, the other half being that which forms the pyranose or furanose ring (1,2).

The upper limit to the number of saccharide (glycosyl) units in an oligosaccharide is undefined. It is often stated to be 10, sometimes 20. There are 2 monosaccharide units in a disaccharide, 3 in a trisaccharide, 4 in a tetrasaccharide, 10 in a decasaccharide, and so on. Chains containing more than 20 units are generally considered to be polymer molecules, that is, polysaccharides. With the exception of only a few naturally occurring oligosaccharides, the glycosyl (generally pyranosyl, occasionally furanosyl) units in oligo- and polysaccharides are joined together by glycosidic linkages in a head-to-tail fashion. As a result, each oligosaccharide has one, and only one, reducing end, that is, an end with a carbonyl or hemiacetal group (Fig. 5), but because most oligosaccharides are produced by depolymerization of polysaccharides, and because polysaccharides are often branched and may have many nonreducing ends, oligosaccharides may have more than one nonreducing end.

Depolymerization of starch molecules produces a series of branched (see "Polysaccharides") and unbranched oligosaccharides known chemically as maltooligosaccharides (Fig. 5) and, perhaps, larger fragments, depending on the extent of depolymerization. Commercially, these mixtures are known as, and labeled, maltodextrins. A range of maltodextrin products with different properties are produced. Mixtures of starch depolymerization products with smaller average degrees of polymerization, but large enough to allow conversion to a dry powder form (rather than a syrup) are known as, and labeled, corn syrup solids, glucose syrup solids, dried glucose syrup, or glucose syrup, dried.

The primary exception to the head-to-tail linkage arrangement is found in the structure of the disaccharide sucrose (1,2) (see the article SUGAR: SUCROSE), in which the linkage is from hemiacetal group to hemiacetal (hemiketal) group (Fig. 6), leaving the molecule with no free or potentially free carbonyl group. Therefore, sucrose is classified as a nonreducing sugar. Tri- and tetrasaccharides related to sucrose (raffinose, and stachyose, respectively)

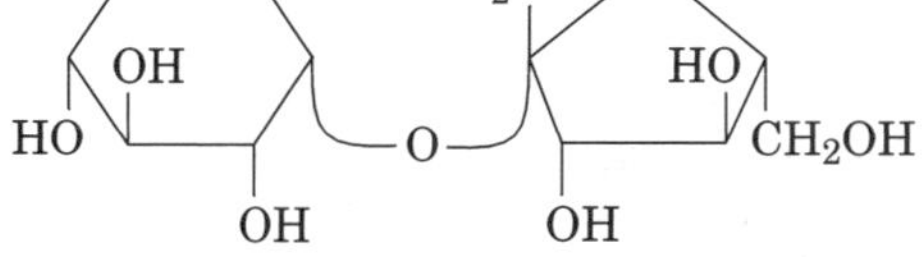

Figure 6. Sucrose. The left-hand portion is an α-D-glucopyranosyl unit. The right-hand portion is a β-D-fructofuranosyl unit.

Figure 7. α,α-Trehalose.

Table 2. Classification of Selected Food Polysaccharides by Source

Source	Examples
	Algal (seaweed extracts)
	Agars, algins, carrageenans, furcellaran
	Higher plants
Insoluble	Cellulose
Extract	Pectins
Seeds	Corn starches, rice starches, wheat starch, guar gum, locust bean gum, psyllium seed gum
Tubers and roots	Potato starch, tapioca starch, konjac mannan
Exudates	Gum arabics, gum tragacanth
	Microorganisms (fermentation gums)
	Xanthans, gellan, curdlan, dextrans
	Derived
From cellulose	Carboxymethylcelluloses (CMC), hydroxypropylcelluloses (HPC), hydroxypropylmethylcelluloses (HPMC), methylcelluloses (MC)
From starch	Starch acetates, starch 1-octenylsuccinates, starch phosphates, starch adipates, starch succinates, hydroxypropylstarches, dextrins
	Synthetic
	Polydextrose

are found primarily in beans. Added sucrose in a food product most often is indicated on the ingredient label as sugar.

The other exception to head-to-tail linkages is that of α, α-trehalose (Fig. 7). Also in this disaccharide, the two monosaccharide (α-D-glucopyranosyl) units are linked reducing end to reducing end.

All glycosidic linkages are susceptible to hydrolysis in the presence of acid and heat. That of sucrose is especially so. This hydrolysis to produce an equimolar mixture of D-glucose and D-fructose is called inversion and the product invert sugar. Invert sugar and invert sugar syrup may be used as label designations.

Unique structures that do involve head-to-tail linkages are those of the cyclodextrins (cycloamyloses), which as their name implies, are cyclic structures with neither a reducing or nonreducing end. Also, as their name implies, they are formed from starch. The commercial one is β-cyclodextrin, which is a ring composed of seven α-D-glucopyranosyl units linked ($1 \rightarrow 4$); that is, a structure like that of Figure 5 in which $n = 5$, and the ends are joined in the same type of linkage joining the other units.

Table 3. Classification of Selected Food Polysaccharides by Structure

Classification schemes	Examples
	By shape
Linear	Algins, starch amyloses, carrageenans, cellulose, konjac mannan, pectins
	Branched
Short branches on a linear backbone	Guar gum, locust bean gum, xanthan
Branch-on-branch structures	Starch amylopectins, gum arabics, gum tragacanth (tragacanthin)
	By different kinds of monosaccharide units
One	Starch amylopectins, starch amylose, cellulose
Two	Algins, carrageenans, galactomannans, konjac mannan, pectins
Three	Gellan, xanthan
Four	Gum arabics
Five	Gum tragacanth (tragacanthin)
	By charge
Neutral	Starch amylopectins, starch amyloses, cellulose, galactomannans, konjac mannan
Anionic (acidic)	Algins, carrageenans, gellan, gum arabics, gum tragacanth (tragacanthin), pectins, xanthan

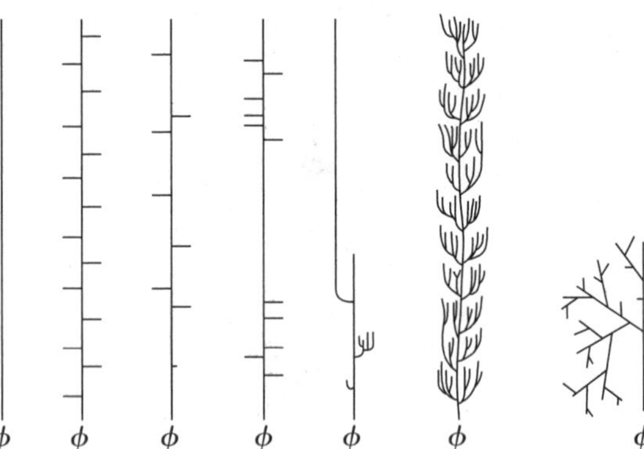

Figure 8. General structural features of polysaccharides. ϕ, Reducing end unit.

Reactions of hydroxyl groups of oligosaccharides are generally not practiced. There are a few exceptions. Derivatives of β-cyclodextrins have been made to improve

Table 4. Label Designations of Food Polysaccharides

Chemical or generic name of food gum	Approved label designations (in addition to chemical or generic name)
Agar	Agar–agar
Alginic acid	
Sodium or potassium alginate	Algin
Propylene glycol alginate	Algin derivative
Carboxymethylcellulose	Sodium carboxymethylcellulose
	Sodium carboxymethyl cellulose
	Carboxymethylcellulose
	Carboxymethyl cellulose
	CMC
	Sodium CMC
	Cellulose gum
κ-carrageenan	Carrageenan
	Chondrus extract
	Irish moss extract
λ,ι-carrageenan	Carrageenan
Cellulose, powdered	
Furcellaran	Danish agar
Gellan	Gellan gum
Guar gum	
Gum arabic	Arabic gum
	Gum acacia
	Acacia gum
Gum ghatti	Ghatti gum
Gum karaya	Karaya gum
Gum tragacanth	Tragacanth gum
Hydroxypropylcellulose	Hydroxypropyl cellulose
	Modified cellulose
Hydroxypropylmethylcellulose	Hydroxypropyl methylcellulose
	Hydroxypropyl methyl cellulose
	HPMC
	Modified cellulose
Konjac mannan	Konjac gum
	Konjac flour
	Yam
	Yam flour
Locust bean gum	Carob gum
Methylcellulose	Methyl cellulose
	MC
	Modified cellulose
Microcrystalline cellulose	Cellulose microcrystalline
	Cellulose gel
Pectin	
Polydextrose	
Psyllium gum	Gum psyllium
Corn, waxy maize, waxy corn, high-amylose corn, potato, tapioca, cassava, wheat, arrowroot, rice, waxy rice starch	Starch
Starch acetate, starch phosphate, starch octenylsuccinate, crosslinked starch, oxidized starch, hydroxypropylstarch, thin-boiling starch	Food starch-modified
	Modified food starch
Dextrins	
Xanthan	Xanthan gum

Note: Approved for the United States.

its solubility. Sucrose is monoesterified with fatty acids to produce food-grade emulsifiers/surfactants and more completely esterified (6–8 fatty acyl groups) to produce the fat substitute olestra. The modified carbohydrate, sorbitan, is also fatty acyl esterified and/or etherified with poly(ethylene glycol)/poly(oxyethylene) chains to produce food-approved emulsifiers/surfactants.

POLYSACCHARIDES

Polysaccharides are polymers of monosaccharide units, which may be substituted with other groups such as ester, ether, and/or cyclic acetal moieties (1). Polysaccharides may be attached to proteins, in which case the substance is referred to as a protein-polysaccharide (when they originate from plants) or a proteoglycan (when they originate from animals), or to a lipid, in which case the substance is referred to as a lipopolysaccharide. Certain polysaccharides of plants, namely, hemicelluloses, may be covalently bound to lignin.

Polysaccharides of foods, either naturally occurring or added as ingredients, are classified in several different ways (3,4), by source (Table 2) and by structure (Table 3) being the most common ways. They may also be classified by composition (3) and by imparted functionalities (4). Water-soluble polysaccharides of foods, especially those added as ingredients, are known as food gums and hydrocolloids (see the article GUMS).

One feature to note from Table 3 is that polysaccharides, unlike proteins, can be branched. Because each monosaccharide unit of a polysaccharide has several hydroxyl groups that can be involved in glycosidic linkages, each unit can potentially be involved in multiple linkages. Therefore, branching can, and does, occur. Branched polysaccharide structures can have several shapes, including essentially linear structures with a very few, very long branches; linear structures decorated with short branches, either regularly spaced, irregularly spaced, or in clusters; branch-on-branch structures with branches in clusters (as in starch amylopectin) (1) (see the article STARCH); and bushlike, branch-on-branch structures, with and without decoration with short branches (Fig. 8). Each polysaccharide has one and only one reducing end unit but may have many nonreducing end units.

The chemistry of polysaccharides, like that of oligosaccharides, is primarily the chemistry of hydroxyl groups and the chemistry of glycosidic linkages; in the case of polysaccharides containing uronic acid units, the chemistry of carboxyl groups may also be involved (3). Hydroxyl groups may be reacted to form ethers (practiced with cellulose [1] and starches [1], primarily corn, waxy maize, and potato starches). Hydroxyl groups may also be reacted to form esters (practiced with starch) (1). Some naturally occurring ester groups are removed by saponification during processing of the food gums konjac mannan (acetate ester groups) and gellan (acetate and glycolate ester groups). Oxidants will convert hydroxyl groups into carbonyl functions; carboxyl groups may also be introduced during oxidation.

Oxidation is usually effected at alkaline pH values. Under these conditions, β-eliminations may result in depolymerization (1). Glycosidic bonds are also cleaved by acid-catalyzed hydrolysis (practiced with starches and gums in general). Starches are also depolymerized by enzyme-catalyzed hydrolyses in the formation of corn syrups and other hydrolytic products. Under low-moisture conditions, heating in the presence of an acid results in both glycosidic bond hydrolysis and transfer, the latter reaction resulting in a more highly branched product (production of dextrins from a starch).

Some of the uronic acid carboxyl groups of algins are esterified in the production of propylene glycol alginate (1). Some naturally occurring methyl ester groups of a high-methoxyl pectin are saponified during its conversion into a low-methoxyl pectin. In other low-methoxyl pectins, the methyl carboxylate groups are converted into carboxamide groups by treatment with ammonia (1).

Allowed label designations of food gums are given in Table 4.

BIBLIOGRAPHY

1. R. L. Whistler and J. N. BeMiller, *Carbohydrate Chemistry for Food Scientists*, Eagen Press, St. Paul, Minn., 1997.
2. J. N. BeMiller and R. L. Whistler, "Carbohydrates," in O. R. Fennema, ed., *Food Chemistry*, Marcel Dekker, New York, 1996.
3. J. N. BeMiller, "Structure–Property Correlations of Non-starch Food Polysaccharides," in H. N. Cheng, G. Cote, and I. C. Baianu, eds., *Application of Polymers in Foods*, Marcel Dekker, New York, 1999.
4. J. N. BeMiller, "Classification, Structure, and Chemistry of Polysaccharides in Foods," in S. S. Cho and M. Dreher, eds., *Handbook of Dietary Fiber and Functional Foods*, Marcel Dekker, New York, 1999.

J. N. BEMILLER
Purdue University
West Lafayette, Indiana

CARBOHYDRATES: FUNCTIONALITY AND PHYSIOLOGICAL SIGNIFICANCE

Carbohydrates are the most important source of food energy in the world. They provide between 40 to 80% of total food energy intake, depending on location, cultural considerations, or economic status. Despite the importance of dietary carbohydrates, the human body only comprises about 1%. Following digestion and absorption, available carbohydrates may be (*1*) used to meet immediate energy needs of tissue cells, (*2*) converted to glycogen and stored in the liver and muscle for later energy needs, and (*3*) converted to fat as a larger reserve of energy.

In addition to the role that carbohydrates play as an energy source, they have a number of other roles as well. Sugars are used as sweeteners to make food more palatable and to assist in preservation. Starches and gums provide important textural properties to processed foods such as baked goods and desserts. Diets high in carbohydrates may reduce individual propensity to obesity, and there is growing evidence that such diets may also provide some protection against various noncommunicable human diseases and conditions.

DESCRIPTION

The major class of dietary carbohydrates (degree of polymerization) are sugars (1–2), oligosaccharides (3–9), and polysaccharides (>9) (1). Sugars comprise monosaccharides, disaccharides, and polyols (sugar alcohols). Oligosaccharides include malto-oligosaccharides, mainly from the hydrolysis of starch, and other oligosaccharides such as fructo-oligosaccharides. Polysaccharides may be divided into starch (α-glucans) and nonstarch polysaccharides, of which the major components are the polysaccharides of plant cell walls (eg, cellulose, hemicellulose, and pectin). The term *complex carbohydrates* is often used in the United States to describe starch, dietary fiber, and nondigestible oligosaccharides (2). One of the major developments in the understanding of the importance of carbohydrates in health over the last 20 years has been the discovery of resistant starch, which is defined as "starch and starch degradation products not absorbed in the small intestine of healthy humans" (1).

FOOD FUNCTIONALITY

Carbohydrates have a number of important roles in food functionality including effects on texture, taste, and stability.

Sugars

Sugars contribute primarily to the sweetness of foods (3). Other important properties of sugars include preservation, fermentation, browning, and viscosity. Concentrated sugar solutions act as a preservative by controlling water activity. Yeast cells in fermented bread and in alcoholic beverages metabolize sugars, under anaerobic conditions, to produce carbon dioxide or ethanol. Carmelization reactions occur when sugars are heated. Sugars are able to increase boiling point and depress the freezing point of a solution. Polyols impart bulking or humectant properties and lower energy content than sugars.

Oligosaccharides

Oligosaccharides provide unique health benefits (4,5). More than half of the "functional foods" on the Japanese market contain prebiotic oligosaccharides as an active component, with the aim of promoting favorable gut microflora. Oligosaccharides such as polydextrose can provide a 1 kcal/g bulking agent providing the body and texture of sugar for energy controlled foods.

Starches

Starch is the major component in flours and commodities such as rice, corn, wheat, and beans (6). One of the major advantages of granular starches is that, being insoluble in cold water, they can be easily dispersed and then rendered functional by heating, which causes the granules to swell, absorb water, and affect viscosity. The behavior of starch depends on the temperature and the concentration. The gelatinization of starch occurs at a critical temperature of between 56 and 68°C depending on the starch type. The rheological properties of starch paste have important food applications. Pregelatinized starches can be used in food products for reduced preparation and processing requirements. Starch retrogradation accompanies the aging of cooked foods such as breads, cereals, puddings, or gravies. With bread, the staling process involves hardening of the starch by retrogradation and the redistribution of the water from the crumb to the crust.

Polysaccharide Gums

Polysaccharide gums are widely used in food processing as thickeners and gelling agents (3). These functions allow the development of a wide variety of unique processed food products such as dietetic jams, jellies, salad dressings, and sauces. There are a variety of gum types derived from (*1*) plant materials such as seaweed, seeds, roots, or tree exudates; (*2*) microbial biosynthesis; and (*3*) chemical modification of natural polysaccharides. Thickening and gelling of foods by gums is a common practice in processed foods to achieve a certain texture, mouth-feel, and body. Most gums have the ability to increase viscosity and many gums can form gels that result from intermolecular associations. Gums tend to be effective at low levels, such as less than 1% of formulation weight.

BASIC PHYSIOLOGICAL EFFECTS

Energy Source

Dietary carbohydrates are generally given an energy value of 4 kcal/g (17 kJ/g), although where carbohydrates are expressed as monosaccharides, the value of 3.75 kcal/g (15.7 kJ/g) is used. However, a number of carbohydrates are only partially or not digested in the small intestine and are fermented in the large bowel to short-chain organic acids. These include the nondigestible oligosaccharides, resistant starch, and nonstarch polysaccharides. This process of fermentation is less efficient than absorption in the small intestine. Although energy values vary for different carbohydrates, a value of 2 kcal/g (8 kJ/g) is a reasonable average figure for carbohydrates that reach the colon (1,4).

Glucose and Insulin

The digestion of carbohydrates starts in the mouth, where salivary α-amylase initiates starch hydrolysis (7). The starch fragments formed are maltose, some glucose, and dextrins. The α-amylase hydrolysis of starch is completed by the pancreatic amylase active in the small intestine. Disaccharides also need to be broken down to monosaccharides to be absorbed. The final hydrolysis is accomplished by "disaccharidases" attached to the intestinal brush border membrane. Disaccharidase deficiencies occur in some populations, causing malabsorption and intolerance, such as in the case of lactose. Lactose, a disaccharide of glucose and galactose, is the principal sugar in milk. At birth, lactase activity is high in the brush border of the small intestine of infants, but declines after weaning, so that most non-Caucasian populations of the world have low activity in adult life (1,8).

Glucose and galactose are transported actively against a concentration gradient into the intestinal mucosal cells by a sodium-dependent transporter (1,7). Fructose undergoes facilitated transport by another mechanism. When absorbed into circulation, carbohydrates cause an elevation of blood glucose concentration. Fructose and galactose have to be converted to glucose mainly in the liver and therefore produce a slower rate of glucose elevation. Insulin, necessary for glucose uptake by body cells, is secreted as a response to blood glucose elevation but is modified by many factors, such as neural and endocrine stimuli.

Plasma Lipids

There is concern that very high carbohydrate diets at the expense of fat might result in a decrease in high density lipoprotein and a corresponding increase in plasma triglycerides. However, this does not appear to be an issue when the increased carbohydrates are in the form of vegetables, fruits, and processed cereal over prolonged periods.

Polysaccharides like oat β-glucan and those of psyllium have been repeatedly shown to lower plasma cholesterol levels in individuals with elevated levels; slight to moderate change of plasma levels are normal (1,9). Proposed mechanisms include impaired bile acid and cholesterol reabsorption through physical entrapment in the small intestine, or inhibitory effects on cholesterol synthesis by products of lower bowel fermentation.

Bowel Function

It has long been recognized that the nonstarch polysaccharide portion of dietary fiber is the principal dietary component affecting bowel regularity (10). This occurs through increases in bowel content bulk and speeding up of intestinal transit time. The degree of effect depends on the chemical and physical properties of the polysaccharides and the extent of fermentation. One important development in recent years has been the demonstration that specific dietary carbohydrates selectively stimulate the growth of individual groups or species of bacteria. For example, fructo-oligosaccharides specifically stimulate growth of bifidobacteria that may protect the host from invasion by pathogenic species and have other health benefits. Foods used to selectively stimulate the growth of gut bacteria are known as prebiotics.

HEALTH MAINTENANCE AND DISEASE PREVENTION

Although the amount of carbohydrate required to avoid ketosis is relatively small (about 50 g/day), carbohydrate provides the majority of the energy in the diets of most people (1). This is desirable not only for providing easily available energy for oxidative metabolism, but also carbohydrate-containing foods are good vehicles for micronutrients and phytochemicals. Carbohydrates are important in maintaining glycemic homeostasis and for gastrointestinal integrity and function. Diets high in carbohydrates, as compared with diets high in fat, reduce the likelihood of developing obesity and related conditions. An optimum diet should consist of at least 55% of the total energy coming from carbohydrates obtained from a variety of food sources. However, carbohydrates intakes at or above 75% of total energy could have significant adverse effects on nutritional status by the exclusion of adequate quantities of protein, fat, and other essential nutrients.

Energy Balance and Weight Maintenance

In adults, it is very important that the energy ingested be matched to the amount of energy expended (1,11). Maintenance of energy balance is necessary to avoid obesity and associated chronic diseases such as diabetes and cardiovascular disease. Positive energy balance (weight gain) occurs when total energy intake exceeds total energy expenditure, regardless of the composition of the excess energy. However, composition of the diet may affect to what extent positive energy balance occurs.

The composition of the diet may affect the body's ability to maintain energy (1,11). Especially, diets containing at least 55% of energy from a variety of carbohydrate sources, as compared with high-fat diets, reduce the probability of body fat accumulation. Substantial data suggest those diets high in fat tend to promote consumption of more total energy than diets high in carbohydrates. This effect may be due to the low energy density of high-carbohydrate diets, because total volume of food consumed appears to provide an important satiety cue. Although there are no data to suggest that different types of carbohydrates affect total energy differently, the composition of the diet may affect the proportion of excess energy stored as body fat. The body has a large fat storage capacity, and excess dietary fat is stored very efficiently in adipose tissue. Alternatively the body's capacity to store carbohydrates is very limited, and excess dietary carbohydrates are not efficiently stored as body fat. Instead, excess carbohydrate tends to be oxidized readily, resulting in indirect fat accumulation by reductions in fat oxidation. Although the overall contribution of de novo lipogenesis from carbohydrates is small, it may increase with appreciable carbohydrates overfeeding, insulin resistance, and with extremely high intakes of sucrose or fructose.

There is considerable controversy surrounding the extent to which sugars and starch promote obesity (1,11). There is no direct evidence to implicate either form of carbohydrate in the etiology of obesity, based on data derived from studies in Western countries. In spite of that, it is important to state that excess energy in any form promotes body fat accumulation. Excess consumption of low-fat foods, although not as obesity producing as excess consumption of high-fat foods, leads to obesity if energy expenditure (such as physical activity) is not increased.

Physical Activity

Maintaining regular physical activity reduces the likelihood of creating positive energy balance or weight gain, regardless of diet composition (1,12). There is general agreement that the combination of high-carbohydrate diets and regular physical activity is the best way to avoid positive energy balance. The increased energy needs of people with high physical activity can be efficiently met by dietary carbohydrates. In many developing countries, the major challenge is to meet the daily energy needs of high physical labor. The importance of carbohydrates in the diet becomes more critical as the amount of and the intensity of physical activity increases.

There is substantial evidence that supplemental carbohydrates can improve performance for elite, endurance-trained athletes (1,12,13). A high-carbohydrate diet including meals and snacks such as energy bars have been shown to enhance performance during long-distance cycling and running. There is, however, no clear evidence that carbohydrate supplements and snacks would improve the performance for the majority of people who engage in recreational physical activity of lower intensity and duration. On the other hand, carbohydrate intake after exercise can help to quickly replenish depleted glycogen stores.

Dental Caries

The incidence of dental caries is influenced by a number of factors (1). Foods containing sugars and starch may be easily hydrolyzed by α-amylase and bacteria in the mouth, which can increase the risk of caries as a result of excess production of organic acids. Starches with a high glycemic index produce more pronounced changes in plaque pH than low glycemic index starch, especially when combined with sugars. However, the impact of carbohydrates on caries is dependent on the food, frequency of consumption, degree of oral hygiene performed, availability of fluoride, salivary function, and genetic factors.

Behavior

There is emerging evidence that food intake may have important effects on behavior (1,14). Although providing breakfast to children who do not typically eat breakfast can increase cognitive performance, it is less clear that overall composition of the diet can affect behavior. Although it is often suggested that sugar consumption may lead to hyperactivity in children, there is no evidence to support the claim that refined sugars have a significant influence on child hyperactivity and related behaviors. Because glucose is an essential energy source for the central nervous system, carbohydrate has been suggested to play a role in memory and cognitive function. Although there appears to be a relationship between glucose levels and memory processing, the clinical significance of this effect has yet to be determined.

Blood Glucose and Diabetes Management

The rise of blood glucose in normal and diabetic people after meals varies markedly and depends on many factors, including the source of the carbohydrate, its method of preparation, and composition of the total meal (1,15). Classification of carbohydrates as simple or complex does not predict their effects on blood glucose or insulin. Rapidly absorbed carbohydrates, which promote large blood glucose responses, may be in the form of both sugars and starches. Sugars added to foods have no different effects on blood glucose from those of sugars alone. The natural sugars from fruits and fruit juices raise blood glucose approximately as much as does sucrose and less than do most refined starchy carbohydrate foods.

Consuming a wide range of carbohydrate foods is now regarded as acceptable in the nutrition management of people who already have non-insulin-dependent diabetes (1). It is suggested that between 60 and 70% of total energy should be derived from a mix of carbohydrates and monounsaturated fats. Carbohydrates should mainly come from a wide range of appropriately processed cereals, vegetables, and fruit. Sucrose and other sugars have not been directly implicated in the etiology of diabetes, and recommendations concerning intake relate primarily to the avoidance of all energy-dense foods in order to reduce obesity. Most recommendations for the management of diabetes permit modest (30–50 g/day) intakes of sucrose and other added sugars in the diabetic dietary plan.

Blood Lipids and Cardiovascular Disease Risk

The cornerstone of dietary advice aimed at reducing coronary heart disease risk is to increase the intake of carbohydrate-rich foods, especially cereals, vegetables, and fruits rich in nonstarch polysaccharides, at the expense of fat (1). There is increasing evidence that antioxidants have a protective effect against coronary heart disease, and complex carbohydrates such as fruits and vegetables tend to be rich sources of antioxidant nutrients and food components. Cereal foods rich in nonstarch polysaccharides tend to have a protective effect, and there is not evidence that sucrose consumption increases risk. Certain nonstarch polysaccharides such as β-glucan have been shown to have an appreciable effect on lowering plasma cholesterol when consumed in naturally occurring foods, enriched foods, or dietary supplements. Many carbohydrate-rich plant foods are rich in potassium, which may help to reduce the risk of hypertension.

Gastrointestinal Illnesses and Cancer Risk Management

Minimally refined carbohydrate staple foods may be a good source of dietary fiber or oligosaccharides and phytochemicals, which might mitigate the risk of colorectal and other cancers (1). The process of complex carbohydrate (eg, dietary fiber, resistant starch, oligosaccharides) fermentation in the large intestine may protect the colorectal area against the genetic damage that may lead to cancer through a range of mechanisms that include (*1*) the dilution of potential carcinogens; (*2*) the reduction of products of protein fermentation through the stimulation of bacterial growth; (*3*) pH effects through the production of butyric acid; (*4*) maintenance of the gut mucosal barrier; and (*5*) effects of bile degradation. High intake of prebiotics such as fructo-oligosaccharides may facilitate the colonization of bifidobacteria and lactobacilli in the large gut, this may reduce the risk of acute infective gastrointestinal illnesses.

Inherited Conditions

There are a number of inherited conditions having significant implications for restricting dietary intake in infants and children (1). These include rare conditions such as fructose intolerance, galactosemia, and sucrase deficiencies. Though rare diseases, their early detection and careful dietary management is important in avoiding severe handicap or pathology.

BIBLIOGRAPHY

1. Food and Agricultural Organization, *Carbohydrates in Human Nutrition*, FAO Food and Nutrition Paper 66, Rome, Italy, 1998.

2. International Life Sciences Committee, "Complex Carbohydrates. The Science and the Label," *Nutr. Rev.* **53**, 186–193 (1995).
3. P. Chinachoti, "Carbohydrates: Functionality in Foods," *Am. J. Clin. Nutr.* **61**, 922S–929S (1995).
4. F. R. J. Bornet, "Undigestible Sugars in Food Products," *Am. J. Clin. Nutr.* **59**, 763S–769S (1994).
5. S. K. Figdor and H. H. Rennhard, "Caloric Utilization and Disposition of Polydextrose in the Rat," *J. Agric. Food Chem.* **29**, 1181 (1981).
6. D. J. Mauro, "An Update on Starch," *Cereal Foods World* **41**, 776–780 (1996).
7. B. Szepesi, "Carbohydrates," in *Present Knowledge in Nutrition*, ILSI Press, Washington, D.C., 1996.
8. L. V. Muir, K. D. Sanders, and W. P. Bishop, "Gastrointestinal Disease," in *Present Knowledge in Nutrition*, ILSI Press, Washington, D.C., 1996.
9. A. S. Truswell and A. C. Beynen, "Dietary Fibre and Plasma Lipids: Potential for Prevention and Treatment of Hyperlipidemias," in *Dietary Fibre—A Component of Food*, Springer-Verlag, New York, 1992.
10. N. W. Read and M. A. Eastwood, "Gastrointestinal Physiology and Function," in *Dietary Fibre—A Component of Food*, Springer-Verlag, New York, 1992.
11. J. O. Hill and B. J. Rolls, *Carbohydrates and Weight Management*, ILSI NA Monograph, ILSI Press, Washington, D.C., 1998.
12. E. F. Coyle, "Substrate Utilization During Exercise in Active People," *Am. J. Clin. Nutr.* **61**, 968S–979S, (1995).
13. W. M. Sherman, "Metabolism of Sugars and Physical Activity," *Am. J. Clin. Nutr.* **62**, 228S–241S (1995).
14. J. W. White and M. Wolraich, "Effect of Sugar on Behavior and Mental Performance," *Am. J. Clin. Nutr.* **62**, 242S–249S (1995).
15. T. M. S. Wolever and J. B. Miller, "Sugars and Blood Glucose Control," *Am. J. Clin. Nutr.* **62**, 212S–227S (1995).

Mark L. Dreher
Mead Johnson Nutritionals, Bristol-Myers Squibb
Evansville, Indiana

CAROTENOID PIGMENTS

Carotenoids as natural pigments are unique in that they are found in both plants and animals. Animals cannot make the basic 40 carbon structure but are capable of making changes in the hydrocarbon rings and chain. Table 1 shows the distribution of the various plant pigments in food. The carotenoid pigments are important from a food science point of view because of the color they impart to various food products and the fact that a number of the pigments have vitamin A activity.

The carotenoids are chemically and biochemically related to the more general class of compounds known collectively as terpenes and terpenoids. These are compounds composed of repeating isoprenoid like units. Mevalonic aid has been shown to be a precursor for the biosynthesis of the terponoids.

The carotenoids have traditionally been classified as carotenes (hydrocarbons) and xanthophylls (containing oxygen in some form). Generally, the hydrocarbons (lycopene, β-carotene, etc) are produced first and the xanthophylls are produced at a later stage. However, it has recently been shown that some fish can reverse the process and make retinol from oxygenated carotenoids.

There are four periods of carotenoid research (1). During the first period (1800s) pigments were characterized by measurements of light absorption. During the second period (1900–1927) attempts were made to define the empirical formulas of the isolated carotenoids. The third period (1928–1949) was a time in which some carotenes with provitamin A activity were discovered. Vitamin A and some of the carotenes, including β-carotene, were synthesized. The fourth, the most recent, period witnessed a virtual explosion in the number of known carotenoids. The instrumentation that has made this possible was developed during this time (2). There were about 80 known carotenoids, and of these perhaps only 35 had established structures. Today there are greater than 600 known carotenoids.

Today there is much interest in the carotenoids because they are natural pigments and also because of their proposed role in medical applications.

Every three years a formal group meets in Europe, the United States, and Japan, and plenary and session lectures are published (2–9). A book on the technical aspects of carotenoids (10) and a book on vitamin A deficiency (11) has also been published.

The carotenoids are the ultimate sources of vitamin A in the diet, and the lack of retinol or provitamin A has led to nutritional blindness. A number of countries have been listed as having a potentially serious vitamin A deficiency problem (12). In some well-studied Southeast Asian countries it has been estimated that half a million children will go blind each year because of a lack of vitamin A. When one extrapolates to other countries in Asia, Africa, and Latin America, the mortality and morbidity rates are staggering.

A body of evidence suggests that some of the carotenoids themselves have medical applications (13). These include the deposition of carotenoid pigments in the skin of people who do not develop protective pigmentation and protection against certain cancers. Although we have known of the protection of green plants from photodynamic destruction by the carotenoids, the protection of animals from neoplastic transformation is a recent discovery.

CAROTENOID NOMENCLATURE

Trivial names were used because the structures were not established and the number was small. With more than 600 carotenoids of proven structure now known there is still a need for trivial names, but there is also a need for more precise names that follow a chemical convention.

Under trivial names the carotenoids are generally grouped as carotenes (hydrocarbons) and the oxygen-containing compounds as xanthophylls. Some of the carotenes' names are preceded by a Greek letter (eg, β, ϵ), and others have names denoting the source (lycopene tomato). Some are a combination such as β-zeacarotene and α-zeacarotene. Here the ring double-bond position and the source (maize) are indicated.

Table 1. Distribution of Natural Coloring Matter

Food	Heme	Carotenoids	Chlorophyll	Betalains	Anthocyanins	Flavonoids	Caramel	Melanina
Red meat	X	—	—	—	—	—	—	—
Fish	X	X	—	—	—	—	—	X
Eggs	—	X	—	—	—	—	—	—
Crustaceans	—	X	—	—	—	—	—	—
Dairy products	—	X	—	—	—	—	—	—
Green vegetables	—	X	X	—	X	X	—	—
Root vegetables	—	X	X	X	X	X	—	—
Fruits	—	X	X	—	X	X	—	—
Cereals	—	X	X	—	—	—	X	—

Xanthophylls are pigments that contain hydroxyl, carbonyl, epoxide, ester, and carboxylic acid groups. The convention today is to indicate the source and add the suffix xanthin (eg, tunaxanthin). Some of the older trivial names such as lutein denoted their yellow color. Structural features are indicated by the prefix iso, as in isozeaxanthin; the hydroxyl group in zeaxanthin is in the three position, and it is in the four position in isozeaxanthin *cis*-Lycopene would indicate a departure from the expected all-trans structure.

In recent years an attempt has been made to limit the number of trivial names by building on a known name such as *β*-carotene. Thus astaxanthin might be 3,3′dihydroxy-4,4′-diketo-*β*-carotene.

The systematic method of carotenoid nomenclature uses the IUPAC convention (14–15) (Fig. 1). The advantage is that any carotenoid can be exactly named, including chiral centers. The disadvantage is that the name for a complicated compound can take two or three lines, whereas the trivial name is a single work (eg, fucoxanthin). In the IUPAC system the central chain is assumed to be unsaturated and all-trans. Any departure from this structure is specified. Thus *β*-carotene is *β*, *β*-carotene. The term nor indicates that a methyl is missing, and seco indicates a break in the ring structure such as 5,6, SECO, *β*, *β*-carotene-5,6-dione (semi-*β*-carotenone). The change in the hydrogenation level to form saturated positions as well as acetylenes and allenes can be designated by hydro and dehydro.

Oxygenated derivatives of carotenoid hydrocarbons are named by use of suffixes and prefixes according to the rules of general chemical nomenclature. The principal group is chosen and is cited by use of a suffix, and the other groups are cited as prefixes. The sequence is as follows: carboxylic acid, ester of carotenoid acid, aldehyde, ketone, alcohol, ester of carotenoid alcohol, and epoxide. Apo-carotenoids are found naturally as well as occurring synthetically. The apo designates where the parent compound has been

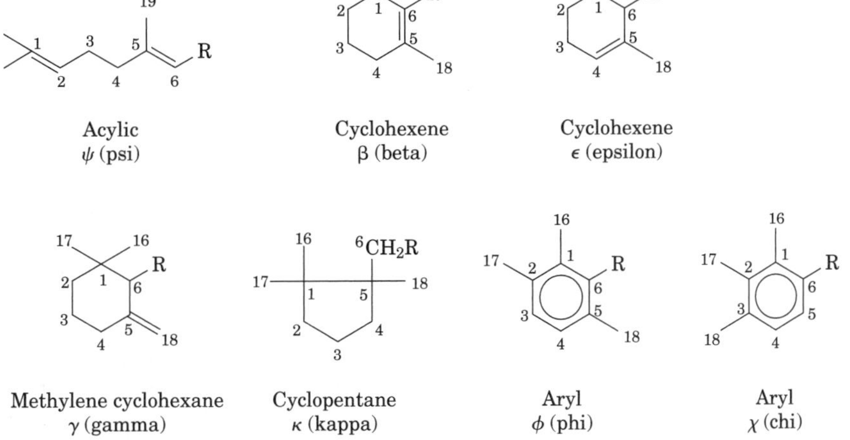

β-Carotene

Acylic
ψ (psi)

Cyclohexene
β (beta)

Cyclohexene
ϵ (epsilon)

Methylene cyclohexane
γ (gamma)

Cyclopentane
κ (kappa)

Aryl
ϕ (phi)

Aryl
χ (chi)

Figure 1. IUPAC rings.

cleaved. Thus β-apo-8′-carotenal should be designated 8′-apo-β-carotene-8′-al.

There are rules for carotenoids with more than 40 carbons, retro double-bond structures, designated chiral centers, and the use of the primes. This area is covered in more detail elsewhere (14). Pfander gives the structure and reference for more than 600 carotenoids (16).

BIOSYNTHESIS

In 1950 a biosynthetic pathway for the conversion of phytoene to lycopene to β-carotene was proposed (17). While the structures of several of these compounds were not proven at that time, it nevertheless was the first sequential pathway proposal.

The 1950s and 1960s were intense years in the polyene biosynthetic field because steroid and carotenoid biogenesis followed similar pathways. Thus when mevalonic acid (MVA) was discovered it was a breakthrough for both groups. The use of stereospecific [14C]-and [3H] mevalonic acids incorporated into cell-free systems provided a valuable tool for studying biosynthetic conversions. Acetyl-CoA may be considered the starting point in terpene biosynthesis. As can be seen in Figure 2, two acetyl-CoA molecules condense to form acetoacetyl-CoA. This in turn condenses to form the branched six-carbon acid β-hydroxy-β-methylglutaryl CoA (HMG-CoA). Through a series of reactions, HMG-CoA can also be formed from leucine (18). HMG-CoA is reduced from the dicarboxylic acid to a monoacidic-monohydroxyl compound, MVA. For most systems, MVA once formed cannot be converted back to HMG-CoA. MVA in the presence of ATP is converted to MVAP and then to MVAPP. With a third ATP, MVAPP is converted to isopentenyl pyrophosphate (IPP). This step involves a phosphorylation and a decarboxylation. IPP is the key intermediate to the terpenoids, which include rubber, sterols, bile acids, squalene, some sex hormones, ubiquinones, essential oils, the phytol side chain of chlorophyll, and vitamins E and K. It is isomerized to form dimethylallylpyrophosphate (DMAPP). IPP and DMAPP condense to form the C_{10} compound geranyl PP (GPP). The 10-carbon unit is further reacted with IPP to form the C_{15} farnesyl (FPP) and the C_{20} geranylgeranyl (GGPP). FPP can be dimerized to form squalene and the C20 GGPP, to form phytoene. The formation of phytoene follows a rather complicated pathway. By analogy to squalene, lycopersene with a saturated 15-15′ position was proposed as the first carotenoid and the precursor to phytoene. More recent studies have shown that the unlikely reduction to form lycopersene from GGPP and the oxidation to form phytoene from lycopersene does not occur; see Figure 2. Plants are able to form phytoene and the higher carotenoids, but animals can only make changes to preformed carotenoids.

More recently a new pathway to IPP has been reported (18). This pathway does not involve mevalonic acid, but follows five carbon intermediates. *B*-carotene has been shown to be formed by this pathway. It remains to be seen how significant these findings are.

There is a branch of neurosporene leading into either lycopene or α- or β-zeacarotenes. The data on the cyclization of neurosporene or lycopene seem to favor lycopene as the major pathway. In the red tomato, synthesis of β-zeacarotene rules out lycopene as the sole precursor to β-carotene because it is difficult to explain the higher temperature-selective inhibition of lycopene (19, Fig. 2). When tomatoes were treated with DMSO at various stages of maturity (20,21), it was found that the acyclic but not the cyclic carotenoids were inhibited. It was proposed that parallel pathways exist for phytoene to lycopene and phytoene to β-carotene. By the use of a mutant tomato it was found that the DMSO inhibited the formation of lycopene at one stage and both pathways at an earlier stage of maturity. Also, the β-carotene is found in the chloroplast, whereas the lycopene is found crystallized in the chromoplasts.

Research has been reported on the use of cell-free homogenates. However, the interpretation of the results of these cell-free systems was difficult because the cell was no longer intact. A number of inhibitors were used. These included mainly nitrogen-containing compounds. One compound, 2-(4-chlorophenyl)triethylamine (CPTA), inhibits the cyclization of lycopene and thus lycopene accumulates. When washing out the CPTa, β-carotene is formed at the expense of lycopene. A number of other inhibitors cause an increase in β-zeacarotene and a decrease in β-carotene. Still other compounds have been found to stimulate β-carotene and β-zeacarotene synthesis. Clearly both pathways are operating and it is possible to "prove" a pathway by the choice of inhibitor, stimulant, or bioregulator (22).

The great deal of information on the biosynthesis of the carotenoids has come from the use of $^{14}C/^{3}H$ dual-labeled MVA. By use of MVA with ^{3}H in the two and four positions and ^{14}C in the two position, it was found that the α- and β-rings are formed independently. Eight carbons and a number of 3H would be found in the entire molecule. Two ^{3}H atoms are found in position four of the β-ring and one in α-carotene. However, if the ϵ-ring is converted to the β-ring, one 3H would be lost. This is not found experimentally. Likewise, the 4-^{3}H from MVA is in the six position; 3H must be lost in the formation of β-carotene but not in the direct formation of α-carotene. Because the 6′-^{3}H is retained in α-carotene, the β-ring is not converted to the ε-ring (23). The nonrandomized dimethyl groups in torularhodin were proved with double labeled MVA. This result suggested that the interconversion of IPP and DMAPP did not result in the randomization of the C-2 hydrogen from MVA (24).

The ring formation in the red pepper to form capsanthin and capsorubin has been suggested to follow a pathway through a postulated epoxide intermediate. It is generally thought that the insertion of oxygen occurs at a later stage in the biosynthetic pathway, and that the epoxide, hydroxyl, and one-carboxyl oxygen came from molecular oxygen.

The view had long been held that the direction of biosynthesis was from saturated to unsaturated, from hydrocarbon to the xanthyphylls. Recently, however, it has been shown that some fish can convert astaxanthin to retinol. Bacteria have also been shown to form a series of 45- and 50-carbon carotenoids. Lycopene, leutein, and cellulose

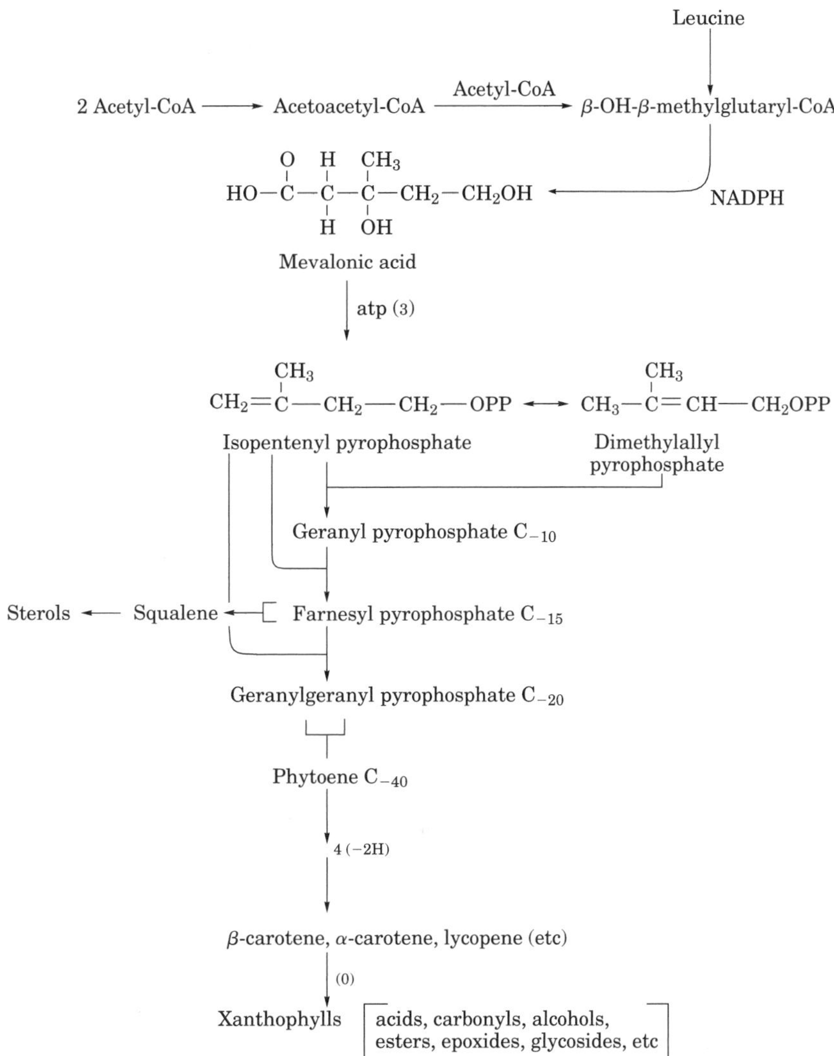

Figure 2. Pathway for terpene biosynthesis.

were fed to rats, and it was found that the conversion of β-carotene to vitamin A was inhibited (25).

OCCURRENCE IN NATURE

As indicated in Table 1 the carotenoids are widely distributed in foods and are the most widespread group of pigments in nature. They are present without exception in photosynthetic tissue; occur with no definite pattern in nonphotosynthetic tissues such as roots, flower petals, seeds, fruits, vegetables; and are found sporadically in the Kingdom Protista, including the (fungi-yeast mushrooms and bacteria). The red, yellow, and orange pigments in the skin, flesh, shell, or exoskeleton of some animal species are due to these pigments. These would include the lobster, shrimp, salmon, goldfish, and flamingo. People who consume large amounts of tomato juice may turn orange or red because of an intake of the pigments. The carotenoids, which cannot be made by animals, are the precursors to vitamin A, but unlike vitamin A, are not toxic in large doses (26).

Generally the concentration of the carotenoids in various tissues is low but can vary widely. The cornea of the eye narcissus contains 18% β-carotene (dry), whereas the concentration in foods is much lower. It has been estimated that the annual production of carotenoids is in on the order of 10^8 tons. The major pigment is fucoxanthin, the pigment in marine brown algae. Other major pigments include those found in green leaves, mainly lutein, violaxanthin, and neoxanthin. By contrast, β-carotene occurs widely but in smaller amounts. It is of interest that (note: zea is major pigment in some algae) 3,3′-dihydroxy-α-carotene (lutein) is a major pigment, whereas 3,3′-dihydroxy-β-carotene (zeaxanthin) is a minor pigment in nature. In contrast, the major hydroxarbon carotenoid found in nature is β-carotene, whereas α-carotene is a minor pigment. A typical percentage profile of the xanthophylls of a green plant could be lutein, 40%; violaxanthin, 34%; neoxanthin, 19%; cryptoxanthin, 4%; and zeaxanthin, 2%. β-carotene, lycopene, and capsanthin can be high in some tissues such as the sweet potato, tomato, and red pepper. By contrast, the carrot is a major course of α-carotene where the concentration can be 30 to 40%.

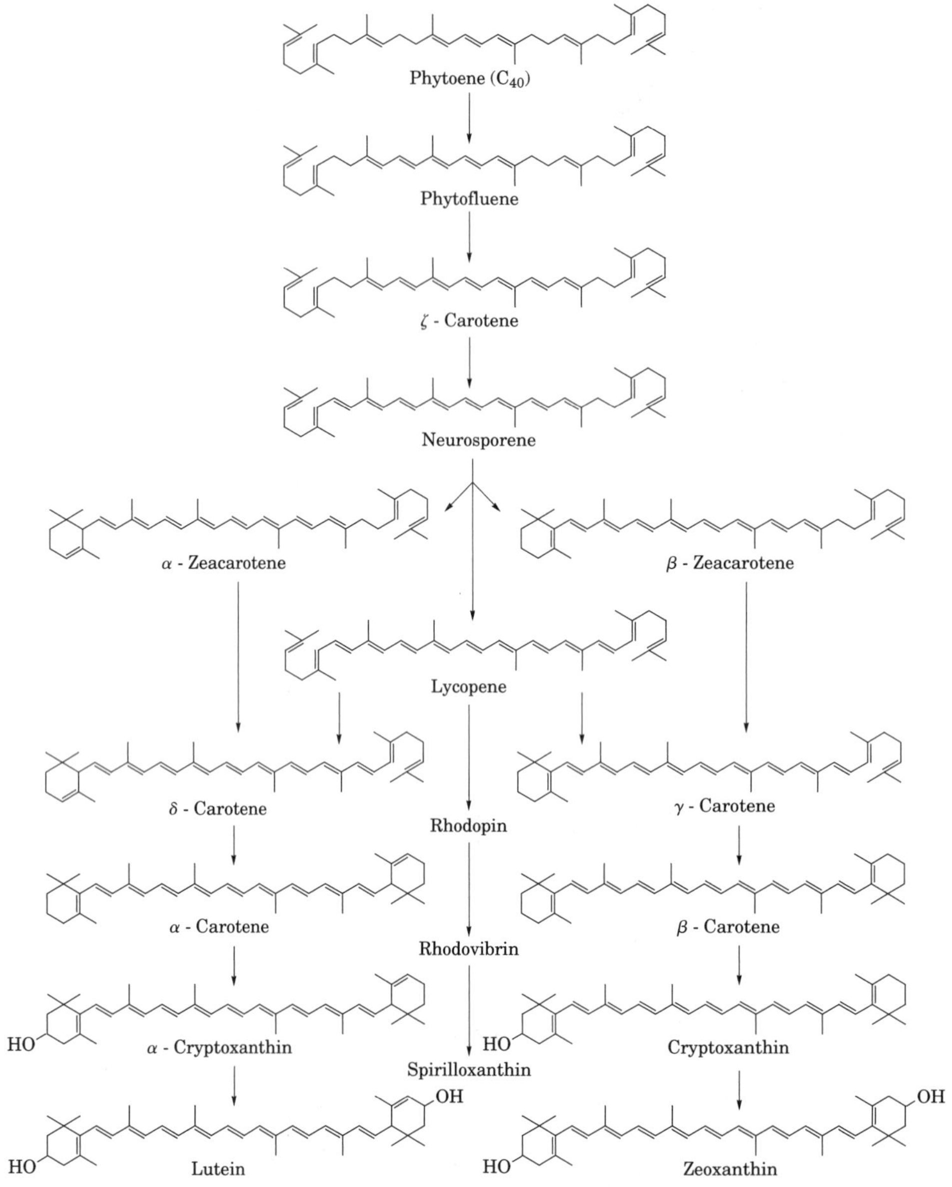

Phytoene (C_{40})

Phytofluene

ζ - Carotene

Neurosporene

α - Zeacarotene

β - Zeacarotene

Lycopene

δ - Carotene

Rhodopin

γ - Carotene

α - Carotene

Rhodovibrin

β - Carotene

HO α - Cryptoxanthin

HO Spirilloxanthin

Cryptoxanthin

OH

HO Lutein

HO Zeoxanthin OH

Figure 3. Common carotenoids.

The level of carotenoids is often directly related to the quality of food. The lack of pigments as in white butter or salmon, for example, is a quality defect (27). However, flour is bleached to destroy the carotenoids, and β-carotene is often removed from red palm oil to make it colorless.

The acyclic pigments phytoene, phytofluene, zeta-carotene, neurosporene, and lycopene are often found in carotenoid-producing systems such as higher and lower plants. They are not often found in animal tissue. Of these, lycopene is the most common and may occur in large amounts in the tomato, watermelon, orange, pink grapefruit, and some apricots. Further reactions in products of lycopene lead to the closing of one or both rings. However in some oranges (Valencia) 3,4-didehydrolycopene is found. There have been a number of acyclic xanthophylls isolated mainly from the purple photosynthetic bacteria. The 1,2-epoxide has been isolated from tomatoes as a minor pigment. Because phytofluene is fluorescent, it must be accounted for in the fluorometric analysis of vitamin A (25).

The alicyclic carotenoids (eg, β-carotene) are very common in higher plants, bacteria, fungi, algae, and animals. It has been estimated, on the basis of structure, that some 50 to 60 alicyclic compounds have potential vitamin A activity. Apricots are an excellent source of β-carotene (60%), whereas the level in tomatoes (10%) and peaches (10%) is much lower. Papaya, if it is yellow/orange, is a good source, whereas the red flesh papaya contains mainly lycopene. Where the flesh is green, some plastid pigments would be present and thus some β-carotene would be present. Red grapes, pears, figs, red apples, and beet root, although brightly colored, are not good sources of β-carotene.

δ-Carotene has been isolated from certain tomato varieties where apo-polycis-isomers of β, α, and ϵ-carotene, and lycopene have been found.

Vegetables may be excellent sources of the provitamin A carotenoids. Spinach may have up to 4.0 mg/100 g β-carotene and carrots may have as high as 6.0 mg/100 g of β-carotene. Yellow corn (maize) contains small amounts of β-carotene and β-zeacarotene. Sweet potatoes may be purple, white, or bright orange. The two former potatoes are low in carotenoids and the latter may be greater than 9 mg/100 g. Squash, broccoli, peas, and pumpkin contain relatively large amounts of carotenes. Most vegetable oils are slightly yellow indicating a small amount of carotenoids. The red palm oil produced in Brazil, Southeast Asia, and parts of Africa may contain large amounts of β- and α-carotene (50 mg/100 g total carotenoids). Egg yolks are a rich source of carotenoids (lutein and zeaxanthin) but relatively poor sources of β-carotene. Green peppers are a better source of carotene than are the red peppers. Wheat flour is low in carotene and this amount is further lowered in the bleaching process.

Hydroxylated carotenoids are very common in plants, where lutein and zeaxanthin are to a lesser extent, β-cryptoxanthin are major pigments. Lutein is a common pigment in freshwater fish, whereas tunaxanthin (3,3′-dihydroxy ϵ-carotene) is common in marine fish. Peaches are a good source of β-cryptoxanthin. Many fruits contain xanthophyll esters. In the persimmon the carotenoid alcohol is β-cryptoxanthin. Corn varieties are selected for their high content of zeaxanthin and lutein. These pigments are fed in order to pigment the skin and egg yolks.

The xanthophylls with hydroxyl or keto substitution in the four position are common in various animal tissues. Echinenone (4-keto-β-carotene) is widely found in marine invertebrates and algae and canthaxanthin (4,4′-keto-β-carotene). The usual epoxides are in the five, six position and these are easily converted to the 5,8-epoxide with acid. Where a keto group is in the four position, in-chain epoxides are formed (eg, 9,10-epoxides of canthaxanthin). The 1,2-epoxides of acylic pigments have been reported in tomato leaves. The allenic seaweed pigment fucoxanthin is thin. The cyclopentyl ketones are found in red peppers and are probably formed by a pinacolic rearrangement of the epoxides of zeaxanthin.

The acetylenic carotenoids are found in diatoms, *Euglena*, giant scallops, mussels, and starfish. A number of apo-carotenoids are formed in some fruit, bacteria, orange peel, and fungi. The most well known apo-compound is bixin, the pigment in annatto seeds. Crocin is an apo-compound occurring in saffron. In a sense, retinol can be considered a degraded β-carotene. The carotenoid acid, torularhodin, is a common pigment in Rhodotorula yeast. Canthaxanthin is found in edible mushrooms, blue-green algae, trout, brine shrimp, and flamingos.

Astaxanthin is the pigment of the salmon and the invertebrates, annelida and crustacea. Recently it has been isolated from red yeast and Adonis flowers (28). Shrimp raised in intensive culture are often blue because of a protein complex with astaxanthin. This complex extends the chromophore resulting in a blue color. This has been mistakenly described as a disease, the condition, however, results in a lower price for the shrimp, particularly on the Japanese market. When cooked, the complex is broken and the red color is seen. The lobster can be green, black, red, or blue because of various protein-astaxanthin complexes all of which are broken with heat, yielding a red pigment. When treated with alkali, astaxanthin is converted to astacene.

Aromatic carotenoids have recently been isolated from photosynthetic bacteria and some sea sponges. Carotenoid epoxides are fairly common in nature. The so-called violaxanthin cycle is an example of the natural occurrence of epoxidation.

$$\text{zeaxanthin} \rightleftharpoons \text{antheraxanthin} \rightleftharpoons \text{violaxanthin}$$

EFFECT OF AGRICULTURAL PRACTICES

A number of agricultural practices, particularly those with plants, directly affect the carotenoid content. Thus the nutrient content of edible plants can vary severalfold to as much as 20 times (29). The major changes are caused by genetic manipulation. However, a number of factors such as sunlight, rainfall, topography, soils, location, season, fertilization of soils, and maturity may affect carotene content. Thus the nutritional tables give an average value with an understood large variation. A number of countries have developed nutritional tables that, when compared, show a great variation for the same fruit or vegetable. In the past, genetic manipulation has been concerned with the color, flavor, texture, and yield. Today there is more of an effort being made to produce foods that are nutritionally superior. Thus, interest is being shown in developing not only the color but also the quality of the color (eg, more β-carotene). No agricultural practice has been as effective as genetic manipulation. High β-carotene tomatoes have been developed but because of the yellow color were not popular. Sweet potatoes can vary from 0 to 7 mg/100 g of carotene and the carotene from carrots can be increased five times. Generally light has a positive effect on carotene synthesis. Fruit grown in the dark, however, will develop some color (30). Temperature has an effect, but this depends on the product. The most pronounced effect is on the ripening of the tomato. In temperatures below 27°C the synthesis of lycopene is favored, whereas a yellow tomato results at a higher temperature. Gamma irradiation also affects the formation of carotene (31).

Most vegetables and fruits increase in carotene during the ripening stage. The carotene content of the carrot increases until a level in maturity is reached. The β-carotene level falls off upon further growth. In the case of the red pepper, the conversion from green to red lowers the carotene content. Because of cultural practices and shipping constraints, fruits may be picked green and thus the full color may not develop.

ANALYSIS

A universal extraction and analysis scheme for the carotenoids is not possible due to the wide variation in carotenoid structure. Thus, numerous methods are suggested for the separation of carotenoids from such things as foodstuffs and blood plasma. The classic methods used by most investigators rely on some method of extraction and a chromatographic separation of the carotenoids (32,33). The methods require some sort of size reduction to increase surface area followed by extraction with a suitable solvent or solvent mixture. Often a bipolar solvent is used to extract nonpolar carotenoids from polar tissue. Ethyl alcohol and acetone are examples of bipolar solvents, and $CHCl_3$/MeoH is an example of a bipolar solvent mixture. The extraction may be followed by saponification, removal of alkali, transfer to a developing solvent, and chromatography under vacuum or pressure. The column may be developed with a gradient of solvents, and the pigments are separated by elution or are cut from the column and extracted. Often rechromatography is required of some bands, followed by verifying chemical reactions. The quantity of each purified fraction is determined by spectroscopy. The method requires several hours, and in some cases artifacts (*cis-trans* isomers, epoxides) are formed due to the long exposure to solvents, absorbents, light, and oxygen. Although the method does give a separation of provitamin A precursors from inactive carotenoids, it is time-consuming and requires skilled technical operators. Thus, the method may not be suitable for analysis of the large number of food items for a table of food compositions. A much simpler method has been developed that chromatographically separates the carotenes from the oxygenated compounds, but it may not separate individual carotenes, their *cis* isomers or carotenoid esters (34). Therefore, this method tends to overestimate provitamin A activity, especially if β-carotene represents only a small part of the total. High-performance liquid chromatography (HPLC) is clearly the method of choice, and the more recent values in the nutritional tables are generally determined by this method. A number of reports have appeared that show the general overestimation of the Association of Official Analytical Chemists (AOAC) method. A modified AOAC method and chromatographic procedures that separated the individual components were used to estimate the provitamin A content of Clingstone peaches (35).

The U.S. RDA for 100 g was much higher by the AOAC method than by the complete separation (60% vs 11% of the RDA for raw, 12% vs 6% for canned).

The comparison of the AOAC method with a stepwise gradient, a saponification step, and HPLC has been published. The AOAC method and HPLC were comparable for green vegetables. For the carrot and pumpkin, the AOAC method was higher for both vegetables than the gradient elution and HPLC method. Where fruit xanthophylls were esterified, these compounds were chromatographed with β-carotene, and thus the estimation was higher in the AOAC method when the fruit extract was not saponified (36).

A number of papers (37) have shown that the open column packet with 40 to 50 μm gives good separation of vitamin A compounds as well as carotenoid compounds. The procedure used a pipette for the injector. Nitrogen gas or vacuum is used in place of a pump.

EFFECT OF FOOD PROCESSING

The carotenoids are altered or partly destroyed by acids, usually but not always stable in inorganic bases, can be destroyed by some enzymes (eg, lipoxygenase), and are usually bleached by light. Most carotenoids are fat soluble and thus are not subject to leaching losses. Generally they are fairly stable to the heat involved in canning but are rapidly lost on dehydration because of oxidation.

Although the most stable form of the carotenoids is the all-*trans* form, the *cis*-geometrical isomer can form upon heating, especially in the presence of H^+. The *cis*-isomer carotenoids are naturally present in the tangerine tomato. It has been shown that if β-carotene activity was set at 100%, the neo-β-carotene vitamin A activity was 38%, and the neo-α-carotene activity was 13%. The transformation from *trans* to *cis* results in a loss of extinction and a shift in absorption maxima to shorter (2–5 mm). Cooking causes the formation of *cis* isomers from all-*trans* β-carotene (38). Broccoli lost 13%, spinach lost 7%, and the sweet potato lost 32% of the all-*trans* isomer. A 15% loss of vitamin A was reported, mainly due to a *cis-trans* isomerization of α- and β-carotene during canning. High-temperature short-time (HTST) processing was shown to have less effect on the carotene in the sweet potato than conventional cooking (35).

Heat processing produced enough acid in pineapple to cause the formation of *cis* isomers. This could result in a lighter yellow color.

The formation of 5,6-and 5,8-epoxides of β-carotene in model systems has been studied. Diphenylamine, a free-radical inhibitor stopped the loss of β-carotene. It is generally thought that epoxide formation is the initial step in the degradation of the carotenoid. Several reports have been published that show the complexity of the reactions. Stored hydrogenated oils and enzymatic-coupled oxidation of carotenoids result in the formation of epoxides and other products.

A number of studies have reported an apparent increase in the level of carotenes in various canned products over the fresh product. There are two reasons for this surprising result. Where soluble solids are leached out of the products during cooking and storage, the fat-soluble carotenes appear to increase in amount. It has also been observed that the fresh control may have an active lipoxygenase-like enzyme that bleaches the carotenoids. When antioxidants were added to the fresh sample before extraction, the vitamin A value in the fresh product increased.

Dehydration and size reduction of fruits and vegetables increase surface area and, if products are not protected from light and oxygen, result in poor stability of the carotenoid pigments. The shelf life of the carotenoids in the dry state is shorter than in the wet state. Different dehydration methods result in varying losses of carotenes. Retention of carotene in fresh carrots varies; fresh carrots retain more than (100%), blanched (95%), freeze-dried (89%), air-dried (80%), and puff-dried (12%). Carotenoid oxidation in foods is generally associated with unsaturated fatty acids and is usually autocatalytic. The oxidation may or may not be enzymatic in nature and is directly related to the available water (A_w), oxygen, heat, and certain metals. Generally antioxidents are effective in slowing the reaction. The decomposition products from a simulated deoderization of red palm oil were isolated. Beta 15^1- and $\beta^{14'}$-apo-carotenals and 13 apo-carotenone were isolated from the reaction mixture and a pathway for the reaction was suggested.

CAROTENOIDS AS FOOD COLORS

It was the color of the carrot that first interested researchers 160 years ago to start work on carotenoids. Many years later it was again the color that was the means of identifying compounds separated by chromatography. It is the same today as it has always been; color is one of the means by which we identify the quality of food. People often reject the unfamiliar. Most pigment changes such as the changes in chlorophyll on thermal processing, the bleaching of anthocyanins, the browning of fresh meat, or the off-coloring of flavonoids do not in themselves affect the product. In case of the destruction of carotene, loss or change in color would result in poorer quality, but also may result in loss of vitamin A activity. β-Carotene has 11 double bonds in conjugation, which is the reason for its yellow-orange color. Any disruption in the chromophore, such as that which is depicted in Figure 2 would result in bleaching.

CONVERSION OF CAROTENOIDS TO VITAMIN A

The carotenoids of fruits, vegetables, and animal products are generally fat soluble and are associated with the lipid fractions. In some cases they may be in the crystalline state, as lycopene is in the tomato. During the digestion process, esterases, lipases, and proteases act on the carotenoids, which are solubilized by the bile salts. Within the mucosal cell β-carotene and the other provitamin A, carotenoids are split by carotenoid dioxygenase(s) resulting in an oxidative cleavage. In theory a molecule of β-carotene should yield two molecules of retinal. Retinol is formed by reduction of retinal. It is esterified and transported to the liver for storage. There is some difference of opinion as to whether there is a central cleavage enzyme or whether there is also a random-splitting enzyme that acts on various places on the chain. The carotenes may also be absorbed, carried by low-density lipoproteins and deposited in the skin, fat, and various organs. The extent of accumulation, rate of accumulation, time to reach saturation, and turnover time were different for each tissue. Liver adrenal and ovary have high concentrations of β-carotene.

Animals differ in the efficiency of the splitting reaction. The rat is generally regarded as very efficient. The mink and the fox are relatively inefficient in this reaction. Man is average with a ratio of about 6:1. The cat does not contain the enzymes necessary to utilize provitamin A carotenes.

Generally, for a compound to be active it must have an unsaturated central chain and an unsubstituted β-ring. On the basis of structure out of the greater than 600 carotenoids listed, the provitamin A compounds would be between 50 and 60. Vitamin A conversion factors include:

$$\text{RE} = \mu\text{g retinol} + \frac{\mu\text{g}\beta\text{-carotene}}{6} + \frac{\mu\text{g other provitamin A compounds}}{12}$$

$$1\ \text{IU} = 0.3\ \mu\text{g retinol}$$
$$0.6\ \mu\text{g}\ \beta\text{-carotene}$$
$$1.2\ \mu\text{g other provitamin A carotenoids.}$$

α-Carotene on the preceding basis would be half as active as β-carotene. The metabolism of α-carotene has been shown to result in the liver storage of retinol and α-vitamin A. The latter compound does not combine with retinol binding protein (RBP). It can induce hypervitaminosis A because it is not delivered to the tissues. There is some confusion in regard to the capacity of the diepoxide to serve as a vitamin A source. The compound is listed as active *in vivo* but not active *in vitro*. Recently, radioactive β-carotene-5,6,5′,6′-diepoxide and luteochrome (5,6,5′,8′-β-carotene-diepoxide) were fed to rats and C^{14} retinol was not detected in the liver. The absorption or conversion to retinal is enhanced by bile salts, proteins, lipids, and zinc. It is not clear what the effect of various nonprovitamin A compounds have on the bioavailability of the provitamin A compounds. Lycopene is reported to enhance carotene uptake. It was found that nonprovitamin A compounds may decrease the utilization of provitamin A pigments, whereas β-cryptoxanthin and α-carotene promote utilization.

Because of the uncertainty of some of the artificial pigments and because of the delisting of several (eg, Red No. 2 in the United States, Red No. 40 in the United Kingdom), interest has increased for using carotenoids as food colorants.

We can consider the use of carotenoids as following a direct or an indirect route. As a case of the former, margarine might be colored with β-carotene. Examples of the latter might be the coloring of poultry skin, or eggs, or salmon flesh with the appropriate pigment. Some confusion exists regarding natural and artificial pigments. The infrared fingerprints for synthetic and natural β-carotene are identical. Many people prefer the term nature-identical to describe the synthetic carotenes.

One of the most common natural coloring preparations is obtained by extracting the seeds of the *Bixa orellana* tree, which grows in the tropics. The apo-compound bixin is the main component of the oil-soluble annatto preparations, and nor-bixin is the 6,6′-diapo-dicarboxylic acid (bixin has one methyl ester). Bixin is most stable at pH 8 and shows a diminished stability in the 4 to 8 pH range. In general, annatto extracts have a good shelf life. Annatto oil applications include the coloring of butter, bakery products, and salad oil. Combined with paprika, annatto oils are used to color cheese. Water-soluble preparations may be used in ice cream.

Saffron consists of the dried stigmas of *Crocus sativus*. It contains crocin, the digentiobioside of crocetin. The preparation is yellow in color. Saffron is also used as a spice and is widely accepted in such foods as soups, meat products, and cheese. Formerly, it was used in cakes and other bakery products.

Tomato extracts are red because of the pigment lycopene. The color of the product would be red to orange depending on the medium. Lycopene is not stable, and its use is limited.

Carrot extracts, carrot oil, and red palm oil contain large amounts of α- and β-carotene. These were used in fat-based products. They have the advantage of being natural but have limited use because of the lower price of synthetic β-carotene. Recently, hypersaline algae have been cultured for their β-carotene content.

Oleoresin paprika is an oil extract of paprika. The main carotenoids are capsanthin and capsorubin. The color of products colored by paprika vary from red to pinkish yellow. Oleoresin paprika can be used in salad dressings, sauces, meat products, and processed cheeses. The indirect coloring of trout with oleoresin paprika resulted in a yellow rather than a red hue of the flesh.

The synthetic pigments have a number of advantages over the natural product extracts. Although the principal advantage is price, it should not be overlooked that the large variation in quality is not a problem with the synthetic FDC dyes. The major synthetic carotenoids that are marketed include β-carotene, β-apo-8′-carotenal, and canthaxanthin. All occur naturally, and only canthaxanthin does not have vitamin A activity.

β-Carotene is widely encountered in nature. In foods it has been used in butter, margarine, salad oil, popcorn, baked goods, confections and candy, eggnog, coffee whiteners, juices, and soups to name a few. β-apo-8′-carotenal has been isolated from oranges and several natural sources. However the main source is the chemical synthesis. It is also yellow, and its uses are similar to those for β-carotene.

Canthaxanin (rosanthin) (4,4′-diketo-β-carotene) is used where a red color is desired. It was first isolated from the mushroom, but it is also a widely distributed animal pigment. Many crustaceans, especially brine shrimp, and birds, notably the flamingo, contain canthaxanthin. One of its main uses is as a pigmenting agent for salmon. Although a red color is obtained, it has been observed that the pigment tends to cook out. Recently canthaxanthin has been used in simulated meats, shrimp, crab, and lobster products.

Synthetic astaxanthin is now available and is approved for use in Europe and the United States. This is the major pigment in a number of fish and shellfish species. The color tends to be more stable than canthaxanthin. When shrimp are raised in intensive culture, often the blue carotenoprotein is found. This can be corrected by feeding more green algae or astaxanthin. Marigold flowers contain mainly lutein and are fed to chickens to supply yellow-orange pigments for the skin and egg yolks.

The crystalline carotenoids are not absorbed or metabolized by animals. This problem has been solved by the development of special water-soluble forms marketed to satisfy various product needs. Two approaches have been primarily employed: the production of oil suspensions of micropulverized crystals and the development of emulsions or beadlet forms containing the carotenoid in supersaturated solution or finely colloidal forms in liquid or dry products.

CAROTENOIDS BY CHEMICAL SYNTHESIS

The total synthesis of β-carotene was reported in 1950. Since that date β-carotene, retinol, *cis*-retinoic acid, canthaxanthin, β-apo-8′-carotenal, and more recently astaxanthin are produced by commercial chemical synthesis. A great many other carotenoids have been synthesized that at the moment have no commercial significance.

The Roche synthesis of β-carotene starts from β-ionone and is related to the vitamin A synthesis. The β-ionone may be prepared from citral, a constituent of lemongrass oil, or prepared synthetically from acetylene as a starting material. In the β-carotene procedure, β-ionone is converted to C-14 aldehyde, which, through further chain-lengthening steps, is converted to C-19 aldehydes. These are joined through a Grignard reaction to form the C-40 diol, which, through allylic rearrangement and dehydration, is converted to 15,15-dehydro-β-carotene. A partial hydrogenation and rearrangement yields *trans*-β-carotene.

Crystalline carotenoids produced by chemical synthesis are of high purity and uniform color. The vast majority of carotenoids, when pure, can be obtained in crystalline form. In general, the melting points of crystalline carotenoids are fairly high and quite sharp for the pure compounds. The synthesis of the carotenoid series of compounds employs a variety of organic synthetic techniques.

The literature on the synthesis of the various carotenoids is extensive, and the details are certainly beyond this chapter.

FUNCTION OF CAROTENOIDS

It is tempting to propose a single universal function for the carotenoids since they are found in such diverse tissues. Failing this, it is tempting to ascribe some function, wherever a carotenoid is found. Although the function of the carotenoids has been proven in some cases, their universal function, if any, remains to be determined. Where the function has been proven, it generally is with some aspect of the light-absorbing property of the carotenoid pigment. The critical role of these pigments in photosynthesis has been the best documented. Phototropism and phototaxis responses of plants seem to be related to the light absorption spectra of the carotenoids. Functions in animals are related to the antioxidant and to the light-absorbing property of these pigments. The function in vitamin A formation has been discussed. Because these pigments are found in many reproductive tissues, it has been suggested that they have a role in reproduction. The results in this area are less convincing than those in other areas.

In the photosynthetic process the carotenoids have been shown to be active in the light-gathering process (eg, they absorb light at wavelengths not absorbed by chlorophyll). In *Chlorella*, other green algae, and higher plants, the light absorbed by the carotenoids was used at low efficiency. In the diatom and brown algae the energy transfer is comparable to chlorophyll where the main pigment in fucoxanthin. In the photosynthetic process two photo systems are involved. In general, more carotenes are found in photosystem I and xanthophylls in photosystem II.

The role of the carotenoids in photoprotection of photosynthetic bacteria has been well documented. A mutant, which only produced the colorless polyenes phytoene and phytofloene, was compared with the wild type *Rhodopseu-*

dowonas sphaeroides. The wild type showed good growth under all conditions of light and oxygen. However, the mutant could only grow under anaerobic-light or aerobic-dark conditions.

Under light with O_2, massive killing resulted at room temperature and bacterial chlorophyll was destroyed. If the mutant was grown in the dark with air, chlorophyll disappeared. Where chlorophyll was missing, light and O_2 have no effect.

Diphenylamine (DPA) inhibits the formation of colored carotenoids. When the wild type blue-green bacteria is treated with DPA, it is destroyed like the mutant. At low temperatures, killing proceeds in the mutant but bacterial chlorophyll is not destroyed.

Corynebacterium poinsettiae is not photosynthetic, and both the wild type and the carotenoid-less mutant can tolerate high intensities of light in the presence of air. When the exogenous dye, toluidine blue, is added as a photosensitizing pigment, the carotenoid-less mutant is killed. The mechanism involves the simultaneous interaction of visible light, a photosensitizing dye, and O_2. Generally, the carotenoids are effective for visible light but have no effect in ultraviolet, gamma, or X radiation.

The reactions are listed as follows:

$$\text{CHL} + h_v \rightarrow {}^*\text{CHL excited state}$$

$$^*\text{CHL} + \rightarrow \text{photosynthesis or } {}^3\text{CHL triplet-excited state intersystem crossing}$$

$$^3\text{CHL} + {}^3\text{O}_2 \rightarrow {}^1\text{CHL} + {}^1\text{O}_2 \quad \text{or } {}^3\text{CHL} + {}^1\text{CAR} \rightarrow {}^1\text{CHL} + {}^3\text{CAR}$$

$$^3\text{CAR}\ {}^1\text{CAR} \rightarrow \text{harmless decay}$$

$$^1\text{O}_2 + \text{CAR CAR O}_2 \text{ CAR can be regenerated} \quad \text{or } {}^1\text{O}_2 + \text{A A O}_2 \text{ photodynamic action}$$

$$^1\text{O}_2 + {}^1\text{CAR}\ {}^3\text{O}_2 + {}^3\text{CAR}$$

$$^3\text{CAR} \rightarrow {}^1\text{CAR harmless decay}$$

As can be seen in these reactions, carotenoids may protect photosynthetic bacteria at various levels by quenching the singlet-excited state of O_2 or the triplet-excited state of chlorophyll. The ground states of oxygen would be $^3O_2$ and for CHL the triplet state. The carotenoids may be the preferred substrate for oxidation or may act in quenching reactive species.

Phototropism is a response of higher plants and some fungi that results in the plant turning to the light. Rival claims for either β-carotene or riboflavin as the active compound have been made based on the action spectra. More recent reports seem to rule out β-carotene as the mediator in the response.

Phototaxis is a response in which an organism such as *Euglena* can move toward the light. The action spectra does not match the spectra of the carotenoids.

The involvement of the carotenoids in reproduction appear to be coincidental to the process of sexual reproduction and of no known significance to the process. The major pigment of the brine shrimp is canthaxanthin; however, the female *Artemia* converts the all-*trans* canthaxanthin to *cis* canthaxanthin during the time of sexual activity.

In animals the major function of carotenoids is as a precursor to the formation of vitamin A. It is assumed that in order to have vitamin A activity a molecule must have one-half of the structure similar to that of β-carotene. Recently it has been shown that astaxanthin can be converted to zeaxanthin in trout where the fish is sufficient in vitamin A. Tritiated astaxanthin was converted to retinol in strips of duodenum or inverted sacks of trout intestines. Astaxanthin, canthaxanthin, and zeaxanthin can be converted to vitamin A and A_2 in guppies.

We have become increasingly aware of a mounting body of evidence that suggests that the carotenoids can function in medical applications apart from their role as vitamin A precursors. The symptoms of erythropoietic *Protoporphyria* can be relieved by large doses of β-carotene (38). This pigment was deposited in the skin of patients who lacked skin pigmentation.

As of late some carotenoids have been claimed to be effective in the treatment of cancer, particularly cancer of epithelial origin. Other carotenoids such as canthaxanthin and phytoene with no vitamin A activity were, in cases, found to be as effective as β-carotene.

Some of the first studies were based on epidemiological studies. In one such study it was shown that the participants who consumed diets rich in carotenoids developed fewer lung cancers whether or not they smoked. This study assumed that the active ingredients are the carotenoids contained in fruits and vegetables.

More recently a number of animal studies have shown that β-carotene or other carotenoids can prevent or slow down the growth of skin cancer and other cancerous tumors. Review have been published on the role of the carotenoids in relation to cancer. Human studies are presently being conducted (39,40).

BIBLIOGRAPHY

1. O. Isler, "Introduction," in O. Isler, ed., *Carotenoids*, Birkhauser Verlag, Basel, 1971, pp. 11–27.
2. P. Karrer and E. Jucker, *Carotenoids*, Birkhauser Verlag, Basel, 1948.
3. *Plenary and Session Lectures, First International Symposium on Carotenoids Other Than Vitamin A-1*, Butterworth, London, 1967, p. 278.
4. *Plenary and Session Lectures, Second International Symposium on Carotenoids Other Than Vitamin A*, Butterworth, Kent, United Kingdom, 1969.
5. *Plenary and Session Lectures, Third International Symposium on Carotenoids Other Than Vitamin A*, Butterworth, Kent, United Kingdom, 1973, p. 130.
6. B. L. C. Weeden, ed., *Plenary and Session Lectures, Fourth International Symposium on Carotenoids—4*, Pergamon Press, Oxford, United Kingdom, 1976, p. 243.
7. T. W. Goodwin, ed., *Plenary and Session Lectures, Fifth International Symposium on Carotenoids—5*, Pergamon Press, Oxford, United Kingdom, 1979, p. 886.
8. G. Britton and T. W. Goodwin, eds., *Plenary and Session Lectures, Sixth International Symposium on Carotenoids—6*, Pergamon Press, Oxford, United Kingdom, 1982.
9. G. Britton, ed., *Plenary and Session Lectures, Seventh International Symposium on Carotenoids—7*. Pergamon Press, Oxford, United Kingdom, 1985.

10. N. I. Krinsky, M. M. Mathews-Roth, and R. F. Taylor, *Carotenoids*, Plenum, New York, 1989.
11. J. C. Bauernfeind, ed., *Carotenoids as Colorants and Vitamin A Precursors. Technological and Nutritional Applications*, Academic Press, Orlando, Fla., 1981, p. 938.
12. J. C. Bauernfiend, ed., *Vitamin A Deficiency and Its Control*, Academic Press, Orlando, Fla., 1986, p. 530.
13. H. Gerster, "Anticarcinogenic Effect of Common Carotenoids," *Int. J. Vitam. Nutr. Res.* **63**, 93–121 (1993).
14. Anonymous, "Tentative Rules for the Nomenclature of Carotenoids," in O. Isler, ed., *Carotenoids*, Birkhauser Verlag, Basel, 1971, pp. 851–864.
15. "Nomenclature of Carotenoids (Rules Approved 1974)," *Journal of Pure and Applied Chemistry* **41**, 406–431 (1975).
16. H. Pfander, *Key to Carotenoids*, Birkhauser, Boston, Mass., 1987.
17. J. W. Porter and R. E. Lincoln, "I. Lycopersicon Selections Containing a High Content of Carotenes and Colorless Polyenes: II. The Mechanism of Carotene Biosynthesis," *Arch. Biochem. Biophys.* **27**, 390–403 (1950).
18. H. K. Lichtenthaler et al., "Biosynthesis of Isoprenoids in Higher Plant Chloroplasts Proceeds via a Mevalonate-Independent Pathway," *FEBS Lett.* **400**, 271–274 (1997).
19. G. Britton, "Biosynthesis of Carotenoids," in T. W. Goodwin, ed., *Chemistry and Biochemistry of Plant Pigments*, Vol. 1, 2nd ed., Academic Press, Orlando, Fla., 1976, pp. 202–327.
20. L. C. Raymundo, A. E. Griffiths, and K. L. Simpson, "Effect of Dimethyl Sulfoxide (DMSO) on the Biosynthesis of Carotenoids in Detached Tomatoes," *Phytochemistry* **6**, 1527–1532 (1967).
21. L. C. Raymundo, A. E. Griffiths, and K. L. Simpson, "Biosynthesis of Carotenoids in the Tomato Fruit," *Phytochemistry* **9**, 1239–1245 (1970).
22. M. Elahi et al., "Effect of CPTA Analogs and Other Nitrogenous Compounds on the Biosynthesis of the Carotenoids in *Phycomyces blakesleeanus* Mutants," *Phytochemistry* **14**, 133–381 (1975).
23. R. J. H. Williams, G. Britton, and T. W. Goodwin, "Biosynthesis of Cyclic Carotenes," *Biochem. J.* **105**, 99–105 (1967).
24. R. E. Tefft, T. W. Goodwin, and K. L. Simpson, "Aspects of the Stereospecificity of Torularhodin Biosynthesis," *Biochem. J.* **117**, 921–927 (1970).
25. S.-I. Huang, Ph.D. Thesis, "Inhibitory Effects of Cellulose and Nonprovitamin A Carotenoids on Beta-Carotene Utilization in Rats," University of Rhode Island, 1995.
26. T. W. Goodwin, *The Biochemistry of the Carotenoids*, Vol. 1, 2nd. ed., Academic Press, Orlando, Fla., 1980, p. 377.
27. J. Ostrander et al., "Sensory Testing of Pen Reared Salmon and Trout," *J. Food Sci.* **41**, 386–390 (1975).
28. T. Kamata and K. L. Simpson, "Study of Astaxanthin Diester Extracted from *Adonis aestivalis*," *Comp. Biochem. Physiol. B* **86**, 587–596 (1987).
29. R. S. Harris, "Effects of Agricultural Practices on the Composition of Foods," in R. S. Harris and E. Karmas, eds., *Nutritional Evaluation of Food Processing*, 2nd ed., AVI Publishing, Westport, Conn., 1957, pp. 33–57.
30. L. C. Raymundo, C. O. Chichester, and K. L. Simpson, "Light Dependent Carotenoid Synthesis in the Tomato Fruit," *J. Agric. Food Chem.* **24**, 59–64 (1976).
31. C. N. Villegas et al., "The Effect of Gamma Irradiation on the Biosynthesis of Carotenoids in the Tomato Fruit," *Plant Physiol.* **50**, 694–697 (1972).
32. B. H. Davies, "Analytical Methods—Carotenoids," in T. W. Goodwin, ed., *Chemistry and Biochemistry of Plant Pigments*, Vol. 2, 2nd ed., Academic Press, Orlando, Fla., 1976, pp. 38–165.
33. E. DeRitter and A. E. Purcell, "Carotenoid Analytical Methods," in J. C. Bauernfiend, ed., *Carotenoids as Colorants and Vitamin A Precursors Technological and Nutritional Applications*, Academic Press, Orlando, Fla., 1981, pp. 915–823.
34. S. E. Gebhardt, E. R. Elkins, and J. Humphrey, "Comparison of Two Methods of Determining the Vitamin A of Clingstone Peaches," *J. Agric. Food Chem.* **25**, 628–631 (1977).
35. D. B. Rodriguez-Amaya, "Assessment of Provitamin A Contents of Foods—The Brazilian Experience," *J. Food Composition and Analysis* **9**, 96–230 (1998).
36. K. L. Simpson, "Chemical Changes in Natural Food Pigments," in T. Richardson and J. W. Finley, eds., *Chemical Changes in Food During Processing*, AVI Publishing, Westport, Conn., 1985, pp. 409–441.
37. L. Wenhong et al., "Rapid Determination of Blood Serum Retinol by Reverse Phase Open Column Chromatography," *Int. J. Vitam. Nutr. Res.* **63**, 82–86 (1993).
38. M. M. Mathews-Roth et al., "Beta-Carotene as a Photoprotective Agent in Erythropoietic Protoporphyria," *N. Engl. J. Med.* **282**, 1231–1234 (1970).
39. R. Peto et al., "Can Dietary B-Carotene Materially Reduce Human Cancer Rates?" *Nature (London)* **290**, 201–208 (1981).
40. M. M. Mathews-Roth, "Carotenoids and Cancer Prevention—Experimental and Epidemiological Studies," *Pure Appl. Chem.* **51**, 717–722 (1985).

KENNETH L. SIMPSON
University of Rhode Island
Kingston, Rhode Island

See also COLORANTS: CAROTENOIDS.

CENTRIFUGES: PRINCIPLES AND APPLICATIONS

Centrifuges are used widely in food processing to separate liquids from solids, liquids from liquids, and even to separate two immiscible liquids from the accompanying solids. Centrifuges can be blanketed with inert gas and operate under high temperatures and pressures, as well as discharge centrate under pressure to retard foam and oxidation. They rinse the mother liquor from solids. They can be cleaned in place, and some can be steam sterilized. Centrifuges may also be used to separate large particles from similar small particles when they are operated as classifiers.

Centrifuges were first used in the 1880s to separate milk from cream. They use rotational acceleration to separate heavier phases from lighter phases, similar to a gravity-settling basin. The separating force is measured in units called *g*'s, which are units of acceleration; $1\,g = 980$ cm/s.^{2}

Centrifuges usually generate at least 1,200 *g* for low-speed machines and up to 63,000 *g* for small tubular centrifuges. From a process standpoint, a higher *g* force results in a higher capacity for a given centrifuge; capacity is measured by volume throughput, degree of solids concentration, clarity of the liquid, and entrainment of one phase in another. Also, because the length of time the fluid is held under high *g* affects the separation, the volume of

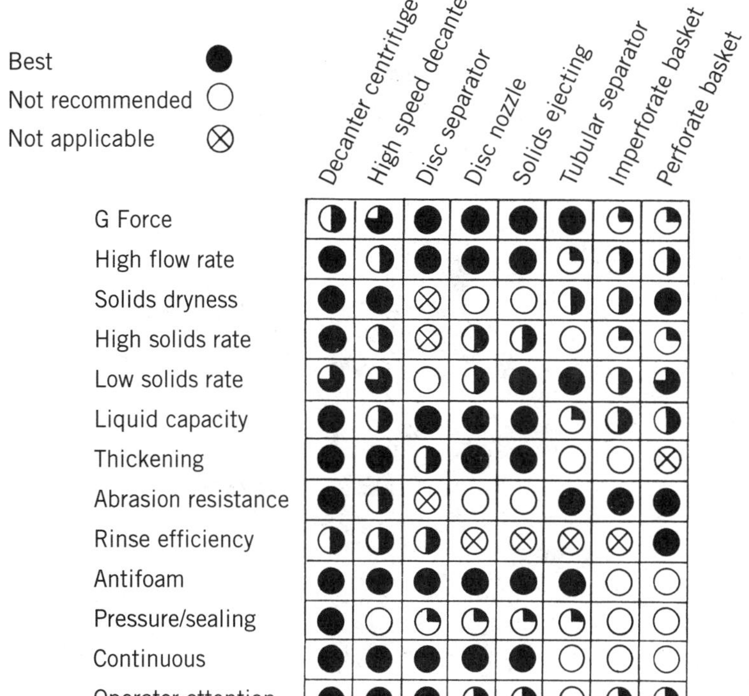

Figure 1. Properties of different centrifuge types.

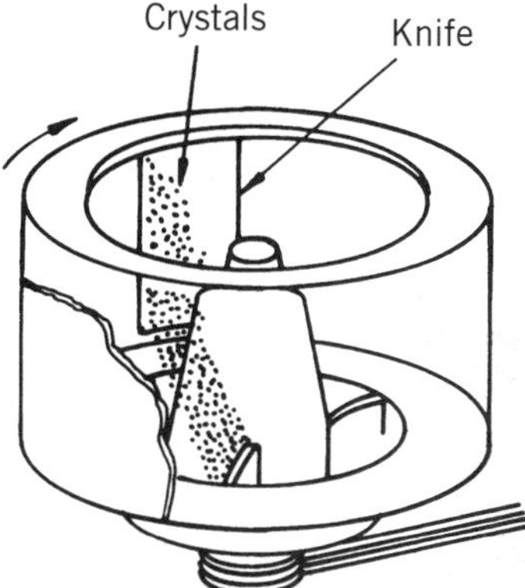

Figure 2. Basket centrifuge: imperforate bowl. Low g force (500–1,000 g) and intermittent feed limit these to applications where moderate solids dryness is needed. Capacity: 100–200 L/min.

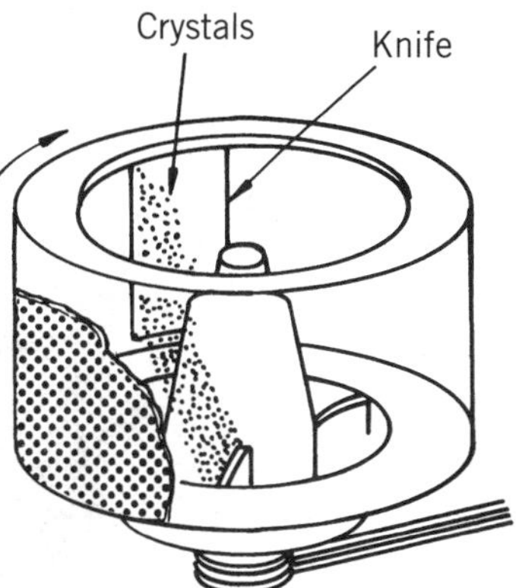

Figure 3. Basket centrifuge: perforate bowl. Same as Figure 2 except that the liquid filters through the solids. Excellent rinsing capability. Very dry cakes if the particle size is large enough to filter well.

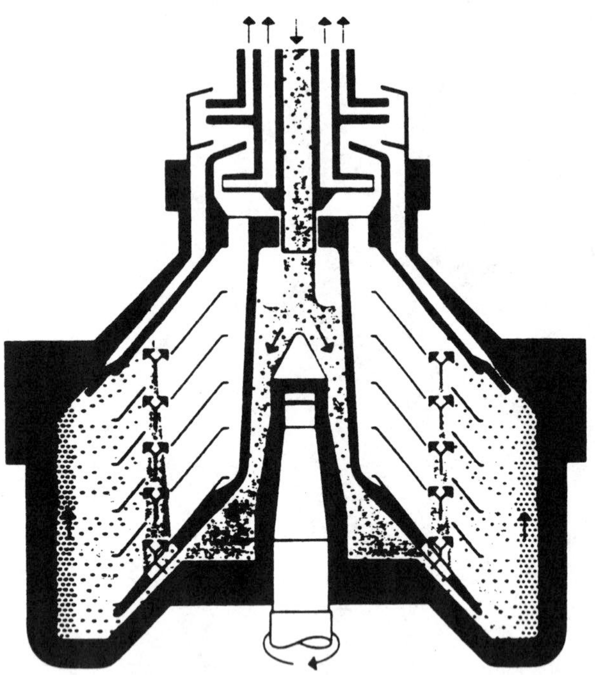

Figure 4. Disc stack centrifuge. The solid bowl disc centrifuge is a solid bowl version liquid-liquid separator operating at 5,000–10,000 *g* with some models as high as 15,000 *g*. They can handle a wide range of rates up to 2,400 L/min.

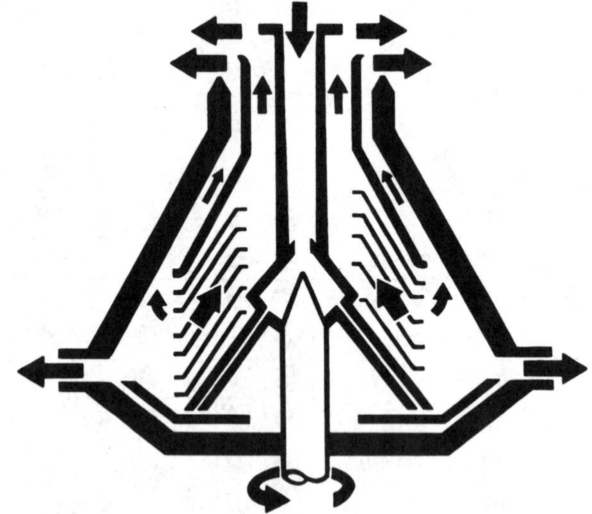

Figure 5. Disc stack centrifuge: nozzle version. Same as Figure 4 except that the solids are continuously discharged as a pumpable slurry. Rates as high as 5,000 L/min are possible.

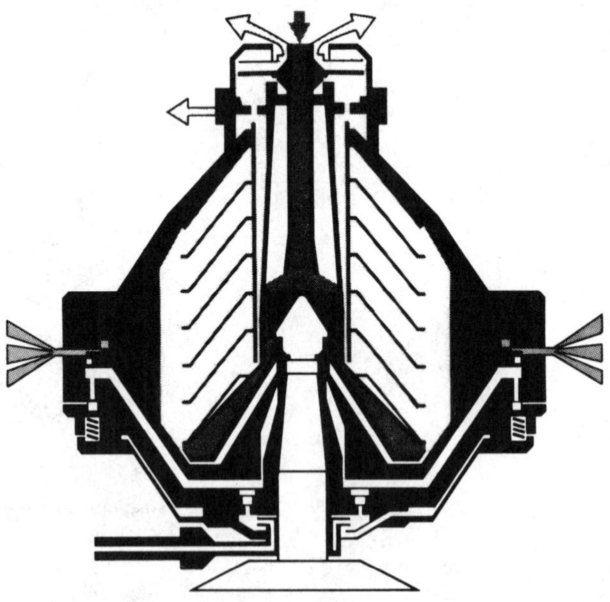

Figure 6. Disc stack centrifuge: solids ejecting version. Used where the solids volume is low, the bowl opens up intermittently to dump the solids as pumpable slurry. Maximum rates 40 L/min solids, 2,500 L/min liquid; *g* force 5,000–10,000 *g*.

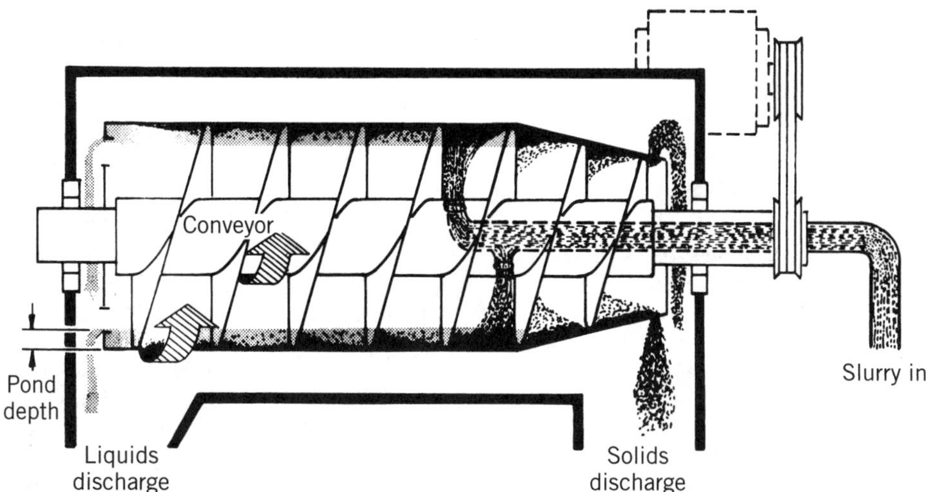

Figure 7. Decanter centrifuges. Solid bowl decanter centrifuges develop 1,000–3,000 g. They are used where large volumes of solids are present and the solid phase must be as dry as possible. Rates vary from 8 to 4,000 L/min. Very high speed decanters from 5–10,000 g are available but rates are lower, 10–200 L/min.

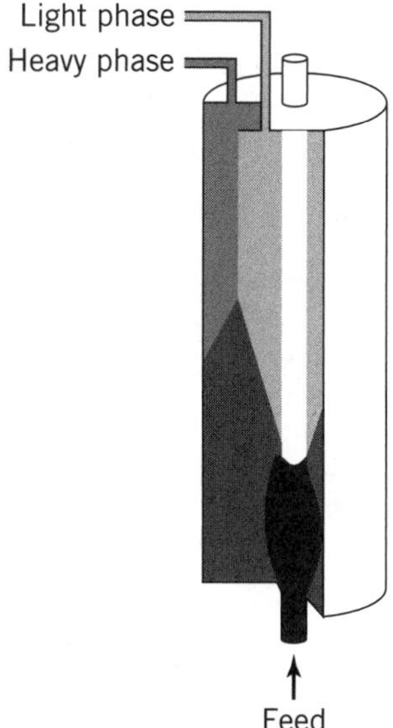

Figure 8. Tubular centrifuges. Manual disassembly to remove solids limits these to applications with little or no solids in the feed. As liquid-liquid separators, they develop much higher g force than most other centrifuges; maximum feed rates are 40–80 L/min.

fluid in the centrifuge directly affects its capacity. Physically, the separation follows Stokes' law for the settling velocity of a particle in a fluid, except that the acceleration of the rotational field is substituted for the earth's gravitational field.

STOKES' LAW

$$v_g = \frac{\Delta\varphi d w^2 r}{18\mu}$$

where V_g = settling velocity, w = rotational speed, r = radial position of the particle, d = diameter of particle, $\Delta\varphi$ = difference in the specific gravity of the liquid and solid, and μ = viscosity of fluid. Stokes' law is used to design separation systems because the faster the settling velocity of a particle, the greater the chance of capture will be. Generally the temperature and the particle size are, to some extent, variables in a system.

For example, in rendering the fat from animal tissues, the amount of fat separated from the tissue increases with increasing temperature. Above 45°C, the residual tissue changes character, and its value as a meat additive drops. For edible rendering, the separation is made at 45°C, for the maximum fat yield with high-value solids. Inedible rendering is done at 90–95°C where the fat yield is at a maximum, because the value of the inedible solids is quite low.

Stokes' law can be generalized as follows. Separation is improved if

1. The difference in specific gravity between the phases is large.
2. The particle size is large.
3. The viscosity is low.
4. The centrifuge speed is high.
5. The feed rate to the centrifuge is low.

Scale-up to estimate between different types of centrifuges is difficult, but within a class (eg, all disk centrifuges) a reasonably accurate scale-up can be developed. Centrifuges are usually scaled up based on g times volume or g times area. More detailed scale-up is usually left to centrifuge specialists.

SCREENING THE VARIOUS CENTRIFUGE TYPES

The goals of the separation must be defined. Centrifuges in general can:

Separate liquids from liquids.
Separate liquids from solids.
Separate solids from immiscible liquids.
Classify solids by size or density.
Rinse mother liquid from solids.

In brewing, the spent grains must be as dry as possible, the solids rate is high, and the solids are somewhat abrasive. Based on Figure 1, the following relationships can be seen.

	Solids dryness	High flow rate	Abrasion resistance
Decanter centrifuge	●	●	●
High-speed decanter	●	◑	◑
Perforate basket	●	◑	●

The decanter centrifuge is the clear choice followed by the perforate basket and, a distant third, the high speed decanter. An inquiry of decanter manufacturers would provide much data and reference accounts. The perforate basket centrifuge, in fact, is a filtration device and is not suitable for the application. Most manufacturers have laboratories where tests of new applications can be run at nominal cost. Some manufacturers also rent portable equipment to test on location.

CENTRIFUGE TYPES AND CONCLUSION

The centrifuge has wide and growing applications in the food industry for both liquid-liquid and solid-liquid separations. Advances in both speed and capacity have been made in the last few years, which increase yields and capacities. There is extensive literature on both theory and applications to aid the design engineer in equipment selection (Figs. 2–8).

PETER L. MONTAGNE
Sharples, Inc.
Warminster, Pennsylvania

CEREALS, NUTRIENTS, AND AGRICULTURAL PRACTICES

CEREAL GRAIN PRODUCTION AND COMPOSITION

Wheat, rice, and corn were the three largest cereal grains produced in the world in 1995/1996 and 1996/1997 (Table 1). Average chemical compositions of these and other commonly consumed grains are listed in Table 2. All cereal grains contain starch as the principal component. Vitamin, mineral, fatty acids, and amino acid contents of cereal grains have been published (3). Content and nutritive value of cereal proteins depend both on seed heredity and environment during cultivation and harvest.

Table 1. Production of Cereal Grains, Million Metric Tons

	World		United States	
Grain	1995/1996	1996/1997	1995/1996	1996/1997
Wheat	536	583	59.4	62.1
Corn	514	591	187	236
Rice	551	565	7.9	7.8
Barley	142	153	7.8	8.7
Sorghum	55.2	68.2	11.7	20.4
Oats	28.7	30.6	2.4	2.2
Rye	21.9	22.2	0.26	0.23

Source: Refs. 1 and 2. Each year such as 1995/1996 covered a 12-month period.

Table 2. Average Composition of Cereal Grains (As-is)

	Barley (pearled)	Corn (field)	Oats (oatmeal)	Rice (brown)	Rye	Sorghum	Wheat[a]
Constituents (%)							
Water	10.1	10.4	8.8	10.4	11.0	9.2	13.1
Protein	9.9	9.4	16.0	7.9	14.8	11.3	12.6
Fat	1.2	4.7	6.3	2.9	2.5	3.3	1.5
Starch	57.5	65.0	50.3	69.2	57.1	65.1	55.7
Total dietary fiber	15.6	7.3	10.6	3.5	14.6	7.6	12.2
Ash	1.1	1.2	1.9	1.5	2.0	1.6	1.6
Energy value, Kcal/100 g	352	365	384	370	335	339	327

Source: Refs. 3–6.
[a]Hard red winter.

WHEAT

Dry and Wet Milling

Wheat (*Triticum vulgare*) products from dry milling hard wheat are farina, straight grade flour, patent flour, first clear flour, second clear flour, germ, bran, and shorts. Similar flour milling processes for soft and durum wheats provide products for many foods that differ from those incorporating hard wheat products, although durum farina is called semolina. A small amount of the world's wheat is processed to starch and gluten (about 80% of the total wheat flour protein) by wet milling. The baking industry is the largest user of gluten. Alkaline extraction procedures can also yield protein and starch from whole wheat (7) and mill feeds (8). Total food uses for wheat was 24 million metric tons in the United States for 1996/1997.

Nutrient Composition of Wheat Products

White wheat flour, the principal refined product of wheat milling, is the major ingredient in many foods: hard wheat flour for breads and noodles; soft wheat flour for cakes, pastries, quick breads, crackers, and snack foods; and durum semolina for spaghetti, macaroni, and other pasta products. The nutritional value of wheat foods depends mainly on the chemical composition of the flours used in their preparation. The average percentage composition of hard wheat milled products is given in Table 3. Wheat averages nearly 13% protein (the range is 7–22%). The lysine content of wheat protein varies between 2.2 and 4.2%, with a mean value of approximately 3.0%. Successful breeding of wheat for better protein quality and quantity could have an important impact on the nutrient content of wheat food products. Celiac disease, a rare allergic reaction, is an ingestion intolerance to wheat gluten (the major protein of wheat) that interferes with the absorption of nutrients.

CORN

Dry and Wet Milling

Corn (*Zea mays* L.) is processed to provide food ingredients, industrial products, feeds, alcoholic beverages, and fuel ethanol. Wet millers produce starch; modified starch products, including dextrose and syrups (by starch hydrolysis); feed products; and oil (10). Dry millers produce hull, germ, and endosperm fractions that vary in particle size and fat content. Endosperm products—grits, meal, and flour—are the primary products used by the food processors and in consumer markets (11).

Corn germ flour prepared from a commercial dry-milled fraction appears to be a promising fortifying ingredient for the food industry (12). Protein concentrates from normal and high-lysine corns (13) and from defatted dry-milled corn germ (14) are obtained by alkaline extraction of corn or corn germ and then precipitating the extracted proteins with acid.

Nutrient Composition of Corn Products

Typical yields and analyses of dry-milled products from corn have been published (11,15). About 95% of the corn produced in the United States in 1996 was yellow dent. Specialty corns such as white, waxy, hard endosperm/food grade, high-oil, high-lysine, and high-amylose are also commercially available. High-oil corn, which has about 7% oil compared with about 4.5% oil for dent corn, is the fastest growing market product among all corns in the United

Table 3. Average Percentage Composition of Dry-Milled Hard Wheat Products (As-is)

Constituent, %	Wheat	Farina	Patent flour	First clear flour	Second clear flour	Germ	Shorts	Bran
Moisture	13.1	14.2	13.9	13.4	12.4	10.5	13.5	14.1
Ash	1.6	0.4	0.4	0.7	1.2	4.0	4.1	6.0
Protein	12.6	10.3	11.0	12.7	13.5	30.0	16.0	14.5
Total dietary fiber	12.2	2.6	2.7	—[a]	—[a]	15.1	—[a]	40.5
Fat	1.5	0.8	0.9	1.3	1.3	10.0	4.5	3.3

Source: Refs. 3, 4, and 9.
[a]Not determined.

States from 1993 to 1997. High-oil corn also has higher protein and higher essential amino acids compared with regular dent corn.

RICE

Rice (*Oryza sativa* L.) consists of 18 to 28% hull and the edible portion (brown rice). After the hull is removed, the resulting brown rice is further processed by abrasive milling, which removes 6 to 10% of the weight as bran polish. The remainder is known as white or milled rice. Conventional and Japanese rice milling systems consist of precleaning, dehulling, rough rice separation, and whitening (16); the chemical composition of rice and its fractions have been listed (17). Brown rice and white rice are usually consumed as food after cooking. Rice bran, because of rancidity problems, was used exclusively as animal feed until recently. Stabilized rice bran is now commercially available in the United States, and perhaps 1% of rice bran in the United States, is used as human food.

Rice is a staple diet item in half of the world's population, but only a relatively small proportion of adverse food reactions are from rice proteins. The main allergen of rice is the 16-kDa globulin, and rice mutants containing low levels of the 16-kDa allergenic protein were obtained by γ-ray irradiation or ethyl methanesulfonate treatment (18). Another method to make hypoallergenic rice is to digest the allergenic protein with a proteolytic enzyme, Actinase (19).

OATS

Processing

Oat (*Avena sativa* L.) processing includes cleaning, dehulling, steaming, and flaking (20). Typical products from milling oats are rolled oats, oat hulls, feed oats, mixed grains and seeds, and fines. Oats are dry milled into break flour, reduction flour, shorts flours, shorts, bran, and hulls (21). Weight distribution and protein contents of dry-milled oat fractions, sequential solvent extraction of proteins from milled fractions, and amino acid compositions of milled fractions and oat extracts were also determined (21). Several commercial high-protein oat cultivars are available. Protein concentrates have been prepared from oat groats (dehulled oats) by wet-milling procedures (22). Air classification (separation of particles by size in a stream of air) of oat groats can concentrate protein in the fine fraction and concentrate β-glucan, a polysaccharide that can lower elevated serum cholesterol in humans, in the coarse fraction (23). Amylodextrins with soluble β-glucan contents as high as 10% were prepared from milled oat flours and brans by α-amylases treatment (24). These amylodextrins are used as fat substitutes in bakery goods, dairy products such as cheese, and beverages. Estimated production of oat amylodextrins was around 5 million lb in 1998.

Nutrient Composition of Oat Products

The nutritional quality of oat protein is good compared with other cereals; rolled oats have a protein efficiency ratio (PER) of 2.2 compared with the milk protein casein with a PER of 2.5. Oat groats used for food contain 11 to 15% protein. However, some dehulled oats from the Near East have protein contents of up to 25%.

BARLEY

Conventional roller milling of dehulled barley (*Hordeum vulgare* L.) gives four principal products: flour, tailings flour, shorts, and bran (25). The yields, protein contents, and amino acid composition of roller-milled barley fractions have been reported (26). Barley protein concentrates from normal and high-protein, high-lysine varieties were prepared by an alkaline extraction method (27). Air classification of flour from high-protein, high β-glucan barleys can concentrate protein in the fine fraction and concentrate β-glucan in the coarse fraction (28).

A hulless high β-glucan barley with 17% β-glucan on dry basis in flour has been reported (28). A high-lysine and high-protein barley variety, Hiproly, has been described (29). Many other high-lysine barleys have also been developed since 1970 (30). Food uses of barley include parched grain, pearled grain, flour, and ground grain. About half of the total U.S. barley crop is used for malting and the remainder is used for feed.

SORGHUM

Sorghum is a major feed grain in the United States. Most of the sorghum that is consumed throughout the world as food is prepared directly from the whole grain. Dry milling of sorghum (*Sorghum bicolor* [L.] Moench) grain ranges from cracking to debranning and degermination, which yield refined bran, germ, meal, and grit fractions (31).

Normal sorghum has the lowest lysine content per 100 g protein among the cereal grains. However, two floury lines of Ethiopian origin were exceptionally high in lysine at relatively high levels of protein (32). Another high-lysine sorghum, *P721 opaque*, was produced by treating normal grain with a chemical mutagen (33). An alkaline extraction process gives protein concentrates and starch from ground normal and high-lysine sorghum (34). Air classification of flour and hard endosperm from high-lysine sorghum yielded fractions with higher protein contents than the starting flour or hard endosperm (35).

RYE

Rye (*Secale cereale* L.) can be milled into flours and meals for bread, crackers, and snack foods. Rye milling is similar to wheat dry milling in general. Ground meat products may contain rye flours as fillers. However, labels for meat with grains must indicate the inclusion of these grains.

DISCLOSURE STATEMENT

Names are necessary to report factually on available data; however, the U.S. Department of Agriculture (USDA) neither guarantees nor warrants the standard of the product, and the use of the name by USDA implies no approval of the product to the exclusion of others that may also be suitable. All programs and services of the USDA are offered on a nondiscriminatory basis without regard to race, color, national origin, religion, sex, age, marital status, or handicap.

BIBLIOGRAPHY

1. U.S. Department of Agriculture, *Agricultural Statistics 1997*, U.S. Government Printing Office, Washington, D.C., 1997.
2. U.S. Department of Agriculture, *Agricultural Statistics 1998*, U.S. Government Printing Office, Washington, D.C., 1998.
3. D. L. Drake, S. E. Gebhardt, and R. H. Matthews, "Composition of Foods: Cereal Grains and Pasta, Raw, Processed, Prepared," in *Agricultural Handbook*, Number 8-20, U.S. Department of Agriculture, Washington, D.C., 1989.
4. R. H. Matthews and P. R. Pehrsson, *Provisional Table on the Dietary Fiber Content of Selected Foods*, HNIS/PT-106, U.S. Department of Agriculture, Human Nutrition Information Service, Washington, D.C., 1988.
5. R. A. Olson and K. J. Frey, eds., *Nutritional Qualities of Cereal Grains*, American Society of Agronomy, Crop Science Society of America, and Soil Science Society of America, Madison, WS., 1987.
6. K. E. Bach Knudsen and L. Munck, "Dietary Fibre Contents and Compositions of Sorghum and Sorghum-Based Foods," *J. Cereal Sci.* **3**, 153–164 (1985).
7. Y. V. Wu and K. R. Sexson, "Preparation of Protein Concentrate from Normal and High-Protein Wheats," *J. Agric. Food Chem.* **23**, 903–905 (1975).
8. R. M. Saunders et al., "Preparation of Protein Concentrates from Wheat Shorts and Wheat Mill Run by Wet Alkaline Process," *Cereal Chem.* **52**, 93–101 (1975).
9. Y. V. Wu and G. E. Inglett, "Effects of Agricultural Practices, Handling, Processing, and Storage on Cereals," in E. Karmas and R. S. Harris, eds., *Nutritional Evaluation of Food Processing*, 3rd ed., AVI/van Nostrand Reinhold, New York, 1988, pp. 101–118.
10. J. B. May, "Wet Milling: Process and Products," in S. A. Watson and P. E. Ramstad, eds., *Corn: Chemistry and Technology*, American Association of Cereal Chemists, St. Paul, Minn., 1987, pp. 377–397.
11. R. J. Alexander, "Corn Dry Milling: Processes, Products, and Applications," in S. A. Watson and P. E. Ramstad, eds., *Corn: Chemistry and Technology*, American Association of Cereal Chemists, St. Paul, Minn., 1987, pp. 351–376.
12. C. W. Blessin et al., "Composition of Three Food Products Containing Defatted Corn Germ Flour," *J. Food Sci.* **38**, 602–606 (1973).
13. Y. V. Wu and K. R. Sexson, "Protein Concentrates from Normal and High-Lysine Corns by Alkaline Extraction: Preparation," *J. Food Sci.* **41**, 509–511 (1976).
14. H. C. Nielsen et al., "Corn Germ Protein Isolate—Preliminary Studies on Preparation and Properties," *Cereal Chem.* **50**, 435–443 (1973).
15. S. A. Watson and P. E. Ramstad, eds., *Corn: Chemistry and Technology*, American Association of Cereal Chemists, St. Paul, Minn., 1987.
16. H. T. L. van Ruiten, "Rice Milling: An Overview," in B. O. Juliano, ed., *Rice: Chemistry and Technology*, 2nd ed., American Association of Cereal Chemists, St. Paul, Minn., 1985, pp. 349–388.
17. B. O. Juliano and D. B. Bechtel, "The Rice Grain and its Gross Composition," in B. O. Juliano, ed., *Rice: Chemistry and Technology*, 2nd ed., American Association of Cereal Chemists, St. Paul, Minn., 1985, pp. 17–57.
18. T. Nishio and S. Iida, "Mutants Having a Low Content of 16-kDa Allergenic Protein in Rice (*Oryza sativa* L.)," *Theoret. Appl. Genet.* **86**, 317–321 (1993).
19. M. Watanabe et al., "Production of Hypoallergenic Rice by Enzymatic Decomposition of Constituent Proteins," *J. Food Sci.* **55**, 781–783 (1990).
20. D. Deane and E. Commers, "Oat Cleaning and Processing," in F. H. Webster, ed., *Oats: Chemistry and Technology*, American Association of Cereal Chemists, St. Paul, Minn., 1986, pp. 371–412.
21. Y. V. Wu et al., "Oats and Their Dry-Milled Fractions: Protein Isolation and Properties of Four Varieties," *J. Agric. Food Chem.* **20**, 757–761 (1972).
22. J. E. Cluskey et al., "Oat Protein Concentrates from a Wet-Milling Process: Preparation," *Cereal Chem.* **50**, 475–481 (1973).
23. Y. V. Wu and A. C. Stringfellow, "Enriched Protein- and β-Glucan Fractions from High-Protein Oats by Air Classification," *Cereal Chem.* **72**, 132–134 (1995).
24. G. E. Inglett, "Amylodextrins Containing β-Glucan from Oat Flours and Bran," *Food Chem.* **47**, 133–136 (1993).
25. Y. Pomeranz, H. Ke, and A. B. Ward, "Composition and Utilization of Milled Barley Products. I. Gross Composition of Roller-Milled and Air-Separated Fractions," *Cereal Chem.* **48**, 47–58 (1971).
26. G. S. Robbins and Y. Pomeranz, "Composition and Utilization of Milled Barley Products. III. Amino Acid Composition," *Cereal Chem.* **49**, 240–246 (1972).
27. Y. V. Wu, K. R. Sexson, and J. E. Sanderson, "Barley Protein Concentrate from High-Protein, High-Lysine Varieties," *J. Food Sci.* **44**, 1580–1583 (1979).
28. Y. V. Wu, A. C. Stringfellow, and G. E. Inglett, "Protein- and β-Glucan Enriched Fractions from High-Protein, High β-Glucan Barleys by Sieving and Air Classification," *Cereal Chem.* **71**, 220–223 (1994).
29. A. Hagberg and K. E. Karlsson, "Breeding for High Protein Content and Quality in Barley," in *Symposium: New Approaches to Breeding for Improved Plant Protein*, edited by the Joint FAO/IAEA Division of Atomic Energy in Food and Agriculture, International Atomic Energy Agency, Vienna, 1968, pp. 17–21.
30. R. S. Bhatty, "Nonmalting Uses of Barley," in A. W. MacGregor and R. S. Bhatty, eds., *Barley: Chemistry and Technology*, American Association of Cereal Chemists, St. Paul, Minn., 1993, pp. 355–417.
31. D. A. V. Dendy, *Sorghum and Millets, Chemistry and Technology*, American Association of Cereal Chemists, St. Paul, Minn., 1995.
32. R. Singh and J. D. Axtell, "High Lysine Mutant Gene HL That Improves Protein Quality and Biological Value of Grain Sorghum," *Crop Sci.* **13**, 535–539 (1973).
33. D. P. Mohan and J. D. Axtell, "Diethyl Sulfate Induced High Lysine Mutants in Sorghum," *Proceedings of the Ninth Biennial Grain Sorghum Research and Utilization Conference*, Lubbock, Tex., March 4–6, 1975.
34. Y. V. Wu, "Protein Concentrate from Normal and High-Lysine Sorghums: Preparation, Composition, and Properties," *J. Agric. Food Chem.* **26**, 305–309 (1978).
35. Y. V. Wu and A. C. Stringfellow, "Protein Concentrate from Air Classification of Flour and Horny Endosperm from High-Lysine Sorghum," *J. Food Sci.* **48**, 304–305 (1981).

GENERAL REFERENCES

W. Bushuk, ed., *Rye: Production, Chemistry and Technology*, American Association of Cereal Chemists, St. Paul, Minn., 1976.

R. C. Hoseney, *Principles of Cereal Science and Technology*, 2nd ed., American Association of Cereal Chemists, St. Paul, Minn., 1994.

Y. Pomeranz, ed., *Wheat: Chemistry and Technology*, 3rd ed., American Association of Cereal Chemists, St. Paul, Minn., 1988.

Y. Victor Wu
U.S. Department of Agriculture, Agricultural Research Service
Peoria, Illinois

CEREALS SCIENCE AND TECHNOLOGY

Cereal grains supply the most calories per acre than do other food sources, can be stored safely for a long time, and can be processed into many acceptable products. They are adapted to a variety of soil and climatic conditions, and can be cultivated both on a large scale mechanically with a small amount of labor, and on a garden scale almost entirely by human labor. They are excellent sources of energy and relatively good sources of inexpensive protein, certain minerals, and vitamins. More than two-thirds of the world's cultivated area is planted with grain crops. Most developing countries rely on cereals as food sources. Cereals provide more than half of the calories consumed for human energy, and in many developing countries they provide more than two-thirds of the total diet. As feeds, cereals also contribute greatly to the production of animal proteins (1–9). Total world production of cereal grains (except for sorghum where data were not available) and their production in various parts of the world are shown in Table 1.

GRAINS

Wheat

Wheat grows on almost every kind of arable land, from sea level to elevations of 3000 m, in regions where water is sufficient or areas that are relatively arid; it also thrives in well-drained loams and clays. Wheat is a basic food throughout the world. In developing countries, wheat consumption is increasing and generally accompanies rising living standards. The increase in wheat consumption comes at the expense of the more costly rice or the less costly barley or sorghum. Wheat is the most widely cultivated grain crop in temperate areas. Winter wheat may be multicropped prior to soybeans while spring wheat is often grown in rotation with barley and oats to reduce loss of soil fertility (5,10–13).

Wheat provides about one-fifth of all calories consumed by humans. It accounts for nearly 30% of worldwide grain production and for over 50% of the world trade in grains. It is harvested somewhere in the world nearly every month of the year. The United States, Canada, Australia, Argentina, and France are the main wheat exporters. Wheat production in North America and Western Europe has declined relative to that of the rest of the world since 1950. These two areas accounted for 41% of world production in 1950 but only 36% in 1951. Eastern Europe and Asia accounted for 48% of the total world production in 1950 and 56% in 1981.

The main wheat-producing areas in Asia are China, India, Turkey, and Pakistan. China produces more wheat than any other nation. In fact, wheat is second only to rice as a source of human food in China. Japan is a large wheat importer; Argentina and Australia are major wheat exporters. The United States is the fourth largest wheat-producing country behind China, the former USSR, and India.

Hexaploid wheats are the most widely grown around the world. They can be classified according to:

1. Growth habitat: spring wheats are sown in the spring and harvested in late summer. Winter wheats are sown in the fall and harvested early the next summer. For winter wheats to produce seed it must be vernalized (be subjected to freezing temperature for several days). Winter wheats are preferred by growers because of a generally higher yield potential.
2. Bran color: white (colorless), red, purple or black.
3. Kernel hardness: hard or soft; and vitreousness: glassy-vitreous or mealy–starchy–floury. In many parts of the world all bread wheats (hard and soft)

Table 1. Production of Cereal Grains, in 10^3 Metric Tons, During 1996–1997 in Various Production Areas

Production area	Wheat	Rye	Rice	Corn	Oats	Barley
North America	95,367	538	6930	266,214	6714	24,578
South America	22,276	50	11,337	56,550	865	1,285
W. Europe	99,970	5,765	1,600	34,969	7,313	52,937
E. Europe	26,300	6,151	42	25,315	2,526	9,708
F. Soviet Union	64,309	9,456	839	4,824	10,322	29,765
Africa	21,881	—	9,704	40,513	164	8,101
Asia	229,106	250	347,667	161,190	902	20,350
Oceania	23,786	20	1,006	516	1,747	7,017
Total World	*583,007*	*22,233*	*379,125*	*590,091*	*30,553*	*153,743*

Source: USDA-NASS Agricultural Statistics 1998. Data for sorghum were not available.

are called soft, and the term hard is reserved for the very hard durum wheats. In the United States the term hard is used for bread wheats; soft is reserved for wheats used to produce cookies, cakes, and so on; and durum for those tetraploid wheats used to produce alimentary pastes.

4. Geographic region of growth.
5. End use properties: protein content, physical dough properties, bread-making quality.
6. Variety.
7. Composite designations (14–17).

Corn

Corn (called maize in much of the world) is now the most popular grain for use as animal feed in temperate regions. In many tropical areas, it is a basic human food. There are five major types of corn: flint, dent, floury, pop, and sweet. Dent varieties, grown most widely in the United States, provide high yields and are used to a large extent as feed grain. Flint corn, with a somewhat higher food value, is common in Europe and South America. It is valued for its physical properties and as a raw material in the production of certain Central American foods (tortilla and arepa). Pop, sweet, and floury corn are relatively minor food crops (18–21).

In corn, the starch contains about 25% amylose and 75% amylopectin. Mutants of corn are grown in which the starch is almost entirely amylopectin (called waxy) or 70 to 80% amylose (amylomaize types). These starches vary widely in their physical properties (gelatinization characteristics, water-binding capacity, gel properties, etc) and usefulness in various foods and industrial applications.

Maize is generally deficient in the basic amino acids. However, high-lysine and high-lysine-tryptophan genetic mutants have improved the nutritional properties of corn and are now available as dent or flint corn. In the future those mutants are likely to be acceptable for the production of maize-based foods.

The United States accounts for more than half of the total world corn production and about 80% of the annual world corn exports. In the United States, most corn is used as feed or seed or in the production of alcoholic beverages. Only about 10% is used as food. This includes the production of starch, corn syrups, breakfast cereals, and various foods. Livestock feed accounts for almost 85% of the total U.S. domestic use. The main importers of U.S. corn are Western Europe and Japan, primarily for livestock feed.

Rice

Rice is the staple food of about half of the human race, providing more than one-fifth of the total food calories consumed by the people of the world. Most rice is produced in the Far East and is primarily consumed within the borders of the country of origin. Asia, which has nearly 60% of the world population, produces and consumes about 90% of the world rice production. The United States produces less than 2% of the world crop, but accounts for about 30% of the world rice trade (22,23).

Rice varieties vary in their kernel shape and properties. They are classified as long, medium, and short grain. Long-grain rice accounts for about 50%, medium-grain rice for 40%, and short-grain rice for about 10% of the U.S. rice production. California is practically the exclusive producer of short-grain rice. California and Louisiana lead in medium-grain rice production. Growers in Arkansas, Texas, and Mississippi produce mostly long-grain rice.

Barley

Barley is a winter-hardy and drought-resistant grain. It matures more rapidly than do wheat, oats, or rye and is used mainly as feed for livestock and in malting and distilling industries. The two main types of barley, two-row and six-row, differ in the arrangement of grains in the ear. The former predominates in Europe and parts of Australia; the latter is more resistant to extreme temperatures and is grown in North America, India, and the Middle East (24,25).

Although barley is grown throughout the world, production is concentrated in the northern latitudes. Since 1950 the world production of barley tripled with most of its increase in Europe. The former USSR is the largest barley producer. Barley is adaptable to a variety of conditions and is produced commercially in 36 states in the United States. In the United States, the use of barley as a livestock feed is declining, while the use for processing into malt is increasing.

Oats

Oats are grown most successfully in cool, humid climates and on neutral to slightly acidic soil. The bulk of the crop is consumed as animal feed (primarily for horses). A covered cereal, the oat hull must first be removed before further processing. The resulting product is called a groat. Heavy oats (with high groat to hull ratios) are processed into rolled oats, breakfast foods, and oatmeal (26).

The world production of oats is about one-tenth of the world production of wheat and slightly over one-fourth of the world production of barley. The chief oat-producing states in the United States are in the north-central Corn Belt and farther northward. About 90% of the oat is left on the farm as a feed, and only a small proportion is processed as food or used for industrial products. The use of oats in food processing is increasing, especially in breakfast cereals.

Sorghum

Sorghum is the fifth most widely grown grain in the world. However, sorghum production is less than 5% of the total grain production. There has been a small increase in the area planted and harvested with sorghum, but the increase in production has been substantial because of increases in yield. This has resulted from the development of high-yielding hybrids. The primary sorghum-producing regions are Asia, Africa, and North America. The primary sorghum-producing countries are the United States, China, India, Argentina, Nigeria, and Mexico. These countries produce more than 75% of the world total, with the United States producing about 30%.

More than 50% of the worldwide production of sorghum is destined for human consumption. However, in the United States, sorghum is grown almost entirely as a feed grain for local use or for export. Sorghum is produced under a wide range of climatic conditions. It tolerates limited moisture conditions and adapts to high temperature conditions (27,28).

The nucellar tissue of some varieties of sorghum contains pigments that complicate production of acceptable and white sorghum starch. Varieties with no nucellar tissue persisting to maturity could be used for starch production, but the ready availability of corn and other problems with the wet-milling of sorghum has discouraged such attempts. Waxy (starch) sorghums are potentially promising for special food uses. Some sorghums are rich in condensed tannins. This reduces both their palatability and feed value.

Rye

Rye is characterized by its good resistance to cold, pests, and diseases, but it cannot compete with wheat or barley on good soils and under improved cultivation practices. Approximately 90% of the world's rye production is in Europe (mainly in Poland and Germany) and in the former USSR. Rye is considered a grain for use in bread in Europe, but its proportion in mixed wheat-rye bread is decreasing. Much of the rye is used as a feed grain and a small proportion in the distilling industry (29).

PHYSICAL PROPERTIES AND STRUCTURE

Optimal utilization of cereal grains requires knowledge of their structure and composition. The practical implications of kernel structure and composition are numerous (30–32). They relate to the various stages of grain production, harvest, storage, marketing, and use.

Some physical properties of cereal grains are listed in Table 2 (33). Table 3 summarizes approximate grain size and the proportions of the principal parts comprising the mature kernels of different cereals (34).

Kernel Structure

The cereal grain is a one-seeded fruit (called a caryopsis), in which the fruit coat adheres to the seed at maturity. As the fruit ripens, the pericarp (fruit wall) becomes firmly attached to the wall of the seed proper. The pericarp, seed coats, nucellus, and aleurone cells form the bran. The embryo comprises only a small part of the seed. The bulk of the seed is taken up by the endosperm, which is a food reservoir for the germinating plant.

The floral envelopes (modified leaves known as lemma and palea), or chaffy parts, within which the caryopsis develops, persist to maturity in the grass family. If the chaffy structures envelop the caryopsis so closely that they remain attached to it when the grain is threshed (as with rice and most varieties of oats and barley), the grain is considered to be covered. However, if the caryopsis readily separates from the floral envelopes when the grain is threshed, as with common wheats, rye, hull-less barleys, and the common varieties of corn, these grains are considered to be naked.

Wheat. The structure of the wheat kernel is shown in Figure 1 (35). The dorsal or back side of the wheat grain is rounded, while the ventral side has a deep groove or crease along the entire longitudinal axis. At the apex or small end (stigniatic end) of the grain is a cluster of short, fine hairs known as brush hairs. The pericarp, or dry fruit coat, consists of four layers: the epidermis, hypodermis, cross cells, and tube cells. The remaining tissues of the grain are the inner bran (seed coat and nucellar tissue), endosperm, and embryo (germ). The aleurone layer consists of large rectangular, thick-walled, starch-free cells. Botanically, the aleurone is the outer layer of the endosperm, but as it tends to remain attached to the outer layers during wheat milling. It is shown in the diagram as the innermost bran layer.

The embryo (germ) consists of the plumule and radical, which are connected by the mesocotyl. The outer layer of the scutellum, the epithelium, may function as both a secretory or an absorptive organ. In a well-filled wheat kernel, the germ comprises about 2 to 3% of the kernel, the bran 13 to 17%, and the endosperm the remainder. The inner bran layers (aleurone) are high in protein, whereas

Table 2. Some Physical Properties of Cereal Grains

Name	Length (mm)	Width (mm)	Grain mass (mg)	Bulk density (kg/m^3)	Unit density (kg/m^3)
Rye	4.5–10	1.5–3.5	21	695	—
Sorghum	3–5	2.5–4.5	23	1360	—
Paddy rice	5–10	1.5–5	27	575–600	1370–1400
Oats	6–13	1–4.5	22	356–520	1360–1390
Wheat	5–8	2.5–4.5	37	790–825	1400–1435
Barley	8–14	1–4.5	37	580–660	1390–1400
Corn	8–17	5–15	285	745	1310
Bulirush millet	3–4	2–3	11	760	1322
Wild rice	8–20	0.5–2	22	388–775	—
Tef	1–1.5	0.5–1	0.3	880	—
Findi	1	1–1.5	0.4	790	—
Finger millet	1.5	1.5	—	—	—

Source: Ref. 31.

Table 3. Approximate Grain Size and Proportions of the Principal Parts Comprising the Mature Kernel of Different Cereals

Cereal	Grain mass (mg)	Embryo (%)	Scutellum (%)	Pericarp (%)	Aleurone (%)	Endosperm (%)
Barley	36–45	1.85	1.53	18.3	79.0	
Bread wheat	30–45	1.2	1.54	7.9	6.7–7.0	81–84
Durum wheat	34–46	1.6		12.0	86.4	
Maize	150–600	1.15	7.25	5.5		82
Oats	15–23	1.6	2.13	28.7–41.4		55.8–68.3
Rice	23–27	2–3	1.5	1.5	4–6	89–94
Rye	15–40	1.8	1.73	12.0	85.1	
Sorghum	8–50	7.8–12.1		7.3–9.3		80–85
Triticale	38–53	3.7		14.4	81.9	

Source: Ref. 32

the outer bran (pericarp, seed coats, and nucellus) is high in cellulose, hemicelluloses, and minerals. The germ is high in proteins, lipids, sugars, and minerals; the endosperm consists largely of starch granules embedded in a protein matrix. The structure of rice, barley, oats, rye, and triticale are similar in structure to wheat, except that rice has no crease.

Rice. Rice is a covered cereal. In the threshed grain (termed rough or paddy rice), the kernel is enclosed in a tough siliceous hull that renders it unsuitable for human consumption. Once this hull is removed during processing, the kernel (or caryopsis), comprised of the pericarp (outer bran) and the seed proper (inner bran, endosperm, and germ), is known as brown rice or sometimes as unpolished rice. Brown rice is in little demand as a food. Unless stored under favorable conditions, it tends to become rancid and is more subject to insect infestation than are the various forms of milled rice. When brown rice is milled further, the bran and germ are removed to produce the purified endosperm marketed as white rice or polished rice. Milled rice is classified according to size as head rice (whole endosperm) and various classes of broken rice, known as second head, screenings, and brewer's rice, in order of decreasing size.

Corn. The corn grain is the largest of all cereals (Fig. 2) (35). The kernel is flattened, wedge shaped, and broader at the apex than at its attachment to the cob. The aleurone cells contains relatively high levels of protein and oil. They also contain the pigments that make certain varieties appear blue, black, or purple. Each corn kernel contains two types of starchy endosperms, horny (vitreous) and floury (opaque). The horny endosperm is tightly packed. The starch granules in this region are polyhedral in shape, In dent corn varieties, horny endosperm is located on the sides and back of the kernel and bulges toward the center at the sides. The floury endosperm fills the crown (upper part) of the kernel and extends downward to surround the germ. The starch granules in the floury endosperm are ellipsoidal in shape. As dent corn matures, the endoperm shrinks causing an indentation at the top of the kernel. In a typical dent corn, the pericarp comprises about 6%, the germ, 11%, and the endosperm, 83% of the kernel by weight. Flint corn varieties contain a higher ratio of horny to floury endosperm than do dent corn.

Barley. The husks of barley are cemented to the kernel and remain attached after threshing. The husks protect the kernel from mechanical injury during commercial malting, strengthen the texture of steeped barley, and contribute to more uniform germination of the kernels. The husks are also important as a filtration bed in the separation of extract components during mashing and contribute to the flavor and astringency of beer. The main types of cultivated covered barley, differ in the arrangement of grains in the ear, two-rowed and six-rowed ear. The axis of the barley ear has nodes throughout its length. The nodes alternate from side to side. In the six-rowed types of barley, three kernels develop on each node, one central kernel and two lateral kernels. In the two-rowed barley, the lateral kernels are sterile and only the central kernels develop.

The kernel of covered barley consists of the caryopsis and the flowering glumes (or husks). The husks consist of two membranous sheaths that completely enclose the caryopsis. During development of the growing barley, a cementing substance causing adherence is secreted by the caryopsis within the first two weeks after pollination. The husk in cultivated malting barley amounts to about 8 to 15% of the grain. The proportion varies according to type, variety, grain size, and climatic conditions. Large kernels have less husk than small kernels. The husk in two-rowed barley is generally lower than that in six-rowed barley. As in wheat, the caryopsis of barley is a one-seeded fruit in which the outer pericarp layers enclose the aleurone, the starchy endosperm, and the germ. The aleurone layer in barley is at least two cell layers thick; in other cereal grains (rice excepted), it is one cell layer thick.

Oats. The common varieties of oats have the fruit (caryopsis) enveloped by a hull composed of the floral envelopes. Naked or hull-less oat varieties are known but not grown extensively. In light thin oats, hulls may comprise as much as 45% of the grain. In very heavy or plump oats, the hull may represent only 20%. The hull normally makes up about 30% of the grain. Oat kernels, obtained by removing hulls, are called groats.

Sorghum. Sorghum kernels are generally spherical to flattened spheres, have a kernel mass of 20 to 30 mg, and may be white, yellow, brown, or red. Sorghum has both vitreous and opaque regions within the kernel similar to that of corn. Sorghum grains contain polyphenolic com-

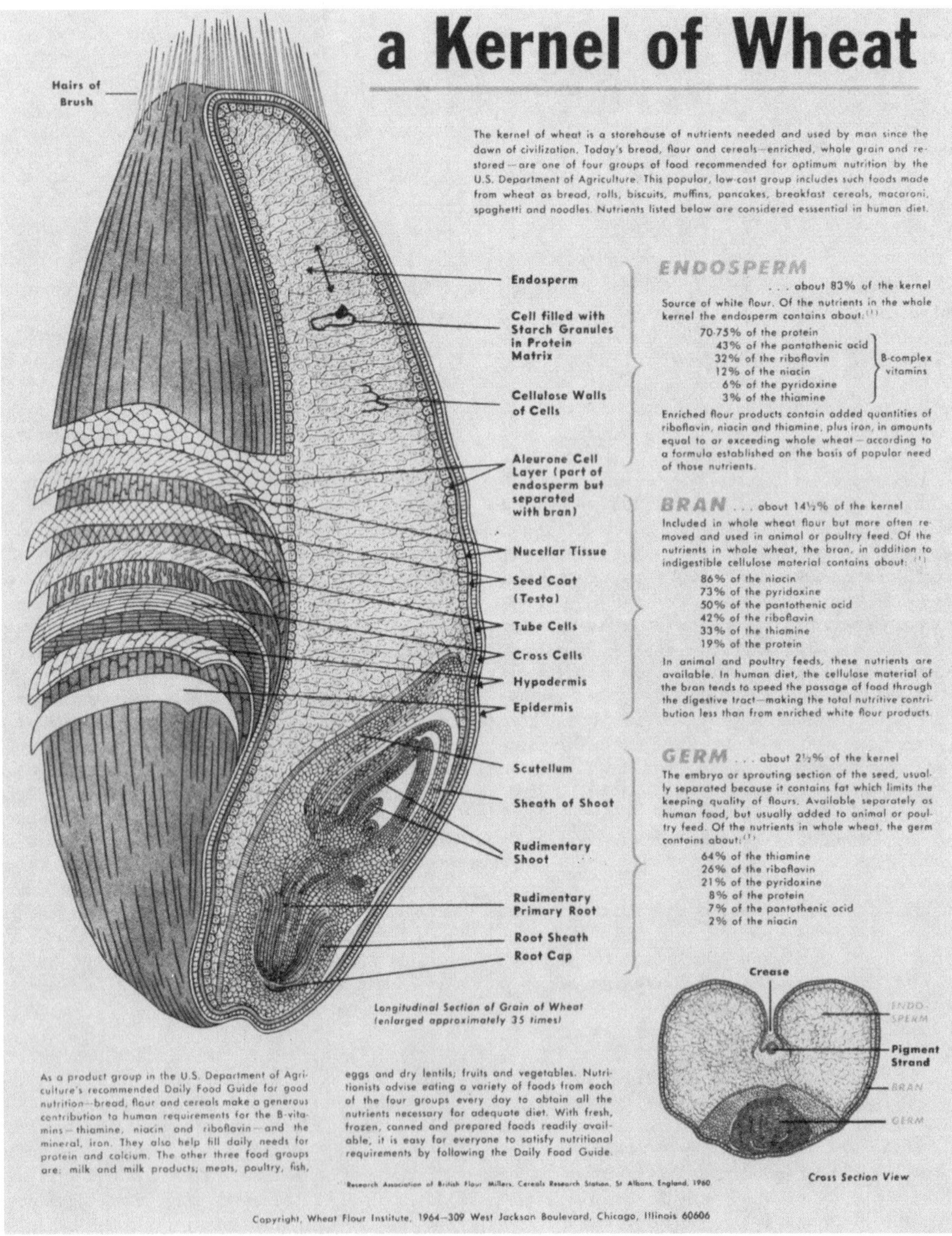

Figure 1. A wheat kernel. Longitudinal section (enlarged ca 12×). *Source:* Ref. 35.

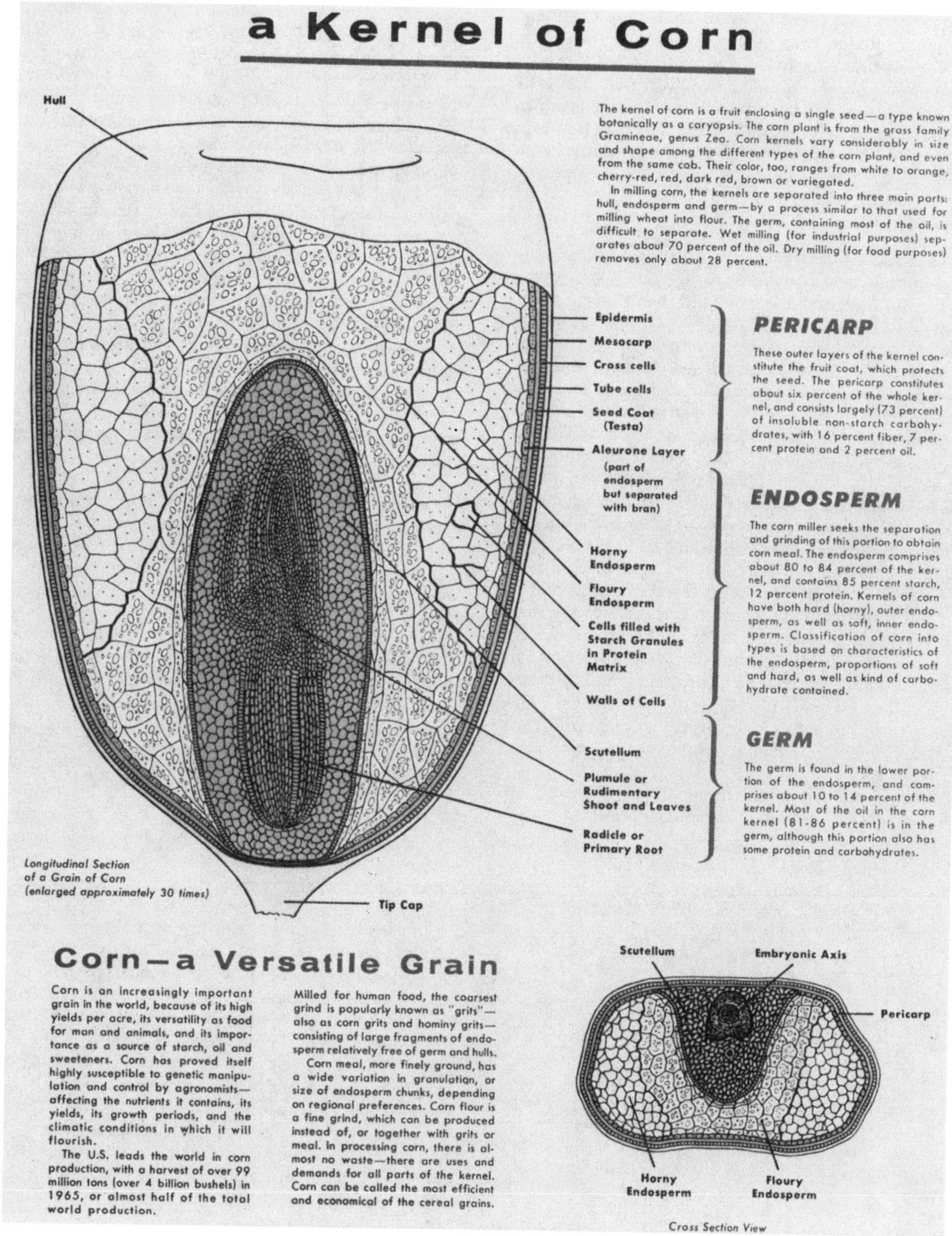

a Kernel of Corn

Hull

The kernel of corn is a fruit enclosing a single seed—a type known botanically as a caryopsis. The corn plant is from the grass family Gramineae, genus Zea. Corn kernels vary considerably in size and shape among the different types of the corn plant, and even from the same cob. Their color, too, ranges from white to orange, cherry-red, red, dark red, brown or variegated.

In milling corn, the kernels are separated into three main parts: hull, endosperm and germ—by a process similar to that used for milling wheat into flour. The germ, containing most of the oil, is difficult to separate. Wet milling (for industrial purposes) separates about 70 percent of the oil. Dry milling (for food purposes) removes only about 28 percent.

Epidermis
Mesocarp
Cross cells
Tube cells
Seed Coat (Testa)
Aleurone Layer (part of endosperm but separated with bran)

PERICARP

These outer layers of the kernel constitute the fruit coat, which protects the seed. The pericarp constitutes about six percent of the whole kernel, and consists largely (73 percent) of insoluble non-starch carbohydrates, with 16 percent fiber, 7 percent protein and 2 percent oil.

Horny Endosperm
Floury Endosperm
Cells filled with Starch Granules in Protein Matrix
Walls of Cells

ENDOSPERM

The corn miller seeks the separation and grinding of this portion to obtain corn meal. The endosperm comprises about 80 to 84 percent of the kernel, and contains 85 percent starch, 12 percent protein. Kernels of corn have both hard (horny), outer endosperm, as well as soft, inner endosperm. Classification of corn into types is based on characteristics of the endosperm, proportions of soft and hard, as well as kind of carbohydrate contained.

Scutellum
Plumule or Rudimentary Shoot and Leaves
Radicle or Primary Root

GERM

The germ is found in the lower portion of the endosperm, and comprises about 10 to 14 percent of the kernel. Most of the oil in the corn kernel (81-86 percent) is in the germ, although this portion also has some protein and carbohydrates.

Longitudinal Section of a Grain of Corn (enlarged approximately 30 times)

Tip Cap

Corn—a Versatile Grain

Corn is an increasingly important grain in the world, because of its high yields per acre, its versatility as food for man and animals, and its importance as a source of starch, oil and sweeteners. Corn has proved itself highly susceptible to genetic manipulation and control by agronomists—affecting the nutrients it contains, its yields, its growth periods, and the climatic conditions in which it will flourish.

The U.S. leads the world in corn production, with a harvest of over 99 million tons (over 4 billion bushels) in 1965, or almost half of the total world production.

Milled for human food, the coarsest grind is popularly known as "grits"—also as corn grits and hominy grits—consisting of large fragments of endosperm relatively free of germ and hulls.

Corn meal, more finely ground, has a wide variation in granulation, or size of endosperm chunks, depending on regional preferences. Corn flour is a fine grind, which can be produced instead of, or together with grits or meal. In processing corn, there is almost no waste—there are uses and demands for all parts of the kernel. Corn can be called the most efficient and economical of the cereal grains.

Scutellum
Embryonic Axis
Pericarp
Horny Endosperm
Floury Endosperm

Cross Section View

Figure 2. A corn kernel. Longitudinal section (enlarged ca 8×). *Source:* Ref. 35.

pounds, primarily in the outer kernel layers and to some degree in the endosperm. These compounds are colored and impart various colors to certain foods from sorghums. Condensed tannins, present only in some sorghum lines, are polyphenolic compounds that interact with and precipitate protein. Those tannins impart a bird resistance to the sorghum. This may be important if the sorghum is grown in small plots in a forest (such as parts of Africa). However, the tannins also reduce nutritional value and germinability.

Composition

The chemical composition of the dry matter of different cereal grains, as with other foods of plant origin, varies widely (36). A proximate analysis of the principal cereal grains is summarized in Table 4. Variations are encountered in the relative amounts of proteins, lipids, carbohydrates, pigments, vitamins, and ash. Mineral elements that are present and the quantities of them also vary widely. As a food group, cereals are characterized by relatively low protein and high carbohydrate contents. The carbohydrates consist of starch (90% or more of the total), pentosans, and simple sugars (monosaccharides).

The various components are not distributed uniformly through the different kernel structures. The hulls and pericarp are high in cellulose, pentosans, and ash; the germ is high in lipid and rich in proteins, sugars, and ash components. The endosperm contains the starch and is lower in protein content than the germ and, in some cereals, than bran. It is also low in crude fat and ash. As a group, cereals are low in nutritionally important calcium. Its concentration and that of other ash components is greatly reduced by the milling processes used to prepare refined flours and grits. In these processes, hulls, germ, and bran, which are the structures rich in minerals and vitamins, are in large part removed.

All cereal grains contain vitamins of the B group, but all are completely lacking in vitamin C (unless the grain is sprouted) and vitamin D. Yellow corn differs from white corn and the other cereal grains in containing carotenoid pigments (principally cryptoxanthin, with smaller quantities of carotenes). These are convertible by the body to vitamin A. Wheat also contains yellow pigments, but they are almost entirely xanthophylls, which are not precursors of vitamin A. The oils of the embryos of cereal grains are rich sources of vitamin E. The relative distribution of vitamins in the kernel is not uniform, although the endosperm invariably contains the lowest concentration.

Table 4. Approximate Analysis of Important Cereal Grains, Percentage of Dry Mass

Cereal	Nitrogen	Protein	Fat	Fiber	Ash	NFE
Barley (grain)	1.2–2.2	11	2.1	6.0	3.1	—
(kernel)	1.2–2.5	9	2.1	2.1	2.3	78.8
Maize	1.4–1.9	10	4.7	2.4	1.5	72.2
Millet	1.7–2.0	11	3.3	8.1	3.4	72.9
Oats (grain)	1.5–2.5	14	5.5	11.8	3.7	—
(kernel)	1.7–3.9	16	7.7	1.6	2.0	68.2
Rice (brown)	1.4–1.7	8	2.4	1.8	1.5	77.4
(milled)	—	—	0.8	0.4	0.8	—
Rye	1.2–2.4	10	1.8	2.6	2.1	73.4
Sorghum	1.5–2.3	10	3.6	2.2	1.6	73.0
Triticale	2.0–2.8	14	1.5	3.1	2.0	71.0
Wheat (bread)	1.4–2.6	12	1.9	2.5	1.4	71.7
(durum)	2.1–2.4	13	—	—	1.5	70.0
Wild rice	2.3–2.5	14	0.7	1.5	1.2	74.4

Note: Typical or average figure. NFE = nitrogen free extract (an approximate measure of the starch content).
Source: Ref. 32.

The protein contents of wheat and barley are important quality indices for the manufacture of various foods. For example, the quantity and quality of wheat protein is critical in bread making. Cereal grains contain water-soluble proteins (albumins), salt-soluble proteins (globulins), aqueous alcohol-soluble proteins (prolamins), and acid- and alkali-soluble proteins (glutelins). The prolamins are characteristic of the grass family, and together with the glutelins, comprise the bulk of the proteins of cereal grains. The following are names given to prolamins in proteins of the cereal grains: gliadins in wheat, hordeins in barley, zeins in maize, avenins in oats, kaffirins in grain sorghum, and secalins in rye.

The various proteins are not distributed uniformly in the kernel. Thus, the proteins isolated from the inner endosperm of wheat consist chiefly of prolamins (gliadin) and glutelins (glutenin). The embryo proteins consist of nucleoprotein, an albumin (leucosin), and globulins, whereas in wheat bran a prolamin predominates with smaller quantities of albumins and globulins. In the presence of water, the wheat endosperm proteins, gliadin, and glutenin form a tenacious colloidal complex known as gluten. Gluten is mostly responsible for the superiority of wheat over the other cereals in the manufacture of leavened products because it makes possible the formation of a dough that retains the carbon dioxide produced by yeast or chemical leavening agents. The gluten proteins collectively contain 17.55% nitrogen. Consequently in estimating the crude protein content of wheat and wheat products from the determination of total nitrogen, the factor 5.7 is normally employed rather than the customary value of 6.25. The latter value is based on the assumption that proteins contain an average of 16% nitrogen.

In general, cereal proteins are not as high in biological value as are those of certain legumes, nuts, or animal products. Most cereals are low in lysine and tryptophan. Oats are a notable exception to this generalization. The biological value of whole cereal grains is greater than that of refined milled products, which consist chiefly of the endosperm. Studies have shown that adults, if satisfying their calorie needs, will not be protein deficient on nearly 100% cereal diets. This may not be true for young children or pregnant women. In general, American and Western European diets normally include animal products as well as cereals and, thus, satisfy protein requirements many times over. Under these conditions, the different proteins tend to supplement each other.

The predominant form of carbohydrate in cereals is starch, which is the primary source of calories provided by the grains. Most of the carbohydrates are in the starchy endosperm.

Table 5. FDA Cereal Enrichment Standards

Product	Thiamine (mg/lb)	Riboflavin (mg/lb)	Niacin (mg/lb)	Iron (mg/lb)	Calcium (mg/lb)	Vitamin D (1U/lb)
Flour[a]	2.9	1.8	24	20	(960)	
Self-rising flour	2.9	1.8	24	20	960	
Corn grits	2.0–3.0	1.2–1.8	16–24	13–26	(500–750)	(250–1000)
Corn meal	2.0–3.0	1.2–1.8	16–24	13–26	(500–750)	(250–1000)
Rice	2.0–4.0	1.2–2.4	16–32	13–26	(500–1000)	(250–1000)
Macaroni products	4.0–5.0	1.7–2.2	27–34	13–16.5	(500–625)	(250–1000)
Noodle products	4.0–5.0	1.7–2.2	27–34	13–16.5	(500–625)	(250–1000)
Bread[a]	1.8	1.1	15	12.5	(600)	
U.S. RDA (mg or IU/day)	1.5	1.7	20	18	1000	400

[a]In addition, flour is to contain 0.7 mg/lb and bread 0.43 mg/lb of folic acid. The figures in this table are the nutrient levels to be present in the final products as set forth by the Food and Drug Administration (21 CFR 101.9, 137 and 139). Figures in parentheses are optional. When figures are shown as a range, these indicate minimum/maximum values. When one figure is shown, it is the minimum required. Calicum is added by the milling company. It cannot be included in the enrichment product. Enriched pasta products are normally made from flours enriched to these levels. Macaroni products with fortified protein must have the maximum figures but no Vitamin D. If enriched bread is manufactured using at least 62.5% of enriched flour, the bread will meet standards for enriched bread.

U.S. RDA (recommended daily allowances) as published in the 1986 21 CFR 104.20.

The lipids in cereals are relatively rich in the essential fatty acid linoleic acid. Saturated fatty acids (mainly palmitic) represent less than 25% of the total fatty acids of most grains.

In summary, cereal grains are a diverse yet primary source of nutrients. Their high starch contents make them major sources of calories. They also contribute to human needs for proteins, lipids, vitamins, and minerals. Vitamins and minerals lost during milling to produce refined food products can be, and in many countries are, replaced by nutrient fortification (Table 5). The composition of cereal grains and their milled products makes them uniquely suited for producing wholesome, nutritional, and consumer-acceptable foods.

STANDARDS AND CLASSIFICATION

At the turn of the century, many organizations in the United States and in several other grain-producing countries tried to develop uniform grain standards. The demand for uniform grades and application of the standards resulted in the introduction during 1903 to 1916 of 26 bills calling for the same in the U.S. Congress.

Those bills and the extensive hearings that accompanied them culminated in the U.S. Grain Standard Act, which was passed August 11, 1916, and amended in 1940 to include soybeans. This act provides in part for the establishment of official grain standards, the federal licensing and supervision of the work of grain inspectors, and the mechanisms for filing appeals concerning grades assigned by licensed inspectors. The official U.S. standards for grain cover wheat, corn, barley, oats, rye, sorghum, flaxseed, soybeans, mixed grain, and triticale.

According to U.S. grain standards, wheat is divided into seven classes on the basis of type and color: hard red spring, durum, red durum, hard red winter, soft red winter, white, and mixed. The division of wheat into classes and subclasses is according to suitability of different wheats for specific uses. Hard red spring, hard red winter, and hard white wheats are valued for the production of flours used in making yeast-leavened breads. Soft red winter, soft white, and white club wheats are especially suited for making chemically leavened baked goods (cookies, cakes). Amber durum wheat is prized for the manufacture of semolina used to produce alimentary pastes (macaroni, vermicelli, etc). Red durum is used primarily as a poultry and livestock feed.

The wheat in each subclass is sorted into a number of grades on the basis of quality and condition. The tests generally used to determine the grade of grain and its conformity to the official U.S. grain standards include plumpness, soundness, cleanliness, dryness, purity of type, and the general condition of the grain (37,38).

Other cereals are graded in a similar manner. Corn is divided into three classes: yellow, white, and mixed. These class designations apply to dent-type corn only. If the corn consists of at least 95% flint varieties, the word flint is used. If the corn contains more than 5% but less than 95% flint varieties, the corn is designated as flint and dent. Waxy corn is corn of any class consisting of at least 95% waxy corn. For many years moisture content was a grading factor for corn and sorghum. Especially at the beginning of the harvest, moisture content is the single most important quality factor of corn (high being problematic). Broken corn and foreign material (BCFM) is determined by sieving through a 12/64 in. (4.76 mm) sieve.

Rough rice is divided into three classes: long, medium, and short. There are separate standards for rough, brown, and milled rice. All rough rice is graded on the basis of milled rice yields.

Barley is divided into three classes (with several subclasses): six-rowed (malting, blue malting, and barley), two-rowed (two-rowed malting and two-rowed), and barley. Plumpness is determined by sieving. The designation malting requires inclusion of only varieties that have been approved as suitable for malting purposes. Special grades and designations of oats are: heavy or extra heavy (test weights of 48.9–51.5 kg/hL and >51.5 kg/hL, respectively), thin, bleached, bright, ergoty, garlicky, smutty, tough (moisture between 14 and 16%), and weevily.

BIBLIOGRAPHY

1. Canadian International Grains Institute, *Grain and Oilseeds, Handling, Marketing, Processing*, 2nd ed., Canadian International Grains Institute, Winnipeg, Canada, 1975.
2. R. C. Hoseney, *Principles of Cereal Science and Technology*, American Association of Cereal Chemists, St. Paul, Minn., 1986.
3. D. W. Kent-Jones and A. J. Amos, *Modern Cereal Chemists*, 6th ed., Food Trade Press, London, United Kingdom, 1967.
4. R. A. Olson and K. J. Frey, eds., *Nutritional Quality of Cereal Grains*, American Society of Agronomy, Madison, Wis., 1987.
5. Y. Pomeranz, *Modern Cereal Science and Technology*, VCH Publishers, New York, 1987.
6. Y. Pomeranz, ed., *Industrial Uses of Cereals*, American Association of Cereal Chemists, St. Paul, Minn., 1973.
7. Y. Pomeranz, ed., *Advanced Cereal Science and Technology*, Vols. I–X, American Association of Cereal Chemists, St. Paul, Minn., 1976–1990.
8. Y. Pomeranz, ed., *Cereals '78: Better Nutrition for the World's Millions*, American Association of Cereal Chemists, St. Paul, Minn., 1978.
9. Y. Pomeranz and L. Munck, eds., *Cereals: A Renewable Resource, Theory and Practice*, American Association of Cereal Chemists, St. Paul, Minn., 1981.
10. H. Hanson, N. E. Bourlaug, and R. G. Anderson, *Wheat in the Third World*, Westview Press, Boulder, Colo., 1982.
11. E. G. Heyne, ed., *Wheat and Wheat Improvement*, 2nd ed., American Society of Agronomy, Madison, Wis., 1987.
12. Y. Pomeranz, ed., *Wheat is Unique*, American Association of Cereal Chemists, St. Paul, Minn., 1989.
13. Y. Pomeranz, ed., *Wheat Chemistry and Technology*, 3rd ed., American Association of Cereal Chemists, St Paul, Minn., 1988.
14. N. L. Kent, *Technology of Cereals With Special Reference to Wheat*, Pergamon, Oxford, United Kingdom, 1960.
15. I. D. Morton, ed., *Cereals in a European Context*, VCH Publishers, New York, 1987.
16. W. R. Akroyd and J. Doughty, *Wheat in Human Nutrition*, FAO Nutritional Studies No. 23, FAO, Rome, Italy, 1970.
17. J. Holas and J. Kratochvil, eds., *Progress in Cereal Chemistry and Technology*, 2 Elsevier Science, New York, 1983.
18. R. W. Jugenheimer, *Corn: Improvement, Seed Production and Uses*, John Wiley & Sons, New York, 1976.
19. P. J. Barnes, ed., *Lipids in Cereal Technology*, Academic Press, Orlando, Fla., 1983.
20. M. N. Leath, L. H. Meyer, and L. D. Hill, *U.S. Corn Industry*, Report No. 479, U.S. Department of Agriculture, Economic Research Service, Washington, D.C., 1982.
21. S. A. Watson and P. E. Ramstad, eds., *Corn: Chemistry and Technology*, American Association of Cereal Chemists, St. Paul, Minn., 1987.
22. B. O. Juliano, ed., *Rice, Chemistry and Technology*, 2nd ed., American Association of Cereal Chemists, St. Paul, Minn., 1987.
23. B. S. Luh, ed., *Rice: Production and Utilization*, AVI Publishing, Westport, Conn., 1980.
24. W. G. Heid, *U.S. Barley Industry*, Agr. Econ. Report No. 395, U.S. Department of Agricultural Economics, Statistics and Cooperative Service, Washington, D.C., 1978.
25. D. C. Rasmusson, ed., *Barley*, American Society of Agronomy Madison, Wis., 1985.
26. F. H. Webster, ed., *Oats, Chemistry and Technology*, American Association of Cereal Chemists, St. Paul, Minn., 1986.
27. H. Doggett, *Sorghum*, Longmans, Green, and Co., London, United Kingdom, 1970.
28. L. W. Rooney, D. S. Murthy, and J. V. Mertui, eds., *International Symposium on Sorghum Grain Quality*, Proceedings, INCRASAT Patancheru, A. P. India, 1982.
29. W. Bushuk, ed., *Rye: Production, Chemistry, and Technology*, American Association of Cereal Chemists, St. Paul, Minn., 1987.
30. *Yearbook of Agriculture*, U.S. Government Printing Office, Washington, D.C., 1985.
31. Y. Pomeranz, *Functional Properties of Food Components*, Academic Press, Orlando, Fla., 1985.
32. J. E. Kruger, D. R. Lineback, and C. E. Stauffer, eds., *Enzymes and Their Role in Cereal Technology*, American Association of Cereal Chemists, St. Paul, Minn., 1987.
33. H. G. Muller, "Some Physical Properties of Cereals and Their Products as Related to Potential Industrial Utilization," in Y. Pomeranz, ed., *Industrial Uses of Cereals*, American Association of Cereal Chemists, St. Paul, Minn., 1973, pp. 20–50.
34. D. H. Simmonds, "Structure, Composition, and Biochemistry of Cereal Grains," in Y. Pomeranz, ed., *Cereals '78: Better Nutrition for the World's Millions*, American Association of Cereal Chemists, St. Paul, Minn., 1978, pp. 105–137.
35. *From Wheat to Flour*, Wheat Flour Institute, Chicago, Ill., 1965.
36. B. Holland, I. D. Unwin, and D. H. Buss, *Cereals and Cereal Products*, Royal Society of Chemistry, Ministry of Agriculture, Fisheries, and Food, London, United Kingdom, 1988.
37. C. M. Christensen, ed., *Storage of Cereal Grains and Their Products*, 3rd ed., American Association of Cereal Chemists, St. Paul, Minn., 1982.
38. J. L. Multon, ed., *Preservation and Storage of Grains, Seeds, and Their By Products*, Lavoisier Publishing, New York, 1988.

R. Carl Hoseney
R & R Research Services
Manhattan, Kansas

See also Wheat science and technology.

CHEESE

ORIGIN

The first use of cheese as food is not known. References to cheeses throughout history are widespread. "Cheese is an art that predates the biblical era" (1). The origin of cheese has been dated to 6000–7000 B.C. and the worldwide number of varieties has been estimated at 500, with an annual production of more than 12 million tons growing at a rate of about 4% (2). This food served as a source of vital nourishment in huts and castles and on journeys; armies also used cheese. It also was a way of preserving food during periods of shortages. A plausible tracing of the origin and development of cheese, as a food, around the world has been published (1).

CLASSIFICATION

Cheeses have been classified in several ways. Several attempts to classify the varieties of cheese have been made. One suggestion consists of scheme that divides cheeses into the following superfamilies based on the coagulating agent.

1. *Rennet Cheeses*. Cheddar, brick, Muenster
2. *Acid Cheeses*. Cottage, Quarg, cream
3. *Heat–Acid*. Ricotta, sapsago
4. *Concentration—Crystallization*. Mysost

A more simple but incomplete scheme, would be to classify cheeses as follows.

1. *Very Hard*. Parmesan, Romano
2. *Hard*. Cheddar, Swiss
3. *Semisoft*. Brick, Muenster, blue, Havarti
4. *Soft*. Bel Paese, Brie, Camembert, feta
5. *Acid*. Cottage, baker's, cream, ricotta

A broad look at cheeses might divide them into two large categories, ripened and fresh.

Ripened

Cheeses can be ripened by adding selected enzymes or microorganisms (bacteria or molds) to the starting milk, to the newly made cheese curds, or to the surface of a finish cheese. The cheese is then ripened (cured) under conditions controlled by one or more of the following elements: temperature, humidity, salt, and time.

Depending on the style of cheese, the ripening can be principally carried out on the cheese surface or the interior. The selection of organisms, the appropriate enzymes, and ripening regime determine the texture and flavor of each cheese type.

Fresh

These cheeses do not undergo curing and are generally the result of acid coagulation of the milk. The composition, as well as processing steps, provides the specific product texture, while the bacteria used to provide the acid usually generate the characteristic flavor of the cheese.

CHEESEMAKING—GENERAL STEPS

1. Milk is clarified by filtration or centrifugation.

2. Depending on the composition of the final cheese, the fat content is standardized in a special centrifuge (separator).

3. Depending on the variety, the milk is pasteurized (generally at about 72°C (161°F) for 16 s) or undergoes any equivalent heat treatment that renders the milk phosphatase negative. Phosphatase is inactivated in the same range as the tuberculin bacilli. The phosphatase test insures that the milk is free of tuberculin bacilli and other pathogens.

4. Before pumping the milk to the cheese vat, the milk fat can be homogenized, if desired. The extent of this homogenization will depend on the cheese variety being made (Fig. 1).

5. A bacterial starter culture is added to the milk, which is at 30–36°C, depending on the cheese being made. This inoculated milk is generally ripened (held at this temperature) for 30–60 min to allow the lactic organisms to multiply sufficiently for their enzyme systems to convert some of the milk sugar (lactose) to lactic acid.

6. After ripening, a milk-coagulating enzyme is added with stirring. The milk in the vat is then allowed to set in a quiescent state.

7. The resulting coagulum is usually ready to be cut 20–30 min after the enzyme addition. For most fresh cheeses little or no coagulating enzyme is added. Coagulation is allowed to occur by the development of lactic acid alone.

8. When the coagulum is firm enough to be cut, a combination of vertical and horizontal stainless steel knives (wires) are pulled lengthwise and crosswise through the coagulum in a rectangular vat. If an automatic enclosed circular vat is used, the cutting is programmed to insure a curd-size range by the speed and timing of the automatic knives.

9. For most varieties a heal time is allowed where no stirring is done for 5–15 min after cutting. This allows the newly cut curd particles to firm up slightly and begin syneresis (excluding whey from the particle during the shrinking process).

10. Following the heal period, heat is applied to the jacket of the vat and gentle stirring begins. For most ripened cheeses the curds are cooked in the whey until the temperature of the mixture reaches 37–41°C, depending on the variety. For some cheeses, for example, Swiss and Parmesan, this temperature can be as high as 53°C. The cheese is generally cooked for 30 min. Fresh cheeses, such as cottage, cream, and Neufchâtel, are cooked at temperatures as high as 51.5–60°C to promote syneresis and provide product stability.

11. Generally, when the cook temperature is reached, a 45–60 min period is allowed to promote curd firming and the releasing of whey. During this stir-out time the contents of the vat are agitated somewhat vigorously.

12. From this point the whey is drained and the treatment of the new curd depends on the nature of the final product.

For cheeses such as Cheddar, Monterey Jack, Colby, brick, and mozzarella the curd can be handled in a number of ways. Cheddar can be made by allowing the curd to settle in a rectangular vat, then draining the whey from the curd. When the whey is removed, the curd is trenched down the middle of the vat, lengthwise. The curd is hand cut, turned over, and then piled, at intervals, one slab on another. During this time acid continues to develop and whey continues to be exuded from the curd. When the proper pH is reached (about 5.1–5.3) the curd is milled by machine into finger-size pieces and slowly dry-salted to

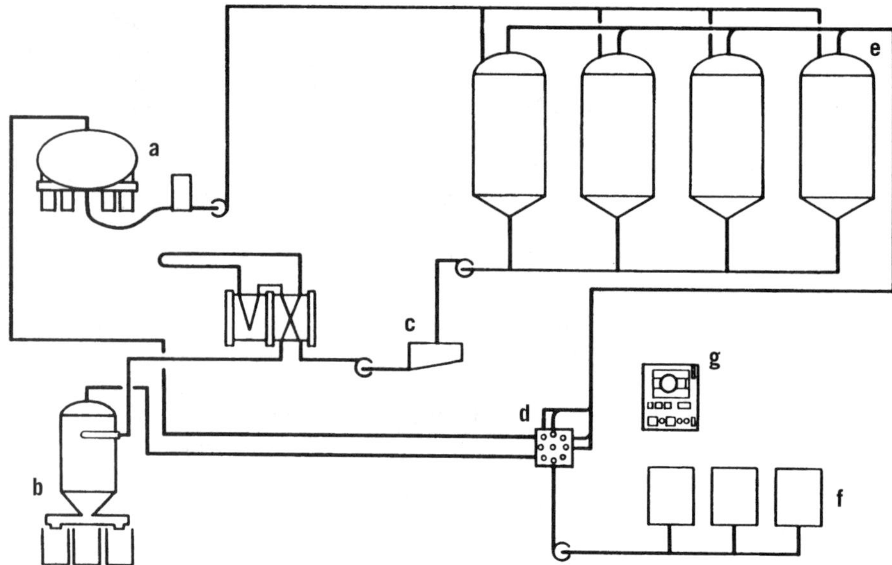

Figure 1. Natural cheese system flow key. a, Receiving pump/air eliminator; b, cyclone filter; c, balance tank; d, flow direction control; e, cheese vat; f, CIP system; g, CIP control panel. *Source:* Courtesy of Sherping Systems, Winsted, Minn.

provide about 1.6–1.8% salt in the final cheese. The salted curds are then packed into 40-lb stainless-steel hoops with porous netting lining the hoops or into 600–700-lb boxes. These containers are generally pressed at 15–20 psi to remove more whey over several hours and then placed into a vacuum chamber while under pressure to fuse the curd particles into a solid mass.

Several methods are used to arrive at a solid cheese block. One process provides for pumping the curds into a tower that allows whey drainage and compacting of the curds by their own weight. At the bottom of the tower a portion of the curd is cut off to fit into a 40-lb hoop, which travels to a press station and, finally, to a packaging station.

In another method, curd and whey are pumped from the vat to a draining-and-matting conveyer (DMC) under controlled temperature and curd depth. When the curd has reached the proper pH, the mat is cut and is automatically salted and transferred onto another conveyer where it is finally carried to a boxing operation (Fig. 2).

There is also a stirred-curd procedure used for mild- and medium-flavored Cheddar as well as Colby, brick, mozzarella, etc. In this method the curd and whey are pumped to a drain table or automatic-finishing vat (AFV) where the whey is drained and the curds are continually stirred to prevent matting. When the proper pH is achieved the cubes are salted and then boxed as previously described (Fig. 3).

If stirred curd Cheddar is intended for use in process cheese it can be packed into 500-lb drums and used after a short curing period. It should be noted that cheeses such as Monterey Jack and Colby, should not be placed under vacuum to preserve a slight open texture.

Blue cheese is made in a similar manner up to this point, but the mold (*Penicillium* sp.) can be added to the starting milk or to the drained curds. This cheese is usually formed in 5-lb hoops to promote an open texture. When the cheese is removed from the hoops it is dry-salted several times over a number of days and then punctured to create air channels in which the mold grows; this develops the characteristic blue veining. This cheese may be salt brined instead to obtain a 4% salt content in the final cheese. The cheese is cured for 60–120 days at 13°C and 95% humidity to achieve full veining and flavor.

Swiss, Parmesan, and other varieties are handled somewhat differently at the whey drainage step. For years these cheeses were manufactured in Europe in wheel-shaped hoops; 20-lb wheels were used for Parmesan and 175–225-lb hoops, for Swiss. The principles are the same so the description here will pertain to most U.S. products. For Swiss cheese the curds and whey are pumped to a large stainless-steel universal vat, which has been equipped with porous side plates, where the curd is allowed to settle evenly at the bottom. Large stainless-steel plates are used to press the curd at a precise depth of the curd mass. This pressing under whey results in a tightly fused cheese, required later for eye development. The whey is drained and the huge curd block is kept under pressure for about 16 h while it continues to develop acidity. Then the 3,000–3,500-lb block is cut into sections of about 180 lb and immersed into 2°C saturated brine for 24 h. The surface of the blocks are then dried. The blocks are packaged and boxed. These boxes are stacked and banded to help keep the block shape as the cheese cures in a hot room at 20–25°C. When the eyes have formed (about 18–28 days) the blocks are transferred to a room at about 2°C to prevent further eye development and held for at least 60 days before cutting into retail sizes. The cheese can be held for longer periods if stronger Swiss flavor is desired.

For Parmesan cheese, the curd is stirred while draining off the whey and then placed into round 20-lb hoops. After pressing, the round cheeses are allowed to surface dry before brining. If salt has not been added during the stirring of the curds, additional brining time is required. The cheeses are held at 13–16°C until the proper weight is reached (moisture reduced) and then vacuum packed in

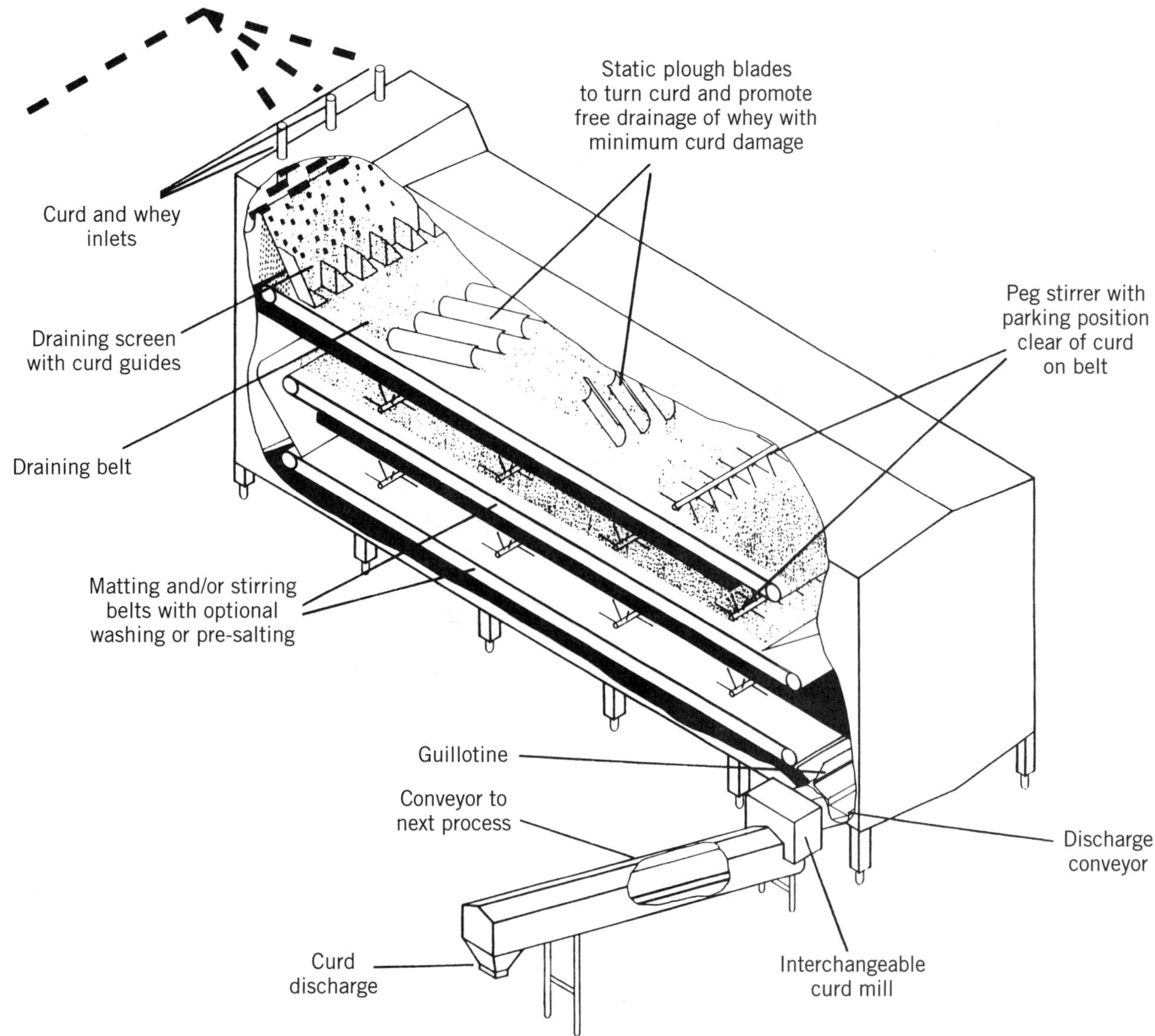

Figure 2. Cheese curd draining conveyor. *Source:* Courtesy of Sherping Systems, Winsted, Minn.

plastic bags with a tight seal. Some factories wax the cheese. This cheese requires a minimum of a 10-mo cure time by federal regulation. Most of the U.S. Parmesan is grated for use as a retail product.

Mozzarella and provolone are among those cheeses referred to as pasta filata varieties. Provolone has some lipase enzymes added to produce its characteristic flavor and sometimes is sold naturally smoked or with smoke flavor added. These cheeses have traditionally been further cooked, after whey drainage, in hot water and stretched by hand with subsequent shaping into the familiar forms. Today, however, most mozzarella and provolone are mechanically cooked under water and further stretched to provide texture and then shaped in molding machines. The cheeses are then brined to the proper salt level.

Cheeses such as Camembert, Brie, Muenster, and Limburger are made in smaller equipment and formed into smaller shapes to allow surface salting or brining and treatment of the cheese surface with selected bacteria or molds to provide the characteristic flavors, and textures.

Fresh cheeses such as cottage, queso fresco, and cream are mostly acid in nature and, following cooking of the curds, are treated differently. Cottage curds are washed with filtered water and then combined with a cream dressing. Queso fresco is pressed into loaves after salting and eaten fresh after slicing. Cream cheese curd is separated centrifugally and then standardized in regard to fat, salt, and stabilizers. Stabilizers are added to prevent moisture release. It can then be packaged in small loaves, hot or chilled.

RECENT DEVELOPMENTS

A number of cheese varieties have been manufactured from milk concentrated by ultrafiltration (3,4). Patents

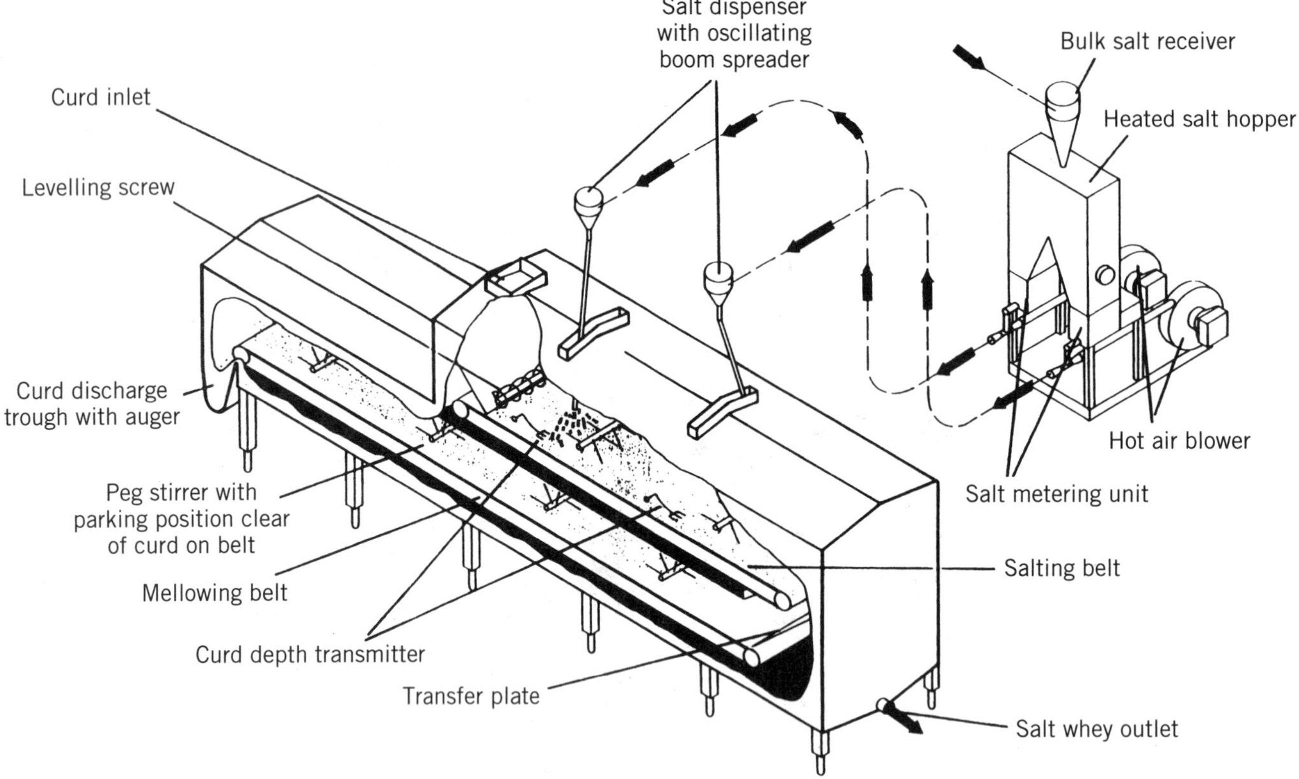

Figure 3. Dual salt applicating and cheese curd mellowing system. Two stage continuous salting and mellowing of milled or granular curd. *Source:* Courtesy of Sherping Systems, Winsted, Minn.

have been issued (5) and some procedures have been reduced to practice, but after more than 20 years of research in this area only a minimal amount of the total world production of cheese is made with ultrafiltered milk.

Bactofugation used to remove microbes and spores from cheese milk has sparked some interest, especially in Europe (6). Microfiltration may find significant application in special milk treatment prior to, or during, product manufacture. Significant developments in recombinant biotechnology and fermentation technology may provide specificity previously not available for milk coagulation and cheese flavor development. New cultures, which are resistant to phage, may be on the horizon. New cheese-making technology may allow previously unattainable processes and product control to become reality.

Cheese Starter Cultures

Starter cultures are organisms that ferment lactose to lactic acid and other products. These include streptococci, leuconostocs, lactobacilli, and streptococcus thermophilus. Starter cultures also include propionibacteria, brevibacteria, and mold species of Penicillium. The latter organisms are used in conjunction with lactic acid bacteria for a particular characteristic of cheese, eg, the holes in Swiss cheese are due to propionibacteria, and the yellowish color and typical flavor of brick cheese is due to brevibacterium linens. Blue cheese and Brie derive their characteristics from the added blue and white molds, respectively. Lactic acid-producing bacteria have several functions:

1. Acid production and coagulation of milk.
2. Acid gives firmness to the coagulum, which influences cheese yield.
3. Developed acidity determines the residual amount of animal rennet influencing cheese ripening; more acid curd binds more rennet.
4. The rate of acid development influences dissociation of colloidal calcium phosphate, which then influences proteolysis during manufacture and the rheological properties of cheese.
5. Acid development and production of other antimicrobials control the growth of certain nonstarter bacteria and pathogens in cheese.
6. Acid development contributes to proteolysis and flavor production in cheese.

Types of Culture

There are some impending changes in the nomenclature of starter culture organisms; the nomenclature used here follows present standards (7). Mesophilic cultures are used for cheese types that do not exceed 40°C during cheese making (8). These starters are propagated at 21–23°C and include *Streptococcus lactis* ssp. *lactis, S. lactis* ssp. *cremoris, S. lactis* ssp. *diacetylactis* and *Leuconostoc cremoris*. These are used for Cheddar, Gouda, brick, Muenster, cream, cottage, and Quarg cheese types, leuconstoc and *S. lactis* ssp. *diacetylactis* ferment citrate to produce carbon dioxide and diacetyl to characterize Edam, Gouda, and cream cheese.

Thermophilic cultures are used in cheese types where curd is cooked to 45–56°C. These starters are propagated at 40–45°C and include *S. thermophilus*, sometimes fecal streptococci, *Lactobacillus helveticus, L. delbrueckii* ssp. *lactis*, and *L. delbrueckii* ssp. *bulgaricus*. These are used for Swiss, Emmentaler, Gruyère, Parmesan, Romano, mozzarella, and Gorgonzola cheeses (9).

Propagation

Cultures are propagated in milk or a medium containing milk components and other nutrients and are heat treated (85°–90°C for 45 min) to render milk free of contaminants. The medium is then cooled to incubation temperature before inoculation. This processing is accomplished in a jacketed stainless-steel vessel provided with agitation and supplied with sterile air pressure. Culture inoculum is available from suppliers in frozen or freeze-dried form. It consists of appropriate mixture of culture strains and is free of contamination. After inoculation, culture is allowed to grow in quiescent state for 12–18 h. At the end of growth the medium is acidic with a pH of 4.5 or lower, depending on the culture. Ripened culture cell population is about 10^9 cfu/g. The ripened culture is cooled to at least 5°C. This form of propagation is called bulk-starter propagation. The culture may be used immediately or held for several days. The growth medium may contain buffering salts or the acid produced may be neutralized and controlled by the addition of ammonia, potassium hydroxide, or sodium hydroxide under controlled conditions. The latter is called pH-controlled propagation.

Starter cultures are susceptible to bacteriophages (viruses), which destroy the cultures and hamper cheese production. Cultures are also sensitive to antibiotics and bacteriocins (proteinaceus bacterial inhibitors). Propagated cultures are used at 1–5% of cheese milk. Highly concentrated (10^{10} cfu/g) frozen cultures are available for direct addition to cheese milk (350 g/5,000 lb). These are called direct vat set (DVS). This form is convenient, but expensive.

Coagulation

Coagulation of milk is essential to cheese making. The coagulation entraps fat and other components of milk. Most proteolytic enzymes can cause milk to coagulate. Rennet (chymosin, EC 3.4.23.4) is widely used for milk coagulation in cheese making (10). It is extracted from the fourth stomach of young calves. Commercial rennets may include blends of chymosin and pepsin (bovine or other animals). Microbial rennets with similar functionality prepared from *M. miehei* are known as Marzyme, Hannilase, Rennilase, and Fromase. Preparations from *M. pusillus* and *Endothia parasiticus* are called Emporase and Sure Curd, respectively. Proteases from plants are known to coagulate milk but are not used in commercial cheese making. Coagulation of milk is also effected by lowering pH to about 4.6 in quiescent state either by fermentation of lactose to lactic acid or by hydrolysis of gamma—delta lactones to produce cottage, Quarg, ricotta, and cream cheese.

Animal and microbial rennets are aspartic proteases with optimum activity under acidic conditions. Their molecular weights range from 30,000 to 38,000. The primary stage of rennet coagulation involves partial hydrolysis of *K*-casein, the principal stabilizing factor of milk protein, at Phe-105 to Met-106 bond. Destabilization of the residual protein (*para*-casein) occurs in the presence of Ca^{2+} at temperatures >18°C in the secondary, nonenzymatic stage of rennet coagulation. About 30% of calf rennet is retained in the curd before pressing and only 5–8% after pressing (11). The amount of rennet retained is governed by the pH of the curd. Microbial rennets are retained to a lesser extent (3–5%).

During coagulation, linkages between casein micelles are formed and many micelles are joined by bridges, but later these appear to contract, bringing the micelles into contact and causing partial fusion. The firmness of curd is due to an increase in both the number and strength of linkages between micelles. It is suggested that phosphoryl side chains of casein specially β-casein are involved. These may be linked by Ca^{2+} bridges. In cheese, αs-1-casein plays a structural role (12).

Casein aggregation and fusion continues after coagulum cutting and throughout cheese making and early stages of cheese ripening. This process of rennet coagulation of milk is fundamental to the conversion of milk to cheese curd and its manipulation is essential for the control of cheese manufacture (Table 1). Casein aggregation leads to curd formation and syneresis. Increase in acidity and temperature and the addition of calcium chloride accelerates the rate of curd formation and its syneresis. On the other hand, cold storage of milk, homogenization, higher pasteurization temperature, and increased fat content impede the two processes. Acid coagulation of milk, as in cottage-cheese making results in less aggregation of casein than is seen with rennet. During cottage-cheese making the pH of milk casein drops to its isoelectric point resulting in a soft, fragile gel, firm enough to be cut. For large-curd cottage cheese, a very small amount of rennet is added. Essentially all added rennet is inactivated at the cooking temperature of curd (57–60°C).

MICROBIOLOGICAL AND BIOCHEMICAL CHANGES DURING CHEESE MAKING AND CHEESE RIPENING

During cheese manufacture, milk is dehydrated to a given composition, characteristic of a family of cheeses. The distinguishing features of a cheese within a family are a function of smaller but significant deviations from set practices. The following pairs of cheeses represent good examples of the effect of manufacturing variations: Edam and Gouda, brick and Limburger, mozzarella and provolone, and Cheddar and Colby. Rennet curds have unique properties that can be handled to produce high- and low-moisture cheeses without hardening. Limited, but essential, proteolysis of milk protein by rennet enzymes augments the shift of microbial populations that, in turn, are pressed into summoning those metabolic activities that must transform the milk component simply to survive. In doing so, the chemical entities generated interact among

Table 1. MHk Coagulants for Cheese Making

Coagulant	Supplier/designation			
	Hansen's	Marschall's	Pfizer	SANOFI
Calf	Standard	Calf rennet Cheese rennet	Calf rennet	American rennet[a]
Bovine	Bovine			Beef rennet
Porcine			Maria Set	Pepsin
Calf-porcine	50-50	Chymoset	Econozyme	Quick Set[b]
Bovine-porcine	B-P			
Calf-bovine-porcine				Trizyme
Mucor miehei (modified thermolabile equivalent to Novo Rennilase XL)	Hannilase (HL)		Mor Curd	
Mucor miehie (modified extra thermolabile equivalent to Novo Rennilase XL)	Superlase	Marzyme Supreme	Mor Curd Pius	
Endothia parasiticus			Sure Curd	
Mucor pusillus				Emporase[c]
Microbial, animal	C and P (Chees-zyme and Pep-zyme blend)	M-P (50-50 Hog and *M. miehei*)		Regalase (porcine and Emporase) Beemase (bovine and Emporase)

[a]DFL sells three gracies of calf rennet: PD300 (highest chymosin), SF100 (intermediate chymosin), and EL400 (lowest chymosin).
[b]DFL sells three grades of Quick Set.
[c]DFL sells three grades of Emporane.

themselves and with the microbial population to result in a more flavorful and preserved milk. Both casein and milk fat hydrolysis is needed for cheese flavor development, but the rate and extent of such hydrolysis must be controlled to maintain cheese identity. Different cheese types are discussed below with these comments in mind.

Lactic starter cultures are added to vat milk to give about 10^6–10^7 cfu/mL. The amount and type of starter may vary significantly depending on the type of cheese and characteristics desired. In most cheese types an overnight pH range is 4.95–5.3.

Cheddar-Type Cheese

In Cheddar, Caciocavallo, and Colby types of cheese, about 30–40% of the added culture cells are lost in whey. The cells trapped in the rennet coagulum rapidly multiply and ferment lactose to lactic acid. The population of starter organisms may reach in excess of 5×10^8 cfu/g in curd before salting. These bacterial numbers include a 10-fold increase due to milk solids concentration. The cultures multiply only slightly during coagulation and cooking, but growth and acid production accelerate after whey is removed and continues through cheddaring as the starter cells are concentrated in the curd. Acid production will continue until the lactose is depleted. At the time of milling the curd may have a pH of 5.3 and titratable acidity of whey at 0.57% or higher.

The milling operation involves cutting the curd into small pieces to facilitate uniform salt distribution and whey drainage. Enough salt is applied to the curd in two or three applications to yield a salt-in-moisture phase ($S/S + M$) of the cheese of about 4–5%. Due to steady acid production throughout cheese making the casein curd is demineralized with respect to calcium and phosphorus. The rate and extent of acid production governs the residual chymosin, minerals, moisture, and the cheese texture and structure. Higher acid in cheese reflects higher starter populations, possibly resulting in bitter cheese due to total proteinase activity from starter cells.

Cheddar-type curd is dry salted. Addition of salt to the curd is the last step in the manufacture of cheese. It controls microbial growth and activity and the final pH of hooped cheese. At hooping Cheddar cheese may contain 0.6% lactose. Its fermentation to lactic acid depends on $S/S + M$. Culture activity is inhibited by $S/S + M$ of >5% (13). Salt also helps to expel moisture. Proteolytic activity of chymosin and pepsin on αs-1-casein is stimulated by salt, whereas β-casein proteolysis is inhibited by $S/S + M$ of 5%. Salt appears to affect physical changes in cheese protein, which influence cheese texture, protein solubility and probably protein conformation. Cheddar-type cheese is internally ripened by chymosin in concert with starter proteinases and adventitious lactobacilli. For normal ripening a high starter population must lyse to release proteinases and peptidases affecting cheese flavor, body, and texture development. In most cheese types there is a shift in the population of starter organisms to lactobacilli and in some cases pediococci. It has been demonstrated that in young cheese there are inhibitory and stimulatory factors for lactobacilli. As the cheese ripens, most of the inhibition vanishes. These stimulatory factors for lactobacilli appear to originate primarily in αs-casein (14). In Cheddar cheese, β-casein is not extensively degraded (15). Studies with starters lacking proteinase indicate the importance of starter proteinases (16). Proteolysis and increasing concentration of free amino acids add to the savory taste and provide substrate for sulfur compounds. Many flavor compounds are chemical interactions of microbially derived substrates under conditions of low pH and low oxidation–reduction potential. Numerous compounds, such as hydrocarbons, alcohols, aldehydes, ketones, acids, esters, lactones, and sulfurs are important in cheese flavor. Hydrogen sulfide and methanethiol along with phenylacetaldehyde, phenylacetic acid, and phenethanol are considered key compounds of good Cheddar flavor (17). Lactic acid bacteria are not lipolytic but small increases in butyric and other fatty acids appear desirable.

Swiss, Emmentaler, and Gruyère

These cheese types are made with thermophilic streptococci and lactobacilli to which propionibacteria are added for eye formation. During the early phases of cheese making *S. thermophilus* multiplies rapidly and utilizes the glucose moiety of lactose to produce *L*-lactate, leaving behind galactose. Lactobacilli start vigorous acid production after whey drainage when the curd temperature drops to 46–49°C. Acid production and starter bacteria number are greater toward the periphery where the cheese has cooled down. Lactic acid fermentation in cheese is complete within 24 h with no detectable sugars.

Changes During Ripening. Increase in starter organisms population and acidity are halted when the cheese is brined. This is due to exhaustion of fermentable sugar and the low brine temperature. Cheese is cured in a hot room (20°–25°C) for 18–40 days. During this period the added starter organisms disappear and the niche is occupied by propionibacteria, adventitious lactobacilli, and fecal streptococci. Propionibacteria ferment lactate to propionic acid and other flavor compounds according to the reaction: $3CH_3CHOHCOOH \rightarrow 2CH_3CH_2COOH + CH_3COOH + CO_2 + H_2O$.

Eye Formation. The quality of Swiss cheese is judged by the size and distribution of eyes. Swiss cheese eyes are essentially due to carbon dioxide production, diffusion, and accumulation in the cheese body (18). The number and size of eyes depend on carbon dioxide pressure, diffusion rate, body texture, and temperature of cheese. The bulk of protein breakdown products are derived from αs-casein. β-Casein is not greatly proteolysed in Swiss cheese (15). Bacterial proteinases and plasmin (milk proteinase) are considered important in Swiss cheese ripening.

Flavor. Cheese flavor is derived in part from cheese milk. However, the characteristic flavor of Swiss-type cheese comes from the microbial transformation of milk components. These contain water-soluble volatiles (acetic acid, propionic acid, butyric acid, and diacetyl), which give the basic sharpness and general cheesy notes (17). Water-soluble nonvolatile amino acids (especially proline), peptides, lactic acid, and salts provide mainly sweet notes. Oil-soluble fractions (short-chain fatty acids) other than the water-soluble volatiles are important to flavor. Nutty flavor is attributed to alkylpyrazines. Several compounds, eg, ketones, aldehydes, esters, lactones, and sulfur-containing compounds are important. Due to the activity of certain strains of lactobacilli and fecal streptococci, biogenic amines are sometimes found in Swiss cheese.

Camembert Cheese

This is a soft cheese with 48–52% moisture. Milk inoculated with 1–5% mesophilic lactic starter is renneted at about 30°C. When the acidity reaches 0.20–0.23% the curd is dipped, molded, and held at 21°C in rooms with 85–90% humidity until it reaches pH 4.5–4.6. During storage, yeast and *Geotrichum candidum* appear on the surface. After about a week at about 10°C *Penicillium camemberti* covers the surface and deacidifies the cheese in 15–20 days. This is followed by colonization of the cheese surface by *B. linens* and micrococci. During ripening, the surface pH rises to about 7.0. β-Casein is not extensively degraded. Appearance of some γ-casein suggests plasmin activity. There is αs-casein degradation but less compared to Cheddar cheese. In Camembert, due to extensive deamination, ammonia constitutes 7–9% of the total *N*. Free fatty acids found in large quantities contribute to the basic flavor of cheese and serve as the precursors of methylketones and secondary alcohols. Aldehydes, methylketones, 1-alkanols, 2-alkanols, esters (C_2, 4, 6, 8, 10-ethyl 2-phenylethylacetate), phenol, *p*-cresol, lactones, hydrogen sulfide, methanethiol, methyldisulfide and other sulfur compounds along with anisoles, amines, and other compounds constitute the volatile compounds.

Blue Cheese

Blue cheese and its relatives, Roquefort, Stilton, and Gorgonzola are characterized by peppery, piquant flavors produced by the mold *Penicillium roqueforti*. Milk for cheese making may be homogenized to promote lipolysis in cheese. An acid curd with pH 4.5–4.6 is produced with mesophilic lactic starter and sometimes include *S. lactics* ssp. *diacetylactis*. Gorgonzola is made with *S. thermophilus* and *Lactobacillus bulgaricus*. The blue mold is added to the milk or curd. The molded curd is dry salted, and salt is rubbed on cheese to attain 4–5% salt. After about a week the cheese heads are pierced with needles to aerate the cheese and let carbon dioxide escape and air enter. This promotes mold growth.

Due to high acid and salt, starter population declines rapidly and the blue mold grows throughout the cheese. Salt-tolerant *B. linens* and micrococci appear on the cheese surface in two to three weeks. Due to deacidification of the cheese and extensive proteolysis, the cheese pH rises from 4.5 to 4.7 at 24 h to a maximum of 6.0–6.25 in two to three months.

Molds are both proteolytic and lipolytic, resulting in extensive proteolysis and lipolysis of cheese. In blue cheese both αs- and β-casein are degraded. There is a large accumulation of free amino acids due to extracellular acidic and alkaline endopeptidases (19). *B. linens* and micrococci are also proteolytic and contribute to the overall proteolysis. Occurrence of citrulline, ornithine, γ-aminobutylic acid, and biogenic amines reflect amino acid break-down products. Other volatile compounds such as ammonia, aldehydes, acids, alcohols, and methanethiol are also produced from amino acid metabolism.

Mold-ripened cheeses have a high free fatty acid content, 27–45 meq of acid per 100 g of fat. *Penicillium roqueforti* produces two lipases, one active at pH 7.5–8.0 or higher and the acid lipase active at pH 6.0–6.5. The fatty acids produced are converted into methylketones via ketoacyl–coenzyme A and the β-keto acids (20). Thiohydrolases are also present in cheese, resulting in the accumulation of 2-heptanone. The rate of fatty acid release governs the rate of methylketone formation (21).

Brick Cheese

Brick cheese is representative of a large group of cheeses (Limburger, Muenster, Tilsiter, Bel Paese, and Trappist) ripened with the aid of surface flora made of yeast, micrococci, and coryneform bacteria (*B. linens*). A culture of *S. lactis* ssp. *lactis, S. lactis* ssp. *cremoris*, and sometimes in combination with *S. thermophilus* is used to cause a cheese pH of 5.0–5.2. After brining for one or more days cheese is held at about 15°C (60°F) in rooms with a 90–95% RH where yeast (*Mycoderma*) appear on the surface in two or three days followed by micrococci and then *B. linens*. Organisms from the genus *Arthrobacter* have also been isolated from the cheese surface. Cheese ripens in four to eight weeks depending on the moisture and intensity of cheese flavor desired. Film-wrapped cheese has more rounded flavor than waxed cheese.

Growth of yeast lowers the acidity on the surface, making it suitable for micrococci and *B. linens* colonization (22). Sometimes *G. candidum* may also be present. αs-Casein is always hydrolyzed but β-casein disappearance was seen in Muenster and not in brick. Yeasts found on Limburger cheese synthesize considerable amounts of pantothenic acid, niacin, and riboflavin (23). Pantothenic acid and *para*-aminobenzoic acid are required by *B. linens* Thus explaining the population sequence. Liberated free amino acids are much higher on the surface than in the interior of cheese. Activities of micrococci appear essential to the development of typical flavor of brick or Limburger cheese. These proteolytic organisms are able to convert methionine to methanethiol. Acetyl methyl disulfide is characteristic of Limburger flavor.

Edam and Gouda

These Dutch cheese varieties are made with *S. lactis* ssp. *lactis, S. lactis* ssp. *cremoris, Leuconstoc cremoris* and/or *L. lactis. Leuconstoc* and *S. lactis* ssp. *diacetylactis* ferment citrate to produce carbon dioxide, diacetyl, and other compounds. The starter organisms reach $\sim 10^9$ cfu/g in curd with a pH of about 5.7. The remaining lactose in curd is fermented in an uncoupled state to lower cheese pH to 5.2. Molded cheese forms are placed in 14% brine at 14°C for three to seven days. After brief drying, the cheese is coated with two to three coats of plastic emulsion, which permits slow moisture evaporation. Carbon dioxide production is essential to eye formation. Ripening changes involve proteolysis by rennet enzymes, enzymes of starter bacteria, and, to a small extent, native milk proteinase. Rennet is responsible for extensive degradation of αs-casein to large molecular weight water-soluble peptides. This proteolysis is significant in determining the body characteristics of cheese. Starter organism production of low molecular weight peptide fractions and amino acids are stimulated by rennet action.

Lipolysis occurs to a small degree mainly due to enzymes of starter bacteria liberating free fatty acids from monoglycerides and diglycerides formed by milk or microbial lipases. In well-made Edam and Gouda lactobacilli do not exceed 10^5–10^6/g or certain defects in the cheese are noticed. Lactic acid, carbon dioxide, diacetyl, aldehydes, ketones, alcohols, esters, and organic acids in proper balance are important to flavor. Anethole and bismethylthiomethane are important odor compounds of Gouda cheese (24).

Mozzarella and Provolone

These cheese types may contain 45–60% moisture. *S. thermophilus*, thermophilic *Lactobacillus* sp. or mesophilic lactics with *S. thermophilus* combination starters are used. Sometimes citric acid or other acids are used in place of lactic starters.

The cultures used generally cleave the galactose moiety of lactose in cheese, which, if not utilized, result in dark brown cheese color on baking. In fresh brined cheese there may be 10^8 cfu/g of starter organisms in 1:1 ratio of streptococci and lactobacilli. αs-Casein is proteolysed to a lesser degree by rennet enzymes compared to other cheese types, while β-casein is largely intact. This level of rennet proteolysis of milk protein is sufficient to give the melt and stretch characteristics to cheese during hot water kneading at about 57°C. The molded cheese is brined. The pH of cheese should be 5.1–5.4. There is little lipolysis and fatty acids liberation in traditional mozzarella cheese (25). Cheese made with mesophilic starter and *S. thermophilus* tends to have greater protein and fat hydrolysis during storage. Such cheese is difficult to shred and has atypical flavor when aged.

Mozzarella and provolone are manufactured in a similar manner. The former is consumed fresh while the latter may be ripened at 12.5°C for three to four weeks and then stored at 4.5°C for 6–12 months for grating. The ripened cheese have mainly *Lactobacillus casei* and its subspecies. Sometimes leuconstocs are detected in cheese, particularly if no starter was used.

Parmesan and Romano

Parmesan and Romano cheese are made with *S. thermophilus, Lactobacillus bulgaricus*, or other species of thermophilic lactobacilli. In addition to rennet, pregastric estrases, or rennet paste may be added to cheese milk for their lipolytic activity. The curd is cooked to 51–54°C, when the whey acidity reaches about 0.2% it is hooped (packed) in round forms. Sometimes salt is added to the curd, which slows the starter and regulates moisture. The cheese at pH 5.1–5.3 is placed in 24% brine for several days.

Compared to other cheese types, the starter populations in the fresh cheese is low. Throughout cheese ripening, 12–24 months, cheese flora seldom exceeds 10^5 cfu/g. Fecal streptococci and salt-tolerant lactobacilli predominate. In these hard grating cheeses αs- and β-caseins are not overly proteolysed compared to other cheese varieties. This is attributed to high cooking temperatures of curd, which inactivates the coagulant, and the high salt in the moisture, which discourages growth of adventitious flora. In ripened cheeses quality varies from location to location. Volatile free fatty acid and nonvolatile free fatty acid (C_4 through C_{18}) concentrations are high in these cheeses, particularly in Romano (25). Butyric acid and minor branched chain fatty acids that occur in milk appear to contribute to the piquant flavor of Parmesan. For a more balanced flavor a concomitant increase in free amino acids (glutamic acid, aspartic acid, valine, and alanine) has been noted. Too high a free fatty acid level in cheese gives a strong, soapy, undesirable flavor.

Cottage Cheese and Cream Cheese

These are cultured to the isoelectric point of casein (pH ~4.6) followed by heating to separate the cheese solids. Cottage cheese curd is blended with a cultured or uncultured cream to no more than 4% fat. Cream cheese is cultured with lactics and citrate-fermenting cultures. Diacetyl, ethanol, acetone, lactic acid, acetic acid, and other less-characterized compounds are responsible for the flavor of these products.

ACCELERATED CHEESE RIPENING

One of the major costs of cheese is the expense of curing time before desired flavor develops. While some maturation time is inevitable, there are systems available where ripening time is shortened by speeding up proteolysis and lipolysis to generate flavor and modify texture. Elevated temperature (13°C or higher) curing offers the simplest approach to speed up ripening of otherwise normal cheese. Cheese intended for this type of curing must not contain measurable levels of heterofermentative lactobacilli or leuconostocs, because an open-texture defect and off-flavors will develop (26).

Microbial proteinases and gastric esterases have been used with little success to achieve acceptable cheese with uniformity. Activity of these exogenous enzymes is unregulated and may contribute to the detriment of cheese quality. Several unproven systems are available from culture houses.

Additions of partially inactivated starter organisms have been used with mixed results. Presently, this is not economical. Most of the proprietary systems investigated cause a minor to major deviation from characteristic flavor, body, and texture of cheese.

CHEESE DEFECTS

Cheese varieties exist because manufacturing and curing are carried out to preserve characteristic flavor, body, and texture of a given cheese. Attributes of one cheese may prove to be defects in another.

Even though certain molds are used for manufacturing and curing of cheese, their growth on most cheeses is undesirable (27). Gas production with putrid, unclean odors due to nonstarter bacteria constitute a defect in cheese. Moisture accumulation on the surface of hard cheeses permits rind rot due to yeasts, molds, and proteolytic bacteria growth, sometimes accompanied by discoloration. Milk itself can at times contribute weedy, feedy, cowy, barny, and related flavors coming from the cow. Psychrotrophic organism growth in milk can generate rancidity due to lipolysis and bitterness due to proteases from these organisms. Specific defects are discussed below.

Swiss Cheese

Swiss is graded by the size, number, and distribution of eyes; there are many defects of the eyes (18). *Blind* or cheese with no eyes may be due to inhibition of carbon dioxide production of proprionibacteria by pH of less than 5.0, prolonged brining (high salt), and antibiotics or other inhibitors in milk or cheese.

Overset (pinny or too many eyes) is generally caused by improper acidity, high moisture, entrapment of air in curd, and other factors that prevent knitting of curd before gas production. Early gas is caused by *Enterobacter aerogenes*. Flavor defects, such as stink spots, white spots (28), and excessive amine production, are caused by *Clostridium tyrobutyricum, S. faecalis* ssp. *liquefaciens*, and strains of *Lactobacillus buchenerii*, respectively.

Cheddar Cheese

Excessive acidity, below pH 5.0 at three or more days, causes short, crumbly body and a sour, often bitter flavor. Normal body and flavor development may not occur in this cheese. On the other hand, too little acid or pH above 5.3 at three or four days may lead to corky, pasty body with off-flavor development. Open-texture defect may be caused by lactobacillus fermentum, L. brevis, and leuconostocs. Raw milk and rennet preparations have been linked to these defects.

Edam and Gouda

Edam and Gouda are subject to spoilage similar to Cheddar. In these cheeses, brine has been known to contaminate the cheese surface with gas-forming lactobacilli. These may cause the occurrence of phenols, putrid odor, and sulfidelike flavors and excessive production of carbon dioxide. Butyric acid fermentation and carbon dioxide production are due to *C. tyrobutyricum*. Growth of propionibacteria may cause a sweet taste and an open texture due to carbon dioxide.

Brick Cheese

Excessive acidity is a major defect. Coliform bacteria can grow during draining and salting, causing early gas, which if excessive can yield a spongy condition. Sometimes, due to lack of surface smear growth, typical flavor does not develop.

Blue Cheese

Blue cheese may not have sufficient mold growth, leading to lack of flavor and general development of cheese attributes. Excessive growth of mold may cause mustiness and loss of flavor. Mold-ripened cheeses are sensitive to bitterness development due to excessive acidity in the curd followed by abundant mycelium growth and proteases. Surface discoloration by other molds is not uncommon. Excessive slime and undesirable flavors occur due to excessive humidity in curing rooms.

Mozzarella

Mozzarella cheese is generally shredded before use. Softening of this cheese, due to *Lactobacillus casei* strain, is a problem during curing. It is also prone to softening and gas production by thermoresistant strains of *Leuconostoc*.

Cottage Cheese

Cottage cheese has a short shelf life generally due to *Pseudomonas*, yeasts, and molds. Sometimes bitterness and excessive acid development due to surviving starter culture is noticed during refrigerated storage (Tables 2, 3).

For total dollar annual sales and a breakout by cheese type see Tables 4 and 5.

PROCESS CHEESE

Process cheese is the food produced by grinding and blending various cheese types into a uniform, pliable mass with the aid of heat. Water, emulsifying salts, color, etc are usually added and the hot plastic mass is cooled into the desired form.

Process cheeses were prepared as early as 1895 in Europe, but the use of emulsifying salts was not widely practiced until 1911 when Gerber and Co. of Switzerland invented process cheese. A patent issued to J. L. Kraft in 1916 marked the origin of the process cheese industry in America and describes the method of heating natural cheese and its emulsification with alkaline salts.

Process cheeses in the United States generally fall into one of the following categories (29).

1. Pasteurized blended cheese. Must conform to the standard of identity and is subject to the requirements prescribed by pasteurized process cheese except
   - A mixture of two or more cheeses may include cream or Neufchâtel.
   - None of the ingredients prescribed or permitted for pasteurized process cheese is used.
   - The moisture content is not more than the arithmetic average of the maximum moisture prescribed by the definitions of the standards of identity for the varieties of cheeses blended.
   - The word process is replaced by the word blended.
2. Pasteurized process cheese.
   - Must be heated at no less than 65.5°C for no less than 30 s. If a single variety is used the moisture content can be no more than 1% greater than that prescribed by the definition of that variety, but in no case greater than 43%, except for special provisions for Swiss, Gruyère, or Limburger.
   - The fat content must not be less than that prescribed for the variety used or in no case less than 47% except for special provisions for Swiss or Gruyère.
   - Further requirements refer to minimum percentages of the cheeses used.
3. Pasteurized process cheese food.
   - Required heat treatment minimum is the same as pasteurized process cheese.
   - Moisture maximum is 44%; fat minimum is 23%.
   - A variety of percentages are prescribed.
   - Optional dairy ingredients may be used, such as cream, milk, skim milk, buttermilk, and cheese whey.
   - May contain any approved emulsifying agent.
   - The weight of the cheese ingredient is not less than 51% of the weight of the finished product.
4. Pasteurized process cheese spread.
   - Moisture is more than 44% but less than 60%.
   - Fat minimum is 20%.
   - Is a blend of cheeses and optional dairy ingredients and is spreadable at 21°C.
   - Has the same heat treatment minimum as pasteurized process cheese.
   - Cheese ingredients must constitute at least 51%.
   - A variety of percentages are prescribed.

Advantages of Process Cheese

- Compact, dense body.
- Smooth texture.

Table 2. General Cheese Composition Ranges

Variety	Moisture (%)	Fat (dry basis, %)	Salt (%)
Blue cheese	43.0–45.0	51.0–54.0	3.0–4.0
Brick	40.0–42.0	51.0–54.0	1.5–1.7
Cheddar	36.5–38.0	51.0–55.0	1.5–1.8
Colby	38.0–39.5	51.5–55.0	1.5–1.8
Cream cheese	53.0–54.5	34.0–36.0	1.0–1.2
Edam	35.0–38.0	41.0–47.0	1.6–2.0
Gorgonzola	35.0–38.0	51.0–53.0	3.0–4.0
Granular-stirred curd (CM)[a]	33.0–35.0	51.0–54.0	1.8–2.0
Limburger	46.0–48.0	51.0–54.0	2.0–2.1
Monterey Jack	40.5–42.5	51.0–54.0	1.5–1.7
Mozzarella	54.0–58.0	46.0–50.0	1.8–2.4
LM mozzarella[b]	45.0–48.0	46.0–50.0	1.7–2.2
PS mozzarella[c]	54.0–58.0	34.0–39.0	1.7–2.2
LMPS mozzarella[d]	46.0–50.0	34.0–39.0	1.6–1.9
Muenster	41.5–43.5	51.0–54.0	1.5–1.7
Neufchâtel	61.0–63.0	21.0–25.0	0.8–1.0
Parmesan	29.5–31.5	36.0–39.0	3.5–4.0
Provolone	41.0–44.0	46.0–48.0	1.8–2.2
Romano	32.5–33.5	39.0–41.0	3.5–4.0
Skim curd	47.0–50.0	11.0–15.0	0.9–1.1
Swiss	37.0–39.5	46.0–48.5	0.2–0.6

[a] Cheese for manufacturing.
[b] Low moisture.
[c] Part skim.
[d] Low moisture, part skim.

Table 3. Code of Federal Regulations Cheese Composition Standards

Cheese type	Legal maximum moisture (%)	Legal minimum fat (dry basis, %)	Legal minimum age
Asiago, fresh	45	50	60 days
Asiago, soft	45	50	60 days
Asiago, medium	35	45	6 months
Asiago, old	32	42	1 year
Blue cheese	46	50	60 days
Brick cheese	44	50	—
Caciocavello Siciliano	40	42	90 days
Cheddar	39	50	—
Low-sodium Cheddar	(Same as cheddar but less than 96 mg of sodium per pound of cheese)		
Colby	40	50	—
Low-sodium Colby	(Same as colby but less than 96 mg of sodium per pound of cheese)		
Cottage cheese (curd)	80	0.5	—
Cream cheese	55	33	—
Washed curd	42	50	—
Edam	45	40	—
Gammelost	52	(skim milk)	—
Gorganzola	42	50	90 days
Gouda	45	46	—
Granular-stirred curd	39	50	—
Hard grating	34	32	6 months
Hard cheese	39	50	—
Gruyère	39	45	90 days
Limburger	50	50	—
Monterey Jack	44	50	—
High-moisture Monterey Jack	44–50	50	—
Mozzarella and scamorza	52–60	45	—
Low-moisture mozzarella and scamorza	45–52	45	—
Part-skim mozzarella and scamorza	52–60	30–45	—
Low-moisture, part-skim mozzarella	45–52	30–45	—
Muenster	46	50	—
Neufchâtel	65	20–33	—
Nuworld	46	50	60 days
Parmesan and reggismo	32	32	10 months
Provolone	45	45	—
Soft-ripened cheese	–	50	—
Romano	34	38	5 months
Roquefort (sheep's milk)	45	50	60 days
Samaoe	41	45	60 days
Sapsago	38	(skim milk)	5 months
Semisoft cheese	39–50	50	—
Semisoft, part-skim cheese	50	45–50	—
Skim-milk cheese for manufacturing	50	(skim milk)	—
Swiss and Emmentaler	41	43	60 days

- Improved slicing properties without crumbling.
- Smooth melting without phase separation.
- Flavor ranges depending on the cheese used.
- Can control organoleptic and physical properties.
- Versatile in use attributes (cooking cheese, dips, sauces, snacks, etc).
- Shelf stability (nonrefrigeration, fixed properties).
- Can take on various shapes and forms.

Basic Emulsification Systems for Cheese Processing

Citrates

- Trisodium citrate (most common, used for slices).
- Tripotassium citrate (used in reduced-sodium formulations, promotes bitterness).
- Calcium citrate (poor emulsification).

Orthophosphates

- Disodium phosphate and trisodium phosphate (most common, used for loaf and slices).
- Sodium aluminum phosphate (used in slices).
- Dicalcium phosphate and tricalcium phosphate (poor emulsification, used for calcium ion fortification).
- Monosodium phosphate (acid taste, open texture).

Table 4. Manufacturers' Sales of Cheese

Year	Total ($, millions)	Annual percent change
1967	1,751.8	—
1972	3,094.6	12.1%[a]
1973	3,644.4	17.8
1974	4,504.7	23.6
1975	4,900.5	8.8
1976	5,764.1	17.6
1977	6,073.6	5.4
1978	6,688.5	10.1
1979	7,903.6	18.2
1980	9,415.9	19.1
1981	10,188.0	8.2
1982	10,170.0	−0.2
1983	10,561.7	3.9
1984	10,492.1	−0.7
1985	10,707.5	2.1
1986	11,378.3	6.3
1987	11,232.5	−1.3
1988[b]	11,388.8	1.4
1997[b]	17,644.8	4.6[a]

Source: Ref. 31.
[a]Average annual growth.
[b]Estimate.

Condensed Phosphates

- Sodium tripolyphosphate (nonmelting).
- Sodium hexametaphosphate (used to restrict melts).
- Tetrasodium pyrophosphate and sodium acid pyrophosphate (minimal usage).

General Rules for Emulsifier Selection

Citrates. Used where firm-textured, close-knit products are desired, such as slices, and when products are refrigerated.

Phosphates. Used where shelf stability is required, pH is to be controlled for melting, and other physical properties.

Optional Dairy Ingredients

- Milk fat.
- Milk and skim milk.
- Nonfat dry milk (NFDM).
- Whey (lots of lactose, some sweetness).
- Whey protein.
- Buttermilk.
- Skim-milk cheese.

Other Optional Ingredients

- Acidifying agents.
- Salt.

Table 5. Manufacturers' Sales of Cheese by Type

	Natural cheese		Process cheese and related products		Cottage cheese		Other cheese[a]	
	Sales ($, millions)	Percent change	Sales ($, millions)	Percent change	Sales ($, millions)	Percent change	Sales ($, millions)	Percent change
1967	829.2	—	562.5	—	218.0	—	142.1	—
1972	1,400.0	11.0[c]	1,134.1	15.1[a]	340.9	9.4[c]	219.6	9.1[b]
1973	1,705.9	21.9	1,363.5	20.2	405.6	19.0	169.4	−22.9
1974	2,458.7	44.1	1,496.6	9.8	456.0	12.4	93.4	−44.9
1975	2,668.7	8.5	1,654.4	10.5	508.7	11.6	68.7	−26.4
1976	3,267.9	22.5	1,859.7	12.4	530.7	4.3	105.8	54.0
1977	2,727.2	−16.5	2,518.5	35.4	545.6	2.8	282.3	66.8
1978	3,104.1	13.8	2,681.4	6.5	588.5	7.9	314.5	11.4
1979	3,949.3	27.2	2,822.0	5.2	729.3	23.9	403.0	28.1
1980	4,821.1	22.1	3,303.4	17.1	840.9	15.3	450.5	11.8
1981	5,225.6	8.4	3,567.9	8.0	856.5	1.9	538.0	19.4
1982	5,625.6	7.7	3,194.3	−10.5	683.2	−20.2	666.9	24.0
1983	5,824.0	3.5	3,325.4	4.1	693.8	1.6	719.5	7.9
1984	5,617.3	−3.5	3,390.1	1.9	748.3	7.9	736.4	2.3
1985	5,664.6	0.8	3,552.6	4.8	738.3	−1.3	752.0	2.1
1986	6,289.8	11.0	3,548.9	−0.1	725.1	−1.8	814.5	8.3
1987	6,208.0	−1.3	3,463.7	−2.4	731.6	0.9	829.2	1.8
1988[c]	6,294.9	1.4	3,529.5	1.9	722.8	−1.2	841.6	1.5
1997[c]	9,826.9	4.7[b]	5,482.8	4.7[b]	1,083.9	3.9[b]	1,251.2	4.2[b]

Source: Ref. 31.
[a]Includes cheese substitution.
[b]Average annual growth.
[c]Estimate.

- Artificial coloring.
- Spices or flavorings (other than cheese flavors).
- Mold inhibitors.
- Enzyme-modified cheese (EMC).
- Gums.
- Sweetening agents.

Processing Equipment

- Grinders.
- Transfer of cheese from grinder to blender.
- Blenders.
- Scale hoppers and auger cart.
- Batch cooking.
- Continuous cooking.
- Aseptic processing.
- Surge tanks.
- Flash tanks, scraped surface heat exchange (SSHE).
- Line screens.
- Shear pumps and silverson mixers.
- Specialty equipment.

Forms of Packaging

- Can and cup.
- Jar.
- Pouch.
- Loaf.
- Slice.

Methods of Cooling

- Chill roll.
- Hot pack.
- Fast cooling.
  Blast coolers.
  Water spray tunnels.
- Slow cooling.
  Cased product.
  Restricted melt properties.

CHEESE ANALOGUES: HISTORY

Cheese analogues are products that resemble natural or process cheeses. They are intended to have the appearance, taste, texture, and nutrition of their counterpart cheeses, but are made without butterfat. The procedures by which they are made are also quite similar to those used for traditional cheeses.

Throughout history, substitute food products have been developed as the result of shortages or the opportunity to reduce costs. With cheese analogues, the impetus was the opportunity to replace valuable butterfat with less-expensive fats made from vegetable oils. Early natural cheese analogues were made by skimming butterfat from whole milk, replacing it with some other fat, then following traditional cheese-making procedures. These products, called filled cheeses, were first made around the turn of the century (30).

In the early 1970s technology was developed in which process cheese analogues could be made by combining dried milk protein, hydrogenated vegetable oil in place of butterfat, emulsifying salts, and other ingredients, then cooking the mixture. These products simulated process American and mozzarella cheeses.

DEFINITIONS

Filled Cheese

A product that simulates a natural cheese, made by substituting the butterfat in fluid whole milk with some other fat, then following traditional natural cheese-making procedures. Typical products are similar to mild Cheddar, Colby, or similar cheeses. A filled cream cheese can also be made by replacing butterfat with vegetable fat following conventional procedures.

Process Cheese Analogue

A product that simulates a process cheese, made by substituting the butterfat in a formulation, combining it with a protein source, emulsifying salts, and other ingredients and cooking the mixture. Typical products are simulations of pasteurized process cheese, pasteurized process cheese food, or pasteurized process cheese spread. Imitation mozzarella cheese is also made with this method.

Cheese analogues are classified as imitation or substitutes depending on their functional and compositional characteristics. An imitation cheese need only resemble the cheese it is designed to replace. Thus the choice of ingredients that can be used is quite extensive. However, a product can generally be called a substitute cheese only if it is nutritionally equivalent to the cheese it is replacing and meets the minimum compositional requirements for that cheese. Thus substitute cheeses will often have higher protein levels than imitation cheeses and be fortified with vitamins and minerals.

COMMERCIAL USE

Since the introduction of these products in the 1970s, cheese analogue consumption in the United States grew to about 208 million lb in 1988, or about 2.2% of the total cheese markets (31). The main incentive was a lower price than natural cheeses. In 1980, for example, the average price of cheese analogues was $0.47/lb less than natural cheese (5). This was due to the use of vegetable oil in analogues, rather than butterfat, as well as imported casein products as the source of dairy protein.

The principal use of cheese analogues has been in sandwich slices or as an ingredient in food products manufactured by food-processing companies or food service operators. In these applications, the analogues can be used as extenders of natural cheeses. Common uses are in frozen pizzas and in food service cheese shreds and cheese sauces. The United States retail market has not grown substantially, because cheese analogues have a lower quality than natural cheeses. Nonetheless, analogues of pasteurized process American cheese and mozzarella shreds are being sold in the grocery stores.

INGREDIENTS IN CHEESE ANALOGUES

Formulations

Cheese analogues all contain a protein source, a fat component, and a variety of minor components chosen to provide the desired flavor, texture, color, nutrition, and keeping quality. Table 6 lists ingredient statements provided by these types of analogue product: a filled Cheddar type, an imitation process cheese, and an imitation cream cheese. The formulations are considerably more complex than for natural cheeses.

PROTEIN SOURCES

Casein Products

The most critical component in cheese analogues is the protein. The cheese analogues that have had commercial success are all based on casein, the protein in milk. In some cases a soluble casein salt, or caseinate, is used. One of three forms of casein is generally used. For each, the starting material is skim milk (Table 7).

Vegetable Proteins

Several patents have been issued for, and papers written about, the use of vegetable proteins such as soy or cottonseed protein in place of caseins (16,32). However, these products have not been commercially successful to date because of many technical problems. These include such important characteristics as flavor, melting properties, gel strength, color, translucence, and mouth feel (such as creaminess or chewiness). The vegetable proteins simply do not function as milk proteins without further modifications, and these developments remain in the laboratory. However, some soybean protein isolates have been used as an extender in cheese analogues to replace up to 30% of the casein.

Table 7. Composition of Casein(ates)

	Acid casein (lactic)	Sodium caseinate	Rennet casein
Protein ($N \times 6.35$)			
Dry basis, %	96.7	94.7	90.6
As is, %	87.3	90.4	80.6
Ash, %	1.8	3.8	7.8
Moisture, %	9.7	4.5	11.0
Fat, %	1.2	1.1	0.5
Lactose, %	0.1	0.1	0.1
pH (5% at 20°C)	4.8	6.7	7.1

Vegetable Fats

Fats are essential ingredients in cheese analogues, for they are replacing butterfat in the products. The fats utilized become emulsified and intermixed with protein and water and, therefore, contribute to melting and textural characteristics. The fats also must behave in the mouth something like butterfat, which does melt in a characteristic way and influences creaminess, flavor, and other factors. Consequently, fats most often utilized in cheese analogues are partially hydrogenated or mixtures of hydrogenated and unhydrogenated oils that stimulate butterfat. These fats are similar to vegetable shortenings and can be made from soybean, corn, cottonseed, palm, or coconut oils or blends of these oils. These blends often have a melting point of about 35–47°C (33).

Table 6. Ingredients in Cheese Analogues

Kraft Golden Image Imitation Cheddar Cheese	Kraft Lunchwagon Imitation Pasteurized Process Cheese	American Whipped Products King Smoothee Imitation Cream Cheese
Pasteurized part-skim milk	Whey	Water
Liquid and partially hydrogenated corn oil	Partially hydrogenated soybean oil	Partially hydrogenated vegetable oil (may contain one or more of the following: coconut oil, soybean oil, corn oil)
Cheese culture	Sodium caseinate	Corn syrup solids
Salt	Granular cheese	Sodium caseinate
Enzymes	Food starch, modified	Tapioca flour
Apocarotenal (color)	Sodium citrate	Monodiglycerides
Vitamin A palmitate	Monodiglycerides	Salt
	Potassium chloride	Lactic acid
	Sorbic acid as a preservative	Guar gum
	Magnesium oxide	Monopotassium phosphate
	Zinc oxide	Artificial flavor
	Vitamin B_{12}	Citric acid
	Apocarotenal (color)	Carageenan
	Vitamin A palmitate	Artificial color
	Calcium pantothenate	Sorbic acid (as a preservative)
	Riboflavin	

Table 8. Cheese Analogues Characteristics

Advantages	Disadvantages
Cost of raw materials	Flavor
Keeping quality	No U.S. large sources of casein
Nutritional equivalency or superiority	Unnatural image
Varying functional properties	

Other Ingredients

The principal ingredients in cheese analogues are protein, fat, and water. Thereafter, ingredients are added to solubilize the protein, add flavor and color, and provide nutritional supplements. A typical formula for an imitation pasteurized process American cheese is presented below (34).

Water	45.20
Acid casein	25.00
Hydrogenated soybean oil	20.60
Adipic acid	1.30
Na–Ca hydroxide	0.80
Color–flavor blend	0.45
Dry ingredients	
Salt	1.50
Tapioca flour	1.50
Tricalcium phosphate	1.10
Disodium phosphate	1.00
Vitamin mix	0.50
Cheese flavor	0.50
Amino acids	0.31
Dipotassium phosphate	0.14
Sorbic acid	0.10
	100.00%

As can be seen, the formulation of cheese analogues is quite complex.

CHARACTERISTICS

Cheese analogues have both advantages and disadvantages compared to their natural counterparts. To date, the biggest problem is flavor. Casein products have a characteristic off-flavor that needs to be masked, in some manner. Likewise, vegetable proteins do not have characteristic, clean dairy flavors. But there are some offsetting advantages (Table 8).

Analogues generally have better keeping quality than natural cheeses because hydrogenated vegetable oil is less likely to develop rancidity than butterfat. The absence of microbial cultures and their enzyme systems creates stable flavor and texture. Analogues made with vegetable oil also have less cholesterol than natural cheeses and can be formulated to have less saturated fat and fewer calories. The varying functional properties can be incorporated by the utilization of different ingredients.

BIBLIOGRAPHY

1. A. Carlson, G. Hill, and N. F. Olson, *Biotechnology and Bioengineering*, Vol. 29, John Wiley & Sons, Inc., New York, 1987, p. 582.
2. P. F. Fox, *Cheese: Chemistry, Physics and Microbiology*, Vol. 2, Elsevier Applied Science Publishers, Ltd., Barking, UK, 1987.
3. M. L. Green, "Effect of Milk Pretreatment and Making Conditions on the Properties of Cheddar Cheese from Milk Concentrated by Ultrafiltration," *Journal of Dairy Research* **52**, 555–564 (1985).
4. U.S. Pat. 3,914,435 (Oct. 21, 1975), J. L. Maubois and G. Mocquot (to Institut National de la Recherche Agronomique-France).
5. G. Sillen, "Modern Bactofuges in the Dairy Industry," *Dairy Industries International* **52**, 27–29 (1987).
6. D. D. Duxbury, "Breakthrough for Cheese—100% Chymosin Rennet," *Food Processing* **50**, 19–22 (1989).
7. P. H. A. Sneath, ed., *Bergey's Manual of Systematic Bacteriology*, Vol. 2, Williams and Wilkins, Baltimore, Md., 1984.
8. C. Daly, "The Use of Mesophilic Cultures in the Dairy Industry," *Antonie Van Leeuwenhock* **49**, 297–312 (1983).
9. J. Auclair and J. P. Accolas, "Use of Thermophilic Lactic Starters in the Dairy Industry," *Antonie Van Leeuwenhock* **49**, 313–326 (1983).
10. M. L. Green, "Review of the Progress of Dairy Science: Milk Coagulants," *Journal of Dairy Research* **44**, 159–188 (1977).
11. C. A. Ernstrom, "Enzyme Survival during Cheesemaking," *Dairy & Ice Cream Field* **159**, 43–46 (1976).
12. D. J. McMahon and R. J. Brown, "Enzymic Coagulation of Casein Micelles: A Review," *Journal of Dairy Science* **67**, 919–929 (1984).
13. R. C. Lawrence and J. Gilles, "Factors That Determine the pH of Young Cheddar Cheese," *New Zealand Journal of Dairy Science Technology* **17**, 1–14 (1982).
14. K. R. Nath and R. A. Ledford, "Growth Response of *Lactobacillus casei* var *casei* to Proteolysis Products Occurring in Cheese During Ripening," *Journal of Dairy Science* **56**, 710–715 (1973).
15. R. A. Ledford, A. C. O'Sullivan, and K. R. Nath, "Residual Casein Fractions in Ripened Cheese Determined by Polyacrylamide-Gel Electrophoresis," *Journal of Dairy Science* **49**, 1098–1101 (1966).
16. J. Stadhouders, L. Toepoal, and T. M. Wouters, "Cheesemaking With prt$^-$ and prt$^+$ Variants of *N*-Streptococci and Their Mixtures. Phage Sensitivity, Proteolysis and Flavor Development during Ripening," *Netherlands Milk Dairy Journal* **42**, 183–193 (1988).
17. D. J. Manning and H. E. Nursten, "Flavor of Milk and Milk Products," in P. F. Fox, ed., *Developments in Dairy Chemistry*, Vol. 3, Elsevier Applied Science Publishers, Ltd., Barking, UK, 1985.
18. G. W. Reinbold, *Swiss Cheese Varieties, Pfizer Cheese Monographs*, Vol. 5, Pfizer, Inc. New York, 1972.
19. P. Trieu-cuot and J. C. Gripon, "A Study of Proteolysis during Camembert Cheese Ripening Using Isoelectric Focusing and Two Dimensional Electrophoresis," *Journal of Dairy Research* **49**, 501–510 (1982).
20. J. E. Kinsella and D. Hwang, "Biosynthesis of Flavors by *Pencillium roqueforti*," *Biotechnology and Bioengineering* **18**, 927–938 (1976).
21. R. D. King and G. H. Clegg, "The Metabolism of Fatty Acids, Methyl Ketones and Secondary Alcohols by *Pencillium roqueforti* in Blue Cheese Slurries," *Journal of the Science of Food and Agriculture* **30**, 197–202 (1979).
22. W. L. Langhus, W. Y. Price, H. H. Sommer, and W. C. Frazier, "The "Smear" of Brick Cheese and Its Relation to Flavor Development," *Journal of Dairy Science* **28**, 827–838 (1945).

23. M. Purko, W. O. Nelson, and W. A. Wood, "The Associative Action between Certain Yeasts and *Bacterium linens*," *Journal of Dairy Science* **34**, 699–705 (1951).

24. D. Sloot and P. D. Harkor, "Volatile Trace Components in Gouda Cheese," *Journal of Agriculture and Food Chemistry* **23**, 356–357 (1975).

25. A. H. Woo and R. C. Lindsay, "Concentration of Major Free Fatty Acids and Flavor Development in Italian Cheese Varieties," *Journal of Dairy Science* **67**, 960–968 (1984).

26. T. Conner, "Advances in Accelerated Ripening of Cheese," *Cultured Dairy Products Journal* **23**, 21–25 (1988).

27. E. M. Foster, F. E. Nelson, R. N. Doetsch, and J. C. Olson, Jr., *Dairy Microbiology*, Prentice Hall, Englewood Cliffs, N.J., 1957.

28. K. R. Nath and B. J. Kostak, "Etiology of White Spot Defect in Swiss Cheese Made from Pasteurized Milk," *Journal of Food Protection* **49**, 718–723 (1986).

29. *Code of Federal Regulations*, Title 21 §100.120.

30. S. K. Gupta, G. R. Patil, and A. A. Patel, "Fabricated Dairy Products," *Indian Dairyman* **39**, 199–208 (1987).

31. U.S. Department of Commerce, Bureau of the Census, International Trade Administration, reprinted in *U.S. Industrial Outlook 1989—Food, Beverages, and Tobacco*, U.S. Government Printing Office, Washington, D.C., Jan. 1989.

32. U.S. Pat. 4,303,691 (Dec. 1, 1981), R. E. Sand and R. E. Johnson (to Anderson, Clayton & Co.).

33. U.S. Pat. 4,822,623 (Apr. 18, 1989), J. L. Middleton (to Universal Foods Corp.).

34. U.S. Pat. 3,922,374 (Nov. 25, 1975), R. J. Bell, J. D. Whynn, G. T. Denton, R. E. Sand, and D. L. Cornelius (to Anderson Clayton & Co.).

GENERAL REFERENCES

V. K. Babayan and J. R. Rosenau, "Medium Chain Triglyceride Cheese," *Food Technol.* **45**, 111–114 (1991).

A. Baer et al., "Microplate Assay of Free Amino Acids in Swiss Cheeses," *Lebensm.-Wiss. Technol.* **29**, 58–62 (1996).

P. Bican and A. Spahni, "Proteolysis in Swiss-type Cheeses: A High Performance Liquid Chromatography Study," *International Dairy Journal* **3**, 73–84 (1993).

P. Bican, A. Spahni, and J. Schaller, "Partial Characterization of Peptides From Emmentaler Cheese," *Food Biotechnology* **8**, 229–241 (1994).

J. O. Bosset et al., "Occurrence of Terpenes and Aliphatic Hydrocarbons in Swiss Gruyere and Etivaz Alpine Cheese Using Dynamic Headspace GC-MS Analysis of Their Volatile Compounds," *Schweizerische Milchwirtschafliche Forschung* **23**, 37–42 (1994).

J. O. Bosset, M. Cullomb, and R. Sieber, "The Aroma Composition of Swiss Gruyere Cheese IV. The Acidic Volatile Components and Their Changes in Content During Ripening," *Lebensm.-Wiss. Technol.* **26**, 581–592 (1993).

S. E. Brummel and K. Lee, "Soluble Hydrocolloids Enable Fat Reduction in Process Cheese Spreads," *J. Food Sci.* **55**, 1290–1292, 1307 (1990).

L. Cabezas, P. Alvarez Martin, and M. D. Cabezudo, "Proteolysis in Gruyere de Comte Cheese Accentuating the Effect of Traditional Salting," *Revista Espanola de Ciencia y Tecnologia de Alimentos* **33**, 501–516 (1993).

S. L. Caldwell et al., "Development of Lactose-Positive *Pediococcus* Species Cheese Starter Cultures," *Journal of Dairy Science* **78**(Suppl. 1), 108 (1995).

E. Carrera et al., "The Effect of Milk Coagulant on the Formation of Hydrophobic Peptides in Cheese Manufactured From Cow, Ewe or Goat Milk," *Journal of Dairy Science* **78**(Suppl. 1), 143 (1995).

F. L. Davies and B. A. Law, eds., *Advances in Microbiology and Biochemistry of Cheese and Fermented Milk*, Elsevier Applied Science Publishers, Ltd., Barking, United Kingdom, 1986.

K. Elsborg, "Cheese Cultures Adapted to Specific Technologies," *European Dairy Magazine* No. 4, 44–45 (1997).

P. F. Fox, *Cheese, Chemistry, Physics and Microbiology*, Vol. 1, Chapman and Hall, London, United Kingdom, 1993.

P. F. Fox, *Cheese: Chemistry, Physics and Microbiology*, Vol. 2, Elsevier Applied Science Publishers, Ltd., Barking, United Kingdom, 1987.

S. K. Garg and B. N. Johri, "Rennet: Current Trends and Future Research," *Food Rev. Int.* **10**, 313–355 (1994).

M. Grand et al., "Development of Ripening Flora on the Surface of Gruyere Cheese," *Schweizerische Milchwirtschaftliche Forschung* **21**, 3–5 (1992).

R. Grappin and J. L. Berdague, "Ripening and Quality of Gruyere de Comte Cheese, VIII. Synthesis and Conclusions," *Lait* **69**, 183–196 (1989).

Y. H. Hong, "Influences of Ingredients and Melting Temperatures on the Physicochemical Properties of Process Cheese," *Korean Journal of Food Science and Technology* **21**, 710–713 (1989).

S. Issanchou, I. Lesschaeve, and E. P. Koester, "Screening Individual Ability to Perform Descriptive Analysis of Food Products: Basic Statements and Application to a Camembert Descriptive Panel," *Journal of Sensory Studies* **10**, 349–368 (1995).

M. Kalab et al., "Structure, Meltability, and Firmness of Process Cheese Containing White Cheese," *Food Structure* **10**, 193–201 (1991).

T. Khieokhachee, "Kinetic Studies of Cheese Starter Cultures," *Dissertation Abstracts International B* **56**, 3522–3523 (1996).

F. A. M. Klaver and M. Twigt, "Improvement of Eye Formation in Cheese," *Voedingsmiddelentechnologie* **26**, 11–14 (1993).

F. Kosikowski, *Cheese and Fermented Milk Foods*, 2d ed. rev., F. V. Kosikowski & Associates Publishers, Brooktondale, New York, 1982.

J. Kubickova and W. Grosch, "Quantification of Potent Odorants in Camembert Cheese and Calculation of Their Odour Activity Values," *International Dairy Journal* **8**, 17–23 (1998).

J. Kubickova and W. Grosch, "Evaluation of Flavour Compounds of Camembert Cheese," *International Dairy Journal* **8**, 11–16 (1998).

B. Kupiec and B. Revell, "Specialty and Artisanal Cheese Today: The Product and the Consumer," *British Food Journal* **100**, 236–243 (1998).

S. Lubbers, N. Cayot, and C. Taisant, "Blue Cheese Taste Intensification in Process Cheese Products," *Science des Ailments* **17**, 393–402 (1997).

S. Marchesseau et al., "Influence of pH on Protein Interactions and Microstructure of Process Cheese," *Journal of Dairy Science* **80**, 1483–1489 (1997).

F. G. Martlery and V. L. Crow, "Open Texture in Cheese: The Contributions of Gas Production by Microorganisms and Cheese Manufacturing Practices," *Journal of Dairy Research* **63**, 489–507 (1996).

P. Palich, W. Derengiewicz, and J. Switka, "Studies on the Shelf Life of Modified Camembert Cheese," *Acta Alimentaria* **19**, 321–329 (1990).

D. B. Perry, J. MacMahon, and C. J. Oberg, "Manufacture of Low Fat Mozzarella Cheese Using Exopolysaccharide-Producing Starter Cultures," *Journal of Dairy Science* **81**, 536–566 (1998).

L. M. Popplewell and J. R. Rosenau, "Incorporation of Milk Proteins Harvested by Direct Acidification Into Process Cheese Products," *Journal of Food Process Engineering* **11**, 203–220 (1989).

M. Preininger and W. Grosch, "Evaluation of Key Odorants of the Neutral Volatiles of Emmentaler Cheese by Calculation of Odour Activity Values," *Lebensm.-Wiss. Technol.* **27**, 237–244 (1994).

S. Rapposch, "Occurrence and Significance of Nonstarter Lactic Acid Bacteria in Cheese," *Deutsche Milchwirtschaft* **48**, 838–841 (1997).

M. Rosenberg, M. McCarthy, and R. Kauten, "Evaluation of Eye Formation and Structural Quality of Swiss-Type Cheese by Magnetic Resonance Imaging," *Journal of Dairy Science* **75**, 2083–2091 (1992).

J. E. Schlesser, S. J. Schmidt, and R. Speckman, "Characterization of Chemical and Physical Changes in Camembert Cheese During Ripening," *Journal of Dairy Science* **75**, 1753–1760 (1992).

E. K. Shumaker and W. L. Wendorff, "Factors Affecting Pink Discoloration in Annato-Colored Pasteurized Process Cheese," *J. Food Sci.* **63**, 828–831 (1998).

R. Smeets, "Enzyme-Coagulation," *Dairy Technology-Paper* **5**, 14–18 (1995).

M. J. Sousa and F. X. Malcata, "Comparison of Plant and Animal Rennets in Terms of Microbiological, Chemical, and Proteolysis Characteristics of Ovine Cheese," *J. Agric. Food Chem.* **45**, 74–81 (1997).

M. Sutheerawattananonda et al., "Fluorescence Image Analysis of Process Cheese Manufactured With Trisodium Citrate and Sodium Chloride," *Journal of Dairy Science* **80**, 620–627 (1997).

F. Vance et al., "Autolysis of *Lactobacillus helveticus* and *Propionbacterium freudenreichii* in Swiss Cheeses: First Evidence by Using Species-Specific Lysis Markers," *Journal of Dairy Research* **65**, 609–620 (1998).

R. Warme, H. D. Belitz, and W. Grosch, "Evaluation of Taste Compounds of Swiss Cheese (Emmentaler)," *Z. Lebensm.-Unters. Forsch.* **203**, 230–235 (1996).

B. H. Webb, A. H. Johnson, and J. A. Alford, *Fundamentals of Dairy Chemistry*, 2d ed., AVI Publishing Co., Westport, Conn., 1974.

G. H. Wilster, *Practical Cheesemaking*, 11th ed., O.S.C. Cooperative Association, Corvalis, Ore., 1959.

P. Zanerl, "Detection of Clostridia Responsible for Cheese Defects in Raw Milk and Cheese," *Deutsche Milkchwirtschaft* **44**, 936, 938, 940 (1993).

K. Rajinder Nuath
John T. Hines
Ronald D. Harris
Kraft, General Foods, Inc.
Glenview, Illinois

CHEESE RHEOLOGY

The word rheology in the most literal sense means the study of flow. At first sight it appears an unlikely pursuit to apply to the study of cheese. However, one may extend the concept of the flow of a material to include the idea of any change in its shape, under the action of an external agency, which is not instantaneous, and which is not entirely recoverable. These are characteristics of flow (1), and the application to cheese becomes immediately obvious. In fact, in the history of rheology applied to food materials, many of the pioneers found dairy products, and particularly cheese, very suitable materials for their experiments. This followed naturally from the fact that cheese-making was originally a craft industry. Long before rheology developed as a science, the skilled cheesemaker used various tactile tests to estimate the progress of the development of body in the curd and the firmness of the final cheese, using memory as a guide to the consistency of the results. The rheologist has, in more recent years, attempted to apply instrumental techniques and universally adopted physical units to these measurements.

At the same time as the science of rheology was developing, and often in parallel with it, the methods of testing using the human senses have been developed into a separate field of study now known as texture studies (2–4). It is necessary for the complete food rheologist to keep abreast of these developments since food is judged, in the final analysis, by its deformation in the mouth during mastication, a purely sensory judgment. Food scientists devote much effort to correlating the physical and sensory observations. It may well be that in endeavoring to seek parity between instrumental rheological measurements and properties judged subjectively one is asking the wrong question and a question to which there can be no final and complete answer. The real questions which can be asked are: What can instrumental measurements tell us about the properties and in particular the structure of the material on which those properties depend? How can this information be useful to the cheese-maker? How may the instrumental measurements be used as a guide to the consumer's assessment?

FIRST PRINCIPLES

Before proceeding to the detailed study of the application of rheology to cheese, it is pertinent to consider a few basic principles. These will appear obvious to the reader educated in a physical science, but may be less obvious to those from other walks of life. One is taught that matter exists in one of three distinct states: gas, liquid, and solid. This context is mainly concerned with only the latter two. Both are characterized by the fact that any sample has a well-defined volume. Solids have a definite shape which can be altered only by the application of an external agency. Liquids, on the other hand, have no characteristic shape: they take up the shape of the vessel containing them.

The property, then, which distinguishes a solid is its rigidity and this term has been formally adopted by physicists as a measure of the effort required to change the sample's shape. The precise definition will be given later. If the material is a true solid, it will follow that this rigidity is invariant for the material. Furthermore, time does not enter into this, so it is understood that the change is instantaneous and once the effort is removed, the sample will recover its original shape spontaneously. This is, in plain language, the physicist's concept of an elastic solid.

In contrast with the solid, the characteristic by which a layman distinguishes a liquid is its fluidity. This, again, has a precise definition, but in practice it is more convenient to use the inverse concept, which is known as the viscosity. This is a measure of the relation between the effort applied to the liquid and the rate at which it flows. As before, once the effort has been withdrawn, the flow ceases, but the liquid remains where it is—there is no recovery. It is the rate of flow which spontaneously reverts to its original value. This describes the physicist's concept of a viscous liquid.

It is important to keep clearly in mind that the fundamental difference in behavior between solids and liquids involves the dimension of time. Time does not enter at all into the description of the behavior of an elastic solid—only spatial dimensions are involved—whereas time is equally as important as the spatial dimensions in the description of fluid behavior.

So far, only ideal materials have been described. These are seldom encountered in practice and are only of interest to rheologists as reference points, the simplest extremes of material behavior, one may say the "black and white" of classical physics. It is that extensive "gray" area between that is the rheologist's domain. Everyone knows that a piece of cheese, once squeezed, may recover its original shape only slowly and probably not completely. Indeed, some of the softer cheeses may, if sufficient effort be applied, be spread and only recover very little. This leads to a definition of a third category of materials. Any material falling within this gray area, possessing at the same time some of the characteristics of elastic and of viscous behavior, is known as a viscoelastic material.

Yet another class of materials may be mentioned in passing, since they may be encountered in everyday parlance. This is the plastic material. Plasticity may be defined as that property of a material whereby it remains rigid until a certain minimum force is applied, whereupon it deforms. Once this force is removed or falls below the critical value, the material again becomes rigid but remains in the deformed state. There is a simple mathematical model for this type of behavior (5) and the rheologist may be particularly interested in the transition from solid to fluid behavior, known as the yield point; however, the ideal plastic is probably as rare as the ideal solid or liquid.

DEFINITIONS

Before the rheological measurements which have been made on cheese can be discussed it is necessary to define a few of the terms used by rheologists. In the preceding paragraphs the words effort and force have been used in their everyday connotation. The term used by the rheologist is stress, denoted by the Greek letter σ. This is formally defined as the force measured in newtons (N) divided by the area in square meters (m^2) over which it is applied, and is measured in pascals (Pa). A convenient aide-mémoire for the size of these units is that a newton is approximately the weight, under the action of gravity, of a medium-sized apple. The stress may be applied normal to the surface, as in Figure 1a, or tangentially, as in Figure 1b. In either case it is equal to the force divided by the area over which it is applied.

In the case of Figure 1a, which depicts a weight resting on the upper surface of a sample, the deformation which

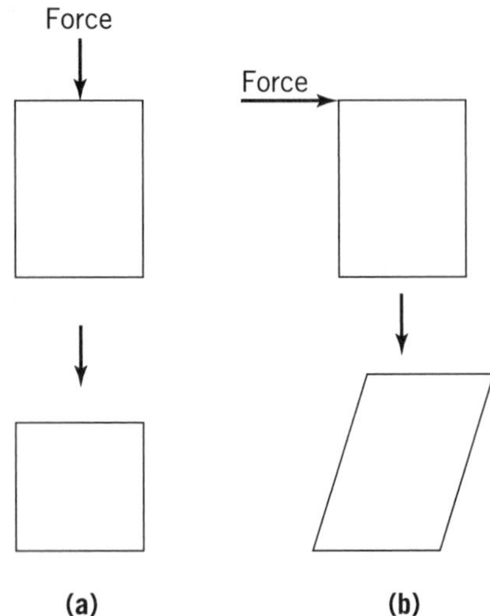

Figure 1. Application of stress to a sample: (**a**) compressive strain; and (**b**) shear.

occurs is a change (Δh) in the height (h) of the sample. The fractional change in height, $\Delta h/h$, defines the strain, which is denoted by the Greek letter ϵ. There is a practical difficulty here: the height is continuously changing during the deformation, so there is some uncertainty over what value should be used for h (6). For small deformations this is relatively unimportant, but when large deformations occur, such as is usual when examining cheese, it is necessary to be clear on this point. For the sake of clarity, the ratio of the change in height to the original height will be referred to as the fractional compression and expressed as a percentage. When the change is related to the varying height, the formula for the strain becomes

$$\epsilon = \ln(h_0/h_1) \tag{1}$$

This is sometimes called the true strain and will be given in decimal form.

If the stress is applied tangentially, as in Figure 1b, the resulting strain is described as a shear. It is defined as the distance through which one plane has traveled relative to another divided by the distance between them and is given the symbol γ. Both the strain and the shear are the quotients of two lengths and are therefore dimensionless numbers.

With these definitions of the physical quantities stress, strain, and shear, it is now possible to define the properties of some ideal materials. The rigidity of a solid, hitherto used loosely, can now be defined formally as the ratio of the tangential stress to the shear produced and is given the symbol n:

$$n = \sigma/\gamma \tag{2}$$

The dimensions of n are evidently the same as the dimensions of the stress, but it is usual to express the modulus of rigidity in newtons per square meter and to reserve the use of pascals for stresses only, thereby preserving an easily recognizable distinction between them.

The second property of an elastic solid which may be defined can be seen by referring again to Figure 1a. This is the modulus of elasticity and is the ratio of the stress to the strain. If the material is ideal, it will be isotropic; that is, the properties will be the same in all directions. During compression the volume will remain constant, so that while the height of the sample is changing in the direction of the applied stress, the dimensions in the two directions at right angles will change simultaneously. At any point within the sample only one-third of the stress will be used in causing the change in height. Another third will be used in each of the two other directions normal to this, causing the sample to expand laterally. Hence, for any ideal elastic material the modulus of elasticity will be three times the modulus of rigidity. However, if the material is compressible, some of the stress will be used up in compressing it and the factor 3 relating the modulus of elasticity to that of rigidity falls toward 2. The importance of this is that some authors express their results in terms of elasticity and some in terms of rigidity. An accurate comparison is possible only when some information is available concerning the compressibility of the material. In practice the ideal incompressible material is as unlikely to exist as any other ideal material. Almost all cheese is somewhat compressible, but in general not so compressible that any serious error is introduced by using the factor 3 to convert from one unit to the other.

In the case of liquids there is only one material constant that is of interest. The rate of flow or deformation is now given by the quotient of the strain and the time taken to reach it if the motion is constant, or in the unsteady state by the first derivative with respect to time of the strain, $\dot{\epsilon}$. The ratio of the applied stress to the rate of flow is called the viscosity and this is denoted by the Greek letter η, so that

$$\eta = \sigma/\dot{\epsilon} \tag{3}$$

The unit of viscosity is the pascal-second. A rough guide to its magnitude is that the viscosity of water at room temperature is very nearly 1/1000 Pa-s.

MODELS OF RHEOLOGICAL BEHAVIOR

In order to provide a theoretical framework within which to discuss the determination of the physical quantities just described it is necessary to consider some models for a viscoelastic body. There are two ways in which this may be visualized. Viscoelasticity may be considered either to be a property intermediate between viscosity and elasticity, or the result of a combination of viscous and elastic behavior occurring simultaneously in the material. In the first case, recalling that the elastic modulus is a function of spatial dimensions only, whereas viscosity is a function also of the first derivative of the strain with respect to time, the suggestion is that viscoelasticity is a function of a fractional derivative, the fraction lying between zero and unity. Although the concept is simple to describe, the mathematical equipment required to convert measurements which are always made in a conventional three-dimensional space and real time to this hypothetical continuum where space and time are no longer independent is formidable. The concept is now of historical interest only, but readers may find some early writers used it to discuss the rheology of cheese.

The alternative hypothesis, that both elastic and viscous behavior occur simultaneously, leads to a number of simple models (7), which require only elementary mathematics. It should be emphasized at this stage that these are only mathematical models, whose purpose is to codify the experimental observations; they cannot be taken to be real physical structures. It will be shown later that it is sometimes possible to relate elements in a model to the existence of particular structural elements within a material. The simplest model for an elastic solid is a spring whose strain or compression ϵ_s is always proportional to the stress. Writing G for the constant of proportionality to distinguish it from the rigidity or elasticity of any real material,

$$\epsilon_s = \sigma/G \tag{4}$$

In a similar manner a viscous liquid may be modeled by an infinitely long dashpot whose displacement is always inversely proportional to the viscosity

$$\dot{\epsilon} = \sigma/\eta \tag{5}$$

There are two ways in which one spring and one dashpot can be combined to give a compound structure. If they are placed in tandem as shown in Figure 2a, the stress is continuous throughout the model and the total displacement is given by the sum of those of the elements:

$$\epsilon = \sigma/G + \sigma t/\eta \tag{6}$$

The first term of the right-hand side expresses the elastic component and indicates that immediately on applying the stress there will be some deformation; thereafter (second term) the deformation will increase with time t. On removal of the stress, the elastic component recovers immediately while the strain due to the dashpot remains.

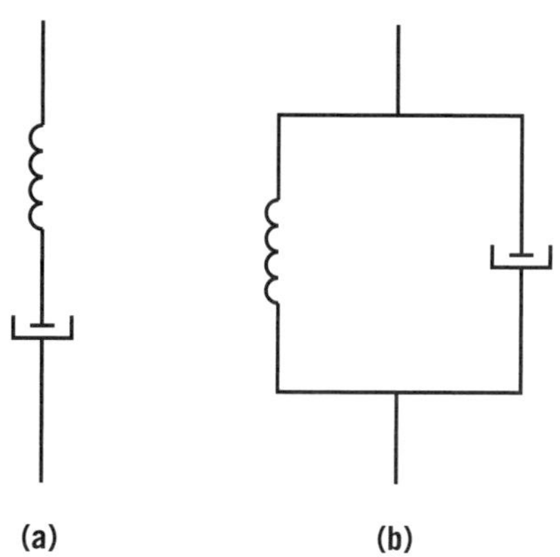

Figure 2. Viscoelastic models: (**a**) Maxwell body, and (**b**) Voigt body.

This model is called a Maxwell body. A complete creep and recovery curve of this model is given in Figure 3a. It is already apparent that a two-component model represents the rheological behavior of a sample of cheese better than the assumption that it is either a solid or a fluid.

If the two components are combined in the other possible way, so that they share the stress but the strain is common to both, we arrive at a formula for a Voigt body, sometimes known as a Kelvin body:

$$\sigma = G\epsilon + \frac{\eta\dot{\epsilon}}{t} \tag{7}$$

Transforming this equation so that it may be compared with equation 6 gives

$$\epsilon = \frac{\sigma}{\eta}\left[1 - \exp\left(-\frac{G}{\eta}t\right)\right] \tag{8}$$

The creep and recovery curve of this model is given in Figure 3b. Again, there is some resemblance between this curve and the behavior observed with cheese. The deformation and the recovery are not instantaneous in this case and this is more in line with practical experience, but the model predicts a complete recovery, which is not always observed.

The student of electric circuit theory will be aware of some familiarity with the Voigt and Maxwell models. If one replaces the spring with a capacitor and the dashpot with a resistor, the two models become the two alternative representations of a leaky dielectric. Just as in electrical network theory these binary combinations can be built up into more complicated networks by adding more elements, so the Voigt and Maxwell bodies can be extended. One popular combination is of a Voigt body placed end-to-end with a Maxwell body. This is known as a Burgers body and it does indeed give a creep and recovery curve in which both the deformation and the recovery are retarded to different degrees and in which the recovery is incomplete.

Although the process of building up more complicated models may be continued indefinitely, the expressions for the behavior rapidly become cumbersome. As the number of elements increases it becomes disproportionately difficult to analyze the data, and the accuracy of the parameters extracted decreases (8). Indeed, the precision of any experimental measurements made on cheese may well be insufficient for more than two or three parameters to be identified with any pretensions of accuracy. Only if the rheological behavior can be linked unequivocally to the structure of the sample may the use of complex models be justified. Otherwise, the model should be kept as simple as possible.

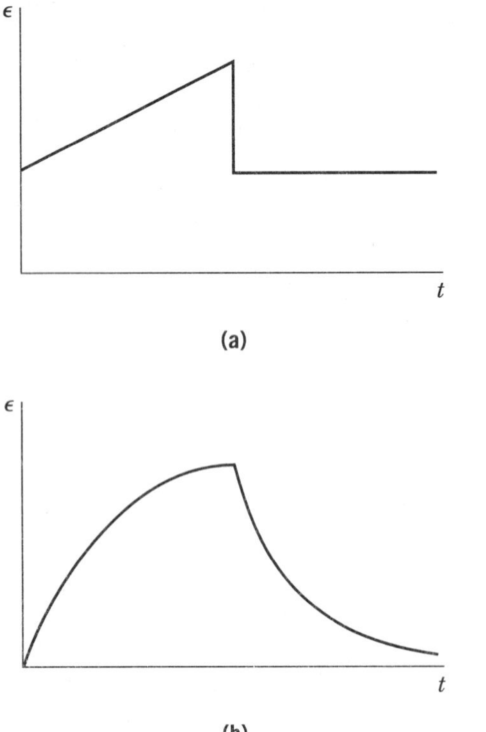

Figure 3. Creep recovery curve: (**a**) Maxwell body; and (**b**) Voigt body.

EMPIRICAL INSTRUMENTAL TESTS

The earliest applications of rheology to cheese, even before rheology was conceived as a scientific discipline, consisted of making subjective estimations of some of the more obvious properties and trying to describe these in plain language. Often the cheese-maker or the grader, though skillful and competent in judgment, would find it difficult to define precisely the terms being used. One can see that confusion can arise when one tries to distinguish between terms such as consistency and body, firmness and hardness, chewiness and meatiness. The grader's mind may be clear about what is understood by each, but communicating the precise meaning to the outside world is less easy. The difficulty is further compounded by the fact that these words do not translate precisely into other languages, making international communication more difficult. To some extent the development of texture studies, particularly in the United States, has clarified the semantic problem (9), but subjective assessment can never match the precision which can be attained with sophisticated instrumentation. Much of the earlier instrumental rheology sought to replace, or at least to reinforce, subjective judgment with numerical quantities which could be measured reproducibly by any competent technical assistant (10).

The earliest instruments were empirical (3) and, in general, intended to give some indication of the firmness or similar qualities. To this end, they were designed to simulate what the scientist perceived to be the mechanical action of the hand or fingers of the expert during examination of the cheese. They were generally unsophisticated, not too expensive, and made no pretence of measuring any precisely defined physical property. Because of this, their measurements cannot be analyzed and expressed in terms of fundamental rheological parameters as defined above. Nevertheless, some of them may still perform a useful function if their limitations are understood, because they possess the undoubted merit that their measurements are easily reproducible between different laboratories and different times.

One such instrument became known as the Ball Compressor (11). In the course of examining a cheese, one of the actions of any grader is to press either a thumb or a finger into the surface and, from the reaction which is felt, gain an impression of the firmness or springiness of the cheese. The Ball Compressor might in fact be described as a mechanical thumb. It consists of a metal hemisphere which is initially placed on the surface of a whole cheese and allowed to sink into the cheese under the action of a load; the depth of indentation can be measured and recorded after a fixed time. The load can then be removed and the recovery of the cheese observed. The instrument is shown diagrammatically in Figure 4.

Making a number of simplifying assumptions, it is possible to convert a reading obtained with the Ball Compressor into a modulus G (12,13), which may be considered to be the equivalent of the modulus of rigidity, using the formula

$$G = 3M/16(RD^3)^{1/2}. \tag{9}$$

where M is the applied load, R the radius, and D the depth of the indentation. This formula has been arrived at assuming that the sample is a homogeneous isotropic elastic solid, the load is static, and there is no friction between the sample and the surface of the hemisphere. It is clear that every one of these conditions is violated. Cheese is far from homogeneous, particularly if it has a rind or has been salted by immersion in brine, causing the properties to vary from the surface to the center. It is not isotropic: the method of manufacture of some varieties, such as Cheddar, on which many of the earlier experiments were carried out, or Mozzarella, is intended to produce an oriented structure. Some of the objections, however, would appear to be much less serious in the case of other cheeses where uniformity of texture is the aim of the manufacturer. Cheese is not an elastic solid; it is viscoelastic. The indentation is not instantaneous and may not be completed within the duration of the experiment. Finally, if the load is removed,

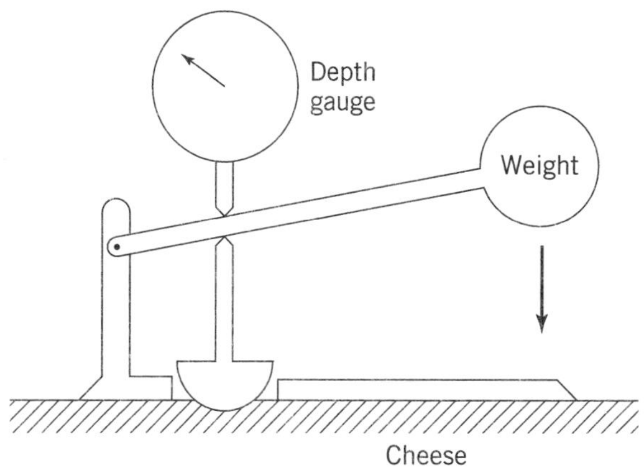

Figure 4. The Ball Compressor (diagrammatic).

some indentation remains, showing that some of the energy has been dissipated in causing flow and not stored up as in an elastic solid. In spite of all these limitations, if the test proceeds at least until the indentation becomes so slow that it has virtually ceased, the use of equation 9 gives an indication of the magnitude of the firmness in readily understandable physical units. This enables the reader to compare measurements given in the early literature with those made more recently by more sophisticated means.

The Ball Compressor has been discussed at some length, not only because of its historical importance, but because it has the merit of being a nondestructive means of testing and does not require an operator to undergo any specialized training. Also, it is used on the whole cheese, whereas almost every other rheological test requires that a sample be cut from the whole. Using the Ball Compressor, it has been shown that in a whole cheese such as a Cheddar or a Cheshire weighing some 25 kg there are considerable variations in the firmness over the surface of the cheese (14) and that the firmness differs on the upper and lower sides. This difference is influenced by the frequency with which the cheese is turned in the store during its maturing period, and the time which has elapsed since it was last turned. The implication of this is that it is not possible, whatever instrumental measurement is made, to assign a single number to any property of the cheese, nor is it possible to assess the properties of the whole cheese by means of a measurement made at one local point in that cheese.

Clearly there is a need for some form of nondestructive test and for many purposes there is no reason why it should not be empirical, provided that the implications are understood. The Ball Compressor has the merits of cheapness and simplicity, but the time taken to obtain a representative reading limits its use to the research laboratory. The problem of devising a suitable test is as yet unsolved. It is unfortunate that the application of ultrasonic techniques, which have proved so useful in the nondestructive testing of many engineering materials, have proved unrewarding with cheese (15,16). This is because the dimensions of the cracks and other inhomogeneities in cheese are commensurate with, or sometimes even larger than, the wavelength of the ultrasound and large-scale scattering takes place. It is difficult for a pulse to penetrate the body of the cheese and both the velocity of propagation and the attenuation are more influenced by the scattering than by the properties of the bulk.

One other empirical test deserves to be mentioned on account of its simplicity. This is the penetrometer (17–19). It is not quite nondestructive, but very nearly so, since it only requires that a needle be driven into the body of the cheese; no separate sampling is required. A penetrometer test may take one of several forms. For example, a needle may be allowed to penetrate under the action of a fixed load (17,18), or it may be forced into the cheese at a predetermined rate (19) and the required force measured.

Whichever method is used, let the specific actions be considered. As the needle penetrates the cheese, that part of the cheese immediately ahead of it is ruptured and forced apart. If the needle is thin, the deformation normal to its axis is small, so that the force required to accomplish this may be ignored. On the other hand, the progress of the needle is retarded by the adhesion of its surface to the cheese through which it passes. This may be expected to increase with the progress of the penetration until a point is reached when the restraining force matches the applied load and further penetration ceases. If a suitable diameter for the needle and a suitable weight have been chosen, this test may be completed in a few seconds. The test will be more useful for cheese whose body is reasonably homogeneous on the macroscopic scale, as, for example, the Dutch cheeses. With cheeses such as Cheddar or the blue cheeses the heterogeneities are generally on too large a scale and the penetration becomes irregular. The needle may pass through weaknesses in the structure or even cracks and so give rise to the impression that the cheese is less firm than it really is, or the point of the needle may attempt to follow a line of weakness, not necessarily vertical, and as a result there will be additional lateral forces acting on the needle and its penetration will be arrested prematurely.

Measurements made with a penetrometer cannot be converted theoretically to any well-defined physical unit. Both the cohesive forces within the cheese and the adhesive forces between the cheese and the surface of the needle are a consequence of the forces binding the structure together. One may infer that these are related to its viscoelastic properties but there is no simple theory which attempts to establish these relations. It has been shown experimentally (20) that there is a statistically significant correlation between firmness as measured by the Ball Compressor and by penetration, but this differs among different types of cheese (21). It has also been shown experimentally that a curvilinear relation exists between the resistance to penetration and an elastic modulus of some Swiss cheeses, calculated from the results of a compression experiment (Fig. 5). While this gives some idea of the magnitude of the forces involved, it has little practical application since it refers only to one series of experiments on only one type of cheese. In the absence of a similar experimentally determined relation for the cheese in which one is particularly interested, penetrometer measurements can only be regarded as empirical.

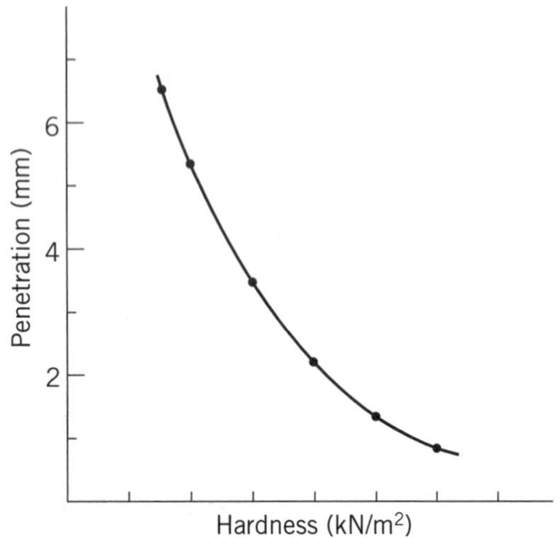

Figure 5. Penetrometer readings compared with compression modulus.

Another test of a similar nature is the use of a cone to penetrate the sample (22). This test was originally developed for use with high-viscosity lubricants. A loaded cone with its apex pointed vertically downward and initially just making contact with the upper surface of the sample is allowed to penetrate the sample until equilibrium is reached. If a suitable combination of the apical angle of the cone and the load is used, the equilibrium is usually reached within a few seconds and the penetration is deep enough to be measured comfortably. The test assumes that the sample behaves as a plastic or Bingham body; that is, when acted upon by small stresses it behaves as a solid, but once a critical stress, the yield stress, is exceeded, it flows as a viscous liquid.

In fact, the test works in reverse. At the commencement the stress is infinite, since the area of application, which is the tip of the cone, is zero. As the cone penetrates, the material is caused to flow laterally and the rate of penetration is controlled by this lateral flow. As the penetration proceeds, the cross-sectional area increases proportionally with the square of the depth of penetration and the stress decreases correspondingly. When the stress no longer exceeds the critical or yield point, motion ceases and an equilibrium is established. It is a matter of contention among rheologists whether a yield value really exists. Some maintain that there is always some flow, albeit minimal, however low the stress. For practical purposes, however, the cone soon appears to be stationary and the depth of penetration can be recorded. This is a simple, quick test and, while it is not entirely nondestructive, it does only a little damage near the surface of the sample. The vertical stress Y may be calculated from the penetration and the angle of the cone:

$$Y = (Mg/\pi h^2)\cot^2\alpha/2 \tag{10}$$

where α is the apical angle of the cone, h the penetration depth, M the applied load, and g the gravitational constant. This is greater than the yield stress since it does not take into account the stresses involved in causing the sample to flow laterally. An estimate of the yield stress may be obtained by multiplying the equilibrium vertical stress by the factor 1/2 sin α (23). It is only an estimate, though, because it takes no account of any friction between the surface of the cone and the cheese.

Another form of penetrometer-type test is that sometimes referred to as the puncture test (24). In place of a needle or cone, a rod is driven into the sample and the resistance to its motion measured. The mechanism of the deformation of the sample is more complicated in this case. At least four principal factors are involved. In the first place, the sample ahead of the rod is compressed. Second, the rod must cut through the sample at its leading edge: hence the name puncture, and this requires a force. Third, there is the frictional resistance between the surface of the rod and the cheese, and finally, there is the force required to set up the lateral flow within the sample. If the sample is large enough compared with the dimensions of the rod so that any compression in directions perpendicular to the motion may be ignored, once a dynamic equilibrium has been set up the first two factors will be constant and the third will increase linearly with the penetration. By using rods of different cross-sectional areas and keeping the perimeter the same it is possible to separate these effects (25,26).

Instead of performing this test on an extensive sample, such as a whole cheese, it is more often carried out on a small sample contained within a rigid box (27–29). When done in this manner, additional forces are called into play. The lateral forces on the rod as the sample is compressed between the rod and the walls of the container now become important. In addition, the compressive force on the leading face of the rod is no longer constant as the distance between this face and the bottom of the container decreases. If the sample is shallow, it is arguable (27,28,30) that this may indeed be the largest single contributory force, and, by ignoring the others in comparison with it, one can calculate a modulus from the ratio of the measured stress to the compression. It is obvious that this will be an overestimate of any true modulus which could have been derived in a simple unrestrained test and that the excess will be arbitrary and unknown. In spite of this the test is useful because it gives an indication of the magnitude of the rigidity of the sample and allows comparisons to be made between similar types of cheese.

PHYSICAL MEASUREMENTS

It has already been pointed out that rheological phenomena take place in real time. The basic rheological experiment consists in applying a known stress for a known time and observing the strain which ensues (31). In this context it is immaterial whether one refers to normal stresses, which result in strains, or to tangential stresses, which result in shears. For the purpose of the argument strain and shear are interchangeable. The generic mathematical equation which connects the variables is

$$\sigma = f(\epsilon, t)$$

This is the equation of a surface in a three-dimensional diagram. The material is said to be linearly viscoelastic if, for any value of t, the line representing the variation of strain with stress is a straight one. For many substances this is approximately true, at least while the strains and stresses are small. Cheeses vary in this respect. For a linear viscoelastic material the two-dimensional graph of strain versus time will contain all the required rheological information.

The basic graph is the creep curve. If the experiment is prolonged indefinitely, some equilibrium will ultimately be established. If the material is essentially a solid, but one whose deformation is retarded, the strain will eventually become constant. There are obviously two parameters which can be used to denote the material behavior, one based on the ultimate deformation and the other on the rate at which it is (asymptotically) established. On the other hand, if the material is more akin to a liquid, the final equilibrium will be a dynamic one in which a constant rate of strain is established. As before, two parameters are needed to describe the behavior: one, the ultimate flow; the other, the rate of attaining it. The experiment may also be conducted rather differently. Instead of allowing it to proceed ad infinitum, the stress may be removed at a predetermined point. Should the material possess any elastic characteristics, some of the energy used in the deformation will be stored within it; on removal of the stress the strain will decrease again and once more eventually reach an equilibrium. This part of the graph is known as the recovery curve. Again there are two points of particular interest—the amount of recovery and the rate.

So far the description refers to what may be called an ideal viscoelastic material and is an oversimplification of real-life experience. If the material is not linearly viscoelastic, the creep and recovery curve can still be obtained, but a single curve will not include all the rheological information about the sample. A series of such curves is necessary, giving the full three-dimensional representation. It has also been assumed so far that the properties of the material are not altered in any way by the action of the strain itself. In the case of cheese it is very evident that at large strains the whole structure breaks down and cannot recover. Indeed, the stress and strain at the point of onset of this breakdown may be a useful quantity when describing the properties of any cheese, as will be seen later. Even at lower strains there may well be unobserved changes in the structure. These could occur during the early part of a cycle of measurement, in which case they may be revealed by the failure of the recovery curve to follow the expected pattern. However, they may occur during handling of the cheese at any stage between manufacture and sampling. A complete rheological description of a sample really needs not only the experimental curve or curves but a statement of the history of the straining of the sample up to the moment of measurement. This is most likely to be unavailable and it is necessary to presume that the sample is in good condition and to allow that variations in its history are included within the natural variation among samples.

It is often more convenient from the point of view of practical instrumentation to carry out the compression experiment in a different way. Instead of applying a fixed stress and observing the reaction, the sample is subjected to a known rate of strain and the build-up of the stress observed. Again, this may be prolonged indefinitely, if the operation of the instrument will allow it. If the straining is stopped at any point, it is not usually practicable to apply a reverse rate of strain. In this case, the strain is usually held constant and the relaxation of the stress observed. Again, a complete cycle of deformation and relaxation contains all the available information about the rheological properties of the sample. Quite sophisticated instruments have been developed which enable this type of test to be carried out with precision and then record the results, making this now the most widely used test. In principle, the operation of these instruments is simple. The sample is compressed between parallel plates at a known fixed rate and the force required is monitored.

However, before discussing the results of any measurements made on these instruments it is important to consider the exact mode of operation. As the measurement is usually carried out, the sample is in the form of a small cylinder, although other right prisms may be used. This is placed with its axis vertical on one plate and is then compressed along that axis. Since the compression is linear, the height of the sample at any instant after the compression has begun is given by

$$h = h_0(1 - at)$$

where h_0 is the original height of the sample and a denotes the rate at which the two places approach each other, ie,

$$a = -dh/dt$$

Using equation 1 for the strain leads to

$$\epsilon = -\ln(1 - at) \tag{11}$$

If the instrument's recorder treats time (or its equivalent linear distance traveled) as one axis, then this is not a true strain axis. As the compression increases from 0 to 100% the true strain increases from zero to infinity. In fact, whatever definition of strain is used (6), other than the linear compression, the time axis is a distortion of the strain axis. Second, as a sample is compressed its volume remains constant, so that the cross-sectional area increases as the height decreases. Assuming that the plates are perfectly smooth, so that there is no lateral friction between them and the end surfaces of the sample, a purely viscous, or equally a perfectly elastic solid, sample would deform uniformly and the cross-sectional area A at any height would vary precisely inversely with the height, ie,

$$A = A_0(h_0/h) \tag{12}$$

A viscoelastic material does not deform so simply, but in such a way that the lateral movement is greatest near the ends, so that a concave shape results (32), as shown in Figure 6. In this case the stress is not uniform throughout the sample. When there is some friction between the plates and the sample, the lateral movement of the end layers is restricted and a rotational couple is set up within the sample, using the perimeter of the end surfaces as fulcrum. If the sample is homogeneous or at least cohesive, the effect of this is that the middle of the sample will spread outward, giving rise to barrel-shaped distortion (Fig. 6c). When the rotational forces which develop are sufficient to cause fractures to develop, even though they may be undetected by eye, then with the weakened structure the sample may spread out near the ends, and what is known as mushroom distortion results (33) (Fig. 6). Both barrel and mushroom distortion may occur simultaneously, and a shape such as Figure 6d may be the outcome.

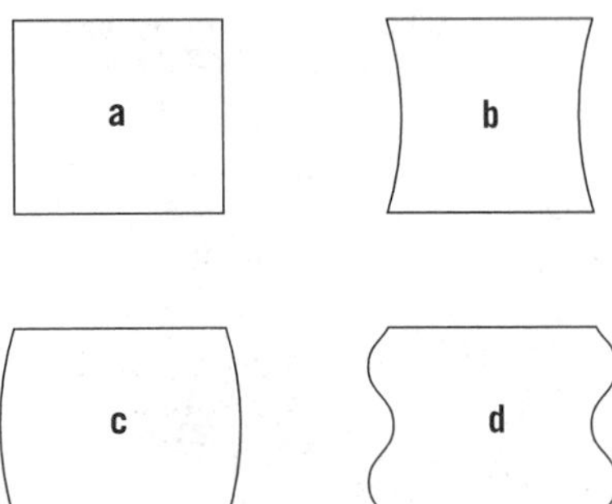

Figure 6. Compression of a cylinder: (**a**) ideal; (**b**) concave; (**c**) barrel; and (**d**) complex.

Also, during the compression some liquid, which may be free fat or water, or both, may be squeezed out. This might be thought to lubricate the sample ends to some extent; but it might also ensure closer contact between what is inevitably the slightly rough cut surface, the degree of roughness depending largely upon the type of cheese, and the smooth plates of the instrument. In practice it has been found (34) that the ends of the cheese do not expand as much as would be predicted by the use of equation 12; yet on the other hand it would be quite erroneous to assume that the cross section remained at its original value. The instrument does not measure stress, but the total force transmitted by the end surface to the measuring transducer. The stress within the sample is obtained by dividing the force F by the cross-sectional area. In the case of a perfectly lubricated sample this is given by

$$\sigma = F/A = (F/A_0)(1 - at) \tag{13}$$

When barrel distortion occurs, the calculation of the true stress is less simple. If one assumes that the material behaves as a perfect elastic solid and that the ends of the sample are firmly bonded to the instrument plates so that they cannot expand laterally, the barreling can be shown to assume a parabolic profile. The actual correction to be applied depends on the height and radius of the uncompressed sample as well as the degree of compression. It is somewhat greater than the correction for a lubricated sample. A force–compression curve is thus not a true representation of a stress–strain curve. Near the origin it is close enough, but the distortion increases as the compression progresses.

If the sample behaves as one of the simple models, it is possible to predict the shape of the stress–strain curve. In the case of the Voigt body the solution is straightforward. The stress is the sum of the elastic and the viscous contributions (eq. 7). From equations 1 and 11

$$\dot{\epsilon} = ae^{\epsilon}$$

so that

$$\sigma = G\epsilon + a\eta e^{\epsilon} \tag{14}$$

This is the equation of a curve in which the intercept on the σ axis is proportional to the viscous component and the rate of compression, and the initial slope is proportional to the elastic component. The curve is concave upward throughout. On substituting for σ and ϵ, equation 14 becomes

$$F = [A_0/(1 - at)][a\eta/(1 - at) - G \ln(1 - at)] \tag{15}$$

This is not so convenient to analyze but there is still an intercept on the F axis proportional to the viscous component and the rate of compression. The solution for the Maxwell body is more complicated. Making the same substitutions as before leads to an infinite series

$$\sigma = Ge^{(G/a\eta)e-\epsilon}\left[\epsilon + \sum_{1}^{\infty} \frac{(-G/a\eta)^{i}(1 - e^{-\epsilon i})}{\mathrm{i.i!}}\right] \quad (16)$$

At the commencement of the compression the stress is zero and the initial slope is proportional to the elastic component. Initially the curve is convex upward, but as the compression proceeds a point of inflexion is reached, depending on the ratio of the viscous to the elastic components and the rate of compression. Finally, the curve becomes asymptotic to a line through the origin, with a slope given by the elastic constant.

If the stress–strain curve obtained for a particular sample can be recognized as being similar to one of these two patterns, the appropriate model can be identified and the material constants evaluated. It is seldom that such a simple fit occurs. A material as heterogeneous as cheese is unlikely to conform to the behavior of a two-element model. More sophisticated models may give a better representation, but adding further elements only makes the theory more complex and the analysis that much more difficult. One word of caution must be entered here. The Voigt body is characterized by the finite intercept on the stress axis. The converse is not necessarily true. A finite intercept also arises in the case of any material possessing a yield value, since this is the stress which must be overcome before any deformation takes place.

In the usual way in which compression tests are carried out, the compression is allowed to proceed to a point far beyond that at which any simple theory may be expected to apply. Often it reaches 80%, a true strain of 1.609. Long before this is reached, any structure which may have been present in the original sample of cheese can be seen to have broken down. If it is carried to this extent, only the early part of the test may be considered as measuring the rheological properties of the original sample, while the latter part becomes a test of the mechanical strength of the structure. A typical compression curve is shown in Figure 7. It is often difficult to decide whether there is an intercept on the stress axis, ie, whether there is an instantaneous build-up of stress at the moment the compression begins or a rapid development of stress rising from zero but in a finite though brief time. Nevertheless, it is essential that the distinction be made if the correct model is to be assigned. The instrument itself necessarily has a finite, even if very rapid, acceleration from rest, and the response of the recording mechanism also takes a finite time, so that even if the stress were instantaneous, it would appear as a very steep rise (35).

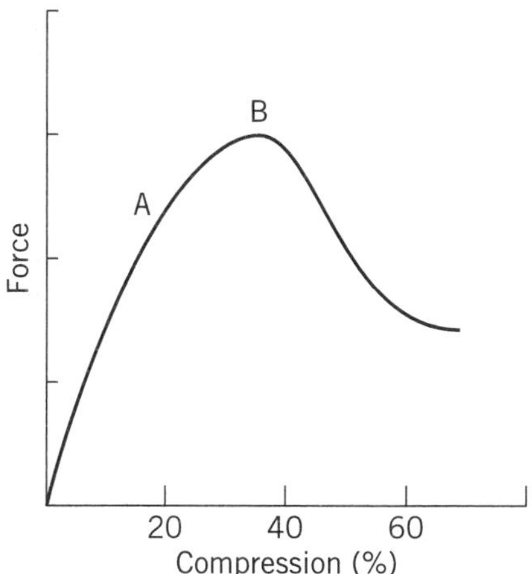

Figure 7. Typical force–compression curve.

However, much more serious than any limitations in the instrument is the problem of attaining perfection in the shape of the sample. Usually the dimensions of the sample and the rate of compression are chosen so that the strain rate at the commencement is of the order of 0.01 s^{-1}. In a sample, say 20 mm in height, a lack of parallelism of the two end surfaces of only 0.2 mm means that for the whole of the first second of compression only part of the sample is being compressed. This will inevitably appear on the record as a flow, whatever the real properties of the cheese.

Assuming that any doubts about the shortcomings of either sample or instrumentation may be satisfactorily allayed, it is pertinent to consider the principal features of the compression curve. In the early stages, after the initial rise there is a smooth increase of stress with the strain. Should the compression be halted at this stage, the cheese would recover either completely or in part and a repeat of the curve could then be obtained, the new curve having the same general shape as the original, showing that the internal structure of the cheese had remained more or less intact, although there may have been some rearrangement. Eventually a point is reached, the point A in Figure 7, where the structure begins to break down and the slope becomes noticeably less. The conventional view is that this is the point at which cracks in the structure appear and the cracks then spread spontaneously (18,36). In the case of a hard cheese these cracks may be evident to the eye, but they may at first be so small and so localized that they do not become visible until the compression has proceeded somewhat further. If the cheese has a very homogeneous structure, it is to be expected that these cracks will develop throughout the structure at about the same time and the change of slope will be clearly defined. In a more heterogeneous cheese they may appear over a range of strains and the change of slope will be much more diffuse. Once this region has been passed, the cracks continue to develop at an increasing rate and become more evident, until at point B the rate at which the structure breaks down overtakes the rate of build-up of stress through further compression and a peak is reached. This value is obviously dependent on a balance between the spontaneous failure of the structure and the effect of increasing deformation applied by the instrument to any residual structure. It is a convenient parameter to determine and may be used as a measure of firmness or hardness of the cheese (18). Beyond this point the stress may continue to fall if the failure of the structure becomes catastrophic, until the fragments become compacted into a new arrangement which can take up the stress and this now rises again.

If the compression test is carried on beyond the point at which it is reasonable to attempt to interpret it in terms of an acceptable model, and hence to evaluate any of the customary rheological parameters, it becomes purely empirical. One of the points on the curve most frequently used is the peak (18,37) (point B). This is a kind of yield point and one can obtain from it two useful parameters, the stress required to cause catastrophic breakdown of the structure and the amount of deformation that the cheese will stand before this breakdown occurs. The two are not entirely independent. In a dynamic measurement such as this on an essentially viscoelastic material, the rate of strain as well as the strain itself influences the moment at which the breakdown occurs. For reliable comparisons, all measurements should be made at the same rate of straining. Fortunately, many workers have found it convenient to use similar sizes and rates of compression, so that at least approximate comparisons can be made between their results. The stress measured at point B is in fact sometimes called a yield value, but it should not be confused with the yield stress of a plastic material determined by other methods such as a cone penetrometer. The yield at point B is not a material constant defining the strength of the sample; breakdown has already been occurring at least since point A, and maybe earlier. The peak value at B only indicates the maximum to which the stress rises before the collapse of structure overtakes the build-up of stress in what remains of that structure.

The other point frequently used is the stress at 80% compression. The damage to the structure when the compression reaches this extent is generally so complete that it is unrealistic to regard the stress as applying to the original sample. This is particularly true in the case of hard cheeses such as the typical English and Grana varieties, which crumble long before this degree of compression is reached. Nevertheless, compressions of this order and much greater arise during mastication (38). The stress at 80% can therefore give some indication of the consumer's response to the firmness of the cheese, although it must be said that in mastication the rate of compression is also much greater.

Measurements have been made on four types of hard cheese, Cheddar, Cheshire, Double Gloucester (36), and Leicester (39), which confirm that the value of the stress at the peak point B is dependent on the rate at which the compression is carried out. It was shown that over at least a twentyfold range of the factor a in equation 11, the rate at which the plates approached each other, the peak stress was linearly related to the fourth root of a. This is shown in Figure 8, where straight lines have been drawn through the points for each cheese type. If one bears in mind that this is a destructive test, so that each measurement had to be made on a different sample, the fit of the lines to the experimental points is quite acceptable. This result is, however, purely empirical; as far as can be seen there is no theoretical justification for it and it is limited to those types of cheese which crumble on breakdown. But there is a practical benefit. Using this finding, it is reasonable to reduce measurements made at any practical rate of compression to a standard rate so that the results of workers in different laboratories may be compared.

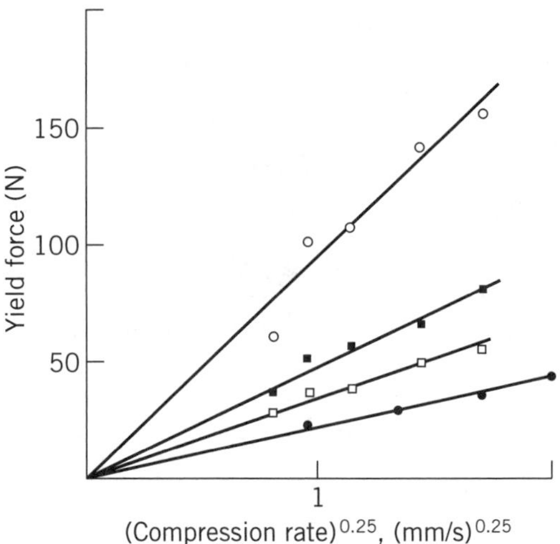

Figure 8. Effect of rate of compression on stress at yield point ●, Double Gloucester; □, Cheshire; ■, Leicester; and ○, Cheddar.

The investigation was carried one step further on Double Gloucester cheese (36). Not only were the peak values found to obey this fourth-root law but so were the stresses at other degrees of compression; the results are shown in Figure 9. The fact that the relation between stress and rate of deformation is more or less constant before, at, and beyond the yield point may be considered to lend support to the hypothesis that the processes taking place within the cheese are similar throughout the compression. The stress at any point is the result of a balance between the rate of collapse of structure, ie, the spread of cracks, and the build-up of stress in that structure which remains. If it is assumed that the basic framework within the cheese has at least some structural strength, there will be some build-up of stress before any cracks appear, and the hypothesis predicts that the stress—strain curve will show an initial instantaneous rise due to viscoelastic deformation before

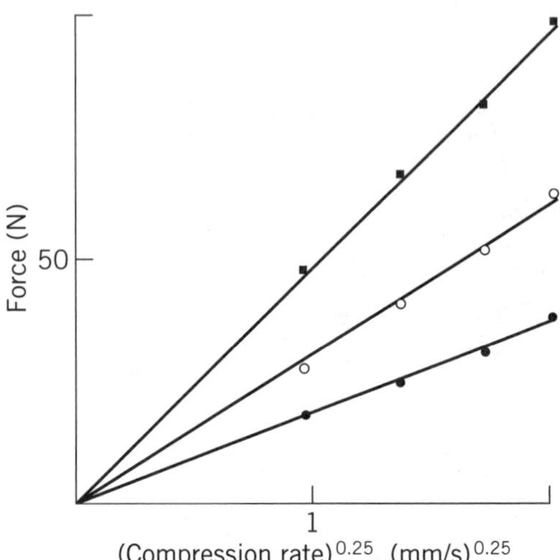

Figure 9. Effect of rate of compression on stress at various compressions: ●, 10%; ○, 40%; and ■, 70%.

any cracks develop, followed by a rising curve, convex upward, as the discrete minute cracks increase, with a pronounced change of slope as they begin to coalesce.

There are alternative explanations which may be advanced for the shape of the curves. In the early stages the curve rises from the origin and is often convex upward. This is characteristic of a Maxwell body. The interpretation could be that cheese behaves as an elastic fluid with a very high viscosity term. Neither the elasticity or the viscosity are readily obtained from the curves, but it is possible to calculate an apparent elasticity from the slope. Sometimes the slope at the origin is taken; more usually, the slope over the middle portion of the rise.

On the other hand, it is arguable that the structure of cheese is basically a solid one, but that even under very small stresses minute cracks begin to appear in that structure (40,41), even though these are far too small to be observed by the naked eye and may be disguised to some extent by the fact that some liquid component from the fat could flow into some of the opening interstices. Experiments on cheese analogs (42) at very low strains have confirmed that the structure does indeed break down well before the compression reaches 1%. It is possible to analyze the curves further on the basis of this hypothesis. Suppose that the cheese has an initial rigid structure giving rise to an elasticity E_0. Then at the instant at which the compression commences the slope of the stress—strain curve $d\sigma/d\epsilon$ is equal to E_0. If the cheese breaks down continuously by the appearance of cracks, infinitessimal at first but becoming gradually more widespread and larger, the strength of the cheese is progressively reduced, so that at any subsequent instant the elasticity $E = d\sigma/d\epsilon$ is less than E_0. If the breakdown of the structure is consequent upon the extent of the strain, the equation may be written

$$E = d\sigma/d\epsilon = E_0[1 - f(\epsilon)] \qquad (17)$$

This is the equation of a curve through the origin with an initial slope E_0 and subsequently convex upward, as is usually observed.

Pursuing this a little further, the function $f(\epsilon)$ is a distribution function of the breaking strains of the interparticulate junctions within the cheese. As a trial hypothesis, one may postulate that the distribution function is linear up to the point at which all the junctions are broken. Replacing $f(\epsilon)$ by a constant c and integrating equation 17 one gets

$$\sigma = E_0\epsilon - c\epsilon^2(\epsilon < \epsilon_{\text{critical}}) \qquad (18)$$

This is the equation of a parabola with its apex upward and a maximum stress of $\sigma = E_0^2/4c$ at a strain of $\epsilon = E_0/2c$. Eventually the situation is reached, at $\epsilon_{\text{critical}}$, where all the structure is more or less completely destroyed and individual crumbs may move more or less independently as in the flow of a powder. If the cheese is spreadable, the stress required to maintain this flow may be expected to be constant. On the other hand, some further compaction may take place and the stress begin to rise again. In general, the distribution would not be expected to be a linear one. While the argument remains the same, leading to a convex upward curve with a peak and a subsequent trough, algebraic analysis is more involved. However, if the constant c is replaced by a general expression, such as a series expansion in ϵ, the slope at the origin is still given by E_0 but the parabolic form becomes distorted.

When studying individual curves it is not always possible to determine the slope at the origin with any confidence, particularly since this is the region where the response time of the recorder is most likely to introduce its own distortion. However, rewriting equation 18 as

$$\sigma/\epsilon = E_0 - c\epsilon$$

it is possible to construct fresh curves of σ/ϵ versus ϵ and these may be easier to extrapolate to zero strain and thereby estimate the value of E_0. The (negative) slope of this curve is then the distribution function of c. In a few cheeses c has been found to be more or less constant up to strains approaching unity (ie, about 60% compression) but with most cheeses the value of c decreases as the strain increases, indicating that the rate of breakdown of structure is actually greatest at the lowest strains. Figure 10 shows a few typical derived curves.

RELAXATION

If the compression of the sample is halted before the sample is completely broken up and then held, the stress which has built up during that compression will decay. The two binary models show characteristically different types of decay behavior. In the Voigt body the stress at any instant is due partly to the strain and partly to the rate of straining. At the instant of arresting the motion the strain rate falls abruptly to zero, so that its contribution to the stress disappears, leaving only the stress due to the amount of strain, which then remains constant. In the Maxwell body the stress decays progressively, ie, relaxes, falling exponentially to zero if followed indefinitely according to the equation

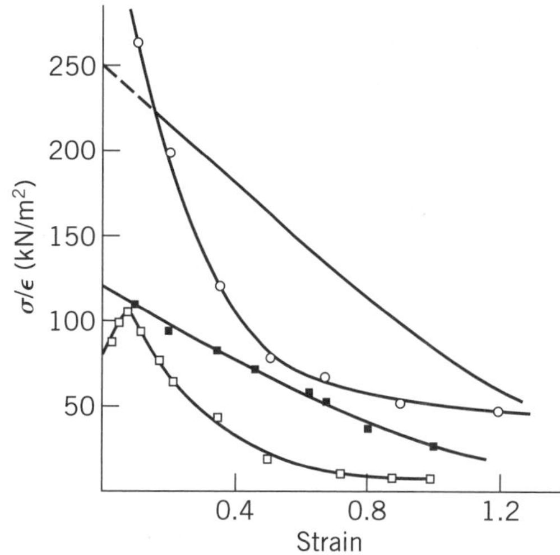

Figure 10. Derived curves of apparent modulus versus strain.

$$\sigma = \sigma_0 \exp(-Gt/\eta) \qquad (19)$$

where σ_0 is the stress at the instant of halting. The ratio ν/G has the dimensions of time and is known as the relaxation time. This is readily determined from the experimental curve, either by redrawing it in logarithmic form or as the time for the stress to fall to $1/e$ of its starting value. The relaxation curve by itself does not allow the elastic and viscous components to be obtained separately. When the simple Maxwell model is evidently insufficient to describe the relaxation behavior, an analysis of the curve may become more difficult. The sample may be described by a more complex model created by combining several simple binary models. There are computational procedures which enable the various constants of the model to be evaluated in such a case, but the precision falls off rapidly as the number of constants increases. The rheologist may sometimes be interested in determining several relaxation times and moduli for a particular sample, especially if these can be assigned to recognizable structural elements in the material, but more often the proliferation of constants does little to clarify the behavior of cheese.

There is an empirical treatment which is sometimes useful in analyzing relaxation data (43,44), particularly in cases where the precision of the data may be in some doubt. If one takes Y_t as the fraction of the stress which has decayed in time t, ie,

$$Y_t = (\sigma_0 - \sigma_t)/\sigma_0$$

where σ_0 is the stress at the commencement of the relaxation and σ_t the stress after time t, it has been found that many complex viscoelastic materials relax in such a way that the relation

$$1/Y_t = k_1 + k_2/t \qquad (20)$$

holds to a fair degree of approximation; k_1 and k_2 may easily be found. Then $1/k_2$ is the extent to which the stress eventually will decay, while the ratio k_2/k_1 is a measure of the rate of decay. For a perfectly elastic body or a Voigt body $1/k_2$ is zero; there is no relaxation. For a liquid or a Maxwell body it is unity. For more complex models it lies somewhere in between. This is a useful device for deciding whether a simple model will suffice. For a Cheddar cheese $1/k_2$ was found to be in the region of 0.8, so a simple model is not adequate for describing the relaxation behavior of this cheese (43). An alternative treatment has been suggested (45), which, it is claimed, sometimes fits experimental data more closely. This may be derived by considering a binary model of the Maxwell type, but one in which the viscosity associated with the dashpot is not constant but follows a power-law variation with stress. The resulting expression is

$$(\sigma_0/\sigma)^{n-1} = 1 + k(n - 1)t \qquad (21)$$

where n is the exponent of the power law. This is an interesting suggestion: One of the consequences of the power-law variation of viscosity is that it becomes infinite at zero stress, and a Maxwell body with an infinitely viscous dashpot is in fact a solid. Thus this is a model for a material which is solid while at rest but becomes progressively more fluid as the stress increases. This is not too different from the description of cheese given above: a solid material whose structure breaks down progressively when subject to strain.

RHEOLOGY AND STRUCTURE

From the foregoing considerations it will be evident that the rheological properties of any material are consequent upon its structure. The three major constituents of cheese, casein, fat, and water, each contribute to the structure and therefore to the rheological properties in their own specific way. At normal room temperatures the casein is solid, the fat is an intimate mixture of solid and liquid fractions, giving it what may be described as plastic properties, while the water is liquid. The casein forms an open meshlike structure (46–48). Within this mesh is entrapped the fat, which had its origin as the fat globules of the milk. The water is more precisely the aqueous phase, since dissolved in it are the soluble constituents of the milk serum together with any salt which has been added during the cheese-making. Some of the water is bound to the protein and therefore largely immobilized; the remainder is free and fills the interstices between the casein matrix and the fat. This structure is common to all types of cheese. The differences among various types of cheese are brought about by the influence of different manufacturing regimes on that structure. In addition to the characteristic differences arising from different procedures, adventitious variations may also occur within the same type of cheese. These reflect differences in the original milk or in the conditions during the manufacturing or maturing processes.

It is the casein within the cheese which is responsible for its solid nature. The primary structure is a three-dimensional cage whose sides consist of chains of casein molecules (49,50). This provides a structure of considerable inherent rigidity. The chains are not linear (51), but have an irregular, somewhat gnarled structure. This may be deformed elastically under the action of external forces and this elasticity modifies the rigidity of the cage.

During the clotting process these chains have been formed by the joining together of individual casein particles in the serum. This serum surrounds the fat globules so that each cage may be expected to encase at least one globule or cluster of globules (52). The distribution of sizes of these cells will be controlled by the distribution and size of the fat globules. For instance, when the milk is first homogenized, the cheese made from it will have a more uniform distribution of cells than cheese made from fresh milk (53). The complete cheese curd at this stage consists of an aggregate of these cells of casein plus fat and the whole is pervaded by the aqueous phase (52).

If a force is applied to such a structure, the deformation will be primarily controlled by the rigidity of the cage, modified by any elasticity in its structural members. In the absence of the fat and water this would behave simply as any other open-type structure and its deformation would be characterized by a single modulus of rigidity or elasticity. However, the deformation of any of the cells is limited by the fat within it. At very low temperatures the fat would be solid and this would only add to the rigidity. At the normal temperatures at which cheese is matured and used the fat has both solid and liquid constituents and has its own peculiar rheological properties. Any deformation of the casein matrix would also require the fat to deform. At the same time, the water between the fat and the casein acts as a lubricant. As a result, the rigidity of the fat is added to that of the casein in a complex manner and it is this which gives rise to the peculiar viscoelastic properties of the cheese.

Even this is not the whole picture. The final cheese is not merely a continuous aggregate of cells as just described. During manufacture the curd is cut into small pieces at least once to allow any excess serum to drain away. As the serum drains away, the casein matrix shrinks on to the fat globules, making a more compact whole. The granules so formed may then be further distorted, as in the cheddaring process, or may be allowed to take up a random distribution, as in a Cheshire cheese. The final cheese mass is an aggregation of these granules which forms a secondary structure having its own set of rheological properties. Even this may be further modified by subsequent processes such as milling, which gives rise to a tertiary structure, and by pressing, which distorts the whole (54).

This rudimentary account of the factors contributing to the rheological properties applies to any cheese. During the course of manufacture, and subsequently during ripening, the basic structure may be modified by mechanical or thermal treatment, or the casein itself may be acted upon by bacteria and any residual enzymes. These agencies may change the organization of the structure or they may cause contiguous fat globules to coalesce. Finally, water may be lost by evaporation from the surface.

Before discussing the effect of the structure of the cheese on its rheological properties it is appropriate to consider differences which may occur within a single cheese or between cheeses from the same batch which might be expected to be alike. A cheese which matures in contact with the air will develop a pronounced rind and may show considerable variation throughout its body. The surface layers lose moisture more rapidly than the inner portions. This results in a difference in composition, which may be reflected in the rheological properties. It may also affect the progress of the maturation process itself. Another source of variation in some cheeses, particularly some of the larger varieties with long maturation periods, is that they are turned at intervals: The top and bottom layers are subjected to alternate low and high compressive forces due to their own weight, while the middle will have a more uniform treatment. This again may be reflected in the nature of the maturation. An otherwise uniform cheese will show a distribution of firmness as in Figure 11. A practical consequence of this is that measurements made on or near the surface of a large cheese will not be characteristic of the main body of that cheese.

Differences between cheeses may be expected to be greater and the magnitude of the variations will depend to some extent on the method of measurement. Penetration methods, where the instrument only acts locally on a limited quantity of the sample, give greater differences than compression testing in which a larger volume of the sample is involved. Variations of the order of 10% of the mean value of a parameter have been found on different samples from the same batch in a compression test (37); this probably indicates the limit of reproducibility which one may reasonably expect when making measurements on cheese. Differences between batches may be considerably larger. As an extreme example, one experimenter measuring eight different lots of cream cheese obtained readings which ranged from 54 to 251 units (55). Before leaving the topic of variation, one other source which must always be borne in mind is that between laboratories nominally making similar measurements. Because this is not a problem unique to either rheology or cheese, it will be disregarded in what follows.

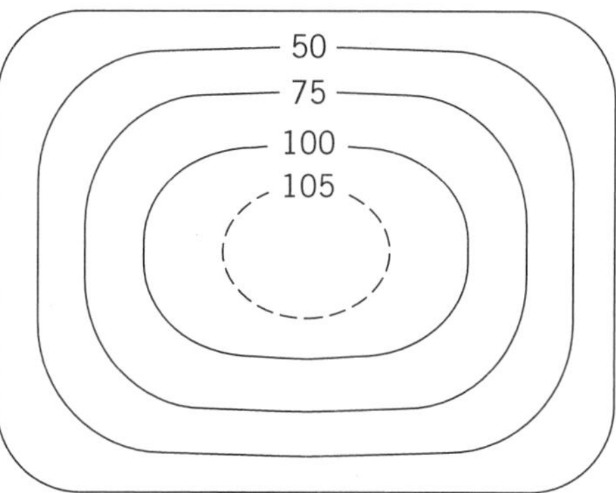

Figure 11. Typical distribution of firmness throughout a mature cheese.

Although it is the interaction of the properties of the principal constituents which gives cheese its viscoelasticity, it is profitable to consider some of the features associated with each of the constituents. First, consider the casein, which, it has been pointed out, gives the cheese its solid appearance. Because the casein forms chains within the spaces around the fat globules, there must necessarily be a minimum amount of casein below which any continuous structure cannot exist. This will depend on the number, size, and size distribution of the fat globules and on the size and size distribution of the casein micelles themselves. Once the quantity of casein required for this minimum structure has been exceeded, any additional casein will serve to strengthen the branches and the junctions. It is to be expected that, irrespective of the type of cheese, there will be a general relationship between the amount of casein present in the cheese and its firmness. Figure 12 shows this relation for some ten different types of hard cheese (29). Naturally, since the data refer to cheeses of very different provenance, there is considerable scatter, but it is possible to draw a regression line through the points and this indicates that about 25% of the weight of the cheese must be casein in order to provide a rigid framework and that above this limit more protein only strengthens it.

The requirement that there be sufficient casein to build a structure around the fat has been clearly shown in mea-

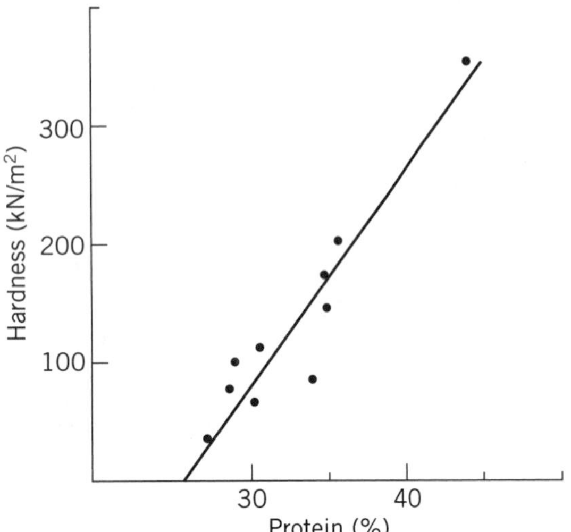

Figure 12. Relation between firmness of cheeses and total protein content.

surements on two other varieties of cheese. In some Meshanger cheese (56) it was found that, unless the casein occupied more than about 37% of the volume of the cheese not occupied by the fat, it had virtually no rigidity and behaved like a soft paste. In a similar way, it has been shown quite dramatically (Fig. 13) that in Mozzarella cheese (57) the protein must exceed about 42% of the nonaqueous part of the cheese for any rigidity to exist. Again, this corresponds to a minimum casein content of about 25% for a rigid structure.

Although it has been claimed that the casein matrix gives rigidity to the cheese, there is still a theoretical question which has not been answered. Is the matrix continuous throughout the whole cheese, so that it may be treated as a solid body, or do the dislocations in the structure which have been introduced by cutting allow it to flow, albeit imperceptibly, however small the stress? The evidence is inconclusive. In theory, an examination of the stress–strain curves at very low strains should decide the issue. If there is a continuous structure, there may be elastic deformation but no flow until a finite stress is reached which is sufficient to rupture some of that structure. Force–compression curves should indicate whether the cheese is of the Maxwell body or the Voigt body type. Commercially available instruments are generally not precise enough in this region, for they are not designed for this purpose. More precise measurements on cheese analogs have already been mentioned as showing that these began to break down at very small strains (42). Measurements of the recovery after compression of a few real cheeses (58) showed that even the smallest strains applied broke down some structure in a Mozzarella cheese, but a Cheddar cheese and a Muenster were able to recover completely (Fig. 14).

Cheddar cheese has also been studied by means of a relaxation experiment (44). Samples were compressed under different stresses, and hence at different rates, to the same ultimate deformation and the subsequent relaxation of the stress was observed. Using Peleg's treatment (43) it was found that the value of $1/k_2$ in equation 20 was dependent on the stress applied. The greater the stress, the more the structure broke down, although only the same strain was reached; the relation was curvilinear, particularly at the lowest stresses (Fig. 15). Extrapolating this curve back to zero stress should indicate whether any breakdown occurred. A zero value of $1/k_2$ would indicate a solid structure. Unfortunately, any extrapolation of this curve would be speculative. It is possible that it would lead to a positive intercept on the vertical axis. If this were so, it would lead to the conclusion that there was no continuous structure throughout that cheese, but to establish this with confidence would require measurements at much lower stresses.

Another series of experiments which could throw light on this question has been carried out (59). A small sinusoidal vibration was applied to one surface of the cheese sample and the stress transmitted through it observed. If

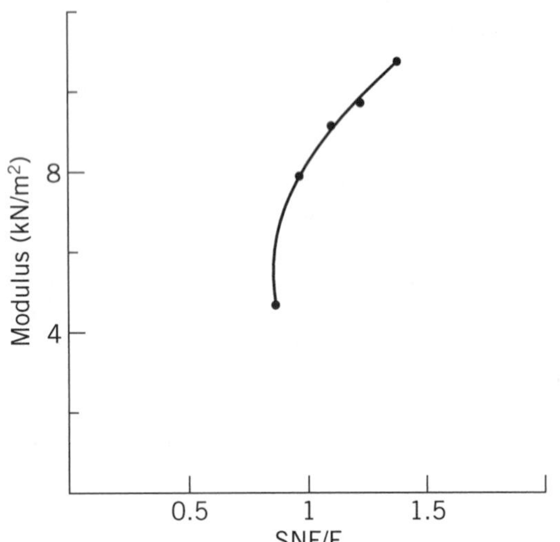

Figure 13. Relation between firmness and protein content in a single cheese. SNF/F = ratio of nonfat to fat solids.

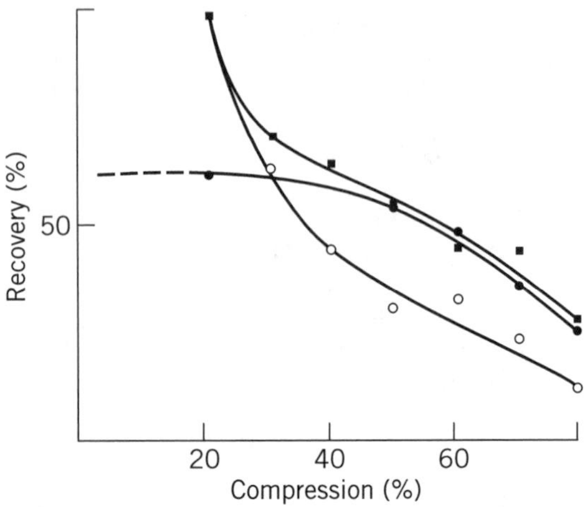

Figure 14. Relation between recovery of cheese and previous compression: ○, Cheddar; ●, Mozzarella; and ■, Muenster.

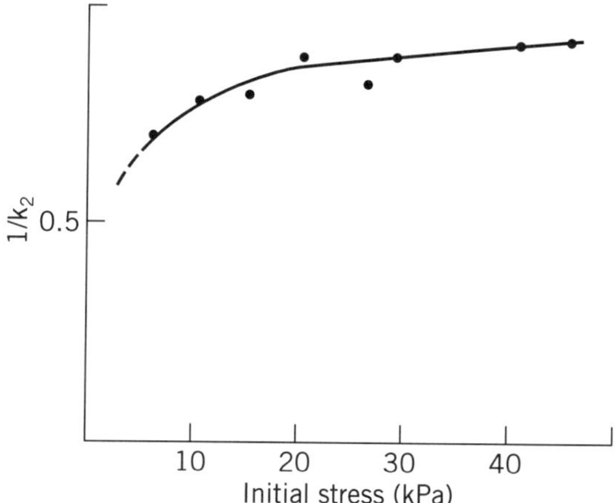

Figure 15. Amount of relaxation after previous stress.

one makes the assumption that the cheese behaves as a Voigt body, a modulus of rigidity can be calculated from the inphase component of the signal received and a viscosity from the out-of-phase component. For a true Voigt body both of these should be independent of the frequency. In fact, neither was constant with either Cheddar or Gouda cheese, even when the strain was as low as 0.04. The inference is that even at this strain some failure in the rigid structure had taken place. Either internal cracks or slip planes had developed within the cheese (42). However, everyday experience suggests that cheese should have some rigid structure. The stress on the lower layers of a large cheese, such as a Parmigiano or a Cheddar, is of the order of 3 to 4 kPa; yet such a cheese retains its shape more or less indefinitely.

Summarizing the previous paragraphs, it appears that the rheological role of the casein is to provide a continuous elastic framework within the individual granules. It is likely that where the casein chains lie in the granule surface and are contiguous with chains in neighboring granules they may be held together by physicochemical bonds which develop during maturation and give rise to some rigidity in the aggregation (54). However, these bonds will be sparser than those within the original network, since they form only where chance contacts occur. This results in a weaker secondary structure. These bonds may be broken when a positive strain is applied, giving a Maxwell body type of response, but are strong enough to preserve rigidity under the cheese's own weight. Some further support is given to this view by the fact that individual granules have a much higher modulus than the whole cheese (60).

The role of the other major constituents is more clearly documented than is that of the casein. The fat derived from the original milk is very roughly one third of the total mass. Its rheological properties are very sensitive to temperature changes and, as is to be expected, this sensitivity is imparted to the whole cheese. At 5°C (around normal refrigeration temperatures) many of the glycerides in the milk fat are solid. The proportion of solid fat decreases with rise of temperature, particularly sharply in the 12–15°C region, which is close to the ripening temperature for many varieties. Above this region the proportion of solid decreases further until at around 35°C almost all is liquid. This is near the temperature that a small portion will rapidly attain if it is placed in the mouth and chewed. The ratio of the solid to liquid fat components is the principal factor influencing the rheological properties of the fat (61). Although factors such as breed and species of animal, pasture, and herd management have some effect on this ratio, temperature is by far the most important. Fat also supercools readily, so that the thermal history is almost as important as the absolute value of the temperature itself. During maturation and storage cheese is usually kept at a lower temperature than that at which it was made. Originally most of the glycerides will have been liquid and at the lower temperature these will slowly solidify. Most of the change will take place in the first day or two, but solidification will continue progressively, maybe for several weeks before final equilibrium is reached.

Measurements made on Cheddar, Cheshire (39), Emmentaler (62), Gouda (63), and Russian (64) cheeses at a range of temperatures, using different instruments, all show the same general trend (Fig. 16). It may be mentioned that different instrumental methods give rise to somewhat different results. In general, penetrometers appear to indicate a rather steeper dependence of firmness on temperature than compression tests. Nevertheless, bearing in mind the difference in origin of the cheeses and the limitations in accuracy of the individual measurements, already discussed, this curve shows clearly the influence of the (test) temperature on the firmness of the cheeses.

Differences in the glyceride composition of the original milk fat are also reflected in the final cheese. Using the iodine number as an indication of the ratio of solid to liquid fat at any one temperature, the firmness of Swiss cheese was shown to vary seasonally with the solidity of the fat (65) (Fig. 17). The lower the iodine number, the greater the degree of saturation of the glycerides and the higher the

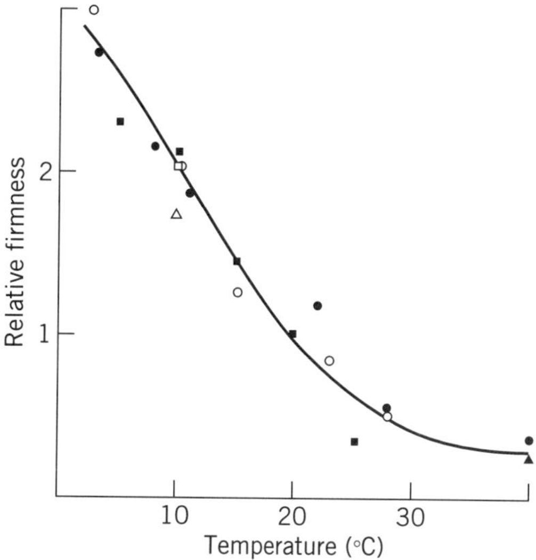

Figure 16. Variation of firmness with temperature: ●, Cheddar; ○, Cheshire; ■, Emmental; □, Gouda; and ▲, Russian.

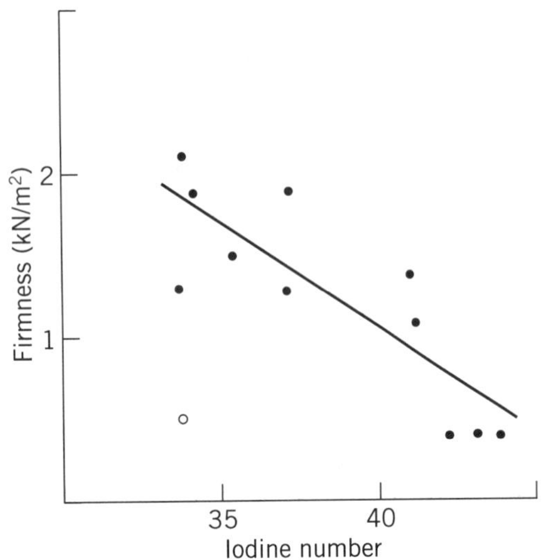

Figure 17. Variation of firmness with saturation of glycerides.

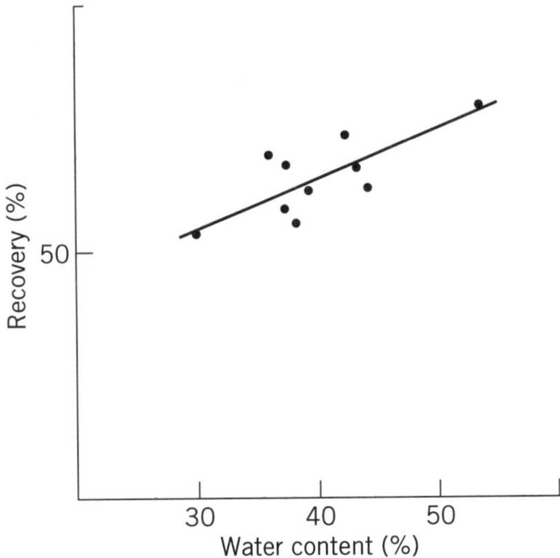

Figure 18. Relation between elastic recovery and moisture content.

proportion of solid fat. In these particular experiments the cheese produced in December appeared to be anomalous. However, even including the December result, the relation was statistically significant: excluding it, it was highly so and this regression has been drawn on the figure.

The water serves two functions, one purely physical, in that it is a low viscosity liquid, the other in its action as a carrier for the substances dissolved in it which may undergo chemical reactions with the fat or casein. It has already been stated that the water occupies the space not taken up by the casein matrix and the fat. Provided that there is a sufficient quantity, it takes up all of that space. Only near the outer surface, for instance, in the rind, is there likely to be any deficiency. In the case of Swiss-type cheeses, which contain eyes filled with gas, the main body of the cheese is still replete with water. The rheological consequence of this arrangement is that the water acts as a low-viscosity lubricant between the surfaces of the fat and the casein. The greater the space between these, which is another way of saying the higher the water content of the cheese, the easier the flow of water within these spaces. Hence there should be less restraint on the movement of the casein mesh around the enclosed fat. This freer movement should manifest itself as a lesser resistance to the deformation of the whole cheese and as a greater recovery after deformation.

Taking the latter point first, a study was made of the elastic recovery of samples of some ten varieties of cheese after they had been compressed (29), but only up to a point at which no visible damage to the structure had occurred (Fig. 18). The cheese with the lowest water content, Parmigiano, showed the least recovery, while a Mozzarella with over 50% water showed the most. From the figure it is seen that while there was considerable scatter about the regression line, as was to be expected with cheeses of different origins, a general trend was well established.

The lubrication effect can be more clearly demonstrated with a single variety and within a single cheese. In a freshly made cheese the water is distributed more or less evenly throughout the whole mass. During the maturation period of those cheeses which are allowed to ripen in air, water evaporates from the surface and is only slowly replaced by more water migrating from the interior. A moisture gradient is set up within the cheese. Measurements made on Emmentaler cheese (62) (Fig. 19) showed a very close relation between the water content of successive layers and the firmness as determined by a penetrometer. Similar results have been obtained with Edam cheese (66). In that series of experiments a close relation was observed between the water content of the cheeses and the firmness, although they were made in two batches at different temperatures (Fig. 20).

In its other role as carrier of active ingredients the presence of water is principally seen in the effect on the

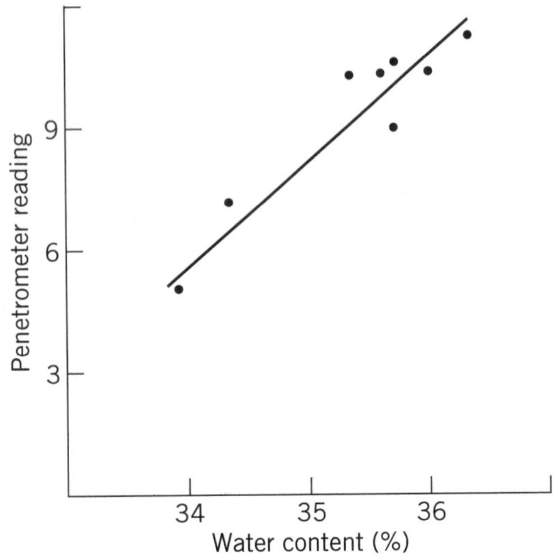

Figure 19. Relation between firmness and moisture content in a single cheese.

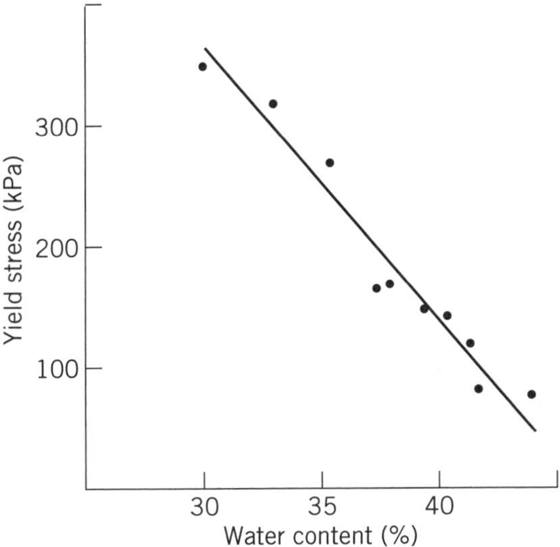

Figure 20. Relation between quasi-modulus and moisture content in a single cheese variety.

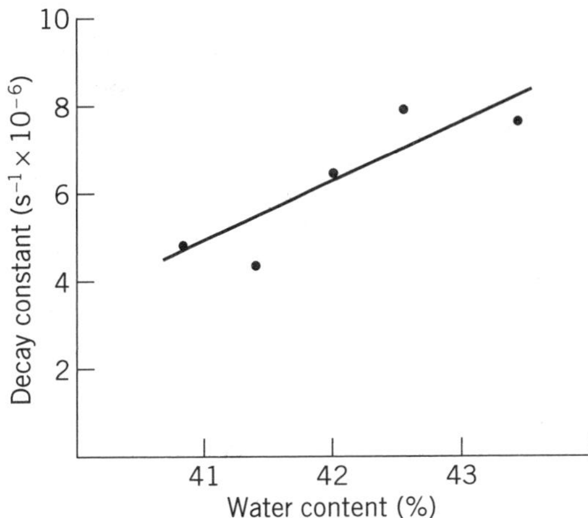

Figure 22. Relation between rate of casein breakdown and moisture content.

changes which take place in the casein network. An illustration of this is found in Meshanger cheese (56). This cheese has the usual well-organized casein structure when freshly made. As the ripening proceeds, the structure breaks down until it becomes a more or less amorphous loose mass of casein submicelles. Figure 21 shows the relation found between an arbitrary measurement of the firmness and the water content of cheeses from some ten batches during ripening. In the same series of experiments the rate of degradation of the casein was determined for some of the batches over the first few weeks of life. The rate of degradation, calculated as a first-order reaction, showed good agreement with the water content (Fig. 22).

Although protein, fat, and water constitute by far the greatest part of any cheese mass, other constituents cannot be ignored. Salt, when added dry before the cheese is pressed, may sometimes appear as crystals embedded in the fat–protein mass. Then it is not usually present in large enough quantities to make any sensible contribution to the rheological properties of the whole. Almost all of the salt is present only in solution in the water. There its effect on the properties of the whole cheese is minimal. Any serious contribution of the salt to the rheological properties of the whole cheese is by indirect action (67). A high concentration of salt increases the osmotic pressure, diverting a significant quantity of water from the structural bonds of the casein network (68,69). This may be seen in the effect on the ripening of some samples of Mozzarella cheese (70). When the salt content was low (0.27%) the modulus decreased from 120 to 45 N/m^2 in five weeks, but when salt was present in excess of 1% no change was observed within the same period.

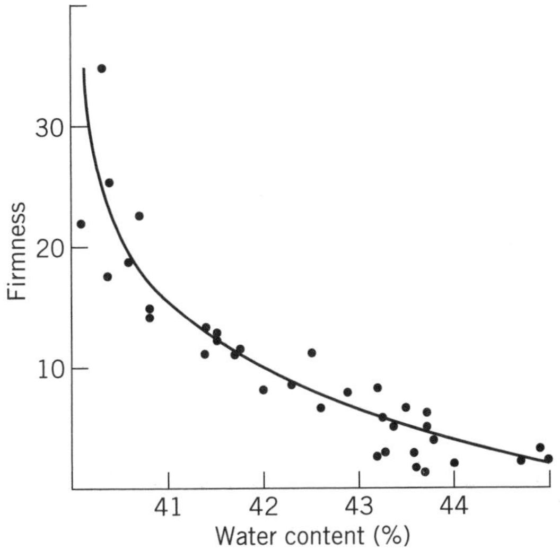

Figure 21. Relation between firmness and moisture content.

In cheese where the salt is added by immersion in brine there is a further consideration. The inward diffusion of the brine results in a concentration gradient being set up within the cheese. This has a twofold effect. First, the presence of the salt simply excludes some of the water; this in turn reinforces the moisture gradient simultaneously arising from the evaporation of water from the surface layers. Probably more important is the fact that the diffusion of the salt into the cheese is a slow process (71). In the early stages of ripening the salt concentration in the inner regions is unlikely to be sufficient to limit the protein degradation described above. The presence of more water and an enhanced proteolytic activity affect the rheological properties in the same sense, giving rise to a weaker structure and therefore a less firm cheese. It is therefore probable that the variation in firmness within a single cheese, which, as described earlier, is closely related to the moisture distribution, is accentuated by bringing.

Some cheeses also contain a significant quantity of gas. In the Swiss-type cheeses this is concentrated in the eyes; the remainder of the cheese has a close waxy texture and is virtually free of any entrapped gas. In hard and crumbly cheeses the gas may simply be air which has infiltrated into cracks which have developed in the mass. The significance of either eyes or cracks is that measurements made on the whole cheese or on its surface are not a fair representation of the properties of the main body of the cheese. Hidden cracks or eyes will give rise to irregularities when measurements are made by any form of penetrometer and probably account for a substantial amount of the observed scatter. In compression measurements they will give rise to premature breakdown. They also account for the fact that difficulties are encountered when any attempt is made to assess the rheological properties of cheese by studying the propagation of ultrasound through it. The velocity of propagation is more influenced by the scattering at the discontinuities than by the properties of the cheese mass.

SOME EXPERIMENTAL RESULTS

Most of the rheological methods described earlier have been used at one time or another for the study of cheese. Problems associated with penetrometers and tests on the surface have already been indicated. The most satisfactory tests are made on samples cut out specifically from the cheese mass. In this case the sample can be inspected to insure that there are at least no major cracks or other inhomogeneities in it and the dimensions can be precisely determined. The force–compression test using small prepared samples has been widely used for routine measurements. However, in spite of its widespread use, there has been no consensus of opinion on the most suitable operating conditions. Sample size and shape, rate of compression, and even temperature have all varied. A most interesting feature is that, notwithstanding these differences and the number of cheeses which have been examined, the force–compression curves show a remarkably common pattern. This underlines its general usefulness.

There are several important considerations to be taken into account when deciding on the operating conditions. In the first place, the sample should be large enough to be representative of the whole (72); the more heterogeneous the cheese, the larger the sample. On the other hand, it should not be so large that it may contain hidden cracks or irregularities. These two requirements are almost mutually exclusive. As a compromise most workers have taken samples with linear dimensions of 10 to 25 mm. The preferred shape has usually been a cylinder. It is easier to prepare a rectangular sample with precise dimensions but the symmetry of the cylindrical shape helps to minimize the development of irregular cracks during compression (63). A wide range of compression rates has been used: from 2 mm/min up to 100 mm/min (3.33×10^{-5} to 1.66×10^{-3} m/s). The faster rates tend to obscure the true behavior at the onset of compression unless a recorder of very short response time is available. The strain rate at any instant depends on both the rate of compression and the height of the sample at that instant.

Table 1 summarizes the principal measurements that have been reported on hard cheeses using the force–compression test. In order to produce this table and make the results from different workers comparable, the original curves from the references cited have been transformed into stress—strain curves. As previously mentioned, the instrument in its usual commercially available form produces curves of force versus linear travel. While linear travel is easily converted to true strain (eq. 11), the stress correction requires a knowledge of the actual shape at any instant during compression. Often this is unknown, or at least unreported. Unless special precautions have been taken, the behavior of the sample will be intermediate between that obtaining in the lubricated and in the bonded situations. For the present purposes the simpler correction applying to the lubricated condition has been used throughout, although it is realized that it will not quite compensate for the barreling which occurs. The values were then read from the transformed curves. The modulus given in the column headed E_0 is from the slope at the origin, ie, it refers to the undeformed sample. The entries in the columns headed yield refer to the stress and strain at the peak of the curve.

Table 1. Principal Measurements of Hard Cheeses Using the Force–Compression Test

Cheese	Modulus, E_c (kN/m^2)	Yield point σ_y (kPa)	Yield point ϵ_y	80% Stress, σ_{30} (kPa)	Ref.
American loaf	70	25	0.72	46	58
Bel Paese	300	44	1.1		37
Caerphilly	1100	92	0.16	67	36
Cheddar	280	41	0.34	50	22
Cheddar		53	0.25		39
Cheddar		63	0.24		47
Cheddar (mild)	160	40	0.60	78	58
Cheddar (strong)	400	58	0.22	65	58
Cheddar	320	45	0.21	64	73
Cheddar (green)	75	54	0.80	37	47
Cheddar (mature)	170	23	0.20	22	47
Double Gloucester	1600+	139	0.24	75	36
Edam	660	214	0.60		36
German loaf	300	18	ca 1.1		33
Gouda	240	98	0.72	52	36
Lancashire	2000+	105	0.23		36
Leicester	400	56	0.31	38	32
Montasio	40	85	0.80		37
Mozzarella	300	(no peak)			34
Mozzarella	80	14	0.57		58
Muenster	40	16	0.72	59	58
Parmigiano	700	132	0.28		37
Provolone	120	37	0.60		37
Silano	230	26	ca 0.30		57
Processed	195	57	0.39	30	36
Processed	440	71	0.27		37
Processed	60	9	0.21		37

Table 2 summarizes a number of results which have been obtained by other methods where it has been possible to calculate a modulus from the data. In order to compare the two tables, it will be recalled that Table 1 lists the values of an elastic modulus. It must be most nearly comparable with any quasi-modulus determined by the puncture test or by simple compression. The yield stress measured by a cone penetrometer is more likely to be comparable with the yield value in the force–compression

Table 2. Modulus Measurements Using Various Methods

Cheese	Modulus (kN/m^2)	Method	Ref.
Brie	1.3	Extruder	74
Sbrinz	41	Extruder	74
Chanakh	58	Cone	69
Emmental	40	Cone	69
Emmental	3.5	Cone	22
Gouda	44	Cone	69
Kostroma	54	Cone	69
Lori	40	Cone	69
Cheddar	195	Punch	29
Edam	340	Punch	29
Emmental	220	Punch	29
Gouda	270	Punch	29
Kachkaval	120	Punch	75
Mozzarella	170	Punch	29
Mozzarella	22	Punch	74
Muenster	120	Punch	29
Parmigiano	550	Punch	29
Provolone	130	Punch	29
Cheddar	270–400	Compression	76
Mozzarella	80	Compression	70
Kachkaval	60–100	Ball Compressor	77

test, since both are influenced by partially broken-down structure. It must be pointed out, however, that lack of adequate documentation makes it difficult to make meaningful comparisons between the results from different laboratories.

VARIATION WITH AGE

The development of firmness during the ripening and subsequent storage of cheese has always been of particular interest to both the practical cheese-maker and the rheologist. During this period all three principal constituents undergo change. Moisture evaporates from the surface of the cheese if this is exposed to the air; additional moisture migrates from the interior to replace that lost and some of this also evaporates. Eventually a moisture gradient is set up, but even the center will be drier than it was originally. The protein in the matrix undergoes progressive change as the available water is reduced and the residual enzymes and bacteria continue to act on it (46,47,75). Furthermore, on the purely physical plane, as the water content decreases, the matrix may shrink or collapse under pressure so that voids do not form and the protein becomes denser. Some of the glycerides in the fat slowly crystallize, resulting in a more solid mass of fat.

These changes which take place in the structural components of the cheese are reflected in force–compression curves. Figure 23 shows curves for two cheeses, Cheddar (47) and Gouda (63), when green and when mature. During aging the strain sustainable by the cheese before breaking down at the yield point (point B in Fig. 7) decreased more or less exponentially with time in both cases. Figure 24 shows this for one batch of Cheddar cheese. At the same time the stress at the yield point diminished. The strength of the matrix had clearly been reduced through aging. In

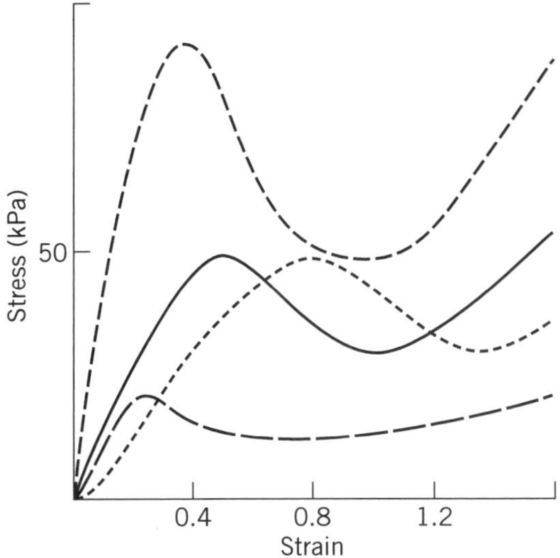

Figure 23. Force–compression curves for young and mature cheeses: — —, Cheddar (mature); - - -, Cheddar (young); —, Gouda (young); and – –, Gouda (mature).

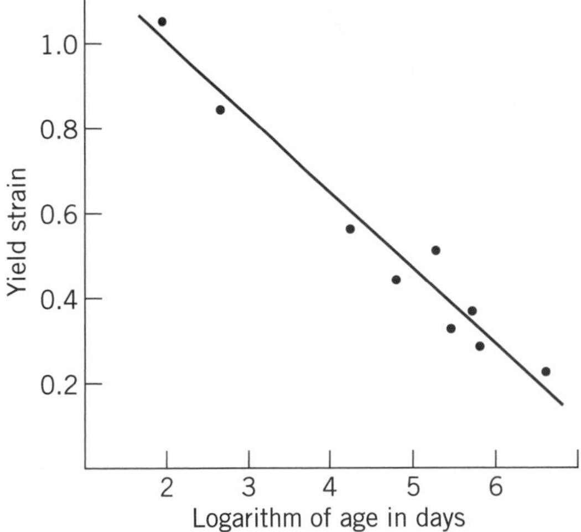

Figure 24. Relation between age and strain at yield point for Cheddar cheese.

the Gouda cheese the behavior was rather different. Although the yield strain decreased, the stress at this point, ie, the ultimate strength of the matrix, had increased. It is only to be expected that as the actual balance between the different mechanisms of aging of the components varies from one cheese to another, so the paths of change in the whole cheese will vary. Figure 25 shows some of the results that have been reported (47,78,79).

Not only does the firmness of cheese change with age, but also its springiness. The degradation of the protein, the solidification of the fat, and the reduction of the water available to lubricate the relative motion all tend to reduce the springiness. Springiness is not a simple rheological concept. It implies not only the ability to recover from compression or indentation but also that this will be immedi-

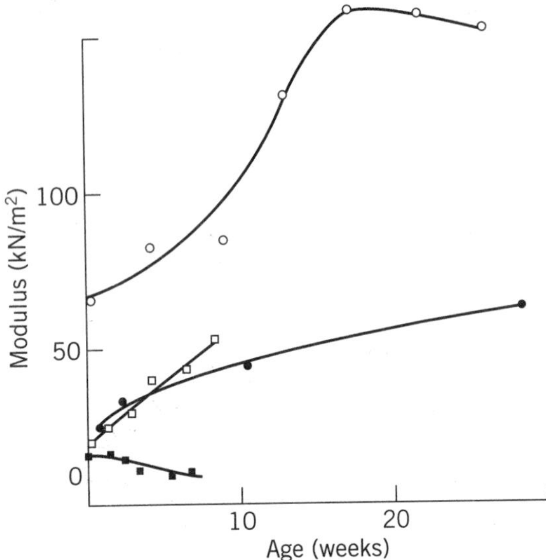

Figure 25. Variation of firmness with age: ●, Cheddar; ○, Kachkaval; □, Tamismi and Russian; and ■, unspecified French cheese.

ately evident. It does not show itself directly in a force–compression curve; a recovery curve is more appropriate for this purpose. Nevertheless, this decrease of springiness has been remarked upon consistently using subjective observations on a number of cheeses (27,29,47,49,80).

SUBJECTIVE ASSESSMENTS

How the rheological properties measured by instruments, as just described, relate to the grader's or the consumer's assessments of the cheese remains to be considered. The emphasis in this article has been on the measurement of the firmness of cheese which arises from its structure. It should be stressed that meaningful comparisons can only be made between instrumental and subjective methods if the subjective terminology is unambiguous and the instrumental measurements precise. Early experiments on single varieties of cheese showed that the subjective assessment of firmness, which is a fairly simple concept, correlates very highly with measurements made with simple apparatus, such as the Ball Compressor (81). But a simple correlation by itself does not indicate to what extent the instrumental measurement should be used to predict a typical user's appraisal. Nor can the results of experiments on a single variety be extrapolated to other types of cheese without verification. Some authors have arbitrarily assigned a specific consumer quality to a particular instrumental reading. For instance, the quasi-modulus calculated from a simple compression test (27), the yield value given by a cone penetrometer (22), and the stress at 80% compression have all been proposed as measures of firmness. The recovery after a limited compression (29,82), arbitrarily decided upon, has been used as an indication of springiness. These intuitive opinions may help visualize the significance of an instrumental measurement and are probably adequate for internal comparisons within a single investigation, but they lack scientific rigor.

The force–compression test, giving as it does a characteristic curve with several readily identifiable features, provides a number of potentially useful parameters. In Figure 26 the values of hardness as assessed by a panel, of a number of different cheeses, versus the stress measured at 80% compression are plotted (73). A logarithmic regression line has been drawn through the points. The standard deviation about this regression line was no greater than that due to the differences among the panel. In this case, then, using the regression line, the instrumental reading can be safely used to predict that panel's assessments. In the same series of experiments the instrumental measurements were carried further. As soon as the compression had reached 80% the compressing plates were withdrawn and the recovered height of the sample was measured. At the same time the springiness was assessed subjectively. As has already been pointed out, springiness involves not only the extent of recovery but also its speed; possibly all individuals will have, subconsciously, their own ideas of the relative contribution of the two to the final judgment. As was to be expected, there was poorer agreement among the panel members in this part of the experiment. Even so, there was a highly significant correlation between the measured recovery and the average panel assessment of springiness.

The relation between user acceptance and instrumental rheological measurement is an ongoing study (82). The few examples cited above show that, if the proper parameter is chosen and the terms clearly understood, rheological measurements can successfully predict users' assessments.

CONCLUSION

Although the study of cheese rheology is as old as rheology itself, only a few cheese varieties have been investigated. These have been drawn almost exclusively from among the

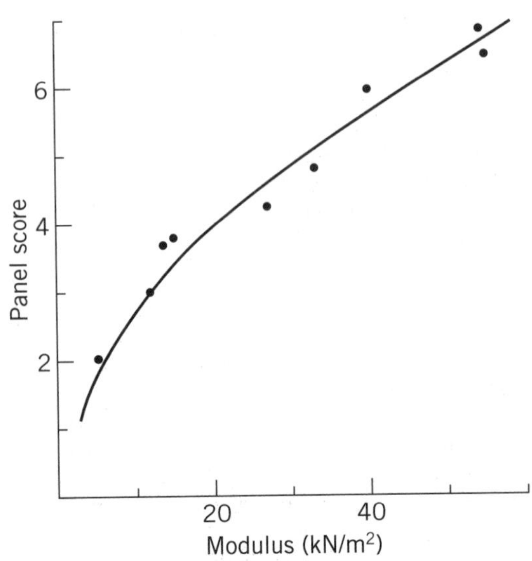

Figure 26. Comparison of subjective and objective measurements of firmness.

firmer varieties. The softer cheeses, of which there are many, have received scant attention. They pose rather different rheological problems because they are less like solids and more akin to stiff pastes or creams; they should, however, be nonetheless interesting to rheologists.

It has been shown, by bringing together diverse observations on the firmer cheeses from a number of workers, that it is possible to build up plausible theories to relate the behavior of the finished cheese to the structure developed during its manufacture. Mathematical models may be posited for the rheological behavior, but with a material as heterogeneous as cheese, these, if they are to be simple enough to be easily handled, are unlikely to give more than an approximate representation of the observed pattern. Using these, or directly from experimental measurements, material constants may be extracted to characterize the cheese. A knowledge of these may be helpful to the cheesemaker, both as a means of controlling the production regime and as a guide to the acceptability of the product.

DISCLOSURE STATEMENT

This article previously appeared in *"Cheese: Chemistry, Physics and Microbiology,"* R. F. Fox, ed., Elsevier Applied Science, New York, 1987. It is printed here with some revision by permission of the publishers.

BIBLIOGRAPHY

1. M. Reiner, *Twelve Lectures in Theoretical Rheology*, North Holland Publishing Co., Amsterdam, 1949, p. 19.
2. A. S. Szczniak, *Journal of Food Science* **28**, 385 (1963).
3. M. Baron, *The Mechanical Properties of Cheese and Butter*, United Trade Press, London, 1952, p. 29.
4. R. Jowitt, *Journal of Texture Studies* **5**, 351 (1974).
5. R. Houwink, *Elasticity, Plasticity and Structure of Matter*, Dover Publications, New York, 1958, p. 5.
6. M. Peleg, *Journal of Texture Studies* **15**, 317 (1984).
7. J. H. Prentice, *Measurements in the Rheology of Foodstuffs*, Elsevier Applied Science Publishers, London, 1984, p. 32.
8. R. B. Bird, *Annual Review of Fluid Mechanics* **8**, 13 (1976).
9. A. S. Szczniak and M. C. Bourne, *Journal of Texture Studies* 1, 52 (1969).
10. J. E. Caffyn, *Dairy Industries* **10**, 257 (1946).
11. J. E. Caffyn and M. Baron, *The Dairyman* **64**, 345 (1947).
12. N. N. Mohsenin, *Physical Properties of Plant and Animal Materials*, Gordon & Breach Scientific Publishers, New York, 1970, p. 288.
13. N. N. Mohsenin, C. T. Morrow, and Y. M. Young, *Proceedings of the 5th International Congress on Rheology*, Vol. 2, 1970, p. 647.
14. C. P. Cox and M. Baron, *Journal of Dairy Research* **22**, 386 (1955).
15. A. D. Konoplev, P. E. Krashenin, and V. P. Tabachnikov, *Trudy Vsesoyuznogo Nauchno-Issledovatel'skogo Instituta masl. i. syr. Prom.* **17**, 40 (1947).
16. S. Poulard, J. Roucou, G. Durrange, and J. Manry, *Rheologica Acta* **13**, 761 (1974).
17. M. Baron, *Dairy Industries* **14**, 146 (1954).
18. P. Eberhard and E. Flückiger, *Schweizerische Milchzeitung* **104**, 24 (1978).
19. V. P. Tabachnikov, V. Ya. Borkov, V. B. Ilyushkin, and I. I. Tetereva, *Trudy Vsesoyuznogo Nauchno-issledovalel'skogo Instituta masl. i. syr. Prom.* **27**, 61 (1979).
20. R. Harper and M. Baron, *British Journal of Applied Physics* **2**, 35 (1951).
21. W. G. Wearmouth, *Dairy Industries* **19**, 213 (1954).
22. E. Flückiger and E. Siegenthaler, *Schweizerische Milchzeitung Wissenschaftliche Beilage* **89**, 707 (1963).
23. F. J. Mottram, *Laboratory Practice* **10**, 767 (1961).
24. M. C. Bourne, in P. Sherman, ed., *Food Texture and Rheology*, Academic Press, London, 1979, p. 95.
25. J. M. deMan, *Journal of Texture Studies* **1**, 114 (1969).
26. B. S. Kamel and J. M. deMan, *Lebensmittel-Wissenschaft und Technologie* **8**, 123 (1975).
27. R. Davidov and N. Barabanshchikov, *Molochnaya Promyshlennost* **9**, 27 (1950).
28. V. P. Tabachnikov, *Trudy Vsesoyuznogo Nauchno-issledovatel'skogo Instituta masl. i. syr. Prom.* **17**, 84 (1974).
29. A. H. Chen, J. W. Larkin, C. J. Clarke, and W. E. Irvine, *Journal of Dairy Science* **62**, 901 (1979).
30. R. Ramanauskas, S. Urbene, L. Galginaitye, and P. Matsulis, *Trudy Litovskii Filial Vsesoyuznogo Nauchno-Issledovatel'skogo Instituta Maslodel'noi Syrodel'noi Promyshlennosti* **13**, 64 (1979).
31. Ref. 7, p. 9.
32. E. J. V. Carter and P. Sherman, *Journal of Texture Studies* **9**, 311 (1978).
33. P. Sherman and G. Atkin, *Advances in Rheology* **4**, 133 (1976).
34. E. M. Casirighi, E. B. Bagley, and D. D. Christianson, *Journal of Texture Studies* **16**, 281 (1985).
35. P. W. Voisey and M. Kloek, *Journal of Texture Studies* **6**, 489 (1975).
36. P. Shama and P. Sherman, *Journal of Texture Studies* **4**, 344 (1973).
37. P. Masi, in *Physical Properties of Foods*, Vol. 2, Elsevier Applied Science Publishers, London, 1987, p. 383.
38. M. M. Bryan and D. Kilcast, *Journal of Texture Studies* **17**, 221 (1986).
39. E. Dickinson and I. C. Goulding, *Journal of Texture Studies* **11**, 51 (1980).
40. N. V. Polak, *Dissertation Abstracts International B* **42**, 3178 (1982).
41. R. Jowitt, in P. Sherman, ed., *Food Texture and Rheology*, Academic Press, London, 1979, p. 146.
42. R. J. Marshall, *Journal of the Science of Food and Agriculture* **50**, 237 (1990).
43. M. Peleg, *Journal of Food Science* **44**, 277 (1979).
44. M. Peleg, *Journal of Rheology* **24**, 451 (1980).
45. B. Launay and S. Buré, *DECHEMA Monographien* **77**, 137 (1974).
46. A. M. Kimber, B. E. Brooker, D. G. Hobbs, and J. H. Prentice, *Journal of Dairy Research* **41**, 389 (1974).
47. H. K. Creamer and N. P. Olson, *Journal of Food Science* **47**, 631 (1982).
48. M. V. Taranto, P. J. Wan, S. L. Chen, and K. C. Rhee, *Scanning Electron Microscopy*, 273 (1979).
49. M. Green, A. Turvey, and D. G. Hobbs, *Journal of Dairy Research* **48**, 343 (1981).
50. T. van Vliet and P. Walstra, *Netherlands Milk and Dairy Journal* **39**, 115 (1985).
51. J. H. Prentice, *Measurements in the Rheology of Foodstuffs*, Elsevier Applied Science Publishers, London, 1984, p. 157.

52. M. L. Green and A. I. Grandison, in P. F. Fox, ed., *Cheese: Chemistry, Physics and Microbiology*, Vol. 1, Elsevier Applied Science Publishers, London, 1987, p. 97.
53. D. B. Emmons, M. Kalab, E. Larmond, and K. J. Lorne, *Journal of Texture Studies* **11**, 15 (1980).
54. M. Kalab, *Milchwissenschaft* **32**, 449 (1977).
55. Z. D. Roundy and W. V. Price, *Journal of Dairy Science* **24**, 135 (1941).
56. L. de Jong, *Netherlands Milk and Dairy Journal* **32**, 1 (1978).
57. P. Masi and F. Addeo, *Advances in Rheology* **4**, 161 (1976).
58. E. M. Imoto, C-H. Lee, and C. K. Rha, *Journal of Food Science* **44**, 343 (1979).
59. S. Taneya, T. Izutsu, and T. Sone, in P. Sherman, ed., *Food Texture and Rheology*, Academic Press, London, 1979, p. 367.
60. E. P. Suchkovae, A. M. Maslov, and N. G. Alekseev, *Izvestiya Vysshikh Uchebnykh Zavedenii Pishchevaya Tekhnologiya* **4**, 113 (1985).
61. H. Jonsson and K. Anderson, *Milchwissenschaft* **31**, 593 (1976).
62. C. Steffen, *Schweizerische Milchwirtschaftliche Forschung* **5**, 43 (1976).
63. J. Culioli and P. Sherman, *Journal of Texture Studies* **7**, 353 (1976).
64. N. Ya. Dykalo and V. P. Tabachnikov, *Trudy Vsesoyoznogo Nauchno-Issledovatel'shoko Instituta masl. i. syr. Prom.* **77**, 85 (1979).
65. E. Flückiger, P. Walser, and H. Hanni, *Deutsche Molkerei Zeitung* **96**, 1524 (1975).
66. C. W. Raadsveld and H. Mulder, *Netherlands Milk and Dairy Journal* **3**, 117 (1949).
67. T. P. Guinee and P. Fox, in P. F. Fox, ed., *Cheese: Chemistry, Physics and Microbiology*, Elsevier Applied Science Publishers, London, 1987, p. 251.
68. R. Ramanauskas, *Proceedings of the 20th International Dairy Congress* 1978, p. E 265.
69. G. G. Khachatryan, K. Zh. Dilanyan, V. P. Tabachnikov, and I. I. Tetereva, *Proceedings of the 19th International Dairy Congress* 1974, 1E, p. 717.
70. M. A. Cervantes, D. B. Lund, and N. F. Olson, *Journal of Dairy Science* **66**, 204 (1983).
71. T. J. Guerts, P. Walstra, and H. Mulder, *Netherlands Milk and Dairy Journal* **34**, 229 (1980).
72. M. Peleg, *Journal of Food Science* **42**, 649 (1977).
73. C-H. Lee, E. M. Imoto, and C. K. Rha, *Journal of Food Science* **43**, 579 (1978).
74. C. S. T. Yang and M. V. Taranto, *Journal of Food Science* **42**, 906 (1982).
75. R. Stefanovic, *DECHEMA Monographien* **77**, 211 (1973).
76. J. C. Weaver and M. Kroger, *Journal of Food Science* **43**, 579 (1978).
77. G. Szabo, *Proceedings of the 19th International Dairy Congress* 1974, 1E, p. 505.
78. S. D. Sakharov, V. P. Tabachnikov, V. K. Nebert, and P. F. Krasheninin, *Trudy Vsesoyuznogo Nauchno-Issledovalel'skogo Instituta masl. i. syr. Prom.* **18**, 29 (1975).
79. M. Ostojic, D. Miocinovic, and G. Niketic, *Mljekarstvo* **32**, 139 (1982).
80. B. A. Nikolaev and R. M. Abdullina, *Trudy Vologogradskaya* moloch. Inst. **30**, 27 (1969).
81. N. Baron and R. Harper, *Dairy Industries* **15**, 407 (1950).
82. A. S. Szczesniak, *Journal of Texture Studies* **18**, 1 (1987).

GENERAL REFERENCES

G. F. Amantea, B. J. Skura and S. Nakai, "Culture Effects on Ripening Characteristics and Rheological Behavior of Cheddar Cheese," *J. Food Sci.* **51**, 912–918 (1986).

E. B. Bagley and D. D. Christianson, "Measurement and Interpretation of Rheological Properties of Foods," *Food Technol.* **41**, 96–99 (1987).

O. H. Campanella et al., "Elongational Viscosity Measurements of Melting Process Cheese," *J. Food Sci.* **52**, 1249–1251 (1987).

S. Cavella, S. Chemin, and P. Masi, "Objective Measurement of the Stretchability of Mozzarella Cheese," *J. Texture Stud.* **23**, 185–194 (1992).

Y. S. Chang et al., "Viscoelasticity of Cheese," *J. Chem. Educ.* **63**, 1077–1078 (1986).

N. Y. Farke and P. F. Fox, "Objective Indices of Cheese Ripening," *Trends Food Sci. Technol.* **1**, 37–40 (1990).

P. Fox, *Cheese Chemistry, Physics and Microbiology*, Vol. 1, Chapman and Hall, London, United Kingdom, 1994, pp. 303–340.

H. C. Goh and P. Sherman, "Influence of Surface Friction on the Stress Relaxation of Gouda Cheese," *J. Texture Stud.* **18**, 389–404 (1987).

T. Izutsu et al., "Rheological Properties of Mozzarella Cheese Curds," *Nippon Kagaku Kaishi* **1990**, 621–627 (1990).

F. R. Jack and A. Patterson, "Texture of Hard Cheeses," *Trends Food Sci. Technol.* **3**, 160–164 (1992).

J. Korolczuk, "Rheological Properties of Fresh Cheeses. I. Stress Evolution During Compression," *Milchwissenschaft* **50**, 674–678 (1995).

J. Koroluczuk and M. Mahaut, "Viscometric Studies on Acid Type Cheese Texture," *J. Texture Stud.* **20**, 169–178 (1989).

J. Korolczuk and M. Mahaut, "Relaxation Studies of Acid Type Cheese Texture by a Constant Speed Cone Penetrometric Method," *J. Texture Stud.* **21**, 107–122 (1990).

E. Kucukoner, "Effect of Various Commercial Fat Replacers on the Physicochemical Properties and Rheology of Low Fat Cheddar Cheese," *Dissertation Abstracts International B* **57**, 4113–4114 (1997).

R. J. Marshall, "Composition, Structure, Rheological Properties, and Sensory Texture of Processed Cheese Analogues," *J. Sci. Food Agric.* **50**, 237–252 (1990).

P. Masi, "Characteristics of History-Dependent Stress-Relaxation Behavior of Cheeses," *J. Texture Stud.* **19**, 373–388 (1989).

E. J. Nolan, V. H. Helsinger, and J. J. Shieh, "Dynamic Rheological Properties of Natural and Imitation Mozzarella Cheese," *J. Texture Stud.* **20**, 179–189 (1989).

A. A. Omar, A. A. Baky Abdel and S. A. Metwally, "Rheological Properties and Microstructure of Ripened Camembert Cheese," *Egyptian Journal of Dairy Science* **26**, 175–192 (1998).

M. Sutheerawattananonda and E. D. Bastian, "Monitoring Process Cheese Meltability Using Dynamic Stress Rheometry," *Journal of Texture Studies* **29**, 169–183 (1998).

M. H. Tunick et al., "Cheddar and Cheshire Cheese Rheology," *J. Dairy Sci.* **73**, 1671–1675 (1990).

M. H. Tunick et al., "Effects of Composition and Storage on the Texture of Mozarella Cheese," *Netherland Milk Dairy Journal* **45**, 117–125 (1991).

M. H. Tunick et al., "Rheology and Microstructure of Low-Fat Mozzarella Cheese," *International Dairy Journal* **3**, 649–662 (1993).

J. H. Prentice
Axminster
Devon, United Kingdom

CHERRIES. See FRUITS, TEMPERATE.

CHILLED FOODS

The chilled-food market has expanded since the advent of freezing technology in the 1940s. Although foods were sold chilled at that time, these were generally raw product such as fish and meats. The current needs of society for time-saving and easily prepared meals has led to a wider range of products being offered to the consumer, including many ethnic dishes. Much of this chilled product development has coincided with expansion in the technology of chilling, modified atmosphere packaging, and an understanding of the microbiological safety issues during storage. The addition of microbiological hurdles has allowed shelf-life extension to many products, which has helped to meet the needs for an economic distribution system. Most supermarkets now offer chilled prepared foods that are fully cooked and ready to eat. The development of a food service industry both in vending and restaurants has allowed large companies to quickly exploit the refrigerated product market and to make the most of such opportunities. This article summarizes the issues and developments in the area of chilled foods related to production, safety, and market trends from different perspectives.

MARKET FOR CHILLED FOODS

The global market for chilled foods was estimated at more than $7 billion in 1996. Although other regions such as Canada and Southeast Asia have significant chilled-food markets, the global market is dominated by sales in the United States ($3.4 billion), Europe ($2.0 billion), and Japan ($590 million) (1). The market share in these regions could be attributed to a modern lifestyle in which less time is set aside for cooking at home as well as a good infrastructure for chilled-food distribution. As the types of available products, distribution of the products, and trends in each of the three largest chilled-food markets differ significantly from the others, these markets will be addressed separately.

UNITED STATES

Available Products

The chilled-food market in the United States can be divided into seven main product categories: precut vegetables and salads, lunches and snacks, fully cooked poultry, pasta, pizza, sauces, and fully cooked entrees and dinners (1). A breakdown of the major product types in the United States chilled-food market can be found in Figure 1.

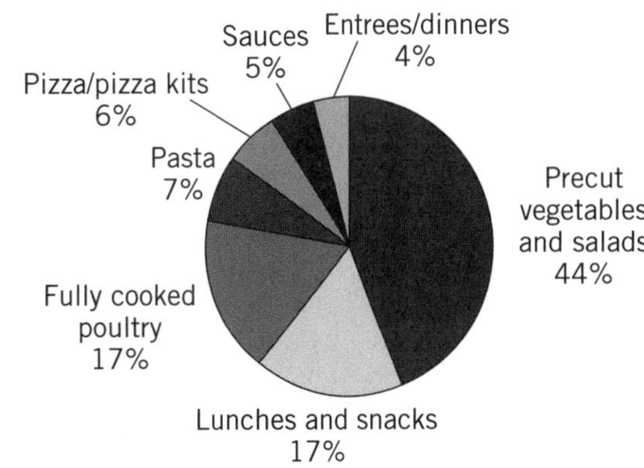

Figure 1. Available product types and market share of chilled foods in the United States, 1996.

Although many food companies may distribute precut vegetables and salads, these items are also prepared, packaged, and stored at the retail level in supermarkets or convenience stores. Prepared salads may also include dressings and/or seasonings, croutons, or breadsticks. Lunches and snacks include small kits for making a sandwich or snack for one person. These kits may include cold cuts, tuna or chicken salads, or cheese and crackers; some kits targeted specifically for children's lunches may include chips or dessert items and a beverage. Fully cooked poultry includes parts (legs, breasts, etc) as well as quarter, half, and whole birds. The majority of chilled poultry marketed in the United States is chicken, with turkey receiving an increasing market share. Most fully cooked, chilled poultry is roasted or fried and may be breaded, seasoned (barbecue, cajun, spicy hot, etc), or marinated. Chilled pasta is not typically fully cooked and usually requires some preparation. The main selling points to this particular category are freshness, variety, and reduced preparation time. Some chilled pasta products are filled with cheese or meat, some may be made with nonstandard flours (corn, bran, or pumpkin), and some may be flavored (1). Chilled pizza products fall into two main segments: kits and completely assembled uncooked or partially cooked pies (1). A pizza kit usually includes the uncooked crust, sauce, cheese, spices, and possibly other toppings. Completely assembled pizzas may be uncooked or partially cooked, but the consumer is usually required to finish the cooking. Chilled sauces tend to be upscale products positioned on gourmet recipes. Italian pasta sauces are the most popular of this category, followed by ethnic and traditional sauces. Examples may include alfredo, pesto, spicy marinara with clams, Thai with lemongrass, Mexican mole, barbecue, and seafood sauces. Fully cooked entrees and dinners are mostly cooked and meant to be heated and served. Examples may include lasagna, meatloaf, and fajitas as well as complete turkey dinners with side dishes for the holiday season (1).

Current and Future Market Trends

The retail market for chilled prepared foods in the United States advanced from $1.8 billion to $3.4 billion during the years 1992 to 1996, an increase of 87.6% (1). Most of the growth has been earned by precut vegetables/salads and lunches/snacks, with fully cooked poultry, pasta, and pizza earning small advances and chilled sauces and entrees/dinners declining slightly. The total chilled-food market is expected to increase to more than $5 billion by the year 2001; this is demonstrated in Figure 2.

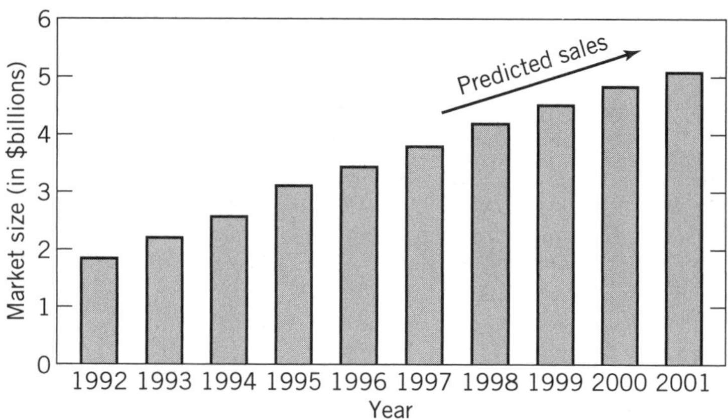

Figure 2. Actual and predicted market for chilled foods in the United States, 1992–2001.

Most growth in the U.S. chilled-food market is predicted to occur in the precut vegetables/salads and lunches/snacks areas. Other chilled categories are expected to remain stable or decline somewhat. Despite the large market growth between 1992 and 1996, most chilled foods are marketed as line extensions of successful frozen or shelf-stable company brands. Although food companies are aware of the consumer demand for fresh, convenient foods, most do not provide proper market support or distribution channels for chilled products. Refrigerated food marketing strategies in the past have included trying to capitalize on already successful frozen or shelf-stable brands rather than pushing the products on their own merits. Most consumers purchase chilled prepared foods because they are perceived as being fresher and more convenient than similar frozen or shelf-stable products. Future trends in the chilled-food area include achieving fresher products through improved distribution practices, improving the market by focusing on the convenience of chilled foods, and developing a greater variety of value-added products to offset premium price levels of chilled foods (1).

EUROPEAN AND JAPANESE MARKETS

Available Products

The European chilled-food market can be divided into seven main product categories: salad lunches, meat dishes, seafood dishes, pasta-based dishes, pizza and pizza toppings, ethnic foods, and other foods such as cheese and dairy products (1). The types of products in each of the preceding categories will vary greatly throughout Europe due to the breadth of cultural diversity. Nonseafood salad lunches account for 21% of the total chilled prepared-food sales, closely followed by meat dishes, seafood dishes, pasta-based dishes, and pizza and pizza toppings. Figure 3 illustrates the major product types and market share of each product category in the European chilled-food market.

The chilled-food market in Japan is comprised of three main product categories: Japanese-style foods, other Asian dishes, and Western dishes (1). Chilled foods in Japan tend to be marketed as snacks or luncheon replacements rather than main meal replacements.

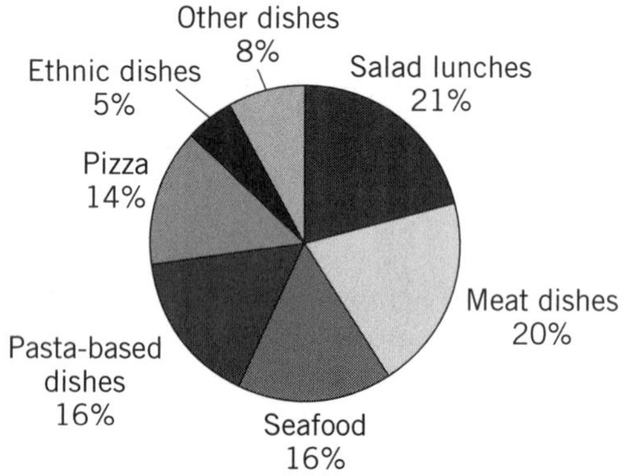

Figure 3. Available product types and market share of chilled foods in the European Community, 1996.

Current and Future Market Trends

Sales of chilled prepared foods in Europe increased from \$1.5 billion to more than \$2 billion during the period 1992 to 1996 (1); actual and forecasted sales figures for the European chilled-food market can be seen in Figure 4.

The growth in the chilled-food market in Europe is not as dramatic as that in the United States over the same time period. It should be noted, however, that chilled-food marketers experienced a drop in unit prices and profit margins due to the marketing of larger product portions in order to gain or expand market share (1). Growth of the European chilled-food market during the period 1997 to 2001 is also expected to be relatively slow; sales of chilled foods in Europe are expected to increase to only \$2.4 billion by the year 2001. Nevertheless, chilled foods account for 19% of the total prepared-foods market in Europe, and this position is expected to increase to 22% by 2001. Scandinavia and the United Kingdom account for more than 50% of the total chilled-food market in Europe, with Germany and France accounting for another 27%. These markets are seen as relatively mature, but greater growth is expected in countries such as Spain, Italy, and Eastern Europe, where chilled-food sales are still relatively small. The chilled prepared-food market in Japan increased from \$433 million to \$587 million during the period 1992 to 1996. Chilled-food sales in Japan are expected to increase

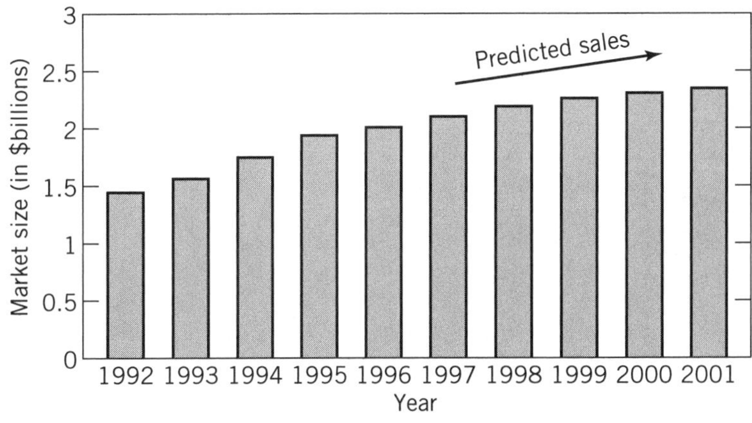

Figure 4. Actual and predicted market for chilled foods in Europe, 1992–2001.

to more than $840 million by 2001 (1); the current and future chilled-food market trend for Japan can be seen in Figure 5.

Most of this market growth is due to the expanding presence of supermarkets, although much of the available chilled foods is distributed through neighborhood convenience stores. The chilled-food market in Japan is seen as relatively new and expanding by market research companies, as chilled-food distribution systems have only recently been opened to Western companies (1). Home delivery of food to the elderly and the expanding presence of automatic vending machines are expected to contribute significantly to the Japanese chilled-food market growth.

Consumers in both Europe and Japan view the nutritional value and health benefits of chilled foods as being superior to similar frozen or shelf-stable products. A demand for superior quality is also evident in both Europe and Japan, as consumers are willing to pay a premium for chilled foods with authentic home-cooked or restaurant-quality taste. Convenient and functional packaging of chilled foods is given as much accord by Japanese customers as the nutritional quality of the food. Consumers in both Europe and Japan are also becoming more price conscious in light of the 1990 European recession and the recent economic troubles in Asia (1).

PROCESSING OF CHILLED FOODS

Chilled-food processing lines are designed to produce quality finished products as well as to control the presence and growth of microorganisms in finished products. A particular production method may be used in concert with another to produce a product of superior organoleptic and microbiological quality. The following procedures are common throughout the industry in the production of chilled foods.

Controlled Product Handling During Processing

Product handling during all phases of chilled-food production is crucial from a microbiological safety and quality standpoint. A typical processing diagram for chilled foods indicating separation of production zones can be found in Figure 6.

Separate zones for raw receiving, preprocess preparation, product assembly and cooking, pasteurization, chilling, and packaging are usually identified and maintained in chilled-food operations (2). Physical separation of different production zones is necessary to prevent cross-contamination between raw and cooked or processed product. Other aspects of hygienic process design, such as clean rooms and air handling systems, may be incorporated into chilled-food processing to reduce the risk of microbial contamination. For example, clean rooms may be maintained in the assembly and packaging of already-processed constituents, and air handling systems may be used to filter particles 0.5 μm or smaller from product assembly areas to minimize postprocess microbial contamination. A rigorous cleaning protocol for machines and the factory environment is also to be included as part of a good manufacturing practice for the manufacture of chilled-food products (3).

Pasteurization

Pasteurization involves heating food products to eliminate specific types of microorganisms. Chilled products can be pasteurized before packaging and clean-filled (hot fill hold) or filled and pasteurized in-pack (IPP). Cooking processes may be considered as part of the pasteurization process, but care must be taken to prevent contamination during latter processing steps. For chilled products having an expected shelf life of 10 to 14 days, a minimal heat process to eliminate infectious pathogens is recommended, especially if the product is ready to eat. An example of a typical process for these product types is 70°C per 2-min hold, which will eliminate 10^6 *Listeria monocytogenes* per gram of product. Chilled products having an expected shelf life of more than two weeks generally require a more severe process to eliminate spore-forming psychrotrophic pathogens. Typical processing conditions for these extended shelf-life products are at least 90°C per 10-min hold to eliminate 10^6 nonproteolytic, psychrotrophic *Clostridium botulinum* per gram of product.

Pasteurization processes are defined in terms of *P*-value, which is the time a product is held at a certain processing temperature (typical value for nonproteolytic *C. botulinum* would be written as $P_{90^\circ C} = 10$). Conventional

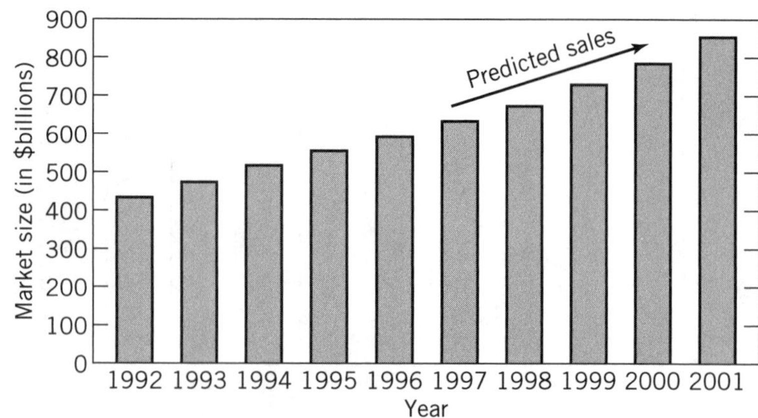

Figure 5. Actual and predicted market for chilled foods in Japan, 1992–2001.

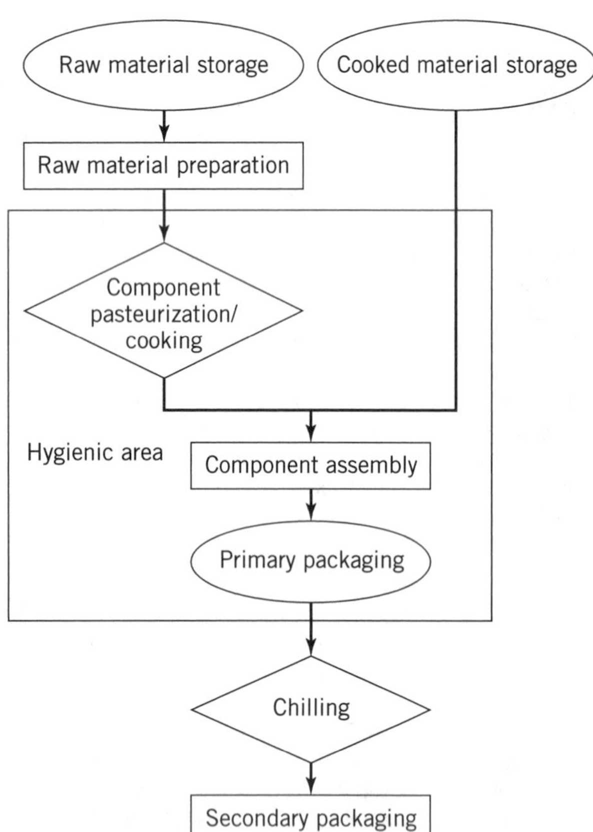

Figure 6. Typical process diagram for the manufacture of chilled foods.

thermal processing z-value, which is used to calculate the increase or decrease in temperature necessary for a 10-fold difference in the reduction of microorganisms, may also be used to describe pasteurization processes. P-values are usually used, however, because unlike commercial sterilization processes, both z-value and heat resistance of the organism may increase during pasteurization (4). The heat resistance of microorganisms generally increases in food products having a lower water activity and decreases in food products having a lower pH.

Examples of pasteurization methods include heating mixes and particles suspended in sauces in jacketed vessels, heating liquid or pumpable ingredients via heat exchangers, heating solid particles in atmospheric ovens, frying, and heating packaged products in water baths or retorts for in-pack pasteurization. Any method used must provide a critical minimum heat treatment, which must be defined by heat penetration analyses and thermal death time experiments. Factors affecting the pasteurization process include the initial temperature of the ingredients, amount of solid particulates, equipment heating rate, and rate of circulation/agitation. These parameters, as well as temperature profiles of the finished product during processing, are usually Hazard Analysis Critical Control Points (HACCPs) and must be accurately and reliably controlled and maintained.

INTRINSIC PRODUCT FACTORS

Chilled-product formulation may have a significant effect on the process applied to a particular chilled food. Several factors may be incorporated into the formulation of chilled foods that may increase the overall keeping quality of the product and subsequently decrease the amount of processing necessary to achieve a certain shelf life. One factor that reduces the amount of heat necessary to provide a particular shelf life is product pH. Microorganisms tend to be more heat sensitive in products with a lower pH, and lower pH may also enhance the keeping quality in combination with chilled storage temperatures, as the optimum pH range for the growth of most foodborne microorganisms is between 5 and 8. An example of utilizing the synergistic effect between storage temperature and product pH is in the prevention of *C. botulinum* type E outgrowth. A product pH of 5.5 is necessary to prevent the outgrowth of *C. botulinum* type E spores at 15.5°C, but if the storage temperature is maintained at 5°C, the optimum pH for control of outgrowth rises to 6.4. Product pH may be lowered using acidic ingredients like tomatoes or by the addition of organic acids such as lactic, citric, or acetic acid. The type of acid selected depends both on sensory and shelf-life requirements. In general, acetic and lactic acids have greater antimicrobial activity than citric acid.

Another factor that may be used to control the possible growth of microorganisms in chilled products, especially nonproteolytic *C. botulinum*, is water activity (a_w). Water activity values below 0.85 may prevent or delay the outgrowth of microorganisms during shelf storage of chilled products, although products with a lower a_w may require a more severe initial heat process. Most chilled products have an a_w above 0.90. Preservatives such as potassium sorbate or sodium benzoate may also be added to chilled foods to control the growth of certain microorganisms during shelf storage. Use of these preservatives is primarily to control the growth of yeasts and molds, and the activity of these substances may depend on product pH. Potassium sorbate may also delay the outgrowth of some spore formers and lactic acid bacteria. Nitrites and sodium chloride have been used in chilled meats, poultry, and fish to protect against *C. botulinum*. The factors just listed may be used in concert for chilled-product safety and quality while improving the overall product quality and appearance. They assist by reducing the pasteurization process necessary for the control of microbial growth. The development of products with these intrinsic factors is known as "Hurdle Technology" and is widely practiced in the food industry.

CHILLING TECHNIQUES

Products distributed in the chilled state must be sufficiently cooled in a reasonable time after cooking and processing. European Economic Community (EEC) Directive 77/99 requires that foods for chilled distribution must be cooled to 10°C or less within two hours of processing and maintained at or below that temperature for the duration of product shelf storage. The U.S. Food and Drug Administration (FDA) requires similar cooling conditions (cooled to less than 7.2°C in less than two hours after processing and maintained at or below that temperature during product shelf storage) for chilled foods.

Several techniques are currently used in the industry that allow for the processed products to be chilled according to the preceding requirements. One of these is the use of air-blast coolers, which circulate chilled air at high velocities (up to 16 m/s for smaller product sizes) to effect a rapid cooling rate. This rapid rate is important for sensitive foods that have been heat processed prior to cooling. The design of these units provides rapid and even chilling, and coolers may also be designed for humidity control in unwrapped product applications. Safety issues such as *L. monocytogenes* contamination can arise in systems for chilling unwrapped products if conditions such as temperature and humidity are not controlled. Chilling tunnels can also provide good product chilling characteristics by altering the airflow methods in different tunnel sections. Products cooled by this method are usually carried through the tunnel on trays or pallets. Direct-contact refrigerants such as liquid nitrogen and dry ice (carbon dioxide) shavings may also be used to rapidly chill heat-processed products. This equipment is usually designed to prevent products from freezing. Immersion hydrocooling and ice bank chilling are two other common methods of rapidly cooling chilled products; these methods provide more control over the chilling process, as they do not allow for the product to freeze. Hydrocooling is carried out by spraying or immersion of the product and can be included as part of a continuous process. Ice bank chilling is primarily used with vegetables, and systems may be designed to maintain a high humidity to prevent moisture loss. After rapid chilling, products may be held in lower-velocity air system cold rooms before packaging or distribution.

Some products may be prepared frozen and held until they reach the retail or food service market. At the retail outlet, the products are thawed at 4 to 6°C and maintained at this temperature. This approach is referred to as "freeze-thaw" and allows greater flexibility in the distribution chain as well as longer product shelf life and improved safety. Sale of such products requires that the consumer be informed that the product was "previously frozen." Another approach, often referred to as Latent Zone chilling, uses temperatures of −2 to 0°C for the preparation and storage of products. Such temperatures have been shown to extend the shelf life of a number of products for up to six weeks. The concept was developed to provide more efficient production of prepared foods.

PACKAGING OF CHILLED FOODS

There are a number of methods for packaging chilled foods, and the diversity is evident when the chilled-product areas in supermarkets are viewed. A series of requirements for chilled-food packaging from both the consumer and manufacturer point of view are listed in Table 1.

Foods undergo numerous changes including chemical, physical, and microbiological changes during storage. Food packaging acts as a protective unit against such changes by providing a barrier against contamination from outside. These properties have been described as active, intelligent, or interactive and may delay or control many of the undesirable changes to the product during the storage period.

Packaging materials are chosen based on the properties necessary to maintain product safety or quality as well as package integrity. The most common materials are paper, variable types of polymers or films, aluminum, or combinations of each as laminates. Lamination of these materials provides strength, improved heat resistance for the package (eg, from direct microwaving), better sealing properties, and more efficient barriers against moisture loss.

Table 1. Principal Package Requirements for Chilled Foods

Manufacturer basic requirements	Consumers' requirements
Nontoxic	Easy opening
Withstand factory process conditions	Resealable
Maintain integrity during transit	Shelf life/nutritional/ingredient information
Control moisture/gas losses	Contents clearly visible
Maintain seal integrity	Minimal handling required during preparation

Rigid polymer materials have become a popular choice for chilled products; these include polyethylene (PE), polyethylene terephthalate (PET), polyvinyl chloride (PVC), and polystyrene (PS). The need for lower-cost packaging materials such as the flexible polymers or packaging films consists of multiple layers of laminates. Ethylene vinyl alcohol (EVOH), for example, is the film of choice for improving the oxygen barrier properties. A good overview of chilled food packaging can be found in Reference 5.

Modified Atmosphere Packaging

Chilled products may be packaged under a modified atmosphere to allow extended shelf storage (6). This type of approach allows products to be shipped longer distances and enables a reduction in economic losses due to spoilage. Modified atmosphere packaging (MAP) depends on replacing air in a package with different gas mixtures to regulate microbial activity within the food. The composition of gases in the pack does not remain constant and will change during storage as a result of chemical, enzymatic, and microbial activity of the product. There are three parameters to consider for the successful use of MAP in chilled foods: the packaging film/container, the gas or gas mixtures being used, and the storage temperature. The film must have high barrier properties against moisture loss and low rates of transmission for the gas mixture being used. The role of the gases is quite distinct; carbon dioxide is inhibitory to both bacteria and yeast and molds and acts by extending the lag phase and, therefore, the generation time. The concentration of the gas and the age, type, and load of the initial bacterial population influence this effect. Low temperatures increase the inhibitory action of carbon dioxide. Use of high levels of this gas can affect sensory properties of a food, producing off-odors and pigment, taste, and texture changes. It may also cause package "collapse" as a result of the carbon dioxide being absorbed by the food. Nitrogen has little or no effect on bacterial growth unless the residual oxygen levels are extremely low (0.2%) and usually acts as a filler for the package. The presence of oxygen in MAP retards anaerobic growth and can improve or stabilize color within the pack. The application of MAP alone for chilled foods may not always provide shelf-life extension but should be considered as a part of a hurdle system. Some commonly used gas mixtures for MAP are shown in Table 2.

Advantages and limitations of MAP in chilled foods are listed in Table 3. Further discussions on MAP technology can be found in References 7 and 8. The desired atmospheres within a package can also be achieved by use of sachet technology, an alternative to gas flushing. This approach referred to as active packaging involves placing a sachet of a gas generator or absorber directly into the package before sealing. Currently there are two kinds of MAP sachets available—oxygen absorbers and carbon dioxide generators. The most common oxygen absorbers work on the principle of iron being converted to ferric hydroxide by oxygen and water vapor. This technology is easily applied during the manufacturing stage, and the atmosphere generated can be maintained during storage. Another application of sachet technology is the use of an ethanol pack. A sachet containing ethanol is placed in a package, and the released vapor acts as an antimicrobial agent.

DISTRIBUTION OF CHILLED FOODS

Chilled products must be distributed under temperature-controlled conditions matching the previously described EEC or FDA guidelines. Once the chilled products have been prepared, a temperature-controlled means of transport must be used to ensure that critical product handling temperatures are maintained. Road transports and intermodal freight containers are generally used to transport chilled food products from factory to warehouse to retail outlet. Road transports may include temperature-controlled trailers that may be run independently of the tractor units or trucks with temperature-controlled storage areas. Intermodal freight containers have self-contained temperature control units and may be transported via train or on the bed of a trailer. Temperature-controlled road transports or intermodal freight containers must be capable of not only maintaining a chilled environment in warmer months but also of maintaining chilled products in a refrigerated state during colder seasons.

Chilled food products are generally maintained in exclusive temperature-controlled warehouses or in temperature-controlled sections of other warehouses via the use of cold room techniques previously described. Once the chilled products have been distributed to the retail outlet, refrigerated display units are generally used to maintain product temperature for the remainder of product shelf

Table 2. Commonly used Gas Mixtures in MAP for Microbial Stability in Chilled Products

Chilled product	Target organisms	CO_2 (%)	Nitrogen (%)	Oxygen (%)
Pasta	Molds, *S. aureus*	>35	>20	—
Meats	Spoilage organisms	—	100	—
Salads	Spoilage organisms	20	—	—
Fish	*C. botulinum*	50–60	—	25
Sandwiches	Spoilage organisms	—	100	—
Pizza	Molds, lactics	50	50	0–10
Poultry	Spoilage organisms	>40	60	—
Cottage cheese	*Listeria*	35	—	—
Soft Cheeses	*Listeria*, molds	—	100	—
Hard Cheeses	*Listeria*, molds	—	100	—

Table 3. Advantages and Limitations of MAP in Chilled Foods

Advantages	Limitations
Shelf-life extension	Does not inactivate pathogens or prevent growth
Restrict growth of aerobic organisms	Shelf-life extension depends on initial population of microorganisms
Delays growth of some facultative organisms	Requires more costly packaging
May improve flavor	Can lead to lower line speeds
Adjunct to other processes	Does not control the growth of lactic acid bacteria

storage. Refrigerated display units may be configured in a variety of ways, and the most common types are freestanding bins without doors and upright shelf units with protective glass doors. Several sources have noted that retail display is the aspect of distribution in which a control of storage conditions is most lacking (9). Proposed regulations in the United States are more stringent in that the temperature requirement has been lowered from 6 to 4°C for refrigerated storage of chilled products. The goal of this regulation if to provide a greater margin of safety and quality to chilled products, which should benefit both the consumer and the processing company.

TEMPERATURE CONTROL AND MONITORING

In the production of chilled foods the focus of any production strategy is either to eliminate or minimize the growth of pathogens or spoilage organisms. However, this focus must be balanced with quality issues such as color, flavor, and the desired shelf life. A number of options will allow such a balance, but all of these depend on temperature control.

A key factor for successful production of chilled foods is the constant awareness of controlling the temperature from the production and manufacturing site to the retail or food service markets and the consumer. The rate of quality loss for chilled products, for example, can increase up to eight times for each 10°C increase in temperature, and the penalty for mishandling of products can be severe both in terms of human illness and product loss (10).

A number of temperature-measuring devices are available for monitoring products during production or transporting or in the retail marketplace. An excellent overview of this area has been published in Reference 11. The most common types consist of single readout units, sensors such as semiconductors or thermocouples, recording charts, data loggers, and chemical indicators. The single-unit devices such as the dial and stick thermometer, often seen in supermarket chill cabinets, lack accuracy in determining the air temperature. Changes to electronic readout versions of these bimetal units have not added significantly to their accuracy. A much more accurate unit is the "point and shoot" infrared thermometers. These handheld guns measure radiation temperatures of a product surface, which is translated to a digital readout. Since the units measure surface temperature, the actual temperature of a food product may need to be confirmed by other means. Similarly the liquid crystal devices seen on the front surfaces of chill cabinets suffer the same problem since they measure the surface temperature of the cabinet.

These single-unit devices are now being replaced with more accurate remote systems. These systems can be linked to a computer for data storage and temperature control. Such systems can give temperature displays from a number of sites at once and may be linked to an alarm device. The temperature-measuring devices in these units may be either thermocouples or thermistors. The accuracy of these sensors ranges between 0.3 and 1.5°C. The thermistor, a device that measures changes in resistance with temperature, is preferable to a thermocouple because of its accuracy.

Data loggers may also be used to monitor refrigerated transport and storage of chilled foods. These units can be programmed to measure and record temperatures at regular intervals over a period of time. When linked to a computer, these data can then be retrieved, and the temperature history during storage is revealed. Current developments with these units have reduced the size of these loggers so that they can easily be placed inside small packages. Temperature monitoring and control is an important issue, especially for chilled food where it remains the primary hurdle against spoilage and foodborne disease. An indication of this is that chilled-product mishandling and abuse due to incorrect storage temperatures contributed to 43% of foodborne disease outbreaks in the United States in 1996. A temperature-monitoring approach that has been successful in the pharmaceutical industry is the use of time-temperature indicators (TTI). These indicators measure temperature or temperature changes based on chemical or biological reactions. The results of these reactions are displayed in the form of color changes. In chilled foods the best type of TTI is the one that integrates both time and temperature. Such a product is available on the market, but current costs for these units prevent many companies from using them on individual packs. Consumer preference with TTI, especially with chilled products, is a "go/no-go" indicator. Systems indicating how much shelf life is remaining in a product are also preferred by consumers (10). Future developments should be able to overcome cost and ease-of-use issues.

Distribution and Storage Temperature Studies

Several studies on refrigeration temperatures maintained at retail outlets and homes of consumers revealed several interesting facts (12). Studies carried out in the U.S. market have shown that the temperature of storage of chilled products varies within a store and by product type. Supermarket surveys have shown that storage temperatures and fluctuations within the raw meat cabinets are better controlled than in the delicatessen areas. A summary of this information can be found in Figure 7.

Comprehensive studies have observed that many refrigerated foods both at the retail and consumer level exceeded 10°C (13). This was particularly true for delicatessen items, where 13% of these items were above 12°C.

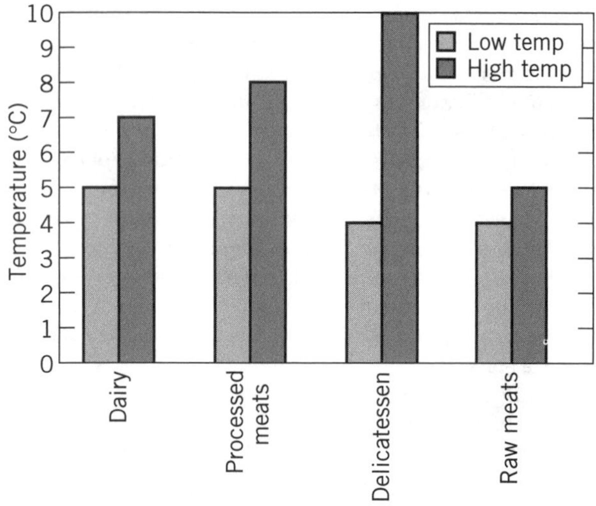

Figure 7. Temperature distribution of chilled products at the retail market.

Twenty percent of home refrigerators surveyed were also operating above 10°C (14). Factors such as shopping time and chilled-product temperature change from store to home have also been studied. In the United States, the variation in shopping time was found to be from 30 to 45 min to over 1 h and 30 min. Product temperatures during transit by consumers were shown to rise from 5°C to 15°C in some instances.

MICROBIOLOGY OF CHILLED FOODS

Other factors apart from temperature control influence microbial growth. These are moisture/water activity (a_w), acidity (pH), packaging, the use of preservatives, and type of initial processing. The most commonly used combinations are pH, a mild heat process, modified atmosphere packaging, and preservatives. These are combined with the use of chilled temperatures for better safety and shelf life of the product.

Microbiological Safety

Significant foodborne pathogens in chilled foods include both infectious and toxigenic organisms. An assessment of the ability of these organisms to grow in such products is critical for an evaluation of hazards. Organisms such as *Staphylococcus aureus*, *Bacillus cereus*, and proteolytic *C. botulinum* are unable to grow below 7°C, but under temperature abuse conditions they should be considered potential hazards. Spore formers such a *B. cereus* and *C. botulinum* are also likely to survive mild pasteurization treatments often given to chilled products, so the outgrowth of these pathogens is controlled by temperature alone. Organisms important to the safety of chilled foods, as well as the parameters for their growth, are listed in Table 4.

The major microbiological concern in chilled foods is pathogens such as *Listeria* and nonproteolytic *C. botulinum*. Recent outbreaks in the United States of *Listeria* in chilled meats emphasize the importance of controlling this pathogen in a chilled-food processing environment. Ready-to-eat deli meats produced by one large company were associated with at least 62 listeriosis cases caused by serotype 4b strain. At the same time, reports of recalls due to contamination by *Listeria* from other ready-to-eat meat producers were also published. The organism is ubiquitous and can be found in raw products such as vegetables, meat, and dairy products. It typically can be isolated from several locations in a factory environment and can contaminate chilled foods as a postprocess contaminant (15). Because the organism cannot be eliminated from food processing plants, strict management of the processing environment is necessary to minimize the probability of cross-contamination. This includes monitoring of ingredients, surfaces, and the environment. This infectious pathogen has the capacity to grow at refrigerated temperatures, with a generation or doubling time at 4°C of about 24 h. Thus chilled foods that do not receive a terminal heat process prior to consumption by the consumer can contain levels of *Listeria* capable of causing listeriosis. Listeriosis is characterized by a range of symptoms from flulike illness to miscarriages, septicemia, and death, depending on the sensitivity of the person. The infectious dose has yet to be determined but the current literature suggests <100 organisms per gram. This organism has limited heat resistance, and a process of 70°C per 2-min hold will ensure the elimination of this organism from the food.

Most serotypes of *C. botulinum* are of little significance in chilled foods with the exception of type E and nonproteolytic type B strains. Such serotypes are capable of growing at chilled temperatures and produce a neurotoxin. Incubation periods under anaerobic conditions of up to one month are required for toxin production at 4°C in meat/fish-containing products. The spores are capable of being destroyed with milder heat treatments (eg, 85°C for 15–30 min) than their proteolytic counterpart strains, which require temperatures above 100°C for any measurable death. Other public heath–related organisms that are capable of growing at refrigerated temperatures include *Yersinia enterocolitica* and *Aeromonas hydrophila*. The more important of these two is *Y. enterocolitica*, as this has been clearly implicated in foodborne outbreaks with chilled products (16). Like *Listeria*, this organism has been isolated from foods such as meats, dairy products, and vegetables. Species of *A. hydrophila* have been associated with chilled salads and other uncooked foods.

B. cereus is a spore-forming organism that functions both as a spoilage organism and a foodborne pathogen. While it is reported that psychrotrophic strains of *B. cereus* have been isolated from refrigerated products, most strains are unable to grow below 8°C. The spores are capable of surviving pasteurization temperatures associated with milk and can germinate and produce toxins under the right conditions. Sources of *B. cereus* include spices, herbs, dairy products, and meats.

Infectious agents such as *Salmonella, Campylobacter*, and *Escherichia coli* 0157:H7, although unable to grow below 7°C, are potential hazards in chilled foods due to their low infectious doses. Such organisms must be eliminated

Table 4. Microorganisms Important in Chilled Foods with Respect to Safety

Microorganism	Minimum growth temperature (°C)	Minimum a_w for growth	Minimum pH for growth
Listeria monocytogenes	0	0.92	4.4
Yersinia enterocolitica	−1	0.93	4.2
Clostridium botulinum (nonproteolytic)	3.3	0.96	4.6
Salmonella	7	0.94	3.8
Escherichia coli 0157:H7	7	0.95	4.4
Bacillus cereus	8	0.93	4.4
Clostridium botulinum (proteolytic)	10	0.93	4.6
Staphylococcus aureus	7	0.86	4.5

in ready-to-eat products. Some viruses such as the Norwalk virus have been associated with oysters and have caused foodborne outbreaks when the product has been eaten raw.

Microbiological Spoilage

Many of the spoilage organisms associated with chilled foods are adapted to their environment (17). At 0°C, for example, the lag time for many organisms is usually about two to three days before initiating growth. Once this occurs, the generation time for these psychrotrophic bacteria at 4°C is in the region of 10 to 12 h. For a chilled food product to achieve maximum shelf life with optimum quality, the initial level of these psychrotrophic spoilage microorganisms must be minimized. The principal groups of spoilage organisms associated with chilled foods are *Pseudmonas* spp., lactic acid bacteria, members of the coliform group, yeast, and molds. These groups are widely distributed in nature and therefore present both in raw ingredients and the factory environment. They are good competitors and capable of producing off-flavors or odors, especially when bacterial populations reach about 1 million cells per gram of food.

Each of the microbial spoilage groups contributes certain attributes to a food product during growth. *Pseudomonas* species cause slime formation, color changes within the pack, and, as a result of their proteolytic metabolism, strong amine odors. These changes occur in an aerobic environment and can be controlled by removing the oxygen from the food. Other Gram-negative groups contribute less drastic changes but will contribute to off-odors and acid flavors. The lactic acid group has a unique role in spoilage since lactic acid is produced as a result of their metabolism. In addition to lactic acid, other antimicrobial agents that are capable of controlling the growth of certain pathogens and competitive microflora may be produced. Lactic acid bacteria are also one of the more persistent groups in chilled foods because they tend to be resistant to most of the preservatives used and can grow under anaerobic conditions. Yeasts and molds are also a problem in chilled foods, and this type of spoilage is easily visible to the consumer. Yeasts cause spoilage problems in chilled products because they are facultative with respect to oxygen requirements, but they can be inhibited by a preservative such as sorbic acid or growth delayed by high carbon dioxide levels in the pack. Molds require aerobic conditions for growth and can be controlled by use of oxygen scavengers in the pack or modified atmosphere packaging.

Nonmicrobial Issues Pertaining to Chilled-Product Safety and Quality

Several toxicological issues relate to the safety of chilled foods. Natural toxicants exist usually at low levels in several types of raw products. These include biogenic amines, glycoalkaloids, phytotoxins, nutritional inhibitors, antivitamins, and antigens that can stimulate anaphylactic shock in sensitive people. Fortunately, many such agents are modified during processing as a result of culling, washing, and heating. Potatoes, for example, can produce glycoalkaloids such as solamine, which is a potent cholinesterase that acts on the central nervous system and interferes with nerve impulses. These amines are produced by stress conditions in the potato and accumulate close to the skin. The process of culling and peeling reduces human exposure to this alkaloid. Lectins associated with legumes such as kidney beans have been associated with toxicosis when eaten without prior treatment. When such products are soaked overnight and subsequently cooked, the toxic lectins are inactivated. An important toxin associated with fish products is caused by algal growth. The algal toxins gain access to the food chain through ingestion by shellfish. These toxins can accumulate in marine life at high levels without adverse effects. It is particularly dangerous to consume such shellfish during these periods of high algal concentration in coastal waters when the nutrient levels and temperature reach an optimum level. Depending on the type of toxin present, consumption by humans of contaminated shellfish causes a number of symptoms including vomiting, disorientation, numbness, and paralysis.

Other chemicals arising from the environment can occur at low levels in foods and should be accounted for when considering safety of a product. The most common and closely monitored contaminants are heavy metals, pesticides, antibiotic residues, and incidental additives from packaging. Toxic metals are of a particular concern, as problems arise as a result of using sludge from sewage farms and waste from settling pools for fertilizing crops. The dissemination of antibiotics from livestock into the food chain has been demonstrated in milk and meats. Although the levels are low, there is a concern with respect to the development of resistant strains of bacteria entering the food chain. See Reference 18 for further information.

Quality Control Plans

Control plans for chilled food include shelf-life testing and the use of predictive microbiology, HACCP plans, and risk assessment to ensure that all products meet local and national regulations.

Shelf-Life Testing. The shelf life of a refrigerated product is, in a sense, a record of its production and applied processes. Products with a long shelf life have to be prepared with more stringent control over ingredients and environment as well as the use of more severe heat treatment or the application of multihurdle technology. Product shelf life can be defined as the period in which the product, when stored under proper chilled conditions, is safe and wholesome to eat. In determining shelf life at chilled temperatures, both chemical and microbiological parameters have to be considered. Likewise, there should be a balance between spoilage and safety with respect to microbiological issues. Chemical issues usually involve texture changes, color degradation, syneresis or separation of water from the product, and flavor changes. Texture changes usually occur as a result of migration of water to or from an ingredient or as a result of enzymatic activity. Color degradation of pigments such as chlorophyll or carotenes, for example, can occur from exposure to high temperatures or pH changes. Sensory testing is an important part of shelf-life evaluation for determining the optimum product quality and how long it can be maintained during storage.

Microbiological shelf life testing usually consists of testing products at both standard chilled temperatures (4–5°C) and at abuse temperatures (10–15°C) over the claimed shelf life. The abuse temperatures are used to determine the margin of safety built into the product by allowing for the growth of certain pathogens if present. Several sampling periods and storage trials are necessary to get a true picture of the microbiological profile of the product. Shelf life is then designated based on data from these studies. In some instances, challenge studies are incorporated in shelf-life studies, especially if new formulations or technology is used. The food is challenged with the pertinent pathogens over the shelf-life standard and abuse temperatures (19). The data give some idea as to the ability of the growth characteristics of the pathogen in that food. Mathematical modeling of data can be used for such studies as a basis for experimental design. These models may aid in predicting growth conditions of selected pathogens under a given set of condition at different temperatures. Using such data can be timesaving and limit the number of tests that need to be performed. Current predictive models also include a number of variables associated with the intrinsic properties of the food such as a_w, pH, and preservative systems.

Needless to say, most of the data used in the establishment of predictive models has been derived from both studies and cannot completely replace standard challenge experiments. Both shelf-life and challenge experiments require technical expertise and adequate resources to carry out the task successfully (20).

Hazard Analysis. Data from shelf-life and challenge studies focus on the safety and quality issues of the product during storage. This information can be tied into the HACCP plan during the manufacturing of the chilled products (21). In essence, this approach examines the safety issues likely to arise during the complete food chain (production to consumer) and lists hazards that are deemed critical. Such hazards are recognized as critical control points (CCP), and control measures are put in place to control the identified hazard during the production, distribution, and sale of the product. The system is science based, applying a systematic approach to the issue of safety (and sometimes quality) for food production. Briefly, the approach uses seven steps that include hazard identification, setting of standards for CCPs, corrective action to be taken, and auditing. The application of a HACCP program requires a team approach that draws on expertise in disciplines such as engineering, product development, microbiology, chemistry, and public health. The HACCP system can also provide other benefits such as reducing end product testing, aid inspection programs and promote international trade.

Risk Assessment. The decision on what is a hazard or whether it is critical is sometimes unclear, and risk assessment can help link relevant steps in the HACCP plan to public health concerns (20). Risk assessment is the scientific basis for the estimation of risks, either qualitative or quantitative and has been used successfully for environmental issues for many years. When a person is exposed to a hazard, it is possible to estimate the severity and likelihood of harm resulting from that exposure. One of the steps in a HACCP study is the establishment of critical limits for the CCPs. These critical limits would more meaningful if they were established by quantitative means. The steps in risk assessment would allow this by determining (*1*) hazard identification, (*2*) assessment of the exposure to the hazard, (*3*) dose response or negative effects of such exposure, and (*4*) estimation of the magnitude of the public health problem.

Thus, the overall objective is to generate and obtain quantitative information of the microbial risks associated with the manufacturing and distribution of the product. This is achieved by assessing the frequency of pathogens in raw materials, the survival or growth patterns during the process, storage, and distribution. This data, along with the dose response for the pathogen, will provide information for a risk assessment to be determined. For further reading see Reference 22.

Legislation

Many countries have standards or guidelines that cover relevant legislation for chilled foods. The FDA in the United States has requirements under Good Manufacturing Practices for the production of chilled foods (CFR Part 110). Included in this law is the requirement that raw materials and ingredients be free from pathogenic microorganisms or be treated in such a way "so that they no longer contain levels that would cause the product to be adulterated." This regulation also has provisions for maintaining the food at 7.2°C or below "as appropriate for the food." It also requires that chilled products be handled in such a way as to protect the product from recontamination. The FDA has also developed the Food Code as a model to assist other state agencies in the area of food service, retail, and vending. It provides guidance for thawing, cold storage, and reheating of potentially hazardous foods. The Code also provides information on labeling requirements and shelf life at specific temperatures for chilled foods. Although this code was intended as a guideline, several states have adopted it into legislation. The U.S. Department of Agriculture (USDA) provides general processing guidelines for meat and poultry products as well as requirements related to the sanitation of the plant. It also requires any manufacturer of chilled foods under its jurisdiction have a sanitation standard operating procedures (9CFR 416) and a HACCP plan for all processed products.

The Food Act of 1990 provides the main provisions in the United Kingdom for the safety and hygiene controls of chilled foods (23). It introduces the concept of "due diligence" by stating that all reasonable precautions and control measures (eg, HACCP plans) must be in place in food preparation. As with other European countries the European Community Directives on Food Hygiene (93/43/EEC) are implemented as part of the regulations on hygiene.

The Food Sanitation Law governs food regulations in Japan and covers a full range of topics including additives and labeling and matters pertaining to chilled foods.

Codes of practice or industrial guides for chilled foods also exist in many countries and these lay out guidelines for the manufacturing of refrigerated products (24). Typical is the Chilled Foods Association (25) chilled-foods guidelines, which covers hygiene requirements, HACCP considerations, lethal rates for selected pathogens, and other relevant codes of practice. The outstanding feature of any legislation for chilled products is the temperature requirements for storage, and this often forms the only basis for local or provincial legislation.

BIBLIOGRAPHY

1. *The International Market for Chilled Foods*, Packaged Facts, New York, 1996.
2. M. H. Brown and G. W. Gould, "Processing," in C. Dennis and M. Stringer, eds., *Chilled Foods, A Comprehensive Guide*, Ellis Horwood Limited, Chichester, United Kingdom, 1992, pp. 111–146.
3. M. Waite-Wright, "Paper 1 Chilled Foods—The Manufacturer's Responsibility," *Food Science and Technology Today* **4**, 223–227 (1990).
4. B. M. McKay and C. M. Derrick, "Changes in Heat Resistance of *Salmonella typhimurium* During Heating at Rising Temperatures," *Lett. Appl. Microbiol.* **3**, 1316 (1987).
5. B. P. F. Day, "Chilled Food Packaging," in C. Dennis and M. Stringer, eds., *Chilled Foods, A Comprehensive Guide*, Ellis Horwood Limited, Chichester, United Kingdom, 1992, pp. 147–163.
6. J. M. Farber, "Microbiological Aspects of Modified Atmosphere Packaging Technology—A Review," *Journal of Food Protection* **54**, 58–70 (1991).
7. C. D. Gill and G. Molin, "Modified Atmospheres and Vacuum Packaging," in N. J. Russel and G. W. Gould, eds., *Food Preservatives*, Blackie and Sons, London, United Kingdom, 1991, pp. 172–199.
8. T. P. Labuza, B. Fu, and P. S. Taokis, "Prediction of Shelf Life and Safety of Minimally Processed CAP/MAP Chilled Foods: A Review," *Journal of Food Protection* **55**, 741–750 (1992).
9. A. L. Brody, "Chilled Foods Distribution Needs Improvement," *Food Technol.* **51**, 120 (1997).
10. M. Sherlock and T. P. Labuza, "Consumer Perceptions of Consumer Time–Temperature Indicators for Use on Refrigerated Products," *J. Dairy Sci.* **75**, 3167–3176 (1992).
11. M. L. Woolfe, "Temperature Monitoring and Measurement," in C. Dennis and M. Stringer, eds., *Chilled Foods, A Comprehensive Guide*, Ellis Horwood, Chichester, United Kingdom, 1992, pp. 77–109.
12. G. E. Skinner and J. W. Larkin, "Conservative Predictions of Time to *Clostridium botulinum* Toxin Formation for Use With Time–Temperature Indicators to Ensure the Safety of Foods," *Journal of Food Protection* **61**, 1154–1160 (1998).
13. R. W. Daniels, "Applying HACCP to New Generation Refrigerated Foods at Retail and Beyond," *Food Technol.* **45**, 122–124 (1991).
14. S. J. Van Garde and M. J. Woodburn, "Food Discard Practices of Households," *Journal of American Dieticians Association* **87**, 322–327 (1987).
15. L. J. Cox et al., "Listeria species in Food Processing, Non-food Processing and Domestic Environments," *Food Microbiol.* **6**, 49–61 (1989).
16. S. A. Rose, "Chilled Foods," in Y. H. Hui, ed., *Encyclopedia of Food Science and Technology*, Vol. 1., John Wiley and Sons, New York, 1992, pp. xx–xx.
17. V. N. Scott, "Interaction of Factors to Control Microbial Spoilage of Refrigerated Foods," *Journal of Food Protection* **52**, 431–435 (1989).
18. H. M. Brown, "Non-Microbiological Factors Affecting Quality and Safety," in C. Dennis and M. Stringer, eds., *Chilled Foods, A Comprehensive Guide*, Ellis Horwood, Chichester, United Kingdom, 1992, pp. 261–288.
19. S. Notermans et al., "A User's Guide to Microbial Challenge Testing for Ensuring the Safety and Stability of Food Products," *Food Microbiol.* **10**, 145–157 (1993).
20. W. B. McNab, "A Literature Review Linking Microbial Risk Assessment, Predictive Microbiology and Dose-Response Modeling," *Dairy, Food and Environmental Sanitation* **17**, 405–416 (1997).
21. A. C. Baird-Parker, "Use of HACCP by the Chilled Food Industry," *Food Control* **5**, 167–170 (1994).
22. L. Jaykus, "The Application of Quantitative Risk Assessment to Microbial Food Safety Risks," *Crit. Rev. Microbiol.* **22**, 279–293 (1996).
23. *Guidelines for the Handling of Chilled Food*, 2nd ed., Institute of Food Science and Technology, London, United Kingdom, 1990.
24. *Guidelines for the Development, Production, Distribution and Handling of Refrigerated Foods*, National Food Processors Association, Washington, D.C., 1989.
25. *Technical Handbook for the Chilled Food Industry*, Chilled Foods Association, Atlanta, Ga., 1990.

David Collins-Thompson
John M. Siddle
Nestlé R&D Center, Inc.
New Milford, Connecticut

CHLOROPHYLL. See COLORANTS: CHLOROPHYLLS.

CHOCOLATE AND COCOA

In the United States, chocolate and coca are standardized by the U.S. Food and Drug Administration under the Federal Food, Drug, and Cosmetic Act. The current definitions and standards resulted from prolonged discussions between the U.S. chocolate industry and the Food and Drug Administration (FDA). The definitions and standards originally published in the *Federal Register* of December 6, 1944, have been revised only slightly.

The FDA announced in the *Federal Register* of January 25, 1989 a proposal to amend the U.S. chocolate and cocoa standards of identity. The proposed amendments respond principally to a citizen petition submitted by the Chocolate Manufacturers Association (CMA) and, to the extent practicable, will achieve consistency with the Codex standards. The proposed amendments would allow for the use of nutritive carbohydrate sweeteners, neutralizing agents, and emulsifiers; reduce slightly the minimum milkfat content and eliminate the nonfat milk solids-to-milkfat ratios in certain cocoa products including milk chocolate; update the language and format of the standards; and provide for optional ingredient labeling requirements. FDA has also received a proposal to establish a new standard of identity for white chocolate. Comments regarding the proposal amendments are under review by FDA, and a final ruling is expected to be issued in the near future.

WHITE CHOCOLATE

There is at present no standard of identity in the United States for white chocolate.

White chocolate has been defined by the European Economic Community (EEC) Directive 75/155/EEC as free of coloring matter and consisting of cocoa butter (not less than 20%); sucrose (not more than 55%); milk or solids obtained by partially or totally dehydrated whole milk, skimmed milk, or cream (not less than 14%); and butter or butter fat (not less than 3.5%).

COCOA BEANS

The cocoa bean is the basic raw ingredient in the manufacture of all cocoa products. The beans are converted to chocolate liquor, the primary ingredient from which all chocolate and cocoa products are made. Figure 1 depicts the conversion of cocoa beans to chocolate liquor, and in turn to the chief chocolate and cocoa products manufactured in the United States, ie, cocoa powder, cocoa butter, and sweet and milk chocolate.

Significant amounts of cocoa beans are produced in about 30 different localities. These areas are confined to latitudes 20° north or south of the equator. Although cocoa trees thrive in this very hot climate, young trees require the shade of larger trees such as banana, coconut, and palm for protection.

Fermentation (Curing)

Prior to shipment from producing countries, most cocoa beans undergo a process known as curing, fermenting, or sweating. These terms are used rather loosely to describe a procedure in which seeds are removed from the pods, fermented, and dried. Unfermented beans, particularly from Haiti and the Dominican Republic, are used in the United States.

Commercial Grades

Most cocoa beans imported into the United States are one of about a dozen commercial varieties that can be generally classified as Criollo or Forastero. Criollo beans have a light color, a mild, nutty flavor, and an odor somewhat like sour wine. Forastero beans have a strong, somewhat bitter flavor and various degrees of astringency. The Forastero varieties are more abundant and provide the basis for most chocolate and cocoa for formulations. The main varieties of cocoa beans imported into the United States, usually named for the country or port of origin, are Ivory Coast, Accra (Ghana), Lagos, Nigeria, Fernando Po, and Sierra Leone (from Africa); Bahia (Brazil), Arriba (Ecuador), and Venezuelan (from South America); Malaysia, New Guinea, Indonesia, and Samoa (in the Pacific); and Sanchez (Dominican Republic), Grenada, and Trinidad (in the West Indies).

Blending

Most chocolate and cocoa products consist of blends of beans chosen for flavor and color characteristics.

Production

Worldwide cocoa bean production ranged from 2.3–2.5-million t and cocoa bean production was stagnant between 1988 and 1993. Indonesia was the only country significantly expanding production. Production in Brazil and Malaysia actually dropped.

Consumption

Worldwide cocoa bean consumption increased by 14% between 1988 and 1993 from approximately 2.1 million t in the 1988–1989 crop year to almost 2.4 million t today. North America and Western Europe increased grind by approximately 26% over this time period, whereas in Russia and Eastern Europe grind dropped by 46%.

Marketing

Most of the cocoa beans and products imported into the United States are done so by New York and London trade houses. The New York Sugar, Coffee, and Cocoa Exchange provides a mechanism by which both chocolate manufacturers and trade houses can hedge their cocoa bean transactions.

CHOCOLATE LIQUOR

Chocolate liquor is the solid or semisolid food prepared by finely grinding the kernel or nib of the cocoa bean. It is also

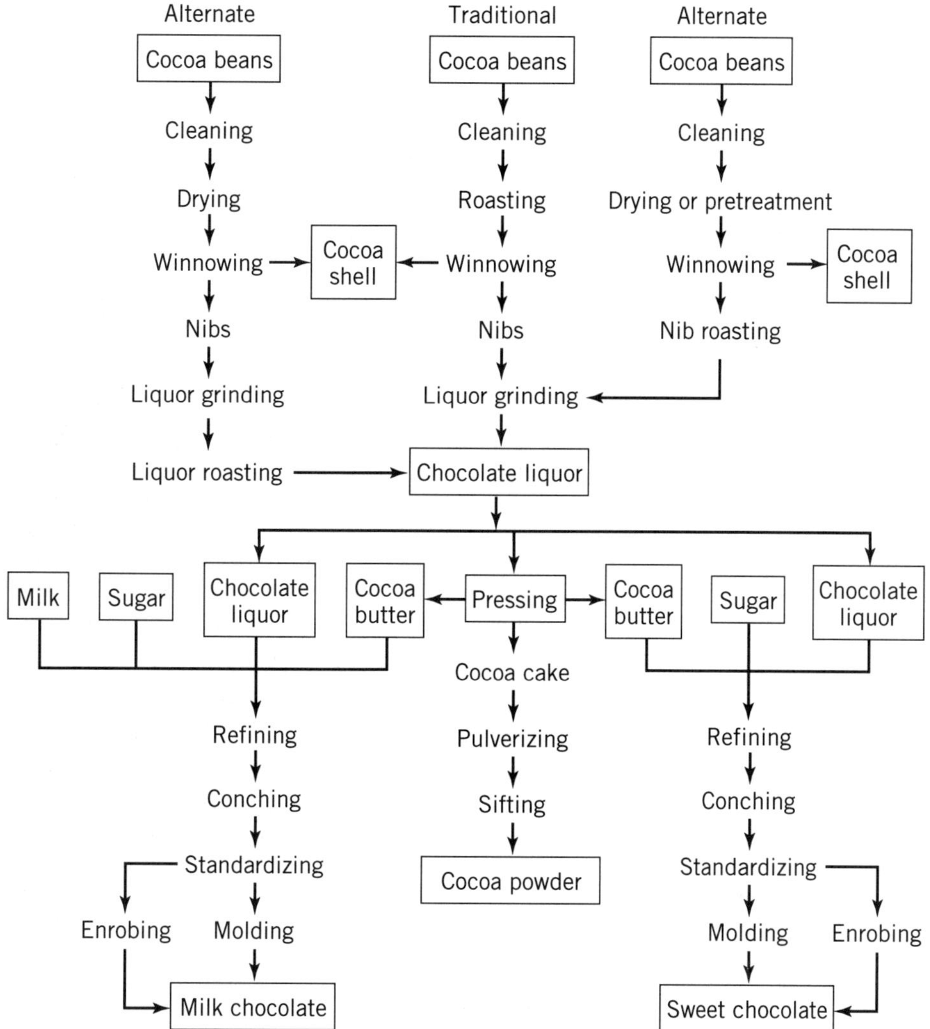

Figure 1. Flow diagram of chocolate and cocoa production.

commonly called chocolate, unsweetened chocolate, baking chocolate, or cooking chocolate. In Europe chocolate liquor is often called chocolate mass or cocoa mass.

COCOA POWDER

Cocoa powder (cocoa) is prepared by pulverizing the remaining material after part of the fat (cocoa butter) is removed from chocolate liquor. The U.S. chocolate standards define three types of cocoas based on their fat content. These are breakfast, or high fat cocoa, containing not less than 22% fat; cocoa, or medium fat cocoa, containing less than 22% fat but more than 10%; and low fat cocoa, containing less than 10% fat.

Cocoa powder production today is an important part of the cocoa and chocolate industry because of increased consumption of chocolate-flavored products. Cocoa powder is the basic flavoring ingredient in most chocolate-flavored cookies, biscuits, syrups, cakes, and ice cream. It is also used extensively in the production of confectionery coatings for candy bars.

COCOA BUTTER

Cocoa butter is the common name given to the fat obtained by subjecting chocolate liquor to hydraulic pressure. It is the main carrier and suspending medium for cocoa particles in chocolate liquor and for sugar and other ingredients in sweet and milk chocolate.

The FDA has not legally defined cocoa butter, and no standard exists for this product under the U.S. Chocolate Standards. For the purpose of enforcement, the FDA defines cocoa butter as the edible fat obtained from cocoa beans either before or after roasting. Cocoa butter as defined in the *U.S. Pharmacopeia* is the fat obtained from the roasted seed of *Theobroma cacao Linne*.

Composition and Properties

Cocoa butter is a unique fat with specific melting characteristics. It is a solid at room temperature (20°C), starts to soften around 30°C, and melts completely just below body temperature. Its distinct melting characteristic makes cocoa butter the preferred fat for chocolate products.

Cocoa butter is composed mainly of glycerides of stearic, palmitic, and oleic fatty acids. The triglyceride structure of cocoa butter has been determined as tri-saturated, 3%; mono-unsaturated (oleo-distearin), 22%; oleo-palmitostearin, 57%; oleo-dipalmitin, 4%; di-unsaturated (stearodiolein), 6%; palmitodiolein, 7%; and tri-unsaturated, triolein, 1%.

Although there are actually six crystalline forms of cocoa butter, four basic forms are generally recognized as alpha, beta, beta prime, and gamma.

Substitutes and Equivalents.

In the past 25 years, many fats have been developed to replace part or all of the added cocoa butter in chocolate-flavored products. These fats fall into two basic categories commonly known as cocoa butter substitutes and cocoa butter equivalents.

Cocoa butter substitutes and equivalents differ greatly with respect to their method of manufacture, source of fats, and functionality; they are produced by several physical and chemical processes. Cocoa butter substitutes are produced from lauric acid fats such as coconut, palm, and palm kernel oils by fractionation and hydrogenation; from domestic fats such as soy, corn, and cotton seed oils by selective hydrogenation; or from palm kernel stearines by fractionation. Cocoa butter equivalents can be produced from palm kernel oil and other specialty fats such as shea and illipe by fractional crystallization; from glycerol and selected fatty acids by direct chemical synthesis; or from edible beef tallow by acetone crystallization.

In the early 1990s, the most frequently used cocoa butter equivalent in the United States was derived from palm kernel oil but a synthesized product was expected to be available in the near future.

SWEET AND MILK CHOCOLATE

Most chocolate consumed in the United States is consumed in the form of milk chocolate and sweet chocolate. Sweet chocolate is chocolate liquor to which sugar and cocoa butter have been added. Milk chocolate contains these same ingredients and milk or milk solids.

U.S. definitions and standards for chocolate are quite specific. Sweet chocolate must contain at least 15% chocolate liquor by weight and must be sweetened with sucrose or mixtures of sucrose, dextrose, and corn syrup solids in specific ratios. Semisweet chocolate and bittersweet chocolate, though often referred to as sweet chocolate, must contain a minimum of 35% chocolate liquor. The three products, sweet chocolate, semisweet chocolate, and bittersweet chocolate, are often simply called chocolate or dark chocolate to distinguish them from milk chocolate.

Sweet chocolate can contain milk or milk solids (up to 12% max), nuts, coffee, honey, malt, salt, vanillin, and other spices and flavors as well as a number of specified emulsifiers. Many different kinds of chocolate can be produced by careful selection of bean blends, controlled roasting temperatures, and varying amounts of ingredients and flavors.

The most popular chocolate in the United States is milk chocolate. The U.S. Chocolate Standards state that milk chocolate shall contain no less than 3.66 wt % of milk fat and not less than 12 wt % of milk solids. In addition, the ratio of nonfat milk solids to milk fat must not exceed 2.43:1 and the chocolate liquor content must not be less than 10% by weight.

Production

The main difference in the production of sweet and milk chocolate is that in the production of milk chocolate, water must be removed from the milk. Many milk chocolate producers in the United States use spray-dried milk powder. Others condense fresh whole milk with sugar, and either dry it, producing milk crumb, or blend it with chocolate liquor and then dry it, producing milk chocolate crumb. These crumbs are mixed with additional chocolate liquor, sugar, and cocoa butter later in the process. Milk chocolates made from crumb typically have a more caramelized milk flavor than those made from spray-dried milk powder.

THEOBROMINE AND CAFFEINE

Chocolate and cocoa products, like coffee, tea, and cola beverages, contain alkaloids. The predominant alkaloid in cocoa and chocolate products is theobromine, though caffeine is also present in smaller amounts. Concentrations of both alkaloids vary depending on the origin of the beans. Published values for the theobromine and caffeine content of chocolate vary widely because of natural differences in cocoa beans and differences in analytical methodology.

NUTRITIONAL PROPERTIES OF CHOCOLATE PRODUCTS

Chocolate and cocoa products supply proteins, fats, carbohydrates, vitamins, and minerals. The Chocolate Manufacturers' Association of the United States (McLean, Virginia) completed a nutritional analysis from 1973 to 1976 of a wide variety of chocolate and cocoa products representative of those generally consumed in the United States.

ECONOMIC ASPECTS

Chocolate consumption (wholesale Dollar value) on a global basis was approximately $23 billion in 1992. In the United States, Hershey, Mars, and Nestlé control about 70% of the market.

The leading chocolate companies continue to pursue a global confectionery business strategy with an increase in the early 1990s of confectionery business activity in the Eastern Bloc countries, Russia, China, and South America. Generally, as per capita income increases, chocolate consumption increases and sugar consumption decreases. Consumer demographics, the declining child population, and the increase in consumer awareness of health issues play important roles in the economics of chocolate consumption. Chocolate confectionery business trends during the early 1990s include product down-sizing leading to snack size finger foods, increased emphasis on specialty chocolates with concentration on dessert chocolates, and chocolate brand equity spread into beverages, baked goods, frozen novelties, and even sugar confections.

GENERAL REFERENCES

W. T. Clarke, *The Literature of Cacao*, ACS, Washington, D.C., 1954.

L. R. Cook, *Chocolate Production and Use*, Magazines for Industry, Inc., New York, 1972.

B. W. Minifie, *Chocolate, Cocoa, and Confectionery: Science and Technology*, AVI, Westport, Conn., 1970.

E. M. Chatt, in Z. J. Kertesz, ed., *Economic Crops*, Vol. 3, Interscience Publishers, Inc., New York, 1953, p. 185.

B. L. Zoumas
J. F. Smullen
Hershey Foods Corporation
University Park, Pennsylvania

CITRUS. See Fruits, semi-tropical; Fruits, tropical.

CLEANING-IN-PLACE (CIP)

From the time of its early use in the 1950s, the practice of cleaning-in-place (CIP) for cleaning of plants processing potable liquids and other products such as ice cream and butter has become widespread and is now considered an established cleaning technique (1,2). The technique of CIP stands for cleaning of the tanks, pipelines, processing equipment, and process lines by circulation of water and chemical solutions (hereafter referred to as solution) through them. The term CIP or cleaning-in-place emphasizes that the technique does not require dismantling of pipelines or equipment, which was the case with manual cleaning. Manual cleaning was extremely time-consuming and expensive, and often the level of hygiene (bacteriologic cleanliness) achieved through it was low and inconsistent (2). Introduction of CIP, which became inevitable in the face of economic pressures to increase throughput, increasing cost of labor, scarcity of labor, and technical developments by equipment manufacturers and detergent chemists, alleviated the problems associated with manual cleaning (2,3). Three forms of energy—chemical, kinetic, and thermal—are generally needed for any cleaning operation. In comparison to manual cleaning, higher temperatures and stronger chemical (detergent) concentrations can be used in CIP. Solution temperatures up to 88°C and detergent pH up to 13 can be used in CIP (4). The equipment design and the properties of the material to be cleaned impose limits on temperature and strength of solution. Lower temperatures may have to be used if product residue (soil on product contact surface) becomes harder to clean at higher temperatures. Manual brushing, which contributes a great deal in manual cleaning, is totally eliminated in CIP and, in some sense, is replaced by the kinetic energy of turbulent flow in pipelines or impingement of jets in vessels (1,4). The circulation time is also a factor in cleaning, and an increase in the time, within a limit, improves the cleaning achieved. Circulation time from 5 min to 1 h is used in practice (5). One of the most important though less noticeable advantages of CIP over manual cleaning is the fact that CIP provides freedom in plant design from the severe limitation of keeping the plant manually cleanable and hence helps in development of new processes and ideas (3). The use of CIP has also made it possible and convenient to automate the cleaning operations in a plant where it was impossible in the case of manual cleaning. The sequence or cycle of operations in any cleaning are the same regardless of cleaning technique (5). A normal cleaning sequence consists of prerinsing with water, washing with detergent solution, and postrinsing with clean water. In addition, there may be a disinfection (sanitizing) stage followed by a final water rinse. Prerinsing is an important stage. It should be started as soon as possible and should continue until the discharging water is free from product residue (soil).

REQUIREMENTS FOR CIP

General Requirements

Only a plant that is properly designed to be cleaned by CIP can be cleaned by CIP efficiently (4). The design of a plant that is to be cleaned in place and the design of the equipment and system that is to apply the CIP technique should be compatible. The CIP system should be designed as an integral part of the processing plant; modifying the plant later for CIP may pose problems (4). All product contact surfaces must be accessible to the pre- and post-rinse water and solution. The material that comes in contact with cleaning solution must be able to withstand the solution at its concentration and temperature. For this reason, most of the construction material is stainless steel.

Requirements for Pipelines

There should be no crevice condition, especially in the pipe joints. Crevice condition can lead to bacterial trap, which may be impossible to clean by CIP (6). As far as possible, pipelines should have welded joints. Proper drainage of prerinse water is important in CIP system to avoid any dilution or cooling of solution. To provide the drainage, all piping should have a minimum fall (pitch or slope) of 1:100. Pipe work should also have good support to prevent the pipes from sagging. Any sags in the pipeline would prevent complete drainage and put strain on the pipe joints. Uncleanable dead pockets must be avoided inside the pipeline. This may not be achieved, if CIP is added to the plant as an afterthought instead of being an integral part (2). A mean flow rate of 1.5 m/s is normally recommended, but a mean solution velocity of 1.0 m/s may be sufficient in some cases. In practice, there is not much gain in exceeding the mean velocity beyond 2.0 m/s (4). Volume flow rate would depend on the diameter of the pipe. Higher flow rates create a higher hydraulic pressure drop, hence power requirements of circulating pumps could be considerably larger. Abrupt changes in the diameter of pipes that disturb the flow and reduce the cleaning efficiency should be avoided (6).

Requirements for the Vessels and Tanks

The cleaning-in-place of storage tanks or vessels is performed by spraying the cleaning solution onto the surface of the vessel through pressure spray devices located inside the vessel. The spray devices may be rotating, oscillating, or fixed. The fixed device, which is in the form of a perforated ball, is used most commonly. The fixed or static spray ball device does not have any moving parts and therefore gives trouble-free operation with minimal maintenance. Rotating or oscillating devices may wear and give a distorted spray pattern. They may also jam or stick in one place with the consequence of incomplete cleaning of the vessel. A single rotating or oscillating jet, however, can clean a larger-sized vessel than can a single static spray ball. The installation of the spray device should result in a spray pattern that must always cover all parts of the vessel, including probes, agitators, and areas shadowed by them. If required, more than one static spray ball should be used for total coverage in the vessel. Suitable filters should be used to prevent blockage of the spray device. Spray devices should be run at designed pressure and throughput. Too high a pressure can cause atomization of the solution and too low a pressure will reduce the force of jet impingement—both resulting in unsatisfactory cleaning. Permanently installed spray devices are commonly used, but removable spray devices may be preferred in certain special circumstances. Adequate venting of vessels is extremely important to avoid the collapse of the vessel due to a vacuum created during in-place-cleaning when a cold-water rinse immediately follows a wash period with hot detergent. The vessel and the supply and return CIP lines to it should have adequate drainage, otherwise undrained prerinse water can dilute and cool the solution, hence resulting in unsatisfactory cleaning.

Cleaning Circuits

Factory installations as a whole are generally divided into a number of circuits that can be cleaned at different times by CIP technique. It is a usual practice to group pipelines, vessels, and special equipment such as heat exchangers and evaporators into different cleaning circuits because of their different cleaning requirements with respect to flow rates, pressures, and chemicals (1). The product residue deposit (soil) should be of the same kind in a circuit, and all components of the circuit must be available for cleaning at the same time. Hot and cold lines of the plant may be placed in different circuits. After the introduction of automation, it has become practical to clean some parts of the plant while production continues in the adjacent areas. In such circumstances, it is necessary to prevent contamination of product by the cleaning solution. In recent years, double seat valves with the internal leakage drains being used to safeguard against such contamination (2).

TYPES OF CIP SETS/SYSTEMS

CIP sets can be categorized as centralized, local, and satellite or decentralized. In a centralized system, the various CIP circuits are connected by a network of pipes to one or two central CIP stations or units, which consist of all necessary equipment for storage and monitoring of cleaning fluids (water for rinse and solution). Large capacity water and detergent tanks (13,600 to 27,700 L) are used in centralized CIP (4). The system may work well except in the case of large processing plants where long pipe runs require large pump capacities and excessive energy use owing to heat losses. Also, the longer pipelines may contain some water after the prerinse operation that can dilute the detergent solution (7). Overcoming this problem increases chemical consumption. Another disadvantage of a centralized system is the fact that total reliance for cleaning the plant is placed on one or two central stations or units (4). In case of failure or malfunctioning of the central units, the whole plant may have to be shut down. In contrast to the centralized system, the local system requires a greater number of smaller tanks and pumps and shorter pipe runs. The system is more reliable. If one local unit fails or malfunctions, the cleaning of the plant not served by that local unit can still be achieved. The local system would require a greater number of heating units and detergent-strength (concentration) controllers (4). The satellite system is a combination of central and local systems. In the satellite system, there are central solution tanks and the local units draw the required volume of solution using properly sized pipes from the central tanks. The heating of the solution is arranged locally (4). Large modern food-processing plants generally employ a satellite system (7) because of the advantages of saving energy, water, and detergent.

The CIP systems are also classified as single-use, reuse, or multiuse systems, depending on whether the same cleaning solution is used for one, many, or few cycles of cleaning (2). The single-use system is most suited for cleaning heavy soil loads (such as in thermal processing equipment) or cleaning of small plants (8). The system uses the minimum amount of detergent needed for cleaning a circuit and discharges the solution to the sewer after one cleaning cycle is over. The system may be wasteful of detergent and energy and can cause effluent problems (4). However, it is simple in installation and operation. By contrast, the reuse system is complex in installation and operation. The reuse system provides for the reclamation and reuse of the cleaning solution and final rinse water—the latter is used as prerinse for the next cleaning cycle. More detergent may be added to the solution between the cleaning cycles to counteract the loss of detergency due to expending chemical energy to remove the soil. The solution is discharged to the sewer when it becomes very dirty. The reuse systems require greater capital expenditure but offer savings on volume of water and detergent solutions and energy used (4). Greater benefits can be derived from the reuse systems in plants that have light soil loads and use large-diameter pipe circuits (especially in the brewing industry) (8). The multiuse system is a compromise between the single-use and reuse systems. The final rinse water and solution are used for a few cleaning cycles before discharging to the sewer. In terms of capital and operating costs and complexity of the installation and operation, it is between single-use and reuse systems. Many factors: type and size of plant to be cleaned, soil loading, range and type of chemicals available, pressure drop in pipeline circuits, and pressure and throughput required for vessel spray device, should be considered before selecting the type and size of the CIP system.

CONTROL SYSTEM

Control systems for operating CIP must try to perform CIP operations in a precise sequence. In the early CIP control systems, pumps and valves were controlled manually and time was kept using a stopwatch. Overdependence on the human element prevented the precise sequencing in early control systems (4,9). As the CIP control system evolved with time, the use of electromechanical relays became common. The relays were used to perform the automatic sequencing, but the system had an inherent ability to fail at the wrong time (9). The present-day CIP control systems are microprocessor-based. The development of inexpensive and reliable microprocessors has accelerated the use of microprocessor-based control systems in CIP. Control systems for CIP have improved a great deal in the past 40 years from the unreliable to a very reliable and flexible microprocessor-based controls where cycle time, chemical concentration, and temperature are fully adjustable to suit individual cleaning needs (10).

The center of any microprocessor-based control system is a process controller containing a central processing unit (CPU), memory, and an input/output (I/O) interface. Input—output devices provide the operator access to the control system. The CPU accesses the operator-oriented equipment (such as pushbuttons, pilot lights, rotary switches, thumbwheel switches, LED readouts, CRT screens, and printers) and process-oriented equipment (such as pump motors, valve solenoids, pressure sensors, level probes, flow meters, and temperature switches) through the I/O devices. An operating logic, which may originate as a verbal description or an algorithm, describes the interaction and coordination among the operator-oriented equipment, the process-oriented equipment, and the control system. The operating logic or computer program can be prepared from the algorithm. The CPU executes the operating logic or the program. Because the controller's actions are determined by the computer program, any configuration of equipment and/or program sequence can be implemented and controlled by such a control system. The application of microprocessor-based control systems has helped in making the modern food-processing plants and associated CIP systems completely automatic (7,9,10). With precisely controlled operations, a more sophisticated CIP unit design has become possible, resulting in increased savings on energy, water, and cleaning chemicals (10). The microprocessor-based control systems have provided new opportunities for modifying and improving CIP operations and concepts.

Even very reliable modern control systems can malfunction; therefore, regular routine checks and maintenance are necessary (1,4).

ENERGY AND COST CONSIDERATIONS

Heat energy required for CIP operations constitutes 99% of the energy requirement (11). Heat energy needed is related to cleaning temperature and hence energy savings can be achieved by reducing the cleaning temperature. However, this would necessitate an increase in the detergency requirement of the solution and/or cleaning time. The most energy efficient method, therefore, may not be the most cost effective method. Cost and energy calculations are required in several plants for combinations of time, temperature, and strength of detergent to achieve a standard level of bacteriologic cleanliness before any meaningful conclusion can be reached (11).

Type of plant to be cleaned, CIP system used, mechanical components, control system, building construction, and plant installation would all influence the capital cost associated with CIP (12). However, it is the operating cost that becomes the major factor in making decisions about the CIP system to be used (12). Cleaning chemicals, water supply, required heat energy, operating labor, effluent discharge, maintenance, and electric power would all contribute to the operating cost. The satellite or decentralized system may be the best option in economic terms (12).

BIBLIOGRAPHY

1. D. A. Timperley, "Cleaning in Place (CIP)," *Journal of the Society of Dairy Technology* **42**, 32–33 (1989).
2. N. P. B. Sharp, "CIP System Design and Philosophy," *Journal of the Society of Dairy Technology* **38**, 17–21 (1985).
3. J. R. Franklin, "The Concept of CIP," in proceedings of a seminar in Melbourne, Australia, June 3–4, 1980.
4. W. Barron, "A Practical Look at CIP," *Dairy Industries International* **49**, 34–34 (1984).
5. R. K. Guthrie, *Food Sanitation*, AVI Publishing Co., Inc., Westport, Conn., 1980.
6. J. Farmer, "Engineering Design for CIP," in proceedings of a seminar in Melbourne, Australia, June 3–4, 1980.
7. "Cleaning of Dairy Equipment," *Dairy Handbook*, Alfa-Laval, S-221 03 Lund, Sweden, pp. 307–318.
8. P. F. Davis, "Single Use, Re-Use and Multi-Use CIP systems," in proceedings of a seminar in Melbourne, Australia, June 3–4, 1980.
9. A. L. Foreshew, "Control Systems for Operating CIP," in proceedings of a seminar in Melbourne, Australia, June 3–4, 1980.
10. J. Hyde, "State-of-the-Art CIP/Sanitation Systems," *Dairy Record* **84**, 101–105 (1983).
11. C. C. Sillett, "Energy Aspects of Cleaning-in-Place," *Journal of the Society of Dairy Technology* **35**, 87–91 (1982).
12. G. F. Taylor, "The Economics of CIP," in proceedings of a seminar in Melbourne, Australia, June 3–4, 1980.

D. S. Jayas
P. Shatadal
S. Cenkowski
University of Manitoba
Winnipeg, Manitoba, Canada

COCONUTS. See Nuts in the Supplement section.

COFFEE

Coffee, originally a wild plant from East Africa, was presumably first cultivated by the Arabians in about 575 A.D. (1). By the sixteenth century, it had become a popular drink in Egypt, Syria, and Turkey. The name coffee derives from the Turkish pronunciation *kahveh* of the Arabian word *gahweh*, signifying an infusion of the bean. Coffee was introduced as a beverage in Europe early in the seventeenth century and its use spread quickly. In 1725 the first coffee plant in the Western Hemisphere was planted on Martinique in the West Indies. Its cultivation expanded rapidly, and its consumption soon gained the wide acceptance it enjoys today.

MODERN COFFEE PRODUCTION

Commercial coffees are grown in tropical and subtropical climates at altitudes up to roughly 1800 m; the best grades are grown at high elevations. Most individual coffees from different producing areas possess characteristic flavors. Commercial roasters obtain preferred flavors by blending varieties before or after roasting. Colombian and washed Central American coffees are generally characterized as mild, winey-acid, and aromatic; Brazilian coffees as heavy body, moderately acid, and aromatic; and African robusta coffees as heavy body, neutral, slightly acid, and slightly aromatic. Premium coffee blends contain higher percentages of Colombian and Central American coffees.

ECONOMIC IMPORTANCE OF COFFEE

Coffee, a significant factor in international trade for about 185 years, is among the leading agricultural products in international trade along with wheat, corn, and soybeans. The total world exportable production of green coffee in the 1997–1998 growing season is approximately 81.7 million bags (Table 1). This compares with a total production of about 107.5 million bags, the difference being internal consumption. Table 2 shows imports and consumption data.

The International Coffee Organization and Association of Coffee Producing Companies assign export quotas to their members to help stabilize coffee pricing. Composite U.S. prices for green coffees in 1997 and 1998 ranged between \$2.20/kg and \$3.97/kg. For Colombians, the range was \$3.22/kg to \$5.82/kg, whereas for robustas it was \$1.49/kg to \$2.11/kg. These compare with a peak of about \$6.60/kg in April 1977 following a disastrous frost in Brazil in 1975.

Table 1. Green Coffee Production and Exportable Product 1997–1998

	Million of 60-kg bags[a]	
Principal countries	Production	Exportable product
Brazil	35.8	23.3
Colombia	11.0	9.4
Indonesia	7.0	4.6
Vietnam	5.8	5.5
Mexico	5.6	4.6
Ivory Coast	4.1	4.0
Uganda	3.8	3.7
Ethiopia	3.7	2.1
India[b]	3.5	2.6
Guatemala	3.1	2.8
Honduras	2.6	2.4
Costa Rica	2.2	1.9
El Salvador	2.0	1.8
Total	*107.5*	*81.7*

[a]1996–1997 figures used because 1997–1998 data were not yet available.
Source: Ref. 2.

Table 2. Imports 1997 and Consumption 1995–1996 of Green Coffee

	Million bags[a]	
Principal countries	Imports	Consumption
United States	18.1[b]	18.1
Germany	13.6	9.7
France	6.7	5.6
Japan	5.6	6.4
Italy	5.6	4.9
Spain	3.6	3.0
Netherlands	3.3	2.4
UK	3.1	2.4
Total	*82.8*	

[a]60-kg bags.
[b]Total 1997 U.S. coffee imports including green, roasted, and soluble coffee were equivalent to 20.2 million bags of green coffee.
Source: Ref. 4.

PROCESSING AND PACKAGING

Green Coffee Processing

The coffee plant is a relatively small tree or shrub, often controlled to a height of 2 to 3 m, belonging to the family Rubiaceae. *Coffea arabica* accounts for 70%, *Coffea robusta*, 29%, and *Coffea liberica* and others, 1% of world production. Arabustas, hybrid coffees that combine many disease-resistant characteristics of robustas with many favorable flavor qualities of arabicas, are also produced. Each species includes several varieties. After the spring rains, the plant produces white flowers. About six months later, the flowers are replaced by fruit, approximately the size of a small cherry. The ripe fruit is red or purple. The outer portion of the fruit is removed by curing; yellowish or light green seeds, the coffee beans, remain. They are covered with a tough parchment and a silvery skin known as the spermoderm. Each cherry normally contains two coffee beans.

Curing is effected by either the dry or wet method. The dry method produces so-called natural coffees; the wet method, washed coffees. The latter coffees are usually more uniform and of higher quality.

Dry curing is used in most of Brazil and in other countries where water is scarce in the harvesting season. The ripe cherries are spread on open drying ground and turned frequently to permit thorough drying by the sun and wind.

Sun drying usually takes two to three weeks depending on weather conditions. Some producing areas use, in addition to the sun, hot air, indirect steam, and other machine-drying devices. When the coffee cherries are thoroughly dry, they are transferred to hulling machines that remove the skin, pulp, parchment shell, and silver skin in a single operation.

In wet curing, freshly picked coffee cherries are fed into a tank for initial washing. Stones and other foreign material are removed. The cherries are then transferred to depulping machines that remove the outer skin and most of the pulp. However, some pulp mucilage clings to the parchment shells that encase the coffee beans. Fermentation tanks, usually containing water, remove the last portions of this pulp. Fermentation may last from 12 h to several days. Because prolonged fermentation may cause development of undesirable flavors and odors in the beans, some operators use enzymes to accelerate the process.

The beans are subsequently dried either in the sun, in mechanical dryers, or in combination. Machine drying continues to gain popularity, in spite of higher costs, because it is faster and independent of weather conditions. When the coffee is thoroughly dried, the parchment is broken by rollers and removed. Further rubbing removes the silver skin to produce ordinary green unroasted coffee containing about 12 to 14% moisture.

Coffee prepared by either the wet or dry method is machine graded into large, medium, and small beans by sieves, oscillating tables, and airveyors. Damaged beans and foreign matter are removed by handpicking, machine separators, electronic color sorters, or a combination of these techniques. Commercial coffee is graded according to the number of imperfections present—black beans, damaged beans, stones, pieces of hull, or other foreign matter. Processors also grade coffee by color, roasting characteristics, and cup quality of the beverage.

Chemical Composition of Green Coffee

Coffee varies in composition according to the type of plant, region from which it comes, altitude, soil, and method of handling the beans. As shown in Table 3, differences are greater between species, for example, arabica versus robusta (African), than within the same species grown in different regions, for example, Colombian versus Brazilian arabicas.

Lower oil, trigonelline, and sucrose contents are typical of robusta beans, as is a higher caffeine content. Green coffee contains little reducing sugar but a considerable quantity of carbohydrate polymers. The polymers are mainly mannose with varying percentages of glucose, arabinose, and galactose. Table 4 summarizes the polysaccharide composition of different sources of green coffee (4).

Effects of Roasting on Major Components

Green coffee has no desirable taste or aroma; these are developed by roasting. Many complex physical and chemical changes occur during roasting, including the obvious change in color from green to brown, and a large increase in bean volume. As the roast nears completion, strong exothermic reactions produce a rapid rise in temperature, usually accompanied by a sudden expansion, or puffing, of the beans, with a volume increase of 50 to 100%. However, this behavior varies widely among coffee varieties because of differences in composition and physical structure.

Table 5 shows the most significant and well-established chemical changes that occur in green coffee as a result of roasting. The principal water-soluble constituents of green coffee are protein, sucrose, chlorogenic acid, and ash, which together account for 70 to 80% of the water-soluble solids. Most sucrose disappears early in the roast. Reducing sugars are apparently formed first and then react rapidly so that the total amount of sugar decreases as the roast nears completion. The sugar reactions, dehydration and polymerization, form high molecular weight water-soluble and water-insoluble materials. Formation of carbon dioxide and other volatile substances as well as the loss of reaction-formed water account for most of the dry-weight roasting loss. These losses range from 2 to 5% of green bean weight for light-roast coffees used in the United States and northern Europe to up to 8 to 11% for dark roasts used in southern Europe.

Roasting essentially insolubilizes the proteins, which constitute 10 to 12% of green coffee, and 20 to 25% of the fraction soluble in cold water. The flavor and aroma of roasted coffee are probably due in large part to breakdown and interaction of the amino acids derived from these proteins. Analyses of the amino acids present after acid hydrolysis in both green and the corresponding roasted coffee show marked decreases in arginine, cysteine, lysine, serine, and threonine in Colombian and Angola robusta types after roasting. The amounts of glutamic acid and leucine for both coffee types, and in the robusta, phenylalanine, proline, and valine increase with roasting (5). Cysteine is the probable source of the many sulfur compounds found in the coffee aroma.

About 15 to 40% of the trigonelline is decomposed during roasting. Trigonelline is a probable source of niacin, which reportedly increases during roasting, and of potent

Table 3. Typical Analyses of Green Coffee Types

Variety	H_2O	Oil	Total nitrogen	Ash	Caffeine	Chlorogenic acid[a]	Trigonelline	Protein	Reducing sugar	Sucrose	Total carbohydrate
African robusta	11.5	7.0	2.5	3.8	2.06	4.7	0.76	11.4	0.40	4.2	35.0
Colombian arabica	13.0	13.7	2.1	3.4	1.10	4.1	0.94	10.5	0.17	7.2	34.1
Brazilian arabica	11.0	14.3	2.2	4.1	1.01	4.1	1.24	11.1	0.27	7.1	32.0

[a]Chlorogenic acid values vary somewhat with the method of analysis used.
Source: Ref. 3.

Table 4. Polysaccharide Analysis of Green Coffee Beans Polysaccharide Content (wt%)

Variety	Arabinose	Mannose	Glucose	Galactose	Total
India robusta	4.1	21.9	7.8	14.0	48.2
Ivory coast robusta	4.0	22.4	8.7	12.4	48.3
Sierra Leone robusta	3.8	21.7	8.0	12.9	46.9
El Salvador arabica	3.6	22.5	6.7	10.7	43.5
Colombia arabica	3.4	22.2	7.0	10.4	43.0
Ethiopia arabica	4.0	21.3	7.8	11.9	45.0

Note: Polysaccharide content is expressed as anhydro-sugars. Wt% is on dry basis.

Table 5. Average Composition of Green and Roasted Coffee

Constituents	Green, percent dry basis	Roasted, percent dry basis[a]
Hemicelluloses	23.0	24.0
Cellulose	12.7	13.2
Lignin	5.6	5.8
Fat	11.4	11.9
Ash	3.8	4.0
Caffeine	1.2	1.3
Sucrose	7.3	0.3
Chlorogenic acid	7.6	3.5
Protein (based on nonalkaloid nitrogen)	11.6	3.1
Trigonelline	1.1	0.7
Reducing sugars	0.7	0.5
Unknown	14.0	31.7
Total	*100.0*	*100.0*

[a]Not corrected for dry-weight roasting loss.

aromatic nitrogen ring compounds, such as pyridine, found in roasted coffee aroma. However, pyrazines, oxazoles, and thiazoles, also components of the coffee aroma, are probably products of protein breakdown. Caffeine is relatively stable, and only small amounts are lost by sublimation during roasting.

The chlorogenic acids—3-caffeoylquinic acid (cholorogenic acid), 4-caffeoylquinic acid (cryptochlorogenic acid), and 5-caffeoylquinic acid (neochlorogenic acid)—occur at least in part on potassium-caffeine-chlorogenate complexes. They decompose in direct relationship to the degree of roast. Table 6 shows the changes that occur in these acids during roasting. Apparently, chlorogenic acids modify and control reactions that occur during the roast and are particularly important to the decomposition of sucrose.

Glycerides of linoleic and palmitic acids, along with some glycerides of stearic and oleic acids, make up the 7 to 16% fat content of coffee. Some cleavage of glycerides and some loss of unsaponifiables occur during roasting. Table 7 details these losses (7).

Aroma

Combinations of advanced chromatography and mass spectra analytical techniques have advanced the identification of volatile flavor components in roasted coffee to more than 800. Although present in minute quantities, they are extremely significant to the balance of flavor in a cup of coffee. A historical account of this work has been published (8). A partial summary of the volatile components by chemical class is in Table 8.

Freshly roasted ground coffee rapidly loses its fresh character when exposed to air, and within a few weeks develops a noticeable stale flavor. The mechanism for the development of staling is not known but is believed to be caused by an oxidation reaction that can be catalyzed by increasing levels of moisture (9).

Table 6. Changes in Chlorogenic Acids during Roasting by Percentage

	Santos				Colombians			
Acid	Green	Light roast	Medium roast	Dark roast	Green	Light roast	Medium roast	Dark roast
Chlorogenic acid	5.56	2.90	1.96	1.11	3.77	2.74	2.16	0.93
Neochlorogenic acid	0.88	1.59	1.02	0.63	0.60	1.53	1.16	0.49
Isochlorogenic acid	0.41		0.24					

Source: Ref. 6.

Table 7. Characteristics of Oil from Green and Roasted Coffee by Percentage

	Green			Roasted		
Variety	Oil	FFA[a]	Unsaponifiables	Oil	FFA[a]	Unsaponifiables
Santos	12.77	0.78	6.56	16.05	2.70	6.10
Robusta (Indonesia)	9.07	1.00	6.40	11.27	1.99	5.65

[a]As oleic; FFA = free fatty acid.

Table 8. Aromatic Components of Roasted Coffee

Hydrocarbons	72
Alcohols	20
Aldehydes	29
Ketones	70
Acids	22
Esters	30
Lactones	9
Amines	22
Sulfur thiols	13
Phenols	44
Furans[a]	115
Thiophenes	30
Pyrroles[a]	74
Oxazoles[a]	29
Thiazoles	27
Pyridines[a]	16
Pyrazines[a]	82
Others	11

[a]Includes benzoderivatives.

ROASTING TECHNOLOGY

The main processing steps in the manufacture of roasted coffee are blending, roasting, grinding, and packaging.

Green coffee is shipped in bags weighing from 60 to 70 kg. Prior to processing, the green coffee is dumped and cleaned of string, lint, dust, hulls, and other foreign matter. Coffees from different varieties or sources are usually blended before or after roasting.

Roasting is usually carried out batchwise by contacting beans with hot combustion gases in rotating cylinders. Continuous roasting using long cylinders fitted with helical conveyor flights is also employed. The beans absorb heat at a fairly uniform rate, and most moisture is removed during the first two-thirds of this period. As the temperature of the coffee increases rapidly during the last few minutes, the beans swell and unfold with a noticeable cracking sound, like that of popping corn, indicating a reaction change from endothermic to exothermic. This stage is known as development of the roast. The final bean temperature, 200 to 230°C, is determined by the blend, variety, or flavor development desired. A water or air quench terminates the roasting reaction. Most, but not all, of any added water is evaporated from the heat of the beans.

Theoretically, about 700 kJ of heat are needed to roast a kilogram of coffee beans. However, the hot gas-recirculation rate determines the thermal efficiency achieved. Older roasters that do not use recirculation may have an efficiency rate as low as 25%, requiring as much as 2800 kJ. Roasters with high levels of gas recirculation have an efficiency rate of 75% or more, requiring about 930 kJ. Volatile organic compounds in roaster discharge gases and smoke have to be reduced to tolerable levels. Some reduction can be effected by partial recycling of roaster gas, but to achieve adequate reduction, discharged roaster gas must be passed through an afterburner.

Most roasters are equipped with controls that regulate the temperature of the hot gas used and automatically quench the roast when a chosen end-of-roast bean temperature is reached. The chosen end-of-roast bean temperature correlates well with the desired color of ground roast coffee as measured by photometric reflectance. Roasts are stopped by diverting the heating gas, injecting a controlled amount of cold water, and blowing room-temperature air through the beans. Control systems introduced in the 1990s provide improved control of roasting reactions and reproducible development of coffee flavor by controlling bean temperature versus time behavior during roasting (10).

Conventional roasting by hot combustion gases in rotating cylinders requires 8 to 15 min. Fast roasting technologies patented in the 1980s develop roast flavor in under 5 min, sometimes in 1 to 3 min (11,12).

Fast roasting, whether carried out in a batch or continuous roaster, is achieved by increasing rates of heat transfer into beans. This may be done by using much more heated gas than used conventionally. The heated gas may flow upward through a bed of beans, fluidizing it. The bed may also be fluidized by hot gas reflected upward from a solid surface after issuing from downward-pointed jets. Rapidly rotating paddles and centrifugal devices that throw beans through heated gas are also used to improve heat transfer to beans. Recirculative lifting of beans by heated gas in a spouted bed is also used. When much larger quantities of gas are used to transfer heat to beans, the gas temperature can be lower, that is, 250 to 310°C compared with 425 to 550°C.

Fast-roasted coffees are less dense, retain more chlorogenic acid, free acid, moisture, and carbon dioxide than conventionally roasted coffees and release 20% more soluble solids during home brewing. It is further claimed that the lower roasting temperature results in higher aroma retention.

Air must be circulated through the beans to remove excess heat before the finished and quenched roasted coffee is conveyed to storage bins. Residual foreign matter, such as stones and tramp iron, which may have passed through the initial green coffee cleaning operation, must be removed before grinding. This is accomplished by an air-lift adjusted to such a high velocity that the roasted coffee beans are carried over into bins above the grinders, and heavier impurities are left behind. The coffee beans flow by airveying or by gravity to mills where they are ground to the desired particle size.

Small and intermediate-size coffee roasters, after virtually disappearing from the United States, have staged a marked comeback. Now, some 1200 to 1500 small, custom roasters roast roughly 12% of the coffee consumed in the United States.

Grinding

Roasted coffee beans are ground to improve extraction efficiency in the preparation of the beverage. Particle size distributions ranging from about 1100 μm average (very coarse) to about 500 μm average (very fine) are tailored by the manufacturer to the various kinds of coffee makers used in households, hotels, restaurants, and institutions.

Finer grinds are provided for espresso coffee (300 to 400 μm), and Turkish coffee (<100 μm).

Most coffee is ground in mills that use multiple steel cutting rolls to produce the most desirable uniform particle size distribution. After passing through cracking rolls, the broken beans are fed between two more rolls, one of which is cut or scored longitudinally, the other, circumferentially. The paired rolls operate at controlled speeds to cut, rather than crush, the coffee particles. A second pair of more finely scored rolls, installed below the main grinding rolls and running at higher speeds, is used for finer grinds. A normalizer section is then used to distribute particles uniformly.

Packaging

Most roasted and ground coffee sold directly to consumers in the United States is vacuum packed in metal cans; 0.45, 0.9, or 1.35 kg are optional, although other sizes have been used to a limited extent. After roasting and grinding, the coffee is conveyed, usually by gravity, to weighing-and-filling machines that achieve the proper fill by tapping or vibrating. A loosely set cover is partially crimped. The can then passes into the vacuum chamber, maintained at about 3.3 kPa (25 mmHg) absolute pressure, or less. The cover is clinched to the can cylinder wall, and the can passes through an exit valve or chamber. This process removes 95% or more of the oxygen from the can. Polyethylene snap caps for reclosure are placed on the cans before they are stacked in cardboard cartons for shipping. A case usually contains 10.9 kg of coffee, and a production packing line usually operates at a rate of 250 to 350 0.45-kg cans per minute.

Though slight losses in fresh roasted character occur due mostly to chemical reactions with the residual oxygen in the can and previous exposure to oxygen prior to packing (9), vacuum-packed coffee retains high quality for at least one year.

Coffee vacuum packed in flexible, bag-in-box packages has gained wide acceptance in Europe and the United States. The inner liner, usually a preformed pouch of plastic-laminated foil, is placed in a paperboard carton that helps shape the bag into a hard brick form when vacuum is applied (13). The carton also protects the package from physical damage during handling and shipping. This type of package provides a barrier to moisture and oxygen as good as that of a metal can.

Inert gas flush packing in plastic-laminated pouches, although less effective than vacuum packing, can remove or displace 80 to 90% of the oxygen in the package. These packages offer satisfactory shelf life and are sold primarily to institutions.

Some coffee in the United States, and an appreciable amount in Europe, is distributed as whole beans, which are ground in the stores or by consumers in their homes. Whole-bean roasted coffee remains fresh longer than unprotected ground coffee and retains its fresh roasted flavor for several days longer than ground.

Roasted whole beans are often packed in bags that contain a one-way valve to allow escape of carbon dioxide gas produced during roasting and prevent air from entering package.

Modified Coffees

Coffee substitutes, which include roasted chicory, chickpeas, cereal, fruit, and vegetable products, have been used in all coffee-consuming countries. Although consumers in some locations prefer the noncoffee beverages, they are generally used as lower-cost beverage sources, rather than as coffee.

Chicory is harvested as fleshy roots, which are dried, cut to a uniform size, and roasted. Chicory contains no caffeine and, on roasting, develops an aroma compatible with that of coffee. It gives a high yield, about 70%, of water-soluble solids with boiling water and can also be extracted and dried in an instant form. Chicory extract has a darker color than does normal coffee brew (14). The growing technology for the processing and use of roasted cereals and chicory is evidenced by the introduction on the market of coffees extended with these materials.

Roughly 2% of the coffee sold in the United States contains added flavors, for example, vanilla, chocolate, and nut flavors. These coffees are manufactured by spraying flavors dissolved in alcohol on roasted beans and drying the beans.

BREWING

Roast and ground coffee is brewed by a wide variety of methods. In most cases, hot water is caused to percolate though a shallow bed of ground coffee supported by a porous surface that retains grounds. Steam-bubble-induced flow may be used to repeatedly apply hot water or hot partly brewed coffee to a bed of relatively coarse grounds, or near-boiling water may be poured on a bed of finer grounds.

Near-boiling water is forced through beds of very fine grounds by a pump when espresso coffee is brewed in modern machines. Steam may be passed through the brew or mixtures of brewed coffee and milk to prepare foamed beverages such as cappucino or latté. Very finely ground, darkly roasted coffee, low water:coffee ratios, and demitasse cups are used when espresso is brewed in Italy. Higher water:coffee ratios and larger-sized cups are often used for espresso made in the United States.

Ground coffee is also cooked in or mixed with boiling water and separated from the resulting brew by decantation, filtration, or sedimentation. Coarse grounds may be used, as sometimes done in Scandinavia, where the practice is decreasing. Superfine grounds are used when Turkish coffee is brewed. Compared with percolation, brewing by mixing grounds with boiling water releases much more caffestol, which elevates serum cholesterol (15).

Roasted coffee contains colloidal material that does not pass through intact cell walls. Very fine and superfine coffees contain many ruptured cells and consequently release colloidal matter during brewing, producing brewed coffee with a mouth-feel different from that produced by more coarsely ground coffee.

Gourmet coffee bars that sell brewed coffee prepared by both standard and espresso methods from custom-roasted, high-quality beans have increased in number in the United

States from 500 in 1991 to roughly 12,000 to 15,000 in 1998.

INSTANT COFFEE

Instant coffee is the dried water-extract of roasted, ground coffee. Although used in army rations during the Civil War, instant coffee did not become a popular consumer item until after World War II. Improvements in manufacturing methods and product quality as well as a demand for convenience foods accounted for its rise in popularity. Some of the soluble coffee consumed in the United States is now manufactured in coffee-growing countries. This may be part of the reason why only 41% as much instant coffee was made in the United States in 1997 as in 1977.

Most soluble coffee blends contain Brazilian, Central American, Colombian, and African robusta coffees. Beans for instant coffee are blended and roasted as for regular coffee but are ground more coarsely to avoid excessive pressure drop and flow maldistribution during subsequent extraction (16). The roasted, ground coffee is charged into columns called percolators through which hot water is pumped to produce a concentrated coffee extract. The extracted solubles are dried, usually by spray or freeze drying, and the final powder is packaged in glass jars at rates of up to 200 jars per minute.

Extraction

Commercial extraction equipment and conditions have been designed to obtain the maximum soluble yield and acceptable flavor. In most processes, the water-soluble components in roasted coffee are first extracted at temperatures slightly lower than 100°C and atmospheric pressure. Additional solubles are then created by hydrolytically converting hemicelluloses and other insoluble components of the roasted coffee into water-soluble materials by pressure extraction at higher temperatures.

The factors influencing extraction efficiency and product quality are (*1*) grind of coffee, (*2*) temperature of water fed to the extractors and temperature profile through the system, (*3*) percolation time, (*4*) ratio of coffee to water, (*5*) premoistening or wetting of the ground coffee, (*6*) design of extraction equipment, and (*7*) flow rate of extract through the percolation columns (16).

Cylindrical percolators with height-to-diameter ratios ranging from 7:1 to 4:1 are common. They are usually operated in series as semicontinuous units of 5 to 10 percolators, with the water flowing countercurrent to the coffee. The ground coffee may be steamed or wetted with water or coffee extract; this supposedly improves extraction (17). Feed water temperatures range from 154 to 182°C, and unless the columns are heated, the temperature drops so that the extract effluent will have cooled to 60 to 82°C. The effluent extract temperature is further reduced by water cooling in plate heat exchangers to minimize flavor and aroma loss prior to drying.

The extract is removed from the percolators and stored in insulated tanks until dried. The extract solubles yield is calculated from the extract weight and soluble solids concentration as measured by specific gravity or refractive index. Yield is controlled directly by adjusting the weight of soluble solids removed and depends primarily on the properties of the coffee, operating temperatures, and percolation time. Soluble yields of 24 to 48% on a roasted coffee basis are possible. Robusta coffees give yields about 10% higher than arabica coffees (16).

High solubles concentrations are desirable to reduce evaporative load in drying and provide good flavor retention during drying. Water inflows that provide extract concentrations in the 20 to 30% soluble solids range are frequently used. More dilute extracts are obtained if greater water inflow is used, but lipophilic flavors and aromas that reside largely in coffee oil in the beans are extracted more efficiently. Some processors concentrate solubles by vacuum evaporation prior to drying. The concentrated percolate may also be clarified by centrifuging prior to drying so that the dry product will be completely free of insoluble fine particles.

The flavor of instant coffee can be enhanced by recovering and returning some of the natural aroma lost during processing. Aroma constituents are liberated and can be collected subsequently by condensation when roasted coffee is ground, when roasted coffee is steamed in a vented extraction column, or when percolate is concentrated. Many patents have been issued dealing with the separation, collection, and transfer of aroma from roasted coffee to instant coffee.

Drying

The following factors are important criteria for good instant coffee drying processes: (*1*) minimum loss or degradation of flavor and aroma, (*2*) free-flowing particles of desired uniform size and shape, (*3*) suitable bulk density for packaging requirements, (*4*) desirable product color, and (*5*) moisture content below 4.5%. Operating costs, product losses, capital investment, and other economic aspects must be considered in selecting the drying process.

Spray Drying

Most instant coffee is spray dried. Concentrated coffee extract is atomized, usually by pumping it through pressure nozzles. Nozzles and nozzle combinations are used that provide the desired particle size, product bulk density, and flow capacity based on the properties of the extract. The resulting stream of drops, joined by concurrently flowing heated air, flows downward into a tall cylindrical drying chamber. The air transfers heat to the drops, causing water evaporation and progressive drying of the drops. The air progressively cools and becomes more and more humid as drying progresses. Most processors prefer to use low inlet-air temperatures (200–260°C) for best flavor quality, and use enough air to provide outlet-air temperatures between 107 and 121°C. Spray dryers are usually constructed of stainless steel and must be provided with adequate dust collection systems, such as cyclones or bag filters (16).

The particles of dried instant coffee are collected from the conical bottom of the spray dryer and from cyclones through rotary valves and are conveyed to bulk storage or packaging bins. Processors may screen the dry product to obtain a uniform particle-size distribution.

Agglomeration

At one time most instant coffee was marketed as small, spherically shaped particles. Now it is marketed mostly in a granular form. Instant coffee granules are produced by moistening small spray-dried particles with steam, thereby fusing clusters of particles together. These clusters are then dried in a tower similar to a spray-drying tower using a low level of heat. Particle fusion and drying may also be carried out on a continuous belt.

Freeze Drying

Freeze-dried instant coffee is made by freezing coffee extract, grinding the frozen extract into granules and then converting the ice and bound-water content of the granules into water vapor by sublimation under high vacuum. Absolute pressures of 67 Pa (500 μmHg), or lower, are used. The heat used to induce sublimation is radiantly transferred from hot plates to the outer surface of beds of granules and then conductively transferred in the bed and granules to sublimation sites. Freeze drying occurs at much lower temperatures than spray drying. Heat input is controlled to give maximum end-point temperatures between 38 and 49°C (18). Drying times are much longer than for spray drying (18). Freeze-dried coffees use high-quality coffee and have better retention of volatile aromatics than do spray-dried coffees and are thus considered quality instant coffees.

Packaging

In the United States, instant coffee for the consumer market is usually packaged in glass jars containing from 56 to 340 g of coffee. Larger units for institutional, hotel, restaurant, and vending-machine use are packaged in bags and pouches of plastic or paper. In Europe, instant coffee is packaged in glass jars and, frequently, plug-closure metal containers with foil liners.

Protective packaging is primarily required to prevent moisture pickup. The flavor quality of regular instant coffee changes very little during storage. However, the powder is hygroscopic, and moisture pickup can cause caking and flavor impairment. Moisture content should be kept below 5%.

Many instant coffee producers in the United States incorporate natural coffee aroma fixed in expelled coffee oil in the powder. These highly volatile and chemically unstable flavor components necessitate inert-gas packing to prevent aroma deterioration and staling from exposure to oxygen.

DECAFFEINATED COFFEE

Decaffeinated coffee was first developed on a commercial basis in Europe about 1900. The basic process is described in a 1908 patent (19). Green coffee beans are moisturized by steam or water to a moisture content of at least 20%. The added water and heat separate the caffeine from its natural complexes and aid its transport through the cell wall to the surface of the beans. Solvents are then used to remove the caffeine from the wet beans.

Up to the 1980s man-made organic solvents were commonly used. The caffeine is removed either by direct contact of solvent with the beans or by contact with a secondary water system that has previously removed the caffeine from the beans (13). In either case additional steaming or stripping is used to remove solvent from the beans. The beans are dried to their original moisture content of about 10 to 12% prior to roasting.

In the 1980s decaffeination processes were commercialized making use of solvents that occur in nature or can be made from substances that occur in nature. The use of these processes are the basis of positioning a coffee product as naturally decaffeinated.

In a 1970 patent, Studiengesellschaft Kohle of Mulheim, Germany, showed that dense supercritical carbon dioxide is a very specific solvent for caffeine. Subsequent patents (20) describe use of this technology to decaffeinate coffee in a semicontinuous commercial process. Caffeine selectively transfers from wetted green coffee beans to supercritical carbon dioxide and then transfers from the carbon dioxide to water, thereby regenerating the carbon dioxide for reuse and recovering the caffeine.

Fats and oils, including oil from roasted coffee, (21) and edible esters, including ethyl acetate, which is present in coffee, are also used to selectively extract caffeine from coffee. Direct water contact with green beans is used in another process. This contact removes caffeine as well as some noncaffeine solids. The caffeine is adsorbed on activated carbon or an ion exchange resin, and the noncaffeine solids, containing flavor precursors, are reabsorbed on green beans prior to drying and roasting.

In all of the preceding decaffeination processes, prewetting of the green beans is necessary, and drying afterward is needed prior to roasting. These steps, in addition to caffeine removal, cause changes in the beans that affect roast flavor development.

The degree of decaffeination as claimed on the product is based on the caffeine content of the starting material and the time-temperature process conditions used by the manufacturer to achieve a desired end point.

Roasted decaffeinated coffee is vacuum packed as ground coffee or whole beans for consumer use. Roasted and ground decaffeinated coffee is made into instant coffee by methods previously described. Decaffeinated coffee represents about 9% of the coffee consumed in 1997 in the United States, declining from a peak of 18% in 1987 (22).

BIBLIOGRAPHY

This article was adapted and updated from the article "Coffee" by A. Stefanucci, W. P. Clinton, and M. Hamell, in M. Grayson, ed., *Kirk-Othmer Encyclopedia of Chemical Technology*, Vol. 6, 3rd ed., John Wiley & Sons, New York, 1979, pp. 511–522 and from an earlier version of this article by W. Clinton in *The Wiley Encyclopedia of Food Science and Technology*.

1. W. A. Ukers, All About Coffee, 2nd ed., *Tea & Coffee Trade Journal*, New York, 1935, pp. 1–3.
2. NTIS, U.S. Department of Commerce, *Tropical Products (Coffee, Cocoa, Spices, Essential Oils)*, June 1998.

3. A. Stefanucci and K. Sloman, *Internal Report*, Technical Center, General Foods Corp.
4. A. G. W. Bradbury and D. J. Halliday, "Polysaccharides in Green Coffee," paper presented at the Association Scientifique Internationale du Cafe, 12th colloquium, Montreux, Switzerland, 1987, p. 266.
5. H. Thaler and R. Gaigal, *Zeitschrift fuer Lebensmittel-Untersuchung und-Forschung* **120**, 449–454 (1963).
6. J. R. Feldman, W. S. Ryder, and T. J. Kung, *J. Agric. Food Chem.* **17**(4), 733–739 (1969).
7. J. Carisano and L. Gariboldi, *J. Sci. Food Agric.* **15**, 619–622 (1964).
8. I. Flament, "Research on the Aroma of Coffee," paper presented at the A.S.I.C. 12th Colloquium, Montreux, Switzerland, 1987, p. 146.
9. W. Clinton, "Evaluation of Stored Coffee Products," paper presented at the A.S.I.C. 9th Colloquium, London, 1980, p. 273.
10. Praxis Werke Inc., *Logofile*, 1995.
11. U.S. Patent 4,169,164 (Sept. 25, 1979), M. Hubbard (to Hills Bros. Coffee Co.).
12. U.S. Patent 4,737,376 (Apr. 12, 1988), L. Brandlein et al. (to General Foods Corp.).
13. U.S. Patent 2,309,092 (Jan. 26, 1943), N. E. Berry and R.H. Walters (to General Foods Corp.).
14. R. J. Clarke and R. Macrae, eds., *Coffee*, Vol. 5, *Related Beverages*, Elsevier Applied Science, Barking, UK.
15. M. B. Katan and R. Ungert, "The Cholesterol Elevating Factor from Coffee Beans," 16th A.S.I.C. Colloquium, Kyoto, 1995, p. 49.
16. M. Sivetz and N. W. Desrosier, *Coffee Processing Technology*, Avi Publishing, Westport, Conn., 1979.
17. U.S. Patent 3,549,380 (Dec. 22, 1970), J.M. Patel et al. (to Procter & Gamble).
18. U.S. Patent 3,438,784 (Apr. 15, 1969), W.P. Clinton and coworkers (to General Foods Corp.).
19. U.S. Patent 897,763 (Sept. 1, 1908), J. F. Meyer and coworkers (to Kaffee-Handels-Aktien-Gesellschaft).
20. U.S. Patent 4,820,537 (Apr. 8, 1989), S. Katz (to General Foods Corp.).
21. U.S. Patent 4,465,699 (Aug. 14, 1984), F.A. Pagliaro and coworkers (to Nestlé, SA).
22. National Coffee Organization, "Coffee Drinking Study, U.S.A.," Winter, 1997.

GENERAL REFERENCES

R. J. Clarke and R. Macrae, eds., *Coffee*, Vols. 1–6, Elsevier Applied Science Publishers, Barking, UK, 1985–1988.

M. N. Clifford and K. C. Wilson, eds., *Coffee: Botany, Biochemistry and Production of Beans and Beverages*, Croom Helm, London, 1985

B. Rothfos, *Coffee Consumption*, Gordian-Max Rieck GmbH, Hamburg, 1986

M. Sivetz and N. W. Desrosier, *Coffee Processing Technology*, Avi Publishing, Westport, Conn., 1979.

Proceedings of International Scientific Colloquium on Coffee (held every two or three years since 1963), Assoc. Scientifique Internationale du Cafe, Paris, France.

HENRY G. SCHWARTZBERG
University of Massachusetts
Amherst, Massachusetts

COLLOID MILLS

A colloid mill is a device used in the preparation of emulsions and dispersions. The name implies that it can generate colloidal size droplets (1–1,000 nm) for the disperse phase, but in reality the colloid mill produces emulsions with droplets in the size range of 1 to 25 μm. It is sometimes used for dispersing solid particulates throughout the continuous phase, but in most cases it does not grind particles, rather it deagglomerates and disperses the solids. In the early years of the colloid mill, one author suggested that it be called a dispersion mill to reflect its function better (1). However, the term colloid mill is still used to describe this type of device.

HISTORY

The first mention of a colloid mill came from a Russian colloid chemist, Von Weimarn, who in 1910 proposed using a mechanical device to generate shearing action to produce colloidal dispersions (1–5). In Hamburg, another Russian engineer named Plauson developed and patented in 1921 a colloid mill. Plauson modified a hammer mill "... so that the beater arms impinged on a liquid surface. Further, he operated the rotor at very high peripheral speed to increase the number of blows in a given time" (2). The Plauson mill used excessive horsepower, was intermittent in its operation, and was not very successful on a plant scale (1).

After the introduction of the Plauson mill, many different designs of colloid mill were introduced. In 1928, Clayton, in his book *The Theory of Emulsions and Their Technical Treatment*, described 22 different colloid mill designs, many protected by patents (2). The colloid mill became popular because of its high volume output and low maintenance costs.

GENERAL DESCRIPTION OF DESIGN

There are many different colloid mills commercially available, but each share some common elements. Colloid mills usually consist of a rotor and a stator (Fig. 1). The rotor is a disk that can have many different forms and surfaces, but it is most commonly conical. This disk rotates at a high speed, usually between 1,500 and 20,000 rpm depending on the particular design. The speed of the rotor is more commonly in the intermediate range of 3,000 to 10,000 rpm for smaller machines and roughly half that range for larger machines. The rotor may be directly connected to the motor, or it may be driven by a gear or drive belt system. The stator is a fixed surface that is close to the rotor. However, colloid mills have been designed with many variations of these elements, such as two rotors spinning in opposite directions close to each other; and rotors and stators with slots, channels, flutes, teeth, projections, corrugations, steps, and intercalating ridges forming a sinuous passage (2). The diameter of the rotor may range from 1 to 21 inches (2.54–53.3 cm). In operation some colloid mills require a feed pump, whereas others are self-pumping.

Flowing between the rotor and stator is a thin film of liquid that is subjected to a very high shear field. The thick-

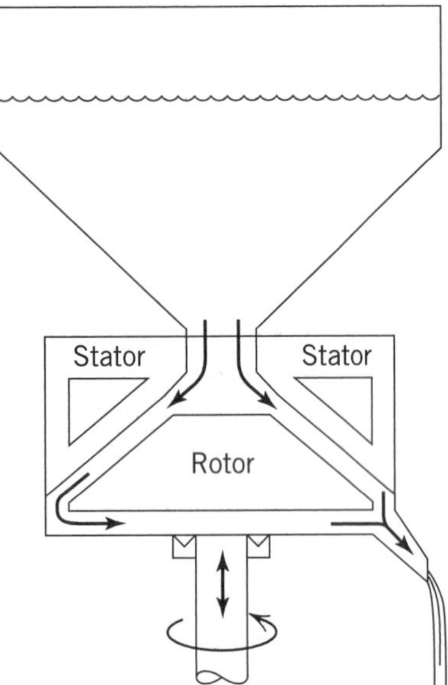

Figure 1. Basic components of a colloid mill: a rotor attached to a rotating shaft and a stator close to the rotor with a small gap between them.

ness of this layer of liquid is typically between 0.001 and 0.050 inches (0.0254–1.27 mm). It is characteristic of a colloid mill that the intensity of the shear field can be adjusted so as to achieve the desired process result. This adjustment is most often made by varying the thickness of the gap between the rotor and the stator.

THEORY OF OPERATION

The prevailing theory of the action in a colloid mill is shear. Fluid shear occurs in a layer of liquid between two planes, one stationary and the other mobile. There exists a gradient of velocity between these two planes. At the mobile plane the velocity is at its maximum, and at the stationary plane it is at its minimum (zero velocity). This means there are parallel layers of liquid moving past each other at different velocities subjecting dispersed particles to a disrupting force. The rate of change in velocity over the distance between these two planes is the shear rate. A common method of estimating the shear rate in a colloid mill is given as (8)

$$\gamma = n\pi D_r/h$$

where n is the number of revolutions per second, Dr is the rotor diameter, and h is the gap width. By this equation, a colloid mill with a rotor diameter of 2 inches (50.8 mm), a speed of 20,000 rpm, and a gap of 0.010 inches (0.254 mm) produces a shear rate of 2.09×10^5 sec^{-1}. At a gap of 0.002 inches (0.0508 mm) the shear rate is 1.05×10^6 sec^{-1}. A rotor diameter of 5 inches (127 mm) at 3,600 rpm with a gap of 0.005 inches (0.127 mm) has a shear rate of 1.88×10^5 sec.^{-1}. These examples illustrate the high shear rate generated in a colloid mill, and how this shear rate varies with operating conditions.

Some authors have suggested that other mechanisms besides shear are important in a colloid mill. Turbulence and cavitation may occur in mills with corrugated rotors and stators (6). Some researchers believe that the dispersion is caused by the collision of particles against each other at the surface of the rotor. Particles are set into rotation by the high shear field in the gap between the rotor and stator. A lift force pushes the rotating particles against the rotor. The comminution or emulsification occurs by "autogenous grinding empowered by the rotational kinetic energy of particles in suspension" (7). The rotational energy spectrum varies with "particle size, fluid viscosity, gap size, mill speed, and other parameters in the system" (7).

A recent paper describing the mechanism of emulsification in a colloid mill emphasizes the importance of residence time in the working area (8). Two models for droplet breakup are presented. One model is capillary breakup, where a droplet is stretched into an elongated form that is then broken into many droplets. The other model is a binary breakup where a cascade of events is needed to complete the reduction of the droplet. There was good agreement between experimental observation and the capillary model. One significant parameter in these models is residence time in the gap. Because these models require a certain time for breakup of the droplets, the residence time is an important factor in understanding scale-up and variations in colloid mill designs.

Residence time in a colloid mill is important because the flow rate in a mill is dependent on several factors. The flow rate depends on the rheological properties of the product, the clearance between the rotor and stator, the in-feed pressure to the mill, the shape of the rotor and stator, the amount of restriction to flow at the discharge from the mill, and the speed of the rotor. Some factors that reduce the flow rate of the mill are high viscosity, reduced clearance between the rotor and stator, restriction to flow at the discharge, and a flow path that is axial or tends radially inward. Factors that increase flow through the mill are increased in-feed pressure, large clearance between the rotor and stator, increased speed of the rotor, low viscosity fluid, no restriction to flow at the discharge, and a flow path that tends radially outward.

The temperature rise in a product flowing through a mill depends on the residence time in the working area. The longer the fluid is in the working area, the greater the temperature rise. Therefore, all the factors that increase residence time most likely increase temperature of the product.

Figure 2 shows the consequences of varying these factors for mayonnaise. Mayonnaise was processed in a laboratory-sized colloid mill. The flow rate, the product viscosity, and the in-feed pressure were measured during the test. When the in-feed pressure is increased, the flow rate through the mill is also increased. However, as the in-feed pressure is increased, the product viscosity is decreased because the residence time for the product is reduced and less work is done on the product.

PRODUCTS

The most common food applications of the colloid mill are the processing of mayonnaise and salad dressing.

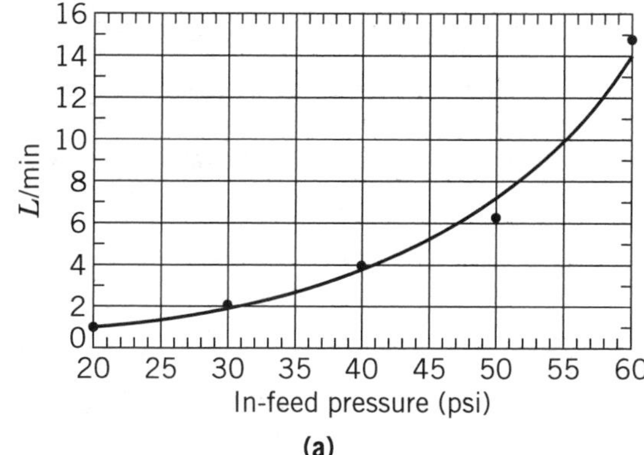

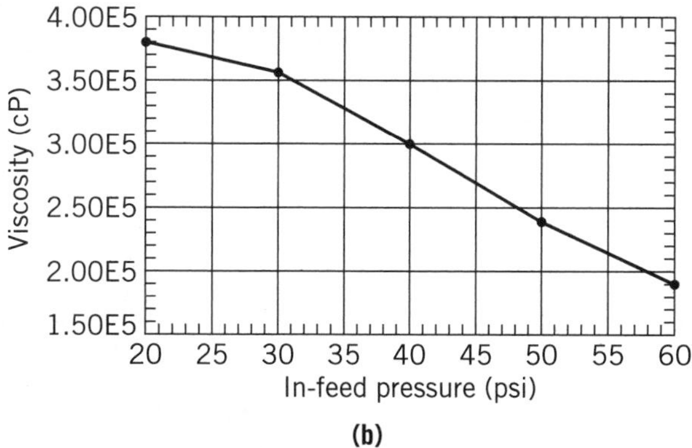

Figure 2. (**a**) Flow rate of a mayonnaise through a colloid mill increases with in-feed pressure. The gap between the rotor and the stator was 0.010 inches (0.254 mm). Density was 920 g/L. (**b**) Viscosity is reduced when the in-feed pressure is increased, causing an increase in flow rate and reduction in residence time. *Source:* From the research laboratory APV Americas-Homogenizers.

Mayonnaise contains 65% or more vegetable oil. The colloid mill produces the best combination of fluid processing conditions to make a good emulsion at this oil level. If this high oil level emulsion is processed on a homogenizer, then the results are poor. The homogenizer makes too small a droplet size, which leads to instability and separation of the emulsion because the natural emulsifier in mayonnaise cannot stabilize the large interfacial area generated. The colloid mill produces a droplet size range that makes a stable, viscous product. On the other hand, a salad dressing at 30% oil can be made on a homogenizer or a colloid mill. Because of the high viscosity usually associated with a salad dressing, a small oil droplet size is not necessary, and the energy level of a colloid mill is adequate to make this type of emulsion.

There are many other products made on the colloid mill. In the cosmetic industry colloid mills are used for making creams and lotions. In the chemical industry they are used for preparing bituminous emulsions, silicone oil emulsions, grease, and pigment dispersions. In many of these applications, the colloid mill works well because of high dispersed phase concentration in the product, high viscosity components, and moderate droplet size requirements.

BIBLIOGRAPHY

1. P. M. Travis, *Mechanochemistry and the Colloid Mill*, Chemical Catalog Co., New York, 1928.
2. W. Clayton, *The Theory of Emulsions and Their Technical Treatment*, 3rd ed., P. Blakiston's Son & Co., Philadelphia, Penn. 1935.
3. E. A. Hauser, *Colloidal Phenomena*, MIT Press, Cambridge, Mass., 1954.
4. E. K. Fischer, *Colloidal Dispersions*, John Wiley & Sons, New York, 1950.
5. J. W. McBain, *Colloid Science*, D. C. Heath, Boston, 1950.
6. P. Walstra, "Formation of Emulsions," in P. Becher, ed., *Encyclopedia of Emulsion Technology*, Vol. 1, Marcel Dekker, New York, 1983, pp. 57–127.
7. A. G. King and S. T. Keswani, "Colloid Mills: Theory and Experiment," *J. Am. Ceram. Soc.* **77**, 769–777 (1994).
8. J. A. Wieringa et al., "Droplet Breakup Mechanisms During Emulsification in Colloid Mills at High Dispersed Phase Volume Fraction," *Trans. Inst. Chem. Eng.* **74**, 554–562 (1996).

William D. Pandolfe
APV Americas-Homogenizers
Wilmington, Massachusetts

COLOR AND FOOD

IMPORTANCE OF FOOD COLOR

The role that color plays in the reaction to food is so automatic that it may be taken for granted. This does not make that role any less important it merely decreases awareness. The color of a food has important considerations for both the consumer and the technologist. These considerations are quite different yet they are interrelated as shown in Figure 1.

The psychological effects of color have long been recognized. A room decorated in red exudes warmth and may increase pulse and respiration rates and even increase vivacity. Whereas a room decorated in blue or green is cool and peaceful encouraging concentration and relaxation. But darkening that blue can turn it into a subdued and even depressing atmosphere. The color of food also has its psychological aspects but these are less of mood swings than they are of learned associations. The color of a food not only sends a message of expectation but can also provide clues as to the condition of that food; a yellow peach is ripe, brown strawberry jam is old. Food of an unnatural color raises a barrier that most people have difficulty overcoming. Witness the red-fleshed potato that cooks up blue. Most people given the chance will avoid tasting it even when assured it tastes similar to the familiar white varieties. In contrast a recently introduced yellow-fleshed variety, reminiscent of buttery mashed potatoes, is gaining in popularity. In a very real sense the appearance of a food acts as a gate keeper and the old saying "We eat with our eyes" is not far off the mark. Experience and memory play important roles in food assessment. This is easily demonstrated by mismatching gelatin samples for color and flavor. Even a familiar flavor such as orange is difficult to

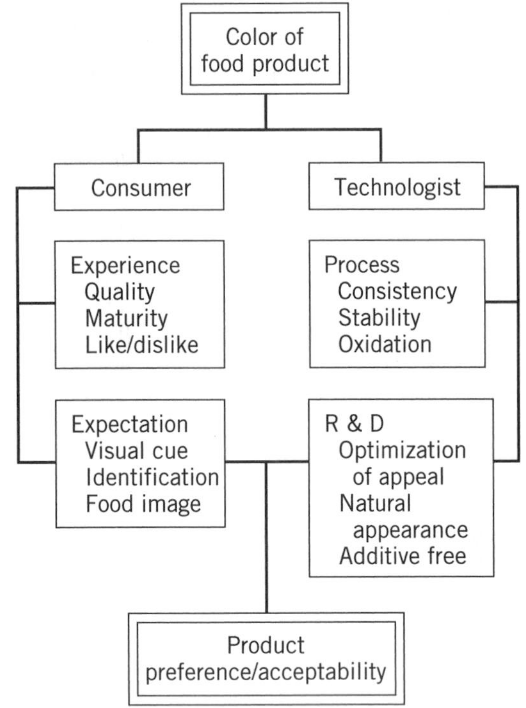

Figure 1. Food color considerations for the consumer and for the food technologist.

recognize when colored red. This is because red is associated with red fruits, thus a red-colored orange gelatin sends the wrong visual cue. This is of such proportion that it is difficult to rely on the messages sent by the taste buds and olfactory receptor.

In practice consumers do not isolate their sensory perceptions of a food but combine them in obtaining a total assessment for a food. This association is sufficiently strong that when asked to separate them there may still be a strong influence of one sensory perception on another. Yet the visual sense is both sophisticated and sensitive, it is only that experience has confirmed what the eye sees, which has caused the interaction of the senses to become unconscious.

The technologist has an interest in the color of food for several reasons. One is the need to maintain a uniformity of color over production runs. A second is the avoidance of color changes brought about by chemical reactions occurring during processing or storage life of the product. A third is the optimizing of color and appearance in relation to consumer preference. A fourth is the maintenance of a color in accordance with consumer experience and expectations. The need to maintain color uniformity over production runs varies with the food and especially with the packaging used. For example, the need is greater for fruit juices packaged in clear containers than for those packaged in individual drinking boxes where very little of the juice is actually seen. The color problems of processed green vegetables are examples of color changes arising from unstable pigment reactions. The retailing practices of fresh meat, particularly beef, have been limited for many years by the need to maintain meat pigments in the bright cherry red of oxymyoglobin because of the strong effect on consumer acceptability.

For many new products the product developer must consider the eye appeal of the product, and this should be given serious consideration in consumer acceptability ratings in any optimization procedure undertaken. The need to be aware of dynamic consumer reactions in regard to color and appearance is paramount. What is preferred in the color of a product is influenced by experience and availability and changes over time. The replacement of nitrites for curing meat products is complicated by the fact that the nitrites contribute to the characteristic color of cured meat. A satisfactory replacement has yet to be found. To support the deletion of nitrites from the process, consumers must exhibit sufficient concern about their diets that they would be willing to accept brown cured meat products or at least products that appeared different from those to which they are accustomed.

COLOR THEORY

A thorough review of color theory is beyond the scope of this article; however, several references have been published (1–5). The portion of the electromagnetic spectrum that comprises visible light falls between 380 and 750 nm, bordered by ultraviolet light on the low end and infrared light on the high end. The visible spectrum is represented by the colors seen in the rainbow with blue existing at wavelengths less than 480 nm; green roughly 480–560 nm; yellow, 560–590; orange 590–630; and red, at wavelengths longer than 630 nm (1). Purple is achieved by mixing blue and red and is considered to be a nonspectral color. If white light is dispersed by a prism a spectrum is obtained representing all the visible colors at appropriate wavelengths. The relative power or energy (power multiplied by time) emitted at these wavelengths can be plotted to produce the spectral power distribution curve of the light source. A group of light sources, black bodies, change from black through red to white when heated, the color produced being dependent on the temperature reached. This is referred to as color temperature. Tungsten filaments are close approximations of black bodies, although their color temperature is not exactly equal to their actual temperatures. Real daylight and fluorescent lights do not approximate black bodies.

When light falls on an object it may be reflected, absorbed, or transmitted or a combination of these may occur. When light passes through a material essentially unchanged it is transmitted. Wherever light is slowed down, as occurs at the boundary of two materials, the light changes speed and the direction of the light beam changes slightly (refractive index). The change in direction is dependant on the wavelength and accounts for the dispersion of light into a spectrum by a prism. The change in refractive index results in some light being reflected from the object. Light that is absorbed by the material is lost as visible light. If light is absorbed completely the object appears black and opaque. If some of the light is not absorbed but transmitted the object appears colored and transparent. Lambert's law, which states that the fraction of light that is absorbed by a substance is independent of the intensity of the incident light, is always true in the absence of light scattering. Beer's law states that the light absorption is directly proportional to the number of molecules through which the light passes. In practice, Beer's law is often not exactly observed, possibly due to chemical changes in some of the absorbing molecules.

When light interacts with matter it may travel in many directions. The sky is blue because of light scattered by molecules of air. Scattering caused by larger particles produces the white of clouds, smoke, etc. Scattering results in light being diffusely reflected from an object. Translucent materials transmit part of the light and scatter part of the light, if no transmittance occurs, the object is said to be opaque. The color of the object is dependent on the amount and kind of scattering and the absorption occurring. The amount of light that is scattered is dependent on the difference between the refractive indexes of the two materials. The boundary between two materials having the same refractive index cannot be seen because no light scattering occurs. Large particles scatter more light than small particles until the particle size approaches the size of the wavelength, at which point scattering decreases. Many foods are not completely reflecting or transmitting materials, which introduces an element of empiricism into an attempt to measure the color (6). With these foods absorption and scattering are factors affecting visual judgments. The absorption coefficient K and the scatter coefficient S in the Kubelka-Munk equations have been used in attempts to deal with this (7,8).

The chemist tends to think of color rather simplistically as being determined by the amount of light absorbed at a specified wavelength. But that is not how color is seen, for it is the light that is reflected from an object that determines color. The appearance of an object may vary over the entire object because of the angle of viewing and the light falling on the object. Appearance attributes have been divided into two categories: color, and geometric or spatial attributes (3). Color refers to the light reflected from the object to provide a portion of the spectrum as well as white, gray, black, or any intermediate. Geometric attributes result from the spatial distribution of the light from the object and are responsible for the variation in perceived light over a surface of uniform color, such as gloss or texture. The mode in which the eye is operating affects the visual evaluation. There are three modes: the illuminant mode, in which the stimulus is seen as a source of light; the object mode, in which the stimulus is an illuminated object; and the aperture mode, in which the stimulus is seen as light. The aperture mode may be thought of as a lighted window where first the light is seen, but at a closer range the object mode takes over and the contents of the room can be seen. The aperture mode views an object through a smallish aperture removing the effects of spatial distribution of light. This method is usually used for visual color-matching experiments. However, it is the object mode that is of practical importance in assessing the appearance of a food. The everyday evaluations of color, haze, clarity, gloss, and opacity are made by observing what the object does to the light falling on it.

The human eye is an incredibly sensitive and discriminating sensor; it can detect up to 10,000,000 different colors (6). The basic units of sight are the eye, the nervous system, and the brain. The cone light receptors located in the fovea of the retina of the eye are responsible for the ability to see color. It is generally agreed that there are three types of cone receptor responding to red, green, or blue and that these responses are converted in nerve-signal-switching areas within the eye and the optic nerve to opponent-color signals as proposed by Muller. Three opponent-color systems, black-white, red-green, and yellow-blue, were first proposed by Hering (3). The other type of light receptor, the rods, are also located in the retina; they increase in density as the distance from the fovea increases. The rods are responsible for black-and-white vision, the ability to see in dim light. Rods do not contribute to color vision. Approximately 8% of the population, predominately men, have abnormal color vision. For all humans the mode of presentation including viewing angle, source of illumination, and background, affect the color perceived by the viewer.

COLOR MEASUREMENT

Tristimulus Colorimetry

It is possible to match colors with a simple laboratory setup. Three projectors with a red, green, and blue filter, respectively, can be focused with the image superimposed on a screen. A color to be matched is projected on the screen by a fourth projector. An operator can match the desired color by determining the amounts of red, green, and blue required for a match. Unfortunately, not all colors can be matched by the red, green, and blue primary colors, so the researchers were given free rein to choose other primaries. They chose X (roughly corresponding to red), Y (roughly corresponding to green), and Z (roughly corresponding to blue). X Y and Z are unreal in that they cannot be produced directly in the laboratory, but they are very useful mathematically. This concept was accepted by the Commission Internationale de Eclairage (CIE) and became known as the CIE system. These tristimulus (three-stimuli) values for the equal-energy spectrum were used to define the 2° 1931 standard observer and were designated as $\bar{x}$, $\bar{y}$, and $\bar{z}$, which are special cases of X, Y, and Z; the 2° refers to the angle of observation. Later systems of measurement were developed with a 10° angle of viewing and are thus called 10° 1964 standard observers. The definition of the XYZ system offered a number of advantages. First, all colors lie within the primary coordinates. Second, the Y value contained all the lightness, and third, many of the coordinates for orange and red colors lie along a straight line, which is a great advantage in color formulation.

The XYZ values, sometimes presented as Y,x,y or xyz as described below, provided a measure of the color of a sample when viewed under a standard source of illumination by a standard observer. The CIE organization defined a number of standard sources of illumination. CIE Source A represented incandescent light and a color temperature of 2,854 K; CIE Source B simulated noon sunlight; and CIE Source C simulated overcast-sky daylight. These light sources were supplemented by the CIE in 1965, and illuminant D_{65} with a color temperature of 6,500 K is now widely used. In 1931 the CIE adopted the standard observer as having color vision representative of the average of that of the population having normal color vision. These data were used to obtain CIE standard-observer curves for the visible spectrum for the tristimulus values r, g, and b, for a set of red, green, and blue primaries. The red, green, and blue data were transformed into XYZ values and provided the basis for the standard observer curves shown in Figure 2 (1). The definition of the standard observer curves made it possible to design a simple colorimeter. Figure 3 shows white light falling on a sample to be measured. The reflected light is passed through a glass filter to determine the X component, the Y component, and the Z component, respectively. The signal from each filter is measured electronically and constitutes a tristimulus reading for the sample. The accuracy of the reading depends on the ability of the manufacturer to produce a filter/photocell combination that duplicates the response of the human eye, that is, the standard observer curves. They have been very successful in this aspect, and several commercial colorimeters available today are based on this principle.

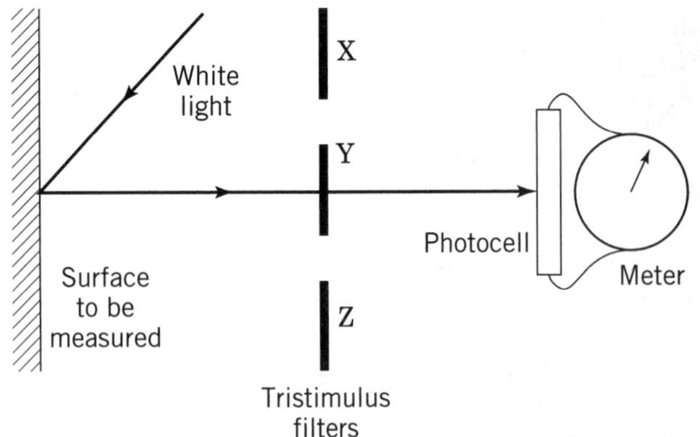

Figure 3. A simple tristimulus colorimeter. The light reflected from the sample passes through the three tristimulus filters, X, Y, and Z, and the signals are recorded by the photocell. The combined outputs from the filter/photocell combination duplicate the standard observer curves x,y, and z and represent the color of the sample.

Spectrophotometry

The CIE values of a color can be calculated by multiplying the energy spectrum of the illuminant by the reflectance, or transmission, spectrum of the sample and the standard observer curves. They can be mathematically represented by the following integral equations:

$$X = \int_{380}^{750} \mathrm{RE}\bar{x}dx$$

$$Y = \int_{380}^{750} \mathrm{RE}\bar{y}dy$$

$$Z = \int_{380}^{750} \mathrm{RE}\bar{z}dz$$

where R is the sample spectrum, E is the source light spectrum, and $\bar{x}$, $\bar{y}$, $\bar{z}$ are the standard observer curves.

Many of the more expensive colorimeters available today record a spectrum from the sample as the initial signal and calculate CIE values from it. This approach allows considerable flexibility because the signal can be programmed to produce data for any illuminant, several types of sample presentation, a number of mathematical color systems, special color scales, and many types of color difference calculations. Interestingly, the early spectrophotometers designed for color measurement 70 years ago all used this approach, but the calculations for CIE values were tedious. Mechanical integrators helped, but they were expensive. This led to the development of the much simpler tristimulus colorimeters about fifty years ago. Today with the development of electronic calculation, the cost of calculation is simply not a significant consideration, and spectrophotometers are coming back into fashion.

Interestingly, a tristimulus colorimeter can also be used as a chemical absorptimeter for analytical purposes (9,10).

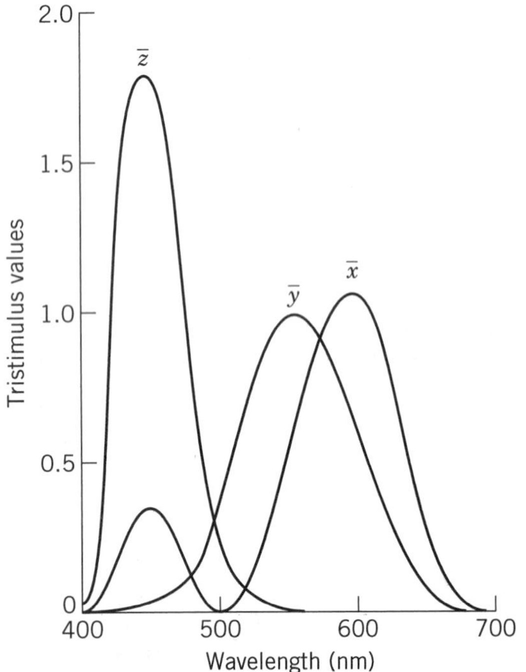

Figure 2. Curves for 2° standard observer expressed as tristimulus values $\bar{X}$, $\bar{Y}$, and $\bar{Z}$.

Data Presentation

Spectrophotometers usually present data in XYZ units but there are other conventional ways. Chromaticity coordinates (xyz) calculated by the following equations are often used to plot a chromaticity diagram (Fig. 4). The convention Yxy is also used, where Y represents CIE Y expressed as a percent of $X + Y + Z$:

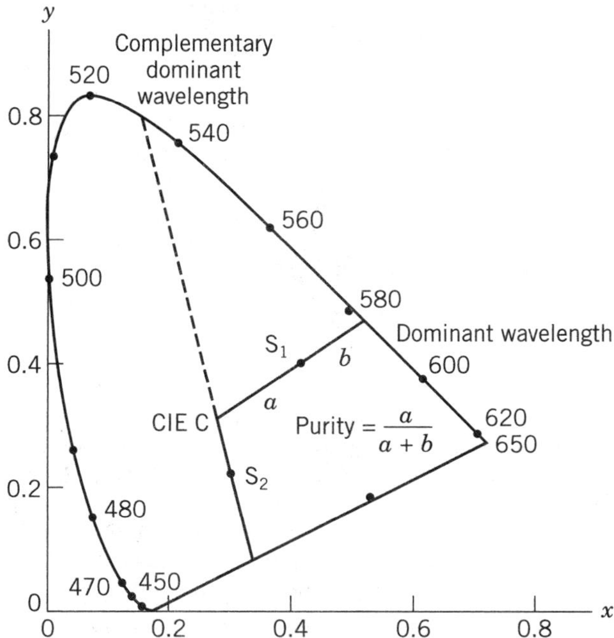

Figure 4. CIE chromaticity diagram showing dominant wavelength and purity.

$$x = \frac{X}{X + Y + Z}$$

$$y = \frac{Y}{X + Y + Z}$$

$$z = \frac{Z}{X + Y + Z}$$

By convention x and y are plotted on the diagram that features the horseshoe-shaped spectrum locus. Dominant wavelength is obtained by drawing a line from the illuminant through the sample to the spectrum locus as shown in Figure 4 and represents visual hue. Purity is obtained by the equation $a/(a + b)$ and represents visual chroma or saturation. Samples located nearer the spectrum locus are more saturated. Colors placing below the illuminant at the horseshoe base are considered nonspectral colors, and a complimentary dominant wavelength is calculated by drawing a line from the sample through the illuminant to the spectrum locus. Purity is determined as for spectral colors. The horseshoe spectrum locus represents two dimensions of color and the third is represented by Y as shown in Figure 5. The plane of the horseshoe rises as the color becomes lighter. Spectrum colors on the spectrum locus have a low lightness factor.

The Hunter L, a, and b color scale is an opponent-type system and has the advantage of being more uniform than the CIE system. It has proved popular for measuring food products. This scale was developed in 1958, allowing computation and direct readout from the dials of the colorimeter (3). Readout is as L_L, a_L, and b_L, and the functions are related to the CIE X, Y, and Z as

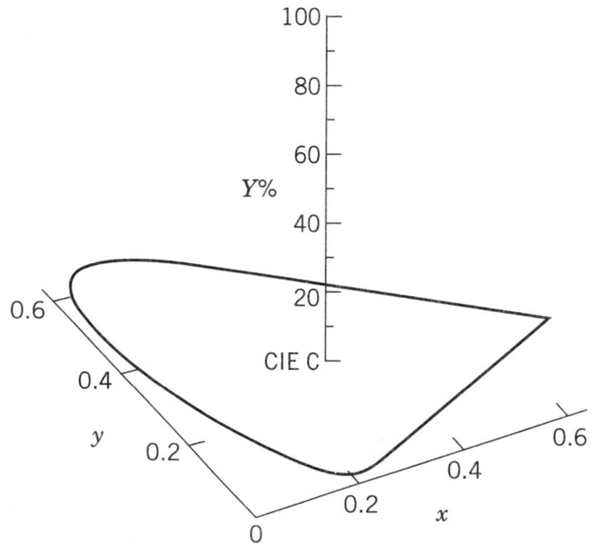

Figure 5. Third dimension of color represented by Y function rising from the plane of the chromaticity diagram.

$$L_L = 10(Y)^{1/2}$$

$$a_L = \frac{17.5(X - Y)}{(Y)^{1/2}}$$

$$b_L = \frac{7.0(Y - Z)}{(Y)^{1/2}}$$

The a and b chromaticity dimensions are expanded so that intervals on the scale for a and b approximate the visual correspondences to those of the 100-unit lightness scale L_L. This color scale is based on Hering's theory that the cone receptors in the eye are coded for light–dark, red–green, and yellow–blue signals. It is argued that a color cannot be both red and green at the same time; therefore, a is used to represent these; $-a$ represents greenness and $+a$ represents redness. Similarly, $-b$ represents blueness and $+b$ represents yellowness. Lightness is represented by L_L (Fig. 6). Hue is calculated as the angle going counterclockwise from $+a$ and hue angle is calculated as $\tan^{-1} b/a$. Chroma is calculated as $(a^2 + b^2)^{1/2}$. Color difference

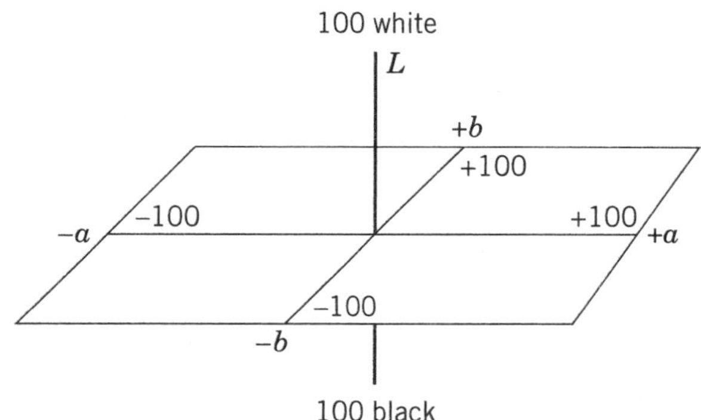

Figure 6. Hunter L a b color scale.

is calculated as $\Delta E = [(L_1 - L_2)^2 + (a_1 - a_2)^2 + (b_1 - b_2)^2]^{1/2}$, but this value gives no indication in which dimension the difference lies. The Hunter color scale was first developed for use with reflectance and later adapted for transmission measurements. Confusion arises with luminosity of dark liquids such as dark fruit juices. This prompted adaptation and revision of scales for use at low luminosity levels (6). These scales are useful when the ratio of absorbance to transmittance is high.

The three-dimensionality of color is an important concept for measuring the color of a food. Examination of the wavelength distribution along the spectrum locus makes it clear that color is not equally visually spaced with this system. The elliptical nature of the color spaces prove troublesome when trying to match or establish specifications for color. Much attention has been devoted to providing more uniform color spaces and the CIELAB (1) system with parameters L*a*b* was designed to do this. Another color system was called CIELCH with parameters L*C*H* (9). The CIELAB system seems to be gaining more prominance. Color-matching computers are now available and are much used in textile and other industries. It may be important to determine whether a color difference exceeds the just-noticeable difference (JND). Computers have removed the computational labor of the more accurate calculations of differences.

SAMPLE PRESENTATION

Most instruments treat foods subjected to measurement as having opaque, matte, and uniform surfaces. In practice, foods seldom meet these requirements but deviate from ideal conditions of flatness, uniformity, opaqueness, diffusion, and specularity in ways that affect the measurement. This introduces an element of empiricism into any attempt to instrumentally assess the color of a food and sample preparation, and presentation becomes important. Color measurements are made to obtain repeatable numbers that correspond to visual assessments of the color of the food. Repeatability requires standardized procedures for sample preparation (9).

Many foods require a container of optical material thus introducing an element of gloss in contrast to the ideal matte surface. A uniform sample works best but grinding, mixing, milling, and blending are all processes that affect the light scattering properties of the sample. This may be to the extent that the sample no longer represents visual evaluation of the product, ie, crushed potato chips. The presence of water or other liquids in the mixture decreases the light scattering of the sample. Foods that consist of pieces that do not fit tightly together result in light from the source becoming trapped and thus reducing the light scattering. These foods are best measured with an aperture larger than the incident light beam to maximize collection of the light that has been trapped and diffused beyond the normal-size aperture. Liquids that depend on the transmission of light require optical cells of a thickness that will maximize the color difference. Translucent samples measured by reflection also pose the problem of light being trapped because it is reflected from within the food not from the surface. Light so trapped is not collected to be included in the measurement and a large aperture is useful.

Because most foods do not represent ideal conditions it is recommended that a second measurement be made after turning the sample 90° and the two readings be averaged. In devising a presentation technique that works well for the measurement it is well to remember to check that it retains a true relationship with the visual assessment of the product in its original form.

DATA INTERPRETATION

For quick assessments the simplest evaluation that will do the job is desired. Three-dimensional data may be considered too complex and difficult to deal with. It is important that any color measurement used to assess a food relates to what the eye sees and this should go further to determining those aspects of color that are important to consumers' assessment of quality. Any reduction of data needs to take these factors into account. Consumer assessments are generally obtained in less-controlled conditions, which are useful in determining the breadth of the problem. An intermediate stage under more—controlled conditions may be necessary in translating consumer results into instrumental measurements. However, using company personnel, closely involved with the technical side of the operation, as a substitute for consumers is risky because they are not representative of the consuming population.

Generally it is advisable to go beyond the basic CIE X, Y, and Z or Hunter L, a, and b instrument readouts to get a true picture of the color of the food. Lightness, hue, and chroma or saturation dimensions should be examined to obtain an indication of how the color of the samples measured differ. If the samples differ in only one dimension that dimension may be enough to provide a useful assessment of the product. However it should be remembered that lightness varies with both hue and chroma.

Some caution is required in using color difference calculations. The general ΔE calculation is affected by the equation selected for the calculation and by the geometry of the instrument on which the data were collected. Therefore, a suitable equation should be selected and used consistenly. In addition, the ΔE calculation does not provide information about the nature of the difference. Calculating differences for the components as either CIELAB ΔL^*, Δa^*, and Δb^* or ΔH for hue, ΔL, or ΔC for chroma, is likely to be more informative. If these are to be used for establishing tolerances they must be checked against visual judgments. It is important to note that while numbers may be convenient, color acceptions or rejections in practice are not number based but made on visual assessments.

The availability of computer-equipped instruments has removed much of the tedium from calculating color-measurement results and made readily available data in different forms. This should not be allowed to lull researchers into a false sense of security. It is still important to use well-prepared, representative samples for measurement. It still takes time to think about the problem, to evaluate, and to understand the results obtained.

COLOR MEASUREMENT OF SPECIFIC FOODS

An extensive review of the color measurement of various foods has been published (10,11). This discussion highlights treatments of color measurement that represent newer developments or deal with conditions that present real problems in color measurement in attempting to relate measured color to what the eye sees.

Orange juice is a food for which color is considered important enough to represent a large proportion of the grading system yet it is difficult to measure because of translucency that may be compounded by the presence of particles. One approach to color measurement of this type of product has been the use of Kubelka-Munk equations developed for reflectance at a single wavelength (7), use with colorimeters is strictly an empirical adaptation. It has been observed that the effect of different rates of transmittance change relative to concentrations; <1.0 produced large changes in measured lightness, hue, and chroma (12). The falloff in lightness and chroma observed at higher concentrations resulted from smaller change in transmittance at concentrations >2.0. It was concluded that for strongly colored scattering materials in dilute suspension, instrumental measurement was inadequate because it did not measure what the human vision perceived as appearance (12). This is because the instrument measures intensity of back-scattered light over a limited angle whereas human vision is stimulated by internally scattered light emerging multidirectionally from the suspension in addition to that which is reflected directly. For purposes of grading, the color of orange juice is measured using the Citrus Colorimeter (3). This instrument has two scales: citrus redness (CR); and a subsidiary citrus yellow (CY), which are used to calculate a color-score equivalent to a visual color score.

The color of fresh meat is of interest because of the importance of the degree of oxygenation of the surface to consumer acceptability and also because of the incidence of dark cutting beef and pale, soft exudate pork. More traditional approaches have been to determine relative concentrations of the myoglobin, oxymyoglobin, and metmyoglobin pigments spectrophotometrically at appropriate wavelengths (13). The problem of measuring meat color is made difficult because of the variability in concentration of the heme pigment myoglobin, the condition of the cut surface that may have undergone a degree of desiccation, the chemical state of the myoglobin, and the light-scattering properties of the muscle pigments (12). The muscle of a freshly slaughtered animal is dark and translucent in appearance becoming lighter and more opaque as the pH falls and the glycogen is converted to lactic acid. Because of the degree of translucency inherent in fresh meat, the Kubelka-Munk equations have been explored as a means of measuring meat color (12,13). The scatter coefficient has been used to measure color as a condition of fresh meat; however, the range of scatter is too large to estimate pigment concentration accurately (12). Scattering data were supplemented with data for L, hue angle, and chroma to produce typical values for conditions of fresh meat. It has been pointed out that derived formulas and multiple regression equations for expressing meat color suffer in accuracy because most people tend to think of how the viewed samples differ from a mental image of ideal meat color (14).

A mathematical approach rather than a visual approach has been used for estimating physical color parameters L^*, a^*, and b^* for red and tawny ports based on consumer data (15). Consumers ratio-scaled 15 blends of port for redness, brownness, and intensity of color and then provided ratings for their ideal port. Assessors were separated on the basis of those who preferred tawny and those who preferred ruby port. Deviations of the samples from the ideal were regressed against L^*, a^*, b^*, hue angle, and chroma. Three simultaneous equations including port preference, hue angle, and chroma were established for relating sensory information with physical information. These data were further treated by the method of inverse simultaneous estimation to obtain estimates for color parameters for ideal ports. However, as with all ideal assessments, caution must be exercised because of the large degree of error. In this case the ideal fell outside the range of blends, a common occurrence with ideal data. Although the study does demonstrate that appropriate use of mathematics and data collection can increase efficiency in relating sensory and physical color data.

ROLE OF COLORANTS

Because the color of a food is used by consumers to aid in forming a judgment about that food, it is not surprising that food processors and manufacturers consider colorants important adjuncts. The use of colorants to enhance the appearance of a food is not a recent development. Spices and condiments have been used since early civilization. However, the use of colorants has not been without its more unscrupulous aspects such as the propensity to use them to mask inferior products in the eighteenth and nineteenth centuries. The use of copper sulfate or copper fittings in making pickles to obtain an appetizing green appearance has lingered to more recent times. Candies, wines, liquors, and even flours have all been subjected to harmful coloring practices (16). As a result the use of colorants in foods came under scrutiny in the late nineteenth century, and the use of food colors has been controlled at the federal level in the United States since an act of Congress in 1886 allowed butter to be colored. This act recognized the need to provide the marketplace with products having both an attractive and a consistent appearance. By 1900 a wide range of foods such as jellies, catsup, alcoholic beverages, milks, ice creams, candy, sausages, pastries, etc were allowed to be colored. Both the substances allowed and the amounts specific to a food have been regulated where coloring has been permitted.

Colorants allowed in foods include those from both natural and synthetic sources. Natural colorants as a rule are more expensive, less stable, and possess lower tinctorial power. In addition, they are frequently present as mixtures in the host materials and vary with season, region, variety, etc. Disadvantages associated with natural colorants have been identified as low yields; color instability resulting from effects of pH, light, heat, and freezing; and possible association with other properties that may be undesirable (17). However, they do present the advantage of being perceived as safer than synthetic colors. Pigments to be used as colorants must be extracted from the host materials and prepared in some form, such as a dried powder or beadlets, which increases the ease of use in a food product (16). A breakthrough in the use of natural pigments occurred with the synthesization of the carotenoids β-carotene, β-apo-8′-carotenal, and canthaxanthin. But even here considerable confusion exists in the interpretations by regulatory bodies. In the United States these synthetic duplications of natural extracts must be identified as an artificial color, whereas in Canada they must be labeled a natural color.

The safety of synthetic colorants continues to be questioned. In 1938 the FDA evaluated synthetic colorants and determined that 15 colorants met the criteria for use in food, drugs, and cosmetics, and these were identified as FD&C colors. Since then the list of FD&C colors has been reduced to seven. The most controversial deletion was FD&C Red No. 2, or amaranth, which was banned in the United States based on concerns regarding teratogenic and reproduction effects raised by USSR studies in 1970 and later confirmed by an in-house FDA study. This colorant was also under fire at the time as contributing to behavioral problems in children. However the Canadian Health Protection Branch, having assessed the same data, retained amaranth as a colorant on the basis of lack of carcinogenic evidence and inadequate experimental control used in the FDA study (16). Sweden, Denmark, the FRG, Japan, and nine European Economic Community countries also retained it. In contrast the FDA gave approval in 1974 for FD&C Red No. 40, and it is now the only general-purpose red color certified in the United States. Canada's Health Protection Branch did not allow FD&C Red Dye No. 40 known as Allura Red, for use in that country. A result of these discrepancies in regulations between countries is an added complexity between trading partners and to global trade.

A quick assessment of the commodity use of synthetic colors indicates that a large proportion are used in fruits and fruit juices and nonalcoholic beverages. These are foods where water solubility and stability are requirements and suitable natural colorants have yet to be developed.

EFFECT OF COLOR ON SENSORY PERCEPTIONS

The effect of color on sensory perceptions has been well demonstrated. Color can influence people's perception of the basic tastes and affect their ability to distinguish threshold levels (18). Presenting the basic tastes in colored solutions raised the thresholds of the test subjects. Thresholds for sour presented in green or yellow solutions were higher than those presented in red solutions, which in turn were higher than those presented in colorless solutions. This is not entirely accounted for by color association with certain foods, for example, both green and yellow sour and nonsour foods are common. In the case of bitter taste, green and yellow presentations resulted in higher concentrations being required for detection than for colorless or red presentations. It was concluded that red color was not associated with bitter flavor (18). In contrast to sour and bitter, sweet thresholds were lower for green-colored solutions. Thresholds for colored salt solutions were no different than those for colorless solutions. This was attributed to the wide variety of foods associated with a salty taste.

The levels of basic tastes in most foods are well above threshold levels and taste-color associations vary with specific foods. Color was shown to be a distraction in correctly identifying the sweeter sample in a paired comparison test of pear nectar, but it was not confirmed that the color green lowers the perception of sweetness, a result obtained in an earlier study (19). It was concluded that the effect of color varied considerably between individuals. The intensity of red color has been shown to affect a subject's perception of sweetness in beverages and solutions (20). A red color in fruit-flavored beverages was associated with an increase in perceived sweetness although the sucrose level was constant. Sweetness of darker red solutions was reported to be perceived as being 2–10% greater than the uncolored reference when actual sucrose concentrations were 1% less. There is an interrelationship between color and flavor, for example, the introduction of blue color to cherry- and strawberry-flavored drinks markedly decreased perception of tartness and flavor (21). The addition of yellow color decreased perception of sweetness to a lesser extent, and the addition of red increased sweetness. This study introduces the possibility of using color to reduce the concentration of sucrose required in a beverage. The perception of sweetness seems to be more affected by color than the perception of salt. It has been observed that color had no influence on salt perception of chicken broth colored red or yellow (22). Neither was flavor preference affected by color.

An investigation into the effect of color on aroma, flavor, and texture of foods, showed that appropriately colored foods were perceived to contain greater intensity of aroma and flavor and as having a higher quality, with aroma being the most affected (23). Perception of texture was not affected by color. Another study found that fruit-flavored beverages were correctly identified when their color was visible but when their color could not be seen dramatic reductions in correct identification occurred (24). Only 70% of the tasters correctly identified grape; 50%, lemon-lime; 30%, cherry; and 20%, orange. When the study was expanded to include combinations of color, flavor, colorless, and flavorless in beverages, flavor identification was color cued, and in the case of the colorless beverages a spread of flavors was identified. Flavorless beverage colored red was identified as strawberry by 22% of the tasters; orange-colored, as orange by 22%; and green-colored, as lime by 26%. Identification as having no flavor ranged from 41% for colored beverages to 48% for the colorless beverage. These same researchers observed that when white cake was varied in lemon flavor and yellow color that flavor intensity was most affected by color when no lemon flavoring was added, but the effect was nonlinear. Intensity of flavor was most affected between no added color and the lowest level. Color has been shown to have a strong influence on the assessment of aroma and flavor for port wines (25). These ratings were shown to be influenced by visual cues when these were made available. Examination of assessments by individual assessors showed assessors differed in the degree of this influence. These researchers reported that trained panelists were able to remove color bias from evaluation of Bordeaux wines when instructed to do so.

That color plays a role in the acceptability of many foods cannot be denied. One study varied the color of cherry- and orange-flavored beverages above and below normal levels. It was observed that the overall liking peaked around the normal color (24). It was also noted that overall acceptability was more closely related to flavor than to color acceptability, suggesting that color cannot substitute for flavor. This same study demonstrated a significant color-flavor interaction for overall acceptability of lemon-flavored cakes. More intense flavors were liked less at higher color concentrations. As flavor levels increased, overall acceptability peaked at lower color levels. These findings suggest that intensities reach a point where they exceed the accustomed expectation and are, therefore, liked less. The interrelationships between pleasantness, color, and sucrose concentration have been observed (16). The interaction was evident in the lower pleasantness rating given to the darker red intensity beverage. This may have resulted because the beverage was perceived as sweeter and, therefore, as less pleasant or the red color may not have been associated with the tasters' expectations of a desirable red color. One study examined the perceived pleasantness of beverages varying in sweetness and color and concluded that the most intense color or sweetest flavor did not always produce the optimum product formulation (21). Color and appearance have been shown to have important influences on overall acceptability of wines but the degree varies with the individual assessor and the study (25).

MARKETING IMPLICATIONS

Any research and development technologist should be aware of the marketing implications related to the color of a food as they affect product development and quality control. The technologist must bear in mind that it is what is perceived by the consuming public that controls sales not necessarily the truth of the perception.

A 1985 survey conducted by *Good Housekeeping* magazine, asked respondents to indicate whether nutrition or appetite appeal was the more-important selection criterion for food purchases; 45% stated appetite appeal (26). Color certainly plays a considerable role in appetite appeal and this is one reason that foods are colored. Yet some consumers consider the addition of color as perpetrating deception and in Europe there have been attempts to ban the use of colorants (17). It has been reported that the lack of artificial colorings and the 3presence of natural color rank high as positive food attributes, whereas the use of artificial color was ranked as negative (26). However, 35–40% of respondents considered artificial color as somewhat acceptable. More than 50% of the respondents considered coal tar color derivatives to be unacceptable. Apparently, the use of colorants for cosmetic reasons, once considered to enhance acceptability, may now be a detractor because of the pressure to reduce the use of additives in processed foods as a result of prevailing consumer attitudes and safety concerns. It should also be remembered that experiences change over time and that this influences acceptability. The combination of experience and prevailing consumer concerns can combine to bring about changes in acceptability. The Spy variety of apple was once prized as a cooking apple but it has gradually been superseded by other varieties for processing and retail sales. When cooked, the Spy apple produces a very intense yellow color. This bright yellow color is less preferred by today's consumers than the duller browner color of the cooked product of more familiar varieties. Part of this is undoubtedly that the intense color is unexpected and unfamiliar but the lower preference may also result from the fact that the intense yellow, although natural, has become suspect.

Although within the marketing profession there has been much exploration of the effects of color, particularly in the packaging, on the sales of food products, little research of this nature has been published. Package design and color are important factors in attracting a buyer at the point of purchase. That packaging becomes part of a familiar product has been witnessed by any shopper who has had to search for an often-purchased product because the color of the package has been changed. Marketers are concerned about product image and thus both the product and the package must contribute to this image. Both can become dated and in an industry that experiences a short life for many of its products this can be of concern for the long-lived products. Black has not generally been considered a desirable color for food packages but with the popularity of art deco design, sophisticated black packages and labels are beginning to appear. Marketers are quick to take advantage of color association with consumer concerns such as that of green with environmental concerns. One Canadian food retailer has capitalized on this to develop an array of G.R.E.E.N. products, which have been promoted as being environmental friendly. This sort of use of color with food is usually beyond the purvey of the technologist, but such marketing decisions may result in formulation requests to research and development.

It can be concluded that color has an important association with food, which extends far beyond the technologist's concern with measurement. An examination of all the facets as they affect any product is becoming increasingly important in coping with the greater degree of sophistication and competitiveness occurring in the food industry.

BIBLIOGRAPHY

1. F. W. Billmeyer, Jr., and M. Saltzman, *Principles of Color Technology*, 2nd ed., John Wiley & Sons, Inc., New York, 1981.
2. D. B. Judd and G. Wyszecki, *Color in Business, Science, and Industry* John Wiley & Sons, Inc., New York, 1963.
3. R. S. Hunter, *The Measurement of Appearance*, John Wiley & Sons, Inc., New York, 1975, pp. **44**, 81, 167.
4. J. B. Hutchings, *Food Colour and Appearance*, Blackie Academic & Professional, Glasgow, Scotland, 1994.
5. R. G. Kuehni, *Color: An Introduction to Practice and Principles*, John Wiley & Sons, New York, 1997.

6. F. J. Francis, "Colour and Appearances as Dominating Sensory Properties of Foods," in, G. G. Birch, J. G. Brennan, and K. J. Parker, eds., *Sensory Properties of Foods*, Elsevier Applied Science Publishers, Ltd., Barking, UK, 1977, p. 27.
7. E. A. Gullett, F. J. Francis, and F. M. Clydesdale, "Colorimetry of Foods: Orange Juice," *Journal of Food Science* **37**, 389 (1972).
8. D. B. MacDougall, "Colour in Meat" in G. G. Birch, J. G. Brennan, and K. J. Parker, eds., *Sensory Properties of Foods*, Applied Science Publishers Ltd., 1977, p. 59.
9. F. J. Francis, "Color Analysis," in S. S. Nielson, ed., *Food Analysis*, 2nd ed., Aspen Publishers, Gaithersburg, Md., 1998, pp. 599–612.
10. F. J. Francis, "Color Measurement and Interpretation," in D. Y. C. Fung and R. F. Matthews, eds., *Instrumental Methods for Quality Assurance in Foods*, Marcel Dekker, Inc. New York, 1991, pp. 189–210.
11. F. J. Francis and F. M. Clydesdale, *Food Colorimetry: Theory and Applications*, AVI Publishing Co., Inc., Westport, Conn., 1975.
12. D. B. MacDougall, "Instrumental Assessment of the Appearance of Foods," in A. A. Williams and R. K. Atkin, eds., *Sensory Quality in Foods and Beverages*, Ellis Horwood Ltd., 1983, p. 121.
13. F. M. Clydesdale and F. J. Francis, "Color Measurement of Foods. 28. The Measurement of Meat Color," *Food Product Development* **5**, 87 (1971).
14. B. A. Eagerman, F. M. Clydesdale, and F. J. Francis, "Determination of Fresh Meat Color by Objective Methods," *Journal of Food Science* **42**, 707 (1977).
15. A. A. Williams, C. R. Baines, and M. S. Finnie, "Optimization of Colour in Commercial Port Blends," *Journal of Food Technology* **21**, 451 (1986).
16. Institute of Food Technologists' Expert Panel on Food Safety & Nutrition and the Committee on Public Information, *Food Colors. Scientific Status Summary, Institute of Food Technologists*, Chicago, Ill., 1980.
17. D. Blenford, "Natural Food Colours," *Food Flavourings, Ingredients and Processing* **7**, 19 (1985).
18. J. A. Maga, "Influence of Color on Taste Thresholds," *Chemical Senses and Flavor* **1**, 115 (1974).
19. R. M. Pangborn and B. Hansen, "The Influence of Color on Discrimination of Sweetness and Sourness in Pear-Nectar," *American Journal of Psychology* **76**, 315 (1963).
20. J. Johnson and F. M. Clydesdale, "Perceived Sweetness and Redness in Colored Sucrose Solutions," *Journal of Food Science* **47**, 747 (1982).
21. J. L. Johnson, E. Dzendolet, R. Damon, M. Sawyer, and F. M. Clydesdale, "Psychophysical Relationships Between Perceived Sweetness and Color in Cherry-Flavored Beverages," *J. Food Protect.* **45**, 601 (1982).
22. S. R. Gifford and F. M. Clydesdale, "The Psychophysical Relationship Between Color and Sodium Chloride Concentrations in Model Systems," *Journal of Food Protection* **49**, 977 (1986).
23. C. M. Christensen, "Effects of Color on Aroma, Flavor and Texture Judgements of Foods," *Journal of Food Science* **48**, 787 (1983).
24. C. N. DuBose, A. V. Cardello, and O. Maller, "Effects of Colorants and Flavorants on Identification, Perceived Flavor Intensity, and Hedonic Quality of Fruit-Flavored Beverages and Cake," *Journal of Food Science* **45**, 1393 (1980).
25. A. A. Williams, S. P. Langron, C. F. Timberlake, and J. Bakker, "Effect of Colour on the Assessment of Ports," *Journal of Food Technology* **19**, 659 (1984).
26. K. W. McNutt, M. E. Powers, and A. E. Sloan, "Food Colors, Flavors and Safety: A Consumer Viewpoint," *Food Technology* **40**, 72 (1986).

F. J. Francis
Editor-in-Chief
University of Massachusetts
Amhert, Massachusetts

See also Colorants section.

COLORANTS

F. J. Francis
Editor-in-Chief
University of Massachusetts
Amhert, Massachusetts

COLORANTS: INTRODUCTION

The appreciation of color dates back to antiquity, and recorded history is replete with descriptions of the use of colorants in everyday life (1). The art of making colored candy is depicted in paintings in Egyptian tombs as far back as 1500 B.C. Pliny the Elder described the artificial coloration of wines in 400 B.C. Spices and condiments were colored at least 500 years ago. Colorants for use in cosmetics were probably more widespread than those in foods or at least were better documented. Archaeologists have pointed out that Egyptian women used green copper ores as eye shadow as early as 5000 B.C. Henna was used to dye hair, carmine to redden lips, and kohl, an arsenic compound, to blacken eyebrows. It was common practice in India thousands of years ago to color faces yellow with saffron or to dye feet red with henna. The Romans used white lead and chalk on their faces and blue dyes on their hair and beards. More recently, in Britain, sugar was imported in the twelfth century in attractive red and violet colors, probably from the colorants Madder and Kermes for the reds, and Tyrian purple for the violet colors.

Early applications of colorants were usually associated with religious festivals and public celebrations. Obviously they could be added to foods, but in the Middle Ages, the trade guilds policed their members and protected the quality of their products. However, when the industrial revolution evolved with its massive social and population changes and much wider food distribution quality safeguards were weakened and adulteration became rampant. Wines and confectionery were particularly suspect. Ancient history contains many examples of the concern about food adulteration and its regulation. We can go back to the laws of Moses, the Roman statutes, and down through the Middle Ages. England seems to stand out in the laws enacted by Parliament in the Middle Ages. For example, in 1266 Parliament prohibited a number of important food staples if they were so adulterated as to be "not wholesome for man's body." This unambiguous statement could form the basis of food regulation today.

In 1920 Accum published a book describing practices such as the coloring of tea leaves with verdigris (copper acetate), Gloucester cheese contaminated with red lead, pickles boiled with a penny to make them green, confectionery with added red lead and copper, and many other appalling practices (2). Hassell in 1857 published a book on adulteration of foods that really raised the public conscience, and things began to change (3). Nearly every country passed laws dealing with the adulteration of foods, and colorants were a major concern since they could be used to conceal inferior food and in some cases were actually dangerous. Until the middle of the nineteenth century, the colorants used in cosmetics, drugs, and foods were those of natural origin such as those from animals, plants, and minerals. In 1856 when Sir William Henry Perkins discovered the first synthetic organic dyestuff, mauve, the supply situation changed drastically. The German dyestuff industry synthesized a number of "coal tar" dyes, which apparently were utilized very quickly. For example, French wines were colored with fuchsine as early as 1860. In the United States, colorants were allowed to be added to butter in 1886 and to cheeses in 1896. By 1900 Americans were eating a wide variety of colored foods such as jellies, cordials, ketchup, butter, cheese, candy, ice cream, wine, noodles, sausage, confectionery, and baked goods. Many of these practices constituted economic fraud but were not actually dangerous. But some were Marmion described a classic horror story in 1860 in which a druggist gave a caterer copper arsenite to use in making a green pudding for a public dinner and two people died (4). At the turn of the century, there were nearly 700 synthetic colorants available, and obviously very little control was exercised over the type and purity of the colorants offered for use in foods. Because of public concerns over these practices, the American food manufacturers decided to police their own industry. One example was a list, published in 1899 by the National Confectioners Association, of 21 colorants that they considered unfit for addition to foods. However, it became apparent that some form of government control was needed, and, in 1900, the Bureau of Chemistry in the U.S. Department of Agriculture (USDA) was given funds to study the problem. A series of Food Inspection Decisions (FID) resulted. One FID in 1904 declared a food to be adulterated "if it be colored, powdered or polished with intent to deceive or to make the article appear to be of better quality than it really is." Another FID in 1905 required the labeling of such products as artificially colored imitation chocolate. Another FID in 1906 concerned a coal tar dye considered to be unsafe for use in food. In effect, it stopped the importation of macaroni colored with Martius Yellow. At this time, the (USDA) launched a monumental task to determine which colorants were safe for use in food and what restrictions should be placed on their use. This effort led to the regulations for the use of synthetic colorants in the United States today.

The development of the regulatory aspects of food colorants is one small portion of the much bigger picture concerning food additives and food quality in general. One of the driving forces for food regulation down through the years has been food safety. This aspect has been brought to public attention, sometimes in rather stormy fashion, by a series of books such as *The Jungle* by Upton Sinclair in 1906. Public laws to minimize obviously unpalatable practices concerned with food sources, processing, handling, ingredients, and so on date back seriously only about 150 years. Much of the criticism has been directed toward the artificial colorants and very little toward the naturally occurring colorants. Perhaps this is partially due to the naive belief that humans have somehow become conditioned to tolerate certain compounds after a long history of use. But it is unlikely that 5000 years is enough to create significant genetic changes in humans. It is true, however, that history has allowed us to identify acute hazards and eliminate them from our diet, but long-term or lifetime exposures are another story. Regardless, much of the regulatory activity has centered around food safety.

The development of food laws in the United States followed in concept those in Europe (5) and are governed by the Food and Drug Acts. The first one of importance was in 1906, and it has been modified several times. The 1938 modification was one of the more important developments, and the 1938 principles are still in effect. The 1938 Act established the term FD& C colors with the initials standing for Food, Drug, and Cosmetic. There is also a D&C classification for colorants allowed in drugs and cosmetics. The Act specified a list of colorants allowed in foods, and presumably they would be held to a higher standard of purity than those available for use in industry. Colorants used in food would have to be identified by their FD&C name, but colorants used for other purposes in industry could still be referred to by their common name. The Society of Dyers and Colorists in England and the American Association of Textile Chemists and Colorists in the United States developed a Color Index (CI) system. Each colorant was given a CI number and name. Colorants are also identified by their Chemical Abstract Service (CAS) Registry code number. The U.S. Food and Drug Administration (FDA) Center for Food Safety and Applied Nutrition (CFSAN) uses a CAS-like code for those substances without a number assigned by CAS.

The 1960 Color Additive Amendment to the Act specified two groups of colorants: certified color additives and color additives exempt from certification. The first group listed seven synthetic FD&C colorants that were certified to comply with the purity specifications required by the FDA. The second group has 26 colorants (6,7) for which the FDA does not require that each batch be analyzed and certified (Table 2 in COLORANTS: FOOD, DRUG AND COSMETIC COLORANTS). Many of the colorants that would be called "natural" in other countries are on the list of 26. The FDA does not accept the concept of "natural" versus "synthetic" and requires that compounds in each group be subjected to the same standards of safety. Actually, in practice, the safety studies required for some of the natural colorants were considerably less than those required for the synthetics. This approach is understandable if one accepts the concept of the decision tree approach for determining safety as suggested by the Food Safety Council (8). The Nutritional Labeling and Education Act of 1990 mandated that certified color additives be specifically declared by their individual names, but the requirements for exempt colorants were left unchanged. Exempt colorants can still be declared generically as "artificial color" or any other specific or generic name for the colorant. But the term "natural" is prohibited because it may lead the consumer to believe that the color is derived from the food itself. There is no such thing as a natural colorant in the United States. Detailed accounts of the safety of the noncertified colorants have recently been published (6,9).

The emphasis on food safety possibly has had another influence, namely the worldwide trend toward the use of colorants derived from natural sources. In one survey of food colorant patents in the 15-year period from 1969 to 1984, there were 356 patents on colorants from natural sources and 71 on synthetics (10). The ratio is even more lopsided when one considers that 21 of the 71 were on colorants linked to a polymer background (the Dynapol concept), and this approach is no longer being considered. There is little effort worldwide being devoted to the development of novel synthetic colorants, but considerable effort is being devoted to the development of colorants from natural sources.

Food colorants are one small portion of the much larger concept of food additives. The FDA has made it easy to determine the status of a food additive by publishing a book entitled *Everything Added to Food in the United States* (11). The information in the book was derived from the files of the FDA and contains 2922 total food additives of which 1755 are regulated food additives, including direct, secondary direct, color, and Generally Recognized as Safe (GRAS) additives. The book also contains administrative and chemical information on 1167 additional substances. The book is arranged alphabetically, and each compound is identified by its CAS number. The appropriate section in Title 21 of the U.S. Code of Federal Regulations for each substance is also listed. For example, Paragraph (e) of CFR 21-178.3297 provides a description and a long list of compounds that "may be safely used as colorants in the manufacture of articles or components of articles intended for use in producing, manufacturing, packing, processing, preparing, treating, packaging, transporting or holding food, subject to the provisions and definitions set forth in this section. . . . The term colorant means a dye, pigment, or other substance that is used to impart color or to alter the color of a food-contact material, but that does not migrate into the food in amounts that will contribute to that food any color apparent to the naked eye. For the purposes of this section, the term 'colorant' includes substances such as optical brighteners and fluorescent whiteners, which may not themselves be colored, but whose use is intended to affect the color of a food-contact material." The preceding section is very general, but the sections on individual colorants are more specific. For example, talc has only one entry (21-182.90) and FD&C Red No. 40 has three (21-74.340, 21-74.1340, and 21-74.2340). The data in *Everything Added to Food in the United States* are also listed in a much broader publication entitled *Food Additives Toxicology, Regulation, and Properties*, edited by F. M. Clydesdale (12). This publication is a CD-ROM playable on Windows software.

Food colorants are an important aspect of the formulation of foods. Walford summarized the reasons for adding colorants to food today:

1. To give an attractive appearance by replacing the natural colour destroyed during processing or anticipated storage conditions.
2. To give colour to those processed foods such as soft drinks, confectionery and ice creams, which otherwise would have little or no colour.
3. To supplement the intensity of the natural colour where this is perceived to be weak.
4. To ensure batch to batch uniformity where raw materials of different source and varying colour intensity have been used. (1)

With the increased emphasis on increasing food production for an ever-increasing population, and the necessity to make food appealing and acceptable, food colorants will remain a major concern. They already are an important part of the economy since Pszczola estimated that the worldwide market for colorants in 1998 was about 939 million U.S. dollars (13). Of this 65% was represented by natural colorants.

BIBLIOGRAPHY

1. J. Walford, "Historical Development of Food Coloration," in J. Walford, ed., *Development of Food Colours*-1, Applied Science, London, 1980, pp. 1–25.
2. F. Accum, *A Treatise on the Adulteration of Food and Culinary Poisons*, London, 1820.
3. A. H. Hassell, *Food and Its Adulterants: Comprising the Reports of the Analytical Sanitary Commission of the Lancet 1857*, London, 1857.
4. D. M. Marmion, *Handbook of U.S. Colorants for Foods, Drugs and Cosmetics*, 2nd ed., Wiley-Interscience, New York, 1984.
5. L. E. Parker, "Regulatory Approaches to Food Coloration," in J. Walford, ed., *Developments in Food Colours-2*, Applied Science, London, 1984, pp. 1–22.

6. J. B. Hallagan, D. C. Allen, and J. F. Borzellaca, "The Safety and Regulatory Status of Food, Drug and Cosmetic Color Additives Exempt from Certification," *Food Chem. Toxicol.* **33**, 515–528 (1995).
7. F. J. Francis, *Handbook of Food Colorants*, Eagen Press, St. Paul, Minn., 1999.
8. *A Proposed Food Safety Evaluation Process. Food Safety Council. Final Report*, Nutrition Foundation, Washington D.C., 1982.
9. F. J. Francis, "Safety of Natural Food Colorants," in G. A. F. Hendry and J. D. Houghton, eds., *Natural Food Colorants*, Blackie, Glasgow, Scotland, 1996, pp. 112–130.
10. F. J. Francis, *Handbook of Food Colorant Patents*, Food and Nutrition Press, Westport, Conn., 1986.
11. U.S. Food and Drug Administration, *Everything Added to Food in the United States*, CRC Press, Boca Raton, Fla., 1993.
12. F. M. Clydesdale, *Food Additives: Toxicology, Regulation, and Properties* [CD ROM], CRC Press, Boca Raton, Fla. 1997.
13. D. E. Pszczola, D. E. "Natural Colors Pigments of Imagination," *Food Technol.* **52**, 70–72 (1998).

COLORANTS: ANTHOCYANINS

The anthocyanins are probably the best known of the natural pigments. They are ubiquitous in the plant kingdom and are responsible for many of the orange, red, blue, violet, and magenta colors in plants. Their very visibility, combined with their role as taxonomic markers, has attracted the efforts of many research workers in the past 75 years. Their use as a colorant goes back to antiquity. The Romans used highly colored berries to augment the color of wine. The American Indians added cranberries to a "trail-stable" product called pemmican. Pemmican is a high-calorie mixture of dried meat strips, fat, and berries. The cranberries did provide an attractive color and flavor, and probably the highly acid cranberries improved storage stability. Certainly the acid would remove the possibility of botulism. Cranberries also contain the well-known preservative benzoic acid. Pemmican may be one of the earliest "fast foods" in the United States.

In view of the ubiquity of the anthocyanins and their high tinctorial power it comes as no surprise that many sources have been suggested as colorants. Francis listed more than 40 plants as potential sources with their pigment profiles and methods of extraction (1,2). Francis also listed 49 patents on anthocyanin sources (3). However, despite the large number of potential sources, only two have had commercial success, grapes and, more recently, red cabbage. Colorants from grapes have been available for nearly 120 years primarily from press cake as a by-product of the wine industry. Grapes are the world's largest fruit crop for processing. The Food and Agriculture Organization (FAO) 1989 estimate of annual production was 60,000,000 metric tons, of which about 80% was used for wine making. This ensures a limitless source of inexpensive raw material for colorant production.

The anthocyanins are chemically similar to the much larger group of flavonoids and are usually included in the general classification of flavonoids. Anthocyanins are usually orange-red in color, whereas the other flavonoids are usually colorless to yellow.

CHEMICAL COMPOSITION

The composition of any colorant containing anthocyanins will reflect the composition of the raw material. At present it is grapes and red cabbage, but other sources may become available in the future. About 275 anthocyanins are known currently and are part of about 5000 flavonoid compounds of similar chemical structure (4). The anthocyanins are composed of an aglycone (anthocyanidin), sugar, and perhaps organic acids. Twenty-two aglycones are known, of which 18 occur naturally. Only 6 are important in foods, and the structure of these is shown in Figure 1. Free aglycones occur very rarely in plants and are nearly always combined with sugars. One reason for this is that the sugars stabilize the molecule. In order of relative abundance, the sugars are glucose, rhamnose, galactose, xylose, arabinose, and glucuronic acid. Anthocyanins may also be acylated, which adds a third component to the molecule. One or more molecules of *p*-coumaric, ferulic, caffeic, malonic, or acetic acids may be esterified to the sugar molecule. The aglycones in grapes are cyanidin, peonidin, malvidin, petunidin, and delphinidin with the organic acids acetic, coumaric, and caffeic (5). The only sugar present is glucose. Grapes usually have a very complex anthocyanin profile with the Concord variety having 31, the greatest number in any single variety. The 15 anthocyanins in red cabbage contain only cyanidin and glucose with the acids ferulic and coumaric.

Plants that contain anthocyanins also invariably contain flavonoids, and the distribution of flavonoids is more widespread than the anthocyanins. In view of the large number of flavonoids (5000), the pigment profile is usually characteristic of the plant family. A number of other polyphenolic compounds are also present in grapes (6). The polyphenolic compounds in grapes, in view of their importance in wines, have received much more research attention than those in red cabbage but the latter can be assumed to also have a complex polyphenol profile. A typical compound from each group of compounds is shown in Figure 2. The composition of grape extracts is further complicated by polymers produced by association with catechins and flavonoids with or without the interaction with acetaldehyde. Actually most of the tinctorial power in enocyanin preparations is due to the complexed anthocyanins. The overall conclusion is that it is impossible to specify accurately the composition of enocyanin preparations.

Figure 1. The anthocyanidin nucleus. Pelargonidin (4′ = OH); cyanidin (3′,4′ = OH); delphinidin (3′,4′,5′ = OH); peonidin (4′ = OH, 3′ = OMe); petunidin (4′,5′ = OH, 3′ = OMe); malvidin (4′ = OH, 3′,5′ = OMe); all other substitution positions = H.

Figure 2. Typical components of grape colorants extracts. (**a**) A red anthocyanin, malvidin-3,5-diglucoside; (**b**) a yellow flavonoid, quercetin-3-rhamnoglucoside, (rutin); (**c**) a component of tannin catechin; (**d**) a phenolic acid, caffeic acid; (**e**) a stilbene phytoalexin, resveritol.

Commercial specifications are usually confined to tinctorial power, total acidity, %solids, %ash, heavy metals, sulfur dioxide, tannins, and alcohol. The tinctorial strength of grape extracts is often expressed as the absorbance of a 1% solution in a 1-cm cell at 520 nm in citrate buffer at pH 3. The tinctorial power of red cabbage extracts is sometimes expressed as the absorbance of a 10% solution in a 1-cm cell measured in citrate buffer at pH 3 (eg, E 10%, 1 cm = 190).

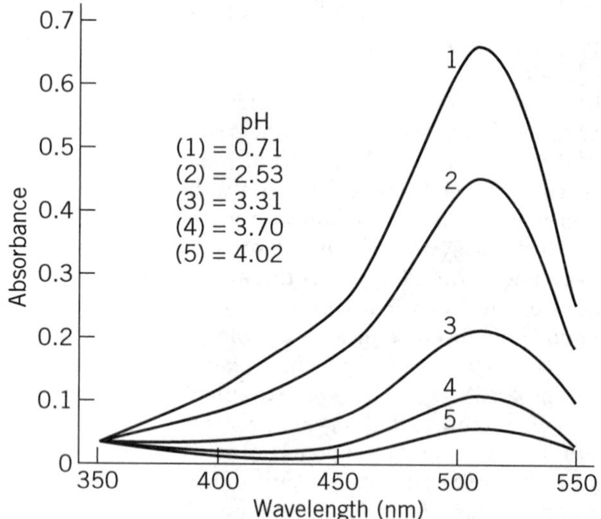

Figure 3. Absorption spectra of cyanidin-3- rhamnoside in buffer solutions at pH values of 0.71–4.02. The concentration of the pigment is 1.6×10^2 g/L.

COMMERCIAL PREPARATION

The most common commercial method involves treatment of the skins with water containing up to 3000 ppm of sulfur dioxide or its equivalent in bisulfite or metabisulfite. After 48 to 72 h, the liquid is removed from the skins, filtered, desulfured, and concentrated. Sometimes a fermentation step is allowed prior to the extraction, in which case the alcohol is removed during subsequent processing. The final product is a particle-free liquid of high tinctorial power. The concentrated liquid may also be dried to produce a soluble dry powder. The presence of sulfur dioxide in the extracting medium results in increased extraction of the pigments as well as increased stability of the final product. The sulfite-anthocyanin complex is colorless, so the sulfite must be removed prior to the final concentration and filtration step. Addition of acid makes the extraction step more efficient, but addition of mineral acid requires neutralization at some point. The addition of tartaric acid can be removed by addition of potassium hydroxide since the resulting precipitate of potassium hydrogen tartrate can be easily removed by filtration. Purification of the crude extract can be accomplished by passing the extract through an ion-exchange resin bed with the added advantage that the colorant eluate can be fractionated. The first portion of

A: Quinoidal base (blue)

$+H^+$

AH^+: Flavylium cation (red)

$+H_2O$ $-H^+$

C: Chalcone (colorless)

B: Carbinol pseudo-base (colorless)

Figure 4. Structural changes in anthocyanins with pH. Malvidin-3-glucoside at 25°C.

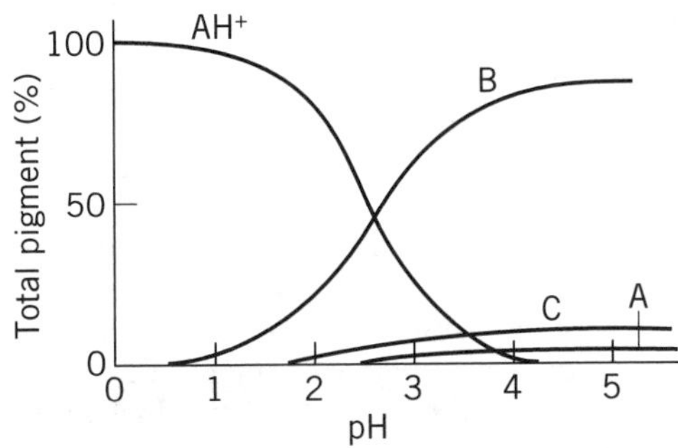

Figure 5. Effect of pH on the distribution of anthocyanin structures of malvidin-3-glucoside. A, B, C, and AH^+ refer to the forms in Figure 4.

the eluate will be redder than the bluer trailing portion, because the higher molecular weight compounds will go through the resin bed at a slower rate. Treatment with resins produces a purer product but at a higher price. The FDA-approved procedures for extraction are specified in the CFR Section 73.170 for Grape Skin Extract.

Extraction of grape skins is much more efficient with acidified methanol or ethanol, but this requires another step to remove the alcohol. Wineries are usually familiar with handling alcohol, but other potential colorant producers seem reluctant to put an alcohol recovery unit in a food plant. Regardless, alcohol extraction does not seem to have commercial appeal.

Current emphasis is on attempts to make colorants containing anthocyanins more stable. Considerable research emphasis has been on the copigmentation or association of anthocyanins with themselves or with a series of other compounds such as flavonoids, polysaccharides, proteins, tannins, and other polyphenolic compounds.

SYNTHETIC COMPOUNDS

Studies on synthetic compounds related to the anthocyanins have shown that methyl or phenyl moieties substituted at the C-4 position in the molecule are virtually resistant to nucleophilic attack at the C-2 position. This observation led to the suggestion that a number of synthetic analogues would prove to be very stable (7). The first natural C-4 substituted anthocyanin reported was purpurinidin fructo-glucoside isolated from willow bark, but shortage of raw material has precluded its commercial development (8). A second group of naturally occurring C-4 substituted anthocyanins was reported by Cameira dos Santos et al. in red wine (9). The new pigments contain a vinylphenol group attached to positions 4 and 5 of the grape anthocyanins and were named anthocyanin-3-glucoside adducts. Sarni-Manchado et al. prepared the vinylphenol adducts of five of the major anthocyanins of wine, namely, delphinidin-3-glucoside, cyanidin-3-glucoside, petunidin-3-glucoside, peonidin-3-glucoside, and malvidin-3-glucoside and tested their stability at 55°C in a model beverage at pH 3 and 5 (10). The adduct pigments were much more stable than the original pigments, particularly at pH 5. The adducts also were unaffected by SO_2. The color of the adducts was slightly more orange than the corresponding anthocyanins. The authors did not state the concentration of the adducts found in red wine. Chemical synthesis of a number of C-4 substituted analogues of anthocyanins is feasible but obviously would require testing for safety and legislative approval.

COMMERCIAL APPLICATIONS

In view of the large body of research on the appearance, stability, analysis, and plant breeding of anthocyanins, that occur naturally in fruit and vegetable products, it is not surprising that anthocyanin colorants would be used to enhance the aesthetic appeal of existing plant products or formulated substitutes. Actually, one of the first applications of enocyanin, a generic term for colorants from grapes, was to improve the color of wine. In the United States, fruit drinks are the biggest market. Colorants containing anthocyanins have been suggested for beverages, jellies, jams, ice cream, yogurt, gelatin desserts, canned fruits, fruit sauces, candy and confections, bakery fillings, toppings, drink-mix crystals, pastries, cosmetics, and pharmaceuticals.

Anthocyanin colorants have their greatest color intensity at pH values between 2 and 4. Figure 3 shows the changes in absorption of cyanidin-3-rhamnoglucoside as the pH is changed from 0.71 to 4.02. Figure 4 shows the chemical changes that accompany a pH change Figure 5 shows the amounts of each pigment present at each pH value. It is possible to use an anthocyanin colorant to color a product with a pH value above 4, but the color will be bluish and considerably more colorant will be required. The reduction in color intensity as the pH is raised, known as the pH effect, is much more obvious with preparations that contain a high proportion of monomeric anthocyanins and less in preparations with a higher content of polymerized or degraded pigments, but all preparations show the pH effect to some degree. The pigments are also susceptible to degradation by light, oxygen, iron, copper, tin, ascorbic acid, and sulfur dioxide. Applications are also restricted to those with minimal heat treatment, but despite these apparent problems with stability, this group of colorants has met with good commercial success.

BIBLIOGRAPHY

1. F. J. Francis, "Miscellaneous Colorants," in G. A. F. Hendry and J. D. Houghton, eds., *Natural Food Colorants*, Blackie, Glasgow, Scotland, 1992, pp. 242–272.
2. F. J. Francis, "Polyphenols as Natural Food Colorants," in A. Scalbert, ed., *Polyphenolic Phenomena*, INRA, Versailles, France 1993, pp. 209–220.
3. F. J. Francis, *Handbook of Food Colorant Patents*, Food and Nutrition Press, Westport, Conn., 1986.
4. J. B. Harborne, *The Flavonoids: Advances in Research Since 1980*, Chapman and Hall, London, 1988.
5. G. Mazza, "Anthocyanins in Grapes and Grape Products," Crit. Rev. Food Sci. Nutr. **35**, 341–371 (1995).
6. F. J. Francis, *Handbook of Food Colorants*, Eagen Press, St. Paul, Minn., 1998.
7. D. Amic, J. Baranac, and V. Vukadinovic, "Reactivity of Some Flavylium Cations and Corresponding Anhydrobases," *J. Agric. Food Chem.* **38**, 936–940 (1990).
8. P. Bridle, K. G. Scott, and C. F. Timberlake, "Antocyanins in *Salix* Species. New Anthocyanin in *Salix purpurea* Bark," *Phytochemistry* **12**, 1103–1106 (1973).
9. H. Fulcrand et al., "A New Class of Wine Pigments Generated by Reaction Between Pyruvic Acid and Grape Anthocyanins," *Phytochem.* **47**, 141–147 (1998).
10. P. Sarni-Manchado et al., "Stability and Color of Unreported Wine Anthocyanin-Derived Pigments," *J. Food Sci.* **61**, 938–941 (1996).

COLORANTS: ANTHRAQUINONES

COCHINEAL AND CARMINE

Cochineal extract is a very old colorant with references as far back as 5000 B.C. when the Egyptian women used it to color their lips. It was introduced to the Western world by Cortez when he found it in Mexico in 1518. The Aztecs had been using it for many years, and the native Mexicans were cultivating the cochineal insect on the aerial parts of cactus, *Opuntia* and *Nopalea* species, particularly *N. coccinelliferna*. The Spaniards guarded the secret of cochineal jealously, and by 1700 as much as 500,000 pounds of cochineal were being shipped to Spain from Mexico each year. This is impressive when one considers that it takes 50,000 to 70,000 insects to produce 1 lb of colorant. Other areas, including the East and West Indies, Central and South America, Palestine, India, Persia, Europe, and Africa, developed the ability to produce cochineal. The cochineal trade peaked in 1870 and then declined rapidly due to the introduction of the synthetic colorants in 1856. Peru today is the major supplier with an annual production of about 400 tons. This constitutes 85% of the world production with Mexico and the Canary Islands sharing the remaining 15%.

American cochineal is the most significant form today, but there are other sources. Similar pigments are produced by insects from the families *Coccoidea* and *Aphidoidea* and have various historical names. Armenian Red is obtained from the insect *Porphyrophera hameli*, which grows on the roots and stems of several grasses in Ajerbaizan and Armenia. Polish cochineal is obtained from the insects *Margaroides polonicus* or *P. pomonica* found on the roots of *Scleranthus perennis*, a grass found in Central and Eastern Europe.

Cochineal and its derivatives are staging a comeback today as a food colorant because of their superior technological properties and the influence of the "natural" trend.

Chemistry

Cochineal extract (CI Natural Red 4, CI No. 75470, EEC No. E 120) is extracted from the bodies of female insects, particularly *Dactylopius coccus* Costa, just prior to egg laying, at which time the insects may contain as much as 22% of their dry weight as pigment. One may wonder at the biological significance of this, but presumably it is a protection from predators. Historically the insects were extracted with hot water, and the colorants were known as "simple" extracts of cochineal. Later methods involved extraction with aqueous alcohol. After removal of the alcohol, the preparation, called cochineal extract, contains approximately 2 to 4% carminic acid (Fig. 1) as the main colorant

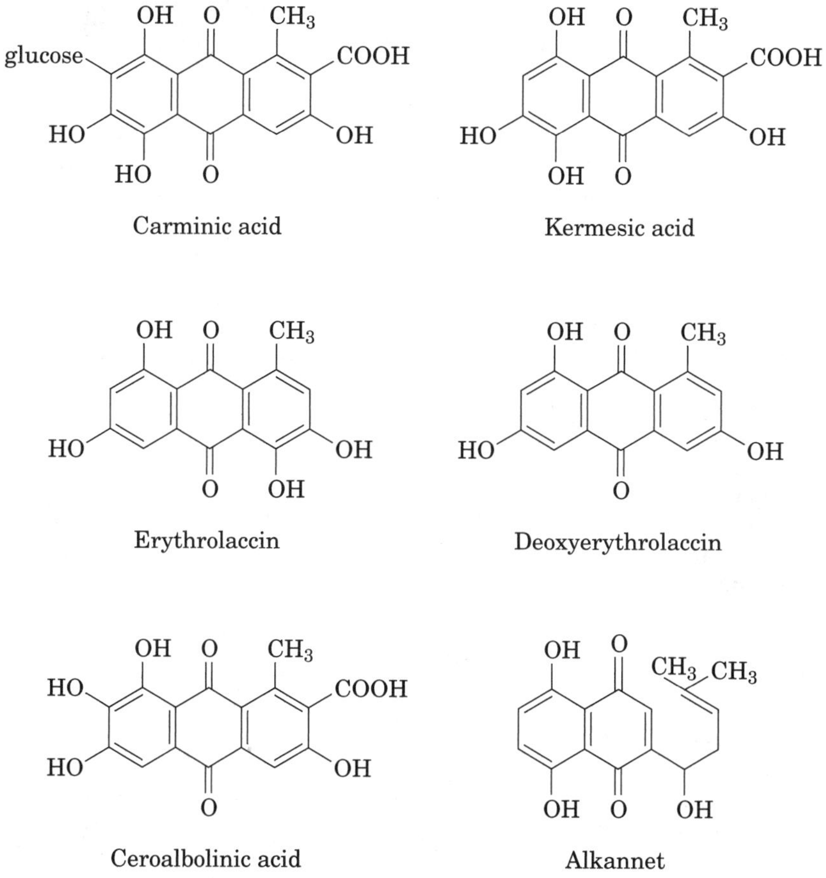

Figure 1. Structures of some anthraquinones.

compound. More recently, purification methods including proteinase enzyme treatments have produced colorants sometimes known as the "carmines of cochineal." The word *carmine* has been used a general term for this class of anthraquinones, but the more usual meaning of the term carmine (CAS Reg. No. 1390-65-4) denotes a magnesium or aluminum lake of carminic acid. Carminic acid (Fig. 1) is the pigment in cochineal, and the lakes usually contain about 50% of carminic acid. The content of carminic acid is usually the way of specifying the strength, but Schul (1) suggested that the price of carmine should be based on its color strength, not on the carminic acid content, and provided details of a method. The FDA-approved procedures are specified in the CFR Section 73.100 for Carmine.

Solutions of carminic acid at pH 4 show a pale yellow color and actually have little intrinsic color at pH values below 7, but they do complex with metals to produce stable brilliant red hues. Complexes with tin and aluminum produce the most desirable hues, and nearly all commercial preparations contain aluminum. A range of hues from "strawberry" to "black currant" can be produced by adjusting the ratio of carminic acid to aluminum. The color is essentially independent of the pH value, being red at pH 4 and bluish-red at pH 10. Aluminum lakes are soluble in alkaline media and insoluble in acids. Carmine is very stable to heat and light, resistant to oxidation, and not affected by sulfur dioxide. The presence of other metal ions may shift the color slightly toward the blue.

Carminic acid is usually available as an aqueous solution with a colorant content about 2 to 5%. It may also be spray-dried. Formulations may also contain propylene glycol citric acid, and sodium citrate. Carmine is usually supplied as an alkaline solution with a carminic acid content of 2 to 7%. Traditionally, ammonia was used as an alkanizing agent, but recently formulations with potassium hydroxide, spray-dried with maltodextrin as a carrier, have become available. They may also contain sodium hydroxide, ammonium hydroxide, and glycerol. The intensity of carmine lakes is almost twice that of carminic acid; thus, they are more cost-effective as colorants.

Applications

Carmine is considered to be technologically a very good food colorant. Its only significant limitation is with products at low pH values. It is ideally suited for comminuted meat products such a sausages, processed poultry products, surimi, and red marinades. Other important uses are in jams and preserves, gelatin desserts, baked goods, confections, icings, toppings, dairy products and soft drinks. The level of usage varies with the product and is usually 0.05 to 1.0%. Cochineal and carmine are permitted as food colorants in the United States.

KERMES

Kermes is a well-known red colorant in eastern Europe. It is derived from the insects *Kermes ilicis* or *Kermococcus vermilis*, which are found in the aboveground portions of several species of oak, particularly *Quercus coccifera*, the "Kermes Oak." The pigment in kermes is kermisic acid, the aglycone of carminic acid (Fig. 1). It also occurs as an isomer ceroalbolinic acid (Fig. 1). The pigment is obviously closely related to carminic acid, and the properties are very similar. Colorants from kermes are not permitted in the United States.

LAC

Lac is a red colorant obtained from the insect *Laccifera lacca* found on the trees *Schleichera oleosa, Ziziphus mauritiana*, and *Butea monosperma*, which grow in India and Malaysia. The lac insects are better known as a source of shellac. The lac pigments are a complex mixture of anthraquinones (Fig. 2). The mixture also contains the closely related pigments erythrolaccin and deoxyerythrolaccin (Fig. 2). The chemistry and applications of lac are similar to cochineal. Colorants from lac are not permitted in the United States.

ALKANNET

Alkannet is a red pigment (Fig. 1) extracted from the roots of *Alkanna tinctoria* Taush and *Alchusa tinctoria* Lom from southern Europe. The pigment is almost insoluble in water but readily soluble in organic solvents. It has been used in Europe to color confectionery, ice cream, and wines, but it is not permitted in the United States.

The anthraquinones and napthoquinones have been the subject of considerable research interest lately. Twelve patents were issued in the 1963 to 1984 era (2), and a large number of new compounds in this class have been reported (3). Interest as food colorants stems mainly from their stability and high tinctorial strength. They were the chromophores of choice in the "linked polymer" concept of colorants pioneered by the Dynapol Company, but this concept is no longer being commercialized (2).

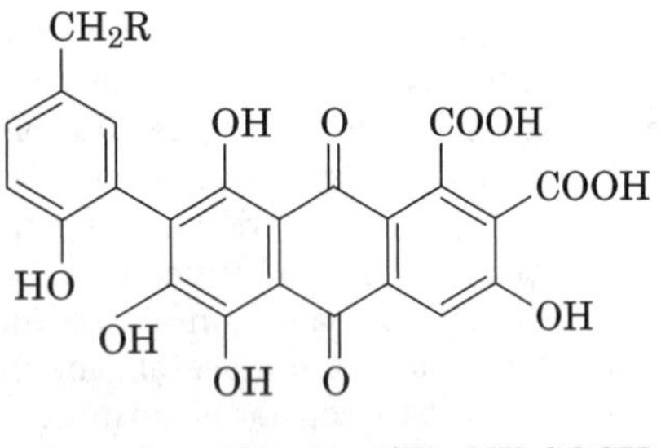

Laccaic acid A: R = $CH_2.NH.COCH_3$
Laccaic acid B: R = CH_2OH
Laccaic acid C: R = $CH.NH_2.COOH$
Laccaic acid E: R = $CH_2.NH_2$

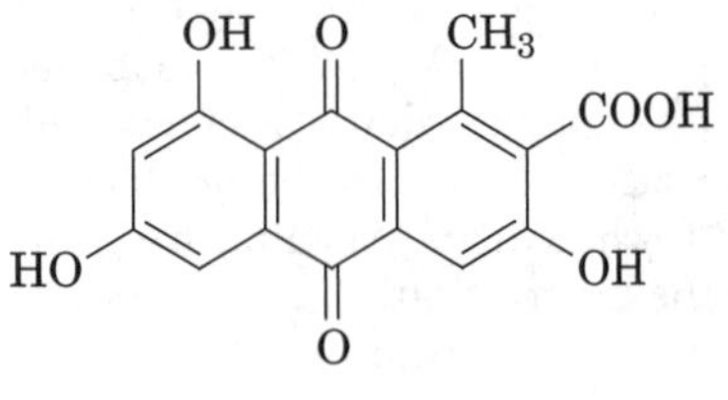

Laccaic acid D

Figure 2. Structures of the anthraquinone pigments in lac.

BIBLIOGRAPHY

1. I. J. Schul, *Proceedings of the First International Symposium on Natural Colorants*, Amherst, Mass., November 7–10, 1993, The Hereld Co, Norwalk, Conn.
2. F. J. Francis, *Handbook of Food Colorant Patents*, Food ands Nutrition Press, Westport, Conn., 1986.
3. F. J. Francis, *Lesser Known Food Colorants*, in G. A. F. Hendry and J. D. Houghton, eds., *Natural Food Colorants*, Blackie Publishing, Glasgow, Scotland, 1996, pp. 310–341.

COLORANTS: BETALAINS

The red roots of beetroot, *Beta vulgaris*, have been known for centuries as an attractive food and as a means of imparting a desirable red color to other foods. Extracts of red beets as colorants are a relatively recent development, but the concept dates back well over 100 years. The berries of pokeweed, *Phytolacca americana*, are intensely colored with betalains and provided some insurance for the wine industry against years with poorly colored grapes. The addition of pokeberry juice to wine was forbidden in France in 1892, primarily because the juice also contains a purgatory and emetic saponin called phytolaccatoxin. The red pigment was formerly called phytolaccanin until it was established that the pigment was identical with betanin in beets (1,2). A method to remove the phytolaccatoxin from pokeberry juice was also published. There was considerable interest in the United States in the development of colorants from beets in the 1980s, but, unfortunately, it coincided with the unfolding of the role of nitrites in the formation of the toxic nitrosamines. Beets are notorious accumulators of nitrates and nitrites in the growing period, and the necessity of reducing the nitrite level was one more hurdle.

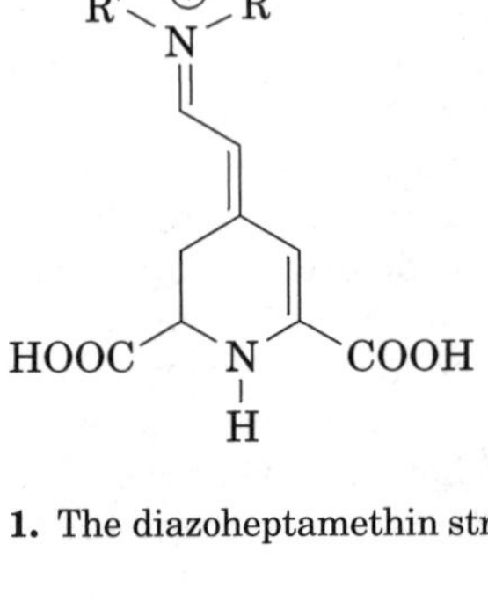

Figure 1. The diazoheptamethin structure.

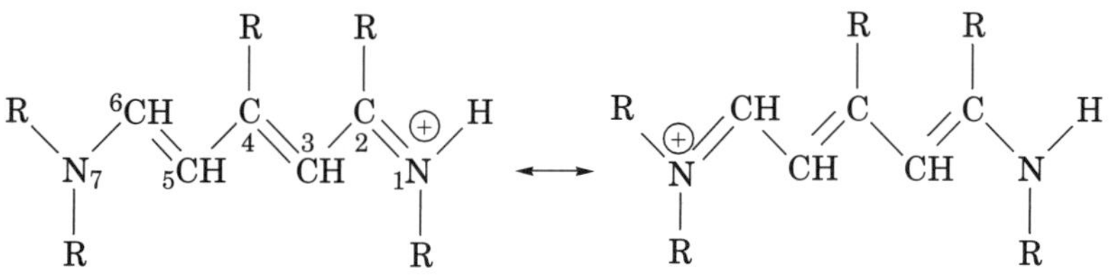

Figure 2. Betacyanin resonating structures.

CHEMISTRY

The betalain group contains about 50 red pigments called betacyanins and 20 yellow pigments termed betaxanthins. The betacyanins are all derivatives of the diazoheptamethin structure shown in Figure 1. The color is due to the resonating structures in Figure 2. If the R or R′ group extends the resonance, the pigment is red as in betanin (Fig. 3). If the resonance is not extended, the pigment is yellow as in vulgaxanthin (Fig. 4). The betacyanins contain the aglycones, betanidin, isobetanidin, and the rarer forms neobetanidin and 2-decarboxybetanidin (Fig. 5). The aglycones are combined usually with glucose and much less frequently with the disaccharide sugars sophorose and rhamnose. The betacyanin molecule may also be acylated with sulfuric, malonic, 3-hydroxy-3-methylglutaric, citric, *p*-coumaric, ferulic, caffeic, and synapic acids. A number of other betalains have been reported, and it is likely that more will be isolated, but those that are most important from a food colorant point of view are betanin for the red colors and vulgaxanthin-1 and vulgaxanthin-2 for the yellow colors.

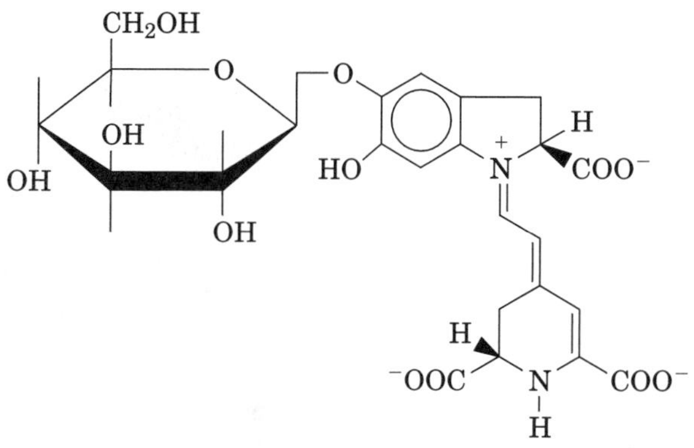

Figure 3. The structure of betanin.

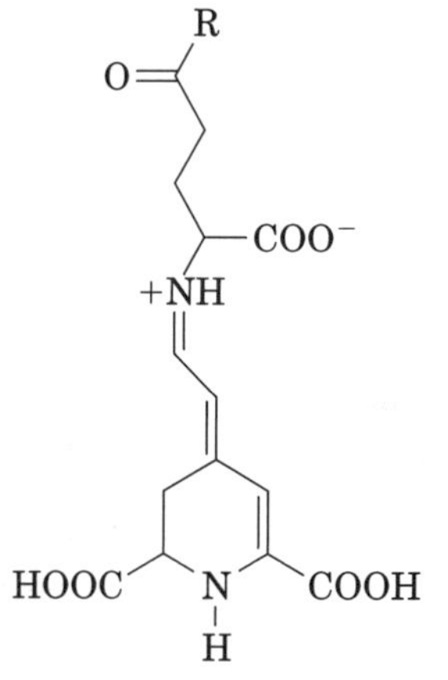

Figure 4. Structures of vulgaxanthin-1 (R = NH_2) and vulgaxanthin-2 (R = OH).

OCCURRENCE

The betalains are confined to 10 closely related families of the order *Caryophyllales*. The only foods containing betacyanins are the red beet, *B. vulgaris*; chard, *B. vulgaris*; cactus fruit, *Opuntia ficus-indica*; and pokeberries, *P. americana*. Pokeberries are not a normal food source, but the leaves are eaten as a green vegetable. Betalains are also found in a number of flowers and the poisonous mushroom, *Amanita muscaria*, but again these are not normal food sources. Their importance as a food colorant derives solely from extracts of red beets. Red beets contain 75 to 95% betanin with the remainder isobetanin, prebetanin, and isoprebetanin. The last two are the sulfate monoesters of betanin and isoprebetanin (Fig. 5), respectively. Beets contain both the red betacyanins and the yellow betaxanthins, but cultivars are available with different ratios of the red and yellow pigments. Cultivars also are available that contain only the yellow betaxanthins. Cultivars of red beets with a wide ratio of red to yellow pigments have been developed. Thus it is possible by selecting appropriate cultivars of red beets and/or blending with extracts of yellow beets to provide a range of red to yellow colorants.

The anthocyanin/flavonoid and betacyanin/betaxanthin groups are mutually exclusive in the plant kingdom. Anthocyanins and betacyanins have never been reported in the same plant.

PREPARATION

Commercial preparations from red beets are currently restricted in most countries to two products. Liquid concentrates can be prepared by pressing blanched ground beets, filtering, and concentrating under vacuum to 60 to 65% total solids. Powders can be prepared by spray-drying the liquid concentrate. The Food and Drug Administration (FDA) approved extraction procedures are described in the 21 CFR Section 73.400 for Beet Powder. Beet juice contains considerable sugar, so a yeast fermentation is sometimes included to reduce the sugar content. The alcohol produced in the fermentation step would have to be removed. Beet juice usually exhibits a beetlike taste and odor due to the

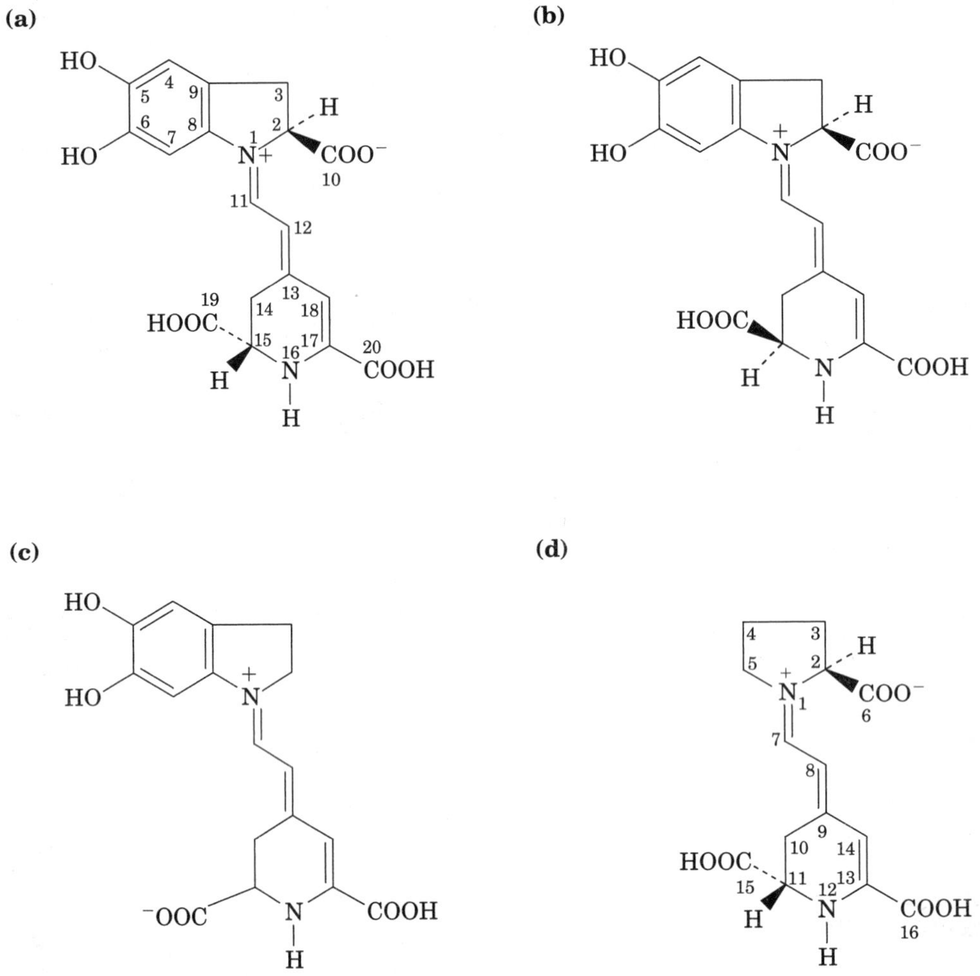

(a)

(b)

(c)

(d)

Figure 5. Structures of betanidin (a), isobetanidin (b), 2-decarboxybetanidin (c), and indicaxanthin (d).

presence of geosmin. Fermentation with *Aspergillus niger, Candida utilis*, and *Saccharomyces oviformis* has been reported to produce a product with improved stability and also to remove the geosmin. Traditionally, hydraulic or rotary presses have been used, but they recover only about 50% of the pigment. Recoveries as high as 90% can be accomplished by continuous diffusion processes. A number of resin and other absorption processes such as Dowex 50 W followed by polyamide column chromatography with methanol as the eluent, polyacrylamide (Bio-Gel P-6), and numerous others (3) are available for purification of the pigments, but they are not usually permitted by legislation. *In vitro* cell suspensions have been reported to yield high levels of pigment production, and they have a number of advantages. The high growth rates of cultures makes pigment production efficient. The characteristic odor of geosmin is not produced. A number of pigments can be produced selectively as desired. Economic commercialization of cell culture production systems has yet to be demonstrated. But there is a need for more highly purified and concentrated colorants.

Commercial beet powders usually contain 0.4 to 1.0% pigment expressed as betanin, 80% sugar, 8% ash, and 10% protein together with citric and/or ascorbic acid as a preservative. The tinctorial power is usually expressed as % betanin. Betanin is identified as EEC No. E 162 or CAS Reg. No. 7659-95-2.

APPLICATIONS

Betanin preparations are water soluble with high tinctorial strength. They are relatively unchanged in color from pH 3 to 7 but are violet at pH values below 3 and bluer at pH values above 7. The absorption spectra of betanin at three pH values are shown in Figure 6 (4). A shift in dominant wavelength toward the left would indicate a yellower color and to the right a bluer color. The spectrum at pH 9 does show a small shift in color toward the bluish-red. Both the red and yellow pigments are thermolabile with and without the presence of oxygen and are also degraded by light. Metal cations such as iron, copper, tin, and aluminum accelerate the degradation. With these limitations, beetroot preparations may be used ideally to color products with short shelf life, which are packaged to reduce exposure to light, oxygen, and high humidity; do not receive

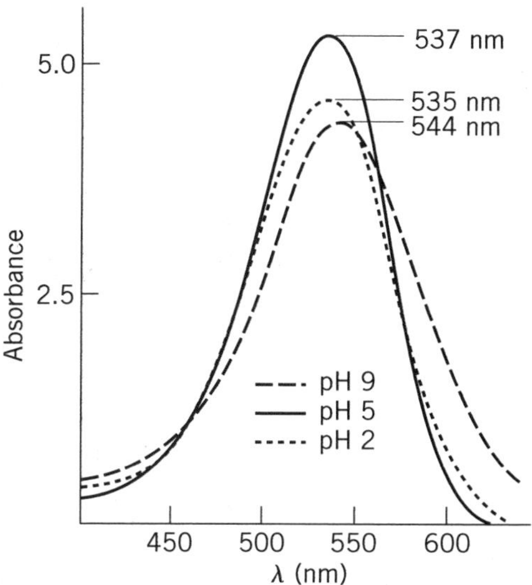

Figure 6. Spectra of aqueous betanin solutions at pH 2, 5, and 9.

extended or high heat treatment; and are marketed in the dry state (3). Beet extracts have been suggested for dairy products such as ice cream and yogurt, salad dressings, frostings, cake mixes, gelatin desserts, meat substitutes, poultry meat sausages, gravy mixes, soft drinks, powdered drink mixes, marshmallow candies, hard candies, and fruit chews. Betanin colorants can also be blended with other colorants to provide desired color matches. Extracts from beets are the only betalain colorants permitted in the United States today.

BIBLIOGRAPHY

1. M. W. Driver and F. J. Francis, "Stability of Phytolaccanin, Betanin, and FD & C Red No. 2 in Dessert Gels," *J. Food Sci.* **44**, 518–520 (1979).
2. M. W. Driver and F. J. Francis, "Purification of Phytolaccanin by Removal of Phytolaccatoxin from *Phytolaccca americana*," *J. Food Sci.* **44**, 52–523 (1979).
3. R. L. Jackman and J. L. Smith, "Anthocyanins and Betalains," in G. A. F. Hendry and J. D. Houghton, eds., *Natural Food Colorants*, Blackie Publishers, Glasgow, Scotland, 1992.
4. J. H. von Elbe, "Stability of Betalains as Food Colors," *Food Technol.* **29**, 42–46 (1975).

COLORANTS: CARAMEL

The heating of sugar preparations to create brown, flavorful, and pleasant-smelling products has been practiced in home cooking for centuries. The sauces or candies are known as caramels, but commercial practices to prepare caramel colorants began in Europe about 1850. The first caramel colorants were prepared by heating sugars in an open pan, but in view of their popularity, modifications were soon introduced. Caramel colorants that were compatible with various food products were desired, and essentially compatibility consists of the absence of flocculation and precipitation in the food. These undesirable effects result from colloidal interactions of charged macromolecular components of caramel with a particular food. Hence, the net ionic charge of the caramel color macromolecules at the pH of the product in which it is used is a prime determinant of compatibility (1).

Caramel (CI Natural Brown 10, EEC No. E 150) is a brown colorant obtained by heating sugars. The official FDA definition is "The color additive caramel is the dark brown liquid or solid resulting from carefully controlled heat treatment of the following food-grade carbohydrates: dextrose, invert sugar, lactose, malt syrup, molasses, starch hydrolysates and fractions thereof, or sucrose."

COMMERCIAL PREPARATION

An appropriate amount of carbohydrate is added to a reaction vessel at 50 to 70°C, the temperature is raised to 100°C, and the reactants are added. After appropriate heating the mass is cooled and filtered, and the pH and specific gravity are adjusted with acids, alkalies, or water to meet customer specifications. The carbohydrate hydrolysates may be obtained from sucrose, corn, wheat, and tapioca. High glucose content is desirable because caramelization only occurs through the monosaccharide. The type and concentration of reactant determines the properties of the caramel colorant. Caramels can be prepared by using carbonates, hydroxides, ammonium compounds, sulfites, and both sulfite and ammonium compounds. There is a direct relationship between reactant concentration and color intensity. Too little produces an underreacted product, and too much produces a product that is too viscous or even solid.

CHEMISTRY

In 1980 the Joint Expert Committee on Food Additives (JECFA) of the Food and Agriculture Organization (FAO)/World Health Organization (WHO) recommended that further information on the chemical properties of caramel be obtained to establish a suitable classification and specification system. This reaction is common to a number of natural extracts and also to some manufactured colorants. For example, with the colorants from grapes, no manufacturer has attempted to provide a chemical profile of the components of the extract. This is understandable in view of the complexity of the chemical profile. A somewhat similar situation occurs with a synthesized product such as FD&C Yellow No. 5. When it was approved for food use in 1916, the manufacturers wanted to produce as pure a product as possible. The impurities they had to cope with were unreacted ingredients and about three compounds produced by side reactions. In the usual process of chemical synthesis, the desired reaction does not normally go to 100% completion with no side reactions, and it is the job of the manufacturer to remove the unwanted by-products. But the increasing sophistication of analytical procedures that can routinely detect one part per billion, or even as low as one molecule, has allowed researchers to detect more impurities. Today, FD&C Yellow No. 5 has been shown to have as many as 17 components present, some in exceedingly low concentrations (2). The situation with other FD&C colorants is probably similar. The same situation exists with synthetic β-carotene. Modern analytical instrumentation has made it possible to identify the impurities in chemicals produced with rigid process control as illustrated with FD&C Yellow No. 5 and β-carotene. The situation with caramel and colorants produced from grapes is different in that it is not possible to specify in detail the structures of all the compounds present.

The question that arises with knowledge of the impurities is, What do they mean from a safety point of view? It also emphasizes that the chemical profile of a given additive in routine manufacture be the same as that in the batches used for toxicology studies. It is economically unfeasible to remove all impurities in food additives; thus, some form of risk assessment is unavoidable. Fortunately all four products mentioned earlier have been given a clean bill of health.

The JECFA wanted data on caramel for the reasons just described. The International Technical Caramel Committee (ITCA) undertook an extensive research program to provide this information. This resulted in the grouping of the caramel formulations into four classes depending on the net ionic charge and the presence of reactants.

Class	Charge	Reactants
I	−	No ammonium or sulfite compounds
II	−	Sulfite compounds
III	+	Ammonium compounds
IV	+	Both sulfite and and ammonium compounds

It is obvious that heating carbohydrates in the presence of a reactant would produce a wide range of chemical compounds. Complete characterization is not feasible; thus, a series of profiles were developed using 157 samples from 11 manufacturers in seven countries. Since Caramel Color IV accounts for 70% of all caramel colors manufactured, it was chosen as a prototype for this ambitious undertaking. The profiles developed were HPLC screening, size fractionation by ultrafiltration to provide low molecular weight, intermediate molecular weight and high molecular weight groupings, and subfractionation by cellulose chromatography. Each fraction was examined by a series of sophisticated analytical approaches. These data combined with 11 physical characteristics confirmed that the four groupings provided real and reproducible classifications (3).

Color is a physical parameter of interest to the formulator. It could, of course, be specified in any of the conventional tristimulus color scales, but it is possible to specify the color accurately by a simpler method (4). The color of caramel is due to a large number of chemical chromophores; thus, different formulations do not show markedly different spectral curves and do not change shape with concentration. Curves of different concentrations can be superimposed on each other by plotting log absorbency against concentration. This is the "optical signature" of a colorant. When log absorbency is plotted against wavelength for caramel solutions, the result is a straight line that moves up or down depending on the concentration. The slope of the line will change according to the hue of the solution. This makes it possible to determine two colorimetric indices, the Hue Index and the Tinctorial Power, from two simple absorbency measurements.

The Hue Index is:

$$10 \log(A_{0.51}/A_{0.61})$$

where 0.51 is the absorbency at 0.51 microns, and 0.61 is the absorbency at 0.51 microns.

The Tinctorial Power is:

$$K^{0.56} = \frac{A_{0.56}}{cb}$$

where 0.56 is the absorbency at 0.565 microns, c is the concentration (g/L), and b is the cell thickness (cm). Both the Hue Index and the Tinctorial Power should reflect the visual appearance over the whole visual spectrum. The preceding equations are reproduced as they appeared in the original publication, but today the term nanometers would be used instead of microns.

One of the major considerations in this research was the safety aspect of the caramel colorants. This program resulted in 11 papers being published in the same 1992 Vol. 30 issue of the journal *Food Chemical Toxicology*, and 7 of them were on toxicology. Caramel was given a clean bill of health and the JECFA assigned an Acceptable Daily Intake (ADI) of 0 to 200 mg/kg/day.

APPLICATIONS

Caramel color is freely soluble in water but insoluble in most organic solvents. In concentrated form the colorant has a characteristic burnt taste, which is not detectable at the levels normally used in foods and beverages. Commercial preparations vary from 50 to 70% total solids with a range of pH values depending on the type of caramel. More than 80% of the caramel produced in the United States is used to color soft drinks, particularly colas and root beers. Type I is designed for use with distilled spirits, desserts, and spice blends. Type II is used with liqueurs. Type III is suitable for baked goods, beer, and gravies. Type IV is used in soft drinks, pet foods, and soups.

BIBLIOGRAPHY

1. D. V. Myers and J. C. Howell, "Characterization and Specifications of Caramel Colours: An Overview," *Food Chem. Toxicol.* **30**, 359–363 (1992).
2. J. E. Kassner, "Modern Technologies in the Manufacturer of Certified Food Colors," *Food Technol.* **41**, 74–76 (1987).
3. B. H. Licht, J. Orr, and D. V. Myer, "Characterization of Caramel Color IV," *Food Chem. Toxicol.* **30**, 365–373 (1992).
4. R. T. Linner, "Caramel Coloring: A New Method of Determining its Color Hue and Tinctorial Power," *Amer. Soft Drink J.*, 26–30 (1971).

COLORANTS: CAROTENOIDS

The carotenoids are one of the largest group of pigments produced in nature, second only to the chlorophylls. They are very widespread with more than 100,000,000 tons produced annually in nature. Most of this amount is produced in the form of fucoxanthin in algae in the ocean and the three main carotenoids of green leaves: lutein, violaxanthin, and neoxanthin (Fig. 1). Other pigments predominate in certain plants, such as lycopene in tomatoes (Fig. 1), capsanthin and capsorubin (Fig. 2) in red peppers, and bixin in annatto (Fig. 2). Colorant preparations have been made from all of these (1), and obviously the composition of the colorant extracts reflects the profile of the starting material. The chemistry and occurrence of the carotenoids is well described in the article entitled COLORANTS: CAROTENOIDS. Carotenoids are probably the best known of the food colorants derived from natural sources (2).

ANNATTO

Annatto (CI Natural Orange 4, CI No. 75120, EEC No. E 160b) is found in the outer layers of the seeds of the shrub *Bixa orellana*, a tropical plant grown in South America, India, East Africa, the Philippines, and the Carribbean. Peru and Brazil are the dominant sources of supply. Annatto is one of the oldest colorants and has been used in antiquity for coloring foods, cosmetics, and textiles. It has been used for more than a hundred years in the United States and Europe primarily as a colorant for dairy products. The colorant is prepared by leaching, with gentle mechanical friction, of the seeds with various solvents, including vegetable oil, fats, and alkali aqueous and alcoholic solutions. Depending on the application, the crude extract may be refined by precipitation with acids and/or recrystallization. Spray-dried powders are also available in both water-soluble and oil-soluble forms. The Food and Drug Administration (FDA) approved procedures for extraction are specified in the 21 CFR section 73.30 for Annatto Extract.

The pigments in annatto are a mixture of bixin, the monomethyl ester of a dicarboxylic carotenoid, and norbixin, the dicarboxylic derivative of the same carotenoid (Fig. 2). Both bixin and norbixin normally occur in the *cis* form, but small amounts of the more stable *trans* form are formed on heating. A yellow degradation product termed C_{17} yellow pigment (Fig. 2) is also produced on heating. The *cis* forms are redder than the *trans* forms or the C_{17} compound; thus, a source of red and yellow pigments are available. The carboxylic portion of the molecule contributes to the solubility in water, and the ester form contributes to oil solubility. This has provided flexibility in applications, and annatto has been used in a wide variety of applications.

Annatto is available in both water-soluble and oil-soluble liquids and powders. The oil-soluble form is somewhat unstable under oxidative conditions, and degradation is increased by exposure to light and is catalyzed by metals. Addition of antioxidants, such as ascorbic acid, tocopherols, and polyphenols, helps to minimize oxidative degradation. Annatto shows more stability to exposure to air than other carotenoids and is moderately stable to heat. Little change in color occurs with pH changes, but products with a low pH may show a pinkish tinge due to isomerization of the pigments.

LYCOPENE

Lycopene occurs as the major (85–90%) pigment (Fig. 1) in red tomatoes, *Lycopersicum esculentum*. The other pigments are β-carotene (10–15%) with small quantities of about 10 other carotenoids. In spite of the fact that large quantities of lycopene are available in the waste from the tomato-processing industry, colorants containing lycopene were not commercially available. This was probably due to the belief that lycopene was susceptible to degradation by oxidation and light. But recently a combination of better manufacturing practices, and the development of a tomato cultivar particularly high in lycopene led to the commercialization of lycopene as a food colorant. Preparation of lycopene extracts from tomatoes is relatively simple, involving an alkali saponification and extraction with a mixture of solvents such as acetone and hexane. The acetone can be removed by washing with water and the hexane by vacuum treatment, leaving a relatively pure lycopene extract. The extract would, of course, contain the other carotenoids in the tomato. Commercialization probably was also helped by the health aspects because it may be an efficient *in vivo* radical scavenger. Lycopene preparations are being marketed as a neutraceutical and as a food colorant. Colorant preparations containing lycopene are currently not allowed in the United States.

PAPRIKA

Paprika is a deep red pungent powder prepared from the ground dried pods of the sweet pepper *Capsicum annum*. Paprika is produced in many warm countries around the world, and several areas have developed products with specific characteristics such as the Hungarian paprika and the Spanish paprika. The same peppers are used in salads and as a source of pimento. The other principal type of red pepper (also *C. annum*), typically called cayenne pepper or cayenne, is usually much more pungent in flavor. Both types are highly pigmented. The red pepper *C. frutescens* is the source of the highly colored and very pungent Tabasco sauce.

Paprika contains capsanthin and capsorubin (Fig. 2) occurring mainly as the lauric acid esters. Smaller quantities of about 20 other carotenoids are also present. Specifications do not attempt to describe the pigment profile but usually specify color strength by the American Spice Trade Association (ASTA) Color Value. This is essentially the absorption at 460 nm in acetone measured against a cobalt solution or a glass standard used as a reference (3).

Paprika oleoresin (EEC No. E 160c) is an orange-red oil-soluble extract from *C. annum*. The FDA-approved procedures are specified in 21 CFR Section 73.345 for Paprika Oleoresin. The dried and ground red peppers are extracted with a volatile solvent, usually a chlorinated hydrocarbon

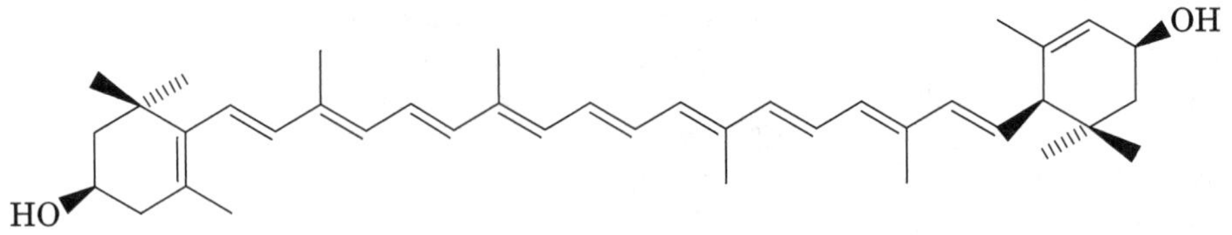

Lutein

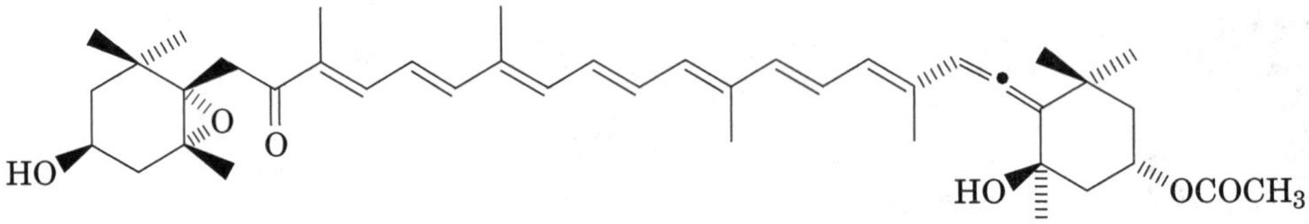

Fucoxanthin

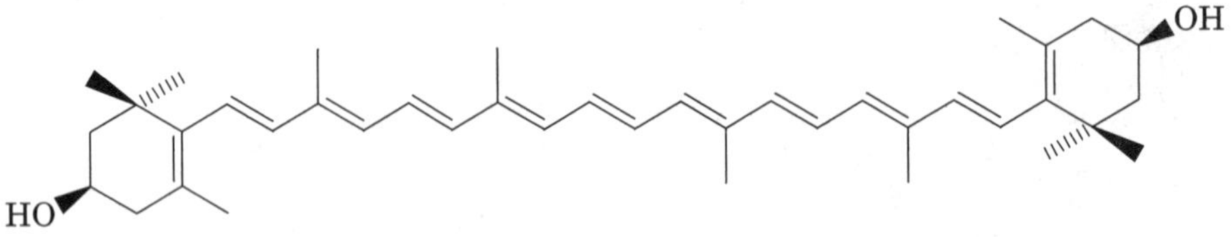

Zeaxanthin

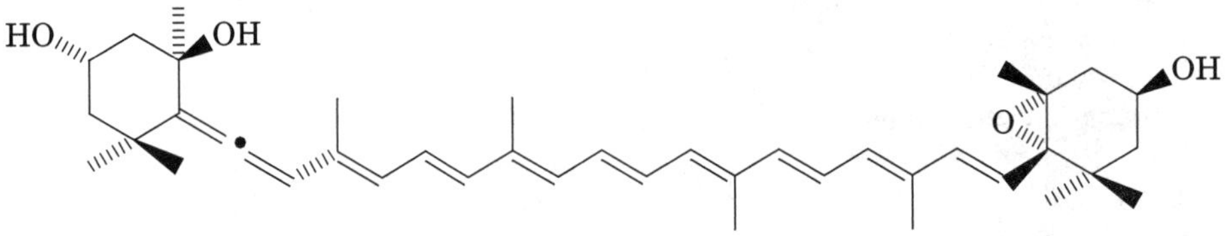

Neoxanthin

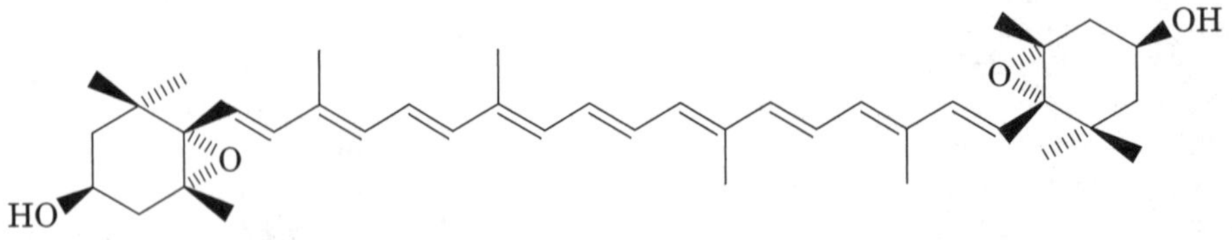

Violaxanthin

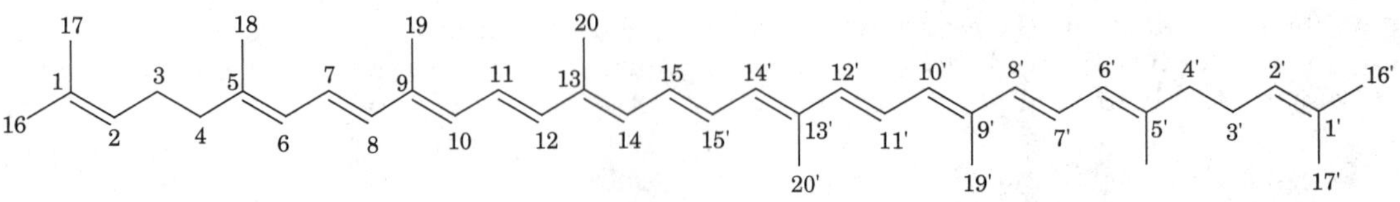

Lycopene

Figure 1. Selected structures for the carotenoids in plants.

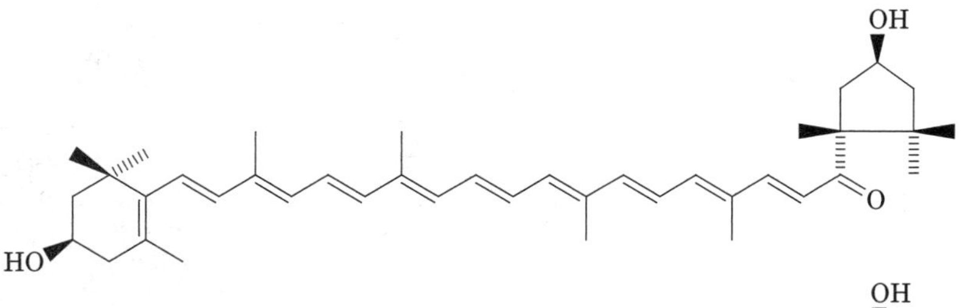

Capsanthin

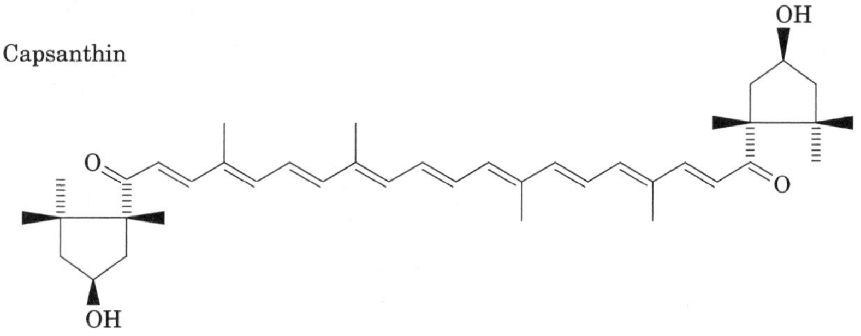

Capsorubin

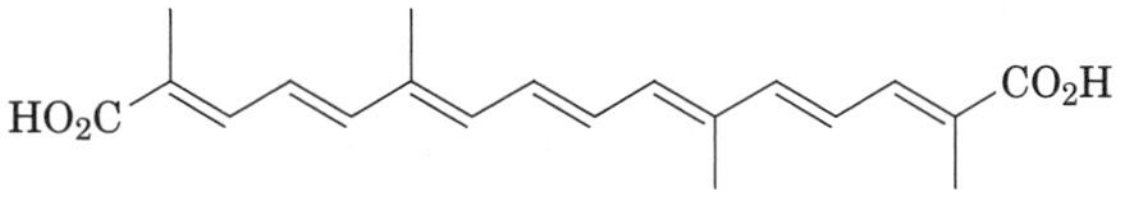

Crocetin

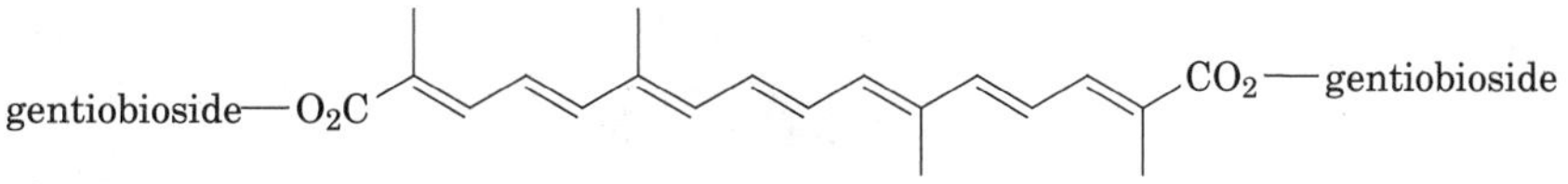

Crocin

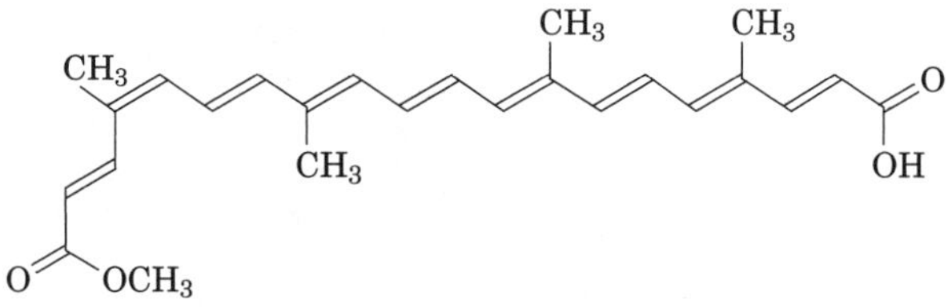

Bixin

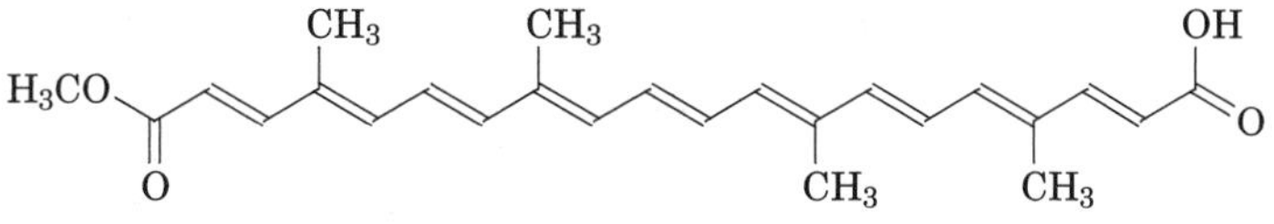

Trans-bixin

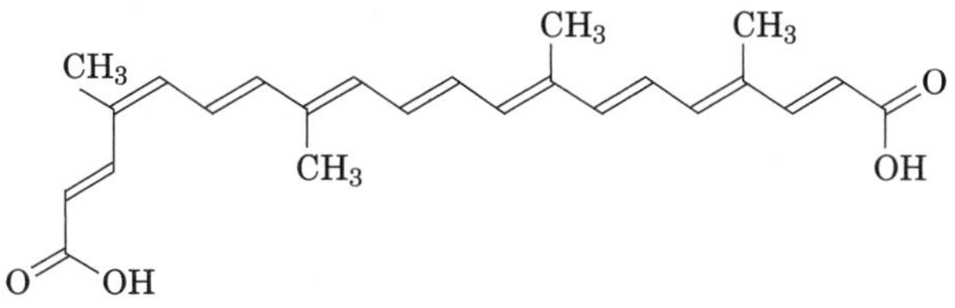

Norbixin

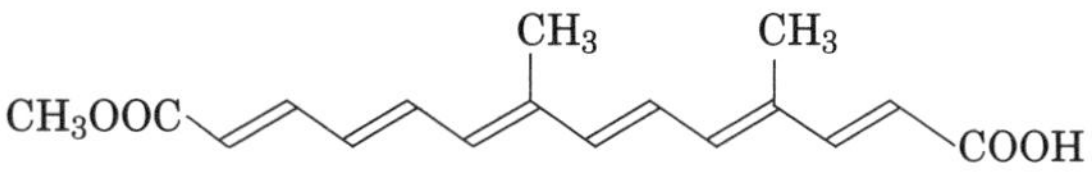

C_{17} Yellow Pigment

Figure 2. The carotenoids of paprika (capsanthin and capsorubin), saffron (crocetin and crocin), and annatto (bixin, transbixin, norbixin, and C_{17} yellow pigment).

followed by removal of the solvent. Preparations can also be made by extraction with a vegetable oil to provide an oil-soluble colorant. Paprika and paprika oleoresin add both color and flavor to a product; thus, the applications are usually limited to products in which both characteristics are desirable. For example, the recent rise in demand for tomato products in the form of pizza, salsa, and so on has increased the demand for paprika. Most of the spice extract used in Europe is used to flavor meat products, soups, and sauces. Smaller quantities are used in salad dressings, snacks, processed cheese, confectionery, and baked goods.

SAFFRON

Saffron (CI Natural Yellow 6, CI No. 75100) is a very old colorant dating back to the twenty-third century B.C. It has come to be known as the "gourmet spice" because of its high price, but it also provides both spice and colorant. Saffron consists of the dried stigmas of the flowers of the crocus bulb, *Crocus sativus*, grown primarily in North Africa, Spain, Switzerland, Austria, Greece, and France. The high price is due to the fact that it takes about 150,000 flowers to produce 1 kg of colorant.

The pigments in saffron are chemically similar to those in annatto (Fig. 2). They are crocetin, a dicarboxylic carotenoid, together with its gentiobioside ester. Gentiobioside is a diglucoside with a β 1–6 linkage. The sugar portion confers solubility in water and makes the colorant very flexible in its applications to a variety of food and pharmaceutical products. The same pigments occur in a number of other plants but *C. sativus* is the only commercial source. The fruits of the Cape Jasmine, *Gardenia jasminoides*, produce the same pigments and have been suggested as an alternate source. But the gardenia fruits supply only the colorant, not the spice flavor, so the sources are quite different. Saffron extracts also contain β-carotene, zeaxanthin, and traces of several other carotenoids.

Saffron preparations are fairly stable to light, oxidation, microbiological attack, and changes in pH. Overall, technically it is a good colorant with high tinctorial strength. Its strength is usually judged by its carotenoid content as measured by the absorption of an aqueous extract at 440 nm, but Alonso et al. (4) suggested that tristimulus methods were more appropriate for classifying samples of saffron.

TAGETES

Tagetes erecta L. (Aztec marigold) is an annual herb that grows from 3 to 4 tall in temperate climates. Colorants are available in three forms: dried ground flower petals, oleoresin extracts, and purified oleoresin extracts. The first two are used primarily in the poultry industry and the third in the food industry. The principal producers are Mexico, Peru, the United States, Spain, and India (5).

Tagetes petals contain up to 80% lutein with smaller amounts of zeaxanthin, cryptoxanthin, β-carotene, and about 14 other carotenoids. The lutein compounds exist as dipalmitate, dimyristate, myristate-palmitate, palmitate-stearate, and distearate esters. Because of the seasonality of the crop, the mature marigold flowers are collected and the petals separated prior to storage in large bins containing up to 80 tons. The petals are removed as needed, pressed, and then dried to less than 10% moisture. They can then be ground and sold as tagetes meal. The ground petals can also be extracted with a number of solvents, but hexane seems to be preferred. After removal of the solvent, a brown oleoresin is obtained, which can be incorporated directly into poultry feed. The resin can be further purified by saponification and sold as Saponified Marigold Extract. This product is suitable for incorporation into poultry feed. The product can be further purified to produce a dry powder suitable for use as a food colorant.

Colorants from tagetes are available in a variety of forms. Formulations for animal feeds are usually ground dried petals, oleoresins, or saponified oleoresins. Food colorants are available in a number of forms such as purified lutein esters in oil-soluble or water-soluble dispersible systems, spray-dried emulsions, gum-based emulsions, and emulsifier-based emulsions. They show good stability to heat, light, pH changes, and sulfur dioxide. They are susceptible to oxidation, which can be minimized through the addition of antioxidants such as ethoxyquin, ascorbic acid, tocopherols, BHA and BHT, and encapsulation. The strength of tagetes extracts is usually measured as the absorption of a 1%/1 cm solution in an appropriate solvent and is often quantified by calculations using the specific extinction coefficient of lutein or lutein esters.

Tagetes colorants provide yellow to orange colors suitable for use in pastas, vegetable oils, margarine, mayonnaise, salad dressings, baked goods, confectionery, dairy products, ice cream, yogurts, citrus juices, and mustard. Tagetes meal and its extracts are approved only as colorants in poultry feed in the United States.

SYNTHETIC CAROTENOIDS

Plant extracts of carotenoids have been used for centuries as food and cosmetic colorants, so it was only natural that synthetic carotenoids would become available as colorants. The original synthesis of β-carotene was reported in 1950, and commercial production followed in 1954. It was followed by β-apo-8′-carotenal in 1962 and canthaxanthin in 1964. The methyl and ethyl esters of β-apo-8′-carotenoic acid, citranaxanthin, and astaxanthin followed (Fig. 3). All carotenoids by virtue of their structure are susceptible to degradation by oxidation, light, and heat. In addition, they are insoluble in water and almost insoluble in oil. But appropriate commercial preparations are currently available for almost any type of food in the yellow to red range.

A variety of approaches are used to prepare appropriate formulations. These involve reducing particle size for suspensions in oil and stabilization with antioxidants. Three approaches are used to develop water-soluble products. These are formulation of colloidal suspensions, emulsification of oily solutions, and dispersion in suitable colloids. They can be stabilized by a wide variety of additives in the protein, carbohydrate, and lipid categories and stabilized

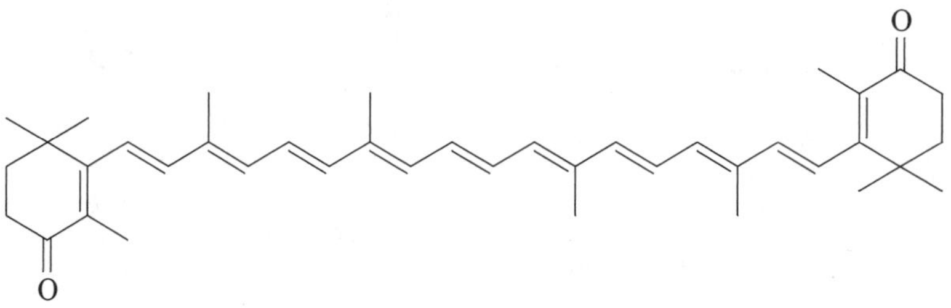

Canthaxanthin

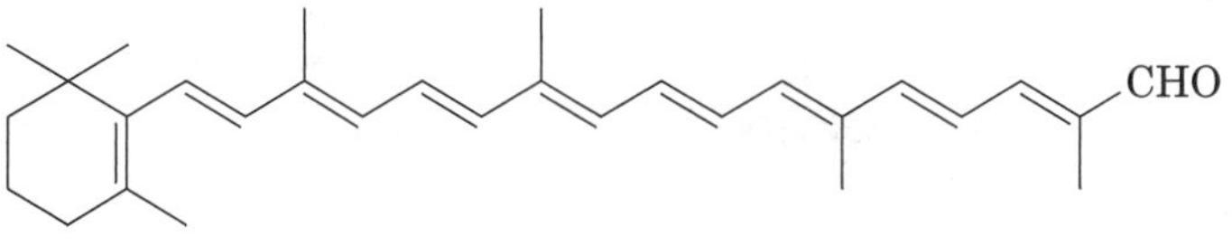

β-apo-8'-carotenal

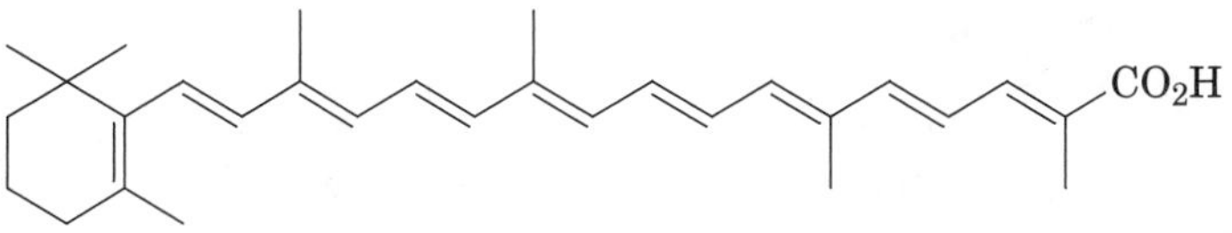

β-apo-8'-carotenoic acid

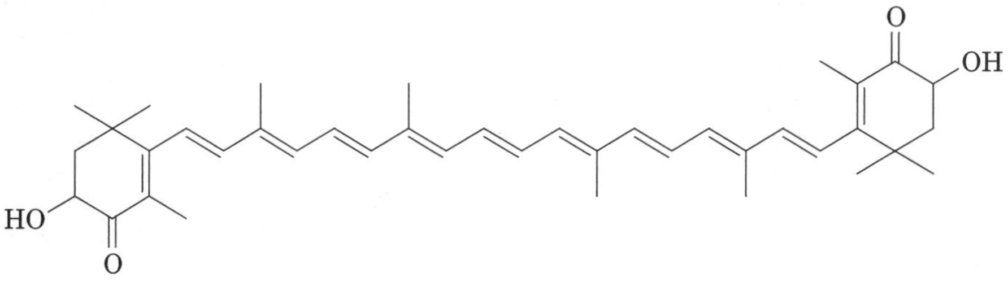

Astaxanthin

Figure 3. Carotenoids synthesized for commercial colorant applications. *β*-Apo-8′-carotenoic acid is usually in the form of the methyl or ethyl esters. *β*-Carotene was also synthesized.

with antioxidants (6). Carotenoid colorants are appropriate for margarines, oils, fats, shortenings, fruit juices, beverages, dry soups, canned soups, dairy products, milk substitutes, coffee whiteners, dessert mixes, preserves, syrups, confectionery, salad dressings, meat products, pasta, egg products, baked goods, and many others.

MISCELLANEOUS CAROTENOID EXTRACTS

The red oil obtained from the fruit of the palm tree *Elaeis guineensis* is very highly pigmented with about 500 mg carotenoids per kilogram of oil. The carotenoid mixture is very complex, containing mainly *β*-carotene and about 20 other carotenoids. The oil can be used as an ingredient to color margarine and other oil-based products, but its chief use is as a cooking oil due to its distinctive flavor or, after refining, as a general purpose edible oil. Xanthophyll pastes are well known in Europe and consist of extracts of alfalfa (lucerne), nettles, or broccoli. Unless they are saponified they will be green due to their high content of chlorophyll. Many concentrated xanthophyll pastes contain as much as 30% carotene with the major pigments as lutein (45%), *β*-carotene (25%), violaxanthin (15%), neoxanthin (15%), and a number of minor pigments. Highly concentrated extracts from alfalfa became available some years ago as by-products of the Pro-Xan process for preparing protein concentrates from alfalfa. The carotenoid extracts were widely promoted as colorant additives for poultry feed, but the Pro-Xan process was never commercially successful. Extracts from citrus peels have been suggested as a means of augmenting the natural color of orange juice since the more highly colored juices command a premium price, but addition of colorants from citrus is not permitted in the United States at the present time. Citrus peel extracts contain *β*-carotene, cryptoxanthin, antheroxanthin, violaxanthin, lycopene and β-apo-8′-carotenol together with a number of others. About 115 carotenoid pigments have been reported in citrus. Colorants from carrots usually contain about 80% β-carotene and up to 20% α-carotene plus small amounts of several others. Some of the high pigment strains of carrots used for colorant extracts also contain lycopene. Astaxanthin is a desirable addition to the diet of salmon and trout in aquaculture because of its ability to impart a desirable red color to the flesh. The usual sources of astaxanthin are the by-products of the lobster and shrimp processing industry, but the demand exceeds the supply. This has led to an interest in growing the red yeast *Phaffia rhodozyma* as a raw material for a concentrated extract. But unfortunately *Phaffia* produces the wrong optical isomer of astaxanthin for optimal accumulation in the flesh of salmon. Hinostroza et al. (7) reported that carotenoids from three sources, synthetic astaxanthin, crabs (*Pleurocodes planiples*), and *P. rhodozyma*, were deposited in trout muscle at the rate of 7.8%, 5.0%, and 4.0%, respectively. The FDA has approved the addition of *Phaffia* products to fish food. There is an ongoing interest to develop other plant sources to compete with the nature-like synthetic carotenoids. Several mutants of the carotogenic molds *Blakeslea trispora* and *Phycomyces blakesleeanus* produce high concentrations of β-carotene, but recent interest has shifted to the microalgae. Species of *Dunaliella* can accumulate up to 10% dry weight of β-carotene. Ponds in Australia originally developed to produce salt by evaporation of salt water are now being used to grow *Dunaliella*. Other installations are in Hawaii. Regardless, both sources command a premium because of their "natural" association. Interestingly, the same ponds can be used to grow *Haematococcus* species, which accumulate astaxanthin. A gene from *H. pluvialis* has been transferred to tobacco plants to provide a plant source of astaxanthin. In view of the widespread occurrence of the carotenoid pigments, it is not surprising that other plants and animals would be suggested as potential sources of colorants. These include krill, chlorella, shrimp, algae, bacteria, molds, and a variety of plant sources (8).

BIBLIOGRAPHY

1. J. C. Bauernfeind, *Carotenoids as Colorants and Vitamin A Precursors*, Academic Press, New York, 1981.
2. F. J. Francis, *Handbook of Food Colorants*, Eagen Press. St. Paul, Minn., 1999.
3. D. M. Marmion, *Handbook of US Colorants*, John Wiley & Sons, New York, 1991.
4. G. L. Alonzo et al., "Color Analysis of Saffron," *Proc. 2nd Int. Symp. on Natural Colorants*, Puerto de Acapulco, Mexico, January 23–26, 1996.
5. J. Verghese, "Focus on Xanthophylls from *Tagetes erecta*, L.—The Giant Natural Colour Complex," *Proc. 2nd International Symposium on Natural Colorants*, Puerto de Acapulco, Mexico, January 23–26, 1996.
6. F. J. Francis, "Carotenoids as Colorants," *World of Ingredients*, 34–38 (1995).
7. G. C. Hinostoza et al., "Pigmentation of Rainbow Trout *Oncorhynchus mykiss* by Astaxanthin from Red Crab *Pleuroncodes planipes* in Comparison with Synthetic Astaxanthin and *Phaffia rhodozyma* Yeast," *Proc. 2nd Int. Symp. on Natural Colorants*, Puerto de Acapulco, Mexico, January 23–26, 1996.
8. F. J. Francis, *Handbook of Food Colorant Patents*, Food and Nutrition Press, Westport, Conn., 1986.

See also CAROTENOID PIGMENTS.

COLORANTS: CARTHAMIN

Carthamin (CI Natural Red 26; CI 75140) is a very old colorant. Its use as a textile colorant dates back to antiquity under a variety of names such as Spanish saffron, African saffron, American saffron, thistle saffron, false saffron, bastard saffron, Dyer's saffron, and so on. More modern names are carthemone, carthamic acid, and safflor red. Carthamin is a yellow to red preparation from safflower flowers, *Carthemus tinctorius*, of the family Compositae, which is cultivated extensively in Europe and America.

Carthamin is classified biochemically as a chalcone in the yellow flavonoid group of pigments. The flavonoids are a very large group of plant pigments with more than 4000 compounds, but only carthamin has been suggested as a colorant. Carthamin contains three chalcones: the red carthamin, safflor Yellow A, and safflor Yellow B (Fig. 1). Fresh yellow flower petals contain precarthamin, which oxidizes to form the red carthamin. Carthamin, under acid conditions, equilibrates to two isomers: red carthamin and yellow isocarthamin (1). The earlier patents on carthamin involved simple aqueous extraction of the petals or crushing the petals prior to extraction to allow for oxidation (2). The three main pigments can be purified by passage through a styrene resin bed. Purification and stabilization can be greatly enhanced by absorption of the pigments on cellulose powder. Apparently cellulose has a great affinity for carthamin and this is known as the "Saito effect." "The effect is so strong that the carthamin may be retained for more than a thousand years without appreciable change to the coloration" (3). No storage data were provided to substantiate this claim. The cellulose absorption method was adapted to large-scale production using a methanolic extract and a cellulose column in acid media. Another method using a cellulose derivative diethylaminoethylcellulose yielded pure precarthamin, carthamin, safflor Yellow A and safflor Yellow B. Tissue culture approaches have been successful in producing the safflower pigments as well as some novel closely related pigments (4,5).

Carthamin has been suggested as a colorant for pineapple juice, yogurt, butter, liqueurs, confectionery and so on. The yellow to red range of colors gives it some flexibility. Currently it is not approved for use in foods in the United States.

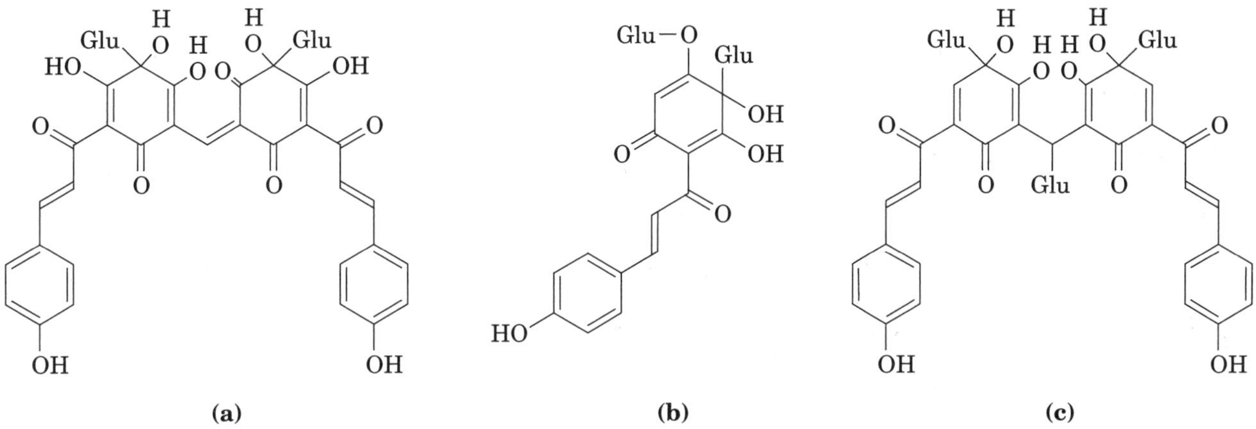

Figure 1. Pigments of carthamin (**a**) carthamin; (**b**) Safflor Yellow A; (**c**) Safflor Yellow B.

BIBLIOGRAPHY

1. K. Saito and A. Fukushima, "Effect of External Conditions on the Stability of Enzymatically Synthesized Carthamin," *Acta. Soc. Bot. Pol.* **55**, 639–651 (1986).
2. F. J. Francis, *Handbook of Food Colorant Patents*, Food and Nutrition Press, Westport, Conn., 1986.
3. K. Saito and A. Fukushima, "On the Mechanism of the Stable Red Colors of Cellulose-bound Carthamin," *Food Chem.* **29**, 161–175 (1988).
4. F. J. Francis, "Less Common Natural Colorants," in G. A. F. Hendry and J. D. Houghton, eds., *Natural Food Colorants*, 2nd ed., Blackie Academic and Professional, Glasgow, Scotland, 1996, pp. 310–342.
5. F. J. Francis, "Miscellaneous Colorants," in G. A. F. Hendry and J. D. Houghton, eds., *Natural Food Colorants*, Blackie Academic and Professional, Glasgow, Scotland, 1992, pp. 242–272.

COLORANTS: CHLOROPHYLLS

The chlorophylls are a group of naturally occurring pigments present in all photosynthetic plants including the algae and some bacteria. They are in greater abundance than any other organic pigment produced in nature. Hendry estimated annual production at about 1,100,000,000 tons as compared with carotenoids at 100,000,000 tons, with about 75% being produced in aquatic, primarily marine, environments (1). Obviously, as a source of raw material for food colorants, supply is no problem.

CHEMISTRY

Five chlorophylls and five bacteriochlorophylls are known in nature. Only two, chlorophylls a and b (Fig. 1) are important as source material for food colorants. Chlorophylls a and b have a complex structure, but they differ only by a -CH_3 and a -CHO group on carbon 7. The numbering system for chlorophylls is shown in Figure 2 (1). The chlorophyll a molecule is easily changed in the presence of acids to remove the magnesium ion, resulting in pheophytin a. The phytyl group is easily removed to produce chlorophyllide a. If both magnesium and phytyl groups are removed, the compound is called pheophorbide a. A similar series of reactions occurs with chlorophyll b. Several other degradative reactions also occur, but the change to pheophytin is the major change that occurs when products containing chlorophyll are heated. The trivial names chlorophyll, chlorophyllide, pheophytin, and pheophorbide were developed historically and have been retained even though the chemical nomenclature has been changed by the IUB-IUPAC Joint Commission on Nomenclature in 1980. A good description of the new nomenclature and the chemistry of chlorophylls has been provided by Hendry (1).

The chlorophylls have a bright green color, whereas the pheophytins are olive green; thus, major efforts to maintain the attractive green color in vegetables during processing have been to retain the magnesium in the molecule. Pretreatment and treatment during processing in alkaline solutions containing magnesium (the Blair Process) were successful in retaining the green color for a short time after thermal processing. Enzymatic treatments to convert the chlorophyll to chlorophyllide were attempted because the chlorophyllides have a color similar to the chlorophylls and are more stable, and they did help to maintain the color. A combination of the two processes with high-temperature short-time (HTST) processing also maintained the color a little longer. But the overall effect was that the retention of attractive color was only a few weeks and not long enough to provide an economic incentive, so the attempts were abandoned. The same problems with stability were experienced with attempts to make colorants containing chlorophyll.

PREPARATION OF COLORANTS

The land plants that contain chlorophylls a and b have been the source of all chlorophyll-containing colorants. The order of stability of the chlorophylls is c > b > a, but chlorophyll c has been ignored as a source of green colorants. It does occur in the brown seaweeds used as a commercial source of alginates and also in single-celled phytoplankton. Algae are already being grown as a source of carotenoids and conceivably could also be grown for chlorophyll content if the increased stability was sufficient to provide an economic motivation. The plants currently being used include

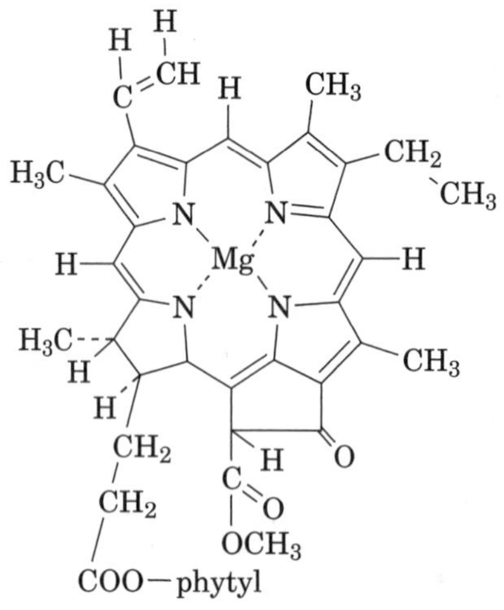

Chlorophyll a

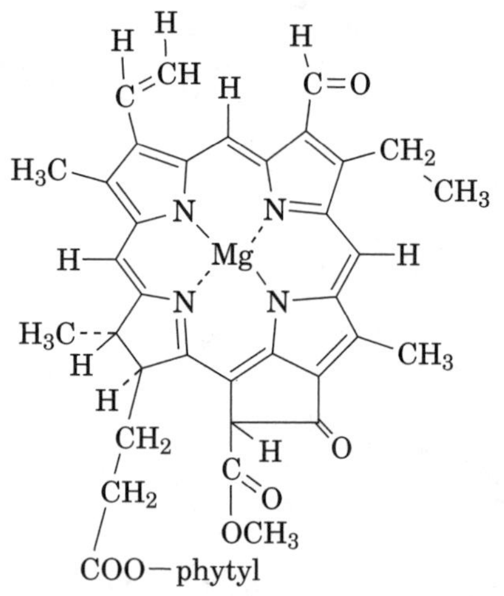

Chlorophyll b

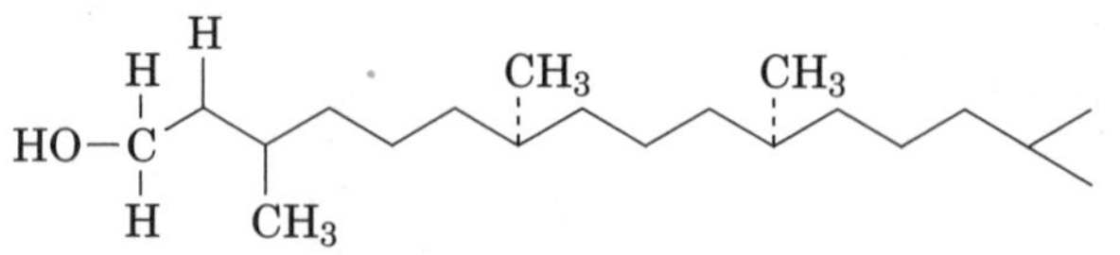

Phytol

Figure 1. Formulas for chlorophylls a and b and the phytol moiety.

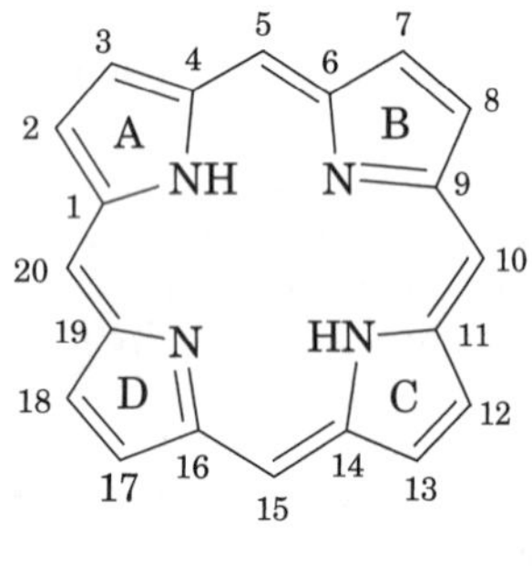

Figure 2. Numbering system for the chlorophylls.

alfalfa (lucerne), *Medicago sativa*; nettles, *Urtica dioica*; and a series of pasture grasses such as fescue. In a living cell, within the chloroplasts, the chlorophylls are complexed with but not covalently bound to one of a series of polypeptides. The chlorophyll–polypeptide or chlorophyll–protein complexes are closely associated with carotenoids and tocopherols (vitamin E) and are actively involved in the processes of photosynthesis. Preparation of a colorant involves recovery and purification of an appropriate form of chlorophyll.

Land plants are usually available for only a few weeks of the year; thus, the chlorophyll colorants are usually made from bulk harvested and dried plants. The dried plants are ground and extracted with acetone or a chlorinated hydrocarbon solvent and washed, and the solvent is removed. A yield of 20% of a mixture of chlorophylls, pheophytins, and other degraded chlorophyll compounds is usually obtained. The dry residue after removal of the solvent can be further purified by treatment with a water-immiscible solvent to obtain an oil-soluble preparation known as metal-free pheophytin. The dry residue can also be acidified in the presence of copper salts to produce an oil-soluble copper pheophytin. It is not commercially practical to produce a colorant containing chlorophyll itself because of the instability of the magnesium compound. The dry residue can also be saponified to replace the phytyl group with sodium or potassium to form water-soluble compounds. After further purification, the product may be marketed as a water-soluble, metal-free, gray-green product. Acidification in the presence of copper salts converts the pigments to more-stable green pigments known as copper chlorophyllins.

APPLICATIONS

The copper-substituted derivatives of pheophytin are relatively stable to light and mineral acids and are more stable than the metal-free pheophytins and pheophorbides. The major portion of the estimated $15 million (in 1994 U.S. dollars) in annual sales of colorants derived from chlorophyll is in the water-soluble forms (1). The major portion of both oil-soluble and water-soluble forms is used in dairy products, edible oils, soups, chewing gum, sugar confections, and drinks. A minor proportion is used in cosmetics and toiletries.

Chlorophyll colorants are widely used in Europe and other parts of the world but are not permitted in the United States.

BIBLIOGRAPHY

1. G. A. F. Hendry, "Chlorophylls and Chlorophyll Derivatives," in G. A. F. Hendry and J. D. Houghton, eds., *Natural Food Colorants*, Blackie Publishers, Glasgow, Scotland, 1996, pp. 131–156.

COLORANTS: FOOD, DRUG, AND COSMETIC COLORANTS

The synthetic colorants are the most obvious of the long list of colorants and probably the most important in terms of scope of applications. They are also the most controversial. The history of synthetic colorants dates back to the discovery of the first synthetic dye, mauve, by Sir William Henry Perkin in 1856. Since then more than 700 colorants have been available to the paint, plastic, textile, and food industry. In 1907 about 80 colorants were offered for use in foods (1), and obviously few, if any, had been tested for safety. At that time, Dr. Bernard Hesse, a German dye expert employed by the U.S. Department of Agriculture (USDA), was asked to study the colorants available for use in foods. He concluded that only 16 of the 80 colorants were probably more or less harmless. Most of Hesse's information was embodied into the Food and Drug Act of 1906. This Act together with FID No. 76 in July 1907 put an end to the indiscriminate use of colorants in food. The new legislation required that only colorants of known chemical structure and that had been tested for safety could be used. The Act also set up a system for certification of synthetic organic food colorants designed for use in foods. The certification of each batch included proof of identity and the levels of impurities. During the next three decades, 10 more colorants were added to the list. In 1938 the federal Food, Drug, and Cosmetic Act of 1938 was enacted. It stated that the use of any uncertified coal-tar color in any food, drug, or cosmetic shipped in interstate commerce was forbidden. The Act also created three categories of coal-tar colors:

1. *FD&C colors*. Those certifiable for use in coloring foods, drugs, and cosmetics.
2. *D&C colors*. Dyes and pigments considered safe in drugs and cosmetics when in contact with mucous membranes or when ingested.
3. *Ext. D&C colors*. Those colorants that, because of their oral toxicity, were not certifiable for use in products intended for ingestion but were considered safe for use in products externally applied.

Each colorant had to be identified by its specific FD&C, D&C, or Ext. D&C name.

SAFETY OF COLORANTS

The passage of the 1938 Act drew public attention to the question of safety of colorants and led to the publication of the Service and Regulatory Announcement, Food, Drugs, and Cosmetics No. 3 in 1940. This document listed specific colorants that could be used together with purity specifications and regulations pertaining to manufacture and sale. In 1950 a new dilemma developed, which was precipitated by two developments. First, a number of children became sick after reportedly eating popcorn and candy colored with excessive amounts of colorant. Second, the Food and Drug Administration (FDA) launched a new round of toxicity tests with animals fed much higher levels of colorants and for a longer time than any tests previously conducted. The results were unfavorable for FD&C Orange No. 1, FD&C Orange No. 2, and FD&C Red No. 32. The FDA based its conclusions on the interpretation of the 1938 Act, which states: "The Secretary shall promulgate regulations providing for the listing of coal-tar colors which are harmless and suitable for food." The FDA interpreted "harmless" to mean harmless at any level. The manufacturers argued that this interpretation was too strict and the FDA should set safe limits. The Supreme Court held that the FDA did not have the authority to set limits, and several more colorants were delisted including FD&C Yellows Nos. 1 to 4. It was obvious that the law was unworkable and, through the efforts of the Certified Color Industry and FDA, a new law was developed called the Color Additive Amendment of 1960. The Amendment allowed for current use of colorants pending completion of testing, and, equally important, it authorized the secretary of health, education, and welfare to establish limits of use, thereby eliminating the "harmless per se" interpretation. Another

Table 1. FD&C Colorants Approved for Use in the United States

FDA name	Common name	CI number	Year listed	Chemical class	Hue
FD&C Blue No. 1	Brilliant Blue	42090	1929	Triphenylmethane	Greenish blue
FD&C Blue No. 2	Indigotine	73015	1907	Indigoid	Deep blue
FD&C Green No. 3	Fast Green	42053	1927	Triphenylmethane	Bluish green
FD&C Yellow No. 5	Tartrazine	19140	1916	Pyrazolone	Lemon yellow
FD&C Yellow No. 6	Sunset Yellow	15985	1929	Monoazo	Reddish yellow
FD&C Red No. 3	Erythrosine	43430	1907	Xanthine	Bluish pink
FD&C Red No. 40	Allura Red	16035	1971	Monoazo	Yellowish red
Orange B[a]	Orange B		1966	Monoazo	
Citrus Red No 2[b]	Citrus Red No. 2		1959	Monoazo	

[a]Allowed only on the surfaces of sausages and frankfurters at concentrations up to 150 ppm by weight.
[b]Allowed only on the skins of oranges, not intended for processing, at concentrations up to 2 ppm by weight of the whole fruit.

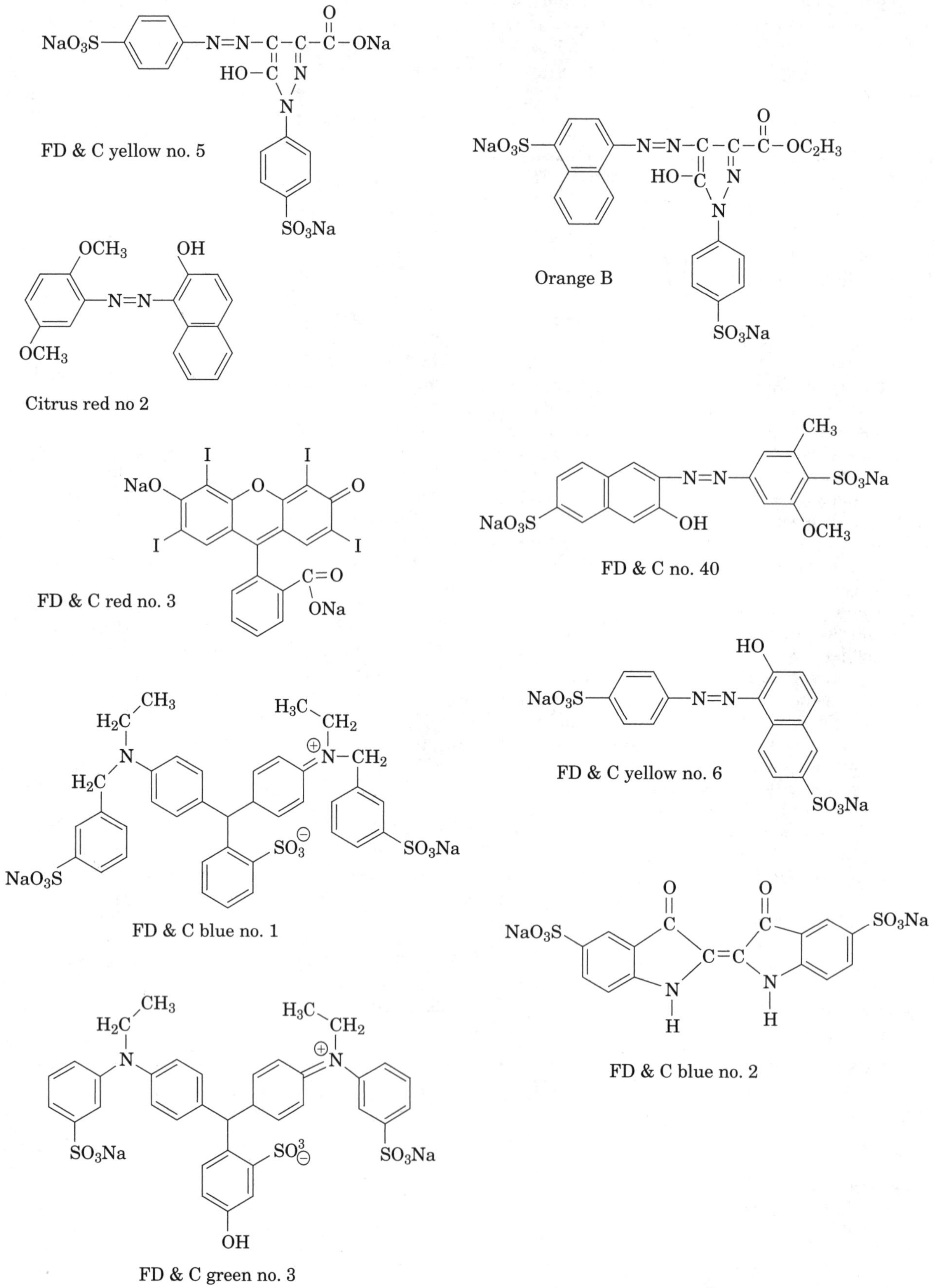

FD & C yellow no. 5

Orange B

Citrus red no 2

FD & C red no. 3

FD & C no. 40

FD & C yellow no. 6

FD & C blue no. 1

FD & C blue no. 2

FD & C green no. 3

Figure 1. Structure of 9 FD&C colorants

feature, known as the "Delany Clause," was also inserted. Briefly, the clause states that no additive shall be added to food "if it is found, after tests which are appropriate for the evaluation of safety of food additives, to induce cancer in man or animals." Unfortunately Congressman Delany did not appreciate the ingenuity of modern chemists to push back the limits of analytical sensitivity. Today, chemists can find almost anything in anything. The clause is clearly unworkable and is currently being replaced by some form of "de minimis," which comes from the term "de minimis curat lex." Freely translated, this means that the law does not concern itself with trifles. Another feature of the 1960 Amendment allowed the secretary of health, education, and welfare to determine which colorants had to be certified and which would be exempt from certification. In view of the expense and time required for adequate testing, some of the colorants had "provisional" listing for a number of years, but this category for straight FD&C colorants has expired at the present time. Because of the expense involved, only commercial colorants of significant economic value were tested, so a number of colorants were delisted by default. Actually this should not be a hardship, because the seven currently allowed colorants (Table 1) should be sufficient in view of the opportunities for blending. Interestingly, Hesse, around 1900, recommended seven colorants for general use in food. Of the seven, only two (FD&C Red No. 3 and FD&C Blue No. 2) remain on the current list of approved colorants. Table 1 shows a list of seven colorants approved for general use, and two approved for restricted use, together with the FD&C name, common name, CI number, year listed, and chemical class. FD&C Blue No. 2 was provisionally listed until final approval in 1987. Similarly, FD&C Yellow No. 6 was finally approved in 1986. Formulae for the nine colorants are shown in Figure 1. The regulatory status of 26 colorants exempt from certification is shown in Table 2.

APPLICATIONS IN FOODS

The seven colorants listed in Table 1 should be sufficient to enable a formulator to create any desired color by mixing

Table 2. Regulatory Status of Colorants Exempt from Certification in the United States

Name of color additive	U.S. food use limit	E. C. status	JECFA ADI[a] (mg/kg/bw)
Annatto	GMP[b]	E 160b	0–0.065 (calculated as bixin)
Dehydrated beets	GMP	E 162	Not evaluated
Ultramarine blue	Salt for animal feed up to 0.5% by weight	Not listed	None
Canthaxanthin	Not to exceed 30 mg/pd of solid/semisolid food or pint of liquid food or 4.41 mg/kg of chicken feed	E 161g	None
Caramel	GMP	E 150	0–200
β-apo-8′-carotenal	Not to exceed 15 mg/pd of solid or semisolid food or 15 mg/pt of liquid food	E 160a	0–5
β-Carotene	GMP	E 160a	0–5
Cochineal extract or carmine	GMP	E 120	0–5
Toasted partially defatted cooked cottonseed flour	GMP	Not listed	Not evaluated
Ferrous gluconate	GMP for ripe olives only	Not listed	Not evaluated
Grape color extract	GMP for nonbeverage foods	E 163	Not evaluated
Grape skin extract (Enocianina)	GMP for beverages	E 163	0–2.5
Synthetic iron oxide	Pet food up to 0.25%	E 172	0–2.5
Fruit juice	GMP	Not listed	Not evaluated
Vegetable juice	GMP	Not listed	Not evaluated
Dried algal meal	GMP for chicken feed	Not listed	Not evaluated
Tagetes (Aztec marigold) meal and extract	GMP for chicken feed	Not listed	Not evaluated
Carrot oil	GMP	Not listed	Not evaluated
Corn endosperm oil	GMP for chicken feed	Not listed	Not evaluated
Paprika	GMP	E 160c	None allocated
Paprika oleoresin	GMP	E 160c	Self-limiting as spice
Riboflavin	GMP	E 101	0–0.5
Saffron	GMP	Not listed	Food ingredient
Titanium dioxide	Not to exceed 1% by weight of food	E 171	Not specified
Turmeric	GMP	E 100 (curcumin)	Temporary ADI not extended
Turmeric oleoresin	GMP	E 100 (curcumin)	Temporary ADI not extended

[a]JECFA refers to the Joint Expert Committee on Food Additives set up by the World Health Organization and the Food and Agriculture Organization of the United Nations. The term ADI refers to Acceptable Daily Intake as usually determined by the JECFA.
[b]GMP = Good manufacturing practice.

colorants as discussed in COLOR AND FOOD. But each colorant probably will behave differently in each food system; therefore prediction from a knowledge of each colorant is difficult. The blending will have to be done under actual food conditions. The stability of the FD&C colorants under acid, alkaline, light, heat, and oxidation/reduction conditions, and the presence of sugars, acids, diluents, solubility, and so on discussed by Francis (2). The FD&C colorants are appropriate for a wide variety of foods, particularly those requiring a water-soluble colorant.

LAKES

Lakes are colorants prepared by precipitating a soluble colorant onto an insoluble base or substratum. A variety of bases such as titanium dioxide, zinc oxide, talc, calcium carbonate, and aluminum benzoate are approved for D&C lakes, but only aluminum, magnesium, and calcium salts are approved in the United States for manufacturing FD&C lakes. The process of manufacture is fairly simple. First, a substratum is prepared by adding sodium carbonate to a solution of aluminum sulfate. Next a certified colorant is added to the slurry, then aluminum sulfate is added to convert the colorant to an aluminum salt, which is absorbed onto the surface of the alumina. The slurry is then washed, dried, and ground to a fine powder. The colored powder can be marketed as is or mixed with a diluent such as hydrogenated vegetable oil, coconut oil, propylene glycol, glycerol, sugar syrup, or other media appropriate for food consumption or for printing food wrappers. Diluents for both pure colorants or lakes have to be either on the Generally Recognized As Safe (GRAS) list or on the list approved by the FDA Code of Federal Regulations (1–3).

Lake colorants have many advantages, primarily because they are insoluble in most solvents including water. They have a high opacity, are easily incorporated into dry media, and have superior stability to light and heat. They are effective colorants for candy and pill coatings since they do not require removal of water prior to processing. They are particularly effective for coloring hydrophobic foods such as fats and oils, confectionery products, bakery products, salad dressings, chocolate substitutes, and other products in which the presence of water is undesirable. Lakes made from some water-soluble FD&C colorants may show some bleeding when formulated in foods outside the pH range 3.5 to 9, but few foods are in this pH range. Lakes have been used to color food packaging materials including lacquers, plastic films, and inks from which straight water-soluble colorants would be soon leached.

BIBLIOGRAPHY

1. D. C. Marmion, *Handbook of U.S. Colorants for Foods, Drugs, and Cosmetics*, 2nd ed., John Wiley and Sons, New York, 1984.
2. D. C. Marmion, *Handbook of U.S. Colorants for Foods, Drugs, and Cosmetics,* 3rd ed., John Wiley and Sons, New York, 1991.
3. F. J. Francis, *Handbook of Food Colorants*, Eagen Press, St. Paul, Minn., 1999.

COLORANTS: HAEMS

Haemoglobin from blood is the only commercially important colorant in the haem group. The term haem, or heme, is used to describe iron derivatives of the cyclic tetrapyrrole protoporphin with the structure shown in Figure 1. The numbering system is similar to that for chlorophyll (see COLORANTS: CHLOROPHYLLS, Fig. 2). This structure is basic to a number of compounds such as chlorophyll, important in photosynthesis; haemoglobin, in oxygen transport; cytochrome, in energy transport; and a number of other enzymes. Haemoglobin is probably the best known, and it is composed of a haem group associated with a protein. Haemoglobin forms a loose association with oxygen as the basis for oxygen transport and carbon dioxide removal and is remarkably stable *in vivo* but not *in vitro*. This introduces some difficulty in using haems as colorants.

Free haem can be easily extracted from blood by treatment with acidified organic solvents. Preparation of haem from blood has been known for centuries and can be accomplished by the addition of whole blood to glacial acetic acid saturated with salt at 100°C. Crystalline haem precipitates out, but it is not stable enough to function as a food colorant. More stable compounds can be produced by allowing other ligands such as carbon monoxide, nitrous oxide, hydroxides, and cyanide to displace the oxygen from the central iron atom. A number of patents exist (1), and they fall into three broad areas: (*1*) treatment of whole blood with a variety of reagents, (*2*) stabilization of the color with a variety of ligands, and (*3*) purification of haem derivatives. Treatment of whole blood to remove the haemoglobin has considerable appeal because the iron is in a very nutritionally available form as a dietary supplement and the protein has a high biological value. It would be still another advantage if the preparation could be made stable enough to act as a food colorant. A number of the patents claim

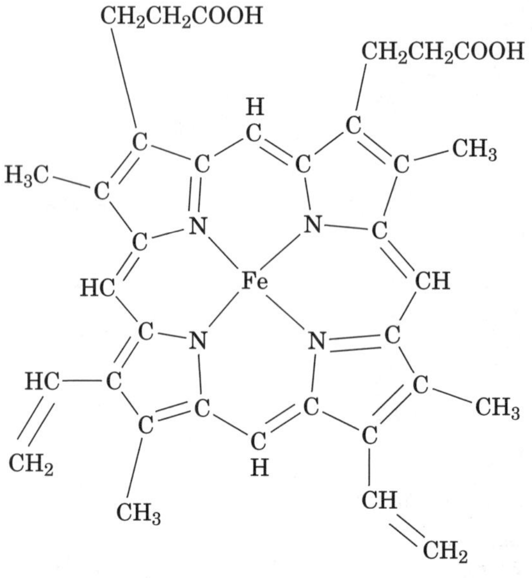

Figure 1. Structure of haem.

that the preparations are suitable for coloring products that normally contain meat, meat analogues, sausages, and so on.

The use of whole blood as a food ingredient is a very old custom. The Europeans are famous for "blood pudding," which may be a delectable food item but is black in color not red. Haemoglobin is denatured in the cooking process. Houghton reported an interesting application of naturally occurring haem pigments, namely, the bile pigments bilirubin and biliverdin (2). They occur in gallstones and hair balls and are in demand in Chinese medicine as an aphrodisiac.

Blood is permitted as a food ingredient if it has been collected and processed in an appropriate sanitary manner, but colorants derived from blood are not permitted in the United States.

BIBLIOGRAPHY

1. F. J. Francis, *Handbook of Food Colorant Patents*, Food and Nutrition Press, Westport, Conn., 1986.
2. J. D. Houghton, "Haems and Bilins," in G. A. F. Hendry and J. D. Houghton, eds., *Natural Food Colorants*, Blackie Publishers, Glasgow, Scotland, 1996, pp. 157–196.

COLORANTS: IRIDOIDS

Colorants from saffron have enjoyed good technological success and gustatory appeal, but their high price has led to searches for other sources with the same pigments. It became apparent that the same pigments, but not the flavor, could be obtained in much larger quantities from the fruits of the gardenia or Cape Jasmine plant. The fruits of gardenia, *Gardenia jasminoides*, contain three major groups of pigments: crocins (see COLORANTS: CAROTENOIDS, Fig. 1), iridoids, and flavonoids. The structures of nine iridoid pigments are shown in Figure 1. This group also comprises a series of flavonoid compounds. The structures of five flavonoids isolated from *G. fosbergii* are shown

Geniposide

Shanzhiside

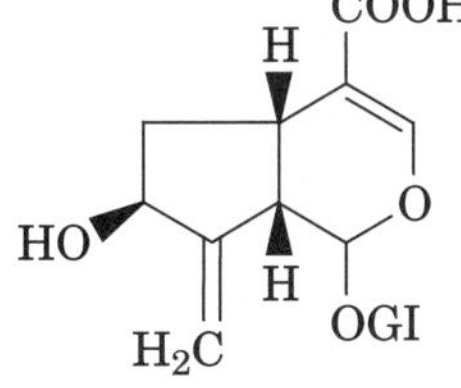

Gardoside

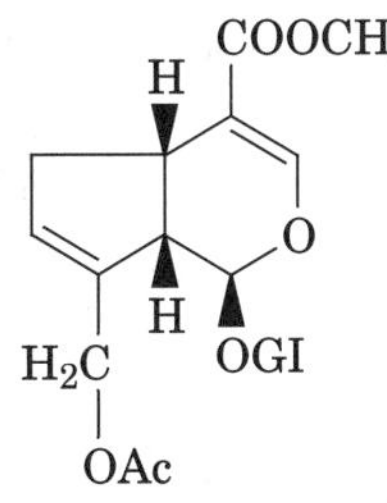

Acetylgeniposide

Methyldeacetylasperuloside

Gardenoside

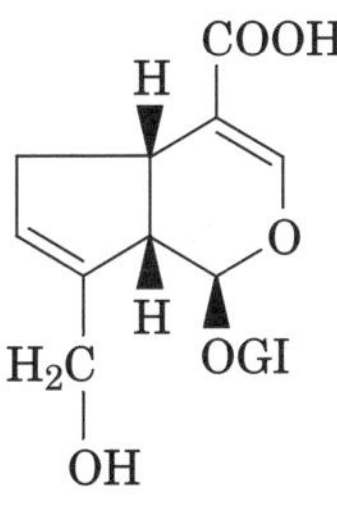

Geniposidic acid

Genipingentiobioside

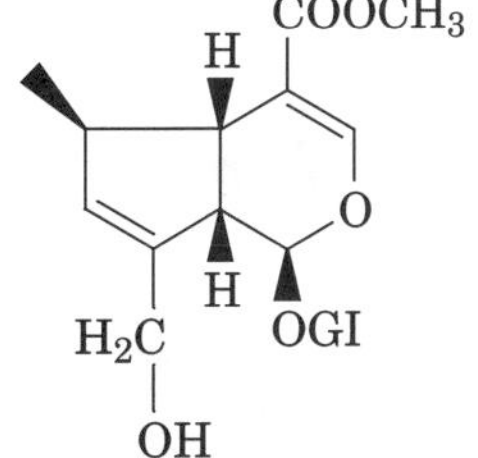

Scandoside methylester

Figure 1. Structures of nine iridoid pigments in *G. jasminoides*. Gl = glucose.

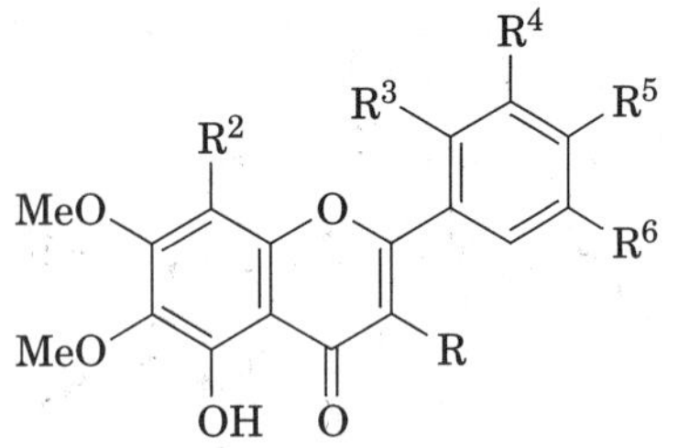

Flavonoid compounds

Compound	R^1	R^2	R^3	R^4	R^5	R^6
1	H	H	H	OMe	OMe	OMe
2	OMe	H	H	OH	OMe	OH
3	H	H	OMe	OMe	H	OH
4	OMe	H	H	OMe	OMe	OH
5	OMe	OMe	H	H	OH	H

Figure 2. Structures of five flavonoid pigments in *G. fosbergii*.

in Figure 2. This is a different species from *G. jasminoides*, but the flavonoid compounds in closely related species tend to be similar. The flavonoids would contribute a pale yellow color and the carotenoids an orange color.

The iridoid pigments constitute an interesting situation as a food colorant because a range of colors from green to yellow, red and blue, can be produced (1). The patent literature suggests extraction of the fruit with water, treatment with enzymes having β-glycosidic or proteolytic (bromelain) activity, followed by reaction with primary amines from amino acids or a protein such as soy. The reaction conditions such as time, pH, temperature, oxygen content, degree of polymerization, and conjugation of the primary amino groups allow a range of colorants to be produced. Several patents involve culturing extracts of gardenia with microorganisms such as *Bacillus subtilis, Aspergillus japonicus*, or a species of *Rhizopus*. Several patents involve hydrolysis of the iridoid glycoside geniposide by the action of β-glucosidase to produce genipin. The genipin can be reacted with taurine to produce a blue colorant. Other amino acids such as glycine, alanine, leucine, phenylalanine, and tyrosine can be reacted to produce brilliant blue colorants. They are claimed to be stable for two weeks at 40°C in 40% ethanol. Four greens, two blues, and one red have been commercialized in Japan.

Colorant preparations from gardenia have been suggested for use with candies, sweets, colored ices, noodles, imitation crab, fish eggs, glazed chestnuts, beans, dried fish substitutes, liqueurs, baked goods, and so on. In view of the wide range of colors and apparent stability, colorants from gardenia appear to have good potential. They are not approved for use in the United States.

BIBLIOGRAPHY

1. F. J. Francis, "Less Common Food Colorants," in G. A. F. Hendry and J. D. Houghton, eds., *Natural Food Colorants*, Blackie Publishers, Glasgow, Scotland, 1996, pp. 310–342.

COLORANTS: MONASCUS

Monascus products have been well known in south and east oriental countries for centuries. Wong and Koehler listed their use as a Chinese medicine as far back as 1590 (1). The genus *Monascus* includes several fungal species that will grow on a number of carbohydrate substrates, especially steamed rice. Traditionally, *Monascus* species were grown on rice, and the whole mass was eaten as an attractive red food. It could also be dried, ground, and incorporated into other foods.

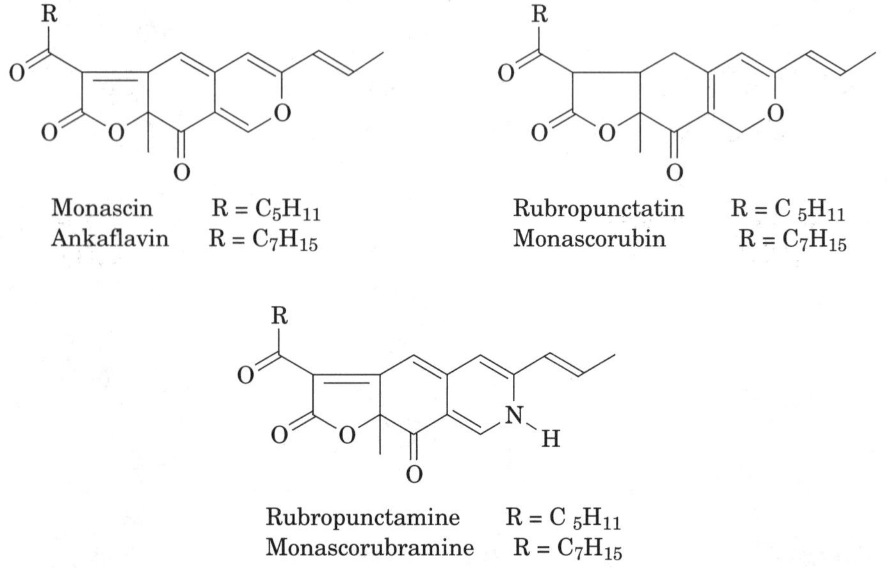

Monascin $R = C_5H_{11}$
Ankaflavin $R = C_7H_{15}$

Rubropunctatin $R = C_5H_{11}$
Monascorubin $R = C_7H_{15}$

Rubropunctamine $R = C_5H_{11}$
Monascorubramine $R = C_7H_{15}$

Figure 1. Pigments of *Monascus*.

CHEMISTRY

Monascus species produce six pigments with the structures shown in Figure 1. Monascin and ankaflavin are yellow, rubropunctatin and monascorubin are red, and rubropunctamine and monascorubramine are purple. The red and yellow pigments are considered to be normal secondary metabolites of fungal growth, and the purple pigments may be formed by enzymatic modification of the red and yellow pigments. The red pigments are very reactive and readily react with amino groups via a ring opening and a Schiff rearrangement to form water-soluble compounds such as that in Figure 2 (2). The *Monascus* pigments have been reacted with amino groups, polyamino acids, amino alcohols, chitin amines or hexamines, proteins, peptides and amino acids, sugar amino acids, browning reaction products, aminoacetic acid, and amino benzoic acid (3). Obviously a number of derivatives are possible, and they have been shown to have greater water solubility, thermostability, and photostability than the parent compounds.

The *Monascus* pigments are readily soluble in ethanol and only slightly soluble in water. Ethanolic solutions are orange at pH 3 to 4, red at 5 to 6 and purplish red at 7 to 9. The pigments fade under prolonged exposure to light, but the pigments in 70% alcohol are more stable than in water. The pigments are very stable to temperatures up to 100°C in neutral or alkaline conditions.

COMMERCIAL PRODUCTION

Monacus colorants are currently being produced in Japan, China and Taiwon. Traditionally, *Monascus* has been grown on solid cereal substrates such as rice and wheat and the entire mass used as a food or a food ingredient. But it became obvious that the fungus could be grown in liquid or fluid solid state media and optimized for pigment production. Much research has been devoted to the determination of the optimal conditions for pigment production on a wide variety of cereal substrates. It is possible to maximize the production of either the red or yellow pigments to produce a range of red to yellow colors. For example, Tadao et al. reported that *M. anka* grown on bread produced only yellow pigments (4). Criticism was directed at the earlier production methods because the fungus produced an antibiotic, and this is considered to be an undesirable food ingredient. Recent strain selections and control of growth conditions have eliminated the antibiotic problem. Han and Mudgett (1992) concluded that *M. purpureus* ATCC 16365 was the most appropriate organism for solid-state fermentation and confirmed that it had no antibiotic activity (5). *Monascus* produces a range of compounds normal to growth metabolism such as alcohol, enzymes, coenzymes, monascolins that modify fat metabolism, antihypertensive agents, and flocculents. This is a strain selection problem, growth optimization, and finally a removal problem for a food colorant.

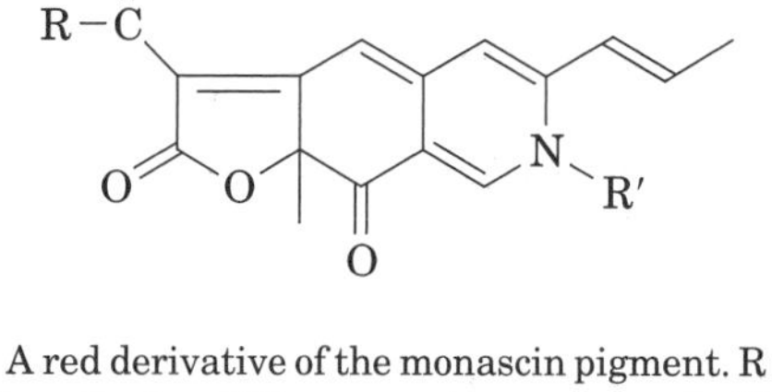

R−C(=O), O, O, O, N, R′

Figure 2. A red derivative of the monascin pigment. R represents an aliphatic radical and R′ represents a compound of the formula HN-R, which represents an amino sugar, a polymer of an amino sugar, or an amino alcohol. *Source:* Ref. 2.

APPLICATIONS

Monascus colorants offer considerable advantages since they can be produced in any quantity on inexpensive substrates. Their range of colors from yellow to red and their stability in neutral media is a real advantage. Sweeny et al. suggested *Monascus* colorants for processed meats, marine products, jam, ice cream, and tomato ketchup (6). They should be appropriate for alcoholic beverages such as saki and also for koji, soy sauce, and kamboko. There is considerable interest in the *Monascus* group, with 38 patents issued in the years 1969 to 1985 (3). Currently *Monascus* colorants are not permitted in the United States.

BIBLIOGRAPHY

1. H. C. Wong and P. E. Koehler, "Mutants for Monascus-pigment Production," *J. Food Sci.* **46**, 956–957 (1981).
2. U.S. Pat. 3,993,789, H. R. Moll and D. R. Farr.
3. F. J. Francis, *Handbook of Food Colorant Patents*, Food and Nutrition Press, Westport, Conn., 1986.
4. Jap. Pat. 81,006,263 (1981), H. Tadao et al.
5. O. H. Han and R. E. Mudgett, "Effects of Oxygen and Carbon Dioxide on Monascus Growth and Pigment Production in Solid State Fermentation," *Biotechnol. Prog.* **8**, 5–12 (1992).
6. J. G. Sweeny et al., "Photoprotection of the Red Pigments of *Monascus anka* in Aqueous Media by 1.4,6-Trihydroxy-naphthalene," *Agric. Food Chem.* **29**, 1189–1193 (1981).

COLORANTS: PHYCOBILINS

Phycobilins are found in abundance as components of the blue-green, red, and cryptomonad algae. They are colored, fluorescent, water-soluble pigment-protein complexes classified into three major groups according to color: the phycoerythrins, which are red in color with a bright orange fluorescence, and the phycocyanins and allophycocyanins, which are both blue and fluorescent red. The range of color and the apparent stability of the pigments appear to make this group attractive as food colorants.

The structures for the two chromophores of the cyanobilins, termed phycocyanobilin and phycoerythrobilin, are shown in Figure 1. Phycocyanins and allophycocyanins share the same chromophore, and the differences in color are due to the different protein groups. Figure 1 shows a bilin written in the conventional linear form and also in acyclic, possibly truer, form, showing its relationship to other pigments such as chlorophyll and haemoglobin. The

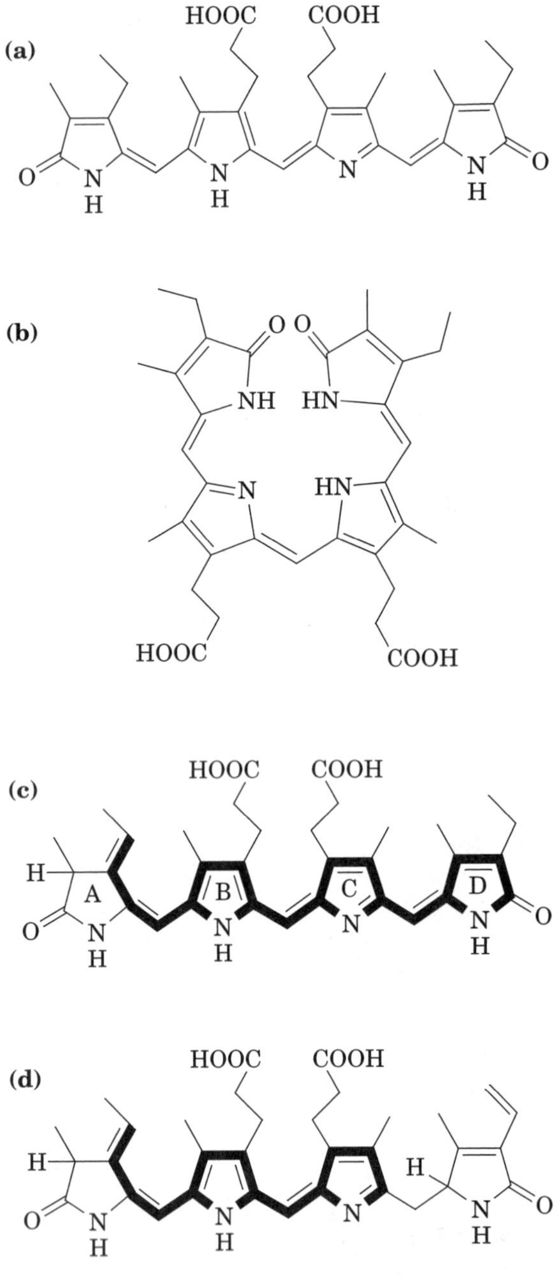

Figure 1. Basic structure of the phycobilins. (**a**) is a bilin in the conventional linear form. (**b**) is the same structure in a cyclic form to show its relationship with the porphyrins. (**c**) is a blue phycocyanobilin pigment and (**d**) is a red phycoerythrobilin pigment.

attachment of the bilin chromophore to its protein is very stable due to covalent bonds between the ethylidene group on position 3 (see COLORANTS: CHLOROPHYLLS, Fig. 2) of the bilin with a cysteine group in the protein. A second covalent bond may be on position 18. The major function of the bilin pigments is to act as light absorbers in the energy transport system.

Phycobilin preparations can be obtained by simply freeze-drying an algal cell suspension, producing a highly colored powder. The chlorophyll and carotenoid components can be removed by centrifuging after breaking the algal cells. The centrifugate is a brightly colored solution of water-soluble protein of which 40% may be phycobilin. Further purification by precipitation with ammonium sulphate and ion-exchange yields almost pure phycobilin. The free chromophore can be obtained by extended refluxing in methanol. An alternative to extensive purification would be to grow an organism such as *Cyanidium caldarium*, which, on addition of 5-aminolevulinic acid to the growth medium, yields pure phycocyanobilin (1). Another approach is to use the existing technology for growth of *Spirulina platensis* in open ponds. This organism has been used for eons in Africa and Mexico, probably because of its high protein content and digestibility. It is important in health food stores as a dietary supplement. The cells are simply harvested and dried. Another unicellular organism, the red algae *Porphyridium cruentum* can also be grown in open ponds or in tubular reactors for production of phycobilin and phycoerythrin. A number of patents (2) exist for the extraction, stabilization, purification, and use of phycocyanins from *Spirulina* and *Aphanotheca nidulans*, and three firms are marketing the product.

The presence of a protein in the pigment suggests that the colorant would be used for products requiring a minimum of heat treatment. But the fact that it takes 16 hours in boiling methanol to separate the protein from the chromophore suggests that applications could include products with mild heat treatment. Suggested applications for phycocyanin colorants include chewing gums, frozen confections, soft drinks, dairy products, sweets, and ice cream. No patents have been filed for the use of phycoerythrin, but it seems like a good candidate for a red colorant.

There are several industrial applications for phycobilins such as fluorescent tracers in biochemical research, fluorescence-activated cell sorting, fluorescence microscopy, and so on (1). Existence of other markets will certainly help to obtain the critical volume necessary for production of food colorants. Colorants containing phycobilins are currently not permitted in the United States.

BIBLIOGRAPHY

1. J. D. Houghton, "Haems and Bilins," in G. A. F. Hendry and J. D. Houghton, eds., *Natural Food Colorants*, Blackie Publishers, Glasgow, Scotland, 1996, pp. 157–196.
2. F. J. Francis, *Handbook of Food Colorant Patents*, Food and Nutrition Press, Westport, Conn., 1986.

COLORANTS: POLYPHENOLS

The major brown food colorant is obviously caramel, but other sources are available. A brown colorant can be prepared from an appropriate mixture of FD&C colorants or from extracts of cacao or tea.

CACAO

Cacao, *Theobroma cacao*, has been suggested as a potential source of brown colorants. The cocao, pods, beans, shells, cotyledons, husks, and stems all contain a complex mixture of polyphenols such as cyanidin glycosides, leucoanthocyanins, (-)-epicatechins, quercitin glycosides, and the acyl acids *p*-coumaric and gentisic. Colorants prepared from cacao are likely to contain a complex mixture of leucoanthocyanins, flavonoid polymers, and catechin-type polymers.

The suggested extraction methods involve hot acid or alkaline water or ethanol followed by filtration and concentration. A more reddish extract can be obtained by prior roasting of the beans followed by extraction with hot aqueous alkaline solutions. The colorant is suitable for most foods where a brown color is desired. Cacao products are permitted as food ingredients in the United States but not as a source of colorants.

TEA

Extracts of tea, *Thea sinensis*, have been used as a brown colorant for centuries. The polyphenols in tea comprise a very complex mixture including the glucosides and rhamnoglucosides of myricetin, quercetin and kaempferol, di-*C*-glycosylapigenins, 7-*O*-glucosylisovitexin, epicatechin, epigallocatechins, epicatechin gallate, gallic acid, chlorogenic acid, ellagic acid, coumarylquinic acid, and many related compounds. In black tea, they may act as precursors to the poorly defined pigments thearubin and theaflavin. Black tea also contains theaflavin gallate, digalloylbisepigallocatechin, triacetinidin, flavanotropolone, and flavanotropolone gallate.

Preparation of extracts of tea is usually a simple extraction of the leaves and stems with warm water or ethanol, followed by filtration and concentration. Extracts of tea are legal food ingredients in the United States but specific colorants prepared from tea products are not. This may appear to be a fine distinction, but it depends on whether the colorant is classified as an ingredient or an additive.

COLORANTS: TURMERIC

Turmeric has been used as a spice and colorant for thousands of years and is today still one of the principal ingredients of curry powder. Turmeric is produced in many tropical countries, including India, China, Pakistan, Haiti, Peru, and the East Indies. Turmeric (CI Natural Yellow 3, CI No. 75300, EEC No. E 100), also called tumeric or curcuma, is the dried ground rhizomes of several species of *Curcuma longa*, a perennial herb of the Zingiberaceae family native to northern Asia.

Turmeric is a bright yellow powder with a characteristic odor and taste and contains three major pigments: curcumin, demethoxycurcumin, and bisdemethoxycurcumin (Fig. 1). These pigments together with the flavor compounds turmerone, cineol, zingeroni, and phellandrene constitute the turmeric and turmeric oleoresin of commerce. The oleoresin is prepared by extraction of dried rhizomes with one or a combination of solvents, including methanol, ethanol, isopropanol, ethyl acetate, hexane, methylene chloride, ethylene dichloride, and trichloroethylene, and removal of the solvent. Other methods include extraction with ether, removal of the ether, and suspension of the residue in vegetable oil. Powdered turmeric preparations consisting of the powdered rhizome or the oleoresin standardized with maltodextrin usually contain 8 to 95% cucurmin. Liquid concentrates of oleoresin suspended in ethanol and/or propylene glycol with a polysorbate emulsifier usually contain 0.5 to 25% curcumin. A variety of other carriers such as edible oils and fats, and monoglycerides are also available (1). The FDA-approved extraction procedures are specified in the CFR Section 73.615 for Turmeric.

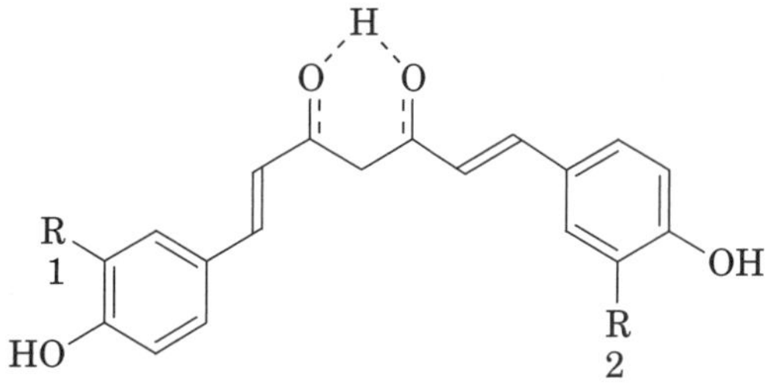

H
O O
R 1
HO
OH
R 2

Curcumin	$R_1 = R_2 = OCH_3$
Demethoxycurcumin	$R_1 = H$ $R_2 = OCH_3$
Bisdemethoxycurcumin	$R_1 = R_2 = H$

Figure 1. Pigments of turmeric.

Turmeric and turmeric oleoresins are both unstable to light and alkaline conditions; thus, several patents involve formulation with citric, gentisic, gallic, and tannic acids. Polyphosphate, sodium citrate, waxy maize, and an emulsifier are sometimes added. Both powder and liquids are susceptible to degradation by oxidation. Both show good tinctorial strength with turmeric usually used in the 0.2 to 60 ppm and the oleoresin in the 2 to 640 ppm range. The tinctorial strength is usually expressed as % turmeric even though preparations from different geographical regions differ in the relative content of curcumin and demethoxycurcumin. Both turmeric and the oleoresin produce bright yellow to greenish yellow shades and are sometimes used as a replacement for FD&C Yellow No. 5. Curcumin is insoluble in water, but water-soluble complexes can be made by complexing with heavy metals such as stannous chloride and zinc chloride to produce an intense orange colorant. Curcumin colorants can also be prepared by absorbing the pigments on finely divided celllose.

The major applications of turmeric are to color cauliflower in pickles and as an ingredient in mustard and curry powder. It is also used alone or in combination with other colorants such as annatto in spices, ice cream, cheeses, baked goods, confectionery, cooking oils, and salad dressings.

BIBLIOGRAPHY

1. F. J. Francis, *Handbook of Food Colorant Patents*, Food and Nutrition Press, Westport, Conn., 1986.

COLORANTS: MISCELLANEOUS COLORANTS

INORGANIC COLORANTS

A number of miscellaneous preparations may be added to foods for various reasons. Some, such as titanium dioxide and carbon black, are obvious colorants, but others, such as zinc oxide and calcium carbonate, are nutritional supplements. Calcium carbonate is a pH adjuster. In both cases their contribution to color is minimal. Still others, such as talc, are indirect additives used in processing or packaging where they do contribute incidentally to appearance but it is not their primary function. There is a long list of compounds in this category as described in CFR 21-178.3297.

Titanium Dioxide

Titanium dioxide (CI Pigment White, CI No. 77891, CAS Reg. No. 1 3463-67-7, EEC No. E 171, Titanic Earth) is a large industrial commodity with current world production estimated at nearly 4 million tons, but only a small percentage is used as a food colorant. Commercial TiO_2 is obtained from the naturally occurring mineral ilmenite ($FeTiO_2$) and occurs in three crystalline forms, anatase, brookite, and rutile, with anatase as the most common form in industry. The only form allowed in food is synthetically produced anatase.

TiO_2 is the whitest pigment known today with a hiding power five times greater than its closest rival zinc oxide. This is the reason for its importance in paint formulations. TiO_2 is a very stable compound with excellent stability toward light, oxidation, pH changes, and microbiological attack. Water-dispersible and oil-dispersible forms are available usually in water, glycerol, propylene glycol, sugar syrups, vegetable oils, or polyglycerol esters of fatty acids. Xanthan gum may be added as a stabilizer and potassium sorbate, citric acid, and methyl and ethyl parabens as preservatives. It is used in confectionery, baked goods, cheeses, icings, and numerous pharmaceuticals and cosmetics up to 1% of the weight of the final product.

Carbon Black

Carbon black is a large volume industrial commodity, but food usage is very small. Food-grade carbon black is derived from vegetable material, usually peat, by complete combustion to residual carbon. The powder colorant has a very small particle size, usually less than 5 nm, and consequently is very difficult to handle. Therefore, it is usual for carbon black to be sold to the food industry in the form of a viscous paste with the colorant suspended in glucose syrup. Little safety data are available, and in the United States in the 1970s when the GRAS list was being reviewed, toxicological data were requested in view of the theoretical possibility of contamination with heterocyclic amines. Apparently the cost of obtaining the data was higher than the entire annual sales of food-grade carbon black so the tests were never done. Carbon black is not currently permitted in the United States.

Carbon black is a very stable and technologically a very effective colorant. It is widely used in Europe and other countries in sugar confectionery.

Ultramarine Blue

Ultramarine blue (CI No. 77007, E 180) is a synthetic blue pigment of rather indefinite composition. The group of ultramarine pigments are aluminosulphosilicates with empirical formulas approximated as $Na_2Al_6Si_6O_{24}S_3$. Ultramarine blue is well known as a cosmetic and is intended to resemble the colorants produced from the naturally occurring semiprecious gem lapis lazuli.

The basic ingredients for the production of the ultramarines are kaolin (China clay), silica, sulfur, soda ash, and sodium sulfate plus a reducing agent such as rosin or charcoal pitch. Depending on the processing conditions, ultramarine green, ultramarine blue, ultramarine violet, and ultramarine red can be produced. All four are important as cosmetic colorants, but the only food use is in salt intended for animal use at the 0.5% w/w level.

Iron Oxides

The iron oxides represent a group of synthetic colorants. The iron oxides are known under a variety of names such as CI Pigment Black 11 and CI Pigment Browns 6 and 7 (CI No. 77499), CI Pigment Yellows 42 and 43 (CI No. 77492), and CI Pigment Reds 101 and 102 (CI No. 77491). The chemical composition varies with the method of manufacture but can be represented by the empirical formula $FeO.xH_2O,Fe_2O_3.xH_2O$ or some combination. Most are produced from ferrous sulfate $FeSO_4.7H_2O$, and the most common forms are yellow hydrated oxides (ochre) and the brown, red, and black oxides.

Iron oxides are very stable compounds insoluble in most solvents but usually soluble in acids. Their main uses are in cosmetics and drugs, but they are allowed (CI Nos. 77491, 77492, 77499) in dog and cat food at levels up to 0.25% by weight of the finished food.

Talc

Talc (CI Pigment White, CI No. 77018, CAS Reg. No. 14807-96-6) is a naturally occurring magnesium silicate, $3MgO.4SiO_2H_2O$, sometimes containing a small amount of aluminum silicate. It is a large industrial commodity produced in many countries, particularly France, Italy, India, and the United States. The lumps are known as soapstone or steatite and the fine powders as talc, tateum, or French chalk. Its major uses are as a dusting powder in medicine; as a white filler in paints, varnishes, and rubber; and as a lubricant in molds for manufacturing. It is used as a release agent in the pharmaceutical and baking industries as well as in oatings for rice grains.

Zinc Oxide

Zinc oxide (CI Pigment White, CI No. 77947, CAS Reg. No. 1314-13-2) is a white or yellowish-white odorless amorphous powder. It is insoluble in water but soluble in most mineral acids and alkali hydroxides. Zinc oxide is the most important white powder used in the cosmetic industry. It has the advantages of brightness, ability to provide opacity without blue undertones, and it has antiseptic and healing effects. It does not have the hiding power of titanium dioxide but sometimes is used as a whitener in wrappers for food. Zinc oxide is also added as a nutrient dietary supplement.

Calcium Carbonate

Calcium carbonate (CI Pigment White 18, CI No. 77220, EEC No. E 170, CAS Reg. No. 471-34-1) is a fine white powder prepared by precipitating $CaCO_3$ from solution. Calcium carbonate occurs naturally as limestone and marble, but the impurities make it unacceptable as a food ingredient. Calcium carbonate has a minor role in foods as a whitener. Its major roles are as a pH adjuster, dietary nutrient supplement primarily for cereals, dough conditioner, firming agent, and yeast food. There are no limitations on use other than Good Manufacturing Practices.

Silver

Silver (EEC No. E 174) is a fine crystalline powder prepared by the reaction of silver nitrate with ferrous sulfate in the presence of nitric, phosphoric, or sulfuric acid. Polyvinyl alcohol is sometimes added to prevent the agglomeration of crystals and the formation of amorphous silver. Its primary use is as a fingernail polish colorant. Colloidal silver is sometimes used in the food industry as a bactericidal aid in purifying solutions.

Silver dragees are small silver-coated candy balls used as an ornament and ingredient in confectionery and baked goods.

Silicon Dioxide

Silicon dioxide is an amorphous material produced synthetically by either a vapor-phase hydrolysis process yielding fumed (colloidal) silica, or by a wet process that yields precipitated silica, silica gel, or hydrous silica. Fumed silica occurs as a hygroscopic, white, fluffy, nongritty powder of very fine particle size. The wet-process silicas occur as hygroscopic, white, fluffy powders or white microcellular beads or granules.

The colorant uses of silicon dioxide are minimal, but it is used widely in foods as an anticaking agent, defoaming agent, conditioning agent, and chillproofing agent in malt beverages.

Gold

Gold is allowed in some alcoholic beverages, usually in the form of flakes.

ORGANIC COLORANTS

Riboflavin

Riboflavin (EEC No. E 101) is a yellow pigment, in the vitamin B group, found in plant and animal cells. Riboflavin is usually added to food as an essential nutrient, but it is also an effective colorant. Together with riboflavin-5′-phosphate, it is a well-known colorant in Europe, but only riboflavin is allowed in the United States. In pure form it is a yellow orange crystalline powder, soluble in water and alcohol and insoluble in ether and chloroform. Riboflavin is somewhat unstable to light and oxidation and is unstable in alkaline media. It is stable to heat. For colorant purposes, riboflavin is confined mainly to cereals, dairy products, and sugar-coated tablets. Riboflavin produces an attractive greenish-yellow color with an intense green fluorescence.

Corn Endosperm Oil

Corn endosperm oil (CAS No 977010 506) is obtained by extraction of corn gluten with isopropyl alcohol and hexane. It is a reddish brown liquid containing fats, fatty acids, sitosterols, carotenoid pigments, and their degradation products. It is added as a nutrient, but it also contributes a yellow/orange color to feedstuffs.

Dried Algal Meal

Dried algal meal is a dry mixture of algal cells produced by the growth of *Spongiococccum* spp. on molasses or cornsteep liquor. It contains a maximum of 0.3% ethoxyquin.

Algae, Brown Extract

Brown extract algae (CAS No. 977026 928) are produced by extraction of the seaweeds *Macrocystis* and *Laminaria* spp.

Algae, Red Extract

Red extract algae (CAS No. 977090 042) are produced by extraction of the seaweeds *Porphyra* spp. *Gloiopeltis furcata*, and *Rhodomenia palmata*. All three algal products are added primarily for nutrient reasons, but they do add to the color.

Cottonseed Products

The FDA recognizes four products from cottonseed kernels:

Cottonseed flour (CAS No 977050 546). It is partially defatted and cooked.

Cottonseed flour (CAS No. 977043 778). It is partially defatted, cooked, and roasted.

Cottonseed kernels (CAS No. 997043 778). Glandless raw kernels.

Cottonseed kernels (CAS No. 997043 78. Glandless roasted kernels.

All four are nutrients, but they do contribute to the color.

Shellac

Shellac is obtained from the resinous secretions of the insect *Lassifer lacca*. Shellac is used primarily in foods as a coating agent, surface-finishing agent, and glaze in baked goods and some fruits, vegetables, and nuts. Shellac is not primarily a colorant, but it does affect the appearance of a product.

Octopus Ink

The secretions of the octopus and squid are a complex mixture of melanoidin polymers. It is an effective black colorant for pasta for special occasions in some ethnic groups, particularly the Portuguese.

COMPUTER APPLICATIONS IN THE FOOD INDUSTRY

BACKGROUND AND DEFINITIONS

The use of computers has become so pervasive in our professional and personal lives that a basic description of computer terms and functions is generally unnecessary. From laptop computers to mainframes the speed, power, and memory of personal computers continues to increase. Computer manufacturers and software developers compete fiercely for market share. New products and systems are constantly emerging or are being revised. Hundreds of books are being printed on computer-related topics every year. Rapidly changing dynamics of the computer industry requires constant updating to stay informed.

Current computer developments must be considered within the context of the entire information marketplace. The information marketplace includes the collection of computers, communications, software, and services engaged in all business and personal transactions. Today's computers successfully support many applications such as spreadsheets and databases. However, the larger problem involves the exchange of information between computers. A lack of standardization of data forms at different sites and with different machines and software packages makes communication difficult. Some degree of standardization may occur as large companies try to impose their standards on the marketplace in order to achieve dominance. The development of tools to encode and decode information and transactions is necessary to take advantage of widespread information exchange. A current area of great interest and growth is information exchange via the Internet. Although the information available through the Internet is extensive, Web browsers can require a great deal of time for searches, and often with uncertain results. The Web has great potential for information exchange, but it is still quite chaotic. Search engines will need to be improved. Security for confidential and financial electronic transactions is uncertain. Many companies provide general information about their operation on a Web page. Most suppliers and trade organizations use Web pages to advertise their products or to report on their activities. Web pages can offer attractive graphics and provide access to company literature and personal contacts for further information or sample requests.

Many computer applications have incorporated some manifestation of artificial intelligence. This term includes devices and applications that exhibit human intelligence and behavior, including robots, expert systems, voice recognition, natural and foreign language processing (1). Expert systems use a knowledge base of human expertise to solve problems. The knowledge base is commonly in the form of if-then-else rule statements, which interact with the user through an inference engine. Expert systems are often used for diagnostic or troubleshooting applications. Knowledge-based expert systems are often embedded as tools within larger programs, somewhat losing their identity. Additional applications widely used include object-oriented programs and neural networks. Object-oriented programs use modules of software to allow their reuse and interchange between programs. Neural networks somewhat mimic biological neurons in that loosely related data can be trained to reflect patterns in a process. Once trained the pattern can predict a particular output for specified input variables, just as a nerve stimulus generates a specific response. Neural networks are being used for more sophisticated process control schemes.

FOOD INDUSTRY APPLICATIONS

Production

Computers have long been a part of agriculture. Much early statistical work began with agricultural experimentation, generating an early need for mathematical computational power. Computers continue to play an important role in agriculture, both in research as well as for farm management. Models of crop growth and yield have been developed to predict the effects of land use and other management practices on yields, profitability, and the environment (2). Dairy cattle nutrition regimens and culling practices are adjusted according to computer records of growth and milk production. Computer-based tools are available to help farmers predict profitability investments (3) and to simulate production and population growth.

Product/Process Development

One of the earliest applications for computers in food processing was for linear programming to obtain optimal use of resources. When a variety of ingredients at varying costs is available to choose from, least-cost formulation programs are widely used. Given certain constraints such as moisture, salt, and fat content, the lowest cost ingredients are selected. More sophisticated programs can consider additional factors such as raw materials, equipment and time requirements, scheduling, and finished goods inventory. A modular neural network system has been developed to optimize the selection of starches for use in processed foods (4). Types of starch along with variables such as source, cost, flavor impact, function, and pH were incorporated into the program. The modular arrangement consisted of two networks trained independently on half of the starches, and then coupled in the final network. A faster learning period resulted. A program based on matrices has been developed that combines food component data with information on process control, inventory management, costing, and quality (5). The information is used to track inventories and is presented in a form ready for label printing. Random centroid optimization has been used to develop a process foods formulation program (6). Compared with other formula optimization techniques it was found to work more efficiently. Accommodating up to 20 factors it was found to be more useful when formulating with constraints.

Product development can be expensive in terms of personnel, ingredients, and plant time. Modeling or simulation programs that predict behaviors given certain conditions can reduce the time and expense of actual trials. Commercial tools are available for developing process models (The MathWorks, Inc., Natick, MA). Numerous reports of food process models have been prepared. Computer-based mathematical models were developed to predict the quality of apple juice during storage (7). Models were based on mass transfer of product through the package and accompanying ascorbic acid oxidation. Computer simulation was used to develop mathematical models for moisture transfer in different packaging systems (8). The models were successfully used in two component bakery mixes. A computer program for calculation of temperature during continuous heating of a particulate food product has been developed (9). The program is useful in predicting the effect of processing on product quality in an aseptic system comparing surface heat transfer and heat conduction inside the particles. Specialized computer packages have been developed for modeling sugar processing (10). Modeling activities include simulation, prediction, optimization, design, evaluation, and control. Optimal conditions for thermal processing have been determined using neural networks (11). Input variables included can size, thermal diffusion, and sterilization temperatures.

Statistical programs such as principal components analysis (PCA) and analysis of variance (ANOVA) have been integrated into programs specific for food product development (12). Statistical and experimental design tools are commonly integrated into packages for product development work. Computer-aided design/computer-aided manufacture (CAD/CAM), along with Latin squares experimental design have been used for label design (13). Consumer response to label changes was identified. Computerized systems have the potential to improve the accuracy of sensory evaluation as it relates to consumer preference and competitor analysis (14). Continual improvements in dietary consumption databases can be helpful in a better understanding of consumer behavior (15).

Quality Assurance

Laboratory automation systems (LAS) and laboratory information management systems (LIMS) have been developed to reduce errors and increase efficiency of laboratory analyses. Rapid access to previous analyses, data manipulation, and report preparation are valuable features of these systems. LIMS continue to evolve as with all computer software. Vendors range from dedicated, smaller companies to business units of larger instrument manufacturers such as Hewlett-Packard (Wilmington, DE) and Beckman Instruments (Fullerton, CA). Continuing improvements include user interfaces, sample scheduling, data storage and retrieval, instrument interfacing, and Web accessibility. Trade journals abound with descriptions of LIMS. Additional information is available through the Internet.

Statistical Quality Control (SQC) charts are widely used to monitor process control and product quality. Important factors to consider when selecting SQC software include the ability to handle process and lab data and the ability to exchange data with corporate or plantwide information systems. Computer programs have been developed for designing hazard analysis and critical control points (HACCP) systems. HACCP software can quickly access critical control points along a process.

Sophisticated computer-based tools for product analysis and identification are available. Computer-controlled automated sampling procedures have been developed to predict noodle discoloration over time (16). Computer vision technology can be used as a quality control process for gauging, verification, flaw determination, recognition, and locating (17). Machine vision and image processing have been used to identify corn kernel shape (18). Grain kernels, nuts, and crackers have also been examined using computerized image analysis and classified with a neural network program (19). Snack and chips with varying morphological and texture features as input and sensory attributes as output were used to train a neural network for quality analysis. The program was validated with predicted output compared with a sensory evaluation panel (20). Computer-controlled shape and size grading and filleting in the seafood industry have been developed (21). Computer-simulated models of flavor loss due to volatilization during heating or baking have been used to improve quality (22). Flavor was related to product-to-air partition coefficients of the flavor components and texture of the foods.

An artificial neural network has been developed to relate residual chlorine in stored water to various water quality and operational parameters (23). A fault analysis program has been developed to diagnose problems at all levels of cake production in order to optimize quality (24). Hybrid object-oriented and rule-based systems have been used to design safe food processes (25). The system relates food composition and process parameters with microbiological safety and food quality.

Processing

Process control systems monitor signals from various sensors, perform analyses on the information, and provide the proper control. Control is commonly accomplished with a programmable logic controller (PLC). Data can be stored in and accessed from the PLC by other hardware for report preparation. The typical function of a PLC is to keep certain parameters constant. More advanced process control systems adjust continuously to provide the best possible settings for the existing process and materials. Most food processing operations have at least some and many a large amount of computer-controlled automation. For example, a milk processing plant has a number of steps in a process. Each step contains a series of logic statements that may control valves, pumps, or thermal devices. To start up the pasteurizer, valves allowing water into the balance tank open, a pump turns on, steam or hot water are applied to the heating section. Local control devices maintain water and temperature levels. A flow diversion valve recycles water until the pasteurization temperature is reached. Once the temperature is reached the operator is cued to switch to milk. The operation is stepped up to another set of valve openings and pump activations, allowing milk into the balance tank. During pasteurization, if the holding tube temperature falls below the legal minimum, milk flow is diverted back to the balance tank. Additional sets of instructions are given for shutdown and cleaning. Neural networks have been used to control process parameters, inspect systems, and monitor machine performance. Incorporation of automation and computers has been credited with increased productivity, reduced waste, more effective cleaning, and improved quality assurance.

Integrated computer systems have been developed in which the food processing operation acts as part of a supply chain. In these systems materials flow from suppliers, through the processor, and on to distribution. Information flows in both directions. Computer integrated manufacturing (CIM) has incorporated business plans and information strategies with business applications software. Electronic data interchange, use of bar codes, automated data collection and retrieval, scheduling, and process control are brought together in a CIM environment. Statistical process control (SPC) programs are commonplace in the food processing industry. Automated production lines require sensors such as thermocouples, flowmeters, and colorimeters to determine when adjustments are required. Data collected from the sensors is used by SPC programs to display the immediate and historical state of the process. Feedback can be sent directly to controllers for changing process variables when required.

Computer automation in which a machine is made to duplicate some labor intensive human operation has involved the implementation of robots. Robots are able to withstand harsh working and cleaning conditions. Labor savings is another robotics benefit. Robots have been extensively used in packaging and palletizing, relieving humans of much physically demanding work. From robotics to control charts, there has been a proliferation of software for the process industries. Mathematical models have been used to determine optimal locations and the method of producing and transporting products. Effective use is made of machine capacity, manpower, raw materials, and warehouse space. Mathematical models and rule-based systems are useful for planning and are able to show the effects of minute by minute changes. For example, given characteristics such as vessel size the program indicates processing time, sequence of products, selection of equipment, and costs. Fuzzy logic programs have been useful for monitoring complex food processes (26). Fuzzy logic control offers a combination of objective process data with subjective human operator input to achieve performance output such as reduced quality defects. In the past raw process data would be modified and reduced before entering into an operational database. With low cost disk space and inexpensive database servers, more often raw data is being saved for further analysis. New techniques are then used to extract information from this data to improve process control and other process optimization. Data can be used with an expert diagnostic system for troubleshooting problems. Neural network technology can also be used to recognize patterns in data unsuitable for conventional statistical techniques.

Manufacturing execution systems (MES) claim to merge process control systems and management information systems (MIS). They integrate plant floor operations with production planning (27). Using current data MES guides, these systems initiate, respond to, and report on plant activities as they occur. Rapid responses can then be made to changing conditions. MES provides tools to integrate other programs such as SPC and supervisory control and data acquisition (SCADA) and then integrate them with planning systems such as manufacturing resource planning (MRP II). As these various computer-based control systems often overlap. Promoters claim MES as an information hub that links to all these other manufacturing systems. Enterprise resource planning (ERP) systems, which already integrate material resource planning with other upper-level business systems (such as financial, sales, and marketing information; cost management; human resources data; and supply-chain management) requires close integration with MES. Object-oriented program technology will help in this integration. Object-oriented programs are based on modular programs that can be modified without affecting each other, reducing costs to customize programs for new business needs.

Warehouse and Distribution

Computer programs have been developed to monitor suppliers and supplies. These systems provide enhanced communication within the plant and between the plant and suppliers. Simulation software has been developed to predict temperature changes under various conditions during storage or transit (28). The predictions can be useful in developing HACCP programs. It is becoming more common for suppliers to be automatically linked to customer inventory databases so that they can keep the customer supplied automatically. This offers advantages of stable, long-term business for the supplier and a never-failing supply of raw materials, parts, and other resources for the customer.

Administrative Operations

Business applications for computers such as word processing, databases, spreadsheets, and presentation graphics are commonly used in the food industry. Human resource management, including health and benefits, accounting, scheduling, training, strategic planning, and all business-related operations in a company, is very dependent upon standard computer programs. Graphic presentation and word processing programs have become very sophisticated with features such as powerful computational functions and data linking operations. Plants and administrative facilities can be closely linked for communications regardless of location. Data from remote and varied locations can be combined for analysis and report generation. International locations and contacts have become more accessible. Plant security systems, which often use key cards for entry and tracking of employee arrivals and departures, are becoming more sophisticated. Biometric recognition has used fingerprints, hand geometry, and eye scans for identification (29).

Food Safety

Implementation of HACCP programs has been greatly facilitated with the use of computer programs. Systems for recall in the event of a product failure require the speed and power of electronic coding and tracking in order to be effective. Microbial modeling programs have created databases and calculated growth, survival, thermal-death-time, and time-to-growth model for various pathogens. Microbial behavior in food is largely determined by factors such as pH, temperature, water activity, and atmosphere. Although limitations of models must be considered, they have been useful in estimating microbial behavior, developing HACCP programs, assisting in planning laboratory experimentation, and educating nonmicrobiologists. In addition, changes in composition or processing can be quickly evaluated, assisting in development of safer products. A limitation of predictive models occurs with outliers. Samples with high contamination, the first pathogenic cells to adjust to a new environment, or the organism with greatest thermal resistance may elude the model.

FUTURE DEVELOPMENTS

In the highly competitive and often saturated markets in which food companies operate, growth is not likely to occur through price increase or geographical expansion. Rather, growth depends on being the low-cost supplier and having access to competitive information. Integration of computer-based tools for sales and marketing management, manufacturing, maintenance, logistics, supply and inventory data, and financial management will be the goal of computer systems suppliers. Use of the Internet and other forms of network communication will continue to have a major impact on information exchange. Competitive intelligence is becoming a critical business strategy. Increased surveillance will cause security of electronic data to become increasingly important. With the proliferation of software, control systems, and networking tools, food processing industries are able to operate more efficiently and competitively. Proper implementation, monitoring, and continual training are key factors for success.

BIBLIOGRAPHY

1. A. Freedman, *The Computer Desktop Encyclopedia*, AMACOM, New York, 1996.
2. P. K. Thornton, H. W. G. Booltink, and J. J. Stoorvogel, "A Computer Program for Geostatistical and Spatial Analysis of Crop Model," *Agronomy Journal* **89**, 620–627 (1997).
3. J. A. A. M. Verstegen et al., "Economic Value of Management Information Systems in Agriculture: A Review of Evaluation Approaches," *Computers and Electronics in Agriculture* **13**, 273–288 (1995).
4. Y. W. Huang et al., "Modular Neural Networks for Identification of Starches in Manufacturing Food Products," *Biotechnol. Prog.* **9**, 401–410 (1993).
5. C. Barcenas and J. P. Norback, "A System for Tracking Food Components for Labeling and Other Purposes," *Food Technol.* **47**, 97–100, 102 (1993).
6. J. Dou, S. Toma, and S. Nakai, "Random-centroid Optimization for Food Formulation," *Food Research International* **26**, 27–37 (1993).
7. J. N. Kim et al., "Computer Simulation Model for Diffusion of Oxygen into a Packaged Liquid Food System with Simultaneous Oxidation of Ascorbic Acid," *IFT Annual Meeting: Book of Abstracts* **129**, (1996).
8. G. C. P. Rangarao, U. V. Chetana, and P. Veerraju, "Mathematical Model for Computer Simulation of Moisture Transfer in Multiple Package Systems," *Lebensmittel-Wissenschaft und Technologie* **28**, 38–42 (1995).
9. C. Skjoldebrand and T. Ohlsson, "A Computer Simulation Program for Evaluation of the Continuous Heat Treatment of Particulate Food Products: II," *Journal of Food Engineering* **20**, 167–187 (1993).
10. P. Thompson, "The Engineers' Toolbox. A Review of Modeling and Analysis Tools and their Application," *International Sugar Journal* **99**, 427–432 (1997).
11. S. S. Sablani, H. S. Ramaswamy, and S. O. Prasher, "A Neural Network Approach for Thermal Processing Applications," *Journal of Food Processing and Preservation* **19**, 283–301, (1995).
12. S. Schonkopf et al., "Computer-aided Product Development in the Food Industry," *Food Technol.* **50**, 69–70 (1996).
13. R. P. Hamlin and K. J. Leith, "The Use of CADCAM and the Latin Square to Isolate and Investigate Individual Consumer Cues in Food Products and Their Packaging," *Food Technologist* **25**, 32–38 (1995).
14. H. McIlveen and G. Armstrong, "Sensory Analysis and the Food Industry: Can Computers Improve Credibility?" *Nutr. Food Sci.* **1**, 36–40 (1996).
15. G. M. Price et al., "Measurement of Diet in a Large National Survey: Comparison of Computerized and Manual Coding of Records in Household Measures," *Journal of Human Nutrition and Dietetics* **8**, 417–428 (1995).
16. H. Corke, L. Au-Yeung, and X. Chen, "An Automated System for the Continuous Measurement of Time-dependent Changes in Noodle Color," *Cereal Chem.* **74**, 356–358 (1997).
17. S. Gunasekaran, "Computer Vision Technology for Food Quality Assurance," *Trends Food Sci. Technol.* **7**, 245–256 (1996).
18. B. Ni, M. R. Paulsen, and J. F. Reid, "Corn Kernel Crown Shape Identification Using Image Processing," *Transactions of the ASAE* **40**, 833–838 (1997).
19. K. Ding, "Food Shape and Microstructure Evaluation with Computer Vision," *Dissertations Abstracts International B* **56**, 3872–3873 (1996).
20. M. S. Sayeed, A. D. Whittaker, and N. D. Kehtarnavaz, "Snack Quality Evaluation Method Based on Image Features and Neural Network Prediction," *Transactions of the ASAE* **38**, 1239–1245 (1995).
21. B. Scudder, "Computers Working for Fish," *Seafood International* **10**, 37–39 (1995).
22. K. B. Roos and E. Graf, "Nonequilibrium Partition Model for Predicting Flavor Retention in Microwave and Convection Heated Foods." *J. Agric. Food Chem.* **43**, 2204–2211 (1995).
23. J. B. Serodes and M. J. Rodriguez, "Predicting Residual Chlorine Evolution in Storage Tanks Within Distribution Systems: Application of a Neural Network Approach," *Aqua* **45**, 57–66 (1996).
24. L. S. Young, "Getting the Product Right," *Food Manufacture* **70**, 43 (1995).
25. M. Schellekens et al., "Computer Aided Microbial Safety Design of Food Processes," *Int. J. Food Microbiol.* **24**, 1–9 (1994).

26. Q. Zhang, J. B. Litchfield, and J. Bentsman, "Fuzzy Logic Control for Food Processes," *AIChE Symposium Series* **89**, 90–97 (1993).
27. C. E. Morris, "MES Boosts Manufacturing Flexibility," *Food Engineering* **69**, 59–66 (1997).
28. C. Alasalvar et al., "Prediction of Time-Temperature Profiles of Atlantic Salmon (*Salmo salar*) During Chilled Transport Using the MAILPROF Computer Program," *IFT Annual Meeting: Book of Abstracts* 80 (1996).
29. R. W. Hardin, "Biometric Recognition: Photonics Ushers in a New Age of Security," *Photonics Spectra* **31**, 88–100 (1997).

ROBERT L. OLSEN
Schreiber Foods, Inc.
Tempe, Arizona

CONTROLLED ATMOSPHERES FOR FRESH FRUITS AND VEGETABLES

Controlled-atmosphere (CA) storage is a technique for maintaining the quality of fresh fruits and vegetables in an atmosphere that differs from normal air with respect to the concentrations of oxygen (O_2), carbon dioxide (CO_2), and/or nitrogen (N_2). The desired compositions of the atmosphere for storing commodities are usually obtained by initially increasing CO_2 or decreasing O_2 levels in a gas-tight storage room or container. Sometimes, the addition of carbon monoxide (CO) or removal of ethylene (C_2H_4) may also be beneficial.

Modified atmosphere (MA) is a condition similar to CA, but with less or no active control of the gas concentrations. In MA, the O_2 level is reduced and the CO_2 level is increased at a rate determined by the respiration rate of the commodity, the storage temperature, and the permeability of the container and film wrap to the gases. Judicious selection of the commodity, the package dimensions, and the package material will insure establishment and maintenance of the desired atmosphere under specified storage temperatures.

TYPES OF CA STORAGE

During the past 60 years, tremendous progress has been made in the technology of CA storage. The commercial application of CA during transit and storage has received considerable attention since 1960, resulting in the development of different methods of establishing and maintaining CA. These include regular CA; short-term, high-CO_2 treatment; rapid CA; low-oxygen CA; low-ethylene CA; and low-pressure, or hypobaric, storage.

Short-Term High-CO_2 Treatment

Short-term, high-CO_2 treatment was originally developed to maintain the firmness of Golden Delicious apples (1). Subsequently, it was found that pears and other fruits and vegetables also benefit from this treatment (2,3). This treatment involves the exposure of fruit to 10–20% CO_2 for four to seven days prior to adjustment of the atmosphere to regular CA concentrations. Carbon dioxide injury of the fruit skin may occur if moisture has condensed on the surface of the fruit. This high-CO_2 treatment gives excellent results in maintaining the quality of Golden Delicious apples and Anjou pears (1,2,4).

Rapid CA

Rapid CA is a strategy that shortens the time between harvest and establishment of the desirable CA conditions (5). The faster the CA conditions are attained after harvest, the better the fruit quality can be maintained, providing that the cooling rate in storage is not adversely affected by the rapid loading of the room. To achieve the objectives of rapid CA, the storage room should be filled and sealed within three days or less of harvest.

Low-Oxygen CA

Low-oxygen CA has recently received increased attention; not only because it markedly retards fruit softening, but also because it greatly reduces the development of storage scald and breakdown of apples and pears (6). In regular CA, the recommended O_2 concentrations are usually 2% or higher. It has been found that O_2 levels between 1 and 1.5% are even more effective in extending the storage life of some fruits and vegetables (7). Careful monitoring to maintain the precise O_2 level is essential to avoid damage due to anaerobic respiration.

Low-Ethylene CA

In low-ethylene CA, ethylene is scrubbed from the CA room to improve storage quality of the fruit. Removal of ethylene from the storage atmosphere results in retardation of ripening, retention of flesh firmness, and reduction of the incidence of superficial scald of apples (8). It is generally recognized that the concentration of ethylene should be maintained below 1 ppm to obtain the beneficial effect of low-ethylene CA. Storage life of several apple varieties such as Empire and Bramley's Seedling can be extended by this technique (9,10).

Low-Pressure Hypobaric, Storage

Low-pressure, or hypobaric, storage consists of storing fruits and vegetables at below-normal atmospheric pressure. Enhanced diffusion of gases under reduced-pressure facilitates the loss of CO_2 and ethylene from the commodity and reduces the O_2 gradient between the inside and outside of the commodity. The partial pressures of O_2 is directly related to the absolute pressure of air (11). Thus the O_2 concentration is equivalent to 0.55% at 20 mm Hg. Ethylene inside the fruit is also reduced proportionally. Therefore, this storage technique combines the advantages of low O_2 storage and low-ethylene storage. Ripening can be inhibited and storage life prolonged when fruit are stored under hypobaric conditions.

BENEFICIAL EFFECTS OF CA STORAGE

Beneficial effects of CA storage include reduction of respiration, scald, decay, discoloration, and internal breakdown, inhibition of ethylene production and ripening, and retention of firmness, flavor, and nutritional quality.

The rate of respiration of fresh fruits and vegetables has been shown to be reduced by low O_2 or high CO_2 (12). The lower respiration rate indicates that CA has an inhibitory effect on the overall metabolic activities of stored commodities. A slower rate of utilization of carbohydrates, organic acids, and other reserves usually leads to prolonging the life of the produce.

Ethylene production of fresh fruits and vegetables is suppressed by low O_2 and/or elevated concentrations of CO_2. The biosynthesis of ethylene in plant tissues requires the presence of O_2 (13,14). When O_2 is absent or when plant tissue is under low O_2 atmosphere, ethylene biosynthesis is inhibited. The time of onset of ethylene production in apples is inversely related to the O_2 concentration in storage, and the maximum rate of production is directly related to the O_2 level (15). High CO_2 inhibits ethylene action on ripening (16), which in turn inhibits the autocatalytic production of enzymes involved in ethylene biosynthesis, including 1-aminocyclopropane-1-carboxylic acid (ACC) synthase and ethylene-forming enzymes (17). These inhibitory effects on ethylene action and production by CA consequently lead to the delay of the ripening process.

The loss of organic acids in apples and pears during storage is reduced by CA (18,19). This reduced loss of organic acids in CA fruits is probably due to an increase in CO_2 fixation, an inhibition of respiratory metabolism, and a lower consumption of acids under CA (19,20).

A slower decline in carbohydrates under CA has been reported in sugar beet, Chinese cabbage, apricots, and peaches (21,22,23). CA storage is also beneficial in the retention of ascorbic acid and amino acids in several fresh fruits and vegetables (3,22,24,25).

EFFECTS ON PHYSIOLOGICAL DISORDERS

Scald is one of the most serious physiological disorders in apples, pears, and many other fruits during and after storage. Storage of fruits in CA, particularly in low O_2 atmosphere, has been shown to reduce the susceptibility of fruits to scald (6,26). Russet spotting of lettuce can be caused by exposure to ethylene or warm shipping or storage temperatures (27). This physiological disorder can be substantially reduced by low O_2 atmosphere (28,29). The development and spread of necrotic spots on the outer leaves of cabbage was largely prevented by low O_2 atmosphere but not by high-CO_2 treatment (30). The incidence and severity of vein streaking of cabbage leaves was also reported to be reduced in CA storage (31).

Controlled atmosphere reduces chilling injury in certain sensitive crops, while it aggravates or has no effect in other crops. Holding zucchini squash in low O_2 alleviated chilling injury during storage at 2.5°C (32,33). Conditioning grapefruit with short-term prestorage treatment of high CO_2 (40%) at 21°C reduced subsequent brown staining and rind pitting, two symptoms of chilling injury, at 1°C (34). Intermittent exposure of unripe avocados to 20% CO_2 reduced chilling injury at 4°C (35). Addition of CO_2 to the storage atmospheres was also effective in reducing chilling-induced internal breakdown and retaining ripening ability of peaches (36,37). Controlled atmosphere has also been reported to reduce the severity of chilling injury symptoms in okra (38). However, increasing the level of CO_2 or reducing O_2 concentration can accentuate symptoms of chilling injury in cucumbers, bell peppers, and tomatoes (39,40).

Some other physiological disorders have also been reported to be aggravated by CA conditions. The development of a brown discoloration in and around the core and adjacent cortex tissue of apples and pears has been associated with high CO_2 (41,42). White core inclusions in kiwifruit can also be induced by elevated CO_2 in combination with ethylene (43). The severity of brown stain on lettuce has also been found to increase with increasing CO_2 levels (44).

CONCLUSION

Controlled atmosphere storage has withstood the test of time and has been proven to be effective for maintaining quality and extending the storage life of a number of horticultural crops. The rate of deterioration of fresh produce is often greatly retarded by CA storage. Each type of fruit or vegetable has its own specific requirement and tolerance for atmosphere modification. The optimum CA condition for each commodity is continually being revised and improved. Table 1 presents the most recent recommendations for O_2 and CO_2 levels, the most suitable storage temperatures, feasible storage periods, and pertinent remarks for some commodities.

The maintenance of CA storage requires continual monitoring of the gases and temperature to prevent any deviation from the recommended conditions. Injuries induced by low O_2 or high CO_2 often lead to tissue discoloration or off-flavor and loss of market value. Proper maturity, the internal condition of the fruit at harvest, and the speed at which storage atmospheres are established are some of the key factors affecting the success of CA storage.

It should be emphasized that proper CA storage should always be accompanied by good temperature control. CA storage is considered to be a supplement—not a substitute—for proper refrigeration and careful handling.

Table 1. Summary of Controlled-Atmosphere Storage Requirements for Fruits and Vegetables

Commodity Cultivar	Beneficial concentration (%) O_2	CO_2	Suitable temperature °C	°F	Approximate storage period	Remarks	Ref.
Fruits of temperate zone							
Apple (*Malus domestica*)							
Boskoop	1.5–2	<1.5	4	39	5–7 mo	Sensitive to low temperature breakdown	45
Bramley's Seedling	2	6	3–4	37–39	7 mo	CA reduces bitter pit	45
Cortland	2–3	5	2	36	4–6 mo		46
Cox Orange Pippin	1–3	<1	3	37	4–6 mo	2% O_2 first week, then 1.00–1.25% O_2	45
Empire	1.5–3	1–5	2	36	5–7 mo		45
Fuji	2–2.5	1–2	0	32	7–8 mo	CA reduces scald	45
Gala	1–2	1–5	0	32	5 mo		45
Golden Delicious	1–3	1–5	0	32	7–11 mo	Rapid CA is beneficial	45
Granny Smith	1–2	1–3	0	32	7–9 mo	CA reduces scald	45
Jonathan	1–3	1–6	0–3	32–37	4–7 mo	CA reduces Jonathan spot	45
McIntosh	1.5–3	1–5	3	37	7–9 mo		45
Newtown	3	5–8	2–4	36–39	8 mo	Susceptible to low temperature injury	
Northern Spy	2–3	2–3	0	32	8 mo		46
Red Delicious	1–3	1–3	0	32	8–10 mo	Susceptible to scald	45
Rome Beauty	2–3	2–5	0	32	6–8 mo		45
Spartan	1.5–2.5	1–2	0	32	6–8 mo		45
Stayman	2–3	2–5	0	32	7–8 mo		46
Worcester Pearmain	3	5	1	34	6 mo		45
Apricot (*Prunus armeniaca*)	2–3	2–3	−0.5–0	31–32	7 wk	CA delays ripening	47
Blackberry (*Rubus* sp.)	5–10	15–20	−0.5–0	31–32	1 wk	Prompt cooling is important	47
Black currant (*Ribes nigrum*)	—	25–50	2	36	4 wk	50% first week, then 25% CO_2	48
Blueberry (*Vaccinium* sp.)	5–10	15–20	−0.5–0	31–32	2–3 wk	Prompt cooling is important	47
Cherry, sweet (*Prunus avium*)	3–10	10–15	−1–0	30–32	4 wk	High CO_2 reduces decay	47
Cranberry (*Vaccinium macrocarpon*)	1–2	0–5	3	37	2–4 mo		47
Fig (*Ficus carica*)	5–10	15–20	−1–0	30–32	2 wk		47
Grape (*Vitis vinifera*)	2–5	1–3	−1–0	30–32	1–6 mo		47
Kiwifruit (*Actinidia chinensis*)	1–2	3–5	0	32	3–5 mo	CA delays ripening	47
Nectarine (*Prunus persica*)	1–2	3–5	−0.5–0	31–32	6–9 wk	CA reduce internal breakdown	47
Peach (*Prunus persica*)	1–2	3–5	−0.5–0	31–32	6–9 wk	Cultivars differ in response	47
Pear, European (*Pyrus communis*)							
Anjou	0.5–2	0.5–2	−1	30	7–9 mo	CA reduce scald	49
Bartlett	1–2	1–2	−1	30	3–5 mo	Rapid cooling recommended	49
Bosc	1–3	0.5–1	−1	30	4–6 mo	Optimum maturity is critical	49
Comice	2	2–3	−1	30	5–7 mo		49
Conference	2	2	−1	30	4–6 mo		49
Packham's Triumph	2–3	1–2	−0.5	31	8 mo	Optimum maturity is critical	49
Passe Crassane	3–4	5–7	−1	30	6–7 mo	Tolerant to high CO_2	49

Pear, Asian (*Pyrus serotina* and *Pyrus breischneideri*)							
Nijiseiki (20th Century)	3	1	0	32	9–12 mo		49
Tsu Li	1–2	3	0	32	6–8 mo		49
Ya Li	3–4	2	0–3	32–37	6–8 mo		49
Persimmon (*Diospyros kaki*)	3–5	5–8	−1–0	30–32	4 mo		47
Plum (*Prunus domestica*)	1–2	0–5	−0.5–0	31–32	4–5 mo	CA delays ripening	47
Rasberry (*Rubus idaeus*)	5–10	15–20	−0.5–0	31–32	1 wk	Prompt cooling is important	47
Strawberry (*Fragaria* sp.)	5–10	15–20	−0.5–0	31–32	1 wk	Commercial use during transport	47
Subtropical and tropical fruits							
Avocado (*Persea americana*)	2–5	3–10	10	50	3–6 wk	CA reduces chilling injury	47
Banana (*Musa* spp)	2–5	2–5	14	58	1–6 mo	CA delays ripening	47
Grapefruit (*Citrus paradisi*)	3–10	5–10	13	55	6–8 wk	CA reduces pitting	47
Lemon (*Citrus limon*)	5–10	0–10	13	55	1–6 mo	CA reduces decay	47
Lime (*Citrus aurantifolia*)	5–10	0–10	13	55	6–8 wk	CA retards degreening	47
Mango (*Mangifera indica*)	3–5	5–10	13	55	5 wk	CA delays ripening	47
Olive (*Olea europaea*)	2–3	0–1	7	45	2 mo		47
Orange (*Citrus sinensis*)	5–10	0–5	7	45	8–12 wk	Storage life varys with cultivars	47
Papaya (*Carica papaya*)	3–5	5–10	12	54	2–3 wk		47
Passion fruit (*Passiflora edulis*)	5	5	7–10	45–50	6 wk		50
Pineapple (*Ananas comosus*)	2–5	5–10	10	50	4 wk		47
Vegetables							
Artichoke (*Cynara scolymus*)	2–3	2–3	0	32	1 mo	CA decreases discoloration	51
Asparagus (*Asparagus officinalis*)	Air	10–14	2	36	3 wk	High CO_2 is beneficial	51
Bean, lima (*Phaseolus limensis*)	>5	10–35	5–7	40–45	7–10 d	Shelled only	52
Bean, snap (*Phaseolus vulgaris*)	2–3	4–7	8	46	2 wk	CA reduces color loss	51
Broccoli (*Brassica oleracea italica*)	1–2	5–10	0	32	1 mo	CA maintain green color	51
Brussels sprouts (*Brassica oleracea gemmifera*)	1–2	5–7	0	32	3–5 wk	CA reduces yellowing	51
Cabbage (*Brassica oleracea, capitata*)	2–3	3–6	0	32	6–8 mo	Large scale commercial use	51
Cantaloupe (*Cucumis melo*)	3–5	10–20	8	46	2–3 mo	CA reduces ripening	51
Cauliflower (*Brassica oleracea botrytis*)	2–3	3–4	0	32	1 mo		51
Celery (*Apium graveolens*)	1–4	3–5	0	32	3 mo		51
Chinese cabbage (*Brassica campestris*)	1–2	0–5	0	32	4–5 mo	CA reduces leaf abscission	51
Corn, sweet (*Zea mays*)	2–4	5–10	0	32	2 wk	CA reduces sugar loss	51
Cucumber (*Cucumis sativus*)	1–4	0	12	54	3 wk	CA reduces yellowing	51
Leek (*Allium porrum*)	1–6	5–10	0	32	4 mo		51
Lettuce, head (*Lactuca sativa*)	1–3	0	0	32	3–4 wk		51
Mushroom (*Agaricus bisporus*)	Air	10–15	0	32	1–2 wk	CO_2 retards cap opening	51
Onion, dry (*Allium cepa*)	0–1	0–5	0	32	8 mo	Tolerant to low O_2	51
Onion, green (*Allium cepa*)	1	5	0	32	2 mo		53
Parsley (*Petroselinum crispum*)	8–10	8–10	0	32	3 mo		51
Pepper, sweet (*Capsicum annuum*)	2–5	0	12	54	3 wk		51
Radish (*Raphanus sativus*)	1–2	2–3	0	32	4 mo		51
Spinach (*Spinacia oleracea*)	7–10	5–10	0	32	3 wk		51
Tomato (*Lycopersicon esculentum*)	3–5	2–3	12	54	4–6 wk		

Note: The approximate storage periods are compiled by the authors based on all the information available.

BIBLIOGRAPHY

1. H. M. Couey and K. L. Olsen, "Storage Response of Golden Delicious Apples Following High Carbon Dioxide Treatment," *Journal of the American Society of Horticulture Science* **100**, 148–150 (1975).
2. C. Y. Wang and W. M. Mellenthin, "Effect of Short Term High CO_2 treatment on Storage of d'Anjou Pear," *Journal of the American Society of Horticulture Science* **100**, 492–495 (1975).
3. C. Y. Wang, "Effect of Short Term High CO_2 Treatment on the Market Quality of Stored Broccoli," *Journal of Food Science* **44**, 1478–1482 (1979).
4. H. M. Couey and T. R. Wright, "Effect of a Prestorage CO_2 Treatment on the Quality of d'Anjou Pears After Regular or controlled Atmosphere Storage," *HortScience* **12**, 244, 245 (1977).
5. O. L. Lau, "Storage Responses of Four Apple Cultivars to a 'Rapid CA' Procedure in Commercial Controlled Atmosphere Facilities," *Journal of the American Society of Horticulture Science* **108**, 530–533 (1983).
6. W. M. Mellenthin, P. M. Chen, and S. B. Kelly, "Low Oxygen Effects on Dessert Quality, Scald Prevention, and Nitrogen Metabolism of d'Anjou Pear Fruit During Long Term Storage," *Journal of the American Society of Horticulture Science* **105**, 522–527 (1980).
7. D. G. Richardson and M. Meheriuk, eds., *Controlled Atmospheres for Storage and Transport of Perishable Agricultural Commodities*, Timber Press, Beaverton, Oreg., 1982.
8. F. W. Liu, "Conditions for Low Ethylene CA Storage of Apple: A Review," in S. M. Blankenship, ed., *Controlled Atmospheres for Storage and Transport of Perishable Agricultural Commodities*, Horticulture Report 126, North Carolina State University, Raleigh, 1985.
9. G. D. Blanpied, "Low Ethylene CA Storage for Empire Apples," in Ref. 8.
10. C. J. Dover, "Effects of Ethylene Removal during Storage of Bramley's Seedling," in Ref. 8.
11. D. R. Dilley, P. L. Irwin, and M. W. McKee, "Low Oxygen, Hypobaric Storage and Ethylene Scrubbing," in Ref. 7.
12. J. B. Biale, "Respiration of Fruits," in W. Ruhland, ed., *Encyclopedia of Plant Physiology*, Vol. 12, Springer-Verlag, New York, 1960.
13. E. Hansen, "Quantitative Study of Ethylene Production in Relation to Respiration of Pears," *Botanical Gazette* **103**, 543–558 (1942).
14. D. O. Adams, and S. F. Yang, "Ethylene Biosynthesis: Identification of 1-Aminocyclopropane-1-Carboxylic Acid as an Intermediate in the Conversion of Methionine to Ethylene," *Proceedings of the National Academy of Sciences of the United States of America* **76**, 170–174 (1979).
15. M. Knee, "Physiological Responses of Apple Fruits to Oxygen Concentrations," *Annals of Applied Biology* **96**, 243–253 (1980).
16. S. P. Burg and E. A. Burg, "Molecular Requirements for the Biological Activity of Ethylene," *Plant Physiology* **42**, 144–152 (1967).
17. S. F. Yang, "Biosynthesis and Action of Ethylene," *HortScience* **20**, 41–45 (1985).
18. P. H. Li and E. Hansen, "Effects of Modified Atmosphere Storage on Organic Acid and Protein Metabolism of Pears," *Proceedings of the American Society for Horticultural Science* **85**, 100–111 (1964).
19. D. A. Kollas, "Preliminary Investigation of the Influence of Controlled Atmosphere Storage on the Organic Acid of Apples," *Nature (London)* **204**, 758–759 (1964).
20. N. Allentoff, W. R. Phillips, and F. B. Johnston, "A ^{14}C Study of Carbon Dioxide Fixation in the Apple. II. Rates of Carbon Dioxide Fixation in the Detached McIntosh Apple," *Journal of the Science of Food and Agriculture* **5**, 234–238 (1954).
21. M. T. Wu, B. Singh, J. C. Theurer, L. E. Olson, and D. K. Salunkhe, "Control of Sucrose Loss in Sugarbeet During Storage by Chemicals and Modified Atmosphere and Certain Associated Physiological Changes," *Journal of the American Society of Sugar Beet Technology* **16**, 117–127 (1970).
22. C. Y. Wang, "Postharvest Responses of Chinese Cabbage to High CO_2 Treatment and Low O_2 Storage," *Journal of the American Society for Horticultural Science* **108**, 125–129 (1983).
23. B. N. Wankier, D. K. Salunkhe, and W. F. Campbell, "Effects of CA Storage on Biochemical Changes in Apricot and Peach Fruits," *Journal of the American Society for Horticultural Science* **95**, 604–609 (1970).
24. J. N. McGill, A. I. Nelson, and M. P. Steinberg, "Effects of Modified Storage Atmospheres on Ascorbic Acid and Other Quality Characteristics of Spinach," *Journal of Food Science* **31**, 510–517 (1966).
25. T. Suzuki, N. Kubo, S. Haginuma, S. Tamura, and H. Yamamoto, "Changes in Free Amino Acid Content During Controlled Atmosphere Storage of Lettuce, Chinese Cabbage and Cucumber," *Report of National Food Research Institute (Japan)* **38**, 46–55 (1981).
26. C. R. Little, H. J. Taylor, and F. McFarlane, "Postharvest and Storage Factors Affecting Superficial Scald and Core Flush of Granny Smith Apples," *HortScience* **20**, 1080–1082 (1985).
27. W. J. Lipton and J. K. Stewart, *An Illustrated Guide to the Identification of Some Market Disorders of Head Lettuce*, U.S. Department of Agriculture, Marketing Research Report, U.S. Government Printing Office, Washington, D.C., **950** (1972).
28. W. J. Lipton, *Market Quality and Rate of Respiration of Head Lettuce Held in Low Oxygen Atmospheres*, U.S. Department of Agriculture, Marketing Research Report, U.S. Government Printing Office, Washington, D.C., **777** (1967).
29. C. S. Parsons, J. E. Gates, and D. H. Spalding, "Quality of Some Fruits and Vegetables after Holding in Nitrogen Atmospheres," *Proceedings of the American Society for Horticultural Science* **84**, 549–556 (1964).
30. H. Bohling and H. Hansen, "Storage of White Cabbage (*Brassica oleracea* var. Capitata) in Controlled Atmospheres," *Acta Horticulturae* **62**, 49–54 (1977).
31. L. S. Beard, "Effect of CA on Several Storage Disorders of White Cabbage," in Ref. 8.
32. F. Mencarelli, W. J. Lipton, and S. J. Peterson, "Responses of Zucchini Squash to Storage in Low O_2 Atmospheres at Chilling and Nonchilling Temperatures," *Journal of the American Society for Horticultural Science* **108**, 884–890 (1983).
33. C. Y. Wang and Z. L. Ji, "Effect of Low Oxygen Storage on Chilling Injury and Polyamines in Zucchini Squash," *Scientia Horticulture* **39**, 1–7 (1989).
34. T. T. Hatton and R. H. Cubbedge, "Conditioning Florida Grapefruit to Reduce Chilling Injury During Low Temperature Storage," *Journal of the American Society for Horticultural Science* **107**, 57–60 (1982).
35. P. Marcellin and A. Chaves, "Effects of Intermittent High CO_2 Treatment on Storage Life of Avocado Fruits in Relation to Respiration and Ethylene Production," *Acta Horticulture* **138**, 155–162 (1983).

36. R. E. Anderson, C. S. Parsons, and W. L. Smith, *Controlled Atmosphere Storage of Eastern-Grown Peaches and Nectarines*, U.S. Department of Agriculture, Marketing Research Report, U.S. Government Printing Office, Washington, D.C.
37. N. L. Wade, "Physiology of Cold Storage Disorders of Fruits and Vegetables," in J. M. Lyons, D. Graham, and J. K. Raison, eds., *Low Temperature Stress in Crop Plants*, Academic Press, Orlando, Fla., 1979.
38. Y. Ilker and L. L. Morris, "Alleviation of Chilling Injury of Okra," *HortScience* **10**, 324 (1975).
39. I. L. Eaks, "Effect of Modified Atmosphere on Cucumbers at Chilling and Nonchilling Temperatures," *Proceedings of the American Society for Horticultural Science* **67**, 473–478 (1956).
40. A. A. Kader, "Biochemical and Physiological Basis for Effects of Controlled and Modified Atmospheres on Fruits and Vegetables," *Food Technology (Chicago)* **40**, 99–104 (1986).
41. R. M. Smock and A. Van Doren, "Controlled Atmosphere Storage of Apples," *Cornell University Agricultural Experimental Station Bulletin* **762** (1941).
42. E. Hansen, "Reactions of Anjou Pears to Carbon Dioxide and Oxygen Content of the Storage Atmosphere," *Proceedings of the American Society for Horticultural Science* **69**, 110–115 (1957).
43. M. L. Arpaia, F. G. Mitchell, A. A. Kader, and G. Mayer, "Effects of 2% O_2 and Varying Concentrations of CO_2 with or without C_2H_4 on the Storage Performance of Kiwifruit," *Journal of the American Society for Horticultural Science* **110**, 200–203 (1985).
44. P. E. Brecht, A. A. Kader, and L. L. Morris, "The Effect of Composition of the Atmosphere and Duration of Exposure on Brown Stain of Lettuce," *Journal of the American Society for Horticultural Science* **98**, 536–538 (1973).
45. M. Meheriuk, "Controlled Atmosphere Recommendations for Apples," in J. Fellman, ed., *Proceedings of the Fifth International CA Research Conference*, Washington State University, Wenatchee, Washington, 1989.
46. R. E. Hardenburg, A. E. Watada, and C. Y. Wang, *The Commercial Storage of Fruits, Vegetables, and Florist and Nursery Stocks*, U.S. Department of Agriculture, Agricultural Handbook **66** (1986).
47. A. A. Kader, "A Summary of CA Requirements and Recommendations for Fruits Other than Pome Fruits," in Ref. 45.
48. W. H. Smith, "The Use of Carbon Dioxide in the Transport and Storage of Fruits and Vegetables," *Advances in Food Research* **12**, 95–146 (1963).
49. D. G. Richardson and M. Meheriuk, "Controlled Atmosphere Recommendations for Pears," in Ref. 45.
50. J. S. Pruthi, "Physiology, Chemistry, and Technology of Passion Fruit," *Advances in Food Research* **12**, 203–282 (1963).
51. M. E. Saltveit, "A Summary of Requirements and Recommendations for the Controlled and Modified Atmosphere Storage of Harvested Vegetables," in Ref. 45.
52. A. L. Ryall and W. J. Lipton, *Handling, Transportation, and Storage of Fruits and Vegetables*, Vol. 1, AVI Publishing Co., Westport, Conn., 1979.
53. H. W. Hruschka, *Storage and Shelf Life of Packaged Green Onions*, U.S. Department of Agriculture, Marketing Research Report, U.S. Government Printing Office, Washington, D.C. **1015** (1974).

GENERAL REFERENCES

D. Batram, "CA Storage Key to Growth of State's Apple Industry," *Good Fruit Grower* **47**, 44–46 (1996).

P. Bertolini et al., "Effects of Controlled Atmosphere Storage on the Physiological Disorders and Quality of Conference Pears," *Italian Journal of Food Science* **9**, 303–312 (1997).

G. Briones Lopez et al., "Storage of Common Mushroom Under Controlled Atmospheres," *International Journal of Food Science and Technology* **27**, 493–505 (1992).

D. M. Burmeister and D. R. Dilley, "A Scald-like Controlled Atmosphere Storage Disorder of Empire Apples—A Chilling Injury Induced by CO2," *Postharvest Biology and Technology* **6**, 1–7 (1995).

S. R. Drake, "Elevated Carbon Dioxide Storage of Anjou Pears Using Ourge-Controlled Atmosphere," *HortScience* **29**, 299–301 (1994).

S. R. Drake and T. A. Eisele, "Quality of Gala Apples as Influenced by Harvest Maturity, Storage Atmosphere and Concomitant Storage With Bartlett Pears," *Journal of Food Quality* **20**, 41–51 (1997).

I. U. Elgar, D. M. Burmeister and C. B. Watkins, "Storage and Handling Effects on a CO2-related Internal Browning Disorder of Braeburn Apples," *HortScience* **33**, 719–722 (1998).

Y. Gariepy, G. S. V. Raghavan, and J. A. Munroe, "Long-Term Storage of Leek Stalks Under Regular and Controlled Atmospheres," *International Journal of Refrigeration* **17**, 140–147 (1994).

G. A. Gonzalez et al., "Low Oxygen Treatment Before Storage in Normal or Modified Atmosphere Packaging of Mangoes to Extend Shelf Life," *Journal of Food Science and Technology India* **34**, 399–404 (1997).

M. Hansen, "New Process Extends Storage Life of Peaches," *Good Fruit Grower* **48**, 27 (1997).

P. L. Hurst et al., "Biochemical Responses of Asparagus to Controlled Atmosphere Storage at 20 Degree," *J. Food Biochem.* **20**, 463–472 (1997).

H. Izumi, A. E. Watada and W. Douglas, "Optimum O2 or CO2 Atmosphere for Storing Broccoli Florets at Various Temperatures," *J. Am. Soc. Hortic. Sci.* **121**, 127–131 (1996).

A. King and Y. Joon Yong, "Effects of Low Temperature and CA on Quality Changes and Physiological Characteristics of Chilling During Storage of Squash (*Curcubitai imoschatai*)," *Journal of the Korean Society for Horticultural Sciences* **39**, 402–407 (1998).

R. E. Lill and V. K. Corrigan, "Asparagus Responds to Controlled Atmospheres in Warm Conditions," *International Journal of Food Science and Technology* **31**, 117–121 (1996).

S. Meir et al., "Further Studies on the Controlled Atmosphere Storage of Avocadoes," *Postharvest Biology and Technology* **5**, 323–330 (1995).

A. Noomhorn and N. Tiasuwan, "Controlled Atmosphere Storage of Mango Fruit, *Mangifera indica L. cv. Rad*," *Journal of Food Processing and Preservation* **19**, 271–281 (1995).

T. Palma et al., "Respiratory Behavior of Cherimoya (*Annona cherimola Mill.*) Under Controlled Atmospheres," *HortScience* **28**, 647–649 (1993).

B. Ratanchinakorn, A. Klieber, and D. H. Simons, "Effect of Short-Term Controlled Atmospheres and Maturity on Ripening and Eating Quality of Tomatoes," *Postharvest Biology and Technology* **11**, 149–154 (1997).

S. Y. Rogers and N. R. Knowles, "Effects of Storage Temperature and Atmosphere on Saskatoon (*Amelancher alnifolia Nutt.*) Fruit Quality, Respiration and Ethylene Production," *Postharvest Biology and Technology* **13**, 183–190 (1998).

T. Saray, "Controlled Atmosphere Storage of Vegetable: The Possibilities," *Food Technology International Europe*, 69–73 (1994).

J. Stow, "The Effects of Storage Atmosphere on the Keeping Quality of Idared Apples," *Journal of Horticultural Science* **70**, 587–595 (1995).

T. K. Thomas and E. M. Sfakiotakis, "Effect of Low-Oxygen Atmosphere on Storage Behaviour of Kiwifruit," *Advances in Horticultural Science* **11**, 137–141 (1997).

A. B. Truter, J. C. Combrink, and S. A. Burger, "Control of Superficial Scald in Granny Smith Apples by Ultra-Low Levels of Oxygen as an Alternative to Diphenylamine," *Journal of Horticultural Science* **69**, 581–587 (1994).

R. K. Voltz et al., "Prediction of Controlled Atmosphere-Induced Flesh Browning in Fuji Fruit," *Postharvest Biology and Technology* **13**, 97–102 (1998).

H. S. Yong and L. A. Kyung, "Physiological Characteristics of Chilling Injury and CA Effect on its Reduction During Cold Storage of Pepper Fruit," *Journal of the Korean Society for Horticultural Sciences* **38**, 478–482 (1997).

H. S. Yong, I. I. Kim, and L. C. Jae, "Effects of Harvest Maturity and Storage Environments on the Incidence of Watercore, Flesh Browning, and Quality in Fuji Apples," *Journal of the Korean Society for Horticultural Sciences* **39**, 569–573 (1998).

P. Younmoon et al., "Preharvest Factors Affecting the Incidence of Physiological Disorders During CA Storage of Fuji Apples," *Journal of the Korean Society for Horticultural Sciences* **38**, 725–729 (1997).

CHIEN YI WANG
ALLEY E. WATADA
USDA/ARS
Beltsville, Maryland

See also PACKAGING: PART IV—CONTROLLED/MODIFIED ATMOSPHERE/VACUUM FOOD PACKAGING.

CORN. See CEREALS SCIENCE AND TECHNOLOGY.

CORROSION AND FOOD PROCESSING

MATERIALS OF CONSTRUCTION

The "Stainless Steels"

Many people refer to this range of alloys as austenitic stainless steels, 18/8s, 18/10s, etc, without a full appreciation of what is meant by the terminology. It is worth devoting a few paragraphs to explain the basic metallurgy of stainless steels.

It was in 1913 that Harry Brearley discovered that the addition of 11% chromium to carbon steel would impart a good level of corrosion and oxidation resistance, and by 1914 these corrosion resisting steels had become commercially available. It was Brearley who pioneered the first commercial use of these steels for cutlery, and it was also he who coined the name "stainless steels." For metallurgical reasons, which are outside the scope of this article, these were known as ferritic steels because of their crystallographic structure. Unfortunately, they lacked the ductility to undergo extensive fabrication and furthermore, they could not be welded. Numerous workers tried to overcome these deficiencies by the addition of other alloying elements and to produce a material where the ferrite was transformed to austenite (another metallurgical phase) that was stable at room temperature. Soft stainless steels that were ductile both before and after welding were developed in Sheffield, England (then the heart of the British steel industry) exploiting scientific work undertaken in Germany. This new group of steels was based on an 18% chromium steel to which nickel was added as a second alloying element. These were termed the austenitic stainless steels. The general relationship between chromium and nickel necessary to maintain a fully austenitic structure is shown diagrammatically in Figure 1. It will be seen that the optimum combination is 18% chromium, 8% nickel—hence the terminology 18/8s.

Probably the next major advance in the development of stainless steels was the discovery that relatively small additions of molybdenum had a pronounced effect on the corrosion resistance, greatly enhancing the ability to withstand the effects of mineral acids and other corrodents such as chloride solutions. Needless to say, from these early developments, there has been tremendous growth in production facilities and the number of grades of stainless steel available. Table 1 lists some of the more commonly available grades, while Figure 2 illustrates how the basic 18/8 composition is modified to enhance specific physical or chemical properties.

In spite of the plethora of stainless steels available, grades 304 and 316 have, and continue to be, the workhorses for fabrication of dairy and food processing equipment.

Although 316 stainless steel offers excellent resistance to a wide range of chemical and nonchemical environments, it does not offer immunity to all. In the case of the food industry, these are notably anything containing salt, especially low-pH products. There was, therefore, a demand by industry to develop more corrosion-resistant materials and these are finding increasing use in the food industry for certain specific processing operations.

Super Stainless Steels and Nickel Alloys

The super stainless steels are a group of alloys that have enhanced levels of chromium, nickel, and molybdenum, compared to the conventional 18/8s. The major constituent is still iron; hence the classification under the "steel" title.

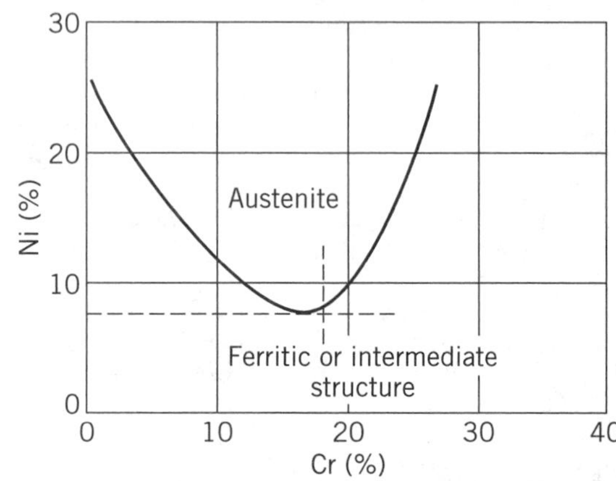

Figure 1. Graph showing the various combinations of chromium and nickel that form austenitic stainless steels. *Source:* Ref. 1.

Table 1. Composition of Some of the More Commonly Used Austenitic Stainless Steels

Alloy	UNS no.	Composition (%)						
		Carbon	Manganese	Silicon	Chromium	Nickel	Molybdenum	Others
302	S30200	0.15	2.00	1.00	16.0–18.0	6.0–8.0		Sulfur 0.030 Phosphorus 0.045
304	S30400	0.08	2.00	1.00	18.0–20.0	8.0–10.5		Sulfur 0.030 Phosphorus 0.045
304L	S30403	0.03	2.00	1.00	18.0–20.0	8.0–12.0		Sulfur 0.030 Phosphorus 0.045
316	S31600	0.08	2.00	1.00	16.0–18.0	10.0–14.0	2.0–3.0	Sulfur 0.030 Phosphorus 0.045
316L	S31603	0.03	2.00	1.00	16.0–18.0	10.0–14.0	2.0–3.0	Sulfur 0.030 Phosphorus 0.045
317	S31700	0.08	2.00	1.00	18.0–20.0	11.0–15.0	3.0–4.0	Sulfur 0.030 Phosphorus 0.045
317L	S31703	0.03	2.00	1.00	18.0–20.0	11.0–15.0	3.0–4.0	Sulfur 0.030 Phosphorus 0.045
321	S32100	0.08	2.00	1.00	17.0–19.0	9.0–12.0		Sulfur 0.030 Ti $\nless$ 5 × Carbon
347	S34700	0.08	2.00	1.00	17.0–19.0	9.0–13.0		Sulfur 0.030 Cb + Ta $\nless$ 10 × Carbon

Note: Unless indicated otherwise, all values are maxima.

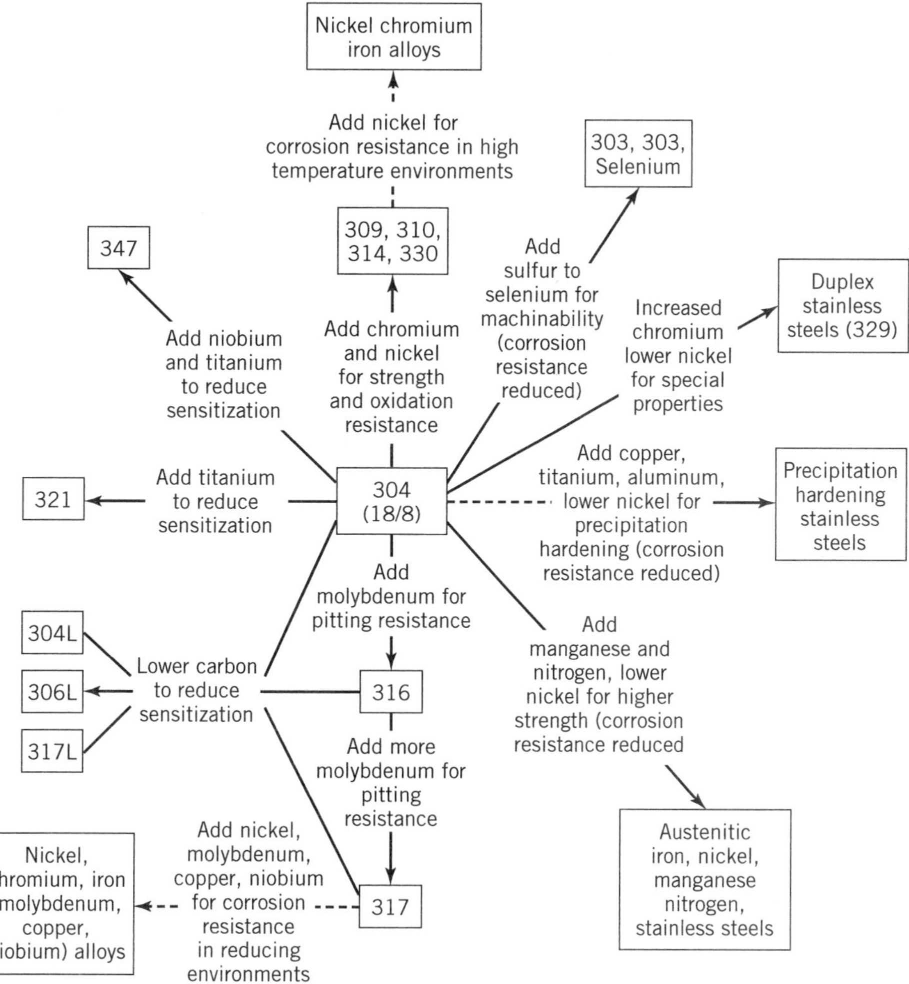

Figure 2. Outline of some compositional modifications of 18/8 austenitic stainless steel to produce special properties. *Source:* Ref. 2.

Still further increases in the three aforementioned alloying elements result in the nickel alloys. (The classification of an alloy is generally under the heading of the major constituent.)

There are a large number of these alloys but those of primary interest to the food industry are shown in Table 2 together with their composition. In general terms, it will be noted that the increase in nickel content is accompanied by an increase in chromium and molybdenum. As stated previously, this element is particularly effective in promoting corrosion resistance.

Just like insurance, you get only what you pay for, and generally speaking, the higher the corrosion resistance, the more expensive the material. In fact, the differential between type 304 stainless steel and a high-nickel alloy may be as much as 20 times, depending on the market prices for the various alloying elements that fluctuate widely with the supply and demand position.

Aluminum

High-purity grades of aluminum (±99.5%) and its alloys still are preferred for some food and pharmaceutical applications because of the reasonable corrosion resistance of the metal. This resistance is attributable to the easy and rapid formation of a thin, continuous, adherent oxide film on exposed surfaces. This oxide film, in turn, exhibits a good corrosion resistance to many foodstuffs, and it is reported that fats, oils, sugar, and some colloids have an inhibitory or sealing effect on these films (3).

As aluminum salts formed by corrosion are colorless, tasteless, and claimed to be nontoxic, the metal is easy to clean, inexpensive, and light and has a high thermal conductivity. It still is used quite extensively in certain areas of food manufacture and distribution. However, in recent years, the claim of nontoxicity is being questioned as a high dietary incidence has been implicated in Alzheimer's disease (senile dementia) with compounds of aluminum (aluminosilicates) being found in the brain tissue of sufferers (4). However, the case is far from proven, and it is not clear if the increased levels of aluminosilicates are due to a high intake of aluminum per se or other factors such as a dietary deficiency of calcium.

For many years, aluminum was used extensively for containment vessels in the diary and brewing industry, and it was Richard Seligman who founded the then Aluminium Plant and Vessel Company (now APV plc) to exploit the technique of welding this material for the fabrication of fermenting vessels in the brewing industry (5). Many of these original vessels are still in use in some of the smaller, privately owned breweries in the United Kingdom.

Although large fermenting vessels and storage tanks now tend to be fabricated from stainless steels, there is still widespread use of aluminum for beer kegs, beer cans, and a miscellany of small-scale equipment where the resistance of aluminum is such that it imparts no change or modification of flavor, even after prolonged storage.

While still used for holding vessels and some equipment when processing cider, wines, and perry, prolonged contact is inadvisable because of the acidity of the sulfites employed as preservatives for these products—inadvisable, that is, unless the surface of the metal has been modified by anodizing or has been protected with a lacquer.

In the manufacture of preserves, aluminum is still employed for boiling pans, the presence of sugar appearing to inhibit any corrosion. In the field of apiculture, it has even been used for making prefabricated honeycombs, which the bees readily accepted.

Extensive use is made of aluminum and the alloys in the baking industry for baking tins, kneading troughs, handling equipment, etc.

In other areas of food manufacture and preparation, the use of aluminum extends virtually over the whole field of activity—butter, margarine, table oils, and edible fats, meat and meat products, fish and shellfish, certain sorts of vinegar, mustards, spices; the list is almost endless.

No mention has so far been made of the application of this metal in the dairy industry, and indeed it still has limited application mostly in the field of packaging, eg, bottle caps, wrapping for cheese, butter, carton caps for yogurt, cream.

It will be appreciated that the uses of aluminum in the food industry so far mentioned have tended to be for equipment used in batch operation, hand utensils, and packaging. There are probably three major factors that have mitigated against its more widespread use, not only in the dairy industry but in brewing and many other branches of food processing.

1. Modern, highly automated plants operating on a continuous or semicontinuous basis employ a wide va-

Table 2. Composition of Some of the More Commonly Used Wrought Super Stainless Steels and Nickel Alloys

		Composition (%)						
Alloy	UNS no.	Carbon	Silicon	Manganese	Chromium	Nickel	Molybdenum	Others
904L	N08904	0.02	0.70	2.0	19.0–21.0	24.0–26.0	4.2–4.7	Cu 1.2–1.7
Avesta 254 SMO	S31254	0.02	0.80	2.0	19.5–20.5	17.5–18.5	6.0–6.5	Cu 0.5–1.0, N 0.18–0.22
Incoloy 825	N08825	0.05	0.50	1.0	19.5–23.5	38.0–46.0	2.5–3.5	Al 0.2, Ti 0.6–1.2
Hastelloy G-30	N06030	0.03	0.08	1.5	28.0–31.5	Bal.	4.0–6.0	Co 5.0, Cu 1.0–2.4, Cb + Ta 0.3–1.5, Fe 13.0–17.0
Inconel 625	N06625	0.10	0.50	0.50	23.0–28.0	Bal.	8.0–10.0	Co 1.0, Fe 5.0, Al 0.4, Ti 0.4
Hastelloy C-276	N10276	0.02	0.08	1.0	14.5–16.5		15.0–17.0	W 3.0–4.5, V 0.35

Note: Unless indicated otherwise, all values are maxima.

riety of materials of construction. Because of the position in the electrochemical series (to be discussed later), aluminum and its alloys are susceptible to galvanic corrosion when coupled with other metals.
2. The commercial availability of stainless steels and their ease of fabrication, strength, ease of maintenance, appearance, and proven track record of reliability.
3. The fact that since modern plants operate on a semicontinuous basis with much higher levels of fouling, cleaning regimes require strongly alkaline detergents to which aluminum has virtually zero corrosion resistance.

Copper and Tinned Copper

Copper and tinned copper were used extensively in former times because of their excellent thermal conductivity (8 times that of stainless steel), ductility, ease of fabrication, and reasonable level of corrosion resistance. However, the demise of copper as a material of construction is largely attributable to the toxic nature of the metal and its catalytic activity in the development of oxidative rancidity in fats and oils. Even at the sub-part-per-million level, copper in vegetable oils and animal fats rapidly causes the development of off-flavors. In equipment where high levels of liquid turbulence are encountered (eg, plate heat exchanger or high-velocity pipe lines) copper is subject to erosion. Nevertheless, there is an area of the beverage industry where copper is still the only acceptable material of construction, ie, pot stills for Scotch and Irish whiskey production. It is also used in the distillation of the spirits such as rum and brandy. Much old copper brewing equipment such as fermenting vessels and wort boilers is still in use throughout the world, and an interesting observation is that even though the wort boilers in modern breweries are fabricated from stainless steel, they are still known as "coppers" and UK craftsmen fabricating stainless steel are still known as coppersmiths.

Titanium

There are certain areas of the food industry, especially in equipment involving heat transfer, where stainless steels are just not capable of withstanding the corrosive effects of salty, low-pH environments. Food processors are increasingly accepting the use of titanium as an alternative, in the full knowledge that it offers corrosion immunity to the more aggressive foodstuffs and provides a long-term solution to what was an on-going problem with stainless steels. Titanium is a light metal, the density of which is almost half that of stainless steels. Although relatively expensive (6–7 times the cost of stainless steel), being a low-density material offsets this price differential for the raw material by almost half. It is ductile and fabricable using normal techniques, although welding it does require a high degree of expertise.

Other Metals

Tin, in the form of tin plate, is used extensively in the canning industry, where its long-term corrosion resistance to a wide range of food acids makes it a material par excellence for this purpose.

Cadmium, used as a protective coating for carbon steel nuts and bolts, was favored at one time. However, the high toxicity of the cadmium compounds has come under increasing scrutiny from many health regulatory bodies and now cadmium-plated bolting is not permitted in food factories. Indeed, Denmark and Sweden have totally banned the import of cadmium-plated components into their countries, and many other countries are likely to follow suit.

Lead and lead-containing products are generally not acceptable for food contact surfaces, although some codes of practice permit the use of lead-containing solder for capillary pipeline joints on water supplies and service lines.

SELECTING MATERIALS OF CONSTRUCTION

Designing equipment is a multidiscipline exercise involving mechanical engineers, materials–corrosion engineers, stressing experts, draftsmen, etc. The corrosion engineer has an important role in this team effort, namely, to ensure the materials specified will offer a corrosion resistance that is just adequate for all the environmental conditions likely to be encountered during normal operation of the equipment. A piece of equipment that prematurely fails by corrosion is as badly designed as one in which the materials have been overspecified. Unfortunately, all too often the functional requirements for a piece of equipment are analyzed in a somewhat arbitrary manner and all too often, the basic cost of the material tends to outweigh other equally important considerations (6).

Figure 3 shows the primary criteria that must be considered in the initial selection process.

- *Corrosion Resistance.* For any processing operation, there will be a range of materials that will offer a corrosion resistance that is adequate (or more than adequate) for a particular job. When considering corrosion resistance, the operational environment is the obvious one, but the other point must be whether the material will also offer corrosion resistance to the chemicals used for cleaning and sanitizing.
- *Cost.* Many of the materials originally considered will be eliminated on the grounds of their high cost. For example, there is no point in considering a high-nickel alloy when a standard 300 series stainless steel at a lower cost will be perfectly satisfactory.

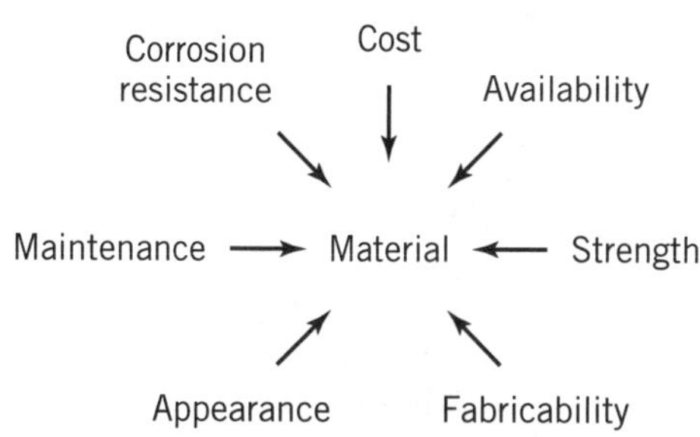

Figure 3. Materials selection criteria.

- *Availability*. Availability is a less obvious feature of the material selection process. Many steel producers will require a minimum order of, say, 3 tons for a nonstandard material. Clearly, the equipment manufacturer is not going to buy this large quantity when the job that is to be done may only require the use of one ton of material.
- *Strength*. Strength is a factor that is taken into account at the design stage but, as with all the others, cannot be considered in isolation. For example, many of the new stronger stainless steels, although more expensive on a ton-for-ton basis than conventional stainless steels, are less expensive when considered on a strength/cost ratio.
- *Fabricability*. There is little point in considering materials that are either unweldable (or unfabricable) or can be welded only under conditions more akin to a surgical operating theater than a general engineering fabrication shop.
- *Appearance*. Appearance may or may not be an important requirement. Equipment located outside must be resistant to environmental weathering and therefore may require the application of protective sheathing, which could double the basic material cost.
- *Maintenance*. Is the equipment to be essentially maintenance-free or is some maintenance, such as periodic repainting, tolerable? How long will the equipment operate without the need for major servicing?

When all these interrelated criteria have been considered, the long list of possible starters will have been reduced to maybe one or two. Also, somewhere through the selection process some of those materials initially rejected because, for example, of their high cost, may have to be reconsidered because of other factors.

TYPES OF CORROSION

Defining Corrosion

Before embarking on a discussion of the various forms of corrosion, it is worthwhile considering exactly what corrosion is. There are several definitions of corrosion. For example, Fontant (7) defines it as extractive metallurgy in reverse using the diagram, shown in Figure 4, to illustrate the point.

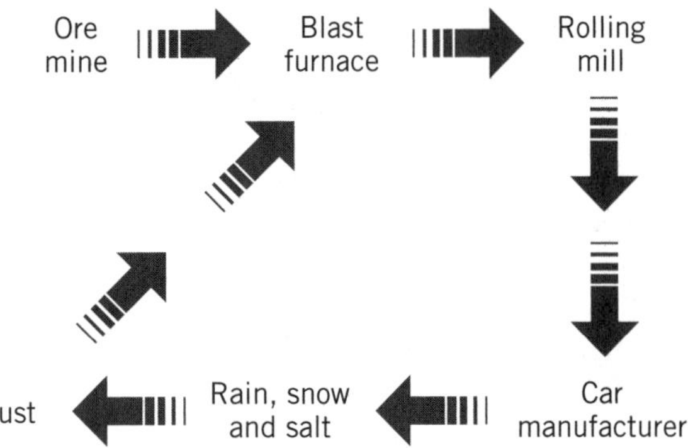

Figure 4. Extractive metallurgy in reverse.

A more general and descriptive definition is "it is the deterioration or destruction of a material through interaction with its environment." This covers all materials of construction including rubber and plastics as well as metal. However, the primary object of this article is to deal with corrosion of metals, in particular stainless steels, and how this corrosion can be classified.

There are two basic forms of corrosion—wet corrosion and dry corrosion. Dry corrosion is concerned with the oxidation of metals at high temperature and clearly outside the scope of this text. Wet corrosion occurs in aqueous solutions or in the presence of electrolytes and is an electrochemical process. It should be noted that the "aqueous" component of the system may be present in only trace quantities (eg, present as moisture); the classical example is the corrosion of steel by chlorine gas. In fact, steel is not corroded by chlorine since steel is the material used for storing liquid chlorine. However, in the presence of even trace quantities of moisture, chlorine rapidly attacks steel and, for that matter, most metals.

The corrosion of metals involves a whole range of factors. These may be chemical, electrochemical, biological, metallurgical, or mechanical, acting singly or conjointly. Nevertheless, the main parameter governing corrosion of metals is related to electrochemistry. Electrochemical principles therefore are the basis for a theoretical understanding of the subject. In fact, electrochemical techniques are now the standard method for investigating corrosion although the standard "weight loss" approach still provides invaluable data. It is not proposed to discuss in depth the electrochemical nature of corrosion but should further information be required, several excellent texts are available (7,8).

Forms of Corrosion

Wet corrosion can be classified under any of eight headings, namely:

- Galvanic or bimetallic corrosion
- Uniform or general attack
- Crevice corrosion
- Pitting corrosion
- Intergranular corrosion
- Stress corrosion cracking
- Corrosion fatigue
- Selective corrosion (castings and free-matching stainless steels)

Galvanic Corrosion. When two dissimilar metals (or alloys) are immersed in a corrosive or conductive solution, an electrical potential or potential difference usually exists between them. If the two metals are electrically connected, then, because of this potential difference, a flow of current occurs. As the corrosion process is an electrochemical phenomenon and dissolution of a metal involves electron flow, the corrosion rates for the two metals is affected. Generally, the corrosion rate for the least corrosion resistant is enhanced while that of the more corrosion resistant is diminished. In simple electrochemical terms, the least resistant metal has become anodic and the more resistant cathodic. This, then, is galvanic or dissimilar metal corrosion.

The magnitude of the changes in corrosion rates depends on the so-called electrode potentials of the two metals; the greater the difference, the greater the enhancement or diminution of the corrosion rates. It is possible to draw up a table of some commercial alloys that ranks them in order of their electrochemical potential. Such a table is known as the galvanic series. A typical one as shown in Table 3 is based on work undertaken by the International Nickel Company (now INCO Ltd.) at their Harbor Island, NC, test facility. This galvanic series relates to tests in unpolluted seawater, although different environments could produce different results and rankings. When coupled, individual metals and alloys from the same group are unlikely to show galvanic effects that will cause any change in their corrosion rates.

The problem of dissimilar metal corrosion, being relatively well understood and appreciated by engineers, is usually avoided in plant construction and in the author's experience, few cases have been encountered. Probably the most common form of unintentional galvanic corrosion is on service lines where brass fittings are used on steel pipelines—the steel suffering an increase in corrosion rate at the bimetallic junction.

One of the worst bimetallic combinations is aluminum and copper. An example of this is in relation to aluminum milk churns used to transport whey from Gruyere cheese manufacture (in Switzerland), where copper is used for the cheesemaking vats and the whey picks up traces of this metal. The effect on the aluminum churns which are internally protected with lacquer that gets worn away through mechanical damage is pretty catastrophic.

Another, somewhat unique, example of galvanic corrosion is related to a weld repair on a 304 stainless-steel storage vessel. Welding consumables containing molybdenum had been employed to effect the repair, and although it is most unusual for the potential difference between molybdenum and nonmolybdenum containing stainless steels to be sufficient to initiate galvanic corrosion, the environmental factors in this particular case were obviously such that corrosion was initiated (Fig. 5). As stated, this is somewhat unique and it is not uncommon for 316 welding consumables to be used for welding 304 stainless steel with no adverse effects. As a practice, however, it is to be deprecated and the correct welding consumables should always be employed.

Not all galvanic corrosion is bad; indeed, galvanic corrosion is used extensively to protect metal and structures by the use of a sacrificial metal coating. A classic example is the galvanizing of sheet steel and fittings, the zinc coating being applied not so much because it does not corrode, but because it does. When the galvanizing film is damaged, the zinc galvanically protects the exposed steel and inhibits rusting. Similarly, sacrificial anodes are fitted to domestic hot water storage tanks to protect the tank.

Table 3. The Galvanic Series of Some Commercial Metals and Alloys in Clean Seawater

	Platinum
	Gold
	Graphite
	Titanium
Noble or cathodic	Silver
	Chlorimet 3 (62 Ni, 18 Cr, 18 Mo)
	Hastelloy C (62 Ni, 17 Cr, 15 Mo)
	18/8 Mo stainless steel (passive)
	18/8 stainless steel (passive)
	Chromium stainless steel 11–30% Cr (passive)
	Inconel (passive) (80 Ni, 13 Cr, 7 Fe)
	Nickel (passive)
	Silver solder
	Monel (70 Ni, 30 Cu)
	Cupronickels (60–90 Cu, 40–10 Ni)
	Bronzes (Cu-Sn)
	Copper
	Brasses (Cu-Zn)
	Chlorimet 2 (66 Ni, 32 Mo, 1 Fe)
	Hastelloy B (60 Ni, 30 Mo, 6 Fe, 1 Mn)
	Inconel (active)
	Nickel (active)
	Tin
	Lead
	Lead-tin solders
	18/8 Mo stainless steel (active)
	18/stainless steel (active)
	Ni-resist (high-Ni cast iron)
	Chromium stainless steel, 13% Cr (active)
	Cast iron
Active or anodic	Steel or iron
	2024 aluminum (4.5 Cu, 1.5 Mg, 0.6 Mn)
	Cadmium
	Commercially pure aluminum (1100)
	Zinc
	Magnesium and magnesium alloys

Uniform or General Attack. As the name implies, this form of corrosion occurs more or less uniformly over the whole surface of the metal exposed to the corrosive envi-

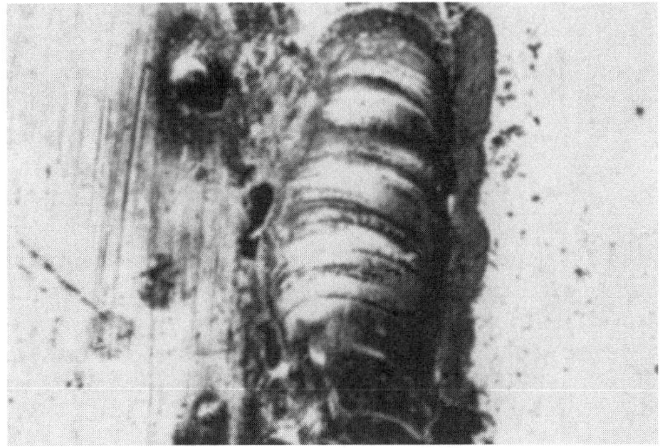

Figure 5. Galvanic corrosion of 304 stainless steel initiated by a 316 weld deposit. Note the large pit associated with the weld splatter.

ronment. It is the most common form of corrosion encountered with the majority of metals, a classic example being the rusting of carbon steel. Insofar as the corrosion occurs uniformly, corrosion rates are predictable and the necessary corrosion allowances built into any equipment. In the case of stainless steels, this form of corrosion is rarely encountered. Corrodents likely to produce general attack of stainless steel are certain mineral acids, some organic acids, and high-strength caustic soda at concentrations and temperatures well in excess of those ever likely to be found in the food industry. The same remark applies to cleaning acids such as nitric, phosphoric, and citric acids, but not for sulfuric or hydrochloric acids, both of which can cause rapid, general corrosion of stainless steels. Hence, they are not recommended for use, especially where corrosion would result in a deterioration of the surface finish of process equipment.

The behavior of both 304 and 316 stainless steels when subjected to some of the more common acids that are encountered in the food industry is graphically illustrated in Figure 6. These isocorrosion graphs, ie, lines that define the conditions of temperature and acid concentration that will produce a constant corrosion rate expressed in mils (0.001 in.) or mm loss of metal thickness per year, are used extensively by corrosion engineers in the material selection process when the form of corrosion is general attack. *They are of no value whatsoever when the corrosion mode is one of the other forms that will be defined, such as pitting or crevice corrosion.*

Crevice Corrosion. This form of corrosion is an intense local attack within crevices or shielded areas on metal surfaces exposed to corrosive solutions. It is characteristically encountered with metals and alloys that rely on a surface oxide film for corrosion protection, eg, stainless steels, titanium, aluminum.

The crevices can be inherent in the design of the equipment (eg, plate heat exchangers) or inadvertently created by bad design. Although crevice corrosion can be initiated at metal-to-metal surfaces (see Fig. 7), it is frequently encountered at metal to nonmetallic sealing faces. Any nonmetallic material that is porous and used, for example,as a gasket, is particularly good (or bad!) for initiating this form of attack. Fibrous materials that have a strong wicking action are notorious in their ability to initiate crevice attack. Similarly, materials that have poor stress relaxation characteristics, ie, have little or no ability to recover their original shape after being deformed, are also crevice creators, as are materials that tend to creep under the influence of applied loads and/or at elevated temperatures. Although used for gasketing, PTFE suffers both these deficiencies. On the other hand, elastomeric materials are particularly good insofar as they exhibit elastic recovery and have the ability to form a crevice-free seal. However, at elevated temperatures, many rubbers harden and in this condition, suffer the deficiencies of many nonelastomeric gasketing materials.

Artificial crevices can also be created by the deposition of scale from one of the process streams to which the metal is exposed. It is necessary, therefore, to maintain food processing equipment in a scale-free condition, especially on surfaces exposed to service fluids such as services side hot/cold water, cooling brines, which tend to be overlooked during plant cleaning operations.

Much research work has been done on the geometry of crevices and the influence of this on the propensity for the initiation of crevice corrosion (9). However, in practical terms, crevice corrosion usually occurs in openings a few tenths of a millimeter or less and rarely is encountered where the crevice is greater than 2 mm (0.08 in.).

Until the 1950s, crevice corrosion was thought to be due to differences in metal ion or oxygen concentration within the crevice and its surroundings. While these are factors in the initiation and propagation of crevice corrosion, they are not the primary cause. Current theory supports the view that through a series of electrochemical reactions and the geometrically restricted access into the crevice, migration of cations, chloride ions in particular, occurs. This alters the environment within, with a large reduction in pH and an increase in the cations by a factor of as much as 10. The pH value can fall from a value of, say, 7 in the surrounding solution to as low as pH 2 within the crevice. As corrosion is initiated, it proceeds in an autocatalytic manner with all the damage and metal dissolution occurring within the crevice and little or no metal loss outside. The confined and autocatalytic nature of crevice corrosion results in significant loss of metal under the surface of site of initiation. As a result, deep and severe undercutting of the metal occurs (see Fig. 8). The time scale for initiation of crevice corrosion can vary from a few hours to several months and, once initiated, can progress very rapidly. Stopping the corrosion process can be extremely difficult as it is necessary to remove all the trapped reactants and completely modify the occluded environment. The difficulty of attaining this will be appreciated by reference to Figure 8, where the entrance to the corroded region is only 0.5 mm (0.020 in.).

While methods for combating the onset of crevice corrosion can be deduced from the foregoing text, a reiteration of some of the more important precautions is not out of place, viz:

- Good-quality, crevice-free welded joints are always preferable to bolted joints
- Good equipment design (well-designed gasket sealing faces) that avoids unintentional crevices and does not permit the development of stagnant regions
- Frequent inspection of equipment and removal of surface deposits
- Use of good-quality rubber gaskets rather than absorbent packings
- Good gasket maintenance; replacement when hardened or damaged

However, certain pieces of equipment are by virtue of their design highly creviced. In such cases, it is necessary to recognize the potential corrosion risk and select the materials of construction that will resist the initiation of crevice corrosion by the environment. Similarly, cleaning and sanitizing regimes must be developed to avoid the onset of attack.

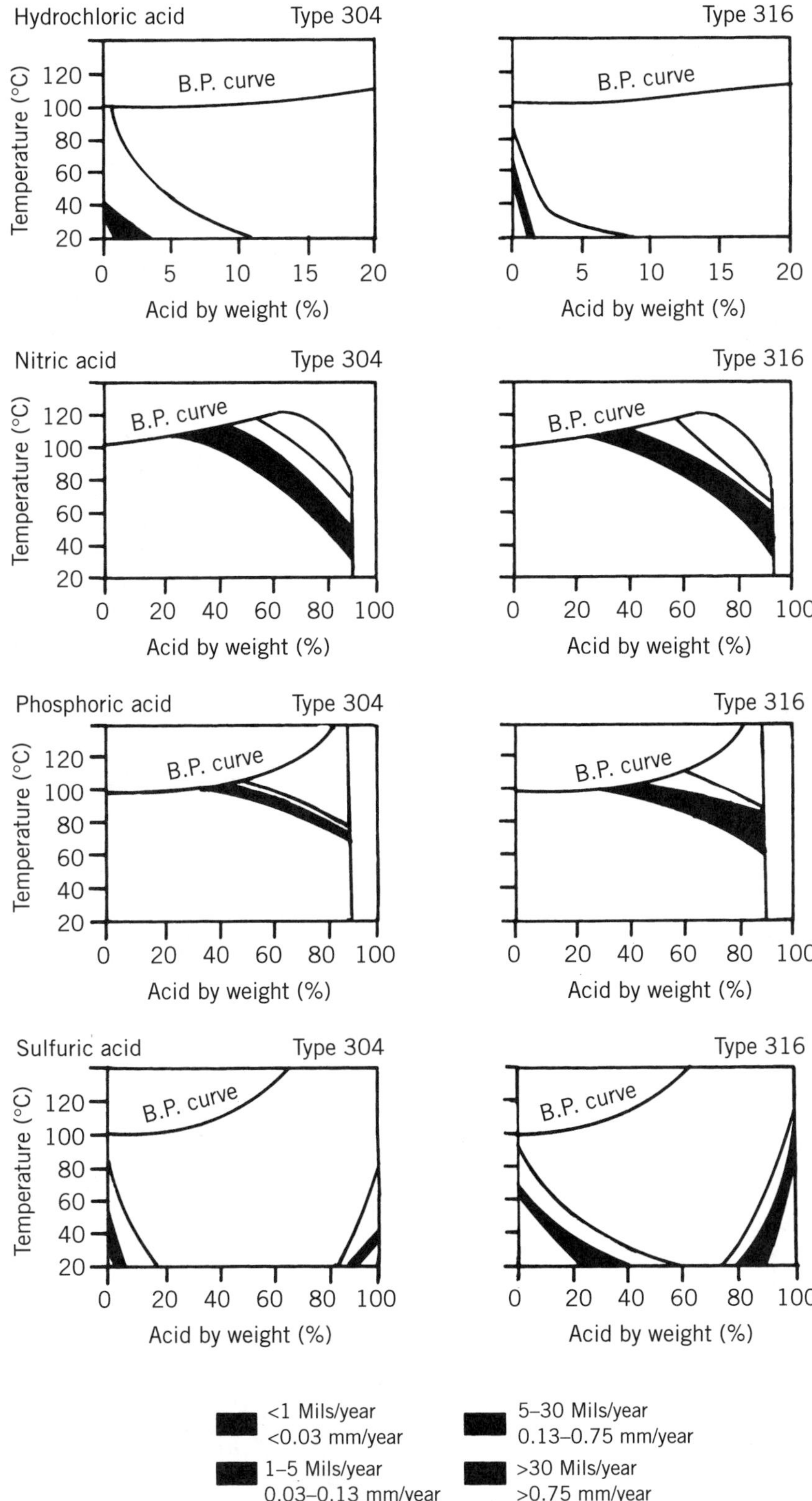

Figure 6. Corrosion resistance of 304 and 316 stainless steels to mineral acids. *Source:* Reproduced by permission of British Steel Plc.

In the case of stainless steels, although there are several ionic species that will initiate the attack, by far the most common are solutions containing chloride. The presence of salt in virtually all foodstuffs highlights the problem. Low pH values also enhance the propensity for initiation of attack.

Other environmental factors such as temperature and the oxygen or dissolved air content of the process stream all play a role in the corrosion process.

Because the presence of oxygen is a prerequisite for the onset of crevice corrosion (and many other forms of attack), in theoretical terms complete removal of oxygen from a

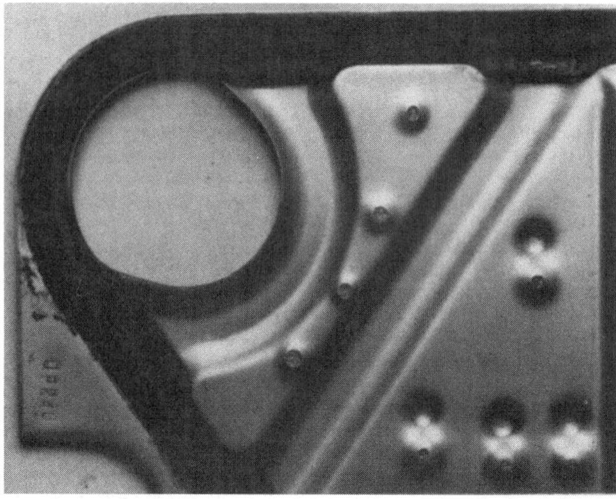

Figure 7. Crevice corrosion at the interplate contact points of a heat-exchanger plate.

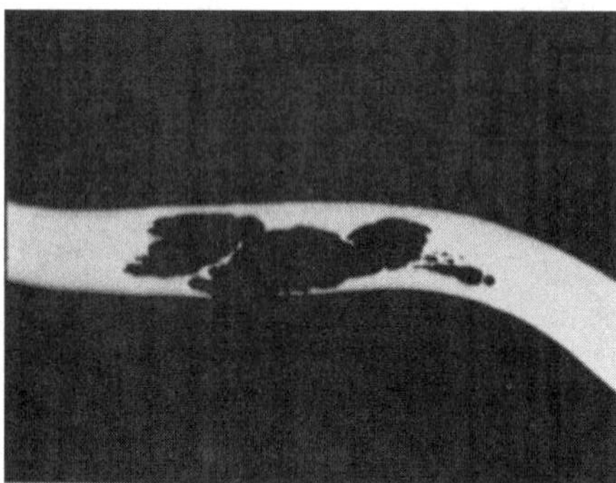

Figure 8. Photomicrograph of a section through a site of crevice corrosion. Note the deep undercutting which is typical of chloride-induced attack on stainless steel.

process stream will inhibit corrosion. In practice, however, this is difficult to achieve. Only in equipment where complete and effective deaeration occurs, such as a multiple effect evaporator operating under reduced pressure, will the beneficial effect of oxygen removal be achieved.

Stainless steels containing molybdenum (316, 317) have a much higher resistance to crevice corrosion than do alloys without this element (304, 321, 347). The higher the molybdenum content, the greater the corrosion resistance. For particularly aggressive process streams, titanium is often the only economically viable material to offer adequate corrosion resistance.

Pitting Corrosion. As the name implies, pitting is a form of corrosion that leads to the development of pits on a metal surface. It is a form of extremely localized but intense attack, insidious insofar as the actual loss of metal is negligible in relation to the total mass of metal that may be affected. Nevertheless, equipment failure by perforation is the usual outcome of pitting corrosion. The pits can be small and sporadically distributed over the metal surface (Fig. 9) or extremely close together, close enough, in fact, to give the appearance of the metal having suffered from general attack.

In the case of stainless steels, environments that will initiate crevice corrosion will also induce pitting. As far as the food industry is concerned, it is almost exclusively caused by chloride containing media, particularly at low pH values.

Many theories have been developed to explain the cause of initiation of pitting corrosion (10), and the one feature they have in common is that there is a breakdown in the passive oxide film. This results in ionic migration and the development of an electrochemical cell. There is, however, no unified theory that explains the reason for the film breakdown. Evans (11) for example, suggests that metal dissolution at the onset of pitting may be due to a surface scratch, an emerging dislocation or other defects, or random variations in solution composition. However, propagation of the pit proceeds by a mechanism similar to that occurring with crevice corrosion. Like crevice corrosion, the pits are often undercut and on vertical surfaces may assume an elongated morphology due to gravitational effects (Fig. 10).

The onset of pitting corrosion can occur in a matter of days but frequently requires several months for the development of recognizable pits. This makes the assessment of the pitting propensity of a particular environment very difficult to determine, and there are no short-cut laboratory testing techniques available. Methods and test solutions are available to rank alloys; the best known and most frequently quoted is ASTM Standard G48 (12), which employs 6% ferric chloride solution. Another chemical method involving ferric chloride determines the temperature at

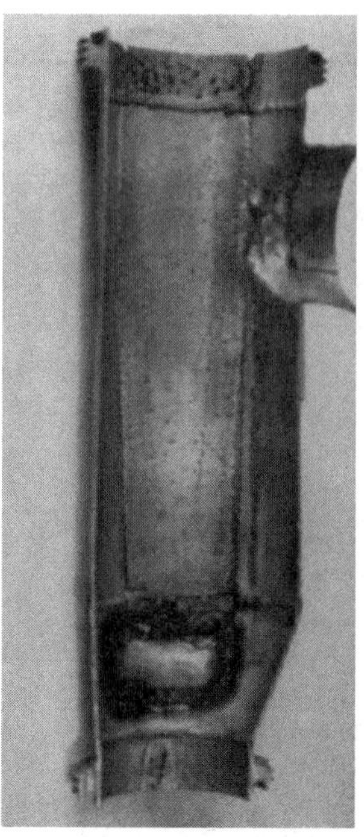

Figure 9. Pitting corrosion of a stainless-steel injector caused by the presence of hydrochloric acid in the steam supply.

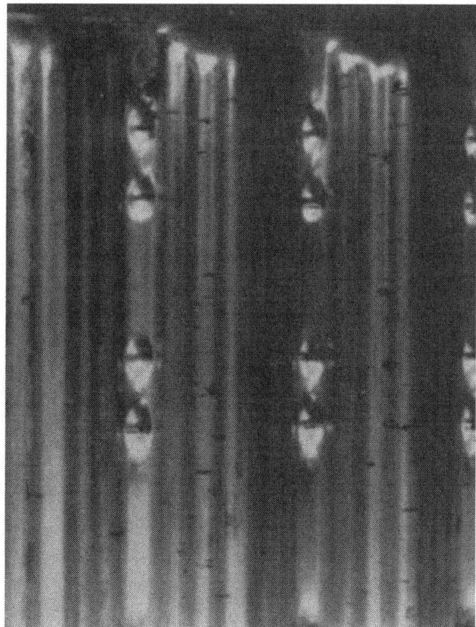

Figure 10. Elongated pitting attack on a 316 stainless-steel heat exchanger plate.

which the solution will cause pitting within a 24-h test period, the results being expressed as the critical pitting temperature or CPT (13). However, as stated, both of these methods are used to rank the susceptibility of a range of alloys rather than define the performance of a material in a service environment. Electrochemical methods have also been used.

As with crevice corrosion, alloy composition has a profound effect on the resistance of a material to pitting attack. Greene and Fontana (14) summarized the effect of various elements as shown in Table 4.

Intergranular Corrosion. A fact not often appreciated is that metals and alloys have a crystalline structure. However, unlike crystalline solids such as sugar or salt, metallic crystals can be deformed or bent without fracturing; in other words, they are ductile. In the molten state, the atoms in a metal are randomly distributed but on cooling and solidification, they become arranged in crystalline form. Because crystallization occurs at many points in the solidification process, these crystals or grains are randomly orientated and the region where they meet are grain boundaries. In thermodynamic terms, the grain boundaries are more susceptible to corrosion attack because of their higher free energy, although in practice the difference in free energy of the grain boundaries and the main crystals or grains in a homogeneous alloy are too small to be significant. However, when the metal or alloy has a heterogeneous structure, preferential attack at or adjacent to the grain boundaries can occur. This is intergranular corrosion as shown in Figure 11.

When austenitic stainless steels are heated to and held in the temperature range of 600–900°C (1100–1650°F), the material becomes sensitized and susceptible to grain boundary corrosion. It is generally agreed that this is due to chromium combining with carbon to form chromium carbide, which is precipitated at the grain boundaries. The net effect is that the metal immediately adjacent to the grain boundaries is denuded of chromium and instead of having a composition of, say, 18% chromium and 8% nickel, it may assume an alloy composition where the chromium content is reduced to 9% or even lower. As such, this zone depleted in chromium bears little similarity to the main metal matrix and has lost one of the major alloying elements on which it relied for its original corrosion resistance. Indeed, the lowering of corrosion resistance in this zone is so great that sensitized materials are subject to attack by even mildly corrosive environments.

As supplied from the steel mills, stainless steels are in the so-called solution annealed condition, ie, the carbon is in solution and does not exist as grain boundary chromium carbide precipitates. During fabrication where welding is involved, the metal adjacent to the weld is subjected to temperatures in the critical range (600–900°C/1100–1650°F) where sensitization can occur. As such, therefore, this zone may be susceptible to the development of intergranular carbide precipitates. Because the formation of chromium carbides is a function of time, the longer the dwell time in the critical temperature zone, the greater the propensity for carbide formation. Hence, the problem is greatest with thicker metal sections due to the thermal mass and slow cooling rate.

By heating a sensitized stainless steel to a temperature of 1050°C (1950°F), the carbide precipitates are taken into solution and by rapidly cooling or quenching the steel from this temperature, the original homogeneous structure is reestablished and the original corrosion resistance restored.

The first stainless steels were produced with carbon contents of up to 0.2% and as such, were extremely susceptible to sensitization and in-service failure after welding. In consequence, the carbon levels were reduced to 0.08%, which represented the lower limit attainable with steel making technology then available. Although this

Table 4. The Effect of Alloying on Pitting Resistance of Stainless-Steel Alloys

Element	Effect on pitting resistance	Element	Effect on pitting resistance
Chromium	Increases	Molybdenum	Increases
Nickel	Increases	Nitrogen	Increases
Titanium/Niobium	No effect in media other than ferric chloride	Sulfur (and selenium)	Decreases
Silicon	Decrease or increase depending on the absence or presence of molybdenum	Carbon	Decreases if present as grain boundary precipitates

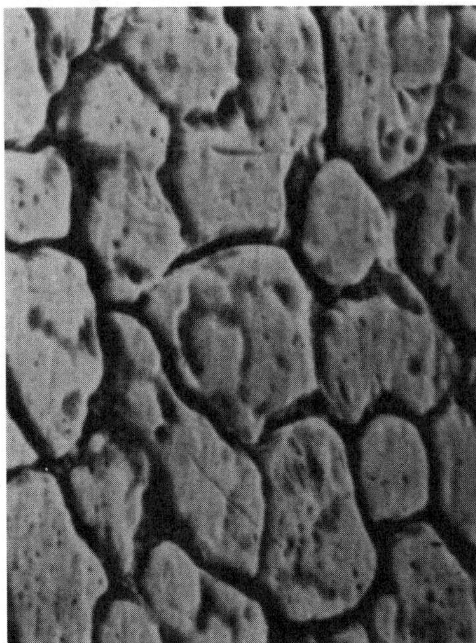

Figure 11. Scanning electron micrograph of the surface of sensitized stainless steel showing preferential attack along the grain boundaries.

move alleviated the problem, it was not wholly successful, particularly when welding thicker sections of the metal. Solution annealing of the fabricated items was rarely a practical proposition and there was a need for a long-term solution. It was shown that titanium or niobium (columbium) had a much greater affinity for the carbon than chromium and by additions of either of these elements, the problem was largely overcome. The titanium or niobium carbides that are formed remain dispersed throughout the metal structure rather than accumulating at the grain boundaries.

Grade 321 is a type 304 (18 Cr, 8 Ni) with titanium added as a stabilizing element, while grade 347 contains niobium. By far, the most commonly used is 321, grade 347 being specified for certain chemical applications.

Modern steelmaking techniques such as AOD (air-oxygen decarburization) were developed to reach even lower levels of carbon, typically less than 0.03%, to produce the "L" grades of stainless steel. These are commercially available and routinely specified where no sensitization can be permitted. With these advances in steel making technology, even the standard grades of stainless steels have typically carbon levels of 0.04–0.05% and, generally speaking, are weldable without risk of chromium carbide precipitates at metal thicknesses of up to 6 mm (1/4 in.). Above this figure or where multipass welding is to be employed, the use of a stabilized or "L" grade is always advisable.

Stress Corrosion Cracking. One of the most insidious forms of corrosion encountered with the austenitic stainless steels is stress corrosion cracking (SCC). The morphology of this type of failure is invariably a fine filamentous crack that propagates through the metal in a transgranular mode. Frequently, the crack is highly branched as shown in Figure 12, although sometimes it can assume a single-crack form. Factors such as metal structure, environment, and stress level have an effect on crack morphology. The disturbing feature of SCC is that there is virtually no loss of metal and frequently, it is not visible by casual inspection and is only apparent after perforation occurs. Some claim that as much as 50% of the failures of stainless steel are attributable to this cause.

Another characteristic of SCC in stainless steels is that once detected, repair by welding is extremely difficult. Crack propagation frequently occurs below the surface of the metal, and any attempt to weld repair results in the crack opening up and running ahead of the welding torch. The only practical method of achieving a satisfactory repair is to completely remove the affected area with a 15–25 cm (6–9 in.) allowance all around the area of visible damage and replace the section. Even then, there is no guarantee that the damaged zone has been entirely removed. In most cases, there are three prerequisites for the initiation of SCC.

- *Tensile Stress*. This may be either residual stress from fabricating operations or applied through the normal operating conditions of the equipment. Furthermore, it has been observed that a corrosion pit can act both as a stress raiser and a nucleation site for SCC.

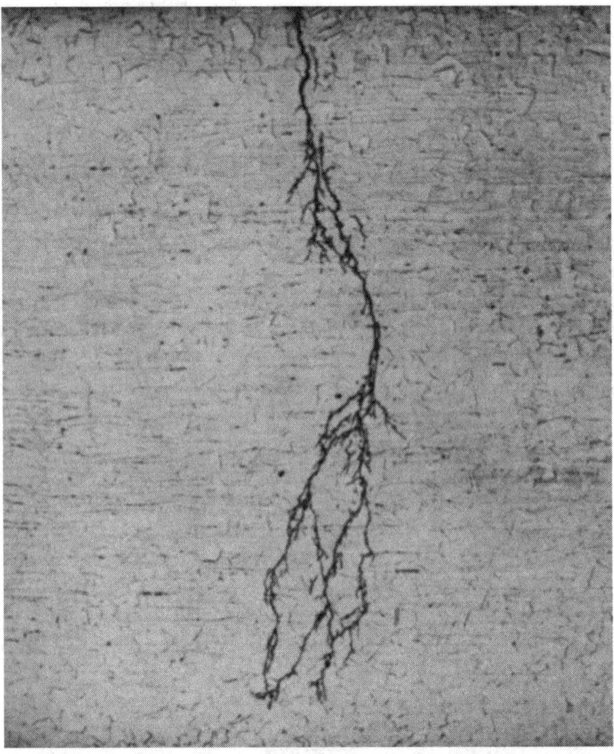

Figure 12. Photomicrograph of a typical stress corrosion crack showing its highly branched morphology and transgranular propagation.

- *Corrosive Species.* Although there are a number of ionic compounds that will act as the corrodent, in the food industry this invariably is the chloride ion. High-strength caustic soda at elevated temperatures will also induce SCC, but the concentrations and temperatures required are well in excess of those ever likely to be encountered. Furthermore, the crack morphology is inter- rather than transgranular. pH also plays a role and generally speaking, the lower the pH the greater the propensity for SCC.
- *Temperature.* It generally is regarded by many that a temperature in excess of 60°C (140°F) is required for this type of failure, although the author has seen examples occurring at 50°C (122°F) in liquid glucose storage vessels.

In the absence of any one of these prerequisites, the initiation of SCC is eliminated. Therefore, it is worth considering the practical approach to its elimination from equipment.

Figure 13 is a diagrammatic representation of the effect of stress on "time-to-failure." As will be seen, by reducing the stress level below a certain critical point, the "time-to-failure" can be increased by several orders of magnitude. On small pieces of equipment, residual stress from manufacturing operations can be removed by stress-relief annealing, which, in the case of stainless steels, is the same as solution annealing. For large pieces of equipment such as storage vessels this approach is clearly impractical. Applied stress is very much a function of the operational conditions of the equipment and only by reducing the stress level by increasing the thickness of the metal can this be reduced. However, this too is a somewhat impractical and uneconomic approach. Some (15) claim that by placing the surface of the metal under compressive rather than tensile stress, by shot peening with glass beads, the problem of SCC can be minimized or eliminated. This too is not a practical proposition for many items of food processing equipment.

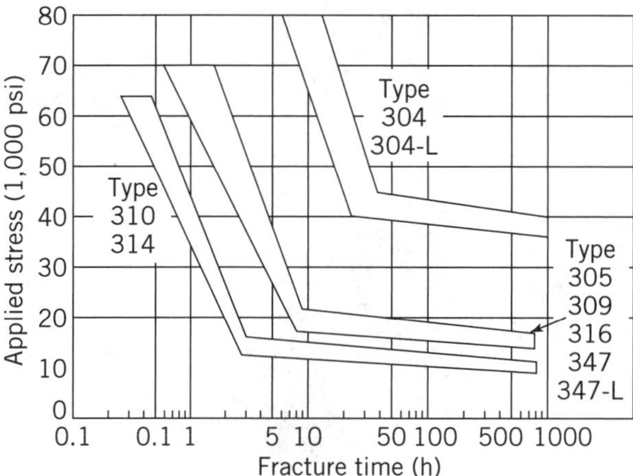

Figure 13. Composite curves illustrating the relative resistance to stress corrosion cracking of some commercial stainless steel in a specific test solution.

As for the matter of corrosive species, it is questionable if anything can be done about elimination. With foodstuffs, for example, this invariably will be the chloride ion, a naturally occurring or essential additive.

Similarly, little can be done in respect to the temperature as this is going to be essential to the processing operation.

As shown by Copson (16), the tendency for iron-chromium-nickel alloys to fail by SCC in a specific test medium (boiling 42% magnesium chloride solution) is related to the nickel content of the alloy. Figure 14 shows this effect, and it is unfortunate that stainless steels with a nominal 10% nickel have the highest susceptibility to failure. Increasing the nickel content of the alloy results in a significant increase in the time-to-failure, but of course, this approach incurs not only the increased cost of the nickel but also the added penalty of having to increase the chromium to maintain a balanced metallurgical structure. The more effective approach is by reducing the nickel content of the alloy.

A group of stainless steels have been developed that exploit this feature, and although their composition varies from producer to producer, they have a nominal composition of 20/22% chromium, 5% nickel. Molybdenum may or may not be present, depending on the environment for which the alloy has been designed. These alloys differ from the austenitic stainless steels insofar as they contain approximately 50% ferrite; hence their designation, austenitic-ferritic, or more commonly, duplex stainless steels. It is only with the advent of modern steel making technology, particularly in relation to the lower carbon levels that can be achieved, that these alloys have become a

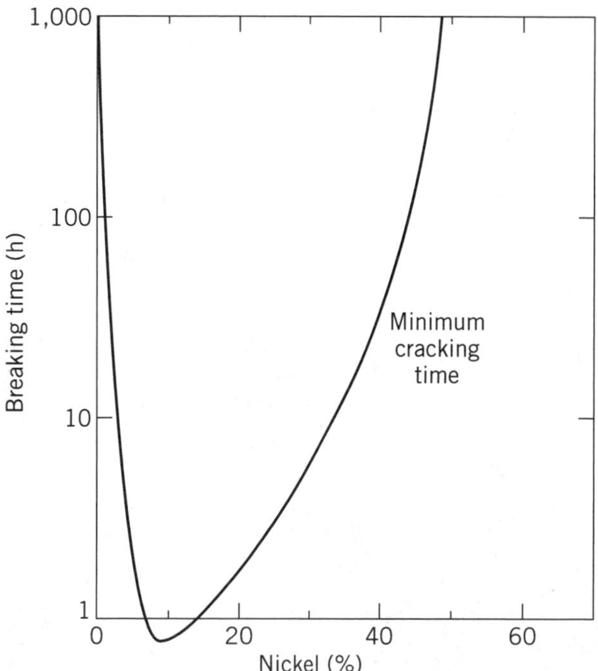

Figure 14. Stress corrosion cracking of iron, chromium, nickel alloys—the Copson curve. Data points have been omitted for clarity.

commercially viable proposition. Because of their low carbon content, typically 0.01–0.02%, the original problems associated with welding ferritic stainless steels and chromium carbide precipitation have been overcome. The alloys are almost twice as strong as the austenitic stainless steels and are ductile and weldable. From a general corrosion standpoint, they are comparable with, or marginally superior to, their 300 series equivalent; but from an SCC standpoint in test work and from field experience, they offer a resistance orders of magnitude better.

Also now available are fully ferritic stainless steels such as grade 444, which contains 18% chromium and 2% molybdenum. This alloy contains carbon at the 0.001% level and therefore does not suffer the problems of welding that were encountered with the original ferritic steels. Furthermore, stabilizing elements such as titanium and niobium are also alloying additions that minimize the tendency for intergranular chromium carbide formation. The main disadvantage of these materials is their susceptibility to grain growth during welding (Fig. 15), which makes them extremely sensitive to fracture even at room temperature. Welding sections thicker than 3 mm (1/8 in.) are not regarded as a practical proposition, and, therefore, their use tends to be limited to tubing.

Corrosion Fatigue. Fatigue is not a form of corrosion in the accepted sense as there is no loss of metal, but can be associated with other forms of localized attack. Because pure fatigue is an *in vacuo* phenomenon, a more correct term is corrosion fatigue or environmental cracking, which is the modern expression and takes into account cracking where the corrosive factor has played a major role on the crack morphology.

The primary cause of corrosion fatigue is the application of fluctuating pressure loads to components that, while of adequate design to withstand normal operating pressures eventually fail under the influence of cyclic loading. The components can be of extremely rigid construction such as a homogenizer block or of relatively light construction such as pipework. There are many potential sources of the fluctuating pressure, the most common of which are positive-displacement pumps (eg, homogenizer or metering pumps), rapid-acting on-off valves that will produce transient pressure peaks, frequent stop-start operations, dead-ending of equipment linked to a filling machine, etc.

Generally speaking, fatigue cracks are straight, without branching and without ductile metal distortion of the material adjacent to the crack. The one characteristic of fatigue cracks is that the crack face frequently has a series of conchoidal markings that represent the stepwise advance of the crack front (Fig. 16). Although, as stated, corrosion fatigue cracks are generally straight and unbranched, where fatiguing conditions pertain in a potentially corrosive environment, the influence of the corrosive component may be superimposed on the cracks. This can lead to branching of the cracks, and in the extreme case, the crack may assume a highly branched morphology which is almost indistinguishable from stress corrosion cracking. It is only when the crack face is examined under high-power magnification that it is possible to categorize the failure mode. An example of this is shown in Figure 17, which has all the features of stress corrosion cracking but the scanning electron micrograph (Fig. 18) clearly shows the stepwise progression of the crack front.

The site for initiation of corrosion fatigue is frequently a discontinuity in the metal section of the component. This may be, for example, a sharp change in diameter of a shaft where inadequate radiusing of the diametric change results in a high cycle stress level through shaft rotation and thus an initiation point for fatigue. A corrosion pit, which under cyclic loading will have a high stress-level association, can also act as a nucleation site (epicenter) for fatigue failure. Because of the corrosive element, it is sometimes difficult to establish whether the pit and associated corrosion were the initiating mechanism for the cracking or the result of corrosion superimposed on a crack, the frac-

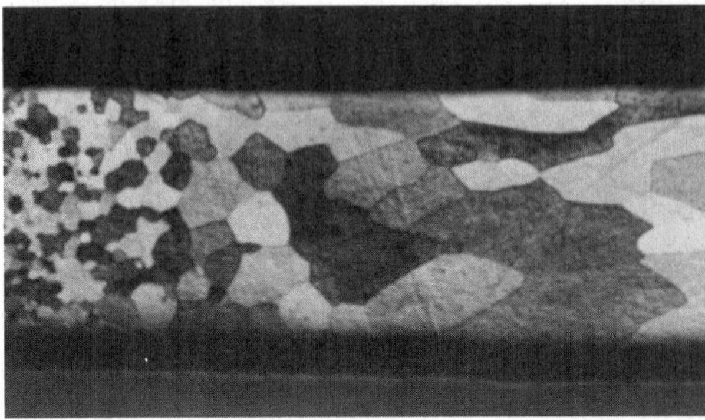

Figure 15. Photomicrograph of a weld deposit on a ferritic stainless steel. Compare the size of grains in the weld with those in the parent material on the left.

Figure 16. Fatigue cracking of a 1-in.-diameter stainless steel bolt showing the characteristic conchoidal striations on the crack face.

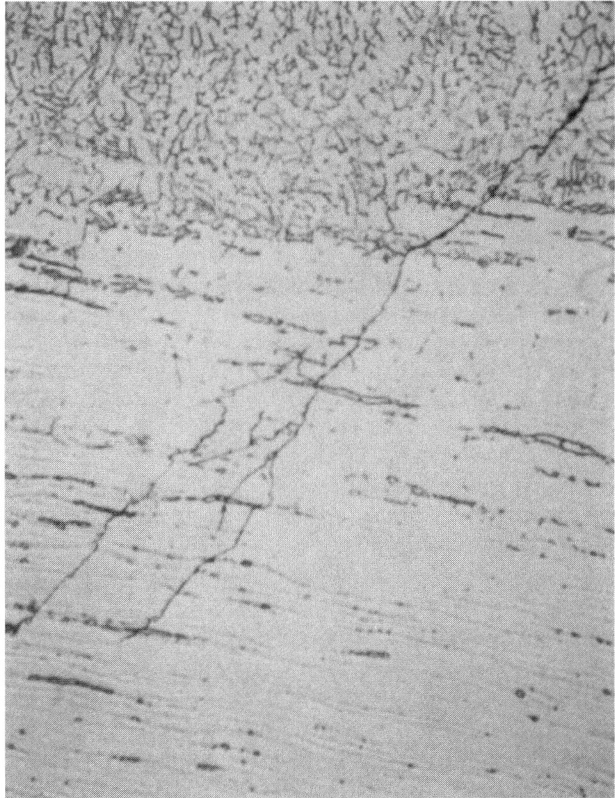

Figure 17. Photomicrograph of a fatigue crack showing all the features of stress corrosion cracking. Compare this with Figure 12.

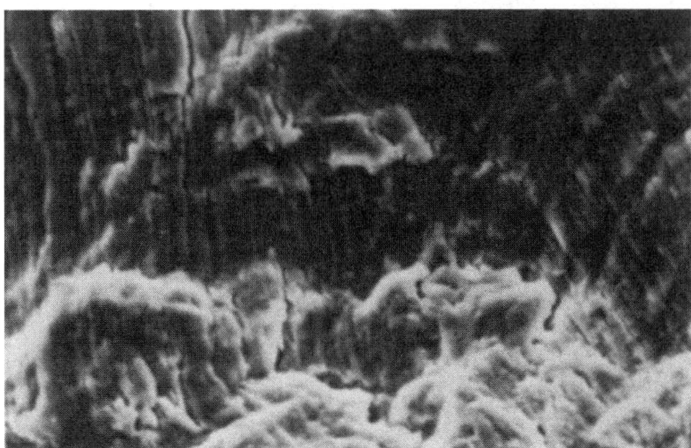

Figure 18. Scanning electron micrograph of the fracture face of the crack shown in Figure 17. Note the stepwise progression of the crack.

ture face of which will be in an "active" state and therefore more susceptible to corrosion processes.

The whole subject of fatigue and corrosion fatigue is complex. However, as far as food processing equipment is concerned, avoidance of fatigue failure is best achieved by avoiding pulsing and pressure peaks. This requires the use of well-engineered valving systems and avoiding the use of positive-displacement pumps. Where this is impractical, provision should be made to incorporate pulsation dampers to smooth out the pressure peaks and minimize the risk of fatigue failures.

Selective Corrosion ***Corrosion of Castings.*** There are a number of stainless-steel components found in food processing plants such as pipeline fittings and pump impellers that are produced as castings rather than fabricated from wrought material—notably, the cast equivalents of grade 304 (CF8) and 316 (CF8M). Although the cast and wrought materials have similar, but not identical, compositions with regard to their chromium and nickel contents, metallurgically they have different structures. Whereas the wrought materials are fully austenitic, castings will contain some ferrite or more terminologically correct, δ (delta) ferrite in the basic austenitic matrix. The ferrite is necessary to permit welding to the castings, to avoid shrinkage cracking during cooling from the casting temperature, and to act as nucleation sites for the precipitation of chromium carbides, which will invariably be present as it will not always be possible to solution anneal the cast components.

The nominal ferrite level is usually 5–12%. Below 5%, cracking problems may be experienced while above 12%, the ferrite tends to form a continuous network rather than remain as isolated pools.

Because the crystallographic structures of the ferrite and austenite differ (austenite is a face-center cubic and the ferrite, body-center cubic), the ferrite has, thermodynamically speaking, a higher free energy, which renders it more susceptible to attack, particularly in low-pH chloride-containing environments such as tomato ketchup and glucose syrups. Although the problem is not so severe when the ferrite occurs as isolated pools, when present as a continuous network, propagation of the corrosion occurs along the ferrite, with the austenite phase being relatively unaffected (Fig. 19). Because the products of corrosion are not leached out from the corrosion site and are more voluminous than the metal, corroded castings frequently assume a blistered or pockmarked appearance. The common environments encountered in the industry that produce preferential ferrite attack are the same as those causing

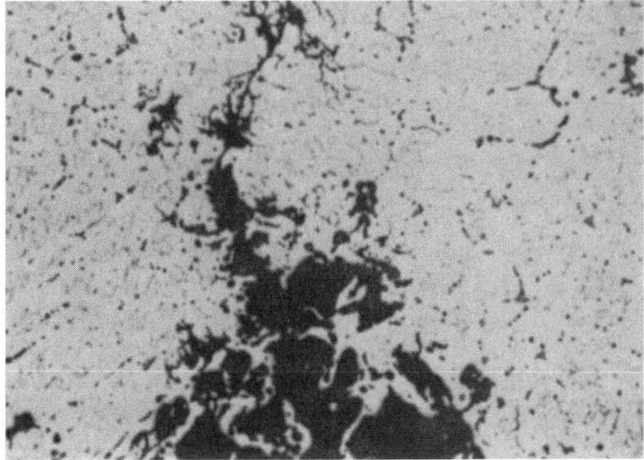

Figure 19. Photomicrograph of a section of cast 316 (CF8M) showing preferential attack of the ferrite phase.

stress corrosion cracking. Therefore, this form of damage is frequently also present (Fig. 20).

Depending on the method used to make the casting, the surface of the castings can be chemically modified, which reduces the corrosion resistance. Small components are usually cast by either the shell molding process or produced as investment castings. In the shell molding process, the sand forming the mold is bonded together with an organic resin that carbonizes when the hot metal is poured. This results in the metal adjacent to the mold having an enhanced carbon level with the formation of intergranular carbide precipitates and hence, a susceptibility to intergranular attack and other forms such as crevice, pitting, and stress corrosion cracking. Methods of overcoming this include solution annealing or machining off the carburized skin of metal.

With investment castings, the mold is made of zircon sand (zirconium silicate), and fired at a high temperature to remove all traces of organic material and wax that is used as a core in the moldmaking process. They do not, therefore, have this carburized layer and offer a much better resistance to surface corrosion.

Free-Machining Stainless Steels. Stainless steels are notoriously difficult to machine, especially turning, not so much because they are hard, but because the swarf tends to form as continuous lengths that clog the machine and to weld to the tip of the machine tools. One method of overcoming this is to incorporate a small amount, typically 0.2%, of sulfur or selenium in the alloy. These elements react with the manganese to form manganese sulfide or selenide, which, being insoluble in the steel, forms as discrete pools in castings or as elongated, continuous stringers in, say, wrought bar. The effect of the sulfide inclusions is to cause the material to form chips rather than long strings of swarf when being machined. Both manganese sulfide and selenide have virtually zero corrosion resistance to dilute mineral acids or other corrosive media. Thus, the free-cutting variants have a much lower corrosion resistance than their designation would imply. Indeed, some believe that the addition of sulfur to a type 316 material will offset the beneficial effect of the alloying addition of molybdenum. As stated, the sulfide inclusions will occur in castings as discrete pools, and therefore there will not be a continuous corrosion path. However, one of the products of corrosion in acidic media will be hydrogen sulfide, which has a profound effect on the corrosivity of even dilute mineral acids, causing attack of the austenitic matrix.

In the case of wrought materials, in particular bar stock, the sulfide inclusions are present as semicontinuous stringers and can suffer so-called end-grain attack in mildly corrosive media.

Stainless-steel nuts and bolts, which are produced on automatic thread-cutting machines, are invariably made from free-cutting materials. Figure 21 illustrates the difference in corrosion resistance of a bolt made from this and a nonfree machining 316. Both bolts were exposed to the same mildly acidic environment.

When specifying materials of construction, quite clearly due cognizance of this difference must be recognized. Any components turned from bar stock which are likely to come into contact with potentially corrosive environments should always be specified in 316 and not in the free machining, sulfur-containing variant.

CORROSION OF SPECIFIC ENVIRONMENTS

From a corrosion standpoint, the environments likely to be encountered in the food industry that may cause premature equipment failure may be classified under main headings:

- Noncorrosive
- Mildly corrosive
- Highly corrosive
- Service fluids
- Alkaline detergents
- Acidic detergents
- Sanitizing agents

Noncorrosive Foodstuffs

In general terms, natural foodstuffs such as milk, cream, natural fruit juices, and whole egg do not cause corrosion

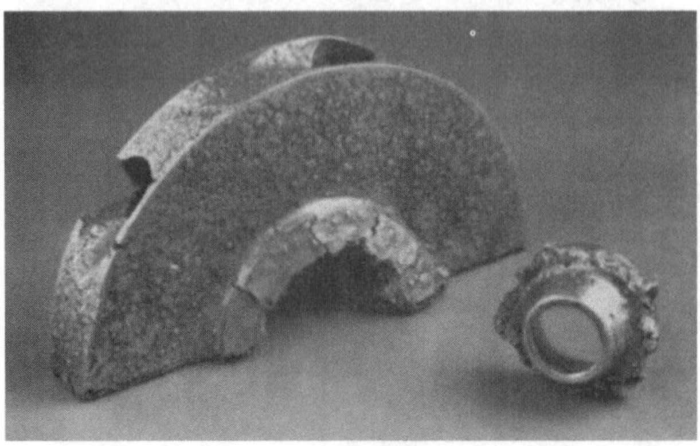

Figure 20. Corroded pump impeller (above) and valve body (below) caused by tomato ketchup. Note the pocklike corrosion sites.

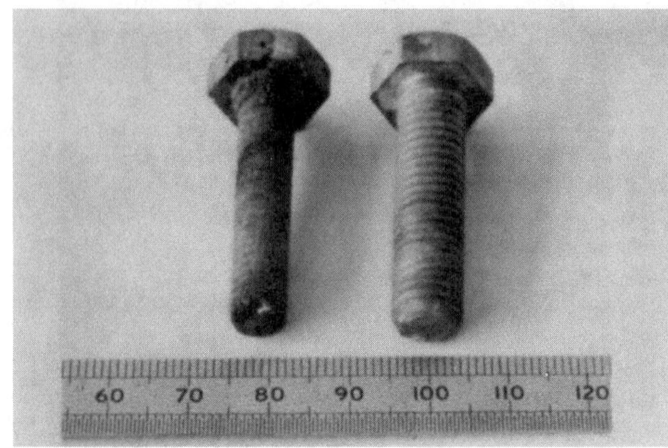

Figure 21. Free machining and nonfree machining bolts after immersion in a mildly acidic environment.

problems with 304 or 316 stainless steels. Prepared foodstuffs to which there is no added salt such as yogurt, beer, ice cream, wines, spirits, and coffee also fall within this classification. For general storage vessels, pipelines, pumps, fittings or valves, grade 304 is perfectly satisfactory. However, for plate heat exchangers that are highly creviced and therefore prone to crevice corrosion, grade 316 frequently is employed. This offers a higher degree of protection against some of the more acidic products such as lemon juice, which may contain small quantities of salt and also provides a higher level of integrity against corrosion by service liquids and sanitizing agents.

It is quite common to use sulfur dioxide or sodium bisulfite for the preservation of fruit juices and gelatin solutions and in such cases, storage vessels always should be constructed from 316. Although the sulfur dioxide is noncorrosive at ambient temperature in the liquid phase, as a gas contained in significant quantities within the head space in a storage tank it tends to dissolve in water droplets on the tank wall. In the presence of air, the sulfurous acid that forms is oxidized to sulfuric acid at a concentration high enough to cause corrosion of 304 but not of 316.

Mildly Corrosive Foodstuffs

This category of foodstuffs covers products containing relatively low levels of salt and where pH values are below 7. Examples include glucose/fructose syrups and gelatin, the production of which may involve the use of hydrochloric acid. For storage vessels, pipelines, fittings, and pumps, grade 316 has established a good track record, and boiling pans in this grade of steel are perfect for long and satisfactory service. The corrosion hazards increase however in processing operations involving high temperatures and where the configuration of the equipment is such as to contain crevices, especially when the product contains dissolved oxygen. For example, multistage evaporators operating on glucose syrup will usually have the first stage, where temperatures may approach 100°C (212°F), constructed in a super stainless steel such as 904L. Subsequent effects where temperatures are lower and where the product has been deoxygenated may be fabricated in grade 316 stainless steel.

As previously indicated, it is common practice to use sulfur dioxide as a preservative in dilute gelatin solutions during storage prior to evaporation. In some cases, excess hydrogen peroxide will be added to neutralize the sulfur dioxide immediately before concentration. This can have a catastrophic effect on the 300 series stainless steels and indeed on even more highly alloyed metals such as 904L, because of the combined effect of the chlorides present with the excess hydrogen peroxide. Because of this, it is a more acceptable practice to make the peroxide addition after rather than before evaporation.

Gelatin for pharmaceutical end use is subject to UHT treatment to ensure sterility. This will involve heating the gelatin solutions to 135°C (285°F) and holding at that temperature for a short period of time. Plate heat exchangers are used extensively for this duty. Although plates made from 316 stainless steel give a reasonable life of typically 2–3 years, a corrosion resistant alloy with an enhanced-level molybdenum is to be preferred.

Highly Corrosive Foodstuffs

The list of foodstuffs falling in this category is almost endless—gravies, ketchups, pickles, salad dressings, butter, and margarine—in fact, anything to which salt has been added at the 1–3% level or even higher. Also within this category must be included cheese salting brine and other brines used in the preservation of foodstuffs that undergo pasteurization to minimize bacterial growth on food residues remaining in the brine. Although these brines are usually too strong to support the growth of common organisms, salt resistant strains (halophiles) are the major problem.

Low-pH products containing acetic acid are particularly aggressive from a corrosion standpoint, but selection of materials for handling these products depends to a great extent on the duty involved.

When trying to define the corrosion risk to a piece of equipment handling potential corrodents, several factors come into play. While temperature, oxygen content, chloride content, and pH are the obvious ones, less obvious and equally important is contact time. All three main forms of corrosion induced in stainless steels (crevice, pitting, and stress corrosion cracking) have an induction period before the onset of corrosion. This can vary from a few hours to several months depending on the other operative factors. In a hypothetical situation where stainless steel is exposed to a potentially corrosive environment, removal of the steel and removal of the corrodents will stop the induction and the status quo is established. On repeating the exposure, the induction period is the same. In other words, the individual periods the steel spends in contact with the corrodent are not cumulative and each period must be taken in isolation.

When the contact period is short, temperatures are low and a rigorous cleaning regime is implemented at the end of each processing period, 316 stainless steel will give excellent service. However, where temperatures are high and contact periods are long, the corrosion process may be initiated. This is especially so in crevices such as the interplate contact points on a plate heat exchanger where, albeit at a microscopic level, corrodents and corrosion products are trapped in pits or cracks. Geometric factors may prevent the complete removal of this debris during cleaning and under such circumstances, the corrosion process will be ongoing.

Because of the perishable nature of foodstuffs, storage is rarely for prolonged periods or at high temperatures, and regular, thorough cleaning tends to be the norm. The one exception to this is buffer storage vessels for holding "self-preserving" ketchup and sauces. For such duties, an alloy such as 904L, Avesta 254 SMO, or even Inconel 625 may be required.

While all the foregoing applies to general equipment, the one exception is plate heat exchangers. Their highly creviced configuration and the high temperatures employed render them particularly susceptible. Plates made from grade 316 have a poor track record on these types of duties. Even the more highly alloyed materials do not offer complete immunity. The only reasonably priced material that is finding increased usage in certain areas of food processing is titanium.

The fact that butter and margarine have been included in this group of corrodents requires comment. Both these foodstuffs are emulsions containing typically 16% water and 2% salt. A fact not often appreciated is that the salt is dissolved in the water phase, being insoluble in the oil. From a corrosion standpoint, therefore, the margarine or butter may be regarded as a suspension of 12% salt solution and as such is very corrosive to 316 stainless steel at the higher processing temperatures. The only mitigating feature that partly offsets their corrosivity is the fact that the aqueous salty phase is dispersed in an oil rather than the reverse and that the oil does tend to preferentially wet the steel surface and provide some degree of protection. However, the pasteurizing heat exchanger in margarine rework systems invariably has titanium plates as the life of 316 stainless steel is limited and has been known to be as little as 6 weeks.

Corrosion by Service Fluids

Steam. Because it is a vapor and free from dissolved salts, steam is not corrosive to stainless steels. Although sometimes contaminated with traces of rust from carbon steel steam lines, in the author's experience no case of corrosion due to industrial boiler steam ever has been encountered.

Water. The quality and dissolved solids content of water supplies varies tremendously, with the aggressive ionic species, chloride ions, being present at levels varying between zero, as found in the lakeland area of England, to several hundred parts per million, as encountered in coastal regions of Holland. It is also normal practice to chlorinate potable water supplies to kill pathogenic bacteria with the amount added dependent on other factors such as the amount of organic matter present. However, most water supply authorities aim to provide water with a residual chlorine content of 0.2 ppm at the point of use. Well waters also vary in composition depending on the geographical location, especially in coastal regions where the chloride content can fluctuate with the rise and fall of the tide.

What constitutes a "good" water? From a general user viewpoint, the important factor is hardness, either temporary hardness caused by calcium and magnesium bicarbonates that can be removed by boiling, or permanent hardness caused by calcium sulfate that can be removed by chemical treatment. While hardness is a factor, chloride content and pH probably are the most important from a corrosion standpoint.

What can be classed as a noncorrosive water supply from a stainless-steel equipment user? Unfortunately, there are no hard-and-fast rules that will determine whether corrosion of equipment will occur. As has been repeatedly stated throughout this article, so many factors come into play. The type of equipment, temperatures, contact times, etc all play a role in the overall corrosion process. Again, as has been stated before, the most critical items of equipment are those with inherent crevices—evaporators and plate heat exchangers among others. Defining conditions of use for this type of equipment will be the regulatory factor. Even then, it is virtually impossible to define the composition of a "safe" water, but as a general guideline, water with less than 100 ppm chloride is unlikely to initiate crevice corrosion of type 316 stainless steel while a maximum level of 50 ppm should be used with type 304.

Cooling tower water systems are frequently overlooked as a potential source of corrodents. It must be appreciated that a cooling tower is an evaporator, and although the supply of makeup water may contain only 25 ppm chloride, over a period of operation this can increase by a factor of 10 unless there is a routine bleed on the pond.

Water scale deposits formed on heat-transfer surfaces should always be removed as part of the routine maintenance schedule. Water scale deposits can accumulate chloride and other soluble salts that tend to concentrate, producing higher levels in contact with the metal than indicated by the water composition. Furthermore, water scale formed on a stainless-steel surface provides an ideal base for the onset of crevice corrosion.

As stated previously, potable water supplies usually have a residual free chlorine content of 0.2 ppm. Where installations have their own private wells, chlorination is undertaken on site. In general terms, the levels employed by the local water authorities should be followed and overchlorination avoided. Levels in excess of 2 ppm could initiate crevice corrosion.

Cooling Brines. Depending on the industry, these can be anything from glycol solutions, sodium nitrate/carbonate, or calcium chloride. It is the latter which are used as a 25% solution that can give rise to corrosion of stainless steel unless maintained in the ideal condition, especially when employed in the final chilling section of plate heat exchangers for milk and beer processing. However, by observing certain precautions, damage can be avoided.

The corrosion of stainless steels by brine can best be represented as shown in Figure 22. It will be noted that an exponential rise in corrosion rate with reducing pH occurs in the range pH 12-7. The diminution in number of pits occurring in the range pH 6-4 corresponds with the change in mode of attack, ie, from pitting to general corrosion. It will be seen that ideally the pH of the solution should be maintained in the region 14-11. However, calcium chloride brine undergoes decomposition at pH values higher than 10.6:

$$CaCl_2 + 2NaOH \rightarrow Ca(OH)_2\downarrow + 2NaCl$$

Scale deposition occurs and heat-transfer surfaces become fouled with calcium hydroxide scale. Furthermore, the scale that forms traps quantities of chloride salts that cannot be effectively removed and remain in contact with the equipment during shutdown periods. This is particularly important in equipment such as plate heat exchangers that are subjected to cleaning and possibly hot-water sterilization cycles at temperatures of 80°C (176°F) or higher.

The other aspect, nonaeration, is equally important. Air contains small quantities of carbon dioxide that form a slightly acidic solution when dissolved in water. This has

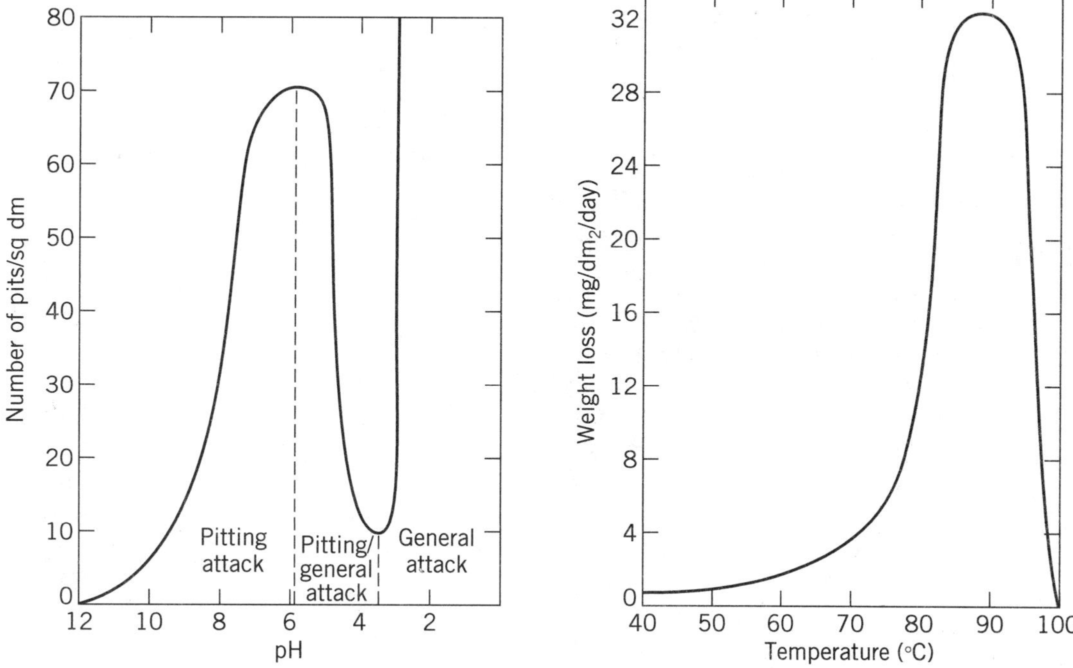

Figure 22. The effect of pH and temperature on the pitting of stainless steel by brine.

the effect of neutralizing the buffering action of any alkaline components in the brine:

$$2NaOH + CO_2 \rightarrow Na_2CO_3 + H_2O$$

$$Na_2CO_3 + CaCl_2 \rightarrow 2NaCl + CaCO_3\downarrow$$

or

$$Ca(OH)_2 + CO_2 \rightarrow CaCO_3\downarrow + H_2O$$

Therefore, the pH of the brine decreases and assumes a value of about pH 6.5, which is the region where pitting incidence is highest. Furthermore, scale deposits of calcium carbonate are laid down on heat-transfer surfaces, creating the problems referred to above.

The precautions to be observed when using brine circuits are

- Ensure correct pH control and maintain in the range pH 9.5–10.
- Eliminate aeration. In particular, make certain that the brine return discharge line is below the surface in the storage tank during running and that the method of feeding brine from the tank does not cause vortexing with resultant air entrainment. Baudelot type evaporators cause aeration of the chilling liquor and should never be used on brine circuits.
- When cleaning and sterilizing the brine section of a pasteurizer, flush out all brine residues until the rinse water is free of chloride. As an added precaution, it is advisable to form a closed circuit and circulate a 1/4–1/2% caustic soda or sodium metasilicate solution to ensure that any brine residues are rendered alkaline.
- In plate heat exchangers and similar equipment, make sure that stainless-steel brine section components remain free of scale.
- When shutting down the plant after a cleaning run, it is advisable to leave the section full with a dilute (1/4%) caustic solution. Before startup, this should be drained and residues flushed out prior to reintroducing brine.

When operating conditions prevailing in a plant do not permit such a disciplined cleaning, operating, and shutdown procedure, only two materials can be considered for the brine section of a plate heat exchanger. These are Hastelloy C-276 or titanium. The corrosion resistance of both of these materials is such that cleaning and sanitizing of the product side of the heat exchanger can be carried out without removing the brine. Although both are more expensive than stainless steels, especially Hastelloy C-276, the flexibility of plant operation which their use permits could offset their premium price.

Alkaline Detergents

Supplied to food processing plants either as bulk shipments of separate chemicals or as carefully preformulated mixtures, the composition of alkaline detergent formulations can vary widely in accordance with individual preference or the cleaning job to be done. The detergents do, however, generally include some or all of the following compounds:

Sodium hydroxide
Sodium polyphosphate
Sodium metasilicate
Sodium carbonate

Additionally, it is not uncommon to find that a selection of sequestering agents such as EDTA and any of the many available wetting agents also may be present in the formulations.

None of these compounds are corrosive to stainless steels at the concentrations and temperatures used by the food industry for cleaning. Indeed, 316 stainless steel is unaffected by concentrations of sodium hydroxide as high as 20% at temperatures up to 160°C (320°F). They can, therefore, be used with impunity at their usual maximum concentration of 5%, even in ultra-high-temperature (UHT) operation where temperatures can rise to 140°C (284°F).

Companies have reported that some of these preformulated alkaline detergents cause discoloration of the equipment.

The discoloration starts as a golden yellow, darkening to blue through mauve and eventually black. It has been established that this discoloration is caused by the EDTA sequestering agent, which complexes with traces of iron in the water. It then decomposes under certain conditions of pH and temperature to form an extremely fine film of hydrated iron oxide, the coloration being interference colors that darken as the film thickness increases. Although the film is not aesthetically pleasing, it is in no way deleterious and removing it by conventional cleaning agents is virtually impossible.

Some alkaline detergents are compounded with chlorine release agents such as sodium hypochlorite, salts of di- or trichlorocyanuric acid that form a solution containing 200–300 ppm available chlorine at their usage strength. Although the high alkalinity reduces the corrosivity of these additives, generally speaking they should not be employed on a regular basis at temperatures exceeding 70°C (160°F).

Acidic Detergents

Alkaline detergents will not remove the inorganic salts such as milkstone and beer-stone deposits frequently found in pasteurizers. For this, an acidic detergent is required and selection must be made with regard to their interaction with the metal. As was shown earlier, sulfuric and hydrochloric acids will cause general corrosion of stainless steels. Although it could be argued that sulfuric acid can be employed under strictly controlled conditions because stainless steels, especially grade 316, have a very low corrosion rate, its use could result in a deterioration of surface finish. This, in corrosion terms, is an extremely low rate but from an aesthetic viewpoint, is undesirable.

Acids such as phosphoric, nitric, and citric, when used at any concentration likely to be employed in a plant cleaning operation, have no effect on stainless steels and can be used with impunity. Three cautionary notes are worthy of mention:

- It is always preferable to use alkaline cleaning before the acid cycle to minimize the risk of interacting the acid with any chloride salts and, therefore, minimize the formation of hydrochloric acid.
- It is inadvisable to introduce an acid into a UHT sterilizing plant when it is at full operating temperature (140°C/285°F) as part of a "clean-on-the-run" regime.
- Nitric acid, which is a strong oxidizing agent, will attack certain types of rubber used as gaskets and seals. As a general guideline, concentrations should not exceed 1% and temperature 65°C (150°F), although at lower concentrations, the upper recommended temperature is 90°C (195°F).

Another acid that is finding increasing use in the food industry for removing water scale and other acid-soluble scales is sulfamic acid. Freshly prepared solutions of up to 5% concentration are relatively innocuous to stainless steels but problems may arise when CIP systems incorporating recovery of detergents and acids are employed. Sulfamic acid will undergo hydrolysis at elevated temperatures to produce ammonium hydrogen sulfate

$$NH_2SO_2OH + H_2O \rightarrow NH_4HSO_4$$

which behaves in much the same way as sulfuric acid. In situations where the use of this acid is contemplated, prolonged storage of dilute solutions at elevated temperature is inadvisable, although at room temperature the hydrolysis is at a low rate.

Sanitizing Agents

Terminology for the process of killing pathogenic bacteria varies from country to country. In Europe, disinfection is preferred; in America, sanitizing. Regardless, the term should not be confused with sterilization, which is the process of rendering equipment free from all live food spoilage organisms including yeasts, mold, thermophilic bacteria, and most importantly, spores. Sterilization with chemicals is not considered to be feasible and the only recommended procedure involves the circulation of pressurized hot water at a temperature of not less than 140°C (285°F).

For sanitization, while hot water (or steam) is preferred, chemical sanitizers are extensively used. These include noncorrosive compounds such as quaternary ammonium salts, anionic compounds, aldehydes, amphoterics, and potentially corrosive groups of compounds that rely on the release of halogens for their efficacy. By far, the most popular sanitizer is sodium hypochlorite (chloros), and this is probably the one material that has caused more corrosion in food plants than any other cleaning agent. For a detailed explanation of the corrosion mechanism, the reader is referred to an article by Boulton and Sorenson (17) that describes a study of the corrosion of 304 and 316 stainless steels by sodium hypochlorite solutions. It is important, therefore, if corrosion is to be avoided, that the conditions under which it is used are strictly controlled. For equipment manufactured from grade 316 stainless steel, the recommended conditions are:

- Maximum concentration—150 ppm available chlorine
- Maximum contact time—20 min
- Maximum temperature—room temperature that is well in excess of the minimum conditions established by Tastayre and Holley to kill *Pseudomonas aeruginosa* (18).

In addition, several other precautions must be observed:

- Before introducing hypochlorite, equipment should be thoroughly clean and free of scale deposits. Organic residues reduce the bactericidal efficiency of the disinfectant and offer an artificial crevice in which stagnant pools of hypochlorite can accumulate.
- It is imperative that acidic residues be removed by adequate rinsing before introducing hypochlorite solutions. Acid solutions will react with hypochlorite to release elementary chlorine, which is extremely corrosive to all stainless steels.
- The equipment must be cooled to room temperature before introducing hypochlorite. In detergent cleaning runs, equipment temperature is raised to 80–85°C (176–185°F), and unless it is cooled during the rinsing cycle, a substantial increase in temperature of the disinfectant can occur. An important point, frequently overlooked, is that a leaking steam valve can cause a rise in the temperature of equipment even though it theoretically is shut off.
- After sanitizing, the solution should be drained and the system flushed with water of an acceptable bacteriological standard. This normally is done by using a high rinse rate, preferably greater than that used in the processing run.

While these comments relate specifically to the sanitizing of plate heat exchangers, similar precautions must be taken with other creviced equipment. Examples include manually operated valves that should be slackened and the plug raised to permit flushing of the seating surface. Pipeline gaskets also should be checked frequently to make sure that they are in good condition and not excessively hardened. Otherwise they will fail to form a crevice-free seal over their entire diameter. Where it is not possible to completely remove hypochlorite residues such as in absorbent gland packing materials, hot water is preferred.

All the foregoing relate specifically to sodium hypochlorite solutions but other sanitizing agents that rely on halogen release, such as di- and trichlorocyanuric acid, should also be used under strictly controlled conditions, including such factors as pH.

Iodophors also are used for sanitizing equipment. These are solutions of iodine in nonionic detergents and contain an acid such as phosphoric to adjust the pH into the range at which they exhibit bactericidal efficacy. This group of sanitizers is employed where hot cleaning is not necessary or on lightly soiled surfaces such as milk road tankers, and farm tanks. Extreme caution should be exercised with this group for, although used at low concentrations (50 ppm), prolonged contact with stainless steel can cause pitting and crevice corrosion. Furthermore, in storage vessels that have been partially filled with iodophor solutions and allowed to stand overnight, pitting corrosion in the head space has been observed as a result of iodine vaporizing from the solution and condensing as pure crystals on the tank wall above the liquid line. Another factor is that iodine can be absorbed by some rubbers. During subsequent processing operations at elevated temperatures, the iodine is released in the form of organic iodine compounds, especially into fatty foods, which can cause an antiseptic taint. The author knows of one dairy that used an iodophor solution to sanitize a plate pasteurizer to kill an infection of a heat resistant spore-forming organism. The following day, there were over 2000 complaints of tainted milk. CIP cleaning cycles did not remove the antiseptic smell from the rubber seals, and complete replacement with new seals was the only method of resolving the problem.

Another sanitizing agent that is assuming increasing popularity, especially in the brewing industry because of its efficacy against yeasts, is peracetic acid. As such, peracetic acid will not cause corrosion of 304 or 316 stainless steels, and the only precautionary measure to be taken is to use a good quality water containing less than 50 ppm of chloride ions for making up the solutions to their usage concentration. Because of the strongly oxidizing nature of some types of peracetic acid solutions, deterioration of some types of rubber may occur. A recent survey undertaken by the IDF for the use of peracetic acid in the dairy industry (19) found few corrosion problems reported. The general consensus of opinion was that it permitted greater flexibility in the conditions of use, compared with sodium hypochlorite, without running the risk of damage to equipment.

For comprehensive information on the cleaning of food processing equipment, albeit primarily written for the dairy industry, the reader is referred to the British Standards Institute publication BS 5305 (20).

CORROSION BY INSULATING MATERIALS

Energy conservation now is widely practiced by all branches of industry, and the food industry is no exception. For example, in the brewing industry, wort from the wort boilers is cooled to fermentation temperature, and the hot water generated in the process is stored in insulated vessels (hot-liquor tanks) for making up the next batch of wort. An area of corrosion science that is receiving increased attention is the subject of corrosion initiated in stainless steels by insulating materials. At temperatures in excess of 60°C (140°F) these can act as a source of chlorides that will induce stress corrosion cracking and pitting corrosion of austenitic stainless steels.

Among the insulating materials that have been used for tanks and pipework are

Foamed plastics—polyurethane, polyisocyanurate, phenolic resins, etc

Cellular and foamed glass

Mineral fiber—glass wool, rock wool

Calcium silicate

Magnesia
Cork

All insulating materials contain chlorides to a lesser (10 ppm) or greater (1.5%) extent. The mineral based insulants such as asbestos may contain them as naturally occurring water soluble salts such as sodium or calcium chloride. The organic foams, on the other hand, may contain hydrolysable organochlorine materials used as blowing agents (to form the foam), fire retardants such as chlorinated organophosphates, or chlorine containing materials present as impurities. Insulation manufacturers are becoming increasingly aware of the potential risk of chlorides in contact with stainless steels and are making serious efforts to market a range of materials that are essentially chloride-free.

A point not frequently appreciated is that even though many of the insulating materials contain high levels of chloride, in isolation they are not corrosive.

Corrosion, which is an electrochemical process, involves ionic species, and in the absence of a solvent (water) the chloride or salts present in the insulation cannot undergo ionization to give chloride ions. Therefore, they are essentially noncorrosive. Similarly, where organochlorine compounds are present, water is necessary for hydrolysis to occur with the formation of hydrochloric acid or other ionisable chloride compounds.

The main problem, therefore, is not so much the insulant but the interaction of the insulant with potential contaminants to release corrosive species. Under ideal conditions, that is, if the insulating material could be maintained perfectly dry, the chloride content would not be a critical factor in the material specification process. Unfortunately, it must be acknowledged that even in the best regulated installation, this ideal is rarely (if ever) achieved and therefore, thought must be given to recommended guidelines.

Any material that is capable of absorbing water must be regarded as a potential source of chloride. Although the chloride content of the insulant may be extremely low (25 ppm), under extractive conditions when the insulation becomes wet and concentration effects come into play, even this material may cause the initiation of stress corrosion cracking. The chloride content of the contaminating water cannot be ignored. Even a "good-quality" water with a low chloride content (30 ppm), if being continuously absorbed by the insulant, forms a potential source of chlorides that, through evaporation, can reach very significant levels and initiate the corrosion mechanism.

The more absorptive the insulant, the greater the risk, and materials such as calcium silicate and certain types of foams must be regarded as least desirable. It is interesting to note that one of the least absorptive insulants from those listed is cork, and in the experience of the author, there have been no reported cases of this insulant causing problems with stress corrosion cracking of stainless steels. Of course, it could be argued that the bitumen used as an adhesive to stick the cork to the vessel walls has to be applied so thickly that the bitumen forms an impermeable barrier preventing contact of the contaminated water with the stainless-steel substrate. Irrespective of the protection mechanism, the net effect is that cork has an extremely good "track record." Unfortunately, due to the cost, cork is now rarely used.

No hard-and-fast rules can be laid down for specifying the acceptable maximum tolerable chloride content of an insulating material. Any specification must take into account what is commercially available as well as all the other factors such as price, flammability, and ease of application. To specify zero chloride is obviously impractical and even a figure of 10 ppm may be difficult to achieve in commercially available products. As a general compromise, 25 ppm maximum is considered to be technically and commercially feasible while minimizing the potential source level of chloride corrodent.

The primary function of the insulant is to provide a thermal barrier between the outside of the vessel and the environment. It does not provide a vapor or moisture barrier, and provision of such protection must be regarded as equally important in the insulating process. As mentioned previously, the use of bitumen as an adhesive for cork insulation must provide an extremely good water barrier, although there are a wide range of products marketed specifically for this purpose. These include specially developed paints of undefined composition and zinc free silicone alkyd paints as well as silicone lacquer. These silicone based products are particularly appropriate because of their inherent low water permeability and also their stability at elevated temperatures. Aluminum foil has also been successfully used as a water-vapor barrier between the stainless steel and the insulant. There is little doubt that the foil provides an extremely good barrier but at laps between the sheets of foil, ingression of water can occur unless sealing is complete, the achievement of which is most unlikely. Furthermore, it is likely that there will be tears in the foil, providing yet another ingression path for water. However, it is believed that aluminum foil has a role additional to that of a barrier. As it is, from an electrochemical aspect, "anodic" to stainless steel, it will provide galvanic protection of the steel in areas where there is a "holiday" in the foil, thus inhibiting corrosion mechanisms.

When insulation material becomes wet, the insulating efficiency shows a dramatic fall-off. In fact, the thermal conductivity of wet insulation will approach that of the wetting medium, and it is ironic that the thermal conductivity of water is among the highest known for liquids. It is imperative that from the standpoint of preventing moisture ingression to minimize corrosion risks and maintain insulation efficiency, the insulation is externally protected from rain and water. There is a variety of materials available for this such as aluminum sheeting, plastic-coated mild steel, and spray applied polyurethane coatings. Much of the value of the protection will be lost unless particular attention is taken to maintaining a weather-tight seal at the overlaps in individuals sheets of cladding. There are many semiflexible caulking agents that can be employed for this. One that is particularly effective is the RTV silicon rubber, which exhibits extremely good weather resistance and long-term reliability. Nevertheless, maintenance is required, especially around areas of discontinuity such as flanged connections and manway doors. Operators are beginning to realize that routine maintenance work on in-

Table 5. Some Synthetic Rubbers

Rubber	Common trade names	Basic structure
Polychloroprene	Neoprene Perbunan C Butachlor Nairit	Poly (2-chlor-1,8 butadiene)
SBR (styrene butadiene rubber)	Buna S Plioflex Intol Krylene	Copolymer of styrene and 1,4 butadiene
Nitrile	Buna N Chemigum N Paracril Perbunan Hycar	Copolymer of acrylonitrile and 1,4 butadiene
EPDM (ethylene propylene diene methylene)	Nordel Royalene Vistalon Dutral Keltan Intolan	Copolymer of ethylene and propylene with a third monomer such as ethylidene norbonene, cyclopentadiene
Butyl	GR-1 Bucar Socabutyl	Copolymer of isobutylene and isoprene
Silicone	Silastic Silastomer Sil-O-Flex	Poly(dimethyl siloxane), poly(methyl vinyl siloxane), etc.
Fluoroelastomer	Viton Technoflon Fluorel	Copolymer of hexafluoropropylene and vinylidene fluoride

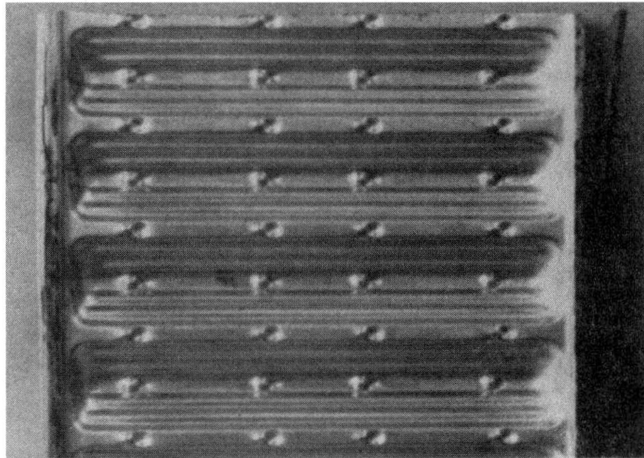

Figure 23. Catastrophic failure of the gasket groove of a heat-exchanger plate by stress corrosion cracking caused by the use of a polychloroprene based adhesive.

sulation is as important as that on all other items of plant and equipment.

It will be appreciated from all the foregoing that insulation of a vessel or pipeline is a composite activity with many interactive parameters. It is impossible to lay down specification rules, as each case much be viewed in the light of requirements. These notes must, therefore, be regarded as guidelines rather than dogma, and for further information, an excellent publication by the American Society for Testing Materials (21) is essential reading.

CORROSION OF RUBBERS

Many involved in the field of corrosion would not regard the deterioration of rubber as a corrosion process. The author's opinion differs from this viewpoint as "rubber undergoes deterioration by interaction with its environment."

General

Rubber and rubber components form an essential part of food processing equipment—joint rings on pipelines, gaskets on heat exchangers, and plate evaporators. Although natural rubber was the first material to be used for manufacturing these components, nowadays they are made almost exclusively from one of the synthetic rubbers listed in Table 5.

Unlike metals and alloys, which have a strictly defined composition, the constituents used in the formulation of rubbers are rarely stipulated. More often, they will reflect the views and idiosyncrasies of the formulator on how to achieve the desired end product. The important constituents of a rubber are

- *Basic Polymer*. Largely determines the general chemical properties of the finished product.
- *Reinforcing Fillers*. These are added to improve the mechanical properties and will invariably be one of the grades of carbon black—or if a white rubber is required, mineral fillers such as clays or calcium silicate.
- *Vulcanizing Agents*. These cross-link the basic polymer and impart rubberlike properties that are maintained at elevated temperatures.
- *Antioxidants*. To stabilize the rubber against oxidative degradation, hardening or softening, after prolonged operating periods at elevated temperatures.
- *Processing Aids*. Which facilitate the molding of the rubber.
- *Plasticizers*. To modify the mechanical properties.

A complicating factor to be considered when formulating rubber for food-contact surfaces is the acceptability (or nonacceptability) of the compounding ingredients. Some countries, notably Germany and the USA, have drawn up lists of permitted ingredients (22,23), while other countries regulate the amount of material that can be extracted from the finished article by various test media.

Invoking these regulations may impose limits on the in-service performance of a rubber component, which could be a compromise, exhibiting desirable properties inferior to those achievable were it for a nonfood application. For example, the resistance to high pressure steam of some rubbers can be enhanced by using lead oxide as an ingredient. Obviously, such materials could not be contemplated for any food contact application.

Table 6. Performance of Rubbers in Some Environments Found in the Food and Beverage Industries

	SBR	Medium nitrile	Natural rubber	Polychloroprene	Butyl	EPDM	Silicone[a]
			Products				
Whole milk	E	E	E	E	E	E	E
Beer, wines, spirits	G-F	E	F	E	E	E	E
Fats, oils, cream	F	E	F	G	P	G	E
Sauces	E	E	F[b]	E	E[b]	E	E
Salad dressings	F	E	F	G	P	G	E
Fruit drinks and juices[d]	E	E	G	E	E	E	E
			Cleaning Agents				
Sodium hydroxide[c]	G	E	G	G	E	E	E
Sodium carbonate[f]	E	E	E	E	E	E	E
Sodium hypochlorite[k]	G	G	G	G	E	E	E
Nitric acid[h]	F	F	P	F	G	G	G
Phosphoric acid[i]	E	E	E	E	E	E	E
Quaternary ammonium compounds[e]	E	E	E	E	G	G	G

[a]Depending on the type of basic polymer.
[b]Depending on the fat/oil content.
[c]Used as an aqueous 1% solution.
[d]The performance of a rubber will be affected by the presence of essential oils.
[e]All strengths up to 5% at maximum operating temperatures of the rubber.
[f]Sodium carbonate and other detergent/additives, eg, sodium phosphate, silicate.
[g]Sodium hypochlorite as used at normal sterilizing concentration—150 mg/L.
[h]Nitric acid as used at normal cleaning strength of 1/2-1%.
[i]Up to 5% strength.

Table 7. Suggested Limits of Concentration and Temperature for Peracetic Acid in Contact with Rubber

Peracetic acid concentration (active ingredient)	Temp. (°C)	Medium nitrile	Butyl (resin-cured)	EPDM	Silicon	Fluoro-elastomer
0.05% (500 mg/L)	20	R[a]	R	R	R	R
	60	R	R	R	R	R
	85	(?)	R	R	R	(?)[b]
0.10% (1 g/L)	20	R	R	R	R	R
	60	NR[c]	R	R	NR	R
	85	NR	(?)	R	NR	NR
0.25% (2.5 g/L)	20	R	R	R	(?)	R
	60	NR	R	R	NR	R
	85	NR	NR	R	NR	NR
0.5% (5 g/L)	20		R	R	NR	R
	60	NR	R	R	NR	R
	85	NR	NR	R	NR	NR

[a]R—Little effect on the rubber.
[b](?)—Possibly some degradation.
[c]NR—Not recommended as significant degradation may occur.

Corrosion by Rubber

The majority of rubbers and formulating ingredients have no effect on stainless steels even under conditions of high temperature and moisture. There are two notable exceptions, namely, polychloroprene and chlorosulfonated polyethylene. Both of these contain chlorine, which under the influence of temperature and moisture, undergo hydrolysis to produce small quantities of hydrochloric acid. In contact with stainless steel, this represents a serious corrosion hazard causing the three main forms of attack.

When specifying a rubber component, it is easy to avoid these two polymers but a fact not often appreciated is that many of the rubber adhesives are produced from one of these polymers. That is the reason why manufacturers of heat exchangers and other equipment, which necessitates sticking the rubber onto metal, specify what type of adhesive should be used. Many DIY adhesives and contact adhesives are formulated from polychloroprene, and it is not unknown for a maintenance engineer having to stick gaskets in a heat exchanger to get the supply of rubber cement from the local hardware shop. The results can be catastrophic, as shown in Figure 23. Similarly, many self adhesive tapes used as a polychloroprene-based adhesive and direct contact of these with stainless steels should be avoided.

Corrosion of Rubber by Environments

From the standpoint of food processing, the environments likely to interact with rubber are classified under the following headings:

- Foodstuffs containing no fat or a low level of fat, eg, milk
- Fatty products: butter, cream, cooking oils, shortenings
- Alkaline detergents
- Acid detergents
- Sanitizing agents

Unlike the corrosion of metals, which is associated with oxidation and loss of metal, rubber deterioration usually takes other forms. When a rubber is immersed in a liquid, it absorbs that liquid or substances present in it to a greater or lesser degree. The amount of absorption determines whether the rubber is compatible with the environment. The absorption will be accompanied by changes in mass, volume, hardness, and tensile strength. For example, immersing and oil resistant rubber in vegetable oil may produce a change in volume of only 2–3%, whereas a non-oil-resistant rubber may swell by 150% or more. Such a volumetric change will be accompanied by a large reduction in the tensile strength and a high degree of softening.

Broadly, speaking, a rubber should not exhibit a volumetric or weight change greater than 10% nor a hardness change of more than 10 degrees (International Rubber Hardness Degrees—IRHD or Shore A) to be classed as compatible. For general guidance, data presented in Tables 6 and 7 (19) indicate the compatibility of rubbers with some food industry environments. But for more information, the reader should refer to one of the national or international test procedures (24–26).

BIBLIOGRAPHY

Adapted from: C. T. Cowan, "Materials Science and Corrosion Prevention for Food Processing Equipment," *APV Corrosion Handbook*, March 1990, 44 pp.
Copyrighted APV Crepaco, Inc. Used with permission.

1. N. Warren, *Metal Corrosion in Boats*, London Stanford Maritime, 1980.
2. A. J. Sedricks, *Corrosion of Stainless Steels*, Wiley, New York, 1979.
3. The British Aluminium Co., Ltd., *Aluminium in the Chemical and Food Industries*, 1951. (London: The British Aluminium Co., Ltd.)
4. G. Ferry, *New Scientist*, 27 Feb. 1986, 23.
5. G. A. Dummett, *From Little Acorns, A History of the APV Company*, Hutchison Benham, London, 1981.
6. C. T. Cowan, "Corrosion Engineering with New Materials," *The Brewer* 163–168 (1985).
7. M. G. Fontana, *Corrosion Engineering*, McGraw-Hill, New York, 1986.
8. L. L. Shrier, *Corrosion*, Newnes-Butterworth, London, 1976.
9. J. W. Oldfield and B. Todd, Transactions of the Institute of Marine Engineering (C), 1984, Conference 1, 139.
10. J. W. Oldfield, *Test Techniques for Pitting and Crevice Corrosion of Stainless Steels and Nickel Based Alloys in Chloride-Containing Environments*, NiDi Technical Series 10 016, 1987.
11. U. R. Evans, *Corrosion* **7**, 238 (1981).
12. ASTM G48-76, *Standard Methods for Pitting and Crevice Corrosion Resistance of Stainless Steels and Related Alloys by Use of Ferric Chloride Solution*, American Society for Testing Materials, Philadelphia.
13. R. J. Brigham, *Materials Performance* **13**, 29 (Nov. 1974).
14. N. D. Greene and M. G. Fontana, *Corrosion* **15**, 25t (1959).
15. M. Woelful and R. Mulhall, *Metal Progress*, 57–59 Sept. 1982.
16. M. R. Copson, "Effect of Composition on SCC of Some Alloys Containing Nickel," in T. Rhodin, ed., *Physical Metallurgy of Stress Corrosion Fracture*, Interscience Publishers, New York, 1989.
17. L. H. Boulton and M. M. Sorensen, *New Zealand Journal of Dairy Science & Technology* **23**, 37–49 (1988).
18. G. M. Tastayre and R. A. Holley, Publication 1806/B. Agriculture Canada, Ottawa, 1986.
19. IDF Bulletin 236, *Corrosion by Peracetic Acid Solutions*, International Dairy Federation, 41 Square Vergote, 1040 Brussels, Belgium.
20. BS 5305 Code of practice for cleaning and disinfecting of plant and equipment used in the dairying industry, British Standards Institute, London.
21. ASTM Special Technical Publication 880. Corrosion of Metals under Thermal Insulation. American Society for Testing Materials, Philadelphia.
22. Sonderdruck aus Bundesgesundheitsblatt, 22.1,79, West Germany.
23. USA *Code of Federal Regulations*. Title 21. Food and Drugs Section 177.2600. Rubber Articles Intended for Repeated Use.
24. BS 903 Part A16, Resistance of Vulcanized Rubbers to Liquids, British Standards Institution, London.
25. ASTM D. 471, Testing of Rubbers and Elastomers. Determination of Resistance to Liquids, Vapors, and Gases, American Society for Testing Materials, Philadelphia.
26. ISO 1817, Vulcanized Rubbers—Resistance to Liquids—Methods of Test.

APV Crepaco, Inc.
Lake Mills, Wisconsin

CRABS AND CRAB PROCESSING

Crabs play an important part in the U.S. fishing industry. Ex-vessel landings and values usually rank low in quantity (Fig. 1) and high in value (Fig. 2). The quantity and value of the major crab fishes in the United States in 1996 are shown in Table 1 (1). Variations in processing methods for crabs are related to the species and the seasonal characteristics that affect yield, ease of meat removal, and color. Speed in handling the product from time of harvesting to freezing or canning is a most important process requirement (2,3). Quality control procedures generally emphasize sanitation, food regulatory requirements, and compliance with end-product specifications of the processor (4).

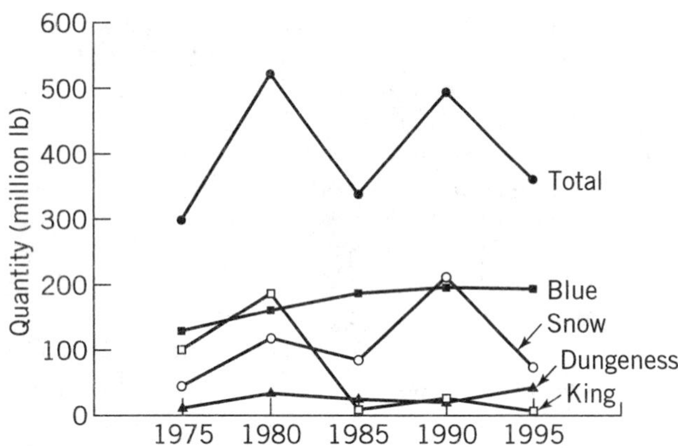

Figure 1. Landings of U.S. crab species.

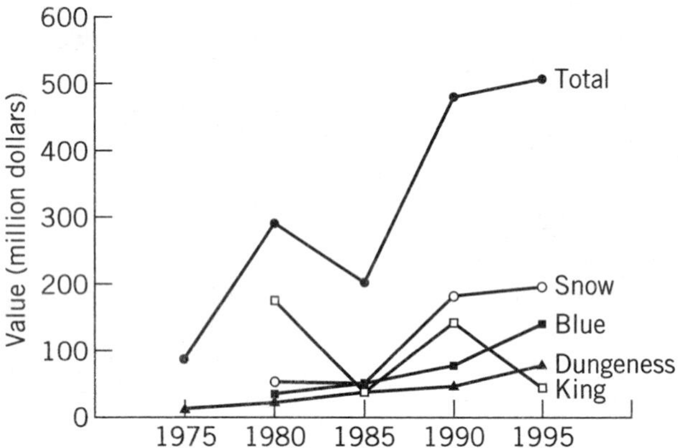

Figure 2. Value of U.S. crab species.

Table 1. Landings and Values of U.S. Crab Species in 1996

	Quantity (million lb)	Value (million $)	Average ex-vessel price ($/lb)
Blue	219.0	147.1	0.67
Chesapeake	68.3		
South Atlantic	86.6		
Gulf	56.9		
Middle Atlantic	7.2		
Dungeness	65.0	87.9	1.35
Washington	27.5		
Alaska	5.9		
Oregon	19.3		
California	12.3		
King			
Alaska	21.0	62.6	2.98
Snow			
Alaska	67.9	93.2	1.37

KING CRAB

Three species of king crab harvested in the North Pacific and Bering Sea are of commercial importance. The most important species is red king crab (*Paralithodes camtschatica*). Blue king crab (*P. platypus*) is caught in commercial quantities primarily near the Pribilof Islands. Brown or golden king crab (*Lithodes aequispina*) are found in deeper waters at the edge of the continental shelf.

King crabs are actually not "true crabs" like dungeness, tanner, or blue crab, but are more closely related to an order of hermit crabs. The legs of true crabs are jointed forward, whereas the legs of king crabs are jointed to fold behind its body (5).

Harvesting

The world's most extensive king crab populations dwell in Alaskan waters, where 15-year-old male king crabs may grow to a weight of 24 lb and a leg span of 6 ft. Most of the commercial catch consists of 7- or 8-year-old crabs that weigh about 6 lb. They are captured by means of large pots consisting of iron frames about 6 ft square and 3 ft high covered with synthetic webbing. These pots may weigh between 300 and 700 lb apiece. The pots are baited with fresh fish (herring is the most preferred) and then dropped to the bottom at 60 to 300 ft depth with a heavy line and marker buoy. The pots are lifted every few days and the male crabs of legal size are kept. The crabs are held alive on the vessel in a well with circulating seawater. At the docks or on the deck of a processor vessel, the live crabs are transferred to seawater tanks to await processing. Weak crabs may be sorted out and processed immediately, but generally all dead and injured crabs are discarded.

Processing

The crab is butchered by use of a stationary iron blade. The back, or carapace, is removed and the crab is split into halves, each with legs attached, called sections. A quick shake will jar out viscera that cling to the body cavities. Revolving nylon brushes are generally used to clean off gills and viscera.

Batch cooking is common, although a continuous cooker requires less labor, and cooking is done without the delay of filling a basket. It also has automatic steam controls and a fixed rate of travel through the tank, ensuring identical cooking. For crab planned for frozen sales, the sections are cooked for 22 to 28 min in either boiling fresh water or seawater. Cook times will vary slightly, but cooking should be continued to ensure that the internal temperature of the crab meat reaches at least 88°C (190 °F).

For canned or retorted products, crab need only be cooked to make meat removal easy. A two-stage cook of 20 min at 60°C (140 °F), rinsing the sections well with a spray of water followed by cooking in boiling water for 12 min, has been proposed (6). This prevents a blue discoloration in canned Kegani crab since the crabmeat is coagulated during the precook while the blood (hemocyanin) can be washed free of the meat. Regardless of cooking method, cooked crab meat should be picked and thermally processed as rapidly as possible to ensure an excellent canned product (7). Normally, canned crabmeat is packed in 307 by 113 C-enamel cans, sealed in a vacuum, retorted for 50 min at 116°C (240°F), then water cooled (8).

For freezing, the sections are cooled in cold running water after cooking. Ideally, chilled water is used to cool the sections to at least 4°C (40°F). Usually, sections are then frozen either in brine or air-blast freezers. The majority of Alaskan king crab production is brine-frozen. The sections are packed with the legs stacked in the center of baskets, shoulders out, and held in liquid brine (22–23%) at a temperature of −18 to −20°C (0 to −5°F) for 30 to 60 min. Internal temperature should reach at least 15°F. The sections are then removed from the brine and dipped in cold fresh water to rinse the salt from the section and to form a protective glaze on the surface of the crab. Blast freezing freezes the product in air by blowing refrigerated air (less than −30°C or −20°F) over it. Proponents of blast freezing contend the method results in a product with a lower sodium content. Brine freezing, if carefully controlled, will be similar to blast freezing, and some feel it results in a moister product.

Product Forms

Almost all of Alaska's king crab is sold frozen. Frozen king crab is packed in a variety of product forms, including both in-shell products and extracted crab meat. The most common king crab products include the following.

Bulk Clusters (Sections). These are packs of shoulders, legs, and claws, commonly packed in 80-lb random weight cartons. The pack should average about 75% legs and 25% shoulder-body meat by weight.

Legs and Claws. These packages are available in a number of forms, including both whole legs and claws or split legs and claws. King crab have six walking legs, two of which have claws. Packs of legs and claws should be in equivalent proportion to the live crab and should include the whole leg from shoulder to walking tip or claw. As a general rule, the larger the leg the higher the price.

King Crab Meat. The body, leg, and claw meats are removed by shaking or by blowing with water under pressure. After the shoulder-body meats are removed, each leg is broken just below the uppermost joint and the two connective tendons that run through the leg meat are removed by pulling the remaining leg sections free. This step is repeated for each leg segment, allowing the leg meats to be removed intact. For larger industrial processes, the legs are divided at the joints, in some cases, by use of a band saw, and fed into two large rubber rollers. The rollers are adjusted for proper clearance and are continuously rotated so that the shell is squeezed as it passes through and the meat is forced out. Meat yield averages about 20% but may be 26% or more by weight of the live crabs (9,10).

The meats are washed, sorted, inspected for shell and debris, and packed into cartons or trays for freezing. King crab meat is available in a number of forms including:

Merus meat. The red-colored merus meat is the largest segment of a king crab's leg and the most expensive. It is available in both blocks and individually quick frozen (IFQ) in poly bags.

Fancy meat. The most common pack of king crab meat and consists of a mixture of merus meat (25–35%), shoulder-body meat (40–50%), and red leg meat (30–35 %). The blocks of meat are packed in triple layers beginning with the merus meat, then white meat, finally red leg meat, and then frozen. The higher the percentage of leg meat in the block, the higher the price.

Salad meat. Salad meat is a cheaper frozen pack of small chunks of both red and white meat.

Rice meat. The cheapest form of king crab meat is rice meat, which consists of uniformly shredded body and leg meats.

King crab that is adequately protected from dehydration can be held for six months or more at −18 to −23°C (0 to −10°F).

SNOW CRAB

Commonly referred to as "tanner crab," the two crab species, *Chionoecetes bairdi* and *C. opilio*, were relatively unknown in the United States until the king crab stocks began to decline. Today, these crab species are marketed as snow crab and have emerged as the most valued of the crabs harvested in the United States (1). *C. bairdi*, the larger and thus more desirable of the two snow crab species, is caught only in the North Pacific and Bering Sea. The smaller species, *C. opilio*, is caught in the North Atlantic as well. Pacific *C. opilio* is caught primarily by Alaskan fishermen while Atlantic *C. opilio*, which is slightly smaller, is caught by Canadians. Canadian snow crab may be marketed as queen crab (5).

Snow crabs are members of the spider crab family. In the North Pacific, they occur broadly from the southeast coast of Alaska westward along the coast throughout the Aleutian Islands and the Bering Sea. Snow crabs are caught in the winter and spring by the same boats that fish for king crab in the fall. Although only half the size of king crabs, snow crabs are still larger than most other species of crab. The largest, *C. bairdi*, occasionally exceed 3 lb and have a leg span of 2 ft.

Alaskan snow crab are harvested and processed in approximately the same manner as king crab and by the same fishers and processors. Snow crab is almost always marketed frozen and the product forms are similar to king crab; that is, clusters, claws, legs, and meats (fancy, merus, salad, and shredded). The yield of picked meats from snow crab is slightly lower than for king crab and averages 17%, ranging from 12 to 25% (9,11). Like fancy meat from king crab, frozen blocks of fancy snow crab meat consist of three layers: red meat (45%), white meat (35–45%), and shreds (10–20%). Snow crab has the same handling characteristics as king crab. Like king crab, care should be taken to ensure the product is well protected by proper glazing and packaging in order to prevent dehydration and subsequent quality loss during frozen storage.

A snow crab product that is popular in the Japanese market is called green crab, which is raw frozen snow crab sections. It is very important to handle, process, and freeze the snow crab quickly and carefully because of the tendency for a bluish to black discoloration to develop in the sections. Although there is no general agreement on the cause(s) for the discoloration (12), it may be the result of the enzymatically mediated oxidation and polymerization of phenolic compounds naturally present in crab (2). Thus antioxidants, sulfites, and chelating agents have been used to block the bluing discoloration. However, speed or quick processing is the best assurance in preventing the bluing, which will still occur and even develop at a faster rate after the frozen raw crab section is thawed. If crab are not cooked properly to destroy these enzymes, then the crab should be further processed quickly or consumed soon after thawing.

DUNGENESS CRAB

Dungeness (*Cancer magister*) crabs inhabit sandy and grassy bottoms along the Pacific Coast from California to the Aleutian Islands in Alaska. They are found at various depths below the intertidal zone but are usually harvested commercially offshore in depths from 12 to 120 ft. Dungeness crabs can grow to a weight of more than 3 lb; however, the commercial catch usually ranges between 1.25 and 2.25 lb. They are harvested with circular Iron pots covered with wire mesh, 3 ft in diameter and constructed with two entrance tunnels on the sides. Pots are hauled up every day or two, the legal-size male crabs are placed in seawater wells in the boat either dry or flooded, and the crabs are delivered alive at the plant. Only live, vigorous crabs are processed (13).

Product Forms

Dungeness crabs are marketed in a variety of forms, although the current trend is more toward whole crabs, both live and cooked.

Live Crab. Live Dungeness are commonly air freighted in 50-lb wetlock boxes (14). As with snow crab, the live crab may be packed in sawdust that has been saturated with water.

Whole-Cooked Crab. For Dungeness crab planned for fresh sales, whole crab are cooked for 20 to 25 min in salted (4–5%) boiling water in stainless steel batch or continuous cookers. Adding citric acid (0.1–0.2%) to the saltwater may enhance the color of the cooked crab by bleaching the surface of the shell, creating a whiter color. The hot crabs are cooled in cold water and, if the crab are to be sold fresh, packed in ice immediately. Properly handled, the shelf life should be up to seven days. If not sold fresh, the crab is frozen as soon after cooking as possible. Brine freezing has become very popular, but any rapid freezing system can be as good if care is taken to avoid any loss of quality from dehydration.

Meat. Dungeness crab sections intended for picking are prepared in the same manner as king crab and generally cooked in fresh (unsalted) boiling water for 10 to 12 min in either stainless steel batch or continuous cookers. After cooling the sections in cold water, the meat is removed from the crab section by hand. The steps in meat extraction have been well documented (15). Generally, while holding the legs of a section with one hand, the body is pressed with the free hand against a stainless steel table and then the key bone is removed. This step frees the body meat, which can be dislodged from the shell by striking the section and hand vigorously against the side of an aluminum or stainless steel pan. Then, the shell associated with the body is broken free of the legs just below the joint of each leg. Using a small metal mallet and anvil the top section of each leg is hit to crack the shell. The leg meat is then removed by striking the leg against the pan. This step is repeated at each joint to remove the leg meat. Body and leg meat are usually kept separate. The meat is placed in a strong salt brine (25% sodium chloride solution) to separate the bits of shell from the meat. The meat floats to the top of the tank and is conveyed out of the tank, where it is immediately inspected under ultraviolet light (shell pieces will fluoresce), then washed with cold water sprays to remove excess salt.

The body and leg meats are weighed separately and packed in about equal proportions in #10 C-enamel cans holding 5 lb net weight of drained crab meat or in #2 cans holding 1 lb of meat. The cans are sealed under a low vacuum, frozen in a blast freezer, and stored at −18 to −23°C (0 to −10°F) or colder. Frozen Dungeness crab meat stores well only if protected by suitable packaging against dehydration and oxidation and is stored at −23°C (−10°F) or lower. Under these conditions it stores well for six months.

BLUE CRAB

Blue crabs (*Callinectes sapidus*) are found along the Atlantic and Gulf Coasts from Massachusetts to Texas. Essentially a shallow water crab, blue crabs live in bays, sounds, and channels near the mouths of coastal rivers. Blue crab, when fully grown, average 5 to 7 in. across the carapace. The top of the shell is generally green and the bottom is a whitish shade. The name blue crab comes from the coloring of the legs, of which there are five pairs. The hind legs, flattened at the outer ends, are used for swimming. At times, using their hind swimming legs, blue crab can swim beautifully through the water with great speed and ease.

Harvesting

Catching methods depend on season, regional preferences, and state regulations. Crabs are taken by crab pots, trotlines, dredges, scrapes, and dip nets. The trotline is a long length of rope with pieces of bait attached at intervals. It is laid on the bottom, ends anchored and marked with buoys. When the fisherman collects his catch, he runs his boat along the line, forcing it to pass over a roller attached to the boat. As the boat moves forward, the crabs cling to the bait until they reach the surface where they are caught with a dip net and placed in a basket or barrel. Fishing boats are usually small. Fishing is usually a one- or two-person operation, employing inexpensive gear. The pots used for blue crab are cubical in shape, 2 ft on each side, and covered with chicken wire. The pot consist of two parts: the crab enter the lower or bait compartment and then pass into the upper or trap compartment through a slit in the partition. The pots are lifted daily, and the crabs are removed from the top compartment.

Processing

At the plant, the crabs are weighed and dumped into baskets for cooking in vertical retorts. Cooking conditions vary based on type of equipment used, but most plants use steam at 121°C (250°F) for 9 to 20 minutes (4,16). A few plants use boiling water for 15 to 20 min. After cooling, many processors deback the cooked crabs, and then wash and refrigerate the crabs overnight before picking the meat.

The meat is picked primarily by hand and divided into three categories: the lump meat, which is the large muscle controlling the swimming legs; the regular or flake meat, which is the remainder of the meat from the body; and claw meat. Total meat yields consist of 50%, flake meat; 25%, lump meat; and 25%, claw meat and varies from 12 to 17% of the weight of the whole crab.

Blue crab meat does not freeze and store well; therefore, only a small volume of frozen meat is produced. Most of the meat is packed in hermetically sealed cans or plastic bags and sold in the fresh chilled form. To extend the marketing period for the chilled product from several days to several weeks, a heat pasteurization process was developed and has been used successfully for 1-lb sealed cans of meat, which must be stored and distributed at 0°C (32°F) or as near to that as practicable (16).

Also, a popular and important product is the soft-shell crab, a product common only in the blue crab industry. Soft-shelled crabs are crabs that have just molted. They are considered a delicacy and bring higher prices. The entire body of a soft-shelled crab may be eaten after cooking. Because the optimal period for shipping soft crabs is so short, crab fishermen of the Atlantic and Gulf Coasts are by necessity close observers of the molting cycle. As the molt approaches, the peelers move from their normal foraging areas to shallow beds of marine grasses near the shoreline. The peelers are collected in pots or with trawl-like devices called scrapes and held in peeler floats, large floating containers. Immediately after molting, the now soft crab is removed, graded for size, and shipped in a refrigerated container to market. Occasionally, soft-shelled crabs are frozen. For freezing, the crabs may be held in refrigerated storage for two to three days, then are killed, eviscerated, washed, wrapped individually with parchment, packed in a carton, and frozen (4).

Handling Tips for Crab

A common problem with processed crab is a discoloration called bluing, although the actual color may range from light blue to blue-gray to black. Despite almost a century of study, general agreement on the causes of or cures for bluing in crab has not been reached (12). However, a scheme for bluing relating the discoloration to the enzymatically mediated oxidation and polymerization of phenolic compounds naturally present in crab has been proposed and can be used to illustrate the importance of quick and proper handling of crab (2).

Only crabs that are alive and in good condition should be processed. Crabs that have been held in a cooler (1–3°C) even for up to four days are edible if properly cooked. However, the chances of bluing occurring in the cooked crab are greatly increased even if proper cooking times and temperatures are used because the compounds causing the bluing have already been formed (2). Certainly, care must be taken to properly cook the crab, since underprocessing will not destroy the enzymes responsible for the bluing. Molting is a factor in the bluing reaction, since the concentration of phenolic compounds responsible for bluing is very high as they are involved in the formation of the new shell (sclerotization). So, during times of molting, extra care must be used in handling and processing crab to prevent bluing.

If the crab meat is to be thermal processed, the cooking time and temperatures of the raw crab are not as critical. However, as soon as the crab is cooked, the meat from the shell must be removed and thermal processed. Regardless of how long the live crab are cooked, any delay in thermal processing the crab meat will result in a higher incidence of bluing. Using frozen crab sections to produce canned crab meat will also increase the risk of bluing. Whole crab or crab sections should be thoroughly cooked before freezing to reduce the risk of bluing. Finally, avoid any contact of copper or iron with the crab since these metals can greatly intensify the bluing discoloration.

BIBLIOGRAPHY

1. M. C. Holliday and B. K. O'Bannon, eds., *Fisheries of the United States, 1988*, Current Fishery Statistics No. 8800, U.S. Commerce, National Oceanic and Atmospheric Administration, National Marine Fisheries Service, Washington, D.C., 1990.
2. J. K. Babbitt, "Blueing Discoloration of Dungeness Crabmeat," in R. E. Martin et al., eds., *Chemistry and Biochemistry of Marine Food Products*, AVI Publishing, Westport, Conn., 1982, p. 423.
3. E. B. Dewberry, "Speed is the Essence," *Food Processing Industry* **39**, 50–53 (1970).
4. J. A. Dassow, "Preparation for Freezing and Freezing of Shellfish," in D. K. Tressler and C. F. Evers, eds., *The Freezing Preservation of Foods*, 4th ed., Vol. 3. AVI Publishing, Westport, Conn., 1968, p. 266.
5. P. Redmayne, "Crab: King, Snow, and Dungeness," *Seafood Leader* **2**, 72 (1982).
6. I. Osakabe, "Low Cook Produces High Quality in Kegani Crab," *Pacific Fisherman* **55**, 48 (1957).
7. J. K. Babbitt, D. K. Law, and D. L. Crawford, "Effect of Precooking on Copper Content, Phenolic Content and Blueing of Canned Dungeness Crabmeat," *J. Food Sci.* **40**, 649 (1975).
8. National Food Processors Association, *Processes for Low-Acid Canned Foods in Metal Containers*, Bulletin 26-L, NFPA, Washington, D.C., 1976.
9. C. Crapo, B. Paust, and J. K. Babbitt, *Recoveries and Yields from Pacific Fish and Shellfish*, Marine Advisory Bulletin No. 37, Alaska Sea Grant College Program, Fairbanks, Alaska, 1988.
10. R. A. Krzeczkowski, R. D. Tenney, and C. Kelley, "Alaska King Crab: Fatty Acid Composition, Carotenoid Index and Proximate Analysis," *J. Food Sci.* **36**, 604 (1974).

11. R. A. Krzeczkowski and F. E. Stone, "Amino Acid, Fatty Acid and Proximate Composition of Snow Crab. (*Chionoecetes bairdi*)," *J. Food Sci.* **39**, 386 (1974).
12. D. D. Boon, "Discoloration in Processed Crabmeat. A Review," *J. Food Sci.* **40**, 756 (1975).
13. J. K. Babbitt, *Improving the Quality of Commercially Processed Dungeness Crab*, Oregon State University Extension Marine Advisory Program SG65, Corvallis, Oreg., 1981.
14. H. J. Barnett, R. W. Nelson, and P. J. Hunter, "Shipping Live Dungeness Crabs by Air to Retail Market," *Commercial Fisheries Review* **31**, 21 (1969).
15. B. Engesser, *Crab Meat Extraction* [training film], College of Engineering, Oregon State University, Corvallis, Oreg., 1972.
16. T. M. Miller, N. B. Webb, and F. B. Thomas, *Technical Operations Manual for the Blue Crab Industry*, University of North Carolina Sea Grant Publication UNC-SG-74-12, Special Scientific Report No. 28, Raleigh, N.C., 1974.

JERRY K. BABBITT
National Marine Fisheries Service
Kodiak, Alaska

CRANBERRIES. See FRUITS, TEMPERATE.

CULTURAL NUTRITION

A LITERAL MELTING POT

Most of us can easily answer the question: What are your favorite foods? Generally, we respond with tastes developed during childhood. Basic food habits usually prove resistant to change.

Despite their food habits, Americans have a wide variety of food tastes. This variety is due partly to the significant mobility of Americans, partly to a wide variety of choices, and partly to such trends as eating at restaurants and eating ethnic. It is considered stylish to sample the cuisines of other countries.

At home, though, we usually prefer less variation. Our food patterns tend to follow family history—and especially what our parents liked and disliked. Where an ethnic community is established, food habits can become readily apparent. Especially in major cities, it is not uncommon to find neighborhood grocery stores that have shelves of food catering to ethnic tastes of a significant segment of the surrounding community.

Personal food patterns become especially important when we are old and when we are ill. Success in encouraging such patients to eat often depends on including their special preferences in their diets. This article addresses food patterns of people in the United States of different regions, ethnic origins, cultural backgrounds, religious backgrounds, and life-styles.

REGIONAL DIFFERENCES

For many years, obvious regional food patterns could be found in the United States. Those pronounced differences have faded considerably as techniques for processing, storing, and transporting foods have advanced. National advertising has had its effects, as has the mobility of our population.

In certain parts of the country, though, remnants of regional food preference persist. Such preferences often reflect availability that has spawned popularity. For example, seafood enjoys popularity in coastal areas because of ready availability, lower cost that reflects lower transportation expense, and local industry advertising. Preference is thus influenced by climate, geography, and economics.

In California, the state's agricultural richness affects preference, as do ethnic demographics—characteristics of human populations. California's large quantities of fruits and vegetables, the popularization of the salad as a part of the meal, and the large numbers of Mexican-Americans, Orientals, and Italians all combine to give California its distinct cuisine.

In the southern states, corn, fish, and rice are often used. Hot breads, such as biscuits or corn bread, accompany meals. Green leafy vegetables are cooked with a piece of fatback (pork back fat), a regional practice that yields a consumable liquid known as pot liquor. Before mechanical refrigeration was developed, milk and cheese were consumed only in small quantities; storage and shipping were not possible. The national consumption of dairy products has shown an increase since the 1940s and 1950s.

ETHNIC, CULTURAL, AND RELIGIOUS INFLUENCES

Each ethnic, cultural, and religious group has certain characteristic food patterns or preferences. Sometimes these characteristics remain unique to the group, but more often they become a part of our nation. In that regard the United States can be described as a melting pot, both figuratively and literally. The remaining part of this article explores such influences.

Europeans

Many of the foods we have come to think of as typically American were in fact brought to us from European countries. Norwegian, Swedish, and Danish immigrants brought a greater use of milk, numerous cheeses, cream, and butter. In their native countries, Western Europeans relied on fish, shellfish, and vegetables. Especially popular were potatoes, dark breads, eggs, cheese, beef, pork, and poultry.

Central Europeans favored potatoes, rye flour, wheat, pork, sausage, and cabbage. Cabbage was especially favored and was eaten raw, cooked, or salted as sauerkraut. Common vegetables included turnips, carrots, squash, onion, beans, and greens.

Italian in the United States have perpetuated many favorite foods of their ancestors and have concurrently made them basics of the American diet. The huge sale of pizza and pasta products and the number of Italian restaurants confirm this fact. There are nearly as many pizza stands in the United States as there are hamburger and fried chicken businesses. Daily, the typical Italian diet features pasta made from hard-wheat dough. Its innumerable forms are seasoned with sauces, onions, tomatoes, cheese, peppers, and meat. Bread usually accompanies the meal.

Differences exist between northern and southern Italian foods. The northern area of Italy is more industrialized, and meat, root vegetables, and dairy products are more popular there. Southern Italy's preference tends toward fish, spices, and olive oil. The southern Italian diet can benefit from more green vegetables, eggs, fruits, meat, and milk. The last two food items can especially benefit the nutrient intake of children.

Experienced dietitians and nurses know that hospitalization can be especially difficult for Italians who have a strong sense of family and a strong preference for traditional foods. Institutional diets that cater to Italian patients can both reduce feelings of isolation and increase nutritional intake.

Native Americans

It may come as a surprise that more than half the plant foods you eat—including corn, potatoes, squash, pumpkin, tomatoes, peppers, beans, wild rice, cranberries, and cocoa—come from North, Central, and South American Indians. The Native American diet also used acorns, wild fruit, fish, wild fowl, small and large game, and seafood. The food preservation methods that produce jerky and pemmican were introduced to the frontier Americans by the Native Americans. Corn, which has become a world staple, is a significant example of their culture's contribution.

Diet varied from tribe to tribe depending on geographic location. Historically, the natural resources of the tribe's home area determined such occupations as hunting, fishing, agriculture, and herding. Native Americans who lived near the coast used large amounts of seafood. Those in the Northwest were and still are great fishermen, and from Alaska to the Columbia River one of the most prized items is salmon. A favorite method of preserving fish and meat is smoking; racks of drying fish can be seen along the river banks where Native Americans live. Along the New England coast, shellfish is a favorite, especially clams. A favored method of cooking is open-air baking of clams and fish over hot coals. Sometimes seafood is baked in the sand. The item is wrapped in seaweed or similar materials and placed in sand kept hot by a fire.

In the past, the Plains Indians depended more on game, fowl, and freshwater fish. Native vegetables and fruits were generously used and cultivated if the group was a stationary, agricultural one. Many vegetables that originated in North and South America have become part of the American and world diet. Because little food storage was possible, nutritional life was characterized by the extremes of bounty and scarcity. Despite such fluctuations in availability, the historical Native American diet was undoubtedly more nutritional than is their diet today.

With the introduction of trading-post products such as lard, sugar, coffee, and canned meat and canned milk, the American Indian diet began to assimilate European food customs, leaving little today that can be distinguished as a Native American food pattern.

Native American food specialties that can be found today include clambakes, fry bread (biscuit dough fried in fat), and wajupi, a pudding made from berries, sugars, and cornstarch. Northern California Indians have revived the use of acorns in preparing soup and flour, methods of preparing eels, and stews combining meat, vegetables, and berries. Often these traditional foods play a part in tribal ceremonies. As cultural renewal has taken place among Native Americans, and as the Native American heritage is becoming better understood, interest in traditional Native American foods is being revived.

The nutritional weakness of the Native American diet is a low intake of milk by children after weaning. Vitamin C intake is also low, and the greater use of fruits, vegetables, and lean meat will improve the diet.

The present generation of Native Americans may be consuming too much sugar and soft drinks. Sound nutritional education is needed. Extensive alcohol abuse among some adult males has also hurt the nutritional status of these people.

Black Americans

Among middle- and upper-income blacks, food selections are similar to those of their white counterparts. Among poor families, many of whom have come from or still live in the southern states, the traditional eating pattern is determined to a large extent by the cost and availability of foods. However, there are many food items preferred by black families, rich or poor.

The popularization of the term *soul food* indicates the effort to recognize the uniqueness of black American food habits. One characteristic black food habit is the use of hominy grits, especially for breakfast with some form of pork. Corn bread, muffins, or biscuits are preferred to yeast bread. There are differences between the dietary habits of southern blacks and northern blacks, and one of these differences is in the customary use of greens.

Among southern blacks, fresh greens such as mustard, turnip, and collard are popular and are usually served with pork such as fatback or salt pork. In the north, fresh greens are less available, and a much wider variety of frozen greens are used. Other vegetable favorites include fresh corn, lima beans, cabbage, sweet potatoes, and squash. The use of fresh fruit emphasizes oranges, watermelon, and peaches (Table 1).

Black-eyed peas and corn are well liked, as indicated earlier; a mixture of black-eyed peas and rice called Hoppin John is a favorite southern dish. A favorite method of preparing green vegetables is to boil them with a small amount of fat pork; the resultant juice (pot liquor) is also used in the meal. Potatoes (white and sweet) and yams are popular, and sweet potatoes are often made into pies.

Carbohydrate is generally obtained from grits, potatoes, and rice, though black-eyed peas and dried beans are also used. The last-named provide some protein, as does fried fish, especially when locally caught. Other meats frequently used include poultry, pork (both cured and fresh), and wild game. All parts of an animal are used, especially among poor families. Dishes made from a slaughtered pig, for instance, include chitlings (the small intestines, cleaned and fried crisp); hog maws (the stomach lining); boiled pig's feet; and neckbone, jaws, snout, and head,

Table 1. Characteristic Black Americans' Food Choices

Protein foods	Milk and milk products	Grain products	Vegetables	Fruits	Other
Meat	Milk	Rice	Broccoli	Apples	Salt pork
Beef	Fluid	Corn bread	Cabbage	Bananas	Carbonated beverages
Pork and ham	Evaporated, in coffee	Hominy grits	Carrots	Grapefruit	Fruit drinks
Sausage	Buttermilk	Biscuits	Corn	Grapes	Gravies
Pig's feet, ears, etc.	Cheese	Muffins	Green beans	Nectarines	
Bacon	Cheddar	White bread	Greens	Oranges	
Luncheon meat	Cottage	Dry cereal	Mustard	Plums	
Organ meats	Ice cream	Cooked cereal	Collard	Tangerines	
Poultry		Macaroni	Kale	Watermelons	
Chicken		Spaghetti	Spinach		
Turkey		Crackers	Turnips, etc.		
Fish			Lima beans		
Catfish			Okra		
Perch			Peas		
Red snapper			Potatoes		
Tuna			Pumpkins		
Salmon			Sweet potatoes		
Sardines			Tomatoes		
Shrimp			Yams		
Eggs					
Legumes					
Kidney beans					
Red beans					
Pinto beans					
Black-eyed peas					
Nuts					
Peanuts					
Peanut butter					

Source: Ref. 1.

which are cooked and made into scrapple. Frying, stewing, and barbecuing are the favored methods of preparation. Sweets such as molasses, pastries, cakes, and candy are especially popular.

Milk consumption, which is often low, should be encouraged, especially among children. Buttermilk and ice cream are well liked by black Americans, although there is little use of cheeses. The use of more citrus fruits should be encouraged.

As the economic status of black people improves, they eat more meat and their diet improves generally. Because their typical diet is high in fat and carbohydrate and low in protein, iron, and vitamin C, an increase in consumption of lean meat, milk, and citrus fruits will help balance their nutrient intake.

It is in preparation that the essence of the term soul food can be found. It is not necessarily the unique foods but rather the care in preparation that is emphasized. Food preparation offers the opportunity to minister to the physical health and well-being of those who will consume the food. The opportunity to bring happiness, health, and love through food preparation lies at the figurative and literal heart of the customary black dietary habits.

Jews

Orthodox Jews follow Old Testament and rabbinical dietary laws. Conservative Jews distinguish between meals served in the home and those outside it. Reformed Jews do not observe dietary regulations. The country of origin can further influence a Jewish family's dietary practices.

Foods used according to strict Jewish laws are referred to as Kosher. Those who follow those dietary laws eat only animals that are designated as clean and are killed in a ritualistic manner. The method of slaughter minimizes pain and maximizes blood drainage. Blood, the symbol of life, is strictly avoided by soaking the meat in cold water, draining it, salting it, and rinsing it three times. Permitted meats include poultry, fish with fins and scales, and quadruped animals that chew the cud and have divided hooves. (Thus, pork cannot be eaten.) The hindquarter of quadruped animals must have the hip sinew or the thigh vein removed.

Kosher laws additionally require separation of meat and milk. Milk and its products must be excluded from meals involving meat but can be consumed before the meat meal. If a milk meal is to be consumed, meat and its products are excluded from the milk meal and for 6 hours thereafter. Usually two milk meals and one meat meal are eaten each day. A Kosher household will maintain two sets of dishes, utensils, and cooking supplies, one for meat meals and one for milk meals.

Certain foods, viewed as neutral, are referred to as pareve. Fruits, uncooked vegetables, eggs, and clean fish, because they are recognized as neutral, can be consumed with either meat or milk meals.

No food can be cooked or heated on Saturday, the Sabbath, so the evening meal preceding the Sabbath tends to the most substantial of the week. Both chicken and fish are served at that meal. Any foods eaten on the Sabbath must be cooked on a preceding day.

On Jewish holidays, symbolic foods are eaten. For example, Passover, a spring festival, commemorates the flight of the Jews from Egypt. Passover celebration lasts 8 days, and only unleavened bread, called matzo, is permitted. An Orthodox household will, therefore, use separate utensils for preparing unleavened bread to avoid any contact with leavening. Other holidays include Rosh Hashanah (Jewish New Year), Sukkoth (fall harvest), Chanukkah (the festival of lights), Purim (the arrival of spring), and Yom Kippur (the day of atonement) when fasting occurs.

Mexican-Americans

People of Mexican descent make up a large part of the population of the Southwest. They therefore reside close to their native country and maintain a close tie to traditional Mexican foods. A lack of refrigeration in such warmer climates creates a scarcity of meat, milk, eggs, vegetables, and fruits. For lower-income families, the expense of these nutrient-rich foods can be prohibitive.

Items basic to the Mexican diet include dry beans, chili peppers, tomatoes, and corn. The corn is ground, soaked in lime water, and baked on a griddle to make tortillas, the popular flat breads that can be rolled and filled with ground beef and vegetables to make tacos. Tacos provide nutrients from all four food groups, although the user of wheat tortillas reduces the calcium obtained from lime-soaked corn tortillas. Many foods are fried, and beans are even refried until they absorb all the fat.

Nutritionists encourage retention of the basic Mexican diet with only minor modifications. Because of iron, vitamin A, and calcium deficiencies, use of lean meat is recommended in taco fillings, dark green and yellow vegetables, fruits, and milk—dried if fresh milk is too expensive—are also recommended. The milk is especially important for children. One Mexican dietary practice that is particularly encouraged is the practice of limiting the amount of sweets consumed (Table 2).

Puerto Ricans

The Puerto Rican diet has similarities to the dietary patterns of other Caribbean Islands but is noteworthy because so many Puerto Ricans emigrate to the United States.

Traditional emphasis is on beans, rice, and starchy root vegetables (and plantains) known as viandas. Rice and beans eaten together are a high-quality, complementary protein combination. The beans are often boiled and then cooked with sofrito, a mixture made from tomatoes, green peppers, onions, garlic, salt pork, lard, and herbs. Viandas are good sources of B vitamins, iron calories, and, in some cases, vitamin C. Salt codfish is more popular than fresh fish. Chicken, beef, and pork are favored. Coffee with 2 to 5 oz of milk per cup—cafe con leche—is drunk several times a day, contributing to an otherwise low consumption of milk. Fruits prove an especially popular food owing to availability, with such native fruits as papaya, mango, and acerola (the West Indian cherry, which is extremely high in vitamin C) being joined by bananas, oranges, and pineapple.

The mainstays of the Puerto Rican diet may be hard to find or expensive in the United States. And because Puerto Ricans come from an agrarian background and thus often lack the job skills for a highly industrialized society, they often find themselves in the lowest socioeconomic classes in the United States, living in crowded urban conditions with poor cooking and refrigeration facilities and unable to feed their families as well as they did in Puerto Rico.

Table 2. Characteristic Mexican-American Food Choices

Protein foods	Milk and milk products	Grain products	Vegetables	Fruits	Other
Meat	Milk	Rice	Avocados	Apples	Salsa
Beef	Fluid	Tortillas	Cabbage	Apricots	(tomato-pepper-onion
Pork	Flavored	Corn	Carrots	Bananas	relish)
Lamb	Evaporated	Flour	Chilies	Guavas	Chili sauce
Tripe	Condensed	Oatmeal	Corn	Lemons	Guacamole
Sausage (chorizo)	Cheese	Dry cereals	Green beans	Mangos	Lard (manteca)
Bologna	American	Cornflakes	Lettuce	Melons	Pork cracklings
Bacon	Monterey jack	Sugar coated	Onions	Oranges	Fruit drinks
Poultry	Hoop	Noodles	Peas	Peaches	Kool-aid
Chicken	Ice cream	Spaghetti	Potatoes	Pears	Carbonated
Eggs		White bread	Prickly pear	Prickly pear	beverages
Legumes		Sweet bread	cactus leaf	cactus fruit	Beer
Pinto beans		(pan dulce)	(nopales)	(tuna)	Coffee
Pink beans			Spinach	Zapote	
Garbanzo beans			Sweet potatoes	(or sapote)	
Lentils			Tomatoes		
Nuts			Zucchini		
Peanuts					
Peanut butter					

Source: Ref. 1.

Consequently, malnutrition is not uncommon among their children. Also, Puerto Rican children born in the United States may have adopted favorite mainland foods such as hamburgers, hot dogs, canned spaghetti, and cold cereals instead of the traditional Puerto Rican food prepared by their mothers. If these children have picked up the mainland habit of snacking on nutrient-light foods, they may be more poorly nourished than people adhering to the relatively inexpensive but nutritionally adequate traditional diet of Puerto Rico.

Nutritionists encourage Puerto Ricans living in northern mainland cities to become familiar with different fruits, which are inexpensive when they are in season or canned; to use more milk, cheese, and inexpensive cuts of meat; and to substitute canned tomatoes for fresh tomatoes, which are expensive when not in season.

Middle Eastern People

Middle East people—Greeks, Iranians, Arabs, Turks, Armenians, and Lebanese—tend to be farmers. Dietary emphasis is on crops and animals raised—cattle, sheep, goats, chickens, ducks, geese, grains, fruits, and vegetables. Lamb holds the position of the favored meat, and wheat products, grains, and rice are the major energy sources. Popular dairy foods, derived from sheep, goats, and camels, are sour milk, including yogurt, fermented milk, and sour cream.

Lentils and beans may be boiled or stewed with tomatoes, onions, and olive oil; they may be eaten alone or mixed with other foods. A favorite combination is seasoned chick-peas mixed with bulgur (wheat that has been steamed, cracked, and fried) and spices and then fried in fat.

Traditional vegetables include okra, squash, tomatoes, onions, leeks, peppers, spinach, brussel sprouts, cabbage, peas, green beans, dandelion greens, eggplant, artichokes, and olives. Grape leaves, used either fresh or canned, are stuffed with rice, bulgur, meat, and seasoning.

Fruits grown in these warm climates are used extensively. Some common ones are dates, figs, melons, cherries, oranges, apricots, and raisins.

Common Middle Eastern cereals include rice, wheat, and barley. Typical breads are flat, thin, and round and may be baked outdoors. Olive oil and butter from sheep's and goat's milk are used generously. Turkish coffee, a favorite beverage in the Middle East, is a strong, dark drink containing the crushed coffee bean that is served with a

Table 3. Characteristic Chinese Food Choices

Protein foods	Milk and milk products	Grain products	Vegetables	Fruits	Other
Meat	Flavored milk	Rice	Bamboo shoots	Apples	Soy sauce
Pork	Milk (cooking)	Noodles	Beans	Bananas	Sweet and
Beef	Ice cream	White bread	Green	Figs	sour sauce
Organ meats		Barley	Yellow	Grapes	Mustard sauce
Poultry		Millet	Bean sprouts	Kumquats	Ginger
Chicken			Bok choy	Loquats	Plum sauce
Duck			Broccoli	Mangos	Red bean paste
Fish			Cabbage	Melons	Tea
White fish			Carrots	Oranges	Coffee
Shrimp			Celery	Peaches	
Lobster			Chinese cabbage	Pears	
Oyster			Corn	Persimmons	
Sardines			Cucumbers	Pineapples	
Eggs			Eggplant	Plums	
Legumes			Greens	Tangerines	
Soybeans			Collard		
Soybean curd (tofu)			Chinese broccoli		
Black beans			Mustard		
Nuts			Kale		
Peanuts			Spinach		
Almonds			Leeks		
Cashews			Lettuce		
			Mushrooms		
			Peppers		
			Potatoes		
			Scallions		
			Snow peas		
			Sweet potatoes		
			Taro		
			Tomatoes		
			Water chestnuts		
			White radishes		
			White turnips		
			Winter melons		

Source: Ref. 1.

generous amount of sugar. The Middle Eastern diet is a good one, contributing adequate nutrients. But Middle Eastern people in the United States may have difficulty obtaining the foods if they cannot afford them, as they are usually expensive.

Asians

The United States is also home to many people of Asian heritage. The diets of the Chinese, Japanese, and Filipino are discussed below.

Chinese. China is a vast country with many different regional foods. In the United States, most of the Chinese on the West Coast come from southern China and adhere to a Cantonese diet. This type of cookery uses little fat and subtle seasonings. Beef, pork, poultry, and all kinds of seafoods are well liked. In Chinese cooking, all parts of the animal are utilized, including organs, blood, and skin. Rice is the predominant cereal used by the Cantonese, both at home and in the United States. Northern Chinese use more bread, noodles, and dumplings, which are prepared from wheat, corn, and millet (Table 3).

The Chinese diet is varied, containing eggs, fish, meat, soybeans, and a great variety of vegetables. Many green, leafy vegetables unfamiliar to Americans, such as leaves of the radish and shephard's purse, are enjoyed. Sprouts of bamboo and beans are incorporated into some dishes, giving a distinctive flavor and texture. Many Chinese recipes call for mushrooms and nuts. Eggs from ducks, hens, and pigeons are widely used—fresh, preserved, and pickled. Soy sauce is used both in preparation and serving. The high salt content of this flavorable condiment is a problem for Chinese patients whose salt intake may be restricted by the physician.

Chinese cookery is unique in many ways. Food is quickly cooked over the heat source, usually cut into small pieces so that the short cooking period is adequate. Vegetables retain their crispness, flavor, and practically all their nutrients; this method of cookery, which has been widely adopted, is quite beneficial. When vegetables are cooked in water, the liquid is also consumed.

Probably the greatest weakness, in the Chinese diet is a low intake of milk and milk products, a consequence of lactose intolerance. However, a higher consumption of meat and the frequent use of soybeans prevent calcium and protein deficiency. In addition, for undefined reasons, Chinese children born in the United States have a higher tolerance for milk.

Japanese. Japanese people who have emigrated to the United States, especially the older people, tend to follow

Table 4. Characteristic Japanese Food Choices

Protein foods	Milk and milk products	Grain products	Vegetables	Fruits	Other
Meat	Milk	Rice	Bamboo shoots	Apples	Soy sauce
Beef	Ice cream	Rice crackers	Bok choy	Apricots	Nori paste (used to season rice)
Pork	Cheese	Noodles (whole wheat noodle called soba)	Broccoli	Bananas	Bean thread (konyaku)
Poultry		Spaghetti	Burdock root	Cherries	Ginger (shoga; dried form called denishoga)
Chicken		White bread	Cabbage	Grapefruits	Tea
Turkey		Oatmeal	Carrots	Grapes	Coffee
Fish		Dry cereals[b]	Cauliflower	Lemons	
Tuna			Celery	Limes	
Mackerel			Cucumbers	Melons	
Sardines (dried form called mezashi)			Eggplants	Oranges	
Sea bass			Green beans	Peaches	
Shrimp			Gourd (kampyo)	Pears	
Abalone			Mushrooms	Persimmons	
Squid			Mustard greens	Pineapples	
Octopus			Napa cabbage	Pomegranates	
Eggs			Peas	Plums (dried pickled plums called umeboshi)	
Legumes			Peppers	Strawberries	
Soybean curd (tofu)			Radishes (white radish called daikon; pickled white radish called takawan)	Tangerines	
Soybean paste (miso)			Snow peas		
Soybeans			Spinach		
Red beans (azuki)			Squash		
Lima beans			Sweet potatoes		
Nuts			Taro (Japanese sweet potato)		
Chestnuts (kuri)			Tomatoes		
			Turnips		
			Water chestnuts		
			Yams		

Source: Ref. 1.
[a]Nisei only.

the food habits of their homeland rather closely. Typical of the Japanese food pattern is much use of fish, both fresh and saltwater. Methods of preparation vary. Fish may be eaten raw or deep-fat fried, dried, or salted. Many kinds of vegetables may be prepared with meat or eaten separately. Protein intake comes from meat, fish, eggs, legumes, nuts, and, most popular of all, soybean curd (tofu), which is sold in many American grocery stores. Eggs are prepared in many ways and are well liked.

The basic cereal is rice, although since World War II wheat has been consumed. Milk and cheese are limited in the traditional Japanese diet, mainly because of lactose intolerance. Japanese born in the United States have a greater tolerance for dairy products such as milk and cheese, and they also eat more fruits. It is important that the Japanese in this country eat enriched, converted, or whole-grain rice, which contains more nutrients than unenriched rice with the husks removed. To avoid excessive nutrients loss, they should also be told to refrain from washing the rice repeatedly before preparing it (Table 4).

Filipinos. Filipino food patterns are similar to those of the Chinese and Japanese. The basic cereal is rice, and the principal sources of protein are fish, meat, eggs, legumes, and nuts. Meat is prepared by roasting, frying, or boiling; the fish used may be dried.

A large variety of vegetables and fruits is found in the diet. Vegetables are usually boiled or pan-fried, and some are used as salad dishes. Most fruits are eaten raw, but they are sometimes used in cooking (Table 5).

Again, because of lactose intolerance, Filipinos have a low intake of milk and milk products. The consumption of good calcium and protein sources should be encouraged, especially among the young. Growing children may also have a problem with low caloric intake. Larger servings of food should be encouraged, especially more meat; food preparers should also be instructed about the use of enriched rice without prewashing.

The Eskimos: Meeting of Two Cultures

For years, the Eskimos of the far north have maintained their culture on a diet consisting almost entirely of meat and fish. This culture has fascinated nutritionists, as it runs counter to all that is known about the necessity of a balanced diet.

Caribou, whale, and seals from the staples of the traditional diet, with walrus, fish, and birds being of minor importance. Edible plants and berries are scarce. Although it had been believed that Eskimos consumed the stomach contents of the animals, that assumption is debatable. There is agreement the Eskimo diet is very high in protein and fat and very low in carbohydrate.

Table 5. Characteristic Filipino Food Choices

Protein foods	Milk and milk products	Grain products	Vegetables	Fruits	Other
Meat	Milk	Rice	Bamboo shoots	Apples	Soy sauce
Pork	Flavored	Cooked cereals	Beets	Bananas	Coffee
Beef	Evaporated	Farina	Cabbage	Grapes	Tea
Goat	Cheese	Oatmeal	Carrots	Guavas	
Deer	Gouda	Dry cereals	Cauliflower	Lemons	
Rabbit	Cheddar	Pastas	Celery	Limes	
Variety meats		Rice noodles	Chinese celery	Mangos	
Poultry		Wheat noodles	Eggplants	Melons	
Chicken		Macaroni	Endive	Oranges	
Fish		Spaghetti	Green beans	Papayas	
Sole			Leeks	Pears	
Bonito			Lettuce	Pineapples	
Herring			Mushrooms	Plums	
Tuna			Okra	Pomegranates	
Mackerel			Onions	Rhubarb	
Crab			Peppers	Strawberries	
Mussels			Potatoes	Tangerines	
Shrimp			Pumpkins		
Squid			Radishes		
Eggs			Snow peas		
Legumes			Spinach		
Black beans			Sweet potatoes		
Chick peas			Tomatoes		
Black-eyed peas			Water chestnuts		
Lentils			Watercress		
Mung beans			Yams		
Lima beans					
White kidney beans					
Nuts					
Peanuts					
Pili nuts					

Source: Ref. 1.

The high protein content provides calories essential to an active life and additionally contributes to blood glucose, an essential metabolic fuel for the nervous system. What carbohydrate has been available has been obtained through synthesis of glycogen from muscles. A by-product of the traditional diet has been the virtual nonexistence of diabetes. This disease has, however, been on the increase since the introduction of sugar to the culture.

A unique characteristic of Eskimo nutrition is the utilization of fatty acids from oils as energy. Tissues can adapt to the utilization of ketone bodies instead of glucose as energy source. The ketone bodies form in the liver and are characteristic of high-fat, low-carbohydrate diets. Eskimos thus provide a human example of a unique metabolic phenomenon.

As we saw in the discussion of vitamins, we should expect scurvy among Eskimos. But their culture has escaped scurvy because they eat fish and meat either raw or only slightly cooked. What little vitamin C is available is not destroyed by oxidation in the cooking process.

Vitamins A, D, E, and K are obtained in sufficient quantities from the traditional Eskimo diet, as are B vitamins. The diet is low in calcium, a problem that surfaces in bone fragility among Eskimos. Children are somewhat protected by a traditionally long nursing period.

As nontraditional dietary habits have gained popularity in the Eskimo culture, our problems have become theirs. Heart disease, diabetes, tooth decay, and hypertension are now on the increase.

PEOPLE AND THEIR LAND

As you have probably determined from these discussions of culture and nutrition, our food habits are very much influenced by our geographic location. Relationships between people and their land can be profound, and they have given rise to fields of knowledge ranging from cultural geography to literary analysis focusing on agrarian mythology. Although it has been said that we are what we eat, perhaps we should also say that we eat where we are.

BIBLIOGRAPHY

This article has been adapted from Y. H. Hui, *Principles and Issues in Nutrition*, Jones and Bartlett Publishers, Inc., Boston, 1985. Used with permission.

1. *Nutrition During Pregnancy and Health*, California Department of Health, Sacramento, Calif., 1975.

GENERAL REFERENCES

L. Bakery, ed., *The Psychobiology of Human Food Selection*, Van Nostrand Reinhold, New York, 1982.

L. K. Brown and K. Mussell, eds., *Ethnic and Regional Foodways in the United States. The Performance of Group Identity*, University of Tennessee Press, Knoxville.

C. A. Bryant et al., 1985, *The Cultural Feast: An Introduction to Food and Society*, West Publishing Co., St. Paul, Minn., 1985.

P. Farb and G. Armelagos, *Consuming Passions: The Anthropology of Eating*, Houghton Mifflin, Boston, 1980.

P. Fieldhouse, *Food and Nutrition: Customs and Culture*, Methuen, New York, 1985.

Food and Nutrition Board, *What is America Eating?* National Academy Press, Washington, D.C., 1986.

Food and Nutrition Board, National Research Council, National Academy of Sciences, *Recommended Dietary Allowances*, 10th ed., National Academy Press, Washington, D.C., 1989.

N. Jerome et al., eds., *Nutritional Anthropology*, Redgrave Publishing, Pleasantville, N.Y., 1980.

F. E. Johnston, ed., *Nutritional Anthropology*, Alan R. Liss, New York, 1987.

P. G. Kittler and K. Sucher, *Food and Culture in America*, Van Nostrand Reinhold, New York, 1989.

A. W. Logue, *The Psychology of Eating and Drinking*, W. H. Freeman, New York, 1986.

J. M. Newman, *Melting Pot: An Annotated Bibliography and Guide to Food and Nutrition Information for Ethnic Groups in America*, Garland, New York, 1986.

W. Root and R. DeRochement, *Eating in America: A History*, Eco Press, New York, 1976.

D. Sanjur, *Social and Cultural Perspectives in Nutrition*, Prentice-Hall, Englewood Cliffs, N.J., 1982.

Y. H. Hui
American Food and Nutrition Center
Cutten, California

CULTURED MILK PRODUCTS

Cultured milk products are produced by the lactic fermentation of milk using various bacteria cultures. Some products may also have other fermentations taking place, for example, alcohol. These fermentations lead to the coagulation of milk and the production of typical cultured milk product flavor.

Fermented milk products originated in the Near East and then spread to parts of southern and eastern Europe. The earliest forms of fermented products were developed accidentally by nomadic tribes who carried milk from cows, sheep, camels, or goats. Under warm storage conditions, milk coagulated or clabbered due to the production of acid end products by lactic bacteria. Fortunately the predominant bacteria were lactic types and, therefore, helped to preserve the product by suppressing spoilage and pathogenic bacteria. Humans evidently enjoyed the refreshing tart taste of their discovery and began to handle milk so that this preserving action would be encouraged.

Milk and curdled milk products are mentioned throughout history dating back as far as 4000 B.C.: "He asked for water, and she gave him milk; in a bowl fit for nobles she brought him curdled milk" (Old Testament, Judges 5:25). There is also remarkable pictorial evidence that the custom of keeping milk in containers for later consumption was already a craft systematically practiced by the Sumerians around 2900 B.C. (1). Through applied scientific principles and advances in manufacturing technology, these early products have developed into a highly diversified group of foods that are popular throughout the world.

Actual figures for world consumption of fermented milk products are not known. According to the International Dairy Federation (IDF), an estimated 1 trillion lb of cow's milk is produced per year that is manufactured into cultured milk products. The northern European countries consume the greatest amount of fermented milk products on a per capita basis with Asian countries and the United States showing rapid increases. Worldwide increases in consumption are leading to advances in the fermented product manufacturing industry. One reason for the continued increase in product consumption is the successful development of functional foods/neutraceuticals that contain health-promoting probiotic cultures.

Throughout the world there are great differences in cultured products. These differences are due to variations in the cultures used and manufacturing principles. Products are most often classified as traditional or nontraditional; however, products can be classified according to culture medium (milk and cream), manufacturing procedure, further processing (packaging and addition of fruits, vegetables, meat, fish, or grains), end use (baking and consumption), and microbial action (type of bacteria or yeast and temperature). Traditional fermented milk products have a long history and are known and made all over the world. Their manufacture is crude and relies on ill-defined procedures and nonstandard cultures that lead to inconsistencies in product characteristics (1). The production of nontraditional products is based on sound scientific principles and leads to the creation of products with consistent characteristics. Cultures and manufacturing methods have been standardized to produce the highest quality possible. Despite some differences, most modern cultured products use the following basic manufacturing steps: (*1*) culture or starter selection and preparation; (*2*) milk processing and treatment (eg, pasteurization to kill undesirable bacteria, standardization of composition by separation of fat and addition of ingredients, and homogenization); (*3*) inoculation with bacterial culture; (*4*) incubation for culture growth and acid production; (*5*) agitation to break the coagulum; (*6*) cooling to stop bacterial growth and acid production; and (*7*) packaging. Table 1 lists the world's principal cultured milk products including type, location, and bacterial culture used.

MICROORGANISMS

A culture (starter) is a controlled bacterial population that is added to milk or milk products to produce acid and flavorful substances that characterize cultured milk products. Cultures may be a single strain of bacteria or a combination of cultures (mixed strain). Mixed strain cultures may be used to enhance the production of specific flavors or characteristics and, therefore, must be compatible and balanced. Some cultures will be antagonistic to each other while others will act in a symbiotic relationship.

The proper production of lactic acid is critical in fermented product manufacture. Lactic acid is not only re-

Table 1. Fermented Foods and Geographical Location of Production

Name	Type	Location	Bacteria
Acidophilus	Dairy	Europe, North America	*Lactobacillus acidophilus* *Bifidobacterium bifidum*
Bulgarian buttermilk	Dairy	Europe	*Lactobacillus bulgaricus*
Buttermilk	Dairy	North America, Europe, Middle East, North Africa, Indian subcontinent, Oceania	*Lactococcus cremoris, Lc. Lacticus, Lc. Lacticus* ssp *diacetylactis Leuconostoc cremoris*
Filmjolk	Dairy	Europe	*Lc. Cremoris, Lc. Lacticus, Lc. Lacticus diacetylactis, Lc. Cremoris, Alcaligenes viscosus, Geatrichus candidum*
Flummery	Cereal and dairy	Europe, South Africa	Naturally present lactic bacteria
Ghee	Dairy and miscellaneous	Indian Subcontinent, Middle East, South Africa, Southeast Asia	*Streptococcus, Lactobacillus, and Leuconostoc* spp
Junket	Dairy	Europe	*Streptococcus and Lactobacillus* spp.
Kefir	Dairy	Middle East, Europe, North Africa	*Streptococcus, Lactobacillus, and Leuconostoc* spp. *Candida kefyr, Kluyveromyces fragilis*
Kishk	Dairy and cereal	North Africa, Middle East, Europe, Indian subcontinent, East Asia,	*Streptococcus, Lactobacillus, and Leuconostoc* spp
Kolatchen	Cereal and dairy	Middle East, Europe	*Lc. Lacticus, Lc. Lacticus* ssp. *diacetylactis, Lc. Cremoris Saccharomyces cerevisiae*
Koumiss	Dairy	Europe, Middle East, East Asia	*Lc. Lacticus, Lb. bulgaricus, Candida Kefyr, Torulopsis*
Kurut	Dairy	North Africa, Middle East, Indian subcontinent, East Asia	*Lactobacillus and Streptococcus* spp. *Saccharomyces lactis, Penicillium*
Lassi	Dairy	Indian subcontinent, East Asia, Middle East, North Africa, South Africa, Europe	*Str. thermophilus, Lb. bulgaricus,* sometimes yeast
Prokllada	Dairy	Europe	*Streptococcus and Lactobacillus* spp
Sour cream	Dairy	Europe, North America, Indian subcontinent, Middle East	*Lc. cremoris, Lc. lacticus, Lc. lacticus* ssp. *diacetylactis*
Yakult	Dairy	East Asia	*Lactobacillus casei*
Yogurt	Dairy	Worldwide	*Str. thermophilus, Lb. bulgaricus*

Source: Ref. 2.

sponsible for the refreshing tart flavor of cultured products but is responsible for the destabilization of the milk protein structure (casein micelle) that allows the milk protein to coagulate, thus contributing to the product's body and texture characteristics. The culture may also contribute other flavorful compounds (Table 2). The major flavor compounds, other than lactic acid, encountered in fermented milk products are acetaldehyde, diacetyl, ethanol, carbon dioxide, and acetic acid. These flavor compounds are responsible for the unique flavor characteristics of various products and great care is taken to promote their production in certain products. For example, proper yogurt flavor depends on the production of acetaldehyde while proper cultured buttermilk flavor depends on the production of diacetyl. Products such as kefir and koumiss also undergo alcoholic fermentation with final levels of ethanol reaching 1 to 2.5%.

NUTRITION

The primary nutritional benefits of cultured milk products are due to their compositional makeup. All milk products are excellent sources of high-quality proteins (casein and whey), calcium, and vitamins (especially riboflavin and other B vitamins). Early in the twentieth century Eli Metchnikoff (1908) proposed in his book, *Prolongation of Life Optimistic Studies*, that the consumption of cultured milk products such as yogurt result in the prolongation of life. Metchnikoff based his claims on the ability of lactic bacteria to prevent putrefactive processes in the digestive tract (4). Although lactic cultures do produce antimicrobial effects from the production of acids, hydrogen peroxide, and antibiotics such as nisin, scientific evidence does not support Metchnikoff's claims of prolonged life due to the consumption of cultured products. However, there is growing scientific evidence that the addition of alternative organisms to cultured products such as *Lactobacillus acidophilus* and *Bifidobacterium* sp. may have therapeutic effects in the lower digestive system (5). Many of these organisms are believed to be able to pass through the upper digestive system and then produce beneficial effects in the lower intestine. Recent research suggest that the consumption of these probiotic cultures in cultured milk products could aid in the prevention of intestinal infections, diarrheal diseases, hypercholesterolaemia, lactose digestion, and cancer.

Table 2. Starter Cultures and Their Principal Metabolic Products

Starter organisms	Important metabolic products
Mesophilic bacteria (optimum growth temperature of 20–35°C)	
Lc. cremoris	Lactic acid
Lc. lactis	Lactic acid
Lc. lactis ssp. diacetylactis	Lactic acid, diacetyl
Leuc. cremoris	Lactic acid, diacetyl
Lb. acidophilus	Lactic acid
Lb. breves	CO_2, acetic acid, lactic acid
Lb. casei	Lactic acid
Thermmophilic bacteria (optimum growth temperature of >35°C)	
Str. thermophilus	Lactic acid, acetaldehyde
Lb. bulgaricus	Lactic acid, acetaldehyde
Lb. lactis	Lactic acid
Bifidobacteria	Lactic acid, acetic acid
Yeast	
Saccharomyces cerevisiae	Ethanol, CO_2
Candida (Torula) kefir	Ethanol, CO_2
Kluveromyces fragilis	Acetaldehyde, CO_2

Note: Lc. = *Lactococcus, Leuc.* = *Leuconostoc, Lb.* = *Lactobacillus, Str.* = *Streptococcus.*
Source: Ref. 3.

BIBLIOGRAPHY

1. M. Kroger, J. A. Kurmann, and J. L. Rasic, "Fermented Milks—Past, Present, and Future," *Food Technol.* **43**, 92–99 (1989).
2. G. Campbell-Platt, *Fermented Foods of the World*, Butterworth, Kent, United Kingdom, 1987.
3. B. A. Law, *Microbiology and Biochemistry of Cheese and Fermented Milk*, Blackie Academic and Professional, London, United Kingdom, 1997.
4. E. Metchnikoff, *The Prolongation of Life Optimistic Studies*, G. P. Putnam's Sons, New York, 1908.
5. K. Kailasapathy and S. Rybka, "*L. acidophilus* and *Bifidobacterium spp.*—Their Therapeutic Potential and Survival in Yogurt," *Aust. J. Dairy Tech.* **52**, 28–35 (1997).

JOHN U. MCGREGOR
Louisiana State University Agricultural Center
Baton Rouge, Louisiana

CYCLOSPORA. See FOODBORNE DISEASES.

D

DAIRY FLAVORS

Dairy products are very ancient foods and are one of the most important classes of foods in our diet. The basic dairy product, milk, is not a very stable food. Today's practice allows us to heat process it to keep it stable from microbiological attack for various periods of time depending on the nature of the processing conditions to which the milk is subjected. The commercial practice of heat processing (pasteurization) has only been in place for the last 60 or so years. In ancient times milk was allowed to spoil in a controlled way so that it would remain eatable.

The development of ghee (butter) from buffalo milk, an early practice of controlling the spoilage of milk, has been traced to India of perhaps about 4,000 years ago. The method of butter making was to appear in Europe at a much later date. The first commercial creamery in the United States was built by Alanson Slaughter in 1881 in Orange County, New York.

Cheese from milk also has a prehistoric origin. Cheese was an excellent way to preserve the important nutrients of milk. Because of the many ways of stabilizing milk into cheese, many forms of cheese with a variety of flavor and taste profiles have been developed. Recently, a USDA bulletin indicated over 800 varieties and the (United States Food and Drug Administration) U.S. FDA has established standards of identity for 30 different type of cheeses. Typically, cheeses are named after the immediate region where they were originally produced.

At first the manufacture of dairy products, including milk, milk products, butter, and cheese, was based on historical practice and the art of the producers. It took the great scientific revelation of Louis Pasteur working in France between 1857 and 1876 to relate the manufacturing process to the new world of the microorganism and fermentation. The use of pure culture techniques began around 1870, and the dairy and cheese industry was transformed from an industry of inconsistent product quality and quantity. The various organisms responsible for the development of the flavor, taste, and texture of dairy and cheese products were discovered and used in day-to-day operation of the manufacturing plants. More recent scientific advances in the area of organic, analytic, and biochemistry have allowed for the identification of those chemical components that develop during the manufacturing steps and contribute to the flavor profiles that we associate with these products.

Besides the well-known nutritional properties of these products, they are very popular owing to their unique flavor, taste, and textual attributes.

OVERVIEW OF DAIRY PRODUCTS

Dairy products may be defined as food products based on milk developed through a variety of manufacturing methods and fluid milk itself. There are several animal sources of milk, including the cow, water buffalo, goat, and sheep, that are commonly used sources.

In the 10 year period from 1987 to 1997, the production of milk has increased by approximately 15 billion lb to 155 billion lb, with 60% of the milk used to create other dairy products. The percent use of milk products to make other dairy products increased by 2% during that period. Total cheese production, excluding cottage cheese, was 7.33 billion lb in 1997, an annual increase of 2%. Wisconsin was the leading state for cheese product, with 29% of the total production. By far the most popular types of cheese are the American varieties, which include mainly cheddar. Cottage cheese is produced by far more plants in the United States (332) and represents production of 1.2 billion lb. Butter production has been declining; in 1997 production was recorded at 1.2 billion lb, a decrease of 2% from 1996. California, at 27% of butter production, was slightly ahead of Wisconsin, at 26%. Table 1 represents the statistics for production of various dairy products in the United States during 1997.

In discussing the nature of the flavors of these products it is natural to organize the discussion based on the types of products manufactured. Based on the major groups of products, there are four distinct areas that will be discussed. They are fluid milk and milk products (dried and canned), milk-derived products (yogurt and similar types of fermented milk products and ice cream), butter, and cheeses.

Milk and Milk Products

These products are minimally processed. Typically they may be reduced in fat, have the fat suspended (homogenized), or be heat treated. The heat-processing step may vary in its time/temperature relationship to assure a safe

Table 1. Dairy Product Production in the United States for 1997

Products	Plants	Total volume in 1997 (in 1,000 pounds)	Growth rate (%)
Butter	91	1,151,250	2.0
Cheeses			2.0
American	215	3,285,203	
Swiss	42	207,583	
Muenster	42	100,191	
Brick	21	8,535	
Limburger	1	746	
Italian types	156	2,880,404	
Cream & Neufchatel	32	614,921	
Cottage Cheese	332	1,164,172	
Canned milk	9	580,965	
Skim milk	51	997,802	
Dry Milk (nonfat)	47	1,217,562	15.0
Yogurt	104	1,574,050	

Source: Ref. 1.

product. Although the heat treatment can have some influence on the taste and flavor of fresh milk, there is a much greater relationship to the time of year, nature of the animals' feed, and the species of animal from which the milk is derived.

Milk is a very complex substance based on fats, sugars (mainly lactose), proteins (such as lactalbumin), and salts (such as calcium phosphate). The taste of milk is mainly affected by the change in this mixture and the degree to which the fat is emulsified (homogenized) in a continuous phase with the aqueous phase. The sensory perception determines the physical nature of the emulsion, with a slightly salty/sweet taste from the salts and lactose. Although some slight flavor changes can be detected in even the mildest of heat-treated milk, the flavor defects are generally considered to be notes associated with good quality, pasteurized milk.

Of the various constituents of the milk, the lipid (fat) fraction has the greatest effect on milk's flavor. It is also the precursor of many chemical components considered important in all types of dairy flavors. The fatty acid composition of the milk lipids is also very complex and unique among food products. More than 60 fatty acids have been reported in cow's milk (2). Quantities of butyric acid and caproic, caprilic, decanoic (fatty acids with 6, 8, and 10 carbon atoms) acids are unique to milk and of great importance in the development of flavors in products based on milk. Oxidation of these lipids gives rise to some key flavor components. For example, the oxidation of unsaturated octadecadienoic acid leads to the formation of 4-*cis*-heptenal, which has been identified as the cream-like flavor component in milk and butter (3). The free fatty acid distribution of milk and butter is shown in Table 2.

The heat treatment of milk is divided into two types of processing: pasteurization in which the milk is heated for 15 s at 72°C and ultra high-temperature treatment (uht), where it is heated at 135–150°C for a few seconds. The flavor changes that occur during the heat treatment may be formed from three distinct chemical mechanisms:

- Degradation of thermally labile precursor found in milk.
- Reaction between the sugars and proteins (Maillard reactions).
- The release or formation of sulfur-containing components.

Table 2. Free Fatty Acids in Milk

Fatty acid (carbon no.: unsaturation)	Fat (%)		
	Fresh milk	Rancid milk	Butter
4:0	7.23	6.03	1.32
6:0	3.40	3.42	0.65
8:0	1.37	2.07	0.83
10:0	5.16	3.98	1.41
12:0	4.89	3.98	4.22
14:0	10.02	9.58	11.02
16:0	23.61	26.09	22.69
18:0	9.98	13.65	12.04
18:1	29.78	27.05	38.40
18:2	4.55	4.14	4.66
18:3	—	—	2.69

Source: Ref. 4.

Degradation of Precursors. The triglycerides (lipids) in milk fat are formed from complex as well as simple fatty acids of many types, including substituted acids such β-keto acids and hydroxy acids. These acids originate in the diet of the animal and also may be the result of normal metabolism. Methyl ketones are formed from the β-keto acids (5). Although they are formed in fresh milk, maximum yield occurs at prolonged elevated heating (140°C for 3 h). The liberation of methyl ketones is generally below their individual odor thresholds, yet they contribute to the aroma of milk because of their synergistic interaction due to their combined concentration (6). Methyl ketones are formed from the decarboxylation of the β-keto acids, whereas ring closure of δ- and τ-hydroxy acids form the corresponding lactones.

The Maillard Reaction. Reaction between the lactose and the milk protein can lead to many different compounds of flavor significance. These reactions are referred to as Maillard reactions. Discussion of the Maillard reaction is complex and outside the scope of this article, but its basic chemical pathway is the reaction between aldose and ketone sugars and α-amino acids resulting in the formation of aldosylamines or ketosylamines. These materials then undergo further rearrangement, resulting in the creation of many key flavoring materials, such as acetyls, furans, pyrroles, aldols, and pyrones. The normal heat treatment of milk does not appear to produce enough of the Maillard reaction product to be a flavor contributor. However, the cooked flavor of milk is noted when it is heated to above 75°C (7). At that point many Maillard reaction products start to contribute to the flavor of the milk. Further heating above the cooked-flavor range produces a caramelization flavor, due to the intense amount of Maillard reaction products being formed and the liberation of many sulfur-containing chemicals.

Formation of Sulfur-Containing Compounds. Dimethyl sulfide is produced from the *S*-methyl methionine sulfonium salt (from the vegetable matter in the animal's diet) that is decomposed on heating (8). At high concentrations this chemical produces a malty or cowy flavor defect in the milk (9). The flavor from the caramelization of milk has found great use in various culinary schools of fine food preparation.

Flavor Defects in Milk. Caramelization is a wanted defect in many food applications; yet, there are some defects that are not wanted. Extended storage of dry milk causes the stale off-flavor where the release to the fatty acid contributes to the further development of methyl ketones above their threshold value, but below the quantity seen in certain cheeses; for example, the family of blue cheeses (10).

The oxidation of milk causes the cardboardy or cappy defect that is evaluated as the milk being metallic, tallowy, oily, or fishy. The oxidation of the unsaturated fatty acids leads to the methyl ketones and series of aldehydes. Very low levels of *cis*-4-heptenal, hexenal, and 1,3 octenone contribute to the oxidized flavor defect (11).

The lipolysis of the milk by lipases remaining active after the heat treatment can cause rancidity of the milk. This is due to the increase of both the short-chain fatty acids and the increased acidity of the milk. In normal milk the total free acid content is about 360 mg/kg, but in rancid milk it rises to 500–1500 mg/kg (4).

One final major flavor defect is known as the sunlight defect. It is believed to be produced from the degradation of methional (a common amino acid found in milk) to several sulfur-containing compounds, including mathanethiol, dimethyl sulfide, and dimethyl disulfide. These contribute to a flavor characterized as burnt or cabbagey. The extent of the flavor defect is related to the length and strength of exposure to light (12).

Table 4. Cultured Milk Products

Product type	Fermentation	Organism
Yogurt (U.S.)	Moderate acid	*Streptococcus thermophilus* *Lactobacillus bulgaricus*
Buttermilk (U.S.)	Moderate acid	*Streptococcus cremoris* *Streptococcus lactis*
Acidophilus milk (U.S.)	High acid	*Lactobacillus acidophilus*
Bulgarican milk (EUR)	High acid	*Lactabacillus bulgaricus*
Yakult (Japan)	Moderate acid	
Dahi (India)	Moderate acid	*Streptococcus diacetyllactic*
Leben (Egypt)	High acid	*Streptococci, Lactobacilli, yeast*
Kifur (USSR)	High alcohol	*Lactobacillus caucascus* *Leuconostoc* spp

Milk's Basic Flavor. Milk's basic flavor, then, is due mainly to the taste components found in the milk as noted in Table 3. The fats, sugars, salts, and protein all contribute to give milk a subtle, but enjoyable, flavor and taste. Unless the milk has been mistreated or subjected to high temperature for prolonged time, no significantly predominate flavor is noted. In contrast, products made from milk have characteristic and, in some cases, strongly predominate aroma and tastes.

Cultured Dairy Products

Among the first products of milk to be developed were cultured products. Cultured products are produced throughout the world, as seen in Table 4. Originally, fermentation occurred naturally to milk set aside for later use. The spoilage organisms in the milk produced the proper conditions for the preservation of the cultured product. The lowering of the pH of the milk by the organisms allows for the preservation of the product and contributes to its characteristic aroma and taste.

The preparation of these cultured products became an art that survives today. The consumption of these products varies around the world, with the northern European countries consuming the greatest amount (~20 kg/y/person); and the United States consuming the least (~1.2 kg/y/person) (14). However, the United States is a 300-fold increase over the 1966 figure. This, no doubt, reflects the healthy eating trend in the United States prompted by television commercials and magazine ads listing the health benefits of cultured milk products.

Table 3. Fat Content of Cow's Milk

Type of fats	Range of occurrence (%)
Triglycerides	97.0–98.0
Diglycerides	0.25–0.48
Monoglycerides	0.015–0.038
Keto acid glycerides	0.85–1.28
Aldehydo glycerides	0.011–0.015
Glyceryl ethers	0.011–0.023
Free Fatty Acids	0.10–0.44
Phospholipids	0.2–1.0
Cerebrosides	0.013–0.066
Sterols	0.12–0.4

Source: Ref. 13.

The basic fermentation is due to lactic acid bacteria. Historically, this was fermentation done by a mixed or unknown culture or back-slop, but modern technology has developed the pure culture with its predictable performance potentials.

The major flavor and taste component in these products is the lactic acid produced by the fermentation of the sugars by the bacteria. Although this flavor is considered acceptable in many societies, it is rejected completely in others. To increase the commercialization of these products, manufacturers have introduced products that contain added sugars, fruit preserves, and flavors. These innovations have also contributed to increased consumption, particularly in the United States.

During the fermentation the lactose in the milk is transformed via pyruvic acid to lactic acid. The small amount of lactose or its hydrolysis products galactose and glucose is below the threshold for detecting sweetness, but the lactic acid may account for more than 2% of the mass of the product and, by that, contribute a significant flavor and taste effect. The fermentation pathway may favor both enantimer formation (L(+) and D(−) lactic acid), as depicted in Figure 1. The preferred metabolite, in flavor terms, is the L(+) or a harsh off-flavor will be observed (15).

The basic microorganisms used are the lactobacilli, streptococci, and leuconostocs. The major difference in their flavor is the amount of lactic acid produced and the perception of the acid taste. One exception is the product from the USSR called Kifur, which undergoes a secondary fermentation to create ethyl alcohol. This product is also known as the champagne of milk (16).

Yogurt Flavor. Yogurt flavor is also influenced by acetaldehyde, another product of lactose from the fermentation of pyruvic acid by *Lactobacillus bulgaricus*. In yogurt the acetaldehyde level differs greatly with the type of microorganism(s) used (17). There is no agreement on the acetaldehyde values that give an optimum flavor effect (18).

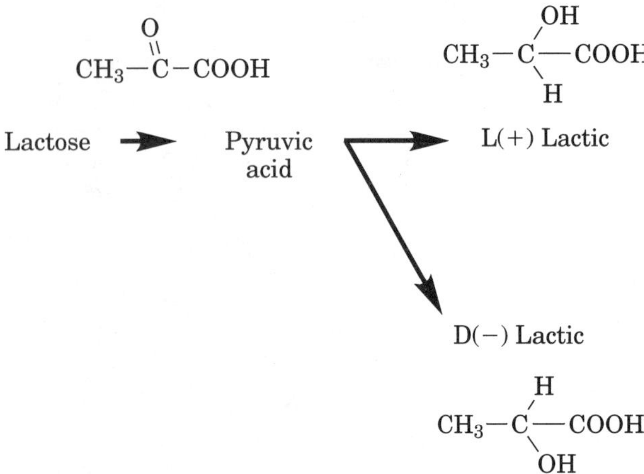

Figure 1. Lactic acid fermentation.

The value of 23–41 ppm is given in some reports, whereas the value of 13–16 ppm is considered by others to be the optimum level (19,20).

One further compound of flavor interest in cultured milk products is diacetyl. It is formed from the citrate present in milk by the action of Streptococci, Leuconostocs, and other organisms that can metabolize citrate. (21) This component strongly contributes to the acid cream butter flavor usually at level of 1–2 ppm (22). It is diacetyl that is a significant contributor to butter's flavor.

Butter and Cream

We have discussed the nature of milk flavor and the cultured products derived from milk. The production of butter from milk involves the concentration of the milk fat in the milk by mechanical separation. The resultant product is a water in oil emulsion that has a mild characteristic flavor. During this process the exogenous enzymes (those enzymes existing outside the cell) can break down the triglycerides to free fatty acids; with improper handling this may occur at a high level, leading to a flavor effect known as rancidity. It is the short-chain free fatty acids (eg, butyric acid) that are responsible for the development of rancidity in butter. The role of other free fatty acids directly on flavor seems small, except the deca- and dodecanoic acids, which appear in sweet cream butter at levels exceeding their flavor threshold (23). Because the lipases responsible for the hydrolysis of the triglycerides are very heat stable, their activity continues even at low temperature during refrigerated or frozen storage.

Rancidity. Rancidity is the major flavor defect in butter. The most useful index of rancidity is related to the amount of butyric acid found in the butter. A rancidity score developed by the University of Wisconsin correlates sensory data with individual free fatty acids concentration by stepwise regression and discriminatory analysis to index the relative rancid flavor intensities in butter (24). This is a very useful flavor index for the evaluation of butter.

Lactone Formation. Lactones have also been considered a major flavor attribute in butter and cream. They are generated from the small amount of δ-hydroxy acids occurring in milk fat. On hydrolysis these compounds may spontaneously lactonize to form the corresponding δ-lactones. The τ-lactones also may be formed from unsaturated C18 fatty acids via hydration and β-oxidation. The amount of lactones formed is related to the type of feed, the season, the type of breed, and the stage of lactation of the animal (25). Estimates of the amount of lactone that can create a flavor effect is reported at 7–85 ppm (26,27).

Heated-Milk Defects. Heated milk fat increases the level of lactones found in butter (28). Although this is considered a defect in butter or cream, it is a desired effect when cooking. The flavor developed is characterized as a coconut-type note, which has been observed in heated butter used to flavor popcorn. Several patents have been issued to commercial companies for the use of lactones or lactone precursors for the flavoring of butter analogues, including margarine.

Imitation Products. Although margarine is not a dairy product, it bears mention here as it is produced in great quantity throughout the world as a substitute for butter. The process goes back to the Napoleonic times as a commercial product. The flavors used in these products generally are composed of diacetyl and lactones or lactone precursors for the butter flavor impression. To various degrees of sophistication, all commercially available margarine has these materials added to the vegetable fat/water emulsion.

Ice Cream. One other product produced from milk and butter is ice cream. Ice cream represents a congealed product produced from cream, milk, skim milk, condensed milk, butter, or a combination of these ingredients, plus the addition of sugar, flavorings, fruits, stabilizers and colors. Again, the basic contribution of the dairy ingredients to the flavor is the fat and lactose, with those aromatic components mentioned in the foregoing discussion and associated with butter or cream flavor. Typically, the aromatic portion of the flavor is greatly modified by the addition of flavors and extracts.

Most of these basic dairy products, like butter, have flavor profiles that are simple. The real flavor masterpieces of the dairy industry are the many varieties of cheese, with their extremes in flavor types and intensities.

Cheese

The many varieties of cheese represent many centuries of the cultivation of the art of cheese-making. Many different, probably accidentally developed, flavors, tastes, and textures have been developed. The products of cheese-making are the result of many types of microorganisms (yeast, bacteria, and mold) working together with the basic milk source (typically cow's, goat's or sheep's milk) and the knowledge of the control of fermentation. The variety of methods is great and, so too, as has been noted, are the varieties of cheese produced throughout the world. We will not attempt to discuss all the types known, but focus our flavor discussion on those well-known cheeses and the basic science of their formation.

Chemistry. The basic chemistry or biochemistry of cheese-making is the slow degradation of the macromolecules (protein and lipid material) to small components of flavor (aroma and taste) value. Proteolysis (breakdown of the protein by enzymes) during cheese ripening and aging has a major influence on peptide formation. Peptides have various taste characteristics based on their polarity. The basic taste sensations of sour/sweet and salt/bitter and the flavor-enhancing property known as umami (the flavor-enhancer taste associated with monosodium glutamate, MSG) are all taste attributes of peptides and amino acids.

It has been observed that the acid- (sour), hydrophobic- (bitter), and glutamic/aspartic-rich (umami) peptides are the key taste peptides and amino acids in aged cheese (29). The hydrophilic (sweet) and salt-based peptides contribute very little, as they are overwhelmed by the basic taste attributes of the cheese. The natural sweetness from lactose and the saltiness due to the salt (sodium chloride) added to all cheese to produce a selective environment and the chemistry needed to develop the specific cheese character play a more significant flavor role than do the peptides that have salty or sweet taste character.

The salt content of selected cheeses is listed in Table 5. Note that the Roquefort or blue-cheese types have the highest salt content and those of the Swiss types have the lowest. This is reflected in the basic salt taste of these cheeses. As we will see, the volatile chemicals formed from the amino acid, peptides, lipids, and sugar are the backbone of the characterizing flavors for the various cheese.

Types of Cheeses. The family of cheeses generally recognized are

- Fresh soft low-fat cheese
- Cream/Neufchâtel
- Gouda/Muenster
- Port du Salut
- Brie/Camembert
- Blue cheeses
- Goat cheeses
- Emmentaler
- Cheddar
- Parmesan/Romano
- Whey/ricotta

Processing. The basic processing involves the progressive dehydration and working of the coagulated and separated curds of milk to form a homogeneous compact mass.

Table 5. Salt Content of Certain Cheeses

Cheese variety	Salt (NaCl) (% w/w)
Roquefort	4.1–5.0
Feta	2.4–4.4
Parmesan	2.1–3.5
Edam	1.7–1.8
Cheddar	1.6–1.7
Emmentaler	0.9–1.0

The initial coagulation is achieved by using an added enzyme, rennin, or by lactic acid fermentation, or both, depending on the desired character of the final product. The green cheese is then ripened via the addition of certain microorganisms that will accomplish the breakdown of the protein, lipids, and sugars to the small molecular weight components that contribute to the cheesey flavors found in the final product.

Flavor Attributes of Cheese by Families. The less cheesey flavored cheeses are those with the least amount of ripening: fresh soft low-fat, higher fat cream/Neufchâtel, and the whey/ricotta type. Their flavors are primarily due to the components contributing to milk, cream, and butter flavor. They should be low in flavor intensity or they would be considered to have a flavor defect.

The next group in terms of flavor strength consist of the Swiss (Emmentaler), German (Muenster), and Dutch (Gouda) cheeses. The importance of peptides was recognized in studies with Swiss cheese (30). Results indicated that the small peptide that interacted with calcium and magnesium gives the cheese a sweet flavor, with the small free peptides and amino acid contributing a slight brothy-nutty flavor. An important reaction in Emmentaler is the conversion of pyruvic acid to propionic acid by propionibacterium. Propionic acid is found in Emmentaler in levels of up to 1%, and it is considered the basis of the sweet taste of that cheese (31). Gouda has a defect called catty taint that is due to the formation of 4-mercato-4-methylpentaone-2 (32).

Cheddar cheese is by far the most consumed cheese in the United States. It is produced by a unique process, cheddaring, introduced in 1857 to repress the growth of spoilage organisms during cheese-making. Cheddaring involves the piling and repiling of blocks of warm cheese curd in the cheese vat for ca 2 h. During this period the lactic acid increases rapidly to a point where coliform bacteria are destroyed.

The basic taste of cheddar is due to the peptide fraction and, in particular, to the umami and bitter components. The development of acetic acid during ripening contributes to the sharpness of aged cheddar. The release of the acetic acid and amino acids continue during the ripening process. The medium chain fatty acids (C6 to C18) that are released during the ripening process have also been shown to be important in cheddar cheese flavor character (33).

Many other studies over a 10-year period indicated that it is necessary to produce the right level of acidity to provide a reducing environment in the finished cheese so that there may be a release of sulfur-containing compounds. The lost of these compounds, in particular, methyl mercaptan, results in a cheese with little cheddar character (34–36). Methanethiol has also been shown to be a significant factor in aged cheddar (37–41). Methanethiol has a very fecal-like aroma with a threshold of detection reported at 0.02 ppb (42).

It has been suggested that acetylpyrazine and 2-methoxy-3-ethylpyrazine are important in aged cheddar (43). The possible microbial origins of pyrazines became evident when it was reported that dimethylpyrazine and 2-methoxy-isopropylpyrazine were isolated from milk con-

Table 6. Summary of the Flavor Chemistry of Dairy Products

Product	Process	Chemistry	Major flavor contributors
Milk	Homogenization and mild heat	Natural components	Balance of fats, salts, and lactose (milk sugar)
Defects	Cooked >57°C	Maillard reactions	Aldehydes and pyrazines
	Prolonged cooking	Caramelization	Sulfur-containing components
	Oxidation	Oxidation	Aldehydes
	Sunlight exposure	Sulfur reactions	Methanethiol
	Rancidity	Lipolysis	Butyric acid
Cream	Fat separation	Enzymolysis	Methyl ketones (low level) diacetyl and lactones
Butter	Further fat separation	Enzymolysis	Diacetyl and butyric acid
Cultured products	Fermentation	Lactic acid formation	L(+) Lactic acid
Yogurt	Fermentation		Acetaldehyde
Cheeses	Salting		NaCl
	Aging	Proteolysis	Peptide formation
Swiss types	Fermentation	Propionibacteria	Propionic acid
Cheddar types	Cheddaring	Lactic acid fermentation	Lactic and acetic acid
		Maillard reactions	Acetyl pyrazine and other pyrazines
Blue types	Fermentation	Penicillium	Methyl ketones (high amounts)

taminated by *Pseudomonas taetrolens* (44). Further alkylpyrazines have been isolated from mold ripened cheese using *Penicillium caseicolum* cultures (45). The overall flavor of good cheddar is a balance between the free fatty acids and the other flavor components noted above.

The blue cheese family has been found to contain from 13 to 37 times the amount of free fatty acids found in cheddar. This is due to the lipolysis of the fat by the lipases generated from the use of the mold *Penicillium roqueforti* (46). Roquefort is made from sheep milk and does not have the high levels of free fatty acids seen in the blue cheese made from cow's milk. Part of the flavor difference between the two is due to the lower concentration of propionic acid (4:0) in Roquefort and the relative larger amount of C8:0 and C10:0 free fatty acids.

The major aroma character of the blue cheeses is due to the high levels of methyl ketones found in the cheese. These components have been isolated in milk and are assumed to be the product of the free fatty acid. The importance of the methyl ketone in creating an imitation blue cheese flavor has been well reported (47–52).

Secondary alcohol may be produced from the methyl ketones when they are generated in high amounts driving the equilibrium from the methyl ketones to the secondary alcohols. Although the secondary alcohols have a similar flavor profile to the methyl ketones, their intensity and quantity in the cheese is less, therefore, making only a small contribution to the complete flavor (52).

CONCLUSION

The flavors of dairy products are extremely varied and represent a great range of chemistry, biochemistry, and microbiology. A summary of the major dairy products and their associated significant flavor values is given in Table 6. It summarizes the nature of the components contributing to the flavor and taste characters of the particular material. Science has yet to identify completely the components responsible for the overall subtle aromas and taste found in dairy products, but Table 6 represents a good basic start to an understanding of the complexities of dairy flavors.

BIBLIOGRAPHY

1. United States Department of Agriculture, *Dairy Industry Report*, National Agricultural Statistics Service, April 1998.
2. S. F. Herb, P. Magidma, F. E. Lubby, and R. W. Riemenscheider, "Fatty Acids of Cow Milk (II). Compounds by Gas-Liquid Chromatography Aided by Fractionation," *Journal of the American Oil Chemists' Society* **39**, 142–149 (1962).
3. P. H. Haverkamp-Begemann and J. C. Koster, "4 Cis-heptenal—cream flavor component of butter," *Nature* **202**, 552–523 (1964).
4. J. A. Knitner and E. A. Day, "Major Free Fatty Acids in Milk," *Journal of Dairy Science* **48**, 1557–1581 (1965).
5. S. Patton and G. W. Kurtz, "A Note on The Thiobarbituric Acid Test for Milk Lipid Oxidation," *Journal of Dairy Science* **38**, 901 (1955).
6. J. E. Langler and E. A. Day, "Development and Flavor Properties of Methyl Ketone in Milk Fat," *Journal of Dairy Science* **47**, 1291–1296 (1964).
7. D. V. Josephson and F. J. Doan, "Cooked Flavor in Milk, Its Sources and Significance," *Milk Dealer* **29**, 35–40 (1939).
8. T. W. Keenan and R. C. Lindsay, "Evidence for a Dimethyl Sulfide Precursor in Milk," *Journal of Dairy Science* **51**, 122–128 (1968).
9. S. Patton, D. A. Forss and E. A. Day, "Methyl Sulfide and the Flavor of Milk," *Journal of Dairy Science* **39**, 1469–1473 (1956).
10. O. N. Parks and S. Patton, "Volatile Carbonyl Compounds in Stored Dry Whole Milk," *Journal of Dairy Science* **44**, 1–9 (1961).
11. H. T. Badings and R. Neeter, "Recent Advances in the Study of Aroma of Milk and Dairy Products," *Netherlands Milk and Dairy Journal* **34**, 9–30 (1980).
12. R. L. Bradley, Jr., "Effect of Light on Alteration of Nutritional Value and Flavor of Milk—A Review," *Journal of Food Protection* **43**, 314–320 (1980).
13. F. E. Kurtz, *Fundamentals of Dairy Chemistry*, The AVI Publishing Co., Inc., Westport, Conn. 1965, pp. 135–180.

14. International Dairy Foundation, *Bulletin 119*, International Dairy Foundation, Brussels, Belgium, 1980.
15. G. Kielwein and U. Dawn, "Occurrence and Significance of D(−)-Lactic Acid in Fermented Milk Products with Special Consideration of Yogurt Based Mixed Products," *Deutsche Molkerei Zeitung* **100**, 290–293 (1979).
16. K. M. Shahani, B. A. Friend, and R. C. Chandan, "Natural Antibiotic Activity of Lactobacillus acidophilus and bulgaricus," *Cult. Dairy Prod. J.* **18**, 1519 (1983).
17. R. K. Robinson, A. Y. Tamine, and L. W. Chubb, "Yogurt-Perceived Regulatory Trends in Europe," *Dairy Ind. Int.* **41**, 449–451 (1976).
18. M. Groux, "Component of Yogurt Flavor," *Le Lait* **53**, 146–153 (1973).
19. H. T. Badings and R. Neeter, "Recent Advances in the Study of Aroma Compounds of Milk and Dairy Products," *Netherlands Milk and Dairy Journal* **34**, 9–30 (1980).
20. M. L. Green and D. J. Manning, "Development of Texture and Flavor in Cheese and Other Fermented Products, *Journal of Dairy Research* **49**, 737–748 (1982).
21. E. B. Collins, "Biosynthesis of Flavor Compounds by Microorganisms," *Journal of Dairy Science* **55**, 1022–1028 (1972).
22. V. K. Steinholt, A. Svensen, and G. Tufto, "The Formation of Citrate in Milk," *Dairy Research Institute* **155**, 1–9 (1971).
23. E. H. Ramshaw, "Volatile Components of Butter and their Relevance to Its Desired Flavor," *Australian Journal of Dairy Technology* **29**, 110–115 (1974).
24. A. H. Woo and R. C. Lindsay, "Statistical Correlation of Quantitative Flavor Intensity Assessments of Free Fatty Acid Measurements for Routine Detection and Prediction of Hydrolytic Rancidity of Flavor in Butter," *Journal of Food Science* **48**, 1761–1766 (1983).
25. P. S. Dimick, N. J. Walker, and S. Patton, "Occurrence and Biochemical Origin of Aliphatic Lactones in Milk Fat—A review," *Journal of Agricultural Food Chemistry* **17**, 649–655 (1969).
26. J. E. Kinsella, S. Patton, and P. S. Dimick, "The Flavor Potential of Milk Fat. A Review of Its Chemical Nature and Biochemical Origins," *Journal of the American Oil Chemistry Society* **44**, 449–454 (1967).
27. R. Ellis and N. P. Wong, "Lactones in Butter, Butter Oil, and Margarine," *Journal of the American Oil Chemistry Society* **52**, 252–255 (1975).
28. G. Urbach, "The Effect of Different Feeds on the Lactone and Methyl Ketone Precursors of Milk Fat," *Lebensmittel Wissenschaft Technologie* **15**, 62–69 (1982).
29. K. H. Ney, "Recent Advances in Cheese Flavor Research," in G. Charalambous and G. Inglett, eds., *The Quality of Foods and Beverages*, Vol. 1, Academic Press Inc., New York, 1981, pp. 389–435.
30. S. L. Biede, A Study of the Chemical and Flavor Profiles of Swiss Cheese, Doctoral Dissertation, Department of Food Science, Iowa State University, Ames, Iowa, 1977.
31. K. H. Ney, "Bitterness and Gel Permeation Chromatography of Enzymatic Protein Hydrolysates," *Fette Seifen Anstrichmittel* **80**, 323–325 (1978).
32. H. T. Badings, "Causes of Ribes Flavor in Cheese," *Journal of Dairy Science* **50**, 1347–1351 (1967).
33. D. D. Bill and E. A. Day, "Determination of the Major Free Fatty Acids of Cheddar Cheese," *Journal of Dairy Science* **47**, 733–738 (1964).
34. T. J. Kristoffersen and I. A. Gould, "Cheddar Cheese Flavor, II., Changes in Flavor Quality and Ripening Products of Commercial Cheddar Cheese During Controlled Curing," *Journal of Dairy Science* **43**, 1202–1209 (1960).
35. B. A. Law, "New Methods for the Controlling Spoilage of Milk and Milk Products," *Perfumer & Flavorist* **7**, 9–13 (1983).
36. T. J. Kristoffersen, I. A. Gould, and G. A. Puruis, "Cheddar Cheese Flavor. III. Active Sulfhydryl Group Production During Ripening," *Journal of Dairy Science* **47**, 599–603 (1964).
37. L. M. Libbey and E. A. Day, "Methyl Mercaptan as a component of Cheddar Cheese," *Journal of Dairy Science* **46**, 859 (1963).
38. D. J. Manning, "Cheddar Cheese Micro Flavor Studies II. Relative Flavor Contribution of Individual Volatile Components," *Journal of Dairy Research* **46**, 531–539 (1979).
39. D. J. Manning, "Headspace Analysis of Hard Cheese," *Journal of Dairy Research* **46**, 539–545 (1979).
40. D. J. Manning, "Cheddar Cheese Aroma—The Effects of Selectively Removing Specific Classes of Compound from Cheese Headspace," *Journal of Dairy Research* **44**, 357–361 (1977).
41. D. J. Manning, "Sulfur Compounds in Relation to Cheddar Cheese Flavor," *Journal of Dairy Research* **41**, 81–87 (1974).
42. R. G. Buttery, D. G. Guadaghi, L. C. Ling, R. M. Seifert, and W. Lipton, "Additional Volatile Components of Cabbage, Broccoli and Cauliflower," *Journal of Agricultural Food Chemistry* **24**, 829–832 (1976).
43. W. A. McGugan, "Cheddar Cheese Flavor, A Review of Current Progress," *Journal of Agricultural Food Chemistry* **23**, 1047–1050 (1975).
44. M. E. Morgan, "The Chemistry of Some Microbially Included Flavor Detects in Milk and Dairy Foods," *Biotechnology and Bioengineering* **18**, 953–965 (1976).
45. T. O. Bosset and R. Liardon, "The Aroma of Swiss Gruyere Cheese, III. Relative Changes in the Content of Alkaline and Neutral Volatile Components During Ripening," *Lebensmittel Wissenchaft Technologie* 18–85, 178 (1985).
46. D. Anderson and E. A. Day, "Quantitative Analysis of the Major Free Fatty Acids in Blue Cheese," *Journal of Dairy Science* **48**, 248–249 (1965).
47. B. K. Dwivedi and J. E. Kinsella, "Continuous Production of Blue-Type Cheese Flavor by Submerged Fermentation of Penicillium Roquefort," *Journal of Food Science* **39**, 620–622 (1974).
48. R. C. Jolly and F. V. Kosikowski, "Flavor Development in Pasteurized Milk Blue Cheese by Animal and Microbial Lipase Preparations," *Journal of Dairy Science* **58**, 846–852 (1975).
49. M. Godinho and P. F. Fox, "Ripening of Blue Cheese. Influence of Salting Rate on Lipolysis and Carbonyl Formation," *Milchwissenschaft* **36**, 476–478 (1981).
50. T. Kanisawa and H. Itoh, "Production of Methyl Ketone Mixture from Lipolyzed Milk Fat by Fungi Isolated from Blue Cheese," *Nippon Shokuhin Kogyo* **31**, 483–487 (1984).
51. J. H. Nelson, "Production of Blue Cheese Flavor via Submerged Fermentation by Penicillium Roqueforti," *Journal of Agricultural Food Chemistry* **18**, 567–569 (1970).
52. E. A. Day and D. F. Anderson, "Gas Chromatographic and Mass Spectral Identification of Natural Components of the Aroma Fraction of Blue Cheese," *Journal of Agricultural Food Chemistry* 13–14, 2 (1965).

Charles H. Manley
Takasago International Corporation, USA, Inc.
Teterboro, New Jersey

DAIRY INGREDIENTS: APPLICATIONS IN MEAT, POULTRY, AND SEAFOODS

Meat and dairy products have several things in common. Both are appreciated from organoleptic and nutritional points of view. They have been traditionally regarded as two of the main protein sources for humans. In a modern diet, meat contributes about 35% of the protein intake, and milk about 25% (1).

Meat and dairy industries, furthermore, have a common interest in maintaining a positive image for protein of animal origin. However, in recent years there has been criticism about the fat content of meat and dairy products, and in some countries nutritional councils have recommended substituting vegetable protein for animal protein in the diet. In response to this criticism, meat and dairy manufacturers have succeeded in introducing low-fat products. A whole new range of meat and dairy foods is available, enabling the consumer to decide on the fat level desired in the diet. In addition to providing protein and fat, meat and dairy products contribute to a great extent to the intake of micronutrients. Meat is very important for the intake of iron and certain vitamins such as thiamin and riboflavin, while dairy foods are critical for the intake of calcium and vitamins.

Liquid or solid dairy products are widely used in the meat industry. They are mostly used with the objective to improve taste or eye appeal, eg, cheese topping on burgers, cordon bleu entrees, and cheese franks. Fresh dairy cream is often used in pâtés or liver products and in many seafood products. These applications are not within the scope of this article on functionality. The two most important ingredients from milk that find application in meat products are milk sugar (lactose) and predominantly milk protein. Milk protein causes functional advantages, while lactose is mostly used for taste improvement.

This article reviews and updates the application of these milk derivates in meat, poultry, and fish products. Scientific backgrounds will be highlighted.

PROCESSED MEAT PRODUCTS

Processed meat products are all meat-based products that need more than just cutting from the carcass. It is estimated that ca 25% of the worldwide meat production is being processed (Table 1). In these processed meats dairy ingredients can be used as functional ingredients.

Table 1. Production of Meat in Different Areas of the World (in 1,000 tons) 1987

Area	Total meat[a]
European community	29,985
United States	26,787
USSR	18,997
Total world	159,669

Source: Ref. 2.

[a]Beef, veal, pork, poultry, and sheep expressed as gross carcass weight.

Meat products can be classified in many different ways. Usually they are categorized as noncomminuted, coarsely comminuted, and finely comminuted. The degree of comminution determines to a large extent the way of applying dairy ingredients (see Table 2).

DAIRY INGREDIENTS IN MEAT PRODUCTS

The two most important ingredients from milk that find application in meat are predominantly milk protein and to some extent the milk sugar or lactose.

Lactose

Water binding in coarse and noncomminuted meats such as pumped ham is improved by extraction and swelling of meat proteins. For optimum extraction of protein from meat, usually phosphate and salt are used. The concentration of these ingredients is limited for organoleptic reasons. The effective salt and phosphate concentration on water and so the ionic strength can be increased by increasing the total solids of the brine mixture. For this purpose lactose is often used. Lactose has advantageous effects on

Taste. Lactose has the capacity of masking the bitter aftertaste of salts and phosphates.

Stability. Lactose improves water binding and sliceability and enhances the cured color.

Yield. Lactose is therefore used in liver products, cooked hams, and cooked sausages. In sterilized products the addition of lactose is less desirable because of the browning reaction.

Lactose as an integrated part of whey powder or skim milk powder (smp) is being used in many meat products.

Table 2. Classification of Meat Product

Degree of comminution	Examples
None	Hams (cooked, smoked, dry), bacon, cooked loins, shoulders, reformed hams
Coarse	(Semi)dry sausages (salami, cervelat), fresh sausages (bratwurst), hamburgers, meat patties, meatballs, chicken nuggets
Finely	Frankfurters, hot dogs, polony, pariser, bologna, mortadella, liver sausages/spreads, meat loaves

Table 3. Ingredients in Final Product

Ingredient	Percent of final product
Phosphates	0.3
Nitrited salt	1.8
Lactose	1.2
Powdered egg white	0.3
Milk protein	Varying from 0–15
Sodium ascorbate	0.05

The protein from smp also contributes to some extent to the stability, but as this is present in micellar form, it is not optimum for emulsifying properties. Also, it is known that calcium may negatively influence the binding properties of meat proteins (3). Therefore Ca-reduced smp has found quite some application in meat products.

Milk Protein

The application of milk protein in the form of sodium caseinate has found broad acceptance in the meat industry (4). The functional properties of milk protein can be explained from its molecular structure (5) (Fig. 1).

Due to the high proline and the low sulfuric amino acids content, caseinates will have a random coil structure with a low percent helix. As a consequence, caseinates will show no heat gelation and denaturation and will have a high viscosity in solution.

Caseinates have a high electrical charge and have several very hydrophobic groups. This makes them perfect emulsifiers, with a strong preference for interfaces fat/water or air/water. The high charge makes them perfectly soluble in water.

Caseinates build strong, flexible membranes that will hardly be influenced by heat. Whey proteins show a different pattern (Fig. 2). They are low in proline and have many S-S bonds, leading to a globular, strongly folded and organized structure. Whey proteins are sensitive to heat. During heating they unfold and, depending on pH and concentration, they will build intermolecular disulfide bonds resulting in gelation. The globular form is also the reason for the low viscosity in solution.

Whey proteins have a rather high hydrophobicity, which is more evenly distributed than in caseinate. Because of this structure, the application possibilities of whey proteins in meat products are very limited and will not be discussed in this article.

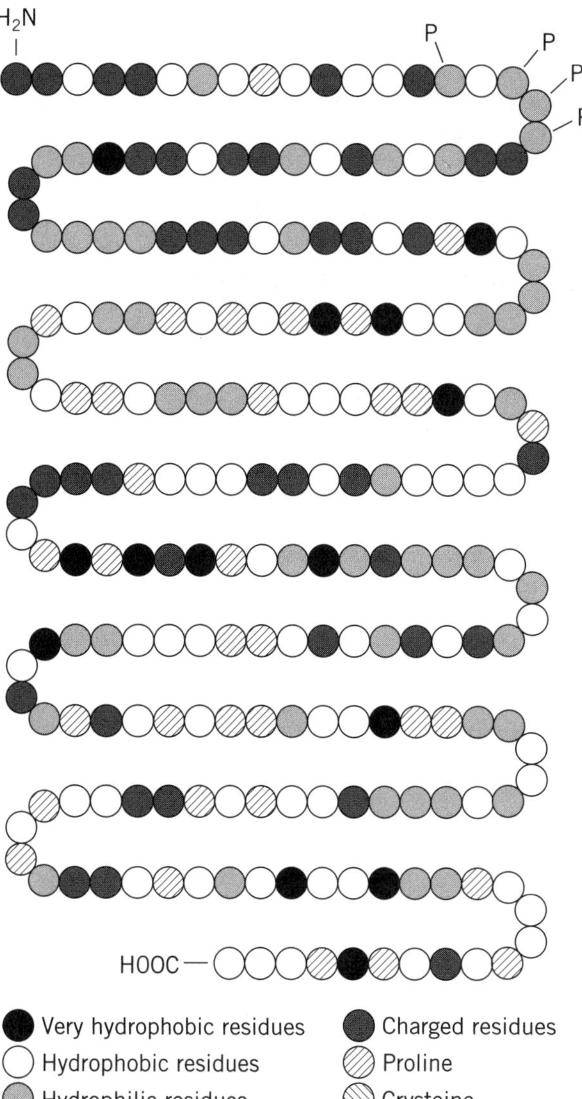

Figure 1. Amino acid sequence of β-casein A.

FINELY COMMINUTED MEAT PRODUCTS

The various finely comminuted meat systems, such as frankfurters, Vienna sausages, mortadellas, luncheon meat, chicken sausages, liver sausages, and pâtés, are commonly indicated as meat emulsions (6).

These products are generally prepared by chopping or grinding meat with water or ice, salt, and frequently phos-

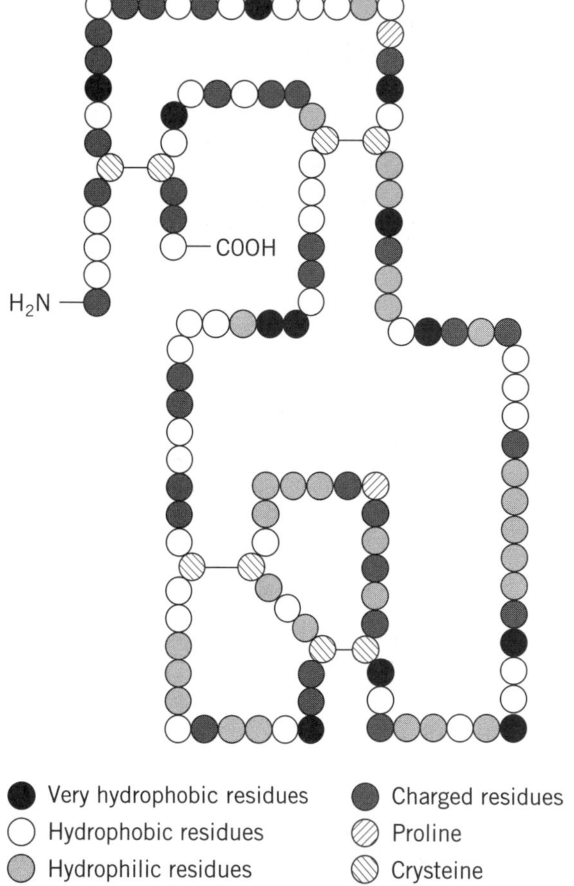

Figure 2. Amino acid sequence of α-lactalbumin B.

phates to a kind of cold meat slurry, then forming the matrix in which the animal fatty tissue and possibly connective tissue material is dispersed.

In general, the final temperature during processing does not exceed 15–18°C. Often flavorings, binders, or other additives are mixed in. Having this in mind, it will be clear that hardly any of the abovementioned finely comminuted products will show structures that are equal to true and simple emulsions as defined in physical chemistry.

The stability of finely comminuted meat products is—apart from production technology, formulation, the use of binders, etc—mainly determined by the quality of the meat cuts used (6,7).

Structure of Lean Meat

Lean meat generally contains about 20% protein, which can be divided into:

30–35% Sarcoplasmic proteins.

50–55% Myofibrillar or structural proteins (myosin, actin).

15–20% Stromal proteins (collagen, elastin).

The essential unit of all muscle is the fiber (8) (Fig. 3). Each fiber is surrounded by a membrane, the sarcolemma, and consists of a great number of parallel ordered myofibrils (8). In turn, each myofibril represents a similarly parallel-ordered structure of very thin protein threads, the actin- and the myosin filaments (Fig. 4).

Within the sarcolemma, surrounding the myofibrils, is the meat juice or sarcoplasma. This meat juice contains the water soluble meat proteins (wsp), consisting of enzymes, the red meat color myoglobin, and structured bodies. It is easily extracted on pressing, freezing/thawing, or chopping the meat. The myofibrillar proteins in warm slaughtered meat are present as free actin- and myosin filaments and are soluble in brine. Therefore, they are called the salt soluble proteins (ssp). In chilled and matured meat, however, the actin and myosin have gone into a reaction to form the complex actomyosin (7,8). This actomyosin is only partly soluble in brine, depending on the condition of the meat, age of the animal, (pre)slaughter conditions, anatomic location, mechanical treatment, etc.

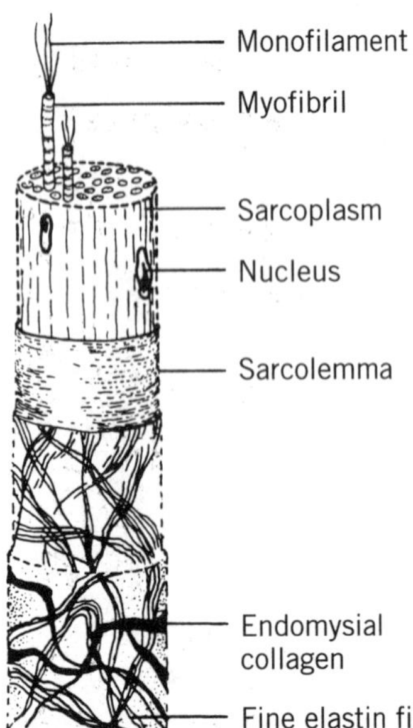

Figure 3. Sketch of a muscle fiber.

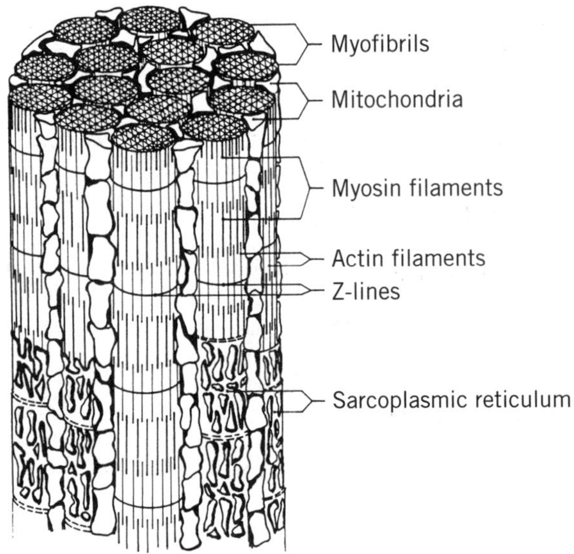

Figure 4. Section of a muscle fiber.

The nonsoluble part of the myofibrillar proteins is able to swell (take up water). In this postrigor condition the water- and fat-holding capacity of the meat are considerably lower than in warm slaughtered meat.

Meat Emulsions

As indicated in the introduction, hardly any of the abovementioned finely comminuted meat products is a true emulsion. In meat emulsion there are at least four different phase systems at the same time:

1. A true aqueous solution of salt, phosphates, nitrite, sugars, etc and a colloidal solution (sol) of sarcoplasmic (water-soluble) proteins and salt-solubilized myofibrillar proteins. This solution is the continuous phase of three different dispersed systems at a time.
2. A suspension of undissolved proteins and fatty tissue.
3. An emulsion containing emulsified fat droplets.
4. A foam; during the comminution process some air is corporated.

This system has to be stabilized against heat treatments such as pasteurization or sterilization, against smoking, cooling, frying, reheating, vacuum treatments, etc—whatever physical stress is put upon the products during conserving, storing, and commercializing.

These stabilization properties are important, not only from the standpoint of production practices, but also from the standpoint of cost and quality characteristics, such as texture, consistency, bite, tenderness, juiciness, appearance, and palatability of the finished sausages.

The mechanism for the binding and stabilization phenomenon still is not completely understood, and there are several conflicting viewpoints expressed in the literature (8–15). It is, however, generally agreed on that the myofibrillar proteins are mainly responsible for this stability (8–10,16,17).

Stability of the Aqueous Matrix

The stability of the matrix is, apart from the effect of binders, determined by the water holding and gelling capacity of the meat proteins, and, more in particular, of the myofibrillar part. In relation to the stability of fresh sausage batters, the water-binding capacity of the nonsolubilized myofibrillar proteins is decisive. Changes in the net-charge of the proteins, resulting in either attraction or repulsion of the filaments, are responsible for shrinkage and swelling of the protein network and run parallel with the decrease or increase, respectively, in water holding (18).

This mechanism of water binding may be compared to that of a sponge. The more salt-soluble proteins are solubilized, the better the water binding will be.

Stability of the Suspended and Emulsified Fat

Animal fatty tissue consists of a cellular network in which the fat is enclosed. The network is built up of connective tissue and water. As long as these cell walls are intact, only minor fat separation will take place. However, on chopping the fatty tissue an increasing number of fat cells is damaged and more and more fat is set free. This free fat should be stabilized to prevent fat separation from the product. The origin of the fat (pork, beef, mutton, poultry, etc) and the anatomic location determine the amount of free fat during comminution and as a consequence the application of the fatty tissue.

Most beef, mutton, and chicken fat, as well as pork flare fat are hard to stabilize in finely comminuted meat products, unless they are pre-emulsified with a nonmeat protein such as sodium caseinate (13,19,20).

Pork fats other than flare fat (back fat, shoulder fat) are softer and consequently can be used directly in the production of the meat emulsion, because there is more fat in a liquid state.

Particularly in finely comminuted meat products, the interaction between the free fat and the meat proteins plays an important role in the stabilization of both fat and water (14).

The salt-soluble myofibrillar meat proteins have excellent emulsifying properties and are quickly preferentially absorbed in the fat/water interface (21). The sarcoplasmic proteins are relatively unimportant in this respect. In addition to their gelling capacity and their role in water binding, this emulsification is another reason to aim at a sufficient extraction of these myofibrillar proteins during first stage of chopping.

However, when the myofibrillar meat protein enters the fat/water interface, the protein structure is altered and, as a result, the myofibrillar protein is no longer capable of gel formation and water binding (8). This means that the consumption of ssp for the emulsification of free fat goes at the expense of the water binding.

Milk Proteins in Meat Emulsions

Milk protein, type sodium caseinate, is a perfect emulsifying protein, which is very strongly attracted by the fat/water interface (12). If this type of milk protein gets the opportunity to surround the free-fat particles during sausage manufacture and before the myofibrillar (ssp) proteins do this, the latter are saved for denaturation in the interface.

Research has proved that milk protein, type sodium caseinate, indeed is better and more quickly absorbed in the oil/water interface when emulsification takes place when both milk proteins and meat proteins are present in the continuous phase (Fig. 5).

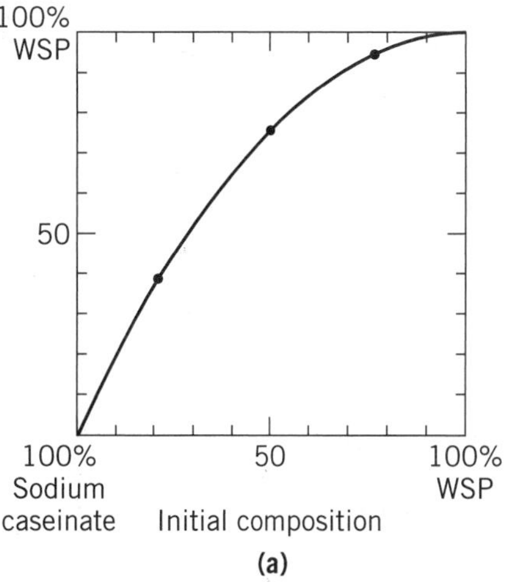

(a)

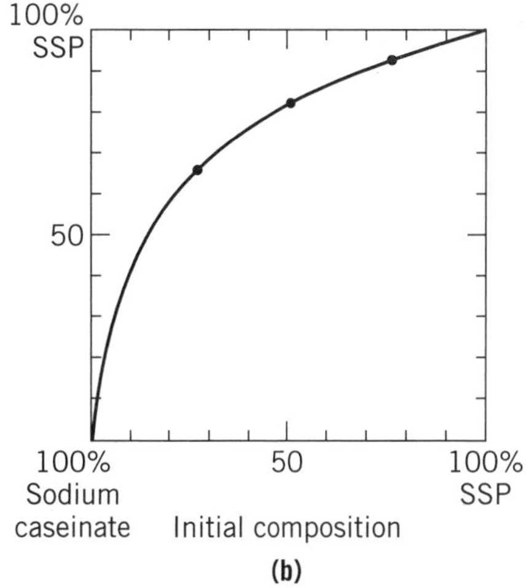

(b)

Figure 5. The preferential adsorption of milk protein, sodium caseinate over water soluble protein (WSP) and salt soluble protein (SSP), in a fat/water/protein emulsion; (**a**) the initial composition of the protein soluble mixtures is shown on the abscissae; (**b**) the composition of the protein solution after removal of the emulsified fat is shown on the ordinates.

In this way the abovementioned loss in gelling capacity is avoided. As milk protein is doing the fat binding, the meat proteins can use their full power in water binding and texture formation. In this way sodium caseinate contributes directly to a better fat binding and indirectly to an improved water binding and texture formation (20–22).

Milk proteins can be applied in three ways:

1. Addition of the powder at the beginning of the comminution process. When milk protein is used in this way, it is important that the addition takes place just before the addition of water/ice. In that way the protein is optimally hydrated; this hydration is very important for its functionality.
2. Addition as a jelly. The milk protein is predissolved in water in a bowl chopper or colloid mill. A common ratio is 1 part milk protein dissolved in 6 parts water.
3. In the form of a pre-emulsion. These pre-emulsions consist of milk protein, fat, and water. Technologically difficult to stabilize fats can be used perfectly by this pre-emulsifying technique (eg, beef suet and flare fat). A common ratio of milk protein:fat:water is 1:5:5. Addition levels of these emulsions to meat emulsions can vary from 10 to 25% maximum.

Temperature control during the production of finely comminuted meat products is very important; when temperatures at the end of comminution exceed 15–18°C, the stability of the final product is affected very negatively. This means that there is little tolerance in processing.

A unique feature of milk proteins is their capacity to widen the processing tolerance as far as temperatures are concerned. This is illustrated in Figure 6.

SECTIONED AND FORMED MEAT PRODUCTS

Principles of Production

Sectioned and formed products are made by mechanically working meat pieces to disrupt the normal muscle cell structure (16). This produces a creamy, tacky exudate on the surface of the meat pieces. When the product is heated during thermal processing, this exudate binds the meat pieces together; during this heating process the solubilized salt-soluble proteins (mainly myosin) are denatured and form a gel (23–27). The mechanical action of tumbling and massaging primarily affects external tissues to produce the surface exudate. Some internal tissue disruption also occurs, which explains the enhanced tenderness, brine penetration and distribution, and improved water-holding capacity. The protein exudate is not only produced by the mechanical action during massaging and tumbling, but also by the synergistic effect of addition of salt combined with alkaline phosphates, which improves yields and maximizes myofibrillar protein solubilization.

Optimum product quality is achieved with the addition of brine, which produces a final product containing 2 to 3% salt and 0.3 to 0.5% phosphate.

Dairy Ingredients

In the production of reformed hams both lactose and milk proteins (sodium caseinate) are very useful ingredients. Lactose can be used in levels varying from 0.5 to 2%. When lactose is used, it serves as a water binder, thereby increasing the yield. It has the capacity of masking the bitter aftertaste of phosphates, giving the final product a more delicate flavor. Milk protein or milk protein hydrolysates can be used in levels varying from 0.8 to 1.6% by dissolving them in the brines used for injection of the meat pieces. Milk protein strongly affects the binding between the meat pieces, thereby an improved sliceability is obtained and they have a positive effect on yield. The effects on yield after cooking as function of milk protein percentage in the final product is presented in Figure 7.

INTERACTION OF MILK PROTEINS WITH MEAT PROTEINS

The results of application research indicate that there is a specific interaction between milk protein and meat proteins. To study this effect we have to make a classification of proteins present in meat. Meat protein can be divided

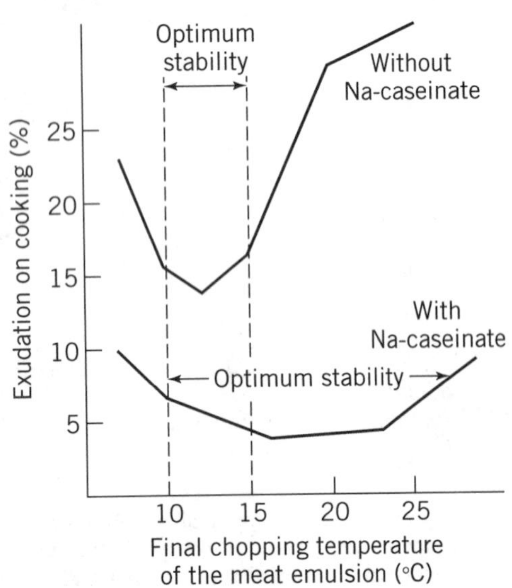

Figure 6. Influence of Na-caseinate on the stability of a meat emulsion in dependence of the final chopping temperature.

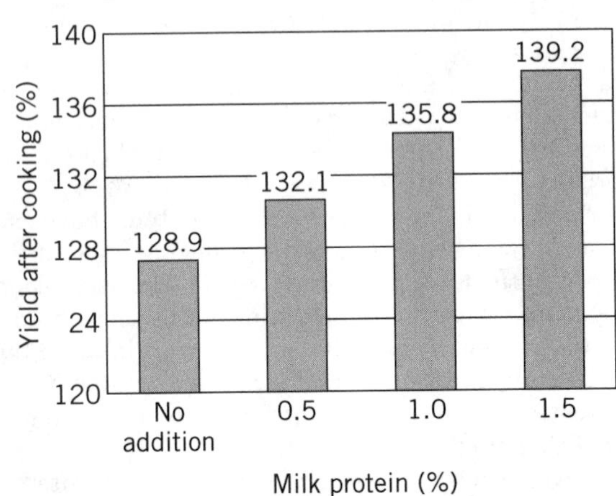

Figure 7. Milk proteins in reformed cooked ham (60% injection).

into different groups according to their specific function and solubility in different solvents (Table 4).

To study the effect of interaction between milk protein and salt-soluble meat proteins, lean meat was extracted according to the extraction procedure shown in Figure 8. Sixty-three percent lean meat (*M. semimembranosis*) is extracted with 2% salt and 35% ice/water in the bowl chopper. After dilution with 200% brine (2% salt), the lean meat slurry is centrifuged.

After centrifugation there are three distinct layers in the centrifuge tube:

Upper layer: contains wsp (sarcoplasmic proteins) and the salt-soluble myofibrillar proteins.

K-fraction: contains nonsolubilized and swollen myofibrillar proteins, mainly actomyosin.

R-fraction: contains connective tissue materials.

The gellification of wsp/ssp proteins was tested in the following solution:

3% ssp/wsp proteins.

3% NaCl.

2% milk protein.

The equipment for these tests is the gelograph (Fig. 9).

The principle of this gelograph is as follows: A protein solution that is placed in a waterbath is slowly heated (1°C/min). During this heat treatment the viscosity/gel strength of the solution is continuously measured by means of an oscillating needle; this equipment operates in a nondestructive way; this means the gel structure that is formed during the heating of the solution is not destroyed because of minimal oscillating of the needle. In this way it is possible to imitate the gelling of meat proteins in pasteurized meat products (eg, a cooked ham).

An example is a product containing 60% lean meat, which contains about 12% total meat protein. Roughly 50% of this protein is myofibrillar proteins. This is about 6%. Suppose that 50% of total myofibrillar proteins are extracted during massaging/tumbling. This means that the concentration of ssp will be around 3%. This is the same concentration as used in the model experiments. Results of gelograph experiments are presented in Figures 10 and 11.

Figure 10 shows the gel strength of a 3% wsp/ssp as a function of temperature. The temperature starts at 20°C (68°F) and is increased at a rate of 1°C/min; during this temperature increase the gel strength is measured as miligels. At 44°C (110°F) there is a sudden increase in gel strength; this is caused by the aggregation of myosin molecules (24). At higher temperatures the viscosity decreases

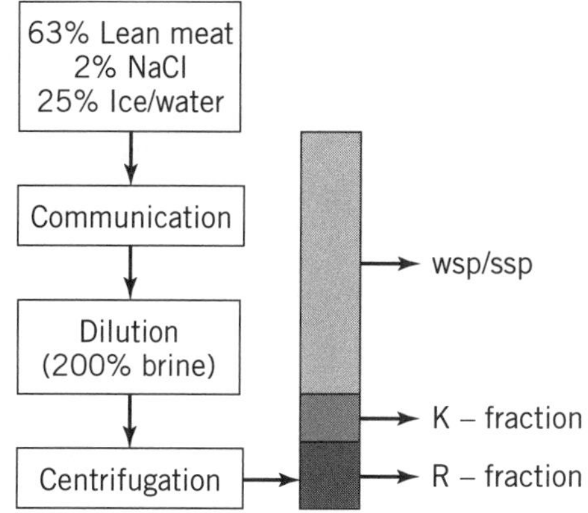

Figure 8. Fractionation of meat proteins.

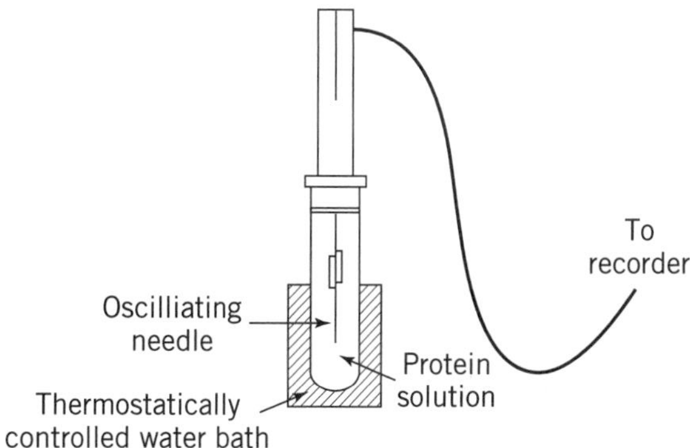

Figure 9. The gelograph.

again. The reason for this phenomenon is not yet exactly known (23,26,27).

At a temperature of 57°C there is again an increase in gel strength. With further heating of the solution to 75°C the gel strength reaches a constant value of ±150 m gels. Cooling down the solution results in a final gel strength of ±200 m gels. The same experiments are done in the presence of 2% milk proteins and 2% soy protein isolate. The results are presented in Figure 12. A conclusion from this figure—addition of 2% milk protein (which has no gellifying capacity itself) results in a much higher gel strength of a heated meat protein solution.

This result supports the theory that there is a synergistic effect of milk proteins on the gellifying capacity of salt-soluble meat proteins, either milk proteins play a role in the cross-linking of myosin molecules or they absorb a significant amount of water, which results in a higher net concentration of myosin molecules. Other nonmeat proteins

Table 4. Classification of Meat Proteins

Protein	Percent	Role	Solubility	Gellifying properties	Emulsifying properties
Sarcoplasmic	30–35	Metabolic	Water	Weak	Weak
Stromal	15–20	Connective	Insoluble	No	Weak
Myofibrillar	50–55	Contractile	Salt soluble	Strong	Strong

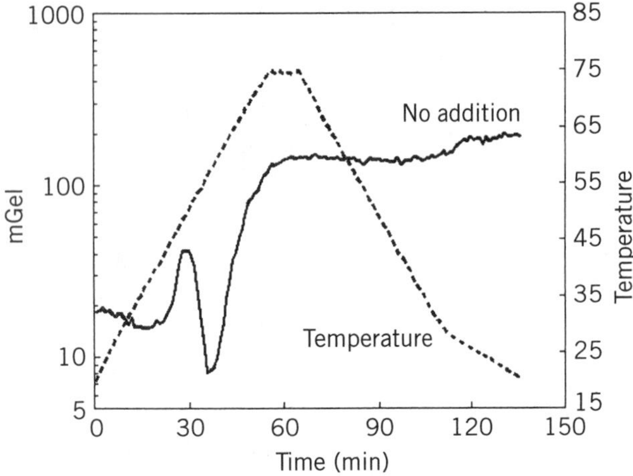

Figure 10. Results of an experiment on gelograph meat protein (no addition).

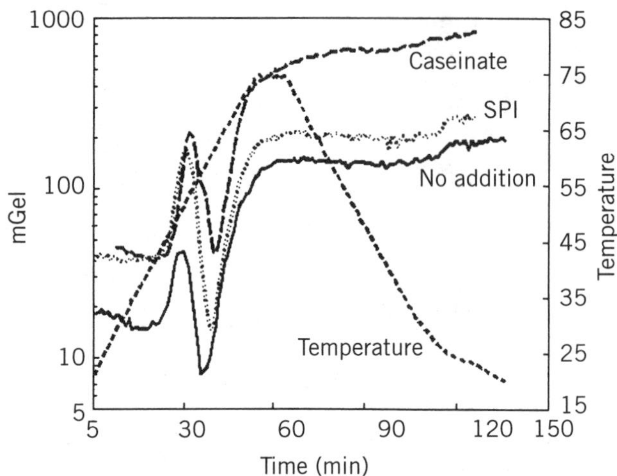

Figure 12. Results of an experiment on gelograph meat protein (no addition, caseinate, SPI).

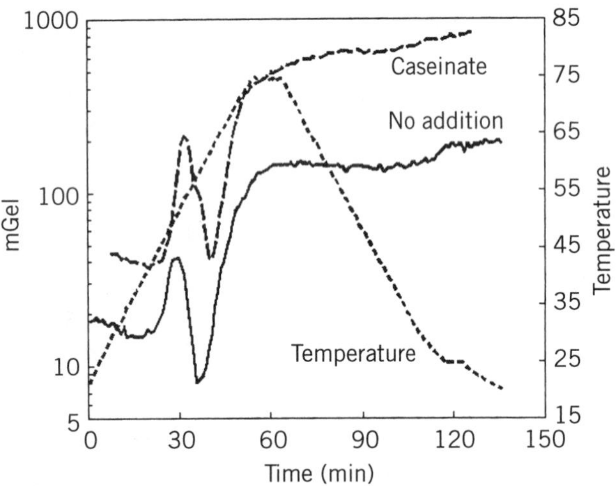

Figure 11. Results of an experiment on gelograph meat protein (no addition, caseinate).

only partially have this specific function (eg, soy protein isolate) or do not have it at all.

POULTRY

Poultry Consumption

Whereas the consumption of red meat is stable or even decreasing, poultry consumption has grown impressively over the last decade. The main reasons for the increase of consumption of poultry are a healthy image (lean, low in fat and cholesterol), increase in the amount of further-processed poultry meat (convenience), and relatively cheap source of animal protein.

The consumption of poultry per capita in different areas of the world is given in Table 5.

Table 5. Per Capita Consumption of Poultry, 1982–1987 (kg)

Area	1982	1983	1984	1985	1986	1987
United States	29.3	29.9	30.5	32.0	33.1	35.6
European community	15.5	15.6	15.6	15.9	16.2	16.9
USSR	9.6	9.4	9.5	10.3	10.7	11.2
Japan	11.1	11.4	11.8	12.2	12.9	13.6

Source: Ref. 2.

Poultry Rolls and Poultry Bolognas

Poultry rolls and poultry bolognas are in appearance very similar to their red meat counterparts. They can be manufactured from materials such as trimmings, skin, fat, and mechanically deboned poultry (mdp). The basic difference between a poultry roll and a poultry bologna is that the bologna sausage contains curing salt and is finely emulsified (20). Poultry rolls normally have a certain ratio of fine meat emulsion to meat chunks.

It is important to create an even color appearance within the roll, ie, the color of the surrounding fine emulsion must match the color of the meat chunks. To obtain this even and white color, it is recommended that milk proteins be used in both basic components: part in the fine emulsion and part mixed into the meat chunks.

Some chicken or turkey bolognas are made from 100% mechanically deboned poultry. From a technologic point of view the poultry meat emulsion is not very different from the red meat emulsion. Sometimes poultry bolognas can have a dry mouthfeel. This can be corrected by the addition of approximately 10% chicken skin/fat emulsion. Such a pre-emulsion is made of 1 part milk protein (sodium caseinate), 5 parts chicken skin, 5 parts chicken fat, 6 parts water, and 2 parts ice.

Without the assistance of such a prestabilized fat/skin emulsion, fatting out often occurs when the total fat content reaches 20% or higher. The addition of milk protein as a dry powder to such products (addition level generally between 1.0 and 1.5%) improves the consistency of the final product as well; generally firmer and better sliceable products are obtained.

Chicken Nuggets and Patties

Since their introduction in the early 1980s, these poultry items have become very popular. Initially, a chicken nugget was a solid piece of slightly marinated breast meat that was battered and breaded (20).

Increased demand has forced the poultry processor to use such parts as thigh meat, trimming, mechanically deboned meat, and chicken skin. The target in combining all these ingredients in a nugget or a patty is to reach the appearance, flavor, and bite of the original product as closely as possible. Important in the stabilization of the final product is the binding of the meat pieces, the water holding capacity, and the stability of the fat.

Again, some basic rules in meat technology are very important: for optimal binding of the meat pieces, it is necessary to grind the meat before use; this creates relatively small meat pieces with a large surface area. During the mixing stage, when salt and phosphates are added, it is important that sufficient meat proteins are extracted and become hydrated in the continuous phase (28–30). On frying, this continuous phase will gellify and cement the meat pieces together. When milk protein is being used in these systems, they boost the meat protein gelation, resulting in higher yields and better consistency of the final products.

Products based on lean meat exclusively tend to be dry and tough after frying. By incorporation of prestabilized chicken fat/skin emulsion (usually 10%), flavor, consistency, and juiciness can be improved.

Whole Muscle Poultry Products

One of the most popular food items in this group of poultry products is breast of turkey. Essentially the manufacturing process comes very close to the production process of reformed hams. Important differences are: lower injection levels are applied (up to 25%); lower salt concentrations are being used; no curing ingredients are being used, because a white color is desired; and the tumbling process is generally much shorter and less intense than that practiced in ham manufacture.

For the production of breast of turkey both milk protein (type sodium caseinate) and milk protein concentrates are used frequently. Usage levels vary from 0.8 to 1.5% (calculated on the final product). Yield increases from 3.0 to 6.0% can be reached. At the same time the consistency, binding, and sliceability of the product are improved. One of the most important characteristics of milk proteins in comparison to, eg, carbohydrate-based stabilizers is the fact that milk proteins do not affect the original flavor and palatability of the final product.

SEAFOOD

Since the growing interest in food and health in general, seafood items have become very popular over the last 10 years. The major reasons for the impressive increase in seafood consumption are sea foods are low in fat, high in protein, contain high percentages of polyunsaturated fatty acids, and most sea foods are relatively cheap.

Although fish and meat are two completely different sources of protein, there is a striking similarity from a technologic point of view. Muscle fibers in seafood are almost identical to fibers in red meat (31). A remarkable difference is the very low content of collagen protein in fishmeat. This is the main reason for its softer texture and bite compared to red meats.

Because of the similarity between red meat and fishmeat structure a lot of processing technologies from the red meat industry can be transferred to processed seafood. This is especially true for fish sausages, fish nuggets, and fish burgers.

Table 6. Water-Packed Tuna

Component	Percent of total
Steam-cooked tuna	75
Brine	25
Brine formula	
Water	86
Salt	4
Milk protein hydrolysate	10

Canned Tuna

Canned tuna, but eg, also canned salmon, mackerel, and sardines, has been on the market for years as preserved fish with a long shelf life. For a good heat transfer and maintenance of the natural appearance and taste during sterilization, it is absolutely necessary that the fish is canned with a liquid as heat transfer medium. Until recently the liquid used was usually oil, but as a result of changes in the food consumption pattern and health consciousness of people, an aqueous solution has been applied more frequently in recent times. The switch from oil to an aqueous solution is naturally associated with some changes in product characteristics, and it is apparent in practice that milk protein hydrolysate is a perfect ingredient for stabilizing the canned product in respect to flavor, juiciness, and yield.

In the preparation of canned tuna and other species of fish the starting point is usually precooked fish. After steam-cooking the fish is cooled and bones, skin, and the dark fish flesh are removed carefully and the fish is suitable for canning. A formula for water-packed tuna is given in Table 6.

Milk Protein Functionality

A light color, mild flavor, juicy texture, and good water penetration during sterilization are extremely important characteristics of premium quality canned tuna. Milk protein hydrolysate promotes the juiciness of the fish structure, gives a mild taste, and has an excellent effect on the water retention of the tuna during sterilization. Adding 10% milk hydrolysate, as indicated in the brine formula, improves the drain weight of the tuna by 6 to 8%. This means that milk proteins not only effect the quality of the final product, but have important economic advantages as well (30).

BIBLIOGRAPHY

1. R. Hermus and H. Albers, "Meat and Meat Products in Nutrition," *Proceedings of 32nd European Meeting of Meat Research Workers*, Gent, Belgium, 1986.

2. ZMP, Bilanz 1987, Zentrale Markt- und Preisberichtstelle für Erzeugnisse der Land-, Forst- und Ernährungswirtschaft GmbH, Bonn, FRG, 1988.
3. F. M. W. Visser, "Dairy Ingredients in Meat Products", *Proceedings of the International Dairy Federation Seminar: Dairy Ingredients in the Food Industry*, Luxembourg, May 19–21, 1981.
4. S. S. Gulayev-Zaitsev, "The Use of Caseinates in Foods," in Organizing Committee of the XXII International Dairy Congress, ed., *Milk the Vital Force*, D. Reidel Publishing Company, Dordrecht, Holland, 1986.
5. P. Walstra and R. Jenness, *Dairy Chemistry and Physics*, John Wiley & Sons, Inc. New York, 1984.
6. P. V. Farrant, "Muscle Proteins in Meat Technology", in P. F. Fox and J. J. Cordon, eds., *Food Proteins*, Applied Science Publishers Ltd., New York, 1982.
7. R. Hamm, *Kolloidchemie des Fleisches*, Paul Parey, Berlin and Hamburg, 1972.
8. J. Schut, "Meat Emulsions'" in S. Friberg, ed., *Food Emulsions*, Marcel Dekker Inc., New York, 1976.
9. P. M. Smith, "Meat Proteins: Functional Properties in Comminuted Meat Products," *Food Technology* **4**, 116–121 (1988).
10. N. L. King and J. J. MacFarlane, "Muscle Proteins," in A. M. Pearson and T. R. Dutson, eds., *Advances in Meat Research, Restructured Meat and Poultry Products*, Vol. 3, Van Nostrand Reinhold Co., New York, 1987.
11. J. Schut, F. Visser, and F. Brouwer, "Microscopical Observations During Heating of Meat Protein Fractions and Emulsions Stabilized by These," *Proceedings of 25th European Meeting of Meat Research Workers*, Budapest, Hungary, Aug. 27–31, 1979, Paper 6.12.
12. J. Schut, "Zur Emulgierbarkeit von Schlachtfetten bei der Herstellung von Brühwurst," *Fleischwirtschaft* **48**, 1201–1204 (1968).
13. J. Schut and F. Brouwer, "The Influence of Milk Proteins on the Stability of Cooked Sausages," *Proceedings of 21st European Meeting of Meat Research Workers*, Aug. 31–Sept. 5, Bern, Switzerland, 1975, pp. 80–82.
14. J. Schut, F. Visser, and F. Brouwer. "Fat Emulsification in Finely Comminuted Sausage: A Model System," *Proceedings of 24th European Meeting of Meat Research Workers*, Kulmbach, FRG, 1978, paper W 12.
15. S. A. Ackerman, C. Swift, R. Carroll, and W. Townsend, "Effects of Types of Fat and of Rates and Temperatures of Comminution on Dispersion of Lipids in Frankfurters," *Journal of Food Science* **36**, 266–269 (1971).
16. L. Knipe, "Sectioned and Formed Meat Products," *Proceedings Annual Sausage and Processed Meat Short Course*, Ames, Iowa, 1982, Res. Paper, 2113 E-7.
17. P. A. Morrisey, D. M. Mulvihill, and E. O'Neill, "Functional Properties of Muscle Proteins", in B. J. F. Hudson, ed., *Development in Food Proteins*, Vol. 5, Elsevier Applied Science, London, 1987.
18. J. E. Kinsella, "Relationships between Structure and Functional Properties of Food Proteins," in P. F. Fox and J. J. Gordon, eds., *Food Proteins*, Applied Science Publishers Ltd., New York, 1982.
19. R. Hamm and J. Grambowska, "Proteinlöslichkeit und Wasserbindung unter den in Brühwurstbräten gegebenen Bedingungen, Part II," *Fleischwirtschaft* **58**, 1345–1348 (1978).
20. H. Hoogenkamp, *Milk Protein. The Complete Guide to Meat, Poultry, and Sea Food*, DMV Campina bv., Veghel, Holland, 1989.
21. J. Schut and F. Brouwer, "Preferential Adsorption of Meat Proteins during Emulsification," *Proceedings of 17th European Meeting of Meat Research Workers*, Bristol, UK, 1971.
22. A. S. Bawa, H. Z. Orr, and W. R. Usborne, "Evaluation of Synergistic Effects Obtained in Emulsion Systems for the Production of Wieners, *Journal of Food Science Technology* **5**, 285–290 (1988).
23. A. Asghar, K. Samejima, and T. Yasui, "Functionality of Muscle Protein in Gelation Mechanisms of Structured Meat Products," *CRC, Critical Reviews in Food Science and Nutrition* **22**, 27–106 (1985).
24. E. A. Foegeding and T. C. Lanier, "The Contribution of Non Muscle Proteins to Texture fo Gelled Muscle Protein Foods," in R. D. Philips and J. W. Fixley, eds., *Protein Quality and the Effects of Processing* Marcel Dekker Inc., New York, 1989.
25. B. Egelandsdal, K. Fretheim, and O. Honbitz, "Dynamic Rheological Measurements on Heat Induced Myosin Gels: An Evaluation of the Methods Suitable for the Filamentous Gels," *Journal of the Science of Food and Agriculture* **37**, 944–954, 1986.
26. M. K. Knight, "Interaction of Proteins in Meat Products: Part III, Functions of Salt Soluble Protein, Insoluble Myofibrillar Protein and Connective Tissue Protein in Meat Products," Research report no. 630, Leatherhead Food R. A., Leatherhead, UK, 1988.
27. G. R. Ziegler and J. C. Acton, "Mechanisms of Gel Formation by Proteins of Muscle Tissue," *Food Technology* **5**, 77–82 (1984).
28. K. Samejima, M. Ishioroshi, and T. Yasui, "Relative Roles of the Head and Tail Portions of the Molecule in Heat Induced Gelation of Myosin," *Journal of Food Science* **46**, 1412–1418 (1981).
29. J. A. Dudziak, E. A. Foegeding, and J. A. Knopp, "Gelation and Thermal Transitions in Post Rigor Turkey Myosin/Actomyosin Suspensions," *Journal of Food Science* **53**, 1278–1281 (1988).
30. D. G. Huber and J. H. Regenstein, "Emulsion Stability Studies of Myosin and Exhaustively Washed Muscle from Adult Chicken Breast Muscle," *Journal of Food Science* **53**, 1282–1293 (1988).
31. J. Jongsma, *Milk Proteins in Processed Sea Food*, DMV Campina bv., Veghel, Holland, 1987.

GENERAL REFERENCES

M. Anese and R. Gormley, "Retaining Quality in Frozen Fish Mince," *Farm & Foods* **5**, 13–15 (1995).

M. Anese and R. Gormley, "Effects of Dairy Ingredients on Some Chemical, Physico-chemical and Functional Properties of Minced Fish During Freezing and Frozen Storage," *Lebensm.-Wiss. Technol.* **29**, 151–157 (1996).

P. Baardseth et al., "Dairy Ingredients Effects on Sausage Sensory Properties Studies by Principal Component Analysis," *J. Food Sci.* **57**, 822–828 (1992).

B. C. Beuschel, J. A. Partridge, and D. M. Smith, "Insolubilized Whey Protein Concentrate and/or Chicken Salt-soluble Protein Gel Properties," *J. Food Sci.* **57**, 852–855 (1992).

A. M. Bhoyar et al., "Effect of Extending Restructured Chicken Steaks With Milk Co-precipitates," *Journal of Food Science and Technology India* **35**, 90–92 (1998).

H. K. Cheng, E. Kolbe, and M. English, "A Nonlinear Programming Technique to Develop Least Cost Formulations of Surimi Products," *Journal of Food Process Engineering* **20**, 179–186 (1997).

S. Cunningham, "Marketing of Dairy Ingredients. Part 1," *World of Ingredients* Jan/Feb., 22–23, 25, 27 (1995).
S. Cunningham, "Marketing of Dairy Ingredients. Part II," *World of Ingredients* March/April, 38–41 (1995).
B. N. Dobson and D. P. Cornforth, "Nonfat Dry Milk Inhibits Pink Discoloration in Turkey Rolls," *Poultry Science* **71**, 1943–1946 (1992).
B. N. Dobson et al., "Instrument for Measuring Bind Strength of Restructured and Emulsion-Type Meat Products," *J. Texture Stud.* **24**, 303–310 (1993).
M. R. Ellekjaer, T. Naes, and P. Baardseth, "Milk Proteins Affect Yield and Sensory Quality of Cooked Sausages," *J. Food Sci.* **61**, 660–666 (1996).
S. B. El-Magoli, S. Larola, and P. M. T. Hansen, "Ultrastructure of Low-Fat Ground Beef Patties With Added Whey Protein Concentrate," *Food Hydrocolloids* **9**, 291–306 (1995).
S. B. El-Magoli, S. Larola, and P. M. T. Hansen, "Flavor and Texture Characteristics of Low Fat Ground Beef Patties Formulated With Whey Protein Concentrate," *Meat Science* **42**, 179–193 (1996).
J. B. German and C. J. Dillard, "Fractionated Milk Fat: Composition, Structure, and Functional Properties," *Food Technol.* **52**, 33–34, 36–38 (1998).
T. P. Guinee and M. O. Corcoran, "Expanded Use of Cheese in Processed Meat Products," *Farm & Food* **4**, 25–28 (1994).
A. Hendrickx, "New Protein Additive for Meat and Poultry Products: Milk-Based Functional Protein Product," *Fleischerei* **43**, 880, 883–884 (1992).
C. K. Hsu and E. Kolbe, "The Market Potential of Whey Concentrate as a Functional Ingredient in Surimi Seafoods," *Journal of Dairy Science* **79**, 2146–2151 (1996).
L. King, "Whey Protein Concentrates as Ingredients," *Food Technology Europe* **3**, 88–89 (1996).
A. Konno and M. Kimura, "Effects of Addition of Whey Protein Concentrates on the Water-Retention Properties and Texture of Ham," *Journal of Japanese Society of Food Science and Technology* **40**, 427–432 (1993).
H. A. M. Lemmers, "The Use of Dairy Ingredients in Further Processed Poultry Products," *Fleischerei* **42**, III–VI (1991).
P. H. Lyons et al., "The Influence of Whey Protein/Carrageenan Gels and Tapioca Starch on the Textural Properties of Low Fat Pork," *Meat Science* **51**, 43–52 (1999).
L. Mancini, "Dairy Ingredients, The natural Choice for Versatility," *Food Engineering* **64**, 73–75 (1992).
A. McCord, A. B. Smith and E. E. O'Neill, "Heat-Induced Gelation Properties of Salt-Soluble Muscle Proteins as Affected by Nonmeat Proteins," *J. Food Sci.* **63**, 580–583 (1998).
L. G. Phillips, M. J. Davis and J. E. Kinsella, "The Effects of Various Milk Proteins on the Foaming Properties of Egg White," *Food Hydrocolloids* **3**, 163–174 (1989).
E. Pitts, "The European Market for Dairy Ingredients," *Journal of the Society for Dairy Technology* **48**, 79–85 (1995).
K. Piyachomkwan and M. H. Penner, "Inhibition of Pacific Whiting Surimi-Associated Protease by Whey Protein Concentrate," *J. Food Prot.* **18**, 341–353 (1995).
C. Ramesh, *Dairy-Based Ingredients*, Eagan Press, Minn., 1997.
D. M. Smith and A. J. Rose, "Properties of Chicken Salt-Soluble Protein and Whey Protein Concentrate Gels as Influenced by Sodium Tripolyphosphate," *Poultry Science* **74**, 169–175 (1995).
S. Solberg, "Effective Product Development for Sausages," *Livsmedelsteknik* **38**, 42–43 (1996).
H. Y. Tan and D. M. Smith, "Dynamic Rheological Properties and Microstructure of Partially Insolubilized Whey Protein Concentrate and Chicken Breast Salt-Soluble Protein Gels," *J. Agric. Food Chem.* **41**, 1372–1378 (1993).
L. Thorsen and T. Bergersen, "Quantitation of Milk Proteins in Heat-Treated Comminuted Meat Products," *InformMAT* **2**, 62–64 (1989).
P. Walkenstrom and A. M. Hermansson, "Fine-Stranded Mixed Gels of Whey Proteins and Gelatin," *Food Hydrocolloids* **10**, 51–62 (1996).
P. Walkenstrom and A. M. Hermansson, "High-Pressure Treated Mixed Gels of Gelatin and Whey Proteins," *Food Hydrocolloids* **11**, 195–208 (1997).
P. Walkenstrom and A. M. Hermansson, "Mixed Gels of Gelatin and Whey Proteins, Formed by Combining Temperature and High Pressure," *Food Hydrocolloids* **11**, 457–470 (1997).
V. C. Weerasinghe et al., "Whey Protein Concentrate as a Proteinase Inhibitor in Pacific Whiting Surimi," *J. Food Sci.* **61**, 367–371 (1996).
J. N. de Wit, "Nutritional and Functional Characteristics of Whey Proteins in Food Products," *Journal of Dairy Science* **81**, 597–608 (1998).
O. Zorba et al., "The Possibility of Using Fluid Whey in Comminuted Meat Products: Capacity and Viscosity of the Model Emulsions Prepared Using Whey and Muscle Protein," *Z. Lebensm.-Unters. Forsch.* **200**, 425–427 (1995).

MARTIEN VAN DEN HOVEN
Creamy Creations
Rijkevoort, The Netherlands

See also DAIRY INGREDIENTS FOR FOODS; WHEY: COMPOSITION, PROPERTIES, PROCESSING, AND USES.

DAIRY INGREDIENTS FOR FOODS

Milk and its products have been consumed by many generations of people of Western, Middle Eastern, Indian, and some African cultures. The nutritional value of dairy materials, properly handled, has been proven by widespread use, and their safety is assured by processing procedures that have been introduced and refined since the time of Pasteur. The great bulk of dairy products today recognize traditions of simple processes and physical separations, avoiding chemicals and thereby preserving the natural properties and the very desirable connotations of traditional products.

The more recently developed food-processing industries have generated requirements for sophisticated performance of dairy materials as ingredients for both familiar and fabricated foods. Milk production does not require the use of pesticides and hormones. Well-managed herds can produce milk with contaminant levels far below the requirements of regulatory authorities. Active components that directly inhibit the processes of digestion, metabolism, and assimilation of nutrients are not present in milk.

The nature of milk and its components is probably better characterized than the properties of any other basic food. Consequently, the ability to transform milk and its individual components into appropriate food ingredients is extensive, as demonstrated by Figure 1.

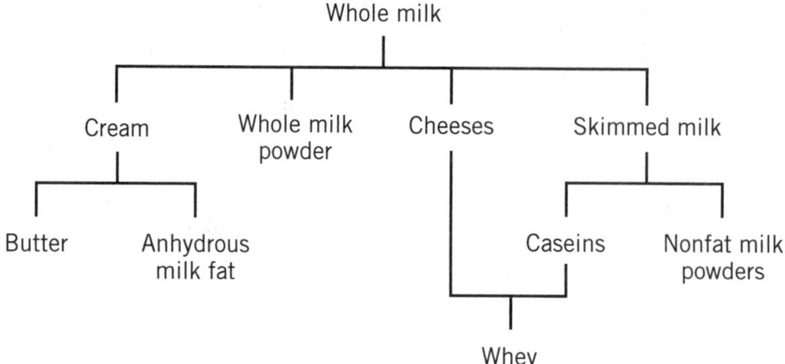

Figure 1. The cascade of dairy materials.

FORMS OF DAIRY INGREDIENTS

Dairy ingredients are provided in many different forms, which can be grouped according to the effect of the process on the products.

1. Compositionally complete or only partially rearranged ingredients (e.g., removal of water and/or fat).

Liquid milk and cream are available, but it is normally more convenient for the food processors to use products from which the water has been removed (as in the manufacture of whole milk powder and butter). The centrifugal removal of cream results in the manufacture of nonfat milk powder. These processes have been industrialized on a very large scale in order to minimize costs. Although they are inherently simple, sophisticated knowledge and process control are required to achieve consistent product performance as may be specified.

2. Products in which a desired component of milk is prepared (e.g., milkfat, protein).

Complete removal of water and nonfat solids from cream will result in anhydrous milkfat that is usable in the same manner as other oils but that is prepared without requiring a complex series of extractive procedures in its manufacture. The major protein of milk, casein, is obtained by simple isoelectric or enzymic precipitation followed by subsequent washing with water and drying.

3. Products obtained by the reassembly, or rearrangement, of the individual components.

At its simplest level the compositional adjustment of milk ingredients is readily achieved. For example, the control of fat levels in cheese products and milk powders is widely practiced; increased levels of whey solids in infant-food products are provided because they are nutritionally desirable; the recombination of powdered ingredients at locations that are far distant from the point of milk production has provided international availability of milk and its products.

4. Products in which the ingredients have been modified for specific purposes.

Judicious application of heat modifies the properties of milk components and can result in a range of milk powders tailored to satisfy specific industrial requirements (1). Addition of food-grade alkali to casein results in its solubilization (2). Physical modification by processes such as homogenization, to provide homogeneity, and agglomeration of powder particles to provide improved dispersibility, is of considerable value. More extensive modifications of dairy materials have been intensively researched. New processes are becoming available, and products that result from these newer technologies hold considerable promise in meeting the increasingly sophisticated requirements of food formulators. Transformations of milk components are being achieved. Lactose can be converted to lactulose (3), protein is being transformed into its hydrolysates (4), and milkfat can be deodorized and decolorized.

Extensively modified materials have been prepared, but the market has not yet taken up all the possibilities; the great bulk of commercial usage in the food industry lies with products that are well known and are based on simple processes.

THE INGREDIENT RANGE

The composition of dairy ingredients varies both naturally (because of their biological origin) and by design (through the manufacturing procedures) but can, in general terms, be set out as in Table 1.

Control of Component Ratios and Performance

A wide variety of performance requirements are imposed by the needs of food formulation. Products that contain various combinations of milk components are available in traditional, widely used forms and also in forms that have been recently modified for food-industry use. The composition of some available product groups is provided in Table 2.

Table 2 reviews the products that have been traditionally available and that are widely recognized. More recently, the specific needs of the food industry have been met by the development of milk powders for milk recombination plants. Recombiners require particular properties such as heat stability for the manufacture of recombined evaporated milks. Similarly, nonfat milk powders that provide manageable viscosity characteristics in the produc-

Table 1. Typical Composition of Dairy Products

Product	Content, % dry basis				Moisture content (%)
	Fat	Protein	Lactose	Mineral	
Milk	32	27	35	5	87
Cheese (cheddar)	53	38	2[a]	6[b]	33
Products with concentrated fat levels					
Cream	87	5	7	1	55
Butter	98	1	1	<0.5	16
Anhydrous milkfat (AMF)	100	<0.1	<0.1	<0.1	<0.1
Products with concentrated protein levels					
Nonfat dry milk	1	40	50	9	3.5
Casein (acid)	1	96	<0.1	2	11
Caseinate	1	95	<0.1	4	4
Whey powder	1	15	76	8	3.4
Whey protein concentrate (WPC)[a]	3	37	54	6	4
Whey protein concentrate[b]	5	79	12	4	4
Whey protein isolate	1	96		3	4
Isolated products					
Lactose			100		5
Minerals	1	8	1	83	10

[a]As lactate.
[b]Includes salt.

Table 2. Combinations of Milk Components

Product class	Performance required by food formulator	Process of manufacture	Reference
Whole milk	A natural liquid	Pasteurized for safety	5
	A stable liquid	Sterilized for microbial stability	5
	A stable liquid with dairy flavor	UHT[a] sterilized and asceptically packaged to protect from oxygen and light	6
Compositionally altered milks	A powder for dry-blend purposes	Dehydrate by roller or spray drier	1
	Fat reduced or zero fat contribution	Remove milk fat centrifugally	7
	Protein enrichment	Protein added before drying	
	Sweetened and stable	Sugar addition, appropriate heat, crystallization concentration	8
	Cocoa crumb as chocolate ingredient	Codried cocoa and sweetened condensed milk	9
Modified milk	Reduced mineral levels (especially for infant foods) or mineral profile adjustment	Ion exchange and/or electrodialysis (of whey generally)	10
	Heat-stable powder	Mineral adjustment and appropriate heat control	11
	A powder for nutritional improvement in bread without loaf volume depression	Appropriate heat control	12
	Sweet but without sugar addition (especially for dietetic uses)	Hydrolysis of lactose by lactase	13
	A powder with high dispersibility	Agglomeration plus surface-coating with hydrophilic materials	14

[a]UHT, ultra-high temperature.

tion of sweetened condensed milks are also required. Manufacture by the appropriate use of heat and by control of the mineral balance of the milks during processing are the essential manufacturing processes. An improvement in the dispersibility of powdered products has been achieved by progressively improved drier design so that powder particles can be agglomerated without damage to the flavor or nutritional properties of the product. More recently, effective coating of the particles with natural materials that impart hydrophillic properties has been achieved. The reduction of mineral levels by the processes of ion exchange and electrodialysis are widely practiced (15,16). This has allowed, in particular, the development of infant foods with low osmotic pressure and with managed mineral profiles that appropriately meet nutritional needs for nutrient balance.

Powders that contain milk fat are manufactured by means that ensure that stale flavors, due to oxidation developed during storage, do not arise in less than 12 months. Fat-containing powders that are intended for dispersion in water require particular care in manufacture (17).

High-fat powders, such as cream and butter powders which are particularly useful in the bakery industry, require appropriate formulation to ensure that free-flow properties are retained without loss of functional performance (18).

PRODUCTS IN WHICH MILKFAT IS THE DOMINANT COMPONENT

The use of cream, butter, and anhydrous milkfat (AMF), represent around 30% of the milkfat that is produced on dairy farms worldwide. The 6.8 million tons used annually means that milkfat is the third most widely used fat in worldwide commerce. The nature of fat on presentation to the manufacturing process is pure and wholesome so that the use of extensive clean-up processes and chemical preservative procedures are unnecessary. Milkfat has a desirable and prestigious flavor, so the maintenance of that flavor is of special importance. For this reason, refrigerated storage of the products of milkfat is common. Because of a significant level of naturally occurring materials that have antioxidant activity, and the widespread use of stainless steel, which prevents contamination of milk with prooxidants, extended shelf life can be achieved without recourse to added antioxidents.

The manufacturing processes that are used are inherently simple physical procedures such as churning and dehydration. The product range is shown in Figure 2.

Rearrangement of the textural properties of milkfat is achieved in a number of ways (19):

1. Mechanical and thermal procedures are often used simultaneously in scraped surface heat exchangers and pin-working machines and result in plasticization. Butter-type products with controlled physical properties are the result.

2. Processes known as dry fractionation consist of crystallization from melted fat after controlled cooling followed by separation of the liquid from solid phase by vacuum filtration or centrifugation without coming into contact with any contaminating process aid. These processes yield products that contain increased solid fat levels that are particularly useful for croissants and puff pastry (20).

3. Compounding of milkfat with vegetable oil to modify spreadability properties of spreadable products results in a number of products that are successful at the retail level and are attractive to the catering industry. Commonly, these products employ 50% milkfat, 40% liquid vegetable oil to increase the proportion of fat liquid at 5°, and 10% hard stock to provide standup properties.

The intensity of the yellow color of milkfat is, to some extent, controllable. The natural color of milk fat is carotene, a precurser of vitamin A, and is not implicated in any concern for health. Selection of the milkfat origin according to breed of cow, feed source to the herd, and season of production will provide a range of color properties. More extensive color modification can be achieved by use of steam deodorization processes that are commonly employed in the oils industry. It yields a white milkfat that is finding extensive use in the manufacture of white cheeses and coffee whiteners and holds promise in other foods.

Experimental investigation into the extensive modification of milkfat by complex processes such as interesterification and hydrogenation has not, so far, yielded products having adequately improved benefits to justify the cost of the processes. Such approaches would simultaneously harm both the distinctive flavor and the naturalness of milkfat (Table 3).

MILK PROTEINS

The products that contain milk proteins are derived from skimmed (nonfat) milk (Fig. 3). The proteins of milk have a well-balanced amino acid profile, having been designed by nature for the rapid early phase of growth that is necessary for infant mammals. By comparison with a reference protein, which has been described by a FAO/WHO

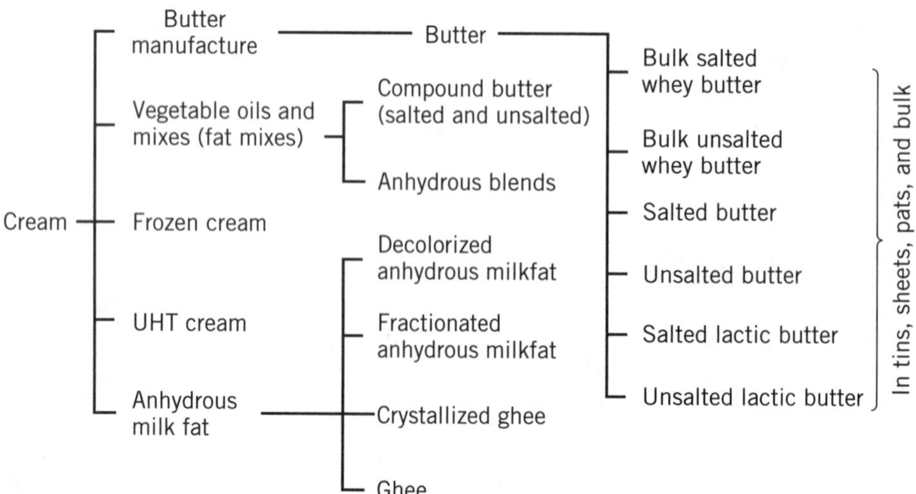

Figure 2. The products derived from cream.

Table 3. Milk Fat Ingredients

Product class	Performance requirement by food formulator	Process of manufacture	Reference
Whole milk			
Cream	Natural flavor and liquid, whippable	Centrifugal separation	7
Butter	Solid or spreadable form	Churning	21
	Spreadability	Ammix process	22
Component: Anhydrous milk fat	Strong dairy flavor with no interfering components	Dehydrate and wash	23
Rearrangement of composition	Control of melting characteristics while retaining natural flavor performance	Selection of natural properties	24
		Fractional crystallization and separation	25
		Inclusion of nondairy oils into churning process (compound butter)	
	Greater spreadability or increased levels of polyunsaturated fatty acids	Inclusion of polyunsaturated oils before churning or by recombining	26
		Introduce protected polyunsaturated oils to feed of the cows	27
Modifications	Light color	Destruction of carotene by steam deodorization	
	Deep color	Select milkfat by origin	28
	Melting-point control	Fractional crystallization and separation	29

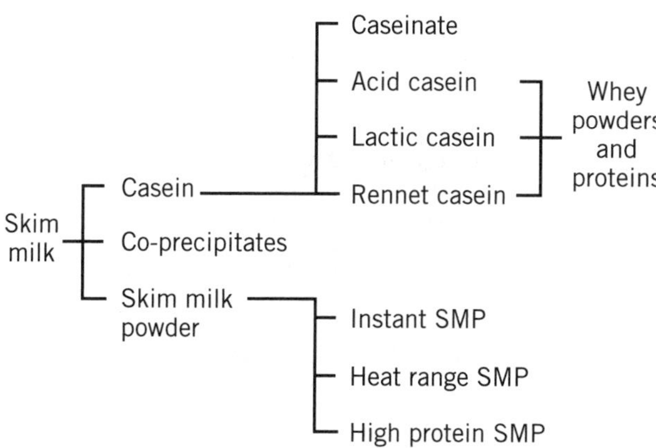

Figure 3. The products derived from skim milk.

group of experts, milk proteins meet all the requirements for essential amino acids. Whey proteins are particularly rich in the sulfur-containing amino acids (methionine and cystine), so they serve a valuable role complementing the amino acid profile of proteins of vegetable origin in food formulations that require high nutritional quality (30).

A physiological role for milk proteins in human nutrition is based on the activity of the immunoglobulins, lactoferrin, and other proteins that are present in small amounts (31). The particular amino acid profile of whey proteins has been proposed as a fundamental property that improves immunocompetence of humans (32).

The proteins in milk are recoverable by long-practiced separation procedures such as by very specific enzymatic precipitation (as in the case of rennet casein), heat precipitation (as in lactalbumin), and by isoelectric precipitation (as in acid caseins). The most widely used of these is the last-named (33).

Recent technological advances have resulted in the manufacture of soluble whey protein concentrates that are produced by the membrane process of ultrafiltration of whey. Functional whey protein concentrates that contain up to about 80% protein are commercially available.

Products that provide the combined benefits of casein and whey protein have been available in the form of coprecipitates, which are manufactured by control of temperature, acidity level, and calcium level (33). More recently developed technology using shifts in pH has resulted in new products that combine strong functionality with high PER levels (34,35).

Specific functional performance has been engineered into caseinates, which have been solubilized from acid casein by reaction with alkali (2). Arising from a precise understanding of the influence of minerals, a growing comprehension of the modifications achievable by controlled management of enzymatic reaction a number of highly specific protein products have been created. Also of considerable importance to the food formulator is the fact that the protein component of a food should not contribute unwelcome flavors. Adsorption technology has been used to enhance the already bland flavor of casein products. The use of casein (and caseinate) as an ingredient in foods is very extensive, as it provides strong functional performance, high levels of protein relative to milk, competitive cost, and strong nutritive quality (36).

Similar product development has been applied to whey protein concentrates (WPCs) with the result that enhanced gelation performance and products with unusual heat stability have been achieved. The selection of an appropriately functional WPC for a particular use in food formulation can offer substantial benefits (37).

Very high purity levels of whey protein have been achieved on the industrial scale by ion exchange processes that produce isolates, but the cost penalty that is incurred means that only the very highest-priced market applications can be considered for such products.

The enzymatic hydrolysis of proteins may be used to produce products of low or zero allergenicity. This technique is receiving intensive scientific attention and is likely to result in products that will have considerable impact in improving the health of infants, and possibly others, who have specific nutritional requirements (Table 4).

CHEESE AND CULTURED PRODUCTS

Cheese and cultured milk products generally are consumed in their native form, but they also can provide a nutritionally valuable ingredient resource with especially useful flavor and compositional characteristics that are well regarded by consumers (48). The cheesemaker can provide a graduated series of flavor, texture, and compositional properties to suit a variety of needs. Consequently cheese types vary widely, as shown in Figure 4.

Natural cheeses are manufactured by controlled fermentation, which influences both acidity and flavor, and which in combination with controlled composition, influences texture (49). Texture control extends from gels, through spreadable pastes, to firmly structured semisolids. Processed cheese, which is made by melting the natural cheese and then dissolving the protein so that a new emulsion is established, extends the range of textural properties and forms of presentation that can be offered.

High-melt properties that can provide intact cheese pieces in, eg, heated sausages have been achieved. Stretch properties as may be required for pizza, and melt charac-

Table 4. Milk Protein

Product class	Performance requirement of food formulation	Process of manufacture	Reference
	Whole proteins		
Coprecipitate	Balanced amino acid profile	Precipitated under managed conditions of heat, acid, and calcium concentration	38
Isolates	Strong functional performance	pH shifts and isoelectric precipitation	34,35
	Components		
Casein-acid	Broad functional performance, bland, nutritionally strong	Isoelectric precipitation, multiple washing procedure	33,36
Casein-rennet	Nutritionally strong, including bioavailable calcium, stable bland flavor, stretch properties for analogue cheese	Enzymic precipitation and multiple washing	39
Whey protein			
Lactalbumin	Amino acid balance complementary to vegetable proteins No interfering function	Heat precipitation and spray dried or ring dried	40
WPC[a] low protein level	Milk powder alternative at economical price	Ultrafiltration to around 35% protein content	
WPC[a], high protein level	Soluble, bland, heat setting	Ultrafilter and diafilter to around 75% protein content	41
Isolates	Pure proteins	Ionic adsorption	42
	Compounding		
	Specific functional performance	Introduce vegetable proteins in blends or by interreaction	43
	Modification		
Caseinates			
Sodium	Solubility, natural emulsification, foam, fat binding, water holding Bland flavor	Treat with sodium alkali and spray dry Carbon asorption	36,44
Calcium	Colloidal protein, viscosity control	Treat with calcium alkali and spray dry	45
Potassium	Soluble and functional with low sodium	Treat with potassium alkali	
Various	Special performance (e.g., high foam, low viscosity, gelation)	Various proprietary procedures	46
Whey proteins, WPC[a]	Special performance (e.g., strong gels, heat stability)	Various proprietary procedures	47
Hydrolysates	Nutritional performance, reduced allergenicity	Enzymic modification	4

[a]WPC, whey protein concentrate.

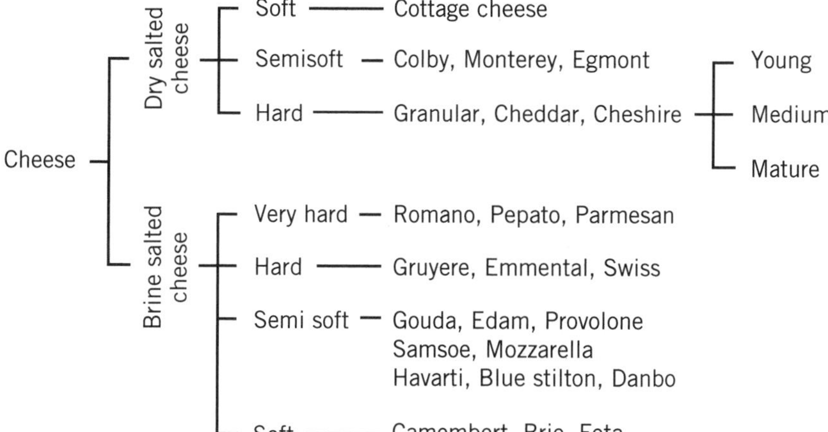

Figure 4. Cheese types.

teristics as may be required for crust appearance, can be built-in during the manufacturing process.

Cheese flavors are often desired characteristics, and they range widely. They include fresh and carbohydrate-derived flavor as in lactic cottage cheese and in creamy harvati; protein-derived flavor as in matured cheddar and camembert; and fat-derived flavors as in parmesan and blue cheeses. Further flavor extension by inclusion of natural materials such as spices and herbs can be used with both natural and processed cheeses. Flavor contributions are particularly significant to cheese sauces, dressings, and parmesan toppings for European-style foods.

Particulate forms of cheese, prepared by grating, dicing, or drying, provide convenience for incorporation into food formulations. Varied flavor intensity (such as by enzyme treatment) can also be built into powdered cheeses. Powdered yogurts are available, for coating and as formulation ingredients. These can be used, along with cream cheese, cottage cheese, and quarg for a range of cheese-cakes and similar products.

New technologies, such as the use of ultrafiltration of milk before cheesemaking, have already yielded commercially successful forms of feta cheese with increased shelf life and promise to further extend the range of product composition that will be created (Table 5).

LACTOSE

The sugar of milk is lactose, a disaccharide that has one of the lowest levels of sweetness contributed by a carbohydrate. Because it is approximately 25% as sweet as sucrose, it can be used in food formulations without dominating the natural flavors of other food components. Its solubility and sweetness vary with temperature and concentration, and these properties may require control in some forms of use (55).

Lactose is prepared by the processes of concentration and multiple crystallization followed by management of crystal size and properties. Lactose is available both at high levels of purity (including pharmaceutical quality) and, at lower cost, as the major component of dry whey powder or dry whey permeates (55).

When used in the baking industry, it acts as a reducing sugar to promote the Maillard reaction, which increases the browning of the crust. At higher temperatures it will caramelize and so contribute to flavor.

In formulated powdered products lactose crystals are slow to take up moisture and consequently minimize the likelihood of caking and lumping. This same property is also a valuable aid to tableting.

Nutritionally, lactose has been shown to promote calcium and phosphorous absorption, which is especially useful in infant feeding preparations (56). It protects against destabilization of the caseinate complex during the drying process. Its presence can also maintain biological activity in the preparation of enzyme products. In brewing and baking applications, it is not fermented by conventional strains of yeast so its contribution to sweetness and color is not destroyed by the biological processes of those industries.

The hydrolysis of lactose yields a sweet syrup that contains glucose and galactose, which may have nutritional advantages in some dietary applications. Lactose-hydrolyzed syrups from permeates and wheys are commercially available and are being used in confectionery and ice cream (Table 6).

THE NUTRITIONALLY VALUABLE MINOR COMPONENTS OF MILK

The new technologies that have arisen from the biotechnology revolution are permitting the extraction of a number of biologically active materials. Some of these biologically active components are present in large proportions in the colostrum milk but decline to low levels during the lactational period. The extraction of lactoferrin, which is an iron-binding protein, has been achieved commercially by sophisticated procedures of chromatography. This protein has been shown to be inhibitory to a number of pathogenic microorganisms, plays a role in the immunological process, and also provides a highly available source of iron. The enzyme lactoperoxidase has also been extracted from dairy sources and is capable of acting to destroy microbes in a

Table 5. Cheeses

Product Class	Performance requirement of food formulation	Process of manufacture	Reference
	Whole		
Natural	Flavor and texture selection, strong nutritional contribution Stretch and melt control	Fermentation, controlled acidity and heat, controlled maturation	50
Powders	Flavor enhancement	Selection of type and maturity, emulsify and spray dry	51
Particulate	Sprinkling (onto pizza)	Shredding or dicing	
	Compositionally modified		
Skim milk cheese	Protein source	Fat-reduced milk	
Cream cheese	Fat source flavor contribution and texture	Cream addition, culturing	
	Modified cheese		
	Flavor enhanced	Enzymic modification	53
	Stretch control	Acidity management	
	Processed cheese		
	Specific texture requirements, flavor modification	Emulsification. Use of nondairy ingredients	53
	Specific shape	e.g., Slices	54

Table 6. Lactose

Product class	Performance requirement	Process	Reference
Component	Low sweetness, reduction of water activity, enhanced browning. Promotion of calcium and phosphorus absorption (especially in infant feeding), dispersing agent	Concentration crystallization and particle size control	55,56,57,59
Modified	Sweet syrup	Hydrolysis by lactase, concentration	13,60
Transformed	New materials	Various	55,61

completely safe manner that would provide an alternative to the use of antimicrobial agents.

Natural antibodies are also present in very low concentration in unprocessed milk, but traditional procedures have resulted in their inactivation. Procedures are now available that allow for preparation of protein products having biological activity, and there is evidence that disease control is achievable (62). Currently these products are targeted toward the problems of animal health. Scientific enquiry is targeted toward the improvement of human health. Wide-ranging implications for a new range of foods that have physiological functions exist and may provide a new wave of the future (63,64).

The recognition of the value of calcium naturally present in milk and cheese in meeting the nutritional needs of aging people, particularly women, has resulted in a demand for pure mineral ingredients of natural origin, fine particle size, and clean flavor. The natural calcium content of milk is being recovered by precipitation processes for direct incorporation into foods (Table 7).

PREPARED INGREDIENTS FOR FOODS

Sophisticated food ingredients have been specifically designed to provide physical functional performance for particular food systems. Strong functional properties are provided by dairy-based ingredients.

Casein protein has an open structure and has separate areas of hydrophobic and hydrophyllic nature along the protein molecule. Consequently, it diffuses readily to interfaces and has powerful emulsion-forming end stabilizing properties. Whey proteins unfold on heating and so, similarly, expose hydrophillic areas of the polypeptide chains. They also gel and can provide natural thickening and stabilizing function. The properties of caseinates and coprecipitates can be tailored by control of ionic content to provide a range of viscosities, solubilities, foaming and whipping (67), and surface activity (68). Measurements of the function of milk proteins have proved difficult to relate to the performance of these proteins in the complex environment of food systems (69), but considerable progress has been made (70,71). Manufacturing techniques that control the function of proteins are being practiced (72).

The functional properties of milkfat vary because of the effects on the secretory process in the mammary glands of cows by the breed of the cow and by the type of feed that she receives and that changes seasonally. Levels of technological control by temperature control during processing, texturizing by physical means, and fractionation by controlled crystallization minimize these variations (19).

Table 7. Nutritionally Valuable Components in Minor Concentrations

Product class	Performance requirement of food formulations	Process of manufacture	Reference
Minerals	Natural, origin, low flavor, bioavailable	Precipitation	
Biologically active			
Lactoferrin	Microbiological inhibition by iron binding. Nutritionally available form of iron	Sophisticated chromatography	65
Lactoperoxidase	Natural antimicrobial system	Sophisticated chromatography	65
Immunoglobulins	Natural antibody active against a wide range of diseases	Specialized ultrafiltration	66

Milkfats tailor-made for specific purposes are available (73). The primary role of milkfat, when incorporated into a food, is the contribution that it makes to flavor, texture, and consumer acceptability (74).

The benefits that dairy ingredients provide to food formulators differ according to the particular needs of each industry. For confectionery products, milkfat is compatible with cocoa butter in that it becomes part of the continuous fat phase and it contributes to smooth flavor and texture of milk chocolate. Caramel flavor is best developed from sweetened condensed milk by utilizing the browning properties of lactose and protein. Proteins, especially caseins, enhance moisture retention by candy and control the quantity of free and bound water (75). Hydrolyzed milk proteins act as whipping agents for frappés and marshmallows (76).

For bakery products, milk powders improve crust color, resilience, and structural strength of cakes. Whey powders improve tenderness and shortness. Whey powder and skim milk powder, when used in dough for cookies and biscuits, reduces the tendency of the dough to tear and produces smooth, even browning. Milk proteins at levels up to 20% improve the nutritional value of cookies end biscuits because they contain high levels of the essential amino acids that are deficient in soy and wheat flour. Butter imparts distinctive flavor to butter cookies and croissants (77,78).

For meat products, milk proteins offer improved appearance, increased yield, and economy. The powerful emulsion capacity of sodium caseinate is particularly valuable in comminuted meat products. Whey protein concentrates of high gel strength are important in reformed ham. Lactose will mask salts, phosphates, and bitter aftertaste while providing a low sweetness profile in liver products, cooked hams, and cooked sausages (79,80).

A very wide variety of proprietary dairy ingredients are manufactured for convenience of use in the food industry. Specific attention is paid to the physical form of powders to ensure dispersibility, mixibility, and low dust levels. Butter is available as flakes and in sheet form. Cheese may be shredded, diced, or sliced. Flavor enhancement, particularly for cheese and milk powder, is offered (81).

CONCLUSION

The remarkably wide range of products that has been obtained from milk demonstrates the intensity of technological attention that has been devoted to dairy materials over a long period of time. The increasing interest in progressively more sophisticated materials that will be required by the food industry is opening up an increasing range of possibilities for the future. While the opportunities for sophisticated materials are exciting, the great bulk of dairy ingredient purchases are based on the recognition of value for money and a clear understanding of the increasing range of uses for products that can be produced by relatively simple processes from a pure and natural raw material.

ACKNOWLEDGMENTS

The authors gratefully acknowledge the valuable assistance of the staff of the New Zealand Dairy Research Institute during the preparation of this paper.

BIBLIOGRAPHY

This article has been adapted and used with permission from K. J. Kirkpatrick and R. M. Fedwick, "Manufacture and General Properties of Dairy Ingredients," *Food Technology* **41**, 58–65 (1987).

1. M. E. Knipschildt, "Drying of Milk and Milk Products." In R. K. Robinson, ed., *Modern Dairy Technology*, vol. 1, Elsevier Applied Science Publishers, London, 1986, p. 131.
2. C. Towler, "Conversion of Casein Curd to Sodium Caseinate," *New Zealand Journal of Dairy Sciences and Technology* **11**, (1976).
3. U.S. Pat. 4,273,922 (1981) K. B. Hicks.
4. R. J. Knights, "Processing and Evaluation of the Antigenicity of Protein Hydrolysates," in F. Lifshitz, ed., *Nutrition for Special Needs in Infancy*, Marcel Dekker, New York, 1985.
5. M. J. Lewis, "Advances in the Heat Treatment of Milk," in Ref. 1, p. 131.
6. International Dairy Federation, *New Monograph on UHT Milk*, Document No. 133, International Dairy Federation, Brussels, Belgium, 1981.
7. C. Towler, "Developments in Cream Separation and Processing," in Ref. 1.
8. B. H. Webb, "Condensed Products," in B. H. Webb and E. O. Whittier, eds., *By-products from Milk*, 2nd ed., AVI Publishing Co., Westport, Conn., 1970, p. 83.
9. B. H. Webb, 1970b. "Miscellaneous Products," in Ref. 8, p. 285.
10. J. G. Zadow, "Utilization of Milk Components: Whey," in Ref. 1, p. 93.
11. K. J. Kirkpatrick, "Raw Material Selection for Recombined Evaporated Milk Products." In *Proceedings of IDF Seminar on Recombination of Milk and Milk Products*, Document No. 142. International Dairy Federation, Brussels, 1982, p. 91.
12. E. J. Guy, "Bakery Products," in Ref. 8, p. 197.
13. R. R. Mahoney, "Modification of Lactose and Lactose-containing Dairy Products with Beta-Galactosidase," *Developments in Dairy Chemistry* **3**, 69 (1985).

14. H. G. Kessler, "Drying-Instantizing," in *Food Engineering and Dairy Technology*, Verlag A. Kessler, Germany, 1981.
15. B. T. Batchelder, "Electrodialysis Applications in Whey Processing," in *Proceedings of International Whey Conferences*. Whey Products Institute and International Dairy Federation, 1986.
16. H. Jonsson and S-O. Arph, "Ion Exchange for Demineralization of Cheese Whey," in Ref. 15.
17. U.S. Pat. 4,737,369 (1988) I. A. Suzuka, and K. Mori.
18. F. G. Kieseker, J. G. Zadow, and B. Aitkin, "Further Developments in the Manufacture of Powdered Whipping Creams," *Australian Journal of Dairy Technology* **34**, 112 (1979).
19. E. Frede, "Technological and Analytical Aspects of Milk Fat Modification," in *Conference Proceedings Food Ingredients Europe 1989*, Expoconsult Publishers, Maarsen, The Netherlands, 1989, pp. 55–61.
20. L. Eyres, "Milkfat Product Development," *Lipid Technology* **1**, 12 (1989).
21. R. A. Wilbey, "Production of Butter and Dairy-based Spreads," Ref. 1, p. 93.
22. H. T. Truong and D. S. Munroe, "The Quality of Butter Produced by the "Ammix" Process," *Brief Communications*, 21st International Dairy Congress, vol. 1, p. 341.
23. A. Fjaervoll, "Anhydrous Milkfat, Manufacturing Techniques and Future Applications," *Dairy Industries* **35**, 424 (1970).
24. M. W. Taylor and R. Norris, "The Physical Properties of Dairy Spreads," *New Zealand Journal of Dairy Science and Technology* **12**, 166 (1977).
25. Austr. Pat. 431,955 (1968), I. T. H. Olsson.
26. M. M. Chrysam, "Table Spreads and Shortenings" in T. H. Applewhite, ed., *Bailey's Industrial Oil and Fat Products*, vol. 3, 1985, p. 1.
27. A. D. Fogerty and A. R. Johnson, "Influence of Nutritional Factors on the Yield and Content of Milkfat. Protected Polyunsaturated Fat in the Diet," Document No. 125, International Dairy Federation, Brussels, 1980, p. 96.
28. A. R. Keen, "Seasonal Variation in the Colour of Milkfat from Selected Herds and Two Dairy Plants," *New Zealand Journal of Dairy Science and Technology* **19**, 263 (1984).
29. A. E. Thomas, "Fractionation and Winterization: Processes and Products," in Ref. 26, p. 1.
30. E. Renner, "Milk Proteins," in *Milk and Dairy Products in Human Nutrition*, W-GmbH, Volkswirtschaftlicher Verlag. Munich, Germany, 1983, pp. 90–115.
31. L. Hambraeus, "Importance of Milk Proteins in Human Nutrition: Physiological Aspects," in T. E. Galesloot and B. J. Tinvbergen, eds. *Milk Proteins 1984*, Pudoc, Wageningen, Germany, 1985, p. 63.
32. G. Bounous and P. A. L. Kongshavn, "Influence of Protein Type in Nutritionally Adequate Diets on the Development of Immunity," in M. Friedman, ed., *Absorption and Utilization of Amino Acids*, vol. 2, CRC Press, Inc., Boca Raton, Fla., 19, pp. 219–233.
33. C. R. Southward and N. J. Walker, "Casein Caseinates and Milk Protein Coprecipitates," in *CRC Handbook of Processing and Utilization in Agriculture*, vol. 1, 1982, p. 445.
34. U.S. Pat. 4,376,072 (1982), P. B. Connolly.
35. U.S. Pat. 4,519,945 (1985), H. A. W. E. M. Ottenhof.
36. C. R. Southward, "Uses of Casein and Caseinates," in P. F. Fox, ed., *Developments in Dairy Chemistry*, vol. 4, *Functional Milk Proteins*, Applied Science Publishers, London, 1989, pp. 173–244.
37. M. J. Rockell, "Selecting the Correct WPC for Your Food Applications via a Knowledge of the Functional Properties of WPC's," in *Food Ingredients Europe, Conference Proceedings*, Expaconsult Publishers, Maarssen, The Netherlands, 1989, p. 51.
38. L. L. Muller, "Manufacture of Casein, Caseinates and Coprecipitates," in Ref. 36, vol. 1, 1982, p. 315.
39. C. R. Southward and N. J. Walker, "The Manufacture and Industrial Use of Casein," *New Zealand Journal of Dairy Technology*, (1980).
40. B. P. Robinson, J. L. Short, and K. R. Marshall, "Traditional Lactalbumin—Manufacture, Properties, and Uses," *New Zealand Journal of Dairy Science and Technology* **11**, 114 (1976).
41. K. R. Marshall, "Proteins, Industrial Isolation of Milk Proteins: Whey Proteins" in Ref. 36, vol. 1, 1982, p. 339.
42. D. E. Palmer, "Recovery of Protein from Food Factory Wastes by Ion Exchange," in P. E. Fox and J. J. Congson, eds., *Food Proteins*, Applied Science Publishers, New York, 1981, p. 341.
43. U.S. Pat. 4,486,343 (1984), N. J. Walker and P. B. Connolly.
44. A. Bergman, "Continuous Production of Spray Dried Sodium Caseinate," *Journal of the Society of Dairy Technology* **25**, 89 (1972).
45. J. Roeper, "Preparation of Calcium Caseinate from Casein Curd," *New Zealand Journal of Dairy Science and Technology* **12**, 182 (1977).
46. U.S. Pat. 4,126,607 (1978), W. C. Easton.
47. J. E. Kinsella, "Proteins from Whey: Factors Affecting Functional Behaviour and Uses," *Proceedings of IDF Seminar, Atlanta, Ga.*, International Dairy Federation, Brussels, Belgium, 1986, p. 87.
48. J. C. Dillon, "Cheese in the Diet," in Ref. 50, pp. 499–511.
49. M. E. Johnson, "Cheese Chemistry," in *Fundamentals of Dairy Chemistry*, 3rd ed., by Van Nostrand Reinhold Co., New York, 1988, pp. 634–654.
50. A. Eck, *Cheesemaking, Science and Technology*, Lavoisier Publishing, Inc., New York, 1986.
51. T. I. Hedrick, "Spray Drying of Cheese," in *Proceedings of Second Marschall International Cheese Conference*, 1981, p. 76.
52. L. Talbott, "The Use of Enzyme Modified Cheeses for Flavouring Processed Cheese Products," in Ref. 51, p. 81.
53. A. Meyer, *Processed Cheese Manufacture, Benckiser-Knapsack GMBH, 1970.*
54. J. H. Carne, "Sliced Process Cheese Production," in *Proceedings from the First Marschall International Cheese Conference*, 1979, p. 305.
55. V. H. Holsinger, "Lactose," in Ref. 49, pp. 279–342.
56. S. G. Coton, T. R. Poynton, and D. Ryder, "Utilisation of Lactose in the Food Industry," *Bulletin of International Dairy Federation*, Document 147, p. 23 (1982).
57. E. Renner, "Lactose," in Ref. 30, pp. 154–171.
58. T. A. Nickerson, "Lactose" in Ref. 8, p. 356.
59. P. A. Morrissey, "Lactose: Chemical and Physico-Chemical Properties," *Developments in Dairy Chemistry* **3**, 1 (1985).
60. J. Rothwell, "Uses for Dairy Ingredients in Ice Cream and Other Frozen Desserts," *Journal of the Society of Dairy Technology* **37**, 119 (1984).
61. L. A. W. Thelwall, "Developments in the Chemistry and Chemical Modification of Lactose," *Developments in Dairy Chemistry* **3**, 35 (1985).
62. C. O. Tacket, G. Losonosky, H. Link, Y. Hoang, P. Guesry, H. Hilpert, and M. M. Levine, "Protection by Milk Immunoglobulin Concentrate Against Oral Challenge with Enteroxigenic

Escherichia coli," *New England Journal of Medicine* **318**, 1240 (1988).

63. S. Dosako, "Application of Milk Proteins to Functional Food," *Japan Food Science* **27**, 25–34 (1988).
64. A. S. Goldman, "Immunologic Supplementation of Cow's Milk Formulations," in *Bulletin No 244*, International Dairy Federation, Brussels, Belgium, 1990.
65. B. Reiter, "The Biological Significance of the Non-immunoglobulin Protective Proteins in Milk," *Developments in Dairy Chemistry* **3**, 281 (1985).
66. Jap. Pat. 60-75433 (1985) Y. Minami.
67. H. W. Modler, "Properties of Non-fat Dairy Ingredients—A Review," *Journal of Dairy Science* **68**, 2195–2205 (1985).
68. J. Leman and J. E. Kinsella, "Surface Activity, Film Formulation, and Emulsifying Properties of Milk Proteins," *Critical Reviews in Food Science and Nutrition* **28**, 115–138 (1989).
69. P. D. Patel and J. C. Fry, *The Search for Standard Methods for Assessing Protein Functionality*.
70. C. V. Morr, "Utilization of Milk Proteins as Starting Materials for Other Foodstuffs," *Journal of Dairy Research* **46**, 369–376 (1979).
71. C. V. Morr, "Functional Properties of Milk Proteins and Their Use as Food Ingredients," Chapter 12, p. 375, in Ref. 36, vol. 1, Proteins, p. 375.
72. H. W. Modler and J. D. Jones, "Selected Processes to Improve the Functionality of Dairy Ingredients," *Food Technology* **41**, 114–117 (1987).
73. D. Illingworth, J. C. Lloyd, and R. Norris, "Tailor-made Fats from Milkfat—the New Zealand Experience," in *Fats for the Future II, International Conference of New Zealand Institute of Chemistry*, IUPAC, 1989.
74. M. I. Gurr and P. Walstra, "Fat Content of Dairy Foods in Relation to Sensory Properties and Consumer Acceptability," *International Dairy Federation Bulletin*, No. 244, 44–46 (1990).
75. L. B. Campbell and S. J. Pavlasek, "Dairy Products as Ingredients in Chocolates and Confections," *Food Technology* **41**, 78–85 (1987).
76. L. Munksgaard and R. Ipsen, "Confectionery Products," in *Dairy Ingredients for the Food Industry 3*, International Dairy Federation, Brussels, Belgium, 1989.
77. W. B. Sanderson, "Cakes, Cookies, & Biscuits," in *Dairy Ingredients for the Food Industry*, International Dairy Federation, Brussels, Belgium, 1985.
78. R. O. Cocup and W. B. Sanderson, "Functionality of Dairy Ingredients," *Bakery Products Food Technology* **41**, 86–90 (1987).
79. J. M. G. Lankveld, "Meat and Meat Products," in *Dairy Ingredients for the Food Industry 2*, International Dairy Federation, Brussels, Belgium, 1987.
80. van deu Hoven, "Functionality of Dairy Ingredients in Meat Products," *Food Technology* **41**, 72–77 (1987).
81. "Product Update—Ingredients from and for Dairy Products," *Food Technology* **43**, 108–122 (1989).

GENERAL REFERENCES

"Research Efforts Aim To Find New Uses for Milkfat Fractions in Food Ingredients," *Canadian Dairy* **75**, 13 (1996).

N. Dhanapati et al., "Effects of Mechanical Agitation, Heating and pH on the Structure of Bovine Alpha Lactalbumin," *Animal Science and Technology* **68**, 545–554 (1997).

D. A. Dionysius and J. M. Milne, "Antibacterial Peptides of Bovine Lactoferrin: Purification and Characterization," *Journal of Dairy Science* **80**, 667–674 (1997).

M. A. Drake et al., "Rheological Characteristics of Milkfat and Milkfat-Blend Sucrose Polyesters," *Food Research International* **27**, 477–481 (1994).

M. A. Drake et al., "Milkfat Sucrose Polyesters as Fat Substitutes in Cheddar-Type Cheeses," *J. Food Sci.* **59**, 326–327, 365 (1994).

D. D. Drohan et al., "Milk Protein–Carrageenan Interactions," *Food Hydrocolloids* **11**, 101–107 (1997).

S. El Shibiny, N. M. Shahein, and M. El Shiekh, "Preparation and Functional Properties of UF Whole Milk Protein Isolates," *Bulletin of the International Dairy Federation* No. 311, 28–30 (1996).

M. J. Facon, "Antibacterial Factors in Cow's Milk and Colostrum: Immunoglobulins and Lactoferrin," *Dissertation Abstracts International B* **56**, 4077 (1996).

L. O. Figura, "Lactose With Tailor-Made Properties," *Lebensmitteltechnik* **25**, 43–45 (1993).

J. B. German and C. J. Dillard, "Fractionated Milk Fat: Composition, Structure and Functional Properties," *Food Technol.* **52**, 33–34, 36–38 (1998).

J. Getler, A. Nielsen, and J. Sprogo, "Powdered Milk Protein Concentrate," *Scandinavian Dairy Information* **10**, 42–43 (1996).

J. Getler, A. Nielsen, and J. Sprogo, "Functional Process for Milk Protein Concentrate (MPC)," *Dairy Industries International* **62**, 25, 27 (1997).

M. S. Haddadin, S. A. Ibrahim, and R. K. Robinson, "Preservation of Raw Milk by Activation of the Natural Lactoperoxidase Systems," *Food Control* **7**, 149–152 (1996).

L. Hambraeus, "Lactoferrin—A Milk Protein in the News," *Livsmedelsteknik* **36**, 42 (1994).

W. R. Harmer and C. Wijesundera, "Heat Stability of Milkfat in Relation to Vegetable Oils," *Australian Journal of Dairy Technology* **51**, 108–111 (1996).

A. G. Khramtsov et al., "Application of Lactulose in Food for Children, Dietetic and Clinical Nutrition," *Voprosy Pitaniya* No. 2, 25–26 (1997).

M. Kosempel, A. McAloon, and L. Roth, "Simulated Scale-Up and Cost Estimate of a Process for Alkaline Isomerization of Lactose to Lactulose Using Boric Acid as Complexation Agent," *J. Chem. Technol. Biotechnol.* **68**, 229–235 (1997).

P. Lee and H. E. Swaisgood, "Modification of Milkfat Physical Properties by Immobilized Fluorescens Lipase," *J. Agric. Food Chem.* **45**, 3343–3349 (1998).

V. V. Mistry and J. B. Pulgar, "Physical and Storage Properties of High Milk Protein Powder," *International Dairy Journal* **6**, 195–203 (1996).

D. L. Moreau and M. Rosenberg, "Storage Stability of Anhydrous Milkfat Microencapsulated in Whey Proteins," *Journal of Dairy Science* **78** (Suppl. 1), 114 (1995).

R. L. Motion, "Hydrolysates of Milk Protein," *Proceedings of the Nutrition Society of New Zealand* **17**, 56–63 (1992).

M. S. Nam et al., "Lactoferrin: A Review," *Korean Journal of Dairy Science* **18**, 289–298 (1996).

I. Nor-Aini et al., "Chemical and Physical Properties of Shortenings Based on Palm Oil and Milkfat," *ASEAN Food Journal* **9**, 141–146 (1994).

I. Nor-Aini et al., "Physical Characteristics of Shortenings Based on Modified Palm Oil, Milkfat and Low Melting Milkfat Fraction," *Fett Wissenschaft Technologie* **97**, 253–260 (1995).

C. I. Onwulata, R. P. Konstance, and V. H. Holsinger, "Flow Properties of Encapsulated Milkfat Powders as Affected by Flow Agent," *J. Food Sci.* **61**, 1211–1215 (1996).

C. I. Onwulata, P. W. Smith, and V. H. Holsinger, "Flow and Compaction of Spray-Dried Powders of Anhydrous Butteroil and High Melting Milkfat Encapsulated in Disaccharides," *J. Food Sci.* **60**, 836–840 (1995).

C. I. Onwulata et al., "Particle Structures of Encapsulated Milkfat Powders," *Lebensm. Wiss. Technol.* **29**, 163–172 (1996).

M. Papalois et al., "Australian Milkfat Survey—Physical Properties," *Australian Journal of Dairy Technology* **51**, 114–117 (1996).

D. Rousseau, A. R. Hill, and A. G. Marangoni, "Rheological Behaviour of Modified Milkfat," *Journal of Dairy Science* **78** (Suppl. 1), 103 (1995).

V. K. S. Shukla, "Milkfat and Its Applications," *World of Ingredients*, 30–31, 33 (Jan./Feb. 1995).

W. Strohmaier, "Lactulose, an Innovative Food Ingredient—Physiological Aspects," *Food Ingredients Europe Conference Proceedings*, 69–72 (1997).

C. Versteeg et al., "New Fractionated Milkfat Products," *Australian Journal of Dairy Technology* **49**, 57–61 (1994).

X. Ye, "Studies on Minor Proteins in Bovine Milk," *Journal of the Faculty of Applied Biological Science Hiroshima University* **34**, 194–195 (1995).

ROBIN M. FENWICK
K. J. KIRKPATRICK
New Zealand Dairy Board
Wellington, New Zealand

See also DAIRY INGREDIENTS: APPLICATIONS IN MEAT, POULTRY, AND SEAFOODS; WHEY: COMPOSITION, PROPERTIES, PROCESSING, AND USES.

DEHYDRATION

Dehydration (or drying) is defined as the application of heat under controlled conditions to remove the majority of the water normally present in a food by evaporation (or in the case of freeze-drying by sublimation). This definition excludes other unit operations that remove water from foods (eg, mechanical separations, membrane concentration, evaporation, and baking) as these normally remove much less water than dehydration. The main purpose of dehydration is to extend the shelf life of foods by a reduction in water activity. This inhibits microbial growth and enzyme activity, but the product temperature is usually insufficient to cause inactivation. The reduction in weight and bulk of food reduces transport and storage costs and, for some types of food, provides greater variety and convenience for the consumer. Drying causes deterioration of both the eating quality and the nutritive value of the food. The design and operation of dehydration equipment aim to minimize these changes by selection of appropriate drying conditions for individual foods. Examples of commercially important dried foods are sugar, coffee, milk, potato, flour (including bakery mixes), beans, pulses, nuts, breakfast cereals, tea, and spices.

THEORY

Dehydration involves the simultaneous application of heat and removal of moisture from foods. Factors that control the rates of heat and mass transfer are described elsewhere in the encyclopedia. Dehydration by heated air or heated surfaces is described in this article. Microwave, dielectric, radiant, and freeze-drying are described in other entries.

Psychrometrics

The capacity of air to remove moisture from a food depends on the temperature and the amount of water vapor already carried by the air. The content of water vapor in air is expressed as either absolute humidity—the mass of water vapor per unit mass of dry air (in kilograms per kilogram), termed moisture content in Fig. 1)—or relative humidity (RH) (in percent)—the ratio of the partial pressure of water vapor in the air to the pressure of saturated water vapor at the same temperature, multiplied by 100. Psychrometry is the study of the interrelationships of the temperature and humidity of air. These properties are most conveniently represented on a phychrometric chart (Fig. 1).

The temperature of the air, measured by a thermometer bulb, is termed the dry-bulb temperature. If the thermometer bulb is surrounded by a wet cloth, heat is removed by evaporation of the water from the cloth and the temperature falls. This lower temperature is called the wet-bulb temperature. The difference between the two temperatures is used to find the relative humidity of air on the psychrometric chart. An increase in air temperature, or reduction in RH, causes water to evaporate from a wet surface more rapidly and therefore produces a greater fall in temperature. The dew point is the temperature at which air becomes saturated with moisture (100% RH). Adiabatic cooling lines are the parallel straight lines sloping across the chart, which show how absolute humidity decreases as the air temperature increases.

Sample Problems 1. Using the psychrometric chart (Fig. 1), calculate the following.

1. The absolute humidity of air that has 50% RH and a dry-bulb temperature of 60°C.
2. The wet-bulb temperature under these conditions.
3. The RH of air having a wet-bulb temperature of 45°C and a dry-bulb temperature of 75°C.
4. The dew point of air cooled adiabatically from a dry-bulb temperature of 55°C and 30% RH.
5. The change in RH of air with a wet-bulb temperature of 39°C, heated from a dry-bulb temperature of 50°C to a dry-bulb temperature of 86°C.
6. The change in RH of air with a wet-bulb temperature of 35°C, cooled adiabatically from a dry-bulb temperature of 70°C to 40°C.

Solution to Sample Problems 1.

1. 0.068 kg/kg of dry air. Find the intersection of the 60°C and 50% RH lines and then follow the chart horizontally right to read the absolute humidity.
2. 47.5°C. From the intersection of the 60°C and 50% RH lines, extrapolate left parallel to the wet-bulb lines to read the wet-bulb temperature.

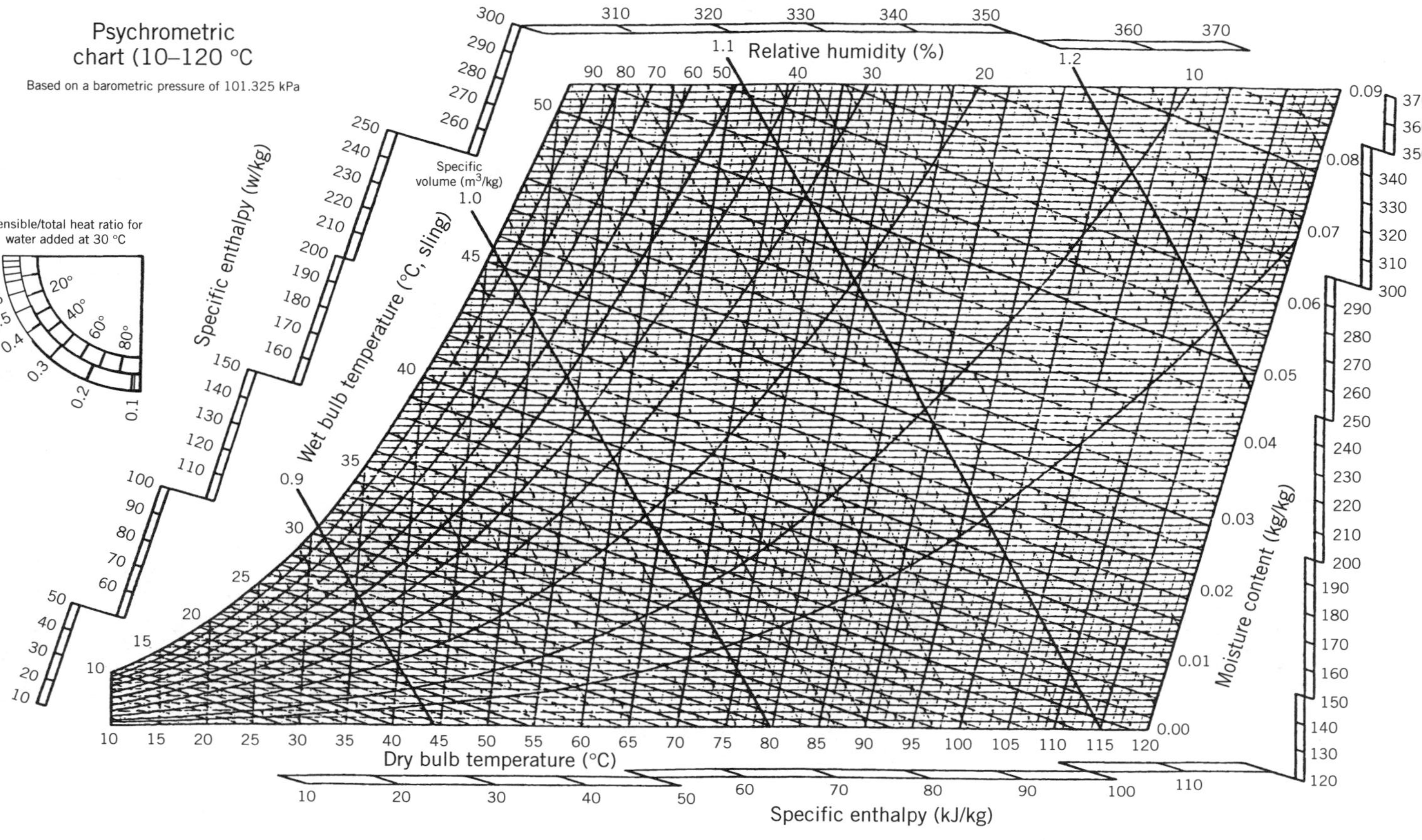

Figure 1. Psychometric chart (10–120°C based on barometric pressure of 101.325 kPa. *Source:* Courtesy of Chartered Institution of Building Services Engineers.

3. 20%. Find the intersection of the 45°C and 75%C lines and follow the sloping RH line upward to read the % RH.
4. 36°C. Find the intersection of the 55°C and 30% RH lines and follow the wet-bulb line left until the RH reaches 100%.
5. 50–100%. Find the intersection of the 39°C wet-bulb and the 50°C dry-bulb temperatures and follow the horizontal line to the intersection with the 86°C dry-bulb line; read the sloping RH line at each intersection (this represents the changes that take place when air is heated before being blown over food.
6. 10–70%. Find the intersection of the 35°C wet-bulb and 70°C dry-bulb temperatures and follow the wet-bulb line left until the intersection with the 40°C dry-bulb line; read sloping RH line at each intersection (this represents the changes taking place as the air is used to dry food; the air is cooled and becomes more humid as it picks up moisture from the food).

Water Activity

Deterioration of foods by microorganisms can take place rapidly, whereas enzymatic and chemical reactions take place more slowly during storage. In either case, water is the single most important factor controlling the rate of deterioration. The moisture content of foods can be expressed either on a wet-weight basis:

$$m = \frac{\text{mass of water}}{\text{mass of sample}} \times 100 \tag{1}$$

$$m = \frac{\text{mass of water}}{\text{mass of water} + \text{solids}} \times 100 \tag{2}$$

or a dry-weight basis (1):

$$M = \frac{\text{mass of water}}{\text{mass of solids}} \tag{3}$$

The dry-weight basis is more commonly used for processing calculations, whereas the wet-weight basis is frequently quoted in food composition tables. It is important, however, to note which system is used when expressing a result. Dry-weight basis is used throughout this article unless otherwise stated.

A knowledge of the moisture content alone is not sufficient to predict the stability of foods. Some foods are unstable at a low moisture content (eg, peanut oil deteriorates if the moisture content exceeds 0.6%), whereas other foods are stable at relatively high moisture contents (eg, potato starch is stable at 20% moisture) (2). It is the availability of water for microbial, enzymatic, or chemical activity that determines the shelf life of a food, and this is measured by the water activity (A_w) of a food. Examples of unit operations that reduce the availability of water in foods include those that physically remove water (dehydration, evaporation, and freeze-drying or freeze concentration) and those that immobilize water in the food (eg, by the use of humectants in intermediate-moisture foods and by formation of ice crystals in freezing). Examples of the moisture content and A_w of foods are shown in Table 1. The effect of reduced A_w on food stability is shown in Table 2.

Water in food exerts a vapor pressure. The size of the vapor pressure depends on

1. The amount of water present.
2. The temperature.
3. The concentration of dissolved solutes (particularly salts and sugars) in the water.

Water activity is defined as the ratio of the vapor pressure of water in a food to the saturated vapor pressure of water at the same temperature:

$$A_W = \frac{P}{P_0} \tag{4}$$

where P (Pa) is vapor pressure of the food and P_0 (Pa) the vapor pressure of pure water at the same temperature. A_w is related to the moisture content by the Brunauer-Emmett-Teller (BET) equation

$$\frac{A_W}{M(1 - A_W)} = \frac{1}{M_1C} + \frac{C - 1}{M_1C} A_W \tag{5}$$

where A_W is the water activity, M the moisture as percentage dry weight, M_1 the moisture (dryweight basis) of a monomolecular layer, and C a constant (2).

A proportion of the total water in a food is strongly bound to specific sites (eg, hydroxyl groups of polysaccharides, carbonyl and amino groups of proteins, the hydrogen bonding). When all sites are (statistically) occupied by adsorbed water the moisture content is termed the BET monolayer value. Typical examples include gelatin (11%), starch (11%), amorphous lactose (6%), and whole spray-dried milk (3%). The BET monolayer value therefore represents the moisture content at which the food is most stable. At moisture contents below this level, there is a higher rate of lipid oxidation and, at higher moisture contents, Maillard browning and then enzymatic and microbiological activities are promoted.

The movement of water vapor from a food to the surrounding air depends on both the moisture content and composition of the food and the temperature and humidity of the air. At a constant temperature the moisture content of food changes until it comes into equilibrium with water vapor in the surrounding air. The food then neither gains nor loses weight on storage under those conditions. This is called the equilibrium moisture content of the food, and the relative humidity of the storage atmosphere is known as the equilibrium relative humidity. When different values of relative humidity versus equilibrium moisture content are plotted, a curve known as a water sorption isotherm is obtained (Fig. 2).

Each food has a unique set of sorption isotherms at different temperatures. The precise shape of the sorption isotherm is caused by differences in the physical structure, chemical composition, and extent of water binding within the food, but all sorption isotherms have a characteristic shape, similar to that shown in Figure 2. The first part of

Table 1. Moisture Content and Water Activity of Foods

Food	Moisture content (%)	Water activity	Degree of protection required
Ice (0°C)	100	1.00[a]	
Fresh meat	70	0.985	
Bread	40	0.96	Package to prevent moisture loss
Ice (−10°C)	100	0.91[a]	
Marmalade	35	0.86	
Ice (−20°C)	100	0.82[a]	
Wheat flour	14.5	0.72	
Ice (−50°C)	100	0.62[a]	
Raisins	27	0.60	Mininum protection or no packaging required
Macaroni	10	0.45	
Cocoa powder		0.40	
Boiled sweets	3.0	0.30	
Biscuits	5.0	0.20	Package to prevent moisture uptake
Dried milk	3.5	0.11	
Potato crisps	1.5	0.08	

Source: Refs. 2–4.
[a]

Table 2. The Importance of Water Activity in Foods

A_W	Phenomenon	Examples
1.00		Highly perishable fresh foods
0.95	Pseudomonads, bacillus, *Clostridium perfringens,* and some yeasts inhibited	Foods with 40% sucrose or 7% salt; cooked sausages, bread
0.90	Lower limit for bacterial growth (general), salmonella, *Vibrio parahemolyticus, Clostridium botulinum,* lactobacillus, and some yeasts and fungi inhibited	Foods with 55% sucrose, 12% salt; cured ham, medium-age cheese. Intermediate-moisture foods (A_W = 0.90 − 0.55)
0.85	Many yeasts inhibited	Foods with 65% sucrose, 15% salt; salami, mature cheese, margarine
0.80	Lower limit for enzyme activity and growth of most fungi; *Staphlococcus aureus* inhibited	Flour, rice (15–17% water) fruit cake, sweetened condensed milk, fruit syrups, fondant
0.75	Lower limit for halophilic bacteria	Marzipan (15–17% water), jams
0.70	Lower limit for growth of most xerophilic fungi	
0.65	Maximum velocity of Maillard reactions	Rolled oats (10% water), fudge, molasses, nuts
0.60	Lower limit for growth of osmophilic or xerophilic yeasts and fungi	Dried fruits (15–20% water), toffees, caramels (8% water), honey
0.55	Deoxyribonucleic acid becomes disordered (lower limit for life to continue)	
0.50		Dried foods, spices, noodles
0.40	Minimum oxidation velocity	Whole egg powder (5% water)
0.30		Crackers, bread crusts (3–5% water)
0.25	Maximum heat resistance of bacterial spores	
0.20		Whole milk powder (2–3% water), dried vegetables (5% water), cornflakes (5% water)

the curve, to point *A*, represents monolayer water, which is very stable, unfreezable, and not removed by drying. The second, relatively straight part of the curve (*AB*) represents water absorbed in multilayers within the food and solutions of soluble components. The third portion (above point *B*) is free water condensed within the capillary structure or in the cells of a food. It is mechanically trapped within the food and held only by weak forces. It is easily removed by drying and easily frozen, as indicated by the steepness of the curve. Free water is available for microbial growth and enzyme activity, and a food that has a moisture content above point *B* on the curve is likely to be susceptible to spoilage.

The sorption isotherm indicates the A_w at which a food is stable and allows predictions of the effect of changes in moisture content on A_w and hence on storage stability. It is used to determine the rate and extent of drying, the optimum frozen storage temperatures, and the moisture-barrier properties required in packaging materials.

The rate of change in A_w on a sorption isotherm differs according to whether moisture is removed from a food (desorption) or it is added to dry food (adsorption) (Fig. 2). This is termed a hysteresis loop. The difference is large in some foods, eg, rice, and is important for example in determining the protection required against moisture uptake.

Figure 2. Water sorption isotherm.

Mechanism of Drying

When hot air is blown over a wet food, heat is transferred to the surface, and latent heat of vaporization causes water to evaporate. Water vapor diffuses through a boundary film of air and is carried away by the moving air (Fig. 3). This creates a region of lower water vapor pressure at the surface of the food, and a water vapor pressure gradient is established from the moist interior of the food to the dry air. This gradient provides the driving force for water removal from the food.

Water moves to the surface by the following mechanisms:

1. Liquid movement by capillary forces.
2. Diffusion of liquids, caused by differences in the concentration of solutes in different regions of the food.
3. Diffusion of liquids, which are adsorbed in layers at the surfaces of solid components of the food.
4. Water vapor diffusion in air spaces within the food caused by vapor pressure gradients.

Foods are characterized as hygroscopic and nonhygroscopic. Hygroscopic foods are those in which the partial pressure of water vapor varies with the moisture content.

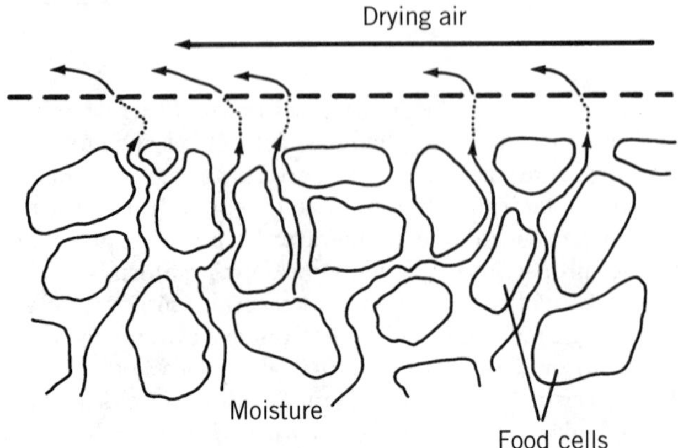

Figure 3. Movement of moisture during drying.

Nonhygroscopic foods have a constant water vapor pressure at different moisture contents. The difference is found by using sorption isotherms.

When food is placed in a dryer, there is a short initial settling down period as the surface heats up to the wet-bulb temperature (*AB* in Fig. 4a). Drying then commences and, provided that water moves from the interior of the food at the same rate as it evaporates from the surface, the surface remains wet. This is known as the constant-rate period and continues until a certain critical moisture content is reached (*BC* in Figs. 4a and b). In practice, however, different areas of the surface of the food dry out at different rates and, overall, the rate of drying declines gradually during the constant-rate period. Thus the critical point is not fixed for a given food and depends on the amount of food in the dryer and the rate of drying.

The three characteristics of air that are necessary for successful drying in the constant rate period are

1. A moderately high dry-bulb temperature.
2. A low RH.
3. A high air velocity.

The boundary film of air surrounding the food acts as a barrier to the transfer of both heat and water vapor during

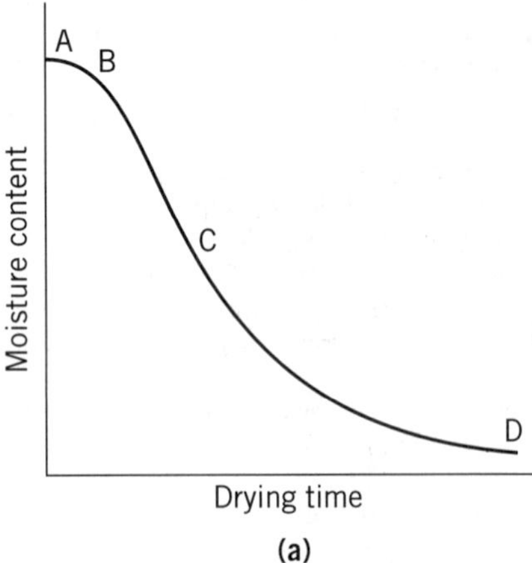

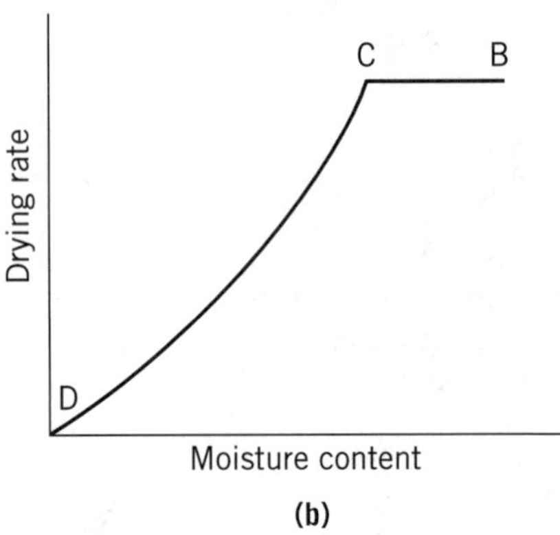

Figure 4. (**a**) and (**b**) Drying curves. The temperature and humidity of the drying air are constant and all heat is supplied to the surface by convection.

drying. The thickness of the film is determined primarily by the air velocity. If this is too low, water vapor leaves the surface of the food and increases the humidity of the surrounding air, to cause a reduction in the water vapor pressure gradient and the rate of drying (Similarly, if the temperature of the drying air falls or the humidity rises, the rate of evaporation falls and drying slows).

When the moisture content of the food falls below the critical moisture content, the rate of drying slowly decreases until it approaches zero at the equilibrium moisture content (ie, the food comes into equilibrium with the drying air). This is known as the falling-rate period. Nonhygroscopic foods have a single falling-rate period (*CD* in Fig. 4a and b), whereas hygroscopic foods have two periods. In the first period, the plane of evaporation moves inside the food, and water diffuses through the dry solids to the drying air. It ends when the plane of evaporation reaches the center of the food and the partial pressure of water falls below the saturated water vapor pressure. The second period occurs when the partial pressure of water is below the saturated vapor pressure and drying is by desorption.

During the falling-rate period, the rate of water movement from the interior of the food to the surface falls below the rate at which water evaporates to the surrounding air. The surface therefore dries out. This is usually the longest period of a drying operation and, in some foods, eg, grain drying, where the initial moisture content is below the critical moisture content, the falling-rate period is the only part of the drying curve to be observed. During the falling-rate period, the factors that control the rate of drying change. Initially, the important factors are similar to those in the constant-rate period, but gradually the rate of mass transfer becomes the controlling factor. This depends mostly on the temperature of the air and the thickness of the food. It is unaffected by both the RH of the air (except in determining the equilibrium moisture content) and the velocity of the air. The air temperature is therefore controlled during the falling rate period, whereas the air velocity and temperature are more important during the constant-rate period. In practice, foods may differ from these idealized drying curves owing to shrinkage, changes in the temperature and rate of moisture diffusion in different parts of the food, and changes in the temperature and humidity of the drying air.

The surface temperature of the food remains close to the wet-bulb temperature of the drying air until the end of the constant-rate period, owing to the cooling effect of the evaporating water. During the falling-rate period the amount of water evaporating from the surface gradually decreases but, as the same amount of heat is being supplied by the air, the surface temperature rises until it reaches the dry-bulb temperature of the drying air. Most heat damage to food therefore occurs in the falling rate period.

Calculation of Drying Rate

The rate of drying depends on the properties of the dryer (the dry-bulb temperature, RH and velocity of the air, and the surface heat transfer coefficient) and the properties of the food (the moisture content, surface-to-volume ratio, and the surface temperature and rate of moisture loss). The size of food pieces has an important effect on the drying rate in both the constant- and the falling-rate periods. In the constant-rate period, smaller pieces have a larger surface area available for evaporation, whereas, in the falling-rate period, smaller pieces have a shorter distance for moisture to travel through the food. Other factors that influence the rate of drying include:

1. The fat content of the food (higher fat contents generally result in slower drying, as water is trapped within the food).
2. The method of preparation of the food (cut surfaces lose moisture more quickly than losses through skin.
3. The amount of food placed into a dryer in relation to its size (in a given dryer, faster drying is achieved with smaller quantities of food).

The rate of heat transfer is found using

$$Q = h_s A(\theta_a - \theta_s) \tag{6}$$

The rate of mass transfer is found using

$$-m_c = K_g A(H_s - H_a) \tag{7}$$

Because during the constant-rate period, an equilibrium exists between the rate of heat transfer to the food and the rate of mass transfer in the form of moisture loss from the food, these rates are related by

$$-m_c = \frac{h_c A}{\lambda}(\theta_a - \theta_s) \tag{8}$$

where Q (J/s) is the rate of heat transfer, h_c W/m^2/°K the surface heat transfer coefficient for convective heating, A (m^2) the surface area available for drying, θ_a (°C) the average dry-bulb temperature of drying air, θ_s (°C) the average wet-bulb temperature of drying air, m_c (kg/s) the change in mass with time (drying rate), K_g (kg/m^2/s) the mass transfer coefficient, H_s (kilograms of moisture per kilogram dry air) the humidity at the surface of the food (saturation humidity), H_a (kilograms of moisture per kilogram dry air) the humidity of air, and λ (J/kg) the latent heat of evaporation at the wet-bulb temperature.

The surface heat transfer coefficient (h_c) is related to the mass flow rate of air using the following equations: for parallel airflow,

$$h_c = 14.3G^{0.8} \tag{9}$$

and for perpendicular airflow,

$$h_c = 24.2G^{0.37} \tag{10}$$

Where G (kg/m^2/s) is the mass flow rate of air.

For a tray of food in which water evaporates only from the upper surface, the drying time is found using

$$-m_c = \frac{h_c}{\rho\lambda x}(\theta_a - \theta_s) \tag{11}$$

where ρ (kg/m^3) is the bulk density of food and x (m) the thickness of the bed of food.

The drying time in the constant-rate period is found using

$$t = \frac{\rho\lambda x(M_i - M_c)}{h_c(\theta_a - \theta_s)} \tag{12}$$

where t (s) is the drying time, M_i (kg/kg of dry solids) the initial moisture content, and M_c (kg/kg of dry solids) the critical moisture content.

For water evaporating from a spherical droplet in a spray dryer, the drying time is found using

$$t = \frac{r^2\rho_1\lambda}{3h_c(\theta_a - \theta_s)} \frac{M_i - M_f}{1 + M_i} \tag{13}$$

where ρ_1 (kg/m^3) is the density of the liquid, r (m) the radius of the droplet, M_f (kg/kg of dry solids) the final moisture content.

In the falling-rate period, moisture gradients change throughout the food and the temperature slowly increases from the wet-bulb temperature to the dry-bulb temperature as the food dries. The following equation is used to calculate the drying time from the start of the falling-rate period to the equilibrium moisture content using a number of assumptions concerning, eg, the nature of moisture movement and the absence of shrinkage of the food:

$$t = \frac{\rho x(M_c - M_e)}{K_g(P_s - P_a)} \ln\left(\frac{M_c - M_e}{M - M_e}\right) \tag{14}$$

where M_e (kg/kg of dry solids) is the equilibrium moisture content, M (kg/kg of dry solids) the moisture content at time t from the start of the falling-rate period, P_s (mmHg) the saturated water vapor pressure at the wet-bulb temperature, and P_a (mmHg) the partial water vapor pressure.

The velocity of the air needed to achieve fluidization of spherical particles is calculated using

$$v_f = \frac{(\rho_s - \rho)g}{\mu} \frac{d^2\varepsilon^3}{180(1 - \varepsilon)} \tag{15}$$

where v_f (m/s) is the fluidization velocity, ρ_s (kg/m^3 the density of the solid particles, ρ (kg/m^3) the density of the fluid, g (m/s^2) the acceleration due to gravity, μ (N · s/m^2) the viscosity of the fluid, d (m) the diameter of the particles, ϵ the voidage of the bed.

Formulas for foods of other shapes are described in Ref. 5. The minimum air velocity needed to convey particles is found using:

$$v_e = \sqrt{\left[\frac{4d(\rho_s - \rho)}{3C_d\rho}\right]} \tag{16}$$

where v_e (m/s) is the minimum air velocity and C_d (= 0.44 for Re = 500–200,000) the drag coefficient.

Sample Problem 2. Peas with an average diameter of 6 mm and a density of 880 kg/m^3 are dried in a fluidized-bed dryer. The minimum voidage is 0.4 and the cross-sectional area of the bed is 0.25 m^2. Calculate the minimum air velocity needed to fluidize the bed if the air density is 0.96 kg/m^3 and the air viscosity is 2.15×10^{-5} N · s/m^2.

Solution of Sample Problem 2. From equation 15

$$v_F = \frac{(880 - 0.96)9.81}{2.15 \times 10^{-8}} \frac{(0.006)^2(0.4)^3}{180(1 - 0.4)} = 8.5 \text{ m/s}^1$$

Derivations of these equations are described in Refs. 6–8.

Sample Problem 3. A conveyor dryer is required to dry peas from an initial moisture content of 78% to 16% moisture (wet-weight basis) in a bed 10 cm deep with a voidage of 0.4. Air at 85°C with a relative humidity of 10% is blown perpendicularly through the bed at 0.9 m/s. The dryer belt measures 0.75 m wide and 4 m long. Assuming that drying takes place from the entire surface of the peas and there is no shrinkage, calculate the drying time and energy consumption in both the constant and the falling-rate periods. (Additional data: The equilibrium moisture content of the peas is 9%, the critical moisture content 300% (dry-weight basis), the average diameter 6 mm, the bulk density 610 kg/m^3, the latent heat of evaporation 2300 kJ/kg, the saturated water vapor pressure at wet-bulb temperature 61.5 mmHg, and the mass transfer coefficient 0.015 kg/m^2/s).

Solution to Sample Problem 3. In the constant-rate period, from equation 10:

$$h_c = 24.2(0.9)^{0.37} = 23.3 \text{ W/m}^2/^\circ\text{K}$$

From Fig. 1 for θ_a = 85°C and RH = 10%,

$$\theta_s = 42^\circ\text{C}$$

To find the area of the peas,

$$\begin{aligned} \text{Volume of a sphere} &= \frac{4}{3}\pi r^3 = 4 \times 3.142(0.003)^3 \\ &= 339 \times 10^{-9} \text{ m}^3 \end{aligned}$$

$$\text{Volume of the bed} = 0.75 \times 4 \times 0.1 = 0.3 \text{ m}^3$$

$$\text{Volume of peas in the bed} = 0.3(1 - 0.4) = 0.18 \text{ m}^3$$

$$\begin{aligned} \text{Number of peas} &= \frac{\text{volume of peas in bed}}{\text{volume each pea}} \\ &= \frac{0.18}{339 \times 10^{-9}} = 5.31 \times 10^5 \end{aligned}$$

$$\begin{aligned} \text{Area of a sphere} &= 4\pi\pi r^2 = 4 \times 3.142(0.003)^2 \\ &= 113 \times 10^{-6} \text{ m}^2 \end{aligned}$$

and

$$\begin{aligned} \text{Total area of peas} &= 5.31 \times 10^5 \times 113 \times 10^{-6} \\ &= 60 \text{ m}^2 \end{aligned}$$

From equation 8:

$$\text{Drying rate} = \frac{23.3 \times 60}{2.3 \times 10^6}(85 - 42) = 0.026 \text{ kg/s}$$

From a mass balance

$$\text{Volume of peas} = 0.18 \text{ m}^3$$

$$\text{Bulk density} = 610 \text{ kg/m}^{-3}$$

Therefore,

$$\text{Mass of peas} = 0.18 \times 610 = 109.8 \text{ kg}$$

$$\text{Initial solids content} = 109.8 \times 0.22 = 24.15 \text{ kg}$$

Therefore,

$$\text{Initial mass water} = 109.8 - 24.15 = 85.6 \text{ kg}$$

After constant-rate period, solids remain constant and

$$\text{Mass of water} = 96.6 - 24.15 = 72.45 \text{ kg}$$

Therefore,

$$(85.6 - 72.45) = 13.15 \text{ kg water lost}$$

at a rate of 0.026 kg/s

$$\text{Drying time} = \frac{13.15}{0.026} = 505 \text{ s} = 8.4 \text{ min}$$

Therefore,

$$\text{Energy required} = 0.026 \times 2.3 \times 10^5 = 6 \times 10^4 \text{ J/s} = 60 \text{ kW}$$

In the falling-rate period,

$$\text{RH} = \frac{P_A}{P_o} \times 100$$

$$10 = \frac{P}{61.5} \times 100$$

Therefore,

$$P = 6.15 \text{ mmHg}$$

The moisture values are

$$M_c = \frac{75}{25} = 3$$

$$M_f = \frac{16}{84} = 0.19$$

$$M_e = \frac{9}{91} = 0.099$$

From Equation 14,

$$t = \frac{(3 - 0.099)610 \times 0.1}{0.015(61.5 - 6.15)} \ln\left(\frac{3 - 0.099}{0.19 - 0.099}\right) = 737.7 \text{ s} = 12.3 \text{ min}$$

From a mass balance, at the critical moisture content, 96.6 kg contains 25% solids = 24.16 kg. After drying in the falling-rate period, 84% solids = 24.16 kg. Therefore,

$$\text{Total mass} = \frac{100}{84} \times 24.16 = 28.8 \text{ kg}$$

and

$$\text{Mass loss} = 96.6 - 28.8 = 67.8 \text{ kg}$$

Thus,

$$\text{Average drying rate} = \frac{67.8}{737.7} = 0.092 \text{ kg/s}$$

$$\text{average energy required} = 0.092 \times 2.3 \times 10^6 = 2.1 \times 10^5 \text{ J/s} = 210 \text{ kW}$$

Drying Using Heated Surfaces

Heat is conducted from a hot surface, through a thin layer of food, and moisture is evaporated from the exposed surface. The main resistance to heat transfer is the thermal conductivity of the food. Knowledge of the rheological properties of the food is necessary to determine the thickness of the layer of food and the way in which it should be applied to the heated surface. Additional resistances to heat transfer arise if the partly dried food lifts off the hot surface. Equation 17 is used in the calculation of drying rates.

$$Q = UA(\theta_a - \theta_b) \tag{17}$$

U is the overall heat transfer coefficient (W/m^2/K), θ_a is the temperature of hot surface (°C), and θ_b is the temperature of food (°C).

Sample Problem 4. A single-drum dryer 0.7 m in diameter and 0.85 m long operates at 150°C and is fitted with a doctor blade to remove food after three-fourths of a revolution. It is used to dry a 0.6-mm layer of 20% w/w solution of gelatin, preheated to 100°C, at atmospheric pressure. Calculate the speed of the drum required to produce a product with a moisture content of 4 kg of solids per kilogram of water. (Additional data: The density of gelatin feed is 1020 kg/m^3, and the overall heat transfer coefficient is 1200 W/m^2/K; assume that the critical moisture content of the gelatin is 450%, dry weight basis.)

Solution to Sample Problem 4. First,

$$\text{Drum area} = \pi dl = 3.142 \times 0.7 \times 0.85 = 1.87 \text{ m}^2$$

Therefore,

$$\text{Mass of food on the drum} = (1.87 \times 0.75)0.0006 \times 1020 = 0.86 \text{ kg}$$

From a mass balance (initially the food contains 80% moisture and 20% solids),

$$\text{Mass of solids} = 0.86 \times 0.2 = 0.172 \text{ kg}$$

After drying, 80% solids = 0.172 kg. Therefore,

$$\text{Mass of dried food} = \frac{100}{80} \times 0.172 = 0.215 \text{ kg}$$

$$\text{Mass loss} = 0.86 - 0.215 = 0.645 \text{ kg}$$

From equation 17,

$$Q = 1200 \times 1.87\,(150 - 100) = 1.12 \times 10^5 \text{ J/s}$$

$$\text{Drying rate} = \frac{1.12 \times 10^5}{2.257 \times 10^6} \text{ kg/s} = 0.05 \text{ kg/s}$$

and

$$\text{Residence time required} = \frac{0.645}{0.05} = 13 \text{ s}$$

As only three-quarters of the drum surface is used, 1 rev should take (100/75) × 13 = 17.3 s. Therefore, speed = 3.5 rev/min.

EQUIPMENT

Most commercial dryers are insulated to reduce heat losses, and they recirculate hot air to save energy. Many designs have energy-saving devices that recover heat from the exhaust air or automatically control the air humidity (9,10). Computer control of dryers is increasingly sophisticated (11) and also results in important savings in energy. The criteria for selection of drying equipment and potential applications are described in Table 3. The relative costs of different drying methods are reported as follows (12): forced-air drying, 198; fluidized-bed drying, 315; drum drying, 327; continuous vacuum drying, 1840; freeze-drying, 3528. Relative energy consumption (in kWh/kg of water removed) are as follows: roller drying, 1.25; pneumatic drying, 1.8; spray drying, 2.5; fluidized-bed drying, 3.5 (13).

Hot-Air Dryers

Bin Dryers (Deep-Bed Dryers). Bin dryers are cylindrical or rectangular containers fitted with a mesh base. Hot air passes up through a bed of food at relatively low speeds (eg, 0.5 $m^3/s/m^2$ of bin area). These dryers have a high capacity and low capital and running costs. They are mainly used for finishing (to 3–6% moisture content) after initial drying in other types of equipment. Bin dryers improve the operating capacity of initial dryers by taking the food when it is in the falling-rate period, when moisture removal is most time consuming. The deep bed of food permits variations in moisture content to be equalized and acts as a store to smooth out fluctuations in the product flow between drying stages and packaging. However, the dryers may be several meters high, and it is therefore important that foods are sufficiently strong to withstand compression at the base and to retain an open structure to permit the passage of hot air through the bed.

Cabinet Dryers (Tray Dryers). These dryers consist of an insulated cabinet fitted with shallow mesh or perforated trays, each of which contains a thin (2–6 cm deep) layer of food. Hot air is circulated through the cabinet at 0.5–5 m/s per square meter tray area. A system of ducts and baffles is used to direct air over and/or through each tray, to promote uniform air distribution. Additional heaters may be placed above or alongside the trays to increase the rate of drying. Tray dryers are used for small-scale production (1–20 tons/day) or for pilot-scale work. They have low capital and maintenance costs, but compared to dryers that have more sophisticated control they produce more variable product quality.

In developing countries the high capital investment for sophisticated dryers often cannot be justified, but there is a need for better-quality products than those produced by sun or solar drying. A small dryer consisting of a 60 kW gas or kerosine heater/blower unit and a cabinet fitted with 15 mesh trays has been developed to meet this need (14). In operation, air is passed across the food in each tray. Trays are loaded at the top of the dryer and unloaded at the base using metal fingers to move the tray stack. The dryer therefore operates semicontinuously and with counter-current airflow. Typical tray loadings are 5 kg, and the tray change cycle is 15–20 minutes.

Conveyor Dryers (Belt Dryers). Continuous conveyor dryers are up to 20 m long and 3 m wide. Food is dried on a mesh belt in beds 5–15 cm deep. The airflow is initially directed upward through the bed of food and then downward in later stages to prevent dried food from blowing out of the bed. Two- or three-stage dryers (Fig. 5) mix and repile the partly dried shrunken food into deeper beds (to 15–25 cm and 250–900 cm in three-stage dryers). This improves uniformity of drying and saves floor space. Foods are dried to 10–15% moisture content and then transferred to bin dryers for finishing. This equipment has good control over drying conditions and high production rates. It is used for large-scale drying of foods (eg, fruits and vegetables are dried in 2–3.5 h at up to 5.5 tons/h). It has independently controlled drying zones and is automatically loaded and unloaded, which reduces labor costs. As a result, it has largely replaced the tunnel dryer in many applications.

A second application of conveyor dryers is foam-mat drying in which liquid foods (eg, fruit juices) are formed into a stable foam by the addition of a stabilizer and aeration with nitrogen or air. The foam is spread on a perforated belt to a depth of 2–3 mm and dried rapidly in two stages by parallel and then counter-current airflows (Table 4). Foam drying is approximately three times faster than drying a similar thickness of liquid. The thin, porous mat

Table 3. Characteristics of Dryers

	Characteristics of the food											
			Initial moisture content		Size of pieces				Drying rate required		Final moisture content required	
Type of dryer	Solid	Liquid	Moderate to high	Low	Heat-sensitive	Small	Intermediate to large	Mechanically strong	Moderate to fast	Slow	Moderate	Low
Bin	*			*			*	*		*		*
Cabinet	*		*				*		*	*	*	
Conveyor	*		*				*		*	*	*	
Drum		*	—	—		—	—	—	*			*
Foam mat		*	—	—	*	—	—	—	*			*
Fluid bed	*		*			*		*	*		*	
Kiln	*		*				*			*	*	
Pneumatic	*			*		*		*	*			*
Rotary	*		*			*		*	*			*
Spray		*	—	—	*	—	—	—	*		*	*
Trough	*		*		*	*		*	*		*	
Tunnel	*		*				*		*	*	*	*
Vacuum band		*	—	—	*	—	—	—	*		*	*
Vacuum shelf	*	*	*		*	*			*	*	*	*
Radiant	*			*		*			*			*
Microwave or dielectric	*			*		*			*	*		*
Solar (sun)	*		*				*				*	

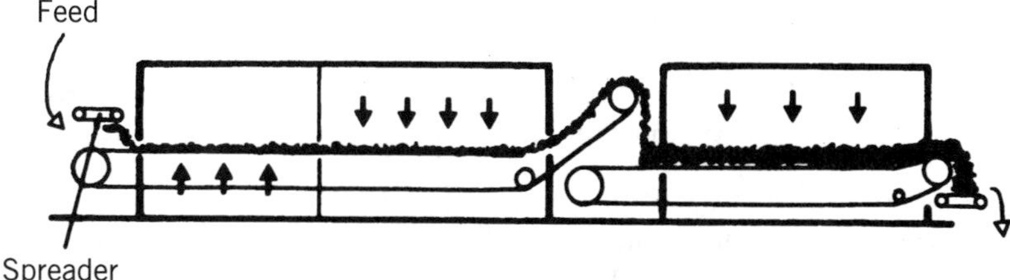

Figure 5. Two-stage conveyor dryer. *Source:* Courtesy of Proctor and Schwartz Inc.

Table 4. Advantages and Limitations of Parallel Flow, Counter-Current Flow, Center-Exhaust, and Cross-Flow Drying

Type of air flow	Advantages	Limitations
Parallel or cocurrent type: Food → Airflow →	Rapid initial drying. Little shrinkage of food. Low bulk density. Less heat damage to food. No risk of spoilage	Low moisture content difficult to achieve as cool, moist air passes over dry food
Counter-current type: Food → Airflow ←	More economical use of energy. Low final moisture content as hot air passes over dry food	Food shrinkage and possible heat damage. Risk of spoilage from warm moist air meeting wet food
Center-exhaust type: Food → Airflow → ↑ ←	Combined benefits of parallel and counter-current dryers but less than cross-flow dryers	More complex and expensive then single-direction air flow
Cross-flow type: Food → Airflow ↑ ↓	Flexible control of drying conditions by separately controlled heating zones, giving uniform drying and high drying rates	More complex and expensive to buy, operate, and maintain

of dried food is ground to a free-flowing powder that has good rehydration properties. The rapid drying and low product temperatures result in high-quality product. However, a large surface area is required for high production rates, so capital costs are therefore high.

Fluidized-Bed Dryers. Metal trays with mesh or perforated bases contain a bed of particulate foods up to 15 cm deep. Hot air is blown through the bed (Fig. 6), causing the food to become suspended and vigorously agitated (fluidized). The air thus acts as both the drying and the fluidiz-

Figure 6. Fluidized-bed drying. *Source:* Courtesy of Petrie and McNaught Ltd.

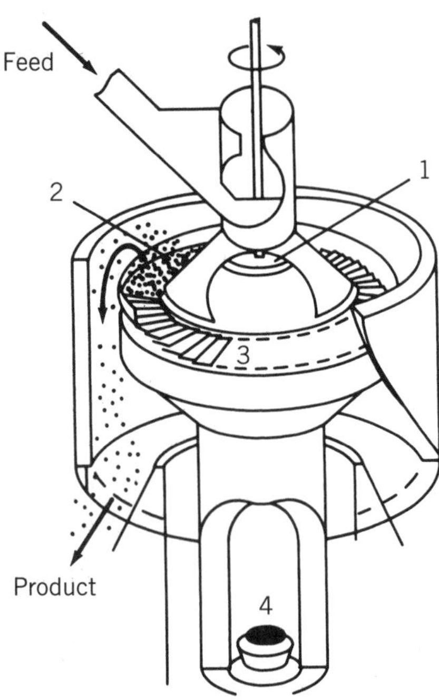

Figure 7. Torbed dryer: 1, rotating disc distributor to deliver raw material evenly into processing chamber; 2, rotating bed of particles; 3, fixed blades with hot gas passing through at high velocity; 4, burner assembly. *Source:* Courtesy of Torftech Ltd.

ing medium and the maximum surface area of food is made available for drying. A sample calculation of the air speed needed for fluidization is described in sample problem 2. Dryers may be batch or continuous in operation; the latter are often fitted with a vibrating base to help move the product. Continuous cascade systems, in which food is discharged under gravity from one tray to the next, employ up to six dryers for high production rates.

Fluidized-bed dryers are compact and have good control over drying conditions, relatively high thermal efficiencies, and high drying rates. In batch operation, products are mixed by fluidization, this leads to uniform drying. In continuous dryers, there is a greater range of moisture content in the dried product; bin dryers are therefore used for finishing. Fluidized-bed dryers are limited to small particulate foods that are capable of being fluidized without excessive mechanical damage (eg, peas, diced or sliced vegetables, grains, powders, or extruded foods). These considerations also apply to fluidized-bed freeze-dryers and freezers.

A development of the fluidized-bed dryer, named the Torbed dryer, has potential applications for drying particulate foods. A fluidized bed of particles is made to rotate around a torus-shaped chamber, by hot air blown directly from a burner (Fig. 7). The dryer has very high rates of heat and mass transfer and substantially reduced drying times. It is likely that some products (eg, vegetable pieces) would require a period of equilibration to allow moisture redistribution before final drying. The dryer operates semicontinuously under microprocessor control and is suitable for agglomeration and puff drying in addition to roasting, cooking, and coating applications.

Kiln Dryers. These dryers are two-story buildings in which a drying room with a slatted floor is located above a furnace. Hot air and the products of combustion from the furnace pass through a bed of food up to 20 cm deep. These dryers have been used traditionally for drying apple rings or slices in the United States, and hops or malt in Europe. There is limited control over drying conditions, and drying times are relatively long. High labor costs are incurred by the need to turn the product regularly, and by manual loading and unloading. However, the dryers have a large capacity and are easily constructed and maintained at low cost.

Pneumatic Dryers. In pneumatic dryers, powders or particulate foods are continuously dried in vertical or horizontal metal ducts. A cyclone separator is used to remove the dried product. The moist food (usually less than 40% moisture) is metered into the ducting and suspended in hot air. In vertical dryers the airflow is adjusted to classify the particles; lighter and smaller particles, which dry more rapidly, are carried to a cyclone more rapidly than are heavier and wetter particles that remain suspended to receive the additional drying required. For longer residence times the ducting is formed into a continuous loop (pneumatic ring dryers) and the product is recirculated until it is adequately dried. High-temperature short-time ring dryers are used to expand the starch-cell structure in potatoes or carrots to give a rigid, porous structure, which enhances subsequent conventional drying and rehydration rates. Calculation of air velocities needed for pneumatic drying is described in equation 16.

Pneumatic dryers have relatively low capital costs, high drying rates and thermal efficiencies, and close control over drying conditions. They are often used after spray drying to produce foods that have a lower moisture content than normal (eg, special milk or egg powders and potato granules). In some applications the simultaneous transportation and drying of the food may be a useful method of materials handling.

Rotary Dryers. A slightly inclined rotating metal cylinder is fitted internally with flights to cause the food to cascade through a stream of hot air as it moves through the dryer. Airflow may be parallel or counter-current (Table 4). The agitation of the food and the large area of food exposed to the air produce high drying rates and a uniformly dried product. The method is especially suitable for foods that tend to mat or stick together in belt or tray dryers. However, the damage caused by impact and abrasion in the dryer restrict this method to relatively few foods (eg, sugar crystals and cocoa beans).

Spray Dryers. A fine dispersion of preconcentrated foods is first atomized to form droplets (10–200 μm in diameter) and sprayed into a current of heated air at 150–300°C in a large drying chamber. The feed rate is controlled to produce an outlet air temperature of 90–100°C, which corresponds to a wet-bulb temperature (and product temperature) of 40–50°C. Complete and uniform atomization is necessary for successful drying, and one of the following types of atomizer is used.

1. *Centrifugal Atomizer*. Liquid is fed to the center of a rotating bowl (with a peripheral velocity of 90–200 m/s). Droplets, 50–60 μm in diameter, are flung from the edge of the bowl to form a uniform spray (Fig. 8a).
2. *Pressure Nozzle Atomizer*. Liquid is forced at a high pressure (700–2000 kPa) through a small aperture. Droplet sizes are 180–250 μm. Grooves on the inside of the nozzle cause the spray to form into a cone shape and therefore to use the full volume of the drying chamber.
3. *Two-Fluid Nozzle Atomizer*. Compressed air creates turbulence, which atomizes the liquid (Fig. 8b). The operating pressure is lower than the pressure nozzle, but a wider range of droplet sizes is produced.

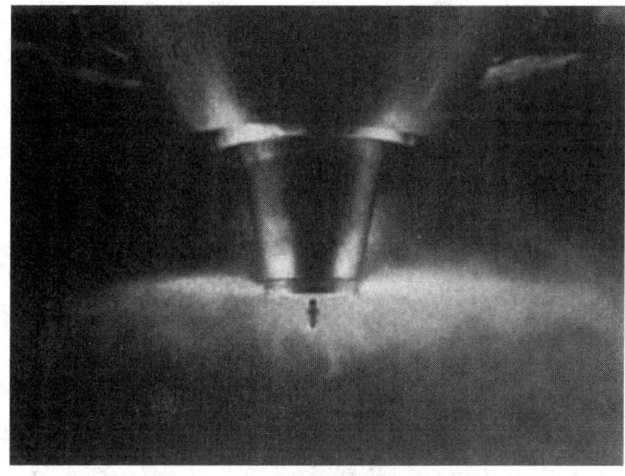

(a)

(b)

Figure 8. Atomizers: (**a**) centrifugal atomizer; (**b**) two-fluid nozzle atomizer (15). *Source:* Courtesy of Elsevier Applied Science.

Both types of nozzle atomizer are susceptible to blockage by particulate foods, and abrasive foods gradually widen the apertures and increase the average droplet size. Studies of droplet drying, including methods for calculating changes in size, density, and trajectory of the droplets are reported in Refs. 10, 16, and 17.

Rapid drying takes place (1–10 s) because of the very large surface area of the droplets. The temperature of the product remains at the wet-bulb temperature of the drying air, and there is minimum heat damage to the food. Airflow may be co- or counter-current (Table 4). The dry powder is collected at the base of the dryer and removed by a screw conveyor or a pneumatic system with a cyclone separator. There are a large number of designs of atomizer, drying chamber, air heating, and powder collecting systems (10,18). The variations in design arise from the different requirements of the very large variety of food materials that are spray dried—eg, milk, egg, coffee, cocoa, tea, potato, ground chicken, ice cream mix, butter, cream, yogurt and cheese powder, coffee whitener, fruit juices, meat and yeast extracts, encapsulated flavors (19), and wheat and corn starch products. Spray dryers may also be fitted with fluidized bed facilities to finish powders taken from the drying chamber.

Spray dryers vary in size from small pilot-scale models for low-volume high-value products (eg, enzymes and flavors) to large commercial models capable of producing 80,000 kg of dried milk per day (20) (Fig. 9). The main advantages are rapid drying, large-scale continuous production, low labor costs, and simple operation and maintenance. The major limitations are high capital costs and the requirement for a relatively high feed moisture content to ensure that the food can be pumped to the atomizer. This results in higher energy costs (to remove the moisture) and higher volatile losses. Conveyor band dryers and fluidized bed dryers are beginning to replace spray dryers, as they are more compact and energy efficient (21).

The bulk density of powders depends on the size of the dried particles and on whether they are hollow or solid.

Figure 9. Spray dryer. *Source:* Courtesy of De Melkindustrie Veghel.

This is determined by the nature of the food and the drying conditions (eg, the uniformity of droplet size, temperature, solids content, and degree of aeration of the feed liquid). Instant powders are produced by either agglomeration or non-agglomeration methods. Agglomeration is achieved by remoistening particles in low-pressure steam in an agglomerator, and then redrying. Fluidized-bed, jet, disc, cone, or belt agglomerators are described in Ref. 22. Alternatively, straight-through agglomeration is achieved directly during spray drying. A relatively moist powder is agglomerated and dried in an attached fluidized bed dryer. Nonagglomeration methods employ a binding agent, (eg, lecithin), to bind particles. This method was previously used for foods with a relatively high fat content, (eg, whole milk powder) but agglomeration procedures have now largely replaced this method (23).

Trough Dryers (Belt-Trough Dryers). Small, uniform pieces of food, (eg, peas or diced vegetables) are dried in a mesh conveyor belt that hangs freely between rollers to form the shape of a trough. Hot air is blown through the bed of food, and the movement of the conveyor mixes and turns it to bring new surfaces continually into contact with the drying air. The mixing action moves food away from the drying air, and this allows time for moisture to move from the interior of the pieces to the dry surface. The moisture is then rapidly evaporated when the food again contacts the hot air. The dryer operates in two stages, to 50–60% moisture and then to 15–20% moisture. Foods are finished in bin dryers. These dryers have high drying rates (eg, 55 min for diced vegetables, compared with 5 h in a tunnel dryer), high energy efficiencies, good control, and minimal heat damage to the product. However, they are not suitable for sticky foods.

Tunnel Dryers. Thin layers of food are dried on trays, which are stacked on trucks programmed to move semicontinuously through an insulated tunnel. Different designs use one of the types of air flow described in Table 4. Food is finished in bin dryers. Typically, a 20-m tunnel contains 12–15 trucks with a total capacity of 5000 kg of food. This ability to dry large quantities of food in a relatively short time (5–16 h) made tunnel drying widely used, especially in the United States. However, the method has now been largely superseded by conveyor drying and fluidized-bed drying as a result of their higher energy efficiency, reduced labor costs, and better product quality.

Sun and Solar Drying. Sun drying (without drying equipment) is the most widely practiced agricultural processing operation in the world; more than 250,000,000 tons of fruits and grains are dried by solar energy per annum. In some countries foods are simply laid out on roofs or other flat surfaces and turned regularly until dry. More sophisticated methods (solar drying) collect solar energy and heat air, which in turn is used for drying. Solar dryers are classified into three categories (4):

1. Direct natural-circulation dryers (a combined collector and drying chamber).
2. Direct dryers with a separate collector.
3. Indirect forced-convection dryers (separate collector and drying chamber).

Both solar and sun drying are simple, inexpensive technologies, in terms of both capital input and operating costs. Energy inputs and skilled labor are not required. Sun drying is therefore the preferred option in developing countries that have a suitable dry season after harvesting food crops such as paddy or maize. The major disadvantages of sun drying are poor control over drying conditions; lower drying rates than in artificial dryers; dependence on sunlight, which causes cessation of drying operations at night or during rain; and contamination of the dried product by dust, etc. Each of these factors contributes to a more variable and generally lower quality product than that produced by artificial dryers.

Solar dryers aim to improve product quality by providing greater control over drying conditions, protection from rain or dust, and higher drying rates. However, they have a relatively small capacity for drying the large bulk of crops at harvest time when compared to sun drying. In addition, the higher capital investment may not result in a higher income from improved quality of the dried crop. More valuable crops such as herbs and spices offer better potential for solar drying owing to the smaller quantities involved and the increased income from improved quality. However, the dependence of solar dryers on sunlight and the better control over drying conditions achieved in artificial (fuel-fired) dryers again limit the potential for solar drying.

Heated-Surface Dryers. Dryers in which heat is supplied to the food by conduction have two main advantages over hot-air drying: (*1*) It is not necessary to heat large volumes of air before drying commences, and the thermal efficiency is therefore high; (*2*) Drying may be carried out in the absence of oxygen to protect components of foods that are easily oxidized.

Typically, heat consumption is 2000–3000 kJ/kg of water evaporated compared with 4000–10,000 kJ/kg of water evaporated for hot-air dryers. However, foods have low thermal conductivities that become lower as the food dries. There should therefore be a thin layer of food to conduct heat rapidly, without causing heat damage. Foods may shrink during drying and lift off the hot surface, therefore introducing an additional barrier to heat transfer. Careful control is necessary over the rheological properties of the feed slurry to minimize shrinkage and to determine the thickness of the feed layer.

Drum Dryers (Roller Dryers). Slowly rotating hollow steel drums are heated internally by pressurized steam to 120–170°C. A thin layer of food is spread uniformly over the outer surface by dipping, spraying, spreading, or auxiliary feed rollers. Before the drum has completed 1 rev (within 20 s/3 min), the dried food is scraped off by a doctor blade that contacts the drum surface uniformly along its length. Dryers may have a single drum (Fig. 10a) or double drums (Fig. 10b) or twin drums. The single drum is widely used as it has greater flexibility, a larger proportion of the drum area available for drying, easier access for maintenance, and no risk of damage caused by metal objects falling between the drums.

Drum dryers have high drying rates and high energy efficiencies. They are suitable for slurries in which the particles are too large for spray drying. However, the high capital cost of the machined drums and heat damage to sensitive foods from high drum temperatures have caused a move to spray drying for many bulk dried foods. Drum drying is used to produce potato flakes, precooked cereals, molasses, some dried soups, and fruit purees, and whey or distillers' solubles for animal feed formulations.

Developments in drum design to improve the sensory and nutritional qualities of dried food include the use of auxiliary rolls to remove and reapply food during drying, the use of high-velocity air to increase the drying rate, and the use of chilled air to cool the product. Drums may be enclosed in a vacuum chamber to dry food at lower temperatures, but the high capital cost of this system restricts its use to high-value heat-sensitive foods.

Vacuum Band and Vacuum Shelf Dryers. A food slurry is spread or sprayed onto a steel belt (or band) that passes over two hollow drums within a vacuum chamber at 1–70 mmHg. The food is dried by the first steam-heated drum, and then by steam-heated coils or radiant heaters located over the band. The dried food is cooled by the second water-cooled drum and removed by a doctor blade. Vacuum shelf dryers consist of hollow shelves in a vacuum chamber. Food is placed in thin layers on flat metal trays that are carefully made to ensure good contact with the shelves. A particular vacuum of 1–70 mmHg is drawn in the chamber and steam or hot water is passed through the shelves to dry the food.

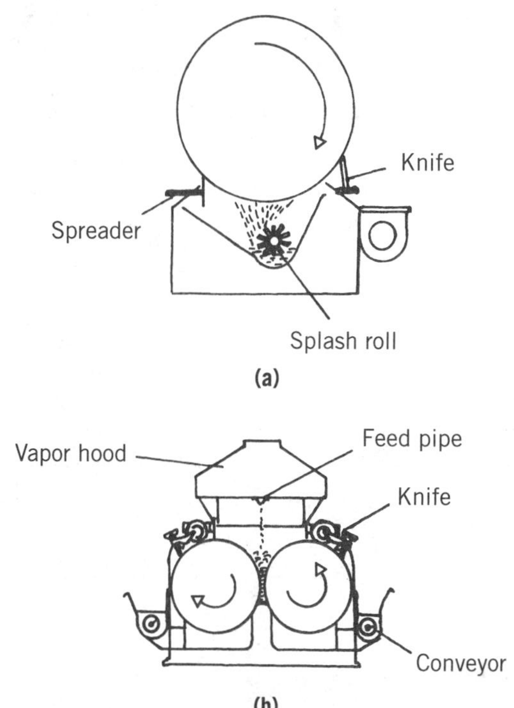

Figure 10. Drum dryers: (**a**) single drum; (**b**) double drum. *Source:* Courtesy of APV Mitchell Ltd.

Rapid drying and limited heat damage to the food make both methods suitable for drying heat-sensitive foods. However, care is necessary to prevent the dried food from burning onto trays in vacuum shelf dryers, and shrinkage reduces the contact between the food and heated surfaces of both types of equipment. Both have relatively high capital and operating costs and low production rates.

Vacuum-band shelf dryers are used to produce puff-dried foods. Explosion puff drying involves partially drying food to a moderate moisture content and then sealing it into a pressure chamber. The pressure and temperature in the chamber are increased and then instantly released. The rapid loss of pressure causes the food to expand and develop a fine porous structure. This permits faster final drying and rapid rehydration. Sensory and nutritional qualities are well retained. The technique was first applied commercially to breakfast cereals and now includes a range of fruit and vegetable products.

EFFECTS ON FOODS

The effect of A_w on microbiological and selected biochemical reactions is shown in Fig. 11 and Table 2. Almost all microbial activity is inhibited below $A_w = 0.6$; most fungi

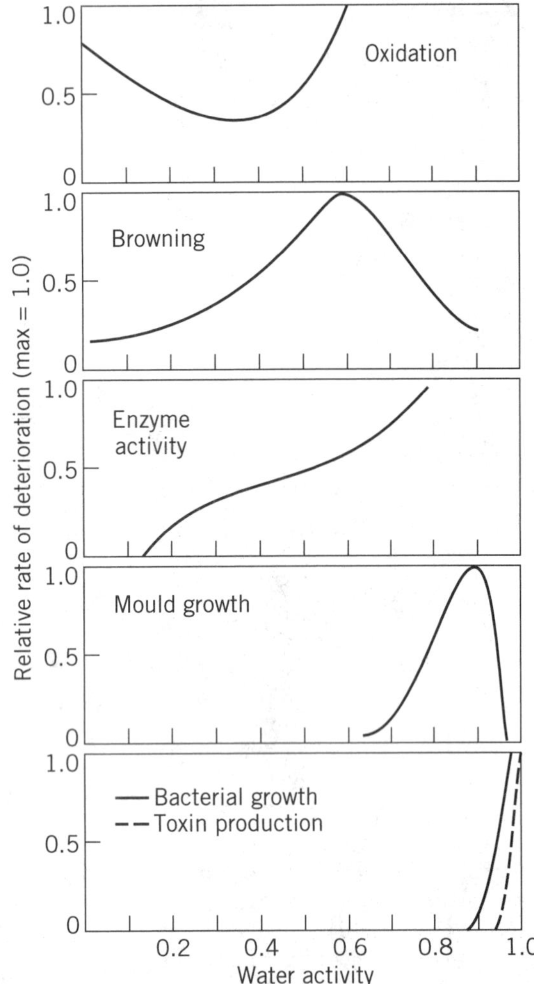

Figure 11. Effect of water activity on microbial, enzymic and chemical changes to foods (7). *Source:* Courtesy of Marcel Dekker.

are inhibited below $A_w = 0.7$; most yeasts are inhibited below $A_w = 0.8$; and most bacteria below $A_w = 0.9$. The interaction of A_w with temperature, pH, oxygen, and carbon dioxide, or chemical preservative has an important effect on the inhibition of microbial growth. When any one of the other environmental conditions is suboptimal for a given microorganism, the effect of reduced A_w is enhanced (Fig. 11). This permits the combination of several mild control mechanisms that result in the preservation of food without substantial loss of nutritional properties or sensory properties (Table 5).

Enzymatic activity virtually ceases at A_w values below the BET monolayer value. This is due to the low substrate mobility and its inability to diffuse to the reactive site on the enzyme. Chemical changes are more complex. The two most important things that occur in foods that have a low A_w are Maillard browning and oxidation of lipids. The A_w that causes the maximum rate of browning varies with different foods. However, in general, a low A_w restricts the mobility of reactants and browning is reduced. At a higher A_w, browning reaches maximum. Water is a product of the condensation reaction in browning, and at higher moisture levels, browning is inhibited by end-product inhibition. At high moisture contents, water dilutes the reactants and the rate of browning falls (Fig. 11).

Oxidation of lipids occurs at low A_w values owing to the action of free radicals. Above the BET monolayer value, antioxidants and chelating agents (which sequester trace metal catalysts) become soluble and reduce the rate of oxidation. At higher A_w values the catalytic activity of metals is reduced by hydration and the formation of insoluble hydroxides but, at high A_w values, metal catalysts become soluble and the structure of the food swells to expose more reactive sites (Fig. 11).

Texture

Changes to the texture of solid foods are an important cause of quality deterioration. The nature and extent of pretreatments (eg, the addition of calcium chloride to blancher water), the type and extent of size reduction, and peeling each affect the texture of rehydrated fruits and vegetables. In foods that are adequately blanched, loss of texture is caused by gelatinization of starch, crystallization of cellulose, and localized variations in the moisture content during dehydration, which set up internal stresses. These rupture, compress, and permanently distort the relatively rigid cells, to give the food a shrunken, shrivelled appearance. On rehydration the product absorbs water more slowly and does not regain the firm texture associated with the fresh material. There are substantial variations in the degree of shrinkage with different foods.

Drying is not commonly applied to meats in many countries owing to the severe changes in texture compared with other methods of preservation. These are caused by aggregation and denaturation of proteins and a loss of water-holding capacity, which leads to toughening of muscle tissue.

The rate and temperature of drying have a substantial effect on the texture of foods. In general, rapid drying and high temperatures cause greater changes than do moderate rates of drying and lower temperatures. As water is removed during dehydration, solutes move from the interior of the food to the surface. The mechanism and rate of movement are specific for each solute and depend on the type of food and the drying conditions used. Evaporation of water causes concentration of solutes at the surface. High air temperatures (particularly with fruits, fish, and meats) cause complex chemical and physical changes to the surface and the formation of a hard, impermeable skin. This is termed case hardening. It reduces the rate of drying and produces a food with a dry surface and a moist interior. It is minimized by controlling the drying conditions to prevent excessively high moisture gradients between the interior and the surface of the food.

In powders, the textural characteristics are related to bulk density and the ease with which they are rehydrated. These properties are determined by the composition of the food, the method of drying, and the particle size of the product. Low-fat foods (eg, fruit juices, potatoes, and coffee) are more easily formed into free-flowing powders than are whole milk or meat extracts. Powders are instantized by

Table 5. Interaction of A_W, pH, and Temperature in Selected Foods

Food	pH	A_W	Shelf life	Notes
Fresh meat	>4.5	>0.95	Days	Preserve by chilling
Cooked meat	>4.5	0.95	Weeks	Ambient storage when packaged
Dry sausage	>4.5	<0.90	Months	Preserved by salt and low A_W
Fresh vegetables	>4.5	>0.95	Weeks	Stable while respiring
Pickles	<4.5	0.90	Months	Low pH maintained by packaging
Bread	>4.5	>0.95	Days	Preserved by heat and low A_W in crust
Fruitcake	>4.5	<0.90	Weeks	Preserved by heat and low A_W
Milk	>4.5	>0.95	Days	Preserved by chilling
Yogurt	<4.5	<0.95	Weeks	Preserved by low pH and chilling
Dried milk	>4.5	<0.90	Months	Preserved by low A_W

treating individual particles so that they form free-flowing agglomerates of aggregates, in which there are relatively few points on contact (Fig. 12). The surface of each particle is easily wetted when the powder is rehydrated, and particles sink below the surface to disperse rapidly through the liquid. These characteristics are respectively termed wettability, sinkability, dispersibility, and solubility. For a powder to be considered instant, it should complete these four stages within a few seconds.

The convenience of instantized powders outweighs the additional expense of production, packaging, and transport for retail products. However, many powdered foods are used as ingredients in other processes, and these are required to possess a high bulk of density and a wider range of particle sizes. Small particles fill the spaces between larger ones and thus exclude air to promote a longer storage life. The characteristics of some powdered foods are described in Table 6.

Table 6. Bulk Density and Moisture Content of Selected Powdered Foods

Food	Bulk density (kg m^3)	Moisture content (%)
Cocoa	480	3–5
Coffee (ground)	330	7
Coffee (instant)	330	2.5
Coffee creamer	470	3
Cornstarch	560	12
Egg, whole	340	2–4
Milk, powdered, skimmed	640	2–4
Milk, instant, skimmed	550	2–4
Salt, granulated	960	0.2
Sugar, granulated	800	0.5
Wheat flour	450	12

Source: Refs. 24 and 25.

Flavor and Aroma

Heat not only vaporizes water during drying but also causes loss of volatile components from the food. The extent of volatile loss depends on the temperature and solids concentration of the food and on the vapor pressure of the volatiles and their solubility in water vapor. Volatiles that have a high relative volatility and diffusivity are lost at an early stage in drying. Fewer volatile components are lost at later stages. Control of drying conditions during each stage of drying minimizes losses. Foods that have a high economic value due to their characteristic flavors (eg, herbs and spices) are dried at lower temperatures.

A second important cause of aroma loss is oxidation of pigments, vitamins, and lipids during storage. The open, porous structure of dried food allows access of oxygen. The rate of deterioration is determined by the storage temperature and the water activity of the food.

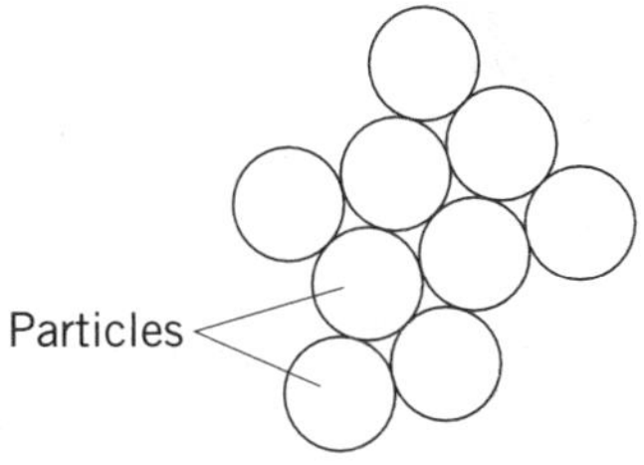

Figure 12. Agglomerated powder.

In dried milk the oxidation of lipids produces rancid flavors owing to the formation of secondary products, including δ-lactones. Most fruits and vegetables contain only small quantities of lipid, but oxidation of unsaturated fatty acids to produce hydroperoxides, which react further by polymerization, dehydration, or oxidation to produce aldehydes, ketones, and acids, causes rancid and objectionable odors. Some foods, (eg, carrots), may develop an odor of violets produced by the oxidation of carotenes to β-ionone. These changes are reduced by vacuum- or gas-packing, low storage temperatures, exclusion of ultraviolet or visible light, maintenance of low moisture contents, addition of synthetic antioxidants, or preservation of natural antioxidants.

The technical enzyme glucose oxidase is also used to protect dried foods from oxidation. A package that is permeable to oxygen but not to moisture and that contains glucose and the enzyme is placed on the dried food inside a container. Oxygen is removed from the headspace during storage. Milk powders are stored under an atmosphere of nitrogen with 10% carbon dioxide. The carbon dioxide is absorbed into the milk and creates a small partial vacuum in the headspace. Air diffuses out of the dried particles and is removed by regassing after 24 h. Flavor changes, due to oxidative or hydrolytic enzymes, are prevented in fruits by the use of sulfur dioxide, ascorbic acid, or citric acid, by pasteurization of milk or fruit juices, and by blanching of vegetables.

Other methods used to retain flavors in dried foods include

1. Recovery of volatiles and their return to the product during drying.
2. Mixing recovered volatiles with flavor-fixing compounds, which are then granulated and added back to the dried product, eg, dried meat powders.
3. Addition of enzymes, or activation of naturally occurring enzymes, to produce flavors from flavor precursors in the food (eg, onion and garlic are dried under conditions that protect the enzymes that release characteristic flavors). Maltose or maltodextrin are used as a carrier material when drying flavor compounds.

Color

Drying changes the surface characteristics of food and hence alters the reflectivity and color. Chemical changes to carotenoid and chlorophyll pigments are caused by heat and oxidation during drying. In general, longer drying times and higher drying temperatures produce greater pigment losses. Oxidation and residual enzyme activity cause browning during storage. This is prevented by improved blanching methods and treatment of fruits with ascorbic acid or sulfur dioxide. For moderately sulfured fruits and vegetables the rate of darkening during storage is inversely proportional to the residual sulfur dioxide content. However, sulfur dioxide bleaches anthocyanins, and residual sulfur dioxide is an important cause of color deterioration in stored dried fruits and vegetables.

The rate of Maillard browning in stored milk and fruit products depends on the water activity of the food and the temperature of storage. The rate of darkening increases markedly at high drying temperatures, when the moisture content of the product exceeds 4–5%, and at storage temperatures above 38°C (26).

Nutritive Value

Large differences in reported data on the nutritive value of dried foods are due to wide variations in the preparation procedures, the drying temperature and time, and the storage conditions. In fruits and vegetables, losses during preparation usually exceed those caused by the drying operation. For example, losses of vitamin C during preparation of apple flakes are reported to be 8% during slicing, 62% from blanching, 10% from pureeing, and 5% from drum drying (27).

Vitamins have different solubilities in water, and, as drying proceeds, some (eg, riboflavin) become supersaturated and precipitate from solution. Losses are therefore small (Table 7). Others, (eg, ascorbic acid) are soluble until the moisture content of the food falls to very low levels and react with solutes at higher rates as drying proceeds. Vitamin C is also sensitive to heat and oxidation. Short drying times, low temperatures, and low moisture and oxygen levels during storage are necessary to avoid large losses. Thiamin is also heat sensitive, but other water-soluble vitamins are more stable to heat and oxidation, and losses during drying rarely exceed 5–10% (excluding blanching losses).

Oil-soluble nutrients (eg, essential fatty acids and vitamins A, D, E, and K) are mostly contained within the dry matter of the food, and they are not therefore concentrated during drying. However, water is a solvent for heavy-metal catalysts that promote oxidation of unsaturated nutrients. As water is removed, the catalysts become more reactive, and the rate of oxidation accelerates (Fig. 12). Fat-soluble vitamins are lost by interaction with the peroxides produced by fat oxidation. Losses during storage are reduced by low oxygen concentrations and storage temperatures and by exclusion of light.

The biological value and digestibility of proteins in most foods does not change substantially. However, milk proteins are partially denatured during drum drying, and this results in a reduction in solubility of the milk powder, aggregation, and loss of clotting ability. A reduction in biological value of 8–30% is reported, depending on the temperature and residence time (30). Spray drying does not affect the biological value of milk proteins. At high storage temperatures and at moisture contents above approximately 5%, the biological value of milk protein is decreased by Maillard reactions between lysine and lactose. Lysine is heat sensitive, and losses in whole milk range from 3–10% in spray drying and 5–40% in drum drying (31).

The importance of nutrient losses during processing depends on the nutritional value of a particular food in the diet. Some foods, eg, bread and milk, are an important source of nutrients for large numbers of people. Vitamin losses are therefore more significant in these foods than in those that either are eaten in small quantities or have low concentrations of nutrients.

In industrialized countries, the majority of the population achieve an adequate supply of nutrients from the mixture of foods that is eaten. Losses due to processing of one component of the diet are therefore insignificant to the long-term health of an individual. In one example, complete meals that initially contained 16.5 μg of vitamin A lost 50% on canning and 100% after storage for 18 months. Although the losses appear to be significant, the original meal contained only 2% of the recommended daily allowance (RDA), and the extent of loss is therefore of minor importance. The same meal contained 9 mg of thiamin and lost 75% after 18 months' storage. The thiamin content is 10 times the RDA, so adequate quantities therefore remained. Possible exceptions are the special dietary needs of preterm infants, pregnant women, and the elderly. In these groups there may be either a special need for certain nutrients or a more restricted diet than normal. These special cases are discussed in detail in Refs. 32, 33 and 34.

Reported vitamin losses during processing give an indication of the severity of each unit operation. However, such data should be treated with caution. Variation in nutrient losses between cultivars or varieties can exceed differences caused by alternative methods of processing. Growth conditions, or handling and preparation procedures before processing, also cause substantial variation in nutrient loss. Data on nutritional changes cannot be di-

Table 7. Vitamin Losses in Selected Dried Foods

Food	Loss (%)						
	Vitamin A	Thiamin	Vitamin B_2	Niacin	Vitamin C	Folic acid	Biotin
Fruits[a]	6	55	0	10	56		
Fig (sun-dried)	—	48	42	37	—	—	—
Whole milk (spray-dried)	—	—	—	—	15	10	10
Whole milk (drum-dried)	—	—	—	—	30	10	10
Pork		50–70					
Vegetables[b]	5	<10	<10				

Source: Refs. 28 and 29.
[a]Fruits: mean loss from fresh apple, apricot, peach, and prune.
[b]Vegetables: mean loss from peas, corn, cabbage, and beans (drying stage only).

rectly applied to individual commercial operations, because of differences in ingredients, processing conditions, and equipment used by different manufacturers.

REHYDRATION

Rehydration is not the reverse of drying. Texture changes, solute migration, and volatile losses are each irreversible. Heat reduces the degree of hydration of starch and the elasticity of cell walls and coagulates proteins to reduce their water-holding capacity. The rate and extent of rehydration may be used as an indicator of food quality; those foods that are dried under optimum conditions suffer less damage and rehydrate more rapidly and completely than poorly dried foods.

NOMENCLATURE

A	Area
A_w	Water activity
C	Constant, drag coefficient
d	Diameter
G	Mass flow rate of air
g	Acceleration due to gravity (9.81 m/s^2)
H	Humidity
h	Surface heat transfer coefficient
K	Mass transfer coefficient
L	Length
M	Moisture content (dry-weight basis)
m	Mass, mass flow rate, moisture content (wet-weight basis)
P	Pressure, vapor pressure
Q	Rate of heat transfer
r	Radius
Re	Reynolds Number
t	Time
U	Overall heat transfer coefficient
v_f	Velocity, air velocity needed for fluidization
x	Thickness, depth
ϵ	Voidage of fluidized bed
θ	Temperature
λ	Latent heat
μ	Viscosity
τ	Constant = 3.142
ρ	Density

ACKNOWLEDGMENTS

Grateful acknowledgment is made for information supplied by the following: APV Mitchell Dryers Ltd, Carlisle, Cumbria, CA5 5DU, UK; Petrie and McNaught Ltd, Rochdale, UK; Zschokke Wartmann Ltd, Stahlrain, CH-5200 Brugg, Switzerland; Proctor and Schwartz Inc., Glasgow, Scotland; Torftech Ltd, Mortimer, Reading, Berkshire, UK; Unilever, London EC4P 4BQ, UK; Chartered Institute of Building Services Engineers, 222 Balham High Road London, SW12 985, UK; De Melkindustrie Veghel.

BIBLIOGRAPHY

Reproduced with permission from P. Fellows, *Food Processing Technology*, Ellis Horwood Limited, Chichester, UK, 1988.

1. M. J. Lewis, *Physical Properties of Foods and Food Processing Systems*, Ellis Horwood, Chichester, West Sussex, UK, 1987.
2. C. Van den Berg, "Water Activity," in D. MacCarthy, ed., *Concentration and Drying of Foods*, Elsevier Applied Science, Barking, Essex, UK, 1986, pp. 11–36.
3. J. A. Troller and J. H. B. Christian, *Water Activity and Food*, Academic Press, London, 1978.
4. B. Brenndorfer, L. Kennedy, C. O. Oswin-Bateman, and D. S. Trim. *Solar Dryers*, Commonwealth Science Council, Commonwealth Secretariat, Pall Mall, London, 1985.
5. D. Kunii and O. Leverspiel, *Fluidisation Engineering*, John Wiley & Sons, New York, 1969.
6. M. Loncin and R. L. Merson, *Food Engineering—Principles and Selected Applications*, Academic Press, New York, 1979.
7. M. Karel, "Dehydration of Foods," in O. R. Fennema, ed., *Principles of Food Science. Part 2. Physical Principles of Food Preservation*, Marcel Dekker, New York, 1975, pp. 309–357.
8. C. W. Hall, *Dictionary of Drying*, Marcell Dekker, New York, 1979.
9. P. E. Zagorzycki, "Automatic Humidity Control of Dryers," *Chemical Engineering Progress*, 66–70 (April 1983).
10. K. Masters, *Spray Drying*. Leonard Hill, London, 1972, pp. 15, 144, 160, 230, 545–586.
11. K. Grikitis, "Dryer Spearheads Dairy Initiative," *Food Process* (1986).

12. S. F. Sapakie and T. A. Renshaw, "Economics of Drying and Concentration of Foods," in B. M. McKenna, ed., *Engineering and Food*, Vol. 2, Elsevier Applied Science, London, 1984, pp. 927–938.
13. C. Tragardh, "Energy and Energy Analysis in Some Food Processing Industries," *Lebensmittel Wissenschaft Technologie* **14**, 213–217 (1981).
14. A. Axtell, A. B. Bush, and M. Molina, *Try Drying It. A Case Study of Small Industry Dryers*, IT Publications, London, 1991.
15. K. Masters, "Recent Developments in Spray Drying," in S. Thorne, ed., *Developments in Food Preservation*, Vol. 2, Applied Science, London, 1983, pp. 95–121.
16. S. E. Charm, *Fundamentals of Food Engineering*, 3rd ed., AVI Publishing Co., Westport, Conn., 1978, pp. 298–408.
17. P. J. A. M. Kerkhof and W. J. A. H. Schoeber, "Theoretical Modelling of the Drying Behaviour of Droplets in Spray Dryers," in A. Spicer, ed., *Advances in Preconcentration and Dehydration of Foods*, Applied Science, London, 1974, pp. 349–397.
18. O. G. Kjaergaard, "Effects of Latest Developments on Design and Practice of Spray Drying. In Ref. 17, pp. 321–348.
19. H. B. Heath, "The Flavour Trap," *Food Flavour Ingredients* **7**, 21, 23, 25 (1985).
20. M. Byrne, "The £8M Dryer," *Food Manufacture*, 67, 69 (September 1986).
21. J. C. Ashworth, "Developments in Dehydration," *Food Manufacture*, 25–27, 29 (December 1981).
22. H. Schubert, (1980) "Processing and Properties of Instant Powdered Foods," in P. Linko, Y. Malkki, J. Olkku, and J. Larinkari, eds., *Food Process Engineering*, Vol. 1, *Food Processing Systems*, Applied Science, London, 1980, pp. 675–684.
23. U.S. Pat. 4,490,403 (1983) J. Pisecky, J. Krag, and Ib. H. Sorensen.
24. B. K. Watt and A. L. Merrill, *Composition of Foods*, Agriculture Handbook 8, U.S. Department of Agriculture, Washington D.C., 1975.
25. M. Peleg, "Physical Characteristics of Food Powders," in M. Peleg and E. B. Bagley, eds., *Physical Properties of Foods*, AVI Publishing Co., Westport, Conn., 1983, pp. 293–323.
26. C. H. Lea, "Chemical Changes in the Preparation and Storage of Dehydrated Foods," in *Proceedings of Fundamental Aspects of Thermal Dehydration of Foodstuffs*, Aberdeen, March 25–27, 1958, Society of Chemical Industry, London pp. 178–194.
27. F. Escher and H. Neukom, "Studies on Drum-drying Apple Flakes," *Trav. Chim. Aliment. Hyg.* **61**, 339–348 (1970) (in German).
28. B. A. Rolls, "Effects of Processing on Nutritive Value of Food: Milk and Milk Products, in M. Rechcigl, ed., *Handbook of the Nutritive Value of Processed Foods*, Vol. 1, CRC Press, Boca Raton, Fla., 1982, pp. 383–389.
29. D. H. Calloway, "Dehydrated Foods," *Nutrition Reviews* **20**, 257–260 (1962).
30. W. Fairbanks and H. H. Mitchell, "The Nutritive Value of Skim-milk Powders with Special Reference to the Sensitivity of Milk Proteins to Heat," *Journal of Agricultural Research* **51**, 1107–1121 (1935).
31. B. A. Rolls and J. W. G. Porter, "Some Effects of Processing and Storage on the Nutritive Value of Milk and Milk Products," *Proceedings of the Nutrition Society* **32**, 9–15 (1973).
32. A. E. Bender, "The Nutritional Aspects of Food Processing," in A. Turner, ed., *Food Technology International Europe*, Sterling, London, 1987, pp. 273–275.
33. R. R. Watson, *Nutrition for the Aged*. CRC Press, Boca Raton, Fla., 1986.
34. D. Francis, *Nutrition for Children*, Blackwell, Oxford, UK, 1986.

GENERAL REFERENCES

References 7, 10, and 12 are good general references.

C. W. Hall, *Dictionary of Drying*, Marcel Dekker, New York, 1979.

O. G. Spicer Kjaergaard, "Effects of Latest Developments on Design and Practice of Spray Drying," in A. Spicer, ed., *Advances in Preconcentration and Dehydration of Foods*. Applied Science, London, 1974, pp. 321–348.

P. Fellows
Intermediate Technology Development Group
Rugby, United Kingdom

DETERGENTS

A detergent is any material that cleanses or provides surface active properties. Detergents assist in the removal of unwanted soils from surfaces. The use of detergents in food-processing facilities is an important component of their sanitation program. Food-processing plants perform regular sanitation within their operations for two primary reasons: to provide a clean and aesthetically appealing environment and to eliminate microbial organisms that cause food spoilage and health problems for the consumer. Detergents selected for use in food-processing areas should be products that are free rinsing, leaving no residual toxins that might contaminate the food. Detergents used in food-processing areas should also leave no residual odor, after rinsing, that might serve to conceal malodors caused by microbial activity in those areas.

The Food and Drug Administration (FDA) publishes a listing of materials that are acceptable for use in the food environment (1). These include ingredients for food and substances present on food contact surfaces, along with parameters for their use. The U.S. Department of Agriculture (USDA) follows the FDA guidelines when approving detergents and other products acceptable for use in meat- and poultry-processing plants.

TYPES OF DETERGENT

Detergents may be either solvent-based or water-soluble products. Water-soluble detergents may be acidic, neutral, or alkaline. Solvent-based detergents are not widely used in food-processing areas. Many solvents have toxic hazards associated with them, possess strong persistent odors and are not readily biodegradable. The few solvent detergents that are acceptable in food-processing plants tend to be either readily biodegradable or are volatile. Most volatile solvents are also flammable. Because of all the liabilities that are associated with solvent-based detergents, their use is limited to specialty applications where other detergents do not provide adequate cleaning. Acid detergents are generally used to remove mineral deposits such as hard-water scale. These products are also effective in removing coagulated protein films. Specialized applications use acid detergents for microbial control. A product formulation that combines an acid with an anionic surfactant makes an effective sanitizer; as do iodophors, which are acid detergents that release iodine in solution. Neutral detergents (pH between 5.0 and 9.0) are generally synergistic surfactant blends. Some neutral products also contain solvents to complement the cleaning action of the surfactants. Because of their pH neutrality these detergents are widely used for hard scrubbing. Neutral detergents tend to be most effective on oils, greases, and fats because of their excellent ability to emulsify and disperse these types of soil. A specialized application in this category is the use of quaternary ammonium chloride formulations as sanitizers. Alkaline detergents are effective on organic soils including fats, proteins, and carbohydrates. These detergents are used extensively to clean food-processing facilities. They work by solubilizing, saponifying, peptizing, or hydrolyzing the soils to be removed.

DETERGENT PROPERTIES

Detergents assist the cleaning process by performing one or more of the following: wetting or penetration, dispersing or deflocculating, suspension, emulsification, peptizing, dissolving, chelating or sequestering, rinsing, and sanitizing (2,3). Wetting refers to the detergent's ability to penetrate through and come in contact with all soils and equipment surfaces, thus loosening the soil and facilitating its removal. A dispersant promotes the breakdown of large soil particles into smaller particles and inhibits the formation of larger soil particles by holding them uniformly distributed in the detergent bath. When soil particles are uniformly distributed throughout the detergent bath, without any precipitation to the bottom or flotation to the surface, they are in suspension. An emulsifier breaks down fat and oil into tiny particles and disburses them uniformly throughout the water phase where they are held in suspension. (This is an oil-in-water emulsion.) Peptizing is the partial solubilization and colloidal dispersion of protein by an alkaline detergent. When a soil is converted into a form that is completely soluble in water it will dissolve. Chelation is the process of chemically removing minerals such as calcium, magnesium, or iron from water. This is done by forming a soluble complex with the mineral thus preventing their precipitation from the detergent bath. This enhances the rinsability of the detergent solution and prevents interference of those minerals with the detergent's cleaning ability. Rinsing is the ability to flush the detergent solution and soil from the surfaces being cleaned without leaving any residue. Sanitizing reduces the total microbial population present on the cleaned surfaces. Sanitizers work most efficiently on cleaned surfaces free of any soil. The ideal detergent would possess the following characteristics: rapid and complete solubility in water and complete removal of all soil from the surfaces being cleaned. It would rinse free of all residue and provide germicidal activity. Finally it would be inexpensive to use and safe to handle. To date no universal product possessing all these characteristics exists; hence the plethora of products available in the marketplace.

DETERGENT COMPONENTS

Most detergents are blends of surfactants and builders, although they may be strictly surfactant blends or blends of builders. Surfactants are classed as anionic, nonionic, or cationic. Anionic surfactants are negatively charged molecules and are the most widely used in cleaning. Nonionic surfactants have no charge and are excellent oil and grease solubilizers and emulsifiers. Cationic surfactants are positively charged molecules and are not used to any significant degree as cleaners. Their use is primarily in specialty applications such as fabric softeners and sanitizers.

In aqueous solution, most solid soils are negatively charged minerals (4). If cationic surfactants are present, an electrostatic attraction occurs that is not favorable to soil removal. The negative charge of the anionic surfactants, on the other hand, creates an electrostatic repulsion that facilitates removal of these soils and prevents their redeposition. Uncharged oily soils are most readily removed by solubilization in micelles or by emulsification with nonionic surfactants (4). The solubilization process is most effective when the nonionic surfactant concentration is above the critical micelle concentration (CMC). Also the effectiveness can be further optimized by maintaining a temperature just below the cloud point of the surfactant. Builders are all the other ingredients that are formulated into a detergent to enhance the performance of the surfactants. Detergents that contain no surfactants rely on saponification to produce surfactants *in situ*. Some builders found in detergents are alkalies, acids, phosphates, silicates, chelants, chlorine, defoamers, antiredeposition agents, abrasives, and corrosion inhibitors (2,3,5). By a judicious selection of builders, detergents can be formulated to enhance one or more of the following properties: emulsification, saponification, hydrolysis, removal of mineral deposits, wetting and penetration, dispersing, antiredeposition, peptizing, water softening and antiscaling, defoaming, and sanitizing.

CLEANING

Regular cleaning is essential to maintaining sanitary conditions in food-processing facilities. A good cleaning program should schedule regular cleaning of all interior building surfaces (floors, walls, and ceilings), all processing equipment, removal of garbage, trash, and other wastes, leaving clean rinsed surfaces that are free of any chemical residues that might contaminate the food (5). The basic steps of the cleaning process are the physical removal of the gross soil load, detergent cleaning, rinsing, and sanitizing. The physical removal of gross soil is accomplished by such methods as squeegeeing, shoveling, scraping, rinsing, or any other mechanical means appropriate to the soil and the surface being cleaned. By reducing the soil load the detergents can work more efficiently. The detergent cleaning process loosens the remaining soil. Rinsing flushes the soil from the surface and prevents its redeposition. Sanitizing inhibits microbiological growth on the cleaned surfaces. Three classes of soil have been characterized: soils bound to the surface by oils or greases, soils absorbed onto the surface, and scale deposits. These soils are separated from surfaces by three possible mechanisms: mechanical action, chemical alteration of the nature of the soil, or reduction of the energy binding the soil to the surface by reduction of the surface tension through wetting and penetration (3). The procedure used for cleaning and the type of detergent selected depend on a number of factors: the type and condition of the soil, the type of surface being cleaned, the condition and temperature of the water supply, the cleaning equipment available, the time alloted for cleanup, and the cost of the sanitation program. The ideal water used in sanitation would be free of microorganisms, clear, colorless, noncorrosive, and free of minerals. Other factors to effective cleaning are time, temperature, concentration of detergent, and mechanical action. In general, cleaning is enhanced with longer time, higher temperature, higher concentrations of detergent, and increased mechanical action. However, this generalization needs to be tempered by practical considerations and limitations of the clean-up project. For example, protein starts to coagulate above 55°C (3). In general the following criteria are used to select an appropriate detergent for cleaning:

Heavy soils, cooked on soils	High alkaline cleaner
Fats and proteins	Chlorinated cleaner
Hand scrubbing, soft metals	Light duty or neutral cleaner
Remove mineral deposits, brighten equipment	Acid cleaner

There are four primary procedures used to apply the detergent cleaners with an infinite variety of variations on these methods. Cleaning in place (CIP) is used primarily on enclosed equipment and uses pumps to circulate the detergent solution through the equipment. This process normally entails the use of an alkaline cleaner followed by an acid cleaner. These cleaners typically are low-foaming detergents, because foam hinders the cleaning process when the solutions are pumped through pipes and spray balls to clean large tanks and other parts of the system. Foaming and high-pressure spraying techniques are the most common methods for cleaning the exposed surfaces of equipment, conveyors, walls, and floors. Foaming is a modification of the high-pressure application where compressed air is injected into the detergent stream to produce foam. Foam is popular for use on vertical surfaces because it holds the detergent onto those surfaces for several minutes before draining off. Soak tank cleaning is used on trays, molds and other portable equipment that can be readily removed to one location for cleaning by soaking in a detergent bath. Boil out is a special soak tank procedure where the equipment to be cleaned is filled with a detergent solution and heat is applied until cleaning is effected. Hand scrubbing is used where none of the other procedures are appropriate, or to supplement inadequate cleaning by those methods. This technique combines the chemical action of the detergent with the mechanical action of wiping or scouring the soil from the surface.

DISPOSAL

Disposal of detergents, as with all chemicals, comes under the purview of federal, state, and local laws and regulations. Food processors should maintain close contact with their waste-disposal facilities to keep abreast of current requirements for disposal in their respective locales. Generally, spent detergents can be flushed down the drain. However, care does need to be taken to ensure that this waste solution does comply with the requirements of the local wastewater treatment plant. Usually their primary concerns are the pH and the phosphate levels in the discharged waste. Avoid using detergents containing phosphates where stringent phosphate disposal levels exist. The pH can be readily neutralized, if necessary, before sending the solution out to the wastewater treatment plant. In the event of a detergent spill, several options exist. One would be to neutralize and dilute the detergent with water and dispose as wastewater. Another would be to contain the spill, absorb onto an appropriate substrate, if a liquid, and dispose in a landfill in accordance with all applicable laws and regulations. Certain solvent cleaners present special disposal problems unique to them, and special care must be taken to comply with the disposal regulations for those products.

FOOD SAFETY

In general, food products must be segregated from the clean-up area to avoid contamination. There are a few specialized cases where detergents are used to process edible food products. These special applications are stringently regulated to ensure the safety and wholesomeness of the resulting food product. However, where this is not the case, the food either must be removed from the clean-up area or otherwise covered or segregated to protect it from any possible contamination during the clean-up process. Also, the containers of detergents should be stored in an area segregated from the food-processing and storage areas, again to avoid the possibility of contamination.

WORKER SAFETY

Many detergents used to clean food processing plants are strongly alkaline, strongly acid, release chlorine, or present other safety hazards to the clean-up personnel. These personnel should familiarize themselves with the products they are using and the safety hazards associated with them by studying the material safety data sheet (MSDS), the product label, and any other information the detergent supplier has made available. Pay special attention to the first aid measures that have been recommended. Acids and alkalies can cause both chemical and thermal burns. Chlorine causes respiratory distress. Many chemicals, such as quaternary ammonium chloride compounds, can be sensitizing agents or allergens. Even mild detergents, because of their ability to solubilize oils and fats, can cause dryness, chapping, and cracking of the skin. Appropriate, safety measures need to be taken by clean-up personnel to avoid injury. The purpose of safety equipment is to ensure a safe air supply for breathing and to protect the skin from direct contact with any hazardous materials being used. Generally the first aid step to take, after eye or skin exposure, is to rinse with copious quantities of water; for breathing difficulties, remove to fresh air. See a doctor if distress persists. Furthermore, in the area of eye safety, contact lenses should never be worn when working with potentially hazardous chemical products. In the event of any chemical splattering in the eye, the contact lens makes it very difficult to adequately flush the eye. When mixing chemical products such as detergents, they should always be added to water. Furthermore, use cold water where the chemical addition produces heat of reaction as, for example, adding a strong alkaline detergent to water, which can generate sufficient heat to boil water. A healthy, clean, and safe working environment should be the goal of every food-processing plant.

BIBLIOGRAPHY

1. *Code of Federal Regulations*, Title 21 §172, U.S. Government Printing Office, Washington, D.C., 1986.
2. M. E. Parker and J. H. Litchfield, *Food Plant Sanitation*, Reinhold Publishing Corp., New York, 1962.
3. N. G. Marriot, *Principles of Food Sanitation*, 2nd ed., Van Nostrand Reinhold, Co., Inc., New York, 1989.
4. D. Myers, *Surfactant Science and Technology*, VCH Publishers, Inc., New York, 1988, pp. 315–316.
5. R. K. Guthrie, *Food Sanitation*, 2nd ed., AVI Publishing Co., Inc., Westport, Conn., 1980.

GENERAL REFERENCES

M. Agosti et al., "In-Field Comparative Evaluation of the Antimicrobial Effects of Plain Detergents and Biocides-Supplemented Detergents," *Industrie Alimentari* **33**, 631–635 (1994).

E. Andueza, "Contributions of Detergents to Sustainable Development in the Dairy Industry," *Alimentacion Equipos y Tecnologia* **17**, 57–59 (1998).

"Developing a Total Cleaning/Sanitation Program," *Food Processing* **48**, 112–113 (1987).

M. J. Banner, "Keeping it Clean," *Beverage World* **106**, 41–49 (1987).

W. G. Cutler and E. Kissa, eds., *Detergency Theory and Technology*, Marcel Dekker, New York, 1987.

E. A. Erten and R. F. Ellis, "Efficient Cleaning Practices Safeguard Product Quality," *Food Processing* **49**, 104–106 (1988).

C. Grimmett, "Developments in Detergents and Disinfectants," *Brewing & Distilling International* **23**, 18–20 (1992).

P. Helisto and T. Korpela, "Effects of Detergents on Activity of Microbial Lipases as Measured by the Nitrophenyl Alkanoate Esters Method," *Enzyme Microb. Technol.* **23**, 113–117 (1998).

D. R. Karsa, *Industrial Applications of Surfactants*, The Royal Society of Chemistry, Burlington House, London, 1987.

F. J. C. Lacroix et al., "*Salmonella typhirium* TnphoA Mutants With Increased Sensitivity to Biological and Chemical Detergents," *Research in Microbiology* **146**, 659–670 (1995).

M. Lucchese, "Ecological Aspects of the Use of Detergents and Sanitizers in the Wine and Food Industries. Effects on Effluents and Purification Aspects," *Vini d'Italia* **34**, 53–60 (1992).

M. H. C. R. Passos and A. Y. Kuaye, "Cleaning of Milk Deposit on Heat Exchange Surface by Alkaline and Acid Detergents," *Cienca e Tecnologia de Alimentos* **14**, 105–112 (1994).

Sanitation Operations Manual, National Restaurant Association, 1979.

A. M. Sjoeberg, G. Wirtanen, and T. S. Mattila, "Biofilm and Residue Investigations of Detergents on Surfaces of Food Processing Equipment," *Food and Bioproducts Processing* **73**(C1), 17–21 (1995).

State Water Quality Board, *Detergent Report, A Study of Detergents in California*, California Department of Water Resources, California Department of Public Health, 1965.

G. K. York, "A Guide to Food Plant Sanitizers," *Food Processing* **48**, 118–120 (1987).

TERRY L. MCANINCH
Birko Corporation
Westminster, Colorado

DISINFECTANTS

DEFINITIONS

Disinfection originally referred to the destruction of the germs of disease. In modern use, a disinfectant is an agent capable of destroying a wide range of microorganisms, but not necessarily bacterial spores. Sterilization is the process of destroying all forms of life, including bacterial spores. Sanitation is a more general term referring to those factors that improve the general cleanliness of our living environment and aid in the preservation of health. A sanitizer is a chemical substance that reduces numbers of microorganisms to an acceptable level. This term is widely used in the United States and it is virtually synonymous with the term disinfectant. In the food industry, sanitation and sanitizers are used to refer to processes and agents, respectively, for cleaning inanimate (work) surfaces. The term hygiene (from *Hygieia*, the Greek goddess of health) is frequently used to refer to personal cleanliness, hence hand hygiene. A soil is an organic or inorganic residue on equipment and other work surfaces. A soiled surface is cleaned by washing with a detergent. Detergency does not imply anything more than assisting in cleaning a surface; microorganisms are reduced only to the extent that they are physically removed from the surface (1).

Disinfectants may be characterized on the basis of their mode and range of antimicrobial action. A biocide is a disinfectant that kills all forms of life, whereas a biostat is a disinfectant that inhibits the growth of all forms of life. Many biocides are biostats at low concentrations. Bactericides and bacteriostats are agents active against bacteria. The prefixes fungi-, spori-, and viru- indicate action against fungi, spores, and viruses, respectively. The term germicide is sometimes used to describe disinfectants that destroy disease-causing organisms.

A sanitation program refers to the schedule and procedures for cleaning and sanitizing a food-processing plant or food-handling facility, as well as requirements for personal hygiene. Procedures for sanitation of inanimate surfaces and equipment involve a two- or three-step process. In each case the initial step is removal of gross soil. In a three-step process, the initial rinse is followed by a detergent wash, rinse, and sanitizing. In a two-step process a combined detergent–sanitizing agent is used, followed by a rinse.

CHEMICAL DISINFECTING AGENTS

Lister used phenol (carbolic acid) as a germicide in 1867. Although it is the parent compound of chemical disinfection, its use today is limited to substituted phenols, eg, the *bis*-phenols used in germicidal soaps. Both chlorine (in the form of hypochlorite) and phenol were used to deodorize waste materials in the early 1800s, before Pasteur established the germ theory of infection and putrefaction. The use of disinfectant chemicals began in clinical surroundings in the late nineteenth century. In 1908 the first large-scale use of chlorine (chloride of lime) in water purification started in Chicago, and its use for this purpose spread rapidly. Yet disinfectants were not readily accepted in food production until the 1940s, when hypochlorite treatment was permitted in the dairy industry as an alternative to steam sterilization. The range of antimicrobial chemical agents has been comprehensively reviewed (2).

Heavy metal ions such as Hg^{2+}, Cu^{2+} and Ag^{+} are universal biocides, but they are generally too toxic to use on food equipment. Alcohols such as ethanol and isopropanol and others with chain length up to 8–10 carbon atoms are biocidal except for bacterial spores. Water is essential for the antimicrobial action of alcohols, and 60–70% (v/v) is the most effective concentration. The alkylating agents formaldehyde and ethylene oxide are active gaseous disinfectants with a broad spectrum of biocidal activity, including bacterial spores. The sporicidal activity is probably due to their ability to penetrate into the spores and the fact that they do not require water for their action. Formaldehyde is too toxic for use with food equipment because it is absorbed onto surfaces as a reversible polymer (paraformaldehyde) that slowly depolymerizes to give free formaldehyde as a food contaminant. Ethylene oxide is the most reliable substance for gaseous disinfection of dry surfaces, especially heat-sensitive materials such as plastics (3).

Halogens

The halogens include fluorine, chlorine, bromine, and iodine, all of which are extremely active oxidizing agents. Fluorine is the most reactive, but it is too toxic, irritant and corrosive to be used as a disinfectant (2). Bromine is also too toxic and irritant for widespread use, but it has some applications in mixed halogen compounds (4). Chlorine is an effective disinfectant (Table 1) through the action of hypochlorous acid (HOCl). Hypochlorous acid is unstable and it is normally stabilized by adding caustic soda to form sodium hypochlorite (NaOCl). HOCl can be formed by any of the following reactions:

Chlorine

$$Cl_2 + H_2O \underset{OH^-}{\overset{H^+}{\rightleftharpoons}} HOCl + H^+ + Cl^-$$

Hypochlorite

$$NaOCl + H_2O \underset{OH^-}{\overset{H^+}{\rightleftharpoons}} HOCl + Na^+ + OH^-$$

Chloramine

$$NH_2Cl + 2H_2O \underset{OH^-}{\overset{H^+}{\rightleftharpoons}} HOCl + NH_4^+ + OH^-$$

At acid pH, solutions have greater germicidal activity because greater amounts of HOCl are formed; however, for product stability a pH > 8 is used. Sodium hypochlorite (NaOCl) in commercial liquid form contains 10–14% available chlorine. Calcium hypochlorite [$Ca(OCl)_2$] is a powder with about 30% available chlorine. Chlorine-based sanitizers are normally used at concentrations that yield 100–200 ppm available chlorine with contact times of 3–30 min. Organic chlorines such as chloramine-T and isocyanuric acid derivatives are more stable than hypochlorites in the presence of organic matter, they are less irritant and less toxic, and they release chlorine more slowly.

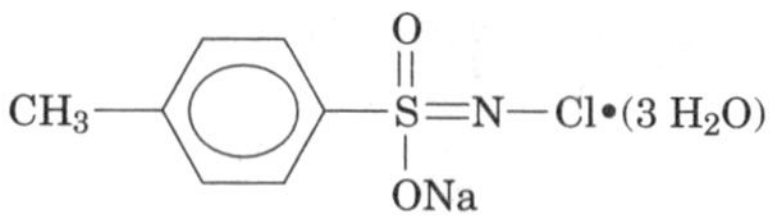

$$CH_3-C_6H_4-S(=O)(ONa)=N-Cl\cdot(3\ H_2O)$$

Chloramine-T

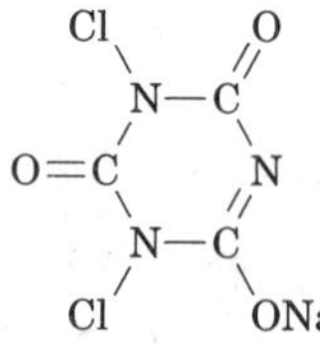

(Structure: ring with Cl, O, N–C, O=C, N, N–C, Cl, ONa)

Isocyanuric acid derivatives

Chloramines have 25–30% available chlorine, but they are weaker bactericides than hypochlorites except that they are more active at pH > 10. Isocyanuric acid products are active over the pH range 6–10. Both of these organic chlorines can be formulated as detergent–sanitizers, but they are relatively expensive (1).

Chlorine dioxide is another form of chlorine that has antimicrobial properties. It is considered to have the same antimicrobial activity as hypochlorite. Principal advantages are that it is not as readily inactivated by organics as hypochlorite, and it retains activity at higher pH values than does hypochlorite. Additionally, chlorine dioxide does not react with ammonia, it can break down phenolics to remove phenolic tastes and odors from water, and it does not form trihalomethanes with nitrogenous organic components of water. Because chlorine dioxide is highly reactive, it cannot be produced and shipped in bulk. Therefore, it is generated at the site of use, usually by combining separate solutions of chlorine and sodium chlorite at carefully controlled reaction conditions. Chlorine dioxide decahydrate can be prepared as a commercial product, but it must be refrigerated because it decomposes with rising temperatures. Under certain conditions it can be explosive. Chlorine dioxide, although it is more expensive than hypochlorite, has been used for treatment of water supplies, in plant chlorination of wastewater treatment, and chlorination of recycled or recirculated waters where slime buildup may occur (5).

Iodine is too insoluble in water to be used as a disinfectant; however, if it is complexed with a nonionic surface

Table 1. Advantages and Disadvantages of Different Classes of Sanitizers

	Hypochlorite	Iodophor	QAC	Acid–Anionics
Antimicrobial spectrum				
Gram-positive bacteria	+	+	+	+
Gram-negative bacteria	+	+	−	+
Spores	+	−	−	−
Viruses	+	−	−	+
Low toxicity	+	+	+	
Nonirritating to skin	−	+	+	
Noncorrosive	−[a]	−	+	
Nonstaining	+	−[b]	+	+
Stability during storage	−	+	+	+
in hard water	+	+	−	+
with organic matter	−	±	+	+
Use at temperatures	+	−	+	+
Mineral–protein films removed	−	+	−	+
Leaves no residues	+	±	−	−
Leaves no flavors or odors	−	−	+	+
Detergent–sanitizer	−	±	+	+
Self-indicating	±[c]	+	−	−
Low cost	+	±	−	

Note: Advantages attribute = −. disadvantageous attribute = −.
[a]Only a problem if misused.
[b]Only some plastic materials and painted walls; staining of skin is temporary.
[c]By odor of available chlorine.

active agent, the resulting iodophor can be used as a detergent–sanitizer (Table 1). Diatomic iodine (I_2) is highly bactericidal; hypoiodous acid (HOI) and the hypoiodite ion (IO^-) are less bactericidal; the iodate ion (IO_3^-) is inactive.

$$I_2 + H_2O \underset{H^+}{\overset{OH^-}{\rightleftharpoons}} HOI + H^+ + I^- \underset{H^+}{\overset{OH^-}{\rightleftharpoons}} I^- + IO_3^-$$

Iodophors are generally used at 12.5–25.0 ppm; at these concentrations they are self-indicating, ie, the presence of available iodine is indicated by the amber color of the solution. They must be used at $< 50°C$ to avoid the release of toxic iodine vapor. Iodophors are formulated with an acid, usually phosphoric acid, because they are more active at pH 3–5. They also solubilize and remove mineral and protein films from food-processing equipment (4).

Surface Active Agents (Surfactants)

These compounds are classified in three categories: (*1*) cationic, (*2*) anionic, and (*3*) nonionic or amphoteric. All three categories have antimicrobial activity to varying degrees, depending on the specifics of molecular moiety and formulation with other ingredients to produce disinfectant or sanitizing compounds. Quaternary ammonium compounds (QACs), or quats, are cationic wetting agents or surfactants. The general chemical configuration of a QAC is

$$\left[\begin{array}{ccc} R_1 & & R_2 \\ & N & \\ R_4 & & R_3 \end{array}\right]^+ H^-$$

where R_1, R_2, R_3, and R_4 are covalently bound organic groups such as alkyl, methyl, benzyl, and cetylbenzyl. The H^- is a halogen, most often chlorine or occasionally bromide. These are synthesized compounds that are formed by reacting tertiary amines with quarterizing agents such as alkyl or cetyl halides. A wide array of groups may be substituted, and this may result in the formation of quaternaries with widely variable properties. The cation or positively charged proton of the molecule is microbiologically active and lyophobic. The negatively charged halide is lyophilic. Because QACs are actively cationic, they are not compatible with detergents or other compounds that are anionic in solution. The chemical configuration for alkyldimethylbenzylammonium chloride, one of the frequently employed quarternaries, is

$$\left[(C_{12}\text{-}C_{18})\text{—}\underset{CH_3}{\overset{CH_3}{N}}\text{—}CH_2\text{—}C_6H_{11}\right]^+ Cl^-$$

Because of the great diversity of chemical composition with QACs, great variation in the antimicrobial activity may occur. Generalization is difficult with these compounds. QACs principally are antibacterial in activity. Often they are more effective against gram-positive bacteria than gram-negative bacteria, especially at alkaline pH. The antifungal activity of quaternaries is variable and depends on the moiety of the molecule of the specific QAC employed and the specific species of fungal organism(s) of concern. Some quaternaries are very effective fungicides, while others may be mediocre at best. Proper selection of these products for antifungal activity is necessary. The QACs are not considered to be effective agents for controlling either bacterial spores or fungal spores. They are variable as viricides, being moderately active against lipophilic viruses but not effective against hydrophilic viruses. In addition to molecular configuration, other factors have an influence on the antimicrobial activity of QACs. As stated previously, quaternaries at alkaline pH generally are more effective against gram-positive than gram-negative bacteria. However, there are complications, conflicts, and resistances related to the pH of solutions. Some gram-negative bacteria, such as *Pseuomonas* and *Escherichia*, show resistance to QACs at alkaline pH but may be much less resistant to neutral or slightly acid quaternary solutions. The effect of organic residues on antimicrobial effectiveness of QACs can be variable. To a large extent, the degree depends on the specific quaternary compound and the type of organic soil that may be present. In general, QACs are more stable to the presence of organic residues than hypochlorite and about equal to that of the iodophors. The activity of quaternary ammonium compounds may be affected by the presence of calcium and magnesium water-hardness salts in solutions. This effect varies with the quaternary compound employed. The QACs that are formulated with chelating agents, such as EDTA and/or sequesterants, have good tolerance to high levels of water-hardness salts. There is good agreement that the effectiveness of QACs increases as the solution temperature increases and that these compounds are more stable than both hypochlorite and iodophors at high temperatures (6).

The quaternaries have found use as sanitizing solutions for dairy and food equipment, skin antiseptics, and hospital sanitizing applications. Quaternary compounds may be combined with nonionic wetting agents and other detergent enhancers into detergent–sanitizer formulations for specific applications. If QACs are to be on sanitizing food-contact surfaces, they must be of a type and in a form that has FDA approval according to provisions of 21 CFR (7). Because of the great variability inherent to these products, it is imperative that they be selected after consultation with QAC manufacturers or suppliers who have been advised of the prevailing conditions under which the quaternary will be expected to perform. After this has been done, the selected product(s) should be used only under the explicit instructions that have been given.

The acid–anionic products are mixtures of anionic surfactant and an acid. Most frequently phosphoric acid is the acid employed. The chemical structure of the anionic surfactants usually is in the form of an alkyl aryl sulfonate. Some examples of these are dodecylbenzene sulfonic acid, sodium lauryl sulfate, sodium dioctylsulfosuccinate, and 1-octane sulfonate. For a more comprehensive list of anionics that have been approved for use in sanitizer solutions for application to food contact surfaces, refer to 21 CFR (7).

The antimicrobial activity of the acid–anionic sanitizers is most pronounced at low pH levels, with the pH range of 3.0 to 1.5 being the most optimal. At the recommended-use concentrations, for specific applications, they provide excellent antimicrobial activity against vegetative bacterial cells and most yeasts. Both bacterial spores and fungal spores show resistance to these products. At pH values above 3.5, the effectiveness of the acid–anionic sanitizers against gram-negative bacteria is reduced, and this becomes more pronounced with increasing pH. Water hardness at up to 1,000 ppm as $CaCO_3$ has not been shown to affect the antimicrobial activity of acid–anionics, but if the water-hardness level should be high enough to cause a rise in pH of a sanitizing solution above 3.5, then reduced sanitizing effectiveness could be anticipated. Because of their acidic reaction in solution, the acid–anionic surfactant sanitizers have found important applications for sanitizing both circulation cleaned-in-place (CIP) equipment and spray-cleaned tanks and vessels in dairy and beverage processing plants where mineral residuals may affect the sanitizing efficiency of the other types of cleaners. The acid–anionic sanitizers are more costly than hypochlorite, but they are less expensive than either the iodophors or the quaternary ammonium sanitizers (8).

Amphoteric surfactants are disinfectants that have been used in European and some other countries for more than 40 years. They have not received FDA approval in the United States for use as sanitizers on food contact surfaces. Most of the commercial amphoteric products are sold under the trade name Tego. They are termed amphoteric or ampholytic because in water solution they yield anions, cations, and zwitterions (dual-charged molecules). The chemical structure of amphoteric surfactants is a molecule composed of an amino acid, most frequently glycine, substituted with a long-chain alkyl amine group. Examples of the most frequently employed amphoterics include dodecylglycine, dodecylaminoethylglycine, and dodecyldiaminoethylglycine. The amphoteric surfactants have been purported to be bactericidal, fungicidal, and viricidal and effective in the presence of soil, lipids, and proteins. The contact times required for effective microbial inactivation are longer than that for other more frequently used sanitizers. Because of their surfactant properties, the amphoteric sanitizers may adsorb on surfaces to form a film that is resistant to rinsing by water. This is an advantageous property when residual antimicrobial activity is desired but may be a disadvantage from the standpoint of possible sanitizer residuals in foods. The effect of organic residues on the antimicrobial of amphoteric surfactant sanitizers are contradictory. This has created a problem for their gaining acceptance and approval as sanitizers. The amphoteric sanitizers have been used as hand-wash disinfectants, floor-wash disinfectants, and food plant equipment sanitizers in European and some other countries (9).

The Peroxides

The two peroxides that are most frequently used for disinfection and sanitizing are hydrogen peroxide and peroxyacetic acid (peracetic acid). Hydrogen peroxide at 3% concentration has been employed as a topical antiseptic and wound irrigation agent for many years. More recently, processes have been developed that allow the production of stable, highly concentrated solutions that may contain up to 90% hydrogen peroxide. This has resulted in the availability of hydrogen peroxide, of various grades and purity, that could be used for microbial control in the food, cosmetic and electronic industries. Hydrogen peroxide has shown effective antimicrobial activity against both non-spore-bearing and spore-bearing bacteria, yeasts, molds, and viruses. It is very effective against the anaerobes because they lack catalase enzyme, which hydrolyses hydrogen peroxide. Also, gram-negative bacteria are more easily inactivated by hydrogen peroxide than are gram-positive bacteria. Among the factors known to affect the antimicrobial activity of hydrogen peroxide are the concentration of chemical, pH, temperature, and presence of metallic salts. Hydrogen peroxide follows the classical relationship that the higher the concentration of chemical antimicrobial in solution, the more rapid the rate of microbial inactivation. Hydrogen peroxide is most active at acidic conditions; its activity is slower at neutrality and becomes progressively less active as the alkalinity of the solution increases. Temperature affects the action of hydrogen peroxide. In general, higher solution temperatures can result in lower concentrations and shorter contact times to obtain microbial inactivation. The presence of certain metallic salts or ions increases the antimicrobial effectiveness of hydrogen peroxide. These include copper, chromium, iron, and molybdenum. A good review on the antimicrobial activity of hydrogen peroxide has been presented by Turner (10).

In industrial applications, hydrogen peroxide has FDA approval for use as a sterilant of containers and equipment used in aseptic packaging of foods (7). For treatment of packaging rooms, 30 to 40% hydrogen peroxide for up to 30 minutes or more contact time at room temperature may be required; whereas at a temperature of 60 to 71°C, the sterilization contact time for containers may be reduced to seconds. Because of the potential hazard of hydrogen peroxide, it is important to insure that all necessary safety practices are in effect for handling and using this material to prevent serious harm to personnel. Hydrogen peroxide has shown promise as an agent for use in vapor-phase sterilization applications (11).

Peroxyacetic acid (peracetic acid) is an organic peroxide that has been used for practical microbial control. Peroxyacetic acid is a strong oxidizing agent that is soluble in water and has a characteristic pungent, vinegarlike odor. Concentrated solutions of peroxyacetic acid are highly unstable, particularly at high temperatures and in the presence of heavy metallic ions. Current commercially available products contain stabilizing ingredients, and they are safe to use. Typically these contain from 4 to 40% peroxyacetic acid. As with all chemicals, they must be handled, stored, and used according to specific label declarations and product safety data sheets for the product. Peroxyacetic acid is superior in action to hydrogen peroxide as an antimicrobial and has a broad spectrum of activity against both gram-positive and gram-negative bacteria, spore-bearing bacteria, and microbial spores. It has both fungicidal and viricidal activity at recommended conditions of use. The FDA allows the use of peroxyacetic acid sanitizing solutions on food-contact surfaces at a concentration of 100 to 200 mg/L (ppm) (7). The antimicrobial activity of peroxyacetic acid is affected by such factors as concentration, pH, and contact time. As is true with other chemical antimicrobials, higher concentrations of peroxyacetic acid in solution are more cidal than are lower concentrations, provided that all other factors are constant. With regard to pH, acid conditions are most optimal to inactivation of microbes, and as the pH increases above pH 7 to 8, there is decreasing cidal activity for a constant concentration of peroxyacetic acid in solution. As might be expected, the temperature of the solution affects the killing power of peroxyacetic acid at a given concentration. Increasing temperature improves the germicidal effect, but peroxyacetic acid retains some antibacterial activity at a temperature as low as 7°C. Because of this, peroxyacetic acid has an advantage over most other commonly used chemical sanitizers for certain applications. Water-hardness minerals do not have a great effect on peroxyacetic acid, and it is not affected as adversely by organic residuals as are other oxidative sanitizers (12,13). As with all hazardous chemicals, necessary precautions are required when using this product. It is essential that personnel be made aware of these and that labeling directions and product safety instructions be followed precisely.

Phenols

Many phenolic compounds are strong germicides, but their potential for use in the food industry is limited by their odor and the possibility of causing off-flavors in foods. Their action involves cell lysis. Depending on concentration, phenols are either bactericidal or bacteriostatic. They have only limited activity against viruses, and they are not sporicidal. Halogenation of phenols increases their activity 3–30 times. Substitution in the *para* position is more effective than in the *ortho* position, substitution with two halogen atoms is more effective than one, and bromophenols are more active than chlorophenols (14). Chloroxylenol or *para*-chloro-*meta*-xylenol (PCMX), the active ingredient of Dettol, is a phenolic compound that can be used as a skin germicide.

Table 2 provides information on comparative information on some of the characteristics of the most common chemical sanitizers that have FDA (7) approval for no-rinse application on food-contact surfaces. Because of the wide variations in the formulation of commercial products, there may be anomalies from that normally expected. In all instances, the information and recommendations given by the manufacturers or supplies of commercial proprietary sanitizer products should be sought and followed explicitly for any application.

Many halogenated *bis*-phenols (or phenylphenols) have considerable activity against bacteria and fungi, but they have low activity against pseudomonads (2). They are used as clinical disinfectants and in germicidal soaps, eg, hexachlorophene.

Hexachlorophene is not very volatile and lacks the unpleasant odor of phenols. It is more active against Gram-positive than Gram-negative bacteria. It is bacteriostatic for *Staphylococcus aureus* at extremely low concentrations (0.05 μg/mL) and requires a suitable quenching agent to inactivate residues of the disinfectant so that its bactericidal activity can be accurately assessed (2). It was widely used as a skin antiseptic marketed as pHisoHex and in a wide range of over-the-counter (OTC) pharmaceutical and personal hygiene products. It can be absorbed through in-

Table 2. Comparative Characteristics of Different Sanitizers

	Hypoclorite	Iodophors	Quarternary ammoniums	Acid anionics	Peroxyacetic acid
Biocidal spectrum					
Gram-positive bacteria	+ +	+ +	+ +	+ +	+ +
Gram-negative bacteria	+ +	+ +	+	+	+ +
Bacterial spores	+ +	+	±	±	+
Yeasts	+	+ +	+	+	+
Molds	+ +	+ +	+	+	+ +
Viruses	+ +	+	±	±	+
pH effectiveness					
Acid	± (unstable)	+ +	±	+ +(<3.5)	+ +
Neutral	+ +	±	+	−	+ +
Alkaline	+	−	+	−	+
Stability to					
Organic residues	−	±	+	+	±
Hard water	+ +	+	±	+	+
Noncorrosive	−	±	+ +	±	±
Skin irritant	+ +	−	−	±	±
Residual activity	−	+	+ +	+	−
High temperature stability	+ +	−	+	+	+ +
Use solution stability	−	+	+ +	+ +	+
Maximum use concentration by FDA[a]	200 mg/l[b]	25 mg/l[b]	200 mg/l[b]	200–430 mg/l[b] (depends on type)	100–200 mg/l[b]

Note: Relative properties: + +, definite; + moderate; ±, variable, −, very low to none.
[a]According to 21CFR, 178:1010; no rinsing after application.
[b]mg/L = milligrams per liter or ppm.

flamed and infant skin with the possibility of serious systemic toxicity (15). As a result, OTC products have generally been limited to 0.75% hexachlorophene. Higher concentrations are sold with medical prescription.

Restrictions on the use of hexachlorophene resulted in the development of a range of other disinfectants for use in germicidal soaps. For example, Irgasan DP 300, also known as triclosan, 2-4-4′trichlor-2′-hydroxy diphenyl ether is active against Gram-positive and Gram-negative organisms.

Other Antimicrobial Agents

Chlorhexidine is one of a family of N^1,N^5-substituted biguanides (2). Its use was licensed in the UK and Canada, but not in the United States.

Chlorhexidine is active against Gram-positive and Gram-negative bacteria; it has limited antifungal activity, but it is not active against acid-fast bacilli, bacterial spores, or viruses. However, some resistant pseudomonads can contaminate aqueous solutions of this compound. It is not compatible with anionic compounds. The digluconate salt is freely soluble in water; however, the diacetate and dihydrochloride salts are only poorly soluble. Alcoholic chlorhexidine or a 4% chlorhexidine detergent (Hibiscrub) are highly effective skin germicides (16).

The salicylanilides and carbanilides are families of antimicrobial chemicals (17). One of the more popular disinfectants among these chemicals is 3,4,4′-trichlorocarbanilide. (TTC). It has been incorporated into soaps to give an antimicrobial product.

PHYSICAL DISINFECTING AGENTS

In addition to chemical compounds, heat and ultraviolet (uv) light can be used to disinfect food contact surfaces. Heat, particularly moist heat, is the most reliable and most widely used method of destroying all forms of microbes (2). The efficacy of heat treatment with steam or hot water (80°C or above) depends on the temperature achieved and the time of exposure. Temperatures above 60°C progressively kill microbial cells. Vegetative cells may be killed or injured. Injured cells can recover in an appropriate environment, for example, in foods. Heat injury has been described in two ways: the extension of the lag phase of growth (18) and the inability to grow on selective media, for example, on salt-containing media for *Staphylococcus aureus* (19) or violet red bile agar for coliform bacteria (20). Because heat cannot be universally applied it is necessary to rely on chemical agents for sanitation programs.

Radiation is an alternative to gaseous disinfection. Gamma (cobalt-60) and electron-accelerated β-ray (<10 MeV) irradiation have good penetrating ability and, therefore, have the potential for use not only in disinfection, but also in food processing. Ultraviolet radiation has a wavelength between 210 and 328 nm, with maximum antimicrobial activity between 240 and 280 nm. Ultraviolet radiation is low energy, so it does not penetrate foods. It is absorbed by glass and plastics, but it can be used for surface disinfection (2). *Micrococcus radiodurans* and bacterial and mold spores are highly resistant to ionizing and uv irradiation. The primary effect of radiation on living cells is by action on DNA. Radiation resistance is genetically determined. Sublethally damaged cells can recover by photoreactivation or by excision and recombination events, which give rise to mutations (2).

EVALUATION OF DISINFECTANTS

A number of laboratory tests have been developed to evaluate the efficacy of disinfectants (2). The Rideal–Walker and Chick–Martin methods of determining the phenol coefficient with *Salmonella typhi* were developed in the early 1900s. Although the Chick–Martin test includes an organic soil, the phenol coefficient tests are artificial in concept, have poor reproducibility and phenol is an unreliable control disinfectant (1). The improved Kelsey–Sykes capacity test attempted to resolve the deficiencies of the phenol tests by using at least four organisms in preliminary screening tests, the most resistant organism being selected for further testing. The disinfectants are prepared in a standardized hard water, with or without a standard soil. Recovery broths are prepared with a neutralizer against the disinfectant. This test also has its shortcomings, but it is used as the official test for disinfectants in the UK (21).

Several tests have been designed to simulate in-use conditions by preparing an air-dried film of microorganisms on an appropriate surface, such as stainless steel, with or without an organic soil. The Lisboa test was developed for testing sanitizers for dairy equipment. Stainless steel tubes are contaminated and disinfected using standardized procedures and neutralized to inactivate any residual disinfectant; the number of surviving organisms is then determined (1). Various modifications of this in-use testing have been developed (2,22). Methods of the Association of Official Analytical Chemists (AOAC) include use-dilution carrier techniques with *Salmonella choleraesuis, Staphylococcus aureus*, or *Pseudomonas aeruginosa* contaminated onto polished stainless-steel cylinders. Tests for fungicidal and sporicidal activity are also specified (23).

Tests for evaluating skin germicides under in-use conditions have been even more difficult to standardize (24), and there are no official methods comparable to the AOAC use-dilution technique for evaluating disinfectants for use on inanimate surfaces. Multiple-basin techniques measure the rate of mechanical removal of microorganisms from the skin. Testing of skin disinfectants must take into account the transient and residual microflora of the skin, the time of exposure, methods of sampling the skin after washing, and methods of contaminating skin with a transient microflora appropriate to the use environment (25). Depending on the germicidal agent used, it may be necessary to neutralize the rinse solution. With skin disinfectants it is appropriate to test for a residual (or substantive) effect, ie, improved results with repetitive use of the same agent compared with nongermicidal soaps. Occlusion and expanded flora tests (in which agents are applied topically to the skin, covered with a plastic film, and sealed with impermeable plastic tape) represent alternative approaches to testing skin disinfectants (26).

Comparisons of antimicrobial soaps by various standardized techniques have given results that question the in-use efficacy of most products tested. Studies have indicated that under the specific test conditions, only an iodophor product (0.75%) and 4% chlorhexidine gluconate significantly reduced the number of microorganisms released from hands (25,27,28). Other products, including intermediate strength iodophor products, 2% chlorhexidine, Irgasan DP 300, *para*-chloro-*meta*-xylenol (PCMX), trichlorocarbanilide or tribromosalicylanilide, under in-use conditions of testing were no better than a nongermicidal soap.

MODULAR SYSTEMS OF DISINFECTION

Food industries that process liquid products, such as milk, beer, and soft drinks, use CIP systems for disinfection. The principles are similar to those for manual cleaning, but mechanical force generated by the velocity of liquid flow through the system is relied on to remove food soils. CIP offers many advantages over manual sanitizing programs, including lower labor costs, more economic operation, better sanitation, faster cleanup and reuse of equipment, less dismantling and reassembly, and greater safety of use and operation (1). The CIP systems can be based on single or multiple use of cleaning solutions. Multiple-use systems are usually automatic, and solutions are recovered according to a preset program and stored in holding tanks for reuse. A typical CIP program is as follows (1):

1. Prerinse (5 min) with cold water.
2. Alkali detergent wash (15 min at 80°C).
3. Intermediate rinse (3 min) with cold water from mains.
4. Cold hypochlorite solution (10 min).
5. Final rinse (3 min) with cold water from mains.

For large tanks that would be uneconomical to fill with cleaning fluids, a permanent or portable spray system is fitted to the vessel. The spraying devices should allow every part of the inside of the vessel to come in contact with the cleaning solutions. Good drainage must be insured to avoid accumulation of fluids and residues on equipment surfaces.

Mechanical aids for plant sanitation include pressurized steam, high-pressure water jets, compressed air, and ultrasound. All of these cleaning aids require specialized equipment and have specific uses. High-pressure steam can remove debris and sterilize. High-pressure water jets also remove debris from inaccessible parts of machinery; however, such inaccessibility should be avoided in the design of food equipment and machinery. Compressed air removes dry powder, dust, and soil, but it spreads, rather than eliminates, the soil. Vacuum cleaners are preferable to compressed air for removing dry solids and dust. Ultrasound is used to clean small or sensitive items of equipment that are otherwise difficult to clean. It requires immersion of the objects in an ultrasound tank for exposure to ultrasonic vibrations, which remove soils by cavitation.

Foam sanitation is an efficient system for cleaning walls, floors, equipment with large contact surfaces, and immovable food-handling equipment. A foaming agent is added to the detergent formulation to produce a thick, long-lasting foam. This gives the cleaning agent a long contact time with the soiled surfaces. This form of cleaning requires a special pressure-generating system. The foam must be removed and the bactericidal agent applied. Foam systems give a good visible awareness of the sanitation process.

LEGISLATION

The food legislation of most countries requires that food is handled in a sanitary manner. The details and standards differ between countries. In the United States the principal federal legislation governing the handling of foods is the Food, Drug and Cosmetic Act of 1938, as amended. This covers all foods except meats and poultry, which are covered by various food inspection acts. Regulations promulgated through provisions of these acts are published in the *Code of Federal Regulations* (CFR), which is revised annually as needed. Emphasis is on good manufacturing practices and hazard analysis and critical control point (HACCP) standards for the handling, preparation, processing, and storage of food. Each state may enact its own laws and establish its own regulations for food sanitation control that apply to its territorial jurisdiction. In most cases, the state laws and regulations are in substantive agreement with the federal counterpart. The U.S. Department of Agriculture has responsibility for meat, poultry, rabbit, and egg products inspection programs in the United States. Under the Federal Insecticide, Fungicide, and Rodenticide Act of 1978, sanitizers for use in food establishments are defined as insecticides and are regulated by the U.S. Environmental Protection Agency (EPA). Antimicrobial agents for applications on humans are regulated by the Food and Drug Administration.

In Canada food legislation is enacted through the federal Food and Drugs Act of 1953, as amended. The Health Protection Branch (HPB) of Health and Welfare Canada is responsible for administering the act. Other agencies also have responsibility for food inspection, including Agriculture Canada, and Fisheries and Oceans Canada. Disinfectants for use in food premises are deemed to be drugs and require a Drug Identification Number (DIN) issued by HPB.

In the UK, food legislation is based on the Food and Drugs Act of 1955, which is the enabling legislation for regulations governing food hygiene and is controlled by the Department of Health and Social Security or the Ministry of Agriculture, Fisheries and Food (MAFF). Enforcement of the regulations is the responsibility of Environmental Health Officers employed by elected local authorities.

In Europe, the EEC has mandatory standards that automatically become law in member countries. In a broader sense, the Codex Alimentarius program of the Food and Agriculture and World Health Organizations (FAO/WHO) has the intention of establishing international agreements on food standards that would safeguard health and encourage good handling practices for foods.

BIBLIOGRAPHY

1. P. R. Hayes, *Food Microbiology and Hygiene*, Elsevier Applied Science Publishers, Ltd., Barbing, UK, 1985.
2. A. D. Russell, W. B. Hugo, and E. A. J. Ayliffe, eds., *Principles and Practice of Disinfection, and Sterilization*, Blackwell Scientific Publications, London, 1982.
3. C. R. Phillips, "Gaseous Sterilization," in S. S. Block, ed., *Disinfection, Sterilization and Preservation*, 2nd. ed., Lea and Febiger, Philadelphia, 1977.
4. J. R. Trueman, "The Halogens" in W. B. Hugo, ed., *Inhibition and Destruction of the Microbial Cell*, Academic Press, Orlando, Fla., 1971, pp. 137–183.
5. M. A. Bernarde et al., "Efficiency of Chlorine Dioxide as a Bactericide," *Appl. Microbiol.* **13**, 776–780 (1965).
6. J. J. Merianos, "Quaternary Ammonium Antimicrobial Compounds," in S. S. Block, ed., *Disinfection, Sterilization and Preservation*, 4th ed., Lea and Febiger, Philadelphia, 1991, pp. 225–255.
7. CFR, *Code of Federal Regulations*, Title 21, Parts 178.1005-1010, Office of Federal Register, U.S. Government Printing Office, Washington, DC, April 1998.
8. G. R. Dychdala and J. A. Lopes, "Surface Active Agents: Acid-Anionic Compounds," in S. S. Block, ed., *Disinfection, Sterilization and Preservation*, 4th ed., Lea and Febiger, Philadelphia, 1991, pp. 256–262.
9. S. S. Block, "Peroxygen Compounds," in S. S. Block, ed., *Disinfection, Sterilization and Preservation*, 4th ed., Lea and Febiger, Philadelphia, 1991, pp. 167–181.
10. F. J. Turner, "Hydrogen Peroxide and Other Oxidant Disinfectants" in S. S. Block, ed., *Disinfection, Sterilization and Preservation*, 3rd ed., Lea and Febiger, Philadelphia, 1983. Penn., 1983, pp. 240–250.
11. N. A. Klapes and D. Vesley, "Vapor-phase Hydrogen Peroxide as a Surface Decontaminant and Sterilant," *Appl. Environ. Microbiol.* **56**, 503–506 (1990).
12. M. G. C. Baldry, "The Bactericidal, Fungicidal and Sporocidal Properties of Hydrogen Peroxide and Peracetic Acid," *J. Appl. Bact.* **54**, 417–423 (1983).
13. M. G. C. Baldry and J. A. L. Fraser, "Disinfection with Peroxygens," in K. R. Payne, ed., *Industrial Biocides*, John Wiley and Sons, New York, 1988, pp. 91–116.
14. R. F. Prindle, "Phenolic Compounds," in S. S. Block, ed., *Disinfection, Sterilization and Preservation*, 3rd ed., Lea and Febiger, Philadelphia, 1983, pp. 197–224.
15. R. A. Chilcote et al., "Hexachlorophene Storage in a Burn Patient Associated with Encephalopathy," *Pediatrics* **59**, 457–459 (1977).
16. E. J. L. Lowbury and H. A. Lilly, "Use of 4% Chlorohexidine Detergent Solution (Hibiscrub) and Other Methods of Skin Disinfection," *Br. Med. J.* **1**, 510–515 (1973).
17. H. C. Stecker, "The Salicylanides and Carbanilides" in S. S. Block, ed., *Disinfection, Sterilization and Preservation*, 2nd ed., Lea and Febiger, Philadelphia, 1977, pp. 282–300.
18. H. Jackson and M. Woodbine, "The Effect of Sublethal Heat Treatment on the Growth of *Staphylococcus aureus*," *J. Appl. Bact.* **26**, 152–158 (1963).
19. M. E. Stiles and L. D. Witter, "Thermal Inactivation, Heat Injury and Recovery of *Staphylcoccus aureus*," *J. Dairy Sci.* **48**, 677–681 (1965).
20. L. A. Roth, M. E. Stiles, and L. F. L. Clegg, "Reliability of Selective Media for the Enumeration and Estimation of *Escherichia coli*," *Can. Inst. Food Sci. Tech. J.* **6**, 230–234 (1973).
21. B. Croshaw, "Disinfectant Testing—With Particular Reference to the Rideal-Walker and Kelsey-Sykes Tests," in C. H. Collins et al., ed. *Disinfectants: Their Use and Evaluation of Effectiveness*, Academic Press, Orlando, Fla., 1981, pp. 1–15.
22. R. M. Blood, J. S. Abbiss, and B. Jarvis, "Assessment of Two Methods for Testing Disinfectants and Sanitizers for Use in the Meat Processing Industry," in C. H. Collins et al., eds., *Disinfectants: Their Use and Evaluation of Effectiveness*, Academic Press, Orlando, Fla., 1981, pp. 17–31.
23. Association of Official Analytical Chemists, *Official Methods of Analysis of the Association of Official Analytical Chemists*, 15th ed., AOAC, Arlington, Va., Vol. 1, 1990, pp. 135–142.

24. G. A. J. Ayliffe, J. R. Babb, and Ha. A. Lilly, "Tests for Hand Disinfection," in C. H. Collins et al., eds., *Disinfectants: Their Use and Evaluation of Effectiveness*, Academic Press, Orlando, Fla., 1981, pp. 37–44.
25. A. Z. Sheena and M. E. Stiles, "Efficacy of Germicidal Hand Wash Agents in Hygienic Hand Disinfection," *J. Food Protection* **45**, 713–720 (1982).
26. R. R. Marples and A. M. Kligmen, "Methods for Evaluation Topical Antibacterial Agents on Human Skin," *Antimicrob. Agents and Chemotherapy* **2**, 8–15 (1974).
27. A. Z. Sheena and M. E. Stiles, "Efficacy of Germicidal Hand Wash Agents Against Transient Bacteria Inoculated Onto Hands," *J. Food Protection* **46**, 722–727 (1983).
28. A. Z. Sheena and M. E. Stiles, "Low Concentration Iodophor for Hand Hygiene," *J. Hygiene* **94**, 269–277 (1985).

DAVID A. EVANS
University of Massachusetts
Amherst, Massachusetts

DISTILLATION: TECHNOLOGY AND ENGINEERING

Although the use of distillation dates back in recorded history to about 50 BC the first truly industrial exploitation of this separation process did not occur until the 12th century, when it was used in the production of alcoholic beverages. By the 16th century, distillation also was being used in the manufacture of vinegar, perfumes, oils, and other such products.

As recently as 200 years ago, distillation stills were small, of the batch type, and usually operated with little or no reflux. With experience, however, came new developments. Tray columns appeared on the scene in the 1820s, along with feed preheating and the use of internal reflux. By the latter part of that century, considerable progress had been made by Hausbrand in Germany and Sorel in France, who developed mathematical relations that turned distillation from an art into a well-defined technology.

Distillation today is a widely used operation in the petroleum, chemical petrochemical, beverage, and pharmaceutical industries. It is important not only for the development of new products but also in many instances for the recovery and reuse of volatile liquids. Pharmaceutical manufacturers, for example, use large quantities of solvents, most of which can be recovered by distillation with a substantial savings in cost and a reduction in pollution.

Although one of the most important unit operations, distillation unfortunately is also one of the most energy-intensive operations. It easily is the largest consumer of energy in petroleum and petrochemical processing and as such, must be approached with conservation in mind. It is a specialized technology in which the correct design of equipment is not always a simple task.

DISTILLATION TERMINOLOGY

To provide a better understanding of the distillation process, the following briefly explains the terminology most often encountered.

Solvent Recovery

The term solvent recovery often has been a somewhat vague label applied to the many and very different ways in which solvents can be reclaimed by industry.

One approach when an impure solvent contains both soluble and insoluble particles is to evaporate the solvent from the solids. This requires the use of a small forced-circulation-type evaporator that combines a heat exchanger, external separator, and vacuum system with a special orifice that causes back pressure in the exchanger and arrests vaporization until the liquid flashes into the separator. Although this will recover a solvent, it will not separate solvents if two or more are present.

A further technique is available to handle an airstream that carries solvents. By chilling the air by means of vent condensers or refrigeration equipment, the solvents can be removed from the condenser. Solvents also can be recovered by using extraction, adsorption, absorption, and distillation methods.

Solvent Extraction

Essentially a liquid/liquid process where one liquid is used to extract another from a secondary stream, solvent extraction generally is performed in a column somewhat similar to a normal distillation column. The primary difference is that the process involves two liquids instead of liquid and vapor.

During the process, the lighter (ie, less dense) liquid is charged to the base of the column and rises through packing or trays while the more dense liquid descends. Mass transfer occurs, and a component is extracted from one stream and passed to the other. Liquid/liquid extraction sometimes is used when the breaking of an azeotrope is difficult or impossible by distillation techniques.

Carbon Adsorption

The carbon adsorption technique is used primarily to recover solvents from dilute air or gas streams. In principle, a solvent-ladened airstream is passed over activated carbon and the solvent is adsorbed into the carbon bed. When the bed becomes saturated, steam is used to desorb the solvent and carry it to a condenser. In such cases as toluene, for example, recovery of the solvent can be achieved simply by decanting the water/solvent two-phase mixture that forms in the condensate. Carbon adsorption beds normally are used in pairs so that the airflow can be diverted to the secondary bed when required.

On occasion, the condensate is in the form of a moderately dilute miscible mixture. The solvent then has to be recovered by distillation. This would apply especially to ethyl alcohol, acetone-type solvents.

Absorption

When carbon adsorption cannot be used because certain solvents either poison the activated carbon bed or create so much heat that the bed can ignite, absorption is an alternative technique. Solvent is recovered by pumping the solvent-ladened airstream through a column countercurrently to a water stream that absorbs the solvent. The air from the top of the column essentially is solvent-free whereas the dilute water/solvent stream discharged from the column bottom usually is concentrated in a distillation column. Absorption also can be applied in cases where an oil rather than water is used to absorb certain organic solvents from an airstream.

Azeotropes

During distillation, some components form an azeotrope at a certain stage of the fractionation and require a third component to break the azeotrope and achieve a higher percentage of concentration. In the case of ethyl alcohol and water, for example, a boiling mixture containing less than 96% by weight ethyl alcohol produces a vapor richer in alcohol than in water and is readily distilled. At the 96% by weight point, however, the ethyl alcohol composition in the vapor remains constant, ie, the same composition as the boiling liquid. This is known as the azeotrope composition. Further concentration requires use of a process known as azeotropic distillation. Other common fluid mixtures that form azeotropes are formic acid/water, isopropyl alcohol/water, and isobutanol/water.

Azeotropic Distillation

In a typical azeotropic distillation procedure, a third component such as benzene, isopropyl ether, or cyclohexane is added to an azeotropic mixture such as ethyl alcohol/water to form a ternary azeotrope. Because the ternary azeotrope is richer in water than the binary ethyl alcohol/water azeotrope, water is carried over the top of the column. The azeotrope, when condensed, forms two phases. The organic phase is refluxed to the column whereas the aqueous phase is discharged to a third column for recovery of the entraining agent.

Certain azeotropes such as the *n*-butanol/water mixture can be separated in a two-column system without the use of a third component. When condensed and decanted, this type of azeotrope forms two phases. The organic phase is fed back to the primary column, and the butanol is recovered from the bottom of the still. The aqueous phase, meanwhile, is charged to the second column, with the water being taken from the column bottom. The vapor from top of both columns is condensed, and the condensate is run to a common decanter (Fig. 1).

Extractive Distillation

This technique is somewhat similar to azeotropic distillation in that it is designed to perform the same type of task. In azeotropic distillation, the azeotrope is broken by carrying over a ternary azeotrope at the top of the column. In extractive distillation, a very high boiling compound is added and the solvent is removed at the base of the column.

Stripping

In distillation terminology, stripping refers to the recovery of a volatile component from a less volatile substance. Again, referring to the ethyl alcohol/water system, stripping is done in the first column below the feed point where the alcohol enters at about 10% by weight and the resulting liquid from the column base contains less than 0.02% alcohol by weight. This is known as the stripping section of the column. This technique does not increase the con-

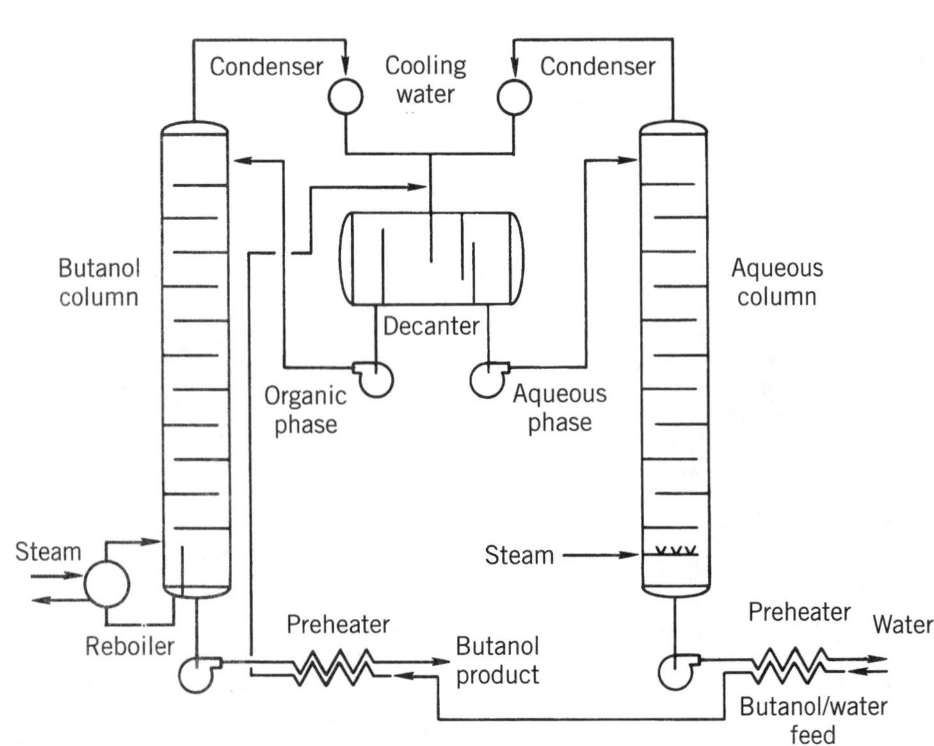

Figure 1. System for recovering butanol from butanol/water mixture.

centration of the more volatile component but rather, decreases its concentration in the less volatile component.

A stripping column also can be used when a liquid such as water contaminated by toluene cannot be discharged to a sewer. For this pure stripping duty, the toluene is removed within the column while vapor from the top is decanted for residual toluene recovery and refluxing of the aqueous phase.

Rectification

For rectification or concentration of the more volatile component, a top section of column above the feed point is required. By means of a series of trays and reflux back to the top of the column, a solvent such as ethyl alcohol can be concentrated to more than 95% by weight.

Batch Distillation

When particularly complex or small operations require recovery of the more volatile component, there are batch-distillation systems of various capacities. Essentially a rectification-type process, batch distillation involves pumping a batch of liquid feed into a tank where boiling occurs. Vapor rising through a column above the tank combines with reflux coming down the column to effect concentration. This approach is not very effective for purifying the less-volatile component.

For many applications, batch distillation requires considerable operator intervention or, alternatively, a significant amount of control instrumentation. Although it is more energy intensive than a continuous system, steam costs generally are less significant on a small operation. Furthermore, it is highly flexible and a single batch column can be used to recover many different solvents.

Continuous Distillation

The most common form of distillation used by the chemical, petroleum, and petrochemical industries is the continuous-mode system. In continuous distillation, feed constantly is charged to the column at a point between the top and bottom trays. The section above the feed point rectifies the more volatile component while the column section below the feed point strips out the more volatile component from the less volatile liquid. In order to separate N components with continuous distillation, a minimum of $N - 1$ distillation columns is required.

Turndown

The turndown ratio of a column is an indication of the operating flexibility. If a column, for example, has a turndown ratio of 3, it means that the column can be operated efficiently at 33% of the design maximum throughput.

SYSTEM COMPONENTS: COLUMNS, INTERNALS, INSTRUMENTATION, AND AUXILIARY EQUIPMENT

The following briefly defines the many components required for a distillation system and the many variations in components that are available to meet different process conditions.

Column Shells

A distillation column shell can be designed for use as a free-standing module or for installation within a supporting steel structure. Generally speaking, unless a column is of very small diameter, a self-supporting column is more economical. This holds true even under extreme seismic 3 conditions.

There are distillation columns built of carbon steel, 304 stainless steel, 316 stainless steel, Monel, titanium, and Incoloy 825. Usually, it is more economical to fabricate columns in a single piece without shell flanges. This technique not only simplifies installation but also eliminates danger of leakage during operation. Columns more than 80 feet long have been shipped by road without transit problems.

Although columns of more than 3-ft diameter normally have been transported without trays to prevent dislodgement and possible damage, recent and more economical techniques have been devised for factory installation of trays with the tray manways omitted. After the column has been erected, manways are added and, at the same time, the fitter inspects each tray.

With packed columns of 20-in. diameter or less that use high-efficiency metal mesh packing, the packing can be installed before shipment. Job-site packing, however, is the norm for larger columns. This prevents packing from bedding down during transit and leaving voids that would reduce operating efficiency. Random packing always is installed after delivery except for those rare occasions when a column can be shipped in a vertical position. Access platforms and interconnecting ladders designed to Occupational Safety and Health Administration (OSHA) standards also are supplied for on-site attachment to free-standing columns.

Installation usually is simple because columns are fitted with lifting lugs and, at the fabrication stage, a template is drilled to match support holes in the column base ring. With these exact template dimensions, supporting bolts can be preset for quick and accurate coupling as the column is lowered into place.

Column Internals

During recent years, the development of sophisticated computer programs and new materials has led to many innovations in the design of trays and packings for more efficient operation of distillation columns. In designing systems for chemical, petroleum, and petrochemical use, full advantage can be taken of available internals to assure optimum distillation performance.

Tray Devices

Although there are perhaps five basic distillation trays suitable for industrial use, there are many design variations of differing degrees of importance and a confusing array of trade names applied to their products by tray manufacturers. The most modern and commonly used devices are sieve, dual-flow, valve, bubble-cap, and baffle trays—each with its advantages and preferred usage. Of these, the sieve- and valve-type trays currently are most often specified.

For a better understanding of tray design, Figure 2 defines and locates typical tray components. The material of construction usually is 14 gauge, with modern trays adopting the integral truss design, which simplifies fabrication. A typical truss tray is shown in Figure 3. For columns less than 3 ft in diameter, it is not possible to assemble the truss trays in the column. Trays therefore must be preassembled on rods into a cartridge section for loading into the column. Figure 4 shows this arrangement in scale-model size.

The hydraulic design of a tray is a most important factor. The upper operating limit generally is governed by the flood point, although, in some cases, entrainment also can restrict performance. By forcing some liquid to flow back up the column, entrainment reduces concentration gradients and lowers efficiency. A column also can flood by downcomer backup when tray design provides insufficient down-comer area or when the pressure drop across the tray is high. When the down-comer is unable to handle all the liquid involved, the trays start to fill and pressures increase. This also can occur when a highly foaming liquid is involved. Flooding associated with high tray pressure drops and small tray spacing takes place when the required liquid seal is higher than the tray spacing. Downcomer design also is particularly important at high operating pressures due to a reduction in the difference between vapor and liquid densities.

The lower limit of tray operation, meanwhile, is influenced by the amount of liquid weeping from one tray to the next. Unlike the upward force of entrainment, weeping liquid flows in the normal direction and considerable amounts can be tolerated before column efficiency is significantly affected. As the vapor rate decreases, however, a point eventually is reached when all the liquid is weeping and there is no liquid seal on the tray. This is known as the dump point, below which there is a severe drop in efficiency.

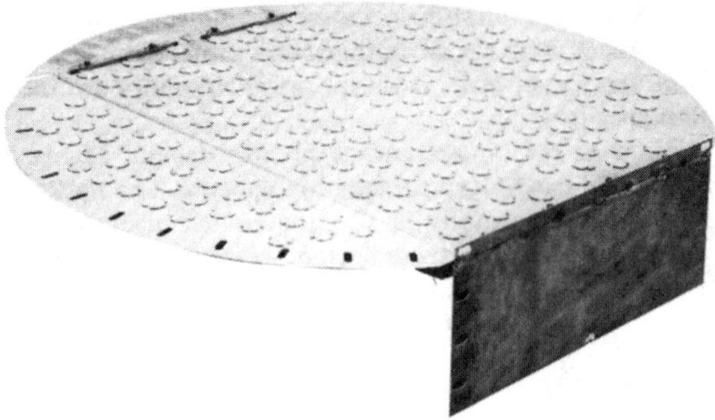

Figure 3. Typical tray of integral truss design.

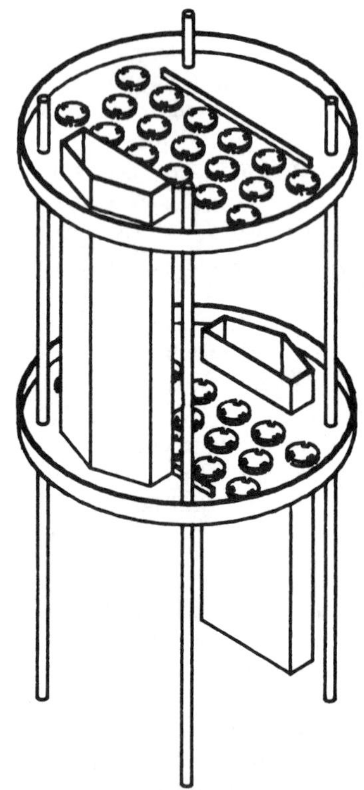

Figure 4. Cartridge tray assembly.

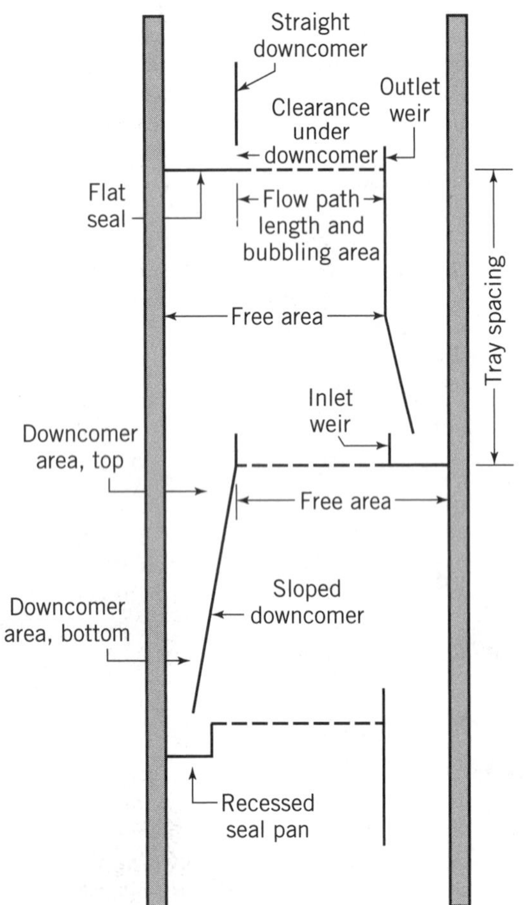

Figure 2. Tray component terminology.

Sieve Tray. The sieve tray is a low-cost device that consists of a perforated plate that usually has holes of 1/16 in. to 1 in. diameter, a down-comer, and an outlet weir. Although inexpensive, a correctly designed sieve tray can be comparable to other styles in vapor and liquid capacities, pressure drop, and efficiency. For flexibility, however, it is inferior to valve and bubble-cap trays and sometimes is unacceptable for low liquid loads when weeping has to be minimized.

Depending on process conditions and allowable pressure drop, the turndown ratio of a sieve tray can vary from 1.5 to 3 and occasionally higher. Ratios of 5, as sometimes claimed, can be achieved only when the tray spacing is large, available pressure drop is very high, liquid loadings are high, and the system is nonfoaming. For many applications, a turndown of 1.5 is acceptable.

It also is possible to increase the flexibility of a sieve tray for occasional low throughput operation by maintaining a high reboil and increasing the reflux ratio. This may be economically desirable when the low throughput occurs for a small fraction of the operating time. Flexibility, likewise, can be increased by the use of blanking plates to reduce the hole area. This is particularly useful for initial operation when it is proposed to increase the plant capacity after a few years. There is no evidence to suggest that blanked-off plates have inferior performance to unblanked plates of similar hole area.

Dual-Flow Tray. The dual-flow tray is a high hole areas sieve tray without a down-comer. The liquid passes down through the same holes through which the vapor rises. Because no down-comer is used, the cost of the tray is lower than that of a conventional sieve tray.

In recent years, use of the dual-flow tray has declined somewhat because of difficulties experienced with partial liquid/vapor bypassing of the two phases, particularly in larger diameter columns. The dual-flow column also has a very restricted operating range and a reduced efficiency because there is no cross flow of liquid.

Valve Tray. Although the valve tray dates back to the rivet type first used in 1922, many design improvements and innumerable valve types have been introduced in recent years. A selection of modern valves as illustrated provide the following advantages (Fig. 5):

1. Throughputs and efficiencies can be as high as sieve or bubble-cap trays.
2. Very high flexibility can be achieved and turndown ratios of 4 to 1 are easily obtained without having to resort to large pressure drops at the high end of the operating range.
3. Special valve designs with venturi-shaped orifices are available for duties involving low pressure drops.
4. Although slightly more expensive than sieve trays, the valve tray is very economical in view of its numerous advantages.
5. Because an operating valve is continuously in movement, the valve tray can be used for light-to-moderate fouling duties. Valve trays can be successfully used on brewery effluent containing waste beer, yeast, and other materials with fouling tendencies.

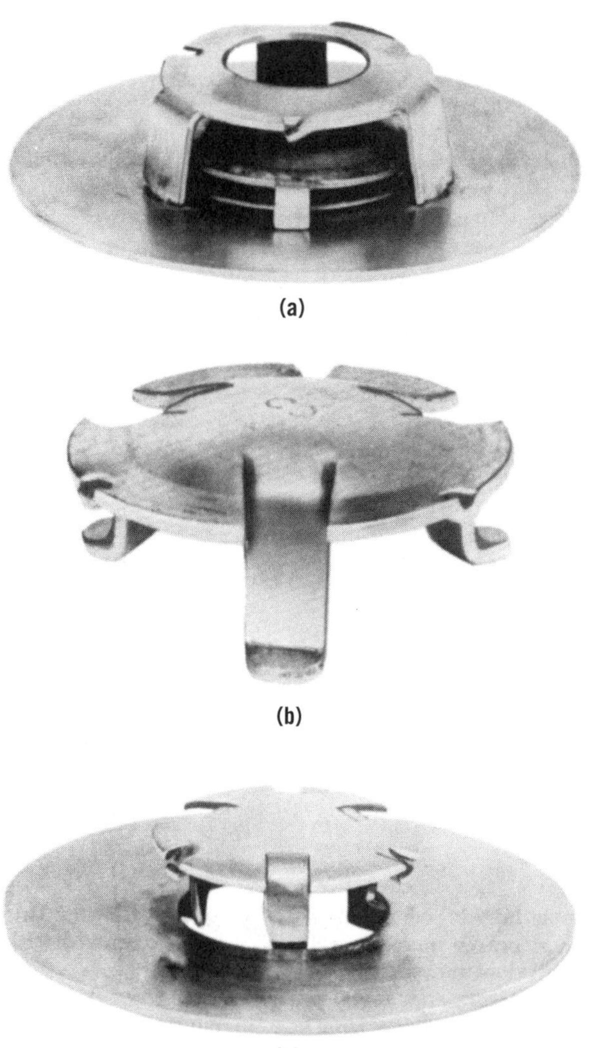

Figure 5. (**a**) Special two-stage valve with lightweight orifice cover for complete closing; (**b**) and (**c**) two typical general purpose valves which may be used in all types of services.

Bubble-Cap Tray. Although many bubble-cap columns still are in operation, bubble-cap trays rarely are specified today because of high cost factors and the excellent performance of the modern valve-type tray. The bubble cap, however, does have a good turndown ratio and is good for low liquid loads.

Baffle Tray. Baffle trays are arranged in a tower in such a manner that the liquid flows down the column by splashing from one baffle to the next lower baffle. The ascending gas or vapor, meanwhile, passes through this curtain of liquid spray.

Although the baffle-type tray has a low efficiency, it can be useful in applications where the liquid contains a high fraction of solids.

Packings

For many types of duties, particularly those involving small-diameter columns, packing is the most economical tower internal. One advantage is that most packing can be purchased from stock on a cubic-foot basis. In addition, the mechanical design and fabrication of a packed column is quite simple. Disadvantages of packing include its unsuitability for fouling duties, breakage of ceramic packing, and, according to some reports, less predictive performance, particularly at low liquid loads or high column diameters.

The most widely used packing is the random packing, usually Rashig rings, Pall rings, and ceramic saddles. These are available in various plastics, a number of different metals and, with the exception of Pall rings, in ceramic materials. Although packings in plastic have the advantage of corrosion resistance, the self-wetting ability of some plastic packing, such as fluorocarbon polymers, sometimes is poor, particularly in aqueous systems. This considerably increases the HETP when compared with equivalent ceramic rings.

High-efficiency metal mesh packing as shown in Figure 6 has found increasing favor in industry during recent years. One type uses a woven wire mesh that becomes self-wetting because of capillary forces. This helps establish good liquid distribution as the liquid flows through the packing in a zig-zag pattern.

Figure 6. Segment of high-efficiency metal mesh packing.

If properly used, high efficiency packings can provide HETP values in the range 6–12 in. This can reduce column heights, especially when a large number of trays is required. Such packings, however, are very expensive, and each application must be studied in great detail.

With random and, in particular, high-efficiency packing, considerable attention must be given to correct liquid distribution. Certain types of high-efficiency packing are extremely sensitive to liquid distribution and should not be used in columns more than 2 ft in diameter. Positioning of these devices and the design of liquid distribution and redistribution are important factors that should be determined only by experts.

Instrumentation

One of the most important aspects of any distillation system is the ability to maintain the correct compositions from the columns by means of proper controls and instrumentation. Although manual controls can be supplied, this approach rarely is used today in the United States. Manual control involves the extensive use of rotameters and thermometers, which, in turn, involves high labor costs, possible energy wastage, and, at times, poor quality control. Far better control is obtained through the use of pneumatic or electronic control systems.

Pneumatic Control Systems. The most common form of distillation column instrumentation is the pneumatic-type analogue control system. Pneumatic instruments have the advantage of being less expensive than other types and, because there are no electrical signals required, there is no risk of an electrical spark. One disadvantage is the need to ensure that the air supply has a very low dew point (usually −40°F) to prevent condensation in the loops.

Electronic Control Systems. Essentially, there are three types of electronic control systems:

1. Conventional electronic instruments.
2. Electronic systems with all field devices explosion proof.
3. Intrinsically safe electronic systems.

The need to have a clear understanding of the differences is important. Most distillation duties involve at least one flammable liquid that is being processed in both the vapor and liquid phases. Because there always is the possibility of a leak of liquid or vapor, particularly from pump seals, it is essential for complete safety that there be no source of ignition in the vicinity of the equipment. Although many instruments, such as controllers and alarms, can be located in a control room removed from the process, all local electronic instruments must be either explosion proof or intrinsically safe.

With explosion-proof equipment, electrical devices and wiring are protected by boxes or conduit that will contain any explosion that may occur. With intrinsically safe equipment, barriers limit the transmission of electrical energy to such a low level that it is not possible to generate a spark. As explosion-proof boxes and conduits are not required, wiring costs are reduced.

For any intrinsically safe system to be accepted for insurance purposes, FM (Factory Mutual) or CSA (Canadian Standards Association) approval usually must be obtained. This approval applies to a combination of barriers and field devices. Therefore, when a loop incorporates such instruments from different manufacturers, it is essential to ensure that approval has been obtained for the combination of instruments.

Auxiliary Equipment

In any distillation system, the design of auxiliary equipment such as the reboiler, condenser, preheaters, and product coolers is as important as the design of the column itself.

Reboiler. Although there are many types of reboilers, the shell-and-tube thermosyphon reboiler is used most frequently. Boiling within the vertical tubes of the exchanger produces liquid circulation and eliminates the need for a pump. A typical arrangement is shown in Figure 7.

For certain duties, particularly when the bottoms liquid has a tendency to foul heat transfer surfaces, it is desirable to pump the liquid around the heat exchanger. Because boiling can be suppressed by use of an orifice plate at the outlet of the unit, fouling is reduced. The liquid being pumped is heated under pressure and then is flashed into the base of the column where vapor is generated.

An alternative approach is the use of a plate heat exchanger as a forced-circulation reboiler. With this technique, the very high liquid turbulent flow that is induced within the heat exchanger through the use of multiple corrugated plates holds fouling to a minimum. Meanwhile, the superior rates of heat transfer that are achieved reduce the surface area required for the reboiler.

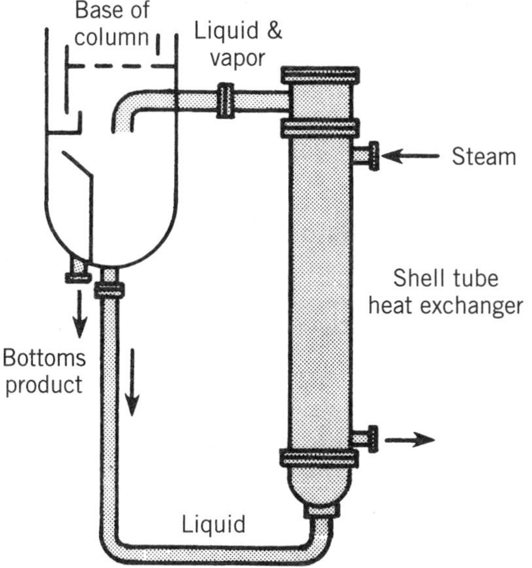

Figure 7. Typical shell and tube thermosyphon reboiler arrangement.

Condensers. Because most distillation column condensers are of shell-and-tube design, the processor has the option of condensing on either the shell or tube side. From the process point of view, condensation on the shell side is preferred because there is less subcooling of condensate and a lower pressure drop is required. These are important factors in vacuum duties. Furthermore, with cooling water on the tube side, any fouling can be removed more easily.

Tube-side condensation, on the other hand, can be more advantageous whenever process fluid characteristics dictate the use of more expensive, exotic materials. Capital cost of the unit then may be cut by using a carbon steel shell.

Preheaters/Coolers. The degree to which fluids are aggressive to metals and gasketing materials generally determines the selection of plate or shell-and-tube preheaters and product coolers. If fluids are not overly aggressive toward gasket materials, a plate heat exchanger is an extremely efficient preheater because a very close temperature approach may be achieved. Added economy is realized by using heat from the top and bottoms product for all necessary preheating.

Very aggressive duties normally are handled in a number of tubular exchangers arranged in series to generate a good mean temperature difference. The use of multiple tubular units obviously is more expensive than a single plate heat exchanger but is unavoidable for certain solutions such as aromatic compounds.

Vent Condenser. It is normal practice on distillation systems to use a vent condenser after the main condenser to minimize the amount of volatiles being driven off into the atmosphere. Usually of the shell-and-tube type, the vent condenser will have about one-tenth the area of the main unit and will use a chilled water supply to cool the noncondensible gases to about 45–50°F.

Pumps. Because most distillation duties involve fluids that are highly flammable and have a low flash point, it is essential that explosion-proof (Class 1, Group D, Division 1) pump motors be supplied. Centrifugal pumps generally are specified as they are reliable and can provide the necessary head and volumetric capacity at moderate costs.

PACKAGED DISTILLATION SYSTEMS

For distillation systems of moderate size, it often can be economical to fabricate and supply columns, heat exchangers, tanks, pumps, and other elements as a fully preassembled package. This technique was used for the distillation unit as pictured in Figure 8 during fabrication as a complete, ready to transport system. For this particular project, which was designed to separate both ethyl alcohol and isopropyl alcohol from water, three columns of relatively small diameter were positioned with associated components and instrumentation on a prefabricated structure. Although insulation normally would not be supplied in order to preclude the possibility of damage during shipment, major elements of this project were insulated at the cus-

Figure 8. Packaged distillation system during fabrication.

tomer's request and were trucked 600 miles to the job site without problems. Final size of the package was 65′ × 12′ × 8′.

The packaged approach proved to be beneficial in a number of ways. Factory fabrication and installation of piping was far more economical than field finishing, while erection time for the system was reduced considerably with significant savings in local labor costs.

An alternative approach for larger systems is the fabrication of equipment modules. For one large batch-distillation unit, the column itself can be supplied as one module, the batch tank and reboiler as the second, and the heat exchangers, decanter, pumps, and auxiliary items as the third. This modular construction before shipment was successful despite the fact that the batch tank was more than 8 ft in diameter.

For systems involving columns in excess of 6-ft diameter, prepackaging generally is not feasible. Components of such systems normally have to be installed and piped at the job site.

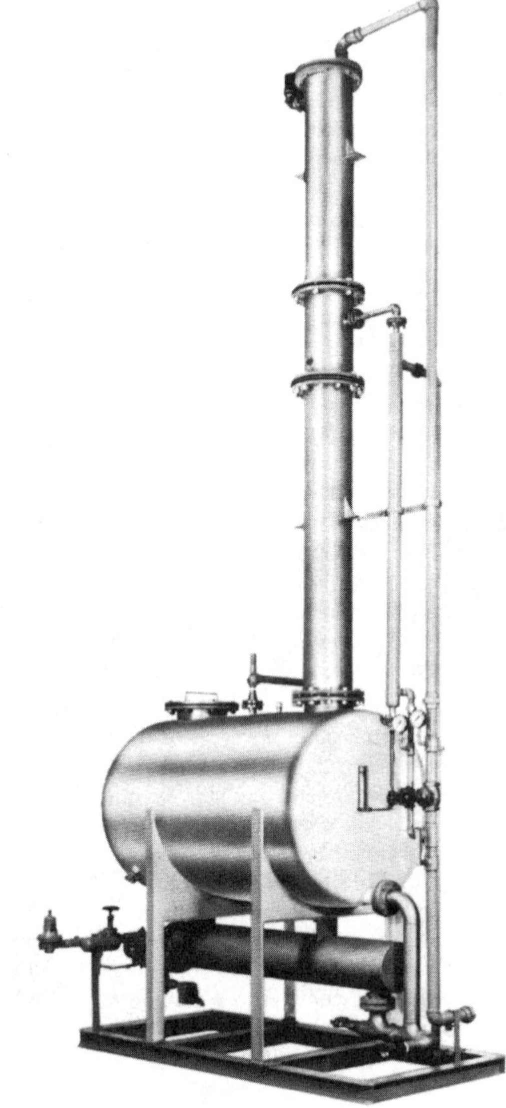

Figure 9. Batch distillation/solvent recovery silhoutte mask.

BATCH DISTILLATION

With process plants becoming increasingly larger, there has been a tendency for the chemical and petroleum industries to focus attention on the use of continuous distillation because this approach becomes progressively more economical as the scale of operation grows. As a result, batch distillation has become a somewhat neglected unit operation. There are, however, many areas where batch distillation can be used more economically than continuous. Batch sizes are available from about 300 to more than 5000 gal. A 300 gal unit is shown in Figure 9.

Advantages

The main advantage of batch distillation is its flexibility. A single unit can, by changing reflux ratios and boil-up rates, be used for many different systems. It also is possible to separate more than two components in the same column, whereas with continuous distillation, at least $N - 1$ separate columns are required to separate N components.

A batch distillation process is very simple to control, and there is no need to balance the feed and draw-off, as is the case with the continuous distillation approach.

Batch distillation is particularly useful when applied to feeds containing residues that have a tendency to foul surfaces as these residues remain in the still and cannot contaminate the rectification column internals. For aqueous systems-handling materials that leave very heavy fouling, it is possible to heat with direct steam injection into the still and thus alleviate problems of buildup on the heat transfer surfaces.

COLUMN DESIGN

Before designing a batch column, it obviously is desirable to have as much detailed information on the system as is possible. Data on vapor liquid equilibria (VLE), vapor and liquid densities, liquid viscosity, and the boiling temperatures of the components are essential if a column is to be properly designed. Failure to have accurate VLE data means that it is necessary to run a small-scale experiment in order to characterize the system. In addition, the customer must identify the product feed composition, the required composition of the residue and distillate, the batch size, and the batch time. Having acquired this data, the vendor finally can commence design work.

To illustrate the design principles, it is useful to study the case of binary mixtures. A typical system is shown in Figure 10. Once the VLE data have been established, it becomes a straightforward task to calculate the number of theoretical stages and reflux ratios.

There are two main techniques for operating a batch column. One is to work with constant reflux ratio during the complete run. The effect of this method is charted in Figure 11. As the composition of the more volatile component (MVC) in the still, x_w, decreases, the fraction of MVC in the top product decreases. For example, to obtain a set composition of 90% in the total amount of top product collected, it will always be necessary to collect initially at a higher composition of about 95% to compensate for a composition below specification at the end of the run. The advantage of constant reflux is that control and operation are very simple.

The second method is to increase the reflux ratio during the run in order to maintain a steady top-product composition. This is shown in Figure 12, where the increase in the slope of the operating line is obtained by increasing the reflux ratio. The gradient of the operating line (L/V) is obtained from the enrichment equation given below.

$$Y_n = \frac{L}{V} x_{n+1} + \frac{D}{V_n} x_d$$

This equation assumes constant molal overflow. Y_n is the mole fraction of the MVC in the vapor leaving the nth stage, x_{n+1} is the mole fraction of the MVC in the liquid arriving at the nth stage, x_d is the mole fraction of the MVC in the top product, D is the molal flow rate of top product, V is the molal vapor flow rate in the column, and L is the molal liquid flow rate in the column. The reflux ratio R is given by

$$R = \frac{L/V}{1 - (L/V)}$$

With the VLE data and the top and bottom compositions, it then is possible to calculate graphically by the

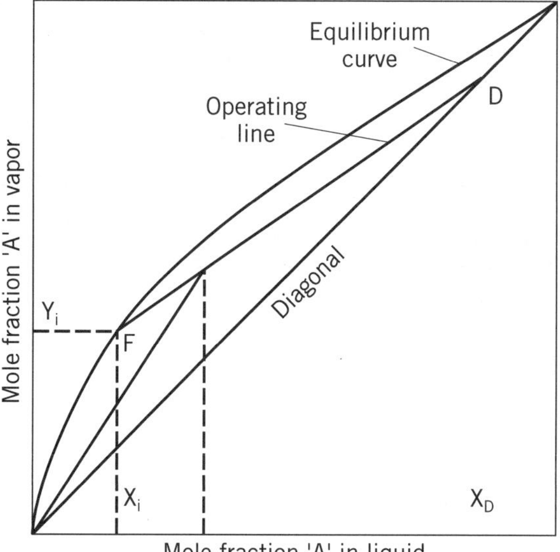

Figure 10. A typical system illustrating the design principles.

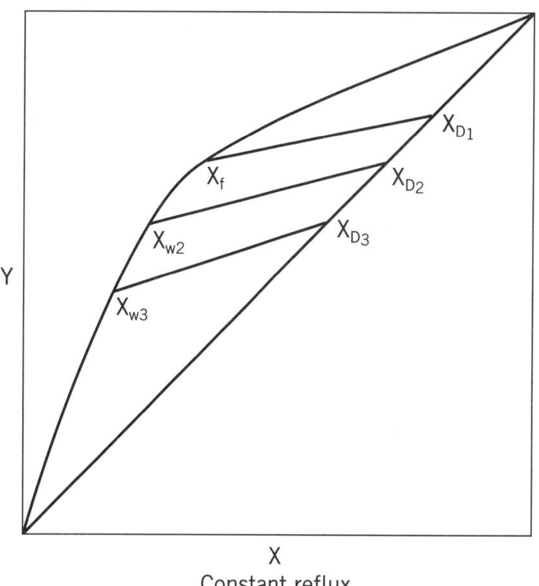

Figure 11. A chart illustrates the effect of the method of working with constant reflux ratio during the complete run. One of two

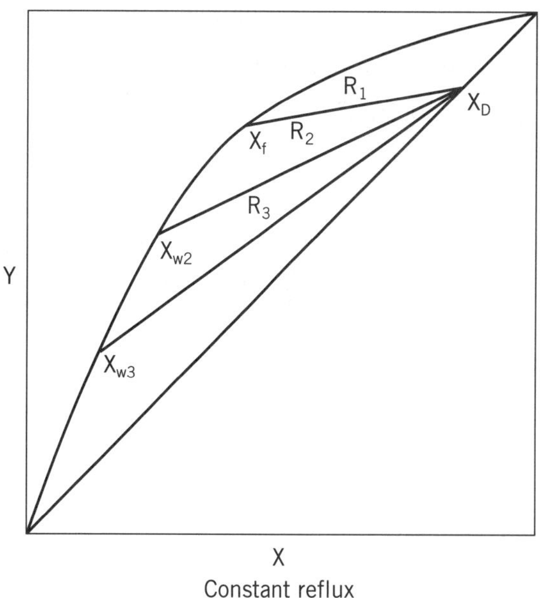

Figure 12. The second method is to increase the reflux ratio during the run in order to maintain a steady top product composition.

Table 1. Computer Program Modes

Mode	Specified	Program calculation
1	Top-product and still composition	Minimum number of theoretical stages Minimum reflux ratio
2	Top-product still composition, reflux ratio	Number of theoretical stages
3	Top-product and still composition Number of theoretical stages	Reflux ratio
4	Top-product composition required Initial and final still composition Still charge Batch time Number of theoretical stages	Reflux ratio at stat and end of batch Boiling rate Product quality
5	Average top-product composition Initial and final still composition Still charge Batch time Number of theoretical stages	Reflux ratio required constant throughout batch Boil-up rate Product quality

McCabe and Thiele procedure the minimum reflux ratio, minimum number of theoretical stages, and other such parameters. In batch distillation in particular, these procedures are tedious and time-consuming because, of course, the composition of the liquid in the still changes with time and it is necessary to repeat the calculations many times. Obviously, the procedures become even more time-consuming with multicomponent systems.

For this type of design work, computer programs can be used to enable the engineer to produce an efficient design very quickly. One program is extremely flexible and operates in a different series of modes, as is charted in Table 1. Naturally, the VLE data have to be specified for all modes. The program further assists in the determination of the number of theoretical stages. Other modes also are available to provide different permutations of operation.

To help determine the sizing of the column diameter, various programs can be used to incorporate different proprietary methods, including those of Fractionation Research Inc. (FRI) of California.

Column Internals

The choice of internals for the column depends mainly on the product being processed and the size of the column to be used. To meet virtually every parameter, packed columns as well as sieve, bubble, and ballast trays are available. As a general rule, sieve trays are not used frequently in batch columns because the turndown ratio of most trays of this type is only about 1.5 to 1. This reduces one of the main advantages of batch columns, mainly flexibility. Usually, small batch columns are packed because the efficiency of trays of less than 2-ft diameter often decreases rapidly. Ballast trays, although expensive, are often used for larger columns because they are efficient and have turndown ratios of up to 9 to 1.

Control

The control of batch columns is very simple, and, therefore, required instrumentation usually is quite inexpensive. Constant reflux operation can be handled by a ratio controller or by using a timed deflection of condensate. For operation with variable reflux, the control of the reflux ratio must be tied to some property of the top product that undergoes a sufficiently large change in value for change in composition. Manual control of this type of reflux operation is not feasible.

Numerous batch distillation systems have been supplied to customers who require the separation of many different components. To simplify the operation of these systems, there are proprietary computers specifically designed for process control. For example, use of certain microprocessors enables the operator to switch from one type of separation to another without having to make many manual adjustments on the control panel. In addition, the computer can be programmed to automatically take care of all necessary tails cuts as well as to clean the system during interims between the various separations.

Batch Systems/Continuous Distillation

In a number of duties, it is essential that the base product contain only very small amounts of one component. For example, in the recovery of solvents from water, it is mandated that the water may retain only trace quantities of solvents before being pumped to waste. Because one disadvantage of a batch system is that there is no stripping section, it is necessary to boil the batch pot for relatively long periods of time in order to reduce the residual solvent to trace quantities.

To resolve this problem, a number of batch distillation systems with the capability of being operated in the continuous mode have been supplied. By incorporating one, two, or three feed points on the column, the feed can be pumped directly to the column instead of to the batch tank. Furthermore, a separate small holdup/reboiler is furnished in certain cases so that the batch tank and its associated reboiler need not be used for the continuous operation.

BIBLIOGRAPHY

This article was adapted from and is used with permission of A. Cooper, *APV Distillation Handbook*, 3rd ed., Lake Mills, Wisc., 1987. Copyrighted APV Crepaco, Inc.

APV Crepaco, Inc.
Lake Mills, Wisconsin

DISTILLED BEVERAGE SPIRITS

HISTORY

The history of distilled spirits goes back into antiquity. Scientists have unearthed pottery in Mesopotamia depicting fermentation scenes dating back to 4200 B.C., a small wooden model of a brewery from about 2000 B.C. is on display at the Metropolitan Museum of Art in New York City, and Aristotle mentions a wine that produces a spirit. The first real distiller was probably a Greek-Egyptian alchemist who in the first or second century A.D., in an attempt to transmute base metal into gold, boiled some wine in a crude still. The discovery of ardent spirits that resulted from this effort was looked on with awe. It was kept a secret for centuries.

The technique of distillation probably came from the Egyptians who had been interested in alchemy since the pre-Christian era. At a later time the Arabians gained this knowledge from the Egyptians. Distillation was introduced into Western Europe either through Spain about A.D. 1150, or by the crusaders who learned about it from the Moslems in the 12th and 13th centuries. Distilled spirits were probably known in Ireland and Scotland before the 12th century, but actually it was not until then that there is a recorded history of distilled spirits in Europe. The first written evidence is in the description by Master Salernus who died in A.D. 1167. However, for another three centuries distilled spirits were regarded only as a rare and costly medicine called *aqua vitae*, "the water of life."

The first treatise on distillation was written by a French chemist, Arnold de Villeneuve, sometime before 1311, and was printed in 1478 in Venice. A Spaniard, Raymond Lully, was also instrumental in spreading the knowledge of distillation through Europe in the late 13th century. *Liber de Arte Distillandi*, by Hieronimus Brunswick, a well-known author of medical works, was printed in 1500 in Strasbourg. A more comprehensive book by Ryff, another medical author, appeared in 1556 in Frankfurt am Main. Both works contain elaborate chapters on herbs and their distillates with indications of their uses as medicines.

Monks had been producing fermented liquors on a substantial scale since A.D. 800, and were the first to practice the new art of distillation because churchmen were the only one capable of reading and understanding the treatises on distillation. Gradually the knowledge of *aqua vitae* spread widely, and during the 16th and 17th centuries the production of distilled spirits became a full-scale industry. The popularity of distilled spirits in Europe grew during this era and may best be traced through court records.

The word *whisky* comes from the Gaelic word *uisgebeatha*, or *usquebaugh*, as the Irish called it, meaning "the water of life." *Usquebaugh*, supposedly the Celtic form of whisky, was actually a cordial made with aniseed, cloves, nutmeg, ginger, caraway seeds, raisins, licorice, sugar, and saffron. The real whiskey of the Irish was called potheen, reputedly a formidable drink, full of heavy-bodied flavors resulting from simple distillations.

The early methods of distillation used the alembic, a simple closed container to which heat was applied. The vapors were transferred through a tube to a cooling chamber in which they were condensed. Alembics made of copper, iron, or tin were preferred. Other metals were said to have an adverse effect on the distillate. Occasionally, glass and potter's earth were used. From this developed stills consisting of clay or brick fireboxes into which the copper pots of the still were fitted. Direct heat was applied to the body or cucurbit containing the fermented mixture, and the vapors rose into the head, passed through a pipe (the crane neck) to the worm tub, where a copper coil was immersed in a barrel of cool water. Variations and improvements of this technique resulted in pot stills, some of which are still in use in the production of malt whiskies, notably in Scotland, and brandies, in France.

During the latter part of the 18th century, great strides were made in the development of distillation equipment, mostly in France. Because brandy (from fruit) was the principal distillate in France, the French had an advantage over the British and Germans, who were both hampered by the thick mashes obtained from grains. Argand invented the preheater, and in 1801, Edouard Adam designed the prototype of a charge still. He used egg-shaped vessels to hold the alcoholic liquid through which vapors from the kettle passed to the condenser. In the early 19th century several stills were patented in France and England, and in 1830 Aeness Coffey, in Dublin, developed his continuous still. The Coffey still (Fig. 1) is composed of two columns: in one, the fermented mash is stripped of its alcohol and, in the other, the vapors are rectified to a high proof (94–96%). Almost all of the fundamentals of distillation had been recognized and incorporated into the Coffey still. Later developments do not differ fundamentally from the stills of Cellier-Blumenthal and Coffey.

The early history of distillation in the United States is as poorly documented as it is elsewhere. No one knows who made the first rum, colonial New England's most important product; no one knows when or where the first rye whisky was made, or with any certainty who made the first bourbon in Kentucky. The Indians may possibly have known distillation. They had fermented drinks from maple syrup, corn, ground acorns, chestnuts, and chinquapins. Columbus reported an Indian drink made from the marrow of maguey. The fermented sap of maguey makes pulque, and distilled pulque becomes mezcal. Whether or not Columbus tasted mezcal is conjectural. However, the Spanish and Portugese brought the still.

The first recorded beverage spirits made from grains (corn and rye) were distilled on Staten Island in 1640 by William Kieft, the director general of the Dutch Colony of New Netherland. During the early colonial days, fermented beverages made from sugar-bearing fruits and vegetables were very popular. Pumpkins, maple sugar, parsnips, peaches, pears, apples, currants, grapes, and elderberries provided a ready source. The more aristocratic colonists preferred imported wines. However, in the 18th century colonial drinking customs changed to distilled drinks, especially rum. Rum was made in Barbados as early as 1650. In colonial America, the earliest reference is in the records of the General Court of Massachusetts in May 1657. By 1750, there were 63 distilleries. One of the first acts of the Continental Congress was to establish a

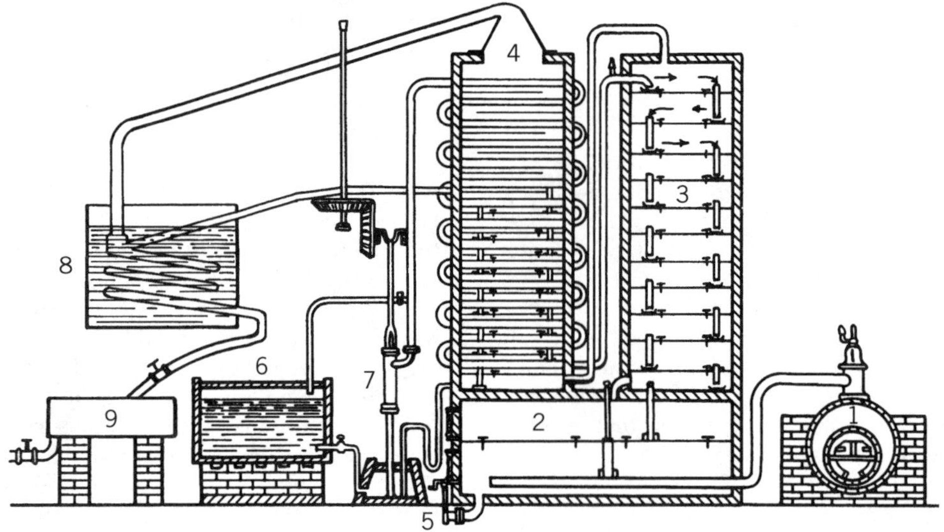

Figure 1. Coffey still. 1, Boiler; 2, spent mash chamber; 3, stripping column (analyzer); 4, rectifying column; 5, residual mash (stillage) outlet; 6, fermented mash (feed); 7, feed pump; 8, condenser; 9, product tank.

ration of rum for soldiers and sailors. Brandy, which had been made in the colonies as early as 1650, and gin, which was popular with the poorer classes in England, never achieved the popularity of rum in this country. Nevertheless, rum was not a native drink, because the raw materials had to be imported. But rum starts with an inexpensive by-product of the sugar industry, molasses, and so could be the least-expensive distilled spirit. There are sufficient sugars left after most of the cane sugar has been removed to serve as the source of alcohol for rum.

With the decline of the three-cornered trade, the revolution, and the movement of settlers from the coastal areas inland, the rum industry slowly declined. Americans turned to rye, a successful crop in Pennsylvania and Maryland, and to corn, which grew in the South and West. George Washington is reported to have made rye whisky. His distillery had an excellent reputation for making fine liquor under the supervision of James Anderson, a Scotsman. In 1798, Washington's net profit was £83 with an inventory of 587 L (155 gal) still unsold.

Canada's first distillery produced rum from molasses in Quebec in 1769, but whisky did not develop until the 19th century. By 1850, there were some 200 distilleries operating in Ontario alone. Today there are only 27 of which 2, Joseph E. Seagram & Sons, Ltd., and Hiram Walker & Sons, Ltd., have been in business continuously since 1857 and 1858, respectively.

The Whisky Rebellion in 1791 resulted in the rise of Kentucky as the greatest whisky-producing state in the union. In 1810, Kentucky which had achieved statehood in 1792 with over 70,000 inhabitants, many of them Scotch-Irish and German farmer immigrants with distiller experience, had 2,000 stills out of 14,000 in the United States producing well over 7,500,000 L (2 million gal) of whisky annually. Many counties of Kentucky claim the first production of bourbon. However, it is generally agreed that the first bourbon whisky (genuine, old-fashioned, handmade, sour-mash bourbon) was made by a Baptist minister, the Reverend Elijah Craig, in 1789 at Georgetown in Scott County which was part of Bourbon County, one of the Virginia counties that made up the Kentucky territory, from which Craig's corn whisky got its name. In order to distinguish this corn whisky from Pennsylvania rye, it was called Kentucky bourbon, with a mashing formula of at least 51% corn grains. Until 1865, few Kentucky distilleries produced more than 160,000 L (1,000 bbl) per year. Most of them were small, with an annual production of only 8,000–80,000 L (50–500 bbl).

The method used in Kentucky of making whisky, with the exception of malt preparation, was in most respects similar to that used in Scotland and Ireland for hundreds of years. Ground corn and rye meal were scalded in tubs somewhat larger than barrels, stirred with paddles, and allowed to cool and sour overnight. Then malt made from rye, corn, or barley was added for conversion of the starch to a fermentable grain sugar. Yeast was added and the mash allowed to ferment for 72–96 h. A simple, single-chambered copper still was used to separate the spirits from the mash. When redistilled, the product was called double distilled. In 1819, New Orleans received some 750,000 L (200,000 gal) per month of this product. By the time it had floated down the Ohio and Mississippi on flatboats, the hot sun and boat movement had aged it.

Before the Civil War, not much attention was paid to aging, even though it was recognized that whisky left in charred oak barrels took on a golden color and some mellowness. Originally, whisky was sold in its natural white state or artificially colored to resemble the amber glint of brandy. No one bothered with brand names, whisky was whisky, as everyone knew, and not too much was made of the wide variation in palatability. The hunter or riverman who drank raw white whisky was not particular about quality.

After the Civil War, the rise in taxes made storage in bond desirable. Many family names, always important in Kentucky lore, now became associated with distinctive whiskies. As the industry grew, it engaged in bitter controversy over what was whisky. As a result, in 1909 during President Taft's administration, whisky was finally defined as any volatile liquor distilled from grain, and standards of identity based on current manufacturing processes for various whisky types, such as rye and bourbon, were established (1). In 1906, the Pure Food and Drug Act was passed requiring a statement of manufacturing process and materials on the label. By 1920 when prohibition began, blended whiskies comprised 70% of the whisky market.

Prohibition brought with it evils that were greater than those it was designed to prevent. During Prohibition, consumption of spirits increased from 530 to 750 million L (140 to 200 million gal) annually. In 1930, the Prohibition Enforcement Bureau estimated that the production of moonshine was more than 3 billion L (800 million gal) per year. In December 1933, Prohibition was repealed and the distilling industry began to catch up with developments in bacteriology, chemistry, engineering, and sanitation. Many of the small producers were absorbed by the major organizations, and capital was provided for new equipment and inventory.

Per capita consumption continued to decline; it was 10.26 L (2.71 gal) in 1864, 6.47 L (1.71 gal) in 1922–1930, and only 4.96 L (1.31 gal) in 1960. However, the growth in population accounted for an increase in total consumption from 550 million L (145 million wine gal) in 1940 to over 1.6 billion L (420 million gal) in 1976. In World War II the industry offered its facilities to the government for war production, even before the United States entered the war. From Pearl Harbor to V-J Day, almost 3 billion L (750 million gal) of 190° proof (95%) alcohol was produced for the war effort in 127 distilling plants.

TAXATION, GOVERNMENT REGULATIONS

The distilling industry has always been taxed heavily, not only in the United States, but throughout the world. On the one hand, governments have laid an unduly heavy burden on the legal producers of distilled beverage spirits. On the other hand, dishonest men have evaded legal obligations and brought the industry into disrepute. Moreover, in many instances, dry interests have attempted to further their cause by advocating high taxes. For example, in England, by 1730, the laws were so complicated and onerous that they all but destroyed the industry, and in 1743 Parliament completely revised the regulations. Whenever taxes were too high, illicit distilling flourished.

Alcoholic beverages were subject to regulations as early as the Babylonian Code of Hammurabi, ca 2000 B.C., which contained provisions for the quality, sale, and use of fermented liquors. In the Magna Carta, a clause provided a standard of measurement for the sale of ale and wine. With the increase in consumption of distilled spirits in Europe, governments became increasingly interested in the tax revenues, and in 1643, Parliament imposed the first tax on distilled spirits.

The first liquor tax in the United States (2 guilders on each half-vat of beer) was imposed in 1640 by William Kieft, director general of New Netherland. The Molasses Act of 1733 and the Sugar Act of 1763, imposing a heavy tax on French and Dutch rum and molasses and leaving the more expensive British products free, provoked protest and action against its enforcement.

In 1791, Alexander Hamilton, in testing the strength of the new federal government, imposed an excise tax to be collected by revenue officers assigned to each district. This tax was fiercely contested and was repealed in 1800 during Jefferson's administration and, except for the years between 1812 and 1817, as a war measure, there was no further excise tax on domestic beverage spirits until 1862.

In 1975, the combined federal excise tax and the average state tax (32 licensed states) amounted to $13.13 per proof gal; federal taxes alone, $10.50 per proof gal. Table 1 lists the federal excise tax rates for various years. The federal excise tax was raised in 1991 to $13.50 per proof gal, ie, a gallon of liquid containing 50% by volume of ethyl alcohol (100° proof). Prior to that, and since 1985, it had been $12.50. per proof gal.

Even though the metric system for containers has been adopted, determination of the excise tax remains on a tax-gallon basis. The Bureau of Alcohol, Tobacco, and Firearms has established an official conversion factor: 1 L = 0.26417 U.S. gal (Treasury Decision A.F.F. 39, Jan. 21, 1977).

In the United States, the revenue from excise taxes on distilled spirits has become a substantial part of income realized by the three levels of government: federal, state, and local. In 1863, with a $0.20 rate, the federal government collected over $5 million; in 1960, $3,090 million; and in 1975 the total revenue amounted to $6,277 million, of which 60% represented the federal government's share, collected at a rate of $10.50 per tax gallon. In 1997, the federal government collected $3.6 billion from distilled spirits and $3.4 billion from beer and $0.6 billion from wine. In addition, the states collected $1.9 billion from beer alone. State and local governments collected $7.5 billion in 1988, of which $3.1 billion came from distilled spirits and $4.4 billion from wine and beer.

Supervision over the production of distilled beverage spirits is maintained by the Bureau of Alcohol, Tobacco,

Table 1. Federal Excise Tax Rates on Distilled Spirits

Year	Rate[a] (dollars)	Year	Rate[a] (dollars)
1812–1817	0.02 (0.09)	1934 (Jan.)	0.53 (2.00)
1862	0.05 (0.20)	1938 (July)	0.59 (2.25)
1846 (July)	0.16 (0.60)	1940 (July)	0.79 (3.00)
1864 (Dec.)	0.53 (2.00)	1941 (Oct.)	1.06 (4.00)
1894	0.29 (1.10)	1942 (Nov.)	1.59 (6.00)
World War I	0.61 (2.30)	1944 (Apr.)	2.38 (9.00)
1919	1.69 (6.40)	1951 (Nov.)	2.77 (10.50)
1933 (Dec.)	1.29 (1.10)	1985 (Oct.)	3.30 (12.50)
		1991 (Jan.)	3.56 (13.50)

[a]Per liter (gal).

and Firearms, Department of the Treasury, which succeeded the Alcohol and Tobacco Tax Division (ATTD) of the Internal Revenue Service in July 1972. (The ATTD was called the Alcohol Tax Unit from the repeal of prohibition to 1952.) This organization's enforcement agents are charged with the responsibility of eliminating illicit operations.

Prior to Prohibition, revenue agents and deputy collectors investigated illegal liquor operations, made arrests, seizures, etc, along with income tax and other miscellaneous work. On January 16, 1920, the effective date of Prohibition, federal prohibition agents took over the duties. An act of Congress created the Bureau of Prohibition in the Treasury Department on April 1, 1927. The Prohibition Reorganization Act, effective July 1, 1930, terminated the Bureau of Prohibition and created the Bureau of Industrial Alcohol in the Treasury Department, responsible for the permissive provisions of the act, and the Bureau of Prohibition in the Department of Justice, responsible for the enforcement of the penal provisions of the National Prohibition Act. These two were merged on December 6, 1933, as the Alcohol Tax Unit, Internal Revenue Department, and on March 15, 1952 it was named the Alcohol and Tobacco Tax Division of the Internal Revenue Service.

PRODUCTION AND CONSUMPTION

In 1810, according to government records, Pennsylvania produced 24.5 million L (6.5 million gal); Indiana 83,000 L (22,000 gal); and Kentucky more than 7.5 million L (2 million gal). In 1917, apparently in anticipation of Prohibition, over 1.1 billion L (300 million gal) of beverage spirits were produced, of which 225 million L (60 million gal) were whisky. In the year 1930, the Prohibition Enforcement Bureau estimated that the illicit production amounted to 3 billion L (800 million gal). Table 2 shows distilled spirits entering trade channels from 1955 to 1995.

STANDARDS OF IDENTITY

Distilled alcoholic beverages are usually characterized by their geographical origin, type of material used in production, and standard of quality, as evaluated by organoleptic analysis (taste and bouquet). Secretiveness was a way of life for ancient distillers; producers today operate according to their own methods. As it is not possible to correlate taste and bouquet with chemical composition, evaluation becomes a matter of organoleptic analysis.

Over the centuries legends, traditions, and to some extent political influences were important factors in the production and identification of potable distilled liquors. Since most governments had an economic interest because of the great source of revenue realized, they established standards for distilled alcoholic beverages, generally in keeping with the customs prevailing in their country. As a result, geographical identification has become accepted, and each country respects the identity and exclusiveness of the other's products.

Within each category of product wide variations in flavor can be caused by (*1*) types of material and their proportions, (*2*) methods of preparation, (*3*) selection of yeast types, (*4*) fermentation conditions, (*5*) distillation processes, (*6*) maturation techniques, and (*7*) blending. Since the alcoholic and water components are insignificant factors in flavor intensity or palatability, distillers are primarily interested in the more flavorful constituents, the so-called congeners, substances that are generated with the alcohol during the fermentation process and in the course of maturation. To produce a palatable product, it is, therefore, necessary to select the proper configuration of these constituents (congeners). In consideration of items 1–7 above, this cannot be accomplished by production techniques alone; therefore, the majority of alcoholic beverages are blended to provide uniformity, balanced bouquet, and palatability. To illustrate the variations in flavor constituents an analysis of some types is given in Table 3.

Whiskies

Although brandy, rum, and gin are substantial items in world markets, whiskies are by far the leading distilled alcoholic beverages, with those from Canada, Scotland, and the United States accounting for most of the sales. Irish whiskey, a distinctive product of Ireland, although not enjoying a large volume of sales, does have good distribution as a specialty item. The Irish use the spelling whiskey, the Scotch and Canadians use whisky, and U.S. citizens use both, although U.S. regulations use the spelling whisky.

Canadian. Canadian whisky is manufactured in compliance with the laws of Canada regulating the manufacture of whisky for consumption in Canada and containing no distilled spirits less than three years old. Canadian whiskies are premium products usually bottled at six years of age or more, and because they are blended, they are not designated as straight whiskies. They are light bodied and, although delicate in flavor, they nevertheless retain a distinct positive character. The Canadian government exercises the customary rigid controls in matters pertaining to labeling and in collection of the excise tax. However, it sets no limitations as to grain formulas, distilling proofs, or special types of cooperage for the maturation of whisky.

The major cereal grains (corn, rye, and barley malt) are used, and their proportions in the mashing formula remains a distiller's trade secret; otherwise, the process is substantially the same as is found in the distilleries of the United States. Because Canadian distillers are not faced with artificial proof restrictions in their distillation procedures, they are able to operate batch and continuous distillation systems under conditions that are optimum for the separation and selection of desirable congeners.

White oak casks (189 L, 50 U.S. gal) are used in the maturation process. A substantial amount of Canadian whiskies are stored in preused cooperage from other whiskies, which should lower the cost. Again, the proportions of new and matured cooperage used for maturation are each distiller's trade secret.

Scotch. Although Scotch whisky has enjoyed a worldwide reputation for its unique smoky flavor and high stan-

Table 2. U.S. distilled spirits entering trade channels, including exports and imports in 1000 wine gallons

	in 1000 wine gallons								
	1955	1960	1965	1970	1975	1981	1985	1990	1995
				Domestic whiskey					
Bonded	12,868	9,394	7,482	6,004	3,993	2,276	1,128	721	327
Straight	44,838	58,939	69,417	77,895	69,531	63,288	48,933	44,193	39,143
Blend of straights	1,249	792	577	751	1,855	1,789	952	3,131	3,154
Blend of netural spirits	81,493	74,074	74,054	—	674	35,398	26,571	18,640	13,869
Other	(−448)	(−118)	—	73,936	58,484	4,153	3,747	878	1,322
				Imported whiskey					
Scotch	12,284	20,584	32,523	51,911	55,275	47,656	40,271	27,889	20,902
Canadian	9,157	12,551	19,636	33,535	48,425	54,639	49,788	54,607	36,240
Irish and others	13	78	104	231	271	821	632	787	535
Belgian	—	—	141	—	—	—	—	—	—
Other	—	72	2	—	—	—	—	—	—
				Gin					
Domestic	20,446	22,000	29,627	35,122	38,916	38,295	31,494	26,836	24,153
Imported	291	1,149	2,498	3,772	4,817	5,872	6,139	5,326	5,045
				Rum					
Puerto Rican	1,856	2,611	4,650	9,049	13,601	26,881	27,077	29,962	18,564
Virgin Island	16	112	136	248	1,314	3,122	3,084	2,739	3,457
Other domestic	662	804	1,625	2,909	1,936	1,218	2,342	2,580	5,902
Imported	181	219	108	141	706	1,357	1,526	2,572	1,784
				Brandy					
Domestic	3,726	5,300	8,263	11,795	12,388	14,541	14,470	13,353	11,642
Imported	893	1,163	1,142	1,764	3,035	6,769	7,862	6,391	6,332
				Vodka					
	6,968	19,405	31,157	50,172	86,775	100,617	98,672	91,995	91,996
				Cordials					
Domestic	6,173	9,156	14,755	20,505	26,323	26,059	32,039	26,881	26,952
Imported	455	812	1,247	2,405	4,769	13,704	16,176	16,274	18,107
Totals	*203,306*	*239,373*	*300,116*	*384,453*	*448,550*	*465,511*	*431,447*	*403,358*	*364,238*

Table 3. Congeneric Content of Various Types of Distilled Alcoholic Beverages, mg/L (ppm) at 80° proof (40%)

	Whisky					
				Bourbon		
Component	U.S.	Canadian	Scotch	Straight	Blended	Cognac brandy
Fusel oil[a]	664	464	1,144	1,624	1,560	1,544
Total acids[b]	240	160	120	552	540	288
Esters[c]	136	112	136	448	344	328
Aldehydes[d]	21.6	23.2	36	54.4	43.2	60.8
Furfural	2.6	0.88	0.88	3.6	7.2	5.4
Total solids	896	776	1,016	1,440	1,272	5,584
Tannins	168	144	64	416	384	200
Total congeners, wt/vol %	0.116	0.085	0.160	0.292	0.309	0.239

Note: Determinations were made according to the official methods of analysis of the Association of Office Agricultural Chemists.

[a]Determined by the Komarowsky colorimetric method.

[b]As acetic acid.

[c]As ethyl acetate.

[d]As acetaldehyde.

dard of quality, consumers know very little about it. Not much information is revealed by government regulations, which only specify the use of cereal grains and a minimum requirement for aging in oak casks; in Britain,

> spirits described as Scotch Whisky shall not be deemed to correspond to that description unless they have been obtained by distillation in Scotland from a mash of cereal grain saccharified by the diastase of malt and have been matured in warehouse in cask for a period of at least three years.

In the United States,

> Scotch Whisky is a distinctive product of Scotland, manufactured in Scotland in compliance with the laws of Great Britain regulating the manufacture of Scotch Whisky for consumption in Great Britain, and containing no distilled spirits less than three years old.

This minimum age requirement is greatly exceeded by the Scotch distillers. For example, nothing under 4 years of age is included in their exports to the United States, and for the most part, 6-, 8-, 10-, and 12-yr minimum ages are featured in their brands. Most Scotch brands are blends of grain whiskies and numerous distinctive malt whiskies are produced by more than 100 distilleries in four major areas of Scotland. Malt whiskies are characterized by their location.

Even though there are many distilleries and, no doubt, slight variations in their production methods (that is how single whiskies acquire the characteristics attributable to each specific plant) there are definite processes generally used. A basic knowledge of these traditional methods, still in use today, is needed to fully appreciate the quality concept inherent in Scotch whisky.

As in Canada, no government limitations are placed on production and maturation techniques. The Scotch distillers are guided by their production experience developed over many centuries and by consumer reactions.

The outstanding taste characteristic of Scotch, its subtle smoky flavor, is due to the techniques used in the production of malt whiskies. Malted barley, dried over peat fires, is the only grain ingredient in the mash. The kind and the amount of peat used in the fires determines the intensity of flavor in the final product. The aroma of the burning peat, known as peat reek, is absorbed by the barley malt. This smoky flavor is carried through to the final distillate and becomes a characteristic of single malt whisky. Peat is heather, fern, and evergreen that have been subjected to aging and compression processes over the centuries.

The dried malted barley is ground to a grist and allowed to hydrolyze in a mash tub. After conversion is completed, the liquid portion, or wort, is drained off, cooled, and placed in a fermenter. After fermentation, the separation of malt whisky from the fermented wort takes place in a batch distillation system, a copper kettle with a "worm," or spiral tube, leading from its head. The size and shape of these pot stills exert a definite influence on the character of the whiskies. Another critical factor is the selection of the product during that portion of the distillation cycle that will produce the desired flavors. The first portion in the cycle is referred to as foreshots and the last as feints (heads and tails in the United States). The middle portion, after further distillation, becomes high wines and is subsequently reduced to maturation proof for storage in oak casks. The final distillation proof is in the 140–160° (70–80%) range.

The grain whiskies used in Scotch brands are produced in a manner similar to the production techniques used in Canada and in the United States. Corn (referred to in the UK as maize), rye, and barley malt are the ingredients. The proportions again depend on the individual distiller. Because delicate flavors are desired, the distillation proof is around 180–186° (90–93%). The distillation system is basically a Coffey still composed of two columns.

The grain whiskies are generally aged in matured oak casks of 190-L capacities not unlike U.S. and Canadian barrels. Some malt whiskies acquire other distinctive qualities by being matured in oak casks that were previously used for sherry.

Whereas the materials, geographic location, and production processes are responsible for the uniqueness of Scotch whisky, it is the skill of the blender that achieves the quality of the final product. As many as 20 and sometimes more, different malt and grain whiskies are married to produce a brand. Of course, the formulae are well-guarded secrets.

Irish. Irish whiskey is manufactured either in the Republic of Ireland or in Northern Ireland in compliances with their laws regulating the manufacture of Irish whiskey for home consumption and containing no distilled spirits less than three years old. Like Scotch, Irish whiskeys are blends of grain and malt whiskeys. Unlike Scotch, Irish whiskey does not have the unique smoky taste because the barley malt is not dried over peat fires. Because whiskey has been known in Ireland as long as in Scotland, it is not surprising that their production techniques are closely related. One variation, however, is the use of some small grain, mostly barley, in addition to barley malt in the production of malt whiskey. The production process involves four basic steps: brewing (mashing), fermentation, distillation, and maturation. Mashing takes place in a Kieve (mash tun), a circular metal vessel with two bottoms; the upper is perforated, which permits the wort (converted grain starch or maltose) to be filtered to the underback and sent on to the washbacks (fermenters) where the inoculation with and action by yeast produces the whiskey. The wash (fermented mash) is then distilled in a pot still (batch), producing a low wine of about 100° proof (50%), which in turn is redistilled to produce first shots (fore shots) from which the final product or whiskey of about 140–150° proof (70–75%) is distilled. Irish whiskey brands are generally considered to be more flavorful and heavier bodied than Scotch blended whiskies.

United States. Distilled spirits for beverage purposes in the United States are characterized specifically as to type, materials, composition, distillation proofs, maturation proofs, storage containers, and the extent of the aging period. The federal government also requires that a detailed statement of the production process be filed for each type and any subsequent improvements or changes must be filed and approved before being placed into operation. In addition, a generalized application of the regulations is made to establish the identity of products where the intensity of flavor may not conform to an arbitrary organoleptic evaluation based on chemical analysis. As a result, in spite of extensive manufacturing facilities and know-how available for the production of a wide range of whiskies, the distiller is restrained within narrow limits and does not enjoy the degree of latitude available to the Canadian and Scotch distillers. U.S. regulations, by the Bureau of Alcohol, Tobacco, and Firearms of the Treasury Department, are very specific and more limiting than those of Canada or the UK. Title 27, *Code of Federal Regulations*, Subpart C, paragraph 5.22 *et seq.* sets the standards of identity for all distilled spirits. Those that are a factor in the U.S. market are given here.

> *Whisky* is an alcoholic distillate from a fermented mash of grain produced at less than 190° proof in such manner that the distillate possesses the taste, aroma, and characteristics generally attributed to whisky, stored in oak containers, and bottled at not less than 80° proof.
>
> *Neutral Spirits* or *alcohol* are distilled spirits produced from any material at or above 190° proof.

The requirement of distilling above 190° proof (95%) gives a distiller an opportunity to take advantage of the physical relationship that exists between water, alcohol, and congeners that are produced during fermentation. The sophisticated equipment utilized by progressive distillers permits a technique known as selective distillation, whereby the distiller can remove all congeners or retain those that are deemed desirable.

> *Grain spirits* are neutral spirits distilled from a fermented mash of grain and stored in oak containers.

Because grain spirits have delicate flavors, they must be stored in oak barrels previously seasoned through the storage of whisky or grain neutral spirits. It is important that the barrel is compatible with the flavor intensity of the grain neutral spirits; otherwise, the woody character of the barrel would overwhelm the light flavor of the grain neutral spirits and thus prevent proper development during aging. Grain neutral spirits are produced on continuous and batch distillation systems. Each produces a number of distillates having a low flavor intensity that, when stored in barrels, develop in flavor in the same manner as whisky (4).

The batch system, related to the pot still, is simpler in concept and offers an opportunity to use the heart-of-the-run principle for the production of grain neutral spirits. Sometimes referred to as a time-cycle distillation system, it involves three stages: (*1*) the heads (aldehydes) are removed, (*2*) the product is removed, and (*3*) the residual distillate (tails) remaining in the kettle is removed for subsequent redistillation in the continuous system. The batch system is composed of a large kettle with a capacity of up to 190,000 L (50,000 wine gal) with a vapor pipe leading to a product-concentrating column having as many as 55 bubble-cap plates. The large capacity of the kettle is important in maintaining product uniformity. Straight whisky, produced in the normal manner in the whisky column, is pumped into the kettle, indirect steam heat is applied through a coil within the kettle, and the alcoholic vapors rise into the product column where they are refined. The grain neutral spirits thus produced are reduced in proof with deionized water to between 110 and 160° proof (55 and 80%), put in oak barrels, and placed in government bonded warehouses for storage.

> *Vodka* is neutral spirits so distilled, or so treated after distillation with charcoal or other materials, as to be without distinctive character, aroma, taste, or color.

This definition has come under fire recently because some vodkas on the market claim to have some flavor. Since the 1950s drinking patterns in the United States have become more diversified. Vodka has moved up from negligible sales in 1949 to 36% in 1996. The fact that it can be mixed with any flavored substance seems to be the reason for its wide acceptance.

> *Rye whisky, bourbon whisky, or malt whisky* is whisky produced at not more than 160° proof from a fermented mash of not less than 51% rye, corn, or malted barley respectively, and stored at not more than 125° proof in charred new oak containers.
>
> *Corn whisky* is whisky produced at not more than 160° proof from a fermented mash of not less than 80% corn. It may or may not be stored in oak containers, but only in used or uncharred ones.

In producing bourbon whisky, eg, the mashing formula must contain at least 51% corn and the remaining ingredients (49%) generally are proportioned between rye grains and barley malt. Each distiller selects a preferred formula. Very little bourbon with 49% small grains is produced. The most popular proportions are 60% corn, 28% rye, 12% barley malt (referred to as 40% small grains), 70-18-12 (30% small grains), and 75-13-12 (25% small grains). Although some bourbons use as much as 15% barley malt, the general practice in the industry is to use 12% barley malt for all bourbon production.

Because rye and barley malt produce more intensive flavors than corn, the formula with the greater small-grains proportion will produce a bourbon with more body, provided, of course, that the same distillation techniques are used.

> *Straight whisky* may be any of the whiskies in the preceding two paragraphs that have been stored in the prescribed oak containers for two years or more. A straight whisky may further be identified as *bottled in bond*, provided it is at least four years of age, bottled at 100° proof (50%), and distilled at one plant by the same proprietor. A bottled-in-bond whisky may contain homogeneous mixtures of whiskies, provided they represent one season, or if consolidated with other seasons, the mixture shall be the distilling season of the youngest spirits contained therein, and shall consist of not less than 10% of spirits of each such season.

Light whisky is whisky produced in the United States at more than 160° proof, and stored in used or uncharred new oak containers; and also includes mixtures of such whiskies. If light whisky is mixed with less than 20% by volume of 100° proof (50%) straight whisky the mixture shall be designated Blended Light Whisky.

Blended whisky is a mixture which contains straight whisky or a blend of straight whiskies at not less than 20 percent on a proof gallon basis, and, separately, or in combination, whisky or neutral spirits. A blended whisky containing not less than 51 percent on a proof gallon basis of one of the types of straight whisky shall be further designated by that specific type of straight whisky; for example, "blended rye whisky."

Scotch whisky is whisky which is a distinctive product of Scotland, manufactured in Scotland in compliance with the laws of the United Kingdom regulating the manufacture of Scotch whisky for consumption in the United Kingdom: such product is a mixture of whiskies, such mixture is "blended Scotch whisky."

Irish whisky is whisky which is a distinctive product of Ireland, manufactured either in the Republic of Ireland or in Northern Ireland, in compliance with their laws regulating the manufacture of Irish whisky for home consumption: if such product is a mixture of whiskies, such mixture is "blended Irish whiskey."

Canadian whisky is whisky which is a distinctive product of Canada, manufactured in Canada in compliance with the laws of Canada regulating the manufacture of Canadian whisky for consumption in Canada: if such product is a mixture of whiskies, such mixture is "blended Canadian whisky."

Gins

Gin is a product obtained by original distillation from mash, or by redistillation of distilled spirits, or by mixing neutral spirits, with or over juniper berries and other aromatics, or with or over extracts derived from infusions, percolations, or maceration of such materials, and includes mixtures of gin and neutral spirits. It shall derive its main characteristic flavor from juniper berries and be bottled at not less than 80° proof.

France de La Boe, a 17th-century professor of medicine at Leyden University, The Netherlands, is credited with being the originator of the botanical-flavored spirits known as gin. Because his product's primary flavor was due to the essential oils extracted from juniper berries, he gave it the French name *jenièvre*, which appeared later as the Dutch *geneva* and finally as the English *gin*. Gin produced exclusively by original distillation or by redistillation may be further designated as distilled. Gin derives its main characteristic flavor from juniper berries. In addition to juniper berries, other botanicals may be used, including angelica root; anise; coriander; caraway seeds; lime, lemon and orange peel; licorice; calamus; cardamom; cassia bark; orris root; and bitter almonds. The use and proportion of any of these botanicals in the gin formula is left to the producer, and the character and quality of the gin depends to a great extent on the skill of the craftsman in formulating the recipe. The more skilled producers formulate their aromatic ingredients on the basis of the essential oil content in the raw materials to assure a greater degree of product uniformity.

To expose the essential oils, the ingredients are reduced to a granular form and then immersed directly into the kettle (pot), which is filled with grain neutral spirits at approximately 100° proof (50%). A vapor-phase extraction may also be used. In this case, the botanical mixture is placed on trays or in baskets in the head of the kettle where the alcoholic vapors passing by extract the essential oils and rise to the condenser.

It is important that the grain spirits be as neutral as possible (devoid of congeners) to avoid undesirable flavors. In addition to the kettle, some gin stills have a refinement section (as many as six plates) above the kettle for flavor stability and enrichment. Indirect steam heat is applied and the various essential oils are distilled over during the entire distillation cycle. The first (heads) and last (tails) portions of the cycle are not included in the product. Only the heart of the run is used, representing approximately an 85% recovery of the original alcohol concentration and varying with the type of product desired. Some distillers, to avoid thermal decomposition of the delicate flavors and to acquire a degree of softness, conduct the distillation under reduced pressure at a temperature of about 57°C. London Dry Gin, for example, is produced in this manner.

British and Canadian regulations permit and recognize the use of maturation techniques for gin. Gins stored in special oak casks acquire a pale golden hue and a unique dryness of flavor. Although distillers are permitted to store gins in the United States for further flavor development, the federal government does not permit any reference to aging to appear on the label.

Holland Gin, characterized by its high flavor intensity derived mostly from juniper berries and cereal grains (corn, rye, and barley malt), is produced by immersing the botanical mixture directly into the grain mash before distillation or by extracting the essential oils from the botanical mixture with the heavy distillate (high wines) from a fermented mash of grain, consisting of corn, rye, and barley malt. Consequently, the flavors produced during fermentation become flavor components of the final product. Compound gin is a mixture of grain spirits and essential oil extracts from botanicals. It does not undergo any distillation procedure.

Brandies

Brandy is an alcoholic distillate from the fermented juice, mash, or wine of fruit, or from the residue thereof, produced at less than 190° proof in such manner that the distillate possesses the taste, aroma, and characteristics generally attributed to the product, and bottled at not less than 80° proof.

The most important category of brandy is fruit brandy, distilled solely from the juice or mash of whole, sound, ripe fruit or from standard grape, citrus, or other fruit wine. Brandy derived exclusively from one variety of fruit is so designated. However, a fruit brandy derived exclusively from grapes may be designated as brandy without further qualification, and unless the product is specifically identified, the term brandy always means grape brandy. Brandy is subject to a distillation limitation of 170° proof (85%). If distilled over 170° proof (85%), it must be further identified as neutral brandy. A minimum of two years of maturation in oak casks is required, otherwise the term immature must be included in the designation. Although the age is not indicated on the label, brandies are normally aged from three to eight years.

Brandies are produced in batch or continuous distillation systems. The pot still or its variation is universally used in France, whereas in the United States both systems are employed. The batch system produces a more flavorful product, the continuous system a lighter, more delicate flavor.

The history of brandy can be said to be the history of distillation, because in the distant past it was the distillation of wine in crude stills that produced *aqua vitae*. In the ensuing evolution, many areas of Europe and of the United States became renowned for their brandies. Perhaps the most popular brandy comes from the Cognac region of France, in the Department of Charente and Charente Inférieure. As such, it enjoys an exclusive identity, Cognac, under which no other brandy may be labeled. Cognac is produced in the traditional pot stills by small farmers and sold to the bottlers who age the brandies in limousin oak casks. When the brandies reach maturity, they are skillfully blended for marketing under their own brand name.

Cognac is a blend of some Grande Champagne, Petite Champagne, Borderies, and Fins Bois, the proportions of each being a well-kept secret. To further characterize Cognac, the bottle is labeled: E, especial; F, fine; V, very; O, old; S, superior; P, pale; X, extra; C, Cognac; eg, VSOP means very superior old pale, and is considered to be a better-quality product.

Another well-known brandy of Frances is Armagnac, produced in southern France. Armagnac is distilled from wines in a continuous system using two pot stills in series. Armagnac is considered to be more heavy-bodied and drier than Cognac. Brandies are distilled in almost every wine region of France; they are called *eau de vie*, exported as French Brandy, never as Cognac.

In the United States, California produces almost all of the grape brandy. Generally, it is a well-integrated operation, the cultivation of the grapes, the making of the wine, the distilling, aging, bottling, and the marketing of the brandy being done by the same firm. Usually a continuous multicolumn distillation system is employed. Of the total U.S. brandy consumption of approximately 56 million L (15 million gal) California brandies account for over 80%.

Blended applejack is accorded a special classification as a mixture that contains at least 20% of apple brandy (applejack) on a proof basis, stored in oak containers for not less than two years, and not more than 80% of neutral spirits on a proof gallon basis and bottled at not less than 80° proof (40%). Another class of beverage spirits, flavored brandy, is a brandy to which natural flavoring materials have been added with or without the addition of sugar; it is bottled at no less than 70° proof (35%). The name of the predominant flavor appears as part of the designation, ie, blackberry flavored brandy; cherry flavored brandy, etc. Such a flavored brandy may contain up to 12.5% of wine derived from the particular fruit corresponding to the labeled flavor.

Certain areas in Europe and in South America are well known for their specialty brandies, such as Spanish brandies, distilled from Jerez sherry wine; the fragrant, fruity Portuguese brandies, distilled from port wine; the pleasant and flowery muscat bouquet of Pisco brandy from Peru; Kirschwasser brandy, with its almond undertone flavor, distilled from a fermented mash of small black cherries, which grow along the Rhine Valley in Germany and Switzerland; and Slivovitz, the plum brandy, which is produced in Hungary and in the Balkan countries.

Cognac or Cognac (grape) brandy, is grape brandy distilled in the Cognac region of France, which is entitled to be so designated by the laws and regulations of the French Government.

Rums

Rum is an alcoholic distillate from the fermented juice of sugar cane, sugar cane syrup, sugar cane molasses, or other sugar cane by-products, produced at less than 190° proof in such manner that the distillate possesses the taste, aroma and characteristics generally attributed to rum, and bottled at not less than 80° proof.

Blackstrap molasses is the most common raw material for the manufacture of rum. Otherwise, the same basic factors that produce different whiskies are responsible for the flavor varieties of rums. The type of yeast, fermentation environment, distillation techniques and systems, the maturation conditions, and not least the blending skill all contribute to the final character and quality of rum. Blackstrap molasses varies in composition according to origin owing to the environment and to some extent on the processing of cane (a greater recovery of sugar is usually reflected in a lower concentration of residual sugar in the molasses). A typical composition of Puerto Rican blackstrap molasses is as follows:

pH	4.3
Density (% solids)	89°Brix
Nonfermentable solids	28.7%
Sucrose	33.8%
Reducing sugars	26.9%
Total sugars	60.7%

Rums are characterized as light bodied, of which the Puerto Ricans are the best known, and full bodied, which come from Jamaica and certain other islands of the West Indies. Light rums are distilled on multicolumn continuous distillation systems over a proof range of 160–180° (80–90%). They are matured in oak casks that are reused for rum storage. Age may, but need not, be stated on the labels.

Rum has been produced in Jamaica for more than 200 yr as some plantation crop records indicate. A typical small plantation would produce a ton of sugar and 41,000 L (9,000 imperial gal) of rum. Today, only a few large facilities exist and the law requires that a distillery may only be operated in conjunction with a sugar refinery. As in every area where distilled spirits are produced, terminology and local practice play a major role in determining the identity and the flavor characteristics of the product; eg, molasses acid results from a natural fermentation that produces alcohol and subsequently an organic acid mixture. Some of this is used in the wash (fermenting mash) for the production of certain flavorful rums. Dunder is the term used for stillage, the material left after the beverage spirits are removed by distillation. Dunder is also used as an ingredient in the fermenting mash, as much as 40 vol %, to provide buffering action, and flavor development.

Jamaican and other full-bodied rums are distilled between 140 and 160° proof (70 and 80%) in pot stills. They are matured in large casks of 422.4 L (111.6 gal) called puncheons. Unlike the light-bodied rums, which use cultured yeast for inoculation, the Jamaican rums rely on natural fermentation, sometimes referred to as wild fermentation. In this method the mash is inoculated by the yeast that is present in the air and in the raw material. Time of fermentation may vary from 2 to as much as 11 days, depending on the desired flavor characteristics.

Puerto Rican rums are generally labeled as white or gold label. The latter is a little more amber in color and has a more pronounced flavor. Although the rums produced in various areas are not considered distinctive types, they do retain their local characteristics and their names may not be applied to rum produced in any other place than the particular region indicated in the name.

Venezuela and to some extent Mexico are major producers of rum. The former requires a minimum two-year aging period and in the latter the amount of production is closely regulated by the government and keyed to the availability of cane and the need for industrial alcohol. In all other respects, the light-bodied rums are similar to those produced in Puerto Rico. A limited amount of more flavorful rums for use in blending is produced in both countries. Continuous and batch distillation systems are used and thus offer a capability of producing rums with varying degrees of flavor intensities and individual characteristics.

Tequila

Tequila, a distinctive product of Mexico, is an alcoholic distillate produced in Mexico from the fermented juice of the heads of *Agave Tequilana Weber* (blue variety), with or without additional fermentable substances, distilled in such a manner that it possess the taste, aroma, and characteristics generally attributed to tequila. This is Mexico's most popular distilled spirits drink.

The mezcal azul (blue mezcal), the primary source for tequila, is cultivated and usually propagated from 2-yr-old sprouts obtained from 7-yr-old mezcal plants. After 8–12 yr the plants are matured; the trimmed heads, referred to as pine apples because of their appearance and weighing 36–59 kg (80–130 lb) each, are transported to the distillery. In the State of Jalisco, up to 20,000 t of mezcal heads are harvested annually. The heads contain (unsteamed): moisture, 62%; total solids, 38%; fiber, 11%; inulin, 20%; and ash, 2.5%, and they have a pH of 5.5.

The juice from the mezcal heads is first extracted in masonry ovens with a capacity of approximately 40 t for 9–24 h at approximately 93°C. The length of the steaming period is critical for the acid hydrolysis of the inulin to monosaccharides. During a 12-h cooling period, some additional juice (*mieles de escurrido*, drained molasses) is recovered. The mezcal heads are now dark brown, soft in texture, with a taste similar to maple syrup. Residual juice is removed by shredding the steamed heads, compressing the strips between roller mills, and finally washing the strips (bagasse) to recover all of the sugary syrup.

The mezcal juice from the steaming ovens, the roller mills, and the bagasse washes is pumped into fermenters of 3,800–7,500 L (1,000–2,000 gal) capacity made of masonry or of local pine wood. Nitrogen nutrients are added to facilitate fermentation. The Mexican government also permits the addition of piloncillo, a brown sugar, up to 30 wt % of the fermentable sugar in the mezcal heads after steaming. After a fermentation period of about 42 h, the alcoholic concentration is 4.5 vol% of the fermenter mash. When piloncillo is not used, the alcoholic yield is lower and the fermentation takes longer.

The fermented mash is pumped to a copper pot still of about 1,100 L (300 gal) capacity provided with a steam coil within the kettle for heating and a condenser for cooling the vapors. The intermediate of the first distillation called ordinario is collected at 28° proof (14%) and redistilled in a slightly larger pot still. This distillation cycle can be controlled to yield a product of approximately 106° proof (53%). The residual distillate from this process is combined with the fermented mash, starting a new cycle in the primary distillation system.

Tequila, as consumed in Mexico, is unaged and usually bottled at 80–86° proof (40–43%). However, some producers do age tequila in seasoned 190-L (50-gal) white oak casks imported from the United States. In aging, tequila becomes golden in color and acquires a pleasant mellowness without altering its inherent taste. Tequila aged one year is identified as Anejo; aged as much as two to four years, it is identified as Muy Anejo. The annual production of tequila in Mexico is around 20 million L (5.2 million U.S. gal), the major portion of which is produced by 28 plants located in the county of Tequila.

Cordials and Liqueurs

Cordials and liqueurs are the same, with the former term being American and the latter European. They are obtained by mixing or redistilling neutral spirits, brandy, gin, or other distilled spirits with or over fruits, flowers, plants, or pure juices therefrom, or other natural flavoring materials, or with extracts derived from infusions, percolations, or maceration of such materials. Cordials must contain a minimum of 2.5 wt% of sugar or dextrose, or a combination of both. If the added sugar and dextrose are less than 10 wt% then it may be designated as dry. Synthetic or imitation flavoring materials cannot be included in United States cordials, nor can they be designated as distilled or compound.

Cordials were known in ancient Egypt and Athens, but commercial production was started in the Middle Ages when alchemists, physicians, and monks, among others, were searching for an elixir of life. From this activity many well-known cordials were developed, such as Benedictine and Chartreuse, both derived from aromatic plant flavors and bearing the names of the monasteries where they were first prepared.

A great variety of cordials are available encompassing a wide spectrum of flavors from fruits, peels, leaves, roots, herbs, and seeds. Organoleptic attainment, however, becomes a matter of experience and skill in the selection of botanicals and in the extraction and formulation of flavors. Although these elements are carefully guarded secrets, the producer must rely on three basic processes, namely maceration, percolation, and distillation, or any combination thereof. Maceration involves the steeping of the raw materials in the spirits, usually in a vat, to impart the desired aroma, flavor, and color. The liquid is then drawn off and provides the base for further processing. Percolation is accomplished by recirculating the spirits through a percolator containing the raw materials. As the spirits seep down through the raw material, the desired constituents are extracted, which will give the proper aroma, flavor intensity, and color. The distillation method is similar to that used in gin production. The ingredients are either immersed in the beverage spirits or placed in trays or pans in the head of the still. The rising vapors extract the essential flavors, which are then condensed and discharged as a colorless liquid. This distillate contains the basic flavor that is used for further processing.

Cordials are characterized and marketed according to their generic names; eg, anisette (aniseed), crème de menthe (peppermint), triple sec (citrus fruit peel), slow gin (sloe berries), and by their trade names (proprietary brands) of which Benedictine and Chartreuse are well-known examples.

THE MANUFACTURING PROCESS

Any material rich in carbohydrates is a potential source of ethyl alcohol, which for industrial purposes is obtained by the fermentation of materials containing sugar (molasses), or a substance convertible into sugar, such as the starches. In the production of distilled spirits for beverage purposes, however, cereal grains are the principal types of raw material used. Any reference to alcohol in beverages is always to ethyl alcohol (C_2H_5OH). Although other alcohols may be present, they are referred to as higher alcohols, fusel oils, or by their specific name.

The chemical composition of grain varies considerably and depends to a large extent on environmental factors such as climatic conditions and the nature of the soil. Another variable is the malt (sprouted or germinated grain) used. Malt is generally understood to be germinated barley, unless it is further qualified as rye malt, wheat malt, etc. The purpose of malting is the development of the amylases, the active ingredients in malt. Amylases are enzymes of organic origin, which change grain starch into the sugar, maltose. Besides providing the means of converting the grain starch into sugar, the malt also contributes to the final flavor and aroma of the distillate. Malting techniques may produce a malt of such unusual character that it may indeed furnish the outstanding characteristics of the final product, as in Scotch whisky. Figure 2 shows the process flow sheet of a modern beverage spirits plant.

Grain Handling and Milling

The beverage distilling industry utilizes premium cereal grains. Each distiller supplements government grain standards with personal specifications, especially in regard to the elimination of grain with objectionable odors, which may have developed during storage or kiln drying at the elevators. Hybrid corn, usually of the readily available dent variety and plump rye, developed from Polish strains (Rosen), are used for beverage alcohol production. Modern distilleries use Airveyor unloading systems, others the traditional power shovel in conjunction with screw conveyors and bucket elevators. Even though the grain has been subjected to a cleaning process at the elevator, it is passed over receiving separators, a series of vibrating screens which sift out the foreign materials. Air jets and dust collectors remove any light material and magnetic separators remove metallic substances.

Milling breaks the outer cellulose protective wall around the kernel and exposes the starch to the cooking and conversion processes. Distillers require an even, coarse meal without flour. Milling is accomplished by three methods: (*1*) in roller mills, using pairs of corrugated rolls (breaks), run sharp to sharp (projections facing projections); (*2*) in hammer mills, where a series of revolving hammers within a close-fitting casing and rotating at 1,800–3,600 rpm shear the grain to a meal, which is removed by suction through a screen, different for various types of grain; and (*3*) in attrition mills (not widely used) where the grain is ground by two counterrotating disks (1,200–2,000 rpm). This mill is not entirely satisfactory because of excessive flour produced. A three-break roller mill (three pairs of corrugated rolls, arranged vertically) with rolls 23 cm in diameter × 76 cm long has a capacity of 3.5 m^3 (100 bu) of corn per hour; a hammer mill 61 cm in width, 10.6 m^3 (300 bu) per hour; and an attrition mill with 41 cm grinding plates, 3.5 m^3 (100 bu) per hour, on the basis that 25% of the ground corn remains above the no. 12 mesh (1.68 mm) screens.

Mashing

The mashing process, consists of cooking, ie, gelatinization of starch, and conversion (saccharification), ie, changing starch to grain sugar (maltose). Cooking can be carried out at atmospheric or higher pressure in a batch or a continuous system. For whisky production, batch cooking at atmospheric pressure is widely used although some batch pressure cooking is practiced. For grain neutral spirits production, both batch and continuous systems are used under pressure. After cooling conversion is accomplished in the cooking vessel by the addition of barley malt meal to the cooked grain. Some distillers pump the mash immediately to a converter for the necessary holding time and thus make the cooking vessel available for the next cook. The converted mash is cooled and pumped to the fermenters.

Distillers vary mashing procedures, but generally conform to basic principles, especially in the maintenance of sanitary conditions. The cooking and conversion equipment is provided with direct or indirect steam, propeller or rake-type agitation, and cooling coils or a barometric condenser. Mashing procedures for rye, corn, and malt grains are described below.

Rye. In the preparation of a bourbon mash, rye is generally subjected to the corn-cooking process. However, rye

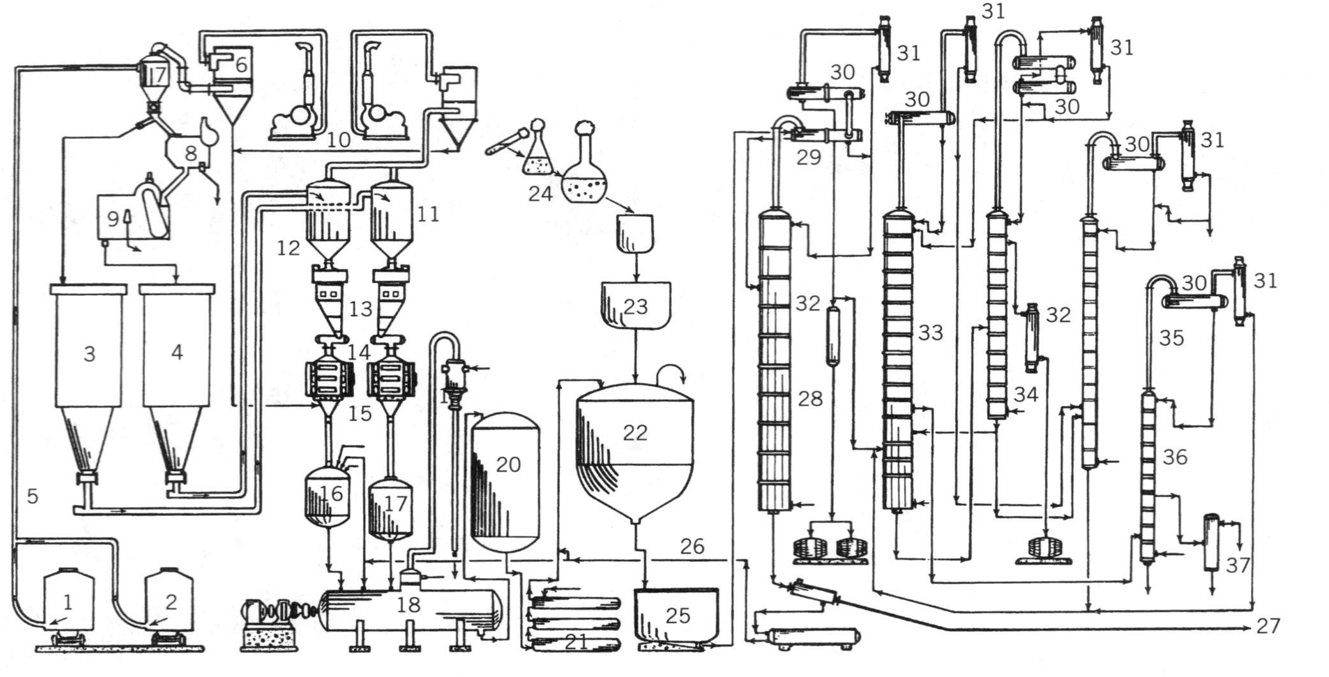

1. Cereal grains
2. Malt
3. Malt bin
4. Cereal grain bin
5. Unloading elevator
6. Dust filter
7. Collector
8. Scalperator
9. Millerator
10. Reclaiming exhauster
11. Malt receiver
12. Cereal grain receiver
13. Automatic scale
14. Mill feeder
15. Roller mills
16. Precooker
17. Malt infusion
18. Cooker
19. Barometric condenser
20. Converter
21. Mash coolers
22. Fermenter
23. Final yeast propagator
24. Yeast culture and intermediate yeast propagator
25. Fermented-mash holding vessel
26. Stillage return system
27. Stillage flow to recovery system
28. Whisky separating column
29. Heat exchanger
30. Dephlegamtor
31. Vent condenser
32. Product cooler
33. Selective distillation column
34. Product concentrating column
35. Aldehyde concentrating column
36. Fusel oil concentrating column
37. Fusel oil decanter

Figure 2. Material process flow, modern beverage spirits plant.

undergoes liquefication at a much lower temperature than corn, which avoids thermal decomposition of critical grain constituents adversely affecting the final flavor of the distillate. For that reason, many distillers mash rye separately.

Water is drawn at the rate of 35 L/m^3 (28 gal/bu) and rye and malt meal are added. The mash is slowly heated to 54°C and held for approximately 30 min. Proteolytic enzymes, active at 43–46°C, aid in reducing the viscosity, and the optimum temperature for β-amylase is 54°C. The mash is then heated to 63–67°C and held for 30–45 min to ensure maximum conversion. The mash (pH 6.0) is then cooled to the fermenting temperature 20–22°C. This process of converting small grains is called infusion mashing.

Corn. Although the starch in corn grains converts rather easily, higher cooking temperatures are necessary to make the starch available. Usually malt is not added at the beginning, but to reduce viscosity, premalt of 0.5% may be added before cooking, preferably at around 66°C. Thin stillage (the residual dealcoholized fermented mash from the whisky distillation process) is added by some producers to adjust pH to 5.2–5.4. For cookers operating at atmospheric pressure, a mashing ratio of 95–115 L (25–30 gal) of slurry (grain, water, and stillage mixture) per 0.03 m^3 (1 bu) and a holding time of 30 min at 100°C are preferable. The mash is cooled to 67°C and malt is added. Primary conversion, the saccharification taking place during conversion, is in the order of 70–80% of the available starches. The remainder of the conversion to fermentable sugar takes place during the fermentation process and is referred to as secondary conversion. For batch cooking under pressure only 65–83 L (17–22 gal) of water are drawn, and the maximum temperature is 120–152°C. In continuous pressure cooking, water is drawn at a ratio of 30 L/m^3 (24 gal/bu) of meal and sufficient thin stillage is added to adjust the pH to 5.2–5.4. The mash is pumped through the continuous pressure cooker, where it is exposed to temperatures of 170–177°C for 2–6 min, and then into a flash chamber where it is cooled immediately by vacuum to the malting (conversion) temperature of 63°C. A malt slurry is continuously introduced and the mixture proceeds through the water cooling system to the fermenters.

Fermentation

In fermentation the grain sugars (largely maltose), produced by the action of malt enzymes (amylases) on gelatinized starch, are converted into nearly equal parts of ethyl alcohol and carbon dioxide. This is accomplished by the enzymes in yeast. Yeast multiples by budding, and a new cell is produced about every 70 min. Although yeasts of several genera are capable of some degree of fermentation, *Saccharomyces cerevisiae* is almost exclusively used by the distilling industry. It has the ability to reproduce prolifically under normal growth conditions found in distilleries, has a high fermentation rate and efficiency, and can tolerate relatively high alcohol concentrations (up to 15–16 vol %). A great variety of strains exist and the characteristics of each strain are evidenced by the type and amount of congeners the yeast is capable of producing.

Alcoholic fermentation is represented by the following reactions:

$$\underset{\text{maltose}}{C_{12}H_{22}O_{11}} \xrightarrow{\text{maltase}} \underset{\text{dextrose}}{C_6H_{12}O_6} \rightarrow \underset{\text{ethyl alcohol}}{2C_2H_5OH} + \underset{\text{carbon dioxide}}{2CO_2}$$

A fermentation efficiency of 95% is obtained based on the sugar available. Of the starch converted to grain sugar and subsequently subjected to fermentation, 5–6% is consumed in side reactions. The extent and type of these reactions depend on: (1) yeast strain characteristics, (2) the composition of the wort, and (3) fermentation conditions such as the oxidation—reduction potential, temperature, and degree of interference by bacterial contaminants.

Secondary products formed by these side reactions largely determine the characteristics and organoleptic qualities of the final product. In the production of whisky, the secondary products (known as congeners) formed and retained during the subsequent operations include a number of aldehydes, esters, higher alcohols (fusel oils), some fatty acids, phenolics or aromatics, and a great many unidentified trace substances (Table 3). In the production of grain neutral spirits, the congeners are removed from the distillate in a complex multicolumn distillation system. Some distillers, however, retain a small portion of the low-boiling esters in the distillate when the grain neutral spirits are to be matured. Fermentation of grain mashes is initiated by the inoculation of the set mash with 2–3 vol % of ripe yeast prepared separately (see below) and followed by three distinct phases.

1. Prefermentation involves rapid multiplication of yeast from an initial 4–8 million/mL to a maximum of 125–130 million/mL of the liquid and an increasing rate of fermentation.
2. Primary fermentation is a rapid rate of fermentation, as indicated by the vigorous "boiling" of the fermenting mash, caused by escape of carbon dioxide. During this phase secondary conversion takes place, ie, the changing of dextrins to fermentable substances.
3. Secondary fermentation is a slow and decreasing rate of fermentation. Conversion of the remaining dextrins, which are difficult to hydrolyze, takes place.

The degree of conversion, agitation of the mash, and temperature directly affect the fermentation rate. Fermenter mash set at a concentration of 144 L (38 gal) of mash per bushel of grain 25.4 kg (56 lb) will be fermented to completion in two to five days, depending on the set and control temperatures. The set temperature (temperature of the mash at the time of inoculation) is largely determined by the available facilities for cooling the fermenting mash. If cooling facilities are adequate, temperatures of 27–30°C may be employed; otherwise, the set temperature must be low enough to ensure that the temperature will not exceed 32°C during fermentation. When no cooling facilities are provided, the inoculation temperature must be below 21°C. Excessive temperatures during the prefermentation phase retard yeast growth and stimulate the development of bacterial contaminants which are likely to produce undesirable flavors.

In the production of sour mash whisky U.S. federal regulations require that a minimum of 25 vol % of the fermenting mash must be stillage (cooled, screened liquid recovered from the base discharge of the whisky separating column, pH 3.8–4.1). In addition to producing a heavier-bodied whisky, this procedure provides the distiller with an economical means of adjusting the setting pH (4.8–5.2) to inhibit bacterial development. It also provides buffering action during the fermentation cycle, which is important because secondary conversion does not take place if the fermenting pH drops below 4.1 in the immediate stages. Thin stillage also provides a means of diluting the cooker mash to the proper fermenter mash concentration, 3,220–3,870 L/m^3 (30–36 gal/bu) of grain for the production of spirits, and 4,080–4,850 L/m^3 (38–45 gal/bu) for making whisky. This concentration gives about a 12–16% soluble solids in the fermenting liquid, within the range used in beer fermentations but lower than that in wine fermentation.

Preparation of the yeast involves a stepwise propagation, first on a laboratory scale and then on a plant scale to produce a sufficient quantity of yeast for stocking the main mash in the fermenters. A strain of yeast is usually carried in a test tube containing a solid medium (agar slant). A series of daily transfers, beginning with the removal of some yeast from the solid medium, are made into successively larger flasks containing liquid media—diamalt (commercial malt extract) diluted to 15–20° Balling, malt extract, or strained sour yeast mash—until the required amount of inoculum is available for the starter yeast mash, called a dona. After one day's fermentation, the dona is added to a yeast mash normally composed of barley malt and rye grains, and representing approximately 2–3.5 wt % of the total grain mashed for each fermenter.

The yeast mash is generally prepared by the infusion mashing method and then soured (acidified) to a pH of 3.9–4.1 by a 4–8-h fermentation at 41–54°C with *Lactobacillus delbrucki*, which ferments carbohydrates to lactic acid. Satisfactory souring can be induced with an inoculum of approximately 0.25% of culture per volume of mash. The water-to-meal ratio of 2,580–3,000 L/m^3 (24–28 gal/bu) attains a yeast mash balling of approximately 21°. Before inoculation with yeast, the soured mash is pasteurized to 71–87°C to curtail bacterial activity, then cooled to the setting temperature of 20–22°C. The sour mash medium offers an optimum condition for yeast growth and also has an inhibitory effect on bacterial contamination. In 16 h the yeast cell count reaches 150–250 million/mL. Some distillers use the sweet yeast method for yeast development. In this instance the lactic acid souring is not included and the inoculation temperature is usually above 26.6°C to insure rapid yeast growth.

Distillation

Distillation separates, selects, and concentrates the alcoholic products of yeast fermentation from the fermented grain mash, sometimes referred to as fermented wort or distillers beer. In addition to the alcohol and the desirable secondary products (congeners), the fermented mash contains solid grain particles, yeast cells, water-soluble proteins, mineral salts, lactic acid, fatty acids, and traces of glycerol and succinic acid. Although a great number of different distillation processes are available, the most common systems used in the United States are (1) the continuous whisky separating column, with or without an auxiliary doubler unit for the production of straight whiskies; (2) the continuous multicolumn, system used for the production of grain neutral spirits; and (3) the batch rectifying column and kettle unit, used primarily in the production of grain neutral spirits that are subsequently stored in barrels for maturation purposes. In the batch system, the heads and tails fractions are separated from the product resulting from the middle portion of the distillation cycle.

Although most modern plants have various capacity whisky stills available, a whisky separating column is usually incorporated into the multicolumn system, thus acquiring a greater range of distillation selectivity, ie, the removal or retention of certain congeners. For example, absorptive distillation involving the addition of water to the upper section of a column in the whisky distillation system is a method of controlling the level of heavier components in a product. In the beverage distillation industry, stills and auxiliary piping are generally fabricated of copper, although stainless steel is also used. All piping that conveys finished products is tin-lined copper, stainless steel, or glass.

The whisky column, a cylindrical shell that is divided into sections and may contain from 14 to 21 perforated plates, spaced 56–61 cm apart. The perforations are usually 1–1.25 cm in diameter and take up about 7–10% of the plate area. The vapors from the bottom of the still pass through the perforations with a velocity of 6–12 m/s. The fermented mash is introduced near the top of the still, and passes from plate to plate through down pipes until it reaches the base where the residual mash is discharged. The vapor leaving the top of the still is condensed and forms the product. Some whisky stills are fitted with entrainment removal chambers and also with bubble-cap plate sections (wine plates) at the top to permit operation at higher distillation proofs. Because whisky stills made of copper, especially the refinement section, supply a superior product, additional copper surface in the upper section of the column may be provided by a demister, a flat disk of copper mesh. The average whisky still uses approximately 1.44–1.80 kg/L (12–15 lb of steam/proof gal) of beverage spirits distilled. Steam is introduced at the base of the column through a sparger. Where economy is an important factor, a calandria is employed as the source of indirect heat. The diameter of the still, number of perforated and bubble-cap plates, capacity of the doubler, and proof of distillation are the critical factors that largely determine the characteristics of a whisky.

The basic continuous distillation system for the production of grain neutral spirits usually consists of a whisky separating column, an aldehyde column (selective distillation column), a product concentrating column (sometimes referred to as an alcohol or rectifying column, from which the product is drawn), and a fusel oil concentrating column. In addition, some distillers, to secure a greater degree of refinement and flexibility, may include an aldehyde concentrating column (heads concentrating column) or a fusel oil stripping column. Bubble-cap plates are used throughout the system (except in the whisky column, which may have some bubble-cap plates).

This distillation system offers a wide range of flexibility for the refinement of distilled beverage spirits. Figure 3 shows a five-column, continuous distillation system for the production of grain neutral spirits. A fermented mash (generally 90% corn and 10% barley malt) with an alcohol concentration of approximately 7 vol % is pumped into the whisky column somewhere between the 13th and 19th perforated plate for stripping. The residual mash is discharged at the base and pumped to the feed recovery plant; the overhead distillate [ranging in proof from 105 to 135° (52.5 to 67.5%)] is fed to the selective distillation column (also called the aldehyde column), which has over 75 bubble-cap plates. The main stream [10–20° proof (5–10%)] from the selective distillation column is pumped to the product concentrating column. A heads draw (aldehydes and esters) from the condenser is pumped to the heads concentrating column (also called the aldehyde concentrating column), and a fusel oil and ester draw is pumped to the fusel oil concentrating column. The product is withdrawn from the product concentrating column.

Some accumulation of heads, at the top of the product concentrating column, are removed at the condenser, and transferred to the aldehyde concentrating column, where the heads from the system are removed for disposal. The fusel oil concentrating column removes the fusel oil from the system. Figures 3, 4, and 5 show the distribution and concentration of alcohols and congeners through the selective distillation column, the product concentrating column, and the heads concentrating column.

By-Products. The discharge from the base of the whisky column is called stillage and contains in solution and in suspension substances derived from grain (except the starch, which has been fermented), and from the mashing and fermentation processes. The suspended solids are recovered by screening and then subjected to a pressing and a drying operation (dehydrating), usually in rotary, steam-tube dryers. The liquid portion, called thin stillage, is concentrated by a multieffect evaporator to a syrup with a solid content of 30–35%. This concentrate can be mixed and dried with the previously screened-out solids, or it can be dried separately, on rotary-drum dryers. These by-products are known as distillers' dried grains and distillers' solubles and are used by the feed industry to fortify

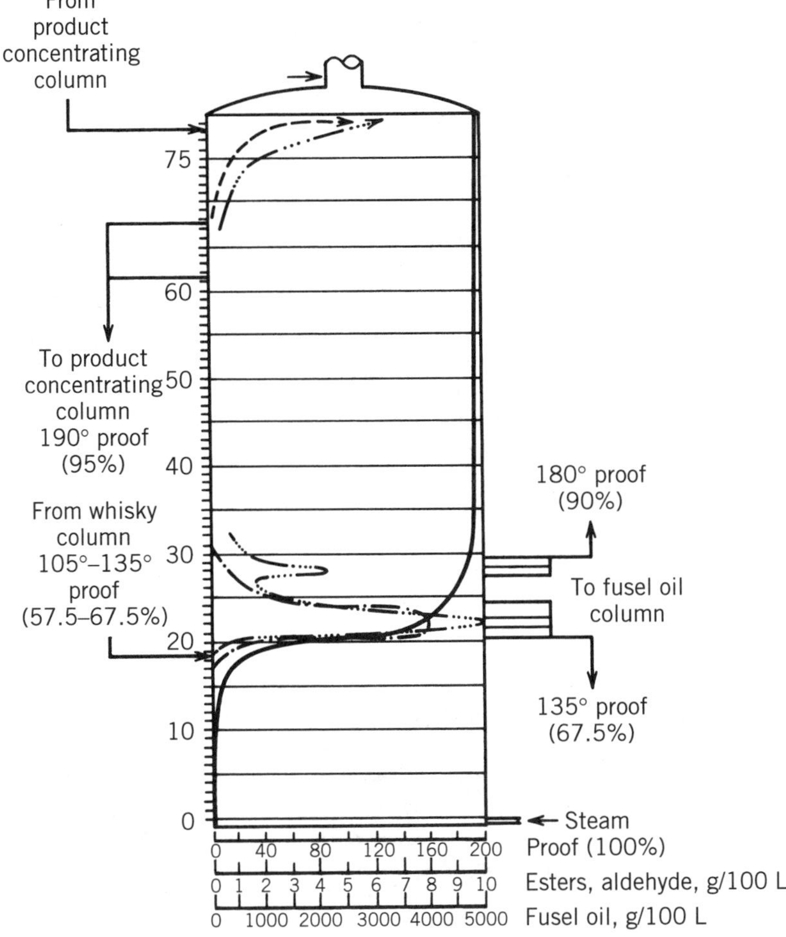

Figure 3. Selective distillation column.

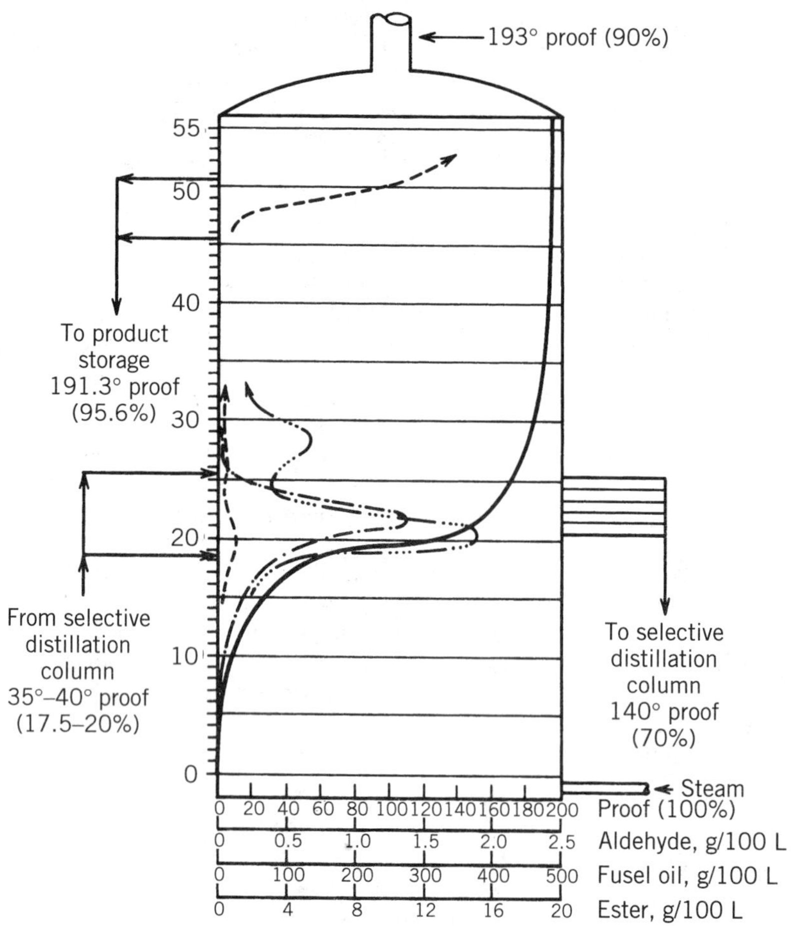

Figure 4. Product concentrating column.

dairy, poultry, and swine formula feeds. Distillers' feeds are rich in proteins (24–35%), fat (8%), choline, niacin, and other B-complex vitamins and contain vital fermentation growth factors that produce good growth responses in livestock and poultry.

Maturation

In the United States, the final phase of the whisky production process, called maturation, is the storage of beverage spirits (which at this point are colorless and rather pungent in taste) in new, white oak barrels, whose staves and headings are charred. The duration (years) of storage in the barrel depends on the time it takes a particular whisky to attain the desirable ripeness or maturity. The staves of the barrel may vary in thickness from 2 to 2.85 m. The outside dimensions of a 190-L (50-gal) barrel are approximately: height, 0.88 m (34.5 in.) and diameter at the head, 0.54 m. Whisky storage warehouses vary in construction from brick and mortar types, single and multiple floors (as many as six) having capacities up to 100,000 barrels of 190-L (50-gal) capacity, to wooden sheet-metal-covered buildings (called iron clads), generally not exceeding a capacity of 30,000 barrels of 190-L (50-gal) capacity. It is customary to provide natural ventilation. The whisky in storage is subject to critical factors that determine the character of the final product. The thickness of the stave, depth of the char (controlled by regulating the duration of the firing, 30–50 s), temperature and humidity, entry proof, and, finally, the length of storage impart definite and intended changes in the aromatic and taste characteristics of a whisky. These changes are caused by three types of reaction occurring simultaneously and continually in the barrel: (*1*) extraction of complex wood constituents by the liquid, (*2*) oxidation of components originally in the liquid and of material extracted from the wood, and (*3*) reaction between the various organic substances present in the liquid, leading to the formation of new congeners.

Comprehensive studies of changes occurring during the maturation of whisky have been published by a number of investigators (4–11). Although these reports contain detailed information on the concentration of various congeners throughout the course of the maturation period, little is revealed of the interrelationships between various congeners except for the nonvolatile groups, such as solids, nonvolatile acids, tannin, and color. Notable among the observed changes are increases in the concentrations of acids, esters, and solids. These studies reveal that throughout the maturation period a linear relationship exists between the increase in acids and esters and the increase in dissolved solids; this suggests that acid and ester formation is dependent on some precursor that is extracted from the barrel at the same rate as the bulk of the solid fraction (9).

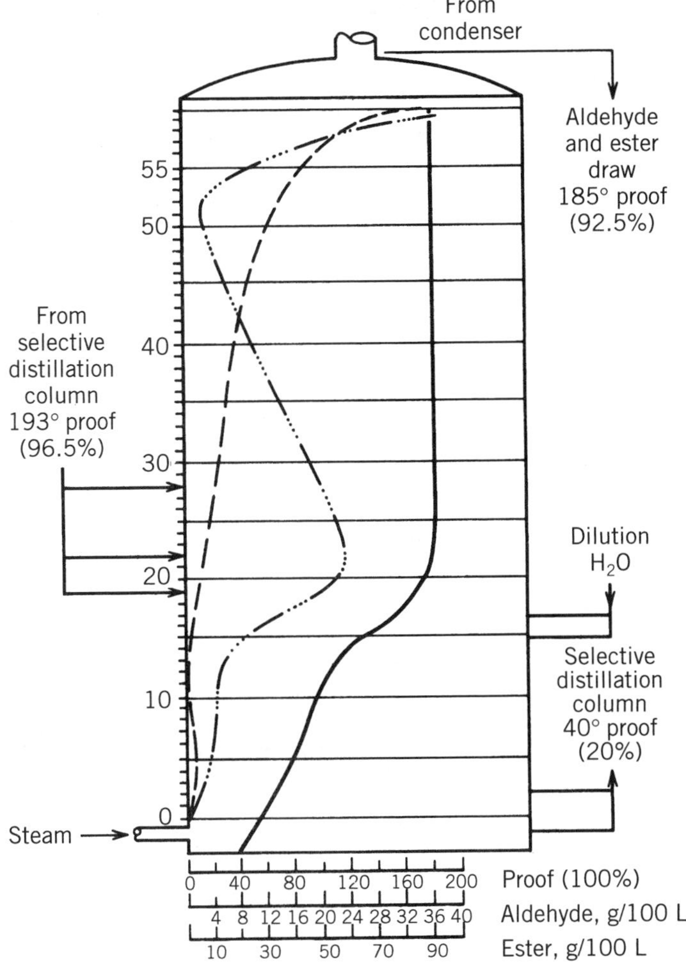

Figure 5. Aldehyde (heads) concentrating column.

In 1962 the U.S. Treasury Department increased the maximum allowable entry proof (storage in barrels) for straight whisky from 110 to 125° (55 to 62.5%). The industry engaged in a study in December 1985 to determine the flavor development, chemical composition, and evaporation loss rates of distillates at six different proofs from 109 to 155° (54.5 to 77.5%) over a planned 8-yr period, and the study was continued an additional 4 yr. Evaporation losses averaged approximately 3.0% a year over the 12-year period.

In addition to chemical analysis, they applied regressive analysis for most congeners listed in Table 4. Regression analysis illustrates the functional relationship between congeners. Except for fusel oil, there was an appreciable continuing development of congeners through 12 yr. Esters, the most important flavor constituent in whisky, developed linearly over this time period. Later studies, using radioactive carbon for direct monitoring, determined the congeners derived from ethanol during whisky maturation. Esters (ethyl acetate) are formed in a system that is affected by changes in concentration of acetic acid, ethanol, and water, ie, a loss of water by evaporation; an increase in acetic acid results in an increase of ethyl acetate.

The maturing progress of a type of bourbon as measured by the principal ingredients is illustrated in Tables 4 and 5. It is evident that the congeners amount to only about 0.5–0.75% of the total weight. Yet it is this small fraction that determines the quality of the final product. Thus matured whiskies vary widely in taste and aroma because of the wide variation in congener concentration. For example, esters may vary from 8.0 to 18.0 g/100 L, and aldehydes from 0.7 to 7.6 g/100 L. Nevertheless, there is no correlation between chemical analysis and quality, ie, taste and aroma. Only the consumer can detect the fine variations and thus evaluate the quality of whiskies.

ANALYSIS

Analytical results are expressed in terms of one component for each chemical class; ie, acetic acid for acids, acetaldehyde for aldehydes, etc (Tables 4 and 5). Individual constituents can, of course, be determined by more refined techniques. Until the advent of gas chromatography, tedious chemical procedures were generally required for separation of the congeners. Distillation, countercurrent distribution, and other physical methods affected concen-

Table 4. Chemical Analysis[a] of Bourbon Whiskies Matured at Six Proofs from 109 to 155° (54.5 to 77.5%), g/100 L at 100° Proof (50%)

Age (yr)	Proof	Color[a]	pH[b]	Solids	Fixed acids	Volatile acids	Esters	Fusel oil	Aldehydes	Furfural	Tannins
0	109	nd[c]	nd	0	0	1.6	4.8	125	0.3	0.06	0
1	106	0.666	4.2	94	6	29	11	134	3.1	1.02	31
	112	0.678	4.2	99	7	27	12	134	3.1	1.03	31
	122	0.686	4.3	90	7	23	11	132	3.1	0.95	29
	132	0.652	4.3	70	6	21	11	128	2.9	0.84	26
	143	0.688	4.2	71	5	21	11	130	2.6	0.78	25
	155	0.652	4.2	63	5	18	11	129	2.6	0.75	23
2	107	0.793	4.1	124	7	31	16	147	3.5	1.16	37
	113	0.810	4.2	105	7	32	18	141	3.5	1.10	37
	123	0.848	4.3	107	7	27	16	143	3.2	1.12	36
	133	0.785	4.1	103	6	28	17	136	2.9	0.96	32
	143	0.807	4.1	92	6	23	16	145	2.5	0.82	30
	152	0.764	4.2	85	5	22	15	142	2.3	0.80	29
3	106	1.036	4.1	149	9	36	21	148	4.5	1.12	45
	112	1.013	4.2	145	9	36	21	146	4.2	1.14	43
	123	1.031	4.1	130	8	32	22	151	3.8	1.09	41
	132	0.928	4.2	118	7	29	19	153	3.5	0.89	36
	142	0.917	4.1	107	7	27	21	142	3.2	0.78	35
	151	0.845	4.1	88	6	25	21	149	2.8	0.71	31
4	106	1.086	4.1	177	12	42	23	148	5.2	1.16	52
	112	1.081	4.1	162	11	40	22	145	5.3	1.16	49
	122	1.081	4.0	152	10	33	21	152	4.9	1.11	45
	131	1.000	4.0	137	9	32	22	148	5.1	0.96	42
	140	0.977	4.1	124	8	28	20	146	4.2	0.88	40
	150	0.910	4.1	110	8	26	20	148	4.0	0.84	36
5	106	1.229	4.1	194	12	43	30	157	6.1	1.20	49
	112	1.260	4.1	182	12	41	31	157	6.0	1.19	50
	122	1.180	4.0	164	10	34	26	155	5.7	1.14	45
	131	1.086	4.1	144	9	30	27	157	5.4	1.02	40
	140	1.102	4.0	131	8	30	28	155	5.3	0.96	38
	149	1.004	4.1	114	7	27	28	159	4.9	0.94	34
6	107	1.215	4.1	186	12	40	33	159	5.6	1.25	49
	113	1.284	4.1	185	12	40	33	163	5.5	1.15	49
	123	1.284	4.0	176	11	34	32	159	5.1	1.30	47
	132	1.155	4.0	153	10	33	32	163	5.1	1.05	42
	141	1.155	4.0	138	10	28	31	163	4.8	1.00	40
	150	1.056	4.0	122	8	27	29	166	4.7	1.00	36
7	107	1.456	4.1	214	13	51	37	157	6.1	1.34	57
	113	1.409	4.1	209	13	49	35	160	6.1	1.34	54
	123	1.409	4.1	190	12	46	36	155	5.7	1.29	51
	132	1.284	4.0	169	10	37	35	164	5.8	1.15	45
	141	1.215	4.0	152	10	32	33	162	5.2	1.08	42
	149	1.137	4.0	135	8	29	35	162	5.3	1.03	40
8	107	1.509	4.1	223	13	50	42	148	6.8	1.38	56
	113	1.523	4.1	218	13	49	43	159	6.8	1.35	57
	122	1.409	4.1	198	12	38	38	155	6.6	1.29	52
	131	1.276	4.0	175	10	34	37	153	5.9	1.13	60
	140	1.284	4.0	57	10	32	40	151	6.1	1.06	60
	149	1.168	4.0	147	9	27	39	152	5.4	1.08	58
9	108	1.509	4.0	225	14	50	48	164	8.7	1.45	57
	113	1.538	4.1	226	13	50	50	160	7.9	1.40	58
	122	1.432	4.0	204	12	38	43	171	7.8	1.27	53
	131	1.284	4.1	179	9	37	42	155	6.2	1.14	45
	140	1.301	4.0	164	9	33	45	159	6.7	1.06	44
	148	1.168	4.1	143	8	28	41	160	6.4	1.06	40
10	108	1.509	4.1	240	15	57	57	182	8.3	1.37	58
	113	1.482	4.1	231	13	52	54	173	8.4	1.33	55
	124	1.523	4.0	206	12	41	54	168	8.2	1.27	53
	132	1.367	4.0	183	11	41	41	175	8.1	1.15	47
	140	1.377	4.0	171	10	37	37	176	7.1	1.03	43
	147	1.215	4.0	152	9	32	32	169	7.1	1.00	41

Table 4. Chemical Analysis[a] of Bourbon Whiskies Matured at Six Proofs from 109 to 155° (54.5 to 77.5%), g/100 L at 100° Proof (50%) *(continued)*

Age (yr)	Proof	Color[a]	pH[b]	Solids	Fixed acids	Volatile acids	Esters	Fusel oil	Aldehydes	Furfural	Tannins
11	108	1.699	4.0	252	14	60	60	166	8.2	1.37	64
	113	1.721	4.0	242	13	56	56	168	7.7	1.37	62
	123	1.658	3.9	219	12	49	57	179	7.8	1.34	59
	131	1.509	3.9	192	12	42	54	171	7.7	1.21	53
	139	1.398	3.9	181	10	38	51	178	6.7	1.08	50
	146	1.319	4.0	162	9	34	51	166	6.7	1.05	47
12	107	1.770	4.0	270	21	68	65	189	9.1	1.41	67
	112	1.745	4.0	258	19	52	62	187	9.3	1.37	65
	123	1.699	3.9	231	17	46	59	187	8.5	1.30	61
	130	1.538	3.9	211	15	41	62	187	9.4	1.20	55
	138	1.409	3.9	190	14	36	58	184	8.0	1.09	52
	145	1.319	3.9	168	10	33	54	185	7.8	1.07	47

Note: Analyses performed annually on samples reduced to 100° proof (50%). Acids as acetic acid, esters are ethyl acetate, aldehydes as acetaldehyde, and tannins as tannic acid (11).

[a]Absorbance at 430 nm at 100° proof (50%).

[b]At 100° proof (50%).

[c]Not determined.

Table 5. Changes Taking Place in Bourbon Whisky Stored in Wood

Age[a] (yr)	Range	Proof	Extract[a]	Acids[a]	Esters[a]	Aldehydes[a]	Furfural[a]	Fusel oil[a]	Color
New	Average	101.0	26.5	10.0	18.4	3.2	0.7	100.9	0.0
	Maximum	104.0	161.0	29.1	53.2	7.9	2.0	171.3	0.0
	Minimum	100.0	4.0	1.2	13.0	1.0	Trace	71.3	0.0
								42.0	
1	Average	101.8	99.4	41.1	28.6	5.8	1.6	110.1	7.1
	Maximum	103.0	193.0	55.3	55.9	8.6	7.9	173.4	10.9
	Minimum	100.0	61.0	24.7	17.2	2.7	Trace	58.0	5.4
			54.0	10.4	10.4			42.8	4.6
2	Average	102.2	126.8	45.6	40.0	8.4	1.6	110.1	8.6
	Maximum	104.0	214.0	61.7	59.8	12.0	9.1	197.1	11.8
	Minimum	100.0	81.0	25.5	24.4	5.9	0.4	86.2	6.9
			78.0	23.5	11.2			42.8	5.7
			78.0	23.5	11.2			42.8	5.7
4	Average	104.3	151.9	58.4	53.5	11.0	1.9	123.9	10.8
	Maximum	108.0	249.0	73.0	80.6	22.0	9.6	237.1	14.8
	Minimum	100.0	101.0	40.0	28.2	6.9	0.8	95.0	8.6
			92.0	40.0	13.8			43.5	7.4
6	Average	107.9	185.1	67.1	64.0	/9	1.8	135.3	13.1
	Maximum	116.0	287.0	81.0	83.9	23.3	9.5	240.0	17.5
	Minimum	102.0	132.0	53.6	36.4	7.7	0.9	98.1	12.0
			127.0	45.0	17.9				9.8
8	Average	111.1	210.3	76.4	65.6	12.9	2.1	143.5	14.2
	Maximum	124.0	326.0	91.4	93.6	28.8	10.0	241.8	20.9
	Minimum	102.0	152.0	64.1	37.7	8.7	1.0	110.0	12.3
			141.0	53.7	22.1			47.6	10.5

Source: Ref. 5.

[a]Grams per 100 L of 100° proof (50%) spirit.

tration of various congener groups. These concentrates reacted to form chemical derivatives that were then subjected to further separation. Paper and column chromatography have been used extensively for this work. After separation of the components, identification is possible by infrared and mass spectrometry.

With the high sensitivity (flame and β-ray ionization detection) and efficiency offered by gas chromatographic techniques, a qualitative profile of alcoholic distillates can be obtained without prior treatment of the sample. Flame ionization uses a hydrogen flame for the combustion of organic substances to produce electrons and negative ions, which are collected on an anode. The resulting electrical current is proportional to the amount of material burned. β-Ray ionization uses beta particles emitted from a source such as strontium-90 to ionize the carrier gas and its components. The measure of the electrical current resulting from the collection of electrons on the anode is used in the determination or detection of the substance. This may augment sensory evaluations. A positive identification of the separated congeners can be made by infrared and mass spectrometers. Nmr may offer possibilities in the identification of congeners present in alcoholic beverages.

PACKAGING

Because distilled spirits are packaged under federal government supervision and the product is then distributed to the various states in compliance with laws and regulations, many factors must be considered normally not involved in glass packaging. First, the product represents a high value, because it includes the federal tax at the time of bottling. In the United States, a bottle of whisky retailing at $5.44 is taxed at $3.11. Obviously, care must be taken to avoid losses. Second, in addition to dealing with cases, bottles, cartons, closures, and labels, the distiller must apply to each bottle a federal strip stamp indicating that the excise tax has been paid and also apply bottle stamps of the decal type to indicate identification or tax payment in the seven states that require these in their system of control. With the heavy investment in product taxes and the necessity of applying state stamps, the distiller's ability to build up a substantial case goods inventory to provide immediate service to his customers in those states having this requirement is limited. Third, because this is a licensed industry, in addition to the normal record keeping necessary for efficient operations and control, federal and state records are required.

Federal regulations prescribe and limit the standards of fill (size of containers). On October 1, 1976, the adoption of the metric system permitted distillers to incorporate the new sizes into their operations as circumstances allow. Once the metric size is adopted for a brand, the distiller may not revert to the former standard.

Now, only the metric size will be permitted, such as 1.75 L (59.2 fl oz); 1.00 L (33.8 fl oz); 750 mL (25.4 fl oz); 500 mL (16.9 fl oz); 200 mL (6.8 fl oz); and 50 mL (1.7 fl oz). Individual states likewise limit the number of sizes that can be distributed within their borders and need not adopt all the sizes made available by the federal government.

The most common packaging operation utilized in the industry is the straight-line system with a line speed from 120 to 200 bottles/min. Along with the general progress in packaging, the distilling industry is moving in the direction of automatic-line operation with variable frequency control systems keyed to the fill and labeling operations. This is only possible, however, when volume on certain sizes is substantial. Most distillers are faced with mechanical equipment changes, requiring approximately four hours per line, to handle various sizes of bottles. In addition to material specifications for quality control on all supplies, built-in quality-control inspection systems are included on the bottling lines. Likewise, the quality-control department, independent of bottling operations, takes random samples for evaluation purposes.

Table 6. U.S. Commercial Exports of Liquor by Class 1996 (in wine gal.)

Class and type	
Domestic whiskey	
Bonded	0
Straight	8,293,519
Blended straights	34,814
Distilled over 160/p	2,355
Blended whiskey	158,576
Subtotal	8,489,264
Imported whiskey	
Scotland and N. Ireland	31,515
Bottled in U.K.	0
Bottled in U.S.	31,616
Canadian	110,562
Bottled in Canada	0
Bottled in U.S.	110,562
Ireland and others	0
Subtotal	142,077
Total all whiskey	*8,631,341*
Gin	
Domestic	553,942
Imported	0
Subtotal	553,942
Vodka	
Domestic	6,008,700
Imported	0
Subtotal	6,008,700
Rum	
Puerto Rico	1,034,908
Virgin Islands	89,834
Other domestic	244,789
Subtotal dom. rum	1,369,531
Subtotal imp. rum	0
Subtotal all run	1,369,531
Brandy	
Domestic	260,337
Imported	0
Subtotal	260,337
Cordials and Specs.	
Domestic	578,408
Imported	0
Subtotal	578,408
Cocktails and mixed drinks	
Domestic	110,167
Imported	0
Subtotal	110,167
Tequila	
Subtotal	308,068
Non-whiskey	
Domestic	8,881,075
Imported	308,058
Total	*9,183,133*

Table 6. U.S. Commercial Exports of Liquor by Class 1996 (in wine gal.) *(continued)*

Class and type	
N.E.S.	
Domestic	14,630,239
Imported	0
Subtotal	14,630,239
Total Dom. Distilled Spirits	*32,000,574*
Total Imp. Distilled Spirits	*450,135*
Grand Total	*32,450,713*

EXPORTS OF DISTILLED SPIRITS

The United States is well known throughout the world for its world-class whiskies and these are exported to many lands. Tastes differ widely in other countries. In some, like Germany and Japan the major export by far is bourbon whisky. In others, like Belgium, it is vodka. In all, bourbon, the prototypical U.S. whisky, is the largest distilled spirit export. Table 6 shows exports of distilled spirits in 1996.

GLOSSARY

Balling. A measure of the sugar concentration in a grain mash, expressed in degrees and approximating percent by weight of the sugar in solution.

Bushel. A distillers bushel of any cereal gain is 25.4 kg (56 lb).

Congeners. The flavor constituents in beverage spirits that are responsible for its flavor and aroma and that result from the fermentation, distillation, and maturation processes.

Feints. The third fraction of the distillation cycle derived from the distillation of low wines in a pot still. This term is also used to describe the undesirable constituents of the wash which are removed during the distillation of grain whisky in a continuous patent still (Coffey). These are mostly aldehydes and fusel oils.

Foreshots. The first fraction of the distillation cycle derived from the distillation of low wines in a pot still.

Fusel oil. An inclusive term for heavier, pungent-tasting alcohols produced during fermentation. Fusel oil is composed of approximately 80% amyl alcohols, 15% butyl alcohols, and 5% other alcohols.

Grain whisky. An alcoholic distillate from a fermented wort derived from malted and unmalted barley and maize (corn), in varying proportions, and distilled in a continuous patent still (Coffey).

Heads. A distillate containing a high percentage of low-boiling components such as aldehydes.

High wines. An all-inclusive term for beverage spirit distillates that have undergone complete distillation.

Low wines. The term for the initial product obtained by separating (in a pot still) the beverage spirits and congeners from the wash. Low wines are subjected to at least one more pot still distillation to attain a greater degree of refinement in the malt whisky.

Malt whisky. An alcoholic distillate made from a fermented wort derived from malted barley only, and distilled in pot stills. It is the second fraction (heart of the run) of the distillation process.

Proof. In Canada, the UK, and the United States, the alcoholic concentration of beverage spirits is expressed in terms of proof. The United States statutes define this standard as follows: proof spirit is held to be that alcoholic liquor that contains one-half its volume of alcohol of a specific gravity of 0.7939 at 15.6°C; ie, the figure for proof is always twice the alcoholic content by volume. For example, 100° proof means 50% alcohol by volume. In the UK as well as Canada, proof spirit is such that at 10.6°C weighs exactly twelve-thirteenths of the weight of an equal bulk of distilled water. A proof of 87.7° would indicate an alcohol concentration of 50%. A conversion factor of 1.142 can be used to change British proof to U.S. proof.

Proof gallon. A U.S. gallon of proof spirits or the alcoholic equivalent thereof; ie, a U.S. gallon of 231 in.^{3} (3,785 cm^3) containing 50% of ethyl alcohol by volume. Thus a gallon of liquor at 120° proof is 1.2 proof gal; a gallon at 86° proof is 0.86 proof gal. A British and Canadian proof gallon is an imperial gallon of 277.4 in.^{3} (4,546 cm^3) at 100° proof (57.1% of ethyl alcohol by volume). An imperial gallon is equivalent to 1.2 U.S. gal. To convert British proof gallons to U.S. proof gallons, multiply by the factor 1.37. Since excise taxes are paid on the basis of proof gallons, this term is synonymous with tax gallons.

Single whisky. The whisky, either grain or malt, produced by one particular distillery. Blended Scotch whisky is not a single whisky.

Spirits. Distilled spirits including all whiskies, gin, brandy, rum, cordials, and others made by a distillation process for nonindustrial use.

Tails. A residual alcoholic distillate.

Wash. The liquid obtained by fermenting wort with yeast. It contains the beverage spirits and congeners developed during fermentation.

Wine gallon. Measure of actual volume; U.S. gallon (3.7845118 L) contains 231 in.^{3} (3,785 cm^3); British (Imperial) gallon contains 277.4 in.^{3} (4,546 cm^3).

Wort. The liquid drained off the mash tun and containing the soluble sugars and amino acids derived from the grains.

BIBLIOGRAPHY

1. *Proceedings Before and By Direction of the President Concerning the Meaning of the Term Whisky*, U.S. Government Printing Office, Washington, D.C., 1909.
2. *Business Week* (Feb. 17, 1962).
3. *Business Week* (Mar. 21, 1977).
4. S. Baldwin et al., *Journal of Agricultural and Food Chemistry* **15**, 381 (1967).
5. C. A. Crampton and L. M. Tolman, *Journal of the American Chemical Society* **30**, 97 (1908).
6. A. J. Liebmann and M. Rosenblatt, *Industrial and Engineering Chemistry* **35**, 994 (1943).

7. A. J. Liebmann and B. Scherl, *Industrial and Engineering Chemistry* **41**, 534 (1949).
8. P. Valaer and W. H. Frazier, *Industrial and Engineering Chemistry* **28**, 92 (1936).
9. M. C. Brockmann, *Journal of the Association of Official Agricultural Chemists* **33**, 127 (1950).
10. G. H. Reazin et al., *Journal of the Association of Official Agricultural Chemists* **59**, 770 (1976).
11. S. Baldwin and A. A. Andreasen, *Journal of the Association of Official Agricultural Chemists* **53**, 940 (1974).
12. Bureau of the Census, U.S. Department of Commerce, *Distilled Spirits Council of the U.S.*, Aug. 1989.

GENERAL REFERENCES

Annual Statistical Revue (1976), Distilled Spirits Council of the US, Inc., Washington, D.C., 1977.

A. Barnard, *The Whiskey Distilleries of the United Kingdom*, London, 1887.

H. Barron, *Distillation of Alcohol*, Joseph E. Seagram & Sons, Inc., New York, 1944.

C. S. Boruff and L. A. Rittschol, *Agric. Food Chem.* **7**, 630 (1959).

Canada, Its History, Products, and Natural Resources, Department of Agriculture of Canada, Ottawa, 1906.

A. Cooper, *The Complete Distiller*, London, 1760.

H. G. Crowgey, *Kentucky Bourbon, The Early Years of Whiskey Making*, The University Press of Kentucky, Lexington, 1971.

Distillers Feed Research, Distillers Feed Research Council, Cincinnati, Ohio, 1971.

G. A. DeBecze, "Alcoholic Beverages, Distilled," *Encyclopedia of Industrial Chemical Analysis*, Vol. 4, Interscience Division of John Wiley & Sons, Inc., New York, 1967, pp. 462–494.

The Excise Act, 1934, Department of National Revenue, Ottawa, Canada, 1947, Chap. 52, pp. 24–25.

G. Foth, *Handbuch der Spiritusfabrikation*, Verlag Paul Parey, Hamburg, Germany, 1929.

G. Foth, *Die Praxis des Brennereibetriebs*, Verlag Paul Parey, Hamburg, Germany, 1935.

A. Herman, E. M. Stallings, and H. F. Willkie, "Chemical Engineering Developments in Grain Distillery," *Trans. Am. Inst. Chem. Eng.* **31**, 1942.

A. J. Liebmann, "Alcoholic Beverages," in R. E. Kirk and D. F. Othmer, eds., *Encyclopedia of Chemical Technology*, Vol. 1, Interscience Publishers, New York, 1947, pp. 228–303.

A. McDonald, *Whisky*, Glasgow, UK, 1934.

Methods of Analysis, 12th ed., Association of Official Agricultural Chemists, New York, 1975.

G. W. Packowski, "Alcoholic Beverages, Distilled" in A. Standen, ed., *Kirk-Othmer Encyclopedia of Chemical Technology*, Vol. 1, 2nd ed., Interscience Division of John Wiley & Sons, Inc., New York, 1963, pp. 501–531.

G. W. Packowski, "Distilled Beverage Spirits," in M. Grayson, ed., *Kirk-Othmer Encyclopedia of Chemical Technology*, Vol. 3, 3rd ed., Wiley Interscience, New York, 1978, pp. 830–869.

D. R. Peryam, *Ind. Qual. Control* **11**, 17 (1950).

Regulations, Distilleries and Their Products, Circular ED 203, Department of National Revenue, Ottawa, Canada, Mar. 30, 1961.

Regulations under the Federal Alcohol Administration Act, Alcohol, Tobacco Products and Firearms (27 CFR) ATF P 5100.8 (12/77), U.S. Government Printing Office, Washington, D.C.

J. Samuelson, *The History of Drink*, London, 1880.

E. D. Unger, H. F. Willkie, and H. C. Blankmyer, "The Development and Design of a Continuous Cooking and Mashing System for Cereal Grains," *Trans. Am. Inst. Chem. Eng.* **40**, 1944.

H. F. Willkie, *Beverage Spirits in America—A Brief History*, The Newcomen Society in North America, Downington, Pa., 1949.

H. F. Willkie and J. A. Proschaska, *Fundamentals of Distillery Practice*, Joseph E. Seagram & Sons, Inc., New York, 1943.

F. B. Wright, *Distillation of Alcohol*, London, 1918.

H. Wustenfeld, *Trinkbranntweine und Liquöre*, Berlin, Germany, 1931.

C. L. Yaws, J. R. Hopper, "Methanol, Ethanol, Propanol and Butanol, Physical and Thermodynamic Properties," *Chemical Engineering* 119 (June 7, 1976).

JOSEPH L. OWADES
Consultant
Sonoma, California

DRY MILK

HISTORY

The development of the dry milk industry apparently stems from the days of Marco Polo in the thirteenth century. Reports are that Marco Polo encountered "sun dried milk" on his journeys through Mongolia and that from this beginning dry milk products evolved.

Through early pioneering scientists, such as Nicholas Appert and Gail Borden, basic processing methods were developed for drying milk products. Martin Ekenberg and Lewis Merrill have been acknowledged as developers of the first commercial roller- and spray-process drying systems, respectively, in the United States.

Since the initial development of commercial drying systems, significant technological advances have been made resulting in such widely recognized and used dry milk products as nonfat dry milk, dry whole milk, and dry buttermilk, which may be manufactured by roller- or spray-dryers (now mostly by the latter) or by more unique processes such as foam or freeze drying. These same dry milk products also may be processed in such a manner as to make them readily soluble and thus "instantized."

PROCESSING

The steps in a typical dry milk processing operation are as follows:

- Receipt of fresh, high-quality milk from modern dairy farms, delivered in refrigerated, stainless steel bulk tankers.
- Separation of the milk (if nonfat dry milk is to be manufactured) to remove milkfat, which commonly is churned into butter. If dry whole milk is to be manufactured, the separation step is omitted but may be replaced by clarification.
- Pasteurization by a continuous high-temperature short-time (HTST) process whereby every particle of milk is subjected to a heat treatment of at least 71.7°C for 15 sec.
- Holding the pasteurized milk at an elevated temperature for an extended period of time (76.7°C for

25–30 min.)—a step used only in the manufacture of high-heat nonfat dry milk, which commonly is used as an ingredient in yeast-raised bakery products and meat products.

- Condensing the milk by removing water in an evaporator or vacuum pan until a milk solids content of 40% is reached.
- Delivery of the condensed product to the dryer.

Commercial U.S. drying processes are of two types—roller (drum) and spray. The former currently is used only to a limited extent. In this process, two large rollers, usually steam-heated internally and located adjacent and parallel to each other, revolve in opposite directions contacting a reservoir of either pasteurized fluid or condensed milk. During rotation, the fluid milk product dries on the hot roller surface. After approximately three-quarters of a revolution, a carefully positioned, sharp stationary knife detaches the milk product, now in the form of a thin dry sheet. The dry milk next is conveyed by an auger to a hammermill where it undergoes a physical treatment to convert it into uniformly fine particles which then are packaged, usually in 25-kg polyethylene-lined multiwall bags.

Two basic configurations of spray dryers presently are in use, these being horizontal (box) and vertical (tower) dryers. In both, pasteurized fluid milk, which has been condensed to a total solids of 40% or greater, is fed under pressure to a spray nozzle, or an atomizer, where the dispersed liquid then comes into contact with a current of filtered, heated air. The droplets of condensed milk are dried almost immediately and fall to the bottom of the fully enclosed, stainless steel drying chamber. The dry milk product continuously is removed from the drying chamber, transported through a cooling and collecting system, and finally conveyed into a hopper for packaging—usually into 25-kg bags, or in tote bins.

Table 1. Yearly U.S. Production of Dry Milks[a]

Year	Nonfat dry milk	Dry whole milk	Dry buttermilk
1960	1,818.6	98.0	86.4
1961	2,019.8	81.7	89.0
1962	2,230.3	86.1	86.4
1963	2,106.1	91.0	87.5
1964	2,177.2	87.6	92.0
1965	1,988.5	88.6	87.4
1966	1,579.8	94.4	76.2
1967	1,678.7	74.3	72.6
1968	1,594.4	79.8	70.4
1969	1,452.3	70.2	66.5
1970	1,444.4	68.9	59.5
1971	1,417.6	72.2	51.7
1972	1,223.5	75.2	49.5
1973	916.6	78.0	43.3
1974	1,020.0	67.7	45.3
1975	1,001.5	63.1	42.8
1976	926.2	78.1	46.3
1977	1,106.6	69.4	53.2
1978	920.4	74.6	47.6
1979	908.7	85.3	44.7
1980	1,160.7	82.7	43.9
1981	1,314.3	92.7	43.8
1982	1,400.5	102.2	38.6
1983	1,499.9	111.2	46.5
1984	1,160.8	119.6	43.5
1985	1,390.0	118.9	51.5
1986	1,284.1	122.4	65.7
1987	1,056.8	145.9	55.6
1988	978.5	172.3	58.7
1989	874.7	175.8	60.5
1990	879.2	175.1	55.9
1991	877.5	106.8	60.0
1992	872.1	168.3	61.1
1993	954.5	153.8	51.0
1994	1,215.6	166.8	52.3
1995	1,233.0	171.3	54.8
1996	1,061.6	134.4	48.8
1997	1,217.6	122.1	49.3
1998	1,121.3	138.7	49.3

[a]In millions of pounds.

THE PRODUCT

The primary dehydrated dairy products manufactured domestically are nonfat dry milk, dry whole milk, and dry buttermilk.

Nonfat dry milk is the product resulting from the removal of fat and water from milk. It contains lactose, milk proteins, and milk minerals in the same relative proportions as in the fresh milk from which it was made. Nonfat dry milk contains not over 5% by weight of moisture. The fat content is not over 1.5% by weight unless otherwise indicated.

Dry whole milk is the product resulting from the removal of water from milk and contains not less than 26% milkfat and not more than 4% moisture. Dry whole milk with milkfat contents of 26 and 28.5% commonly are produced.

Table 2. Approximate Composition and Food Value of Dry Milks

Constituents	Nonfat dry milk	Dry whole milk	Dry buttermilk
Protein ($N \times 6.38$) (%)	36.0	26.0	34.0
Lactose (milk sugar) (%)	51.0	38.0	48.0
Fat (%)	0.7	26.75	5.0
Moisture (%)	3.0	2.25	3.0
Minerals (ash) (%)	8.2	6.0	7.9
Calcium (%)	(1.31)	(0.97)	(1.3)
Phosphorus (%)	(1.02)	(0.75)	(1.0)
Vitamin A (IU/lb)	165.0	4,950.0	2,300.0
Riboflavin (mg/lb)	9.2	6.7	14.0
Thiamin (mg/lb)	1.6	1.2	1.2
Niacin (mg/lb)	4.2	3.1	4.5
Niacin equivalents[a] (mg/lb)	42.2	30.6	40.6
Pantothenic acid (mg/lb)	15.0	13.0	14.0
Pyridoxine (mg/lb)	2.0	1.5	2.0
Biotin (mg/lb)	0.2	0.185	0.2
Choline (mg/lb)	500.0	400.0	500.0
Energy (cal/lb)	1,630.0	2,260.0	1,700.0

[a]Includes contribution of tryptophan.

Dry buttermilk is the product resulting from the removal of water from liquid buttermilk derived from the manufacture of butter. It contains not less than 4.5% milkfat and not more than 5% moisture.

All of these products may be processed in such a way that their dispersing and reliquefaction properties are substantially improved. When this is accomplished, the products are called instantized. "Single-pass" and "agglomerating" processes are used to make instant dry milk products, with the primarily method being the latter.

Table 1 reflects the annual domestic production of nonfat dry milk, dry whole milk, and dry buttermilk during the period 1960 to 1998. Standards for dry milk products initially were developed by the dairy industry in 1929. Since that time, additional product standards have been developed—and revised—both by industry and government agencies. These standards are based on various general and specific product characteristics, which serve as a basis for determining overall quality of the dry milk products. Table 2 shows the approximate composition and food value of the three most commonly manufactured dry milk products.

UTILIZATION

An industry-wide survey of end uses for dry milk products is conducted annually by the American Dairy Products Institute. Data compiled in the survey are published under the title *Dry Milk Products—Utilization and Production Trends*. Included in the census data are end-use markets for nonfat dry milk, dry whole milk, and dry buttermilk. For the past several years the largest markets for nonfat dry milk have been dairy, prepared dry mixes, confectionery, and bakery. These market areas accounted for 781 million pounds in 1998—approximately 89% of the domestic sales. The primary market for dry whole milk is confectionery. In 1998, this market area accounted for 88 million pounds of dry whole milk—over 66% of the domestic sales for this product. Dry buttermilk is used in three primary markets—dairy, bakery, and prepared dry mixes. In 1998, these markets used 39.6 million pounds of dry buttermilk—representing 89% of domestic sales.

GENERAL REFERENCES

American Dairy Products Institute, *Standards for Grades of Dry Milks Including Methods of Analysis*, Bulletin 916 (Rev.), American Dairy Products Institute, Chicago, Ill., 1990.

American Dairy Products Institute, *Dry Milk Products—Utilization and Production Trends, 1997*, Bulletin 1000, American Dairy Products Institute, Chicago, Ill., 1999.

C. E. Beardslee, *Dry Milks—The Story of an Industry*, American Dry Milk Institute, Chicago, Ill., 1948.

C. W. Hall and T. I. Hedrick, *Drying of Milk and Milk Products*, 2nd ed., AVI Publishing, Westport, Conn., 1971.

WARREN S. CLARK, JR.
American Dairy Products Institute
Chicago, Illinois

DRYERS: TECHNOLOGY AND ENGINEERING

Throughout the food processing industries, there are many and varied requirements for thermal drying. Some involve the removal of moisture or volatiles from various food ingredients or products that differ in both chemical and physical characteristics. Others involve the drying of solutions or liquid suspensions and different approaches to the problem. To assist manufacturers in arriving at a reasonably accurate first assessment of the type, size, and cost of element for a particular duty, this article describes the most widely used types of both batch and continuous dryers in the food industries and gives an indication of approximate sizes and capital costs for typical installations.

Three basic methods of heat transfer are used in industrial dryers in varying degrees of prominence and combinations, specifically, convection, conduction, and radiation.

In the chemical processing industry, the majority of dryers employ forced convection and continuous operation. With the exception of the indirectly heated rotary dryer and the film drum dryer, units in which heat is transferred by conduction are suitable only for batch use. This limitation effectively restricts them to applications involving somewhat modest production runs.

Radiant, or "infrared," heating is rarely used in drying materials such as fine chemicals or pigments. Its main application is in such operations as the drying of surface coatings on large plane surfaces since for efficient utilization, it generally is true that the material being irradiated must have a sight of the heat source or emitter. There is, however, in all the dryers considered here a radiant component in the heat-transfer mechanism.

Direct heating is used extensively in industrial drying equipment where much higher thermal efficiencies are exhibited than with indirectly heated dryers. This is because there are no heat exchanger losses and the maximum heat release from the fuel is available for the process. However, this method is not always acceptable, especially where product contamination cannot be tolerated. In such cases, indirect heating must be used.

With forced-convection equipment, indirect heating employs a condensing vapor such as steam in an extended surface tubular heat exchanger or in a steam jacket where conduction is the method of heat transfer. Alternative systems that employ proprietary heat-transfer fluids also can be used. These enjoy the advantage of obtaining elevated temperatures without the need for high-pressure operation as may be required with conventional steam heating. This may be reflected in the design and manufacturing cost of the dryer. Furthermore, in addition to the methods listed above, oil- or gas-fired indirect heat exchangers also can be used.

In general, dryers are either suitable for batch or continuous operation. A number of the more common types are listed in Table 1, where an application rating based on practical considerations is given. In the following review, some of the factors likely to influence selection of the various types are discussed for particular applications.

BATCH DRYERS

It will be apparent that batch operated equipment usually is related to small production runs or to operations requir-

Table 1. Product Classification and Dryer Types as an Aid to Selection

	Evaporation rate, lb/ft^2 · h mean rate = E_{av}	Fluids, liquid suspension	Pastes, dewatered cake	Powders	Granules, pellets, extrudates	Operation
Forced convection (cross-airflow)	0.15–0.25					
	E_{av} = 0.2	Poor	Fair	Fair	Good	Batch
Forced convection (throughflow)	1.0–2.0					
	E_{av} = 1.5	—	—	—	Good	Batch
Agitated pan (sub-atmospheric)	1.0–5.0					
	E_{av} = 3.0	Fair	Fair	Fair	Poor	Batch
Agitated pan (atmospheric)	1.0–5.0					
	E_{av} = 3.0	Fair	Fair	Fair	Poor	Batch
Double-cone tumbler (sub atmospheric)	1.0–3.0					
	E_{av} = 2.0	—	Poor	Fair	Poor	Batch
Fluidized bed (throughflow)	2–50					
	E_{av} = 26	—	—	Good	Good	Continuous
Conveyor band (throughflow)	2.0–10.0					
	E_{av} = 6.0	—	Fair	—	Good	Continuous
Rotary (indirect)	1.0–3.0[a]					
	E_{av} = 2.0	—	Poor	Good	Fair	Continuous
Rotary (direct)	2.0–6.0[a]					
	E_{av} = 4.0	—	Fair	Fair	Good	Continuous
Film drum (atmospheric)	3.0–6.0					
	E_{av} = 4.5	Good	Fair	—	—	Continuous
Pulumatic or flash	50–250					
	E_{av} = 150	—	Fair	Good	Fair	Continuous
Spray	7.0–33.0					
	E_{av} = 20.0	Good	—	—	—	Continuous

[a]Evaporation rates for rotary dryers are expressed in lb/ft^3 · h.

ing great flexibility. As a result, the batch-type forced-convection unit certainly finds the widest possible application of any dryer used today.

The majority of designs employ recirculatory air systems incorporating large-volume, low-pressure fans that, with the use of properly insulated enclosures, usually provide thermal efficiencies in the region of 50–60%. However, in special applications of this type of dryer that call for total air rejection, this figure is somewhat lower and is largely related to the volume and temperature of the exhaust air. Capital investment is relatively low, as are installation costs. Furthermore, by using the fan systems, both power requirements and operating costs also are minimal. Against these advantages, labor costs can be high.

In such a plant, the drying cycles are extended, with 24–45 hours being quite common in certain cases. This is a direct result of the low evaporative rate, which normally is in the region of 0.15–0.25 lb/ft^2 h.

Following the recent trend and interest shown in preforming feedstock and in particular with regard to the design of extruding and tray-filling equipment for dewatered cakes, it is now possible to obtain the maximum benefit of enhanced evaporative rates by using through-air circulation dryers when handling preformed materials. Figure 1 shows how a high-performance dryer can product 1950 pounds of dried material in a 24-h period at a terminal figure of 0.5% moisture when handling a preformed filter cake having an initial moisture content of 58%. The very great improvement in performance can readily be seen from the curve, in which it is clear that the corresponding number of conventional two-truck recirculatory units

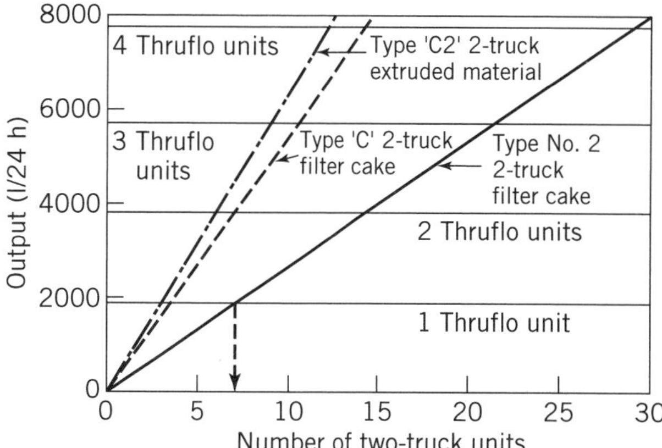

Figure 1. Comparative performance curves for Thruflo and conventional units.

would be between seven and eight for the same duty. The advantage is more apparent when it is seen that respective floor areas occupied are 55 ft^2 for the Thruflo dryer pictured in Figure 2 and 245 ft^2 in the case of conventional units using transverse airflow.

Reference to the drying curves for the processing of materials in solid or filter cake form or, in fact, in the case of wet powders, clearly indicates that the ultimate rate-governing factor is the rate of diffusion of moisture from the wet mass. This becomes increasingly so during the falling rate period of drying. This situation, however, can be

Figure 2. Thruflo dryer.

improved by preforming the product in order to increase the effective surface area presented to heat and mass transfer. The logical extension of this technique is to total dispersion drying (flash or pneumatic dryers, fluid beds, etc) where discrete particles can be brought into contact with the hot gas. This produces rapid heat transfer with correspondingly short drying times.

Batch-type fluidized-bed dryers have, therefore, superseded forced-convection units in many cases, notably in the drying of pharmaceuticals. These machines generally are available in a range of standard sizes with batch capacities of 50–200 lb, although much larger units are made for special applications.

When considering this type of dryer, it is important to ensure that the feed material can be fluidized in both initial and final conditions. It also should be remembered that standard fan arrangements are not equally suitable for a variety of materials of different densities. Therefore, it is necessary to determine accurately the minimum fluidizing velocity for each product.

If the feedback is at an acceptable level of moisture content for fluidization, this type of dryer provides many advantages over a batch-type unit. Simplified loading and unloading results in lower labor costs, high thermal efficiencies are common, and the drying time is reduced to minutes as opposed to hours in conventional units. Current developments of this type of equipment now include techniques for the simultaneous evaporation of water and the granulation of solids. This makes the units ideal for use in the pharmaceutical field.

The various batch dryers referred to operate by means of forced convection, the transfer of thermal energy being designed to increase the vapor pressure of the absorbed moisture while the circulated air scavenges the overlying vapor. Good conditions thus are maintained for continued effective drying. Alternatively and where the material is thermosensitive—implying low temperatures with consequently low evaporative rates—some improvement can be effected by the use of subatmospheric dryers, ie, by reducing the vapor pressure. Several different configurations are in use and all fall into the category of conduction-type dryers. The most usual type of heating is by steam, although hot water or one of the proprietary heat-transfer fluids can be used.

Two particular types are the double-cone dryer (Fig. 3) with capacities of $\leq$400 ft^3 and the agitated-pan dryer not normally larger than 8 ft diameter, where average evaporative rates per unit wetted area usually are in the region of 4 lb/ft^2 h. These units are comparatively simple to operate and, when adequately insulated, are thermally quite efficient, although drying times can be extended. They are especially suitable for applications involving solvent recovery and will handle powders and granules moderately well. They do, however, suffer from the disadvantage with some materials that the tumbling action in double-cone dryers and the action of the agitator in agitated-pan machines can produce a degree of attrition in the dried product that may prove unacceptable.

Similarly, quite large rotary vacuum dryers are used for pigment pastes and other such materials, especially where organic solvents present in the feedstock have to be recovered. These units normally are jacketed and equipped with an internal agitator that constantly lifts and turns the material. Heat transfer here is entirely by conduction from the wall of the dryer and from the agitator. Owing to the nature of their construction, initial cost is high relative to capacity. Installation costs also are considerable. In general, these dryers find only limited application.

CONTINUOUS DRYERS

For the drying of liquids of liquid suspensions, two types of dryers can be used: film drum dryers for duties in the region of 600 lb/h for a larger dryer of about 4 ft diameter × 10 ft face length or large spray dryers (as in Fig. 4) with drying rates of approximately 22,000 lb/h. Where tonnage production is required, the drum dryer is at a disadvan-

Figure 3. Double-cone vacuum dryer.

Figure 4. Conical section of a large spray dryer.

tage. However, the thermal efficiency of the drum dryer is high in the region of 1.3–1.5 lb steam/lb of water evaporated and for small to medium production runs, it does have many applications.

Drum dryers usually are steam-heated, although work has been done involving the development of units for direct gas or oil heating. Completely packaged and capable of independent operation, these dryers can be divided into two broad classifications, ie, single-drum and double-drum.

Double-drum machines normally employ a "nip" feed device with the space between the drums capable of being adjusted to provide a means of controlling the film thickness. Alternatively, and in the case of the single-drum types, a variety of feeding methods can be used to apply material to the drum. The most usual is the simple "dip" feed. With this arrangement, good liquor circulation in the trough is desirable in order to avoid increasing the concentration of the feed by evaporation. Again, for special applications, single-drum dryers use top roller feed. While the number of rolls is related to the particular application and the material being handled, in general this method of feeding is used for pasty materials such as starches. Where the feed is very mobile, rotating devices such as spray feeds are used.

It will be seen from the Figure 5 drawings that there are a number of different feeding arrangements for drum dryers, all of which have a particular use. In practice, these variants are necessary owing to the differing characteristics of the materials to be dried and to the fact that no universally satisfactory feeding device has yet been developed. This again illustrates the need for testing, not only in support of theoretical calculations for the determination of the best dryer size but also to establish where a satisfactory film can be formed.

It must be emphasized that the method of feeding the product to the dryer is of paramount importance to selection or design. There are, of course, certain materials that are temperature-sensitive to such a degree that their handling would preclude the use of an atmospheric drum dryer. In such cases, special subatmospheric equipment may provide the answer, although the capital cost in relation to output generally would restrict its use to premium grade products.

As an alternative, the spray dryer offers an excellent solution to many drying problems. Many materials that would suffer from thermal degradation if dried by other methods often can be handled by spray drying owing to the rapid flash evaporation and its accompanying cooling effect. The continuous method of operation also lends itself to large outputs and with the correct application of control equipment, to low labor cost as well.

SPRAY DRYERS

Fundamentally, the spray-drying process is a simple one. However, the design of an efficient spray-drying plant requires considerable expertise along with access to large-scale test facilities, particularly where particle size and bulk density requirements in the dried product are critical. The sizing of spray dryers on a purely thermal basis is a comparatively simple matter since the evaporation is entirely a function of the Δt across the dryer. Tests on pilot-scale equipment are not sufficient in the face of such imponderables as possible wall buildup, bulk density, and particle size predictions. Atomization of the feed is of prime importance to efficient drying and three basic feed devices are used extensively:

1. Single-fluid nozzle or pressure type
2. Two-fluid nozzle or pneumatic type
3. Centrifugal (spinning disk)

The single-fluid nozzle produces a narrow spray of fine particles. While a multiplicity of nozzles of this type are used in tonnage plants to obtain the desired feedrate, because of the high pressures employed (up to 7000 psig), excessive wear can result, particularly with abrasive products, as an alternative, the two-fluid nozzle with external mixing is used for a variety of abrasive materials. This system generally is limited to small-capacity installations. Normally, the feed is pumped at about 25 psig merely to induce mobility while the secondary fluid is introduced at 50–100 psig, thus producing the required atomization.

Centrifugal atomization achieves dispersion by centrifugal force; the feed liquor is pumped to a spinning disk. This system is suitable for and generally used on large productions. When stacked or multiple disks are employed, feed rates of 40,000–60,000 lb/h are not uncommon.

Many spray dryer configurations are in current use along with a variety of airflow patterns. The nature of the chamber geometry selected is strictly related to the system

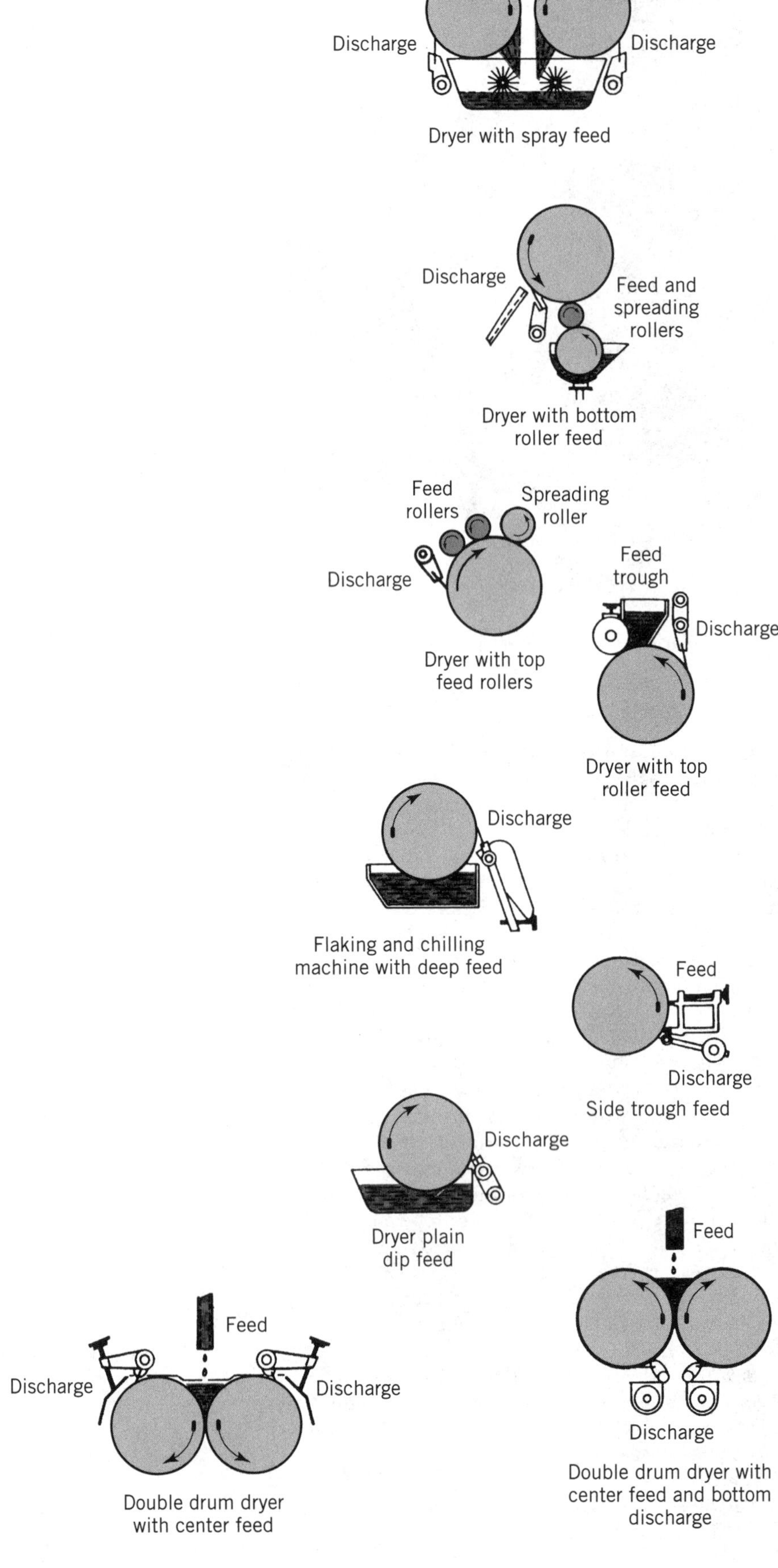

Figure 5. Feeding arrangements for drum dryers.

of atomization used. An example of this is the tower configuration designed to accommodate the inverted jet of the two-fluid nozzle whereas the cylinder and cone of the more usual configuration are designed for the spray pattern produced by a disk-type atomizer (Fig. 6).

The product collection systems incorporated in spray-drying installations are many and varied and can constitute a substantial proportion of the total capital investment. In some cases, this can be as high as 20–25% of the installed plant cost. It also must be remembered that to be suitable for spray drying, the feed must be in a pumpable condition. Therefore, consideration must be given to the upstream process, ie, whether there is any need to reslurry in order to make the feed suitable for spray drying.

It generally is accepted that mechanical dewatering is less costly than thermal methods and while the spray dryer exhibits quite high thermal efficiencies, it often is at a disadvantage relative to other drying systems because of the greater absolute weight of water to be evaporated owing to the nature of the feed. It is, however, interesting to consider this point further. A classic case for comparison is provided by a thixotropic material that may be handled either in a spray dryer using a mechanical disperser doing work on a filtered cake to make it amenable to spray drying or alternatively, drying the same feedstock on a continuous-band dryer. In the latter instance, the cake is fed to an extruder and suitably preformed prior to being deposited on the conveyor band.

The operating costs presented in Table 2 are based on requirements for thermal and electrical energy only. No consideration is given to labor costs for either type of plant since these are likely to be approximately the same. Probably the most obvious figure emerging from this comparison is the 20% price differential in favor of the continuous-band dryer. Furthermore, while energy costs favor spray drying, they are not significantly lower.

It is, of course, impossible to generalize since the economic viability of a drying process ultimately depends on the cost per pound of the dried product and, as mentioned previously, the spray dryer usually has a greater amount of water to remove by thermal methods than other types. In the particular case illustrated, the spray dryer would have an approximate diameter of 21 ft for the evaporation of 6000 lb/h. If, however, the feed solids were reduced to 30% by dilution, the hourly evaporation rate would increase to 14,000 lb and the chamber diameter would be about 30 ft with corresponding increases in thermal input and air volume. The system would, as a result, also require larger fans and product collection systems. The overall thermal efficiency would remain substantially constant at 76% with reducing feed solids. Installed plant costs, however, increase proportionally with increasing dryer size necessary for the higher evaporation involved in producing a dried output equivalent to that shown in Table 2.

Spray drying does have many advantages, particularly with regard to the final product form. This is especially so where pressing grade materials are required, ie, in the production of ceramics and dust-free products such as dyestuffs. It is certain that with the introduction of new ge-

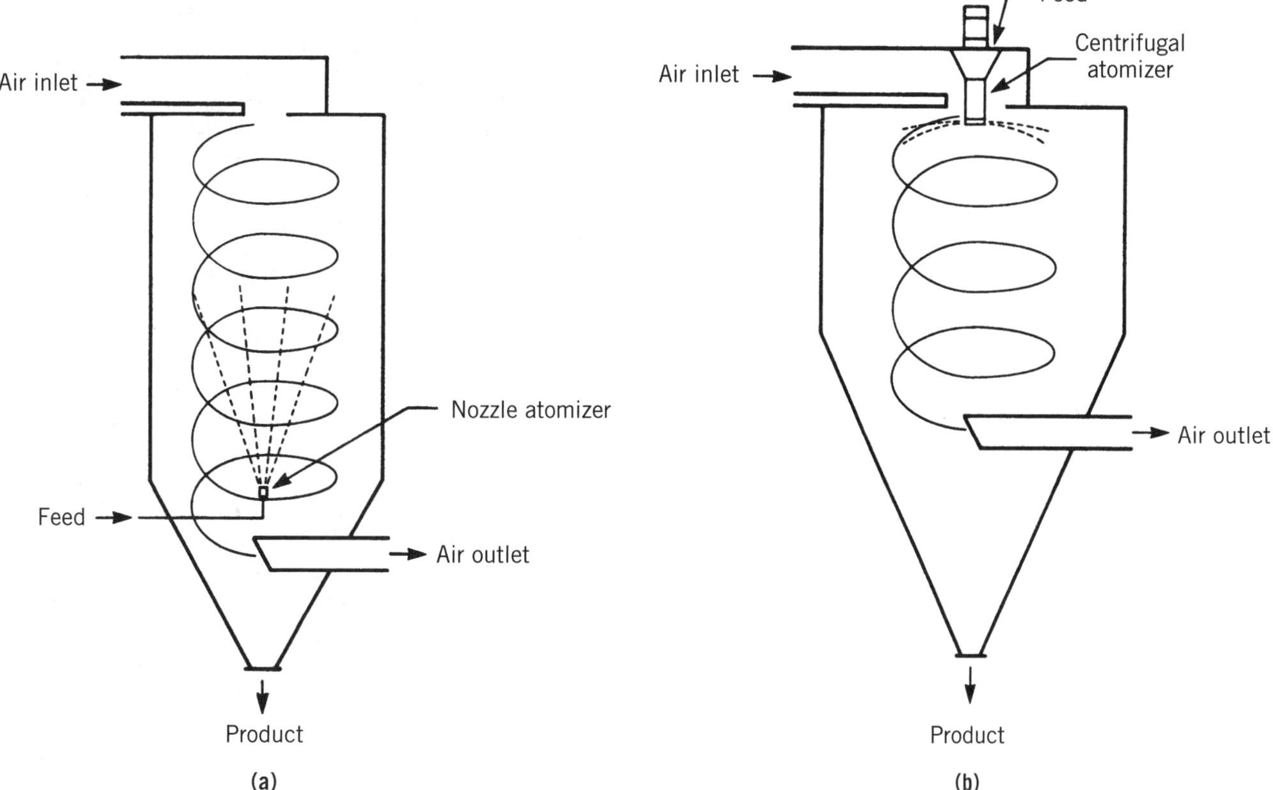

Figure 6. Alternative configurations of spray dryers showing (**a**) tall form type and (**b**) conical.

Table 2. Comparison between Direct-Fired Band Dryer and Spray-Dryer Costs in the Processing of Titanium Dioxide

	Spray dryer	Band dryer
Initial moisture content, W/W	50%	50%
Final moisture content, W/W	0.5%	0.5%
Feed, lb/h	12,000	12,000
Evaporation lb/h	5970	5970
Output, lb/h	6030	6030
Inlet air temperature, °F	1150	393
Exhaust air temperature, °F	250	300
Direct gas firing	Gas CV = 1000 Btu/ft^2	Gas CV = 1000 Btu/ft^2
Hourly operating cost gas and electricity	$39.00	$42.26
Thermal efficiency	76.0%	69.5%
Installed horsepower	149	154
Floor area occupied, ft^2	650	2100
Installed cost	$465,000	$330,000
Energy cost per pound of dried product	$0.065	$0.070

ometries and techniques there will be further development into areas such as foods and in the production of powders that may be easily reconstituted.

ROTARY DRYERS

Another type of dryer that is very much in evidence in the chemical and process industries is the continuous rotary dryer. This machine generally is associated with tonnage production and as a result of its ability to handle products having a considerable size variation, can be used to dry a wide range of materials. The principal sources of thermal energy are oil, gas, and coal. While typical inlet temperatures for direct-fired dryers using these fuels is in the order of 1200°F, in certain instances they may be as high as 1500–1600°F depending largely on the nature of the product handled. Where feed materials are thermosensitive, steam heating from an indirect heat exchanger also is used extensively. These dryers are available in a variety of designs but in general, can be divided into two main types: those arranged for direct heating and those designed for indirect heating. As seen from Figure 7, certain variants do exist. For example, the direct—indirect dryer uses both systems simultaneously.

Where direct heating is used, the product of combustion are in intimate contact with the material to be dried while in the case of the indirect system, the hot gases are arranged to circulate around the dryer shell. Heat transfer then is by conduction and radiation through the shell.

With the indirect—direct system, hot gases first pass down a central tube coaxial with the dryer shell and return through the annular space between the tube and the shell. The material being cascaded in this annulus picks up heat from the gases and also by conduction from direct contact with the central tube. This design is thermally very efficient. Again, while there are a number of proprietary designs employing different systems of airflow, in the main these dryers are of two types: parallel and countercurrent flow. With parallel flow, only high-moisture-content material comes into contact with the hot gases, and, as a result, higher evaporative rates can be achieved than when using countercurrent flow.

In addition, many thermosensitive materials can be dried successfully by this method. Such an arrangement lends itself to the handling of pasty materials since the rapid flashing off of moisture and consequent surface drying limits the possibility of wall buildup or agglomeration within the dryer. On the other hand, countercurrent operation normally is used where a low terminal moisture content is required. In this arrangement, the high-temperature gases are brought into contact with the product immediately prior to discharge where the final traces of moisture in the product must be driven off.

In both these types, however, gas velocities can be sufficiently high to produce product entrainment. They therefore would be unsuitable for low-density or fine-particle materials such as carbon black. In such cases, the indirect-fired function-type dryer is more suitable since the dryer shell usually is enclosed in a brick housing or outer steel jacket into which the hot gases are introduced. As heat transfer is entirely by conduction, conventional flighting and cascading of the material is not used. Rather, the inside of the shell is fitted with small lifters designed to gently turn the product while at the same time maintaining maximum contact with the heated shell.

Another type of indirectly heated dryer that is particularly useful for fine particles or heat-sensitive materials is the steam-tube unit. This dryer can be of the fixed-tube variety equipped with conventional lifting flights designed to cascade the product through a nest of square section tubes, or alternatively, a central rotating tube nest can be used. Figure 8 shows a fixed-tube rotary dryer, which normally has an electrical vibrator fitted to the tube nest in order to eliminate the possibility of bridging of the product with consequent loss of heat-transfer surface. Since the heat exchanger is positioned within the insulated shell in this type of dryer, the air rejection rate is extremely low and thermal efficiencies are high. In general, this design is suitable only for free-flowing materials.

A considerable amount of work has been done on the development of various types of lifting flights, all designed to produce a continuous curtain of material over the cross section of the dryer shell. Other special configurations involve cruciform arrangements to produce a labyrinth path. The object is to give longer residence times where this is

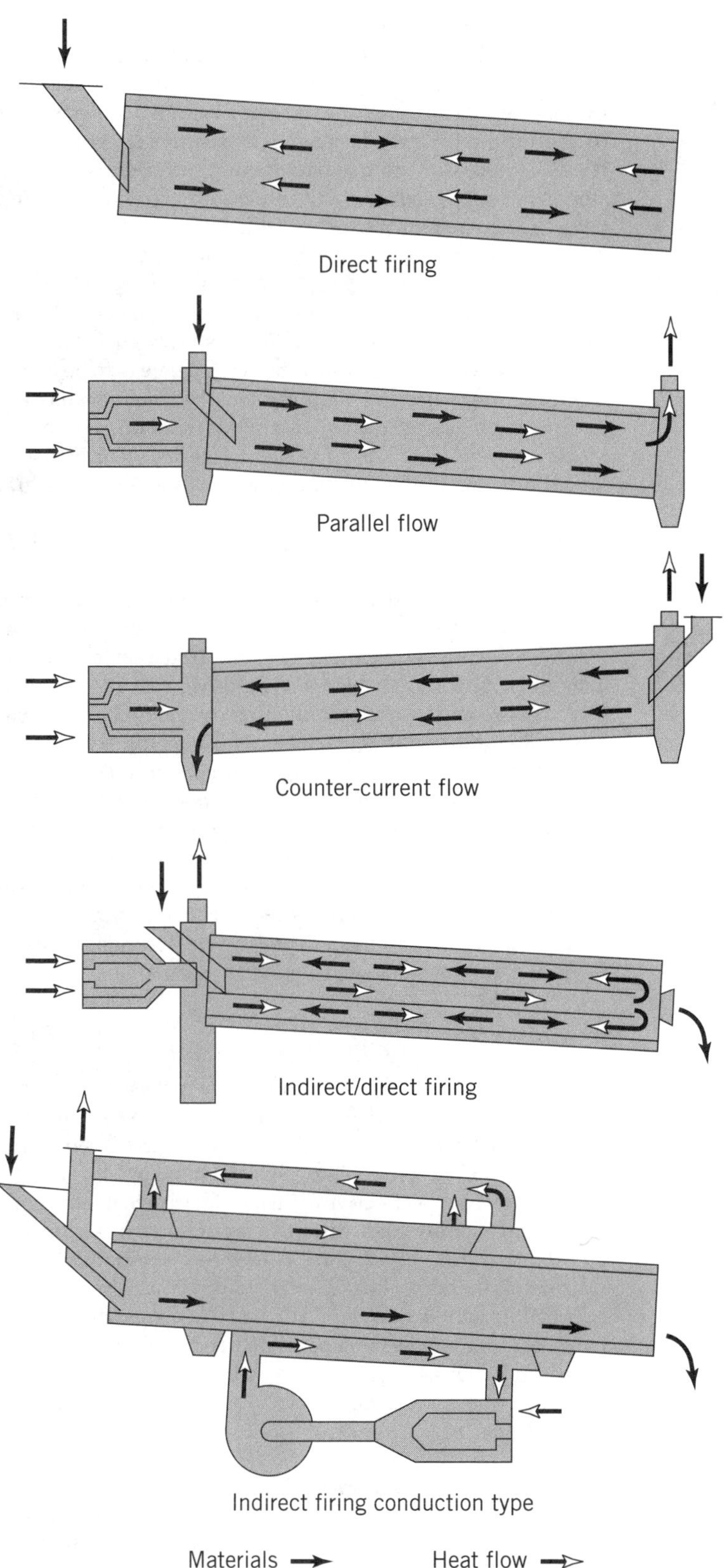

Figure 7. Typical rotary dryer arrangements.

Figure 8. Fixed-tube rotary dryer.

necessary. When the diffusional characteristics of the material or other process considerations call for extended residence times, these machines no doubt will continue to find application.

PNEUMATIC DRYERS

Where total dispersion of the product in a heated gas stream can be achieved with a significant increase in evaporative rates, pneumatic or continuous fluid-bed dryers are preferred. The capital cost of these alternatives generally is lower and maintenance is limited to such components as circulating fans and rotary valves. When considering these two types of dryers, it is convenient to examine them together since both share similar characteristics. Both employ forced convection with dispersion of the feedstock, and as a result of the intimate contact between the drying medium and the wet solids, both exhibit much higher drying rates than do any of the other dryers mentioned previously.

In a fluidized-bed dryer, the degree of dispersion and agitation of the wet solids is limited, whereas in a pneumatic dryer, the degree of dispersion is total and the material is completely entrained in the gas stream. This often is turned to advantage as the drying medium is used as a vehicle for the partially dried product. Other operations such as product classification also can be carried out where required. A further feature of fluid-bed and flash dryers is that the method of operation allows many temperature-sensitive materials to be dried without thermal degradation due to the rapid absorption of the latent heat of vaporization. This generally permits high-rate drying, whereas in other types of dryers lower temperatures would be necessary and correspondingly larger and more costly equipment would be required.

A very good degree of temperature control can be achieved in fluid-bed dryers and the residence time of the material can be varied either by the adjustment of the discharge weir or by the use of multistage units. Similarly, the residence time in the flash dryer can be adjusted by the use of variable cross-sectional area and, therefore, variable velocity. In addition, multiple-effect columns can be incorporated to give an extended path length or continuous recirculatory systems employing both air and product recycle can be used as illustrated in Figure 9.

Generally speaking, the residence time in fluidized-bed dryers is measured in minutes and in the pneumatic dryer in seconds. Both dryers feature high thermal efficiencies, particularly where the moisture content of the wet feed is sufficiently high to produce a significant drop between inlet and outlet temperatures. While the condition of the feed in the pneumatic dryer is somewhat less critical than that in the fluid-bed dryer because it is completely entrained, it still is necessary to use backmixing techniques on occasion in order to produce a suitable feed. A variety of feeding devices are used with these machines.

In fluid-bed dryers, special attention must be paid to the nature of the proposed feed since one condition can militate against another. To some extent, this is reflected in the range of variation in the figures given in Table 1 for evaporative rates. If a large or heavy particle is to be dried, the fluidizing velocities required may be considerable and involve high power usage. In such circumstances, if the moisture content is low and the surface/mass ratio also is low, the thermal efficiency and evaporation would be low. This would make selection of a fluid-bed dryer completely unrealistic and probably would suggest a conventional rotary dryer for the application.

Another case is that in which the minimum fluidizing velocity is so low that a dryer of very large surface dimensions is necessary to obtain the required thermal input. This also occurs in problems of fluid-bed cooling and usually is overcome by the introduction or removal of thermal energy by additional heating or cooling media through an extended-surface heat exchanger immersed in the bed.

With both types of dispersion dryer many configurations are available. While the power requirements of each usually is well in excess of other dryers because of the use of high-efficiency product recovery systems, the smaller size of the fluid-bed dryer compared with conventional rotaries and the fact that the flash dryer can be arranged to fit in limited floor space makes them very attractive.

BAND DRYERS

When selecting a dryer, it always is necessary to consider the final product form. When the degree of product attrition common to pneumatic and fluidized-bed dryer operation is unacceptable, continuous-band or "apron" dryers can provide an effective solution. These are widely used where moderately high rates of throughput on a continuous basis are called for. The most commonly used continuous-band dryer is the single-pass machine employing through-air circulation. Alternatively, and where there is limited floor space or a possible need for long residence time, multi-pass units are used with the conveyors mounted one above the other. In similar circumstances, another special type of multi-deck dryer can be used that employs a system of tilting trays so that the product is supported on both the normal working and the inside of the return run of the conveyor band. This arrangement considerably increases the residence time within the dryer and is particularly useful where the product has poor diffusional characteristics.

The method of airflow employed in these dryers is either vertically downward through the material and the sup-

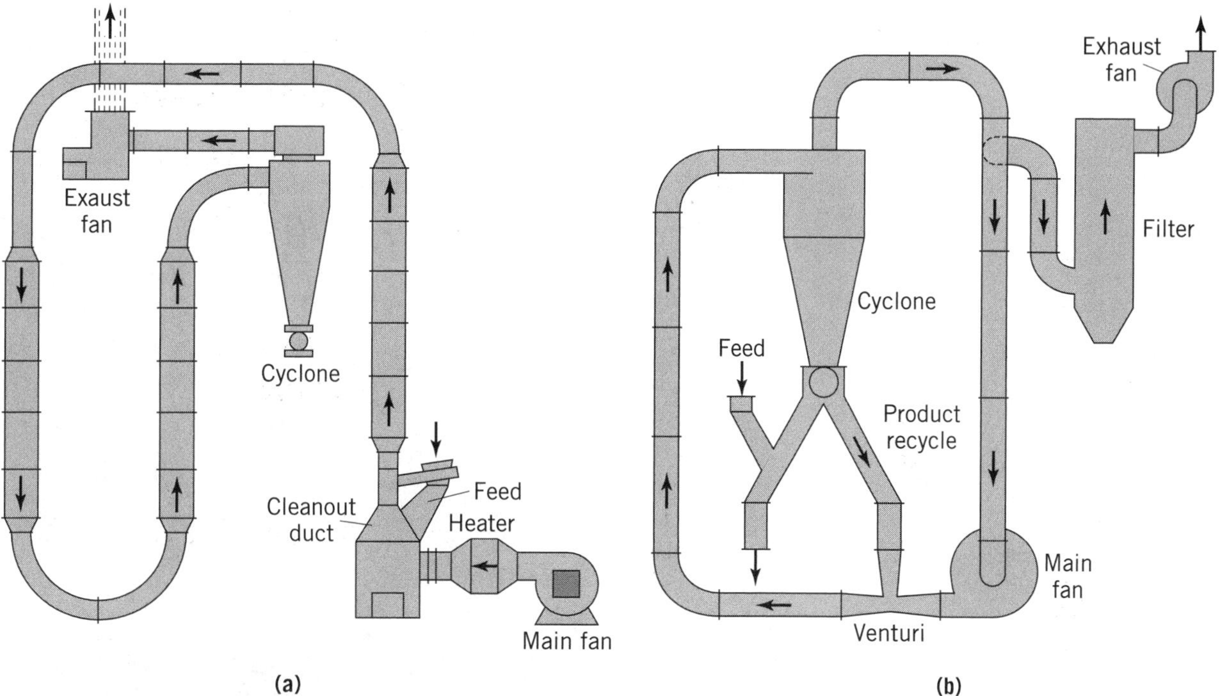

Figure 9. Multipass (**a**) and air recycle (**b**) arrangements in flash dryers.

porting band or alternatively upward. Sometimes a combination of the two may be dictated by the nature of the wet feed. It occasionally happens that extruded materials have a tendency to coalesce when deposited on the band, in which case one or more sections at the wet end of the dryer may be arranged for upward airflow to reduce the effect. Wherever possible, through-air circulation is used as opposed to transverse airflow. This results in greatly increased evaporative rates as may be seen from Table 1.

An illustration of the relatively high performance of a band dryer operating on this system as compared with a unit having transverse airflow can be cited in a case involving the processing of a 70% moisture content filter cake. When this material is dried in a conventional unit, the cycle time is in the region of 28 h. This is reduced to 55 min in the through-circulation band dryer largely as a result of using an extruder–performer designed to produce a dimensionally stable bed of sufficient porosity to permit air circulation through the feed.

In view of this, transverse airflow usually is used only where the type of conveyor necessary to support the product does not allow through-flow or where the product form is not suitable for this method of airflow. The most usual method of heating is by steam through heat exchangers mounted in the side plenums or above the band, although direct oil and gas firing sometimes is used. In such cases, the products of combustion normally are introduced to a hot well or duct at an elevated temperature from where they are drawn off and mixed with circulating air in each zone or section of the dryer.

Another alternative with direct firing is to use a series of small individual burners positioned so that each serves one or more zones of the dryer. Typical single-pass dryers of modular construction are illustrated in Figures 10, 11, and 12.

With this type and size of dryer, the average product throughput is about 5600 lb/h and involves an evaporation of 1600 lb/h moisture. It is not unusual, however, to find equipment with evaporative capacities of 3000 lb/h. Such outputs involve quite a large band area with correspondingly large floor area requirements. Various types of feeding arrangements are available to spread or distribute the wet product over the width of the band. Here again, the nature of the feed is an important prerequisite for efficient drying. Steam-heated, finned drums have been used as a means of producing a partially dried, preformed feed. While the amount of predrying achieved is reflected in increased output for a given dryer size or, alternatively, enables a smaller dryer to be used, these items are usually much more costly than many of the mechanical extruders that are available.

Generally, these extruders operate with rubber covered rollers moving over a perforated die plate with feed in the form of pressed cakes or more usually, as the discharge from a rotary vacuum filter. Others of the pressure type employ a gear pump arrangement, with extrusion taking place through a series of individual nozzles, while some use screw feeds that usually are set up to oscillate in order to obtain effective coverage of the band. Alternative designs include rotating cam blades or conventional bar-type granulators, although the latter often produce a high proportion of fines because of the pronounced shearing effect. This makes the product rather unsuitable because of the entrainment problems that can occur.

Each of the types available is designed to produce continuous–discontinuous extrudates or granules, with

Figure 10. Continuous conveyor band dryer arrangement for direct gas firing.

Figure 11. Multistage band dryer.

the grid perforations spaced to meet product characteristics. In selecting the proper type of extruder, it is essential to carry out tests on semiscale equipment as no other valid assessment of suitability can be made.

As a further illustration of the desirability of using a preforming technique, tests on a designated material exhibited a mean evaporative rate of 1.9 lb $ft^2 \cdot h$ when processed in filter-cake form without preforming. When extruded, however, the same material being dried under identical conditions gave a mean evaporative rate of 3.8 lb $ft^2 \cdot h$. This indicates, of course, that the effective band area required when working on extruded material would be only 50% of that required in the initial test. Unfortunately, the capital cost is not halved as might be expected since the feed and delivery ends of the machine housing the drive and terminals remain the same and form an increased proportion of the cost of the smaller dryer. While the cost of the extruder also must be taken into account in the comparison, cost reduction still would be about 15%. Of course, other advantages result from the installation of the smaller dryer. These include reduced radiation and convection losses and a savings of approximately 40% in the floor area occupied.

This type of plant does not involve high installation costs, and both maintenance and operating labor requirements are minimal. Since they generally are built on a zonal principle, with each zone having an integral heater and fan, a good measure of process control can be achieved.

Figure 12. Band dryer and extruder with dye stuffs.

Furthermore, they provide a high degree of flexibility because of the provision of variable speed control on the conveyor.

SELECTION

The application ratings given in Table 1 in which are listed the approximate mean evaporative rates for products under a generic classification are based on APV experience over a number of years in the design and selection of drying equipment. It should be appreciated, however, that drying rates vary considerably in view of the variety of materials and their widely differing chemical and physical characteristics. Furthermore, drying conditions such as temperature and the moisture range over which the material is to be dried have a very definite effect on the actual evaporative rate. It is important, therefore, when using the figures quoted that an attempt is made to assess carefully the nature of the product to be handled and the conditions to which it may be subjected in order to achieve greater accuracy. With these factors in mind, it is hoped that the foregoing observations on drying techniques along with the appropriate tables and curves will provide a basis for making an assessment of the type, size, and cost of drying equipment.

In making a preliminary assessment for dryer selection, there are a number of further points to consider:

1. What is the nature of the upstream process? Is it feasible to modify the physical properties of the feed, eg, mechanical dewatering to reduce the evaporative load?
2. Does the quantity to be handled per unit time suggest batch or continuous operation?
3. From a knowledge of the product, select the type(s) of dryer that it appears would handle both the wet feedstock and the dried product satisfactorily and relate this to the equipment having the highest application rating in Table 1.
4. From a knowledge of the required evaporative duty, ie, the total mass of water to be evaporated per unit time and by the application of the approximate E_{av} figure given in Table 1, estimate the size of the dryer.
5. Having established the size of the dryer on an area or volumetric basis, refer to the appropriate curve in Figures 13 and 14 and establish an order of cost for the particular type of unit.

Although a great deal of fundamental work has been carried out into the mechanics of drying, which provides a basis for recommendations, it is most desirable for pilot plant testing to be done. This is necessary not only to support theoretical calculations but also to establish whether a particular dryer will handle the product satisfactorily. In the final analysis, it is essential to discuss the drying application with the equipment manufacturer who has the necessary test facilities to examine the alternatives objectively and has the correlated data and experience from field trials to make the best recommendation.

EFFICIENT ENERGY UTILIZATION IN DRYING

It generally is necessary to employ thermal methods of drying in order to reach what is termed a "commercially dry"

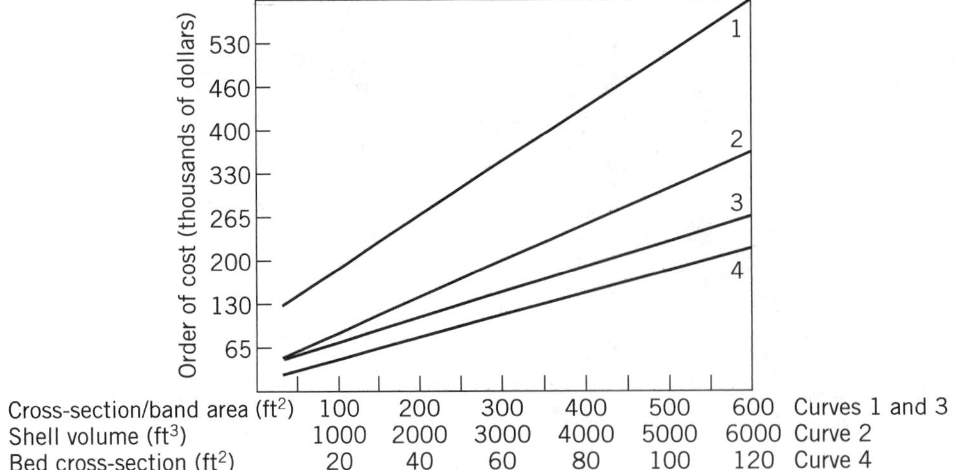

Figure 13. Approximate capital costs of (1) spray dryers, (2) rotary dryers (direct-oil-fired, mild steel, excluding product collection system), (3) continuous-band dryers steam-heated with roller extruder), (4) continuous-fluid-bed dryers (direct-oil-fired with primary cyclone).

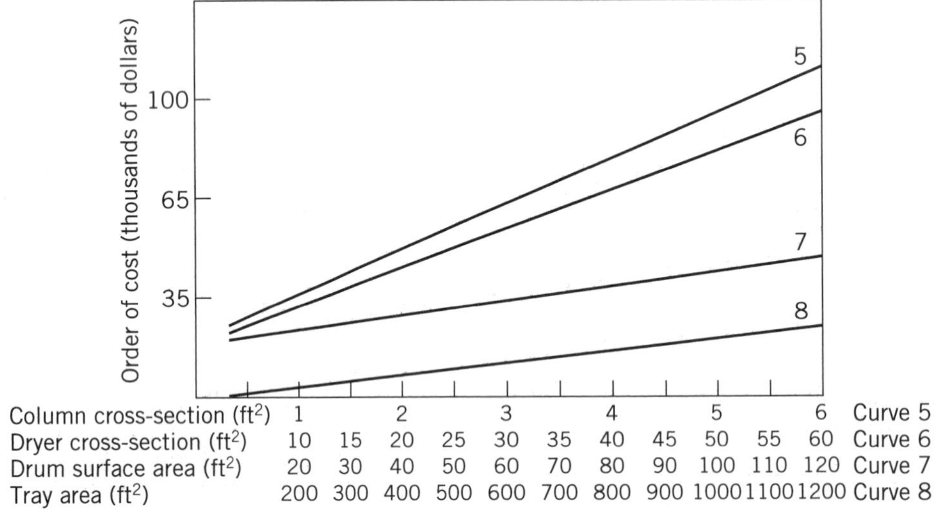

Figure 14. Approximate capital costs of (5) pneumatic dryers (direct-fired, stainless steel, including primary cyclone), (6) double-cone dryers (stainless steel, excluding vacuum equipment), (7) drum dryers (cast-iron drums, dip feed), (8) tray dryers (steam/electric-heated).

condition. As a result, drying forms an important part of most food and chemical processes and accounts for a significant proportion of total fuel consumption.

The rapid escalation of fuel costs over the past 10 yr, together with the prevailing uncertainty of future availability, cost, and possible supply limitations, highlights the continued need to actively engage in the practice of energy conservation. Some of the factors affecting dryer efficiency and certain techniques designed to reduce the cost of the drying operation are discussed in the next few sections.

In making an appraisal of factors affecting dryer efficiency, it is useful to draw up a checklist of those items that have a significant bearing on both operation and economy. It also is appropriate prior to examining various options to emphasize that in the final analysis, the primary concern is the "cost per unit weight" of the dried product. This single fact must largely govern any approach to dryer selection and operation. Additionally, there is an increasing necessity to consider the unit operation of drying in concert with other upstream processes such as mechanical dewatering and preforming techniques in a "total energy" evaluation.

In considering which factors have a bearing on dryer efficiency and what can be done to maximize that efficiency, the following aims should be kept in mind:

1. Maximum temperature drop across the dryer system indicating high energy utilization. This implies maximum inlet and minimum outlet temperature.

2. Employ maximum permissible air recirculation, ie, reduce to an absolute minimum the quantity of dryer exhaust, having due regard for humidity levels and possible condensation problems.
3. Examine the possibility of countercurrent drying, ie, two-stage operation with exhaust gases from a final dryer being passed to a predryer, or alternatively, preheating of incoming air through the use of a heat exchanger located in the exhaust gases.
4. Utilize "direct" heating wherever possible in order to obtain maximum heat release from the fuel and eliminate heat-exchanger loss.
5. Reduce radiation and convection losses by means of efficient thermal insulation.

While the above clearly are basic requirements, there are a number of other areas where heat losses occur in practice, including sensible heat of solids. Furthermore, other opportunities exist for improving the overall efficiency of the process. These are related to dryer types, methods of operation, or possibly the use of a combination of drying approaches to obtain optimum conditions.

Types of Dryers

Considering the requirements for high inlet temperature, the "flash" or pneumatic dryer offers great potential for economic drying. This stems from the simultaneous flash cooling effect that results from the rapid absorption of the latent heat of vaporization and allows the use of high inlet temperatures without thermal degradation of the product. This type of dryer also exhibits extremely high evaporative rate characteristics, but the short gas-solids contact time can in certain cases make it impossible to achieve a very low terminal moisture. However, a pneumatic dryer working in conjunction with a rotary or a continuous fluidized-bed dryer provides sufficient residence time for diffusion of moisture to take place. Such an arrangement combines the most desirable features of two dryer types and provides a compact plant and conceivably, the optimum solution.

As a further example, it is common practice in the process industries to carry out pretreatment of filter-cake materials using extruders or preformers prior to drying. The primary object is to increase the surface area of the product in order to produce enhanced rates of evaporation and smaller and more efficient drying plants. It is interesting to examine the improvements in energy utilization in a conveyor-band dryer resulting from a reduction in overall size of the dryer simply because of a change in the physical form of the feed material.

Consider a typical case. As a result of preforming a filter cake, the evaporative rate per unit area increases by a factor of 2, eg, 1.9 to 3.8 lb/ft^2 h. This permits the effective dryer size to be reduced to half that required for the non-preformed material and, for a plant handling 1 ton/h of a particular product at a solids content of 60%, the radiation and convection losses from the smaller dryer enclosure show an overall reduction of some 140,000 Btu/h. Although it is necessary to introduce another processing item into the line to carry out the pretreatment, it can be shown that the reduction in the number of dryer sections and savings in horsepower more than offset the power required for the extruder. This differential is approximately 15 kW and the overall saving in total energy is approximately 10% ie, 466 kW compared with 518 kW. Furthermore, the fact that the dryer has appreciably smaller overall dimensions provides an added bonus in better utilization of factory floor space.

Drying Techniques

If, as stated, the prime concern is the cost per pound of dried product, then major savings can be achieved by reducing the amount of water in the feedstock to a minimum prior to applying thermal methods of drying. Again, since it generally is accepted that the mechanical removal of water is less costly than thermal drying, it follows that considerable economies can be made when there is a substantial amount of water that can be readily removed by filtration or centrifuging. This approach, however, may involve changing the drying technique. For example, whereas a liquid suspension or mobile slurry would require a spray dryer for satisfactory handling of the feed, the drying of a filter cake calls for a different type of dryer and certainly presents a totally different materials handling situation.

In approaching a problem by two alternative methods, the overall savings in energy usage may be considerable as illustrated in Figure 15, which is a plot of feed moisture content versus dryer heat load. As may be seen, the difference in thermal energy used in (A) drying from a moisture content of 35% down to 0.1%, or alternatively (B) drying the same material from 14% down to 0.1% is 24.3 $\times$ 10^6

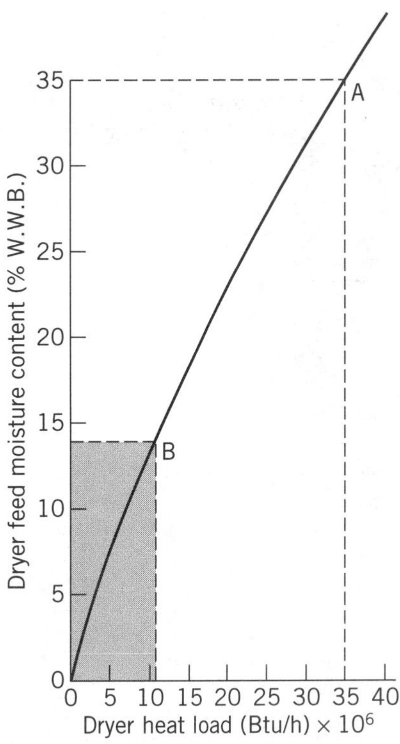

Figure 15. Thermal energy required for drying versus feed moisture content (duty as in Table 1).

Btu. Route A involves the use of a spray dryer in which the absolute weight of water in the feedstock is considerably greater than in route B, which involves the use of a thermoventuri dryer. Both types, incidentally, come under the classification of dispersion dryers and generally operate over similar temperature ranges. As a result, their efficiencies are substantially the same.

It will be apparent from Table 3 that the spray dryer is at a considerable disadvantage because of the requirements for a pumpable feed. This highlights the need to consider the upstream processes and, where a particular "route" to produce a dry product requires a slurried feed, to consider whether a better alternative would be to dewater mechanically and use a different type of dryer.

Comparing these two alternative methods for drying a mineral concentrate, the spray dryer uses a single-step atomization of a pumpable slurry having an initial moisture content of 35% and employs thermal drying techniques alone down to a final figure of 0.1%. The alternative method commences with the same feedstock at 35% moisture content but employs a rotary vacuum filter to mechanically dewater to 14%. From that point, a pneumatic dryer handles the cake as the second of two stages and thermally dries the product to the same final moisture figure.

While the difference in thermal requirements for the alternative routes already has been noted, it also is necessary to take into account the energy used for mechanical dewatering in accordance with assumptions outlined in the footnote to Table 3. From reference to this chart, it will be evident that the two-stage system requires approximately one-third of the energy needed by the single-stage operation but that the savings are not limited to operating costs. There also are significant differences in the basic air volumes required for the two dryers, which means that ancillaries such as product collection–gas cleaning equipment will be smaller and less costly in the case of the pneumatic dryer. The same applies to combustion equipment, fans, and similar items. Actually, the chart shows that capital cost savings are in the nature of 50% in favor of the filter and pneumatic drying system.

This illustration amply demonstrates the need for a more detailed consideration of drying techniques than perhaps has been the case in the past. It also points to the desirability of examining a problem on a "total energy" basis rather than taking the drying operation in isolation. Such a full evaluation approach often will prove it advantageous to change technology, ie, to use a different type of dryer than the one that possibly has evolved on the basis of custom and practice.

Operating Economies

Looking at dryer operations in terms of improving efficiency, it is interesting to see the savings that accrue if, for example, a pneumatic dryer is used on a closed-circuit basis, ie, with the recycle of hot gases instead of total rejection. Briefly, the "self-inertizing" pneumatic dryer consists of a closed loop as shown in Figure 16 with the duct system sealed to eliminate the ingress of ambient air. This means that the hot gases are recycled with only a relatively small quantity rejected at the exhaust and a correspondingly small amount of fresh air admitted at the burner. In practice, therefore, oxygen levels of the order of 5% by volume are maintained. This method of operation raises a number of interesting possibilities. It permits the use of elevated temperatures and provides a capability to dry products that under normal conditions would oxidize. The result is an increase in thermal efficiency. Furthermore, the amount of exhaust gas is only a fraction of the quantity exhausted by conventional pneumatic dryers. This is clearly important wherever there may be a gaseous effluent problem inherent in the drying operation.

A comparison of a self-inertizing versus a conventional pneumatic dryer is detailed in Table 4 with the many advantages clearly apparent. Drying with closed-circuit

Table 3. Comparison of Operating Conditions and Energy Utilization of Pneumatic-and-Spray Dryers Processing Concentrates

Parameter	Pneumatic dryer	Spray dryer
Feed rate	88,200 lb/h	88,200 lb/h
Filtrate rate	21,500 lb/h	—
Evaporative rate	9200 lb/h	30,700 lb/h
Production rate	57,500 lb/h	57,500 lb/h
Initial moisture content	14%	35%
Final moisture content	0.1%	0.1%
Air inlet temperature, °F	752	752
Air outlet temperature, °F	230	230
Total thermal input	21.43×10^6 Btu/h	66.2×10^6 Btu/h
Basic air volume (BAV) at NTP	28,100 NCFM	86,700 NCFM
Fuel consumption	1080 lb/h	3290 lb/h
Total dryer horsepower	187 kW	530 kW
Total filter horsepower	295 kW	—
Total system horsepower	482 kW	530 kW
Total thermal input expressed as kilowatts	6280 kW	19,400 kW
Total energy input to system	6762 kW	19,930 kW

Note: An assumed volumetric flow of 3 ft^3/min ft^2 of fitter area with approximately 1 hp10 ft^3 min has been used as being typical of the flow rates and energy requirements for the filtration or mechanical dewatering equipment.

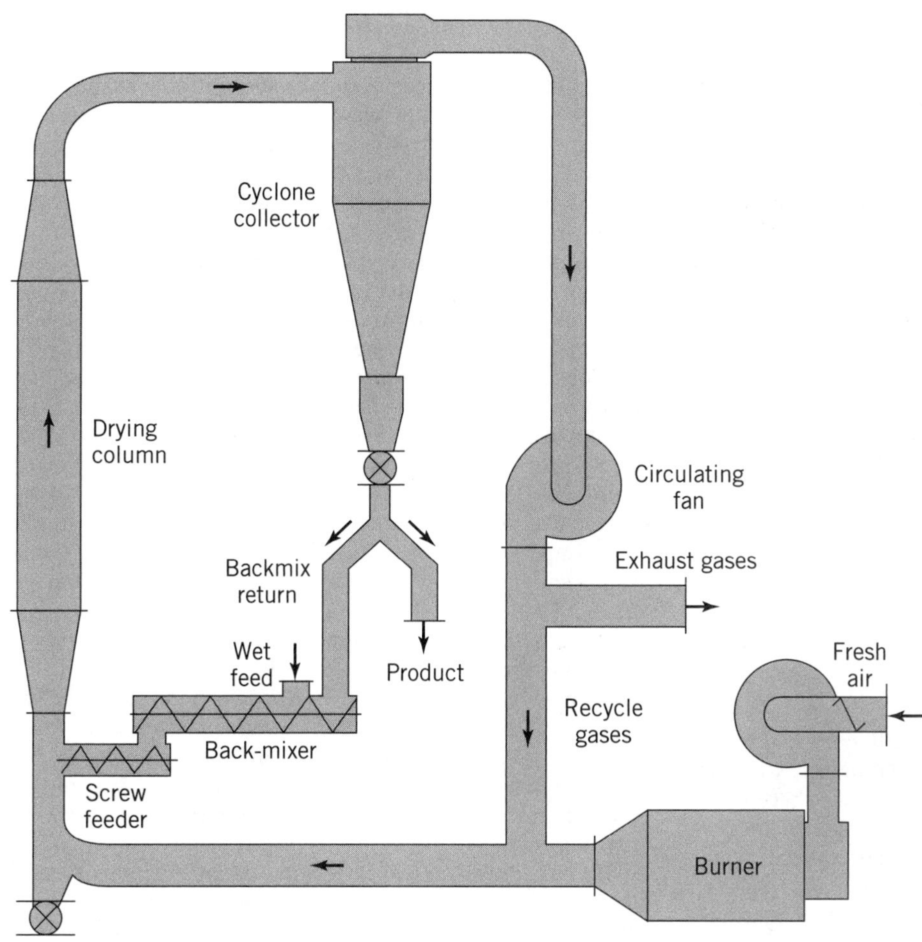

Figure 16. Self-inertizing pneumatic dryer with backmix facility.

Table 4. Comparison of Self-Inertizing Thermoventuri Dryer Versus Total Rejection Thermoventuri Dryer

Parameter	Closed circuit	Total rejection
Evaporation, lb/h	1,250	1,250
IMC, % wwb	85	85
FMC, % wwb	10	10
Feed, lb/h	1,500	1,500
Air inlet temperature, °F	932	662
Air outlet temperature, °F	302	248
Exhaust dry airflow, lb/h	1,368	16,400
Exhaust losses, Btu/h	79,700	744,000
Heat losses, Btu/h	120,000	150,000
Total heat input, Btu/h	1,680,400	2,375,700
Efficiency, %	87.9	62.4

operation is tending toward superheated vapor drying, and in practice 40–50% of the gases in circulation are water vapor. This suggests that the specific heat of the gas will be approximately 0.34 Btu/lb °F compared with about 0.24 Btu/lb °F in a conventional total rejection dryer. Since the mass of gas for a given thermal capacity is appreciably less than in a conventional dryer and, as previously mentioned, oxygen levels are low, much higher operating temperatures may be used. This permits a reduction in the size of the closed-circuit dryer. For the case given in Table 4 and comparing the two dryers on the basis of their cross-sectional areas, the total rejection dryer would have rather more than double the area of the closed-circuit system. There are obvious limitations to the use of such a technique, but the advantages illustrated are very apparent, especially the savings in thermal energy alone, viz, 695,000 Btu/h representing about 29.2% of the total requirements of the conventional drying system.

Effect of Changing Feed Rates

Since it has a significant bearing on efficiency, another factor of major importance to consider when designing dryers of this type is the possible effect of reducing feed rate. What variation in quantity of feed is likely to occur as a result of operational changes in the plant upstream of the dryer and as a result, what turndown ratio is required of the dryer? With spray dryers and rotary dryers, the mass airflow can be varied facilitating modulation of the dryer when operating at reduced throughputs.

This, however, is not the case with pneumatic and true fluidized-bed dryers since the gases perform a dual function of providing the thermal input for drying and acting as a vehicle for transporting the material. Since the mass flow has to remain constant, the only means open for modulating these dryers is to reduce the inlet temperature. This clearly has an adverse effect on thermal efficiency. It therefore is of paramount importance to establish realistic production requirements. This will avoid the inclusion of excessive scale-up factors or oversizing of drying equipment and thereby maximize operating efficiency.

Figure 17 illustrates the effects on thermal efficiency of either increasing the evaporative capacity by increased inlet temperatures or alternatively, reducing the inlet temperature with the exhaust temperature remaining constant at the level necessary to produce an acceptable dry product. While the figure refers to the total rejection case of the previous illustration where design throughput corresponds to an efficiency of 62.4%, the curve shows that if the unit is used at only 60% of design, dryer efficiency falls to 50%. The converse, of course, also is true.

In this brief presentation, an attempt has been made to highlight some of the factors affecting efficiency in drying operations and promote an awareness of where savings can be made by applying new techniques. While conditions differ from one drying process to another, it is clear that economies can and should be made.

SPRAY DRYING: NEW DEVELOPMENTS, NEW ECONOMICS

In the field of spray drying, the past 10 years have witnessed many new developments initiated mainly to meet three demands from food and dairy processors—better energy efficiency, improved functional properties of the finished powder, and production of new, specialized dried products. This has been accompanied by progress made in the automation of drying systems and in stricter requirements for the reduction of environmental impacts resulting from the operation of spray dryers.

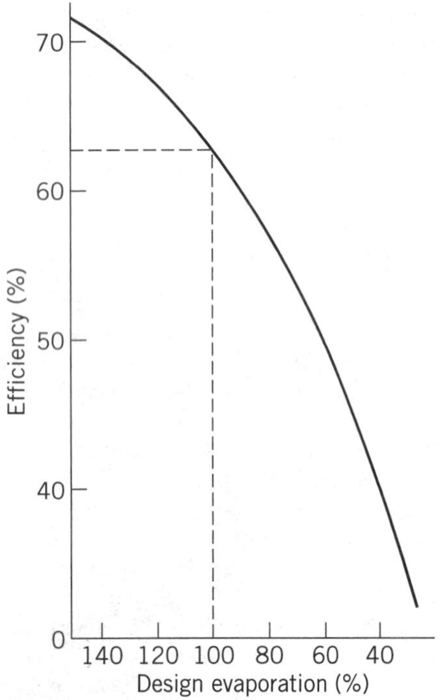

Figure 17. Variation of efficiency with dryer turndown.

While demands for energy efficient drying may have become somewhat relaxed recently with the stabilizing of energy costs, the fact remains that spray dryers and evaporators are among the most energy-intensive processing systems and the future of energy costs is still uncertain.

The extra functionality that may be required of the powders is to a large extent determined by the drying conditions, including pre- and posttreatment of the concentrate and final powder. Desirable characteristics are dust-free and good reconstitution properties that involve carefully controlled drying conditions, often with added agglomeration and lecithination steps.

The interest in the production of high-value specialty dried products instead of straight commodity products has been accentuated by the government's reduced support to dairy producers, whereby there will be less milk solids available. Processors naturally are looking into ways to better use these available supplies by producing products with improved functional, nutritional, and handling properties and, coincidentally, products with higher profit margins.

This is particularly noticeable in the utilization of whey and whey product derivatives. A larger percentage of the total whey produced today is handled and processed so that it will be suitable as food ingredients for human consumption. It also is finding new applications by being refined into whey protein concentrate and lactose through crystallization, hydrolyzation, and separation by ultrafiltration.

To handle the increase in plant size and complexity and to meet demands for closer adherence to exact product specifications, dryer instrumentation has been improved and operation gradually has evolved from manual to fully automated. Today, it is normal for milk drying plants to be controlled by such microprocessor-based systems as the APV ACCOS, with automation covering the increasingly complex startup, shutdown, and CIP procedures as well as sophisticated automatic adjustment of the dryer operating parameters to variations in ambient-air conditions.

Finally, more stringent demands governing environmental aspects of plant operations have resulted in that virtually all spray dryers today have to be equipped with exhaust air pollution control devices such as wet scrubbers or bag filters.

Principle and Approaches

Spray drying basically is accomplished in a specially designed chamber by atomizing feed liquid into a hot-air stream. Evaporation of water from the droplets takes place almost instantaneously, resulting in dried particles being carried with the spent drying air. The powder is subsequently separated from the air in cyclone separators or bag filters and collected for packaging. The powder particles being protected during the drying phase by the evaporative cooling effect results in very low heat exposure to the product, and the process is thus suitable also for heat-sensitive materials.

A number of different spray dryer configurations are available, distinguished by the type of feed atomization device and the drying chamber design.

Atomization Techniques

Proper atomization is essential for satisfactory drying and for producing a powder of prime quality. To meet varying parameters, APV offers a selection of atomizing systems: centrifugal, pneumatic nozzles, or pressure nozzles.

Centrifugal Atomization. With centrifugal or spinning-disk atomization (Fig. 18), liquid feed is accelerated to a velocity in excess of 800 ft/s to produce fine droplets that mix with the drying air. Particle size is controlled mainly be liquid properties and wheel speed. There are no vibrations, little noise, and small risk of clogging. Furthermore, the system allows maximum flexibility in feed rate, provides capacities in excess of 200 tons/h and operates with low power consumption.

Steam Injection. To produce a product with significantly increased bulk density and fewer fines, the APV steam injection technique (Fig. 19) has been refined to a point that allows its use with centrifugal atomizers in large drying operations.

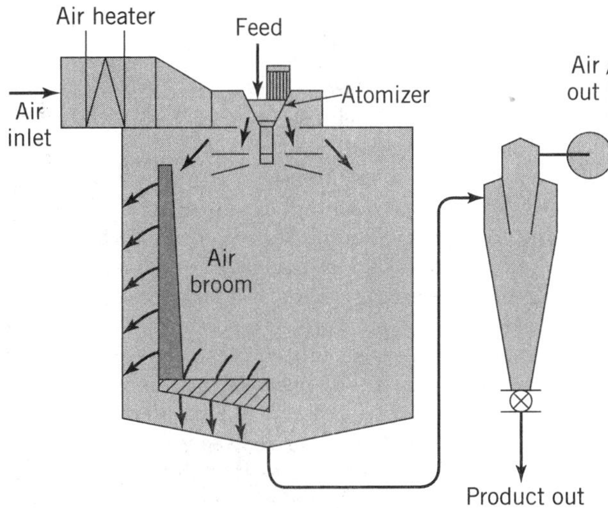

Figure 18. Flat-bottomed drying chamber incorporates rotating powder discharger and can be equipped with an air broom.

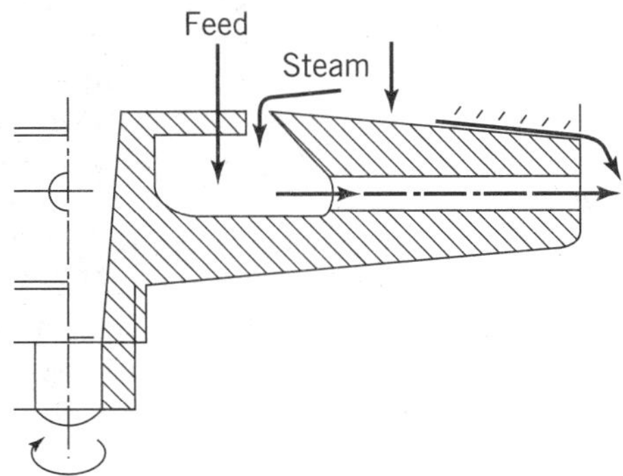

Figure 19. Steam is added around and into the atomizing disk to minimize air in the atomized liquid droplet.

During centrifugal atomization, air around the rotating wheel may be entrapped within the atomized droplets.

When heat is transferred from the drying air into the feed droplet, water evaporates and diffuses out through the surface, at the same time creating a hard shell around the particle and some hollow spaces inside. If the droplet already contains some air bubbles, this incorporated air will expand to fill out the created vacuoles and the particle will not shrink much during the drying process, resulting in porous particles as shown in Figure 20 microphoto of spray-dried autolyzed yeast.

In the steam injection process the air atmosphere around the spinning disk is replaced with a steam atmosphere, thus reducing the amount of air incorporated in the droplets. This means that there is no air that will expand within the particle and the created vacuoles will collapse, thus shrinking the particle and resulting in the type of dense, void-free particles illustrated in the Figure 21 microphoto. Control of the amount of steam injected permits a precise adjustment in powder bulk density. Furthermore, reduction of air-exposed surfaces often reduces product oxidation and prolongs powder shelf life.

Pressure Nozzles. With the pressure nozzle system (Fig. 22), liquid feed is atomized when forced under high pressure through a narrow orifice. This approach offers great versatility in the selection of the spray angle, direction of the spray, and positioning of the atomizer within the chamber. It also allows cocurrent, mixed-current, or countercurrent drying, with the production of powders having par-

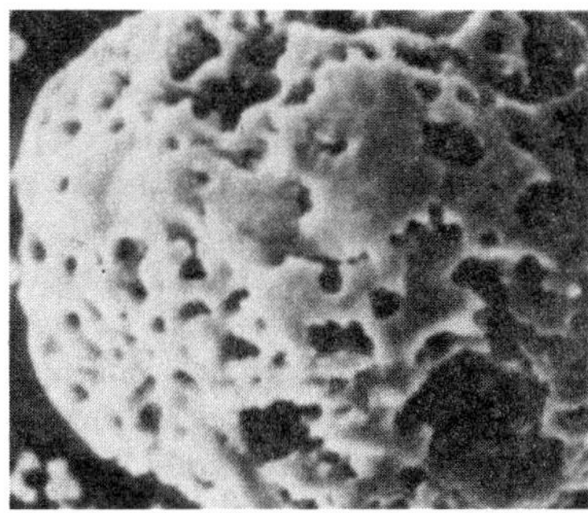

Figure 20. Spray-dried autolyzed yeast (2000× magnification). Particle is hollow and filled with crevices.

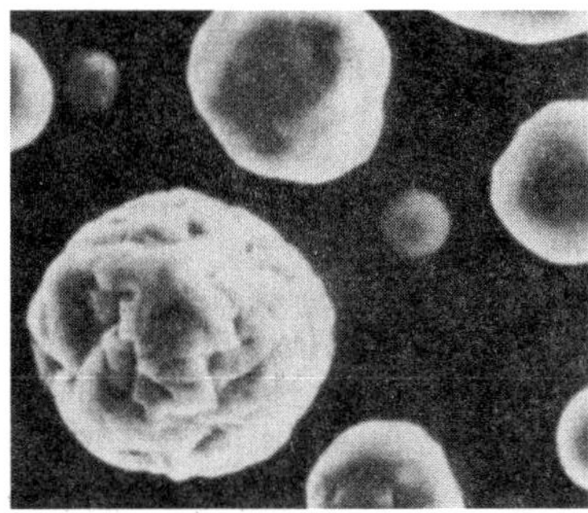

Figure 21. Spray-dried autolyzed yeast with steam injection (2000× magnification). Particle is solid and essentially void-free.

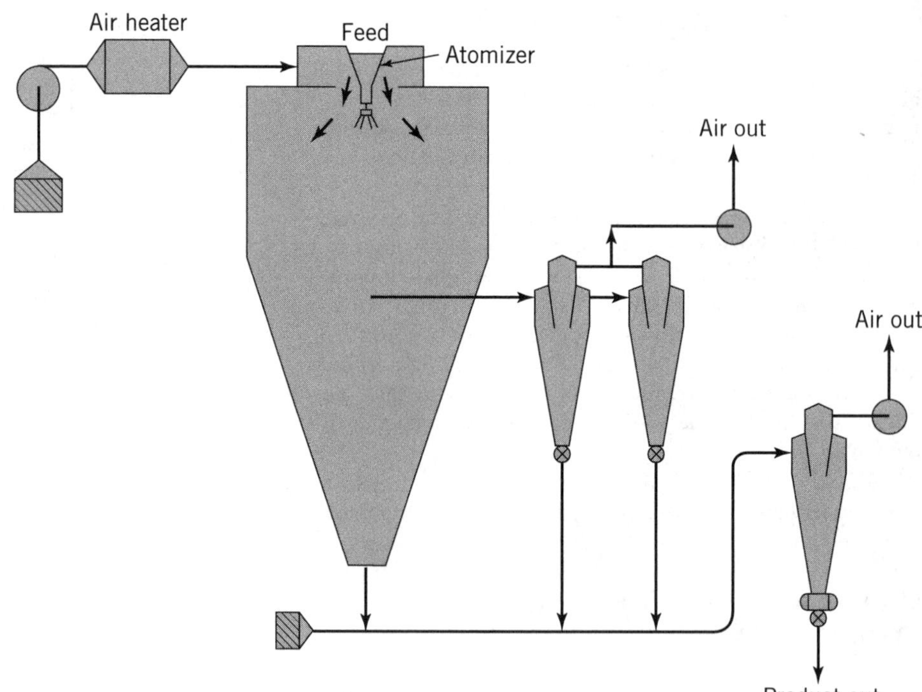

Figure 22. Typical spray dryer with conical chamber; arranged with high-pressure atomizer and air broom.

ticularly narrow particle size distribution and/or coarse characteristics. Since the particle size is dependent on the feed rate, dryers with pressure nozzles are somewhat limited relative to changing product characteristics and operating rates.

Pneumatic (Two-Fluid) Nozzles. This type of nozzle uses compressed air to accomplish the atomization of the feed product. The advantage with this type of atomizer is that it allows a greater flexibility in feed capacity than what can be obtained with pressure nozzles. However, particle size distribution is not as good as what can be obtained with the centrifugal or high-pressure nozzle-type atomizers, and because of the large quantities of compressed air required, operating costs tend to be quite high.

Single-Stage Drying

Defined as a process in which the product is dried to its final moisture content within the spray-drying chamber alone, the single-stage dryer design is well known throughout the industry, although some difference in airflow patterns and chamber design exist.

As illustrated, the drying air is drawn through filters, heated to the drying temperature, typically by means of a direct-combustion natural-gas heater, and enters the spray drying chamber through an air distributor located at the top of the chamber. The feed liquid enters the chamber through the atomizer, which disperses the liquid into a well-defined mist of very fine droplets. The drying air and droplets are very intimately mixed, causing a rapid evaporation of water. As this happens, the temperature of the air drops, as the heat is transferred to the droplets and used to supply the necessary heat of evaporation of the liquid. Each droplet thus is transformed into a powder particle. During this operation, the temperature of the particles will not increase much owing to the cooling effect resulting from water evaporation.

The rate of drying will be very high at first but then declines as the moisture content of the particles decreases. When the powder reaches the bottom of the drying chamber, it has attained its final moisture and normally is picked up in a pneumatic cooling system to be cooled down to a suitable bagging or storage temperature. Powder that is carried with the air is separated in cyclone collectors. As the powder is cooled in the pneumatic system, it also is subjected to attrition. This results in a powder that is fine, dusting, and has relatively poor redissolving properties.

Single-stage dryers are made in different configurations. The basic ones are described below.

Box Dryers. A very common type of dryer in the past, the drying air enters from the side of a boxlike drying chamber. Atomization is from a large number of high-pressure nozzles also mounted in the side of the chamber. Dried powder will collect on the flat floor of the unit, from which it is removed by moving scrapers. The exhaust air is filtered through filter bags that also are mounted in the chamber. In some instances, cyclones will be substituted for the filter bags.

Tall-Form Dryers. These are very tall towers in which the airflow is parallel to the chamber walls. The atomization is by high-pressure nozzles. This type of dryer is used mainly for the production of straightforward, commodity-type dried powders.

Wide-Body Dryers. While the height requirement is less than for the tall-form type, the drying chamber is considerably wider. The atomizing device can be either centrifugal or nozzle type. The product is sprayed as an umbrella-shaped cloud, and the airflow follows a spiral path inside the chamber. As shown in Figs. 18 and 22, this type of dryer is designed either with a flat or conical bottom.

Flat-Bottom Dryer. The flat-bottom dryer is not a new development, but the design has gained renewed interest in recent years.

The chamber has a nearly flat bottom from which the powder is removed continuously by means of a rotating pneumatic powder discharger. The dryer has the obvious advantage of considerably reduced installed height requirements compared to cone-bottom dryers, and it is also easy for operators to enter the chamber for cleaning and inspection. Another advantage is that the rotating powder discharger provides a positive powder removal from the chamber and a well-defined powder residence time. This has been shown to be important in the processing of heat-sensitive products such as enzymes and flavor compounds. Depending on the product involved, a pneumatic powder cooling system also may be installed.

Typically, this type of dryer is used for egg products, blood albumin, tanning agents, ice cream powder and toppings. The flat- and bottom chamber, incidentally, may be provided with an air broom, which is indicated by the colored section in Figure 18. By blowing tempered air onto the chamber walls while rotating, this device blows away loose powder deposits and cools the chamber walls to keep the temperature below the sticking point of certain products. Some items with which the air bottom technique has been successful are for fruit and vegetable pulp and juices, meat extracts, and blood.

Cone-Bottom Dryer. Figure 22, meanwhile, shows a conical-bottom chamber arrangement with a side air outlet, high-pressure nozzle atomization, and pneumatic product transport beneath the chamber. This single-stage dryer is very well suited to making relatively large particles of dairy products or proteins for cattle feed. If the product will not withstand pneumatic transport, it may be taken out unharmed directly from the chamber bottom.

Multiple-Stage Dryers

The best way to reduce energy usage in spray drying is, of course, to try and reduce the specific energy consumption of the process. With advances in atomizer and air distributor designs it has been possible with many products to operate with higher dryer inlet temperature and lower outlet temperature. While this procedure substantially cuts energy needs and does not harm most heat-sensitive products, care must be taken and proper balance struck. The nature of the product usually defines the upper limit, ie, the denaturation of milk protein or discoloring of other products. A higher inlet temperature requires close control of the airflow in the spray-drying chamber, and particularly around the atomizer. Furthermore, it must be noted that a lower outlet temperature will increase the humidity of the powder.

Generally speaking, the drying process can be divided into two phases: (1) the constant-rate drying period when drying proceeds quickly and when surface moisture and moisture within the particles that can move by capillary action are extracted and (2) the falling-rate period when diffusion of water to the particle surface becomes the determining factor. Since the rate of diffusion decreases with the moisture content, the time required to remove the last few percent of moisture in the case of single-stage drying takes up the major part of the residence time within the dryer. The residence time of the powder thus is essentially the same as that of the air and is limited to between 15 and 30 s. As the rate of water removal is decreasing toward the end of the drying process, the outlet air temperature has to be kept fairly high in order to provide enough driving force to finish the drying process within the available air residence time in the chamber.

In multiple-stage drying, the residence time is increased by separating the powder from the main drying air and subjecting it to further drying under conditions where the powder residence time can be varied independently of the airflow. Technically, this is done either by suspension in a fluidized bed or by retention on a moving belt. Since a longer residence time can be allowed during the falling-rate period of the drying, it is possible and desirable to reduce the drying air outlet temperature. Enough time to complete the drying process can be made available under more lenient operating conditions.

The introduction of this concept has led to higher thermal efficiencies. Fifteen years ago, the typical inlet and outlet temperatures of a milk spray dryer were 360–205°F. Today, the inlet temperature is often above 430°F, with outlet temperatures down to about 185°F.

Two-Stage Drying. In a typical two stage drying process as shown in Figure 23, powder at approximately 7% moisture is discharged from the primary drying chamber to an APV fluid bed for final drying and cooling. The fact that the powder leaves the spray dryer zone at a relatively high moisture content means that either the outlet air tem-

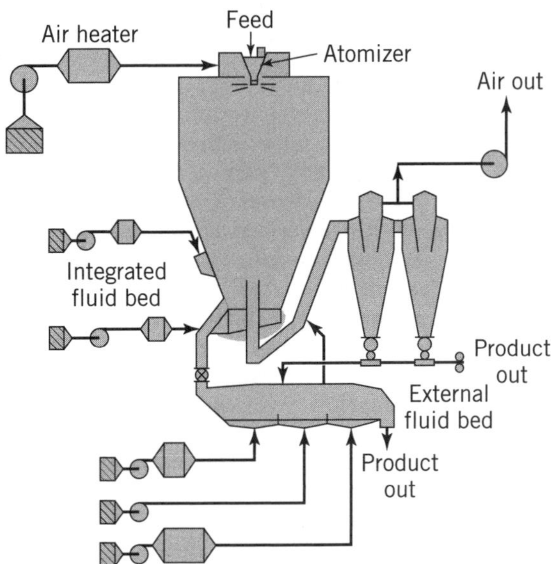

Figure 23. Two-stage drying with external fluid bed.

perature can be lowered or the inlet air temperature increased. Compared to single-stage drying, this will result in better thermal efficiency and higher capacity from the same size drying chamber. As the product is protected by its surrounding moisture in the spray-drying phase, there normally are no adverse effects on product quality resulting from the higher inlet air temperature.

The outlet air from the chamber leaves through a side air outlet duct, and the powder is discharged at the bottom of the chamber into the fluid bed. This prolongs the drying time from about 22 s in a typical single-stage dryer to more than 10 min, thus allowing for low-temperature use in the fluid bed.

In the development of this type of drying system, an initial difficulty was to provide a means to reliably fluidize the semimoist powders. This stems from the fact that milk powder products, and especially whey-based ones, show thermoplastic behavior. This makes them difficult to fluidize when warm and having high moisture content, a problem that was largely overcome by the development of a new type of vibrating fluid bed.

The vibrating fluid bed has a well-defined powder flow and typically is equipped with different air supply sections, each allowing a different temperature level for a optimum temperature profile. The last section normally is where the product is cooled to bagging and storage temperature.

While the specific energy consumption in the fluid-bed process may be relatively high, the evaporation is minor compared with the spray-drying process and the total energy use therefore is 15–20% lower.

Figure 24 shows the specific energy costs as a function of the water evaporated in the fluid bed or the residual moisture in the powder from the spray drying to the fluid-bed process. The curves are only shown qualitatively because the absolute values depend on energy prices, inlet temperatures, required residual moisture content, and other such parameters. The total cost curve shows a minimum that defines the quantity of water to be evaporated in the fluid bed to minimize the energy costs.

The introduction of two-stage drying also created the potential for a general improvement in powder quality, especially as far as dissolving properties are concerned. The slower and more gentle the drying process produces more solid particles of improved density and solubility and opens up the possibility of producing agglomerated powders in a straight-through process. If desired, the fluid bed can be designed for rewet instantizing or powder agglomeration and for the addition of a surface-active agent such as lecithin.

Three-Stage Drying. This advanced drying concept basically is an extension of two-stage drying in which the second drying stage is integrated into the spray-drying chamber with final drying conducted in an external third stage (Fig. 25). The design allows a higher moisture content from the spray drying zone than is possible in a two-stage unit and results in an even lower outlet air temperature. An added advantage is that it improves the drying conditions for several difficult products. This is accomplished by spray drying the powder to high moisture content but at the same time avoiding any contact with metal

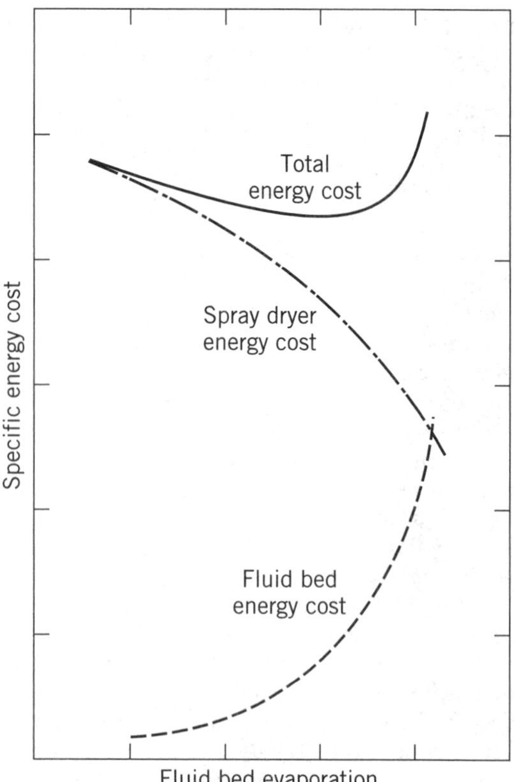

Figure 24. Specific energy cost as a function of the water evaporated in fluid bed.

surfaces by handling the powder directly on a fluidized powder layer in the integrated fluid bed. Moist powder particles thus are surrounded by already dry particles, and any tendency to stick to the chamber walls will be reduced.

In this drying concept, it is not necessary to vibrate the chamber fluid bed, although often the powders entering from the spray drying zone are difficult to fluidize because of their high moisture content and thermoplastic and hygroscopic characteristics. The fluidization characteristics are improved as the wet powder from the spray-drying zone is mixed and coated with dryer powder in the integrated fluid bed. The fluidization is further assisted by the use of a special type of perforated plate with directional airflow. The stationary fluid bed operates with high fluidizing velocities and high bed depths, both optimized to the product being processed.

The air outlet of the integrated fluid-bed system is located in the middle of the fluid bed at the bottom of the chamber. This forms an annulus around the air outlet duct and creates an aerodynamically clean design, completely eliminating the mechanical obstructions found in older two-point discharge designs. The dryer is equipped with tangential air inlets, so called Wall Sweeps, for conditioned air. These are important to the operation of the dryer in two ways: they will cool the wall and remove powder that may have a tendency to accumulate on the wall, and they also serve to stabilize the airflow within the chamber. The dryer thus operates with a very steady and well-defined airflow pattern that minimizes the amount of wall buildup.

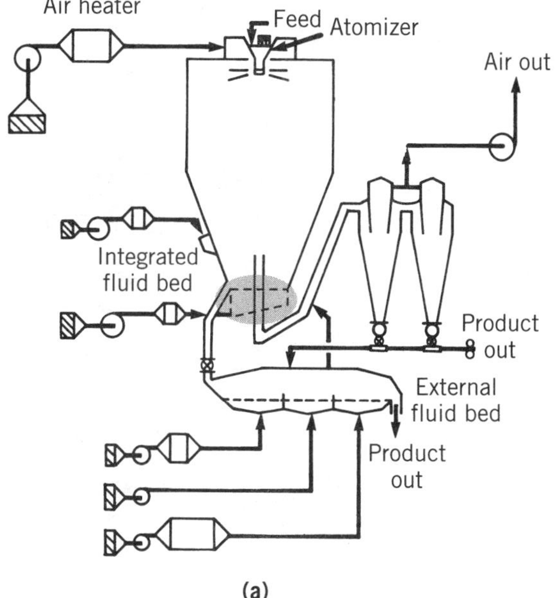

(a)

(b)

Figure 25. (**a,b**) Three-stage dryer combines conical drying chamber, integrated stationary fluid bed with directional air flow and external vibrating fluid bed. Provides optimum conditions for producing nondusty, hygroscopic, and high-fat-content products.

Atomization can be either by pressure nozzles or by a centrifugal-type atomizer.

The three-stage system is exceptionally suitable for the production on nondusty, hygroscopic, and high-fat-content products. The dryer can produce either high-bulk-density powder by returning the fines to the external fluid bed, or powder with improved wettability by straight-through agglomeration with the fines reintroduced into an atomizing zone. The concept also allows the addition of liquid into the internal fluid bed, thus opening the path to the production of very sophisticated agglomerated and instantized products, ie, excellent straight-through lecithinated powders. Whey powder from such a system shows improved quality because of the higher moisture level present when the powder enters the integrated fluid bed, providing good conditions for lactose after crystallization within the powder.

A three-stage dryer also offers a high production capacity in a small equipment volume. The specific energy requirement is about 10% less than for a two-stage dryer.

Spray-Bed Dryer

While the concept of a three-stage drying system with an integrated fluid evolved from traditional dryer technology, it served to spur the development of a modified technique referred to by APV as the "Spray Bed Dryer." This machine is characterized by use of an integrated fluid bed at the bottom of the drying zone but with drying air both entering and exiting at the top of the chamber. Atomization can be with either nozzles or a centrifugal atomizer.

The chamber fluid bed is vigorously agitated by a high fluidization velocity. Particles from the spray-drying zone enter the fluid bed with a moisture content as high as 10–15% depending on the type of product and are dried in the bed to about 5%. Final drying and cooling take place in an external fluid bed.

The structure of the powder produced in the Spray Bed differs considerably from the conventional. It is coarse, consists of large agglomerates, and consequently has low bulk density but exhibits excellent flowability. The dryer is very suited to the processing of complicated products having high contents of fat, sugar, and protein. Two of the many possible variations of this type drying are shown in Figures 26 and 27.

Spray-drying systems divided into two or more stages undoubtedly will be a characteristic of almost all future dryer installations. The advantages resulting from this technology will provide dairy and food processors with the necessary flexibility and energy efficiency required to meet today's uncertain market and whatever changes will be called for in the future.

Pollution-Control Devices

Virtually, all plants have a system to ensure clean exhaust air while collecting powder at an efficiency of about 99.5%, and may be supplied with secondary pollution-control collection equipment if necessary. In most new spray dryers this is accomplished using bag filter collectors after the primary collection device. In the case of such products as acid whey where the use of bag collectors is not practical, high-efficiency scrubbers are used. Some multipurpose plants have both plus an elaborate system of ducts and dampers to switch between the devices as different products are dried.

An added benefit from the use of sanitary bag collectors is an improvement in the yield of sale product. Although this increase typically is not more than 0.5%, it still can be of substantial value over a year of operation.

Heat-Recovery Equipment

Although it is possible to reduce the direct heating energy consumption by the use of multiple-stage drying, optimum thermal conditions generally require that heat recuperators be used as well. A few approaches are shown in Figure 28.

Despite the many recuperators available for recovering heat from drying air, only a few are suited for the spray-

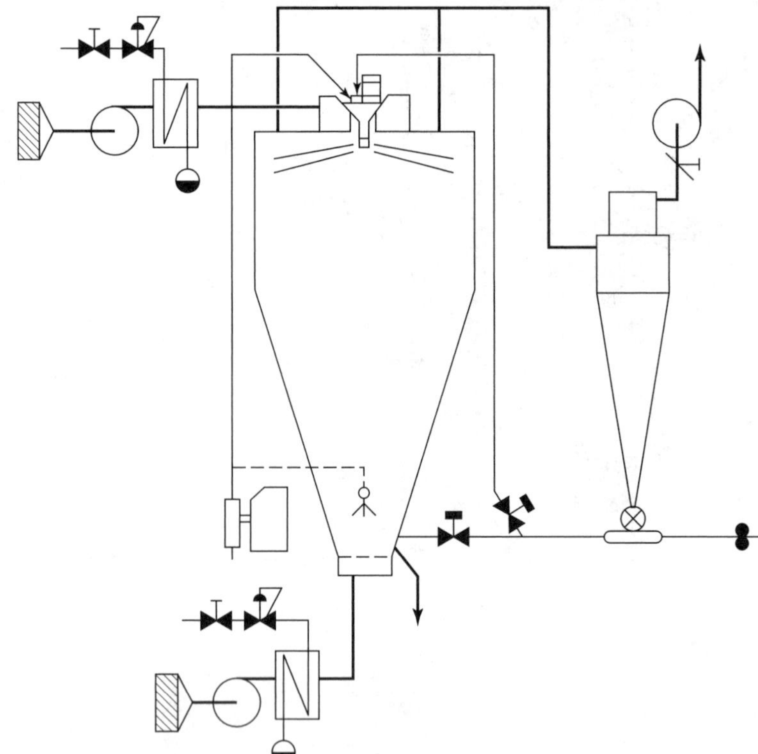

Figure 26. One variation of APV Crepaco Spray Bed Dryer arrangements (see Fig. 27 for other variation). Basic dryer with centrifugal atomizer and integrated fluid bed, which is agitated by high fluidation velocity.

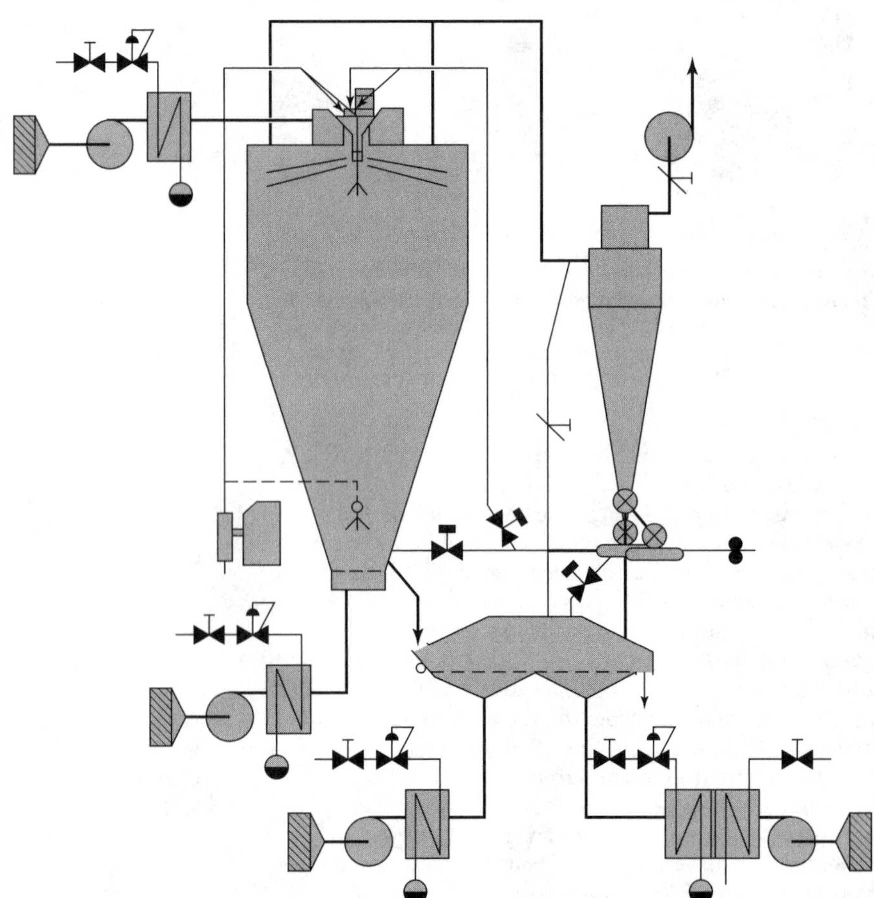

Figure 27. Spray bed dryer with added external fluid bed for final powder drying and cooling.

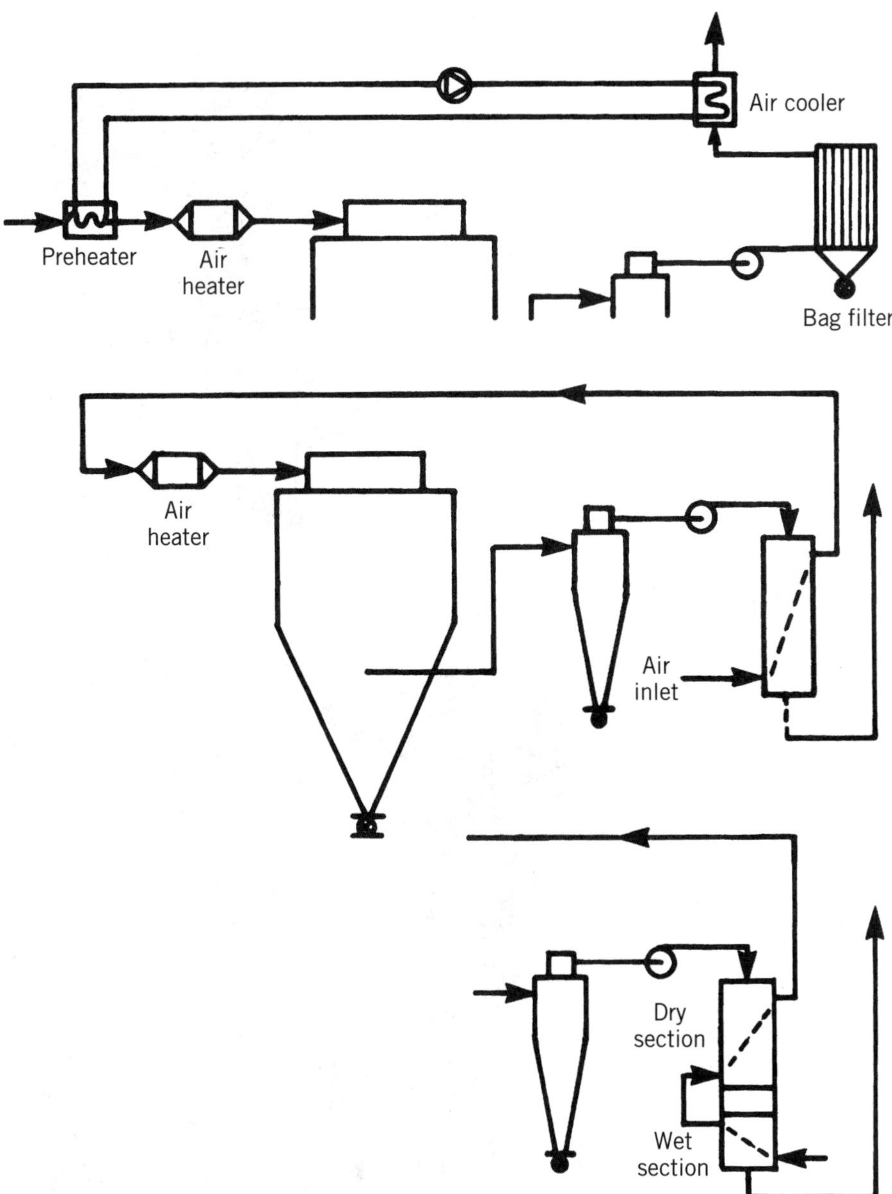

Figure 28. Spray-drying plant with (top) liquid coupled recuperator, (middle) single-stage recuperator, and (bottom) two-stage recuperator.

drying process. This is due to the dust-laden air that is involved in the process and that tends to contaminate heat exchange surfaces. To be effective, therefore, recuperators for spray-dryer use should have the following properties:

- Modular system
- High thermal efficiency
- Low pressure drop
- Automatic cleaning
- Large temperature range
- Stainless steel
- Large capacity
- Low price

Two such recuperators are the air-to-air tubular and the air-to-liquid plate designs.

Air-to-Air Heat Recuperators. For optimum flexibility, this heat exchanger is of modular design, each module consisting of 804 tubes welded to an end plate. Available in various standard lengths, the modules can be used for counterflow, cross-flow, or two-stage counterflow as sketched in Figure 29. While the first two designs are based on well-known principles, the two-stage counterflow (patent pending) system is an APV development. This recuperator consists of two sections, a long dry element and a shorter wet section. During operation, hot waste air is passed downward through the dry module tubes and cooled to just above the dew point. It then goes to the wet section where latent heat is recovered. The inner surfaces of these shorter tubes are kept wet be recirculating water. The system, therefore, not only recovers free and latent heat from the waste air but also scrubs the air and reduces the powder emission by more than 80%. Effectiveness and

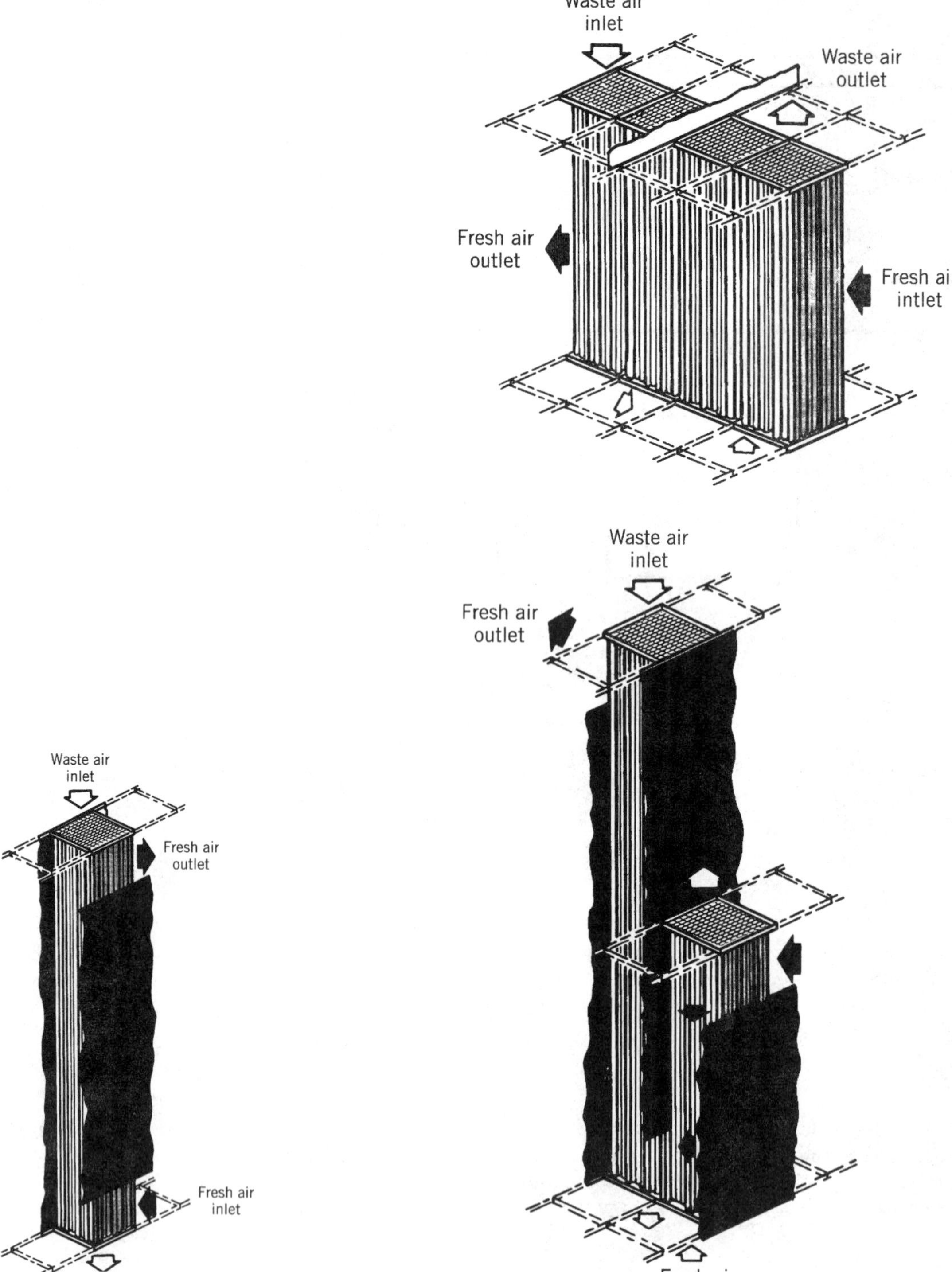

Figure 29. Air-to-air tubular heat recuperators.

economy are such that the system replaces a conventional recuperator-scrubber combination.

Air-to-Liquid Recuperator. In cases where an air-to-air recuperator is impractical or uneconomical because of space limitations or the length of the distance between the outlet and inlet air to be preheated, a liquid heat-recovery system is available. This plate recuperator as shown in Figure 30 uses various standard plates as the base with waste air being cooled in counterflow by circulating a water/glycol mixture. At high temperature, circulating thermal oils are used. Heat recovered by this means gen-

Figure 30. Thermoplate for air-to-liquid plate recuperator.

erally is recycled to preheat the drying air but can be used for other purposes such as heating water, clean-in-place (CIP) liquids, or buildings.

Other Heat-Recovery Methods. While many spray-drying plants are equipped with bag filters to minimize emissions to the environment and recover valuable powder, the filtration system can be coupled with a finned-tube recuperator for heat-recovery purposes. This type of recuperator is particularly effective when the air is not dust-loaded. It is compact, very flexible and normally, very inexpensive.

Some degree of heat recovery can also be accomplished by using available low-temperature waste streams to preheat the drying air through a finned-tube heat exchanger. Examples of such sources of waste heat include evaporator condensate and scrubber liquids.

SPRAY DRYERS: FLUID-BED AGGLOMERATION

Most powderlike products produced by spray drying or grinding are dusting, exhibit poor flow characteristics, and are difficult to rehydrate.

It is well known, however, that agglomeration in most instances will improve the redispersion characteristics of a powder. Added benefits of the agglomerated powder is that it exhibits improved flowability and is nondusting. All of these are characteristics for which the demand has increased in recent years.

Depending on the application or the area of industry where the process is being used, the process sometimes is also referred to as granulation or instantizing.

Instant Powders

Powders with particle size less than about 100 μm typically tend to form lumps when mixed with water and require strong mechanical stirring in order to become homogeneously dispersed or dissolved in the liquid. What is happening is that as water aided by capillary forces penetrates into the narrow spaces between the particles, the powder will start to dissolve. As it does, it will form a thick, gel-like mass that resists further penetration of water. Thus, lumps will be formed that contain dry powder in the middle and, if enough air is locked into them, will float on the surface of the liquid and resist further dispersion.

In order to produce a more readily dispersible product, the specific surface of the powder has to be reduced and the liquid needs to penetrate more evenly around the particles. In an agglomerated powder with its open structure, the large passages between the individual powder particles will assist in quickly displacing the air and allow liquid to penetrate before an impenetrable gel layer is formed. The powder thus can disperse into the bulk of the liquid, after which the final dissolution can take place.

Although there always is some degree of overlap between them, the reconstitution of an agglomerated product can be considered as consisting of the following steps:

1. Granular particles are wetted as they touch the water surface.
2. Water penetrates into the pores of the granule structure.
3. The wetted particles sink into the water.
4. The granules disintegrate into their original smallest particles, which disperse in the water.
5. The small dispersed particles dissolve in the water.

It is important to realize that it is the total time required for all these steps that should be the criterion in evaluating a product's instant properties. It is not unusual to see products characterized only on their wettability. This neglects the importance of the dispersion and disso-

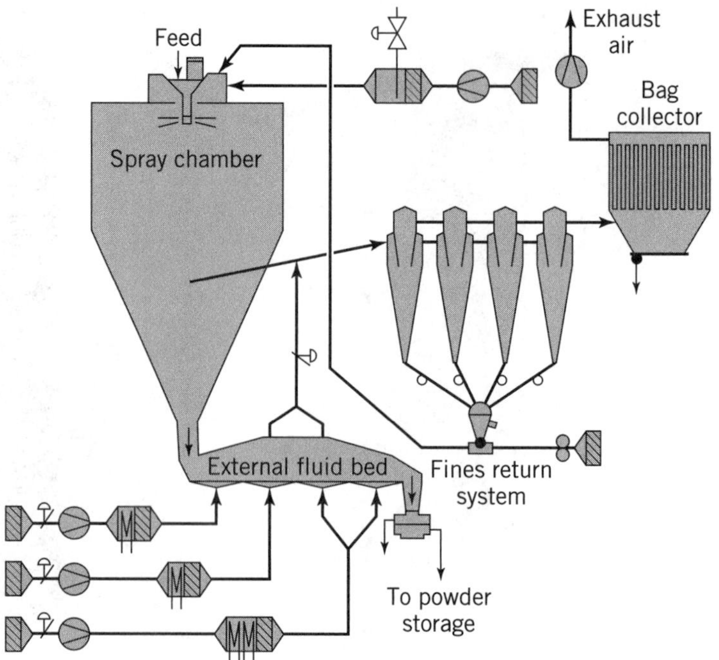

Figure 31. Straight-through agglomeration.

lution steps, the time for which may vary considerably with different agglomeration methods.

For powders that are produced by spray drying, there are a number of ways in which the agglomeration can be accomplished in the spray dryer itself. This often is referred to as the "straight-through" process and is illustrated in Figure 31. Note that fines powder from the cyclone is conveyed up to the atomizer, where it is introduced into the wet zone surrounding the spray cloud. Cluster formation will occur between the semimoist freshly produced particles and the recycled fines. The agglomerated product then is removed from the bottom of the drying chamber, cooled, and packaged. This method produces a degree of agglomeration that is sufficient for many applications.

An alternative approach to agglomeration is referred to as the "rewet method." This is characterized by processing an already existing fine dry powder into an agglomerate using fluidized-bed technology.

The Agglomeration Mechanism

Two particles can be made to agglomerate if they are brought into contact and at least one of them has a sticky surface. This condition can be obtained by one or a combination of the following means:

1. Droplet humidification whereby the surface of the particles is uniformly wetted by the application of a finely dispersed liquid.
2. Steam humidification whereby saturated steam injected into the powder causes condensation on the particles.
3. Heating—for the thermoplastic materials.
4. Addition of binder media, ie, a solution that can serve as an adhesive between the particles.

The steam condensation method usually cannot provide enough wetting without adversely heating the material and is used less frequently on newer systems.

After having been brought into a sticky state, the particles are contacted under such conditions that a suitable, stable agglomerate structure can be formed. The success of this formation will depend on such physical properties as product solubility and surface tension as well as on the conditions that can be generated in the process equipment.

For most products, possible combinations of moisture and temperature can be established as shown in Figure 32. Usually, the window for operation is further narrowed down by the specifications for product characteristics. Once the agglomerate structure is created, the added mois-

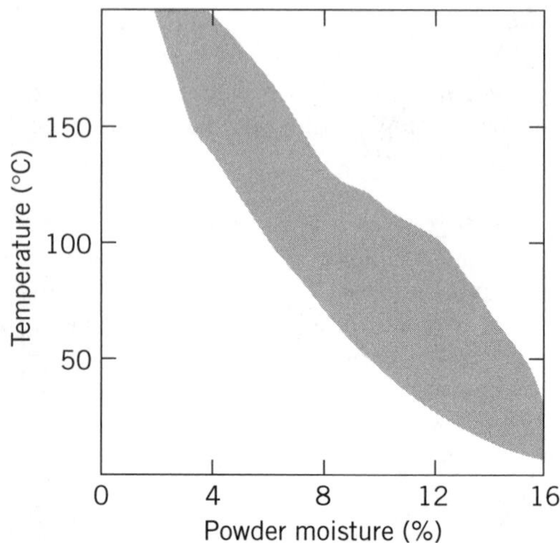

Figure 32. Typical combinations of conditions for agglomeration.

ture is dried off and the powder cooled below its thermoplastic point.

Agglomeration Equipment

While slightly different in equipment design and operation, most commercially available agglomeration processes fundamentally are the same. Each relies on the formation of agglomerates by the mechanism already described. This is followed by final drying, cooling, and size classification to eliminate the particle agglomerates that are either too small or too large. Generally, designs involve a re-wet chamber followed by a belt or a fluid bed for moisture removal. Such a system is shown in Figure 33.

It is obvious that this system is quite sensitive to even very minor variations in powder or liquid rates. A very brief reduction in powder feed rate will result in overwetting of the material with consequent deposit formation in the chamber. Conversely, a temporary reduction in liquid rate will result in sufficiently wetted powder and, therefore, weak agglomerates. Many designs rely on the product impacting the walls of the agglomeration chamber to build up agglomerate strength. Other designs include equipment for breaking large lumps into suitably sized agglomerates before the final drying. Obviously, deposit formation always will be a concern in agglomeration equipment as the process depends on the creation of conditions where the material becomes sticky.

Typically, equipment designs are very complicated, probably reflecting the fact that agglomeration actually is a complicated process. Despite the complexity of the process, however, it is possible to carry out agglomeration by means of comparatively simple equipment that involves the use of a fluidized bed for the rewetting and particle contact phase. This approach provides the following advantages:

1. There is sufficient agitation in the bed to obtain a satisfactory distribution of the binder liquid on the particle surfaces and to prevent lump formation.
2. Agglomerate characteristics can be influenced by varying operating parameters such as the fluidizing velocity, rewet binder rate, and temperature levels.
3. The system can accept some degree of variation of the feed rate of powder and liquid as the product level in the fluid bed always is constant, controlled by an overflow weir. Thus, the rewetting section will not be emptied of powder. Even during a complete interruption of powder flow, the fluidized material will remain in the rewet section as a stabilizing factor in the process.
4. By using fluid bed drying and cooling of the formed agglomerates, it is possible to combine the entire agglomeration process into one continuously operating unit.
5. Startup, shutdown, and operation of the fluid bed agglomerator are greatly simplified owing the stabilizing effect of the powder volume in the rewet zone.

Proper implementation of a fluid-bed agglomeration system requires a detailed knowledge of the fluidization technology itself. Fluidization velocities, bed heights, airflow patterns, residence time distribution, and the mechanical design of vibrating equipment must be known.

Features of Fluid-Bed Agglomeration

Figure 34 shows a typical agglomerator system where the process is implemented through the use of a vibrated, continuous fluid bed.

The powder is fed into the agglomerator by a volumetric screw feeder. As a result of the previously mentioned stabilizing effect of the material already in the fluid bed, the

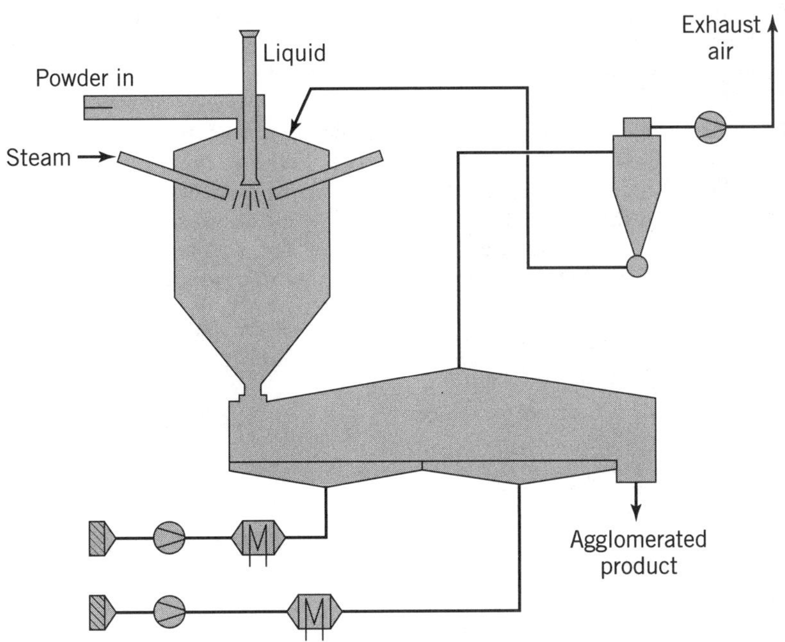

Figure 33. Typical agglomeration system.

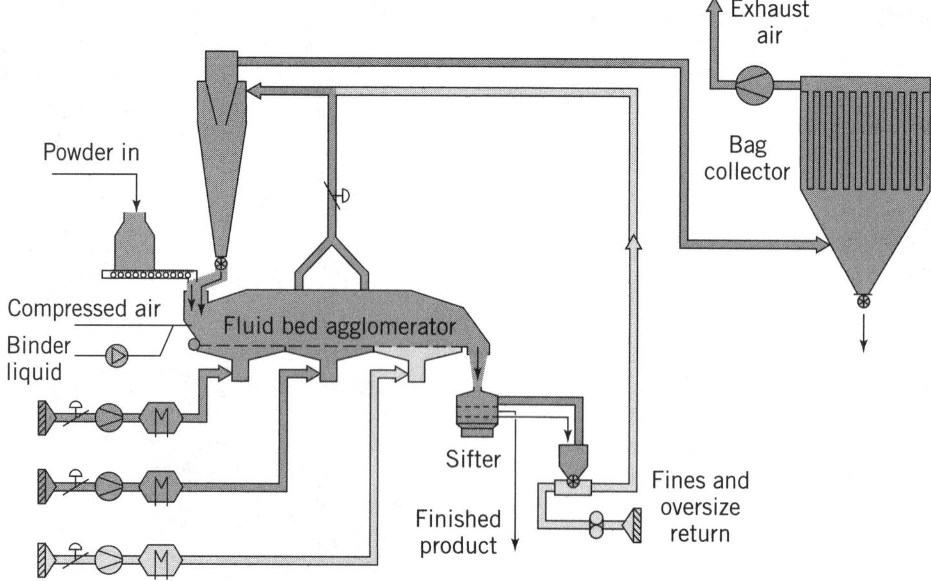

Figure 34. Fluid bed rewet agglomeration system.

Table 5. Frequently Used Binders

For food products	For chemical products
Malto dextrines	Lignosulfonates
Gum arabic	Poly(vinyl alcohol) (PVA)
Starch	Any of the food product binders
Gelatin	
Molasses	
Sugar	

reproducibility of a volumetric feeder is satisfactory and there is no need for a complicated feed system such as a loss-in-weight or similar type.

The fluid bed unit is constructed with several processing zones, each with a separate air supply system. The first section is the rewet and agglomeration section where agglomerates are formed. Here, the powder is fluidized with heated air in order to utilize its thermoplastic characteristics.

The binder liquid almost always is water or a water-based solution, whereas steam, as already explained, rarely is used. The binder liquid is sprayed over the fluidized layer using two fluid nozzles driven by compressed air. For large systems, numerous nozzles are used. Powder deposits are minimized by accurate selection of spray nozzle angles and nozzle position patterns. Powder movement is enhanced by the vibration of the fluid-bed unit itself and by the use of a special perforated air distribution plate with directional air slots. A proper detailed design is vital in order to have a trouble-free operation.

From the agglomeration zone, the powder automatically will flow into the drying area where the added moisture is removed by fluidization with heated air. In some instances, more than one drying section is required, and in such cases, these sections are operated at successively lower drying temperatures in order to reduce thermal exposure of the more heat-sensitive dry powder.

The final zone is for cooling where either ambient or cooled air is used to cool the agglomerates to a suitable packaging temperature.

During processing, air velocities are adjusted so that fine, unagglomerated powder is blown off the fluidized layer. The exhaust air is passed through a cyclone separator for removal and return of entrained powder to the inlet of the agglomerator. When there are high demands for a narrow particle size distribution, the agglomerated powder is passed through a sifter, where the desired fraction is removed while over- and undersize material is recycled into the process.

As with all rewet agglomeration equipment, the operation must be performed within certain operating parameters. Overwetting will lead to poor product quality, while underwetted powder will produce fragile agglomerates and an excessive amount of fines. However, fluid-bed agglomeration does offer a great degree of flexibility in controlling the final result of the process. The characteristics of the formed agglomerate can be influenced by operating conditions such as binder liquid rate, fluidizing velocity, and temperature. Typically, the fluid-bed rewet method will produce agglomerated products with superior redispersion characteristics.

As indicated by this partial list, this method has been used successfully with a number of products.

Dairy products
Infant formula
Calf milk replacer
Flavor compounds
Corn syrup solids
Sweeteners
Detergents
Enzymes
Fruit extracts

Malto dextrines
Herbicides
Egg albumin

In most cases, the agglomeration can be accomplished using only water as a rewet medium. This applies to most dairy products and to malto dextrine-based flavor formulations. For some products, increased agglomerate size and strength has been obtained by using a solution of the material itself as the binder liquid. In the case of relatively water-insoluble materials, a separate binder material has been used, but it must be one that does not compromise the integrity of the final product. The addition of the binder material may have a beneficial effect on the end product at times. This is seen, for example, in flavor compounds when a pure solution of malto dextrine or gum arabic may further encapsulate the volatile flavor essences and create better shelf life. In other instances, the added binder can become part of the final formulation as is the case with some detergents.

For some materials, the addition of a binder compounds is an unavoidable inconvenience. At such times, the selected binder must be as neutral as possible and must be added in small quantities so that the main product is not unnecessarily diluted. An example of this is herbicide formulations, which often have a very well defined level of active ingredients.

For products containing fat, the normal process often is combined with a step by which the agglomerates are coated with a thin layer of surface-active material, usually lecithin. This is done by mounting an extra set of spray nozzles near the end of the drying section where the surfactant is applied.

Variations of the fluid-bed rewet technology have been developed whereby the system serves as a mixer for several dry and wet products. An example of this may be seen in the APV fluid mix process for the continuous production of detergent formulations from metered inputs of the dry and wet ingredients. This provides an agglomerated end product that is produced with minimum energy input when compared to traditional approaches.

Spray-Bed Dryer Agglomeration

While the fluid-bed rewet agglomeration method produces an excellent product that is, in most respects, superior to that made directly by the straight-through process, a new generation of spray dryers has evolved that combines fluid-bed agglomeration with spray drying. These are referred to as "spray-bed" dryers. The concept was developed from spray dryers having a fluid bed integrated into the spray chamber itself and is depicted in Figure 35. What distinguishes the spray-bed dryer is that it has the drying air both entering and leaving at the top of the chamber. Atomization can be with nozzles or by a centrifugal atomizer.

During operation, the chamber fluid bed is vigorously agitated by a high fluidization velocity, and as the particles from the spray-drying zone, they enter the fluid bed with a very high moisture content and agglomerate with the powder in the bed. Fines carried upward in the dryer by the high fluidizing velocity have to pass through the spray cloud, thus forming agglomerates at this point as well. Material from the integrated fluid bed is taken to an external fluid bed for final drying and cooling.

The spray-bed dryer is a highly specialized unit that can produce only agglomerated powder. It is best suited for small to medium-sized plants since the efficiency of the agglomeration process unfortunately decreases as the plant increases in size. This is because the spray zone becomes too far removed from the fluid-bed zone as the size of the dryer increases.

Agglomerates from the spray-bed dryer exhibit excellent characteristics. They are very compact and show high agglomerate strength and good flowability.

Considerations and Conclusions

While the agglomeration process improves the redispersion, flowability, and nondustiness of most fine powders, it invariably decreases the bulk density. The comparison in Table 6 clearly shows that agglomeration improves the powder wettability and dispersibility. Individual powder particles with a mean diameter of less than 100 μm are converted into agglomerates ranging in size from 250 to 400 μm, with the rewet method being able to produce the coarser agglomerate. The powder bulk density will decrease from about 43 lb/ft^3 to approximately 28 lb/ft^3. Use of the rewet method will expose the product to one additional processing step that, in this case, will somewhat affect the proteins and result in a slightly poorer solubility.

Since fluid bed agglomeration can be operated as an independent process, it can be used with already existing power-producing equipment. It offers great flexibility and ease of operation, and provides a convenient way to add functionality, nondustiness, and value to a number of products.

SPIN FLASH DRYERS

Background Information

While mechanical dewatering of a feed slurry is significantly less expensive than thermal drying, this process results in a paste or filter cake that cannot be spray dried and can be difficult to handle in other types of dryers. The spin flash dryer is one option available for continuous powder production from pastes and filter cakes without the need for grinding.

Powders generally are produced by some form of drying operation. There are several generic types of dryers, but all must involve the evaporation of water, which can take anywhere from 1000 to 2500 Btu/lb depending on dryer type. The most common of these dryers probably is the spray dryer because of its ability to produce a uniform powder at relatively low temperatures. However, by its definition, a spray dryer requires a fluid feed material to allow its atomization device to be employed. Generally, there is a maximum viscosity limitation in the range of 1000–1500 SSU (Saybolt Seconds Universal viscosity unit) (see Fig. 36).

Figure 37 illustrates the amount of water that must be evaporated to produce 1 lb of bone dry powder from a range

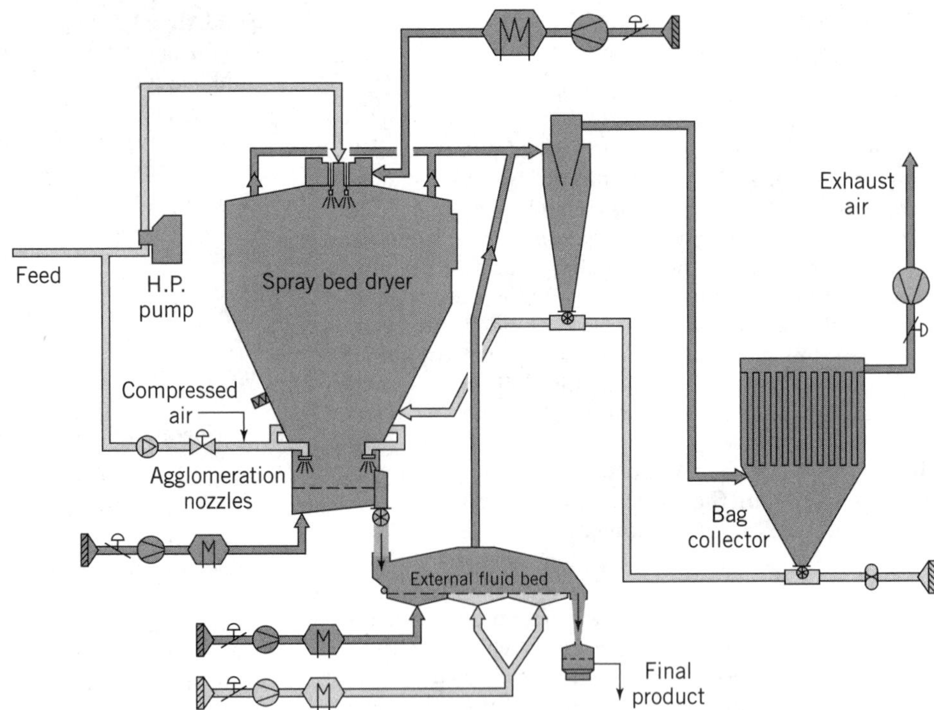

Figure 35. Spray bed-type agglomerating spray dryer.

Table 6. Reconstitutability and Physical Structure of Different Types of Skim Milk Powder

	Ordinary spray dried powder	Integrated fluid-bed agglomeration	Rewet agglomerated powder
Wettability, s	>1000	<20	<10
Dispersibility: %	60–80	92–98	92–98
Insolubility index	<0.10	<0.10	<0.20
Average particle size, μm	<100	>250	>400
Density, lb/ft^3	40–43	28–34	28–31

of different feed solids. It can clearly be seen that even a 5% increase in total solids will reduce the water evaporation and, hence, the dryer operating costs by about 20%. If this water removal can be done mechanically by, for example, filtration or centrifuging, the cost will be infinitely lower than that required to heat and evaporate the same water. The direct energy cost can be calculated as equivalent to 3–8 Btu/lb, compared to an average 1500 Btu/lb for evaporation. This increase in solids, however, inevitably will result in an increase in viscosity that may exceed the limitations of a spray dryer.

Options available for drying these higher-viscosity feed materials are listed in Table 7. The spin flash dryer is among the newest of the dryer options and has the capability of drying most materials ranging from a dilatent fluid to a cohesive paste.

Description

The spin flash dryer was developed and introduced in 1970 in response to a demand by the chemical industry to produce a uniform powder on a continuous basis from high-viscosity fluids, cohesive pastes, and sludges.

The spin flash dryer can be described as an agitated fluid bed. As shown in Figure 38, the unit consists primarily of a drying chamber (9), which is a vertical cylinder with an inverted conical bottom, an annular air inlet (7), and an axially mounted rotor (8). The drying air enters the air heater (4), is typically heated by a direct-fired gas burner (5), and enters the hot-air inlet plenum (6) tangentially. This tangential inlet, together with the action of the rotor, causes a turbulent whirling gas flow in the drying chamber.

The wet feed material, typically filter cake, is dropped into the feed vat (1) where the low-speed agitator (2) breaks up the cake to a uniform consistency and gently presses it down into the feed screw (3). Both agitator and feed screw are provided with variable speed drives.

In the case of a dilatent fluid feed, the agitated vat and screw would be replaced with a progressive cavity pump and several liquid injection ports at the same elevation as the feed screw.

As the feed material is extruded off the end of the screw into the drying chamber, it becomes coated in dried powder. The powder-coated lumps then fall into the fluid bed and are kept in motion by the rotor. As they dry, the friable

Figure 36. Typical APV Anhydro spin flash dryer with general dryer characteristics. Dryer characteristics: drying method—direct gas contact; flow—cocurrent; food material—dilatent fluids, cohesive paste, filter cake, moist granules; drying medium—air, inert gas, low-humidity waste gas; inlet temperature—up to 1800°F; capacity—up to 10 tons per hour of final product; product residence time—5–500 s.

surface material is abraded by a combination of attrition in the bed and the mechanical action of the rotor. Thus, a balanced fluidized bed is formed that contains all intermediate phases between raw material and finished product.

The dryer and lighter particles become airborne in the drying airstream and rise up the walls of the drying chamber, passing the end of the feed screw and providing, in effect, a continuous backmixing action within the heart of the dryer. At the top of the chamber, they must pass through the classification orifice, which can be sized to prevent the larger particles from passing on to the bag collector. These larger lumps tend to fall back into the fluid bed to continue drying.

Table 7. Dryer Options for High-Viscosity Materials

Direct suspension dryers
Pneumatic or flash dryers
Spin flash dryers
Fluid-bed dryers
Direct nonsuspension dryers
Tray dryers
Tunnel dryers
Belt dryers
Rotary dryers
Indirect dryers
Screw conveyor dryers
Vacuum pan dryers
Steam tube rotary dryers

Air exiting from the bag collector (10) passes through the exhaust fan (12) and is clean enough for use in a heat-recovery system. Dried powder is discharged continuously from the bottom of the bag collector through the discharge valve (11).

Two important features make the spin flash dryer suitable for products that tend to be heat-sensitive: (1) the dry powder is carried away as soon as it becomes light enough and therefore is not reintroduced into the hot-air zone, and (2) the fluid bed consists mainly of moist powder, which constantly sweeps the bottom and lower walls of the drying chamber and keeps them at a temperature lower than the dryer air outlet temperature. In addition to this self-cooling capacity, the lower edge of the drying chamber directly above the hot-air inlet can be provided with an auxiliary cooling ring.

Figure 39 illustrates the very rapid reduction in air temperature that occurs as a result of the high heat-transfer rate obtained in the fluid bed.

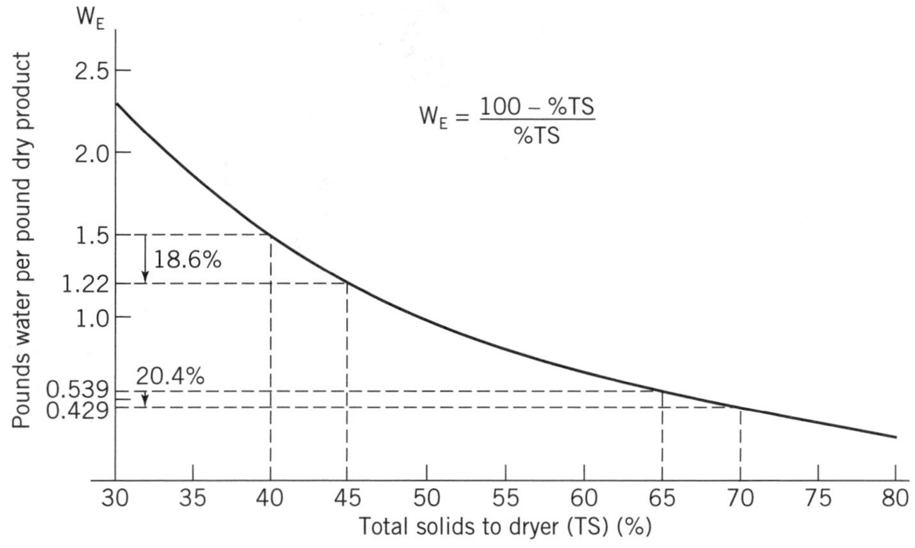

Figure 37. Water evaporation/total solids ratio.

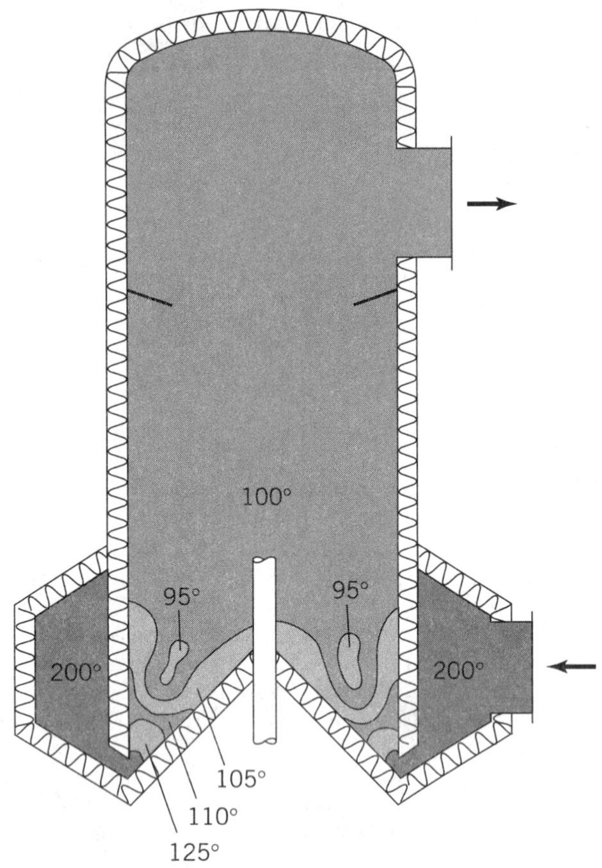

Figure 38. Fluid bed provides rapid air temperature reduction.

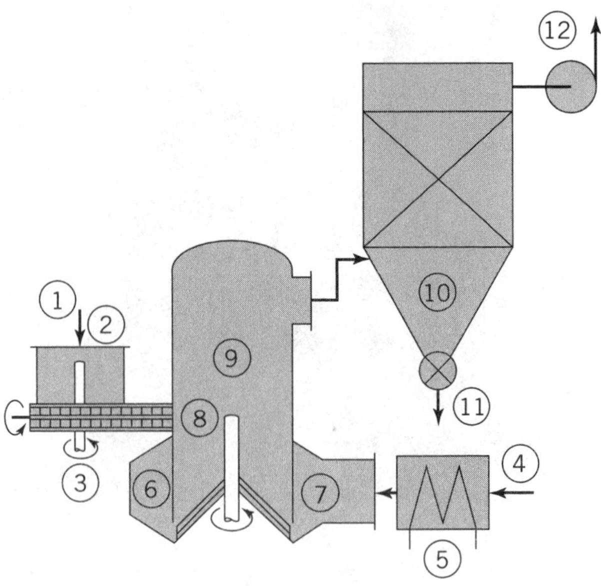

Figure 39. Standard spin flash dryer configuration.

Operating Parameters

Inlet temperature of the drying air introduced into the chamber is dependent on the particular characteristics of the product being dried but generally would be similar to that used on a spray dryer for the same product.

Outlet temperature is selected by test work to provide the desired powder moisture and is controlled by the speed of the feed screw. Since the spin flash dryer produces a finer particle size than does a spray dryer, it has been found that a slightly lower outlet temperature may be used to obtain

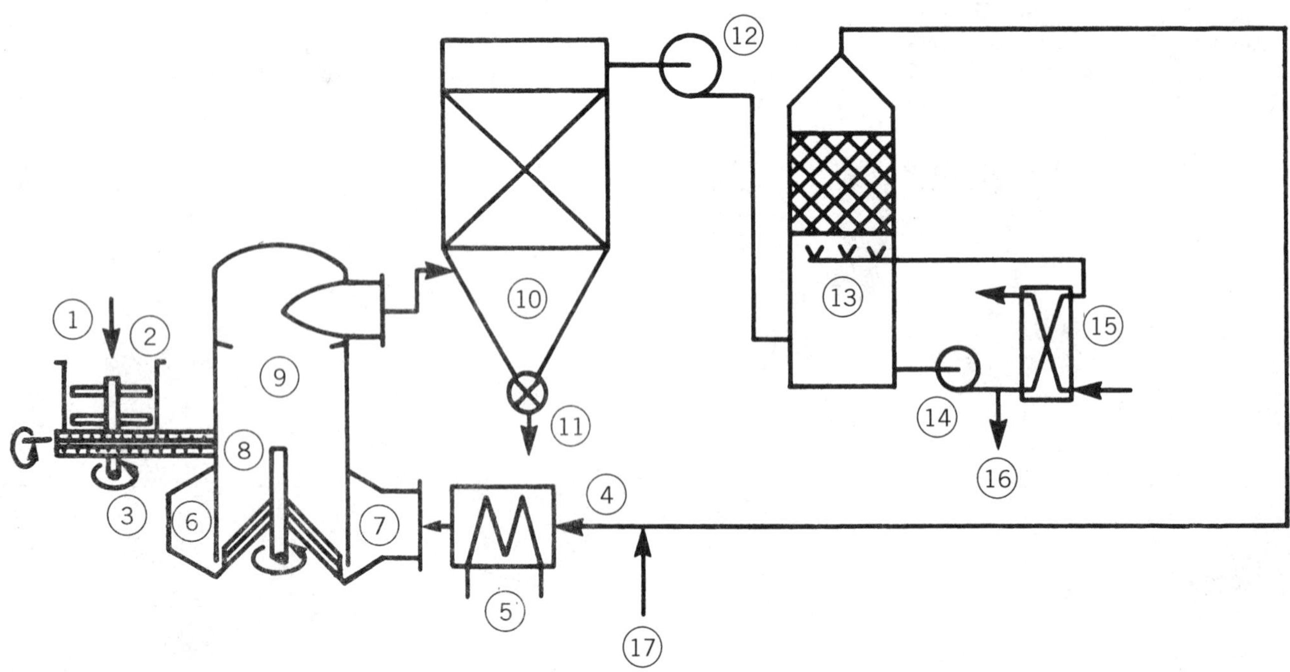

Figure 40. Spin flash dryer in closed-cycle arrangement.

Figure 41. Spin flash dryer being prepared for shipment.

the same powder moisture. This provides an increase in thermal efficiency.

Air velocity through the cross section of the drying chamber is an important design factor and is determined in part by the final particle size that is required. A lower velocity will tend to reduce the final dried particle size carried out of the chamber. The major factor, however, is the stability of the very complex bed that must neither settle back into the air distributor nor blow out of the top of the chamber. Once the maximum velocity has been determined by test work for a given product, the diameter of a drying chamber can be selected to provide the desired water evaporation rate.

Capacity can be adjusted to suit the output from the preceding process equipment, which may be difficult to control and slow in response time. This is achieved by a cascade control from a feed vat level sensor to the inlet temperature controller set-point. The feed vat sometimes can be oversized to accept the batch discharge from a preceding filter press while allowing the dryer to operate continuously.

Closed-Cycle Drying

Once a decision has been made to increase feed solids prior to drying, the small size and lower airflow requirements of the spin flash dryer make it practical to design the system as a closed-cycle dryer with nitrogen as the drying medium. This type of system can be used to dry a solvent-based powder, allowing complete recovery of the solvent.

The simplified Figure 40 schematic shows a possible configuration of a closed-cycle spin flash dryer. The operating process as described earlier, is extended, with exhaust gas from the baghouse being scrubbed and cooled in a condenser (13) using cooled solvent from an external plate heat exchanger (15) as the scrubbing medium. Recovered solvent is bled off at (16), downstream from the scrubber recirculation pump (14) at a controlled rate based on scrubber liquid level.

Table 8. Products Dried on Spin Flash Equipment

	Unit	Sludge of various metal hydroxides	Dolomite	Yellow iron oxide	Alumina	Aluminum silicate	Calcium carbonate with binder	Calcium carbonate (pure)	Ferrite	Tartrazine (azo dye)	Food yellow (color index 15985)	Calcium stearate	Polyvanadate	Nickel catalyst on diatomaceous earth	Barium sulfate	Titanium dioxide
Inlet air temperature	F	482	590	617	482	842	536	968	932	410	437	212	914	662	1112	1292
Outlet air temperature	F	185	293	212	194	212	194	338	257	203	203	126	320	311	275	257
Temperature of the feed	F	60	60	55	65	60	55	32	55	50	70	70	55	70	60	60
Total solids in feed	%	70	34	35	29	20	58	67	62	50	28	43	35	33	78	65
Residual moisture	%	4.5	0.4	0.6	12.5	5.5	0.3	0	1.0	5.0	9	0.32	2.7	3	0.1	0.5
Mean particle size	μm	40	15	5	70	20	50	5	15	10	10	16	20	15	5	3
Bulk density	gr/cm^3	0.8	0.45	0.3	0.4	0.2	0.45	0.8	0.9	0.7	0.3	0.14	0.76	0.32	1.6	0.6

Table 9. Size and Cost Comparison: Spray Dryer vs. Spin Flash Dryer

Space requirements	Spray dryer type III-K no. 70/71	Spin flash type III no. 59
	Space requirements	
Heater type (gas)	Direct-fired	Direct-fired
Chamber diameter, ft	14′0″	2′7-1/2″
Building floor area, ft^2	650	325
Building height, ft	46′0″	16′6″
Building volume, ft^3	24,700	5,300
	Capacity	
Powder, lb/h	880	880
	Performance data	
Feed solids, %	30	45
Feed rate, lb/h	2,995	1,951
Water evaporation, lb/h	2,115	1,071
Powder moisture, %	0.4	0.4
Gas consumption, SCF/h[a] (1000 Btu/SCF)	4,416	2,192
Power consumption (kWh)	40	30
	Investment cost, U.S. $	
Building	75,000	20,000
Dryer equipment	174,000	147,000
Baghouse	22,000	18,000
Installation	35,000	22,000
	$306,000	$207,000
	Assumptions	
Gas cost, $/M Btu $4.32		
Electricity cost, $/kWh $0.07		
Salaries per hour $15.00		
Manpower required = 1 operator		
	Variable operating costs per hour	
Wages	$15.00	$15.00
Gas	19.08	9.47
Electricity	2.80	2.10
Total	$36.88	$26.57
Cost per lb of powder	$0.042	$ 0.03

Note: It can be seen that the investment in a spin flash dryer plant is about 32% lower than a spray dryer for the same capacity and that spin flash operating costs are approximately 29% lower. These figures, however, do not include the capital investment in filtration equipment.

[a]Standard cubic feet.

The drying chamber would be maintained at a pressure slightly higher than ambient using a pressurized nitrogen purge (17). The heater (5) would use either a steam coil or a thermal fluid system with an external heater.

A less expensive alternative to the closed-cycle approach is a "Lo-Ox" (low-oxygen) system where a low excess air burner is used in a direct-fired heater. The products of combustion are recirculated through the condenser, and the surplus gas is vented to atmosphere. The oxygen level in such a system can be controlled to less than 5%. (See also Figure 41.)

Spin-Flash-Dried Products

While Table 8 charts some typical products being dried on spin flash equipment, it should be noted that in some cases the inlet temperatures are limited by the heat source used and not by the spin flash process.

Other products test dried on a laboratory scale spin flash dryer include the following:

Product	Feed Solids, %	Powder Moisture, %
Iron oxide	70	2
Effluent treatment sludge	22	40
Lignin	49	15
Gum	40	13
Chitosan gel	6	40
Crab meat paste	30	9
Dutched cocoa cake	70	2

Cost Benefits

In comparison to a spray dryer, the spin flash dryer has a much shorter residence time and consequently is considerably smaller and requires less building space. Its ability to dry to even higher solids than a spray dryer results in operating cost savings. Table 9 shows a detailed size and cost comparison based on actual test drying of yellow iron oxide.

Conclusion

Despite its obvious size and cost advantages, there are many instances when a spin flash dryer cannot replace a spray dryer. Typically, such cases occur when a free-flowing spherical particle of a particular size range is required or when agglomeration is needed.

There are, however, many situations in both the food and chemical industries where the particular capabilities of the spin flash dryer to produce powders from paste warrant careful consideration of its use.

PARTIAL LIST OF PRODUCTS PROCESSED ON VARIOUS TYPES OF APV DRYERS

Spray Dryer

Albumen
Alcoholic extracts
Alfalfa
Alginates
Amino acids

Baby food
Bananas
Beef extract
Beer wort
Bile extract
Bouillon
Brain
Butter
Buttermilk

Calcium carbonate
Calcium salts
Carbon, active
Carboxymethylcellulose
Carrageenan
Casein hydrolysate
Caseinates
Catalysts
Cellulose acetate
Cellulose hydrate
Cheese
Chelates
Chlorophyll
Chocolate milk
Citrates
Colors
Cornstarch
Corn steep liquor
Corn syrup
Cream

Dextranes
Dextrin maltose
Dextrose
Diastase
Diatomaceous earth
Dietetic products
Distillers waste

Egg white
Egg yolk
Emulsifiers
Enzymes

Fat-enriched milk
Fatty alcohol sulfonates
Fish meat, hydrolyzed
Fish pulp
Fish solubles
Flavors
Fruit juice
Fruit pulp fullers earth

Garlic
Gelatin
Glands
Glauber salt
Glazes
Glucoheptonate
Glucose
Gluten
Glyceryl monostearate
Gum arabic

Hormones
Hydrolysates
Hypochlorites

Ice cream mix
Lactates
Lactose
Licorice
Lignin
Liver extract

Magnesium carbonate
Magnesium oxide
Magnesium salts
Magnesium sulfate
Malted milk
Malt extract
Mango
Meat extract
Milk-cocoa
Molasses with filler

Nitrates

Olive paste

Pancreas
Papain
Peanut milk
Pectin
Penicillin
Pepsin
Peptones
Phosphates
Pigments
Potassium sulfate
Potatoes
Potato waste liquor
Proteins

Quaternary salts
Rennet

Saponin
Seaweed extract
Senna
Sequestering agents
Silica-alumina gels
Silicates
Skim milk
Sodium adipate
Sodium phenate
Sorbate
Sorbose
Soybean milk
Soybean protein
Soy flour
Starch products
Stearic acid
Stearyl-tartrate
Sulfonates

Tannin extract
Tapioca
Tea extract
Thiamin
Thymus
Tomato
Vegetable extracts
Vegetable proteins

Wetting agents
Whey
Whey nonhygroscopic
Whole egg
Whole milk
Yeast
Yeast autolysate
Yeast hydrolysate

Zinc ammonium chloride
Zinc chromate
Zinc stearate

Film Drum Dyers

Agar
Apple pectin
Avocado pulp

Baby food
Beef, comminuted
Beef extract
Beef protein (hyrolyzed)
Beer concentrate
Blood
Buttermilk

Calcium acetate
Calcium carbonate
Calcium propionate
Caramel
Cattle food
Cellulose
Cereal (baby)
Chicken offal
Cider less
Corn syrup
Manure
Meat extract
Meat pieces
Milk crumb
Milk (skimmed)
Dextrine

Effluent
Essence

Fat
Fatty alcohol sulfates
Fermentation waste
Ferric ammonium syrup
Ferrous sulfate
Fish food
Fish stick liquor
Flavors
Flour

Gelatin
Glauber's salts

Herbicide
Hydrolyzed protein
Maize gluten
Mango flake
Single-cell protein
Sorghum and maize
Starch
Stearate
Stearic acid

Molasses
Oat flour
Onion pulp
Palm oil residue
Pectin
Phenylglycine casein
Potato (waste)
Press liquors
Protein (hyrolyzed)
Rice starch
Seaweed extract
Semolina
Steep liquor
Stick liquor
Sugar solution
Oats
Tapioca starch
Vegetable extract
Vegetable protein
Wax
Whale solubles
Wheat starch
Whey
Yeast cream
Yeast extract

Thermoventuri Dryer

Alginates
Alginic acid
Bread crumbs
Cocoa residue
Cornstarch
Coumarin
Dark grains
Demerara sugar
Dextrose
Fishmeal
Flour
Glucose starch
Groundnut residue
Lactose
Maize starch
Organic acids
Pectin
Potato
Spent grains
Sphagnum moss
Starches
Stearates
Sulfur waste
Tartaric acid
Wheat germ
Wheat gluten
Wheat starch
Yeast

Coveyor-Band Dryer

Almonds
Apples
Apricots
Assorted biscuits
Beans
Bran
Cabbage
Carrots
Celery
Chocolate milk crumb
Coconuts
Corn grit
Bread crumbs
Breakfast cereals
Filter board
Food concentrates
French fries
Garlic
Gelatin
Grapes
Iced biscuits
Jellies
Meat
Mints
Molding powders
Nougat
Nuts
Onions
Dog biscuits
Dog food
Pastilles
Peanuts
Peas
Pectin
Peppers
Pet food
Plums
Potatoes
Puffed cereals
Rice
Sausage rusk
Seaweed
Soya protein
Sugar dough
Sugar- and/or honey-coated cereals
Turnip
Wheat flakes

BIBLIOGRAPHY

1. D. Noden, "Industrial Dryers Selection, Sizing, and Costs," *CPI Digest* **8**(1), 6–23, (1985).
2. D. Noden, "Efficient Energy Utilization in Drying," *dfi News* **7**(4), 8–14, (1985).
3. B. Bjarekull, "Spray Drying New Developments, New Economics," *dfi News* **8**(2), 6–13 (1985).
4. B. Bjarekull, "Fluid Bed Agglomerization for the Production of Free-Flowing Dustless Powders," *dfi News* **8**(2), 6–13 (1985).
5. S. G. Gibson, "Spin Flash Dryers for Continuous Powder Production from Pastes and Filter Cakes," *APV Dryer Handbook*, Oct. 1985, pp. 36–42.

APV CREPACO, INC.
Lake Mills, Wisconsin

E

***E. COLI* 0157.** See FOODBORNE DISEASES.

EDIBLE FILMS AND COATINGS

Edible films and coatings are generally defined as thin layers of edible material that are applied on (or within) food products to extend shelf life by functioning as solute, gas, and vapor barriers. Also, they can modify product appearance, improve structural integrity, act as carriers of food additives (eg, antimicrobials, antioxidants, and flavorings) and, in certain instances, partially or completely replace synthetic packaging materials. The terms films and coatings are often used interchangeably. Generally, coatings are formed on foods after application by dipping, brushing, spraying, or tumbling, whereas stand-alone films are preformed by casting or extrusion and then placed on (or within) foods. Relevant commercial applications range from wax coatings for fresh produce, to extruded collagen casings for sausages, to sugar coatings for bakery and confectionery products, to water-soluble pouches for premeasured food additives. Interest in development, characterization, and application of edible films and coatings has recently intensified, resulting in voluminous peer-reviewed and patent literature on the subject. Several protein, polysaccharide, and lipid materials derived from plants and animals are known film-forming ingredients with current or potential edible packaging applications. Typically, protein- and polysaccharide-based films and coatings require incorporation of low molecular weight edible plasticizers such as glycerin, sorbitol, polyethylene glycol, propylene glycol, and fatty acids. The present summary is intended as a reference for various types of edible film-forming materials and not as an edible packaging applications guide.

PROTEIN FILMS

Collagen

One of the earliest uses of edible films is the use of natural casings made from animal intestines to restructure comminuted meat. Natural casings have the disadvantages of low mechanical strength, inconsistent quality, and limited availability. Therefore, natural casings have been largely supplanted by (artificial) collagen casings fabricated from swollen collagen fibers. Sausage casings hold and shape the meat batter while the meat protein is being heat-set. They have no real barrier properties; they allow moisture to pass when needed, and allow smoke and flavorings to enter the sausage. Compared with natural intestine casings, fabricated edible collagen casings offer improved uniformity, processibility, and sanitation. However, edible collagen casings are somewhat tough and tend to split when grilled.

The currently used extrusion process for fabricating collagen casings from regenerated bovine hide corium was developed in the late 1950s. The process involves dehairing hides by alkaline or acid treatment, decalcifying hide coria; grinding into small pieces; acidifying to produce a swollen slurry (4–5% solids); homogenizing; extrusion forming into tubular casings (8–10% solids); feeding extrudates into coagulating baths of brine or ammonium; washing casings free of salts; adding plasticizers and cross-linking agents; and drying (1). Perpendicular shear is applied to the collagen fibers as they exit the extruder to disrupt their laminar alignment to produce a more woven fiber structure. A collagen casing extrusion process developed in Germany in the 1930s employs collagen dispersions of higher solids contents (>12%) (1). Such dispersions require substantially higher extrusion pressures. This method has had limited acceptance outside Europe.

As an alternative to preformed collagen casings, a technology where the collagen casing is coextruded around the sausage meat batter was developed in the 1980s (2). The collagen dispersion is prepared *in situ* and pumped to a special extrusion nozzle. The meat emulsion is pumped through an inner opening in the nozzle and the collagen dispersion is applied to the emerging meat batter through counterrotating concentric cones that orient collagen fibers into a woven alignment. The continuous encased sausage rope is passed through a brine bath to dehydrate and set the collagen casing. The sausage rope is crimped into links, and the sausage is dried and smoked to further set the collagen casing. Coextruded sausage casings are more tender than preformed casings and, therefore, have better eating qualities, yet they remain more durable during grilling. The high initial cost of the coextrusion equipment has limited the acceptance of this method to large processing plants. Besides sausage casings, collagen-based edible films for use on netted roasts, boneless hams, fish fillets, and roast beef also have been commercialized since the late 1980s (3). Such films (trade name Coffi) can reduce cook shrink, increase product juiciness, and allow for easy removal of elastic stretch netting after heat processing. Reportedly, both refrigerated and frozen/thawed round beef steaks wrapped in collagen films prior to retail packaging (permeable film overwrap) or vacuum packaging exhibited substantially less fluid exudate than unwrapped steaks (4).

Gelatin

Gelatin is obtained by partial hydrolysis of collagen through alkaline or acidic treatments. It is an important hydrocolloid used in a number of food, photographic, pharmaceutical, and technical applications (5). In particular, gelatin is used widely in pharmaceutical and dietary supplement industries as an encapsulating agent (edible barrier) in hard (two-piece) capsules, soft capsules (softgels), and microcapsules (usually as coacervate with anionic polysaccharides). However, the film-forming properties of gelatin that have made it indispensable to the pharmaceutical and photographic industries have not been utilized by the food industry. This is mainly due to gelatin being too easily hydrated and solubilized to form an effective barrier in foods.

Cereal Proteins

Zein, the prolamin fraction of corn proteins has been studied extensively for its ability to form glossy, grease-resistant films that are not water-soluble. Protective coatings based on zein are commercially available for shelled nuts and confections (6) and have shown potential for extending the shelf life of tomatoes (7) and cooked turkey (8). Also, due to their grease resistance, zein-based coatings can reduce oil uptake by deep-fat fried foods (9). Wheat gluten also possesses good film-forming properties due to its cohesiveness and elasticity. At low moisture contents, wheat gluten films exhibited 10-fold lower aroma permeability than low-density polyethylene films (10). Typically, wheat gluten films are cast from aqueous alcohol solutions under alkaline (11) or acidic (12) conditions. The use of organic solvents (eg, ethanol or acetone) in zein and wheat gluten film-forming solutions has hindered the application of such films. Recently, a method was patented for preparing film-forming colloidal dispersions from dilute aqueous acid solutions of zein and wheat gluten (13). Also, zein-based moldable resins have been rolled into sheets (14). The relationship of wheat gluten (and other cereal proteins) to celiac disease, a dietary intolerance, should be considered when developing edible films, as well as allergies associated with various proteins. The film-forming properties of other cereal proteins, such as sorghum kafirin (15) and rice bran protein (16), have recently been investigated.

Oilseed Proteins

Soy protein-lipid films have traditionally been fabricated in East Asia on the surface of soy milk heated in open, shallow pans (6). Films are successively formed on and removed from the soy milk surface until protein in the solution is exhausted. Such films, known as "yuba" in Japan, can be used as meat and vegetable wrappers prior to cooking. Similar films have been prepared from peanut and cottonseed "milk." Also, heat-catalyzed protein polymerization through surface dehydration has been reported for heated solutions of soy protein isolates (17). This method is not suitable for large-scale commercial film production. Films have been cast from aqueous film-forming solutions of soy protein isolates (18). Soy protein films are generally less water resistant than zein or wheat gluten films.

Milk Proteins

Transparent and flavorless cast films have been prepared from both whey proteins and caseins (19). Reportedly, cross-linking with transglutaminase reduced water resistance and increased mechanical strength of milk protein films (20). Transglutaminase also catalyzed the formation of heterologous biopolymers from whey protein and 11S soy protein fraction (21). Besides film formation by casting, a modified wet spinning process was recently proposed for the continuous preparation of casein films (22). Edible sausage casings from casein have been prepared (23). Whey protein coatings have shown potential as oxygen barriers for dry roasted peanuts (24).

Other Proteins

Egg white (albumen) is a complex protein system with ovalbumin, ovotransferrin, ovomucoid, lysozyme, globulins, and ovomucin as the principal fractions. Films cast from aqueous alkaline solutions of desugared dried egg white (25) were clearer and more transparent than wheat gluten, zein, or soy protein films. Egg white films have limited water resistance and may find use as water-soluble packages (eg, pouches for premeasured food ingredients). Water-insoluble, transparent films with tensile strength close to those of polyethylene films have been developed from fish myofibrillar proteins (26). Other proteins that have received limited attention as film-formers are pea protein, feather keratin, and blood protein.

POLYSACCHARIDE FILMS

Alginates

Alginates are salts of alginic acid, a linear copolymer of D-mannuronic acid and L-guluronic acids that is extracted from brown seaweeds. The ability of alginates to gel in the presence of divalent or trivalent cations is utilized in forming alginate films. Calcium cations, which are more effective gelling agents than other polyvalent cations, "bridge" adjacent alginate chains via ionic interactions, after which interchain hydrogen bonding occurs (27). The application of alginate-based coatings on meat and seafood products has been discussed in several published studies and patent disclosures (3). A two-step procedure was used for forming such coatings. Application of aqueous sodium alginate solutions on food products, by dipping or spraying, was followed by treatment with a calcium salt solution (typically calcium chloride) to induce gelation (fixing). Calcium alginate coatings have reduced moisture losses from coated foods effectively, mainly due to their high moisture content. The coating acts as a sacrificing agent when moisture evaporates from the gel prior to any notable desiccation of the enrobed food (27). A bitter taste imparted on coated foods by calcium chloride has been a concern with alginate coatings. Patents (28,29) have been awarded for a fresh meat coating process in which a sodium alginate-oligosaccharide solution and a calcium chloride-thickening gum solution were successively applied on fresh meat by spraying or dipping. Based on these patents, an edible alginate-based coating for meats and other foods was developed and marketed (trade name Flavor-Tex) in the 1970s. Flavor-Tex's formulation included sodium alginate with maltodextrin in the first solution and calcium chloride with carboxymethyl cellulose in the second solution. Beef cuts coated with Flavor-Tex and stored at 5°C had reduced moisture loss but similar surface microbial counts compared with uncoated samples (30). Alginate coatings also extended shelf life and preserved texture of coated mushrooms by reducing the rate of water evaporation (31). In an active packaging application, lactic or acetic acids immobilized in calcium alginate gels were more effective in reducing viable counts of *Listeria monocytogenes* attached to lean beef surfaces after six days at 5°C than were acid treatments without alginate (32).

Carrageenan

Carrageenans are sulfated polysaccharides extracted from red seaweeds. The κ- and ι-carrageenan fractions form gels in the presence of monovalent and divalent cations. Locust bean gum and other mannans often are added to carrageenan gels to prevent syneresis and increase elasticity. Protective carrageenan coatings have been used to extend the shelf lives of meats and fish (3). A patented process described dipping precooked meat into an aqueous calcium carrageenan dispersion prior to freezing (33). The formed coatings extended product shelf life by preventing moisture loss and, thus, "freezer burn." Similar to alginate films, carrageenan films probably reduce moisture loss by serving as a reservoir of moisture that is sacrificed. A carrageenan-based active coating containing sorbic acid created a pH gradient on the surface of an intermediate moisture cheese analog product, thus enhancing the antimicrobial activity of sorbate (34).

Cellulose Derivatives

Water-soluble cellulose ethers including methyl cellulose (MC), hydroxypropyl cellulose (HPC), hydroxypropylmethyl cellulose (HPMC), and carboxymethyl cellulose (CMC) possess good film-forming properties. Their relative hydrophilicities increase in the order of HPC < MC < HPMC < CMC. Ethyl cellulose (EC) is another film-forming cellulose ether, which is water-insoluble. As discussed next, cellulose ethers often are used as matrices in composite films. Aqueous-based HPMC solutions are widely used in the pharmaceutical industry for tablet coating. MC and HPMC have been used as batter ingredients to reduce oil uptake during deep-fat frying of battered and breaded foods. Meatballs prepared from ground chicken breast and coated with HPMC absorbed up to 33.7% less fat than uncoated samples during deep-fat frying (35). HPMC coatings also reduced the number of viable *Salmonella montevideo* cells on the surface of tomatoes (36). Water-soluble, edible pouches from HPMC are used commercially to deliver preweighed minor manufacturing ingredients such as dough conditioners, leavening agents, vitamins, and minerals. Nature-Seal, a proprietary fresh produce coating formulation based on cellulose ethers and emulsifiers is used for delaying ripening of climacteric fruits (37).

Pectin

Pectins are composed primarily of the methyl esters of linear chains of 1,4-α-D-galacturonic acid units. They are extracted mainly from citrus peel and apple pomace. High-methoxyl (HM) and low-methoxyl (LM) pectins have degree of esterification above or below 50%, respectively. HM pectins gel at a soluble solids concentration of 55 to 80% and at acidic pH (2.8–3.7), whereas LM pectins gel in presence of divalent cations similarly to alginates (38). Applications of LM pectin coatings on oily or sticky food products such as shelled nuts and candied fruit have been proposed (39). A recent patent describes film casting from pectin or pectin/gelatinized starch mixtures (40).

Starch and Starch Derivatives

Starch, a glucose polymer, is composed of linear amylose and branched amylopectin. Both fractions can form films although amylose films have better mechanical and barrier properties than amylopectin films (41). Reportedly, tensile strength of nonplasticized starch films increased continuously as amylose content increased from 0 to 100% (42). As early as the 1950s, transparent films cast from water-butanol solutions of gelatinized amylose were evaluated for their functional properties (43). Extruded edible film wraps based on hydroxypropylated high amylose starch were commercially available (trade name Ediflex) in the late 1960s for application on frozen foods. High amylose starch coatings reduced moisture loss, maintained firmness, and reduced decay in coated, refrigerated strawberries (44). Also, various commercial modified starches, dextrins, and maltodextrins improved the quality of coated precooked dehydrated pinto beans (45).

Chitosan

Chitin, a glucosamine polymer, and chitosan (deacetylated chitin) are produced industrially from crustacean shell waste. Although abundantly available, chitin and chitosan have limited commercial use. They are approved in Japan and in Canada for various food applications but have not received approval by the FDA in the United States. Free-standing chitosan films can be cast from acetic, formic, or dilute hydrochloric acid solutions. For example, glycerin-plasticized films were cast from 3% w/w solutions of chitosan in 1% acetic acid (46). Protective chitosan coatings substantially reduced water loss in bell peppers and cucumbers (47). Also, chitosan coatings, due to the inherent antimicrobial properties of chitosan, inhibited growth of several spoilage microorganisms on raw refrigerated shrimp (48). Water-soluble *N,O*-carboxymethyl chitosan (NOCC) is produced by reacting chitin with chloroacetic acid under alkaline conditions. A commercially available NOCC coating (Nutri-Save) reportedly had some success as a selectively permeable postharvest coating for fresh fruit (38).

LIPIDS AND RESINS

Acetylated Glycerides

Acetylated glycerides can be obtained either by reacting glycerides with acetic anhydride or through the catalyzed interesterification of a fat or oil with triacetin (49). They solidify into stable α-form crystals that are waxy and relatively impermeable to moisture migration. Distilled acetylated monoglyceride coatings have been used in edible packaging applications since the 1960s. Claims have been made in patent disclosures that such coatings preserved meat freshness (50,51). However, acetylated monoglyceride coatings (trade name Dermatex) did not affect significantly the physical and sensory characteristics (eg, moisture, surface discoloration, and overall appearance scores) of vacuum-packaged beef steaks and roasts stored at 2°C for seven to eight weeks (52). More recently, acetylated monoglyceride (trade name Myvacet) coatings substantially reduced moisture loss and delayed the onset of lipid oxidation in frozen salmon pieces stored at −23°C (53). Certain application and sensory problems are associated with acetylated monoglyceride coatings including the tendency of highly saturated acetylated glycerides to crack and flake during storage. An acidic, bitter aftertaste also has been attributed to acetylated glycerides.

Waxes

Coating fresh produce with waxes to extend postharvest shelf life has a long history of use. Both natural waxes (eg, carnauba, beeswax, candelilla, and rice bran) and synthetic waxes (eg, paraffin and oxidized polyethylene) are used for this purpose. For example, commercial wax coating formulations include Primafresh 31 (carnauba wax), Shild-Brite PR-190 (natural waxes), Apple-Britex 559 (carnauba wax), and Fresh Wax 625 (oxidized polyethylene wax) (37). Generally, waxes are substantially more resistant to moisture transport than most other lipid or nonlipid edible films (27). Waxes applied to perishable produce replace natural surface waxes of the cuticle, prevent moisture loss, and improve surface luster. Application of wax coatings on the commodity surface may be accomplished by dipping, spraying, dripping, or controlled dropping (54). Waxes, by themselves, however, do not reduce decay. In fact, waxes can restrict oxygen and carbon dioxide gas exchange and cause uneven ripening or decay (55,56). For this reason, fungicides and growth regulators often are incorporated into waxes to control spoilage.

Resins

Resins are acidic substances that are secreted by many trees and shrubs in response to injury. Shellac resin is the alcohol soluble exudate of the insect *Laccifer lacca* and is the only commercially used natural resin of animal origin. Regular bleached shellac containing the natural shellac wax, or refined, dewaxed bleached shellac, which yields clearer solutions, are available. Shellac coatings are extensively used for candies (eg, candy corn), shelled nuts, and tablets (eg, multivitamins). Shellac-based enteric release coatings for solid dosage forms are utilized in the pharmaceutical industry (57). Fresh produce coatings from shellac or mixtures of shellac and waxes, such as carnauba wax, are commercially available (37,58). Besides shellac, other resins such as coumarone indene (petroleum-derived) and wood rosin are permitted for use in food coatings (59).

COMPOSITE FILMS

Combining various biopolymers is an obvious approach to tailor the functional properties of edible films and coatings to specific applications. Generally, polysaccharide and protein films provide limited resistance to moisture transmission due to the inherent hydrophilicity of such materials and to the notable amounts of incorporated hydrophilic plasticizers needed to impart adequate film flexibility. In contrast, hydrophobic lipids are effective moisture barriers. Composite films (prepared from emulsions or by lamination) made of lipids and polysaccharides or proteins combine the satisfactory structural and oxygen barrier properties of polysaccharide or protein films with the good moisture barrier characteristics of lipids. For example, bicomponent films prepared from cellulose ethers and lipids (eg, waxes, acetylated monoglycerides, and fatty acids) effectively restricted moisture migration in simulated food products containing major components differing substantially in water activity (60). Coatings comprised of MC or HPMC and fatty acids also may be used to control diffusion of antimicrobial agents (eg, potassium sorbate) into foods (61). Recently, highly refined cellulose dispersions made from fibrous agricultural by-products were copulverized at high pressure with various polysaccharides (guar gum, alginate, carrageenan, pectin, and xanthan gum) and coconut oil to form emulsions, which were cast into films (62). Such films had substantially lower water vapor permeability than those of polysaccharide-based films reported in the literature. Coatings comprised of sucrose esters of fatty acids and the sodium salt of CMC (trade names Semperfresh and Pro-long) effectively reduced ripening rate of fresh produce (63,64). Edible coating compositions for multicomponent food products combining shellac, HPC, and, optionally, fatty acids have been patented (65). Bicomponent chitosan-lauric acid films had 50% lower water vapor permeability than chitosan films (66). Composite protein-lipid films had lower water vapor permeability values than control protein films (19,67). Sodium caseinate-stearic acid coatings reduced white blush on peeled carrots by alleviating surface dehydration (68). Besides lipid-polysaccharide or lipid-protein films, composite films combining proteins and polysaccharides, such as starch, also have been studied recently (69,70). Blending various protein and polysaccharide film-formers with inexpensive starch can substantially reduce the cost of edible film formulations (71).

BIBLIOGRAPHY

1. L. L. Hood, "Collagen in Sausage Casings," in A. M. Pearson, T. R. Dutson, and A. J. Bailey, eds., *Advances in Meat Research*, Vol. 4, Van Nostrand Reinhold, New York, 1987, pp. 109–129.
2. J. W. Smits, "The Sausage Coextrusion Process," in B. Krol, P. S. van Roon, and J. H. Houben, eds., *Trends in Modern Meat Technology*, Pudoc, Wageningen, The Netherlands, 1985, pp. 60–62.
3. A. Gennadios, M. A. Hanna, and L. B. Kurth, "Application of Edible Coatings on Meats, Poultry and Seafoods: A Review," *Lebensm.-Wiss. Technol.* **30**, 337–350 (1997).
4. M. M. Farouk, J. F. Price, and A. M. Salih, "Effect of an Edible Collagen Film Overwrap on Exudation and Lipid Oxidation in Beef Round Steak," *J. Food Sci.* **55**, 1510–1512, 1563 (1990).
5. P. I. Rose, "Gelatin," in H. F. Mark et al., eds., *Encyclopedia of Polymer Science and Engineering*, 2nd ed., Vol. 7, John Wiley & Sons, New York, 1987.

6. A. Gennadios et al., "Edible Coatings and Films Based on Proteins," in J. M. Krochta, E. A. Baldwin, and M. Nisperos-Carriedo, eds., *Edible Coatings and Films to Improve Food Quality*, Technomic Publishing Company, Lancaster, Penn., 1994, pp. 201–277.
7. H. J. Park, M. S. Chinnan, and R. L. Shewfelt, "Edible Coating Effects on Storage Life and Quality of Tomatoes," *J. Food Sci.* **59**, 568–570 (1994).
8. T. J. Herald et al., "Corn Zein Packaging Materials for Cooked Turkey," *J. Food Sci.* **61**, 415–417, 421 (1996).
9. U.S. Pat. 5,217,736 (June 8, 1993), R. D. Feeney, G. Haralampu, and A. Gross (to Opta Food Ingredients, Inc.).
10. F. Debeaufort and A. Voilley, "Aroma Compound and Water Vapor Permeability of Edible Films and Polymeric Packaging," *J. Agric. Food Chem.* **42**, 2871–2875 (1994).
11. U.S. Pat. 3,653,925 (April 4, 1972), C. A. Anker, G. A. Foster, Jr., and M. A. Loader (to General Mills, Inc.).
12. N. Gontard, S. Guilbert, and J.-L. Cuq, "Edible Wheat Gluten Films: Influence of the Main Process Variables on Film Properties Using Response Surface Methodology," *J. Food Sci.* **57**, 190–195, 199 (1992).
13. U.S. Pat. 5,736,178 (April 7, 1998), R. B. Cook and M. L. Shulman (to Opta Food Ingredients, Inc.).
14. H.-M. Lai, G. W. Padua, and L. S. Wei, "Properties and Microstructure of Zein Sheets Plasticized With Palmitic and Stearic Acids," *Cereal Chem.* **74**, 83–90 (1997).
15. R. A. Buffo, C. L. Weller, and A. Gennadios, "Films From Laboratory-Extracted Sorghum Kafirin," *Cereal Chem.* **74**, 473–475 (1997).
16. R. Gnanasambandam, N. S. Hettiarachchy, and M. Coleman, "Mechanical and Barrier Properties of Rice Bran Films," *J. Food Sci.* **62**, 395–398 (1997).
17. S. Okamoto, "Factors Affecting Protein Film Formation," *Cereal Foods World* **23**, 256–262 (1978).
18. A. H. Brandenburg, C. L. Weller, and R. F. Testin, "Edible Films and Coatings From Soy Protein," *J. Food Sci.* **58**, 1086–1089 (1993).
19. T. H. McHugh and J. M. Krochta, "Milk Protein-Based Edible Films and Coatings," *Food Technol.* **48**(1), 97–103 (1994).
20. H. Chen, "Functional Properties and Applications of Edible Films Made of Milk Proteins," *J. Dairy Sci.* **78**, 2563–2583 (1995).
21. M. Yildirim and N. S. Hettiarachchy, "Biopolymers Produced by Cross-linking Soybean 11S Globulin With Whey Proteins Using Transglutaminase," *J. Food Sci.* **62**, 270–275 (1997).
22. A. Frinault et al., "Preparation of Casein Films by a Modified Wet Spinning Process," *J. Food Sci.* **62**, 744–747 (1997).
23. U.S. Pat. 5,681,517 (October 28, 1997), W. Metzger (to Doxa GmbH).
24. J. I. Maté and J. M. Krochta, "Whey Protein Coating Effect on the Oxygen Uptake of Dry Roasted Peanuts," *J. Food Sci.* **61**, 1202–1206, 1210 (1996).
25. A. Gennadios et al., "Mechanical and Barrier Properties of Egg Albumen Films," *J. Food Sci.* **61**, 585–589 (1996).
26. B. Cuq, N. Gontard, and S. Guilbert, "Proteins as Agricultural Polymers for Packaging Production," *Cereal Chem.* **75**, 1–9 (1998).
27. J. J. Kester and O. R. Fennema, "Edible Films and Coatings: A Review," *Food Technol.* **40**(12), 47–59 (1986).
28. U.S. Pat. 3,395,024 (July 30, 1968), R. D. Earle (to R. D. Earle).
29. U.S. Pat. 3,991,218 (November 9, 1976), R. D. Earle and D. H. McKee (to Food Research, Inc.).
30. S. K. Williams, J. L. Oblinger, and R. L. West, "Evaluation of a Calcium Alginate Film for Use on Beef Cuts," *J. Food Sci.* **43**, 292–296 (1978).
31. A. Nussinovitch and N. Kampf, "Shelf-Life Extension and Conserved Texture of Alginate-Coated Mushrooms (*Agaricus bisporus*)," *Lebensm.-Wiss. Technol.* **26**, 469–475 (1993).
32. G. R. Siragusa and J. S. Dickson, "Inhibition of *Listeria monocytogenes* on Beef Tissue by Application of Organic Acids Immobilized in a Calcium Gel," *J. Food Sci.* **57**, 293–296 (1992).
33. U.S. Pat. 4,196,219 (April 1, 1980), C. P. Shaw, J. L. Secrist, and J. M. Tuomy (to The United States of America as represented by the Secretary of the Army).
34. J. A. Torres, J. O. Bouzas, and M. Karel, "Microbial Stabilization of Intermediate Moisture Food Surfaces. II. Control of Surface pH," *J. Food Proc. Preserv.* **9**, 93–106 (1985).
35. V. M. Balasubramaniam et al., "The Effect of Edible Film on Oil Uptake and Moisture Retention of a Deep-Fat Fried Poultry Product," *J. Food Proc. Engr.* **20**, 17–29 (1997).
36. R. Zhuang et al., "Inactivation of *Salmonella montevideo* on Tomatoes by Applying Cellulose-Based Edible Films," *J. Food Prot.* **59**, 808–812 (1996).
37. M. O. Nisperos and E. A. Baldwin, "Edible Coatings for Whole and Minimally Processed Fruits and Vegetables," *Food Austr.* **48**, 27–31 (1996).
38. M. O. Nisperos-Carriedo, "Edible Coatings and Films Based on Polyssacharides," in J. M. Krochta, E. A. Baldwin, and M. O. Nisperos-Carriedo, eds., *Edible Coatings and Films To Improve Food Quality*, Technomic Publishing Company, Lancaster, Penn., 1994, pp. 305–355.
39. H. A. Swenson et al., "Pectinate and Pectate Coatings. II. Application to Nuts and Fruit Products" *Food Technol.* **7**, 232–235 (1953).
40. U.S. Pat. 5,451,673 (September 19, 1995), M. L. Fishman and D. R. Coffin (to The United States of America as represented by the Secretary of Agriculture).
41. Å. Rindlav-Westling et al., "Structure, Mechanical and Barrier Properties of Amylose and Amylopectin Films," *Carbohydr. Polym.* **36**, 217–224 (1998).
42. D. Lourdin, G. Della Valle, and P. Colonna, "Influence of Amylose Content on Starch Films and Foams," *Carbohydr. Polym.* **27**, 261–270 (1995).
43. J. C. Rankin et al., "Permeability of Amylose Film to Moisture Vapor, Selected Organic Vapors, and the Common Gases," *Ind. Engr. Chem. (Chem. & Engr. Data Series)* **3**, 120–123 (1958).
44. M. A. García, M. N. Martino, and N. E. Zaritzky, "Starch-Based Coatings: Effect on Refrigerated Strawberry (*Fragaria ananassa*) Quality," *J. Sci. Food Agric.* **76**, 411–420 (1998).
45. H. L. Su and K. C. Chang, "Dehydrated Precooked Pinto Bean Quality as Affected by Cultivar and Coating Biopolymers," *J. Food Sci.* **60**, 1330–1332 (1995).
46. B. L. Butler et al., "Mechanical and Barrier Properties of Edible Chitosan Films as Affected by Composition and Storage," *J. Food Sci.* **61**, 953–955, 961 (1996).
47. A. El Ghaouth et al., "Use of Chitosan Coating to Reduce Water Loss and Maintain Quality of Cucumber and Bell Pepper Fruits," *J. Food Proc. Preserv.* **15**, 359–368 (1992).
48. B. K. Simpson et al., "Utilization of Chitosan for Preservation of Raw Shrimp," *Food Biotechnol.* **11**, 25–44 (1997).
49. R. O. Feuge, "Acetoglycerides—New Fat Products of Potential Value to the Food Industry," *Food Technol.* **9**, 314–318 (1955).
50. U.S. Pat. 3,851,077 (November 26, 1974), M. Stemmler and H. Stemmler (to M. Stemmler & H. Stemmler).

51. U.S. Pat. 4,137,334 (January 30, 1979), C. Heine, R. Wüst, and B. Kamp (to Henkel KgaA).
52. D. B. Griffin et al., "Physical and Sensory Characteristics of Vacuum Packaged Beef Steaks and Roasts Treated With an Edible Acetylated Monoglyceride," *J. Food Prot.* **50**, 550–553 (1987).
53. Y. M. Stuchell and J. M. Krochta, "Edible Coatings on Frozen King Salmon: Effect of Whey Protein Isolate and Acetylated Monoglycerides on Moisture Loss and Lipid Oxidation," *J. Food Sci.* **60**, 28–31 (1995).
54. L. A. Grant and J. Burns, "Application of Coatings," in J. M. Krochta, E. A. Baldwin, and M. Nisperos-Carriedo, eds., *Edible Coatings and Films To Improve Food Quality*, Technomic Publishing Co., Lancaster, Penn., 1994, pp. 189–200.
55. L. A. Risse et al., "Volatile Production and Decay During Storage of Cucumbers Waxed, Imazalil-Treated, and Film-Wrapped," *HortScience.* **22**, 274–276 (1987).
56. C. H. Mannheim and T. Soffer, "Permeability of Different Wax Coatings and Their Effect on Citrus Fruit Quality," *J. Agric. Food Chem.* **44**, 919–923 (1996).
57. F. Specht et al., "The Application of Shellac as an Acidic Polymer for Enteric Coating," *Pharm. Technol.* **23**(3), 146, 148, 150, 152, 154 (1999).
58. R. D. Hagenmaier and R. A. Baker, "Layered Coatings to Control Weight Loss and Preserve Gloss of Citrus Fruit," *Postharv. Biol. Technol.* **30**, 296–298 (1995).
59. E. Hernandez, "Edible Coatings From Lipids and Resins," in J. M. Krochta, E. A. Baldwin, and M. Nisperos-Carriedo, eds., *Edible Coatings and Films To Improve Food Quality*, Technomic Publishing Co., Lancaster, Penn., 1994, pp. 279–303.
60. O. Fennema, I. G. Donhowe, and J. J. Kester, "Edible Films: Barriers to Moisture Migration in Frozen Foods," *Food Austr.* **45**, 521–525 (1993).
61. F. Vojdani and J. A. Torres, "Potassium Sorbate Permeability of Methylcellulose and Hydroxypropyl Methylcellulose Coatings: Effect of Fatty Acids," *J. Food Sci.* **55**, 841–846 (1990).
62. R. R. Ruan, L. Xu, and P. L. Chen, "Water Vapor Permeability and Tensile Strength of Cellulose-Based Composite Edible Films," *Appl. Engr. Agric.* **14**, 411–413 (1998).
63. F. H. Motlagh and P. C. Quantick, "Effect of Permeable Coatings on the Storage Life of Fruits. I. Pro-long Treatment of Limes (*Citrus aurantifolia* cv. Persian)," *Int. J. Food Sci. Technol.* **23**, 99–105 (1988).
64. C. R. Santerre, T. F. Leach, and J. N. Cash, "The Influence of the Sucrose Polyester, Semperfresh™, on the Storage of Michigan Grown 'McIntosh' and 'Golden Delicious' Apples," *J. Food Proc. Preserv.* **13**, 293–305 (1989).
65. U.S. Pat. 4,661,359 (April 28, 1987), J. Seaborne and C. Egberg (to General Mills, Inc.).
66. D. W. S. Wong et al., "Chitosan-Lipid Films: Microstructure and Surface Energy," *J. Agric. Food Chem.* **40**, 540–544 (1992).
67. N. Gontard et al., "Water Vapor Permeability of Edible Bilayer Films of Wheat Gluten and Lipids," *Int. J. Food Sci. Technol.* **30**, 49–56 (1995).
68. R. J. Avena-Bustillos et al., "Application of Casein-Lipid Edible Film Emulsions to Reduce White Blush on Minimally Processed Carrots," *Postharv. Biol. Technol.* **4**, 319–329 (1994).
69. N. Parris et al., "Water Vapor Permeability and Solubility of Zein/Starch Hydrophilic Films Prepared from Dry Milled Corn Extract," *J. Food Engr.* **32**, 199–207 (1997).
70. I. Arvanitoyannis and C. G. Biliaderis, "Physical Properties of Polyol-Plasticized Edible Films Made From Sodium Caseinate and Soluble Starch Blends," *Food Chem.* **62**, 333–342 (1998).
71. M. L. Fishman, "Edible and Biodegradable Polymer Films: Challenges and Opportunities," *Food Technol.* **51**(2), 16 (1997).

ARISTIPPOS GENNADIOS
Banner Pharmacaps, Inc.
High Point, North Carolina
CURTIS L. WELLER
MILFORD A. HANNA
University of Nebraska
Lincoln, Nebraska

EEL

Eel is a popular food fish in Europe and the Far East, especially in Japan. In some countries, however, eel is not so popular because of its snakelike appearance. In spite of this, the demand for eels has increased considerably during the last two decades.

Annual world production of eels is about 200,000 t. Twenty-five percent of these are wild eels captured mainly in Europe, North America, and Oceania, whereas the remaining 75% are cultured eels.

Japan pioneered the culture of eels more than 150 years ago. Its eel culture technique is now one of the most advanced in the world. Taiwan, which adopted Japanese culture methods, follows closely. In annual eel output, Taiwan, however, surpasses Japan, making it the world's largest supplier of eels (1). Meanwhile, several countries are also developing their own eel culture industry. Fluctuating natural eel stocks and the pollution of the eel's habitat have made world eel production unstable. The culture of eels may be the best way of ensuring the adequate supply of the fish.

BIOLOGY

Species and Characteristics

Freshwater eels are widely distributed throughout the world, mostly in the areas around the Atlantic and Indo-Pacific oceans (Fig. 1) (2). A total of 16 species and six subspecies are recorded, namely, *Anguilla anguilla* (European eel), *A. rostrata* (American eel), *A. japonica* (Japanese eel), *A. marmorata, A. bicolor* (subspecies: *A. bicolor pacifica* and *A. bicolor bicolor*), *A borneensis, A. celebesensis, A, ancestralis, A. obscura, A. australis* (subspecies: *A. australis australis* and *A. australis schmidti*), *A. megastoma, A. mossambica, A. nebulosa* (subspecies *A. nebulosa labiata* and *A. nebulosa nebulosa*), *A. reinhardti, A. interioris*, and *A. dieffenbachi*. However, *A. celebesensis* and *A. ancestralis* are classified as synonyms (3). There are several characteristics that distinguish the world's eel species, including dorsal fin length, vertebral number, color, size, head shape, habitat, and distribution (Table 1).

Dorsal Fin Length. Eels are generally divided into two types according to the proportion of the length from the

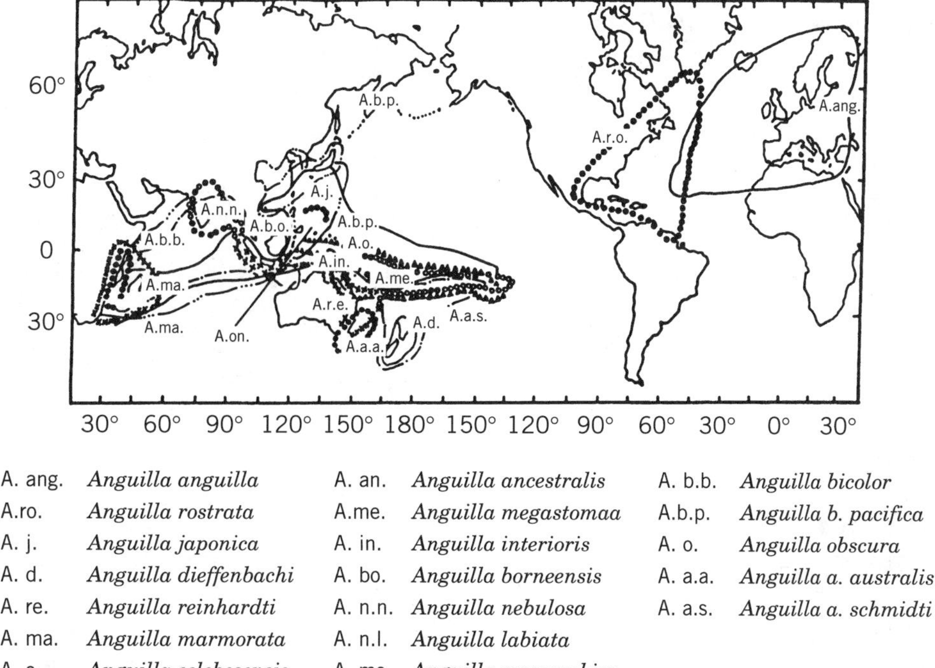

Figure 1. Natural distribution of freshwater eels. *Source:* Ref. 2.

anterior base of the dorsal fin to the anus and the total body length. The long-fin type has a proportion of 7–17%, while the short-fin type has a proportion of 0–5%. Thirteen of the eel species belong to the long-fin category, three species belong to the short-fin type.

Vertebral Number. Another taxonomic criterion for eel types is the number of vertebrae. Eels have 100–119 vertebrae; the number varies among the different species. The number even varies within each species, which increases with higher latitudes.

Color. Eels can be mottled or plain. There are seven mottled and nine plain species.

Size. There is a considerable difference between female and male eels. The growth of female eels in inland water is much more superior than that of male eels. The females also stay longer in inland water before returning to the sea. Eels can grow at maximum sizes of 27 kg and 200 cm.

Head Shape. Eels are also classified according to the shape of their head. Those with a narrow head and thin, narrow lips are classified as narrow-head type while those with broad head and thick, broad lips are classified as broad-head type (4,5).

Distribution

Of the 16 species of eels, 14 are distributed in the Indo-Pacific areas. Some of these eels inhabit the tropical zone, while others abound in the temperate zone. Eels in the Indo-Pacific Ocean are distributed in both the southern and northern hemispheres. The northernmost boundary is at about 45° N in Hokkaido, Japan, while the southernmost limit is at about 50° S in the Auckland Islands, New Zealand. The two other species, the European eel and the American eel, inhabit the Atlantic Ocean and abound in the temperate zones of Europe and North America, respectively (Fig. 1).

LIFE HISTORY

The first successful attempt to understand the amazing life cycle and unusual traits of the eel has generally been accredited to two men who devoted most of their lives to studying eels: Danish biologist Johannes Schmidt and his student Vilhelm Ege. The discovery of a leptocephalus by Schmidt in 1904 started a 35-year study that helped establish a solid foundation to the ecological study of eels. Nevertheless, many questions concerning the life history of eels still remain unanswered. Although the life history of three species, European eel, American eel, and the Japanese eel are well investigated, little is known about the other species.

Eels are catadromous migrants, that is, they spawn at sea and spend most of their lives in inland waters. The spawning ground and migration route vary with each species.

European Eel

The European eel spawns at the Sargasso Sea, 22–30° N, 48–65° W, in the western Atlantic Ocean from early spring

Table 1. Major Details of the World's *Anguilla* Species

Dorsal fin length	Vertebral number	Color[a]	Maximum size of females kg	Maximum size of females cm	Species	Zone[b]		Distribution[c]
Long	110–119	P	6.0	125[d]	*A. anguilla*	Temp	ATL	Europe, North Africa
Long	103–111	P	6.0	125[d]	*A. rostrata*	Temp	ATL	United States, Canada, Greenland
Long	112–119	P	6.0	125[d]	*A. japonica*	Temp	PAC	Japan, Korea, Taiwan, People's Republic of China
Long	109–116	P	20.0	150	*A. dieffenbachi*	Temp	PAC	New Zealand
Long	104–110	M	18.0	170	*A. reinhardti*	Trop	PAC	Australia, New Caledonia
Long	100–110	M	27.0	200	*A. marmorata*[e]	Trop	PAC	Africa, Madagascar, Indonesia, People's Republic of China, Japan, Pacific Islands, Sumatra
Long	101–107	M			*A. celebesensis*[f]	Trop	PAC	Philippines
Long	101–106	M			*A. ancestralis*[f]	Trop	PAC	Celebes
Long	108–116	M	22.0	190	*A. megastoma*	Trop	PAC	New Caledonia
Long	104–108	M			*A. interioris*	Trop	PAC	New Guinea, New Caledonia
Long	103–108	P	2.0	90	*A. borneensis*	Trop	PAC	Brunei
Long		M	10.0	150	*A. nebulosa*	—	—	—
Long	106–112				*A. n. nebulosa*	Trop	IND	Sri Lanka, Burma, Sumatra
Long	107–115				*A. n. labiata*	Trop	IND	South Africa
Long	100–106	P	5.0	125	*A. mossambica*	Trop	IND	Africa
Short		P	3.0	110	*A. bicolor*	—	—	—
	106–115	P			*A. b. bicolor*	Trop	IND	South Africa, Madagascar, Sri Lanka, Burma, Sumatra, Australia
Short	103–111				*A. b. pacifica*	Trop	PAC	Brunei
Short	101–107	P			*A. obscura*	Trop	PAC	New Caledonia
Short		P	2.5	95	*A. australis*			
Short	109–116				*A. a. australis*	Trop	PAC	Australia
Short	108–115				*A. a. schmidti*	Temp	PAC	New Caledonia, New Zealand

Source: Refs. 5 and 9.
[a]Color abbreviations are P = plain, M = mottled.
[b]Zone abbreviations are Trop = tropical, Temp = temperate.
[c]Distribution abbreviations are ATL = Atlantic Ocean, IND = Indian Ocean, PAC = Pacific Ocean.
[d]*A. anguilla. A. rostrata,* and *A. japonica* are closely related (9).
[e]*A. marmorata* is the most widely distributed species (9).
[f]Synonymous (3).

through early summer. The prelarvae grow to about 25 mm in June. They then attenuate in shape, and gradually grow into transparent, leaf-shaped leptocephalus. The leptocephali drift with the current away from the Sargasso Sea to the coasts of Europe in about three years. The leptocephalus can attain a body length of 50 mm after a year and about 75 mm at two years old. They metamorphose into slim transparent elvers (Fig. 2) at 2.5 years old. The elvers then head for the coast and enter estuaries.

Both the female and male European eels stay in rivers until they grow to 6–12 and 9–19 years old, respectively. When they become sexually mature, the eels migrate downstream and back to the Sargasso Sea to spawn. After spawning, the adults die.

American Eel

The American eel also spawns in the Sargasso Sea. The elvers only take about 1.5 years to reach the coastal area of North America (6).

Japanese Eel

The spawning ground of the Japanese eel is estimated to be within the eastern part of Taiwan and the southern part of the Ryukyu islands, which may extend to the northeastern part of the Philippines. The spawning season is estimated to be between June and July. At the latter part of their development, the leptocephali drift with the Kuroshio current from the spawning ground and metamorphose into elvers when they reach the estuaries of Taiwan,

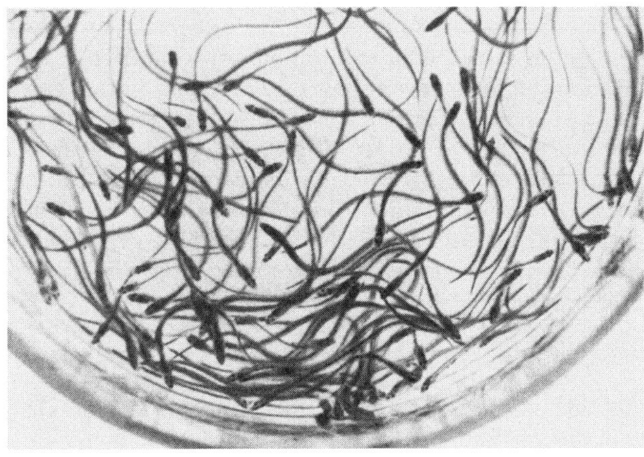

Figure 2. Elvers.

the People's Republic of China, Korea, and Japan after about 4–5 months. The migration time is estimated by the daily growth rings in the otolith. Using this index, the peak of elvers upstream migration is between December and January. The elvers enter the rivers and live and feed on small fish, shrimp, and aquatic insects for about 5–20 years until they reach adult size. They mature during autumn and winter (September–November). The pectoral fin, dorsal and ventral part of the mature males turn blackish and silvery. While mature males can weigh only about 500 g on maturity, the females can reach 2–3 kg. Sexually mature eels then migrate down the river into the ocean to spawn. Figure 3 illustrates the estimated spawning ground and life cycle of the Japanese eel (7). The Japanese eel is the most popular and most commonly eaten eel, particularly in Japan. Thus the following discussion focuses on the Japanese eel.

ECOLOGY OF ELVERS

The amount of catch for elvers in coastal waters seems to be correlated with temperature. The maximum catch of elvers in the upstream areas has often been found during or several days after daily seawater temperature reaches its lowest in winter. The lowest water temperature ever recorded in an eel habitat is 15–16°C.

Maximum catch has also been observed to occur at the same time when salinity has leveled off and the flow of seawater has reached its maximum. The elvers also become most active during this period at night.

The biological rhythm of elver activity was found to follow the lunar cycle in the coastal waters. The peak catches occur only once a month, about the time of the new moon. On the other hand, a semilunar rhythm of the elvers in the rivers was observed, namely, two peak catches occurred in each lunar cycle, one around full moon, the other around new moon. The semilunar rhythm of upstream elvers also coincided with the spring tide (8).

CULTURE TECHNIQUES

Species Cultured

Of the different species of eels in the world, three are valued for its economic importance, namely, the European eel,

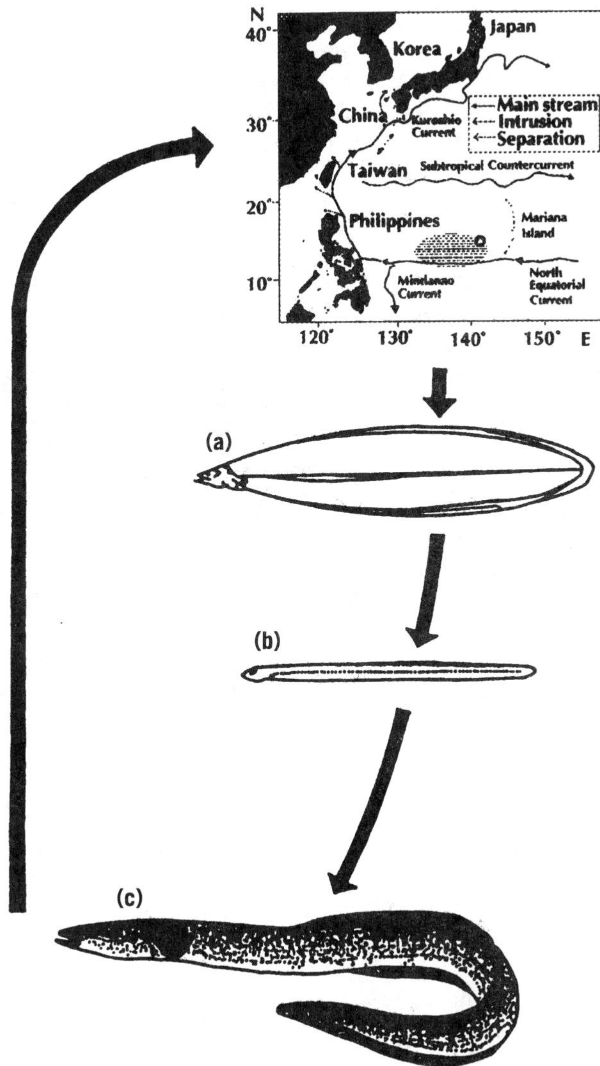

Figure 3. Estimated spawning ground and life cycle of the Japanese eel: (**a**) Leptocephalus, 5.5 cm T.L., drifted by current; (**b**) elver (ascending) migrates from estuary to river; (**c**) adult (descending) migrates from river to spawning ground. *Source:* Ref. 7.

the American eel, and the Japanese eel. From a marketing point of view, these species are similar in shape, growth, and behavior (8).

Elver Collection

The collection of the elvers is the starting point of eel culture. Attempts to propagate eels artificially have not been successful to date. In 1974, the Japanese succeeded in inducing artificial spawning of the eel and hatching the eel eggs. But attempts to rear the elver during the first three weeks of its life have not been as successful. Thus eel fry are still sourced entirely from the wild-caught elvers.

A growing concern within the eel culture industry in recent years is the diminishing number of large seaward running eels, as well as the elvers swimming upstream. This has been attributed mainly to pollution in estuaries, rivers, and streams and the overfishing of elvers. To remedy this, captured elvers are now widely used not only for culturing eels for consumption, but also for restocking in the lower reaches of rivers and streams.

Elvers are usually captured at nighttime. Light is used to attract them (Fig. 4) and they are caught in scoop nets in shallow water, in a fine-mesh net set across the width of the river, or in trap nets set in the estuaries (10).

Elvers are quite hardy; nevertheless, great care is taken when handling them to prevent mortality due to injury. They are not handled directly. They are placed in boxes lined with wet muslin or hung in mesh cages in the river. These boxes or cages are taken to eel farms within a few hours. The high demand coupled with a supply shortage has resulted in a phenomenal increase in the price of elvers.

CULTURE SYSTEMS

Eels can be cultured both extensively and intensively. Most eel farms, however, use the intensive culture system. This culture system is characterized by high stocking densities, stringent water-quality management, additional inputs such as formulated diets to increase fish production, and aggressive disease prevention.

Elver Rearing

There are various sizes and types of tanks being used in stocking elvers. The most commonly used, however, is the circular concrete tank, about 5 m in diameter and 60 cm in depth. Tanks are usually built under greenhouses and heated to above 25°C by thermostatically controlled electric immersion heaters (10). This is to protect the elvers from low temperatures.

The average stocking density of intensive culture is 1–2 kg of elvers/3.3m^2. Sorting and restocking are usually done after 20–30 days, when a significant difference in the size of eels is observed. The average density used for the second stocking phase is 1 kg of fish/3.3 m^2. Frequent sorting and restocking will bring about higher survival rate. These will also help maintain uniform growth and provide better feed conversion rates.

Figure 4. Offshore collection using boat and light to attract elvers.

The elvers have to be trained to take formulated feeds. The common method is to lower baskets containing tubifex worms or minced meat of oyster and clam twice a day, in the early morning and in the late evening under a wooden enclosure lit by a 20–40 W lamp. Elvers do not feed at temperatures below 13°C, thus temperatures must be kept above 13°C. Feeding time is gradually shifted to daytime.

Initially, elvers may not eat. In the long run, however, they learn to eat and eventually, attain rapid growth. After 100–150 days, a size of 100–200 fingerlings/kg is reached.

Feeding Management

Eels are carnivorous fish and thus require high animal protein input. The more animal protein that can be provided, the greater the eels' weight will be. Most eel farmers feed their eels with formulated feeds that are chiefly made of fish meal with added carbohydrates, minerals, vitamins, and other ingredients. Due to the high cost of these feeds, most eel farms supplement these with raw fish such as mackerel, sardine, and anchovy. Trash fish or scraps that are minced using electric grinders are also used widely as feeds. Eels, however, convert formulated feeds more efficiently than they convert raw fish.

Feeding is done once a day, at about 7:00 A.M. during summer and about 10:00 A.M. in winter. The usual feeding rate is 5–15% of the total weight of the eel for minced trash fish and 1.0–3.5% for formulated feeds (11). Nevertheless, factors such as the previous day's feeding condition and water and air temperature as well as water quality are considered when determining the feeding rate. Ideally, feeds should be consumed within 20–30 min.

Precautions must be taken to insure that no unnecessary organic debris are allowed to fall into the ponds. Thus eel farmers do not just broadcast the feeds into the ponds. Instead, a special feeding platform, usually shaded, within the pond is designated for feeding, and the eels learn to swim toward the area each feeding time. Feeding areas are preferably located where the oxygen level is high as this encourages the eels to eat well. Formulated feed is lowered on a perforated tray suspended just under the surface and lifted after feeding is done (Fig. 5). Raw fish, dipped for a few minutes in boiling water to soften the skin, are threaded through the eyes and also lowered and raised once eels have eaten all the flesh. Thus the only organic matter that goes into the pond is that which the eels eat.

Pond and Water Management

Growout ponds are either rectangular or square. In Japan, adult eels are raised in several types of ponds that make use of efficient drainage with water temperatures of about 28°C. Until recently, these eel ponds were between 5,000–20,000 m^2 in area. However, the practice of intensive culture has reduced pond size to about 500–1,000 m^2, and is expected to be further reduced in the future (10).

In Taiwan, two types of eel ponds are used, namely, the hard pond and the soft pond (Fig. 6). The hard pond has a concrete or red brick bottom and dikes and is small (1,000–1,650 m^2 and water depth of 0.8–1.2 m). The soft pond, on the other hand, has a mud bottom and concrete or brick dikes and is generally large (0.6–1.0 ha and water depth

Figure 5. Eels feeding on formulated feed paste from a perforated tray lowered into the pond. The tray is lifted out once the eels stop feeding to avoid scraps fouling the water.

Figure 6. Soft ponds.

of 2.0–3.0 m). Recently, some farmers have reverted back to using earthen ponds (Fig. 7).

Most eel farms have many ponds of different sizes because eels of different sizes must be sorted to maintain uniform growth. The eels that grow fast are separated from those that grow slow and are transferred to larger ponds.

Figure 7. Earthen ponds.

Pond water quality is also an important aspect of eel culture because it could affect the health and growth of the cultured eels. A well-maintained pond could spell the difference between healthy and fast-growing eels and poor quality ones.

Pond water color can also be used to gauge water quality. The most favorable color is green because it can show the presence of zooplanktons, phytoplanktons, ideal pH value, high dissolved oxygen concentration, and other factors affecting water quality.

Eel ponds require large amounts of fresh water and are thus generally located in areas with good water supply both quantitatively and qualitatively. Water is kept free from pollution and within pH values of 6.5–8.0. Temperature is kept above 13°C because eels stop eating below this temperature. Oxygen, which is considered the most critical water characteristic, must be maintained above 1 mL/L because eels cannot sustain life below this level.

To ensure and improve the water quality in eel ponds, various methods may be adopted. These would include allowing more water to flow into the pond to increase the rate of water exchange, installing aerators to increase oxygen supply, and liming to improve water quality. Bottom sediments are also removed after harvest.

Disease Problems, Prevention, and Treatment

The eel's body surface is covered by thick mucus. Thus the eel generally is highly resistant to diseases, unless its skin is injured or the skin or gill is infected by parasites (12).

Various fungi and bacteria can infect eels. The most common of these are fungus disease, red disease, branchionephritis, swollen intestine disease, gill disease, and anchor worm disease.

Diseases usually infect eels during the changes in seasons. These may be caused by a wide variety of factors such as poor pond water quality and inefficient feeding management.

Prevention rather than medication should be the program followed in disease management. Prevention of diseases would include efficient feeding management and proper water management. Moreover, eels are handled with care to avoid skin damage, which is the primary cause of infections. In case of disease outbreak, medicines are added to the food or to the pond water.

Harvesting and Marketing

When eels reach marketable size, they are harvested daily or once every few days, usually at feeding time. The ideal marketable size is 5–6 fish/kg for the Japanese market and more than 6 fish/kg for other markets.

There are several ways of eel harvesting in the ponds. The common practice is by placing a net below the feeding platform. Other methods include draining the pond and catching the eels in long net bags or by drawing a seine across the pond.

Harvested eels are sorted into different sizes. Smaller ones are put back to the pond for further growth. The harvested eels are then starved for about two days by holding them in baskets in front of the water inlet of the pond or in midpond, or stacking perforated plastic baskets under showers of trickling water. They are then packed and sent to the market. Starvation removes undigested substances inside the intestine. This minimizes the risk of meat contamination with gastroenteric bacteria during cutting and deboning in processing plants.

Eels, when marketed live or quick frozen and glazed, fetch fair prices. When they are transported to the market live, eels are packed in containers with conditions that may vary, depending on the transport time and distance (Fig. 8).

CONSUMPTION AND PROCESSING

Different countries have different preferences and eating habits, and thus have different ways of preparing and eating eels. In Europe, the main eel consumers are the Germans, Dutch, Danes, and Swedes. These consumers commonly prefer smoked eel, which ranks as an expensive and luxurious food. The British also eat eels, but prefer the jellied type.

In the Far East, the main eel-eating country is Japan, where it is a custom to eat roasted eel (*kabayaki*) especially on the *Ushinohi* (a special eel-eating day near the end of July). The Japanese believe that the eel is nutritious and can make up for overexhaustion during the summer. Eel is also consumed widely in Taiwan, the People's Republic of China, Korea, and Hong Kong.

Preprocessing

Before eels are processed, they are graded according to size (Fig. 9). One way of preparing them is by putting them in a deep container and sprinkling them with salt. This treatment removes their slime until they die of asphyxiation. Another way is by putting them in fresh water and stunning them with an electric shock (10). Or they may simply be put in cold storage overnight. In this way, eel activities can be slowed down, and thus they can easily be handled for gutting or deboning.

Figure 8. Eels readied for the market. Eels are packed in polyethylene bags inflated with oxygen. (Note oxygen tank and corrugated cardboard boxes in background.)

Figure 9. Eels being graded according to size.

Cutting Methods

There are generally two types of cutting, namely, *Kansai* style (cut abdominally) and *Kanto* style (cut dorsally). Newly cut eels are usually cleaned of slime by washing them in cold water and scraping. The eels are then gutted. This is done by slitting from the throat through either the belly (*Kansai* style) or up toward the back (*Kanto* style) to the tail, about 2.5 cm beyond the anus. The guts are then emptied and the backbone and head removed. The eel is washed thoroughly to remove traces of slime and blood. Machines specifically designed to eviscerate eels efficiently are also available.

Preparations

Fresh Eels. Fresh eels are cooked in several ways, such as braising and steaming. Steaming, however, is the more popular method. In both Taiwan and Mainland China, eel is either steamed with various vegetables and mushrooms or stewed with Chinese herbs, such as medlar, lovage, and dates. This herb-filled soup is considered as a revitalizing tonic soup for frail or disabled persons.

Smoked Eels. After the eels' are cut, they are brined in a saltwater solution, then hot smoked. In this process, the eels are dried and smoked. To allow uniform drying throughout the thickness of the fish, the temperature during smoking is increased gradually. The smoking process may vary from country to country. The finished product is then packed either in boxes or in cans (10).

Jellied Eels. Jellied eels is a traditional British way of preparing eels. First, gutted, cleaned, and cut eels are placed into boiling water and then simmered until the flesh becomes tender. Cooking time depends primarily on size. The cooked pieces marinated with hot liquor are then poured into large bowls containing gelatine dissolved in a small amount of water. The amount of gelatine solution is determined by considering the condition of the eels and their natural capacity to gel. Experience is thus necessary to get the right recipe. Once the mixture has cooled, the pieces in jelly are packed into cartons for immediate fresh consumption. Shelf life can reach two weeks at chilled temperatures. There are various other recipes for preparing jellied eels (10).

Roasted Eels. Roasted eel is a favorite Japanese cuisine. The Japanese prepare roasted eels in several ways, such as, *kabayaki* (roasted eel with seasoning), *shirayaki* (roasted eel without seasoning), *kimoyaki* (roasted eel viscera with seasoning), and *capitalyaki* (roasted eel head with seasoning). Until recently, these preparations were only available in Japanese specialty restaurants. Domestic production, however, has not kept up with the large demand for roasted eels in Japan. Japan thus imports frozen roasted eel from Taiwan, where the roasted eel processing industry has boomed in recent decades, and the People's Republic of China, where the industry is in its initial stages.

Live eels are processed into frozen roasted eel in processing plants. The ideal size of the raw material is between five and six eels per kilogram. Within this size range, the smaller size commands a higher price.

In the cut fillet style, the eel is cut crosswise into three equal sections, with the tail cut lengthwise into two. If the weight is not enough, each half of the tail is stretched together with one of the other sections by using bamboo sticks. This step is usually done using a stretching machine. Stretching is done to keep the meat flat during cooking. For the whole fillet style, the fish is not cut into sections after the degutting process; they are, however, also stretched.

After cutting, the fillets are arranged on a conveyor and pass through single-sided or double-sided roasting machines. The fillet may be roasted either seasoned or unseasoned (Figs. 10 and 11). In single-sided machines, the inside portion is roasted first, then the eel is turned and the other side is roasted. Liquefied petroleum gas is used as fuel. The appropriate roasting time is about 3–5 min. Roasting indicators are the color of the meat (it should become evenly light scorched), the scorch bubbles (some should appear on the skin; under 3% of the total area), and the central temperature of the meat (it should reach 78°C). The most delicate part in the whole procedure would be in the precooling step because this is where the fillet becomes most susceptible to recontamination. It is thus necessary to control the bacterial drop rate, until the total plate count is under three colony-forming units/min.

Figure 10. Frozen roasted eel without seasoning, cut fillet style.

Figure 11. Roasted eel dish with seasoning, cut whole fillet style.

Some manufacturers use precooling tunnels and spray cold air into the surface of the fillet: a shorter precooling time is maintained to minimize recontamination. The use of either individual quick freezing or contact freezing equipment is popular. In these types of equipment, the temperature drops to −18°C within 30 min. After the central temperature has reached −30°C to −35°C, the frozen fillet is then removed from the pan or conveyor, then packed and stored at a temperature of −20°C (13). Comparing the weight of the processed eel (eviscerated but not yet seasoned and roasted) to the total weight of the raw material, the yield for cut and whole fillet may reach as much as 58–60% and 68–70%, respectively.

Sanitary standards for the products include negative amounts of coliforms, total bacterial count of under 3,000/g, and negative presence of residues of contaminated chemicals in the meat, such as malachite green, methylene blue, oxolinic acid, antibiotics, nitrofurans, sulfamides, insecticides, and herbicides.

SKIN PROCESSING

Eel is not only used for food but also for other uses. Eel skin, for example, is processed into leather (Figs. 12 and 13). Although there is no fur on the eel skin, there are scales that are covered by follicles dispersed in the corium (14). The major steps in eel skin processing are liming, decoloring and removal of scales, removal of excess fat, tanning using alum, and glazing. Considering that the fat content of the eel skin is more than 25%, which is even higher than that of cattle, the removal of excess fat and scales are the more important steps in eel skin processing.

FUTURE DEVELOPMENT

The eel is a highly valued fish. Compared with other fish, the eel is rich in vitamins A and E, and calcium. The eel is recognized as one of the most nutritious cultured fishes. The fish is high in energy content, and calorific values,

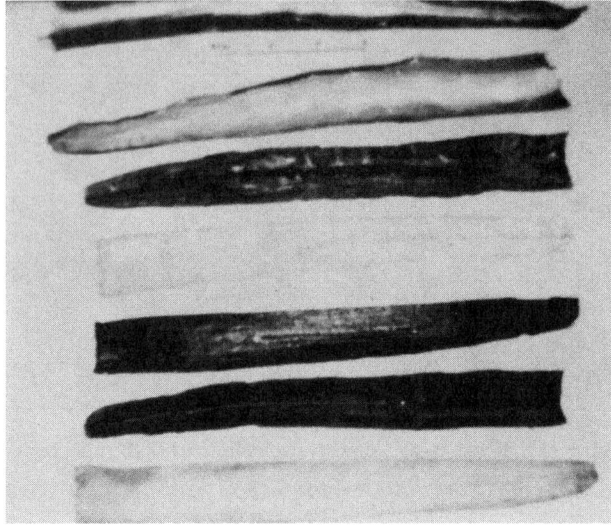

Figure 12. Tanned eel skin stained with different kinds of dye.

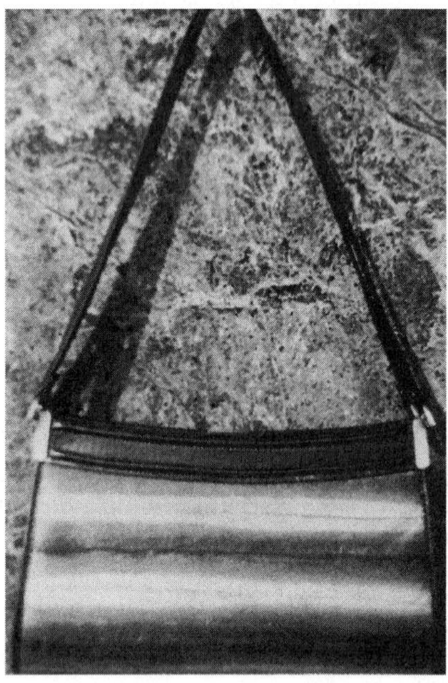

Figure 13. A shoulder bag made from tanned eel skin.

such that it is eaten to create appetite and stamina especially in hot and humid days. Its health benefits make the eel a valuable fish.

Eel skin, when processed, can be used in leathercrafts. More research, however, must be undertaken in finding ways to efficiently tan its skin to enable it to compete with other leathers.

Interest in the culture of eel has reached a high level in recent years because of its versatility and popularity, and high demand as gourmet food particularly in Japan, where it commands a good price. In eel-eating countries such as Japan, Taiwan, and some parts of Europe (Germany, Denmark, and the UK) enterprising activities concerning eels are in progress. These activities include research into new techniques of eel culture in cold climate, ie, Northern Europe.

Elvers used in eel culture are still sourced from the wild, and are diminishing in numbers. Thus investigations on eel propagation in captivity should be made. Attempts albeit unsuccessful have been made to propagate elvers artificially. The challenge for researchers now is to look for efficient and commercially successful ways to propagate them artificially. Researchers may gain hints from the lessons of past investigations and look for ways to unlock the mystery of the long life history of the eel. The fish's potentials are so great that the search will surely be worthwhile.

BIBLIOGRAPHY

1. C. C. Huang, "Taiwan Becomes World Leader in Exports of Eel," *Almanac of Food Industries in Taiwan, R.O.C.*, 56–57 (1990).
2. "Fisheries in Japan, Eels," *Jap. Mar. Prod.* 1977.
3. D. H. J. Castle and G. R. Williamson, "On the Validity of the Freshwater Eel Species, *Anguilla ancestralis* Ege, from Celebes," *Copeia* **2**, 569–570 (1974).
4. L. Bertin, *Eels: A Biological Study*, Cleaver-Hume Press, London, 1956.
5. I. Matsui, *Eel Study: Biology*, Vol. 1, Kosei-Sha Kosei-Kaku, Tokyo, 1972 (in Japanese).
6. F. W. Tesch, *The Eel: Biology and Management of Anguillid Eels*, Chapman and Hall, London, 1977.
7. K. Tsukamoto, "Breeding Places of Freshwater Eels," in O. Tabeta, ed., *Early Life History and Prospects of Seed Production of the Japanese Eel* Anguilla japonica (in Japanese), Suisan Gaku Ser. 107, Koseisha Koseikaku, Japan, 1996, pp. 11–21.
8. I. C. Liao et al., "Investigation on *Anguilla japonica leptocephali* by Fishery Researcher I," in K. Aida and K. Tsukamoto, eds., *Studies on the Life Cycle of Eels* (in Japanese), Kaiyo Monthly Special Issue 18, 1999, pp. 27–33.
9. W. N. Tzeng, "Immigration Timing and Activity Rhythms of the Eel, *Anguilla japonica*, Elvers in the Estuary of Northern Taiwan, with Emphasis on Environmental Influences," *Bull. Jap. Soc. Fish. Oceanogr.* 47–48, 11–28 (1985).
10. A. Usui, *Eel Culture*, Fishing News (Books) Ltd., Surrey, UK, 1974.
11. T. P. Chen, *Aquaculture Practices in Taiwan*, Fishing News (Books) Ltd., Surrey, UK, 1976, pp. 17–28.
12. S. Egusa, *Epidemic Diseases of Fish*, Kosei-Sha Kosei-Kaku, Tokyo, Japan, 1978 (in Japanese).
13. H. C. Chen, "Frozen Roasted Eel Processing Industry in Taiwan," *in* J. L. Chuang, B. S. Pun and G. C. Chen, eds., *Fishery Product of Taiwan, JCRR Fisheries Series* **25B**, Joint Commission on Rural Reconstruction. Taiwan, R.O.C., 1977, pp. 20–26.
14. Y. S. Lai and Y. Y. Kuo, "Primary Report on the Eel Skin Tanning," *Bulletin of the Taiwan Fisheries Research Institute* **32**, 439–441 (1980) (in Chinese with English abstract).

I CHIU LIAO
Taiwan Fisheries Research Institute
Keelung, Taiwan

EGGS AND EGG PRODUCTS

DEVELOPMENT OF THE EGG INDUSTRY

Eggs have been a human food source since the beginning of human residence on earth. The first usage of eggs was probably the taking of eggs from nests of wild birds. In primitive societies this continues to be significant source of high-quality protein, especially during the egg production season of wild birds. As civilization developed, the birds were domesticated and were kept in pens and shelters. In Western societies the chicken became the primary source of eggs. The growth of the egg industry in selected countries from 1965 to 1995 is shown in Table 1. The egg production in China amounted to almost 50% of total egg production in the world in 1995 and is continuing to increase at a rate greater than world average.

Until relatively recent times the keeping of chickens on farms was generally only a sideline to the main business of farm operation. Commercial egg farms became a primary source of farm income only in the last century. Large poultry farm operations for egg production did not become common until the last half of the twentieth century. The change in size of egg production units in the United States is shown in the data presented in Table 2. Data for recent years are not available in a form as presented in Table 2. However, a 1995 survey indicated that less than 1% of all eggs produced in the United States were from flocks of less than 1000 layers. In 1997 there were less than 350 egg-producing farms with more than 75,000 laying hens.

Table 2. Changes in Size of Laying Flocks in the United States from 1959 to 1974

	Percentage of all eggs sold		
Size of flock	1959	1964	1974
Less than 400	26.78	10.87	3.64
400 to 1,599	22.58	12.64	1.80
1,600 to 3,199	12.83	7.39	1.14
More than 3,200	37.81	69.10	93.43
More than 20,000		30.10	67.74
3,200 to 9,999			9.59
10,000 to 19,999			16.10
20,000 to 49,999			24.02
50,000 to 99,999			13.14
More than 100,000			30.58

Source: Ref. 2.

EGG FORMATION AND STRUCTURE

The egg consists of four distinct parts: shell, shell membranes, albumen or white, and yolk. A schematic drawing of an egg is shown in Figure 1. The yolk is formed in the ovary. During embryonic development of the female chick, tiny ovules appear on the ovary, that may later develop into yolks of eggs to be produced by that hen. The number of immature ovules present in the chick at the time of hatching is far in excess of the number of eggs that will be produced by the hen during her lifetime.

Approximately 14 days prior to the laying of an egg, one of the rudimentary ovules in the ovary starts to increase in size. During the next 13 days the yolk develops in its follicular membrane. A suture line forms in the membrane

Table 1. Egg Production in Selected Countries

Country	1965	1975	1985	1995
United States	65,692	64,379	68,250	74,280
Canada	5,194	5,339	5,855	5,792
Mexico	5,000	7,446	18,092	21,200
Argentina	2,880	3,480	3,150	NA
Brazil	8,124	5,000	9,000	16,065
Venezuela	508	1,721	2,736	NA
France	9,220	13,120	14,910	16,991
Germany	16,110[a]	20,050[a]	18,746[a]	13,847[a]
Italy	9,990	11,400	10,900	12,017
The Netherlands	4,206	5,320	10,051	9,970
Spain	6,320	10,152	10,164	9,983
United Kingdom	7,840	13,861	13,117	10,644
Poland	6,264	8,013	8,631	6,500
USSR	29,000	57,700	77,000	43,220[b]
Japan	18,625	29,798	35,700	42,167
Australia	2,196	3,384	3,825	NA
Turkey	1,448	2,597	5,700	8,000
China	NA	NA	NA	335,340
Republic of Korea	NA	NA	NA	8,317
Taiwan	NA	NA	NA	6,237
Thailand	NA	NA	NA	7,700

Note: All values in millions of eggs.
[a]Includes East and West Germany.
[b]Includes only Russia and Ukraine of former USSR.
Source: Ref. 1 (1967, 1977, 1987, 1997).

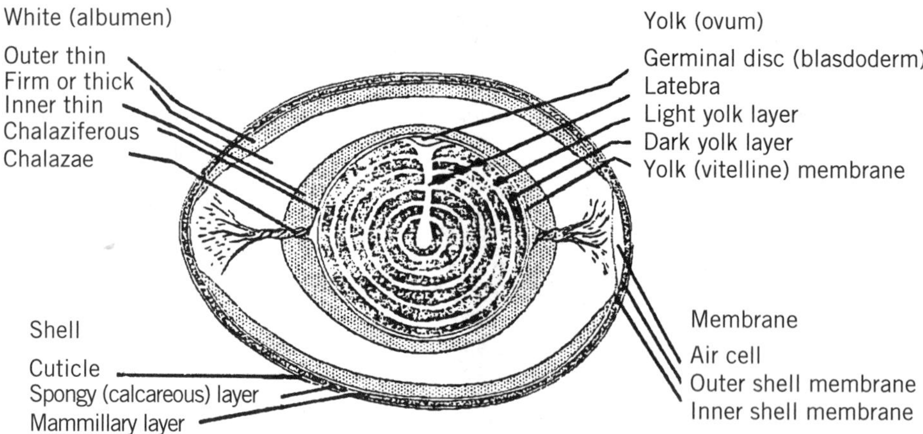

Figure 1. A schematic of the parts of an egg. *Source:* Ref. 4.

so that the matured yolk can be dropped into the infundibulum of the oviduct without the rupture of any blood vessels in the follicular membrane. A blood spot in an egg is generally caused by incomplete formation of the suture line.

The oviduct consists of five identifiable sections: infundibulum, magnum, isthmus, uterus, and vagina. The infundibulum, or funnel, collects the yolk, or ovum, and by peristaltic action moves it on to the magnum region, where the thick albumen is secreted and laid down in layers on the yolk. In the isthmus the shell membranes are formed around the yolk and thick albumen. The membranes, two in number, only loosely fit the enclosed material. Most of the time that the egg is in the oviduct is spent in the uterine section where thin albumen fills the shell membrane sack and the shell is laid down to form the rigid egg protective layer. The shell color is determined by the breed of the hen. Generally, hens with red earlobes produce brown-shelled eggs, and white-shelled egg producers have white earlobes. The total time spent in the oviduct is usually about 24 hs. Total time for the formation of the egg from the start of rapid growth of the ovule to time of laying varies from 12 to 15 days.

COMPOSITION OF THE EGG

The egg is composed of approximately 10% shell, 30% yolk, and 60% white or albumen. The egg is a very good source of high-quality protein and many minerals and vitamins. The exact levels of the minerals and vitamins is determined by the level of each in the diet of the hen. An average chemical composition of eggs, including shells, is summarized in Table 3.

The protein of egg white is complete; it contains all of the essential amino acids in well-balanced proportions. The white is made up mainly of the proteins ovalbumin, conalbumin or ovotransferrin, ovomucin, ovoglobulins, ovomucoid, and lysozyme. The structure of the thick white is the result of a complex of ovomucin and lysozyme.

The important yolk proteins are ovovitellin and ovolivetin. The lipid materials in the egg are all in the yolk. The fatty acid composition can easily be modified by changing the fatty acid makeup in the feed. In most commercial eggs where the hens are fed a corn and soybean meal diet the fatty acids are about one-third saturated and two-thirds mono- and polyunsaturated. Yolk color is controlled by the level of pigments, mostly xanthophyll, in the feed of the hens.

Some elements of the medical profession have emphasized the role of dietary cholesterol in human cardiovascular problems. The negative effect on egg consumption that this emphasis on cholesterol has had led to a number of research attempts to reduce the cholesterol content of eggs. Genetically it has been possible to achieve only slight reductions. Selection of hens for small yolk size has met with some success. A review on altering cholesterol by feeding (5) concluded that dietary modifications resulted in only minor changes in the cholesterol content of egg yolk. Another report (6) states that including 1.5 to 3.0% of menhaden oil in the laying hen's ration results in a temporary reduction of about 50% in the cholesterol concentration in the yolk. The fatty acid composition of the eggs was also modified in that the eicosopentenoic acid and docosahexanoic acid (omega-3 fatty acids) content of the yolks was significantly increased when the fish oil was included in the hen's ration. Levels of fish oil in excess of 3% in the ration resulted in fishy flavored eggs.

The relative ease with which egg composition can be modified has led to the development of a variety of "designer eggs." It has been reported that all vitamins and minerals content of eggs, except for choline and zinc, can be adjusted significantly by dietary modification for the hens (7).

PRODUCTION PRACTICES

As shown in Table 2, the majority of eggs at this time are produced on large egg production farms. With this change, the level of mechanization used in the production of eggs has increased. Over 95% of all laying hens in the United States and over 75% of all laying hens in the world are kept in cages (8). Legislation has been introduced in many

Table 3. Chemical Composition of the Egg

	Water (%)	Protein (%)	Fat (%)	CHO (%)	Ash (%)
Edible portion					
Whole egg	75.33	12.49	10.02	1.22	0.94
White	88.00	10.52	0	1.03	0.64
Yolk	48.80	16.76	30.87	1.78	1.77

	Calcium carbonate (%)	Magnesium carbonate (%)	Calcium phosphate (%)	Organic matter (%)
Shell	94.0	1.0	1.0	1.0

Source: Ref. 3 for edible portion; Ref. 4 for shell.

countries and passed in a few that would ban the use of cages for laying hens due to activities of animal welfare groups. Full details on production practices are given by North and Bell (8).

SHELL EGG PROCESSING

In a modern egg production unit hens are housed in cages with sloping floors so that eggs will roll from the cage onto an egg gathering belt. The eggs are transported by conveyor belts to a separate unit of the facility where they are washed, inspected for defects by passing them over an intense light source (candling), sized, and packed into cartons for retail or onto trays for bulk packaging. Throughout the operation individual eggs are not touched by any workers. The cartons or trays are packed into master containers by hand.

Some countries ban the washing of eggs to be sold at retail. In others, including the United States and Canada, almost all eggs are washed in a detergent solution at a temperature of about 45°C (113°F) and then sanitized with an approved sanitizing agent, frequently a chlorine-based compound. All of the washing should be done with water spray with or without brushes depending on the equipment.

After the eggs are washed and dried, they are conveyed to the candling area of the processing line. Here there is a strong light source under the eggs so that any defects such as cracked shells, dirty or stained shells, or inclusions in the egg might be detected and the defective eggs removed from the conveyor. Candling of eggs is used to segregate eggs into the several grades that are found in the marketplace. The retail grades in the United States are AA, A, and B. Quality factors used in grading include the shell (cleanliness and soundness), air cell (size), white (clarity and firmness), yolk (visibility of the outline), and freedom from defects. All grades of eggs are required to have a sound, not cracked, shell. Other quality factors are judgment calls, with smallness of the air cell, freedom from stains or dirt on the shell, firmness of the white, and dimness of the yolk outline being considered most desirable. For eggs that do not meet the grade standards two additional groupings are provided: dirty and checked or cracked. Details on the grading standards have been published (9).

There are six size classifications of eggs in the United States. Table 4 gives minimum weight requirements for each of the sizes. On the processing line the sizes are segregated by passing the eggs over a series of scales set so that the jumbo eggs are removed from the line first and the peewees are removed last.

After the eggs have been mechanically placed in cartons or on trays, many processors spray the tops of the eggs with a food grade mineral oil to preserve the interior quality of the egg during movement through market channels. The oil partially seals the pores of the shell to reduce loss of gases from the egg and to reduce the rate of pH change of the albumen. As a further aid in maintaining quality, eggs should be handled at refrigerated temperatures of less than 7°C (45°F) throughout market channels, and consumers are advised to keep eggs in the refrigerator after they purchase them. With good handling throughout market channels and in the home, eggs will remain high quality for several weeks and in a usable condition for several months. The factors that have the greatest effect on quality loss in eggs are time, temperature, humidity, and handling.

During the last decade a potential health risk in the usage of eggs has appeared. A new strain of *Salmonella enteritidis* is capable of infecting the reproductive system of the hen, so that some eggs are produced with the *S. enteritidis* in the egg at the time of production (10). The frequency of finding an egg positive for *S. enteritidis* in market eggs has been estimated to be between 1 in 10,000 and 1 in 20,000. Most eggs are heated to over 60°C (140°F)

Table 4. Weight Classes of Shell Eggs in the United States

Size name	Minimum ounces per dozen	Minimum average ounces per egg	Minimum average grams per egg
Jumbo	30	2.5	70.0
Extra Large	27	2.25	63.8
Large	24	2.0	56.7
Medium	21	1.75	49.6
Small	18	1.5	42.5
Peewee	15	1.25	35.4

Source: Ref. 4.

prior to eating, so the pathogen would be killed by cooking. Handling of eggs at refrigerated temperatures inhibits the increase in numbers of bacteria during marketing and holding in the home. With this threat to food safety from shell eggs, research has led to the development of systems for pasteurization of intact shell eggs to destroy all potential pathogens in the egg or on its shell (11).

CONVERSION OF SHELL EGGS TO LIQUID PRODUCT

An ever-increasing number of eggs are being broken at commercial egg breaking plants. The early growth of the egg products industry has been reviewed (12). In the first half of the twentieth century most of the egg breaking plants were located in the north central and western plains states. Since 1960 many of the new breaking plants have been built in the southeastern states. The yield of liquid products from a case (30 dozen) of large eggs is shown in Table 5. Data in this table also include the yields of solids that might be obtained if the egg products were dehydrated by either spray drying, pan drying, or freeze drying.

The growth of the egg products industry is summarized in Table 6. The shift from dried to liquid since 1970 was likely the result of mandatory pasteurization of all egg products since 1971 (14). The shift to liquid from 1988 to 1990 was likely the result of ultrapasteurization and aseptic packaging of liquid eggs that will give more than 4 weeks usable shelf life for the liquid product (15). The rapid expansion since 1986 is at least partially attributable to the potential incidence of *S. enteritidis* in the shell eggs. The savings in costs of freezing and in time saved by not having to thaw the products before using in food formulations are evident.

The pasteurization of liquid eggs was practiced for a number of years prior to the passage of the law making it mandatory for all egg products in the United States. The times and temperatures required for pasteurization in several countries are listed in Table 7. All of these values are for liquid whole egg. Details of pasteurization have been discussed (17).

In the processing of egg white it is necessary to remove the reducing sugars before the liquid can be dried. In early work on dehydration of egg white the process included a natural fermentation by whatever organisms were present in the egg white. A patent was issued in 1931 covering the use of lactic acid bacteria for egg white fermentation. The next development was the use a yeast, *Saccharomyces apiculatus*, and then the use of an enzyme, glucose oxidase. Each of these methods is still in use in some parts of the world.

The natural egg products of white and yolk are also supplied to the food processing industries in a number of blends and modifications often with the addition of sodium chloride or sugar. The chemical composition of some of the egg products is shown in Table 8. The difference in composition between pure yolk and commercial yolk should be emphasized. The difference is due to the inclusion of a significant amount of albumen with yolk in the commercial product.

Table 5. Yields of Liquid Egg and Egg Solids per 30-Dozen Case of Large Size Eggs

	Liquid		Solids	
	Kilograms	Pounds	Kilograms	Pounds
White	13.8	30.4	1.67	3.68
Yolk	6.1	13.5	3.12	6.87
Whole egg	19.9	43.9	4.79	10.55

Source: Ref. 13.

Table 6. Processed Egg Product Production in the United States (in millions of pounds)

Year	Liquid	Frozen	Dried
1960	44	362	176
1965	44	368	216
1970	110	357	281
1974	301	367	73
1977	394	348	75
1980	422	331	81
1982	460	339	80
1984	528	322	76
1986	632	337	89
1988	797	371	106
1990	932	384	105
1992	1,276	418	130
1994	1,689	428	133
1996	2,033	406	139

Source: Ref. 1.

NUTRITIONAL AND FUNCTIONAL PROPERTIES OF EGGS

Complete nutritional data on eggs and egg products prepared in several ways have been summarized (3). An abbreviated listing of nutrients in whole egg, yolk, and white is given in Table 9. As mentioned earlier, the content of various trace minerals, vitamins, and fatty acids in the egg can be modified significantly by varying the diet of the hen. The various uses for eggs in food products have been reviewed (19). The functional properties include whipping, emulsifying, coagulation, flavor, color, and nutrition. The foam produced during whipping should give a large volume and be stable. The whipping time required to get a maximum foam volume will vary for eggs of different qualities. Eggs with a high percentage of thick albumen require a longer time for whipping but the foam formed will be more

Table 7. Minimum Pasteurization Requirements for Whole Egg Products

Country	Temperature °C	Temperature °F	Time, minutes	Reference
United Kingdom	64.4	148	2.5	15
Poland	66.1–67.8	151–154	3.0	16
China (PRC)	63.3	146	2.5	16
Australia	62.5	144.5	2.5	16
Denmark	65–69	149–156.5	1.5–3.0	16
United States	60	140	3.5	13

Source: Ref. 16.

Table 8. Approximate Composition of Selected Egg Products (per 100 g)

Product	Calories	Composition				
		Water	Protein	Lipid	CHO	Ash
		Frozen eggs				
Whole egg	158	74.57	12.14	11.15	1.20	0.94
White	49	88.07	10.14	Trace	1.23	0.56
Yolk, pure[a]	377	48.20	16.10	34.10	—	1.69
Yolk, commercial[b]	323	55.04	14.52	28.65	0.36	1.43
Yolk, sugared	323	50.82	12.92	25.50	9.49	1.27
		Dehydrated eggs				
Whole egg	594	4.14	45.83	41.81	4.77	3.45
Whole egg, stabilized	615	1.87	48.17	43.95	2.38	3.63
White, stabilized, flakes	351	14.62	76.92	0.04	4.17	4.25
White, stabilized, powder	376	8.54	82.40	0.04	4.47	4.55
Yolk	687	4.65	30.52	61.28	0.39	3.16

[a]Contains approximately 17% white.
Source: Refs. 3 and 18.

stable. Cakes made using very low quality eggs generally do not have the volume of similar cakes made with higher-quality eggs. When egg white is dried, there is generally a loss in whipping properties. To overcome the decrease in foam volume, whipping aids such as triethyl citrate are frequently added to the dried egg white.

PRODUCTION OF EGG-RICH CONVENIENCE FOODS

Reference 20 outlines procedures for the manufacture of a number of value-added egg products, some of which will be described here.

Hard-Cooked Eggs

The preparation of hard-cooked eggs has been a function of kitchens for many years. In about 1970 the production of hard-cooked eggs for sale to salad bars, restaurants, and commercial food caterers was begun. The original products were cooked in hot water, partially cooled, peeled, and stored in a solution of citric acid and sodium benzoate. The concentration of the citric acid was initially 2% but has been lowered over the last 28 years to from 0.5 to 0.8%. The lower levels of acid produce eggs with softer cooked whites, much closer to freshly cooked eggs.

Cooking of hard-cooked eggs is now being done either in steam or hot water. Peeling of eggs was primarily a hand operation aided by equipment to crack the shells. Equipment has been developed that strips the shell from a cracked hard cooked egg so that efficiencies of production have been significantly increased. With improvements in packaging technology, some hard-cooked eggs are now marketed as a dry-packed, refrigerated product. It is estimated that almost 1% of all eggs are now being marketed as hard-cooked product.

Deviled Eggs

One use of hard-cooked eggs is the production of deviled eggs. A hard-cooked egg is cut in half through the middle either longitudinally or horizontally, and the yolk is removed. The yolk is removed and mixed with a salad dressing and other ingredients to produce a filling for the depression left in the white. Deviled eggs are a favorite for hospitality suites or buffets.

Scotch Eggs

This product was developed years ago but received very little publicity until relatively recently. A hard-cooked egg is wrapped in a thin layer of sausage meat. The sausage is then cooked in a deep fat fryer. In producing this item the sausage used should be very low in fat to get better adhesion of the surface of the hard-cooked egg. Scotch eggs are a snack food item frequently sold from refrigerated vending machines and are also used for hors d'oeuvres by cutting the finished product into quarters or halves.

Diced Egg Products

Diced eggs were first marketed as a means of utilizing hard-cooked eggs that were not perfectly smooth. With the popularity of hard-cooked diced eggs on salad bars, the demand was greater than the quantity available from the imperfect hard-cooked eggs. At this point innovative producers separated yolks and whites as liquids and steam-cooked trays of the two liquids. The cooked yolk and white were then diced. The product thus prepared gave yolk cubes that held together because of the residual albumen in commercially separated liquid yolk. The diced egg products have been marketed as individually quick frozen material in most instances. However, where market channels are short, nonfrozen, controlled atmosphere packaging is also being used.

Scrambled Eggs

Scrambled eggs are prepared as fully cooked, freeze-dried products for the camper; as fully cooked, frozen items for the microwaveable meal; and as frozen prepared liquid

Table 9. Distribution of Nutrients in Large Chicken Eggs (per egg, edible portion)

	Whole egg	Yolk[a]	White
Weight per egg, g	50	17	33
Water, g	37.28	8.29	29.06
Calories	79	63	16
Protein, g	6.07	2.79	3.35
Lipid, g	5.58	5.60	Trace
Carbohydrate, g	0.60	0.04	0.41
Fiber, g	0	0	0
Ash, g	0.47	0.29	0.18
	Minerals		
Calcium, mg	28	26	4
Iron, mg	1.04	0.95	0.01
Magnesium, mg	6	3	3
Phosphorus, mg	90	86	4
Potassium, mg	65	15	45
Sodium, mg	69	8	50
Zinc, mg	0.72	0.58	0.01
	Vitamins		
Ascorbic acid, mg	0	0	0
Thiamin, mg	0.044	0.043	0.002
Riboflavin, mg	0.150	0.074	0.094
Niacin, mg	0.031	0.012	0.029
Pantothenic acid, mg	0.864	0.753	0.080
Vitamin B_6, mg	0.060	0.053	0.001
Folacin, μg	32	26	5
Vitamin B_{12}, μg	0.773	0.647	0.021
Vitamin A, R.E.	78	94	0
	Lipids		
Saturated fatty acids, g	1.67	1.68	0
14:0	0.02	0.02	0
16:0	1.23	1.24	0
18:0	0.43	0.43	0
Monounsaturated FA, g	2.23	2.24	0
16:1	0.19	0.19	0
18:1	2.04	2.05	0
Polyunsaturated FA, g	0.72	0.73	0
18:2	0.62	0.62	0
18:3	0.02	0.02	0
20:4	0.05	0.05	0
Cholesterol, mg	213	213	0
	Amino acids		
Tryptophan, g	0.097	0.041	0.051
Threonine, g	0.298	0.151	0.149
Isoleucine, g	0.380	0.160	0.204
Leucine, g	0.533	0.237	0.291
Methionine, g	0.196	0.171	0.130
Cystine, g	0.145	0.050	0.083
Phenylalanine, g	0.343	0.121	0.210
Tyrosine, g	0.253	0.120	0.134
Valine, g	0.437	0.170	0.251
Arginine, g	0.388	0.193	0.195
Histidine, g	0.147	0.067	0.076
Alanine, g	0.354	0.140	0.215
Aspartic acid, g	0.602	0.233	0.296
Glutamic acid, g	0.773	0.341	0.467
Glycine, g	0.202	0.084	0.125
Proline, g	0.241	0.116	0.126
Serine, g	0.461	0.231	0.247

Note: Shell is 12% of weight of egg.
[a]Fresh yolk includes a small proportion of white.
Source: Ref. 3

mixes, packaged in cook-in-bag film. The basic formula for scrambled eggs consists of about 70% whole liquid egg and 30% milk with seasonings of salt and pepper. The most significant form of scrambled eggs is the frozen prepared mix packaged in film. This product is used extensively in hospitals and nursing home feeding because the cooked product remains hot for an extended period in the film bag. As some dieticians desire a reduced-cholesterol egg product for nursing home feeding, it is possible to use a blend of whites and yolk with only enough yolk for color as the egg source instead of whole liquid egg.

Omelets

The major difference between omelets and scrambled eggs is that in a true omelet no milk is included in the formula. Water is used instead of the milk. Omelets are produced for distribution as frozen product either fully cooked or as a premix. The formulation of omelet mixes frequently includes diced ham, bacon bits, mushrooms, onions, green or red peppers, and a number of other vegetables.

Crepes, Pancakes, and Waffles

Each of these products has a formula including flour and eggs. The crepe mixture is usually richer in eggs than the other two. Each of these products is fully cooked and frozen for distribution. There are also dry mixes for preparation of the products in the home.

Snack Foods from Eggs

A number of egg-rich snack foods have been proposed but, thus far, none have become significant users of eggs. A breaded, fried egg white ring that looks much like an onion ring was patented (21), and a procedure for producing an egg jerky flavored to taste like dried meat jerky was suggested (22). A cookie formulation was prepared so that each cookie contained one egg equivalent. Another snack food that is quite common is yogurt. It is possible to make a yogurt-type product using egg albumen as a partial replacement for the milk. In a sensory comparison of normal yogurt and the egg-substituted product, the latter was judged superior for mouth-feel and smoothness.

A number of egg-rich drinks have been produced. The drink sold in greatest quantity is eggnog. This is produced as a dried product as well as the refrigerated liquid drink. A number of fruit juices have been mixed with eggs to produce drinks of orange juice and egg, apple juice and egg, and cranberry juice and egg. When preparing any of the drinks in quantity, it is recommended that pasteurized liquid eggs be used. The prepared drinks should be kept at temperatures below 5°C (41°F).

MODIFYING THE COMPOSITION OF EGGS BY PROCESSING

The composition of egg products can be modified rather easily by utilizing yolk, white, or whole egg in various amounts. Much interest has been shown in reducing the amount of cholesterol in eggs. The cholesterol can be removed from the yolk by using supercritical extraction techniques with carbon dioxide as the extracting solvent. Much of the research has been with dried egg yolk, but it is possible to remove more than 80% of the cholesterol from the liquid yolk by supercritical extraction techniques (23).

EGGS AS A SOURCE OF PHARMACEUTICAL PRODUCTS

Lysozyme makes up 3.5% of the egg white. A cation exchange system consisting of a macroporous, weak acid resin can be used to recover most of the lysozyme (24). The system can be designed as a continuous operation. It was found that the residual lysozyme-free egg white possessed superior whipping, gelling, and emulsifying properties compared with native egg albumen. Lysozyme-free albumen has been approved for food product usage by federal agencies in the United States, providing the labeling is appropriate. Extracted lysozyme may be used in pharmaceutical products or as an antibacterial agent in the preservation of food products.

Fertile eggs are used extensively in the production of vaccines. Some recent work has shown promise for the egg to be used as a means of producing antibodies by manipulation of diet and management of the hen.

BIBLIOGRAPHY

1. USDA, *Agricultural Statistics*, U.S. Government Printing Office, Washington, D.C., 1961–1997.
2. W. J. Stadelman, and O. J. Cotterill, *Egg Science and Technology*, 3rd ed., AVI Publishing Westport, Conn., 1986.
3. USDA, "Composition of Foods, Dairy and Egg Products—Raw, Processed, Prepared," *Agricultural Handbook 8.1, USDA, ARS*, U.S. Government Printing Office, Washington, D.C., 1976; revised, 1989.
4. USDA, *Egg Grading Manual, USDA, AMS, Agriculture Handbook 75*, U.S. Government Printing Office, Washington, D.C., 1983
5. E. C. Naber, "The Effect of Nutrition on the Composition of Eggs," *Poultry Sci.* **58**, 518–528 (1979).
6. R. L. Adams et al., "Introduction of Omega-3 Polyunsaturated Fatty Acids into Eggs," *Abstracts of Papers of Southern Poultry Science Society*, Atlanta, Ga. (1989).
7. W. J. Stadelman and D. E. Pratt, "Factors Influencing Composition of the Hen's Egg," *World's Poultry Sci. J.* **45**, 247–266 (1989).
8. M. A. North, and D. D. Bell, *Commercial Chicken Production Manual*, 4th ed., Chapman & Hall, New York, 1990.
9. USDA, "Regulations Governing the Grading of Shell Eggs and United States Standards, Grades and Weight Class for Shell Eggs," *USDA, AMS, Poultry Division, 7 CFR, Part 56*, 1987.
10. R. G. Board and R. Fuller, *Microbiology of the Avian Egg*, Chapman and Hall, London.
11. W. J. Stadelman et al., "Pasteurization of Eggs in the Shell," *Poultry Sci.* **75**, 1122–1125 (1996).
12. J. W. Koudele, and E. C. Heinsohn, "The Egg Products Industry of the United States. I. Historical Highlights, 1900–1959," *Kansas Agric. Exp. Station Bull.* 423, 1960, pp. 1–48.
13. O. J. Cotterill and G. S. Geiger, "Egg Product Yield Trends from Shell Eggs," *Poultry Sci.* **56**, 1027–1031 (1977).
14. USDA, "Regulations Governing the Inspection of Eggs and Egg Products," *USDA, AMS, Poultry Division, 7 CFR, Part 59*, 1984.
15. H. R. Ball, Jr., et al., "Functions and Shelflife of Ultrapasteurized, Acseptically Packaged Whole Egg," *Poultry Sci.* (Suppl. 1), 63 (1985).
16. "The Liquid Egg (Pasteurization) Regulations 1963," *Statutory Instruments, No. 1503*, H. M. Stationery Office, London, 1963.
17. F. E. Cunningham, "Egg Product Pasteurization," in W. J. Stadelman and O. J. Cotterill, eds., *Egg Science and Technology*, 4th ed., Haworth Publishing, Binghamton, N. Y., 1995.
18. O. J. Cotterill, and J. L. Glauert, "Nutrient Values for Shell, Liquid/Frozen and Dehydrated Eggs Derived by Linear Regression Analysis and Conversion Factors," *Poultry Sci.* **58**, 131–134 (1979).
19. Sheng-Chin Yang, and R. E. Baldwin, "Functional Properties of Eggs in Foods," in W. J. Stadelman and O. J. Cotterill, eds., *Egg Science and Technology*, 4th ed., Haworth Publishing, Binghamton, N. Y., 1995.
20. W. J. Stadelman et al., *Egg and Poultry Meat Processing*, Ellis Horwood, Chichester, England, 1988.
21. U.S. Pat. 4,421,770 (Dec. 20, 1983), J. M. Wicker and F. E. Cunningham (to Kansas State University).
22. U.S. Pat. 4,537,788 (August 28, 1985), V. Proctor and F. E. Cunningham (to Kansas State University).
23. G. Zeidler, G. Pasin, and A. King, "Removing Cholesterol from Liquid Egg Yolk by Carbon Dioxide-Supercritical Fluid Extraction," in J. S. Sim and S. Nakai, eds., *Egg Uses and Processing Technologies—New Developments*, CAB International, Oxon, U.K.
24. E. Li-Chan et al., "Lysozyme Separation from Egg White by Cation Exchange Column Chromatography," *J. Food Sci.* **53**, 425–427 (1986).

WILLIAM J. STADELMAN
Purdue University
West Lafayette, Indiana

ELASTINS AND MEAT LIGAMENTS

Elastin and collagen are the principal components of connective tissue. They form a network that is responsible for the transmission of tension and the structure of muscle and other tissues. Elastin forms a lesser proportion of connective tissue than collagen and is not soluble during heating. The ratio of collagen to elastin in connective tissue is dependent on the tissue and its location. This ratio affects the tissue's mechanical properties and its physiological functions (1). Elastin is present in muscle only in small amounts, less than 3% of the total connective tissue. *Musculus semitendinosus*, however, contains more elastin, up to 37% of the total connective tissue. Elastin normally forms fibers and lamellae, is abundant in elastic ligaments, and elastic blood vessels, and is found to a small extent in the skin, lungs, and other organs.

Elastic tissue is often referred to as yellow connective tissue because of its color. It contains the elastin fibers with filamentous, refractive, and fluorescent with a blue-white appearance under ultraviolet light. Because of the large number of nonpolar amino acids in the structure and their hydrophobic nature, the elastin fibers stain poorly with acid or basic dyes but do stain selectively with phenolic dyes such as orcein.

MORPHOLOGY

The basic morphological component of elastin is filament made up of a linear sequence of globular structures with a probable maximum diameter of 9–10 nm. Dehydrated elastin may be 4–6 nm in diameter (2). The freeze-etching technique reveals elastin's structure as a regular, three-dimensional network of filaments. The filaments cross their neighbors every 8–12 globular subunits, giving rise to a network with a large, less-dense central core (3). The filaments of elastin can be disordered and swollen when immersed in 25% glycerol for 8–10 days. The swollen elastin shows a much higher affinity for stains.

CHEMISTRY AND BIOCHEMISTRY

Elastin can be determined from the amount of desmosine or isodesmosine found in the tissue. Its amino acid composition is essentially identical as extracted from various tissues and are not affected by the age of the animal (4). However, compositions differ from animal to animal. Elastin contains 1–2% hydroxyproline. Almost 95% of its amino acids are nonpolar (Table 1). The sulfur amino acids, tryptophan, and tyrosine are present in small amounts or nonexistent. In contrast to collagen, elastin does not contain hydroxylysine. Elastin has about one-third of its amino acids as glycine, but their arrangement in the elastin molecule is unknown. Desmosine and isodesmosine, unique cyclic polymers, are made up of four lysine residues, which connect two to four elastin molecules. This reaction requires the oxidative deamination of three lysine ϵ-amino groups which give an intermediate followed by condensation with the ϵ-amino group of a fourth lysyl residue, and is catalyzed by a copper-containing enzyme, lysine oxidase. Therefore, the synthesis of elastin can be blocked by inhibition of this enzyme through dietary copper depletion, or by β-aminopropionitrile, the active lathyrogen of the sweet pea. The synthesis is age dependent; the amount of desmosine and isodesmosine increase and the lysine content decreases with advancing age. The elasticity of elastin is due to the exposure of nonpolar sequences of its molecules to water during the stretching, followed by spontaneous refolding of the molecule on removal of the applied force.

Elastin can be isolated as the residue remaining after the extraction of tissue with 0.1 *N* NaOH at 98°C for 45 min (4). Elastin is heat stable to a temperature of 140–150°C and is insoluble in a wide range of hydrogen bond breaking solvents at temperatures up to 100°C. It must be processed with harsh enzymatic or chemical treatments in order to solubilize it. This includes KOH/ETOH solubilized κ-2-elastin, oxalic acid solubilized α-elastin, elastase solubilized elastin, and various salt-soluble forms of elastin (9).

Using immunological techniques, two serologically distinct elastin antibody fractions in antiserum prepared against bovine ligamentum nuchae elastin were identified (9). One is a species-specific fraction that bound only to bovine elastin. The other population of antibodies has affinity for elastin from porcine aorta, bovine ligamentum nuchae and aorta, rabbit and hamster lung, and human aorta. The elastin radioimmunoassay can be modified to detect bovine or other elastins, depending on the choice of radiolabeled antigen, in meat products. This assay also permits the study and quantification of elastin synthesis and degradation in a wide range of meat-animal species without having to raise specific antibodies to each one.

Table 1. Amino Acid Compositions of Elastins from Selected Tissues of Various Meat Animals

	Residues per 1000 residues			
Amino Acid	Avian, insol[a]	Bovine, semitendinosus[b]	Sheep, vascular tissue[c]	Yellowfin tuna, dark muscle[d]
Asp	2	7.4	1.8	66.0
Hyp	22	12.4	0.0	6.0
Thr	3	8.2	9.2	51.0
Ser	5	8.3	9.0	53.0
Glu	12	17.1	20.4	134.6
Pro	128	110.9	105.3	65.1
Gly	352	325.9	241.1	128.5
Ala	176	223.1	288.7	91.7
Val	175	145.8	111.7	55.9
Cys/2	1	—	—	17.6
Met	—	—	—	22.3
Ile	19	26.5	17.5	38.2
Leu	47	61.4	57.5	82.9
Tyr	12	8.4	13.4	31.7
Phe	23	29.8	34.5	29.9
Hyl	22	12.4	—	0.0
Lys	4	5.3	30.7	62.2
His	1	0.6	13.6	14.8
Arg	5	5.7	6.8	48.0
isodes	3	2.1[a]	29.3[b]	0.5[a]
Des	3	1.2[a]	7.2[b]	0.3[a]

Source: [a]Ref. 4, [b]Ref. 5, [c]Ref. 6, and [d]Ref. 7.

The emission spectrum of elastin is similar to that of Type I collagen with peak excitation near 370 nm. Macroscopic ultraviolet fluorometry was used to measure the gristle contents, elastin, and collagen Type I of beef (10).

Fluorescence emissions were measured with a monochromator and a photomultiplier tube. Intact tendons and elastic ligaments had a strong fluorescence emission peak around 440–450 nm and only weak fluorescence around 510 nm. The 510:450 nm ratio was correlated to the amount of gristle in comminuted meats.

INDUSTRIAL APPLICATIONS

The visible gristle in meat is from a number of sources, including tendons (particularly from myotendon junctions), ligaments (such as the ligamentum nuchae in rib roasts), perimysium (particularly in the pennate extensors and flexors of the lower limbs), fasciae (such as the lumbodorsal fascia of the longissimus muscle), and intramuscular vessels (particularly arteries). All these sources consist mainly of Type I collagen or elastin. Elastin-contained gristle retains its tensile strength after typical cooking procedures, such as roasting and broiling (11,12). Fresh meat cuts containing gristle are generally perceived as lower quality products, because of the unpleasant sensation of gristle. However, the elastin concentration is not consistently related to variations in the tenderness of muscles, at least in the bovine species (13), nor is the total concentration of connective tissue components (collagen and elastin). Tenderness is actually more related to soluble collagen and the overall fiber arrangement in the meat.

Aging is the industrial practice of storing meat carcasses above the freezing point for a certain period of time to improve the tenderness. The changes are the result of the physical breakage of muscle and collagen protein fibers by rigor and enzymatic reactions. However, aging has no apparent effect on the structure of elastin tissue and its physical properties.

Tissues with high elastin content, such as elastic ligaments, blood vessels, and lung, are removed in the slaughtering operation. They are processed as offals with other organs and by-products in the rendering industry as a protein source for animal feed. In this application, they may be rendered at high or low temperature, depending on the planned use of the fat. Fat is separated from the solids by centrifugation. The solids are dried and batched with other protein sources for optimum amino acid content and distribution, and then used in animal feed. The make-up of the feed is determined by a linear program based computer program which takes account of the amino acids, fat, carbohydrate, vitamin, and mineral content required in feed formulations.

In the ready-to-eat processed meat industry, elastin tissues and meat ligaments retain their tensile strength and are not gelatinized under most heat-processing conditions. For this reason, elastin tissues in meat are removed, with other gristle, by mechanical means, by hand or mechanical gristle removers. Grinders equipped with gristle and bone removers are widely used in the processed meat industry for coarse-ground, nonemulsified products. In this case, the meat is put through a grinder and the bone chips and gristle are separated from the ground meat. For large-muscle products, such as ham, the meat must be hand trimmed. The separated gristle with elastin and collagen tissue is discarded or chopped fine and used in emulsified products. It is used as a filler rather than functional proteins for economic reasons.

Meat cuts with a high gristle, collagen, and elastin content are not used in large-muscle products. They are used most often in emulsified products to minimize the toughness problem of the gristle. However, elastin and collagen tissues from gristle do not bind water or function as emulsifiers in meat emulsion; therefore, emulsion breakdown occurs more often in these products. To solve these problems, a computer-based, least-cost formulations program, which is widely used in the processed meat industry, has incorporated the connective tissue in individual meat cuts as a quality constraint in the computer simulation of product formulations. But no effort has been made to separate the elastin or ligament contribution as a constraint in the computer program, because of the low content of elastin in most meat cuts.

Unlike collagen, elastin does not swell at mild acid conditions and has no film-forming properties. It is not used in gelatin production because of its poor water solubility and gelation properties. It cannot be substituted for collagen in regenerated collagen casing, or collagen film and tissue. Elastin also has little use in the processed meat industry because it functions poorly as a binder and emulsifier. Because it is found in only small amounts in meats, because it is uneconomical to separate it from other meat tissues, and because it is poor nutritionally, elastin's technological development for use in foods and other industries has been hindered.

BIBLIOGRAPHY

1. R. J. Bagshaw, E. Weit, and R. H. Cox, "Aortic Connective Tissue Content in White Leghorn Females," *Poultry Science* **65**, 403 (1986).
2. C. Fornieri, I. P. Ronchetti, A. C. Edman, and M. Sjostrom, "Contribution of Cryotechniques to the Study of Elastin Ultrastructure," *Journal of Microscopy* (Oxford) **126**, 97 (1982).
3. M. Morocutti, M. Raspanti, P. Govoni, A. Kadar, and A. Ruggeri, "Ultrastructural Aspects of Freeze-Fractured and Etched Elastin," *Connective Tissue Research* **18**, 55 (1988).
4. H. R. Cross, G. C. Smith, and Z. L. Carpenter, "Quantitative Isolation and Partial Characterization of Elastin in Bovine Muscle Tissue," *Journal of Agricultural and Food Chemistry* **21**, 716 (1973).
5. J. A. Foster "Elastin Structure and Biosynthesis: An Overview," in, *Structural and Contractile Proteins. Part A. Extracellular Matrix*, L. W. Cunningham and D. W. Frederiksen, eds. Academic Press, New York, 1982, pp. 559–570.
6. J. R. Bendall, "The Elastin Content of Various Muscles of Beef Animals," *Journal of the Science of Food and Agriculture* **18**, 553 (1967).
7. S. Kanoh, T. Suzuki, K. Maeyama, T. Takewa, S. Watabe, and K. Hashimoto, "Comparative Studies on Ordinary and Dark Muscles of Tuna Fish," *Bulletin of the Japanese Society of Scientific Fisheries (NIHON SUISAN GAEFFAI-SHI)* **52**(10), 1807 (1986).

8. J. N. Manning, P. F. Davis, N. S. Greenhill, A. J. Sigley, "Salt Soluble Cross-linked Elastin: Formation and Composition of Fibers," *Connective Tissue Research* **13**(4), 313 (1985).
9. R. P. Mecham and G. Lange, "Measurement by Radioimmunoassay of Soluble Elastins from Different Animal Species," *Connective Tissue Research* **7**, 247 (1980).
10. H. J. Swatland, "Fiber-Optic Reflectance and Autofluorescence of Bovine Elastin and Differences between Intramuscular and Extramuscular Tendon," *Journal of Animal Science* **65**, 158 (1987).
11. A. M. Pearson and F. W. Tauber, *Processed Meats*, AVI Publishing Co., Inc., Westport, Conn. 1984.
12. P. P. Purslow, "The Physical Basis of Meat Texture: Observations on the Fracture Behavior of Cooked Bovine *M. Semidendinosus*," *Meat Science* **12**, 39 (1985).
13. H. R. Cross, Z. L. Carpenter, and G. C. Smith, "Effects of Intramuscular Collagen and Elastin on Bovine Muscle Tenderness," *Food Science* **38**, 998 (1973).

Rudy R. Lin
Swift-Eckrich, Inc.
Downers Grove, Illinois

EMULSIFIER TECHNOLOGY IN FOODS

Most foods consumed by humans are emulsions, in that they contain either fats or oils dispersed in an aqueous phase, or they contain an aqueous phase dispersed in a fat or oil. An example of the prior emulsion would be milk; an example of the latter would be butter.

A food emulsion may be described as a multiple-phase system that contains at least two components that are normally immiscible. Oil and water are examples of such incompatible components. The water or aqueous phase may contain many other ingredients that are truly soluble in the aqueous phase such as sugars and salts. It may also contain colloidally dispersed components such as proteins and carbohydrate polymers such as cellulose. The fat or oil phase may contain other lipids such as phospholipids and glycolipids, and certain vitamins, colors, and sterols.

The preparation by humans of food emulsions not found in nature has been considered a culinary art and has been practiced as an art for centuries. A perfect example of an artful culinary emulsion has been mayonnaise. Such emulsions did not require long-term stability because they were consumed within a short period of time in the home or restaurant. Such problems as freeze/thaw stability, long-term storage, the physical rigors of transportation, and high degree of aeration did not concern the chef of bygone eras. His emulsions would be considered crude by today's standards. His emulsion—a sauce, for example—would be acceptable if stable for only a few hours, as compared with the months of stability required today in commercially produced emulsions such as salad dressings.

Milk and salad dressings are two of the primary examples of one type of emulsion—the oil-in-water emulsion. The oil or lipid phase is also known as the disperse, discontinuous, or internal phase. It is dispersed in the aqueous serum phase, also known as the external or continuous phase.

Butter and margarine are obvious examples of the opposite type of emulsion, the water-in-oil emulsion. The aqueous phase is dispersed in the fat or oil and is now known as the discontinuous or internal phase, as the lipid has become the continuous or external phase.

A gaseous phase can also exist and may be external, as in the case of an aerosol, or internal, as in whipped topping. Dual emulsions are commonly encountered; a cake batter is a good example of the latter. Air or leavening gasses are dispersed in the fat, which is, in turn, dispersed in the aqueous phase of the oil-in-water cake batter.

The viscosity of a food emulsion is a function primarily of the external phase. It is generally lower when the external phase is aqueous, as in the case of milk, and higher when the external phase is lipid, as in the case of margarine. Several other factors can affect emulsion viscosity. In the case of an oil-in-water type of emulsion, packing or clumping of the internal phase can affect yield value and viscosity. An example of such an emulsion is mayonnaise. The internal dispersed phase (oil) occupies over 70% of its volume, closely approaching its theoretical maximum when the oil globules are of uniform spherical shape. Such an emulsion exhibits little or no flow properties, and its yield value and viscosity are such as to lead the causal observer to assume it may be a water-in-oil emulsion.

Oil-in-water emulsions can contain higher than theoretical limits of lipid internal phase (74%) if globules are misshapen due to compression, and viscosity can be further affected by hydrogen bonding and van der Waals forces. Surface charges supplied by components such as phospholipids can reduce viscosity by imparting repellency.

When the internal phase of an oil-in-water emulsion is low, as in the case of dairy products, ingredients such as colloids, including cellulose, starch, and algins, may be added to increase viscosity. Homogenization of the lipid phase to attain more surface area per unit weight of disperse phase will increase viscosity. Indeed, when homogenization of dairy foods was introduced in the United States in the 1930s, it so improved perceived richness due to increased surface area and resultant viscosity that an attempt was made to ban the process on the basis that it deceived the consumer. As an example, using typical homogenization procedures, milk with an average 8 μ fat globule is converted to 512 globules of 1 μ providing an eightfold increase in surface area.

Most commercial food emulsions contain the disperse phase in small globules in the range of 0.1 to 0.2 μ. With water-in-oil emulsions such as margarine, the viscosity, or plasticity, is a function primarily of the crystalline solids contribution of the fat, as well as its melting point.

The aqueous phase included in the margarine lipid phase is dispersed using a scraped surface crystallizer, such as the votator, so that the water particles are as small as possible. Large particles explode into steam when the margarine is used for cooking with such violence as to cause spattering. A properly manufactured margarine will have the dispersed water particles in the 1 to 5 μ range.

Most oil-in-water emulsions such as dairy-type frozen desserts, because of very efficient modern homogenizers, possess fat globules with a diameter of less than 0.5 μ. The size of the fat globule will affect the light reflected to the viewers eyes in a simple oil in water emulsion. In fact, the size of the oil globule can be estimated from the appearance of the emulsion as given in Table 1.

The physical stability of an emulsion, that is, the tendency of the two phases not to separate, takes place in accordance with Stokes Law:

$$V = 2gr^2 \frac{d_1 - d_2}{9\eta}$$

where V is the velocity of the sphere, r the radius of the sphere, η the viscosity of the continuous phase, d_1 the density of the continuous phase and d_2, the density of the disperse phase, and g the force of gravity. In an oil-in-water emulsion, the oil globules of lower density will tend to rise through the more dense aqueous phase. They will rise more rapidly if they coalesce into larger globules, and their separation will be slowed by increased viscosity.

When an oil is finely divided in a aqueous phase, a tension is created due to the increased surface energy created by division. The oil tends to coalesce to reduce this tension by returning to the lowest free energy state. An emulsifier (surface active agent) will reduce this tension by adsorption at the interface between the two antagonistic phases, thus reducing the free energy (amphiphilic).

Some surfactants further stabilize food emulsions by inhibiting coalescence due to contributing an electrical charge (usually negative) at the fat globule surface. The likely charged globules repel one another, further reducing the opportunity for coalescence.

Most food emulsions produced today contain added surface active agents (emulsifiers). Modern food processing procedures are designed to produce food with long-term shelf life and to withstand the rigors of transportation, including temperature extremes and mechanical abuse. Added surfactants contribute significantly to long-term emulsion stability. As an example, salad dressings that would normally separate in a matter of hours are stable and acceptable to the consumer after months of storage.

Most surfactants used in food processing are esters of fatty acids, and although found in nature, they are in most cases not present in the food after processing in sufficient quantity to provide the desired emulsion stability. Emulsifiers are characterized by the fact that they are amphoteric; that is, they posses both an oil-loving and a water-loving moiety within the same molecule. Since they are neither completely lipophilic (oil loving) nor hydrophilic (water loving) they proceed to the interface between both components and orient themselves in a structure making the fat and water more compatible.

Table 1. Estimation of the Globule Size

Globule size (μm)	Appearance
>1	Milky white
0.1–1.0	Blue white
0.05–0.1	Gray, semitransparent
<0.05	Transparent

Source: Ref. 1.

Food emulsifiers can be manufactured by the esterification of a polyol such as glycerine or sorbitol with a source of fatty acid. Glycerine can be directly esterified with a fatty acid or, as is usually more economical, glycerine can be transesterified with a fat (glycerolysis). Both procedures will produce a surface active glyceryl monoester such as glyceryl monostearate. The resulting monoester will be surface active because it will possess the fat-loving stearic acid moiety and the water-loving hydroxyl groups of the unesterified polyol carbons within the same molecule (Fig. 1).

Although glyceryl monoesters are the most commonly used surfactants in the food industry and are found naturally occurring in most oil-bearing food (milk, nuts, olives, wheat, etc), other alcohols and aldehydes may be used to supply the hydroxyl groups. When a shortage of glycerine developed during World War II, it was found that a monoester of sorbitol produced by hydrogenation of glucose could be substituted in most cases. Not only were such esters found to be equivalent to those derived from glycerol, they were usually superior. Sorbitol differs from glycerine in that it contains six water-loving hydroxyl groups instead of three and is thus more hydrophilic (Fig. 2).

Other alcohols and aldehydes such as propylene glycol and sucrose may be used to supply the hydrophilic moiety of the surfactant. The hydrophilic character and functional value may be further enhanced by further esterification with organic acids such as lactic, citric, or acetic to produce the diester. The esters can be made even more hydrophilic by ethoxylation (Fig. 3) or by esterification with polymers such as polyglycerol. Because of its amphiphilic character, the surfactant will orient itself to the interface between the lipid and aqueous phase. The more hydrophilic the emul-

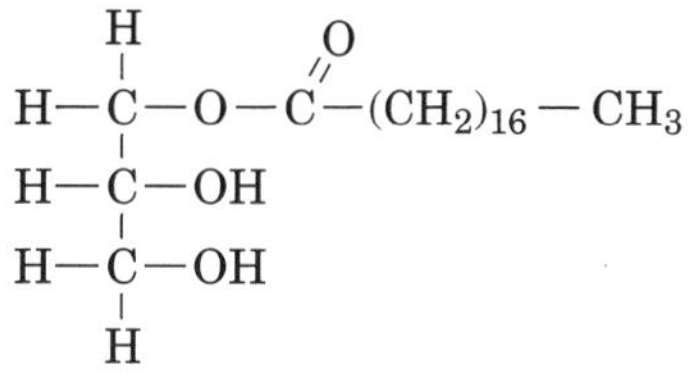

H O
H—C—O—C—$(CH_2)_{16}$—CH_3
H—C—OH
H—C—OH
H

Figure 1. Glyceryl monostearate.

Glycerine
H
H—C—O
H—C—O
H—C—O
H

Sorbitol
H
H—C—OH
H—C—OH
OH—C—H
H—C—OH
H—C—OH
H—C—OH
H

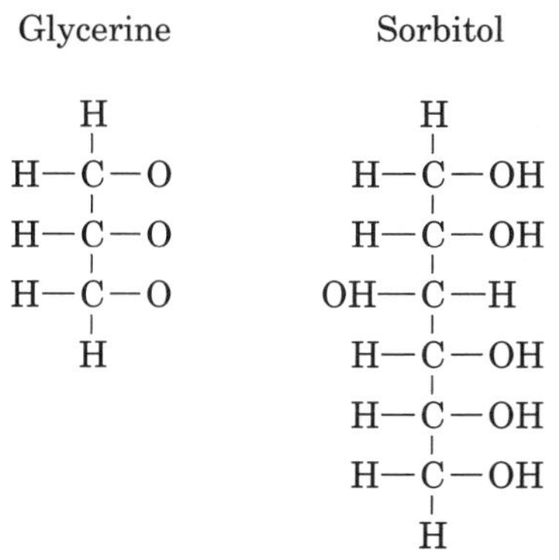

Figure 2. Structures of glycerine and sorbitol.

$$
\begin{array}{l}
\quad CH_2 \\
\quad HCO(C_2H_4O)_wH \qquad O \\
H(OC_2H_4)_xOCH \\
\quad HC \\
\quad HCO(C_2H_4O)_yH \\
\quad CH_2O(C_2H_4O)_zOCR
\end{array}
$$

Figure 3. Structure of polysorbate 60 ethoxylated sorbitan monostearate.

sifier is, the greater will be the reduction in interfacial tension (Fig. 4). Its function is therefore threefold:

1. Reduction of interfacial tension
2. Formation of a cohesive film, encapsulating the fat or oil globule
3. Imparting an electrical charge to further aid in preventing coalescence through repellency

Although the functional value of a surfactant in forming stable emulsions is most valuable, these amphiphiles have other valuable physical and chemical attributes useful in food technology. Surfactants improve the texture and shelf life of starch-containing foods by forming complexes with starch components such as amylose, modifying the rheological properties of protein-containing foods such as wheat doughs, and improving the consistency and textural qualities of foods containing fats and other lipids by modifying their crystalline behavior (size and polymorphism) (2).

The reduction of surface tension improves the ease with which oil and water will mix. Reduction of surface tension will require less energy input during agitation when dispersing the oil through the aqueous phase. However, when it is realized that surface energy is increased by a factor of 10^6 in a fine emulsion and that the use of a surfactant will reduce surface tension by a factor of 20 to 25, it is apparent that the reduction of surface tension alone may be insufficient to provide a stable emulsion.

The formation of a cohesive film around the oil globule and between the interface of the oil and aqueous phase will make a major contribution to emulsion stability (Fig. 5). The α-tending surfactants, which are nonpolymorphic, are particularly effective encapsulating agents. These include propylene glycol monostearate, lactylated and acetylated monoglycerides, and sorbitan tristearate. These surfactants form a crystalline protective coating around the oil globule with great cohesive strength (4).

When a surfactant is adsorbed on the oil globule surface, the lipophilic hydrocarbon chain is oriented into the oil, with the polar moiety, for example, the OH group at

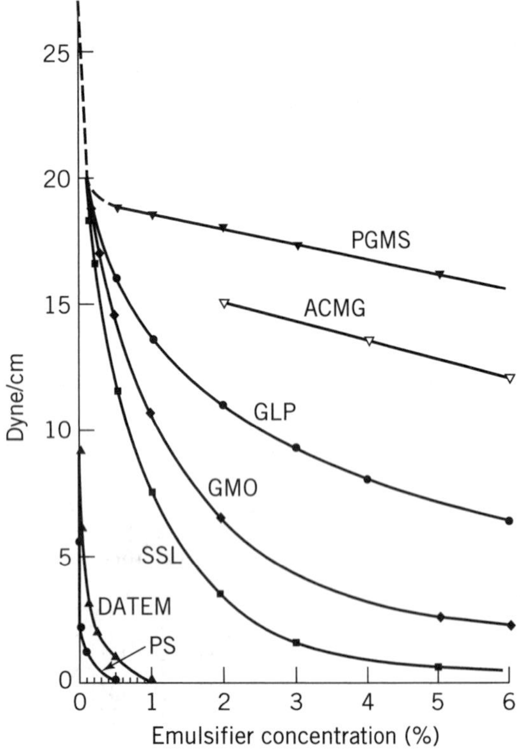

Figure 4. Interfacial tension between soya bean oil and water of 50°C. Measured by the Du Nouy technique. Surfactants tested are propylene glycol monostearate (PGMS), acetylated monoglycerides (ACMG), lactated glyceryl monopalmitate (GLP), distilled monoglyceride from sunflower oil (GMO), sodium stearyl-2-lactylate (SSL), diacetyltartaric acid ester of monoglycerides (DATEM), and polysorbate 60 (PS). *Source:* Ref. 2.

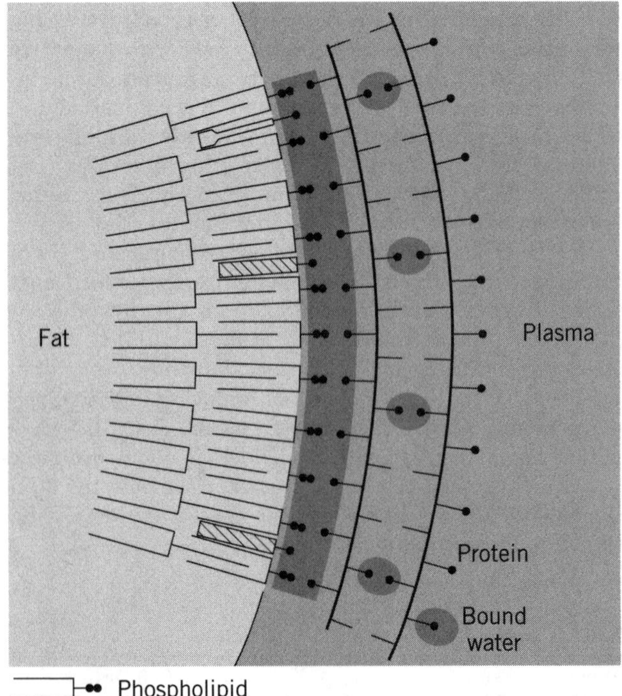

Figure 5. Fat globule surface of milk. *Source:* Ref. 3.

the interface with the aqueous phase. The globule surface is now relatively more hydrophilic with a vastly reduced tendency to coalesce with adjoining fat globules. The tendency to coalesce is further reduced when the fatty acid chains of the lipophilic moiety are high melting because the oil globule surface now takes on a melting point approaching that of the surfactant.

If the surfactant imparts an electrical charge to the surface of the oil globule, it further aids in contributing to emulsion stability because the likely charged globules tend to repel each other, further reducing the opportunity for coalescence. In addition, repellency aids in slowing the separation of the phases (migration of the internal phase). The presence of a surface electrical charge improves the stability of an emulsion when frozen because the emulsified globules are less likely to coalesce following compression and rupture of the surface film, which occurs as a result of concentration into the reduced unfrozen aqueous phase.

Most surfactants added to improve food emulsions are nonionic and do not impart an electrical charge to the globule surface. Lecithin is an exception. Lecithin, with its component phosphatides, although amphoteric, is relatively anionic.

The relative propensity of a surfactant for oil or water is known as its hydrophile–lipophile balance, or HLB. The HLB value may be expressed as the weight percentage of the hydrophilic moiety of the surfactant molecule divided by 5 to reduce the magnitude of the numbers with which we work. For example, if 50% of the molecular weight is hydrophilic, the HLB value would be 50 ÷ 5, or 10. The HLB values of some surfactants used in food emulsification are given in Table 2. HLB values will vary significantly depending on monoester and fatty acid content. The surfactants with lower HLB values tend to induce the formation of water-in-oil emulsions, such as margarine; those with higher values (>10) tend to induce the formation of oil-in-water emulsions. Some relationships between HLB value and the application of the surfactant to food emulsions are given in Table 3. It should be remembered that food emulsions usually contain a significant quantity of naturally occurring surfactants such as phospholipids; the added emulsifiers are merely supplementing the activity of the naturally occurring ones. HLB may be estimated from the solubility of the surfactant, as seen in Table 4.

Obviously, surfactants with high HLB values will tend to disperse more readily in aqueous media while those with low HLB values will be more easily dispersed in the lipid phase. However, there are important other factors that significantly affect surfactant dispersibility in any phase; most surfactants, especially those most effective when used in food processing, are lipophilic and solids at room temperature. As such, they disperse in aqueous media with difficulty. When first introduced as cake emulsifiers, the lipophilic monoglycerides were dispersed in the shortening, which was then added to the batter or cake mix. Monoglycerides and other lipophilic surfactants such as the sorbitan esters, when used in dairy products, were melted into the mix during pasteurization and dispersed to the interface of the fat and water on a molecular basis with the aid of hydrophilic surfactants naturally occurring in the cream phase. When lipophilic or high-melting surfactants such as monoglycerides were required for inclusion in a bread dough at room temperature, it was found necessary to "hydrate" them into a paste emulsion form with water to disperse them in dough or include less effective unsaturated fluid monoesters to form a soft, dispersible plastic.

Table 2. Calculated or Determined HLB Number of Some Surfactants Used in Foods

Chemical name	HLB number
Glycerol monooleate	2.8
Propylene glycol monostearate	1.8
Glycerol monostearate	3.7
Lecithin	4.2
Sorbitan monostearate	4.7
Glycerol monostearate, self-emulsifying	5.5
Polyoxyethylene (20) sorbitan tristearate	10.5
Polyoxyethylene (20) sorbitan monostearate	14.9
Polyoxyethylene (20) sorbitan monooleate	15.0
Polyoxyethylene (40) stearate	16.9

Table 3. Relationship Between HLB and Surfactant End Use

HLB number	Application
4–6	Emulsifiers for w/o systems
7–9	Wetting agents
8–18	Emulsifiers for o/w systems
13–15	Detergents
15–18	Solubilizers

Table 4. Estimation of HLB by Water Solubility

Application in water	HLB range
Not dispersible	1–4
Poor dispersibility	3–8
Milky dispersion after vigorous agitation	6–8
Stable milky dispersion	8–10
Translucent to clear dispersion	10–13
Clear	13+

The first monoglycerides used for the production of hydrates were fully saturated mono/diglycerides. They formed stable emulsions in paste form with about 75% water. Small quantities (ca 0.5%) of hydrophilic coemulsifiers such as sodium stearate or lecithin were added to improve emulsion stability.

When monoglycerides were later distilled to provide a higher monoester content of about 90%, it was found that such products formed gel-like lumps with water, making dispersion in the aqueous phase impossible (4). It was found that distilled monoglycerides and some other surfactants formed mesophases with water and that temperature and water content had to be carefully controlled to obtain adequate dispersion. Some mesophases are dispersible in the aqueous phase, others are more compatible with the lipid phase.

Surface active agents, when dispersed in an aqueous phase, can exist in essentially six mesomorphic or liquid crystalline phases. These phases are anhydrous crystals, fluid isotropic, neat lamellar, gel, dispersion, hexagonal and cubic or viscous isotropic (Fig. 6). When a liquid crystal results from heating the anhydrous crystal, it is known as a thermotropic mesomorph; when resulting from a crystallization from solvent such as water, it is known as a lyotropic mesomorph.

When existing as a pure crystal, the polar heads of the surfactant adjoin one another head to head, and the rigid hydrocarbon tails also adjoin. When heat is applied in the presence of water, the hydrocarbon tails liquify because the weak energy of the van der Waals forces between them is overcome. However, the hydrogen bonding between the polar heads remains, resulting in a semiliquid crystalline structure. The contraction of the disordered hydrocarbon chains opens a gap between the polar heads, allowing water to enter. A lamellar mesophase is formed. When this phase is cooled below the temperature where the hydrocarbon chains resolidify (known as the Krafft Point), a gel phase is formed with the water remaining between the polar groups. The gel is metastable; the water is eventually expelled, and a coagel of fine β crystals dispersed in water is formed.

When some surfactant crystals, such as monoglycerides, are heated to relatively high temperature with little water, the water is dispersed in the fluid lipid as micellar aggregates, cylinders at low water content, discs at higher. This phase is known as fluid isotropic.

When some surfactants are heated to higher temperature in the presence of water, they may form hexagonal cylindrical aggregates. Two types may be formed. The hexagonal I phase consists of cylinders with the polar heads on the outside and the hydrocarbon chains oriented inward as a core. Such phases are infinitely dilutable in aqueous media. The hexagonal II cylinder is the reverse of the first, with the polar moiety in the interior, surrounding a core of water, with the hydrocarbon chains oriented to the exterior. This phase is found only at low water content, usually less than 30%. At higher water content, this phase will separate from the aqueous media as a viscous mesophase.

Another lipophilic phase, somewhat associated with the hexagonal II phase, is the liquid crystalline cubic phase, frequently encountered with unsaturated monoglycerides. It is viscous and isotropic. Larsson (6) has described a cubic phase of monoglyceride as a polygon aggregate of water cylinders enclosed in a matrix of hexagonal II configura-

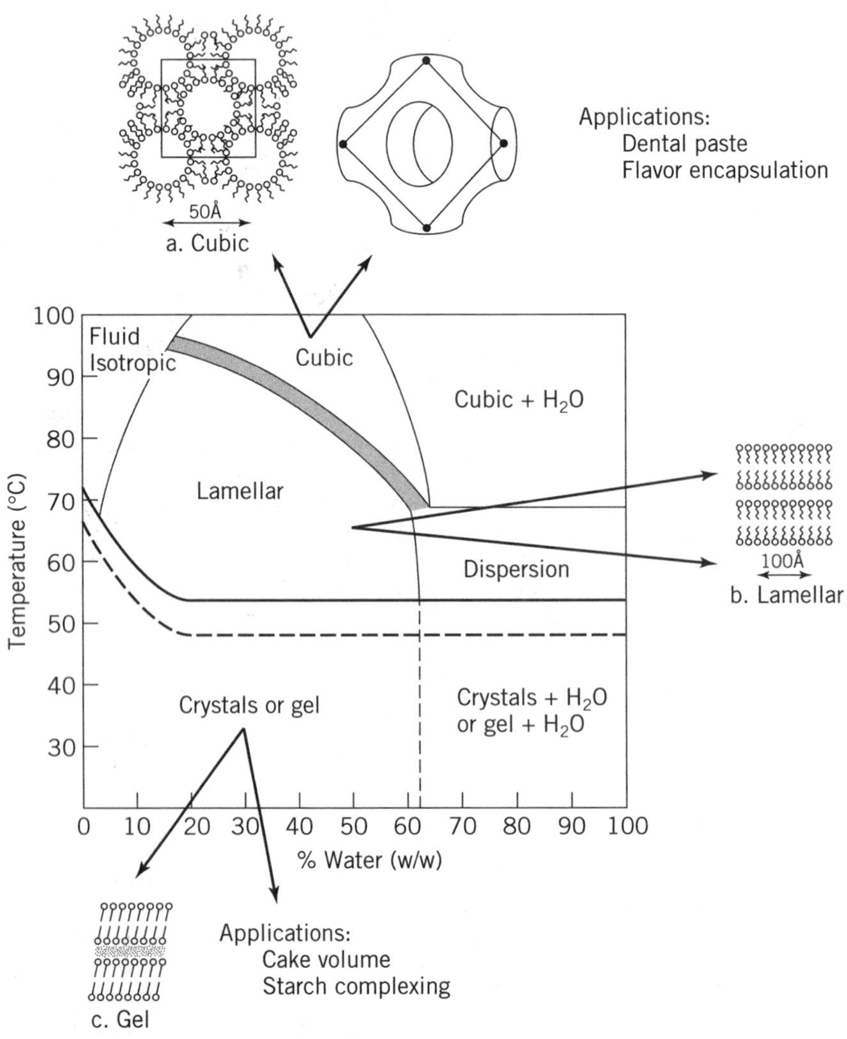

Figure 6. Phase diagram of glyceryl monopalmitate depicting molecular packing of cubic, lamellar, and gel mesophases. *Source:* Ref. 5.

tion, associated as a three-dimensional cubic lattice (Fig. 6). It can absorb up to about 40% water. When dispersed in excess water, surrounded by a lamellar phase, Larsson (7) has referred to the phase as a "cubosome."

When the lamellar mesophase is further diluted with water at the proper temperature, a dispersion phase is formed. The dispersion phase is not merely a dilution of the neat or lamellar phase. The particles are spheres of bimolecular layers alternating with water, and the hollow center can be filled with an aqueous phase. The lamellar bilayer is not stable because the hydrophobic edges resist the aqueous media, and the lamellar bilayer rolls up to form a sphere, with the polar groups external, protecting the internal hydrophobic tails ("liposome"). The physical form of a surfactant drastically affects its functionality. In some cases, the surfactant is completely nonfunctional due to nondispersal on a molecular basis. In other cases, its functional value can be doubled simply by changing the physical form to provide full dispersion.

Although there had been some reference to the use of monodiglycerides in margarine as early as 1921 (4), the first significant commercial use of a surfactant in food was the inclusion of lecithin to improve the viscosity of molten chocolate in the early 1930s (8).

Lecithin was found to lower the viscosity of chocolate, reducing the level of expensive cocoa butter normally required to do so. Lecithin was found to form a monolayer over the hydrophilic constituents of chocolate (sugar, milk, etc) with the lipophilic hydrocarbon tail extended into the molten cocoa fat, facilitating the flow of the vehicle by reducing friction (9).

The incorporation of sorbitan monostearate in chocolate inhibited the formation of bloom (10). Bloom is the result of migration to the surface of chocolate of unstable polymorphs of cocoa butter. As the polymorphs of fat migrate, they leave behind the cocoa fibers, which impart color and ultimately resolidify on the surface of chocolate as light-colored blotches. Sorbitan monostearate forms a monolayer on chocolate nonfat solids, impeding the capillary migration of the unstable cocoa fat polymorph to the surface, thus, inhibiting bloom (11).

About 1933, the so-called superglycerinated shortenings were introduced in the United States for use in the production of cakes (12,13). These contained a significant amount of monoglycerides, for example, 3%. These were included either by production *in situ* by alcoholisis of the fat with excess glycerine during refining or by direct addition of a monodiglyceride.

Shortenings with added surface-active monoglycerides were found to impart greater structural stability to the cakes, allowing for the inclusion of higher ratios of sugar to flour. Cakes with improved texture, volume, and symmetry, as well as keeping quality, resulted from the use of such shortening.

In 1968 the following definition for cake was proposed (14): Cake is a protein foam stabilized with gelatinized wheat starch and containing fat, emulsifiers, mineral salts, and flour and aerated principally by gases evolved by chemical reaction *in situ*.

It is well known that oil is an antifoam that tends to destroy any foam structure, including cake, by weakening the protein film. Plastic shortenings are composed principally of an oil fraction, usually 70 to 75%, blended with solid fats for solidity at room temperature. One would expect the shortening to act as an antifoam, and indeed it does, unless properly encapsulated within an emulsifier. The α-tending surfactants such as propylene glycol monostearate and glyceryl lactopalmatate are exceptionally effective emulsifiers, especially when liquid oil is used as the shortening, because of their ability to form a crystalline membrane around the oil globules, preventing the oil from migrating into the protein lamellae (15). Monoglycerides are also very effective encapsulating agents.

The dispersion of the emulsifier in the batter is most important from the standpoint of obtaining functional improvement. Dispersion by inclusion within the shortening is adequate if the shortening is fully dispersed. Complete dispersion of a plastic fat in an aqueous medium may require extensive art and still might not produce optimum results.

Wren (16) has described six possible physical states for the inclusion of surface active lipids (Table 5). Inclusion of the surfactant in shortening or margarine would represent the anhydrous state. However, it is often necessary to include the emulsifier in the aqueous phase. As an example, the traditional sponge cake formula contains no shortening. To obtain the benefits of an emulsifier, it must be added as an aqueous dispersion. Krog (17) compared the effectiveness of six different mesophases of a distilled glyceryl monostearate in the aqueous state (Table 6). The best results were obtained with the dispersion phase, followed closely by the crystalline gel. Poorest results were obtained with the coagel of β crystals in water. There is no doubt that the dispersion phase was more effective because it was more intimately dispersed in batter. The viscous isotropic cubic phase, poorly dispersible, provided very poor results as is but provided vastly improved results when dispersed in water.

Hydrophilic surfactants (eg, polysorbate 60) are added to cake batters to further reduce surface tension, improving the dispersibility of the lipid phase (eg, shortening) and

Table 5. Possible Physical States of a Surface-Active Lipid

Anhydrous	Polymorphic crystal
	Mixed crystal
	Liquid
	Solution in oil
Aqueous	Molecular solution
	Micellar solution
	Liquid crystalline phase
	Gel phase
	Coagel (crystals + water)
Lipoprotein	Insoluble
	Soluble
Inclusion compound with amylose	
Other macromolecular complex	
Surface film between air–oil–water–solid	"Simple" (monolayer)
	Lipoprotein
	Other complex
	Liquid crystalline phase

Source: Ref. 15.

Table 6. The Effect of Various Aqueous Preparations of DGMS in Sponge Cake

			Specific volume (cm^3/kg)	
Type of DGMS preparations added	DGMS–water proportion	Temperature of preparation (°C)	Cake batter	Cake
Neat, lamellar phase	60:40	65	1140	3400
Viscous isotropic cubic phase	60:40	80	1112	3320
Viscous isotropic + water	10:90	80	1560	5060
Dispersion (pH 7.0)	10:90	65	2970	6900
Gel, α crystalline	10:90	25	2700	6000
Coagel, β crystals in water	10:90	25	1020	2400

Source: Ref. 16.

promoting aeration of the batter. They are usually added in hydrated paste or gel form (25–50% solids) to aid in dispersion. Naturally occurring surfactants such as phospholipids from egg also contribute to emulsification of the batter.

As early as 1925, it had been confirmed that the addition of eggs improved the whipping quality of ice cream mixes (18). It was assumed that the albumin protein fraction of egg was the principal aerating agent. In 1928 it was demonstrated that egg albumin had no beneficial effect on the ice cream mix and that the improved aerating quality was being contributed by the yolk (19). It was later determined that the improvement was due to the emulsification imparted by phospholipids and lipoprotein in the yolk (20,21). Inclusion of yolk in the mix was also shown to increase the electrical charge (anionic) carried by the fat globule (22).

In 1936 a patent was granted on the use of glyceryl monostearate as an emulsifier and whipping aid for ice cream (23). It was found that the monoester was as effective at 0.1 to 0.2% as egg yolk was at 0.5%. It was also found that the use of the monoester provided an ice cream that was drier and stiffer on freezing and thus extruded and packaged with greater facility.

Surfactants based on saturated fatty acids were found to be more effective as aerating agents when used as ice cream emulsifiers, as compared with those based on unsaturated fatty acids (Figs. 7 and 8). The more hydrophilic ethoxylated surfactants are more effective than those of lower HLB value. Those approved for use in food contain an average of 20 mol of ethylene oxide and have an HLB in the 14 to 16 range.

In the production of whippable emulsions, such as ice cream or whipped toppings, it is desirable to obtain good volume and stiffness. Contrary to published hypothesis, it was demonstrated that a partial breakdown of the emulsion during whipping to produce agglomerated (but not coalesced) fat globules provides maximum product stiffness (24,25). It was found that the more hydrophilic ethoxylated surfactants based on saturated fatty acids such as stearic provided maximum whippability, whereas those based on unsaturated fatty acids such as oleic provided maximum dryness and stiffness to the aerated product. A turbidimetic procedure was used to prove that surfactants actually accelerated the breakdown of the ice cream mix emulsion to provide agglomerated fat globules (Fig. 9) (24).

Combining a lipophilic saturated surfactant such as glyceryl monostearate with a hydrophilic unsaturated surfactant such as ethoxylated sorbitan monooleate resulted in a synergistic effect in frozen aerated desserts whereby the combination produced improved whippability and stiffness (26).

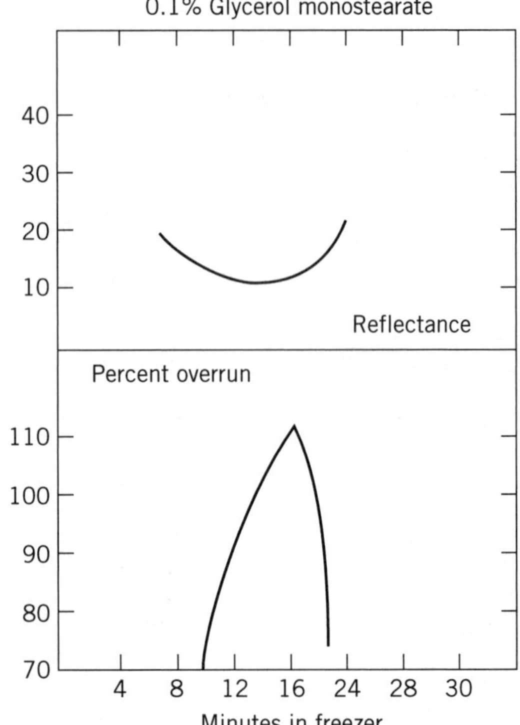

Figure 7. Performance characteristics, ice cream emulsifier glyceryl monostearate.

As with cake batter foams, α-tending surfactants form strong protective films around oil globules and are especially effective in promoting agglomeration of fat globules in dairy-type emulsions. The α-tending surfactants are adsorbed onto the globule surface, interfering with strong protein bonding and making the protective protein film more easily swept off the surface during whipping, resulting in increased agglomeration of globules with resulting firm texture and stiffness in whipped foams (eg, toppings).

Margarine is probably the simulated dairy product most widely consumed. It is a water-in-oil emulsion usually containing 80% fat and 20% aqueous phase. A lipophilic emulsifier system of 0.5 to 1.0% is used to stabilize the emul-

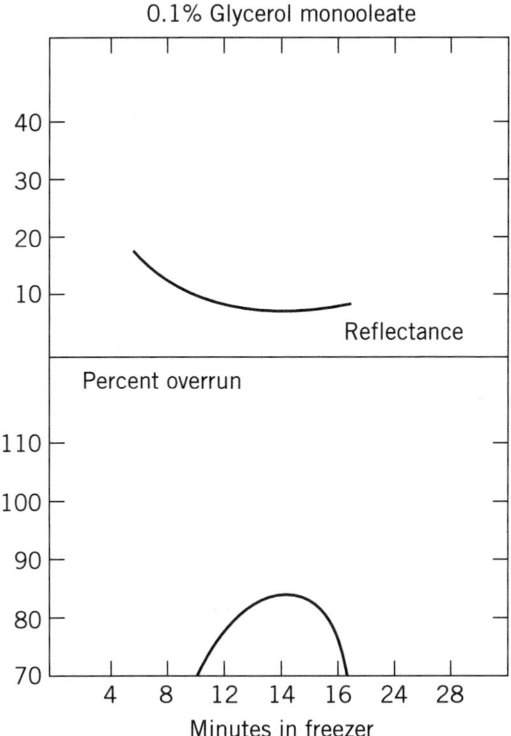

Figure 8. Performance characteristics ice cream emulsifier glyceryl monooleate.

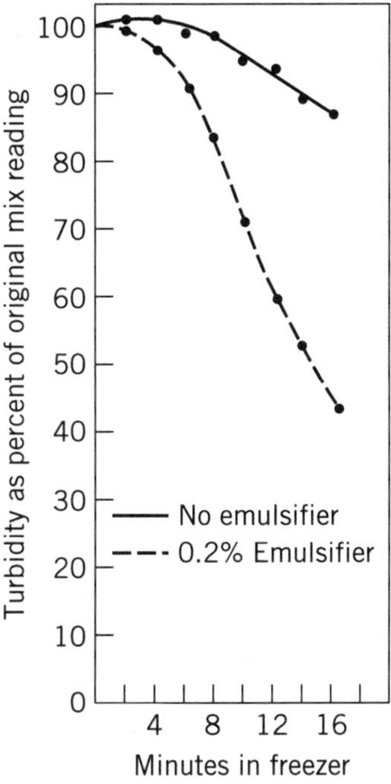

Figure 9. Effect of emulsifier on fat destabilization in the freezer. *Source:* Ref. 24.

sion, preventing leakage of the aqueous phase (weeping) and reducing spattering when the margarine is used for frying.

Surface-active agents with a high degree of unsaturation are usually preferred for margarine, especially the low-fat/high-aqueous types as they more tenaciously incorporate the high-level water into the oil phase.

Another category of simulated dairy products with a large market is milk replacers for young animals. Such emulsions are usually spray-dried in such a manner as to retain the emulsion potential intact, and the emulsion reforms instantly on rehydration.

The largest consumption of surfactants within the food industry is by the baking industry. Aside from the use of surfactants for cake-batter emulsification previously discussed, surfactants are used in yeast-raised products such as bread and rolls for the purpose of reducing the rate of crumb firming, the principal factor associated with staling, and to strengthen gluten (dough conditioning).

As long ago as 1852, the French chemist Boussingault (27) proved that bread did not stale because of moisture loss simply by hermetically sealing it in cans, where the crumb firmed as rapidly as that of unsealed loaves. It is now generally conceded that bread crumb firms primarily owing to the retrogradation of the gelatinized wheat starch. The linear amylose fraction (Fig. 10) retrogrades almost immediately after the baked product cools, thereby supplying physical structure to the baked food. It is not further involved in an ongoing firming. The branched amylopectin fraction (Fig. 10), however, continues to retrograde slowly over a period of days, causing the crumb to become firmer with aging.

Certain surfactants can retard crumb firming because they are able to form complexes with the glucose polymers of starch. Although these surfactants will form complexes with both the straight-chain amylose and the branched-chain amylopectin, it is now understood that it is most important to complex with the amylopectin since it is responsible for the progressive firming of crumb. It has been shown that the starch fractions will form a helix around the hydrocarbon chain of surfactants possessing the proper steric configuration (Fig. 11). The inside of the helix is lipophilic owing to the CH groups, and the exterior is hydrophilic owing to the presence of OH groups. Straight hydrocarbon chains such as stearic acid can accommodate the core of the helix with its diameter of 4.5 to 6.0 A. On the other hand, unsaturated fatty acids of the *cis* configuration cannot enter the helix, as they are not straight.

The complexing ability of a number of surfactants has been evaluated (28); some of these are given in Table 7. A correlation has been established between the ability of a surfactant to complex with amylose and the ability to also complex with amylopectin and retard crumb firming (29,30). The starch components are not as prone to retrograde, owing to interference with hydrogen bonding when the helical clathrates form. For complex formation to take place, the surfactant must possess not only the proper steric configuration but must be dispersible on a molecular basis in the aqueous phase of the dough or batter.

Surfactants such as monoglycerides are used because of their starch-complexing ability in other foods containing

Amylose

1,6 Linkage

Amylopectin

Figure 10. Structure of amylose and amylopectin.

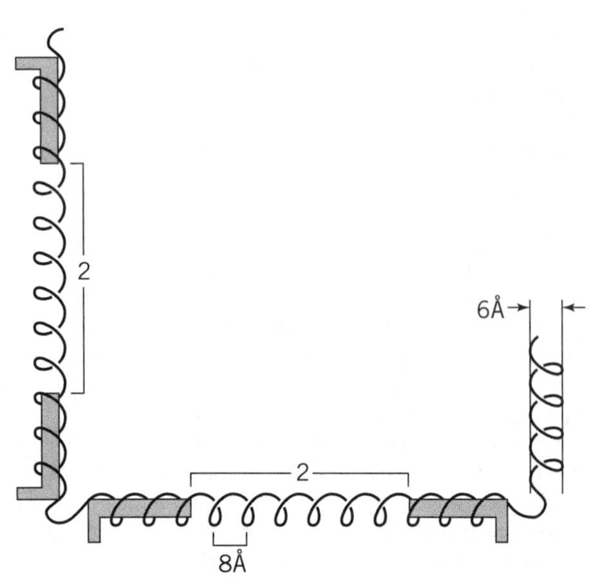

Figure 11. Schematic diagram of glyceryl monopalmitate/amylose complex. a = "Free" space.

Table 7. Complex Formation Between Amylose and Food Emulsifiers

Materials	Amylose complexing index
Distilled monoglycerides, 90–92% 1 monoester	
Lard, hydrogenated	92
Lard, unhydrogenated, 45% monoolein	35
Soya bean oil, hydrogenated, 85% monostearin	87
Soya bean oil, unhydrogenated, 55% monolinolein	28
Monodiglycerides, 45% monoester	28
Organic acid esters of monoglycerides	
Lactic acid esters	22
Succinic acid esters	63
Diacetylated tartaric acid esters	49
Distilled propylene glycol monostearate, 90%	15
Sucrose monostearate (commercial sample)	26
Polyoxylethylene-sorbitan-(20)-monostearate (commercial sample)	32
Stearoyl-2-lactylate	79
No stearoyl-2-lactylate	72
Ca stearoyl-2-lactylate	65
Na stearyl-fumarate	67

Source: Ref. 27.

starch, such as pasta, instant potato products, and starch-based desserts. By complexing with the starch, surfactants improve the texture, and cohesive strength and prevent stickiness in cooked pasta; they also, for the same reason, prevent lumpiness and stickiness in rehydrated instant potatoes and starch-based desserts, which would normally occur with the presence of free amylose.

Certain surfactants that are either anionic or ethoxylated have the ability to react with gluten and other proteins, thereby improving their film-forming ability and cohesive strength. They are known as dough strengtheners or dough conditioners.

Their ability to strengthen protein is the result of the surfactant forming a complex between glutinous protein fractions, together with nonglutinous protein and native polar lipids (31).

Because of the improved cohesive strength of the gluten, the dough has improved tolerance to mixing and increased tolerance to the addition of nonwheat proteins. These changes result in improved loaf volume and symmetry, better crumb texture and cell structure, and greater resistance to staling. The best known of these dough conditioners are the sodium and calcium salts of stearoyl lactylic acid, the ethoxylated glyceryl and sorbitan monostearates, and datem esters. Hydrophilic lecithins have also been shown to possess gluten-strengthening properties (32).

Surfactants are used to stabilize salad-dressing emulsions and to emulsify flavor oils. They are used in chewing gum and certain other confections such as caramel to prevent wrapper sticking and sticking to false dentures. Surfactants find use as extrusion aids in snack foods and meat analogues. Surfactants based on unsaturated fatty acids such as glyceryl monooleate are used as antifoams in sugar processing and confectionery manufacture.

The market for surfactants in foods is now about 400,000,000 lb annually in the United States. About 275,000,000 lb are consumed in foods in Europe and 180,000,000 in Japan.

BIBLIOGRAPHY

1. M. J. Lynch and W. C. Griffin, "Food Emulsions," in K. J. Lissant, ed., *Emulsions and Emulsion Technology*, Marcel Dekker, New York, 1974, p. 249.
2. N. Krog, *J. Am. Oil Chem. Soc.* **54**, 124 (1977).
3. N. King, *The Milk Fat Globule Membrane*, Commonwealth Agr. Bureau, Farnham Royal, Bucks, United Kingdom, 1955.
4. N. Krog, "Food Emulsifiers," in S. Friberg, ed., *Food Emulsions*, Marcel Dekker, New York, 1976, pp. 67–139.
5. E. Boyle, *Food Technol.* **51**, 52 (1997).
6. K. Larsson, *Chem. Phys. Lipids* **9**, 181 (1972).
7. K. Larsson, *J. Phys. Chem.* **93**, 7304 (1989).
8. German Patent 530187 (May 15, 1930), Hansea Muhle, A.G.
9. W. Duck, *Pennsylvania Manufacturing Conference Progress Report 10*, Pennsylvania Manufacturing Confectioners Association, Lancaster, Pa., 1955.
10. N. Easton et al., *J. Food Technol.* **6**, 21 (1952).
11. W. H. Knightly and J. W. DuRoss, *Proceedings of the 19th Annual Production Conference*, Pennsylvania Manufacturing Confectioners Assn. (Sect. 15), Lancaster, Pa., 1965.
12. U.S. Patent 2,132,393–2,132,398 (October 11, 1938), N. S. Goeth, A. S. Richardson, and V. M. Votaw (to Proctor and Gamble).
13. U.S. Patent 2,132,406 (October 11, 1938), A. K. Epstein and B. R. Harris (to Proctor and Gamble).
14. W. H. Knightly, "Surface Active Lipids in Foods," *Society of Chemical Industry Monograph* **32**, 141–157 (1968).
15. I. S. Shepherd and R. W. Yoell, "Cake Emulsions," in S. Friberg, ed., *Food Emulsions*, Marcel Dekker, New York, 1976, pp. 270–273.
16. J. J. Wren, "Surface Active Lipids in Foods," *Society of Chemical Industry Monograph* **32**, 158–170 (1968).
17. N. Krog, "The Effect of Various Mesomorphic Phases of Monoglyceride-Water Systems in Starch Products," Paper No. 268, AACC-ACS, 1968 Joint Meeting, Washington D.C., March 31–April 4, 1968.
18. H. F. DePew and S. W. Dyer, *Creamery and Milk Plant Monthly* **14**, 87 (1925).
19. D. A. Pettee, *Ice Cream Manufacturing* **2**, 56 (1929).
20. C. C. Walts and C. D. Dahle, *J. Agric. Res.* **47**, 967–977 (1933).
21. C. D. Dahle and D. V. Josephson, *Ice Cream Review* **20**, 60 (1937).
22. G. C. North, Ph.D. Thesis, University of Wisconsin, 1931.
23. U.S. Patent 2,065,398 (Dec. 22, 1936), W. D. Roth (to Industrial Patents Corp.).
24. P. G. Keeney, *Ice Cream Field* **71**, 20 (1958).
25. W. H. Knightly, *Ice Cream Trade Journal* **55**, 24 (1959).
26. U.S. Patent 3,124,464 (March 10, 1964), W. H. Knightly and G. P. Lensack (to Atlas Chemical Industries, Inc.).
27. J. B. Boussingault, *Annals of Chemistry and Physics* **36**, 470 (1852).
28. N. Krog, *Die Staerke* **23**, 206 (1971).
29. N. Krog and B. N. Jensen, *J. Food Technol.* **5**, 77 (1970).
30. N. Krog et al., *Cereal Foods World* **34**, 281–285 (1989).
31. O. K. Chung, C. C. Tsen, and R. J. Robinson, *Cereal Chem.* **58**, 220 (1981).
32. W. H. Knightly, "Lecithin in Baking Applications," in B. F. Szuhaj, ed., *Lecithins—Sources, Manufacture and Uses*, American Oil Chemists Society, Champaign, Ill., 1989, pp. 174–196.

William H. Knightly
Emulsion Technology, Inc.
Wilmington, Delaware

See also Emulsifiers, stabilizers, and thickeners.

EMULSIFIERS, STABILIZERS, AND THICKENERS

The food additives that will be discussed in this article fall into three functional categories. Although the definition of each function will be clarified in the discussion, the definitions given in 21 *Code of Federal Regulations* are as follows:

> §170.3(o)(8). Emulsifiers and emulsifier salts: Substances which modify surface tension in the component phase of an emulsion to establish a uniform dispersion or emulsion.

§170.3(n)(28). Stabilizers and thickeners: Substances used to produce viscous solutions or dispersions, to impart body, improve consistency, or stabilize emulsions, including suspending and bodying agents, setting agents, jellying agents, and bulking agents, etc.

§170.3(o)(29). Surface-active agents: Substances used to modify surface properties of liquid food components for a variety of effects, other than emulsifiers, but including solubilizing agents, dispersants, detergents, wetting agents, rehydration enhancers, whipping agents, foaming agents, and defoaming agents, etc.

Generally emulsifiers and surface-active agents are relatively small molecules (molecular weight less than 1,000 Da), while stabilizers and thickeners are polymers such as gums and proteins. There are exceptions to this; calcium stearate is listed as a GRAS (generally recognized as safe) stabilizer and thickener while gum ghatti is approved for use as a GRAS emulsifier.

FUNCTIONS OF EMULSIFIERS, STABILIZERS, AND THICKENERS

As shown in Figure 1 surfactants have a lipophilic (fat-loving) portion and a hydrophilic (water-loving) portion; for this reason they are sometimes called amphiphilic (both-loving) compounds. The lipophilic part of food surfactants is usually a long-chain fatty acid obtained from a food grade fat or oil. The hydrophilic portion is either nonionic (glycerol is shown in Fig. 1), anionic (negatively charged, such as lactate), or amphoteric, carrying both positive and negative charges (phosphorylcholine is shown in Fig. 1). Cationic (positively charged) surfactants are usually bactericidal and somewhat toxic; they are not used as food additives. Examples of the three types indicated are a monoglyceride (nonionic), stearoyl lactylate (anionic), and lecithin (amphoteric). The nonionic surfactants are relatively insensitive to pH and salt concentration in the aqueous phase, while the functionality of the ionic types may be markedly influenced by pH and ionic strength.

Emulsifiers

The formation of an oil-in-water emulsion is outlined in Figure 2. If 100 mL of pure vegetable oil plus 500 mL of water are mixed vigorously to obtain an emulsion in which the average diameter of the oil globules is 1 μm, 600 m^2 (slightly more than 6,400 ft^2) of oil-water interface is generated. Using a purified vegetable oil, the interfacial tension (γ) versus water is about 3×10^{-6} J/cm^2. To form the emulsion, 18 J of the energy input is converted into interfacial energy. The addition of 1% glycerol monostearate (GMS) to the oil phase will lower γ to about 3×10^{-7} J/cm^2 so that the excess interfacial energy is only 1.8 J. This excess interfacial energy is the driving force behind coalescence of the oil globules and also the force resisting subdivision of oil droplets during mixing (each division increases the amount of interface present in the system). In a series of emulsification experiments in which the amount of mixing energy was constant and γ was changed by adding emulsifier, it was found that the average oil droplet diameter paralleled γ; that is, as more emulsifier was added, γ decreased and so did the average droplet size. The total excess interfacial energy was roughly constant.

Stabilizers

The phenomenon just described is the promotion of emulsion formation; this is not the same as stabilization of emulsions. The difference is seen by reference to Figure 2. After the emulsion is formed, if it is allowed to stand, the oil droplets will rise to the top or cream (assuming the volume ratio of oil to water is low enough for flotation to occur). The rate of creaming is inversely related to droplet diameter and to the viscosity of the aqueous phase: large droplets rise faster than small droplets, and faster in water than in a viscous gum solution. Emulsifiers that promote formation of smaller diameter drops and additives that increase viscosity give emulsions in which the rate of separation is slower, so in that sense only the emulsion is stabilized.

When two oil droplets make contact (facilitated by the creaming process), they may either clump (stick together, but retain their individual identity) or coalesce into one larger droplet, reducing total surface area and total excess interfacial energy. Clumped droplets may be readily redispersed by simply stirring the system; inverting a bottle of creamed (nonhomogenized) milk a few times redistributes the clumped milkfat droplets. In coalescence the emulsifier

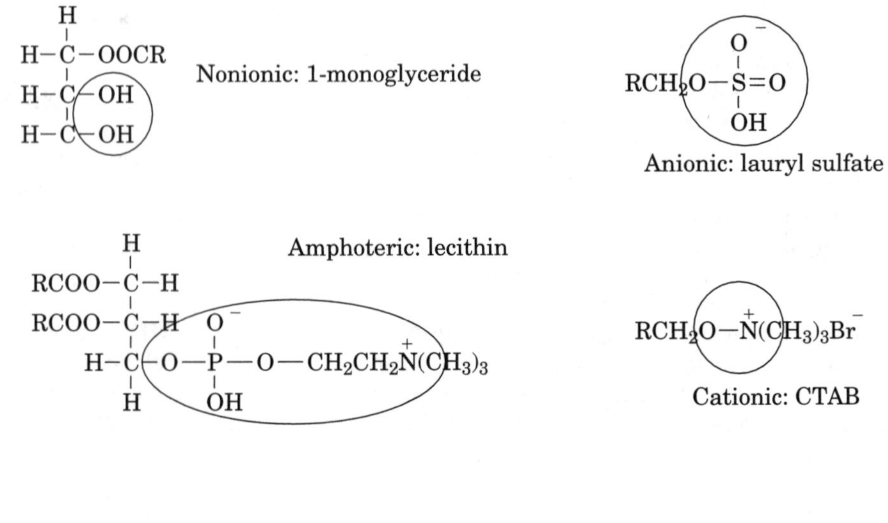

Nonionic: 1-monoglyceride

Anionic: lauryl sulfate

Amphoteric: lecithin

Cationic: CTAB

Figure 1. Generalized structure of amphiphilic molecules. R represents a lipophilic hydrocarbon (a fatty acid or fatty alcohol). The hydrophilic groups are glyceryl, (nonionic), sulfate (anionic), phosphorylcholine (amphoteric) and quaternary ammonium (cationic).

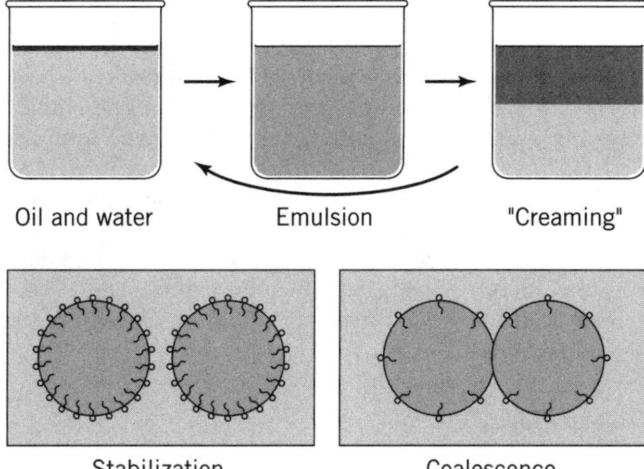

Figure 2. States of emulsion formation and breakdown. In the upper figures the dark area represents oil, the clear areas water. In the lower figures, shaded circles represent oil surrounded by water.

used may be efficient at reducing γ even at a low interfacial concentration but does not prevent the oil droplets from touching and coalescing during creaming.

A true emulsion stabilizer prevents coalescence (Fig. 2). In essence, the thin layer of water between the oil droplets is stabilized by various mechanisms. If the surfactant is anionic, then the surfaces of both oil droplets carry a negative charge and they are mutually repelled by electrostatic effects. This sort of stabilization is sensitive to ionic strength, and a high salt concentration will suppress the electrostatic repulsion, promote contact and coalescence, and lead to rapid emulsion breakdown.

A second kind of stabilization is shown by surfactants in which the hydrophilic portion is quite large; for example, the polyoxyethylene chain of the Tweens or ethoxylated monoglyceride. In this case the chain is anchored at the surface of the oil droplet by the lipophilic tail, but it is strongly hydrated and generates a layer of bound water around the droplet, preventing contact and coalescence. This functionality is relatively insensitive to salt concentration.

A third kind of stabilization is due to simple steric hindrance of contact. The alpha-tending emulsifiers such as propylene glycol monostearate (PGMS) form an actual solid layer at the oil-water interface (1,2). This film physically prevents the contents of oil droplets from coalescing even though their surfaces may be touching. Gums such as gum arabic and gum ghatti stabilize oil-in-water emulsions by a similar mechanism, forming a film of adsorbed polymer around the oil droplet, while some water-soluble proteins perform the same function in mixtures of ground meat and fat for making sausage.

It should be noted that emulsion stabilization is not directly related to the ability to lower interfacial tension. For example, 1% GMS in the oil phase lowers γ to 3×10^{-7} J/cm^2 but has little effect on coalescence rate, whereas 12% PGMS yields a γ of 9×10^{-7} J/cm^2 but gives an emulsion with excellent long-term stability.

Foam Stabilizers

The generation and stabilization of a foam is subject to the same thermodynamic energy considerations as emulsions; lowering the interfacial (surface) tension favors foam formation. Small surfactant molecules dissolved in the aqueous phase will promote foaming, and the stability of the foam is dependent on the maintenance of a film of water between air bubbles. The lipophilic portion of the surfactant enters the gas phase (rather than an oil phase) but in all other respects the situation is analogous to that where both phases are liquid (ie, emulsions).

Proteins are amphiphilic molecules in that many amino acids (eg, leucine, isoleucine, and valine) contain hydrophobic (lipophilic) side chains while others (glutamic, aspartic, lysine, and arginine) have ionic, hence hydrophilic, side chains. Normally, proteins such as egg albumin in solution are folded in such a way that the hydrophobic side chains are buried in the interior of the molecule in a nonpolar environment, while the hydrophilic side chains are on the surface of the molecule and interact with the polar aqueous environment. Introduction of air bubbles into the solution presents a new possibility for the lowest energy state of the protein, namely for it to unfold with the hydrophobic side chains entering the air phase and the hydrophilic chains remaining in the water phase. The portion of the proteins located in the aqueous phase hold water, preventing it from draining away from this region, hence stabilizing the foam. Whipping aids enhance the ability of the protein to unfold at the air-water interface; the energetics of protein unfolding become more favorable and the ease of foam formation increases. Solutes that increase the viscosity of the water phase, for example, gums, slow the rate of draining and thus increase foam stability.

Thickeners

Thickeners may function in four different ways: (*1*) by stabilizing an emulsion with a high volume percentage of internal phase, (*2*) by increasing the viscosity of the external phase, (*3*) by forming an elastic network in the external phase, and (*4*) by removing a portion of the external phase. An example of the first kind is mayonnaise, where egg yolk lipoprotein stabilizes an oil-in-water emulsion in which the internal phase (oil) represents more than 70% of the total volume. As the particle size of the oil droplets decreases, the relative amount of water that is immobilized around the surface increases, and the amount of mobile external phase (free water, in a nonthermodynamic sense) decreases, contributing to the high viscosity and body of the final product.

Certain gums form aqueous solutions that have a high viscosity (Table 1). When such a solution is used as the water component of a food formula, the viscosity of the final product reflects the viscosity of the solution. In a simple salad dressing formulation, increased viscosity in the aqueous phase slows down the flotation rate of the oil droplets formed during shaking and (in this sense) stabilizes the emulsion. Commercial batters (ie, cake, cake donut, and fish breading) require a certain viscosity for optimum functionality; gums may be used to impart and control batter viscosity.

Table 1. Properties of Some Gums

Gum	1% solution Viscosity, mPa·s	*21 CFR*
	Low viscosity gums	
Arabic	2–5	184.1330
Ghatti	4–10	184.1333
Larch	2–10	172.610
	Medium-viscosity gums	
Sodium alginate	25–800	184.724
Tragacanth	200–500	184.1351
Xanthan	800–1400	172.695
	High-viscosity gums	
Guar	2000–3500	184.1339
Karaya	2500–3500	184.1349
Locust bean	3000–3500	184.1343
	Cellulose gums	
Sodium carboxymethylcellulose	50–5000	182.1745
Hydroxypropyl methylcellulose	20–50,000	172.874
Methylcellulose	10–2000	182.1480
	Gel formers	
Agar	Gel	184.1115
Calcium alginate	Gel	184.1187
Carrageenan	Gel	172.620
Furcelleran	Gel	172.655
Pectin	Gel	184.1588
Gellan	Gel	172.665

Gums and many modified starches in aqueous solution form gels under the proper conditions (Table 1). Gelation is usually dependent on some additional factor (temperature change and presence of divalent cations), and these properties are used during production to generate the desired degree of elasticity in the final food product. The elastic gel network prevents separation of dispersed oil and solids (eg, fruit) as well as giving a pleasing texture to the food. In most food products the external phase is water or an aqueous solution and the internal phase is a mixture of solid material or emulsified lipid. The viscosity of the product depends in part on the volume ratio of internal to external phases. Insoluble fibrous materials can physically adsorb several times their weight in water, so the addition of a small percentage of a material such as cellulose (either alpha or microcrystalline) or hemicellulose (cereal bran) will markedly reduce the amount of water, raising the internal to external phases ratio and hence the viscosity. As an example, a sauce might contain 20% solids (80% water) for an internal to external ratio of 20:80, or 0.25. Adding 5% powdered cellulose will convert about 30% of the water to the solid phase via adsorption, raising the internal to external ratio to 55:50, or 1.1, producing a marked difference in the viscosity of the sauce.

TYPES OF EMULSIFIER

Emulsifiers may be divided into four classes as mentioned earlier: (*1*) nonionic, (*2*) anionic, (*3*) amphoteric, and (*4*) cationic. The actual item of commerce is seldom exactly like the organic chemical structures but is usually a mixture of similar compounds derived from natural raw materials. As a simple comparison, dextrose is one single, relatively pure chemical entity, described by the formula $C_6H_{12}O_6$. On the other hand, glycerol monostearate (GMS) is made from a hydrogenated natural fat or oil and the saturated fatty acid composition may well be something like 1% C_{12} 2% C_{14} 30% C_{16} 65% C_{18} and 2% C_{20}. In addition, the monoglyceride will be approximately 91% 1-monoglyceride and 9% 2-monoglyceride, which represents the chemical equilibrium.

Nonionic Emulsifiers

Monoglycerides and Derivatives. The manufacture of monoglycerides and derivatives used by the food industry were estimated to be more than 200 million lb in 1981 (3–5). The use of monoglycerides in food products first began in the 1930s when superglycerinated shortening became commercially available. Glycerine was added to ordinary shortening along with a small amount of alkaline catalyst, the mixture was heated causing some interesterification of triglyceride with the glycerine, and the catalyst was removed by neutralization and washing with water. The resulting emulsified shortening contained about 3% monoglyceride and was widely used for making cakes, particularly with high sugar levels. Subsequent use of monoglyceride to retard staling (crumb firming) in bread used plastic monoglyceride made by altering the ratio of glycerine to fat to achieve a higher final concentration of monoglyceride, with most of the remainder being diglyceride (commercial mono- and diglycerides may contain from 42 to 60% α-monoglyceride). Later developments in monoglyceride technology include (*1*) distilled monoglyceride, consisting of a minimum of 90% monoglyceride; (*2*) hydrated monoglyceride, which contains roughly 25% monoglyceride, 3% sodium stearoyl lactylate (SSL), and 72% water, forming a lamellar mesophase for better water dispersibility; and (*3*) powdered distilled monoglyceride in which the composition of the original feedstock fat is balance between saturated and unsaturated fatty acids so that the resulting powder hydrates fairly rapidly during mixing in an aqueous system such as bread dough.

The monoglyceride structure shown in Figure 3 is for 1-monostearin, also called α-monostearin. If the fatty acid is esterified at the middle hydroxyl, the compound is 2-monostearin, or β-monostearin. In technical specifications manufacturers usually give the monoglyceride content of their product as a percentage of α-monoglyceride. The routine analytical method for monoglyceride (6) detects only the 1-isomer; quantitation of the 2-isomer is much more tedious. The total monoglyceride content of a product is about 10% higher than the reported α-monoglyceride content. In a practical sense, however, when the functionality and cost effectiveness of various products are being compared, the α-monoglyceride content is a useful number because for all products it equals about 91% of the total monoglyceride present.

The fatty acid composition of monoglyceride reflects the makeup of the triglyceride fat from which it is made. Com-

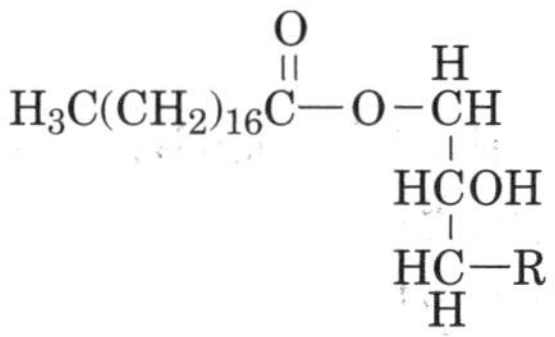

$$H_3C(CH_2)_{16}\overset{\overset{O}{\|}}{C}-O-\overset{H}{\underset{|}{C}}H$$
$$HCOH$$
$$\underset{H}{HC}-R$$

R	Emulsifier
$-OH$	Glycerol monostearate (GMS)
$-O(CH_2CH_2O)_nH$	Polyoxyethylene monoglyceride (EMG)
$-O-\overset{O}{\|}C\overset{H}{C}(OH)CH_3$	Lactylated monoglyceride (LacMG)
$-O-\overset{O}{\|}CCH_3$	Acetylated monoglyceride (AcMG)
H	Propylene glycol monostearate (PGMS)

Figure 3. Nonionic emulsifiers based on glycerol esters and derivatives.

mercial GMS, for example, may contain as little as 65% stearate, if made from fully hydrogenated lard, or as much as 87% stearate, if made from fully hydrogenated soybean oil. The other principal saturated fatty acid will be palmitic, and because hydrogenation is practically never carried out to the extent that all unsaturation is removed (iodine value of zero), a small percentage of unsaturated (oleic and elaidic) acid is also usually present. Iodine values for powdered distilled monoglycerides (monoglyceride in a beaded form that hydrates readily when incorporated in a bread dough) are in the range of 19 to 36, and for plastic monoglycerides a typical range is 65 to 75. The unsaturated fatty acids are a mixture of oleic and linoleic and the *trans* isomers of these acids.

The product of the manufacture of ethoxylated monoglyceride (EMG) is somewhat random in structure. Monoglyceride is treated with ethylene oxide gas under pressure in the presence of alkaline catalyst at elevated temperatures. Ethylene oxide polymerizes via a series of ether linkages and also forms ether bonds with the free hydroxyl groups on the monoglyceride. The average number of ethylene oxide units per monoglyceride molecule is about 20 units ($n = 20$ in Fig. 3), distributed between two chains if the monoglyceride is doubly substituted. Chains may be attached to hydroxyls at both the two and three positions of the monoglyceride, although many more chains will be located at the α (three) position than at the β (two) position because of the difference in their chemical reactivities. The exact distribution of polymer chain lengths and distribution between α and β positions are functions of reaction conditions, for example, catalyst type and concentration, gas pressure, temperature, agitation, and length of reaction time.

The second group of monoglyceride derivatives, the α-tending emulsifiers (lactylated monoglyceride, acetylated monoglyceride, and PGMS), find their main use in cake production. These emulsifiers are dissolved in the shortening phase of the cake formulation and contribute to the emulsification of the shortening in the water phase as well as promoting incorporation of air into the fat phase. The particular property of these emulsifiers that makes them valuable in liquid shortening cakes is that they form a solid film at the oil water interface, not only stabilizing the emulsion but also keeping the lipid phase from preventing air incorporation (protein-stabilized foam formation) during cake batter mixing.

Sorbitan Derivatives. When the sugar alcohol sorbitol is heated with stearic acid in the presence of a catalyst, two reactions occur: sorbitol cyclizes to form the five-membered sorbitan ring and the remaining primary hydroxyl group is esterified by the acid. The resulting sorbitan monostearate (Fig. 4) is oil soluble, has a rather low hydrophilic lipophilic balance (HLB) value, and is the only one of the many sorbitan esters explicitly approved for food use in the United States (sorbitan tristearate is marketed as a crystal modifier, as a self-affirmed GRAS material). Other sorbitan esters of importance are the monooleate and the tristearate. Any of the three esters may be reacted with ethylene oxide to give polyoxyethylene derivatives, as indicated in Figure 4 which are water soluble and have a relatively high HLB. The monostearate derivative is known as Polysorbate 60, the tristearate is Polysorbate 65, and the monooleate is Polysorbate 80. The remarks made in connection with EMG regarding the length and location of the polyoxyethylene chains apply to these compounds. The average number of oxyethylene monomers per sorbitan ester molecule is 20 ($n = 20$), and in the case of the monoesters, chains of varying length (but with about 20 ethylene oxide units in aggregate) may be attached to more than one hydroxyl group of the sorbitan ring (triester has only one hydroxyl group available for derivatization).

Sorbitan monostearate is used in applications where fat is the continuous phase, that is, in fat-based confectionery

Sorbitan monostearate

Polyoxyethylene (20) sorbitan monostearate (Polysorbate 60)

Polyoxyethylene (20) sorbitan tristearate (Polysorbate 65)

Figure 4. Nonionic emulsifiers based on sorbitan esters.

coatings, in bakery icings, and in coffee creamers. The polyoxyethylene derivatives have found wider usage. Polysorbate 60 is used in a variety of applications where a high-HLB emulsifier is required, and is used in fluid oil cake shortening systems (4,5,7), generally in combination with GMS and PGMS. The various sorbitan derivatives are often used in combination to obtain a specific desired HLB; the regulations regarding the permitted levels of each emulsifier in the final product are rather complicated: details can be found in 21 *CFR*.

Esters of Polyhydric Compounds. Polyglycerol esters (Fig. 5) have a variety of applications as emulsifiers in the food industry (3,8). The polyglycerol portion is synthesized by heating glycerol in the presence of an alkaline catalyst; ether linkages are formed between the primary hydroxyls of glycerol. In Figure 5 n may take any value, but for food emulsifiers the most common ones are $n = 1$ (triglycerol), $n = 4$ (hexaglycerol), $n = 6$ (octaglycerol), and $n = 8$ (decaglycerol) (in all cases n is an average value for the molecules present in the commercial preparation). The polyglycerol backbone is then esterified to varying extents, either by direct reaction with a fatty acid or by interesterification with a triglyceride fat. Again, the number of acid groups esterified to a polyglycerol molecule varies around some central value, so an octaglycerol octaoleate really should be understood as (approximately octa)glycerol (approximately octa)oleate ester. By good control of feedstocks and reaction conditions manufacturers do manage to keep the properties of their various products relatively constant from batch to batch.

The HLB of these esters depends on the length of the polyglycerol chain (the number of hydrophilic hydroxyl groups present) and the degree of esterification. As ex-

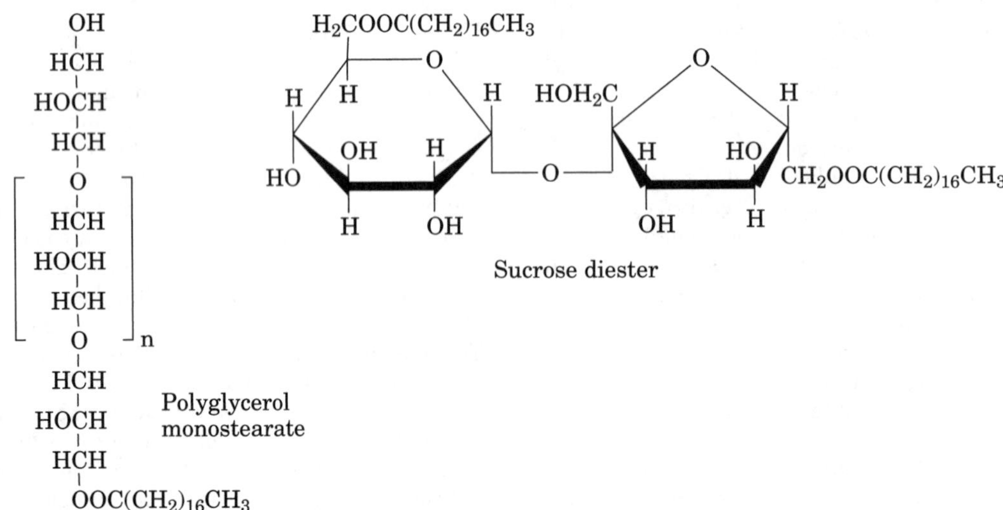

Figure 5. Nonionic emulsifiers based on polyhydric hydrophilic groups.

amples, decaglycerol monostearate has an HLB of 14.5, while triglycerol tristearate has an HLB of 3.6. Intermediate species have intermediate HLB values, and any desired value may be obtained by appropriate blending. The wide range of possible compositions and HLB values make these materials versatile emulsifiers for food applications.

Sucrose has eight free hydroxyl groups that are potential sites for esterification to fatty acids. Compounds containing six or more fatty acids per sucrose molecule have been approved for use as noncaloric fat substitutes under the name Olestra; this material acts like a triglyceride fat and has no surfactant properties. Compounds containing one to three fatty acid esters (Fig. 5) act as emulsifiers and are approved for food use in that capacity (9). They are manufactured by the following steps. First, an emulsion is made of fatty acid methyl ester in a concentrated aqueous sucrose solution. The water is then removed under vacuum at elevated temperature. An alkaline catalyst is added and the temperature of the dispersion is slowly raised to 150°C under vacuum, distilling off methanol formed on transesterification. Finally, the reaction mixture is cooled and purified. The degree of esterification is controlled by the reaction conditions, especially the sucrose to methyl ester ratio, and the final product is a mixture of esters. The HLB value of a particular product is lower (more lipophilic) as the degree of esterification increases, as would be expected.

Anionic Emulsifiers

Monoglyceride Derivatives. A large number of anionic derivatives of monoglyceride have been synthesized and tested, but the two shown in Figure 6 are the most widely used. Others listed in Table 2 include stearyl monoglyceridyl citrate, succistearin (the succinate monoester of propylene glycol monostearate) monoglyceride citrate), and monoglyceride phosphate. Succinyl monoglyceride (SMG) is manufactured by reacting succinic anhydride with monoglyceride, forming the succinate monoester. Diacetyltartrate ester of monoglyceride (DATEM) is made by mixing monoglyceride, tartaric acid, and acetic anhydride in a defined ratio and reacting under closely controlled conditions. The product is a mixture of the possible esters, with the one shown in Figure 6 forming the largest proportion. The most usual application of these compounds is as dough conditioners (ie, surfactants added to bread dough), which increase the strength and elasticity of gluten in the dough. Bread and other yeast-leavened baked foods made with the addition of dough conditioners typically have larger volume and finer internal texture than products made without such conditioners.

Other Anionic Surfactants. In addition to SMG and DATEM, some other anionic surfactants that have been tried as dough conditioners are shown in Figure 7. Currently SSL is the one most widely used in the United States; sodium stearyl fumarate (10) has been offered but did not find acceptance, and sodium lauryl sulfate is used only as a whipping agent with egg whites. Lactic acid, having both a carboxylic acid and a hydroxyl function on the same molecule, readily forms an ester with itself. When stearic acid is heated with polylactic acid under the proper reaction conditions and then neutralized with sodium hydroxide, a product having the structure shown in Figure 7 is obtained. The monomer lactylic acid shown represents the predominant product; the lactylic dimer is also present, as well as lactylic trimers and tetramers (11). As with all compounds based on commercial stearic acid derived from fats, some percentage of the fatty acid is palmitic, with small portions of myristic and arachidic acids also found. Although SSL is readily water soluble, the calcium salt is practically insoluble. In this respect it mimics a soap; for example, sodium stearate is water soluble but calcium stearate is oil soluble. Either form may be used, depending on the details of the intended application. The free (unneutralized) form is also approved for certain uses.

Stearyl fumarate is a half ester of fumaric acid with stearyl alcohol (octadecanol). Although it might be expected to have dough-strengthening properties similar to SSL, this was not found to be so in practice and the product was not a commercial success.

The anionic structure shown in Figure 1 is sodium dodecyl sulfate (SDS), a sulfate ester of the C_{12} alcohol dodecanol. Commercially lauryl alcohol is produced by reduction of coconut oil. The alcohol portion of sodium lauryl sulfate is a mixture of chain lengths; the approximate composition is 8% C_8, 7% C_{10}, 48% C_{12}, 20% C_{14}, 10% C_{16}, and small amounts of longer chains. The most common use of

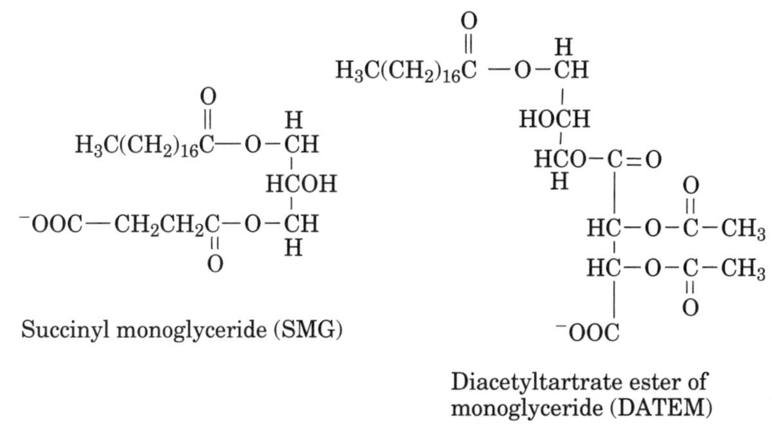

```
          O
          ||       H
H3C(CH2)16C—O−CH
                 |
               HCOH
                 |
−OOC—CH2CH2C—O−CH
            ||     H
            O
```

Succinyl monoglyceride (SMG)

```
              O
              ||       H
H3C(CH2)16C —O−CH
                  |
                HOCH
                  |
                HCO−C=O
                H     |      O
                      |      ||
                    HC−O−C−CH3
                      |
                    HC−O−C−CH3
                      |      ||
                      |      O
                   −OOC
```

Diacetyltartrate ester of monoglyceride (DATEM)

Figure 6. Anionic derivatives of monoglycerides.

Table 2. Regulatory Status of Emulsifiers

Emulsifier	US *21 CFR*	Canadian[a]	EU No.[b]
Mono- and diglycerides (GRAS)[c]	182.4505	M.4, M.5	E 471
Succinyl monoglyceride	172.830		
Lactylated monoglyceride	172.852	L.1	E 472
Acetylated monoglyceride	172.828	A.2	E 472
Monoglyceride citrate	172.832		E 472
Monoglyceride phosphate (GRAS)	182.4521	A.94, C.7	
Stearyl monoglyceride citrate	172.755		E 472
Diacetyl-tartrate ester of monoglyceride (GRAS)	182.4101	A.3	E 472
Polyoxyethylene monoglyceride	172.834		
Polyoxyethylene (8) stearate		P.5	
Propylene glycol monoester	172.854	P.14	E 477
Lactylated propylene glycol monoester	172.850		
Succinylated propylene glycol monoester (succistearin)	172.765		
Sorbitan monostearate	172.842	S.18	E 491
Sorbitan tristearate		S.18B	
Polysorbate 60	172.836	P.3	E 435
Polysorbate 65	172.838	P.4	E 436
Polysorbate 80	172.840	P.2	E 433
Calcium stearoyl lactylate	172.844		E 482
Sodium stearoyl lactylate	172.846	S.15A	E 481
Stearoyl lactylic acid	172.848	L.1A	
Stearyl tartrate			E 483
Stearyl monoglyceridyl citrate	172.755	S.19	
Sodium stearyl fumarate	172.826		
Sodium lauryl sulfate	172.822		
Dioctyl sodium sulfosuccinate	172.810		
Polyglycerol esters	172.854	P.1A	E 475
Sucrose esters	172.859	S.20	E 473
Sucrose glycerides			E 474
Lecithin (GRAS)	184.1400	L.2	E 322
Hydroxylated lecithin	172.814	H.1	E 322
Triethyl citrate (GRAS)	182.1911		

[a]Canadian Food and Drug Regulations Table IV, Div. 16.
[b]European Parliament and Council Directive No 95/2/EC - 20 Feb 95
[c]Generally recognized as safe.

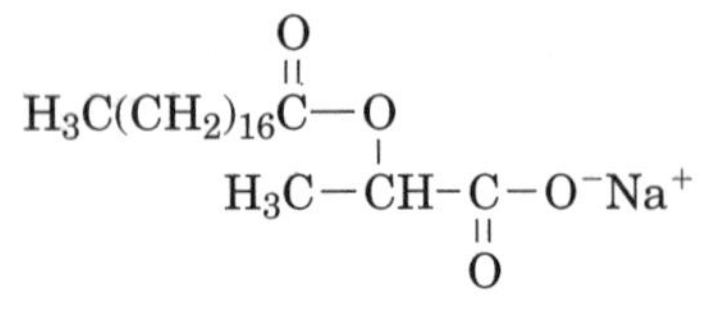

$$H_3C(CH_2)_{16}C(=O)-O-CH(CH_3)-C(=O)-O^-Na^+$$

Sodium stearoyl lactylate (SSL)

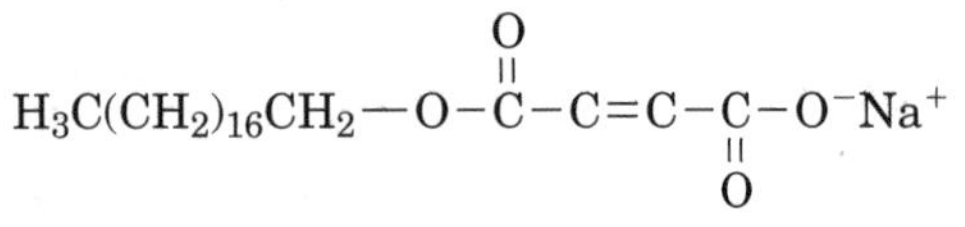

$$H_3C(CH_2)_{16}CH_2-O-C(=O)-C{=}C-C(=O)-O^-Na^+$$

Sodium stearyl fumarate

Figure 7. Sodium salts of two anionic emulsifiers. The third one, SDS, is shown in Fig. 1.

sodium lauryl sulfate is as a whipping aid. It is added to liquid egg whites at a maximum concentration of 0.0125%, or to egg white solids at a level of 0.1%. It promotes the unfolding of egg albumin at the air-water interface and foam stabilization as discussed earlier. (Triethyl citrate, not strictly an emulsifier, is also used as a whipping aid.)

Amphoteric Emulsifiers

Lecithin. During the purification of crude oil extracted from soybean, safflower, or corn germ a phosphatide-rich gum is obtained that is treated and purified to give the various commercial lecithin products available to the food processor today (12–14). The dark crude material is bleached to give a more acceptable light brown color. Treatment with up to 1.5% hydrogen peroxide gives the product known as single-bleached lecithin, and further addition of up to 0.5% benzoyl peroxide yields double-bleached lecithin (14). Reaction with even higher levels of hydrogen peroxide plus lactic acid hydroxylates unsaturated fatty acid side chains at the double bond (yielding, eg; dihydroxystearic acid from oleic acid) giving hydroxylated lecithin, which is more dispersible in cold water than the other types and is more effective as an emulsifier for oil-in-water emulsions

(15). Normally some vegetable oil is added to the lecithin to reduce the viscosity to 7.5 to 10 Pa·s; this is called standardized fluid lecithin and contains roughly one-third oil and two-thirds phosphatides.

Figure 8 shows the structure of the main surface active components of lecithin. The phosphatidyl group is a phosphate ester of diglyceride. The fatty acid composition of the diglyceride is similar to that of the basic oil (16) so a number of different fatty acids are found, not just the stearic and oleic acids depicted. There is little phosphatidylserine present in soybean lecithin, and the other three derivatives are found in approximately equal amounts. Phosphatidyl ethanolamine (PE) and phosphatidyl choline (PC) are amphoteric surfactants, whereas phosphatidyl inositol (PI) is anionic. The HLB values of the three species are varied, with PC having a high, PE an intermediate, and PI a low value. The HLB of the natural blend is around 9 to 10, and emulsifier mixtures having this value will tend to form either oil-in-water or water-in-oil emulsions, although neither one is highly stable. On the other hand, intermediate HLB emulsifiers are excellent wetting agents and this is a principal application for lecithin.

The emulsifying properties of lecithin can be improved by ethanol fractionation (15). PC is soluble in ethanol, PI is rather insoluble, and PE is partially soluble. Adding lecithin to ethanol gives a soluble and an insoluble fraction. The phosphatide compositions of the two are (*1*) ethanol soluble—60%, PC; 30%, PE; 2%, PI; and (*2*) ethanol insoluble—4%, PC; 29%, PE; 55%, PI (17). The soluble fraction is effective in promoting and stabilizing oil-in-water emulsions, whereas the insoluble portion promotes and stabilizes water-in-oil emulsions. At least one company (Lucas Meyer) is using this process to produce industrial food-grade emulsifiers.

Cationic Emulsifiers

Quaternary Ammonium Surfactants. Cationic emulsifiers are not used as food additives. Some of them, such as the quaternary ammonium compound cetyltrimethylammonium bromide (CTAB) (Fig. 1), have bactericidal properties and are used for sanitation of food-processing equipment. The structure is shown to indicate the chemical nature of this group of surfactants and to complete the description of classes of emulsifiers and surfactant.

USES OF EMULSIFIERS

The regulatory status of these food additives falls into one of three groups. The first group, Section 172 additives, includes those additives that have been examined and judged to be allowable for certain uses at concentrations not to exceed a set level. As an example, stearyl monoglyceridyl citrate (21 *CFR* 172.755) is allowed for use to stabilize shortenings containing emulsifiers. To use this additive as a wetting agent in a dry gelatin dessert mix, the manufacturer would have to submit a petition to the Food and Drug Administration (FDA) seeking clearance for this application and could not sell it for this purpose until such clearance was given. Several of the additives listed in Section 172 are cleared for general use in foods and therefore may be used in any food where they are functional, at a level not to exceed that amount that produces the desired functional effects.

Sections 182 and 184 of 21 *CFR* deal with additives that are generally recognized as safe (GRAS) and thus can be used in any food. In some cases the regulation allows use at any level that is consistent with good manufacturing practices, while for other additives a tolerance or maximum use level is established. Section 182 (Substances Generally Recognized as Safe) lists additives that are recognized as safe based on a history of use without reported safety problems. If an additive undergoes a specific detailed evaluation and its safety is affirmed, it is then moved to Section 184 (Direct Food Substances Affirmed as Generally Recognized as Safe).

The relevant U.S., Canadian, and European Union regulatory references are listed in Table 2 for most food emulsifiers. In many cases the maximum permissible amount of emulsifier is specified. In other instances, the manufacturer is expected to use no more than the amount necessary to provide the desired technological effect.

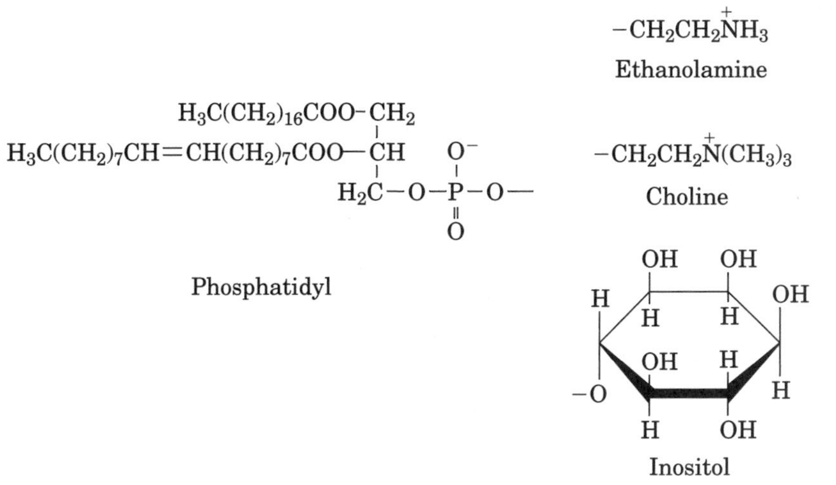

$$H_3C(CH_2)_{16}COO{-}CH_2$$
$$H_3C(CH_2)_7CH{=}CH(CH_2)_7COO{-}CH \quad O^-$$
$$H_2C{-}O{-}P(=O){-}O{-}$$

Phosphatidyl

$$-CH_2CH_2\overset{+}{N}H_3$$

Ethanolamine

$$-CH_2CH_2\overset{+}{N}(CH_3)_3$$

Choline

Inositol (ring substituents: OH, OH, H, OH, H, H, OH, H, H, –O, H, OH)

Figure 8. Surface active components of commercial lecithin. Phosphatidyl ethanolamine and phosphatidyl choline are amphoteric emulsifiers, while phosphatidyl inositol is anionic.

Many emulsifiers have functions that are not directly related to emulsification. For example, monoglycerides are widely used to retard firming (staling) of bread and similar baked goods. In this use the monoglyceride forms a complex with gelatinized starch in the bread crumb decreasing the rate at which it recrystallizes. Lecithin is added to melted chocolate intended for coating confectionery goods. The lecithin decreases the viscosity of the chocolate, improving coverage of the confectionery piece at lower use levels of the chocolate. Some emulsifiers (succistearin and polyglycerol esters) are used in oils to inhibit or modify fat crystal formation when the oils are cooled. The emulsifier deposits on the surface of the fat microcrystals (nuclei) and interferes with normal crystal growth.

CONCLUSION

Amphiphilic compounds (surfactants and emulsifiers) promote the formation of emulsions by lowering the interfacial free energy (surface tension γ) between the two phases, thus facilitating the subdivision of the internal phase into droplets of smaller diameter. They may be nonionic (monoglyceride), anionic (sodium lauryl sulfate), amphoteric (phosphatidyl choline), or cationic (cetyltrimethylammonium bromide); cationic surfactants are not used as food additives.

Emulsifiers stabilize emulsions by preventing the contact between the droplets that would lead to coalescence. This may occur by three mechanisms: (*1*) establishing electrostatic charges on the droplet surfaces, (*2*) generating a layer of bound water (in oil-in-water emulsions) around the surface of the droplet, and (*3*) forming a solid film at the interface.

Materials that stabilize foams usually perform both functions. They lower the surface tension between the liquid and air phases and also stabilize the thin liquid film between the occluded air bubbles. For food purposes, effective foaming agents are usually proteins (egg white and gelatin), but certain whipping agents can enhance this functionality.

Compounds that act as thickeners or bodying agents function in one of two ways. They may increase the viscosity of the aqueous phase, or they may form a gel network within that phase. Viscosity builders are usually gums or modified starches, while gelling agents can be gums, modified starches, or proteins. Some solid materials (eg, microcrystalline cellulose) increase the viscosity in a formulated product by adsorbing water and thus increasing the ratio of internal phase (solids, oil) to external phase (mobile water).

BIBLIOGRAPHY

1. J. C. Wootton et al., "The Role of Emulsifiers in the Incorporation of Air into Layer Cake Batter Systems," *Cereal Chem.* **44**, 333–343 (1967).
2. C. E. Stauffer, "The Interfacial Properties of Some Propylenc Glycol Monoesters," *J. Colloid Interface Sci.* **27**, 625–633 (1968).
3. H. Birnbaum, "The Monoglycerides: Manufacture, Concentration, Derivatives and Applications," *Bakers Digest* **55**, 6–18 (1981).
4. W. H. Knightly, "Surfactants in Baked Foods: Current Practice and Future Trends," *Cereal Foods World* **33**, 405–412 (1988).
5. D. T. Rusch, "Emulsifiers: Uses in Cereal and Bakery Foods," *Cereal Foods World* **26**, 111–115 (1981).
6. American Association of Cereal Chemists, *Approved Methods of the American Association of Cereal Chemists*, 8th ed., AACC, St. Paul, Minn., 1983, Method 58–45.
7. D. I. Hartnett, "Cake Shortenings," *J. Am. Oil Chem. Soc.* **54**, 557–560 (1977).
8. V. K. Babayan, "Polyglycerol Esters: Unique Additives for the Bakery Industry," *Cereal Foods World* **27**, 510–512 (1982).
9. C. E. Walker, "Food Applications of Sucrose Esters," *Cereal Foods World* **29**, 286–289 (1984).
10. B. A. Brachfeld, J. J. Geminder, and C. P. Hetzel, "Sodium Stearyl Fumarate: A New Dough Improver," *Bakers Digest* **40**, 53–58, 86 (1966).
11. J. B. Lauridsen, "Food Surfactants, Their Structure and Polymorphism," *AOCS Short Course on Physical Chemistry of Fats and Oils*, May 11–14, 1986, Hawaii.
12. W. J. Wolf and D. J. Sessa, "Lecithin," in M. S. Peterson and A. H. Johnson, eds., *Encyclopedia of Food Technology and Food Science*, AVI Publishing, Westport, Conn., 1978, pp. 461–467.
13. C. R. Scholfield, "Composition of Soybean Lecithin," *J. Am. Oil Chem. Soc.* **58**, 889–892 (1981).
14. B. F. Szuhaj, "Lecithin Production and Utilization," *J. Am. Oil Chem. Soc.* **60**, 258A–261A (1983).
15. W. Van Nieuwenhuyzen, "Lecithin Production and Properties," *J. Am. Oil Chem. Soc.* **53**, 425–427 (1976).
16. J. Stanley, "Production and Utilization of Lecithin," in K. S. Markley, ed., *Soybeans and Soybean Products*, Interscience Publishers, New York, 1951, pp. 593–647.
17. O. L. Brekke, "Oil Degumming and Soybean Lecithin," in D. R. Erickson et al., eds., *Handbook of Soy Oil Processing and Utilization*, American Soybean Association, St. Louis, Mo., 1980, pp. 71–88.

CLYDE E. STAUFFER
Technical Food Consultants
Cincinnati, Ohio

See also EMULSIFIER TECHNOLOGY IN FOODS.

ENCAPSULATION TECHNIQUES

DEFINITION AND TECHNOLOGY DEVELOPMENT

Encapsulate is most commonly defined in its verb form as the act of enclosing or when used as a noun, it is synonymous with capsule. Microencapsulate applies to enclosure, or encasement in microcapsules.

Terms that are often used interchangeably with encapsulate are coated, or protected, eg, coated leavening, protected salt, and coated acids. All three terms are used when referring to practically any core material that is encased or enclosed in an outer shell. However, there are subtle distinctions that often dictate which term is more properly used. Very small capsules, perhaps 100 μ and less, are practically always called capsules or microcapsules. The adjectives coated or protected usually imply a cheaper, more crudely manufactured capsule. Coated or protected particles are generally larger than 100 μ.

In the late 1800s a pharmacist named Upjohn patented a process still referred to as pan coating (1). Relatively large solid particles were tumbled in a cylindrically shaped mixer as liquid coating material was sprayed on. Typically a sugar in water solution was the liquid used and the process involved alternately spraying and evaporating water, until through deposition of sugar, the rounded polished capsules achieved the desired size, shape, and shell thickness. The process was originally applied to drugs and pharmaceuticals for the purpose of taste masking—sugar coated pills—and is still being used extensively. Its application has expanded into several well-known brands of candies and confections.

Most texts establish pioneer work in microencapsulation as having been done in the 1930s by the National Cash Register Company (Dayton, Ohio) (2). Using a chemical process often termed coacervation, National Cash Register developed a means of occluding a colorless dye in a cross-linked gelatin shell. The particles were deposited (less than 20 μ in a cross section) in a thin layer on the underside of a sheet of paper in contact with a second sheet impregnated with a colorless reagent. Pressure exerted on the laminate by the point of a pen or pencil fractured the microencapsulate releasing the reagent-sensitive dye, producing a colored image. The process, which became commercial in 1954, produced the world's first carbonless duplicating paper. In 1981 production of carbonless copy business forms exceeded 500,000 tons per year (3,4).

The National Cash Register invention and the new coacervation process created interest during the mid-1950s in encapsulation in a wide variety of fields from agriculture to cosmetics and toiletries, from electronic parts to food processing (5). The coacervation process itself was examined and, for some applications, was found technically suitable and economical. For many others, of the myriad potential uses that were conceived, coacervation had technical limitations and was much too costly. However, the interest in coacervation spawned development of new encapsulation technologies; some of which have been found entirely suitable for food ingredient applications.

SPRAY DRYING

Spray drying is probably the oldest encapsulation technique and the most often used for preparation of dry, stable, food additives, particularly flavors. Spray-dried flavors have been available since the mid-1930s (6). The process is economical, flexible, and adaptable to commonly used, readily available processing equipment. Generally it is accomplished in three stages. A flavor, commonly an oil, although water-soluble flavors are also fixed using this technique, is mixed with an edible, food-grade polymeric material, such as gelatin, vegetable gum, modified starch, dextrin, or nongelling proteins often in the ratio of one part flavor to four parts fixatives (5). An emulsifier is added and the mixture is homogenized to produce an oil-in-water emulsion with a small micelle structure (7). The emulsion is atomized using one of several different methods, introducing an aerosol into a column of heated air in a drying chamber. The small droplets develop a spherical shape with the flavor oil gravitating to the center, or core and the aqueous–hydrocolloidal phase forming the outside coating. Rapid evaporation of water from the coating's surface maintains a temperature below 100°C (212°F) even in air columns of much higher temperatures. Fortunately the dwell time (exposure to heat in the chamber) is relatively short, usually only a few seconds (6).

Water-soluble flavors (or other food ingredients) are also fixed or entrapped in edible hydrocolloids but the particles do not have a clearly defined core and coating. Rather the spray-dried particle tends to be homogeneous.

Spray-dried particles are usually produced in particle sizes less than 100 μ, an optimum range for easy water solubility, or dispersability. Where necessary spray dried particles may readily be agglomerated or granulated for dry mixes where greater uniformity of particle size is desirable.

Before the introduction of spray drying, dry flavors were produced by plating out liquid flavors on dry substrates, typically sucrose, dextrose, starches, salt, or gelatin, depending principally on the composition of the mix being flavored. Although this crude method provided good flavor rendition and was obviously inexpensive (really nothing more than mixing), it provided no protection against loss of volatile components and oxygen-sensitive fractions were readily degraded (5).

With the advent of spray-dried flavors, oxygen degradation was significantly reduced as was loss of volatiles through evaporation. The colloidal, polymeric coatings provided at least some measurable protection. Some commercial products began to be referred to as locked-in flavors or protected. Even encapsulated began to be used to describe some spray-dried flavors in the early 1950s.

While spray drying offered a step forward in obtaining stability, the process created a problem in obtaining good flavor rendition (from wet to dry). Air, hot enough to instantly vaporize water, is going to drive off at least some low-boiling molecules, and large volumes of hot air will cause some instant oxidation particularly with sensitive terpene-rich citrus oils. Some answers to the problems associated with spray drying flavors came from the flavor chemist who began tailoring liquid flavors using less oxygen-sensitive molecules and fortifying liquids with excess volatiles to allow for rounding off during drying.

Spray drying remains the most common method for producing dry flavors. Some highly sophisticated flavor labs believe emphatically that they can accomplish anything with spray dehydration that can be done with more recently developed encapsulation techniques. However, the inadequacies of the process are generally recognized and it is the desire for superior dry flavors that have spurred the search for new encapsulation techniques and spread these concepts into many other food ingredient applications.

SOLVENT DEHYDRATION

This encapsulation method involves the preparation of a flavor emulsion or solution with the same polymeric hydrocolloidal coatings used in making spray-dried flavors. The emulsions are homogenized and atomized directly into polar solvents such as isopropenol, ethanol, glycerin, or polyglycols, which act to dehydrate the aerosol particles. Resulting microcapsules are recovered by filtration and vacuum dried at low temperatures (8). The technique offers the benefit of tighter protected flavor capsules, somewhat higher payloads than spray drying, and excellent flavor rendition. On the down side processing equipment is expensive and specialized and processing costs are significantly higher than with spray drying.

ENCAPSULATION BY EXTRUSION

Encapsulation of flavors by extrusion incorporated the basic techniques of solvent dehydration with a novel method for providing excellent protection to the core materials. The process involves the preparation of a low-moisture (5–10%) melt (100°–130°C) of low DE malto dextrin, sugar, and modified edible starch. The flavor to be encapsulated is combined along with an emulsifier and optionally an antioxidant, and mixed with a melt under extreme agitation. The molten emulsion is extruded through holes about 1/64 in. into a cold, isopropanol bath. The melt solidifies and is broken into small rods by agitation. The flavor particles are recovered by filtration, or centrifugation, often mixed with a flow agent and packaged for sale. The literature discusses many variations in the process. Some are designed to increase the payload (typically 8–10% flavor); often flavor capsules are given additional solvent washes to remove traces of surface oils further improving the exceptional stability this process affords to deterioration by oxidation. As such it is particularly useful with citrus-flavored oils (9).

Development of today's encapsulation by extrusion solvent dehydration began more than 30 years ago. It really comes from a process developed in 1956 involving adding flavor oils to a molten solution of sucrose and dextrose, cooling the solution to a hard slab, and subsequently grinding the rock-candy-like product to the desired particle size (10,11). From the development of pulverized rock-candy flavor to the current state-of-the-art flavor encapsulation by extrusion, significant processing and product improvements have been made by a number of inventors covered by at least six important patents. At one point there were three companies in the United States selling what was advertised as extruded flavor capsules. The volume, however, although significant never reached the size that the quality of these encapsulates would seem to merit. High process cost (perhaps as much as six times the cost of spray drying) is the major factor. Improvements in spray-drying technology, particularly improvements in formulating flavors designed for spray dehydration have narrowed the quality gap. When a dry flavor is required spray drying is still the technology of choice unless the need for the superior quality of extruded capsules is great enough to warrant the added costs.

AIR SUSPENSION COATING

Perhaps more commonly termed fluid-bed processing, this technology probably accounts for the second largest commercial production of encapsulates for the food industry—second only to spray dehydration. Solid particles to be spray coated are suspended in a column of moving air of a controlled temperature. The coating material, which may be molten or dissolved in an evaporable solvent, can be a wide range of water-soluble or insoluble polymeric material, including starches, emulsifiers, dextrins, protein derivatives, and lipids atomized through spray nozzles into the air chamber and deposited on the particles of suspended core material. The movement from the bottom of the chamber through the aerosol to the top of the chamber is random allowing for a rather uniform coating of core material. The constant flow of air allows the molten lipid coating material impinging on the particle to cool and harden, or in the case of a solubilized coating, the solvent to evaporate. The partially coated capsule falls to the bottom of the chamber and the cycle is continued until the parameters for complete or near complete encapsulation have been achieved. The finished product is removed from the chamber, cooled, or perhaps put through a final drying procedure prior to packaging (2,4,12).

Fluid-bed encapsulation has achieved outstanding success as applied to food ingredients, probably because of its flexibility in being suitable for many different core materials. For a wide variety of coatings, it is cost-effective for many ingredient applications, although it is not inexpensive. Credit for the development process in the 1950s is generally given to Wurster, who, while a professor of pharmacy at the University of Wisconsin, used the technique for coating pharmaceutical tablets (13). The Wurster process has become popular, aided perhaps by the commercial availability of suitable solid particle fluidizing equipment. There have been many modifications, mostly centering around different methods to dispense coatings to achieve improved encapsulation or special product characteristics. In principle, however, they all adhere to the basic concept of Wurster, spraying aerosol droplets to impinge on and coat a solid particle (12).

CENTRIFUGAL EXTRUSION

The Southwest Research Institute (SWRI) is responsible for developing several generations of a unique encapsulating concept extending back into the 1960s. Two tubes are arranged concentrically to receive a flow of core material through the center tube and coating material through the outer orifice (the liquids are pumpled). The tubes converge in a common orifice (nozzle) through which is extruded a liquid column within a liquid column. The SWRI equipment actually includes multiple nozzles mounted on a rotating shaft so that they spin around a verticle axis of the shaft. Centrifugal force acts to break off the liquid shaft into small particles as they emerge from the nozzles. The process relies on surface tension to form the spherical core and for the outer liquid to form a continuous coating around the core.

Particles are collected in either solids (starches) or liquid solvents, which tend to cushion the impact and protect the particles and may also perform additional functions. Starches absorb excess moisture and coatings. Gelatin coatings may be hardened or solubility reduced through emersion in solvents containing suitable cross-linking agents (2,14).

COACERVATION

Coacervation is a colloidal chemical phenomenon first used as a microencapsulation technique in the early 1940s by Green of the National Cash Register Corp. It is considered by many to be the first true microencapsulation process.

Simple coacervation involves dispersing a colloid, eg, gelatin, in an appropriate solvent, eg, water. A core material, eg, a hydrophobic citrus oil, is dispersed in the mixture with agitation. By one or more method, such as the addition of sodium sulfate and lowering the temperature, the solubility of gelatin in water is reduced, creating a two-phase system. The colloid-rich phase appears as an amorphous cloud in the colloid-poor, more aqueous phase. If left to stand, the minute droplets would coalesce forming a separate liquid layer. However, under proper conditions (and most literature refers to coacervation as an art as well as a science), the coalescence of the polymeric colloid occurs around the suspended core particles of citrus oils, creating small, still unstable microcapsules. Final steps in the process include adding a suitable cross-linking, or hardening, agent such as glutaraldehyde; adjusting the pH; and subsequent collecting, washing, and drying of the now-stable citrus oil encapsulate.

Coacervation as described is also referred to as aqueous-phase separation, or oil-in-water encapsulation. Complex coacervation is possible only at pH values below the isoelectric point of gelatin. It is at these pH values that gelatin becomes positively charged, but gum arabic continues to be negatively charged. A typical complex coacervation process begins with the suspension, or emulsification, of core material in either gelatin or gum arabic solution. Then the gelatin or gum arabic solution (whichever was not used to suspend the core material) is added to the system while mixing continuously. The pH is adjusted to 3.8–4.3, the system is cooled to 5°C (41°F) and the gelled coacervate capsule walls are insolubilized with glutaraldehyde, or another hardening agent. Microcapsules are collected, washed, and dried.

It is possible to microencapsulate hydrophilic core material in oil-soluble coatings. A polar core is dispersed in an organic, nonpolar solvent at an elevated temperature. The coating material is then dissolved in the solvent. Encapsulation is achieved by lowering the temperature and allowing the polymeric coating material to emerge as a separate coacervate phase microencapsulating the core particles. The coating gradually solidifies and remains insoluble in cold solvent. This process is called either water-in-oil microencapsulation, or organic-phase separation. It is used typically in the pharmaceutical industry for encapsulation with ethylcellulose. However, ethylcellulose does not have general approval for use in the food industry.

Coacervation is an efficient, but expensive, microencapsulation technique. When small particle sizes are required it is probably the only process that can produce submicron-size particles. With typical payloads in the range of 85–95%, it might be expected that the process would be economical to allow for its application in the food industry for myriad ingredients, such is not the case. Aside from a few specialized flavor applications there are no current uses for encapsulated food ingredients by the coacervation process; costs are much too high (2–4).

INCLUSION COMPLEXATION

B-Cyclodextrin is a cyclic glucose polymer consisting of seven glucopyranose units linked in a $1 \rightarrow 4$ position. The molecule is a doughnut-shaped structure with a hollow center, which enables it to form complexes with many flavors, colors, and vitamins. The *B*-cyclodextrin molecule forms inclusion complexes in the presence of water with compounds that can fit dimensionally within its cavity. Water molecules within the nonpolar center of the *B*-cyclodextrin molecules are readily substituted by less-polar guest molecules. The resulting complex precipitates from the solution and is recovered by filtration and a subsequent drying step. Core material ranges from 6 to 15% (w/w) (15). The odorless, crystalline complexes will tenaciously bind the guest molecules up to temperatures of 200°C (392°F) without decomposition. The same complexes easily release the bound molecule in the temperature and moisture conditions of the mouth (16).

Inclusion complexes are stable to oxygen, light, and radiation, and even onion and garlic oil complexes are almost odorless. Cyclodextrins—while approved for food applications in Eastern Europe, where they were developed, and in Japan—are not approved for use in foods in Western Europe, or in the United States (17,18).

SPRAY CHILLING

Material to be encapsulated, usually dry solids of a relatively small particle size, are suspended in molten coating material, most commonly fractionated and hydrogenated vegetable oils. The mixture is atomized through spray nozzles into a chamber of tempered air (usually refrigerated) where the droplets solidify into small spherical particles. Spray chilling is applied principally to solid food additives including vitamins and minerals, and also to spray-dried flavors to retard volatilization during thermal processing. Spray-chilled products are commonly used in dry mixes. The process is relatively inexpensive. Spray-chilled capsules provide a controlled release of core material. For systems in which a more positive barrier or shell is required, other encapsulation techniques may be preferable (3,19,20).

BASIC CRITERIA FOR ENCAPSULATING FOOD INGREDIENTS

Economics

When microencapsulation was in its infancy in the early 1970s and food technologists were searching for scraps of information about the concept from any source they could find, there was no lack of apparently good ideas for developing new encapsulated ingredients. Few were so foolish to ignore the cost of encapsulating, but most inventive minds operated on the basis that the real problem was to develop the technology, ie, to make the encapsulate and adapt it to the system. It was generally known that once a prototype is made, the reduction in costs is just a matter of time.

Since 1970 the development of encapsulation technology has been startling; however, the cost of encapsulating has not decreased, but has kept pace with inflation. Some relatively crude, lower—cost methods of encapsulation have been developed, but these have been limited in application where a poor encapsulate was good enough.

Most FDA-approved food ingredients have been encapsulated on a laboratory or small pilot-plant basis, but the real world exists outside the laboratory and there commercial viability of an encapsulate revolves around economics in its broadest sense.

Encapsulated ingredient projects begin with a need or a problem. Sometimes the problem is a matter of poor shelf life. More often a project begins with the concept of an encapsulate that would allow the development of an entirely new product that cannot be available without it. Whatever the need for encapsulation, the wisest first step is not to analyze the various methods of encapsulation that might work, but to look at the project, really a processed food system, in its entirety. Assume that the encapsulate is available and technically satisfactory in all respects. Estimate the cost of the encapsulate, ie, ingredient, coating, and both minimum and maximum processing costs.

Then make a judgment: Under the best technical and economic conditions should this project be taken to commercialization? The answer may be no, and for a variety of reasons having nothing to do with encapsulation. The maze of approvals and evaluations that are prerequisites for taking new processes or products to commercialization can be formidable. But assume that in the best-case scenario the answer is yes. It is then just as important to take a look at the worst case, ie, a product that is more costly than desirable and technical requirements that are not quite fulfilled. A definite yes answer in this case makes the project much more attractive, but a no answer, although perhaps disappointing, at least more clearly defines what is really necessary. Projects having long odds may be abandoned, avoiding an investment and a high-risk gamble.

The evaluation of projects prior to embarking on costly research and development is obviously necessary for all new ideas regardless of whether encapsulation is required. However, there is particular wisdom in carefully evaluating encapsulation projects. All too often problems are solved technically with an encapsulate that never becomes commercial for reasons that may have been obvious from the outset. This is particularly true of systems where low-cost food ingredients have been encapsulated. When a formulator must work with a \$0.75/lb item that in its raw form costs only \$0.10/lb, it is amazing how many other solutions to the problem can be developed in place of encapsulation.

Most encapsulation companies charge for development and preparation of encapsulated prototypes; not to make a profit or even to recover expenses, but primarily to provide that extra insurance that a project has been well thought out and will have some reasonable chance of becoming commercial. Once technical objectives are met, commitment to any financial investment requires consideration and approvals at a relatively high managerial level.

ENCAPSULATED INGREDIENTS—COMMERCIAL TYPES AND APPLICATIONS

The encapsulated food ingredients that are commonly discussed in the literature may represent just the tip of the iceberg. There are more than 20 companies advertising services in encapsulation, granulation, and agglomeration (21). Much of their work is done on a custom basis, and volumes and applications are unknown. There is also a significant amount of encapsulation being done by companies for captive use on a proprietary basis. However, there has developed since the late 1960s a major market for encapsulated ingredients, in addition to the dry flavor business, that by that time had already been established.

Flavors

In 1971 it was estimated that more than 15,000,000 lb of dry flavors were produced domestically (5). This figure has at least tripled. The primary reason for flavor encapsulation is still to provide a dry form for liquid materials, but as technology has advanced, encapsulated flavors have been developed to provide outstanding flavor rendition, excellent protection from degradation through oxidation, and reduced or controlled volatility to the extent that significant economies can be achieved by reducing high temperature losses through evaporation of expensive flavoring ingredients. Technology has been applied to flavors as diverse as citrus oils, menthol, oleoresins of spices and herbs, cinnamaldehyde, peppermint, natural and artificial cheese and seafood flavors, and many other (2,5).

Acidulants

Encapsulated acidulants were first offered commercially in the early 1970s. Initially dry, granular acids including fumaric, malic, citric, tartaric, and adipic were encapsulated to improve stability of acid-sensitive ingredients in dry mixes. By encapsulating acids with coatings such as hydrogenated vegetable oils, or malto dextrin, it was possible to reduce contact between acids and starches in puddings and pie fillings, thereby reducing potential prehydrolysis and significantly extending shelf life. An attendant benefit with encapsulated citric acid was that it reduced hygroscopicity. Core acids were released with addition of water or milk to a mix or, in the case of products employing hydrogenated vegetable oil coatings, on heating (2,5,22).

Encapsulated acids have been used to prevent undesirable discoloration of acid-sensitive food colors and dyes, degradation of aspartame, control of dusting, as means of regulating the gelling effect of acidulants on acid-sensitive pectins in fruit-based products, and alteration of bulk density. The most recent, and by now the largest application, for encapsulated acidulants is in the meat industry where the use of encapsulated citric acid, lactic acid, and glucono-*delta*-lactone, a gluconic acid precursor, provide a means of direct acidification of processed meat. In the manufacture of fermented sausage the industry has long relied on naturally occurring microorganisms or on *Lactobacillus* starter cultures whose activity on carbohydrates produces lactic acid, thereby lowering the pH to a level adequate to ensure protection against pathogenic microorganisms. The fermentation process can take as long as four to five days, and although it has been used for hundreds of years, it is not without problems, primarily the length of fermentation time and the variability from batch to batch. The potential for direct acidification was obvious, but not easily accomplished. Contact of ground meat with acid causes rapid separation of fat from protein, discoloration, and a change in texture from a soft uniform emulsion to hard, brittle, crumbly particles that are no longer usable. Use of encapsulated acidulants having hydrogenated vegetable oil coatings with appropriate melting points offer a mechanism for releasing acidulants into meat slowly and uniformly throughout the mass as temperature is elevated during cooking. The fermentation step is eliminated. The concept has been expanded to meat systems where acidification is possible to replace high-temperature retorting with short-term pasteurization (23–26). The desirability of using lactic acid as the acidulant by some technologists has extended the creativity of the encapsulators to include liquid lactic acid, which is first absorbed on solid calcium lactate particles prior to tight encapsulation in hydrogenated vegetable oil.

Dough-Conditioning Agents

Encapsulation technology has been applied to conditioners for protection and to control activity until the optimum point in the baking cycle is reached. Ascorbic acid, which is used as an oxidizing agent in strengthening and conditioning bread doughs, is unstable in the presence of water and degrades rapidly. Encapsulated ascorbic acid can be added to typical prefermenting broths with minor loss of potency, through the fermentation stage. During baking the hydrogenated vegetable oil coating melts, releasing the ascorbic acid for its conditioning–strengthening effect at the stage where it is most effective (27).

Leavening

Shelf life of chemically leavened doughs is restricted by reaction prior to proofing or baking. Dry, packaged mixes lose leavening activity particularly in high-moisture systems. The same is true of chemically leavened frozen doughs. Refrigerated and microwaveable frozen doughs offer unique leavening problems in that activity needs to be controlled, not simply prevented. Refrigerated doughs require some leavening initially when the dough is formed, but ideally no more reaction through the packaging stage. Once in the closed containers and held under suitable proofing conditions, rapid leavening is desirable to purge the seal and provide maximum can pressure. Microwaveable frozen products require initial protection from reaction, with rapid, intense leavening during the very brief time that a rise may occur within the microwaving process.

Encapsulated sodium bicarbonate with precise release characteristics in combination with selected leavening acids offers a commercial means of providing protection, and optimizing activity in a wide variety of baking systems. Encapsulation has also been applied to leavening acids, particularly glucono-*delta*-lactone for similar applications.

Protection of Yeast-Leavened Baked Goods

Sensitivity of yeast cells to a variety of ingredients used in baked products has created a highly specialized need for encapsulates. Even at low flavor levels acidulants, eg, citric, lactic, and acetic acids, will reduce the viability of yeast and may further affect the cellular structure of dough by reducing the total effect of yeast leavening. Preservatives such as sorbic acid and calcium proprionate, and even salt may have a similar killing effect on yeast as on the molds they are intended to inhibit. Citric, lactic, and acidic acids encapsulated in hydrogenated vegetable oil (HVO) are being used on a commercial basis to prevent the effect on yeast of low pH, as is HVO-coated sorbic acid. Other encapsulated additives that affect yeast negatively in the raw form have been produced and evaluated on a developmental basis with promising results. Potentially dried yeast cells themselves may one day be commercially available in an encapsulated form. Work has been done along this line, but apparently with limited success. The coating may be extremely useful in maintaining the viability of the cells, but encapsulation itself can have an inhibiting effect on yeast, and certainly the benefits of encapsulation would have to be major to warrant the added cost.

Vitamins and Minerals

A variety of dry mixes are fortified with vitamins and minerals including cereals, infant formulas, pet foods, and beverages. Encapsulation with both lipids and water soluble coatings are used to reduce off-flavors, permit controlled release in the digestive tract, improve flow properties, facilitate tableting, and enhance stability, particularly of vitamins, to extremes of moisture and temperature. Iron is encapsulated as are other minerals to control bulk density, reduce stratification in combination with lighter mix components, and to retard catalytic oxidative rancidity.

One use for an encapsulated vitamin provides such a major product improvement with an accompanying economic advantage that it merits special comment. The aquaculture industry has grown enormously in the last decade. The key to maintaining the health of pond-grown fish is in proper feeding. A number of nutrient supplements are added to fish rations, but one stands out prominently in any discussion of the applications for microencapsulation technology. Fish must ingest vitamin C in an adequate dosage for survival. Fish along with humans, apes, guinea pigs, and the fruit bat lack the enzymes required to synthesize ascorbic acid from protein consumed in their diet. Using raw ascorbic acid for feed fortification requires 5–10 times the level recommended in the normal diet of fish because oxygenated water is used in all three of the basic processes for producing fish feed, which rapidly reduces the potency of ascorbic acid. This reduction is accelerated with temperature and pressure elevation, which is required in all three processing techniques.

Encapsulation of vitamin C in a HVO coating prevents contact of the oxygen-sensitive ascorbic acid with water. The half-life of raw ascorbic acid in fish feed is measured in days. The half-life of the digestible, bioavailable, encapsulated version of vitamin C is measured in months (28–30).

Sodium Chloride

There has been a significant, although not major market, for coated salt as a food ingredient since the late 1960s, primarily in the meat industry where it is used in ground pork sausage where contact with raw salt will affect texture and turns the bright pink color to a dull bluish gray. Fat-coated salt, even when crudely applied, extends shelf life and the improvement is worth the additional cost at least to many manufacturers. As technology has improved so has the quality of HVO-encapsulated salt. Although still restricted from becoming a principal ingredient in the food industry because of the economics involved, applications have been extended to baked products to protect yeast and to retard the stiffening effect induced by raw salt through water absorption. Use in comminuted meat products has also increased to control changes in viscosity and texture produced by water absorption. Maintenance of a free-flowing viscosity of ground beef allows the cost of encapsulation to be offset by savings provided by increased throughput.

FUTURE DEVELOPMENT

A new and better technique to microencapsulate food ingredients may be developed in the next 10 years, and perhaps the new process will be more economical than the current state of the art. However, if the available technology stays the same, the market for encapsulates in the food industry will certainly keep growing at an accelerating rate. The benefits are significant.

The industry is becoming accustomed to what encapsulation can do and as the confidence level grows, so do the opportunities. What is perhaps equally important is knowing what the new technology can do and knowing what it can't do. Fortunately the food industry no longer looks at encapsulation as some far out technology that might be considered if all else fails. The industry has matured to the extent that products are tailored to fit specific applications. It is interesting to observe that with one common core ingredient—citric acid—one of the major ingredient producers offers 11 different grades, the other producer offers 8. Each one differs in variables that are fundamental to encapsulation, eg, payload, particle size, coatings, water solubility, or fat coatings. Other important differences include the coatings, melting points, origins, and costs. Each product is designed to be useful in a specific food ingredient system.

While commercialization of encapsulation technology covers many diverse fields within the broad term Food Industry, the majority of products and the greatest volume is focused on two segments, ie, meat processing and baking. It is interesting to observe that in both, encapsulation has been treated not as a tool to be tried and abandoned if it doesn't perform adequately the first time, but rather as a part of the answer to a food-processing problem, albeit an important part. Most initial attempts to solve problems with encapsulates are only partially successful. It is in the fine tuning, not only of the encapsulates to fit a system but, where possible, of the system's design to fit what can be achieved through encapsulation. The meat and baking industries have been doing this. Other industries such as the dairy and microwaveable entrée industry would seem to be candidates for a similar developmental approach.

A significant part of encapsulation technology is in the selection or development of optimum edible Food and Drug Administration approved coatings to enclose and protect and still release under the desired conditions. There is room for new coating materials, particularly ones that will have higher melting points or will remain intact under higher temperatures than what is currently available. Edible coatings having release points of ca 74°C (165°F) would be particularly interesting.

BIBLIOGRAPHY

1. R. J. Versic, "Flavor Encapsulation, An Overview," *ACS Symposium Flavor Encapsulation, Series 370*, ACS, Washington, D.C., 1988, Chap. 1, p. 5.
2. J. D. Dziezak, "Microencapsulation and Encapsulated Ingredients," *Food Technology* **42**(4), 137 (Apr. 1988).
3. R. J. Versic, "Coacervation for Flavor Encapsulation," in Ref. 1, Chap. 14, pp. 126–131.
4. R. E. Sparks, "Microencapsulation," in M. Grayson, ed., *Kirk-Othmer Encyclopedia of Chemical Technology*, Vol. 15, 3rd ed., John Wiley & Sons, Inc., New York, 1981.
5. R. E. Graves, "Uses for Microencapsulation in Food Additives," *Cereal Science Today* **17**(4), 107, 1972.
6. G. A. Reineccius, "Spray-Drying of Food Flavors," in Ref 1, Chap. 7.
7. J. Brenner, "The Essence of Spray Dried Flavors: The State of the Art," *Perfumer and Flavorist* **8**, 40 (Apr.–May 1983).
8. R. Zilberboium, I. J. Kopelman, and Y. Talmon, "Microencapsulation by a Dehydrating Liquid: Retention of Paprika Oleoresin and Aromatic Esters," *Food Science* **1**(5), 1301 (1986).
9. S. J. Risch, "Encapsulation of Flavors by Extrusion," in Ref. 1, pp. 103–109.
10. U.S. Pat. 2,809,895 (Oct. 15, 1957), H. E. Swisher (to Sunkist Growers, Inc.).
11. T. H. Schultz, K. P. Dimick, B. Makower, "Incorporation of Natural Fruit Flavors into Fruit Juice Powder," *Food Technology* **10**, 57–60 (1956).
12. D. M. Jones, "Controlling Particle Size and Release Properties," in Ref. 1, pp. 158–176.
13. U.S. Pat. 2,648,609 (Aug. 11, 1953), D. E. Wurster (to Wisconsin Alumni Research Foundation).
14. R. E. Lyle, D. J. Mangold, and W. W. Harlow, "The Making of Microcapsules," *Technology Today* **3**(3), 13–16 (Sept. 1904).
15. H. B. Heath and G. A. Reineccius, "Flavor Production," in *Flavor Chemistry and Technology*, AVI Publishing Co., Inc., Westport, Conn., 1986.

16. J. S. Pagington, "Molecular Encapsulation with *B*-Cyclodextrin Food Flavor Ingredients," *Process Packaging* **7**(9) 50 (Sept. 1985).
17. G. A. Reineccius and S. J. Risch, "Encapsulation of Artificial Flavors with *B*-Cyclodextrin Food Flavor Ingredients," *Process Packaging* **8**(9), 1 (Aug.–Sept. 1986).
18. L. Szente and J. Szejti, "Stabilization of Flavors by Cyclodextrins," in Ref. 1, Chap. 16, pp. 148–157.
19. J. A. Bakan and J. S. Anderson, "Microencapsulation," in L. Lachman, H. A. Lieberman, and J. L. Kanig, eds., *The Theory and Practice of Industrial Pharmacy*, Lea & Febiger, Philadelphia, 1970, p. 384.
20. D. Blenford, "Fully Protected Food Flavor Ingredients," *Process Packaging* **8**(8), 43 (July 1986).
21. H. Weiss, president, Balchem Corp., unpublished data.
22. L. E. Werner, "Encapsulated Food Acids," *Cereal Food World* **25**(3), 102 (1980).
23. R. E. Graves, "Sausage Fermentation: New Ways to Control Acidulations of Meat," *National Provisioner* **198**(21), 32–34, 49 (May 1988).
24. J. Bacus, "Fermenting Meat, Part I," *Meat Processing* **24**(2), 26–31.
25. J. Bacus, "Fermenting Meat, Part II," *Meat Processing* **24**(3), 32–35 (Mar. 1986).
26. G. J. Jedlicka, "Chemical Acidulation of Semi-Dry Sausage." *paper presented at American Meat Institute Convention*, New Orleans, 1984.
27. U.S. Pat. 3,959,496 (May 25, 1976) S. Jackel and R. V. Diachuk (to Baker Research and Development Co.).
28. C. Andres, "Encapsulated Ingredients," *Food Processing* **38**(12), **44** (Nov. 1977).
29. L. A. Gorton, "Encapsulation Stabilizes Vitamin C in Cupcakes," *Baking Industry* **148**(1805), 20–21 (Jan. 1981).
30. C. J. Pacifico, "A Novel Approach to Vitamin C Nutrition for Fish Rations," *Aquaculture Today* **2**(4), 20–22 (1989).

Bob Graves
Herb Weiss
Balchem Corporation
Slate Hill, New York

ENERGY USAGE IN FOOD PROCESSING PLANTS

There is a continuing concern about the energy demand by our food system. It is estimated that 17% of U.S. energy consumption is attributed to the food system (1). This figure includes energy used for production through processing, distribution, out-of-home preparation, and in-home preparation. The food industry requires energy for a variety of equipment such as gas fired ovens; dryers; steam boilers; electrical motors; refrigeration units; and heating, ventilation, and air-conditioning systems.

Natural gas is the predominant source of energy used by the U.S. food industry. In the last few decades about 50% of the gross energy used in food processing was from natural gas; 15%, from fuel oil, 13%, from electricity; and about 22%, from propane, butane, other petroleum products, coal, and some renewable energy sources (2). Within the food industry, the principal types of energy use include direct fuel use, steam, and electricity. Nearly 50% of energy use is in the form of direct fuel use. Almost 30% of energy is used to process steam and 10% to heat water. Almost 67% of electrical energy consumption is for generating mechanical power to operate conveyors, pumps, compressors, and other machinery. Refrigeration equipment consumes about 17% of electricity; lights consume about 10%; and heating, ventilation, and air-conditioning use approximately 4% of electricity.

Between 1973 and 1986, the cost of energy escalated dramatically. This trend promoted the installation of heat recovery equipment to conserve energy (3). It also focused the attention of scientists and engineers on the economic feasibility of alternative energy sources in various food-processing situations like cheese processing (4), food dehydration (5), and water heating (6). Energy consumption patterns of various food industries are briefly discussed.

MILK PROCESSING

Considerable variations in energy requirements for milk processing have been reported for different plants. Fuel requirements for producing pasteurized bottled milk ranges from 0.25 to 2.65 MJ/L of milk (7). The energy used for processing cheese during the regular plant operation is approximately 2.13 kJ/kg. Excluding drying and evaporation, energy required to produce 1 kg of cheese is in the range of 3.37 to 17.53 MJ (4). Some of this variation is due to a lack of concern to energy conservation, but there are also operational factors that may explain variations among plants.

Considering energy allocation to individual unit operations separately in a typical processing of pasteurized bottled milk, most of the energy is used in pasteurization step (approximately 23% of electricity and 35% of fuel). In the production of pasteurized bottled milk, bottle washing consumes approximately 58% of the fuel energy and 6% of the electricity used in the process.

In butter production pasteurization consumes approximately 30% of the fuel energy. The churning process and cold storage of butter consume approximately 50% of the electricity required for butter processing.

In yogurt production energy for heating and cooling per kilogram of product averages 1,146 kJ/kg (8). The heating of the base yogurt from 10 to 87.8°C uses 80% of the energy, and the electrical equipment used in product handling and packaging consumes only 6%. The manufacturing process of sour cream is similar to that for yogurt, except the amount of heating is considerably less because the highest product temperature is 22.2°C for sour cream compared to 87.8°C for yogurt. In packaging, the shrink wrap machine is not used for holding the sour cream cartons. The thermal energy ratio is 273 kJ/kg of sour cream and packaging uses less electrical energy primarily because of the lack of the shrink wrap operation. Product handling and packaging uses 49 kJ/kg of sour cream compared to 78 kJ/kg of yogurt. Total energy, thermal plus electrical, is 287 kJ/kg for sour cream compared to 1,224 kJ/kg for yogurt. The above values of energy to manufacture sour cream and yogurt exclude the following: the energy required to pasteurize the raw milk, the energy required to pump the base material into the processing vat, and the energy required in temporary cold storage prior to shipping. Yogurt in the processing vat is cooled from 87.8 to 42.2°C partially with well water. Measurements of the energy taken from the yogurt by the well water showed an average of 212 kJ/L of yogurt in 4.16 m^3 batch. The cooling could be considered free, except for the cost of the water and pumping energy (8).

Spray drying is the common means of converting fluid feedstocks into solids in the form of powders, granules or agglomerates. It is widely used for dehydration in the manufacture of a wide variety of food, pharmaceutical, and chemical products. Among the foods that have been successfully spray dried are milk, whey, cheese, coffee whitener, eggs, soups, baby foods, and fruits. Energy costs constitute a significant fraction of the operating costs. A number of methods have been suggested to decrease energy related costs during the spray-drying process by using insulation to decrease heat losses to the environment and heat recovery from exhaust gases (8). Typically, in the manufacture of spray-dried milk powder, air heating for the spray-drying operation requires between 50 and 80% of total fuel consumption and approximately 30–35% of total electricity consumption in the whole process. The variation in the primary fuel input results from heat recovery steps.

Cleaning in place (CIP) in milk processing plants is one of the operations that takes significant input of energy. In the manufacturing of Cheddar cheese, CIP consumes approximately 31% of the total fuel energy and 18% of the total electricity required for the whole process. In other processing operations such as milk pasteurization, creamery butter, and acid casein processing, the fuel consumption in CIP operation does not exceed 3% of total fuel used in the whole process, and the use of electricity is up to approximately 1% of total electric energy used for the whole process. In spray-dried milk powder production, CIP operation costs approximately 5–8% of the total fuel consumption and 4–5% of the total electricity used in milk powder processing.

LIQUID FOOD CONCENTRATION

Evaporation of liquid foods is most commonly accomplished using multieffect evaporators. The word *effect* indicates vapor flow in the evaporator. The total evaporation costs depend on the steam consumption, the specific area of the evaporator, and the specific product losses (9). The energy efficiency of an evaporator is usually expressed by thermal or steam economy, which is defined as a ratio of quantity of water evaporated to quantity of steam consumed (10). The steam economy of an evaporator increases with the increase in number of evaporator effect, because subsequent effects use vapor from the previous effect as the heating medium. As an example, a single-effect evaporator takes 1.5 kilogram of steam to evaporate 1 kg of water, whereas a double-effect evaporator only consumes 0.75 kg of steam to evaporate the same amount of water (11).

In conventional multieffect evaporators, those without mechanical or thermal recompression, the steam evaporated from the first effect is used as the heating medium for the next effect and so on. A mechanical vapor recompression evaporator is very much like any other evaporator except that steam is recycled. A compressor is installed, which takes the vapor off an effect and pressurizes it to a higher pressure for use as the heating medium. The major energy expense in such a system is electricity. The operating economy of a mechanical vapor recompression unit is equivalent to an 8–20-effect evaporator. All vaporization is carried out at the same temperature. Steam from the boiler and water are only needed for startup, which accounts for the tremendous savings.

The dairy industry carries out the evaporation of whole milk, skim milk, buttermilk, and whey. The steam economy of two-stage falling film plate evaporator with mechanical vapor recompression, when used to concentrate buttermilk and skim milk to 25% solids, was reported to be 6, which was claimed by the evaporator manufacture (12) to be equivalent to an 8–30-effect conventional evaporator, depending on the cost of fuel. The estimated fuel savings were expected to be enough to amortize the rotary compressor in less than one year (12). A 7-effect mechanical vapor recompression evaporator can remove 17.5 kg of water from the feed for each kilogram of steam supplied to the turbine, which gives the steam economy equal to 17.5 (13) while concentrating whey, skim milk, or whole milk.

Frozen concentrated orange juice (FCOJ) is consumed widely in the United States, European Common Market countries, Japan, Venezuela, and other countries. São Paulo, Brazil, and Florida are the two leading producers of FCOJ. Temperature accelerated short-time evaporator (TASTE) evaporators are used extensively for concentrating orange juice (14).

Energy consumption in a four-effect, seven-stage (the word *stage* indicates the flow of orange juice in the evaporator) TASTE evaporator with a capacity of evaporation of 18.140 kg H_2O/h ranged between 840 and 1,000 kJ/kg H_2O evaporated (15). By introducing one effect and one stage after the fourth stage, the magnitude of energy consumption was reduced to 570–640 kJ/kg of water evaporated. In most orange juice evaporators, the flow of orange juice and the steam pressure are controlled manually to obtain concentrated juice of the desired degree Brix (percent sucrose equals degree Brix). By means of automatic control, the energy consumption in a six-effect, eight-stage evaporator with a capacity to evaporate 9,070 kg H_2O/h was reduced by 6.7%

The data on energy consumption during the 1978–1981 processing season in a concentrated orange juice plant in Brazil were presented for a plant with a capacity to produce 70,000 t/yr of 65° Brix of frozen concentrated orange juice and 86,000 t/yr of citrus pulp pellets made from orange residues. The plant had on-site storage facilities for 17,600 t/yr of frozen orange juice at −8°C. At a separate location away from the plant, 13,650 t/yr of concentrated orange juice were stored at −25°C. For the plant as a whole, electricity was accounted for only 10% of the total energy consumed in the plant. Most of the steam generated was used to concentrate orange juice in evaporators; a small portion (16%) was used in the pelletizing operation of citrus pulp (16).

Evaporators are also used to produce tomato products with varying degrees of concentration. The primary products are classified as purées (solids content from 11% to 22%) and paste (solids content from 28% to 45%). A number of energy audits on tomato product evaporators have been performed (17). Daily average performance data are given for single-, double-, and triple-effect evaporators. Average daily steam economies for the single-effect evaporator averaged 0.84 compared to a theoretical average of 0.95. Two double-effect evaporators had average daily steam economies from 0.79 to 2.03. The average daily steam economy measured was 1.45 while the theoretical average was 1.91. As expected, the triple-effect evaporators showed the best steam economies ranging from 1.66 to 3.06. The lower ranges of steam economy were caused by frequent fouling of the heat exchangers by the burning product. The recent evaporators installed in tomato processing plants have four effects. In the multieffect evaporators the energy cost is directly proportional to the temperature difference across each effect. If the temperature difference is cut in half, twice as many effects can be put in, cutting the energy consumption approximately in half. It will also approximately double the equipment cost (18).

Beets are approximately 15% sugar of which about 88% is extracted giving 132 kg sugar/t of beets. The various sugar beet processors in California provided data that showed that for each ton of beets, 3.19 GJ of primary energy is consumed directly in processing (19). Dried pulp, a by-product of sugar beet processing, accounted for large amounts of energy used. Pulp drying requires 0.73 GJ/ton of sugar beets. This value is nearly 23% of the total direct energy consumed during processing of sugar beets. The evaporators used in the sugar concentrating process consumed nearly 44% of the total direct energy required.

Energy requirements were monitored during the production of a beet colorant on a pilot scale (20). Colorant production involved centrifugation, concentration of ferment, vacuum concentration, and spray drying. The first operation was juice extraction. The beets were blanched, abrasion peeled, and washed. They were then dried and immediately comminuted. The product was pumped to a centrifugal separator for separation. The clarified juice was sent to a plate heat and exchanger pasteurized at 88°C for 2 min, then cooled to 2°C. Extraction of the juice from 1 t of beets used 69.3 ± 0.2 kWh of electricity and 226 ± 13 kg of steam. It took 95 ± 5 min and yielded 550 ± 45 kg of juice at 7% solids. Then the juice was vacuum concentrated to 14% total solids. This used 343 ± 14 kg of steam and 5.1 ± 0.2 kWh of electricity.

The second operation in the production of beet colorant was the fermentation of the clarified beet juice. The temperature was maintained at 30 ± 2°C during the continuous fermentation. The juice extracted from 1 t of beets used 10.2 ± 0.2 kWh of electricity. The fermentation removed about 80% of the solids (20).

In the third operation, the fermented juice was centrifically separated. The clarified juice was vacuum concentrated in a single-effect falling film evaporator to 30% soluble solids, cooled to 16°C, and spray dried. The yeast cell slurry was also spray dried. The vacuum evaporation used 396 ± 14 kg of steam and 4.6 ± 0.2 kWh of electricity to concentrate juice from 1 t of beets. Spray drying the juice took 29.6 ± 1.5 m^3 of natural gas and 2.8 ± 0.2 kWh of electricity per ton of beets. Spray drying of yeast cells took 26.6 ± 1.5 m^3 of natural gas and 2.5 ± 0.2 kWh of electricity per ton of beets.

FOOD BLANCHING

Blanching is an important unit operation in the pretreatment of fruits and vegetables. It is used prior to freezing, canning, and dehydration. The blanching process is a relatively fast process. Blanching time is a function of piece size, heating medium, and temperature as well as material packing in the blancher. The most common heating media are steam and water. Design of a blancher has a significant influence on the energy used. The open steam blanchers are approximately 10% less efficient than the sealed units (21). Steam losses from the unsealed entrance and exit account for almost 80% of the energy input to the blancher. The steam blancher with end seals is similar in design to the conventional open steam blancher. Water sprays are positioned inside the ends of the blancher to liquify steam that might escape. They also serve to help cool the product at the exit. In such a design the steam losses are up to approximately 51% of the thermal energy input to the blancher. The thermal energy required to blanch spinach using a conventional steam blancher is 2.12 MJ/kg of product (21). Spinach processed in a blancher with hydrostatic seals requires only 0.95 MJ/kg of product (21). In a commercial blancher, at a spinach processing plant, energy used is 6.5 MJ of natural gas and fuel oil and 0.072 MJ of electricity per 1 kg of raw spinach (22).

Energy consumption data for three commercial water blanchers (tubular, screw conveyor, and water tank) indicate that type of product and type of blancher dictate the energy usage (21,23). For lima beans processed in a tubular water blancher, thermal energy usage is 0.543 MJ/kg of processed beans and thermal efficiency is 44.5%, the highest of all three water blanchers tested (23).

The screw conveyor water blancher processed cauliflower cuts. Thermal energy consumption averages 0.91 MJ/kg of product and its thermal efficiency is 31.2%. Energy accounting data on a commercial blancher of a water tank type at the spinach processing plant indicate that 31% of the energy input goes to heating the product and 69% is lost through incomplete condensation of steam, hot water discharge to a drain, and heat losses by convection and radiation (22).

POULTRY PROCESSING

Poultry processing consists of several operations: live bird holding, hanging, slaughtering, scalding, defeathering, eviscerating and chilling, grading, cutup, and packaging. All processes require some energy either as fossil fuel or electricity. Electricity is used for conveyors, refrigeration, lighting, air-conditioning, pumps, and the mechanical drives. Fossil fuels are used for space heating, production of steam or hot water, and feather singeing. The electrical demand for three selected poultry processing plants indicates that the refrigeration and other mechanical drives account for more than 80% of the total calculated electrical use. Lighting requires 6.8% and fans 7.8% of the total electricity. The rest of the total electrical energy is used for heating and air-conditioning of the plants (24). Calculating the electrical use by major area in those three processing plants, the average electrical use in the eviscerating and chilling areas together is 6.7% of the total. This percentage does not include electricity used for ice making and water chilling. Packing and shipping uses 1.6% of the total electricity, which is about 3.8 kWh/1,000 head (total electricity consumption is 237.5 kWh/1,000 head); but this does not include refrigeration for storage areas. Offal and waste handling equipment and waste treatment use 12.7% of the total electricity, and shops and services account for 11.9% of the total electricity use (24).

In another study (25), the electrical energy use per 1,000 broiler equivalents processed was 212 kWh. The cost of ice making, used for a water chiller, and in transportation accounted for approximately 45% of the total electricity consumption. It was noted that 7.2% of the electricity was used in clean-up operations.

FISH PROCESSING

The fresh and frozen packaged fish industry consumes 0.6–0.8% of the energy required by the food-processing industry (26). The fish plant manufactures fillets; dressed, gilled, and puffed fish; and ground fish scrap, which is sold as animal feed. Electrical energy consumption makes up 80% of the plant's total energy requirements (26). Natural gas is used for space heating purposes. Electrical energy is primarily used to operate the refrigeration equipment for the chilling and freezing of whole fish and processed fish products. In addition, it is used to a lesser extent to power motor drives and air-conditioning equipment, to generate hot water for cleaning purposes and to supply lighting. The average energy used ranges between 3.93 and 4.89 MJ/kg of fish product (26).

Ammonia reciprocating compressors are the largest single electricity consumers in the ice production plant. Electrical energy is the main source of energy in that production, representing approximately 85% of its energy requirements. The reported annual energy use in the ice industry was between 581 and 697 kJ/kg in the mid-1970s. This ratio has decreased in recent years due to more energy-efficient equipment and a shift to higher production rates of fragmentary ice and is reported as 380–420 kJ/kg of ice (26).

CANNING

The energy use for canning of tomato products was obtained from the daily amount of tomato received by the plant and the total daily energy consumption (2). Tomato processing consists of several operations. At the receiving station the tomatoes are removed from the gondolas with the use of water. The water containing field dirt of the tomatoes is pumped to a mud-settling tank. The tomatoes are conveyed in a hydraulic flume for additional washing and initial inspection. Water is recirculated in the hydraulic flume. The majority of the electrical energy at the receiving station was consumed by the pump conveying water to the mud-settling tank. The energy use of the receiving operation was 0.32 Wh/kg of raw tomatoes. Using data from six days of processing a thermal energy intensity value of 1,251.4 kJ/kg of tomatoes was calculated. This value represents processing of tomatoes into three products: tomato juice, canned peeled tomatoes, and tomato paste. The electrical energy intensity value is 0.025 kWh/kg of raw tomatoes.

Production of tomato juice involves crushing tomatoes, heating them rapidly to inactivate enzymes, filling juice into cans, and retorting the cans. Most operations consume electricity. Hot-break heaters and retorts use steam. Combining the thermal energy and electrical energy data to a similar base unit of kilojoules, the hot-break heaters and retorts account for almost 97% of the energy consumed in the processing line and the energy intensity of tomato juice production was 1,086.2 kJ/kg of raw tomatoes (2).

Processing of canned, peeled tomatoes requires lye-bath peelers to facilitate peel removal and retorts to sterilize the canned product and conveying equipment. Lyebath peelers and retorts consume steam. These two operations account for almost 99% of the total energy consumed (2). All other equipment operates with electricity. The energy intensity is 1,300.2 kJ/kg of raw tomatoes.

In tomato paste production, several operations such as sorting, pulping, and finishing are similar to tomato juice production. Steam is used in vertical heat exchangers to preheat tomato juice. A large quantity of steam is used in the evaporator. The evaporator and the heat exchanger used in preconcentration account for the majority of steam consumption. The energy intensity of tomato paste production in the cannery industry is 1,307.2 kJ/kg of raw tomatoes (2).

As a consequence of the seasonal vegetable or fruit production, most of the energy consumption is also seasonal. Energy is also consumed during the nonprocessing season, which adds to the energy cost of processed fruits and vegetables. A study of four vegetable canneries located in Western, New York (27) producing apple sauce, canned beans and carrots, indicate that, on the average, blanching and sterilization consume 9% (theoretical estimate); lighting and electric motors, 21%; and movement of vehicles, 4.1%. The rest of the energy is lost in different ways. A significant portion, 22.6%, is lost in steam generation; 21% is lost to the environment after steam generation; and 22.3% is spent on other activities (losses to start up and operation of process equipment prior to actual processing and inefficiencies of process equipment).

Peach canning involves receiving peaches in bins, dumping fruit into a water tank and then elevating it out with a conveyor. The peaches are then graded for size, rinsed with water and pitted. After pit removal, the peach halves are oriented cup down on a belt conveyor and conveyed to a lye-bath peeler. After exposure to caustic solution for a predetermined time, the fruit are again rinsed. The fruit are allowed to orientate cup up; are reinspected, are size-graded, and filled into cans along with syrup. The canned fruit are then heated in retorts to achieve the desired sterilization (2). Within various operations, transport of fruit and waste consume the most electricity. It was determined that 23% of the total electrical energy consumption occurs in dry conveying whereas 38% is consumed in pumping water to convey fruit and waste products (2). Retorting and lye-bath peeling are the most energy-consumptive unit operations, accounting for 89% and 10%, respectively, of total energy consumption. The energy used by the three lye-bath peelers is 135.7 kJ/kg of canned peaches.

There are four energy-intensive operations in citrus packing: degreening–precooling, wash–wax–drying, holding, and storage. Energy analysis of four citrus packing plants in California showed that electricity and natural gas are two major types of energy source (28). Orange and lemon packing differ primarily in the duration of storage required before shipment. Oranges are shipped as soon as they are harvested and packed in a plant. Storage is used mainly to chill and hold the fruit until shipment. This holding time does not exceed a few days. Lemons might be stored from two weeks to six months. Another important distinction between an orange and a lemon packing plant is in the degreening operation. Lemons degreen naturally in storage, oranges are put into a chamber with air containing ethylene at 18–21°C (28).

In the unload–dump unit, field bins of fruit are conveyed from a receiving area to the dump station where the fruit drop into a water tank. Fruit are conveyed in a water flume to the wash–was–dry operation, which consists of four steps: manual sorting and removal of rotten fruit, washing and rinsing, waxing and surface moisture removal by heated air, and manual sorting of fruit for juice processing. The above sequence is followed when water-based waxes are used. In cases when solvent-based waxes are used, the surface moisture is removed both prior to and after waxing.

The total energy intensities for orange and lemon packing vary between 0.64 and 1.03 MJ/kg (28). Electrical energy intensities are about the same in citrus packing plants (0.55–0.82 MJ/kg). Natural gas intensities, however, vary from 0.07 to 0.3 MJ/kg. Another study on consumption of energy in lemon, orange and grapefruit packing plants indicates that total energy intensities for plants have a range of 362.6–702.4 kJ/kg (29). Electrical power intensities vary from 163.3 to 583.3 kJ/kg. Natural gas use intensities for citrus packing vary from 53.8 to 159.8 kJ/kg. The possible reasons for the differences in energy used for packing are different harvest times, the refrigerated holding costs, and storage time.

BAKING INDUSTRY

The largest consumers of energy in a bakery are the ovens, heating boilers, steam generators, and refrigerators. The differences among the energy balances of the bakeries can be large. They are influenced, for example, by the size of the bakery, production structure, amount of production, location, and apparatus.

Based on energy measurement in two bakeries in Sweden, energy consumption of 13.96 MJ/kg of bread for a bakery with a capacity of 250,000 kg of bread per year and 4.88 MJ/kg bread for another bakery with a capacity of 3,500,000 kg of bread per year was reported (30). An investigation conducted in the United States (31) reported an energy consumption for the baking industry of 7.26 MJ/kg bread baked. This is based on measurements on a bakery with a capacity of 35,000 kg of bread per day. In another study, an energy consumption of three bakeries on an average was found to be 6.99 MJ/kg bread baked (32).

The difference in the energy consumption figures may be due to several factors including the size of the bakery. A small baker uses more energy per unit production than a large one. If the bakery has many different products, this will also cause an increase in the energy use. The transportation costs are also important. It was shown (32) that 13–21% of the total energy use in the bakeries was for the delivery vans. Thus the distribution range causes differences in energy use between different bakeries.

A German oven manufacturer reported that the energy use of their oven (a tunnel oven with convective heat transfer) was as low as 0.6 MJ/kg baked bread. This low figure is probably explained by an optimal use of the oven and the convective heat transfer. Much less energy consumption was mostly due to less ventilation.

Considering the energy distribution in bread making, the heating of pans and lids used 26% of the energy. The rest of the energy use lost either through ventilation (31%), exhaust gases (13%), or radiation and convection losses from the walls (30%) (33). These figures show that there are some major energy-conserving opportunities in the baking industry. These steps include minimization of the ventilation of the oven, use of materials with lower heat capacities in the pans and lids and use of heat exchangers to recover heat from the hot exhaust gases.

The British-style crumpet is a small, round bakery product made of unsweetened batter and cooked on a griddle. Traditionally, the product is marketed in the frozen form. In 1982, a modified-atmosphere technique was introduced to package crumpets (34). The aim of modified-atmosphere packaging was to reduce energy consumption without adversely affecting quality during storage and marketing of the product. The production of the crumpets involves mixing, pumping, baking, cooling, packaging, freezing, and storage. Electricity for mixing and pumping of batter and power for running associated motors of the griddle, refrigeration units, and packaging machines together total 92.1 kWh for every 1,000 kg of crumpets produced (35). Thermal energy requirements for process heating (baking) and cooling of 1,000 kg crumpets are 1,050 and 160 MJ, respectively. The results (35) indicate that the freezing process requires 440 MJ, an additional 29% energy to produce and cool the crumpets.

A major advantage of modified-atmosphere packaging over frozen crumpets is that the former requires no special conditions for storage, whereas, for the latter, refrigeration is necessary. Storage of 1,000 kg of crumpets uses 7.4 kWh of electricity and 642.8 MJ of refrigeration for 30 days in storage. In terms of total thermal equivalents, frozen storage adds approximately 43% more to the consumption of process energy for crumpets above that of the modified-atmosphere packaging system.

BIBLIOGRAPHY

1. *Energy Consumption in the Food System* report XIV, 13392-007-001, Federal Energy Administration, Washington, D.C., 1975.
2. R. P. Singh, "Energetics of an Industrial Food System," in R. P. Singh, ed., *Energy in Food Processing*, Vol. 1, Elsevier Science Publishing Co., Inc., New York, 1986.
3. D. P. Donhowe, C. H. Amundson, and C. G. Hill, Jr., "Performance of Heat Recovery System for a Spray Dryer," *Journal of Food Process Engineering* **12**, 13–32 (1989).
4. R. K. Singh, D. B. Lund, and F. H. Buelow, "Applications of Solar Energy in Food Processing I: Cheese Processing," *Transactions of the American Society of Agricultural Engineers* **26**, 1562–1569 (1983).
5. R. K. Singh, D. B. Lund, and F. H. Buelow, "Application of Solar Energy in Food Processing II: Food Dehydration," *Transactions of the American Society of Agricultural Engineers* **26**, 1569–1574 (1983).
6. R. K. Singh, D. B. Lund, and F. H. Buelow, "Application of Solar Energy in Food Processing IV: Effect of Collector Type and Hot Water Storage Volume on Economic Feasibility," *Transactions of the American Society of Agricultural Engineers* **26**, 1580–1583 (1983).
7. B. Elsy, "Survey of Energy and Water Usage in Liquid Milk Processing," *Milk Industry* **82**(10), 18–23 (1980).
8. G. H. Brusewitz and R. P. Singh, "Energy Accounting and Conservation in the Manufacture of Yogurt and Sour Cream," *Transactions of the American Society of Agricultural Engineers* **24**, 533–536 (1981).
9. S. Bouman, D. W. Brinkman, P. de Jong, and R. Waalewijn, "Multistage Evaporation in the Dairy Industry: Energy Savings, Product Losses and Cleaning," in S. Bruin, ed., *Preconcentration and Drying of Food Materials*, Elsevier Science Publishing Co., Inc., New York, 1988, pp. 51–60.
10. A. H. Zaida, S. C. Sarma, P. D. Grover, and D. R. Heldman, "Milk Concentration by Direct Contact Heat Exchange," *Journal of Food Process Engineering* **9**, 63–79 (1986).
11. D. P. Lubelski, S. L. Clark, and M. R. Okos, "Process Modifications to Reduce Energy Usage," in *Energy Management and Membrane Technology in Food and Dairy Processing, Proceedings from the Special Food Engineering Symposium Held in Conjunction with Food and Dairy Expo 83*, American Society of Agricultural Engineers, St. Joseph, Mich., 1983, pp. 7–13.
12. "MVR Evaporator Provides $60,000 Annual Fuel Cost Savings," *Food Process* **40**(1), 92 (1979).
13. P. Standford, "A Milestone in Evaporation Systems," *Dairy Rec.* **84**(7), 89–90, 92, 94, 96, 98 (1983).
14. C. S. Chen, "Citrus Evaporator Technology," *Transactions of the American Society of Agricultural Engineers* **25**, 1457–1463 (1982).
15. C. S. Chen, R. D. Carter, and B. S. Buslig, "Energy Requirements for the TASTE Citrus Juice Evaporator," in *Energy Use Features*, Vol. 4, Pergamon Press, New York, 1979, p. 1841.
16. J. Filho, A. Vitali, C. P. Viegas, and M. A. Rao, "Energy Consumption in a Concentrated Orange Juice Plant," *J. Food Process Eng.* **7**, 77–89 (1984).
17. T. R. Rumsey, T. T. Conant, T. Fortis, E. P. Scott, L. D. Pederson, and W. W. Rose, "Energy Use in Tomato Paste Evaporation," *Journal of Food Process Engineering* **7**, 111–121 (1984).
18. R. W. Cook, "Multiple Effect Evaporation," in R. F. Matthews, ed., *Energy Conservation and Its Relation to Materials Handling in the Food Industry*, Proceedings of the 15th Annual Short Course for the Food Industry, 1975, pp. 78–82.
19. P. K. Avlani, R. P. Singh, and W. J. Chancellor, "Energy Consumption in Sugar Beet Production and Processing in California," *Transactions of the American Society of Agricultural Engineers* **23**, 783–787, 782 (1980).
20. J. E. Block, C. H. Amundson, and J. H. Von Elbe, "Energy Requirements of Beet Colorant Production," *Journal of Food Process Engineering* **5**, 67–75 (1981).
21. E. P. Scott, P. A. Carroad, T. R. Rumsey, J. Horn, J. Buhlert, and W. W. Rose, "Energy Consumption in Steam Blanchers," *Journal of Food Process Engineering* **5**, 77–88 (1981).
22. M. S. Chhinnan, R. P. Singh, L. D. Pedersen, P. A. Carroad, W. W. Rose, and N. L. Jacob, "Analysis of Energy Utilization in Spinach Processing," *Transactions of the American Society of Agricultural Engineers* **23**, 503–507 (1980).
23. T. R. Rumsey, E. P. Scott, and P. A. Carroad, "Energy Consumption in Water Blanching," *Journal of Food Science* **47**, 295–298 (1981).
24. W. K. Whitehead and W. L. Shupe, "Energy Requirements for Processing Poultry," *Transactions of the American Society of Agricultural Engineers* **22**, 889–893 (1979).
25. L. E. Carr, "Identifying Broiler Processing Plant High Electrical Use Area," *Transactions of the American Society of Agricultural Engineers* **24**, 1054–1057 (1981).
26. L. G. Enriquez, G. J. Flick, and W. H. Mashburn, "An Energy Use Analysis of a Fresh and Frozen Fish Processing Company," *Journal of Food Process Engineering* **8**, 213–230 (1986).
27. W. Vergara, M. A. Rao and W. K. Jordan, "Analysis of Direct Energy Usage in Vegetable Canneries," *Transactions of the American Society of Agricultural Engineers* **21**, 1246–1249 (1978).
28. M. Naughton, R. P. Singh, P. Hardt, and T. R. Rumsey, "Energy Use in Citrus Packing Plants," *Transactions of the American Society of Agricultural Engineers* **22**, 188–192 (1979).
29. L. P. Mayou and R. P. Singh, "Energy Use Profiles in Citrus Packing Plants in California," *Transactions of the American Society of Agricultural Engineers* **23**, 234–236, 241 (1980).
30. C. Trägårdh and co-workers, "Energy Relations in Some Swedish Food Industries," in P. Linko and co-workers, eds., *Food Process Engineering*, Elsevier Applied Science Publishers, Ltd., Barking, UK, 1980.
31. L. A. Johnson and W. J. Hoover, "Energy Use in Baking Bread," *Bakers Digest* **51**, 58 (1977).
32. G. A. Beech, "Energy Use in Bread Baking," *Journal of the Science of Food and Agriculture* **31**, 289 (1980).
33. A. Christensen, and R. P. Singh, "Energy Consumption in the Baking Industry," in B. M. McKenna, ed., *Engineering and Food*, Vol. 2, Elsevier Applied Science Publishers, Ltd., Barking, UK, 1983, pp. 965–973.
34. B. Ooraikul, "Gas-packaging for a Bakery Product," *Canadian Institute of Food Science Technology Journal* **15**, 313 (1982).
35. N. Y. Aboagye, B. Ooraikul, R. Lawrence, and E. D. Jackson, "Energy Costs in Modified Atmosphere Packaging and Freezing Processes as Applied to a Baked Product," in M. Le Maguer and P. Jelen, eds., *Food Engineering and Process Applications*, Vol. 2, Elsevier Applied Science Publishers, Ltd., Barking, UK, 1986, pp. 417–425.

S. Cenkowski
D. S. Jayas
University of Manitoba
Winnipeg, Manitoba
Canada

ENTERAL FORMULAS AND FEEDING SYSTEMS

Using a feeding tube to deliver nutrients directly into the stomach or upper gastrointestinal (GI) tract can be a lifeline for people who are unable to consume an adequate oral diet. Nutrient administration by vein (parenteral nutrition) is another option. Recent research results showing the superior physiologic benefits of enteral feeding over parenteral and favorable third party reimbursement have encouraged clinicians to use enteral tube feeding whenever possible, reserving parenteral nutrition for those with totally dysfunctional GI tracts or in whom enteral feeding tubes cannot be placed (1). Further, the availability of specially designed enteral feeding tubes that can be placed at the bedside, during surgery, or percutaneously using radiologic, endoscopic, or laparoscopic techniques has simplified feeding tube placement and management. These advances plus the availability of a wide variety of standard and condition-specific enteral formulas have made it possible to enterally feed many patients who previously would not have been considered candidates for enteral nutrition support (Table 1).

ENTERAL FORMULAS

Enteral formulas can be categorized as standard or condition-specific, based on their nutrient composition. Generally, the caloric distribution of standard enteral formulas are consistent with recommendations for a healthy diet for the general population (2). Carbohydrate contributes 50 to 60% of calories, protein contributes 12 to 20% and fat about 20 to 30% of calories. Standard enteral formulas are designed to also provide 100% of recommended intakes (RDI) of vitamins and minerals in 1 to 2 liters. They are comprised of intact nutrients similar to those occurring in a usual oral diet (Table 2) that require digestion and absorption just as solid food would. Standard enteral formulas are also called polymeric formulas to indicate that they contain long-chained macronutrients that require digestion prior to absorption. The most widely used standard enteral formulas also contain fiber.

Initially fiber was added to help moderate bowel function in chronically tube-fed patients. Dietary fiber increases stool bulk as a result of the water-holding capacity of the fiber, the osmotic effect of the short-chain fatty acids produced by bacterial fermentation, and the bacterial mass itself (3). Dietary fiber also moderates transit time; that is, the time taken by undigested particles to move through the GI tract. An increase in dietary fiber hastens transit time in people with slow transit and can prolong it in people with too rapid a transit rate (3). Because of this moderating effect, adding fiber to the diet can be effective for both diarrhea and constipation. To supply fiber in enteral products, soy polysaccharide was found to be the most compatible with liquids and well tolerated by patients.

Table 1. Some Conditions When Tube Feeding Can Be Used

Dysphagia
Upper GI tract obstruction
Malabsorption syndromes
GI disease
Critical illness or injury
Pancreatitis
Short bowel syndrome
Glucose intolerance
Renal dysfunction
Cancer
Pulmonary dysfunction

Table 2. Common Components of Enteral Solutions

Carbohydrates	Proteins	Lipids
Maltodextrin	Casein	Corn oil
Corn syrup	Whey	Canola oil
Sucrose	Soy protein isolate	Safflower oil
Fructose	Egg white solids	Soybean oil
Cereal solids	Skim milk	Marine oil
Fruit/vegetable purees	Pea protein	Fractionated coconut oil
		Butterfat

In 1987, the Federation of American Societies of Experimental Biology prepared a review of the known physiological effects of fiber (4). Based on their findings, they recommended an intake of dietary fiber of 10 to 13 g/Cal comprised of 70 to 75% insoluble fiber and 25 to 30% soluble fiber. This was felt to best represent the types of fiber that would be found in a healthy diet containing a variety of foods. Within the past several years, new fibers have become available for use in enteral nutrition products. A recent advance in enteral formulations then has been the addition of fiber blends that provide both insoluble and soluble fibers at these recommended levels. These fiber blends may include soy fiber, oat fiber, carboxymethylcellulose, hydrolyzed guar gum, gum arabic, or pectin. Some indigestible oligosaccharides, such as the fructooligosaccharides, although not technically classified as a fiber, have recently been added to enteral products because they are completely soluble, physically compatible with the product components, and provide fiberlike benefits (5).

People being fed standard enteral formulas must have adequate GI function to process nutrients and fiber. People who have malabsorption syndromes or other medical conditions (eg, short bowel syndrome, pancreatitis, inflammatory bowel disease) that compromise the gut's digestive capability may benefit from use of partially hydrolyzed or elemental formulas. The caloric distribution of these formulas is similar to that of standard enteral formulas. The nutrient sources, however, are partially hydrolyzed, rendering them smaller and easier to digest. For example, amino acids and/or peptides may be used instead of intact proteins, glucose polymers instead of complex carbohydrates. Medium chain triglycerides, fatty acids that contain 8- or 10-carbon fatty acids, often comprise at least part of the lipid component of hydrolyzed formulas. Because they are well absorbed and transported directly into the portal blood, they bypass the complex steps required for digestion and absorption of long-chain triglycerides, thus overcoming problems associated with inability to absorb long-chain fatty acids (6).

Standard formulas are available in caloric densities ranging from 1.0 to 2.0 kcal/mL. Caloric requirements or medical condition dictates the caloric concentration of the formula to be used. For example, a relatively stable adult could be provided with an adequate daily intake of protein, essential fatty acids, minerals, electrolytes, vitamins, and water in about 2,000 mL of a 1 kcal/mL product. Contrast this with an elderly patient with congestive heart failure who has a much lower daily calorie requirement and needs fluid restriction, but still must have a full complement of protein, vitamins, and minerals. In this case, a more concentrated formula would be desirable.

Generally the amount of water contained in a formula is inversely related to its caloric density. Most 1.0 kcal/mL formulas are approximately 80% water; 2.0 kcal/mL formulas may contain as little as 50% water. Ordinarily, most people need about 1 mL water/kcal fed (7) or a minimum of 1,500 mL of water unless medically contraindicated for people who need fluid restrictions (8). In most cases, all tube-fed patients need to have water in addition to formula.

Standard enteral formulas meet the metabolic needs of most patients. There are some conditions, however, when special formulations are needed. Some of these special formulations, called condition- or disease-specific formulas, are discussed in the following section.

CONDITION-SPECIFIC NUTRITIONALS

Hypermetabolic Stress or Trauma

Severe injury causes abrupt changes in metabolism. Critically ill patients experience a metabolic response to injury that is divided into what is known as an ebb phase and a flow phase (9). During the ebb phase, blood pressure, cardiac output, body temperature, and oxygen consumption are reduced, usually in conjunction with hemorrhage. This depression in metabolic rate may be a protective mechanism during this period of hemodynamic instability.

As blood volume stabilizes, the ebb phase is replaced by the flow phase. The acute response is a hypermetabolic state characterized by increased cardiac output and urinary nitrogen losses, altered glucose metabolism, and accelerated catabolism. Levels of the catabolic stress hormones, glucocorticoids (10,11), glucagon (12–14), and catecholamines (15–16) are increased. These hormonal conditions favor the breakdown of muscle tissue to provide amino acids for gluconeogenesis and hepatic protein synthesis. The physiologic consequence is a rapid loss of nitrogen and lean body tissue after severe injury. The levels of catabolic hormones decrease gradually during the adaptive flow phase. Recovery continues throughout this response, which is associated with the predominance of anabolic processes such as wound healing.

Historically, parenteral nutrition predominated in the care of critically ill patients because of the routine availability of central venous access and standardized techniques. Health care professionals preferred parenteral nutrition because it bypassed the GI tract, which they assumed to be totally dysfunctional in the critically ill. However, evidence indicates that enteral nutrition support confers benefits not seen with parenteral nutrition such as reduced incidence of infection and maintenance of gut structure and function. As a result, clinicians are increasingly advocating tube feeding for critically ill patients, whenever possible.

Recent nutrition research has focused on the clinical benefits of enterally feeding hypermetabolic patients. Several trials have reported significant decreases in infections and/or wound healing complications (17–20), hospital length of stay (17,20,21), and multiple organ failure (19) in patients fed specialized enteral formulas compared with patients fed standard enteral formulas. These specialized enteral formulas contain ingredients shown to enhance the immune response, improve wound healing, reduce free radical formation, and alter the inflammatory response. Included are the amino acids arginine (22) and glutamine (23) as conditionally essential nutrients during trauma, omega-3 fatty acids (24), and antioxidants (25–27).

Pulmonary Disease

Patients with chronic obstructive pulmonary disease (COPD) develop various degrees of respiratory muscle fatigue, hypoventilation, carbon dioxide retention (hypercapnia), and oxygen depletion (hypoxemia). In the majority of patients who experience difficulty in breathing, the obstruction of airflow is due to chronic bronchitis or emphysema (28). Typically, these patients experience marked weight loss, which may result both from decreased caloric intake (29,30) and from increased energy expenditure, presumably because of increased work of breathing (29). Because the oxidation of fat results in less carbon dioxide production than the oxidation of a equal caloric load of carbohydrate, formulas high in fat and low in carbohydrate have been developed for pulmonary patients. Such formulas have been shown to benefit patients with COPD, leading to an increase in general activity (31).

Acute respiratory distress syndrome (ARDS) is an often fatal form of acute lung injury that results from direct (eg, drowning) or indirect (eg, sepsis, trauma, pneumonia) injury to the lungs. Patients with ARDS have distinctive nutritional requirements due to their hypermetabolic status and presence of pulmonary inflammation. ARDS patients typically have elevated energy needs due to trauma and/or sepsis (32). However, overfeeding must be avoided as it may contribute to respiratory failure by increasing minute ventilation and dead space ventilation (33). On the other hand, inadequate nutrition can actually prolong the need for use of a ventilator if the respiratory muscle mass is broken down to provide energy and protein substrates for the body (34,35). This further contributes to increased complication rates, mortality, and prolonged length of hospital stay among patients on mechanical ventilation (34).

To address the nutritional needs of these patients, a formula was designed containing ingredients that could help control or reduce lung damage from inflammation and oxidation. Preclinical studies have shown that high-fat, low-carbohydrate diets containing omega-3 fatty acids from fish oil and γ-linolenic acid from borage oil rapidly modulate the fatty acid composition of inflammatory cell membranes (36–38), reduce the synthesis of important proinflammatory mediators of lung injury (36,39,40), attenuate lung permeability (36,40) and improve cardiopulmonary hemodynamics and respiratory gas exchange in models of ARDS (36,39). In a multicenter clinical trial, patients receiving this formula experienced a significant improvement in gas exchange and clinical outcome (41).

Cancer

Cachexia or wasting as a result of altered metabolism or decreased intake is a hallmark of cancer. The presence of a malignant tumor can cause a number of physiologic changes that can affect nutritional status. Some tumors cause an increase in basal metabolic rate (42), and changes in glucose (43) and protein (44) metabolism have been reported. Fluid and electrolyte imbalances (45), malabsorption (45), and anorexia (46) are other reported problems. There also can be serious nutritional consequences of cancer therapy: surgery, chemotherapy, or radiation therapy can cause nausea and vomiting, which can sometimes be treated with antiemetic drugs (45,47–49). Creative approaches are needed to maintain food intake, such as offering small high-calorie meals to anorexics or pureed foods to patients with painful swallowing (50,51). If oral intake is not satisfactory, tube feeding may be necessary; patients with obstructions of the nasopharynx or the esophagus can be fed in different sites (see "Delivery Systems").

The maintenance of adequate dietary intake in cancer patients does have benefits. Oral supplementation and tube feeding can help prevent weight loss during therapy (52,53), which is strongly related to the positive outcome of therapy. Aggressive tube feeding during chemotherapy can improve hematopoietic status (54), which may be related to more active chemotactic and phagocytic function (55). Peptide elemental diets have been shown to provide protection against radiation damage to the intestinal mucosa (56).

Diabetes

Although controversy continues to exist regarding the optimal diet for individuals with diabetes, most health care diabetes specialists use the nutrition recommendations issued by the American Diabetes Association. These latest guidelines (57) specify that protein supply 10 to 20% of total caloric intake, saturated fatty acids less than 10% of total calories, and polyunsaturated fatty acids less than 10% of total calories. Thus, 60 to 70% of total calories remain to be divided between carbohydrate (CHO) and monounsaturated fatty acid (MUFA) intake. The distribution is individualized based on nutritional assessment and treatment goals. Nevertheless, individuals who are at a healthy weight and have normal lipid levels are encouraged to use the nutrition recommendations of the National Cholesterol Education Program, in which individuals over 2 years of age are advised to limit fat intake to ≤30% of total calories (58,59). The rationale for these recommendations is the desire to lower saturated fat and cholesterol intake to reduce cardiovascular risk (60), and this is often difficult to achieve in a "typical" diet without lowering total fat intake. Most standard enteral formulas now available comply with these recommendations. However, tube feeding or dietary supplementation with standard liquid medical nutritional products can compromise the metabolic control of patients with abnormal glucose tolerance. This occurs because rapid gastric emptying of liquid formulas and efficient absorption of nutrients supplied by these products result in a rapid elevation of blood glucose. Standard enteral formulas empty from the stomach at least twice as fast as an isocaloric solid-food meal (61) and have been shown to produce a peak blood glucose response equivalent to that seen when an equivalent solution of pure glucose is fed (62). Rapid gastric emptying of formulas and rapid absorption of nutrients can complicate metabolic control of hyperglycemic patients. A number of clinical studies have shown that high-carbohydrate, low-fat diets may be inappropriate for many patients with diabetes (63–65). Moreover, recent research has demonstrated that substitution of MUFA for CHO can significantly improve glycemic control, lower serum triglycerides, and maintain or increase high density lipoprotein (HDL)-cholesterol, and raise the HDL-cholesterol to low density lipoprotein (LDL)-cholesterol ratio (66–68). Therefore, carbohydrate-reduced, high-monounsaturated fat formula seems a reasonable choice for patients with abnormal glucose tolerance.

Renal Disease

Once an initial injury to the kidneys causes chronic renal failure, renal disease typically progresses toward end-stage renal disease until dialysis or kidney transplantation are required, even if the cause of the initial damage is mitigated. Accordingly, nutritional management of people with renal disease depends on the level of residual renal function. The goals of nutritional management are to optimize nutritional status, minimize the burden of the diet on the person with renal failure, and to the extent possible, modify the nutrient profile to compensate for the nutritional and metabolic abnormalities associated with the loss of renal function (69). For persons with residual renal function, there is an additional goal of delaying progression of renal disease (69).

One of many features of chronic renal failure is a gradual increase in toxic nitrogenous wastes, the by-products of protein and amino acid metabolism. Protein restriction is advocated to optimize nutritional status (69) and help compensate for the metabolic effects of the loss of renal function (70–72). Although somewhat controversial, a moderate protein restriction also may limit the rate of disease progression (73), though nutritional status must be monitored closely to prevent undernutrition (74). Typical recommendations for protein intake are 0.7 to 0.8 g protein/kg/d for normally nourished people with chronic renal failure (69). Protein requirements are elevated in persons with end-stage renal disease on either hemodialysis or peritoneal dialysis due to increased nitrogen loss during dialysis. Typical recommendations are 1.0 to 1.5 g protein/kg/d for stable persons receiving dialytic therapy (69,75,76).

Energy requirements of people with chronic or end-stage renal disease are comparable to those of healthy adults, approximately 35 kcal/kg/d (76–78). Inadequate energy intake is one of the primary causes of wasting in this population (77). In addition to protein and energy, loss of renal function alters metabolism and excretion of many other nutrients. Recommendations for dietary fiber and fat intakes correspond to those for normal populations (69). Fluid and electrolyte restrictions typically are required for people treated with hemodialysis and are sometimes necessary during peritoneal dialysis (69). With the loss of renal function, phosphate excretion is impaired, necessitating both dietary phosphate restriction and administration of phosphate binders, typically in the form of calcium salts. Enteral products used in end-stage renal disease should contain nutrient levels tailored to meet the distinctive requirements of this population (79). Supplemental levels of vitamins A, C, and D typically are limited to RDI levels due to abnormalities in their metabolism (hypervitaminosis A (69,80–83), oxalosis (84–87), and loss of the ability to activate vitamin D (88), respectively). In contrast, supranormal levels of folic acid (89–91) and pyridoxine (92–94) are required. Metabolism of taurine (95,96), carnitine (97–99), selenium (100–102), and magnesium (69) is altered by loss of renal function. Deficiencies of the first three may develop, and supplements containing levels found in typical diets may be used. In contrast, hypermagnesemia is common and recommended intakes are approximately half of the RDI (69). Clinical research demonstrates that people on dialysis tolerate sole-source enteral nutrition using disease-tailored products and that these products may offer some biochemical advantages over standard nutritionals in the management of this population (79).

Pediatrics

Until 1988, the only enteral products available for feeding to children were infant formulas and adult nutritional products. Neither were seen as nutritionally ideal for this population. Infant formulas were too dilute for older children. Products for adults were perceived as too concentrated in some nutrients (eg, protein, sodium) and too low in others when the concentrations of protein and sodium were diluted to acceptable levels. This stimulated the development of products specifically for children aged 1 to 10 years, designed to provide complete, balanced nutrition for the support of both growth and tissue maintenance as well as the increased nutrient demands during acute and chronic illness.

Most of the products on the market have been designed for both oral and tube feeding. They have a caloric density of 1.0 Kcal/mL. The caloric distribution for pediatric products are typically 10 to 12% of calories from protein, 25 to 44% from fat, and 44 to 63% from carbohydrate. Like adult formulas, these formulas are generally lactose-free and gluten-free. They consist of intact milk protein sources, blends of vegetable oils, hydrolyzed starches, and sucrose, and some have added fiber. All have been formulated to meet the nutrient standards set forth by the National Academy of Sciences–National Research Council Recommended Dietary Allowances (RDAs) (7). These RDAs differ from the RDIs used for adult nutritionals in that they are more age-specific. In most cases, certain nutrients have been increased above these standards to account for increased needs during stress and illness. Recently, elemental formulas with hydrolyzed proteins or free amino acids have been developed for children who have severe food allergies or conditions resulting in maldigestion or malabsorption.

For most cases of failure to thrive, poor appetite, or when weight gain is desired, oral supplementation is usually sufficient. Conditions that require tube feeding in children are similar to those that necessitate tube feeding in adults (eg, neurologic disease, burns, severe infections, anorexia from disease, cancer, esophageal injury or obstruction, trauma, and surgery). Guidelines for attending to the nutritional needs of these children are available (103,104). The feedings may be for days, weeks, or years, depending on the nature and severity of the condition. When a child cannot receive adequate nutrition by tube or has impaired GI function, parenteral nutrition (nutrition infused by the vein) is required.

DELIVERY SYSTEMS

Nasoenteric tube feeding dates back as far as the fifteenth century; modern use began in the early 1950s with the development of fine polyethylene tubes and feeding pumps (105). In the 1960s, the development of total parenteral nutrition (TPN), which was viewed as more sophisticated and problem free, contributed to reduced use of enteral tube feeding in the 1970s and early 1980s. However, tube feeding is now resurging because of a number of factors. Enteral nutrition is less expensive than TPN (106) and results in fewer serious complications (107). Research shows that enteral feeding maintains gut mass, integrity, and function better than parenteral nutrition (1). The use of small-bore tubes made of soft biocompatible materials has led to increased patient tolerance of nasally placed tubes.

Most feeding tubes are now made of polyurethane or silicone, which does not disintegrate or become brittle over time and therefore does not need frequent replacement. Feeding tubes are mostly easily inserted into the stomach by way of the nasopharaynx. If the patient is at high risk for aspiration, the tube can be passed further down into the duodenum or jejunum, although it is sometimes difficult to get the tube through the pylorus (105). If the transversal route is not possible or long-term feeding is desired (more than six weeks) tubes can be inserted directly into the esophagus or into the stomach or duodenum through the abdominal wall. (Tube placements are depicted in Figure 1.) Cervical pharyngostomies, or more rarely, esophagostomies are performed when the access site has been obtained during surgery for head and neck cancer or surgical repair of the maxillofacial area (108). Feeding gastrostomies can be placed surgically or percutaneously (109); the advantage of percutaneous endoscope placement is that it can be done as an outpatient procedure by gastroenterologist with local anesthesia. Feeding jejunostomies are usually performed in conjunction with other intestinal surgery, facilitating early nutrition support. The diameter of feeding tubes is described in French (F) units (F unit = 0.33 mm). Large-diameter tubes ($\geq$16 F) tend to cause

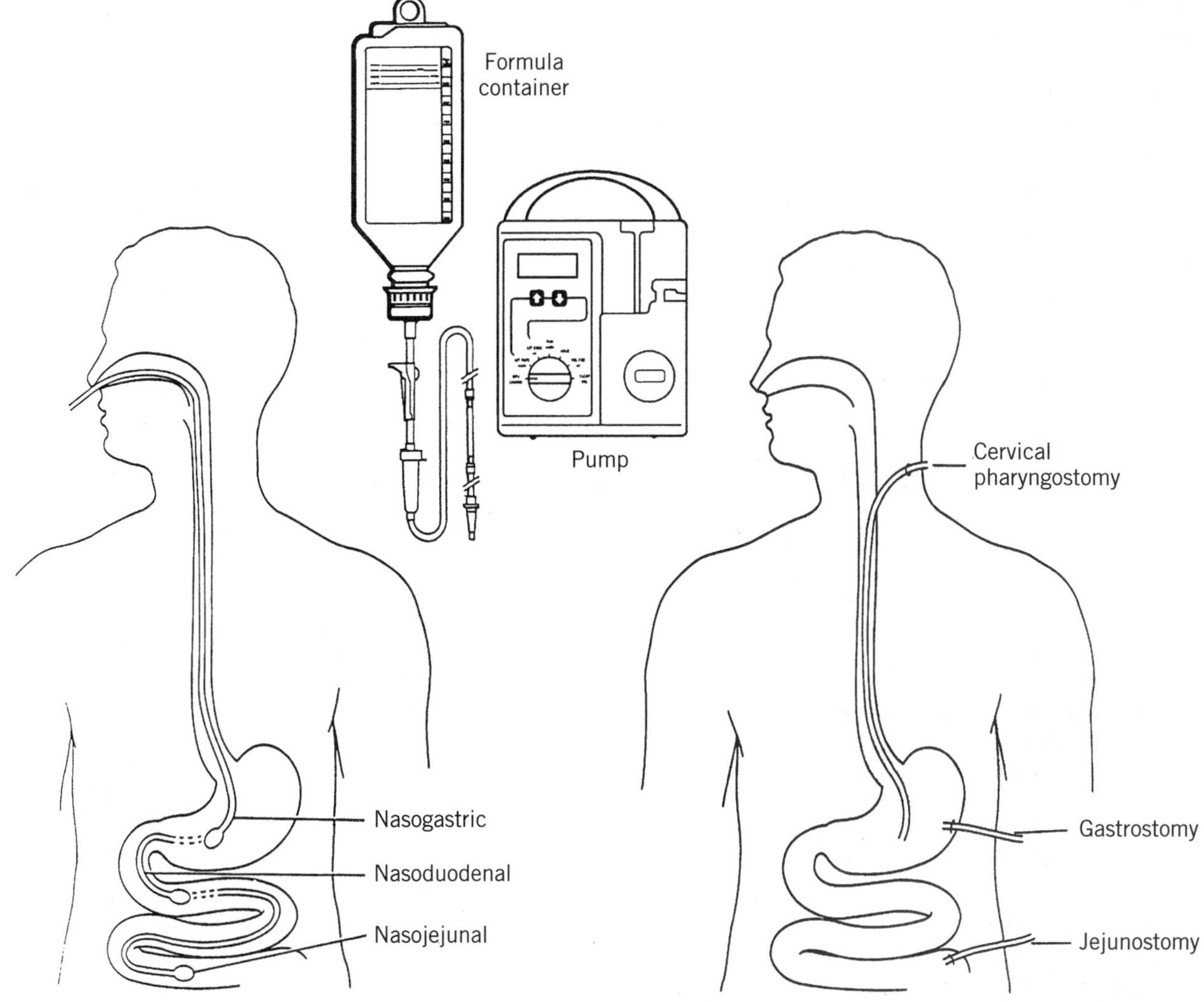

Figure 1. Tube feeding sites and devices. *Source:* Courtesy of Applied Therapeutics, Inc., Vancouver, Washington.

pressure necrosis in the nose as well as inflammation of throat and esophagus. They may compromise the function of the esophageal sphincter, increasing the potential for pulmonary aspiration. Large tubes may also interfere with swallowing, discouraging oral consumption of liquid supplements. Most adults tolerate nasal tubes ranging between 8 and 12 F. High viscosity formulas or those supplemented with fiber require at least a 10 F tube unless a feeding pump is being used. Extremely small caliber tubes (<6 F) require very low viscosity formulas such as elemental formulas.

Once the tube is in place, formula is usually administered at a slow rate and increased in increments depending on the condition of the patient until the patient's needs are met. Formula can be poured into 500 or 1,000 mL rigid containers or bags. To provide the safest environment for tube-fed patients, formulas are available in commercially prefilled containers that decrease the risk of bacterial contamination from handling and reduce staff time. The formula flows from the container through a feeding set, by gravity or by a pump (a small portable pump is shown in Figure 1), into the feeding tube (110). Formula can be delivered by a continuous or intermittent administration schedule. A continuous schedule is preferred when tube feeding is initiated, when patients have not been fed for several days, when patients are critically ill, and when duodenal or jejunal feeding sites are used (105,107). Intermittent feedings given five to eight times a day can simulate a meal pattern. Ambulatory tube-fed patients frequently use an intermittent or 16 to 18 h continuous feeding schedule because it permits more freedom of movement than does 24-h continuous feeding. Intermittent feeding may also be preferable in patients with diabetes because it simplifies insulin dosing. Most often delivery is controlled by means of an enteral feeding pump thus ensuring an even and controlled flow of formula, but it is also possible to infuse by gravity or with the use of a syringe (bolus feeding).

Enteral feeding pumps have been improved greatly since first being introduced. Enteral feeding pumps work via one of two mechanisms; rotary peristaltic or cassette technology. The rotary peristaltic pump works by looping the tubing of the administration set around the pumping mechanism and pressure against the tubing forces the formula out in a constant sequence of motions. The delivery is regulated by a drop-counting sensor. The cassette pump has a cassette that holds a specific volume of fluid that works with a pumping mechanism to discharge precise volumes at a specified rate. The cassette fits easily into place and eliminates the need to loop tubing for proper pump operation. Pumps have very sophisticated features and alarms. There are pumps that automatically flush the feeding tube at set intervals; this has been proven to reduce tube clogging. Safety features and alarms vary by pump manufacturer; however, basic safety features and alarms include memory, free-flow alarm, no flow alarm, and low battery alarm. Additional features may include select rate alarm, select run alarm, system self-check, lockout, and hold.

Enteral feeding pumps are used to control the rate of feeding delivery and provide increased safety in providing enteral nutrition support. Controlling the rate of formula delivery is imperative in the following patient care situations: (*1*) when feeding into the small bowel, (*2*) when feeding patients who are fluid restricted (eg, cardiac, renal), (*3*) when feeding patients intragastrically who have a history of gastroesophageal reflux or aspiration, or (*4*) when feeding patients who may have delayed gastric emptying secondary to their underlying disease (eg, diabetes). Enteral feeding pumps can have very positive outcomes as well as lower overall cost of patient care.

PROCESSING TECHNOLOGY

Enteral nutrition can be a patient's sole source of nutrition for extended periods. Therefore, high standards of care in the manufacturing process are necessary to assure consistent nutrient content and product sterility. Enteral nutrition manufacturers maintain rigid specifications for ingredient quality, processing conditions, and package integrity. The combination of specifications and documentation comprise a thorough quality assurance program to demonstrate the batch-to-batch consistency of the finished product. The specific processing parameters and quality assurance programs used in the manufacture of enteral formulas are proprietary, but are described in general terms in this section.

Ingredients

The ingredients for enteral nutritional products are selected following examination of attributes such as nutritional quality, safety for use, functional properties (eg, viscosity), and processing ease (eg, dispersibility in water). Following receipt of ingredients at the manufacturing facility, a rigid testing, storage, and handling program are used to assure ingredient quality. The testing program includes identity, microbiological and infestation testing, organoleptic testing, and nutritional potency. Storage conditions are carefully monitored to assure that extremes of temperature and moisture do not damage the ingredients. Water quality in enteral nutritional processing is assured through use of treated potable water supplies and water quality monitoring programs.

Ingredients are added during the manufacturing process at various points dictated by factors such as manufacturing convenience and assurance of distribution of the ingredient into the product. For example, it is most convenient to add carbohydrate sources (eg, maltodextrins) to the water phase to assure dispersion. Incorporation of ingredients into the product varies from use of direct addition to use of pumps such as triblenders.

Liquids

The ingredients in the oil and water phases are blended together in a concentrated mixture, heated, and the pH adjusted to stabilize the protein (Fig. 2). This mixture is subjected to a combination of processes which are similar to typical dairy processes: homogenization, pasteurization, UHTST (ultrahigh-temperature short-time sterilization), HTST (high temperature short time sterilization), deaeration, and clarification. Heat-labile nutrients and flavors are added after these processing steps. Samples are taken for measurement of pH and total solids at this point, and the mix is adjusted to final total solids and pH.

Shelf-stable, liquid enteral products are commercially sterilized using either retort or aseptic processes. In the retort process, the final product is filled in either metal, glass, or plastic containers and sealed, normally under a slight vacuum. The filled containers are subjected to rapid heating with superheated steam. In most cases, the containers are rotated or shaken to assure even distribution of heat throughout the product. In the aseptic filling process, the product undergoes a UHTST treatment just prior to filling. The containers and lids are independently treated with a sterilant (eg, hydrogen peroxide) to assure sterility. The product is filled in the package in a sterile filling zone in the filler. Aseptic packages include plastic, glass, aluminum, steel, and fiberboard containers.

In the case of both retort and aseptic sterilization processes, the manufacturer must establish that the product and package receive treatment that is adequate to reduce pathogenic microorganisms in the finished product to a nonviable level for the shelf life of the product. Typically, the manufacturer subjects the product and package to testing protocols to assure sterilization processes are adequate. Before commercial release, the product is incubated at typical storage temperatures to assure that no microbial contamination occurs postprocess.

Powders

Powdered products are sometimes the preferred form for certain products or specific settings. For a given volume of final reconstituted liquid product, powders have less bulk and weigh less than processed liquid products and, therefore, are easier and less expensive to store and transport. Powders generally have a longer shelf life than liquid forms. Some products are available only as powders because certain components cannot be exposed to thermal processing or are not stable for long periods in aqueous solution.

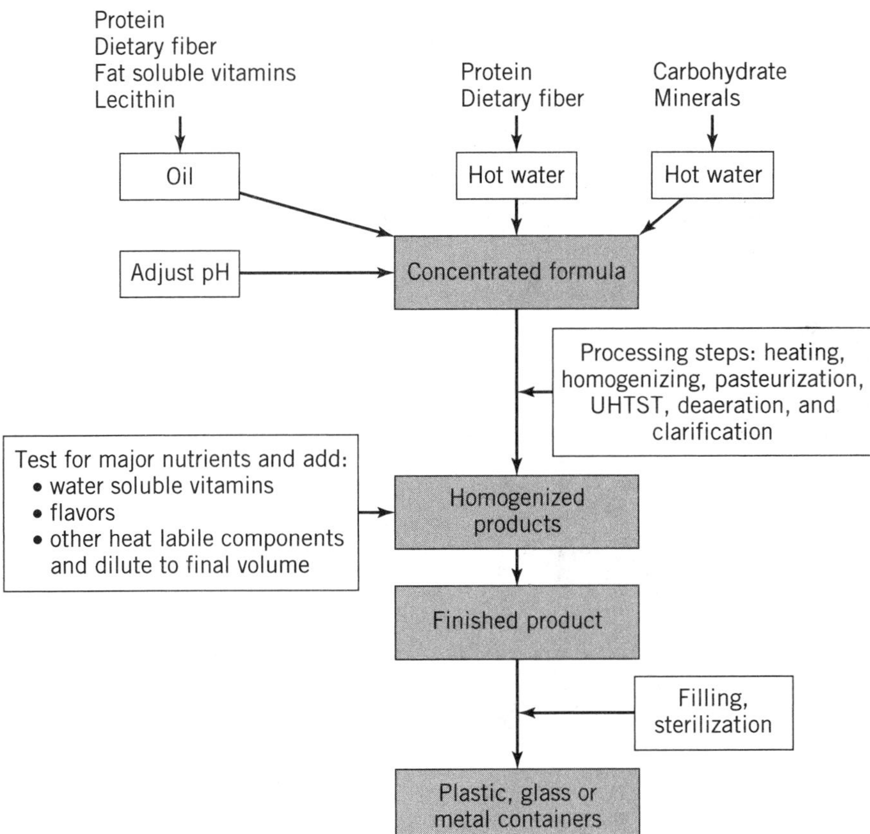

Figure 2. Formulation and processing of a liquid enteral formula.

Two methods of manufacturing are commonly used for powdered enteral products: spray-drying and dry-blending, either separately or in combination. Spray-dried product is UHTST or HTST processed as a concentrated liquid prior to the drying step. In the drying step over 95% of the water is removed to bring the powder moisture content to a level that does not promote microbiological growth. Dry-blending is a process where dry ingredients are mixed in a chamber by rotating ribbons or plows. Dry-blending is desirable when there is heat- or moisture-sensitive ingredients in the product formulation. Occasionally, base powders are spray-dried, followed by dry-blending of sensitive ingredients.

The process of spray-drying or dry-blending can produce a powder that does not disperse easily in liquid. The process of agglomeration is used to improve powder dispersability in liquid by binding small, dense particles together to provide a larger surface area to "wet."

Packaging

Enteral products are packaged in glass, metal (steel and aluminum), fiberboard, and plastic packages. Packaging materials are selected for ease and cost of manufacturing, convenience for the customer, and ability to protect the product from environmental conditions such as light, moisture, and oxygen. Glass and metal containers offer excellent protection from moisture or oxygen. Plastic containers are normally multilayer with one or more of the layers possessing properties that act as a barrier to oxygen, moisture, or light. The product label, outer carton, and shipping containers are also designed to provide protection to the product and help increase the package's durability. Packages are subjected to physical tests to assure that the rigors of shipping and handling do not adversely affect the integrity of the package's hermetic seal.

Product Shelf Life

A minimum of a one year room temperature storage shelf life for enteral nutrition products is normally the requirement of institutional health care providers for reasons of convenience and storage. Assurance of the product's nutritional and physical quality and the package's durability under normal storage conditions is accomplished by shelf life studies prior to product introduction. Ongoing product monitoring programs provide assurance that the product and package meet specifications throughout the product's life cycle.

REGULATORY ACTIVITIES

Enteral diets that are formulated simply as adult meal replacements are considered to be foods by the U.S. FDA and are not regulated any differently than standard foods. Webster's Dictionary defines food as "material consisting essentially of protein, carbohydrate, and fat used in the body of an organism to sustain growth, repair, and vital processes, and to furnish energy; also such food together with supplementary substances (as minerals, vitamins, and condiments)." Like regular foods, enteral diets are formulated with items that are generally recognized as safe (GRAS) or items approved as food additives by the FDA.

The FDA has a role in approving or advising food manufacturers regarding the health messages for conventional foods. There are two opposing views about the potential effects of regulating such messages: if health claims are not allowed, then the public may not be informed of the potential benefit of consuming certain foods. On the other hand, allowing health messages on foods opens the door for false or exaggerated claims to be made. Recently, the policy of the FDA has been changing toward the allowance of certain health messages. Regulations or food labeling are detailed in two sections of the Code of Federal Regulations (111).

The development of more specialized enteral formulas for specific diseases has placed these formulas under increasingly intense secrutiny by the FDA. In 1988, the orphan drug amendment created a new category of food called medical food, which was defined as follows: "The term medical food means a food which is formulated to be consumed or administered enterally under the supervision of a physician and which is intended for the specific dietary management of a disease or condition for which distinctive nutritional requirements based on recognized scientific principles, are established by medical evaluation" (112). Implicit in this definition is the notion that disease conditions can change an individual's nutritional requirements or food intake leading to a deficiency or overabundance of a nutrient. Although the amendment did issue the definition of a medical food under the current federal regulations that implement the Food, Drug, and Cosmetic Act, there is no procedural framework for the evaluation of medical food products. This contrasts with the detailed regulations for the development and approval of new drugs and food additives as well as the regulations for the formulation and manufacture of infant formulas specified in the Infant Formula Act of 1980 (113).

ACKNOWLEDGMENTS

The authors would like to express their appreciation to Pamela Anderson, Sheila Campbell, David Cockram, Lisa Craig, Joyce Hall, Jeffrey Nelson, and Sean Moore for their invaluable contributions.

BIBLIOGRAPHY

1. ASPEN Board of Directors, "Guidelines for Use of Parenteral and Enteral Nutrition in Adult and Pediatric Patients," *JPEN, J. Parenter. Enteral Nutr.* **17**, 1SA–52SA (1993).
2. A. Chait et al., "Rationale of the Diet-Heart Statement of the American Heart Association. Report of the Nutrition Committee," *Circulation* **88**, 3008–3029 (1993).
3. J. H. Cummings, "Consequences of the Metabolism of Fiber in the Human Large Intestine," in G. V. Vahouny and D. Kritchevsky, eds., *Dietary Fiber in Health and Disease*, Plenum Press, New York, 1982, pp. 9–22.
4. Federation of American Societies for Experimental Biology, *Physiological Effects and Health Consequences of Dietary Fiber*. Washington, D.C., U.S. Dept. of Health and Human Services, 1987, pp. 160–163.
5. J. E. Spiegel et al., "Safety and Benefits of Fructooligosaccharides as Food Ingredients," *Food Technol.* **48**, 85–89 (1994).
6. A. C. Bach and V. K. Babayan, "Medium-Chain Trigylcerides: An Update," *Am. J. Clin. Nutr.* **36**, 950–962 (1982).
7. National Research Council, *Recommended Dietary Allowances*, 10th ed., Washington, D.C., National Academy Press, 1989.
8. S. Mobarhan and L. S. Trumbore, "Nutritional Problems of the Elderly," *Clin. Geriatr. Med.* **7**, 191–214 (1991).
9. D. Cuthbertson and W. J. Tilstone, "Metabolism During The Post-Injury Period," *Adv. Clin. Chem.* **12**, 1–55 (1969).
10. J. W. Bane, R. E. McCaa, and C. S. McCaa, "The Pattern of Aldosterone and Cortisone Blood Levels in Thermal Burn Patients," *J. Trauma* **14**, 605–611 (1974).
11. G. M. Vaughn et al., "Cortisol and Corticotrophin in Burned Patients," *J. Trauma* **22**, 263–273 (1982).
12. D. W. Wilmore et al., "Hyperglucanemia After Burns," *Lancet* **1**, 73–75 (1974).
13. J. M. Shuck et al., "Dynamics of Insulin and Glucagon Secretions in Severely Burned Patients," *J. Trauma* **17**, 706–713 (1977).
14. R. R. Wolfe et al., "Glucose Metabolism in Severely Burned Patients," *Metabolism* **28**, 1031–1039 (1979).
15. D. W. Wilmore et al., "Catecholamines: Mediator of the Hypermetabolic Response to Thermal Injury," *J. Trauma* **15**, 697–703 (1975).
16. K. N. Frayn, P. F. Maycock, and H. B. Stoner, "The Relationship of Plasma Catecholamines to Acute Metabolic and Hormonal Responses to Injury in Man," *Circ. Shock* **16**, 229–240 (1985).
17. J. M. Daly et al., "Enteral Nutrition With Supplemental Arginine, RNA, and Omega-3 Fatty Acids in Patients After Operation: Immunologic, Metabolic and Clinical Outcome," *Surgery* **112**, 56–67 (1992).
18. R. O. Brown et al., "Comparison of Specialized and Standard Enteral Formulas in Trauma Patients," *Pharmacology* **14**, 314–320 (1994).
19. F. A. Moore et al., "Clinical Benefits of an Immune-Enhancing Diet for Early Postinjury Enteral Feeding," *J. Trauma* **37**, 607–615 (1994).
20. J. M. Daly et al., "Enteral Nutrition During Multimodality Therapy in Upper Gastrointestinal Cancer Patients," *Ann. Surg.* **221**, 327–338 (1995).
21. K. A. Kudsk et al., "A Randomized Trial of Isonitrogenous Enteral Diets After Severe Trauma: An Immune-Enhancing Diet Reduces Septic Complications," *Ann. Surg.* **224**, 531–543 (1996).
22. A. Barbul, "Arginine: Biochemistry, Physiology and Therapeutic Implications," *J. Parenter. Enteral Nutr.* **10**, 227–238 (1986).
23. R. J. Smith and D. W. Wilmore, "Glutamine Nutrition and Requirements," *J. Parenter. Enteral Nutr.* **14**, 94S–99S (1990).
24. M. M. Gottschlich, "Selection of Optimal Lipid Sources in Enteral and Parenteral Nutrition," *Nutr. Clin. Prac.* **7**, 152–165 (1992).
25. C. L. Pirani and S. M. Levenson, "Effect of Vitamin C Deficiency on Healed Wounds," *Proc. Soc. Exp. Biol. Med.* **82**, 95 (1953).
26. T. W. Goodwin, "Metabolism, Nutrition and Function of Carotenoids," *Annu. Rev. Nutr.* **6**, 273–297 (1986).

27. E. G. Maderazo et al., "A Randomized Trial of Replacement Antioxidant Vitamin Therapy for Neutrophil Locomotory Dysfunction in Blunt Trauma," *J. Trauma* **31**, 1142–1150 (1991).
28. R. S. Mitchell and T. L. Petty, "Chronic Obstructive Pulmonary Disease (COPD)," in R. S. Mitchell and T. L. Petty, eds., *Synopsis of Clinical Pulmonary Disease*, 3rd ed, CV Mosby, St. Louis, Mo., 1982, pp. 97–113.
29. S. E. Brown and R. W. Light, "What Is Know Known About Protein-Energy Depletion: When COPD Patients Are Malnourished," *Journal of Respiratory Diseases*, 36–50 (May 1983).
30. R. J. Browning and A. M. Olsen, "The Functional Gastrointestinal Disorders of Pulmonary Emphysema," *Mayo Clinic Proceedings* **26**, 537–543 (1961).
31. J. D. Frankfort et al., "Comparison of Ensure-HN Plus and Pulmocare on Maximum Exercise Performance in Patients With Chronic Airflow Obstruction (CAO)," *American Review of Respiratory Diseases* **135**, A361 (1987).
32. F. B. Cerra et al., "Applied Nutrition in ICU Patients: A Consensus Statement of the American College of Chest Physicians," *Chest* **111**, 769–778 (1997).
33. P. Amene et al., "Hypercapnia During Total Parenteral Nutrition With Hypertonic Dextrose," *Crit. Care Med.* **15**, 171–172 (1987).
34. S. A. McClave, "The Consequences of Overfeeding and Underfeeding," *J. Resp. Care Pract.* **10**, 57–64 (1997).
35. K. Ideno et al., "Nutrition in the ICU," *J. Resp. Care Pract.* **9**, Apr/May, 41–50 (1996).
36. M. J. Karlstad et al., "The Anti-inflammatory Role of Gamma-Linolenic and Eicosapentanoic Acids in Acute Lung Injury," in Y. S. Huang and D. E. Mills, eds., *Gamma-Linolenic Acid: Biochemistry and Role in Medicine and Nutrition*, AOCS Press, Champaign, Ill., 1996, pp. 137–167.
37. J. D. Palombo et al., "Rapid Modulation of Lung and Liver Macrophage Phospholipid Fatty Acids in Endotoxemic Rats by Continuous Enteral Feeding with n-3 and Gamma-Linolenic Fatty Acids," *Am. J. Clin. Nutr.* **63**, 208–219 (1996).
38. J. D. Palombo et al., "Cyclic vs Continuous Enteral Feeding with n-3 and Gamma-Linolenic Fatty Acids: Effect on Modulation of Phospholipid Fatty Acids in Rat Lung and Liver Macrophages," *J. Parenter. Enteral Nutr.* **21**, 123–132 (1997).
39. M. J. Murray et al., "Select Dietary Fatty Acids Attenuate Cardiopulmonary Dysfunction During Acute Lung Injury in Pigs," *Am. J. Physiol.* **269**, H2090–H2099 (1995).
40. P. Mancuso et al., "Dietary Fish Oil and Fish and Borage Oil Attentuate Proinflammatory Eicosanoids in Bronchoalveolar Lavage Fluid and Pulmonary Neutrophil Accumulation in Endotoxic Rats," *Crit. Care Med.* **25**, 1198–1206 (1997).
41. J. E. Gadek et al., "Specialized Enteral Nutrition Improves Clinical Outcomes in Patients With or at Risk of Acute Respiratory Distress Syndrome (ARDS): A Prospective, Blinded, Randomized, Controlled Multicenter Trial," *Am. J. Resp. Crit. Care Med.* **157**, A677 (1998).
42. J. Warnold, K. Lundolm, and T. Scherstein, "Energy Balance and Body Composition in Cancer Patients," *Cancer Res.* **38**, 1801–1907 (1976).
43. C. P. Holroyde et al., "Altered Glucose Metabolism in Metastatic Carcinoma," *Cancer Res.* **35**, 3710–3714 (1975).
44. M. F. Brennan and M. E. Burt, "Nitrogen Metabolism in Cancer Patients," *Cancer Treatment Report* **65** (Suppl 5), 67–68 (1981).
45. M. E. Shils, "Nutritional Problems Induced by Cancer," *Med. Clin. North Am.* **63**, 1009–1025 (1979).
46. A. Theologides, "Anorexia in Cancer: Another Speculation on Its Pathogenesis," *Nutr. Cancer* **2**, 133–135 (1981).
47. J. G. Harris, "Nausea, Vomiting and Cancer Treatment," *Cancer* **28**, 194–201 (1978).
48. J. Medoff, "A Double-Blind Evaluation of the Anti-Emetic Efficacy of Benzquinamide, Prochlorperazine, and Trimethobenzamide in Office Practice," *Current Theoretical Research* **12**, 706–710 (1970).
49. R. A. Harrington et al., "Metoclopramide: An Updated Review of Its Pharmacological Properties and Clinical Use," *Drugs* **25**, 451–494 (1983).
50. U.S. Department of Health, Education, and Welfare, *Eating Hints: Recipes and Tips for Better Nutrition During Cancer Treatment*, National Institutes of Health, Rockville, Md., 1980.
51. E. H. Rosenbaum et al., *Nutrition for the Cancer Patient*, Bull Publishing, Palo Alto, Calif., 1980.
52. B. C. Walike and J. W. Walike, "Relative Lactose Intolerance. A Clinical Study of Tube-Fed Patients," *J. Am. Med. Assoc.* **238**, 948–951 (1977).
53. H. L. Greene et al., "Nasogastric Tube Feeding at Home: A Method for Adjunctive Nutritional Support of Malnourished Patients," *Am. J. Clin. Nutr.* **34**, 1131–1138 (1981).
54. D. A. Lipschitz and C. O. Mitchell, "Enteral Hyperalimentation and Hemoatopoietic Toxicity Caused by Chemotherapy of Small Cell Lung Cancer [Abstract]," *J. Parenter. Enteral Nutr.* **4**, 593 (1980).
55. L. H. Brubaker et al., "Serum Opsonic Defect in Malnourished Cancer Patients and Improvement Following Nutritional Therapy [Abstract]," *Clin. Res.* **27**, 774A (1979).
56. A. H. McArdle et al., "Prophylaxis Against Radiation Injury," *Arch. Surg.* **121**, 879–885 (1986).
57. American Diabetes Association, "Position Statement. Nutrition Recommendations and Principles for People with Diabetes Mellitus," *Diabetes Care* **21**(Suppl 1), S32–S35 (1998).
58. Expert Panel on Detection, Evaluation, and Treatment of High Blood Cholesterol in Adults, "Summary of the Second Report of the National Cholesterol Education Program (NCEP) Expert Panel on Detection, Evaluation, and Treatment of High Blood Cholesterol in Adults. (Adult Treatment Panel II)," *JAMA, J. Am. Med. Assoc.* **269**, 3015–3023 (1993).
59. Expert Panel on Blood Cholesterol Levels in Children and Adolescents, "Treatment Recommendations of the National Cholesterol Education Program. Report of the Expert Panel on Blood Cholesterol Levels in Children and Adolescents," *Pediatrics* **89**(Suppl) 525–584 (1992).
60. U.S. Department of Health and Human Services, *The Surgeon General's Report on Nutrition and Health. Summary and Recommendations*, U.S. Government Printing Office, Washington, D.C., 1988.
61. C. Stevens, D. L. Costill, and B. Maxwell, "Impact of Carbohydrate (CHO) Source and Osmolality on Gastric Emptying Rates of Liquid Nutritionals [Abstract]," *J. Parenter. Enteral Nutr.* **3**, 32 (1979).
62. K. A. Cashmere et al., "Serum Endocrine and Glucose Response Elicited from Ingestion of Enteral Feedings [Abstract]," *Federation Proceedings* **40**, 440A (1981).
63. A. Peters, M. Davidson, and R. M. Isaac, "Lack of Glucose Elevation After Simulated Tube Feeding With a Low-Carbohydrate, High-Fat Enteral Formula in Patients with Type 1 Diabetes," *Am. J. Med.* **87**, 178–182 (1989).
64. W. Sturmer et al., "Favorable Glycaemic Effects of a New Balanced Liquid Diet for Enteral Nutrition. Results of a Short Term Study in 30 Type II Diabetic Patients," *Clin. Nutr.* **13**, 221–227 (1994).

65. L. D. Craig et al., "Use of a Reduced-Carbohydrate, Modified-Fat Enteral Formula for Improving Metabolic Control and Clinical Outcomes in Long-Term Care Residents With Type 2 Diabetes: Results of a Pilot Trial," *Nutrition* **14**, 529–534 (1998).
66. L. V. Campbell et al., "The High-Monounsaturated Fat Diet as a Practical Alternative for NIDDM," *Diabetes Care* **17**, 177–182 (1994).
67. A. Garg et al., "Effects of Varying Carbohydrate Content of Diet in Patients With Non-Insulin Dependent Diabetes Mellitus," *J. Am. Med. Assoc.* **271**, 1428–1428 (1994).
68. B. Gumbiner, C. C. Low, and P. D. Reaven, "Effects of a Monounsaturated Fatty Acid-Enriched Hypocaloric Diet on Cardiovascular Risk Factors in Obese Patients with Type 2 Diabetes," *Diabetes Care* **21**, 9–15 (1998).
69. J. D. Kopple, "Nutrition, Diet, and the Kidney," in M. E. Shils, J. A. Olson, and M. Shike, eds., *Modern Nutrition in Health and Disease*, 8th ed., Lea & Febiger, Philadelphia, Penn. 1994, pp. 1102–1134.
70. W. E. Mitch and B. J. Maroni, "Nutritional Considerations in the Treatment of Patients With Chronic Uremia," *Miner. Electrolyte Metab.* **24**, 285–289 (1998).
71. E. Lusvarghi et al., "Natural History of Nutrition in Chronic Renal Failure," *Nephrol Dial Transplant* **11**(Suppl 9), 75–84 (1996).
72. W. E. Mitch, "Robert H. Herman Memorial Award in Clinical Nutrition Lecture, 1997: Mechanisms Causing Loss of Lean Body Mass in Kidney Disease," *Am. J. Clin. Nutr.* **67**, 359–366 (1998).
73. "Effects of Dietary Protein Restriction on the Progression of Moderate Renal Disease in the Modification of Diet in Renal Disease Study," *J. Am. Soc. Nephrol.* **7**, 2616–2626 (1996).
74. J. S. Park et al., "Protein Intake and the Nutritional Status in Patients With Predialysis Chronic Renal Failure on Unrestricted Diet," *Korean J. Intern. Med.* **12**, 115–121 (1997).
75. M. J. Blumenkrantz et al., "Metabolic Balance Studies and Dietary Protein Requirements in Patients Undergoing Ambulatory Peritoneal Dialysis," *Kidney Int.* **21**, 849–861 (1982).
76. L. A. Slomowitz et al., "Effect of Energy Intake on Nutritional Status in Maintenance Hemodialysis Patients," *Kidney Int.* **35**, 704–711 (1989).
77. B. Schneeweiss et al., "Energy Metabolism in Acute and Chronic Renal Failure (see comments)," *Am. J. Clin. Nutr.* **52**, 596–601 (1990).
78. J. D. Kopple, "McCollum Award Lecture, 1996: Protein-Energy Malnutrition in Maintenance Dialysis Patients," *Am. J. Clin. Nutr.* **65**, 1544–1557 (1997).
79. D. B. Cockram et al., "Safety and Tolerance of Medical Nutritional Products as Sole Sources of Nutrition for People on Hemodialysis," *J. Ren. Nutr.* **8**, 25–33 (1998).
80. N. Cano et al., "Prealbumin-Retinol Binding Protein-Retinol Complex in Hemodialysis Patients," *Am. J. Clin. Nutr.* **47**, 664–667 (1988).
81. J. Kelleher et al., "Vitamin A and Its Transport Proteins in Patients with Chronic Renal Failure Receiving Maintenance Haemodialysis and After Renal Transplantation," *Clin. Sci.* **65**, 619–626 (1983).
82. F. R. Smith and D. S. Goodman, "The Effects of Diseases of the Liver, Thyroid and Kidneys on the Transport of Vitamin A in Human Plasma," *J. Clin. Invest.* **50**, 2426–2436 (1971).
83. H. Yatzidis, P. Digenis, and P. Fountas, "Hypervitaminosis A Accompanying Advanced Chronic Renal Failure," *Br. Med. J.* **3**, 352–353 (1975).
84. M. A. Allman et al., "Vitamin Supplementation of Patients Receiving Hemodialysis," *Med. J. Aust.* **150**, 130–133 (1989).
85. C. Pru, J. Eaton, and C. Kjellstrand, "Vitamin C Intoxication and Hyperoxalemia in Chronic Hemodialysis Patients," *Nephron* **39**, 112–116 (1985).
86. P. Balcke et al., "Ascorbic Acid Aggravates Secondary Hyperoxalemia in Patients on Chronic Hemodialysis," *Ann. Intern. Med.* **101**, 344–345 (1984).
87. W. R. Salyer and D. Keren, "Oxalosis as a Complication of Chronic Renal Failure," *Kidney Int.* **4**, 61–66 (1973).
88. K. Satomura et al., "Renal 25-hydroxyvitamin D3-1-hydroxylase in Patients With Renal Disease," *Kidney Int.* **34**, 712–716 (1988).
89. J. C. Jeanette and I. D. Goldman, "Inhibition of the Membrane Transport of Folates by Anions Retained in Uremia," *J. Lab. Clin. Med.* **86**, 834–843 (1975).
90. V. A. Skoutakis et al., "Folic Acid Dosage for Chronic Hemodialysis Patients," *Clin. Pharmacol. Ther.* **18**, 200–204 (1975).
91. V. W. Dennis and K. Robinson, "Homocysteinemia and Vascular Disease in End-Stage Renal Disease," *Kidney Int. Suppl.* **57**, S11–S17 (1996).
92. E. A. Ross et al., "Vitamin B6 Requirements of Patients on Chronic Peritoneal Dialysis," *Kidney Int.* **19**, 694–704 (1989).
93. J. D. Kopple et al., "Daily Requirement for Pyridoxine Supplements in Chronic Renal Failure," *Kidney Int.* **19**, 694–704 (1981).
94. A. G. Bostom et al., "High Dose B-Vitamin Treatment of Hyperhomocysteinemia in Dialysis Patients", *Kidney Int.* **49**, 147–152 (1996).
95. M. E. Suliman et al., "Total, Free and Protein-Bound Sulphur Amino Acids in Uraemic Patients," *Nephrol. Dial. Transplant* **12**, 2332–2338 (1997).
96. M. E. Suliman, B. Anderstam, and J. Bergstrom, "Evidence of Taurine Depletion and Accumulation of Cysteinesulfinic Acid in Chronic Dialysis Patients," *Kidney Int.* **50**, 1713–1717 (1996).
97. Y. Sakurauchi et al., "Effects of L-Carnitine Supplementation on Muscular Symptoms in Hemodialyzed Patients," *Am. J. Kidney Dis.* **32**, 258–264 (1998).
98. A. S. Alhomida et al., "Influence of Sex and Chronic Haemodialysis Treatment on Total, Free and Acyl Carnitine Concentrations in Human Serum," *Int. Urol. Nephrol.* **29**, 479–487 (1997).
99. B. de los Reyes et al., "L-Carnitine Normalizes the Reduced Carnitine Palmitoyl Transferase Activity in Red Cells From Haemodialysis Patients [letter]," *Nephrol. Dial. Transplant* **12**, 1300–1301 (1997).
100. T. H. Lin et al., "Trace Elements and Lipid Peroxidation in Uremic Patients on Hemodialysis," *Biol. Trace Elem. Res.* **51**, 277–283 (1996).
101. B. Dworkin et al., "Diminished Blood Selenium Levels in Renal Failure Patients on Dialysis: Correlations With Nutritional Status," *Am. J. Med. Sci.* **293**, 6–12 (1987).
102. G. Kallistratos et al., "Selenium and Haemodialysis: Serum Selenium Levels in Healthy Persons, Non-Cancer and Cancer Patients With Chronic Renal Failure," *Nephron* **41**, 217–222 (1985).
103. P. M. Queen and C. E. Lang eds. *Handbook of Pediatric Nutrition*, Aspen Publishers, Gaithersburg Md., 1993.
104. S. B. Baker, R. D. Baker, Jr., and A. Davis, eds. *Pediatric Enteral Nutrition*, Chapman and Hall, New York, 1994.

105. J. L. Rombeau and D. O. Jacobs, "Nasoenteric Tube Feeding," in J. L. Rombeau and M. D. Caldwell, eds., *Enteral and Tube Feeding*, W. B. Saunders, Philadelphia, Penn. 1984. pp. xx–xx.
106. A. H. McArdle et al., "A Rationale for Enteral Feeding as the Preferable Route for Hyperalimentation," *Surgery* **90**, 613–623 (1981).
107. T. Jones, "Enteral Feeding: Techniques of Administration," *Gut* **27**, 47–49 (1986).
108. C. Page, R. Andrassay, and J. Sandler, "Techniques in Delivery of Liquid Diet: Short-Term and Long-Term," in M. Deitel, ed., *Nutrition in Clinical Surgery*, Williams & Wilkins, Baltimore, Md., 1985, pp. 60–87.
109. J. Rombeau et al., "Feeding by Tube Enterostomy," in J. L. Rombeau and M. D. Caldwell, eds., *Enteral and Tube Feeding*, W.B. Saunders, Philadelphia, Penn., 1984, pp. xx–xx.
110. L. Y. Young and M. A. Joda-Kimble, eds., *Applied Therapeutics: The Clinical Use of Drugs*, 4th ed., Applied Therapeutics, Vancouver, Wash., 1988.
111. *Regulations for Food Labeling: Conventional Foods*, 21 CFR PART 101, *Foods for Special Dietary Use*, 21 CFR Part 105.
112. *Definition of Medical Food*, Orphan Drug Amendments of 1988, 21 USC 360 e(b)(3).
113. Infant Formula Act of 1980, Public Law 96-359.

DEBRA ROONEY
DAVID DEIS
Abbott Laboratories
Columbus, Ohio

ENZYME ASSAYS FOR FOOD SCIENTISTS

REASONS FOR ASSAYING ENZYMES

In designing enzyme assays it is important to determine the level of sophistication required of the data obtained. One may be led astray by answers that are too simplistic; conversely, excessive time and effort may be expended in obtaining highly precise data to meet "quick-and-dirty" needs. Understanding the basis of various kinds of assays will help in choosing the proper level of assay sophistication to avoid these two kinds of errors.

Characterization of Enzymes for Applications

In industrial food applications the usual requirement of an enzyme is that it produce the desired functionality for the minimum cost. This often implies that offerings by alternate suppliers are assayed to find the best enzyme source for the process in hand. The characterizing factors are (*1*) rate (activity per gram of enzyme); (*2*) pH optimum; (*3*) temperature optimum; (*4*) stability under conditions of use; and (*5*) presence (or absence) of potentially deleterious side activities.

Rate. By rate, the food processor usually means the amount of modification obtained during the time allowed for enzyme action in the process. This may be different from the initial rate of conversion of substrate to product as defined by an enzymologist. An assay that measures the latter rate may be misleading if the modification occurs during extended incubation in the process.

pH, Temperature. The pH and temperature optimum curves published by suppliers usually confound the true influence of these factors on enzyme catalytic properties with the effect on enzyme stability. Enzyme denaturation is influenced by numerous factors, is usually irreversible, and occurs with first-order kinetics. Optimum curves are constructed using assays that measure the amount of substrate modification over a period of time and represent a summation of true rate effects plus denaturation during that time. They should be used with caution.

Stability. The presence of substrate stabilizes enzyme against denaturation; thus, the real optimum of interest is the stability of enzyme under the conditions of use: time, pH, temperature, substrate concentration, inorganic ions, organic solvents, and so on. An assay used to screen enzymes for a particular application should mimic as nearly as possible the actual use conditions, to give a reliable estimate of cost benefit.

Side activities. Preparing a pure isolated enzyme is cost-prohibitive for food applications. All commercial enzymes contain some activities other than those declared on the label; for example, fungal amylase usually contains some proteolytic activity, a bacterial protease may also contain some xylanase, and so on. These side activities may present a problem when the enzyme is used in certain food processing situations, and assays should be applied that will detect them, as well as the main enzyme of interest.

Characterization of Enzymes in Raw Materials

Specificity. Often enzymes with similar activities (ie, proteases) are obtained from different sources and assays are used to characterize them in terms of units per gram. If the different enzymes have varying specificity requirements, the results may be misleading. For example, if two proteases have specificities corresponding to elastase and trypsin, an assay based on azocollagen substrate will give a much higher value for the former enzyme, while one based on casein will favor the latter. If the protein to be modified is something quite different, for example, wheat gluten, neither assay will give a reliable comparison. Substrate specificity may be of major importance, as with proteases, or a negligible factor, as with lipoxygenase. It is best to assume that it is important until the contrary is established.

If an assay is used to monitor production of enzyme from biofermentation, specificity may be overlooked. Here the ideal is the quickest possible assay consistent with reasonable accuracy, making the assumption that enzyme specificity is constant. (This also depends on the conditions of biofermentation remaining constant, particularly the purity of the microbiological innoculum being used.) A quick assay based on azocollagen may suffice even if the protease being made is similar to trypsin in its specificity.

Characterization of Enzyme Rate Parameters

For a study made for the purpose of establishing basic enzymatic parameters (catalytic rate constant, affinity for substrate, inhibitor binding, etc) the assay must provide appropriate velocity estimates. In general this means the catalytic rate at time zero, that is, when the enzyme and substrate are first combined. Assays involving incubation for a fixed length of time present some difficulties for this purpose. Progress curves, in which the concentration of product is measured at intervals during the incubation, are valid and are not used as often as they might be to determine enzymatic parameters.

THEORETICAL ASPECTS OF ENZYME ASSAYS

Properly judging the nature of the desired assay requires a certain amount of theoretical understanding. This does not have to be in great depth; the requisite fundamentals are easily grasped and applied. The effort will be repaid by improved enzyme assays for routine work and process design (1).

Assay Characteristics

A sound assay, regardless of the level of sophistication required, will have (*1*) linear dependence on enzyme concentration, (*2*) adequate consideration of pH and temperature effects, (*3*) appropriate accuracy, (*4*) adequate sensitivity, and (*5*) speed and ease of performance. Unfortunately, many assays emphasize the last factor at the expense of the other four.

Enzyme Linearity. Enzyme linearity is paramount. A curved plot of assay response versus amount of enzyme used indicates that some chemical, physical, and/or enzymological factors have been overlooked. In fixed-time assays the formation of product is often not strictly linear with time because of substrate depletion or product inhibition and this nonlinearity is more pronounced at higher enzyme concentrations. Occasionally the chemical reaction used to quantitate the amount of product formed is not stoichiometric, leading to a nonlinear plot of, for example, spectrophotometric absorbance versus enzyme amount. Numerous nonlinear assays found in the literature indicate an incomplete understanding of the system being used for the assay.

Many assays are linear over a limited range of activity. These may be used (particularly if they meet the requirement for ease of use) if care is taken to ensure that measurements are made only within the linear range.

Temperature, pH. Elevated temperatures and extremes of pH contribute to enzyme denaturation during the assay. Temperature optimum curves are always due to this phenomenon, and activity decreases at high or low pH also may be due to enzyme instability. The pH also may affect enzyme catalytic activity, influencing the ionization state of the active site and possibly the substrate (2–4). A sound assay will take these factors into account.

Accuracy, Sensitivity. The required levels of accuracy (more properly, precision) and sensitivity should be carefully considered. Measurements of α-amylase activity may have a coefficient of variation of 2% or of 10%, depending on whether the assay is a replicated colorimetric one using a modified substrate (5) or a viscometric one using gelatinized ground grain (6). For standardizing a purified amylase for use in bread production, the former would be appropriate, while for identifying bins of wheat that has been subjected to sprouting, the latter method is quite adequate. Likewise, assays at almost any level of sensitivity may be constructed. Using a fluorescent substrate, picomolar concentrations of trypsin may be assayed (7), while in monitoring the production of microbial protease a simple protein-based assay (8) will do the job, although it is some five orders of magnitude less sensitive.

Convenience. Speed and ease of performance should be the last factors considered. These are important in many industrial contexts, but a fast, easy, inadequate assay will only result in the rapid generation of much useless data. After linearity, precision, enzyme stability, and sensitivity are established, then steps may be taken to increase output.

Initial Rates. Most assays measure the rate of formation of product from substrate, that is, $d[P]/dt$. This measurement may be of the initial rate, $d[P]/dt$, at the initiation of the reaction, or of the amount of product formed during a fixed time of incubation of substrate with enzyme. Fixed-time assays are convenient, in that a large number of samples may be run simultaneously. Unfortunately, they are also more prone to complications leading to the nonlinearity mentioned above. Accurate initial rate measurements are not contaminated by effects such as substrate depletion, product inhibition, or enzyme denaturation. However, they are not always easy to obtain. A relatively simple method of finding the initial rate is the following. Set up the assay system, take samples at various times t, and measure product concentration $[P]$ at each time. Usually there is a slight downward curve of the plot of $[P]$ versus t (Fig. 1) that makes the determination of the tangent ($d[P]/dt$) at zero time difficult. Using a simple program (Basic, or even most spreadsheet programs today are capable of this), fit the data points with a least squares quadratic curve:

$$[P] = a + b \times t + c \times t^2$$

The first derivative gives the desired initial rate with reasonable accuracy:

$$d[P]/dt = b + 2c \times t; \text{ and } d[P]/dt_{(t=0)} = b.$$

Determining Rate Parameters

Michaelis-Menten Parameters. In designing assays it is useful to know the maximum rate obtainable with a given amount of enzyme, V_{max}, and the concentration of substrate that gives half that rate, K_M. These are the fundamental parameters in the Michaelis-Menten (M-M) rate

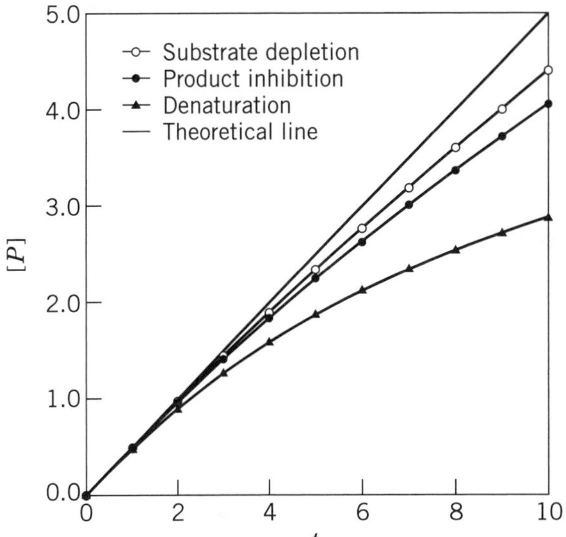

Figure 1. Nonlinearity of product formation with time, due to various factors. Initial rate for the theoretical reaction is 0.5. Fitting the nonlinear rates with a second degree polynomial, the calculated initial rates are: Substrate depletion, 0.51; Product inhibition, 0.51; Enzyme denaturation, 0.44. If the curves are fitted to a third degree polynomial, the calculated rates are 0.50, 0.50, and 0.49, respectively. *Source:* Ref. 1, p. 93.

equation: $v = V_{max}[S]/(K_M + [S])$, where $[S]$ is substrate concentration and v is the actual rate $d[P]/dt$. The usual procedure is to measure v at several concentrations $[S]$ and then calculate V_{max} and K_M, either by applying a computer program (9) such as HYPER (Fig. 2a) or by fitting a straight line to one of the linear transforms of the M-M equation. The usual transform is the double reciprocal or Lineweaver-Burk plot: $1/v = 1/V_{max} + (K_M/V_{max})(1/[S])$, where $1/v$ is plotted versus $1/[S]$ (Fig. 2b). From statistical considerations, this is the least desirable transform to use (11).

A better plot is the Hanes plot of $[S]/v$ versus $[S]$ (Fig. 2c): $[S]/v = K_M/V_{max} + (1/V_{max})[S]$. The points are spaced along the x axis at the same intervals as in the M-M plot rather than being crowded together near the y axis. The larger experimental errors inherent in the smaller values of v (at low $[S]$) have less influence on the linear least squares regression line. The Hanes plot should be used in treating v, $[S]$ data graphically; the use of the unsatisfactory Lineweaver-Burk plot should be discontinued.

The M-M equation is a differential equation, that is, v equals $d[P]/dt$ at the instantaneous value of $[S]$, usually taken as the initial substrate concentration. If v is only available as $[P]/t$ from a fixed-time assay, then the value taken for $[S]$ for the above calculations should be the average of the substrate concentration at the beginning and end of the incubation period, $([S]_0 + [S]_t)/2$. This approximation gives estimates of V_{max} and K_M that are much closer to the true values than if the initial value of $[S]$ is used (12).

Inhibitors

Two types of enzyme inhibitors are of interest to food scientists: (*1*) low-affinity inhibitors and (*2*) high-affinity inhibitors. The former are effective in the millimolar to micromolar concentration range and readily dissociate from the enzyme; an example is inorganic phosphate inhibiting phytase. The latter are effective in the nanomolar to picomolar concentration range and are bound tightly to the enzyme; an example is soybean trypsin inhibitor. For low-affinity inhibitors the dissociation constant K_i is a useful parameter to know; for high-affinity inhibitors the amount present is usually of more concern.

Inhibition Model. A general equilibrium model of inhibition is shown in Figure 3. If the parameter α is very large, so that the species EIS does not exist, the inhibition is termed competitive (inhibitor competes with S for the enzyme). The effect, shown in Figure 4a, is that V_{max} is unchanged, and K_M increases. If $\alpha = 1$ and $\beta = 0$, EIS is formed but does not proceed to product P, noncompetitive inhibition results. As shown in Figure 4a, V_{max} decreases and K_M is unchanged. If α is greater than 1 but not extremely large, mixed inhibition occurs. These types are diagnosed by comparing the Hanes plots in the absence and presence of inhibitor (Figure 4b).

Inhibitor Constant. If the inhibition is competitive, K_M is determined from the ratio of K_{app} (the apparent value of K in the presence of inhibitor of concentration $[I]$) to K_M (no inhibitor present): $K_{app}/K_M = 1 + [I]/K_i$. If inhibition is noncompetitive, the ratio of true maximum velocity to apparent maximum velocity in the presence of inhibitor gives $V_{max}/V_{app} = 1 + [I]/K_i$. Determining K_i, α, and β in the cases of partial and mixed inhibition is too complex to be discussed here.

High-Affinity Inhibitors. Measuring the concentration of high-affinity inhibitors (eg, soy trypsin inhibitor) is relatively straightforward (13). Trypsin is mixed with aliquots of soy meal extract and after a brief incubation (for formation of the inhibitor-enzyme complex) the amount of uninhibited enzyme remaining is measured by a simple assay. The rate of reaction is plotted versus the size of the extract aliquot; the straight line through the data obtained at lower levels of inhibition intersects the x axis at a point where the amount of enzyme equals the amount of inhibitor present (Fig. 5) (14). In the example shown, 1.2 mL of soy meal extract contained a molar amount of trypsin inhibitor equal to the number of moles of trypsin used in each assay tube. If the absolute amount of trypsin is established by a titration assay, the amount of soy trypsin inhibitor can be expressed in absolute molar units rather than arbitrary trypsin inhibitor units (15). Conversely, if the absolute concentration of high-affinity inhibitor is known, this method serves to measure the molar amount of enzyme present.

Endogenous Inhibitors. Endogenous inhibitors may be present in crude extracts of materials containing the enzyme being assayed, resulting in a marked nonlinearity in the assay (Fig. 6a). The uninhibited rate may be found as follows (1). Let $[e]$ be the amount of enzyme in the largest aliquot of extract used in making the plot of Figure 6. X is the fraction of that aliquot used for each of the other points.

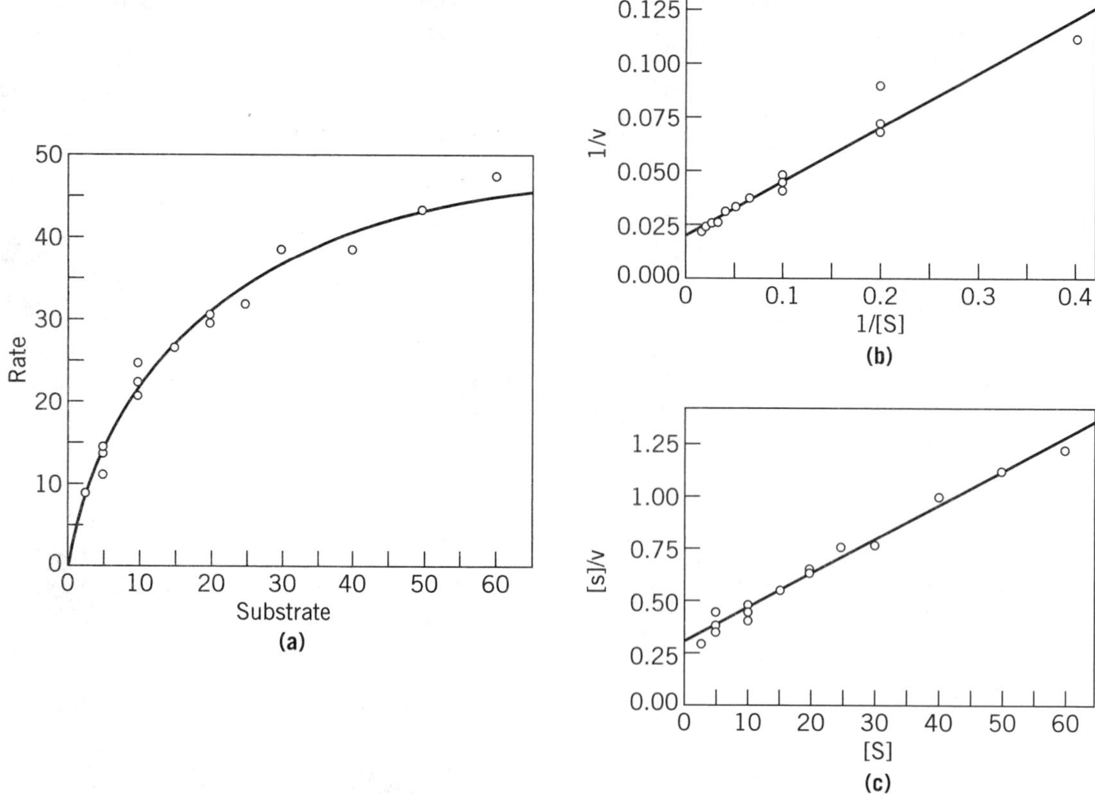

Figure 2. Determining V_{max} and K_M from experimental data using three different methods: (a) HYPER computer program: $V_{max} = 58.55$, $K_M = 17.44$, Std. Dev. 1.65; (b) Lineweaver-Burk plot: $V_{max} = 49.60$, $K_M = 12.85$, Std. Dev. 2.54; (c) Hanes plot: $V_{max} = 58.96$, $K_M = 17.56$, Std. Dev. 1.66. *Source:* Data from Ref. 10.

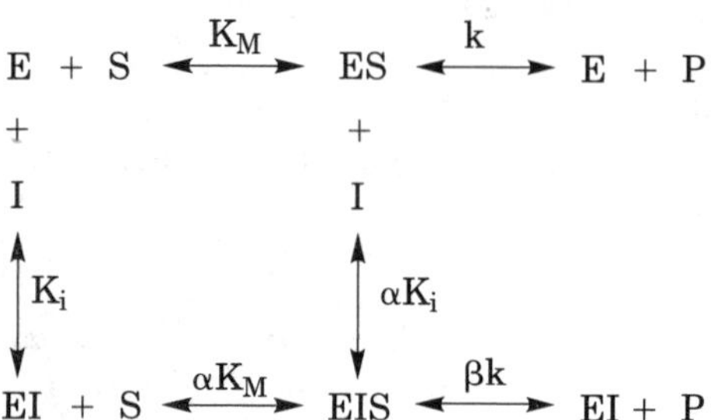

Figure 3. General equilibrium model for reversible enzyme inhibition. *Source:* Ref. 1, p. 38.

The measured rate at each point is v_i. A plot is made of X/v_i versus X (Fig. 6b). From the y-axis intercept calculate the uninhibited rate; that is, 1/intercept equals the true rate due to enzyme concentration [e]. This method is useful for comparing enzyme amounts from different sources, and during enzyme purification until the inhibitor has been removed.

SPECIFIC ENZYME ASSAYS

Several assays for enzymes most commonly measured by food scientists will be described. These are only examples of the wide range of ingenious methods that have been reported for following the rate of product formation by food-related enzymes. In most cases factors such as pH, temperature, activating ions, time of reaction, and detection methods may be adjusted to fit the specific needs of the project in hand. Thus, these should be considered as starting points for designing assays to meet particular requirements, not the only way to measure the enzyme activity under investigation.

Proteases and Peptidases

Substrates. Protein substrates for proteases are most often hemoglobin or casein and must be completely soluble in buffer. Casein for protease assays is designated "nach Hammarsten," while hemoglobin "for protease assay" is usually of high quality. Casein precipitates below pH 6, so it is used at neutral to alkaline pH. Hemoglobin must be denatured before use, either by treatment with acid (if the assay is at acidic pH) (16) or urea (neutral to alkaline pH assay) (17). Gelatin, sometimes used in viscometric assays, is quite heterogeneous, so lot-to-lot reproducibility is a concern.

Proteins may be derivatized to fit assay needs. Diazotized protein allows measurement of solubilized peptide with a visible-range colorimeter (18). If the amino groups freed during hydrolysis are quantitated using, for example,

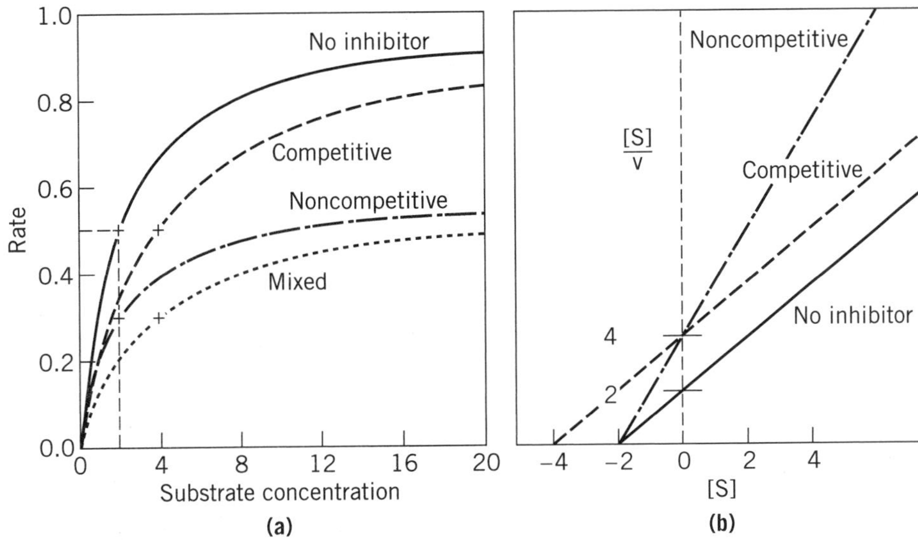

Figure 4. Hyperbolic and Hanes plots showing different modes of inhibition.

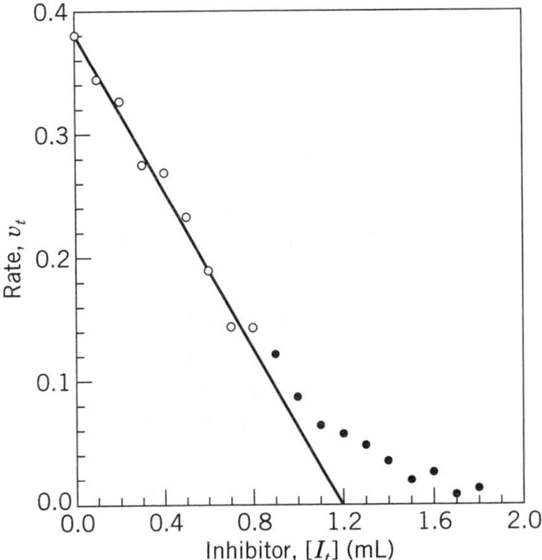

Figure 5. Plot of enzyme rates in presence of high-affinity soy trypsin inhibitor. *Source:* Data from Ref. 14.

TNBS (trinitrobenzenesulfonic acid), the ϵ-amino groups of lysine give a high blank value that may be removed by making the succinyl (19) or *N,N*-dimethyl (20) derivative of the protein. The latter is preferable for trypsinlike proteases.

Small synthetic molecules are also useful for assaying proteases. These give a change in spectrophotometric absorbance as they are hydrolyzed, so a continuous assay with its advantages is possible. Table 1 lists a number of small molecule substrates. Most of these are applicable to serine and/or sulfhydryl proteases, with two exceptions: FAGLA is a substrate for metalloprotease (neutral protease), and Z-Gly-Phe is a carboxypeptidase substrate. Acidic protease may be assayed using a chromogenic peptide. Aminopeptidase is usually assayed using an amino acid derivative such as L-leucine-β-naphthylamide.

Protease Assays. Most assays with protein substrates involve incubation for a set length of time, stopping the reaction with TCA (trichloroacetic acid), and measuring the amount of soluble peptide. Buffered TCA (0.11 MTCA, 18 g/L; 0.22 M sodium acetate, 18 g/L; and 0.33 M acetic acid, 19.8 g/L) gives superior enzyme linearity compared

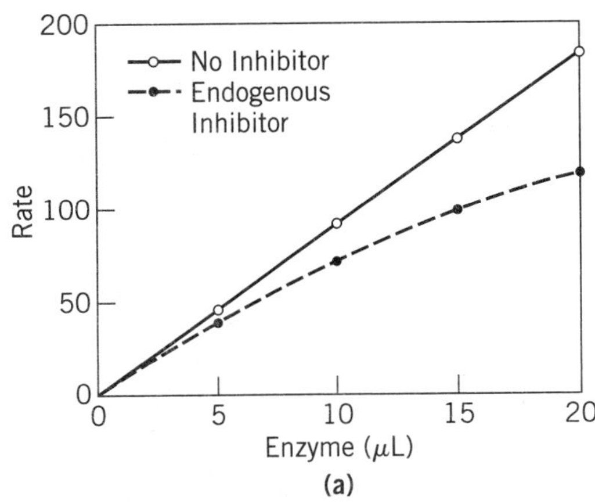

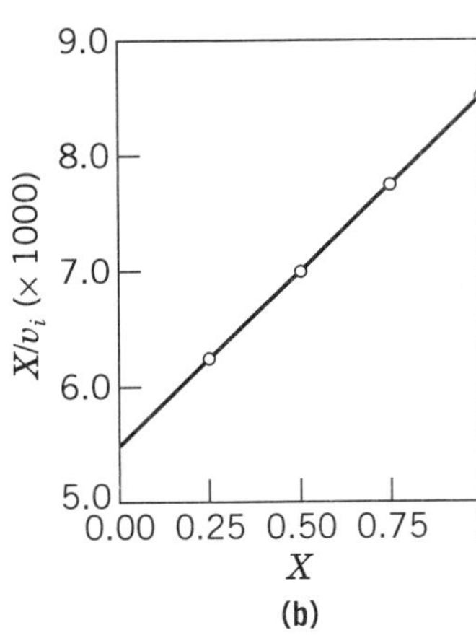

Figure 6. Method for finding uninhibited rate in the presence of endogenous inhibitor. *Source:* Ref. 1, p. 109.

Table 1. Absorbance Changes for Some Synthetic Substrates

Substrate	Acronym	λ, nm	Δε, mM	Reference
N-acetyl-L-tyrosine ethyl ester	ATEE	237	+0.23	21
		275	−0.17	21
N-benzoyl-L-arginine ethyl ester	BAEE	253	+1.15	21
N-benzoyl-L-arginine-*p*-nitroanilide	BAPA	410	+8.8	22
N-benzoyl-L-tyrosine ethyl ester	BTEE	254	+1.03	23
		256	+0.96	24
N-tosyl-L-Arginine ethyl ester · HCL	TAME	244	+0.80	23
		247	+0.41	25
3-(2-furylacryloyl)-glycyl-L-leucinamide	FAGLA	345	−0.32	26
N-carbobenzoxy-glycyl-L-phenylalanine	Z-Gly-Phe	224	−1.0	27
N-carbobenzoxy-L-tyrosine-*p*-nitrophenyl ester	Z-Tyr-pNP	400	+18.8	23

with a simple aqueous TCA solution (8). Quantitation may be direct (absorbance of filtrate at 275 nm) or colorimetric (Folin-Lowry [28], bicinchoninic acid [29], or TNBS [30]). An example of a casein-based assay is the following (8). To 5 mL casein (12 mg/mL in 0.03 M phosphate buffer, pH 7.5), add 1 mL enzyme solution and after 10 min reaction add 5 mL of buffered TCA. After 30 min the mixture is filtered and absorbance at 275 nm is measured. For Folin-Lowry quantitation, to 1 mL of filtrate add 1 mL alkaline buffer (1 M Na_2CO_3, 0.25 M NaOH), 0.4 mL copper reagent (0.1% $CuSO_4 \cdot 5H_2O$, 0.2% NaK tartrate), mix, and allow to stand 10 min. Then add 0.75 mL diluted phenol reagent (Folin-Ciocalteau reagent diluted with 3 volumes H_2O), mix, wait 10 min, and measure absorbance at 700 nm versus an appropriate reagent blank. A plot of log absorbance versus log protein is linear over the range 3 to 400 μg of protein (eg, bovine serum albumin standard) (31).

An assay using TNBS to measure freed amino groups (30) uses *N,N*-dimethyl casein (20) (1 mg/mL) in buffer as substrate. To 1 mL in reaction tube, add 0.1 mL enzyme solution, incubate for the desired time, and heat the tube briefly in boiling water to stop the reaction. Add 1 mL 0.4 M phosphate buffer, pH 8.2 (containing 0.25% sodium dodecyl sulfate), 2 mL freshly made TNBS (1 mg/mL in water), and incubate in the dark 1 h at 60°C. Add 4 mL 0.1 N HCl, cool to room temperature, and measure absorbance at 340 nm. The reagent blank contains only buffer and TNBS; a complete assay with zero incubation time gives the correction for free amino groups in substrate and enzyme.

The absorbance change during the reaction of synthetic substrates with enzyme is recorded on a stripchart recorder, giving a continuous trace from which the initial velocity or the first-order reaction rate is obtained. Two assays will be described as examples.

1. TAME a substrate for trypsin (23). The reaction solution is 1.04 mM TAME (0.394 mg/mL) in 0.04 M Tris buffer, pH 8.1, 10 mM in CaCl. To 2.9 mL, add 0.1 mL enzyme solution and record the increase in absorbance at 244 nm.
2. FAGLA is a useful substrate for metalloproteases such as thermolysin (26). To 100 mL 0.05 M phosphate buffer, pH 7.2, add 76.9 mg FAGLA to give a 2.5 mM solution. This is well below the K_M of thermolysin for this substrate, so the reaction is kinetically first order. In the cuvette place 1.5 mL substrate, 1.4 mL buffer, and 0.1 mL enzyme, and measure the decrease in absorbance at 345 nm from 0.96 (zero time) toward an infinite-time value of 0.56. Plot log(absorbance − 0.56) versus time; the slope divided by 2.303 gives the first-order rate for the reaction, which is directly proportional to the concentration of enzyme. Preferably, a simple Basic computer program (1) may also be used to calculate the first-order rate constant.

Peptidase Assays. Carboxypeptidase is assayed by incubation with an *N*-acylated dipeptide, followed by measurement of the amino group freed (32). Substrate solution is 1 mM dipeptide (eg, 35.6 mg Z-Phe-Gly per 100 mL) in 0.025 M phosphate buffer, pH 7.2. To 200 μL add 50 μL enzyme, incubate for 1 h, stop by heating in boiling water, and measure free amino groups by reaction with TNBS as described previously. Carboxypeptidase may also be assayed continuously using Z-Gly-Phe in 0.05 M Tris buffer, pH 7.5, following the decrease in absorbance at 224 nm (27).

Aminopeptidase assays usually employ the β-naphthylamide of an amino acid. After incubation the freed β-naphthylamine is quantitated by diazotization (33). Substrate stock solution is 2 mM amino acid naphthylamide in 0.01 N HCI. Buffer is 0.025 M phosphate, pH 7.2. Color reagent is the diazonium salt Fast Garnet GBC (1 mg/mL) in 1M acetate buffer, pH 4.2, containing 10% (v/v) Tween 20. To the assay tube, add 1.6 mL buffer, 0.2 mL substrate, 0.2 mL enzyme, and incubate 3 h. Then add 1 mL color reagent, allow 5 min for color development, and read absorbance at 525 nm.

Carbohydrases

Substrates. Starch, the substrate for amylases, consists of a linear polymer, amylose, and a highly branched polymer, amylopectin. The ratio of these two components varies in starches obtained from different plant sources. They may be separated by treating a solution of gelatinized starch with thymol (34). In certain assays for α-amylase the use of one or the other fraction is preferable. If the assay involves colorimetric quantitation of freed reducing groups, the blank due to reducing ends of the substrate chain may be removed by reduction with $NABH_4$ (5).

Soluble cellulose substrates for cellulase are the carboxymethyl, hydroxypropyl, or hydroxyethyl derivatives (35). These are available as commercial products but should be characterized before use to ensure comparability between lots. Insoluble complexes of dyes with starch (36,37) or cellulose (38) are also available; enzyme activity is assayed by measuring the amount of dye that is solubilized during an incubation period.

Small synthetic substrates, usually the *p*-nitrophenylglycosides, are available for some glycosidases (39). Specific oligosaccharide derivatives have been used for assays of α-amylase and β-amylase (40,41).

Amylase Assays. For α-amylase (5), use 1 mL of a 1% solution of reduced starch in 0.02 M acetate buffer, pH 4.7, containing 1 mM $CaCl_2$. Add 1 mL enzyme solution, incubate for the time desired, then add 4 mL each of Reagent A and Reagent B of the neocuproine system (42). Heat in boiling water 12 min, make up to 25 mL, and read absorbance at 450 nm. Reagent A: in 600 mL water, dissolve 40 g anhydrous Na_2CO_3, 10 g glycine, and 0.45 g $CuSO_4 \cdot 5H_2O$. Make up to 1 L. Reagent B: dissolve 0.12 g neocuproine (2,9-dimethyl-1,10-phenanthroline-HCl) in 100 mL water. Store in a brown bottle.

Amylopectin forms a complex with I_2 which absorbs at 570 nm. As it is hydrolyzed by α-amylase the color decreases, the basis for a fixed end-point assay frequently used in cereals-related industries (43). A continuous assay based on this phenomenon is the following (44). Suspend 1 g soluble starch in 10 mL water, then slowly add this to 50 mL boiling water. Boil gently for 2 min, cool, and make up to 100 mL volume. To 2.5 mL buffered iodine solution (32 mg I_2, 330 mg KI, and 23.4 mg NaCl in 100 mL of 0.01 M phosphate buffer, pH 7.0), add 10 μL enzyme followed by 10 μL starch. Record the decrease in absorbance at 570 nm. This assay was developed for pancreatic α-amylase; for cereal α-amylases, the buffer system would be 0.02 M acetate, pH 4.7, 1 mM in $CaCl_2$.

α-Amylase is an *endo* glycosidase, hydrolyzing internal bonds in α-1,4 linked glucose polymers (40). A colorimetric assay using a blocked substrate takes advantage of this fact. The substrate is maltoheptaose, with a *p*-nitrophenyl group attached to the reducing end and a blocking agent (45) at the nonreducing end. α-Amylase cleaves this molecule in the middle. The assay mixture also contains glucoamylase and α-glucosidase, which combine to hydrolyze the short maltose (or maltotriose) derivative, freeing the chromogen, *p*-nitrophenol. The reaction is stopped by the addition of Tris buffer at pH 10, and the absorbance due to ionized *p*-nitrophenol is read at 410 nm.

β-Amylase hydrolyzes soluble starch to yield maltose, which is quantitated by colorimetric reaction with DNS (dinitrosalicylic acid) reagent (46). A 1% soluble starch solution is prepared as already described, except that 10 mL 0.16 M acetate buffer, pH 4.8, is added before adjusting to 100 mL total volume. Mix 0.5 mL substrate with 0.5 mL enzyme solution, incubate at 25°C for 3 min, then add 1 mL DNS reagent (50). Heat in boiling water 5 min, cool, add 10 mL water, and read absorbance at 540 nm. DNS reagent: in 500 mL water dissolve 10 g NAOH, 10 g 3,5-dinitrosalicylic acid, 2 g phenol, 0.5 g Na_2SO_3, and 200 g NaK tartrate, then make up to 1 L volume.

A chromogenic substrate, *p*-nitrophenyl maltopentaose, is used to assay this enzyme (41). β-Amylase removes two maltose units from the substrate, and α-glucosidase in the assay mixture then frees the chromogen. The reaction is stopped with alkaline buffer, and absorbance is read at 410 nm.

Cellulase Assays. Cellulase activity may be monitored using the DNS reagent (47). Dissolve 11.4 g CMC (carboxymethylcellulose) in 700 mL water with stirring, then add 7 g citric acid monohydrate, 19.6 g sodium citrate dihydrate, 0.1 g merthiolate, and 0.1 g glucose and make up to 1 L. To 1 mL substrate, add 1 mL enzyme solution, incubate 20 min at 50°C, add 3 mL DNS reagent, heat 15 min in boiling water, cool, and read absorbance at 540 nm. Cellulase activity may also be assayed by following the decrease in viscosity of a solution of CMC during incubation with enzyme (48,49). This requires a great deal of care and preparation and will not be discussed here.

Glycosidase. As an example of assays using nitrophenyl glycosides, that for β-galactosidase is given (50). In the reaction tube, combine 50 μL 0.1 M citrate buffer, pH 4.3, 50 μL 0.4% bovine serum albumin in water, 50 μL 20 mM pNP-β-galactoside in water, and 50 μL enzyme solution. After 30 min at 37°C, add 2 mL 0.25 M glycine, pH 10, and read the absorbance at 410 nm.

Xylanase. Interest has been growing recently in the use of xylanases in the food and feed industries. Xylans are pentose polymers, present primarily in cell walls of plants (an older name for them, based on this origin, is hemicellulose). Both soluble and insoluble types are found. The substrate for xylanases is usually derived either from a cereal grain (rye is the most abundant source) or wood (beech or larch gums). Assays may involve reducing-sugar techniques, viscometry, or chromogenic substrates (51). The chromogenic substrates are available in both soluble and insoluble (cross-linked) versions.

Xylanases, like amylases, may have either *endo* or *exo* specificities (similar to α- and β-amylases). Recently the two types have been separated and shown to have different kinds of improving actions in baked foods. Unfortunately, the specific assays that differentiate them have not been published. Using a soluble xylan as substrate, the *endo*-xylanase would show a rapid decrease in solution viscosity with a relatively slow increase in reducing groups, while the *exo*-xylanase would display the opposite behavior. The comparison of results from a viscometric assay and a reducing-sugar assay would readily differentiate the two types.

Ester Hydrolases

Lipase. Lipase acts on esters at the interface between water and oil (52); the substrate is a liquid vegetable oil (olive oil, soy oil) purified by percolation over a column of activated alumina or silica gel. It is emulsified with the help of vegetable gum or bile salt. The release of fatty acid is monitored by addition of base to keep the pH constant (continuous titrimetry, or pH stat, method) or by extraction and colorimetric measurement of the copper soap (53). For the latter assay, make substrate emulsion by adding 2 mL vegetable oil solution (1% in absolute EtOH) to 100 mL 0.025 M Tris buffer, pH 8.8, containing 0.6% sodium deoxycholate. To 5 mL substrate, add 0.1 mL enzyme, incubate 5 min, add 1 mL 1 N HCI and 10 mL isooctane, shake, and allow phase separation. To 5 mL of the upper (organic solvent) layer, add 1 mL copper reagent (5% cupric acetate in water, pH adjusted to 6.1 with pyridine), shake vigorously for 90 s, and allow the phases to separate. Read the absorbance at 715 nm of the upper phase.

Esterase. The best assay for esterase activity is the pH stat method. The substrate in water or salt solution is adjusted to the desired pH, enzyme is added, and base is continually added to keep the pH constant. The plot of base consumption versus time gives the rate of the reaction. For a spectrophotometric assay the system is lightly buffered at the desired pH, an indicator dye is included, and the change in absorption is measured. An assay for pectin methylesterase (54) uses 0.5% citrus pectin in water adjusted to pH 7.5, an indicator/buffer system of 0.01% bromthymol blue in 3 mM phosphate buffer, pH 7.5, and enzyme in the same buffer. To 2 mL of pectin solution, add 0.15 mL indicator/buffer and 0.83 mL water. Record absorbance at 620 nm briefly to establish a baseline, then add 20 μL enzyme and record the rate of absorbance change due to esteratic action. The system (without enzyme) is titrated with galacturonic acid to establish the correspondence between absorbance change and acid concentration.

Phosphatase. Phosphatase activity is monitored by colorimetric quantitation of P_i (inorganic phosphate) formed (55). An example is the assay for acid phosphatase using glucose-6-phosphate (G-6-P) (56). Substrate is 0.2 M maleate buffer, pH 6.7, containing 60 mM G-6-P, 8 mM KF, and 8 mM EDTA. To 1 mL, add 1 mL enzyme extract, incubate 20 min at 37°C, and add 1 mL cold 10% TCA to stop. To 1 mL, add 2 mL P_i reagent, hold 20 min at 45°C, and read absorbance at 820 nm. Reagent A: 10% (w/v) ascorbic acid in water. Reagent B: 4.2 g ammonium molybdate tetrahydrate in 1 L 1 N H_2SO_4. P_i reagent: 1 part A plus 6 parts B, made fresh daily, and kept in an ice bath until used.

Oxidases

Polyphenoloxidase. Polyphenoloxidase enzymes (eg, tyrosinase) catalyze two reactions: (*1*) oxygenation of a phenol to *o*-diphenol (cresolase) and (*2*) oxidation of a diphenol to *o*-quinone (catecholase). Substrates commonly used for assays are *p*-cresol (4-methylphenol), L-tyrosine (4-hydroxy phenylalanine), and *p*-coumaric acid (4-hydroxycinnamic acid). The corresponding diphenols are 4-methylcatechol, L-Dopa, and caffeic acid. Rates are monitored by absorbance changes at specific wavelengths as oxygenation or oxidation occurs. An assay system that differentiates the two kinds of activities is the following (57). For cresolase activity, mix 1.5 mL 0.05 M acetate, pH 4.8, 0.4 mL 1 mM *p*-cresol, 0.1 mL 1 mM 4-methylcatechol and 10 μL enzyme. Record increase in absorbance at 291 nm. The change is due only to oxygenation; oxidation of diphenol to quinone produces no absorbance change at 291 nm. For catecholase activity, mix 1.6 mL buffer, 0.4 mL 1 mM 4-methylcatechol, and 10 μL enzyme, recording the decrease in absorbance at 280 nm.

Lipoxygenase. Native lipoxygenase requires activation by hydroperoxydecadienoic acid, the product of its reaction with linoleic acid and oxygen; there is a lag period in the reaction if enzyme is simply mixed with a solution of substrate (58). The delay is removed by including a small amount of product in the stock substrate solution (168 mg linoleic acid and 14 mg hydroperoxylinoleic acid in 100 mL ethanol) (59). For assay, to 2.9 mL 0.2 M borate buffer, pH 9.0, add 50 μL substrate stock, followed by 50 μL enzyme solution. Mix and record the increase in absorbance at 235 nm.

Ascorbic Acid Oxidase (AAO). A simple spectrophotometric assay for AAO is based on the loss of absorbance at 265 nm when ascorbate is oxidized to dehydroascorbate (60,61). To 2.8 mL buffer (0.025 M citrate, 0.05 M phosphate, pH 5.6), add 100 μL substrate solution (0.05% ascorbate, 1% NA_2EDTA, neutralized), 50 μL 1% bovine serum albumin solution, and finally 50 μL enzyme solution. Record the rate of disappearance of absorbance at 265 nm.

Catalase, Peroxidase. For peroxidase, the reaction mixture is 0.01 M phosphate, pH 7.0, containing 25 mg/L H_2O_2 and 2.5 g/L pyrogallol (62). After adding enzyme, record the increase at 430 nm due to the formation of purpurogallin. The reaction of catalase with H_2O_2 is a first-order reaction (63). Hydrogen peroxide has a broad absorption band in the far ultraviolet; molar absorbance is 120 at 200 nm and 30 at 250 nm. To 2 mL assay solution (0.05 M phosphate, pH 7.0, containing 2 g/L H_2O_2), add 1 mL diluted catalase solution and record the ultraviolet absorbance. Plot log absorbance versus time; the slope divided by 2.303 equals the first-order rate constant, which is directly proportional to the amount of catalase in the assay (or use a computer program to calculate the rate constant).

Other Enzymes

Lysozyme. Hydrolysis of bacterial cell walls by lysozyme causes the cells to lyse; a turbid suspension of cells will slowly clear. The turbidimetric assay (64) uses a suspension of *Micrococcus lysodeikticus* (0.5 g/L) in 0.06 M phosphate, pH 6.2. To 1.5 mL, add 0.5 mL 0.3 M NaCl and 1 mL enzyme, place the cuvette in a colorimeter, and record % transmittance (not absorbance) at 540 nm. A plot of %T versus time is linear from about 10 to 40%, and the rate %T/min is proportional to enzyme in the range of 0 to 10 μg.

BIBLIOGRAPHY

1. C. E. Stauffer, *Enzyme Assays for Food Scientists*, Van Nostrand Reinhold, New York, 1989.

2. K. F. Tipton and H. B. F. Dixon, "Effects of pH on Enzymes," *Methods Enzymol.* **63**, 183–233 (1979).
3. W. W. Cleland, "The Use of pH Studies to Determine Chemical Mechanism of Enzyme-Catalyzed Reaction," *Methods Enzymol.* **87**, 390–404 (1982).
4. R. L. Van Etten, "Human Prostatic Acid Phosphatase: A Histidine Phosphatase," *Ann. N.Y. Acad. Sci.* **390**, 27–51 (1982).
5. D. H. Strumeyer, "A Modified Starch for Use in Amylase Assays," *Anal. Biochem.* **19**, 61–71 (1967).
6. H. Perten, "Application of the Falling Number Method for Evaluating Alpha-Amylase Activity," *Cereal Chem.* **41**, 127–140 (1964).
7. M. Monsigny, C. Kieda, and T. Maillet, "Assay for Proteolytic Activity Using a New Fluorogenic Substrate (Peptidyl-3amino-9-ethylcarbazole)," *EMBO J.* **1**, 303–306 (1982).
8. B. Hagihara et al., "Crystalline Bacterial Proteinase. 1. Preparation of Crystalline Proteinase of Bacillus subtilis," *J. Biochem.* **45**, 185–194 (1958).
9. W. W. Cleland, "Statistical Analysis of Enzyme Kinetic Data," *Methods Enzymol.* **63**, 103–137 (1979).
10. J.-M. Caillat and R. Drapron, "La lipase du blé. Caracteristiques de son action en milieux aqueux et peu hydrate," *Annales de Technologie Agricole* **23**, 273–286 (1974).
11. G. L. Atkins and I. A. Nimmo, "Current Trends in the Estimation of Michaelis-Menten Parameters," *Anal. Biochem.* **104**, 1–9 (1980).
12. H.-J. Lee and I. B. Wilson, "Enzymic Parameters: Measurement of V and K_m" *Biochim. Biophys. Acta* **242**, 519–522 (1971).
13. M. L. Kakade et al., "Determination of Trypsin Inhibitor Activity of Soy Products: A Collaborative Analysis of an Improved Procedure," *Cereal Chemistry* **51**, 376–382 (1974).
14. G. E. Hamerstrand, L. T. Black, and J. D. Glover, "Trypsin Inhibitors in Soy Products: Modification of the Standard Analytical Procedure," *Cereal Chemistry* **58**, 42–45 (1981).
15. C. E. Stauffer, "Measuring Trypsin Inhibitor in Soy Meal: Suggested Improvements in the Standard Method," *Cereal Chemistry* **67**, 296–302 (1990).
16. American Association of Cereal Chemists (AACC), *Approved Methods of the American Association of Cereal Chemists*, 8th ed., AACC, St. Paul, Minn., 1983, Method 22-62.
17. M. L. Anson, "The Estimation of Pepsin, Trypsin, Papain and Cathepsin with Hemoglobin," *J. Gen. Physiol.* **22**, 79–89 (1938).
18. R. M. Tomarelli, J. Charney, and M. L. Harding, "The Use of Azoalbumin as a Substrate in the Colorimetric Determination of Peptic and Tryptic Activity," *J. Lab. Clin. Med.* **34**, 428–433 (1949).
19. C. Schwabe, "A Fluorescent Assay for Proteolytic Activity," *Anal. Biochem.* **53**, 484–490 (1973).
20. Y. Lin, G. E. Means, and R. E. Feeney, "The Action of Proteolytic Enzymes on *N,N*-Dimethyl Proteins," *J. Biol. Chem.* **244**, 789–793 (1969).
21. G. W. Schwert and Y. Takenaka, "A Spectrophotometric Determination of Trypsin and Chymotrypsin," *Biochim. Biophys. Acta* **16**, 570–575 (1955).
22. B. F. Erlanger, N. Kowosky, and W. Cohen, "The Preparation and Properties of Two New Chromogenic Substrates of Trypsin," *Arch. Biochem. Biophys.* **95**, 271–278 (1961).
23. B. C. W. Hummel, "A Modified Spectrophotometric Determination of Chymotrypsin, Trypsin and Thrombin," *Can. J. Biochem. Physiol.* **37**, 1393–1399 (1959).
24. K. A. Walsh and P. E. Wilcox, "Serine Proteases," *Methods Enzymol.* **19**, 31–41 (1970).
25. K. A. Walsh, "Trypsinogens and Trypsins of Various Species," *Methods Enzymol.* **19**, 41–63 (1970).
26. J. Feder, "A Spectrophotometric Assay for Neutral Protease," *Biochem. Biophys. Res. Commun.* **32**, 326–332 (1968).
27. P. H. Petra, "Bovine Procarboxypeptidase and Carboxypeptidase A," *Methods Enzymol.* **19**, 460–503 (1970).
28. O. H. Lowry et al., "Protein Measurement With the Folin Phenol Reagent," *J. Biol. Chem.* **193**, 265–275 (1951).
29. P. K. Smith et al., "Measurement of Protein Using Bicinchoninic Acid," *Anal. Biochem.* **150**, 76–85 (1985).
30. J. Adlers-Nissen, "Determination of the Degree of Hydrolysis of Food Protein Hydrolysates by Trinitrobenzenesulfonic Acid," *J. Agric. Food Chem.* **27**, 1256–1262 (1979).
31. C. E. Stauffer, "A Linear Standard Curve for the Folin-Lowry Determination of Protein," *Anal. Biochem.* **69**, 646–648 (1975).
32. J. E. Kruger and K. Preston, "The Distribution of Carboxypeptidases in Anatomical Tissues of Developing and Germinating Wheat Kernels," *Cereal Chem.* **54**, 167–174 (1977).
33. J. E. Kruger and K. R. Preston, "Changes in Aminopeptidases of Wheat Kernels During Growth and Maturation," *Cereal Chem.* **55**, 360–372 (1978).
34. H. V. Street and J. R. Close, "Determination of Amylase Activity in Biological Fluids," *Clinica Chimica Acta* **1**, 256–268 (1956).
35. J. J. Child, D. E. Eveleigh, and A. S. Sieben, "Determination of Cellulase Activity Using Hydroxyethylcellulose as Substrate," *Can. J. Biochem.* **51**, 39–43 (1973).
36. M. Ceska, E. Hultman, and B. Ingleman, "The Determination of Alpha-amylase," *Experientia* **25**, 555–556 (1969).
37. B. Klein, J. A. Foreman, and R. L. Searcy, "The Synthesis and Utilization of Cibachron Blue-Amylose: A New Chromogenic Substrate for Determination of Amylase Activity," *Anal. Biochem.* **31**, 412–425 (1969).
38. T. K. Ng and J. G. Zeikus, "A Continuous Spectrophotometric Assay for the Determination of Cellulase Solubilizing Activity," *Analytical Biochemist* **103**, 42–50 (1980).
39. S. Matsubara, T. Ikenaka, and S. Akabori, "Studies on Taka-amylase A. VI. On the α-Maltosidase Activity of Takaamylase A," *J. Biochem.* **46**, 425–431 (1959).
40. B. V. McCleary and H. Sheehan, "Measurement of Cereal α-Amylase: A New Assay Procedure," *J. Cereal Sci.* **6**, 237–251 (1987).
41. B. V. McCleary and R. Codd, "Measurement of β-Amylase in Cereal Flours and Commercial Enzyme Preparations," *J. Cereal Sci.* **9**, 17–33 (1989).
42. S. Dygert et al., "Determination of Reducing Sugar With Improved Precision," *Anal. Biochem.* **13**, 367–374 (1965).
43. American Association of Cereal Chemists (AACC), Approved Methods of the American Association of Cereal Chemists, 8th ed., AACC, St. Paul, Minn., 1983, Method 22-01.
44. L. B. Marshall and G. D. Christian, "A Rapid Spectrophotometric Method for lodimetric α-Amylase Assay," *Anal. Chim. Acta* **100**, 223–228 (1978).
45. U.S. Pat. 4 649 108 (March 10, 1987) H. E. Blair (to Genzyme Corporation).
46. P. Bernfeld, "Amylases, α and β," *Methods Enzymol.* **1**, 149–158 (1955).
47. G. L. Miller et al., "Measurement of Carboxymethyl-cellulase Activity," *Anal. Biochem.* **1**, 127–132 (1960).

48. M. A. Hulme, "Viscometric Determination of Carboxymethylcellulase in Standard International Units," *Arch. Biochem. Biophys.* **147**, 49–54 (1971).

49. K. Manning, "Improved Viscometric Assay for Cellulase," *J. Biochem. Biophys. Methods* **5**, 189–202 (1981).

50. J. J. Distler and G. W. Jourdian, "The Purification and Properties of α-Galactosidase From Bovine Testes," *J. Biol. Chem.* **248**, 6772–6780 (1973).

51. B. V. McCleary, "Measurement of *endo*-1,4-β-D-Xylanase", in J. Visser, ed., *Xylans and Xylanases*, Elsevier Science Publishers B. V., Amsterdam, 1992, pp. 161–170.

52. L. Sarda and P. Desnuelle, "Action de la lipase pancreatique sur les esters en emulsion," *Biochim. Biophys. Acta* **30**, 513–521 (1958).

53. D. Y. Kwon and J. S. Rhee, "A Simple and Rapid Colorimetric Method for Determination of Free Fatty Acids for Lipase Assay," *J. Am. Oil Chem. Soc.* **63**, 89–92 (1986).

54. A. E. Hagerman and P. J. Austin, "Continuous Spectrophotometric Assay for Plant Pectin Methylesterase," *J. Agric. Food Chem.* **34**, 440–444 (1986).

55. P. S. J. Chen, T. Y. Toribara, and H. Warner, "Microdetermination of Phosphorus," *Anal. Chem.* **28**, 1756–1758 (1956).

56. A. Rossi et al., "Properties of Acid Phosphatase From Scutella of Germinating Maize Seeds," *Phytochemistry* **20**, 1823–1826 (1981).

57. M. H. Keyes and F. E. Semersky, "A Quantitative Method for the Determination of the Activities of Mushroom Tyrosinase," *Arch. Biochem. Biophys.* **148**, 256–261 (1972).

58. B. Axelrod, T. M. Cheesbrough, and S. Laakso, "Lipoxygenase from Soybeans," *Methods Enzymol.* **71**, 441–451 (1981).

59. M. J. Gibian and R. A. Galaway, "Steady-State Kinetics of Lipoxygenase Oxygenation of Unsaturated Fatty Acids," *Biochemistry* **15**, 4209–4214 (1976).

60. E. Racker, "Spectrophotometric Measurements of the Metabolic Formation and Degradation of Thiol Esters and Enediol Compounds," *Biochim. Biophys. Acta* **9**, 577–578 (1952).

61. T. Tono and S. Fujita, "Determination of Ascorbic Acid by Spectrophotometric Method Based on Difference Spectra. II. Spectrophotometric Determination Based on Difference Spectra of L-Ascorbic Acid in Plant and Animal Foods," *Agric. Biol. Chem.* **46**, 2953–2959 (1982).

62. B. Chance and A. C. Maehly, "Assay of Catalases and Peroxidases," *Methods Enzymol.* **2**, 764–775 (1955).

63. R. M. J. Parry, R. C. Chandan, and K. H. Shahani, "A Rapid and Sensitive Assay of Muramidase," *Proc. Soc. Exp. Biol. Med.* **119**, 384–386 (1965).

CLYDE E. STAUFFER
Technical Food Consultants
Cincinnati, Ohio

ENZYMES IN FOOD PRODUCTION

The use of enzymes in the formulation and processing of food products is frequently thought of in terms of the addition of commercially available enzymes from an industrial source, that is, using enzymes as additives from an external or exogenous source. However, it should be recognized that virtually all food materials are, or were, living organisms and, as such, may contain a number of enzymes normally utilized in the metabolic pathways of the organism. These endogenous enzymes can play important roles in food processing. In some cases these enzymes provide an improving function, whereas in others their effects can be deleterious either to the processing itself and/or to the final product characteristics. This article will provide a background describing endogenous enzymes from a number of food sources and briefly describe how these enzymes may influence processing of the food material.

GENERAL

The amount and type of endogenous enzyme present in the fruit of a plant, whether a cereal grain seed or a fruit such as an apple or orange, depends, to a large extent, on the maturation stage of the plant. As a seed or fruit ripens, its enzyme content changes. As ripening proceeds, more hydrolytic enzymes are produced. In the case of a cereal grain, these enzymes begin the process of breaking down carbohydrate and protein stored in the seed in order to provide energy and amino acids necessary for the synthesis of new plant parts. While necessary for plant growth, these enzymes can have quite detrimental effects when flour from germinated wheat kernels is used in the production of foods, particularly in baked goods.

In ripening fruits, analogous enzymes begin a similar process of cellular breakdown that leads to characteristic loss of firmness normally associated with overripe, less desirable produce. The freshness of fruits and vegetables is often judged by the color and texture of the produce such as, for example, the bright green color and firm texture of fresh peas, beans, or peppers. When many products such as these ripen, their color changes to red, orange, yellow, brown, or even black. This is due, to a large degree, to endogenous enzymes, some of which degrade the green chlorophyll, thus changing the color of the product. Other enzymes such as lipoxygenase attack triglycerides forming free radicals and hydroperoxides. These free radicals can also lead to loss of green color due to oxidative degradation of chlorophyll and also changes in reds and orange colors due to reactions with carotenoid compounds. Oxidative reactions from these compounds can also lead to nutritional damage in the food product through detrimental action on vitamins as well as some proteins. Other color changes are also found to result from enzymatic browning reactions involving polyphenol oxidase (see "Browning Reactions—Color and Flavor") (1).

Texture is also affected by endogenous enzymes that hydrolyze cellular material, most of which is composed of various forms of carbohydrate material. Most commonly found in plants, these include cellulose, hemicellulose, starch, pectic material, and lignin. These carbohydrates are involved in cell structure of most plants and are responsible for the firm texture in young, preripe fruits and vegetables.

Meat color is also influenced by endogenous enzymes involved in oxidation–reduction reactions that alter the oxidation state of several forms of myoglobin as well as the amount of available oxygen. Changes in the oxidation state of oxymyoglobin and deoxymyoglobin can lead to production of metmyoglobin, which is responsible for the brown coloration in meat (1).

BAKED GOODS

Although a number of cereal grains may be used in the production of baked goods, the primary source for such products is wheat. Wheat is utilized in the production of baked goods such as breads, cakes, rolls, and pastries. The enzymes of greatest interest for this discussion will be those involved in protein and carbohydrate metabolism, specifically, the amylases, pentosanases, and proteases. Amylases are starch-degrading enzymes that act on the linkages between the glucose monomers making up the starch polymer. In starch, these linkages are either α-1,4 or α-1,6 glucosidic bonds connecting adjacent glucose residues. β-Amylase (EC 3.22.1.2) is an exo-type hydrolase that is able to hydrolyze only the penultimate α-1,4 bond, each cleavage producing a molecule of maltose. α-Amylase (EC 3.2.1.1) is an endo-type hydrolase that, while specific for the α-1,4 glucosidic bond, is capable of hydrolyzing such bonds anywhere along the internal portion of the glucose chain, producing small amounts of glucose as well as many different glucose oligomers. The difference between the two enzymes is important because they have different effects on the rheology of a starch system. β-Amylase has relatively little effect on the molecular weight of the starch molecule because it removes only one maltose at a time from starch molecules having molecular weights as large as several million. α-Amylase, however, can drastically alter the molecular weight because it can hydrolyze anywhere along the chain. That is, it can hydrolyze an internal α-1,4 glucosidic bond about in the middle of the glucose chain and immediately reduce the molecular weight by half. This action can greatly affect (decrease) the viscosity of the starch system.

Under ideal harvest conditions, wheat is in a dormant storage stage and contains comparatively few active enzymes. There may be significant amounts of β-amylase, but α-amylase levels are quite low. Thus, the amount of potentially detrimental endogenous starch-degrading enzyme in the sound wheat kernel is low. However, it is not unusual, particularly in the Pacific Northwest, that harvest conditions are not ideal. The most serious consequence from the standpoint of enzyme content is when harvest is preceded by humid, rainy conditions. In this environment, the seed breaks its dormancy and begins to grow. This phenomenon is commonly referred to as "sprouting" in the grain industry. This growth requires that the stored carbohydrate and protein be utilized for energy and synthesis of new plant parts. Thus, enzymes such as those which hydrolyze carbohydrate and protein material are synthesized and act on the stored material in the plant seed. It is the plant seed that is then processed into a food material such as wheat flour.

Whether the endogenous enzymes are a factor in utilizing the food material depends on a number of factors. First, how extensive was the sprouting? It may range from slight, in which case little enzyme was synthesized, to extensive, in which case a large amount of carbohydrase and proteinase enzymes have probably been formed. Second, what are the processing conditions for the manufacture of the food material? In general, high-moisture products such as breads will be much more susceptible to the action of these hydrolytic enzymes than will low-moisture products such as crackers. In bread processing, the dough may contain approximately 35 to 40% moisture, and the final baked product retains much of that moisture. Since processing involves a lengthy proofing period as well as the bake time, the presence of such high levels of water (most with a water activity [a_w] of 0.95–0.98) may lead to extensive hydrolysis if significant levels of endogenous enzymes are present.

In the dough stage, only that portion of the starch (both amylose and amylopectin) that is physically damaged in the milling process is susceptible to amylolytic attack. Depending on the type of wheat, the amount of starch damage may vary from about 3 to 8%. The amylase enzymes can attack the damaged starch and will release some water that was associated with the starch; however, the effect on the dough consistency will be relatively minor. However, the starch will gelatinize during the baking process as the internal temperature of the loaf reaches about 60 to 65°C (104°F). Both amylase enzymes may then rapidly degrade the gelatinized starch. However, it should be noted that endogenous amylase enzymes are relatively unstable above 65°C, and so are less effective as the temperature increases. The overall effect of the starch-degrading enzymes will depend on the amount of endogenous enzyme present and how long it has to act on the gelatinized starch.

It was thought for many years that α-amylase was the enzyme from sprouted wheat that was most responsible for the detrimental effects observed during baking of high-moisture bread products, but this is not the case. The most damaging of the hydrolytic enzymes in such a process are the proteases. Sprouted wheat contains proteolytic enzymes that can act on the protein fraction throughout the dough stage and into the baking stage. The result of excessive hydrolysis by proteolytic enzymes is a loss of gluten structure. This is manifest as a slack dough that has lost the viscoelasticity characteristic of a well-developed wheat flour dough. The finished product will have poor volume and texture characteristics. If hydrolysis proceeds to an extreme, all gluten structure may be lost, resulting in a very sticky, slack semiliquid. Some effects of these enzymes may be observed even when the degree of sprouting is relatively low. An assay for this type of enzyme would serve as a good diagnostic indicator of the quality of the flour used in an industrial food process.

In low-moisture baked products such as crackers where the final moisture content of the product is only about 3%, endogenous enzymes usually have less potential for detrimental effects. Because the processing conditions require less time compared with breads, and water bakeout in the oven during baking is more rapid, starch-degrading enzymes usually have relatively little effect. The rapid loss of water in the initial stage of baking leads to more rapid increases in the temperature of the dough piece, thus tending to inactivate the enzymes faster. In addition, because the available water is driven off rapidly, the starch gelatinization temperature increases such that relatively little starch is gelatinized in most cracker products. Thus, the amylolytic enzymes have much less of an opportunity to hydrolyze the starch in its most susceptible condition. Proteolytic enzymes may have an effect even in low-moisture products, but this will depend on proof time (long or short), since they will normally be rapidly inactivated in the baking process.

Unsprouted wheat flour also contains endogenous protease enzyme(s). These enzymes are activated only at relatively low pH (<4.0) and so are not normally active in most baked product processes. However, processing of fermented products such as soda crackers utilize long periods during which lactobacilli-based fermentation leads to low pH in the sponge stage. These low-pH endogenous proteases can be an important processing factor under these conditions, softening the dough and helping to "mellow" the final cracker dough-handling characteristics. Many chemically leavened cracker doughs have relatively basic pH values, so the endogenous low-pH proteases will not be activated.

Lipoxygenase enzymes in cereals, especially soybeans and wheat, may have significant consequences, both good and adverse. The desirable effect is that of bleaching the carotenoid pigments, resulting in whiter products made from the flour. This is a desirable attribute in products such as white bread and rolls. The amount of the lipoxygenase enzymes is greater in soy compared with wheat; thus, wheat flour may be supplemented with soy flour to achieve the bleaching effect. The downside to this activity is that too much lipoxygenase activity may lead to oxidation of native flour lipids, producing an undesirable rancid flavor/aroma characteristic. The type I lipoxygenase enzyme acts only on free fatty acids while type II lipoxygenase can act on triglycerides as well.

The pentosan fraction of the flour is an important functional participant in the formation of the dough. Though normally only about 2 to 3% of the flour, it is capable of absorbing up to 25% of the moisture in the dough. Hydrolysis of the pentosan fraction would have a significant effect on dough viscosity. However, at present, while pentosanase activity has been detected in germinated wheat, the levels appear to be quite low and of relatively little practical significance in terms of processing effects. Exogenous pentosanases, available commercially, may be used to greatly alter dough viscosity by destroying the water-holding capacity of the pentosan network.

Flour may also be used in nonbaking applications. For example, flour is often added to soup as a thickening agent. Endogenous enzymes, which may be present in that flour, are normally inactivated during the retorting step in the manufacture of the soup. If the flour is sound (unsprouted) containing relatively low levels of hydrolytic enzymes, the enzymes have little effect prior to retorting, and the desired thickening effect is maintained in the canned soup for long periods. However, if a sprouted wheat flour is used in this application, the endogenous enzymes can, prior to retorting, hydrolyze both starch and protein components, producing a thin, watery soup consistency.

FRESH FRUITS AND VEGETABLES

As mentioned in the "General" section, most plants contain various forms of carbohydrate, either as part of their cellular structure or as components of their "flesh." These include starch, cellulose, hemicellulose, pectins, and lignins. Several endogenous pectin-degrading enzymes contribute to the softening of many fruits and vegetables. Pectin methylesterase (EC 3.1.1.11) hydrolyzes the methyl ester from the methylated galactose units of pectin to form pectic acid and methanol. In the presence of sufficient calcium ion (Ca^{+2}) a complex is formed that can actually lead to an increase in firming of the plant tissue. However, this reaction can also lead to some problems in the fruit juice industry (see "Fruit juices"). Another enzyme, polygalacturanase, splits the α-1,4 glycosidic bond between adjacent galacturonosyl units. There are actually several forms of this enzyme, an endo-acting form (EC 3.2.1.15) and two exo-acting forms, EC 3.2.1.67 and EC 3.2.1.82 (2), which release D-galacturonic acid. These enzymes lead to extensive degradation of the substrate carbohydrate, leading to significant loss of structural integrity and a concomitant loss of firmness in the product. There is also a pectate lyase (EC 4.2.2.10) found, so far, only in microorganisms, which also splits the galacturonosyl chain at the α-1,4 glycosidic linkage leading to a loss of structural integrity. Although not strictly an enzyme endogenous to the plant, the microorganisms containing the enzyme may often be found to be present, thereby affecting the processing conditions in the absence of added commercial enzymes.

Another common plant cell wall constituent is cellulose. Cellulases (EC 3.2.1.4) are found in many plants, primarily in trees, but also in fruits and vegetables. This enzyme acts on the β-1,4 glucosidic bonds found in cellulose. Though cellulose is closely related to starch in being a polymer of glucose monomers, amylase enzymes cannot hydrolyze cellulose, and cellulases have no effect on starch. This is due to the different geometry of the β-1,4 bond in cellulose as compared with the α-1,4 bond found in starch. The loss of firmness attributable to the enzyme found in fruits and vegetables is still unconfirmed, but cellulases of microbial origin, often found with the plant, are known to be effective at hydrolyzing plant cellulose.

Pentosans, sometimes referred to as hemicellulose, are polymers of xylose, arabinose, or both. Generally the material is an arabinoxylan with β-1,4 linkages between the backbone xylose units. These polymers are commonly found in higher plants and are particularly important in cereal products because the pentosan polymer is capable of absorbing very large amounts of water, thus affecting formulations. There seems to be relatively little endogenous pentosanase enzyme active in cereals.

BEVERAGES

Alcoholic

Beer is a virtually universal libation of quite ancient origin. The methods used today are cleaner and more efficient, but the fundamental approach remains similar to that developed thousands of years ago. By accident or design, brewing beer represents one of the first systematic uses of enzyme technology in the production of a food product.

The basic goal of brewing is to convert carbohydrate to alcohol and protein to peptides and amino acids. Most often, the source of these substrates is the seed of the barley plant, though other seeds are also used. In many areas of Africa, for example, grain sorghum seed is utilized in the brewing process. As described for wheat kernels, these seeds also store energy as starch and amino acids as proteins. The first step in the brewing process involves "malting" of the barley seed. This is a process of controlled germination resulting in the synthesis of hydrolytic enzymes. The barley kernels are steeped in water under controlled temperature to a moisture content of about 45% and then allowed to germinate, usually on large floor areas where good aeration is maintained by turning and mixing the grain. After a suitable period at controlled temperature (about 15–20°C) the germinated grain is dried at 35 to 40°C for about 12 hours and subsequently ground to a course powder (3). This material then contains much increased levels of hydrolytic enzymes, primarily amylase and protease activity, which are essential to the brewing process.

As described previously in the baking section, the amylase enzymes (both alpha and beta) hydrolyze starch-producing sugars (maltose and glucose) and some oligosaccharides. With the addition of yeast (normally a strain of *Saccharomyces*) in a process known as "pitching," the glucose is converted to alcohol and CO_2 according to the well-known equation:

$$C_6H_{12}O_6 \rightarrow 2CO_2 + 2CH_3CH_2OH$$

Yeast utilizes a maltase enzyme to convert the maltose to glucose. Thus, this step in the brewing process is facilitated through the action of endogenous enzymes from both the barley seed as well as a series of enzymatic steps utilizing enzymes endogenous to the yeast.

The same basic process of converting carbohydrate to alcohol is utilized in the production of distilled spirits. Such products have an ethanol content of about 20% and higher (beer having about 3–6% and wine having about 10–15% alcohol). While the variety of such alcoholic beverages is quite large, those derived from cereals will be the focus of this discussion. Many different cereal grains, including wheat, rye, barley, corn, and others, are used in the production of alcohol spirits. The source grains are ground, mixed with water, and heated to affect the gelatinization of the starch. The conversion of the starch to sugar is then accomplished by the addition of a malt containing high levels of saccharifying enzymes (α- and β-amylase). The "mash" is mixed and stirred under controlled conditions producing high levels of fermentable maltose and glucose. The mash is heated to inactivate the enzymes, cooled and fermented. This mixture is then distilled to produce the elevated alcohol content characteristic of such beverages. Again it is the use of the endogenous malt enzymes that are central to the production of these products.

Fruit Juices

Fruit juices are another category of beverage in which endogenous enzymes may play a role in the production process. Unlike cereals, the polysaccharides of fruits are primarily cellulose, hemicellulose, and pectic material. These compounds tend to cause a number of production problems, most of which are addressed by the addition of exogenous enzymes. This is the usual approach because most fruits intended for human consumption are harvested prior to achieving biological ripeness in order to maintain the characteristics (color and firmness) desired by the consumer. The native enzymes, which would be expressed with advanced ripening, are not available. Thus, the problems of clouding and poor filtration associated with the presence of pectic material and cellulose and hemicellulosic compounds are normally treated with microbially derived enzymes added as processing aids.

One interesting problem, normally characteristic of orange juice, is the loss of cloud stability in the juice. Orange juice is normally produced by pressing the juice from the fruit. Rather than being clear, the juice is meant to be an opaque liquid. The insoluble material that accompanies the juice as the fruit is pressed contributes desirable color, opaqueness, and flavor characteristics. Loss of this cloudiness is associated with the action of an endogenous enzyme, a pectin methylesterase (EC 3.1.1.11), which desterifies the pectin to pectic acid. The ionized pectic acid is then able to form a complex with normally present calcium ions (Ca^{+2}), which results in a gel-like complex. This complex will precipitate over time as an undesirable layer of sediment leaving a clarified juice above it. There are several approaches to preventing this troublesome condition. One involves the heating of the juice to a temperature sufficient to inactivate the offending pectin methylesterase. This approach, however, has a deleterious effect on the flavor components of the juice and thus represents an approach of last resort. A more common and less detrimental solution is the addition of an exogenous polygalacturonase (EC 3.2.1.15), which depolymerizes the pectic acid prior to significant formation of the calcium complex (2).

Although the formation of this calcium–pectic acid complex presents problems for the orange juice industry, it can be used to advantage in other circumstances. For example, many vegetables tend to lose their firmness after harvest, particularly after canning. The addition of calcium chloride to canned tomatoes results in firming of the fruit. The firming has been attributed to the formation of the calcium–pectic acid complex. This phenomenon has been used to achieve desirable firming in many vegetables as well, including potatoes, peppers, and beans.

DAIRY FOODS

Cheeses are derived from the milk of cows, goats, and other ruminant animals. The textural characteristics of cheeses are determined, to a large extent, by the methods of curd manufacturing, but the flavors and aromas of cheeses are developed primarily in the ripening process. Cheese ripening is an enzymatic process involving enzymes endogenous to the milk, to the ruminants from which the milk is taken, and to microorganisms that are an essential part of the cheese-making process.

Milk protein is composed primarily from a set of proteins collectively called casein. These proteins form an association in milk to form micelles that remain suspended. These suspended micelles, along with fat globules and calcium phosphate, give milk its characteristic opaque white color. The first step in making cheese is the formation of a curd. Usually, the curd results from treatment of whole milk with acid or with enzyme to precipitate the casein fractions. Traditionally, this curd formation was accomplished using rennin (EC 3.4.23.4), an enzyme from the stomach of ruminants. The enzyme has a restricted specificity such that the hydrolysis proceeds to the extent necessary to precipitate the milk protein without causing extensive further protein hydrolysis. A commercially available substitute for rennin derived from microorganisms is called rennet. Other commercially available proteases may be used, but many cause too much hydrolysis, resulting in the formation of "bitter peptides" that contribute undesirable flavor characteristics. Protease treatment accomplishes the concentration of the milk protein into a dense curd, but it also initiates reactions that contribute to flavor development. This flavor development is enhanced by protease enzymes from the milk itself, primarily plasmin (EC 3.4.21.7), as well as enzymes endogenous to microorganisms used in the cheese process. These microorganisms may be bacterial and/or fungal depending on the specific cheese.

A number of processes occur during ripening, all enzyme mediated, which contribute to textural and flavor characteristics. They can be described by three categories of enzymatic activity. A more detailed description of each of these stages may be found in Reference 4. First, probably the most important enzymatic process involves proteolysis of the milk protein by rennin, plasmin, and microorganism-derived enzymes. Proteolysis results in the formation of peptides, amino acids, thiols, and other compounds, all of which contribute significantly to the flavor of the cheese.

A second lipolytic enzymatic process also contributes to the essence of cheeses. Although some lipolysis may lead to a perceived rancidity in some cheeses, the free fatty acids released by lipase (EC 3.1.1.3) activity do influence flavor positively as well. For the most part, the lipases involved are endogenous to microorganisms, although there is some evidence for participation of milk lipases as well. Esterases (EC 3.1.1.1) are also thought to contribute to the process by hydrolyzing mono- and diglycerides to ester compounds that are known to provide various flavor and aroma attributes.

A third process involves glycolysis, in which the lactose in milk is enzymatically converted to lactic and other acids through microorganism metabolism. The acids contribute significantly to the character of the cheese. In modern cheese making, these enzymatic reactions are accelerated by the addition of exogenous esterases and lipases to produce stronger flavors in a shorter period of time while providing addition control over the process.

ENDOGENOUS ENZYMES AS PROCESSING INDICATORS

Enzymes present in food materials may be used as diagnostic indicators of processing steps that would be difficult to assess otherwise. The following are several examples of the use of endogenous enzymes as indicators of the effectiveness of a processing step, often involving heat treatments to inactivate organisms or enzyme activity.

Pasteurization of Milk

The purpose of pasteurization is to destroy potentially harmful microorganisms in the milk through a process of rapid heating to about 85°C for a very short time (several seconds). It is sometimes difficult to determine whether the processing has effectively eliminated the targeted microorganisms. One way to assess the effectiveness of the treatment is to test the viability of the organisms. This, however, is a lengthy process not well suited for remedial heating. It happens that endogenous milk alkaline phosphatase (EC 3.1.3.1) is heat inactivated in a temperature range similar to that required to destroy the potentially harmful organisms. Thus, a simple colorimetric test (5) for alkaline phosphatase can quickly indicate whether the temperature of the milk has been raised sufficiently to inactivate the enzyme and, by inference, the microorganisms.

Blanching of Vegetables

Vegetables are often placed in boiling water to inactivate certain hydrolytic enzymes that would otherwise cause detrimental effects in terms of flavor and/or texture. Although certainly less a problem in the home, the proper degree of heating required to inactivate these enzymes, while not adversely affecting texture is critical to industries involved in the longer-term storage of such products in cans or bottles. Inadequate blanching in this case can result in seriously degraded product by the time the consumer opens the container for consumption of the product. Thus, it is important to know when a sufficient amount of heating has been achieved. A convenient way to assess the blanching process is to look at the activity of endogenous peroxidase (EC 1.11.1.7). While complete inactivation of peroxidase is not desirable, the proper degree of blanching is achieved when about 5 to 10% of the peroxidase activity remains. The assay method usually utilized for this test involves the use of gulacol and is based on the colorimetric determination of the enzymatic product tetragulacol (6). This methodology has been used to assess the degree of blanching in products such as peas, corn, beans, and other vegetables.

Oat Rancidity

Oats contain lipolytic enzymes, the action of which can lead rapidly to rancidity. In this case, as discussed for vegetables, heat inactivation of those lipolytic enzymes is often used to extend the shelf life of oats and products formulated with oats. Since the assay of the lipolytic enzymes is problematic, endogenous oat peroxidase activity serves as a more convenient indicator of sufficient thermal inactivation of the lipolytic enzymes.

Soybean Treatment

Soybeans are very high in protein and are therefore used to supplement other foodstuffs that are lower in protein quantity as well as quality. However, soybean meal is usually heated to alter the bitter flavor and to inactivate some antinutritional components such as trypsin inhibitors as well as lipoxygenase activity. Applying too much or too little heat in this process can produce undesirable effects. The endogenous urease enzyme (EC 3.5.1.5) is used as a convenient assay (7) to indicate the correct degree of thermal treatment.

BROWNING REACTIONS—COLOR AND FLAVOR

Maillard Browning

Several types of chemical reactions can affect color and flavor in food products. The Maillard reaction is a nonenzymatic reaction between a sugar and an amino function, often one on an amino acid. Although the reaction itself can proceed without the participation of an enzyme, the reaction is influenced by the concentration of the reactants, which in turn may be influenced by the presence of exoproteases and exo-amylases. Thus, endogenous enzymes such as β-amylase, a carboxypeptidase, or a leucineaminopeptidase can contribute to Maillard browning in food products, baked goods in particular. These reactions lead to significant color formation (evidenced by the color of the crust on a loaf of bread) as well as many flavor characteristics.

Enzymatic Browning

Another source of color in foods comes from a reaction usually referred to as enzymatic browning. This reaction is mediated by polyphenol oxidase (EC 1.10.3.1). This enzyme is able to interact with a large number of phenolic compounds and thus had been referred to by several names including tyrosinase, catecholase, polyphenolase, and cresolase. The basic oxidation reaction normally results in the formation of an unstable quinone, which then proceeds through a number of steps to form melanins. Melanins are brown to black in color and are responsible for undesirable spots on many fruits and vegetables such as bananas, apples, mushrooms, and potatoes. Because of the widespread presence of polyphenol oxidase and the damage it causes in many food products, numerous methods have been developed to try to prevent this reaction from occurring. Clearly, eliminating as much oxygen as possible will help. In addition, reducing agents such as ascorbic acid, sodium sulfite, and thioi compounds can also help by reacting with the quinone compound. These reactions do not, however, inactivate the enzyme itself, so when the oxidizing agents are depleted, the browning reaction may still proceed. Under the proper conditions, ascorbic acid can inactivate polyphenol oxidase reacting with an active-site histidine residue. Likewise, thiol compounds can also affect the enzyme activity by chelating the required metal cofactor (Cu^{++}) (1).

Other Flavor Reactions

Members of the plant genus *Allium*, which includes onions, shallots, garlic, and leeks, have strong, often irritating redolence that is the result of endogenous enzymatic activity. The characteristic odor of these plants is generally unnoticeable in the whole, undamaged bulb because the enzyme involved, allinase, is compartmentalized and unable to catalyze the odor-causing reaction. When tissue damage occurs, such as slicing the onion or garlic bulb, the enzyme is released and proceeds to react, producing the thiol compounds responsible for the offending (to some) aroma. For more detailed discussion of similar reactions in related food materials, the reader is referred to Reference 8.

BIBLIOGRAPHY

1. J. R. Whitaker, "Enzymes," in O. R. Fenema, ed., *Food Chemistry*, 3rd ed., Marcel Dekker, New York, 1996, pp. 431–530.
2. W. Pilnik and G. J. Voragen, "The Significance of Endogenous and Exogenous Pectic Enzymes. In Fruit and Vegetable Processing," in P. F. Fox, ed., *Food Enzymology*, Vol. 1. Elsevier Applied Science, London, United Kingdom, pp. 303–333.
3. T. Godfrey, "Brewing," in T. Godfrey and J. Reichelt, eds., *Industrial Enzymology—The Application of Enzymes in Industry*, The Nature Press, New York, 1983, pp. 221–259.
4. P. F. Fox and J. Law, "Enzymology of Cheese Ripening," *Food Biotechnol.* **5**, 239–262 (1991).
5. D. A. Schiemann and M. H. Brodsky, "Studies of Scharer's Original Method for Alkaline Phosphatase in Milk With a Modification Utilizing an Organic Buffer," *J. Milk Food Technol.* **39**, 191–195 (1976).
6. P. Varoquaux et al., "Automatic Measurement of Heat Destruction and Regeneration of Peroxide," (in French) *Lebensm. Wiss. Technol.* **8**, 60–63 (1975).
7. K. Brocklehurst, "Electrochemical Assays: The pH-Stat," in R. Eisenthal and M. J. Danson, eds. *Enzyme Assays: A Practical Approach*, Oxford University Press, Oxford, U.K., 1992, pp. 191–216.
8. R. C. Lindsay, "Flavors," in O. R. Fenema, ed., *Food Chemistry*, 3rd ed., Marcel Dekker, New York, 1996, pp. 723–765.

GENERAL REFERENCES

H.-D. Belitz and W. Grosch, *Food Chemistry*, Springer-Verlag, New York, 1987.

S. Schwimmer, *Sourcebook of Food Enzymology*, AVI Publishing, Westport, Conn., 1981.

C. Zapsalis and R. A. Beck, *Food Chemistry and Nutritional Biochemistry*, John Wiley & Sons, New York, 1985.

PAUL R. MATHEWSON
Food Technology Resource Group
Park City, Utah

See also ENZYMOLOGY.

ENZYMOLOGY

Enzymes can be defined as proteins with very specific powers of catalysis. The enhancement in reaction rate is very high and may be as much as 10^{14}-fold in some instances. Although all enzymes are proteins, some require an additional small molecule, called a coenzyme or cofactor (cofactors can be metal ions or small organic molecules), to function. The role of enzymes in cells means that growth, maturation, storage, processing, and consumption–digestion of food all depend on various enzyme activities. In plants, enzymes are responsible for changes associated with ripening, including alterations in color, flavor, and texture. Enzymatic changes continue after harvesting of plants and also after the death of animals, for example, in the conversion of muscle to meat. These changes affect subsequent food quality. In many cases the action of endogenous enzymes is arrested or controlled by processing and storage.

Enzymes are also widely used to provide desirable changes in food quality attributes through changes in the chemical structure of food components. Enzymes have significant advantages over chemical catalysts, of which the most important are specificity and the ability to work at moderate temperatures. As a consequence, side reactions are minimized, and undesirable changes caused by harsh conditions are averted. Enzymes have traditionally been obtained from their natural sources: animals, plants, and microorganisms. Advances in molecular biology have led to the cloning and expression of enzymes from many sources into microorganisms (primarily yeasts and fungi) that are efficient producers and regarded as safe hosts. The value of enzymes used in food processing is already several hundred million dollars per annum and will likely expand as new applications are discovered or improved enzymes become available, for example, by genetic engineering to improve operating characteristics such as the pH-activity profile or thermostability.

KINETICS

The activity of an enzyme is determined by many factors, including enzyme, substrate, and cofactor concentrations; ionic strength; pH; and temperature. For conversion of substrate (S) to product (P) by an enzyme (E) the reaction scheme can be simply represented as

$$E + S \overset{k_s}{\rightleftharpoons} ES \overset{k_{cat}}{\rightarrow} E + P$$

where ES is the enzyme–substrate complex, K_s is the dissociation constant, and K_{cat} is the turnover number/rate constant for the breakdown of ES.

The initial reaction velocity (V) is then given by the Michaelis–Menton equation

$$V = \frac{k_{cat}(E)(S)}{K_m + (S)}$$

where K_m is the substrate concentration at which V equals one-half the maximum velocity (V_{max}). Integration of the equation with respect to time gives

$$V_{max} = K_m \ln \frac{(S)}{(S_t)} + [(S) - (S_t)]$$

where S and S_t are the substrate concentrations at zero time and time t, respectively. This equation is particularly useful in industrial situations where a reaction is allowed to proceed to near completion or equilibrium.

APPLICATIONS IN FOOD PROCESSING

α-Amylase (EC 3.2.1.1) catalyzes random hydrolysis of α 1–4 linkages in amylose and amylopectin to form straight- and branched-chain dextrins, oligosaccharides, and monosaccharides. Bacteria α-amylase is used to partially hydrolyze or "thin" gelatinized starch, often prior to further degradation to glucose syrups. The enzyme from *Bacillus* spp. is very thermostable and is used at 85 to 105°C. The enzyme can also be used to hydrolyze starch in sugar cane juice and in the brewing mash.

Fungal α-amylases, derived from *Aspergillus niger* and *A. oryzae* are much less heat stable. They produce large amounts of maltose and maltotriose and some glucose. Their principal applications are production of maltose syrups used in jam and confectionery, as brewing aids to improve fermentability of the mash and in removal of starch haze in beer, and in supplementation of endogenous α-amylase in bread flour to enhance the rate of fermentation by yeast and reduce dough viscosity, thereby improving loaf volume. Fungal α-amylase has largely replaced malt α-amylase previously used in these applications.

β-Amylase (EC 3.2.1.2) splits off β-maltose from the nonreducing ends of starch molecules. In the case of amylopectin, this produces β-limit dextrins since the enzyme cannot bypass the β-1-6 branch points. Complete hydrolysis of liquified starch produces about 80% maltose and 20% dextrins. Unlike fungal α-amylase, β-amylase does not produce maltotriose. The enzyme is produced from cereal or microbial sources and is used for production of maltose syrup from starch, and in both brewing and baking to produce maltose for fermentation by yeast to CO_2 and to alcohol.

Amyloglucosidase (glucoamylase) (EC 3.2.1.3) is an exoamylase that catalyzes the stepwise hydrolysis of α 1–4 linkages in starch, thereby releasing glucose molecules from the nonreducing end. It is used mainly to produce glucose syrups from liquified starch previously treated with α-amylase. Since the α 1–6 linkages in amylopectin are also slowly hydrolyzed, the final conversion to glucose reaches 95 to 97% w/w with the remainder being mostly maltose and higher saccharides. The glucose produced is usually used as a syrup, crystallized out, or converted to fructose by glucose isomerase. The enzyme can also be used to hydrolyze residual oligosaccharides in high-fructose corn syrup and in the analysis of the starch content of foods. Glucoamylase is produced commercially from *Aspergillus* or *Rhizopus* spp. It is typically used to process liquified starch at 60°C. Immobilization of the enzyme has been widely studied to provide the benefits of a continuous process. However, the soluble enzyme is relatively cheap so it continues to be used mostly in that form.

Catalase (EC 1.11.1.6) specifically catalyzes the decomposition of hydrogen peroxide to water and oxygen. In the dairy industry, a low concentration of H_2O_2 (up to 0.05% w/w) is used to "cold pasteurize" milk destined for cheese making, and to preserve milk and whey in some instances where refrigeration is not practical. The H_2O_2 is then destroyed by catalase. Commercial sources include beef liver and *A. niger*. The enzyme is frequently used in conjunction with glucose oxidase (see later) to remove H_2O_2 produced by that enzyme.

Cellulase (EC 3.2.1.4) is the name given to a complex of several enzymes that, acting together, hydrolyze cellulose into β-dextrins and glucose. The native structure of cellulose is a major impediment to enzyme action, so pretreatment by milling, or with alkali or steam, is often necessary to make the substrate accessible. The fungus *Trichoderma reesie* is a good source of cellulase. It contains (*1*) endocellulases that randomly split internal β 1–4 linkages to produce dextrins and cellobiose, (*2*) exocellulases that act from the nonreducing end of the polymer to produce cellobiose, and (*3*) cellobiase (or β-glucosidase) that converts cellobiose to glucose. The latter is especially important since it relieves end-product inhibition of the other enzymes by cellobiose. Cellulase is used primarily to turn cellulosic wastes into glucose that can be fermented to ethanol. It is also used on a small scale to degrade cellulose in foodstuffs.

β-Glucanase (EC 3.2.1.6) hydrolyzes β 1–3 or β 1–4 bonds in β-D-glucans. The products are oligosaccharides and glucose. The enzyme is used primarily in brewing, where it is added, along with other enzymes, to malted barley at the mash stage. The β-glucanase degrades and solubilizes barley gums (β-glucan polymers), including those contributed by dying yeast cells, which would otherwise increase the viscosity of the wort. These residual glucans can contribute to haze problems in the final product. Commercial sources of β-glucanase include *Bacillus subtilis* and *A. niger*.

Glucose isomerase (EC 5.1.3.5) catalyzes conversion of glucose to fructose, thereby increasing sweetness and value. The enzyme is actually a xylose isomerase, which also acts on glucose and requires magnesium as a cofactor. The reaction is readily reversible and at equilibrium, a fructose/glucose ratio of 52/48 is achieved. However, in practice it is uneconomical to take the conversion beyond 42% fructose, when the syrup is isosweet with glucose on a solids basis. The product (high-fructose corn syrup) is produced at a rate of several million tons per annum and has displaced sucrose and glucose syrups from many products, especially in soft drinks in the United States.

The enzyme is produced commercially from several microbial sources including *Bacillus, Actinoplanes*, and *Streptomyces* spp. and is readily immobilized by a variety of methods (1).

Glucose oxidase (EC 1.1.3.4) catalyses the reaction:

$$\text{glucose} + O_2 \rightarrow \text{gluconic acid} + H_2O_2.$$

It is often used in conjunction with catalase, which breaks down H_2O_2, thereby sparing the glucose oxidase from denaturation as well as providing oxygen for use by glucose oxidase. Commercial preparations are derived from *A. niger* or *Penicillum* spp. Its main uses are to remove glucose from foods and as an antioxidant to prevent changes in color and flavor, particularly during food storage. It is used for removal of glucose from egg whites and whole eggs. Significant browning and off-flavor development due to Maillard reactions occurs if eggs are not desugared prior to drying. Other applications include removal of either dissolved or headspace oxygen from citrus drinks, canned soft drinks, beer, and wine, thereby preventing oxidative deterioration, and as an antioxidant in mayonnaise and production of gluconic acid. It is often the method of choice for assay of glucose since it is highly specific and sensitive.

Invertase (β-fructofuranosidase, EC 3.2.1.6) hydrolyzes sucrose into an equimolar mixture of glucose and fructose. Although the same result can be achieved by acid hydrolysis, the syrup produced by enzyme action is more pure and free from discoloration. Invert sugar syrup has several advantages over sucrose syrup: it is slightly sweeter, it does not crystallize at higher concentrations, and the sweetness intensity is stable in acidic foods. Until the advent of high-fructose corn syrup, it was used extensively to replace sucrose in jams and confectionery.

Other applications of invertase include its use in the production of liquid or soft centers in chocolate-coated sucrose candies, in recovery of scrap candy, in artificial honey, and as a humectant to hold moisture in foods. Invertase is produced commercially from yeasts, usually *Saccharomyces* or *Candida* spp., and is relatively cheap to produce. Consequently, it is used as a soluble enzyme and there is little commercial impetus for its application in an immobilized form.

Lactase (β-galactosidase, EC 3.2.1.23) hydrolyzes lactose into glucose and galactose. Small quantities of oligosaccharides containing galactose may also be formed as byproducts. Compared with lactose, the main products are, in combination, three to four times more soluble, about twice as sweet, easier to ferment, and directly absorbed from the intestine. Applications of lactase take advantage of these changes (2).

Hydrolysis of lactose in milk makes the milk more digestible for those who are lactose intolerant (especially infants) due to low levels of intestinal β-galactosidase. It also prevents crystallization of lactose in concentrated or frozen milk products, such as ice cream. Cheese whey is a major disposal problem, mostly because of its high lactose content. After separating the valuable whey protein, the lactose can be hydrolyzed by soluble or immobilized lactase. The product is then concentrated to give a stable, sweet syrup, which may be used in a variety of foods or fermented to ethanol (3).

Lactases produced commercially from yeasts such as *Kluyveromyces marxianus* have a neutral pH optimum and are used in milk processing. Lactases from molds such *A. niger* or *A. oryzae* have an acid pH optimum and are more suitable for whey processing. Lactoses are readily immobilized on a variety of supports for industrial use and for analysis of lactose in dairy products.

Lipases (EC 3.1.1.3) hydrolyze ester linkages of triglycerides to give free fatty acids, diglycerides, monoglycerides, and eventually glycerol. The enzyme is widely distributed in food tissues and belongs to the general class of esterases. Lipase action has traditionally been regarded as a problem in food science because the fatty acids produced may be unpalatable or susceptible to oxidation. This is a particular problem in stored cereals such as wheat and rice and in milk, where the release of short-chain fatty acids from milk fat by endogenous lipase leads directly to off-flavors. However, it is recognized that lipase action is also responsible for some of the desirable flavor in matured cheeses. Consequently, impure preparations (containing both lipase and esterase activities) are produced from molds such as *Mucor, Rhizopus*, and *Aspergillus* spp. for use in accelerated cheese ripening, often in conjunction with proteases. Concentrated cheese flavors produced in this fashion can be used in a variety of snack products. Lipase can also be used in directed transesterification of fats to improve functional properties and uses.

Pectinases are a group of enzymes that act on various pectic substances (pectins) in higher plants. There are three major types of pectinase: polygalacturonase (PG), pectin lyase (transeliminase) (PL), and pectin esterase (PE). PG (EC 3.2.1.15) splits glycosidic bonds within (endo-) or at the end of (exo-) the pectin molecule. Endo-PG action leads to a large decrease in viscosity of pectin solutions. PL (EC 4.2.2.10) also splits endoglycosidic bonds but the transelimination reaction yields a C^4-C^5 double bond on the nonreducing end. PE (EC 3.1.1.11) cleaves methanol from carboxyl groups, yielding low-methoxy pectin and polygalacturonic acid.

Since pectin is a major structural element in and between plant cell walls, alteration of the size or esterification of pectins can change the texture of fruits and vegetables during ripening and storage. Endogenous pectinases may soften texture during ripening. However, degradation of pectin can be inhibited by antisense RNA technology and this has been successfully exploited in tomatoes (4). Microbial pectinases are often responsible for postharvest rotting and decay. However, fungal pectinases, mostly from *Aspergillus* spp. are widely used as processing aids, mainly for extraction and clarification of fruit juices.

Polyphenol oxidase (PPO) (EC 1.10.3.1) is also known as phenolase tyrosinase, or catecholase. The enzyme is widely distributed in plants and fungi and in some mammals. In fruits and vegetables it is separated from phenolic substrates in the intact tissue. However, on cutting or other injury and exposure to oxygen, phenolase activity results in rapid browning due to polymerization of the quinones to give brown pigments (melanins).

Enzymatic browning is a major problem in handling and storage of fresh produce. However, in foods such as tea, coffee, and cocoa and in dried fruits such as raisins and dates, it improves color and flavor. Methods for control of enzymatic browning include: inactivation of the enzyme by heat, sulfites, or proteases; use of acidulants to lower pH and inhibit the enzyme; and use of chelators to remove the copper that is essential for activity and exclusion (or removal) of oxygen by appropriate packaging or glucose oxidase activity. Polyphenol oxidase activity in shrimp and other crustacea leads to blackening during harvest and storage. This can be prevented by use of the substrate analogue 4-hexylresorcinol (5).

Proteases hydrolyze peptide bonds in proteins and polypeptides to produce smaller peptides and amino acids. Many proteases also have esterase activity. These enzymes vary widely in their substrate specificity and optimum pH range. Enzymatic hydrolysis of protein is employed in a number of industries to create changes in product taste, texture, and appearance as well as in waste recovery. Plant proteases such as papain and ficin have broad substrate specificity and good thermal stability. They are used to tenderize meat, to chillproof beer, and also to recover protein hydrolyzate from scrap fish and bones. Animal proteases are generally more specific than the plant proteases. Trypsin has been used to inhibit the development of an oxidized flavor in stored milk and to solubilize heat-denatured whey protein (6). Pepsin is used primarily as an extender for calf rennet. Fungal proteases from *A. niger* and *A. oryzae* are use to modify gluten in bread flour, so as to reduce dough viscosity and improve color, texture and loaf volume. Bacterial proteases from *Bacillus* spp. are widely used in biscuit, cookie, and cracker dough since hydrolysis of gluten yields very elastic doughs that can be spread thinly without rupture. Other uses of microbial proteases include modification of soy and other food proteins to give improved functionality, decolorization of red blood cells to facilitate use of waste blood plasma protein from animal slaughter and production, and modification of gelatin from collagen. Recently there has been interest in using proteases in reverse to synthesize peptides. An example of this is the use of the enzyme thermolysin to produce the dipeptide sweetner aspartame (7).

Pullulanase (EC 3.2.1.41) and *isoamylase* (3.2.1.68) act specifically on the α 1–6 bonds of amylopectin in liquified starch to produce maltose and maltotriose. These enzymes improve the yield of mono- and disaccharides from starch when used in conjunction with other amylases. Pullulanase is produced commercially from *Klebsiella pneumonia* and is used in brewing to remove limit dextrins and allow the production of specialty beers such as "high alcohol" or "low calorie."

Rennet is obtained from the fourth stomach of unweaned calves and contains several enzymes including pepsin and chymosin (EC 3.44.3). Chymosin is an acid protease used in cheese making to coagulate the casein in milk. It catalyzes very specific and limited proteolysis of *k*-casein, thereby destabilizing the casein micelle and causing subsequent formation of curd. Calf rennet is expensive and is increasingly being replaced by microbial proteases with a high ratio of milk-clotting activity to general proteolytic activity. Suitable proteases have been obtained from *Endothia parasitica, Mucor meihei* and *M. pusillus*. Microbial proteases are now used in about one-half of all cheese production worldwide. Their main disadvantage is that they tend to be heat stable and cause proteolysis in whey products. However, a second generation of modified microbial rennets with lower stability is now available. The gene for chymosin has now been successfully cloned and expressed in several microorganisms (8). Products from this source may well replace both traditional and microbial rennets in the near future.

BIBLIOGRAPHY

1. S. A. Barker and G. S. Petch, "Enzymatic Processes for High-Fructose Corn Syrup," in A. Laskin, ed., *Enzymes and Immobilized Cells in Biotechnology*, Benjamin/Cummings, Menlo Park, Calif., 1985, pp. 93–105.
2. R. R. Mahoney, "Lactose: Enzymatic Modification," in P. F. Fox, ed., *Advanced Dairy Chemistry*, Vol. 3, Chapman and Hall, London, United Kingdom, 1997, pp. 77–105.
3. J. Barry, "A New Source for Alcohol," *Food Manufacture* **58**, 63–67 (1983).
4. G. Tucker, "Improvement of Tomato Fruit Quality and Processing Characteristics by Genetic Engineering," *Food Science and Technology Today* **7**, 103–108 (1993).
5. A. J. McEvily, R. Iyengar, and W. S. Otwell, "Sulfite Alternative Presents Shrimp Melanosis," *Food Technol.* **45**, 80–86 (1991).
6. J. C. Monti and R. Jost, "Solubilization of Cheese Whey Protein by Trypsin and a Process to Recover the Active Enzyme from the Digest," *Biotechnol. Bioeng.* **20**, 1173–1185 (1981).
7. K. Nakanishi and R. Matsuno, "Enzymatic Synthesis of Aspartame," in R. D. King and P. S. J. Cheetham, eds., *Food Biotechnology*, Vol. 2, Elsevier Applied Science, London, United Kingdom, 1988, pp. 219–249.
8. D. Jackson, "Cost Reduction in Food Processing Using Biotechnology," in S. K. Harlander and T. B. Lubuza eds., *Biotechnology in Food Processing*, Noyes Press, Park Ridge, N.J., 1986, pp. 285–295.

GENERAL REFERENCES

P. S. J. Cheetham, "The Application of Enzymes in Industry," in A. Wiseman, ed., *Handbook of Enzyme Biotechnology*, 2nd ed., Ellis Horwood, Chichester, United Kingdom, 1985, pp. 274–373.

P. F. Fox, ed., *Food Enzymology*, Vols. 1 and 2, Elsevier Applied Science, London, United Kingdom, 1991.

T. Nagodawithana and G. Reed, eds., *Enzymes in Food Processing*, 3d ed., Academic Press, San Diego, Calif., 1993.

G. A. Tucker and L. F. J. Woods, eds., *Enzymes in Food Processing*, 2nd ed., Blackie Academic and Professional, London, United Kingdom, 1995.

RAYMOND R. MAHONEY
University of Massachusetts
Amherst, Massachusetts

See also ENZYMES IN FOOD PRODUCTION.

EVAPORATED MILK

HISTORY

Evaporated milk, like other processed canned foods, originated with the experiments of the French scientist Nicholas Appert. Appert, whose work on food preservation began in 1795, was the first to evaporate milk by boiling it in an open container and preserve it by heating the product in a sealed container. Fifty years later, Louis Pasteur laid the scientific foundation for heat preservation by demonstrating that food spoilage could be caused by bacteria and other microorganisms.

Patents for the preservation of milk after evaporation in a vacuum were granted to Gail Borden by the United States and England in 1856. Although these patents applied to concentrating milk without the addition of sugar, Borden's first commercial process was for the manufacture of sweetened condensed milk. The original vacuum equipment developed by Borden is now on display at the Smithsonian Institution.

Borden produced sweetened condensed milk at Wassaic, New York, in 1861. By 1865, new plants were opened in Brewster, New York, and Elgin, Illinois. In 1866, Charles A. Page and his three brothers built Europe's first commercial sweetened condensed milk plant in Switzerland. They later built plants in England and in the United States.

In 1884, a U.S. patent was issued for "an apparatus for preserving milk," and in 1885, the world's first commercial evaporated milk plant was opened in a converted wool factory in Highland, Illinois, where evaporated cream was manufactured and sold.

The first advertisement for evaporated milk appeared in 1893, calling the product "a perfect instant food." In 1894, the first recipe booklet for evaporated milk was distributed, and by 1895, the product was popular both in western mining areas of the United States, where fresh milk was scarce, and in the South, where there was little refrigeration. Consumers quickly recognized the value of evaporated milk as a safe, wholesome, and convenient food as well as a nutritional ingredient for cooking that had the added benefit of storage stability.

THE PRODUCT

Evaporated milk is a canned whole milk concentrate to which a specified quantity of vitamin D has been added and to which vitamin A may be added. It conforms to the Food and Drug Administration (FDA) Standard of Identity (21 CFR 131.130), having a minimum of 6.5% milkfat, 16.5% of nonfat milk solids, 23.0% total milk solids, and 25 IU vitamin D per fluid ounce. Related evaporated milk products are evaporated nonfat milk, evaporated lowfat milk, evaporated filled milk, and evaporated goat milk. Standards of Identity do not exist for these related evaporated milk products. Their typical compositions are the following:

Evaporated nonfat milk	Less than 0.5% milkfat, 20.0% total milk solids, 25 IU vitamin D and 125 IU vitamin A per fluid ounce added
Evaporated lowfat milk	2.0% milkfat, 18% nonfat milk solids, vitamins A and D added
Evaporated filled milk	6.0% vegetable fat, 17.5% nonfat milk solids, vitamins A and D added
Evaporated goat milk	Not less than 7.0% milkfat and 15.0% nonfat milk solids, vitamin D added

Table 1 shows the nutritional content of evaporated, evaporated nonfat, and evaporated lowfat milks.

Table 1. Nutritional Content per One-Half Cup Evaporated Milk

	Original	Nonfat	Lowfat
Calories	170	100	110
Protein (g)	8	9	8
Carbohydrates (g)	12	14	12
Fat (g)	10	Less than 1	3
Percentage of U.S. Recommended Daily Allowance (U.S. RDA)			
Protein	20	20	20
Vitamin A	4	10	10
Vitamin C	[a]	[a]	[a]
Thiamine	2	2	2
Riboflavin	20	20	20
Niacin	[a]	[a]	[a]
Calcium	30	30	30
Iron	[a]	[a]	[a]
Vitamin D	25	25	25
Phosphorus	25	25	25

[a]Contains less than 2% of the U.S. RDA of these nutrients.

PRODUCT PROCESSING

A typical processing scheme for evaporated milk begins with high quality, fresh whole milk, which is standardized to produce the exact composition desired in the final product (addition or removal of cream or nonfat milk). The product is heated, concentrated under reduced pressure in an evaporator, homogenized, and cooled. Vitamins (A or D, or both) are added, and the final composition is verified. Final standardization is accomplished, when necessary. After cans are filled and sealed, they are sterilized in a three-phase continuous system consisting of preheater, retort, and cooler, cooled; labeled; and packed for shipment. In the United States, evaporated milk is packed in 5-, 12-, 20-, and 97-fluid ounce lead-free cans.

INDUSTRY PRODUCTION

Table 2 reflects U.S. production of evaporated milk and related products during the period 1975 to 1998. The data are based on manufacturers' production reported to the American Dairy Products Institute. Before 1983, data were not collected on the production of related evaporated milk products.

Table 2. Evaporated Milk Production

Year	Evaporated milk	Evaporated milk and related products[a]
1975	20,796,000	
1976	20,362,000	
1977	18,347,000	
1978	17,340,000	
1979	17,110,000	
1980	15,500,000	
1981	15,840,000	
1982	15,050,000	
1983	14,300,000	16,612,000
1984	13,700,000	15,892,000
1985	14,360,000	16,537,000
1986	12,996,000	14,881,000
1987	12,966,000	14,759,000
1988	13,357,000	15,164,000
1989	12,451,000	14,193,000
1990	13,169,000	14,895,000
1991	11,313,000	12,776,000
1992	12,153,000	13,595,000
1993	11,561,000	13,114,000
1994	10,585,000	11,771,000
1995	10,847,000	12,462,000
1996	9,853,000	11,271,000
1997	10,656,000	12,204,000
1998	10,275,000	11,843,000

Notes: Figures reported in cases; 48-tall equivalent. Product weight/case since 1985: 40 lb.

[a]Includes evaporated milk plus evaporated nonfat milk, evaporated lowfat milk, and evaporated filled milk. Data not available before 1983.

GENERAL REFERENCES

C. W. Hall and T. I. Hedrick, *Drying of Milk and Milk Products*, 2nd ed., AVI, Westport, Conn., 1971.

E. H. Parfitt, "The Development of the Evaporated Milk Industry in the United States," *Journal of Dairy Science* **39**, 838 (1956).

R. Seltzer, *The Dairy Industry in America*, Magazines for Industry, New York, 1976.

U.S. Food and Drug Administration, Department of Health and Human Services, "Evaporated Milk," *Code of Federal Regulations* **21**, 131.130 (1999).

WARREN S. CLARK, JR.
American Dairy Products Institute
Chicago, Illinois